# Be sure to keep up-to-date in 2011!

Register now and we'll notify you by email each quarter when the newest cost index is available free online.

Please provide your name, address, and email below and return this card by mail or
Fax: 1-800-632-6732.

To register online:
**www.reedconstructiondata.com/rsmeans/chgnotice/842339**

Name _____

email _____

Title _____

Company _____

Street _____

City/Town _____ State/Prov. _____ Zip/Postal Code _____

RS**Means**

RSMeans
Quarterly Updates
700 Longwater Drive
Suite 100
Norwell, MA 02061
USA

2194478  C

# RSMeans
# Building
# Construction
# Cost Data

## 69TH ANNUAL EDITION

# 2011

**RSMeans**
A division of Reed Construction Data
*Construction Publishers & Consultants*
700 Longwater Drive
Norwell, MA 02061
USA

**1-800-334-3509**

Copyright© 2010 by RSMeans
A division of Reed Construction Data
All rights reserved.

Printed in the United States of America
ISSN 0068-3531
ISBN 978-1-936335-03-9

**Senior Editor**
Phillip R. Waier, PE

**Contributing Editors**
Christopher Babbitt
Ted Baker
Barbara Balboni
Robert A. Bastoni
John H. Chiang, PE
Gary W. Christensen
David G. Drain, PE
Cheryl Elsmore
Robert J. Kuchta
Robert C. McNichols
Melville J. Mossman, PE
Jeannene D. Murphy
Stephen C. Plotner
Eugene R. Spencer
Marshall J. Stetson

**Cover Design**
Paul Robertson
Wayne Engebretson

**President & CEO**
Iain Melville

**Vice President, Product & Development**
Richard Remington

**Vice President of Finance & Operations**
David C. Walsh

**Vice President, Data Aquisition & Fulfillment**
Jane Rafeedie

**Chief Customer Officer**
Lisa Fiondella

**Director, RSMeans**
Daimon Bridge

**Vice President, Marketing & Business**
Steven Ritchie

**Engineering Manager**
Bob Mewis, CCC

**Product Manager**
Andrea Sillah

**Production Manager**
Michael Kokernak

**Production Coordinator**
Jill Goodman

**Production**
Paula Reale-Camelio
Sheryl A. Rose
Jonathan Forgit
Debbie Panarelli
Mary Lou Geary
Caroline Sullivan

**Technical Support**
Chris Anderson
Tom Dion
Roger Hancock
Gary L. Hoitt
Genevieve Medeiros
Kathryn S. Rodriguez
Sharon Proulx

014621 ✓

This book is printed using soy-based printing ink.
This book is recyclable.

PRINTED WITH SOY INK

# Related RSMeans Products and Services

This book is aimed primarily at commercial and industrial projects or large multi-family housing projects costing $3,500,000 and up. For civil engineering structures such as bridges, dams, highways, or the like, please refer to *RSMeans Heavy Construction Cost Data*.

The engineers at RSMeans suggest the following products and services as companion information resources to *RSMeans Building Construction Cost Data*:

## Construction Cost Data Books
*Assemblies Cost Data 2011*
*Square Foot Costs 2011*

## Reference Books
*ADA Compliance Pricing Guide*
*Estimating Building Costs*
*RSMeans Estimating Handbook*
*Green Building: Project Planning & Estimating*
*How to Estimate with Means Data and CostWorks*
*Plan Reading & Material Takeoff*
*Project Scheduling & Management for Construction*
*Repair & Remodeling Estimating Methods*
*Universal Design*

## Seminars and In-House Training
Unit Price Estimating
*Means CostWorks®* Training
Means Data for Job Order Contracting (JOC)
Practical Project Management for Construction Professionals
Scheduling with MSProject for Construction Professionals
Mechanical & Electrical Estimating

## RSMeans on the Internet
Visit RSMeans at **www.reedconstructiondata.com/rsmeans** for a one-stop portal for the most reliable and current resources available on the market. Here you can request or download **FREE** estimating software demos, visit our bookstore for convenient ordering, and learn more about new publications and companion products.
Also visit **www.reedconstructiondata.com** for RSMeans business solutions, seminars, training, and more.

## RSMeans Electronic Data
Get the information found in RSMeans cost books electronically on *Means CostWorks®* CD or on the Web at **MeansCostWorks.com**.

## RSMeans Business Solutions
Cost engineers and analysts conduct facility life cycle and benchmark research studies and predictive cost modeling, as well as offer consultation services for conceptual estimating, real property management, and job order contracting. Research studies are designed with the application of proprietary cost and project data from Reed Construction Data and RSMeans' extensive North American databases. Analysts offer building product manufacturers qualitative and quantitative market research, as well as time/motion studies for new products, and Web-based dashboards for market opportunity analysis. Clients are from both the public and private sectors, including federal, state, and municipal agencies, corporations, institutions, construction management firms, hospitals, and associations.

## RSMeans for Job Order Contracting (JOC)
Best practice job order contracting (JOC) services support the owner in the development of their program. RSMeans develops all required contracting documents, works with stakeholders, and develops qualified and experienced contractor lists. Means JOCWorks™ software is the best of class in the functional tools that meet JOC specific requirements. RSMeans data is organized to meet JOC cost estimating requirements.

## Construction Costs for Software Applications
Over 25 unit price and assemblies cost databases are available through a number of leading estimating and facilities management software providers (listed below). For more information see the "Other RSMeans Products" pages at the back of this publication.
MeansData™ is also available to federal, state, and local government agencies as multi-year, multi-seat licenses.

- 4Clicks-Solutions, LLC
- ArenaSoft Estimating
- Beck Technology
- BSD – Building Systems Design, Inc.
- CMS – Construction Management Software
- Corecon Technologies, Inc.
- CorVet Systems
- Hard Dollar Corporation
- MC²
- Parsons Corporation
- Sage Timberline Office
- UDA Technologies, Inc.
- US Cost, Inc.
- VFA – Vanderweil Facility Advisers
- WinEstimator, Inc.
- R & K Solutions
- Maxwell Systems
- Tririga

# Table of Contents

# Foreword

## Our Mission

Since 1942, RSMeans has been actively engaged in construction cost publishing and consulting throughout North America.

Today, more than 65 years after RSMeans began, our primary objective remains the same: to provide you, the construction and facilities professional, with the most current and comprehensive construction cost data possible.

Whether you are a contractor, owner, architect, engineer, facilities manager, or anyone else who needs a reliable construction cost estimate, you'll find this publication to be a highly useful and necessary tool.

With the constant flow of new construction methods and materials today, it's difficult to find the time to look at and evaluate all the different construction cost possibilities. In addition, because labor and material costs keep changing, last year's cost information is not a reliable basis for today's estimate or budget.

That's why so many construction professionals turn to RSMeans. We keep track of the costs for you, along with a wide range of other key information, from city cost indexes . . . to productivity rates . . . to crew composition . . . to contractor's overhead and profit rates.

RSMeans performs these functions by collecting data from all facets of the industry and organizing it in a format that is instantly accessible to you. From the preliminary budget to the detailed unit price estimate, you'll find the data in this book useful for all phases of construction cost determination.

## The Staff, the Organization, and Our Services

When you purchase one of RSMeans' publications, you are, in effect, hiring the services of a full-time staff of construction and engineering professionals.

Our thoroughly experienced and highly qualified staff works daily at collecting, analyzing, and disseminating comprehensive cost information for your needs. These staff members have years of practical construction experience and engineering training prior to joining the firm. As a result, you can count on them not only for accurate cost figures, but also for additional background reference information that will help you create a realistic estimate.

The RSMeans organization is always prepared to help you solve construction problems through its variety of data solutions, including online, CD, and print book formats, as well as cost estimating expertise available via our business solutions, training, and seminars.

Besides a full array of construction cost estimating books, RSMeans also publishes a number of other reference works for the construction industry. Subjects include construction estimating and project and business management, special topics such as green building and job order contracting, and a library of facility management references.

In addition, you can access all of our construction cost data electronically using *Means CostWorks*® CD or on the Web using **MeansCostWorks.com.**

What's more, you can increase your knowledge and improve your construction estimating and management performance with an RSMeans construction seminar or in-house training program. These two-day seminar programs offer unparalleled opportunities for everyone in your organization to become updated on a wide variety of construction-related issues.

RSMeans is also a worldwide provider of construction cost management and analysis services for commercial and government owners.

In short, RSMeans can provide you with the tools and expertise for constructing accurate and dependable construction estimates and budgets in a variety of ways.

## Robert Snow Means Established a Tradition of Quality That Continues Today

Robert Snow Means spent years building RSMeans, making certain he always delivered a quality product.

Today, at RSMeans, we do more than talk about the quality of our data and the usefulness of our books. We stand behind all of our data, from historical cost indexes to construction materials and techniques to current costs.

If you have any questions about our products or services, please call us toll-free at 1-800-334-3509. Our customer service representatives will be happy to assist you. You can also visit our Web site at **www.rsmeans.com**

# How the Book is Built: An Overview

The Construction Specifications Institute (CSI) and Construction Specifications Canada (CSC) have produced the 2010 edition of MasterFormat™, a system of titles and numbers used extensively to organize construction information.

All unit price data in the RSMeans cost data books is now arranged in the 50-division MasterFormat™ 2010 system.

## A Powerful Construction Tool

You have in your hands one of the most powerful construction tools available today. A successful project is built on the foundation of an accurate and dependable estimate. This book will enable you to construct just such an estimate.

For the casual user the book is designed to be:

- quickly and easily understood so you can get right to your estimate.
- filled with valuable information so you can understand the necessary factors that go into the cost estimate.

For the regular user, the book is designed to be:

- a handy desk reference that can be quickly referred to for key costs.
- a comprehensive, fully reliable source of current construction costs and productivity rates so you'll be prepared to estimate any project.
- a source book for preliminary project cost, product selections, and alternate materials and methods.

To meet all of these requirements, we have organized the book into the following clearly defined sections.

## New: Quick Start

See our new "Quick Start" instructions on the following page to get started right away.

## Estimating with RSMeans Unit Price Cost Data

Please refer to these steps for guidance on completing an estimate using RSMeans unit price cost data.

## How To Use the Book:
### The Details

This section contains an in-depth explanation of how the book is arranged . . . and how you can use it to determine a reliable construction cost estimate. It includes information about how we develop our cost figures and how to completely prepare your estimate.

### Unit Price Section

All cost data has been divided into the 50 divisions according to the MasterFormat system of classification and numbering. For a listing of these divisions and an outline of their subdivisions, see the Unit Price Section Table of Contents.

*Estimating tips are included at the beginning of each division.*

### Reference Section

This section includes information on Equipment Rental Costs, Crew Listings, Historical Cost Indexes, City Cost Indexes, Location Factors, Reference Tables, Change Orders, Square Foot Costs, and a listing of Abbreviations.

**Equipment Rental Costs:** This section contains the average costs to rent and operate hundreds of pieces of construction equipment.

**Crew Listings:** This section lists all of the crews referenced in the book. For the purposes of this book, a crew is composed of more than one trade classification and/or the addition of power equipment to any trade classification. Power equipment is included in the cost of the crew. Costs are shown both with bare labor rates and with the installing contractor's overhead and profit added. For each, the total crew cost per eight-hour day and the composite cost per labor-hour are listed.

**Historical Cost Indexes:** These indexes provide you with data to adjust construction costs over time.

**City Cost Indexes:** All costs in this book are U.S. national averages. Costs vary because of the regional economy. You can adjust costs by CSI Division to over 700 locations throughout the U.S. and Canada by using the data in this section.

**Location Factors:** You can adjust total project costs to over 900 locations throughout the U.S. and Canada by using the data in this section.

**Reference Tables:** At the beginning of selected major classifications in the Unit Price section are reference numbers shown in a shaded box. These numbers refer you to related information in the Reference Section. In this section, you'll find reference tables, explanations, estimating information that support how we develop the unit price data, technical data, and estimating procedures.

**Change Orders:** This section includes information on the factors that influence the pricing of change orders.

**Square Foot Costs:** This section contains costs for 59 different building types that allow you to make a rough estimate for the overall cost of a project or its major components.

**Abbreviations:** A listing of abbreviations used throughout this book, along with the terms they represent, is included in this section.

### Index

A comprehensive listing of all terms and subjects in this book will help you quickly find what you need when you are not sure where it occurs in MasterFormat.

## The Scope of This Book

This book is designed to be as comprehensive and as easy to use as possible. To that end we have made certain assumptions and limited its scope in two key ways:

1. We have established material prices based on a national average.
2. We have computed labor costs based on a 30-city national average of union wage rates.

For a more detailed explanation of how the cost data is developed, see "How To Use the Book: The Details."

## Project Size/Type

The material prices in Means data cost books are "contractor's prices." They are the prices that contractors can expect to pay at the lumberyards, suppliers/distributers warehouses, etc. Small orders of speciality items would be higher than the costs shown, while very large orders, such as truckload lots, would be less. The variation would depend on the size, timing, and negotiating power of the contractor. The labor costs are primarily for new construction or major renovation rather than repairs or minor alterations.

*With reasonable exercise of judgment, the figures can be used for any building work.*

v

# Absolute Essentials for a Quick Start

If you feel you are ready to use this book and don't think you will need the detailed instructions that begin on the following page, this Absolute Essentials for a Quick Start page is for you. These steps will allow you to get started estimating in a matter of minutes.

## 1 Scope

Think through the project that you will be estimating, and identify the many individual work tasks that will need to be covered in your estimate.

## 2 Quantify

Determine the number of units that will be required for each work task that you identified.

## 3 Pricing

Locate individual Unit Price line items that match the work tasks you identified. The Unit Price Section Table of Contents that begins on page 1 and the Index in the back of the book will help you find these line items.

## 4 Multiply

Multiply the Total Incl O&P cost for a Unit Price line item in the book by your quantity for that item. The price you calculate will be an estimate for a completed item of work performed by a subcontractor. Keep adding line items in this manner to build your estimate.

## 5 Project Overhead

Include project overhead items in your estimate. These items are needed to make the job run and are typically, but not always, provided by the General Contractor. They can be found in Division 1. An alternate method of estimating project overhead costs is to apply a percentage of the total project cost.

Include rented tools not included in crews, waste, rubbish handling, and cleanup.

## 6 Estimate Summary

Include General Contractor's markup on subcontractors, General Contractor's office overhead and profit, and sales tax on materials and equipment.

Adjust your estimate to the project's location by using the City Cost Indexes or Location Factors found in the Reference Section.

---

*Editors' Note: We urge you to spend time reading and understanding the supporting material in the front of this book. An accurate estimate requires experience, knowledge, and careful calculation. The more you know about how we at RSMeans developed the data, the more accurate your estimate will be. In addition, it is important to take into consideration the reference material in the back of the book such as Equipment Listings, Crew Listings, City Cost Indexes, Location Factors, and Reference Numbers.*

# Estimating with RSMeans Unit Price Cost Data

Following these steps will allow you to complete an accurate estimate using RSMeans Unit Price cost data.

## 1 Scope Out the Project

- Identify the individual work tasks that will need to be covered in your estimate.
- The Unit Price data in this book has been divided into 50 Divisions according to the CSI MasterFormat 2010—their titles are listed on the back cover of your book.
- Think through the project and identify those CSI Divisions needed in your estimate.
- The Unit Price Section Table of Contents on page 1 may also be helpful when scoping out your project.
- Experienced estimators find it helpful to begin with Division 2 and continue through completion. Division 1 can be estimated after the full project scope is known.

## 2 Quantify

- Determine the number of units required for each work task that you identified.
- Experienced estimators include an allowance for waste in their quantities. (Waste is not included in RSMeans Unit Price line items unless so stated.)

## 3 Price the Quantities

- Use the Unit Price Table of Contents, and the Index, to locate individual Unit Price line items for your estimate.
- Reference Numbers indicated within a Unit Price section refer to additional information that you may find useful.
- The crew indicates who is performing the work for that task. Crew codes are expanded in the Crew Listings in the Reference Section to include all trades and equipment that comprise the crew.
- The Daily Output is the amount of work the crew is expected to do in one day.
- The Labor-Hours value is the amount of time it will take for the crew to install one unit of work.

- The abbreviated Unit designation indicates the unit of measure upon which the crew, productivity, and prices are based.
- Bare Costs are shown for materials, labor, and equipment needed to complete the Unit Price line item. Bare costs do not include waste, project overhead, payroll insurance, payroll taxes, main office overhead, or profit.
- The Total Incl O&P cost is the billing rate or invoice amount of the installing contractor or subcontractor who performs the work for the Unit Price line item.

## 4 Multiply

- Multiply the total number of units needed for your project by the Total Incl O&P cost for each Unit Price line item.
- Be careful that your take off unit of measure matches the unit of measure in the Unit column.
- The price you calculate is an estimate for a completed item of work.
- Keep scoping individual tasks, determining the number of units required for those tasks, matching each task with individual Unit Price line items in the book, and multiply quantities by Total Incl O&P costs.
- An estimate completed in this manner is priced as if a subcontractor, or set of subcontractors, is performing the work. The estimate does not yet include Project Overhead or Estimate Summary components such as General Contractor markups on subcontracted work, General Contractor office overhead and profit, contingency, and location factor.

## 5 Project Overhead

- Include project overhead items from Division 1 – General Requirements.
- These items are needed to make the job run. They are typically, but not always, provided by the General Contractor. Items include, but are not limited to, field personnel, insurance, performance bond, permits, testing, temporary utilities, field office and storage facilities, temporary scaffolding and platforms, equipment mobilization and demobilization, temporary roads and sidewalks, winter protection, temporary barricades and fencing, temporary security, temporary signs, field engineering and layout, final cleaning and commissioning.
- Each item should be quantified, and matched to individual Unit Price line items in Division 1, and then priced and added to your estimate.
- An alternate method of estimating project overhead costs is to apply a percentage of the total project cost, usually 5% to 15% with an average of 10% (see General Conditions, page ix).
- Include other project related expenses in your estimate such as:
  - Rented equipment not itemized in the Crew Listings
  - Rubbish handling throughout the project (see 02 41 19.23)

# 6 Estimate Summary

- Include sales tax as required by laws of your state or country.
- Include the General Contractor's markup on self-performed work, usually 5% to 15% with an average of 10%.
- Include the General Contractor's markup on subcontracted work, usually 5% to 15% with an average of 10%.
- Include General Contractor's main office overhead and profit.
- RSMeans gives general guidelines on the General Contractor's main office overhead (see section 01 31 13.50 and Reference Number R013113-50).
- RSMeans gives no guidance on the General Contractor's profit.
- Markups will depend on the size of the General Contractor's operations, his projected annual revenue, the level of risk he is taking on, and on the level of competition in the local area and for this project in particular.
- Include a contingency, usually 3% to 5%, if appropriate.

- Adjust your estimate to the project's location by using the City Cost Indexes or the Location Factors in the Reference Section.
- Look at the rules on the pages for How to Use the City Cost Indexes to see how to apply the Indexes for your location.
- When the proper Index or Factor has been identified for the project's location, convert it to a multiplier by dividing it by 100, and then multiply that multiplier by your estimated total cost. The original estimated total cost will now be adjusted up or down from the national average to a total that is appropriate for your location.

*Editors' Notes:*

1) *We urge you to spend time reading and understanding the supporting material in the front of this book. An accurate estimate requires experience, knowledge, and careful calculation. The more you know about how we at RSMeans developed the data, the more accurate your estimate will be. In addition, it is important to take into consideration the reference material in the back of the book such as Equipment Listings, Crew Listings, City Cost Indexes, Location Factors, and Reference Numbers.*

2) *Contractors who are bidding or are involved in JOC, DOC, SABER, or IDIQ type contracts are cautioned that workers' compensation insurance, federal and state payroll taxes, waste, project supervision, project overhead, main office overhead, and profit are not included in bare costs. The coefficient or multiplier must cover these costs.*

# How to Use the Book: The Details

## What's Behind the Numbers? The Development of Cost Data

The staff at RSMeans continually monitors developments in the construction industry in order to ensure reliable, thorough, and up-to-date cost information. While overall construction costs may vary relative to general economic conditions, price fluctuations within the industry are dependent upon many factors. Individual price variations may, in fact, be opposite to overall economic trends. Therefore, costs are constantly tracked and complete updates are published yearly. Also, new items are frequently added in response to changes in materials and methods.

## Costs—$ (U.S.)

All costs represent U.S. national averages and are given in U.S. dollars. The RSMeans City Cost Indexes can be used to adjust costs to a particular location. The City Cost Indexes for Canada can be used to adjust U.S. national averages to local costs in Canadian dollars. No exchange rate conversion is necessary.

**G** The processes or products identified by the green symbol in our publications have been determined to be environmentally responsible and/or resource-efficient solely by the RSMeans engineering staff. The inclusion of the green symbol does not represent compliance with any specific industry association or standard.

## *Material Costs*

The RSMeans staff contacts manufacturers, dealers, distributors, and contractors all across the U.S. and Canada to determine national average material costs. If you have access to current material costs for your specific location, you may wish to make adjustments to reflect differences from the national average. Included within material costs are fasteners for a normal installation. RSMeans engineers use manufacturers' recommendations, written specifications, and/or standard construction practice for size and spacing of fasteners. Adjustments to material costs may be required for your specific application or location. Material costs do not include sales tax.

## *Labor Costs*

Labor costs are based on the average of wage rates from 30 major U.S. cities. Rates are determined from labor union agreements or prevailing wages for construction trades for the current year. Rates, along with overhead and profit markups, are listed on the inside back cover of this book.

- If wage rates in your area vary from those used in this book, or if rate increases are expected within a given year, labor costs should be adjusted accordingly.

Labor costs reflect productivity based on actual working conditions. In addition to actual installation, these figures include time spent during a normal weekday on tasks such as, material receiving and handling, mobilization at site, site movement, breaks, and cleanup. Productivity data is developed over an extended period so as not to be influenced by abnormal variations and reflects a typical average.

## *Equipment Costs*

Equipment costs include not only rental, but also operating costs for equipment under normal use. The operating costs include parts and labor for routine servicing such as repair and replacement of pumps, filters, and worn lines. Normal operating expendables, such as fuel, lubricants, tires, and electricity (where applicable), are also included. Extraordinary operating expendables with highly variable wear patterns, such as diamond bits and blades, are excluded. These costs are included under materials. Equipment rental rates are obtained from industry sources throughout North America—contractors, suppliers, dealers, manufacturers, and distributors.

Rental rates can also be treated as reimbursement costs for contractor-owned equipment. Owned equipment costs include depreciation, loan payments, interest, taxes, insurance, storage and major repairs.

Equipment costs do not include operators' wages; nor do they include the cost to move equipment to a job site (mobilization) or from a job site (demobilization).

**Equipment Cost/Day**——The cost of power equipment required for each crew is included in the Crew Listings in the Reference Section (small tools that are considered as essential everyday tools are not listed out separately). The Crew Listings itemize specialized tools and heavy equipment along with labor trades. The daily cost of itemized equipment included in a crew is based on dividing the weekly bare rental rate by 5 (number of working days per week) and then adding the hourly operating cost times 8 (the number of hours per day). This Equipment Cost/Day is shown in the last column of the Equipment Rental Cost pages in the Reference Section.

**Mobilization/Demobilization**——The cost to move construction equipment from an equipment yard or rental company to the job site and back again is not included in equipment costs. Mobilization (to the site) and demobilization (from the site) costs can be found in the Unit Price Section. If a piece of equipment is already at the job site, it is not appropriate to utilize mob./demob. costs again in an estimate.

## *Overhead and Profit*

Total Cost including O&P for the *Installing Contractor* is shown in the last column on the Unit Price and/or the Assemblies pages of this book. This figure is the sum of the bare material cost plus 10% for profit, the bare labor cost plus total overhead and profit, and the bare equipment cost plus 10% for profit. Details for the calculation of Overhead and Profit on labor are shown on the inside back cover and in the Reference Section of this book. (See the "How to Use the Unit Price Pages" for an example of this calculation.)

## *General Conditions*

Cost data in this book is presented in two ways: Bare Costs and Total Cost including O&P (Overhead and Profit). General Conditions, or General Requirements, of the contract should also be added to the Total Cost including O&P when applicable. Costs for General Conditions are listed in Division 1 of the Unit Price Section and the Reference

Section of this book. General Conditions for the *Installing Contractor* may range from 0% to 10% of the Total Cost including O&P. For the *General* or *Prime Contractor*, costs for General Conditions may range from 5% to 15% of the Total Cost including O&P, with a figure of 10% as the most typical allowance. If applicable, the Assemblies and Models sections of this book use costs that include the installing contractor's overhead and profit (O&P).

## Factors Affecting Costs

Costs can vary depending upon a number of variables. Here's how we have handled the main factors affecting costs.

**Quality**——The prices for materials and the workmanship upon which productivity is based represent sound construction work. They are also in line with U.S. government specifications.

**Overtime**——We have made no allowance for overtime. If you anticipate premium time or work beyond normal working hours, be sure to make an appropriate adjustment to your labor costs.

**Productivity**——The productivity, daily output, and labor-hour figures for each line item are based on working an eight-hour day in daylight hours in moderate temperatures. For work that extends beyond normal work hours or is performed under adverse conditions, productivity may decrease. (See the section in "How To Use the Unit Price Pages" for more on productivity.)

**Size of Project**——The size, scope of work, and type of construction project will have a significant impact on cost. Economies of scale can reduce costs for large projects. Unit costs can often run higher for small projects.

**Location**——Material prices in this book are for metropolitan areas. However, in dense urban areas, traffic and site storage limitations may increase costs. Beyond a 20-mile radius of large cities, extra trucking or transportation charges may also increase the material costs slightly. On the other hand, lower wage rates may be in effect. Be sure to consider both of these factors when preparing an estimate, particularly if the job site is located in a central city or remote rural location. In addition, highly specialized subcontract items may require travel and per-diem expenses for mechanics.

**Other Factors**——

- season of year
- contractor management
- weather conditions
- local union restrictions
- building code requirements
- availability of:
  - adequate energy
  - skilled labor
  - building materials
- owner's special requirements/restrictions
- safety requirements
- environmental considerations

**Unpredictable Factors**——General business conditions influence "in-place" costs of all items. Substitute materials and construction methods may have to be employed. These may affect the installed cost and/or life cycle costs. Such factors may be difficult to evaluate and cannot necessarily be predicted on the basis of the job's location in a particular section of the country. Thus, where these factors apply, you may find significant but unavoidable cost variations for which you will have to apply a measure of judgment to your estimate.

## Rounding of Costs

In general, all unit prices in excess of $5.00 have been rounded to make them easier to use and still maintain adequate precision of the results. The rounding rules we have chosen are in the following table.

| Prices from . . . | Rounded to the nearest . . . |
|---|---|
| $.01 to $5.00 | $.01 |
| $5.01 to $20.00 | $.05 |
| $20.01 to $100.00 | $.50 |
| $100.01 to $300.00 | $1.00 |
| $300.01 to $1,000.00 | $5.00 |
| $1,000.01 to $10,000.00 | $25.00 |
| $10,000.01 to $50,000.00 | $100.00 |
| $50,000.01 and above | $500.00 |

## How Subcontracted Items Affect Costs

A considerable portion of all large construction jobs is usually subcontracted. In fact, the percentage done by subcontractors is constantly increasing and may run over 90%. Since the workers employed by these companies do nothing else but install their particular product, they soon become expert in that line. The result is, installation by these firms is accomplished so efficiently that the total in-place cost, even adding the general contractor's overhead and profit, is no more, and often less, than if the principal contractor had handled the installation himself/herself. Companies that deal with construction specialties are anxious to have their product perform well and, consequently, the installation will be the best possible.

## Contingencies

The allowance for contingencies generally provides for unforeseen construction difficulties. On alterations or repair jobs, 20% is not too much. If drawings are final and only field contingencies are being considered, 2% or 3% is probably sufficient, and often nothing need be added. Contractually, changes in plans will be covered by extras. The contractor should consider inflationary price trends and possible material shortages during the course of the job. These escalation factors are dependent upon both economic conditions and the anticipated time between the estimate and actual construction. If drawings are not complete or approved, or a budget cost is wanted, it is wise to add 5% to 10%. Contingencies, then, are a matter of judgment. Additional allowances for contingencies are shown in Division 1.

## Important Estimating Considerations

The "productivity," or daily output, of each craftsman includes mobilization and cleanup time, break time and plan layout time, as well as an allowance to carry stock from the storage trailer or location on the job site up to 200' into the building and to the first or second floor. If material has to be transported over greater distances or to higher floors, an additional allowance should be considered by the estimator. An allowance has also been included in the piping and fittings installation time for leak checking and minor tightening. Equipment installation time includes the following applicable items: positioning, leveling and securing the unit in place, connecting all associated piping, ducts, vents, etc., which shall have been estimated separately, connecting to an adjacent power source, filling/bleeding, startup, adjusting the controls up and down to ensure proper response, setting the integral controls/valves/regulators/thermostats for proper operation (does not include external building type control systems, DDC systems, etc.), explaining/training owner's operator, and

warranty. A reasonable breakdown of the labor costs is as follows:

| | | |
|---|---|---|
| 1. Movement into building, installation/ setting of equipment | 35% |
| 2. Connection to piping/duct/power, etc. | 25% |
| 3. Filling/flushing/cleaning/touchup, etc. | 15% |
| 4. Startup/running adjustments | 5% |
| 5. Training owner's representative | 5% |
| 6. Warranty/call back/service | 15% |

Note that cranes or other lifting equipment are not included on any lines in the Mechanical divisions. For example, if a crane is required to lift a heavy piece of pipe into place high above a gym floor, or to put a rooftop unit on the roof of a four-story building, etc., it must be added. Due to the potential for extreme variation—from nothing additional required to a major crane or helicopter—we feel that including a nominal amount for "lifting contingency" would be useless and detract from the accuracy of the estimate. When using equipment rental from RSMeans, remember to include the cost of the operator(s).

## Estimating Labor-Hours

The labor-hours expressed in this publication are based on Average Installation time, using an efficiency level of approximately 60%–65% (see item 7), which has been found reasonable and acceptable by many contractors. The book uses this national efficiency average to establish a consistent benchmark. For bid situations, adjustments to this efficiency level should be the responsibility of the contractor bidding the project. The unit labor-hour is divided in the following manner. A typical day for a journeyman might be:

| | | | |
|---|---|---|---|
| 1. | Study Plans | 3% | 14.4 min. |
| 2. | Material Procurement | 3% | 14.4 min. |
| 3. | Receiving and Storing | 3% | 14.4 min |
| 4. | Mobilization | 5% | 24.0 min. |
| 5. | Site Movement | 5% | 24.0 min. |
| 6. | Layout and Marking | 8% | 38.4 min. |
| 7. | Actual Installation | 64% | 307.2 min. |
| 8. | Cleanup | 3% | 14.4 min. |
| 9. | Breaks, Nonproductive | 6% | 28.8 min. |
| | | 100% | 480.0 min. |

If any of the percentages expressed in this breakdown do not apply to the particular work or project situation, then that percentage or a portion of it may be deducted from or added to labor-hours.

## Final Checklist

Estimating can be a straightforward process provided you remember the basics. Here's a checklist of some of the steps you should remember to complete before finalizing your estimate. Did you remember to . . .

- factor in the City Cost Index for your locale?
- take into consideration which items have been marked up and by how much?
- mark up the entire estimate sufficiently for your purposes?
- read the background information on techniques and technical matters that could impact your project time span and cost?
- include all components of your project in the final estimate?
- double check your figures for accuracy?
- call RSMeans if you have any questions about your estimate or the data you've found in our publications? Remember, RSMeans stands behind its publications. If you have any questions about your estimate . . . about the costs you've used from our books . . . or even about the technical aspects of the job that may affect your estimate, feel free to call the RSMeans editors at 1-800-334-3509.

# Unit Price Section

## Table of Contents

1

# Table of Contents (cont.)

# Table of Contents (cont.)

# How to Use the Unit Price Pages

The following is a detailed explanation of a sample entry in the Unit Price Section. Next to each bold number below is the described item with the appropriate component of the sample entry following in parentheses. Some prices are listed as bare costs; others as costs that include overhead and profit of the installing contractor. In most cases, if the work is to be subcontracted, the general contractor will need to add an additional markup (RSMeans suggests using 10%) to the figures in the column "Total Incl. O&P."

## 1 Section Number/Title (03 30/Cast-In-Place Concrete)

Use the Unit Price Section Table of Contents to locate specific Level 2 Sections. The sections are classified according to the CSI MasterFormat 2004 system.

## 2 Line Numbers (03 30 53.40 3920)

Each unit price line item has been assigned a unique 12-digit code based on the CSI MasterFormat classification.

```
┌──────── MasterFormat Level 1 (Division 03)
│ ┌────── MasterFormat Level 2 (03 30)
│ │ ┌──── MasterFormat Level 3 (03 30 53)
│ │ │
03 30 53.40   3920
      │         │
      └──── MasterFormat Level 4 (03 30 53.40)
                │
                └──── RSMeans 12-Digit Line Number
```

## 3 Description (Concrete In Place, etc.)

Each line item is described in detail. Sub-items and additional sizes are indented beneath the appropriate line items. The first line or two after the main item (in boldface) may contain descriptive information that pertains to all line items beneath this boldface listing.

## 4 Reference Number Information

**R033105 -20** You'll see reference numbers shown in shaded boxes at the beginning of some sections. These refer to related items in the Reference Section, visually identified by a vertical gray bar on the page edges.

The relation may be an estimating procedure that should be read before estimating, or technical information.

The "R" designates the Reference Section. The numbers refer to the MasterFormat 2004 classification system.

**It is strongly recommended that you review all reference numbers that appear within the section in which you are working.**

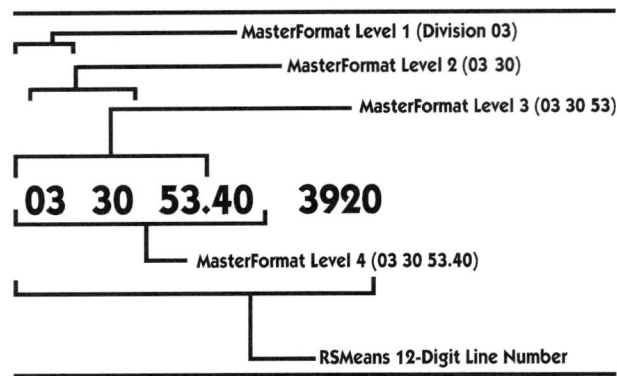

## 03 30 Cast-In-Place Concrete

### 03 30 53 – Miscellaneous Cast-In-Place Concrete

| 03 30 53.40 Concrete In Place | | Crew | Daily Output | Labor-Hours | Unit | Material | 2011 Bare Costs Labor | Equipment | Total | Total Incl O&P |
|---|---|---|---|---|---|---|---|---|---|---|
| 0010 | **CONCRETE IN PLACE** | R033105-10 | | | | | | | | |
| 0020 | Including forms (4 uses), reinforcing steel, concrete, placement, | R033105-20 | | | | | | | | |
| 0050 | and finishing unless otherwise indicated | R033105-50 | | | | | | | | |
| 0300 | Beams (3500 psi), 5 kip per L.F., 10' span | R033105-65 | C-14A | 15.62 | 12.804 | C.Y. | 291 | 555 | 49 | 895 | 1,225 |
| 3825 | 1 C.Y. to 5 C.Y. | | | 43 | 2.605 | | 185 | 108 | .55 | 293.55 | 370 |
| 3850 | Over 5 C.Y. | | | 75 | 1.493 | | 171 | 61.50 | .31 | 232.81 | 283 |
| 3900 | Footings, strip (3000 psi), 18" x 9", unreinforced | | C-14L | 40 | 2.400 | | 119 | 96.50 | .58 | 216.08 | 279 |
| 3920 | 18" x 9", reinforced | | C-14C | 35 | 3.200 | | 141 | 132 | .67 | 273.67 | 360 |
| 3925 | 20" x 10", unreinforced | | C-14L | 45 | 2.133 | | 116 | 85.50 | .51 | 202.01 | 260 |
| 3930 | 20" x 10", reinforced | | C-14C | 40 | 2.800 | | 134 | 116 | .59 | 250.59 | 325 |
| 3935 | 24" x 12", unreinforced | | C-14L | 55 | 1.745 | | 114 | 70 | .42 | 184.42 | 233 |
| 3940 | 24" x 12", reinforced | | C-14C | 48 | 2.333 | | 132 | 96.50 | .49 | 228.99 | 294 |

## Crew (C-14C)

The "Crew" column designates the typical trade or crew used to install the item. If an installation can be accomplished by one trade and requires no power equipment, that trade and the number of workers are listed (for example, "2 Carpenters"). If an installation requires a composite crew, a crew code designation is listed (for example, "C-14C"). You'll find full details on all composite crews in the Crew Listings.

- For a complete list of all trades utilized in this book and their abbreviations, see the inside back cover.

### Crews

| Crew No. | Bare Costs | | Incl. Subs O&P | | Cost Per Labor-Hour | |
|---|---|---|---|---|---|---|
| Crew C-14C | Hr. | Daily | Hr. | Daily | Bare Costs | Incl. O&P |
| 1 Carpenter Foreman (out) | $45.05 | $360.40 | $69.10 | $552.80 | $41.31 | $63.60 |
| 6 Carpenters | 43.05 | 2066.40 | 66.05 | 3170.40 | | |
| 2 Rodmen (reinf.) | 48.40 | 774.40 | 77.35 | 1237.60 | | |
| 4 Laborers | 34.35 | 1099.20 | 52.70 | 1686.40 | | |
| 1 Cement Finisher | 40.85 | 326.80 | 59.50 | 476.00 | | |
| 1 Gas Engine Vibrator | | 23.20 | | 25.52 | .21 | .23 |
| 112 L.H., Daily Totals | | $4650.40 | | $7148.72 | $41.52 | $63.83 |

## Productivity: Daily Output (35.0)/ Labor-Hours (3.20)

The "Daily Output" represents the typical number of units the designated crew will install in a normal 8-hour day. To find out the number of days the given crew would require to complete the installation, divide your quantity by the daily output. For example:

| Quantity | ÷ | Daily Output | = | Duration |
|---|---|---|---|---|
| 100 C.Y. | ÷ | 35.0/ Crew Day | = | 2.86 Crew Days |

The "Labor-Hours" figure represents the number of labor-hours required to install one unit of work. To find out the number of labor-hours required for your particular task, multiply the quantity of the item times the number of labor-hours shown. For example:

| Quantity | x | Productivity Rate | = | Duration |
|---|---|---|---|---|
| 100 C.Y. | x | 3.20 Labor-Hours/ C.Y. | = | 320 Labor-Hours |

## Unit (C.Y.)

The abbreviated designation indicates the unit of measure upon which the price, production, and crew are based (C.Y. = Cubic Yard). For a complete listing of abbreviations, refer to the Abbreviations Listing in the Reference Section of this book.

## Bare Costs:
### Mat. (Bare Material Cost) (141)

The unit material cost is the "bare" material cost with no overhead or profit included. Costs shown reflect national average material prices for January of the current year and include delivery to the job site. No sales taxes are included.

### Labor (132)

The unit labor cost is derived by multiplying bare labor-hour costs for Crew C-14C by labor-hour units. The bare labor-hour cost is found in the Crew Section under C-14C. (If a trade is listed, the hourly labor cost—the wage rate—is found on the inside back cover.)

| Labor-Hour Cost Crew C-14C | x | Labor-Hour Units | = | Labor |
|---|---|---|---|---|
| $41.31 | x | 3.20 | = | $132.00 |

### Equip. (Equipment) (.67)

Costs for equipment included in each crew are listed in the Crew Listings in the Reference Section. Costs for small tools that are considered as essential everyday tools are covered in the "Overhead" column on the "Installing Contractor's Overhead and Profit" table that lists labor trades, base rates and markups. The unit equipment cost displayed here is derived by multiplying the bare equipment hourly cost in the Crew Listing by the unit Labor-Hours value.

| Equipment Cost Crew C-14C | x | Labor-Hour Units | = | Equip. |
|---|---|---|---|---|
| $.21 | x | 3.20 | = | $.67 |

### Total (273.67)

The total of the bare costs is the arithmetic total of the three previous columns: mat., labor, and equip.

| Material | + | Labor | + | Equip. | = | Total |
|---|---|---|---|---|---|---|
| $141 | + | $132 | + | $.67 | = | $273.67 |

## Total Costs Including O&P

This figure is the sum of the bare material cost plus 10% for profit; the bare labor cost plus total overhead and profit (per the inside back cover or, if a crew is listed, from the crew listings); and the bare equipment cost plus 10% for profit.

| Material is Bare Material Cost + 10% = 141 + 14.1 | = | $155.10 |
|---|---|---|
| Labor for Crew C-14C = Labor-Hour Cost (63.60) x Labor-Hour Units (3.20) | = | $203.50 |
| Equip. is Bare Equip. Cost + 10% = .67 + .07 | = | $ .74 |
| Total (Rounded) | = | $ 360 |

# Estimating Tips

## 01 20 00 Price and Payment Procedures

- Allowances that should be added to estimates to cover contingencies and job conditions that are not included in the national average material and labor costs are shown in section 01 21.

- When estimating historic preservation projects (depending on the condition of the existing structure and the owner's requirements), a 15%–20% contingency or allowance is recommended, regardless of the stage of the drawings.

## 01 30 00 Administrative Requirements

- Before determining a final cost estimate, it is a good practice to review all the items listed in Subdivisions 01 31 and 01 32 to make final adjustments for items that may need customizing to specific job conditions.

- Requirements for initial and periodic submittals can represent a significant cost to the General Requirements of a job. Thoroughly check the submittal specifications when estimating a project to determine any costs that should be included.

## 01 40 00 Quality Requirements

- All projects will require some degree of Quality Control. This cost is not included in the unit cost of construction listed in each division. Depending upon the terms of the contract, the various costs of inspection and testing can be the responsibility of either the owner or the contractor. Be sure to include the required costs in your estimate.

## 01 50 00 Temporary Facilities and Controls

- Barricades, access roads, safety nets, scaffolding, security, and many more requirements for the execution of a safe project are elements of direct cost. These costs can easily be overlooked when preparing an estimate. When looking through the major classifications of this subdivision, determine which items apply to each division in your estimate.

- Construction Equipment Rental Costs can be found in the Reference Section in section 01 54 33. Operators' wages are not included in equipment rental costs.

- Equipment mobilization and demobilization costs are not included in equipment rental costs and must be considered separately in section 01 54 36.50.

- The cost of small tools provided by the installing contractor for his workers is covered in the "Overhead" column on the "Installing Contractor's Overhead and Profit" table that lists labor trades, base rates and markups and, therefore, is included in the "Total Incl. O&P" cost of any Unit Price line item. For those users who are constrained by contract terms to use only bare costs, there are two line items in section 01 54 39.70 for small tools as a percentage of bare labor cost. If some of those users are further constrained by contract terms to refrain from using line items from Division 1, you are advised to cover the cost of small tools within your coefficient or multiplier.

## 01 70 00 Execution and Closeout Requirements

- When preparing an estimate, thoroughly read the specifications to determine the requirements for Contract Closeout. Final cleaning, record documentation, operation and maintenance data, warranties and bonds, and spare parts and maintenance materials can all be elements of cost for the completion of a contract. Do not overlook these in your estimate.

# Reference Numbers

Reference numbers are shown in shaded boxes at the beginning of some major classifications. These numbers refer to related items in the Reference Section. The reference information may be an estimating procedure, an alternate pricing method, or technical information.

*Note:* Not all subdivisions listed here necessarily appear in this publication.

# 01 11 Summary of Work

## 01 11 31 – Professional Consultants

### 01 11 31.10 Architectural Fees

| | | | Crew | Daily Output | Labor-Hours | Unit | Material | 2011 Bare Costs Labor | Equipment | Total | Total Incl O&P |
|---|---|---|---|---|---|---|---|---|---|---|---|
| 0010 | **ARCHITECTURAL FEES** | R011110-10 | | | | | | | | | |
| 0020 | For new construction | | | | | | | | | | |
| 0060 | Minimum | | | | | Project | | | | | 4.90% |
| 0090 | Maximum | | | | | | | | | | 16% |
| 0100 | For alteration work, to $500,000, add to new construction fee | | | | | | | | | | 50% |
| 0150 | Over $500,000, add to new construction fee | | | | | | | | | | 25% |
| 2000 | For "Greening" of building | G | | | | | | | | | 3% |

### 01 11 31.20 Construction Management Fees

| | | | | | | | | | | | |
|---|---|---|---|---|---|---|---|---|---|---|---|
| 0010 | **CONSTRUCTION MANAGEMENT FEES** | | | | | | | | | | |
| 0020 | $1,000,000 job, minimum | | | | | Project | | | | | 4.50% |
| 0050 | Maximum | | | | | | | | | | 7.50% |
| 0300 | $50,000,000 job, minimum | | | | | | | | | | 2.50% |
| 0350 | Maximum | | | | | | | | | | 4% |

### 01 11 31.30 Engineering Fees

| | | | | | | | | | | | |
|---|---|---|---|---|---|---|---|---|---|---|---|
| 0010 | **ENGINEERING FEES** | R011110-30 | | | | | | | | | |
| 0020 | Educational planning consultant, minimum | | | | | Project | | | | | .50% |
| 0100 | Maximum | | | | | " | | | | | 2.50% |
| 0200 | Electrical, minimum | | | | | Contrct | | | | | 4.10% |
| 0300 | Maximum | | | | | | | | | | 10.10% |
| 0400 | Elevator & conveying systems, minimum | | | | | | | | | | 2.50% |
| 0500 | Maximum | | | | | | | | | | 5% |
| 0600 | Food service & kitchen equipment, minimum | | | | | | | | | | 8% |
| 0700 | Maximum | | | | | | | | | | 12% |
| 0800 | Landscaping & site development, minimum | | | | | | | | | | 2.50% |
| 0900 | Maximum | | | | | | | | | | 6% |
| 1000 | Mechanical (plumbing & HVAC), minimum | | | | | | | | | | 4.10% |
| 1100 | Maximum | | | | | | | | | | 10.10% |
| 1200 | Structural, minimum | | | | | Project | | | | | 1% |
| 1300 | Maximum | | | | | " | | | | | 2.50% |

### 01 11 31.50 Models

| | | | | | | | | | | | |
|---|---|---|---|---|---|---|---|---|---|---|---|
| 0010 | **MODELS** | | | | | | | | | | |
| 0020 | Cardboard & paper, 1 building, minimum | | | | | Ea. | 710 | | | 710 | 780 |
| 0050 | Maximum | | | | | | 1,625 | | | 1,625 | 1,775 |
| 0100 | 2 buildings, minimum | | | | | | 945 | | | 945 | 1,050 |
| 0150 | Maximum | | | | | | 2,150 | | | 2,150 | 2,350 |
| 0200 | Plexiglass and metal, basic layout | | | | | SF Flr. | .06 | | | .06 | .07 |
| 0210 | Including equipment and personnel | | | | | " | .31 | | | .31 | .34 |
| 0300 | Site plan layout, minimum | | | | | Ea. | 1,350 | | | 1,350 | 1,500 |
| 0350 | Maximum | | | | | " | 2,275 | | | 2,275 | 2,500 |

### 01 11 31.75 Renderings

| | | | | | | | | | | | |
|---|---|---|---|---|---|---|---|---|---|---|---|
| 0010 | **RENDERINGS** Color, matted, 20" x 30", eye level, | | | | | | | | | | |
| 0020 | 1 building, minimum | | | | | Ea. | 1,975 | | | 1,975 | 2,175 |
| 0050 | Average | | | | | | 2,800 | | | 2,800 | 3,100 |
| 0100 | Maximum | | | | | | 4,500 | | | 4,500 | 4,950 |
| 1000 | 5 buildings, minimum | | | | | | 3,950 | | | 3,950 | 4,325 |
| 1100 | Maximum | | | | | | 7,875 | | | 7,875 | 8,675 |
| 2000 | Aerial perspective, color, 1 building, minimum | | | | | | 2,800 | | | 2,800 | 3,100 |
| 2100 | Maximum | | | | | | 7,875 | | | 7,875 | 8,675 |
| 3000 | 5 buildings, minimum | | | | | | 5,675 | | | 5,675 | 6,225 |
| 3100 | Maximum | | | | | | 11,200 | | | 11,200 | 12,400 |

# 01 21 Allowances

## 01 21 16 – Contingency Allowances

| 01 21 16.50 Contingencies | Crew | Daily Output | Labor-Hours | Unit | Material | 2011 Bare Costs Labor | Equipment | Total | Total Incl O&P |
|---|---|---|---|---|---|---|---|---|---|
| 0010 | **CONTINGENCIES**, Add to estimate | | | | | | | | | |
| 0020 | Conceptual stage | | | | Project | | | | | 20% |
| 0050 | Schematic stage | | | | | | | | | 15% |
| 0100 | Preliminary working drawing stage (Design Dev.) | | | | | | | | | 10% |
| 0150 | Final working drawing stage | | | | | | | | | 3% |

## 01 21 55 – Job Conditions Allowance

### 01 21 55.50 Job Conditions

| | | Crew | Daily Output | Labor-Hours | Unit | Material | 2011 Bare Costs Labor | Equipment | Total | Total Incl O&P |
|---|---|---|---|---|---|---|---|---|---|---|
| 0010 | **JOB CONDITIONS** Modifications to applicable | | | | | | | | | |
| 0020 | cost summaries | | | | | | | | | |
| 0100 | Economic conditions, favorable, deduct | | | | Project | | | | 2% | 2% |
| 0200 | Unfavorable, add | | | | | | | | 5% | 5% |
| 0300 | Hoisting conditions, favorable, deduct | | | | | | | | 2% | 2% |
| 0400 | Unfavorable, add | | | | | | | | 5% | 5% |
| 0500 | General Contractor management, experienced, deduct | | | | | | | | 2% | 2% |
| 0600 | Inexperienced, add | | | | | | | | 10% | 10% |
| 0700 | Labor availability, surplus, deduct | | | | | | | | 1% | 1% |
| 0800 | Shortage, add | | | | | | | | 10% | 10% |
| 0900 | Material storage area, available, deduct | | | | | | | | 1% | 1% |
| 1000 | Not available, add | | | | | | | | 2% | 2% |
| 1100 | Subcontractor availability, surplus, deduct | | | | | | | | 5% | 5% |
| 1200 | Shortage, add | | | | | | | | 12% | 12% |
| 1300 | Work space, available, deduct | | | | | | | | 2% | 2% |
| 1400 | Not available, add | | | | | | | | 5% | 5% |

## 01 21 57 – Overtime Allowance

### 01 21 57.50 Overtime

| | | Crew | Daily Output | Labor-Hours | Unit | Material | 2011 Bare Costs Labor | Equipment | Total | Total Incl O&P |
|---|---|---|---|---|---|---|---|---|---|---|
| 0010 | **OVERTIME** for early completion of projects or where | R012909-90 | | | | | | | | |
| 0020 | labor shortages exist, add to usual labor, up to | | | | Costs | | 100% | | | |

## 01 21 61 – Cost Indexes

### 01 21 61.10 Construction Cost Index

| | | Crew | Daily Output | Labor-Hours | Unit | Material | 2011 Bare Costs Labor | Equipment | Total | Total Incl O&P |
|---|---|---|---|---|---|---|---|---|---|---|
| 0010 | **CONSTRUCTION COST INDEX** (Reference) over 930 zip code locations in | | | | | | | | | |
| 0020 | the U.S. and Canada, total bldg. cost, min. (Guymon, OK) | | | | % | | | | | 67.30% |
| 0050 | Average | | | | | | | | | 100% |
| 0100 | Maximum (New York, NY) | | | | | | | | | 133% |

### 01 21 61.20 Historical Cost Indexes

| | | Crew | Daily Output | Labor-Hours | Unit | Material | 2011 Bare Costs Labor | Equipment | Total | Total Incl O&P |
|---|---|---|---|---|---|---|---|---|---|---|
| 0010 | **HISTORICAL COST INDEXES** (See Reference Section) | | | | | | | | | |

### 01 21 61.30 Labor Index

| | | Crew | Daily Output | Labor-Hours | Unit | Material | 2011 Bare Costs Labor | Equipment | Total | Total Incl O&P |
|---|---|---|---|---|---|---|---|---|---|---|
| 0010 | **LABOR INDEX** (Reference) For over 930 zip code locations in | | | | | | | | | |
| 0020 | the U.S. and Canada, minimum (Guymon, OK) | | | | % | | 31.10% | | | |
| 0050 | Average | | | | | | 100% | | | |
| 0100 | Maximum (New York, NY) | | | | | | 166.30% | | | |

### 01 21 61.50 Material Index

| | | Crew | Daily Output | Labor-Hours | Unit | Material | 2011 Bare Costs Labor | Equipment | Total | Total Incl O&P |
|---|---|---|---|---|---|---|---|---|---|---|
| 0010 | **MATERIAL INDEX** (Reference) For over 930 zip code locations in | | | | | | | | | |
| 0020 | the U.S. and Canada, minimum (Elizabethtown, KY) | | | | % | 91% | | | | |
| 0040 | Average | | | | | 100% | | | | |
| 0060 | Maximum (Ketchikan, AK) | | | | | 136.90% | | | | |

## 01 21 63 – Taxes

### 01 21 63.10 Taxes

| | | Crew | Daily Output | Labor-Hours | Unit | Material | 2011 Bare Costs Labor | Equipment | Total | Total Incl O&P |
|---|---|---|---|---|---|---|---|---|---|---|
| 0010 | **TAXES** | R012909-80 | | | | | | | | |
| 0020 | Sales tax, State, average | | | | % | 5.04% | | | | |
| 0050 | Maximum | R012909-85 | | | | 8.25% | | | | |

# 01 21 Allowances

## 01 21 63 – Taxes

| 01 21 63.10 Taxes | | Crew | Daily Output | Labor-Hours | Unit | Material | 2011 Bare Costs | | Total | Total Incl O&P |
|---|---|---|---|---|---|---|---|---|---|---|
| | | | | | | | Labor | Equipment | | |
| 0200 | Social Security, on first $106,800 of wages | | | | % | | 7.65% | | | |
| 0300 | Unemployment, combined Federal and State, minimum | | | | | | .80% | | | |
| 0350 | Average | | | | | | 6.20% | | | |
| 0400 | Maximum | | | | | | 13.16% | | | |

# 01 31 Project Management and Coordination

## 01 31 13 – Project Coordination

### 01 31 13.20 Field Personnel

| | | Crew | Daily Output | Labor-Hours | Unit | Material | 2011 Bare Costs | | Total | Total Incl O&P |
|---|---|---|---|---|---|---|---|---|---|---|
| | | | | | | | Labor | Equipment | | |
| 0010 | **FIELD PERSONNEL** | | | | | | | | | |
| 0020 | Clerk, average | | | | Week | | 410 | | 410 | 630 |
| 0100 | Field engineer, minimum | | | | | | 975 | | 975 | 1,500 |
| 0120 | Average | | | | | | 1,265 | | 1,265 | 1,950 |
| 0140 | Maximum | | | | | | 1,450 | | 1,450 | 2,225 |
| 0160 | General purpose laborer, average | | | | | | 1,350 | | 1,350 | 2,075 |
| 0180 | Project manager, minimum | | | | | | 1,800 | | 1,800 | 2,750 |
| 0200 | Average | | | | | | 2,075 | | 2,075 | 3,175 |
| 0220 | Maximum | | | | | | 2,375 | | 2,375 | 3,650 |
| 0240 | Superintendent, minimum | | | | | | 1,750 | | 1,750 | 2,675 |
| 0260 | Average | | | | | | 1,925 | | 1,925 | 2,950 |
| 0280 | Maximum | | | | | | 2,200 | | 2,200 | 3,375 |
| 0290 | Timekeeper, average | | | | | | 1,130 | | 1,130 | 1,725 |

### 01 31 13.30 Insurance

| | | Crew | Daily Output | Labor-Hours | Unit | Material | 2011 Bare Costs | | Total | Total Incl O&P |
|---|---|---|---|---|---|---|---|---|---|---|
| | | | | | | | Labor | Equipment | | |
| 0010 | **INSURANCE** R013113-40 | | | | | | | | | |
| 0020 | Builders risk, standard, minimum | | | | Job | | | | | .24% |
| 0050 | Maximum R013113-50 | | | | | | | | | .64% |
| 0200 | All-risk type, minimum | | | | | | | | | .25% |
| 0250 | Maximum R013113-60 | | | | | | | | | .62% |
| 0400 | Contractor's equipment floater, minimum | | | | Value | | | | | .50% |
| 0450 | Maximum | | | | " | | | | | 1.50% |
| 0600 | Public liability, average | | | | Job | | | | | 2.02% |
| 0800 | Workers' compensation & employer's liability, average | | | | | | | | | |
| 0850 | by trade, carpentry, general | | | | Payroll | | 16.07% | | | |
| 0900 | Clerical | | | | | | .51% | | | |
| 0950 | Concrete | | | | | | 13.33% | | | |
| 1000 | Electrical | | | | | | 5.81% | | | |
| 1050 | Excavation | | | | | | 9.10% | | | |
| 1100 | Glazing | | | | | | 12.89% | | | |
| 1150 | Insulation | | | | | | 12.30% | | | |
| 1200 | Lathing | | | | | | 8.59% | | | |
| 1250 | Masonry | | | | | | 12.66% | | | |
| 1300 | Painting & decorating | | | | | | 11.07% | | | |
| 1350 | Pile driving | | | | | | 18% | | | |
| 1400 | Plastering | | | | | | 11.43% | | | |
| 1450 | Plumbing | | | | | | 7.04% | | | |
| 1500 | Roofing | | | | | | 29.20% | | | |
| 1550 | Sheet metal work (HVAC) | | | | | | 8.94% | | | |
| 1600 | Steel erection, structural | | | | | | 36.30% | | | |
| 1650 | Tile work, interior ceramic | | | | | | 8.45% | | | |
| 1700 | Waterproofing, brush or hand caulking | | | | | | 6.43% | | | |
| 1800 | Wrecking | | | | | | 30.85% | | | |
| 2000 | Range of 35 trades in 50 states, excl. wrecking, min. | | | | | | 2.36% | | | |

# 01 31 Project Management and Coordination

## 01 31 13 – Project Coordination

| 01 31 13.30 Insurance | Crew | Daily Output | Labor-Hours | Unit | Material | 2011 Bare Costs Labor | Equipment | Total | Total Incl O&P |
|---|---|---|---|---|---|---|---|---|---|
| 2100 Average | | | | Payroll | | 14.10% | | | |
| 2200 Maximum | | | | ↓ | | 155.90% | | | |

| 01 31 13.40 Main Office Expense | Crew | Daily Output | Labor-Hours | Unit | Material | 2011 Bare Costs Labor | Equipment | Total | Total Incl O&P |
|---|---|---|---|---|---|---|---|---|---|
| 0010 **MAIN OFFICE EXPENSE** Average for General Contractors    R013113-50 | | | | | | | | | |
| 0020     As a percentage of their annual volume | | | | | | | | | |
| 0125 Annual volume under 1 million dollars | | | | % Vol. | | | | 17.50% | |
| 0145     Up to 2.5 million dollars | | | | | | | | 8% | |
| 0150     Up to 4.0 million dollars | | | | | | | | 6.80% | |
| 0200     Up to 7.0 million dollars | | | | | | | | 5.60% | |
| 0250     Up to 10 million dollars | | | | | | | | 5.10% | |
| 0300     Over 10 million dollars | | | | ↓ | | | | 3.90% | |

| 01 31 13.50 General Contractor's Mark-Up | Crew | Daily Output | Labor-Hours | Unit | Material | 2011 Bare Costs Labor | Equipment | Total | Total Incl O&P |
|---|---|---|---|---|---|---|---|---|---|
| 0010 **GENERAL CONTRACTOR'S MARK-UP** on Change Orders | | | | | | | | | |
| 0200 Extra work, by subcontractors, add | | | | % | | | | | 10% |
| 0250     By General Contractor, add | | | | | | | | | 15% |
| 0400 Omitted work, by subcontractors, deduct all but | | | | | | | | | 5% |
| 0450     By General Contractor, deduct all but | | | | | | | | | 7.50% |
| 0600 Overtime work, by subcontractors, add | | | | | | | | | 15% |
| 0650     By General Contractor, add | | | | ↓ | | | | | 10% |

| 01 31 13.60 Installing Contractor's Main Office Overhead | Crew | Daily Output | Labor-Hours | Unit | Material | 2011 Bare Costs Labor | Equipment | Total | Total Incl O&P |
|---|---|---|---|---|---|---|---|---|---|
| 0010 **INSTALLING CONTRACTOR'S MAIN OFFICE OVERHEAD**    R013113-50 | | | | | | | | | |
| 0020 As percent of direct costs, minimum | | | | % | | | | | 5% |
| 0050     Average | | | | | | | | | 13% |
| 0100     Maximum | | | | ↓ | | | | | 30% |

| 01 31 13.80 Overhead and Profit | Crew | Daily Output | Labor-Hours | Unit | Material | 2011 Bare Costs Labor | Equipment | Total | Total Incl O&P |
|---|---|---|---|---|---|---|---|---|---|
| 0010 **OVERHEAD & PROFIT** Allowance to add to items in this | | | | | | | | | |
| 0020     book that do not include Subs O&P, average | | | | % | | | | | 25% |
| 0100 Allowance to add to items in this book that | | | | | | | | | |
| 0110     do include Subs O&P, minimum | | | | % | | | | | 5% |
| 0150         Average | | | | | | | | | 10% |
| 0200         Maximum | | | | | | | | | 15% |
| 0300 Typical, by size of project, under $100,000 | | | | | | | | 30% | |
| 0350     $500,000 project | | | | | | | | 25% | |
| 0400     $2,000,000 project | | | | | | | | 20% | |
| 0450     Over $10,000,000 project | | | | ↓ | | | | 15% | |

| 01 31 13.90 Performance Bond | Crew | Daily Output | Labor-Hours | Unit | Material | 2011 Bare Costs Labor | Equipment | Total | Total Incl O&P |
|---|---|---|---|---|---|---|---|---|---|
| 0010 **PERFORMANCE BOND**    R013113-80 | | | | | | | | | |
| 0020 For buildings, minimum | | | | Job | | | | | .60% |
| 0100     Maximum | | | | " | | | | | 2.50% |

# 01 32 Construction Progress Documentation

## 01 32 13 – Scheduling of work

| 01 32 13.50 Scheduling | Crew | Daily Output | Labor-Hours | Unit | Material | 2011 Bare Costs Labor | 2011 Bare Costs Equipment | Total | Total Incl O&P |
|---|---|---|---|---|---|---|---|---|---|
| 0010 **SCHEDULING** | | | | | | | | | |
| 0020 Critical path, as % of architectural fee, minimum | | | | % | | | | | .50% |
| 0100 Maximum | | | | " | | | | | 1% |
| 0300 Computer-update, micro, no plots, minimum | | | | Ea. | | | | 455 | 500 |
| 0400 Including plots, maximum | | | | " | | | | 1,450 | 1,600 |
| 0600 Rule of thumb, CPM scheduling, small job ($10 Million) | | | | Job | | | | | .05% |
| 0650 Large job ($50 Million +) | | | | | | | | | .03% |
| 0700 Including cost control, small job | | | | | | | | | .08% |
| 0750 Large job | | | | | | | | | .04% |

## 01 32 33 – Photographic Documentation

| 01 32 33.50 Photographs | Crew | Daily Output | Labor-Hours | Unit | Material | 2011 Bare Costs Labor | 2011 Bare Costs Equipment | Total | Total Incl O&P |
|---|---|---|---|---|---|---|---|---|---|
| 0010 **PHOTOGRAPHS** | | | | | | | | | |
| 0020 8" x 10", 4 shots, 2 prints ea., std. mounting | | | | Set | 475 | | | 475 | 520 |
| 0100 Hinged linen mounts | | | | | 530 | | | 530 | 580 |
| 0200 8" x 10", 4 shots, 2 prints each, in color | | | | | 415 | | | 415 | 460 |
| 0300 For I.D. slugs, add to all above | | | | | 5.30 | | | 5.30 | 5.85 |
| 0500 Aerial photos, initial fly-over, 6 shots, 1 print ea., 8" x 10" | | | | | 845 | | | 845 | 925 |
| 0550 11" x 14" prints | | | | | 1,025 | | | 1,025 | 1,125 |
| 0600 16" x 20" prints | | | | | 1,200 | | | 1,200 | 1,325 |
| 0700 For full color prints, add | | | | | 40% | | | | 40% |
| 0750 Add for traffic control area | | | | | 305 | | | 305 | 335 |
| 0900 For over 30 miles from airport, add per | | | | Mile | 5.45 | | | 5.45 | 6 |
| 1000 Vertical photography, 4 to 6 shots with | | | | | | | | | |
| 1010 different scales, 1 print each | | | | Set | 1,125 | | | 1,125 | 1,225 |
| 1500 Time lapse equipment, camera and projector, buy | | | | | 845 | | | 845 | 930 |
| 1550 Rent per month | | | | | 260 | | | 260 | 286 |
| 1700 Cameraman and film, including processing, B.&W. | | | | Day | 1,375 | | | 1,375 | 1,525 |
| 1720 Color | | | | " | 1,375 | | | 1,375 | 1,525 |

# 01 41 Regulatory Requirements

## 01 41 26 – Permit Requirements

| 01 41 26.50 Permits | Crew | Daily Output | Labor-Hours | Unit | Material | 2011 Bare Costs Labor | 2011 Bare Costs Equipment | Total | Total Incl O&P |
|---|---|---|---|---|---|---|---|---|---|
| 0010 **PERMITS** | | | | | | | | | |
| 0020 Rule of thumb, most cities, minimum | | | | Job | | | | | .50% |
| 0100 Maximum | | | | " | | | | | 2% |

# 01 45 Quality Control

## 01 45 23 – Testing and Inspecting Services

| 01 45 23.50 Testing | Crew | Daily Output | Labor-Hours | Unit | Material | 2011 Bare Costs Labor | 2011 Bare Costs Equipment | Total | Total Incl O&P |
|---|---|---|---|---|---|---|---|---|---|
| 0010 **TESTING** and Inspecting Services | | | | | | | | | |
| 0015 For concrete building costing $1,000,000, minimum | | | | Project | | | | 4,725 | 5,200 |
| 0020 Maximum | | | | | | | | 38,000 | 41,800 |
| 0050 Steel building, minimum | | | | | | | | 4,725 | 5,200 |
| 0070 Maximum | | | | | | | | 14,800 | 16,300 |
| 0100 For building costing, $10,000,000, minimum | | | | | | | | 30,100 | 33,100 |
| 0150 Maximum | | | | | | | | 48,200 | 53,000 |
| 0200 Asphalt testing, compressive strength Marshall stability, set of 3 | | | | Ea. | | | | 145 | 165 |
| 0220 Density, set of 3 | | | | | | | | 86 | 95 |
| 0250 Extraction, individual tests on sample | | | | | | | | 136 | 150 |

| 01 45 23.50 Testing | | Crew | Daily Output | Labor-Hours | Unit | Material | 2011 Bare Costs Labor | Equipment | Total | Total Incl O&P |
|---|---|---|---|---|---|---|---|---|---|---|
| 0300 | Penetration | | | | Ea. | | | | 41 | 45 |
| 0350 | Mix design, 5 specimens | | | | | | | | 182 | 200 |
| 0360 | Additional specimen | | | | | | | | 36 | 40 |
| 0400 | Specific gravity | | | | | | | | 41 | 45 |
| 0420 | Swell test | | | | | | | | 64 | 70 |
| 0450 | Water effect and cohesion, set of 6 | | | | | | | | 182 | 200 |
| 0470 | Water effect and plastic flow | | | | | | | | 64 | 70 |
| 0600 | Concrete testing, aggregates, abrasion, ASTM C 131 | | | | | | | | 136 | 150 |
| 0650 | Absorption, ASTM C 127 | | | | | | | | 42 | 46 |
| 0800 | Petrographic analysis, ASTM C 295 | | | | | | | | 775 | 850 |
| 0900 | Specific gravity, ASTM C 127 | | | | | | | | 50 | 55 |
| 1000 | Sieve analysis, washed, ASTM C 136 | | | | | | | | 59 | 65 |
| 1050 | Unwashed | | | | | | | | 59 | 65 |
| 1200 | Sulfate soundness | | | | | | | | 114 | 125 |
| 1300 | Weight per cubic foot | | | | | | | | 36 | 40 |
| 1500 | Cement, physical tests, ASTM C 150 | | | | | | | | 320 | 350 |
| 1600 | Chemical tests, ASTM C 150 | | | | | | | | 245 | 270 |
| 1800 | Compressive test, cylinder, delivered to lab, ASTM C 39 | | | | | | | | 12 | 13 |
| 1900 | Picked up by lab, minimum | | | | | | | | 14 | 15 |
| 1950 | Average | | | | | | | | 18 | 20 |
| 2000 | Maximum | | | | | | | | 27 | 30 |
| 2200 | Compressive strength, cores (not incl. drilling), ASTM C 42 | | | | | | | | 36 | 40 |
| 2250 | Core drilling, 4" diameter (plus technician) | | | | Inch | | | | 23 | 25 |
| 2260 | Technician for core drilling | | | | Hr. | | | | 45 | 50 |
| 2300 | Patching core holes | | | | Ea. | | | | 22 | 24 |
| 2400 | Drying shrinkage at 28 days | | | | | | | | 236 | 260 |
| 2500 | Flexural test beams, ASTM C 78 | | | | | | | | 59 | 65 |
| 2600 | Mix design, one batch mix | | | | | | | | 259 | 285 |
| 2650 | Added trial batches | | | | | | | | 120 | 132 |
| 2800 | Modulus of elasticity, ASTM C 469 | | | | | | | | 164 | 180 |
| 2900 | Tensile test, cylinders, ASTM C 496 | | | | | | | | 45 | 50 |
| 3000 | Water-Cement ratio curve, 3 batches | | | | | | | | 141 | 155 |
| 3100 | 4 batches | | | | | | | | 186 | 205 |
| 3300 | Masonry testing, absorption, per 5 brick, ASTM C 67 | | | | | | | | 45 | 50 |
| 3350 | Chemical resistance, per 2 brick | | | | | | | | 50 | 55 |
| 3400 | Compressive strength, per 5 brick, ASTM C 67 | | | | | | | | 68 | 75 |
| 3420 | Efflorescence, per 5 brick, ASTM C 67 | | | | | | | | 68 | 75 |
| 3440 | Imperviousness, per 5 brick | | | | | | | | 87 | 96 |
| 3470 | Modulus of rupture, per 5 brick | | | | | | | | 86 | 95 |
| 3500 | Moisture, block only | | | | | | | | 32 | 35 |
| 3550 | Mortar, compressive strength, set of 3 | | | | | | | | 23 | 25 |
| 4100 | Reinforcing steel, bend test | | | | | | | | 55 | 61 |
| 4200 | Tensile test, up to #8 bar | | | | | | | | 36 | 40 |
| 4220 | #9 to #11 bar | | | | | | | | 41 | 45 |
| 4240 | #14 bar and larger | | | | | | | | 64 | 70 |
| 4400 | Soil testing, Atterberg limits, liquid and plastic limits | | | | | | | | 59 | 65 |
| 4510 | Hydrometer analysis | | | | | | | | 109 | 120 |
| 4530 | Specific gravity, ASTM D 354 | | | | | | | | 44 | 48 |
| 4600 | Sieve analysis, washed, ASTM D 422 | | | | | | | | 55 | 60 |
| 4700 | Unwashed, ASTM D 422 | | | | | | | | 59 | 65 |
| 4710 | Consolidation test (ASTM D2435), minimum | | | | | | | | 250 | 275 |
| 4715 | Maximum | | | | | | | | 430 | 475 |
| 4720 | Density and classification of undisturbed sample | | | | | | | | 73 | 80 |

# 01 45 Quality Control

## 01 45 23 – Testing and Inspecting Services

### 01 45 23.50 Testing

| | | Crew | Daily Output | Labor-Hours | Unit | Material | 2011 Bare Costs Labor | Equipment | Total | Total Incl O&P |
|---|---|---|---|---|---|---|---|---|---|---|
| 4735 | Soil density, nuclear method, ASTM D2922 | | | | Ea. | | | | 35 | 38.50 |
| 4740 | Sand cone method ASTM D1556 | | | | | | | | 27 | 30 |
| 4750 | Moisture content, ASTM D 2216 | | | | | | | | 9 | 10 |
| 4780 | Permeability test, double ring infiltrometer | | | | | | | | 500 | 550 |
| 4800 | Permeability, var. or constant head, undist., ASTM D 2434 | | | | | | | | 227 | 250 |
| 4850 | Recompacted | | | | | | | | 250 | 275 |
| 4900 | Proctor compaction, 4" standard mold, ASTM D 698 | | | | | | | | 123 | 135 |
| 4950 | 6" modified mold | | | | | | | | 68 | 75 |
| 5100 | Shear tests, triaxial, minimum | | | | | | | | 410 | 450 |
| 5150 | Maximum | | | | | | | | 545 | 600 |
| 5300 | Direct shear, minimum, ASTM D 3080 | | | | | | | | 320 | 350 |
| 5350 | Maximum | | | | | | | | 410 | 450 |
| 5550 | Technician for inspection, per day, earthwork | | | | | | | | 320 | 350 |
| 5650 | Bolting | | | | | | | | 400 | 440 |
| 5750 | Roofing | | | | | | | | 335 | 370 |
| 5790 | Welding | | | | | | | | 370 | 405 |
| 5820 | Non-destructive metal testing, dye penetrant | | | | Day | | | | 310 | 340 |
| 5840 | Magnetic particle | | | | | | | | 310 | 340 |
| 5860 | Radiography | | | | | | | | 450 | 495 |
| 5880 | Ultrasonic | | | | | | | | 310 | 340 |
| 6000 | Welding certification, minimum | | | | Ea. | | | | 91 | 100 |
| 6100 | Maximum | | | | " | | | | 250 | 275 |
| 7000 | Underground storage tank | | | | | | | | | |
| 7500 | Volumetric tightness test ,<=12,000 gal | | | | Ea. | | | | 435 | 480 |
| 7510 | <=30,000 gal | | | | " | | | | 615 | 675 |
| 7600 | Vadose zone (soil gas) sampling, 10-40 samples, min. | | | | Day | | | | 1,375 | 1,500 |
| 7610 | Maximum | | | | " | | | | 2,275 | 2,500 |
| 7700 | Ground water monitoring incl. drilling 3 wells, min. | | | | Total | | | | 4,550 | 5,000 |
| 7710 | Maximum | | | | " | | | | 6,375 | 7,000 |
| 8000 | X-ray concrete slabs | | | | Ea. | | | | 182 | 200 |
| 9000 | Thermographic testing, for bldg envelope heat loss, average 2,000 S.F. [G] | | | | " | | | | | 500 |

# 01 51 Temporary Utilities

## 01 51 13 – Temporary Electricity

### 01 51 13.80 Temporary Utilities

| | | Crew | Daily Output | Labor-Hours | Unit | Material | 2011 Bare Costs Labor | Equipment | Total | Total Incl O&P |
|---|---|---|---|---|---|---|---|---|---|---|
| 0010 | **TEMPORARY UTILITIES** | | | | | | | | | |
| 0100 | Heat, incl. fuel and operation, per week, 12 hrs. per day | 1 Skwk | 100 | .080 | CSF Flr | 27.50 | 3.55 | | 31.05 | 36 |
| 0200 | 24 hrs. per day | " | 60 | .133 | | 52.50 | 5.90 | | 58.40 | 67 |
| 0350 | Lighting, incl. service lamps, wiring & outlets, minimum | 1 Elec | 34 | .235 | | 2.73 | 11.85 | | 14.58 | 20.50 |
| 0360 | Maximum | " | 17 | .471 | | 5.80 | 23.50 | | 29.30 | 41.50 |
| 0400 | Power for temp lighting only, per month, min/month 6.6 KWH | | | | | | | | .92 | 1.01 |
| 0430 | Average/month 11.8 KWH | | | | | | | | 1.65 | 1.82 |
| 0450 | Maximum/month 23.6 KWH | | | | | | | | 3.30 | 3.63 |
| 0600 | Power for job duration incl. elevator, etc., minimum | | | | | | | | 47 | 51.50 |
| 0650 | Maximum | | | | | | | | 110 | 121 |
| 1000 | Toilet, portable, see Equip. Rental 01 54 33 in Reference Section | | | | | | | | | |

# 01 52 Construction Facilities

## 01 52 13 – Field Offices and Sheds

### 01 52 13.20 Office and Storage Space

| | 01 52 13.20 Office and Storage Space | Crew | Daily Output | Labor-Hours | Unit | Material | 2011 Bare Costs Labor | Equipment | Total | Total Incl O&P |
|---|---|---|---|---|---|---|---|---|---|---|
| 0010 | **OFFICE AND STORAGE SPACE** | | | | | | | | | |
| 0020 | Trailer, furnished, no hookups, 20' x 8', buy | 2 Skwk | 1 | 16 | Ea. | 8,550 | 710 | | 9,260 | 10,500 |
| 0250 | Rent per month | | | | | 146 | | | 146 | 160 |
| 0300 | 32' x 8', buy | 2 Skwk | .70 | 22.857 | | 12,300 | 1,025 | | 13,325 | 15,100 |
| 0350 | Rent per month | | | | | 193 | | | 193 | 213 |
| 0400 | 50' x 10', buy | 2 Skwk | .60 | 26.667 | | 23,300 | 1,175 | | 24,475 | 27,500 |
| 0450 | Rent per month | | | | | 288 | | | 288 | 315 |
| 0500 | 50' x 12', buy | 2 Skwk | .50 | 32 | | 26,000 | 1,425 | | 27,425 | 30,800 |
| 0550 | Rent per month | | | | | 360 | | | 360 | 395 |
| 0700 | For air conditioning, rent per month, add | | | | | 41.50 | | | 41.50 | 45.50 |
| 0800 | For delivery, add per mile | | | | Mile | 4.60 | | | 4.60 | 5.05 |
| 1000 | Portable buildings, prefab, on skids, economy, 8' x 8' | 2 Carp | 265 | .060 | S.F. | 85 | 2.60 | | 87.60 | 97.50 |
| 1100 | Deluxe, 8' x 12' | " | 150 | .107 | " | 95 | 4.59 | | 99.59 | 112 |
| 1200 | Storage boxes, 20' x 8', buy | 2 Skwk | 1.80 | 8.889 | Ea. | 2,175 | 395 | | 2,570 | 2,975 |
| 1250 | Rent per month | | | | | 71.50 | | | 71.50 | 78.50 |
| 1300 | 40' x 8', buy | 2 Skwk | 1.40 | 11.429 | | 3,025 | 505 | | 3,530 | 4,100 |
| 1350 | Rent per month | | | | | 94 | | | 94 | 103 |
| 5000 | Air supported structures, see Section 13 31 13.13 | | | | | | | | | |

### 01 52 13.40 Field Office Expense

| | | Crew | Daily Output | Labor-Hours | Unit | Material | 2011 Bare Costs Labor | Equipment | Total | Total Incl O&P |
|---|---|---|---|---|---|---|---|---|---|---|
| 0010 | **FIELD OFFICE EXPENSE** | | | | | | | | | |
| 0100 | Office equipment rental average | | | | Month | 200 | | | 200 | 220 |
| 0120 | Office supplies, average | | | | " | 86 | | | 86 | 94.50 |
| 0125 | Office trailer rental, see Section 01 52 13.20 | | | | | | | | | |
| 0140 | Telephone bill; avg. bill/month incl. long dist. | | | | Month | 81 | | | 81 | 89 |
| 0160 | Lights & HVAC | | | | " | 152 | | | 152 | 167 |

# 01 54 Construction Aids

## 01 54 09 – Protection Equipment

### 01 54 09.50 Personnel Protective Equipment

| | | Crew | Daily Output | Labor-Hours | Unit | Material | 2011 Bare Costs Labor | Equipment | Total | Total Incl O&P |
|---|---|---|---|---|---|---|---|---|---|---|
| 0010 | **PERSONNEL PROTECTIVE EQUIPMENT** | | | | | | | | | |
| 0015 | Hazardous waste protection | | | | | | | | | |
| 0020 | Respirator mask only, full face, silicone | | | | Ea. | 242 | | | 242 | 266 |
| 0030 | Half face, silicone | | | | | 36 | | | 36 | 39.50 |
| 0040 | Respirator cartridges, 2 req'd/mask, dust or asbestos | | | | | 2.34 | | | 2.34 | 2.57 |
| 0050 | Chemical vapor | | | | | 2.66 | | | 2.66 | 2.92 |
| 0060 | Combination vapor and dust | | | | | 7 | | | 7 | 7.70 |
| 0100 | Emergency escape breathing apparatus, 5 min | | | | | 465 | | | 465 | 510 |
| 0110 | 10 min | | | | | 500 | | | 500 | 550 |
| 0150 | Self contained breathing apparatus with full face piece, 30 min | | | | | 1,750 | | | 1,750 | 1,925 |
| 0160 | 60 min | | | | | 2,925 | | | 2,925 | 3,225 |
| 0200 | Encapsulating suits, limited use, level A | | | | | 950 | | | 950 | 1,050 |
| 0210 | Level B | | | | | 244 | | | 244 | 269 |
| 0300 | Over boots, latex | | | | Pr. | 6.35 | | | 6.35 | 7 |
| 0310 | PVC | | | | | 21.50 | | | 21.50 | 24 |
| 0320 | Neoprene | | | | | 41.50 | | | 41.50 | 46 |
| 0400 | Gloves, nitrile/PVC | | | | | 21 | | | 21 | 23.50 |
| 0410 | Neoprene coated | | | | | 24 | | | 24 | 26.50 |

### 01 54 09.60 Safety Nets

| | | Crew | Daily Output | Labor-Hours | Unit | Material | 2011 Bare Costs Labor | Equipment | Total | Total Incl O&P |
|---|---|---|---|---|---|---|---|---|---|---|
| 0010 | **SAFETY NETS** | | | | | | | | | |
| 0020 | No supports, stock sizes, nylon, 4" mesh | | | | S.F. | 1.10 | | | 1.10 | 1.21 |

# 01 54 Construction Aids

## 01 54 09 – Protection Equipment

### 01 54 09.60 Safety Nets

| | | Crew | Daily Output | Labor-Hours | Unit | Material | 2011 Bare Costs Labor | Equipment | Total | Total Incl O&P |
|---|---|---|---|---|---|---|---|---|---|---|
| 0100 | Polypropylene, 6" mesh | | | | S.F. | 1.59 | | | 1.59 | 1.75 |
| 0200 | Small mesh debris nets, 1/4" & 3/4" mesh, stock sizes | | | | | .74 | | | .74 | .81 |
| 0220 | Combined 4" mesh and 1/4" mesh, stock sizes | | | | | 2.05 | | | 2.05 | 2.26 |
| 0300 | Monthly rental, 4" mesh, stock sizes, 1st month | | | | | .50 | | | .50 | .55 |
| 0320 | 2nd month rental | | | | | .25 | | | .25 | .28 |
| 0340 | Maximum rental/year | | | | | 1.15 | | | 1.15 | 1.27 |

## 01 54 16 – Temporary Hoists

### 01 54 16.50 Weekly Forklift Crew

| | | Crew | Daily Output | Labor-Hours | Unit | Material | 2011 Bare Costs Labor | Equipment | Total | Total Incl O&P |
|---|---|---|---|---|---|---|---|---|---|---|
| 0010 | **WEEKLY FORKLIFT CREW** | | | | | | | | | |
| 0100 | All-terrain forklift, 45' lift, 35' reach, 9000 lb. capacity | A-3P | .20 | 40 | Week | | 1,750 | 2,225 | 3,975 | 5,050 |

## 01 54 19 – Temporary Cranes

### 01 54 19.50 Daily Crane Crews

| | | Crew | Daily Output | Labor-Hours | Unit | Material | 2011 Bare Costs Labor | Equipment | Total | Total Incl O&P |
|---|---|---|---|---|---|---|---|---|---|---|
| 0010 | **DAILY CRANE CREWS** for small jobs, portal to portal | | | | | | | | | |
| 0100 | 12-ton truck-mounted hydraulic crane | A-3H | 1 | 8 | Day | | 370 | 870 | 1,240 | 1,525 |
| 0200 | 25-ton | A-3I | 1 | 8 | | | 370 | 1,025 | 1,395 | 1,675 |
| 0300 | 40-ton | A-3J | 1 | 8 | | | 370 | 1,250 | 1,620 | 1,925 |
| 0400 | 55-ton | A-3K | 1 | 16 | | | 695 | 1,750 | 2,445 | 2,950 |
| 0500 | 80-ton | A-3L | 1 | 16 | | | 695 | 2,425 | 3,120 | 3,700 |
| 0600 | 100-ton | A-3M | 1 | 16 | | | 695 | 2,400 | 3,095 | 3,650 |

### 01 54 19.60 Monthly Tower Crane Crew

| | | Crew | Daily Output | Labor-Hours | Unit | Material | 2011 Bare Costs Labor | Equipment | Total | Total Incl O&P |
|---|---|---|---|---|---|---|---|---|---|---|
| 0010 | **MONTHLY TOWER CRANE CREW**, excludes concrete footing | | | | | | | | | |
| 0100 | Static tower crane, 130' high, 106' jib, 6200 lb. capacity | A-3N | .05 | 176 | Month | | 8,175 | 21,700 | 29,875 | 36,100 |

## 01 54 23 – Temporary Scaffolding and Platforms

### 01 54 23.60 Pump Staging

| | | Crew | Daily Output | Labor-Hours | Unit | Material | 2011 Bare Costs Labor | Equipment | Total | Total Incl O&P |
|---|---|---|---|---|---|---|---|---|---|---|
| 0010 | **PUMP STAGING**, Aluminum      R015423-20 | | | | | | | | | |
| 0200 | 24' long pole section, buy | | | | Ea. | 395 | | | 395 | 435 |
| 0300 | 18' long pole section, buy | | | | | 305 | | | 305 | 335 |
| 0400 | 12' long pole section, buy | | | | | 206 | | | 206 | 227 |
| 0500 | 6' long pole section, buy | | | | | 108 | | | 108 | 119 |
| 0600 | 6' long splice joint section, buy | | | | | 80.50 | | | 80.50 | 88.50 |
| 0700 | Pump jack, buy | | | | | 139 | | | 139 | 153 |
| 0900 | Foldable brace, buy | | | | | 56.50 | | | 56.50 | 62.50 |
| 1000 | Workbench/back safety rail support, buy | | | | | 74.50 | | | 74.50 | 82 |
| 1100 | Scaffolding planks/workbench, 14" wide x 24' long, buy | | | | | 695 | | | 695 | 760 |
| 1200 | Plank end safety rail, buy | | | | | 370 | | | 370 | 405 |
| 1250 | Safety net, 22' long, buy | | | | | 335 | | | 335 | 365 |
| 1300 | System in place, 50' working height, per use based on 50 uses | 2 Carp | 84.80 | .189 | C.S.F. | 6.35 | 8.10 | | 14.45 | 19.40 |
| 1400 | 100 uses | | 84.80 | .189 | | 3.16 | 8.10 | | 11.26 | 15.95 |
| 1500 | 150 uses | | 84.80 | .189 | | 2.12 | 8.10 | | 10.22 | 14.80 |

### 01 54 23.70 Scaffolding

| | | Crew | Daily Output | Labor-Hours | Unit | Material | 2011 Bare Costs Labor | Equipment | Total | Total Incl O&P |
|---|---|---|---|---|---|---|---|---|---|---|
| 0010 | **SCAFFOLDING**      R015423-10 | | | | | | | | | |
| 0015 | Steel tube, regular, no plank, labor only to erect & dismantle | | | | | | | | | |
| 0090 | Building exterior, wall face, 1 to 5 stories, 6'-4" x 5' frames | 3 Carp | 8 | 3 | C.S.F. | | 129 | | 129 | 198 |
| 0200 | 6 to 12 stories | 4 Carp | 8 | 4 | | | 172 | | 172 | 264 |
| 0301 | 13 to 20 stories | 5 Clab | 8 | 5 | | | 172 | | 172 | 264 |
| 0460 | Building interior, wall face area, up to 16' high | 3 Carp | 12 | 2 | | | 86 | | 86 | 132 |
| 0560 | 16' to 40' high | | 10 | 2.400 | | | 103 | | 103 | 159 |
| 0800 | Building interior floor area, up to 30' high | | 150 | .160 | C.C.F. | | 6.90 | | 6.90 | 10.55 |
| 0900 | Over 30' high | 4 Carp | 160 | .200 | " | | 8.60 | | 8.60 | 13.20 |
| 0906 | Complete system for face of walls, no plank, material only rent/mo | | | | C.S.F. | 36 | | | 36 | 39.50 |

| 01 54 23.70 Scaffolding | Crew | Daily Output | Labor-Hours | Unit | Material | 2011 Bare Costs Labor | Equipment | Total | Total Incl O&P |
|---|---|---|---|---|---|---|---|---|---|
| 0908    Interior spaces, no plank, material only rent/mo | | | | C.C.F. | 3.42 | | | 3.42 | 3.76 |
| 0910  Steel tubular, heavy duty shoring, buy | | | | | | | | | |
| 0920    Frames 5' high 2' wide | | | | Ea. | 93 | | | 93 | 102 |
| 0925     5' high 4' wide | | | | | 105 | | | 105 | 116 |
| 0930     6' high 2' wide | | | | | 106 | | | 106 | 117 |
| 0935     6' high 4' wide | | | | | 124 | | | 124 | 136 |
| 0940    Accessories | | | | | | | | | |
| 0945     Cross braces | | | | Ea. | 18 | | | 18 | 19.80 |
| 0950     U-head, 8" x 8" | | | | | 21.50 | | | 21.50 | 24 |
| 0955     J-head, 4" x 8" | | | | | 15.80 | | | 15.80 | 17.40 |
| 0960     Base plate, 8" x 8" | | | | | 17.60 | | | 17.60 | 19.35 |
| 0965     Leveling jack | | | | | 38 | | | 38 | 41.50 |
| 1000  Steel tubular, regular, buy | | | | | | | | | |
| 1100    Frames 3' high 5' wide | | | | Ea. | 71 | | | 71 | 78 |
| 1150     5' high 5' wide | | | | | 83 | | | 83 | 91.50 |
| 1200     6'-4" high 5' wide | | | | | 104 | | | 104 | 114 |
| 1350     7'-6" high 6' wide | | | | | 179 | | | 179 | 197 |
| 1500    Accessories cross braces | | | | | 20.50 | | | 20.50 | 22.50 |
| 1550     Guardrail post | | | | | 18.60 | | | 18.60 | 20.50 |
| 1600     Guardrail 7' section | | | | | 9.90 | | | 9.90 | 10.90 |
| 1650     Screw jacks & plates | | | | | 27 | | | 27 | 29.50 |
| 1700     Sidearm brackets | | | | | 43 | | | 43 | 47.50 |
| 1750     8" casters | | | | | 35.50 | | | 35.50 | 39 |
| 1800     Plank 2" x 10" x 16'-0" | | | | | 51 | | | 51 | 56 |
| 1900     Stairway section | | | | | 325 | | | 325 | 360 |
| 1910     Stairway starter bar | | | | | 36.50 | | | 36.50 | 40 |
| 1920     Stairway inside handrail | | | | | 66 | | | 66 | 72.50 |
| 1930     Stairway outside handrail | | | | | 98 | | | 98 | 108 |
| 1940     Walk-thru frame guardrail | | | | | 47 | | | 47 | 51.50 |
| 2000  Steel tubular, regular, rent/mo. | | | | | | | | | |
| 2100    Frames 3' high 5' wide | | | | Ea. | 5 | | | 5 | 5.50 |
| 2150     5' high 5' wide | | | | | 5 | | | 5 | 5.50 |
| 2200     6'-4" high 5' wide | | | | | 5.15 | | | 5.15 | 5.65 |
| 2250     7'-6" high 6' wide | | | | | 7 | | | 7 | 7.70 |
| 2500    Accessories, cross braces | | | | | 1 | | | 1 | 1.10 |
| 2550     Guardrail post | | | | | 1 | | | 1 | 1.10 |
| 2600     Guardrail 7' section | | | | | 1 | | | 1 | 1.10 |
| 2650     Screw jacks & plates | | | | | 2 | | | 2 | 2.20 |
| 2700     Sidearm brackets | | | | | 2 | | | 2 | 2.20 |
| 2750     8" casters | | | | | 8 | | | 8 | 8.80 |
| 2800     Outrigger for rolling tower | | | | | 3 | | | 3 | 3.30 |
| 2850     Plank 2" x 10" x 16'-0" | | | | | 6 | | | 6 | 6.60 |
| 2900     Stairway section | | | | | 40 | | | 40 | 44 |
| 2940     Walk-thru frame guardrail | | | | | 2.50 | | | 2.50 | 2.75 |
| 3000  Steel tubular, heavy duty shoring, rent/mo. | | | | | | | | | |
| 3250     5' high 2' & 4' wide | | | | Ea. | 5 | | | 5 | 5.50 |
| 3300     6' high 2' & 4' wide | | | | | 5 | | | 5 | 5.50 |
| 3500    Accessories, cross braces | | | | | 1 | | | 1 | 1.10 |
| 3600     U - head, 8" x 8" | | | | | 1 | | | 1 | 1.10 |
| 3650     J - head, 4" x 8" | | | | | 1 | | | 1 | 1.10 |
| 3700     Base plate, 8" x 8" | | | | | 1 | | | 1 | 1.10 |
| 3750     Leveling jack | | | | | 2 | | | 2 | 2.20 |
| 5700  Planks, 2" x 10" x 16'-0", labor only to erect & remove to 50' H | 3 Carp | 72 | .333 | | | 14.35 | | 14.35 | 22 |

## 01 54 23 – Temporary Scaffolding and Platforms

### 01 54 23.70 Scaffolding

| | | Crew | Daily Output | Labor-Hours | Unit | Material | 2011 Bare Costs Labor | Equipment | Total | Total Incl O&P |
|---|---|---|---|---|---|---|---|---|---|---|
| 5800 | Over 50' high | 4 Carp | 80 | .400 | Ea. | | 17.20 | | 17.20 | 26.50 |
| 6000 | Heavy duty shoring for elevated slab forms to 8'-2" high, floor area | 4 Carp | 16 | 2 | C.S.F. | | 86 | | 86 | 132 |
| 6100 | Labor only to erect & dismantle | | | | | | | | | |
| 6110 | Materials only, rent.mo | | | | " | 29.50 | | | 29.50 | 32.50 |
| 6500 | To 14'-8" high | | | | | | | | | |
| 6600 | Labor only to erect & dismantle | 4 Carp | 10 | 3.200 | C.S.F. | | 138 | | 138 | 211 |
| 6610 | Materials only, rent/mo | | | | " | 43 | | | 43 | 47.50 |

### 01 54 23.75 Scaffolding Specialties

| | | Crew | Daily Output | Labor-Hours | Unit | Material | 2011 Bare Costs Labor | Equipment | Total | Total Incl O&P |
|---|---|---|---|---|---|---|---|---|---|---|
| 0010 | **SCAFFOLDING SPECIALTIES** | | | | | | | | | |
| 1200 | Sidewalk bridge, heavy duty steel posts & beams, including | | | | | | | | | |
| 1210 | parapet protection & waterproofing | | | | | | | | | |
| 1220 | 8' to 10' wide, 2 posts | 3 Carp | 15 | 1.600 | L.F. | 35 | 69 | | 104 | 144 |
| 1230 | 3 posts | " | 10 | 2.400 | " | 53 | 103 | | 156 | 218 |
| 1500 | Sidewalk bridge using tubular steel | | | | | | | | | |
| 1510 | scaffold frames, including planking | 3 Carp | 45 | .533 | L.F. | 5.60 | 23 | | 28.60 | 41 |
| 1600 | For 2 uses per month, deduct from all above | | | | | 50% | | | | |
| 1700 | For 1 use every 2 months, add to all above | | | | | 100% | | | | |
| 1900 | Catwalks, 20" wide, no guardrails, 7' span, buy | | | | Ea. | 145 | | | 145 | 160 |
| 2000 | 10' span, buy | | | | | 203 | | | 203 | 223 |
| 3720 | Putlog, standard, 8' span, with hangers, buy | | | | | 75 | | | 75 | 82.50 |
| 3730 | Rent per month | | | | | 10 | | | 10 | 11 |
| 3750 | 12' span, buy | | | | | 113 | | | 113 | 124 |
| 3755 | Rent per month | | | | | 15 | | | 15 | 16.50 |
| 3760 | Trussed type, 16' span, buy | | | | | 260 | | | 260 | 286 |
| 3770 | Rent per month | | | | | 20 | | | 20 | 22 |
| 3790 | 22' span, buy | | | | | 310 | | | 310 | 345 |
| 3795 | Rent per month | | | | | 30 | | | 30 | 33 |
| 3800 | Rolling ladders with handrails, 30" wide, buy, 2 step | | | | | 237 | | | 237 | 261 |
| 4000 | 7 step | | | | | 725 | | | 725 | 800 |
| 4050 | 10 step | | | | | 1,000 | | | 1,000 | 1,100 |
| 4100 | Rolling towers, buy, 5' wide, 7' long, 10' high | | | | | 1,350 | | | 1,350 | 1,500 |
| 4200 | For 5' high added sections, to buy, add | | | | | 207 | | | 207 | 228 |
| 4300 | Complete incl. wheels, railings, outriggers, | | | | | | | | | |
| 4350 | 21' high, to buy | | | | Ea. | 2,250 | | | 2,250 | 2,475 |
| 4400 | Rent/month = 5% of purchase cost | | | | " | 113 | | | 113 | 124 |

### 01 54 23.80 Staging Aids

| | | Crew | Daily Output | Labor-Hours | Unit | Material | 2011 Bare Costs Labor | Equipment | Total | Total Incl O&P |
|---|---|---|---|---|---|---|---|---|---|---|
| 0010 | **STAGING AIDS** and fall protection equipment | | | | | | | | | |
| 0100 | Sidewall staging bracket, tubular, buy | | | | Ea. | 47.50 | | | 47.50 | 52 |
| 0110 | Cost each per day, based on 250 days use | | | | Day | .19 | | | .19 | .21 |
| 0200 | Guard post, buy | | | | Ea. | 22.50 | | | 22.50 | 25 |
| 0210 | Cost each per day, based on 250 days use | | | | Day | .09 | | | .09 | .10 |
| 0300 | End guard chains, buy per pair | | | | Pair | 31 | | | 31 | 34 |
| 0310 | Cost per set per day, based on 250 days use | | | | Day | .19 | | | .19 | .21 |
| 1000 | Roof shingling bracket, steel, buy | | | | Ea. | 10.10 | | | 10.10 | 11.10 |
| 1010 | Cost each per day, based on 250 days use | | | | Day | .04 | | | .04 | .04 |
| 1100 | Wood bracket, buy | | | | Ea. | 17.10 | | | 17.10 | 18.80 |
| 1110 | Cost each per day, based on 250 days use | | | | Day | .07 | | | .07 | .08 |
| 2000 | Ladder jack, aluminum, buy per pair | | | | Pair | 108 | | | 108 | 119 |
| 2010 | Cost per pair per day, based on 250 days use | | | | Day | .43 | | | .43 | .48 |
| 2100 | Steel siderail jack, buy per pair | | | | Pair | 122 | | | 122 | 134 |
| 2110 | Cost per pair per day, based on 250 days use | | | | Day | .49 | | | .49 | .54 |
| 3000 | Laminated wood plank, 2" x 10" x 16', buy | | | | Ea. | 51 | | | 51 | 56 |

# 01 54 Construction Aids

## 01 54 23 – Temporary Scaffolding and Platforms

| 01 54 23.80 Staging Aids | | Crew | Daily Output | Labor-Hours | Unit | Material | 2011 Bare Costs Labor | Equipment | Total | Total Incl O&P |
|---|---|---|---|---|---|---|---|---|---|---|
| 3010 | Cost each per day, based on 250 days use | | | | Day | .20 | | | .20 | .22 |
| 3100 | Aluminum scaffolding plank, 20" wide x 24' long, buy | | | | Ea. | 895 | | | 895 | 985 |
| 3110 | Cost each per day, based on 250 days use | | | | Day | 3.58 | | | 3.58 | 3.94 |
| 4000 | Nylon full body harness, lanyard and rope grab | | | | Ea. | 195 | | | 195 | 214 |
| 4010 | Cost each per day, based on 250 days use | | | | Day | .78 | | | .78 | .86 |
| 4100 | Rope for safety line, 5/8" x 100' nylon, buy | | | | Ea. | 70 | | | 70 | 77 |
| 4110 | Cost each per day, based on 250 days use | | | | Day | .28 | | | .28 | .31 |
| 4200 | Permanent U-Bolt roof anchor, buy | | | | Ea. | 35 | | | 35 | 38.50 |
| 4300 | Temporary (one use) roof ridge anchor, buy | | | | " | 28.50 | | | 28.50 | 31 |
| 5000 | Installation (setup and removal) of staging aids | | | | | | | | | |
| 5010 | Sidewall staging bracket | 2 Carp | 64 | .250 | Ea. | | 10.75 | | 10.75 | 16.50 |
| 5020 | Guard post with 2 wood rails | " | 64 | .250 | | | 10.75 | | 10.75 | 16.50 |
| 5030 | End guard chains, set | 1 Carp | 64 | .125 | | | 5.40 | | 5.40 | 8.25 |
| 5100 | Roof shingling bracket | | 96 | .083 | | | 3.59 | | 3.59 | 5.50 |
| 5200 | Ladder jack | ↓ | 64 | .125 | | | 5.40 | | 5.40 | 8.25 |
| 5300 | Wood plank, 2" x 10" x 16' | 2 Carp | 80 | .200 | | | 8.60 | | 8.60 | 13.20 |
| 5310 | Aluminum scaffold plank, 20" x 24' | " | 40 | .400 | | | 17.20 | | 17.20 | 26.50 |
| 5410 | Safety rope | 1 Carp | 40 | .200 | | | 8.60 | | 8.60 | 13.20 |
| 5420 | Permanent U-Bolt roof anchor (install only) | 2 Carp | 40 | .400 | | | 17.20 | | 17.20 | 26.50 |
| 5430 | Temporary roof ridge anchor (install only) | 1 Carp | 64 | .125 | ↓ | | 5.40 | | 5.40 | 8.25 |

## 01 54 26 – Temporary Swing Staging

| 01 54 26.50 Swing Staging | | Crew | Daily Output | Labor-Hours | Unit | Material | 2011 Bare Costs Labor | Equipment | Total | Total Incl O&P |
|---|---|---|---|---|---|---|---|---|---|---|
| 0010 | SWING STAGING, 500 lb cap., 2' wide to 24' long, hand operated | | | | | | | | | |
| 0020 | steel cable type, with 60' cables, buy | | | | Ea. | 4,050 | | | 4,050 | 4,450 |
| 0030 | Rent per month | | | | " | 405 | | | 405 | 445 |
| 0600 | Lightweight (not for masons) 24' long for 150' height, | | | | | | | | | |
| 0610 | manual type, buy | | | | Ea. | 4,150 | | | 4,150 | 4,575 |
| 0620 | Rent per month | | | | | 415 | | | 415 | 455 |
| 0700 | Powered, electric or air, to 150' high, buy | | | | | 13,300 | | | 13,300 | 14,700 |
| 0710 | Rent per month | | | | | 935 | | | 935 | 1,025 |
| 0780 | To 300' high, buy | | | | | 13,500 | | | 13,500 | 14,800 |
| 0800 | Rent per month | | | | | 945 | | | 945 | 1,025 |
| 1000 | Bosun's chair or work basket 3' x 3.5', to 300' high, electric, buy | | | | | 8,625 | | | 8,625 | 9,500 |
| 1010 | Rent per month | | | | ↓ | 605 | | | 605 | 665 |
| 2200 | Move swing staging (setup and remove) | E-4 | 2 | 16 | Move | | 785 | 54.50 | 839.50 | 1,425 |

## 01 54 36 – Equipment Mobilization

| 01 54 36.50 Mobilization | | Crew | Daily Output | Labor-Hours | Unit | Material | 2011 Bare Costs Labor | Equipment | Total | Total Incl O&P |
|---|---|---|---|---|---|---|---|---|---|---|
| 0010 | MOBILIZATION (Use line item again for demobilization) R015433-10 | | | | | | | | | |
| 0015 | Up to 25 mi haul dist (50 mi RT for mob/demob crew) | | | | | | | | | |
| 0020 | Dozer, loader, backhoe, excav., grader, paver, roller, 70 to 150 H.P. | B-34N | 4 | 2 | Ea. | | 69 | 116 | 185 | 233 |
| 0100 | Above 150 HP | B-34K | 3 | 2.667 | | | 91.50 | 259 | 350.50 | 425 |
| 0300 | Scraper, towed type (incl. tractor), 6 C.Y. capacity | | 3 | 2.667 | | | 91.50 | 259 | 350.50 | 425 |
| 0400 | 10 C.Y. | | 2.50 | 3.200 | | | 110 | 310 | 420 | 505 |
| 0600 | Self-propelled scraper, 15 C.Y. | | 2.50 | 3.200 | | | 110 | 310 | 420 | 505 |
| 0700 | 24 C.Y. | | 2 | 4 | | | 138 | 390 | 528 | 640 |
| 0900 | Shovel or dragline, 3/4 C.Y. | | 3.60 | 2.222 | | | 76.50 | 216 | 292.50 | 355 |
| 1000 | 1-1/2 C.Y. | ↓ | 3 | 2.667 | | | 91.50 | 259 | 350.50 | 425 |
| 1100 | Small equipment, placed in rear of, or towed by pickup truck | A-3A | 8 | 1 | | | 33.50 | 14.60 | 48.10 | 67 |
| 1150 | Equip up to 70 HP, on flatbed trailer behind pickup truck | A-3D | 4 | 2 | | | 67 | 57 | 124 | 165 |
| 2000 | Crane, truck-mounted, up to 75 ton, (driver only, one-way) | 1 Eqhv | 7.20 | 1.111 | | | 51.50 | | 51.50 | 77 |
| 2100 | Crane, truck-mounted, over 75 ton | A-3E | 2.50 | 6.400 | | | 259 | 46.50 | 305.50 | 440 |
| 2200 | Crawler-mounted, up to 75 ton | A-3F | 2 | 8 | ↓ | | 325 | 400 | 725 | 920 |

# 01 54 Construction Aids

## 01 54 36 – Equipment Mobilization

| 01 54 36.50 Mobilization | Crew | Daily Output | Labor-Hours | Unit | Material | 2011 Bare Costs Labor | Equipment | Total | Total Incl O&P |
|---|---|---|---|---|---|---|---|---|---|
| 2300     Over 75 ton | A-3G | 1.50 | 10.667 | Ea. | | 430 | 595 | 1,025 | 1,300 |
| 2500     For each additional 5 miles haul distance, add | | | | | | 10% | 10% | | |
| 3000     For large pieces of equipment, allow for assembly/knockdown | | | | | | | | | |
| 3001     For mob/demob of vibrofloatation equip, see Section 31 45 13.10 | | | | | | | | | |
| 3100     For mob/demob of micro-tunneling equip, see Section 33 05 23.19 | | | | | | | | | |
| 3200     For mob/demob of pile driving equip, see Section 31 62 19.10 | | | | | | | | | |
| 3300     For mob/demob of caisson drilling equip, see Section 31 63 26.13 | | | | | | | | | |

## 01 54 39 – Construction Equipment

### 01 54 39.70 Small Tools

| | | | | Unit | Material | Labor | Equipment | Total | Total Incl O&P |
|---|---|---|---|---|---|---|---|---|---|
| 0010 **SMALL TOOLS** | R013113-50 | | | | | | | | |
| 0020   As % of contractor's bare labor cost for project, minimum | | | | Total | | .50% | | | |
| 0100     Maximum | | | | " | | 2% | | | |

# 01 55 Vehicular Access and Parking

## 01 55 23 – Temporary Roads

### 01 55 23.50 Roads and Sidewalks

| | Crew | Daily Output | Labor-Hours | Unit | Material | Labor | Equipment | Total | Total Incl O&P |
|---|---|---|---|---|---|---|---|---|---|
| 0010 **ROADS AND SIDEWALKS** Temporary | | | | | | | | | |
| 0050   Roads, gravel fill, no surfacing, 4" gravel depth | B-14 | 715 | .067 | S.Y. | 4 | 2.43 | .45 | 6.88 | 8.60 |
| 0100     8" gravel depth | " | 615 | .078 | " | 8 | 2.83 | .52 | 11.35 | 13.70 |
| 1000   Ramp, 3/4" plywood on 2" x 6" joists, 16" O.C. | 2 Carp | 300 | .053 | S.F. | 1.34 | 2.30 | | 3.64 | 5 |
| 1100     On 2" x 10" joists, 16" O.C. | " | 275 | .058 | " | 1.92 | 2.50 | | 4.42 | 5.95 |

# 01 56 Temporary Barriers and Enclosures

## 01 56 13 – Temporary Air Barriers

### 01 56 13.60 Tarpaulins

| | Crew | Daily Output | Labor-Hours | Unit | Material | Labor | Equipment | Total | Total Incl O&P |
|---|---|---|---|---|---|---|---|---|---|
| 0010 **TARPAULINS** | | | | | | | | | |
| 0020   Cotton duck, 10 oz. to 13.13 oz. per S.Y., minimum | | | | S.F. | .48 | | | .48 | .53 |
| 0050     Maximum | | | | | .60 | | | .60 | .66 |
| 0100   Polyvinyl coated nylon, 14 oz. to 18 oz., minimum | | | | | .48 | | | .48 | .53 |
| 0150     Maximum | | | | | .68 | | | .68 | .75 |
| 0200   Reinforced polyethylene 3 mils thick, white | | | | | .15 | | | .15 | .17 |
| 0300     4 mils thick, white, clear or black | | | | | .20 | | | .20 | .22 |
| 0400     5.5 mils thick, clear | | | | | .30 | | | .30 | .33 |
| 0500       White, fire retardant | | | | | .35 | | | .35 | .39 |
| 0600     7.5 mils, oil resistant, fire retardant | | | | | .40 | | | .40 | .44 |
| 0700     8.5 mils, black | | | | | .53 | | | .53 | .58 |
| 0710   Woven polyethylene, 6 mils thick | | | | | .35 | | | .35 | .39 |
| 0730   Polyester reinforced w/integral fastening system 11 mils thick | | | | | 1.07 | | | 1.07 | 1.18 |
| 0740   Mylar polyester, non-reinforced, 7 mils thick | | | | | 1.17 | | | 1.17 | 1.29 |

### 01 56 13.90 Winter Protection

| | Crew | Daily Output | Labor-Hours | Unit | Material | Labor | Equipment | Total | Total Incl O&P |
|---|---|---|---|---|---|---|---|---|---|
| 0010 **WINTER PROTECTION** | | | | | | | | | |
| 0100   Framing to close openings | 2 Clab | 750 | .021 | S.F. | .39 | .73 | | 1.12 | 1.55 |
| 0200   Tarpaulins hung over scaffolding, 8 uses, not incl. scaffolding | | 1500 | .011 | | .25 | .37 | | .62 | .84 |
| 0250   Tarpaulin polyester reinf. w/integral fastening system 11 mils thick | | 1600 | .010 | | .80 | .34 | | 1.14 | 1.41 |
| 0300   Prefab fiberglass panels, steel frame, 8 uses | | 1200 | .013 | | .85 | .46 | | 1.31 | 1.64 |

# 01 56 Temporary Barriers and Enclosures

## 01 56 16 – Temporary Dust Barriers

### 01 56 16.10 Dust Barriers, Temporary

| | | Crew | Daily Output | Labor-Hours | Unit | Material | 2011 Bare Costs Labor | Equipment | Total | Total Incl O&P |
|---|---|---|---|---|---|---|---|---|---|---|
| 0010 | **DUST BARRIERS, TEMPORARY** | | | | | | | | | |
| 0020 | Spring loaded telescoping pole & head, to 12', erect and dismantle | 1 Clab | 240 | .033 | Ea. | | 1.14 | | 1.14 | 1.76 |
| 0025 | Cost per day (based upon 250 days) | | | | Day | .22 | | | .22 | .24 |
| 0030 | To 21', erect and dismantle | 1 Clab | 240 | .033 | Ea. | | 1.14 | | 1.14 | 1.76 |
| 0035 | Cost per day (based upon 250 days) | | | | Day | .39 | | | .39 | .42 |
| 0040 | Accessories, caution tape reel, erect and dismantle | 1 Clab | 480 | .017 | Ea. | | .57 | | .57 | .88 |
| 0045 | Cost per day (based upon 250 days) | | | | Day | .05 | | | .05 | .05 |
| 0060 | Foam rail and connector, erect and dismantle | 1 Clab | 240 | .033 | Ea. | | 1.14 | | 1.14 | 1.76 |
| 0065 | Cost per day (based upon 250 days) | | | | Day | .10 | | | .10 | .11 |
| 0070 | Caution tape | 1 Clab | 384 | .021 | C.L.F. | 2.67 | .72 | | 3.39 | 4.03 |
| 0080 | Zipper, standard duty | | 60 | .133 | Ea. | 8 | 4.58 | | 12.58 | 15.85 |
| 0090 | Heavy duty | | 48 | .167 | " | 9.50 | 5.75 | | 15.25 | 19.25 |
| 0100 | Polyethylene sheet, 4 mil | | 37 | .216 | Sq. | 3.34 | 7.45 | | 10.79 | 15.05 |
| 0110 | 6 mil | | 37 | .216 | " | 5.25 | 7.45 | | 12.70 | 17.15 |
| 1000 | Dust partition, 6 mil polyethylene, 1" x 3" frame | 2 Carp | 2000 | .008 | S.F. | .28 | .34 | | .62 | .84 |
| 1080 | 2" x 4" frame | " | 2000 | .008 | " | .30 | .34 | | .64 | .86 |

## 01 56 23 – Temporary Barricades

### 01 56 23.10 Barricades

| | | Crew | Daily Output | Labor-Hours | Unit | Material | 2011 Bare Costs Labor | Equipment | Total | Total Incl O&P |
|---|---|---|---|---|---|---|---|---|---|---|
| 0010 | **BARRICADES** | | | | | | | | | |
| 0020 | 5' high, 3 rail @ 2" x 8", fixed | 2 Carp | 20 | .800 | L.F. | 4.93 | 34.50 | | 39.43 | 58.50 |
| 0150 | Movable | " | 30 | .533 | " | 4.10 | 23 | | 27.10 | 39.50 |
| 0300 | Stock units, 6' high, 8' wide, plain, buy | | | | Ea. | 435 | | | 435 | 480 |
| 0350 | With reflective tape, buy | | | | " | 525 | | | 525 | 580 |
| 0400 | Break-a-way 3" PVC pipe barricade | | | | | | | | | |
| 0410 | with 3 ea. 1' x 4' reflectorized panels, buy | | | | Ea. | 305 | | | 305 | 335 |
| 0500 | Plywood with steel legs, 32" wide | | | | | 72 | | | 72 | 79 |
| 0600 | Telescoping Christmas tree, 9' high, 5 flags, buy | | | | | 122 | | | 122 | 134 |
| 0800 | Traffic cones, PVC, 18" high | | | | | 10.55 | | | 10.55 | 11.65 |
| 0850 | 28" high | | | | | 17.85 | | | 17.85 | 19.65 |
| 1000 | Guardrail, wooden, 3' high, 1" x 6", on 2" x 4" posts | 2 Carp | 200 | .080 | L.F. | 1.11 | 3.44 | | 4.55 | 6.50 |
| 1100 | 2" x 6", on 4" x 4" posts | " | 165 | .097 | | 2.08 | 4.17 | | 6.25 | 8.70 |
| 1200 | Portable metal with base pads, buy | | | | | 19.30 | | | 19.30 | 21 |
| 1250 | Typical installation, assume 10 reuses | 2 Carp | 600 | .027 | | 2.50 | 1.15 | | 3.65 | 4.51 |
| 1300 | Barricade tape, polyethylene, 7 mil, 3" wide x 500' long roll | | | | Ea. | 25 | | | 25 | 27.50 |
| 5000 | Barricades, see Section 01 54 33.40 | | | | | | | | | |

## 01 56 26 – Temporary Fencing

### 01 56 26.50 Temporary Fencing

| | | Crew | Daily Output | Labor-Hours | Unit | Material | 2011 Bare Costs Labor | Equipment | Total | Total Incl O&P |
|---|---|---|---|---|---|---|---|---|---|---|
| 0010 | **TEMPORARY FENCING** | | | | | | | | | |
| 0020 | Chain link, 11 ga, 5' high | 2 Clab | 400 | .040 | L.F. | 4.50 | 1.37 | | 5.87 | 7.05 |
| 0100 | 6' high | | 300 | .053 | | 5.30 | 1.83 | | 7.13 | 8.65 |
| 0200 | Rented chain link, 6' high, to 1000' (up to 12 mo.) | | 400 | .040 | | 6.30 | 1.37 | | 7.67 | 9.05 |
| 0250 | Over 1000' (up to 12 mo.) | | 300 | .053 | | 3.29 | 1.83 | | 5.12 | 6.45 |
| 0350 | Plywood, painted, 2" x 4" frame, 4' high | A-4 | 135 | .178 | | 5.60 | 7.35 | | 12.95 | 17.30 |
| 0400 | 4" x 4" frame, 8' high | " | 110 | .218 | | 10.45 | 9 | | 19.45 | 25 |
| 0500 | Wire mesh on 4" x 4" posts, 4' high | 2 Carp | 100 | .160 | | 8.55 | 6.90 | | 15.45 | 19.95 |
| 0550 | 8' high | " | 80 | .200 | | 13 | 8.60 | | 21.60 | 27.50 |

## 01 56 29 – Temporary Protective Walkways

### 01 56 29.50 Protection

| | | Crew | Daily Output | Labor-Hours | Unit | Material | 2011 Bare Costs Labor | Equipment | Total | Total Incl O&P |
|---|---|---|---|---|---|---|---|---|---|---|
| 0010 | **PROTECTION** | | | | | | | | | |
| 0020 | Stair tread, 2" x 12" planks, 1 use | 1 Carp | 75 | .107 | Tread | 4.08 | 4.59 | | 8.67 | 11.55 |

# 01 56 Temporary Barriers and Enclosures

## 01 56 29 – Temporary Protective Walkways

### 01 56 29.50 Protection

| 01 56 29.50 Protection | | Crew | Daily Output | Labor-Hours | Unit | Material | 2011 Bare Costs Labor | Equipment | Total | Total Incl O&P |
|---|---|---|---|---|---|---|---|---|---|---|
| 0100 | Exterior plywood, 1/2" thick, 1 use | 1 Carp | 65 | .123 | Tread | 1.74 | 5.30 | | 7.04 | 10.05 |
| 0200 | 3/4" thick, 1 use | | 60 | .133 | | 2.45 | 5.75 | | 8.20 | 11.50 |
| 2200 | Sidewalks, 2" x 12" planks, 2 uses | | 350 | .023 | S.F. | .68 | .98 | | 1.66 | 2.26 |
| 2300 | Exterior plywood, 2 uses, 1/2" thick | | 750 | .011 | | .29 | .46 | | .75 | 1.02 |
| 2400 | 5/8" thick | | 650 | .012 | | .35 | .53 | | .88 | 1.19 |
| 2500 | 3/4" thick | | 600 | .013 | | .41 | .57 | | .98 | 1.33 |

## 01 56 32 – Temporary Security

### 01 56 32.50 Watchman

| | | Crew | Daily Output | Labor-Hours | Unit | Material | 2011 Bare Costs Labor | Equipment | Total | Total Incl O&P |
|---|---|---|---|---|---|---|---|---|---|---|
| 0010 | **WATCHMAN** | | | | | | | | | |
| 0020 | Service, monthly basis, uniformed person, minimum | | | | Hr. | | | | 25 | 27.50 |
| 0100 | Maximum | | | | | | | | 45.50 | 50 |
| 0200 | Person and command dog, minimum | | | | | | | | 31 | 34 |
| 0300 | Maximum | | | | | | | | 54.50 | 60 |
| 0500 | Sentry dog, leased, with job patrol (yard dog), 1 dog | | | | Week | | | | 290 | 320 |
| 0600 | 2 dogs | | | | " | | | | 390 | 430 |
| 0800 | Purchase, trained sentry dog, minimum | | | | Ea. | | | | 1,375 | 1,500 |
| 0900 | Maximum | | | | " | | | | 2,725 | 3,000 |

# 01 58 Project Identification

## 01 58 13 – Temporary Project Signage

### 01 58 13.50 Signs

| | | Crew | Daily Output | Labor-Hours | Unit | Material | 2011 Bare Costs Labor | Equipment | Total | Total Incl O&P |
|---|---|---|---|---|---|---|---|---|---|---|
| 0010 | **SIGNS** | | | | | | | | | |
| 0020 | High intensity reflectorized, no posts, buy | | | | S.F. | 26.50 | | | 26.50 | 29.50 |

# 01 71 Examination and Preparation

## 01 71 23 – Field Engineering

### 01 71 23.13 Construction Layout

| | | Crew | Daily Output | Labor-Hours | Unit | Material | 2011 Bare Costs Labor | Equipment | Total | Total Incl O&P |
|---|---|---|---|---|---|---|---|---|---|---|
| 0010 | **CONSTRUCTION LAYOUT** | | | | | | | | | |
| 1100 | Crew for layout of building, trenching or pipe laying, 2 person crew | A-6 | 1 | 16 | Day | | 690 | 69.50 | 759.50 | 1,125 |
| 1200 | 3 person crew | A-7 | 1 | 24 | | | 1,125 | 69.50 | 1,194.50 | 1,800 |
| 1400 | Crew for roadway layout, 4 person crew | A-8 | 1 | 32 | | | 1,475 | 70 | 1,545 | 2,300 |

### 01 71 23.19 Surveyor Stakes

| | | Crew | Daily Output | Labor-Hours | Unit | Material | 2011 Bare Costs Labor | Equipment | Total | Total Incl O&P |
|---|---|---|---|---|---|---|---|---|---|---|
| 0010 | **SURVEYOR STAKES** | | | | | | | | | |
| 0020 | Hardwood, 1" x 1" x 48" long | | | | C | 58 | | | 58 | 64 |
| 0100 | 2" x 2" x 18" long | | | | | 66 | | | 66 | 72.50 |
| 0150 | 2" x 2" x 24" long | | | | | 110 | | | 110 | 121 |
| 0200 | 2" x 2" x 30" long | | | | | 74 | | | 74 | 81.50 |

# 01 74 Cleaning and Waste Management

## 01 74 13 – Progress Cleaning

| 01 74 13.20 Cleaning Up | Crew | Daily Output | Labor-Hours | Unit | Material | 2011 Bare Costs Labor | Equipment | Total | Total Incl O&P |
|---|---|---|---|---|---|---|---|---|---|
| 0010 **CLEANING UP** | | | | | | | | | |
| 0020 After job completion, allow, minimum | | | | Job | | | | | .30% |
| 0040 Maximum | | | | " | | | | | 1% |
| 0050 Cleanup of floor area, continuous, per day, during const. | A-5 | 24 | .750 | M.S.F. | 1.70 | 25.50 | 1.79 | 28.99 | 43.50 |
| 0100 Final by GC at end of job | " | 11.50 | 1.565 | " | 2.71 | 53.50 | 3.73 | 59.94 | 89 |
| 0200 Rubbish removal, see Section 02 41 19.23 | | | | | | | | | |

# 01 76 Protecting Installed Construction

## 01 76 13 – Temporary Protection of Installed Construction

| 01 76 13.20 Temporary Protection | Crew | Daily Output | Labor-Hours | Unit | Material | 2011 Bare Costs Labor | Equipment | Total | Total Incl O&P |
|---|---|---|---|---|---|---|---|---|---|
| 0010 **TEMPORARY PROTECTION** | | | | | | | | | |
| 0020 Flooring, 1/8" tempered hardboard, taped seams | 2 Carp | 1500 | .011 | S.F. | .38 | .46 | | .84 | 1.12 |
| 0030 Peel away carpet protection | 1 Clab | 3200 | .003 | " | .11 | .09 | | .20 | .25 |

# 01 91 Commissioning

## 01 91 13 – General Commissioning Requirements

| 01 91 13.50 Building Commissioning | Crew | Daily Output | Labor-Hours | Unit | Material | 2011 Bare Costs Labor | Equipment | Total | Total Incl O&P |
|---|---|---|---|---|---|---|---|---|---|
| 0010 **BUILDING COMMISSIONING** | | | | | | | | | |
| 0100 Basic building commissioning, minimum | | | | % | | | | | .25% |
| 0150 Maximum | | | | | | | | | .50% |
| 0200 Enhanced building commissioning, minimum | | | | | | | | | .50% |
| 0250 Maximum | | | | | | | | | 1% |

# Division Notes

| | CREW | DAILY OUTPUT | LABOR-HOURS | UNIT | 2011 BARE COSTS | | | | TOTAL INCL O&P |
|---|---|---|---|---|---|---|---|---|---|
| | | | | | MAT. | LABOR | EQUIP. | TOTAL | |
| | | | | | | | | | |
| | | | | | | | | | |
| | | | | | | | | | |
| | | | | | | | | | |
| | | | | | | | | | |
| | | | | | | | | | |
| | | | | | | | | | |
| | | | | | | | | | |
| | | | | | | | | | |
| | | | | | | | | | |
| | | | | | | | | | |
| | | | | | | | | | |
| | | | | | | | | | |
| | | | | | | | | | |
| | | | | | | | | | |
| | | | | | | | | | |
| | | | | | | | | | |
| | | | | | | | | | |
| | | | | | | | | | |
| | | | | | | | | | |
| | | | | | | | | | |
| | | | | | | | | | |

# Estimating Tips

## 02 30 00 Subsurface Investigation

In preparing estimates on structures involving earthwork or foundations, all information concerning soil characteristics should be obtained. Look particularly for hazardous waste, evidence of prior dumping of debris, and previous stream beds.

## 02 41 00 Demolition and Structure Moving

The costs shown for selective demolition do not include rubbish handling or disposal. These items should be estimated separately using RSMeans data or other sources.

- Historic preservation often requires that the contractor remove materials from the existing structure, rehab them, and replace them. The estimator must be aware of any related measures and precautions that must be taken when doing selective demolition and cutting and patching. Requirements may include special handling and storage, as well as security.

- In addition to Subdivision 02 41 00, you can find selective demolition items in each division. Example: Roofing demolition is in Division 7.

## 02 42 10 Building Deconstruction

This section provides costs for the careful dismantling and recycling of most of low-rise building materials.

## 02 56 13 Containment of Hazardous Waste

This section addresses on-site hazardous waste disposal costs.

## 02 80 00 Hazardous Material Disposal/ Remediation

This subdivision includes information on hazardous waste handling, asbestos remediation, lead remediation, and mold remediation. See reference R028213-20 and R028319-60 for further guidance in using these unit price lines.

## 02 91 10 Monitoring Chemical Sampling, Testing Analysis

This section provides costs for on-site sampling and testing hazardous waste.

# Reference Numbers

Reference numbers are shown in shaded boxes at the beginning of some major classifications. These numbers refer to related items in the Reference Section. The reference information may be an estimating procedure, an alternate pricing method, or technical information.

*Note:* Not all subdivisions listed here necessarily appear in this publication.

# 02 21 Surveys

## 02 21 13 – Site Surveys

### 02 21 13.09 Topographical Surveys

| 02 21 13.09 Topographical Surveys | Crew | Daily Output | Labor-Hours | Unit | Material | 2011 Bare Costs Labor | Equipment | Total | Total Incl O&P |
|---|---|---|---|---|---|---|---|---|---|
| 0010 **TOPOGRAPHICAL SURVEYS** | | | | | | | | | |
| 0020 Topographical surveying, conventional, minimum | A-7 | 3.30 | 7.273 | Acre | 18.20 | 340 | 21 | 379.20 | 565 |
| 0100 Maximum | A-8 | .60 | 53.333 | " | 55.50 | 2,450 | 116 | 2,621.50 | 3,900 |

### 02 21 13.13 Boundary and Survey Markers

| 02 21 13.13 Boundary and Survey Markers | Crew | Daily Output | Labor-Hours | Unit | Material | 2011 Bare Costs Labor | Equipment | Total | Total Incl O&P |
|---|---|---|---|---|---|---|---|---|---|
| 0010 **BOUNDARY AND SURVEY MARKERS** | | | | | | | | | |
| 0300 Lot location and lines, large quantities, minimum | A-7 | 2 | 12 | Acre | 32.50 | 565 | 35 | 632.50 | 930 |
| 0320 Average | " | 1.25 | 19.200 | | 51.50 | 900 | 55.50 | 1,007 | 1,500 |
| 0400 Small quantities, maximum | A-8 | 1 | 32 | ↓ | 68.50 | 1,475 | 70 | 1,613.50 | 2,375 |
| 0600 Monuments, 3' long | A-7 | 10 | 2.400 | Ea. | 30.50 | 113 | 6.95 | 150.45 | 212 |
| 0800 Property lines, perimeter, cleared land | " | 1000 | .024 | L.F. | .03 | 1.13 | .07 | 1.23 | 1.82 |
| 0900 Wooded land | A-8 | 875 | .037 | " | .05 | 1.67 | .08 | 1.80 | 2.69 |

### 02 21 13.16 Aerial Surveys

| 02 21 13.16 Aerial Surveys | Crew | Daily Output | Labor-Hours | Unit | Material | 2011 Bare Costs Labor | Equipment | Total | Total Incl O&P |
|---|---|---|---|---|---|---|---|---|---|
| 0010 **AERIAL SURVEYS** | | | | | | | | | |
| 1500 Aerial surveying, including ground control, minimum fee, 10 acres | | | | Total | | | | | 4,200 |
| 1510 100 acres | | | | | | | | | 8,400 |
| 1550 From existing photography, deduct | | | | ↓ | | | | | 1,400 |
| 1600 2' contours, 10 acres | | | | Acre | | | | | 155 |
| 1650 20 acres | | | | | | | | | 75 |
| 1800 50 acres | | | | | | | | | 30 |
| 1850 100 acres | | | | | | | | | 25 |
| 2000 1000 acres | | | | | | | | | 25 |
| 2050 10,000 acres | | | | ↓ | | | | | 25 |
| 2150 For 1' contours and | | | | | | | | | |
| 2160 dense urban areas, add to above | | | | Acre | | | | | 20% |
| 3000 Inertial guidance system for | | | | | | | | | |
| 3010 locating coordinates, rent per day | | | | Ea. | | | | | 4,300 |

# 02 32 Geotechnical Investigations

## 02 32 13 – Subsurface Drilling and Sampling

### 02 32 13.10 Boring and Exploratory Drilling

| 02 32 13.10 Boring and Exploratory Drilling | Crew | Daily Output | Labor-Hours | Unit | Material | 2011 Bare Costs Labor | Equipment | Total | Total Incl O&P |
|---|---|---|---|---|---|---|---|---|---|
| 0010 **BORING AND EXPLORATORY DRILLING** | | | | | | | | | |
| 0020 Borings, initial field stake out & determination of elevations | A-6 | 1 | 16 | Day | | 690 | 69.50 | 759.50 | 1,125 |
| 0100 Drawings showing boring details | | | | Total | | 300 | | 300 | 375 |
| 0200 Report and recommendations from P.E. | | | | | | 700 | | 700 | 875 |
| 0300 Mobilization and demobilization, minimum | B-55 | 4 | 6 | ↓ | | 204 | 231 | 435 | 565 |
| 0350 For over 100 miles, per added mile | | 450 | .053 | Mile | | 1.81 | 2.05 | 3.86 | 5.05 |
| 0600 Auger holes in earth, no samples, 2-1/2" diameter | | 78.60 | .305 | L.F. | | 10.40 | 11.75 | 22.15 | 29 |
| 0650 4" diameter | | 67.50 | .356 | | | 12.10 | 13.70 | 25.80 | 33.50 |
| 0800 Cased borings in earth, with samples, 2-1/2" diameter | | 55.50 | .432 | | 20.50 | 14.70 | 16.65 | 51.85 | 63.50 |
| 0850 4" diameter | ↓ | 32.60 | .736 | | 32.50 | 25 | 28.50 | 86 | 105 |
| 1000 Drilling in rock, "BX" core, no sampling | B-56 | 34.90 | .458 | | | 17.85 | 37 | 54.85 | 67.50 |
| 1050 With casing & sampling | | 31.70 | .505 | | 20.50 | 19.65 | 40.50 | 80.65 | 96.50 |
| 1200 "NX" core, no sampling | | 25.92 | .617 | | | 24 | 49.50 | 73.50 | 91 |
| 1250 With casing and sampling | ↓ | 25 | .640 | ↓ | 25 | 25 | 51.50 | 101.50 | 122 |
| 1400 Borings, earth, drill rig and crew with truck mounted auger | B-55 | 1 | 24 | Day | | 815 | 925 | 1,740 | 2,275 |
| 1450 Rock using crawler type drill | B-56 | 1 | 16 | " | | 625 | 1,300 | 1,925 | 2,375 |
| 1500 For inner city borings add, minimum | | | | | | | | | 10% |
| 1510 Maximum | | | | | | | | | 20% |

# 02 32 Geotechnical Investigations

## 02 32 19 – Exploratory Excavations

| 02 32 19.10 Test Pits | | Crew | Daily Output | Labor-Hours | Unit | Material | 2011 Bare Costs Labor | 2011 Bare Costs Equipment | Total | Total Incl O&P |
|---|---|---|---|---|---|---|---|---|---|---|
| 0010 | **TEST PITS** | | | | | | | | | |
| 0020 | Hand digging, light soil | 1 Clab | 4.50 | 1.778 | C.Y. | | 61 | | 61 | 93.50 |
| 0100 | Heavy soil | " | 2.50 | 3.200 | | | 110 | | 110 | 169 |
| 0120 | Loader-backhoe, light soil | B-11M | 28 | .571 | | | 23 | 12.50 | 35.50 | 48.50 |
| 0130 | Heavy soil | " | 20 | .800 | ↓ | | 32 | 17.45 | 49.45 | 67 |
| 1000 | Subsurface exploration, mobilization | | | | Mile | | | | 6 | 7.50 |
| 1010 | Difficult access for rig, add | | | | Hr. | | | | 230 | 290 |
| 1020 | Auger borings, drill rig, incl. samples | | | | L.F. | | | | 20 | 25 |
| 1030 | Hand auger | | | | | | | | 25 | 30 |
| 1050 | Drill and sample every 5', split spoon | | | | ↓ | | | | 25 | 30 |
| 1060 | Extra samples | | | | Ea. | | | | 30 | 40 |

# 02 41 Demolition

## 02 41 13 – Selective Site Demolition

### 02 41 13.15 Hydrodemolition

| | | | Crew | Daily Output | Labor-Hours | Unit | Material | Labor | Equipment | Total | Total Incl O&P |
|---|---|---|---|---|---|---|---|---|---|---|---|
| 0010 | **HYDRODEMOLITION** | R024119-10 | | | | | | | | | |
| 0015 | Hydrodemolition, concrete pavement, 4000 PSI, 2" depth | | B-5 | 500 | .112 | S.F. | | 4.23 | 2.41 | 6.64 | 9.10 |
| 0120 | 4" depth | | | 450 | .124 | | | 4.70 | 2.67 | 7.37 | 10.10 |
| 0130 | 6" depth | | | 400 | .140 | | | 5.30 | 3.01 | 8.31 | 11.35 |
| 0410 | 6000 PSI, 2" depth | | | 410 | .137 | | | 5.15 | 2.93 | 8.08 | 11.10 |
| 0420 | 4" depth | | | 350 | .160 | | | 6.05 | 3.44 | 9.49 | 13 |
| 0430 | 6" depth | | | 300 | .187 | | | 7.05 | 4.01 | 11.06 | 15.10 |
| 0510 | 8000 PSI, 2" depth | | | 330 | .170 | | | 6.40 | 3.65 | 10.05 | 13.75 |
| 0520 | 4" depth | | | 280 | .200 | | | 7.55 | 4.30 | 11.85 | 16.25 |
| 0530 | 6" depth | | ↓ | 240 | .233 | ↓ | | 8.80 | 5 | 13.80 | 18.90 |

### 02 41 13.17 Demolish, Remove Pavement and Curb

| | | | Crew | Daily Output | Labor-Hours | Unit | Material | Labor | Equipment | Total | Total Incl O&P |
|---|---|---|---|---|---|---|---|---|---|---|---|
| 0010 | **DEMOLISH, REMOVE PAVEMENT AND CURB** | R024119-10 | | | | | | | | | |
| 5010 | Pavement removal, bituminous roads, 3" thick | | B-38 | 690 | .058 | S.Y. | | 2.25 | 1.65 | 3.90 | 5.20 |
| 5050 | 4" to 6" thick | | | 420 | .095 | | | 3.70 | 2.71 | 6.41 | 8.60 |
| 5100 | Bituminous driveways | | | 640 | .063 | | | 2.43 | 1.78 | 4.21 | 5.65 |
| 5200 | Concrete to 6" thick, hydraulic hammer, mesh reinforced | | | 255 | .157 | | | 6.10 | 4.46 | 10.56 | 14.10 |
| 5300 | Rod reinforced | | | 200 | .200 | ↓ | | 7.75 | 5.70 | 13.45 | 18 |
| 5400 | Concrete, 7" to 24" thick, plain | | | 33 | 1.212 | C.Y. | | 47 | 34.50 | 81.50 | 110 |
| 5500 | Reinforced | | ↓ | 24 | 1.667 | " | | 64.50 | 47.50 | 112 | 150 |
| 5600 | With hand held air equipment, bituminous, to 6" thick | | B-39 | 1900 | .025 | S.F. | | .92 | .10 | 1.02 | 1.51 |
| 5700 | Concrete to 6" thick, no reinforcing | | | 1600 | .030 | | | 1.09 | .12 | 1.21 | 1.79 |
| 5800 | Mesh reinforced | | | 1400 | .034 | | | 1.24 | .14 | 1.38 | 2.05 |
| 5900 | Rod reinforced | | ↓ | 765 | .063 | ↓ | | 2.27 | .25 | 2.52 | 3.75 |
| 6000 | Curbs, concrete, plain | | B-6 | 360 | .067 | L.F. | | 2.50 | .89 | 3.39 | 4.77 |
| 6100 | Reinforced | | | 275 | .087 | | | 3.27 | 1.17 | 4.44 | 6.25 |
| 6200 | Granite | | | 360 | .067 | | | 2.50 | .89 | 3.39 | 4.77 |
| 6300 | Bituminous | | ↓ | 528 | .045 | ↓ | | 1.70 | .61 | 2.31 | 3.25 |

### 02 41 13.33 Minor Site Demolition

| | | | Crew | Daily Output | Labor-Hours | Unit | Material | Labor | Equipment | Total | Total Incl O&P |
|---|---|---|---|---|---|---|---|---|---|---|---|
| 0010 | **MINOR SITE DEMOLITION** | R024119-10 | | | | | | | | | |
| 0015 | No hauling, abandon catch basin or manhole | | B-6 | 7 | 3.429 | Ea. | | 128 | 46 | 174 | 246 |
| 0020 | Remove existing catch basin or manhole, masonry | | | 4 | 6 | | | 225 | 80 | 305 | 430 |
| 0030 | Catch basin or manhole frames and covers, stored | | | 13 | 1.846 | | | 69 | 24.50 | 93.50 | 132 |
| 0040 | Remove and reset | | ↓ | 7 | 3.429 | | | 128 | 46 | 174 | 246 |
| 0100 | Roadside delineators, remove only | | B-80 | 175 | .183 | | | 6.75 | 3.54 | 10.29 | 14.15 |
| 0110 | Remove and reset | | " | 100 | .320 | ↓ | | 11.80 | 6.20 | 18 | 25 |

# 02 41 Demolition

## 02 41 13 – Selective Site Demolition

### 02 41 13.33 Minor Site Demolition

| | | Crew | Daily Output | Labor-Hours | Unit | Material | Labor | 2011 Bare Costs Equipment | Total | Total Incl O&P |
|---|---|---|---|---|---|---|---|---|---|---|
| 0800 | Guiderail, corrugated steel, remove only | B-80A | 100 | .240 | L.F. | | 8.25 | 2.17 | 10.42 | 15.05 |
| 0850 | Remove and reset | " | 40 | .600 | " | | 20.50 | 5.40 | 25.90 | 37.50 |
| 0860 | Guide posts, remove only | B-80B | 120 | .267 | Ea. | | 9.80 | 1.88 | 11.68 | 16.95 |
| 0870 | Remove and reset | B-55 | 50 | .480 | | | 16.35 | 18.45 | 34.80 | 45.50 |
| 0900 | Hydrants, fire, remove only | B-21A | 5 | 8 | | | 340 | 87 | 427 | 610 |
| 0950 | Remove and reset | " | 2 | 20 | ▼ | | 855 | 217 | 1,072 | 1,550 |
| 1000 | Masonry walls, block, solid | B-5 | 1800 | .031 | C.F. | | 1.18 | .67 | 1.85 | 2.53 |
| 1200 | Brick, solid | | 900 | .062 | | | 2.35 | 1.34 | 3.69 | 5.05 |
| 1400 | Stone, with mortar | | 900 | .062 | | | 2.35 | 1.34 | 3.69 | 5.05 |
| 1500 | Dry set | ▼ | 1500 | .037 | ▼ | | 1.41 | .80 | 2.21 | 3.02 |
| 1600 | Median barrier, precast concrete, remove and store | B-3 | 430 | .112 | L.F. | | 4.08 | 4.70 | 8.78 | 11.35 |
| 1610 | Remove and reset | " | 390 | .123 | | | 4.50 | 5.20 | 9.70 | 12.55 |
| 2900 | Pipe removal, sewer/water, no excavation, 12" diameter | B-6 | 175 | .137 | | | 5.15 | 1.83 | 6.98 | 9.80 |
| 2930 | 15"-18" diameter | B-12Z | 150 | .160 | | | 6.15 | 10.75 | 16.90 | 21 |
| 2960 | 21"-24" diameter | | 120 | .200 | | | 7.70 | 13.45 | 21.15 | 26.50 |
| 3000 | 27"-36" diameter | ▼ | 90 | .267 | | | 10.25 | 17.90 | 28.15 | 35.50 |
| 3200 | Steel, welded connections, 4" diameter | B-6 | 160 | .150 | | | 5.60 | 2 | 7.60 | 10.75 |
| 3300 | 10" diameter | " | 80 | .300 | | | 11.25 | 4.01 | 15.26 | 21.50 |
| 3500 | Railroad track removal, ties and track | B-13 | 330 | .170 | ▼ | | 6.30 | 2.27 | 8.57 | 12.10 |
| 3600 | Ballast | B-14 | 500 | .096 | C.Y. | | 3.48 | .64 | 4.12 | 6 |
| 3700 | Remove and re-install, ties & track using new bolts & spikes | | 50 | .960 | L.F. | | 35 | 6.40 | 41.40 | 60 |
| 3800 | Turnouts using new bolts and spikes | ▼ | 1 | 48 | Ea. | | 1,750 | 320 | 2,070 | 3,000 |
| 4000 | Sidewalk removal, bituminous, 2-1/2" thick | B-6 | 325 | .074 | S.Y. | | 2.76 | .99 | 3.75 | 5.30 |
| 4050 | Brick, set in mortar | | 185 | .130 | | | 4.86 | 1.73 | 6.59 | 9.30 |
| 4100 | Concrete, plain, 4" | | 160 | .150 | | | 5.60 | 2 | 7.60 | 10.75 |
| 4200 | Mesh reinforced | ▼ | 150 | .160 | ▼ | | 6 | 2.14 | 8.14 | 11.45 |
| 4300 | Slab on grade removal, plain | B-5 | 45 | 1.244 | C.Y. | | 47 | 26.50 | 73.50 | 101 |
| 4310 | Mesh reinforced | | 33 | 1.697 | | | 64 | 36.50 | 100.50 | 138 |
| 4320 | Rod reinforced | ▼ | 25 | 2.240 | | | 84.50 | 48 | 132.50 | 182 |
| 4400 | For congested sites or small quantities, add up to | | | | | | | | 200% | 200% |
| 4450 | For disposal on site, add | B-11A | 232 | .069 | | | 2.75 | 4.53 | 7.28 | 9.15 |
| 4500 | To 5 miles, add | B-34D | 76 | .105 | ▼ | | 3.62 | 7.90 | 11.52 | 14.20 |

### 02 41 13.60 Selective Demolition Fencing

| | | Crew | Daily Output | Labor-Hours | Unit | Material | Labor | 2011 Bare Costs Equipment | Total | Total Incl O&P |
|---|---|---|---|---|---|---|---|---|---|---|
| 0010 | **SELECTIVE DEMOLITION FENCING** R024119-10 | | | | | | | | | |
| 1600 | Fencing, barbed wire, 3 strand | 2 Clab | 430 | .037 | L.F. | | 1.28 | | 1.28 | 1.96 |
| 1650 | 5 strand | " | 280 | .057 | | | 1.96 | | 1.96 | 3.01 |
| 1700 | Chain link, posts & fabric, 8' to 10' high, remove only | B-6 | 445 | .054 | | | 2.02 | .72 | 2.74 | 3.86 |
| 1750 | Remove and reset | " | 70 | .343 | ▼ | | 12.85 | 4.58 | 17.43 | 24.50 |

## 02 41 16 – Structure Demolition

### 02 41 16.13 Building Demolition

| | | Crew | Daily Output | Labor-Hours | Unit | Material | Labor | 2011 Bare Costs Equipment | Total | Total Incl O&P |
|---|---|---|---|---|---|---|---|---|---|---|
| 0010 | **BUILDING DEMOLITION** Large urban projects, incl. 20 mi. haul R024119-10 | | | | | | | | | |
| 0011 | No foundation or dump fees, C.F. is vol. of building standing | | | | | | | | | |
| 0020 | Steel | B-8 | 21500 | .003 | C.F. | | .11 | .13 | .24 | .31 |
| 0050 | Concrete | | 15300 | .004 | | | .16 | .18 | .34 | .44 |
| 0080 | Masonry | | 20100 | .003 | | | .12 | .14 | .26 | .33 |
| 0100 | Mixture of types, average | ▼ | 20100 | .003 | | | .12 | .14 | .26 | .33 |
| 0500 | Small bldgs, or single bldgs, no salvage included, steel | B-3 | 14800 | .003 | | | .12 | .14 | .26 | .33 |
| 0600 | Concrete | | 11300 | .004 | | | .16 | .18 | .34 | .44 |
| 0650 | Masonry | | 14800 | .003 | | | .12 | .14 | .26 | .33 |
| 0700 | Wood | ▼ | 14800 | .003 | | | .12 | .14 | .26 | .33 |
| 0750 | For buildings with no interior walls, deduct | | | | ▼ | | | | 50% | |
| 1000 | Demoliton single family house, one story, wood 1600 S.F. | B-3 | 1 | 48 | Ea. | | 1,750 | 2,025 | 3,775 | 4,900 |

# 02 41 Demolition

## 02 41 16 – Structure Demolition

### 02 41 16.13 Building Demolition

| | | Crew | Daily Output | Labor-Hours | Unit | Material | 2011 Bare Costs Labor | Equipment | Total | Total Incl O&P |
|---|---|---|---|---|---|---|---|---|---|---|
| 1020 | 3200 S.F. | B-3 | .50 | 96 | Ea. | | 3,500 | 4,050 | 7,550 | 9,775 |
| 1200 | Demoliton two family house, two story, wood 2400 S.F. | | .67 | 71.964 | | | 2,625 | 3,025 | 5,650 | 7,325 |
| 1220 | 4200 S.F. | | .38 | 128 | | | 4,675 | 5,375 | 10,050 | 13,100 |
| 1300 | Demoliton three family house, three story, wood 3200 S.F. | | .50 | 96 | | | 3,500 | 4,050 | 7,550 | 9,775 |
| 1320 | 5400 S.F. | | .30 | 160 | | | 5,850 | 6,725 | 12,575 | 16,300 |
| 5000 | For buildings with no interior walls, deduct | | | | | | | | 50% | |

### 02 41 16.15 Explosive/Implosive Demolition

| | | Crew | Daily Output | Labor-Hours | Unit | Material | 2011 Bare Costs Labor | Equipment | Total | Total Incl O&P |
|---|---|---|---|---|---|---|---|---|---|---|
| 0010 | **EXPLOSIVE/IMPLOSIVE DEMOLITION** R024119-10 | | | | | | | | | |
| 0011 | Large projects, | | | | | | | | | |
| 0020 | No disposal fee based on building volume, steel building | B-5B | 16900 | .003 | C.F. | | .11 | .13 | .24 | .31 |
| 0100 | Concrete building | | 16900 | .003 | | | .11 | .13 | .24 | .31 |
| 0200 | Masonry building | | 16900 | .003 | | | .11 | .13 | .24 | .31 |
| 0400 | Disposal of material, minimum | B-3 | 445 | .108 | C.Y. | | 3.94 | 4.54 | 8.48 | 11 |
| 0500 | Maximum | " | 365 | .132 | " | | 4.80 | 5.55 | 10.35 | 13.40 |

### 02 41 16.17 Building Demolition Footings and Foundations

| | | Crew | Daily Output | Labor-Hours | Unit | Material | 2011 Bare Costs Labor | Equipment | Total | Total Incl O&P |
|---|---|---|---|---|---|---|---|---|---|---|
| 0010 | **BUILDING DEMOLITION FOOTINGS AND FOUNDATIONS** R024119-10 | | | | | | | | | |
| 0200 | Floors, concrete slab on grade, | | | | | | | | | |
| 0240 | 4" thick, plain concrete | B-9 | 500 | .080 | S.F. | | 2.78 | .39 | 3.17 | 4.69 |
| 0280 | Reinforced, wire mesh | | 470 | .085 | | | 2.96 | .42 | 3.38 | 5 |
| 0300 | Rods | | 400 | .100 | | | 3.48 | .49 | 3.97 | 5.90 |
| 0400 | 6" thick, plain concrete | | 375 | .107 | | | 3.71 | .52 | 4.23 | 6.25 |
| 0420 | Reinforced, wire mesh | | 340 | .118 | | | 4.09 | .57 | 4.66 | 6.90 |
| 0440 | Rods | | 300 | .133 | | | 4.63 | .65 | 5.28 | 7.80 |
| 1000 | Footings, concrete, 1' thick, 2' wide | B-5 | 300 | .187 | L.F. | | 7.05 | 4.01 | 11.06 | 15.10 |
| 1080 | 1'-6" thick, 2' wide | | 250 | .224 | | | 8.45 | 4.81 | 13.26 | 18.15 |
| 1120 | 3' wide | | 200 | .280 | | | 10.60 | 6 | 16.60 | 22.50 |
| 1140 | 2' thick, 3' wide | | 175 | .320 | | | 12.10 | 6.85 | 18.95 | 26 |
| 1200 | Average reinforcing, add | | | | | | | | 10% | 10% |
| 1220 | Heavy reinforcing, add | | | | | | | | 20% | 20% |
| 2000 | Walls, block, 4" thick | 1 Clab | 180 | .044 | S.F. | | 1.53 | | 1.53 | 2.34 |
| 2040 | 6" thick | | 170 | .047 | | | 1.62 | | 1.62 | 2.48 |
| 2080 | 8" thick | | 150 | .053 | | | 1.83 | | 1.83 | 2.81 |
| 2100 | 12" thick | | 150 | .053 | | | 1.83 | | 1.83 | 2.81 |
| 2200 | For horizontal reinforcing, add | | | | | | | | 10% | 10% |
| 2220 | For vertical reinforcing, add | | | | | | | | 20% | 20% |
| 2400 | Concrete, plain concrete, 6" thick | B-9 | 160 | .250 | | | 8.70 | 1.22 | 9.92 | 14.70 |
| 2420 | 8" thick | | 140 | .286 | | | 9.95 | 1.39 | 11.34 | 16.80 |
| 2440 | 10" thick | | 120 | .333 | | | 11.60 | 1.63 | 13.23 | 19.55 |
| 2500 | 12" thick | | 100 | .400 | | | 13.90 | 1.95 | 15.85 | 23.50 |
| 2600 | For average reinforcing, add | | | | | | | | 10% | 10% |
| 2620 | For heavy reinforcing, add | | | | | | | | 20% | 20% |
| 4000 | For congested sites or small quantities, add up to | | | | | | | | 200% | 200% |
| 4200 | Add for disposal, on site | B-11A | 232 | .069 | C.Y. | | 2.75 | 4.53 | 7.28 | 9.15 |
| 4250 | To five miles | B-30 | 220 | .109 | " | | 4.15 | 8.90 | 13.05 | 16.05 |

## 02 41 19 – Selective Demolition

### 02 41 19.13 Selective Building Demolition

| | | Crew | Daily Output | Labor-Hours | Unit | Material | 2011 Bare Costs Labor | Equipment | Total | Total Incl O&P |
|---|---|---|---|---|---|---|---|---|---|---|
| 0010 | **SELECTIVE BUILDING DEMOLITION** | | | | | | | | | |
| 0020 | Costs related to selective demolition of specific building components | | | | | | | | | |
| 0025 | are included under Common Work Results (XX 05) | | | | | | | | | |
| 0030 | in the component's appropriate division. | | | | | | | | | |

### 02 41 19.16 Selective Demolition, Cutout

| | | Daily Output | Labor-Hours | Unit | Material | 2011 Bare Costs Labor | 2011 Bare Costs Equipment | Total | Total Incl O&P |
|---|---|---|---|---|---|---|---|---|---|
| 0010 | **SELECTIVE DEMOLITION, CUTOUT** R024119-10 | | | | | | | | |
| 0020 | Concrete, elev. slab, light reinforcement, under 6 C.F. | B-9 | 65 | .615 | C.F. | | 21.50 | 3 | 24.50 | 36.50 |
| 0050 | Light reinforcing, over 6 C.F. | | 75 | .533 | " | | 18.55 | 2.60 | 21.15 | 31.50 |
| 0200 | Slab on grade to 6" thick, not reinforced, under 8 S.F. | | 85 | .471 | S.F. | | 16.35 | 2.30 | 18.65 | 27.50 |
| 0250 | 8 – 16 S.F. | | 175 | .229 | " | | 7.95 | 1.12 | 9.07 | 13.45 |
| 0255 | For over 16 S.F. see Line 02 41 16.17 0400 | | | | | | | | | |
| 0600 | Walls, not reinforced, under 6 C.F. | B-9 | 60 | .667 | C.F. | | 23 | 3.25 | 26.25 | 39 |
| 0650 | 6 – 12 C.F. | " | 80 | .500 | " | | 17.40 | 2.44 | 19.84 | 29 |
| 0655 | For over 12 C.F. see Line 02 41 16.17 2500 | | | | | | | | | |
| 1000 | Concrete, elevated slab, bar reinforced, under 6 C.F. | B-9 | 45 | .889 | C.F. | | 31 | 4.34 | 35.34 | 52.50 |
| 1050 | Bar reinforced, over 6 C.F. | | 50 | .800 | " | | 28 | 3.90 | 31.90 | 47 |
| 1200 | Slab on grade to 6" thick, bar reinforced, under 8 S.F. | | 75 | .533 | S.F. | | 18.55 | 2.60 | 21.15 | 31.50 |
| 1250 | 8 – 16 S.F. | | 150 | .267 | " | | 9.25 | 1.30 | 10.55 | 15.65 |
| 1255 | For over 16 S.F. see Line 02 41 16.17 0440 | | | | | | | | | |
| 1400 | Walls, bar reinforced, under 6 C.F. | B-9 | 50 | .800 | C.F. | | 28 | 3.90 | 31.90 | 47 |
| 1450 | 6 – 12 C.F. | " | 70 | .571 | " | | 19.85 | 2.79 | 22.64 | 33.50 |
| 1455 | For over 12 C.F. see Lines 02 41 16.17 2500 and 2600 | | | | | | | | | |
| 2000 | Brick, to 4 S.F. opening, not including toothing | | | | | | | | | |
| 2040 | 4" thick | B-9 | 30 | 1.333 | Ea. | | 46.50 | 6.50 | 53 | 78 |
| 2060 | 8" thick | | 18 | 2.222 | | | 77 | 10.85 | 87.85 | 130 |
| 2080 | 12" thick | | 10 | 4 | | | 139 | 19.50 | 158.50 | 235 |
| 2400 | Concrete block, to 4 S.F. opening, 2" thick | | 35 | 1.143 | | | 39.50 | 5.60 | 45.10 | 67 |
| 2420 | 4" thick | | 30 | 1.333 | | | 46.50 | 6.50 | 53 | 78 |
| 2440 | 8" thick | | 27 | 1.481 | | | 51.50 | 7.25 | 58.75 | 87 |
| 2460 | 12" thick | | 24 | 1.667 | | | 58 | 8.15 | 66.15 | 98 |
| 2600 | Gypsum block, to 4 S.F. opening, 2" thick | | 80 | .500 | | | 17.40 | 2.44 | 19.84 | 29 |
| 2620 | 4" thick | | 70 | .571 | | | 19.85 | 2.79 | 22.64 | 33.50 |
| 2640 | 8" thick | | 55 | .727 | | | 25.50 | 3.55 | 29.05 | 43 |
| 2800 | Terra cotta, to 4 S.F. opening, 4" thick | | 70 | .571 | | | 19.85 | 2.79 | 22.64 | 33.50 |
| 2840 | 8" thick | | 65 | .615 | | | 21.50 | 3 | 24.50 | 36.50 |
| 2880 | 12" thick | | 50 | .800 | | | 28 | 3.90 | 31.90 | 47 |
| 3000 | Toothing masonry cutouts, brick, soft old mortar | 1 Brhe | 40 | .200 | V.L.F. | | 6.95 | | 6.95 | 10.45 |
| 3100 | Hard mortar | | 30 | .267 | | | 9.30 | | 9.30 | 13.90 |
| 3200 | Block, soft old mortar | | 70 | .114 | | | 3.98 | | 3.98 | 5.95 |
| 3400 | Hard mortar | | 50 | .160 | | | 5.55 | | 5.55 | 8.35 |
| 6000 | Walls, interior, not including re-framing, | | | | | | | | | |
| 6010 | openings to 5 S.F. | | | | | | | | | |
| 6100 | Drywall to 5/8" thick | 1 Clab | 24 | .333 | Ea. | | 11.45 | | 11.45 | 17.55 |
| 6200 | Paneling to 3/4" thick | | 20 | .400 | | | 13.75 | | 13.75 | 21 |
| 6300 | Plaster, on gypsum lath | | 20 | .400 | | | 13.75 | | 13.75 | 21 |
| 6340 | On wire lath | | 14 | .571 | | | 19.65 | | 19.65 | 30 |
| 7000 | Wood frame, not including re-framing, openings to 5 S.F. | | | | | | | | | |
| 7200 | Floors, sheathing and flooring to 2" thick | 1 Clab | 5 | 1.600 | Ea. | | 55 | | 55 | 84.50 |
| 7310 | Roofs, sheathing to 1" thick, not including roofing | | 6 | 1.333 | | | 46 | | 46 | 70.50 |
| 7410 | Walls, sheathing to 1" thick, not including siding | | 7 | 1.143 | | | 39.50 | | 39.50 | 60 |

### 02 41 19.18 Selective Demolition, Disposal Only

| | | Daily Output | Labor-Hours | Unit | Material | 2011 Bare Costs Labor | 2011 Bare Costs Equipment | Total | Total Incl O&P |
|---|---|---|---|---|---|---|---|---|---|
| 0010 | **SELECTIVE DEMOLITION, DISPOSAL ONLY** R024119-10 | | | | | | | | |
| 0015 | Urban bldg w/salvage value allowed | | | | | | | | | |
| 0020 | Including loading and 5 mile haul to dump | | | | | | | | | |
| 0200 | Steel frame | B-3 | 430 | .112 | C.Y. | | 4.08 | 4.70 | 8.78 | 11.35 |
| 0300 | Concrete frame | | 365 | .132 | | | 4.80 | 5.55 | 10.35 | 13.40 |
| 0400 | Masonry construction | | 445 | .108 | | | 3.94 | 4.54 | 8.48 | 11 |

# 02 41 Demolition

## 02 41 19 – Selective Demolition

| 02 41 19.18 Selective Demolition, Disposal Only | | Crew | Daily Output | Labor-Hours | Unit | Material | 2011 Bare Costs Labor | Equipment | Total | Total Incl O&P |
|---|---|---|---|---|---|---|---|---|---|---|
| 0500 | Wood frame | B-3 | 247 | .194 | C.Y. | | 7.10 | 8.20 | 15.30 | 19.80 |

### 02 41 19.19 Selective Demolition, Dump Charges

| | | | | | | | | | | |
|---|---|---|---|---|---|---|---|---|---|---|
| 0010 | **SELECTIVE DEMOLITION, DUMP CHARGES** | R024119-10 | | | | | | | | |
| 0020 | Dump charges, typical urban city, tipping fees only | | | | | | | | | |
| 0100 | Building construction materials | | | | Ton | 82 | | | 82 | 90 |
| 0200 | Trees, brush, lumber | | | | | 70 | | | 70 | 77 |
| 0300 | Rubbish only | | | | | 70 | | | 70 | 77 |
| 0500 | Reclamation station, usual charge | | | | | 82 | | | 82 | 90 |

### 02 41 19.21 Selective Demolition, Gutting

| | | | | | | | | | | |
|---|---|---|---|---|---|---|---|---|---|---|
| 0010 | **SELECTIVE DEMOLITION, GUTTING** | R024119-10 | | | | | | | | |
| 0020 | Building interior, including disposal, dumpster fees not included | | | | | | | | | |
| 0500 | Residential building | | | | | | | | | |
| 0560 | Minimum | B-16 | 400 | .080 | SF Flr. | | 2.79 | 1.26 | 4.05 | 5.65 |
| 0580 | Maximum | " | 360 | .089 | " | | 3.10 | 1.41 | 4.51 | 6.30 |
| 0900 | Commercial building | | | | | | | | | |
| 1000 | Minimum | B-16 | 350 | .091 | SF Flr. | | 3.19 | 1.45 | 4.64 | 6.45 |
| 1020 | Maximum | " | 250 | .128 | " | | 4.46 | 2.02 | 6.48 | 9.10 |

### 02 41 19.23 Selective Demolition, Rubbish Handling

| | | | | | | | | | | |
|---|---|---|---|---|---|---|---|---|---|---|
| 0010 | **SELECTIVE DEMOLITION, RUBBISH HANDLING** | R024119-10 | | | | | | | | |
| 0020 | The following are to be added to the demolition prices | | | | | | | | | |
| 0400 | Chute, circular, prefabricated steel, 18" diameter | B-1 | 40 | .600 | L.F. | 53 | 21 | | 74 | 90.50 |
| 0440 | 30" diameter | " | 30 | .800 | " | 47.50 | 28 | | 75.50 | 95.50 |
| 0600 | Dumpster, weekly rental, 1 dump/week, 6 C.Y. capacity (2 Tons) | | | | Week | 300 | | | 300 | 330 |
| 0700 | 10 C.Y. capacity (4 Tons) | | | | | 475 | | | 475 | 525 |
| 0725 | 20 C.Y. capacity (8 Tons) | R024119-20 | | | | 630 | | | 630 | 695 |
| 0800 | 30 C.Y. capacity (10 Tons) | | | | | 750 | | | 750 | 825 |
| 0840 | 40 C.Y. capacity (13 Tons) | | | | | 875 | | | 875 | 965 |
| 2000 | Load, haul, dump and return, 50' haul, hand carried | 2 Clab | 24 | .667 | C.Y. | | 23 | | 23 | 35 |
| 2005 | Wheeled | | 37 | .432 | | | 14.85 | | 14.85 | 23 |
| 2040 | 51' to 100' haul, hand carried | | 16.50 | .970 | | | 33.50 | | 33.50 | 51 |
| 2045 | Wheeled | | 25 | .640 | | | 22 | | 22 | 33.50 |
| 2080 | Over 100' haul, add per 100 L.F., hand carried | | 35.50 | .451 | | | 15.50 | | 15.50 | 24 |
| 2085 | Wheeled | | 54 | .296 | | | 10.20 | | 10.20 | 15.60 |
| 2120 | In elevators, per 10 floors, add | | 140 | .114 | | | 3.93 | | 3.93 | 6 |
| 2130 | Load, haul, dump and return, up to 50' haul, incl. up to 5 rsr stair, hand | | 23 | .696 | | | 24 | | 24 | 36.50 |
| 2135 | Wheeled | | 35 | .457 | | | 15.70 | | 15.70 | 24 |
| 2140 | 6 – 10 riser stairs, hand carried | | 22 | .727 | | | 25 | | 25 | 38.50 |
| 2145 | Wheeled | | 34 | .471 | | | 16.15 | | 16.15 | 25 |
| 2150 | 11 – 20 riser stairs, hand carried | | 20 | .800 | | | 27.50 | | 27.50 | 42 |
| 2155 | Wheeled | | 31 | .516 | | | 17.75 | | 17.75 | 27 |
| 2160 | 21 – 40 riser stairs, hand carried | | 16 | 1 | | | 34.50 | | 34.50 | 52.50 |
| 2165 | Wheeled | | 24 | .667 | | | 23 | | 23 | 35 |
| 2170 | 100' haul, incl. 5 riser stair, hand carried | | 15 | 1.067 | | | 36.50 | | 36.50 | 56 |
| 2175 | Wheeled | | 23 | .696 | | | 24 | | 24 | 36.50 |
| 2180 | 6 – 10 riser stair, hand carried | | 14 | 1.143 | | | 39.50 | | 39.50 | 60 |
| 2185 | Wheeled | | 21 | .762 | | | 26 | | 26 | 40 |
| 2190 | 11 – 20 riser stair, hand carried | | 12 | 1.333 | | | 46 | | 46 | 70.50 |
| 2195 | Wheeled | | 18 | .889 | | | 30.50 | | 30.50 | 47 |
| 2200 | 21 – 40 riser stair, hand carried | | 8 | 2 | | | 68.50 | | 68.50 | 105 |
| 2205 | Wheeled | | 12 | 1.333 | | | 46 | | 46 | 70.50 |
| 2210 | Over 100' haul, add per 100 L.F., hand carried | | 35.50 | .451 | | | 15.50 | | 15.50 | 24 |
| 2215 | Wheeled | | 54 | .296 | | | 10.20 | | 10.20 | 15.60 |

# 02 41 Demolition

## 02 41 19 – Selective Demolition

### 02 41 19.23 Selective Demolition, Rubbish Handling

| | | Crew | Daily Output | Labor-Hours | Unit | Material | 2011 Bare Costs Labor | Equipment | Total | Total Incl O&P |
|---|---|---|---|---|---|---|---|---|---|---|
| 2220 | For each additional flight of stairs, up to 5 risers, add | 2 Clab | 550 | .029 | Flight | | 1 | | 1 | 1.53 |
| 2225 | 6 – 10 risers, add | | 275 | .058 | | | 2 | | 2 | 3.07 |
| 2230 | 11 – 20 risers, add | | 138 | .116 | | | 3.98 | | 3.98 | 6.10 |
| 2235 | 21 – 40 risers, add | ↓ | 69 | .232 | ↓ | | 7.95 | | 7.95 | 12.20 |
| 3000 | Loading & trucking, including 2 mile haul, chute loaded | B-16 | 45 | .711 | C.Y. | | 25 | 11.25 | 36.25 | 50.50 |
| 3040 | Hand loading truck, 50' haul | " | 48 | .667 | | | 23 | 10.55 | 33.55 | 47 |
| 3080 | Machine loading truck | B-17 | 120 | .267 | | | 9.80 | 5.30 | 15.10 | 20.50 |
| 5000 | Haul, per mile, up to 8 C.Y. truck | B-34B | 1165 | .007 | | | .24 | .43 | .67 | .84 |
| 5100 | Over 8 C.Y. truck | " | 1550 | .005 | ↓ | | .18 | .33 | .51 | .63 |

### 02 41 19.25 Selective Demolition, Saw Cutting

| | | | Crew | Daily Output | Labor-Hours | Unit | Material | 2011 Bare Costs Labor | Equipment | Total | Total Incl O&P |
|---|---|---|---|---|---|---|---|---|---|---|---|
| 0010 | **SELECTIVE DEMOLITION, SAW CUTTING** | R024119-10 | | | | | | | | | |
| 0015 | Asphalt, up to 3" deep | | B-89 | 1050 | .015 | L.F. | .36 | .59 | .35 | 1.30 | 1.67 |
| 0020 | Each additional inch of depth | | " | 1800 | .009 | | .12 | .34 | .21 | .67 | .88 |
| 1200 | Masonry walls, hydraulic saw, brick, per inch of depth | | B-89B | 300 | .053 | | .44 | 2.05 | 2.18 | 4.67 | 5.95 |
| 1220 | Block walls, solid, per inch of depth | | " | 250 | .064 | | .44 | 2.46 | 2.61 | 5.51 | 7.05 |
| 2000 | Brick or masonry w/hand held saw, per inch of depth | | A-1 | 125 | .064 | | .35 | 2.20 | .51 | 3.06 | 4.32 |
| 5000 | Wood sheathing to 1" thick, on walls | | 1 Carp | 200 | .040 | | | 1.72 | | 1.72 | 2.64 |
| 5020 | On roof | | " | 250 | .032 | ↓ | | 1.38 | | 1.38 | 2.11 |

### 02 41 19.27 Selective Demolition, Torch Cutting

| | | | Crew | Daily Output | Labor-Hours | Unit | Material | 2011 Bare Costs Labor | Equipment | Total | Total Incl O&P |
|---|---|---|---|---|---|---|---|---|---|---|---|
| 0010 | **SELECTIVE DEMOLITION, TORCH CUTTING** | R024119-10 | | | | | | | | | |
| 0020 | Steel, 1" thick plate | | 1 Clab | 360 | .022 | L.F. | .24 | .76 | | 1 | 1.44 |
| 0040 | 1" diameter bar | | " | 210 | .038 | Ea. | | 1.31 | | 1.31 | 2.01 |
| 1000 | Oxygen lance cutting, reinforced concrete walls | | | | | | | | | | |
| 1040 | 12" to 16" thick walls | | 1 Clab | 10 | .800 | L.F. | | 27.50 | | 27.50 | 42 |
| 1080 | 24" thick walls | | " | 6 | 1.333 | " | | 46 | | 46 | 70.50 |

# 02 42 Removal and Salvage of Construction Materials

## 02 42 10 – Building Deconstruction

### 02 42 10.10 Estimated Salvage Value or Savings

| | | | Crew | Daily Output | Labor-Hours | Unit | Material | 2011 Bare Costs Labor | Equipment | Total | Total Incl O&P |
|---|---|---|---|---|---|---|---|---|---|---|---|
| 0010 | **ESTIMATED SALVAGE VALUE OR SAVINGS** | | | | | | | | | | |
| 0015 | Excludes material handling, packaging, container costs and | | | | | | | | | | |
| 0020 | transportation for salvage or disposal | | | | | | | | | | |
| 0050 | All Items in Section 02 42 10.10 are credit deducts and not costs | | | | | | | | | | |
| 0100 | Copper Wire Salvage Value | G | | | | Lb. | | | | 1.30 | 1.30 |
| 0110 | Disposal Savings | G | | | | | | | | .03 | .03 |
| 0200 | Copper Pipe Salvage Value | G | | | | | | | | 1.34 | 1.34 |
| 0210 | Disposal Savings | G | | | | | | | | .03 | .03 |
| 0300 | Steel Pipe Salvage Value | G | | | | | | | | .06 | .06 |
| 0310 | Disposal Savings | G | | | | | | | | .03 | .03 |
| 0400 | Cast Iron Pipe Salvage Value | G | | | | | | | | .06 | .06 |
| 0410 | Disposal Savings | G | | | | | | | | .03 | .03 |
| 0500 | Steel Doors or Windows Salvage Value | G | | | | | | | | .06 | .06 |
| 0510 | Aluminum | G | | | | | | | | .66 | .66 |
| 0520 | Disposal Savings | G | | | | | | | | .03 | .03 |
| 0600 | Aluminum Siding Salvage Value | G | | | | | | | | .49 | .49 |
| 0630 | Disposal Savings | G | | | | | | | | .03 | .03 |
| 0640 | Wood Siding ( no lead or asbestos) | G | | | | C.Y. | | | | 10.90 | 10.90 |
| 0800 | Clean Concrete Disposal Savings | G | | | | Ton | | | | 60 | 60 |
| 0850 | Asphalt Shingles Disposal Savings | G | | | | " | | | | 60 | 60 |
| 1000 | Wood wall framing clean salvage value | G | | | | M.B.F. | | | | 50 | 50 |

# 02 42 Removal and Salvage of Construction Materials

## 02 42 10 – Building Deconstruction

### 02 42 10.10 Estimated Salvage Value or Savings

| | | | Crew | Daily Output | Labor-Hours | Unit | Material | 2011 Bare Costs Labor | Equipment | Total | Total Incl O&P |
|---|---|---|---|---|---|---|---|---|---|---|---|
| 1010 | Painted | G | | | | M.B.F. | | | | 40 | 40 |
| 1020 | Floor framing | G | | | | | | | | 50 | 50 |
| 1030 | Painted | G | | | | | | | | 40 | 40 |
| 1050 | Roof framing | G | | | | | | | | 50 | 50 |
| 1060 | Painted | G | | | | | | | | 40 | 40 |
| 1100 | Wood beams salvage value | G | | | | | | | | 50 | 50 |
| 1200 | Wood framing and beams disposal savings | G | | | | Ton | | | | 60 | 60 |
| 1220 | Wood sheating and sub-base flooring | G | | | | | | | | 60 | 60 |
| 1230 | Wood wall paneling (1/4 inch thick) | G | | | | | | | | 60 | 60 |
| 1300 | Wood panel 3/4-1 inch thick low salvage value | G | | | | S.F. | | | | .50 | .50 |
| 1350 | high salvage value | G | | | | " | | | | 2 | 2 |
| 1400 | Disposal savings | G | | | | Ton | | | | 60 | 60 |
| 1500 | Flooring tongue and groove 25/32 inch thick low salvage value | G | | | | S.F. | | | | .50 | .50 |
| 1530 | High salvage value | G | | | | " | | | | 1 | 1 |
| 1560 | Disposal savings | G | | | | Ton | | | | 60 | 60 |
| 1600 | Drywall or sheet rock salvage value | G | | | | | | | | 20 | 20 |
| 1650 | Disposal savings | G | | | | | | | | 60 | 60 |

### 02 42 10.20 Deconstruction of Building Components

| | | | Crew | Daily Output | Labor-Hours | Unit | Material | 2011 Bare Costs Labor | Equipment | Total | Total Incl O&P |
|---|---|---|---|---|---|---|---|---|---|---|---|
| 0010 | **DECONSTRUCTION OF BUILDING COMPONENTS** | | | | | | | | | | |
| 0012 | Buildings one or two stories only | | | | | | | | | | |
| 0015 | Excludes material handling, packaging, container costs and | | | | | | | | | | |
| 0020 | transportation for salvage or disposal | | | | | | | | | | |
| 0050 | Deconstruction of Plumbing Fixtures | | | | | | | | | | |
| 0100 | Wall hung or countertop lavatory | G | 2 Clab | 16 | 1 | Ea. | | 34.50 | | 34.50 | 52.50 |
| 0110 | Single or double compartment kitchen sink | G | | 14 | 1.143 | | | 39.50 | | 39.50 | 60 |
| 0120 | Wall hung urinal | G | | 14 | 1.143 | | | 39.50 | | 39.50 | 60 |
| 0130 | Floor mounted | G | | 8 | 2 | | | 68.50 | | 68.50 | 105 |
| 0140 | Floor mounted water closet | G | | 16 | 1 | | | 34.50 | | 34.50 | 52.50 |
| 0150 | Wall hung | G | | 14 | 1.143 | | | 39.50 | | 39.50 | 60 |
| 0160 | Water fountain, free standing | G | | 16 | 1 | | | 34.50 | | 34.50 | 52.50 |
| 0170 | Wall hung or deck mounted | G | | 12 | 1.333 | | | 46 | | 46 | 70.50 |
| 0180 | Bathtub, steel or fiberglass | G | | 10 | 1.600 | | | 55 | | 55 | 84.50 |
| 0190 | Cast iron | G | | 8 | 2 | | | 68.50 | | 68.50 | 105 |
| 0200 | Shower, single | G | | 6 | 2.667 | | | 91.50 | | 91.50 | 141 |
| 0210 | Group | G | | 7 | 2.286 | | | 78.50 | | 78.50 | 120 |
| 0300 | Deconstruction of Electrical Fixtures | | | | | | | | | | |
| 0310 | Surface mount incandescent fixtures | G | 2 Clab | 48 | .333 | Ea. | | 11.45 | | 11.45 | 17.55 |
| 0320 | Fluorescent, 2 lamp | G | | 32 | .500 | | | 17.20 | | 17.20 | 26.50 |
| 0330 | 4 lamp | G | | 24 | .667 | | | 23 | | 23 | 35 |
| 0340 | Strip Fluorescent, 1 lamp | G | | 40 | .400 | | | 13.75 | | 13.75 | 21 |
| 0350 | 2 lamp | G | | 32 | .500 | | | 17.20 | | 17.20 | 26.50 |
| 0400 | Recessed drop-in fluorescent fixture, 2 lamp | G | | 27 | .593 | | | 20.50 | | 20.50 | 31 |
| 0410 | 4 lamp | G | | 18 | .889 | | | 30.50 | | 30.50 | 47 |
| 0500 | Deconstruction of appliances | | | | | | | | | | |
| 0510 | Cooking stoves | G | 2 Clab | 26 | .615 | Ea. | | 21 | | 21 | 32.50 |
| 0520 | Dishwashers | G | " | 26 | .615 | " | | 21 | | 21 | 32.50 |
| 0600 | Deconstruction of millwork and trim | | | | | | | | | | |
| 0610 | Cabinets, wood | G | 2 Carp | 40 | .400 | L.F. | | 17.20 | | 17.20 | 26.50 |
| 0620 | Countertops | G | | 100 | .160 | " | | 6.90 | | 6.90 | 10.55 |
| 0630 | Wall paneling, 1 inch thick | G | | 500 | .032 | S.F. | | 1.38 | | 1.38 | 2.11 |
| 0640 | Ceiling trim | G | | 500 | .032 | L.F. | | 1.38 | | 1.38 | 2.11 |
| 0650 | Wainscoting | G | | 500 | .032 | S.F. | | 1.38 | | 1.38 | 2.11 |

## 02 42 10 – Building Deconstruction

| 02 42 10.20 Deconstruction of Building Components | Crew | Daily Output | Labor-Hours | Unit | Material | 2011 Bare Costs Labor | Equipment | Total | Total Incl O&P |
|---|---|---|---|---|---|---|---|---|---|
| 0660 | Base, 3/4" to 1" thick | G 2 Carp | 600 | .027 | L.F. | | 1.15 | | 1.15 | 1.76 |
| 0700 | Deconstruction of doors and windows | | | | | | | | | |
| 0710 | Doors, wrap, interior, wood, single, no closers | G 2 Carp | 21 | .762 | Ea. | 7.50 | 33 | | 40.50 | 59 |
| 0720 | Double | G | 13 | 1.231 | | 15 | 53 | | 68 | 98 |
| 0730 | Solid core, single, exterior or interior | G | 10 | 1.600 | | 7.50 | 69 | | 76.50 | 114 |
| 0740 | Double | G | 8 | 2 | | 15 | 86 | | 101 | 149 |
| 0810 | Windows, wrap, wood, single | | 21 | .762 | | 7.50 | 33 | | 40.50 | 59 |
| 0812 | with no casement or cladding | G | | | | | | | | |
| 0820 | with casement and/or cladding | G 2 Carp | 18 | .889 | Ea. | 7.50 | 38.50 | | 46 | 67 |
| 0900 | Deconstruction of interior finishes | | | | | | | | | |
| 0910 | Drywall for recycling | G 2 Clab | 1775 | .009 | S.F. | | .31 | | .31 | .47 |
| 0920 | Plaster wall, first floor | G | 1775 | .009 | | | .31 | | .31 | .47 |
| 0930 | Second floor | G | 1330 | .012 | | | .41 | | .41 | .63 |
| 1000 | Deconstruction of roofing and accessories | | | | | | | | | |
| 1010 | Built-up roofs | G 2 Clab | 570 | .028 | S.F. | | .96 | | .96 | 1.48 |
| 1020 | Gutters, facia and rakes | G " | 1140 | .014 | L.F. | | .48 | | .48 | .74 |
| 2000 | Deconstruction of wood components | | | | | | | | | |
| 2010 | Roof sheeting | G 2 Clab | 570 | .028 | S.F. | | .96 | | .96 | 1.48 |
| 2020 | Main roof framing | G | 760 | .021 | L.F. | | .72 | | .72 | 1.11 |
| 2030 | Porch roof framing | G | 445 | .036 | | | 1.24 | | 1.24 | 1.90 |
| 2040 | Beams 4" x 8" | G B-1 | 375 | .064 | | | 2.24 | | 2.24 | 3.44 |
| 2050 | 4" x 10" | G | 300 | .080 | | | 2.80 | | 2.80 | 4.30 |
| 2055 | 4" x 12" | G | 250 | .096 | | | 3.36 | | 3.36 | 5.15 |
| 2060 | 6" x 8" | G | 250 | .096 | | | 3.36 | | 3.36 | 5.15 |
| 2065 | 6" x 10" | G | 200 | .120 | | | 4.20 | | 4.20 | 6.45 |
| 2070 | 6" x 12" | G | 170 | .141 | | | 4.94 | | 4.94 | 7.60 |
| 2075 | 8" x 12" | G | 126 | .190 | | | 6.65 | | 6.65 | 10.25 |
| 2080 | 10" x 12" | G | 100 | .240 | | | 8.40 | | 8.40 | 12.90 |
| 2100 | Ceiling joists | G 2 Clab | 800 | .020 | | | .69 | | .69 | 1.05 |
| 2150 | Wall framing, interior | G | 1230 | .013 | | | .45 | | .45 | .69 |
| 2160 | Sub-floor | G | 2000 | .008 | S.F. | | .27 | | .27 | .42 |
| 2170 | Floor joists | G | 2000 | .008 | L.F. | | .27 | | .27 | .42 |
| 2200 | Wood siding (no lead or asbestos) | G | 1300 | .012 | S.F. | | .42 | | .42 | .65 |
| 2300 | Wall framing, exterior | G | 1600 | .010 | L.F. | | .34 | | .34 | .53 |
| 2400 | Stair risers | G | 53 | .302 | Ea. | | 10.35 | | 10.35 | 15.90 |
| 2500 | Posts | G | 800 | .020 | L.F. | | .69 | | .69 | 1.05 |
| 3000 | Deconstruction of exterior brick walls | | | | | | | | | |
| 3010 | Exterior brick walls, first floor | G 2 Clab | 200 | .080 | S.F. | | 2.75 | | 2.75 | 4.22 |
| 3020 | Second floor | G | 64 | .250 | " | | 8.60 | | 8.60 | 13.20 |
| 3030 | Brick chiminey | G | 100 | .160 | C.F. | | 5.50 | | 5.50 | 8.45 |
| 4000 | Deconstruction of concrete | | | | | | | | | |
| 4010 | Slab on grade, 4" thick, plain concrete | G B-9 | 500 | .080 | S.F. | | 2.78 | .39 | 3.17 | 4.69 |
| 4020 | Wire mesh reinforced | G | 470 | .085 | | | 2.96 | .42 | 3.38 | 5 |
| 4030 | Rod reinforced | G | 400 | .100 | | | 3.48 | .49 | 3.97 | 5.90 |
| 4110 | Foundation wall, 6" thick, plain concrete | G | 160 | .250 | | | 8.70 | 1.22 | 9.92 | 14.70 |
| 4120 | 8" thick | G | 140 | .286 | | | 9.95 | 1.39 | 11.34 | 16.80 |
| 4130 | 10" thick | G | 120 | .333 | | | 11.60 | 1.63 | 13.23 | 19.55 |
| 9000 | Deconstruction process, support equipment as needed | | | | | | | | | |
| 9010 | Daily use, portal to portal, 12-ton truck-mounted hydraulic crane crew | G A-3H | 1 | 8 | Day | | 370 | 870 | 1,240 | 1,525 |
| 9020 | Daily use, skid steer and operator | G A-3C | 1 | 8 | | | 350 | 261 | 611 | 805 |
| 9030 | Daily use, backhoe 48 H. P., operator and labor | G " | 1 | 8 | | | 350 | 261 | 611 | 805 |

# 02 42 Removal and Salvage of Construction Materials

## 02 42 10 – Building Deconstruction

| 02 42 10.30 Deconstruction Material Handling | | Crew | Daily Output | Labor-Hours | Unit | Material | 2011 Bare Costs Labor | Equipment | Total | Total Incl O&P |
|---|---|---|---|---|---|---|---|---|---|---|
| 0010 | **DECONSTRUCTION MATERIAL HANDLING** | | | | | | | | | |
| 0012 | Buildings one or two stories only | | | | | | | | | |
| 0100 | Clean and stack brick on pallet | G | 2 Clab | 1200 | .013 | Ea. | | .46 | | .46 | .70 |
| 0200 | Haul 50' and load rough lumber up to 2" x 8" size | G | | 2000 | .008 | " | | .27 | | .27 | .42 |
| 0210 | Lumber larger than 2" x 8" | G | | 3200 | .005 | B.F. | | .17 | | .17 | .26 |
| 0300 | Finish wood for recycling stack and wrap per pallet | G | | 8 | 2 | Ea. | 60 | 68.50 | | 128.50 | 171 |
| 0350 | Light fixtures | | | 6 | 2.667 | | 108 | 91.50 | | 199.50 | 260 |
| 0375 | Windows | | | 6 | 2.667 | | 102 | 91.50 | | 193.50 | 253 |
| 0400 | Miscellaneous materials | | ▼ | 8 | 2 | ▼ | 30 | 68.50 | | 98.50 | 138 |
| 1000 | See Section 02 41 19.23 for bulk material handling | | | | | | | | | | |

# 02 43 Structure Moving

## 02 43 13 – Structure Relocation

### 02 43 13.13 Building Relocation

| | | Crew | Daily Output | Labor-Hours | Unit | Material | 2011 Bare Costs Labor | Equipment | Total | Total Incl O&P |
|---|---|---|---|---|---|---|---|---|---|---|
| 0010 | **BUILDING RELOCATION** | | | | | | | | | |
| 0011 | One day move, up to 24' wide | | | | | | | | | |
| 0020 | Reset on new foundation, patch & hook-up, average move | | | | | Total | | | | | 11,200 |
| 0040 | Wood or steel frame bldg., based on ground floor area | G | B-4 | 185 | .259 | S.F. | | 9 | 2.32 | 11.32 | 16.35 |
| 0060 | Masonry bldg., based on ground floor area | G | " | 137 | .350 | | | 12.15 | 3.14 | 15.29 | 22 |
| 0200 | For 24' to 42' wide, add | | | | | ▼ | | | | | 15% |
| 0220 | For each additional day on road, add | G | B-4 | 1 | 48 | Day | | 1,675 | 430 | 2,105 | 3,025 |
| 0240 | Construct new basement, move building, 1 day | | | | | | | | | | |
| 0300 | move, patch & hook-up, based on ground floor area | G | B-3 | 155 | .310 | S.F. | 9.25 | 11.30 | 13.05 | 33.60 | 42 |

# 02 56 Site Containment

## 02 56 13 – Waste Containment

### 02 56 13.10 Containment of Hazardous Waste

| | | Crew | Daily Output | Labor-Hours | Unit | Material | 2011 Bare Costs Labor | Equipment | Total | Total Incl O&P |
|---|---|---|---|---|---|---|---|---|---|---|
| 0010 | **CONTAINMENT OF HAZARDOUS WASTE** | | | | | | | | | |
| 0020 | OSHA hazard level C | | | | | | | | | |
| 0030 | OSHA Hazard level D decrease labor and equipment, deduct | | | | | | | -45% | -45% | | |
| 0035 | OSHA Hazard level B increase labor and equipment, add | | | | | | | 22% | 22% | | |
| 0040 | OSHA Hazard level A increase labor and equipment, add | | | | | | | 71% | 71% | | |
| 0100 | Excavation of contaminated soil & waste | | | | | | | | | |
| 0105 | Includes one respirator filter and two disposable suits per work day | | | | | | | | | |
| 0110 | 3/4 C.Y. excavator to 10 feet deep | | B-12F | 51 | .314 | B.C.Y. | 1.64 | 12.70 | 12.55 | 26.89 | 35 |
| 0120 | Labor crew to 6' deep | | B-2 | 19 | 2.105 | | 11 | 73 | | 84 | 124 |
| 0130 | 6' - 12' deep | | " | 12 | 3.333 | ▼ | 17.40 | 116 | | 133.40 | 197 |
| 0200 | Move contaminated soil/waste upto 150' on-site with 2.5 C.Y. loader | | B-10T | 300 | .040 | L.C.Y. | .28 | 1.67 | 1.33 | 3.28 | 4.28 |
| 0210 | 300' | | " | 186 | .065 | " | .45 | 2.69 | 2.14 | 5.28 | 6.90 |
| 0300 | Secure burial cell construction | | | | | | | | | |
| 0310 | Various liner and cover materials | | | | | | | | | |
| 0400 | Very low density polyethelene (VLDPE) | | | | | | | | | |
| 0410 | 50 mil top cover | | B-47H | 4000 | .008 | S.F. | .37 | .36 | .05 | .78 | 1.02 |
| 0420 | 80 mil liner | | " | 4000 | .008 | " | .37 | .36 | .05 | .78 | 1.02 |
| 0500 | Chlorosulfunated polyethylene | | | | | | | | | |
| 0510 | 36 mil hypalon top cover | | B-47H | 4000 | .008 | S.F. | 1.18 | .36 | .05 | 1.59 | 1.91 |
| 0520 | 45 mil hypalon liner | | " | 4000 | .008 | " | 1.24 | .36 | .05 | 1.65 | 1.97 |
| 0600 | Polyvinyl chloride (PVC) | | | | | | | | | |
| 0610 | 60 mil top cover | | B-47H | 4000 | .008 | S.F. | .65 | .36 | .05 | 1.06 | 1.33 |

# 02 56 Site Containment

## 02 56 13 – Waste Containment

| 02 56 13.10 Containment of Hazardous Waste | Crew | Daily Output | Labor-Hours | Unit | Material | 2011 Bare Costs Labor | 2011 Bare Costs Equipment | Total | Total Incl O&P |
|---|---|---|---|---|---|---|---|---|---|
| 0620 | 80 mil liner | B-47H | 4000 | .008 | S.F. | .76 | .36 | .05 | 1.17 | 1.45 |
| 0700 | Rough textured H.D. polyethylene (HDPE) | | | | | | | | | |
| 0710 | 40 mil top cover | B-47H | 4000 | .008 | S.F. | .37 | .36 | .05 | .78 | 1.02 |
| 0720 | 60 mil top cover | | 4000 | .008 | | .39 | .36 | .05 | .80 | 1.04 |
| 0722 | 60 mil liner | | 4000 | .008 | | .35 | .36 | .05 | .76 | 1 |
| 0730 | 80 mil liner | ↓ | 3800 | .008 | ↓ | .42 | .38 | .06 | .86 | 1.10 |
| 1000 | 3/4" crushed stone, 6" deep ballast around liner | B-6 | 30 | .800 | L.C.Y. | 38 | 30 | 10.70 | 78.70 | 99.50 |
| 1100 | Hazardous waste, ballast cover with common borrow material | B-63 | 56 | .714 | | 15 | 26 | 2.71 | 43.71 | 59 |
| 1110 | Mixture of common borrow & topsoil | | 56 | .714 | | 19.50 | 26 | 2.71 | 48.21 | 64 |
| 1120 | Bank sand | | 56 | .714 | | 15 | 26 | 2.71 | 43.71 | 59 |
| 1130 | Medium priced clay | | 44 | .909 | | 25 | 33 | 3.45 | 61.45 | 81.50 |
| 1140 | Mixture of common borrow & medium priced clay | ↓ | 56 | .714 | ↓ | 20 | 26 | 2.71 | 48.71 | 64.50 |

# 02 58 Snow Control

## 02 58 13 – Snow Fencing

### 02 58 13.10 Snow Fencing System

| | | Crew | Daily Output | Labor-Hours | Unit | Material | 2011 Bare Costs Labor | 2011 Bare Costs Equipment | Total | Total Incl O&P |
|---|---|---|---|---|---|---|---|---|---|---|
| 0010 | **SNOW FENCING SYSTEM** | | | | | | | | | |
| 7001 | Snow fence on steel posts 10' O.C., 4' high | B-1 | 500 | .048 | L.F. | 2.02 | 1.68 | | 3.70 | 4.80 |

# 02 65 Underground Storage Tank Removal

## 02 65 10 – Underground Tank and Contaminated Soil Removal

### 02 65 10.30 Removal of Underground Storage Tanks

| | | | Crew | Daily Output | Labor-Hours | Unit | Material | 2011 Bare Costs Labor | 2011 Bare Costs Equipment | Total | Total Incl O&P |
|---|---|---|---|---|---|---|---|---|---|---|---|
| 0010 | **REMOVAL OF UNDERGROUND STORAGE TANKS** | R026510-20 | | | | | | | | | |
| 0011 | Petroleum storage tanks, non-leaking | | | | | | | | | | |
| 0100 | Excavate & load onto trailer | | | | | | | | | | |
| 0110 | 3000 gal. to 5000 gal. tank | G | B-14 | 4 | 12 | Ea. | | 435 | 80 | 515 | 755 |
| 0120 | 6000 gal. to 8000 gal. tank | G | B-3A | 3 | 13.333 | | | 485 | 315 | 800 | 1,100 |
| 0130 | 9000 gal. to 12000 gal. tank | G | " | 2 | 20 | ↓ | | 730 | 475 | 1,205 | 1,650 |
| 0190 | Known leaking tank, add | | | | | % | | | | 100% | 100% |
| 0200 | Remove sludge, water and remaining product from tank bottom | | | | | | | | | | |
| 0201 | of tank with vacuum truck | | | | | | | | | | |
| 0300 | 3000 gal. to 5000 gal. tank | G | A-13 | 5 | 1.600 | Ea. | | 70 | 147 | 217 | 266 |
| 0310 | 6000 gal. to 8000 gal. tank | G | | 4 | 2 | | | 87 | 184 | 271 | 330 |
| 0320 | 9000 gal. to 12000 gal. tank | G | ↓ | 3 | 2.667 | ↓ | | 116 | 245 | 361 | 445 |
| 0390 | Dispose of sludge off-site, average | | | | | Gal. | | | | 6 | 6.60 |
| 0400 | Insert inert solid CO$_2$ "dry ice" into tank | | | | | | | | | | |
| 0401 | For cleaning/transporting tanks (1.5 lbs./100 gal. cap) | G | 1 Clab | 500 | .016 | Lb. | 1.87 | .55 | | 2.42 | 2.90 |
| 0403 | Insert solid carbon dioxide, 1.5 lbs./100 gal. | G | " | 400 | .020 | " | 1.87 | .69 | | 2.56 | 3.11 |
| 0503 | Disconnect and remove piping | G | 1 Plum | 160 | .050 | L.F. | | 2.68 | | 2.68 | 4 |
| 0603 | Transfer liquids, 10% of volume | G | " | 1600 | .005 | Gal. | | .27 | | .27 | .40 |
| 0703 | Cut accessway into underground storage tank | G | 1 Clab | 5.33 | 1.501 | Ea. | | 51.50 | | 51.50 | 79 |
| 0813 | Remove sludge, wash and wipe tank, 500 gal. | G | 1 Plum | 8 | 1 | | | 53.50 | | 53.50 | 80 |
| 0823 | 3,000 gal. | G | | 6.67 | 1.199 | | | 64.50 | | 64.50 | 96 |
| 0833 | 5,000 gal. | G | | 6.15 | 1.301 | | | 69.50 | | 69.50 | 104 |
| 0843 | 8,000 gal. | G | | 5.33 | 1.501 | | | 80.50 | | 80.50 | 120 |
| 0853 | 10,000 gal. | G | | 4.57 | 1.751 | | | 94 | | 94 | 140 |
| 0863 | 12,000 gal. | G | ↓ | 4.21 | 1.900 | ↓ | | 102 | | 102 | 152 |
| 1020 | Haul tank to certified salvage dump, 100 miles round trip | | | | | | | | | | |
| 1023 | 3000 gal. to 5000 gal. tank | | | | | Ea. | | | | 710 | 780 |

# 02 65 Underground Storage Tank Removal

## 02 65 10 – Underground Tank and Contaminated Soil Removal

| 02 65 10.30 Removal of Underground Storage Tanks | | Crew | Daily Output | Labor-Hours | Unit | Material | 2011 Bare Costs Labor | Equipment | Total | Total Incl O&P |
|---|---|---|---|---|---|---|---|---|---|---|
| 1026 | 6000 gal. to 8000 gal. tank | | | | Ea. | | | | 810 | 890 |
| 1029 | 9,000 gal. to 12,000 gal. tank | | | | ↓ | | | | 950 | 1,050 |
| 1100 | Disposal of contaminated soil to landfill | | | | | | | | | |
| 1110 | Minimum | | | | C.Y. | | | | 125 | 140 |
| 1111 | Maximum | | | | " | | | | 400 | 440 |
| 1120 | Disposal of contaminated soil to | | | | | | | | | |
| 1121 | bituminous concrete batch plant | | | | | | | | | |
| 1130 | Minimum | | | | C.Y. | | | | 70 | 77 |
| 1131 | Maximum | | | | " | | | | 110 | 120 |
| 1203 | Excavate, pull, & load tank, backfill hole, 8,000 gal. + | G | B-12C | .50 | 32 | Ea. | | 1,300 | 2,450 | 3,750 | 4,650 |
| 1213 | Haul tank to certified dump, 100 miles rt, 8,000 gal. + | G | B-34K | 1 | 8 | | | 275 | 780 | 1,055 | 1,275 |
| 1223 | Excavate, pull, & load tank, backfill hole, 500 gal. | G | B-11C | 1 | 16 | | | 640 | 320 | 960 | 1,325 |
| 1233 | Excavate, pull, & load tank, backfill hole, 3,000 – 5,000 gal. | G | B-11M | .50 | 32 | | | 1,275 | 700 | 1,975 | 2,700 |
| 1243 | Haul tank to certified dump, 100 miles rt, 500 gal. | G | B-34L | 1 | 8 | ↓ | | 350 | 171 | 521 | 710 |
| 2010 | Decontamination of soil on site incl poly tarp on top/bottom | | | | | | | | | |
| 2011 | Soil containment berm, and chemical treatment | | | | | | | | | |
| 2020 | Minimum | G | B-11C | 100 | .160 | C.Y. | 7.05 | 6.40 | 3.20 | 16.65 | 21 |
| 2021 | Maximum | G | " | 100 | .160 | | 9.15 | 6.40 | 3.20 | 18.75 | 23.50 |
| 2050 | Disposal of decontaminated soil, minimum | | | | | | | | | 125 | 140 |
| 2055 | Maximum | | | | | ↓ | | | | 400 | 440 |

# 02 81 Transportation and Disposal of Hazardous Materials

## 02 81 20 – Hazardous Waste Handling

### 02 81 20.10 Hazardous Waste Cleanup/Pickup/Disposal

| | | Crew | Daily Output | Labor-Hours | Unit | Material | 2011 Bare Costs Labor | Equipment | Total | Total Incl O&P |
|---|---|---|---|---|---|---|---|---|---|---|
| 0010 | **HAZARDOUS WASTE CLEANUP/PICKUP/DISPOSAL** | | | | | | | | | |
| 0100 | For contractor rental equipment, i.e., Dozer, | | | | | | | | | |
| 0110 | Front end loader, Dump truck, etc., see 01 54 33 Reference Section | | | | | | | | | |
| 1000 | Solid pickup | | | | | | | | | |
| 1100 | 55 gal. drums | | | | Ea. | | | | 225 | 250 |
| 1120 | Bulk material, minimum | | | | Ton | | | | 170 | 190 |
| 1130 | Maximum | | | | " | | | | 575 | 635 |
| 1200 | Transportation to disposal site | | | | | | | | | |
| 1220 | Truckload = 80 drums or 25 C.Y. or 18 tons | | | | | | | | | |
| 1260 | Minimum | | | | Mile | | | | 3.50 | 3.85 |
| 1270 | Maximum | | | | " | | | | 6.40 | 7.05 |
| 3000 | Liquid pickup, vacuum truck, stainless steel tank | | | | | | | | | |
| 3100 | Minimum charge, 4 hours | | | | | | | | | |
| 3110 | 1 compartment, 2200 gallon | | | | Hr. | | | | 120 | 135 |
| 3120 | 2 compartment, 5000 gallon | | | | " | | | | 170 | 190 |
| 3400 | Transportation in 6900 gallon bulk truck | | | | Mile | | | | 7 | 7.70 |
| 3410 | In teflon lined truck | | | | " | | | | 9 | 9.90 |
| 5000 | Heavy sludge or dry vacuumable material | | | | Hr. | | | | 125 | 140 |
| 6000 | Dumpsite disposal charge, minimum | | | | Ton | | | | 125 | 140 |
| 6020 | Maximum | | | | " | | | | 400 | 440 |

## 02 82 13 – Asbestos Abatement

| 02 82 13.39 Asbestos Remediation Plans and Methods | Crew | Daily Output | Labor-Hours | Unit | Material | 2011 Bare Costs Labor | Equipment | Total | Total Incl O&P |
|---|---|---|---|---|---|---|---|---|---|
| 0010 **ASBESTOS REMEDIATION PLANS AND METHODS** | | | | | | | | | |
| 0100 Building Survey-Commercial Building | | | | Ea. | | | | 2,000 | 2,200 |
| 0200 Asbestos Abatement Remediation Plan | | | | " | | | | 1,200 | 1,325 |

### 02 82 13.41 Asbestos Abatement Equip.

| | Crew | Daily Output | Labor-Hours | Unit | Material | 2011 Bare Costs Labor | Equipment | Total | Total Incl O&P |
|---|---|---|---|---|---|---|---|---|---|
| 0010 **ASBESTOS ABATEMENT EQUIP.** R028213-20 | | | | | | | | | |
| 0011 Equipment and supplies, buy | | | | | | | | | |
| 0200 Air filtration device, 2000 C.F.M. | | | | Ea. | 870 | | | 870 | 955 |
| 0250 Large volume air sampling pump, minimum | | | | | 355 | | | 355 | 390 |
| 0260 Maximum | | | | | 520 | | | 520 | 570 |
| 0300 Airless sprayer unit, 2 gun | | | | | 4,800 | | | 4,800 | 5,300 |
| 0350 Light stand, 500 watt | | | | | 46 | | | 46 | 50.50 |
| 0400 Personal respirators | | | | | | | | | |
| 0410 Negative pressure, 1/2 face, dual operation, min. | | | | Ea. | 25.50 | | | 25.50 | 28 |
| 0420 Maximum | | | | | 31.50 | | | 31.50 | 35 |
| 0450 P.A.P.R., full face, minimum | | | | | 107 | | | 107 | 118 |
| 0460 Maximum | | | | | 195 | | | 195 | 215 |
| 0470 Supplied air, full face, incl. air line, minimum | | | | | 187 | | | 187 | 206 |
| 0480 Maximum | | | | | 385 | | | 385 | 425 |
| 0500 Personnel sampling pump, minimum | | | | | 221 | | | 221 | 244 |
| 1500 Power panel, 20 unit, incl. G.F.I. | | | | | 550 | | | 550 | 610 |
| 1600 Shower unit, including pump and filters | | | | | 1,175 | | | 1,175 | 1,275 |
| 1700 Supplied air system (type C) | | | | | 5,325 | | | 5,325 | 5,875 |
| 1750 Vacuum cleaner, HEPA, 16 gal., stainless steel, wet/dry | | | | | 1,175 | | | 1,175 | 1,300 |
| 1760 55 gallon | | | | | 1,925 | | | 1,925 | 2,100 |
| 1800 Vacuum loader, 9 – 18 ton/hr | | | | | 90,000 | | | 90,000 | 99,000 |
| 1900 Water atomizer unit, including 55 gal. drum | | | | | 230 | | | 230 | 253 |
| 2000 Worker protection, whole body, foot, head cover & gloves, plastic | | | | | 8.70 | | | 8.70 | 9.60 |
| 2500 Respirator, single use | | | | | 29 | | | 29 | 32 |
| 2550 Cartridge for respirator | | | | | 6.85 | | | 6.85 | 7.55 |
| 2570 Glove bag, 7 mil, 50" x 64" | | | | | 17.25 | | | 17.25 | 19 |
| 2580 10 mil, 44" x 60" | | | | | 5.40 | | | 5.40 | 5.95 |
| 2590 6 mil, 44" x 60" | | | | | 5.40 | | | 5.40 | 5.95 |
| 3000 HEPA vacuum for work area, minimum | | | | | 1,375 | | | 1,375 | 1,500 |
| 3050 Maximum | | | | | 1,425 | | | 1,425 | 1,575 |
| 6000 Disposable polyethylene bags, 6 mil, 3 C.F. | | | | | 2.30 | | | 2.30 | 2.53 |
| 6300 Disposable fiber drums, 3 C.F. | | | | | 16.90 | | | 16.90 | 18.60 |
| 6400 Pressure sensitive caution labels, 3" x 5" | | | | | 3.16 | | | 3.16 | 3.48 |
| 6450 11" x 17" | | | | | 7.05 | | | 7.05 | 7.75 |
| 6500 Negative air machine, 1800 C.F.M. | | | | | 820 | | | 820 | 905 |

### 02 82 13.42 Preparation of Asbestos Containment Area

| | Crew | Daily Output | Labor-Hours | Unit | Material | 2011 Bare Costs Labor | Equipment | Total | Total Incl O&P |
|---|---|---|---|---|---|---|---|---|---|
| 0010 **PREPARATION OF ASBESTOS CONTAINMENT AREA** | | | | | | | | | |
| 0100 Pre-cleaning, HEPA vacuum and wet wipe, flat surfaces | A-9 | 12000 | .005 | S.F. | .02 | .25 | | .27 | .41 |
| 0200 Protect carpeted area, 2 layers 6 mil poly on 3/4" plywood | " | 1000 | .064 | | 2 | 3.01 | | 5.01 | 6.85 |
| 0300 Separation barrier, 2" x 4" @ 16", 1/2" plywood ea. side, 8' high | 2 Carp | 400 | .040 | | 3 | 1.72 | | 4.72 | 5.95 |
| 0310 12' high | | 320 | .050 | | 3 | 2.15 | | 5.15 | 6.60 |
| 0320 16' high | | 200 | .080 | | 2 | 3.44 | | 5.44 | 7.50 |
| 0400 Personnel decontam. chamber, 2" x 4" @ 16", 3/4" ply ea. side | | 280 | .057 | | 4 | 2.46 | | 6.46 | 8.15 |
| 0450 Waste decontam. chamber, 2" x 4" studs @ 16", 3/4" ply ea. side | | 360 | .044 | | 4 | 1.91 | | 5.91 | 7.35 |
| 0500 Cover surfaces with polyethylene sheeting | | | | | | | | | |
| 0501 Including glue and tape | | | | | | | | | |
| 0550 Floors, each layer, 6 mil | A-9 | 8000 | .008 | S.F. | .05 | .38 | | .43 | .64 |
| 0551 4 mil | | 9000 | .007 | | .04 | .33 | | .37 | .56 |

## 02 82 13 - Asbestos Abatement

### 02 82 13.42 Preparation of Asbestos Containment Area

| | 02 82 13.42 Preparation of Asbestos Containment Area | Crew | Daily Output | Labor-Hours | Unit | Material | 2011 Bare Costs Labor | Equipment | Total | Total Incl O&P |
|---|---|---|---|---|---|---|---|---|---|---|
| 0560 | Walls, each layer, 6 mil | A-9 | 6000 | .011 | S.F. | .05 | .50 | | .55 | .83 |
| 0561 | 4 mil | ↓ | 7000 | .009 | ↓ | .04 | .43 | | .47 | .70 |
| 0570 | For heights above 12', add | | | | | | 20% | | | |
| 0575 | For heights above 20', add | | | | | | 30% | | | |
| 0580 | For fire retardant poly, add | | | | | 100% | | | | |
| 0590 | For large open areas, deduct | | | | | 10% | 20% | | | |
| 0600 | Seal floor penetrations with foam firestop to 36 sq. in. | 2 Carp | 200 | .080 | Ea. | 7.60 | 3.44 | | 11.04 | 13.70 |
| 0610 | 36 sq. in. to 72 sq. in. | | 125 | .128 | | 15.25 | 5.50 | | 20.75 | 25 |
| 0615 | 72 sq. in. to 144 sq. in. | | 80 | .200 | | 30.50 | 8.60 | | 39.10 | 46.50 |
| 0620 | Wall penetrations, to 36 square inches | | 180 | .089 | | 7.60 | 3.83 | | 11.43 | 14.25 |
| 0630 | 36 sq. in. to 72 sq. in. | | 100 | .160 | | 15.25 | 6.90 | | 22.15 | 27.50 |
| 0640 | 72 sq. in. to 144 sq. in. | ↓ | 60 | .267 | ↓ | 30.50 | 11.50 | | 42 | 51 |
| 0800 | Caulk seams with latex | 1 Carp | 230 | .035 | L.F. | .15 | 1.50 | | 1.65 | 2.47 |
| 0900 | Set up neg. air machine, 1-2k C.F.M./25 M.C.F. volume | 1 Asbe | 4.30 | 1.860 | Ea. | | 87.50 | | 87.50 | 135 |

### 02 82 13.43 Bulk Asbestos Removal

| | | Crew | Daily Output | Labor-Hours | Unit | Material | 2011 Bare Costs Labor | Equipment | Total | Total Incl O&P |
|---|---|---|---|---|---|---|---|---|---|---|
| 0010 | **BULK ASBESTOS REMOVAL** | | | | | | | | | |
| 0020 | Includes disposable tools and 2 suits and 1 respirator filter/day/worker | | | | | | | | | |
| 0100 | Beams, W 10 x 19 | A-9 | 235 | .272 | L.F. | .83 | 12.80 | | 13.63 | 20.50 |
| 0110 | W 12 x 22 | | 210 | .305 | | .93 | 14.30 | | 15.23 | 23 |
| 0120 | W 14 x 26 | | 180 | .356 | | 1.08 | 16.70 | | 17.78 | 27 |
| 0130 | W 16 x 31 | | 160 | .400 | | 1.22 | 18.80 | | 20.02 | 30.50 |
| 0140 | W 18 x 40 | | 140 | .457 | | 1.39 | 21.50 | | 22.89 | 34.50 |
| 0150 | W 24 x 55 | | 110 | .582 | | 1.77 | 27.50 | | 29.27 | 44 |
| 0160 | W 30 x 108 | | 85 | .753 | | 2.29 | 35.50 | | 37.79 | 57 |
| 0170 | W 36 x 150 | | 72 | .889 | ↓ | 2.70 | 41.50 | | 44.20 | 67.50 |
| 0200 | Boiler insulation | | 480 | .133 | S.F. | .48 | 6.25 | | 6.73 | 10.20 |
| 0210 | With metal lath, add | | | | % | | | | 50% | |
| 0300 | Boiler breeching or flue insulation | A-9 | 520 | .123 | S.F. | .37 | 5.80 | | 6.17 | 9.35 |
| 0310 | For active boiler, add | | | | % | | | | 100% | |
| 0400 | Duct or AHU insulation | A-10B | 440 | .073 | S.F. | .22 | 3.42 | | 3.64 | 5.55 |
| 0500 | Duct vibration isolation joints, up to 24 sq. in. duct | A-9 | 56 | 1.143 | Ea. | 3.47 | 53.50 | | 56.97 | 87 |
| 0520 | 25 sq. in. to 48 sq. in. duct | | 48 | 1.333 | | 4.05 | 62.50 | | 66.55 | 101 |
| 0530 | 49 sq. in. to 76 sq. in. duct | | 40 | 1.600 | ↓ | 4.86 | 75 | | 79.86 | 121 |
| 0600 | Pipe insulation, air cell type, up to 4" diameter pipe | | 900 | .071 | L.F. | .22 | 3.34 | | 3.56 | 5.40 |
| 0610 | 4" to 8" diameter pipe | | 800 | .080 | | .24 | 3.76 | | 4 | 6.05 |
| 0620 | 10" to 12" diameter pipe | | 700 | .091 | | .28 | 4.29 | | 4.57 | 6.95 |
| 0630 | 14" to 16" diameter pipe | | 550 | .116 | ↓ | .35 | 5.45 | | 5.80 | 8.85 |
| 0650 | Over 16" diameter pipe | | 650 | .098 | S.F. | .30 | 4.62 | | 4.92 | 7.50 |
| 0700 | With glove bag up to 3" diameter pipe | | 200 | .320 | L.F. | 6.05 | 15.05 | | 21.10 | 29.50 |
| 1000 | Pipe fitting insulation up to 4" diameter pipe | | 320 | .200 | Ea. | .61 | 9.40 | | 10.01 | 15.15 |
| 1100 | 6" to 8" diameter pipe | | 304 | .211 | | .64 | 9.90 | | 10.54 | 16 |
| 1110 | 10" to 12" diameter pipe | | 192 | .333 | | 1.01 | 15.65 | | 16.66 | 25 |
| 1120 | 14" to 16" diameter pipe | | 128 | .500 | ↓ | 1.52 | 23.50 | | 25.02 | 38 |
| 1130 | Over 16" diameter pipe | | 176 | .364 | S.F. | 1.11 | 17.10 | | 18.21 | 27.50 |
| 1200 | With glove bag, up to 8" diameter pipe | | 75 | .853 | L.F. | 6.55 | 40 | | 46.55 | 69 |
| 2000 | Scrape foam fireproofing from flat surface | | 2400 | .027 | S.F. | .08 | 1.25 | | 1.33 | 2.03 |
| 2100 | Irregular surfaces | | 1200 | .053 | | .16 | 2.50 | | 2.66 | 4.05 |
| 3000 | Remove cementitious material from flat surface | | 1800 | .036 | | .11 | 1.67 | | 1.78 | 2.70 |
| 3100 | Irregular surface | | 1000 | .064 | | .14 | 3.01 | | 3.15 | 4.80 |
| 4000 | Scrape acoustical coating/fireproofing, from ceiling | | 3200 | .020 | | .06 | .94 | | 1 | 1.52 |
| 5000 | Remove VAT and mastic from floor by hand | | 2400 | .027 | | .08 | 1.25 | | 1.33 | 2.03 |
| 5100 | By machine | A-11 | 4800 | .013 | ↓ | .04 | .63 | .01 | .68 | 1.02 |

## 02 82 13 – Asbestos Abatement

### 02 82 13.43 Bulk Asbestos Removal

| | | Crew | Daily Output | Labor-Hours | Unit | Material | 2011 Bare Costs Labor | Equipment | Total | Total Incl O&P |
|---|---|---|---|---|---|---|---|---|---|---|
| 5150 | For 2 layers, add | | | | % | | | | 50% | |
| 6000 | Remove contaminated soil from crawl space by hand | A-9 | 400 | .160 | C.F. | .49 | 7.50 | | 7.99 | 12.15 |
| 6100 | With large production vacuum loader | A-12 | 700 | .091 | " | .28 | 4.29 | 1.05 | 5.62 | 8.10 |
| 7000 | Radiator backing, not including radiator removal | A-9 | 1200 | .053 | S.F. | .16 | 2.50 | | 2.66 | 4.05 |
| 8000 | Cement-asbestos transite board | 2 Asbe | 1000 | .016 | | .14 | .75 | | .89 | 1.31 |
| 8100 | Transite shingle siding | A-10B | 750 | .043 | | .21 | 2.01 | | 2.22 | 3.33 |
| 8200 | Shingle roofing | " | 2000 | .016 | | .08 | .75 | | .83 | 1.25 |
| 8250 | Built-up, no gravel, non-friable | B-2 | 1400 | .029 | | .08 | .99 | | 1.07 | 1.61 |
| 8260 | Bituminous flashing | 1 Rofc | 300 | .027 | | .08 | .97 | | 1.05 | 1.71 |
| 8300 | Asbestos millboard | 2 Asbe | 1000 | .016 | | .08 | .75 | | .83 | 1.25 |
| 9000 | For type B (supplied air) respirator equipment, add | | | | % | | | | | 10% |

### 02 82 13.44 Demolition In Asbestos Contaminated Area

| | | Crew | Daily Output | Labor-Hours | Unit | Material | 2011 Bare Costs Labor | Equipment | Total | Total Incl O&P |
|---|---|---|---|---|---|---|---|---|---|---|
| 0010 | **DEMOLITION IN ASBESTOS CONTAMINATED AREA** | | | | | | | | | |
| 0200 | Ceiling, including suspension system, plaster and lath | A-9 | 2100 | .030 | S.F. | .09 | 1.43 | | 1.52 | 2.31 |
| 0210 | Finished plaster, leaving wire lath | | 585 | .109 | | .33 | 5.15 | | 5.48 | 8.30 |
| 0220 | Suspended acoustical tile | | 3500 | .018 | | .06 | .86 | | .92 | 1.39 |
| 0230 | Concealed tile grid system | | 3000 | .021 | | .06 | 1 | | 1.06 | 1.62 |
| 0240 | Metal pan grid system | | 1500 | .043 | | .13 | 2 | | 2.13 | 3.24 |
| 0250 | Gypsum board | | 2500 | .026 | | .08 | 1.20 | | 1.28 | 1.95 |
| 0260 | Lighting fixtures up to 2' x 4' | | 72 | .889 | Ea. | 2.70 | 41.50 | | 44.20 | 67.50 |
| 0400 | Partitions, non load bearing | | | | | | | | | |
| 0410 | Plaster, lath, and studs | A-9 | 690 | .093 | S.F. | .85 | 4.36 | | 5.21 | 7.70 |
| 0450 | Gypsum board and studs | " | 1390 | .046 | " | .14 | 2.16 | | 2.30 | 3.49 |
| 9000 | For type B (supplied air) respirator equipment, add | | | | % | | | | | 10% |

### 02 82 13.45 OSHA Testing

| | | Crew | Daily Output | Labor-Hours | Unit | Material | 2011 Bare Costs Labor | Equipment | Total | Total Incl O&P |
|---|---|---|---|---|---|---|---|---|---|---|
| 0010 | **OSHA TESTING** | | | | | | | | | |
| 0100 | Certified technician, minimum | | | | Day | | | | 200 | 220 |
| 0110 | Maximum | | | | | | | | 300 | 330 |
| 0120 | Industrial hygienist, minimum | | | | | | | | 300 | 330 |
| 0130 | Maximum | | | | | | | | 400 | 440 |
| 0200 | Asbestos sampling and PCM analysis, NIOSH 7400, minimum | 1 Asbe | 8 | 1 | Ea. | 2.75 | 47 | | 49.75 | 75.50 |
| 0210 | Maximum | | 4 | 2 | | 3 | 94 | | 97 | 148 |
| 1000 | Cleaned area samples | | 8 | 1 | | 2.63 | 47 | | 49.63 | 75.50 |
| 1100 | PCM air sample analysis, NIOSH 7400, minimum | | 8 | 1 | | 30.50 | 47 | | 77.50 | 107 |
| 1110 | Maximum | | 4 | 2 | | 3.20 | 94 | | 97.20 | 149 |
| 1200 | TEM air sample analysis, NIOSH 7402, minimum | | | | | | | | 90 | 120 |
| 1210 | Maximum | | | | | | | | 360 | 450 |

### 02 82 13.46 Decontamination of Asbestos Containment Area

| | | Crew | Daily Output | Labor-Hours | Unit | Material | 2011 Bare Costs Labor | Equipment | Total | Total Incl O&P |
|---|---|---|---|---|---|---|---|---|---|---|
| 0010 | **DECONTAMINATION OF ASBESTOS CONTAINMENT AREA** | | | | | | | | | |
| 0100 | Spray exposed substrate with surfactant (bridging) | | | | | | | | | |
| 0200 | Flat surfaces | A-9 | 6000 | .011 | S.F. | .36 | .50 | | .86 | 1.17 |
| 0250 | Irregular surfaces | | 4000 | .016 | " | .31 | .75 | | 1.06 | 1.50 |
| 0300 | Pipes, beams, and columns | | 2000 | .032 | L.F. | .56 | 1.50 | | 2.06 | 2.94 |
| 1000 | Spray encapsulate polyethylene sheeting | | 8000 | .008 | S.F. | .31 | .38 | | .69 | .92 |
| 1100 | Roll down polyethylene sheeting | | 8000 | .008 | " | | .38 | | .38 | .58 |
| 1500 | Bag polyethylene sheeting | | 400 | .160 | Ea. | .77 | 7.50 | | 8.27 | 12.45 |
| 2000 | Fine clean exposed substrate, with nylon brush | | 2400 | .027 | S.F. | | 1.25 | | 1.25 | 1.94 |
| 2500 | Wet wipe substrate | | 4800 | .013 | | | .63 | | .63 | .97 |
| 2600 | Vacuum surfaces, fine brush | | 6400 | .010 | | | .47 | | .47 | .73 |
| 3000 | Structural demolition | | | | | | | | | |
| 3100 | Wood stud walls | A-9 | 2800 | .023 | S.F. | | 1.07 | | 1.07 | 1.66 |
| 3500 | Window manifolds, not incl. window replacement | | 4200 | .015 | | | .72 | | .72 | 1.11 |

# 02 82 Asbestos Remediation

## 02 82 13 – Asbestos Abatement

| 02 82 13.46 Decontamination of Asbestos Containment Area | Crew | Daily Output | Labor-Hours | Unit | Material | 2011 Bare Costs Labor | Equipment | Total | Total Incl O&P |
|---|---|---|---|---|---|---|---|---|---|
| 3600 | Plywood carpet protection | A-9 | 2000 | .032 | S.F. | | 1.50 | | 1.50 | 2.32 |
| 4000 | Remove custom decontamination facility | A-10A | 8 | 3 | Ea. | 15 | 141 | | 156 | 235 |
| 4100 | Remove portable decontamination facility | 3 Asbe | 12 | 2 | " | 12.75 | 94 | | 106.75 | 159 |
| 5000 | HEPA vacuum, shampoo carpeting | A-9 | 4800 | .013 | S.F. | .05 | .63 | | .68 | 1.03 |
| 9000 | Final cleaning of protected surfaces | A-10A | 8000 | .003 | " | | .14 | | .14 | .22 |

### 02 82 13.47 Asbestos Waste Pkg., Handling, and Disp.

| | | Crew | Daily Output | Labor-Hours | Unit | Material | Labor | Equipment | Total | Total Incl O&P |
|---|---|---|---|---|---|---|---|---|---|---|
| 0010 | **ASBESTOS WASTE PACKAGING, HANDLING, AND DISPOSAL** | | | | | | | | | |
| 0100 | Collect and bag bulk material, 3 C.F. bags, by hand | A-9 | 400 | .160 | Ea. | 2.30 | 7.50 | | 9.80 | 14.15 |
| 0200 | Large production vacuum loader | A-12 | 880 | .073 | | .80 | 3.42 | .83 | 5.05 | 7.10 |
| 1000 | Double bag and decontaminate | A-9 | 960 | .067 | | 2.30 | 3.13 | | 5.43 | 7.35 |
| 2000 | Containerize bagged material in drums, per 3 C.F. drum | " | 800 | .080 | | 16.90 | 3.76 | | 20.66 | 24.50 |
| 3000 | Cart bags 50' to dumpster | 2 Asbe | 400 | .040 | | | 1.88 | | 1.88 | 2.90 |
| 5000 | Disposal charges, not including haul, minimum | | | | C.Y. | | | | 60 | 66 |
| 5020 | Maximum | | | | " | | | | 350 | 385 |
| 5100 | Remove refrigerant from system | 1 Stpi | 40 | .200 | Lb. | | 10.80 | | 10.80 | 16.15 |
| 9000 | For type B (supplied air) respirator equipment, add | | | | % | | | | | 10% |

### 02 82 13.48 Asbestos Encapsulation With Sealants

| | | Crew | Daily Output | Labor-Hours | Unit | Material | Labor | Equipment | Total | Total Incl O&P |
|---|---|---|---|---|---|---|---|---|---|---|
| 0010 | **ASBESTOS ENCAPSULATION WITH SEALANTS** | | | | | | | | | |
| 0100 | Ceilings and walls, minimum | A-9 | 21000 | .003 | S.F. | .27 | .14 | | .41 | .52 |
| 0110 | Maximum | | 10600 | .006 | | .42 | .28 | | .70 | .90 |
| 0200 | Columns and beams, minimum | | 13300 | .005 | | .27 | .23 | | .50 | .65 |
| 0210 | Maximum | | 5325 | .012 | | .47 | .56 | | 1.03 | 1.39 |
| 0300 | Pipes to 12" diameter including minor repairs, minimum | | 800 | .080 | L.F. | .37 | 3.76 | | 4.13 | 6.20 |
| 0310 | Maximum | | 400 | .160 | " | 1.04 | 7.50 | | 8.54 | 12.75 |

# 02 83 Lead Remediation

## 02 83 19 – Lead-Based Paint Remediation

### 02 83 19.21 Lead Paint Remediation Plans and Methods

| | | Crew | Daily Output | Labor-Hours | Unit | Material | Labor | Equipment | Total | Total Incl O&P |
|---|---|---|---|---|---|---|---|---|---|---|
| 0010 | **LEAD PAINT REMEDIATION PLANS AND METHODS** | | | | | | | | | |
| 0100 | Building Survey-Commercial Building | | | | Ea. | | | | 2,000 | 2,200 |
| 0200 | Lead Abatement Remediation Plan | | | | | | | | 1,200 | 1,325 |
| 0300 | Lead Paint Testing, AAS Analysis | | | | | | | | 50 | 55 |
| 0400 | Lead Paint Testing, X-Ray Fluorescence | | | | | | | | 50 | 55 |

### 02 83 19.23 Encapsulation of Lead-Based Paint

| | | Crew | Daily Output | Labor-Hours | Unit | Material | Labor | Equipment | Total | Total Incl O&P |
|---|---|---|---|---|---|---|---|---|---|---|
| 0010 | **ENCAPSULATION OF LEAD-BASED PAINT** | | | | | | | | | |
| 0020 | Interior, brushwork, trim, under 6" | 1 Pord | 240 | .033 | L.F. | 2.46 | 1.25 | | 3.71 | 4.57 |
| 0030 | 6" to 12" wide | | 180 | .044 | | 3.28 | 1.67 | | 4.95 | 6.10 |
| 0040 | Balustrades | | 300 | .027 | | 1.98 | 1 | | 2.98 | 3.67 |
| 0050 | Pipe to 4" diameter | | 500 | .016 | | 1.19 | .60 | | 1.79 | 2.20 |
| 0060 | To 8" diameter | | 375 | .021 | | 1.57 | .80 | | 2.37 | 2.92 |
| 0070 | To 12" diameter | | 250 | .032 | | 2.36 | 1.20 | | 3.56 | 4.38 |
| 0080 | To 16" diameter | | 170 | .047 | | 3.47 | 1.77 | | 5.24 | 6.45 |
| 0090 | Cabinets, ornate design | | 200 | .040 | S.F. | 2.97 | 1.50 | | 4.47 | 5.50 |
| 0100 | Simple design | | 250 | .032 | " | 2.36 | 1.20 | | 3.56 | 4.38 |
| 0110 | Doors, 3' x 7', both sides, incl. frame & trim | | | | | | | | | |
| 0120 | Flush | 1 Pord | 6 | 1.333 | Ea. | 30.50 | 50 | | 80.50 | 108 |
| 0130 | French, 10 – 15 lite | | 3 | 2.667 | | 6.05 | 100 | | 106.05 | 156 |
| 0140 | Panel | | 4 | 2 | | 36.50 | 75 | | 111.50 | 151 |
| 0150 | Louvered | | 2.75 | 2.909 | | 33.50 | 109 | | 142.50 | 199 |
| 0160 | Windows, per interior side, per 15 S.F. | | | | | | | | | |

## 02 83 19 – Lead-Based Paint Remediation

### 02 83 19.23 Encapsulation of Lead-Based Paint

| | | Crew | Daily Output | Labor-Hours | Unit | Material | 2011 Bare Costs Labor | Equipment | Total | Total Incl O&P |
|---|---|---|---|---|---|---|---|---|---|---|
| 0170 | 1 to 6 lite | 1 Pord | 14 | .571 | Ea. | 21 | 21.50 | | 42.50 | 55 |
| 0180 | 7 to 10 lite | | 7.50 | 1.067 | | 23 | 40 | | 63 | 85 |
| 0190 | 12 lite | | 5.75 | 1.391 | | 31 | 52 | | 83 | 112 |
| 0200 | Radiators | | 8 | 1 | | 73.50 | 37.50 | | 111 | 137 |
| 0210 | Grilles, vents | | 275 | .029 | S.F. | 2.15 | 1.09 | | 3.24 | 3.99 |
| 0220 | Walls, roller, drywall or plaster | | 1000 | .008 | | .59 | .30 | | .89 | 1.10 |
| 0230 | With spunbonded reinforcing fabric | | 720 | .011 | | .68 | .42 | | 1.10 | 1.37 |
| 0240 | Wood | | 800 | .010 | | .74 | .38 | | 1.12 | 1.37 |
| 0250 | Ceilings, roller, drywall or plaster | | 900 | .009 | | .68 | .33 | | 1.01 | 1.25 |
| 0260 | Wood | | 700 | .011 | | .84 | .43 | | 1.27 | 1.56 |
| 0270 | Exterior, brushwork, gutters and downspouts | | 300 | .027 | L.F. | 1.98 | 1 | | 2.98 | 3.67 |
| 0280 | Columns | | 400 | .020 | S.F. | 1.47 | .75 | | 2.22 | 2.73 |
| 0290 | Spray, siding | | 600 | .013 | " | .99 | .50 | | 1.49 | 1.83 |
| 0300 | Miscellaneous | | | | | | | | | |
| 0310 | Electrical conduit, brushwork, to 2" diameter | 1 Pord | 500 | .016 | L.F. | 1.19 | .60 | | 1.79 | 2.20 |
| 0320 | Brick, block or concrete, spray | | 500 | .016 | S.F. | 1.19 | .60 | | 1.79 | 2.20 |
| 0330 | Steel, flat surfaces and tanks to 12" | | 500 | .016 | | 1.19 | .60 | | 1.79 | 2.20 |
| 0340 | Beams, brushwork | | 400 | .020 | | 1.47 | .75 | | 2.22 | 2.73 |
| 0350 | Trusses | | 400 | .020 | | 1.47 | .75 | | 2.22 | 2.73 |

### 02 83 19.26 Removal of Lead-Based Paint

| | | Crew | Daily Output | Labor-Hours | Unit | Material | 2011 Bare Costs Labor | Equipment | Total | Total Incl O&P |
|---|---|---|---|---|---|---|---|---|---|---|
| 0010 | **REMOVAL OF LEAD-BASED PAINT** R028319-60 | | | | | | | | | |
| 0011 | By chemicals, per application | | | | | | | | | |
| 0050 | Baseboard, to 6" wide | 1 Pord | 64 | .125 | L.F. | .66 | 4.69 | | 5.35 | 7.70 |
| 0070 | To 12" wide | | 32 | .250 | " | 1.32 | 9.40 | | 10.72 | 15.40 |
| 0200 | Balustrades, one side | | 28 | .286 | S.F. | 1.32 | 10.75 | | 12.07 | 17.35 |
| 1400 | Cabinets, simple design | | 32 | .250 | | 1.32 | 9.40 | | 10.72 | 15.40 |
| 1420 | Ornate design | | 25 | .320 | | 1.32 | 12 | | 13.32 | 19.25 |
| 1600 | Cornice, simple design | | 60 | .133 | | 1.32 | 5 | | 6.32 | 8.90 |
| 1620 | Ornate design | | 20 | .400 | | 5.05 | 15 | | 20.05 | 28 |
| 2800 | Doors, one side, flush | | 84 | .095 | | 1 | 3.58 | | 4.58 | 6.40 |
| 2820 | Two panel | | 80 | .100 | | 1.32 | 3.76 | | 5.08 | 7 |
| 2840 | Four panel | | 45 | .178 | | 1.32 | 6.70 | | 8.02 | 11.35 |
| 2880 | For trim, one side, add | | 64 | .125 | L.F. | .66 | 4.69 | | 5.35 | 7.70 |
| 3000 | Fence, picket, one side | | 30 | .267 | S.F. | 1.32 | 10 | | 11.32 | 16.30 |
| 3200 | Grilles, one side, simple design | | 30 | .267 | | 1.32 | 10 | | 11.32 | 16.30 |
| 3220 | Ornate design | | 25 | .320 | | 1.32 | 12 | | 13.32 | 19.25 |
| 4400 | Pipes, to 4" diameter | | 90 | .089 | L.F. | 1 | 3.34 | | 4.34 | 6.05 |
| 4420 | To 8" diameter | | 50 | .160 | | 2 | 6 | | 8 | 11.10 |
| 4440 | To 12" diameter | | 36 | .222 | | 3 | 8.35 | | 11.35 | 15.70 |
| 4460 | To 16" diameter | | 20 | .400 | | 4 | 15 | | 19 | 27 |
| 4500 | For hangers, add | | 40 | .200 | Ea. | 2.54 | 7.50 | | 10.04 | 13.95 |
| 4800 | Siding | | 90 | .089 | S.F. | 1.17 | 3.34 | | 4.51 | 6.25 |
| 5000 | Trusses, open | | 55 | .145 | SF Face | 1.85 | 5.45 | | 7.30 | 10.15 |
| 6200 | Windows, one side only, double hung, 1/1 light, 24" x 48" high | | 4 | 2 | Ea. | 25.50 | 75 | | 100.50 | 139 |
| 6220 | 30" x 60" high | | 3 | 2.667 | | 34 | 100 | | 134 | 187 |
| 6240 | 36" x 72" high | | 2.50 | 3.200 | | 41 | 120 | | 161 | 223 |
| 6280 | 40" x 80" high | | 2 | 4 | | 51 | 150 | | 201 | 279 |
| 6400 | Colonial window, 6/6 light, 24" x 48" high | | 2 | 4 | | 51 | 150 | | 201 | 280 |
| 6420 | 30" x 60" high | | 1.50 | 5.333 | | 68 | 200 | | 268 | 370 |
| 6440 | 36" x 72" high | | 1 | 8 | | 102 | 300 | | 402 | 555 |
| 6480 | 40" x 80" high | | 1 | 8 | | 102 | 300 | | 402 | 555 |
| 6600 | 8/8 light, 24" x 48" high | | 2 | 4 | | 51 | 150 | | 201 | 280 |

# 02 83 Lead Remediation

## 02 83 19 – Lead-Based Paint Remediation

### 02 83 19.26 Removal of Lead-Based Paint

| | | Crew | Daily Output | Labor-Hours | Unit | Material | 2011 Bare Costs Labor | Equipment | Total | Total Incl O&P |
|---|---|---|---|---|---|---|---|---|---|---|
| 6620 | 40" x 80" high | 1 Pord | 1 | 8 | Ea. | 102 | 300 | | 402 | 555 |
| 6800 | 12/12 light, 24" x 48" high | | 1 | 8 | | 102 | 300 | | 402 | 555 |
| 6820 | 40" x 80" high | ↓ | .75 | 10.667 | ↓ | 136 | 400 | | 536 | 745 |
| 6840 | Window frame & trim items, included in pricing above | | | | | | | | | |

# 02 85 Mold Remediation

## 02 85 16 – Mold Remediation Preparation and Containment

### 02 85 16.40 Mold Remediation Plans and Methods

| | | Crew | Daily Output | Labor-Hours | Unit | Material | 2011 Bare Costs Labor | Equipment | Total | Total Incl O&P |
|---|---|---|---|---|---|---|---|---|---|---|
| 0010 | **MOLD REMEDIATION PLANS AND METHODS** | | | | | | | | | |
| 0020 | Initial inspection, average 3 bedroom home | | | | Total | | | | 250 | 275 |
| 0030 | Average 5 bedroom home | | | | | | | | 400 | 440 |
| 0040 | Testing, air sample each | | | | | | | | 125 | 138 |
| 0050 | Swab sample | | | | | | | | 125 | 138 |
| 0060 | Tape sample | | | | | | | | 125 | 140 |
| 0070 | Post remediation air test | | | | | | | | 125 | 140 |
| 0080 | Mold abatement plan, average 3 bedroom home | | | | | | | | 1,200 | 1,300 |
| 0090 | Average 5 bedroom home | | | | | | | | 1,600 | 1,750 |
| 0100 | Packup & removal of contents, average 3 bedroom home, excl storage | | | | | | | | 8,000 | 8,800 |
| 0110 | Average 5 bedroom home, excl storage | | | | ↓ | | | | 15,000 | 16,500 |
| 0600 | For demolition in mold contaminated areas, see Section 02 85 33.50 | | | | | | | | | |
| 0610 | For personal protection equipment, see Section 02 82 13.41 | | | | | | | | | |

### 02 85 16.50 Preparation of Mold Containment Area

| | | Crew | Daily Output | Labor-Hours | Unit | Material | 2011 Bare Costs Labor | Equipment | Total | Total Incl O&P |
|---|---|---|---|---|---|---|---|---|---|---|
| 0010 | **PREPARATION OF MOLD CONTAINMENT AREA** | | | | | | | | | |
| 0100 | Pre-cleaning, HEPA vacuum and wet wipe, flat surfaces | A-9 | 12000 | .005 | S.F. | .02 | .25 | | .27 | .41 |
| 0300 | Separation barrier, 2" x 4" @ 16", 1/2" plywood ea. side, 8' high | 2 Carp | 400 | .040 | | 3 | 1.72 | | 4.72 | 5.95 |
| 0310 | 12' high | | 320 | .050 | | 3 | 2.15 | | 5.15 | 6.60 |
| 0320 | 16' high | | 200 | .080 | | 2 | 3.44 | | 5.44 | 7.50 |
| 0400 | Personnel decontam. chamber, 2" x 4" @ 16", 3/4" ply ea. side | | 280 | .057 | | 4 | 2.46 | | 6.46 | 8.15 |
| 0450 | Waste decontam. chamber, 2" x 4" studs @ 16", 3/4" ply each side | ↓ | 360 | .044 | ↓ | 4 | 1.91 | | 5.91 | 7.35 |
| 0500 | Cover surfaces with polyethylene sheeting | | | | | | | | | |
| 0501 | Including glue and tape | | | | | | | | | |
| 0550 | Floors, each layer, 6 mil | A-9 | 8000 | .008 | S.F. | .06 | .38 | | .44 | .65 |
| 0551 | 4 mil | | 9000 | .007 | | .04 | .33 | | .37 | .57 |
| 0560 | Walls, each layer, 6 mil | | 6000 | .011 | | .09 | .50 | | .59 | .87 |
| 0561 | 4 mil | ↓ | 7000 | .009 | ↓ | .07 | .43 | | .50 | .74 |
| 0570 | For heights above 12', add | | | | | | 20% | | | |
| 0575 | For heights above 20', add | | | | | | 30% | | | |
| 0580 | For fire retardant poly, add | | | | | 100% | | | | |
| 0590 | For large open areas, deduct | | | | | 10% | 20% | | | |
| 0600 | Seal floor penetrations with foam firestop to 36 sq. in. | 2 Carp | 200 | .080 | Ea. | 7.60 | 3.44 | | 11.04 | 13.70 |
| 0610 | 36 sq. in. to 72 sq. in. | | 125 | .128 | | 15.25 | 5.50 | | 20.75 | 25 |
| 0615 | 72 sq. in. to 144 sq. in. | | 80 | .200 | | 30.50 | 8.60 | | 39.10 | 46.50 |
| 0620 | Wall penetrations, to 36 square inches | | 180 | .089 | | 7.60 | 3.83 | | 11.43 | 14.25 |
| 0630 | 36 sq. in. to 72 sq. in. | | 100 | .160 | | 15.25 | 6.90 | | 22.15 | 27.50 |
| 0640 | 72 sq. in. to 144 sq. in. | ↓ | 60 | .267 | ↓ | 30.50 | 11.50 | | 42 | 51 |
| 0800 | Caulk seams with latex caulk | 1 Carp | 230 | .035 | L.F. | .15 | 1.50 | | 1.65 | 2.47 |
| 0900 | Set up neg. air machine, 1-2k C.F.M./25 M.C.F. volume | 1 Asbe | 4.30 | 1.860 | Ea. | | 87.50 | | 87.50 | 135 |

## 02 85 33 – Removal and Disposal of Materials with Mold

### 02 85 33.50 Demolition in Mold Contaminated Area

| | | Crew | Daily Output | Labor-Hours | Unit | Material | 2011 Bare Costs Labor | Equipment | Total | Total Incl O&P |
|---|---|---|---|---|---|---|---|---|---|---|
| 0010 | **DEMOLITION IN MOLD CONTAMINATED AREA** | | | | | | | | | |

# 02 85 Mold Remediation

## 02 85 33 – Removal and Disposal of Materials with Mold

| 02 85 33.50 Demolition in Mold Contaminated Area | Crew | Daily Output | Labor-Hours | Unit | Material | 2011 Bare Costs Labor | Equipment | Total | Total Incl O&P |
|---|---|---|---|---|---|---|---|---|---|
| 0200 | Ceiling, including suspension system, plaster and lath | A-9 | 2100 | .030 | S.F. | .09 | 1.43 | | 1.52 | 2.31 |
| 0210 | Finished plaster, leaving wire lath | | 585 | .109 | | .33 | 5.15 | | 5.48 | 8.30 |
| 0220 | Suspended acoustical tile | | 3500 | .018 | | .06 | .86 | | .92 | 1.39 |
| 0230 | Concealed tile grid system | | 3000 | .021 | | .06 | 1 | | 1.06 | 1.62 |
| 0240 | Metal pan grid system | | 1500 | .043 | | .13 | 2 | | 2.13 | 3.24 |
| 0250 | Gypsum board | | 2500 | .026 | | .08 | 1.20 | | 1.28 | 1.95 |
| 0255 | Plywood | | 2500 | .026 | | .08 | 1.20 | | 1.28 | 1.95 |
| 0260 | Lighting fixtures up to 2' x 4' | | 72 | .889 | Ea. | 2.70 | 41.50 | | 44.20 | 67.50 |
| 0400 | Partitions, non load bearing | | | | | | | | | |
| 0410 | Plaster, lath, and studs | A-9 | 690 | .093 | S.F. | .85 | 4.36 | | 5.21 | 7.70 |
| 0450 | Gypsum board and studs | | 1390 | .046 | | .14 | 2.16 | | 2.30 | 3.49 |
| 0465 | Carpet & pad | | 1390 | .046 | | .14 | 2.16 | | 2.30 | 3.49 |
| 0600 | Pipe insulation, air cell type, up to 4" diameter pipe | | 900 | .071 | L.F. | .22 | 3.34 | | 3.56 | 5.40 |
| 0610 | 4" to 8" diameter pipe | | 800 | .080 | | .24 | 3.76 | | 4 | 6.05 |
| 0620 | 10" to 12" diameter pipe | | 700 | .091 | | .28 | 4.29 | | 4.57 | 6.95 |
| 0630 | 14" to 16" diameter pipe | | 550 | .116 | | .35 | 5.45 | | 5.80 | 8.85 |
| 0650 | Over 16" diameter pipe | | 650 | .098 | S.F. | .30 | 4.62 | | 4.92 | 7.50 |
| 9000 | For type B (supplied air) respirator equipment, add | | | | % | | | | | 10% |

# 02 91 Chemical Sampling, Testing, and Analysis

## 02 91 10 – Monitoring, Sampling, Testing, and Analysis

| 02 91 10.10 Monitoring, Chemical Sampling, Testing, and Analysis | Crew | Daily Output | Labor-Hours | Unit | Material | 2011 Bare Costs Labor | Equipment | Total | Total Incl O&P |
|---|---|---|---|---|---|---|---|---|---|
| 0010 | **MONITORING, CHEMICAL SAMPLING, TESTING, AND ANALYSIS** | | | | | | | | | |
| 0015 | Field Sampling of waste | | | | | | | | | |
| 0100 | Field samples, sample collection, sludge | 1 Skwk | 32 | .250 | Ea. | | 11.10 | | 11.10 | 17.05 |
| 0110 | Contaminated soils | " | 32 | .250 | " | | 11.10 | | 11.10 | 17.05 |
| 0200 | Vials and bottles | | | | | | | | | |
| 0210 | 32 oz. clear wide mouth jar (case of 12) | | | | Ea. | 41.50 | | | 41.50 | 46 |
| 0220 | 32 oz. Boston round bottle (case of 12) | | | | | 29 | | | 29 | 32 |
| 0230 | 32 oz. HDPE bottle (case of 12) | | | | | 17.60 | | | 17.60 | 19.35 |
| 0300 | Laboratory analytical services | | | | | | | | | |
| 0310 | Laboratory testing 13 metals | | | | Ea. | 190 | | | 190 | 209 |
| 0312 | 13 metals + mercury | | | | | 223 | | | 223 | 245 |
| 0314 | 8 metals | | | | | 160 | | | 160 | 176 |
| 0316 | Mercury only | | | | | 37.50 | | | 37.50 | 41.50 |
| 0318 | Single metal (only Cs, Li, Sr, Ta) | | | | | 20 | | | 20 | 22 |
| 0320 | Single metal (excludes Hg, Cs, Li, Sr, Ta) | | | | | 17.50 | | | 17.50 | 19.25 |
| 0400 | Hydrocarbons standard | | | | | 90 | | | 90 | 99 |
| 0410 | Hydrocarbons fingerprint | | | | | 163 | | | 163 | 179 |
| 0500 | Radioactivity gross alpha | | | | | 150 | | | 150 | 165 |
| 0510 | Gross alpha & beta | | | | | 150 | | | 150 | 165 |
| 0520 | Radium 226 | | | | | 85 | | | 85 | 93.50 |
| 0530 | Radium 228 | | | | | 125 | | | 125 | 138 |
| 0540 | Radon | | | | | 145 | | | 145 | 160 |
| 0550 | Uranium | | | | | 75 | | | 75 | 82.50 |
| 0600 | Volatile organics without GC/MS | | | | | 140 | | | 140 | 154 |
| 0610 | Volatile organics including GC/MS | | | | | 185 | | | 185 | 204 |
| 0630 | Synthetic organic compounds | | | | | 1,000 | | | 1,000 | 1,100 |
| 0640 | Herbicides | | | | | 213 | | | 213 | 234 |
| 0650 | Pesticides | | | | | 135 | | | 135 | 149 |
| 0660 | PCB's | | | | | 135 | | | 135 | 149 |

# Estimating Tips

## General

- Carefully check all the plans and specifications. Concrete often appears on drawings other than structural drawings, including mechanical and electrical drawings for equipment pads. The cost of cutting and patching is often difficult to estimate. See Subdivision 03 81 for Concrete Cutting, Subdivision 02 41 19.16 for Cutout Demolition, Subdivision 03 05 05.10 for Concrete Demolition, and Subdivision 02 41 19.23 for Rubbish Handling (handling, loading and hauling of debris).

- Always obtain concrete prices from suppliers near the job site. A volume discount can often be negotiated, depending upon competition in the area. Remember to add for waste, particularly for slabs and footings on grade.

## 03 10 00 Concrete Forming and Accessories

- A primary cost for concrete construction is forming. Most jobs today are constructed with prefabricated forms. The selection of the forms best suited for the job and the total square feet of forms required for efficient concrete forming and placing are key elements in estimating concrete construction. Enough forms must be available for erection to make efficient use of the concrete placing equipment and crew.

- Concrete accessories for forming and placing depend upon the systems used. Study the plans and specifications to ensure that all special accessory requirements have been included in the cost estimate, such as anchor bolts, inserts, and hangers.

- Included within costs for forms-in-place are all necessary bracing and shoring.

## 03 20 00 Concrete Reinforcing

- Ascertain that the reinforcing steel supplier has included all accessories, cutting, bending, and an allowance for lapping, splicing, and waste. A good rule of thumb is 10% for lapping, splicing, and waste. Also, 10% waste should be allowed for welded wire fabric.

- The unit price items in the subdivision for Reinforcing In Place include the labor to install accessories such as beam and slab bolsters, high chairs, and bar ties and tie wire. The material cost for these accessories is not included; they may be obtained from the Accessories Division.

## 03 30 00 Cast-In-Place Concrete

- When estimating structural concrete, pay particular attention to requirements for concrete additives, curing methods, and surface treatments. Special consideration for climate, hot or cold, must be included in your estimate. Be sure to include requirements for concrete placing equipment, and concrete finishing.

- For accurate concrete estimating, the estimator must consider each of the following major components individually: forms, reinforcing steel, ready-mix concrete, placement of the concrete, and finishing of the top surface. For faster estimating, Subdivision 03 30 53.40 for Concrete-In-Place can be used; here, various items of concrete work are presented that include the costs of all five major components (unless specifically stated otherwise).

## 03 40 00 Precast Concrete
## 03 50 00 Cast Decks and Underlayment

- The cost of hauling precast concrete structural members is often an important factor. For this reason, it is important to get a quote from the nearest supplier. It may become economically feasible to set up precasting beds on the site if the hauling costs are prohibitive.

## Reference Numbers

Reference numbers are shown in shaded boxes at the beginning of some major classifications. These numbers refer to related items in the Reference Section. The reference information may be an estimating procedure, an alternate pricing method, or technical information.

*Note:* Not all subdivisions listed here necessarily appear in this publication.

# 03 01 Maintenance of Concrete

## 03 01 30 – Maintenance of Cast-In-Place Concrete

| 03 01 30.62 Concrete Patching | Crew | Daily Output | Labor-Hours | Unit | Material | 2011 Bare Costs Labor | 2011 Bare Costs Equipment | Total | Total Incl O&P |
|---|---|---|---|---|---|---|---|---|---|
| 0010 **CONCRETE PATCHING** | | | | | | | | | |
| 0100 Floors, 1/4" thick, small areas, regular grout | 1 Cefi | 170 | .047 | S.F. | 1.15 | 1.92 | | 3.07 | 4.06 |
| 0150 Epoxy grout | " | 100 | .080 | " | 7.65 | 3.27 | | 10.92 | 13.15 |
| 2000 Walls, including chipping, cleaning and epoxy grout | | | | | | | | | |
| 2100 1/4" deep | 1 Cefi | 65 | .123 | S.F. | 7.50 | 5.05 | | 12.55 | 15.55 |
| 2150 1/2" deep | | 50 | .160 | | 15.05 | 6.55 | | 21.60 | 26 |
| 2200 3/4" deep | ↓ | 40 | .200 | ↓ | 22.50 | 8.15 | | 30.65 | 37 |

# 03 05 Common Work Results for Concrete

## 03 05 05 – Selective Concrete Demolition

### 03 05 05.10 Selective Demolition, Concrete

| | Crew | Daily Output | Labor-Hours | Unit | Material | 2011 Bare Costs Labor | 2011 Bare Costs Equipment | Total | Total Incl O&P |
|---|---|---|---|---|---|---|---|---|---|
| 0010 **SELECTIVE DEMOLITION, CONCRETE**  R024119-10 | | | | | | | | | |
| 0012 Excludes saw cutting, torch cutting, loading or hauling | | | | | | | | | |
| 0050 Break up into small pieces, minimum reinforcing | B-9 | 24 | 1.667 | C.Y. | | 58 | 8.15 | 66.15 | 98 |
| 0060 Average reinforcing | | 16 | 2.500 | | | 87 | 12.20 | 99.20 | 146 |
| 0070 Maximum reinforcing | ↓ | 8 | 5 | ↓ | | 174 | 24.50 | 198.50 | 294 |
| 0150 Remove whole pieces, up to 2 tons per piece | E-18 | 36 | 1.111 | Ea. | | 53.50 | 29 | 82.50 | 124 |
| 0160 2 – 5 tons per piece | | 30 | 1.333 | | | 64.50 | 34.50 | 99 | 148 |
| 0170 5 – 10 tons per piece | | 24 | 1.667 | | | 80.50 | 43 | 123.50 | 186 |
| 0180 10 – 15 tons per piece | ↓ | 18 | 2.222 | | | 107 | 57.50 | 164.50 | 247 |
| 0250 Precast unit embedded in masonry, up to 1 C.F. | D-1 | 16 | 1 | | | 39 | | 39 | 58.50 |
| 0260 1 – 2 C.F. | | 12 | 1.333 | | | 52 | | 52 | 78 |
| 0270 2 – 5 C.F. | | 10 | 1.600 | | | 62.50 | | 62.50 | 93.50 |
| 0280 5 – 10 C.F. | ↓ | 8 | 2 | ↓ | | 78 | | 78 | 117 |
| 0990 For hydrodemolition see Section 02 41 13.15 | | | | | | | | | |

## 03 05 13 – Basic Concrete Materials

### 03 05 13.20 Concrete Admixtures and Surface Treatments

| | Crew | Daily Output | Labor-Hours | Unit | Material | 2011 Bare Costs Labor | 2011 Bare Costs Equipment | Total | Total Incl O&P |
|---|---|---|---|---|---|---|---|---|---|
| 0010 **CONCRETE ADMIXTURES AND SURFACE TREATMENTS** | | | | | | | | | |
| 0040 Abrasives, aluminum oxide, over 20 tons | | | | Lb. | 1.48 | | | 1.48 | 1.63 |
| 0050 1 to 20 tons | | | | | 1.63 | | | 1.63 | 1.80 |
| 0070 Under 1 ton | | | | | 1.70 | | | 1.70 | 1.87 |
| 0100 Silicon carbide, black, over 20 tons | | | | | 2.15 | | | 2.15 | 2.37 |
| 0110 1 to 20 tons | | | | | 2.49 | | | 2.49 | 2.74 |
| 0120 Under 1 ton | | | | ↓ | 2.59 | | | 2.59 | 2.85 |
| 0200 Air entraining agent, .7 to 1.5 oz. per bag, 55 gallon drum | | | | Gal. | 10.55 | | | 10.55 | 11.60 |
| 0220 5 gallon pail | | | | | 14.50 | | | 14.50 | 15.95 |
| 0300 Bonding agent, acrylic latex, 250 S.F. per gallon, 5 gallon pail | | | | | 18.85 | | | 18.85 | 20.50 |
| 0320 Epoxy resin, 80 S.F. per gallon, 4 gallon case | | | | ↓ | 55.50 | | | 55.50 | 61.50 |
| 0400 Calcium chloride, 50 lb. bags, TL lots | | | | Ton | 610 | | | 610 | 670 |
| 0420 Less than truckload lots | | | | Bag | 19.05 | | | 19.05 | 21 |
| 0500 Carbon black, liquid, 2 to 8 lbs. per bag of cement | | | | Lb. | 6.85 | | | 6.85 | 7.55 |
| 0600 Colors, integral, 2 to 10 lb. per bag of cement, minimum | | | | | 2.52 | | | 2.52 | 2.77 |
| 0610 Average | | | | | 3.46 | | | 3.46 | 3.81 |
| 0620 Maximum | | | | ↓ | 5.50 | | | 5.50 | 6.05 |
| 0920 Dustproofing compound, 250 S.F./gal., 5 gallon pail | | | | Gal. | 7.40 | | | 7.40 | 8.10 |
| 1010 Epoxy based, 125 S.F./gal., 5 gallon pail | | | | " | 53 | | | 53 | 58.50 |
| 1100 Hardeners, metallic, 55 lb. bags, natural (grey) | | | | Lb. | .71 | | | .71 | .78 |
| 1200 Colors | | | | | .87 | | | .87 | .96 |
| 1300 Non-metallic, 55 lb. bags, natural grey | | | | | .45 | | | .45 | .50 |
| 1320 Colors | | | | ↓ | .75 | | | .75 | .82 |

## 03 05 13 – Basic Concrete Materials

| 03 05 13.20 Concrete Admixtures and Surface Treatments | | Crew | Daily Output | Labor-Hours | Unit | Material | 2011 Bare Costs Labor | Equipment | Total | Total Incl O&P |
|---|---|---|---|---|---|---|---|---|---|---|
| 1550 | Release agent, for tilt slabs, 5 gallon pail | | | | Gal. | 14.90 | | | 14.90 | 16.35 |
| 1570 | For forms, 5 gallon pail | | | | | 11.10 | | | 11.10 | 12.20 |
| 1590 | Concrete release agent for forms, 100% biodegradeable, zero VOC, 5 gal pail [G] | | | | | 16.75 | | | 16.75 | 18.45 |
| 1595 | 55 gallon drum [G] | | | | | 14 | | | 14 | 15.40 |
| 1600 | Sealer, hardener and dustproofer, epoxy-based, 125 S.F./gal., 5 gallon, min | | | | | 53 | | | 53 | 58.50 |
| 1620 | 5 gallon pail, maximum | | | | | 58.50 | | | 58.50 | 64 |
| 1630 | Sealer, solvent-based, 250 S.F./gal., 55 gallon drum | | | | | 16.75 | | | 16.75 | 18.40 |
| 1640 | 5 gallon pail | | | | | 24 | | | 24 | 26.50 |
| 1650 | Sealer, water based, 350 S.F., 55 gallon drum | | | | | 18.75 | | | 18.75 | 20.50 |
| 1660 | 5 gallon pail | | | | | 20 | | | 20 | 22 |
| 1900 | Set retarder, 100 S.F./gal., 1 gallon pail | | | | | 17.80 | | | 17.80 | 19.60 |
| 2000 | Waterproofing, integral 1 lb. per bag of cement | | | | Lb. | 1.60 | | | 1.60 | 1.76 |
| 2100 | Powdered metallic, 40 lbs. per 100 S.F., minimum | | | | | 1.90 | | | 1.90 | 2.09 |
| 2120 | Maximum | | | | ▼ | 2.65 | | | 2.65 | 2.92 |
| 3000 | For integral colors, 2500 psi (5 bag mix) | | | | | | | | | |
| 3100 | Red, yellow or brown, 1.8 lb. per bag, add | | | | C.Y. | 22.50 | | | 22.50 | 25 |
| 3200 | 9.4 lb. per bag, add | | | | | 118 | | | 118 | 130 |
| 3400 | Black, 1.8 lb. per bag, add | | | | | 31 | | | 31 | 34.50 |
| 3500 | 7.5 lb. per bag, add | | | | | 130 | | | 130 | 143 |
| 3700 | Green, 1.8 lb. per bag, add | | | | | 49.50 | | | 49.50 | 54.50 |
| 3800 | 7.5 lb. per bag, add | | | | ▼ | 206 | | | 206 | 226 |
| 6000 | Concrete ready-mix additives, recycled coal fly ash, mixed at plant [G] | | | | Ton | 56 | | | 56 | 61.50 |
| 6010 | Recycled blast furnace slag, mixed at plant [G] | | | | " | 87 | | | 87 | 95.50 |

### 03 05 13.25 Aggregate

| | | Crew | Daily Output | Labor-Hours | Unit | Material | 2011 Bare Costs Labor | Equipment | Total | Total Incl O&P |
|---|---|---|---|---|---|---|---|---|---|---|
| 0010 | **AGGREGATE** R033105-20 | | | | | | | | | |
| 0100 | Lightweight vermiculite or perlite, 4 C.F. bag, C.L. lots [G] | | | | Bag | 19.25 | | | 19.25 | 21 |
| 0150 | L.C.L. lots R033105-40 [G] | | | | " | 21.50 | | | 21.50 | 23.50 |
| 0250 | Sand & stone, loaded at pit, crushed bank gravel | | | | Ton | 21.50 | | | 21.50 | 23.50 |
| 0350 | Sand, washed, for concrete R033105-50 | | | | | 16.60 | | | 16.60 | 18.25 |
| 0400 | For plaster or brick | | | | | 16.60 | | | 16.60 | 18.25 |
| 0450 | Stone, 3/4" to 1-1/2" | | | | | 18.60 | | | 18.60 | 20.50 |
| 0470 | Round, river stone | | | | | 30 | | | 30 | 33 |
| 0500 | 3/8" roofing stone & 1/2" pea stone | | | | | 24 | | | 24 | 26.50 |
| 0550 | For trucking 10 miles (5 mile R/T), add to the above | B-34B | 117 | .068 | | | 2.35 | 4.33 | 6.68 | 8.35 |
| 0600 | 30 miles (15 mi R/T), add to the above | " | 72 | .111 | ▼ | | 3.82 | 7.05 | 10.87 | 13.55 |
| 0850 | Sand & stone, loaded at pit, crushed bank gravel | | | | C.Y. | 30 | | | 30 | 33 |
| 0950 | Sand, washed, for concrete | | | | | 23 | | | 23 | 25.50 |
| 1000 | For plaster or brick | | | | | 23 | | | 23 | 25.50 |
| 1050 | Stone, 3/4" to 1-1/2" | | | | | 35.50 | | | 35.50 | 39 |
| 1055 | Round, river stone | | | | | 33.50 | | | 33.50 | 37 |
| 1100 | 3/8" roofing stone & 1/2" pea stone | | | | | 39.50 | | | 39.50 | 43.50 |
| 1150 | For trucking 10 miles (5 mi R/T), add to the above | B-34B | 78 | .103 | | | 3.53 | 6.50 | 10.03 | 12.50 |
| 1200 | 30 miles (15 mi R/T), add to the above | " | 48 | .167 | ▼ | | 5.75 | 10.55 | 16.30 | 20.50 |
| 1310 | Quartz chips, 50 lb. bags | | | | Cwt. | 19.20 | | | 19.20 | 21 |
| 1330 | Silica chips, 50 lb. bags | | | | | 10.55 | | | 10.55 | 11.60 |
| 1410 | White marble, 3/8" to 1/2", 50 lb. bags | | | | ▼ | 15.60 | | | 15.60 | 17.15 |
| 1430 | 3/4" | | | | Ton | 90 | | | 90 | 99 |

### 03 05 13.30 Cement

| | | Crew | Daily Output | Labor-Hours | Unit | Material | 2011 Bare Costs Labor | Equipment | Total | Total Incl O&P |
|---|---|---|---|---|---|---|---|---|---|---|
| 0010 | **CEMENT** R033105-20 | | | | | | | | | |
| 0240 | Portland, Type I/II, TL lots, 94 lb bags | | | | Bag | 10.70 | | | 10.70 | 11.80 |
| 0250 | LTL/LCL lots R033105-30 | | | | " | 12.45 | | | 12.45 | 13.70 |
| 0300 | Trucked in bulk, per cwt | | | | Cwt. | 7.90 | | | 7.90 | 8.70 |

# 03 05 Common Work Results for Concrete

## 03 05 13 – Basic Concrete Materials

### 03 05 13.30 Cement

| | | Crew | Daily Output | Labor-Hours | Unit | Material | 2011 Bare Costs Labor | Equipment | Total | Total Incl O&P |
|---|---|---|---|---|---|---|---|---|---|---|
| 0400 | Type III, high early strength, TL lots, 94 lb bags | R033105-40 | | | | Bag | 12.50 | | | 12.50 | 13.75 |
| 0420 | L.T.L. or L.C.L. lots | | | | | | 12.40 | | | 12.40 | 13.65 |
| 0500 | White, type III, high early strength, T.L. or C.L. lots, bags | R033105-50 | | | | | 25 | | | 25 | 27.50 |
| 0520 | L.T.L. or L.C.L. lots | | | | | | 25 | | | 25 | 27.50 |
| 0600 | White, type I, T.L. or C.L. lots, bags | | | | | | 23.50 | | | 23.50 | 25.50 |
| 0620 | L.T.L. or L.C.L. lots | | | | | | 24.50 | | | 24.50 | 27 |

### 03 05 13.80 Waterproofing and Dampproofing

| | | Crew | Daily Output | Labor-Hours | Unit | Material | 2011 Bare Costs Labor | Equipment | Total | Total Incl O&P |
|---|---|---|---|---|---|---|---|---|---|---|
| 0010 | **WATERPROOFING AND DAMPPROOFING** | | | | | | | | | | |
| 0050 | Integral waterproofing, add to cost of regular concrete | | | | | C.Y. | 9.60 | | | 9.60 | 10.55 |

### 03 05 13.85 Winter Protection

| | | Crew | Daily Output | Labor-Hours | Unit | Material | 2011 Bare Costs Labor | Equipment | Total | Total Incl O&P |
|---|---|---|---|---|---|---|---|---|---|---|
| 0010 | **WINTER PROTECTION** | | | | | | | | | | |
| 0012 | For heated ready mix, add, minimum | | | | | C.Y. | 4.25 | | | 4.25 | 4.68 |
| 0050 | Maximum | | | | | " | 5.30 | | | 5.30 | 5.85 |
| 0100 | Temporary heat to protect concrete, 24 hours, minimum | 2 Clab | 50 | .320 | M.S.F. | 525 | 11 | | 536 | 595 |
| 0150 | Maximum | " | 25 | .640 | " | 685 | 22 | | 707 | 790 |
| 0200 | Temporary shelter for slab on grade, wood frame/polyethylene sheeting | | | | | | | | | |
| 0201 | Build or remove, minimum | 2 Carp | 10 | 1.600 | M.S.F. | 268 | 69 | | 337 | 400 |
| 0210 | Maximum | " | 3 | 5.333 | " | 325 | 230 | | 555 | 705 |
| 0500 | Electrically heated pads, 110 volts, 15 watts per S.F., buy | | | | S.F. | 5.90 | | | 5.90 | 6.50 |
| 0600 | 20 watts per S.F., buy | | | | | 7.85 | | | 7.85 | 8.65 |
| 0710 | Electrically, heated pads, 15 watts/S.F., 20 uses, minimum | | | | | .21 | | | .21 | .23 |
| 0800 | Maximum | | | | | .35 | | | .35 | .39 |

# 03 11 Concrete Forming

## 03 11 13 – Structural Cast-In-Place Concrete Forming

### 03 11 13.20 Forms In Place, Beams and Girders

| | | | Crew | Daily Output | Labor-Hours | Unit | Material | 2011 Bare Costs Labor | Equipment | Total | Total Incl O&P |
|---|---|---|---|---|---|---|---|---|---|---|---|
| 0010 | **FORMS IN PLACE, BEAMS AND GIRDERS** | R031113-40 | | | | | | | | | |
| 0500 | Exterior spandrel, job-built plywood, 12" wide, 1 use | | C-2 | 225 | .213 | SFCA | 2.50 | 8.95 | | 11.45 | 16.45 |
| 0550 | 2 use | R031113-60 | | 275 | .175 | | 1.31 | 7.30 | | 8.61 | 12.70 |
| 0600 | 3 use | | | 295 | .163 | | 1 | 6.80 | | 7.80 | 11.55 |
| 0650 | 4 use | | | 310 | .155 | | .81 | 6.50 | | 7.31 | 10.85 |
| 1000 | 18" wide, 1 use | | | 250 | .192 | | 2.19 | 8.05 | | 10.24 | 14.75 |
| 1050 | 2 use | | | 275 | .175 | | 1.21 | 7.30 | | 8.51 | 12.60 |
| 1100 | 3 use | | | 305 | .157 | | .88 | 6.60 | | 7.48 | 11.05 |
| 1150 | 4 use | | | 315 | .152 | | .72 | 6.40 | | 7.12 | 10.60 |
| 1500 | 24" wide, 1 use | | | 265 | .181 | | 2 | 7.60 | | 9.60 | 13.85 |
| 1550 | 2 use | | | 290 | .166 | | 1.13 | 6.95 | | 8.08 | 11.90 |
| 1600 | 3 use | | | 315 | .152 | | .80 | 6.40 | | 7.20 | 10.70 |
| 1650 | 4 use | | | 325 | .148 | | .65 | 6.20 | | 6.85 | 10.20 |
| 2000 | Interior beam, job-built plywood, 12" wide, 1 use | | | 300 | .160 | | 3.02 | 6.70 | | 9.72 | 13.65 |
| 2050 | 2 use | | | 340 | .141 | | 1.46 | 5.90 | | 7.36 | 10.70 |
| 2100 | 3 use | | | 364 | .132 | | 1.21 | 5.55 | | 6.76 | 9.85 |
| 2150 | 4 use | | | 377 | .127 | | .98 | 5.35 | | 6.33 | 9.30 |
| 2500 | 24" wide, 1 use | | | 320 | .150 | | 2.04 | 6.30 | | 8.34 | 11.90 |
| 2550 | 2 use | | | 365 | .132 | | 1.15 | 5.50 | | 6.65 | 9.70 |
| 2600 | 3 use | | | 385 | .125 | | .81 | 5.25 | | 6.06 | 8.90 |
| 2650 | 4 use | | | 395 | .122 | | .66 | 5.10 | | 5.76 | 8.50 |
| 3000 | Encasing steel beam, hung, job-built plywood, 1 use | | | 325 | .148 | | 2.43 | 6.20 | | 8.63 | 12.15 |
| 3050 | 2 use | | | 390 | .123 | | 1.33 | 5.15 | | 6.48 | 9.35 |
| 3100 | 3 use | | | 415 | .116 | | .97 | 4.85 | | 5.82 | 8.50 |

# 03 11 Concrete Forming

## 03 11 13 – Structural Cast-In-Place Concrete Forming

### 03 11 13.20 Forms In Place, Beams and Girders

| | | Crew | Daily Output | Labor-Hours | Unit | Material | 2011 Bare Costs Labor | Equipment | Total | Total Incl O&P |
|---|---|---|---|---|---|---|---|---|---|---|
| 3150 | 4 use | C-2 | 430 | .112 | SFCA | .79 | 4.68 | | 5.47 | 8.05 |
| 3500 | Bottoms only, to 30" wide, job-built plywood, 1 use | | 230 | .209 | | 3.42 | 8.75 | | 12.17 | 17.20 |
| 3550 | 2 use | | 265 | .181 | | 1.91 | 7.60 | | 9.51 | 13.75 |
| 3600 | 3 use | | 280 | .171 | | 1.37 | 7.20 | | 8.57 | 12.55 |
| 3650 | 4 use | | 290 | .166 | | 1.11 | 6.95 | | 8.06 | 11.85 |
| 4000 | Sides only, vertical, 36" high, job-built plywood, 1 use | | 335 | .143 | | 3.46 | 6 | | 9.46 | 13 |
| 4050 | 2 use | | 405 | .119 | | 1.91 | 4.97 | | 6.88 | 9.70 |
| 4100 | 3 use | | 430 | .112 | | 1.39 | 4.68 | | 6.07 | 8.70 |
| 4150 | 4 use | | 445 | .108 | | 1.13 | 4.52 | | 5.65 | 8.20 |
| 4500 | Sloped sides, 36" high, 1 use | | 305 | .157 | | 3.41 | 6.60 | | 10.01 | 13.85 |
| 4550 | 2 use | | 370 | .130 | | 1.90 | 5.45 | | 7.35 | 10.45 |
| 4600 | 3 use | | 405 | .119 | | 1.36 | 4.97 | | 6.33 | 9.10 |
| 4650 | 4 use | | 425 | .113 | | 1.11 | 4.74 | | 5.85 | 8.45 |
| 5000 | Upstanding beams, 36" high, 1 use | | 225 | .213 | | 4.14 | 8.95 | | 13.09 | 18.25 |
| 5050 | 2 use | | 255 | .188 | | 2.30 | 7.90 | | 10.20 | 14.65 |
| 5100 | 3 use | | 275 | .175 | | 1.67 | 7.30 | | 8.97 | 13.10 |
| 5150 | 4 use | | 280 | .171 | | 1.36 | 7.20 | | 8.56 | 12.55 |

### 03 11 13.25 Forms In Place, Columns

| | | | Crew | Daily Output | Labor-Hours | Unit | Material | 2011 Bare Costs Labor | Equipment | Total | Total Incl O&P |
|---|---|---|---|---|---|---|---|---|---|---|---|
| 0010 | **FORMS IN PLACE, COLUMNS** | R031113-40 | | | | | | | | | |
| 0500 | Round fiberglass, 4 use per mo., rent, 12" diameter | | C-1 | 160 | .200 | L.F. | 8.15 | 8.20 | | 16.35 | 21.50 |
| 0550 | 16" diameter | R031113-60 | | 150 | .213 | | 9.70 | 8.70 | | 18.40 | 24 |
| 0600 | 18" diameter | | | 140 | .229 | | 10.85 | 9.35 | | 20.20 | 26.50 |
| 0650 | 24" diameter | | | 135 | .237 | | 13.50 | 9.70 | | 23.20 | 29.50 |
| 0700 | 28" diameter | | | 130 | .246 | | 15.10 | 10.05 | | 25.15 | 32 |
| 0800 | 30" diameter | | | 125 | .256 | | 15.75 | 10.45 | | 26.20 | 33.50 |
| 0850 | 36" diameter | | | 120 | .267 | | 21 | 10.90 | | 31.90 | 39.50 |
| 1500 | Round fiber tube, recycled paper, 1 use, 8" diameter | G | | 155 | .206 | | 1.42 | 8.45 | | 9.87 | 14.50 |
| 1550 | 10" diameter | G | | 155 | .206 | | 1.97 | 8.45 | | 10.42 | 15.10 |
| 1600 | 12" diameter | G | | 150 | .213 | | 2.30 | 8.70 | | 11 | 15.95 |
| 1650 | 14" diameter | G | | 145 | .221 | | 3.33 | 9 | | 12.33 | 17.50 |
| 1700 | 16" diameter | G | | 140 | .229 | | 4.06 | 9.35 | | 13.41 | 18.80 |
| 1720 | 18" diameter | G | | 140 | .229 | | 4.75 | 9.35 | | 14.10 | 19.60 |
| 1750 | 20" diameter | G | | 135 | .237 | | 6.55 | 9.70 | | 16.25 | 22 |
| 1800 | 24" diameter | G | | 130 | .246 | | 8.40 | 10.05 | | 18.45 | 24.50 |
| 1850 | 30" diameter | G | | 125 | .256 | | 11.40 | 10.45 | | 21.85 | 28.50 |
| 1900 | 36" diameter | G | | 115 | .278 | | 15.05 | 11.40 | | 26.45 | 34 |
| 1950 | 42" diameter | G | | 100 | .320 | | 38 | 13.10 | | 51.10 | 61.50 |
| 2000 | 48" diameter | G | | 85 | .376 | | 44 | 15.40 | | 59.40 | 72 |
| 2200 | For seamless type, add | | | | | | 15% | | | | |
| 3000 | Round, steel, 4 use per mo., rent, regular duty, 14" diameter | G | C-1 | 145 | .221 | L.F. | 15.15 | 9 | | 24.15 | 30.50 |
| 3050 | 16" diameter | G | | 125 | .256 | | 15.50 | 10.45 | | 25.95 | 33 |
| 3100 | Heavy duty, 20" diameter | G | | 105 | .305 | | 17 | 12.45 | | 29.45 | 38 |
| 3150 | 24" diameter | G | | 85 | .376 | | 18.65 | 15.40 | | 34.05 | 44 |
| 3200 | 30" diameter | G | | 70 | .457 | | 21.50 | 18.70 | | 40.20 | 52 |
| 3250 | 36" diameter | G | | 60 | .533 | | 23 | 22 | | 45 | 59 |
| 3300 | 48" diameter | G | | 50 | .640 | | 68 | 26 | | 94 | 115 |
| 3350 | 60" diameter | G | | 45 | .711 | | 42 | 29 | | 71 | 90.50 |
| 4500 | For second and succeeding months, deduct | | | | | | 50% | | | | |
| 5000 | Job-built plywood, 8" x 8" columns, 1 use | | C-1 | 165 | .194 | SFCA | 2.31 | 7.95 | | 10.26 | 14.70 |
| 5050 | 2 use | | | 195 | .164 | | 1.32 | 6.70 | | 8.02 | 11.75 |
| 5100 | 3 use | | | 210 | .152 | | .92 | 6.25 | | 7.17 | 10.55 |
| 5150 | 4 use | | | 215 | .149 | | .76 | 6.10 | | 6.86 | 10.20 |

## 03 11 13 – Structural Cast-In-Place Concrete Forming

### 03 11 13.25 Forms In Place, Columns

| | | Crew | Daily Output | Labor-Hours | Unit | Material | 2011 Bare Costs Labor | Equipment | Total | Total Incl O&P |
|---|---|---|---|---|---|---|---|---|---|---|
| 5500 | 12" x 12" columns, 1 use | C-1 | 180 | .178 | SFCA | 2.19 | 7.25 | | 9.44 | 13.55 |
| 5550 | 2 use | | 210 | .152 | | 1.20 | 6.25 | | 7.45 | 10.85 |
| 5600 | 3 use | | 220 | .145 | | .87 | 5.95 | | 6.82 | 10.05 |
| 5650 | 4 use | | 225 | .142 | | .71 | 5.80 | | 6.51 | 9.70 |
| 6000 | 16" x 16" columns, 1 use | | 185 | .173 | | 2.17 | 7.05 | | 9.22 | 13.25 |
| 6050 | 2 use | | 215 | .149 | | 1.16 | 6.10 | | 7.26 | 10.65 |
| 6100 | 3 use | | 230 | .139 | | .87 | 5.70 | | 6.57 | 9.65 |
| 6150 | 4 use | | 235 | .136 | | .71 | 5.55 | | 6.26 | 9.35 |
| 6500 | 24" x 24" columns, 1 use | | 190 | .168 | | 2.43 | 6.90 | | 9.33 | 13.20 |
| 6550 | 2 use | | 216 | .148 | | 1.33 | 6.05 | | 7.38 | 10.75 |
| 6600 | 3 use | | 230 | .139 | | .97 | 5.70 | | 6.67 | 9.75 |
| 6650 | 4 use | | 238 | .134 | | .79 | 5.50 | | 6.29 | 9.30 |
| 7000 | 36" x 36" columns, 1 use | | 200 | .160 | | 1.77 | 6.55 | | 8.32 | 12 |
| 7050 | 2 use | | 230 | .139 | | 1.01 | 5.70 | | 6.71 | 9.80 |
| 7100 | 3 use | | 245 | .131 | | .71 | 5.35 | | 6.06 | 9 |
| 7150 | 4 use | | 250 | .128 | | .58 | 5.25 | | 5.83 | 8.70 |
| 7400 | Steel framed plywood, based on 50 uses of purchased | | | | | | | | | |
| 7420 | forms, and 4 uses of bracing lumber | | | | | | | | | |
| 7500 | 8" x 8" column | C-1 | 340 | .094 | SFCA | 1.81 | 3.85 | | 5.66 | 7.90 |
| 7550 | 10" x 10" | | 350 | .091 | | 1.58 | 3.74 | | 5.32 | 7.50 |
| 7600 | 12" x 12" | | 370 | .086 | | 1.35 | 3.54 | | 4.89 | 6.90 |
| 7650 | 16" x 16" | | 400 | .080 | | 1.05 | 3.27 | | 4.32 | 6.15 |
| 7700 | 20" x 20" | | 420 | .076 | | .93 | 3.11 | | 4.04 | 5.80 |
| 7750 | 24" x 24" | | 440 | .073 | | .67 | 2.97 | | 3.64 | 5.30 |
| 7755 | 30" x 30" | | 440 | .073 | | .86 | 2.97 | | 3.83 | 5.50 |
| 7760 | 36" x 36" | | 460 | .070 | | .75 | 2.84 | | 3.59 | 5.20 |

### 03 11 13.30 Forms In Place, Culvert

| | | | | | | | | | | |
|---|---|---|---|---|---|---|---|---|---|---|
| 0010 | **FORMS IN PLACE, CULVERT** | R031113-40 | | | | | | | | |
| 0015 | 5' to 8' square or rectangular, 1 use | | C-1 | 170 | .188 | SFCA | 3.10 | 7.70 | | 10.80 | 15.20 |
| 0050 | 2 use | R031113-60 | | 180 | .178 | | 1.83 | 7.25 | | 9.08 | 13.15 |
| 0100 | 3 use | | | 190 | .168 | | 1.40 | 6.90 | | 8.30 | 12.10 |
| 0150 | 4 use | | | 200 | .160 | | 1.19 | 6.55 | | 7.74 | 11.35 |

### 03 11 13.35 Forms In Place, Elevated Slabs

| | | | | | | | | | | |
|---|---|---|---|---|---|---|---|---|---|---|
| 0010 | **FORMS IN PLACE, ELEVATED SLABS** | R031113-40 | | | | | | | | |
| 1000 | Flat plate, job-built plywood, to 15' high, 1 use | R031113-60 | C-2 | 470 | .102 | S.F. | 3.16 | 4.28 | | 7.44 | 10.05 |
| 1050 | 2 use | | | 520 | .092 | | 1.74 | 3.87 | | 5.61 | 7.85 |
| 1100 | 3 use | | | 545 | .088 | | 1.26 | 3.69 | | 4.95 | 7.05 |
| 1150 | 4 use | | | 560 | .086 | | 1.03 | 3.59 | | 4.62 | 6.65 |
| 1500 | 15' to 20' high ceilings, 4 use | | | 495 | .097 | | 1.03 | 4.07 | | 5.10 | 7.40 |
| 1600 | 21' to 35' high ceilings, 4 use | | | 450 | .107 | | 1.29 | 4.47 | | 5.76 | 8.25 |
| 2000 | Flat slab, drop panels, job-built plywood, to 15' high, 1 use | | | 449 | .107 | | 3.59 | 4.48 | | 8.07 | 10.85 |
| 2050 | 2 use | | | 509 | .094 | | 1.97 | 3.95 | | 5.92 | 8.20 |
| 2100 | 3 use | | | 532 | .090 | | 1.43 | 3.78 | | 5.21 | 7.40 |
| 2150 | 4 use | | | 544 | .088 | | 1.17 | 3.70 | | 4.87 | 7 |
| 2250 | 15' to 20' high ceilings, 4 use | | | 480 | .100 | | 2.08 | 4.19 | | 6.27 | 8.75 |
| 2350 | 20' to 35' high ceilings, 4 use | | | 435 | .110 | | 2.34 | 4.63 | | 6.97 | 9.70 |
| 3000 | Floor slab hung from steel beams, 1 use | | | 485 | .099 | | 1.99 | 4.15 | | 6.14 | 8.55 |
| 3050 | 2 use | | | 535 | .090 | | 1.54 | 3.76 | | 5.30 | 7.45 |
| 3100 | 3 use | | | 550 | .087 | | 1.40 | 3.66 | | 5.06 | 7.15 |
| 3150 | 4 use | | | 565 | .085 | | 1.32 | 3.56 | | 4.88 | 6.90 |
| 3500 | Floor slab, with 1-way joist pans, 1 use | | | 415 | .116 | | 5.35 | 4.85 | | 10.20 | 13.35 |
| 3550 | 2 use | | | 445 | .108 | | 3.73 | 4.52 | | 8.25 | 11.05 |

# 03 11 Concrete Forming

## 03 11 13 – Structural Cast-In-Place Concrete Forming

### 03 11 13.35 Forms In Place, Elevated Slabs

| | | Crew | Daily Output | Labor-Hours | Unit | Material | 2011 Bare Costs Labor | Equipment | Total | Total Incl O&P |
|---|---|---|---|---|---|---|---|---|---|---|
| 3600 | 3 use | C-2 | 475 | .101 | S.F. | 3.19 | 4.24 | | 7.43 | 10 |
| 3650 | 4 use | | 500 | .096 | | 2.92 | 4.03 | | 6.95 | 9.40 |
| 4500 | With 2-way waffle domes, 1 use | | 405 | .119 | | 5.50 | 4.97 | | 10.47 | 13.65 |
| 4520 | 2 use | | 450 | .107 | | 3.86 | 4.47 | | 8.33 | 11.10 |
| 4530 | 3 use | | 460 | .104 | | 3.32 | 4.38 | | 7.70 | 10.35 |
| 4550 | 4 use | | 470 | .102 | | 3.05 | 4.28 | | 7.33 | 9.90 |
| 5000 | Box out for slab openings, over 16" deep, 1 use | | 190 | .253 | SFCA | 3.51 | 10.60 | | 14.11 | 20 |
| 5050 | 2 use | | 240 | .200 | " | 1.93 | 8.40 | | 10.33 | 14.95 |
| 5500 | Shallow slab box outs, to 10 S.F. | | 42 | 1.143 | Ea. | 9.65 | 48 | | 57.65 | 84 |
| 5550 | Over 10 S.F. (use perimeter) | | 600 | .080 | L.F. | 1.29 | 3.35 | | 4.64 | 6.55 |
| 6000 | Bulkhead forms for slab, with keyway, 1 use, 2 piece | | 500 | .096 | | 1.67 | 4.03 | | 5.70 | 8.05 |
| 6100 | 3 piece (see also edge forms) | | 460 | .104 | | 1.80 | 4.38 | | 6.18 | 8.70 |
| 6200 | Slab bulkhead form, 4-1/2" high, exp metal, w/keyway & stakes G | C-1 | 1200 | .027 | | 1.64 | 1.09 | | 2.73 | 3.47 |
| 6210 | 5-1/2" high G | | 1100 | .029 | | 1.90 | 1.19 | | 3.09 | 3.91 |
| 6215 | 7-1/2" high G | | 960 | .033 | | 2.31 | 1.36 | | 3.67 | 4.63 |
| 6220 | 9-1/2" high G | | 840 | .038 | | 2.40 | 1.56 | | 3.96 | 5.05 |
| 6500 | Curb forms, wood, 6" to 12" high, on elevated slabs, 1 use | | 180 | .178 | SFCA | 1.34 | 7.25 | | 8.59 | 12.65 |
| 6550 | 2 use | | 205 | .156 | | .74 | 6.40 | | 7.14 | 10.60 |
| 6600 | 3 use | | 220 | .145 | | .54 | 5.95 | | 6.49 | 9.70 |
| 6650 | 4 use | | 225 | .142 | | .44 | 5.80 | | 6.24 | 9.40 |
| 7000 | Edge forms to 6" high, on elevated slab, 4 use | | 500 | .064 | L.F. | .17 | 2.62 | | 2.79 | 4.20 |
| 7500 | Depressed area forms to 12" high, 4 use | | 300 | .107 | | .78 | 4.36 | | 5.14 | 7.55 |
| 7550 | 12" to 24" high, 4 use | | 175 | .183 | | 1.06 | 7.50 | | 8.56 | 12.60 |
| 8000 | Perimeter deck and rail for elevated slabs, straight | | 90 | .356 | | 9.70 | 14.55 | | 24.25 | 33 |
| 8050 | Curved | | 65 | .492 | | 13.35 | 20 | | 33.35 | 45.50 |
| 8500 | Void forms, round fiber, 3" diameter G | | 450 | .071 | | .99 | 2.91 | | 3.90 | 5.55 |
| 8550 | 4" diameter G | | 425 | .075 | | 1.46 | 3.08 | | 4.54 | 6.35 |
| 8600 | 6" diameter G | | 400 | .080 | | 2.28 | 3.27 | | 5.55 | 7.50 |
| 8650 | 8" diameter G | | 375 | .085 | | 3.90 | 3.49 | | 7.39 | 9.65 |
| 8700 | 10" diameter G | | 350 | .091 | | 2.52 | 3.74 | | 6.26 | 8.50 |
| 8750 | 12" diameter G | | 300 | .107 | | 3.20 | 4.36 | | 7.56 | 10.20 |

### 03 11 13.40 Forms In Place, Equipment Foundations

| | | | Crew | Daily Output | Labor-Hours | Unit | Material | Labor | Equipment | Total | Total Incl O&P |
|---|---|---|---|---|---|---|---|---|---|---|---|
| 0010 | **FORMS IN PLACE, EQUIPMENT FOUNDATIONS** | R031113-40 | | | | | | | | | |
| 0020 | 1 use | | C-2 | 160 | .300 | SFCA | 2.66 | 12.60 | | 15.26 | 22 |
| 0050 | 2 use | R031113-60 | | 190 | .253 | | 1.47 | 10.60 | | 12.07 | 17.85 |
| 0100 | 3 use | | | 200 | .240 | | 1.07 | 10.05 | | 11.12 | 16.60 |
| 0150 | 4 use | | | 205 | .234 | | .87 | 9.80 | | 10.67 | 16 |

### 03 11 13.45 Forms In Place, Footings

| | | | Crew | Daily Output | Labor-Hours | Unit | Material | Labor | Equipment | Total | Total Incl O&P |
|---|---|---|---|---|---|---|---|---|---|---|---|
| 0010 | **FORMS IN PLACE, FOOTINGS** | R031113-40 | | | | | | | | | |
| 0020 | Continuous wall, plywood, 1 use | | C-1 | 375 | .085 | SFCA | 5 | 3.49 | | 8.49 | 10.85 |
| 0050 | 2 use | R031113-60 | | 440 | .073 | | 2.76 | 2.97 | | 5.73 | 7.60 |
| 0100 | 3 use | | | 470 | .068 | | 2 | 2.78 | | 4.78 | 6.45 |
| 0150 | 4 use | | | 485 | .066 | | 1.63 | 2.70 | | 4.33 | 5.95 |
| 0500 | Dowel supports for footings or beams, 1 use | | | 500 | .064 | L.F. | .74 | 2.62 | | 3.36 | 4.83 |
| 1000 | Integral starter wall, to 4" high, 1 use | | | 400 | .080 | | .79 | 3.27 | | 4.06 | 5.85 |
| 1500 | Keyway, 4 use, tapered wood, 2" x 4" | | 1 Carp | 530 | .015 | | .18 | .65 | | .83 | 1.20 |
| 1550 | 2" x 6" | | | 500 | .016 | | .26 | .69 | | .95 | 1.35 |
| 2000 | Tapered plastic | | | 530 | .015 | | .45 | .65 | | 1.10 | 1.50 |
| 2250 | For keyway hung from supports, add | | | 150 | .053 | | .74 | 2.30 | | 3.04 | 4.34 |
| 3000 | Pile cap, square or rectangular, job-built plywood, 1 use | | C-1 | 290 | .110 | SFCA | 2.33 | 4.51 | | 6.84 | 9.45 |
| 3050 | 2 use | | | 346 | .092 | | 1.28 | 3.78 | | 5.06 | 7.20 |
| 3100 | 3 use | | | 371 | .086 | | .93 | 3.53 | | 4.46 | 6.40 |

## 03 11 13 – Structural Cast-In-Place Concrete Forming

### 03 11 13.45 Forms In Place, Footings

| | | Crew | Daily Output | Labor-Hours | Unit | Material | 2011 Bare Costs Labor | Equipment | Total | Total Incl O&P |
|---|---|---|---|---|---|---|---|---|---|---|
| 3150 | 4 use | C-1 | 383 | .084 | SFCA | .76 | 3.42 | | 4.18 | 6.10 |
| 4000 | Triangular or hexagonal, 1 use | | 225 | .142 | | 2.72 | 5.80 | | 8.52 | 11.90 |
| 4050 | 2 use | | 280 | .114 | | 1.50 | 4.67 | | 6.17 | 8.80 |
| 4100 | 3 use | | 305 | .105 | | 1.09 | 4.29 | | 5.38 | 7.80 |
| 4150 | 4 use | | 315 | .102 | | .88 | 4.15 | | 5.03 | 7.30 |
| 5000 | Spread footings, job-built lumber, 1 use | | 305 | .105 | | 1.79 | 4.29 | | 6.08 | 8.55 |
| 5050 | 2 use | | 371 | .086 | | 1 | 3.53 | | 4.53 | 6.50 |
| 5100 | 3 use | | 401 | .080 | | .72 | 3.26 | | 3.98 | 5.80 |
| 5150 | 4 use | | 414 | .077 | | .58 | 3.16 | | 3.74 | 5.50 |
| 6000 | Supports for dowels, plinths or templates, 2' x 2' footing | | 25 | 1.280 | Ea. | 4.92 | 52.50 | | 57.42 | 86 |
| 6050 | 4' x 4' footing | | 22 | 1.455 | | 9.85 | 59.50 | | 69.35 | 102 |
| 6100 | 8' x 8' footing | | 20 | 1.600 | | 19.70 | 65.50 | | 85.20 | 122 |
| 6150 | 12' x 12' footing | | 17 | 1.882 | | 24 | 77 | | 101 | 144 |
| 7000 | Plinths, job-built plywood, 1 use | | 250 | .128 | SFCA | 2.49 | 5.25 | | 7.74 | 10.80 |
| 7100 | 4 use | | 270 | .119 | " | .82 | 4.85 | | 5.67 | 8.35 |

### 03 11 13.47 Forms In Place, Gas Station Forms

| | | | Crew | Daily Output | Labor-Hours | Unit | Material | 2011 Bare Costs Labor | Equipment | Total | Total Incl O&P |
|---|---|---|---|---|---|---|---|---|---|---|---|
| 0010 | **FORMS IN PLACE, GAS STATION FORMS** | | | | | | | | | | |
| 0050 | Curb fascia, with template, 12 ga. steel, left in place, 9" high | G | 1 Carp | 50 | .160 | L.F. | 11.80 | 6.90 | | 18.70 | 23.50 |
| 1000 | Sign or light bases, 18" diameter, 9" high | G | | 9 | .889 | Ea. | 74.50 | 38.50 | | 113 | 140 |
| 1050 | 30" diameter, 13" high | G | | 8 | 1 | | 118 | 43 | | 161 | 196 |
| 2000 | Island forms, 10' long, 9" high, 3'-6" wide | G | C-1 | 10 | 3.200 | | 330 | 131 | | 461 | 565 |
| 2050 | 4' wide | G | | 9 | 3.556 | | 340 | 145 | | 485 | 600 |
| 2500 | 20' long, 9" high, 4' wide | G | | 6 | 5.333 | | 545 | 218 | | 763 | 935 |
| 2550 | 5' wide | G | | 5 | 6.400 | | 570 | 262 | | 832 | 1,025 |

### 03 11 13.50 Forms In Place, Grade Beam

| | | | Crew | Daily Output | Labor-Hours | Unit | Material | 2011 Bare Costs Labor | Equipment | Total | Total Incl O&P |
|---|---|---|---|---|---|---|---|---|---|---|---|
| 0010 | **FORMS IN PLACE, GRADE BEAM** | R031113-40 | | | | | | | | | |
| 0020 | Job-built plywood, 1 use | | C-2 | 530 | .091 | SFCA | 2.37 | 3.80 | | 6.17 | 8.45 |
| 0050 | 2 use | R031113-60 | | 580 | .083 | | 1.30 | 3.47 | | 4.77 | 6.75 |
| 0100 | 3 use | | | 600 | .080 | | .95 | 3.35 | | 4.30 | 6.20 |
| 0150 | 4 use | | | 605 | .079 | | .77 | 3.33 | | 4.10 | 5.95 |

### 03 11 13.55 Forms In Place, Mat Foundation

| | | | Crew | Daily Output | Labor-Hours | Unit | Material | 2011 Bare Costs Labor | Equipment | Total | Total Incl O&P |
|---|---|---|---|---|---|---|---|---|---|---|---|
| 0010 | **FORMS IN PLACE, MAT FOUNDATION** | R031113-40 | | | | | | | | | |
| 0020 | Job-built plywood, 1 use | | C-2 | 290 | .166 | SFCA | 2.44 | 6.95 | | 9.39 | 13.35 |
| 0050 | 2 use | R031113-60 | | 310 | .155 | | .99 | 6.50 | | 7.49 | 11.05 |
| 0100 | 3 use | | | 330 | .145 | | .65 | 6.10 | | 6.75 | 10.05 |
| 0120 | 4 use | | | 350 | .137 | | .58 | 5.75 | | 6.33 | 9.45 |

### 03 11 13.65 Forms In Place, Slab On Grade

| | | | Crew | Daily Output | Labor-Hours | Unit | Material | 2011 Bare Costs Labor | Equipment | Total | Total Incl O&P |
|---|---|---|---|---|---|---|---|---|---|---|---|
| 0010 | **FORMS IN PLACE, SLAB ON GRADE** | R031113-40 | | | | | | | | | |
| 1000 | Bulkhead forms w/keyway, wood, 6" high, 1 use | | C-1 | 510 | .063 | L.F. | .86 | 2.57 | | 3.43 | 4.88 |
| 1050 | 2 uses | R031113-60 | | 400 | .080 | | .47 | 3.27 | | 3.74 | 5.50 |
| 1100 | 4 uses | | | 350 | .091 | | .28 | 3.74 | | 4.02 | 6.05 |
| 1400 | Bulkhead form for slab, 4-1/2" high, exp metal, incl keyway & stakes | G | | 1200 | .027 | | 1.64 | 1.09 | | 2.73 | 3.47 |
| 1410 | 5-1/2" high | G | | 1100 | .029 | | 1.90 | 1.19 | | 3.09 | 3.91 |
| 1420 | 7-1/2" high | G | | 960 | .033 | | 2.31 | 1.36 | | 3.67 | 4.63 |
| 1430 | 9-1/2" high | G | | 840 | .038 | | 2.40 | 1.56 | | 3.96 | 5.05 |
| 2000 | Curb forms, wood, 6" to 12" high, on grade, 1 use | | | 215 | .149 | SFCA | 1.80 | 6.10 | | 7.90 | 11.35 |
| 2050 | 2 use | | | 250 | .128 | | 1 | 5.25 | | 6.25 | 9.15 |
| 2100 | 3 use | | | 265 | .121 | | .72 | 4.94 | | 5.66 | 8.35 |
| 2150 | 4 use | | | 275 | .116 | | .58 | 4.76 | | 5.34 | 7.95 |
| 3000 | Edge forms, wood, 4 use, on grade, to 6" high | | | 600 | .053 | L.F. | .27 | 2.18 | | 2.45 | 3.64 |
| 3050 | 7" to 12" high | | | 435 | .074 | SFCA | .59 | 3.01 | | 3.60 | 5.25 |
| 3500 | For depressed slabs, 4 use, to 12" high | | | 300 | .107 | L.F. | .60 | 4.36 | | 4.96 | 7.35 |

# 03 11 Concrete Forming

## 03 11 13 – Structural Cast-In-Place Concrete Forming

### 03 11 13.65 Forms In Place, Slab On Grade

| | | Crew | Daily Output | Labor-Hours | Unit | Material | 2011 Bare Costs Labor | Equipment | Total | Total Incl O&P |
|---|---|---|---|---|---|---|---|---|---|---|
| 3550 | To 24" high | C-1 | 175 | .183 | L.F. | .79 | 7.50 | | 8.29 | 12.30 |
| 4000 | For slab blockouts, to 12" high, 1 use | | 200 | .160 | | .63 | 6.55 | | 7.18 | 10.75 |
| 4050 | To 24" high, 1 use | | 120 | .267 | | .80 | 10.90 | | 11.70 | 17.60 |
| 4100 | Plastic (extruded), to 6" high, multiple use, on grade | | 800 | .040 | | 7.90 | 1.64 | | 9.54 | 11.20 |
| 5000 | Screed, 24 ga. metal key joint, see Section 03 15 05.25 | | | | | | | | | |
| 5020 | Wood, incl. wood stakes, 1" x 3" | C-1 | 900 | .036 | L.F. | .68 | 1.45 | | 2.13 | 2.98 |
| 5050 | 2" x 4" | | 900 | .036 | " | .67 | 1.45 | | 2.12 | 2.97 |
| 6000 | Trench forms in floor, wood, 1 use | | 160 | .200 | SFCA | 1.55 | 8.20 | | 9.75 | 14.25 |
| 6050 | 2 use | | 175 | .183 | | .85 | 7.50 | | 8.35 | 12.40 |
| 6100 | 3 use | | 180 | .178 | | .62 | 7.25 | | 7.87 | 11.85 |
| 6150 | 4 use | | 185 | .173 | | .50 | 7.05 | | 7.55 | 11.40 |
| 8760 | Void form, corrugated fiberboard, 4" x 12", 4' long [G] | | 3000 | .011 | S.F. | 2.39 | .44 | | 2.83 | 3.29 |
| 8770 | 6" x 12", 4' long | | 3000 | .011 | | 2.81 | .44 | | 3.25 | 3.76 |
| 8780 | 1/4" thick hardboard protective cover for void form | 2 Carp | 1500 | .011 | | .57 | .46 | | 1.03 | 1.33 |

### 03 11 13.85 Forms In Place, Walls

| | | Crew | Daily Output | Labor-Hours | Unit | Material | 2011 Bare Costs Labor | Equipment | Total | Total Incl O&P |
|---|---|---|---|---|---|---|---|---|---|---|
| 0010 | **FORMS IN PLACE, WALLS** R031113-10 | | | | | | | | | |
| 0100 | Box out for wall openings, to 16" thick, to 10 S.F. | C-2 | 24 | 2 | Ea. | 23 | 84 | | 107 | 154 |
| 0150 | Over 10 S.F. (use perimeter) R031113-40 | " | 280 | .171 | L.F. | 1.94 | 7.20 | | 9.14 | 13.20 |
| 0250 | Brick shelf, 4" w, add to wall forms, use wall area abv shelf | | | | | | | | | |
| 0260 | 1 use R031113-60 | C-2 | 240 | .200 | SFCA | 2.06 | 8.40 | | 10.46 | 15.10 |
| 0300 | 2 use | | 275 | .175 | | 1.13 | 7.30 | | 8.43 | 12.50 |
| 0350 | 4 use | | 300 | .160 | | .82 | 6.70 | | 7.52 | 11.20 |
| 0500 | Bulkhead, wood with keyway, 1 use, 2 piece | | 265 | .181 | L.F. | 1.69 | 7.60 | | 9.29 | 13.50 |
| 0600 | Bulkhead forms with keyway, 1 piece expanded metal, 8" wall [G] | C-1 | 1000 | .032 | | 2.31 | 1.31 | | 3.62 | 4.55 |
| 0610 | 10" wall [G] | | 800 | .040 | | 2.40 | 1.64 | | 4.04 | 5.15 |
| 0620 | 12" wall [G] | | 525 | .061 | | 2.88 | 2.49 | | 5.37 | 7 |
| 0700 | Buttress, to 8' high, 1 use | C-2 | 350 | .137 | SFCA | 2.89 | 5.75 | | 8.64 | 12 |
| 0750 | 2 use | | 430 | .112 | | 1.59 | 4.68 | | 6.27 | 8.95 |
| 0800 | 3 use | | 460 | .104 | | 1.16 | 4.38 | | 5.54 | 8 |
| 0850 | 4 use | | 480 | .100 | | .96 | 4.19 | | 5.15 | 7.50 |
| 1000 | Corbel or haunch, to 12" wide, add to wall forms, 1 use | | 150 | .320 | L.F. | 1.91 | 13.40 | | 15.31 | 22.50 |
| 1050 | 2 use | | 170 | .282 | | 1.05 | 11.85 | | 12.90 | 19.30 |
| 1100 | 3 use | | 175 | .274 | | .76 | 11.50 | | 12.26 | 18.50 |
| 1150 | 4 use | | 180 | .267 | | .62 | 11.20 | | 11.82 | 17.85 |
| 2000 | Wall, job-built plywood, to 8' high, 1 use | | 370 | .130 | SFCA | 2.17 | 5.45 | | 7.62 | 10.75 |
| 2050 | 2 use | | 435 | .110 | | 1.38 | 4.63 | | 6.01 | 8.60 |
| 2100 | 3 use | | 495 | .097 | | 1.01 | 4.07 | | 5.08 | 7.35 |
| 2150 | 4 use | | 505 | .095 | | .82 | 3.99 | | 4.81 | 7 |
| 2400 | Over 8' to 16' high, 1 use | | 280 | .171 | | 2.38 | 7.20 | | 9.58 | 13.65 |
| 2450 | 2 use | | 345 | .139 | | 1.08 | 5.85 | | 6.93 | 10.15 |
| 2500 | 3 use | | 375 | .128 | | .77 | 5.35 | | 6.12 | 9.10 |
| 2550 | 4 use | | 395 | .122 | | .63 | 5.10 | | 5.73 | 8.50 |
| 2700 | Over 16' high, 1 use | | 235 | .204 | | 2.18 | 8.55 | | 10.73 | 15.55 |
| 2750 | 2 use | | 290 | .166 | | 1.20 | 6.95 | | 8.15 | 11.95 |
| 2800 | 3 use | | 315 | .152 | | .87 | 6.40 | | 7.27 | 10.75 |
| 2850 | 4 use | | 330 | .145 | | .71 | 6.10 | | 6.81 | 10.15 |
| 4000 | Radial, smooth curved, job-built plywood, 1 use | | 245 | .196 | | 2.16 | 8.20 | | 10.36 | 15 |
| 4050 | 2 use | | 300 | .160 | | 1.19 | 6.70 | | 7.89 | 11.60 |
| 4100 | 3 use | | 325 | .148 | | .86 | 6.20 | | 7.06 | 10.45 |
| 4150 | 4 use | | 335 | .143 | | .70 | 6 | | 6.70 | 9.95 |
| 4200 | Below grade, job-built plywood, 1 use | | 225 | .213 | | 2.25 | 8.95 | | 11.20 | 16.20 |
| 4210 | 2 use | | 225 | .213 | | 1.24 | 8.95 | | 10.19 | 15.05 |

# 03 11 Concrete Forming

## 03 11 13 – Structural Cast-In-Place Concrete Forming

| 03 11 13.85 Forms In Place, Walls | | Crew | Daily Output | Labor-Hours | Unit | Material | 2011 Bare Costs Labor | Equipment | Total | Total Incl O&P |
|---|---|---|---|---|---|---|---|---|---|---|
| 4220 | 3 use | C-2 | 225 | .213 | SFCA | 1.03 | 8.95 | | 9.98 | 14.85 |
| 4230 | 4 use | | 225 | .213 | | .73 | 8.95 | | 9.68 | 14.50 |
| 4300 | Curved, 2' chords, job-built plywood, 1 use | | 290 | .166 | | 1.76 | 6.95 | | 8.71 | 12.60 |
| 4350 | 2 use | | 355 | .135 | | .97 | 5.65 | | 6.62 | 9.75 |
| 4400 | 3 use | | 385 | .125 | | .70 | 5.25 | | 5.95 | 8.75 |
| 4450 | 4 use | | 400 | .120 | | .57 | 5.05 | | 5.62 | 8.35 |
| 4500 | Over 8' high, 1 use | | 290 | .166 | | .82 | 6.95 | | 7.77 | 11.55 |
| 4525 | 2 use | | 355 | .135 | | .45 | 5.65 | | 6.10 | 9.20 |
| 4550 | 3 use | | 385 | .125 | | .33 | 5.25 | | 5.58 | 8.35 |
| 4575 | 4 use | | 400 | .120 | | .27 | 5.05 | | 5.32 | 8 |
| 4600 | Retaining wall, battered, job-built plyw'd, to 8' high, 1 use | | 300 | .160 | | 1.66 | 6.70 | | 8.36 | 12.15 |
| 4650 | 2 use | | 355 | .135 | | .91 | 5.65 | | 6.56 | 9.70 |
| 4700 | 3 use | | 375 | .128 | | .66 | 5.35 | | 6.01 | 9 |
| 4750 | 4 use | | 390 | .123 | | .54 | 5.15 | | 5.69 | 8.50 |
| 4900 | Over 8' to 16' high, 1 use | | 240 | .200 | | 1.81 | 8.40 | | 10.21 | 14.85 |
| 4950 | 2 use | | 295 | .163 | | .99 | 6.80 | | 7.79 | 11.55 |
| 5000 | 3 use | | 305 | .157 | | .72 | 6.60 | | 7.32 | 10.90 |
| 5050 | 4 use | | 320 | .150 | | .59 | 6.30 | | 6.89 | 10.30 |
| 5500 | For gang wall forming, 192 S.F. sections, deduct | | | | | 10% | 10% | | | |
| 5550 | 384 S.F. sections, deduct | | | | | 20% | 20% | | | |
| 5750 | Liners for forms (add to wall forms), A.B.S. plastic | | | | | | | | | |
| 5800 | Aged wood, 4" wide, 1 use | 1 Carp | 256 | .031 | SFCA | 2.91 | 1.35 | | 4.26 | 5.25 |
| 5820 | 2 use | | 256 | .031 | | 1.60 | 1.35 | | 2.95 | 3.82 |
| 5830 | 3 use | | 256 | .031 | | 1.16 | 1.35 | | 2.51 | 3.34 |
| 5840 | 4 use | | 256 | .031 | | .95 | 1.35 | | 2.30 | 3.10 |
| 5900 | Fractured rope rib, 1 use | | 192 | .042 | | 4.22 | 1.79 | | 6.01 | 7.40 |
| 5925 | 2 use | | 192 | .042 | | 2.32 | 1.79 | | 4.11 | 5.30 |
| 5950 | 3 use | | 192 | .042 | | 1.69 | 1.79 | | 3.48 | 4.61 |
| 6000 | 4 use | | 192 | .042 | | 1.37 | 1.79 | | 3.16 | 4.26 |
| 6100 | Ribbed, 3/4" deep x 1-1/2" O.C., 1 use | | 224 | .036 | | 4.22 | 1.54 | | 5.76 | 7 |
| 6125 | 2 use | | 224 | .036 | | 2.32 | 1.54 | | 3.86 | 4.91 |
| 6150 | 3 use | | 224 | .036 | | 1.69 | 1.54 | | 3.23 | 4.22 |
| 6200 | 4 use | | 224 | .036 | | 1.37 | 1.54 | | 2.91 | 3.87 |
| 6300 | Rustic brick pattern, 1 use | | 224 | .036 | | 2.91 | 1.54 | | 4.45 | 5.55 |
| 6325 | 2 use | | 224 | .036 | | 1.60 | 1.54 | | 3.14 | 4.12 |
| 6350 | 3 use | | 224 | .036 | | 1.16 | 1.54 | | 2.70 | 3.64 |
| 6400 | 4 use | | 224 | .036 | | .95 | 1.54 | | 2.49 | 3.40 |
| 6500 | 3/8" striated, random, 1 use | | 224 | .036 | | 2.91 | 1.54 | | 4.45 | 5.55 |
| 6525 | 2 use | | 224 | .036 | | 1.60 | 1.54 | | 3.14 | 4.12 |
| 6550 | 3 use | | 224 | .036 | | 1.16 | 1.54 | | 2.70 | 3.64 |
| 6600 | 4 use | | 224 | .036 | | .95 | 1.54 | | 2.49 | 3.40 |
| 6850 | Random vertical rustication, 1 use | | 384 | .021 | | 5.55 | .90 | | 6.45 | 7.50 |
| 6900 | 2 use | | 384 | .021 | | 3.04 | .90 | | 3.94 | 4.73 |
| 6925 | 3 use | | 384 | .021 | | 2.21 | .90 | | 3.11 | 3.81 |
| 6950 | 4 use | | 384 | .021 | | 1.80 | .90 | | 2.70 | 3.36 |
| 7050 | Wood, beveled edge, 3/4" deep, 1 use | | 384 | .021 | L.F. | .19 | .90 | | 1.09 | 1.59 |
| 7100 | 1" deep, 1 use | | 384 | .021 | " | .39 | .90 | | 1.29 | 1.81 |
| 7200 | 4" wide aged cedar, 1 use | | 256 | .031 | SFCA | 2.91 | 1.35 | | 4.26 | 5.25 |
| 7300 | 4" variable depth rough cedar | | 224 | .036 | | 4.22 | 1.54 | | 5.76 | 7 |
| 7500 | Lintel or sill forms, 1 use | | 30 | .267 | | 2.62 | 11.50 | | 14.12 | 20.50 |
| 7520 | 2 use | | 34 | .235 | | 1.44 | 10.15 | | 11.59 | 17.15 |
| 7540 | 3 use | | 36 | .222 | | 1.05 | 9.55 | | 10.60 | 15.85 |
| 7560 | 4 use | | 37 | .216 | | .85 | 9.30 | | 10.15 | 15.25 |

# 03 11 Concrete Forming

## 03 11 13 – Structural Cast-In-Place Concrete Forming

### 03 11 13.85 Forms In Place, Walls

| | | Crew | Daily Output | Labor-Hours | Unit | Material | 2011 Bare Costs Labor | Equipment | Total | Total Incl O&P |
|---|---|---|---|---|---|---|---|---|---|---|
| 7800 | Modular prefabricated plywood, based on 20 uses of purchased | | | | | | | | | |
| 7820 | forms, and 4 uses of bracing lumber | | | | | | | | | |
| 7860 | To 8' high | C-2 | 800 | .060 | SFCA | .89 | 2.52 | | 3.41 | 4.84 |
| 8060 | Over 8' to 16' high | | 600 | .080 | | .93 | 3.35 | | 4.28 | 6.20 |
| 8600 | Pilasters, 1 use | | 270 | .178 | | 2.51 | 7.45 | | 9.96 | 14.20 |
| 8620 | 2 use | | 330 | .145 | | 1.38 | 6.10 | | 7.48 | 10.85 |
| 8640 | 3 use | | 370 | .130 | | 1 | 5.45 | | 6.45 | 9.45 |
| 8660 | 4 use | | 385 | .125 | | .81 | 5.25 | | 6.06 | 8.90 |
| 9010 | Steel framed plywood, based on 50 uses of purchased | | | | | | | | | |
| 9020 | forms, and 4 uses of bracing lumber | | | | | | | | | |
| 9060 | To 8' high | C-2 | 600 | .080 | SFCA | .55 | 3.35 | | 3.90 | 5.75 |
| 9260 | Over 8' to 16' high | | 450 | .107 | | .55 | 4.47 | | 5.02 | 7.45 |
| 9460 | Over 16' to 20' high | | 400 | .120 | | .55 | 5.05 | | 5.60 | 8.30 |
| 9475 | For elevated walls, add | | | | | | 10% | | | |
| 9480 | For battered walls, 1 side battered, add | | | | | 10% | 10% | | | |
| 9485 | For battered walls, 2 sides battered, add | | | | | 15% | 15% | | | |

## 03 11 19 – Insulating Concrete Forming

### 03 11 19.10 Insulating Forms, Left In Place

| | | | Crew | Daily Output | Labor-Hours | Unit | Material | 2011 Bare Costs Labor | Equipment | Total | Total Incl O&P |
|---|---|---|---|---|---|---|---|---|---|---|---|
| 0010 | **INSULATING FORMS, LEFT IN PLACE** | | | | | | | | | | |
| 0020 | S.F. is for exterior face, but includes forms for both faces (total R22) | | | | | | | | | | |
| 2000 | 4" wall, straight block, 16" x 48" (5.33 SF) | G | 2 Carp | 90 | .178 | Ea. | 16.45 | 7.65 | | 24.10 | 30 |
| 2010 | 90 corner block, exterior 16" x 38" x 22" (6.67 SF) | G | | 75 | .213 | | 18.80 | 9.20 | | 28 | 34.50 |
| 2020 | 45 corner block, exterior 16" x 34" x 18" (5.78 SF) | G | | 75 | .213 | | 18.95 | 9.20 | | 28.15 | 35 |
| 2100 | 6" wall, straight block, 16" x 48" (5.33 SF) | G | | 90 | .178 | | 16.70 | 7.65 | | 24.35 | 30 |
| 2110 | 90 corner block, exterior 16" x 32" x 24" (6.22 SF) | G | | 75 | .213 | | 20.50 | 9.20 | | 29.70 | 36.50 |
| 2120 | 45 corner block, exterior 16" x 26" x 18" (4.89 SF) | G | | 75 | .213 | | 17.80 | 9.20 | | 27 | 33.50 |
| 2130 | Brick ledge block, 16" x 48" (5.33 SF) | G | | 80 | .200 | | 21 | 8.60 | | 29.60 | 36.50 |
| 2140 | Taper top block, 16" x 48" (5.33 SF) | G | | 80 | .200 | | 19.65 | 8.60 | | 28.25 | 34.50 |
| 2200 | 8" wall, straight block, 16" x 48" (5.33 SF) | G | | 90 | .178 | | 17.25 | 7.65 | | 24.90 | 30.50 |
| 2210 | 90 corner block, exterior 16" x 34" x 26" (6.67 SF) | G | | 75 | .213 | | 22.50 | 9.20 | | 31.70 | 38.50 |
| 2220 | 45 corner block, exterior 16" x 28" x 20" (5.33 SF) | G | | 75 | .213 | | 19.15 | 9.20 | | 28.35 | 35 |
| 2230 | Brick ledge block, 16" x 48" (5.33 SF) | G | | 80 | .200 | | 22 | 8.60 | | 30.60 | 37 |
| 2240 | Taper top block, 16" x 48" (5.33 SF) | G | | 80 | .200 | | 20 | 8.60 | | 28.60 | 35 |

### 03 11 19.60 Roof Deck Form Boards

| | | | Crew | Daily Output | Labor-Hours | Unit | Material | 2011 Bare Costs Labor | Equipment | Total | Total Incl O&P |
|---|---|---|---|---|---|---|---|---|---|---|---|
| 0010 | **ROOF DECK FORM BOARDS** | R051223-50 | | | | | | | | | |
| 0050 | Includes bulb tee sub-purlins @ 32-5/8" O.C. | | | | | | | | | | |
| 0070 | Non-asbestos fiber cement, 5/16" thick | | C-13 | 2950 | .008 | S.F. | 2.49 | .38 | .04 | 2.91 | 3.41 |
| 0100 | Fiberglass, 1" thick | | | 2700 | .009 | | 3.48 | .42 | .04 | 3.94 | 4.56 |
| 0500 | Wood fiber, 1" thick | G | | 2700 | .009 | | 2.04 | .42 | .04 | 2.50 | 2.98 |

## 03 11 23 – Permanent Stair Forming

### 03 11 23.75 Forms In Place, Stairs

| | | | Crew | Daily Output | Labor-Hours | Unit | Material | 2011 Bare Costs Labor | Equipment | Total | Total Incl O&P |
|---|---|---|---|---|---|---|---|---|---|---|---|
| 0010 | **FORMS IN PLACE, STAIRS** | R031113-40 | | | | | | | | | |
| 0015 | (Slant length x width), 1 use | | C-2 | 165 | .291 | S.F. | 4.65 | 12.20 | | 16.85 | 24 |
| 0050 | 2 use | R031113-60 | | 170 | .282 | | 2.65 | 11.85 | | 14.50 | 21 |
| 0100 | 3 use | | | 180 | .267 | | 1.98 | 11.20 | | 13.18 | 19.35 |
| 0150 | 4 use | | | 190 | .253 | | 1.65 | 10.60 | | 12.25 | 18.05 |
| 1000 | Alternate pricing method (1.0 L.F./S.F.), 1 use | | | 100 | .480 | LF Rsr | 4.65 | 20 | | 24.65 | 36 |
| 1050 | 2 use | | | 105 | .457 | | 2.65 | 19.15 | | 21.80 | 32.50 |
| 1100 | 3 use | | | 110 | .436 | | 1.98 | 18.30 | | 20.28 | 30 |
| 1150 | 4 use | | | 115 | .417 | | 1.65 | 17.50 | | 19.15 | 29 |
| 2000 | Stairs, cast on sloping ground (length x width), 1 use | | | 220 | .218 | S.F. | 1.88 | 9.15 | | 11.03 | 16.10 |

# 03 11 Concrete Forming

## 03 11 23 – Permanent Stair Forming

| 03 11 23.75 Forms In Place, Stairs | | Crew | Daily Output | Labor-Hours | Unit | Material | 2011 Bare Costs Labor | Equipment | Total | Total Incl O&P |
|---|---|---|---|---|---|---|---|---|---|---|
| 2025 | 2 use | C-2 | 232 | .207 | S.F. | 1.03 | 8.70 | | 9.73 | 14.45 |
| 2050 | 3 use | | 244 | .197 | | .75 | 8.25 | | 9 | 13.50 |
| 2100 | 4 use | | 256 | .188 | | .61 | 7.85 | | 8.46 | 12.70 |

# 03 15 Concrete Accessories

## 03 15 05 – Concrete Forming Accessories

### 03 15 05.02 Anchor Bolt Accessories

| | | | Crew | Daily Output | Labor-Hours | Unit | Material | Labor | Equipment | Total | Total Incl O&P |
|---|---|---|---|---|---|---|---|---|---|---|---|
| 0010 | **ANCHOR BOLT ACCESSORIES** | | | | | | | | | | |
| 0015 | For anchor bolts set in fresh concrete, see Section 05 05 23.05 | | | | | | | | | | |
| 8150 | Anchor bolt sleeve, plastic, 1" diam. bolts | | 1 Carp | 60 | .133 | Ea. | 7 | 5.75 | | 12.75 | 16.50 |
| 8500 | 1-1/2" diameter | | | 28 | .286 | | 12.40 | 12.30 | | 24.70 | 32.50 |
| 8600 | 2" diameter | | | 24 | .333 | | 15.45 | 14.35 | | 29.80 | 39 |
| 8650 | 3" diameter | | | 20 | .400 | | 28.50 | 17.20 | | 45.70 | 57.50 |

### 03 15 05.12 Chamfer Strips

| | | | Crew | Daily Output | Labor-Hours | Unit | Material | Labor | Equipment | Total | Total Incl O&P |
|---|---|---|---|---|---|---|---|---|---|---|---|
| 0010 | **CHAMFER STRIPS** | | | | | | | | | | |
| 2000 | Polyvinyl chloride, 1/2" wide with leg | | 1 Carp | 535 | .015 | L.F. | .42 | .64 | | 1.06 | 1.45 |
| 2200 | 3/4" wide with leg | | | 525 | .015 | | .69 | .66 | | 1.35 | 1.77 |
| 2400 | 1" radius with leg | | | 515 | .016 | | .30 | .67 | | .97 | 1.36 |
| 2800 | 2" radius with leg | | | 500 | .016 | | .52 | .69 | | 1.21 | 1.63 |
| 5000 | Wood, 1/2" wide | | | 535 | .015 | | .13 | .64 | | .77 | 1.13 |
| 5200 | 3/4" wide | | | 525 | .015 | | .19 | .66 | | .85 | 1.22 |
| 5400 | 1" wide | | | 515 | .016 | | .39 | .67 | | 1.06 | 1.46 |

### 03 15 05.15 Column Form Accessories

| | | | Crew | Daily Output | Labor-Hours | Unit | Material | Labor | Equipment | Total | Total Incl O&P |
|---|---|---|---|---|---|---|---|---|---|---|---|
| 0010 | **COLUMN FORM ACCESSORIES** | | | | | | | | | | |
| 1000 | Column clamps, adjustable to 24" x 24", buy | G | | | | Set | 126 | | | 126 | 139 |
| 1100 | Rent per month | G | | | | | 11.80 | | | 11.80 | 13 |
| 1300 | For sizes to 30" x 30", buy | G | | | | | 128 | | | 128 | 141 |
| 1400 | Rent per month | G | | | | | 12.55 | | | 12.55 | 13.80 |
| 1600 | For sizes to 36" x 36", buy | G | | | | | 134 | | | 134 | 147 |
| 1700 | Rent per month | G | | | | | 13.80 | | | 13.80 | 15.15 |
| 2000 | Bull winch (band iron) 36" x 36", buy | G | | | | | 52 | | | 52 | 57 |
| 2100 | Rent per month | G | | | | | 5.30 | | | 5.30 | 5.85 |
| 2300 | 48" x 48", buy | G | | | | | 53 | | | 53 | 58.50 |
| 2400 | Rent per month | G | | | | | 6.35 | | | 6.35 | 7 |
| 3000 | Chain & wedge type 36" x 36", buy | G | | | | | 72 | | | 72 | 79 |
| 3100 | Rent per month | G | | | | | 8.10 | | | 8.10 | 8.90 |
| 3300 | 60" x 60", buy | G | | | | | 98 | | | 98 | 108 |
| 3400 | Rent per month | G | | | | | 11.65 | | | 11.65 | 12.80 |
| 4000 | Friction collars 2'-6" diam., buy | G | | | | | 735 | | | 735 | 810 |
| 4100 | Rent per month | G | | | | | 72 | | | 72 | 79 |
| 4300 | 4'-0" diam., buy | G | | | | | 845 | | | 845 | 925 |
| 4400 | Rent per month | G | | | | | 93 | | | 93 | 102 |

### 03 15 05.20 Dovetail Anchor System

| | | | Crew | Daily Output | Labor-Hours | Unit | Material | Labor | Equipment | Total | Total Incl O&P |
|---|---|---|---|---|---|---|---|---|---|---|---|
| 0010 | **DOVETAIL ANCHOR SYSTEM** | | | | | | | | | | |
| 0500 | Anchor slot, galv., filled, 24 ga. | G | 1 Carp | 425 | .019 | L.F. | .74 | .81 | | 1.55 | 2.05 |
| 0600 | 20 ga. | G | | 400 | .020 | | 1.57 | .86 | | 2.43 | 3.05 |
| 0625 | 12 ga. | G | | 400 | .020 | | 1.77 | .86 | | 2.63 | 3.27 |
| 0900 | 26 ga. stainless steel, foam filled | G | | 375 | .021 | | 1.18 | .92 | | 2.10 | 2.71 |
| 1200 | Brick anchor, corr., galv., 3-1/2" long, 16 ga. | G | 1 Bric | 10.50 | .762 | C | 26 | 33 | | 59 | 78 |
| 1300 | 12 ga. | G | | 10.50 | .762 | | 37 | 33 | | 70 | 90 |

# 03 15 Concrete Accessories

## 03 15 05 – Concrete Forming Accessories

| 03 15 05.20 Dovetail Anchor System | | Crew | Daily Output | Labor-Hours | Unit | Material | 2011 Bare Costs Labor | Equipment | Total | Total Incl O&P |
|---|---|---|---|---|---|---|---|---|---|---|
| 1500 | Flat, galv., 3-1/2" long, 16 ga. | G | 1 Bric | 10.50 | .762 | C | 38 | 33 | | 71 | 91 |
| 1600 | 12 ga. | G | | 10.50 | .762 | | 53 | 33 | | 86 | 108 |
| 2000 | Cavity wall, corr., galv., 5" long, 16 ga. | G | | 10.50 | .762 | | 33 | 33 | | 66 | 86 |
| 2100 | 12 ga. | G | | 10.50 | .762 | | 47 | 33 | | 80 | 102 |
| 3000 | Furring anchors, corr., galv., 1-1/2" long, 16 ga. | G | | 10.50 | .762 | | 12.80 | 33 | | 45.80 | 63.50 |
| 3100 | 12 ga. | G | | 10.50 | .762 | | 22 | 33 | | 55 | 74 |
| 6000 | Stone anchors, 3-1/2" long, galv., 1/8" x 1" wide | G | | 10.50 | .762 | | 82 | 33 | | 115 | 140 |
| 6100 | 1/4" x 1" wide | G | | 10.50 | .762 | | 94.50 | 33 | | 127.50 | 154 |

| 03 15 05.25 Expansion Joints | | Crew | Daily Output | Labor-Hours | Unit | Material | 2011 Bare Costs Labor | Equipment | Total | Total Incl O&P |
|---|---|---|---|---|---|---|---|---|---|---|
| 0010 | **EXPANSION JOINTS** | | | | | | | | | |
| 0020 | Keyed, cold, 24 ga., incl. stakes, 3-1/2" high | G | 1 Carp | 200 | .040 | L.F. | 1.41 | 1.72 | | 3.13 | 4.19 |
| 0050 | 4-1/2" high | G | | 200 | .040 | | 1.64 | 1.72 | | 3.36 | 4.44 |
| 0100 | 5-1/2" high | G | | 195 | .041 | | 1.90 | 1.77 | | 3.67 | 4.80 |
| 0150 | 7-1/2" high | G | | 190 | .042 | | 2.31 | 1.81 | | 4.12 | 5.30 |
| 0160 | 9-1/2" high | G | | 185 | .043 | | 2.40 | 1.86 | | 4.26 | 5.50 |
| 0300 | Poured asphalt, plain, 1/2" x 1" | | 1 Clab | 450 | .018 | | .65 | .61 | | 1.26 | 1.66 |
| 0350 | 1" x 2" | | | 400 | .020 | | 2.60 | .69 | | 3.29 | 3.91 |
| 0500 | Neoprene, liquid, cold applied, 1/2" x 1" | | | 450 | .018 | | 2.22 | .61 | | 2.83 | 3.38 |
| 0550 | 1" x 2" | | | 400 | .020 | | 8.85 | .69 | | 9.54 | 10.80 |
| 0700 | Polyurethane, poured, 2 part, 1/2" x 1" | | | 400 | .020 | | 1.30 | .69 | | 1.99 | 2.48 |
| 0750 | 1" x 2" | | | 350 | .023 | | 5.20 | .79 | | 5.99 | 6.90 |
| 0900 | Rubberized asphalt, hot or cold applied, 1/2" x 1" | | | 450 | .018 | | .35 | .61 | | .96 | 1.33 |
| 0950 | 1" x 2" | | | 400 | .020 | | 1.40 | .69 | | 2.09 | 2.59 |
| 1100 | Hot applied, fuel resistant, 1/2" x 1" | | | 450 | .018 | | .53 | .61 | | 1.14 | 1.52 |
| 1150 | 1" x 2" | | | 400 | .020 | | 2.10 | .69 | | 2.79 | 3.36 |
| 2000 | Premolded, bituminous fiber, 1/2" x 6" | | 1 Carp | 375 | .021 | | .41 | .92 | | 1.33 | 1.86 |
| 2050 | 1" x 12" | | | 300 | .027 | | 1.91 | 1.15 | | 3.06 | 3.86 |
| 2140 | Concrete expansion joint, recycled paper and fiber, 1/2" x 6" | G | | 390 | .021 | | .36 | .88 | | 1.24 | 1.75 |
| 2150 | 1/2" x 12" | G | | 360 | .022 | | .72 | .96 | | 1.68 | 2.27 |
| 2250 | Cork with resin binder, 1/2" x 6" | | | 375 | .021 | | 1.81 | .92 | | 2.73 | 3.40 |
| 2300 | 1" x 12" | | | 300 | .027 | | 7.55 | 1.15 | | 8.70 | 10.05 |
| 2500 | Neoprene sponge, closed cell, 1/2" x 6" | | | 375 | .021 | | 1.78 | .92 | | 2.70 | 3.37 |
| 2550 | 1" x 12" | | | 300 | .027 | | 6.65 | 1.15 | | 7.80 | 9.05 |
| 2750 | Polyethylene foam, 1/2" x 6" | | | 375 | .021 | | .62 | .92 | | 1.54 | 2.09 |
| 2800 | 1" x 12" | | | 300 | .027 | | 1.99 | 1.15 | | 3.14 | 3.95 |
| 3000 | Polyethylene backer rod, 3/8" diameter | | | 460 | .017 | | .05 | .75 | | .80 | 1.21 |
| 3050 | 3/4" diameter | | | 460 | .017 | | .10 | .75 | | .85 | 1.26 |
| 3100 | 1" diameter | | | 460 | .017 | | .19 | .75 | | .94 | 1.36 |
| 3500 | Polyurethane foam, with polybutylene, 1/2" x 1/2" | | | 475 | .017 | | .95 | .73 | | 1.68 | 2.16 |
| 3550 | 1" x 1" | | | 450 | .018 | | 2.40 | .77 | | 3.17 | 3.81 |
| 3750 | Polyurethane foam, regular, closed cell, 1/2" x 6" | | | 375 | .021 | | .70 | .92 | | 1.62 | 2.18 |
| 3800 | 1" x 12" | | | 300 | .027 | | 2.50 | 1.15 | | 3.65 | 4.51 |
| 4000 | Polyvinyl chloride foam, closed cell, 1/2" x 6" | | | 375 | .021 | | 1.90 | .92 | | 2.82 | 3.50 |
| 4050 | 1" x 12" | | | 300 | .027 | | 6.55 | 1.15 | | 7.70 | 8.95 |
| 4250 | Rubber, gray sponge, 1/2" x 6" | | | 375 | .021 | | 1.60 | .92 | | 2.52 | 3.17 |
| 4300 | 1" x 12" | | | 300 | .027 | | 5.75 | 1.15 | | 6.90 | 8.10 |
| 4500 | Lead wool for joints | | | | | Lb. | 7.85 | | | 7.85 | 8.60 |
| 5000 | For installation in walls, add | | | | | | | 75% | | | |
| 5250 | For installation in boxouts, add | | | | | | | 25% | | | |

## 03 15 05.30 Hangers

| | | | | | | | | | | |
|---|---|---|---|---|---|---|---|---|---|---|
| 0010 | **HANGERS** | | | | | | | | | |
| 0020 | Slab and beam form | | | | | | | | | |

# 03 15 Concrete Accessories

## 03 15 05 – Concrete Forming Accessories

| 03 15 05.30 Hangers | | Crew | Daily Output | Labor-Hours | Unit | Material | 2011 Bare Costs Labor | 2011 Bare Costs Equipment | Total | Total Incl O&P |
|---|---|---|---|---|---|---|---|---|---|---|
| 0500 | Banding iron, 3/4" x 22 ga, 14 L.F. per lb or | | | | | | | | | |
| 0550 | 1/2" x 14 ga, 7 L.F. per lb. | G | | | | Lb. | 1.24 | | | 1.24 | 1.36 |
| 1000 | Fascia ties, coil type, to 24" long | G | | | | C | 320 | | | 320 | 355 |
| 1500 | Frame ties to 8-1/8" | G | | | | | 380 | | | 380 | 415 |
| 1550 | 8-1/8" to 10-1/8" | G | | | | | 400 | | | 400 | 440 |
| 5000 | Snap tie hanger, to 30" overall length, 3000 # | G | | | | | 320 | | | 320 | 350 |
| 5050 | To 36" overall length | G | | | | | 350 | | | 350 | 385 |
| 5100 | To 48" overall length | G | | | | | 410 | | | 410 | 450 |
| 5500 | Steel beam hanger | | | | | | | | | |
| 5600 | Flange to 8-1/8" | G | | | | C | 380 | | | 380 | 415 |
| 5650 | 8-1/8" to 10-1/8" | G | | | | | 400 | | | 400 | 440 |
| 6000 | Tie hangers to 30" overall length, 4000 # | G | | | | | 365 | | | 365 | 400 |
| 6100 | To 36" overall length | G | | | | | 395 | | | 395 | 435 |
| 6500 | Tie back hanger, up to 12-1/8" flange | G | | | | | 1,025 | | | 1,025 | 1,125 |
| 8500 | Wire, black annealed, 9 ga | G | | | | Cwt. | 200 | | | 200 | 220 |
| 8600 | 16 ga | | | | | " | 164 | | | 164 | 180 |

| 03 15 05.35 Inserts | | Crew | Daily Output | Labor-Hours | Unit | Material | 2011 Bare Costs Labor | 2011 Bare Costs Equipment | Total | Total Incl O&P |
|---|---|---|---|---|---|---|---|---|---|---|
| 0010 | **INSERTS** | | | | | | | | | |
| 1000 | Inserts, slotted nut type for 3/4" bolts, 4" long | G | 1 Carp | 84 | .095 | Ea. | 16 | 4.10 | | 20.10 | 24 |
| 2100 | 6" long | G | | 84 | .095 | | 18 | 4.10 | | 22.10 | 26 |
| 2150 | 8" long | G | | 84 | .095 | | 24 | 4.10 | | 28.10 | 33 |
| 2200 | Slotted, strap type, 4" long | G | | 84 | .095 | | 15.70 | 4.10 | | 19.80 | 23.50 |
| 2250 | 6" long | G | | 84 | .095 | | 17.70 | 4.10 | | 21.80 | 26 |
| 2300 | 8" long | G | | 84 | .095 | | 23.50 | 4.10 | | 27.60 | 32.50 |
| 2350 | Strap for slotted insert, 4" long | G | | 84 | .095 | | 8.60 | 4.10 | | 12.70 | 15.75 |
| 4100 | 6" long | G | | 84 | .095 | | 9.30 | 4.10 | | 13.40 | 16.55 |
| 4150 | 8" long | G | | 84 | .095 | | 11.60 | 4.10 | | 15.70 | 19.05 |
| 4200 | 10" long | G | | 84 | .095 | | 14.10 | 4.10 | | 18.20 | 22 |
| 7000 | Loop ferrule type | | | | | | | | | |
| 7100 | 1/4" diameter bolt | G | 1 Carp | 84 | .095 | Ea. | 1.71 | 4.10 | | 5.81 | 8.20 |
| 7350 | 7/8" diameter bolt | G | " | 84 | .095 | " | 7.95 | 4.10 | | 12.05 | 15.05 |
| 9000 | Wedge type | | | | | | | | | |
| 9100 | For 3/4" diameter bolt | G | 1 Carp | 60 | .133 | Ea. | 6.60 | 5.75 | | 12.35 | 16.05 |
| 9800 | Cut washers, black | | | | | | | | | |
| 9900 | 3/4" bolt | G | | | | Ea. | .82 | | | .82 | .90 |
| 9950 | For galvanized inserts, add | | | | | | 30% | | | | |

| 03 15 05.70 Shores | | Crew | Daily Output | Labor-Hours | Unit | Material | 2011 Bare Costs Labor | 2011 Bare Costs Equipment | Total | Total Incl O&P |
|---|---|---|---|---|---|---|---|---|---|---|
| 0010 | **SHORES** | | | | | | | | | |
| 0020 | Erect and strip, by hand, horizontal members | | | | | | | | | |
| 0500 | Aluminum joists and stringers | G | 2 Carp | 60 | .267 | Ea. | | 11.50 | | 11.50 | 17.60 |
| 0600 | Steel, adjustable beams | G | | 45 | .356 | | | 15.30 | | 15.30 | 23.50 |
| 0700 | Wood joists | | | 50 | .320 | | | 13.80 | | 13.80 | 21 |
| 0800 | Wood stringers | | | 30 | .533 | | | 23 | | 23 | 35 |
| 1000 | Vertical members to 10' high | G | | 55 | .291 | | | 12.50 | | 12.50 | 19.20 |
| 1050 | To 13' high | G | | 50 | .320 | | | 13.80 | | 13.80 | 21 |
| 1100 | To 16' high | G | | 45 | .356 | | | 15.30 | | 15.30 | 23.50 |
| 1500 | Reshoring | G | | 1400 | .011 | S.F. | .50 | .49 | | .99 | 1.31 |
| 1600 | Flying truss system | G | C-17D | 9600 | .009 | SFCA | | .39 | .09 | .48 | .70 |
| 1760 | Horizontal, aluminum joists, 6-1/4" high x 5' to 21' span, buy | G | | | | L.F. | 13.40 | | | 13.40 | 14.70 |
| 1770 | Beams, 7-1/4" high x 4' to 30' span | G | | | | " | 15.65 | | | 15.65 | 17.25 |
| 1810 | Horizontal, steel beam, W8x10, 7' span, buy | G | | | | Ea. | 47.50 | | | 47.50 | 52 |
| 1830 | 10' span | G | | | | | 56 | | | 56 | 61.50 |

# 03 15 Concrete Accessories

## 03 15 05 – Concrete Forming Accessories

### 03 15 05.70 Shores

| | | | Crew | Daily Output | Labor-Hours | Unit | Material | 2011 Bare Costs Labor | 2011 Bare Costs Equipment | Total | Total Incl O&P |
|---|---|---|---|---|---|---|---|---|---|---|---|
| 1920 | 15' span | G | | | | Ea. | 96 | | | 96 | 106 |
| 1940 | 20' span | G | | | | ▼ | 135 | | | 135 | 149 |
| 1970 | Steel stringer, W8x10, 4' to 16' span, buy | G | | | | L.F. | 5.60 | | | 5.60 | 6.15 |
| 3000 | Rent for job duration, aluminum joist @ 2' O.C., per mo | G | | | | SF Flr. | .33 | | | .33 | .37 |
| 3050 | Steel W8x10 | G | | | | | .14 | | | .14 | .15 |
| 3060 | Steel adjustable | G | | | | ▼ | .14 | | | .14 | .15 |
| 3500 | #1 post shore, steel, 5'-7" to 9'-6" high, 10000# cap., buy | G | | | | Ea. | 128 | | | 128 | 141 |
| 3550 | #2 post shore, 7'-3" to 12'-10" high, 7800# capacity | G | | | | | 148 | | | 148 | 163 |
| 3600 | #3 post shore, 8'-10" to 16'-1" high, 3800# capacity | G | | | | ▼ | 162 | | | 162 | 178 |
| 5010 | Frame shoring systems, steel, 12000#/leg, buy | | | | | | | | | | |
| 5040 | Frame, 2' wide x 6' high | G | | | | Ea. | 106 | | | 106 | 117 |
| 5250 | X-brace | G | | | | | 18 | | | 18 | 19.80 |
| 5550 | Base plate | G | | | | | 17.60 | | | 17.60 | 19.35 |
| 5600 | Screw jack | G | | | | | 38 | | | 38 | 41.50 |
| 5650 | U-head, 8" x 8" | G | | | | ▼ | 21.50 | | | 21.50 | 24 |

### 03 15 05.75 Sleeves and Chases

| | | | Crew | Daily Output | Labor-Hours | Unit | Material | 2011 Bare Costs Labor | 2011 Bare Costs Equipment | Total | Total Incl O&P |
|---|---|---|---|---|---|---|---|---|---|---|---|
| 0010 | **SLEEVES AND CHASES** | | | | | | | | | | |
| 0100 | Plastic, 1 use, 12" long, 2" diameter | | 1 Carp | 100 | .080 | Ea. | 1.74 | 3.44 | | 5.18 | 7.20 |
| 0150 | 4" diameter | | | 90 | .089 | | 5.10 | 3.83 | | 8.93 | 11.45 |
| 0200 | 6" diameter | | | 75 | .107 | | 9 | 4.59 | | 13.59 | 16.95 |
| 0250 | 12" diameter | | ▼ | 60 | .133 | ▼ | 25.50 | 5.75 | | 31.25 | 37 |

### 03 15 05.80 Snap Ties

| | | | Crew | Daily Output | Labor-Hours | Unit | Material | 2011 Bare Costs Labor | 2011 Bare Costs Equipment | Total | Total Incl O&P |
|---|---|---|---|---|---|---|---|---|---|---|---|
| 0010 | **SNAP TIES**, 8-1/4" L&W (Lumber and wedge) | | | | | | | | | | |
| 0100 | 2250 lb., w/flat washer, 8" wall | G | | | | C | 68 | | | 68 | 75 |
| 0200 | 12" wall | G | | | | | 79 | | | 79 | 87 |
| 0250 | 16" wall | G | | | | | 91 | | | 91 | 100 |
| 0300 | 18" wall | G | | | | | 97 | | | 97 | 107 |
| 0500 | With plastic cone, 8" wall | G | | | | | 61 | | | 61 | 67 |
| 0600 | 12" wall | G | | | | | 67 | | | 67 | 73.50 |
| 0650 | 16" wall | G | | | | | 74 | | | 74 | 81.50 |
| 0700 | 18" wall | G | | | | | 77 | | | 77 | 84.50 |
| 1000 | 3350 lb., w/flat washer, 8" wall | G | | | | | 96 | | | 96 | 106 |
| 1150 | 12" wall | G | | | | | 108 | | | 108 | 119 |
| 1200 | 16" wall | G | | | | | 124 | | | 124 | 136 |
| 1250 | 18" wall | G | | | | | 129 | | | 129 | 142 |
| 1500 | With plastic cone, 8" wall | G | | | | | 96 | | | 96 | 106 |
| 1600 | 12" wall | G | | | | | 108 | | | 108 | 119 |
| 1650 | 16" wall | G | | | | | 124 | | | 124 | 136 |
| 1700 | 18" wall | G | | | | | 129 | | | 129 | 142 |

### 03 15 05.85 Stair Tread Inserts

| | | | Crew | Daily Output | Labor-Hours | Unit | Material | 2011 Bare Costs Labor | 2011 Bare Costs Equipment | Total | Total Incl O&P |
|---|---|---|---|---|---|---|---|---|---|---|---|
| 0010 | **STAIR TREAD INSERTS** | | | | | | | | | | |
| 0015 | Cast iron, abrasive, 3" wide | G | 1 Carp | 90 | .089 | L.F. | 9.65 | 3.83 | | 13.48 | 16.45 |
| 0020 | 4" wide | G | | 80 | .100 | | 12.80 | 4.31 | | 17.11 | 20.50 |
| 0040 | 6" wide | G | | 75 | .107 | | 19.25 | 4.59 | | 23.84 | 28 |
| 0050 | 9" wide | G | | 70 | .114 | | 29 | 4.92 | | 33.92 | 39.50 |
| 0100 | 12" wide | G | ▼ | 65 | .123 | | 38.50 | 5.30 | | 43.80 | 50.50 |
| 0300 | Cast aluminum, compared to cast iron, deduct | | | | | | 10% | | | | |
| 0500 | Extruded aluminum safety tread, 3" wide | G | 1 Carp | 75 | .107 | | 10.95 | 4.59 | | 15.54 | 19.10 |
| 0550 | 4" wide | G | | 75 | .107 | | 14.60 | 4.59 | | 19.19 | 23 |
| 0600 | 6" wide | G | | 75 | .107 | | 22 | 4.59 | | 26.59 | 31 |
| 0650 | 9" wide to resurface stairs | G | ▼ | 70 | .114 | | 33 | 4.92 | | 37.92 | 43.50 |
| 1700 | Cement fill for pan-type metal treads, plain | | 1 Cefi | 115 | .070 | S.F. | 1.83 | 2.84 | | 4.67 | 6.15 |

## 03 15 05 – Concrete Forming Accessories

| 03 15 05.85 **Stair Tread Inserts** | | Crew | Daily Output | Labor-Hours | Unit | Material | 2011 Bare Costs Labor | Equipment | Total | Total Incl O&P |
|---|---|---|---|---|---|---|---|---|---|---|
| 1750 | Non-slip | 1 Cefi | 100 | .080 | S.F. | 2.02 | 3.27 | | 5.29 | 7 |

| 03 15 05.95 **Wall and Foundation Form Accessories** | | | | | | | | | | |
|---|---|---|---|---|---|---|---|---|---|---|
| 0010 | **WALL AND FOUNDATION FORM ACCESSORIES** | | | | | | | | | |
| 2000 | Footings, turnbuckle form aligner [G] | | | | Ea. | 13.50 | | | 13.50 | 14.85 |
| 2050 | Spreaders for footer, adjustable [G] | | | | " | 17.90 | | | 17.90 | 19.70 |
| 3000 | Form oil, coverage varies greatly, minimum | | | | Gal. | 6.80 | | | 6.80 | 7.50 |
| 3050 | Maximum | | | | " | 15.90 | | | 15.90 | 17.50 |
| 3500 | Form patches, 1-3/4" diameter | | | | C | 39 | | | 39 | 43 |
| 3550 | 2-3/4" diameter | | | | " | 68 | | | 68 | 75 |
| 4000 | Nail stakes, 3/4" diameter, 18" long [G] | | | | Ea. | 2.44 | | | 2.44 | 2.68 |
| 4050 | 24" long [G] | | | | | 3.16 | | | 3.16 | 3.48 |
| 4200 | 30" long [G] | | | | | 4.05 | | | 4.05 | 4.46 |
| 4250 | 36" long [G] | | | | | 4.89 | | | 4.89 | 5.40 |

## 03 15 13 – Waterstops

| 03 15 13.50 **Waterstops** | | Crew | Daily Output | Labor-Hours | Unit | Material | 2011 Bare Costs Labor | Equipment | Total | Total Incl O&P |
|---|---|---|---|---|---|---|---|---|---|---|
| 0010 | **WATERSTOPS**, PVC and Rubber | | | | | | | | | |
| 0020 | PVC, ribbed 3/16" thick, 4" wide | 1 Carp | 155 | .052 | L.F. | .95 | 2.22 | | 3.17 | 4.46 |
| 0050 | 6" wide | | 145 | .055 | | 1.79 | 2.38 | | 4.17 | 5.60 |
| 0500 | With center bulb, 6" wide, 3/16" thick | | 135 | .059 | | 1.61 | 2.55 | | 4.16 | 5.70 |
| 0550 | 3/8" thick | | 130 | .062 | | 2.91 | 2.65 | | 5.56 | 7.25 |
| 0600 | 9" wide x 3/8" thick | | 125 | .064 | | 4.48 | 2.76 | | 7.24 | 9.15 |
| 0800 | Dumbbell type, 6" wide, 3/16" thick | | 150 | .053 | | 1.65 | 2.30 | | 3.95 | 5.35 |
| 0850 | 3/8" thick | | 145 | .055 | | 3.27 | 2.38 | | 5.65 | 7.25 |
| 1000 | 9" wide, 3/8" thick, plain | | 130 | .062 | | 4.86 | 2.65 | | 7.51 | 9.40 |
| 1050 | Center bulb | | 130 | .062 | | 6.95 | 2.65 | | 9.60 | 11.65 |
| 1250 | Ribbed type, split, 3/16" thick, 6" wide | | 145 | .055 | | 1.62 | 2.38 | | 4 | 5.40 |
| 1300 | 3/8" thick | | 130 | .062 | | 3.43 | 2.65 | | 6.08 | 7.85 |
| 2000 | Rubber, flat dumbbell, 3/8" thick, 6" wide | | 145 | .055 | | 8.10 | 2.38 | | 10.48 | 12.60 |
| 2050 | 9" wide | | 135 | .059 | | 12.40 | 2.55 | | 14.95 | 17.55 |
| 2500 | Flat dumbbell split, 3/8" thick, 6" wide | | 145 | .055 | | 1.62 | 2.38 | | 4 | 5.40 |
| 2550 | 9" wide | | 135 | .059 | | 3.43 | 2.55 | | 5.98 | 7.70 |
| 3000 | Center bulb, 1/4" thick, 6" wide | | 145 | .055 | | 6.95 | 2.38 | | 9.33 | 11.30 |
| 3050 | 9" wide | | 135 | .059 | | 15.55 | 2.55 | | 18.10 | 21 |
| 3500 | Center bulb split, 3/8" thick, 6" wide | | 145 | .055 | | 10.30 | 2.38 | | 12.68 | 15 |
| 3550 | 9" wide | | 135 | .059 | | 17.80 | 2.55 | | 20.35 | 23.50 |
| 5000 | Waterstop fittings, rubber, flat | | | | | | | | | |
| 5010 | Dumbbell or center bulb, 3/8" thick, | | | | | | | | | |
| 5200 | Field union, 6" wide | 1 Carp | 50 | .160 | Ea. | 21 | 6.90 | | 27.90 | 33.50 |
| 5250 | 9" wide | | 50 | .160 | | 27 | 6.90 | | 33.90 | 40.50 |
| 5500 | Flat cross, 6" wide | | 30 | .267 | | 35 | 11.50 | | 46.50 | 56 |
| 5550 | 9" wide | | 30 | .267 | | 51 | 11.50 | | 62.50 | 73.50 |
| 6000 | Flat tee, 6" wide | | 30 | .267 | | 36 | 11.50 | | 47.50 | 57 |
| 6050 | 9" wide | | 30 | .267 | | 49 | 11.50 | | 60.50 | 71.50 |
| 6500 | Flat ell, 6" wide | | 40 | .200 | | 35 | 8.60 | | 43.60 | 51.50 |
| 6550 | 9" wide | | 40 | .200 | | 45.50 | 8.60 | | 54.10 | 63 |
| 7000 | Vertical tee, 6" wide | | 25 | .320 | | 28 | 13.80 | | 41.80 | 51.50 |
| 7050 | 9" wide | | 25 | .320 | | 39.50 | 13.80 | | 53.30 | 64.50 |
| 7500 | Vertical ell, 6" wide | | 35 | .229 | | 26 | 9.85 | | 35.85 | 43.50 |
| 7550 | 9" wide | | 35 | .229 | | 35 | 9.85 | | 44.85 | 53.50 |

# 03 21 Reinforcing Steel

## 03 21 05 – Reinforcing Steel Accessories

| 03 21 05.10 Rebar Accessories | | Crew | Daily Output | Labor-Hours | Unit | Material | 2011 Bare Costs Labor | Equipment | Total | Total Incl O&P |
|---|---|---|---|---|---|---|---|---|---|---|
| 0010 | **REBAR ACCESSORIES** | | | | | | | | | |
| 0030 | Steel & plastic made from recycled materials | G | | | | | | | | |
| 0100 | Beam bolsters (BB), lower, 1-1/2" high, plain steel | G | | | | C.L.F. | 43 | | | 43 | 47.50 |
| 0102 | Galvanized | G | | | | | 51.50 | | | 51.50 | 57 |
| 0104 | Stainless tipped legs | G | | | | | 345 | | | 345 | 375 |
| 0106 | Plastic tipped legs | G | | | | | 47 | | | 47 | 51.50 |
| 0108 | Epoxy dipped | G | | | | | 90 | | | 90 | 99 |
| 0110 | 2" high, plain | G | | | | | 47 | | | 47 | 51.50 |
| 0120 | Galvanized | G | | | | | 56 | | | 56 | 61.50 |
| 0140 | Stainless tipped legs | G | | | | | 345 | | | 345 | 380 |
| 0160 | Plastic tipped legs | G | | | | | 48 | | | 48 | 53 |
| 0162 | Epoxy dipped | G | | | | | 100 | | | 100 | 110 |
| 0200 | Upper (BBU), 1-1/2" high, plain steel | G | | | | | 84 | | | 84 | 92.50 |
| 0210 | 3" high | G | | | | | 97 | | | 97 | 107 |
| 0300 | Beam bolster with plate (BBP), 1-1/2" high, plain steel | G | | | | | 167 | | | 167 | 184 |
| 0310 | 3" high | G | | | | | 267 | | | 267 | 294 |
| 0500 | Slab bolsters, continuous (SB), 1" high, plain steel | G | | | | | 23 | | | 23 | 25.50 |
| 0502 | Galvanized | G | | | | | 40 | | | 40 | 44 |
| 0504 | Stainless tipped legs | G | | | | | 325 | | | 325 | 355 |
| 0506 | Plastic tipped legs | G | | | | | 30 | | | 30 | 33 |
| 0510 | 2" high, plain steel | G | | | | | 25 | | | 25 | 27.50 |
| 0515 | Galvanized | G | | | | | 51 | | | 51 | 56 |
| 0520 | Stainless tipped legs | G | | | | | 325 | | | 325 | 360 |
| 0525 | Plastic tipped legs | G | | | | | 35 | | | 35 | 38.50 |
| 0530 | For bolsters with wire runners (SBR), add | G | | | | | 32 | | | 32 | 35 |
| 0540 | For bolsters with plates (SBP), add | G | | | | | 64 | | | 64 | 70.50 |
| 0700 | Bag ties, 16 ga., plain, 4" long | G | | | | C | 2 | | | 2 | 2.20 |
| 0710 | 5" long | G | | | | | 3 | | | 3 | 3.30 |
| 0720 | 6" long | G | | | | | 2 | | | 2 | 2.20 |
| 0730 | 7" long | G | | | | | 3 | | | 3 | 3.30 |
| 1200 | High chairs, individual (HC), 3" high, plain steel | G | | | | | 34 | | | 34 | 37.50 |
| 1202 | Galvanized | G | | | | | 41 | | | 41 | 45 |
| 1204 | Stainless tipped legs | G | | | | | 335 | | | 335 | 365 |
| 1206 | Plastic tipped legs | G | | | | | 46 | | | 46 | 50.50 |
| 1210 | 5" high, plain | G | | | | | 50 | | | 50 | 55 |
| 1212 | Galvanized | G | | | | | 55 | | | 55 | 60.50 |
| 1214 | Stainless tipped legs | G | | | | | 350 | | | 350 | 385 |
| 1216 | Plastic tipped legs | G | | | | | 63 | | | 63 | 69.50 |
| 1220 | 8" high, plain | G | | | | | 102 | | | 102 | 112 |
| 1222 | Galvanized | G | | | | | 122 | | | 122 | 135 |
| 1224 | Stainless tipped legs | G | | | | | 400 | | | 400 | 440 |
| 1226 | Plastic tipped legs | G | | | | | 131 | | | 131 | 144 |
| 1230 | 12" high, plain | G | | | | | 238 | | | 238 | 262 |
| 1232 | Galvanized | G | | | | | 285 | | | 285 | 315 |
| 1234 | Stainless tipped legs | G | | | | | 540 | | | 540 | 590 |
| 1236 | Plastic tipped legs | G | | | | | 238 | | | 238 | 262 |
| 1400 | Individual high chairs, with plate (HCP), 5" high | G | | | | | 147 | | | 147 | 162 |
| 1410 | 8" high | G | | | | | 204 | | | 204 | 224 |
| 1500 | Bar chair (BC), 1-1/2" high, plain steel | G | | | | | 22 | | | 22 | 24 |
| 1520 | Galvanized | G | | | | | 31 | | | 31 | 34 |
| 1530 | Stainless tipped legs | G | | | | | 320 | | | 320 | 355 |
| 1540 | Plastic tipped legs | G | | | | | 35 | | | 35 | 38.50 |
| 1550 | Joist chair (JC), 3/4" x 5", plain steel | G | | | | | 28 | | | 28 | 31 |

## 03 21 05 – Reinforcing Steel Accessories

| 03 21 05.10 Rebar Accessories | | Crew | Daily Output | Labor-Hours | Unit | Material | 2011 Bare Costs Labor | Equipment | Total | Total Incl O&P |
|---|---|---|---|---|---|---|---|---|---|---|
| 1580 | Galvanized | G | | | C | 33.50 | | | 33.50 | 37 |
| 1600 | Stainless tipped legs | G | | | | 330 | | | 330 | 360 |
| 1620 | Plastic tipped legs | G | | | | 47 | | | 47 | 51.50 |
| 1630 | Epoxy dipped | G | | | | 104 | | | 104 | 114 |
| 1700 | Continuous high chairs (CHC), legs 8" O.C., 4" high, plain steel | G | | | C.L.F. | 41 | | | 41 | 45 |
| 1705 | Galvanized | G | | | | 49 | | | 49 | 54 |
| 1710 | Stainless tipped legs | G | | | | 340 | | | 340 | 375 |
| 1715 | Plastic tipped legs | G | | | | 54 | | | 54 | 59.50 |
| 1718 | Epoxy dipped | G | | | | 75 | | | 75 | 82.50 |
| 1720 | 6" high, plain | G | | | | 59 | | | 59 | 65 |
| 1725 | Galvanized | G | | | | 71 | | | 71 | 78 |
| 1730 | Stainless tipped legs | G | | | | 360 | | | 360 | 395 |
| 1735 | Plastic tipped legs | G | | | | 82 | | | 82 | 90 |
| 1738 | Epoxy dipped | G | | | | 99 | | | 99 | 109 |
| 1740 | 8" high, plain | G | | | | 87 | | | 87 | 95.50 |
| 1745 | Galvanized | G | | | | 104 | | | 104 | 115 |
| 1750 | Stainless tipped legs | G | | | | 385 | | | 385 | 425 |
| 1755 | Plastic tipped legs | G | | | | 131 | | | 131 | 144 |
| 1758 | Epoxy dipped | G | | | | 150 | | | 150 | 165 |
| 1900 | For continuous bottom wire runners, add | G | | | | 32 | | | 32 | 35 |
| 1940 | For continuous bottom plate, add | G | | | | 151 | | | 151 | 166 |
| 2100 | Paper tubing, 4' lengths, for #4 bar | G | | | | 64 | | | 64 | 70.50 |
| 2120 | For #6 bar | G | | | | 75 | | | 75 | 82.50 |
| 2200 | Screed chair base, 1/2" coil thread diam., 2-1/2" high, plain steel | G | | | C | 244 | | | 244 | 268 |
| 2210 | Galvanized | G | | | | 257 | | | 257 | 283 |
| 2220 | 5-1/2" high, plain | G | | | | 295 | | | 295 | 325 |
| 2250 | Galvanized | G | | | | 310 | | | 310 | 340 |
| 2300 | 3/4" coil thread diam., 2-1/2" high, plain steel | G | | | | 310 | | | 310 | 340 |
| 2310 | Galvanized | G | | | | 415 | | | 415 | 455 |
| 2320 | 5-1/2" high, plain steel | G | | | | 395 | | | 395 | 435 |
| 2350 | Galvanized | G | | | | 530 | | | 530 | 585 |
| 2400 | Screed holder, 1/2" coil thread diam. for pipe screed, plain steel, 6" long | G | | | | 275 | | | 275 | 305 |
| 2420 | 12" long | G | | | | 425 | | | 425 | 465 |
| 2500 | 3/4" coil thread diam. for pipe screed, plain steel, 6" long | G | | | | 395 | | | 395 | 435 |
| 2520 | 12" long | G | | | | 635 | | | 635 | 700 |
| 2700 | Screw anchor for bolts, plain steel, 3/4" diameter x 4" long | G | | | | 405 | | | 405 | 445 |
| 2720 | 1" diameter x 6" long | G | | | | 660 | | | 660 | 730 |
| 2740 | 1-1/2" diameter x 8" long | G | | | | 840 | | | 840 | 925 |
| 2800 | Screw anchor eye bolts, 3/4" x 7" long | G | | | | 3,450 | | | 3,450 | 3,800 |
| 2820 | 1" x 9" long | G | | | | 6,825 | | | 6,825 | 7,525 |
| 2840 | 1-1/2" x 14" long | G | | | | 17,300 | | | 17,300 | 19,000 |
| 2900 | Screw anchor bolts, 3/4" x 9" long | G | | | | 1,200 | | | 1,200 | 1,325 |
| 2920 | 1" x 12" long | G | | | | 2,325 | | | 2,325 | 2,575 |
| 3001 | Slab lifting inserts, single pickup, galv, 3/4" diam., 5" high | G | | | | 1,150 | | | 1,150 | 1,250 |
| 3010 | 6" high | G | | | | 1,175 | | | 1,175 | 1,275 |
| 3030 | 7" high | G | | | | 1,175 | | | 1,175 | 1,300 |
| 3100 | 1" diameter, 5-1/2" high | G | | | | 1,200 | | | 1,200 | 1,325 |
| 3120 | 7" high | G | | | | 1,250 | | | 1,250 | 1,375 |
| 3200 | Double pickup lifting inserts, 1" diameter, 5-1/2" high | G | | | | 2,325 | | | 2,325 | 2,575 |
| 3220 | 7" high | G | | | | 2,575 | | | 2,575 | 2,850 |
| 3330 | 1-1/2" diameter, 5-1/2" high | G | | | | 2,825 | | | 2,825 | 3,125 |
| 3800 | Subgrade chairs, #4 bar head, 3-1/2" high | G | | | | 28 | | | 28 | 31 |
| 3850 | 12" high | G | | | | 40 | | | 40 | 44 |

# 03 21 Reinforcing Steel

## 03 21 05 – Reinforcing Steel Accessories

### 03 21 05.10 Rebar Accessories

| | | Crew | Daily Output | Labor-Hours | Unit | Material | 2011 Bare Costs Labor | Equipment | Total | Total Incl O&P |
|---|---|---|---|---|---|---|---|---|---|---|
| 3900 | #6 bar head, 3-1/2" high | G | | | C | 28 | | | 28 | 31 |
| 3950 | 12" high | G | | | | 40 | | | 40 | 44 |
| 4200 | Subgrade stakes, no nail holes, 3/4" diameter, 12" long | G | | | | 196 | | | 196 | 216 |
| 4250 | 24" long | G | | | | 284 | | | 284 | 310 |
| 4300 | 7/8" diameter, 12" long | G | | | | 277 | | | 277 | 305 |
| 4350 | 24" long | G | | | | 465 | | | 465 | 510 |
| 4500 | Tie wire, 16 ga. annealed steel | G | | | Cwt. | 164 | | | 164 | 180 |
| 4550 | Tie wire holder, plastic case | G | | | Ea. | 40 | | | 40 | 44 |
| 4600 | Aluminum case | G | | | " | 56.50 | | | 56.50 | 62.50 |

### 03 21 05.75 Splicing Reinforcing Bars

| | | Crew | Daily Output | Labor-Hours | Unit | Material | 2011 Bare Costs Labor | Equipment | Total | Total Incl O&P |
|---|---|---|---|---|---|---|---|---|---|---|
| 0010 | **SPLICING REINFORCING BARS**　　R032110-70 | | | | | | | | | |
| 0020 | Including holding bars in place while splicing | | | | | | | | | |
| 0100 | Standard, self-aligning type, taper threaded, #4 bars | G | C-25 | 190 | .168 | Ea. | 4.55 | 6.35 | | 10.90 | 15.30 |
| 0105 | #5 bars | G | | 170 | .188 | | 5.55 | 7.10 | | 12.65 | 17.65 |
| 0110 | #6 bars | G | | 150 | .213 | | 6.40 | 8.05 | | 14.45 | 20 |
| 0120 | #7 bars | G | | 130 | .246 | | 7.45 | 9.30 | | 16.75 | 23.50 |
| 0300 | #8 bars | G | | 115 | .278 | | 12.90 | 10.50 | | 23.40 | 31.50 |
| 0305 | #9 bars | G | C-5 | 105 | .533 | | 14.10 | 25 | 7.15 | 46.25 | 63 |
| 0310 | #10 bars | G | | 95 | .589 | | 15.70 | 28 | 7.90 | 51.60 | 70 |
| 0320 | #11 bars | G | | 85 | .659 | | 16.70 | 31 | 8.80 | 56.50 | 77 |
| 0330 | #14 bars | G | | 65 | .862 | | 25 | 40.50 | 11.50 | 77 | 104 |
| 0340 | #18 bars | G | | 45 | 1.244 | | 38 | 59 | 16.65 | 113.65 | 153 |
| 0500 | Transition self-aligning, taper threaded, #18-14 | G | | 45 | 1.244 | | 40.50 | 59 | 16.65 | 116.15 | 155 |
| 0510 | #18-11 | G | | 45 | 1.244 | | 41.50 | 59 | 16.65 | 117.15 | 157 |
| 0520 | #14-11 | G | | 65 | .862 | | 29.50 | 40.50 | 11.50 | 81.50 | 109 |
| 0540 | #11-10 | G | | 85 | .659 | | 23.50 | 31 | 8.80 | 63.30 | 84.50 |
| 0550 | #10-9 | G | | 95 | .589 | | 22 | 28 | 7.90 | 57.90 | 76.50 |
| 0560 | #9-8 | G | C-25 | 105 | .305 | | 21 | 11.50 | | 32.50 | 41.50 |
| 0580 | #8-7 | G | | 115 | .278 | | 19.55 | 10.50 | | 30.05 | 38.50 |
| 0590 | #7-6 | G | | 130 | .246 | | 19.05 | 9.30 | | 28.35 | 36 |
| 0600 | Position coupler for curved bars, taper threaded, #4 bars | G | | 160 | .200 | | 22 | 7.55 | | 29.55 | 36.50 |
| 0610 | #5 bars | G | | 145 | .221 | | 23 | 8.35 | | 31.35 | 38.50 |
| 0620 | #6 bars | G | | 130 | .246 | | 26 | 9.30 | | 35.30 | 44 |
| 0630 | #7 bars | G | | 110 | .291 | | 38.50 | 11 | | 49.50 | 60.50 |
| 0640 | #8 bars | G | | 100 | .320 | | 40 | 12.10 | | 52.10 | 63.50 |
| 0650 | #9 bars | G | C-5 | 90 | .622 | | 43.50 | 29.50 | 8.30 | 81.30 | 103 |
| 0660 | #10 bars | G | | 80 | .700 | | 47 | 33 | 9.35 | 89.35 | 114 |
| 0670 | #11 bars | G | | 70 | .800 | | 49 | 38 | 10.70 | 97.70 | 125 |
| 0680 | #14 bars | G | | 55 | 1.018 | | 61 | 48 | 13.60 | 122.60 | 158 |
| 0690 | #18 bars | G | | 40 | 1.400 | | 88 | 66 | 18.70 | 172.70 | 221 |
| 0700 | Transition position coupler for curved bars, taper threaded, #18-14 | G | | 40 | 1.400 | | 111 | 66 | 18.70 | 195.70 | 247 |
| 0710 | #18-11 | G | | 40 | 1.400 | | 112 | 66 | 18.70 | 196.70 | 248 |
| 0720 | #14-11 | G | | 55 | 1.018 | | 95 | 48 | 13.60 | 156.60 | 194 |
| 0730 | #11-10 | G | | 70 | .800 | | 54.50 | 38 | 10.70 | 103.20 | 131 |
| 0740 | #10-9 | G | | 80 | .700 | | 52 | 33 | 9.35 | 94.35 | 119 |
| 0750 | #9-8 | G | C-25 | 90 | .356 | | 48 | 13.40 | | 61.40 | 75 |
| 0760 | #8-7 | G | | 100 | .320 | | 44 | 12.10 | | 56.10 | 68 |
| 0770 | #7-6 | G | | 110 | .291 | | 42.50 | 11 | | 53.50 | 65 |
| 0800 | Sleeve type w/grout filler, for precast concrete, #6 bars | G | | 72 | .444 | | 23.50 | 16.80 | | 40.30 | 53 |
| 0802 | #7 bars | G | | 64 | .500 | | 25.50 | 18.90 | | 44.40 | 58.50 |
| 0805 | #8 bars | G | | 56 | .571 | | 29 | 21.50 | | 50.50 | 67 |
| 0807 | #9 bars | G | | 48 | .667 | | 31.50 | 25 | | 56.50 | 76 |

# 03 21 Reinforcing Steel

## 03 21 05 – Reinforcing Steel Accessories

### 03 21 05.75 Splicing Reinforcing Bars

| | | | Crew | Daily Output | Labor-Hours | Unit | Material | 2011 Bare Costs Labor | Equipment | Total | Total Incl O&P |
|---|---|---|---|---|---|---|---|---|---|---|---|
| 0810 | #10 bars | G | C-5 | 40 | 1.400 | Ea. | 41 | 66 | 18.70 | 125.70 | 170 |
| 0900 | #11 bars | G | | 32 | 1.750 | | 49 | 82.50 | 23.50 | 155 | 210 |
| 0920 | #14 bars | G | | 24 | 2.333 | | 79.50 | 110 | 31 | 220.50 | 295 |
| 1000 | Sleeve type w/ferrous filler, for critical structures, #6 bars | G | C-25 | 72 | .444 | | 41 | 16.80 | | 57.80 | 72.50 |
| 1210 | #7 bars | G | | 64 | .500 | | 42 | 18.90 | | 60.90 | 76.50 |
| 1220 | #8 bars | G | | 56 | .571 | | 44 | 21.50 | | 65.50 | 83.50 |
| 1230 | #9 bars | G | C-5 | 48 | 1.167 | | 45 | 55 | 15.60 | 115.60 | 154 |
| 1240 | #10 bars | G | | 40 | 1.400 | | 48 | 66 | 18.70 | 132.70 | 178 |
| 1250 | #11 bars | G | | 32 | 1.750 | | 58 | 82.50 | 23.50 | 164 | 220 |
| 1260 | #14 bars | G | | 24 | 2.333 | | 73 | 110 | 31 | 214 | 288 |
| 1270 | #18 bars | G | | 16 | 3.500 | | 106 | 165 | 47 | 318 | 430 |
| 2000 | Weldable half coupler, taper threaded, #4 bars | G | E-16 | 120 | .133 | | 7.10 | 6.60 | .91 | 14.61 | 20.50 |
| 2100 | #5 bars | G | | 112 | .143 | | 8.40 | 7.10 | .97 | 16.47 | 23 |
| 2200 | #6 bars | G | | 104 | .154 | | 13.35 | 7.60 | 1.05 | 22 | 29.50 |
| 2300 | #7 bars | G | | 96 | .167 | | 15.50 | 8.25 | 1.14 | 24.89 | 33 |
| 2400 | #8 bars | G | | 88 | .182 | | 16.15 | 9 | 1.24 | 26.39 | 35 |
| 2500 | #9 bars | G | | 80 | .200 | | 17.80 | 9.90 | 1.36 | 29.06 | 38.50 |
| 2600 | #10 bars | G | | 72 | .222 | | 18.20 | 11 | 1.51 | 30.71 | 41 |
| 2700 | #11 bars | G | | 64 | .250 | | 19.45 | 12.40 | 1.70 | 33.55 | 45.50 |
| 2800 | #14 bars | G | | 56 | .286 | | 22.50 | 14.15 | 1.95 | 38.60 | 51.50 |
| 2900 | #18 bars | G | | 48 | .333 | | 36.50 | 16.50 | 2.27 | 55.27 | 71.50 |

## 03 21 10 – Uncoated Reinforcing Steel

### 03 21 10.60 Reinforcing In Place

| | | | Crew | Daily Output | Labor-Hours | Unit | Material | 2011 Bare Costs Labor | Equipment | Total | Total Incl O&P |
|---|---|---|---|---|---|---|---|---|---|---|---|
| 0015 | **REINFORCING IN PLACE**, 50-60 ton lots, A615 Grade 60 | R032110-10 | | | | | | | | | |
| 0020 | Includes labor, but not material cost, to install accessories | | | | | | | | | | |
| 0030 | Made from recycled materials | G | | | | | | | | | |
| 0100 | Beams & Girders, #3 to #7 | G | 4 Rodm | 1.60 | 20 | Ton | 900 | 970 | | 1,870 | 2,550 |
| 0150 | #8 to #18 | R032110-20 | G | | 2.70 | 11.852 | | 900 | 575 | | 1,475 | 1,900 |
| 0200 | Columns, #3 to #7 | G | | 1.50 | 21.333 | | 900 | 1,025 | | 1,925 | 2,650 |
| 0250 | #8 to #18 | G | | 2.30 | 13.913 | | 900 | 675 | | 1,575 | 2,075 |
| 0300 | Spirals, hot rolled, 8" to 15" diameter | G | | 2.20 | 14.545 | | 1,400 | 705 | | 2,105 | 2,675 |
| 0320 | 15" to 24" diameter | R032110-40 | G | | 2.20 | 14.545 | | 1,350 | 705 | | 2,055 | 2,600 |
| 0330 | 24" to 36" diameter | G | | 2.30 | 13.913 | | 1,275 | 675 | | 1,950 | 2,475 |
| 0340 | 36" to 48" diameter | R032110-50 | G | | 2.40 | 13.333 | | 1,225 | 645 | | 1,870 | 2,350 |
| 0360 | 48" to 64" diameter | G | | 2.50 | 12.800 | | 1,350 | 620 | | 1,970 | 2,475 |
| 0380 | 64" to 84" diameter | R032110-70 | G | | 2.60 | 12.308 | | 1,400 | 595 | | 1,995 | 2,500 |
| 0390 | 84" to 96" diameter | G | | 2.70 | 11.852 | | 1,475 | 575 | | 2,050 | 2,550 |
| 0400 | Elevated slabs, #4 to #7 | R032110-80 | G | | 2.90 | 11.034 | | 955 | 535 | | 1,490 | 1,900 |
| 0500 | Footings, #4 to #7 | G | | 2.10 | 15.238 | | 855 | 740 | | 1,595 | 2,125 |
| 0550 | #8 to #18 | G | | 3.60 | 8.889 | | 810 | 430 | | 1,240 | 1,575 |
| 0600 | Slab on grade, #3 to #7 | G | | 2.30 | 13.913 | | 855 | 675 | | 1,530 | 2,025 |
| 0700 | Walls, #3 to #7 | G | | 3 | 10.667 | | 855 | 515 | | 1,370 | 1,775 |
| 0750 | #8 to #18 | G | | 4 | 8 | | 855 | 385 | | 1,240 | 1,550 |
| 0900 | For other than 50 – 60 ton lots | | | | | | | | | | |
| 1000 | Under 10 ton job, #3 to #7, add | | | | | Ton | 25% | 10% | | | |
| 1010 | #8 to #18, add | | | | | | 20% | 10% | | | |
| 1050 | 10 – 50 ton job, #3 to #7, add | | | | | | 10% | | | | |
| 1060 | #8 to #18, add | | | | | | 5% | | | | |
| 1100 | 60 – 100 ton job, #3 to #7, deduct | | | | | | 5% | | | | |
| 1110 | #8 to #18, deduct | | | | | | 10% | | | | |
| 1150 | Over 100 ton job, #3 to #7, deduct | | | | | | 10% | | | | |
| 1160 | #8 to #18, deduct | | | | | | 15% | | | | |

# 03 21 Reinforcing Steel

## 03 21 10 – Uncoated Reinforcing Steel

### 03 21 10.60 Reinforcing In Place

| | | Crew | Daily Output | Labor-Hours | Unit | Material | 2011 Bare Costs Labor | Equipment | Total | Total Incl O&P |
|---|---|---|---|---|---|---|---|---|---|---|
| 1200 | Reinforcing in place, A615 Grade 75, add [G] | | | | Ton | 82.50 | | | 82.50 | 90.50 |
| 1220 | Grade 90, add | | | | | 99 | | | 99 | 109 |
| 2000 | Unloading & sorting, add to above | C-5 | 100 | .560 | | | 26.50 | 7.50 | 34 | 50 |
| 2200 | Crane cost for handling, add to above, minimum | | 135 | .415 | | | 19.60 | 5.55 | 25.15 | 37 |
| 2210 | Average | | 92 | .609 | | | 29 | 8.15 | 37.15 | 54 |
| 2220 | Maximum | | 35 | 1.600 | | | 75.50 | 21.50 | 97 | 143 |
| 2400 | Dowels, 2 feet long, deformed, #3 [G] | 2 Rodm | 520 | .031 | Ea. | .37 | 1.49 | | 1.86 | 2.79 |
| 2410 | #4 [G] | | 480 | .033 | | .66 | 1.61 | | 2.27 | 3.31 |
| 2420 | #5 [G] | | 435 | .037 | | 1.03 | 1.78 | | 2.81 | 3.98 |
| 2430 | #6 [G] | | 360 | .044 | | 1.49 | 2.15 | | 3.64 | 5.10 |
| 2450 | Longer and heavier dowels, add [G] | | 725 | .022 | Lb. | .50 | 1.07 | | 1.57 | 2.25 |
| 2500 | Smooth dowels, 12" long, 1/4" or 3/8" diameter [G] | | 140 | .114 | Ea. | .72 | 5.55 | | 6.27 | 9.65 |
| 2520 | 5/8" diameter [G] | | 125 | .128 | | 1.26 | 6.20 | | 7.46 | 11.30 |
| 2530 | 3/4" diameter [G] | | 110 | .145 | | 1.57 | 7.05 | | 8.62 | 12.95 |
| 2600 | Dowel sleeves for CIP concrete, 2-part system | | | | | | | | | |
| 2610 | Sleeve base, plastic, for 5/8" smooth dowel sleeve, fasten to edge form | 1 Rodm | 200 | .040 | Ea. | .57 | 1.94 | | 2.51 | 3.72 |
| 2615 | Sleeve, plastic, 12" long, for 5/8" smooth dowel, snap onto base | | 400 | .020 | | 1.09 | .97 | | 2.06 | 2.75 |
| 2620 | Sleeve base, for 3/4" smooth dowel sleeve | | 175 | .046 | | .54 | 2.21 | | 2.75 | 4.13 |
| 2625 | Sleeve, 12" long, for 3/4" smooth dowel | | 350 | .023 | | .97 | 1.11 | | 2.08 | 2.84 |
| 2630 | Sleeve base, for 1" smooth dowel sleeve | | 150 | .053 | | .64 | 2.58 | | 3.22 | 4.83 |
| 2635 | Sleeve, 12" long, for 1" smooth dowel | | 300 | .027 | | 1.28 | 1.29 | | 2.57 | 3.47 |
| 2700 | Dowel caps, visual warning only, plastic, #3 to #8 | 2 Rodm | 800 | .020 | | .26 | .97 | | 1.23 | 1.84 |
| 2720 | #8 to #18 | | 750 | .021 | | .78 | 1.03 | | 1.81 | 2.51 |
| 2750 | Impalement protective, plastic, #4 to #9 | | 800 | .020 | | 1.41 | .97 | | 2.38 | 3.10 |

### 03 21 10.70 Glass Fiber Reinforced Polymer Bars

| | | Crew | Daily Output | Labor-Hours | Unit | Material | 2011 Bare Costs Labor | Equipment | Total | Total Incl O&P |
|---|---|---|---|---|---|---|---|---|---|---|
| 0010 | **GLASS FIBER REINFORCED POLYMER BARS** | | | | | | | | | |
| 0050 | #2 bar, .043 lb./L.F. | 4 Rodm | 9500 | .003 | L.F. | .35 | .16 | | .51 | .65 |
| 0100 | #3 bar, .092 lb./L.F. | | 9300 | .003 | | .48 | .17 | | .65 | .80 |
| 0150 | #4 bar, .160 lb./L.F. | | 9100 | .004 | | .70 | .17 | | .87 | 1.04 |
| 0200 | #5 bar, .258 lb./L.F. | | 8700 | .004 | | 1.05 | .18 | | 1.23 | 1.44 |
| 0250 | #6 bar, .372 lb./L.F. | | 8300 | .004 | | 1.38 | .19 | | 1.57 | 1.82 |
| 0300 | #7 bar, .497 lb./L.F. | | 7900 | .004 | | 1.73 | .20 | | 1.93 | 2.21 |
| 0350 | #8 bar, .620 lb./L.F. | | 7400 | .004 | | 2.32 | .21 | | 2.53 | 2.88 |
| 0400 | #9 bar, .800 lb./L.F. | | 6800 | .005 | | 2.92 | .23 | | 3.15 | 3.57 |
| 0450 | #10 bar, 1.08 lb./L.F. | | 5800 | .006 | | 3.50 | .27 | | 3.77 | 4.28 |
| 0500 | For Bends, add per bend | | | | Ea. | 1.25 | | | 1.25 | 1.38 |

## 03 21 13 – Galvanized Reinforcing Steel

### 03 21 13.10 Galvanized Reinforcing

| | | Crew | Daily Output | Labor-Hours | Unit | Material | 2011 Bare Costs Labor | Equipment | Total | Total Incl O&P |
|---|---|---|---|---|---|---|---|---|---|---|
| 0010 | **GALVANIZED REINFORCING** | | | | | | | | | |
| 0150 | Add to uncoated reinforcing price for galvanizing | | | | Ton | 575 | | | 575 | 635 |

## 03 21 16 – Epoxy-Coated Reinforcing Steel

### 03 21 16.10 Epoxy-Coated Reinforcing

| | | Crew | Daily Output | Labor-Hours | Unit | Material | 2011 Bare Costs Labor | Equipment | Total | Total Incl O&P |
|---|---|---|---|---|---|---|---|---|---|---|
| 0010 | **EPOXY-COATED REINFORCING** | | | | | | | | | |
| 0100 | Add to uncoated reinforcing price for coating with epoxy | | | | Ton | 350 | | | 350 | 385 |

# 03 22 Welded Wire Fabric Reinforcing

## 03 22 05 – Uncoated Welded Wire Fabric

| 03 22 05.50 Welded Wire Fabric | | Crew | Daily Output | Labor-Hours | Unit | Material | 2011 Bare Costs Labor | Equipment | Total | Total Incl O&P |
|---|---|---|---|---|---|---|---|---|---|---|
| 0010 | **WELDED WIRE FABRIC** ASTM A185    R032205-30 | | | | | | | | | |
| 0030 | Made from recycled materials    [G] | | | | | | | | | |
| 0050 | Sheets | | | | | | | | | |
| 0100 | 6 x 6 - W1.4 x W1.4 (10 x 10) 21 lb. per C.S.F.    [G] | 2 Rodm | 35 | .457 | C.S.F. | 12.50 | 22 | | 34.50 | 49.50 |
| 0200 | 6 x 6 - W2.1 x W2.1 (8 x 8) 30 lb. per C.S.F.    [G] | | 31 | .516 | | 18.90 | 25 | | 43.90 | 61 |
| 0300 | 6 x 6 - W2.9 x W2.9 (6 x 6) 42 lb. per C.S.F.    [G] | | 29 | .552 | | 21.50 | 26.50 | | 48 | 66 |
| 0400 | 6 x 6 - W4 x W4 (4 x 4) 58 lb. per C.S.F.    [G] | | 27 | .593 | | 29.50 | 28.50 | | 58 | 78.50 |
| 0500 | 4 x 4 - W1.4 x W1.4 (10 x 10) 31 lb. per C.S.F.    [G] | | 31 | .516 | | 17.90 | 25 | | 42.90 | 59.50 |
| 0600 | 4 x 4 - W2.1 x W2.1 (8 x 8) 44 lb. per C.S.F.    [G] | | 29 | .552 | | 22.50 | 26.50 | | 49 | 67 |
| 0650 | 4 x 4 - W2.9 x W2.9 (6 x 6) 61 lb. per C.S.F.    [G] | | 27 | .593 | | 36 | 28.50 | | 64.50 | 85.50 |
| 0700 | 4 x 4 - W4 x W4 (4 x 4) 85 lb. per C.S.F.    [G] | ↓ | 25 | .640 | ↓ | 45 | 31 | | 76 | 99 |
| 0750 | Rolls | | | | | | | | | |
| 0800 | 2 x 2 - #14 galv., 21 lb./C.S.F., beam & column wrap    [G] | 2 Rodm | 6.50 | 2.462 | C.S.F. | 44 | 119 | | 163 | 239 |
| 0900 | 2 x 2 - #12 galv. for gunite reinforcing    [G] | " | 6.50 | 2.462 | " | 65 | 119 | | 184 | 262 |

# 03 23 Stressing Tendons

## 03 23 05 – Prestressing Tendons

### 03 23 05.50 Prestressing Steel

| | | Crew | Daily Output | Labor-Hours | Unit | Material | 2011 Bare Costs Labor | Equipment | Total | Total Incl O&P |
|---|---|---|---|---|---|---|---|---|---|---|
| 0010 | **PRESTRESSING STEEL**    R034136-90 | | | | | | | | | |
| 0100 | Grouted strand, post-tensioned in field, 50' span, 100 kip    [G] | C-3 | 1200 | .053 | Lb. | 2.28 | 2.38 | .09 | 4.75 | 6.35 |
| 0150 | 300 kip    [G] | | 2700 | .024 | | 1.04 | 1.06 | .04 | 2.14 | 2.85 |
| 0300 | 100' span, 100 kip    [G] | | 1700 | .038 | | 2.29 | 1.68 | .06 | 4.03 | 5.20 |
| 0350 | 300 kip    [G] | | 3200 | .020 | | 1.98 | .89 | .03 | 2.90 | 3.62 |
| 0500 | 200' span, 100 kip    [G] | | 2700 | .024 | | 2.28 | 1.06 | .04 | 3.38 | 4.21 |
| 0550 | 300 kip    [G] | | 3500 | .018 | | 1.98 | .81 | .03 | 2.82 | 3.48 |
| 0800 | Grouted bars, 50' span, 42 kip    [G] | | 2600 | .025 | | .91 | 1.10 | .04 | 2.05 | 2.77 |
| 0850 | 143 kip    [G] | | 3200 | .020 | | .88 | .89 | .03 | 1.80 | 2.40 |
| 1000 | 75' span, 42 kip    [G] | | 3200 | .020 | | .93 | .89 | .03 | 1.85 | 2.46 |
| 1050 | 143 kip    [G] | ↓ | 4200 | .015 | | .78 | .68 | .03 | 1.49 | 1.96 |
| 1200 | Ungrouted strand, 50' span, 100 kip    [G] | C-4 | 1275 | .025 | | .62 | 1.23 | .02 | 1.87 | 2.67 |
| 1250 | 300 kip    [G] | | 1475 | .022 | | .62 | 1.06 | .02 | 1.70 | 2.40 |
| 1400 | 100' span, 100 kip    [G] | | 1500 | .021 | | .62 | 1.04 | .02 | 1.68 | 2.37 |
| 1450 | 300 kip    [G] | | 1650 | .019 | | .62 | .95 | .02 | 1.59 | 2.22 |
| 1600 | 200' span, 100 kip    [G] | | 1500 | .021 | | .62 | 1.04 | .02 | 1.68 | 2.37 |
| 1650 | 300 kip    [G] | | 1700 | .019 | | .62 | .92 | .02 | 1.56 | 2.17 |
| 1800 | Ungrouted bars, 50' span, 42 kip    [G] | | 1400 | .023 | | .45 | 1.12 | .02 | 1.59 | 2.31 |
| 1850 | 143 kip    [G] | | 1700 | .019 | | .45 | .92 | .02 | 1.39 | 1.99 |
| 2000 | 75' span, 42 kip    [G] | | 1800 | .018 | | .45 | .87 | .02 | 1.34 | 1.91 |
| 2050 | 143 kip    [G] | | 2200 | .015 | | .45 | .71 | .01 | 1.17 | 1.65 |
| 2220 | Ungrouted single strand, 100' slab, 25 kip    [G] | | 1200 | .027 | | .62 | 1.30 | .02 | 1.94 | 2.79 |
| 2250 | 35 kip    [G] | ↓ | 1475 | .022 | ↓ | .62 | 1.06 | .02 | 1.70 | 2.40 |
| 3000 | Slabs on grade, 0.5-inch diam. non-bonded strands, HDPE sheathed, | | | | | | | | | |
| 3050 | attached dead-end anchors, loose stressing-end anchors | | | | | | | | | |
| 3100 | 25' x 30' slab, strands @ 36" O.C., placing | 2 Rodm | 2940 | .005 | S.F. | .59 | .26 | | .85 | 1.07 |
| 3105 | Stressing | C-4A | 3750 | .004 | | | .21 | .01 | .22 | .34 |
| 3110 | 42" O.C., placing | 2 Rodm | 3200 | .005 | | .52 | .24 | | .76 | .96 |
| 3115 | Stressing | C-4A | 4040 | .004 | | | .19 | .01 | .20 | .32 |
| 3120 | 48" O.C., placing | 2 Rodm | 3510 | .005 | | .46 | .22 | | .68 | .85 |
| 3125 | Stressing | C-4A | 4390 | .004 | | | .18 | .01 | .19 | .29 |
| 3150 | 25' x 40' slab, strands @ 36" O.C., placing | 2 Rodm | 3370 | .005 | | .57 | .23 | | .80 | 1 |
| 3155 | Stressing | C-4A | 4360 | .004 | ↓ | | .18 | .01 | .19 | .29 |

# 03 23 Stressing Tendons

## 03 23 05 – Prestressing Tendons

### 03 23 05.50 Prestressing Steel

| | | Crew | Daily Output | Labor-Hours | Unit | Material | 2011 Bare Costs Labor | Equipment | Total | Total Incl O&P |
|---|---|---|---|---|---|---|---|---|---|---|
| 3160 | 42" O.C., placing | 2 Rodm | 3760 | .004 | S.F. | .49 | .21 | | .70 | .87 |
| 3165 | Stressing | C-4A | 4820 | .003 | | | .16 | .01 | .17 | .27 |
| 3170 | 48" O.C., placing | 2 Rodm | 4090 | .004 | | .44 | .19 | | .63 | .78 |
| 3175 | Stressing | C-4A | 5190 | .003 | | | .15 | .01 | .16 | .25 |
| 3200 | 30' x 30' slab, strands @ 36" O.C., placing | 2 Rodm | 3260 | .005 | | .57 | .24 | | .81 | 1.01 |
| 3205 | Stressing | C-4A | 4190 | .004 | | | .18 | .01 | .19 | .31 |
| 3210 | 42" O.C., placing | 2 Rodm | 3530 | .005 | | .51 | .22 | | .73 | .92 |
| 3215 | Stressing | C-4A | 4500 | .004 | | | .17 | .01 | .18 | .29 |
| 3220 | 48" O.C., placing | 2 Rodm | 3840 | .004 | | .46 | .20 | | .66 | .82 |
| 3225 | Stressing | C-4A | 4850 | .003 | | | .16 | .01 | .17 | .27 |
| 3230 | 30' x 40' slab, strands @ 36" O.C., placing | 2 Rodm | 3780 | .004 | | .55 | .20 | | .75 | .94 |
| 3235 | Stressing | C-4A | 4920 | .003 | | | .16 | .01 | .17 | .26 |
| 3240 | 42" O.C., placing | 2 Rodm | 4190 | .004 | | .48 | .18 | | .66 | .83 |
| 3245 | Stressing | C-4A | 5410 | .003 | | | .14 | .01 | .15 | .24 |
| 3250 | 48" O.C., placing | 2 Rodm | 4520 | .004 | | .44 | .17 | | .61 | .75 |
| 3255 | Stressing | C-4A | 5790 | .003 | | | .13 | .01 | .14 | .22 |
| 3260 | 30' x 50' slab, strands @ 36" O.C., placing | 2 Rodm | 4300 | .004 | | .52 | .18 | | .70 | .86 |
| 3265 | Stressing | C-4A | 5650 | .003 | | | .14 | .01 | .15 | .23 |
| 3270 | 42" O.C., placing | 2 Rodm | 4720 | .003 | | .46 | .16 | | .62 | .77 |
| 3275 | Stressing | C-4A | 6150 | .003 | | | .13 | .01 | .14 | .21 |
| 3280 | 48" O.C., placing | 2 Rodm | 5240 | .003 | | .41 | .15 | | .56 | .69 |
| 3285 | Stressing | C-4A | 6760 | .002 | | | .11 | .01 | .12 | .19 |

# 03 24 Fibrous Reinforcing

## 03 24 05 – Reinforcing Fibers

### 03 24 05.30 Synthetic Fibers

| | | | | | Unit | Material | Labor | Equipment | Total | Total Incl O&P |
|---|---|---|---|---|---|---|---|---|---|---|
| 0010 | **SYNTHETIC FIBERS** | | | | | | | | | |
| 0100 | Synthetic fibers, add to concrete | | | | Lb. | 5.05 | | | 5.05 | 5.55 |
| 0110 | 1-1/2 lb. per C.Y. | | | | C.Y. | 7.80 | | | 7.80 | 8.60 |

### 03 24 05.70 Steel Fibers

| | | | | | | Unit | Material | Labor | Equipment | Total | Total Incl O&P |
|---|---|---|---|---|---|---|---|---|---|---|---|
| 0010 | **STEEL FIBERS** | | | | | | | | | | |
| 0150 | Steel fibers, add to concrete | G | | | | Lb. | .77 | | | .77 | .85 |
| 0155 | 25 lb. per C.Y. | G | | | | C.Y. | 19.25 | | | 19.25 | 21 |
| 0160 | 50 lb. per C.Y. | G | | | | | 38.50 | | | 38.50 | 42.50 |
| 0170 | 75 lb. per C.Y. | G | | | | | 59.50 | | | 59.50 | 65 |
| 0180 | 100 lb. per C.Y. | G | | | | | 77 | | | 77 | 84.50 |

# 03 30 Cast-In-Place Concrete

## 03 30 53 – Miscellaneous Cast-In-Place Concrete

### 03 30 53.40 Concrete In Place

| | | | Crew | Daily Output | Labor-Hours | Unit | Material | Labor | Equipment | Total | Total Incl O&P |
|---|---|---|---|---|---|---|---|---|---|---|---|
| 0010 | **CONCRETE IN PLACE** | R033105-10 | | | | | | | | | |
| 0020 | Including forms (4 uses), reinforcing steel, concrete, placement, | R033105-20 | | | | | | | | | |
| 0050 | and finishing unless otherwise indicated | R033105-50 | | | | | | | | | |
| 0300 | Beams (3500 psi), 5 kip per L.F., 10' span | R033105-65 | C-14A | 15.62 | 12.804 | C.Y. | 291 | 555 | 49 | 895 | 1,225 |
| 0350 | 25' span | R033105-70 | " | 18.55 | 10.782 | | 305 | 465 | 41 | 811 | 1,100 |
| 0500 | Chimney foundations (5000 psi), industrial, minimum | R033105-85 | C-14C | 32.22 | 3.476 | | 144 | 144 | .73 | 288.73 | 380 |
| 0510 | Maximum | | " | 23.71 | 4.724 | | 168 | 195 | .99 | 363.99 | 485 |
| 0700 | Columns, square (4000 psi), 12" x 12", minimum reinforcing | | C-14A | 11.96 | 16.722 | | 325 | 725 | 64 | 1,114 | 1,550 |
| 0720 | Average reinforcing | | | 10.13 | 19.743 | | 510 | 855 | 75.50 | 1,440.50 | 1,975 |

## 03 30 53 – Miscellaneous Cast-In-Place Concrete

| 03 30 53.40 Concrete In Place | Crew | Daily Output | Labor-Hours | Unit | Material | 2011 Bare Costs Labor | Equipment | Total | Total Incl O&P |
|---|---|---|---|---|---|---|---|---|---|
| 0740 | Maximum reinforcing | C-14A | 9.03 | 22.148 | C.Y. | 770 | 960 | 84.50 | 1,814.50 | 2,425 |
| 0800 | 16" x 16", minimum reinforcing | | 16.22 | 12.330 | | 259 | 535 | 47 | 841 | 1,150 |
| 0820 | Average reinforcing | | 12.57 | 15.911 | | 435 | 690 | 61 | 1,186 | 1,600 |
| 0840 | Maximum reinforcing | | 10.25 | 19.512 | | 675 | 845 | 74.50 | 1,594.50 | 2,125 |
| 0900 | 24" x 24", minimum reinforcing | | 23.66 | 8.453 | | 221 | 365 | 32.50 | 618.50 | 845 |
| 0920 | Average reinforcing | | 17.71 | 11.293 | | 390 | 490 | 43 | 923 | 1,225 |
| 0940 | Maximum reinforcing | | 14.15 | 14.134 | | 620 | 610 | 54 | 1,284 | 1,700 |
| 1000 | 36" x 36", minimum reinforcing | | 33.69 | 5.936 | | 197 | 257 | 22.50 | 476.50 | 635 |
| 1020 | Average reinforcing | | 23.32 | 8.576 | | 345 | 370 | 33 | 748 | 985 |
| 1040 | Maximum reinforcing | | 17.82 | 11.223 | | 580 | 485 | 43 | 1,108 | 1,425 |
| 1100 | Columns, round (4000 psi), tied, 12" diameter, minimum reinforcing | | 20.97 | 9.537 | | 285 | 415 | 36.50 | 736.50 | 990 |
| 1120 | Average reinforcing | | 15.27 | 13.098 | | 470 | 565 | 50 | 1,085 | 1,450 |
| 1140 | Maximum reinforcing | | 12.11 | 16.515 | | 715 | 715 | 63 | 1,493 | 1,950 |
| 1200 | 16" diameter, minimum reinforcing | | 31.49 | 6.351 | | 262 | 275 | 24.50 | 561.50 | 740 |
| 1220 | Average reinforcing | | 19.12 | 10.460 | | 450 | 455 | 40 | 945 | 1,250 |
| 1240 | Maximum reinforcing | | 13.77 | 14.524 | | 675 | 630 | 55.50 | 1,360.50 | 1,775 |
| 1300 | 20" diameter, minimum reinforcing | | 41.04 | 4.873 | | 264 | 211 | 18.60 | 493.60 | 635 |
| 1320 | Average reinforcing | | 24.05 | 8.316 | | 435 | 360 | 32 | 827 | 1,075 |
| 1340 | Maximum reinforcing | | 17.01 | 11.758 | | 675 | 510 | 45 | 1,230 | 1,575 |
| 1400 | 24" diameter, minimum reinforcing | | 51.85 | 3.857 | | 248 | 167 | 14.75 | 429.75 | 545 |
| 1420 | Average reinforcing | | 27.06 | 7.391 | | 435 | 320 | 28 | 783 | 1,000 |
| 1440 | Maximum reinforcing | | 18.29 | 10.935 | | 665 | 475 | 42 | 1,182 | 1,500 |
| 1500 | 36" diameter, minimum reinforcing | | 75.04 | 2.665 | | 247 | 115 | 10.20 | 372.20 | 460 |
| 1520 | Average reinforcing | | 37.49 | 5.335 | | 415 | 231 | 20.50 | 666.50 | 835 |
| 1540 | Maximum reinforcing | | 22.84 | 8.757 | | 645 | 380 | 33.50 | 1,058.50 | 1,325 |
| 1900 | Elevated slab (4000 psi), flat slab with drops, 125 psf Sup. Load, 20' span | C-14B | 38.45 | 5.410 | | 234 | 234 | 19.85 | 487.85 | 640 |
| 1950 | 30' span | | 50.99 | 4.079 | | 252 | 176 | 14.95 | 442.95 | 565 |
| 2100 | Flat plate, 125 psf Sup. Load, 15' span | | 30.24 | 6.878 | | 215 | 297 | 25 | 537 | 720 |
| 2150 | 25' span | | 49.60 | 4.194 | | 227 | 181 | 15.40 | 423.40 | 545 |
| 2300 | Waffle const., 30" domes, 125 psf Sup. Load, 20' span | | 37.07 | 5.611 | | 233 | 242 | 20.50 | 495.50 | 655 |
| 2350 | 30' span | | 44.07 | 4.720 | | 218 | 204 | 17.30 | 439.30 | 575 |
| 2500 | One way joists, 30" pans, 125 psf Sup. Load, 15' span | | 27.38 | 7.597 | | 278 | 330 | 28 | 636 | 840 |
| 2550 | 25' span | | 31.15 | 6.677 | | 263 | 288 | 24.50 | 575.50 | 760 |
| 2700 | One way beam & slab, 125 psf Sup. Load, 15' span | | 20.59 | 10.102 | | 231 | 435 | 37 | 703 | 965 |
| 2750 | 25' span | | 28.36 | 7.334 | | 222 | 315 | 27 | 564 | 760 |
| 2900 | Two way beam & slab, 125 psf Sup. Load, 15' span | | 24.04 | 8.652 | | 224 | 375 | 32 | 631 | 855 |
| 2950 | 25' span | | 35.87 | 5.799 | | 196 | 251 | 21.50 | 468.50 | 625 |
| 3100 | Elevated slabs, flat plate, including finish, not | | | | | | | | | |
| 3110 | including forms or reinforcing | | | | | | | | | |
| 3150 | Regular concrete (4000 psi), 4" slab | C-8 | 2613 | .021 | S.F. | 1.32 | .82 | .28 | 2.42 | 2.99 |
| 3200 | 6" slab | | 2585 | .022 | | 1.96 | .82 | .29 | 3.07 | 3.71 |
| 3250 | 2-1/2" thick floor fill | | 2685 | .021 | | .85 | .79 | .28 | 1.92 | 2.42 |
| 3300 | Lightweight, 110# per C.F., 2-1/2" thick floor fill | | 2585 | .022 | | 1.14 | .82 | .29 | 2.25 | 2.82 |
| 3400 | Cellular concrete, 1-5/8" fill, under 5000 S.F. | | 2000 | .028 | | .76 | 1.07 | .37 | 2.20 | 2.85 |
| 3450 | Over 10,000 S.F. | | 2200 | .025 | | .73 | .97 | .34 | 2.04 | 2.63 |
| 3500 | Add per floor for 3 to 6 stories high | | 31800 | .002 | | | .07 | .02 | .09 | .13 |
| 3520 | For 7 to 20 stories high | | 21200 | .003 | | | .10 | .03 | .13 | .19 |
| 3540 | Equipment pad (3000 psi), 3' x 3' x 6" thick | C-14H | 45 | 1.067 | Ea. | 40.50 | 45.50 | .51 | 86.51 | 115 |
| 3550 | 4' x 4' x 6" thick | | 30 | 1.600 | | 62 | 68 | .77 | 130.77 | 173 |
| 3560 | 5' x 5' x 8" thick | | 18 | 2.667 | | 111 | 113 | 1.28 | 225.28 | 297 |
| 3570 | 6' x 6' x 8" thick | | 14 | 3.429 | | 150 | 146 | 1.65 | 297.65 | 390 |
| 3580 | 8' x 8' x 10" thick | | 8 | 6 | | 320 | 255 | 2.88 | 577.88 | 750 |
| 3590 | 10' x 10' x 12" thick | | 5 | 9.600 | | 550 | 410 | 4.61 | 964.61 | 1,225 |

# 03 30 Cast-In-Place Concrete

## 03 30 53 – Miscellaneous Cast-In-Place Concrete

| 03 30 53.40 Concrete In Place | | Crew | Daily Output | Labor-Hours | Unit | Material | 2011 Bare Costs Labor | 2011 Bare Costs Equipment | Total | Total Incl O&P |
|---|---|---|---|---|---|---|---|---|---|---|
| 3800 | Footings (3000 psi), spread under 1 C.Y. | C-14C | 28 | 4 | C.Y. | 158 | 165 | .84 | 323.84 | 430 |
| 3825 | 1 C.Y. to 5 C.Y. | | 43 | 2.605 | | 185 | 108 | .55 | 293.55 | 370 |
| 3850 | Over 5 C.Y. | | 75 | 1.493 | | 171 | 61.50 | .31 | 232.81 | 283 |
| 3900 | Footings, strip (3000 psi), 18" x 9", unreinforced | C-14L | 40 | 2.400 | | 119 | 96.50 | .58 | 216.08 | 279 |
| 3920 | 18" x 9", reinforced | C-14C | 35 | 3.200 | | 141 | 132 | .67 | 273.67 | 360 |
| 3925 | 20" x 10", unreinforced | C-14L | 45 | 2.133 | | 116 | 85.50 | .51 | 202.01 | 260 |
| 3930 | 20" x 10", reinforced | C-14C | 40 | 2.800 | | 134 | 116 | .59 | 250.59 | 325 |
| 3935 | 24" x 12", unreinforced | C-14L | 55 | 1.745 | | 114 | 70 | .42 | 184.42 | 233 |
| 3940 | 24" x 12", reinforced | C-14C | 48 | 2.333 | | 132 | 96.50 | .49 | 228.99 | 294 |
| 3945 | 36" x 12", unreinforced | C-14L | 70 | 1.371 | | 111 | 55 | .33 | 166.33 | 206 |
| 3950 | 36" x 12", reinforced | C-14C | 60 | 1.867 | | 127 | 77 | .39 | 204.39 | 259 |
| 4000 | Foundation mat (3000 psi), under 10 C.Y. | | 38.67 | 2.896 | | 192 | 120 | .61 | 312.61 | 395 |
| 4050 | Over 20 C.Y. | | 56.40 | 1.986 | | 169 | 82 | .42 | 251.42 | 310 |
| 4200 | Wall, free-standing (3000 psi), 8" thick, 8' high | C-14D | 45.83 | 4.364 | | 160 | 187 | 16.65 | 363.65 | 480 |
| 4250 | 14' high | | 27.26 | 7.337 | | 192 | 315 | 28 | 535 | 725 |
| 4260 | 12" thick, 8' high | | 64.32 | 3.109 | | 146 | 133 | 11.90 | 290.90 | 380 |
| 4270 | 14' high | | 40.01 | 4.999 | | 155 | 214 | 19.10 | 388.10 | 520 |
| 4300 | 15" thick, 8' high | | 80.02 | 2.499 | | 140 | 107 | 9.55 | 256.55 | 330 |
| 4350 | 12' high | | 51.26 | 3.902 | | 140 | 167 | 14.90 | 321.90 | 425 |
| 4500 | 18' high | | 48.85 | 4.094 | | 156 | 176 | 15.65 | 347.65 | 460 |
| 4520 | Handicap access ramp (4000 psi), railing both sides, 3' wide | C-14H | 14.58 | 3.292 | L.F. | 278 | 140 | 1.58 | 419.58 | 520 |
| 4525 | 5' wide | | 12.22 | 3.928 | | 288 | 167 | 1.89 | 456.89 | 575 |
| 4530 | With 6" curb and rails both sides, 3' wide | | 8.55 | 5.614 | | 287 | 238 | 2.69 | 527.69 | 685 |
| 4535 | 5' wide | | 7.31 | 6.566 | | 292 | 279 | 3.15 | 574.15 | 755 |
| 4650 | Slab on grade (3500 psi), not including finish, 4" thick | C-14E | 60.75 | 1.449 | C.Y. | 117 | 61.50 | .38 | 178.88 | 225 |
| 4700 | 6" thick | " | 92 | .957 | " | 113 | 41 | .25 | 154.25 | 188 |
| 4701 | Thickened slab edge (3500 psi), for slab on grade poured | | | | | | | | | |
| 4702 | monolithically with slab; depth is in addition to slab thickness; | | | | | | | | | |
| 4703 | formed vertical outside edge, earthen bottom and inside slope | | | | | | | | | |
| 4705 | 8" deep x 8" wide bottom, unreinforced | C-14L | 2190 | .044 | L.F. | 3.18 | 1.76 | .01 | 4.95 | 6.20 |
| 4710 | 8" x 8", reinforced | C-14C | 1670 | .067 | | 5.30 | 2.77 | .01 | 8.08 | 10.15 |
| 4715 | 12" deep x 12" wide bottom, unreinforced | C-14L | 1800 | .053 | | 6.55 | 2.14 | .01 | 8.70 | 10.50 |
| 4720 | 12" x 12", reinforced | C-14C | 1310 | .086 | | 10.40 | 3.53 | .02 | 13.95 | 16.90 |
| 4725 | 16" deep x 16" wide bottom, unreinforced | C-14L | 1440 | .067 | | 11.10 | 2.68 | .02 | 13.80 | 16.30 |
| 4730 | 16" x 16", reinforced | C-14C | 1120 | .100 | | 15.70 | 4.13 | .02 | 19.85 | 23.50 |
| 4735 | 20" deep x 20" wide bottom, unreinforced | C-14L | 1150 | .083 | | 16.85 | 3.35 | .02 | 20.22 | 23.50 |
| 4740 | 20" x 20", reinforced | C-14C | 920 | .122 | | 22.50 | 5.05 | .03 | 27.58 | 33 |
| 4745 | 24" deep x 24" wide bottom, unreinforced | C-14L | 930 | .103 | | 24 | 4.14 | .02 | 28.16 | 32.50 |
| 4750 | 24" x 24", reinforced | C-14C | 740 | .151 | | 31.50 | 6.25 | .03 | 37.78 | 44 |
| 4751 | Slab on grade (3500 psi), incl. troweled finish, not incl. forms | | | | | | | | | |
| 4760 | or reinforcing, over 10,000 S.F., 4" thick | C-14F | 3425 | .021 | S.F. | 1.29 | .82 | .01 | 2.12 | 2.64 |
| 4820 | 6" thick | | 3350 | .021 | | 1.89 | .84 | .01 | 2.74 | 3.33 |
| 4840 | 8" thick | | 3184 | .023 | | 2.59 | .88 | .01 | 3.48 | 4.16 |
| 4900 | 12" thick | | 2734 | .026 | | 3.88 | 1.02 | .01 | 4.91 | 5.80 |
| 4950 | 15" thick | | 2505 | .029 | | 4.88 | 1.12 | .01 | 6.01 | 7 |
| 5000 | Slab on grade (3000 psi), incl. textured finish, not incl. forms | | | | | | | | | |
| 5001 | or reinforcing, 4" thick | C-14G | 2873 | .019 | S.F. | 1.29 | .75 | .01 | 2.05 | 2.53 |
| 5010 | 6" thick | | 2590 | .022 | | 2.01 | .83 | .01 | 2.85 | 3.45 |
| 5020 | 8" thick | | 2320 | .024 | | 2.62 | .93 | .01 | 3.56 | 4.28 |
| 5200 | Lift slab in place above the foundation, incl. forms, | | | | | | | | | |
| 5210 | reinforcing, concrete (4000 psi) and columns, minimum | C-14B | 2113 | .098 | S.F. | 6.20 | 4.25 | .36 | 10.81 | 13.75 |
| 5250 | Average | | 1650 | .126 | | 6.75 | 5.45 | .46 | 12.66 | 16.30 |
| 5300 | Maximum | | 1500 | .139 | | 7.30 | 6 | .51 | 13.81 | 17.80 |

# 03 30 Cast-In-Place Concrete

## 03 30 53 – Miscellaneous Cast-In-Place Concrete

| 03 30 53.40 Concrete In Place | Crew | Daily Output | Labor-Hours | Unit | Material | 2011 Bare Costs Labor | Equipment | Total | Total Incl O&P |
|---|---|---|---|---|---|---|---|---|---|
| 5500 | Lightweight, ready mix, including screed finish only, | | | | | | | | |
| 5510 | not including forms or reinforcing | | | | | | | | |
| 5550 | 1:4 (2500 psi) for structural roof decks | C-14B | 260 | .800 | C.Y. | 136 | 34.50 | 2.94 | 173.44 | 205 |
| 5600 | 1:6 (3000 psi) for ground slab with radiant heat | C-14F | 92 | .783 | | 133 | 30.50 | .25 | 163.75 | 192 |
| 5650 | 1:3:2 (2000 psi) with sand aggregate, roof deck | C-14B | 260 | .800 | | 133 | 34.50 | 2.94 | 170.44 | 202 |
| 5700 | Ground slab (2000 psi) | C-14F | 107 | .673 | | 133 | 26 | .22 | 159.22 | 185 |
| 5900 | Pile caps (3000 psi), incl. forms and reinf., sq. or rect., under 10 C.Y. | C-14C | 54.14 | 2.069 | | 154 | 85.50 | .43 | 239.93 | 300 |
| 5950 | Over 10 C.Y. | | 75 | 1.493 | | 147 | 61.50 | .31 | 208.81 | 256 |
| 6000 | Triangular or hexagonal, under 10 C.Y. | | 53 | 2.113 | | 118 | 87.50 | .44 | 205.94 | 264 |
| 6050 | Over 10 C.Y. | ↓ | 85 | 1.318 | | 132 | 54.50 | .28 | 186.78 | 229 |
| 6200 | Retaining walls (3000 psi), gravity, 4' high see Section 32 32 | C-14D | 66.20 | 3.021 | | 136 | 130 | 11.55 | 277.55 | 360 |
| 6250 | 10' high | | 125 | 1.600 | | 129 | 68.50 | 6.10 | 203.60 | 254 |
| 6300 | Cantilever, level backfill loading, 8' high | | 70 | 2.857 | | 148 | 122 | 10.90 | 280.90 | 365 |
| 6350 | 16' high | ↓ | 91 | 2.198 | LF Nose | 141 | 94 | 8.40 | 243.40 | 310 |
| 6800 | Stairs (3500 psi), not including safety treads, free standing, 3'-6" wide | C-14H | 83 | .578 | LF Nose | 5 | 24.50 | .28 | 29.78 | 43.50 |
| 6850 | Cast on ground | | 125 | .384 | " | 4.28 | 16.30 | .18 | 20.76 | 30 |
| 7000 | Stair landings, free standing | | 200 | .240 | S.F. | 4.03 | 10.20 | .12 | 14.35 | 20 |
| 7050 | Cast on ground | ↓ | 475 | .101 | " | 3.29 | 4.29 | .05 | 7.63 | 10.25 |

# 03 31 Structural Concrete

## 03 31 05 – Normal Weight Structural Concrete

### 03 31 05.30 Concrete, Field Mix

| | | Crew | Daily Output | Labor-Hours | Unit | Material | Labor | Equipment | Total | Total Incl O&P |
|---|---|---|---|---|---|---|---|---|---|---|
| 0010 | **CONCRETE, FIELD MIX** | R033105-65 | | | | | | | | |
| 0015 | FOB forms 2250 psi | | | | C.Y. | 85 | | | 85 | 93.50 |
| 0020 | 3000 psi | | | | " | 92 | | | 92 | 101 |

### 03 31 05.35 Normal Weight Concrete, Ready Mix

| | | | | | | | | | | |
|---|---|---|---|---|---|---|---|---|---|---|
| 0010 | **NORMAL WEIGHT CONCRETE, READY MIX**, delivered | R033105-10 | | | | | | | | |
| 0012 | Includes local aggregate, sand, Portland cement, and water | | | | | | | | | |
| 0015 | Excludes all additives and treatments | R033105-20 | | | | | | | | |
| 0020 | 2000 psi | | | | C.Y. | 91.50 | | | 91.50 | 101 |
| 0100 | 2500 psi | R033105-30 | | | | 94 | | | 94 | 103 |
| 0150 | 3000 psi | | | | | 99 | | | 99 | 109 |
| 0200 | 3500 psi | R033105-40 | | | | 99.50 | | | 99.50 | 110 |
| 0300 | 4000 psi | | | | | 103 | | | 103 | 113 |
| 0350 | 4500 psi | R033105-50 | | | | 106 | | | 106 | 116 |
| 0400 | 5000 psi | | | | | 111 | | | 111 | 122 |
| 0411 | 6000 psi | | | | | 127 | | | 127 | 139 |
| 0412 | 8000 psi | | | | | 206 | | | 206 | 227 |
| 0413 | 10,000 psi | | | | | 293 | | | 293 | 320 |
| 0414 | 12,000 psi | | | | | 355 | | | 355 | 390 |
| 1000 | For high early strength cement, add | | | | | 10% | | | | |
| 1010 | For structural lightweight with regular sand, add | | | | | 25% | | | | |
| 1300 | For winter concrete (hot water), add | | | | | 4.25 | | | 4.25 | 4.68 |
| 1400 | For hot weather concrete (ice), add | | | | | 9.35 | | | 9.35 | 10.25 |
| 1410 | For mid-range water reducer, add | | | | | 4.13 | | | 4.13 | 4.54 |
| 1420 | For high-range water reducer/superplasticizer, add | | | | | 6.35 | | | 6.35 | 6.95 |
| 1430 | For retarder, add | | | | | 2.71 | | | 2.71 | 2.98 |
| 1440 | For non-Chloride accelerator, add | | | | | 4.83 | | | 4.83 | 5.30 |
| 1450 | For Chloride accelerator, per 1%, add | | | | | 3.28 | | | 3.28 | 3.61 |
| 1460 | For fiber reinforcing, synthetic (1 lb./C.Y.), add | | | | | 6.65 | | | 6.65 | 7.30 |
| 1500 | For Saturday delivery, add | | | | ↓ | 8.85 | | | 8.85 | 9.70 |

# 03 31 Structural Concrete

## 03 31 05 – Normal Weight Structural Concrete

### 03 31 05.35 Normal Weight Concrete, Ready Mix

| | | Crew | Daily Output | Labor-Hours | Unit | Material | 2011 Bare Costs Labor | Equipment | Total | Total Incl O&P |
|---|---|---|---|---|---|---|---|---|---|---|
| 1510 | For truck holding/waiting time past 1st hour per load, add | | | | Hr. | 87.50 | | | 87.50 | 96.50 |
| 1520 | For short load (less than 4 C.Y.), add per load | | | | Ea. | 112 | | | 112 | 124 |
| 2000 | For all lightweight aggregate, add | | | | C.Y. | 45% | | | | |

### 03 31 05.70 Placing Concrete

| | | Crew | Daily Output | Labor-Hours | Unit | Material | 2011 Bare Costs Labor | Equipment | Total | Total Incl O&P |
|---|---|---|---|---|---|---|---|---|---|---|
| 0010 | **PLACING CONCRETE**  R033105-70 | | | | | | | | | |
| 0020 | Includes labor and equipment to place, strike off and consolidate | | | | | | | | | |
| 0050 | Beams, elevated, small beams, pumped | C-20 | 60 | 1.067 | C.Y. | | 39 | 13.10 | 52.10 | 74 |
| 0100 | With crane and bucket | C-7 | 45 | 1.600 | | | 59.50 | 28.50 | 88 | 121 |
| 0200 | Large beams, pumped | C-20 | 90 | .711 | | | 26 | 8.75 | 34.75 | 49 |
| 0250 | With crane and bucket | C-7 | 65 | 1.108 | | | 41 | 19.65 | 60.65 | 84 |
| 0400 | Columns, square or round, 12" thick, pumped | C-20 | 60 | 1.067 | | | 39 | 13.10 | 52.10 | 74 |
| 0450 | With crane and bucket | C-7 | 40 | 1.800 | | | 67 | 32 | 99 | 136 |
| 0600 | 18" thick, pumped | C-20 | 90 | .711 | | | 26 | 8.75 | 34.75 | 49 |
| 0650 | With crane and bucket | C-7 | 55 | 1.309 | | | 48.50 | 23 | 71.50 | 99 |
| 0800 | 24" thick, pumped | C-20 | 92 | .696 | | | 25.50 | 8.55 | 34.05 | 48.50 |
| 0850 | With crane and bucket | C-7 | 70 | 1.029 | | | 38 | 18.25 | 56.25 | 78 |
| 1000 | 36" thick, pumped | C-20 | 140 | .457 | | | 16.80 | 5.60 | 22.40 | 31.50 |
| 1050 | With crane and bucket | C-7 | 100 | .720 | | | 27 | 12.75 | 39.75 | 54.50 |
| 1400 | Elevated slabs, less than 6" thick, pumped | C-20 | 140 | .457 | | | 16.80 | 5.60 | 22.40 | 31.50 |
| 1450 | With crane and bucket | C-7 | 95 | .758 | | | 28 | 13.45 | 41.45 | 57.50 |
| 1500 | 6" to 10" thick, pumped | C-20 | 160 | .400 | | | 14.70 | 4.92 | 19.62 | 28 |
| 1550 | With crane and bucket | C-7 | 110 | .655 | | | 24.50 | 11.60 | 36.10 | 50 |
| 1600 | Slabs over 10" thick, pumped | C-20 | 180 | .356 | | | 13.10 | 4.37 | 17.47 | 24.50 |
| 1650 | With crane and bucket | C-7 | 130 | .554 | | | 20.50 | 9.85 | 30.35 | 42 |
| 1900 | Footings, continuous, shallow, direct chute | C-6 | 120 | .400 | | | 14.30 | .39 | 14.69 | 22 |
| 1950 | Pumped | C-20 | 150 | .427 | | | 15.70 | 5.25 | 20.95 | 30 |
| 2000 | With crane and bucket | C-7 | 90 | .800 | | | 29.50 | 14.20 | 43.70 | 60.50 |
| 2100 | Footings, continuous, deep, direct chute | C-6 | 140 | .343 | | | 12.25 | .33 | 12.58 | 19 |
| 2150 | Pumped | C-20 | 160 | .400 | | | 14.70 | 4.92 | 19.62 | 28 |
| 2200 | With crane and bucket | C-7 | 110 | .655 | | | 24.50 | 11.60 | 36.10 | 50 |
| 2400 | Footings, spread, under 1 C.Y., direct chute | C-6 | 55 | .873 | | | 31 | .85 | 31.85 | 48.50 |
| 2450 | Pumped | C-20 | 65 | .985 | | | 36 | 12.10 | 48.10 | 68.50 |
| 2500 | With crane and bucket | C-7 | 45 | 1.600 | | | 59.50 | 28.50 | 88 | 121 |
| 2600 | Over 5 C.Y., direct chute | C-6 | 120 | .400 | | | 14.30 | .39 | 14.69 | 22 |
| 2650 | Pumped | C-20 | 150 | .427 | | | 15.70 | 5.25 | 20.95 | 30 |
| 2700 | With crane and bucket | C-7 | 100 | .720 | | | 27 | 12.75 | 39.75 | 54.50 |
| 2900 | Foundation mats, over 20 C.Y., direct chute | C-6 | 350 | .137 | | | 4.91 | .13 | 5.04 | 7.60 |
| 2950 | Pumped | C-20 | 400 | .160 | | | 5.90 | 1.97 | 7.87 | 11.10 |
| 3000 | With crane and bucket | C-7 | 300 | .240 | | | 8.90 | 4.26 | 13.16 | 18.20 |
| 3200 | Grade beams, direct chute | C-6 | 150 | .320 | | | 11.45 | .31 | 11.76 | 17.75 |
| 3250 | Pumped | C-20 | 180 | .356 | | | 13.10 | 4.37 | 17.47 | 24.50 |
| 3300 | With crane and bucket | C-7 | 120 | .600 | | | 22.50 | 10.65 | 33.15 | 45.50 |
| 3500 | High rise, for more than 5 stories, pumped, add per story | C-20 | 2100 | .030 | | | 1.12 | .37 | 1.49 | 2.11 |
| 3510 | With crane and bucket, add per story | C-7 | 2100 | .034 | | | 1.27 | .61 | 1.88 | 2.60 |
| 3700 | Pile caps, under 5 C.Y., direct chute | C-6 | 90 | .533 | | | 19.10 | .52 | 19.62 | 29.50 |
| 3750 | Pumped | C-20 | 110 | .582 | | | 21.50 | 7.15 | 28.65 | 40.50 |
| 3800 | With crane and bucket | C-7 | 80 | .900 | | | 33.50 | 15.95 | 49.45 | 68 |
| 3850 | Pile cap, 5 C.Y. to 10 C.Y., direct chute | C-6 | 175 | .274 | | | 9.80 | .27 | 10.07 | 15.20 |
| 3900 | Pumped | C-20 | 200 | .320 | | | 11.75 | 3.94 | 15.69 | 22 |
| 3950 | With crane and bucket | C-7 | 150 | .480 | | | 17.85 | 8.50 | 26.35 | 36.50 |
| 4000 | Over 10 C.Y., direct chute | C-6 | 215 | .223 | | | 8 | .22 | 8.22 | 12.40 |
| 4050 | Pumped | C-20 | 240 | .267 | | | 9.80 | 3.28 | 13.08 | 18.50 |

# 03 31 Structural Concrete

## 03 31 05 – Normal Weight Structural Concrete

### 03 31 05.70 Placing Concrete

| | | Crew | Daily Output | Labor-Hours | Unit | Material | 2011 Bare Costs Labor | Equipment | Total | Total Incl O&P |
|---|---|---|---|---|---|---|---|---|---|---|
| 4100 | With crane and bucket | C-7 | 185 | .389 | C.Y. | | 14.45 | 6.90 | 21.35 | 29.50 |
| 4300 | Slab on grade, up to 6" thick, direct chute | C-6 | 110 | .436 | | | 15.60 | .42 | 16.02 | 24 |
| 4350 | Pumped | C-20 | 130 | .492 | | | 18.10 | 6.05 | 24.15 | 34 |
| 4400 | With crane and bucket | C-7 | 110 | .655 | | | 24.50 | 11.60 | 36.10 | 50 |
| 4600 | Over 6" thick, direct chute | C-6 | 165 | .291 | | | 10.40 | .28 | 10.68 | 16.10 |
| 4650 | Pumped | C-20 | 185 | .346 | | | 12.75 | 4.26 | 17.01 | 24 |
| 4700 | With crane and bucket | C-7 | 145 | .497 | | | 18.45 | 8.80 | 27.25 | 37.50 |
| 4900 | Walls, 8" thick, direct chute | C-6 | 90 | .533 | | | 19.10 | .52 | 19.62 | 29.50 |
| 4950 | Pumped | C-20 | 100 | .640 | | | 23.50 | 7.85 | 31.35 | 44 |
| 5000 | With crane and bucket | C-7 | 80 | .900 | | | 33.50 | 15.95 | 49.45 | 68 |
| 5050 | 12" thick, direct chute | C-6 | 100 | .480 | | | 17.15 | .47 | 17.62 | 26.50 |
| 5100 | Pumped | C-20 | 110 | .582 | | | 21.50 | 7.15 | 28.65 | 40.50 |
| 5200 | With crane and bucket | C-7 | 90 | .800 | | | 29.50 | 14.20 | 43.70 | 60.50 |
| 5300 | 15" thick, direct chute | C-6 | 105 | .457 | | | 16.35 | .44 | 16.79 | 25.50 |
| 5350 | Pumped | C-20 | 120 | .533 | | | 19.60 | 6.55 | 26.15 | 37 |
| 5400 | With crane and bucket | C-7 | 95 | .758 | ▼ | | 28 | 13.45 | 41.45 | 57.50 |
| 5600 | Wheeled concrete dumping, add to placing costs above | | | | | | | | | |
| 5610 | Walking cart, 50' haul, add | C-18 | 32 | .281 | C.Y. | | 9.70 | 1.72 | 11.42 | 16.80 |
| 5620 | 150' haul, add | | 24 | .375 | | | 12.95 | 2.29 | 15.24 | 22.50 |
| 5700 | 250' haul, add | ▼ | 18 | .500 | | | 17.30 | 3.06 | 20.36 | 30 |
| 5800 | Riding cart, 50' haul, add | C-19 | 80 | .113 | | | 3.89 | 1.15 | 5.04 | 7.20 |
| 5810 | 150' haul, add | | 60 | .150 | | | 5.20 | 1.53 | 6.73 | 9.65 |
| 5900 | 250' haul, add | ▼ | 45 | .200 | ▼ | | 6.90 | 2.04 | 8.94 | 12.85 |

# 03 35 Concrete Finishing

## 03 35 29 – Tooled Concrete Finishing

### 03 35 29.30 Finishing Floors

| | | Crew | Daily Output | Labor-Hours | Unit | Material | 2011 Bare Costs Labor | Equipment | Total | Total Incl O&P |
|---|---|---|---|---|---|---|---|---|---|---|
| 0010 | **FINISHING FLOORS** | | | | | | | | | |
| 0012 | Finishing requires that concrete first be placed, struck off & consolidated | | | | | | | | | |
| 0015 | Basic finishing for various unspecified flatwork | | | | | | | | | |
| 0100 | Bull float only | C-10 | 4000 | .006 | S.F. | | .23 | | .23 | .34 |
| 0125 | Bull float & manual float | | 2000 | .012 | | | .46 | | .46 | .69 |
| 0150 | Bull float, manual float, & broom finish, w/edging & joints | | 1850 | .013 | | | .50 | | .50 | .74 |
| 0200 | Bull float, manual float & manual steel trowel | ▼ | 1265 | .019 | ▼ | | .73 | | .73 | 1.09 |
| 0210 | For specified Random Access Floors in ACI Classes 1, 2, 3 and 4 to achieve | | | | | | | | | |
| 0215 | Composite Overall Floor Flatness and Levelness values up to F35/F25 | | | | | | | | | |
| 0250 | Bull float, machine float & machine trowel (walk-behind) | C-10C | 1715 | .014 | S.F. | | .54 | .02 | .56 | .82 |
| 0300 | Power screed, bull float, machine float & trowel (walk-behind) | C-10D | 2400 | .010 | | | .39 | .04 | .43 | .62 |
| 0350 | Power screed, bull float, machine float & trowel (ride-on) | C-10E | 4000 | .006 | ▼ | | .23 | .06 | .29 | .40 |
| 0352 | For specified Random Access Floors in ACI Classes 5, 6, 7 and 8 to achieve | | | | | | | | | |
| 0354 | Composite Overall Floor Flatness and Levelness values up to F50/F50 | | | | | | | | | |
| 0356 | Add for two-dimensional restraightening after power float | C-10 | 6000 | .004 | S.F. | | .15 | | .15 | .23 |
| 0358 | For specified Random or Defined Access Floors in ACI Class 9 to achieve | | | | | | | | | |
| 0360 | Composite Overall Floor Flatness and Levelness values up to F100/F100 | | | | | | | | | |
| 0362 | Add for two-dimensional restraightening after bull float & power float | C-10 | 3000 | .008 | S.F. | | .31 | | .31 | .46 |
| 0364 | For specified Superflat Defined Access Floors in ACI Class 9 to achieve | | | | | | | | | |
| 0366 | Minimum Floor Flatness and Levelness values of F100/F100 | | | | | | | | | |
| 0368 | Add for 2-dim'l restraightening after bull float, power float, power trowel | C-10 | 2000 | .012 | S.F. | | .46 | | .46 | .69 |
| 0400 | Integral topping and finish, using 1:1:2 mix, 3/16" thick | C-10B | 1000 | .040 | | .10 | 1.48 | .23 | 1.81 | 2.58 |
| 0450 | 1/2" thick | | 950 | .042 | | .27 | 1.56 | .24 | 2.07 | 2.89 |
| 0500 | 3/4" thick | ▼ | 850 | .047 | ▼ | .41 | 1.74 | .27 | 2.42 | 3.36 |

# 03 35 Concrete Finishing

## 03 35 29 – Tooled Concrete Finishing

### 03 35 29.30 Finishing Floors

| | | Crew | Daily Output | Labor-Hours | Unit | Material | 2011 Bare Costs Labor | 2011 Bare Costs Equipment | Total | Total Incl O&P |
|---|---|---|---|---|---|---|---|---|---|---|
| 0600 | 1" thick | C-10B | 750 | .053 | S.F. | .55 | 1.97 | .30 | 2.82 | 3.89 |
| 0800 | Granolithic topping, laid after, 1:1:1-1/2 mix, 1/2" thick | | 590 | .068 | | .31 | 2.51 | .39 | 3.21 | 4.53 |
| 0820 | 3/4" thick | | 580 | .069 | | .46 | 2.55 | .39 | 3.40 | 4.75 |
| 0850 | 1" thick | | 575 | .070 | | .61 | 2.57 | .40 | 3.58 | 4.97 |
| 0950 | 2" thick | | 500 | .080 | | 1.22 | 2.96 | .46 | 4.64 | 6.25 |
| 1200 | Heavy duty, 1:1:2, 3/4" thick, preshrunk, gray, 20 M.S.F. | | 320 | .125 | | .72 | 4.62 | .71 | 6.05 | 8.50 |
| 1300 | 100 M.S.F. | | 380 | .105 | | .41 | 3.89 | .60 | 4.90 | 6.95 |
| 1600 | Exposed local aggregate finish, minimum | 1 Cefi | 625 | .013 | | .21 | .52 | | .73 | .99 |
| 1650 | Maximum | | 465 | .017 | | .64 | .70 | | 1.34 | 1.72 |
| 1800 | Floor abrasives, .25 psf, aluminum oxide | | 850 | .009 | | .43 | .38 | | .81 | 1.03 |
| 1850 | Silicon carbide | | 850 | .009 | | .65 | .38 | | 1.03 | 1.27 |
| 2000 | Floor hardeners, metallic, light service, .50 psf, add | | 850 | .009 | | .51 | .38 | | .89 | 1.12 |
| 2050 | Medium service, .75 psf | | 750 | .011 | | .76 | .44 | | 1.20 | 1.46 |
| 2100 | Heavy service, 1.0 psf | | 650 | .012 | | 1.01 | .50 | | 1.51 | 1.84 |
| 2150 | Extra heavy, 1.5 psf | | 575 | .014 | | 1.52 | .57 | | 2.09 | 2.50 |
| 2300 | Non-metallic, light service, .50 psf | | 850 | .009 | | .23 | .38 | | .61 | .81 |
| 2350 | Medium service, .75 psf | | 750 | .011 | | .34 | .44 | | .78 | 1.01 |
| 2400 | Heavy service, 1.00 psf | | 650 | .012 | | .45 | .50 | | .95 | 1.23 |
| 2450 | Extra heavy, 1.50 psf | | 575 | .014 | | .68 | .57 | | 1.25 | 1.58 |
| 2800 | Trap rock wearing surface for monolithic floors | | | | | | | | | |
| 2810 | 2.0 psf | C-10B | 1250 | .032 | S.F. | .02 | 1.18 | .18 | 1.38 | 2 |
| 3000 | Floor coloring, dusted on, minimum (0.6 psf), add to above | 1 Cefi | 1300 | .006 | | .35 | .25 | | .60 | .75 |
| 3050 | Maximum (1.0 psf), add to above | " | 625 | .013 | | .58 | .52 | | 1.10 | 1.39 |
| 3100 | Colored powder only | | | | Lb. | .58 | | | .58 | .63 |
| 3600 | 1/2" topping using 0.6 psf powdered color | C-10B | 590 | .068 | S.F. | 4.41 | 2.51 | .39 | 7.31 | 9.05 |
| 3650 | 1/2" topping using 1.0 psf powdered color | " | 590 | .068 | | 4.64 | 2.51 | .39 | 7.54 | 9.30 |
| 3800 | Dustproofing, solvent-based, 1 coat | 1 Cefi | 1900 | .004 | | .14 | .17 | | .31 | .40 |
| 3850 | 2 coats | | 1300 | .006 | | .50 | .25 | | .75 | .92 |
| 4000 | Epoxy-based, 1 coat | | 1500 | .005 | | .12 | .22 | | .34 | .45 |
| 4050 | 2 coats | | 1500 | .005 | | .23 | .22 | | .45 | .57 |
| 4400 | Stair finish, float | | 275 | .029 | | | 1.19 | | 1.19 | 1.73 |
| 4500 | Steel trowel finish | | 200 | .040 | | | 1.63 | | 1.63 | 2.38 |
| 4600 | Silicon carbide finish, .25 psf | | 150 | .053 | | .43 | 2.18 | | 2.61 | 3.64 |

### 03 35 29.35 Control Joints, Saw Cut

| | | Crew | Daily Output | Labor-Hours | Unit | Material | 2011 Bare Costs Labor | 2011 Bare Costs Equipment | Total | Total Incl O&P |
|---|---|---|---|---|---|---|---|---|---|---|
| 0010 | **CONTROL JOINTS, SAW CUT** | | | | | | | | | |
| 0100 | Sawcut control joints in green concrete | | | | | | | | | |
| 0120 | 1" depth | C-27 | 2000 | .008 | L.F. | .05 | .33 | .07 | .45 | .60 |
| 0140 | 1-1/2" depth | | 1800 | .009 | | .07 | .36 | .07 | .50 | .69 |
| 0160 | 2" depth | | 1600 | .010 | | .10 | .41 | .08 | .59 | .80 |
| 0180 | Sawcut joint reservoir in cured concrete | | | | | | | | | |
| 0182 | 3/8" wide x 3/4" deep, with single saw blade | C-27 | 1000 | .016 | L.F. | .07 | .65 | .13 | .85 | 1.18 |
| 0184 | 1/2" wide x 1" deep, with double saw blades | | 900 | .018 | | .14 | .73 | .15 | 1.02 | 1.38 |
| 0186 | 3/4" wide x 1-1/2" deep, with double saw blades | | 800 | .020 | | .30 | .82 | .17 | 1.29 | 1.70 |
| 0190 | Water blast joint to wash away laitance, 2 passes | C-29 | 2500 | .003 | | | .11 | .02 | .13 | .20 |
| 0200 | Air blast joint to blow out debris and air dry, 2 passes | C-28 | 2000 | .004 | | | .16 | .01 | .17 | .25 |
| 0300 | For backer rod, see Section 07 91 23.10 | | | | | | | | | |
| 0340 | For joint sealant, see Sections 03 15 05.25 or 07 92 13.20 | | | | | | | | | |
| 0900 | For replacement of joint sealant, see Section 07 01 90.81 | | | | | | | | | |

### 03 35 29.60 Finishing Walls

| | | Crew | Daily Output | Labor-Hours | Unit | Material | 2011 Bare Costs Labor | 2011 Bare Costs Equipment | Total | Total Incl O&P |
|---|---|---|---|---|---|---|---|---|---|---|
| 0010 | **FINISHING WALLS** | | | | | | | | | |
| 0020 | Break ties and patch voids | 1 Cefi | 540 | .015 | S.F. | .03 | .61 | | .64 | .92 |
| 0050 | Burlap rub with grout | | 450 | .018 | | .03 | .73 | | .76 | 1.10 |

# 03 35 Concrete Finishing

## 03 35 29 – Tooled Concrete Finishing

### 03 35 29.60 Finishing Walls

| | | Crew | Daily Output | Labor-Hours | Unit | Material | 2011 Bare Costs Labor | Equipment | Total | Total Incl O&P |
|---|---|---|---|---|---|---|---|---|---|---|
| 0100 | Carborundum rub, dry | 1 Cefi | 270 | .030 | S.F. | | 1.21 | | 1.21 | 1.76 |
| 0150 | Wet rub | ↓ | 175 | .046 | | | 1.87 | | 1.87 | 2.72 |
| 0300 | Bush hammer, green concrete | B-39 | 1000 | .048 | | | 1.74 | .19 | 1.93 | 2.86 |
| 0350 | Cured concrete | " | 650 | .074 | | | 2.68 | .30 | 2.98 | 4.41 |
| 0600 | Float finish, 1/16" thick | 1 Cefi | 300 | .027 | | .29 | 1.09 | | 1.38 | 1.91 |
| 0700 | Sandblast, light penetration | E-11 | 1100 | .029 | | .73 | 1.13 | .18 | 2.04 | 2.86 |
| 0750 | Heavy penetration | " | 375 | .085 | ↓ | 1.46 | 3.31 | .52 | 5.29 | 7.65 |
| 0850 | Grind form fins flush | 1 Clab | 700 | .011 | L.F. | | .39 | | .39 | .60 |

## 03 35 33 – Stamped Concrete Finishing

### 03 35 33.50 Slab Texture Stamping

| | | Crew | Daily Output | Labor-Hours | Unit | Material | 2011 Bare Costs Labor | Equipment | Total | Total Incl O&P |
|---|---|---|---|---|---|---|---|---|---|---|
| 0010 | **SLAB TEXTURE STAMPING** | | | | | | | | | |
| 0050 | Stamping requires that concrete first be placed, struck off, consolidated, | | | | | | | | | |
| 0060 | bull floated and free of bleed water. Decorative stamping tasks include: | | | | | | | | | |
| 0100 | Step 1 - first application of dry shake colored hardener | 1 Cefi | 6400 | .001 | S.F. | .38 | .05 | | .43 | .49 |
| 0110 | Step 2 - bull float | ↓ | 6400 | .001 | | | .05 | | .05 | .07 |
| 0130 | Step 3 - second application of dry shake colored hardener | ↓ | 6400 | .001 | | .19 | .05 | | .24 | .28 |
| 0140 | Step 4 - bull float, manual float & steel trowel | 3 Cefi | 1280 | .019 | | | .77 | | .77 | 1.12 |
| 0150 | Step 5 - application of dry shake colored release agent | 1 Cefi | 6400 | .001 | | .08 | .05 | | .13 | .16 |
| 0160 | Step 6 - place, tamp & remove mats | 3 Cefi | 2400 | .010 | | 1.41 | .41 | | 1.82 | 2.15 |
| 0170 | Step 7 - touch up edges, mat joints & simulated grout lines | 1 Cefi | 1280 | .006 | | | .26 | | .26 | .37 |
| 0300 | Alternate stamping estimating method includes all tasks above | 4 Cefi | 800 | .040 | | 2.07 | 1.63 | | 3.70 | 4.66 |
| 0400 | Step 8 - pressure wash @ 3000 psi after 24 hours | 1 Cefi | 1600 | .005 | | | .20 | | .20 | .30 |
| 0500 | Step 9 - roll 2 coats cure/seal compound when dry | " | 800 | .010 | ↓ | .48 | .41 | | .89 | 1.13 |

## 03 35 43 – Polished Concrete Finishing

### 03 35 43.10 Polished Concrete Floors

| | | Crew | Daily Output | Labor-Hours | Unit | Material | 2011 Bare Costs Labor | Equipment | Total | Total Incl O&P |
|---|---|---|---|---|---|---|---|---|---|---|
| 0010 | **POLISHED CONCRETE FLOORS** | | | | | | | | | |
| 0015 | Processing of cured concrete to include grinding, honing, | | | | | | | | | |
| 0020 | and polishing of interior floors with 22" segmented diamond | | | | | | | | | |
| 0025 | planetary floor grinder (2 passes in different directions per grit) | | | | | | | | | |
| 0100 | Removal of pre-existing coatings, dry, with carbide discs using | | | | | | | | | |
| 0105 | dry vacuum pick-up system, final hand sweeping | | | | | | | | | |
| 0110 | Glue, adhesive or tar | J-4 | 1.60 | 15 | M.S.F. | 19.50 | 580 | 107 | 706.50 | 1,000 |
| 0120 | Paint, epoxy, 1 coat | ↓ | 3.60 | 6.667 | | 19.50 | 258 | 47.50 | 325 | 455 |
| 0130 | 2 coats | ↓ | 1.80 | 13.333 | ↓ | 19.50 | 515 | 95.50 | 630 | 890 |
| 0200 | Grinding and edging, wet, including wet vac pick-up and auto | | | | | | | | | |
| 0205 | scrubbing between grit changes | | | | | | | | | |
| 0210 | 40-grit diamond/metal matrix | J-4A | 1.60 | 20 | M.S.F. | 33 | 750 | 204 | 987 | 1,375 |
| 0220 | 80-grit diamond/metal matrix | | 2 | 16 | | 33 | 600 | 163 | 796 | 1,125 |
| 0230 | 120-grit diamond/metal matrix | | 2.40 | 13.333 | | 33 | 500 | 136 | 669 | 935 |
| 0240 | 200-grit diamond/metal matrix | ↓ | 2.80 | 11.429 | | 33 | 430 | 116 | 579 | 805 |
| 0300 | Spray on dye or stain (1 coat) | 1 Cefi | 16 | .500 | | 271 | 20.50 | | 291.50 | 330 |
| 0400 | Spray on densifier/hardener (2 coats) | " | 8 | 1 | | 279 | 41 | | 320 | 365 |
| 0410 | Auto scrubbing after 2nd coat, when dry | J-4B | 16 | .500 | ↓ | | 17.20 | 9.65 | 26.85 | 37 |
| 0500 | Honing and edging, wet, including wet vac pick-up and auto | | | | | | | | | |
| 0505 | scrubbing between grit changes | | | | | | | | | |
| 0510 | 100-grit diamond/resin matrix | J-4A | 2.80 | 11.429 | M.S.F. | 33 | 430 | 116 | 579 | 805 |
| 0520 | 200-grit diamond/resin matrix | " | 2.80 | 11.429 | " | 33 | 430 | 116 | 579 | 805 |
| 0530 | Dry, including dry vacuum pick-up system, final hand sweeping | | | | | | | | | |
| 0540 | 400-grit diamond/resin matrix | J-4A | 2.80 | 11.429 | M.S.F. | 33 | 430 | 116 | 579 | 805 |
| 0600 | Polishing and edging, dry, including dry vac pick-up and hand | | | | | | | | | |
| 0605 | sweeping between grit changes | | | | | | | | | |
| 0610 | 800-grit diamond/resin matrix | J-4A | 2.80 | 11.429 | M.S.F. | 33 | 430 | 116 | 579 | 805 |

# 03 35 Concrete Finishing

## 03 35 43 – Polished Concrete Finishing

| 03 35 43.10 Polished Concrete Floors | | Crew | Daily Output | Labor-Hours | Unit | Material | 2011 Bare Costs Labor | Equipment | Total | Total Incl O&P |
|---|---|---|---|---|---|---|---|---|---|---|
| 0620 | 1500-grit diamond/resin matrix | J-4A | 2.80 | 11.429 | M.S.F. | 33 | 430 | 116 | 579 | 805 |
| 0630 | 3000-grit diamond/resin matrix | ↓ | 2.80 | 11.429 | | 33 | 430 | 116 | 579 | 805 |
| 0700 | Auto scrubbing after final polishing step | J-4B | 16 | .500 | ↓ | | 17.20 | 9.65 | 26.85 | 37 |

# 03 37 Specialty Placed Concrete

## 03 37 13 – Shotcrete

### 03 37 13.30 Gunite (Dry-Mix)

| | 03 37 13.30 Gunite (Dry-Mix) | Crew | Daily Output | Labor-Hours | Unit | Material | 2011 Bare Costs Labor | Equipment | Total | Total Incl O&P |
|---|---|---|---|---|---|---|---|---|---|---|
| 0010 | **GUNITE (DRY-MIX)** | | | | | | | | | |
| 0020 | Typical in place, 1" layers, no mesh included | C-16 | 2000 | .028 | S.F. | .33 | 1.07 | .10 | 1.50 | 2.07 |
| 0100 | Mesh for gunite 2 x 2, #12 | 2 Rodm | 800 | .020 | | .65 | .97 | | 1.62 | 2.27 |
| 0150 | #4 reinforcing bars @ 6" each way | " | 500 | .032 | | 1.50 | 1.55 | | 3.05 | 4.13 |
| 0300 | Typical in place, including mesh, 2" thick, minimum | C-16 | 1000 | .056 | | 1.31 | 2.13 | .20 | 3.64 | 4.86 |
| 0350 | Maximum | | 500 | .112 | | 1.31 | 4.26 | .40 | 5.97 | 8.30 |
| 0500 | 4" thick, minimum | | 750 | .075 | | 1.96 | 2.84 | .27 | 5.07 | 6.70 |
| 0550 | Maximum | ↓ | 350 | .160 | | 1.96 | 6.10 | .57 | 8.63 | 11.95 |
| 0900 | Prepare old walls, no scaffolding, minimum | C-10 | 1000 | .024 | | | .93 | | .93 | 1.37 |
| 0950 | Maximum | " | 275 | .087 | | | 3.38 | | 3.38 | 4.99 |
| 1100 | For high finish requirement or close tolerance, add, minimum | | | | ↓ | | 50% | | | |
| 1150 | Maximum | | | | | | 110% | | | |

### 03 37 13.60 Shotcrete (Wet-Mix)

| | 03 37 13.60 Shotcrete (Wet-Mix) | Crew | Daily Output | Labor-Hours | Unit | Material | 2011 Bare Costs Labor | Equipment | Total | Total Incl O&P |
|---|---|---|---|---|---|---|---|---|---|---|
| 0010 | **SHOTCRETE (WET-MIX)** | | | | | | | | | |
| 0020 | Wet mix, placed @ up to 12 C.Y. per hour, 3000 psi | C-8C | 80 | .600 | C.Y. | 106 | 22.50 | 5.25 | 133.75 | 157 |
| 0100 | Up to 35 C.Y. per hour | C-8E | 240 | .200 | " | 95.50 | 7.45 | 2.04 | 104.99 | 119 |
| 1010 | Fiber reinforced, 1" thick | C-8C | 1740 | .028 | S.F. | .79 | 1.04 | .24 | 2.07 | 2.71 |
| 1020 | 2" thick | | 900 | .053 | | 1.58 | 2.01 | .47 | 4.06 | 5.30 |
| 1030 | 3" thick | | 825 | .058 | | 2.37 | 2.19 | .51 | 5.07 | 6.50 |
| 1040 | 4" thick | ↓ | 750 | .064 | ↓ | 3.16 | 2.41 | .56 | 6.13 | 7.75 |

# 03 39 Concrete Curing

## 03 39 13 – Water Concrete Curing

### 03 39 13.50 Water Curing

| | 03 39 13.50 Water Curing | Crew | Daily Output | Labor-Hours | Unit | Material | 2011 Bare Costs Labor | Equipment | Total | Total Incl O&P |
|---|---|---|---|---|---|---|---|---|---|---|
| 0010 | **WATER CURING** | | | | | | | | | |
| 0015 | With burlap, 4 uses assumed, 7.5 oz. | 2 Clab | 55 | .291 | C.S.F. | 8.55 | 10 | | 18.55 | 25 |
| 0100 | 10 oz. | " | 55 | .291 | " | 15.40 | 10 | | 25.40 | 32.50 |
| 0400 | Curing blankets, 1" to 2" thick, buy, minimum | | | | S.F. | .32 | | | .32 | .35 |
| 0450 | Maximum | | | | " | .48 | | | .48 | .53 |

## 03 39 23 – Membrane Concrete Curing

### 03 39 23.13 Chemical Compound Membrane Concrete Curing

| | 03 39 23.13 Chemical Compound Membrane Concrete Curing | Crew | Daily Output | Labor-Hours | Unit | Material | 2011 Bare Costs Labor | Equipment | Total | Total Incl O&P |
|---|---|---|---|---|---|---|---|---|---|---|
| 0010 | **CHEMICAL COMPOUND MEMBRANE CONCRETE CURING** | | | | | | | | | |
| 0300 | Sprayed membrane curing compound | 2 Clab | 95 | .168 | C.S.F. | 5.60 | 5.80 | | 11.40 | 15.05 |
| 0700 | Curing compound, solvent based, 400 S.F./gal, 55 gallon lots | | | | Gal. | 16.75 | | | 16.75 | 18.40 |
| 0720 | 5 gallon lots | | | | | 24 | | | 24 | 26.50 |
| 0800 | Curing compound, water based, 250 S.F./gal, 55 gallon lots | | | | ↓ | 18.75 | | | 18.75 | 20.50 |
| 0820 | 5 gallon lots | | | | | 20 | | | 20 | 22 |

### 03 39 23.23 Sheet Membrane Concrete Curing

| | 03 39 23.23 Sheet Membrane Concrete Curing | Crew | Daily Output | Labor-Hours | Unit | Material | 2011 Bare Costs Labor | Equipment | Total | Total Incl O&P |
|---|---|---|---|---|---|---|---|---|---|---|
| 0010 | **SHEET MEMBRANE CONCRETE CURING** | | | | | | | | | |
| 0200 | Curing blanket, burlap/poly, 2-ply | 2 Clab | 70 | .229 | C.S.F. | 14.85 | 7.85 | | 22.70 | 28.50 |

# 03 41 Precast Structural Concrete

## 03 41 05 – Precast Concrete Members

### 03 41 05.10 Precast Beams

| | | Crew | Daily Output | Labor-Hours | Unit | Material | 2011 Bare Costs Labor | Equipment | Total | Total Incl O&P |
|---|---|---|---|---|---|---|---|---|---|---|
| 0010 | **PRECAST BEAMS** R034105-30 | | | | | | | | | |
| 0011 | L-shaped, 20' span, 12" x 20" | C-11 | 32 | 2.250 | Ea. | 2,000 | 107 | 59.50 | 2,166.50 | 2,450 |
| 1000 | Inverted tee beams, add to above, small beams | | | | | 15% | | | | |
| 1050 | Large beams | | | | | 20% | | | | |
| 1200 | Rectangular, 20' span, 12" x 20" | C-11 | 32 | 2.250 | | 1,500 | 107 | 59.50 | 1,666.50 | 1,900 |
| 1250 | 18" x 36" | | 24 | 3 | | 2,725 | 143 | 79.50 | 2,947.50 | 3,325 |
| 1300 | 24" x 44" | | 22 | 3.273 | | 3,775 | 156 | 86.50 | 4,017.50 | 4,550 |
| 1400 | 30' span, 12" x 36" | | 24 | 3 | | 3,425 | 143 | 79.50 | 3,647.50 | 4,100 |
| 1450 | 18" x 44" | | 20 | 3.600 | | 4,925 | 171 | 95.50 | 5,191.50 | 5,825 |
| 1500 | 24" x 52" | | 16 | 4.500 | | 6,575 | 214 | 119 | 6,908 | 7,725 |
| 1600 | 40' span, 12" x 52" | | 20 | 3.600 | | 6,375 | 171 | 95.50 | 6,641.50 | 7,400 |
| 1650 | 18" x 52" | | 16 | 4.500 | | 7,575 | 214 | 119 | 7,908 | 8,825 |
| 1700 | 24" x 52" | | 12 | 6 | | 8,775 | 286 | 159 | 9,220 | 10,300 |
| 2000 | "T" shaped, 20' span, 12" x 20" | | 32 | 2.250 | | 2,600 | 107 | 59.50 | 2,766.50 | 3,100 |
| 2050 | 18" x 36" | | 24 | 3 | | 4,675 | 143 | 79.50 | 4,897.50 | 5,475 |
| 2100 | 24" x 44" | | 22 | 3.273 | | 6,475 | 156 | 86.50 | 6,717.50 | 7,500 |
| 2200 | 30' span, 12" x 36" | | 24 | 3 | | 5,825 | 143 | 79.50 | 6,047.50 | 6,725 |
| 2250 | 18" x 44" | | 20 | 3.600 | | 8,375 | 171 | 95.50 | 8,641.50 | 9,600 |
| 2300 | 24" x 52" | | 16 | 4.500 | | 11,200 | 214 | 119 | 11,533 | 12,800 |
| 2500 | 40' span, 12" x 52" | | 20 | 3.600 | | 10,800 | 171 | 95.50 | 11,066.50 | 12,200 |
| 2550 | 18" x 52" | | 16 | 4.500 | | 12,900 | 214 | 119 | 13,233 | 14,700 |
| 2600 | 24" x 52" | | 12 | 6 | | 14,900 | 286 | 159 | 15,345 | 17,100 |

### 03 41 05.15 Precast Columns

| | | Crew | Daily Output | Labor-Hours | Unit | Material | Labor | Equipment | Total | Total Incl O&P |
|---|---|---|---|---|---|---|---|---|---|---|
| 0010 | **PRECAST COLUMNS** R034105-30 | | | | | | | | | |
| 0020 | Rectangular to 12' high, small columns | C-11 | 120 | .600 | L.F. | 106 | 28.50 | 15.90 | 150.40 | 182 |
| 0050 | Large columns | | 96 | .750 | | 184 | 35.50 | 19.85 | 239.35 | 285 |
| 0300 | 24' high, small columns | | 192 | .375 | | 106 | 17.85 | 9.95 | 133.80 | 157 |
| 0350 | Large columns | | 144 | .500 | | 184 | 24 | 13.25 | 221.25 | 258 |

### 03 41 05.25 Precast Joists

| | | Crew | Daily Output | Labor-Hours | Unit | Material | Labor | Equipment | Total | Total Incl O&P |
|---|---|---|---|---|---|---|---|---|---|---|
| 0010 | **PRECAST JOISTS** R034105-30 | | | | | | | | | |
| 0015 | 40 psf L.L., 6" deep for 12' spans | C-12 | 600 | .080 | L.F. | 17.65 | 3.40 | 1.09 | 22.14 | 26 |
| 0050 | 8" deep for 16' spans | | 575 | .083 | | 29.50 | 3.55 | 1.14 | 34.19 | 39 |
| 0100 | 10" deep for 20' spans | | 550 | .087 | | 51.50 | 3.71 | 1.19 | 56.40 | 63.50 |
| 0150 | 12" deep for 24' spans | | 525 | .091 | | 70.50 | 3.89 | 1.25 | 75.64 | 85 |

## 03 41 13 – Precast Concrete Hollow Core Planks

### 03 41 13.50 Precast Slab Planks

| | | Crew | Daily Output | Labor-Hours | Unit | Material | Labor | Equipment | Total | Total Incl O&P |
|---|---|---|---|---|---|---|---|---|---|---|
| 0010 | **PRECAST SLAB PLANKS** R034105-30 | | | | | | | | | |
| 0020 | Prestressed roof/floor members, grouted, solid, 4" thick | C-11 | 2400 | .030 | S.F. | 5.55 | 1.43 | .79 | 7.77 | 9.40 |
| 0050 | 6" thick | | 2800 | .026 | | 6.50 | 1.22 | .68 | 8.40 | 10 |
| 0100 | Hollow, 8" thick | | 3200 | .023 | | 7.20 | 1.07 | .60 | 8.87 | 10.40 |
| 0150 | 10" thick | | 3600 | .020 | | 7.50 | .95 | .53 | 8.98 | 10.45 |
| 0200 | 12" thick | | 4000 | .018 | | 8 | .86 | .48 | 9.34 | 10.80 |

## 03 41 16 – Precast Concrete Slabs

### 03 41 16.20 Precast Concrete Channel Slabs

| | | Crew | Daily Output | Labor-Hours | Unit | Material | Labor | Equipment | Total | Total Incl O&P |
|---|---|---|---|---|---|---|---|---|---|---|
| 0010 | **PRECAST CONCRETE CHANNEL SLABS** | | | | | | | | | |
| 0335 | Lightweight concrete channel slab, long runs, 2-3/4" thick | C-12 | 1575 | .030 | S.F. | 7.25 | 1.30 | .42 | 8.97 | 10.45 |
| 0375 | 3-3/4" thick | | 1550 | .031 | | 7.55 | 1.32 | .42 | 9.29 | 10.80 |
| 0475 | 4-3/4" thick | | 1525 | .031 | | 8.25 | 1.34 | .43 | 10.02 | 11.60 |
| 1275 | Short pieces, 2-3/4" thick | | 785 | .061 | | 10.90 | 2.60 | .84 | 14.34 | 16.85 |
| 1375 | 3-3/4" thick | | 770 | .062 | | 11.35 | 2.65 | .85 | 14.85 | 17.45 |
| 1475 | 4-3/4" thick | | 762 | .063 | | 12.40 | 2.68 | .86 | 15.94 | 18.65 |

# 03 41 Precast Structural Concrete

## 03 41 16 – Precast Concrete Slabs

| 03 41 16.50 Precast Lightweight Concrete Plank | Crew | Daily Output | Labor-Hours | Unit | Material | 2011 Bare Costs Labor | Equipment | Total | Total Incl O&P |
|---|---|---|---|---|---|---|---|---|---|
| 0010 **PRECAST LIGHTWEIGHT CONCRETE PLANK** | | | | | | | | | |
| 0015 Lightweight plank, nailable, T&G, 2" thick | C-12 | 1800 | .027 | S.F. | 6.65 | 1.13 | .36 | 8.14 | 9.45 |
| 0150 For premium ceiling finish, add | | | | " | 10% | | | | |
| 0200 For sloping roofs, slope over 4 in 12, add | | | | | | 25% | | | |
| 0250 Slope over 6 in 12, add | | | | | | 150% | | | |

## 03 41 23 – Precast Concrete Stairs

| 03 41 23.50 Precast Stairs | Crew | Daily Output | Labor-Hours | Unit | Material | 2011 Bare Costs Labor | Equipment | Total | Total Incl O&P |
|---|---|---|---|---|---|---|---|---|---|
| 0010 **PRECAST STAIRS** | | | | | | | | | |
| 0020 Precast concrete treads on steel stringers, 3' wide | C-12 | 75 | .640 | Riser | 129 | 27 | 8.75 | 164.75 | 193 |
| 0300 Front entrance, 5' wide with 48" platform, 2 risers | | 16 | 3 | Flight | 460 | 128 | 41 | 629 | 745 |
| 0350 5 risers | | 12 | 4 | | 715 | 170 | 54.50 | 939.50 | 1,100 |
| 0500 6' wide, 2 risers | | 15 | 3.200 | | 510 | 136 | 43.50 | 689.50 | 815 |
| 0550 5 risers | | 11 | 4.364 | | 790 | 186 | 59.50 | 1,035.50 | 1,225 |
| 0700 7' wide, 2 risers | | 14 | 3.429 | | 660 | 146 | 47 | 853 | 1,000 |
| 0750 5 risers | | 10 | 4.800 | | 1,075 | 204 | 65.50 | 1,344.50 | 1,575 |
| 1200 Basement entrance stairs, steel bulkhead doors, minimum | B-51 | 22 | 2.182 | | 1,350 | 75.50 | 7.80 | 1,433.30 | 1,625 |
| 1250 Maximum | " | 11 | 4.364 | | 2,275 | 151 | 15.60 | 2,441.60 | 2,750 |

## 03 41 33 – Precast Structural Pretensioned Concrete

| 03 41 33.60 Tees | Crew | Daily Output | Labor-Hours | Unit | Material | 2011 Bare Costs Labor | Equipment | Total | Total Incl O&P |
|---|---|---|---|---|---|---|---|---|---|
| 0010 **TEES** R034105-30 | | | | | | | | | |
| 0020 Quad tee, short spans, roof | C-11 | 7200 | .010 | S.F. | 7.90 | .48 | .26 | 8.64 | 9.80 |
| 0050 Floor | | 7200 | .010 | | 7.90 | .48 | .26 | 8.64 | 9.80 |
| 0200 Double tee, floor members, 60' span | | 8400 | .009 | | 9.40 | .41 | .23 | 10.04 | 11.30 |
| 0250 80' span | | 8000 | .009 | | 12.15 | .43 | .24 | 12.82 | 14.35 |
| 0300 Roof members, 30' span | | 4800 | .015 | | 7.90 | .71 | .40 | 9.01 | 10.35 |
| 0350 50' span | | 6400 | .011 | | 8.65 | .54 | .30 | 9.49 | 10.75 |
| 0400 Wall members, up to 55' high | | 3600 | .020 | | 11.75 | .95 | .53 | 13.23 | 15.10 |
| 0500 Single tee roof members, 40' span | | 3200 | .023 | | 10.70 | 1.07 | .60 | 12.37 | 14.25 |
| 0550 80' span | | 5120 | .014 | | 12.15 | .67 | .37 | 13.19 | 14.90 |
| 0600 100' span | | 6000 | .012 | | 17.80 | .57 | .32 | 18.69 | 21 |
| 0650 120' span | | 6000 | .012 | | 19.50 | .57 | .32 | 20.39 | 23 |
| 1000 Double tees, floor members | | | | | | | | | |
| 1100 Lightweight, 20" x 8' wide, 45' span | C-11 | 20 | 3.600 | Ea. | 3,375 | 171 | 95.50 | 3,641.50 | 4,100 |
| 1150 24" x 8' wide, 50' span | | 18 | 4 | | 3,650 | 191 | 106 | 3,947 | 4,475 |
| 1200 32" x 10' wide, 60' span | | 16 | 4.500 | | 6,175 | 214 | 119 | 6,508 | 7,300 |
| 1250 Standard weight, 12" x 8' wide, 20' span | | 22 | 3.273 | | 1,250 | 156 | 86.50 | 1,492.50 | 1,750 |
| 1300 16" x 8' wide, 25' span | | 20 | 3.600 | | 1,675 | 171 | 95.50 | 1,941.50 | 2,250 |
| 1350 18" x 8' wide, 30' span | | 20 | 3.600 | | 2,125 | 171 | 95.50 | 2,391.50 | 2,750 |
| 1400 20" x 8' wide, 45' span | | 18 | 4 | | 2,400 | 191 | 106 | 2,697 | 3,075 |
| 1450 24" x 8' wide, 50' span | | 16 | 4.500 | | 3,100 | 214 | 119 | 3,433 | 3,900 |
| 1500 32" x 10' wide, 60' span | | 14 | 5.143 | | 5,675 | 245 | 136 | 6,056 | 6,800 |
| 2000 Roof members | | | | | | | | | |
| 2050 Lightweight, 20" x 8' wide, 40' span | C-11 | 20 | 3.600 | Ea. | 2,850 | 171 | 95.50 | 3,116.50 | 3,525 |
| 2100 24" x 8' wide, 50' span | | 18 | 4 | | 3,800 | 191 | 106 | 4,097 | 4,625 |
| 2150 32" x 10' wide, 60' span | | 16 | 4.500 | | 6,275 | 214 | 119 | 6,608 | 7,400 |
| 2200 Standard weight, 12" x 8' wide, 30' span | | 22 | 3.273 | | 1,900 | 156 | 86.50 | 2,142.50 | 2,475 |
| 2250 16" x 8' wide, 30' span | | 20 | 3.600 | | 2,000 | 171 | 95.50 | 2,266.50 | 2,600 |
| 2300 18" x 8' wide, 30' span | | 20 | 3.600 | | 2,225 | 171 | 95.50 | 2,491.50 | 2,850 |
| 2350 20" x 8' wide, 40' span | | 18 | 4 | | 2,275 | 191 | 106 | 2,572 | 2,950 |
| 2400 24" x 8' wide, 50' span | | 16 | 4.500 | | 3,050 | 214 | 119 | 3,383 | 3,850 |
| 2450 32" x 10' wide, 60' span | | 14 | 5.143 | | 5,325 | 245 | 136 | 5,706 | 6,425 |

# 03 41 Precast Structural Concrete

## 03 41 36 – Precast Structural Post-Tensioned Concrete

| 03 41 36.50 Post-Tensioned Jobs | | Crew | Daily Output | Labor-Hours | Unit | Material | 2011 Bare Costs Labor | Equipment | Total | Total Incl O&P |
|---|---|---|---|---|---|---|---|---|---|---|
| 0010 | **POST-TENSIONED JOBS** R034105-30 | | | | | | | | | |
| 0100 | Post-tensioned in place, small job R034136-90 | C-17B | 8.50 | 9.647 | C.Y. | 1,175 | 435 | 51 | 1,661 | 2,025 |
| 0200 | Large job | " | 10 | 8.200 | " | 885 | 370 | 43.50 | 1,298.50 | 1,600 |

# 03 45 Precast Architectural Concrete

## 03 45 13 – Faced Architectural Precast Concrete

### 03 45 13.50 Precast Wall Panels

| 03 45 13.50 Precast Wall Panels | | Crew | Daily Output | Labor-Hours | Unit | Material | 2011 Bare Costs Labor | Equipment | Total | Total Incl O&P |
|---|---|---|---|---|---|---|---|---|---|---|
| 0010 | **PRECAST WALL PANELS** R034513-10 | | | | | | | | | |
| 0050 | Uninsulated 4" thick, smooth gray | | | | | | | | | |
| 0150 | Low rise, 4' x 8' x 4" thick | C-11 | 320 | .225 | S.F. | 21.50 | 10.70 | 5.95 | 38.15 | 48.50 |
| 0210 | 8' x 8', 4" thick | | 576 | .125 | | 21 | 5.95 | 3.31 | 30.26 | 37.50 |
| 0250 | 8' x 16' x 4" thick | | 1024 | .070 | | 21 | 3.35 | 1.86 | 26.21 | 31 |
| 0600 | High rise, 4' x 8' x 4" thick | | 288 | .250 | | 21.50 | 11.90 | 6.60 | 40 | 51.50 |
| 0650 | 8' x 8' x 4" thick | | 512 | .141 | | 21 | 6.70 | 3.72 | 31.42 | 39 |
| 0700 | 8' x 16' x 4" thick | | 768 | .094 | | 21 | 4.47 | 2.48 | 27.95 | 33.50 |
| 0750 | 20' x 10', 6" thick, smooth gray | ▼ | 1400 | .051 | | 35.50 | 2.45 | 1.36 | 39.31 | 45 |
| 0800 | Insulated panel, 2" polystyrene, add | | | | | 1 | | | 1 | 1.10 |
| 0850 | 2" urethane, add | | | | | .79 | | | .79 | .87 |
| 1200 | Finishes, white, add | | | | | 2.73 | | | 2.73 | 3.01 |
| 1250 | Exposed aggregate, add | | | | | 1.77 | | | 1.77 | 1.95 |
| 1300 | Granite faced, domestic, add | | | | | 29 | | | 29 | 32 |
| 1350 | Brick faced, modular, red, add | | | | ▼ | 8.75 | | | 8.75 | 9.65 |
| 2200 | Fiberglass reinforced cement with urethane core | | | | | | | | | |
| 2210 | R20, 8' x 8', minimum | E-2 | 750 | .075 | S.F. | 22 | 3.54 | 2.16 | 27.70 | 32.50 |
| 2220 | Maximum | " | 600 | .093 | " | 33 | 4.42 | 2.70 | 40.12 | 46.50 |

# 03 47 Site-Cast Concrete

## 03 47 13 – Tilt-Up Concrete

### 03 47 13.50 Tilt-Up Wall Panels

| 03 47 13.50 Tilt-Up Wall Panels | | Crew | Daily Output | Labor-Hours | Unit | Material | 2011 Bare Costs Labor | Equipment | Total | Total Incl O&P |
|---|---|---|---|---|---|---|---|---|---|---|
| 0010 | **TILT-UP WALL PANELS** R034713-20 | | | | | | | | | |
| 0015 | Wall panel construction, walls only, 5-1/2" thick | C-14 | 1600 | .090 | S.F. | 4.78 | 3.80 | 1.06 | 9.64 | 12.25 |
| 0100 | 7-1/2" thick | | 1550 | .093 | | 6 | 3.92 | 1.09 | 11.01 | 13.85 |
| 0500 | Walls and columns, 5-1/2" thick walls, 12" x 12" columns | | 1565 | .092 | | 7.20 | 3.88 | 1.08 | 12.16 | 15.05 |
| 0550 | 7-1/2" thick wall, 12" x 12" columns | | 1370 | .105 | ▼ | 8.85 | 4.44 | 1.24 | 14.53 | 17.90 |
| 0800 | Columns only, site precast, minimum | | 200 | .720 | L.F. | 11.90 | 30.50 | 8.45 | 50.85 | 69 |
| 0850 | Maximum | ▼ | 105 | 1.371 | " | 15.05 | 58 | 16.15 | 89.20 | 123 |

# 03 48 Precast Concrete Specialties

## 03 48 43 - Precast Concrete Trim

### 03 48 43.40 Precast Lintels

| | | Crew | Daily Output | Labor-Hours | Unit | Material | 2011 Bare Costs Labor | Equipment | Total | Total Incl O&P |
|---|---|---|---|---|---|---|---|---|---|---|
| 0010 | **PRECAST LINTELS**, smooth gray, prestressed, stock units only | | | | | | | | | |
| 0800 | 4" wide, 8" high, x 4' long | D-10 | 28 | 1.143 | Ea. | 22.50 | 48.50 | 15.50 | 86.50 | 115 |
| 0850 | 8' long | | 24 | 1.333 | | 57.50 | 56.50 | 18.10 | 132.10 | 168 |
| 1000 | 6" wide, 8" high, x 4' long | | 26 | 1.231 | | 33 | 52 | 16.70 | 101.70 | 132 |
| 1050 | 10' long | | 22 | 1.455 | | 88 | 61.50 | 19.75 | 169.25 | 211 |
| 1200 | 8" wide, 8" high, x 4' long | | 24 | 1.333 | | 41 | 56.50 | 18.10 | 115.60 | 150 |
| 1250 | 12' long | | 20 | 1.600 | | 134 | 68 | 21.50 | 223.50 | 274 |
| 1275 | For custom sizes, types, colors, or finishes of precast lintels, add | | | | | 150% | | | | |

### 03 48 43.90 Precast Window Sills

| | | Crew | Daily Output | Labor-Hours | Unit | Material | 2011 Bare Costs Labor | Equipment | Total | Total Incl O&P |
|---|---|---|---|---|---|---|---|---|---|---|
| 0010 | **PRECAST WINDOW SILLS** | | | | | | | | | |
| 0600 | Precast concrete, 4" tapers to 3", 9" wide | D-1 | 70 | .229 | L.F. | 12.25 | 8.90 | | 21.15 | 27 |
| 0650 | 11" wide | | 60 | .267 | | 16 | 10.40 | | 26.40 | 33 |
| 0700 | 13" wide, 3 1/2" tapers to 2 1/2", 12" wall | | 50 | .320 | | 16 | 12.45 | | 28.45 | 36.50 |

# 03 51 Cast Roof Decks

## 03 51 13 - Cementitious Wood Fiber Decks

### 03 51 13.50 Cementitious/Wood Fiber Planks

| | | | Crew | Daily Output | Labor-Hours | Unit | Material | 2011 Bare Costs Labor | Equipment | Total | Total Incl O&P |
|---|---|---|---|---|---|---|---|---|---|---|---|
| 0010 | **CEMENTITIOUS/WOOD FIBER PLANKS** | R051223-50 | | | | | | | | | |
| 0050 | Plank, beveled edge, 1" thick | | 2 Carp | 1000 | .016 | S.F. | 2.32 | .69 | | 3.01 | 3.61 |
| 0100 | 1-1/2" thick | | | 975 | .016 | | 3.02 | .71 | | 3.73 | 4.40 |
| 0150 | T & G, 2" thick | | | 950 | .017 | | 3.54 | .73 | | 4.27 | 5 |
| 0200 | 2-1/2" thick | | | 925 | .017 | | 4.14 | .74 | | 4.88 | 5.70 |
| 0250 | 3" thick | | | 900 | .018 | | 4.75 | .77 | | 5.52 | 6.40 |
| 1000 | Bulb tee, sub-purlin and grout, 6' span, add | | E-1 | 5000 | .005 | | 2.06 | .23 | .02 | 2.31 | 2.66 |
| 1100 | 8' span | | " | 4200 | .006 | | 2.31 | .27 | .03 | 2.61 | 3.03 |

## 03 51 16 - Gypsum Concrete Roof Decks

### 03 51 16.50 Gypsum Roof Deck

| | | Crew | Daily Output | Labor-Hours | Unit | Material | 2011 Bare Costs Labor | Equipment | Total | Total Incl O&P |
|---|---|---|---|---|---|---|---|---|---|---|
| 0010 | **GYPSUM ROOF DECK** | | | | | | | | | |
| 1000 | Poured gypsum, 2" thick | C-8 | 6000 | .009 | S.F. | 1.25 | .36 | .12 | 1.73 | 2.04 |
| 1100 | 3" thick | " | 4800 | .012 | " | 1.87 | .44 | .15 | 2.46 | 2.90 |

# 03 52 Lightweight Concrete Roof Insulation

## 03 52 16 - Lightweight Insulating Concrete

### 03 52 16.13 Lightweight Cellular Insulating Concrete

| | | | Crew | Daily Output | Labor-Hours | Unit | Material | 2011 Bare Costs Labor | Equipment | Total | Total Incl O&P |
|---|---|---|---|---|---|---|---|---|---|---|---|
| 0010 | **LIGHTWEIGHT CELLULAR INSULATING CONCRETE** | R035216-10 | | | | | | | | | |
| 0020 | Portland cement and foaming agent | G | C-8 | 50 | 1.120 | C.Y. | 144 | 42.50 | 14.80 | 201.30 | 238 |

### 03 52 16.16 Lightweight Aggregate Insulating Concrete

| | | | Crew | Daily Output | Labor-Hours | Unit | Material | 2011 Bare Costs Labor | Equipment | Total | Total Incl O&P |
|---|---|---|---|---|---|---|---|---|---|---|---|
| 0010 | **LIGHTWEIGHT AGGREGATE INSULATING CONCRETE** | R035216-10 | | | | | | | | | |
| 0100 | Poured vermiculite or perlite, field mix, | | | | | | | | | | |
| 0110 | 1:6 field mix | G | C-8 | 50 | 1.120 | C.Y. | 222 | 42.50 | 14.80 | 279.30 | 325 |
| 0200 | Ready mix, 1:6 mix, roof fill, 2" thick | G | | 10000 | .006 | S.F. | 1.24 | .21 | .07 | 1.52 | 1.76 |
| 0250 | 3" thick | G | | 7700 | .007 | | 1.85 | .28 | .10 | 2.23 | 2.57 |
| 0400 | Expanded volcanic glass rock, with binder, minimum | G | 2 Carp | 1500 | .011 | | .40 | .46 | | .86 | 1.14 |
| 0450 | Maximum | G | " | 1200 | .013 | | 1.19 | .57 | | 1.76 | 2.19 |

# 03 54 Cast Underlayment

## 03 54 13 – Gypsum Cement Underlayment

### 03 54 13.50 Poured Gypsum Underlayment

| 03 54 13.50 Poured Gypsum Underlayment | Crew | Daily Output | Labor-Hours | Unit | Material | 2011 Bare Costs Labor | Equipment | Total | Total Incl O&P |
|---|---|---|---|---|---|---|---|---|---|
| 0010 **POURED GYPSUM UNDERLAYMENT** | | | | | | | | | |
| 0400 Underlayment, gypsum based, self-leveling 2500 psi, pumped, 1/2" | C-8 | 24000 | .002 | S.F. | .31 | .09 | .03 | .43 | .50 |
| 0500    3/4" | | 20000 | .003 | | .47 | .11 | .04 | .62 | .71 |
| 0600    1" | | 16000 | .004 | | .62 | .13 | .05 | .80 | .94 |
| 1400    Hand placed, 1/2" | C-18 | 450 | .020 | | .31 | .69 | .12 | 1.12 | 1.53 |
| 1500    3/4" | " | 300 | .030 | | .47 | 1.04 | .18 | 1.69 | 2.30 |

## 03 54 16 – Hydraulic Cement Underlayment

### 03 54 16.50 Cement Underlayment

| 03 54 16.50 Cement Underlayment | Crew | Daily Output | Labor-Hours | Unit | Material | 2011 Bare Costs Labor | Equipment | Total | Total Incl O&P |
|---|---|---|---|---|---|---|---|---|---|
| 0010 **CEMENT UNDERLAYMENT** | | | | | | | | | |
| 2510 Underlayment, P.C based self-leveling, 4100 psi, pumped, 1/4" | C-8 | 20000 | .003 | S.F. | 1.35 | .11 | .04 | 1.50 | 1.69 |
| 2520    1/2" | | 19000 | .003 | | 2.71 | .11 | .04 | 2.86 | 3.19 |
| 2530    3/4" | | 18000 | .003 | | 4.06 | .12 | .04 | 4.22 | 4.69 |
| 2540    1" | | 17000 | .003 | | 5.40 | .13 | .04 | 5.57 | 6.20 |
| 2550    1-1/2" | | 15000 | .004 | | 8.10 | .14 | .05 | 8.29 | 9.20 |
| 2560    Hand placed, 1/2" | C-18 | 450 | .020 | | 2.71 | .69 | .12 | 3.52 | 4.17 |
| 2610 Topping, P.C. based self-level/dry 6100 psi, pumped, 1/4" | C-8 | 20000 | .003 | | 2.03 | .11 | .04 | 2.18 | 2.44 |
| 2620    1/2" | | 19000 | .003 | | 4.06 | .11 | .04 | 4.21 | 4.68 |
| 2630    3/4" | | 18000 | .003 | | 6.10 | .12 | .04 | 6.26 | 6.95 |
| 2660    1" | | 17000 | .003 | | 8.15 | .13 | .04 | 8.32 | 9.20 |
| 2670    1-1/2" | | 15000 | .004 | | 12.20 | .14 | .05 | 12.39 | 13.65 |
| 2680    Hand placed, 1/2" | C-18 | 450 | .020 | | 4.06 | .69 | .12 | 4.87 | 5.65 |

# 03 62 Non-Shrink Grouting

## 03 62 13 – Non-Metallic Non-Shrink Grouting

### 03 62 13.50 Grout, Non-Metallic Non-shrink

| 03 62 13.50 Grout, Non-Metallic Non-shrink | Crew | Daily Output | Labor-Hours | Unit | Material | 2011 Bare Costs Labor | Equipment | Total | Total Incl O&P |
|---|---|---|---|---|---|---|---|---|---|
| 0010 **GROUT, NON-METALLIC NON-SHRINK** | | | | | | | | | |
| 0300 Non-shrink, non-metallic, 1" deep | 1 Cefi | 35 | .229 | S.F. | 7.30 | 9.35 | | 16.65 | 21.50 |
| 0350    2" deep | " | 25 | .320 | " | 14.65 | 13.05 | | 27.70 | 35 |

## 03 62 16 – Metallic Non-Shrink Grouting

### 03 62 16.50 Grout, Metallic Non-Shrink

| 03 62 16.50 Grout, Metallic Non-Shrink | Crew | Daily Output | Labor-Hours | Unit | Material | 2011 Bare Costs Labor | Equipment | Total | Total Incl O&P |
|---|---|---|---|---|---|---|---|---|---|
| 0010 **GROUT, METALLIC NON-SHRINK** | | | | | | | | | |
| 0020 Column & machine bases, non-shrink, metallic, 1" deep | 1 Cefi | 35 | .229 | S.F. | 8.65 | 9.35 | | 18 | 23 |
| 0050    2" deep | " | 25 | .320 | " | 17.35 | 13.05 | | 30.40 | 38 |

# 03 63 Epoxy Grouting

## 03 63 05 – Grouting of Dowels and Fasteners

### 03 63 05.10 Epoxy Only

| 03 63 05.10 Epoxy Only | Crew | Daily Output | Labor-Hours | Unit | Material | 2011 Bare Costs Labor | Equipment | Total | Total Incl O&P |
|---|---|---|---|---|---|---|---|---|---|
| 0010 **EPOXY ONLY** | | | | | | | | | |
| 1500 Chemical anchoring, epoxy cartridge, excludes layout, drilling, fastener | | | | | | | | | |
| 1530    For fastener 3/4" diam. x 6" embedment | 2 Skwk | 72 | .222 | Ea. | 4.95 | 9.85 | | 14.80 | 20.50 |
| 1535    1" diam. x 8" embedment | | 66 | .242 | | 7.45 | 10.75 | | 18.20 | 24.50 |
| 1540    1-1/4" diam. x 10" embedment | | 60 | .267 | | 14.85 | 11.85 | | 26.70 | 34.50 |
| 1545    1-3/4" diam. x 12" embedment | | 54 | .296 | | 25 | 13.15 | | 38.15 | 47 |
| 1550    14" embedment | | 48 | .333 | | 29.50 | 14.80 | | 44.30 | 55 |
| 1555    2" diam. x 12" embedment | | 42 | .381 | | 39.50 | 16.90 | | 56.40 | 69.50 |
| 1560    18" embedment | | 32 | .500 | | 49.50 | 22 | | 71.50 | 88.50 |

# 03 81 Concrete Cutting

## 03 81 13 – Flat Concrete Sawing

### 03 81 13.50 Concrete Floor/Slab Cutting

| | | Crew | Daily Output | Labor-Hours | Unit | Material | 2011 Bare Costs Labor | Equipment | Total | Total Incl O&P |
|---|---|---|---|---|---|---|---|---|---|---|
| 0010 | **CONCRETE FLOOR/SLAB CUTTING** | | | | | | | | | |
| 0300 | Saw cut concrete slabs, plain, up to 3" deep | B-89 | 1060 | .015 | L.F. | .18 | .58 | .35 | 1.11 | 1.45 |
| 0320 | Each additional inch of depth | | 3180 | .005 | | .06 | .19 | .12 | .37 | .49 |
| 0400 | Mesh reinforced, up to 3" deep | | 980 | .016 | | .20 | .63 | .38 | 1.21 | 1.58 |
| 0420 | Each additional inch of depth | | 2940 | .005 | | .07 | .21 | .13 | .41 | .53 |
| 0500 | Rod reinforced, up to 3" deep | | 800 | .020 | | .25 | .77 | .46 | 1.48 | 1.94 |
| 0520 | Each additional inch of depth | | 2400 | .007 | | .08 | .26 | .15 | .49 | .65 |

## 03 81 16 – Track Mounted Concrete Wall Sawing

### 03 81 16.50 Concrete Wall Cutting

| | | Crew | Daily Output | Labor-Hours | Unit | Material | 2011 Bare Costs Labor | Equipment | Total | Total Incl O&P |
|---|---|---|---|---|---|---|---|---|---|---|
| 0010 | **CONCRETE WALL CUTTING** | | | | | | | | | |
| 0800 | Concrete walls, hydraulic saw, plain, per inch of depth | B-89B | 250 | .064 | L.F. | .46 | 2.46 | 2.61 | 5.53 | 7.10 |
| 0820 | Rod reinforcing, per inch of depth | " | 150 | .107 | " | .65 | 4.11 | 4.36 | 9.12 | 11.70 |

# 03 82 Concrete Boring

## 03 82 13 – Concrete Core Drilling

### 03 82 13.10 Core Drilling

| | | Crew | Daily Output | Labor-Hours | Unit | Material | 2011 Bare Costs Labor | Equipment | Total | Total Incl O&P |
|---|---|---|---|---|---|---|---|---|---|---|
| 0010 | **CORE DRILLING** | | | | | | | | | |
| 0020 | Reinf. conc slab, up to 6" thick, incl. bit, layout & set up | | | | | | | | | |
| 0100 | 1" diameter core | B-89A | 17 | .941 | Ea. | 3.66 | 37 | 6.45 | 47.11 | 68 |
| 0150 | Each added inch thick in same hole, add | | 1440 | .011 | | .69 | .44 | .08 | 1.21 | 1.51 |
| 0300 | 3" diameter core | | 16 | 1 | | 12.20 | 39.50 | 6.85 | 58.55 | 81.50 |
| 0350 | Each added inch thick in same hole, add | | 720 | .022 | | 1.41 | .88 | .15 | 2.44 | 3.06 |
| 0500 | 4" diameter core | | 15 | 1.067 | | 15.45 | 42 | 7.30 | 64.75 | 89.50 |
| 0550 | Each added inch thick in same hole, add | | 480 | .033 | | 1.73 | 1.31 | .23 | 3.27 | 4.16 |
| 0700 | 6" diameter core | | 14 | 1.143 | | 20 | 45 | 7.85 | 72.85 | 99.50 |
| 0750 | Each added inch thick in same hole, add | | 360 | .044 | | 2.74 | 1.75 | .30 | 4.79 | 6.05 |
| 0900 | 8" diameter core | | 13 | 1.231 | | 33 | 48.50 | 8.45 | 89.95 | 120 |
| 0950 | Each added inch thick in same hole, add | | 288 | .056 | | 3.76 | 2.19 | .38 | 6.33 | 7.90 |
| 1100 | 10" diameter core | | 12 | 1.333 | | 26.50 | 52.50 | 9.15 | 88.15 | 120 |
| 1150 | Each added inch thick in same hole, add | | 240 | .067 | | 3.76 | 2.63 | .46 | 6.85 | 8.65 |
| 1300 | 12" diameter core | | 11 | 1.455 | | 31 | 57.50 | 10 | 98.50 | 133 |
| 1350 | Each added inch thick in same hole, add | | 206 | .078 | | 4.56 | 3.06 | .53 | 8.15 | 10.30 |
| 1500 | 14" diameter core | | 10 | 1.600 | | 37 | 63 | 11 | 111 | 149 |
| 1550 | Each added inch thick in same hole, add | | 180 | .089 | | 6.20 | 3.50 | .61 | 10.31 | 12.85 |
| 1700 | 18" diameter core | | 9 | 1.778 | | 45.50 | 70 | 12.20 | 127.70 | 170 |
| 1750 | Each added inch thick in same hole, add | | 144 | .111 | | 8.80 | 4.38 | .76 | 13.94 | 17.25 |
| 1760 | For horizontal holes, add to above | | | | | | 20% | 20% | | |
| 1770 | Prestressed hollow core plank, 8" thick | | | | | | | | | |
| 1780 | 1" diameter core | B-89A | 17.50 | .914 | Ea. | 2.38 | 36 | 6.25 | 44.63 | 64.50 |
| 1790 | Each added inch thick in same hole, add | | 3840 | .004 | | .48 | .16 | .03 | .67 | .81 |
| 1800 | 3" diameter core | | 17 | .941 | | 4.96 | 37 | 6.45 | 48.41 | 69.50 |
| 1810 | Each added inch thick in same hole, add | | 1920 | .008 | | .86 | .33 | .06 | 1.25 | 1.51 |
| 1820 | 4" diameter core | | 16.50 | .970 | | 6.80 | 38 | 6.65 | 51.45 | 73.50 |
| 1830 | Each added inch thick in same hole, add | | 1280 | .013 | | 1.26 | .49 | .09 | 1.84 | 2.24 |
| 1840 | 6" diameter core | | 15.50 | 1.032 | | 8.35 | 40.50 | 7.10 | 55.95 | 79.50 |
| 1850 | Each added inch thick in same hole, add | | 960 | .017 | | 1.43 | .66 | .11 | 2.20 | 2.71 |
| 1860 | 8" diameter core | | 15 | 1.067 | | 11.85 | 42 | 7.30 | 61.15 | 85.50 |
| 1870 | Each added inch thick in same hole, add | | 768 | .021 | | 2 | .82 | .14 | 2.96 | 3.62 |
| 1880 | 10" diameter core | | 14 | 1.143 | | 15.70 | 45 | 7.85 | 68.55 | 95 |
| 1890 | Each added inch thick in same hole, add | | 640 | .025 | | 2.08 | .98 | .17 | 3.23 | 3.99 |

# 03 82 Concrete Boring

## 03 82 13 – Concrete Core Drilling

| 03 82 13.10 Core Drilling | | Crew | Daily Output | Labor-Hours | Unit | Material | 2011 Bare Costs Labor | Equipment | Total | Total Incl O&P |
|---|---|---|---|---|---|---|---|---|---|---|
| 1900 | 12" diameter core | B-89A | 13.50 | 1.185 | Ea. | 18.35 | 46.50 | 8.15 | 73 | 100 |
| 1910 | Each added inch thick in same hole, add | | 548 | .029 | ↓ | 3.33 | 1.15 | .20 | 4.68 | 5.65 |
| 1999 | Drilling, core, minimum labor/equipment charge | ↓ | 4 | 4 | Job | | 158 | 27.50 | 185.50 | 272 |
| 3010 | Bits for core drill, diamond, premium, 1" diameter | | | | Ea. | 80 | | | 80 | 88 |
| 3020 | 3" diameter | | | | | 206 | | | 206 | 226 |
| 3040 | 4" diameter | | | | | 229 | | | 229 | 252 |
| 3050 | 6" diameter | | | | | 365 | | | 365 | 405 |
| 3080 | 8" diameter | | | | | 585 | | | 585 | 645 |
| 3120 | 12" diameter | | | | | 900 | | | 900 | 990 |
| 3180 | 18" diameter | | | | | 1,400 | | | 1,400 | 1,525 |
| 3240 | 24" diameter | | | | ↓ | 1,700 | | | 1,700 | 1,875 |

## 03 82 16 – Concrete Drilling

### 03 82 16.10 Concrete Impact Drilling

| | 03 82 16.10 Concrete Impact Drilling | Crew | Daily Output | Labor-Hours | Unit | Material | 2011 Bare Costs Labor | Equipment | Total | Total Incl O&P |
|---|---|---|---|---|---|---|---|---|---|---|
| 0010 | **CONCRETE IMPACT DRILLING** | | | | | | | | | |
| 0050 | Up to 4" deep in conc/brick floor/wall, incl. bit & layout, no anchor | | | | | | | | | |
| 0100 | Holes, 1/4" diameter | 1 Carp | 75 | .107 | Ea. | .05 | 4.59 | | 4.64 | 7.10 |
| 0150 | For each additional inch of depth, add | | 430 | .019 | | .01 | .80 | | .81 | 1.24 |
| 0200 | 3/8" diameter | | 63 | .127 | | .05 | 5.45 | | 5.50 | 8.45 |
| 0250 | For each additional inch of depth, add | | 340 | .024 | | .01 | 1.01 | | 1.02 | 1.56 |
| 0300 | 1/2" diameter | | 50 | .160 | | .05 | 6.90 | | 6.95 | 10.60 |
| 0350 | For each additional inch of depth, add | | 250 | .032 | | .01 | 1.38 | | 1.39 | 2.12 |
| 0400 | 5/8" diameter | | 48 | .167 | | .07 | 7.20 | | 7.27 | 11.10 |
| 0450 | For each additional inch of depth, add | | 240 | .033 | | .02 | 1.43 | | 1.45 | 2.22 |
| 0500 | 3/4" diameter | | 45 | .178 | | .10 | 7.65 | | 7.75 | 11.85 |
| 0550 | For each additional inch of depth, add | | 220 | .036 | | .02 | 1.57 | | 1.59 | 2.43 |
| 0600 | 7/8" diameter | | 43 | .186 | | .13 | 8 | | 8.13 | 12.45 |
| 0650 | For each additional inch of depth, add | | 210 | .038 | | .03 | 1.64 | | 1.67 | 2.56 |
| 0700 | 1" diameter | | 40 | .200 | | .14 | 8.60 | | 8.74 | 13.35 |
| 0750 | For each additional inch of depth, add | | 190 | .042 | | .04 | 1.81 | | 1.85 | 2.82 |
| 0800 | 1-1/4" diameter | | 38 | .211 | | .22 | 9.05 | | 9.27 | 14.15 |
| 0850 | For each additional inch of depth, add | | 180 | .044 | | .06 | 1.91 | | 1.97 | 3 |
| 0900 | 1-1/2" diameter | | 35 | .229 | | .34 | 9.85 | | 10.19 | 15.45 |
| 0950 | For each additional inch of depth, add | ↓ | 165 | .048 | ↓ | .08 | 2.09 | | 2.17 | 3.29 |
| 1000 | For ceiling installations, add | | | | | | 40% | | | |

# Estimating Tips

## 04 05 00 Common Work Results for Masonry

- The terms *mortar* and *grout* are often used interchangeably, and incorrectly. Mortar is used to bed masonry units, seal the entry of air and moisture, provide architectural appearance, and allow for size variations in the units. Grout is used primarily in reinforced masonry construction and is used to bond the masonry to the reinforcing steel. Common mortar types are M(2500 psi), S(1800 psi), N(750 psi), and O(350 psi), and conform to ASTM C270. Grout is either fine or coarse and conforms to ASTM C476, and in-place strengths generally exceed 2500 psi. Mortar and grout are different components of masonry construction and are placed by entirely different methods. An estimator should be aware of their unique uses and costs.

- Mortar is included in all assembled masonry line items. The mortar cost, part of the assembled masonry material cost, includes all ingredients, all labor, and all equipment required. Please see reference number R040513-10.

- Waste, specifically the loss/droppings of mortar and the breakage of brick and block, is included in all masonry assemblies in this division. A factor of 25% is added for mortar and 3% for brick and concrete masonry units.

- Scaffolding or staging is not included in any of the Division 4 costs. Refer to Subdivision 01 54 23 for scaffolding and staging costs.

## 04 20 00 Unit Masonry

- The most common types of unit masonry are brick and concrete masonry. The major classifications of brick are building brick (ASTM C62), facing brick (ASTM C216), glazed brick, fire brick, and pavers. Many varieties of texture and appearance can exist within these classifications, and the estimator would be wise to check local custom and availability within the project area. For repair and remodeling jobs, matching the existing brick may be the most important criteria.

- Brick and concrete block are priced by the piece and then converted into a price per square foot of wall. Openings less than two square feet are generally ignored by the estimator because any savings in units used is offset by the cutting and trimming required.

- It is often difficult and expensive to find and purchase small lots of historic brick. Costs can vary widely. Many design issues affect costs, selection of mortar mix, and repairs or replacement of masonry materials. Cleaning techniques must be reflected in the estimate.

- All masonry walls, whether interior or exterior, require bracing. The cost of bracing walls during construction should be included by the estimator, and this bracing must remain in place until permanent bracing is complete. Permanent bracing of masonry walls is accomplished by masonry itself, in the form of pilasters or abutting wall corners, or by anchoring the walls to the structural frame. Accessories in the form of anchors, anchor slots, and ties are used, but their supply and installation can be by different trades. For instance, anchor slots on spandrel beams and columns are supplied and welded in place by the steel fabricator, but the ties from the slots into the masonry are installed by the bricklayer. Regardless of the installation method, the estimator must be certain that these accessories are accounted for in pricing.

## Reference Numbers

Reference numbers are shown in shaded boxes at the beginning of some major classifications. These numbers refer to related items in the Reference Section. The reference information may be an estimating procedure, an alternate pricing method, or technical information.

*Note:* Not all subdivisions listed here necessarily appear in this publication.

## 04 01 20 – Maintenance of Unit Masonry

| 04 01 20.20 Pointing Masonry | | Crew | Daily Output | Labor-Hours | Unit | Material | 2011 Bare Costs Labor | Equipment | Total | Total Incl O&P |
|---|---|---|---|---|---|---|---|---|---|---|
| 0010 | **POINTING MASONRY** | | | | | | | | | |
| 0300 | Cut and repoint brick, hard mortar, running bond | 1 Bric | 80 | .100 | S.F. | .53 | 4.32 | | 4.85 | 7.10 |
| 0320 | Common bond | | 77 | .104 | | .53 | 4.48 | | 5.01 | 7.35 |
| 0360 | Flemish bond | | 70 | .114 | | .56 | 4.93 | | 5.49 | 8 |
| 0400 | English bond | | 65 | .123 | | .56 | 5.30 | | 5.86 | 8.55 |
| 0600 | Soft old mortar, running bond | | 100 | .080 | | .53 | 3.45 | | 3.98 | 5.80 |
| 0620 | Common bond | | 96 | .083 | | .53 | 3.60 | | 4.13 | 6 |
| 0640 | Flemish bond | | 90 | .089 | | .56 | 3.84 | | 4.40 | 6.35 |
| 0680 | English bond | | 82 | .098 | | .56 | 4.21 | | 4.77 | 6.90 |
| 0700 | Stonework, hard mortar | | 140 | .057 | L.F. | .70 | 2.47 | | 3.17 | 4.47 |
| 0720 | Soft old mortar | | 160 | .050 | " | .70 | 2.16 | | 2.86 | 4.01 |
| 1000 | Repoint, mask and grout method, running bond | | 95 | .084 | S.F. | .70 | 3.63 | | 4.33 | 6.20 |
| 1020 | Common bond | | 90 | .089 | | .70 | 3.84 | | 4.54 | 6.50 |
| 1040 | Flemish bond | | 86 | .093 | | .74 | 4.01 | | 4.75 | 6.80 |
| 1060 | English bond | | 77 | .104 | | .74 | 4.48 | | 5.22 | 7.55 |
| 2000 | Scrub coat, sand grout on walls, minimum | | 120 | .067 | | 2.77 | 2.88 | | 5.65 | 7.35 |
| 2020 | Maximum | | 98 | .082 | | 3.84 | 3.52 | | 7.36 | 9.55 |

## 04 01 20.30 Pointing CMU

| | | Crew | Daily Output | Labor-Hours | Unit | Material | Labor | Equipment | Total | Total Incl O&P |
|---|---|---|---|---|---|---|---|---|---|---|
| 0010 | **POINTING CMU** | | | | | | | | | |
| 0300 | Cut and repoint block, hard mortar, running bond | 1 Bric | 190 | .042 | S.F. | .22 | 1.82 | | 2.04 | 2.97 |
| 0310 | Stacked bond | | 200 | .040 | | .22 | 1.73 | | 1.95 | 2.83 |
| 0600 | Soft old mortar, running bond | | 230 | .035 | | .22 | 1.50 | | 1.72 | 2.49 |
| 0610 | Stacked bond | | 245 | .033 | | .22 | 1.41 | | 1.63 | 2.35 |

## 04 01 20.40 Sawing Masonry

| | | Crew | Daily Output | Labor-Hours | Unit | Material | Labor | Equipment | Total | Total Incl O&P |
|---|---|---|---|---|---|---|---|---|---|---|
| 0010 | **SAWING MASONRY** | | | | | | | | | |
| 0050 | Brick or block by hand, per inch depth | A-1 | 125 | .064 | L.F. | | 2.20 | .51 | 2.71 | 3.93 |

## 04 01 30 – Unit Masonry Cleaning

### 04 01 30.20 Cleaning Masonry

| | | Crew | Daily Output | Labor-Hours | Unit | Material | Labor | Equipment | Total | Total Incl O&P |
|---|---|---|---|---|---|---|---|---|---|---|
| 0010 | **CLEANING MASONRY** | | | | | | | | | |
| 0200 | Chemical cleaning, new construction, brush and wash, minimum | D-1 | 1000 | .016 | S.F. | .05 | .62 | | .67 | 1 |
| 0220 | Average | | 800 | .020 | | .08 | .78 | | .86 | 1.26 |
| 0240 | Maximum | | 600 | .027 | | .10 | 1.04 | | 1.14 | 1.67 |
| 0260 | Light restoration, minimum | | 800 | .020 | | .08 | .78 | | .86 | 1.26 |
| 0270 | Average | | 400 | .040 | | .12 | 1.56 | | 1.68 | 2.47 |
| 0280 | Maximum | | 330 | .048 | | .16 | 1.89 | | 2.05 | 3.02 |
| 0300 | Heavy restoration, minimum | | 600 | .027 | | .10 | 1.04 | | 1.14 | 1.67 |
| 0310 | Average | | 400 | .040 | | .15 | 1.56 | | 1.71 | 2.50 |
| 0320 | Maximum | | 250 | .064 | | .19 | 2.49 | | 2.68 | 3.95 |
| 0400 | High pressure water only, minimum | B-9 | 2000 | .020 | | | .70 | .10 | .80 | 1.18 |
| 0420 | Average | | 1500 | .027 | | | .93 | .13 | 1.06 | 1.56 |
| 0440 | Maximum | | 1000 | .040 | | | 1.39 | .20 | 1.59 | 2.34 |
| 2000 | Steam cleaning, minimum | | 3000 | .013 | | | .46 | .07 | .53 | .78 |
| 2020 | Average | | 2500 | .016 | | | .56 | .08 | .64 | .94 |
| 2040 | Maximum | | 1500 | .027 | | | .93 | .13 | 1.06 | 1.56 |
| 4000 | Add for masking doors and windows | 1 Clab | 800 | .010 | | .05 | .34 | | .39 | .59 |
| 4200 | Add for pedestrian protection | | | | Job | | | | | 10% |

### 04 01 30.40 Masonry Building Cleaning

| | | Crew | Daily Output | Labor-Hours | Unit | Material | Labor | Equipment | Total | Total Incl O&P |
|---|---|---|---|---|---|---|---|---|---|---|
| 0010 | **MASONRY BUILDING CLEANING** | | | | | | | | | |
| 0020 | Steam clean building, not incl scaffolding, minimum | B-9 | 3000 | .013 | S.F. | | .46 | .07 | .53 | .78 |
| 0100 | Maximum | | 1500 | .027 | | | .93 | .13 | 1.06 | 1.56 |
| 0300 | Common face brick | | 1750 | .023 | | | .79 | .11 | .90 | 1.34 |

# 04 01 Maintenance of Masonry

## 04 01 30 – Unit Masonry Cleaning

| 04 01 30.40 Masonry Building Cleaning | Crew | Daily Output | Labor-Hours | Unit | Material | 2011 Bare Costs Labor | Equipment | Total | Total Incl O&P |
|---|---|---|---|---|---|---|---|---|---|
| 0400 | Wire cut face brick | B-9 | 1250 | .032 | S.F. | | 1.11 | .16 | 1.27 | 1.88 |

### 04 01 30.60 Brick Washing

| | | Crew | Daily Output | Labor-Hours | Unit | Material | Labor | Equipment | Total | Total Incl O&P |
|---|---|---|---|---|---|---|---|---|---|---|
| 0010 | **BRICK WASHING** R040130-10 | | | | | | | | | |
| 0012 | Acid cleanser, smooth brick surface | 1 Bric | 560 | .014 | S.F. | .05 | .62 | | .67 | .98 |
| 0050 | Rough brick | | 400 | .020 | | .06 | .86 | | .92 | 1.37 |
| 0060 | Stone, acid wash | | 600 | .013 | | .08 | .58 | | .66 | .94 |
| 1000 | Muriatic acid, price per gallon in 5 gallon lots | | | | Gal. | 9.55 | | | 9.55 | 10.50 |

# 04 05 Common Work Results for Masonry

## 04 05 05 – Selective Masonry Demolition

### 04 05 05.10 Selective Demolition

| | | Crew | Daily Output | Labor-Hours | Unit | Material | Labor | Equipment | Total | Total Incl O&P |
|---|---|---|---|---|---|---|---|---|---|---|
| 0010 | **SELECTIVE DEMOLITION** R024119-10 | | | | | | | | | |
| 0200 | Bond beams, 8" block with #4 bar | 2 Clab | 32 | .500 | L.F. | | 17.20 | | 17.20 | 26.50 |
| 0300 | Concrete block walls, unreinforced, 2" thick | | 1200 | .013 | S.F. | | .46 | | .46 | .70 |
| 0310 | 4" thick | | 1150 | .014 | | | .48 | | .48 | .73 |
| 0320 | 6" thick | | 1100 | .015 | | | .50 | | .50 | .77 |
| 0330 | 8" thick | | 1050 | .015 | | | .52 | | .52 | .80 |
| 0340 | 10" thick | | 1000 | .016 | | | .55 | | .55 | .84 |
| 0360 | 12" thick | | 950 | .017 | | | .58 | | .58 | .89 |
| 0380 | Reinforced alternate courses, 2" thick | | 1130 | .014 | | | .49 | | .49 | .75 |
| 0390 | 4" thick | | 1080 | .015 | | | .51 | | .51 | .78 |
| 0400 | 6" thick | | 1035 | .015 | | | .53 | | .53 | .81 |
| 0410 | 8" thick | | 990 | .016 | | | .56 | | .56 | .85 |
| 0420 | 10" thick | | 940 | .017 | | | .58 | | .58 | .90 |
| 0430 | 12" thick | | 890 | .018 | | | .62 | | .62 | .95 |
| 0440 | Reinforced alternate courses & vertically 48" OC, 4" thick | | 900 | .018 | | | .61 | | .61 | .94 |
| 0450 | 6" thick | | 850 | .019 | | | .65 | | .65 | .99 |
| 0460 | 8" thick | | 800 | .020 | | | .69 | | .69 | 1.05 |
| 0480 | 10" thick | | 750 | .021 | | | .73 | | .73 | 1.12 |
| 0490 | 12" thick | | 700 | .023 | | | .79 | | .79 | 1.20 |
| 1000 | Chimney, 16" x 16", soft old mortar | 1 Clab | 55 | .145 | C.F. | | 5 | | 5 | 7.65 |
| 1020 | Hard mortar | | 40 | .200 | | | 6.85 | | 6.85 | 10.55 |
| 1030 | 16" x 20", soft old mortar | | 55 | .145 | | | 5 | | 5 | 7.65 |
| 1040 | Hard mortar | | 40 | .200 | | | 6.85 | | 6.85 | 10.55 |
| 1050 | 16" x 24", soft old mortar | | 55 | .145 | | | 5 | | 5 | 7.65 |
| 1060 | Hard mortar | | 40 | .200 | | | 6.85 | | 6.85 | 10.55 |
| 1080 | 20" x 20", soft old mortar | | 55 | .145 | | | 5 | | 5 | 7.65 |
| 1100 | Hard mortar | | 40 | .200 | | | 6.85 | | 6.85 | 10.55 |
| 1110 | 20" x 24", soft old mortar | | 55 | .145 | | | 5 | | 5 | 7.65 |
| 1120 | Hard mortar | | 40 | .200 | | | 6.85 | | 6.85 | 10.55 |
| 1140 | 20" x 32", soft old mortar | | 55 | .145 | | | 5 | | 5 | 7.65 |
| 1160 | Hard mortar | | 40 | .200 | | | 6.85 | | 6.85 | 10.55 |
| 1200 | 48" x 48", soft old mortar | | 55 | .145 | | | 5 | | 5 | 7.65 |
| 1220 | Hard mortar | | 40 | .200 | | | 6.85 | | 6.85 | 10.55 |
| 1250 | Metal, high temp steel jacket, 24" diameter | E-2 | 130 | .431 | V.L.F. | | 20.50 | 12.50 | 33 | 48 |
| 1260 | 60" diameter | " | 60 | .933 | | | 44 | 27 | 71 | 105 |
| 1280 | Flue lining, up to 12" x 12" | 1 Clab | 200 | .040 | | | 1.37 | | 1.37 | 2.11 |
| 1282 | Up to 24" x 24" | | 150 | .053 | | | 1.83 | | 1.83 | 2.81 |
| 2000 | Columns, 8" x 8", soft old mortar | | 48 | .167 | | | 5.75 | | 5.75 | 8.80 |
| 2020 | Hard mortar | | 40 | .200 | | | 6.85 | | 6.85 | 10.55 |
| 2060 | 16" x 16", soft old mortar | | 16 | .500 | | | 17.20 | | 17.20 | 26.50 |

## 04 05 05 – Selective Masonry Demolition

| 04 05 05.10 Selective Demolition | Crew | Daily Output | Labor-Hours | Unit | Material | 2011 Bare Costs Labor | Equipment | Total | Total Incl O&P |
|---|---|---|---|---|---|---|---|---|---|
| 2100 | Hard mortar | 1 Clab | 14 | .571 | V.L.F. | | 19.65 | | 19.65 | 30 |
| 2140 | 24" x 24", soft old mortar | | 8 | 1 | | | 34.50 | | 34.50 | 52.50 |
| 2160 | Hard mortar | | 6 | 1.333 | | | 46 | | 46 | 70.50 |
| 2200 | 36" x 36", soft old mortar | | 4 | 2 | | | 68.50 | | 68.50 | 105 |
| 2220 | Hard mortar | | 3 | 2.667 | | | 91.50 | | 91.50 | 141 |
| 2230 | Alternate pricing method, soft old mortar | | 30 | .267 | C.F. | | 9.15 | | 9.15 | 14.05 |
| 2240 | Hard mortar | | 23 | .348 | " | | 11.95 | | 11.95 | 18.35 |
| 3000 | Copings, precast or masonry, to 8" wide | | | | | | | | | |
| 3020 | Soft old mortar | 1 Clab | 180 | .044 | L.F. | | 1.53 | | 1.53 | 2.34 |
| 3040 | Hard mortar | " | 160 | .050 | " | | 1.72 | | 1.72 | 2.64 |
| 3100 | To 12" wide | | | | | | | | | |
| 3120 | Soft old mortar | 1 Clab | 160 | .050 | L.F. | | 1.72 | | 1.72 | 2.64 |
| 3140 | Hard mortar | " | 140 | .057 | " | | 1.96 | | 1.96 | 3.01 |
| 4000 | Fireplace, brick, 30" x 24" opening | | | | | | | | | |
| 4020 | Soft old mortar | 1 Clab | 2 | 4 | Ea. | | 137 | | 137 | 211 |
| 4040 | Hard mortar | | 1.25 | 6.400 | | | 220 | | 220 | 335 |
| 4100 | Stone, soft old mortar | | 1.50 | 5.333 | | | 183 | | 183 | 281 |
| 4120 | Hard mortar | | 1 | 8 | | | 275 | | 275 | 420 |
| 5000 | Veneers, brick, soft old mortar | | 140 | .057 | S.F. | | 1.96 | | 1.96 | 3.01 |
| 5020 | Hard mortar | | 125 | .064 | | | 2.20 | | 2.20 | 3.37 |
| 5100 | Granite and marble, 2" thick | | 180 | .044 | | | 1.53 | | 1.53 | 2.34 |
| 5120 | 4" thick | | 170 | .047 | | | 1.62 | | 1.62 | 2.48 |
| 5140 | Stone, 4" thick | | 180 | .044 | | | 1.53 | | 1.53 | 2.34 |
| 5160 | 8" thick | | 175 | .046 | | | 1.57 | | 1.57 | 2.41 |
| 5400 | Alternate pricing method, stone, 4" thick | | 60 | .133 | C.F. | | 4.58 | | 4.58 | 7.05 |
| 5420 | 8" thick | | 85 | .094 | " | | 3.23 | | 3.23 | 4.96 |

## 04 05 13 – Masonry Mortaring

### 04 05 13.10 Cement

| | | Crew | Daily Output | Labor-Hours | Unit | Material | 2011 Bare Costs Labor | Equipment | Total | Total Incl O&P |
|---|---|---|---|---|---|---|---|---|---|---|
| 0010 | **CEMENT** | | | | | | | | | |
| 0100 | Masonry, 70 lb. bag, T.L. lots | | | | Bag | 9.45 | | | 9.45 | 10.35 |
| 0150 | L.T.L. lots | | | | | 10 | | | 10 | 11 |
| 0200 | White, 70 lb. bag, T.L. lots | | | | | 14.85 | | | 14.85 | 16.35 |
| 0250 | L.T.L. lots | | | | | 16.70 | | | 16.70 | 18.35 |

### 04 05 13.20 Lime

| | | Crew | Daily Output | Labor-Hours | Unit | Material | 2011 Bare Costs Labor | Equipment | Total | Total Incl O&P |
|---|---|---|---|---|---|---|---|---|---|---|
| 0010 | **LIME** | | | | | | | | | |
| 0020 | Masons, hydrated, 50 lb. bag, T.L. lots | | | | Bag | 9.10 | | | 9.10 | 10 |
| 0050 | L.T.L. lots | | | | | 9.30 | | | 9.30 | 10.25 |
| 0200 | Finish, double hydrated, 50 lb. bag, T.L. lots | | | | | 9.90 | | | 9.90 | 10.90 |
| 0250 | L.T.L. lots | | | | | 11 | | | 11 | 12.10 |

### 04 05 13.23 Surface Bonding Masonry Mortaring

| | | Crew | Daily Output | Labor-Hours | Unit | Material | 2011 Bare Costs Labor | Equipment | Total | Total Incl O&P |
|---|---|---|---|---|---|---|---|---|---|---|
| 0010 | **SURFACE BONDING MASONRY MORTARING** | | | | | | | | | |
| 0020 | Gray or white colors, not incl. block work | 1 Bric | 540 | .015 | S.F. | .10 | .64 | | .74 | 1.07 |

### 04 05 13.30 Mortar

| | | Crew | Daily Output | Labor-Hours | Unit | Material | 2011 Bare Costs Labor | Equipment | Total | Total Incl O&P |
|---|---|---|---|---|---|---|---|---|---|---|
| 0010 | **MORTAR** R040513-10 | | | | | | | | | |
| 0020 | With masonry cement | | | | | | | | | |
| 0100 | Type M, 1:1:6 mix | 1 Brhe | 143 | .056 | C.F. | 4.88 | 1.95 | | 6.83 | 8.25 |
| 0200 | Type N, 1:3 mix | " | 143 | .056 | " | 4.36 | 1.95 | | 6.31 | 7.70 |
| 2000 | With portland cement and lime | | | | | | | | | |
| 2100 | Type M, 1:1/4:3 mix | 1 Brhe | 143 | .056 | C.F. | 8.80 | 1.95 | | 10.75 | 12.60 |
| 2200 | Type N, 1:1:6 mix, 750 psi | | 143 | .056 | | 6.80 | 1.95 | | 8.75 | 10.35 |
| 2300 | Type O, 1:2:9 mix (Pointing Mortar) | | 143 | .056 | | 7.50 | 1.95 | | 9.45 | 11.15 |

## 04 05 13 – Masonry Mortaring

### 04 05 13.30 Mortar

| | | Crew | Daily Output | Labor-Hours | Unit | Material | 2011 Bare Costs Labor | Equipment | Total | Total Incl O&P |
|---|---|---|---|---|---|---|---|---|---|---|
| 2650 | Pre-mixed, type S or N | | | | C.F. | 4.94 | | | 4.94 | 5.45 |
| 2700 | Mortar for glass block | 1 Brhe | 143 | .056 | ↓ | 10.05 | 1.95 | | 12 | 13.95 |
| 2900 | Mortar for Fire Brick, 80 lb. bag, T.L. Lots | | | | Bag | 25.50 | | | 25.50 | 28 |

### 04 05 13.91 Masonry Restoration Mortaring

| | | | | | | | | | | |
|---|---|---|---|---|---|---|---|---|---|---|
| 0010 | **MASONRY RESTORATION MORTARING** | | | | | | | | | |
| 0020 | Masonry restoration mix | | | | Lb. | 1.97 | | | 1.97 | 2.17 |
| 0050 | White | | | | " | 2.27 | | | 2.27 | 2.50 |

### 04 05 13.93 Mortar Pigments

| | | | | | | | | | | |
|---|---|---|---|---|---|---|---|---|---|---|
| 0010 | **MORTAR PIGMENTS**, 50 lb. bags (2 bags per M bricks)    R040513-10 | | | | | | | | | |
| 0020 | Color admixture, range 2 to 10 lb. per bag of cement, minimum | | | | Lb. | 5.25 | | | 5.25 | 5.80 |
| 0050 | Average | | | | | 7.80 | | | 7.80 | 8.60 |
| 0100 | Maximum | | | | ↓ | 11.10 | | | 11.10 | 12.20 |

### 04 05 13.95 Sand

| | | | | | | | | | | |
|---|---|---|---|---|---|---|---|---|---|---|
| 0010 | **SAND**, screened and washed at pit | | | | | | | | | |
| 0020 | For mortar, per ton | | | | Ton | 16.60 | | | 16.60 | 18.25 |
| 0050 | With 10 mile haul | | | | | 27 | | | 27 | 29.50 |
| 0100 | With 30 mile haul | | | | ↓ | 36 | | | 36 | 39.50 |
| 0200 | Screened and washed, at the pit | | | | C.Y. | 23 | | | 23 | 25.50 |
| 0250 | With 10 mile haul | | | | | 37.50 | | | 37.50 | 41 |
| 0300 | With 30 mile haul | | | | ↓ | 50 | | | 50 | 55 |

### 04 05 13.98 Mortar Admixtures

| | | | | | | | | | | |
|---|---|---|---|---|---|---|---|---|---|---|
| 0010 | **MORTAR ADMIXTURES** | | | | | | | | | |
| 0020 | Waterproofing admixture, per quart (1 qt. to 2 bags of masonry cement) | | | | Qt. | 9.50 | | | 9.50 | 10.45 |

## 04 05 16 – Masonry Grouting

### 04 05 16.30 Grouting

| | | Crew | Daily Output | Labor-Hours | Unit | Material | 2011 Bare Costs Labor | Equipment | Total | Total Incl O&P |
|---|---|---|---|---|---|---|---|---|---|---|
| 0010 | **GROUTING**    R040513-10 | | | | | | | | | |
| 0011 | Bond beams & lintels, 8" deep, 6" thick, 0.15 C.F. per L.F. | D-4 | 1480 | .022 | L.F. | .64 | .85 | .09 | 1.58 | 2.06 |
| 0020 | 8" thick, 0.2 C.F. per L.F. | | 1400 | .023 | | .99 | .89 | .09 | 1.97 | 2.52 |
| 0050 | 10" thick, 0.25 C.F. per L.F. | | 1200 | .027 | | 1.06 | 1.04 | .10 | 2.20 | 2.84 |
| 0060 | 12" thick, 0.3 C.F. per L.F. | ↓ | 1040 | .031 | ↓ | 1.27 | 1.20 | .12 | 2.59 | 3.33 |
| 0200 | Concrete block cores, solid, 4" thk., by hand, 0.067 C.F./S.F. of wall | D-8 | 1100 | .036 | S.F. | .28 | 1.45 | | 1.73 | 2.48 |
| 0210 | 6" thick, pumped, 0.175 C.F. per S.F. | D-4 | 720 | .044 | | .74 | 1.74 | .17 | 2.65 | 3.60 |
| 0250 | 8" thick, pumped, 0.258 C.F. per S.F. | | 680 | .047 | | 1.09 | 1.84 | .18 | 3.11 | 4.16 |
| 0300 | 10" thick, pumped, 0.340 C.F. per S.F. | | 660 | .048 | | 1.44 | 1.90 | .19 | 3.53 | 4.63 |
| 0350 | 12" thick, pumped, 0.422 C.F. per S.F. | | 640 | .050 | | 1.79 | 1.95 | .20 | 3.94 | 5.10 |
| 0500 | Cavity walls, 2" space, pumped, 0.167 C.F./S.F. of wall | | 1700 | .019 | | .71 | .74 | .07 | 1.52 | 1.96 |
| 0550 | 3" space, 0.250 C.F./S.F. | | 1200 | .027 | | 1.06 | 1.04 | .10 | 2.20 | 2.84 |
| 0600 | 4" space, 0.333 C.F. per S.F. | | 1150 | .028 | | 1.41 | 1.09 | .11 | 2.61 | 3.30 |
| 0700 | 6" space, 0.500 C.F. per S.F. | | 800 | .040 | ↓ | 2.12 | 1.56 | .16 | 3.84 | 4.84 |
| 0800 | Door frames, 3' x 7' opening, 2.5 C.F. per opening | | 60 | .533 | Opng. | 10.60 | 21 | 2.10 | 33.70 | 45 |
| 0850 | 6' x 7' opening, 3.5 C.F. per opening | | 45 | .711 | " | 14.80 | 28 | 2.79 | 45.59 | 61 |
| 2000 | Grout, C476, for bond beams, lintels and CMU cores | ↓ | 350 | .091 | C.F. | 4.23 | 3.57 | .36 | 8.16 | 10.40 |

## 04 05 19 – Masonry Anchorage and Reinforcing

### 04 05 19.05 Anchor Bolts

| | | Crew | Daily Output | Labor-Hours | Unit | Material | 2011 Bare Costs Labor | Equipment | Total | Total Incl O&P |
|---|---|---|---|---|---|---|---|---|---|---|
| 0010 | **ANCHOR BOLTS** | | | | | | | | | |
| 0015 | Installed in fresh grout in CMU bond beams or filled cores, no templates | | | | | | | | | |
| 0020 | Hooked, with nut and washer, 1/2" diam., 8" long | 1 Bric | 132 | .061 | Ea. | 1.36 | 2.62 | | 3.98 | 5.40 |
| 0030 | 12" long | | 131 | .061 | | 1.51 | 2.64 | | 4.15 | 5.60 |
| 0040 | 5/8" diameter, 8" long | | 129 | .062 | | 2.96 | 2.68 | | 5.64 | 7.30 |
| 0050 | 12" long | ↓ | 127 | .063 | ↓ | 3.65 | 2.72 | | 6.37 | 8.10 |

## 04 05 19 – Masonry Anchorage and Reinforcing

### 04 05 19.05 Anchor Bolts

| | | Crew | Daily Output | Labor-Hours | Unit | Material | 2011 Bare Costs Labor | 2011 Bare Costs Equipment | Total | Total Incl O&P |
|---|---|---|---|---|---|---|---|---|---|---|
| 0060 | 3/4" diameter, 8" long | 1 Bric | 127 | .063 | Ea. | 3.65 | 2.72 | | 6.37 | 8.10 |
| 0070 | 12" long | ↓ | 125 | .064 | ↓ | 4.56 | 2.76 | | 7.32 | 9.15 |

### 04 05 19.16 Masonry Anchors

| | | Crew | Daily Output | Labor-Hours | Unit | Material | Labor | Equipment | Total | Total Incl O&P |
|---|---|---|---|---|---|---|---|---|---|---|
| 0010 | **MASONRY ANCHORS** | | | | | | | | | |
| 0020 | For brick veneer, galv., corrugated, 7/8" x 7", 22 Ga. | 1 Bric | 10.50 | .762 | C | 9.45 | 33 | | 42.45 | 60 |
| 0100 | 24 Ga. | | 10.50 | .762 | | 8.90 | 33 | | 41.90 | 59.50 |
| 0150 | 16 Ga. | | 10.50 | .762 | | 25 | 33 | | 58 | 77 |
| 0200 | Buck anchors, galv., corrugated, 16 gauge, 2" bend, 8" x 2" | | 10.50 | .762 | | 49 | 33 | | 82 | 104 |
| 0250 | 8" x 3" | | 10.50 | .762 | | 51 | 33 | | 84 | 106 |
| 0660 | Cavity wall, Z-type, galvanized, 6" long, 1/8" diam. | | 10.50 | .762 | | 23 | 33 | | 56 | 74.50 |
| 0670 | 3/16" diameter | | 10.50 | .762 | | 21 | 33 | | 54 | 72.50 |
| 0680 | 1/4" diameter | | 10.50 | .762 | | 39.50 | 33 | | 72.50 | 93 |
| 0850 | 8" long, 3/16" diameter | | 10.50 | .762 | | 23.50 | 33 | | 56.50 | 75.50 |
| 0855 | 1/4" diameter | | 10.50 | .762 | | 46.50 | 33 | | 79.50 | 101 |
| 1000 | Rectangular type, galvanized, 1/4" diameter, 2" x 6" | | 10.50 | .762 | | 67 | 33 | | 100 | 123 |
| 1050 | 4" x 6" | | 10.50 | .762 | | 83.50 | 33 | | 116.50 | 142 |
| 1100 | 3/16" diameter, 2" x 6" | | 10.50 | .762 | | 38 | 33 | | 71 | 91 |
| 1150 | 4" x 6" | | 10.50 | .762 | | 39 | 33 | | 72 | 92.50 |
| 1500 | Rigid partition anchors, plain, 8" long, 1" x 1/8" | | 10.50 | .762 | | 109 | 33 | | 142 | 170 |
| 1550 | 1" x 1/4" | | 10.50 | .762 | | 180 | 33 | | 213 | 248 |
| 1580 | 1-1/2" x 1/8" | | 10.50 | .762 | | 145 | 33 | | 178 | 210 |
| 1600 | 1-1/2" x 1/4" | | 10.50 | .762 | | 295 | 33 | | 328 | 375 |
| 1650 | 2" x 1/8" | | 10.50 | .762 | | 188 | 33 | | 221 | 257 |
| 1700 | 2" x 1/4" | ↓ | 10.50 | .762 | ↓ | 345 | 33 | | 378 | 430 |

### 04 05 19.26 Masonry Reinforcing Bars

| | | Crew | Daily Output | Labor-Hours | Unit | Material | Labor | Equipment | Total | Total Incl O&P |
|---|---|---|---|---|---|---|---|---|---|---|
| 0010 | **MASONRY REINFORCING BARS** R040519-50 | | | | | | | | | |
| 0015 | Steel bars A615, placed horiz., #3 & #4 bars | 1 Bric | 450 | .018 | Lb. | .45 | .77 | | 1.22 | 1.65 |
| 0020 | #5 & #6 bars | | 800 | .010 | | .45 | .43 | | .88 | 1.15 |
| 0050 | Placed vertical, #3 & #4 bars | | 350 | .023 | | .45 | .99 | | 1.44 | 1.98 |
| 0060 | #5 & #6 bars | | 650 | .012 | | .45 | .53 | | .98 | 1.30 |
| 0200 | Joint reinforcing, regular truss, to 6" wide, mill std galvanized | | 30 | .267 | C.L.F. | 15.45 | 11.50 | | 26.95 | 34.50 |
| 0250 | 12" wide | | 20 | .400 | | 17 | 17.25 | | 34.25 | 44.50 |
| 0400 | Cavity truss with drip section, to 6" wide | | 30 | .267 | | 14.80 | 11.50 | | 26.30 | 33.50 |
| 0450 | 12" wide | ↓ | 20 | .400 | ↓ | 15.20 | 17.25 | | 32.45 | 42.50 |

## 04 05 23 – Masonry Accessories

### 04 05 23.13 Masonry Control and Expansion Joints

| | | Crew | Daily Output | Labor-Hours | Unit | Material | Labor | Equipment | Total | Total Incl O&P |
|---|---|---|---|---|---|---|---|---|---|---|
| 0010 | **MASONRY CONTROL AND EXPANSION JOINTS** | | | | | | | | | |
| 0020 | Rubber, for double wythe 8" minimum wall (Brick/CMU) | 1 Bric | 400 | .020 | L.F. | 1.72 | .86 | | 2.58 | 3.19 |
| 0025 | "T" shaped | | 320 | .025 | | 1.05 | 1.08 | | 2.13 | 2.78 |
| 0030 | Cross-shaped for CMU units | | 280 | .029 | | 1.38 | 1.23 | | 2.61 | 3.37 |
| 0050 | PVC, for double wythe 8" minimum wall (Brick/CMU) | | 400 | .020 | | 1.26 | .86 | | 2.12 | 2.69 |
| 0120 | "T" shaped | | 320 | .025 | | .64 | 1.08 | | 1.72 | 2.32 |
| 0160 | Cross-shaped for CMU units | | 280 | .029 | | .88 | 1.23 | | 2.11 | 2.82 |

### 04 05 23.19 Masonry Cavity Drainage, Weepholes, and Vents

| | | Crew | Daily Output | Labor-Hours | Unit | Material | Labor | Equipment | Total | Total Incl O&P |
|---|---|---|---|---|---|---|---|---|---|---|
| 0010 | **MASONRY CAVITY DRAINAGE, WEEPHOLES, AND VENTS** | | | | | | | | | |
| 0020 | Extruded aluminum, 4" deep, 2-3/8" x 8-1/8" | 1 Bric | 30 | .267 | Ea. | 30.50 | 11.50 | | 42 | 51 |
| 0050 | 5" x 8-1/8" | | 25 | .320 | | 40 | 13.80 | | 53.80 | 64.50 |
| 0100 | 2-1/4" x 25" | | 25 | .320 | | 69.50 | 13.80 | | 83.30 | 97 |
| 0150 | 5" x 16-1/2" | | 22 | .364 | | 58.50 | 15.70 | | 74.20 | 87.50 |
| 0200 | 6" x 16-1/2" | | 22 | .364 | | 81 | 15.70 | | 96.70 | 113 |
| 0250 | 7-3/4" x 16-1/2" | ↓ | 20 | .400 | | 69 | 17.25 | | 86.25 | 102 |

# 04 05 Common Work Results for Masonry

## 04 05 23 – Masonry Accessories

| 04 05 23.19 Masonry Cavity Drainage, Weepholes, and Vents | Crew | Daily Output | Labor-Hours | Unit | Material | 2011 Bare Costs Labor | 2011 Bare Costs Equipment | Total | Total Incl O&P |
|---|---|---|---|---|---|---|---|---|---|
| 0400 | For baked enamel finish, add | | | | Ea. | 35% | | | | |
| 0500 | For cast aluminum, painted, add | | | | | 60% | | | | |
| 1000 | Stainless steel ventilators, 6" x 6" | 1 Bric | 25 | .320 | | 126 | 13.80 | | 139.80 | 159 |
| 1050 | 8" x 8" | | 24 | .333 | | 132 | 14.40 | | 146.40 | 168 |
| 1100 | 12" x 12" | | 23 | .348 | | 153 | 15 | | 168 | 191 |
| 1150 | 12" x 6" | | 24 | .333 | | 134 | 14.40 | | 148.40 | 169 |
| 1200 | Foundation block vent, galv., 1-1/4" thk, 8" high, 16" long, no damper | | 30 | .267 | | 24.50 | 11.50 | | 36 | 44.50 |
| 1250 | For damper, add | | | | | 8.20 | | | 8.20 | 9 |

## 04 05 23.95 Wall Plugs

| | | Crew | Daily Output | Labor-Hours | Unit | Material | Labor | Equipment | Total | Total Incl O&P |
|---|---|---|---|---|---|---|---|---|---|---|
| 0010 | **WALL PLUGS** (for nailing to brickwork) | | | | | | | | | |
| 0020 | 26 ga., galvanized, plain | 1 Bric | 10.50 | .762 | C | 31 | 33 | | 64 | 83.50 |
| 0050 | Wood filled | " | 10.50 | .762 | " | 104 | 33 | | 137 | 164 |

# 04 21 Clay Unit Masonry

## 04 21 13 – Brick Masonry

### 04 21 13.13 Brick Veneer Masonry

| | | Crew | Daily Output | Labor-Hours | Unit | Material | Labor | Equipment | Total | Total Incl O&P |
|---|---|---|---|---|---|---|---|---|---|---|
| 0010 | **BRICK VENEER MASONRY**, T.L. lots, excl. scaff., grout & reinforcing  R042110-20 | | | | | | | | | |
| 0015 | Material costs incl. 3% brick and 25% mortar waste | | | | | | | | | |
| 0020 | Standard, select common, 4" x 2-2/3" x 8" (6.75/S.F.) | D-8 | 1.50 | 26.667 | M | 535 | 1,050 | | 1,585 | 2,200 |
| 0050 | Red, 4" x 2-2/3" x 8", running bond | | 1.50 | 26.667 | | 620 | 1,050 | | 1,670 | 2,275 |
| 0100 | Full header every 6th course (7.88/S.F.)  R042110-50 | | 1.45 | 27.586 | | 620 | 1,100 | | 1,720 | 2,325 |
| 0150 | English, full header every 2nd course (10.13/S.F.) | | 1.40 | 28.571 | | 615 | 1,125 | | 1,740 | 2,375 |
| 0200 | Flemish, alternate header every course (9.00/S.F.) | | 1.40 | 28.571 | | 620 | 1,125 | | 1,745 | 2,375 |
| 0250 | Flemish, alt. header every 6th course (7.13/S.F.) | | 1.45 | 27.586 | | 620 | 1,100 | | 1,720 | 2,325 |
| 0300 | Full headers throughout (13.50/S.F.) | | 1.40 | 28.571 | | 615 | 1,125 | | 1,740 | 2,375 |
| 0350 | Rowlock course (13.50/S.F.) | | 1.35 | 29.630 | | 615 | 1,175 | | 1,790 | 2,450 |
| 0400 | Rowlock stretcher (4.50/S.F.) | | 1.40 | 28.571 | | 625 | 1,125 | | 1,750 | 2,375 |
| 0450 | Soldier course (6.75/S.F.) | | 1.40 | 28.571 | | 620 | 1,125 | | 1,745 | 2,375 |
| 0500 | Sailor course (4.50/S.F.) | | 1.30 | 30.769 | | 625 | 1,225 | | 1,850 | 2,525 |
| 0601 | Buff or gray face, running bond, (6.75/S.F.) | | 1.50 | 26.667 | | 620 | 1,050 | | 1,670 | 2,275 |
| 0700 | Glazed face, 4" x 2-2/3" x 8", running bond | | 1.40 | 28.571 | | 1,850 | 1,125 | | 2,975 | 3,750 |
| 0750 | Full header every 6th course (7.88/S.F.) | | 1.35 | 29.630 | | 1,775 | 1,175 | | 2,950 | 3,725 |
| 1000 | Jumbo, 6" x 4" x 12", (3.00/S.F.) | | 1.30 | 30.769 | | 1,775 | 1,225 | | 3,000 | 3,800 |
| 1051 | Norman, 4" x 2-2/3" x 12" (4.50/S.F.) | | 1.45 | 27.586 | | 1,100 | 1,100 | | 2,200 | 2,875 |
| 1100 | Norwegian, 4" x 3-1/5" x 12" (3.75/S.F.) | | 1.40 | 28.571 | | 1,150 | 1,125 | | 2,275 | 2,950 |
| 1150 | Economy, 4" x 4" x 8" (4.50 per S.F.) | | 1.40 | 28.571 | | 940 | 1,125 | | 2,065 | 2,725 |
| 1201 | Engineer, 4" x 3-1/5" x 8", (5.63/S.F.) | | 1.45 | 27.586 | | 630 | 1,100 | | 1,730 | 2,350 |
| 1251 | Roman, 4" x 2" x 12", (6.00/S.F.) | | 1.50 | 26.667 | | 985 | 1,050 | | 2,035 | 2,675 |
| 1300 | S.C.R. 6" x 2-2/3" x 12" (4.50/S.F.) | | 1.40 | 28.571 | | 1,175 | 1,125 | | 2,300 | 3,000 |
| 1350 | Utility, 4" x 4" x 12" (3.00/S.F.) | | 1.35 | 29.630 | | 1,550 | 1,175 | | 2,725 | 3,475 |
| 1360 | For less than truck load lots, add | | | | | 15 | | | 15 | 16.50 |
| 1400 | For battered walls, add | | | | | | 30% | | | |
| 1450 | For corbels, add | | | | | | 75% | | | |
| 1500 | For curved walls, add | | | | | | 30% | | | |
| 1550 | For pits and trenches, deduct | | | | | | 20% | | | |
| 1999 | Alternate method of figuring by square foot | | | | | | | | | |
| 2000 | Standard, sel. common, 4" x 2-2/3" x 8", (6.75/S.F.) | D-8 | 230 | .174 | S.F. | 3.61 | 6.90 | | 10.51 | 14.35 |
| 2020 | Standard, red, 4" x 2-2/3" x 8", running bond (6.75/S.F.) | | 220 | .182 | | 4.19 | 7.25 | | 11.44 | 15.45 |
| 2050 | Full header every 6th course (7.88/S.F.) | | 185 | .216 | | 4.88 | 8.60 | | 13.48 | 18.25 |
| 2100 | English, full header every 2nd course (10.13/S.F.) | | 140 | .286 | | 6.25 | 11.35 | | 17.60 | 24 |
| 2150 | Flemish, alternate header every course (9.00/S.F.) | | 150 | .267 | | 5.55 | 10.60 | | 16.15 | 22 |

# 04 21 Clay Unit Masonry

## 04 21 13 – Brick Masonry

### 04 21 13.13 Brick Veneer Masonry

| | | Crew | Daily Output | Labor-Hours | Unit | Material | 2011 Bare Costs Labor | Equipment | Total | Total Incl O&P |
|---|---|---|---|---|---|---|---|---|---|---|
| 2200 | Flemish, alt. header every 6th course (7.13/S.F.) | D-8 | 205 | .195 | S.F. | 4.42 | 7.75 | | 12.17 | 16.50 |
| 2250 | Full headers throughout (13.50/S.F.) | | 105 | .381 | | 8.30 | 15.15 | | 23.45 | 32 |
| 2300 | Rowlock course (13.50/S.F.) | | 100 | .400 | | 8.30 | 15.90 | | 24.20 | 33 |
| 2350 | Rowlock stretcher (4.50/S.F.) | | 310 | .129 | | 2.81 | 5.15 | | 7.96 | 10.80 |
| 2400 | Soldier course (6.75/S.F.) | | 200 | .200 | | 4.19 | 7.95 | | 12.14 | 16.55 |
| 2450 | Sailor course (4.50/S.F.) | | 290 | .138 | | 2.81 | 5.50 | | 8.31 | 11.35 |
| 2600 | Buff or gray face, running bond, (6.75/S.F.) | | 220 | .182 | | 4.43 | 7.25 | | 11.68 | 15.70 |
| 2700 | Glazed face brick, running bond | | 210 | .190 | | 11.95 | 7.60 | | 19.55 | 24.50 |
| 2750 | Full header every 6th course (7.88/S.F.) | | 170 | .235 | | 13.95 | 9.35 | | 23.30 | 29.50 |
| 3000 | Jumbo, 6" x 4" x 12" running bond (3.00/S.F.) | | 435 | .092 | | 4.87 | 3.66 | | 8.53 | 10.85 |
| 3050 | Norman, 4" x 2-2/3" x 12" running bond, (4.5/S.F.) | | 320 | .125 | | 5.70 | 4.98 | | 10.68 | 13.70 |
| 3100 | Norwegian, 4" x 3-1/5" x 12" (3.75/S.F.) | | 375 | .107 | | 4.15 | 4.25 | | 8.40 | 10.90 |
| 3150 | Economy, 4" x 4" x 8" (4.50/S.F.) | | 310 | .129 | | 4.19 | 5.15 | | 9.34 | 12.30 |
| 3200 | Engineer, 4" x 3-1/5" x 8" (5.63/S.F.) | | 260 | .154 | | 3.54 | 6.10 | | 9.64 | 13.10 |
| 3250 | Roman, 4" x 2" x 12" (6.00/S.F.) | | 250 | .160 | | 5.75 | 6.35 | | 12.10 | 15.90 |
| 3300 | SCR, 6" x 2-2/3" x 12" (4.50/S.F.) | | 310 | .129 | | 5.25 | 5.15 | | 10.40 | 13.45 |
| 3350 | Utility, 4" x 4" x 12" (3.00/S.F.) | | 450 | .089 | | 4.51 | 3.54 | | 8.05 | 10.25 |
| 3400 | For cavity wall construction, add | | | | | | 15% | | | |
| 3450 | For stacked bond, add | | | | | | 10% | | | |
| 3500 | For interior veneer construction, add | | | | | | 15% | | | |
| 3550 | For curved walls, add | | | | | | 30% | | | |

### 04 21 13.14 Thin Brick Veneer

| | | Crew | Daily Output | Labor-Hours | Unit | Material | 2011 Bare Costs Labor | Equipment | Total | Total Incl O&P |
|---|---|---|---|---|---|---|---|---|---|---|
| 0010 | **THIN BRICK VENEER** | | | | | | | | | |
| 0015 | Material costs incl. 3% brick and 25% mortar waste | | | | | | | | | |
| 0020 | On & incl. metal panel support sys, modular, 2-2/3" x 5/8" x 8", red | D-7 | 92 | .174 | S.F. | 8.75 | 6.30 | | 15.05 | 18.80 |
| 0100 | Closure, 4" x 5/8" x 8" | | 110 | .145 | | 8.75 | 5.25 | | 14 | 17.30 |
| 0110 | Norman, 2-2/3" x 5/8" x 12" | | 110 | .145 | | 8.95 | 5.25 | | 14.20 | 17.50 |
| 0120 | Utility, 4" x 5/8" x 12" | | 125 | .128 | | 8.95 | 4.63 | | 13.58 | 16.60 |
| 0130 | Emperor, 4" x 3/4" x 16" | | 175 | .091 | | 10.15 | 3.31 | | 13.46 | 16 |
| 0140 | Super emperor, 8" x 3/4" x 16" | | 195 | .082 | | 10.15 | 2.97 | | 13.12 | 15.55 |
| 0150 | For L shaped corners with 4" return, add | | | | L.F. | 9.25 | | | 9.25 | 10.20 |
| 0200 | On masonry/plaster back-up, modular, 2-2/3" x 5/8" x 8", red | D-7 | 137 | .117 | S.F. | 3.95 | 4.22 | | 8.17 | 10.50 |
| 0210 | Closure, 4" x 5/8" x 8" | | 165 | .097 | | 3.95 | 3.51 | | 7.46 | 9.45 |
| 0220 | Norman, 2-2/3" x 5/8" x 12" | | 165 | .097 | | 4.15 | 3.51 | | 7.66 | 9.65 |
| 0230 | Utility, 4" x 5/8" x 12" | | 185 | .086 | | 4.15 | 3.13 | | 7.28 | 9.15 |
| 0240 | Emperor, 4" x 3/4" x 16" | | 260 | .062 | | 5.35 | 2.22 | | 7.57 | 9.15 |
| 0250 | Super emperor, 8" x 3/4" x 16" | | 285 | .056 | | 5.35 | 2.03 | | 7.38 | 8.85 |
| 0260 | For L shaped corners with 4" return, add | | | | L.F. | 9.25 | | | 9.25 | 10.20 |
| 0270 | For embedment into pre-cast concrete panels, add | | | | S.F. | 14.40 | | | 14.40 | 15.85 |

### 04 21 13.15 Chimney

| | | Crew | Daily Output | Labor-Hours | Unit | Material | 2011 Bare Costs Labor | Equipment | Total | Total Incl O&P |
|---|---|---|---|---|---|---|---|---|---|---|
| 0010 | **CHIMNEY**, excludes foundation, scaffolding, grout and reinforcing | | | | | | | | | |
| 0100 | Brick, 16" x 16", 8" flue | D-1 | 18.20 | .879 | V.L.F. | 21.50 | 34.50 | | 56 | 75 |
| 0150 | 16" x 20" with one 8" x 12" flue | | 16 | 1 | | 34 | 39 | | 73 | 96 |
| 0200 | 16" x 24" with two 8" x 8" flues | | 14 | 1.143 | | 50 | 44.50 | | 94.50 | 122 |
| 0250 | 20" x 20" with one 12" x 12" flue | | 13.70 | 1.168 | | 38.50 | 45.50 | | 84 | 111 |
| 0300 | 20" x 24" with two 8" x 12" flues | | 12 | 1.333 | | 57 | 52 | | 109 | 141 |
| 0350 | 20" x 32" with two 12" x 12" flues | | 10 | 1.600 | | 68 | 62.50 | | 130.50 | 169 |
| 1800 | Metal, high temp. steel jacket, factory lining, 24" diam. | E-2 | 65 | .862 | | 210 | 41 | 25 | 276 | 330 |
| 1900 | 60" diameter | " | 30 | 1.867 | | 765 | 88.50 | 54 | 907.50 | 1,050 |
| 2100 | Poured concrete, brick lining, 200' high x 10' diam. | | | | | 7,275 | | | 7,275 | 8,000 |
| 2800 | 500' x 20' diameter | | | | | 12,700 | | | 12,700 | 14,000 |

# 04 21 Clay Unit Masonry

## 04 21 13 - Brick Masonry

### 04 21 13.18 Columns

| | | Crew | Daily Output | Labor-Hours | Unit | Material | 2011 Bare Costs Labor | 2011 Bare Costs Equipment | Total | Total Incl O&P |
|---|---|---|---|---|---|---|---|---|---|---|
| 0010 | **COLUMNS**, solid, excludes scaffolding, grout and reinforcing   R042110-10 | | | | | | | | | |
| 0050 | Brick, 8" x 8", 9 brick per V.L.F. | D-1 | 56 | .286 | V.L.F. | 5.45 | 11.15 | | 16.60 | 22.50 |
| 0100 | 12" x 8", 13.5 brick per V.L.F. | | 37 | .432 | | 8.15 | 16.85 | | 25 | 34.50 |
| 0200 | 12" x 12", 20 brick per V.L.F. | | 25 | .640 | | 12.10 | 25 | | 37.10 | 51 |
| 0300 | 16" x 12", 27 brick per V.L.F. | | 19 | .842 | | 16.35 | 33 | | 49.35 | 67.50 |
| 0400 | 16" x 16", 36 brick per V.L.F. | | 14 | 1.143 | | 22 | 44.50 | | 66.50 | 91 |
| 0500 | 20" x 16", 45 brick per V.L.F. | | 11 | 1.455 | | 27 | 56.50 | | 83.50 | 115 |
| 0600 | 20" x 20", 56 brick per V.L.F. | | 9 | 1.778 | | 34 | 69.50 | | 103.50 | 142 |
| 0700 | 24" x 20", 68 brick per V.L.F. | | 7 | 2.286 | | 41 | 89 | | 130 | 180 |
| 0800 | 24" x 24", 81 brick per V.L.F. | | 6 | 2.667 | | 49 | 104 | | 153 | 210 |
| 1000 | 36" x 36", 182 brick per V.L.F. | | 3 | 5.333 | | 110 | 208 | | 318 | 430 |

### 04 21 13.30 Oversized Brick

| | | Crew | Daily Output | Labor-Hours | Unit | Material | 2011 Bare Costs Labor | 2011 Bare Costs Equipment | Total | Total Incl O&P |
|---|---|---|---|---|---|---|---|---|---|---|
| 0010 | **OVERSIZED BRICK**, excludes scaffolding, grout and reinforcing | | | | | | | | | |
| 0100 | Veneer, 4" x 2.25" x 16" | D-8 | 387 | .103 | S.F. | 6.45 | 4.11 | | 10.56 | 13.20 |
| 0105 | 4" x 2.75" x 16" | | 412 | .097 | | 5.80 | 3.87 | | 9.67 | 12.15 |
| 0110 | 4" x 4" x 16" | | 460 | .087 | | 5.30 | 3.46 | | 8.76 | 11.05 |
| 0120 | 4" x 8" x 16" | | 533 | .075 | | 6.10 | 2.99 | | 9.09 | 11.25 |
| 0125 | Loadbearing, 6" x 4" x 16", grouted and reinforced | | 387 | .103 | | 9 | 4.11 | | 13.11 | 16.05 |
| 0130 | 8" x 4" x 16", grouted and reinforced | | 327 | .122 | | 9.40 | 4.87 | | 14.27 | 17.65 |
| 0135 | 6" x 8" x 16", grouted and reinforced | | 440 | .091 | | 8.65 | 3.62 | | 12.27 | 15 |
| 0140 | 8" x 8" x 16", grouted and reinforced | | 400 | .100 | | 9.55 | 3.98 | | 13.53 | 16.45 |
| 0145 | Curtainwall/reinforced veneer, 6" x 4" x 16" | | 387 | .103 | | 13.20 | 4.11 | | 17.31 | 20.50 |
| 0150 | 8" x 4" x 16" | | 327 | .122 | | 15.65 | 4.87 | | 20.52 | 24.50 |
| 0155 | 6" x 8" x 16" | | 440 | .091 | | 13.30 | 3.62 | | 16.92 | 20 |
| 0160 | 8" x 8" x 16" | | 400 | .100 | | 15.80 | 3.98 | | 19.78 | 23.50 |
| 0200 | For 1 to 3 slots in face, add | | | | | 15% | | | | |
| 0210 | For 4 to 7 slots in face, add | | | | | 25% | | | | |
| 0220 | For bond beams, add | | | | | 20% | | | | |
| 0230 | For bullnose shapes, add | | | | | 20% | | | | |
| 0240 | For open end knockout, add | | | | | 10% | | | | |
| 0250 | For white or gray color group, add | | | | | 10% | | | | |
| 0260 | For 135 degree corner, add | | | | | 250% | | | | |

### 04 21 13.35 Common Building Brick

| | | Crew | Daily Output | Labor-Hours | Unit | Material | 2011 Bare Costs Labor | 2011 Bare Costs Equipment | Total | Total Incl O&P |
|---|---|---|---|---|---|---|---|---|---|---|
| 0010 | **COMMON BUILDING BRICK**, C62, TL lots, material only   R042110-20 | | | | | | | | | |
| 0020 | Standard, minimum | | | | M | 400 | | | 400 | 440 |
| 0050 | Average (select) | | | | " | 430 | | | 430 | 475 |

### 04 21 13.40 Structural Brick

| | | Crew | Daily Output | Labor-Hours | Unit | Material | 2011 Bare Costs Labor | 2011 Bare Costs Equipment | Total | Total Incl O&P |
|---|---|---|---|---|---|---|---|---|---|---|
| 0010 | **STRUCTURAL BRICK** C652, Grade SW, incl. mortar, scaffolding not incl. | | | | | | | | | |
| 0100 | Standard unit, 4-5/8" x 2-3/4" x 9-5/8" | D-8 | 245 | .163 | S.F. | 5.15 | 6.50 | | 11.65 | 15.45 |
| 0120 | Bond beam | | 225 | .178 | | 9.65 | 7.10 | | 16.75 | 21.50 |
| 0140 | V cut bond beam | | 225 | .178 | | 10.45 | 7.10 | | 17.55 | 22 |
| 0160 | Stretcher quoin, 5-5/8" x 2-3/4" x 9-5/8" | | 245 | .163 | | 10.10 | 6.50 | | 16.60 | 21 |
| 0180 | Corner quoin | | 245 | .163 | | 11.30 | 6.50 | | 17.80 | 22 |
| 0200 | Corner, 45 deg, 4-5/8" x 2-3/4" x 10-7/16" | | 235 | .170 | | 11.70 | 6.80 | | 18.50 | 23 |

### 04 21 13.45 Face Brick

| | | Crew | Daily Output | Labor-Hours | Unit | Material | 2011 Bare Costs Labor | 2011 Bare Costs Equipment | Total | Total Incl O&P |
|---|---|---|---|---|---|---|---|---|---|---|
| 0010 | **FACE BRICK** Material Only, C216, TL lots   R042110-20 | | | | | | | | | |
| 0300 | Standard modular, 4" x 2-2/3" x 8", minimum | | | | M | 515 | | | 515 | 565 |
| 0350 | Maximum | | | | | 685 | | | 685 | 750 |
| 0450 | Economy, 4" x 4" x 8", minimum | | | | | 815 | | | 815 | 895 |
| 0500 | Maximum | | | | | 1,050 | | | 1,050 | 1,150 |
| 0510 | Economy, 4" x 4" x 12", minimum | | | | | 1,200 | | | 1,200 | 1,325 |

## 04 21 13 – Brick Masonry

| 04 21 13.45 Face Brick | | Crew | Daily Output | Labor-Hours | Unit | Material | 2011 Bare Costs Labor | Equipment | Total | Total Incl O&P |
|---|---|---|---|---|---|---|---|---|---|---|
| 0520 | Maximum | | | | M | 1,550 | | | 1,550 | 1,725 |
| 0550 | Jumbo, 6" x 4" x 12", minimum | | | | | 1,475 | | | 1,475 | 1,625 |
| 0600 | Maximum | | | | | 1,875 | | | 1,875 | 2,075 |
| 0610 | Jumbo, 8" x 4" x 12", minimum | | | | | 1,475 | | | 1,475 | 1,625 |
| 0620 | Maximum | | | | | 1,875 | | | 1,875 | 2,075 |
| 0650 | Norwegian, 4" x 3-1/5" x 12", minimum | | | | | 980 | | | 980 | 1,075 |
| 0700 | Maximum | | | | | 1,275 | | | 1,275 | 1,400 |
| 0710 | Norwegian, 6" x 3-1/5" x 12", minimum | | | | | 1,425 | | | 1,425 | 1,575 |
| 0720 | Maximum | | | | | 1,875 | | | 1,875 | 2,050 |
| 0850 | Standard glazed, plain colors, 4" x 2-2/3" x 8", minimum | | | | | 1,625 | | | 1,625 | 1,800 |
| 0900 | Maximum | | | | | 2,125 | | | 2,125 | 2,325 |
| 1000 | Deep trim shades, 4" x 2-2/3" x 8", minimum | | | | | 1,900 | | | 1,900 | 2,100 |
| 1050 | Maximum | | | | | 2,175 | | | 2,175 | 2,400 |
| 1080 | Jumbo utility, 4" x 4" x 12" | | | | | 1,350 | | | 1,350 | 1,500 |
| 1120 | 4" x 8" x 8" | | | | | 1,675 | | | 1,675 | 1,850 |
| 1140 | 4" x 8" x 16" | | | | | 4,925 | | | 4,925 | 5,400 |
| 1260 | Engineer, 4" x 3-1/5" x 8", minimum | | | | | 520 | | | 520 | 570 |
| 1270 | Maximum | | | | | 775 | | | 775 | 850 |
| 1350 | King, 4" x 2-3/4" x 10", minimum | | | | | 460 | | | 460 | 505 |
| 1360 | Maximum | | | | | 560 | | | 560 | 620 |
| 1400 | Norman, 4" x 2-3/4" x 12" | | | | | 460 | | | 460 | 505 |
| 1450 | Roman, 4" x 2" x 12" | | | | | 460 | | | 460 | 505 |
| 1500 | SCR, 6" x 2-2/3" x 12" | | | | | 460 | | | 460 | 505 |
| 1550 | Double, 4" x 5-1/3" x 8" | | | | | 460 | | | 460 | 505 |
| 1600 | Triple, 4" x 5-1/3" x 12" | | | | | 460 | | | 460 | 505 |
| 1770 | Standard modular, double glazed, 4" x 2-2/3" x 8" | | | | | 2,275 | | | 2,275 | 2,500 |
| 1850 | Jumbo, colored glazed ceramic, 6" x 4" x 12" | | | | | 2,400 | | | 2,400 | 2,650 |
| 2050 | Jumbo utility, glazed, 4" x 4" x 12" | | | | | 4,550 | | | 4,550 | 5,000 |
| 2100 | 4" x 8" x 8" | | | | | 5,325 | | | 5,325 | 5,875 |
| 2150 | 4" x 16" x 8" | | | | | 6,200 | | | 6,200 | 6,825 |
| 2170 | For less than truck load lots, add | | | | | 15 | | | 15 | 16.50 |
| 2180 | For buff or gray brick, add | | | | | 16 | | | 16 | 17.60 |

## 04 21 26 – Glazed Structural Clay Tile Masonry

### 04 21 26.10 Structural Facing Tile

| | | Crew | Daily Output | Labor-Hours | Unit | Material | 2011 Bare Costs Labor | Equipment | Total | Total Incl O&P |
|---|---|---|---|---|---|---|---|---|---|---|
| 0010 | **STRUCTURAL FACING TILE**, std. colors, excl. scaffolding, grout, reinforcing | | | | | | | | | |
| 0020 | 6T series, 5-1/3" x 12", 2.3 pieces per S.F., glazed 1 side, 2" thick | D-8 | 225 | .178 | S.F. | 7.95 | 7.10 | | 15.05 | 19.35 |
| 0100 | 4" thick | | 220 | .182 | | 10.95 | 7.25 | | 18.20 | 23 |
| 0150 | Glazed 2 sides | | 195 | .205 | | 14.05 | 8.15 | | 22.20 | 27.50 |
| 0250 | 6" thick | | 210 | .190 | | 16.25 | 7.60 | | 23.85 | 29.50 |
| 0300 | Glazed 2 sides | | 185 | .216 | | 19.30 | 8.60 | | 27.90 | 34 |
| 0400 | 8" thick | | 180 | .222 | | 21.50 | 8.85 | | 30.35 | 37 |
| 0500 | Special shapes, group 1 | | 400 | .100 | Ea. | 6.80 | 3.98 | | 10.78 | 13.40 |
| 0550 | Group 2 | | 375 | .107 | | 11.05 | 4.25 | | 15.30 | 18.55 |
| 0600 | Group 3 | | 350 | .114 | | 14.30 | 4.55 | | 18.85 | 22.50 |
| 0650 | Group 4 | | 325 | .123 | | 29.50 | 4.90 | | 34.40 | 40 |
| 0700 | Group 5 | | 300 | .133 | | 35.50 | 5.30 | | 40.80 | 47 |
| 0750 | Group 6 | | 275 | .145 | | 48.50 | 5.80 | | 54.30 | 62 |
| 1000 | Fire rated, 4" thick, 1 hr. rating | | 210 | .190 | S.F. | 15.75 | 7.60 | | 23.35 | 29 |
| 1300 | Acoustic, 4" thick | | 210 | .190 | " | 34 | 7.60 | | 41.60 | 49 |
| 2000 | 8W series, 8" x 16", 1.125 pieces per S.F. | | | | | | | | | |
| 2050 | 2" thick, glazed 1 side | D-8 | 360 | .111 | S.F. | 9.30 | 4.42 | | 13.72 | 16.85 |
| 2100 | 4" thick, glazed 1 side | | 345 | .116 | | 11.95 | 4.62 | | 16.57 | 20 |

# 04 21 Clay Unit Masonry

## 04 21 26 – Glazed Structural Clay Tile Masonry

| 04 21 26.10 Structural Facing Tile | | Crew | Daily Output | Labor-Hours | Unit | Material | 2011 Bare Costs Labor | Equipment | Total | Total Incl O&P |
|---|---|---|---|---|---|---|---|---|---|---|
| 2150 | Glazed 2 sides | D-8 | 325 | .123 | S.F. | 13.85 | 4.90 | | 18.75 | 22.50 |
| 2200 | 6" thick, glazed 1 side | | 330 | .121 | | 21 | 4.83 | | 25.83 | 30.50 |
| 2250 | 8" thick, glazed 1 side | | 310 | .129 | | 22 | 5.15 | | 27.15 | 31.50 |
| 2500 | Special shapes, group 1 | | 300 | .133 | Ea. | 12.85 | 5.30 | | 18.15 | 22 |
| 2550 | Group 2 | | 280 | .143 | | 17.20 | 5.70 | | 22.90 | 27.50 |
| 2600 | Group 3 | | 260 | .154 | | 18.25 | 6.10 | | 24.35 | 29 |
| 2650 | Group 4 | | 250 | .160 | | 36.50 | 6.35 | | 42.85 | 49.50 |
| 2700 | Group 5 | | 240 | .167 | | 39 | 6.65 | | 45.65 | 53 |
| 2750 | Group 6 | | 230 | .174 | | 84.50 | 6.90 | | 91.40 | 103 |
| 3000 | 4" thick, glazed 1 side | | 345 | .116 | S.F. | 11.45 | 4.62 | | 16.07 | 19.55 |
| 3100 | Acoustic, 4" thick | | 345 | .116 | " | 13.95 | 4.62 | | 18.57 | 22.50 |
| 3120 | 4W series, 8" x 8", 2.25 pieces per S.F. | | | | | | | | | |
| 3125 | 2" thick, glazed 1 side | D-8 | 360 | .111 | S.F. | 8 | 4.42 | | 12.42 | 15.45 |
| 3130 | 4" thick, glazed 1 side | | 345 | .116 | | 9.65 | 4.62 | | 14.27 | 17.55 |
| 3135 | Glazed 2 sides | | 325 | .123 | | 13.20 | 4.90 | | 18.10 | 22 |
| 3140 | 6" thick, glazed 1 side | | 330 | .121 | | 13.80 | 4.83 | | 18.63 | 22.50 |
| 3150 | 8" thick, glazed 1 side | | 310 | .129 | | 20 | 5.15 | | 25.15 | 29.50 |
| 3155 | Special shapes, group I | | 300 | .133 | Ea. | 6.35 | 5.30 | | 11.65 | 14.95 |
| 3160 | Group II | | 280 | .143 | " | 7.30 | 5.70 | | 13 | 16.55 |
| 3200 | For designer colors, add | | | | | 25% | | | | |
| 3300 | For epoxy mortar joints, add | | | | S.F. | 1.58 | | | 1.58 | 1.74 |

## 04 21 29 – Terra Cotta Masonry

### 04 21 29.10 Terra Cotta Masonry Components

| | | Crew | Daily Output | Labor-Hours | Unit | Material | 2011 Bare Costs Labor | Equipment | Total | Total Incl O&P |
|---|---|---|---|---|---|---|---|---|---|---|
| 0010 | **TERRA COTTA MASONRY COMPONENTS** | | | | | | | | | |
| 0020 | Coping, split type, not glazed, 9" wide | D-1 | 90 | .178 | L.F. | 10.40 | 6.95 | | 17.35 | 22 |
| 0100 | 13" wide | | 80 | .200 | | 15.75 | 7.80 | | 23.55 | 29 |
| 0200 | Coping, split type, glazed, 9" wide | | 90 | .178 | | 17.40 | 6.95 | | 24.35 | 29.50 |
| 0250 | 13" wide | | 80 | .200 | | 22.50 | 7.80 | | 30.30 | 36.50 |
| 0500 | Partition or back-up blocks, scored, in C.L. lots | | | | | | | | | |
| 0700 | Non-load bearing 12" x 12", 3" thick, special order | D-8 | 550 | .073 | S.F. | 26 | 2.90 | | 28.90 | 33 |
| 0750 | 4" thick, standard | | 500 | .080 | | 7.10 | 3.18 | | 10.28 | 12.60 |
| 0800 | 6" thick | | 450 | .089 | | 8.70 | 3.54 | | 12.24 | 14.85 |
| 0850 | 8" thick | | 400 | .100 | | 11.05 | 3.98 | | 15.03 | 18.15 |
| 1000 | Load bearing, 12" x 12", 4" thick, in walls | | 500 | .080 | | 7.60 | 3.18 | | 10.78 | 13.15 |
| 1050 | In floors | | 750 | .053 | | 7.60 | 2.12 | | 9.72 | 11.55 |
| 1200 | 6" thick, in walls | | 450 | .089 | | 8.70 | 3.54 | | 12.24 | 14.85 |
| 1250 | In floors | | 675 | .059 | | 8.70 | 2.36 | | 11.06 | 13.10 |
| 1400 | 8" thick, in walls | | 400 | .100 | | 11.05 | 3.98 | | 15.03 | 18.15 |
| 1450 | In floors | | 575 | .070 | | 11.05 | 2.77 | | 13.82 | 16.35 |
| 1600 | 10" thick, in walls, special order | | 350 | .114 | | 23.50 | 4.55 | | 28.05 | 32.50 |
| 1650 | In floors, special order | | 500 | .080 | | 23.50 | 3.18 | | 26.68 | 30.50 |
| 1800 | 12" thick, in walls, special order | | 300 | .133 | | 28 | 5.30 | | 33.30 | 38.50 |
| 1850 | In floors, special order | | 450 | .089 | | 28 | 3.54 | | 31.54 | 36 |
| 2000 | For reinforcing with steel rods, add to above | | | | | 15% | 5% | | | |
| 2100 | For smooth tile instead of scored, add | | | | | 3.74 | | | 3.74 | 4.11 |
| 2200 | For L.C.L. quantities, add | | | | | 10% | 10% | | | |

### 04 21 29.20 Terra Cotta Tile

| | | Crew | Daily Output | Labor-Hours | Unit | Material | 2011 Bare Costs Labor | Equipment | Total | Total Incl O&P |
|---|---|---|---|---|---|---|---|---|---|---|
| 0010 | **TERRA COTTA TILE**, on walls, dry set, 1/2" thick | | | | | | | | | |
| 0100 | Square, hexagonal or lattice shapes, unglazed | 1 Tilf | 135 | .059 | S.F. | 6.35 | 2.40 | | 8.75 | 10.45 |
| 0300 | Glazed, plain colors | | 130 | .062 | | 8.75 | 2.49 | | 11.24 | 13.25 |
| 0400 | Intense colors | | 125 | .064 | | 10.20 | 2.59 | | 12.79 | 15 |

| 04 22 10.11 Autoclave Aerated Concrete Block | | Crew | Daily Output | Labor-Hours | Unit | Material | 2011 Bare Costs Labor | Equipment | Total | Total Incl O&P |
|---|---|---|---|---|---|---|---|---|---|---|
| 0010 | **AUTOCLAVE AERATED CONCRETE BLOCK**, excl. scaffolding, grout & reinforcing | | | | | | | | | |
| 0050 | Solid, 4" x 12" x 24", incl mortar ⒢ | D-8 | 600 | .067 | S.F. | 1.39 | 2.65 | | 4.04 | 5.50 |
| 0060 | 6" x 12" x 24" ⒢ | | 600 | .067 | | 2.08 | 2.65 | | 4.73 | 6.25 |
| 0070 | 8" x 8" x 24" ⒢ | | 575 | .070 | | 2.78 | 2.77 | | 5.55 | 7.20 |
| 0080 | 10" x 12" x 24" ⒢ | | 575 | .070 | | 3.40 | 2.77 | | 6.17 | 7.90 |
| 0090 | 12" x 12" x 24" ⒢ | | 550 | .073 | | 4.16 | 2.90 | | 7.06 | 8.90 |

| 04 22 10.12 Chimney Block | | | | | | | | | | |
|---|---|---|---|---|---|---|---|---|---|---|
| 0010 | **CHIMNEY BLOCK**, excludes scaffolding, grout and reinforcing | | | | | | | | | |
| 0220 | 1 piece, with 8" x 8" flue, 16" x 16" | D-1 | 28 | .571 | V.L.F. | 17.50 | 22.50 | | 40 | 53 |
| 0230 | 2 piece, 16" x 16" | | 26 | .615 | | 15.05 | 24 | | 39.05 | 52.50 |
| 0240 | 2 piece, with 8" x 12" flue, 16" x 20" | | 24 | .667 | | 30 | 26 | | 56 | 72 |

| 04 22 10.14 Concrete Block, Back-Up | | | | | | | | | | |
|---|---|---|---|---|---|---|---|---|---|---|
| 0010 | **CONCRETE BLOCK, BACK-UP**, C90, 2000 psi    R042210-20 | | | | | | | | | |
| 0020 | Normal weight, 8" x 16" units, tooled joint 1 side | | | | | | | | | |
| 0050 | Not-reinforced, 2000 psi, 2" thick | D-8 | 475 | .084 | S.F. | 1.39 | 3.35 | | 4.74 | 6.60 |
| 0200 | 4" thick | | 460 | .087 | | 1.58 | 3.46 | | 5.04 | 6.95 |
| 0300 | 6" thick | | 440 | .091 | | 2.18 | 3.62 | | 5.80 | 7.85 |
| 0350 | 8" thick | | 400 | .100 | | 2.49 | 3.98 | | 6.47 | 8.70 |
| 0400 | 10" thick | | 330 | .121 | | 3.12 | 4.83 | | 7.95 | 10.70 |
| 0450 | 12" thick | D-9 | 310 | .155 | | 3.52 | 6.05 | | 9.57 | 12.95 |
| 1000 | Reinforced, alternate courses, 4" thick | D-8 | 450 | .089 | | 1.71 | 3.54 | | 5.25 | 7.20 |
| 1100 | 6" thick | | 430 | .093 | | 2.31 | 3.70 | | 6.01 | 8.10 |
| 1150 | 8" thick | | 395 | .101 | | 2.62 | 4.03 | | 6.65 | 8.95 |
| 1200 | 10" thick | | 320 | .125 | | 3.25 | 4.98 | | 8.23 | 11 |
| 1250 | 12" thick | D-9 | 300 | .160 | | 3.65 | 6.25 | | 9.90 | 13.35 |

| 04 22 10.16 Concrete Block, Bond Beam | | | | | | | | | | |
|---|---|---|---|---|---|---|---|---|---|---|
| 0010 | **CONCRETE BLOCK, BOND BEAM**, C90, 2000 psi | | | | | | | | | |
| 0020 | Not including grout or reinforcing | | | | | | | | | |
| 0125 | Regular block, 6" thick | D-8 | 584 | .068 | L.F. | 2.32 | 2.73 | | 5.05 | 6.65 |
| 0130 | 8" high, 8" thick | " | 565 | .071 | | 2.50 | 2.82 | | 5.32 | 7 |
| 0150 | 12" thick | D-9 | 510 | .094 | | 3.50 | 3.67 | | 7.17 | 9.35 |
| 0525 | Lightweight, 6" thick | D-8 | 592 | .068 | | 2.55 | 2.69 | | 5.24 | 6.85 |
| 0530 | 8" high, 8" thick | " | 575 | .070 | | 3.09 | 2.77 | | 5.86 | 7.55 |
| 0550 | 12" thick | D-9 | 520 | .092 | | 4.16 | 3.60 | | 7.76 | 9.95 |
| 2000 | Including grout and 2 #5 bars | | | | | | | | | |
| 2100 | Regular block, 8" high, 8" thick | D-8 | 300 | .133 | L.F. | 4.45 | 5.30 | | 9.75 | 12.85 |
| 2150 | 12" thick | D-9 | 250 | .192 | | 6 | 7.50 | | 13.50 | 17.85 |
| 2500 | Lightweight, 8" high, 8" thick | D-8 | 305 | .131 | | 5.05 | 5.20 | | 10.25 | 13.40 |
| 2550 | 12" thick | D-9 | 255 | .188 | | 6.65 | 7.35 | | 14 | 18.30 |

| 04 22 10.18 Concrete Block, Column | | | | | | | | | | |
|---|---|---|---|---|---|---|---|---|---|---|
| 0010 | **CONCRETE BLOCK, COLUMN** or pilaster | | | | | | | | | |
| 0050 | Including vertical reinforcing (4-#4 bars) and grout | | | | | | | | | |
| 0160 | 1 piece unit, 16" x 16" | D-1 | 26 | .615 | V.L.F. | 18.05 | 24 | | 42.05 | 56 |
| 0170 | 2 piece units, 16" x 20" | | 24 | .667 | | 23.50 | 26 | | 49.50 | 65 |
| 0180 | 20" x 20" | | 22 | .727 | | 34.50 | 28.50 | | 63 | 80.50 |
| 0190 | 22" x 24" | | 18 | .889 | | 48.50 | 34.50 | | 83 | 106 |
| 0200 | 20" x 32" | | 14 | 1.143 | | 53 | 44.50 | | 97.50 | 126 |

| 04 22 10.19 Concrete Block, Insulation Inserts | | | | | | | | | | |
|---|---|---|---|---|---|---|---|---|---|---|
| 0010 | **CONCRETE BLOCK, INSULATION INSERTS** | | | | | | | | | |
| 0100 | Styrofoam, plant installed, add to block prices | | | | | | | | | |
| 0200 | 8" x 16" units, 6" thick | | | | S.F. | 1.81 | | | 1.81 | 1.99 |

# 04 22 Concrete Unit Masonry

## 04 22 10 – Concrete Masonry Units

| 04 22 10.19 Concrete Block, Insulation Inserts | Crew | Daily Output | Labor-Hours | Unit | Material | 2011 Bare Costs Labor | Equipment | Total | Total Incl O&P |
|---|---|---|---|---|---|---|---|---|---|
| 0250 | 8" thick | | | | S.F. | 1.81 | | | 1.81 | 1.99 |
| 0300 | 10" thick | | | | | 2.13 | | | 2.13 | 2.34 |
| 0350 | 12" thick | | | | | 2.24 | | | 2.24 | 2.46 |
| 0500 | 8" x 8" units, 8" thick | | | | | 1.49 | | | 1.49 | 1.64 |
| 0550 | 12" thick | | | | | 1.81 | | | 1.81 | 1.99 |

### 04 22 10.23 Concrete Block, Decorative

| | | Crew | Daily Output | Labor-Hours | Unit | Material | Labor | Equipment | Total | Total Incl O&P |
|---|---|---|---|---|---|---|---|---|---|---|
| 0010 | **CONCRETE BLOCK, DECORATIVE**, C90, 2000 psi | | | | | | | | | |
| 0020 | Embossed, simulated brick face | | | | | | | | | |
| 0100 | 8" x 16" units, 4" thick | D-8 | 400 | .100 | S.F. | 3.51 | 3.98 | | 7.49 | 9.80 |
| 0200 | 8" thick | | 340 | .118 | | 4.83 | 4.68 | | 9.51 | 12.35 |
| 0250 | 12" thick | | 300 | .133 | | 6.35 | 5.30 | | 11.65 | 14.95 |
| 0400 | Embossed both sides | | | | | | | | | |
| 0500 | 8" thick | D-8 | 300 | .133 | S.F. | 5.40 | 5.30 | | 10.70 | 13.90 |
| 0550 | 12" thick | " | 275 | .145 | " | 6.85 | 5.80 | | 12.65 | 16.25 |
| 1000 | Fluted high strength | | | | | | | | | |
| 1100 | 8" x 16" x 4" thick, flutes 1 side, | D-8 | 345 | .116 | S.F. | 4.17 | 4.62 | | 8.79 | 11.55 |
| 1150 | Flutes 2 sides | | 335 | .119 | | 5.05 | 4.75 | | 9.80 | 12.70 |
| 1200 | 8" thick | | 300 | .133 | | 6.60 | 5.30 | | 11.90 | 15.20 |
| 1250 | For special colors, add | | | | | .41 | | | .41 | .45 |
| 1400 | Deep grooved, smooth face | | | | | | | | | |
| 1450 | 8" x 16" x 4" thick | D-8 | 345 | .116 | S.F. | 2.72 | 4.62 | | 7.34 | 9.95 |
| 1500 | 8" thick | " | 300 | .133 | " | 4.68 | 5.30 | | 9.98 | 13.10 |
| 2000 | Formblock, incl. inserts & reinforcing | | | | | | | | | |
| 2100 | 8" x 16" x 8" thick | D-8 | 345 | .116 | S.F. | 4.76 | 4.62 | | 9.38 | 12.20 |
| 2150 | 12" thick | " | 310 | .129 | " | 5.95 | 5.15 | | 11.10 | 14.20 |
| 2500 | Ground face | | | | | | | | | |
| 2600 | 8" x 16" x 4" thick | D-8 | 345 | .116 | S.F. | 3.27 | 4.62 | | 7.89 | 10.55 |
| 2650 | 6" thick | | 325 | .123 | | 4.78 | 4.90 | | 9.68 | 12.60 |
| 2700 | 8" thick | | 300 | .133 | | 6.35 | 5.30 | | 11.65 | 14.90 |
| 2750 | 12" thick | D-9 | 265 | .181 | | 5.70 | 7.05 | | 12.75 | 16.85 |
| 2900 | For special colors, add, minimum | | | | | 15% | | | | |
| 2950 | For special colors, add, maximum | | | | | 45% | | | | |
| 4000 | Slump block | | | | | | | | | |
| 4100 | 4" face height x 16" x 4" thick | D-1 | 165 | .097 | S.F. | 3.94 | 3.78 | | 7.72 | 10 |
| 4150 | 6" thick | | 160 | .100 | | 5.45 | 3.90 | | 9.35 | 11.80 |
| 4200 | 8" thick | | 155 | .103 | | 6.05 | 4.02 | | 10.07 | 12.70 |
| 4250 | 10" thick | | 140 | .114 | | 10.90 | 4.46 | | 15.36 | 18.70 |
| 4300 | 12" thick | | 130 | .123 | | 11.40 | 4.80 | | 16.20 | 19.75 |
| 4400 | 6" face height x 16" x 6" thick | | 155 | .103 | | 4.75 | 4.02 | | 8.77 | 11.25 |
| 4450 | 8" thick | | 150 | .107 | | 6.50 | 4.16 | | 10.66 | 13.40 |
| 4500 | 10" thick | | 130 | .123 | | 10 | 4.80 | | 14.80 | 18.20 |
| 4550 | 12" thick | | 120 | .133 | | 10.35 | 5.20 | | 15.55 | 19.20 |
| 5000 | Split rib profile units, 1" deep ribs, 8 ribs | | | | | | | | | |
| 5100 | 8" x 16" x 4" thick | D-8 | 345 | .116 | S.F. | 3.17 | 4.62 | | 7.79 | 10.45 |
| 5150 | 6" thick | | 325 | .123 | | 3.64 | 4.90 | | 8.54 | 11.35 |
| 5200 | 8" thick | | 300 | .133 | | 4.16 | 5.30 | | 9.46 | 12.55 |
| 5250 | 12" thick | D-9 | 275 | .175 | | 4.93 | 6.80 | | 11.73 | 15.65 |
| 5400 | For special deeper colors, 4" thick, add | | | | | 1.15 | | | 1.15 | 1.27 |
| 5450 | 12" thick, add | | | | | 1.19 | | | 1.19 | 1.31 |
| 5600 | For white, 4" thick, add | | | | | 1.15 | | | 1.15 | 1.27 |
| 5650 | 6" thick, add | | | | | 1.14 | | | 1.14 | 1.26 |
| 5700 | 8" thick, add | | | | | 1.20 | | | 1.20 | 1.32 |

| | | Daily Output | Labor-Hours | Unit | Material | 2011 Bare Costs Labor | 2011 Bare Costs Equipment | Total | Total Incl O&P |
|---|---|---|---|---|---|---|---|---|---|
| **04 22 10.23 Concrete Block, Decorative** | | Crew | | | | | | | |
| 5750 | 12" thick, add | | | S.F. | 1.23 | | | 1.23 | 1.35 |
| 6000 | Split face | | | | | | | | |
| 6100 | 8" x 16" x 4" thick | D-8 | 350 | .114 | S.F. | 2.82 | 4.55 | 7.37 | 9.95 |
| 6150 | 6" thick | | 325 | .123 | | 3.28 | 4.90 | 8.18 | 10.95 |
| 6200 | 8" thick | | 300 | .133 | | 3.73 | 5.30 | 9.03 | 12.05 |
| 6250 | 12" thick | D-9 | 270 | .178 | | 4.57 | 6.95 | 11.52 | 15.45 |
| 6300 | For scored, add | | | | | .35 | | .35 | .39 |
| 6400 | For special deeper colors, 4" thick, add | | | | | .67 | | .67 | .73 |
| 6450 | 6" thick, add | | | | | .66 | | .66 | .72 |
| 6500 | 8" thick, add | | | | | .68 | | .68 | .75 |
| 6550 | 12" thick, add | | | | | .70 | | .70 | .77 |
| 6650 | For white, 4" thick, add | | | | | 1.14 | | 1.14 | 1.26 |
| 6700 | 6" thick, add | | | | | 1.13 | | 1.13 | 1.24 |
| 6750 | 8" thick, add | | | | | 1.16 | | 1.16 | 1.28 |
| 6800 | 12" thick, add | | | | | 1.19 | | 1.19 | 1.31 |
| 7000 | Scored ground face, 2 to 5 scores | | | | | | | | |
| 7100 | 8" x 16" x 4" thick | D-8 | 340 | .118 | S.F. | 4.91 | 4.68 | 9.59 | 12.45 |
| 7150 | 6" thick | | 310 | .129 | | 5.60 | 5.15 | 10.75 | 13.90 |
| 7200 | 8" thick | | 290 | .138 | | 6.60 | 5.50 | 12.10 | 15.50 |
| 7250 | 12" thick | D-9 | 265 | .181 | | 8.20 | 7.05 | 15.25 | 19.65 |
| 8000 | Hexagonal face profile units, 8" x 16" units | | | | | | | | |
| 8100 | 4" thick, hollow | D-8 | 340 | .118 | S.F. | 3.12 | 4.68 | 7.80 | 10.50 |
| 8200 | Solid | | 340 | .118 | | 4.03 | 4.68 | 8.71 | 11.50 |
| 8300 | 6" thick, hollow | | 310 | .129 | | 3.49 | 5.15 | 8.64 | 11.55 |
| 8350 | 8" thick, hollow | | 290 | .138 | | 4.73 | 5.50 | 10.23 | 13.45 |
| 8500 | For stacked bond, add | | | | | | 26% | | |
| 8550 | For high rise construction, add per story | D-8 | 67.80 | .590 | M.S.F. | | 23.50 | 23.50 | 35 |
| 8600 | For scored block, add | | | | | 10% | | | |
| 8650 | For honed or ground face, per face, add | | | | Ea. | .38 | | .38 | .42 |
| 8700 | For honed or ground end, per end, add | | | | " | 2.98 | | 2.98 | 3.28 |
| 8750 | For bullnose block, add | | | | | 10% | | | |
| 8800 | For special color, add | | | | | 13% | | | |
| **04 22 10.24 Concrete Block, Exterior** | | | | | | | | | |
| 0010 | **CONCRETE BLOCK, EXTERIOR,** C90, 2000 psi | | | | | | | | |
| 0020 | Reinforced alt courses, tooled joints 2 sides | | | | | | | | |
| 0100 | Normal weight, 8" x 16" x 6" thick | D-8 | 395 | .101 | S.F. | 2.56 | 4.03 | 6.59 | 8.85 |
| 0200 | 8" thick | | 360 | .111 | | 3.26 | 4.42 | 7.68 | 10.25 |
| 0250 | 10" thick | | 290 | .138 | | 4.28 | 5.50 | 9.78 | 12.95 |
| 0300 | 12" thick | D-9 | 250 | .192 | | 4.29 | 7.50 | 11.79 | 15.95 |
| 0500 | Lightweight, 8" x 16" x 6" thick | D-8 | 450 | .089 | | 2.73 | 3.54 | 6.27 | 8.30 |
| 0600 | 8" thick | | 430 | .093 | | 3.47 | 3.70 | 7.17 | 9.35 |
| 0650 | 10" thick | | 395 | .101 | | 4.32 | 4.03 | 8.35 | 10.80 |
| 0700 | 12" thick | D-9 | 350 | .137 | | 6.35 | 5.35 | 11.70 | 14.95 |
| **04 22 10.26 Concrete Block Foundation Wall** | | | | | | | | | |
| 0010 | **CONCRETE BLOCK FOUNDATION WALL,** C90/C145 | | | | | | | | |
| 0050 | Normal-weight, cut joints, horiz joint reinf, no vert reinf | | | | | | | | |
| 0200 | Hollow, 8" x 16" x 6" thick | D-8 | 455 | .088 | S.F. | 2.62 | 3.50 | 6.12 | 8.15 |
| 0250 | 8" thick | | 425 | .094 | | 2.94 | 3.75 | 6.69 | 8.85 |
| 0300 | 10" thick | | 350 | .114 | | 3.57 | 4.55 | 8.12 | 10.80 |
| 0350 | 12" thick | D-9 | 300 | .160 | | 3.98 | 6.25 | 10.23 | 13.75 |
| 0500 | Solid, 8" x 16" block, 6" thick | D-8 | 440 | .091 | | 2.69 | 3.62 | 6.31 | 8.40 |
| 0550 | 8" thick | " | 415 | .096 | | 3.80 | 3.84 | 7.64 | 9.95 |

# 04 22 Concrete Unit Masonry

## 04 22 10 – Concrete Masonry Units

| 04 22 10.26 Concrete Block Foundation Wall | Crew | Daily Output | Labor-Hours | Unit | Material | 2011 Bare Costs Labor | Equipment | Total | Total Incl O&P |
|---|---|---|---|---|---|---|---|---|---|
| 0600 | 12" thick | D-9 | 350 | .137 | S.F. | 5.40 | 5.35 | | 10.75 | 13.95 |

### 04 22 10.28 Concrete Block, High Strength

| | | Crew | Daily Output | Labor-Hours | Unit | Material | Labor | Equipment | Total | Total Incl O&P |
|---|---|---|---|---|---|---|---|---|---|---|
| 0010 | **CONCRETE BLOCK, HIGH STRENGTH** | | | | | | | | | |
| 0050 | Hollow, reinforced alternate courses, 8" x 16" units | | | | | | | | | |
| 0200 | 3500 psi, 4" thick | D-8 | 440 | .091 | S.F. | 2.22 | 3.62 | | 5.84 | 7.90 |
| 0250 | 6" thick | | 395 | .101 | | 2.47 | 4.03 | | 6.50 | 8.75 |
| 0300 | 8" thick | ↓ | 360 | .111 | | 3.18 | 4.42 | | 7.60 | 10.15 |
| 0350 | 12" thick | D-9 | 250 | .192 | | 4.17 | 7.50 | | 11.67 | 15.85 |
| 0500 | 5000 psi, 4" thick | D-8 | 440 | .091 | | 2.43 | 3.62 | | 6.05 | 8.15 |
| 0550 | 6" thick | | 395 | .101 | | 3.05 | 4.03 | | 7.08 | 9.40 |
| 0600 | 8" thick | ↓ | 360 | .111 | | 4.06 | 4.42 | | 8.48 | 11.10 |
| 0650 | 12" thick | D-9 | 300 | .160 | ↓ | 6 | 6.25 | | 12.25 | 15.95 |
| 1000 | For 75% solid block, add | | | | | 30% | | | | |
| 1050 | For 100% solid block, add | | | | | 50% | | | | |

### 04 22 10.30 Concrete Block, Interlocking

| | | Crew | Daily Output | Labor-Hours | Unit | Material | Labor | Equipment | Total | Total Incl O&P |
|---|---|---|---|---|---|---|---|---|---|---|
| 0010 | **CONCRETE BLOCK, INTERLOCKING** | | | | | | | | | |
| 0100 | Not including grout or reinforcing | | | | | | | | | |
| 0200 | 8" x 16" units, 2,000 psi, 8" thick | D-1 | 245 | .065 | S.F. | 2.72 | 2.55 | | 5.27 | 6.80 |
| 0300 | 12" thick | | 220 | .073 | | 4.03 | 2.84 | | 6.87 | 8.70 |
| 0350 | 16" thick | ↓ | 185 | .086 | | 6.05 | 3.37 | | 9.42 | 11.70 |
| 0400 | Including grout & reinforcing, 8" thick | D-4 | 245 | .131 | | 7.35 | 5.10 | .51 | 12.96 | 16.25 |
| 0450 | 12" thick | | 220 | .145 | | 8.85 | 5.70 | .57 | 15.12 | 18.85 |
| 0500 | 16" thick | ↓ | 185 | .173 | ↓ | 11.05 | 6.75 | .68 | 18.48 | 23 |

### 04 22 10.32 Concrete Block, Lintels

| | | Crew | Daily Output | Labor-Hours | Unit | Material | Labor | Equipment | Total | Total Incl O&P |
|---|---|---|---|---|---|---|---|---|---|---|
| 0010 | **CONCRETE BLOCK, LINTELS**, C90, normal weight | | | | | | | | | |
| 0100 | Including grout and horizontal reinforcing | | | | | | | | | |
| 0200 | 8" x 8" x 8", 1 #4 bar | D-4 | 300 | .107 | L.F. | 4.85 | 4.17 | .42 | 9.44 | 12.05 |
| 0250 | 2 #4 bars | | 295 | .108 | | 5.05 | 4.24 | .43 | 9.72 | 12.35 |
| 0400 | 8" x 16" x 8", 1 #4 bar | | 275 | .116 | | 3.33 | 4.55 | .46 | 8.34 | 10.95 |
| 0450 | 2 #4 bars | | 270 | .119 | | 3.54 | 4.63 | .47 | 8.64 | 11.35 |
| 1000 | 12" x 8" x 8", 1 #4 bar | | 275 | .116 | | 6.80 | 4.55 | .46 | 11.81 | 14.80 |
| 1100 | 2 #4 bars | | 270 | .119 | | 7 | 4.63 | .47 | 12.10 | 15.15 |
| 1150 | 2 #5 bars | | 270 | .119 | | 7.25 | 4.63 | .47 | 12.35 | 15.40 |
| 1200 | 2 #6 bars | | 265 | .121 | | 7.50 | 4.72 | .47 | 12.69 | 15.80 |
| 1500 | 12" x 16" x 8", 1 #4 bar | | 250 | .128 | | 10.80 | 5 | .50 | 16.30 | 19.90 |
| 1600 | 2 #3 bars | | 245 | .131 | | 10.80 | 5.10 | .51 | 16.41 | 20 |
| 1650 | 2 #4 bars | | 245 | .131 | | 11 | 5.10 | .51 | 16.61 | 20.50 |
| 1700 | 2 #5 bars | ↓ | 240 | .133 | ↓ | 11.20 | 5.20 | .52 | 16.92 | 20.50 |

### 04 22 10.34 Concrete Block, Partitions

| | | Crew | Daily Output | Labor-Hours | Unit | Material | Labor | Equipment | Total | Total Incl O&P |
|---|---|---|---|---|---|---|---|---|---|---|
| 0010 | **CONCRETE BLOCK, PARTITIONS**, excludes scaffolding    R042210-20 | | | | | | | | | |
| 0100 | Acoustical slotted block | | | | | | | | | |
| 0200 | 4" thick, type A-1 | D-8 | 315 | .127 | S.F. | 3.74 | 5.05 | | 8.79 | 11.70 |
| 0210 | 8" thick | | 275 | .145 | | 4.72 | 5.80 | | 10.52 | 13.90 |
| 0250 | 8" thick, type Q | | 275 | .145 | | 8.30 | 5.80 | | 14.10 | 17.85 |
| 0260 | 4" thick, type RSC | | 315 | .127 | | 5.80 | 5.05 | | 10.85 | 14 |
| 0270 | 6" thick | | 295 | .136 | | 5.80 | 5.40 | | 11.20 | 14.50 |
| 0280 | 8" thick | | 275 | .145 | | 5.80 | 5.80 | | 11.60 | 15.10 |
| 0290 | 12" thick | | 250 | .160 | | 5.80 | 6.35 | | 12.15 | 15.95 |
| 0300 | 8" thick, type RSR | | 275 | .145 | | 5.80 | 5.80 | | 11.60 | 15.10 |
| 0400 | 8" thick, type RSC/RF | | 275 | .145 | | 6.55 | 5.80 | | 12.35 | 15.90 |
| 0410 | 10" thick | | 260 | .154 | | 7 | 6.10 | | 13.10 | 16.90 |
| 0420 | 12" thick | ↓ | 250 | .160 | | 7.55 | 6.35 | | 13.90 | 17.85 |

| 04 22 10.34 Concrete Block, Partitions | | Crew | Daily Output | Labor-Hours | Unit | Material | 2011 Bare Costs Labor | Equipment | Total | Total Incl O&P |
|---|---|---|---|---|---|---|---|---|---|---|
| 0430 | 12" thick, type RSC/RF-4 | D-8 | 250 | .160 | S.F. | 8.95 | 6.35 | | 15.30 | 19.40 |
| 0500 | NRC .60 type R, 8" thick | | 265 | .151 | | 8 | 6 | | 14 | 17.80 |
| 0600 | NRC .65 type RR, 8" thick | | 265 | .151 | | 7.35 | 6 | | 13.35 | 17.05 |
| 0700 | NRC .65 type 4R-RF, 8" thick | | 265 | .151 | | 8.25 | 6 | | 14.25 | 18.05 |
| 0710 | NRC .70 type R, 12" thick | | 245 | .163 | | 8.65 | 6.50 | | 15.15 | 19.25 |
| 1000 | Lightweight block, tooled joints, 2 sides, hollow | | | | | | | | | |
| 1100 | Not reinforced, 8" x 16" x 4" thick | D-8 | 440 | .091 | S.F. | 1.65 | 3.62 | | 5.27 | 7.25 |
| 1150 | 6" thick | | 410 | .098 | | 2.37 | 3.88 | | 6.25 | 8.45 |
| 1200 | 8" thick | | 385 | .104 | | 2.88 | 4.14 | | 7.02 | 9.35 |
| 1250 | 10" thick | | 370 | .108 | | 3.49 | 4.30 | | 7.79 | 10.30 |
| 1300 | 12" thick | D-9 | 350 | .137 | | 4.34 | 5.35 | | 9.69 | 12.75 |
| 2000 | Not reinforced, 8" x 24" x 4" thick, hollow | | 460 | .104 | | 1.21 | 4.07 | | 5.28 | 7.45 |
| 2100 | 6" thick | | 440 | .109 | | 1.71 | 4.25 | | 5.96 | 8.30 |
| 2150 | 8" thick | | 415 | .116 | | 2.12 | 4.51 | | 6.63 | 9.10 |
| 2200 | 10" thick | | 385 | .125 | | 2.56 | 4.86 | | 7.42 | 10.10 |
| 2250 | 12" thick | | 365 | .132 | | 3.17 | 5.15 | | 8.32 | 11.20 |
| 2800 | Solid, not reinforced, 8" x 16" x 2" thick | D-8 | 440 | .091 | | 1.46 | 3.62 | | 5.08 | 7.05 |
| 2900 | 4" thick | | 420 | .095 | | 2.36 | 3.79 | | 6.15 | 8.30 |
| 2950 | 6" thick | | 390 | .103 | | 2.63 | 4.08 | | 6.71 | 9.05 |
| 3000 | 8" thick | | 365 | .110 | | 3.38 | 4.36 | | 7.74 | 10.25 |
| 3050 | 10" thick | | 350 | .114 | | 4.22 | 4.55 | | 8.77 | 11.50 |
| 3100 | 12" thick | D-9 | 330 | .145 | | 6.25 | 5.65 | | 11.90 | 15.35 |
| 4000 | Regular block, tooled joints, 2 sides, hollow | | | | | | | | | |
| 4100 | Not reinforced, 8" x 16" x 4" thick | D-8 | 430 | .093 | S.F. | 1.49 | 3.70 | | 5.19 | 7.20 |
| 4150 | 6" thick | | 400 | .100 | | 2.09 | 3.98 | | 6.07 | 8.25 |
| 4200 | 8" thick | | 375 | .107 | | 2.40 | 4.25 | | 6.65 | 9 |
| 4250 | 10" thick | | 360 | .111 | | 3.03 | 4.42 | | 7.45 | 10 |
| 4300 | 12" thick | D-9 | 340 | .141 | | 3.43 | 5.50 | | 8.93 | 12.05 |
| 4500 | Reinforced alternate courses, 8" x 16" x 4" thick | D-8 | 425 | .094 | | 1.62 | 3.75 | | 5.37 | 7.40 |
| 4550 | 6" thick | | 395 | .101 | | 2.22 | 4.03 | | 6.25 | 8.50 |
| 4600 | 8" thick | | 370 | .108 | | 2.53 | 4.30 | | 6.83 | 9.25 |
| 4650 | 10" thick | | 355 | .113 | | 3.43 | 4.49 | | 7.92 | 10.50 |
| 4700 | 12" thick | D-9 | 335 | .143 | | 3.56 | 5.60 | | 9.16 | 12.30 |
| 4900 | Solid, not reinforced, 2" thick | D-8 | 435 | .092 | | 1.34 | 3.66 | | 5 | 6.95 |
| 5000 | 3" thick | | 430 | .093 | | 1.60 | 3.70 | | 5.30 | 7.30 |
| 5050 | 4" thick | | 415 | .096 | | 1.83 | 3.84 | | 5.67 | 7.75 |
| 5100 | 6" thick | | 385 | .104 | | 2.16 | 4.14 | | 6.30 | 8.60 |
| 5150 | 8" thick | | 360 | .111 | | 3.26 | 4.42 | | 7.68 | 10.25 |
| 5200 | 12" thick | D-9 | 325 | .148 | | 4.88 | 5.75 | | 10.63 | 14 |
| 5500 | Solid, reinforced alternate courses, 4" thick | D-8 | 420 | .095 | | 1.90 | 3.79 | | 5.69 | 7.80 |
| 5550 | 6" thick | | 380 | .105 | | 2.25 | 4.19 | | 6.44 | 8.80 |
| 5600 | 8" thick | | 355 | .113 | | 3.34 | 4.49 | | 7.83 | 10.45 |
| 5650 | 12" thick | D-9 | 320 | .150 | | 4.41 | 5.85 | | 10.26 | 13.60 |

| 04 22 10.38 Concrete Brick | | | | | | | | | | |
|---|---|---|---|---|---|---|---|---|---|---|
| 0010 | **CONCRETE BRICK**, C55, grade N, type 1 | | | | | | | | | |
| 0100 | Regular, 4 x 2-1/4 x 8 | D-8 | 660 | .061 | Ea. | .45 | 2.41 | | 2.86 | 4.12 |
| 0125 | Rusticated, 4 x 2-1/4 x 8 | | 660 | .061 | | .50 | 2.41 | | 2.91 | 4.17 |
| 0150 | Frog, 4 x 2-1/4 x 8 | | 660 | .061 | | .49 | 2.41 | | 2.90 | 4.15 |
| 0200 | Double, 4 x 4-7/8 x 8 | | 535 | .075 | | .79 | 2.98 | | 3.77 | 5.35 |

| 04 22 10.42 Concrete Block, Screen Block | | | | | | | | | | |
|---|---|---|---|---|---|---|---|---|---|---|
| 0010 | **CONCRETE BLOCK, SCREEN BLOCK** | | | | | | | | | |
| 0200 | 8" x 16", 4" thick | D-8 | 330 | .121 | S.F. | 1.97 | 4.83 | | 6.80 | 9.40 |

# 04 22 Concrete Unit Masonry

## 04 22 10 – Concrete Masonry Units

### 04 22 10.42 Concrete Block, Screen Block

| | | Crew | Daily Output | Labor-Hours | Unit | Material | 2011 Bare Costs Labor | Equipment | Total | Total Incl O&P |
|---|---|---|---|---|---|---|---|---|---|---|
| 0300 | 8" thick | D-8 | 270 | .148 | S.F. | 2.99 | 5.90 | | 8.89 | 12.15 |
| 0350 | 12" x 12", 4" thick | | 290 | .138 | | 2.21 | 5.50 | | 7.71 | 10.70 |
| 0500 | 8" thick | ↓ | 250 | .160 | ↓ | 3.06 | 6.35 | | 9.41 | 12.90 |

### 04 22 10.44 Glazed Concrete Block

| | | Crew | Daily Output | Labor-Hours | Unit | Material | 2011 Bare Costs Labor | Equipment | Total | Total Incl O&P |
|---|---|---|---|---|---|---|---|---|---|---|
| 0010 | **GLAZED CONCRETE BLOCK** C744 | | | | | | | | | |
| 0100 | Single face, 8" x 16" units, 2" thick | D-8 | 360 | .111 | S.F. | 8.25 | 4.42 | | 12.67 | 15.75 |
| 0200 | 4" thick | | 345 | .116 | | 8.50 | 4.62 | | 13.12 | 16.30 |
| 0250 | 6" thick | | 330 | .121 | | 9.15 | 4.83 | | 13.98 | 17.30 |
| 0300 | 8" thick | | 310 | .129 | | 9.25 | 5.15 | | 14.40 | 17.85 |
| 0350 | 10" thick | ↓ | 295 | .136 | | 11.10 | 5.40 | | 16.50 | 20.50 |
| 0400 | 12" thick | D-9 | 280 | .171 | | 11.80 | 6.70 | | 18.50 | 23 |
| 0700 | Double face, 8" x 16" units, 4" thick | D-8 | 340 | .118 | | 12.20 | 4.68 | | 16.88 | 20.50 |
| 0750 | 6" thick | | 320 | .125 | | 14.50 | 4.98 | | 19.48 | 23.50 |
| 0800 | 8" thick | | 300 | .133 | ↓ | 15.20 | 5.30 | | 20.50 | 24.50 |
| 1000 | Jambs, bullnose or square, single face, 8" x 16", 2" thick | | 315 | .127 | Ea. | 17.50 | 5.05 | | 22.55 | 27 |
| 1050 | 4" thick | | 285 | .140 | " | 20.50 | 5.60 | | 26.10 | 31 |
| 1200 | Caps, bullnose or square, 8" x 16", 2" thick | | 420 | .095 | L.F. | 14.15 | 3.79 | | 17.94 | 21.50 |
| 1250 | 4" thick | | 380 | .105 | | 15.95 | 4.19 | | 20.14 | 24 |
| 1500 | Cove base, 8" x 16", 2" thick | | 315 | .127 | | 11 | 5.05 | | 16.05 | 19.70 |
| 1550 | 4" thick | | 285 | .140 | | 7.15 | 5.60 | | 12.75 | 16.25 |
| 1600 | 6" thick | | 265 | .151 | | 7.70 | 6 | | 13.70 | 17.45 |
| 1650 | 8" thick | ↓ | 245 | .163 | ↓ | 8.10 | 6.50 | | 14.60 | 18.70 |

# 04 23 Glass Unit Masonry

## 04 23 13 – Vertical Glass Unit Masonry

### 04 23 13.10 Glass Block

| | | Crew | Daily Output | Labor-Hours | Unit | Material | 2011 Bare Costs Labor | Equipment | Total | Total Incl O&P |
|---|---|---|---|---|---|---|---|---|---|---|
| 0010 | **GLASS BLOCK** | | | | | | | | | |
| 0100 | Plain, 4" thick, under 1,000 S.F., 6" x 6" | D-8 | 115 | .348 | S.F. | 25.50 | 13.85 | | 39.35 | 49 |
| 0150 | 8" x 8" | | 160 | .250 | | 15 | 9.95 | | 24.95 | 31.50 |
| 0160 | end block | | 160 | .250 | | 50 | 9.95 | | 59.95 | 70 |
| 0170 | 90 deg corner | | 160 | .250 | | 49.50 | 9.95 | | 59.45 | 69.50 |
| 0180 | 45 deg corner | | 160 | .250 | | 46 | 9.95 | | 55.95 | 65.50 |
| 0200 | 12" x 12" | | 175 | .229 | | 21.50 | 9.10 | | 30.60 | 37 |
| 0210 | 4" x 8" | | 160 | .250 | | 15.30 | 9.95 | | 25.25 | 32 |
| 0220 | 6" x 8" | | 160 | .250 | | 14.70 | 9.95 | | 24.65 | 31 |
| 0300 | 1,000 to 5,000 S.F., 6" x 6" | | 135 | .296 | | 25 | 11.80 | | 36.80 | 45 |
| 0350 | 8" x 8" | | 190 | .211 | | 14.70 | 8.40 | | 23.10 | 29 |
| 0400 | 12" x 12" | | 215 | .186 | | 21 | 7.40 | | 28.40 | 34.50 |
| 0410 | 4" x 8" | | 215 | .186 | | 6.05 | 7.40 | | 13.45 | 17.75 |
| 0420 | 6" x 8" | | 215 | .186 | | 5.75 | 7.40 | | 13.15 | 17.45 |
| 0500 | Over 5,000 S.F., 6" x 6" | | 145 | .276 | | 24.50 | 11 | | 35.50 | 43.50 |
| 0550 | 8" x 8" | | 215 | .186 | | 14.25 | 7.40 | | 21.65 | 27 |
| 0600 | 12" x 12" | | 240 | .167 | | 20.50 | 6.65 | | 27.15 | 32.50 |
| 0610 | 4" x 8" | | 240 | .167 | | 5.65 | 6.65 | | 12.30 | 16.20 |
| 0620 | 6" x 8" | ↓ | 240 | .167 | ↓ | 5.75 | 6.65 | | 12.40 | 16.30 |
| 0700 | For solar reflective blocks, add | | | | | 100% | | | | |
| 1000 | Thinline, plain, 3-1/8" thick, under 1,000 S.F., 6" x 6" | D-8 | 115 | .348 | S.F. | 17.50 | 13.85 | | 31.35 | 40.50 |
| 1050 | 8" x 8" | | 160 | .250 | | 9.80 | 9.95 | | 19.75 | 25.50 |
| 1200 | Over 5,000 S.F., 6" x 6" | | 145 | .276 | | 16.40 | 11 | | 27.40 | 34.50 |
| 1250 | 8" x 8" | | 215 | .186 | | 9.20 | 7.40 | | 16.60 | 21 |
| 1400 | For cleaning block after installation (both sides), add | ↓ | 1000 | .040 | ↓ | .11 | 1.59 | | 1.70 | 2.51 |

# 04 23 Glass Unit Masonry

## 04 23 13 – Vertical Glass Unit Masonry

| 04 23 13.10 Glass Block | | Crew | Daily Output | Labor-Hours | Unit | Material | 2011 Bare Costs Labor | Equipment | Total | Total Incl O&P |
|---|---|---|---|---|---|---|---|---|---|---|
| 4000 | Accessories | | | | | | | | | |
| 4100 | Anchors, 20 ga. galv., 1-3/4" wide x 24" long | | | | Ea. | 3.17 | | | 3.17 | 3.49 |
| 4200 | Emulsion asphalt | | | | Gal. | 8.10 | | | 8.10 | 8.90 |
| 4300 | Expansion joint, fiberglass | | | | L.F. | .57 | | | .57 | .63 |
| 4400 | Steel mesh, double galvanized | | | | " | .94 | | | .94 | 1.03 |

# 04 24 Adobe Unit Masonry

## 04 24 16 – Manufactured Adobe Unit Masonry

### 04 24 16.06 Adobe Brick

| | | | Crew | Daily Output | Labor-Hours | Unit | Material | Labor | Equipment | Total | Total Incl O&P |
|---|---|---|---|---|---|---|---|---|---|---|---|
| 0010 | **ADOBE BRICK**, Semi-stabilized, with cement mortar | | | | | | | | | | |
| 0060 | Brick, 10" x 4" x 14", 2.6/S.F. | G | D-8 | 560 | .071 | S.F. | 3.49 | 2.84 | | 6.33 | 8.10 |
| 0080 | 12" x 4" x 16", 2.3/S.F. | G | | 580 | .069 | | 4.61 | 2.75 | | 7.36 | 9.15 |
| 0100 | 10" x 4" x 16", 2.3/S.F. | G | | 590 | .068 | | 4.37 | 2.70 | | 7.07 | 8.85 |
| 0120 | 8" x 4" x 16", 2.3/S.F. | G | | 560 | .071 | | 3.56 | 2.84 | | 6.40 | 8.20 |
| 0140 | 4" x 4" x 16", 2.3/S.F. | G | | 540 | .074 | | 2.56 | 2.95 | | 5.51 | 7.25 |
| 0160 | 6" x 4" x 16", 2.3/S.F. | G | | 540 | .074 | | 2.59 | 2.95 | | 5.54 | 7.25 |
| 0180 | 4" x 4" x 12", 3.0/S.F. | G | | 520 | .077 | | 2.45 | 3.06 | | 5.51 | 7.30 |
| 0200 | 8" x 4" x 12", 3.0/S.F. | G | | 520 | .077 | | 3.16 | 3.06 | | 6.22 | 8.05 |

# 04 25 Unit Masonry Panels

## 04 25 20 – Pre-Fabricated Masonry Panels

### 04 25 20.10 Brick and Epoxy Mortar Panels

| | | Crew | Daily Output | Labor-Hours | Unit | Material | Labor | Equipment | Total | Total Incl O&P |
|---|---|---|---|---|---|---|---|---|---|---|---|
| 0010 | **BRICK AND EPOXY MORTAR PANELS** | | | | | | | | | |
| 0020 | Prefabricated brick & epoxy mortar, 4" thick, minimum | C-11 | 775 | .093 | S.F. | 9.75 | 4.42 | 2.46 | 16.63 | 21 |
| 0100 | Maximum | " | 500 | .144 | | 10.75 | 6.85 | 3.81 | 21.41 | 28 |
| 0200 | For 2" concrete back-up, add | | | | | 50% | | | | |
| 0300 | For 1" urethane & 3" concrete back-up, add | | | | | 70% | | | | |

# 04 27 Multiple-Wythe Unit Masonry

## 04 27 10 – Multiple-Wythe Masonry

### 04 27 10.20 Cavity Walls

| | | Crew | Daily Output | Labor-Hours | Unit | Material | Labor | Equipment | Total | Total Incl O&P |
|---|---|---|---|---|---|---|---|---|---|---|
| 0010 | **CAVITY WALLS**, brick and CMU, includes joint reinforcing and ties | | | | | | | | | |
| 0200 | 4" face brick, 4" block | D-8 | 165 | .242 | S.F. | 5.80 | 9.65 | | 15.45 | 21 |
| 0400 | 6" block | | 145 | .276 | | 6.25 | 11 | | 17.25 | 23.50 |
| 0600 | 8" block | | 125 | .320 | | 6.45 | 12.75 | | 19.20 | 26 |

### 04 27 10.30 Brick Walls

| | | | Crew | Daily Output | Labor-Hours | Unit | Material | Labor | Equipment | Total | Total Incl O&P |
|---|---|---|---|---|---|---|---|---|---|---|---|
| 0010 | **BRICK WALLS**, including mortar, excludes scaffolding | R042110-20 | | | | | | | | | |
| 0020 | Estimating by number of brick | | | | | | | | | | |
| 0140 | Face brick, 4" thick wall, 6.75 brick/S.F. | | D-8 | 1.45 | 27.586 | M | 610 | 1,100 | | 1,710 | 2,325 |
| 0150 | Common brick, 4" thick wall, 6.75 brick/S.F. | | | 1.60 | 25 | | 495 | 995 | | 1,490 | 2,050 |
| 0204 | 8" thick, 13.50 bricks per S.F. | | | 1.80 | 22.222 | | 510 | 885 | | 1,395 | 1,875 |
| 0250 | 12" thick, 20.25 bricks per S.F. | | | 1.90 | 21.053 | | 515 | 840 | | 1,355 | 1,825 |
| 0304 | 16" thick, 27.00 bricks per S.F. | | | 2 | 20 | | 520 | 795 | | 1,315 | 1,775 |
| 0500 | Reinforced, face brick, 4" thick wall, 6.75 brick/S.F. | | | 1.40 | 28.571 | | 635 | 1,125 | | 1,760 | 2,400 |
| 0520 | Common brick, 4" thick wall, 6.75 brick/S.F. | | | 1.55 | 25.806 | | 515 | 1,025 | | 1,540 | 2,125 |
| 0550 | 8" thick, 13.50 bricks per S.F. | | | 1.75 | 22.857 | | 535 | 910 | | 1,445 | 1,950 |
| 0600 | 12" thick, 20.25 bricks per S.F. | | | 1.85 | 21.622 | | 535 | 860 | | 1,395 | 1,900 |

# 04 27 Multiple-Wythe Unit Masonry

## 04 27 10 – Multiple-Wythe Masonry

| 04 27 10.30 Brick Walls | | Crew | Daily Output | Labor-Hours | Unit | Material | 2011 Bare Costs Labor | Equipment | Total | Total Incl O&P |
|---|---|---|---|---|---|---|---|---|---|---|
| 0650 | 16" thick, 27.00 bricks per S.F. | D-8 | 1.95 | 20.513 | M | 545 | 815 | | 1,360 | 1,825 |
| 0790 | Alternate method of figuring by square foot | | | | | | | | | |
| 0800 | Face brick, 4" thick wall, 6.75 brick/S.F. | D-8 | 215 | .186 | S.F. | 4.12 | 7.40 | | 11.52 | 15.65 |
| 0850 | Common brick, 4" thick wall, 6.75 brick/S.F. | | 240 | .167 | | 3.33 | 6.65 | | 9.98 | 13.60 |
| 0900 | 8" thick, 13.50 bricks per S.F. | | 135 | .296 | | 6.90 | 11.80 | | 18.70 | 25.50 |
| 1000 | 12" thick, 20.25 bricks per S.F. | | 95 | .421 | | 10.40 | 16.75 | | 27.15 | 36.50 |
| 1050 | 16" thick, 27.00 bricks per S.F. | | 75 | .533 | | 14.05 | 21 | | 35.05 | 47.50 |
| 1200 | Reinforced, face brick, 4" thick wall, 6.75 brick/S.F. | | 210 | .190 | | 4.27 | 7.60 | | 11.87 | 16.10 |
| 1220 | Common brick, 4" thick wall, 6.75 brick/S.F. | | 235 | .170 | | 3.48 | 6.80 | | 10.28 | 14 |
| 1250 | 8" thick, 13.50 bricks per S.F. | | 130 | .308 | | 7.20 | 12.25 | | 19.45 | 26.50 |
| 1300 | 12" thick, 20.25 bricks per S.F. | | 90 | .444 | | 10.85 | 17.70 | | 28.55 | 38.50 |
| 1350 | 16" thick, 27.00 bricks per S.F. | | 70 | .571 | | 14.70 | 23 | | 37.70 | 50 |

## 04 27 10.40 Steps

| | | Crew | Daily Output | Labor-Hours | Unit | Material | Labor | Equipment | Total | Total Incl O&P |
|---|---|---|---|---|---|---|---|---|---|---|
| 0010 | **STEPS** | | | | | | | | | |
| 0012 | Entry steps, select common brick | D-1 | .30 | 53.333 | M | 430 | 2,075 | | 2,505 | 3,600 |

# 04 41 Dry-Placed Stone

## 04 41 10 – Dry-Placed Stone

### 04 41 10.10 Rough Stone Wall

| | | | Crew | Daily Output | Labor-Hours | Unit | Material | Labor | Equipment | Total | Total Incl O&P |
|---|---|---|---|---|---|---|---|---|---|---|---|
| 0011 | **ROUGH STONE WALL**, Dry | | | | | | | | | | |
| 0012 | Dry laid (no mortar), under 18" thick | G | D-1 | 60 | .267 | C.F. | 8.80 | 10.40 | | 19.20 | 25.50 |
| 0100 | Random fieldstone, under 18" thick | G | D-12 | 60 | .533 | | 8.80 | 21 | | 29.80 | 41 |
| 0150 | Over 18" thick | G | " | 63 | .508 | | 10.55 | 20 | | 30.55 | 41.50 |
| 0500 | Field stone veneer | G | D-8 | 120 | .333 | S.F. | 9.95 | 13.25 | | 23.20 | 31 |
| 0510 | Valley stone veneer | G | | 120 | .333 | | 9.95 | 13.25 | | 23.20 | 31 |
| 0520 | River stone veneer | G | | 120 | .333 | | 9.95 | 13.25 | | 23.20 | 31 |
| 0600 | Rubble stone walls, in mortar bed, up to 18" thick | G | D-11 | 75 | .320 | C.F. | 10.65 | 13.15 | | 23.80 | 31.50 |

# 04 43 Stone Masonry

## 04 43 10 – Masonry with Natural and Processed Stone

### 04 43 10.05 Ashlar Veneer

| | | Crew | Daily Output | Labor-Hours | Unit | Material | Labor | Equipment | Total | Total Incl O&P |
|---|---|---|---|---|---|---|---|---|---|---|
| 0011 | **ASHLAR VENEER** 4" + or - thk, random or random rectangular | | | | | | | | | |
| 0150 | Sawn face, split joints, low priced stone | D-8 | 140 | .286 | S.F. | 9.50 | 11.35 | | 20.85 | 27.50 |
| 0200 | Medium priced stone | | 130 | .308 | | 11.30 | 12.25 | | 23.55 | 31 |
| 0300 | High priced stone | | 120 | .333 | | 14.50 | 13.25 | | 27.75 | 36 |
| 0600 | Seam face, split joints, medium price stone | | 125 | .320 | | 7.55 | 12.75 | | 20.30 | 27.50 |
| 0700 | High price stone | | 120 | .333 | | 15 | 13.25 | | 28.25 | 36.50 |
| 1000 | Split or rock face, split joints, medium price stone | | 125 | .320 | | 10.80 | 12.75 | | 23.55 | 31 |
| 1100 | High price stone | | 120 | .333 | | 15 | 13.25 | | 28.25 | 36.50 |

### 04 43 10.10 Bluestone

| | | Crew | Daily Output | Labor-Hours | Unit | Material | Labor | Equipment | Total | Total Incl O&P |
|---|---|---|---|---|---|---|---|---|---|---|
| 0010 | **BLUESTONE**, cut to size | | | | | | | | | |
| 0500 | Sills, natural cleft, 10" wide to 6' long, 1-1/2" thick | D-11 | 70 | .343 | L.F. | 13.35 | 14.05 | | 27.40 | 35.50 |
| 0550 | 2" thick | " | 63 | .381 | | 14.60 | 15.65 | | 30.25 | 39.50 |
| 1000 | Stair treads, natural cleft, 12" wide, 6' long, 1-1/2" thick | D-10 | 115 | .278 | | 30 | 11.80 | 3.78 | 45.58 | 55 |
| 1050 | 2" thick | | 105 | .305 | | 30 | 12.90 | 4.14 | 47.04 | 57 |
| 1100 | Smooth finish, 1-1/2" thick | | 115 | .278 | | 15 | 11.80 | 3.78 | 30.58 | 38.50 |
| 1300 | Thermal finish, 1-1/2" thick | | 115 | .278 | | 15 | 11.80 | 3.78 | 30.58 | 38.50 |
| 1350 | 2" thick | | 105 | .305 | | 15 | 12.90 | 4.14 | 32.04 | 40.50 |

## 04 43 10 – Masonry with Natural and Processed Stone

| 04 43 10.45 Granite | Crew | Daily Output | Labor-Hours | Unit | Material | 2011 Bare Costs Labor | Equipment | Total | Total Incl O&P |
|---|---|---|---|---|---|---|---|---|---|
| 0010 **GRANITE**, cut to size | | | | | | | | | |
| 0050 Veneer, polished face, 3/4" to 1-1/2" thick | | | | | | | | | |
| 0150 Low price, gray, light gray, etc. | D-10 | 130 | .246 | S.F. | 26.50 | 10.45 | 3.34 | 40.29 | 48.50 |
| 0220 High price, red, black, etc. | " | 130 | .246 | " | 41.50 | 10.45 | 3.34 | 55.29 | 65.50 |
| 0300 1-1/2" to 2-1/2" thick, veneer | | | | | | | | | |
| 0350 Low price, gray, light gray, etc. | D-10 | 130 | .246 | S.F. | 28.50 | 10.45 | 3.34 | 42.29 | 50.50 |
| 0550 High price, red, black, etc. | " | 130 | .246 | " | 51.50 | 10.45 | 3.34 | 65.29 | 76 |
| 0700 2-1/2" to 4" thick, veneer | | | | | | | | | |
| 0750 Low price, gray, light gray, etc. | D-10 | 110 | .291 | S.F. | 38 | 12.35 | 3.95 | 54.30 | 65 |
| 0950 High price, red, black, etc. | " | 110 | .291 | | 62.50 | 12.35 | 3.95 | 78.80 | 91.50 |
| 1000 For bush hammered finish, deduct | | | | | 5% | | | | |
| 1050 Coarse rubbed finish, deduct | | | | | 10% | | | | |
| 1100 Honed finish, deduct | | | | | 5% | | | | |
| 1150 Thermal finish, deduct | | | | | 18% | | | | |
| 2450 For radius under 5', add | | | | L.F. | 100% | | | | |
| 2500 Steps, copings, etc., finished on more than one surface | | | | | | | | | |
| 2550 Minimum | D-10 | 50 | .640 | C.F. | 91 | 27 | 8.70 | 126.70 | 150 |
| 2600 Maximum | " | 50 | .640 | " | 146 | 27 | 8.70 | 181.70 | 210 |
| 2800 Pavers, 4" x 4" x 4" blocks, split face and joints | | | | | | | | | |
| 2850 Minimum | D-11 | 80 | .300 | S.F. | 12.95 | 12.30 | | 25.25 | 32.50 |
| 2900 Maximum | " | 80 | .300 | | 28.50 | 12.30 | | 40.80 | 50 |
| 4000 Soffits, 2" thick, minimum | D-13 | 35 | 1.371 | | 37.50 | 56.50 | 12.40 | 106.40 | 140 |
| 4100 Maximum | | 35 | 1.371 | | 90 | 56.50 | 12.40 | 158.90 | 198 |
| 4200 4" thick, minimum | | 35 | 1.371 | | 62 | 56.50 | 12.40 | 130.90 | 167 |
| 4300 Maximum | | 35 | 1.371 | | 117 | 56.50 | 12.40 | 185.90 | 228 |

### 04 43 10.50 Lightweight Natural Stone

| 04 43 10.50 Lightweight Natural Stone | | Crew | Daily Output | Labor-Hours | Unit | Material | Labor | Equipment | Total | Total Incl O&P |
|---|---|---|---|---|---|---|---|---|---|---|
| 0011 **LIGHTWEIGHT NATURAL STONE** Lava type | | | | | | | | | | |
| 0100 Veneer, rubble face, sawed back, irregular shapes | G | D-10 | 130 | .246 | S.F. | 6.30 | 10.45 | 3.34 | 20.09 | 26 |
| 0200 Sawed face and back, irregular shapes | G | " | 130 | .246 | " | 6.30 | 10.45 | 3.34 | 20.09 | 26 |

### 04 43 10.55 Limestone

| 04 43 10.55 Limestone | Crew | Daily Output | Labor-Hours | Unit | Material | Labor | Equipment | Total | Total Incl O&P |
|---|---|---|---|---|---|---|---|---|---|
| 0010 **LIMESTONE**, cut to size | | | | | | | | | |
| 0020 Veneer facing panels | | | | | | | | | |
| 0500 Texture finish, light stick, 4-1/2" thick, 5' x 12' | D-4 | 300 | .107 | S.F. | 48 | 4.17 | .42 | 52.59 | 59.50 |
| 0750 5" thick, 5' x 14' panels | D-10 | 275 | .116 | | 52.50 | 4.93 | 1.58 | 59.01 | 66.50 |
| 1000 Sugarcube finish, 2" Thick, 3' x 5' panels | | 275 | .116 | | 35 | 4.93 | 1.58 | 41.51 | 47.50 |
| 1050 3" Thick, 4' x 9' panels | | 275 | .116 | | 38 | 4.93 | 1.58 | 44.51 | 51 |
| 1200 4" Thick, 5' x 11' panels | | 275 | .116 | | 41 | 4.93 | 1.58 | 47.51 | 54 |
| 1400 Sugarcube, textured finish, 4-1/2" thick, 5' x 12' | | 275 | .116 | | 45 | 4.93 | 1.58 | 51.51 | 58.50 |
| 1450 5" thick, 5' x 14' panels | | 275 | .116 | | 53.50 | 4.93 | 1.58 | 60.01 | 67.50 |
| 2000 Coping, sugarcube finish, top & 2 sides | | 30 | 1.067 | C.F. | 68 | 45 | 14.50 | 127.50 | 159 |
| 2100 Sills, lintels, jambs, trim, stops, sugarcube finish, average | | 20 | 1.600 | | 68 | 68 | 21.50 | 157.50 | 201 |
| 2150 Detailed | | 20 | 1.600 | | 68 | 68 | 21.50 | 157.50 | 201 |
| 2300 Steps, extra hard, 14" wide, 6" rise | | 50 | .640 | L.F. | 41.50 | 27 | 8.70 | 77.20 | 96 |
| 3000 Quoins, plain finish, 6" x 12" x 12" | D-12 | 25 | 1.280 | Ea. | 38 | 50.50 | | 88.50 | 118 |
| 3050 6" x 16" x 24" | " | 25 | 1.280 | " | 50.50 | 50.50 | | 101 | 132 |

### 04 43 10.60 Marble

| 04 43 10.60 Marble | Crew | Daily Output | Labor-Hours | Unit | Material | Labor | Equipment | Total | Total Incl O&P |
|---|---|---|---|---|---|---|---|---|---|
| 0011 **MARBLE**, ashlar, split face, 4" + or - thick, random | | | | | | | | | |
| 0040 Lengths 1' to 4' & heights 2" to 7-1/2", average | D-8 | 175 | .229 | S.F. | 17.75 | 9.10 | | 26.85 | 33 |
| 0100 Base, polished, 3/4" or 7/8" thick, polished, 6" high | D-10 | 65 | .492 | L.F. | 16.15 | 21 | 6.70 | 43.85 | 56.50 |
| 0300 Carvings or bas relief, from templates, average | | 80 | .400 | S.F. | 144 | 16.95 | 5.45 | 166.40 | 189 |
| 0350 Maximum | | 80 | .400 | " | 335 | 16.95 | 5.45 | 357.40 | 400 |

# 04 43 Stone Masonry

## 04 43 10 – Masonry with Natural and Processed Stone

### 04 43 10.60 Marble

| | | Crew | Daily Output | Labor-Hours | Unit | Material | 2011 Bare Costs Labor | Equipment | Total | Total Incl O&P |
|---|---|---|---|---|---|---|---|---|---|---|
| 0600 | Columns, cornices, mouldings, etc. | | | | | | | | | |
| 0650 | Hand or special machine cut, average | D-10 | 35 | .914 | C.F. | 144 | 39 | 12.40 | 195.40 | 230 |
| 0700 | Maximum | " | 35 | .914 | " | 310 | 39 | 12.40 | 361.40 | 410 |
| 1000 | Facing, polished finish, cut to size, 3/4" to 7/8" thick | | | | | | | | | |
| 1050 | Average | D-10 | 130 | .246 | S.F. | 23.50 | 10.45 | 3.34 | 37.29 | 45.50 |
| 1100 | Maximum | | 130 | .246 | | 55 | 10.45 | 3.34 | 68.79 | 80 |
| 1300 | 1-1/4" thick, average | | 125 | .256 | | 34.50 | 10.85 | 3.48 | 48.83 | 57.50 |
| 1350 | Maximum | | 125 | .256 | | 68.50 | 10.85 | 3.48 | 82.83 | 95.50 |
| 1500 | 2" thick, average | | 120 | .267 | | 40 | 11.30 | 3.62 | 54.92 | 65 |
| 1550 | Maximum | | 120 | .267 | | 69 | 11.30 | 3.62 | 83.92 | 97 |
| 2200 | Window sills, 6" x 3/4" thick | D-1 | 85 | .188 | L.F. | 8.95 | 7.35 | | 16.30 | 21 |
| 2500 | Flooring, polished tiles, 12" x 12" x 3/8" thick | | | | | | | | | |
| 2510 | Thin set, average | D-11 | 90 | .267 | S.F. | 11.90 | 10.95 | | 22.85 | 29.50 |
| 2600 | Maximum | | 90 | .267 | | 183 | 10.95 | | 193.95 | 217 |
| 2700 | Mortar bed, average | | 65 | .369 | | 10.65 | 15.15 | | 25.80 | 34.50 |
| 2740 | Maximum | | 65 | .369 | | 183 | 15.15 | | 198.15 | 224 |
| 2780 | Travertine, 3/8" thick, average | D-10 | 130 | .246 | | 15.15 | 10.45 | 3.34 | 28.94 | 36 |
| 2790 | Maximum | " | 130 | .246 | | 37 | 10.45 | 3.34 | 50.79 | 60.50 |
| 2800 | Patio tile, non-slip, 1/2" thick, flame finish | D-11 | 75 | .320 | | 15.45 | 13.15 | | 28.60 | 36.50 |
| 2900 | Shower or toilet partitions, 7/8" thick partitions | | | | | | | | | |
| 3050 | 3/4" or 1-1/4" thick stiles, polished 2 sides, average | D-11 | 75 | .320 | S.F. | 46 | 13.15 | | 59.15 | 70 |
| 3201 | Soffits, add to above prices | | | | " | 20% | 100% | | | |
| 3210 | Stairs, risers, 7/8" thick x 6" high | D-10 | 115 | .278 | L.F. | 15.55 | 11.80 | 3.78 | 31.13 | 39 |
| 3360 | Treads, 12" wide x 1-1/4" thick | " | 115 | .278 | " | 22.50 | 11.80 | 3.78 | 38.08 | 47 |
| 3500 | Thresholds, 3' long, 7/8" thick, 4" to 5" wide, plain | D-12 | 24 | 1.333 | Ea. | 17.15 | 52.50 | | 69.65 | 98 |
| 3550 | Beveled | | 24 | 1.333 | " | 19.90 | 52.50 | | 72.40 | 101 |
| 3700 | Window stools, polished, 7/8" thick, 5" wide | | 85 | .376 | L.F. | 15.95 | 14.85 | | 30.80 | 40 |

### 04 43 10.75 Sandstone or Brownstone

| | | Crew | Daily Output | Labor-Hours | Unit | Material | 2011 Bare Costs Labor | Equipment | Total | Total Incl O&P |
|---|---|---|---|---|---|---|---|---|---|---|
| 0011 | **SANDSTONE OR BROWNSTONE** | | | | | | | | | |
| 0100 | Sawed face veneer, 2-1/2" thick, to 2' x 4' panels | D-10 | 130 | .246 | S.F. | 21 | 10.45 | 3.34 | 34.79 | 42.50 |
| 0150 | 4" thick, to 3'-6" x 8' panels | | 100 | .320 | | 21 | 13.55 | 4.35 | 38.90 | 48.50 |
| 0300 | Split face, random sizes | | 100 | .320 | | 14 | 13.55 | 4.35 | 31.90 | 40.50 |
| 0350 | Cut stone trim (limestone) | | | | | | | | | |
| 0360 | Ribbon stone, 4" thick, 5' pieces | D-8 | 120 | .333 | Ea. | 172 | 13.25 | | 185.25 | 209 |
| 0370 | Cove stone, 4" thick, 5' pieces | | 105 | .381 | | 173 | 15.15 | | 188.15 | 213 |
| 0380 | Cornice stone, 10" to 12" wide | | 90 | .444 | | 214 | 17.70 | | 231.70 | 262 |
| 0390 | Band stone, 4" thick, 5' pieces | | 145 | .276 | | 110 | 11 | | 121 | 138 |
| 0410 | Window and door trim, 3" to 4" wide | | 160 | .250 | | 93.50 | 9.95 | | 103.45 | 118 |
| 0420 | Key stone, 18" long | | 60 | .667 | | 98.50 | 26.50 | | 125 | 148 |

### 04 43 10.80 Slate

| | | Crew | Daily Output | Labor-Hours | Unit | Material | 2011 Bare Costs Labor | Equipment | Total | Total Incl O&P |
|---|---|---|---|---|---|---|---|---|---|---|
| 0010 | **SLATE** | | | | | | | | | |
| 0040 | Pennsylvania - blue gray to black | | | | | | | | | |
| 0050 | Vermont - unfading green, mottled green & purple, gray & purple | | | | | | | | | |
| 0100 | Virginia - blue black | | | | | | | | | |
| 0200 | Exterior paving, natural cleft, 1" thick | | | | | | | | | |
| 0500 | 24" x 24", Pennsylvania | D-12 | 120 | .267 | S.F. | 13 | 10.55 | | 23.55 | 30 |
| 0550 | Vermont | | 120 | .267 | | 31 | 10.55 | | 41.55 | 50 |
| 0600 | Virginia | | 120 | .267 | | 20.50 | 10.55 | | 31.05 | 38.50 |
| 1000 | Interior flooring, natural cleft, 1/2" thick | | | | | | | | | |
| 1300 | 24" x 24" Pennsylvania | D-12 | 120 | .267 | S.F. | 7.75 | 10.55 | | 18.30 | 24.50 |
| 1350 | Vermont | | 120 | .267 | | 25 | 10.55 | | 35.55 | 43.50 |
| 1400 | Virginia | | 120 | .267 | | 14.75 | 10.55 | | 25.30 | 32 |

# 04 43 Stone Masonry

## 04 43 10 - Masonry with Natural and Processed Stone

### 04 43 10.80 Slate

| | | Crew | Daily Output | Labor-Hours | Unit | Material | 2011 Bare Costs Labor | Equipment | Total | Total Incl O&P |
|---|---|---|---|---|---|---|---|---|---|---|
| 2000 | Facing panels, 1-1/4" thick, to 4' x 4' panels | | | | | | | | | |
| 2100 | Natural cleft finish, Pennsylvania | D-10 | 180 | .178 | S.F. | 34 | 7.55 | 2.41 | 43.96 | 51 |
| 2110 | Vermont | | 180 | .178 | | 27 | 7.55 | 2.41 | 36.96 | 44 |
| 2120 | Virginia | | 180 | .178 | | 33 | 7.55 | 2.41 | 42.96 | 50.50 |
| 2150 | Sand rubbed finish, surface, add | | | | | 10.15 | | | 10.15 | 11.20 |
| 2200 | Honed finish, add | | | | | 7.35 | | | 7.35 | 8.10 |
| 2500 | Ribbon, natural cleft finish, 1" thick, to 9 S.F. | D-10 | 80 | .400 | | 13.10 | 16.95 | 5.45 | 35.50 | 46 |
| 2700 | 1-1/2" thick | | 78 | .410 | | 17.05 | 17.40 | 5.55 | 40 | 51 |
| 2850 | 2" thick | | 76 | .421 | | 20.50 | 17.85 | 5.70 | 44.05 | 56 |
| 3500 | Stair treads, sand finish, 1" thick x 12" wide | | | | | | | | | |
| 3600 | 3 L.F. to 6 L.F. | D-10 | 120 | .267 | L.F. | 24 | 11.30 | 3.62 | 38.92 | 47 |
| 3700 | Ribbon, sand finish, 1" thick x 12" wide | | | | | | | | | |
| 3750 | To 6 L.F. | D-10 | 120 | .267 | L.F. | 20 | 11.30 | 3.62 | 34.92 | 43 |
| 4000 | Stools or sills, sand finish, 1" thick, 6" wide | D-12 | 160 | .200 | | 11.50 | 7.90 | | 19.40 | 24.50 |
| 4200 | 10" wide | | 90 | .356 | | 17.75 | 14.05 | | 31.80 | 40.50 |
| 4400 | 2" thick, 6" wide | | 140 | .229 | | 18.50 | 9 | | 27.50 | 34 |
| 4600 | 10" wide | | 90 | .356 | | 29 | 14.05 | | 43.05 | 53 |
| 4800 | For lengths over 3', add | | | | | 25% | | | | |

### 04 43 10.85 Window Sill

| | | Crew | Daily Output | Labor-Hours | Unit | Material | 2011 Bare Costs Labor | Equipment | Total | Total Incl O&P |
|---|---|---|---|---|---|---|---|---|---|---|
| 0010 | **WINDOW SILL** | | | | | | | | | |
| 0020 | Bluestone, thermal top, 10" wide, 1-1/2" thick | D-1 | 85 | .188 | S.F. | 16.05 | 7.35 | | 23.40 | 28.50 |
| 0050 | 2" thick | | 75 | .213 | " | 18.75 | 8.30 | | 27.05 | 33 |
| 0100 | Cut stone, 5" x 8" plain | | 48 | .333 | L.F. | 11.90 | 13 | | 24.90 | 32.50 |
| 0200 | Face brick on edge, brick, 8" wide | | 80 | .200 | | 2.53 | 7.80 | | 10.33 | 14.50 |
| 0400 | Marble, 9" wide, 1" thick | | 85 | .188 | | 8.50 | 7.35 | | 15.85 | 20.50 |
| 0900 | Slate, colored, unfading, honed, 12" wide, 1" thick | | 85 | .188 | | 17.10 | 7.35 | | 24.45 | 30 |
| 0950 | 2" thick | | 70 | .229 | | 24 | 8.90 | | 32.90 | 39.50 |

# 04 51 Flue Liner Masonry

## 04 51 10 - Clay Flue Lining

### 04 51 10.10 Flue Lining

| | | Crew | Daily Output | Labor-Hours | Unit | Material | 2011 Bare Costs Labor | Equipment | Total | Total Incl O&P |
|---|---|---|---|---|---|---|---|---|---|---|
| 0010 | **FLUE LINING**, including mortar | | | | | | | | | |
| 0020 | Clay, 8" x 8" | D-1 | 125 | .128 | V.L.F. | 4.97 | 4.99 | | 9.96 | 12.95 |
| 0100 | 8" x 12" | | 103 | .155 | | 7.55 | 6.05 | | 13.60 | 17.40 |
| 0200 | 12" x 12" | | 93 | .172 | | 9.75 | 6.70 | | 16.45 | 21 |
| 0300 | 12" x 18" | | 84 | .190 | | 18.90 | 7.40 | | 26.30 | 32 |
| 0400 | 18" x 18" | | 75 | .213 | | 22.50 | 8.30 | | 30.80 | 37 |
| 0500 | 20" x 20" | | 66 | .242 | | 36 | 9.45 | | 45.45 | 53.50 |
| 0600 | 24" x 24" | | 56 | .286 | | 47 | 11.15 | | 58.15 | 68 |
| 1000 | Round, 18" diameter | | 66 | .242 | | 30.50 | 9.45 | | 39.95 | 47.50 |
| 1100 | 24" diameter | | 47 | .340 | | 67.50 | 13.25 | | 80.75 | 94 |

# 04 54 Refractory Brick Masonry

## 04 54 10 – Refractory Brick Work

| 04 54 10.10 Fire Brick | | Crew | Daily Output | Labor-Hours | Unit | Material | 2011 Bare Costs Labor | Equipment | Total | Total Incl O&P |
|---|---|---|---|---|---|---|---|---|---|---|
| 0010 | **FIRE BRICK** | | | | | | | | | |
| 0012 | Low duty, 2000°F, 9" x 2-1/2" x 4-1/2" | D-1 | .60 | 26.667 | M | 1,375 | 1,050 | | 2,425 | 3,075 |
| 0050 | High duty, 3000°F | " | .60 | 26.667 | " | 2,100 | 1,050 | | 3,150 | 3,875 |

| 04 54 10.20 Fire Clay | | | | | | | | | | |
|---|---|---|---|---|---|---|---|---|---|---|
| 0010 | **FIRE CLAY** | | | | | | | | | |
| 0020 | Gray, high duty, 100 lb. bag | | | | Bag | 59 | | | 59 | 65 |
| 0050 | 100 lb. drum, premixed (400 brick per drum) | | | | Drum | 73 | | | 73 | 80.50 |

# 04 57 Masonry Fireplaces

## 04 57 10 – Brick or Stone Fireplaces

| 04 57 10.10 Fireplace | | Crew | Daily Output | Labor-Hours | Unit | Material | 2011 Bare Costs Labor | Equipment | Total | Total Incl O&P |
|---|---|---|---|---|---|---|---|---|---|---|
| 0010 | **FIREPLACE** | | | | | | | | | |
| 0100 | Brick fireplace, not incl. foundations or chimneys | | | | | | | | | |
| 0110 | 30" x 29" opening, incl. chamber, plain brickwork | D-1 | .40 | 40 | Ea. | 535 | 1,550 | | 2,085 | 2,950 |
| 0200 | Fireplace box only (110 brick) | " | 2 | 8 | " | 176 | 310 | | 486 | 665 |
| 0300 | For elaborate brickwork and details, add | | | | | 35% | 35% | | | |
| 0400 | For hearth, brick & stone, add | D-1 | 2 | 8 | Ea. | 197 | 310 | | 507 | 685 |
| 0410 | For steel angle, damper, cleanouts, add | | 4 | 4 | | 138 | 156 | | 294 | 385 |
| 0600 | Plain brickwork, incl. metal circulator | | .50 | 32 | | 1,025 | 1,250 | | 2,275 | 3,000 |
| 0800 | Face brick only, standard size, 8" x 2-2/3" x 4" | | .30 | 53.333 | M | 535 | 2,075 | | 2,610 | 3,725 |

# 04 71 Manufactured Brick Masonry

## 04 71 10 – Simulated or Manufactured Brick

| 04 71 10.10 Simulated Brick | | Crew | Daily Output | Labor-Hours | Unit | Material | 2011 Bare Costs Labor | Equipment | Total | Total Incl O&P |
|---|---|---|---|---|---|---|---|---|---|---|
| 0010 | **SIMULATED BRICK** | | | | | | | | | |
| 0020 | Aluminum, baked on colors | 1 Carp | 200 | .040 | S.F. | 3.53 | 1.72 | | 5.25 | 6.50 |
| 0050 | Fiberglass panels | | 200 | .040 | | 3.87 | 1.72 | | 5.59 | 6.90 |
| 0100 | Urethane pieces cemented in mastic | | 150 | .053 | | 6.70 | 2.30 | | 9 | 10.85 |
| 0150 | Vinyl siding panels | | 200 | .040 | | 2.67 | 1.72 | | 4.39 | 5.60 |
| 0160 | Cement base, brick, incl. mastic | D-1 | 100 | .160 | | 4.22 | 6.25 | | 10.47 | 14 |
| 0170 | Corner | | 50 | .320 | V.L.F. | 10.55 | 12.45 | | 23 | 30.50 |
| 0180 | Stone face, incl. mastic | | 100 | .160 | S.F. | 10.20 | 6.25 | | 16.45 | 20.50 |
| 0190 | Corner | | 50 | .320 | V.L.F. | 11.25 | 12.45 | | 23.70 | 31 |

# 04 72 Cast Stone Masonry

## 04 72 10 – Cast Stone Masonry Features

| 04 72 10.10 Coping | | Crew | Daily Output | Labor-Hours | Unit | Material | 2011 Bare Costs Labor | Equipment | Total | Total Incl O&P |
|---|---|---|---|---|---|---|---|---|---|---|
| 0010 | **COPING**, stock units | | | | | | | | | |
| 0050 | Precast concrete, 10" wide, 4" tapers to 3-1/2", 8" wall | D-1 | 75 | .213 | L.F. | 17.55 | 8.30 | | 25.85 | 32 |
| 0100 | 12" wide, 3-1/2" tapers to 3", 10" wall | | 70 | .229 | | 17.55 | 8.90 | | 26.45 | 32.50 |
| 0110 | 14" wide, 4" tapers to 3-1/2", 12" wall | | 65 | .246 | | 20.50 | 9.60 | | 30.10 | 37 |
| 0150 | 16" wide, 4" tapers to 3-1/2", 14" wall | | 60 | .267 | | 17.10 | 10.40 | | 27.50 | 34.50 |
| 0250 | Precast concrete corners | | 40 | .400 | Ea. | 28.50 | 15.60 | | 44.10 | 54.50 |
| 0300 | Limestone for 12" wall, 4" thick | | 90 | .178 | L.F. | 15.35 | 6.95 | | 22.30 | 27.50 |
| 0350 | 6" thick | | 80 | .200 | | 23 | 7.80 | | 30.80 | 37 |
| 0500 | Marble, to 4" thick, no wash, 9" wide | | 90 | .178 | | 23 | 6.95 | | 29.95 | 36 |
| 0550 | 12" wide | | 80 | .200 | | 35 | 7.80 | | 42.80 | 50 |

# 04 72 Cast Stone Masonry

## 04 72 10 – Cast Stone Masonry Features

| 04 72 10.10 Coping | Crew | Daily Output | Labor-Hours | Unit | Material | 2011 Bare Costs Labor | Equipment | Total | Total Incl O&P |
|---|---|---|---|---|---|---|---|---|---|
| 0700 | Terra cotta, 9" wide | D-1 | 90 | .178 | L.F. | 5.60 | 6.95 | | 12.55 | 16.55 |
| 0750 | 12" wide | | 80 | .200 | | 9.20 | 7.80 | | 17 | 22 |
| 0800 | Aluminum, for 12" wall | | 80 | .200 | | 13.30 | 7.80 | | 21.10 | 26.50 |

## 04 72 20 – Cultured Stone Veneer

### 04 72 20.10 Cultured Stone Veneer Components

| | | Crew | Daily Output | Labor-Hours | Unit | Material | 2011 Bare Costs Labor | Equipment | Total | Total Incl O&P |
|---|---|---|---|---|---|---|---|---|---|---|
| 0010 | **CULTURED STONE VENEER COMPONENTS** | | | | | | | | | |
| 0110 | On wood frame and sheathing substrate, random sized cobbles, corner stones | D-8 | 70 | .571 | V.L.F. | 12.10 | 23 | | 35.10 | 47.50 |
| 0120 | Field stones | | 140 | .286 | S.F. | 8.80 | 11.35 | | 20.15 | 26.50 |
| 0130 | Random sized flats, corner stones | | 70 | .571 | V.L.F. | 11.90 | 23 | | 34.90 | 47 |
| 0140 | Field stones | | 140 | .286 | S.F. | 10.05 | 11.35 | | 21.40 | 28 |
| 0150 | Horizontal lined ledgestones, corner stones | | 75 | .533 | V.L.F. | 12.10 | 21 | | 33.10 | 45.50 |
| 0160 | Field stones | | 150 | .267 | S.F. | 8.80 | 10.60 | | 19.40 | 25.50 |
| 0170 | Random shaped flats, corner stones | | 65 | .615 | V.L.F. | 12.10 | 24.50 | | 36.60 | 50.50 |
| 0180 | Field stones | | 150 | .267 | S.F. | 8.80 | 10.60 | | 19.40 | 25.50 |
| 0190 | Random shaped/textured face, corner stones | | 65 | .615 | V.L.F. | 12.10 | 24.50 | | 36.60 | 50.50 |
| 0200 | Field stones | | 130 | .308 | S.F. | 8.80 | 12.25 | | 21.05 | 28 |
| 0210 | Random shaped river rock, corner stones | | 65 | .615 | V.L.F. | 12.10 | 24.50 | | 36.60 | 50.50 |
| 0220 | Field stones | | 130 | .308 | S.F. | 8.80 | 12.25 | | 21.05 | 28 |
| 0240 | On concrete or CMU substrate, random sized cobbles, corner stones | | 70 | .571 | V.L.F. | 11.50 | 23 | | 34.50 | 46.50 |
| 0250 | Field stones | | 140 | .286 | S.F. | 8.50 | 11.35 | | 19.85 | 26.50 |
| 0260 | Random sized flats, corner stones | | 70 | .571 | V.L.F. | 11.30 | 23 | | 34.30 | 46.50 |
| 0270 | Field stones | | 140 | .286 | S.F. | 9.75 | 11.35 | | 21.10 | 28 |
| 0280 | Horizontal lined ledgestones, corner stones | | 75 | .533 | V.L.F. | 11.50 | 21 | | 32.50 | 44.50 |
| 0290 | Field stones | | 150 | .267 | S.F. | 8.50 | 10.60 | | 19.10 | 25.50 |
| 0300 | Random shaped flats, corner stones | | 70 | .571 | V.L.F. | 11.50 | 23 | | 34.50 | 46.50 |
| 0310 | Field stones | | 140 | .286 | S.F. | 8.50 | 11.35 | | 19.85 | 26.50 |
| 0320 | Random shaped/textured face, corner stones | | 65 | .615 | V.L.F. | 11.50 | 24.50 | | 36 | 49.50 |
| 0330 | Field stones | | 130 | .308 | S.F. | 8.50 | 12.25 | | 20.75 | 28 |
| 0340 | Random shaped river rock, corner stones | | 65 | .615 | V.L.F. | 11.50 | 24.50 | | 36 | 49.50 |
| 0350 | Field stones | | 130 | .308 | S.F. | 8.50 | 12.25 | | 20.75 | 28 |
| 0360 | Cultured stone veneer, #15 felt weather resistant barrier | 1 Clab | 3700 | .002 | Sq. | 5.65 | .07 | | 5.72 | 6.30 |
| 0370 | Expanded metal lath, diamond, 2.5 lb./S.Y., galvanized | 1 Lath | 85 | .094 | S.Y. | 2.63 | 3.63 | | 6.26 | 8.20 |
| 0390 | Water table or window sill, 18" long | 1 Bric | 80 | .100 | Ea. | 10.60 | 4.32 | | 14.92 | 18.15 |

# 04 73 Manufactured Stone Masonry

## 04 73 20 – Simulated or Manufactured Stone

### 04 73 20.10 Simulated Stone

| | | Crew | Daily Output | Labor-Hours | Unit | Material | 2011 Bare Costs Labor | Equipment | Total | Total Incl O&P |
|---|---|---|---|---|---|---|---|---|---|---|
| 0010 | **SIMULATED STONE** | | | | | | | | | |
| 0100 | Insulated fiberglass panels, 5/8" ply backer | L-4 | 200 | .120 | S.F. | 11.40 | 4.86 | | 16.26 | 19.95 |

# Estimating Tips

## 05 05 00 Common Work Results for Metals

- Nuts, bolts, washers, connection angles, and plates can add a significant amount to both the tonnage of a structural steel job and the estimated cost. As a rule of thumb, add 10% to the total weight to account for these accessories.

- Type 2 steel construction, commonly referred to as "simple construction," consists generally of field-bolted connections with lateral bracing supplied by other elements of the building, such as masonry walls or x-bracing. The estimator should be aware, however, that shop connections may be accomplished by welding or bolting. The method may be particular to the fabrication shop and may have an impact on the estimated cost.

## 05 12 23 Structural Steel

- Steel items can be obtained from two sources: a fabrication shop or a metals service center. Fabrication shops can fabricate items under more controlled conditions than can crews in the field. They are also more efficient and can produce items more economically. Metal service centers serve as a source of long mill shapes to both fabrication shops and contractors.

- Most line items in this structural steel subdivision, and most items in 05 50 00 Metal Fabrications, are indicated as being shop fabricated. The bare material cost for these shop fabricated items is the "Invoice Cost" from the shop and includes the mill base price of steel plus mill extras, transportation to the shop, shop drawings and detailing where warranted, shop fabrication and handling, sandblasting and a shop coat of primer paint, all necessary structural bolts, and delivery to the job site. The bare labor cost and bare equipment cost for these shop fabricated items is for field installation or erection.

- Line items in Subdivision 05 12 23.40 Lightweight Framing, and other items scattered in Division 5, are indicated as being field fabricated. The bare material cost for these field fabricated items is the "Invoice Cost" from the metals service center and includes the mill base price of steel plus mill extras, transportation to the metals service center, material handling, and delivery of long lengths of mill shapes to the job site. Material costs for structural bolts and welding rods should be added to the estimate. The bare labor cost and bare equipment cost for these items is for both field fabrication and field installation or erection, and include time for cutting, welding and drilling in the fabricated metal items. Drilling into concrete and fasteners to fasten field fabricated items to other work are not included and should be added to the estimate.

## 05 20 00 Steel Joist Framing

- In any given project the total weight of open web steel joists is determined by the loads to be supported and the design. However, economies can be realized in minimizing the amount of labor used to place the joists. This is done by maximizing the joist spacing, and therefore minimizing the number of joists required to be installed on the job. Certain spacings and locations may be required by the design, but in other cases maximizing the spacing and keeping it as uniform as possible will keep the costs down.

## 05 30 00 Steel Decking

- The takeoff and estimating of metal deck involves more than simply the area of the floor or roof and the type of deck specified or shown on the drawings. Many different sizes and types of openings may exist. Small openings for individual pipes or conduits may be drilled after the floor/roof is installed, but larger openings may require special deck lengths as well as reinforcing or structural support. The estimator should determine who will be supplying this reinforcing. Additionally, some deck terminations are part of the deck package, such as screed angles and pour stops, and others will be part of the steel contract, such as angles attached to structural members and cast-in-place angles and plates. The estimator must ensure that all pieces are accounted for in the complete estimate.

## 05 50 00 Metal Fabrications

- The most economical steel stairs are those that use common materials, standard details, and most importantly, a uniform and relatively simple method of field assembly. Commonly available A36 channels and plates are very good choices for the main stringers of the stairs, as are angles and tees for the carrier members. Risers and treads are usually made by specialty shops, and it is most economical to use a typical detail in as many places as possible. The stairs should be pre-assembled and shipped directly to the site. The field connections should be simple and straightforward to be accomplished efficiently, and with minimum equipment and labor.

## Reference Numbers

Reference numbers are shown in shaded boxes at the beginning of some major classifications. These numbers refer to related items in the Reference Section. The reference information may be an estimating procedure, an alternate pricing method, or technical information. *Note:* Not all subdivisions listed here necessarily appear in this publication.

# 05 01 Maintenance of Metals

## 05 01 10 – Maintenance of Structural Metal Framing

### 05 01 10.51 Cleaning of Structural Metal Framing

| | | Daily Output | Labor-Hours | Unit | Material | 2011 Bare Costs Labor | Equipment | Total | Total Incl O&P |
|---|---|---|---|---|---|---|---|---|---|
| | | Crew | | | | | | | |
| 0010 | **CLEANING OF STRUCTURAL METAL FRAMING** | | | | | | | | |
| 6125 | Steel surface treatments, PDCA guidelines | | | | | | | | |
| 6170 | Wire brush, hand (SSPC-SP2) | 1 Psst | 400 | .020 | S.F. | .02 | .77 | | .79 | 1.40 |
| 6180 | Power tool (SSPC-SP3) | " | 700 | .011 | | .06 | .44 | | .50 | .86 |
| 6215 | Pressure washing, up to 5000 psi, 5000-15,000 S.F./day | 1 Pord | 10000 | .001 | | | .03 | | .03 | .04 |
| 6220 | Steam cleaning, 600 psi @ 300 F, 1250 – 2500 S.F./day | | 2000 | .004 | | | .15 | | .15 | .22 |
| 6225 | Water blasting, up to 25,000 psi, 1750 – 3500 S.F./day | | 2500 | .003 | | | .12 | | .12 | .18 |
| 6230 | Brush-off blast (SSPC-SP7) | E-11 | 1750 | .018 | | .24 | .71 | .11 | 1.06 | 1.56 |
| 6235 | Com'l blast (SSPC-SP6), loose scale, fine pwder rust, 2.0#/S.F. sand | | 1200 | .027 | | .49 | 1.04 | .16 | 1.69 | 2.43 |
| 6240 | Tight mill scale, little/no rust, 3.0#/S.F. sand | | 1000 | .032 | | .73 | 1.24 | .20 | 2.17 | 3.07 |
| 6245 | Exist coat blistered/pitted, 4.0#/S.F. sand | | 875 | .037 | | .97 | 1.42 | .22 | 2.61 | 3.66 |
| 6250 | Exist coat badly pitted/nodules, 6.7#/S.F. sand | | 825 | .039 | | 1.63 | 1.51 | .24 | 3.38 | 4.53 |
| 6255 | Near white blast (SSPC-SP10), loose scale, fine rust, 5.6#/S.F. sand | | 450 | .071 | | 1.36 | 2.76 | .44 | 4.56 | 6.55 |
| 6260 | Tight mill scale, little/no rust, 6.9#/S.F. sand | | 325 | .098 | | 1.68 | 3.82 | .60 | 6.10 | 8.80 |
| 6265 | Exist coat blistered/pitted, 9.0#/S.F. sand | | 225 | .142 | | 2.19 | 5.50 | .87 | 8.56 | 12.45 |
| 6270 | Exist coat badly pitted/nodules, 11.3#/S.F. sand | | 150 | .213 | | 2.75 | 8.30 | 1.31 | 12.36 | 18.10 |

# 05 05 Common Work Results for Metals

## 05 05 05 – Selective Metals Demolition

### 05 05 05.10 Selective Demolition, Metals

| | | Crew | Daily Output | Labor-Hours | Unit | Material | 2011 Bare Costs Labor | Equipment | Total | Total Incl O&P |
|---|---|---|---|---|---|---|---|---|---|---|
| 0010 | **SELECTIVE DEMOLITION, METALS** | R024119-10 | | | | | | | | |
| 0015 | Excludes shores, bracing, cutting, loading, hauling, dumping | | | | | | | | | |
| 0020 | Remove nuts only up to 3/4" diameter | 1 Sswk | 480 | .017 | Ea. | | .81 | | .81 | 1.43 |
| 0030 | 7/8" to 1-1/4" diameter | | 240 | .033 | | | 1.62 | | 1.62 | 2.86 |
| 0040 | 1-3/8" to 2" diameter | | 160 | .050 | | | 2.43 | | 2.43 | 4.29 |
| 0060 | Unbolt and remove structural bolts up to 3/4" diameter | | 240 | .033 | | | 1.62 | | 1.62 | 2.86 |
| 0070 | 7/8" to 2" diameter | | 160 | .050 | | | 2.43 | | 2.43 | 4.29 |
| 0140 | Light weight framing members, remove whole or cut up, up to 20 lb | | 240 | .033 | | | 1.62 | | 1.62 | 2.86 |
| 0150 | 21 – 40 lb | 2 Sswk | 210 | .076 | | | 3.70 | | 3.70 | 6.55 |
| 0160 | 41 – 80 lb | 3 Sswk | 180 | .133 | | | 6.45 | | 6.45 | 11.45 |
| 0170 | 81 – 120 lb | 4 Sswk | 150 | .213 | | | 10.35 | | 10.35 | 18.30 |
| 0230 | Structural members, remove whole or cut up, up to 500 lb | E-19 | 48 | .500 | | | 24 | 21.50 | 45.50 | 63.50 |
| 0240 | 1/4 – 2 tons | E-18 | 36 | 1.111 | | | 53.50 | 29 | 82.50 | 124 |
| 0250 | 2 – 5 tons | E-24 | 30 | 1.067 | | | 51 | 25 | 76 | 114 |
| 0260 | 5 – 10 tons | E-20 | 24 | 2.667 | | | 127 | 52.50 | 179.50 | 274 |
| 0270 | 10 – 15 tons | E-2 | 18 | 3.111 | | | 147 | 90 | 237 | 350 |
| 0340 | Fabricated item, remove whole or cut up, up to 20 lb | 1 Sswk | 96 | .083 | | | 4.05 | | 4.05 | 7.15 |
| 0350 | 21 – 40 lb | 2 Sswk | 84 | .190 | | | 9.25 | | 9.25 | 16.35 |
| 0360 | 41 – 80 lb | 3 Sswk | 72 | .333 | | | 16.20 | | 16.20 | 28.50 |
| 0370 | 81 – 120 lb | 4 Sswk | 60 | .533 | | | 26 | | 26 | 45.50 |
| 0380 | 121 – 500 lb | E-19 | 48 | .500 | | | 24 | 21.50 | 45.50 | 63.50 |
| 0390 | 501 – 1000 lb | " | 36 | .667 | | | 31.50 | 29 | 60.50 | 85 |

## 05 05 13 – Shop-Applied Coatings for Metal

### 05 05 13.50 Paints and Protective Coatings

| | | Crew | Daily Output | Labor-Hours | Unit | Material | 2011 Bare Costs Labor | Equipment | Total | Total Incl O&P |
|---|---|---|---|---|---|---|---|---|---|---|
| 0010 | **PAINTS AND PROTECTIVE COATINGS** | | | | | | | | | |
| 5900 | Galvanizing structural steel in shop, under 1 ton | | | | Ton | 460 | | | 460 | 505 |
| 5950 | 1 ton to 20 tons | | | | | 400 | | | 400 | 440 |
| 6000 | Over 20 tons | | | | | 375 | | | 375 | 415 |

## 05 05 21 – Fastening Methods for Metal

| 05 05 21.10 Cutting Steel | | Crew | Daily Output | Labor-Hours | Unit | Material | 2011 Bare Costs Labor | Equipment | Total | Total Incl O&P |
|---|---|---|---|---|---|---|---|---|---|---|
| 0010 | **CUTTING STEEL** | | | | | | | | | |
| 0020 | Hand burning, incl. preparation, torch cutting & grinding, no staging | | | | | | | | | |
| 0100 | Steel to 1/2" thick | E-25 | 320 | .025 | L.F. | | 1.26 | .32 | 1.58 | 2.58 |
| 0150 | 3/4" thick | | 260 | .031 | | | 1.56 | .39 | 1.95 | 3.18 |
| 0200 | 1" thick | | 200 | .040 | | | 2.02 | .50 | 2.52 | 4.12 |

### 05 05 21.15 Drilling Steel

| | | Crew | Daily Output | Labor-Hours | Unit | Material | Labor | Equipment | Total | Total Incl O&P |
|---|---|---|---|---|---|---|---|---|---|---|
| 0010 | **DRILLING STEEL** | | | | | | | | | |
| 1910 | Drilling & layout for steel, up to 1/4" deep, no anchor | | | | | | | | | |
| 1920 | Holes, 1/4" diameter | 1 Sswk | 112 | .071 | Ea. | .07 | 3.47 | | 3.54 | 6.20 |
| 1925 | For each additional 1/4" depth, add | | 336 | .024 | | .07 | 1.16 | | 1.23 | 2.11 |
| 1930 | 3/8" diameter | | 104 | .077 | | .08 | 3.73 | | 3.81 | 6.70 |
| 1935 | For each additional 1/4" depth, add | | 312 | .026 | | .08 | 1.24 | | 1.32 | 2.29 |
| 1940 | 1/2" diameter | | 96 | .083 | | .08 | 4.05 | | 4.13 | 7.25 |
| 1945 | For each additional 1/4" depth, add | | 288 | .028 | | .08 | 1.35 | | 1.43 | 2.47 |
| 1950 | 5/8" diameter | | 88 | .091 | | .12 | 4.41 | | 4.53 | 7.95 |
| 1955 | For each additional 1/4" depth, add | | 264 | .030 | | .12 | 1.47 | | 1.59 | 2.73 |
| 1960 | 3/4" diameter | | 80 | .100 | | .15 | 4.86 | | 5.01 | 8.75 |
| 1965 | For each additional 1/4" depth, add | | 240 | .033 | | .15 | 1.62 | | 1.77 | 3.02 |
| 1970 | 7/8" diameter | | 72 | .111 | | .19 | 5.40 | | 5.59 | 9.75 |
| 1975 | For each additional 1/4" depth, add | | 216 | .037 | | .19 | 1.80 | | 1.99 | 3.39 |
| 1980 | 1" diameter | | 64 | .125 | | .20 | 6.05 | | 6.25 | 10.90 |
| 1985 | For each additional 1/4" depth, add | | 192 | .042 | | .20 | 2.02 | | 2.22 | 3.79 |
| 1990 | For drilling up, add | | | | | | 40% | | | |

### 05 05 21.90 Welding Steel

| | | Crew | Daily Output | Labor-Hours | Unit | Material | Labor | Equipment | Total | Total Incl O&P |
|---|---|---|---|---|---|---|---|---|---|---|
| 0010 | **WELDING STEEL**, Structural    R050521-20 | | | | | | | | | |
| 0020 | Field welding, 1/8" E6011, cost per welder, no oper. engr | E-14 | 8 | 1 | Hr. | 4.22 | 50.50 | 13.65 | 68.37 | 109 |
| 0200 | With 1/2 operating engineer | E-13 | 8 | 1.500 | | 4.22 | 72.50 | 13.60 | 90.32 | 142 |
| 0300 | With 1 operating engineer | E-12 | 8 | 2 | | 4.22 | 94 | 13.60 | 111.82 | 174 |
| 0500 | With no operating engineer, 2# weld rod per ton | E-14 | 8 | 1 | Ton | 4.22 | 50.50 | 13.65 | 68.37 | 109 |
| 0600 | 8# E6011 per ton | " | 2 | 4 | | 16.85 | 202 | 54.50 | 273.35 | 435 |
| 0800 | With one operating engineer per welder, 2# E6011 per ton | E-12 | 8 | 2 | | 4.22 | 94 | 13.60 | 111.82 | 174 |
| 0900 | 8# E6011 per ton | " | 2 | 8 | | 16.85 | 375 | 54.50 | 446.35 | 700 |
| 1200 | Continuous fillet, stick welding, incl. equipment | | | | | | | | | |
| 1300 | Single pass, 1/8" thick, 0.1#/L.F. | E-14 | 150 | .053 | L.F. | .21 | 2.70 | .73 | 3.64 | 5.80 |
| 1400 | 3/16" thick, 0.2#/L.F. | | 75 | .107 | | .42 | 5.40 | 1.45 | 7.27 | 11.55 |
| 1500 | 1/4" thick, 0.3#/L.F. | | 50 | .160 | | .63 | 8.10 | 2.18 | 10.91 | 17.40 |
| 1610 | 5/16" thick, 0.4#/L.F. | | 38 | .211 | | .84 | 10.65 | 2.87 | 14.36 | 23 |
| 1800 | 3 passes, 3/8" thick, 0.5#/L.F. | | 30 | .267 | | 1.05 | 13.50 | 3.63 | 18.18 | 29 |
| 2010 | 4 passes, 1/2" thick, 0.7#/L.F. | | 22 | .364 | | 1.48 | 18.40 | 4.96 | 24.84 | 39.50 |
| 2200 | 5 to 6 passes, 3/4" thick, 1.3#/L.F. | | 12 | .667 | | 2.74 | 33.50 | 9.10 | 45.34 | 72.50 |
| 2400 | 8 to 11 passes, 1" thick, 2.4#/L.F. | | 6 | 1.333 | | 5.05 | 67.50 | 18.15 | 90.70 | 145 |
| 2600 | For all position welding, add, minimum | | | | | | 20% | | | |
| 2700 | Maximum | | | | | | 300% | | | |
| 2900 | For semi-automatic welding, deduct, minimum | | | | | | 5% | | | |
| 3000 | Maximum | | | | | | 15% | | | |
| 4000 | Cleaning and welding plates, bars, or rods | | | | | | | | | |
| 4010 | to existing beams, columns, or trusses | E-14 | 12 | .667 | L.F. | 1.05 | 33.50 | 9.10 | 43.65 | 70.50 |

## 05 05 23 – Metal Fastenings

### 05 05 23.05 Anchor Bolts

| | | | | | | | | | | |
|---|---|---|---|---|---|---|---|---|---|---|
| 0010 | **ANCHOR BOLTS** | | | | | | | | | |
| 0015 | Made from recycled materials    G | | | | | | | | | |

## 05 05 23 – Metal Fastenings

| 05 05 23.05 **Anchor Bolts** | | Crew | Daily Output | Labor-Hours | Unit | Material | 2011 Bare Costs Labor | Equipment | Total | Total Incl O&P |
|---|---|---|---|---|---|---|---|---|---|---|
| 0025 | Single bolts installed in fresh concrete, no templates | | | | | | | | | |
| 0030 | Hooked w/nut and washer, 1/2" diameter, 8" long | G 1 Carp | 132 | .061 | Ea. | 1.36 | 2.61 | | 3.97 | 5.50 |
| 0040 | 12" long | G | 131 | .061 | | 1.51 | 2.63 | | 4.14 | 5.70 |
| 0050 | 5/8" diameter, 8" long | G | 129 | .062 | | 2.96 | 2.67 | | 5.63 | 7.35 |
| 0060 | 12" long | G | 127 | .063 | | 3.65 | 2.71 | | 6.36 | 8.15 |
| 0070 | 3/4" diameter, 8" long | G | 127 | .063 | | 3.65 | 2.71 | | 6.36 | 8.15 |
| 0080 | 12" long | G | 125 | .064 | | 4.56 | 2.76 | | 7.32 | 9.25 |
| 0090 | 2-bolt pattern, including job-built 2-hole template, per set | | | | | | | | | |
| 0100 | J-type, incl. hex nut & washer, 1/2" diameter x 6" long | G 1 Carp | 21 | .381 | Set | 4.50 | 16.40 | | 20.90 | 30 |
| 0110 | 12" long | G | 21 | .381 | | 5.10 | 16.40 | | 21.50 | 30.50 |
| 0120 | 18" long | G | 21 | .381 | | 6 | 16.40 | | 22.40 | 31.50 |
| 0130 | 3/4" diameter x 8" long | G | 20 | .400 | | 9.40 | 17.20 | | 26.60 | 37 |
| 0140 | 12" long | G | 20 | .400 | | 11.20 | 17.20 | | 28.40 | 39 |
| 0150 | 18" long | G | 20 | .400 | | 13.95 | 17.20 | | 31.15 | 42 |
| 0160 | 1" diameter x 12" long | G | 19 | .421 | | 19.65 | 18.15 | | 37.80 | 49.50 |
| 0170 | 18" long | G | 19 | .421 | | 23.50 | 18.15 | | 41.65 | 53.50 |
| 0180 | 24" long | G | 19 | .421 | | 28 | 18.15 | | 46.15 | 58.50 |
| 0190 | 36" long | G | 18 | .444 | | 37.50 | 19.15 | | 56.65 | 71 |
| 0200 | 1-1/2" diameter x 18" long | G | 17 | .471 | | 60 | 20.50 | | 80.50 | 97 |
| 0210 | 24" long | G | 16 | .500 | | 71.50 | 21.50 | | 93 | 112 |
| 0300 | L-type, incl. hex nut & washer, 3/4" diameter x 12" long | G | 20 | .400 | | 10.70 | 17.20 | | 27.90 | 38.50 |
| 0310 | 18" long | G | 20 | .400 | | 13.10 | 17.20 | | 30.30 | 41 |
| 0320 | 24" long | G | 20 | .400 | | 15.50 | 17.20 | | 32.70 | 43.50 |
| 0330 | 30" long | G | 20 | .400 | | 19.15 | 17.20 | | 36.35 | 47.50 |
| 0340 | 36" long | G | 20 | .400 | | 21.50 | 17.20 | | 38.70 | 50 |
| 0350 | 1" diameter x 12" long | G | 19 | .421 | | 16.75 | 18.15 | | 34.90 | 46.50 |
| 0360 | 18" long | G | 19 | .421 | | 20.50 | 18.15 | | 38.65 | 50.50 |
| 0370 | 24" long | G | 19 | .421 | | 24.50 | 18.15 | | 42.65 | 55 |
| 0380 | 30" long | G | 19 | .421 | | 28.50 | 18.15 | | 46.65 | 59.50 |
| 0390 | 36" long | G | 18 | .444 | | 32.50 | 19.15 | | 51.65 | 65.50 |
| 0400 | 42" long | G | 18 | .444 | | 39 | 19.15 | | 58.15 | 72.50 |
| 0410 | 48" long | G | 18 | .444 | | 43.50 | 19.15 | | 62.65 | 77.50 |
| 0420 | 1-1/4" diameter x 18" long | G | 18 | .444 | | 30 | 19.15 | | 49.15 | 62.50 |
| 0430 | 24" long | G | 18 | .444 | | 35 | 19.15 | | 54.15 | 68.50 |
| 0440 | 30" long | G | 17 | .471 | | 40.50 | 20.50 | | 61 | 75.50 |
| 0450 | 36" long | G | 17 | .471 | | 45.50 | 20.50 | | 66 | 81.50 |
| 0460 | 42" long | G 2 Carp | 32 | .500 | | 51.50 | 21.50 | | 73 | 89.50 |
| 0470 | 48" long | G | 32 | .500 | | 58.50 | 21.50 | | 80 | 97 |
| 0480 | 54" long | G | 31 | .516 | | 68.50 | 22 | | 90.50 | 109 |
| 0490 | 60" long | G | 31 | .516 | | 75 | 22 | | 97 | 117 |
| 0500 | 1-1/2" diameter x 18" long | G | 33 | .485 | | 43.50 | 21 | | 64.50 | 80 |
| 0510 | 24" long | G | 32 | .500 | | 50.50 | 21.50 | | 72 | 88.50 |
| 0520 | 30" long | G | 31 | .516 | | 57 | 22 | | 79 | 96.50 |
| 0530 | 36" long | G | 30 | .533 | | 65 | 23 | | 88 | 107 |
| 0540 | 42" long | G | 30 | .533 | | 74 | 23 | | 97 | 116 |
| 0550 | 48" long | G | 29 | .552 | | 82.50 | 24 | | 106.50 | 128 |
| 0560 | 54" long | G | 28 | .571 | | 100 | 24.50 | | 124.50 | 148 |
| 0570 | 60" long | G | 28 | .571 | | 110 | 24.50 | | 134.50 | 159 |
| 0580 | 1-3/4" diameter x 18" long | G | 31 | .516 | | 61.50 | 22 | | 83.50 | 102 |
| 0590 | 24" long | G | 30 | .533 | | 72 | 23 | | 95 | 114 |
| 0600 | 30" long | G | 29 | .552 | | 83.50 | 24 | | 107.50 | 128 |
| 0610 | 36" long | G | 28 | .571 | | 95 | 24.50 | | 119.50 | 142 |
| 0620 | 42" long | G | 27 | .593 | | 106 | 25.50 | | 131.50 | 156 |

## 05 05 23 – Metal Fastenings

### 05 05 23.05 Anchor Bolts

| | | Crew | Daily Output | Labor-Hours | Unit | Material | 2011 Bare Costs Labor | 2011 Bare Costs Equipment | Total | Total Incl O&P |
|---|---|---|---|---|---|---|---|---|---|---|
| 0630 | 48" long | G 2 Carp | 26 | .615 | Set | 116 | 26.50 | | 142.50 | 169 |
| 0640 | 54" long | G | 26 | .615 | | 144 | 26.50 | | 170.50 | 199 |
| 0650 | 60" long | G | 25 | .640 | | 155 | 27.50 | | 182.50 | 214 |
| 0660 | 2" diameter x 24" long | G | 27 | .593 | | 96.50 | 25.50 | | 122 | 145 |
| 0670 | 30" long | G | 27 | .593 | | 108 | 25.50 | | 133.50 | 158 |
| 0680 | 36" long | G | 26 | .615 | | 119 | 26.50 | | 145.50 | 172 |
| 0690 | 42" long | G | 25 | .640 | | 132 | 27.50 | | 159.50 | 189 |
| 0700 | 48" long | G | 24 | .667 | | 152 | 28.50 | | 180.50 | 211 |
| 0710 | 54" long | G | 23 | .696 | | 180 | 30 | | 210 | 244 |
| 0720 | 60" long | G | 23 | .696 | | 194 | 30 | | 224 | 259 |
| 0730 | 66" long | G | 22 | .727 | | 207 | 31.50 | | 238.50 | 276 |
| 0740 | 72" long | G | 21 | .762 | | 227 | 33 | | 260 | 300 |
| 1000 | 4-bolt pattern, including job-built 4-hole template, per set | | | | | | | | | |
| 1100 | J-type, incl. hex nut & washer, 1/2" diameter x 6" long | G 1 Carp | 19 | .421 | Set | 6.90 | 18.15 | | 25.05 | 35.50 |
| 1110 | 12" long | G | 19 | .421 | | 8.15 | 18.15 | | 26.30 | 37 |
| 1120 | 18" long | G | 18 | .444 | | 9.95 | 19.15 | | 29.10 | 40.50 |
| 1130 | 3/4" diameter x 8" long | G | 17 | .471 | | 16.70 | 20.50 | | 37.20 | 49.50 |
| 1140 | 12" long | G | 17 | .471 | | 20.50 | 20.50 | | 41 | 53.50 |
| 1150 | 18" long | G | 17 | .471 | | 26 | 20.50 | | 46.50 | 59.50 |
| 1160 | 1" diameter x 12" long | G | 16 | .500 | | 37.50 | 21.50 | | 59 | 74 |
| 1170 | 18" long | G | 15 | .533 | | 44.50 | 23 | | 67.50 | 84 |
| 1180 | 24" long | G | 15 | .533 | | 54 | 23 | | 77 | 94 |
| 1190 | 36" long | G | 15 | .533 | | 73 | 23 | | 96 | 115 |
| 1200 | 1-1/2" diameter x 18" long | G | 13 | .615 | | 118 | 26.50 | | 144.50 | 171 |
| 1210 | 24" long | G | 12 | .667 | | 140 | 28.50 | | 168.50 | 199 |
| 1300 | L-type, incl. hex nut & washer, 3/4" diameter x 12" long | G | 17 | .471 | | 19.25 | 20.50 | | 39.75 | 52 |
| 1310 | 18" long | G | 17 | .471 | | 24 | 20.50 | | 44.50 | 57.50 |
| 1320 | 24" long | G | 17 | .471 | | 29 | 20.50 | | 49.50 | 63 |
| 1330 | 30" long | G | 16 | .500 | | 36 | 21.50 | | 57.50 | 73 |
| 1340 | 36" long | G | 16 | .500 | | 41 | 21.50 | | 62.50 | 78 |
| 1350 | 1" diameter x 12" long | G | 16 | .500 | | 31.50 | 21.50 | | 53 | 67.50 |
| 1360 | 18" long | G | 15 | .533 | | 38.50 | 23 | | 61.50 | 77.50 |
| 1370 | 24" long | G | 15 | .533 | | 47 | 23 | | 70 | 87 |
| 1380 | 30" long | G | 15 | .533 | | 55.50 | 23 | | 78.50 | 96 |
| 1390 | 36" long | G | 15 | .533 | | 63 | 23 | | 86 | 104 |
| 1400 | 42" long | G | 14 | .571 | | 76 | 24.50 | | 100.50 | 121 |
| 1410 | 48" long | G | 14 | .571 | | 85 | 24.50 | | 109.50 | 131 |
| 1420 | 1-1/4" diameter x 18" long | G | 14 | .571 | | 58 | 24.50 | | 82.50 | 101 |
| 1430 | 24" long | G | 14 | .571 | | 68.50 | 24.50 | | 93 | 113 |
| 1440 | 30" long | G | 13 | .615 | | 79 | 26.50 | | 105.50 | 127 |
| 1450 | 36" long | G | 13 | .615 | | 89.50 | 26.50 | | 116 | 139 |
| 1460 | 42" long | G 2 Carp | 25 | .640 | | 101 | 27.50 | | 128.50 | 154 |
| 1470 | 48" long | G | 24 | .667 | | 115 | 28.50 | | 143.50 | 170 |
| 1480 | 54" long | G | 23 | .696 | | 135 | 30 | | 165 | 194 |
| 1490 | 60" long | G | 23 | .696 | | 148 | 30 | | 178 | 208 |
| 1500 | 1-1/2" diameter x 18" long | G | 25 | .640 | | 85 | 27.50 | | 112.50 | 136 |
| 1510 | 24" long | G | 24 | .667 | | 99 | 28.50 | | 127.50 | 153 |
| 1520 | 30" long | G | 23 | .696 | | 112 | 30 | | 142 | 169 |
| 1530 | 36" long | G | 22 | .727 | | 128 | 31.50 | | 159.50 | 189 |
| 1540 | 42" long | G | 22 | .727 | | 146 | 31.50 | | 177.50 | 208 |
| 1550 | 48" long | G | 21 | .762 | | 163 | 33 | | 196 | 230 |
| 1560 | 54" long | G | 20 | .800 | | 198 | 34.50 | | 232.50 | 271 |
| 1570 | 60" long | G | 20 | .800 | | 217 | 34.50 | | 251.50 | 292 |

### 05 05 23.05 Anchor Bolts

| | | Crew | Daily Output | Labor-Hours | Unit | Material | 2011 Bare Costs Labor | Equipment | Total | Total Incl O&P |
|---|---|---|---|---|---|---|---|---|---|---|
| 1580 | 1-3/4" diameter x 18" long | G 2 Carp | 22 | .727 | Set | 121 | 31.50 | | 152.50 | 181 |
| 1590 | 24" long | G | 21 | .762 | | 142 | 33 | | 175 | 207 |
| 1600 | 30" long | G | 21 | .762 | | 165 | 33 | | 198 | 232 |
| 1610 | 36" long | G | 20 | .800 | | 187 | 34.50 | | 221.50 | 259 |
| 1620 | 42" long | G | 19 | .842 | | 210 | 36.50 | | 246.50 | 287 |
| 1630 | 48" long | G | 18 | .889 | | 231 | 38.50 | | 269.50 | 315 |
| 1640 | 54" long | G | 18 | .889 | | 286 | 38.50 | | 324.50 | 375 |
| 1650 | 60" long | G | 17 | .941 | | 310 | 40.50 | | 350.50 | 400 |
| 1660 | 2" diameter x 24" long | G | 19 | .842 | | 191 | 36.50 | | 227.50 | 266 |
| 1670 | 30" long | G | 18 | .889 | | 215 | 38.50 | | 253.50 | 295 |
| 1680 | 36" long | G | 18 | .889 | | 236 | 38.50 | | 274.50 | 320 |
| 1690 | 42" long | G | 17 | .941 | | 263 | 40.50 | | 303.50 | 350 |
| 1700 | 48" long | G | 16 | 1 | | 300 | 43 | | 343 | 395 |
| 1710 | 54" long | G | 15 | 1.067 | | 360 | 46 | | 406 | 465 |
| 1720 | 60" long | G | 15 | 1.067 | | 385 | 46 | | 431 | 495 |
| 1730 | 66" long | G | 14 | 1.143 | | 410 | 49 | | 459 | 530 |
| 1740 | 72" long | G | 14 | 1.143 | | 450 | 49 | | 499 | 570 |
| 1990 | For galvanized, add | | | | Ea. | 75% | | | | |

### 05 05 23.10 Bolts and Hex Nuts

| | | Crew | Daily Output | Labor-Hours | Unit | Material | 2011 Bare Costs Labor | Equipment | Total | Total Incl O&P |
|---|---|---|---|---|---|---|---|---|---|---|
| 0010 | **BOLTS & HEX NUTS**, Steel, A307 | | | | | | | | | |
| 0100 | 1/4" diameter, 1/2" long | G 1 Sswk | 140 | .057 | Ea. | .07 | 2.77 | | 2.84 | 4.98 |
| 0200 | 1" long | G | 140 | .057 | | .08 | 2.77 | | 2.85 | 4.99 |
| 0300 | 2" long | G | 130 | .062 | | .11 | 2.99 | | 3.10 | 5.40 |
| 0400 | 3" long | G | 130 | .062 | | .16 | 2.99 | | 3.15 | 5.50 |
| 0500 | 4" long | G | 120 | .067 | | .18 | 3.24 | | 3.42 | 5.90 |
| 0600 | 3/8" diameter, 1" long | G | 130 | .062 | | .18 | 2.99 | | 3.17 | 5.50 |
| 0700 | 2" long | G | 130 | .062 | | .22 | 2.99 | | 3.21 | 5.55 |
| 0800 | 3" long | G | 120 | .067 | | .29 | 3.24 | | 3.53 | 6 |
| 0900 | 4" long | G | 120 | .067 | | .36 | 3.24 | | 3.60 | 6.10 |
| 1000 | 5" long | G | 115 | .070 | | .45 | 3.38 | | 3.83 | 6.45 |
| 1100 | 1/2" diameter, 1-1/2" long | G | 120 | .067 | | .43 | 3.24 | | 3.67 | 6.15 |
| 1200 | 2" long | G | 120 | .067 | | .49 | 3.24 | | 3.73 | 6.25 |
| 1300 | 4" long | G | 115 | .070 | | .78 | 3.38 | | 4.16 | 6.80 |
| 1400 | 6" long | G | 110 | .073 | | 1.08 | 3.53 | | 4.61 | 7.45 |
| 1500 | 8" long | G | 105 | .076 | | 1.42 | 3.70 | | 5.12 | 8.10 |
| 1600 | 5/8" diameter, 1-1/2" long | G | 120 | .067 | | 1.01 | 3.24 | | 4.25 | 6.80 |
| 1700 | 2" long | G | 120 | .067 | | 1.12 | 3.24 | | 4.36 | 6.95 |
| 1800 | 4" long | G | 115 | .070 | | 1.61 | 3.38 | | 4.99 | 7.70 |
| 1900 | 6" long | G | 110 | .073 | | 2.07 | 3.53 | | 5.60 | 8.55 |
| 2000 | 8" long | G | 105 | .076 | | 3.07 | 3.70 | | 6.77 | 9.95 |
| 2100 | 10" long | G | 100 | .080 | | 3.87 | 3.88 | | 7.75 | 11.10 |
| 2200 | 3/4" diameter, 2" long | G | 120 | .067 | | 1.22 | 3.24 | | 4.46 | 7.05 |
| 2300 | 4" long | G | 110 | .073 | | 1.74 | 3.53 | | 5.27 | 8.15 |
| 2400 | 6" long | G | 105 | .076 | | 2.23 | 3.70 | | 5.93 | 9 |
| 2500 | 8" long | G | 95 | .084 | | 3.35 | 4.09 | | 7.44 | 10.90 |
| 2600 | 10" long | G | 85 | .094 | | 4.39 | 4.57 | | 8.96 | 12.90 |
| 2700 | 12" long | G | 80 | .100 | | 5.15 | 4.86 | | 10.01 | 14.25 |
| 2800 | 1" diameter, 3" long | G | 105 | .076 | | 3.53 | 3.70 | | 7.23 | 10.45 |
| 2900 | 6" long | G | 90 | .089 | | 5.25 | 4.32 | | 9.57 | 13.35 |
| 3000 | 12" long | G | 75 | .107 | | 9.55 | 5.20 | | 14.75 | 19.65 |
| 3100 | For galvanized, add | | | | | 75% | | | | |
| 3200 | For stainless, add | | | | | 350% | | | | |

## 05 05 23 – Metal Fastenings

### 05 05 23.15 Chemical Anchors

| | Crew | Daily Output | Labor-Hours | Unit | Material | 2011 Bare Costs Labor | Equipment | Total | Total Incl O&P |
|---|---|---|---|---|---|---|---|---|---|
| 0010 **CHEMICAL ANCHORS** | | | | | | | | | |
| 0020 Includes layout & drilling | | | | | | | | | |
| 1430 Chemical anchor, w/rod & epoxy cartridge, 3/4" diam. x 9-1/2" long | B-89A | 27 | .593 | Ea. | 8.25 | 23.50 | 4.07 | 35.82 | 49.50 |
| 1435     1" diameter x 11-3/4" long | | 24 | .667 | | 16.05 | 26.50 | 4.57 | 47.12 | 63 |
| 1440     1-1/4" diameter x 14" long | | 21 | .762 | | 29.50 | 30 | 5.25 | 64.75 | 84.50 |
| 1445     1-3/4" diameter x 15" long | | 20 | .800 | | 58 | 31.50 | 5.50 | 95 | 118 |
| 1450     18" long | | 17 | .941 | | 69.50 | 37 | 6.45 | 112.95 | 141 |
| 1455     2" diameter x 18" long | | 16 | 1 | | 84 | 39.50 | 6.85 | 130.35 | 161 |
| 1460     24" long | ↓ | 15 | 1.067 | ↓ | 109 | 42 | 7.30 | 158.30 | 193 |

### 05 05 23.20 Expansion Anchors

| | | Crew | Daily Output | Labor-Hours | Unit | Material | 2011 Bare Costs Labor | Equipment | Total | Total Incl O&P |
|---|---|---|---|---|---|---|---|---|---|---|
| 0010 **EXPANSION ANCHORS** | | | | | | | | | | |
| 0100 Anchors for concrete, brick or stone, no layout and drilling | | | | | | | | | | |
| 0200 Expansion shields, zinc, 1/4" diameter, 1-5/16" long, single | G | 1 Carp | 90 | .089 | Ea. | .42 | 3.83 | | 4.25 | 6.30 |
| 0300     1-3/8" long, double | G | | 85 | .094 | | .50 | 4.05 | | 4.55 | 6.75 |
| 0400     3/8" diameter, 1-1/2" long, single | G | | 85 | .094 | | .72 | 4.05 | | 4.77 | 7 |
| 0500     2" long, double | G | | 80 | .100 | | 1.08 | 4.31 | | 5.39 | 7.80 |
| 0600     1/2" diameter, 2-1/16" long, single | G | | 80 | .100 | | 1.36 | 4.31 | | 5.67 | 8.10 |
| 0700     2-1/2" long, double | G | | 75 | .107 | | 1.65 | 4.59 | | 6.24 | 8.85 |
| 0800     5/8" diameter, 2-5/8" long, single | G | | 75 | .107 | | 2.46 | 4.59 | | 7.05 | 9.75 |
| 0900     2-3/4" long, double | G | | 70 | .114 | | 2.39 | 4.92 | | 7.31 | 10.20 |
| 1000     3/4" diameter, 2-3/4" long, single | G | | 70 | .114 | | 2.74 | 4.92 | | 7.66 | 10.55 |
| 1100     3-15/16" long, double | G | ↓ | 65 | .123 | ↓ | 4.65 | 5.30 | | 9.95 | 13.25 |
| 2100 Hollow wall anchors for gypsum wall board, plaster or tile | | | | | | | | | | |
| 2300     1/8" diameter, short | G | 1 Carp | 160 | .050 | Ea. | .19 | 2.15 | | 2.34 | 3.51 |
| 2400     Long | G | | 150 | .053 | | .22 | 2.30 | | 2.52 | 3.76 |
| 2500     3/16" diameter, short | G | | 150 | .053 | | .34 | 2.30 | | 2.64 | 3.89 |
| 2600     Long | G | | 140 | .057 | | .48 | 2.46 | | 2.94 | 4.30 |
| 2700     1/4" diameter, short | G | | 140 | .057 | | .56 | 2.46 | | 3.02 | 4.39 |
| 2800     Long | G | | 130 | .062 | | .52 | 2.65 | | 3.17 | 4.63 |
| 3000 Toggle bolts, bright steel, 1/8" diameter, 2" long | G | | 85 | .094 | | .18 | 4.05 | | 4.23 | 6.40 |
| 3100     4" long | G | | 80 | .100 | | .23 | 4.31 | | 4.54 | 6.85 |
| 3200     3/16" diameter, 3" long | G | | 80 | .100 | | .21 | 4.31 | | 4.52 | 6.85 |
| 3300     6" long | G | | 75 | .107 | | .31 | 4.59 | | 4.90 | 7.40 |
| 3400     1/4" diameter, 3" long | G | | 75 | .107 | | .27 | 4.59 | | 4.86 | 7.35 |
| 3500     6" long | G | | 70 | .114 | | .45 | 4.92 | | 5.37 | 8.05 |
| 3600     3/8" diameter, 3" long | G | | 70 | .114 | | .67 | 4.92 | | 5.59 | 8.30 |
| 3700     6" long | G | | 60 | .133 | | 1.13 | 5.75 | | 6.88 | 10.05 |
| 3800     1/2" diameter, 4" long | G | | 60 | .133 | | 1.67 | 5.75 | | 7.42 | 10.65 |
| 3900     6" long | G | ↓ | 50 | .160 | ↓ | 2.30 | 6.90 | | 9.20 | 13.10 |
| 4000 Nailing anchors | | | | | | | | | | |
| 4100     Nylon nailing anchor, 1/4" diameter, 1" long | | 1 Carp | 3.20 | 2.500 | C | 12.45 | 108 | | 120.45 | 179 |
| 4200     1-1/2" long | | | 2.80 | 2.857 | | 15.85 | 123 | | 138.85 | 206 |
| 4300     2" long | | | 2.40 | 3.333 | | 17.95 | 144 | | 161.95 | 240 |
| 4400     Metal nailing anchor, 1/4" diameter, 1" long | G | | 3.20 | 2.500 | | 15.20 | 108 | | 123.20 | 182 |
| 4500     1-1/2" long | G | | 2.80 | 2.857 | | 19.95 | 123 | | 142.95 | 211 |
| 4600     2" long | G | ↓ | 2.40 | 3.333 | ↓ | 25 | 144 | | 169 | 248 |
| 5000 Screw anchors for concrete, masonry, | | | | | | | | | | |
| 5100     stone & tile, no layout or drilling included | | | | | | | | | | |
| 5700     Lag screw shields, 1/4" diameter, short | G | 1 Carp | 90 | .089 | Ea. | .31 | 3.83 | | 4.14 | 6.20 |
| 5800     Long | G | | 85 | .094 | | .40 | 4.05 | | 4.45 | 6.65 |
| 5900     3/8" diameter, short | G | | 85 | .094 | | .59 | 4.05 | | 4.64 | 6.85 |
| 6000     Long | G | ↓ | 80 | .100 | ↓ | .73 | 4.31 | | 5.04 | 7.40 |

### 05 05 23.20 Expansion Anchors

| | | Crew | Daily Output | Labor-Hours | Unit | Material | 2011 Bare Costs Labor | Equipment | Total | Total Incl O&P |
|---|---|---|---|---|---|---|---|---|---|---|
| 6100 | 1/2" diameter, short | G | 1 Carp | 80 | .100 | Ea. | .99 | 4.31 | | 5.30 | 7.70 |
| 6200 | Long | G | | 75 | .107 | | 1.35 | 4.59 | | 5.94 | 8.55 |
| 6300 | 5/8" diameter, short | G | | 70 | .114 | | 1.58 | 4.92 | | 6.50 | 9.30 |
| 6400 | Long | G | | 65 | .123 | | 2.12 | 5.30 | | 7.42 | 10.50 |
| 6600 | Lead, #6 & #8, 3/4" long | G | | 260 | .031 | | .16 | 1.32 | | 1.48 | 2.21 |
| 6700 | #10 - #14, 1-1/2" long | G | | 200 | .040 | | .26 | 1.72 | | 1.98 | 2.93 |
| 6800 | #16 & #18, 1-1/2" long | G | | 160 | .050 | | .32 | 2.15 | | 2.47 | 3.65 |
| 6900 | Plastic, #6 & #8, 3/4" long | | | 260 | .031 | | .04 | 1.32 | | 1.36 | 2.07 |
| 7000 | #8 & #10, 7/8" long | | | 240 | .033 | | .04 | 1.43 | | 1.47 | 2.24 |
| 7100 | #10 & #12, 1" long | | | 220 | .036 | | .05 | 1.57 | | 1.62 | 2.46 |
| 7200 | #14 & #16, 1-1/2" long | | | 160 | .050 | | .07 | 2.15 | | 2.22 | 3.38 |
| 8000 | Wedge anchors, not including layout or drilling | | | | | | | | | | |
| 8050 | Carbon steel, 1/4" diameter, 1-3/4" long | G | 1 Carp | 150 | .053 | Ea. | .32 | 2.30 | | 2.62 | 3.87 |
| 8100 | 3-1/4" long | G | | 140 | .057 | | .42 | 2.46 | | 2.88 | 4.23 |
| 8150 | 3/8" diameter, 2-1/4" long | G | | 145 | .055 | | .38 | 2.38 | | 2.76 | 4.06 |
| 8200 | 5" long | G | | 140 | .057 | | .67 | 2.46 | | 3.13 | 4.51 |
| 8250 | 1/2" diameter, 2-3/4" long | G | | 140 | .057 | | .79 | 2.46 | | 3.25 | 4.64 |
| 8300 | 7" long | G | | 125 | .064 | | 1.35 | 2.76 | | 4.11 | 5.70 |
| 8350 | 5/8" diameter, 3-1/2" long | G | | 130 | .062 | | 1.40 | 2.65 | | 4.05 | 5.60 |
| 8400 | 8-1/2" long | G | | 115 | .070 | | 2.97 | 3 | | 5.97 | 7.85 |
| 8450 | 3/4" diameter, 4-1/4" long | G | | 115 | .070 | | 2.09 | 3 | | 5.09 | 6.90 |
| 8500 | 10" long | G | | 95 | .084 | | 4.75 | 3.63 | | 8.38 | 10.80 |
| 8550 | 1" diameter, 6" long | G | | 100 | .080 | | 7.65 | 3.44 | | 11.09 | 13.70 |
| 8575 | 9" long | G | | 85 | .094 | | 9.95 | 4.05 | | 14 | 17.10 |
| 8600 | 12" long | G | | 75 | .107 | | 10.70 | 4.59 | | 15.29 | 18.85 |
| 8650 | 1-1/4" diameter, 9" long | G | | 70 | .114 | | 21.50 | 4.92 | | 26.42 | 31 |
| 8700 | 12" long | G | | 60 | .133 | | 27.50 | 5.75 | | 33.25 | 39 |
| 8750 | For type 303 stainless steel, add | | | | | | 350% | | | | |
| 8800 | For type 316 stainless steel, add | | | | | | 450% | | | | |
| 8950 | Self-drilling concrete screw, hex washer head, 3/16" diam. x 1-3/4" long | G | 1 Carp | 300 | .027 | Ea. | .18 | 1.15 | | 1.33 | 1.96 |
| 8960 | 2-1/4" long | G | | 250 | .032 | | .20 | 1.38 | | 1.58 | 2.33 |
| 8970 | Phillips flat head, 3/16" diam. x 1-3/4" long | G | | 300 | .027 | | .18 | 1.15 | | 1.33 | 1.96 |
| 8980 | 2-1/4" long | G | | 250 | .032 | | .19 | 1.38 | | 1.57 | 2.32 |

### 05 05 23.25 High Strength Bolts

| | | Crew | Daily Output | Labor-Hours | Unit | Material | 2011 Bare Costs Labor | Equipment | Total | Total Incl O&P |
|---|---|---|---|---|---|---|---|---|---|---|
| 0010 | **HIGH STRENGTH BOLTS** R050523-10 | | | | | | | | | | |
| 0020 | A325 Type 1, structural steel, bolt-nut-washer set | | | | | | | | | | |
| 0100 | 1/2" diameter x 1-1/2" long | G | 1 Sswk | 130 | .062 | Ea. | .92 | 2.99 | | 3.91 | 6.30 |
| 0120 | 2" long | G | | 125 | .064 | | .99 | 3.11 | | 4.10 | 6.60 |
| 0150 | 3" long | G | | 120 | .067 | | 1.38 | 3.24 | | 4.62 | 7.20 |
| 0170 | 5/8" diameter x 1-1/2" long | G | | 125 | .064 | | 1.61 | 3.11 | | 4.72 | 7.25 |
| 0180 | 2" long | G | | 120 | .067 | | 1.73 | 3.24 | | 4.97 | 7.60 |
| 0190 | 3" long | G | | 115 | .070 | | 2.14 | 3.38 | | 5.52 | 8.30 |
| 0200 | 3/4" diameter x 2" long | G | | 120 | .067 | | 2.62 | 3.24 | | 5.86 | 8.60 |
| 0220 | 3" long | G | | 115 | .070 | | 3.14 | 3.38 | | 6.52 | 9.40 |
| 0250 | 4" long | G | | 110 | .073 | | 3.84 | 3.53 | | 7.37 | 10.50 |
| 0300 | 6" long | G | | 105 | .076 | | 4.99 | 3.70 | | 8.69 | 12.05 |
| 0350 | 8" long | G | | 95 | .084 | | 9.85 | 4.09 | | 13.94 | 18.05 |
| 0360 | 7/8" diameter x 2" long | G | | 115 | .070 | | 3.66 | 3.38 | | 7.04 | 10 |
| 0365 | 3" long | G | | 110 | .073 | | 4.33 | 3.53 | | 7.86 | 11 |
| 0370 | 4" long | G | | 105 | .076 | | 5.20 | 3.70 | | 8.90 | 12.30 |
| 0380 | 6" long | G | | 100 | .080 | | 6.65 | 3.88 | | 10.53 | 14.15 |
| 0390 | 8" long | G | | 90 | .089 | | 10.55 | 4.32 | | 14.87 | 19.20 |

## 05 05 23 – Metal Fastenings

| 05 05 23.25 High Strength Bolts | | Crew | Daily Output | Labor-Hours | Unit | Material | 2011 Bare Costs Labor | Equipment | Total | Total Incl O&P |
|---|---|---|---|---|---|---|---|---|---|---|
| 0400 | 1" diameter x 2" long | G 1 Sswk | 105 | .076 | Ea. | 4.67 | 3.70 | | 8.37 | 11.70 |
| 0420 | 3" long | G | 100 | .080 | | 5.25 | 3.88 | | 9.13 | 12.65 |
| 0450 | 4" long | G | 95 | .084 | | 5.95 | 4.09 | | 10.04 | 13.75 |
| 0500 | 6" long | G | 90 | .089 | | 7.90 | 4.32 | | 12.22 | 16.30 |
| 0550 | 8" long | G | 85 | .094 | | 13.70 | 4.57 | | 18.27 | 23 |
| 0600 | 1-1/4" diameter x 3" long | G | 85 | .094 | | 10.80 | 4.57 | | 15.37 | 19.90 |
| 0650 | 4" long | G | 80 | .100 | | 2.78 | 4.86 | | 7.64 | 11.65 |
| 0700 | 6" long | G | 75 | .107 | | 15.25 | 5.20 | | 20.45 | 26 |
| 0750 | 8" long | G | 70 | .114 | | 19.40 | 5.55 | | 24.95 | 31.50 |
| 1020 | A490, bolt-nut-washer set | | | | | | | | | |
| 1170 | 5/8" diameter x 1-1/2" long | G 1 Sswk | 125 | .064 | Ea. | 2.02 | 3.11 | | 5.13 | 7.70 |
| 1180 | 2" long | G | 120 | .067 | | 2.38 | 3.24 | | 5.62 | 8.30 |
| 1190 | 3" long | G | 115 | .070 | | 2.89 | 3.38 | | 6.27 | 9.15 |
| 1200 | 3/4" diameter x 2" long | G | 120 | .067 | | 3.47 | 3.24 | | 6.71 | 9.50 |
| 1220 | 3" long | G | 115 | .070 | | 4.11 | 3.38 | | 7.49 | 10.45 |
| 1250 | 4" long | G | 110 | .073 | | 4.79 | 3.53 | | 8.32 | 11.50 |
| 1300 | 6" long | G | 105 | .076 | | 7.05 | 3.70 | | 10.75 | 14.30 |
| 1350 | 8" long | G | 95 | .084 | | 12 | 4.09 | | 16.09 | 20.50 |
| 1360 | 7/8" diameter x 2" long | G | 115 | .070 | | 5.25 | 3.38 | | 8.63 | 11.70 |
| 1365 | 3" long | G | 110 | .073 | | 6.15 | 3.53 | | 9.68 | 13.05 |
| 1370 | 4" long | G | 105 | .076 | | 7.65 | 3.70 | | 11.35 | 15 |
| 1380 | 6" long | G | 100 | .080 | | 10.80 | 3.88 | | 14.68 | 18.75 |
| 1390 | 8" long | G | 90 | .089 | | 15.75 | 4.32 | | 20.07 | 25 |
| 1400 | 1" diameter x 2" long | G | 105 | .076 | | 6.20 | 3.70 | | 9.90 | 13.40 |
| 1420 | 3" long | G | 100 | .080 | | 7.55 | 3.88 | | 11.43 | 15.15 |
| 1450 | 4" long | G | 95 | .084 | | 8.75 | 4.09 | | 12.84 | 16.80 |
| 1500 | 6" long | G | 90 | .089 | | 11.75 | 4.32 | | 16.07 | 20.50 |
| 1550 | 8" long | G | 85 | .094 | | 18.70 | 4.57 | | 23.27 | 28.50 |
| 1600 | 1-1/4" diameter x 3" long | G | 85 | .094 | | 13.70 | 4.57 | | 18.27 | 23 |
| 1650 | 4" long | G | 80 | .100 | | 15.70 | 4.86 | | 20.56 | 26 |
| 1700 | 6" long | G | 75 | .107 | | 21.50 | 5.20 | | 26.70 | 33 |
| 1750 | 8" long | G | 70 | .114 | | 28 | 5.55 | | 33.55 | 41 |

| 05 05 23.30 Lag Screws | | | | | | | | | | |
|---|---|---|---|---|---|---|---|---|---|---|
| 0010 | **LAG SCREWS** | | | | | | | | | |
| 0020 | Steel, 1/4" diameter, 2" long | G 1 Carp | 200 | .040 | Ea. | .09 | 1.72 | | 1.81 | 2.74 |
| 0100 | 3/8" diameter, 3" long | G | 150 | .053 | | .26 | 2.30 | | 2.56 | 3.81 |
| 0200 | 1/2" diameter, 3" long | G | 130 | .062 | | .48 | 2.65 | | 3.13 | 4.59 |
| 0300 | 5/8" diameter, 3" long | G | 120 | .067 | | .99 | 2.87 | | 3.86 | 5.50 |

| 05 05 23.35 Machine Screws | | | | | | | | | | |
|---|---|---|---|---|---|---|---|---|---|---|
| 0010 | **MACHINE SCREWS** | | | | | | | | | |
| 0020 | Steel, round head, #8 x 1" long | G 1 Carp | 4.80 | 1.667 | C | 3.73 | 72 | | 75.73 | 114 |
| 0110 | #8 x 2" long | G | 2.40 | 3.333 | | 5.75 | 144 | | 149.75 | 226 |
| 0200 | #10 x 1" long | G | 4 | 2 | | 5.80 | 86 | | 91.80 | 138 |
| 0300 | #10 x 2" long | G | 2 | 4 | | 7.80 | 172 | | 179.80 | 273 |

| 05 05 23.40 Machinery Anchors | | | | | | | | | | |
|---|---|---|---|---|---|---|---|---|---|---|
| 0010 | **MACHINERY ANCHORS**, heavy duty, incl. sleeve, floating base nut, | | | | | | | | | |
| 0020 | lower stud & coupling nut, fiber plug, connecting stud, washer & nut. | | | | | | | | | |
| 0030 | For flush mounted embedment in poured concrete heavy equip. pads. | | | | | | | | | |
| 0200 | Stud & bolt, 1/2" diameter | G E-16 | 40 | .400 | Ea. | 68.50 | 19.80 | 2.72 | 91.02 | 114 |
| 0300 | 5/8" diameter | G | 35 | .457 | | 76 | 22.50 | 3.11 | 101.61 | 127 |
| 0500 | 3/4" diameter | G | 30 | .533 | | 87.50 | 26.50 | 3.63 | 117.63 | 147 |
| 0600 | 7/8" diameter | G | 25 | .640 | | 95.50 | 31.50 | 4.36 | 131.36 | 166 |

### 05 05 23.40 Machinery Anchors

| | | Crew | Daily Output | Labor-Hours | Unit | Material | 2011 Bare Costs Labor | Equipment | Total | Total Incl O&P |
|---|---|---|---|---|---|---|---|---|---|---|
| 0800 | 1" diameter | G E-16 | 20 | .800 | Ea. | 100 | 39.50 | 5.45 | 144.95 | 186 |
| 0900 | 1-1/4" diameter | G ↓ | 15 | 1.067 | ↓ | 133 | 53 | 7.25 | 193.25 | 249 |

### 05 05 23.50 Powder Actuated Tools and Fasteners

| | | Crew | Daily Output | Labor-Hours | Unit | Material | Labor | Equipment | Total | Total Incl O&P |
|---|---|---|---|---|---|---|---|---|---|---|
| 0010 | **POWDER ACTUATED TOOLS & FASTENERS** | | | | | | | | | |
| 0020 | Stud driver, .22 caliber, buy, minimum | | | | Ea. | 104 | | | 104 | 114 |
| 0100 | Maximum | | | | " | 370 | | | 370 | 405 |
| 0300 | Powder charges for above, low velocity | | | | C | 7 | | | 7 | 7.70 |
| 0400 | Standard velocity | | | | | 7.70 | | | 7.70 | 8.50 |
| 0600 | Drive pins & studs, 1/4" & 3/8" diam., to 3" long, minimum | G 1 Carp | 4.80 | 1.667 | | 4.02 | 72 | | 76.02 | 114 |
| 0700 | Maximum | G " | 4 | 2 | ↓ | 12.85 | 86 | | 98.85 | 146 |

### 05 05 23.55 Rivets

| | | Crew | Daily Output | Labor-Hours | Unit | Material | Labor | Equipment | Total | Total Incl O&P |
|---|---|---|---|---|---|---|---|---|---|---|
| 0010 | **RIVETS** | | | | | | | | | |
| 0100 | Aluminum rivet & mandrel, 1/2" grip length x 1/8" diameter | G 1 Carp | 4.80 | 1.667 | C | 6.60 | 72 | | 78.60 | 117 |
| 0200 | 3/16" diameter | G | 4 | 2 | | 11.20 | 86 | | 97.20 | 144 |
| 0300 | Aluminum rivet, steel mandrel, 1/8" diameter | G | 4.80 | 1.667 | | 8.75 | 72 | | 80.75 | 120 |
| 0400 | 3/16" diameter | G | 4 | 2 | | 14 | 86 | | 100 | 147 |
| 0500 | Copper rivet, steel mandrel, 1/8" diameter | G | 4.80 | 1.667 | | 7.70 | 72 | | 79.70 | 119 |
| 0800 | Stainless rivet & mandrel, 1/8" diameter | G | 4.80 | 1.667 | | 24 | 72 | | 96 | 137 |
| 0900 | 3/16" diameter | G | 4 | 2 | | 39.50 | 86 | | 125.50 | 176 |
| 1000 | Stainless rivet, steel mandrel, 1/8" diameter | G | 4.80 | 1.667 | | 20.50 | 72 | | 92.50 | 133 |
| 1100 | 3/16" diameter | G | 4 | 2 | | 28 | 86 | | 114 | 163 |
| 1200 | Steel rivet and mandrel, 1/8" diameter | G | 4.80 | 1.667 | | 7.50 | 72 | | 79.50 | 118 |
| 1300 | 3/16" diameter | G ↓ | 4 | 2 | ↓ | 11.10 | 86 | | 97.10 | 144 |
| 1400 | Hand riveting tool, minimum | | | | Ea. | 55 | | | 55 | 60.50 |
| 1500 | Maximum | | | | | 257 | | | 257 | 283 |
| 1600 | Power riveting tool, minimum | | | | | 450 | | | 450 | 495 |
| 1700 | Maximum | | | | ↓ | 2,450 | | | 2,450 | 2,700 |

### 05 05 23.70 Structural Blind Bolts

| | | Crew | Daily Output | Labor-Hours | Unit | Material | Labor | Equipment | Total | Total Incl O&P |
|---|---|---|---|---|---|---|---|---|---|---|
| 0010 | **STRUCTURAL BLIND BOLTS** | | | | | | | | | |
| 0100 | 1/4" diameter x 1/4" grip | G 1 Sswk | 240 | .033 | Ea. | 1.24 | 1.62 | | 2.86 | 4.22 |
| 0150 | 1/2" grip | G | 216 | .037 | | .95 | 1.80 | | 2.75 | 4.23 |
| 0200 | 3/8" diameter x 1/2" grip | G | 232 | .034 | | 1.75 | 1.67 | | 3.42 | 4.89 |
| 0250 | 3/4" grip | G | 208 | .038 | | 1.83 | 1.87 | | 3.70 | 5.30 |
| 0300 | 1/2" diameter x 1/2" grip | G | 224 | .036 | | 5.15 | 1.73 | | 6.88 | 8.70 |
| 0350 | 3/4" grip | G | 200 | .040 | | 5.60 | 1.94 | | 7.54 | 9.60 |
| 0400 | 5/8" diameter x 3/4" grip | G | 216 | .037 | | 8.25 | 1.80 | | 10.05 | 12.30 |
| 0450 | 1" grip | G ↓ | 192 | .042 | ↓ | 9.05 | 2.02 | | 11.07 | 13.50 |

### 05 05 23.80 Vibration and Bearing Pads

| | | Crew | Daily Output | Labor-Hours | Unit | Material | Labor | Equipment | Total | Total Incl O&P |
|---|---|---|---|---|---|---|---|---|---|---|
| 0010 | **VIBRATION & BEARING PADS** | | | | | | | | | |
| 0300 | Laminated synthetic rubber impregnated cotton duck, 1/2" thick | 2 Sswk | 24 | .667 | S.F. | 64 | 32.50 | | 96.50 | 128 |
| 0400 | 1" thick | | 20 | .800 | | 126 | 39 | | 165 | 208 |
| 0600 | Neoprene bearing pads, 1/2" thick | | 24 | .667 | | 24.50 | 32.50 | | 57 | 84 |
| 0700 | 1" thick | | 20 | .800 | | 49 | 39 | | 88 | 123 |
| 0900 | Fabric reinforced neoprene, 5000 psi, 1/2" thick | | 24 | .667 | | 10.75 | 32.50 | | 43.25 | 69 |
| 1000 | 1" thick | | 20 | .800 | | 21.50 | 39 | | 60.50 | 92 |
| 1200 | Felt surfaced vinyl pads, cork and sisal, 5/8" thick | | 24 | .667 | | 27 | 32.50 | | 59.50 | 87 |
| 1300 | 1" thick | | 20 | .800 | | 49 | 39 | | 88 | 123 |
| 1500 | Teflon bonded to 10 ga. carbon steel, 1/32" layer | | 24 | .667 | | 48 | 32.50 | | 80.50 | 110 |
| 1600 | 3/32" layer | | 24 | .667 | | 72 | 32.50 | | 104.50 | 136 |
| 1800 | Bonded to 10 ga. stainless steel, 1/32" layer | | 24 | .667 | | 85 | 32.50 | | 117.50 | 151 |
| 1900 | 3/32" layer | ↓ | 24 | .667 | ↓ | 110 | 32.50 | | 142.50 | 178 |
| 2100 | Circular machine leveling pad & stud | | | | Kip | 6 | | | 6 | 6.60 |

## 05 05 23 – Metal Fastenings

| 05 05 23.85 Weld Shear Connectors | | Crew | Daily Output | Labor-Hours | Unit | Material | 2011 Bare Costs Labor | Equipment | Total | Total Incl O&P |
|---|---|---|---|---|---|---|---|---|---|---|
| 0010 | **WELD SHEAR CONNECTORS** | | | | | | | | | |
| 0020 | 3/4" diameter, 3-3/16" long | G | E-10 | 960 | .017 | Ea. | .51 | .83 | .34 | 1.68 | 2.39 |
| 0030 | 3-3/8" long | G | | 950 | .017 | | .53 | .83 | .34 | 1.70 | 2.44 |
| 0200 | 3-7/8" long | G | | 945 | .017 | | .57 | .84 | .35 | 1.76 | 2.49 |
| 0300 | 4-3/16" long | G | | 935 | .017 | | .60 | .85 | .35 | 1.80 | 2.54 |
| 0500 | 4-7/8" long | G | | 930 | .017 | | .67 | .85 | .35 | 1.87 | 2.64 |
| 0600 | 5-3/16" long | G | | 920 | .017 | | .70 | .86 | .35 | 1.91 | 2.68 |
| 0800 | 5-3/8" long | G | | 910 | .018 | | .70 | .87 | .36 | 1.93 | 2.70 |
| 0900 | 6-3/16" long | G | | 905 | .018 | | .77 | .88 | .36 | 2.01 | 2.80 |
| 1000 | 7-3/16" long | G | | 895 | .018 | | .96 | .89 | .36 | 2.21 | 3.02 |
| 1100 | 8-3/16" long | G | | 890 | .018 | | 1.05 | .89 | .37 | 2.31 | 3.12 |
| 1500 | 7/8" diameter, 3-11/16" long | G | | 920 | .017 | | .82 | .86 | .35 | 2.03 | 2.82 |
| 1600 | 4-3/16" long | G | | 910 | .018 | | .88 | .87 | .36 | 2.11 | 2.90 |
| 1700 | 5-3/16" long | G | | 905 | .018 | | 1 | .88 | .36 | 2.24 | 3.05 |
| 1800 | 6-3/16" long | G | | 895 | .018 | | 1.12 | .89 | .36 | 2.37 | 3.19 |
| 1900 | 7-3/16" long | G | | 890 | .018 | | 1.24 | .89 | .37 | 2.50 | 3.34 |
| 2000 | 8-3/16" long | G | | 880 | .018 | | 1.36 | .90 | .37 | 2.63 | 3.49 |

### 05 05 23.87 Weld Studs

| | | | Crew | Daily Output | Labor-Hours | Unit | Material | Labor | Equipment | Total | Total Incl O&P |
|---|---|---|---|---|---|---|---|---|---|---|---|
| 0010 | **WELD STUDS** | | | | | | | | | | |
| 0020 | 1/4" diameter, 2-11/16" long | G | E-10 | 1120 | .014 | Ea. | .33 | .71 | .29 | 1.33 | 1.93 |
| 0100 | 4-1/8" long | G | | 1080 | .015 | | .31 | .73 | .30 | 1.34 | 1.97 |
| 0200 | 3/8" diameter, 4-1/8" long | G | | 1080 | .015 | | .36 | .73 | .30 | 1.39 | 2.03 |
| 0300 | 6-1/8" long | G | | 1040 | .015 | | .47 | .76 | .31 | 1.54 | 2.20 |
| 0400 | 1/2" diameter, 2-1/8" long | G | | 1040 | .015 | | .34 | .76 | .31 | 1.41 | 2.06 |
| 0500 | 3-1/8" long | G | | 1025 | .016 | | .41 | .77 | .32 | 1.50 | 2.17 |
| 0600 | 4-1/8" long | G | | 1010 | .016 | | .48 | .78 | .32 | 1.58 | 2.28 |
| 0700 | 5-5/16" long | G | | 990 | .016 | | .59 | .80 | .33 | 1.72 | 2.42 |
| 0800 | 6-1/8" long | G | | 975 | .016 | | .64 | .81 | .33 | 1.78 | 2.51 |
| 0900 | 8-1/8" long | G | | 960 | .017 | | .90 | .83 | .34 | 2.07 | 2.82 |
| 1000 | 5/8" diameter, 2-11/16" long | G | | 1000 | .016 | | .58 | .79 | .33 | 1.70 | 2.40 |
| 1010 | 4-3/16" long | G | | 990 | .016 | | .72 | .80 | .33 | 1.85 | 2.56 |
| 1100 | 6-9/16" long | G | | 975 | .016 | | .94 | .81 | .33 | 2.08 | 2.84 |
| 1200 | 8-3/16" long | G | | 960 | .017 | | 1.26 | .83 | .34 | 2.43 | 3.22 |

### 05 05 23.90 Welding Rod

| | | | Crew | Daily Output | Labor-Hours | Unit | Material | Labor | Equipment | Total | Total Incl O&P |
|---|---|---|---|---|---|---|---|---|---|---|---|
| 0010 | **WELDING ROD** | | | | | | | | | | |
| 0020 | Steel, type 6011, 1/8" diam., less than 500# | | | | | Lb. | 2.11 | | | 2.11 | 2.32 |
| 0100 | 500# to 2,000# | | | | | | 1.90 | | | 1.90 | 2.09 |
| 0200 | 2,000# to 5,000# | | | | | | 1.79 | | | 1.79 | 1.96 |
| 0300 | 5/32" diameter, less than 500# | | | | | | 2.23 | | | 2.23 | 2.45 |
| 0310 | 500# to 2,000# | | | | | | 2.01 | | | 2.01 | 2.21 |
| 0320 | 2,000# to 5,000# | | | | | | 1.89 | | | 1.89 | 2.08 |
| 0400 | 3/16" diam., less than 500# | | | | | | 2.26 | | | 2.26 | 2.49 |
| 0500 | 500# to 2,000# | | | | | | 2.04 | | | 2.04 | 2.24 |
| 0600 | 2,000# to 5,000# | | | | | | 1.92 | | | 1.92 | 2.11 |
| 0620 | Steel, type 6010, 1/8" diam., less than 500# | | | | | | 2.20 | | | 2.20 | 2.42 |
| 0630 | 500# to 2,000# | | | | | | 1.98 | | | 1.98 | 2.18 |
| 0640 | 2,000# to 5,000# | | | | | | 1.86 | | | 1.86 | 2.05 |
| 0650 | Steel, type 7018 Low Hydrogen, 1/8" diam., less than 500# | | | | | | 2.12 | | | 2.12 | 2.33 |
| 0660 | 500# to 2,000# | | | | | | 1.91 | | | 1.91 | 2.10 |
| 0670 | 2,000# to 5,000# | | | | | | 1.80 | | | 1.80 | 1.97 |
| 0700 | Steel, type 7024 Jet Weld, 1/8" diam., less than 500# | | | | | | 2.26 | | | 2.26 | 2.49 |
| 0710 | 500# to 2,000# | | | | | | 2.04 | | | 2.04 | 2.24 |

## 05 05 23 – Metal Fastenings

### 05 05 23.90 Welding Rod

| | | Crew | Daily Output | Labor-Hours | Unit | Material | 2011 Bare Costs Labor | Equipment | Total | Total Incl O&P |
|---|---|---|---|---|---|---|---|---|---|---|
| 0720 | 2,000# to 5,000# | | | | Lb. | 1.92 | | | 1.92 | 2.11 |
| 1550 | Aluminum, type 4043 TIG, 1/8" diam., less than 10# | | | | | 4.40 | | | 4.40 | 4.84 |
| 1560 | 10# to 60# | | | | | 3.96 | | | 3.96 | 4.36 |
| 1570 | Over 60# | | | | | 3.72 | | | 3.72 | 4.09 |
| 1600 | Aluminum, type 5356 TIG, 1/8" diam., less than 10# | | | | | 4.47 | | | 4.47 | 4.92 |
| 1610 | 10# to 60# | | | | | 4.03 | | | 4.03 | 4.43 |
| 1620 | Over 60# | | | | | 3.79 | | | 3.79 | 4.17 |
| 1900 | Cast iron, type 8 Nickel, 1/8" diam., less than 500# | | | | | 27.50 | | | 27.50 | 30 |
| 1910 | 500# to 1,000# | | | | | 24.50 | | | 24.50 | 27 |
| 1920 | Over 1,000# | | | | | 23 | | | 23 | 25.50 |
| 2000 | Stainless steel, type 316/316L, 1/8" diam., less than 500# | | | | | 7.60 | | | 7.60 | 8.35 |
| 2100 | 500# to 1000# | | | | | 6.85 | | | 6.85 | 7.55 |
| 2220 | Over 1000# | | | | | 6.45 | | | 6.45 | 7.10 |

# 05 12 Structural Steel Framing

## 05 12 23 – Structural Steel for Buildings

### 05 12 23.05 Canopy Framing

| | | | Crew | Daily Output | Labor-Hours | Unit | Material | 2011 Bare Costs Labor | Equipment | Total | Total Incl O&P |
|---|---|---|---|---|---|---|---|---|---|---|---|
| 0010 | **CANOPY FRAMING** | | | | | | | | | | |
| 0020 | 6" and 8" members, shop fabricated | G | E-4 | 3000 | .011 | Lb. | 1.35 | .52 | .04 | 1.91 | 2.45 |

### 05 12 23.10 Ceiling Supports

| | | | Crew | Daily Output | Labor-Hours | Unit | Material | 2011 Bare Costs Labor | Equipment | Total | Total Incl O&P |
|---|---|---|---|---|---|---|---|---|---|---|---|
| 0010 | **CEILING SUPPORTS** | | | | | | | | | | |
| 1000 | Entrance door/folding partition supports, shop fabricated | G | E-4 | 60 | .533 | L.F. | 22.50 | 26 | 1.82 | 50.32 | 73 |
| 1100 | Linear accelerator door supports | G | | 14 | 2.286 | | 102 | 112 | 7.80 | 221.80 | 320 |
| 1200 | Lintels or shelf angles, hung, exterior hot dipped galv. | G | | 267 | .120 | | 15.35 | 5.90 | .41 | 21.66 | 28 |
| 1250 | Two coats primer paint instead of galv. | G | | 267 | .120 | | 13.30 | 5.90 | .41 | 19.61 | 25.50 |
| 1400 | Monitor support, ceiling hung, expansion bolted | G | | 4 | 8 | Ea. | 355 | 390 | 27.50 | 772.50 | 1,125 |
| 1450 | Hung from pre-set inserts | G | | 6 | 5.333 | | 385 | 262 | 18.20 | 665.20 | 900 |
| 1600 | Motor supports for overhead doors | G | | 4 | 8 | | 181 | 390 | 27.50 | 598.50 | 925 |
| 1700 | Partition support for heavy folding partitions, without pocket | G | | 24 | 1.333 | L.F. | 51 | 65.50 | 4.55 | 121.05 | 178 |
| 1750 | Supports at pocket only | G | | 12 | 2.667 | | 102 | 131 | 9.10 | 242.10 | 355 |
| 2000 | Rolling grilles & fire door supports | G | | 34 | .941 | | 44 | 46 | 3.21 | 93.21 | 134 |
| 2100 | Spider-leg light supports, expansion bolted to ceiling slab | G | | 8 | 4 | Ea. | 146 | 196 | 13.65 | 355.65 | 520 |
| 2150 | Hung from pre-set inserts | G | | 12 | 2.667 | " | 158 | 131 | 9.10 | 298.10 | 415 |
| 2400 | Toilet partition support | G | | 36 | .889 | L.F. | 51 | 43.50 | 3.03 | 97.53 | 137 |
| 2500 | X-ray travel gantry support | G | | 12 | 2.667 | " | 176 | 131 | 9.10 | 316.10 | 435 |

### 05 12 23.15 Columns, Lightweight

| | | | Crew | Daily Output | Labor-Hours | Unit | Material | 2011 Bare Costs Labor | Equipment | Total | Total Incl O&P |
|---|---|---|---|---|---|---|---|---|---|---|---|
| 0010 | **COLUMNS, LIGHTWEIGHT** | | | | | | | | | | |
| 1000 | Lightweight units (lally), 3-1/2" diameter | | E-2 | 780 | .072 | L.F. | 6.65 | 3.40 | 2.08 | 12.13 | 15.40 |
| 1050 | 4" diameter | | " | 900 | .062 | " | 8.10 | 2.95 | 1.80 | 12.85 | 15.85 |
| 5800 | Adjustable jack post, 8' maximum height, 2-3/4" diameter | G | | | | Ea. | 42 | | | 42 | 46.50 |
| 5850 | 4" diameter | G | | | | " | 67.50 | | | 67.50 | 74 |

### 05 12 23.17 Columns, Structural

| | | | Crew | Daily Output | Labor-Hours | Unit | Material | 2011 Bare Costs Labor | Equipment | Total | Total Incl O&P |
|---|---|---|---|---|---|---|---|---|---|---|---|
| 0010 | **COLUMNS, STRUCTURAL** | R051223-10 | | | | | | | | | |
| 0015 | Made from recycled materials | G | | | | | | | | | |
| 0020 | Shop fab'd for 100-ton, 1-2 story project, bolted connections | | | | | | | | | | |
| 0800 | Steel, concrete filled, extra strong pipe, 3-1/2" diameter | | E-2 | 660 | .085 | L.F. | 37 | 4.02 | 2.46 | 43.48 | 50.50 |
| 0830 | 4" diameter | | | 780 | .072 | | 41.50 | 3.40 | 2.08 | 46.98 | 53.50 |
| 0890 | 5" diameter | | | 1020 | .055 | | 49.50 | 2.60 | 1.59 | 53.69 | 60 |
| 0930 | 6" diameter | | | 1200 | .047 | | 65.50 | 2.21 | 1.35 | 69.06 | 77 |
| 0940 | 8" diameter | | | 1100 | .051 | | 65.50 | 2.41 | 1.47 | 69.38 | 77.50 |

### 05 12 23.17 Columns, Structural

| | | Crew | Daily Output | Labor-Hours | Unit | Material | 2011 Bare Costs Labor | Equipment | Total | Total Incl O&P |
|---|---|---|---|---|---|---|---|---|---|---|
| 1100 | For galvanizing, add | | | | Lb. | .20 | | | .20 | .22 |
| 1300 | For web ties, angles, etc., add per added lb. | 1 Sswk | 945 | .008 | | 1.13 | .41 | | 1.54 | 1.97 |
| 1500 | Steel pipe, extra strong, no concrete, 3" to 5" diameter [G] | E-2 | 16000 | .004 | | 1.13 | .17 | .10 | 1.40 | 1.63 |
| 1600 | 6" to 12" diameter [G] | | 14000 | .004 | | 1.13 | .19 | .12 | 1.44 | 1.69 |
| 1700 | Steel pipe, extra strong, no concrete, 3" diameter x 12'-0" [G] | | 60 | .933 | Ea. | 138 | 44 | 27 | 209 | 257 |
| 1750 | 4" diameter x 12'-0" [G] | | 58 | .966 | | 202 | 45.50 | 28 | 275.50 | 330 |
| 1800 | 6" diameter x 12'-0" [G] | | 54 | 1.037 | | 385 | 49 | 30 | 464 | 540 |
| 1850 | 8" diameter x 14'-0" [G] | | 50 | 1.120 | | 685 | 53 | 32.50 | 770.50 | 875 |
| 1900 | 10" diameter x 16'-0" [G] | | 48 | 1.167 | | 985 | 55.50 | 34 | 1,074.50 | 1,200 |
| 1950 | 12" diameter x 18'-0" [G] | | 45 | 1.244 | | 1,325 | 59 | 36 | 1,420 | 1,600 |
| 3300 | Structural tubing, square, A500GrB, 4" to 6" square, light section [G] | | 11270 | .005 | Lb. | 1.13 | .24 | .14 | 1.51 | 1.80 |
| 3600 | Heavy section [G] | | 32000 | .002 | " | 1.13 | .08 | .05 | 1.26 | 1.44 |
| 4000 | Concrete filled, add | | | | L.F. | 4.03 | | | 4.03 | 4.43 |
| 4500 | Structural tubing, sq, 4" x 4" x 1/4" x 12'-0" [G] | E-2 | 58 | .966 | Ea. | 186 | 45.50 | 28 | 259.50 | 315 |
| 4550 | 6" x 6" x 1/4" x 12'-0" [G] | | 54 | 1.037 | | 305 | 49 | 30 | 384 | 450 |
| 4600 | 8" x 8" x 3/8" x 14'-0" [G] | | 50 | 1.120 | | 660 | 53 | 32.50 | 745.50 | 850 |
| 4650 | 10" x 10" x 1/2" x 16'-0" [G] | | 48 | 1.167 | | 1,225 | 55.50 | 34 | 1,314.50 | 1,475 |
| 5100 | Structural tubing, rect, 5" to 6" wide, light section [G] | | 8000 | .007 | Lb. | 1.13 | .33 | .20 | 1.66 | 2.02 |
| 5200 | Heavy section [G] | | 12000 | .005 | | 1.13 | .22 | .14 | 1.49 | 1.76 |
| 5300 | 7" to 10" wide, light section [G] | | 15000 | .004 | | 1.13 | .18 | .11 | 1.42 | 1.66 |
| 5400 | Heavy section [G] | | 18000 | .003 | | 1.13 | .15 | .09 | 1.37 | 1.59 |
| 5500 | Structural tubing, rect, 5" x 3" x 1/4" x 12'-0" [G] | | 58 | .966 | Ea. | 180 | 45.50 | 28 | 253.50 | 305 |
| 5550 | 6" x 4" x 5/16" x 12'-0" [G] | | 54 | 1.037 | | 281 | 49 | 30 | 360 | 425 |
| 5600 | 8" x 4" x 3/8" x 12'-0" [G] | | 54 | 1.037 | | 410 | 49 | 30 | 489 | 565 |
| 5650 | 10" x 6" x 3/8" x 14'-0" [G] | | 50 | 1.120 | | 660 | 53 | 32.50 | 745.50 | 850 |
| 5700 | 12" x 8" x 1/2" x 16'-0" [G] | | 48 | 1.167 | | 1,225 | 55.50 | 34 | 1,314.50 | 1,450 |
| 6800 | W Shape, A992 steel, 2 tier, W8 x 24 [G] | | 1080 | .052 | L.F. | 29.50 | 2.46 | 1.50 | 33.46 | 38.50 |
| 6850 | W8 x 31 [G] | | 1080 | .052 | | 38.50 | 2.46 | 1.50 | 42.46 | 48 |
| 6900 | W8 x 48 [G] | | 1032 | .054 | | 59.50 | 2.57 | 1.57 | 63.64 | 71.50 |
| 6950 | W8 x 67 [G] | | 984 | .057 | | 83 | 2.70 | 1.65 | 87.35 | 97.50 |
| 7000 | W10 x 45 [G] | | 1032 | .054 | | 55.50 | 2.57 | 1.57 | 59.64 | 67.50 |
| 7050 | W10 x 68 [G] | | 984 | .057 | | 84 | 2.70 | 1.65 | 88.35 | 99 |
| 7100 | W10 x 112 [G] | | 960 | .058 | | 139 | 2.76 | 1.69 | 143.45 | 159 |
| 7150 | W12 x 50 [G] | | 1032 | .054 | | 62 | 2.57 | 1.57 | 66.14 | 74 |
| 7200 | W12 x 87 [G] | | 984 | .057 | | 108 | 2.70 | 1.65 | 112.35 | 124 |
| 7250 | W12 x 120 [G] | | 960 | .058 | | 149 | 2.76 | 1.69 | 153.45 | 170 |
| 7300 | W12 x 190 [G] | | 912 | .061 | | 235 | 2.91 | 1.78 | 239.69 | 266 |
| 7350 | W14 x 74 [G] | | 984 | .057 | | 91.50 | 2.70 | 1.65 | 95.85 | 107 |
| 7400 | W14 x 120 [G] | | 960 | .058 | | 149 | 2.76 | 1.69 | 153.45 | 170 |
| 7450 | W14 x 176 [G] | | 912 | .061 | | 218 | 2.91 | 1.78 | 222.69 | 247 |
| 8090 | For projects 75 to 99 tons, add | | | | All | 10% | | | | |
| 8092 | 50 to 74 tons, add | | | | | 20% | | | | |
| 8094 | 25 to 49 tons, add | | | | | 30% | 10% | | | |
| 8096 | 10 to 24 tons, add | | | | | 50% | 25% | | | |
| 8098 | 2 to 9 tons, add | | | | | 75% | 50% | | | |
| 8099 | Less than 2 tons, add | | | | | 100% | 100% | | | |

### 05 12 23.20 Curb Edging

| | | Crew | Daily Output | Labor-Hours | Unit | Material | 2011 Bare Costs Labor | Equipment | Total | Total Incl O&P |
|---|---|---|---|---|---|---|---|---|---|---|
| 0010 | **CURB EDGING** | | | | | | | | | |
| 0020 | Steel angle w/anchors, shop fabricated, on forms, 1" x 1", 0.8#/L.F. [G] | E-4 | 350 | .091 | L.F. | 1.44 | 4.48 | .31 | 6.23 | 9.80 |
| 0100 | 2" x 2" angles, 3.92#/L.F. [G] | | 330 | .097 | | 5.65 | 4.76 | .33 | 10.74 | 14.95 |
| 0200 | 3" x 3" angles, 6.1#/L.F. [G] | | 300 | .107 | | 8.90 | 5.25 | .36 | 14.51 | 19.45 |
| 0300 | 4" x 4" angles, 8.2#/L.F. [G] | | 275 | .116 | | 11.75 | 5.70 | .40 | 17.85 | 23.50 |

## 05 12 23 – Structural Steel for Buildings

| 05 12 23.20 Curb Edging | | Crew | Daily Output | Labor-Hours | Unit | Material | 2011 Bare Costs Labor | Equipment | Total | Total Incl O&P |
|---|---|---|---|---|---|---|---|---|---|---|
| 1000 | 6" x 4" angles, 12.3#/L.F. | G E-4 | 250 | .128 | L.F. | 17.30 | 6.30 | .44 | 24.04 | 30.50 |
| 1050 | Steel channels with anchors, on forms, 3" channel, 5#/L.F. | G | 290 | .110 | | 7.10 | 5.40 | .38 | 12.88 | 17.75 |
| 1100 | 4" channel, 5.4#/L.F. | G | 270 | .119 | | 7.65 | 5.80 | .40 | 13.85 | 19.10 |
| 1200 | 6" channel, 8.2#/L.F. | G | 255 | .125 | | 11.75 | 6.15 | .43 | 18.33 | 24.50 |
| 1300 | 8" channel, 11.5#/L.F. | G | 225 | .142 | | 16.20 | 7 | .49 | 23.69 | 30.50 |
| 1400 | 10" channel, 15.3#/L.F. | G | 180 | .178 | | 21.50 | 8.70 | .61 | 30.81 | 39.50 |
| 1500 | 12" channel, 20.7#/L.F. | G | 140 | .229 | | 28.50 | 11.20 | .78 | 40.48 | 52 |
| 2000 | For curved edging, add | | | | | 35% | 10% | | | |

| 05 12 23.40 Lightweight Framing | | Crew | Daily Output | Labor-Hours | Unit | Material | 2011 Bare Costs Labor | Equipment | Total | Total Incl O&P |
|---|---|---|---|---|---|---|---|---|---|---|
| 0010 | **LIGHTWEIGHT FRAMING** R051223-35 | G | | | | | | | | |
| 0015 | Made from recycled materials | G | | | | | | | | |
| 0200 | For load-bearing steel studs see Section 05 41 13.30 | | | | | | | | | |
| 0400 | Angle framing, field fabricated, 4" and larger   R051223-45 | G E-3 | 440 | .055 | Lb. | .65 | 2.69 | .25 | 3.59 | 5.75 |
| 0450 | Less than 4" angles | G | 265 | .091 | " | .68 | 4.46 | .41 | 5.55 | 9.05 |
| 0460 | 1/2" x 1/2" x 1/8" | G | 200 | .120 | L.F. | .14 | 5.90 | .54 | 6.58 | 11.20 |
| 0462 | 3/4" x 3/4" x 1/8" | G | 160 | .150 | | .38 | 7.40 | .68 | 8.46 | 14.20 |
| 0464 | 1" x 1" x 1/8" | G | 135 | .178 | | .54 | 8.75 | .81 | 10.10 | 16.95 |
| 0466 | 1-1/4" x 1-1/4" x 3/16" | G | 115 | .209 | | 1 | 10.25 | .95 | 12.20 | 20.50 |
| 0468 | 1-1/2" x 1-1/2" x 3/16" | G | 100 | .240 | | 1.22 | 11.80 | 1.09 | 14.11 | 23.50 |
| 0470 | 2" x 2" x 1/4" | G | 90 | .267 | | 2.15 | 13.15 | 1.21 | 16.51 | 26.50 |
| 0472 | 2-1/2" x 2-1/2" x 1/4" | G | 72 | .333 | | 2.77 | 16.40 | 1.51 | 20.68 | 33.50 |
| 0474 | 3" x 2" x 3/8" | G | 65 | .369 | | 3.98 | 18.15 | 1.68 | 23.81 | 38 |
| 0476 | 3" x 3" x 3/8" | G | 57 | .421 | | 4.86 | 20.50 | 1.91 | 27.27 | 44 |
| 0600 | Channel framing, field fabricated, 8" and larger | G | 500 | .048 | Lb. | .68 | 2.36 | .22 | 3.26 | 5.15 |
| 0650 | Less than 8" channels | G | 335 | .072 | " | .68 | 3.53 | .33 | 4.54 | 7.35 |
| 0660 | C2 x 1.78 | G | 115 | .209 | L.F. | 1.20 | 10.25 | .95 | 12.40 | 20.50 |
| 0662 | C3 x 4.1 | G | 80 | .300 | | 2.77 | 14.75 | 1.36 | 18.88 | 30.50 |
| 0664 | C4 x 5.4 | G | 66 | .364 | | 3.65 | 17.90 | 1.65 | 23.20 | 37.50 |
| 0666 | C5 x 6.7 | G | 57 | .421 | | 4.52 | 20.50 | 1.91 | 26.93 | 43.50 |
| 0668 | C6 x 8.2 | G | 55 | .436 | | 5.35 | 21.50 | 1.98 | 28.83 | 46 |
| 0670 | C7 x 9.8 | G | 40 | .600 | | 6.60 | 29.50 | 2.72 | 38.82 | 62.50 |
| 0672 | C8 x 11.5 | G | 36 | .667 | | 7.75 | 33 | 3.03 | 43.78 | 70 |
| 0710 | Structural bar tee, field fabricated, 3/4" x 3/4" x 1/8" | G | 160 | .150 | | .38 | 7.40 | .68 | 8.46 | 14.20 |
| 0712 | 1" x 1" x 1/8" | G | 135 | .178 | | .54 | 8.75 | .81 | 10.10 | 16.95 |
| 0714 | 1-1/2" x 1-1/2" x 1/4" | G | 114 | .211 | | 1.58 | 10.35 | .96 | 12.89 | 21 |
| 0716 | 2" x 2" x 1/4" | G | 89 | .270 | | 2.15 | 13.25 | 1.22 | 16.62 | 27 |
| 0718 | 2-1/2" x 2-1/2" x 3/8" | G | 72 | .333 | | 3.98 | 16.40 | 1.51 | 21.89 | 35 |
| 0720 | 3" x 3" x 3/8" | G | 57 | .421 | | 4.86 | 20.50 | 1.91 | 27.27 | 44 |
| 0730 | Structural zee, field fabricated, 1-1/4" x 1-3/4" x 1-3/4" | G | 114 | .211 | | .51 | 10.35 | .96 | 11.82 | 19.90 |
| 0732 | 2-11/16" x 3" x 2-11/16" | G | 114 | .211 | | 1.20 | 10.35 | .96 | 12.51 | 20.50 |
| 0734 | 3-1/16" x 4" x 3-1/16" | G | 133 | .180 | | 1.82 | 8.90 | .82 | 11.54 | 18.60 |
| 0736 | 3-1/4" x 5" x 3-1/4" | G | 133 | .180 | | 2.48 | 8.90 | .82 | 12.20 | 19.30 |
| 0738 | 3-1/2" x 6" x 3-1/2" | G | 160 | .150 | | 3.73 | 7.40 | .68 | 11.81 | 17.90 |
| 0740 | Junior beam, field fabricated, 3" | G | 80 | .300 | | 3.85 | 14.75 | 1.36 | 19.96 | 31.50 |
| 0742 | 4" | G | 72 | .333 | | 5.20 | 16.40 | 1.51 | 23.11 | 36.50 |
| 0744 | 5" | G | 67 | .358 | | 6.75 | 17.65 | 1.63 | 26.03 | 40 |
| 0746 | 6" | G | 62 | .387 | | 8.45 | 19.05 | 1.76 | 29.26 | 44.50 |
| 0748 | 7" | G | 57 | .421 | | 10.35 | 20.50 | 1.91 | 32.76 | 50 |
| 0750 | 8" | G | 53 | .453 | | 12.40 | 22.50 | 2.06 | 36.96 | 55.50 |
| 1000 | Continuous slotted channel framing system, shop fab, minimum | G 2 Sswk | 2400 | .007 | Lb. | 3.49 | .32 | | 3.81 | 4.41 |
| 1200 | Maximum | G " | 1600 | .010 | | 3.94 | .49 | | 4.43 | 5.20 |
| 1300 | Cross bracing, rods, shop fabricated, 3/4" diameter | G E-3 | 700 | .034 | | 1.35 | 1.69 | .16 | 3.20 | 4.64 |

# 05 12 Structural Steel Framing

## 05 12 23 – Structural Steel for Buildings

### 05 12 23.40 Lightweight Framing

| | | | Crew | Daily Output | Labor-Hours | Unit | Material | 2011 Bare Costs Labor | Equipment | Total | Total Incl O&P |
|---|---|---|---|---|---|---|---|---|---|---|---|
| 1310 | 7/8" diameter | G | E-3 | 850 | .028 | Lb. | 1.35 | 1.39 | .13 | 2.87 | 4.08 |
| 1320 | 1" diameter | G | | 1000 | .024 | | 1.35 | 1.18 | .11 | 2.64 | 3.70 |
| 1330 | Angle, 5" x 5" x 3/8" | G | | 2800 | .009 | | 1.35 | .42 | .04 | 1.81 | 2.27 |
| 1350 | Hanging lintels, shop fabricated, average | G | | 850 | .028 | | 1.35 | 1.39 | .13 | 2.87 | 4.08 |
| 1380 | Roof frames, shop fabricated, 3'-0" square, 5' span | G | E-2 | 4200 | .013 | | 1.35 | .63 | .39 | 2.37 | 2.98 |
| 1400 | Tie rod, not upset, 1-1/2" to 4" diameter, with turnbuckle | G | 2 Sswk | 800 | .020 | | 1.46 | .97 | | 2.43 | 3.33 |
| 1420 | No turnbuckle | G | | 700 | .023 | | 1.41 | 1.11 | | 2.52 | 3.51 |
| 1500 | Upset, 1-3/4" to 4" diameter, with turnbuckle | G | | 800 | .020 | | 1.46 | .97 | | 2.43 | 3.33 |
| 1520 | No turnbuckle | G | | 700 | .023 | | 1.41 | 1.11 | | 2.52 | 3.51 |

### 05 12 23.45 Lintels

| | | | Crew | Daily Output | Labor-Hours | Unit | Material | 2011 Bare Costs Labor | Equipment | Total | Total Incl O&P |
|---|---|---|---|---|---|---|---|---|---|---|---|
| 0010 | **LINTELS** | | | | | | | | | | |
| 0015 | Made from recycled materials | G | | | | | | | | | |
| 0020 | Plain steel angles, shop fabricated, under 500 lb. | G | 1 Bric | 550 | .015 | Lb. | .87 | .63 | | 1.50 | 1.89 |
| 0100 | 500 to 1000 lb. | G | | 640 | .013 | | .84 | .54 | | 1.38 | 1.74 |
| 0200 | 1,000 to 2,000 lb. | G | | 640 | .013 | | .82 | .54 | | 1.36 | 1.71 |
| 0300 | 2,000 to 4,000 lb. | G | | 640 | .013 | | .80 | .54 | | 1.34 | 1.69 |
| 0500 | For built-up angles and plates, add to above | G | | | | | .28 | | | .28 | .31 |
| 0700 | For engineering, add to above | | | | | | .11 | | | .11 | .12 |
| 0900 | For galvanizing, add to above, under 500 lb. | | | | | | .25 | | | .25 | .28 |
| 0950 | 500 to 2,000 lb. | | | | | | .23 | | | .23 | .25 |
| 1000 | Over 2,000 lb. | | | | | | .20 | | | .20 | .22 |
| 2000 | Steel angles, 3-1/2" x 3", 1/4" thick, 2'-6" long | G | 1 Bric | 47 | .170 | Ea. | 12.15 | 7.35 | | 19.50 | 24.50 |
| 2100 | 4'-6" long | G | | 26 | .308 | | 22 | 13.30 | | 35.30 | 44 |
| 2600 | 4" x 3-1/2", 1/4" thick, 5'-0" long | G | | 21 | .381 | | 28 | 16.45 | | 44.45 | 55 |
| 2700 | 9'-0" long | G | | 12 | .667 | | 50 | 29 | | 79 | 98 |

### 05 12 23.60 Pipe Support Framing

| | | | Crew | Daily Output | Labor-Hours | Unit | Material | 2011 Bare Costs Labor | Equipment | Total | Total Incl O&P |
|---|---|---|---|---|---|---|---|---|---|---|---|
| 0010 | **PIPE SUPPORT FRAMING** | | | | | | | | | | |
| 0020 | Under 10#/L.F., shop fabricated | G | E-4 | 3900 | .008 | Lb. | 1.51 | .40 | .03 | 1.94 | 2.40 |
| 0200 | 10.1 to 15#/L.F. | G | | 4300 | .007 | | 1.49 | .36 | .03 | 1.88 | 2.30 |
| 0400 | 15.1 to 20#/L.F. | G | | 4800 | .007 | | 1.46 | .33 | .02 | 1.81 | 2.22 |
| 0600 | Over 20#/L.F. | G | | 5400 | .006 | | 1.44 | .29 | .02 | 1.75 | 2.11 |

### 05 12 23.65 Plates

| | | | Crew | Daily Output | Labor-Hours | Unit | Material | 2011 Bare Costs Labor | Equipment | Total | Total Incl O&P |
|---|---|---|---|---|---|---|---|---|---|---|---|
| 0010 | **PLATES** | R051223-80 | | | | | | | | | |
| 0015 | Made from recycled materials | G | | | | | | | | | |
| 0020 | For connections & stiffener plates, shop fabricated | | | | | | | | | | |
| 0050 | 1/8" thick (5.1 lb./S.F.) | G | | | | S.F. | 5.75 | | | 5.75 | 6.30 |
| 0100 | 1/4" thick (10.2 lb./S.F.) | G | | | | | 11.50 | | | 11.50 | 12.60 |
| 0300 | 3/8" thick (15.3 lb./S.F.) | G | | | | | 17.20 | | | 17.20 | 18.95 |
| 0400 | 1/2" thick (20.4 lb./S.F.) | G | | | | | 23 | | | 23 | 25.50 |
| 0450 | 3/4" thick (30.6 lb./S.F.) | G | | | | | 34.50 | | | 34.50 | 38 |
| 0500 | 1" thick (40.8 lb./S.F.) | G | | | | | 46 | | | 46 | 50.50 |
| 2000 | Steel plate, warehouse prices, no shop fabrication | | | | | | | | | | |
| 2100 | 1/4" thick (10.2 lb./S.F.) | G | | | | S.F. | 7.15 | | | 7.15 | 7.85 |

### 05 12 23.70 Stressed Skin Steel Roof and Ceiling System

| | | | Crew | Daily Output | Labor-Hours | Unit | Material | 2011 Bare Costs Labor | Equipment | Total | Total Incl O&P |
|---|---|---|---|---|---|---|---|---|---|---|---|
| 0010 | **STRESSED SKIN STEEL ROOF & CEILING SYSTEM** | | | | | | | | | | |
| 0020 | Double panel flat roof, spans to 100' | G | E-2 | 1150 | .049 | S.F. | 9 | 2.31 | 1.41 | 12.72 | 15.35 |
| 0100 | Double panel convex roof, spans to 200' | G | | 960 | .058 | | 14.65 | 2.76 | 1.69 | 19.10 | 22.50 |
| 0200 | Double panel arched roof, spans to 300' | G | | 760 | .074 | | 22.50 | 3.49 | 2.13 | 28.12 | 33.50 |

### 05 12 23.75 Structural Steel Members

| | | | Crew | Daily Output | Labor-Hours | Unit | Material | 2011 Bare Costs Labor | Equipment | Total | Total Incl O&P |
|---|---|---|---|---|---|---|---|---|---|---|---|
| 0010 | **STRUCTURAL STEEL MEMBERS** | R051223-10 | | | | | | | | | |
| 0015 | Made from recycled materials | G | | | | | | | | | |

# 05 12 Structural Steel Framing

## 05 12 23 – Structural Steel for Buildings

### 05 12 23.75 Structural Steel Members

| | | Crew | Daily Output | Labor-Hours | Unit | Material | Labor | Equipment | Total | Total Incl O&P |
|---|---|---|---|---|---|---|---|---|---|---|
| | | | | | | | **2011 Bare Costs** | | | |
| 0020 | Shop fab'd for 100-ton, 1-2 story project, bolted connections | | | | | | | | | |
| 0100 | W 6 x 9 | G E-2 | 600 | .093 | L.F. | 11.15 | 4.42 | 2.70 | 18.27 | 22.50 |
| 0120 | x 15 | G | 600 | .093 | | 18.55 | 4.42 | 2.70 | 25.67 | 31 |
| 0140 | x 20 | G | 600 | .093 | | 25 | 4.42 | 2.70 | 32.12 | 37.50 |
| 0300 | W 8 x 10 | G | 600 | .093 | | 12.40 | 4.42 | 2.70 | 19.52 | 24 |
| 0320 | x 15 | G | 600 | .093 | | 18.55 | 4.42 | 2.70 | 25.67 | 31 |
| 0350 | x 21 | G | 600 | .093 | | 26 | 4.42 | 2.70 | 33.12 | 39 |
| 0360 | x 24 | G | 550 | .102 | | 29.50 | 4.82 | 2.95 | 37.27 | 44 |
| 0370 | x 28 | G | 550 | .102 | | 34.50 | 4.82 | 2.95 | 42.27 | 49.50 |
| 0500 | x 31 | G | 550 | .102 | | 38.50 | 4.82 | 2.95 | 46.27 | 53.50 |
| 0520 | x 35 | G | 550 | .102 | | 43.50 | 4.82 | 2.95 | 51.27 | 59 |
| 0540 | x 48 | G | 550 | .102 | | 59.50 | 4.82 | 2.95 | 67.27 | 77 |
| 0600 | W 10 x 12 | G | 600 | .093 | | 14.85 | 4.42 | 2.70 | 21.97 | 27 |
| 0620 | x 15 | G | 600 | .093 | | 18.55 | 4.42 | 2.70 | 25.67 | 31 |
| 0700 | x 22 | G | 600 | .093 | | 27 | 4.42 | 2.70 | 34.12 | 40.50 |
| 0720 | x 26 | G | 600 | .093 | | 32 | 4.42 | 2.70 | 39.12 | 46 |
| 0740 | x 33 | G | 550 | .102 | | 41 | 4.82 | 2.95 | 48.77 | 56.50 |
| 0900 | x 49 | G | 550 | .102 | | 60.50 | 4.82 | 2.95 | 68.27 | 78 |
| 1100 | W 12 x 16 | G | 880 | .064 | | 19.80 | 3.01 | 1.84 | 24.65 | 29 |
| 1300 | x 22 | G | 880 | .064 | | 27 | 3.01 | 1.84 | 31.85 | 37 |
| 1500 | x 26 | G | 880 | .064 | | 32 | 3.01 | 1.84 | 36.85 | 42.50 |
| 1520 | x 35 | G | 810 | .069 | | 43.50 | 3.27 | 2 | 48.77 | 55.50 |
| 1560 | x 50 | G | 750 | .075 | | 62 | 3.54 | 2.16 | 67.70 | 76.50 |
| 1580 | x 58 | G | 750 | .075 | | 72 | 3.54 | 2.16 | 77.70 | 87.50 |
| 1700 | x 72 | G | 640 | .088 | | 89 | 4.14 | 2.53 | 95.67 | 108 |
| 1740 | x 87 | G | 640 | .088 | | 108 | 4.14 | 2.53 | 114.67 | 128 |
| 1900 | W 14 x 26 | G | 990 | .057 | | 32 | 2.68 | 1.64 | 36.32 | 42 |
| 2100 | x 30 | G | 900 | .062 | | 37 | 2.95 | 1.80 | 41.75 | 48 |
| 2300 | x 34 | G | 810 | .069 | | 42 | 3.27 | 2 | 47.27 | 54.50 |
| 2320 | x 43 | G | 810 | .069 | | 53 | 3.27 | 2 | 58.27 | 66.50 |
| 2340 | x 53 | G | 800 | .070 | | 65.50 | 3.32 | 2.03 | 70.85 | 80 |
| 2360 | x 74 | G | 760 | .074 | | 91.50 | 3.49 | 2.13 | 97.12 | 109 |
| 2380 | x 90 | G | 740 | .076 | | 111 | 3.58 | 2.19 | 116.77 | 131 |
| 2500 | x 120 | G | 720 | .078 | | 149 | 3.68 | 2.25 | 154.93 | 172 |
| 2700 | W 16 x 26 | G | 1000 | .056 | | 32 | 2.65 | 1.62 | 36.27 | 42 |
| 2900 | x 31 | G | 900 | .062 | | 38.50 | 2.95 | 1.80 | 43.25 | 49 |
| 3100 | x 40 | G | 800 | .070 | | 49.50 | 3.32 | 2.03 | 54.85 | 62.50 |
| 3120 | x 50 | G | 800 | .070 | | 62 | 3.32 | 2.03 | 67.35 | 76 |
| 3140 | x 67 | G | 760 | .074 | | 83 | 3.49 | 2.13 | 88.62 | 99.50 |
| 3300 | W 18 x 35 | G E-5 | 960 | .083 | | 43.50 | 3.99 | 1.80 | 49.29 | 56.50 |
| 3500 | x 40 | G | 960 | .083 | | 49.50 | 3.99 | 1.80 | 55.29 | 63.50 |
| 3520 | x 46 | G | 960 | .083 | | 57 | 3.99 | 1.80 | 62.79 | 71.50 |
| 3700 | x 50 | G | 912 | .088 | | 62 | 4.20 | 1.90 | 68.10 | 77.50 |
| 3900 | x 55 | G | 912 | .088 | | 68 | 4.20 | 1.90 | 74.10 | 84.50 |
| 3920 | x 65 | G | 900 | .089 | | 80.50 | 4.26 | 1.92 | 86.68 | 98 |
| 3940 | x 76 | G | 900 | .089 | | 94 | 4.26 | 1.92 | 100.18 | 112 |
| 3960 | x 86 | G | 900 | .089 | | 106 | 4.26 | 1.92 | 112.18 | 126 |
| 3980 | x 106 | G | 900 | .089 | | 131 | 4.26 | 1.92 | 137.18 | 153 |
| 4100 | W 21 x 44 | G | 1064 | .075 | | 54.50 | 3.60 | 1.63 | 59.73 | 68 |
| 4300 | x 50 | G | 1064 | .075 | | 62 | 3.60 | 1.63 | 67.23 | 76 |
| 4500 | x 62 | G | 1036 | .077 | | 76.50 | 3.70 | 1.67 | 81.87 | 92.50 |
| 4700 | x 68 | G | 1036 | .077 | | 84 | 3.70 | 1.67 | 89.37 | 101 |
| 4720 | x 83 | G | 1000 | .080 | | 103 | 3.83 | 1.73 | 108.56 | 122 |

## 05 12 23 – Structural Steel for Buildings

| 05 12 23.75 Structural Steel Members | | Crew | Daily Output | Labor-Hours | Unit | Material | 2011 Bare Costs Labor | Equipment | Total | Total Incl O&P |
|---|---|---|---|---|---|---|---|---|---|---|
| 4740 | x 93 | G | E-5 | 1000 | .080 | L.F. | 115 | 3.83 | 1.73 | 120.56 | 136 |
| 4760 | x 101 | G | | 1000 | .080 | | 125 | 3.83 | 1.73 | 130.56 | 146 |
| 4780 | x 122 | G | | 1000 | .080 | | 151 | 3.83 | 1.73 | 156.56 | 175 |
| 4900 | W 24 x 55 | G | | 1110 | .072 | | 68 | 3.45 | 1.56 | 73.01 | 82.50 |
| 5100 | x 62 | G | | 1110 | .072 | | 76.50 | 3.45 | 1.56 | 81.51 | 92 |
| 5300 | x 68 | G | | 1110 | .072 | | 84 | 3.45 | 1.56 | 89.01 | 100 |
| 5500 | x 76 | G | | 1110 | .072 | | 94 | 3.45 | 1.56 | 99.01 | 111 |
| 5700 | x 84 | G | | 1080 | .074 | | 104 | 3.55 | 1.60 | 109.15 | 122 |
| 5720 | x 94 | G | | 1080 | .074 | | 116 | 3.55 | 1.60 | 121.15 | 136 |
| 5740 | x 104 | G | | 1050 | .076 | | 129 | 3.65 | 1.65 | 134.30 | 150 |
| 5760 | x 117 | G | | 1050 | .076 | | 145 | 3.65 | 1.65 | 150.30 | 167 |
| 5780 | x 146 | G | | 1050 | .076 | | 181 | 3.65 | 1.65 | 186.30 | 207 |
| 5800 | W 27 x 84 | G | | 1190 | .067 | | 104 | 3.22 | 1.45 | 108.67 | 121 |
| 5900 | x 94 | G | | 1190 | .067 | | 116 | 3.22 | 1.45 | 120.67 | 135 |
| 5920 | x 114 | G | | 1150 | .070 | | 141 | 3.33 | 1.51 | 145.84 | 162 |
| 5940 | x 146 | G | | 1150 | .070 | | 181 | 3.33 | 1.51 | 185.84 | 206 |
| 5960 | x 161 | G | | 1150 | .070 | | 199 | 3.33 | 1.51 | 203.84 | 226 |
| 6100 | W 30 x 99 | G | | 1200 | .067 | | 123 | 3.19 | 1.44 | 127.63 | 142 |
| 6300 | x 108 | G | | 1200 | .067 | | 134 | 3.19 | 1.44 | 138.63 | 154 |
| 6500 | x 116 | G | | 1160 | .069 | | 144 | 3.31 | 1.49 | 148.80 | 165 |
| 6520 | x 132 | G | | 1160 | .069 | | 163 | 3.31 | 1.49 | 167.80 | 187 |
| 6540 | x 148 | G | | 1160 | .069 | | 183 | 3.31 | 1.49 | 187.80 | 208 |
| 6560 | x 173 | G | | 1120 | .071 | | 214 | 3.42 | 1.55 | 218.97 | 244 |
| 6580 | x 191 | G | | 1120 | .071 | | 236 | 3.42 | 1.55 | 240.97 | 268 |
| 6700 | W 33 x 118 | G | | 1176 | .068 | | 146 | 3.26 | 1.47 | 150.73 | 168 |
| 6900 | x 130 | G | | 1134 | .071 | | 161 | 3.38 | 1.53 | 165.91 | 184 |
| 7100 | x 141 | G | | 1134 | .071 | | 174 | 3.38 | 1.53 | 178.91 | 199 |
| 7120 | x 169 | G | | 1100 | .073 | | 209 | 3.49 | 1.57 | 214.06 | 238 |
| 7140 | x 201 | G | | 1100 | .073 | | 249 | 3.49 | 1.57 | 254.06 | 282 |
| 7300 | W 36 x 135 | G | | 1170 | .068 | | 167 | 3.28 | 1.48 | 171.76 | 191 |
| 7500 | x 150 | G | | 1170 | .068 | | 186 | 3.28 | 1.48 | 190.76 | 211 |
| 7600 | x 170 | G | | 1150 | .070 | | 210 | 3.33 | 1.51 | 214.84 | 238 |
| 7700 | x 194 | G | | 1125 | .071 | | 240 | 3.41 | 1.54 | 244.95 | 272 |
| 7900 | x 231 | G | | 1125 | .071 | | 286 | 3.41 | 1.54 | 290.95 | 325 |
| 7920 | x 262 | G | | 1035 | .077 | | 325 | 3.70 | 1.67 | 330.37 | 365 |
| 8100 | x 302 | G | | 1035 | .077 | | 375 | 3.70 | 1.67 | 380.37 | 420 |
| 8490 | For projects 75 to 99 tons, add | | | | | | 10% | | | | |
| 8492 | 50 to 74 tons, add | | | | | | 20% | | | | |
| 8494 | 25 to 49 tons, add | | | | | | 30% | 10% | | | |
| 8496 | 10 to 24 tons, add | | | | | | 50% | 25% | | | |
| 8498 | 2 to 9 tons, add | | | | | | 75% | 50% | | | |
| 8499 | Less than 2 tons, add | | | | | | 100% | 100% | | | |

## 05 12 23.77 Structural Steel Projects

| | | | Crew | Daily Output | Labor-Hours | Unit | Material | Labor | Equipment | Total | Total Incl O&P |
|---|---|---|---|---|---|---|---|---|---|---|---|
| 0010 | **STRUCTURAL STEEL PROJECTS** | R050516-30 | | | | | | | | | |
| 0015 | Made from recycled materials | G | | | | | | | | | |
| 0020 | Shop fab'd for 100-ton, 1-2 story project, bolted connections | | | | | | | | | | |
| 0200 | Apartments, nursing homes, etc., 1 to 2 stories | R050523-10 G | E-5 | 10.30 | 7.767 | Ton | 2,250 | 370 | 168 | 2,788 | 3,300 |
| 0300 | 3 to 6 stories | G | " | 10.10 | 7.921 | | 2,300 | 380 | 171 | 2,851 | 3,375 |
| 0400 | 7 to 15 stories | R051223-10 G | E-6 | 14.20 | 9.014 | | 2,350 | 430 | 133 | 2,913 | 3,475 |
| 0500 | Over 15 stories | G | " | 13.90 | 9.209 | | 2,425 | 440 | 136 | 3,001 | 3,575 |
| 0700 | Offices, hospitals, etc., steel bearing, 1 to 2 stories | R051223-20 G | E-5 | 10.30 | 7.767 | | 2,250 | 370 | 168 | 2,788 | 3,300 |
| 0800 | 3 to 6 stories | G | E-6 | 14.40 | 8.889 | | 2,300 | 425 | 131 | 2,856 | 3,400 |

# 05 12 Structural Steel Framing

## 05 12 23 – Structural Steel for Buildings

| 05 12 23.77 Structural Steel Projects | | Crew | Daily Output | Labor-Hours | Unit | Material | 2011 Bare Costs Labor | Equipment | Total | Total Incl O&P |
|---|---|---|---|---|---|---|---|---|---|---|
| 0900 | 7 to 15 stories          R051223-25 | G | E-6 | 14.20 | 9.014 | Ton | 2,350 | 430 | 133 | 2,913 | 3,475 |
| 1000 | Over 15 stories | G | ↓ | 13.90 | 9.209 | | 2,425 | 440 | 136 | 3,001 | 3,575 |
| 1100 | For multi-story masonry wall bearing construction, add    R051223-30 | | | | | | | 30% | | | |
| 1300 | Industrial bldgs., 1 story, beams & girders, steel bearing | G | E-5 | 12.90 | 6.202 | | 2,250 | 297 | 134 | 2,681 | 3,125 |
| 1400 | Masonry bearing | G | " | 10 | 8 | ↓ | 2,250 | 385 | 173 | 2,808 | 3,325 |
| 1500 | Industrial bldgs., 1 story, under 10 tons, | | | | | | | | | | |
| 1510 | steel from warehouse, trucked | G | E-2 | 7.50 | 7.467 | Ton | 2,700 | 355 | 216 | 3,271 | 3,825 |
| 1600 | 1 story with roof trusses, steel bearing | G | E-5 | 10.60 | 7.547 | | 2,650 | 360 | 163 | 3,173 | 3,725 |
| 1700 | Masonry bearing | G | " | 8.30 | 9.639 | | 2,650 | 460 | 209 | 3,319 | 3,950 |
| 1900 | Monumental structures, banks, stores, etc., minimum | G | E-6 | 13 | 9.846 | | 2,250 | 470 | 146 | 2,866 | 3,450 |
| 2000 | Maximum | G | " | 9 | 14.222 | | 3,725 | 680 | 210 | 4,615 | 5,500 |
| 2200 | Churches, minimum | G | E-5 | 11.60 | 6.897 | | 2,100 | 330 | 149 | 2,579 | 3,025 |
| 2300 | Maximum | G | " | 5.20 | 15.385 | | 2,800 | 735 | 335 | 3,870 | 4,725 |
| 2800 | Power stations, fossil fuels, minimum | G | E-6 | 11 | 11.636 | | 2,250 | 560 | 172 | 2,982 | 3,625 |
| 2900 | Maximum | G | | 5.70 | 22.456 | | 3,375 | 1,075 | 330 | 4,780 | 5,950 |
| 2950 | Nuclear fuels, non-safety steel, minimum | G | | 7 | 18.286 | | 2,250 | 875 | 270 | 3,395 | 4,275 |
| 3000 | Maximum | G | | 5.50 | 23.273 | | 3,375 | 1,125 | 345 | 4,845 | 6,025 |
| 3040 | Safety steel, minimum | G | | 2.50 | 51.200 | | 3,275 | 2,450 | 755 | 6,480 | 8,675 |
| 3070 | Maximum | G | ↓ | 1.50 | 85.333 | | 4,325 | 4,100 | 1,250 | 9,675 | 13,200 |
| 3100 | Roof trusses, minimum | G | E-5 | 13 | 6.154 | | 3,150 | 295 | 133 | 3,578 | 4,125 |
| 3200 | Maximum | G | | 8.30 | 9.639 | | 3,825 | 460 | 209 | 4,494 | 5,225 |
| 3210 | Schools, minimum | G | | 14.50 | 5.517 | | 2,250 | 264 | 119 | 2,633 | 3,050 |
| 3220 | Maximum | G | ↓ | 8.30 | 9.639 | | 3,275 | 460 | 209 | 3,944 | 4,650 |
| 3400 | Welded construction, simple commercial bldgs., 1 to 2 stories | G | E-7 | 7.60 | 10.526 | | 2,300 | 505 | 242 | 3,047 | 3,650 |
| 3500 | 7 to 15 stories | G | E-9 | 8.30 | 15.422 | | 2,650 | 740 | 261 | 3,651 | 4,475 |
| 3700 | Welded rigid frame, 1 story, minimum | G | E-7 | 15.80 | 5.063 | | 2,350 | 243 | 116 | 2,709 | 3,125 |
| 3800 | Maximum | G | " | 5.50 | 14.545 | ↓ | 3,050 | 695 | 335 | 4,080 | 4,925 |
| 3810 | Fabrication shop costs (included in project material cost, above) | | | | | | | | | | |
| 3820 | Mini mill base price, A992 | G | | | | Ton | 770 | | | 770 | 845 |
| 3830 | Mill extra for delivery to shop | | | | | | 240 | | | 240 | 264 |
| 3840 | Shop extra for shop drawings and detailing | | | | | | 270 | | | 270 | 297 |
| 3850 | Shop fabricating and handling | | | | | | 730 | | | 730 | 805 |
| 3860 | Shop sandblasting and primer coat of paint | | | | | | 135 | | | 135 | 149 |
| 3870 | Shop delivery to the job site | | | | | | 105 | | | 105 | 116 |
| 3880 | Total material cost, shop fabricated, primed, delivered | | | | | ↓ | 2,250 | | | 2,250 | 2,475 |
| 3900 | High strength steel mill spec extras: | | | | | | | | | | |
| 3950 | A529, A572 (50 ksi) and A36: same as A992 steel (no extra) | | | | | | | | | | |
| 4000 | Add to A992 price for A572 (60, 65 ksi) | G | | | | Ton | 100 | | | 100 | 110 |
| 4100 | A242 and A588 Weathering | G | | | | " | 85 | | | 85 | 93.50 |
| 4200 | Mill size extras for W-Shapes: 0 to 30 plf: no extra charge | | | | | | | | | | |
| 4210 | Member sizes 31 to 65 plf, deduct | G | | | | Ton | .01 | | | .01 | .01 |
| 4220 | Member sizes 66 to 100 plf, deduct | G | | | | | 8.40 | | | 8.40 | 9.25 |
| 4230 | Member sizes 101 to 387 plf, add | G | | | | ↓ | 58 | | | 58 | 64 |
| 4300 | Column base plates, light, up to 150 lb | G | 2 Sswk | 2000 | .008 | Lb. | 1.24 | .39 | | 1.63 | 2.05 |
| 4400 | Heavy, over 150 lb | G | E-2 | 7500 | .007 | " | 1.29 | .35 | .22 | 1.86 | 2.26 |
| 4600 | Castellated beams, light sections, to 50#/L.F., minimum | G | | 10.70 | 5.234 | Ton | 2,375 | 248 | 152 | 2,775 | 3,175 |
| 4700 | Maximum | G | | 7 | 8 | | 2,600 | 380 | 232 | 3,212 | 3,750 |
| 4900 | Heavy sections, over 50# per L.F., minimum | G | | 11.70 | 4.786 | | 2,475 | 227 | 139 | 2,841 | 3,250 |
| 5000 | Maximum | G | ↓ | 7.80 | 7.179 | | 2,700 | 340 | 208 | 3,248 | 3,775 |
| 5390 | For projects 75 to 99 tons, add | | | | | | 10% | | | | |
| 5392 | 50 to 74 tons, add | | | | | | 20% | | | | |
| 5394 | 25 to 49 tons, add | | | | | | 30% | 10% | | | |
| 5396 | 10 to 24 tons, add | | | | | | 50% | 25% | | | |

124

# 05 12 Structural Steel Framing

## 05 12 23 – Structural Steel for Buildings

| 05 12 23.77 Structural Steel Projects | | Crew | Daily Output | Labor-Hours | Unit | Material | 2011 Bare Costs Labor | Equipment | Total | Total Incl O&P |
|---|---|---|---|---|---|---|---|---|---|---|
| 5398 | 2 to 9 tons, add | | | | Ton | 75% | 50% | | | |
| 5399 | Less than 2 tons, add | | | | | 100% | 100% | | | |

### 05 12 23.80 Subpurlins

| 0010 | **SUBPURLINS** | R051223-50 | | | | | | | | |
|---|---|---|---|---|---|---|---|---|---|---|
| 0015 | Made from recycled materials | G | | | | | | | | |
| 0020 | Bulb tees, shop fabricated, painted, 32-5/8" O.C., 40 psf L.L. | | | | | | | | | |
| 0100 | Type 178, max 8'-9" span, 2.15 plf, 2" high x 1-5/8" wide | G | E-1 | 4200 | .006 | S.F. | 1.56 | .27 | .03 | 1.86 | 2.20 |
| 0200 | Type 218, max 10'-2" span, 3.19 plf, 2-1/8" high x 2-1/8" wide | G | " | 3100 | .008 | | 1.81 | .37 | .04 | 2.22 | 2.65 |
| 1420 | For 24-5/8" spacing, add | | | | | | 33% | 33% | | | |
| 1430 | For 48-5/8" spacing, deduct | | | | | | 50% | 50% | | | |

# 05 14 Structural Aluminum Framing

## 05 14 23 – Non-Exposed Structural Aluminum Framing

### 05 14 23.05 Aluminum Shapes

| 0010 | **ALUMINUM SHAPES** | | | | | | | | | | |
|---|---|---|---|---|---|---|---|---|---|---|---|
| 0015 | Made from recycled materials | G | | | | | | | | | |
| 0020 | Structural shapes, 1" to 10" members, under 1 ton | G | E-2 | 1050 | .053 | Lb. | 2.98 | 2.53 | 1.54 | 7.05 | 9.25 |
| 0050 | 1 to 5 tons | G | | 1330 | .042 | | 2.73 | 1.99 | 1.22 | 5.94 | 7.70 |
| 0100 | Over 5 tons | G | | 1330 | .042 | | 2.61 | 1.99 | 1.22 | 5.82 | 7.60 |
| 0300 | Extrusions, over 5 tons, stock shapes | G | | 1330 | .042 | | 3.10 | 1.99 | 1.22 | 6.31 | 8.15 |
| 0400 | Custom shapes | G | | 1330 | .042 | | 3.10 | 1.99 | 1.22 | 6.31 | 8.15 |

# 05 15 Wire Rope Assemblies

## 05 15 16 – Steel Wire Rope Assemblies

### 05 15 16.05 Accessories for Steel Wire Rope

| 0010 | **ACCESSORIES FOR STEEL WIRE ROPE** | | | | | | | | | | |
|---|---|---|---|---|---|---|---|---|---|---|---|
| 0015 | Made from recycled materials | G | | | | | | | | | |
| 1500 | Thimbles, heavy duty, 1/4" | G | E-17 | 160 | .100 | Ea. | .50 | 4.96 | | 5.46 | 9.30 |
| 1510 | 1/2" | G | | 160 | .100 | | 2.21 | 4.96 | | 7.17 | 11.20 |
| 1520 | 3/4" | G | | 105 | .152 | | 5 | 7.55 | | 12.55 | 18.85 |
| 1530 | 1" | G | | 52 | .308 | | 10.05 | 15.25 | | 25.30 | 38 |
| 1540 | 1-1/4" | G | | 38 | .421 | | 15.45 | 21 | | 36.45 | 54 |
| 1550 | 1-1/2" | G | | 13 | 1.231 | | 43.50 | 61 | | 104.50 | 156 |
| 1560 | 1-3/4" | G | | 8 | 2 | | 89.50 | 99 | | 188.50 | 274 |
| 1570 | 2" | G | | 6 | 2.667 | | 130 | 132 | | 262 | 375 |
| 1580 | 2-1/4" | G | | 4 | 4 | | 177 | 198 | | 375 | 545 |
| 1600 | Clips, 1/4" diameter | G | | 160 | .100 | | 2.54 | 4.96 | | 7.50 | 11.55 |
| 1610 | 3/8" diameter | G | | 160 | .100 | | 2.78 | 4.96 | | 7.74 | 11.80 |
| 1620 | 1/2" diameter | G | | 160 | .100 | | 4.47 | 4.96 | | 9.43 | 13.65 |
| 1630 | 3/4" diameter | G | | 102 | .157 | | 7.25 | 7.75 | | 15 | 21.50 |
| 1640 | 1" diameter | G | | 64 | .250 | | 12.10 | 12.40 | | 24.50 | 35.50 |
| 1650 | 1-1/4" diameter | G | | 35 | .457 | | 19.80 | 22.50 | | 42.30 | 62 |
| 1670 | 1-1/2" diameter | G | | 26 | .615 | | 26.50 | 30.50 | | 57 | 83.50 |
| 1680 | 1-3/4" diameter | G | | 16 | 1 | | 62 | 49.50 | | 111.50 | 156 |
| 1690 | 2" diameter | G | | 12 | 1.333 | | 69 | 66 | | 135 | 193 |
| 1700 | 2-1/4" diameter | G | | 10 | 1.600 | | 102 | 79.50 | | 181.50 | 252 |
| 1800 | Sockets, open swage, 1/4" diameter | G | | 160 | .100 | | 41 | 4.96 | | 45.96 | 54 |
| 1810 | 1/2" diameter | G | | 77 | .208 | | 59 | 10.30 | | 69.30 | 83 |
| 1820 | 3/4" diameter | G | | 19 | .842 | | 91.50 | 41.50 | | 133 | 175 |

# 05 15 Wire Rope Assemblies

## 05 15 16 – Steel Wire Rope Assemblies

| 05 15 16.05 Accessories for Steel Wire Rope | | Crew | Daily Output | Labor-Hours | Unit | Material | 2011 Bare Costs Labor | Equipment | Total | Total Incl O&P |
|---|---|---|---|---|---|---|---|---|---|---|
| 1830 | 1" diameter | G | E-17 | 9 | 1.778 | Ea. | 164 | 88 | | 252 | 335 |
| 1840 | 1-1/4" diameter | G | | 5 | 3.200 | | 227 | 159 | | 386 | 530 |
| 1850 | 1-1/2" diameter | G | | 3 | 5.333 | | 500 | 264 | | 764 | 1,025 |
| 1860 | 1-3/4" diameter | G | | 3 | 5.333 | | 885 | 264 | | 1,149 | 1,450 |
| 1870 | 2" diameter | G | | 1.50 | 10.667 | | 1,350 | 530 | | 1,880 | 2,400 |
| 1900 | Closed swage, 1/4" diameter | G | | 160 | .100 | | 24 | 4.96 | | 28.96 | 35.50 |
| 1910 | 1/2" diameter | G | | 104 | .154 | | 41.50 | 7.60 | | 49.10 | 59.50 |
| 1920 | 3/4" diameter | G | | 32 | .500 | | 62.50 | 25 | | 87.50 | 113 |
| 1930 | 1" diameter | G | | 15 | 1.067 | | 109 | 53 | | 162 | 214 |
| 1940 | 1-1/4" diameter | G | | 7 | 2.286 | | 164 | 113 | | 277 | 380 |
| 1950 | 1-1/2" diameter | G | | 4 | 4 | | 298 | 198 | | 496 | 675 |
| 1960 | 1-3/4" diameter | G | | 3 | 5.333 | | 440 | 264 | | 704 | 950 |
| 1970 | 2" diameter | G | | 2 | 8 | | 855 | 395 | | 1,250 | 1,650 |
| 2000 | Open spelter, galv., 1/4" diameter | G | | 160 | .100 | | 52 | 4.96 | | 56.96 | 66.50 |
| 2010 | 1/2" diameter | G | | 70 | .229 | | 54.50 | 11.35 | | 65.85 | 80 |
| 2020 | 3/4" diameter | G | | 26 | .615 | | 81.50 | 30.50 | | 112 | 144 |
| 2030 | 1" diameter | G | | 10 | 1.600 | | 226 | 79.50 | | 305.50 | 390 |
| 2040 | 1-1/4" diameter | G | | 5 | 3.200 | | 325 | 159 | | 484 | 635 |
| 2050 | 1-1/2" diameter | G | | 4 | 4 | | 685 | 198 | | 883 | 1,100 |
| 2060 | 1-3/4" diameter | G | | 2 | 8 | | 1,200 | 395 | | 1,595 | 2,025 |
| 2070 | 2" diameter | G | | 1.20 | 13.333 | | 1,375 | 660 | | 2,035 | 2,700 |
| 2080 | 2-1/2" diameter | G | | 1 | 16 | | 2,525 | 795 | | 3,320 | 4,200 |
| 2100 | Closed spelter, galv., 1/4" diameter | G | | 160 | .100 | | 43 | 4.96 | | 47.96 | 56.50 |
| 2110 | 1/2" diameter | G | | 88 | .182 | | 46 | 9 | | 55 | 66.50 |
| 2120 | 3/4" diameter | G | | 30 | .533 | | 70 | 26.50 | | 96.50 | 124 |
| 2130 | 1" diameter | G | | 13 | 1.231 | | 149 | 61 | | 210 | 271 |
| 2140 | 1-1/4" diameter | G | | 7 | 2.286 | | 238 | 113 | | 351 | 460 |
| 2150 | 1-1/2" diameter | G | | 6 | 2.667 | | 510 | 132 | | 642 | 795 |
| 2160 | 1-3/4" diameter | G | | 2.80 | 5.714 | | 680 | 283 | | 963 | 1,250 |
| 2170 | 2" diameter | G | | 2 | 8 | | 840 | 395 | | 1,235 | 1,625 |
| 2200 | Jaw & jaw turnbuckles, 1/4" x 4" | G | | 160 | .100 | | 16.35 | 4.96 | | 21.31 | 27 |
| 2250 | 1/2" x 6" | G | | 96 | .167 | | 20.50 | 8.25 | | 28.75 | 37 |
| 2260 | 1/2" x 9" | G | | 77 | .208 | | 27.50 | 10.30 | | 37.80 | 48.50 |
| 2270 | 1/2" x 12" | G | | 66 | .242 | | 31 | 12 | | 43 | 55 |
| 2300 | 3/4" x 6" | G | | 38 | .421 | | 40.50 | 21 | | 61.50 | 81.50 |
| 2310 | 3/4" x 9" | G | | 30 | .533 | | 45 | 26.50 | | 71.50 | 96 |
| 2320 | 3/4" x 12" | G | | 28 | .571 | | 57.50 | 28.50 | | 86 | 114 |
| 2330 | 3/4" x 18" | G | | 23 | .696 | | 69 | 34.50 | | 103.50 | 137 |
| 2350 | 1" x 6" | G | | 17 | .941 | | 78.50 | 46.50 | | 125 | 169 |
| 2360 | 1" x 12" | G | | 13 | 1.231 | | 86 | 61 | | 147 | 203 |
| 2370 | 1" x 18" | G | | 10 | 1.600 | | 129 | 79.50 | | 208.50 | 282 |
| 2380 | 1" x 24" | G | | 9 | 1.778 | | 142 | 88 | | 230 | 310 |
| 2400 | 1-1/4" x 12" | G | | 7 | 2.286 | | 145 | 113 | | 258 | 360 |
| 2410 | 1-1/4" x 18" | G | | 6.50 | 2.462 | | 179 | 122 | | 301 | 410 |
| 2420 | 1-1/4" x 24" | G | | 5.60 | 2.857 | | 241 | 142 | | 383 | 515 |
| 2450 | 1-1/2" x 12" | G | | 5.20 | 3.077 | | 425 | 152 | | 577 | 735 |
| 2460 | 1-1/2" x 18" | G | | 4 | 4 | | 450 | 198 | | 648 | 850 |
| 2470 | 1-1/2" x 24" | G | | 3.20 | 5 | | 610 | 248 | | 858 | 1,100 |
| 2500 | 1-3/4" x 18" | G | | 3.20 | 5 | | 920 | 248 | | 1,168 | 1,450 |
| 2510 | 1-3/4" x 24" | G | | 2.80 | 5.714 | | 1,050 | 283 | | 1,333 | 1,650 |
| 2550 | 2" x 24" | G | | 1.60 | 10 | | 1,425 | 495 | | 1,920 | 2,425 |

## 05 15 16 – Steel Wire Rope Assemblies

### 05 15 16.50 Steel Wire Rope

| | | Crew | Daily Output | Labor-Hours | Unit | Material | 2011 Bare Costs Labor | Equipment | Total | Total Incl O&P |
|---|---|---|---|---|---|---|---|---|---|---|
| 0010 | **STEEL WIRE ROPE** | | | | | | | | | |
| 0015 | Made from recycled materials | G | | | | | | | | |
| 0020 | 6 x 19, bright, fiber core, 5000' rolls, 1/2" diameter | G | | | L.F. | .76 | | | .76 | .84 |
| 0050 | Steel core | G | | | | 1 | | | 1 | 1.10 |
| 0100 | Fiber core, 1" diameter | G | | | | 2.56 | | | 2.56 | 2.82 |
| 0150 | Steel core | G | | | | 2.92 | | | 2.92 | 3.21 |
| 0300 | 6 x 19, galvanized, fiber core, 1/2" diameter | G | | | | 1.12 | | | 1.12 | 1.23 |
| 0350 | Steel core | G | | | | 1.28 | | | 1.28 | 1.41 |
| 0400 | Fiber core, 1" diameter | G | | | | 3.28 | | | 3.28 | 3.61 |
| 0450 | Steel core | G | | | | 3.44 | | | 3.44 | 3.78 |
| 0500 | 6 x 7, bright, IPS, fiber core, <500 L.F. w/acc., 1/4" diameter | G | E-17 | 6400 | .003 | | 1.34 | .12 | | 1.46 | 1.69 |
| 0510 | 1/2" diameter | G | | 2100 | .008 | | 3.27 | .38 | | 3.65 | 4.27 |
| 0520 | 3/4" diameter | G | | 960 | .017 | | 5.90 | .83 | | 6.73 | 7.95 |
| 0550 | 6 x 19, bright, IPS, IWRC, <500 L.F. w/acc., 1/4" diameter | G | | 5760 | .003 | | .84 | .14 | | .98 | 1.16 |
| 0560 | 1/2" diameter | G | | 1730 | .009 | | 1.36 | .46 | | 1.82 | 2.31 |
| 0570 | 3/4" diameter | G | | 770 | .021 | | 2.36 | 1.03 | | 3.39 | 4.42 |
| 0580 | 1" diameter | G | | 420 | .038 | | 4 | 1.89 | | 5.89 | 7.75 |
| 0590 | 1-1/4" diameter | G | | 290 | .055 | | 6.65 | 2.73 | | 9.38 | 12.15 |
| 0600 | 1-1/2" diameter | G | | 192 | .083 | | 8.15 | 4.13 | | 12.28 | 16.30 |
| 0610 | 1-3/4" diameter | G | E-18 | 240 | .167 | | 13 | 8.05 | 4.31 | 25.36 | 33 |
| 0620 | 2" diameter | G | | 160 | .250 | | 16.70 | 12.10 | 6.45 | 35.25 | 46 |
| 0630 | 2-1/4" diameter | G | | 160 | .250 | | 22.50 | 12.10 | 6.45 | 41.05 | 52 |
| 0650 | 6 x 37, bright, IPS, IWRC, <500 L.F. w/acc., 1/4" diameter | G | E-17 | 6400 | .003 | | 1 | .12 | | 1.12 | 1.32 |
| 0660 | 1/2" diameter | G | | 1730 | .009 | | 1.70 | .46 | | 2.16 | 2.68 |
| 0670 | 3/4" diameter | G | | 770 | .021 | | 2.75 | 1.03 | | 3.78 | 4.84 |
| 0680 | 1" diameter | G | | 430 | .037 | | 4.36 | 1.84 | | 6.20 | 8.05 |
| 0690 | 1-1/4" diameter | G | | 290 | .055 | | 6.60 | 2.73 | | 9.33 | 12.10 |
| 0700 | 1-1/2" diameter | G | | 190 | .084 | | 9.40 | 4.17 | | 13.57 | 17.70 |
| 0710 | 1-3/4" diameter | G | E-18 | 260 | .154 | | 14.95 | 7.45 | 3.98 | 26.38 | 33.50 |
| 0720 | 2" diameter | G | | 200 | .200 | | 19.40 | 9.65 | 5.20 | 34.25 | 44 |
| 0730 | 2-1/4" diameter | G | | 160 | .250 | | 25.50 | 12.10 | 6.45 | 44.05 | 55.50 |
| 0800 | 6 x 19 & 6 x 37, swaged, 1/2" diameter | G | E-17 | 1220 | .013 | | 2.20 | .65 | | 2.85 | 3.57 |
| 0810 | 9/16" diameter | G | | 1120 | .014 | | 2.55 | .71 | | 3.26 | 4.06 |
| 0820 | 5/8" diameter | G | | 930 | .017 | | 3.03 | .85 | | 3.88 | 4.84 |
| 0830 | 3/4" diameter | G | | 640 | .025 | | 3.86 | 1.24 | | 5.10 | 6.45 |
| 0840 | 7/8" diameter | G | | 480 | .033 | | 4.87 | 1.65 | | 6.52 | 8.25 |
| 0850 | 1" diameter | G | | 350 | .046 | | 5.95 | 2.26 | | 8.21 | 10.55 |
| 0860 | 1-1/8" diameter | G | | 288 | .056 | | 7.30 | 2.75 | | 10.05 | 12.90 |
| 0870 | 1-1/4" diameter | G | | 230 | .070 | | 8.85 | 3.45 | | 12.30 | 15.85 |
| 0880 | 1-3/8" diameter | G | | 192 | .083 | | 10.20 | 4.13 | | 14.33 | 18.55 |
| 0890 | 1-1/2" diameter | G | E-18 | 300 | .133 | | 12.40 | 6.45 | 3.45 | 22.30 | 28.50 |

### 05 15 16.60 Galvanized Steel Wire Rope and Accessories

| | | Crew | Daily Output | Labor-Hours | Unit | Material | 2011 Bare Costs Labor | Equipment | Total | Total Incl O&P |
|---|---|---|---|---|---|---|---|---|---|---|
| 0010 | **GALVANIZED STEEL WIRE ROPE & ACCESSORIES** | | | | | | | | | |
| 0015 | Made from recycled materials | G | | | | | | | | |
| 3000 | Aircraft cable, galvanized, 7 x 7 x 1/8" | G | E-17 | 5000 | .003 | L.F. | .17 | .16 | | .33 | .47 |
| 3100 | Clamps, 1/8" | G | " | 125 | .128 | Ea. | 1.81 | 6.35 | | 8.16 | 13.20 |

### 05 15 16.70 Temporary Cable Safety Railing

| | | Crew | Daily Output | Labor-Hours | Unit | Material | 2011 Bare Costs Labor | Equipment | Total | Total Incl O&P |
|---|---|---|---|---|---|---|---|---|---|---|
| 0010 | **TEMPORARY CABLE SAFETY RAILING**, Each 100' strand incl. | | | | | | | | | |
| 0020 | 2 eyebolts, 1 turnbuckle, 100' cable, 2 thimbles, 6 clips | | | | | | | | | |
| 0025 | Made from recycled materials | G | | | | | | | | |
| 0100 | One strand using 1/4" cable & accessories | G | 2 Sswk | 4 | 4 | C.L.F. | 191 | 194 | | 385 | 555 |
| 0200 | 1/2" cable & accessories | G | " | 2 | 8 | " | 400 | 390 | | 790 | 1,125 |

# 05 21 Steel Joist Framing

## 05 21 13 – Deep Longspan Steel Joist Framing

| 05 21 13.50 Deep Longspan Joists | | Crew | Daily Output | Labor-Hours | Unit | Material | 2011 Bare Costs Labor | Equipment | Total | Total Incl O&P |
|---|---|---|---|---|---|---|---|---|---|---|
| 0010 | **DEEP LONGSPAN JOISTS** | | | | | | | | | |
| 3010 | DLH series, 40-ton job lots, bolted cross bridging, shop primer | | | | | | | | | |
| 3015 | Made from recycled materials | G | | | | | | | | |
| 3020 | Spans to 144' (shipped in 2 pieces), minimum | G | E-7 | 16 | 5 | Ton | 1,375 | 240 | 115 | 1,730 | 2,050 |
| 3040 | Average | G | | 13 | 6.154 | | 1,475 | 295 | 142 | 1,912 | 2,275 |
| 3100 | Maximum | G | | 11 | 7.273 | | 1,800 | 350 | 167 | 2,317 | 2,750 |
| 3200 | 52DLH11, 26 lb./L.F. | G | | 2000 | .040 | L.F. | 18.55 | 1.92 | .92 | 21.39 | 25 |
| 3220 | 52DLH16, 45 lb./L.F. | G | | 2000 | .040 | | 33.50 | 1.92 | .92 | 36.34 | 41 |
| 3240 | 56DLH11, 26 lb./L.F. | G | | 2000 | .040 | | 19.30 | 1.92 | .92 | 22.14 | 25.50 |
| 3260 | 56DLH16, 46 lb./L.F. | G | | 2000 | .040 | | 34 | 1.92 | .92 | 36.84 | 42 |
| 3280 | 60DLH12, 29 lb./L.F. | G | | 2000 | .040 | | 21.50 | 1.92 | .92 | 24.34 | 28 |
| 3300 | 60DLH17, 52 lb./L.F. | G | | 2000 | .040 | | 38.50 | 1.92 | .92 | 41.34 | 47 |
| 3320 | 64DLH12, 31 lb./L.F. | G | | 2200 | .036 | | 23 | 1.74 | .84 | 25.58 | 29.50 |
| 3340 | 64DLH17, 52 lb./L.F. | G | | 2200 | .036 | | 38.50 | 1.74 | .84 | 41.08 | 46.50 |
| 3360 | 68DLH13, 37 lb./L.F. | G | | 2200 | .036 | | 27.50 | 1.74 | .84 | 30.08 | 34 |
| 3380 | 68DLH18, 61 lb./L.F. | G | | 2200 | .036 | | 45.50 | 1.74 | .84 | 48.08 | 54 |
| 3400 | 72DLH14, 41 lb./L.F. | G | | 2200 | .036 | | 30.50 | 1.74 | .84 | 33.08 | 37.50 |
| 3420 | 72DLH19, 70 lb./L.F. | G | | 2200 | .036 | | 52 | 1.74 | .84 | 54.58 | 61 |
| 3500 | For less than 40-ton job lots | | | | | | | | | |
| 3502 | For 30 to 39 tons, add | | | | | | 10% | | | |
| 3504 | 20 to 29 tons, add | | | | | | 20% | | | |
| 3506 | 10 to 19 tons, add | | | | | | 30% | | | |
| 3507 | 5 to 9 tons, add | | | | | | 50% | 25% | | |
| 3508 | 1 to 4 tons, add | | | | | | 75% | 50% | | |
| 3509 | Less than 1 ton, add | | | | | | 100% | 100% | | |
| 4010 | SLH series, 40-ton job lots, bolted cross bridging, shop primer | | | | | | | | | |
| 4020 | Spans to 200' (shipped in 3 pieces), minimum | G | E-7 | 16 | 5 | Ton | 1,350 | 240 | 115 | 1,705 | 2,025 |
| 4040 | Average | G | | 13 | 6.154 | | 1,525 | 295 | 142 | 1,962 | 2,325 |
| 4060 | Maximum | G | | 11 | 7.273 | | 1,825 | 350 | 167 | 2,342 | 2,775 |
| 4200 | 80SLH15, 40 lb./L.F. | G | | 1500 | .053 | L.F. | 30.50 | 2.56 | 1.23 | 34.29 | 39 |
| 4220 | 80SLH20, 75 lb./L.F. | G | | 1500 | .053 | | 57 | 2.56 | 1.23 | 60.79 | 68.50 |
| 4240 | 88SLH16, 46 lb./L.F. | G | | 1500 | .053 | | 35 | 2.56 | 1.23 | 38.79 | 44 |
| 4260 | 88SLH21, 89 lb./L.F. | G | | 1500 | .053 | | 68 | 2.56 | 1.23 | 71.79 | 80 |
| 4280 | 96SLH17, 52 lb./L.F. | G | | 1500 | .053 | | 39.50 | 2.56 | 1.23 | 43.29 | 49 |
| 4300 | 96SLH22, 102 lb./L.F. | G | | 1500 | .053 | | 78 | 2.56 | 1.23 | 81.79 | 91 |
| 4320 | 104SLH18, 59 lb./L.F. | G | | 1800 | .044 | | 45 | 2.13 | 1.02 | 48.15 | 54.50 |
| 4340 | 104SLH23, 109 lb./L.F. | G | | 1800 | .044 | | 83 | 2.13 | 1.02 | 86.15 | 96.50 |
| 4360 | 112SLH19, 67 lb./L.F. | G | | 1800 | .044 | | 51 | 2.13 | 1.02 | 54.15 | 61 |
| 4380 | 112SLH24, 131 lb./L.F. | G | | 1800 | .044 | | 100 | 2.13 | 1.02 | 103.15 | 115 |
| 4400 | 120SLH20, 77 lb./L.F. | G | | 1800 | .044 | | 58.50 | 2.13 | 1.02 | 61.65 | 69.50 |
| 4420 | 120SLH25, 152 lb./L.F. | G | | 1800 | .044 | | 116 | 2.13 | 1.02 | 119.15 | 133 |
| 6100 | For less than 40-ton job lots | | | | | | | | | |
| 6102 | For 30 to 39 tons, add | | | | | | 10% | | | |
| 6104 | 20 to 29 tons, add | | | | | | 20% | | | |
| 6106 | 10 to 19 tons, add | | | | | | 30% | | | |
| 6107 | 5 to 9 tons, add | | | | | | 50% | 25% | | |
| 6108 | 1 to 4 tons, add | | | | | | 75% | 50% | | |
| 6109 | Less than 1 ton, add | | | | | | 100% | 100% | | |

## 05 21 16 – Longspan Steel Joist Framing

### 05 21 16.50 Longspan Joists

| | | | | | | | | | | |
|---|---|---|---|---|---|---|---|---|---|---|
| 0010 | **LONGSPAN JOISTS** | | | | | | | | | |
| 2000 | LH series, 40-ton job lots, bolted cross bridging, shop primer | | | | | | | | | |

# 05 21 Steel Joist Framing

## 05 21 16 – Longspan Steel Joist Framing

| 05 21 16.50 Longspan Joists | | Crew | Daily Output | Labor-Hours | Unit | Material | 2011 Bare Costs Labor | Equipment | Total | Total Incl O&P |
|---|---|---|---|---|---|---|---|---|---|---|
| 2015 | Made from recycled materials | G | | | | | | | | |
| 2020 | Spans to 96', minimum | G | E-7 | 16 | 5 | Ton | 1,275 | 240 | 115 | 1,630 | 1,950 |
| 2040 | Average | G | | 13 | 6.154 | | 1,400 | 295 | 142 | 1,837 | 2,200 |
| 2080 | Maximum | G | | 11 | 7.273 | | 1,675 | 350 | 167 | 2,192 | 2,600 |
| 2200 | 18LH04, 12 lb./L.F. | G | | 1400 | .057 | L.F. | 8.45 | 2.74 | 1.31 | 12.50 | 15.45 |
| 2220 | 18LH08, 19 lb./L.F. | G | | 1400 | .057 | | 13.40 | 2.74 | 1.31 | 17.45 | 21 |
| 2240 | 20LH04, 12 lb./L.F. | G | | 1400 | .057 | | 8.45 | 2.74 | 1.31 | 12.50 | 15.45 |
| 2260 | 20LH08, 19 lb./L.F. | G | | 1400 | .057 | | 13.40 | 2.74 | 1.31 | 17.45 | 21 |
| 2280 | 24LH05, 13 lb./L.F. | G | | 1400 | .057 | | 9.20 | 2.74 | 1.31 | 13.25 | 16.25 |
| 2300 | 24LH10, 23 lb./L.F. | G | | 1400 | .057 | | 16.25 | 2.74 | 1.31 | 20.30 | 24 |
| 2320 | 28LH06, 16 lb./L.F. | G | | 1800 | .044 | | 11.30 | 2.13 | 1.02 | 14.45 | 17.20 |
| 2340 | 28LH11, 25 lb./L.F. | G | | 1800 | .044 | | 17.65 | 2.13 | 1.02 | 20.80 | 24 |
| 2360 | 32LH08, 17 lb./L.F. | G | | 1800 | .044 | | 12 | 2.13 | 1.02 | 15.15 | 18 |
| 2380 | 32LH13, 30 lb./L.F. | G | | 1800 | .044 | | 21 | 2.13 | 1.02 | 24.15 | 28.50 |
| 2400 | 36LH09, 21 lb./L.F. | G | | 1800 | .044 | | 14.80 | 2.13 | 1.02 | 17.95 | 21 |
| 2420 | 36LH14, 36 lb./L.F. | G | | 1800 | .044 | | 25.50 | 2.13 | 1.02 | 28.65 | 33 |
| 2440 | 40LH10, 21 lb./L.F. | G | | 2200 | .036 | | 14.80 | 1.74 | .84 | 17.38 | 20 |
| 2460 | 40LH15, 36 lb./L.F. | G | | 2200 | .036 | | 25.50 | 1.74 | .84 | 28.08 | 32 |
| 2480 | 44LH11, 22 lb./L.F. | G | | 2200 | .036 | | 15.55 | 1.74 | .84 | 18.13 | 21 |
| 2500 | 44LH16, 42 lb./L.F. | G | | 2200 | .036 | | 29.50 | 1.74 | .84 | 32.08 | 36.50 |
| 2520 | 48LH11, 22 lb./L.F. | G | | 2200 | .036 | | 15.55 | 1.74 | .84 | 18.13 | 21 |
| 2540 | 48LH16, 42 lb./L.F. | G | | 2200 | .036 | | 29.50 | 1.74 | .84 | 32.08 | 36.50 |
| 2600 | For less than 40-ton job lots | | | | | | | | | | |
| 2602 | For 30 to 39 tons, add | | | | | | 10% | | | | |
| 2604 | 20 to 29 tons, add | | | | | | 20% | | | | |
| 2606 | 10 to 19 tons, add | | | | | | 30% | | | | |
| 2607 | 5 to 9 tons, add | | | | | | 50% | 25% | | | |
| 2608 | 1 to 4 tons, add | | | | | | 75% | 50% | | | |
| 2609 | Less than 1 ton, add | | | | | | 100% | 100% | | | |
| 6000 | For welded cross bridging, add | | | | | | | 30% | | | |

## 05 21 19 – Open Web Steel Joist Framing

| 05 21 19.10 Open Web Joists | | Crew | Daily Output | Labor-Hours | Unit | Material | 2011 Bare Costs Labor | Equipment | Total | Total Incl O&P |
|---|---|---|---|---|---|---|---|---|---|---|
| 0010 | **OPEN WEB JOISTS** | | | | | | | | | | |
| 0015 | Made from recycled materials | G | | | | | | | | | |
| 0020 | K series, 40-ton lots, horiz. bridging, spans to 30', shop primer, minimum | G | E-7 | 15 | 5.333 | Ton | 1,125 | 256 | 123 | 1,504 | 1,825 |
| 0050 | Average | G | | 12 | 6.667 | | 1,275 | 320 | 153 | 1,748 | 2,125 |
| 0080 | Maximum | G | | 9 | 8.889 | | 1,525 | 425 | 204 | 2,154 | 2,625 |
| 0130 | 8K1, 5.1 lb./L.F. | G | | 1200 | .067 | L.F. | 3.25 | 3.19 | 1.53 | 7.97 | 10.75 |
| 0140 | 10K1, 5.0 lb./L.F. | G | | 1200 | .067 | | 3.19 | 3.19 | 1.53 | 7.91 | 10.70 |
| 0160 | 12K3, 5.7 lb./L.F. | G | | 1500 | .053 | | 3.63 | 2.56 | 1.23 | 7.42 | 9.75 |
| 0180 | 14K3, 6.0 lb./L.F. | G | | 1500 | .053 | | 3.82 | 2.56 | 1.23 | 7.61 | 9.95 |
| 0200 | 16K3, 6.3 lb./L.F. | G | | 1800 | .044 | | 4.01 | 2.13 | 1.02 | 7.16 | 9.20 |
| 0220 | 16K6, 8.1 lb./L.F. | G | | 1800 | .044 | | 5.15 | 2.13 | 1.02 | 8.30 | 10.50 |
| 0240 | 18K5, 7.7 lb./L.F. | G | | 2000 | .040 | | 4.91 | 1.92 | .92 | 7.75 | 9.70 |
| 0260 | 18K9, 10.2 lb./L.F. | G | | 2000 | .040 | | 6.50 | 1.92 | .92 | 9.34 | 11.45 |
| 0410 | Span 30' to 50', minimum | G | | 17 | 4.706 | Ton | 1,125 | 226 | 108 | 1,459 | 1,725 |
| 0440 | Average | G | | 17 | 4.706 | | 1,250 | 226 | 108 | 1,584 | 1,875 |
| 0460 | Maximum | G | | 10 | 8 | | 1,325 | 385 | 184 | 1,894 | 2,300 |
| 0500 | 20K5, 8.2 lb./L.F. | G | | 2000 | .040 | L.F. | 5.15 | 1.92 | .92 | 7.99 | 9.95 |
| 0520 | 20K9, 10.8 lb./L.F. | G | | 2000 | .040 | | 6.75 | 1.92 | .92 | 9.59 | 11.75 |
| 0540 | 22K5, 8.8 lb./L.F. | G | | 2000 | .040 | | 5.50 | 1.92 | .92 | 8.34 | 10.35 |
| 0560 | 22K9, 11.3 lb./L.F. | G | | 2000 | .040 | | 7.05 | 1.92 | .92 | 9.89 | 12.05 |

## 05 21 19 – Open Web Steel Joist Framing

| 05 21 19.10 Open Web Joists | | Crew | Daily Output | Labor-Hours | Unit | Material | 2011 Bare Costs Labor | Equipment | Total | Total Incl O&P |
|---|---|---|---|---|---|---|---|---|---|---|
| 0580 | 24K6, 9.7 lb./L.F. | G | E-7 | 2200 | .036 | L.F. | 6.05 | 1.74 | .84 | 8.63 | 10.55 |
| 0600 | 24K10, 13.1 lb./L.F. | G | | 2200 | .036 | | 8.20 | 1.74 | .84 | 10.78 | 12.90 |
| 0620 | 26K6, 10.6 lb./L.F. | G | | 2200 | .036 | | 6.65 | 1.74 | .84 | 9.23 | 11.20 |
| 0640 | 26K10, 13.8 lb./L.F. | G | | 2200 | .036 | | 8.65 | 1.74 | .84 | 11.23 | 13.40 |
| 0660 | 28K8, 12.7 lb./L.F. | G | | 2400 | .033 | | 7.95 | 1.60 | .77 | 10.32 | 12.35 |
| 0680 | 28K12, 17.1 lb./L.F. | G | | 2400 | .033 | | 10.70 | 1.60 | .77 | 13.07 | 15.35 |
| 0700 | 30K8, 13.2 lb./L.F. | G | | 2400 | .033 | | 8.25 | 1.60 | .77 | 10.62 | 12.70 |
| 0720 | 30K12, 17.6 lb./L.F. | G | ▼ | 2400 | .033 | ▼ | 11 | 1.60 | .77 | 13.37 | 15.70 |
| 0800 | For less than 40-ton job lots | | | | | | | | | | |
| 0802 | For 30 to 39 tons, add | | | | | | 10% | | | | |
| 0804 | 20 to 29 tons, add | | | | | | 20% | | | | |
| 0806 | 10 to 19 tons, add | | | | | | 30% | | | | |
| 0807 | 5 to 9 tons, add | | | | | | 50% | 25% | | | |
| 0808 | 1 to 4 tons, add | | | | | | 75% | 50% | | | |
| 0809 | Less than 1 ton, add | | | | | | 100% | 100% | | | |
| 1010 | CS series, 40-ton job lots, horizontal bridging, shop primer | | | | | | | | | | |
| 1020 | Spans to 30', minimum | G | E-7 | 15 | 5.333 | Ton | 1,175 | 256 | 123 | 1,554 | 1,875 |
| 1040 | Average | G | | 12 | 6.667 | | 1,300 | 320 | 153 | 1,773 | 2,175 |
| 1060 | Maximum | G | | 9 | 8.889 | ▼ | 1,550 | 425 | 204 | 2,179 | 2,650 |
| 1100 | 10CS2, 7.5 lb./L.F. | G | | 1200 | .067 | L.F. | 4.91 | 3.19 | 1.53 | 9.63 | 12.60 |
| 1120 | 12CS2, 8.0 lb./L.F. | G | | 1500 | .053 | | 5.25 | 2.56 | 1.23 | 9.04 | 11.50 |
| 1140 | 14CS2, 8.0 lb./L.F. | G | | 1500 | .053 | | 5.25 | 2.56 | 1.23 | 9.04 | 11.50 |
| 1160 | 16CS2, 8.5 lb./L.F. | G | | 1800 | .044 | | 5.55 | 2.13 | 1.02 | 8.70 | 10.90 |
| 1180 | 16CS4, 14.5 lb./L.F. | G | | 1800 | .044 | | 9.50 | 2.13 | 1.02 | 12.65 | 15.25 |
| 1200 | 18CS2, 9.0 lb./L.F. | G | | 2000 | .040 | | 5.90 | 1.92 | .92 | 8.74 | 10.80 |
| 1220 | 18CS4, 15.0 lb./L.F. | G | | 2000 | .040 | | 9.80 | 1.92 | .92 | 12.64 | 15.10 |
| 1240 | 20CS2, 9.5 lb./L.F. | G | | 2000 | .040 | | 6.20 | 1.92 | .92 | 9.04 | 11.15 |
| 1260 | 20CS4, 16.5 lb./L.F. | G | | 2000 | .040 | | 10.80 | 1.92 | .92 | 13.64 | 16.20 |
| 1280 | 22CS2, 10.0 lb./L.F. | G | | 2000 | .040 | | 6.55 | 1.92 | .92 | 9.39 | 11.50 |
| 1300 | 22CS4, 16.5 lb./L.F. | G | | 2000 | .040 | | 10.80 | 1.92 | .92 | 13.64 | 16.20 |
| 1320 | 24CS2, 10.0 lb./L.F. | G | | 2200 | .036 | | 6.55 | 1.74 | .84 | 9.13 | 11.10 |
| 1340 | 24CS4, 16.5 lb./L.F. | G | | 2200 | .036 | | 10.80 | 1.74 | .84 | 13.38 | 15.80 |
| 1360 | 26CS2, 10.0 lb./L.F. | G | | 2200 | .036 | | 6.55 | 1.74 | .84 | 9.13 | 11.10 |
| 1380 | 26CS4, 16.5 lb./L.F. | G | | 2200 | .036 | | 10.80 | 1.74 | .84 | 13.38 | 15.80 |
| 1400 | 28CS2, 10.5 lb./L.F. | G | | 2400 | .033 | | 6.85 | 1.60 | .77 | 9.22 | 11.15 |
| 1420 | 28CS4, 16.5 lb./L.F. | G | | 2400 | .033 | | 10.80 | 1.60 | .77 | 13.17 | 15.50 |
| 1440 | 30CS2, 11.0 lb./L.F. | G | | 2400 | .033 | | 7.20 | 1.60 | .77 | 9.57 | 11.50 |
| 1460 | 30CS4, 16.5 lb./L.F. | G | ▼ | 2400 | .033 | ▼ | 10.80 | 1.60 | .77 | 13.17 | 15.50 |
| 1500 | For less than 40-ton job lots | | | | | | | | | | |
| 1502 | For 30 to 39 tons, add | | | | | | 10% | | | | |
| 1504 | 20 to 29 tons, add | | | | | | 20% | | | | |
| 1506 | 10 to 19 tons, add | | | | | | 30% | | | | |
| 1507 | 5 to 9 tons, add | | | | | | 50% | 25% | | | |
| 1508 | 1 to 4 tons, add | | | | | | 75% | 50% | | | |
| 1509 | Less than 1 ton, add | | | | | | 100% | 100% | | | |
| 6200 | For shop prime paint other than mfrs. standard, add | | | | | | 20% | | | | |
| 6300 | For bottom chord extensions, add per chord | G | | | | Ea. | 27.50 | | | 27.50 | 30.50 |
| 6400 | Individual steel bearing plate, 6" x 6" x 1/4" with J-hook | G | 1 Bric | 160 | .050 | " | 6.75 | 2.16 | | 8.91 | 10.70 |

## 05 21 23 – Steel Joist Girder Framing

### 05 21 23.50 Joist Girders

| 0010 | JOIST GIRDERS | | | | | | | | | | |
|---|---|---|---|---|---|---|---|---|---|---|---|
| 0015 | Made from recycled materials | G | | | | | | | | | |

# 05 21 Steel Joist Framing

## 05 21 23 – Steel Joist Girder Framing

| 05 21 23.50 Joist Girders | | Crew | Daily Output | Labor-Hours | Unit | Material | 2011 Bare Costs Labor | Equipment | Total | Total Incl O&P |
|---|---|---|---|---|---|---|---|---|---|---|
| 7000 | Joist girders, 40-ton job lots, shop primer, minimum | G E-5 | 15 | 5.333 | Ton | 1,150 | 256 | 115 | 1,521 | 1,850 |
| 7020 | Average | G | 13 | 6.154 | | 1,275 | 295 | 133 | 1,703 | 2,050 |
| 7040 | Maximum | G ↓ | 11 | 7.273 | ↓ | 1,350 | 350 | 157 | 1,857 | 2,250 |
| 7100 | For less than 40-ton job lots | | | | | | | | | |
| 7102 | For 30 to 39 tons, add | | | | | 10% | | | | |
| 7104 | 20 to 29 tons, add | | | | | 20% | | | | |
| 7106 | 10 to 19 tons, add | | | | | 30% | | | | |
| 7107 | 5 to 9 tons, add | | | | | 50% | 25% | | | |
| 7108 | 1 to 4 tons, add | | | | | 75% | 50% | | | |
| 7109 | Less than 1 ton, add | | | | | 100% | 100% | | | |
| 8000 | Trusses, 40-ton job lots, shop fabricated WT chords, shop primer, average | G E-5 | 11 | 7.273 | Ton | 4,175 | 350 | 157 | 4,682 | 5,375 |
| 8100 | For less than 40-ton job lots | | | | | | | | | |
| 8102 | For 30 to 39 tons, add | | | | | 10% | | | | |
| 8104 | 20 to 29 tons, add | | | | | 20% | | | | |
| 8106 | 10 to 19 tons, add | | | | | 30% | | | | |
| 8107 | 5 to 9 tons, add | | | | | 50% | 25% | | | |
| 8108 | 1 to 4 tons, add | | | | | 75% | 50% | | | |
| 8109 | Less than 1 ton, add | | | | | 100% | 100% | | | |

# 05 31 Steel Decking

## 05 31 13 – Steel Floor Decking

| 05 31 13.50 Floor Decking | | Crew | Daily Output | Labor-Hours | Unit | Material | 2011 Bare Costs Labor | Equipment | Total | Total Incl O&P |
|---|---|---|---|---|---|---|---|---|---|---|
| 0010 | **FLOOR DECKING** R053100-10 | | | | | | | | | |
| 0015 | Made from recycled materials | G | | | | | | | | |
| 5100 | Non-cellular composite decking, galvanized, 1-1/2" deep, 16 gauge | G E-4 | 3500 | .009 | S.F. | 2.39 | .45 | .03 | 2.87 | 3.45 |
| 5120 | 18 gauge | G | 3650 | .009 | | 1.92 | .43 | .03 | 2.38 | 2.90 |
| 5140 | 20 gauge | G | 3800 | .008 | | 1.55 | .41 | .03 | 1.99 | 2.46 |
| 5200 | 2" deep, 22 gauge | G | 3860 | .008 | | 1.35 | .41 | .03 | 1.79 | 2.24 |
| 5300 | 20 gauge | G | 3600 | .009 | | 1.49 | .44 | .03 | 1.96 | 2.44 |
| 5400 | 18 gauge | G | 3380 | .009 | | 1.88 | .46 | .03 | 2.37 | 2.93 |
| 5500 | 16 gauge | G | 3200 | .010 | | 2.37 | .49 | .03 | 2.89 | 3.51 |
| 5700 | 3" deep, 22 gauge | G | 3200 | .010 | | 1.47 | .49 | .03 | 1.99 | 2.53 |
| 5800 | 20 gauge | G | 3000 | .011 | | 1.64 | .52 | .04 | 2.20 | 2.76 |
| 5900 | 18 gauge | G | 2850 | .011 | | 2 | .55 | .04 | 2.59 | 3.21 |
| 6000 | 16 gauge | G ↓ | 2700 | .012 | ↓ | 2.69 | .58 | .04 | 3.31 | 4.03 |

## 05 31 23 – Steel Roof Decking

| 05 31 23.50 Roof Decking | | Crew | Daily Output | Labor-Hours | Unit | Material | 2011 Bare Costs Labor | Equipment | Total | Total Incl O&P |
|---|---|---|---|---|---|---|---|---|---|---|
| 0010 | **ROOF DECKING** | | | | | | | | | |
| 0015 | Made from recycled materials | G | | | | | | | | |
| 2100 | Open type, 1-1/2" deep, Type B, wide rib, galv., 22 gauge, under 50 sq. | G E-4 | 4500 | .007 | S.F. | 1.42 | .35 | .02 | 1.79 | 2.21 |
| 2200 | 50-500 squares | G | 4900 | .007 | | 1.10 | .32 | .02 | 1.44 | 1.80 |
| 2400 | Over 500 squares | G | 5100 | .006 | | 1.02 | .31 | .02 | 1.35 | 1.68 |
| 2600 | 20 gauge, under 50 squares | G | 3865 | .008 | | 1.67 | .41 | .03 | 2.11 | 2.59 |
| 2650 | 50-500 squares | G | 4170 | .008 | | 1.34 | .38 | .03 | 1.75 | 2.16 |
| 2700 | Over 500 squares | G | 4300 | .007 | | 1.21 | .36 | .03 | 1.60 | 2 |
| 2900 | 18 gauge, under 50 squares | G | 3800 | .008 | | 2.15 | .41 | .03 | 2.59 | 3.13 |
| 2950 | 50-500 squares | G | 4100 | .008 | | 1.72 | .38 | .03 | 2.13 | 2.60 |
| 3000 | Over 500 squares | G | 4300 | .007 | | 1.55 | .36 | .03 | 1.94 | 2.37 |
| 3050 | 16 gauge, under 50 squares | G | 3700 | .009 | | 2.90 | .42 | .03 | 3.35 | 3.97 |
| 3060 | 50-500 squares | G ↓ | 4000 | .008 | ↓ | 2.32 | .39 | .03 | 2.74 | 3.27 |

# 05 31 Steel Decking

## 05 31 23 – Steel Roof Decking

### 05 31 23.50 Roof Decking

| | | Crew | Daily Output | Labor-Hours | Unit | Material | 2011 Bare Costs Labor | Equipment | Total | Total Incl O&P |
|---|---|---|---|---|---|---|---|---|---|---|
| 3100 | Over 500 squares [G] | E-4 | 4200 | .008 | S.F. | 2.09 | .37 | .03 | 2.49 | 2.98 |
| 3200 | 3" deep, 22 gauge, under 50 squares [G] | | 3600 | .009 | | 2.10 | .44 | .03 | 2.57 | 3.11 |
| 3250 | 50-500 squares [G] | | 3800 | .008 | | 1.68 | .41 | .03 | 2.12 | 2.61 |
| 3260 | over 500 squares [G] | | 4000 | .008 | | 1.51 | .39 | .03 | 1.93 | 2.38 |
| 3300 | 20 gauge, under 50 squares [G] | | 3400 | .009 | | 2.27 | .46 | .03 | 2.76 | 3.35 |
| 3350 | 50-500 squares [G] | | 3600 | .009 | | 1.81 | .44 | .03 | 2.28 | 2.79 |
| 3360 | over 500 squares [G] | | 3800 | .008 | | 1.63 | .41 | .03 | 2.07 | 2.55 |
| 3400 | 18 gauge, under 50 squares [G] | | 3200 | .010 | | 2.94 | .49 | .03 | 3.46 | 4.14 |
| 3450 | 50-500 squares [G] | | 3400 | .009 | | 2.35 | .46 | .03 | 2.84 | 3.44 |
| 3460 | over 500 squares [G] | | 3600 | .009 | | 2.11 | .44 | .03 | 2.58 | 3.12 |
| 3500 | 16 gauge, under 50 squares [G] | | 3000 | .011 | | 3.88 | .52 | .04 | 4.44 | 5.20 |
| 3550 | 50-500 squares [G] | | 3200 | .010 | | 3.10 | .49 | .03 | 3.62 | 4.32 |
| 3560 | over 500 squares [G] | | 3400 | .009 | | 2.79 | .46 | .03 | 3.28 | 3.93 |
| 3700 | 4-1/2" deep, 20 gauge, over 50 squares [G] | | 2700 | .012 | | 3.64 | .58 | .04 | 4.26 | 5.05 |
| 3800 | 18 gauge [G] | | 2460 | .013 | | 4.69 | .64 | .04 | 5.37 | 6.35 |
| 3900 | 16 gauge [G] | | 2350 | .014 | | 3.51 | .67 | .05 | 4.23 | 5.10 |
| 4100 | 6" deep, 18 gauge, over 50 squares [G] | | 2000 | .016 | | 6.70 | .78 | .05 | 7.53 | 8.85 |
| 4200 | 16 gauge [G] | | 1930 | .017 | | 5 | .81 | .06 | 5.87 | 7 |
| 4300 | 14 gauge [G] | | 1860 | .017 | | 6.45 | .84 | .06 | 7.35 | 8.65 |
| 4500 | 7-1/2" deep, 18 gauge, over 50 squares [G] | | 1690 | .019 | | 7.35 | .93 | .06 | 8.34 | 9.80 |
| 4600 | 16 gauge [G] | | 1590 | .020 | | 5.50 | .99 | .07 | 6.56 | 7.85 |
| 4700 | 14 gauge [G] | | 1490 | .021 | | 7.10 | 1.05 | .07 | 8.22 | 9.75 |
| 4800 | For painted instead of galvanized, deduct | | | | | 5% | | | | |
| 5000 | For acoustical perforated with fiberglass insulation, add | | | | S.F. | 25% | | | | |
| 5100 | For type F intermediate rib instead of type B wide rib, add [G] | | | | | 25% | | | | |
| 5150 | For type A narrow rib instead of type B wide rib, add [G] | | | | | 25% | | | | |

## 05 31 33 – Steel Form Decking

### 05 31 33.50 Form Decking

| | | Crew | Daily Output | Labor-Hours | Unit | Material | 2011 Bare Costs Labor | Equipment | Total | Total Incl O&P |
|---|---|---|---|---|---|---|---|---|---|---|
| 0010 | **FORM DECKING** | | | | | | | | | |
| 0015 | Made from recycled materials [G] | | | | | | | | | |
| 6100 | Slab form, steel, 28 gauge, 9/16" deep, type UFS, uncoated [G] | E-4 | 4000 | .008 | S.F. | 1.07 | .39 | .03 | 1.49 | 1.90 |
| 6200 | Galvanized [G] | | 4000 | .008 | | .95 | .39 | .03 | 1.37 | 1.77 |
| 6220 | 24 gauge, 1" deep , type UF1X, uncoated [G] | | 3900 | .008 | | 1.04 | .40 | .03 | 1.47 | 1.88 |
| 6240 | Galvanized [G] | | 3900 | .008 | | 1.22 | .40 | .03 | 1.65 | 2.08 |
| 6300 | 24 gauge, 1-5/16" deep, type UFX, uncoated [G] | | 3800 | .008 | | 1.10 | .41 | .03 | 1.54 | 1.97 |
| 6400 | Galvanized [G] | | 3800 | .008 | | 1.29 | .41 | .03 | 1.73 | 2.18 |
| 6500 | 22 gauge, 1-5/16" deep, uncoated [G] | | 3700 | .009 | | 1.38 | .42 | .03 | 1.83 | 2.30 |
| 6600 | Galvanized [G] | | 3700 | .009 | | 1.41 | .42 | .03 | 1.86 | 2.33 |
| 6700 | 22 gauge, 2" deep, uncoated [G] | | 3600 | .009 | | 1.82 | .44 | .03 | 2.29 | 2.80 |
| 6800 | Galvanized [G] | | 3600 | .009 | | 1.78 | .44 | .03 | 2.25 | 2.76 |
| 7000 | Sheet metal edge closure form, 12" wide with 2 bends, galv | | | | | | | | | |
| 7100 | 18 gauge [G] | E-14 | 360 | .022 | L.F. | 2.92 | 1.12 | .30 | 4.34 | 5.50 |
| 7200 | 16 gauge [G] | " | 360 | .022 | " | 3.97 | 1.12 | .30 | 5.39 | 6.70 |

## 05 35 13 – Steel Cellular Decking

| 05 35 13.50 Cellular Decking | | Crew | Daily Output | Labor-Hours | Unit | Material | 2011 Bare Costs Labor | Equipment | Total | Total Incl O&P |
|---|---|---|---|---|---|---|---|---|---|---|
| 0010 | **CELLULAR DECKING** | | | | | | | | | |
| 0015 | Made from recycled materials | G | | | | | | | | |
| 0200 | Cellular units, galv, 1-1/2" deep, Type BC, 20-20 gauge, over 15 squares | G | E-4 | 1460 | .022 | S.F. | 5.65 | 1.08 | .07 | 6.80 | 8.20 |
| 0250 | 18-20 gauge | G | | 1420 | .023 | | 6.40 | 1.11 | .08 | 7.59 | 9.10 |
| 0300 | 18-18 gauge | G | | 1390 | .023 | | 6.55 | 1.13 | .08 | 7.76 | 9.30 |
| 0320 | 16-18 gauge | G | | 1360 | .024 | | 7.80 | 1.15 | .08 | 9.03 | 10.75 |
| 0340 | 16-16 gauge | G | | 1330 | .024 | | 8.70 | 1.18 | .08 | 9.96 | 11.75 |
| 0400 | 3" deep, Type NC, galvanized, 20-20 gauge | G | | 1375 | .023 | | 6.20 | 1.14 | .08 | 7.42 | 8.90 |
| 0500 | 18-20 gauge | G | | 1350 | .024 | | 7.50 | 1.16 | .08 | 8.74 | 10.35 |
| 0600 | 18-18 gauge | G | | 1290 | .025 | | 7.45 | 1.22 | .08 | 8.75 | 10.45 |
| 0700 | 16-18 gauge | G | | 1230 | .026 | | 8.40 | 1.28 | .09 | 9.77 | 11.60 |
| 0800 | 16-16 gauge | G | | 1150 | .028 | | 9.15 | 1.37 | .09 | 10.61 | 12.60 |
| 1000 | 4-1/2" deep, Type JC, galvanized, 18-20 gauge | G | | 1100 | .029 | | 8.65 | 1.43 | .10 | 10.18 | 12.15 |
| 1100 | 18-18 gauge | G | | 1040 | .031 | | 8.60 | 1.51 | .10 | 10.21 | 12.25 |
| 1200 | 16-18 gauge | G | | 980 | .033 | | 9.65 | 1.60 | .11 | 11.36 | 13.60 |
| 1300 | 16-16 gauge | G | | 935 | .034 | | 10.55 | 1.68 | .12 | 12.35 | 14.70 |
| 1500 | For acoustical deck, add | | | | | | 15% | | | | |
| 1700 | For cells used for ventilation, add | | | | | | 15% | | | | |
| 1900 | For multi-story or congested site, add | | | | | | | 50% | | | |
| 8000 | Metal deck and trench, 2" thick, 20 gauge, combination | | | | | | | | | | |
| 8010 | 60% cellular, 40% non-cellular, inserts and trench | G | R-4 | 1100 | .036 | S.F. | 11.25 | 1.79 | .10 | 13.14 | 15.55 |

## 05 41 13 – Load-Bearing Metal Stud Framing

### 05 41 13.05 Bracing

| 05 41 13.05 Bracing | | Crew | Daily Output | Labor-Hours | Unit | Material | 2011 Bare Costs Labor | Equipment | Total | Total Incl O&P |
|---|---|---|---|---|---|---|---|---|---|---|
| 0010 | **BRACING**, shear wall X-bracing, per 10' x 10' bay, one face | | | | | | | | | | |
| 0015 | Made of recycled materials | G | | | | | | | | | |
| 0120 | Metal strap, 20 ga x 4" wide | G | 2 Carp | 18 | .889 | Ea. | 16.65 | 38.50 | | 55.15 | 77 |
| 0130 | 6" wide | G | | 18 | .889 | | 25 | 38.50 | | 63.50 | 86 |
| 0160 | 18 ga x 4" wide | G | | 16 | 1 | | 26 | 43 | | 69 | 94.50 |
| 0170 | 6" wide | G | | 16 | 1 | | 38.50 | 43 | | 81.50 | 109 |
| 0410 | Continuous strap bracing, per horizontal row on both faces | | | | | | | | | | |
| 0420 | Metal strap, 20 ga x 2" wide, studs 12" O.C. | G | 1 Carp | 7 | 1.143 | C.L.F. | 46 | 49 | | 95 | 127 |
| 0430 | 16" O.C. | G | | 8 | 1 | | 46 | 43 | | 89 | 117 |
| 0440 | 24" O.C. | G | | 10 | .800 | | 46 | 34.50 | | 80.50 | 104 |
| 0450 | 18 ga x 2" wide, studs 12" O.C. | G | | 6 | 1.333 | | 64 | 57.50 | | 121.50 | 158 |
| 0460 | 16" O.C. | G | | 7 | 1.143 | | 64 | 49 | | 113 | 146 |
| 0470 | 24" O.C. | G | | 8 | 1 | | 64 | 43 | | 107 | 136 |

### 05 41 13.10 Bridging

| 05 41 13.10 Bridging | | Crew | Daily Output | Labor-Hours | Unit | Material | 2011 Bare Costs Labor | Equipment | Total | Total Incl O&P |
|---|---|---|---|---|---|---|---|---|---|---|
| 0010 | **BRIDGING**, solid between studs w/1-1/4" leg track, per stud bay | | | | | | | | | | |
| 0015 | Made from recycled materials | G | | | | | | | | | |
| 0200 | Studs 12" O.C., 18 ga x 2-1/2" wide | G | 1 Carp | 125 | .064 | Ea. | .76 | 2.76 | | 3.52 | 5.05 |
| 0210 | 3-5/8" wide | G | | 120 | .067 | | .91 | 2.87 | | 3.78 | 5.40 |
| 0220 | 4" wide | G | | 120 | .067 | | .98 | 2.87 | | 3.85 | 5.50 |
| 0230 | 6" wide | G | | 115 | .070 | | 1.26 | 3 | | 4.26 | 6 |
| 0240 | 8" wide | G | | 110 | .073 | | 1.57 | 3.13 | | 4.70 | 6.55 |
| 0300 | 16 ga x 2-1/2" wide | G | | 115 | .070 | | .97 | 3 | | 3.97 | 5.65 |
| 0310 | 3-5/8" wide | G | | 110 | .073 | | 1.17 | 3.13 | | 4.30 | 6.10 |
| 0320 | 4" wide | G | | 110 | .073 | | 1.25 | 3.13 | | 4.38 | 6.15 |
| 0330 | 6" wide | G | | 105 | .076 | | 1.60 | 3.28 | | 4.88 | 6.80 |
| 0340 | 8" wide | G | | 100 | .080 | | 2.01 | 3.44 | | 5.45 | 7.50 |

## 05 41 13 – Load-Bearing Metal Stud Framing

### 05 41 13.10 Bridging

| | | Crew | Daily Output | Labor-Hours | Unit | Material | 2011 Bare Costs Labor | Equipment | Total | Total Incl O&P |
|---|---|---|---|---|---|---|---|---|---|---|
| 1200 | Studs 16" O.C., 18 ga x 2-1/2" wide | G 1 Carp | 125 | .064 | Ea. | .97 | 2.76 | | 3.73 | 5.30 |
| 1210 | 3-5/8" wide | G | 120 | .067 | | 1.17 | 2.87 | | 4.04 | 5.70 |
| 1220 | 4" wide | G | 120 | .067 | | 1.25 | 2.87 | | 4.12 | 5.80 |
| 1230 | 6" wide | G | 115 | .070 | | 1.62 | 3 | | 4.62 | 6.40 |
| 1240 | 8" wide | G | 110 | .073 | | 2.01 | 3.13 | | 5.14 | 7 |
| 1300 | 16 ga x 2-1/2" wide | G | 115 | .070 | | 1.24 | 3 | | 4.24 | 5.95 |
| 1310 | 3-5/8" wide | G | 110 | .073 | | 1.50 | 3.13 | | 4.63 | 6.45 |
| 1320 | 4" wide | G | 110 | .073 | | 1.60 | 3.13 | | 4.73 | 6.55 |
| 1330 | 6" wide | G | 105 | .076 | | 2.05 | 3.28 | | 5.33 | 7.30 |
| 1340 | 8" wide | G | 100 | .080 | | 2.57 | 3.44 | | 6.01 | 8.15 |
| 2200 | Studs 24" O.C., 18 ga x 2-1/2" wide | G | 125 | .064 | | 1.41 | 2.76 | | 4.17 | 5.80 |
| 2210 | 3-5/8" wide | G | 120 | .067 | | 1.69 | 2.87 | | 4.56 | 6.25 |
| 2220 | 4" wide | G | 120 | .067 | | 1.81 | 2.87 | | 4.68 | 6.40 |
| 2230 | 6" wide | G | 115 | .070 | | 2.34 | 3 | | 5.34 | 7.15 |
| 2240 | 8" wide | G | 110 | .073 | | 2.91 | 3.13 | | 6.04 | 8 |
| 2300 | 16 ga x 2-1/2" wide | G | 115 | .070 | | 1.79 | 3 | | 4.79 | 6.55 |
| 2310 | 3-5/8" wide | G | 110 | .073 | | 2.17 | 3.13 | | 5.30 | 7.20 |
| 2320 | 4" wide | G | 110 | .073 | | 2.32 | 3.13 | | 5.45 | 7.35 |
| 2330 | 6" wide | G | 105 | .076 | | 2.96 | 3.28 | | 6.24 | 8.30 |
| 2340 | 8" wide | G | 100 | .080 | | 3.72 | 3.44 | | 7.16 | 9.40 |
| 3000 | Continuous bridging, per row | | | | | | | | | |
| 3100 | 16 ga x 1-1/2" channel thru studs 12" O.C. | G 1 Carp | 6 | 1.333 | C.L.F. | 41 | 57.50 | | 98.50 | 133 |
| 3110 | 16" O.C. | G | 7 | 1.143 | | 41 | 49 | | 90 | 121 |
| 3120 | 24" O.C. | G | 8.80 | .909 | | 41 | 39 | | 80 | 105 |
| 4100 | 2" x 2" angle x 18 ga, studs 12" O.C. | G | 7 | 1.143 | | 64 | 49 | | 113 | 146 |
| 4110 | 16" O.C. | G | 9 | .889 | | 64 | 38.50 | | 102.50 | 129 |
| 4120 | 24" O.C. | G | 12 | .667 | | 64 | 28.50 | | 92.50 | 114 |
| 4200 | 16 ga, studs 12" O.C. | G | 5 | 1.600 | | 81.50 | 69 | | 150.50 | 196 |
| 4210 | 16" O.C. | G | 7 | 1.143 | | 81.50 | 49 | | 130.50 | 165 |
| 4220 | 24" O.C. | G | 10 | .800 | | 81.50 | 34.50 | | 116 | 143 |

### 05 41 13.25 Framing, Boxed Headers/Beams

| | | Crew | Daily Output | Labor-Hours | Unit | Material | 2011 Bare Costs Labor | Equipment | Total | Total Incl O&P |
|---|---|---|---|---|---|---|---|---|---|---|
| 0010 | **FRAMING, BOXED HEADERS/BEAMS** | | | | | | | | | |
| 0015 | Made from recycled materials | G | | | | | | | | |
| 0200 | Double, 18 ga x 6" deep | G 2 Carp | 220 | .073 | L.F. | 4.40 | 3.13 | | 7.53 | 9.65 |
| 0210 | 8" deep | G | 210 | .076 | | 4.84 | 3.28 | | 8.12 | 10.40 |
| 0220 | 10" deep | G | 200 | .080 | | 5.90 | 3.44 | | 9.34 | 11.80 |
| 0230 | 12" deep | G | 190 | .084 | | 6.45 | 3.63 | | 10.08 | 12.60 |
| 0300 | 16 ga x 8" deep | G | 180 | .089 | | 5.60 | 3.83 | | 9.43 | 12 |
| 0310 | 10" deep | G | 170 | .094 | | 6.75 | 4.05 | | 10.80 | 13.60 |
| 0320 | 12" deep | G | 160 | .100 | | 7.30 | 4.31 | | 11.61 | 14.65 |
| 0400 | 14 ga x 10" deep | G | 140 | .114 | | 7.75 | 4.92 | | 12.67 | 16.10 |
| 0410 | 12" deep | G | 130 | .123 | | 8.50 | 5.30 | | 13.80 | 17.50 |
| 1210 | Triple, 18 ga x 8" deep | G | 170 | .094 | | 7 | 4.05 | | 11.05 | 13.90 |
| 1220 | 10" deep | G | 165 | .097 | | 8.45 | 4.17 | | 12.62 | 15.70 |
| 1230 | 12" deep | G | 160 | .100 | | 9.25 | 4.31 | | 13.56 | 16.80 |
| 1300 | 16 ga x 8" deep | G | 145 | .110 | | 8.10 | 4.75 | | 12.85 | 16.20 |
| 1310 | 10" deep | G | 140 | .114 | | 9.70 | 4.92 | | 14.62 | 18.20 |
| 1320 | 12" deep | G | 135 | .119 | | 10.55 | 5.10 | | 15.65 | 19.50 |
| 1400 | 14 ga x 10" deep | G | 115 | .139 | | 10.60 | 6 | | 16.60 | 21 |
| 1410 | 12" deep | G | 110 | .145 | | 11.70 | 6.25 | | 17.95 | 22.50 |

## 05 41 13 – Load-Bearing Metal Stud Framing

| 05 41 13.30 Framing, Stud Walls | | Crew | Daily Output | Labor-Hours | Unit | Material | 2011 Bare Costs Labor | Equipment | Total | Total Incl O&P |
|---|---|---|---|---|---|---|---|---|---|---|
| 0010 | **FRAMING, STUD WALLS** w/top & bottom track, no openings, | | | | | | | | | |
| 0020 | Headers, beams, bridging or bracing | | | | | | | | | |
| 0025 | Made from recycled materials | G | | | | | | | | |
| 4100 | 8' high walls, 18 ga x 2-1/2" wide, studs 12" O.C. | G | 2 Carp | 54 | .296 | L.F. | 7.30 | 12.75 | | 20.05 | 27.50 |
| 4110 | 16" O.C. | G | | 77 | .208 | | 5.85 | 8.95 | | 14.80 | 20 |
| 4120 | 24" O.C. | G | | 107 | .150 | | 4.37 | 6.45 | | 10.82 | 14.70 |
| 4130 | 3-5/8" wide, studs 12" O.C. | G | | 53 | .302 | | 8.60 | 13 | | 21.60 | 29.50 |
| 4140 | 16" O.C. | G | | 76 | .211 | | 6.90 | 9.05 | | 15.95 | 21.50 |
| 4150 | 24" O.C. | G | | 105 | .152 | | 5.20 | 6.55 | | 11.75 | 15.75 |
| 4160 | 4" wide, studs 12" O.C. | G | | 52 | .308 | | 9.05 | 13.25 | | 22.30 | 30.50 |
| 4170 | 16" O.C. | G | | 74 | .216 | | 7.25 | 9.30 | | 16.55 | 22.50 |
| 4180 | 24" O.C. | G | | 103 | .155 | | 5.45 | 6.70 | | 12.15 | 16.25 |
| 4190 | 6" wide, studs 12" O.C. | G | | 51 | .314 | | 11.45 | 13.50 | | 24.95 | 33 |
| 4200 | 16" O.C. | G | | 73 | .219 | | 9.20 | 9.45 | | 18.65 | 24.50 |
| 4210 | 24" O.C. | G | | 101 | .158 | | 6.90 | 6.80 | | 13.70 | 18.05 |
| 4220 | 8" wide, studs 12" O.C. | G | | 50 | .320 | | 13.85 | 13.80 | | 27.65 | 36.50 |
| 4230 | 16" O.C. | G | | 72 | .222 | | 11.15 | 9.55 | | 20.70 | 27 |
| 4240 | 24" O.C. | G | | 100 | .160 | | 8.45 | 6.90 | | 15.35 | 19.80 |
| 4300 | 16 ga x 2-1/2" wide, studs 12" O.C. | G | | 47 | .340 | | 8.65 | 14.65 | | 23.30 | 32 |
| 4310 | 16" O.C. | G | | 68 | .235 | | 6.85 | 10.15 | | 17 | 23 |
| 4320 | 24" O.C. | G | | 94 | .170 | | 5.05 | 7.35 | | 12.40 | 16.80 |
| 4330 | 3-5/8" wide, studs 12" O.C. | G | | 46 | .348 | | 10.20 | 14.95 | | 25.15 | 34.50 |
| 4340 | 16" O.C. | G | | 66 | .242 | | 8.10 | 10.45 | | 18.55 | 25 |
| 4350 | 24" O.C. | G | | 92 | .174 | | 6 | 7.50 | | 13.50 | 18.10 |
| 4360 | 4" wide, studs 12" O.C. | G | | 45 | .356 | | 10.80 | 15.30 | | 26.10 | 35.50 |
| 4370 | 16" O.C. | G | | 65 | .246 | | 8.60 | 10.60 | | 19.20 | 25.50 |
| 4380 | 24" O.C. | G | | 90 | .178 | | 6.35 | 7.65 | | 14 | 18.75 |
| 4390 | 6" wide, studs 12" O.C. | G | | 44 | .364 | | 13.50 | 15.65 | | 29.15 | 39 |
| 4400 | 16" O.C. | G | | 64 | .250 | | 10.75 | 10.75 | | 21.50 | 28.50 |
| 4410 | 24" O.C. | G | | 88 | .182 | | 7.95 | 7.85 | | 15.80 | 21 |
| 4420 | 8" wide, studs 12" O.C. | G | | 43 | .372 | | 16.65 | 16 | | 32.65 | 43 |
| 4430 | 16" O.C. | G | | 63 | .254 | | 13.25 | 10.95 | | 24.20 | 31.50 |
| 4440 | 24" O.C. | G | | 86 | .186 | | 9.85 | 8 | | 17.85 | 23 |
| 5100 | 10' high walls, 18 ga x 2-1/2" wide, studs 12" O.C. | G | | 54 | .296 | | 8.75 | 12.75 | | 21.50 | 29 |
| 5110 | 16" O.C. | G | | 77 | .208 | | 6.90 | 8.95 | | 15.85 | 21.50 |
| 5120 | 24" O.C. | G | | 107 | .150 | | 5.10 | 6.45 | | 11.55 | 15.50 |
| 5130 | 3-5/8" wide, studs 12" O.C. | G | | 53 | .302 | | 10.35 | 13 | | 23.35 | 31.50 |
| 5140 | 16" O.C. | G | | 76 | .211 | | 8.20 | 9.05 | | 17.25 | 23 |
| 5150 | 24" O.C. | G | | 105 | .152 | | 6.05 | 6.55 | | 12.60 | 16.70 |
| 5160 | 4" wide, studs 12" O.C. | G | | 52 | .308 | | 10.85 | 13.25 | | 24.10 | 32.50 |
| 5170 | 16" O.C. | G | | 74 | .216 | | 8.60 | 9.30 | | 17.90 | 24 |
| 5180 | 24" O.C. | G | | 103 | .155 | | 6.35 | 6.70 | | 13.05 | 17.25 |
| 5190 | 6" wide, studs 12" O.C. | G | | 51 | .314 | | 13.70 | 13.50 | | 27.20 | 35.50 |
| 5200 | 16" O.C. | G | | 73 | .219 | | 10.90 | 9.45 | | 20.35 | 26.50 |
| 5210 | 24" O.C. | G | | 101 | .158 | | 8.05 | 6.80 | | 14.85 | 19.30 |
| 5220 | 8" wide, studs 12" O.C. | G | | 50 | .320 | | 16.60 | 13.80 | | 30.40 | 39.50 |
| 5230 | 16" O.C. | G | | 72 | .222 | | 13.20 | 9.55 | | 22.75 | 29 |
| 5240 | 24" O.C. | G | | 100 | .160 | | 9.80 | 6.90 | | 16.70 | 21.50 |
| 5300 | 16 ga x 2-1/2" wide, studs 12" O.C. | G | | 47 | .340 | | 10.45 | 14.65 | | 25.10 | 34 |
| 5310 | 16" O.C. | G | | 68 | .235 | | 8.20 | 10.15 | | 18.35 | 24.50 |
| 5320 | 24" O.C. | G | | 94 | .170 | | 5.95 | 7.35 | | 13.30 | 17.80 |
| 5330 | 3-5/8" wide, studs 12" O.C. | G | | 46 | .348 | | 12.35 | 14.95 | | 27.30 | 36.50 |

## 05 41 13 – Load-Bearing Metal Stud Framing

| 05 41 13.30 Framing, Stud Walls | | Crew | Daily Output | Labor-Hours | Unit | Material | 2011 Bare Costs Labor | Equipment | Total | Total Incl O&P |
|---|---|---|---|---|---|---|---|---|---|---|
| 5340 | 16" O.C. | G 2 Carp | 66 | .242 | L.F. | 9.70 | 10.45 | | 20.15 | 26.50 |
| 5350 | 24" O.C. | G | 92 | .174 | | 7.05 | 7.50 | | 14.55 | 19.25 |
| 5360 | 4" wide, studs 12" O.C. | G | 45 | .356 | | 13.05 | 15.30 | | 28.35 | 38 |
| 5370 | 16" O.C. | G | 65 | .246 | | 10.25 | 10.60 | | 20.85 | 27.50 |
| 5380 | 24" O.C. | G | 90 | .178 | | 7.45 | 7.65 | | 15.10 | 19.95 |
| 5390 | 6" wide, studs 12" O.C. | G | 44 | .364 | | 16.30 | 15.65 | | 31.95 | 42 |
| 5400 | 16" O.C. | G | 64 | .250 | | 12.85 | 10.75 | | 23.60 | 30.50 |
| 5410 | 24" O.C. | G | 88 | .182 | | 9.35 | 7.85 | | 17.20 | 22.50 |
| 5420 | 8" wide, studs 12" O.C. | G | 43 | .372 | | 20 | 16 | | 36 | 46.50 |
| 5430 | 16" O.C. | G | 63 | .254 | | 15.80 | 10.95 | | 26.75 | 34 |
| 5440 | 24" O.C. | G | 86 | .186 | | 11.55 | 8 | | 19.55 | 25 |
| 6190 | 12' high walls, 18 ga x 6" wide, studs 12" O.C. | G | 41 | .390 | | 15.95 | 16.80 | | 32.75 | 43.50 |
| 6200 | 16" O.C. | G | 58 | .276 | | 12.55 | 11.90 | | 24.45 | 32 |
| 6210 | 24" O.C. | G | 81 | .198 | | 9.20 | 8.50 | | 17.70 | 23 |
| 6220 | 8" wide, studs 12" O.C. | G | 40 | .400 | | 19.30 | 17.20 | | 36.50 | 47.50 |
| 6230 | 16" O.C. | G | 57 | .281 | | 15.25 | 12.10 | | 27.35 | 35.50 |
| 6240 | 24" O.C. | G | 80 | .200 | | 11.15 | 8.60 | | 19.75 | 25.50 |
| 6390 | 16 ga x 6" wide, studs 12" O.C. | G | 35 | .457 | | 19.10 | 19.70 | | 38.80 | 51 |
| 6400 | 16" O.C. | G | 51 | .314 | | 14.90 | 13.50 | | 28.40 | 37 |
| 6410 | 24" O.C. | G | 70 | .229 | | 10.75 | 9.85 | | 20.60 | 27 |
| 6420 | 8" wide, studs 12" O.C. | G | 34 | .471 | | 23.50 | 20.50 | | 44 | 57 |
| 6430 | 16" O.C. | G | 50 | .320 | | 18.40 | 13.80 | | 32.20 | 41 |
| 6440 | 24" O.C. | G | 69 | .232 | | 13.25 | 10 | | 23.25 | 30 |
| 6530 | 14 ga x 3-5/8" wide, studs 12" O.C. | G | 34 | .471 | | 18.20 | 20.50 | | 38.70 | 51 |
| 6540 | 16" O.C. | G | 48 | .333 | | 14.20 | 14.35 | | 28.55 | 37.50 |
| 6550 | 24" O.C. | G | 65 | .246 | | 10.20 | 10.60 | | 20.80 | 27.50 |
| 6560 | 4" wide, studs 12" O.C. | G | 33 | .485 | | 19.20 | 21 | | 40.20 | 53 |
| 6570 | 16" O.C. | G | 47 | .340 | | 15 | 14.65 | | 29.65 | 39 |
| 6580 | 24" O.C. | G | 64 | .250 | | 10.80 | 10.75 | | 21.55 | 28.50 |
| 6730 | 12 ga x 3-5/8" wide, studs 12" O.C. | G | 31 | .516 | | 25 | 22 | | 47 | 61.50 |
| 6740 | 16" O.C. | G | 43 | .372 | | 19.40 | 16 | | 35.40 | 46 |
| 6750 | 24" O.C. | G | 59 | .271 | | 13.70 | 11.65 | | 25.35 | 33 |
| 6760 | 4" wide, studs 12" O.C. | G | 30 | .533 | | 27 | 23 | | 50 | 64.50 |
| 6770 | 16" O.C. | G | 42 | .381 | | 20.50 | 16.40 | | 36.90 | 48 |
| 6780 | 24" O.C. | G | 58 | .276 | | 14.60 | 11.90 | | 26.50 | 34.50 |
| 7390 | 16' high walls, 16 ga x 6" wide, studs 12" O.C. | G | 33 | .485 | | 24.50 | 21 | | 45.50 | 59 |
| 7400 | 16" O.C. | G | 48 | .333 | | 19.10 | 14.35 | | 33.45 | 43 |
| 7410 | 24" O.C. | G | 67 | .239 | | 13.50 | 10.30 | | 23.80 | 30.50 |
| 7420 | 8" wide, studs 12" O.C. | G | 32 | .500 | | 30.50 | 21.50 | | 52 | 66.50 |
| 7430 | 16" O.C. | G | 47 | .340 | | 23.50 | 14.65 | | 38.15 | 48.50 |
| 7440 | 24" O.C. | G | 66 | .242 | | 16.65 | 10.45 | | 27.10 | 34.50 |
| 7560 | 14 ga x 4" wide, studs 12" O.C. | G | 31 | .516 | | 25 | 22 | | 47 | 61.50 |
| 7570 | 16" O.C. | G | 45 | .356 | | 19.20 | 15.30 | | 34.50 | 44.50 |
| 7580 | 24" O.C. | G | 61 | .262 | | 13.60 | 11.30 | | 24.90 | 32.50 |
| 7590 | 6" wide, studs 12" O.C. | G | 30 | .533 | | 31 | 23 | | 54 | 69.50 |
| 7600 | 16" O.C. | G | 44 | .364 | | 24 | 15.65 | | 39.65 | 50.50 |
| 7610 | 24" O.C. | G | 60 | .267 | | 17.10 | 11.50 | | 28.60 | 36.50 |
| 7760 | 12 ga x 4" wide, studs 12" O.C. | G | 29 | .552 | | 35 | 24 | | 59 | 75 |
| 7770 | 16" O.C. | G | 40 | .400 | | 27 | 17.20 | | 44.20 | 56 |
| 7780 | 24" O.C. | G | 55 | .291 | | 18.70 | 12.50 | | 31.20 | 39.50 |
| 7790 | 6" wide, studs 12" O.C. | G | 28 | .571 | | 44 | 24.50 | | 68.50 | 86 |
| 7800 | 16" O.C. | G | 39 | .410 | | 34 | 17.65 | | 51.65 | 64.50 |
| 7810 | 24" O.C. | G | 54 | .296 | | 23.50 | 12.75 | | 36.25 | 45.50 |

# 05 41 Structural Metal Stud Framing

## 05 41 13 – Load-Bearing Metal Stud Framing

| 05 41 13.30 Framing, Stud Walls | | Crew | Daily Output | Labor-Hours | Unit | Material | 2011 Bare Costs Labor | Equipment | Total | Total Incl O&P |
|---|---|---|---|---|---|---|---|---|---|---|
| 8590 | 20' high walls, 14 ga x 6" wide, studs 12" O.C. | **G** 2 Carp | 29 | .552 | L.F. | 38 | 24 | | 62 | 78.50 |
| 8600 | 16" O.C. | **G** | 42 | .381 | | 29.50 | 16.40 | | 45.90 | 57.50 |
| 8610 | 24" O.C. | **G** | 57 | .281 | | 20.50 | 12.10 | | 32.60 | 41 |
| 8620 | 8" wide, studs 12" O.C. | **G** | 28 | .571 | | 46.50 | 24.50 | | 71 | 89 |
| 8630 | 16" O.C. | **G** | 41 | .390 | | 36 | 16.80 | | 52.80 | 65.50 |
| 8640 | 24" O.C. | **G** | 56 | .286 | | 25 | 12.30 | | 37.30 | 46.50 |
| 8790 | 12 ga x 6" wide, studs 12" O.C. | **G** | 27 | .593 | | 54.50 | 25.50 | | 80 | 99 |
| 8800 | 16" O.C. | **G** | 37 | .432 | | 41.50 | 18.60 | | 60.10 | 74.50 |
| 8810 | 24" O.C. | **G** | 51 | .314 | | 28.50 | 13.50 | | 42 | 52 |
| 8820 | 8" wide, studs 12" O.C. | **G** | 26 | .615 | | 66 | 26.50 | | 92.50 | 113 |
| 8830 | 16" O.C. | **G** | 36 | .444 | | 50.50 | 19.15 | | 69.65 | 85 |
| 8840 | 24" O.C. | **G** | 50 | .320 | | 35 | 13.80 | | 48.80 | 59.50 |

# 05 42 Cold-Formed Metal Joist Framing

## 05 42 13 – Cold-Formed Metal Floor Joist Framing

### 05 42 13.05 Bracing

| | | Crew | Daily Output | Labor-Hours | Unit | Material | 2011 Bare Costs Labor | Equipment | Total | Total Incl O&P |
|---|---|---|---|---|---|---|---|---|---|---|
| 0010 | **BRACING**, continuous, per row, top & bottom | | | | | | | | | |
| 0015 | Made from recycled materials | **G** | | | | | | | | |
| 0120 | Flat strap, 20 ga x 2" wide, joists at 12" O.C. | **G** 1 Carp | 4.67 | 1.713 | C.L.F. | 48.50 | 74 | | 122.50 | 166 |
| 0130 | 16" O.C. | **G** | 5.33 | 1.501 | | 46.50 | 64.50 | | 111 | 151 |
| 0140 | 24" O.C. | **G** | 6.66 | 1.201 | | 45 | 51.50 | | 96.50 | 129 |
| 0150 | 18 ga x 2" wide, joists at 12" O.C. | **G** | 4 | 2 | | 63 | 86 | | 149 | 202 |
| 0160 | 16" O.C. | **G** | 4.67 | 1.713 | | 62 | 74 | | 136 | 182 |
| 0170 | 24" O.C. | **G** | 5.33 | 1.501 | | 61 | 64.50 | | 125.50 | 166 |

### 05 42 13.10 Bridging

| | | Crew | Daily Output | Labor-Hours | Unit | Material | 2011 Bare Costs Labor | Equipment | Total | Total Incl O&P |
|---|---|---|---|---|---|---|---|---|---|---|
| 0010 | **BRIDGING**, solid between joists w/1-1/4" leg track, per joist bay | | | | | | | | | |
| 0015 | Made from recycled materials | **G** | | | | | | | | |
| 0230 | Joists 12" O.C., 18 ga track x 6" wide | **G** 1 Carp | 80 | .100 | Ea. | 1.26 | 4.31 | | 5.57 | 8 |
| 0240 | 8" wide | **G** | 75 | .107 | | 1.57 | 4.59 | | 6.16 | 8.80 |
| 0250 | 10" wide | **G** | 70 | .114 | | 1.96 | 4.92 | | 6.88 | 9.70 |
| 0260 | 12" wide | **G** | 65 | .123 | | 2.23 | 5.30 | | 7.53 | 10.60 |
| 0330 | 16 ga track x 6" wide | **G** | 70 | .114 | | 1.60 | 4.92 | | 6.52 | 9.30 |
| 0340 | 8" wide | **G** | 65 | .123 | | 2.01 | 5.30 | | 7.31 | 10.35 |
| 0350 | 10" wide | **G** | 60 | .133 | | 2.48 | 5.75 | | 8.23 | 11.55 |
| 0360 | 12" wide | **G** | 55 | .145 | | 2.86 | 6.25 | | 9.11 | 12.75 |
| 0440 | 14 ga track x 8" wide | **G** | 60 | .133 | | 2.51 | 5.75 | | 8.26 | 11.55 |
| 0450 | 10" wide | **G** | 55 | .145 | | 3.11 | 6.25 | | 9.36 | 13.05 |
| 0460 | 12" wide | **G** | 50 | .160 | | 3.59 | 6.90 | | 10.49 | 14.50 |
| 0550 | 12 ga track x 10" wide | **G** | 45 | .178 | | 4.54 | 7.65 | | 12.19 | 16.75 |
| 0560 | 12" wide | **G** | 40 | .200 | | 4.66 | 8.60 | | 13.26 | 18.30 |
| 1230 | 16" O.C., 18 ga track x 6" wide | **G** | 80 | .100 | | 1.62 | 4.31 | | 5.93 | 8.40 |
| 1240 | 8" wide | **G** | 75 | .107 | | 2.01 | 4.59 | | 6.60 | 9.25 |
| 1250 | 10" wide | **G** | 70 | .114 | | 2.51 | 4.92 | | 7.43 | 10.30 |
| 1260 | 12" wide | **G** | 65 | .123 | | 2.85 | 5.30 | | 8.15 | 11.30 |
| 1330 | 16 ga track x 6" wide | **G** | 70 | .114 | | 2.05 | 4.92 | | 6.97 | 9.80 |
| 1340 | 8" wide | **G** | 65 | .123 | | 2.57 | 5.30 | | 7.87 | 11 |
| 1350 | 10" wide | **G** | 60 | .133 | | 3.18 | 5.75 | | 8.93 | 12.30 |
| 1360 | 12" wide | **G** | 55 | .145 | | 3.66 | 6.25 | | 9.91 | 13.65 |
| 1440 | 14 ga track x 8" wide | **G** | 60 | .133 | | 3.22 | 5.75 | | 8.97 | 12.35 |
| 1450 | 10" wide | **G** | 55 | .145 | | 3.99 | 6.25 | | 10.24 | 14 |
| 1460 | 12" wide | **G** | 50 | .160 | | 4.60 | 6.90 | | 11.50 | 15.60 |

## 05 42 13 – Cold-Formed Metal Floor Joist Framing

### 05 42 13.10 Bridging

| | | | Crew | Daily Output | Labor-Hours | Unit | Material | 2011 Bare Costs Labor | Equipment | Total | Total Incl O&P |
|---|---|---|---|---|---|---|---|---|---|---|---|
| 1550 | 12 ga track x 10" wide | G | 1 Carp | 45 | .178 | Ea. | 5.80 | 7.65 | | 13.45 | 18.15 |
| 1560 | 12" wide | G | | 40 | .200 | | 5.95 | 8.60 | | 14.55 | 19.75 |
| 2230 | 24" O.C., 18 ga track x 6" wide | G | | 80 | .100 | | 2.34 | 4.31 | | 6.65 | 9.15 |
| 2240 | 8" wide | G | | 75 | .107 | | 2.91 | 4.59 | | 7.50 | 10.25 |
| 2250 | 10" wide | G | | 70 | .114 | | 3.63 | 4.92 | | 8.55 | 11.55 |
| 2260 | 12" wide | G | | 65 | .123 | | 4.13 | 5.30 | | 9.43 | 12.70 |
| 2330 | 16 ga track x 6" wide | G | | 70 | .114 | | 2.96 | 4.92 | | 7.88 | 10.80 |
| 2340 | 8" wide | G | | 65 | .123 | | 3.72 | 5.30 | | 9.02 | 12.25 |
| 2350 | 10" wide | G | | 60 | .133 | | 4.61 | 5.75 | | 10.36 | 13.85 |
| 2360 | 12" wide | G | | 55 | .145 | | 5.30 | 6.25 | | 11.55 | 15.45 |
| 2440 | 14 ga track x 8" wide | G | | 60 | .133 | | 4.65 | 5.75 | | 10.40 | 13.90 |
| 2450 | 10" wide | G | | 55 | .145 | | 5.80 | 6.25 | | 12.05 | 15.95 |
| 2460 | 12" wide | G | | 50 | .160 | | 6.65 | 6.90 | | 13.55 | 17.90 |
| 2550 | 12 ga track x 10" wide | G | | 45 | .178 | | 8.45 | 7.65 | | 16.10 | 21 |
| 2560 | 12" wide | G | | 40 | .200 | | 8.65 | 8.60 | | 17.25 | 22.50 |

### 05 42 13.25 Framing, Band Joist

| | | | Crew | Daily Output | Labor-Hours | Unit | Material | 2011 Bare Costs Labor | Equipment | Total | Total Incl O&P |
|---|---|---|---|---|---|---|---|---|---|---|---|
| 0010 | **FRAMING, BAND JOIST** (track) fastened to bearing wall | | | | | | | | | | |
| 0015 | Made from recycled materials | G | | | | | | | | | |
| 0220 | 18 ga track x 6" deep | G | 2 Carp | 1000 | .016 | L.F. | 1.03 | .69 | | 1.72 | 2.19 |
| 0230 | 8" deep | G | | 920 | .017 | | 1.28 | .75 | | 2.03 | 2.56 |
| 0240 | 10" deep | G | | 860 | .019 | | 1.60 | .80 | | 2.40 | 2.99 |
| 0320 | 16 ga track x 6" deep | G | | 900 | .018 | | 1.30 | .77 | | 2.07 | 2.60 |
| 0330 | 8" deep | G | | 840 | .019 | | 1.64 | .82 | | 2.46 | 3.06 |
| 0340 | 10" deep | G | | 780 | .021 | | 2.03 | .88 | | 2.91 | 3.58 |
| 0350 | 12" deep | G | | 740 | .022 | | 2.33 | .93 | | 3.26 | 3.99 |
| 0430 | 14 ga track x 8" deep | G | | 750 | .021 | | 2.05 | .92 | | 2.97 | 3.66 |
| 0440 | 10" deep | G | | 720 | .022 | | 2.54 | .96 | | 3.50 | 4.27 |
| 0450 | 12" deep | G | | 700 | .023 | | 2.93 | .98 | | 3.91 | 4.73 |
| 0540 | 12 ga track x 10" deep | G | | 670 | .024 | | 3.71 | 1.03 | | 4.74 | 5.65 |
| 0550 | 12" deep | G | | 650 | .025 | | 3.80 | 1.06 | | 4.86 | 5.80 |

### 05 42 13.30 Framing, Boxed Headers/Beams

| | | | Crew | Daily Output | Labor-Hours | Unit | Material | 2011 Bare Costs Labor | Equipment | Total | Total Incl O&P |
|---|---|---|---|---|---|---|---|---|---|---|---|
| 0010 | **FRAMING, BOXED HEADERS/BEAMS** | | | | | | | | | | |
| 0015 | Made from recycled materials | G | | | | | | | | | |
| 0200 | Double, 18 ga x 6" deep | G | 2 Carp | 220 | .073 | L.F. | 4.40 | 3.13 | | 7.53 | 9.65 |
| 0210 | 8" deep | G | | 210 | .076 | | 4.84 | 3.28 | | 8.12 | 10.40 |
| 0220 | 10" deep | G | | 200 | .080 | | 5.90 | 3.44 | | 9.34 | 11.80 |
| 0230 | 12" deep | G | | 190 | .084 | | 6.45 | 3.63 | | 10.08 | 12.60 |
| 0300 | 16 ga x 8" deep | G | | 180 | .089 | | 5.60 | 3.83 | | 9.43 | 12 |
| 0310 | 10" deep | G | | 170 | .094 | | 6.75 | 4.05 | | 10.80 | 13.60 |
| 0320 | 12" deep | G | | 160 | .100 | | 7.30 | 4.31 | | 11.61 | 14.65 |
| 0400 | 14 ga x 10" deep | G | | 140 | .114 | | 7.75 | 4.92 | | 12.67 | 16.10 |
| 0410 | 12" deep | G | | 130 | .123 | | 8.50 | 5.30 | | 13.80 | 17.50 |
| 0500 | 12 ga x 10" deep | G | | 110 | .145 | | 10.25 | 6.25 | | 16.50 | 21 |
| 0510 | 12" deep | G | | 100 | .160 | | 11.35 | 6.90 | | 18.25 | 23 |
| 1210 | Triple, 18 ga x 8" deep | G | | 170 | .094 | | 7 | 4.05 | | 11.05 | 13.90 |
| 1220 | 10" deep | G | | 165 | .097 | | 8.45 | 4.17 | | 12.62 | 15.70 |
| 1230 | 12" deep | G | | 160 | .100 | | 9.25 | 4.31 | | 13.56 | 16.80 |
| 1300 | 16 ga x 8" deep | G | | 145 | .110 | | 8.10 | 4.75 | | 12.85 | 16.20 |
| 1310 | 10" deep | G | | 140 | .114 | | 9.70 | 4.92 | | 14.62 | 18.20 |
| 1320 | 12" deep | G | | 135 | .119 | | 10.55 | 5.10 | | 15.65 | 19.50 |
| 1400 | 14 ga x 10" deep | G | | 115 | .139 | | 11.25 | 6 | | 17.25 | 21.50 |
| 1410 | 12" deep | G | | 110 | .145 | | 12.35 | 6.25 | | 18.60 | 23 |

# 05 42 Cold-Formed Metal Joist Framing

## 05 42 13 – Cold-Formed Metal Floor Joist Framing

| 05 42 13.30 Framing, Boxed Headers/Beams | | Crew | Daily Output | Labor-Hours | Unit | Material | 2011 Bare Costs Labor | Equipment | Total | Total Incl O&P |
|---|---|---|---|---|---|---|---|---|---|---|
| 1500 | 12 ga x 10" deep | G 2 Carp | 90 | .178 | L.F. | 15 | 7.65 | | 22.65 | 28.50 |
| 1510 | 12" deep | G ↓ | 85 | .188 | ↓ | 16.60 | 8.10 | | 24.70 | 30.50 |

### 05 42 13.40 Framing, Joists

| | | | | | | | | | | |
|---|---|---|---|---|---|---|---|---|---|---|
| 0010 | **FRAMING, JOISTS**, no band joists (track), web stiffeners, headers, | | | | | | | | | |
| 0020 | Beams, bridging or bracing | | | | | | | | | |
| 0025 | Made from recycled materials | G | | | | | | | | |
| 0030 | Joists (2" flange) and fasteners, materials only | | | | | | | | | |
| 0220 | 18 ga x 6" deep | G | | | L.F. | 1.35 | | | 1.35 | 1.49 |
| 0230 | 8" deep | G | | | | 1.59 | | | 1.59 | 1.74 |
| 0240 | 10" deep | G | | | | 1.87 | | | 1.87 | 2.06 |
| 0320 | 16 ga x 6" deep | G | | | | 1.65 | | | 1.65 | 1.81 |
| 0330 | 8" deep | G | | | | 1.97 | | | 1.97 | 2.17 |
| 0340 | 10" deep | G | | | | 2.31 | | | 2.31 | 2.54 |
| 0350 | 12" deep | G | | | | 2.61 | | | 2.61 | 2.88 |
| 0430 | 14 ga x 8" deep | G | | | | 2.47 | | | 2.47 | 2.71 |
| 0440 | 10" deep | G | | | | 2.86 | | | 2.86 | 3.14 |
| 0450 | 12" deep | G | | | | 3.24 | | | 3.24 | 3.57 |
| 0540 | 12 ga x 10" deep | G | | | | 4.16 | | | 4.16 | 4.57 |
| 0550 | 12" deep | G | | | ↓ | 4.73 | | | 4.73 | 5.20 |
| 1010 | Installation of joists to band joists, beams & headers, labor only | | | | | | | | | |
| 1220 | 18 ga x 6" deep | | 2 Carp | 110 | .145 | Ea. | | 6.25 | | 6.25 | 9.60 |
| 1230 | 8" deep | | | 90 | .178 | | | 7.65 | | 7.65 | 11.75 |
| 1240 | 10" deep | | | 80 | .200 | | | 8.60 | | 8.60 | 13.20 |
| 1320 | 16 ga x 6" deep | | | 95 | .168 | | | 7.25 | | 7.25 | 11.10 |
| 1330 | 8" deep | | | 70 | .229 | | | 9.85 | | 9.85 | 15.10 |
| 1340 | 10" deep | | | 60 | .267 | | | 11.50 | | 11.50 | 17.60 |
| 1350 | 12" deep | | | 55 | .291 | | | 12.50 | | 12.50 | 19.20 |
| 1430 | 14 ga x 8" deep | | | 65 | .246 | | | 10.60 | | 10.60 | 16.25 |
| 1440 | 10" deep | | | 45 | .356 | | | 15.30 | | 15.30 | 23.50 |
| 1450 | 12" deep | | | 35 | .457 | | | 19.70 | | 19.70 | 30 |
| 1540 | 12 ga x 10" deep | | | 40 | .400 | | | 17.20 | | 17.20 | 26.50 |
| 1550 | 12" deep | | ↓ | 30 | .533 | ↓ | | 23 | | 23 | 35 |

### 05 42 13.45 Framing, Web Stiffeners

| | | | | | | | | | | |
|---|---|---|---|---|---|---|---|---|---|---|
| 0010 | **FRAMING, WEB STIFFENERS** at joist bearing, fabricated from | | | | | | | | | |
| 0020 | Stud piece (1-5/8" flange) to stiffen joist (2" flange) | | | | | | | | | |
| 0025 | Made from recycled materials | G | | | | | | | | |
| 2120 | For 6" deep joist, with 18 ga x 2-1/2" stud | G | 1 Carp | 120 | .067 | Ea. | 1.61 | 2.87 | | 4.48 | 6.15 |
| 2130 | 3-5/8" stud | G | | 110 | .073 | | 1.76 | 3.13 | | 4.89 | 6.75 |
| 2140 | 4" stud | G | | 105 | .076 | | 1.71 | 3.28 | | 4.99 | 6.95 |
| 2150 | 6" stud | G | | 100 | .080 | | 1.86 | 3.44 | | 5.30 | 7.35 |
| 2160 | 8" stud | G | | 95 | .084 | | 1.90 | 3.63 | | 5.53 | 7.65 |
| 2220 | 8" deep joist, with 2-1/2" stud | G | | 120 | .067 | | 1.76 | 2.87 | | 4.63 | 6.35 |
| 2230 | 3-5/8" stud | G | | 110 | .073 | | 1.90 | 3.13 | | 5.03 | 6.90 |
| 2240 | 4" stud | G | | 105 | .076 | | 1.87 | 3.28 | | 5.15 | 7.10 |
| 2250 | 6" stud | G | | 100 | .080 | | 2.04 | 3.44 | | 5.48 | 7.55 |
| 2260 | 8" stud | G | | 95 | .084 | | 2.19 | 3.63 | | 5.82 | 7.95 |
| 2320 | 10" deep joist, with 2-1/2" stud | G | | 110 | .073 | | 2.48 | 3.13 | | 5.61 | 7.55 |
| 2330 | 3-5/8" stud | G | | 100 | .080 | | 2.71 | 3.44 | | 6.15 | 8.30 |
| 2340 | 4" stud | G | | 95 | .084 | | 2.69 | 3.63 | | 6.32 | 8.50 |
| 2350 | 6" stud | G | | 90 | .089 | | 2.91 | 3.83 | | 6.74 | 9.05 |
| 2360 | 8" stud | G | | 85 | .094 | | 2.93 | 4.05 | | 6.98 | 9.45 |
| 2420 | 12" deep joist, with 2-1/2" stud | G | ↓ | 110 | .073 | ↓ | 2.63 | 3.13 | | 5.76 | 7.70 |

# 05 42 Cold-Formed Metal Joist Framing

## 05 42 13 – Cold-Formed Metal Floor Joist Framing

### 05 42 13.45 Framing, Web Stiffeners

| | | | Crew | Daily Output | Labor-Hours | Unit | Material | 2011 Bare Costs Labor | Equipment | Total | Total Incl O&P |
|---|---|---|---|---|---|---|---|---|---|---|---|
| 2430 | 3-5/8" stud | G | 1 Carp | 100 | .080 | Ea. | 2.84 | 3.44 | | 6.28 | 8.40 |
| 2440 | 4" stud | G | | 95 | .084 | | 2.79 | 3.63 | | 6.42 | 8.60 |
| 2450 | 6" stud | G | | 90 | .089 | | 3.05 | 3.83 | | 6.88 | 9.20 |
| 2460 | 8" stud | G | | 85 | .094 | | 3.26 | 4.05 | | 7.31 | 9.80 |
| 3130 | For 6" deep joist, with 16 ga x 3-5/8" stud | G | | 100 | .080 | | 1.86 | 3.44 | | 5.30 | 7.35 |
| 3140 | 4" stud | G | | 95 | .084 | | 1.85 | 3.63 | | 5.48 | 7.60 |
| 3150 | 6" stud | G | | 90 | .089 | | 2.02 | 3.83 | | 5.85 | 8.05 |
| 3160 | 8" stud | G | | 85 | .094 | | 2.14 | 4.05 | | 6.19 | 8.55 |
| 3230 | 8" deep joist, with 3-5/8" stud | G | | 100 | .080 | | 2.06 | 3.44 | | 5.50 | 7.55 |
| 3240 | 4" stud | G | | 95 | .084 | | 2.03 | 3.63 | | 5.66 | 7.80 |
| 3250 | 6" stud | G | | 90 | .089 | | 2.24 | 3.83 | | 6.07 | 8.30 |
| 3260 | 8" stud | G | | 85 | .094 | | 2.41 | 4.05 | | 6.46 | 8.85 |
| 3330 | 10" deep joist, with 3-5/8" stud | G | | 85 | .094 | | 2.82 | 4.05 | | 6.87 | 9.30 |
| 3340 | 4" stud | G | | 80 | .100 | | 2.88 | 4.31 | | 7.19 | 9.75 |
| 3350 | 6" stud | G | | 75 | .107 | | 3.11 | 4.59 | | 7.70 | 10.50 |
| 3360 | 8" stud | G | | 70 | .114 | | 3.26 | 4.92 | | 8.18 | 11.15 |
| 3430 | 12" deep joist, with 3-5/8" stud | G | | 85 | .094 | | 3.07 | 4.05 | | 7.12 | 9.60 |
| 3440 | 4" stud | G | | 80 | .100 | | 3.02 | 4.31 | | 7.33 | 9.95 |
| 3450 | 6" stud | G | | 75 | .107 | | 3.34 | 4.59 | | 7.93 | 10.70 |
| 3460 | 8" stud | G | | 70 | .114 | | 3.59 | 4.92 | | 8.51 | 11.50 |
| 4230 | For 8" deep joist, with 14 ga x 3-5/8" stud | G | | 90 | .089 | | 2.67 | 3.83 | | 6.50 | 8.80 |
| 4240 | 4" stud | G | | 85 | .094 | | 2.72 | 4.05 | | 6.77 | 9.20 |
| 4250 | 6" stud | G | | 80 | .100 | | 2.95 | 4.31 | | 7.26 | 9.85 |
| 4260 | 8" stud | G | | 75 | .107 | | 3.15 | 4.59 | | 7.74 | 10.50 |
| 4330 | 10" deep joist, with 3-5/8" stud | G | | 75 | .107 | | 3.75 | 4.59 | | 8.34 | 11.20 |
| 4340 | 4" stud | G | | 70 | .114 | | 3.72 | 4.92 | | 8.64 | 11.65 |
| 4350 | 6" stud | G | | 65 | .123 | | 4.09 | 5.30 | | 9.39 | 12.65 |
| 4360 | 8" stud | G | | 60 | .133 | | 4.26 | 5.75 | | 10.01 | 13.50 |
| 4430 | 12" deep joist, with 3-5/8" stud | G | | 75 | .107 | | 3.99 | 4.59 | | 8.58 | 11.45 |
| 4440 | 4" stud | G | | 70 | .114 | | 4.06 | 4.92 | | 8.98 | 12 |
| 4450 | 6" stud | G | | 65 | .123 | | 4.40 | 5.30 | | 9.70 | 13 |
| 4460 | 8" stud | G | | 60 | .133 | | 4.71 | 5.75 | | 10.46 | 14 |
| 5330 | For 10" deep joist, with 12 ga x 3-5/8" stud | G | | 65 | .123 | | 3.96 | 5.30 | | 9.26 | 12.50 |
| 5340 | 4" stud | G | | 60 | .133 | | 4.06 | 5.75 | | 9.81 | 13.25 |
| 5350 | 6" stud | G | | 55 | .145 | | 4.48 | 6.25 | | 10.73 | 14.55 |
| 5360 | 8" stud | G | | 50 | .160 | | 4.90 | 6.90 | | 11.80 | 15.95 |
| 5430 | 12" deep joist, with 3-5/8" stud | G | | 65 | .123 | | 4.39 | 5.30 | | 9.69 | 13 |
| 5440 | 4" stud | G | | 60 | .133 | | 4.28 | 5.75 | | 10.03 | 13.50 |
| 5450 | 6" stud | G | | 55 | .145 | | 4.88 | 6.25 | | 11.13 | 14.95 |
| 5460 | 8" stud | G | | 50 | .160 | | 5.60 | 6.90 | | 12.50 | 16.70 |

## 05 42 23 – Cold-Formed Metal Roof Joist Framing

### 05 42 23.05 Framing, Bracing

| | | | Crew | Daily Output | Labor-Hours | Unit | Material | 2011 Bare Costs Labor | Equipment | Total | Total Incl O&P |
|---|---|---|---|---|---|---|---|---|---|---|---|
| 0010 | **FRAMING, BRACING** | | | | | | | | | | |
| 0015 | Made from recycled materials | G | | | | | | | | | |
| 0020 | Continuous bracing, per row | | | | | | | | | | |
| 0100 | 16 ga x 1-1/2" channel thru rafters/trusses @ 16" O.C. | G | 1 Carp | 4.50 | 1.778 | C.L.F. | 41 | 76.50 | | 117.50 | 162 |
| 0120 | 24" O.C. | G | | 6 | 1.333 | | 41 | 57.50 | | 98.50 | 133 |
| 0300 | 2" x 2" angle x 18 ga, rafters/trusses @ 16" O.C. | G | | 6 | 1.333 | | 64 | 57.50 | | 121.50 | 158 |
| 0320 | 24" O.C. | G | | 8 | 1 | | 64 | 43 | | 107 | 136 |
| 0400 | 16 ga, rafters/trusses @ 16" O.C. | G | | 4.50 | 1.778 | | 81.50 | 76.50 | | 158 | 207 |
| 0420 | 24" O.C. | G | | 6.50 | 1.231 | | 81.50 | 53 | | 134.50 | 171 |

# 05 42 Cold-Formed Metal Joist Framing

## 05 42 23 – Cold-Formed Metal Roof Joist Framing

### 05 42 23.10 Framing, Bridging

| 05 42 23.10 Framing, Bridging | | Crew | Daily Output | Labor-Hours | Unit | Material | 2011 Bare Costs Labor | Equipment | Total | Total Incl O&P |
|---|---|---|---|---|---|---|---|---|---|---|
| 0010 | **FRAMING, BRIDGING** | | | | | | | | | |
| 0015 | Made from recycled materials | G | | | | | | | | |
| 0020 | Solid, between rafters w/1-1/4" leg track, per rafter bay | | | | | | | | | |
| 1200 | Rafters 16" O.C., 18 ga x 4" deep | G | 1 Carp | 60 | .133 | Ea. | 1.25 | 5.75 | 7 | 10.20 |
| 1210 | 6" deep | G | | 57 | .140 | | 1.62 | 6.05 | 7.67 | 11.05 |
| 1220 | 8" deep | G | | 55 | .145 | | 2.01 | 6.25 | 8.26 | 11.80 |
| 1230 | 10" deep | G | | 52 | .154 | | 2.51 | 6.60 | 9.11 | 12.90 |
| 1240 | 12" deep | G | | 50 | .160 | | 2.85 | 6.90 | 9.75 | 13.70 |
| 2200 | 24" O.C., 18 ga x 4" deep | G | | 60 | .133 | | 1.81 | 5.75 | 7.56 | 10.80 |
| 2210 | 6" deep | G | | 57 | .140 | | 2.34 | 6.05 | 8.39 | 11.80 |
| 2220 | 8" deep | G | | 55 | .145 | | 2.91 | 6.25 | 9.16 | 12.80 |
| 2230 | 10" deep | G | | 52 | .154 | | 3.63 | 6.60 | 10.23 | 14.15 |
| 2240 | 12" deep | G | | 50 | .160 | | 4.13 | 6.90 | 11.03 | 15.10 |

### 05 42 23.50 Framing, Parapets

| 05 42 23.50 Framing, Parapets | | Crew | Daily Output | Labor-Hours | Unit | Material | Labor | Equipment | Total | Total Incl O&P |
|---|---|---|---|---|---|---|---|---|---|---|
| 0010 | **FRAMING, PARAPETS** | | | | | | | | | |
| 0015 | Made from recycled materials | G | | | | | | | | |
| 0100 | 3' high installed on 1st story, 18 ga x 4" wide studs, 12" O.C. | G | 2 Carp | 100 | .160 | L.F. | 4.56 | 6.90 | 11.46 | 15.55 |
| 0110 | 16" O.C. | G | | 150 | .107 | | 3.89 | 4.59 | 8.48 | 11.35 |
| 0120 | 24" O.C. | G | | 200 | .080 | | 3.21 | 3.44 | 6.65 | 8.85 |
| 0200 | 6" wide studs, 12" O.C. | G | | 100 | .160 | | 5.80 | 6.90 | 12.70 | 16.90 |
| 0210 | 16" O.C. | G | | 150 | .107 | | 4.94 | 4.59 | 9.53 | 12.50 |
| 0220 | 24" O.C. | G | | 200 | .080 | | 4.10 | 3.44 | 7.54 | 9.80 |
| 1100 | Installed on 2nd story, 18 ga x 4" wide studs, 12" O.C. | G | | 95 | .168 | | 4.56 | 7.25 | 11.81 | 16.10 |
| 1110 | 16" O.C. | G | | 145 | .110 | | 3.89 | 4.75 | 8.64 | 11.60 |
| 1120 | 24" O.C. | G | | 190 | .084 | | 3.21 | 3.63 | 6.84 | 9.10 |
| 1200 | 6" wide studs, 12" O.C. | G | | 95 | .168 | | 5.80 | 7.25 | 13.05 | 17.45 |
| 1210 | 16" O.C. | G | | 145 | .110 | | 4.94 | 4.75 | 9.69 | 12.75 |
| 1220 | 24" O.C. | G | | 190 | .084 | | 4.10 | 3.63 | 7.73 | 10.05 |
| 2100 | Installed on gable, 18 ga x 4" wide studs, 12" O.C. | G | | 85 | .188 | | 4.56 | 8.10 | 12.66 | 17.45 |
| 2110 | 16" O.C. | G | | 130 | .123 | | 3.89 | 5.30 | 9.19 | 12.45 |
| 2120 | 24" O.C. | G | | 170 | .094 | | 3.21 | 4.05 | 7.26 | 9.75 |
| 2200 | 6" wide studs, 12" O.C. | G | | 85 | .188 | | 5.80 | 8.10 | 13.90 | 18.80 |
| 2210 | 16" O.C. | G | | 130 | .123 | | 4.94 | 5.30 | 10.24 | 13.60 |
| 2220 | 24" O.C. | G | | 170 | .094 | | 4.10 | 4.05 | 8.15 | 10.70 |

### 05 42 23.60 Framing, Roof Rafters

| 05 42 23.60 Framing, Roof Rafters | | Crew | Daily Output | Labor-Hours | Unit | Material | Labor | Equipment | Total | Total Incl O&P |
|---|---|---|---|---|---|---|---|---|---|---|
| 0010 | **FRAMING, ROOF RAFTERS** | | | | | | | | | |
| 0015 | Made from recycled materials | G | | | | | | | | |
| 0100 | Boxed ridge beam, double, 18 ga x 6" deep | G | 2 Carp | 160 | .100 | L.F. | 4.40 | 4.31 | 8.71 | 11.45 |
| 0110 | 8" deep | G | | 150 | .107 | | 4.84 | 4.59 | 9.43 | 12.40 |
| 0120 | 10" deep | G | | 140 | .114 | | 5.90 | 4.92 | 10.82 | 14.05 |
| 0130 | 12" deep | G | | 130 | .123 | | 6.45 | 5.30 | 11.75 | 15.20 |
| 0200 | 16 ga x 6" deep | G | | 150 | .107 | | 4.96 | 4.59 | 9.55 | 12.50 |
| 0210 | 8" deep | G | | 140 | .114 | | 5.60 | 4.92 | 10.52 | 13.70 |
| 0220 | 10" deep | G | | 130 | .123 | | 6.75 | 5.30 | 12.05 | 15.55 |
| 0230 | 12" deep | G | | 120 | .133 | | 7.30 | 5.75 | 13.05 | 16.85 |
| 1100 | Rafters, 2" flange, material only, 18 ga x 6" deep | G | | | | | 1.35 | | 1.35 | 1.49 |
| 1110 | 8" deep | G | | | | | 1.59 | | 1.59 | 1.74 |
| 1120 | 10" deep | G | | | | | 1.87 | | 1.87 | 2.06 |
| 1130 | 12" deep | G | | | | | 2.15 | | 2.15 | 2.37 |
| 1200 | 16 ga x 6" deep | G | | | | | 1.65 | | 1.65 | 1.81 |
| 1210 | 8" deep | G | | | | | 1.97 | | 1.97 | 2.17 |
| 1220 | 10" deep | G | | | | | 2.31 | | 2.31 | 2.54 |

## 05 42 23 – Cold-Formed Metal Roof Joist Framing

### 05 42 23.60 Framing, Roof Rafters

| | | Crew | Daily Output | Labor-Hours | Unit | Material | 2011 Bare Costs Labor | 2011 Bare Costs Equipment | Total | Total Incl O&P |
|---|---|---|---|---|---|---|---|---|---|---|
| 1230 | 12" deep **G** | | | | L.F. | 2.61 | | | 2.61 | 2.88 |
| 2100 | Installation only, ordinary rafter to 4:12 pitch, 18 ga x 6" deep | 2 Carp | 35 | .457 | Ea. | | 19.70 | | 19.70 | 30 |
| 2110 | 8" deep | | 30 | .533 | | | 23 | | 23 | 35 |
| 2120 | 10" deep | | 25 | .640 | | | 27.50 | | 27.50 | 42.50 |
| 2130 | 12" deep | | 20 | .800 | | | 34.50 | | 34.50 | 53 |
| 2200 | 16 ga x 6" deep | | 30 | .533 | | | 23 | | 23 | 35 |
| 2210 | 8" deep | | 25 | .640 | | | 27.50 | | 27.50 | 42.50 |
| 2220 | 10" deep | | 20 | .800 | | | 34.50 | | 34.50 | 53 |
| 2230 | 12" deep | | 15 | 1.067 | | | 46 | | 46 | 70.50 |
| 8100 | Add to labor, ordinary rafters on steep roofs | | | | | | 25% | | | |
| 8110 | Dormers & complex roofs | | | | | | 50% | | | |
| 8200 | Hip & valley rafters to 4:12 pitch | | | | | | 25% | | | |
| 8210 | Steep roofs | | | | | | 50% | | | |
| 8220 | Dormers & complex roofs | | | | | | 75% | | | |
| 8300 | Hip & valley jack rafters to 4:12 pitch | | | | | | 50% | | | |
| 8310 | Steep roofs | | | | | | 75% | | | |
| 8320 | Dormers & complex roofs | | | | | | 100% | | | |

### 05 42 23.70 Framing, Soffits and Canopies

| | | Crew | Daily Output | Labor-Hours | Unit | Material | 2011 Bare Costs Labor | 2011 Bare Costs Equipment | Total | Total Incl O&P |
|---|---|---|---|---|---|---|---|---|---|---|
| 0010 | **FRAMING, SOFFITS & CANOPIES** | | | | | | | | | |
| 0015 | Made from recycled materials **G** | | | | | | | | | |
| 0130 | Continuous ledger track @ wall, studs @ 16" O.C., 18 ga x 4" wide **G** | 2 Carp | 535 | .030 | L.F. | .84 | 1.29 | | 2.13 | 2.90 |
| 0140 | 6" wide **G** | | 500 | .032 | | 1.08 | 1.38 | | 2.46 | 3.30 |
| 0150 | 8" wide **G** | | 465 | .034 | | 1.34 | 1.48 | | 2.82 | 3.75 |
| 0160 | 10" wide **G** | | 430 | .037 | | 1.67 | 1.60 | | 3.27 | 4.30 |
| 0230 | Studs @ 24" O.C., 18 ga x 4" wide **G** | | 800 | .020 | | .80 | .86 | | 1.66 | 2.20 |
| 0240 | 6" wide **G** | | 750 | .021 | | 1.03 | .92 | | 1.95 | 2.54 |
| 0250 | 8" wide **G** | | 700 | .023 | | 1.28 | .98 | | 2.26 | 2.92 |
| 0260 | 10" wide **G** | | 650 | .025 | | 1.60 | 1.06 | | 2.66 | 3.39 |
| 1000 | Horizontal soffit and canopy members, material only | | | | | | | | | |
| 1030 | 1-5/8" flange studs, 18 ga x 4" deep **G** | | | | L.F. | 1.08 | | | 1.08 | 1.19 |
| 1040 | 6" deep **G** | | | | | 1.36 | | | 1.36 | 1.49 |
| 1050 | 8" deep **G** | | | | | 1.63 | | | 1.63 | 1.80 |
| 1140 | 2" flange joists, 18 ga x 6" deep **G** | | | | | 1.55 | | | 1.55 | 1.70 |
| 1150 | 8" deep **G** | | | | | 1.81 | | | 1.81 | 1.99 |
| 1160 | 10" deep **G** | | | | | 2.14 | | | 2.14 | 2.35 |
| 4030 | Installation only, 18 ga, 1-5/8" flange x 4" deep | 2 Carp | 130 | .123 | Ea. | | 5.30 | | 5.30 | 8.15 |
| 4040 | 6" deep | | 110 | .145 | | | 6.25 | | 6.25 | 9.60 |
| 4050 | 8" deep | | 90 | .178 | | | 7.65 | | 7.65 | 11.75 |
| 4140 | 2" flange, 18 ga x 6" deep | | 110 | .145 | | | 6.25 | | 6.25 | 9.60 |
| 4150 | 8" deep | | 90 | .178 | | | 7.65 | | 7.65 | 11.75 |
| 4160 | 10" deep | | 80 | .200 | | | 8.60 | | 8.60 | 13.20 |
| 6010 | Clips to attach facia to rafter tails, 2" x 2" x 18 ga angle **G** | 1 Carp | 120 | .067 | | .75 | 2.87 | | 3.62 | 5.25 |
| 6020 | 16 ga angle **G** | " | 100 | .080 | | .96 | 3.44 | | 4.40 | 6.35 |

## 05 44 13 – Cold-Formed Metal Roof Trusses

| 05 44 13.60 Framing, Roof Trusses | | Crew | Daily Output | Labor-Hours | Unit | Material | 2011 Bare Costs Labor | Equipment | Total | Total Incl O&P |
|---|---|---|---|---|---|---|---|---|---|---|
| 0010 | **FRAMING, ROOF TRUSSES** | | | | | | | | | |
| 0015 | Made from recycled materials | G | | | | | | | | |
| 0020 | Fabrication of trusses on ground, Fink (W) or King Post, to 4:12 pitch | | | | | | | | | |
| 0120 | 18 ga x 4" chords, 16' span | G | 2 Carp | 12 | 1.333 | Ea. | 50.50 | 57.50 | | 108 | 144 |
| 0130 | 20' span | G | | 11 | 1.455 | | 63 | 62.50 | | 125.50 | 166 |
| 0140 | 24' span | G | | 11 | 1.455 | | 75.50 | 62.50 | | 138 | 179 |
| 0150 | 28' span | G | | 10 | 1.600 | | 88 | 69 | | 157 | 203 |
| 0160 | 32' span | G | | 10 | 1.600 | | 101 | 69 | | 170 | 217 |
| 0250 | 6" chords, 28' span | G | | 9 | 1.778 | | 111 | 76.50 | | 187.50 | 239 |
| 0260 | 32' span | G | | 9 | 1.778 | | 127 | 76.50 | | 203.50 | 256 |
| 0270 | 36' span | G | | 8 | 2 | | 142 | 86 | | 228 | 289 |
| 0280 | 40' span | G | | 8 | 2 | | 158 | 86 | | 244 | 305 |
| 1120 | 5:12 to 8:12 pitch, 18 ga x 4" chords, 16' span | G | | 10 | 1.600 | | 57.50 | 69 | | 126.50 | 170 |
| 1130 | 20' span | G | | 9 | 1.778 | | 72 | 76.50 | | 148.50 | 196 |
| 1140 | 24' span | G | | 9 | 1.778 | | 86.50 | 76.50 | | 163 | 212 |
| 1150 | 28' span | G | | 8 | 2 | | 101 | 86 | | 187 | 243 |
| 1160 | 32' span | G | | 8 | 2 | | 115 | 86 | | 201 | 259 |
| 1250 | 6" chords, 28' span | G | | 7 | 2.286 | | 127 | 98.50 | | 225.50 | 290 |
| 1260 | 32' span | G | | 7 | 2.286 | | 145 | 98.50 | | 243.50 | 310 |
| 1270 | 36' span | G | | 6 | 2.667 | | 163 | 115 | | 278 | 355 |
| 1280 | 40' span | G | | 6 | 2.667 | | 181 | 115 | | 296 | 375 |
| 2120 | 9:12 to 12:12 pitch, 18 ga x 4" chords, 16' span | G | | 8 | 2 | | 72 | 86 | | 158 | 211 |
| 2130 | 20' span | G | | 7 | 2.286 | | 90 | 98.50 | | 188.50 | 250 |
| 2140 | 24' span | G | | 7 | 2.286 | | 108 | 98.50 | | 206.50 | 270 |
| 2150 | 28' span | G | | 6 | 2.667 | | 126 | 115 | | 241 | 315 |
| 2160 | 32' span | G | | 6 | 2.667 | | 144 | 115 | | 259 | 335 |
| 2250 | 6" chords, 28' span | G | | 5 | 3.200 | | 158 | 138 | | 296 | 385 |
| 2260 | 32' span | G | | 5 | 3.200 | | 181 | 138 | | 319 | 410 |
| 2270 | 36' span | G | | 4 | 4 | | 203 | 172 | | 375 | 490 |
| 2280 | 40' span | G | | 4 | 4 | | 226 | 172 | | 398 | 515 |
| 5120 | Erection only of roof trusses, to 4:12 pitch, 16' span | | F-6 | 48 | .833 | | | 33.50 | 13.65 | 47.15 | 66 |
| 5130 | 20' span | | | 46 | .870 | | | 35 | 14.25 | 49.25 | 69 |
| 5140 | 24' span | | | 44 | .909 | | | 36.50 | 14.90 | 51.40 | 72.50 |
| 5150 | 28' span | | | 42 | .952 | | | 38.50 | 15.60 | 54.10 | 75.50 |
| 5160 | 32' span | | | 40 | 1 | | | 40.50 | 16.40 | 56.90 | 79.50 |
| 5170 | 36' span | | | 38 | 1.053 | | | 42.50 | 17.25 | 59.75 | 83.50 |
| 5180 | 40' span | | | 36 | 1.111 | | | 44.50 | 18.20 | 62.70 | 88 |
| 5220 | 5:12 to 8:12 pitch, 16' span | | | 42 | .952 | | | 38.50 | 15.60 | 54.10 | 75.50 |
| 5230 | 20' span | | | 40 | 1 | | | 40.50 | 16.40 | 56.90 | 79.50 |
| 5240 | 24' span | | | 38 | 1.053 | | | 42.50 | 17.25 | 59.75 | 83.50 |
| 5250 | 28' span | | | 36 | 1.111 | | | 44.50 | 18.20 | 62.70 | 88 |
| 5260 | 32' span | | | 34 | 1.176 | | | 47.50 | 19.30 | 66.80 | 93 |
| 5270 | 36' span | | | 32 | 1.250 | | | 50.50 | 20.50 | 71 | 99 |
| 5280 | 40' span | | | 30 | 1.333 | | | 53.50 | 22 | 75.50 | 106 |
| 5320 | 9:12 to 12:12 pitch, 16' span | | | 36 | 1.111 | | | 44.50 | 18.20 | 62.70 | 88 |
| 5330 | 20' span | | | 34 | 1.176 | | | 47.50 | 19.30 | 66.80 | 93 |
| 5340 | 24' span | | | 32 | 1.250 | | | 50.50 | 20.50 | 71 | 99 |
| 5350 | 28' span | | | 30 | 1.333 | | | 53.50 | 22 | 75.50 | 106 |
| 5360 | 32' span | | | 28 | 1.429 | | | 57.50 | 23.50 | 81 | 114 |
| 5370 | 36' span | | | 26 | 1.538 | | | 62 | 25 | 87 | 122 |
| 5380 | 40' span | | | 24 | 1.667 | | | 67 | 27.50 | 94.50 | 132 |

# 05 51 Metal Stairs

## 05 51 13 – Metal Pan Stairs

### 05 51 13.50 Pan Stairs

| 05 51 13.50 Pan Stairs | | Crew | Daily Output | Labor-Hours | Unit | Material | 2011 Bare Costs Labor | Equipment | Total | Total Incl O&P |
|---|---|---|---|---|---|---|---|---|---|---|
| 0010 | **PAN STAIRS**, shop fabricated, steel stringers | | | | | | | | | |
| 0015 | Made from recycled materials | G | | | | | | | | |
| 0200 | Cement fill metal pan, picket rail, 3'-6" wide | G | E-4 | 35 | .914 | Riser | 420 | 45 | 3.12 | 468.12 | 540 |
| 0300 | 4'-0" wide | G | | 30 | 1.067 | | 475 | 52.50 | 3.64 | 531.14 | 620 |
| 0350 | Wall rail, both sides, 3'-6" wide | G | | 53 | .604 | | 320 | 29.50 | 2.06 | 351.56 | 405 |
| 1500 | Landing, steel pan, conventional | G | | 160 | .200 | S.F. | 56 | 9.80 | .68 | 66.48 | 79.50 |
| 1600 | Pre-erected | G | | 255 | .125 | " | 98.50 | 6.15 | .43 | 105.08 | 119 |
| 1700 | Pre-erected, steel pan tread, 3'-6" wide, 2 line pipe rail | G | E-2 | 87 | .644 | Riser | 465 | 30.50 | 18.65 | 514.15 | 580 |

## 05 51 16 – Metal Floor Plate Stairs

### 05 51 16.50 Floor Plate Stairs

| 05 51 16.50 Floor Plate Stairs | | Crew | Daily Output | Labor-Hours | Unit | Material | 2011 Bare Costs Labor | Equipment | Total | Total Incl O&P |
|---|---|---|---|---|---|---|---|---|---|---|
| 0010 | **FLOOR PLATE STAIRS**, shop fabricated, steel stringers | | | | | | | | | |
| 0015 | Made from recycled materials | G | | | | | | | | |
| 0400 | Cast iron tread and pipe rail, 3'-6" wide | G | E-4 | 35 | .914 | Riser | 450 | 45 | 3.12 | 498.12 | 575 |
| 0500 | Checkered plate tread, industrial, 3'-6" wide | G | | 28 | 1.143 | | 281 | 56 | 3.90 | 340.90 | 415 |
| 0550 | Circular, for tanks, 3'-0" wide | G | | 33 | .970 | | 310 | 47.50 | 3.31 | 360.81 | 430 |
| 0600 | For isolated stairs, add | | | | | | | 100% | | | |
| 0800 | Custom steel stairs, 3'-6" wide, minimum | G | E-4 | 35 | .914 | | 420 | 45 | 3.12 | 468.12 | 540 |
| 0810 | Average | G | | 30 | 1.067 | | 560 | 52.50 | 3.64 | 616.14 | 710 |
| 0900 | Maximum | G | | 20 | 1.600 | | 700 | 78.50 | 5.45 | 783.95 | 915 |
| 1100 | For 4' wide stairs, add | | | | | | 5% | 5% | | | |
| 1300 | For 5' wide stairs, add | | | | | | 10% | 10% | | | |

## 05 51 19 – Metal Grating Stairs

### 05 51 19.50 Grating Stairs

| 05 51 19.50 Grating Stairs | | Crew | Daily Output | Labor-Hours | Unit | Material | 2011 Bare Costs Labor | Equipment | Total | Total Incl O&P |
|---|---|---|---|---|---|---|---|---|---|---|
| 0010 | **GRATING STAIRS**, shop fabricated, steel stringers, safety nosing on treads | | | | | | | | | |
| 0015 | Made from recycled materials | G | | | | | | | | |
| 0020 | Grating tread and pipe railing, 3'-6" wide | G | E-4 | 35 | .914 | Riser | 281 | 45 | 3.12 | 329.12 | 390 |
| 0100 | 4'-0" wide | G | " | 30 | 1.067 | " | 365 | 52.50 | 3.64 | 421.14 | 495 |

## 05 51 23 – Metal Fire Escapes

### 05 51 23.25 Fire Escapes

| 05 51 23.25 Fire Escapes | | Crew | Daily Output | Labor-Hours | Unit | Material | 2011 Bare Costs Labor | Equipment | Total | Total Incl O&P |
|---|---|---|---|---|---|---|---|---|---|---|
| 0010 | **FIRE ESCAPES**, shop fabricated | | | | | | | | | |
| 0200 | 2' wide balcony, 1" x 1/4" bars 1-1/2" O.C. | G | 1 Sswk | 5 | 1.600 | L.F. | 49 | 77.50 | | 126.50 | 191 |
| 0400 | 1st story cantilevered stair, standard | G | | .09 | 88.889 | Ea. | 2,050 | 4,325 | | 6,375 | 9,875 |
| 0700 | Platform & fixed stair, 36" x 40" | G | | .17 | 47.059 | Flight | 905 | 2,275 | | 3,180 | 5,025 |
| 0900 | For 3'-6" wide escapes, add to above | | | | | | 100% | 150% | | | |

### 05 51 23.50 Fire Escape Stairs

| 05 51 23.50 Fire Escape Stairs | | Crew | Daily Output | Labor-Hours | Unit | Material | 2011 Bare Costs Labor | Equipment | Total | Total Incl O&P |
|---|---|---|---|---|---|---|---|---|---|---|
| 0010 | **FIRE ESCAPE STAIRS**, shop fabricated | | | | | | | | | |
| 0020 | One story, disappearing, stainless steel | G | 2 Sswk | 20 | .800 | V.L.F. | 220 | 39 | | 259 | 310 |
| 0100 | Portable ladder | | | | | Ea. | 120 | | | 120 | 132 |
| 1100 | Fire escape, galvanized steel, 8'-0" to 10'-4" ceiling | G | 2 Carp | 1 | 16 | | 1,600 | 690 | | 2,290 | 2,800 |
| 1110 | 10'-6" to 13'-6" ceiling | G | " | 1 | 16 | | 2,100 | 690 | | 2,790 | 3,375 |

## 05 51 33 – Metal Ladders

### 05 51 33.13 Vertical Metal Ladders

| 05 51 33.13 Vertical Metal Ladders | | Crew | Daily Output | Labor-Hours | Unit | Material | 2011 Bare Costs Labor | Equipment | Total | Total Incl O&P |
|---|---|---|---|---|---|---|---|---|---|---|
| 0010 | **VERTICAL METAL LADDERS**, shop fabricated | | | | | | | | | |
| 0015 | Made from recycled materials | G | | | | | | | | |
| 0020 | Steel, 20" wide, bolted to concrete, with cage | G | E-4 | 50 | .640 | V.L.F. | 65 | 31.50 | 2.18 | 98.68 | 129 |
| 0100 | Without cage | G | | 85 | .376 | | 36 | 18.45 | 1.28 | 55.73 | 73.50 |
| 0300 | Aluminum, bolted to concrete, with cage | G | | 50 | .640 | | 111 | 31.50 | 2.18 | 144.68 | 181 |
| 0400 | Without cage | G | | 85 | .376 | | 48 | 18.45 | 1.28 | 67.73 | 87 |

# 05 51 Metal Stairs

## 05 51 33 – Metal Ladders

### 05 51 33.16 Inclined Metal Ladders

| | | Crew | Daily Output | Labor-Hours | Unit | Material | 2011 Bare Costs Labor | Equipment | Total | Total Incl O&P |
|---|---|---|---|---|---|---|---|---|---|---|
| 0010 | **INCLINED METAL LADDERS**, shop fabricated | | | | | | | | | |
| 0015 | Made from recycled materials | G | | | | | | | | |
| 3900 | Industrial ships ladder, steel, 24" W, grating treads, 2 line pipe rail | G | E-4 | 30 | 1.067 | Riser | 187 | 52.50 | 3.64 | 243.14 | 305 |
| 4000 | Aluminum | G | " | 30 | 1.067 | " | 206 | 52.50 | 3.64 | 262.14 | 325 |

### 05 51 33.23 Alternating Tread Ladders

| | | Crew | Daily Output | Labor-Hours | Unit | Material | 2011 Bare Costs Labor | Equipment | Total | Total Incl O&P |
|---|---|---|---|---|---|---|---|---|---|---|
| 0010 | **ALTERNATING TREAD LADDERS**, shop fabricated | | | | | | | | | |
| 0015 | Made from recycled materials | G | | | | | | | | |
| 1350 | Alternating tread stair, 56/68°, steel, standard paint color | G | 2 Sswk | 50 | .320 | V.L.F. | 258 | 15.55 | | 273.55 | 310 |
| 1360 | Non-standard paint color | G | | 50 | .320 | | 292 | 15.55 | | 307.55 | 350 |
| 1370 | Galvanized steel | G | | 50 | .320 | | 300 | 15.55 | | 315.55 | 360 |
| 1380 | Stainless steel | G | | 50 | .320 | | 435 | 15.55 | | 450.55 | 505 |
| 1390 | 68°, aluminum | G | | 50 | .320 | | 315 | 15.55 | | 330.55 | 380 |

# 05 52 Metal Railings

## 05 52 13 – Pipe and Tube Railings

### 05 52 13.50 Railings, Pipe

| | | Crew | Daily Output | Labor-Hours | Unit | Material | 2011 Bare Costs Labor | Equipment | Total | Total Incl O&P |
|---|---|---|---|---|---|---|---|---|---|---|
| 0010 | **RAILINGS, PIPE**, shop fab'd, 3'-6" high, posts @ 5' O.C. | | | | | | | | | |
| 0015 | Made from recycled materials | G | | | | | | | | |
| 0020 | Aluminum, 2 rail, satin finish, 1-1/4" diameter | G | E-4 | 160 | .200 | L.F. | 31.50 | 9.80 | .68 | 41.98 | 52.50 |
| 0030 | Clear anodized | G | | 160 | .200 | | 39 | 9.80 | .68 | 49.48 | 61 |
| 0040 | Dark anodized | G | | 160 | .200 | | 44 | 9.80 | .68 | 54.48 | 66.50 |
| 0080 | 1-1/2" diameter, satin finish | G | | 160 | .200 | | 37.50 | 9.80 | .68 | 47.98 | 59.50 |
| 0090 | Clear anodized | G | | 160 | .200 | | 42 | 9.80 | .68 | 52.48 | 64 |
| 0100 | Dark anodized | G | | 160 | .200 | | 46.50 | 9.80 | .68 | 56.98 | 69 |
| 0140 | Aluminum, 3 rail, 1-1/4" diam., satin finish | G | | 137 | .234 | | 48 | 11.45 | .80 | 60.25 | 74 |
| 0150 | Clear anodized | G | | 137 | .234 | | 60 | 11.45 | .80 | 72.25 | 87 |
| 0160 | Dark anodized | G | | 137 | .234 | | 66.50 | 11.45 | .80 | 78.75 | 94 |
| 0200 | 1-1/2" diameter, satin finish | G | | 137 | .234 | | 57.50 | 11.45 | .80 | 69.75 | 84.50 |
| 0210 | Clear anodized | G | | 137 | .234 | | 65.50 | 11.45 | .80 | 77.75 | 93 |
| 0220 | Dark anodized | G | | 137 | .234 | | 71.50 | 11.45 | .80 | 83.75 | 99.50 |
| 0500 | Steel, 2 rail, on stairs, primed, 1-1/4" diameter | G | | 160 | .200 | | 21 | 9.80 | .68 | 31.48 | 41 |
| 0520 | 1-1/2" diameter | G | | 160 | .200 | | 23 | 9.80 | .68 | 33.48 | 43.50 |
| 0540 | Galvanized, 1-1/4" diameter | G | | 160 | .200 | | 29 | 9.80 | .68 | 39.48 | 50 |
| 0560 | 1-1/2" diameter | G | | 160 | .200 | | 32.50 | 9.80 | .68 | 42.98 | 54 |
| 0580 | Steel, 3 rail, primed, 1-1/4" diameter | G | | 137 | .234 | | 31.50 | 11.45 | .80 | 43.75 | 55.50 |
| 0600 | 1-1/2" diameter | G | | 137 | .234 | | 33 | 11.45 | .80 | 45.25 | 57.50 |
| 0620 | Galvanized, 1-1/4" diameter | G | | 137 | .234 | | 44 | 11.45 | .80 | 56.25 | 69.50 |
| 0640 | 1-1/2" diameter | G | | 137 | .234 | | 51.50 | 11.45 | .80 | 63.75 | 77.50 |
| 0700 | Stainless steel, 2 rail, 1-1/4" diam. #4 finish | G | | 137 | .234 | | 94 | 11.45 | .80 | 106.25 | 124 |
| 0720 | High polish | G | | 137 | .234 | | 152 | 11.45 | .80 | 164.25 | 188 |
| 0740 | Mirror polish | G | | 137 | .234 | | 190 | 11.45 | .80 | 202.25 | 230 |
| 0760 | Stainless steel, 3 rail, 1-1/2" diam., #4 finish | G | | 120 | .267 | | 142 | 13.10 | .91 | 156.01 | 180 |
| 0770 | High polish | G | | 120 | .267 | | 235 | 13.10 | .91 | 249.01 | 283 |
| 0780 | Mirror finish | G | | 120 | .267 | | 285 | 13.10 | .91 | 299.01 | 340 |
| 0900 | Wall rail, alum. pipe, 1-1/4" diam., satin finish | G | | 213 | .150 | | 17.95 | 7.35 | .51 | 25.81 | 33.50 |
| 0905 | Clear anodized | G | | 213 | .150 | | 22 | 7.35 | .51 | 29.86 | 37.50 |
| 0910 | Dark anodized | G | | 213 | .150 | | 26.50 | 7.35 | .51 | 34.36 | 42.50 |
| 0915 | 1-1/2" diameter, satin finish | G | | 213 | .150 | | 20 | 7.35 | .51 | 27.86 | 35.50 |
| 0920 | Clear anodized | G | | 213 | .150 | | 25 | 7.35 | .51 | 32.86 | 41 |
| 0925 | Dark anodized | G | | 213 | .150 | | 31 | 7.35 | .51 | 38.86 | 47.50 |

# 05 52 Metal Railings

## 05 52 13 – Pipe and Tube Railings

### 05 52 13.50 Railings, Pipe

| | | | Crew | Daily Output | Labor-Hours | Unit | Material | 2011 Bare Costs Labor | Equipment | Total | Total Incl O&P |
|---|---|---|---|---|---|---|---|---|---|---|---|
| 0930 | Steel pipe, 1-1/4" diameter, primed | G | E-4 | 213 | .150 | L.F. | 12.70 | 7.35 | .51 | 20.56 | 27.50 |
| 0935 | Galvanized | G | | 213 | .150 | | 18.45 | 7.35 | .51 | 26.31 | 34 |
| 0940 | 1-1/2" diameter | G | | 176 | .182 | | 13.10 | 8.90 | .62 | 22.62 | 31 |
| 0945 | Galvanized | G | | 213 | .150 | | 18.50 | 7.35 | .51 | 26.36 | 34 |
| 0955 | Stainless steel pipe, 1-1/2" diam., #4 finish | G | | 107 | .299 | | 75 | 14.65 | 1.02 | 90.67 | 110 |
| 0960 | High polish | G | | 107 | .299 | | 153 | 14.65 | 1.02 | 168.67 | 195 |
| 0965 | Mirror polish | G | | 107 | .299 | | 181 | 14.65 | 1.02 | 196.67 | 226 |
| 2000 | 2-line pipe rail (1-1/2" T&B) with 1/2" pickets @ 4-1/2" O.C., | | | | | | | | | | | |
| 2005 | attached handrail on brackets | | | | | | | | | | | |
| 2010 | 42" high aluminum, satin finish, straight & level | G | E-4 | 120 | .267 | L.F. | 175 | 13.10 | .91 | 189.01 | 217 |
| 2050 | 42" high steel, primed, straight & level | G | " | 120 | .267 | | 107 | 13.10 | .91 | 121.01 | 142 |
| 4000 | For curved and level rails, add | | | | | | | 10% | 10% | | | |
| 4100 | For sloped rails for stairs, add | | | | | | | 30% | 30% | | | |

## 05 52 16 – Industrial Railings

### 05 52 16.50 Railings, Industrial

| | | | Crew | Daily Output | Labor-Hours | Unit | Material | 2011 Bare Costs Labor | Equipment | Total | Total Incl O&P |
|---|---|---|---|---|---|---|---|---|---|---|---|
| 0010 | **RAILINGS, INDUSTRIAL**, shop fab'd, 3'-6" high, posts @ 5' O.C. | | | | | | | | | | | |
| 0020 | 2 rail, 3'-6" high, 1-1/2" pipe | G | E-4 | 255 | .125 | L.F. | 23 | 6.15 | .43 | 29.58 | 37 |
| 0100 | 2" angle rail | G | " | 255 | .125 | | 21 | 6.15 | .43 | 27.58 | 34.50 |
| 0200 | For 4" high kick plate, 10 gauge, add | G | | | | | | 4.80 | | | 4.80 | 5.30 |
| 0300 | 1/4" thick, add | G | | | | | | 6.20 | | | 6.20 | 6.80 |
| 0500 | For curved level rails, add | | | | | | | 10% | 10% | | | |
| 0550 | For sloped rails for stairs, add | | | | | | | 30% | 30% | | | |

# 05 53 Metal Gratings

## 05 53 13 – Bar Gratings

### 05 53 13.10 Floor Grating, Aluminum

| | | | Crew | Daily Output | Labor-Hours | Unit | Material | 2011 Bare Costs Labor | Equipment | Total | Total Incl O&P |
|---|---|---|---|---|---|---|---|---|---|---|---|
| 0010 | **FLOOR GRATING, ALUMINUM**, field fabricated from panels | | | | | | | | | | | |
| 0015 | Made from recycled materials | G | | | | | | | | | | |
| 0110 | Bearing bars @ 1-3/16" O.C., cross bars @ 4" O.C., | | | | | | | | | | | |
| 0111 | Up to 300 S.F., 1" x 1/8" bar | G | E-4 | 900 | .036 | S.F. | 23.50 | 1.74 | .12 | 25.36 | 29 |
| 0112 | Over 300 S.F. | G | | 850 | .038 | | 21.50 | 1.85 | .13 | 23.48 | 27 |
| 0113 | 1-1/4" x 1/8" bar, up to 300 S.F. | G | | 800 | .040 | | 27.50 | 1.96 | .14 | 29.60 | 34 |
| 0114 | Over 300 S.F. | G | | 1000 | .032 | | 25 | 1.57 | .11 | 26.68 | 30.50 |
| 0122 | 1-1/4" x 3/16" bar, up to 300 S.F. | G | | 750 | .043 | | 36 | 2.09 | .15 | 38.24 | 43.50 |
| 0124 | Over 300 S.F. | G | | 1000 | .032 | | 32.50 | 1.57 | .11 | 34.18 | 39 |
| 0132 | 1-1/2" x 3/16" bar, up to 300 S.F. | G | | 700 | .046 | | 41.50 | 2.24 | .16 | 43.90 | 49.50 |
| 0134 | Over 300 S.F. | G | | 1000 | .032 | | 37.50 | 1.57 | .11 | 39.18 | 44.50 |
| 0136 | 1-3/4" x 3/16" bar, up to 300 S.F. | G | | 500 | .064 | | 50 | 3.14 | .22 | 53.36 | 61 |
| 0138 | Over 300 S.F. | G | | 1000 | .032 | | 45.50 | 1.57 | .11 | 47.18 | 53 |
| 0146 | 2-1/4" x 3/16" bar, up to 300 S.F. | G | | 600 | .053 | | 69 | 2.62 | .18 | 71.80 | 80.50 |
| 0148 | Over 300 S.F. | G | | 1000 | .032 | | 62.50 | 1.57 | .11 | 64.18 | 72 |
| 0162 | Cross bars @ 2" O.C., 1" x 1/8", up to 300 S.F. | G | | 600 | .053 | | 30.50 | 2.62 | .18 | 33.30 | 38.50 |
| 0164 | Over 300 S.F. | G | | 1000 | .032 | | 27.50 | 1.57 | .11 | 29.18 | 33.50 |
| 0172 | 1-1/4" x 3/16" bar, up to 300 S.F. | G | | 600 | .053 | | 40.50 | 2.62 | .18 | 43.30 | 50 |
| 0174 | Over 300 S.F. | G | | 1000 | .032 | | 37 | 1.57 | .11 | 38.68 | 43.50 |
| 0182 | 1-1/2" x 3/16" bar, up to 300 S.F. | G | | 600 | .053 | | 45.50 | 2.62 | .18 | 48.30 | 55 |
| 0184 | Over 300 S.F. | G | | 1000 | .032 | | 41.50 | 1.57 | .11 | 43.18 | 48.50 |
| 0186 | 1-3/4" x 3/16" bar, up to 300 S.F. | G | | 600 | .053 | | 55.50 | 2.62 | .18 | 58.30 | 66 |
| 0188 | Over 300 S.F. | G | | 1000 | .032 | | 50.50 | 1.57 | .11 | 52.18 | 58.50 |
| 0200 | For straight cuts, add | | | | | | L.F. | 3.70 | | | 3.70 | 4.07 |

# 05 53 Metal Gratings

## 05 53 13 – Bar Gratings

| 05 53 13.10 Floor Grating, Aluminum | | Crew | Daily Output | Labor-Hours | Unit | Material | 2011 Bare Costs Labor | Equipment | Total | Total Incl O&P |
|---|---|---|---|---|---|---|---|---|---|---|
| 0300 | For curved cuts, add | | | | L.F. | 4.55 | | | 4.55 | 5 |
| 0400 | For straight banding, add | G | | | | 4.75 | | | 4.75 | 5.25 |
| 0500 | For curved banding, add | G | | | | 5.75 | | | 5.75 | 6.35 |
| 0600 | For aluminum checkered plate nosings, add | G | | | | 6.20 | | | 6.20 | 6.80 |
| 0700 | For straight toe plate, add | G | | | | 9.35 | | | 9.35 | 10.30 |
| 0800 | For curved toe plate, add | G | | | | 10.90 | | | 10.90 | 12 |
| 1000 | For cast aluminum abrasive nosings, add | G | | | | 9.10 | | | 9.10 | 10 |
| 1200 | Expanded aluminum, .65# per S.F. | G | E-4 | 1050 | .030 | S.F. | 11.65 | 1.50 | .10 | 13.25 | 15.55 |
| 1400 | Extruded I bars are 10% less than 3/16" bars | | | | | | | | | |
| 1600 | Heavy duty, all extruded plank, 3/4" deep, 1.8 # per S.F. | G | E-4 | 1100 | .029 | S.F. | 30 | 1.43 | .10 | 31.53 | 35.50 |
| 1700 | 1-1/4" deep, 2.9# per S.F. | G | | 1000 | .032 | | 35 | 1.57 | .11 | 36.68 | 41.50 |
| 1800 | 1-3/4" deep, 4.2# per S.F. | G | | 925 | .035 | | 42 | 1.70 | .12 | 43.82 | 49 |
| 1900 | 2-1/4" deep, 5.0# per S.F. | G | | 875 | .037 | | 50 | 1.79 | .12 | 51.91 | 58.50 |
| 2100 | For safety serrated surface, add | | | | | | 15% | | | | |

## 05 53 13.70 Floor Grating, Steel

| | | | Crew | Daily Output | Labor-Hours | Unit | Material | Labor | Equipment | Total | Total Incl O&P |
|---|---|---|---|---|---|---|---|---|---|---|---|
| 0010 | **FLOOR GRATING, STEEL**, field fabricated from panels | | | | | | | | | | |
| 0015 | Made from recycled materials | G | | | | | | | | | |
| 0300 | Platforms, to 12' high, rectangular | G | E-4 | 3150 | .010 | Lb. | 3.20 | .50 | .03 | 3.73 | 4.44 |
| 0400 | Circular | G | " | 2300 | .014 | " | 4 | .68 | .05 | 4.73 | 5.65 |
| 0410 | Painted bearing bars @ 1-3/16" | | | | | | | | | | |
| 0412 | Cross bars @ 4" O.C., 3/4" x 1/8" bar, up to 300 S.F. | G | E-2 | 500 | .112 | S.F. | 7.30 | 5.30 | 3.24 | 15.84 | 20.50 |
| 0414 | Over 300 S.F. | G | | 750 | .075 | | 6.65 | 3.54 | 2.16 | 12.35 | 15.70 |
| 0422 | 1-1/4" x 3/16", up to 300 S.F. | G | | 400 | .140 | | 11.20 | 6.65 | 4.05 | 21.90 | 28 |
| 0424 | Over 300 S.F. | G | | 600 | .093 | | 10.20 | 4.42 | 2.70 | 17.32 | 21.50 |
| 0432 | 1-1/2" x 3/16", up to 300 S.F. | G | | 400 | .140 | | 12.95 | 6.65 | 4.05 | 23.65 | 30 |
| 0434 | Over 300 S.F. | G | | 600 | .093 | | 11.75 | 4.42 | 2.70 | 18.87 | 23.50 |
| 0436 | 1-3/4" x 3/16", up to 300 S.F. | G | | 400 | .140 | | 17.90 | 6.65 | 4.05 | 28.60 | 35.50 |
| 0438 | Over 300 S.F. | G | | 600 | .093 | | 16.25 | 4.42 | 2.70 | 23.37 | 28.50 |
| 0452 | 2-1/4" x 3/16", up to 300 S.F. | G | | 300 | .187 | | 21.50 | 8.85 | 5.40 | 35.75 | 44.50 |
| 0454 | Over 300 S.F. | G | | 450 | .124 | | 19.55 | 5.90 | 3.60 | 29.05 | 35.50 |
| 0462 | Cross bars @ 2" O.C., 3/4" x 1/8", up to 300 S.F. | G | | 500 | .112 | | 13.65 | 5.30 | 3.24 | 22.19 | 27.50 |
| 0464 | Over 300 S.F. | G | | 750 | .075 | | 11.35 | 3.54 | 2.16 | 17.05 | 21 |
| 0472 | 1-1/4" x 3/16", up to 300 S.F. | G | | 400 | .140 | | 17.85 | 6.65 | 4.05 | 28.55 | 35.50 |
| 0474 | Over 300 S.F. | G | | 600 | .093 | | 14.85 | 4.42 | 2.70 | 21.97 | 27 |
| 0482 | 1-1/2" x 3/16", up to 300 S.F. | G | | 400 | .140 | | 20 | 6.65 | 4.05 | 30.70 | 37.50 |
| 0484 | Over 300 S.F. | G | | 600 | .093 | | 16.75 | 4.42 | 2.70 | 23.87 | 29 |
| 0486 | 1-3/4" x 3/16", up to 300 S.F. | G | | 400 | .140 | | 30.50 | 6.65 | 4.05 | 41.20 | 49 |
| 0488 | Over 300 S.F. | G | | 600 | .093 | | 25 | 4.42 | 2.70 | 32.12 | 38 |
| 0502 | 2-1/4" x 3/16", up to 300 S.F. | G | | 300 | .187 | | 29.50 | 8.85 | 5.40 | 43.75 | 53.50 |
| 0504 | Over 300 S.F. | G | | 450 | .124 | | 24.50 | 5.90 | 3.60 | 34 | 41 |
| 0690 | For galvanized grating, add | | | | | | 25% | | | | |
| 0800 | For straight cuts, add | | | | | L.F. | 5.35 | | | 5.35 | 5.90 |
| 0900 | For curved cuts, add | | | | | | 6.75 | | | 6.75 | 7.45 |
| 1000 | For straight banding, add | G | | | | | 5.75 | | | 5.75 | 6.35 |
| 1100 | For curved banding, add | G | | | | | 7.50 | | | 7.50 | 8.25 |
| 1200 | For checkered plate nosings, add | G | | | | | 6.60 | | | 6.60 | 7.25 |
| 1300 | For straight toe or kick plate, add | G | | | | | 11.85 | | | 11.85 | 13.05 |
| 1400 | For curved toe or kick plate, add | G | | | | | 13.40 | | | 13.40 | 14.75 |
| 1500 | For abrasive nosings, add | G | | | | | 8.80 | | | 8.80 | 9.70 |
| 1510 | For stair treads, see Section 05 55 13.50 | | | | | | | | | | |
| 1600 | For safety serrated surface, minimum, add | | | | | | 15% | | | | |
| 1700 | Maximum, add | | | | | | 25% | | | | |

# 05 53 Metal Gratings

## 05 53 13 – Bar Gratings

### 05 53 13.70 Floor Grating, Steel

| | | Crew | Daily Output | Labor-Hours | Unit | Material | 2011 Bare Costs Labor | Equipment | Total | Total Incl O&P |
|---|---|---|---|---|---|---|---|---|---|---|
| 2000 | Stainless steel gratings, close spaced, 1" x 1/8" bars, up to 300 S.F. | G | E-4 | 450 | .071 | S.F. | 106 | 3.49 | .24 | 109.73 | 123 |
| 2100 | Standard spacing, 3/4" x 1/8" bars | G | | 500 | .064 | | 75.50 | 3.14 | .22 | 78.86 | 89 |
| 2200 | 1-1/4" x 3/16" bars | G | | 400 | .080 | | 116 | 3.92 | .27 | 120.19 | 135 |
| 2400 | Expanded steel grating, at ground, 3.0# per S.F. | G | | 900 | .036 | | 7.30 | 1.74 | .12 | 9.16 | 11.25 |
| 2500 | 3.14# per S.F. | G | | 900 | .036 | | 7.85 | 1.74 | .12 | 9.71 | 11.85 |
| 2600 | 4.0# per S.F. | G | | 850 | .038 | | 9.70 | 1.85 | .13 | 11.68 | 14.05 |
| 2650 | 4.27# per S.F. | G | | 850 | .038 | | 10.40 | 1.85 | .13 | 12.38 | 14.85 |
| 2700 | 5.0# per S.F. | G | | 800 | .040 | | 12.35 | 1.96 | .14 | 14.45 | 17.20 |
| 2800 | 6.25# per S.F. | G | | 750 | .043 | | 16 | 2.09 | .15 | 18.24 | 21.50 |
| 2900 | 7.0# per S.F. | G | | 700 | .046 | | 17.70 | 2.24 | .16 | 20.10 | 23.50 |
| 3100 | For flattened expanded steel grating, add | | | | | | 8% | | | | |
| 3300 | For elevated installation above 15', add | | | | | | | 15% | | | |

## 05 53 16 – Plank Gratings

### 05 53 16.50 Grating Planks

| | | Crew | Daily Output | Labor-Hours | Unit | Material | 2011 Bare Costs Labor | Equipment | Total | Total Incl O&P |
|---|---|---|---|---|---|---|---|---|---|---|
| 0010 | GRATING PLANKS, field fabricated from planks | | | | | | | | | | |
| 0020 | Aluminum, 9-1/2" wide, 14 ga., 2" rib | G | E-4 | 950 | .034 | L.F. | 28 | 1.65 | .11 | 29.76 | 34 |
| 0200 | Galvanized steel, 9-1/2" wide, 14 ga., 2-1/2" rib | G | | 950 | .034 | | 15.15 | 1.65 | .11 | 16.91 | 19.70 |
| 0300 | 4" rib | G | | 950 | .034 | | 17.80 | 1.65 | .11 | 19.56 | 22.50 |
| 0500 | 12 gauge, 2-1/2" rib | G | | 950 | .034 | | 26 | 1.65 | .11 | 27.76 | 31.50 |
| 0600 | 3" rib | G | | 950 | .034 | | 21 | 1.65 | .11 | 22.76 | 26 |
| 0800 | Stainless steel, type 304, 16 ga., 2" rib | G | | 950 | .034 | | 41 | 1.65 | .11 | 42.76 | 48 |
| 0900 | Type 316 | G | | 950 | .034 | | 50.50 | 1.65 | .11 | 52.26 | 58.50 |

## 05 53 19 – Floor Grating Frame

### 05 53 19.30 Grating Frame

| | | Crew | Daily Output | Labor-Hours | Unit | Material | 2011 Bare Costs Labor | Equipment | Total | Total Incl O&P |
|---|---|---|---|---|---|---|---|---|---|---|
| 0010 | GRATING FRAME, field fabricated | | | | | | | | | | |
| 0020 | Aluminum, for gratings 1" to 1-1/2" deep | G | 1 Sswk | 70 | .114 | L.F. | 3.44 | 5.55 | | 8.99 | 13.60 |
| 0100 | For each corner, add | G | | | | Ea. | 5.15 | | | 5.15 | 5.70 |

# 05 54 Metal Floor Plates

## 05 54 13 – Floor Plates

### 05 54 13.20 Checkered Plates

| | | Crew | Daily Output | Labor-Hours | Unit | Material | 2011 Bare Costs Labor | Equipment | Total | Total Incl O&P |
|---|---|---|---|---|---|---|---|---|---|---|
| 0010 | CHECKERED PLATES, steel, field fabricated | | | | | | | | | | |
| 0015 | Made from recycled materials | G | | | | | | | | | |
| 0020 | 1/4" & 3/8", 2000 to 5000 S.F., bolted | G | E-4 | 2900 | .011 | Lb. | .66 | .54 | .04 | 1.24 | 1.73 |
| 0100 | Welded | G | | 4400 | .007 | " | .62 | .36 | .02 | 1 | 1.34 |
| 0300 | Pit or trench cover and frame, 1/4" plate, 2' to 3' wide | G | | 100 | .320 | S.F. | 9.10 | 15.70 | 1.09 | 25.89 | 39 |
| 0400 | For galvanizing, add | G | | | | Lb. | .29 | | | .29 | .32 |
| 0500 | Platforms, 1/4" plate, no handrails included, rectangular | G | E-4 | 4200 | .008 | | 3.20 | .37 | .03 | 3.60 | 4.21 |
| 0600 | Circular | G | " | 2500 | .013 | | 4 | .63 | .04 | 4.67 | 5.55 |

### 05 54 13.70 Trench Covers

| | | Crew | Daily Output | Labor-Hours | Unit | Material | 2011 Bare Costs Labor | Equipment | Total | Total Incl O&P |
|---|---|---|---|---|---|---|---|---|---|---|
| 0010 | TRENCH COVERS, field fabricated | | | | | | | | | | |
| 0020 | Cast iron grating with bar stops and angle frame, to 18" wide | G | 1 Sswk | 20 | .400 | L.F. | 259 | 19.40 | | 278.40 | 320 |
| 0100 | Frame only (both sides of trench), 1" grating | G | | 45 | .178 | | 1.26 | 8.65 | | 9.91 | 16.65 |
| 0150 | 2" grating | G | | 35 | .229 | | 2.70 | 11.10 | | 13.80 | 22.50 |
| 0200 | Aluminum, stock units, including frames and | | | | | | | | | | |
| 0210 | 3/8" plain cover plate, 4" opening | G | E-4 | 205 | .156 | L.F. | 25.50 | 7.65 | .53 | 33.68 | 42 |
| 0300 | 6" opening | G | | 185 | .173 | | 30 | 8.50 | .59 | 39.09 | 48.50 |
| 0400 | 10" opening | G | | 170 | .188 | | 37.50 | 9.25 | .64 | 47.39 | 58.50 |
| 0500 | 16" opening | G | | 155 | .206 | | 50 | 10.15 | .70 | 60.85 | 73.50 |
| 0700 | Add per inch for additional widths to 24" | G | | | | | 2 | | | 2 | 2.20 |

# 05 54 Metal Floor Plates

## 05 54 13 – Floor Plates

| 05 54 13.70 Trench Covers | | Crew | Daily Output | Labor-Hours | Unit | Material | 2011 Bare Costs Labor | Equipment | Total | Total Incl O&P |
|---|---|---|---|---|---|---|---|---|---|---|
| 0900 | For custom fabrication, add | | | | L.F. | 50% | | | | |
| 1100 | For 1/4" plain cover plate, deduct | | | | | 12% | | | | |
| 1500 | For cover recessed for tile, 1/4" thick, deduct | | | | | 12% | | | | |
| 1600 | 3/8" thick, add | | | | | 5% | | | | |
| 1800 | For checkered plate cover, 1/4" thick, deduct | | | | | 12% | | | | |
| 1900 | 3/8" thick, add | | | | | 2% | | | | |
| 2100 | For slotted or round holes in cover, 1/4" thick, add | | | | | 3% | | | | |
| 2200 | 3/8" thick, add | | | | | 4% | | | | |
| 2300 | For abrasive cover, add | | | | | 12% | | | | |

# 05 55 Metal Stair Treads and Nosings

## 05 55 13 – Metal Stair Treads

### 05 55 13.50 Stair Treads

| | | | Crew | Daily Output | Labor-Hours | Unit | Material | 2011 Bare Costs Labor | Equipment | Total | Total Incl O&P |
|---|---|---|---|---|---|---|---|---|---|---|---|
| 0010 | **STAIR TREADS** | | | | | | | | | | |
| 0020 | Aluminum grating, 3' long, 1-1/2" x 3/16" rect. bars, 6" wide | G | 1 Sswk | 24 | .333 | Ea. | 59 | 16.20 | | 75.20 | 93.50 |
| 0100 | 12" wide | G | | 22 | .364 | | 87 | 17.65 | | 104.65 | 127 |
| 0200 | 1-1/2" x 3/16" I-bars, 6" wide | G | | 24 | .333 | | 49.50 | 16.20 | | 65.70 | 83 |
| 0300 | 12" wide | G | | 22 | .364 | | 80 | 17.65 | | 97.65 | 119 |
| 0400 | For abrasive nosings, add | G | | | | | 15.55 | | | 15.55 | 17.10 |
| 0500 | For narrow mesh, add | | | | | | 60% | | | | |
| 0600 | Stair treads, not incl. stringers, see also Section 03 15 05.85 | | | | | | | | | | |
| 0700 | Cast aluminum, abrasive, 3' long x 12" wide, 5/16" thick | G | 1 Sswk | 15 | .533 | Ea. | 131 | 26 | | 157 | 190 |
| 0800 | 3/8" thick | G | | 15 | .533 | | 140 | 26 | | 166 | 199 |
| 0900 | 1/2" thick | G | | 15 | .533 | | 153 | 26 | | 179 | 214 |
| 1000 | Cast bronze, abrasive, 3/8" thick | G | | 8 | 1 | | 595 | 48.50 | | 643.50 | 740 |
| 1100 | 1/2" thick | G | | 8 | 1 | | 780 | 48.50 | | 828.50 | 945 |
| 1200 | Cast iron, abrasive, 3' long x 12" wide, 3/8" thick | G | | 15 | .533 | | 116 | 26 | | 142 | 173 |
| 1300 | 1/2" thick | G | | 15 | .533 | | 133 | 26 | | 159 | 192 |
| 1400 | Fiberglass reinforced plastic with safety nosing, | | | | | | | | | | |
| 1500 | 1-1/2" thick, 12" wide, 24" long | | 1 Sswk | 22 | .364 | Ea. | 49.50 | 17.65 | | 67.15 | 85.50 |
| 1600 | 30" long | | | 22 | .364 | | 59 | 17.65 | | 76.65 | 96 |
| 1700 | 36" long | | | 22 | .364 | | 69.50 | 17.65 | | 87.15 | 108 |
| 2000 | Steel grating, painted, 3' long, 1-1/4" x 3/16" bars, 6" wide | G | | 20 | .400 | | 48 | 19.40 | | 67.40 | 87.50 |
| 2010 | 12" wide | G | | 18 | .444 | | 56 | 21.50 | | 77.50 | 99.50 |
| 2100 | Add for abrasive nosing, 3' long, painted | G | | | | | 15.05 | | | 15.05 | 16.55 |
| 2200 | Galvanized | G | | | | | 20 | | | 20 | 22 |
| 2300 | Painting 3' treads, in shop, nonstandard paint | | | | | | 2.67 | | | 2.67 | 2.94 |
| 2400 | Added coats, standard paint | | | | | | 1.14 | | | 1.14 | 1.25 |
| 2500 | Expanded steel, 2-1/2" deep, 9" x 3' long, 18 gauge | G | 1 Sswk | 20 | .400 | | 26.50 | 19.40 | | 45.90 | 63.50 |
| 2600 | 14 gauge | G | " | 20 | .400 | | 32.50 | 19.40 | | 51.90 | 70.50 |

# 05 56 Metal Castings

## 05 56 13 – Metal Construction Castings

| 05 56 13.50 Construction Castings | | Crew | Daily Output | Labor-Hours | Unit | Material | 2011 Bare Costs Labor | Equipment | Total | Total Incl O&P |
|---|---|---|---|---|---|---|---|---|---|---|
| 0010 | **CONSTRUCTION CASTINGS** | | | | | | | | | |
| 0020 | Manhole covers and frames, see Section 33 44 13.13 | | | | | | | | | |
| 0100 | Column bases, cast iron, 16" x 16", approx. 65 lb. | G | E-4 | 46 | .696 | Ea. | 138 | 34 | 2.37 | 174.37 | 215 |
| 0200 | 32" x 32", approx. 256 lb. | G | " | 23 | 1.391 | | 515 | 68 | 4.74 | 587.74 | 690 |
| 0400 | Cast aluminum for wood columns, 8" x 8" | G | 1 Carp | 32 | .250 | | 41 | 10.75 | | 51.75 | 61.50 |
| 0500 | 12" x 12" | G | " | 32 | .250 | | 89 | 10.75 | | 99.75 | 115 |
| 0600 | Miscellaneous C.I. castings, light sections, less than 150 lb. | G | E-4 | 3200 | .010 | Lb. | 2.13 | .49 | .03 | 2.65 | 3.25 |
| 1100 | Heavy sections, more than 150 lb. | G | | 4200 | .008 | | 1.04 | .37 | .03 | 1.44 | 1.83 |
| 1300 | Special low volume items | G | | 3200 | .010 | | 3.70 | .49 | .03 | 4.22 | 4.98 |
| 1500 | For ductile iron, add | | | | | | 100% | | | | |

# 05 58 Formed Metal Fabrications

## 05 58 21 – Formed Chain

### 05 58 21.05 Alloy Steel Chain

| | | | Crew | Daily Output | Labor-Hours | Unit | Material | Labor | Equipment | Total | Total Incl O&P |
|---|---|---|---|---|---|---|---|---|---|---|---|
| 0010 | **ALLOY STEEL CHAIN**, Grade 80, for lifting | | | | | | | | | | |
| 0015 | Self-colored, cut lengths, 1/4" | G | E-17 | 4 | 4 | C.L.F. | 690 | 198 | | 888 | 1,100 |
| 0020 | 3/8" | G | | 2 | 8 | | 875 | 395 | | 1,270 | 1,650 |
| 0030 | 1/2" | G | | 1.20 | 13.333 | | 1,400 | 660 | | 2,060 | 2,700 |
| 0040 | 5/8" | G | | .72 | 22.222 | | 2,125 | 1,100 | | 3,225 | 4,300 |
| 0050 | 3/4" | G | E-18 | .48 | 83.333 | | 2,900 | 4,025 | 2,150 | 9,075 | 12,500 |
| 0060 | 7/8" | G | | .40 | 100 | | 4,850 | 4,825 | 2,600 | 12,275 | 16,500 |
| 0070 | 1" | G | | .35 | 114 | | 6,550 | 5,525 | 2,950 | 15,025 | 19,900 |
| 0080 | 1-1/4" | G | | .24 | 166 | | 13,600 | 8,050 | 4,325 | 25,975 | 33,600 |
| 0110 | Hook, Grade 80, Clevis slip, 1/4" | G | | | | Ea. | 33 | | | 33 | 36.50 |
| 0120 | 3/8" | G | | | | | 43 | | | 43 | 47.50 |
| 0130 | 1/2" | G | | | | | 69.50 | | | 69.50 | 76.50 |
| 0140 | 5/8" | G | | | | | 87.50 | | | 87.50 | 96.50 |
| 0150 | 3/4" | G | | | | | 143 | | | 143 | 157 |
| 0160 | Hook, Grade 80, eye/sling w/hammerlock coupling, 15 Ton | G | | | | | 345 | | | 345 | 380 |
| 0170 | 22 Ton | G | | | | | 1,375 | | | 1,375 | 1,500 |
| 0180 | 37 Ton | G | | | | | 2,300 | | | 2,300 | 2,525 |

## 05 58 23 – Formed Metal Guards

### 05 58 23.90 Window Guards

| | | | Crew | Daily Output | Labor-Hours | Unit | Material | Labor | Equipment | Total | Total Incl O&P |
|---|---|---|---|---|---|---|---|---|---|---|---|
| 0010 | **WINDOW GUARDS**, shop fabricated | | | | | | | | | | |
| 0015 | Expanded metal, steel angle frame, permanent | G | E-4 | 350 | .091 | S.F. | 19.50 | 4.48 | .31 | 24.29 | 29.50 |
| 0025 | Steel bars, 1/2" x 1/2", spaced 5" O.C. | G | " | 290 | .110 | " | 13.50 | 5.40 | .38 | 19.28 | 25 |
| 0030 | Hinge mounted, add | G | | | | Opng. | 39 | | | 39 | 43 |
| 0040 | Removable type, add | | | | | " | 25 | | | 25 | 27.50 |
| 0050 | For galvanized guards, add | | | | | S.F. | 35% | | | | |
| 0070 | For pivoted or projected type, add | | | | | | 105% | 40% | | | |
| 0100 | Mild steel, stock units, economy | G | E-4 | 405 | .079 | | 5.30 | 3.88 | .27 | 9.45 | 13 |
| 0200 | Deluxe | G | | 405 | .079 | | 10.85 | 3.88 | .27 | 15 | 19.10 |
| 0400 | Woven wire, stock units, 3/8" channel frame, 3' x 5' opening | G | | 40 | .800 | Opng. | 143 | 39 | 2.73 | 184.73 | 230 |
| 0500 | 4' x 6' opening | G | | 38 | .842 | | 230 | 41.50 | 2.87 | 274.37 | 330 |
| 0800 | Basket guards for above, add | G | | | | | 196 | | | 196 | 216 |
| 1000 | Swinging guards for above, add | G | | | | | 67 | | | 67 | 73.50 |

## 05 58 25 – Formed Lamp Posts

### 05 58 25.40 Lamp Posts

| | | | Crew | Daily Output | Labor-Hours | Unit | Material | Labor | Equipment | Total | Total Incl O&P |
|---|---|---|---|---|---|---|---|---|---|---|---|
| 0010 | **LAMP POSTS** | | | | | | | | | | |
| 0020 | Aluminum, 7' high, stock units, post only | G | 1 Carp | 16 | .500 | Ea. | 49 | 21.50 | | 70.50 | 87 |

# 05 58 Formed Metal Fabrications

## 05 58 25 – Formed Lamp Posts

| 05 58 25.40 Lamp Posts | | Crew | Daily Output | Labor-Hours | Unit | Material | 2011 Bare Costs Labor | Equipment | Total | Total Incl O&P |
|---|---|---|---|---|---|---|---|---|---|---|
| 0100 | Mild steel, plain | G 1 Carp | 16 | .500 | Ea. | 57 | 21.50 | | 78.50 | 95.50 |

# 05 71 Decorative Metal Stairs

## 05 71 13 – Fabricated Metal Spiral Stairs

| 05 71 13.50 Spiral Stairs | | Crew | Daily Output | Labor-Hours | Unit | Material | 2011 Bare Costs Labor | Equipment | Total | Total Incl O&P |
|---|---|---|---|---|---|---|---|---|---|---|
| 0010 | **SPIRAL STAIRS**, shop fabricated | | | | | | | | | |
| 1810 | Spiral aluminum, 5'-0" diameter, stock units | G E-4 | 45 | .711 | Riser | 510 | 35 | 2.42 | 547.42 | 625 |
| 1820 | Custom units | G | 45 | .711 | | 965 | 35 | 2.42 | 1,002.42 | 1,125 |
| 1900 | Spiral, cast iron, 4'-0" diameter, ornamental, minimum | G | 45 | .711 | | 455 | 35 | 2.42 | 492.42 | 565 |
| 1920 | Maximum | G | 25 | 1.280 | | 620 | 63 | 4.36 | 687.36 | 795 |
| 2000 | Spiral, steel, industrial checkered plate, 4' diameter | G | 45 | .711 | | 455 | 35 | 2.42 | 492.42 | 565 |
| 2200 | Stock units, 6'-0" diameter | G | 40 | .800 | | 550 | 39 | 2.73 | 591.73 | 680 |
| 3110 | Spiral steel, stock units, primed, flat metal tread, 3'-6" dia | G 2 Carp | 1.60 | 10 | Flight | 1,250 | 430 | | 1,680 | 2,000 |
| 3120 | 4'-0" dia | G | 1.45 | 11.034 | | 1,425 | 475 | | 1,900 | 2,300 |
| 3130 | 4'-6" dia | G | 1.35 | 11.852 | | 1,575 | 510 | | 2,085 | 2,500 |
| 3140 | 5'-0" dia | G | 1.25 | 12.800 | | 1,700 | 550 | | 2,250 | 2,725 |
| 3210 | Galvanized, 3'-6" dia | G | 1.60 | 10 | | 1,475 | 430 | | 1,905 | 2,275 |
| 3220 | 4'-0" dia | G | 1.45 | 11.034 | | 1,700 | 475 | | 2,175 | 2,600 |
| 3230 | 4'-6" dia | G | 1.35 | 11.852 | | 1,900 | 510 | | 2,410 | 2,850 |
| 3240 | 5'-0" dia | G | 1.25 | 12.800 | | 2,050 | 550 | | 2,600 | 3,100 |
| 3310 | Checkered plate tread, 3'-6" dia | G | 1.45 | 11.034 | | 1,450 | 475 | | 1,925 | 2,325 |
| 3320 | 4'-0" dia | G | 1.35 | 11.852 | | 1,650 | 510 | | 2,160 | 2,575 |
| 3330 | 4'-6" dia | G | 1.25 | 12.800 | | 1,800 | 550 | | 2,350 | 2,825 |
| 3340 | 5'-0" dia | G | 1.15 | 13.913 | | 1,950 | 600 | | 2,550 | 3,075 |
| 3410 | Galvanized, 3'-6" dia | G | 1.45 | 11.034 | | 1,750 | 475 | | 2,225 | 2,650 |
| 3420 | 4'-0" dia | G | 1.35 | 11.852 | | 1,975 | 510 | | 2,485 | 2,950 |
| 3430 | 4'-6" dia | G | 1.25 | 12.800 | | 2,175 | 550 | | 2,725 | 3,225 |
| 3440 | 5'-0" dia | G | 1.15 | 13.913 | | 2,350 | 600 | | 2,950 | 3,500 |
| 3510 | Red oak tread on flat metal, 3'-6" dia | | 1.35 | 11.852 | | 2,550 | 510 | | 3,060 | 3,600 |
| 3520 | 4'-0" dia | | 1.25 | 12.800 | | 2,825 | 550 | | 3,375 | 3,950 |
| 3530 | 4'-6" dia | | 1.15 | 13.913 | | 3,025 | 600 | | 3,625 | 4,275 |
| 3540 | 5'-0" dia | | 1.05 | 15.238 | | 3,325 | 655 | | 3,980 | 4,650 |

# 05 73 Decorative Metal Railings

## 05 73 16 – Wire Rope Decorative Metal Railings

| 05 73 16.10 Cable Railings | | Crew | Daily Output | Labor-Hours | Unit | Material | 2011 Bare Costs Labor | Equipment | Total | Total Incl O&P |
|---|---|---|---|---|---|---|---|---|---|---|
| 0010 | **CABLE RAILINGS**, with 316 stainless steel 1 x 19 cable, 3/16" diameter | | | | | | | | | |
| 0015 | Made from recycled materials | G | | | | | | | | |
| 0100 | 1-3/4" diameter stainless steel posts x 42" high, cables 4" OC | G 2 Sswk | 25 | .640 | L.F. | 61.50 | 31 | | 92.50 | 123 |

## 05 73 23 – Ornamental Railings

| 05 73 23.50 Railings, Ornamental | | Crew | Daily Output | Labor-Hours | Unit | Material | 2011 Bare Costs Labor | Equipment | Total | Total Incl O&P |
|---|---|---|---|---|---|---|---|---|---|---|
| 0010 | **RAILINGS, ORNAMENTAL**, 3'-6" high, posts @ 6' O.C. | | | | | | | | | |
| 0020 | Bronze or stainless, hand forged, minimum | G 2 Sswk | 24 | .667 | L.F. | 112 | 32.50 | | 144.50 | 180 |
| 0100 | Maximum | G | 18 | .889 | | 167 | 43 | | 210 | 259 |
| 0200 | Aluminum, panelized, minimum | G | 24 | .667 | | 19.85 | 32.50 | | 52.35 | 79 |
| 0300 | Maximum | G | 18 | .889 | | 36 | 43 | | 79 | 116 |
| 0400 | Wrought iron, hand forged, minimum | G | 24 | .667 | | 74.50 | 32.50 | | 107 | 139 |
| 0500 | Maximum | G | 18 | .889 | | 119 | 43 | | 162 | 207 |
| 0550 | Steel, panelized, minimum | G | 24 | .667 | | 18.70 | 32.50 | | 51.20 | 77.50 |

# 05 73 Decorative Metal Railings

## 05 73 23 – Ornamental Railings

### 05 73 23.50 Railings, Ornamental

| | | Crew | Daily Output | Labor-Hours | Unit | Material | 2011 Bare Costs Labor | Equipment | Total | Total Incl O&P |
|---|---|---|---|---|---|---|---|---|---|---|
| 0560 | Maximum | G 2 Sswk | 18 | .889 | L.F. | 24 | 43 | | 67 | 102 |
| 0600 | Composite metal/wood/glass, minimum | | 18 | .889 | | 102 | 43 | | 145 | 188 |
| 0700 | Maximum | | 12 | 1.333 | | 204 | 64.50 | | 268.50 | 340 |

# 05 75 Decorative Formed Metal

## 05 75 13 – Columns

### 05 75 13.10 Aluminum Columns

| | | | Crew | Daily Output | Labor-Hours | Unit | Material | 2011 Bare Costs Labor | Equipment | Total | Total Incl O&P |
|---|---|---|---|---|---|---|---|---|---|---|---|
| 0010 | **ALUMINUM COLUMNS** | | | | | | | | | | |
| 0015 | Made from recycled materials | G | | | | | | | | | |
| 0020 | Aluminum, extruded, stock units, no cap or base, 6" diameter | G | E-4 | 240 | .133 | L.F. | 9.45 | 6.55 | .45 | 16.45 | 22.50 |
| 0100 | 8" diameter | G | | 170 | .188 | | 12.55 | 9.25 | .64 | 22.44 | 31 |
| 0200 | 10" diameter | G | | 150 | .213 | | 17.15 | 10.45 | .73 | 28.33 | 38 |
| 0300 | 12" diameter | G | | 140 | .229 | | 30.50 | 11.20 | .78 | 42.48 | 54 |
| 0400 | 15" diameter | G | | 120 | .267 | | 40 | 13.10 | .91 | 54.01 | 68 |
| 0410 | Caps and bases, plain, 6" diameter | G | | | | Set | 22.50 | | | 22.50 | 24.50 |
| 0420 | 8" diameter | G | | | | | 26.50 | | | 26.50 | 29.50 |
| 0430 | 10" diameter | G | | | | | 37.50 | | | 37.50 | 41 |
| 0440 | 12" diameter | G | | | | | 53 | | | 53 | 58.50 |
| 0450 | 15" diameter | G | | | | | 102 | | | 102 | 112 |
| 0460 | Caps, ornamental, minimum | G | | | | | 278 | | | 278 | 305 |
| 0470 | Maximum | G | | | | | 1,350 | | | 1,350 | 1,475 |
| 0500 | For square columns, add to column prices above | | | | | L.F. | 50% | | | | |
| 0700 | Residential, flat, 8' high, plain | G | E-4 | 20 | 1.600 | Ea. | 95 | 78.50 | 5.45 | 178.95 | 249 |
| 0720 | Fancy | G | | 20 | 1.600 | | 181 | 78.50 | 5.45 | 264.95 | 345 |
| 0740 | Corner type, plain | G | | 20 | 1.600 | | 165 | 78.50 | 5.45 | 248.95 | 325 |
| 0760 | Fancy | G | | 20 | 1.600 | | 320 | 78.50 | 5.45 | 403.95 | 495 |

### 05 75 13.20 Columns, Ornamental

| | | | Crew | Daily Output | Labor-Hours | Unit | Material | 2011 Bare Costs Labor | Equipment | Total | Total Incl O&P |
|---|---|---|---|---|---|---|---|---|---|---|---|
| 0010 | **COLUMNS, ORNAMENTAL**, shop fabricated | R051223-10 | | | | | | | | | |
| 6400 | Mild steel, flat, 9" wide, stock units, painted, plain | G | E-4 | 160 | .200 | V.L.F. | 8.90 | 9.80 | .68 | 19.38 | 28 |
| 6450 | Fancy | G | | 160 | .200 | | 17 | 9.80 | .68 | 27.48 | 37 |
| 6500 | Corner columns, painted, plain | G | | 160 | .200 | | 15.50 | 9.80 | .68 | 25.98 | 35 |
| 6550 | Fancy | G | | 160 | .200 | | 30 | 9.80 | .68 | 40.48 | 51 |

# Estimating Tips

## 06 05 00 Common Work Results for Wood, Plastics, and Composites

- Common to any wood-framed structure are the accessory connector items such as screws, nails, adhesives, hangers, connector plates, straps, angles, and hold-downs. For typical wood-framed buildings, such as residential projects, the aggregate total for these items can be significant, especially in areas where seismic loading is a concern. For floor and wall framing, the material cost is based on 10 to 25 lbs. per MBF. Hold-downs, hangers, and other connectors should be taken off by the piece.

  Included with material costs are fasteners for a normal installation. RSMeans engineers use manufacturer's recommendations, written specifications, and/or standard construction practice for size and spacing of fasteners. Prices for various fasteners are shown for informational purposes only. Adjustments should be made if unusual fastening conditions exist.

## 06 10 00 Carpentry

- Lumber is a traded commodity and therefore sensitive to supply and demand in the marketplace. Even in "budgetary" estimating of wood-framed projects, it is advisable to call local suppliers for the latest market pricing.

- Common quantity units for wood-framed projects are "thousand board feet" (MBF). A board foot is a volume of wood, 1″ x 1′ x 1′, or 144 cubic inches. Board-foot quantities are generally calculated using nominal material dimensions—dressed sizes are ignored. Board foot per lineal foot of any stick of lumber can be calculated by dividing the nominal cross-sectional area by 12. As an example, 2,000 lineal feet of 2 x 12 equates to 4 MBF by dividing the nominal area, 2 x 12, by 12, which equals 2, and multiplying by 2,000 to give 4,000 board feet. This simple rule applies to all nominal dimensioned lumber.

- Waste is an issue of concern at the quantity takeoff for any area of construction. Framing lumber is sold in even foot lengths, i.e., 10′, 12′, 14′, 16′, and depending on spans, wall heights, and the grade of lumber, waste is inevitable. A rule of thumb for lumber waste is 5%–10% depending on material quality and the complexity of the framing.

- Wood in various forms and shapes is used in many projects, even where the main structural framing is steel, concrete, or masonry. Plywood as a back-up partition material and 2x boards used as blocking and cant strips around roof edges are two common examples. The estimator should ensure that the costs of all wood materials are included in the final estimate.

## 06 20 00 Finish Carpentry

- It is necessary to consider the grade of workmanship when estimating labor costs for erecting millwork and interior finish. In practice, there are three grades: premium, custom, and economy. The RSMeans daily output for base and case moldings is in the range of 200 to 250 L.F. per carpenter per day. This is appropriate for most average custom-grade projects. For premium projects, an adjustment to productivity of 25%–50% should be made, depending on the complexity of the job.

## Reference Numbers

Reference numbers are shown in shaded boxes at the beginning of some major classifications. These numbers refer to related items in the Reference Section. The reference information may be an estimating procedure, an alternate pricing method, or technical information.

*Note:* Not all subdivisions listed here necessarily appear in this publication.

### 06 05 05.10 Selective Demolition Wood Framing

| | | Crew | Daily Output | Labor-Hours | Unit | Material | 2011 Bare Costs Labor | Equipment | Total | Total Incl O&P |
|---|---|---|---|---|---|---|---|---|---|---|
| 0010 | **SELECTIVE DEMOLITION WOOD FRAMING**      R024119-10 | | | | | | | | | |
| 0100 | Timber connector, nailed, small | 1 Clab | 96 | .083 | Ea. | | 2.86 | | 2.86 | 4.39 |
| 0110 | Medium | | 60 | .133 | | | 4.58 | | 4.58 | 7.05 |
| 0120 | Large | | 48 | .167 | | | 5.75 | | 5.75 | 8.80 |
| 0130 | Bolted, small | | 48 | .167 | | | 5.75 | | 5.75 | 8.80 |
| 0140 | Medium | | 32 | .250 | | | 8.60 | | 8.60 | 13.20 |
| 0150 | Large | | 24 | .333 | | | 11.45 | | 11.45 | 17.55 |
| 2958 | Beams, 2" x 6" | 2 Clab | 1100 | .015 | L.F. | | .50 | | .50 | .77 |
| 2960 | 2" x 8" | | 825 | .019 | | | .67 | | .67 | 1.02 |
| 2965 | 2" x 10" | | 665 | .024 | | | .83 | | .83 | 1.27 |
| 2970 | 2" x 12" | | 550 | .029 | | | 1 | | 1 | 1.53 |
| 2972 | 2" x 14" | | 470 | .034 | | | 1.17 | | 1.17 | 1.79 |
| 2975 | 4" x 8" | B-1 | 413 | .058 | | | 2.04 | | 2.04 | 3.12 |
| 2980 | 4" x 10" | | 330 | .073 | | | 2.55 | | 2.55 | 3.91 |
| 2985 | 4" x 12" | | 275 | .087 | | | 3.06 | | 3.06 | 4.69 |
| 3000 | 6" x 8" | | 275 | .087 | | | 3.06 | | 3.06 | 4.69 |
| 3040 | 6" x 10" | | 220 | .109 | | | 3.82 | | 3.82 | 5.85 |
| 3080 | 6" x 12" | | 185 | .130 | | | 4.54 | | 4.54 | 6.95 |
| 3120 | 8" x 12" | | 140 | .171 | | | 6 | | 6 | 9.20 |
| 3160 | 10" x 12" | | 110 | .218 | | | 7.65 | | 7.65 | 11.70 |
| 3162 | Alternate pricing method | | 1.10 | 21.818 | M.B.F. | | 765 | | 765 | 1,175 |
| 3170 | Blocking, in 16" OC wall framing, 2" x 4" | 1 Clab | 600 | .013 | L.F. | | .46 | | .46 | .70 |
| 3172 | 2" x 6" | | 400 | .020 | | | .69 | | .69 | 1.05 |
| 3174 | In 24" OC wall framing, 2" x 4" | | 600 | .013 | | | .46 | | .46 | .70 |
| 3176 | 2" x 6" | | 400 | .020 | | | .69 | | .69 | 1.05 |
| 3178 | Alt method, wood blocking removal from wood framing | | .40 | 20 | M.B.F. | | 685 | | 685 | 1,050 |
| 3179 | Wood blocking removal from steel framing | | .36 | 22.222 | " | | 765 | | 765 | 1,175 |
| 3180 | Bracing, let in, 1" x 3", studs 16" OC | | 1050 | .008 | L.F. | | .26 | | .26 | .40 |
| 3181 | Studs 24" OC | | 1080 | .007 | | | .25 | | .25 | .39 |
| 3182 | 1" x 4", studs 16" OC | | 1050 | .008 | | | .26 | | .26 | .40 |
| 3183 | Studs 24" OC | | 1080 | .007 | | | .25 | | .25 | .39 |
| 3184 | 1" x 6", studs 16" OC | | 1050 | .008 | | | .26 | | .26 | .40 |
| 3185 | Studs 24" OC | | 1080 | .007 | | | .25 | | .25 | .39 |
| 3186 | 2" x 3", studs 16" OC | | 800 | .010 | | | .34 | | .34 | .53 |
| 3187 | Studs 24" OC | | 830 | .010 | | | .33 | | .33 | .51 |
| 3188 | 2" x 4", studs 16" OC | | 800 | .010 | | | .34 | | .34 | .53 |
| 3189 | Studs 24" OC | | 830 | .010 | | | .33 | | .33 | .51 |
| 3190 | 2" x 6", studs 16" OC | | 800 | .010 | | | .34 | | .34 | .53 |
| 3191 | Studs 24" OC | | 830 | .010 | | | .33 | | .33 | .51 |
| 3192 | 2" x 8", studs 16" OC | | 800 | .010 | | | .34 | | .34 | .53 |
| 3193 | Studs 24" OC | | 830 | .010 | | | .33 | | .33 | .51 |
| 3194 | "T" shaped metal bracing, studs at 16" OC | | 1060 | .008 | | | .26 | | .26 | .40 |
| 3195 | Studs at 24" OC | | 1200 | .007 | | | .23 | | .23 | .35 |
| 3196 | Metal straps, studs at 16" OC | | 1200 | .007 | | | .23 | | .23 | .35 |
| 3197 | Studs at 24" OC | | 1240 | .006 | | | .22 | | .22 | .34 |
| 3200 | Columns, round, 8' to 14' tall | | 40 | .200 | Ea. | | 6.85 | | 6.85 | 10.55 |
| 3202 | Dimensional lumber sizes | 2 Clab | 1.10 | 14.545 | M.B.F. | | 500 | | 500 | 765 |
| 3250 | Blocking, between joists | 1 Clab | 320 | .025 | Ea. | | .86 | | .86 | 1.32 |
| 3252 | Bridging, metal strap, between joists | | 320 | .025 | Pr. | | .86 | | .86 | 1.32 |
| 3254 | Wood, between joists | | 320 | .025 | " | | .86 | | .86 | 1.32 |
| 3260 | Door buck, studs, header & access., 8' high 2" x 4" wall, 3' wide | | 32 | .250 | Ea. | | 8.60 | | 8.60 | 13.20 |
| 3261 | 4' wide | | 32 | .250 | | | 8.60 | | 8.60 | 13.20 |
| 3262 | 5' wide | | 32 | .250 | | | 8.60 | | 8.60 | 13.20 |

| 06 05 05.10 Selective Demolition Wood Framing | Crew | Daily Output | Labor-Hours | Unit | Material | 2011 Bare Costs Labor | Equipment | Total | Total Incl O&P |
|---|---|---|---|---|---|---|---|---|---|
| 3263 | 6' wide | 1 Clab | 32 | .250 | Ea. | | 8.60 | | 8.60 | 13.20 |
| 3264 | 8' wide | | 30 | .267 | | | 9.15 | | 9.15 | 14.05 |
| 3265 | 10' wide | | 30 | .267 | | | 9.15 | | 9.15 | 14.05 |
| 3266 | 12' wide | | 30 | .267 | | | 9.15 | | 9.15 | 14.05 |
| 3267 | 2" x 6" wall, 3' wide | | 32 | .250 | | | 8.60 | | 8.60 | 13.20 |
| 3268 | 4' wide | | 32 | .250 | | | 8.60 | | 8.60 | 13.20 |
| 3269 | 5' wide | | 32 | .250 | | | 8.60 | | 8.60 | 13.20 |
| 3270 | 6' wide | | 32 | .250 | | | 8.60 | | 8.60 | 13.20 |
| 3271 | 8' wide | | 30 | .267 | | | 9.15 | | 9.15 | 14.05 |
| 3272 | 10' wide | | 30 | .267 | | | 9.15 | | 9.15 | 14.05 |
| 3273 | 12' wide | | 30 | .267 | | | 9.15 | | 9.15 | 14.05 |
| 3274 | Window buck, studs, header & access, 8' high 2" x 4" wall, 2' wide | | 24 | .333 | | | 11.45 | | 11.45 | 17.55 |
| 3275 | 3' wide | | 24 | .333 | | | 11.45 | | 11.45 | 17.55 |
| 3276 | 4' wide | | 24 | .333 | | | 11.45 | | 11.45 | 17.55 |
| 3277 | 5' wide | | 24 | .333 | | | 11.45 | | 11.45 | 17.55 |
| 3278 | 6' wide | | 24 | .333 | | | 11.45 | | 11.45 | 17.55 |
| 3279 | 7' wide | | 24 | .333 | | | 11.45 | | 11.45 | 17.55 |
| 3280 | 8' wide | | 22 | .364 | | | 12.50 | | 12.50 | 19.15 |
| 3281 | 10' wide | | 22 | .364 | | | 12.50 | | 12.50 | 19.15 |
| 3282 | 12' wide | | 22 | .364 | | | 12.50 | | 12.50 | 19.15 |
| 3283 | 2" x 6" wall, 2' wide | | 24 | .333 | | | 11.45 | | 11.45 | 17.55 |
| 3284 | 3' wide | | 24 | .333 | | | 11.45 | | 11.45 | 17.55 |
| 3285 | 4' wide | | 24 | .333 | | | 11.45 | | 11.45 | 17.55 |
| 3286 | 5' wide | | 24 | .333 | | | 11.45 | | 11.45 | 17.55 |
| 3287 | 6' wide | | 24 | .333 | | | 11.45 | | 11.45 | 17.55 |
| 3288 | 7' wide | | 24 | .333 | | | 11.45 | | 11.45 | 17.55 |
| 3289 | 8' wide | | 22 | .364 | | | 12.50 | | 12.50 | 19.15 |
| 3290 | 10' wide | | 22 | .364 | | | 12.50 | | 12.50 | 19.15 |
| 3291 | 12' wide | | 22 | .364 | | | 12.50 | | 12.50 | 19.15 |
| 3400 | Fascia boards, 1" x 6" | | 500 | .016 | L.F. | | .55 | | .55 | .84 |
| 3440 | 1" x 8" | | 450 | .018 | | | .61 | | .61 | .94 |
| 3480 | 1" x 10" | | 400 | .020 | | | .69 | | .69 | 1.05 |
| 3490 | 2" x 6" | | 450 | .018 | | | .61 | | .61 | .94 |
| 3500 | 2" x 8" | | 400 | .020 | | | .69 | | .69 | 1.05 |
| 3510 | 2" x 10" | | 350 | .023 | | | .79 | | .79 | 1.20 |
| 3610 | Furring, on wood walls or ceiling | | 4000 | .002 | S.F. | | .07 | | .07 | .11 |
| 3620 | On masonry or concrete walls or ceiling | | 1200 | .007 | " | | .23 | | .23 | .35 |
| 3800 | Headers over openings, 2 @ 2" x 6" | | 110 | .073 | L.F. | | 2.50 | | 2.50 | 3.83 |
| 3840 | 2 @ 2" x 8" | | 100 | .080 | | | 2.75 | | 2.75 | 4.22 |
| 3880 | 2 @ 2" x 10" | | 90 | .089 | | | 3.05 | | 3.05 | 4.68 |
| 3885 | Alternate pricing method | | .26 | 30.651 | M.B.F. | | 1,050 | | 1,050 | 1,625 |
| 3920 | Joists, 1" x 4" | | 1250 | .006 | L.F. | | .22 | | .22 | .34 |
| 3930 | 1" x 6" | | 1135 | .007 | | | .24 | | .24 | .37 |
| 3940 | 1" x 8" | | 1000 | .008 | | | .27 | | .27 | .42 |
| 3950 | 1" x 10" | | 895 | .009 | | | .31 | | .31 | .47 |
| 3960 | 1" x 12" | | 765 | .010 | | | .36 | | .36 | .55 |
| 4200 | 2" x 4" | 2 Clab | 1000 | .016 | | | .55 | | .55 | .84 |
| 4230 | 2" x 6" | | 970 | .016 | | | .57 | | .57 | .87 |
| 4240 | 2" x 8" | | 940 | .017 | | | .58 | | .58 | .90 |
| 4250 | 2" x 10" | | 910 | .018 | | | .60 | | .60 | .93 |
| 4280 | 2" x 12" | | 880 | .018 | | | .62 | | .62 | .96 |
| 4281 | 2" x 14" | | 850 | .019 | | | .65 | | .65 | .99 |
| 4282 | Composite joists, 9-1/2" | | 960 | .017 | | | .57 | | .57 | .88 |

| 06 05 05.10 Selective Demolition Wood Framing | | Crew | Daily Output | Labor-Hours | Unit | Material | 2011 Bare Costs Labor | Equipment | Total | Total Incl O&P |
|---|---|---|---|---|---|---|---|---|---|---|
| 4283 | 11-7/8" | 2 Clab | 930 | .017 | L.F. | | .59 | | .59 | .91 |
| 4284 | 14" | | 897 | .018 | | | .61 | | .61 | .94 |
| 4285 | 16" | | 865 | .019 | | | .64 | | .64 | .98 |
| 4290 | Wood joists, alternate pricing method | | 1.50 | 10.667 | M.B.F. | | 365 | | 365 | 560 |
| 4500 | Open web joist, 12" deep | | 500 | .032 | L.F. | | 1.10 | | 1.10 | 1.69 |
| 4505 | 14" deep | | 475 | .034 | | | 1.16 | | 1.16 | 1.77 |
| 4510 | 16" deep | | 450 | .036 | | | 1.22 | | 1.22 | 1.87 |
| 4520 | 18" deep | | 425 | .038 | | | 1.29 | | 1.29 | 1.98 |
| 4530 | 24" deep | | 400 | .040 | | | 1.37 | | 1.37 | 2.11 |
| 4550 | Ledger strips, 1" x 2" | 1 Clab | 1200 | .007 | | | .23 | | .23 | .35 |
| 4560 | 1" x 3" | | 1200 | .007 | | | .23 | | .23 | .35 |
| 4570 | 1" x 4" | | 1200 | .007 | | | .23 | | .23 | .35 |
| 4580 | 2" x 2" | | 1100 | .007 | | | .25 | | .25 | .38 |
| 4590 | 2" x 4" | | 1000 | .008 | | | .27 | | .27 | .42 |
| 4600 | 2" x 6" | | 1000 | .008 | | | .27 | | .27 | .42 |
| 4601 | 2" x 8" or 2" x 10" | | 800 | .010 | | | .34 | | .34 | .53 |
| 4602 | 4" x 6" | | 600 | .013 | | | .46 | | .46 | .70 |
| 4604 | 4" x 8" | | 450 | .018 | | | .61 | | .61 | .94 |
| 5400 | Posts, 4" x 4" | 2 Clab | 800 | .020 | | | .69 | | .69 | 1.05 |
| 5405 | 4" x 6" | | 550 | .029 | | | 1 | | 1 | 1.53 |
| 5410 | 4" x 8" | | 440 | .036 | | | 1.25 | | 1.25 | 1.92 |
| 5425 | 4" x 10" | | 390 | .041 | | | 1.41 | | 1.41 | 2.16 |
| 5430 | 4" x 12" | | 350 | .046 | | | 1.57 | | 1.57 | 2.41 |
| 5440 | 6" x 6" | | 400 | .040 | | | 1.37 | | 1.37 | 2.11 |
| 5445 | 6" x 8" | | 350 | .046 | | | 1.57 | | 1.57 | 2.41 |
| 5450 | 6" x 10" | | 320 | .050 | | | 1.72 | | 1.72 | 2.64 |
| 5455 | 6" x 12" | | 290 | .055 | | | 1.90 | | 1.90 | 2.91 |
| 5480 | 8" x 8" | | 300 | .053 | | | 1.83 | | 1.83 | 2.81 |
| 5500 | 10" x 10" | | 240 | .067 | | | 2.29 | | 2.29 | 3.51 |
| 5660 | Tongue and groove floor planks | | 2 | 8 | M.B.F. | | 275 | | 275 | 420 |
| 5682 | 16" OC, 2" x 4" | | 880 | .018 | S.F. | | .62 | | .62 | .96 |
| 5683 | 2" x 6" | | 840 | .019 | | | .65 | | .65 | 1 |
| 5684 | 2" x 8" | | 820 | .020 | | | .67 | | .67 | 1.03 |
| 5685 | 2" x 10" | | 820 | .020 | | | .67 | | .67 | 1.03 |
| 5686 | 2" x 12" | | 810 | .020 | | | .68 | | .68 | 1.04 |
| 5687 | 24" OC, 2" x 4" | | 1170 | .014 | | | .47 | | .47 | .72 |
| 5688 | 2" x 6" | | 1117 | .014 | | | .49 | | .49 | .75 |
| 5689 | 2" x 8" | | 1091 | .015 | | | .50 | | .50 | .77 |
| 5690 | 2" x 10" | | 1091 | .015 | | | .50 | | .50 | .77 |
| 5691 | 2" x 12" | | 1077 | .015 | | | .51 | | .51 | .78 |
| 5795 | Rafters, ordinary, 2" x 4" (alternate method) | | 862 | .019 | L.F. | | .64 | | .64 | .98 |
| 5800 | 2" x 6" (alternate method) | | 850 | .019 | | | .65 | | .65 | .99 |
| 5840 | 2" x 8" (alternate method) | | 837 | .019 | | | .66 | | .66 | 1.01 |
| 5855 | 2" x 10" (alternate method) | | 825 | .019 | | | .67 | | .67 | 1.02 |
| 5865 | 2" x 12" (alternate method) | | 812 | .020 | | | .68 | | .68 | 1.04 |
| 5870 | Sill plate, 2" x 4" | 1 Clab | 1170 | .007 | | | .24 | | .24 | .36 |
| 5871 | 2" x 6" | | 780 | .010 | | | .35 | | .35 | .54 |
| 5872 | 2" x 8" | | 586 | .014 | | | .47 | | .47 | .72 |
| 5873 | Alternate pricing method | | .78 | 10.256 | M.B.F. | | 350 | | 350 | 540 |
| 5885 | Ridge board, 1" x 4" | 2 Clab | 900 | .018 | L.F. | | .61 | | .61 | .94 |
| 5886 | 1" x 6" | | 875 | .018 | | | .63 | | .63 | .96 |
| 5887 | 1" x 8" | | 850 | .019 | | | .65 | | .65 | .99 |
| 5888 | 1" x 10" | | 825 | .019 | | | .67 | | .67 | 1.02 |

| 06 05 05.10 Selective Demolition Wood Framing | Crew | Daily Output | Labor-Hours | Unit | Material | 2011 Bare Costs Labor | Equipment | Total | Total Incl O&P |
|---|---|---|---|---|---|---|---|---|---|
| 5889 | 1" x 12" | 2 Clab | 800 | .020 | L.F. | | .69 | | .69 | 1.05 |
| 5890 | 2" x 4" | | 900 | .018 | | | .61 | | .61 | .94 |
| 5892 | 2" x 6" | | 875 | .018 | | | .63 | | .63 | .96 |
| 5894 | 2" x 8" | | 850 | .019 | | | .65 | | .65 | .99 |
| 5896 | 2" x 10" | | 825 | .019 | | | .67 | | .67 | 1.02 |
| 5898 | 2" x 12" | | 800 | .020 | | | .69 | | .69 | 1.05 |
| 6050 | Rafter tie, 1" x 4" | | 1250 | .013 | | | .44 | | .44 | .67 |
| 6052 | 1" x 6" | | 1135 | .014 | | | .48 | | .48 | .74 |
| 6054 | 2" x 4" | | 1000 | .016 | | | .55 | | .55 | .84 |
| 6056 | 2" x 6" | ▼ | 970 | .016 | | | .57 | | .57 | .87 |
| 6070 | Sleepers, on concrete, 1" x 2" | 1 Clab | 4700 | .002 | | | .06 | | .06 | .09 |
| 6075 | 1" x 3" | | 4000 | .002 | | | .07 | | .07 | .11 |
| 6080 | 2" x 4" | | 3000 | .003 | | | .09 | | .09 | .14 |
| 6085 | 2" x 6" | ▼ | 2600 | .003 | ▼ | | .11 | | .11 | .16 |
| 6086 | Sheathing from roof, 5/16" | 2 Clab | 1600 | .010 | S.F. | | .34 | | .34 | .53 |
| 6088 | 3/8" | | 1525 | .010 | | | .36 | | .36 | .55 |
| 6090 | 1/2" | | 1400 | .011 | | | .39 | | .39 | .60 |
| 6092 | 5/8" | | 1300 | .012 | | | .42 | | .42 | .65 |
| 6094 | 3/4" | | 1200 | .013 | | | .46 | | .46 | .70 |
| 6096 | Board sheathing from roof | | 1400 | .011 | | | .39 | | .39 | .60 |
| 6100 | Sheathing, from walls, 1/4" | | 1200 | .013 | | | .46 | | .46 | .70 |
| 6110 | 5/16" | | 1175 | .014 | | | .47 | | .47 | .72 |
| 6120 | 3/8" | | 1150 | .014 | | | .48 | | .48 | .73 |
| 6130 | 1/2" | | 1125 | .014 | | | .49 | | .49 | .75 |
| 6140 | 5/8" | | 1100 | .015 | | | .50 | | .50 | .77 |
| 6150 | 3/4" | | 1075 | .015 | | | .51 | | .51 | .78 |
| 6152 | Board sheathing from walls | | 1500 | .011 | | | .37 | | .37 | .56 |
| 6158 | Subfloor, with boards | | 1050 | .015 | | | .52 | | .52 | .80 |
| 6160 | Plywood, 1/2" thick | | 768 | .021 | | | .72 | | .72 | 1.10 |
| 6162 | 5/8" thick | | 760 | .021 | | | .72 | | .72 | 1.11 |
| 6164 | 3/4" thick | | 750 | .021 | | | .73 | | .73 | 1.12 |
| 6165 | 1-1/8" thick | ▼ | 720 | .022 | | | .76 | | .76 | 1.17 |
| 6166 | Underlayment, particle board, 3/8" thick | 1 Clab | 780 | .010 | | | .35 | | .35 | .54 |
| 6168 | 1/2" thick | | 768 | .010 | | | .36 | | .36 | .55 |
| 6170 | 5/8" thick | | 760 | .011 | | | .36 | | .36 | .55 |
| 6172 | 3/4" thick | ▼ | 750 | .011 | ▼ | | .37 | | .37 | .56 |
| 6200 | Stairs and stringers, minimum | 2 Clab | 40 | .400 | Riser | | 13.75 | | 13.75 | 21 |
| 6240 | Maximum | " | 26 | .615 | " | | 21 | | 21 | 32.50 |
| 6300 | Components, tread | 1 Clab | 110 | .073 | Ea. | | 2.50 | | 2.50 | 3.83 |
| 6320 | Riser | | 80 | .100 | " | | 3.44 | | 3.44 | 5.25 |
| 6390 | Stringer, 2" x 10" | | 260 | .031 | L.F. | | 1.06 | | 1.06 | 1.62 |
| 6400 | 2" x 12" | | 260 | .031 | | | 1.06 | | 1.06 | 1.62 |
| 6410 | 3" x 10" | | 250 | .032 | | | 1.10 | | 1.10 | 1.69 |
| 6420 | 3" x 12" | ▼ | 250 | .032 | | | 1.10 | | 1.10 | 1.69 |
| 6590 | Wood studs, 2" x 3" | 2 Clab | 3076 | .005 | | | .18 | | .18 | .27 |
| 6600 | 2" x 4" | | 2000 | .008 | | | .27 | | .27 | .42 |
| 6640 | 2" x 6" | ▼ | 1600 | .010 | ▼ | | .34 | | .34 | .53 |
| 6720 | Wall framing, including studs, plates and blocking, 2" x 4" | 1 Clab | 600 | .013 | S.F. | | .46 | | .46 | .70 |
| 6740 | 2" x 6" | | 480 | .017 | " | | .57 | | .57 | .88 |
| 6750 | Headers, 2" x 4" | | 1125 | .007 | L.F. | | .24 | | .24 | .37 |
| 6755 | 2" x 6" | | 1125 | .007 | | | .24 | | .24 | .37 |
| 6760 | 2" x 8" | | 1050 | .008 | | | .26 | | .26 | .40 |
| 6765 | 2" x 10" | ▼ | 1050 | .008 | ▼ | | .26 | | .26 | .40 |

# 06 05 Common Work Results for Wood, Plastics, and Composites

## 06 05 05 – Selective Wood and Plastics Demolition

### 06 05 05.10 Selective Demolition Wood Framing

| | 06 05 05.10 Selective Demolition Wood Framing | Crew | Daily Output | Labor-Hours | Unit | Material | 2011 Bare Costs Labor | Equipment | Total | Total Incl O&P |
|---|---|---|---|---|---|---|---|---|---|---|
| 6770 | 2" x 12" | 1 Clab | 1000 | .008 | L.F. | | .27 | | .27 | .42 |
| 6780 | 4" x 10" | | 525 | .015 | | | .52 | | .52 | .80 |
| 6785 | 4" x 12" | | 500 | .016 | | | .55 | | .55 | .84 |
| 6790 | 6" x 8" | | 560 | .014 | | | .49 | | .49 | .75 |
| 6795 | 6" x 10" | | 525 | .015 | | | .52 | | .52 | .80 |
| 6797 | 6" x 12" | | 500 | .016 | | | .55 | | .55 | .84 |
| 8000 | Soffit, T & G wood | | 520 | .015 | S.F. | | .53 | | .53 | .81 |
| 8010 | Hardboard, vinyl or aluminum | | 640 | .013 | | | .43 | | .43 | .66 |
| 8030 | Plywood | 2 Carp | 315 | .051 | | | 2.19 | | 2.19 | 3.35 |
| 9500 | See Section 02 41 19.23 for rubbish handling | | | | | | | | | |

### 06 05 05.20 Selective Demolition Millwork and Trim

| | 06 05 05.20 Selective Demolition Millwork and Trim | Crew | Daily Output | Labor-Hours | Unit | Material | 2011 Bare Costs Labor | Equipment | Total | Total Incl O&P |
|---|---|---|---|---|---|---|---|---|---|---|
| 0010 | **SELECTIVE DEMOLITION MILLWORK AND TRIM** R024119-10 | | | | | | | | | |
| 1000 | Cabinets, wood, base cabinets, per L.F. | 2 Clab | 80 | .200 | L.F. | | 6.85 | | 6.85 | 10.55 |
| 1020 | Wall cabinets, per L.F. | " | 80 | .200 | " | | 6.85 | | 6.85 | 10.55 |
| 1060 | Remove and reset, base cabinets | 2 Carp | 18 | .889 | Ea. | | 38.50 | | 38.50 | 58.50 |
| 1070 | Wall cabinets | " | 20 | .800 | " | | 34.50 | | 34.50 | 53 |
| 1100 | Steel, painted, base cabinets | 2 Clab | 60 | .267 | L.F. | | 9.15 | | 9.15 | 14.05 |
| 1120 | Wall cabinets | " | 60 | .267 | " | | 9.15 | | 9.15 | 14.05 |
| 1200 | Casework, large area | | 320 | .050 | S.F. | | 1.72 | | 1.72 | 2.64 |
| 1220 | Selective | | 200 | .080 | " | | 2.75 | | 2.75 | 4.22 |
| 1500 | Counter top, minimum | | 200 | .080 | L.F. | | 2.75 | | 2.75 | 4.22 |
| 1510 | Maximum | | 120 | .133 | | | 4.58 | | 4.58 | 7.05 |
| 1550 | Remove and reset, minimum | 2 Carp | 50 | .320 | | | 13.80 | | 13.80 | 21 |
| 1560 | Maximum | " | 40 | .400 | | | 17.20 | | 17.20 | 26.50 |
| 2000 | Paneling, 4' x 8' sheets | 2 Clab | 2000 | .008 | S.F. | | .27 | | .27 | .42 |
| 2100 | Boards, 1" x 4" | | 700 | .023 | | | .79 | | .79 | 1.20 |
| 2120 | 1" x 6" | | 750 | .021 | | | .73 | | .73 | 1.12 |
| 2140 | 1" x 8" | | 800 | .020 | | | .69 | | .69 | 1.05 |
| 3000 | Trim, baseboard, to 6" wide | | 1200 | .013 | L.F. | | .46 | | .46 | .70 |
| 3040 | Greater than 6" and up to 12" wide | | 1000 | .016 | | | .55 | | .55 | .84 |
| 3080 | Remove and reset, minimum | 2 Carp | 400 | .040 | | | 1.72 | | 1.72 | 2.64 |
| 3090 | Maximum | " | 300 | .053 | | | 2.30 | | 2.30 | 3.52 |
| 3100 | Ceiling trim | 2 Clab | 1000 | .016 | | | .55 | | .55 | .84 |
| 3120 | Chair rail | | 1200 | .013 | | | .46 | | .46 | .70 |
| 3140 | Railings with balusters | | 240 | .067 | | | 2.29 | | 2.29 | 3.51 |
| 3160 | Wainscoting | | 700 | .023 | S.F. | | .79 | | .79 | 1.20 |

## 06 05 23 – Wood, Plastic, and Composite Fastenings

### 06 05 23.10 Nails

| | 06 05 23.10 Nails | Crew | Daily Output | Labor-Hours | Unit | Material | 2011 Bare Costs Labor | Equipment | Total | Total Incl O&P |
|---|---|---|---|---|---|---|---|---|---|---|
| 0010 | **NAILS**, material only, based upon 50# box purchase | | | | | | | | | |
| 0020 | Copper nails, plain | | | | Lb. | 10 | | | 10 | 11 |
| 0400 | Stainless steel, plain | | | | | 7.80 | | | 7.80 | 8.60 |
| 0500 | Box, 3d to 20d, bright | | | | | 1.25 | | | 1.25 | 1.38 |
| 0520 | Galvanized | | | | | 1.77 | | | 1.77 | 1.95 |
| 0600 | Common, 3d to 60d, plain | | | | | 1.30 | | | 1.30 | 1.43 |
| 0700 | Galvanized | | | | | 1.56 | | | 1.56 | 1.72 |
| 0800 | Aluminum | | | | | 5.80 | | | 5.80 | 6.40 |
| 1000 | Annular or spiral thread, 4d to 60d, plain | | | | | 1.80 | | | 1.80 | 1.98 |
| 1200 | Galvanized | | | | | 3 | | | 3 | 3.30 |
| 1400 | Drywall nails, plain | | | | | 3.59 | | | 3.59 | 3.95 |
| 1600 | Galvanized | | | | | 2.19 | | | 2.19 | 2.41 |
| 1800 | Finish nails, 4d to 10d, plain | | | | | 1.36 | | | 1.36 | 1.50 |
| 2000 | Galvanized | | | | | 1.79 | | | 1.79 | 1.97 |

## 06 05 23 – Wood, Plastic, and Composite Fastenings

### 06 05 23.10 Nails

| | | Crew | Daily Output | Labor-Hours | Unit | Material | 2011 Bare Costs Labor | Equipment | Total | Total Incl O&P |
|---|---|---|---|---|---|---|---|---|---|---|
| 2100 | Aluminum | | | | Lb. | 5.80 | | | 5.80 | 6.40 |
| 2300 | Flooring nails, hardened steel, 2d to 10d, plain | | | | | 3.10 | | | 3.10 | 3.41 |
| 2400 | Galvanized | | | | | 4 | | | 4 | 4.40 |
| 2500 | Gypsum lath nails, 1-1/8", 13 ga. flathead, blued | | | | | 3.59 | | | 3.59 | 3.95 |
| 2600 | Masonry nails, hardened steel, 3/4" to 3" long, plain | | | | | 2.20 | | | 2.20 | 2.42 |
| 2700 | Galvanized | | | | | 3.75 | | | 3.75 | 4.13 |
| 2900 | Roofing nails, threaded, galvanized | | | | | 1.63 | | | 1.63 | 1.79 |
| 3100 | Aluminum | | | | | 4.80 | | | 4.80 | 5.30 |
| 3300 | Compressed lead head, threaded, galvanized | | | | | 2.47 | | | 2.47 | 2.72 |
| 3600 | Siding nails, plain shank, galvanized | | | | | 2 | | | 2 | 2.20 |
| 3800 | Aluminum | | | | | 5.80 | | | 5.80 | 6.40 |
| 5000 | Add to prices above for cement coating | | | | | .11 | | | .11 | .12 |
| 5200 | Zinc or tin plating | | | | | .14 | | | .14 | .15 |
| 5500 | Vinyl coated sinkers, 8d to 16d | | | | | .65 | | | .65 | .72 |

### 06 05 23.40 Sheet Metal Screws

| | | Crew | Daily Output | Labor-Hours | Unit | Material | 2011 Bare Costs Labor | Equipment | Total | Total Incl O&P |
|---|---|---|---|---|---|---|---|---|---|---|
| 0010 | **SHEET METAL SCREWS** | | | | | | | | | |
| 0020 | Steel, standard, #8 x 3/4", plain | | | | C | 2.92 | | | 2.92 | 3.21 |
| 0100 | Galvanized | | | | | 2.92 | | | 2.92 | 3.21 |
| 0300 | #10 x 1", plain | | | | | 4.03 | | | 4.03 | 4.43 |
| 0400 | Galvanized | | | | | 4.03 | | | 4.03 | 4.43 |
| 0600 | With washers, #14 x 1", plain | | | | | 12.35 | | | 12.35 | 13.60 |
| 0700 | Galvanized | | | | | 12.35 | | | 12.35 | 13.60 |
| 0900 | #14 x 2", plain | | | | | 16.90 | | | 16.90 | 18.60 |
| 1000 | Galvanized | | | | | 16.90 | | | 16.90 | 18.60 |
| 1500 | Self-drilling, with washers, (pinch point) #8 x 3/4", plain | | | | | 7.80 | | | 7.80 | 8.55 |
| 1600 | Galvanized | | | | | 7.80 | | | 7.80 | 8.55 |
| 1800 | #10 x 3/4", plain | | | | | 8.90 | | | 8.90 | 9.80 |
| 1900 | Galvanized | | | | | 8.90 | | | 8.90 | 9.80 |
| 3000 | Stainless steel w/aluminum or neoprene washers, #14 x 1", plain | | | | | 30 | | | 30 | 33 |
| 3100 | #14 x 2", plain | | | | | 40 | | | 40 | 44 |

### 06 05 23.50 Wood Screws

| | | Crew | Daily Output | Labor-Hours | Unit | Material | 2011 Bare Costs Labor | Equipment | Total | Total Incl O&P |
|---|---|---|---|---|---|---|---|---|---|---|
| 0010 | **WOOD SCREWS** | | | | | | | | | |
| 0020 | #8 x 1" long, steel | | | | C | 3.41 | | | 3.41 | 3.75 |
| 0100 | Brass | | | | | 10.50 | | | 10.50 | 11.55 |
| 0200 | #8, 2" long, steel | | | | | 5.25 | | | 5.25 | 5.80 |
| 0300 | Brass | | | | | 19.95 | | | 19.95 | 22 |
| 0400 | #10, 1" long, steel | | | | | 3.65 | | | 3.65 | 4.02 |
| 0500 | Brass | | | | | 14.15 | | | 14.15 | 15.60 |
| 0600 | #10, 2" long, steel | | | | | 5.85 | | | 5.85 | 6.45 |
| 0700 | Brass | | | | | 24.50 | | | 24.50 | 27 |
| 0800 | #10, 3" long, steel | | | | | 9.80 | | | 9.80 | 10.75 |
| 1000 | #12, 2" long, steel | | | | | 8.05 | | | 8.05 | 8.85 |
| 1100 | Brass | | | | | 29.50 | | | 29.50 | 32 |
| 1500 | #12, 3" long, steel | | | | | 11.65 | | | 11.65 | 12.80 |
| 2000 | #12, 4" long, steel | | | | | 21.50 | | | 21.50 | 23.50 |

### 06 05 23.60 Timber Connectors

| | | Crew | Daily Output | Labor-Hours | Unit | Material | 2011 Bare Costs Labor | Equipment | Total | Total Incl O&P |
|---|---|---|---|---|---|---|---|---|---|---|
| 0010 | **TIMBER CONNECTORS** | | | | | | | | | |
| 0020 | Add up cost of each part for total cost of connection | | | | | | | | | |
| 0100 | Connector plates, steel, with bolts, straight | 2 Carp | 75 | .213 | Ea. | 28 | 9.20 | | 37.20 | 44.50 |
| 0110 | Tee, 7 gauge | | 50 | .320 | | 32 | 13.80 | | 45.80 | 56 |
| 0120 | T-Strap, 14 gauge, 12" x 8" x 2" | | 50 | .320 | | 32 | 13.80 | | 45.80 | 56 |
| 0150 | Anchor plates, 7 ga, 9" x 7" | | 75 | .213 | | 28 | 9.20 | | 37.20 | 44.50 |

## 06 05 23 – Wood, Plastic, and Composite Fastenings

| 06 05 23.60 Timber Connectors | | Crew | Daily Output | Labor-Hours | Unit | Material | 2011 Bare Costs Labor | 2011 Bare Costs Equipment | Total | Total Incl O&P |
|---|---|---|---|---|---|---|---|---|---|---|
| 0200 | Bolts, machine, sq. hd. with nut & washer, 1/2" diameter, 4" long | 1 Carp | 140 | .057 | Ea. | .78 | 2.46 | | 3.24 | 4.63 |
| 0300 | 7-1/2" long | | 130 | .062 | | 1.41 | 2.65 | | 4.06 | 5.60 |
| 0500 | 3/4" diameter, 7-1/2" long | | 130 | .062 | | 3.35 | 2.65 | | 6 | 7.75 |
| 0610 | Machine bolts, w/nut, washer, 3/4" diam., 15" L, HD's & beam hangers | | 95 | .084 | | 6.25 | 3.63 | | 9.88 | 12.40 |
| 0800 | Drilling bolt holes in timber, 1/2" diameter | | 450 | .018 | Inch | | .77 | | .77 | 1.17 |
| 0900 | 1" diameter | | 350 | .023 | " | | .98 | | .98 | 1.51 |
| 1100 | Framing anchor, angle, 3" x 3" x 1-1/2", 12 ga | | 175 | .046 | Ea. | 2.40 | 1.97 | | 4.37 | 5.65 |
| 1150 | Framing anchors, 18 gauge, 4-1/2" x 2-3/4" | | 175 | .046 | | 2.40 | 1.97 | | 4.37 | 5.65 |
| 1160 | Framing anchors, 18 gauge, 4-1/2" x 3" | | 175 | .046 | | 2.40 | 1.97 | | 4.37 | 5.65 |
| 1170 | Clip anchors plates, 18 gauge, 12" x 1-1/8" | | 175 | .046 | | 2.40 | 1.97 | | 4.37 | 5.65 |
| 1250 | Holdowns, 3 gauge base, 10 gauge body | | 8 | 1 | | 23 | 43 | | 66 | 91 |
| 1260 | Holdowns, 7 gauge 11-1/16" x 3-1/4" | | 8 | 1 | | 23 | 43 | | 66 | 91 |
| 1270 | Holdowns, 7 gauge 14-3/8" x 3-1/8" | | 8 | 1 | | 23 | 43 | | 66 | 91 |
| 1275 | Holdowns, 12 gauge 8" x 2-1/2" | | 8 | 1 | | 23 | 43 | | 66 | 91 |
| 1300 | Joist and beam hangers, 18 ga. galv., for 2" x 4" joist | | 175 | .046 | | .67 | 1.97 | | 2.64 | 3.76 |
| 1400 | 2" x 6" to 2" x 10" joist | | 165 | .048 | | 1.28 | 2.09 | | 3.37 | 4.61 |
| 1600 | 16 ga. galv., 3" x 6" to 3" x 10" joist | | 160 | .050 | | 2.72 | 2.15 | | 4.87 | 6.30 |
| 1700 | 3" x 10" to 3" x 14" joist | | 160 | .050 | | 4.52 | 2.15 | | 6.67 | 8.25 |
| 1800 | 4" x 6" to 4" x 10" joist | | 155 | .052 | | 2.82 | 2.22 | | 5.04 | 6.50 |
| 1900 | 4" x 10" to 4" x 14" joist | | 155 | .052 | | 4.62 | 2.22 | | 6.84 | 8.50 |
| 2000 | Two-2" x 6" to two-2" x 10" joists | | 150 | .053 | | 3.85 | 2.30 | | 6.15 | 7.75 |
| 2100 | Two-2" x 10" to two-2" x 14" joists | | 150 | .053 | | 4.30 | 2.30 | | 6.60 | 8.25 |
| 2300 | 3/16" thick, 6" x 8" joist | | 145 | .055 | | 59.50 | 2.38 | | 61.88 | 69 |
| 2400 | 6" x 10" joist | | 140 | .057 | | 62 | 2.46 | | 64.46 | 72.50 |
| 2500 | 6" x 12" joist | | 135 | .059 | | 64.50 | 2.55 | | 67.05 | 75 |
| 2700 | 1/4" thick, 6" x 14" joist | | 130 | .062 | | 67.50 | 2.65 | | 70.15 | 78 |
| 2900 | Plywood clips, extruded aluminum H clip, for 3/4" panels | | | | | .22 | | | .22 | .24 |
| 3000 | Galvanized 18 ga. back-up clip | | | | | .17 | | | .17 | .19 |
| 3200 | Post framing, 16 ga. galv. for 4" x 4" base, 2 piece | 1 Carp | 130 | .062 | | 15.05 | 2.65 | | 17.70 | 20.50 |
| 3300 | Cap | | 130 | .062 | | 21 | 2.65 | | 23.65 | 27 |
| 3500 | Rafter anchors, 18 ga. galv., 1-1/2" wide, 5-1/4" long | | 145 | .055 | | .45 | 2.38 | | 2.83 | 4.14 |
| 3600 | 10-3/4" long | | 145 | .055 | | 1.34 | 2.38 | | 3.72 | 5.10 |
| 3800 | Shear plates, 2-5/8" diameter | | 120 | .067 | | 2.19 | 2.87 | | 5.06 | 6.80 |
| 3900 | 4" diameter | | 115 | .070 | | 5.20 | 3 | | 8.20 | 10.30 |
| 4000 | Sill anchors, embedded in concrete or block, 25-1/2" long | | 115 | .070 | | 11.65 | 3 | | 14.65 | 17.45 |
| 4100 | Spike grids, 3" x 6" | | 120 | .067 | | .88 | 2.87 | | 3.75 | 5.35 |
| 4400 | Split rings, 2-1/2" diameter | | 120 | .067 | | 1.80 | 2.87 | | 4.67 | 6.40 |
| 4500 | 4" diameter | | 110 | .073 | | 2.73 | 3.13 | | 5.86 | 7.80 |
| 4550 | Tie plate, 20 gauge, 7" x 3 1/8" | | 110 | .073 | | 2.73 | 3.13 | | 5.86 | 7.80 |
| 4560 | Tie plate, 20 gauge, 5" x 4 1/8" | | 110 | .073 | | 2.73 | 3.13 | | 5.86 | 7.80 |
| 4575 | Twist straps, 18 gauge, 12" x 1 1/4" | | 110 | .073 | | 2.73 | 3.13 | | 5.86 | 7.80 |
| 4580 | Twist straps, 18 gauge, 16" x 1 1/4" | | 110 | .073 | | 2.73 | 3.13 | | 5.86 | 7.80 |
| 4600 | Strap ties, 20 ga., 2 -1/16" wide, 12 13/16" long | | 180 | .044 | | .87 | 1.91 | | 2.78 | 3.90 |
| 4700 | Strap ties, 16 ga., 1-3/8" wide, 12" long | | 180 | .044 | | .87 | 1.91 | | 2.78 | 3.90 |
| 4800 | 21-5/8" x 1-1/4" | | 160 | .050 | | 2.72 | 2.15 | | 4.87 | 6.30 |
| 5000 | Toothed rings, 2-5/8" or 4" diameter | | 90 | .089 | | 1.63 | 3.83 | | 5.46 | 7.65 |
| 5200 | Truss plates, nailed, 20 gauge, up to 32' span | | 17 | .471 | Truss Ea. | 11.85 | 20.50 | | 32.35 | 44 |
| 5400 | Washers, 2" x 2" x 1/8" | | | | Ea. | .37 | | | .37 | .41 |
| 5500 | 3" x 3" x 3/16" | | | | " | .98 | | | .98 | 1.08 |
| 6000 | Angles and gussets, painted | | | | | | | | | |
| 6012 | 7 ga., 3-1/4" x 3-1/4" x 2-1/2" long | 1 Carp | 1.90 | 4.211 | C | 1,025 | 181 | | 1,206 | 1,400 |
| 6014 | 3-1/4" x 3-1/4" x 5" long | | 1.90 | 4.211 | | 1,975 | 181 | | 2,156 | 2,450 |
| 6016 | 3-1/4" x 3-1/4" x 7-1/2" long | | 1.85 | 4.324 | | 3,750 | 186 | | 3,936 | 4,400 |

| 06 05 23.60 Timber Connectors | | Crew | Daily Output | Labor-Hours | Unit | Material | 2011 Bare Costs Labor | Equipment | Total | Total Incl O&P |
|---|---|---|---|---|---|---|---|---|---|---|
| 6018 | 5-3/4" x 5-3/4" x 2-1/2" long | 1 Carp | 1.85 | 4.324 | C | 2,425 | 186 | | 2,611 | 2,925 |
| 6020 | 5-3/4" x 5-3/4" x 5" long | | 1.85 | 4.324 | | 3,850 | 186 | | 4,036 | 4,500 |
| 6022 | 5-3/4" x 5-3/4" x 7-1/2" long | | 1.80 | 4.444 | | 5,675 | 191 | | 5,866 | 6,550 |
| 6024 | 3 ga., 4-1/4" x 4-1/4" x 3" long | | 1.85 | 4.324 | | 2,575 | 186 | | 2,761 | 3,100 |
| 6026 | 4-1/4" x 4-1/4" x 6" long | | 1.85 | 4.324 | | 5,525 | 186 | | 5,711 | 6,350 |
| 6028 | 4-1/4" x 4-1/4" x 9" long | | 1.80 | 4.444 | | 6,200 | 191 | | 6,391 | 7,125 |
| 6030 | 7-1/4" x 7-1/4" x 3" long | | 1.80 | 4.444 | | 4,450 | 191 | | 4,641 | 5,175 |
| 6032 | 7-1/4" x 7-1/4" x 6" long | | 1.80 | 4.444 | | 6,000 | 191 | | 6,191 | 6,900 |
| 6034 | 7-1/4" x 7-1/4" x 9" long | | 1.75 | 4.571 | | 13,400 | 197 | | 13,597 | 15,100 |
| 6036 | Gussets | | | | | | | | | |
| 6038 | 7 ga., 8-1/8" x 8-1/8" x 2-3/4" long | 1 Carp | 1.80 | 4.444 | C | 4,250 | 191 | | 4,441 | 4,975 |
| 6040 | 3 ga., 9-3/4" x 9-3/4" x 3-1/4" long | " | 1.80 | 4.444 | " | 5,875 | 191 | | 6,066 | 6,775 |
| 6101 | Beam hangers, polymer painted | | | | | | | | | |
| 6102 | Bolted, 3 ga., (W x H x L) | | | | | | | | | |
| 6104 | 3-1/4" x 9" x 12" top flange | 1 Carp | 1 | 8 | C | 18,000 | 345 | | 18,345 | 20,300 |
| 6106 | 5-1/4" x 9" x 12" top flange | | 1 | 8 | | 18,800 | 345 | | 19,145 | 21,200 |
| 6108 | 5-1/4" x 11" x 11-3/4" top flange | | 1 | 8 | | 21,400 | 345 | | 21,745 | 24,000 |
| 6110 | 6-7/8" x 9" x 12" top flange | | 1 | 8 | | 19,500 | 345 | | 19,845 | 21,900 |
| 6112 | 6-7/8" x 11" x 13-1/2" top flange | | 1 | 8 | | 22,500 | 345 | | 22,845 | 25,200 |
| 6114 | 8-7/8" x 11" x 15-1/2" top flange | | 1 | 8 | | 24,000 | 345 | | 24,345 | 26,900 |
| 6116 | Nailed, 3 ga., (W x H x L) | | | | | | | | | |
| 6118 | 3-1/4" x 10-1/2" x 10" top flange | 1 Carp | 1.80 | 4.444 | C | 18,800 | 191 | | 18,991 | 21,000 |
| 6120 | 3-1/4" x 10-1/2" x 12" top flange | | 1.80 | 4.444 | | 18,800 | 191 | | 18,991 | 21,000 |
| 6122 | 5-1/4" x 9-1/2" x 10" top flange | | 1.80 | 4.444 | | 19,200 | 191 | | 19,391 | 21,400 |
| 6124 | 5-1/4" x 9-1/2" x 12" top flange | | 1.80 | 4.444 | | 19,200 | 191 | | 19,391 | 21,400 |
| 6128 | 6-7/8" x 8-1/2" x 12" top flange | | 1.80 | 4.444 | | 19,700 | 191 | | 19,891 | 21,900 |
| 6134 | Saddle hangers, glu-lam (W x H x L) | | | | | | | | | |
| 6136 | 3-1/4" x 10-1/2" x 5-1/4" x 6" saddle | 1 Carp | .50 | 16 | C | 14,100 | 690 | | 14,790 | 16,600 |
| 6138 | 3-1/4" x 10-1/2" x 6-7/8" x 6" saddle | | .50 | 16 | | 14,800 | 690 | | 15,490 | 17,400 |
| 6140 | 3-1/4" x 10-1/2" x 8-7/8" x 6" saddle | | .50 | 16 | | 15,600 | 690 | | 16,290 | 18,200 |
| 6142 | 3-1/4" x 19-1/2" x 5-1/4" x 10-1/8" saddle | | .40 | 20 | | 14,100 | 860 | | 14,960 | 16,800 |
| 6144 | 3-1/4" x 19-1/2" x 6-7/8" x 10-1/8" saddle | | .40 | 20 | | 14,800 | 860 | | 15,660 | 17,600 |
| 6146 | 3-1/4" x 19-1/2" x 8-7/8" x 10-1/8" saddle | | .40 | 20 | | 15,600 | 860 | | 16,460 | 18,400 |
| 6148 | 5-1/4" x 9-1/2" x 5-1/4" x 12" saddle | | .50 | 16 | | 16,800 | 690 | | 17,490 | 19,600 |
| 6150 | 5-1/4" x 9-1/2" x 6-7/8" x 9" saddle | | .50 | 16 | | 18,400 | 690 | | 19,090 | 21,300 |
| 6152 | 5-1/4" x 10-1/2" x spec x 12" saddle | | .50 | 16 | | 20,000 | 690 | | 20,690 | 23,100 |
| 6154 | 5-1/4" x 18" x 5-1/4" x 12-1/8" saddle | | .40 | 20 | | 16,800 | 860 | | 17,660 | 19,800 |
| 6156 | 5-1/4" x 18" x 6-7/8" x 12-1/8" saddle | | .40 | 20 | | 18,400 | 860 | | 19,260 | 21,500 |
| 6158 | 5-1/4" x 18" x spec x 12-1/8" saddle | | .40 | 20 | | 20,000 | 860 | | 20,860 | 23,300 |
| 6160 | 6-7/8" x 8-1/2" x 6-7/8" x 12" saddle | | .50 | 16 | | 20,100 | 690 | | 20,790 | 23,200 |
| 6162 | 6-7/8" x 8-1/2" x 8-7/8" x 12" saddle | | .50 | 16 | | 20,800 | 690 | | 21,490 | 23,900 |
| 6164 | 6-7/8" x 10-1/2" x spec x 12" saddle | | .50 | 16 | | 20,100 | 690 | | 20,790 | 23,200 |
| 6166 | 6-7/8" x 18" x 6-7/8" x 13-3/4" saddle | | .40 | 20 | | 20,100 | 860 | | 20,960 | 23,400 |
| 6168 | 6-7/8" x 18" x 8-7/8" x 13-3/4" saddle | | .40 | 20 | | 20,800 | 860 | | 21,660 | 24,100 |
| 6170 | 6-7/8" x 18" x spec x 13-3/4" saddle | | .40 | 20 | | 22,400 | 860 | | 23,260 | 25,900 |
| 6172 | 8-7/8" x 18" x spec x 15-3/4" saddle | | .40 | 20 | | 35,000 | 860 | | 35,860 | 39,800 |
| 6201 | Beam and purlin hangers, galvanized, 12 ga. | | | | | | | | | |
| 6202 | Purlin or joist size, 3" x 8" | 1 Carp | 1.70 | 4.706 | C | 1,875 | 203 | | 2,078 | 2,350 |
| 6204 | 3" x 10" | | 1.70 | 4.706 | | 2,000 | 203 | | 2,203 | 2,500 |
| 6206 | 3" x 12" | | 1.65 | 4.848 | | 2,300 | 209 | | 2,509 | 2,850 |
| 6208 | 3" x 14" | | 1.65 | 4.848 | | 2,450 | 209 | | 2,659 | 3,025 |
| 6210 | 3" x 16" | | 1.65 | 4.848 | | 2,600 | 209 | | 2,809 | 3,175 |
| 6212 | 4" x 8" | | 1.65 | 4.848 | | 1,875 | 209 | | 2,084 | 2,375 |

| 06 05 23.60 Timber Connectors | | Crew | Daily Output | Labor-Hours | Unit | Material | 2011 Bare Costs Labor | Equipment | Total | Total Incl O&P |
|---|---|---|---|---|---|---|---|---|---|---|
| 6214 | 4" x 10" | 1 Carp | 1.65 | 4.848 | C | 2,025 | 209 | | 2,234 | 2,550 |
| 6216 | 4" x 12" | | 1.60 | 5 | | 2,400 | 215 | | 2,615 | 2,950 |
| 6218 | 4" x 14" | | 1.60 | 5 | | 2,550 | 215 | | 2,765 | 3,125 |
| 6220 | 4" x 16" | | 1.60 | 5 | | 2,675 | 215 | | 2,890 | 3,275 |
| 6222 | 6" x 8" | | 1.60 | 5 | | 2,400 | 215 | | 2,615 | 2,975 |
| 6224 | 6" x 10" | | 1.55 | 5.161 | | 2,450 | 222 | | 2,672 | 3,050 |
| 6226 | 6" x 12" | | 1.55 | 5.161 | | 4,200 | 222 | | 4,422 | 4,975 |
| 6228 | 6" x 14" | | 1.50 | 5.333 | | 4,450 | 230 | | 4,680 | 5,250 |
| 6230 | 6" x 16" | ▼ | 1.50 | 5.333 | ▼ | 4,725 | 230 | | 4,955 | 5,525 |
| 6250 | Beam seats | | | | | | | | | |
| 6252 | Beam size, 5-1/4" wide | | | | | | | | | |
| 6254 | 5" x 7" x 1/4" | 1 Carp | 1.80 | 4.444 | C | 7,100 | 191 | | 7,291 | 8,100 |
| 6256 | 6" x 7" x 3/8" | | 1.80 | 4.444 | | 7,950 | 191 | | 8,141 | 9,050 |
| 6258 | 7" x 7" x 3/8" | | 1.80 | 4.444 | | 8,500 | 191 | | 8,691 | 9,650 |
| 6260 | 8" x 7" x 3/8" | ▼ | 1.80 | 4.444 | ▼ | 10,000 | 191 | | 10,191 | 11,300 |
| 6262 | Beam size, 6-7/8" wide | | | | | | | | | |
| 6264 | 5" x 9" x 1/4" | 1 Carp | 1.80 | 4.444 | C | 8,475 | 191 | | 8,666 | 9,625 |
| 6266 | 6" x 9" x 3/8" | | 1.80 | 4.444 | | 11,100 | 191 | | 11,291 | 12,500 |
| 6268 | 7" x 9" x 3/8" | | 1.80 | 4.444 | | 11,200 | 191 | | 11,391 | 12,600 |
| 6270 | 8" x 9" x 3/8" | ▼ | 1.80 | 4.444 | | 13,200 | 191 | | 13,391 | 14,900 |
| 6272 | Special beams, over 6-7/8" wide | | | | | | | | | |
| 6274 | 5" x 10" x 3/8" | 1 Carp | 1.80 | 4.444 | C | 11,500 | 191 | | 11,691 | 12,900 |
| 6276 | 6" x 10" x 3/8" | | 1.80 | 4.444 | | 13,400 | 191 | | 13,591 | 15,000 |
| 6278 | 7" x 10" x 3/8" | | 1.80 | 4.444 | | 14,000 | 191 | | 14,191 | 15,700 |
| 6280 | 8" x 10" x 3/8" | | 1.75 | 4.571 | | 15,000 | 197 | | 15,197 | 16,800 |
| 6282 | 5-1/4" x 12" x 5/16" | | 1.75 | 4.571 | | 11,600 | 197 | | 11,797 | 13,100 |
| 6284 | 6-1/2" x 12" x 3/8" | | 1.75 | 4.571 | | 19,200 | 197 | | 19,397 | 21,500 |
| 6286 | 5-1/4" x 16" x 5/16" | | 1.70 | 4.706 | | 17,100 | 203 | | 17,303 | 19,100 |
| 6288 | 6-1/2" x 16" x 3/8" | | 1.70 | 4.706 | | 22,300 | 203 | | 22,503 | 24,800 |
| 6290 | 5-1/4" x 20" x 5/16" | | 1.70 | 4.706 | | 20,100 | 203 | | 20,303 | 22,400 |
| 6292 | 6-1/2" x 20" x 3/8" | ▼ | 1.65 | 4.848 | ▼ | 26,200 | 209 | | 26,409 | 29,100 |
| 6300 | Column bases | | | | | | | | | |
| 6302 | 4 x 4, 16 ga. | 1 Carp | 1.80 | 4.444 | C | 715 | 191 | | 906 | 1,075 |
| 6306 | 7 ga. | | 1.80 | 4.444 | | 2,625 | 191 | | 2,816 | 3,175 |
| 6308 | 4 x 6, 16 ga. | | 1.80 | 4.444 | | 1,675 | 191 | | 1,866 | 2,125 |
| 6312 | 7 ga. | | 1.80 | 4.444 | | 2,725 | 191 | | 2,916 | 3,300 |
| 6314 | 6 x 6, 16 ga. | | 1.75 | 4.571 | | 1,900 | 197 | | 2,097 | 2,375 |
| 6318 | 7 ga. | | 1.75 | 4.571 | | 3,725 | 197 | | 3,922 | 4,400 |
| 6320 | 6 x 8, 7 ga. | | 1.70 | 4.706 | | 2,900 | 203 | | 3,103 | 3,500 |
| 6322 | 6 x 10, 7 ga. | | 1.70 | 4.706 | | 3,125 | 203 | | 3,328 | 3,725 |
| 6324 | 6 x 12, 7 ga. | | 1.70 | 4.706 | | 3,375 | 203 | | 3,578 | 4,025 |
| 6326 | 8 x 8, 7 ga. | | 1.65 | 4.848 | | 5,700 | 209 | | 5,909 | 6,600 |
| 6330 | 8 x 10, 7 ga. | | 1.65 | 4.848 | | 6,825 | 209 | | 7,034 | 7,850 |
| 6332 | 8 x 12, 7 ga. | | 1.60 | 5 | | 7,425 | 215 | | 7,640 | 8,475 |
| 6334 | 10 x 10, 3 ga. | | 1.60 | 5 | | 7,575 | 215 | | 7,790 | 8,650 |
| 6336 | 10 x 12, 3 ga. | | 1.60 | 5 | | 8,700 | 215 | | 8,915 | 9,900 |
| 6338 | 12 x 12, 3 ga. | ▼ | 1.55 | 5.161 | ▼ | 9,450 | 222 | | 9,672 | 10,700 |
| 6350 | Column caps, painted, 3 ga. | | | | | | | | | |
| 6352 | 3-1/4" x 3-5/8" | 1 Carp | 1.80 | 4.444 | C | 9,600 | 191 | | 9,791 | 10,900 |
| 6354 | 3-1/4" x 5-1/2" | | 1.80 | 4.444 | | 9,600 | 191 | | 9,791 | 10,900 |
| 6356 | 3-5/8" x 3-5/8" | | 1.80 | 4.444 | | 7,850 | 191 | | 8,041 | 8,950 |
| 6358 | 3-5/8" x 5-1/2" | | 1.80 | 4.444 | | 7,850 | 191 | | 8,041 | 8,950 |
| 6360 | 5-1/4" x 5-1/2" | ▼ | 1.75 | 4.571 | | 10,300 | 197 | | 10,497 | 11,600 |

| 06 05 23.60 Timber Connectors | | Crew | Daily Output | Labor-Hours | Unit | Material | 2011 Bare Costs Labor | Equipment | Total | Total Incl O&P |
|---|---|---|---|---|---|---|---|---|---|---|
| 6362 | 5-1/4" x 7-1/2" | 1 Carp | 1.75 | 4.571 | C | 10,300 | 197 | | 10,497 | 11,600 |
| 6364 | 5-1/2" x 3-5/8" | | 1.75 | 4.571 | | 11,100 | 197 | | 11,297 | 12,500 |
| 6366 | 5-1/2" x 5-1/2" | | 1.75 | 4.571 | | 11,100 | 197 | | 11,297 | 12,500 |
| 6368 | 5-1/2" x 7-1/2" | | 1.70 | 4.706 | | 11,100 | 203 | | 11,303 | 12,500 |
| 6370 | 6-7/8" x 5-1/2" | | 1.70 | 4.706 | | 11,600 | 203 | | 11,803 | 13,000 |
| 6372 | 6-7/8" x 6-7/8" | | 1.70 | 4.706 | | 11,600 | 203 | | 11,803 | 13,000 |
| 6374 | 6-7/8" x 7-1/2" | | 1.70 | 4.706 | | 11,600 | 203 | | 11,803 | 13,000 |
| 6376 | 7-1/2" x 5-1/2" | | 1.65 | 4.848 | | 12,100 | 209 | | 12,309 | 13,600 |
| 6378 | 7-1/2" x 7-1/2" | | 1.65 | 4.848 | | 12,100 | 209 | | 12,309 | 13,600 |
| 6380 | 8-7/8" x 5-1/2" | | 1.60 | 5 | | 12,800 | 215 | | 13,015 | 14,300 |
| 6382 | 8-7/8" x 7-1/2" | | 1.60 | 5 | | 12,800 | 215 | | 13,015 | 14,300 |
| 6384 | 9-1/2" x 5-1/2" | ▼ | 1.60 | 5 | ▼ | 17,300 | 215 | | 17,515 | 19,300 |
| 6400 | Floor tie anchors, polymer paint | | | | | | | | | |
| 6402 | 10 ga., 3" x 37-1/2" | 1 Carp | 1.80 | 4.444 | C | 4,600 | 191 | | 4,791 | 5,350 |
| 6404 | 3-1/2" x 45-1/2" | | 1.75 | 4.571 | | 4,825 | 197 | | 5,022 | 5,600 |
| 6406 | 3 ga., 3-1/2" x 56" | ▼ | 1.70 | 4.706 | ▼ | 8,500 | 203 | | 8,703 | 9,650 |
| 6410 | Girder hangers | | | | | | | | | |
| 6412 | 6" wall thickness, 4" x 6" | 1 Carp | 1.80 | 4.444 | C | 2,575 | 191 | | 2,766 | 3,150 |
| 6414 | 4" x 8" | | 1.80 | 4.444 | | 2,875 | 191 | | 3,066 | 3,475 |
| 6416 | 8" wall thickness, 4" x 6" | | 1.80 | 4.444 | | 3,000 | 191 | | 3,191 | 3,575 |
| 6418 | 4" x 8" | ▼ | 1.80 | 4.444 | ▼ | 2,975 | 191 | | 3,166 | 3,575 |
| 6420 | Hinge connections, polymer painted | | | | | | | | | |
| 6422 | 3/4" thick top plate | | | | | | | | | |
| 6424 | 5-1/4" x 12" w/5" x 5" top | 1 Carp | 1 | 8 | C | 33,000 | 345 | | 33,345 | 36,800 |
| 6426 | 5-1/4" x 15" w/6" x 6" top | | .80 | 10 | | 35,100 | 430 | | 35,530 | 39,300 |
| 6428 | 5-1/4" x 18" w/7" x 7" top | | .70 | 11.429 | | 36,900 | 490 | | 37,390 | 41,300 |
| 6430 | 5-1/4" x 26" w/9" x 9" top | ▼ | .60 | 13.333 | ▼ | 39,200 | 575 | | 39,775 | 44,100 |
| 6432 | 1" thick top plate | | | | | | | | | |
| 6434 | 6-7/8" x 14" w/5" x 5" top | 1 Carp | .80 | 10 | C | 40,200 | 430 | | 40,630 | 44,900 |
| 6436 | 6-7/8" x 17" w/6" x 6" top | | .80 | 10 | | 44,600 | 430 | | 45,030 | 49,800 |
| 6438 | 6-7/8" x 21" w/7" x 7" top | | .70 | 11.429 | | 48,700 | 490 | | 49,190 | 54,500 |
| 6440 | 6-7/8" x 31" w/9" x 9" top | ▼ | .60 | 13.333 | ▼ | 53,500 | 575 | | 54,075 | 59,500 |
| 6442 | 1-1/4" thick top plate | | | | | | | | | |
| 6444 | 8-7/8" x 16" w/5" x 5" top | 1 Carp | .60 | 13.333 | C | 50,000 | 575 | | 50,575 | 56,000 |
| 6446 | 8-7/8" x 21" w/6" x 6" top | | .50 | 16 | | 55,000 | 690 | | 55,690 | 61,500 |
| 6448 | 8-7/8" x 26" w/7" x 7" top | | .40 | 20 | | 62,500 | 860 | | 63,360 | 70,000 |
| 6450 | 8-7/8" x 39" w/9" x 9" top | ▼ | .30 | 26.667 | ▼ | 78,000 | 1,150 | | 79,150 | 87,500 |
| 6460 | Holddowns | | | | | | | | | |
| 6462 | Embedded along edge | | | | | | | | | |
| 6464 | 26" long, 12 ga. | 1 Carp | .90 | 8.889 | C | 1,250 | 385 | | 1,635 | 1,950 |
| 6466 | 35" long, 12 ga. | | .85 | 9.412 | | 1,675 | 405 | | 2,080 | 2,450 |
| 6468 | 35" long, 10 ga. | ▼ | .85 | 9.412 | ▼ | 1,325 | 405 | | 1,730 | 2,075 |
| 6470 | Embedded away from edge | | | | | | | | | |
| 6472 | Medium duty, 12 ga. | | | | | | | | | |
| 6474 | 18-1/2" long | 1 Carp | .95 | 8.421 | C | 760 | 365 | | 1,125 | 1,400 |
| 6476 | 23-3/4" long | | .90 | 8.889 | | 885 | 385 | | 1,270 | 1,550 |
| 6478 | 28" long | | .85 | 9.412 | | 905 | 405 | | 1,310 | 1,625 |
| 6480 | 35" long | ▼ | .85 | 9.412 | ▼ | 1,250 | 405 | | 1,655 | 2,000 |
| 6482 | Heavy duty, 10 ga. | | | | | | | | | |
| 6484 | 28" long | 1 Carp | .85 | 9.412 | C | 1,625 | 405 | | 2,030 | 2,400 |
| 6486 | 35" long | " | .85 | 9.412 | " | 1,775 | 405 | | 2,180 | 2,575 |
| 6490 | Surface mounted (W x H) | | | | | | | | | |
| 6492 | 2-1/2" x 5-3/4", 7 ga. | 1 Carp | 1 | 8 | C | 1,850 | 345 | | 2,195 | 2,550 |

| 06 05 23.60 Timber Connectors | Crew | Daily Output | Labor-Hours | Unit | Material | 2011 Bare Costs Labor | Equipment | Total | Total Incl O&P |
|---|---|---|---|---|---|---|---|---|---|
| 6494 | 2-1/2" x 8", 12 ga. | 1 Carp | 1 | 8 | C | 1,125 | 345 | | 1,470 | 1,775 |
| 6496 | 2-7/8" x 6-3/8", 7 ga. | | 1 | 8 | | 4,075 | 345 | | 4,420 | 5,000 |
| 6498 | 2-7/8" x 12-1/2", 3 ga. | | 1 | 8 | | 4,175 | 345 | | 4,520 | 5,125 |
| 6500 | 3-3/16" x 9-3/8", 10 ga. | | 1 | 8 | | 2,800 | 345 | | 3,145 | 3,600 |
| 6502 | 3-1/2" x 11-5/8", 3 ga. | | 1 | 8 | | 5,100 | 345 | | 5,445 | 6,125 |
| 6504 | 3-1/2" x 14-3/4", 3 ga. | | 1 | 8 | | 3,400 | 345 | | 3,745 | 4,250 |
| 6506 | 3-1/2" x 16-1/2", 3 ga. | | 1 | 8 | | 7,775 | 345 | | 8,120 | 9,075 |
| 6508 | 3-1/2" x 20-1/2", 3 ga. | | .90 | 8.889 | | 7,950 | 385 | | 8,335 | 9,325 |
| 6510 | 3-1/2" x 24-1/2", 3 ga. | | .90 | 8.889 | | 10,100 | 385 | | 10,485 | 11,700 |
| 6512 | 4-1/4" x 20-3/4", 3 ga. | | .90 | 8.889 | | 6,850 | 385 | | 7,235 | 8,100 |
| 6520 | Joist hangers | | | | | | | | | |
| 6522 | Sloped, field adjustable, 18 ga. | | | | | | | | | |
| 6524 | 2" x 6" | 1 Carp | 1.65 | 4.848 | C | 500 | 209 | | 709 | 870 |
| 6526 | 2" x 8" | | 1.65 | 4.848 | | 900 | 209 | | 1,109 | 1,300 |
| 6528 | 2" x 10" and up | | 1.65 | 4.848 | | 1,500 | 209 | | 1,709 | 1,975 |
| 6530 | 3" x 10" and up | | 1.60 | 5 | | 1,125 | 215 | | 1,340 | 1,550 |
| 6532 | 4" x 10" and up | | 1.55 | 5.161 | | 1,350 | 222 | | 1,572 | 1,850 |
| 6536 | Skewed 45°, 16 ga. | | | | | | | | | |
| 6538 | 2" x 4" | 1 Carp | 1.75 | 4.571 | C | 800 | 197 | | 997 | 1,175 |
| 6540 | 2" x 6" or 2" x 8" | | 1.65 | 4.848 | | 815 | 209 | | 1,024 | 1,225 |
| 6542 | 2" x 10" or 2" x 12" | | 1.65 | 4.848 | | 935 | 209 | | 1,144 | 1,350 |
| 6544 | 2" x 14" or 2" x 16" | | 1.60 | 5 | | 1,675 | 215 | | 1,890 | 2,150 |
| 6546 | (2) 2" x 6" or (2) 2" x 8" | | 1.60 | 5 | | 1,500 | 215 | | 1,715 | 1,975 |
| 6548 | (2) 2" x 10" or (2) 2" x 12" | | 1.55 | 5.161 | | 1,625 | 222 | | 1,847 | 2,125 |
| 6550 | (2) 2" x 14" or (2) 2" x 16" | | 1.50 | 5.333 | | 2,550 | 230 | | 2,780 | 3,150 |
| 6552 | 4" x 6" or 4" x 8" | | 1.60 | 5 | | 1,250 | 215 | | 1,465 | 1,725 |
| 6554 | 4" x 10" or 4" x 12" | | 1.55 | 5.161 | | 1,475 | 222 | | 1,697 | 1,975 |
| 6556 | 4" x 14" or 4" x 16" | | 1.55 | 5.161 | | 2,275 | 222 | | 2,497 | 2,850 |
| 6560 | Skewed 45°, 14 ga. | | | | | | | | | |
| 6562 | (2) 2" x 6" or (2) 2" x 8" | 1 Carp | 1.60 | 5 | C | 1,725 | 215 | | 1,940 | 2,225 |
| 6564 | (2) 2" x 10" or (2) 2" x 12" | | 1.55 | 5.161 | | 2,400 | 222 | | 2,622 | 3,000 |
| 6566 | (2) 2" x 14" or (2) 2" x 16" | | 1.50 | 5.333 | | 3,525 | 230 | | 3,755 | 4,225 |
| 6568 | 4" x 6" or 4" x 8" | | 1.60 | 5 | | 2,075 | 215 | | 2,290 | 2,600 |
| 6570 | 4" x 10" or 4" x 12" | | 1.55 | 5.161 | | 2,175 | 222 | | 2,397 | 2,750 |
| 6572 | 4" x 14" or 4" x 16" | | 1.55 | 5.161 | | 2,925 | 222 | | 3,147 | 3,575 |
| 6590 | Joist hangers, heavy duty 12 ga., galvanized | | | | | | | | | |
| 6592 | 2" x 4" | 1 Carp | 1.75 | 4.571 | C | 1,175 | 197 | | 1,372 | 1,575 |
| 6594 | 2" x 6" | | 1.65 | 4.848 | | 1,275 | 209 | | 1,484 | 1,725 |
| 6595 | 2" x 6", 16 gauge | | 1.65 | 4.848 | | 1,275 | 209 | | 1,484 | 1,725 |
| 6596 | 2" x 8" | | 1.65 | 4.848 | | 1,950 | 209 | | 2,159 | 2,450 |
| 6597 | 2" x 8", 16 gauge | | 1.65 | 4.848 | | 1,950 | 209 | | 2,159 | 2,450 |
| 6598 | 2" x 10" | | 1.65 | 4.848 | | 2,000 | 209 | | 2,209 | 2,525 |
| 6600 | 2" x 12" | | 1.65 | 4.848 | | 2,450 | 209 | | 2,659 | 3,025 |
| 6602 | 2" x 14" | | 1.65 | 4.848 | | 2,550 | 209 | | 2,759 | 3,125 |
| 6604 | 2" x 16" | | 1.65 | 4.848 | | 2,700 | 209 | | 2,909 | 3,300 |
| 6606 | 3" x 4" | | 1.65 | 4.848 | | 1,625 | 209 | | 1,834 | 2,100 |
| 6608 | 3" x 6" | | 1.65 | 4.848 | | 2,175 | 209 | | 2,384 | 2,725 |
| 6610 | 3" x 8" | | 1.65 | 4.848 | | 2,200 | 209 | | 2,409 | 2,750 |
| 6612 | 3" x 10" | | 1.60 | 5 | | 2,550 | 215 | | 2,765 | 3,150 |
| 6614 | 3" x 12" | | 1.60 | 5 | | 3,075 | 215 | | 3,290 | 3,700 |
| 6616 | 3" x 14" | | 1.60 | 5 | | 3,600 | 215 | | 3,815 | 4,275 |
| 6618 | 3" x 16" | | 1.60 | 5 | | 3,950 | 215 | | 4,165 | 4,675 |
| 6620 | (2) 2" x 4" | | 1.75 | 4.571 | | 1,900 | 197 | | 2,097 | 2,400 |

| 06 05 23.60 Timber Connectors | | Crew | Daily Output | Labor-Hours | Unit | Material | 2011 Bare Costs Labor | Equipment | Total | Total Incl O&P |
|---|---|---|---|---|---|---|---|---|---|---|
| 6622 | (2) 2" x 6" | 1 Carp | 1.60 | 5 | C | 2,300 | 215 | | 2,515 | 2,850 |
| 6624 | (2) 2" x 8" | | 1.60 | 5 | | 2,350 | 215 | | 2,565 | 2,900 |
| 6626 | (2) 2" x 10" | | 1.55 | 5.161 | | 2,550 | 222 | | 2,772 | 3,150 |
| 6628 | (2) 2" x 12" | | 1.55 | 5.161 | | 3,250 | 222 | | 3,472 | 3,925 |
| 6630 | (2) 2" x 14" | | 1.50 | 5.333 | | 3,275 | 230 | | 3,505 | 3,950 |
| 6632 | (2) 2" x 16" | | 1.50 | 5.333 | | 3,300 | 230 | | 3,530 | 4,000 |
| 6634 | 4" x 4" | | 1.65 | 4.848 | | 1,375 | 209 | | 1,584 | 1,850 |
| 6636 | 4" x 6" | | 1.60 | 5 | | 1,525 | 215 | | 1,740 | 2,000 |
| 6638 | 4" x 8" | | 1.60 | 5 | | 1,750 | 215 | | 1,965 | 2,250 |
| 6640 | 4" x 10" | | 1.55 | 5.161 | | 2,150 | 222 | | 2,372 | 2,725 |
| 6642 | 4" x 12" | | 1.55 | 5.161 | | 2,300 | 222 | | 2,522 | 2,875 |
| 6644 | 4" x 14" | | 1.55 | 5.161 | | 2,750 | 222 | | 2,972 | 3,375 |
| 6646 | 4" x 16" | | 1.55 | 5.161 | | 3,000 | 222 | | 3,222 | 3,650 |
| 6648 | (3) 2" x 10" | | 1.50 | 5.333 | | 3,325 | 230 | | 3,555 | 4,000 |
| 6650 | (3) 2" x 12" | | 1.50 | 5.333 | | 3,700 | 230 | | 3,930 | 4,425 |
| 6652 | (3) 2" x 14" | | 1.45 | 5.517 | | 3,975 | 238 | | 4,213 | 4,750 |
| 6654 | (3) 2" x 16" | | 1.45 | 5.517 | | 4,050 | 238 | | 4,288 | 4,825 |
| 6656 | 6" x 6" | | 1.60 | 5 | | 1,775 | 215 | | 1,990 | 2,300 |
| 6658 | 6" x 8" | | 1.60 | 5 | | 1,850 | 215 | | 2,065 | 2,350 |
| 6660 | 6" x 10" | | 1.55 | 5.161 | | 2,200 | 222 | | 2,422 | 2,775 |
| 6662 | 6" x 12" | | 1.55 | 5.161 | | 2,500 | 222 | | 2,722 | 3,100 |
| 6664 | 6" x 14" | | 1.50 | 5.333 | | 3,150 | 230 | | 3,380 | 3,800 |
| 6666 | 6" x 16" | ▼ | 1.50 | 5.333 | ▼ | 3,700 | 230 | | 3,930 | 4,425 |
| 6690 | Knee braces, galvanized, 12 ga. | | | | | | | | | |
| 6692 | Beam depth, 10" x 15" x 5' long | 1 Carp | 1.80 | 4.444 | C | 4,800 | 191 | | 4,991 | 5,600 |
| 6694 | 15" x 22-1/2" x 7' long | | 1.70 | 4.706 | | 5,525 | 203 | | 5,728 | 6,375 |
| 6696 | 22-1/2" x 28-1/2" x 8' long | | 1.60 | 5 | | 5,925 | 215 | | 6,140 | 6,850 |
| 6698 | 28-1/2" x 36" x 10' long | | 1.55 | 5.161 | | 6,175 | 222 | | 6,397 | 7,150 |
| 6700 | 36" x 42" x 12' long | ▼ | 1.50 | 5.333 | ▼ | 6,825 | 230 | | 7,055 | 7,850 |
| 6710 | Mudsill anchors | | | | | | | | | |
| 6714 | 2" x 4" or 3" x 4" | 1 Carp | 115 | .070 | C | 1,225 | 3 | | 1,228 | 1,350 |
| 6716 | 2" x 6" or 3" x 6" | | 115 | .070 | | 1,225 | 3 | | 1,228 | 1,350 |
| 6718 | Block wall, 13-1/4" long | | 115 | .070 | | 850 | 3 | | 853 | 940 |
| 6720 | 21-1/4" long | ▼ | 115 | .070 | ▼ | 127 | 3 | | 130 | 144 |
| 6730 | Post bases, 12 ga. galvanized | | | | | | | | | |
| 6732 | Adjustable, 3-9/16" x 3-9/16" | 1 Carp | 1.30 | 6.154 | C | 980 | 265 | | 1,245 | 1,475 |
| 6734 | 3-9/16" x 5-1/2" | | 1.30 | 6.154 | | 1,275 | 265 | | 1,540 | 1,800 |
| 6736 | 4" x 4" | | 1.30 | 6.154 | | 715 | 265 | | 980 | 1,200 |
| 6738 | 4" x 6" | | 1.30 | 6.154 | | 1,675 | 265 | | 1,940 | 2,225 |
| 6740 | 5-1/2" x 5-1/2" | | 1.30 | 6.154 | | 2,475 | 265 | | 2,740 | 3,125 |
| 6742 | 6" x 6" | | 1.30 | 6.154 | | 1,900 | 265 | | 2,165 | 2,475 |
| 6744 | Elevated, 3-9/16" x 3-1/4" | | 1.30 | 6.154 | | 1,075 | 265 | | 1,340 | 1,575 |
| 6746 | 5-1/2" x 3-5/16" | | 1.30 | 6.154 | | 1,525 | 265 | | 1,790 | 2,075 |
| 6748 | 5-1/2" x 5" | | 1.30 | 6.154 | | 2,275 | 265 | | 2,540 | 2,900 |
| 6750 | Regular, 3-9/16" x 3-3/8" | | 1.30 | 6.154 | | 815 | 265 | | 1,080 | 1,300 |
| 6752 | 4" x 3-3/8" | | 1.30 | 6.154 | | 1,150 | 265 | | 1,415 | 1,650 |
| 6754 | 18 gauge, 5-1/4" x 3-1/8" | | 1.30 | 6.154 | | 1,200 | 265 | | 1,465 | 1,725 |
| 6755 | 5-1/2" x 3-3/8" | | 1.30 | 6.154 | | 1,200 | 265 | | 1,465 | 1,725 |
| 6756 | 5-1/2" x 5-3/8" | | 1.30 | 6.154 | | 1,725 | 265 | | 1,990 | 2,300 |
| 6758 | 6" x 3-3/8" | | 1.30 | 6.154 | | 2,075 | 265 | | 2,340 | 2,675 |
| 6760 | 6" x 5-3/8" | ▼ | 1.30 | 6.154 | ▼ | 2,300 | 265 | | 2,565 | 2,950 |
| 6762 | Post combination cap/bases | | | | | | | | | |
| 6764 | 3-9/16" x 3-9/16" | 1 Carp | 1.20 | 6.667 | C | 465 | 287 | | 752 | 950 |

| 06 05 23.60 **Timber Connectors** | | Crew | Daily Output | Labor-Hours | Unit | Material | 2011 Bare Costs Labor | Equipment | Total | Total Incl O&P |
|---|---|---|---|---|---|---|---|---|---|---|
| 6766 | 3-9/16" x 5-1/2" | 1 Carp | 1.20 | 6.667 | C | 1,050 | 287 | | 1,337 | 1,600 |
| 6768 | 4" x 4" | | 1.20 | 6.667 | | 2,100 | 287 | | 2,387 | 2,750 |
| 6770 | 5-1/2" x 5-1/2" | | 1.20 | 6.667 | | 1,150 | 287 | | 1,437 | 1,725 |
| 6772 | 6" x 6" | | 1.20 | 6.667 | | 3,825 | 287 | | 4,112 | 4,675 |
| 6774 | 7-1/2" x 7-1/2" | | 1.20 | 6.667 | | 4,200 | 287 | | 4,487 | 5,075 |
| 6776 | 8" x 8" | | 1.20 | 6.667 | | 4,350 | 287 | | 4,637 | 5,225 |
| 6790 | Post-beam connection caps | | | | | | | | | |
| 6792 | Beam size 3-9/16" | | | | | | | | | |
| 6794 | 12 ga. post, 4" x 4" | 1 Carp | 1 | 8 | C | 2,650 | 345 | | 2,995 | 3,450 |
| 6796 | 4" x 6" | | 1 | 8 | | 3,550 | 345 | | 3,895 | 4,425 |
| 6798 | 4" x 8" | | 1 | 8 | | 5,225 | 345 | | 5,570 | 6,275 |
| 6800 | 16 ga. post, 4" x 4" | | 1 | 8 | | 1,200 | 345 | | 1,545 | 1,850 |
| 6802 | 4" x 6" | | 1 | 8 | | 2,000 | 345 | | 2,345 | 2,725 |
| 6804 | 4" x 8" | | 1 | 8 | | 3,375 | 345 | | 3,720 | 4,225 |
| 6805 | 18 ga. post, 2-7/8" x 3" | | 1 | 8 | | 3,375 | 345 | | 3,720 | 4,225 |
| 6806 | Beam size 5-1/2" | | | | | | | | | |
| 6808 | 12 ga. post, 6" x 4" | 1 Carp | 1 | 8 | C | 3,475 | 345 | | 3,820 | 4,350 |
| 6810 | 6" x 6" | | 1 | 8 | | 5,475 | 345 | | 5,820 | 6,550 |
| 6812 | 6" x 8" | | 1 | 8 | | 3,775 | 345 | | 4,120 | 4,675 |
| 6816 | 16 ga. post, 6" x 4" | | 1 | 8 | | 1,875 | 345 | | 2,220 | 2,600 |
| 6818 | 6" x 6" | | 1 | 8 | | 2,000 | 345 | | 2,345 | 2,725 |
| 6820 | Beam size 7-1/2" | | | | | | | | | |
| 6822 | 12 ga. post, 8" x 4" | 1 Carp | 1 | 8 | C | 4,800 | 345 | | 5,145 | 5,825 |
| 6824 | 8" x 6" | | 1 | 8 | | 5,050 | 345 | | 5,395 | 6,075 |
| 6826 | 8" x 8" | | 1 | 8 | | 7,575 | 345 | | 7,920 | 8,875 |
| 6840 | Purlin anchors, embedded | | | | | | | | | |
| 6842 | Heavy duty, 10 ga. | | | | | | | | | |
| 6844 | Straight, 28" long | 1 Carp | 1.60 | 5 | C | 1,325 | 215 | | 1,540 | 1,775 |
| 6846 | 35" long | | 1.50 | 5.333 | | 1,625 | 230 | | 1,855 | 2,150 |
| 6848 | Twisted, 28" long | | 1.60 | 5 | | 1,325 | 215 | | 1,540 | 1,775 |
| 6850 | 35" long | | 1.50 | 5.333 | | 1,625 | 230 | | 1,855 | 2,150 |
| 6852 | Regular duty, 12 ga. | | | | | | | | | |
| 6854 | Straight, 18-1/2" long | 1 Carp | 1.80 | 4.444 | C | 755 | 191 | | 946 | 1,125 |
| 6856 | 23-3/4" long | | 1.70 | 4.706 | | 950 | 203 | | 1,153 | 1,350 |
| 6858 | 29" long | | 1.60 | 5 | | 970 | 215 | | 1,185 | 1,400 |
| 6860 | 35" long | | 1.50 | 5.333 | | 1,325 | 230 | | 1,555 | 1,825 |
| 6862 | Twisted, 18" long | | 1.80 | 4.444 | | 755 | 191 | | 946 | 1,125 |
| 6866 | 28" long | | 1.60 | 5 | | 900 | 215 | | 1,115 | 1,325 |
| 6868 | 35" long | | 1.50 | 5.333 | | 1,325 | 230 | | 1,555 | 1,825 |
| 6870 | Straight, plastic coated | | | | | | | | | |
| 6872 | 23-1/2" long | 1 Carp | 1.60 | 5 | C | 1,850 | 215 | | 2,065 | 2,375 |
| 6874 | 26-7/8" long | | 1.60 | 5 | | 2,175 | 215 | | 2,390 | 2,725 |
| 6876 | 32-1/2" long | | 1.50 | 5.333 | | 2,325 | 230 | | 2,555 | 2,900 |
| 6878 | 35-7/8" long | | 1.50 | 5.333 | | 2,425 | 230 | | 2,655 | 3,000 |
| 6890 | Purlin hangers, painted | | | | | | | | | |
| 6892 | 12 ga., 2" x 6" | 1 Carp | 1.80 | 4.444 | C | 1,700 | 191 | | 1,891 | 2,175 |
| 6894 | 2" x 8" | | 1.80 | 4.444 | | 1,850 | 191 | | 2,041 | 2,350 |
| 6896 | 2" x 10" | | 1.80 | 4.444 | | 2,000 | 191 | | 2,191 | 2,500 |
| 6898 | 2" x 12" | | 1.75 | 4.571 | | 2,150 | 197 | | 2,347 | 2,675 |
| 6900 | 2" x 14" | | 1.75 | 4.571 | | 2,300 | 197 | | 2,497 | 2,825 |
| 6902 | 2" x 16" | | 1.75 | 4.571 | | 2,450 | 197 | | 2,647 | 2,975 |
| 6904 | 3" x 6" | | 1.70 | 4.706 | | 1,725 | 203 | | 1,928 | 2,200 |
| 6906 | 3" x 8" | | 1.70 | 4.706 | | 1,875 | 203 | | 2,078 | 2,350 |

| 06 05 23.60 Timber Connectors | | Crew | Daily Output | Labor-Hours | Unit | Material | 2011 Bare Costs Labor | Equipment | Total | Total Incl O&P |
|---|---|---|---|---|---|---|---|---|---|---|
| 6908 | 3" x 10" | 1 Carp | 1.70 | 4.706 | C | 2,000 | 203 | | 2,203 | 2,500 |
| 6910 | 3" x 12" | | 1.65 | 4.848 | | 2,300 | 209 | | 2,509 | 2,850 |
| 6912 | 3" x 14" | | 1.65 | 4.848 | | 2,450 | 209 | | 2,659 | 3,025 |
| 6914 | 3" x 16" | | 1.65 | 4.848 | | 2,600 | 209 | | 2,809 | 3,175 |
| 6916 | 4" x 6" | | 1.65 | 4.848 | | 1,725 | 209 | | 1,934 | 2,225 |
| 6918 | 4" x 8" | | 1.65 | 4.848 | | 1,875 | 209 | | 2,084 | 2,375 |
| 6920 | 4" x 10" | | 1.65 | 4.848 | | 2,025 | 209 | | 2,234 | 2,550 |
| 6922 | 4" x 12" | | 1.60 | 5 | | 2,400 | 215 | | 2,615 | 2,950 |
| 6924 | 4" x 14" | | 1.60 | 5 | | 2,550 | 215 | | 2,765 | 3,125 |
| 6926 | 4" x 16" | | 1.60 | 5 | | 2,675 | 215 | | 2,890 | 3,275 |
| 6928 | 6" x 6" | | 1.60 | 5 | | 2,275 | 215 | | 2,490 | 2,825 |
| 6930 | 6" x 8" | | 1.60 | 5 | | 2,400 | 215 | | 2,615 | 2,975 |
| 6932 | 6" x 10" | | 1.55 | 5.161 | | 2,450 | 222 | | 2,672 | 3,050 |
| 6934 | double 2" x 6" | | 1.70 | 4.706 | | 1,875 | 203 | | 2,078 | 2,350 |
| 6936 | double 2" x 8" | | 1.70 | 4.706 | | 2,000 | 203 | | 2,203 | 2,500 |
| 6938 | double 2" x 10" | | 1.70 | 4.706 | | 2,150 | 203 | | 2,353 | 2,675 |
| 6940 | double 2" x 12" | | 1.65 | 4.848 | | 2,300 | 209 | | 2,509 | 2,850 |
| 6942 | double 2" x 14" | | 1.65 | 4.848 | | 2,450 | 209 | | 2,659 | 3,025 |
| 6944 | double 2" x 16" | | 1.65 | 4.848 | | 2,600 | 209 | | 2,809 | 3,175 |
| 6960 | 11 ga., 4" x 6" | | 1.65 | 4.848 | | 3,400 | 209 | | 3,609 | 4,075 |
| 6962 | 4" x 8" | | 1.65 | 4.848 | | 3,675 | 209 | | 3,884 | 4,350 |
| 6964 | 4" x 10" | | 1.65 | 4.848 | | 3,925 | 209 | | 4,134 | 4,625 |
| 6966 | 6" x 6" | | 1.60 | 5 | | 3,450 | 215 | | 3,665 | 4,125 |
| 6968 | 6" x 8" | | 1.60 | 5 | | 3,700 | 215 | | 3,915 | 4,400 |
| 6970 | 6" x 10" | | 1.55 | 5.161 | | 3,950 | 222 | | 4,172 | 4,700 |
| 6972 | 6" x 12" | | 1.55 | 5.161 | | 4,200 | 222 | | 4,422 | 4,975 |
| 6974 | 6" x 14" | | 1.55 | 5.161 | | 4,450 | 222 | | 4,672 | 5,250 |
| 6976 | 6" x 16" | | 1.50 | 5.333 | | 4,725 | 230 | | 4,955 | 5,525 |
| 6978 | 7 ga., 8" x 6" | | 1.60 | 5 | | 3,750 | 215 | | 3,965 | 4,450 |
| 6980 | 8" x 8" | | 1.60 | 5 | | 4,000 | 215 | | 4,215 | 4,725 |
| 6982 | 8" x 10" | | 1.55 | 5.161 | | 4,250 | 222 | | 4,472 | 5,025 |
| 6984 | 8" x 12" | | 1.55 | 5.161 | | 4,500 | 222 | | 4,722 | 5,300 |
| 6986 | 8" x 14" | | 1.50 | 5.333 | | 4,750 | 230 | | 4,980 | 5,575 |
| 6988 | 8" x 16" | | 1.50 | 5.333 | | 5,000 | 230 | | 5,230 | 5,850 |
| 7000 | Strap connectors, galvanized | | | | | | | | | |
| 7002 | 12 ga., 2-1/16" x 36" | 1 Carp | 1.55 | 5.161 | C | 1,075 | 222 | | 1,297 | 1,550 |
| 7004 | 2-1/16" x 47" | | 1.50 | 5.333 | | 1,500 | 230 | | 1,730 | 2,000 |
| 7005 | 10 ga., 2-1/16" x 72" | | 1.50 | 5.333 | | 1,500 | 230 | | 1,730 | 2,000 |
| 7006 | 7 ga., 2-1/16" x 34" | | 1.55 | 5.161 | | 2,700 | 222 | | 2,922 | 3,325 |
| 7008 | 2-1/16" x 45" | | 1.50 | 5.333 | | 3,500 | 230 | | 3,730 | 4,200 |
| 7010 | 3 ga., 3" x 32" | | 1.55 | 5.161 | | 4,525 | 222 | | 4,747 | 5,350 |
| 7012 | 3" x 41" | | 1.55 | 5.161 | | 4,700 | 222 | | 4,922 | 5,525 |
| 7014 | 3" x 50" | | 1.50 | 5.333 | | 7,175 | 230 | | 7,405 | 8,250 |
| 7016 | 3" x 59" | | 1.50 | 5.333 | | 8,750 | 230 | | 8,980 | 9,975 |
| 7018 | 3-1/2" x 68" | | 1.45 | 5.517 | | 8,875 | 238 | | 9,113 | 10,100 |
| 7030 | Tension ties | | | | | | | | | |
| 7032 | 19-1/8" long, 16 ga., 3/4" anchor bolt | 1 Carp | 1.80 | 4.444 | C | 1,250 | 191 | | 1,441 | 1,675 |
| 7034 | 20" long, 12 ga., 1/2" anchor bolt | | 1.80 | 4.444 | | 1,625 | 191 | | 1,816 | 2,075 |
| 7036 | 20" long, 12 ga., 3/4" anchor bolt | | 1.80 | 4.444 | | 1,625 | 191 | | 1,816 | 2,075 |
| 7038 | 27-3/4" long, 12 ga., 3/4" anchor bolt | | 1.75 | 4.571 | | 2,850 | 197 | | 3,047 | 3,450 |
| 7050 | Truss connectors, galvanized | | | | | | | | | |
| 7052 | Adjustable hanger | | | | | | | | | |
| 7054 | 18 ga., 2" x 6" | 1 Carp | 1.65 | 4.848 | C | 490 | 209 | | 699 | 855 |

## 06 05 23 – Wood, Plastic, and Composite Fastenings

| 06 05 23.60 Timber Connectors | | Crew | Daily Output | Labor-Hours | Unit | Material | 2011 Bare Costs Labor | Equipment | Total | Total Incl O&P |
|---|---|---|---|---|---|---|---|---|---|---|
| 7056 | 4" x 6" | 1 Carp | 1.65 | 4.848 | C | 640 | 209 | | 849 | 1,025 |
| 7058 | 16 ga., 4" x 10" | | 1.60 | 5 | | 940 | 215 | | 1,155 | 1,350 |
| 7060 | (2) 2" x 10" | ↓ | 1.60 | 5 | ↓ | 940 | 215 | | 1,155 | 1,350 |
| 7062 | Connectors to plate | | | | | | | | | |
| 7064 | 16 ga., 2" x 4" plate | 1 Carp | 1.80 | 4.444 | C | 480 | 191 | | 671 | 825 |
| 7066 | 2" x 6" plate | " | 1.80 | 4.444 | " | 620 | 191 | | 811 | 980 |
| 7068 | Hip jack connector | | | | | | | | | |
| 7070 | 14 ga. | 1 Carp | 1.50 | 5.333 | C | 2,525 | 230 | | 2,755 | 3,125 |

### 06 05 23.70 Rough Hardware

| 0010 | ROUGH HARDWARE, average percent of carpentry material | | | | | | | | | |
|---|---|---|---|---|---|---|---|---|---|---|
| 0020 | Minimum | | | | Job | .50% | | | | |
| 0200 | Maximum | | | | | 1.50% | | | | |
| 0210 | In seismic or hurricane areas, up to | | | | ↓ | 10% | | | | |

### 06 05 23.80 Metal Bracing

| 0010 | METAL BRACING | | | | | | | | | |
|---|---|---|---|---|---|---|---|---|---|---|
| 0302 | Let-in, "T" shaped, 22 ga. galv. steel, studs at 16" O.C. | 1 Carp | 580 | .014 | L.F. | .75 | .59 | | 1.34 | 1.74 |
| 0402 | Studs at 24" O.C. | | 600 | .013 | | .75 | .57 | | 1.32 | 1.71 |
| 0502 | 16 ga. galv. steel straps, studs at 16" O.C. | | 600 | .013 | | .97 | .57 | | 1.54 | 1.95 |
| 0602 | Studs at 24" O.C. | ↓ | 620 | .013 | ↓ | .97 | .56 | | 1.53 | 1.92 |

# 06 11 Wood Framing

## 06 11 10 – Framing with Dimensional, Engineered or Composite Lumber

### 06 11 10.01 Forest Stewardship Council Certification

| 0010 | FOREST STEWARDSHIP COUNCIL CERTIFICATION | | | | | | | | | |
|---|---|---|---|---|---|---|---|---|---|---|
| 0020 | For Forest Stewardship Council (FSC) cert dimension lumber, add    G | | | | | 65% | | | | |

### 06 11 10.02 Blocking

| 0010 | BLOCKING | | | | | | | | | |
|---|---|---|---|---|---|---|---|---|---|---|
| 2600 | Miscellaneous, to wood construction | | | | | | | | | |
| 2620 | 2" x 4" | 1 Carp | .17 | 47.059 | M.B.F. | 495 | 2,025 | | 2,520 | 3,650 |
| 2625 | Pneumatic nailed | | .21 | 38.095 | | 495 | 1,650 | | 2,145 | 3,075 |
| 2660 | 2" x 8" | | .27 | 29.630 | | 545 | 1,275 | | 1,820 | 2,550 |
| 2665 | Pneumatic nailed | ↓ | .33 | 24.242 | ↓ | 545 | 1,050 | | 1,595 | 2,200 |
| 2720 | To steel construction | | | | | | | | | |
| 2740 | 2" x 4" | 1 Carp | .14 | 57.143 | M.B.F. | 495 | 2,450 | | 2,945 | 4,325 |
| 2780 | 2" x 8" | " | .21 | 38.095 | " | 545 | 1,650 | | 2,195 | 3,125 |

### 06 11 10.04 Wood Bracing

| 0010 | WOOD BRACING | | | | | | | | | |
|---|---|---|---|---|---|---|---|---|---|---|
| 0011 | Let-in, with 1" x 6" boards, studs 16" O.C. | 1 Carp | 1.50 | 5.333 | C.L.F. | 67.50 | 230 | | 297.50 | 425 |
| 0200 | Studs @ 24" O.C. | " | 2.30 | 3.478 | " | 67.50 | 150 | | 217.50 | 305 |

### 06 11 10.06 Bridging

| 0010 | BRIDGING | | | | | | | | | |
|---|---|---|---|---|---|---|---|---|---|---|
| 0011 | Wood, for joists 16" O.C., 1" x 3" | 1 Carp | 1.30 | 6.154 | C.Pr. | 51 | 265 | | 316 | 460 |
| 0015 | Pneumatic nailed | | 1.70 | 4.706 | | 51 | 203 | | 254 | 365 |
| 0100 | 2" x 3" bridging | | 1.30 | 6.154 | | 49.50 | 265 | | 314.50 | 460 |
| 0105 | Pneumatic nailed | | 1.70 | 4.706 | | 49.50 | 203 | | 252.50 | 365 |
| 0300 | Steel, galvanized, 18 ga., for 2" x 10" joists at 12" O.C. | | 1.30 | 6.154 | | 172 | 265 | | 437 | 595 |
| 0400 | 24" O.C. | | 1.40 | 5.714 | | 185 | 246 | | 431 | 580 |
| 0600 | For 2" x 14" joists at 16" O.C. | | 1.30 | 6.154 | | 185 | 265 | | 450 | 610 |
| 0700 | 24" O.C. | | 1.40 | 5.714 | | 283 | 246 | | 529 | 685 |
| 0900 | Compression type, 16" O.C., 2" x 8" joists | ↓ | 2 | 4 | ↓ | 178 | 172 | | 350 | 460 |

# 06 11 Wood Framing

## 06 11 10 – Framing with Dimensional, Engineered or Composite Lumber

### 06 11 10.06 Bridging

| | | Crew | Daily Output | Labor-Hours | Unit | Material | 2011 Bare Costs Labor | Equipment | Total | Total Incl O&P |
|---|---|---|---|---|---|---|---|---|---|---|
| 1000 | 2" x 12" joists | 1 Carp | 2 | 4 | C.Pr. | 178 | 172 | | 350 | 460 |

### 06 11 10.10 Beam and Girder Framing

| | | Crew | Daily Output | Labor-Hours | Unit | Material | 2011 Bare Costs Labor | Equipment | Total | Total Incl O&P |
|---|---|---|---|---|---|---|---|---|---|---|
| 0010 | **BEAM AND GIRDER FRAMING**  R061110-30 | | | | | | | | | |
| 3500 | Single, 2" x 6" | 2 Carp | .70 | 22.857 | M.B.F. | 525 | 985 | | 1,510 | 2,075 |
| 3505 | Pneumatic nailed | | .81 | 19.704 | | 525 | 850 | | 1,375 | 1,875 |
| 3520 | 2" x 8" | | .86 | 18.605 | | 545 | 800 | | 1,345 | 1,825 |
| 3525 | Pneumatic nailed | | 1 | 16.048 | | 545 | 690 | | 1,235 | 1,650 |
| 3540 | 2" x 10" | | 1 | 16 | | 670 | 690 | | 1,360 | 1,775 |
| 3545 | Pneumatic nailed | | 1.16 | 13.793 | | 670 | 595 | | 1,265 | 1,650 |
| 3560 | 2" x 12" | | 1.10 | 14.545 | | 680 | 625 | | 1,305 | 1,700 |
| 3565 | Pneumatic nailed | | 1.28 | 12.539 | | 680 | 540 | | 1,220 | 1,575 |
| 3580 | 2" x 14" | | 1.17 | 13.675 | | 725 | 590 | | 1,315 | 1,700 |
| 3585 | Pneumatic nailed | | 1.36 | 11.791 | | 725 | 510 | | 1,235 | 1,575 |
| 3600 | 3" x 8" | | 1.10 | 14.545 | | 1,050 | 625 | | 1,675 | 2,125 |
| 3620 | 3" x 10" | | 1.25 | 12.800 | | 1,075 | 550 | | 1,625 | 2,025 |
| 3640 | 3" x 12" | | 1.35 | 11.852 | | 1,100 | 510 | | 1,610 | 1,975 |
| 3660 | 3" x 14" | ↓ | 1.40 | 11.429 | | 1,100 | 490 | | 1,590 | 1,950 |
| 3680 | 4" x 8" | F-3 | 2.66 | 15.038 | | 1,125 | 660 | 246 | 2,031 | 2,500 |
| 3700 | 4" x 10" | | 3.16 | 12.658 | | 1,150 | 555 | 207 | 1,912 | 2,325 |
| 3720 | 4" x 12" | | 3.60 | 11.111 | | 1,375 | 485 | 182 | 2,042 | 2,475 |
| 3740 | 4" x 14" | ↓ | 3.96 | 10.101 | | 1,525 | 440 | 166 | 2,131 | 2,525 |
| 4000 | Double, 2" x 6" | 2 Carp | 1.25 | 12.800 | | 525 | 550 | | 1,075 | 1,425 |
| 4005 | Pneumatic nailed | | 1.45 | 11.034 | | 525 | 475 | | 1,000 | 1,300 |
| 4020 | 2" x 8" | | 1.60 | 10 | | 545 | 430 | | 975 | 1,250 |
| 4025 | Pneumatic nailed | | 1.86 | 8.621 | | 545 | 370 | | 915 | 1,175 |
| 4040 | 2" x 10" | | 1.92 | 8.333 | | 670 | 360 | | 1,030 | 1,275 |
| 4045 | Pneumatic nailed | | 2.23 | 7.185 | | 670 | 310 | | 980 | 1,200 |
| 4060 | 2" x 12" | | 2.20 | 7.273 | | 680 | 315 | | 995 | 1,225 |
| 4065 | Pneumatic nailed | | 2.55 | 6.275 | | 680 | 270 | | 950 | 1,175 |
| 4080 | 2" x 14" | | 2.45 | 6.531 | | 725 | 281 | | 1,006 | 1,225 |
| 4085 | Pneumatic nailed | | 2.84 | 5.634 | | 725 | 243 | | 968 | 1,175 |
| 5000 | Triple, 2" x 6" | | 1.65 | 9.697 | | 525 | 415 | | 940 | 1,225 |
| 5005 | Pneumatic nailed | | 1.91 | 8.377 | | 525 | 360 | | 885 | 1,125 |
| 5020 | 2" x 8" | | 2.10 | 7.619 | | 545 | 330 | | 875 | 1,100 |
| 5025 | Pneumatic nailed | | 2.44 | 6.568 | | 545 | 283 | | 828 | 1,025 |
| 5040 | 2" x 10" | | 2.50 | 6.400 | | 670 | 276 | | 946 | 1,150 |
| 5045 | Pneumatic nailed | | 2.90 | 5.517 | | 670 | 238 | | 908 | 1,100 |
| 5060 | 2" x 12" | | 2.85 | 5.614 | | 680 | 242 | | 922 | 1,125 |
| 5065 | Pneumatic nailed | | 3.31 | 4.840 | | 680 | 208 | | 888 | 1,075 |
| 5080 | 2" x 14" | | 3.15 | 5.079 | | 725 | 219 | | 944 | 1,125 |
| 5085 | Pneumatic nailed | ↓ | 3.35 | 4.770 | ↓ | 725 | 205 | | 930 | 1,125 |

### 06 11 10.12 Ceiling Framing

| | | Crew | Daily Output | Labor-Hours | Unit | Material | 2011 Bare Costs Labor | Equipment | Total | Total Incl O&P |
|---|---|---|---|---|---|---|---|---|---|---|
| 0010 | **CEILING FRAMING** | | | | | | | | | |
| 6400 | Suspended, 2" x 3" | 2 Carp | .50 | 32 | M.B.F. | 660 | 1,375 | | 2,035 | 2,850 |
| 6450 | 2" x 4" | | .59 | 27.119 | | 495 | 1,175 | | 1,670 | 2,350 |
| 6500 | 2" x 6" | | .80 | 20 | | 525 | 860 | | 1,385 | 1,900 |
| 6550 | 2" x 8" | ↓ | .86 | 18.605 | ↓ | 545 | 800 | | 1,345 | 1,825 |

### 06 11 10.14 Posts and Columns

| | | Crew | Daily Output | Labor-Hours | Unit | Material | 2011 Bare Costs Labor | Equipment | Total | Total Incl O&P |
|---|---|---|---|---|---|---|---|---|---|---|
| 0010 | **POSTS AND COLUMNS** | | | | | | | | | |
| 0400 | 4" x 4" | 2 Carp | .52 | 30.769 | M.B.F. | 1,100 | 1,325 | | 2,425 | 3,250 |
| 0420 | 4" x 6" | | .55 | 29.091 | | 1,175 | 1,250 | | 2,425 | 3,225 |
| 0440 | 4" x 8" | ↓ | .59 | 27.119 | ↓ | 1,125 | 1,175 | | 2,300 | 3,025 |

# 06 11 Wood Framing

## 06 11 10 – Framing with Dimensional, Engineered or Composite Lumber

### 06 11 10.14 Posts and Columns

| | | Crew | Daily Output | Labor-Hours | Unit | Material | 2011 Bare Costs Labor | Equipment | Total | Total Incl O&P |
|---|---|---|---|---|---|---|---|---|---|---|
| 0460 | 6" x 6" | 2 Carp | .65 | 24.615 | M.B.F. | 1,200 | 1,050 | | 2,250 | 2,950 |
| 0480 | 6" x 8" | | .70 | 22.857 | | 1,200 | 985 | | 2,185 | 2,825 |
| 0500 | 6" x 10" | | .75 | 21.333 | | 1,300 | 920 | | 2,220 | 2,825 |

### 06 11 10.18 Joist Framing

| | | Crew | Daily Output | Labor-Hours | Unit | Material | 2011 Bare Costs Labor | Equipment | Total | Total Incl O&P |
|---|---|---|---|---|---|---|---|---|---|---|
| 0010 | **JOIST FRAMING** R061110-30 | | | | | | | | | |
| 2650 | Joists, 2" x 4" | 2 Carp | .83 | 19.277 | M.B.F. | 495 | 830 | | 1,325 | 1,825 |
| 2655 | Pneumatic nailed | | .96 | 16.667 | | 495 | 720 | | 1,215 | 1,650 |
| 2680 | 2" x 6" | | 1.25 | 12.800 | | 525 | 550 | | 1,075 | 1,425 |
| 2685 | Pneumatic nailed | | 1.44 | 11.111 | | 525 | 480 | | 1,005 | 1,325 |
| 2700 | 2" x 8" | | 1.46 | 10.959 | | 545 | 470 | | 1,015 | 1,325 |
| 2705 | Pneumatic nailed | | 1.68 | 9.524 | | 545 | 410 | | 955 | 1,225 |
| 2720 | 2" x 10" | | 1.49 | 10.738 | | 670 | 460 | | 1,130 | 1,450 |
| 2725 | Pneumatic nailed | | 1.71 | 9.357 | | 670 | 405 | | 1,075 | 1,350 |
| 2740 | 2" x 12" | | 1.75 | 9.143 | | 680 | 395 | | 1,075 | 1,350 |
| 2745 | Pneumatic nailed | | 2.01 | 7.960 | | 680 | 345 | | 1,025 | 1,275 |
| 2760 | 2" x 14" | | 1.79 | 8.939 | | 725 | 385 | | 1,110 | 1,400 |
| 2765 | Pneumatic nailed | | 2.06 | 7.767 | | 725 | 335 | | 1,060 | 1,325 |
| 2780 | 3" x 6" | | 1.39 | 11.511 | | 1,050 | 495 | | 1,545 | 1,900 |
| 2790 | 3" x 8" | | 1.90 | 8.421 | | 1,050 | 365 | | 1,415 | 1,725 |
| 2800 | 3" x 10" | | 1.95 | 8.205 | | 1,075 | 355 | | 1,430 | 1,725 |
| 2820 | 3" x 12" | | 1.80 | 8.889 | | 1,100 | 385 | | 1,485 | 1,775 |
| 2840 | 4" x 6" | | 1.60 | 10 | | 1,175 | 430 | | 1,605 | 1,950 |
| 2850 | 4" x 8" | | 4.15 | 3.855 | | 1,125 | 166 | | 1,291 | 1,475 |
| 2860 | 4" x 10" | | 2 | 8 | | 1,150 | 345 | | 1,495 | 1,775 |
| 2880 | 4" x 12" | | 1.80 | 8.889 | | 1,375 | 385 | | 1,760 | 2,100 |
| 3000 | Composite wood joist 9-1/2" deep | | .90 | 17.778 | M.L.F. | 1,550 | 765 | | 2,315 | 2,875 |
| 3010 | 11-1/2" deep | | .88 | 18.182 | | 1,725 | 785 | | 2,510 | 3,100 |
| 3020 | 14" deep | | .82 | 19.512 | | 2,400 | 840 | | 3,240 | 3,925 |
| 3030 | 16" deep | | .78 | 20.513 | | 2,975 | 885 | | 3,860 | 4,600 |
| 4000 | Open web joist 12" deep | | .88 | 18.182 | | 2,775 | 785 | | 3,560 | 4,250 |
| 4010 | 14" deep | | .82 | 19.512 | | 2,975 | 840 | | 3,815 | 4,575 |
| 4020 | 16" deep | | .78 | 20.513 | | 2,950 | 885 | | 3,835 | 4,600 |
| 4030 | 18" deep | | .74 | 21.622 | | 3,175 | 930 | | 4,105 | 4,925 |
| 6000 | Composite rim joist, 1-1/4" x 9-1/2" | | .90 | 17.778 | | 1,675 | 765 | | 2,440 | 3,025 |
| 6010 | 1-1/4" x 11-1/2" | | .88 | 18.182 | | 2,150 | 785 | | 2,935 | 3,550 |
| 6020 | 1-1/4" x 14-1/2" | | .82 | 19.512 | | 2,400 | 840 | | 3,240 | 3,950 |
| 6030 | 1-1/4" x 16-1/2" | | .78 | 20.513 | | 2,950 | 885 | | 3,835 | 4,600 |

### 06 11 10.24 Miscellaneous Framing

| | | Crew | Daily Output | Labor-Hours | Unit | Material | 2011 Bare Costs Labor | Equipment | Total | Total Incl O&P |
|---|---|---|---|---|---|---|---|---|---|---|
| 0010 | **MISCELLANEOUS FRAMING** | | | | | | | | | |
| 8500 | Firestops, 2" x 4" | 2 Carp | .51 | 31.373 | M.B.F. | 495 | 1,350 | | 1,845 | 2,625 |
| 8505 | Pneumatic nailed | | .62 | 25.806 | | 495 | 1,100 | | 1,595 | 2,250 |
| 8520 | 2" x 6" | | .60 | 26.667 | | 525 | 1,150 | | 1,675 | 2,325 |
| 8525 | Pneumatic nailed | | .73 | 21.858 | | 525 | 940 | | 1,465 | 2,025 |
| 8540 | 2" x 8" | | .60 | 26.667 | | 545 | 1,150 | | 1,695 | 2,350 |
| 8560 | 2" x 12" | | .70 | 22.857 | | 680 | 985 | | 1,665 | 2,250 |
| 8600 | Nailers, treated, wood construction, 2" x 4" | | .53 | 30.189 | | 540 | 1,300 | | 1,840 | 2,600 |
| 8605 | Pneumatic nailed | | .64 | 25.157 | | 540 | 1,075 | | 1,615 | 2,250 |
| 8620 | 2" x 6" | | .75 | 21.333 | | 615 | 920 | | 1,535 | 2,075 |
| 8625 | Pneumatic nailed | | .90 | 17.778 | | 615 | 765 | | 1,380 | 1,850 |
| 8640 | 2" x 8" | | .93 | 17.204 | | 715 | 740 | | 1,455 | 1,925 |
| 8645 | Pneumatic nailed | | 1.12 | 14.337 | | 715 | 615 | | 1,330 | 1,725 |
| 8660 | Steel construction, 2" x 4" | | .50 | 32 | | 540 | 1,375 | | 1,915 | 2,725 |

# 06 11 Wood Framing

## 06 11 10 – Framing with Dimensional, Engineered or Composite Lumber

### 06 11 10.24 Miscellaneous Framing

| | | Crew | Daily Output | Labor-Hours | Unit | Material | 2011 Bare Costs Labor | Equipment | Total | Total Incl O&P |
|---|---|---|---|---|---|---|---|---|---|---|
| 8680 | 2" x 6" | 2 Carp | .70 | 22.857 | M.B.F. | 615 | 985 | | 1,600 | 2,175 |
| 8700 | 2" x 8" | | .87 | 18.391 | | 715 | 790 | | 1,505 | 2,025 |
| 8760 | Rough bucks, treated, for doors or windows, 2" x 6" | | .40 | 40 | | 615 | 1,725 | | 2,340 | 3,325 |
| 8765 | Pneumatic nailed | | .48 | 33.333 | | 615 | 1,425 | | 2,040 | 2,875 |
| 8780 | 2" x 8" | | .51 | 31.373 | | 715 | 1,350 | | 2,065 | 2,875 |
| 8785 | Pneumatic nailed | | .61 | 26.144 | | 715 | 1,125 | | 1,840 | 2,525 |
| 8800 | Stair stringers, 2" x 10" | | .22 | 72.727 | | 670 | 3,125 | | 3,795 | 5,525 |
| 8820 | 2" x 12" | | .26 | 61.538 | | 680 | 2,650 | | 3,330 | 4,825 |
| 8840 | 3" x 10" | | .31 | 51.613 | | 1,075 | 2,225 | | 3,300 | 4,575 |
| 8860 | 3" x 12" | | .38 | 42.105 | | 1,100 | 1,825 | | 2,925 | 3,975 |
| 8870 | Laminated structural lumber, 1-1/4" x 11-1/2" | | 130 | .123 | L.F. | 2.14 | 5.30 | | 7.44 | 10.50 |
| 8880 | 1-1/4" x 14-1/2" | | 130 | .123 | " | 2.41 | 5.30 | | 7.71 | 10.80 |

### 06 11 10.26 Partitions

| | | Crew | Daily Output | Labor-Hours | Unit | Material | 2011 Bare Costs Labor | Equipment | Total | Total Incl O&P |
|---|---|---|---|---|---|---|---|---|---|---|
| 0010 | **PARTITIONS** | | | | | | | | | |
| 0020 | Single bottom and double top plate, no waste, std. & better lumber | | | | | | | | | |
| 0180 | 2" x 4" studs, 8' high, studs 12" O.C. | 2 Carp | 80 | .200 | L.F. | 3.64 | 8.60 | | 12.24 | 17.20 |
| 0185 | 12" O.C., pneumatic nailed | | 96 | .167 | | 3.64 | 7.20 | | 10.84 | 15 |
| 0200 | 16" O.C. | | 100 | .160 | | 2.98 | 6.90 | | 9.88 | 13.80 |
| 0205 | 16" O.C., pneumatic nailed | | 120 | .133 | | 2.98 | 5.75 | | 8.73 | 12.05 |
| 0300 | 24" O.C. | | 125 | .128 | | 2.31 | 5.50 | | 7.81 | 11 |
| 0305 | 24" O.C., pneumatic nailed | | 150 | .107 | | 2.31 | 4.59 | | 6.90 | 9.60 |
| 0380 | 10' high, studs 12" O.C. | | 80 | .200 | | 4.30 | 8.60 | | 12.90 | 17.95 |
| 0385 | 12" O.C., pneumatic nailed | | 96 | .167 | | 4.30 | 7.20 | | 11.50 | 15.75 |
| 0400 | 16" O.C. | | 100 | .160 | | 3.47 | 6.90 | | 10.37 | 14.35 |
| 0405 | 16" O.C., pneumatic nailed | | 120 | .133 | | 3.47 | 5.75 | | 9.22 | 12.60 |
| 0500 | 24" O.C. | | 125 | .128 | | 2.65 | 5.50 | | 8.15 | 11.35 |
| 0505 | 24" O.C., pneumatic nailed | | 150 | .107 | | 2.65 | 4.59 | | 7.24 | 9.95 |
| 0580 | 12' high, studs 12" O.C. | | 65 | .246 | | 4.96 | 10.60 | | 15.56 | 21.50 |
| 0585 | 12" O.C., pneumatic nailed | | 78 | .205 | | 4.96 | 8.85 | | 13.81 | 19 |
| 0600 | 16" O.C. | | 80 | .200 | | 3.97 | 8.60 | | 12.57 | 17.55 |
| 0605 | 16" O.C., pneumatic nailed | | 96 | .167 | | 3.97 | 7.20 | | 11.17 | 15.35 |
| 0700 | 24" O.C. | | 100 | .160 | | 2.98 | 6.90 | | 9.88 | 13.80 |
| 0705 | 24" O.C., pneumatic nailed | | 120 | .133 | | 2.98 | 5.75 | | 8.73 | 12.05 |
| 0780 | 2" x 6" studs, 8' high, studs 12" O.C. | | 70 | .229 | | 5.80 | 9.85 | | 15.65 | 21.50 |
| 0785 | 12" O.C., pneumatic nailed | | 84 | .190 | | 5.80 | 8.20 | | 14 | 19 |
| 0800 | 16" O.C. | | 90 | .178 | | 4.74 | 7.65 | | 12.39 | 16.95 |
| 0805 | 16" O.C., pneumatic nailed | | 108 | .148 | | 4.74 | 6.40 | | 11.14 | 15 |
| 0900 | 24" O.C. | | 115 | .139 | | 3.69 | 6 | | 9.69 | 13.25 |
| 0905 | 24" O.C., pneumatic nailed | | 138 | .116 | | 3.69 | 4.99 | | 8.68 | 11.70 |
| 0980 | 10' high, studs 12" O.C. | | 70 | .229 | | 6.85 | 9.85 | | 16.70 | 22.50 |
| 0985 | 12" O.C., pneumatic nailed | | 84 | .190 | | 6.85 | 8.20 | | 15.05 | 20 |
| 1000 | 16" O.C. | | 90 | .178 | | 5.55 | 7.65 | | 13.20 | 17.85 |
| 1005 | 16" O.C., pneumatic nailed | | 108 | .148 | | 5.55 | 6.40 | | 11.95 | 15.90 |
| 1100 | 24" O.C. | | 115 | .139 | | 4.22 | 6 | | 10.22 | 13.85 |
| 1105 | 24" O.C., pneumatic nailed | | 138 | .116 | | 4.22 | 4.99 | | 9.21 | 12.30 |
| 1180 | 12' high, studs 12" O.C. | | 55 | .291 | | 7.90 | 12.50 | | 20.40 | 28 |
| 1185 | 12" O.C., pneumatic nailed | | 66 | .242 | | 7.90 | 10.45 | | 18.35 | 24.50 |
| 1200 | 16" O.C. | | 70 | .229 | | 6.30 | 9.85 | | 16.15 | 22 |
| 1205 | 16" O.C., pneumatic nailed | | 84 | .190 | | 6.30 | 8.20 | | 14.50 | 19.55 |
| 1300 | 24" O.C. | | 90 | .178 | | 4.74 | 7.65 | | 12.39 | 16.95 |
| 1305 | 24" O.C., pneumatic nailed | | 108 | .148 | | 4.74 | 6.40 | | 11.14 | 15 |
| 1400 | For horizontal blocking, 2" x 4", add | | 600 | .027 | | .33 | 1.15 | | 1.48 | 2.12 |

# 06 11 Wood Framing

## 06 11 10.26 Partitions

| | | Crew | Daily Output | Labor-Hours | Unit | Material | 2011 Bare Costs Labor | Equipment | Total | Total Incl O&P |
|---|---|---|---|---|---|---|---|---|---|---|
| 1500 | 2" x 6", add | 2 Carp | 600 | .027 | L.F. | .53 | 1.15 | | 1.68 | 2.34 |
| 1600 | For openings, add | ↓ | 250 | .064 | ↓ | | 2.76 | | 2.76 | 4.23 |
| 1700 | Headers for above openings, material only, add | | | | M.B.F. | 545 | | | 545 | 600 |

## 06 11 10.28 Porch or Deck Framing

| | | Crew | Daily Output | Labor-Hours | Unit | Material | 2011 Bare Costs Labor | Equipment | Total | Total Incl O&P |
|---|---|---|---|---|---|---|---|---|---|---|
| 0010 | **PORCH OR DECK FRAMING** | | | | | | | | | |
| 0100 | Treated lumber, posts or columns, 4" x 4" | 2 Carp | 390 | .041 | L.F. | 1.37 | 1.77 | | 3.14 | 4.21 |
| 0110 | 4" x 6" | | 275 | .058 | | 2.46 | 2.50 | | 4.96 | 6.55 |
| 0120 | 4" x 8" | | 220 | .073 | | 4.74 | 3.13 | | 7.87 | 10 |
| 0130 | Girder, single, 4" x 4" | | 675 | .024 | | 1.37 | 1.02 | | 2.39 | 3.07 |
| 0140 | 4" x 6" | | 600 | .027 | | 2.46 | 1.15 | | 3.61 | 4.47 |
| 0150 | 4" x 8" | | 525 | .030 | | 4.74 | 1.31 | | 6.05 | 7.20 |
| 0160 | Double, 2" x 4" | | 625 | .026 | | .75 | 1.10 | | 1.85 | 2.51 |
| 0170 | 2" x 6" | | 600 | .027 | | 1.28 | 1.15 | | 2.43 | 3.16 |
| 0180 | 2" x 8" | | 575 | .028 | | 1.97 | 1.20 | | 3.17 | 4.01 |
| 0190 | 2" x 10" | | 550 | .029 | | 2.65 | 1.25 | | 3.90 | 4.84 |
| 0200 | 2" x 12" | | 525 | .030 | | 3.49 | 1.31 | | 4.80 | 5.85 |
| 0210 | Triple, 2" x 4" | | 575 | .028 | | 1.12 | 1.20 | | 2.32 | 3.08 |
| 0220 | 2" x 6" | | 550 | .029 | | 1.92 | 1.25 | | 3.17 | 4.03 |
| 0230 | 2" x 8" | | 525 | .030 | | 2.96 | 1.31 | | 4.27 | 5.25 |
| 0240 | 2" x 10" | | 500 | .032 | | 3.98 | 1.38 | | 5.36 | 6.50 |
| 0250 | 2" x 12" | | 475 | .034 | | 5.25 | 1.45 | | 6.70 | 7.95 |
| 0260 | Ledger, bolted 4' O.C., 2" x 4" | | 400 | .040 | | .48 | 1.72 | | 2.20 | 3.17 |
| 0270 | 2" x 6" | | 395 | .041 | | .74 | 1.74 | | 2.48 | 3.49 |
| 0280 | 2" x 8" | | 390 | .041 | | 1.07 | 1.77 | | 2.84 | 3.89 |
| 0300 | 2" x 12" | | 380 | .042 | | 1.82 | 1.81 | | 3.63 | 4.78 |
| 0310 | Joists, 2" x 4" | | 1250 | .013 | | .38 | .55 | | .93 | 1.26 |
| 0320 | 2" x 6" | | 1250 | .013 | | .64 | .55 | | 1.19 | 1.56 |
| 0330 | 2" x 8" | | 1100 | .015 | | .99 | .63 | | 1.62 | 2.05 |
| 0340 | 2" x 10" | | 900 | .018 | | 1.33 | .77 | | 2.10 | 2.64 |
| 0350 | 2" x 12" | ↓ | 875 | .018 | | 1.60 | .79 | | 2.39 | 2.97 |
| 0360 | Railings and trim , 1" x 4" | 1 Carp | 300 | .027 | | .53 | 1.15 | | 1.68 | 2.34 |
| 0370 | 2" x 2" | | 300 | .027 | | .35 | 1.15 | | 1.50 | 2.15 |
| 0380 | 2" x 4" | | 300 | .027 | | .36 | 1.15 | | 1.51 | 2.16 |
| 0390 | 2" x 6" | | 300 | .027 | ↓ | .62 | 1.15 | | 1.77 | 2.44 |
| 0400 | Decking, 1" x 4" | | 275 | .029 | S.F. | 1.55 | 1.25 | | 2.80 | 3.63 |
| 0410 | 2" x 4" | | 300 | .027 | | 1.23 | 1.15 | | 2.38 | 3.11 |
| 0420 | 2" x 6" | | 320 | .025 | | 1.34 | 1.08 | | 2.42 | 3.12 |
| 0430 | 5/4" x 6" | ↓ | 320 | .025 | ↓ | 2.09 | 1.08 | | 3.17 | 3.95 |
| 0440 | Balusters, square, 2" x 2" | 2 Carp | 660 | .024 | L.F. | .36 | 1.04 | | 1.40 | 1.99 |
| 0450 | Turned, 2" x 2" | | 420 | .038 | | .48 | 1.64 | | 2.12 | 3.04 |
| 0460 | Stair stringer, 2" x 10" | | 130 | .123 | | 1.33 | 5.30 | | 6.63 | 9.60 |
| 0470 | 2" x 12" | | 130 | .123 | | 1.60 | 5.30 | | 6.90 | 9.90 |
| 0480 | Stair treads, 1" x 4" | | 140 | .114 | | 1.56 | 4.92 | | 6.48 | 9.25 |
| 0490 | 2" x 4" | | 140 | .114 | | .38 | 4.92 | | 5.30 | 7.95 |
| 0500 | 2" x 6" | | 160 | .100 | | .57 | 4.31 | | 4.88 | 7.20 |
| 0510 | 5/4" x 6" | | 160 | .100 | ↓ | .98 | 4.31 | | 5.29 | 7.65 |
| 0520 | Turned handrail post, 4" x 4" | | 64 | .250 | Ea. | 32 | 10.75 | | 42.75 | 51.50 |
| 0530 | Lattice panel, 4' x 8', 1/2" | | 1600 | .010 | S.F. | .68 | .43 | | 1.11 | 1.41 |
| 0535 | 3/4" | | 1600 | .010 | " | 1.02 | .43 | | 1.45 | 1.78 |
| 0540 | Cedar, posts or columns, 4" x 4" | | 390 | .041 | L.F. | 3.01 | 1.77 | | 4.78 | 6 |
| 0550 | 4" x 6" | | 275 | .058 | | 5.75 | 2.50 | | 8.25 | 10.15 |
| 0560 | 4" x 8" | ↓ | 220 | .073 | ↓ | 9.70 | 3.13 | | 12.83 | 15.50 |

# 06 11 Wood Framing

## 06 11 10 – Framing with Dimensional, Engineered or Composite Lumber

### 06 11 10.28 Porch or Deck Framing

| | | Crew | Daily Output | Labor-Hours | Unit | Material | 2011 Bare Costs Labor | Equipment | Total | Total Incl O&P |
|---|---|---|---|---|---|---|---|---|---|---|
| 0800 | Decking, 1" x 4" | 2 Carp | 550 | .029 | L.F. | 1.49 | 1.25 | | 2.74 | 3.56 |
| 0810 | 2" x 4" | | 600 | .027 | | 2.98 | 1.15 | | 4.13 | 5.05 |
| 0820 | 2" x 6" | | 640 | .025 | | 5.25 | 1.08 | | 6.33 | 7.45 |
| 0830 | 5/4" x 6" | | 640 | .025 | | 3.50 | 1.08 | | 4.58 | 5.50 |
| 0840 | Railings and trim, 1" x 4" | | 600 | .027 | | 1.49 | 1.15 | | 2.64 | 3.40 |
| 0860 | 2" x 4" | | 600 | .027 | | 2.98 | 1.15 | | 4.13 | 5.05 |
| 0870 | 2" x 6" | | 600 | .027 | | 5.25 | 1.15 | | 6.40 | 7.55 |
| 0920 | Stair treads, 1" x 4" | | 140 | .114 | | 1.49 | 4.92 | | 6.41 | 9.20 |
| 0930 | 2" x 4" | | 140 | .114 | | 2.98 | 4.92 | | 7.90 | 10.85 |
| 0940 | 2" x 6" | | 160 | .100 | | 5.25 | 4.31 | | 9.56 | 12.40 |
| 0950 | 5/4" x 6" | | 160 | .100 | | 3.50 | 4.31 | | 7.81 | 10.45 |
| 0980 | Redwood, posts or columns, 4" x 4" | | 390 | .041 | | 6.40 | 1.77 | | 8.17 | 9.75 |
| 0990 | 4" x 6" | | 275 | .058 | | 11.20 | 2.50 | | 13.70 | 16.15 |
| 1000 | 4" x 8" | | 220 | .073 | | 20.50 | 3.13 | | 23.63 | 27.50 |
| 1240 | Redwood decking, 1" x 4" | 1 Carp | 275 | .029 | S.F. | 3.97 | 1.25 | | 5.22 | 6.30 |
| 1260 | 2" x 6" | | 340 | .024 | | 7.50 | 1.01 | | 8.51 | 9.80 |
| 1270 | 5/4" x 6" | | 320 | .025 | | 4.74 | 1.08 | | 5.82 | 6.85 |
| 1280 | Railings and trim, 1" x 4" | 2 Carp | 600 | .027 | L.F. | 1.17 | 1.15 | | 2.32 | 3.04 |
| 1310 | 2" x 6" | | 600 | .027 | | 7.50 | 1.15 | | 8.65 | 10 |
| 1420 | Alternative decking, wood/plastic composite, 5/4" x 6"  [G] | | 640 | .025 | | 3 | 1.08 | | 4.08 | 4.95 |
| 1430 | Vinyl, 1-1/2" x 5-1/2" | | 640 | .025 | | 4.15 | 1.08 | | 5.23 | 6.20 |
| 1440 | 1" x 4" square edge fir | | 550 | .029 | | 1.57 | 1.25 | | 2.82 | 3.64 |
| 1450 | 1" x 4" tongue and groove fir | | 450 | .036 | | 1.41 | 1.53 | | 2.94 | 3.90 |
| 1460 | 1" x 4" mahogany | | 550 | .029 | | 2.51 | 1.25 | | 3.76 | 4.68 |
| 1462 | 5/4" x 6" PVC | | 550 | .029 | | 2.72 | 1.25 | | 3.97 | 4.91 |
| 1465 | Framing, porch or deck, alt deck fastening, screws, add | 1 Carp | 240 | .033 | S.F. | | 1.43 | | 1.43 | 2.20 |
| 1470 | Accessories, joist hangers, 2" x 4" | | 160 | .050 | Ea. | .67 | 2.15 | | 2.82 | 4.04 |
| 1480 | 2" x 6" through 2" x 12" | | 150 | .053 | | 1.28 | 2.30 | | 3.58 | 4.93 |
| 1530 | Post footing, incl excav, backfill, tube form & concrete, 4' deep, 8" dia | F-7 | 12 | 2.667 | | 10.65 | 103 | | 113.65 | 170 |
| 1540 | 10" diameter | | 11 | 2.909 | | 15.80 | 113 | | 128.80 | 190 |
| 1550 | 12" diameter | | 10 | 3.200 | | 21 | 124 | | 145 | 213 |

### 06 11 10.30 Roof Framing

| | | Crew | Daily Output | Labor-Hours | Unit | Material | 2011 Bare Costs Labor | Equipment | Total | Total Incl O&P |
|---|---|---|---|---|---|---|---|---|---|---|
| 0010 | **ROOF FRAMING** | | | | | | | | | |
| 5250 | Composite rafter, 9-1/2" deep | 2 Carp | 575 | .028 | L.F. | 1.54 | 1.20 | | 2.74 | 3.53 |
| 5260 | 11-1/2" deep | | 575 | .028 | " | 1.74 | 1.20 | | 2.94 | 3.75 |
| 6070 | Fascia boards, 2" x 8" | | .30 | 53.333 | M.B.F. | 545 | 2,300 | | 2,845 | 4,125 |
| 6080 | 2" x 10" | | .30 | 53.333 | | 670 | 2,300 | | 2,970 | 4,250 |
| 7000 | Rafters, to 4 in 12 pitch, 2" x 6" | | 1 | 16 | | 525 | 690 | | 1,215 | 1,625 |
| 7060 | 2" x 8" | | 1.26 | 12.698 | | 545 | 545 | | 1,090 | 1,450 |
| 7300 | Hip and valley rafters, 2" x 6" | | .76 | 21.053 | | 525 | 905 | | 1,430 | 1,975 |
| 7360 | 2" x 8" | | .96 | 16.667 | | 545 | 720 | | 1,265 | 1,700 |
| 7540 | Hip and valley jacks, 2" x 6" | | .60 | 26.667 | | 525 | 1,150 | | 1,675 | 2,325 |
| 7600 | 2" x 8" | | .65 | 24.615 | | 545 | 1,050 | | 1,595 | 2,225 |
| 7780 | For slopes steeper than 4 in 12, add | | | | | | 30% | | | |
| 7790 | For dormers or complex roofs, add | | | | | | 50% | | | |
| 7800 | Rafter tie, 1" x 4", #3 | 2 Carp | .27 | 59.259 | M.B.F. | 1,600 | 2,550 | | 4,150 | 5,675 |
| 7820 | Ridge board, #2 or better, 1" x 6" | | .30 | 53.333 | | 1,350 | 2,300 | | 3,650 | 5,000 |
| 7840 | 1" x 8" | | .37 | 43.243 | | 1,450 | 1,850 | | 3,300 | 4,450 |
| 7860 | 1" x 10" | | .42 | 38.095 | | 1,450 | 1,650 | | 3,100 | 4,125 |
| 7880 | 2" x 6" | | .50 | 32 | | 525 | 1,375 | | 1,900 | 2,700 |
| 7900 | 2" x 8" | | .60 | 26.667 | | 545 | 1,150 | | 1,695 | 2,350 |
| 7920 | 2" x 10" | | .66 | 24.242 | | 670 | 1,050 | | 1,720 | 2,325 |

# 06 11 Wood Framing

## 06 11 10 – Framing with Dimensional, Engineered or Composite Lumber

### 06 11 10.30 Roof Framing

| | | Crew | Daily Output | Labor-Hours | Unit | Material | 2011 Bare Costs Labor | Equipment | Total | Total Incl O&P |
|---|---|---|---|---|---|---|---|---|---|---|
| 7940 | Roof cants, split, 4" x 4" | 2 Carp | .86 | 18.605 | M.B.F. | 1,100 | 800 | | 1,900 | 2,450 |
| 7960 | 6" x 6" | | 1.80 | 8.889 | | 1,200 | 385 | | 1,585 | 1,900 |
| 7980 | Roof curbs, untreated, 2" x 6" | | .52 | 30.769 | | 525 | 1,325 | | 1,850 | 2,600 |
| 8000 | 2" x 12" | | .80 | 20 | | 680 | 860 | | 1,540 | 2,075 |

### 06 11 10.32 Sill and Ledger Framing

| | | Crew | Daily Output | Labor-Hours | Unit | Material | 2011 Bare Costs Labor | Equipment | Total | Total Incl O&P |
|---|---|---|---|---|---|---|---|---|---|---|
| 0010 | **SILL AND LEDGER FRAMING** | | | | | | | | | |
| 4482 | Ledgers, nailed, 2" x 4" | 2 Carp | .50 | 32 | M.B.F. | 495 | 1,375 | | 1,870 | 2,675 |
| 4484 | 2" x 6" | | .60 | 26.667 | | 525 | 1,150 | | 1,675 | 2,325 |
| 4486 | Bolted, not including bolts, 3" x 8" | | .65 | 24.615 | | 1,050 | 1,050 | | 2,100 | 2,800 |
| 4488 | 3" x 12" | | .70 | 22.857 | | 1,100 | 985 | | 2,085 | 2,700 |
| 4490 | Mud sills, redwood, construction grade, 2" x 4" | | .59 | 27.119 | | 3,400 | 1,175 | | 4,575 | 5,525 |
| 4492 | 2" x 6" | | .78 | 20.513 | | 3,400 | 885 | | 4,285 | 5,100 |
| 4500 | Sills, 2" x 4" | | .40 | 40 | | 495 | 1,725 | | 2,220 | 3,200 |
| 4520 | 2" x 6" | | .55 | 29.091 | | 525 | 1,250 | | 1,775 | 2,500 |
| 4540 | 2" x 8" | | .67 | 23.881 | | 545 | 1,025 | | 1,570 | 2,175 |
| 4600 | Treated, 2" x 4" | | .36 | 44.444 | | 540 | 1,925 | | 2,465 | 3,525 |
| 4620 | 2" x 6" | | .50 | 32 | | 615 | 1,375 | | 1,990 | 2,800 |
| 4640 | 2" x 8" | | .60 | 26.667 | | 715 | 1,150 | | 1,865 | 2,550 |
| 4700 | 4" x 4" | | .60 | 26.667 | | 1,000 | 1,150 | | 2,150 | 2,850 |
| 4720 | 4" x 6" | | .70 | 22.857 | | 1,200 | 985 | | 2,185 | 2,825 |
| 4740 | 4" x 8" | | .80 | 20 | | 1,750 | 860 | | 2,610 | 3,250 |
| 4760 | 4" x 10" | | .87 | 18.391 | | 1,750 | 790 | | 2,540 | 3,150 |

### 06 11 10.34 Sleepers

| | | Crew | Daily Output | Labor-Hours | Unit | Material | 2011 Bare Costs Labor | Equipment | Total | Total Incl O&P |
|---|---|---|---|---|---|---|---|---|---|---|
| 0010 | **SLEEPERS** | | | | | | | | | |
| 0300 | On concrete, treated, 1" x 2" | 2 Carp | .39 | 41.026 | M.B.F. | 1,700 | 1,775 | | 3,475 | 4,575 |
| 0320 | 1" x 3" | | .50 | 32 | | 2,100 | 1,375 | | 3,475 | 4,425 |
| 0340 | 2" x 4" | | .99 | 16.162 | | 540 | 695 | | 1,235 | 1,675 |
| 0360 | 2" x 6" | | 1.30 | 12.308 | | 615 | 530 | | 1,145 | 1,500 |

### 06 11 10.36 Soffit and Canopy Framing

| | | Crew | Daily Output | Labor-Hours | Unit | Material | 2011 Bare Costs Labor | Equipment | Total | Total Incl O&P |
|---|---|---|---|---|---|---|---|---|---|---|
| 0010 | **SOFFIT AND CANOPY FRAMING** | | | | | | | | | |
| 1300 | Canopy or soffit framing, 1" x 4" | 2 Carp | .30 | 53.333 | M.B.F. | 1,600 | 2,300 | | 3,900 | 5,275 |
| 1340 | 1" x 8" | | .50 | 32 | | 1,450 | 1,375 | | 2,825 | 3,725 |
| 1360 | 2" x 4" | | .41 | 39.024 | | 495 | 1,675 | | 2,170 | 3,125 |
| 1400 | 2" x 8" | | .67 | 23.881 | | 545 | 1,025 | | 1,570 | 2,175 |
| 1420 | 3" x 4" | | .50 | 32 | | 850 | 1,375 | | 2,225 | 3,050 |
| 1460 | 3" x 8" | | .60 | 26.667 | | 1,050 | 1,150 | | 2,200 | 2,925 |

### 06 11 10.38 Treated Lumber Framing Material

| | | Crew | Daily Output | Labor-Hours | Unit | Material | 2011 Bare Costs Labor | Equipment | Total | Total Incl O&P |
|---|---|---|---|---|---|---|---|---|---|---|
| 0010 | **TREATED LUMBER FRAMING MATERIAL** | | | | | | | | | |
| 0100 | 2" x 4" | | | | M.B.F. | 540 | | | 540 | 595 |
| 0110 | 2" x 6" | | | | | 615 | | | 615 | 675 |
| 0120 | 2" x 8" | | | | | 715 | | | 715 | 790 |
| 0130 | 2" x 10" | | | | | 775 | | | 775 | 850 |
| 0140 | 2" x 12" | | | | | 850 | | | 850 | 935 |
| 0200 | 4" x 4" | | | | | 1,000 | | | 1,000 | 1,100 |
| 0210 | 4" x 6" | | | | | 1,200 | | | 1,200 | 1,325 |
| 0220 | 4" x 8" | | | | | 1,750 | | | 1,750 | 1,925 |

### 06 11 10.40 Wall Framing

| | | Crew | Daily Output | Labor-Hours | Unit | Material | 2011 Bare Costs Labor | Equipment | Total | Total Incl O&P |
|---|---|---|---|---|---|---|---|---|---|---|
| 0010 | **WALL FRAMING** R061110-30 | | | | | | | | | |
| 5860 | Headers over openings, 2" x 6" | 2 Carp | .36 | 44.444 | M.B.F. | 525 | 1,925 | | 2,450 | 3,500 |
| 5865 | 2" x 6", pneumatic nailed | | .43 | 37.209 | | 525 | 1,600 | | 2,125 | 3,025 |
| 5880 | 2" x 8" | | .45 | 35.556 | | 545 | 1,525 | | 2,070 | 2,950 |

# 06 11 Wood Framing

## 06 11 10 – Framing with Dimensional, Engineered or Composite Lumber

### 06 11 10.40 Wall Framing

| | | Crew | Daily Output | Labor-Hours | Unit | Material | 2011 Bare Costs Labor | 2011 Bare Costs Equipment | Total | Total Incl O&P |
|---|---|---|---|---|---|---|---|---|---|---|
| 5885 | 2" x 8", pneumatic nailed | 2 Carp | .54 | 29.630 | M.B.F. | 545 | 1,275 | | 1,820 | 2,550 |
| 5900 | 2" x 10" | | .53 | 30.189 | | 670 | 1,300 | | 1,970 | 2,725 |
| 5905 | 2" x 10", pneumatic nailed | | .67 | 23.881 | | 670 | 1,025 | | 1,695 | 2,300 |
| 5920 | 2" x 12" | | .60 | 26.667 | | 680 | 1,150 | | 1,830 | 2,500 |
| 5925 | 2" x 12", pneumatic nailed | | .72 | 22.222 | | 680 | 955 | | 1,635 | 2,225 |
| 5940 | 4" x 12" | | .76 | 21.053 | | 1,375 | 905 | | 2,280 | 2,925 |
| 5945 | 4" x 12", pneumatic nailed | | .92 | 17.391 | | 1,375 | 750 | | 2,125 | 2,675 |
| 5960 | 6" x 12" | | .84 | 19.048 | | 1,700 | 820 | | 2,520 | 3,125 |
| 5965 | 6" x 12", pneumatic nailed | | 1.01 | 15.873 | | 1,700 | 685 | | 2,385 | 2,925 |
| 6000 | Plates, untreated, 2" x 3" | | .43 | 37.209 | | 660 | 1,600 | | 2,260 | 3,175 |
| 6005 | 2" x 3", pneumatic nailed | | .52 | 30.769 | | 660 | 1,325 | | 1,985 | 2,750 |
| 6020 | 2" x 4" | | .53 | 30.189 | | 495 | 1,300 | | 1,795 | 2,550 |
| 6025 | 2" x 4", pneumatic nailed | | .67 | 23.881 | | 495 | 1,025 | | 1,520 | 2,125 |
| 6040 | 2" x 6" | | .75 | 21.333 | | 525 | 920 | | 1,445 | 1,975 |
| 6045 | 2" x 6", pneumatic nailed | | .90 | 17.778 | | 525 | 765 | | 1,290 | 1,750 |
| 6120 | Studs, 8' high wall, 2" x 3" | | .60 | 26.667 | | 660 | 1,150 | | 1,810 | 2,475 |
| 6125 | 2" x 3", pneumatic nailed | | .72 | 22.222 | | 660 | 955 | | 1,615 | 2,200 |
| 6140 | 2" x 4" | | .92 | 17.391 | | 495 | 750 | | 1,245 | 1,700 |
| 6145 | 2" x 4", pneumatic nailed | | 1.10 | 14.493 | | 495 | 625 | | 1,120 | 1,500 |
| 6160 | 2" x 6" | | 1 | 16 | | 525 | 690 | | 1,215 | 1,625 |
| 6165 | 2" x 6", pneumatic nailed | | 1.20 | 13.333 | | 525 | 575 | | 1,100 | 1,450 |
| 6180 | 3" x 4" | | .80 | 20 | | 850 | 860 | | 1,710 | 2,250 |
| 6185 | 3" x 4", pneumatic nailed | | .96 | 16.667 | | 850 | 720 | | 1,570 | 2,025 |
| 8200 | For 12' high walls, deduct | | | | | | 5% | | | |
| 8220 | For stub wall, 6' high, add | | | | | | 20% | | | |
| 8240 | 3' high, add | | | | | | 40% | | | |
| 8250 | For second story & above, add | | | | | | 5% | | | |
| 8300 | For dormer & gable, add | | | | | | 15% | | | |

### 06 11 10.42 Furring

| | | Crew | Daily Output | Labor-Hours | Unit | Material | 2011 Bare Costs Labor | 2011 Bare Costs Equipment | Total | Total Incl O&P |
|---|---|---|---|---|---|---|---|---|---|---|
| 0010 | **FURRING** | | | | | | | | | |
| 0012 | Wood strips, 1" x 2", on walls, on wood | 1 Carp | 550 | .015 | L.F. | .25 | .63 | | .88 | 1.23 |
| 0015 | On wood, pneumatic nailed | | 710 | .011 | | .25 | .49 | | .74 | 1.01 |
| 0300 | On masonry | | 495 | .016 | | .25 | .70 | | .95 | 1.34 |
| 0400 | On concrete | | 260 | .031 | | .25 | 1.32 | | 1.57 | 2.30 |
| 0600 | 1" x 3", on walls, on wood | | 550 | .015 | | .34 | .63 | | .97 | 1.34 |
| 0605 | On wood, pneumatic nailed | | 710 | .011 | | .34 | .49 | | .83 | 1.12 |
| 0700 | On masonry | | 495 | .016 | | .34 | .70 | | 1.04 | 1.45 |
| 0800 | On concrete | | 260 | .031 | | .34 | 1.32 | | 1.66 | 2.41 |
| 0850 | On ceilings, on wood | | 350 | .023 | | .34 | .98 | | 1.32 | 1.89 |
| 0855 | On wood, pneumatic nailed | | 450 | .018 | | .34 | .77 | | 1.11 | 1.55 |
| 0900 | On masonry | | 320 | .025 | | .34 | 1.08 | | 1.42 | 2.03 |
| 0950 | On concrete | | 210 | .038 | | .34 | 1.64 | | 1.98 | 2.90 |

### 06 11 10.44 Grounds

| | | Crew | Daily Output | Labor-Hours | Unit | Material | 2011 Bare Costs Labor | 2011 Bare Costs Equipment | Total | Total Incl O&P |
|---|---|---|---|---|---|---|---|---|---|---|
| 0010 | **GROUNDS** | | | | | | | | | |
| 0020 | For casework, 1" x 2" wood strips, on wood | 1 Carp | 330 | .024 | L.F. | .25 | 1.04 | | 1.29 | 1.87 |
| 0100 | On masonry | | 285 | .028 | | .25 | 1.21 | | 1.46 | 2.12 |
| 0200 | On concrete | | 250 | .032 | | .25 | 1.38 | | 1.63 | 2.38 |
| 0400 | For plaster, 3/4" deep, on wood | | 450 | .018 | | .25 | .77 | | 1.02 | 1.44 |
| 0500 | On masonry | | 225 | .036 | | .25 | 1.53 | | 1.78 | 2.62 |
| 0600 | On concrete | | 175 | .046 | | .25 | 1.97 | | 2.22 | 3.29 |
| 0700 | On metal lath | | 200 | .040 | | .25 | 1.72 | | 1.97 | 2.91 |

## 06 12 10 – Structural Insulated Panels

| 06 12 10.10 OSB Faced Panels | | Crew | Daily Output | Labor-Hours | Unit | Material | 2011 Bare Costs Labor | Equipment | Total | Total Incl O&P |
|---|---|---|---|---|---|---|---|---|---|---|
| 0010 | **OSB FACED PANELS** | | | | | | | | | |
| 0100 | Structural insul. panels, 7/16" OSB both faces, EPS insul, 3-5/8" T | G | F-3 | 2075 | .019 | S.F. | 2.72 | .84 | .32 | 3.88 | 4.63 |
| 0110 | 5-5/8" thick | G | | 1725 | .023 | | 3.05 | 1.01 | .38 | 4.44 | 5.35 |
| 0120 | 7-3/8" thick | G | | 1425 | .028 | | 3.33 | 1.23 | .46 | 5.02 | 6.05 |
| 0130 | 9-3/8" thick | G | | 1125 | .036 | | 3.64 | 1.56 | .58 | 5.78 | 7 |
| 0140 | 7/16" OSB one face, EPS insul, 3-5/8" thick | G | | 2175 | .018 | | 2.90 | .80 | .30 | 4 | 4.75 |
| 0150 | 5-5/8" thick | G | | 1825 | .022 | | 3.32 | .96 | .36 | 4.64 | 5.50 |
| 0160 | 7-3/8" thick | G | | 1525 | .026 | | 3.70 | 1.15 | .43 | 5.28 | 6.30 |
| 0170 | 9-3/8" thick | G | | 1225 | .033 | | 4.12 | 1.43 | .54 | 6.09 | 7.30 |
| 0190 | 7/16" OSB - 1/2" GWB faces , EPS insul, 3-5/8" T | G | | 2075 | .019 | | 2.87 | .84 | .32 | 4.03 | 4.80 |
| 0200 | 5-5/8" thick | G | | 1725 | .023 | | 3.29 | 1.01 | .38 | 4.68 | 5.60 |
| 0210 | 7-3/8" thick | G | | 1425 | .028 | | 3.66 | 1.23 | .46 | 5.35 | 6.40 |
| 0220 | 9-3/8" thick | G | | 1125 | .036 | | 4.29 | 1.56 | .58 | 6.43 | 7.75 |
| 0240 | 7/16" OSB - 1/2" MRGWB faces , EPS insul, 3-5/8" T | G | | 2075 | .019 | | 2.87 | .84 | .32 | 4.03 | 4.80 |
| 0250 | 5-5/8" thick | G | | 1725 | .023 | | 3.29 | 1.01 | .38 | 4.68 | 5.60 |
| 0260 | 7-3/8" thick | G | | 1425 | .028 | | 3.66 | 1.23 | .46 | 5.35 | 6.40 |
| 0270 | 9-3/8" thick | G | | 1125 | .036 | | 4.29 | 1.56 | .58 | 6.43 | 7.75 |
| 0300 | For 1/2" GWB added to OSB skin, add | G | | | | | .60 | | | .60 | .66 |
| 0310 | For 1/2" MRGWB added to OSB skin, add | G | | | | | .60 | | | .60 | .66 |
| 0320 | For one T1-11 skin, add to OSB-OSB | G | | | | | 1.39 | | | 1.39 | 1.53 |
| 0330 | For one 19/32" CDX skin, add to OSB-OSB | G | | | | | .70 | | | .70 | .77 |
| 0500 | Structural insulated panel, 7/16" OSB both sides, straw core | | | | | | | | | | |
| 0510 | 4-3/8" T, walls (w/sill, splines, plates) | G | F-6 | 2400 | .017 | S.F. | 6.60 | .67 | .27 | 7.54 | 8.55 |
| 0520 | Floors (w/splines) | G | | 2400 | .017 | | 6.60 | .67 | .27 | 7.54 | 8.55 |
| 0530 | Roof (w/splines) | G | | 2400 | .017 | | 6.60 | .67 | .27 | 7.54 | 8.55 |
| 0550 | 7-7/8" T, walls (w/sill, splines, plates) | G | | 2400 | .017 | | 9.95 | .67 | .27 | 10.89 | 12.25 |
| 0560 | Floors (w/splines) | G | | 2400 | .017 | | 9.95 | .67 | .27 | 10.89 | 12.25 |
| 0570 | Roof (w/splines) | G | | 2400 | .017 | | 9.95 | .67 | .27 | 10.89 | 12.25 |

## 06 12 19 – Composite Shearwall Panels

| 06 12 19.10 Steel and Wood Composite Shearwall Panels | | Crew | Daily Output | Labor-Hours | Unit | Material | 2011 Bare Costs Labor | Equipment | Total | Total Incl O&P |
|---|---|---|---|---|---|---|---|---|---|---|
| 0010 | **STEEL & WOOD COMPOSITE SHEARWALL PANELS** | | | | | | | | | |
| 0020 | Anchor bolts, 36" long (must be placed in wet concrete) | 1 Carp | 150 | .053 | Ea. | 32.50 | 2.30 | | 34.80 | 39 |
| 0030 | On concrete, 2 x 4 & 2 x 6 walls, 7' - 10' high, 360 lbs shear, 12" wide | 2 Carp | 8 | 2 | | 278 | 86 | | 364 | 435 |
| 0040 | 715 lbs shear, 15" wide | | 8 | 2 | | 320 | 86 | | 406 | 480 |
| 0050 | 1860 lbs shear, 18" wide | | 8 | 2 | | 355 | 86 | | 441 | 520 |
| 0060 | 2780 lbs shear, 21" wide | | 8 | 2 | | 365 | 86 | | 451 | 530 |
| 0070 | 3790 lbs shear, 24" wide | | 8 | 2 | | 460 | 86 | | 546 | 635 |
| 0080 | 2 x 6 walls, 11' to 13' high, 1180 lbs shear, 18" wide | | 6 | 2.667 | | 350 | 115 | | 465 | 560 |
| 0090 | 1555 lbs shear, 21" wide | | 6 | 2.667 | | 510 | 115 | | 625 | 735 |
| 0100 | 2280 lbs shear, 24" wide | | 6 | 2.667 | | 620 | 115 | | 735 | 855 |
| 0110 | For installing above on wood floor frame, add | | | | | | | | | |
| 0120 | Coupler nuts, threaded rods, bolts, shear transfer plate kit | 1 Carp | 16 | .500 | Ea. | 48 | 21.50 | | 69.50 | 86 |
| 0130 | Framing anchors, angle (2 required) | " | 96 | .083 | " | 2.40 | 3.59 | | 5.99 | 8.15 |
| 0140 | For blocking see Section 06 11 10.02 | | | | | | | | | |
| 0150 | For installing above, first floor to second floor, wood floor frame, add | | | | | | | | | |
| 0160 | Add stack option to first floor wall panel | | | | Ea. | 53 | | | 53 | 58.50 |
| 0170 | Threaded rods, bolts, shear transfer plate kit | 1 Carp | 16 | .500 | | 57.50 | 21.50 | | 79 | 96.50 |
| 0180 | Framing anchors, angle (2 required) | " | 96 | .083 | | 2.40 | 3.59 | | 5.99 | 8.15 |
| 0200 | For installing stacked panels, balloon framing | | | | | | | | | |
| 0210 | Add stack option to first floor wall panel | | | | Ea. | 53 | | | 53 | 58.50 |
| 0220 | Threaded rods, bolts kit | 1 Carp | 16 | .500 | " | 31 | 21.50 | | 52.50 | 67 |

# 06 13 Heavy Timber Construction

## 06 13 23 – Heavy Timber Framing

| 06 13 23.10 Heavy Framing | Crew | Daily Output | Labor-Hours | Unit | Material | 2011 Bare Costs Labor | Equipment | Total | Total Incl O&P |
|---|---|---|---|---|---|---|---|---|---|
| 0010 **HEAVY FRAMING** | | | | | | | | | |
| 0020 Beams, single 6" x 10" | 2 Carp | 1.10 | 14.545 | M.B.F. | 1,400 | 625 | | 2,025 | 2,500 |
| 0100 Single 8" x 16" | | 1.20 | 13.333 | | 1,975 | 575 | | 2,550 | 3,050 |
| 0200 Built from 2" lumber, multiple 2" x 14" | | .90 | 17.778 | | 725 | 765 | | 1,490 | 1,975 |
| 0210 Built from 3" lumber, multiple 3" x 6" | | .70 | 22.857 | | 1,050 | 985 | | 2,035 | 2,650 |
| 0220 Multiple 3" x 8" | | .80 | 20 | | 1,050 | 860 | | 1,910 | 2,500 |
| 0230 Multiple 3" x 10" | | .90 | 17.778 | | 1,075 | 765 | | 1,840 | 2,350 |
| 0240 Multiple 3" x 12" | | 1 | 16 | | 1,100 | 690 | | 1,790 | 2,250 |
| 0250 Built from 4" lumber, multiple 4" x 6" | | .80 | 20 | | 1,175 | 860 | | 2,035 | 2,625 |
| 0260 Multiple 4" x 8" | | .90 | 17.778 | | 1,125 | 765 | | 1,890 | 2,400 |
| 0270 Multiple 4" x 10" | | 1 | 16 | | 1,150 | 690 | | 1,840 | 2,300 |
| 0280 Multiple 4" x 12" | | 1.10 | 14.545 | | 1,375 | 625 | | 2,000 | 2,475 |
| 0290 Columns, structural grade, 1500f, 4" x 4" | | .60 | 26.667 | | 1,275 | 1,150 | | 2,425 | 3,150 |
| 0300 6" x 6" | | .65 | 24.615 | | 1,300 | 1,050 | | 2,350 | 3,050 |
| 0400 8" x 8" | | .70 | 22.857 | | 1,350 | 985 | | 2,335 | 2,975 |
| 0500 10" x 10" | | .75 | 21.333 | | 1,450 | 920 | | 2,370 | 3,000 |
| 0600 12" x 12" | | .80 | 20 | | 1,525 | 860 | | 2,385 | 3,000 |
| 0800 Floor planks, 2" thick, T & G, 2" x 6" | | 1.05 | 15.238 | | 1,050 | 655 | | 1,705 | 2,150 |
| 0900 2" x 10" | | 1.10 | 14.545 | | 1,050 | 625 | | 1,675 | 2,100 |
| 1100 3" thick, 3" x 6" | | 1.05 | 15.238 | | 1,250 | 655 | | 1,905 | 2,375 |
| 1200 3" x 10" | | 1.10 | 14.545 | | 1,250 | 625 | | 1,875 | 2,325 |
| 1400 Girders, structural grade, 12" x 12" | | .80 | 20 | | 1,525 | 860 | | 2,385 | 3,000 |
| 1500 10" x 16" | | 1 | 16 | | 2,425 | 690 | | 3,115 | 3,725 |
| 2300 Roof purlins, 4" thick, structural grade | | 1.05 | 15.238 | | 1,125 | 655 | | 1,780 | 2,225 |

# 06 15 Wood Decking

## 06 15 16 – Wood Roof Decking

### 06 15 16.10 Solid Wood Roof Decking

| 06 15 16.10 Solid Wood Roof Decking | Crew | Daily Output | Labor-Hours | Unit | Material | 2011 Bare Costs Labor | Equipment | Total | Total Incl O&P |
|---|---|---|---|---|---|---|---|---|---|
| 0010 **SOLID WOOD ROOF DECKING** | | | | | | | | | |
| 0400 Cedar planks, 3" thick | 2 Carp | 320 | .050 | S.F. | 7.70 | 2.15 | | 9.85 | 11.80 |
| 0500 4" thick | | 250 | .064 | | 10.30 | 2.76 | | 13.06 | 15.60 |
| 0700 Douglas fir, 3" thick | | 320 | .050 | | 3.60 | 2.15 | | 5.75 | 7.25 |
| 0800 4" thick | | 250 | .064 | | 4.80 | 2.76 | | 7.56 | 9.55 |
| 1000 Hemlock, 3" thick | | 320 | .050 | | 3.28 | 2.15 | | 5.43 | 6.90 |
| 1100 4" thick | | 250 | .064 | | 4.36 | 2.76 | | 7.12 | 9.05 |
| 1300 Western white spruce, 3" thick | | 320 | .050 | | 3.28 | 2.15 | | 5.43 | 6.90 |
| 1400 4" thick | | 250 | .064 | | 4.36 | 2.76 | | 7.12 | 9.05 |

## 06 15 23 – Laminated Wood Decking

### 06 15 23.10 Laminated Roof Deck

| 06 15 23.10 Laminated Roof Deck | Crew | Daily Output | Labor-Hours | Unit | Material | 2011 Bare Costs Labor | Equipment | Total | Total Incl O&P |
|---|---|---|---|---|---|---|---|---|---|
| 0010 **LAMINATED ROOF DECK** | | | | | | | | | |
| 0020 Pine or hemlock, 3" thick | 2 Carp | 425 | .038 | S.F. | 3.75 | 1.62 | | 5.37 | 6.60 |
| 0100 4" thick | | 325 | .049 | | 5 | 2.12 | | 7.12 | 8.75 |
| 0300 Cedar, 3" thick | | 425 | .038 | | 4.11 | 1.62 | | 5.73 | 7 |
| 0400 4" thick | | 325 | .049 | | 5.25 | 2.12 | | 7.37 | 9 |
| 0600 Fir, 3" thick | | 425 | .038 | | 3.80 | 1.62 | | 5.42 | 6.65 |
| 0700 4" thick | | 325 | .049 | | 5.05 | 2.12 | | 7.17 | 8.85 |

# 06 16 Sheathing

## 06 16 23 – Subflooring

### 06 16 23.10 Subfloor

| | | Crew | Daily Output | Labor-Hours | Unit | Material | 2011 Bare Costs Labor | Equipment | Total | Total Incl O&P |
|---|---|---|---|---|---|---|---|---|---|---|
| 0010 | **SUBFLOOR** R061636-20 | | | | | | | | | |
| 0011 | Plywood, CDX, 1/2" thick | 2 Carp | 1500 | .011 | SF Flr. | .58 | .46 | | 1.04 | 1.34 |
| 0015 | Pneumatic nailed | | 1860 | .009 | | .58 | .37 | | .95 | 1.21 |
| 0100 | 5/8" thick | | 1350 | .012 | | .70 | .51 | | 1.21 | 1.55 |
| 0105 | Pneumatic nailed | | 1674 | .010 | | .70 | .41 | | 1.11 | 1.40 |
| 0200 | 3/4" thick | | 1250 | .013 | | .82 | .55 | | 1.37 | 1.75 |
| 0205 | Pneumatic nailed | | 1550 | .010 | | .82 | .44 | | 1.26 | 1.58 |
| 0300 | 1-1/8" thick, 2-4-1 including underlayment | | 1050 | .015 | | 1.53 | .66 | | 2.19 | 2.69 |
| 0440 | With boards, 1" x 6", S4S, laid regular | | 900 | .018 | | 1.47 | .77 | | 2.24 | 2.78 |
| 0450 | 1" x 8", laid regular | | 1000 | .016 | | 1.55 | .69 | | 2.24 | 2.76 |
| 0460 | Laid diagonal | | 850 | .019 | | 1.55 | .81 | | 2.36 | 2.94 |
| 0500 | 1" x 10", laid regular | | 1100 | .015 | | 1.52 | .63 | | 2.15 | 2.63 |
| 0600 | Laid diagonal | | 900 | .018 | | 1.52 | .77 | | 2.29 | 2.84 |
| 8990 | Subfloor adhesive, 3/8" bead | 1 Carp | 2300 | .003 | L.F. | .09 | .15 | | .24 | .33 |

## 06 16 26 – Underlayment

### 06 16 26.10 Wood Product Underlayment

| | | Crew | Daily Output | Labor-Hours | Unit | Material | 2011 Bare Costs Labor | Equipment | Total | Total Incl O&P |
|---|---|---|---|---|---|---|---|---|---|---|
| 0010 | **WOOD PRODUCT UNDERLAYMENT** R061636-20 | | | | | | | | | |
| 0030 | Plywood, underlayment grade, 3/8" thick | 2 Carp | 1500 | .011 | SF Flr. | .77 | .46 | | 1.23 | 1.55 |
| 0070 | Pneumatic nailed | | 1860 | .009 | | .77 | .37 | | 1.14 | 1.42 |
| 0100 | 1/2" thick | | 1450 | .011 | | .80 | .47 | | 1.27 | 1.61 |
| 0105 | Pneumatic nailed | | 1798 | .009 | | .80 | .38 | | 1.18 | 1.47 |
| 0200 | 5/8" thick | | 1400 | .011 | | 1.01 | .49 | | 1.50 | 1.87 |
| 0205 | Pneumatic nailed | | 1736 | .009 | | 1.01 | .40 | | 1.41 | 1.72 |
| 0300 | 3/4" thick | | 1300 | .012 | | 1.23 | .53 | | 1.76 | 2.16 |
| 0305 | Pneumatic nailed | | 1612 | .010 | | 1.23 | .43 | | 1.66 | 2.01 |
| 0500 | Particle board, 3/8" thick [G] | | 1500 | .011 | | .33 | .46 | | .79 | 1.06 |
| 0505 | Pneumatic nailed [G] | | 1860 | .009 | | .33 | .37 | | .70 | .93 |
| 0600 | 1/2" thick [G] | | 1450 | .011 | | .36 | .47 | | .83 | 1.13 |
| 0605 | Pneumatic nailed [G] | | 1798 | .009 | | .36 | .38 | | .74 | .99 |
| 0800 | 5/8" thick [G] | | 1400 | .011 | | .46 | .49 | | .95 | 1.27 |
| 0805 | Pneumatic nailed [G] | | 1736 | .009 | | .46 | .40 | | .86 | 1.12 |
| 0900 | 3/4" thick [G] | | 1300 | .012 | | .65 | .53 | | 1.18 | 1.53 |
| 0905 | Pneumatic nailed [G] | | 1612 | .010 | | .65 | .43 | | 1.08 | 1.38 |
| 0955 | Particleboard, 100% recycled straw/wheat, 4' x 8' x 1/4" [G] | | 1450 | .011 | S.F. | .26 | .47 | | .73 | 1.02 |
| 0960 | 4' x 8' x 3/8" [G] | | 1450 | .011 | | .39 | .47 | | .86 | 1.16 |
| 0965 | 4' x 8' x 1/2" [G] | | 1350 | .012 | | .53 | .51 | | 1.04 | 1.36 |
| 0970 | 4' x 8' x 5/8" [G] | | 1300 | .012 | | .63 | .53 | | 1.16 | 1.50 |
| 0975 | 4' x 8' x 3/4" [G] | | 1250 | .013 | | .73 | .55 | | 1.28 | 1.65 |
| 0980 | 4' x 8' x 1" [G] | | 1150 | .014 | | .98 | .60 | | 1.58 | 2 |
| 0985 | 4' x 8' x 1-1/4" [G] | | 1100 | .015 | | 1.15 | .63 | | 1.78 | 2.23 |
| 1100 | Hardboard, underlayment grade, 4' x 4', .215" thick [G] | | 1500 | .011 | SF Flr. | .57 | .46 | | 1.03 | 1.33 |

## 06 16 36 – Wood Panel Product Sheathing

### 06 16 36.10 Sheathing

| | | Crew | Daily Output | Labor-Hours | Unit | Material | 2011 Bare Costs Labor | Equipment | Total | Total Incl O&P |
|---|---|---|---|---|---|---|---|---|---|---|
| 0010 | **SHEATHING** R061636-20 | | | | | | | | | |
| 0012 | Plywood on roofs, CDX | | | | | | | | | |
| 0030 | 5/16" thick | 2 Carp | 1600 | .010 | S.F. | .55 | .43 | | .98 | 1.26 |
| 0035 | Pneumatic nailed R061110-30 | | 1952 | .008 | | .55 | .35 | | .90 | 1.14 |
| 0050 | 3/8" thick | | 1525 | .010 | | .47 | .45 | | .92 | 1.21 |
| 0055 | Pneumatic nailed | | 1860 | .009 | | .47 | .37 | | .84 | 1.09 |
| 0100 | 1/2" thick | | 1400 | .011 | | .58 | .49 | | 1.07 | 1.40 |
| 0105 | Pneumatic nailed | | 1708 | .009 | | .58 | .40 | | .98 | 1.26 |
| 0200 | 5/8" thick | | 1300 | .012 | | .70 | .53 | | 1.23 | 1.58 |

# 06 16 Sheathing

## 06 16 36 – Wood Panel Product Sheathing

| 06 16 36.10 Sheathing | Crew | Daily Output | Labor-Hours | Unit | Material | 2011 Bare Costs Labor | 2011 Bare Costs Equipment | Total | Total Incl O&P |
|---|---|---|---|---|---|---|---|---|---|
| 0205 | Pneumatic nailed | 2 Carp | 1586 | .010 | S.F. | .70 | .43 | | 1.13 | 1.44 |
| 0300 | 3/4" thick | | 1200 | .013 | | .82 | .57 | | 1.39 | 1.78 |
| 0305 | Pneumatic nailed | | 1464 | .011 | | .82 | .47 | | 1.29 | 1.62 |
| 0500 | Plywood on walls with exterior CDX, 3/8" thick | | 1200 | .013 | | .47 | .57 | | 1.04 | 1.40 |
| 0505 | Pneumatic nailed | | 1488 | .011 | | .47 | .46 | | .93 | 1.23 |
| 0600 | 1/2" thick | | 1125 | .014 | | .58 | .61 | | 1.19 | 1.58 |
| 0605 | Pneumatic nailed | | 1395 | .011 | | .58 | .49 | | 1.07 | 1.40 |
| 0700 | 5/8" thick | | 1050 | .015 | | .70 | .66 | | 1.36 | 1.78 |
| 0705 | Pneumatic nailed | | 1302 | .012 | | .70 | .53 | | 1.23 | 1.58 |
| 0800 | 3/4" thick | | 975 | .016 | | .82 | .71 | | 1.53 | 1.98 |
| 0805 | Pneumatic nailed | | 1209 | .013 | | .82 | .57 | | 1.39 | 1.77 |
| 0840 | Oriented strand board, 7/16" thick G | | 1400 | .011 | | .33 | .49 | | .82 | 1.12 |
| 0845 | Pneumatic nailed G | | 1736 | .009 | | .33 | .40 | | .73 | .97 |
| 0846 | 1/2" thick G | | 1325 | .012 | | .33 | .52 | | .85 | 1.16 |
| 0847 | Pneumatic nailed G | | 1643 | .010 | | .33 | .42 | | .75 | 1 |
| 0852 | 5/8" thick G | | 1250 | .013 | | .79 | .55 | | 1.34 | 1.72 |
| 0857 | Pneumatic nailed G | | 1550 | .010 | | .79 | .44 | | 1.23 | 1.55 |
| 1000 | For shear wall construction, add | | | | | | 20% | | | |
| 1200 | For structural 1 exterior plywood, add | | | | S.F. | 10% | | | | |
| 1400 | With boards, on roof 1" x 6" boards, laid horizontal | 2 Carp | 725 | .022 | | 1.47 | .95 | | 2.42 | 3.07 |
| 1500 | Laid diagonal | | 650 | .025 | | 1.47 | 1.06 | | 2.53 | 3.24 |
| 1700 | 1" x 8" boards, laid horizontal | | 875 | .018 | | 1.55 | .79 | | 2.34 | 2.91 |
| 1800 | Laid diagonal | | 725 | .022 | | 1.55 | .95 | | 2.50 | 3.16 |
| 2000 | For steep roofs, add | | | | | | 40% | | | |
| 2200 | For dormers, hips and valleys, add | | | | | 5% | 50% | | | |
| 2400 | Boards on walls, 1" x 6" boards, laid regular | 2 Carp | 650 | .025 | | 1.47 | 1.06 | | 2.53 | 3.24 |
| 2500 | Laid diagonal | | 585 | .027 | | 1.47 | 1.18 | | 2.65 | 3.42 |
| 2700 | 1" x 8" boards, laid regular | | 765 | .021 | | 1.55 | .90 | | 2.45 | 3.08 |
| 2800 | Laid diagonal | | 650 | .025 | | 1.55 | 1.06 | | 2.61 | 3.33 |
| 2850 | Gypsum, weatherproof, 1/2" thick | | 1125 | .014 | | .38 | .61 | | .99 | 1.36 |
| 2900 | With embedded glass mats | | 1100 | .015 | | .75 | .63 | | 1.38 | 1.79 |
| 3000 | Wood fiber, regular, no vapor barrier, 1/2" thick | | 1200 | .013 | | .56 | .57 | | 1.13 | 1.50 |
| 3100 | 5/8" thick | | 1200 | .013 | | .71 | .57 | | 1.28 | 1.66 |
| 3300 | No vapor barrier, in colors, 1/2" thick | | 1200 | .013 | | .68 | .57 | | 1.25 | 1.63 |
| 3400 | 5/8" thick | | 1200 | .013 | | .72 | .57 | | 1.29 | 1.67 |
| 3600 | With vapor barrier one side, white, 1/2" thick | | 1200 | .013 | | .55 | .57 | | 1.12 | 1.49 |
| 3700 | Vapor barrier 2 sides, 1/2" thick | | 1200 | .013 | | .77 | .57 | | 1.34 | 1.73 |
| 3800 | Asphalt impregnated, 25/32" thick | | 1200 | .013 | | .28 | .57 | | .85 | 1.19 |
| 3850 | Intermediate, 1/2" thick | | 1200 | .013 | | .20 | .57 | | .77 | 1.10 |

# 06 17 Shop-Fabricated Structural Wood

## 06 17 33 – Wood I-Joists

### 06 17 33.10 Wood and Composite I-Joists

| | | Crew | Daily Output | Labor-Hours | Unit | Material | 2011 Bare Costs Labor | 2011 Bare Costs Equipment | Total | Total Incl O&P |
|---|---|---|---|---|---|---|---|---|---|---|
| 0010 | **WOOD AND COMPOSITE I-JOISTS** | | | | | | | | | |
| 0100 | Plywood webs, incl. bridging & blocking, panels 24" O.C. | | | | | | | | | |
| 1200 | 15' to 24' span, 50 psf live load | F-5 | 2400 | .013 | SF Flr. | 1.76 | .58 | | 2.34 | 2.82 |
| 1300 | 55 psf live load | | 2250 | .014 | | 1.98 | .62 | | 2.60 | 3.13 |
| 1400 | 24' to 30' span, 45 psf live load | | 2600 | .012 | | 2.73 | .54 | | 3.27 | 3.82 |
| 1500 | 55 psf live load | | 2400 | .013 | | 3.38 | .58 | | 3.96 | 4.61 |

# 06 17 Shop-Fabricated Structural Wood

## 06 17 53 – Shop-Fabricated Wood Trusses

| 06 17 53.10 Roof Trusses | Crew | Daily Output | Labor-Hours | Unit | Material | 2011 Bare Costs Labor | Equipment | Total | Total Incl O&P |
|---|---|---|---|---|---|---|---|---|---|
| 0010 **ROOF TRUSSES** | | | | | | | | | |
| 0100 Fink (W) or King post type, 2'-0" O.C. | | | | | | | | | |
| 0200 Metal plate connected, 4 in 12 slope | | | | | | | | | |
| 0210     24' to 29' span | F-3 | 3000 | .013 | SF Flr. | 1.50 | .58 | .22 | 2.30 | 2.78 |
| 0300     30' to 43' span | | 3000 | .013 | | 2.18 | .58 | .22 | 2.98 | 3.53 |
| 0400     44' to 60' span | | 3000 | .013 | | 2.36 | .58 | .22 | 3.16 | 3.72 |
| 0700 Glued and nailed, add | | | | | 50% | | | | |

# 06 18 Glued-Laminated Construction

## 06 18 13 – Glued-Laminated Beams

### 06 18 13.20 Laminated Framing

| | Crew | Daily Output | Labor-Hours | Unit | Material | 2011 Bare Costs Labor | Equipment | Total | Total Incl O&P |
|---|---|---|---|---|---|---|---|---|---|
| 0010 **LAMINATED FRAMING** | | | | | | | | | |
| 0020 30 lb., short term live load, 15 lb. dead load | | | | | | | | | |
| 0200 Straight roof beams, 20' clear span, beams 8' O.C. | F-3 | 2560 | .016 | SF Flr. | 1.99 | .68 | .26 | 2.93 | 3.51 |
| 0300     Beams 16' O.C. | | 3200 | .013 | | 1.44 | .55 | .20 | 2.19 | 2.64 |
| 0500     40' clear span, beams 8' O.C. | | 3200 | .013 | | 3.80 | .55 | .20 | 4.55 | 5.25 |
| 0600     Beams 16' O.C. | | 3840 | .010 | | 3.11 | .46 | .17 | 3.74 | 4.31 |
| 0800     60' clear span, beams 8' O.C. | F-4 | 2880 | .017 | | 6.55 | .72 | .42 | 7.69 | 8.75 |
| 0900     Beams 16' O.C. | " | 3840 | .013 | | 4.87 | .54 | .31 | 5.72 | 6.50 |
| 1100 Tudor arches, 30' to 40' clear span, frames 8' O.C. | F-3 | 1680 | .024 | | 8.50 | 1.04 | .39 | 9.93 | 11.35 |
| 1200     Frames 16' O.C. | " | 2240 | .018 | | 6.70 | .78 | .29 | 7.77 | 8.85 |
| 1400     50' to 60' clear span, frames 8' O.C. | F-4 | 2200 | .022 | | 9.20 | .94 | .55 | 10.69 | 12.15 |
| 1500     Frames 16' O.C. | | 2640 | .018 | | 7.80 | .78 | .46 | 9.04 | 10.30 |
| 1700 Radial arches, 60' clear span, frames 8' O.C. | | 1920 | .025 | | 8.60 | 1.08 | .63 | 10.31 | 11.80 |
| 1800     Frames 16' O.C. | | 2880 | .017 | | 6.60 | .72 | .42 | 7.74 | 8.80 |
| 2000     100' clear span, frames 8' O.C. | | 1600 | .030 | | 8.90 | 1.30 | .76 | 10.96 | 12.60 |
| 2100     Frames 16' O.C. | | 2400 | .020 | | 7.80 | .86 | .50 | 9.16 | 10.45 |
| 2300     120' clear span, frames 8' O.C. | | 1440 | .033 | | 11.85 | 1.44 | .84 | 14.13 | 16.10 |
| 2400     Frames 16' O.C. | | 1920 | .025 | | 10.80 | 1.08 | .63 | 12.51 | 14.25 |
| 2600 Bowstring trusses, 20' O.C., 40' clear span | F-3 | 2400 | .017 | | 5.35 | .73 | .27 | 6.35 | 7.25 |
| 2700     60' clear span | F-4 | 3600 | .013 | | 4.79 | .58 | .34 | 5.71 | 6.50 |
| 2800     100' clear span | | 4000 | .012 | | 6.80 | .52 | .30 | 7.62 | 8.55 |
| 2900     120' clear span | | 3600 | .013 | | 7.25 | .58 | .34 | 8.17 | 9.25 |
| 3000 For less than 1000 B.F., add | | | | | 20% | | | | |
| 3050 For over 5000 B.F., deduct | | | | | 10% | | | | |
| 3100 For premium appearance, add to S.F. prices | | | | | 5% | | | | |
| 3300 For industrial type, deduct | | | | | 15% | | | | |
| 3500 For stain and varnish, add | | | | | 5% | | | | |
| 3900 For 3/4" laminations, add to straight | | | | | 25% | | | | |
| 4100     Add to curved | | | | | 15% | | | | |
| 4300 Alternate pricing method: (use nominal footage of | | | | | | | | | |
| 4310     components). Straight beams, camber less than 6" | F-3 | 3.50 | 11.429 | M.B.F. | 2,950 | 500 | 187 | 3,637 | 4,225 |
| 4400     Columns, including hardware | | 2 | 20 | | 3,175 | 875 | 330 | 4,380 | 5,175 |
| 4600 Curved members, radius over 32' | | 2.50 | 16 | | 3,250 | 700 | 262 | 4,212 | 4,950 |
| 4700     Radius 10' to 32' | | 3 | 13.333 | | 3,225 | 585 | 219 | 4,029 | 4,650 |
| 4900 For complicated shapes, add maximum | | | | | 100% | | | | |
| 5100 For pressure treating, add to straight | | | | | 35% | | | | |
| 5200     Add to curved | | | | | 45% | | | | |
| 6000 Laminated veneer members, southern pine or western species | | | | | | | | | |
| 6050     1-3/4" wide x 5-1/2" deep | 2 Carp | 480 | .033 | L.F. | 3.19 | 1.43 | | 4.62 | 5.70 |

## 06 18 13 – Glued-Laminated Beams

| 06 18 13.20 Laminated Framing | | Crew | Daily Output | Labor-Hours | Unit | Material | 2011 Bare Costs Labor | Equipment | Total | Total Incl O&P |
|---|---|---|---|---|---|---|---|---|---|---|
| 6100 | 9-1/2" deep | 2 Carp | 480 | .033 | L.F. | 4.56 | 1.43 | | 5.99 | 7.20 |
| 6150 | 14" deep | | 450 | .036 | | 7.30 | 1.53 | | 8.83 | 10.40 |
| 6200 | 18" deep | ▼ | 450 | .036 | ▼ | 9.90 | 1.53 | | 11.43 | 13.20 |
| 6300 | Parallel strand members, southern pine or western species | | | | | | | | | |
| 6350 | 1-3/4" wide x 9-1/4" deep | 2 Carp | 480 | .033 | L.F. | 4.56 | 1.43 | | 5.99 | 7.20 |
| 6400 | 11-1/4" deep | | 450 | .036 | | 5.70 | 1.53 | | 7.23 | 8.60 |
| 6450 | 14" deep | | 400 | .040 | | 7.30 | 1.72 | | 9.02 | 10.70 |
| 6500 | 3-1/2" wide x 9-1/4" deep | | 480 | .033 | | 15.85 | 1.43 | | 17.28 | 19.65 |
| 6550 | 11-1/4" deep | | 450 | .036 | | 19.15 | 1.53 | | 20.68 | 23.50 |
| 6600 | 14" deep | | 400 | .040 | | 23.50 | 1.72 | | 25.22 | 28 |
| 6650 | 7" wide x 9-1/4" deep | | 450 | .036 | | 31.50 | 1.53 | | 33.03 | 37.50 |
| 6700 | 11-1/4" deep | | 420 | .038 | | 39.50 | 1.64 | | 41.14 | 46 |
| 6750 | 14" deep | ▼ | 400 | .040 | ▼ | 46.50 | 1.72 | | 48.22 | 54 |
| 8000 | Straight beams | | | | | | | | | |
| 8102 | 20' span | | | | | | | | | |
| 8104 | 3-1/8" x 9" | F-3 | 30 | 1.333 | Ea. | 138 | 58.50 | 22 | 218.50 | 265 |
| 8106 | X 10-1/2" | | 30 | 1.333 | | 161 | 58.50 | 22 | 241.50 | 290 |
| 8108 | X 12" | | 30 | 1.333 | | 184 | 58.50 | 22 | 264.50 | 315 |
| 8110 | X 13-1/2" | | 30 | 1.333 | | 207 | 58.50 | 22 | 287.50 | 340 |
| 8112 | X 15" | | 29 | 1.379 | | 230 | 60.50 | 22.50 | 313 | 370 |
| 8114 | 5-1/8" x 10-1/2" | | 30 | 1.333 | | 265 | 58.50 | 22 | 345.50 | 405 |
| 8116 | X 12" | | 30 | 1.333 | | 300 | 58.50 | 22 | 380.50 | 450 |
| 8118 | X 13-1/2" | | 30 | 1.333 | | 340 | 58.50 | 22 | 420.50 | 490 |
| 8120 | X 15" | | 29 | 1.379 | | 380 | 60.50 | 22.50 | 463 | 530 |
| 8122 | X 16-1/2" | | 29 | 1.379 | | 415 | 60.50 | 22.50 | 498 | 570 |
| 8124 | X 18" | | 29 | 1.379 | | 455 | 60.50 | 22.50 | 538 | 615 |
| 8126 | X 19-1/2" | | 29 | 1.379 | | 490 | 60.50 | 22.50 | 573 | 655 |
| 8128 | X 21" | | 28 | 1.429 | | 530 | 62.50 | 23.50 | 616 | 700 |
| 8130 | X 22-1/2" | | 28 | 1.429 | | 565 | 62.50 | 23.50 | 651 | 745 |
| 8132 | X 24" | | 28 | 1.429 | | 605 | 62.50 | 23.50 | 691 | 785 |
| 8134 | 6-3/4" x 12" | | 29 | 1.379 | | 400 | 60.50 | 22.50 | 483 | 555 |
| 8136 | X 13-1/2" | | 29 | 1.379 | | 450 | 60.50 | 22.50 | 533 | 610 |
| 8138 | X 15" | | 29 | 1.379 | | 500 | 60.50 | 22.50 | 583 | 665 |
| 8140 | X 16-1/2" | | 28 | 1.429 | | 550 | 62.50 | 23.50 | 636 | 720 |
| 8142 | X 18" | | 28 | 1.429 | | 595 | 62.50 | 23.50 | 681 | 775 |
| 8144 | X 19-1/2" | | 28 | 1.429 | | 645 | 62.50 | 23.50 | 731 | 830 |
| 8146 | X 21" | | 27 | 1.481 | | 695 | 65 | 24.50 | 784.50 | 890 |
| 8148 | X 22-1/2" | | 27 | 1.481 | | 745 | 65 | 24.50 | 834.50 | 945 |
| 8150 | X 24" | | 27 | 1.481 | | 795 | 65 | 24.50 | 884.50 | 1,000 |
| 8152 | X 25-1/2" | | 27 | 1.481 | | 845 | 65 | 24.50 | 934.50 | 1,050 |
| 8154 | X 27" | | 26 | 1.538 | | 895 | 67.50 | 25 | 987.50 | 1,125 |
| 8156 | X 28-1/2" | | 26 | 1.538 | | 945 | 67.50 | 25 | 1,037.50 | 1,175 |
| 8158 | X 30" | ▼ | 26 | 1.538 | ▼ | 995 | 67.50 | 25 | 1,087.50 | 1,225 |
| 8200 | 30' span | | | | | | | | | |
| 8250 | 3-1/8" x 9" | F-3 | 30 | 1.333 | Ea. | 207 | 58.50 | 22 | 287.50 | 340 |
| 8252 | X 10-1/2" | | 30 | 1.333 | | 242 | 58.50 | 22 | 322.50 | 380 |
| 8254 | X 12" | | 30 | 1.333 | | 277 | 58.50 | 22 | 357.50 | 420 |
| 8256 | X 13-1/2" | | 30 | 1.333 | | 310 | 58.50 | 22 | 390.50 | 455 |
| 8258 | X 15" | | 29 | 1.379 | | 345 | 60.50 | 22.50 | 428 | 495 |
| 8260 | 5-1/8" x 10-1/2" | | 30 | 1.333 | | 395 | 58.50 | 22 | 475.50 | 550 |
| 8262 | X 12" | | 30 | 1.333 | | 455 | 58.50 | 22 | 535.50 | 615 |
| 8264 | X 13-1/2" | | 30 | 1.333 | | 510 | 58.50 | 22 | 590.50 | 675 |
| 8266 | X 15" | ▼ | 29 | 1.379 | ▼ | 565 | 60.50 | 22.50 | 648 | 740 |

## 06 18 13 – Glued-Laminated Beams

| 06 18 13.20 Laminated Framing | | Crew | Daily Output | Labor-Hours | Unit | Material | 2011 Bare Costs Labor | Equipment | Total | Total Incl O&P |
|---|---|---|---|---|---|---|---|---|---|---|
| 8268 | X 16-1/2" | F-3 | 29 | 1.379 | Ea. | 625 | 60.50 | 22.50 | 708 | 800 |
| 8270 | X 18" | | 29 | 1.379 | | 680 | 60.50 | 22.50 | 763 | 865 |
| 8272 | X 19-1/2" | | 29 | 1.379 | | 735 | 60.50 | 22.50 | 818 | 925 |
| 8274 | X 21" | | 28 | 1.429 | | 795 | 62.50 | 23.50 | 881 | 995 |
| 8276 | X 22-1/2" | | 28 | 1.429 | | 850 | 62.50 | 23.50 | 936 | 1,050 |
| 8278 | X 24" | | 28 | 1.429 | | 905 | 62.50 | 23.50 | 991 | 1,125 |
| 8280 | 6-3/4" x 12" | | 29 | 1.379 | | 595 | 60.50 | 22.50 | 678 | 770 |
| 8282 | X 13-1/2" | | 29 | 1.379 | | 670 | 60.50 | 22.50 | 753 | 855 |
| 8284 | X 15" | | 29 | 1.379 | | 745 | 60.50 | 22.50 | 828 | 935 |
| 8286 | X 16-1/2" | | 28 | 1.429 | | 820 | 62.50 | 23.50 | 906 | 1,025 |
| 8288 | X 18" | | 28 | 1.429 | | 895 | 62.50 | 23.50 | 981 | 1,100 |
| 8290 | X 19-1/2" | | 28 | 1.429 | | 970 | 62.50 | 23.50 | 1,056 | 1,200 |
| 8292 | X 21" | | 27 | 1.481 | | 1,050 | 65 | 24.50 | 1,139.50 | 1,275 |
| 8294 | X 22-1/2" | | 27 | 1.481 | | 1,125 | 65 | 24.50 | 1,214.50 | 1,350 |
| 8296 | X 24" | | 27 | 1.481 | | 1,200 | 65 | 24.50 | 1,289.50 | 1,450 |
| 8298 | X 25-1/2" | | 27 | 1.481 | | 1,275 | 65 | 24.50 | 1,364.50 | 1,525 |
| 8300 | X 27" | | 26 | 1.538 | | 1,350 | 67.50 | 25 | 1,442.50 | 1,600 |
| 8302 | X 28-1/2" | | 26 | 1.538 | | 1,425 | 67.50 | 25 | 1,517.50 | 1,675 |
| 8304 | X 30" | | 26 | 1.538 | | 1,500 | 67.50 | 25 | 1,592.50 | 1,775 |
| 8400 | 40' span | | | | | | | | | |
| 8402 | 3-1/8" x 9" | F-3 | 30 | 1.333 | Ea. | 277 | 58.50 | 22 | 357.50 | 420 |
| 8404 | X 10-1/2" | | 30 | 1.333 | | 325 | 58.50 | 22 | 405.50 | 470 |
| 8406 | X 12" | | 30 | 1.333 | | 370 | 58.50 | 22 | 450.50 | 520 |
| 8408 | X 13-1/2" | | 30 | 1.333 | | 415 | 58.50 | 22 | 495.50 | 570 |
| 8410 | X 15" | | 29 | 1.379 | | 460 | 60.50 | 22.50 | 543 | 620 |
| 8412 | 5-1/8" x 10-1/2" | | 30 | 1.333 | | 530 | 58.50 | 22 | 610.50 | 695 |
| 8414 | X 12" | | 30 | 1.333 | | 605 | 58.50 | 22 | 685.50 | 780 |
| 8416 | X 13-1/2" | | 30 | 1.333 | | 680 | 58.50 | 22 | 760.50 | 865 |
| 8418 | X 15" | | 29 | 1.379 | | 755 | 60.50 | 22.50 | 838 | 945 |
| 8420 | X 16-1/2" | | 29 | 1.379 | | 830 | 60.50 | 22.50 | 913 | 1,025 |
| 8422 | X 18" | | 29 | 1.379 | | 905 | 60.50 | 22.50 | 988 | 1,125 |
| 8424 | X 19-1/2" | | 29 | 1.379 | | 985 | 60.50 | 22.50 | 1,068 | 1,200 |
| 8426 | X 21" | | 28 | 1.429 | | 1,050 | 62.50 | 23.50 | 1,136 | 1,300 |
| 8428 | X 22-1/2" | | 28 | 1.429 | | 1,125 | 62.50 | 23.50 | 1,211 | 1,375 |
| 8430 | X 24" | | 28 | 1.429 | | 1,200 | 62.50 | 23.50 | 1,286 | 1,450 |
| 8432 | 6-3/4" x 12" | | 29 | 1.379 | | 795 | 60.50 | 22.50 | 878 | 990 |
| 8434 | X 13-1/2" | | 29 | 1.379 | | 895 | 60.50 | 22.50 | 978 | 1,100 |
| 8436 | X 15" | | 29 | 1.379 | | 995 | 60.50 | 22.50 | 1,078 | 1,225 |
| 8438 | X 16-1/2" | | 28 | 1.429 | | 1,100 | 62.50 | 23.50 | 1,186 | 1,325 |
| 8440 | X 18" | | 28 | 1.429 | | 1,200 | 62.50 | 23.50 | 1,286 | 1,450 |
| 8442 | X 19-1/2" | | 28 | 1.429 | | 1,300 | 62.50 | 23.50 | 1,386 | 1,550 |
| 8444 | X 21" | | 27 | 1.481 | | 1,400 | 65 | 24.50 | 1,489.50 | 1,650 |
| 8446 | X 22-1/2" | | 27 | 1.481 | | 1,500 | 65 | 24.50 | 1,589.50 | 1,775 |
| 8448 | X 24" | | 27 | 1.481 | | 1,600 | 65 | 24.50 | 1,689.50 | 1,875 |
| 8450 | X 25-1/2" | | 27 | 1.481 | | 1,700 | 65 | 24.50 | 1,789.50 | 1,975 |
| 8452 | X 27" | | 26 | 1.538 | | 1,800 | 67.50 | 25 | 1,892.50 | 2,100 |
| 8454 | X 28-1/2" | | 26 | 1.538 | | 1,900 | 67.50 | 25 | 1,992.50 | 2,200 |
| 8456 | X 30" | | 26 | 1.538 | | 2,000 | 67.50 | 25 | 2,092.50 | 2,325 |

# 06 22 Millwork

## 06 22 13 – Standard Pattern Wood Trim

### 06 22 13.15 Moldings, Base

| | | Crew | Daily Output | Labor-Hours | Unit | Material | 2011 Bare Costs Labor | Equipment | Total | Total Incl O&P |
|---|---|---|---|---|---|---|---|---|---|---|
| 0010 | **MOLDINGS, BASE** | | | | | | | | | |
| 0500 | Base, stock pine, 9/16" x 3-1/2" | 1 Carp | 240 | .033 | L.F. | 2.22 | 1.43 | | 3.65 | 4.64 |
| 0550 | 9/16" x 4-1/2" | | 200 | .040 | | 2.46 | 1.72 | | 4.18 | 5.35 |
| 0561 | Base shoe, oak, 3/4" x 1" | ▼ | 240 | .033 | ▼ | 1.13 | 1.43 | | 2.56 | 3.44 |

### 06 22 13.30 Moldings, Casings

| | | Crew | Daily Output | Labor-Hours | Unit | Material | 2011 Bare Costs Labor | Equipment | Total | Total Incl O&P |
|---|---|---|---|---|---|---|---|---|---|---|
| 0010 | **MOLDINGS, CASINGS** | | | | | | | | | |
| 0090 | Apron, stock pine, 5/8" x 2" | 1 Carp | 250 | .032 | L.F. | 1.22 | 1.38 | | 2.60 | 3.45 |
| 0110 | 5/8" x 3-1/2" | | 220 | .036 | | 1.62 | 1.57 | | 3.19 | 4.18 |
| 0300 | Band, stock pine, 11/16" x 1-1/8" | | 270 | .030 | | .82 | 1.28 | | 2.10 | 2.86 |
| 0350 | 11/16" x 1-3/4" | | 250 | .032 | | 1.06 | 1.38 | | 2.44 | 3.28 |
| 0700 | Casing, stock pine, 11/16" x 2-1/2" | | 240 | .033 | | 1.33 | 1.43 | | 2.76 | 3.66 |
| 0750 | 11/16" x 3-1/2" | ▼ | 215 | .037 | ▼ | 1.62 | 1.60 | | 3.22 | 4.24 |

### 06 22 13.35 Moldings, Ceilings

| | | Crew | Daily Output | Labor-Hours | Unit | Material | 2011 Bare Costs Labor | Equipment | Total | Total Incl O&P |
|---|---|---|---|---|---|---|---|---|---|---|
| 0010 | **MOLDINGS, CEILINGS** | | | | | | | | | |
| 0600 | Bed, stock pine, 9/16" x 1-3/4" | 1 Carp | 270 | .030 | L.F. | 1.02 | 1.28 | | 2.30 | 3.08 |
| 0650 | 9/16" x 2" | | 240 | .033 | | 1.22 | 1.43 | | 2.65 | 3.54 |
| 1200 | Cornice molding, stock pine, 9/16" x 1-3/4" | | 330 | .024 | | 1.02 | 1.04 | | 2.06 | 2.72 |
| 1300 | 9/16" x 2-1/4" | | 300 | .027 | | 1.24 | 1.15 | | 2.39 | 3.12 |
| 2400 | Cove scotia, stock pine, 9/16" x 1-3/4" | | 270 | .030 | | .67 | 1.28 | | 1.95 | 2.70 |
| 2500 | 11/16" x 2-3/4" | | 255 | .031 | | 1.86 | 1.35 | | 3.21 | 4.12 |
| 2600 | Crown, stock pine, 9/16" x 3-5/8" | | 250 | .032 | | 2.30 | 1.38 | | 3.68 | 4.64 |
| 2700 | 11/16" x 4-5/8" | ▼ | 220 | .036 | ▼ | 3.87 | 1.57 | | 5.44 | 6.65 |

### 06 22 13.40 Moldings, Exterior

| | | Crew | Daily Output | Labor-Hours | Unit | Material | 2011 Bare Costs Labor | Equipment | Total | Total Incl O&P |
|---|---|---|---|---|---|---|---|---|---|---|
| 0010 | **MOLDINGS, EXTERIOR** | | | | | | | | | |
| 1500 | Cornice, boards, pine, 1" x 2" | 1 Carp | 330 | .024 | L.F. | .25 | 1.04 | | 1.29 | 1.87 |
| 1700 | 1" x 6" | | 250 | .032 | | .67 | 1.38 | | 2.05 | 2.85 |
| 2000 | 1" x 12" | | 180 | .044 | | 1.46 | 1.91 | | 3.37 | 4.55 |
| 2200 | Three piece, built-up, pine, minimum | | 80 | .100 | | 1.45 | 4.31 | | 5.76 | 8.20 |
| 2300 | Maximum | | 65 | .123 | | 3.64 | 5.30 | | 8.94 | 12.15 |
| 3000 | Corner board, sterling pine, 1" x 4" | | 200 | .040 | | .79 | 1.72 | | 2.51 | 3.51 |
| 3100 | 1" x 6" | | 200 | .040 | | 1.05 | 1.72 | | 2.77 | 3.80 |
| 3350 | Fascia, sterling pine, 1" x 6" | | 250 | .032 | | 1.05 | 1.38 | | 2.43 | 3.27 |
| 3370 | 1" x 8" | | 225 | .036 | | 1.43 | 1.53 | | 2.96 | 3.92 |
| 3400 | Trim, back band, 11/16" x 1-1/16" | | 250 | .032 | | .83 | 1.38 | | 2.21 | 3.02 |
| 3500 | Casing, 11/16" x 4-1/4" | | 250 | .032 | | 2.74 | 1.38 | | 4.12 | 5.10 |
| 3600 | Crown, 11/16" x 4-1/4" | | 250 | .032 | | 3.85 | 1.38 | | 5.23 | 6.35 |
| 3700 | 1" x 4" and 1" x 6" railing w/balusters 4" O.C. | | 22 | .364 | | 22.50 | 15.65 | | 38.15 | 48.50 |
| 3800 | Insect screen framing stock, 1-1/16" x 1-3/4" | | 395 | .020 | | 2.03 | .87 | | 2.90 | 3.57 |
| 4100 | Verge board, sterling pine, 1" x 4" | | 200 | .040 | | .79 | 1.72 | | 2.51 | 3.51 |
| 4200 | 1" x 6" | | 200 | .040 | | 1.05 | 1.72 | | 2.77 | 3.80 |
| 4300 | 2" x 6" | | 165 | .048 | | 2.10 | 2.09 | | 4.19 | 5.50 |
| 4400 | 2" x 8" | ▼ | 165 | .048 | ▼ | 2.86 | 2.09 | | 4.95 | 6.35 |
| 4700 | For redwood trim, add | | | | | 200% | | | | |

### 06 22 13.45 Moldings, Trim

| | | Crew | Daily Output | Labor-Hours | Unit | Material | 2011 Bare Costs Labor | Equipment | Total | Total Incl O&P |
|---|---|---|---|---|---|---|---|---|---|---|
| 0010 | **MOLDINGS, TRIM** | | | | | | | | | |
| 0200 | Astragal, stock pine, 11/16" x 1-3/4" | 1 Carp | 255 | .031 | L.F. | 1.22 | 1.35 | | 2.57 | 3.41 |
| 0250 | 1-5/16" x 2-3/16" | | 240 | .033 | | 2.57 | 1.43 | | 4 | 5.05 |
| 0800 | Chair rail, stock pine, 5/8" x 2-1/2" | | 270 | .030 | | 1.43 | 1.28 | | 2.71 | 3.53 |
| 0900 | 5/8" x 3-1/2" | | 240 | .033 | | 2.26 | 1.43 | | 3.69 | 4.69 |
| 1000 | Closet pole, stock pine, 1-1/8" diameter | | 200 | .040 | | 1.05 | 1.72 | | 2.77 | 3.80 |
| 1100 | Fir, 1-5/8" diameter | | 200 | .040 | | 1.93 | 1.72 | | 3.65 | 4.76 |
| 3300 | Half round, stock pine, 1/4" x 1/2" | | 270 | .030 | ▼ | .24 | 1.28 | | 1.52 | 2.22 |

## 06 22 13 – Standard Pattern Wood Trim

### 06 22 13.45 Moldings, Trim

| | | Crew | Daily Output | Labor-Hours | Unit | Material | 2011 Bare Costs Labor | Equipment | Total | Total Incl O&P |
|---|---|---|---|---|---|---|---|---|---|---|
| 3350 | 1/2" x 1" | 1 Carp | 255 | .031 | L.F. | .62 | 1.35 | | 1.97 | 2.75 |
| 3400 | Handrail, fir, single piece, stock, hardware not included | | | | | | | | | |
| 3450 | 1-1/2" x 1-3/4" | 1 Carp | 80 | .100 | L.F. | 2.03 | 4.31 | | 6.34 | 8.85 |
| 3470 | Pine, 1-1/2" x 1-3/4" | | 80 | .100 | | 2.03 | 4.31 | | 6.34 | 8.85 |
| 3500 | 1-1/2" x 2-1/2" | | 76 | .105 | | 3.04 | 4.53 | | 7.57 | 10.30 |
| 3600 | Lattice, stock pine, 1/4" x 1-1/8" | | 270 | .030 | | .42 | 1.28 | | 1.70 | 2.42 |
| 3700 | 1/4" x 1-3/4" | | 250 | .032 | | .65 | 1.38 | | 2.03 | 2.83 |
| 3800 | Miscellaneous, custom, pine, 1" x 1" | | 270 | .030 | | .35 | 1.28 | | 1.63 | 2.35 |
| 3900 | 1" x 3" | | 240 | .033 | | 1.06 | 1.43 | | 2.49 | 3.37 |
| 4100 | Birch or oak, nominal 1" x 1" | | 240 | .033 | | .47 | 1.43 | | 1.90 | 2.71 |
| 4200 | Nominal 1" x 3" | | 215 | .037 | | 1.40 | 1.60 | | 3 | 4 |
| 4400 | Walnut, nominal 1" x 1" | | 215 | .037 | | .51 | 1.60 | | 2.11 | 3.02 |
| 4500 | Nominal 1" x 3" | | 200 | .040 | | 1.53 | 1.72 | | 3.25 | 4.33 |
| 4700 | Teak, nominal 1" x 1" | | 215 | .037 | | 2.47 | 1.60 | | 4.07 | 5.20 |
| 4800 | Nominal 1" x 3" | | 200 | .040 | | 7.40 | 1.72 | | 9.12 | 10.80 |
| 4900 | Quarter round, stock pine, 1/4" x 1/4" | | 275 | .029 | | .22 | 1.25 | | 1.47 | 2.16 |
| 4950 | 3/4" x 3/4" | | 255 | .031 | ↓ | .60 | 1.35 | | 1.95 | 2.73 |
| 5600 | Wainscot moldings, 1-1/8" x 9/16", 2' high, minimum | | 76 | .105 | S.F. | 11.15 | 4.53 | | 15.68 | 19.25 |
| 5700 | Maximum | ↓ | 65 | .123 | " | 20 | 5.30 | | 25.30 | 30.50 |

### 06 22 13.50 Moldings, Window and Door

| | | Crew | Daily Output | Labor-Hours | Unit | Material | 2011 Bare Costs Labor | Equipment | Total | Total Incl O&P |
|---|---|---|---|---|---|---|---|---|---|---|
| 0010 | **MOLDINGS, WINDOW AND DOOR** | | | | | | | | | |
| 2800 | Door moldings, stock, decorative, 1-1/8" wide, plain | 1 Carp | 17 | .471 | Set | 45 | 20.50 | | 65.50 | 80.50 |
| 2900 | Detailed | | 17 | .471 | " | 87.50 | 20.50 | | 108 | 128 |
| 2960 | Clear pine door jamb, no stops, 11/16" x 4-9/16" | | 240 | .033 | L.F. | 5.15 | 1.43 | | 6.58 | 7.85 |
| 3150 | Door trim set, 1 head and 2 sides, pine, 2-1/2 wide | | 12 | .667 | Opng. | 22.50 | 28.50 | | 51 | 69 |
| 3170 | 3-1/2" wide | | 11 | .727 | " | 27.50 | 31.50 | | 59 | 78.50 |
| 3250 | Glass beads, stock pine, 3/8" x 1/2" | | 275 | .029 | L.F. | .31 | 1.25 | | 1.56 | 2.26 |
| 3270 | 3/8" x 7/8" | | 270 | .030 | | .41 | 1.28 | | 1.69 | 2.41 |
| 4850 | Parting bead, stock pine, 3/8" x 3/4" | | 275 | .029 | | .36 | 1.25 | | 1.61 | 2.32 |
| 4870 | 1/2" x 3/4" | | 255 | .031 | | .45 | 1.35 | | 1.80 | 2.57 |
| 5000 | Stool caps, stock pine, 11/16" x 3-1/2" | | 200 | .040 | | 2.27 | 1.72 | | 3.99 | 5.15 |
| 5100 | 1-1/16" x 3-1/4" | | 150 | .053 | ↓ | 3.56 | 2.30 | | 5.86 | 7.45 |
| 5300 | Threshold, oak, 3' long, inside, 5/8" x 3-5/8" | | 32 | .250 | Ea. | 9 | 10.75 | | 19.75 | 26.50 |
| 5400 | Outside, 1-1/2" x 7-5/8" | ↓ | 16 | .500 | " | 38 | 21.50 | | 59.50 | 75 |
| 5900 | Window trim sets, including casings, header, stops, | | | | | | | | | |
| 5910 | stool and apron, 2-1/2" wide, minimum | 1 Carp | 13 | .615 | Opng. | 32.50 | 26.50 | | 59 | 76 |
| 5950 | Average | | 10 | .800 | | 38 | 34.50 | | 72.50 | 95 |
| 6000 | Maximum | ↓ | 6 | 1.333 | ↓ | 62.50 | 57.50 | | 120 | 157 |

### 06 22 13.60 Moldings, Soffits

| | | Crew | Daily Output | Labor-Hours | Unit | Material | 2011 Bare Costs Labor | Equipment | Total | Total Incl O&P |
|---|---|---|---|---|---|---|---|---|---|---|
| 0010 | **MOLDINGS, SOFFITS** | | | | | | | | | |
| 0200 | Soffits, pine, 1" x 4" | 2 Carp | 420 | .038 | L.F. | .53 | 1.64 | | 2.17 | 3.10 |
| 0210 | 1" x 6" | | 420 | .038 | | .67 | 1.64 | | 2.31 | 3.26 |
| 0220 | 1" x 8" | | 420 | .038 | | .97 | 1.64 | | 2.61 | 3.58 |
| 0230 | 1" x 10" | | 400 | .040 | | 1.21 | 1.72 | | 2.93 | 3.97 |
| 0240 | 1" x 12" | | 400 | .040 | | 1.46 | 1.72 | | 3.18 | 4.25 |
| 0250 | STK cedar, 1" x 4" | | 420 | .038 | | .43 | 1.64 | | 2.07 | 2.99 |
| 0260 | 1" x 6" | | 420 | .038 | | .76 | 1.64 | | 2.40 | 3.36 |
| 0270 | 1" x 8" | | 420 | .038 | | 1.18 | 1.64 | | 2.82 | 3.82 |
| 0280 | 1" x 10" | | 400 | .040 | | 1.68 | 1.72 | | 3.40 | 4.49 |
| 0290 | 1" x 12" | | 400 | .040 | ↓ | 2.49 | 1.72 | | 4.21 | 5.40 |
| 1000 | Exterior AC plywood, 1/4" thick | | 420 | .038 | S.F. | .62 | 1.64 | | 2.26 | 3.20 |
| 1050 | 3/8" thick | ↓ | 420 | .038 | | .77 | 1.64 | | 2.41 | 3.37 |

# 06 22 Millwork

## 06 22 13 – Standard Pattern Wood Trim

| 06 22 13.60 Moldings, Soffits | | Crew | Daily Output | Labor-Hours | Unit | Material | 2011 Bare Costs Labor | Equipment | Total | Total Incl O&P |
|---|---|---|---|---|---|---|---|---|---|---|
| 1100 | 1/2" thick | 2 Carp | 420 | .038 | S.F. | .80 | 1.64 | | 2.44 | 3.40 |
| 1150 | Polyvinyl chloride, white, solid | 1 Carp | 230 | .035 | | 1.96 | 1.50 | | 3.46 | 4.46 |
| 1160 | Perforated | " | 230 | .035 | | 1.96 | 1.50 | | 3.46 | 4.46 |

# 06 25 Prefinished Paneling

## 06 25 13 – Prefinished Hardboard Paneling

### 06 25 13.10 Paneling, Hardboard

| 06 25 13.10 Paneling, Hardboard | | | Crew | Daily Output | Labor-Hours | Unit | Material | 2011 Bare Costs Labor | Equipment | Total | Total Incl O&P |
|---|---|---|---|---|---|---|---|---|---|---|---|
| 0010 | **PANELING, HARDBOARD** | | | | | | | | | | |
| 0050 | Not incl. furring or trim, hardboard, tempered, 1/8" thick | G | 2 Carp | 500 | .032 | S.F. | .38 | 1.38 | | 1.76 | 2.53 |
| 0100 | 1/4" thick | G | | 500 | .032 | | .55 | 1.38 | | 1.93 | 2.72 |
| 0300 | Tempered pegboard, 1/8" thick | G | | 500 | .032 | | .45 | 1.38 | | 1.83 | 2.61 |
| 0400 | 1/4" thick | G | | 500 | .032 | | .64 | 1.38 | | 2.02 | 2.81 |
| 0600 | Untempered hardboard, natural finish, 1/8" thick | G | | 500 | .032 | | .38 | 1.38 | | 1.76 | 2.53 |
| 0700 | 1/4" thick | G | | 500 | .032 | | .46 | 1.38 | | 1.84 | 2.62 |
| 0900 | Untempered pegboard, 1/8" thick | G | | 500 | .032 | | .39 | 1.38 | | 1.77 | 2.54 |
| 1000 | 1/4" thick | G | | 500 | .032 | | .42 | 1.38 | | 1.80 | 2.57 |
| 1200 | Plastic faced hardboard, 1/8" thick | G | | 500 | .032 | | .68 | 1.38 | | 2.06 | 2.86 |
| 1300 | 1/4" thick | G | | 500 | .032 | | .84 | 1.38 | | 2.22 | 3.03 |
| 1500 | Plastic faced pegboard, 1/8" thick | G | | 500 | .032 | | .62 | 1.38 | | 2 | 2.79 |
| 1600 | 1/4" thick | G | | 500 | .032 | | .74 | 1.38 | | 2.12 | 2.92 |
| 1800 | Wood grained, plain or grooved, 1/4" thick, minimum | G | | 500 | .032 | | .58 | 1.38 | | 1.96 | 2.75 |
| 1900 | Maximum | G | | 425 | .038 | | 1.12 | 1.62 | | 2.74 | 3.72 |
| 2100 | Moldings for hardboard, wood or aluminum, minimum | | | 500 | .032 | L.F. | .39 | 1.38 | | 1.77 | 2.54 |
| 2200 | Maximum | | | 425 | .038 | " | 1.09 | 1.62 | | 2.71 | 3.69 |

## 06 25 16 – Prefinished Plywood Paneling

### 06 25 16.10 Paneling, Plywood

| 06 25 16.10 Paneling, Plywood | | | Crew | Daily Output | Labor-Hours | Unit | Material | 2011 Bare Costs Labor | Equipment | Total | Total Incl O&P |
|---|---|---|---|---|---|---|---|---|---|---|---|
| 0010 | **PANELING, PLYWOOD** | R061636-20 | | | | | | | | | |
| 2400 | Plywood, prefinished, 1/4" thick, 4' x 8' sheets | | | | | | | | | | |
| 2410 | with vertical grooves. Birch faced, minimum | | 2 Carp | 500 | .032 | S.F. | .89 | 1.38 | | 2.27 | 3.09 |
| 2420 | Average | | | 420 | .038 | | 1.35 | 1.64 | | 2.99 | 4.01 |
| 2430 | Maximum | | | 350 | .046 | | 1.97 | 1.97 | | 3.94 | 5.20 |
| 2600 | Mahogany, African | | | 400 | .040 | | 2.51 | 1.72 | | 4.23 | 5.40 |
| 2700 | Philippine (Lauan) | | | 500 | .032 | | 1.08 | 1.38 | | 2.46 | 3.30 |
| 2900 | Oak or Cherry, minimum | | | 500 | .032 | | 2.10 | 1.38 | | 3.48 | 4.42 |
| 3000 | Maximum | | | 400 | .040 | | 3.22 | 1.72 | | 4.94 | 6.20 |
| 3200 | Rosewood | | | 320 | .050 | | 4.58 | 2.15 | | 6.73 | 8.35 |
| 3400 | Teak | | | 400 | .040 | | 3.22 | 1.72 | | 4.94 | 6.20 |
| 3600 | Chestnut | | | 375 | .043 | | 4.77 | 1.84 | | 6.61 | 8.05 |
| 3800 | Pecan | | | 400 | .040 | | 2.06 | 1.72 | | 3.78 | 4.91 |
| 3900 | Walnut, minimum | | | 500 | .032 | | 2.75 | 1.38 | | 4.13 | 5.15 |
| 3950 | Maximum | | | 400 | .040 | | 5.20 | 1.72 | | 6.92 | 8.40 |
| 4000 | Plywood, prefinished, 3/4" thick, stock grades, minimum | | | 320 | .050 | | 1.24 | 2.15 | | 3.39 | 4.66 |
| 4100 | Maximum | | | 224 | .071 | | 5.40 | 3.08 | | 8.48 | 10.60 |
| 4300 | Architectural grade, minimum | | | 224 | .071 | | 3.98 | 3.08 | | 7.06 | 9.10 |
| 4400 | Maximum | | | 160 | .100 | | 6.10 | 4.31 | | 10.41 | 13.30 |
| 4600 | Plywood, "A" face, birch, V.C., 1/2" thick, natural | | | 450 | .036 | | 1.89 | 1.53 | | 3.42 | 4.43 |
| 4700 | Select | | | 450 | .036 | | 2.06 | 1.53 | | 3.59 | 4.62 |
| 4900 | Veneer core, 3/4" thick, natural | | | 320 | .050 | | 1.99 | 2.15 | | 4.14 | 5.50 |
| 5000 | Select | | | 320 | .050 | | 2.23 | 2.15 | | 4.38 | 5.75 |
| 5200 | Lumber core, 3/4" thick, natural | | | 320 | .050 | | 2.99 | 2.15 | | 5.14 | 6.60 |

# 06 25 Prefinished Paneling

## 06 25 16 – Prefinished Plywood Paneling

| | 06 25 16.10 Paneling, Plywood | Crew | Daily Output | Labor-Hours | Unit | Material | 2011 Bare Costs Labor | Equipment | Total | Total Incl O&P |
|---|---|---|---|---|---|---|---|---|---|---|
| 5500 | Plywood, knotty pine, 1/4" thick, A2 grade | 2 Carp | 450 | .036 | S.F. | 1.63 | 1.53 | | 3.16 | 4.14 |
| 5600 | A3 grade | | 450 | .036 | | 2.06 | 1.53 | | 3.59 | 4.62 |
| 5800 | 3/4" thick, veneer core, A2 grade | | 320 | .050 | | 2.11 | 2.15 | | 4.26 | 5.60 |
| 5900 | A3 grade | | 320 | .050 | | 2.38 | 2.15 | | 4.53 | 5.90 |
| 6100 | Aromatic cedar, 1/4" thick, plywood | | 400 | .040 | | 2.08 | 1.72 | | 3.80 | 4.93 |
| 6200 | 1/4" thick, particle board | | 400 | .040 | | 1.01 | 1.72 | | 2.73 | 3.75 |

## 06 25 26 – Panel System

| | 06 25 26.10 Panel Systems | Crew | Daily Output | Labor-Hours | Unit | Material | 2011 Bare Costs Labor | Equipment | Total | Total Incl O&P |
|---|---|---|---|---|---|---|---|---|---|---|
| 0010 | **PANEL SYSTEMS** | | | | | | | | | |
| 0100 | Raised panel, eng. wood core w/wood veneer, std., paint grade | 2 Carp | 300 | .053 | S.F. | 12.20 | 2.30 | | 14.50 | 16.90 |
| 0110 | Oak veneer | | 300 | .053 | | 20 | 2.30 | | 22.30 | 25.50 |
| 0120 | Maple veneer | | 300 | .053 | | 24.50 | 2.30 | | 26.80 | 30 |
| 0130 | Cherry veneer | | 300 | .053 | | 29 | 2.30 | | 31.30 | 35.50 |
| 0300 | Class I fire rated, paint grade | | 300 | .053 | | 14.25 | 2.30 | | 16.55 | 19.15 |
| 0310 | Oak veneer | | 300 | .053 | | 23.50 | 2.30 | | 25.80 | 29.50 |
| 0320 | Maple veneer | | 300 | .053 | | 30 | 2.30 | | 32.30 | 36.50 |
| 0330 | Cherry veneer | | 300 | .053 | | 38.50 | 2.30 | | 40.80 | 45.50 |
| 0510 | Beadboard, 5/8" MDF, standard, primed | | 300 | .053 | | 6.60 | 2.30 | | 8.90 | 10.75 |
| 0520 | Oak veneer, unfinished | | 300 | .053 | | 10.05 | 2.30 | | 12.35 | 14.55 |
| 0530 | Maple veneer, unfinished | | 300 | .053 | | 11.20 | 2.30 | | 13.50 | 15.80 |
| 0610 | Rustic paneling, 5/8" MDF, standard, maple veneer, unfinished | | 300 | .053 | | 12.70 | 2.30 | | 15 | 17.45 |

# 06 26 Board Paneling

## 06 26 13 – Profile Board Paneling

| | 06 26 13.10 Paneling, Boards | Crew | Daily Output | Labor-Hours | Unit | Material | 2011 Bare Costs Labor | Equipment | Total | Total Incl O&P |
|---|---|---|---|---|---|---|---|---|---|---|
| 0010 | **PANELING, BOARDS** | | | | | | | | | |
| 6400 | Wood board paneling, 3/4" thick, knotty pine | 2 Carp | 300 | .053 | S.F. | 1.89 | 2.30 | | 4.19 | 5.60 |
| 6500 | Rough sawn cedar | | 300 | .053 | | 2.86 | 2.30 | | 5.16 | 6.65 |
| 6700 | Redwood, clear, 1" x 4" boards | | 300 | .053 | | 4.96 | 2.30 | | 7.26 | 8.95 |
| 6900 | Aromatic cedar, closet lining, boards | | 275 | .058 | | 2 | 2.50 | | 4.50 | 6.05 |

# 06 43 Wood Stairs and Railings

## 06 43 13 – Wood Stairs

| | 06 43 13.20 Prefabricated Wood Stairs | Crew | Daily Output | Labor-Hours | Unit | Material | 2011 Bare Costs Labor | Equipment | Total | Total Incl O&P |
|---|---|---|---|---|---|---|---|---|---|---|
| 0010 | **PREFABRICATED WOOD STAIRS** | | | | | | | | | |
| 0100 | Box stairs, prefabricated, 3'-0" wide | | | | | | | | | |
| 0110 | Oak treads, up to 14 risers | 2 Carp | 39 | .410 | Riser | 85 | 17.65 | | 102.65 | 121 |
| 0600 | With pine treads for carpet, up to 14 risers | " | 39 | .410 | " | 55 | 17.65 | | 72.65 | 87.50 |
| 1100 | For 4' wide stairs, add | | | | Flight | 25% | | | | |
| 1550 | Stairs, prefabricated stair handrail with balusters | 1 Carp | 30 | .267 | L.F. | 75 | 11.50 | | 86.50 | 99.50 |
| 1700 | Basement stairs, prefabricated, pine treads | | | | | | | | | |
| 1710 | Pine risers, 3' wide, up to 14 risers | 2 Carp | 52 | .308 | Riser | 55 | 13.25 | | 68.25 | 81 |
| 4000 | Residential, wood, oak treads, prefabricated | | 1.50 | 10.667 | Flight | 1,100 | 460 | | 1,560 | 1,900 |
| 4200 | Built in place | | .44 | 36.364 | " | 1,750 | 1,575 | | 3,325 | 4,325 |
| 4400 | Spiral, oak, 4'-6" diameter, unfinished, prefabricated, | | | | | | | | | |
| 4500 | incl. railing, 9' high | 2 Carp | 1.50 | 10.667 | Flight | 3,625 | 460 | | 4,085 | 4,675 |

# 06 43 Wood Stairs and Railings

## 06 43 13 – Wood Stairs

| 06 43 13.40 Wood Stair Parts | Crew | Daily Output | Labor-Hours | Unit | Material | 2011 Bare Costs Labor | Equipment | Total | Total Incl O&P |
|---|---|---|---|---|---|---|---|---|---|
| **0010 WOOD STAIR PARTS** | | | | | | | | | |
| 0020 Pin top balusters, 1-1/4", oak, 34" | 1 Carp | 96 | .083 | Ea. | 7.80 | 3.59 | | 11.39 | 14.10 |
| 0030 38" | | 96 | .083 | | 9.85 | 3.59 | | 13.44 | 16.35 |
| 0040 42" | | 96 | .083 | | 12.10 | 3.59 | | 15.69 | 18.80 |
| 0050 Poplar, 34" | | 96 | .083 | | 3.05 | 3.59 | | 6.64 | 8.85 |
| 0060 38" | | 96 | .083 | | 3.62 | 3.59 | | 7.21 | 9.50 |
| 0070 42" | | 96 | .083 | | 8.55 | 3.59 | | 12.14 | 14.95 |
| 0080 Maple, 34" | | 96 | .083 | | 4.90 | 3.59 | | 8.49 | 10.90 |
| 0090 38" | | 96 | .083 | | 6.20 | 3.59 | | 9.79 | 12.30 |
| 0100 42" | | 96 | .083 | | 7.60 | 3.59 | | 11.19 | 13.85 |
| 0130 Primed, 34" | | 96 | .083 | | 2.85 | 3.59 | | 6.44 | 8.65 |
| 0140 38" | | 96 | .083 | | 3.62 | 3.59 | | 7.21 | 9.50 |
| 0150 42" | | 96 | .083 | | 4.42 | 3.59 | | 8.01 | 10.35 |
| 0180 Box top balusters, 1-1/4", oak, 34" | | 60 | .133 | | 10.85 | 5.75 | | 16.60 | 21 |
| 0190 38" | | 60 | .133 | | 12.50 | 5.75 | | 18.25 | 22.50 |
| 0200 42" | | 60 | .133 | | 12.95 | 5.75 | | 18.70 | 23 |
| 0210 Poplar, 34" | | 60 | .133 | | 8.65 | 5.75 | | 14.40 | 18.30 |
| 0220 38" | | 60 | .133 | | 9.50 | 5.75 | | 15.25 | 19.25 |
| 0230 42" | | 60 | .133 | | 10 | 5.75 | | 15.75 | 19.80 |
| 0240 Maple, 34" | | 60 | .133 | | 11.90 | 5.75 | | 17.65 | 22 |
| 0250 38" | | 60 | .133 | | 13.80 | 5.75 | | 19.55 | 24 |
| 0260 42" | | 60 | .133 | | 14.20 | 5.75 | | 19.95 | 24.50 |
| 0290 Primed, 34" | | 60 | .133 | | 6.95 | 5.75 | | 12.70 | 16.45 |
| 0300 38" | | 60 | .133 | | 8 | 5.75 | | 13.75 | 17.60 |
| 0310 42" | | 60 | .133 | | 8.35 | 5.75 | | 14.10 | 18 |
| 0340 Square balusters, cut from lineal stock, pine, 1-1/16" x 1-1/16" | | 180 | .044 | L.F. | 1.32 | 1.91 | | 3.23 | 4.39 |
| 0350 1-5/16" x 1-5/16" | | 180 | .044 | | 1.70 | 1.91 | | 3.61 | 4.81 |
| 0360 1-5/8" x 1-5/8" | | 180 | .044 | | 2.82 | 1.91 | | 4.73 | 6.05 |
| 0370 Turned newel, oak, 3-1/2" square, 48" high | | 8 | 1 | Ea. | 94.50 | 43 | | 137.50 | 170 |
| 0380 62" high | | 8 | 1 | | 138 | 43 | | 181 | 218 |
| 0390 Poplar, 3-1/2" square, 48" high | | 8 | 1 | | 52 | 43 | | 95 | 123 |
| 0400 62" high | | 8 | 1 | | 75 | 43 | | 118 | 149 |
| 0410 Maple, 3-1/2" square, 48" high | | 8 | 1 | | 66 | 43 | | 109 | 139 |
| 0420 62" high | | 8 | 1 | | 93 | 43 | | 136 | 168 |
| 0430 Square newel, oak, 3-1/2" square, 48" high | | 8 | 1 | | 50 | 43 | | 93 | 121 |
| 0440 58" high | | 8 | 1 | | 70 | 43 | | 113 | 143 |
| 0450 Poplar, 3-1/2" square, 48" high | | 8 | 1 | | 49 | 43 | | 92 | 120 |
| 0460 58" high | | 8 | 1 | | 54 | 43 | | 97 | 126 |
| 0470 Maple, 3" square, 48" high | | 8 | 1 | | 50 | 43 | | 93 | 121 |
| 0480 58" high | | 8 | 1 | | 62 | 43 | | 105 | 134 |
| 0490 Railings, oak, minimum | | 96 | .083 | L.F. | 10.40 | 3.59 | | 13.99 | 16.95 |
| 0500 Average | | 96 | .083 | | 12.50 | 3.59 | | 16.09 | 19.25 |
| 0510 Maximum | | 96 | .083 | | 15.50 | 3.59 | | 19.09 | 22.50 |
| 0520 Maple, minimum | | 96 | .083 | | 15 | 3.59 | | 18.59 | 22 |
| 0530 Average | | 96 | .083 | | 15.25 | 3.59 | | 18.84 | 22.50 |
| 0540 Maximum | | 96 | .083 | | 15.50 | 3.59 | | 19.09 | 22.50 |
| 0550 Oak, for bending rail, minimum | | 48 | .167 | | 21 | 7.20 | | 28.20 | 34 |
| 0560 Average | | 48 | .167 | | 22 | 7.20 | | 29.20 | 35 |
| 0570 Maximum | | 48 | .167 | | 23 | 7.20 | | 30.20 | 36.50 |
| 0580 Maple, for bending rail, minimum | | 48 | .167 | | 23 | 7.20 | | 30.20 | 36 |
| 0590 Average | | 48 | .167 | | 25 | 7.20 | | 32.20 | 38 |
| 0600 Maximum | | 48 | .167 | | 27 | 7.20 | | 34.20 | 40.50 |

# 06 43 Wood Stairs and Railings

## 06 43 13 – Wood Stairs

| 06 43 13.40 Wood Stair Parts | | Crew | Daily Output | Labor-Hours | Unit | Material | 2011 Bare Costs Labor | Equipment | Total | Total Incl O&P |
|---|---|---|---|---|---|---|---|---|---|---|
| 0610 | Risers, oak, 3/4" x 8", 36" long | 1 Carp | 80 | .100 | Ea. | 18 | 4.31 | | 22.31 | 26.50 |
| 0620 | 42" long | | 80 | .100 | | 21 | 4.31 | | 25.31 | 29.50 |
| 0630 | 48" long | | 80 | .100 | | 24 | 4.31 | | 28.31 | 33 |
| 0640 | 54" long | | 80 | .100 | | 27 | 4.31 | | 31.31 | 36 |
| 0650 | 60" long | | 80 | .100 | | 30 | 4.31 | | 34.31 | 39.50 |
| 0660 | 72" long | | 80 | .100 | | 36 | 4.31 | | 40.31 | 46 |
| 0670 | Poplar, 3/4" x 8", 36" long | | 80 | .100 | | 11.10 | 4.31 | | 15.41 | 18.80 |
| 0680 | 42" long | | 80 | .100 | | 12.95 | 4.31 | | 17.26 | 21 |
| 0690 | 48" long | | 80 | .100 | | 14.80 | 4.31 | | 19.11 | 23 |
| 0700 | 54" long | | 80 | .100 | | 16.65 | 4.31 | | 20.96 | 25 |
| 0710 | 60" long | | 80 | .100 | | 18.50 | 4.31 | | 22.81 | 27 |
| 0720 | 72" long | | 80 | .100 | | 22 | 4.31 | | 26.31 | 31 |
| 0730 | Pine, 1" x 8", 36" long | | 80 | .100 | | 2.90 | 4.31 | | 7.21 | 9.80 |
| 0740 | 42" long | | 80 | .100 | | 3.38 | 4.31 | | 7.69 | 10.30 |
| 0750 | 48" long | | 80 | .100 | | 3.87 | 4.31 | | 8.18 | 10.85 |
| 0760 | 54" long | | 80 | .100 | | 4.35 | 4.31 | | 8.66 | 11.40 |
| 0770 | 60" long | | 80 | .100 | | 4.83 | 4.31 | | 9.14 | 11.90 |
| 0780 | 72" long | | 80 | .100 | | 5.80 | 4.31 | | 10.11 | 13 |
| 0790 | Treads, oak, no returns, 1-1/32" x 11-1/2" x 36" long | | 32 | .250 | | 32 | 10.75 | | 42.75 | 51.50 |
| 0800 | 42" long | | 32 | .250 | | 37.50 | 10.75 | | 48.25 | 57.50 |
| 0810 | 48" long | | 32 | .250 | | 42.50 | 10.75 | | 53.25 | 63.50 |
| 0820 | 54" long | | 32 | .250 | | 48 | 10.75 | | 58.75 | 69.50 |
| 0830 | 60" long | | 32 | .250 | | 53.50 | 10.75 | | 64.25 | 75 |
| 0840 | 72" long | | 32 | .250 | | 64 | 10.75 | | 74.75 | 87 |
| 0850 | Mitred return one end, 1-1/32" x 11-1/2" x 36" long | | 24 | .333 | | 46 | 14.35 | | 60.35 | 72.50 |
| 0860 | 42" long | | 24 | .333 | | 53.50 | 14.35 | | 67.85 | 81 |
| 0870 | 48" long | | 24 | .333 | | 61.50 | 14.35 | | 75.85 | 89.50 |
| 0880 | 54" long | | 24 | .333 | | 69 | 14.35 | | 83.35 | 98 |
| 0890 | 60" long | | 24 | .333 | | 76.50 | 14.35 | | 90.85 | 107 |
| 0900 | 72" long | | 24 | .333 | | 92 | 14.35 | | 106.35 | 123 |
| 0910 | Mitred return two ends, 1-1/32" x 11-1/2" x 36" long | | 12 | .667 | | 49 | 28.50 | | 77.50 | 98 |
| 0920 | 42" long | | 12 | .667 | | 57 | 28.50 | | 85.50 | 107 |
| 0930 | 48" long | | 12 | .667 | | 65.50 | 28.50 | | 94 | 116 |
| 0940 | 54" long | | 12 | .667 | | 73.50 | 28.50 | | 102 | 125 |
| 0950 | 60" long | | 12 | .667 | | 81.50 | 28.50 | | 110 | 134 |
| 0960 | 72" long | | 12 | .667 | | 98 | 28.50 | | 126.50 | 152 |
| 0970 | Starting step, oak, 48", bullnose | | 8 | 1 | | 192 | 43 | | 235 | 277 |
| 0980 | Double end bullnose | | 8 | 1 | | 345 | 43 | | 388 | 445 |
| 1030 | Skirt board, pine, 1" x 10" | | 55 | .145 | L.F. | 1.21 | 6.25 | | 7.46 | 10.95 |
| 1040 | 1" x 12" | | 52 | .154 | " | 1.46 | 6.60 | | 8.06 | 11.75 |
| 1050 | Oak landing tread, 1-1/16" thick | | 54 | .148 | S.F. | 8.25 | 6.40 | | 14.65 | 18.90 |
| 1060 | Oak cove molding | | 96 | .083 | L.F. | 1.45 | 3.59 | | 5.04 | 7.10 |
| 1070 | Oak stringer molding | | 96 | .083 | " | 2.75 | 3.59 | | 6.34 | 8.55 |
| 1090 | Rail bolt, 5/16" x 3-1/2" | | 48 | .167 | Ea. | 2.75 | 7.20 | | 9.95 | 14.05 |
| 1100 | 5/16" x 4-1/2" | | 48 | .167 | | 2.75 | 7.20 | | 9.95 | 14.05 |
| 1120 | Newel post anchor | | 16 | .500 | | 13.50 | 21.50 | | 35 | 48 |
| 1130 | Tapered plug, 1/2" | | 240 | .033 | | .30 | 1.43 | | 1.73 | 2.53 |
| 1140 | 1" | | 240 | .033 | | .40 | 1.43 | | 1.83 | 2.64 |

## 06 43 16 – Wood Railings

### 06 43 16.10 Wood Handrails and Railings

| | | Crew | Daily Output | Labor-Hours | Unit | Material | Labor | Equipment | Total | Total Incl O&P |
|---|---|---|---|---|---|---|---|---|---|---|
| 0010 | **WOOD HANDRAILS AND RAILINGS** | | | | | | | | | |
| 0020 | Custom design, architectural grade, hardwood, minimum | 1 Carp | 38 | .211 | L.F. | 9.50 | 9.05 | | 18.55 | 24.50 |

# 06 43 Wood Stairs and Railings

## 06 43 16 – Wood Railings

| 06 43 16.10 Wood Handrails and Railings | Crew | Daily Output | Labor-Hours | Unit | Material | 2011 Bare Costs Labor | Equipment | Total | Total Incl O&P |
|---|---|---|---|---|---|---|---|---|---|
| 0100 | Maximum | 1 Carp | 30 | .267 | L.F. | 50 | 11.50 | | 61.50 | 72.50 |
| 0300 | Stock interior railing with spindles 4" O.C., 4' long | | 40 | .200 | | 37 | 8.60 | | 45.60 | 53.50 |
| 0400 | 8' long | | 48 | .167 | | 37 | 7.20 | | 44.20 | 51.50 |

# 06 44 Ornamental Woodwork

## 06 44 19 – Wood Grilles

### 06 44 19.10 Grilles

| | | Crew | Daily Output | Labor-Hours | Unit | Material | 2011 Bare Costs Labor | Equipment | Total | Total Incl O&P |
|---|---|---|---|---|---|---|---|---|---|---|
| 0010 | GRILLES and panels, hardwood, sanded | | | | | | | | | |
| 0020 | 2' x 4' to 4' x 8', custom designs, unfinished, minimum | 1 Carp | 38 | .211 | S.F. | 54 | 9.05 | | 63.05 | 73.50 |
| 0050 | Average | | 30 | .267 | | 62.50 | 11.50 | | 74 | 86.50 |
| 0100 | Maximum | | 19 | .421 | | 71 | 18.15 | | 89.15 | 106 |

## 06 44 33 – Wood Mantels

### 06 44 33.10 Fireplace Mantels

| | | Crew | Daily Output | Labor-Hours | Unit | Material | 2011 Bare Costs Labor | Equipment | Total | Total Incl O&P |
|---|---|---|---|---|---|---|---|---|---|---|
| 0010 | FIREPLACE MANTELS | | | | | | | | | |
| 0015 | 6" molding, 6' x 3'-6" opening, minimum | 1 Carp | 5 | 1.600 | Opng. | 228 | 69 | | 297 | 355 |
| 0100 | Maximum | | 5 | 1.600 | | 410 | 69 | | 479 | 560 |
| 0300 | Prefabricated pine, colonial type, stock, deluxe | | 2 | 4 | | 1,250 | 172 | | 1,422 | 1,675 |
| 0400 | Economy | | 3 | 2.667 | | 530 | 115 | | 645 | 760 |

### 06 44 33.20 Fireplace Mantel Beam

| | | Crew | Daily Output | Labor-Hours | Unit | Material | 2011 Bare Costs Labor | Equipment | Total | Total Incl O&P |
|---|---|---|---|---|---|---|---|---|---|---|
| 0010 | FIREPLACE MANTEL BEAM | | | | | | | | | |
| 0020 | Rough texture wood, 4" x 8" | 1 Carp | 36 | .222 | L.F. | 8.30 | 9.55 | | 17.85 | 24 |
| 0100 | 4" x 10" | | 35 | .229 | " | 10.90 | 9.85 | | 20.75 | 27 |
| 0300 | Laminated hardwood, 2-1/4" x 10-1/2" wide, 6' long | | 5 | 1.600 | Ea. | 110 | 69 | | 179 | 227 |
| 0400 | 8' long | | 5 | 1.600 | " | 150 | 69 | | 219 | 271 |
| 0600 | Brackets for above, rough sawn | | 12 | .667 | Pr. | 10 | 28.50 | | 38.50 | 55 |
| 0700 | Laminated | | 12 | .667 | " | 15 | 28.50 | | 43.50 | 60.50 |

## 06 44 39 – Wood Posts and Columns

### 06 44 39.10 Decorative Beams

| | | Crew | Daily Output | Labor-Hours | Unit | Material | 2011 Bare Costs Labor | Equipment | Total | Total Incl O&P |
|---|---|---|---|---|---|---|---|---|---|---|
| 0010 | DECORATIVE BEAMS | | | | | | | | | |
| 0020 | Rough sawn cedar, non-load bearing, 4" x 4" | 2 Carp | 180 | .089 | L.F. | 1.31 | 3.83 | | 5.14 | 7.30 |
| 0100 | 4" x 6" | | 170 | .094 | | 1.97 | 4.05 | | 6.02 | 8.35 |
| 0200 | 4" x 8" | | 160 | .100 | | 2.63 | 4.31 | | 6.94 | 9.50 |
| 0300 | 4" x 10" | | 150 | .107 | | 3.99 | 4.59 | | 8.58 | 11.45 |
| 0400 | 4" x 12" | | 140 | .114 | | 5.05 | 4.92 | | 9.97 | 13.15 |
| 0500 | 8" x 8" | | 130 | .123 | | 5.25 | 5.30 | | 10.55 | 13.95 |

### 06 44 39.20 Columns

| | | Crew | Daily Output | Labor-Hours | Unit | Material | 2011 Bare Costs Labor | Equipment | Total | Total Incl O&P |
|---|---|---|---|---|---|---|---|---|---|---|
| 0010 | COLUMNS | | | | | | | | | |
| 0050 | Aluminum, round colonial, 6" diameter | 2 Carp | 80 | .200 | V.L.F. | 16 | 8.60 | | 24.60 | 31 |
| 0100 | 8" diameter | | 62.25 | .257 | | 19.30 | 11.05 | | 30.35 | 38 |
| 0200 | 10" diameter | | 55 | .291 | | 23.50 | 12.50 | | 36 | 45 |
| 0250 | Fir, stock units, hollow round, 6" diameter | | 80 | .200 | | 26.50 | 8.60 | | 35.10 | 42 |
| 0300 | 8" diameter | | 80 | .200 | | 31 | 8.60 | | 39.60 | 47.50 |
| 0350 | 10" diameter | | 70 | .229 | | 39.50 | 9.85 | | 49.35 | 58.50 |
| 0400 | Solid turned, to 8' high, 3-1/2" diameter | | 80 | .200 | | 9 | 8.60 | | 17.60 | 23 |
| 0500 | 4-1/2" diameter | | 75 | .213 | | 14 | 9.20 | | 23.20 | 29.50 |
| 0600 | 5-1/2" diameter | | 70 | .229 | | 18.40 | 9.85 | | 28.25 | 35 |
| 0800 | Square columns, built-up, 5" x 5" | | 65 | .246 | | 15.50 | 10.60 | | 26.10 | 33.50 |
| 0900 | Solid, 3-1/2" x 3-1/2" | | 130 | .123 | | 9 | 5.30 | | 14.30 | 18.05 |
| 1600 | Hemlock, tapered, T & G, 12" diam., 10' high | | 100 | .160 | | 53 | 6.90 | | 59.90 | 68.50 |

# 06 44 Ornamental Woodwork

## 06 44 39 – Wood Posts and Columns

### 06 44 39.20 Columns

| | | Crew | Daily Output | Labor-Hours | Unit | Material | 2011 Bare Costs Labor | Equipment | Total | Total Incl O&P |
|---|---|---|---|---|---|---|---|---|---|---|
| 1700 | 16' high | 2 Carp | 65 | .246 | V.L.F. | 79.50 | 10.60 | | 90.10 | 103 |
| 1900 | 10' high, 14" diameter | | 100 | .160 | | 121 | 6.90 | | 127.90 | 144 |
| 2000 | 18' high | | 65 | .246 | | 107 | 10.60 | | 117.60 | 134 |
| 2200 | 18" diameter, 12' high | | 65 | .246 | | 153 | 10.60 | | 163.60 | 185 |
| 2300 | 20' high | | 50 | .320 | | 148 | 13.80 | | 161.80 | 184 |
| 2500 | 20" diameter, 14' high | | 40 | .400 | | 159 | 17.20 | | 176.20 | 202 |
| 2600 | 20' high | | 35 | .457 | | 149 | 19.70 | | 168.70 | 193 |
| 2800 | For flat pilasters, deduct | | | | | 33% | | | | |
| 3000 | For splitting into halves, add | | | | Ea. | 116 | | | 116 | 128 |
| 4000 | Rough sawn cedar posts, 4" x 4" | 2 Carp | 250 | .064 | V.L.F. | 2.98 | 2.76 | | 5.74 | 7.50 |
| 4100 | 4" x 6" | | 235 | .068 | | 5.70 | 2.93 | | 8.63 | 10.75 |
| 4200 | 6" x 6" | | 220 | .073 | | 10.30 | 3.13 | | 13.43 | 16.15 |
| 4300 | 8" x 8" | | 200 | .080 | | 19.30 | 3.44 | | 22.74 | 27 |

# 06 48 Wood Frames

## 06 48 13 – Exterior Wood Door Frames

### 06 48 13.10 Exterior Wood Door Frames and Accessories

| | | Crew | Daily Output | Labor-Hours | Unit | Material | 2011 Bare Costs Labor | Equipment | Total | Total Incl O&P |
|---|---|---|---|---|---|---|---|---|---|---|
| 0010 | **EXTERIOR WOOD DOOR FRAMES AND ACCESSORIES** | | | | | | | | | |
| 0400 | Exterior frame, incl. ext. trim, pine, 5/4 x 4-9/16" deep | 2 Carp | 375 | .043 | L.F. | 5.55 | 1.84 | | 7.39 | 8.90 |
| 0420 | 5-3/16" deep | | 375 | .043 | | 11.35 | 1.84 | | 13.19 | 15.30 |
| 0440 | 6-9/16" deep | | 375 | .043 | | 8.50 | 1.84 | | 10.34 | 12.15 |
| 0600 | Oak, 5/4 x 4-9/16" deep | | 350 | .046 | | 10.70 | 1.97 | | 12.67 | 14.80 |
| 0620 | 5-3/16" deep | | 350 | .046 | | 11.80 | 1.97 | | 13.77 | 15.95 |
| 0640 | 6-9/16" deep | | 350 | .046 | | 14.50 | 1.97 | | 16.47 | 18.95 |
| 1000 | Sills, 8/4 x 8" deep, oak, no horns | | 100 | .160 | | 17 | 6.90 | | 23.90 | 29.50 |
| 1020 | 2" horns | | 100 | .160 | | 17.10 | 6.90 | | 24 | 29.50 |
| 1040 | 3" horns | | 100 | .160 | | 17.10 | 6.90 | | 24 | 29.50 |
| 1100 | 8/4 x 10" deep, oak, no horns | | 90 | .178 | | 20.50 | 7.65 | | 28.15 | 34.50 |
| 1120 | 2" horns | | 90 | .178 | | 21 | 7.65 | | 28.65 | 35.50 |
| 1140 | 3" horns | | 90 | .178 | | 21 | 7.65 | | 28.65 | 35.50 |
| 2000 | Exterior, colonial, frame & trim, 3' opng., in-swing, minimum | | 22 | .727 | Ea. | 315 | 31.50 | | 346.50 | 400 |
| 2010 | Average | | 21 | .762 | | 470 | 33 | | 503 | 565 |
| 2020 | Maximum | | 20 | .800 | | 1,075 | 34.50 | | 1,109.50 | 1,225 |
| 2100 | 5'-4" opening, in-swing, minimum | | 17 | .941 | | 360 | 40.50 | | 400.50 | 455 |
| 2120 | Maximum | | 15 | 1.067 | | 1,075 | 46 | | 1,121 | 1,250 |
| 2140 | Out-swing, minimum | | 17 | .941 | | 370 | 40.50 | | 410.50 | 465 |
| 2160 | Maximum | | 15 | 1.067 | | 1,100 | 46 | | 1,146 | 1,300 |
| 2400 | 6'-0" opening, in-swing, minimum | | 16 | 1 | | 345 | 43 | | 388 | 445 |
| 2420 | Maximum | | 10 | 1.600 | | 1,100 | 69 | | 1,169 | 1,325 |
| 2460 | Out-swing, minimum | | 16 | 1 | | 375 | 43 | | 418 | 480 |
| 2480 | Maximum | | 10 | 1.600 | | 1,300 | 69 | | 1,369 | 1,525 |
| 2600 | For two sidelights, add, minimum | | 30 | .533 | Opng. | 355 | 23 | | 378 | 425 |
| 2620 | Maximum | | 20 | .800 | " | 1,125 | 34.50 | | 1,159.50 | 1,300 |
| 2700 | Custom birch frame, 3'-0" opening | | 16 | 1 | Ea. | 205 | 43 | | 248 | 292 |
| 2750 | 6'-0" opening | | 16 | 1 | | 305 | 43 | | 348 | 400 |
| 2900 | Exterior, modern, plain trim, 3' opng., in-swing, minimum | | 26 | .615 | | 35.50 | 26.50 | | 62 | 79.50 |
| 2920 | Average | | 24 | .667 | | 42 | 28.50 | | 70.50 | 90 |
| 2940 | Maximum | | 22 | .727 | | 50.50 | 31.50 | | 82 | 104 |

# 06 48 Wood Frames

## 06 48 16 – Interior Wood Door Frames

### 06 48 16.10 Interior Wood Door Jamb and Frames

| | | Daily Output | Labor-Hours | Unit | Material | 2011 Bare Costs Labor | Equipment | Total | Total Incl O&P |
|---|---|---|---|---|---|---|---|---|---|
| | | Crew | | | | | | | |
| 0010 | **INTERIOR WOOD DOOR JAMB AND FRAMES** | | | | | | | | |
| 3000 | Interior frame, pine, 11/16" x 3-5/8" deep | 2 Carp | 375 | .043 | L.F. | 5.55 | 1.84 | | 7.39 | 8.90 |
| 3020 | 4-9/16" deep | | 375 | .043 | | 6.50 | 1.84 | | 8.34 | 9.95 |
| 3200 | Oak, 11/16" x 3-5/8" deep | | 350 | .046 | | 4.35 | 1.97 | | 6.32 | 7.80 |
| 3220 | 4-9/16" deep | | 350 | .046 | | 4.50 | 1.97 | | 6.47 | 7.95 |
| 3240 | 5-3/16" deep | | 350 | .046 | | 4.58 | 1.97 | | 6.55 | 8.05 |
| 3400 | Walnut, 11/16" x 3-5/8" deep | | 350 | .046 | | 7.30 | 1.97 | | 9.27 | 11.05 |
| 3420 | 4-9/16" deep | | 350 | .046 | | 7.50 | 1.97 | | 9.47 | 11.25 |
| 3440 | 5-3/16" deep | | 350 | .046 | | 7.90 | 1.97 | | 9.87 | 11.65 |
| 3600 | Pocket door frame | | 16 | 1 | Ea. | 74.50 | 43 | | 117.50 | 148 |
| 3800 | Threshold, oak, 5/8" x 3-5/8" deep | | 200 | .080 | L.F. | 3.83 | 3.44 | | 7.27 | 9.50 |
| 3820 | 4-5/8" deep | | 190 | .084 | | 4.58 | 3.63 | | 8.21 | 10.60 |
| 3840 | 5-5/8" deep | | 180 | .089 | | 7.15 | 3.83 | | 10.98 | 13.70 |

# 06 49 Wood Screens and Exterior Wood Shutters

## 06 49 19 – Exterior Wood Shutters

### 06 49 19.10 Shutters, Exterior

| | | Crew | Daily Output | Labor-Hours | Unit | Material | 2011 Bare Costs Labor | Equipment | Total | Total Incl O&P |
|---|---|---|---|---|---|---|---|---|---|---|
| 0010 | **SHUTTERS, EXTERIOR** | | | | | | | | | |
| 0012 | Aluminum, louvered, 1'-4" wide, 3'-0" long | 1 Carp | 10 | .800 | Pr. | 61 | 34.50 | | 95.50 | 121 |
| 0400 | 6'-8" long | | 9 | .889 | | 123 | 38.50 | | 161.50 | 194 |
| 1000 | Pine, louvered, primed, each 1'-2" wide, 3'-3" long | | 10 | .800 | | 106 | 34.50 | | 140.50 | 170 |
| 1100 | 4'-7" long | | 10 | .800 | | 144 | 34.50 | | 178.50 | 211 |
| 1500 | Each 1'-6" wide, 3'-3" long | | 10 | .800 | | 113 | 34.50 | | 147.50 | 177 |
| 1600 | 4'-7" long | | 10 | .800 | | 158 | 34.50 | | 192.50 | 227 |
| 1620 | Hemlock, louvered, 1'-2" wide, 5'-7" long | | 10 | .800 | | 174 | 34.50 | | 208.50 | 244 |
| 1630 | Each 1'-4" wide, 2'-2" long | | 10 | .800 | | 108 | 34.50 | | 142.50 | 172 |
| 1670 | 4'-3" long | | 10 | .800 | | 130 | 34.50 | | 164.50 | 196 |
| 1690 | 5'-11" long | | 10 | .800 | | 183 | 34.50 | | 217.50 | 254 |
| 1700 | Door blinds, 6'-9" long, each 1'-3" wide | | 9 | .889 | | 184 | 38.50 | | 222.50 | 261 |
| 1710 | 1'-6" wide | | 9 | .889 | | 198 | 38.50 | | 236.50 | 277 |
| 1720 | Hemlock, solid raised panel, each 1'-4" wide, 3'-3" long | | 10 | .800 | | 175 | 34.50 | | 209.50 | 246 |
| 1740 | 4'-3" long | | 10 | .800 | | 222 | 34.50 | | 256.50 | 297 |
| 1770 | 5'-11" long | | 10 | .800 | | 297 | 34.50 | | 331.50 | 380 |
| 1800 | Door blinds, 6'-9" long, each 1'-3" wide | | 9 | .889 | | 335 | 38.50 | | 373.50 | 425 |
| 1900 | 1'-6" wide | | 9 | .889 | | 365 | 38.50 | | 403.50 | 460 |
| 2500 | Polystyrene, solid raised panel, each 1'-4" wide, 3'-3" long | | 10 | .800 | | 41 | 34.50 | | 75.50 | 98 |
| 2700 | 4'-7" long | | 10 | .800 | | 51.50 | 34.50 | | 86 | 110 |
| 4500 | Polystyrene, louvered, each 1'-2" wide, 3'-3" long | | 10 | .800 | | 28 | 34.50 | | 62.50 | 83.50 |
| 4750 | 5'-3" long | | 10 | .800 | | 39.50 | 34.50 | | 74 | 96.50 |
| 6000 | Vinyl, louvered, each 1'-2" x 4'-7" long | | 10 | .800 | | 34.50 | 34.50 | | 69 | 90.50 |
| 6200 | Each 1'-4" x 6'-8" long | | 9 | .889 | | 48 | 38.50 | | 86.50 | 112 |

# 06 51 Structural Plastic Shapes and Plates

## 06 51 13 – Plastic Lumber

| 06 51 13.10 Recycled Plastic Lumber | | Crew | Daily Output | Labor-Hours | Unit | Material | 2011 Bare Costs Labor | Equipment | Total | Total Incl O&P |
|---|---|---|---|---|---|---|---|---|---|---|
| 0010 | **RECYCLED PLASTIC LUMBER** | | | | | | | | | |
| 4000 | Sheeting, recycled plastic, black or white, 4' x 8' x 1/8" | G | 2 Carp | 1100 | .015 | S.F. | 1.56 | .63 | | 2.19 | 2.68 |
| 4010 | 4' x 8' x 3/16" | G | | 1100 | .015 | | 2.33 | .63 | | 2.96 | 3.52 |
| 4020 | 4' x 8' x 1/4" | G | | 950 | .017 | | 3.11 | .73 | | 3.84 | 4.53 |
| 4030 | 4' x 8' x 3/8" | G | | 950 | .017 | | 4.67 | .73 | | 5.40 | 6.25 |
| 4040 | 4' x 8' x 1/2" | G | | 900 | .018 | | 6.20 | .77 | | 6.97 | 8 |
| 4050 | 4' x 8' x 5/8" | G | | 900 | .018 | | 7.70 | .77 | | 8.47 | 9.60 |
| 4060 | 4' x 8' x 3/4" | G | | 850 | .019 | | 9.15 | .81 | | 9.96 | 11.35 |
| 4070 | Add for colors | G | | | | Ea. | 5% | | | 5% | 6% |
| 8500 | 100% recycled plastic, var colors, NLB, 2" x 2" | G | | | | L.F. | 1.41 | | | 1.41 | 1.55 |
| 8510 | 2" x 4" | G | | | | | 2.95 | | | 2.95 | 3.25 |
| 8520 | 2" x 6" | G | | | | | 4.63 | | | 4.63 | 5.10 |
| 8530 | 2" x 8" | G | | | | | 6.35 | | | 6.35 | 6.95 |
| 8540 | 2" x 10" | G | | | | | 9.25 | | | 9.25 | 10.15 |
| 8550 | 5/4" x 4" | G | | | | | 3.02 | | | 3.02 | 3.32 |
| 8560 | 5/4" x 6" | G | | | | | 2.99 | | | 2.99 | 3.29 |
| 8570 | 1" x 6" | G | | | | | 2.28 | | | 2.28 | 2.51 |
| 8580 | 1/2" x 8" | G | | | | | 2.46 | | | 2.46 | 2.71 |
| 8590 | 2" x 10" T & G | G | | | | | 8.60 | | | 8.60 | 9.45 |
| 8600 | 3" x 10" T & G | G | | | | | 14.25 | | | 14.25 | 15.70 |
| 8610 | Add for premium colors | G | | | | | 20% | | | 20% | 22% |

# 06 52 Plastic Structural Assemblies

## 06 52 10 – Fiberglass Structural Assemblies

| 06 52 10.10 Castings, Fiberglass | | Crew | Daily Output | Labor-Hours | Unit | Material | 2011 Bare Costs Labor | Equipment | Total | Total Incl O&P |
|---|---|---|---|---|---|---|---|---|---|---|
| 0010 | **CASTINGS, FIBERGLASS** | | | | | | | | | |
| 0100 | Angle, 1" x 1" x 1/8" thick | 2 Sswk | 240 | .067 | L.F. | 1.44 | 3.24 | | 4.68 | 7.30 |
| 0120 | 3" x 3" x 1/4" thick | | 200 | .080 | | 6 | 3.88 | | 9.88 | 13.45 |
| 0140 | 4" x 4" x 1/4" thick | | 200 | .080 | | 7.70 | 3.88 | | 11.58 | 15.35 |
| 0160 | 4" x 4" x 3/8" thick | | 200 | .080 | | 11.85 | 3.88 | | 15.73 | 19.90 |
| 0180 | 6" x 6" x 1/2" thick | | 160 | .100 | | 23.50 | 4.86 | | 28.36 | 34.50 |
| 1000 | Flat sheet, 1/8" thick | | 140 | .114 | S.F. | 5.05 | 5.55 | | 10.60 | 15.35 |
| 1020 | 1/4" thick | | 120 | .133 | | 9.90 | 6.45 | | 16.35 | 22.50 |
| 1040 | 3/8" thick | | 100 | .160 | | 16.85 | 7.75 | | 24.60 | 32 |
| 1060 | 1/2" thick | | 80 | .200 | | 22 | 9.70 | | 31.70 | 41.50 |
| 2000 | Handrail, 42" high, 2" diam. rails pickets 5' O.C. | | 32 | .500 | L.F. | 51 | 24.50 | | 75.50 | 99 |
| 3000 | Round bar, 1/4" diam. | | 240 | .067 | | .58 | 3.24 | | 3.82 | 6.35 |
| 3020 | 1/2" diam. | | 200 | .080 | | 1.01 | 3.88 | | 4.89 | 7.95 |
| 3040 | 3/4" diam. | | 200 | .080 | | 2.05 | 3.88 | | 5.93 | 9.10 |
| 3060 | 1" diam. | | 160 | .100 | | 3.62 | 4.86 | | 8.48 | 12.60 |
| 3080 | 1-1/4" diam. | | 160 | .100 | | 4.13 | 4.86 | | 8.99 | 13.15 |
| 3100 | 1-1/2" diam. | | 140 | .114 | | 5.05 | 5.55 | | 10.60 | 15.35 |
| 3500 | Round tube, 1" diam. x 1/8" thick | | 240 | .067 | | 1.97 | 3.24 | | 5.21 | 7.85 |
| 3520 | 2" diam. x 1/4" thick | | 200 | .080 | | 6.35 | 3.88 | | 10.23 | 13.85 |
| 3540 | 3" diam. x 1/4" thick | | 160 | .100 | | 10.10 | 4.86 | | 14.96 | 19.70 |
| 4000 | Square bar, 1/2" square | | 240 | .067 | | 3.98 | 3.24 | | 7.22 | 10.10 |
| 4020 | 1" square | | 200 | .080 | | 5.10 | 3.88 | | 8.98 | 12.45 |
| 4040 | 1-1/2" square | | 160 | .100 | | 10.40 | 4.86 | | 15.26 | 20 |
| 4500 | Square tube, 1" x 1" x 1/8" thick | | 240 | .067 | | 2.24 | 3.24 | | 5.48 | 8.15 |
| 4520 | 2" x 2" x 1/8" thick | | 200 | .080 | | 4.34 | 3.88 | | 8.22 | 11.60 |
| 4540 | 3" x 3" x 1/4" thick | | 160 | .100 | | 11.85 | 4.86 | | 16.71 | 21.50 |

# 06 52 Plastic Structural Assemblies

## 06 52 10 – Fiberglass Structural Assemblies

### 06 52 10.10 Castings, Fiberglass

| | | Crew | Daily Output | Labor-Hours | Unit | Material | 2011 Bare Costs Labor | 2011 Bare Costs Equipment | Total | Total Incl O&P |
|---|---|---|---|---|---|---|---|---|---|---|
| 5000 | Threaded rod, 3/8" diam. | 2 Sswk | 320 | .050 | L.F. | 3.45 | 2.43 | | 5.88 | 8.10 |
| 5020 | 1/2" diam. | | 320 | .050 | | 4.06 | 2.43 | | 6.49 | 8.75 |
| 5040 | 5/8" diam. | | 280 | .057 | | 4.49 | 2.77 | | 7.26 | 9.85 |
| 5060 | 3/4" diam. | | 280 | .057 | | 5.10 | 2.77 | | 7.87 | 10.50 |
| 6000 | Wide flange beam, 4" x 4" x 1/4" thick | | 120 | .133 | | 10.55 | 6.45 | | 17 | 23 |
| 6020 | 6" x 6" x 1/4" thick | | 100 | .160 | | 19.95 | 7.75 | | 27.70 | 35.50 |
| 6040 | 8" x 8" x 3/8" thick | ▼ | 80 | .200 | ▼ | 35 | 9.70 | | 44.70 | 55.50 |

### 06 52 10.20 Fiberglass Stair Treads

| | | Crew | Daily Output | Labor-Hours | Unit | Material | 2011 Bare Costs Labor | 2011 Bare Costs Equipment | Total | Total Incl O&P |
|---|---|---|---|---|---|---|---|---|---|---|
| 0010 | **FIBERGLASS STAIR TREADS** | | | | | | | | | |
| 0100 | 24" wide | 2 Sswk | 52 | .308 | Ea. | 39.50 | 14.95 | | 54.45 | 70 |
| 0140 | 30" wide | | 52 | .308 | | 47 | 14.95 | | 61.95 | 78.50 |
| 0180 | 36" wide | | 52 | .308 | | 55.50 | 14.95 | | 70.45 | 87.50 |
| 0220 | 42" wide | ▼ | 52 | .308 | ▼ | 65.50 | 14.95 | | 80.45 | 98.50 |

### 06 52 10.30 Fiberglass Grating

| | | Crew | Daily Output | Labor-Hours | Unit | Material | 2011 Bare Costs Labor | 2011 Bare Costs Equipment | Total | Total Incl O&P |
|---|---|---|---|---|---|---|---|---|---|---|
| 0010 | **FIBERGLASS GRATING** | | | | | | | | | |
| 0100 | Molded, green (for mod. corrosive environment) | | | | | | | | | |
| 0140 | 1" x 4" mesh, 1" thick | 2 Sswk | 400 | .040 | S.F. | 11.15 | 1.94 | | 13.09 | 15.75 |
| 0180 | 1-1/2" square mesh, 1" thick | | 400 | .040 | | 16.60 | 1.94 | | 18.54 | 21.50 |
| 0220 | 1-1/4" thick | | 400 | .040 | | 13.25 | 1.94 | | 15.19 | 18.05 |
| 0260 | 1-1/2" thick | | 400 | .040 | | 24 | 1.94 | | 25.94 | 30 |
| 0300 | 2" square mesh, 2" thick | ▼ | 320 | .050 | ▼ | 22.50 | 2.43 | | 24.93 | 29.50 |
| 1000 | Orange (for highly corrosive environment) | | | | | | | | | |
| 1040 | 1" x 4" mesh, 1" thick | 2 Sswk | 400 | .040 | S.F. | 14.05 | 1.94 | | 15.99 | 18.90 |
| 1080 | 1-1/2" square mesh, 1" thick | | 400 | .040 | | 18.60 | 1.94 | | 20.54 | 24 |
| 1120 | 1-1/4" thick | | 400 | .040 | | 19.30 | 1.94 | | 21.24 | 25 |
| 1160 | 1-1/2" thick | | 400 | .040 | | 21 | 1.94 | | 22.94 | 26.50 |
| 1200 | 2" square mesh, 2" thick | ▼ | 320 | .050 | ▼ | 25 | 2.43 | | 27.43 | 32 |
| 3000 | Pultruded, green (for mod. corrosive environment) | | | | | | | | | |
| 3040 | 1" O.C. bar spacing, 1" thick | 2 Sswk | 400 | .040 | S.F. | 16.65 | 1.94 | | 18.59 | 21.50 |
| 3080 | 1-1/2" thick | | 320 | .050 | | 18.50 | 2.43 | | 20.93 | 25 |
| 3120 | 1-1/2" O.C. bar spacing, 1" thick | | 400 | .040 | | 13.50 | 1.94 | | 15.44 | 18.30 |
| 3160 | 1-1/2" thick | ▼ | 400 | .040 | ▼ | 15.15 | 1.94 | | 17.09 | 20 |
| 4000 | Grating support legs, fixed height, no base | | | | Ea. | 43.50 | | | 43.50 | 48 |
| 4040 | With base | | | | | 39 | | | 39 | 43 |
| 4080 | Adjustable to 60" | | | | ▼ | 59 | | | 59 | 65 |

### 06 52 10.40 Fiberglass Floor Grating

| | | Crew | Daily Output | Labor-Hours | Unit | Material | 2011 Bare Costs Labor | 2011 Bare Costs Equipment | Total | Total Incl O&P |
|---|---|---|---|---|---|---|---|---|---|---|
| 0010 | **FIBERGLASS FLOOR GRATING** | | | | | | | | | |
| 0100 | Reinforced polyester, fire retardant, 1" x 4" grid, 1" thick | E-4 | 510 | .063 | S.F. | 14.25 | 3.08 | .21 | 17.54 | 21.50 |
| 0200 | 1-1/2" x 6" mesh, 1-1/2" thick | | 500 | .064 | | 16.65 | 3.14 | .22 | 20.01 | 24 |
| 0300 | With grit surface, 1-1/2" x 6" grid, 1-1/2" thick | ▼ | 500 | .064 | ▼ | 17 | 3.14 | .22 | 20.36 | 24.50 |

# 06 63 Plastic Railings

## 06 63 10 – Plastic (PVC) Railings

| 06 63 10.10 Plastic Railings | | Crew | Daily Output | Labor-Hours | Unit | Material | 2011 Bare Costs Labor | Equipment | Total | Total Incl O&P |
|---|---|---|---|---|---|---|---|---|---|---|
| 0010 | **PLASTIC RAILINGS** | | | | | | | | | |
| 0100 | Horizontal PVC handrail with balusters, 3-1/2" wide, 36" high | 1 Carp | 96 | .083 | L.F. | 24 | 3.59 | | 27.59 | 32 |
| 0150 | 42" high | | 96 | .083 | | 27 | 3.59 | | 30.59 | 35 |
| 0200 | Angled PVC handrail with balusters, 3-1/2" wide, 36" high | | 72 | .111 | | 28 | 4.78 | | 32.78 | 38 |
| 0250 | 42" high | | 72 | .111 | | 30.50 | 4.78 | | 35.28 | 41 |
| 0300 | Post sleeve for 4 x 4 post | | 96 | .083 | | 12.55 | 3.59 | | 16.14 | 19.30 |
| 0400 | Post cap for 4 x 4 post, flat profile | | 48 | .167 | Ea. | 11.40 | 7.20 | | 18.60 | 23.50 |
| 0450 | Newel post style profile | | 48 | .167 | | 21 | 7.20 | | 28.20 | 34 |
| 0500 | Raised corbeled profile | | 48 | .167 | | 23 | 7.20 | | 30.20 | 36 |
| 0550 | Post base trim for 4 x 4 post | | 96 | .083 | | 12.80 | 3.59 | | 16.39 | 19.60 |

# 06 65 Plastic Simulated Wood Trim

## 06 65 10 – PVC Trim

### 06 65 10.10 PVC Trim, Exterior

| 06 65 10.10 PVC Trim, Exterior | | Crew | Daily Output | Labor-Hours | Unit | Material | 2011 Bare Costs Labor | Equipment | Total | Total Incl O&P |
|---|---|---|---|---|---|---|---|---|---|---|
| 0010 | **PVC TRIM, EXTERIOR** | | | | | | | | | |
| 0100 | Cornerboards, 5/4" x 6" x 6" | 1 Carp | 240 | .033 | L.F. | 7.30 | 1.43 | | 8.73 | 10.25 |
| 0110 | Door/window casing, 1" x 4" | | 200 | .040 | | 1.42 | 1.72 | | 3.14 | 4.20 |
| 0120 | 1" x 6" | | 200 | .040 | | 2.24 | 1.72 | | 3.96 | 5.10 |
| 0130 | 1" x 8" | | 195 | .041 | | 2.96 | 1.77 | | 4.73 | 5.95 |
| 0140 | 1" x 10" | | 195 | .041 | | 3.74 | 1.77 | | 5.51 | 6.80 |
| 0150 | 1" x 12" | | 190 | .042 | | 4.57 | 1.81 | | 6.38 | 7.85 |
| 0160 | 5/4" x 4" | | 195 | .041 | | 2.05 | 1.77 | | 3.82 | 4.97 |
| 0170 | 5/4" x 6" | | 195 | .041 | | 3.24 | 1.77 | | 5.01 | 6.25 |
| 0180 | 5/4" x 8" | | 190 | .042 | | 4.24 | 1.81 | | 6.05 | 7.45 |
| 0190 | 5/4" x 10" | | 190 | .042 | | 5.40 | 1.81 | | 7.21 | 8.75 |
| 0200 | 5/4" x 12" | | 185 | .043 | | 6.60 | 1.86 | | 8.46 | 10.10 |
| 0210 | Fascia, 1" x 4" | | 250 | .032 | | 1.42 | 1.38 | | 2.80 | 3.67 |
| 0220 | 1" x 6" | | 250 | .032 | | 2.24 | 1.38 | | 3.62 | 4.57 |
| 0230 | 1" x 8" | | 225 | .036 | | 2.96 | 1.53 | | 4.49 | 5.60 |
| 0240 | 1" x 10" | | 225 | .036 | | 3.74 | 1.53 | | 5.27 | 6.45 |
| 0250 | 1" x 12" | | 200 | .040 | | 4.57 | 1.72 | | 6.29 | 7.70 |
| 0260 | 5/4" x 4" | | 240 | .033 | | 2.05 | 1.43 | | 3.48 | 4.46 |
| 0270 | 5/4" x 6" | | 240 | .033 | | 3.24 | 1.43 | | 4.67 | 5.75 |
| 0280 | 5/4" x 8" | | 215 | .037 | | 4.24 | 1.60 | | 5.84 | 7.10 |
| 0290 | 5/4" x 10" | | 215 | .037 | | 5.40 | 1.60 | | 7 | 8.40 |
| 0300 | 5/4" x 12" | | 190 | .042 | | 6.60 | 1.81 | | 8.41 | 10.05 |
| 0310 | Frieze, 1" x 4" | | 250 | .032 | | 1.42 | 1.38 | | 2.80 | 3.67 |
| 0320 | 1" x 6" | | 250 | .032 | | 2.24 | 1.38 | | 3.62 | 4.57 |
| 0330 | 1" x 8" | | 225 | .036 | | 2.96 | 1.53 | | 4.49 | 5.60 |
| 0340 | 1" x 10" | | 225 | .036 | | 3.74 | 1.53 | | 5.27 | 6.45 |
| 0350 | 1" x 12" | | 200 | .040 | | 4.57 | 1.72 | | 6.29 | 7.70 |
| 0360 | 5/4" x 4" | | 240 | .033 | | 2.05 | 1.43 | | 3.48 | 4.46 |
| 0370 | 5/4" x 6" | | 240 | .033 | | 3.24 | 1.43 | | 4.67 | 5.75 |
| 0380 | 5/4" x 8" | | 215 | .037 | | 4.24 | 1.60 | | 5.84 | 7.10 |
| 0390 | 5/4" x 10" | | 215 | .037 | | 5.40 | 1.60 | | 7 | 8.40 |
| 0400 | 5/4" x 12" | | 190 | .042 | | 6.60 | 1.81 | | 8.41 | 10.05 |
| 0410 | Rake, 1" x 4" | | 200 | .040 | | 1.42 | 1.72 | | 3.14 | 4.20 |
| 0420 | 1" x 6" | | 200 | .040 | | 2.24 | 1.72 | | 3.96 | 5.10 |
| 0430 | 1" x 8" | | 190 | .042 | | 2.96 | 1.81 | | 4.77 | 6.05 |
| 0440 | 1" x 10" | | 190 | .042 | | 3.74 | 1.81 | | 5.55 | 6.90 |
| 0450 | 1" x 12" | | 180 | .044 | | 4.57 | 1.91 | | 6.48 | 8 |

# 06 65 Plastic Simulated Wood Trim

## 06 65 10 – PVC Trim

| 06 65 10.10 PVC Trim, Exterior | | Crew | Daily Output | Labor-Hours | Unit | Material | 2011 Bare Costs Labor | Equipment | Total | Total Incl O&P |
|---|---|---|---|---|---|---|---|---|---|---|
| 0460 | 5/4" x 4" | 1 Carp | 195 | .041 | L.F. | 2.05 | 1.77 | | 3.82 | 4.97 |
| 0470 | 5/4" x 6" | | 195 | .041 | | 3.24 | 1.77 | | 5.01 | 6.25 |
| 0480 | 5/4" x 8" | | 185 | .043 | | 4.24 | 1.86 | | 6.10 | 7.50 |
| 0490 | 5/4" x 10" | | 185 | .043 | | 5.40 | 1.86 | | 7.26 | 8.80 |
| 0500 | 5/4" x 12" | | 175 | .046 | | 6.60 | 1.97 | | 8.57 | 10.25 |
| 0510 | Rake trim, 1" x 4" | | 225 | .036 | | 1.42 | 1.53 | | 2.95 | 3.91 |
| 0520 | 1" x 6" | | 225 | .036 | | 2.24 | 1.53 | | 3.77 | 4.81 |
| 0560 | 5/4" x 4" | | 220 | .036 | | 2.05 | 1.57 | | 3.62 | 4.66 |
| 0570 | 5/4" x 6" | | 220 | .036 | | 3.24 | 1.57 | | 4.81 | 5.95 |
| 0610 | Soffit, 1" x 4" | 2 Carp | 420 | .038 | | 1.42 | 1.64 | | 3.06 | 4.08 |
| 0620 | 1" x 6" | | 420 | .038 | | 2.24 | 1.64 | | 3.88 | 4.98 |
| 0630 | 1" x 8" | | 420 | .038 | | 2.96 | 1.64 | | 4.60 | 5.80 |
| 0640 | 1" x 10" | | 400 | .040 | | 3.74 | 1.72 | | 5.46 | 6.75 |
| 0650 | 1" x 12" | | 400 | .040 | | 4.57 | 1.72 | | 6.29 | 7.70 |
| 0660 | 5/4" x 4" | | 410 | .039 | | 2.05 | 1.68 | | 3.73 | 4.84 |
| 0670 | 5/4" x 6" | | 410 | .039 | | 3.24 | 1.68 | | 4.92 | 6.15 |
| 0680 | 5/4" x 8" | | 410 | .039 | | 4.24 | 1.68 | | 5.92 | 7.25 |
| 0690 | 5/4" x 10" | | 390 | .041 | | 5.40 | 1.77 | | 7.17 | 8.65 |
| 0700 | 5/4" x 12" | | 390 | .041 | | 6.60 | 1.77 | | 8.37 | 9.95 |

# Division Notes

| | CREW | DAILY OUTPUT | LABOR-HOURS | UNIT | 2011 BARE COSTS | | | | TOTAL INCL O&P |
|---|---|---|---|---|---|---|---|---|---|
| | | | | | MAT. | LABOR | EQUIP. | TOTAL | |
| | | | | | | | | | |
| | | | | | | | | | |
| | | | | | | | | | |
| | | | | | | | | | |
| | | | | | | | | | |
| | | | | | | | | | |
| | | | | | | | | | |
| | | | | | | | | | |
| | | | | | | | | | |
| | | | | | | | | | |
| | | | | | | | | | |
| | | | | | | | | | |
| | | | | | | | | | |
| | | | | | | | | | |
| | | | | | | | | | |
| | | | | | | | | | |
| | | | | | | | | | |
| | | | | | | | | | |
| | | | | | | | | | |
| | | | | | | | | | |
| | | | | | | | | | |
| | | | | | | | | | |

# Estimating Tips

## 07 10 00 Dampproofing and Waterproofing

- Be sure of the job specifications before pricing this subdivision. The difference in cost between waterproofing and dampproofing can be great. Waterproofing will hold back standing water. Dampproofing prevents the transmission of water vapor. Also included in this section are vapor retarding membranes.

## 07 20 00 Thermal Protection

- Insulation and fireproofing products are measured by area, thickness, volume or R-value. Specifications may give only what the specific R-value should be in a certain situation. The estimator may need to choose the type of insulation to meet that R-value.

## 07 30 00 Steep Slope Roofing
## 07 40 00 Roofing and Siding Panels

- Many roofing and siding products are bought and sold by the square. One square is equal to an area that measures 100 square feet.

This simple change in unit of measure could create a large error if the estimator is not observant. Accessories necessary for a complete installation must be figured into any calculations for both material and labor.

## 07 50 00 Membrane Roofing
## 07 60 00 Flashing and Sheet Metal
## 07 70 00 Roofing and Wall Specialties and Accessories

- The items in these subdivisions compose a roofing system. No one component completes the installation, and all must be estimated. Built-up or single-ply membrane roofing systems are made up of many products and installation trades. Wood blocking at roof perimeters or penetrations, parapet coverings, reglets, roof drains, gutters, downspouts, sheet metal flashing, skylights, smoke vents, and roof hatches all need to be considered along with the roofing material. Several different installation trades will need to work together on the roofing system. Inherent difficulties in the scheduling and coordination of various trades must be accounted for when estimating labor costs.

## 07 90 00 Joint Protection

- To complete the weather-tight shell, the sealants and caulkings must be estimated. Where different materials meet—at expansion joints, at flashing penetrations, and at hundreds of other locations throughout a construction project—they provide another line of defense against water penetration. Often, an entire system is based on the proper location and placement of caulking or sealants. The detailed drawings that are included as part of a set of architectural plans show typical locations for these materials. When caulking or sealants are shown at typical locations, this means the estimator must include them for all the locations where this detail is applicable. Be careful to keep different types of sealants separate, and remember to consider backer rods and primers if necessary.

# Reference Numbers

Reference numbers are shown in shaded boxes at the beginning of some major classifications. These numbers refer to related items in the Reference Section. The reference information may be an estimating procedure, an alternate pricing method, or technical information.

*Note:* Not all subdivisions listed here necessarily appear in this publication.

## 07 01 50 – Maintenance of Membrane Roofing

### 07 01 50.10 Roof Coatings

| | | Crew | Daily Output | Labor-Hours | Unit | Material | 2011 Bare Costs Labor | Equipment | Total | Total Incl O&P |
|---|---|---|---|---|---|---|---|---|---|---|
| 0010 | **ROOF COATINGS** | | | | | | | | | |
| 0012 | Asphalt, brush grade, material only | | | | Gal. | 7.90 | | | 7.90 | 8.65 |
| 0200 | Asphalt base, fibered aluminum coating [G] | | | | | 10.75 | | | 10.75 | 11.85 |
| 0300 | Asphalt primer, 5 gallon | | | | ↓ | 6.40 | | | 6.40 | 7.05 |
| 0600 | Coal tar pitch, 200 lb. barrels | | | | Ton | 1,100 | | | 1,100 | 1,200 |
| 0700 | Tar roof cement, 5 gal. lots | | | | Gal. | 14 | | | 14 | 15.40 |
| 0800 | Glass fibered roof & patching cement, 5 gallon | | | | " | 7.40 | | | 7.40 | 8.15 |
| 0900 | Reinforcing glass membrane, 450 S.F./roll | | | | Ea. | 111 | | | 111 | 122 |
| 1000 | Neoprene roof coating, 5 gal, 2 gal/sq | | | | Gal. | 27.50 | | | 27.50 | 30.50 |
| 1100 | Roof patch & flashing cement, 5 gallon | | | | | 8.90 | | | 8.90 | 9.75 |
| 1200 | Roof resurtant, glass fibered, 3 gal/sq | | | | ↓ | 5.35 | | | 5.35 | 5.85 |

## 07 01 90 – Maintenance of Joint Protection

### 07 01 90.81 Joint Sealant Replacement

| | | Crew | Daily Output | Labor-Hours | Unit | Material | 2011 Bare Costs Labor | Equipment | Total | Total Incl O&P |
|---|---|---|---|---|---|---|---|---|---|---|
| 0010 | **JOINT SEALANT REPLACEMENT** | | | | | | | | | |
| 0050 | Control joints in concrete floors/slabs | | | | | | | | | |
| 0100 | Option 1 for joints with hard dry sealant | | | | | | | | | |
| 0110 | Step 1: Sawcut to remove 95% of old sealant | | | | | | | | | |
| 0112 | 1/4" wide x 1/2" deep, with single saw blade | C-27 | 4800 | .003 | L.F. | .02 | .14 | .03 | .19 | .25 |
| 0114 | 3/8" wide x 3/4" deep, with single saw blade | | 4000 | .004 | | .04 | .16 | .03 | .23 | .33 |
| 0116 | 1/2" wide x 1" deep, with double saw blades | | 3600 | .004 | | .08 | .18 | .04 | .30 | .39 |
| 0118 | 3/4" wide x 1-1/2" deep, with double saw blades | ↓ | 3200 | .005 | | .17 | .20 | .04 | .41 | .53 |
| 0120 | Step 2: Water blast joint faces and edges | C-29 | 2500 | .003 | | | .11 | .02 | .13 | .20 |
| 0130 | Step 3: Air blast joint faces and edges | C-28 | 2000 | .004 | | | .16 | .01 | .17 | .25 |
| 0140 | Step 4: Sand blast joint faces and edges | E-11 | 2000 | .016 | | | .62 | .10 | .72 | 1.13 |
| 0150 | Step 5: Air blast joint faces and edges | C-28 | 2000 | .004 | ↓ | | .16 | .01 | .17 | .25 |
| 0200 | Option 2 for joints with soft pliable sealant | | | | | | | | | |
| 0210 | Step 1: Plow joint with rectangular blade | B-62 | 2600 | .009 | L.F. | | .35 | .06 | .41 | .58 |
| 0220 | Step 2: Sawcut to re-face joint faces | | | | | | | | | |
| 0222 | 1/4" wide x 1/2" deep, with single saw blade | C-27 | 2400 | .007 | L.F. | .03 | .27 | .06 | .36 | .49 |
| 0224 | 3/8" wide x 3/4" deep, with single saw blade | | 2000 | .008 | | .06 | .33 | .07 | .46 | .61 |
| 0226 | 1/2" wide x 1" deep, with double saw blades | | 1800 | .009 | | .11 | .36 | .07 | .54 | .73 |
| 0228 | 3/4" wide x 1-1/2" deep, with double saw blades | ↓ | 1600 | .010 | | .22 | .41 | .08 | .71 | .93 |
| 0230 | Step 3: Water blast joint faces and edges | C-29 | 2500 | .003 | | | .11 | .02 | .13 | .20 |
| 0240 | Step 4: Air blast joint faces and edges | C-28 | 2000 | .004 | | | .16 | .01 | .17 | .25 |
| 0250 | Step 5: Sand blast joint faces and edges | E-11 | 2000 | .016 | | | .62 | .10 | .72 | 1.13 |
| 0260 | Step 6: Air blast joint faces and edges | C-28 | 2000 | .004 | ↓ | | .16 | .01 | .17 | .25 |
| 0290 | For saw cutting new control joints, see Section 03 35 29.35 | | | | | | | | | |
| 8910 | For backer rod, see Section 07 91 23.10 | | | | | | | | | |
| 8920 | For joint sealant, see Sections 03 15 05.25 or 07 92 13.20 | | | | | | | | | |

## 07 05 05 – Selective Demolition

### 07 05 05.10 Selective Demo., Thermal and Moist. Protection

| | | | Crew | Daily Output | Labor-Hours | Unit | Material | 2011 Bare Costs Labor | Equipment | Total | Total Incl O&P |
|---|---|---|---|---|---|---|---|---|---|---|---|
| 0010 | **SELECTIVE DEMO., THERMAL AND MOISTURE PROTECTION** | | | | | | | | | | |
| 0020 | Caulking/sealant, to 1" x 1" joint | R024119-10 | 1 Clab | 600 | .013 | L.F. | | .46 | | .46 | .70 |
| 0120 | Downspouts, including hangers | | | 350 | .023 | " | | .79 | | .79 | 1.20 |
| 0220 | Flashing, sheet metal | | | 290 | .028 | S.F. | | .95 | | .95 | 1.45 |
| 0420 | Gutters, aluminum or wood, edge hung | | | 240 | .033 | L.F. | | 1.14 | | 1.14 | 1.76 |
| 0520 | Built-in | | | 100 | .080 | " | | 2.75 | | 2.75 | 4.22 |
| 0620 | Insulation, air/vapor barrier | | ↓ | 3500 | .002 | S.F. | | .08 | | .08 | .12 |

| 07 05 05.10 Selective Demo., Thermal and Moist. Protection | Crew | Daily Output | Labor-Hours | Unit | Material | 2011 Bare Costs Labor | Equipment | Total | Total Incl O&P |
|---|---|---|---|---|---|---|---|---|---|
| 0670 | Batts or blankets | 1 Clab | 1400 | .006 | C.F. | | .20 | | .20 | .30 |
| 0720 | Foamed or sprayed in place | 2 Clab | 1000 | .016 | B.F. | | .55 | | .55 | .84 |
| 0770 | Loose fitting | 1 Clab | 3000 | .003 | C.F. | | .09 | | .09 | .14 |
| 0870 | Rigid board | | 3450 | .002 | B.F. | | .08 | | .08 | .12 |
| 1120 | Roll roofing, cold adhesive | | 12 | .667 | Sq. | | 23 | | 23 | 35 |
| 1170 | Roof accessories, adjustable metal chimney flashing | | 9 | .889 | Ea. | | 30.50 | | 30.50 | 47 |
| 1325 | Plumbing vent flashing | | 32 | .250 | " | | 8.60 | | 8.60 | 13.20 |
| 1375 | Ridge vent strip, aluminum | | 310 | .026 | L.F. | | .89 | | .89 | 1.36 |
| 1620 | Skylight to 10 S.F. | | 8 | 1 | Ea. | | 34.50 | | 34.50 | 52.50 |
| 2120 | Roof edge, aluminum soffit and fascia | ▼ | 570 | .014 | L.F. | | .48 | | .48 | .74 |
| 2170 | Concrete coping, up to 12" wide | 2 Clab | 160 | .100 | | | 3.44 | | 3.44 | 5.25 |
| 2220 | Drip edge | 1 Clab | 1650 | .005 | | | .17 | | .17 | .26 |
| 2270 | Gravel stop | | 1650 | .005 | | | .17 | | .17 | .26 |
| 2370 | Sheet metal coping, up to 12" wide | ▼ | 240 | .033 | ▼ | | 1.14 | | 1.14 | 1.76 |
| 2470 | Roof insulation board, over 2" thick | B-2 | 7800 | .005 | B.F. | | .18 | | .18 | .27 |
| 2520 | Up to 2" thick | " | 3900 | .010 | S.F. | | .36 | | .36 | .55 |
| 2620 | Roof ventilation, louvered gable vent | 1 Clab | 16 | .500 | Ea. | | 17.20 | | 17.20 | 26.50 |
| 2670 | Remove, roof hatch | G-3 | 15 | 2.133 | | | 91.50 | | 91.50 | 140 |
| 2675 | Rafter vents | 1 Clab | 960 | .008 | ▼ | | .29 | | .29 | .44 |
| 2720 | Soffit vent and/or fascia vent | | 575 | .014 | L.F. | | .48 | | .48 | .73 |
| 2775 | Soffit vent strip, aluminum, 3" to 4" wide | | 160 | .050 | | | 1.72 | | 1.72 | 2.64 |
| 2820 | Roofing accessories, shingle moulding, to 1" x 4" | ▼ | 1600 | .005 | | | .17 | | .17 | .26 |
| 2870 | Cant strip | B-2 | 2000 | .020 | | | .70 | | .70 | 1.07 |
| 2920 | Concrete block walkway | 1 Clab | 230 | .035 | ▼ | | 1.19 | | 1.19 | 1.83 |
| 3070 | Roofing, felt paper, 15# | | 70 | .114 | Sq. | | 3.93 | | 3.93 | 6 |
| 3125 | #30 felt | ▼ | 30 | .267 | " | | 9.15 | | 9.15 | 14.05 |
| 3170 | Asphalt shingles, 1 layer | B-2 | 3500 | .011 | S.F. | | .40 | | .40 | .61 |
| 3180 | 2 layers | | 1750 | .023 | " | | .79 | | .79 | 1.22 |
| 3370 | Modified bitumen | | 26 | 1.538 | Sq. | | 53.50 | | 53.50 | 82 |
| 3420 | Built-up, no gravel, 3 ply | | 25 | 1.600 | | | 55.50 | | 55.50 | 85.50 |
| 3470 | 4 ply | | 21 | 1.905 | ▼ | | 66 | | 66 | 102 |
| 3620 | 5 ply | | 1600 | .025 | S.F. | | .87 | | .87 | 1.33 |
| 3720 | 5 ply, with gravel | | 890 | .045 | | | 1.56 | | 1.56 | 2.40 |
| 3725 | Gravel removal, minimum | | 5000 | .008 | | | .28 | | .28 | .43 |
| 3730 | Maximum | | 2000 | .020 | | | .70 | | .70 | 1.07 |
| 3870 | Fiberglass sheet | | 1200 | .033 | | | 1.16 | | 1.16 | 1.78 |
| 4120 | Slate shingles | | 1900 | .021 | ▼ | | .73 | | .73 | 1.12 |
| 4170 | Ridge shingles, clay or slate | | 2000 | .020 | L.F. | | .70 | | .70 | 1.07 |
| 4320 | Single ply membrane, attached at seams | | 52 | .769 | Sq. | | 26.50 | | 26.50 | 41 |
| 4370 | Ballasted | | 75 | .533 | | | 18.55 | | 18.55 | 28.50 |
| 4420 | Fully adhered | ▼ | 39 | 1.026 | ▼ | | 35.50 | | 35.50 | 54.50 |
| 4550 | Roof hatch, 2'-6" x 3'-0" | 1 Clab | 10 | .800 | Ea. | | 27.50 | | 27.50 | 42 |
| 4670 | Wood shingles | B-2 | 2200 | .018 | S.F. | | .63 | | .63 | .97 |
| 4820 | Sheet metal roofing | " | 2150 | .019 | | | .65 | | .65 | .99 |
| 4970 | Siding, horizontal wood clapboards | 1 Clab | 380 | .021 | | | .72 | | .72 | 1.11 |
| 5025 | Exterior insulation finish system | " | 120 | .067 | | | 2.29 | | 2.29 | 3.51 |
| 5070 | Tempered hardboard, remove and reset | 1 Carp | 380 | .021 | | | .91 | | .91 | 1.39 |
| 5120 | Tempered hardboard sheet siding | " | 375 | .021 | ▼ | | .92 | | .92 | 1.41 |
| 5170 | Metal, corner strips | 1 Clab | 850 | .009 | L.F. | | .32 | | .32 | .50 |
| 5225 | Horizontal strips | | 444 | .018 | S.F. | | .62 | | .62 | .95 |
| 5320 | Vertical strips | | 400 | .020 | | | .69 | | .69 | 1.05 |
| 5520 | Wood shingles | | 350 | .023 | | | .79 | | .79 | 1.20 |
| 5620 | Stucco siding | ▼ | 360 | .022 | ▼ | | .76 | | .76 | 1.17 |

# 07 05 Common Work Results for Thermal and Moisture Protection

## 07 05 05 – Selective Demolition

### 07 05 05.10 Selective Demo., Thermal and Moist. Protection

| 07 05 05.10 Selective Demo., Thermal and Moist. Protection | Crew | Daily Output | Labor-Hours | Unit | Material | 2011 Bare Costs Labor | 2011 Bare Costs Equipment | Total | Total Incl O&P |
|---|---|---|---|---|---|---|---|---|---|
| 5670 | Textured plywood | 1 Clab | 725 | .011 | S.F. | | .38 | | .38 | .58 |
| 5720 | Vinyl siding | | 510 | .016 | | | .54 | | .54 | .83 |
| 5770 | Corner strips | | 900 | .009 | L.F. | | .31 | | .31 | .47 |
| 5870 | Wood, boards, vertical | | 400 | .020 | S.F. | | .69 | | .69 | 1.05 |
| 5920 | Waterproofing, protection/drain board | 2 Clab | 3900 | .004 | B.F. | | .14 | | .14 | .22 |
| 5970 | Over 1/2" thick | | 1750 | .009 | S.F. | | .31 | | .31 | .48 |
| 6020 | To 1/2" thick | | 2000 | .008 | " | | .27 | | .27 | .42 |

# 07 11 Dampproofing

## 07 11 13 – Bituminous Dampproofing

### 07 11 13.10 Bituminous Asphalt Coating

| | | Crew | Daily Output | Labor-Hours | Unit | Material | 2011 Bare Costs Labor | 2011 Bare Costs Equipment | Total | Total Incl O&P |
|---|---|---|---|---|---|---|---|---|---|---|
| 0010 | **BITUMINOUS ASPHALT COATING** | | | | | | | | | |
| 0030 | Brushed on, below grade, 1 coat | 1 Rofc | 665 | .012 | S.F. | .20 | .44 | | .64 | .95 |
| 0100 | 2 coat | | 500 | .016 | | .39 | .58 | | .97 | 1.40 |
| 0300 | Sprayed on, below grade, 1 coat, 25.6 S.F./gal. | | 830 | .010 | | .20 | .35 | | .55 | .81 |
| 0400 | 2 coat, 20.5 S.F./gal. | | 500 | .016 | | .38 | .58 | | .96 | 1.39 |
| 0500 | Asphalt coating, with fibers | | | | Gal. | 10.75 | | | 10.75 | 11.85 |
| 0600 | Troweled on, asphalt with fibers, 1/16" thick | 1 Rofc | 500 | .016 | S.F. | .47 | .58 | | 1.05 | 1.48 |
| 0700 | 1/8" thick | | 400 | .020 | | .83 | .73 | | 1.56 | 2.13 |
| 1000 | 1/2" thick | | 350 | .023 | | 2.69 | .84 | | 3.53 | 4.35 |

## 07 11 16 – Cementitious Dampproofing

### 07 11 16.20 Cementitious Parging

| | | Crew | Daily Output | Labor-Hours | Unit | Material | 2011 Bare Costs Labor | 2011 Bare Costs Equipment | Total | Total Incl O&P |
|---|---|---|---|---|---|---|---|---|---|---|
| 0010 | **CEMENTITIOUS PARGING** | | | | | | | | | |
| 0020 | Portland cement, 2 coats, 1/2" thick | D-1 | 250 | .064 | S.F. | .29 | 2.49 | | 2.78 | 4.06 |
| 0100 | Waterproofed Portland cement, 1/2" thick, 2 coats | " | 250 | .064 | " | 3.01 | 2.49 | | 5.50 | 7.05 |

# 07 12 Built-up Bituminous Waterproofing

## 07 12 13 – Built-Up Asphalt Waterproofing

### 07 12 13.20 Membrane Waterproofing

| | | Crew | Daily Output | Labor-Hours | Unit | Material | 2011 Bare Costs Labor | 2011 Bare Costs Equipment | Total | Total Incl O&P |
|---|---|---|---|---|---|---|---|---|---|---|
| 0010 | **MEMBRANE WATERPROOFING** | | | | | | | | | |
| 0012 | On slabs, 1 ply, felt, mopped | G-1 | 3000 | .019 | S.F. | .33 | .64 | .14 | 1.11 | 1.57 |
| 0100 | On slabs, 1 ply, glass fiber fabric, mopped | | 2100 | .027 | | .34 | .91 | .20 | 1.45 | 2.11 |
| 0300 | On slabs, 2 ply, felt, mopped | | 2500 | .022 | | .66 | .76 | .16 | 1.58 | 2.18 |
| 0400 | On slabs, 2 ply, glass fiber fabric, mopped | | 1650 | .034 | | .77 | 1.16 | .25 | 2.18 | 3.04 |
| 0600 | On slabs, 3 ply, felt, mopped | | 2100 | .027 | | .99 | .91 | .20 | 2.10 | 2.82 |
| 0700 | On slabs, 3 ply, glass fiber fabric, mopped | | 1550 | .036 | | 1.03 | 1.23 | .26 | 2.52 | 3.47 |
| 0710 | Asphaltic hardboard protection board, 1/8" thick | 2 Rofc | 500 | .032 | | .40 | 1.17 | | 1.57 | 2.39 |
| 1000 | 1/4" EPS membrane protection board | | 3500 | .005 | | .28 | .17 | | .45 | .59 |
| 1050 | 3/8" thick | | 3500 | .005 | | .32 | .17 | | .49 | .63 |
| 1060 | 1/2" thick | | 3500 | .005 | | .36 | .17 | | .53 | .67 |
| 1070 | Fiberglass fabric, black, 20/10 mesh | | 116 | .138 | Sq. | 18.10 | 5.05 | | 23.15 | 28.50 |
| 1080 | White, 20/10 mesh | | 116 | .138 | " | 18.10 | 5.05 | | 23.15 | 28.50 |

# 07 13 Sheet Waterproofing

## 07 13 53 – Elastomeric Sheet Waterproofing

| 07 13 53.10 Elastomeric Sheet Waterproofing and Access. | Crew | Daily Output | Labor-Hours | Unit | Material | 2011 Bare Costs Labor | Equipment | Total | Total Incl O&P |
|---|---|---|---|---|---|---|---|---|---|
| 0010 | **ELASTOMERIC SHEET WATERPROOFING AND ACCESS.** | | | | | | | | | |
| 0090 | EPDM, plain, 45 mils thick | 2 Rofc | 580 | .028 | S.F. | 1.30 | 1.01 | | 2.31 | 3.11 |
| 0100 | 60 mils thick | | 570 | .028 | | 1.37 | 1.03 | | 2.40 | 3.22 |
| 0300 | Nylon reinforced sheets, 45 mils thick | | 580 | .028 | | 1.33 | 1.01 | | 2.34 | 3.14 |
| 0400 | 60 mils thick | | 570 | .028 | | 1.59 | 1.03 | | 2.62 | 3.46 |
| 0600 | Vulcanizing splicing tape for above, 2" wide | | | | C.L.F. | 48 | | | 48 | 53 |
| 0700 | 4" wide | | | | " | 110 | | | 110 | 121 |
| 0900 | Adhesive, bonding, 60 SF per gal | | | | Gal. | 21 | | | 21 | 23 |
| 1000 | Splicing, 75 SF per gal | | | | " | 31 | | | 31 | 34 |
| 1200 | Neoprene sheets, plain, 45 mils thick | 2 Rofc | 580 | .028 | S.F. | 1.53 | 1.01 | | 2.54 | 3.36 |
| 1300 | 60 mils thick | | 570 | .028 | | 2.61 | 1.03 | | 3.64 | 4.58 |
| 1500 | Nylon reinforced, 45 mils thick | | 580 | .028 | | 1.75 | 1.01 | | 2.76 | 3.61 |
| 1600 | 60 mils thick | | 570 | .028 | | 2.38 | 1.03 | | 3.41 | 4.33 |
| 1800 | 120 mils thick | | 500 | .032 | | 4.74 | 1.17 | | 5.91 | 7.15 |
| 1900 | Adhesive, splicing, 150 S.F. per gal. per coat | | | | Gal. | 31 | | | 31 | 34 |
| 2100 | Fiberglass reinforced, fluid applied, 1/8" thick | 2 Rofc | 500 | .032 | S.F. | 1.90 | 1.17 | | 3.07 | 4.04 |
| 2200 | Polyethylene and rubberized asphalt sheets, 1/8" thick | | 550 | .029 | | .70 | 1.06 | | 1.76 | 2.54 |
| 2400 | Polyvinyl chloride sheets, plain, 10 mils thick | | 580 | .028 | | .17 | 1.01 | | 1.18 | 1.87 |
| 2500 | 20 mils thick | | 570 | .028 | | .28 | 1.03 | | 1.31 | 2.02 |
| 2700 | 30 mils thick | | 560 | .029 | | .39 | 1.04 | | 1.43 | 2.17 |
| 3000 | Adhesives, trowel grade, 40-100 SF per gal | | | | Gal. | 21.50 | | | 21.50 | 23.50 |
| 3100 | Brush grade, 100-250 SF per gal. | | | | " | 17.50 | | | 17.50 | 19.25 |
| 3300 | Bitumen modified polyurethane, fluid applied, 55 mils thick | 2 Rofc | 665 | .024 | S.F. | .78 | .88 | | 1.66 | 2.32 |

# 07 16 Cementitious and Reactive Waterproofing

## 07 16 16 – Crystalline Waterproofing

### 07 16 16.20 Cementitious Waterproofing

| | | Crew | Daily Output | Labor-Hours | Unit | Material | Labor | Equipment | Total | Total Incl O&P |
|---|---|---|---|---|---|---|---|---|---|---|
| 0010 | **CEMENTITIOUS WATERPROOFING** | | | | | | | | | |
| 0020 | 1/8" application, sprayed on | G-2A | 1000 | .024 | S.F. | 2 | .78 | .59 | 3.37 | 4.12 |
| 0050 | 4 coat cementitious metallic slurry | 1 Cefi | 1.20 | 6.667 | C.S.F. | 26 | 272 | | 298 | 425 |

# 07 17 Bentonite Waterproofing

## 07 17 13 – Bentonite Panel Waterproofing

### 07 17 13.10 Bentonite

| | | Crew | Daily Output | Labor-Hours | Unit | Material | Labor | Equipment | Total | Total Incl O&P |
|---|---|---|---|---|---|---|---|---|---|---|
| 0010 | **BENTONITE** | | | | | | | | | |
| 0020 | Panels, 4' x 4', 3/16" thick | 1 Rofc | 625 | .013 | S.F. | 1.10 | .47 | | 1.57 | 1.99 |
| 0100 | Rolls, 3/8" thick, with geotextile fabric both sides | " | 550 | .015 | " | 1.29 | .53 | | 1.82 | 2.31 |
| 0300 | Granular bentonite, 50 lb. bags (.625 C.F.) | | | | Bag | 9.80 | | | 9.80 | 10.80 |
| 0400 | 3/8" thick, troweled on | 1 Rofc | 475 | .017 | S.F. | .49 | .62 | | 1.11 | 1.56 |
| 0500 | Drain board, expanded polystyrene, 1-1/2" thick | 1 Rohe | 1600 | .005 | | .44 | .14 | | .58 | .71 |
| 0510 | 2" thick | | 1600 | .005 | | .58 | .14 | | .72 | .87 |
| 0520 | 3" thick | | 1600 | .005 | | .87 | .14 | | 1.01 | 1.19 |
| 0530 | 4" thick | | 1600 | .005 | | 1.16 | .14 | | 1.30 | 1.51 |
| 0600 | With filter fabric, 1-1/2" thick | | 1600 | .005 | | .51 | .14 | | .65 | .79 |
| 0625 | 2" thick | | 1600 | .005 | | .66 | .14 | | .80 | .95 |
| 0650 | 3" thick | | 1600 | .005 | | .95 | .14 | | 1.09 | 1.27 |
| 0675 | 4" thick | | 1600 | .005 | | 1.24 | .14 | | 1.38 | 1.59 |

# 07 19 Water Repellents

## 07 19 19 – Silicone Water Repellents

| 07 19 19.10 Silicone Based Water Repellents | | Crew | Daily Output | Labor-Hours | Unit | Material | 2011 Bare Costs Labor | Equipment | Total | Total Incl O&P |
|---|---|---|---|---|---|---|---|---|---|---|
| 0010 | **SILICONE BASED WATER REPELLENTS** | | | | | | | | | |
| 0020 | Water base liquid, roller applied | 2 Rofc | 7000 | .002 | S.F. | .86 | .08 | | .94 | 1.09 |
| 0200 | Silicone or stearate, sprayed on CMU, 1 coat | 1 Rofc | 4000 | .002 | | .36 | .07 | | .43 | .52 |
| 0300 | 2 coats | " | 3000 | .003 | | .73 | .10 | | .83 | .96 |

# 07 21 Thermal Insulation

## 07 21 13 – Board Insulation

### 07 21 13.10 Rigid Insulation

| | | | Crew | Daily Output | Labor-Hours | Unit | Material | 2011 Bare Costs Labor | Equipment | Total | Total Incl O&P |
|---|---|---|---|---|---|---|---|---|---|---|---|
| 0010 | **RIGID INSULATION**, for walls | | | | | | | | | | |
| 0040 | Fiberglass, 1.5#/C.F., unfaced, 1" thick, R4.1 | G | 1 Carp | 1000 | .008 | S.F. | .40 | .34 | | .74 | .97 |
| 0060 | 1-1/2" thick, R6.2 | G | | 1000 | .008 | | .58 | .34 | | .92 | 1.17 |
| 0080 | 2" thick, R8.3 | G | | 1000 | .008 | | .64 | .34 | | .98 | 1.23 |
| 0120 | 3" thick, R12.4 | G | | 800 | .010 | | .76 | .43 | | 1.19 | 1.50 |
| 0370 | 3#/C.F., unfaced, 1" thick, R4.3 | G | | 1000 | .008 | | .46 | .34 | | .80 | 1.04 |
| 0390 | 1-1/2" thick, R6.5 | G | | 1000 | .008 | | .86 | .34 | | 1.20 | 1.48 |
| 0400 | 2" thick, R8.7 | G | | 890 | .009 | | .91 | .39 | | 1.30 | 1.59 |
| 0420 | 2-1/2" thick, R10.9 | G | | 800 | .010 | | .96 | .43 | | 1.39 | 1.72 |
| 0440 | 3" thick, R13 | G | | 800 | .010 | | 1.39 | .43 | | 1.82 | 2.19 |
| 0520 | Foil faced, 1" thick, R4.3 | G | | 1000 | .008 | | .83 | .34 | | 1.17 | 1.44 |
| 0540 | 1-1/2" thick, R6.5 | G | | 1000 | .008 | | 1.24 | .34 | | 1.58 | 1.89 |
| 0560 | 2" thick, R8.7 | G | | 890 | .009 | | 1.55 | .39 | | 1.94 | 2.30 |
| 0580 | 2-1/2" thick, R10.9 | G | | 800 | .010 | | 1.82 | .43 | | 2.25 | 2.66 |
| 0600 | 3" thick, R13 | G | | 800 | .010 | | 2 | .43 | | 2.43 | 2.86 |
| 1600 | Isocyanurate, 4' x 8' sheet, foil faced, both sides | | | | | | | | | | |
| 1610 | 1/2" thick | G | 1 Carp | 800 | .010 | S.F. | .29 | .43 | | .72 | .98 |
| 1620 | 5/8" thick | G | | 800 | .010 | | .32 | .43 | | .75 | 1.01 |
| 1630 | 3/4" thick | G | | 800 | .010 | | .33 | .43 | | .76 | 1.02 |
| 1640 | 1" thick | G | | 800 | .010 | | .51 | .43 | | .94 | 1.22 |
| 1650 | 1-1/2" thick | G | | 730 | .011 | | .64 | .47 | | 1.11 | 1.42 |
| 1660 | 2" thick | G | | 730 | .011 | | .79 | .47 | | 1.26 | 1.59 |
| 1670 | 3" thick | G | | 730 | .011 | | 1.78 | .47 | | 2.25 | 2.68 |
| 1680 | 4" thick | G | | 730 | .011 | | 2 | .47 | | 2.47 | 2.92 |
| 1700 | Perlite, 1" thick, R2.77 | G | | 800 | .010 | | .30 | .43 | | .73 | .99 |
| 1750 | 2" thick, R5.55 | G | | 730 | .011 | | .60 | .47 | | 1.07 | 1.38 |
| 1900 | Extruded polystyrene, 25 PSI compressive strength, 1" thick, R5 | G | | 800 | .010 | | .50 | .43 | | .93 | 1.21 |
| 1940 | 2" thick R10 | G | | 730 | .011 | | 1 | .47 | | 1.47 | 1.82 |
| 1960 | 3" thick, R15 | G | | 730 | .011 | | 1.44 | .47 | | 1.91 | 2.30 |
| 2100 | Expanded polystyrene, 1" thick, R3.85 | G | | 800 | .010 | | .29 | .43 | | .72 | .98 |
| 2120 | 2" thick, R7.69 | G | | 730 | .011 | | .58 | .47 | | 1.05 | 1.36 |
| 2140 | 3" thick, R11.49 | G | | 730 | .011 | | .87 | .47 | | 1.34 | 1.68 |

### 07 21 13.13 Foam Board Insulation

| | | | Crew | Daily Output | Labor-Hours | Unit | Material | Labor | Equipment | Total | Total Incl O&P |
|---|---|---|---|---|---|---|---|---|---|---|---|
| 0010 | **FOAM BOARD INSULATION** | | | | | | | | | | |
| 0600 | Polystyrene, expanded, 1" thick, R4 | G | 1 Carp | 680 | .012 | S.F. | .29 | .51 | | .80 | 1.10 |
| 0700 | 2" thick, R8 | G | " | 675 | .012 | " | .58 | .51 | | 1.09 | 1.42 |

## 07 21 16 – Blanket Insulation

### 07 21 16.10 Blanket Insulation for Floors/Ceilings

| | | | Crew | Daily Output | Labor-Hours | Unit | Material | Labor | Equipment | Total | Total Incl O&P |
|---|---|---|---|---|---|---|---|---|---|---|---|
| 0010 | **BLANKET INSULATION FOR FLOORS/CEILINGS** | | | | | | | | | | |
| 0020 | Including spring type wire fasteners | | | | | | | | | | |
| 2000 | Fiberglass, blankets or batts, paper or foil backing | | | | | | | | | | |
| 2100 | 3-1/2" thick, R13 | G | 1 Carp | 700 | .011 | S.F. | .32 | .49 | | .81 | 1.11 |

# 07 21 Thermal Insulation

## 07 21 16 – Blanket Insulation

| 07 21 16.10 Blanket Insulation for Floors/Ceilings | | Crew | Daily Output | Labor-Hours | Unit | Material | 2011 Bare Costs Labor | Equipment | Total | Total Incl O&P |
|---|---|---|---|---|---|---|---|---|---|---|
| 2150 | 6-1/4" thick, R19 | G | 1 Carp | 600 | .013 | S.F. | .42 | .57 | | .99 | 1.34 |
| 2210 | 9-1/2" thick, R30 | G | | 500 | .016 | | .70 | .69 | | 1.39 | 1.83 |
| 2220 | 12" thick, R38 | G | | 475 | .017 | | .85 | .73 | | 1.58 | 2.05 |
| 3000 | Unfaced, 3-1/2" thick, R13 | G | | 600 | .013 | | .28 | .57 | | .85 | 1.19 |
| 3010 | 6-1/4" thick, R19 | G | | 500 | .016 | | .44 | .69 | | 1.13 | 1.54 |
| 3020 | 9-1/2" thick, R30 | G | | 450 | .018 | | .70 | .77 | | 1.47 | 1.94 |
| 3030 | 12" thick, R38 | G | ▼ | 425 | .019 | ▼ | .73 | .81 | | 1.54 | 2.04 |

### 07 21 16.20 Blanket Insulation for Walls

| | | Crew | Daily Output | Labor-Hours | Unit | Material | 2011 Bare Costs Labor | Equipment | Total | Total Incl O&P |
|---|---|---|---|---|---|---|---|---|---|---|
| 0010 | **BLANKET INSULATION FOR WALLS** | | | | | | | | | | |
| 0020 | Kraft faced fiberglass, 3-1/2" thick, R11, 15" wide | G | 1 Carp | 1350 | .006 | S.F. | .28 | .26 | | .54 | .70 |
| 0030 | 23" wide | G | | 1600 | .005 | | .28 | .22 | | .50 | .64 |
| 0060 | R13, 11" wide | G | | 1150 | .007 | | .27 | .30 | | .57 | .76 |
| 0080 | 15" wide | G | | 1350 | .006 | | .27 | .26 | | .53 | .69 |
| 0100 | 23" wide | G | | 1600 | .005 | | .27 | .22 | | .49 | .63 |
| 0110 | R-15, 11" wide | G | | 1150 | .007 | | .41 | .30 | | .71 | .91 |
| 0120 | 15" wide | G | | 1350 | .006 | | .41 | .26 | | .67 | .84 |
| 0130 | 23" wide | G | | 1600 | .005 | | .41 | .22 | | .63 | .78 |
| 0140 | 6" thick, R19, 11" wide | G | | 1150 | .007 | | .37 | .30 | | .67 | .87 |
| 0160 | 15" wide | G | | 1350 | .006 | | .37 | .26 | | .63 | .80 |
| 0180 | 23" wide | G | | 1600 | .005 | | .37 | .22 | | .59 | .74 |
| 0182 | R21, 11" wide | G | | 1150 | .007 | | .51 | .30 | | .81 | 1.02 |
| 0184 | 15" wide | G | | 1350 | .006 | | .51 | .26 | | .77 | .95 |
| 0186 | 23" wide | G | | 1600 | .005 | | .51 | .22 | | .73 | .89 |
| 0188 | 9" thick, R-30, 11" wide | G | | 985 | .008 | | .71 | .35 | | 1.06 | 1.32 |
| 0200 | 15" wide | G | | 1150 | .007 | | .71 | .30 | | 1.01 | 1.24 |
| 0220 | 23" wide | G | | 1350 | .006 | | .71 | .26 | | .97 | 1.17 |
| 0230 | 12" thick, R38, 11" wide | G | | 985 | .008 | | .95 | .35 | | 1.30 | 1.59 |
| 0240 | 15" wide | G | | 1150 | .007 | | .95 | .30 | | 1.25 | 1.51 |
| 0260 | 23" wide | G | | 1350 | .006 | | .95 | .26 | | 1.21 | 1.44 |
| 0410 | Foil faced fiberglass, 3-1/2" thick, R13, 11" wide | G | | 1150 | .007 | | .54 | .30 | | .84 | 1.05 |
| 0420 | 15" wide | G | | 1350 | .006 | | .54 | .26 | | .80 | .98 |
| 0440 | 23" wide | G | | 1600 | .005 | | .54 | .22 | | .76 | .92 |
| 0442 | R15, 11" wide | G | | 1150 | .007 | | .56 | .30 | | .86 | 1.08 |
| 0444 | 15" wide | G | | 1350 | .006 | | .56 | .26 | | .82 | 1.01 |
| 0446 | 23" wide | G | | 1600 | .005 | | .56 | .22 | | .78 | .95 |
| 0448 | 6" thick, R19, 11" wide | G | | 1150 | .007 | | .77 | .30 | | 1.07 | 1.31 |
| 0460 | 15" wide | G | | 1350 | .006 | | .77 | .26 | | 1.03 | 1.24 |
| 0480 | 23" wide | G | | 1600 | .005 | | .77 | .22 | | .99 | 1.18 |
| 0482 | R-21, 11" wide | G | | 1150 | .007 | | .82 | .30 | | 1.12 | 1.36 |
| 0484 | 15" wide | G | | 1350 | .006 | | .82 | .26 | | 1.08 | 1.29 |
| 0486 | 23" wide | G | | 1600 | .005 | | .82 | .22 | | 1.04 | 1.23 |
| 0488 | 9" thick, R-30, 11" wide | G | | 985 | .008 | | .89 | .35 | | 1.24 | 1.52 |
| 0500 | 9" thick, R30, 15" wide | G | | 1150 | .007 | | .89 | .30 | | 1.19 | 1.44 |
| 0550 | 23" wide | G | | 1350 | .006 | | .89 | .26 | | 1.15 | 1.37 |
| 0560 | 12" thick, R-38, 11" wide | G | | 985 | .008 | | .90 | .35 | | 1.25 | 1.53 |
| 0570 | 15" wide | G | | 1150 | .007 | | .90 | .30 | | 1.20 | 1.45 |
| 0580 | 23" wide | G | | 1350 | .006 | | .90 | .26 | | 1.16 | 1.38 |
| 0620 | Unfaced fiberglass, 3-1/2" thick, R-13, 11" wide | G | | 1150 | .007 | | .28 | .30 | | .58 | .77 |
| 0820 | 15" wide | G | | 1350 | .006 | | .28 | .26 | | .54 | .70 |
| 0830 | 23" wide | G | | 1600 | .005 | | .28 | .22 | | .50 | .64 |
| 0832 | R15, 11" wide | G | | 1150 | .007 | | .40 | .30 | | .70 | .90 |
| 0834 | 15" wide | G | ▼ | 1350 | .006 | ▼ | .40 | .26 | | .66 | .83 |

## 07 21 16 – Blanket Insulation

### 07 21 16.20 Blanket Insulation for Walls

| | | Crew | Daily Output | Labor-Hours | Unit | Material | 2011 Bare Costs Labor | Equipment | Total | Total Incl O&P |
|---|---|---|---|---|---|---|---|---|---|---|
| 0836 | 23" wide | G 1 Carp | 1600 | .005 | S.F. | .40 | .22 | | .62 | .77 |
| 0838 | 6" thick, R19, 11" wide | G | 1150 | .007 | | .44 | .30 | | .74 | .94 |
| 0860 | 15" wide | G | 1150 | .007 | | .44 | .30 | | .74 | .94 |
| 0880 | 23" wide | G | 1350 | .006 | | .44 | .26 | | .70 | .87 |
| 0882 | R-21, 11" wide | G | 1150 | .007 | | .46 | .30 | | .76 | .97 |
| 0886 | 15" wide | G | 1350 | .006 | | .46 | .26 | | .72 | .90 |
| 0888 | 23" wide | G | 1600 | .005 | | .46 | .22 | | .68 | .84 |
| 0890 | 9" thick, R30, 11" wide | G | 985 | .008 | | .70 | .35 | | 1.05 | 1.31 |
| 0900 | 15" wide | G | 1150 | .007 | | .70 | .30 | | 1 | 1.23 |
| 0920 | 23" wide | G | 1350 | .006 | | .70 | .26 | | .96 | 1.16 |
| 0930 | 12" thick, R38, 11" wide | G | 985 | .008 | | .73 | .35 | | 1.08 | 1.34 |
| 0940 | 15" wide | G | 1000 | .008 | | .73 | .34 | | 1.07 | 1.33 |
| 0960 | 23" wide | G | 1150 | .007 | | .73 | .30 | | 1.03 | 1.26 |
| 1300 | Mineral fiber batts, kraft faced | | | | | | | | | |
| 1320 | 3-1/2" thick, R12 | G 1 Carp | 1600 | .005 | S.F. | .38 | .22 | | .60 | .75 |
| 1340 | 6" thick, R19 | G | 1600 | .005 | | .42 | .22 | | .64 | .79 |
| 1380 | 10" thick, R30 | G | 1350 | .006 | | .67 | .26 | | .93 | 1.13 |
| 1850 | Friction fit wire insulation supports, 16" O.C. | | 960 | .008 | Ea. | .05 | .36 | | .41 | .61 |

## 07 21 23 – Loose-Fill Insulation

### 07 21 23.10 Poured Loose-Fill Insulation

| | | Crew | Daily Output | Labor-Hours | Unit | Material | 2011 Bare Costs Labor | Equipment | Total | Total Incl O&P |
|---|---|---|---|---|---|---|---|---|---|---|
| 0010 | **POURED LOOSE-FILL INSULATION** | | | | | | | | | |
| 0020 | Cellulose fiber, R3.8 per inch | G 1 Carp | 200 | .040 | C.F. | .56 | 1.72 | | 2.28 | 3.26 |
| 0040 | Ceramic type (perlite), R3.2 per inch | G | 200 | .040 | | 1.95 | 1.72 | | 3.67 | 4.79 |
| 0080 | Fiberglass wool, R4 per inch | G | 200 | .040 | | .42 | 1.72 | | 2.14 | 3.10 |
| 0100 | Mineral wool, R3 per inch | G | 200 | .040 | | .39 | 1.72 | | 2.11 | 3.07 |
| 0300 | Polystyrene, R4 per inch | G | 200 | .040 | | 3.09 | 1.72 | | 4.81 | 6.05 |
| 0400 | Perlite, R2.7 per inch | G | 200 | .040 | | 1.95 | 1.72 | | 3.67 | 4.79 |
| 0700 | Wood fiber, R3.85 per inch | G | 200 | .040 | | .70 | 1.72 | | 2.42 | 3.41 |

### 07 21 23.20 Masonry Loose-Fill Insulation

| | | Crew | Daily Output | Labor-Hours | Unit | Material | 2011 Bare Costs Labor | Equipment | Total | Total Incl O&P |
|---|---|---|---|---|---|---|---|---|---|---|
| 0010 | **MASONRY LOOSE-FILL INSULATION**, vermiculite or perlite | | | | | | | | | |
| 0100 | In cores of concrete block, 4" thick wall, .115 C.F./S.F. | G D-1 | 4800 | .003 | S.F. | .22 | .13 | | .35 | .44 |
| 0200 | 6" thick wall, .175 C.F./S.F. | G | 3000 | .005 | | .34 | .21 | | .55 | .69 |
| 0300 | 8" thick wall, .258 C.F./S.F. | G | 2400 | .007 | | .50 | .26 | | .76 | .94 |
| 0400 | 10" thick wall, .340 C.F./S.F. | G | 1850 | .009 | | .66 | .34 | | 1 | 1.24 |
| 0500 | 12" thick wall, .422 C.F./S.F. | G | 1200 | .013 | | .82 | .52 | | 1.34 | 1.69 |
| 0550 | For sand fill, deduct from above | G | | | | 70% | | | | |
| 0600 | Poured cavity wall, vermiculite or perlite, water repellant | G D-1 | 250 | .064 | C.F. | 1.95 | 2.49 | | 4.44 | 5.90 |
| 0700 | Foamed in place, urethane in 2-5/8" cavity | G G-2A | 1035 | .023 | S.F. | .49 | .76 | .57 | 1.82 | 2.40 |
| 0800 | For each 1" added thickness, add | G " | 2372 | .010 | " | .12 | .33 | .25 | .70 | .95 |

## 07 21 26 – Blown Insulation

### 07 21 26.10 Blown Insulation

| | | Crew | Daily Output | Labor-Hours | Unit | Material | 2011 Bare Costs Labor | Equipment | Total | Total Incl O&P |
|---|---|---|---|---|---|---|---|---|---|---|
| 0010 | **BLOWN INSULATION** Ceilings, with open access | | | | | | | | | |
| 0020 | Cellulose, 3-1/2" thick, R13 | G G-4 | 5000 | .005 | S.F. | .17 | .17 | .06 | .40 | .52 |
| 0030 | 5-3/16" thick, R19 | G | 3800 | .006 | | .25 | .22 | .08 | .55 | .71 |
| 0050 | 6-1/2" thick, R22 | G | 3000 | .008 | | .32 | .28 | .10 | .70 | .90 |
| 0100 | 8-11/16" thick, R30 | G | 2600 | .009 | | .44 | .32 | .11 | .87 | 1.11 |
| 0120 | 10-7/8" thick, R38 | G | 1800 | .013 | | .56 | .47 | .16 | 1.19 | 1.51 |
| 1000 | Fiberglass, 5.5" thick, R11 | G | 3800 | .006 | | .15 | .22 | .08 | .45 | .60 |
| 1050 | 6" thick, R12 | G | 3000 | .008 | | .19 | .28 | .10 | .57 | .74 |
| 1100 | 8.8" thick, R19 | G | 2200 | .011 | | .26 | .38 | .13 | .77 | 1.03 |
| 1200 | 10" thick, R22 | G | 1800 | .013 | | .31 | .47 | .16 | .94 | 1.24 |

# 07 21 Thermal Insulation

## 07 21 26 – Blown Insulation

### 07 21 26.10 Blown Insulation

| | 07 21 26.10 Blown Insulation | | Crew | Daily Output | Labor-Hours | Unit | Material | 2011 Bare Costs Labor | Equipment | Total | Total Incl O&P |
|---|---|---|---|---|---|---|---|---|---|---|---|
| 1300 | 11.5" thick, R26 | G | G-4 | 1500 | .016 | S.F. | .37 | .56 | .20 | 1.13 | 1.49 |
| 1350 | 13" thick, R30 | G | | 1400 | .017 | | .40 | .60 | .21 | 1.21 | 1.59 |
| 1450 | 16" thick, R38 | G | | 1145 | .021 | | .49 | .73 | .26 | 1.48 | 1.95 |
| 1500 | 20" thick, R49 | G | | 920 | .026 | | .62 | .91 | .32 | 1.85 | 2.43 |

## 07 21 27 – Reflective Insulation

### 07 21 27.10 Reflective Insulation Options

| | | | | | | | | | | | |
|---|---|---|---|---|---|---|---|---|---|---|---|
| 0010 | **REFLECTIVE INSULATION OPTIONS** | | | | | | | | | | |
| 0020 | Aluminum foil on reinforced scrim | G | 1 Carp | 19 | .421 | C.S.F. | 14.20 | 18.15 | | 32.35 | 43.50 |
| 0100 | Reinforced with woven polyolefin | G | | 19 | .421 | | 18 | 18.15 | | 36.15 | 48 |
| 0500 | With single bubble air space, R8.8 | G | | 15 | .533 | | 28 | 23 | | 51 | 66 |
| 0600 | With double bubble air space, R9.8 | G | | 15 | .533 | | 29 | 23 | | 52 | 66.50 |

## 07 21 29 – Sprayed Insulation

### 07 21 29.10 Sprayed-On Insulation

| | | | | | | | | | | | |
|---|---|---|---|---|---|---|---|---|---|---|---|
| 0010 | **SPRAYED-ON INSULATION** | | | | | | | | | | |
| 0020 | Fibrous/cementitious, finished wall, 1" thick, R3.7 | G | G-2 | 2050 | .012 | S.F. | .24 | .42 | .06 | .72 | .97 |
| 0100 | Attic, 5.2" thick, R19 | G | | 1550 | .015 | " | .42 | .56 | .08 | 1.06 | 1.39 |
| 0200 | Fiberglass, R4 per inch, vertical | G | | 2050 | .012 | B.F. | .15 | .42 | .06 | .63 | .88 |
| 0210 | Horizontal | G | | 1550 | .015 | " | .15 | .56 | .08 | .79 | 1.10 |
| 0300 | Closed cell, spray polyurethane foam, 2 pounds per cubic foot density | | | | | | | | | | |
| 0310 | 1" thick | G | G-2A | 6000 | .004 | S.F. | .41 | .13 | .10 | .64 | .77 |
| 0320 | 2" thick | G | | 3000 | .008 | | .82 | .26 | .20 | 1.28 | 1.54 |
| 0330 | 3" thick | G | | 2000 | .012 | | 1.23 | .39 | .30 | 1.92 | 2.32 |
| 0335 | 3-1/2" thick | G | | 1715 | .014 | | 1.44 | .46 | .35 | 2.25 | 2.70 |
| 0340 | 4" thick | G | | 1500 | .016 | | 1.64 | .52 | .40 | 2.56 | 3.09 |
| 0350 | 5" thick | G | | 1200 | .020 | | 2.05 | .65 | .49 | 3.19 | 3.86 |
| 0355 | 5-1/2" thick | G | | 1090 | .022 | | 2.26 | .72 | .54 | 3.52 | 4.24 |
| 0360 | 6" thick | G | | 1000 | .024 | | 2.46 | .78 | .59 | 3.83 | 4.63 |

# 07 22 Roof and Deck Insulation

## 07 22 16 – Roof Board Insulation

### 07 22 16.10 Roof Deck Insulation

| | | | | | | | | | | | |
|---|---|---|---|---|---|---|---|---|---|---|---|
| 0010 | **ROOF DECK INSULATION** | | | | | | | | | | |
| 0020 | Fiberboard low density, 1/2" thick R1.39 | G | 1 Rofc | 1000 | .008 | S.F. | .26 | .29 | | .55 | .78 |
| 0030 | 1" thick R2.78 | G | | 800 | .010 | | .46 | .37 | | .83 | 1.12 |
| 0080 | 1 1/2" thick R4.17 | G | | 800 | .010 | | .70 | .37 | | 1.07 | 1.38 |
| 0100 | 2" thick R5.56 | G | | 800 | .010 | | .93 | .37 | | 1.30 | 1.63 |
| 0110 | Fiberboard high density, 1/2" thick R1.3 | G | | 1000 | .008 | | .24 | .29 | | .53 | .75 |
| 0120 | 1" thick R2.5 | G | | 800 | .010 | | .48 | .37 | | .85 | 1.14 |
| 0130 | 1-1/2" thick R3.8 | G | | 800 | .010 | | .72 | .37 | | 1.09 | 1.40 |
| 0200 | Fiberglass, 3/4" thick R2.78 | G | | 1000 | .008 | | .53 | .29 | | .82 | 1.07 |
| 0400 | 15/16" thick R3.70 | G | | 1000 | .008 | | .71 | .29 | | 1 | 1.27 |
| 0460 | 1-1/16" thick R4.17 | G | | 1000 | .008 | | .89 | .29 | | 1.18 | 1.47 |
| 0600 | 1-5/16" thick R5.26 | G | | 1000 | .008 | | 1.20 | .29 | | 1.49 | 1.81 |
| 0650 | 2-1/16" thick R8.33 | G | | 800 | .010 | | 1.27 | .37 | | 1.64 | 2.01 |
| 0700 | 2-7/16" thick R10 | G | | 800 | .010 | | 1.47 | .37 | | 1.84 | 2.23 |
| 1650 | Perlite, 1/2" thick R1.32 | G | | 1050 | .008 | | .34 | .28 | | .62 | .83 |
| 1655 | 3/4" thick R2.08 | G | | 800 | .010 | | .37 | .37 | | .74 | 1.02 |
| 1660 | 1" thick R2.78 | G | | 800 | .010 | | .46 | .37 | | .83 | 1.12 |
| 1670 | 1-1/2" thick R4.17 | G | | 800 | .010 | | .48 | .37 | | .85 | 1.14 |
| 1680 | 2" thick R5.56 | G | | 700 | .011 | | .80 | .42 | | 1.22 | 1.58 |

## 07 22 16 – Roof Board Insulation

| 07 22 16.10 Roof Deck Insulation | | Crew | Daily Output | Labor-Hours | Unit | Material | 2011 Bare Costs Labor | Equipment | Total | Total Incl O&P |
|---|---|---|---|---|---|---|---|---|---|---|
| 1685 | 2-1/2" thick R6.67 | G | 1 Rofc | 700 | .011 | S.F. | .95 | .42 | | 1.37 | 1.75 |
| 1690 | Tapered for drainage | G | | 800 | .010 | B.F. | .73 | .37 | | 1.10 | 1.41 |
| 1700 | Polyisocyanurate, 2#/C.F. density, 3/4" thick | G | | 1500 | .005 | S.F. | .44 | .19 | | .63 | .80 |
| 1705 | 1" thick | G | | 1400 | .006 | | .51 | .21 | | .72 | .91 |
| 1715 | 1-1/2" thick | G | | 1250 | .006 | | .66 | .23 | | .89 | 1.12 |
| 1725 | 2" thick | G | | 1100 | .007 | | .78 | .27 | | 1.05 | 1.30 |
| 1735 | 2-1/2" thick | G | | 1050 | .008 | | 1.01 | .28 | | 1.29 | 1.57 |
| 1745 | 3" thick | G | | 1000 | .008 | | 1.22 | .29 | | 1.51 | 1.83 |
| 1755 | 3-1/2" thick | G | | 1000 | .008 | | 1.95 | .29 | | 2.24 | 2.64 |
| 1765 | Tapered for drainage | G | | 1400 | .006 | B.F. | 1.95 | .21 | | 2.16 | 2.50 |
| 1900 | Extruded Polystyrene | | | | | | | | | | |
| 1910 | 15 PSI compressive strength, 1" thick, R5 | G | 1 Rofc | 1500 | .005 | S.F. | .44 | .19 | | .63 | .80 |
| 1920 | 2" thick, R10 | G | | 1250 | .006 | | .56 | .23 | | .79 | 1.01 |
| 1930 | 3" thick R15 | G | | 1000 | .008 | | 1.15 | .29 | | 1.44 | 1.76 |
| 1932 | 4" thick R20 | G | | 1000 | .008 | | 1.52 | .29 | | 1.81 | 2.16 |
| 1934 | Tapered for drainage | G | | 1500 | .005 | B.F. | .51 | .19 | | .70 | .88 |
| 1940 | 25 PSI compressive strength, 1" thick R5 | G | | 1500 | .005 | S.F. | .61 | .19 | | .80 | .99 |
| 1942 | 2" thick R10 | G | | 1250 | .006 | | 1.15 | .23 | | 1.38 | 1.66 |
| 1944 | 3" thick R15 | G | | 1000 | .008 | | 1.75 | .29 | | 2.04 | 2.42 |
| 1946 | 4" thick R20 | G | | 1000 | .008 | | 2.50 | .29 | | 2.79 | 3.24 |
| 1948 | Tapered for drainage | G | | 1500 | .005 | B.F. | .53 | .19 | | .72 | .90 |
| 1950 | 40 psi compressive strength, 1" thick R5 | G | | 1500 | .005 | S.F. | .47 | .19 | | .66 | .84 |
| 1952 | 2" thick R10 | G | | 1250 | .006 | | .88 | .23 | | 1.11 | 1.36 |
| 1954 | 3" thick R15 | G | | 1000 | .008 | | 1.28 | .29 | | 1.57 | 1.90 |
| 1956 | 4" thick R20 | G | | 1000 | .008 | | 1.70 | .29 | | 1.99 | 2.36 |
| 1958 | Tapered for drainage | G | | 1400 | .006 | B.F. | .67 | .21 | | .88 | 1.09 |
| 1960 | 60 PSI compressive strength, 1" thick R5 | G | | 1450 | .006 | S.F. | .65 | .20 | | .85 | 1.06 |
| 1962 | 2" thick R10 | G | | 1200 | .007 | | 1.25 | .24 | | 1.49 | 1.79 |
| 1964 | 3" thick R15 | G | | 975 | .008 | | 2 | .30 | | 2.30 | 2.70 |
| 1966 | 4" thick R20 | G | | 950 | .008 | | 2.50 | .31 | | 2.81 | 3.26 |
| 1968 | Tapered for drainage | G | | 1400 | .006 | B.F. | .85 | .21 | | 1.06 | 1.29 |
| 2010 | Expanded polystyrene, 1#/C.F. density, 3/4" thick R2.89 | G | | 1500 | .005 | S.F. | .22 | .19 | | .41 | .56 |
| 2020 | 1" thick R3.85 | G | | 1500 | .005 | | .29 | .19 | | .48 | .64 |
| 2100 | 2" thick R7.69 | G | | 1250 | .006 | | .58 | .23 | | .81 | 1.03 |
| 2110 | 3" thick R11.49 | G | | 1250 | .006 | | .87 | .23 | | 1.10 | 1.35 |
| 2120 | 4" thick R15.38 | G | | 1200 | .007 | | 1.16 | .24 | | 1.40 | 1.69 |
| 2130 | 5" thick R19.23 | G | | 1150 | .007 | | 1.45 | .25 | | 1.70 | 2.02 |
| 2140 | 6" thick R23.26 | G | | 1150 | .007 | | 1.74 | .25 | | 1.99 | 2.33 |
| 2150 | Tapered for drainage | G | | 1500 | .005 | B.F. | .54 | .19 | | .73 | .91 |
| 2400 | Composites with 2" EPS | | | | | | | | | | |
| 2410 | 1" fiberboard | G | 1 Rofc | 950 | .008 | S.F. | 1.28 | .31 | | 1.59 | 1.92 |
| 2420 | 7/16" oriented strand board | G | | 800 | .010 | | 1 | .37 | | 1.37 | 1.71 |
| 2430 | 1/2" plywood | G | | 800 | .010 | | 1.20 | .37 | | 1.57 | 1.93 |
| 2440 | 1" perlite | G | | 800 | .010 | | 1.18 | .37 | | 1.55 | 1.91 |
| 2450 | Composites with 1-1/2" polyisocyanurate | | | | | | | | | | |
| 2460 | 1" fiberboard | G | 1 Rofc | 800 | .010 | S.F. | 1.20 | .37 | | 1.57 | 1.93 |
| 2470 | 1" perlite | G | | 850 | .009 | | 1.08 | .34 | | 1.42 | 1.76 |
| 2480 | 7/16" oriented strand board | G | | 800 | .010 | | .84 | .37 | | 1.21 | 1.53 |

# 07 24 Exterior Insulation and Finish Systems

## 07 24 13 – Polymer-Based Exterior Insulation and Finish System

| | 07 24 13.10 Exterior Insulation and Finish Systems | | Crew | Daily Output | Labor-Hours | Unit | Material | 2011 Bare Costs Labor | Equipment | Total | Total Incl O&P |
|---|---|---|---|---|---|---|---|---|---|---|---|
| 0010 | **EXTERIOR INSULATION AND FINISH SYSTEMS** | | | | | | | | | | |
| 0095 | Field applied, 1" EPS insulation | G | J-1 | 390 | .103 | S.F. | 2.09 | 3.86 | .32 | 6.27 | 8.40 |
| 0100 | With 1/2" cement board sheathing | G | | 268 | .149 | | 2.76 | 5.60 | .46 | 8.82 | 11.90 |
| 0105 | 2" EPS insulation | G | | 390 | .103 | | 2.38 | 3.86 | .32 | 6.56 | 8.70 |
| 0110 | With 1/2" cement board sheathing | G | | 268 | .149 | | 3.05 | 5.60 | .46 | 9.11 | 12.20 |
| 0115 | 3" EPS insulation | G | | 390 | .103 | | 2.67 | 3.86 | .32 | 6.85 | 9.05 |
| 0120 | With 1/2" cement board sheathing | G | | 268 | .149 | | 3.34 | 5.60 | .46 | 9.40 | 12.55 |
| 0125 | 4" EPS insulation | G | | 390 | .103 | | 2.96 | 3.86 | .32 | 7.14 | 9.35 |
| 0130 | With 1/2" cement board sheathing | G | | 268 | .149 | | 4.31 | 5.60 | .46 | 10.37 | 13.60 |
| 0140 | Premium finish add | | | 1265 | .032 | | .31 | 1.19 | .10 | 1.60 | 2.22 |
| 0150 | Heavy duty reinforcement add | | | 914 | .044 | | 1.08 | 1.65 | .14 | 2.87 | 3.79 |
| 0160 | 2.5#/S.Y. metal lath substrate add | | 1 Lath | 75 | .107 | S.Y. | 2.37 | 4.11 | | 6.48 | 8.60 |
| 0170 | 3.4#/S.Y. metal lath substrate add | | " | 75 | .107 | " | 2.58 | 4.11 | | 6.69 | 8.85 |
| 0180 | Color or texture change, | | J-1 | 1265 | .032 | S.F. | .83 | 1.19 | .10 | 2.12 | 2.79 |
| 0190 | With substrate leveling base coat | | 1 Plas | 530 | .015 | | .83 | .59 | | 1.42 | 1.79 |
| 0210 | With substrate sealing base coat | | 1 Pord | 1224 | .007 | | .08 | .25 | | .33 | .45 |
| 0370 | V groove shape in panel face | | | | | L.F. | .60 | | | .60 | .66 |
| 0380 | U groove shape in panel face | | | | | " | .79 | | | .79 | .87 |
| 0440 | For higher than one story, add | | | | | | | 25% | | | |

# 07 25 Weather Barriers

## 07 25 10 – Weather Barriers or Wraps

### 07 25 10.10 Weather Barriers

| | | Crew | Daily Output | Labor-Hours | Unit | Material | 2011 Bare Costs Labor | Equipment | Total | Total Incl O&P |
|---|---|---|---|---|---|---|---|---|---|---|
| 0010 | **WEATHER BARRIERS** | | | | | | | | | |
| 0400 | Asphalt felt paper, 15# | 1 Carp | 37 | .216 | Sq. | 5.65 | 9.30 | | 14.95 | 20.50 |
| 0401 | Per square foot | " | 3700 | .002 | S.F. | .06 | .09 | | .15 | .20 |
| 0450 | Housewrap, exterior, spun bonded polypropylene | | | | | | | | | |
| 0470 | Small roll | 1 Carp | 3800 | .002 | S.F. | .21 | .09 | | .30 | .38 |
| 0480 | Large roll | " | 4000 | .002 | " | .13 | .09 | | .22 | .27 |
| 2100 | Asphalt felt roof deck vapor barrier, class 1 metal decks | 1 Rofc | 37 | .216 | Sq. | 19.70 | 7.90 | | 27.60 | 34.50 |
| 2200 | For all other decks | " | 37 | .216 | | 14.05 | 7.90 | | 21.95 | 28.50 |
| 2800 | Asphalt felt, 50% recycled content, 15 lb, 4 sq per roll | 1 Carp | 36 | .222 | | 5.65 | 9.55 | | 15.20 | 21 |
| 2810 | 30 lb, 2 sq per roll | " | 36 | .222 | | 9.45 | 9.55 | | 19 | 25 |
| 3000 | Building wrap, spunbonded polyethylene | 2 Carp | 8000 | .002 | S.F. | .14 | .09 | | .23 | .28 |

# 07 26 Vapor Retarders

## 07 26 10 – Above-Grade Vapor Retarders

### 07 26 10.10 Vapor Retarders

| | | | Crew | Daily Output | Labor-Hours | Unit | Material | 2011 Bare Costs Labor | Equipment | Total | Total Incl O&P |
|---|---|---|---|---|---|---|---|---|---|---|---|
| 0010 | **VAPOR RETARDERS** | | | | | | | | | | |
| 0020 | Aluminum and kraft laminated, foil 1 side | G | 1 Carp | 37 | .216 | Sq. | 9.35 | 9.30 | | 18.65 | 24.50 |
| 0100 | Foil 2 sides | G | | 37 | .216 | | 9.75 | 9.30 | | 19.05 | 25 |
| 0600 | Polyethylene vapor barrier, standard, .002" thick | G | | 37 | .216 | | 1.19 | 9.30 | | 10.49 | 15.60 |
| 0700 | .004" thick | G | | 37 | .216 | | 3.34 | 9.30 | | 12.64 | 17.95 |
| 0900 | .006" thick | G | | 37 | .216 | | 5.25 | 9.30 | | 14.55 | 20 |
| 1200 | .010" thick | G | | 37 | .216 | | 6.70 | 9.30 | | 16 | 21.50 |
| 1300 | Clear reinforced, fire retardant, .008" thick | G | | 37 | .216 | | 10.15 | 9.30 | | 19.45 | 25.50 |
| 1350 | Cross laminated type, .003" thick | G | | 37 | .216 | | 7.10 | 9.30 | | 16.40 | 22 |
| 1400 | .004" thick | G | | 37 | .216 | | 7.80 | 9.30 | | 17.10 | 23 |
| 1800 | Reinf. waterproof, .002" polyethylene backing, 1 side | | | 37 | .216 | | 5.65 | 9.30 | | 14.95 | 20.50 |

# 07 26 Vapor Retarders

## 07 26 10 – Above-Grade Vapor Retarders

| 07 26 10.10 Vapor Retarders | Crew | Daily Output | Labor-Hours | Unit | Material | 2011 Bare Costs Labor | Equipment | Total | Total Incl O&P |
|---|---|---|---|---|---|---|---|---|---|
| 1900 | 2 sides | 1 Carp | 37 | .216 | Sq. | 7.45 | 9.30 | | 16.75 | 22.50 |
| 2400 | Waterproofed kraft with sisal or fiberglass fibers, minimum | | 37 | .216 | | 6.10 | 9.30 | | 15.40 | 21 |
| 2500 | Maximum | | 37 | .216 | | 15.10 | 9.30 | | 24.40 | 31 |

# 07 31 Shingles and Shakes

## 07 31 13 – Asphalt Shingles

### 07 31 13.10 Asphalt Roof Shingles

| | | Crew | Daily Output | Labor-Hours | Unit | Material | 2011 Bare Costs Labor | Equipment | Total | Total Incl O&P |
|---|---|---|---|---|---|---|---|---|---|---|
| 0010 | **ASPHALT ROOF SHINGLES** | | | | | | | | | |
| 0100 | Standard strip shingles | | | | | | | | | |
| 0150 | Inorganic, class A, 210-235 lb./sq. | 1 Rofc | 5.50 | 1.455 | Sq. | 74 | 53 | | 127 | 170 |
| 0155 | Pneumatic nailed | | 7 | 1.143 | | 74 | 42 | | 116 | 151 |
| 0200 | Organic, class C, 235-240 lb./sq. | | 5 | 1.600 | | 72.50 | 58.50 | | 131 | 178 |
| 0205 | Pneumatic nailed | | 6.25 | 1.280 | | 72.50 | 47 | | 119.50 | 158 |
| 0250 | Standard, laminated multi-layered shingles | | | | | | | | | |
| 0300 | Class A, 240-260 lb./sq. | 1 Rofc | 4.50 | 1.778 | Sq. | 97.50 | 65 | | 162.50 | 215 |
| 0305 | Pneumatic nailed | | 5.63 | 1.422 | | 97.50 | 52 | | 149.50 | 194 |
| 0350 | Class C, 260-300 lb/square, 4 bundles/square | | 4 | 2 | | 120 | 73 | | 193 | 254 |
| 0355 | Pneumatic nailed | | 5 | 1.600 | | 120 | 58.50 | | 178.50 | 230 |
| 0400 | Premium, laminated multi-layered shingles | | | | | | | | | |
| 0450 | Class A, 260-300 lb, 4 bundles/sq | 1 Rofc | 3.50 | 2.286 | Sq. | 157 | 83.50 | | 240.50 | 310 |
| 0455 | Pneumatic nailed | | 4.37 | 1.831 | | 157 | 67 | | 224 | 284 |
| 0500 | Class C, 300-385 lb/square, 5 bundles/square | | 3 | 2.667 | | 150 | 97.50 | | 247.50 | 325 |
| 0505 | Pneumatic nailed | | 3.75 | 2.133 | | 150 | 78 | | 228 | 295 |
| 0800 | #15 felt underlayment | | 64 | .125 | | 5.65 | 4.57 | | 10.22 | 13.80 |
| 0825 | #30 felt underlayment | | 58 | .138 | | 9.45 | 5.05 | | 14.50 | 18.75 |
| 0850 | Self adhering polyethylene and rubberized asphalt underlayment | | 22 | .364 | | 70 | 13.30 | | 83.30 | 99 |
| 0900 | Ridge shingles | | 330 | .024 | L.F. | 1.90 | .89 | | 2.79 | 3.57 |
| 0905 | Pneumatic nailed | | 412.50 | .019 | " | 1.90 | .71 | | 2.61 | 3.27 |
| 1000 | For steep roofs (7 to 12 pitch or greater), add | | | | | | 50% | | | |

## 07 31 16 – Metal Shingles

### 07 31 16.10 Aluminum Shingles

| | | Crew | Daily Output | Labor-Hours | Unit | Material | 2011 Bare Costs Labor | Equipment | Total | Total Incl O&P |
|---|---|---|---|---|---|---|---|---|---|---|
| 0010 | **ALUMINUM SHINGLES** | | | | | | | | | |
| 0020 | Mill finish, .019 thick | 1 Carp | 5 | 1.600 | Sq. | 191 | 69 | | 260 | 315 |
| 0100 | .020" thick | " | 5 | 1.600 | | 214 | 69 | | 283 | 340 |
| 0300 | For colors, add | | | | | 20 | | | 20 | 22 |
| 0600 | Ridge cap, .024" thick | 1 Carp | 170 | .047 | L.F. | 2.18 | 2.03 | | 4.21 | 5.50 |
| 0700 | End wall flashing, .024" thick | | 170 | .047 | | 1.78 | 2.03 | | 3.81 | 5.05 |
| 0900 | Valley section, .024" thick | | 170 | .047 | | 2.90 | 2.03 | | 4.93 | 6.30 |
| 1000 | Starter strip, .024" thick | | 400 | .020 | | 1.41 | .86 | | 2.27 | 2.87 |
| 1200 | Side wall flashing, .024" thick | | 170 | .047 | | 1.78 | 2.03 | | 3.81 | 5.05 |
| 1500 | Gable flashing, .024" thick | | 400 | .020 | | 1.41 | .86 | | 2.27 | 2.87 |

### 07 31 16.20 Steel Shingles

| | | Crew | Daily Output | Labor-Hours | Unit | Material | 2011 Bare Costs Labor | Equipment | Total | Total Incl O&P |
|---|---|---|---|---|---|---|---|---|---|---|
| 0010 | **STEEL SHINGLES** | | | | | | | | | |
| 0012 | Galvanized, 26 gauge | 1 Rots | 2.20 | 3.636 | Sq. | 250 | 134 | | 384 | 500 |
| 0200 | 24 gauge | " | 2.20 | 3.636 | | 265 | 134 | | 399 | 515 |
| 0300 | For colored galvanized shingles, add | | | | | 55 | | | 55 | 60.50 |
| 0500 | For 1" factory applied polystyrene insulation, add | | | | | 40 | | | 40 | 44 |

# 07 31 Shingles and Shakes

## 07 31 26 – Slate Shingles

### 07 31 26.10 Slate Roof Shingles

| | | Crew | Daily Output | Labor-Hours | Unit | Material | 2011 Bare Costs Labor | Equipment | Total | Total Incl O&P |
|---|---|---|---|---|---|---|---|---|---|---|
| 0010 | **SLATE ROOF SHINGLES** R073126-20 | | | | | | | | | |
| 0100 | Buckingham Virginia black, 3/16" - 1/4" thick  G | 1 Rots | 1.75 | 4.571 | Sq. | 475 | 168 | | 643 | 805 |
| 0200 | 1/4" thick  G | | 1.75 | 4.571 | | 475 | 168 | | 643 | 805 |
| 0900 | Pennsylvania black, Bangor, #1 clear  G | | 1.75 | 4.571 | | 490 | 168 | | 658 | 820 |
| 1200 | Vermont, unfading, green, mottled green  G | | 1.75 | 4.571 | | 480 | 168 | | 648 | 805 |
| 1300 | Semi-weathering green & gray  G | | 1.75 | 4.571 | | 350 | 168 | | 518 | 660 |
| 1400 | Purple  G | | 1.75 | 4.571 | | 425 | 168 | | 593 | 745 |
| 1500 | Black or gray  G | | 1.75 | 4.571 | | 460 | 168 | | 628 | 785 |
| 1600 | Red  G | | 1.75 | 4.571 | | 1,175 | 168 | | 1,343 | 1,550 |
| 1700 | Variegated purple | | 1.75 | 4.571 | | 415 | 168 | | 583 | 735 |

## 07 31 29 – Wood Shingles and Shakes

### 07 31 29.13 Wood Shingles

| | | Crew | Daily Output | Labor-Hours | Unit | Material | 2011 Bare Costs Labor | Equipment | Total | Total Incl O&P |
|---|---|---|---|---|---|---|---|---|---|---|
| 0010 | **WOOD SHINGLES** | | | | | | | | | |
| 0012 | 16" No. 1 red cedar shingles, 5" exposure, on roof | 1 Carp | 2.50 | 3.200 | Sq. | 215 | 138 | | 353 | 445 |
| 0015 | Pneumatic nailed | | 3.25 | 2.462 | | 215 | 106 | | 321 | 400 |
| 0200 | 7-1/2" exposure, on walls | | 2.05 | 3.902 | | 143 | 168 | | 311 | 415 |
| 0205 | Pneumatic nailed | | 2.67 | 2.996 | | 143 | 129 | | 272 | 355 |
| 0300 | 18" No. 1 red cedar perfections, 5-1/2" exposure, on roof | | 2.75 | 2.909 | | 220 | 125 | | 345 | 435 |
| 0305 | Pneumatic nailed | | 3.57 | 2.241 | | 220 | 96.50 | | 316.50 | 390 |
| 0500 | 7-1/2" exposure, on walls | | 2.25 | 3.556 | | 162 | 153 | | 315 | 415 |
| 0505 | Pneumatic nailed | | 2.92 | 2.740 | | 162 | 118 | | 280 | 360 |
| 0600 | Resquared, and rebutted, 5-1/2" exposure, on roof | | 3 | 2.667 | | 238 | 115 | | 353 | 440 |
| 0605 | Pneumatic nailed | | 3.90 | 2.051 | | 238 | 88.50 | | 326.50 | 395 |
| 0900 | 7-1/2" exposure, on walls | | 2.45 | 3.265 | | 175 | 141 | | 316 | 410 |
| 0905 | Pneumatic nailed | | 3.18 | 2.516 | | 175 | 108 | | 283 | 360 |
| 1000 | Add to above for fire retardant shingles | | | | | 55 | | | 55 | 60.50 |
| 1060 | Preformed ridge shingles | 1 Carp | 400 | .020 | L.F. | 3.60 | .86 | | 4.46 | 5.30 |
| 2000 | White cedar shingles, 16" long, extras, 5" exposure, on roof | | 2.40 | 3.333 | Sq. | 149 | 144 | | 293 | 385 |
| 2005 | Pneumatic nailed | | 3.12 | 2.564 | | 149 | 110 | | 259 | 335 |
| 2050 | 5" exposure on walls | | 2 | 4 | | 149 | 172 | | 321 | 430 |
| 2055 | Pneumatic nailed | | 2.60 | 3.077 | | 149 | 132 | | 281 | 365 |
| 2100 | 7-1/2" exposure, on walls | | 2 | 4 | | 106 | 172 | | 278 | 380 |
| 2105 | Pneumatic nailed | | 2.60 | 3.077 | | 106 | 132 | | 238 | 320 |
| 2150 | "B" grade, 5" exposure on walls | | 2 | 4 | | 136 | 172 | | 308 | 415 |
| 2155 | Pneumatic nailed | | 2.60 | 3.077 | | 136 | 132 | | 268 | 350 |
| 2300 | For 15# organic felt underlayment on roof, 1 layer, add | | 64 | .125 | | 5.65 | 5.40 | | 11.05 | 14.45 |
| 2400 | 2 layers, add | | 32 | .250 | | 11.25 | 10.75 | | 22 | 29 |
| 2600 | For steep roofs (7/12 pitch or greater), add to above | | | | | | 50% | | | |
| 2700 | Panelized systems, No.1 cedar shingles on 5/16" CDX plywood | | | | | | | | | |
| 2800 | On walls, 8' strips, 7" or 14" exposure | 2 Carp | 700 | .023 | S.F. | 5.55 | .98 | | 6.53 | 7.60 |
| 3500 | On roofs, 8' strips, 7" or 14" exposure | 1 Carp | 3 | 2.667 | Sq. | 555 | 115 | | 670 | 785 |
| 3505 | Pneumatic nailed | " | 4 | 2 | " | 555 | 86 | | 641 | 740 |

### 07 31 29.16 Wood Shakes

| | | Crew | Daily Output | Labor-Hours | Unit | Material | 2011 Bare Costs Labor | Equipment | Total | Total Incl O&P |
|---|---|---|---|---|---|---|---|---|---|---|
| 0010 | **WOOD SHAKES** | | | | | | | | | |
| 1100 | Hand-split red cedar shakes, 1/2" thick x 24" long, 10" exp. on roof | 1 Carp | 2.50 | 3.200 | Sq. | 233 | 138 | | 371 | 465 |
| 1105 | Pneumatic nailed | | 3.25 | 2.462 | | 233 | 106 | | 339 | 420 |
| 1110 | 3/4" thick x 24" long, 10" exp. on roof | | 2.25 | 3.556 | | 233 | 153 | | 386 | 490 |
| 1115 | Pneumatic nailed | | 2.92 | 2.740 | | 233 | 118 | | 351 | 435 |
| 1200 | 1/2" thick, 18" long, 8-1/2" exp. on roof | | 2 | 4 | | 159 | 172 | | 331 | 440 |
| 1205 | Pneumatic nailed | | 2.60 | 3.077 | | 159 | 132 | | 291 | 380 |
| 1210 | 3/4" thick x 18" long, 8 1/2" exp. on roof | | 1.80 | 4.444 | | 159 | 191 | | 350 | 470 |

# 07 31 Shingles and Shakes

## 07 31 29 – Wood Shingles and Shakes

### 07 31 29.16 Wood Shakes

| | | Crew | Daily Output | Labor-Hours | Unit | Material | 2011 Bare Costs Labor | Equipment | Total | Total Incl O&P |
|---|---|---|---|---|---|---|---|---|---|---|
| 1215 | Pneumatic nailed | 1 Carp | 2.34 | 3.419 | Sq. | 159 | 147 | | 306 | 400 |
| 1255 | 10" exp. on walls | | 2 | 4 | | 154 | 172 | | 326 | 435 |
| 1260 | 10" exposure on walls, pneumatic nailed | ↓ | 2.60 | 3.077 | | 154 | 132 | | 286 | 370 |
| 1700 | Add to above for fire retardant shakes, 24" long | | | | | 55 | | | 55 | 60.50 |
| 1800 | 18" long | | | | ↓ | 55 | | | 55 | 60.50 |
| 1810 | Ridge shakes | 1 Carp | 350 | .023 | L.F. | 3.60 | .98 | | 4.58 | 5.45 |

# 07 32 Roof Tiles

## 07 32 13 – Clay Roof Tiles

### 07 32 13.10 Clay Tiles

| | | | Crew | Daily Output | Labor-Hours | Unit | Material | 2011 Bare Costs Labor | Equipment | Total | Total Incl O&P |
|---|---|---|---|---|---|---|---|---|---|---|---|
| 0010 | **CLAY TILES** | | | | | | | | | | |
| 0200 | Lanai tile or Classic tile, 158 pc per sq | G | 1 Rots | 1.65 | 4.848 | Sq. | 200 | 178 | | 378 | 515 |
| 0300 | Americana, 158 pc per sq, most colors | G | | 1.65 | 4.848 | | 440 | 178 | | 618 | 780 |
| 0350 | Green, gray or brown | G | | 1.65 | 4.848 | | 440 | 178 | | 618 | 780 |
| 0400 | Blue | G | | 1.65 | 4.848 | | 440 | 178 | | 618 | 780 |
| 0600 | Spanish tile, 171 pc per sq, red | G | | 1.80 | 4.444 | | 325 | 163 | | 488 | 630 |
| 0800 | Buff, green, gray, brown | G | | 1.80 | 4.444 | | 550 | 163 | | 713 | 875 |
| 0900 | Glazed white | G | | 1.80 | 4.444 | | 630 | 163 | | 793 | 965 |
| 1100 | Mission tile, 192 pc per sq, machine scored finish, red | G | | 1.15 | 6.957 | | 740 | 256 | | 996 | 1,250 |
| 1700 | French tile, 133 pc per sq, smooth finish, red | G | | 1.35 | 5.926 | | 675 | 218 | | 893 | 1,100 |
| 1750 | Blue or green | G | | 1.35 | 5.926 | | 870 | 218 | | 1,088 | 1,325 |
| 1800 | Norman black 317 pc per sq | G | | 1 | 8 | | 1,050 | 294 | | 1,344 | 1,650 |
| 2200 | Williamsburg tile, 158 pc per sq, aged cedar | G | | 1.35 | 5.926 | | 660 | 218 | | 878 | 1,100 |
| 2250 | Gray or green | G | ↓ | 1.35 | 5.926 | | 595 | 218 | | 813 | 1,025 |
| 3010 | Clay tile, #15 felt underlayment | | 1 Rofc | 64 | .125 | | 5.65 | 4.57 | | 10.22 | 13.80 |
| 3020 | Clay tile, #30 felt underlayment | | | 58 | .138 | | 9.45 | 5.05 | | 14.50 | 18.75 |
| 3040 | Clay tile, polyethylene and rubberized asph. underlayment | | ↓ | 22 | .364 | ↓ | 70 | 13.30 | | 83.30 | 99 |

## 07 32 16 – Concrete Roof Tiles

### 07 32 16.10 Concrete Tiles

| | | Crew | Daily Output | Labor-Hours | Unit | Material | 2011 Bare Costs Labor | Equipment | Total | Total Incl O&P |
|---|---|---|---|---|---|---|---|---|---|---|---|
| 0010 | **CONCRETE TILES** | | | | | | | | | |
| 0020 | Corrugated, 13" x 16-1/2", 90 per sq, 950 lb per sq | | | | | | | | | |
| 0050 | Earthtone colors, nailed to wood deck | 1 Rots | 1.35 | 5.926 | Sq. | 97.50 | 218 | | 315.50 | 470 |
| 0150 | Blues | | 1.35 | 5.926 | | 101 | 218 | | 319 | 475 |
| 0200 | Greens | | 1.35 | 5.926 | | 101 | 218 | | 319 | 475 |
| 0250 | Premium colors | ↓ | 1.35 | 5.926 | ↓ | 101 | 218 | | 319 | 475 |
| 0500 | Shakes, 13" x 16-1/2", 90 per sq, 950 lb per sq | | | | | | | | | |
| 0600 | All colors, nailed to wood deck | 1 Rots | 1.50 | 5.333 | Sq. | 260 | 196 | | 456 | 610 |
| 1500 | Accessory pieces, ridge & hip, 10" x 16-1/2", 8 lbs. each | " | 120 | .067 | Ea. | 3.64 | 2.45 | | 6.09 | 8.10 |
| 1700 | Rake, 6-1/2" x 16-3/4", 9 lbs. each | | | | | 3.64 | | | 3.64 | 4 |
| 1800 | Mansard hip, 10" x 16-1/2", 9.2 lbs. each | | | | | 3.64 | | | 3.64 | 4 |
| 1900 | Hip starter, 10" x 16-1/2", 10.5 lbs. each | | | | | 11.45 | | | 11.45 | 12.60 |
| 2000 | 3 or 4 way apex, 10" each side, 11.5 lbs. each | | | | ↓ | 13.25 | | | 13.25 | 14.55 |

## 07 32 19 – Metal Roof Tiles

### 07 32 19.10 Metal Roof Tiles

| | | Crew | Daily Output | Labor-Hours | Unit | Material | 2011 Bare Costs Labor | Equipment | Total | Total Incl O&P |
|---|---|---|---|---|---|---|---|---|---|---|---|
| 0010 | **METAL ROOF TILES** | | | | | | | | | |
| 0020 | Accessories included, .032" thick aluminum, mission tile | 1 Carp | 2.50 | 3.200 | Sq. | 815 | 138 | | 953 | 1,100 |
| 0200 | Spanish tiles | " | 3 | 2.667 | " | 560 | 115 | | 675 | 790 |

# 07 33 Natural Roof Coverings

## 07 33 63 – Vegetated Roofing

### 07 33 63.10 Green Roof Systems

| | | Crew | Daily Output | Labor-Hours | Unit | Material | 2011 Bare Costs Labor | Equipment | Total | Total Incl O&P |
|---|---|---|---|---|---|---|---|---|---|---|
| 0010 | **GREEN ROOF SYSTEMS** | | | | | | | | | |
| 0020 | Soil mixture for green roof 30% sand, 55% gravel, 15% soil | G | | | | | | | | |
| 0100 | Hoist and spread soil mixture 4 inch depth up to five stories tall roof | G | B-13B | 4000 | .014 | S.F. | .24 | .52 | .30 | 1.06 | 1.38 |
| 0150 | 6 inch depth | G | | 2667 | .021 | | .36 | .78 | .45 | 1.59 | 2.08 |
| 0200 | 8 inch depth | G | | 2000 | .028 | | .48 | 1.04 | .60 | 2.12 | 2.77 |
| 0250 | 10 inch depth | G | | 1600 | .035 | | .60 | 1.30 | .76 | 2.66 | 3.47 |
| 0300 | 12 inch depth | G | | 1335 | .042 | | .72 | 1.56 | .90 | 3.18 | 4.16 |
| 0310 | Alt. man-made soil mix, hoist & spread, 4" deep up to 5 stories tall roof | G | | 4000 | .014 | | 1.50 | .52 | .30 | 2.32 | 2.77 |
| 0350 | Mobilization 55 ton crane to site | G | 1 Eqhv | 3.60 | 2.222 | Ea. | | 103 | | 103 | 154 |
| 0355 | Hoisting cost to five stories per day (Avg. 28 picks per day) | G | B-13B | 1 | 56 | Day | | 2,075 | 1,200 | 3,275 | 4,500 |
| 0360 | Mobilization or demobilization, 100 ton crane to site driver & escort | G | A-3E | 2.50 | 6.400 | Ea. | | 259 | 46.50 | 305.50 | 440 |
| 0365 | Hoisting cost six to ten stories per day (Avg. 21 picks per day) | G | B-13C | 1 | 56 | Day | | 2,075 | 1,650 | 3,725 | 5,000 |
| 0370 | Hoist and spread soil mixture 4 inch depth six to ten stories tall roof | G | | 4000 | .014 | S.F. | .24 | .52 | .41 | 1.17 | 1.50 |
| 0375 | 6 inch depth | G | | 2667 | .021 | | .36 | .78 | .62 | 1.76 | 2.26 |
| 0380 | 8 inch depth | G | | 2000 | .028 | | .48 | 1.04 | .83 | 2.35 | 3.02 |
| 0385 | 10 inch depth | G | | 1600 | .035 | | .60 | 1.30 | 1.03 | 2.93 | 3.78 |
| 0390 | 12 inch depth | G | | 1335 | .042 | | .72 | 1.56 | 1.24 | 3.52 | 4.52 |
| 0400 | Green roof edging treated lumber 4" x 4" no hoisting included | G | 2 Carp | 400 | .040 | L.F. | 1.37 | 1.72 | | 3.09 | 4.14 |
| 0410 | 4" x 6" | G | | 400 | .040 | | 2.46 | 1.72 | | 4.18 | 5.35 |
| 0420 | 4" x 8" | G | | 360 | .044 | | 4.74 | 1.91 | | 6.65 | 8.15 |
| 0430 | 4" x 6" double stacked | G | | 300 | .053 | | 4.93 | 2.30 | | 7.23 | 8.90 |
| 0500 | Green roof edging redwood lumber 4" x 4" no hoisting included | G | | 400 | .040 | | 6.40 | 1.72 | | 8.12 | 9.70 |
| 0510 | 4" x 6" | G | | 400 | .040 | | 11.20 | 1.72 | | 12.92 | 14.95 |
| 0520 | 4" x 8" | G | | 360 | .044 | | 20.50 | 1.91 | | 22.41 | 25.50 |
| 0530 | 4" x 6" double stacked | G | | 300 | .053 | | 22.50 | 2.30 | | 24.80 | 28 |
| 0550 | Roof membrane and root barrier, not including insulation: | G | | | | | | | | | |
| 0560 | Fluid applied rubber membrane, reinforced, 215 mil thick | G | G-5 | 350 | .114 | S.F. | .26 | 3.79 | .48 | 4.53 | 7.10 |
| 0570 | Root barrier | G | 2 Rofc | 775 | .021 | | .70 | .75 | | 1.45 | 2.03 |
| 0580 | Moisture retention barrier and reservoir | G | " | 900 | .018 | | 2.70 | .65 | | 3.35 | 4.05 |
| 0600 | Planting sedum, light soil, potted, 2-1/4" diameter, two per SF | G | 1 Clab | 420 | .019 | | 4.50 | .65 | | 5.15 | 5.95 |
| 0610 | one per SF | G | " | 840 | .010 | | 2.25 | .33 | | 2.58 | 2.98 |
| 0630 | Planting sedum mat per SF including shipping (4000 SF Minimum) | G | 4 Clab | 4000 | .008 | | 5.40 | .27 | | 5.67 | 6.35 |
| 0640 | Installation sedum mat system (no soil required) per SF (4000 SF minimum) | G | " | 4000 | .008 | | 7.50 | .27 | | 7.77 | 8.65 |
| 0645 | Note: pricing of sedum mats shipped in full truck loads (4000-5000 SF) | G | | | | | | | | | |

# 07 41 Roof Panels

## 07 41 13 – Metal Roof Panels

### 07 41 13.10 Aluminum Roof Panels

| | | Crew | Daily Output | Labor-Hours | Unit | Material | 2011 Bare Costs Labor | Equipment | Total | Total Incl O&P |
|---|---|---|---|---|---|---|---|---|---|---|
| 0010 | **ALUMINUM ROOF PANELS** | | | | | | | | | |
| 0020 | Corrugated or ribbed, .0155" thick, natural | G-3 | 1200 | .027 | S.F. | .96 | 1.15 | | 2.11 | 2.80 |
| 0300 | Painted | | 1200 | .027 | | 1.40 | 1.15 | | 2.55 | 3.28 |
| 0400 | Corrugated, .018" thick, on steel frame, natural finish | | 1200 | .027 | | 1.27 | 1.15 | | 2.42 | 3.14 |
| 0600 | Painted | | 1200 | .027 | | 1.55 | 1.15 | | 2.70 | 3.45 |
| 0700 | Corrugated, on steel frame, natural, .024" thick | | 1200 | .027 | | 1.80 | 1.15 | | 2.95 | 3.72 |
| 0800 | Painted | | 1200 | .027 | | 2.19 | 1.15 | | 3.34 | 4.15 |
| 0900 | .032" thick, natural | | 1200 | .027 | | 2.32 | 1.15 | | 3.47 | 4.29 |
| 1200 | Painted | | 1200 | .027 | | 2.97 | 1.15 | | 4.12 | 5 |
| 1300 | V-Beam, on steel frame construction, .032" thick, natural | | 1200 | .027 | | 2.45 | 1.15 | | 3.60 | 4.44 |
| 1500 | Painted | | 1200 | .027 | | 3.01 | 1.15 | | 4.16 | 5.05 |
| 1600 | .040" thick, natural | | 1200 | .027 | | 2.99 | 1.15 | | 4.14 | 5.05 |
| 1800 | Painted | | 1200 | .027 | | 3.59 | 1.15 | | 4.74 | 5.70 |

# 07 41 Roof Panels

## 07 41 13 – Metal Roof Panels

| 07 41 13.10 Aluminum Roof Panels | Crew | Daily Output | Labor-Hours | Unit | Material | 2011 Bare Costs Labor | Equipment | Total | Total Incl O&P |
|---|---|---|---|---|---|---|---|---|---|
| 1900 | .050" thick, natural | G-3 | 1200 | .027 | S.F. | 3.58 | 1.15 | | 4.73 | 5.70 |
| 2100 | Painted | | 1200 | .027 | | 4.33 | 1.15 | | 5.48 | 6.50 |
| 2200 | For roofing on wood frame, deduct | | 4600 | .007 | | .08 | .30 | | .38 | .55 |
| 2400 | Ridge cap, .032" thick, natural | | 800 | .040 | L.F. | 2.83 | 1.72 | | 4.55 | 5.75 |

### 07 41 13.20 Steel Roofing Panels

| | | Crew | Daily Output | Labor-Hours | Unit | Material | Labor | Equipment | Total | Total Incl O&P |
|---|---|---|---|---|---|---|---|---|---|---|
| 0010 | **STEEL ROOFING PANELS** | | | | | | | | | |
| 0012 | Corrugated or ribbed, on steel framing, 30 ga galv | G-3 | 1100 | .029 | S.F. | 1.68 | 1.25 | | 2.93 | 3.75 |
| 0100 | 28 ga | | 1050 | .030 | | 1.79 | 1.31 | | 3.10 | 3.96 |
| 0300 | 26 ga | | 1000 | .032 | | 1.85 | 1.38 | | 3.23 | 4.13 |
| 0400 | 24 ga | | 950 | .034 | | 2.91 | 1.45 | | 4.36 | 5.40 |
| 0600 | Colored, 28 ga | | 1050 | .030 | | 1.68 | 1.31 | | 2.99 | 3.84 |
| 0700 | 26 ga | | 1000 | .032 | | 1.99 | 1.38 | | 3.37 | 4.28 |
| 0710 | Flat profile, 1-3/4" standing seams, 10" wide, standard finish, 26 ga | | 1000 | .032 | | 3.89 | 1.38 | | 5.27 | 6.35 |
| 0715 | 24 ga | | 950 | .034 | | 4.51 | 1.45 | | 5.96 | 7.15 |
| 0720 | 22 ga | | 900 | .036 | | 5.55 | 1.53 | | 7.08 | 8.50 |
| 0725 | Zinc aluminum alloy finish, 26 ga | | 1000 | .032 | | 3.05 | 1.38 | | 4.43 | 5.45 |
| 0730 | 24 ga | | 950 | .034 | | 3.64 | 1.45 | | 5.09 | 6.20 |
| 0735 | 22 ga | | 900 | .036 | | 4.17 | 1.53 | | 5.70 | 6.90 |
| 0740 | 12" wide, standard finish, 26 ga | | 1000 | .032 | | 3.88 | 1.38 | | 5.26 | 6.35 |
| 0745 | 24 ga | | 950 | .034 | | 5.10 | 1.45 | | 6.55 | 7.80 |
| 0750 | Zinc aluminum alloy finish, 26 ga | | 1000 | .032 | | 4.41 | 1.38 | | 5.79 | 6.95 |
| 0755 | 24 ga | | 950 | .034 | | 3.63 | 1.45 | | 5.08 | 6.20 |
| 0840 | Flat profile, 1" x 3/8" batten, 12" wide, standard finish, 26 ga | | 1000 | .032 | | 3.42 | 1.38 | | 4.80 | 5.85 |
| 0845 | 24 ga | | 950 | .034 | | 4.01 | 1.45 | | 5.46 | 6.60 |
| 0850 | 22 ga | | 900 | .036 | | 4.80 | 1.53 | | 6.33 | 7.65 |
| 0855 | Zinc aluminum alloy finish, 26 ga | | 1000 | .032 | | 3.28 | 1.38 | | 4.66 | 5.70 |
| 0860 | 24 ga | | 950 | .034 | | 3.66 | 1.45 | | 5.11 | 6.25 |
| 0865 | 22 ga | | 900 | .036 | | 4.23 | 1.53 | | 5.76 | 7 |
| 0870 | 16-1/2" wide, standard finish, 24 ga | | 950 | .034 | | 3.95 | 1.45 | | 5.40 | 6.55 |
| 0875 | 22 ga | | 900 | .036 | | 4.42 | 1.53 | | 5.95 | 7.20 |
| 0880 | Zinc aluminum alloy finish, 24 ga | | 950 | .034 | | 3.45 | 1.45 | | 4.90 | 6 |
| 0885 | 22 ga | | 900 | .036 | | 3.85 | 1.53 | | 5.38 | 6.55 |
| 0890 | Flat profile, 2" x 2" batten, 12" wide, standard finish, 26 ga | | 1000 | .032 | | 3.93 | 1.38 | | 5.31 | 6.40 |
| 0895 | 24 ga | | 950 | .034 | | 4.69 | 1.45 | | 6.14 | 7.35 |
| 0900 | 22 ga | | 900 | .036 | | 5.75 | 1.53 | | 7.28 | 8.70 |
| 0905 | Zinc aluminum alloy finish, 26 ga | | 1000 | .032 | | 3.66 | 1.38 | | 5.04 | 6.10 |
| 0910 | 24 ga | | 950 | .034 | | 4.18 | 1.45 | | 5.63 | 6.80 |
| 0915 | 22 ga | | 900 | .036 | | 4.87 | 1.53 | | 6.40 | 7.70 |
| 0920 | 16-1/2" wide, standard finish, 24 ga | | 950 | .034 | | 4.32 | 1.45 | | 5.77 | 6.95 |
| 0925 | 22 ga | | 900 | .036 | | 5.05 | 1.53 | | 6.58 | 7.90 |
| 0930 | Zinc aluminum alloy finish, 24 ga | | 950 | .034 | | 3.91 | 1.45 | | 5.36 | 6.50 |
| 0935 | 22 ga | | 900 | .036 | | 4.45 | 1.53 | | 5.98 | 7.25 |
| 1200 | Ridge, galvanized, 10" wide | | 800 | .040 | L.F. | 3.18 | 1.72 | | 4.90 | 6.10 |
| 1210 | 20" wide | | 750 | .043 | " | 4.93 | 1.83 | | 6.76 | 8.20 |

## 07 41 33 – Plastic Roof Panels

### 07 41 33.10 Fiberglass Panels

| | | Crew | Daily Output | Labor-Hours | Unit | Material | Labor | Equipment | Total | Total Incl O&P |
|---|---|---|---|---|---|---|---|---|---|---|
| 0010 | **FIBERGLASS PANELS** | | | | | | | | | |
| 0012 | Corrugated panels, roofing, 8 oz per S.F. | G-3 | 1000 | .032 | S.F. | 1.74 | 1.38 | | 3.12 | 4 |
| 0100 | 12 oz per S.F. | | 1000 | .032 | | 3.44 | 1.38 | | 4.82 | 5.85 |
| 0300 | Corrugated siding, 6 oz per S.F. | | 880 | .036 | | 1.53 | 1.56 | | 3.09 | 4.06 |
| 0400 | 8 oz per S.F. | | 880 | .036 | | 1.74 | 1.56 | | 3.30 | 4.29 |
| 0500 | Fire retardant | | 880 | .036 | | 3.20 | 1.56 | | 4.76 | 5.90 |

# 07 41 Roof Panels

## 07 41 33 – Plastic Roof Panels

| 07 41 33.10 Fiberglass Panels | | Crew | Daily Output | Labor-Hours | Unit | Material | 2011 Bare Costs Labor | Equipment | Total | Total Incl O&P |
|---|---|---|---|---|---|---|---|---|---|---|
| 0600 | 12 oz. siding, textured | G-3 | 880 | .036 | S.F. | 3.31 | 1.56 | | 4.87 | 6 |
| 0700 | Fire retardant | | 880 | .036 | | 4.40 | 1.56 | | 5.96 | 7.20 |
| 0900 | Flat panels, 6 oz per S.F., clear or colors | | 880 | .036 | | 1.95 | 1.56 | | 3.51 | 4.53 |
| 1100 | Fire retardant, class A | | 880 | .036 | | 3.18 | 1.56 | | 4.74 | 5.90 |
| 1300 | 8 oz per S.F., clear or colors | | 880 | .036 | | 2.15 | 1.56 | | 3.71 | 4.75 |
| 1700 | Sandwich panels, fiberglass, 1-9/16" thick, panels to 20 S.F. | | 180 | .178 | | 23 | 7.65 | | 30.65 | 37 |
| 1900 | As above, but 2-3/4" thick, panels to 100 S.F. | ↓ | 265 | .121 | ↓ | 17 | 5.20 | | 22.20 | 26.50 |

# 07 42 Wall Panels

## 07 42 13 – Metal Wall Panels

### 07 42 13.10 Mansard Panels

| | | Crew | Daily Output | Labor-Hours | Unit | Material | 2011 Bare Costs Labor | Equipment | Total | Total Incl O&P |
|---|---|---|---|---|---|---|---|---|---|---|
| 0010 | **MANSARD PANELS** | | | | | | | | | |
| 0600 | Aluminum, stock units, straight surfaces | 1 Shee | 115 | .070 | S.F. | 3.35 | 3.59 | | 6.94 | 9.15 |
| 0700 | Concave or convex surfaces | | 75 | .107 | " | 3.67 | 5.50 | | 9.17 | 12.40 |
| 0800 | For framing, to 5' high, add | | 115 | .070 | L.F. | 3.67 | 3.59 | | 7.26 | 9.50 |
| 0900 | Soffits, to 1' wide | ↓ | 125 | .064 | S.F. | 1.82 | 3.31 | | 5.13 | 7 |

### 07 42 13.20 Aluminum Siding Panels

| | | Crew | Daily Output | Labor-Hours | Unit | Material | 2011 Bare Costs Labor | Equipment | Total | Total Incl O&P |
|---|---|---|---|---|---|---|---|---|---|---|
| 0010 | **ALUMINUM SIDING PANELS** | | | | | | | | | |
| 0012 | Corrugated, on steel framing, .019 thick, natural finish | G-3 | 775 | .041 | S.F. | 1.47 | 1.78 | | 3.25 | 4.32 |
| 0100 | Painted | | 775 | .041 | | 1.58 | 1.78 | | 3.36 | 4.44 |
| 0400 | Farm type, .021" thick on steel frame, natural | | 775 | .041 | | 1.48 | 1.78 | | 3.26 | 4.33 |
| 0600 | Painted | | 775 | .041 | | 1.58 | 1.78 | | 3.36 | 4.44 |
| 0700 | Industrial type, corrugated, on steel, .024" thick, mill | | 775 | .041 | | 2.06 | 1.78 | | 3.84 | 4.97 |
| 0900 | Painted | | 775 | .041 | | 2.22 | 1.78 | | 4 | 5.15 |
| 1000 | .032" thick, mill | | 775 | .041 | | 2.37 | 1.78 | | 4.15 | 5.30 |
| 1200 | Painted | | 775 | .041 | | 2.89 | 1.78 | | 4.67 | 5.90 |
| 1300 | V-Beam, on steel frame, .032" thick, mill | | 775 | .041 | | 2.67 | 1.78 | | 4.45 | 5.65 |
| 1500 | Painted | | 775 | .041 | | 3.06 | 1.78 | | 4.84 | 6.05 |
| 1600 | .040" thick, mill | | 775 | .041 | | 3.24 | 1.78 | | 5.02 | 6.25 |
| 1800 | Painted | | 775 | .041 | | 3.80 | 1.78 | | 5.58 | 6.90 |
| 1900 | .050" thick, mill | | 775 | .041 | | 3.83 | 1.78 | | 5.61 | 6.90 |
| 2100 | Painted | | 775 | .041 | | 4.57 | 1.78 | | 6.35 | 7.75 |
| 2200 | Ribbed, 3" profile, on steel frame, .032" thick, natural | | 775 | .041 | | 2.31 | 1.78 | | 4.09 | 5.25 |
| 2400 | Painted | | 775 | .041 | | 2.93 | 1.78 | | 4.71 | 5.90 |
| 2500 | .040" thick, natural | | 775 | .041 | | 2.64 | 1.78 | | 4.42 | 5.60 |
| 2700 | Painted | | 775 | .041 | | 3.08 | 1.78 | | 4.86 | 6.10 |
| 2750 | .050" thick, natural | | 775 | .041 | | 3.04 | 1.78 | | 4.82 | 6.05 |
| 2760 | Painted | | 775 | .041 | | 3.49 | 1.78 | | 5.27 | 6.55 |
| 3300 | For siding on wood frame, deduct from above | ↓ | 2800 | .011 | ↓ | .09 | .49 | | .58 | .85 |
| 3400 | Screw fasteners, aluminum, self tapping, neoprene washer, 1" | | | | M | 200 | | | 200 | 220 |
| 3600 | Stitch screws, self tapping, with neoprene washer, 5/8" | | | | " | 150 | | | 150 | 165 |
| 3630 | Flashing, sidewall, .032" thick | G-3 | 800 | .040 | L.F. | 2.77 | 1.72 | | 4.49 | 5.65 |
| 3650 | End wall, .040" thick | | 800 | .040 | | 3.24 | 1.72 | | 4.96 | 6.20 |
| 3670 | Closure strips, corrugated, .032" thick | | 800 | .040 | | .82 | 1.72 | | 2.54 | 3.52 |
| 3680 | Ribbed, 4" or 8", .032" thick | | 800 | .040 | | .82 | 1.72 | | 2.54 | 3.52 |
| 3690 | V-beam, .040" thick | ↓ | 800 | .040 | ↓ | 1.11 | 1.72 | | 2.83 | 3.84 |
| 3800 | Horizontal, colored clapboard, 8" wide, plain | 2 Carp | 515 | .031 | S.F. | 2.17 | 1.34 | | 3.51 | 4.44 |
| 3810 | Insulated | | 515 | .031 | | 2.32 | 1.34 | | 3.66 | 4.60 |
| 4000 | Vertical board & batten, colored, non-insulated | ↓ | 515 | .031 | | 1.65 | 1.34 | | 2.99 | 3.87 |
| 4200 | For simulated wood design, add | | | | ↓ | .10 | | | .10 | .11 |
| 4300 | Corners for above, outside | 2 Carp | 515 | .031 | V.L.F. | 3.19 | 1.34 | | 4.53 | 5.55 |

# 07 42 Wall Panels

## 07 42 13 – Metal Wall Panels

### 07 42 13.20 Aluminum Siding Panels

| | | Crew | Daily Output | Labor-Hours | Unit | Material | 2011 Bare Costs Labor | Equipment | Total | Total Incl O&P |
|---|---|---|---|---|---|---|---|---|---|---|
| 4500 | Inside corners | 2 Carp | 515 | .031 | V.L.F. | 1.39 | 1.34 | | 2.73 | 3.58 |
| 4600 | Sandwich panels, 1" insulation, single story | G-3 | 395 | .081 | S.F. | 7.35 | 3.48 | | 10.83 | 13.35 |
| 4900 | Multi-story | " | 345 | .093 | | 10.30 | 3.99 | | 14.29 | 17.40 |
| 5100 | For baked enamel finish 1 side, add | | | | | .40 | | | .40 | .44 |

### 07 42 13.30 Steel Siding

| | | Crew | Daily Output | Labor-Hours | Unit | Material | 2011 Bare Costs Labor | Equipment | Total | Total Incl O&P |
|---|---|---|---|---|---|---|---|---|---|---|
| 0010 | **STEEL SIDING** | | | | | | | | | |
| 0020 | Beveled, vinyl coated, 8" wide | 1 Carp | 265 | .030 | S.F. | 1.97 | 1.30 | | 3.27 | 4.16 |
| 0050 | 10" wide | " | 275 | .029 | | 2.11 | 1.25 | | 3.36 | 4.24 |
| 0080 | Galv, corrugated or ribbed, on steel frame, 30 gauge | G-3 | 800 | .040 | | 1.32 | 1.72 | | 3.04 | 4.07 |
| 0100 | 28 gauge | | 795 | .040 | | 1.38 | 1.73 | | 3.11 | 4.15 |
| 0300 | 26 gauge | | 790 | .041 | | 1.94 | 1.74 | | 3.68 | 4.78 |
| 0400 | 24 gauge | | 785 | .041 | | 1.95 | 1.75 | | 3.70 | 4.82 |
| 0600 | 22 gauge | | 770 | .042 | | 2.25 | 1.79 | | 4.04 | 5.20 |
| 0700 | Colored, corrugated/ribbed, on steel frame, 10 yr. finish, 28 ga. | | 800 | .040 | | 2.05 | 1.72 | | 3.77 | 4.88 |
| 0900 | 26 gauge | | 795 | .040 | | 2.14 | 1.73 | | 3.87 | 4.98 |
| 1000 | 24 gauge | | 790 | .041 | | 2.49 | 1.74 | | 4.23 | 5.40 |
| 1020 | 20 gauge | | 785 | .041 | | 3.16 | 1.75 | | 4.91 | 6.15 |
| 1200 | Factory sandwich panel, 26 ga., 1" insulation, galvanized | | 380 | .084 | | 5.25 | 3.62 | | 8.87 | 11.30 |
| 1300 | Colored 1 side | | 380 | .084 | | 6.45 | 3.62 | | 10.07 | 12.55 |
| 1500 | Galvanized 2 sides | | 380 | .084 | | 8 | 3.62 | | 11.62 | 14.30 |
| 1600 | Colored 2 sides | | 380 | .084 | | 8.25 | 3.62 | | 11.87 | 14.60 |
| 1800 | Acrylic paint face, regular paint liner | ▼ | 380 | .084 | | 6.25 | 3.62 | | 9.87 | 12.35 |
| 1900 | For 2" thick polystyrene, add | | | | | .96 | | | .96 | 1.06 |
| 2000 | 22 ga, galv, 2" insulation, baked enamel exterior | G-3 | 360 | .089 | | 12.20 | 3.82 | | 16.02 | 19.25 |
| 2100 | Polyvinylidene exterior finish | " | 360 | .089 | | 12.85 | 3.82 | | 16.67 | 19.95 |

# 07 44 Faced Panels

## 07 44 73 – Metal Faced Panels

### 07 44 73.10 Metal Faced Panels and Accessories

| | | Crew | Daily Output | Labor-Hours | Unit | Material | 2011 Bare Costs Labor | Equipment | Total | Total Incl O&P |
|---|---|---|---|---|---|---|---|---|---|---|
| 0010 | **METAL FACED PANELS AND ACCESSORIES** | | | | | | | | | |
| 0400 | Textured aluminum, 4' x 8' x 5/16" plywood backing, single face | 2 Shee | 375 | .043 | S.F. | 3.67 | 2.20 | | 5.87 | 7.35 |
| 0600 | Double face | | 375 | .043 | | 4.99 | 2.20 | | 7.19 | 8.85 |
| 0700 | 4' x 10' x 5/16" plywood backing, single face | | 375 | .043 | | 3.93 | 2.20 | | 6.13 | 7.65 |
| 0900 | Double face | | 375 | .043 | | 5.25 | 2.20 | | 7.45 | 9.15 |
| 1000 | 4' x 12' x 5/16" plywood backing, single face | | 375 | .043 | | 3.33 | 2.20 | | 5.53 | 7 |
| 1300 | Smooth aluminum, 1/4" plywood panel, fluoropolymer finish, double face | | 375 | .043 | | 5.45 | 2.20 | | 7.65 | 9.30 |
| 1350 | Clear anodized finish, double face | | 375 | .043 | | 9.55 | 2.20 | | 11.75 | 13.85 |
| 1400 | Double face textured aluminum, structural panel, 1" EPS insulation | ▼ | 375 | .043 | ▼ | 4.92 | 2.20 | | 7.12 | 8.75 |
| 1500 | Accessories, outside corner | 1 Shee | 175 | .046 | L.F. | 1.90 | 2.36 | | 4.26 | 5.65 |
| 1600 | Inside corner | | 175 | .046 | | 1.34 | 2.36 | | 3.70 | 5.05 |
| 1800 | Batten mounting clip | | 200 | .040 | | .49 | 2.07 | | 2.56 | 3.66 |
| 1900 | Low profile batten | | 480 | .017 | | .61 | .86 | | 1.47 | 1.97 |
| 2100 | High profile batten | | 480 | .017 | | 1.41 | .86 | | 2.27 | 2.85 |
| 2200 | Water table | | 200 | .040 | | 2.12 | 2.07 | | 4.19 | 5.45 |
| 2400 | Horizontal joint connector | | 200 | .040 | | 1.67 | 2.07 | | 3.74 | 4.96 |
| 2500 | Corner cap | | 200 | .040 | | 1.84 | 2.07 | | 3.91 | 5.15 |
| 2700 | H - moulding | ▼ | 480 | .017 | ▼ | 1.25 | .86 | | 2.11 | 2.68 |

# 07 46 Siding

## 07 46 23 – Wood Siding

### 07 46 23.10 Wood Board Siding

| 07 46 23.10 Wood Board Siding | Crew | Daily Output | Labor-Hours | Unit | Material | 2011 Bare Costs Labor | Equipment | Total | Total Incl O&P |
|---|---|---|---|---|---|---|---|---|---|
| 0010 **WOOD BOARD SIDING** | | | | | | | | | |
| 3200 Wood, cedar bevel, A grade, 1/2" x 6" | 1 Carp | 295 | .027 | S.F. | 4.39 | 1.17 | | 5.56 | 6.60 |
| 3300 1/2" x 8" | | 330 | .024 | | 5.85 | 1.04 | | 6.89 | 8 |
| 3500 3/4" x 10", clear grade | | 375 | .021 | | 6.55 | .92 | | 7.47 | 8.60 |
| 3600 "B" grade | | 375 | .021 | | 3.79 | .92 | | 4.71 | 5.60 |
| 3800 Cedar, rough sawn, 1" x 4", A grade, natural | | 220 | .036 | | 3.88 | 1.57 | | 5.45 | 6.65 |
| 3900 Stained | | 220 | .036 | | 4.25 | 1.57 | | 5.82 | 7.10 |
| 4100 1" x 12", board & batten, #3 & Btr., natural | | 420 | .019 | | 4.02 | .82 | | 4.84 | 5.70 |
| 4200 Stained | | 420 | .019 | | 4.25 | .82 | | 5.07 | 5.95 |
| 4400 1" x 8" channel siding, #3 & Btr., natural | | 330 | .024 | | 2.92 | 1.04 | | 3.96 | 4.81 |
| 4500 Stained | | 330 | .024 | | 2.67 | 1.04 | | 3.71 | 4.54 |
| 4700 Redwood, clear, beveled, vertical grain, 1/2" x 4" | | 220 | .036 | | 3.85 | 1.57 | | 5.42 | 6.65 |
| 4750 1/2" x 6" | | 295 | .027 | | 3.85 | 1.17 | | 5.02 | 6.05 |
| 4800 1/2" x 8" | | 330 | .024 | | 4.10 | 1.04 | | 5.14 | 6.10 |
| 5000 3/4" x 10" | | 375 | .021 | | 4.12 | .92 | | 5.04 | 5.95 |
| 5200 Channel siding, 1" x 10", B grade | | 375 | .021 | | 4.12 | .92 | | 5.04 | 5.95 |
| 5250 Redwood, T&G boards, B grade, 1" x 4" | | 220 | .036 | | 2.54 | 1.57 | | 4.11 | 5.20 |
| 5270 1" x 8" | | 330 | .024 | | 4.10 | 1.04 | | 5.14 | 6.10 |
| 5400 White pine, rough sawn, 1" x 8", natural | | 330 | .024 | | 2.20 | 1.04 | | 3.24 | 4.02 |
| 5500 Stained | | 330 | .024 | | 2.60 | 1.04 | | 3.64 | 4.46 |

## 07 46 29 – Plywood Siding

### 07 46 29.10 Plywood Siding Options

| 07 46 29.10 Plywood Siding Options | Crew | Daily Output | Labor-Hours | Unit | Material | 2011 Bare Costs Labor | Equipment | Total | Total Incl O&P |
|---|---|---|---|---|---|---|---|---|---|
| 0010 **PLYWOOD SIDING OPTIONS** | | | | | | | | | |
| 0900 Plywood, medium density overlaid, 3/8" thick | 2 Carp | 750 | .021 | S.F. | 1.25 | .92 | | 2.17 | 2.79 |
| 1000 1/2" thick | | 700 | .023 | | 1.22 | .98 | | 2.20 | 2.85 |
| 1100 3/4" thick | | 650 | .025 | | 1.55 | 1.06 | | 2.61 | 3.34 |
| 1600 Texture 1-11, cedar, 5/8" thick, natural | | 675 | .024 | | 2.43 | 1.02 | | 3.45 | 4.24 |
| 1700 Factory stained | | 675 | .024 | | 1.95 | 1.02 | | 2.97 | 3.72 |
| 1900 Texture 1-11, fir, 5/8" thick, natural | | 675 | .024 | | 1.37 | 1.02 | | 2.39 | 3.08 |
| 2000 Factory stained | | 675 | .024 | | 1.73 | 1.02 | | 2.75 | 3.47 |
| 2050 Texture 1-11, S.Y.P., 5/8" thick, natural | | 675 | .024 | | 1.19 | 1.02 | | 2.21 | 2.88 |
| 2100 Factory stained | | 675 | .024 | | 1.24 | 1.02 | | 2.26 | 2.93 |
| 2200 Rough sawn cedar, 3/8" thick, natural | | 675 | .024 | | 1.20 | 1.02 | | 2.22 | 2.89 |
| 2300 Factory stained | | 675 | .024 | | .96 | 1.02 | | 1.98 | 2.63 |
| 2500 Rough sawn fir, 3/8" thick, natural | | 675 | .024 | | .76 | 1.02 | | 1.78 | 2.41 |
| 2600 Factory stained | | 675 | .024 | | .96 | 1.02 | | 1.98 | 2.63 |
| 2800 Redwood, textured siding, 5/8" thick | | 675 | .024 | | 1.92 | 1.02 | | 2.94 | 3.68 |
| 3000 Polyvinyl chloride coated, 3/8" thick | | 750 | .021 | | 1.11 | .92 | | 2.03 | 2.63 |

## 07 46 33 – Plastic Siding

### 07 46 33.10 Vinyl Siding

| 07 46 33.10 Vinyl Siding | Crew | Daily Output | Labor-Hours | Unit | Material | 2011 Bare Costs Labor | Equipment | Total | Total Incl O&P |
|---|---|---|---|---|---|---|---|---|---|
| 0010 **VINYL SIDING** | | | | | | | | | |
| 3995 Clapboard profile, woodgrain texture, .048 thick, double 4 | 2 Carp | 495 | .032 | S.F. | .87 | 1.39 | | 2.26 | 3.09 |
| 4000 Double 5 | | 550 | .029 | | .87 | 1.25 | | 2.12 | 2.88 |
| 4005 Single 8 | | 495 | .032 | | .94 | 1.39 | | 2.33 | 3.17 |
| 4010 Single 10 | | 550 | .029 | | 1.13 | 1.25 | | 2.38 | 3.16 |
| 4015 .044 thick, double 4 | | 495 | .032 | | .82 | 1.39 | | 2.21 | 3.04 |
| 4020 Double 5 | | 550 | .029 | | .82 | 1.25 | | 2.07 | 2.83 |
| 4025 .042 thick, double 4 | | 495 | .032 | | .82 | 1.39 | | 2.21 | 3.04 |
| 4030 Double 5 | | 550 | .029 | | .82 | 1.25 | | 2.07 | 2.83 |
| 4035 Cross sawn texture, .040 thick, double 4 | | 495 | .032 | | .64 | 1.39 | | 2.03 | 2.84 |
| 4040 Double 5 | | 550 | .029 | | .64 | 1.25 | | 1.89 | 2.63 |
| 4045 Smooth texture, .042 thick, double 4 | | 495 | .032 | | .68 | 1.39 | | 2.07 | 2.88 |

## 07 46 33 – Plastic Siding

| 07 46 33.10 Vinyl Siding | Crew | Daily Output | Labor-Hours | Unit | Material | 2011 Bare Costs Labor | Equipment | Total | Total Incl O&P |
|---|---|---|---|---|---|---|---|---|---|
| 4050 | Double 5 | 2 Carp | 550 | .029 | S.F. | .73 | 1.25 | | 1.98 | 2.73 |
| 4055 | Single 8 | | 495 | .032 | | .94 | 1.39 | | 2.33 | 3.17 |
| 4060 | Cedar texture, .044 thick, double 4 | | 495 | .032 | | .85 | 1.39 | | 2.24 | 3.07 |
| 4065 | Double 6 | | 600 | .027 | | .94 | 1.15 | | 2.09 | 2.80 |
| 4070 | Dutch lap profile, woodgrain texture, .048 thick, double 5 | | 550 | .029 | | .91 | 1.25 | | 2.16 | 2.93 |
| 4075 | .044 thick, double 4.5 | | 525 | .030 | | .87 | 1.31 | | 2.18 | 2.97 |
| 4080 | .042 thick, double 4.5 | | 525 | .030 | | .72 | 1.31 | | 2.03 | 2.81 |
| 4085 | .040 thick, double 4.5 | | 525 | .030 | | .65 | 1.31 | | 1.96 | 2.73 |
| 4090 | Shingle profile, random grooves, double 7 | | 400 | .040 | | 2.71 | 1.72 | | 4.43 | 5.65 |
| 4095 | Triple 5 | | 400 | .040 | | 2.71 | 1.72 | | 4.43 | 5.65 |
| 4100 | Shake profile, 10" wide | | 400 | .040 | | 2.14 | 1.72 | | 3.86 | 5 |
| 4105 | Vertical pattern, .046 thick, double 5 | | 550 | .029 | | 1.27 | 1.25 | | 2.52 | 3.32 |
| 4110 | .044 thick, triple 3 | | 550 | .029 | | 1.31 | 1.25 | | 2.56 | 3.37 |
| 4115 | .040 thick, triple 4 | | 550 | .029 | | 1.25 | 1.25 | | 2.50 | 3.30 |
| 4120 | .040 thick, triple 2.66 | | 550 | .029 | | 1.41 | 1.25 | | 2.66 | 3.48 |
| 4125 | Insulation, fan folded extruded polystyrene, 1/4" | | 2000 | .008 | | .23 | .34 | | .57 | .79 |
| 4130 | 3/8" | | 2000 | .008 | ▼ | .26 | .34 | | .60 | .81 |
| 4135 | Accessories, J channel, 5/8" pocket | | 700 | .023 | L.F. | .37 | .98 | | 1.35 | 1.91 |
| 4140 | 3/4" pocket | | 695 | .023 | | .38 | .99 | | 1.37 | 1.93 |
| 4145 | 1-1/4" pocket | | 680 | .024 | | .59 | 1.01 | | 1.60 | 2.20 |
| 4150 | Flexible, 3/4" pocket | | 600 | .027 | | 1.92 | 1.15 | | 3.07 | 3.87 |
| 4155 | Under sill finish trim | | 500 | .032 | | .39 | 1.38 | | 1.77 | 2.54 |
| 4160 | Vinyl starter strip | | 700 | .023 | | .39 | .98 | | 1.37 | 1.94 |
| 4165 | Aluminum starter strip | | 700 | .023 | | .26 | .98 | | 1.24 | 1.79 |
| 4170 | Window casing, 2-1/2" wide, 3/4" pocket | | 510 | .031 | | 1.22 | 1.35 | | 2.57 | 3.41 |
| 4175 | Outside corner, woodgrain finish, 4" face, 3/4" pocket | | 700 | .023 | | 1.58 | .98 | | 2.56 | 3.25 |
| 4180 | 5/8" pocket | | 700 | .023 | | 1.66 | .98 | | 2.64 | 3.34 |
| 4185 | Smooth finish, 4" face, 3/4" pocket | | 700 | .023 | | 1.58 | .98 | | 2.56 | 3.25 |
| 4190 | 7/8" pocket | | 690 | .023 | | 1.46 | 1 | | 2.46 | 3.14 |
| 4195 | 1-1/4" pocket | | 700 | .023 | | 1.05 | .98 | | 2.03 | 2.67 |
| 4200 | Soffit and fascia, 1' overhang, solid | | 120 | .133 | | 3.38 | 5.75 | | 9.13 | 12.50 |
| 4205 | Vented | | 120 | .133 | | 3.38 | 5.75 | | 9.13 | 12.50 |
| 4207 | 18" overhang, solid | | 110 | .145 | | 4.02 | 6.25 | | 10.27 | 14 |
| 4208 | Vented | | 110 | .145 | | 4.02 | 6.25 | | 10.27 | 14 |
| 4210 | 2' overhang, solid | | 100 | .160 | | 4.65 | 6.90 | | 11.55 | 15.65 |
| 4215 | Vented | | 100 | .160 | | 4.65 | 6.90 | | 11.55 | 15.65 |
| 4217 | 3' overhang, solid | | 100 | .160 | | 5.90 | 6.90 | | 12.80 | 17.05 |
| 4218 | Vented | ▼ | 100 | .160 | ▼ | 5.90 | 6.90 | | 12.80 | 17.05 |
| 4220 | Colors for siding and soffits, add | | | | S.F. | .16 | | | .16 | .18 |
| 4225 | Colors for accessories and trim, add | | | | L.F. | .32 | | | .32 | .35 |

## 07 46 46 – Fiber Cement Siding

| 07 46 46.10 Fiber Cement Siding | Crew | Daily Output | Labor-Hours | Unit | Material | 2011 Bare Costs Labor | Equipment | Total | Total Incl O&P |
|---|---|---|---|---|---|---|---|---|---|
| 0010 | **FIBER CEMENT SIDING** | | | | | | | | | |
| 0020 | Lap siding, 5/16" thick, 6" wide, 4-3/4" exposure, smooth texture | 2 Carp | 415 | .039 | S.F. | 1.16 | 1.66 | | 2.82 | 3.83 |
| 0025 | Woodgrain texture | | 415 | .039 | | 1.16 | 1.66 | | 2.82 | 3.83 |
| 0030 | 7-1/2" wide, 6-1/4" exposure, smooth texture | | 425 | .038 | | 1.04 | 1.62 | | 2.66 | 3.63 |
| 0035 | Woodgrain texture | | 425 | .038 | | 1.04 | 1.62 | | 2.66 | 3.63 |
| 0040 | 8" wide, 6-3/4" exposure, smooth texture | | 425 | .038 | | 1.12 | 1.62 | | 2.74 | 3.72 |
| 0045 | Roughsawn texture | | 425 | .038 | | 1.12 | 1.62 | | 2.74 | 3.72 |
| 0050 | 9-1/2" wide, 8-1/4" exposure, smooth texture | | 440 | .036 | | 1.16 | 1.57 | | 2.73 | 3.68 |
| 0055 | Woodgrain texture | | 440 | .036 | | 1.16 | 1.57 | | 2.73 | 3.68 |
| 0060 | 12" wide, 10-3/8" exposure, smooth texture | ▼ | 455 | .035 | | 1.28 | 1.51 | | 2.79 | 3.73 |

# 07 46 Siding

## 07 46 46 – Fiber Cement Siding

### 07 46 46.10 Fiber Cement Siding

| | | Crew | Daily Output | Labor-Hours | Unit | Material | 2011 Bare Costs Labor | Equipment | Total | Total Incl O&P |
|---|---|---|---|---|---|---|---|---|---|---|
| 0065 | Woodgrain texture | 2 Carp | 455 | .035 | S.F. | 1.28 | 1.51 | | 2.79 | 3.73 |
| 0070 | Panel siding, 5/16" thick, smooth texture | | 750 | .021 | | 1.05 | .92 | | 1.97 | 2.57 |
| 0075 | Stucco texture | | 750 | .021 | | 1.05 | .92 | | 1.97 | 2.57 |
| 0080 | Grooved woodgrain texture | | 750 | .021 | | 1.05 | .92 | | 1.97 | 2.57 |
| 0085 | V - grooved woodgrain texture | | 750 | .021 | ↓ | 1.05 | .92 | | 1.97 | 2.57 |
| 0090 | Wood starter strip | ↓ | 400 | .040 | L.F. | .53 | 1.72 | | 2.25 | 3.22 |

## 07 46 73 – Soffit

### 07 46 73.10 Soffit Options

| | | Crew | Daily Output | Labor-Hours | Unit | Material | 2011 Bare Costs Labor | Equipment | Total | Total Incl O&P |
|---|---|---|---|---|---|---|---|---|---|---|
| 0010 | **SOFFIT OPTIONS** | | | | | | | | | |
| 0012 | Aluminum, residential, .020" thick | 1 Carp | 210 | .038 | S.F. | 1.96 | 1.64 | | 3.60 | 4.68 |
| 0100 | Baked enamel on steel, 16 or 18 gauge | | 105 | .076 | | 5.35 | 3.28 | | 8.63 | 10.95 |
| 0300 | Polyvinyl chloride, white, solid | | 230 | .035 | | 1.96 | 1.50 | | 3.46 | 4.46 |
| 0400 | Perforated | ↓ | 230 | .035 | | 1.96 | 1.50 | | 3.46 | 4.46 |
| 0500 | For colors, add | | | | ↓ | .13 | | | .13 | .14 |

# 07 51 Built-Up Bituminous Roofing

## 07 51 13 – Built-Up Asphalt Roofing

### 07 51 13.10 Built-Up Roofing Components

| | | Crew | Daily Output | Labor-Hours | Unit | Material | 2011 Bare Costs Labor | Equipment | Total | Total Incl O&P |
|---|---|---|---|---|---|---|---|---|---|---|
| 0010 | **BUILT-UP ROOFING COMPONENTS** | | | | | | | | | |
| 0012 | Asphalt saturated felt, #30, 2 square per roll | 1 Rofc | 58 | .138 | Sq. | 9.45 | 5.05 | | 14.50 | 18.75 |
| 0200 | #15, 4 sq per roll, plain or perforated, not mopped | | 58 | .138 | | 5.65 | 5.05 | | 10.70 | 14.60 |
| 0300 | Roll roofing, smooth, #65 | | 15 | .533 | | 9.65 | 19.50 | | 29.15 | 43 |
| 0500 | #90 | | 12 | .667 | | 34.50 | 24.50 | | 59 | 78.50 |
| 0520 | Mineralized | | 12 | .667 | | 34.50 | 24.50 | | 59 | 78.50 |
| 0540 | D.C. (Double coverage), 19" selvage edge | ↓ | 10 | .800 | ↓ | 44 | 29 | | 73 | 96.50 |
| 0580 | Adhesive (lap cement) | | | | Gal. | 4.79 | | | 4.79 | 5.25 |
| 0600 | Steep, flat or dead level asphalt, 10 ton lots, bulk | | | | Ton | 415 | | | 415 | 455 |
| 0800 | Packaged | | | | " | 700 | | | 700 | 770 |

### 07 51 13.13 Cold-Applied Built-Up Asphalt Roofing

| | | Crew | Daily Output | Labor-Hours | Unit | Material | 2011 Bare Costs Labor | Equipment | Total | Total Incl O&P |
|---|---|---|---|---|---|---|---|---|---|---|
| 0010 | **COLD-APPLIED BUILT-UP ASPHALT ROOFING** | | | | | | | | | |
| 0020 | 3 ply system, installation only (components listed below) | G-5 | 50 | .800 | Sq. | | 26.50 | 3.37 | 29.87 | 47.50 |
| 0100 | Spunbond poly. fabric, 1.35 oz./S.Y., 36"W, 10.8 sq./roll | | | | Ea. | 175 | | | 175 | 193 |
| 0200 | 49" wide, 14.6 sq./roll | | | | | 244 | | | 244 | 268 |
| 0300 | 2.10 oz./S.Y., 36" wide, 10.8 sq./roll | | | | | 266 | | | 266 | 293 |
| 0400 | 49" wide, 14.6 sq./roll | | | | ↓ | 360 | | | 360 | 395 |
| 0500 | Base & finish coat, 3 gal./sq., 5 gal./can | | | | Gal. | 5 | | | 5 | 5.50 |
| 0600 | Coating, ceramic granules, 1/2 sq./bag | | | | Ea. | 16.75 | | | 16.75 | 18.45 |
| 0700 | Aluminum, 2 gal./sq. | | | | Gal. | 13.50 | | | 13.50 | 14.85 |
| 0800 | Emulsion, fibered or non-fibered, 4 gal./sq. | | | | " | 4.70 | | | 4.70 | 5.15 |

### 07 51 13.20 Built-Up Roofing Systems

| | | Crew | Daily Output | Labor-Hours | Unit | Material | 2011 Bare Costs Labor | Equipment | Total | Total Incl O&P |
|---|---|---|---|---|---|---|---|---|---|---|
| 0010 | **BUILT-UP ROOFING SYSTEMS** R075113-20 | | | | | | | | | |
| 0120 | Asphalt flood coat with gravel/slag surfacing, not including | | | | | | | | | |
| 0140 | Insulation, flashing or wood nailers | | | | | | | | | |
| 0200 | Asphalt base sheet, 3 plies #15 asphalt felt, mopped | G-1 | 22 | 2.545 | Sq. | 82.50 | 87 | 18.60 | 188.10 | 257 |
| 0350 | On nailable decks | | 21 | 2.667 | | 87.50 | 91 | 19.50 | 198 | 270 |
| 0500 | 4 plies #15 asphalt felt, mopped | | 20 | 2.800 | | 115 | 95.50 | 20.50 | 231 | 310 |
| 0550 | On nailable decks | | 19 | 2.947 | | 103 | 101 | 21.50 | 225.50 | 305 |
| 0700 | Coated glass base sheet, 2 plies glass (type IV), mopped | | 22 | 2.545 | | 86 | 87 | 18.60 | 191.60 | 260 |
| 0850 | 3 plies glass, mopped | | 20 | 2.800 | | 103 | 95.50 | 20.50 | 219 | 295 |
| 0950 | On nailable decks | ↓ | 19 | 2.947 | | 97 | 101 | 21.50 | 219.50 | 298 |

# 07 51 Built-Up Bituminous Roofing

## 07 51 13 – Built-Up Asphalt Roofing

### 07 51 13.20 Built-Up Roofing Systems

| | | Crew | Daily Output | Labor-Hours | Unit | Material | 2011 Bare Costs Labor | Equipment | Total | Total Incl O&P |
|---|---|---|---|---|---|---|---|---|---|---|
| 1100 | 4 plies glass fiber felt (type IV), mopped | G-1 | 20 | 2.800 | Sq. | 125 | 95.50 | 20.50 | 241 | 320 |
| 1150 | On nailable decks | | 19 | 2.947 | | 113 | 101 | 21.50 | 235.50 | 315 |
| 1200 | Coated & saturated base sheet, 3 plies #15 asph. felt, mopped | | 20 | 2.800 | | 92.50 | 95.50 | 20.50 | 208.50 | 284 |
| 1250 | On nailable decks | | 19 | 2.947 | | 86.50 | 101 | 21.50 | 209 | 287 |
| 1300 | 4 plies #15 asphalt felt, mopped | | 22 | 2.545 | | 108 | 87 | 18.60 | 213.60 | 285 |
| 2000 | Asphalt flood coat, smooth surface | | | | | | | | | |
| 2200 | Asphalt base sheet & 3 plies #15 asphalt felt, mopped | G-1 | 24 | 2.333 | Sq. | 87.50 | 79.50 | 17.05 | 184.05 | 248 |
| 2400 | On nailable decks | | 23 | 2.435 | | 81.50 | 83 | 17.80 | 182.30 | 248 |
| 2600 | 4 plies #15 asphalt felt, mopped | | 24 | 2.333 | | 103 | 79.50 | 17.05 | 199.55 | 265 |
| 2700 | On nailable decks | | 23 | 2.435 | | 97 | 83 | 17.80 | 197.80 | 265 |
| 2900 | Coated glass fiber base sheet, mopped, and 2 plies of | | | | | | | | | |
| 2910 | glass fiber felt (type IV) | G-1 | 25 | 2.240 | Sq. | 80.50 | 76.50 | 16.35 | 173.35 | 234 |
| 3100 | On nailable decks | | 24 | 2.333 | | 76 | 79.50 | 17.05 | 172.55 | 235 |
| 3200 | 3 plies, mopped | | 23 | 2.435 | | 97 | 83 | 17.80 | 197.80 | 265 |
| 3300 | On nailable decks | | 22 | 2.545 | | 91 | 87 | 18.60 | 196.60 | 266 |
| 3800 | 4 plies glass fiber felt (type IV), mopped | | 23 | 2.435 | | 114 | 83 | 17.80 | 214.80 | 283 |
| 3900 | On nailable decks | | 22 | 2.545 | | 108 | 87 | 18.60 | 213.60 | 284 |
| 4000 | Coated & saturated base sheet, 3 plies #15 asph. felt, mopped | | 24 | 2.333 | | 86.50 | 79.50 | 17.05 | 183.05 | 247 |
| 4200 | On nailable decks | | 23 | 2.435 | | 81 | 83 | 17.80 | 181.80 | 247 |
| 4300 | 4 plies #15 organic felt, mopped | | 22 | 2.545 | | 102 | 87 | 18.60 | 207.60 | 278 |
| 4500 | Coal tar pitch with gravel/slag surfacing | | | | | | | | | |
| 4600 | 4 plies #15 tarred felt, mopped | G-1 | 21 | 2.667 | Sq. | 176 | 91 | 19.50 | 286.50 | 370 |
| 4800 | 3 plies glass fiber felt (type IV), mopped | " | 19 | 2.947 | " | 146 | 101 | 21.50 | 268.50 | 350 |
| 5000 | Coated glass fiber base sheet, and 2 plies of | | | | | | | | | |
| 5010 | glass fiber felt, (type IV), mopped | G-1 | 19 | 2.947 | Sq. | 150 | 101 | 21.50 | 272.50 | 360 |
| 5300 | On nailable decks | | 18 | 3.111 | | 131 | 106 | 22.50 | 259.50 | 345 |
| 5600 | 4 plies glass fiber felt (type IV), mopped | | 21 | 2.667 | | 202 | 91 | 19.50 | 312.50 | 395 |
| 5800 | On nailable decks | | 20 | 2.800 | | 183 | 95.50 | 20.50 | 299 | 385 |

### 07 51 13.30 Cants

| | | Crew | Daily Output | Labor-Hours | Unit | Material | 2011 Bare Costs Labor | Equipment | Total | Total Incl O&P |
|---|---|---|---|---|---|---|---|---|---|---|
| 0010 | **CANTS** | | | | | | | | | |
| 0012 | Lumber, treated, 4" x 4" cut diagonally | 1 Rofc | 325 | .025 | L.F. | 2 | .90 | | 2.90 | 3.70 |
| 0100 | Foamglass | | 325 | .025 | | 2.23 | .90 | | 3.13 | 3.95 |
| 0300 | Mineral or fiber, trapezoidal, 1" x 4" x 48" | | 325 | .025 | | .19 | .90 | | 1.09 | 1.71 |
| 0400 | 1-1/2" x 5-5/8" x 48" | | 325 | .025 | | .31 | .90 | | 1.21 | 1.84 |

### 07 51 13.40 Felts

| | | Crew | Daily Output | Labor-Hours | Unit | Material | 2011 Bare Costs Labor | Equipment | Total | Total Incl O&P |
|---|---|---|---|---|---|---|---|---|---|---|
| 0010 | **FELTS** | | | | | | | | | |
| 0012 | Glass fibered roofing felt, #15, not mopped | 1 Rofc | 58 | .138 | Sq. | 6.90 | 5.05 | | 11.95 | 15.95 |
| 0300 | Base sheet, #45, channel vented | | 58 | .138 | | 32 | 5.05 | | 37.05 | 44 |
| 0400 | #50, coated | | 58 | .138 | | 16.45 | 5.05 | | 21.50 | 26.50 |
| 0500 | Cap, mineral surfaced | | 58 | .138 | | 45 | 5.05 | | 50.05 | 58 |
| 0600 | Flashing membrane, #65 | | 16 | .500 | | 9.65 | 18.30 | | 27.95 | 41 |
| 0800 | Coal tar fibered, #15, no mopping | | 58 | .138 | | 14 | 5.05 | | 19.05 | 24 |
| 0900 | Asphalt felt, #15, 4 sq per roll, no mopping | | 58 | .138 | | 5.65 | 5.05 | | 10.70 | 14.60 |
| 1100 | #30, 2 sq per roll | | 58 | .138 | | 9.45 | 5.05 | | 14.50 | 18.75 |
| 1200 | Double coated, #33 | | 58 | .138 | | 9.90 | 5.05 | | 14.95 | 19.30 |
| 1400 | #40, base sheet | | 58 | .138 | | 10.90 | 5.05 | | 15.95 | 20.50 |
| 1450 | Coated and saturated | | 58 | .138 | | 9.95 | 5.05 | | 15 | 19.35 |
| 1500 | Tarred felt, organic, #15, 4 sq rolls | | 58 | .138 | | 11.50 | 5.05 | | 16.55 | 21 |
| 1550 | #30, 2 sq roll | | 58 | .138 | | 23 | 5.05 | | 28.05 | 34 |
| 1700 | Add for mopping above felts, per ply, asphalt, 24 lb per sq | G-1 | 192 | .292 | | 8.40 | 9.95 | 2.13 | 20.48 | 28 |
| 1800 | Coal tar mopping, 30 lb per sq | | 186 | .301 | | 16.50 | 10.30 | 2.20 | 29 | 37.50 |
| 1900 | Flood coat, with asphalt, 60 lb per sq | | 60 | .933 | | 21 | 32 | 6.80 | 59.80 | 83.50 |

# 07 51 Built-Up Bituminous Roofing

## 07 51 13 – Built-Up Asphalt Roofing

### 07 51 13.40 Felts

| | | Crew | Daily Output | Labor-Hours | Unit | Material | 2011 Bare Costs Labor | Equipment | Total | Total Incl O&P |
|---|---|---|---|---|---|---|---|---|---|---|
| 2000 | With coal tar, 75 lb per sq | G-1 | 56 | 1 | Sq. | 41.50 | 34 | 7.30 | 82.80 | 111 |

### 07 51 13.50 Walkways for Built-Up Roofs

| | | Crew | Daily Output | Labor-Hours | Unit | Material | Labor | Equipment | Total | Total Incl O&P |
|---|---|---|---|---|---|---|---|---|---|---|
| 0010 | **WALKWAYS FOR BUILT-UP ROOFS** | | | | | | | | | |
| 0020 | Asphalt impregnated, 3' x 6' x 1/2" thick | 1 Rofc | 400 | .020 | S.F. | 1.72 | .73 | | 2.45 | 3.11 |
| 0100 | 3' x 3' x 3/4" thick | " | 400 | .020 | | 4.96 | .73 | | 5.69 | 6.65 |
| 0300 | Concrete patio blocks, 2" thick, natural | 1 Clab | 115 | .070 | | 3.71 | 2.39 | | 6.10 | 7.75 |
| 0400 | Colors | " | 115 | .070 | | 3.75 | 2.39 | | 6.14 | 7.80 |

# 07 52 Modified Bituminous Membrane Roofing

## 07 52 13 – Atactic-Polypropylene-Modified Bituminous Membrane Roofing

### 07 52 13.10 APP Modified Bituminous Membrane

| | | Crew | Daily Output | Labor-Hours | Unit | Material | Labor | Equipment | Total | Total Incl O&P |
|---|---|---|---|---|---|---|---|---|---|---|
| 0010 | **APP MODIFIED BITUMINOUS MEMBRANE**   R075213-30 | | | | | | | | | |
| 0020 | Base sheet, #15 glass fiber felt, nailed to deck | 1 Rofc | 58 | .138 | Sq. | 7.80 | 5.05 | | 12.85 | 17 |
| 0030 | Spot mopped to deck | G-1 | 295 | .190 | | 11.10 | 6.50 | 1.39 | 18.99 | 24.50 |
| 0040 | Fully mopped to deck | " | 192 | .292 | | 15.30 | 9.95 | 2.13 | 27.38 | 36 |
| 0050 | #15 organic felt, nailed to deck | 1 Rofc | 58 | .138 | | 6.55 | 5.05 | | 11.60 | 15.60 |
| 0060 | Spot mopped to deck | G-1 | 295 | .190 | | 9.85 | 6.50 | 1.39 | 17.74 | 23 |
| 0070 | Fully mopped to deck | " | 192 | .292 | | 14.05 | 9.95 | 2.13 | 26.13 | 34.50 |
| 2100 | APP mod., smooth surf. cap sheet, poly. reinf., torched, 160 mils | G-5 | 2100 | .019 | S.F. | .48 | .63 | .08 | 1.19 | 1.67 |
| 2150 | 170 mils | | 2100 | .019 | | .55 | .63 | .08 | 1.26 | 1.75 |
| 2200 | Granule surface cap sheet, poly. reinf., torched, 180 mils | | 2000 | .020 | | .59 | .66 | .08 | 1.33 | 1.84 |
| 2250 | Smooth surface flashing, torched, 160 mils | | 1260 | .032 | | .48 | 1.05 | .13 | 1.66 | 2.43 |
| 2300 | 170 mils | | 1260 | .032 | | .55 | 1.05 | .13 | 1.73 | 2.51 |
| 2350 | Granule surface flashing, torched, 180 mils | | 1260 | .032 | | .59 | 1.05 | .13 | 1.77 | 2.55 |
| 2400 | Fibrated aluminum coating | 1 Rofc | 3800 | .002 | | .11 | .08 | | .19 | .25 |

## 07 52 16 – Styrene-Butadiene-Styrene Modified Bituminous Membrane Roofing

### 07 52 16.10 SBS Modified Bituminous Membrane

| | | Crew | Daily Output | Labor-Hours | Unit | Material | Labor | Equipment | Total | Total Incl O&P |
|---|---|---|---|---|---|---|---|---|---|---|
| 0010 | **SBS MODIFIED BITUMINOUS MEMBRANE** | | | | | | | | | |
| 0080 | SBS modified, granule surf cap sheet, polyester rein., mopped | | | | | | | | | |
| 0600 | 150 mils | G-1 | 2000 | .028 | S.F. | .58 | .96 | .20 | 1.74 | 2.46 |
| 1100 | 160 mils | | 2000 | .028 | | .77 | .96 | .20 | 1.93 | 2.67 |
| 1500 | Glass fiber reinforced, mopped, 160 mils | | 2000 | .028 | | .49 | .96 | .20 | 1.65 | 2.36 |
| 1600 | Smooth surface cap sheet, mopped, 145 mils | | 2100 | .027 | | .49 | .91 | .20 | 1.60 | 2.27 |
| 1700 | Smooth surface flashing, 145 mils | | 1260 | .044 | | .49 | 1.52 | .32 | 2.33 | 3.43 |
| 1800 | 150 mils | | 1260 | .044 | | .48 | 1.52 | .32 | 2.32 | 3.42 |
| 1900 | Granular surface flashing, 150 mils | | 1260 | .044 | | .58 | 1.52 | .32 | 2.42 | 3.53 |
| 2000 | 160 mils | | 1260 | .044 | | .77 | 1.52 | .32 | 2.61 | 3.74 |

# 07 53 Elastomeric Membrane Roofing

## 07 53 16 – Chlorosulfonate-Polyethylene Roofing

### 07 53 16.10 Chlorosulfonated Polyethylene Roofing

| | | Crew | Daily Output | Labor-Hours | Unit | Material | Labor | Equipment | Total | Total Incl O&P |
|---|---|---|---|---|---|---|---|---|---|---|
| 0010 | **CHLOROSULFONATED POLYETHYLENE ROOFING** | | | | | | | | | |
| 0800 | Chlorosulfonated polyethylene-hypalon (CSPE), 45 mils, | | | | | | | | | |
| 0900 | 0.29 P.S.F., fully adhered | G-5 | 26 | 1.538 | Sq. | 185 | 51 | 6.50 | 242.50 | 296 |
| 1100 | Loose-laid & ballasted with stone (10 P.S.F.) | | 51 | .784 | | 193 | 26 | 3.30 | 222.30 | 259 |
| 1200 | Mechanically attached | | 35 | 1.143 | | 186 | 38 | 4.81 | 228.81 | 273 |
| 1300 | Plates with adhesive attachment | | 35 | 1.143 | | 183 | 38 | 4.81 | 225.81 | 269 |

# 07 53 Elastomeric Membrane Roofing

## 07 53 23 – Ethylene-Propylene-Diene-Monomer Roofing

### 07 53 23.20 Ethylene-Propylene-Diene-Monomer Roofing

| | | Daily Output | Labor-Hours | Unit | Material | 2011 Bare Costs Labor | Equipment | Total | Total Incl O&P |
|---|---|---|---|---|---|---|---|---|---|
| 0010 | **ETHYLENE-PROPYLENE-DIENE-MONOMER ROOFING (E.P.D.M.)** | | | | | | | | |
| 3500 | Ethylene-propylene-diene-monomer (EPDM), 45 mils, 0.28 P.S.F. | | | | | | | | |
| 3600 | Loose-laid & ballasted with stone (10 P.S.F.) | G-5 | 51 | .784 | Sq. | 73.50 | 26 | 3.30 | 102.80 | 128 |
| 3700 | Mechanically attached | | 35 | 1.143 | | 65.50 | 38 | 4.81 | 108.31 | 140 |
| 3800 | Fully adhered with adhesive | ↓ | 26 | 1.538 | ↓ | 101 | 51 | 6.50 | 158.50 | 203 |
| 4500 | 60 mils, 0.40 P.S.F. | | | | | | | | |
| 4600 | Loose-laid & ballasted with stone (10 P.S.F.) | G-5 | 51 | .784 | Sq. | 87 | 26 | 3.30 | 116.30 | 143 |
| 4700 | Mechanically attached | | 35 | 1.143 | | 77.50 | 38 | 4.81 | 120.31 | 153 |
| 4800 | Fully adhered with adhesive | ↓ | 26 | 1.538 | | 113 | 51 | 6.50 | 170.50 | 216 |
| 4810 | 45 mil, .28 PSF, membrane only | | | | | 39.50 | | | 39.50 | 43.50 |
| 4820 | 60 mil, .40 PSF, membrane only | | | | ↓ | 50.50 | | | 50.50 | 55.50 |
| 4850 | Seam tape for membrane, 3" x 100' roll | | | | Ea. | 48 | | | 48 | 53 |
| 4900 | Batten strips, 10' sections | | | | | 3.05 | | | 3.05 | 3.36 |
| 4910 | Cover tape for batten strips, 6" x 100' roll | | | | ↓ | 110 | | | 110 | 121 |
| 4930 | Plate anchors | | | | M | 105 | | | 105 | 116 |
| 4970 | Adhesive for fully adhered systems, 60 S.F./gal. | | | | Gal. | 21 | | | 21 | 23 |

## 07 53 29 – Polyisobutylene Roofing

### 07 53 29.10 Polyisobutylene Roofing

| | | Daily Output | Labor-Hours | Unit | Material | 2011 Bare Costs Labor | Equipment | Total | Total Incl O&P |
|---|---|---|---|---|---|---|---|---|---|
| 0010 | **POLYISOBUTYLENE ROOFING** | | | | | | | | |
| 7500 | Polyisobutylene (PIB), 100 mils, 0.57 P.S.F. | | | | | | | | |
| 7600 | Loose-laid & ballasted with stone/gravel (10 P.S.F.) | G-5 | 51 | .784 | Sq. | 181 | 26 | 3.30 | 210.30 | 246 |
| 7700 | Partially adhered with adhesive | | 35 | 1.143 | | 227 | 38 | 4.81 | 269.81 | 320 |
| 7800 | Hot asphalt attachment | | 35 | 1.143 | | 217 | 38 | 4.81 | 259.81 | 305 |
| 7900 | Fully adhered with contact cement | ↓ | 26 | 1.538 | ↓ | 234 | 51 | 6.50 | 291.50 | 350 |

# 07 54 Thermoplastic Membrane Roofing

## 07 54 19 – Polyvinyl-Chloride Roofing

### 07 54 19.10 Polyvinyl-Chloride Roofing (P.V.C.)

| | | Daily Output | Labor-Hours | Unit | Material | 2011 Bare Costs Labor | Equipment | Total | Total Incl O&P |
|---|---|---|---|---|---|---|---|---|---|
| 0010 | **POLYVINYL-CHLORIDE ROOFING (P.V.C.)** | | | | | | | | |
| 8200 | Heat welded seams | | | | | | | | |
| 8700 | Reinforced, 48 mils, 0.33 P.S.F. | | | | | | | | |
| 8750 | Loose-laid & ballasted with stone/gravel (12 P.S.F.) | G-5 | 51 | .784 | Sq. | 102 | 26 | 3.30 | 131.30 | 159 |
| 8800 | Mechanically attached | | 35 | 1.143 | | 94 | 38 | 4.81 | 136.81 | 171 |
| 8850 | Fully adhered with adhesive | ↓ | 26 | 1.538 | ↓ | 127 | 51 | 6.50 | 184.50 | 231 |
| 8860 | Reinforced, 60 mils, .40 P.S.F. | | | | | | | | |
| 8870 | Loose-laid & ballasted with stone/gravel (12 P.S.F.) | G-5 | 51 | .784 | Sq. | 103 | 26 | 3.30 | 132.30 | 161 |
| 8880 | Mechanically attached | | 35 | 1.143 | | 95.50 | 38 | 4.81 | 138.31 | 173 |
| 8890 | Fully adhered with adhesive | ↓ | 26 | 1.538 | ↓ | 128 | 51 | 6.50 | 185.50 | 233 |

## 07 54 23 – Thermoplastic-Polyolefin Roofing

### 07 54 23.10 Thermoplastic Polyolefin Roofing (T.P.O)

| | | Daily Output | Labor-Hours | Unit | Material | 2011 Bare Costs Labor | Equipment | Total | Total Incl O&P |
|---|---|---|---|---|---|---|---|---|---|
| 0010 | **THERMOPLASTIC POLYOLEFIN ROOFING (T.P.O.)** | | | | | | | | |
| 0100 | 45 mils, loose laid & ballasted with stone(1/2 ton/sq.) | G-5 | 51 | .784 | Sq. | 83.50 | 26 | 3.30 | 112.80 | 139 |
| 0120 | Fully adhered | | 25 | 1.600 | | 72.50 | 53 | 6.75 | 132.25 | 176 |
| 0140 | Mechanically attached | | 34 | 1.176 | | 72 | 39 | 4.95 | 115.95 | 149 |
| 0160 | Self adhered | | 35 | 1.143 | | 88 | 38 | 4.81 | 130.81 | 165 |
| 0180 | 60 mil membrane, heat welded seams, ballasted | | 50 | .800 | | 89 | 26.50 | 3.37 | 118.87 | 146 |
| 0200 | Fully adhered | | 25 | 1.600 | | 77 | 53 | 6.75 | 136.75 | 181 |
| 0220 | Mechanically attached | | 34 | 1.176 | | 80.50 | 39 | 4.95 | 124.45 | 159 |
| 0240 | Self adhered | ↓ | 35 | 1.143 | ↓ | 109 | 38 | 4.81 | 151.81 | 188 |

# 07 56 Fluid-Applied Roofing

## 07 56 10 – Fluid-Applied Roofing Elastomers

| 07 56 10.10 Elastomeric Roofing | Crew | Daily Output | Labor-Hours | Unit | Material | 2011 Bare Costs Labor | Equipment | Total | Total Incl O&P |
|---|---|---|---|---|---|---|---|---|---|
| 0010 **ELASTOMERIC ROOFING** | | | | | | | | | |
| 0110 Acrylic rubber, fluid applied, 20 mils thick | G-5 | 2000 | .020 | S.F. | 2.04 | .66 | .08 | 2.78 | 3.43 |
| 0120 50 mils, reinforced | | 1200 | .033 | | 3.16 | 1.11 | .14 | 4.41 | 5.45 |
| 0130 For walking surface, add | | 900 | .044 | | .96 | 1.47 | .19 | 2.62 | 3.72 |
| 0300 Hypalon neoprene, fluid applied, 20 mil thick, not-reinforced | G-1 | 1135 | .049 | | 2.57 | 1.68 | .36 | 4.61 | 6.05 |
| 0600 Non-woven polyester, reinforced | | 960 | .058 | | 2.61 | 1.99 | .43 | 5.03 | 6.65 |
| 0700 5 coat neoprene deck, 60 mil thick, under 10,000 SF | | 325 | .172 | | 5.75 | 5.90 | 1.26 | 12.91 | 17.50 |
| 0900 Over 10,000 SF | | 625 | .090 | | 5.35 | 3.06 | .66 | 9.07 | 11.70 |
| 1300 Vinyl plastic traffic deck, sprayed, 2 to 4 mils thick | | 625 | .090 | | 1.68 | 3.06 | .66 | 5.40 | 7.65 |
| 1500 Vinyl and neoprene membrane traffic deck | | 1550 | .036 | | 1.77 | 1.23 | .26 | 3.26 | 4.29 |

# 07 57 Coated Foamed Roofing

## 07 57 13 – Sprayed Polyurethane Foam Roofing

| 07 57 13.10 Sprayed Polyurethane Foam Roofing (S.P.F.) | Crew | Daily Output | Labor-Hours | Unit | Material | 2011 Bare Costs Labor | Equipment | Total | Total Incl O&P |
|---|---|---|---|---|---|---|---|---|---|
| 0010 **SPRAYED POLYURETHANE FOAM ROOFING (S.P.F.)** | | | | | | | | | |
| 0100 Primer for metal substrate (when required) | G-2A | 3000 | .008 | S.F. | .42 | .26 | .20 | .88 | 1.10 |
| 0200 Primer for non-metal substrate (when required) | | 3000 | .008 | | .16 | .26 | .20 | .62 | .82 |
| 0300 Closed cell spray, polyurethane foam, 3 lb. per C.F. density, 1", R6.7 | | 15000 | .002 | | .55 | .05 | .04 | .64 | .73 |
| 0400 2", R13.4 | | 13125 | .002 | | 1.10 | .06 | .05 | 1.21 | 1.36 |
| 0500 3", R18.6 | | 11485 | .002 | | 1.66 | .07 | .05 | 1.78 | 1.99 |
| 0550 4", R24.8 | | 10080 | .002 | | 2.21 | .08 | .06 | 2.35 | 2.62 |
| 0700 Spray-on silicone coating | | 2500 | .010 | | .92 | .31 | .24 | 1.47 | 1.78 |
| 0800 Warranty 5-20 year manufacturer's | | | | | | | | | .15 |
| 0900 Warranty 20 year, no dollar limit | | | | | | | | | .20 |

# 07 58 Roll Roofing

## 07 58 10 – Asphalt Roll Roofing

| 07 58 10.10 Roll Roofing | Crew | Daily Output | Labor-Hours | Unit | Material | 2011 Bare Costs Labor | Equipment | Total | Total Incl O&P |
|---|---|---|---|---|---|---|---|---|---|
| 0010 **ROLL ROOFING** | | | | | | | | | |
| 0100 Asphalt, mineral surface | | | | | | | | | |
| 0200 1 ply #15 organic felt, 1 ply mineral surfaced | | | | | | | | | |
| 0300 Selvage roofing, lap 19", nailed & mopped | G-1 | 27 | 2.074 | Sq. | 60.50 | 71 | 15.15 | 146.65 | 201 |
| 0400 3 plies glass fiber felt (type IV), 1 ply mineral surfaced | | | | | | | | | |
| 0500 Selvage roofing, lapped 19", mopped | G-1 | 25 | 2.240 | Sq. | 98 | 76.50 | 16.35 | 190.85 | 253 |
| 0600 Coated glass fiber base sheet, 2 plies of glass fiber | | | | | | | | | |
| 0700 Felt (type IV), 1 ply mineral surfaced selvage | | | | | | | | | |
| 0800 Roofing, lapped 19", mopped | G-1 | 25 | 2.240 | Sq. | 108 | 76.50 | 16.35 | 200.85 | 264 |
| 0900 On nailable decks | " | 24 | 2.333 | " | 99.50 | 79.50 | 17.05 | 196.05 | 261 |
| 1000 3 plies glass fiber felt (type III), 1 ply mineral surfaced | | | | | | | | | |
| 1100 Selvage roofing, lapped 19", mopped | G-1 | 25 | 2.240 | Sq. | 98 | 76.50 | 16.35 | 190.85 | 253 |

# 07 61 Sheet Metal Roofing

## 07 61 13 – Standing Seam Sheet Metal Roofing

| 07 61 13.10 Standing Seam Sheet Metal Roofing, Field Fab. | Crew | Daily Output | Labor-Hours | Unit | Material | 2011 Bare Costs Labor | Equipment | Total | Total Incl O&P |
|---|---|---|---|---|---|---|---|---|---|
| 0010 **STANDING SEAM SHEET METAL ROOFING, FIELD FABRICATED** | | | | | | | | | |
| 0400 Copper standing seam roofing, over 10 squares, 16 oz, 125 lb per sq | 1 Shee | 1.30 | 6.154 | Sq. | 735 | 320 | | 1,055 | 1,300 |
| 0600     18 oz, 140 lb per sq | | 1.20 | 6.667 | | 825 | 345 | | 1,170 | 1,425 |
| 0700     20 oz, 150 lb per sq | | 1.10 | 7.273 | | 920 | 375 | | 1,295 | 1,575 |
| 1200 For abnormal conditions or small areas, add | | | | | 25% | 100% | | | |
| 1300 For lead-coated copper, add | | | | | 25% | | | | |

## 07 61 16 – Batten Seam Sheet Metal Roofing

| 07 61 16.10 Batten Seam Sheet Metal Roofing, Field Fabricated | | | | | | | | | |
|---|---|---|---|---|---|---|---|---|---|
| 0010 **BATTEN SEAM SHEET METAL ROOFING, FIELD FABRICATED** | | | | | | | | | |
| 0012 Copper batten seam roofing, over 10 sq, 16 oz, 130 lb per sq | 1 Shee | 1.10 | 7.273 | Sq. | 935 | 375 | | 1,310 | 1,600 |
| 0020 Lead batten seam roofing, 5 lb. per SF | | 1.20 | 6.667 | | 600 | 345 | | 945 | 1,175 |
| 0100 Zinc/copper alloy batten seam roofing, .020 thick | | 1.20 | 6.667 | | 1,100 | 345 | | 1,445 | 1,750 |
| 0200 Copper roofing, batten seam, over 10 sq, 18 oz, 145 lb per sq | | 1 | 8 | | 1,050 | 415 | | 1,465 | 1,775 |
| 0300     20 oz, 160 lb per sq | | 1 | 8 | | 1,175 | 415 | | 1,590 | 1,900 |
| 0500 Stainless steel batten seam roofing, type 304, 28 gauge | | 1.20 | 6.667 | | 625 | 345 | | 970 | 1,200 |
| 0600     26 gauge | | 1.15 | 6.957 | | 630 | 360 | | 990 | 1,250 |
| 0800 Zinc, copper alloy roofing, batten seam, .027" thick | | 1.15 | 6.957 | | 1,550 | 360 | | 1,910 | 2,275 |
| 0900     .032" thick | | 1.10 | 7.273 | | 1,875 | 375 | | 2,250 | 2,650 |
| 1000     .040" thick | | 1.05 | 7.619 | | 2,100 | 395 | | 2,495 | 2,900 |

## 07 61 19 – Flat Seam Sheet Metal Roofing

| 07 61 19.10 Flat Seam Sheet Metal Roofing, Field Fabricated | | | | | | | | | |
|---|---|---|---|---|---|---|---|---|---|
| 0010 **FLAT SEAM SHEET METAL ROOFING, FIELD FABRICATED** | | | | | | | | | |
| 0900 Copper flat seam roofing, over 10 squares, 16 oz, 115 lb per sq | 1 Shee | 1.20 | 6.667 | Sq. | 685 | 345 | | 1,030 | 1,275 |
| 1000     20 oz, 145 lb per sq | | 1.10 | 7.273 | | 855 | 375 | | 1,230 | 1,500 |
| 1100 Lead flat seam roofing, 5 lb per SF | | 1.30 | 6.154 | | 600 | 320 | | 920 | 1,150 |

# 07 65 Flexible Flashing

## 07 65 10 – Sheet Metal Flashing

| 07 65 10.10 Sheet Metal Flashing and Counter Flashing | Crew | Daily Output | Labor-Hours | Unit | Material | Labor | Equipment | Total | Total Incl O&P |
|---|---|---|---|---|---|---|---|---|---|
| 0010 **SHEET METAL FLASHING AND COUNTER FLASHING** | | | | | | | | | |
| 0011 Including up to 4 bends | | | | | | | | | |
| 0020 Aluminum, mill finish, .013" thick | 1 Rofc | 145 | .055 | S.F. | .73 | 2.02 | | 2.75 | 4.16 |
| 0030     .016" thick | | 145 | .055 | | .84 | 2.02 | | 2.86 | 4.28 |
| 0060     .019" thick | | 145 | .055 | | 1.07 | 2.02 | | 3.09 | 4.54 |
| 0100     .032" thick | | 145 | .055 | | 1.25 | 2.02 | | 3.27 | 4.74 |
| 0200     .040" thick | | 145 | .055 | | 2.14 | 2.02 | | 4.16 | 5.70 |
| 0300     .050" thick | | 145 | .055 | | 2.25 | 2.02 | | 4.27 | 5.85 |
| 0325     Mill finish 5" x 7" step flashing, .016" thick | | 1920 | .004 | Ea. | .18 | .15 | | .33 | .45 |
| 0350     Mill finish 12" x 12" step flashing, .016" thick | | 1600 | .005 | " | .62 | .18 | | .80 | .98 |
| 0400     Painted finish, add | | | | S.F. | .34 | | | .34 | .37 |
| 1000     Mastic-coated 2 sides, .005" thick | 1 Rofc | 330 | .024 | | 1.67 | .89 | | 2.56 | 3.32 |
| 1100     .016" thick | | 330 | .024 | | 1.84 | .89 | | 2.73 | 3.50 |
| 1600 Copper, 16 oz, sheets, under 1000 lbs. | | 115 | .070 | | 5.90 | 2.54 | | 8.44 | 10.75 |
| 1700     Over 4000 lbs. | | 155 | .052 | | 5.55 | 1.89 | | 7.44 | 9.25 |
| 1900     20 oz sheets, under 1000 lbs. | | 110 | .073 | | 7.35 | 2.66 | | 10.01 | 12.55 |
| 2000     Over 4000 lbs. | | 145 | .055 | | 7 | 2.02 | | 9.02 | 11.05 |
| 2200     24 oz sheets, under 1000 lbs. | | 105 | .076 | | 9.25 | 2.78 | | 12.03 | 14.85 |
| 2300     Over 4000 lbs. | | 135 | .059 | | 8.80 | 2.17 | | 10.97 | 13.25 |
| 2500     32 oz sheets, under 1000 lbs. | | 100 | .080 | | 11.10 | 2.92 | | 14.02 | 17.05 |
| 2600     Over 4000 lbs. | | 130 | .062 | | 10.55 | 2.25 | | 12.80 | 15.35 |

# 07 65 Flexible Flashing

## 07 65 10 – Sheet Metal Flashing

### 07 65 10.10 Sheet Metal Flashing and Counter Flashing

| | | Crew | Daily Output | Labor-Hours | Unit | Material | 2011 Bare Costs Labor | 2011 Bare Costs Equipment | Total | Total Incl O&P |
|---|---|---|---|---|---|---|---|---|---|---|
| 2700 | W shape for valleys, 16 oz, 24" wide | 1 Rofc | 100 | .080 | L.F. | 12 | 2.92 | | 14.92 | 18.05 |
| 5800 | Lead, 2.5 lb. per SF, up to 12" wide | | 135 | .059 | S.F. | 5.40 | 2.17 | | 7.57 | 9.55 |
| 5900 | Over 12" wide | | 135 | .059 | | 4.43 | 2.17 | | 6.60 | 8.50 |
| 8900 | Stainless steel sheets, 32 ga, .010" thick | | 155 | .052 | | 4 | 1.89 | | 5.89 | 7.55 |
| 9000 | 28 ga, .015" thick | | 155 | .052 | | 5.65 | 1.89 | | 7.54 | 9.35 |
| 9100 | 26 ga, .018" thick | | 155 | .052 | | 7.25 | 1.89 | | 9.14 | 11.15 |
| 9200 | 24 ga, .025" thick | | 155 | .052 | | 8.90 | 1.89 | | 10.79 | 12.95 |
| 9290 | For mechanically keyed flashing, add | | | | | 40% | | | | |
| 9320 | Steel sheets, galvanized, 20 gauge | 1 Rofc | 130 | .062 | S.F. | 1.57 | 2.25 | | 3.82 | 5.45 |
| 9340 | 30 gauge | | 160 | .050 | | 1.46 | 1.83 | | 3.29 | 4.65 |
| 9400 | Terne coated stainless steel, .015" thick, 28 ga | | 155 | .052 | | 7.35 | 1.89 | | 9.24 | 11.25 |
| 9500 | .018" thick, 26 ga | | 155 | .052 | | 8.30 | 1.89 | | 10.19 | 12.25 |
| 9600 | Zinc and copper alloy (brass), .020" thick | | 155 | .052 | | 11.10 | 1.89 | | 12.99 | 15.35 |
| 9700 | .027" thick | | 155 | .052 | | 15.60 | 1.89 | | 17.49 | 20.50 |
| 9800 | .032" thick | | 155 | .052 | | 18.80 | 1.89 | | 20.69 | 23.50 |
| 9900 | .040" thick | | 155 | .052 | | 21 | 1.89 | | 22.89 | 26 |

## 07 65 12 – Fabric and Mastic Flashings

### 07 65 12.10 Fabric and Mastic Flashing and Counter Flashing

| | | Crew | Daily Output | Labor-Hours | Unit | Material | 2011 Bare Costs Labor | 2011 Bare Costs Equipment | Total | Total Incl O&P |
|---|---|---|---|---|---|---|---|---|---|---|
| 0010 | **FABRIC AND MASTIC FLASHING AND COUNTER FLASHING** | | | | | | | | | |
| 1300 | Asphalt flashing cement, 5 gallon | | | | Gal. | 10.80 | | | 10.80 | 11.85 |
| 4900 | Fabric, asphalt-saturated cotton, specification grade | 1 Rofc | 35 | .229 | S.Y. | 2.36 | 8.35 | | 10.71 | 16.50 |
| 5000 | Utility grade | | 35 | .229 | | 1.26 | 8.35 | | 9.61 | 15.30 |
| 5300 | Close-mesh fabric, saturated, 17 oz per S.Y. | | 35 | .229 | | 2.36 | 8.35 | | 10.71 | 16.50 |
| 5500 | Fiberglass, resin-coated | | 35 | .229 | | 1.23 | 8.35 | | 9.58 | 15.25 |
| 8500 | Shower pan, bituminous membrane, 7 oz | | 155 | .052 | S.F. | 2 | 1.89 | | 3.89 | 5.35 |

## 07 65 13 – Laminated Sheet Flashing

### 07 65 13.10 Laminated Sheet Flashing

| | | Crew | Daily Output | Labor-Hours | Unit | Material | 2011 Bare Costs Labor | 2011 Bare Costs Equipment | Total | Total Incl O&P |
|---|---|---|---|---|---|---|---|---|---|---|
| 0010 | **LAMINATED SHEET FLASHING,** Including up to 4 bends | | | | | | | | | |
| 0500 | Fabric-backed 2 sides, .004" thick | 1 Rofc | 330 | .024 | S.F. | 1.41 | .89 | | 2.30 | 3.03 |
| 0700 | .005" thick | | 330 | .024 | | 1.67 | .89 | | 2.56 | 3.32 |
| 0750 | Mastic-backed, self adhesive | | 460 | .017 | | 3.46 | .64 | | 4.10 | 4.87 |
| 0800 | Mastic-coated 2 sides, .004" thick | | 330 | .024 | | 1.41 | .89 | | 2.30 | 3.03 |
| 2800 | Copper, paperbacked 1 side, 2 oz | | 330 | .024 | | 1.25 | .89 | | 2.14 | 2.86 |
| 2900 | 3 oz | | 330 | .024 | | 1.63 | .89 | | 2.52 | 3.27 |
| 3100 | Paperbacked 2 sides, 2 oz | | 330 | .024 | | 1.26 | .89 | | 2.15 | 2.87 |
| 3150 | 3 oz | | 330 | .024 | | 1.62 | .89 | | 2.51 | 3.26 |
| 3200 | 5 oz | | 330 | .024 | | 2.43 | .89 | | 3.32 | 4.15 |
| 3250 | 7 oz | | 330 | .024 | | 3.96 | .89 | | 4.85 | 5.85 |
| 3400 | Mastic-backed 2 sides, copper, 2 oz | | 330 | .024 | | 1.49 | .89 | | 2.38 | 3.12 |
| 3500 | 3 oz | | 330 | .024 | | 1.83 | .89 | | 2.72 | 3.49 |
| 3700 | 5 oz | | 330 | .024 | | 2.68 | .89 | | 3.57 | 4.43 |
| 3800 | Fabric-backed 2 sides, copper, 2 oz | | 330 | .024 | | 1.57 | .89 | | 2.46 | 3.21 |
| 4000 | 3 oz | | 330 | .024 | | 2.04 | .89 | | 2.93 | 3.72 |
| 4100 | 5 oz | | 330 | .024 | | 2.76 | .89 | | 3.65 | 4.52 |
| 4300 | Copper-clad stainless steel, .015" thick, under 500 lbs. | | 115 | .070 | | 5.80 | 2.54 | | 8.34 | 10.65 |
| 4400 | Over 2000 lbs. | | 155 | .052 | | 5.55 | 1.89 | | 7.44 | 9.25 |
| 4600 | .018" thick, under 500 lbs. | | 100 | .080 | | 6.55 | 2.92 | | 9.47 | 12.05 |
| 4700 | Over 2000 lbs. | | 145 | .055 | | 6.20 | 2.02 | | 8.22 | 10.20 |
| 6100 | Lead-coated copper, fabric-backed, 2 oz | | 330 | .024 | | 2.42 | .89 | | 3.31 | 4.14 |
| 6200 | 5 oz | | 330 | .024 | | 2.78 | .89 | | 3.67 | 4.54 |
| 6400 | Mastic-backed 2 sides, 2 oz | | 330 | .024 | | 1.89 | .89 | | 2.78 | 3.56 |
| 6500 | 5 oz | | 330 | .024 | | 2.34 | .89 | | 3.23 | 4.05 |

# 07 65 Flexible Flashing

## 07 65 13 – Laminated Sheet Flashing

| 07 65 13.10 Laminated Sheet Flashing | Crew | Daily Output | Labor-Hours | Unit | Material | Labor | Equipment | Total | Total Incl O&P |
|---|---|---|---|---|---|---|---|---|---|
| | | | | | | 2011 Bare Costs | | | |
| 6700 | Paperbacked 1 side, 2 oz | 1 Rofc | 330 | .024 | S.F. | 1.63 | .89 | | 2.52 | 3.27 |
| 6800 | 3 oz | | 330 | .024 | | 1.92 | .89 | | 2.81 | 3.59 |
| 7000 | Paperbacked 2 sides, 2 oz | | 330 | .024 | | 1.69 | .89 | | 2.58 | 3.34 |
| 7100 | 5 oz | | 330 | .024 | | 2.75 | .89 | | 3.64 | 4.51 |
| 8550 | 3 ply copper and fabric, 3 oz | | 155 | .052 | | 2.32 | 1.89 | | 4.21 | 5.70 |
| 8600 | 7 oz | | 155 | .052 | | 3.25 | 1.89 | | 5.14 | 6.70 |
| 8700 | Lead on copper and fabric, 5 oz | | 155 | .052 | | 2.78 | 1.89 | | 4.67 | 6.20 |
| 8800 | 7 oz | | 155 | .052 | | 5.25 | 1.89 | | 7.14 | 8.95 |
| 9300 | Stainless steel, paperbacked 2 sides, .005" thick | | 330 | .024 | | 3.62 | .89 | | 4.51 | 5.45 |

## 07 65 19 – Plastic Sheet Flashing

### 07 65 19.10 Plastic Sheet Flashing and Counter Flashing

| | | Crew | Daily Output | Labor-Hours | Unit | Material | Labor | Equipment | Total | Total Incl O&P |
|---|---|---|---|---|---|---|---|---|---|---|
| 0010 | PLASTIC SHEET FLASHING AND COUNTER FLASHING | | | | | | | | | |
| 7300 | Polyvinyl chloride, black, .010" thick | 1 Rofc | 285 | .028 | S.F. | .21 | 1.03 | | 1.24 | 1.94 |
| 7400 | .020" thick | | 285 | .028 | | .30 | 1.03 | | 1.33 | 2.04 |
| 7600 | .030" thick | | 285 | .028 | | .39 | 1.03 | | 1.42 | 2.14 |
| 7700 | .056" thick | | 285 | .028 | | .94 | 1.03 | | 1.97 | 2.74 |
| 7900 | Black or white for exposed roofs, .060" thick | | 285 | .028 | | 2 | 1.03 | | 3.03 | 3.91 |
| 8060 | PVC tape, 5" x 45 mils, for joint covers, 100 L.F./roll | | | | Ea. | 128 | | | 128 | 141 |
| 8850 | Polyvinyl chloride, .030" thick | 1 Rofc | 160 | .050 | S.F. | .46 | 1.83 | | 2.29 | 3.55 |

## 07 65 23 – Rubber Sheet Flashing

### 07 65 23.10 Rubber Sheet Flashing and Counterflashing

| | | Crew | Daily Output | Labor-Hours | Unit | Material | Labor | Equipment | Total | Total Incl O&P |
|---|---|---|---|---|---|---|---|---|---|---|
| 0010 | RUBBER SHEET FLASHING AND COUNTERFLASHING | | | | | | | | | |
| 8100 | Rubber, butyl, 1/32" thick | 1 Rofc | 285 | .028 | S.F. | 1.75 | 1.03 | | 2.78 | 3.64 |
| 8200 | 1/16" thick | | 285 | .028 | | 2.38 | 1.03 | | 3.41 | 4.33 |
| 8300 | Neoprene, cured, 1/16" thick | | 285 | .028 | | 1.92 | 1.03 | | 2.95 | 3.82 |
| 8400 | 1/8" thick | | 285 | .028 | | 4 | 1.03 | | 5.03 | 6.10 |

# 07 71 Roof Specialties

## 07 71 19 – Manufactured Gravel Stops and Fascias

### 07 71 19.10 Gravel Stop

| | | Crew | Daily Output | Labor-Hours | Unit | Material | Labor | Equipment | Total | Total Incl O&P |
|---|---|---|---|---|---|---|---|---|---|---|
| 0010 | GRAVEL STOP | | | | | | | | | |
| 0020 | Aluminum, .050" thick, 4" face height, mill finish | 1 Shee | 145 | .055 | L.F. | 6.10 | 2.85 | | 8.95 | 11.05 |
| 0080 | Duranodic finish | | 145 | .055 | | 5.90 | 2.85 | | 8.75 | 10.80 |
| 0100 | Painted | | 145 | .055 | | 6.85 | 2.85 | | 9.70 | 11.80 |
| 0300 | 6" face height | | 135 | .059 | | 6.55 | 3.06 | | 9.61 | 11.90 |
| 0350 | Duranodic finish | | 135 | .059 | | 6.95 | 3.06 | | 10.01 | 12.30 |
| 0400 | Painted | | 135 | .059 | | 8.10 | 3.06 | | 11.16 | 13.55 |
| 0600 | 8" face height | | 125 | .064 | | 8.20 | 3.31 | | 11.51 | 14.05 |
| 0650 | Duranodic finish | | 125 | .064 | | 8 | 3.31 | | 11.31 | 13.80 |
| 0700 | Painted | | 125 | .064 | | 8.15 | 3.31 | | 11.46 | 13.95 |
| 0900 | 12" face height, .080 thick, 2 piece | | 100 | .080 | | 10.95 | 4.13 | | 15.08 | 18.30 |
| 0950 | Duranodic finish | | 100 | .080 | | 10.05 | 4.13 | | 14.18 | 17.30 |
| 1000 | Painted | | 100 | .080 | | 11.90 | 4.13 | | 16.03 | 19.35 |
| 1350 | Galv steel, 24 ga., 4" leg, plain, with continuous cleat, 4" face | | 145 | .055 | | 2.65 | 2.85 | | 5.50 | 7.25 |
| 1360 | 6" face height | | 145 | .055 | | 4 | 2.85 | | 6.85 | 8.70 |
| 1500 | Polyvinyl chloride, 6" face height | | 135 | .059 | | 4.90 | 3.06 | | 7.96 | 10.05 |
| 1600 | 9" face height | | 125 | .064 | | 5.80 | 3.31 | | 9.11 | 11.35 |
| 1800 | Stainless steel, 24 ga., 6" face height | | 135 | .059 | | 12.60 | 3.06 | | 15.66 | 18.50 |
| 1900 | 12" face height | | 100 | .080 | | 26.50 | 4.13 | | 30.63 | 35.50 |
| 2100 | 20 ga., 6" face height | | 135 | .059 | | 14.30 | 3.06 | | 17.36 | 20.50 |

# 07 71 Roof Specialties

## 07 71 19 – Manufactured Gravel Stops and Fascias

| 07 71 19.10 Gravel Stop | Crew | Daily Output | Labor-Hours | Unit | Material | 2011 Bare Costs Labor | Equipment | Total | Total Incl O&P |
|---|---|---|---|---|---|---|---|---|---|
| 2200     12" face height | 1 Shee | 100 | .080 | L.F. | 30 | 4.13 | | 34.13 | 39.50 |

### 07 71 19.30 Fascia

| | Crew | Daily Output | Labor-Hours | Unit | Material | Labor | Equipment | Total | Total Incl O&P |
|---|---|---|---|---|---|---|---|---|---|
| 0010 **FASCIA** | | | | | | | | | |
| 0100     Aluminum, reverse board and batten, .032" thick, colored, no furring incl | 1 Shee | 145 | .055 | S.F. | 5.95 | 2.85 | | 8.80 | 10.85 |
| 0300     Steel, galv and enameled, stock, no furring, long panels | | 145 | .055 | | 4 | 2.85 | | 6.85 | 8.70 |
| 0600       Short panels | | 115 | .070 | | 5.45 | 3.59 | | 9.04 | 11.45 |

## 07 71 23 – Manufactured Gutters and Downspouts

### 07 71 23.10 Downspouts

| | Crew | Daily Output | Labor-Hours | Unit | Material | Labor | Equipment | Total | Total Incl O&P |
|---|---|---|---|---|---|---|---|---|---|
| 0010 **DOWNSPOUTS** | | | | | | | | | |
| 0020     Aluminum 2" x 3", .020" thick, embossed | 1 Shee | 190 | .042 | L.F. | 1.03 | 2.18 | | 3.21 | 4.42 |
| 0100       Enameled | | 190 | .042 | | 1.26 | 2.18 | | 3.44 | 4.68 |
| 0300       Enameled, .024" thick, 2" x 3" | | 180 | .044 | | 1.89 | 2.30 | | 4.19 | 5.55 |
| 0400       3" x 4" | | 140 | .057 | | 2.75 | 2.95 | | 5.70 | 7.50 |
| 0600     Round, corrugated aluminum, 3" diameter, .020" thick | | 190 | .042 | | 1.60 | 2.18 | | 3.78 | 5.05 |
| 0700       4" diameter, .025" thick | | 140 | .057 | | 2.47 | 2.95 | | 5.42 | 7.20 |
| 0900     Wire strainer, round, 2" diameter | | 155 | .052 | Ea. | 3.42 | 2.67 | | 6.09 | 7.80 |
| 1000       4" diameter | | 155 | .052 | | 6.30 | 2.67 | | 8.97 | 11 |
| 1200     Rectangular, perforated, 2" x 3" | | 145 | .055 | | 2.30 | 2.85 | | 5.15 | 6.85 |
| 1300       3" x 4" | | 145 | .055 | | 3.32 | 2.85 | | 6.17 | 7.95 |
| 1500     Copper, round, 16 oz., stock, 2" diameter | | 190 | .042 | L.F. | 6.90 | 2.18 | | 9.08 | 10.90 |
| 1600       3" diameter | | 190 | .042 | | 6.90 | 2.18 | | 9.08 | 10.90 |
| 1800       4" diameter | | 145 | .055 | | 10.20 | 2.85 | | 13.05 | 15.50 |
| 1900       5" diameter | | 130 | .062 | | 14.55 | 3.18 | | 17.73 | 21 |
| 2100     Rectangular, corrugated copper, stock, 2" x 3" | | 190 | .042 | | 7.55 | 2.18 | | 9.73 | 11.60 |
| 2200       3" x 4" | | 145 | .055 | | 9.95 | 2.85 | | 12.80 | 15.25 |
| 2400     Rectangular, plain copper, stock, 2" x 3" | | 190 | .042 | | 6.70 | 2.18 | | 8.88 | 10.65 |
| 2500       3" x 4" | | 145 | .055 | | 9.05 | 2.85 | | 11.90 | 14.30 |
| 2700     Wire strainers, rectangular, 2" x 3" | | 145 | .055 | Ea. | 5.50 | 2.85 | | 8.35 | 10.35 |
| 2800       3" x 4" | | 145 | .055 | | 6.60 | 2.85 | | 9.45 | 11.55 |
| 3000       Round, 2" diameter | | 145 | .055 | | 6.75 | 2.85 | | 9.60 | 11.70 |
| 3100       3" diameter | | 145 | .055 | | 6.75 | 2.85 | | 9.60 | 11.70 |
| 3300       4" diameter | | 145 | .055 | | 12.20 | 2.85 | | 15.05 | 17.70 |
| 3400       5" diameter | | 115 | .070 | | 16.90 | 3.59 | | 20.49 | 24 |
| 3600     Lead-coated copper, round, stock, 2" diameter | | 190 | .042 | L.F. | 34 | 2.18 | | 36.18 | 41 |
| 3700       3" diameter | | 190 | .042 | | 30 | 2.18 | | 32.18 | 36.50 |
| 3900       4" diameter | | 145 | .055 | | 31 | 2.85 | | 33.85 | 39 |
| 4200       6" diameter, corrugated | | 105 | .076 | | 30 | 3.94 | | 33.94 | 39 |
| 4300     Rectangular, corrugated, stock, 2" x 3" | | 190 | .042 | | 12.70 | 2.18 | | 14.88 | 17.25 |
| 4500       Plain, stock, 2" x 3" | | 190 | .042 | | 16.35 | 2.18 | | 18.53 | 21.50 |
| 4600       3" x 4" | | 145 | .055 | | 18.20 | 2.85 | | 21.05 | 24.50 |
| 4800     Steel, galvanized, round, corrugated, 2" or 3" diameter, 28 gauge | | 190 | .042 | | 1.72 | 2.18 | | 3.90 | 5.20 |
| 4900       4" diameter, 28 gauge | | 145 | .055 | | 2.35 | 2.85 | | 5.20 | 6.90 |
| 5100       5" diameter, 28 gauge | | 130 | .062 | | 3.48 | 3.18 | | 6.66 | 8.65 |
| 5200       26 gauge | | 130 | .062 | | 3.60 | 3.18 | | 6.78 | 8.75 |
| 5400       6" diameter, 28 gauge | | 105 | .076 | | 4 | 3.94 | | 7.94 | 10.35 |
| 5500       26 gauge | | 105 | .076 | | 5.20 | 3.94 | | 9.14 | 11.70 |
| 5700     Rectangular, corrugated, 28 gauge, 2" x 3" | | 190 | .042 | | 1.28 | 2.18 | | 3.46 | 4.70 |
| 5800       3" x 4" | | 145 | .055 | | 1.64 | 2.85 | | 4.49 | 6.10 |
| 6000     Rectangular, plain, 28 gauge, galvanized, 2" x 3" | | 190 | .042 | | 1.28 | 2.18 | | 3.46 | 4.70 |
| 6100       3" x 4" | | 145 | .055 | | 2.33 | 2.85 | | 5.18 | 6.85 |
| 6300       Epoxy painted, 24 gauge, corrugated, 2" x 3" | | 190 | .042 | | 2 | 2.18 | | 4.18 | 5.50 |
| 6400       3" x 4" | | 145 | .055 | | 2.50 | 2.85 | | 5.35 | 7.05 |

## 07 71 23 – Manufactured Gutters and Downspouts

### 07 71 23.10 Downspouts

| | | Crew | Daily Output | Labor-Hours | Unit | Material | 2011 Bare Costs Labor | Equipment | Total | Total Incl O&P |
|---|---|---|---|---|---|---|---|---|---|---|
| 6600 | Wire strainers, rectangular, 2" x 3" | 1 Shee | 145 | .055 | Ea. | 12 | 2.85 | | 14.85 | 17.50 |
| 6700 | 3" x 4" | | 145 | .055 | | 14 | 2.85 | | 16.85 | 19.70 |
| 6900 | Round strainers, 2" or 3" diameter | | 145 | .055 | | 2.80 | 2.85 | | 5.65 | 7.40 |
| 7000 | 4" diameter | | 145 | .055 | | 5.20 | 2.85 | | 8.05 | 10 |
| 7800 | Stainless steel tubing, schedule 5, 2" x 3" or 3" diameter | | 190 | .042 | L.F. | 44.50 | 2.18 | | 46.68 | 52.50 |
| 7900 | 3" x 4" or 4" diameter | | 145 | .055 | | 56.50 | 2.85 | | 59.35 | 66.50 |
| 8100 | 4" x 5" or 5" diameter | | 135 | .059 | | 117 | 3.06 | | 120.06 | 133 |
| 8200 | Vinyl, rectangular, 2" x 3" | | 210 | .038 | | 1.94 | 1.97 | | 3.91 | 5.10 |
| 8300 | Round, 2-1/2" | | 220 | .036 | | 1.28 | 1.88 | | 3.16 | 4.25 |

### 07 71 23.20 Downspout Elbows

| | | Crew | Daily Output | Labor-Hours | Unit | Material | 2011 Bare Costs Labor | Equipment | Total | Total Incl O&P |
|---|---|---|---|---|---|---|---|---|---|---|
| 0010 | **DOWNSPOUT ELBOWS** | | | | | | | | | |
| 0020 | Aluminum, 2" x 3", embossed | 1 Shee | 100 | .080 | Ea. | .90 | 4.13 | | 5.03 | 7.25 |
| 0100 | Enameled | | 100 | .080 | | .96 | 4.13 | | 5.09 | 7.30 |
| 0200 | 3" x 4", .025" thick, embossed | | 100 | .080 | | 3.50 | 4.13 | | 7.63 | 10.10 |
| 0300 | Enameled | | 100 | .080 | | 1.75 | 4.13 | | 5.88 | 8.20 |
| 0400 | Round corrugated, 3", embossed, .020" thick | | 100 | .080 | | 2.27 | 4.13 | | 6.40 | 8.75 |
| 0500 | 4", .025" thick | | 100 | .080 | | 4.74 | 4.13 | | 8.87 | 11.45 |
| 0600 | Copper, 16 oz. round, 2" diameter | | 100 | .080 | | 8 | 4.13 | | 12.13 | 15.05 |
| 0700 | 3" diameter | | 100 | .080 | | 10.55 | 4.13 | | 14.68 | 17.85 |
| 0800 | 4" diameter | | 100 | .080 | | 12.40 | 4.13 | | 16.53 | 19.90 |
| 1000 | 2" x 3" corrugated | | 100 | .080 | | 7.25 | 4.13 | | 11.38 | 14.25 |
| 1100 | 3" x 4" corrugated | | 100 | .080 | | 12.15 | 4.13 | | 16.28 | 19.60 |
| 1300 | Vinyl, 2-1/2" diameter, 45° or 75° | | 100 | .080 | | 3.98 | 4.13 | | 8.11 | 10.65 |
| 1400 | Tee Y junction | | 75 | .107 | | 12.90 | 5.50 | | 18.40 | 22.50 |

### 07 71 23.30 Gutters

| | | Crew | Daily Output | Labor-Hours | Unit | Material | 2011 Bare Costs Labor | Equipment | Total | Total Incl O&P |
|---|---|---|---|---|---|---|---|---|---|---|
| 0010 | **GUTTERS** | | | | | | | | | |
| 0012 | Aluminum, stock units, 5" K type, .027" thick, plain | 1 Shee | 120 | .067 | L.F. | 2.58 | 3.44 | | 6.02 | 8.05 |
| 0100 | Enameled | | 120 | .067 | | 2.51 | 3.44 | | 5.95 | 7.95 |
| 0300 | 5" K type type, .032" thick, plain | | 120 | .067 | | 2.74 | 3.44 | | 6.18 | 8.20 |
| 0400 | Enameled | | 120 | .067 | | 3.29 | 3.44 | | 6.73 | 8.80 |
| 0700 | Copper, half round, 16 oz, stock units, 4" wide | | 120 | .067 | | 8.20 | 3.44 | | 11.64 | 14.20 |
| 0900 | 5" wide | | 120 | .067 | | 8.55 | 3.44 | | 11.99 | 14.65 |
| 1000 | 6" wide | | 115 | .070 | | 9.65 | 3.59 | | 13.24 | 16.10 |
| 1200 | K type, 16 oz, stock, 4" wide | | 120 | .067 | | 6.85 | 3.44 | | 10.29 | 12.75 |
| 1300 | 5" wide | | 120 | .067 | | 10.65 | 3.44 | | 14.09 | 16.95 |
| 1500 | Lead coated copper, half round, stock, 4" wide | | 120 | .067 | | 9.45 | 3.44 | | 12.89 | 15.60 |
| 1600 | 6" wide | | 115 | .070 | | 15.60 | 3.59 | | 19.19 | 22.50 |
| 1800 | K type, stock, 4" wide | | 120 | .067 | | 11.35 | 3.44 | | 14.79 | 17.70 |
| 1900 | 5" wide | | 120 | .067 | | 12.55 | 3.44 | | 15.99 | 19.05 |
| 2100 | Stainless steel, half round or box, stock, 4" wide | | 120 | .067 | | 5.90 | 3.44 | | 9.34 | 11.65 |
| 2200 | 5" wide | | 120 | .067 | | 6.35 | 3.44 | | 9.79 | 12.15 |
| 2400 | Steel, galv, half round or box, 28 ga, 5" wide, plain | | 120 | .067 | | 1.98 | 3.44 | | 5.42 | 7.40 |
| 2500 | Enameled | | 120 | .067 | | 1.51 | 3.44 | | 4.95 | 6.85 |
| 2700 | 26 ga, stock, 5" wide | | 120 | .067 | | 1.59 | 3.44 | | 5.03 | 6.95 |
| 2800 | 6" wide | | 120 | .067 | | 2.79 | 3.44 | | 6.23 | 8.25 |
| 3000 | Vinyl, O.G., 4" wide | 1 Carp | 110 | .073 | | 1.08 | 3.13 | | 4.21 | 6 |
| 3100 | 5" wide | | 110 | .073 | | 1.38 | 3.13 | | 4.51 | 6.30 |
| 3200 | 4" half round, stock units | | 110 | .073 | | 1.08 | 3.13 | | 4.21 | 6 |
| 3250 | Joint connectors | | | | Ea. | 2.88 | | | 2.88 | 3.17 |
| 3300 | Wood, clear treated cedar, fir or hemlock, 3" x 4" | 1 Carp | 100 | .080 | L.F. | 7.95 | 3.44 | | 11.39 | 14.05 |
| 3400 | 4" x 5" | " | 100 | .080 | " | 11.55 | 3.44 | | 14.99 | 18 |

# 07 71 Roof Specialties

## 07 71 23 – Manufactured Gutters and Downspouts

| 07 71 23.35 Gutter Guard | Crew | Daily Output | Labor-Hours | Unit | Material | 2011 Bare Costs Labor | Equipment | Total | Total Incl O&P |
|---|---|---|---|---|---|---|---|---|---|
| 0010 **GUTTER GUARD** | | | | | | | | | |
| 0020 6" wide strip, aluminum mesh | 1 Carp | 500 | .016 | L.F. | 2.18 | .69 | | 2.87 | 3.46 |
| 0100 Vinyl mesh | " | 500 | .016 | " | 2.58 | .69 | | 3.27 | 3.90 |

## 07 71 26 – Reglets

### 07 71 26.10 Reglets and Accessories

| 07 71 26.10 Reglets and Accessories | Crew | Daily Output | Labor-Hours | Unit | Material | 2011 Bare Costs Labor | Equipment | Total | Total Incl O&P |
|---|---|---|---|---|---|---|---|---|---|
| 0010 **REGLETS AND ACCESSORIES** | | | | | | | | | |
| 0020 Aluminum, .025" thick, in concrete parapet | 1 Carp | 225 | .036 | L.F. | 1.62 | 1.53 | | 3.15 | 4.13 |
| 0100 Copper, 10 oz. | | 225 | .036 | | 2.62 | 1.53 | | 4.15 | 5.25 |
| 0300 16 oz. | | 225 | .036 | | 5.10 | 1.53 | | 6.63 | 8 |
| 0400 Galvanized steel, 24 gauge | | 225 | .036 | | .94 | 1.53 | | 2.47 | 3.38 |
| 0600 Stainless steel, .020" thick | | 225 | .036 | | 3.15 | 1.53 | | 4.68 | 5.80 |
| 0700 Zinc and copper alloy, 20 oz. | | 225 | .036 | | 2.62 | 1.53 | | 4.15 | 5.25 |
| 0900 Counter flashing for above, 12" wide, .032" aluminum | 1 Shee | 150 | .053 | | 1.63 | 2.75 | | 4.38 | 5.95 |
| 1000 Copper, 10 oz. | | 150 | .053 | | 4.95 | 2.75 | | 7.70 | 9.60 |
| 1200 16 oz. | | 150 | .053 | | 4.80 | 2.75 | | 7.55 | 9.45 |
| 1300 Galvanized steel, .020" thick | | 150 | .053 | | 1.17 | 2.75 | | 3.92 | 5.45 |
| 1500 Stainless steel, .020" thick | | 150 | .053 | | 4.69 | 2.75 | | 7.44 | 9.30 |
| 1600 Zinc and copper alloy, 20 oz. | | 150 | .053 | | 4.78 | 2.75 | | 7.53 | 9.40 |

## 07 71 29 – Manufactured Roof Expansion Joints

### 07 71 29.10 Expansion Joints

| 07 71 29.10 Expansion Joints | Crew | Daily Output | Labor-Hours | Unit | Material | 2011 Bare Costs Labor | Equipment | Total | Total Incl O&P |
|---|---|---|---|---|---|---|---|---|---|
| 0010 **EXPANSION JOINTS** | | | | | | | | | |
| 0300 Butyl or neoprene center with foam insulation, metal flanges | | | | | | | | | |
| 0400 Aluminum, .032" thick for openings to 2-1/2" | 1 Rofc | 165 | .048 | L.F. | 16 | 1.77 | | 17.77 | 20.50 |
| 0600 For joint openings to 3-1/2" | | 165 | .048 | | 16 | 1.77 | | 17.77 | 20.50 |
| 0610 For joint openings to 5" | | 165 | .048 | | 16.40 | 1.77 | | 18.17 | 21 |
| 0620 For joint openings to 8" | | 165 | .048 | | 17.30 | 1.77 | | 19.07 | 22 |
| 0700 Copper, 16 oz. for openings to 2-1/2" | | 165 | .048 | | 24.50 | 1.77 | | 26.27 | 30 |
| 0900 For joint openings to 3-1/2" | | 165 | .048 | | 24.50 | 1.77 | | 26.27 | 30 |
| 0910 For joint openings to 5" | | 165 | .048 | | 25.50 | 1.77 | | 27.27 | 31 |
| 0920 For joint openings to 8" | | 165 | .048 | | 27 | 1.77 | | 28.77 | 33 |
| 1000 Galvanized steel, 26 ga. for openings to 2-1/2" | | 165 | .048 | | 13 | 1.77 | | 14.77 | 17.25 |
| 1200 For joint openings to 3-1/2" | | 165 | .048 | | 13 | 1.77 | | 14.77 | 17.25 |
| 1210 For joint openings to 5" | | 165 | .048 | | 13.90 | 1.77 | | 15.67 | 18.25 |
| 1220 For joint openings to 8" | | 165 | .048 | | 17.30 | 1.77 | | 19.07 | 22 |
| 1300 Lead-coated copper, 16 oz. for openings to 2-1/2" | | 165 | .048 | | 33.50 | 1.77 | | 35.27 | 40 |
| 1500 For joint openings to 3-1/2" | | 165 | .048 | | 40 | 1.77 | | 41.77 | 47 |
| 1600 Stainless steel, .018", for openings to 2-1/2" | | 165 | .048 | | 17.90 | 1.77 | | 19.67 | 22.50 |
| 1800 For joint openings to 3-1/2" | | 165 | .048 | | 17.90 | 1.77 | | 19.67 | 22.50 |
| 1810 For joint openings to 5" | | 165 | .048 | | 18.60 | 1.77 | | 20.37 | 23.50 |
| 1820 For joint openings to 8" | | 165 | .048 | | 20.50 | 1.77 | | 22.27 | 25.50 |
| 1900 Neoprene, double-seal type with thick center, 4-1/2" wide | | 125 | .064 | | 14 | 2.34 | | 16.34 | 19.30 |
| 1950 Polyethylene bellows, with galv steel flat flanges | | 100 | .080 | | 5.40 | 2.92 | | 8.32 | 10.80 |
| 1960 With galvanized angle flanges | | 100 | .080 | | 6 | 2.92 | | 8.92 | 11.45 |
| 2000 Roof joint with extruded aluminum cover, 2" | 1 Shee | 115 | .070 | | 37.50 | 3.59 | | 41.09 | 47 |
| 2100 Roof joint, plastic curbs, foam center, standard | 1 Rofc | 100 | .080 | | 13.70 | 2.92 | | 16.62 | 19.90 |
| 2200 Large | | 100 | .080 | | 18.25 | 2.92 | | 21.17 | 25 |
| 2300 Transitions, regular, minimum | | 10 | .800 | Ea. | 120 | 29 | | 149 | 181 |
| 2350 Maximum | | 4 | 2 | | 150 | 73 | | 223 | 288 |
| 2400 Large, minimum | | 9 | .889 | | 173 | 32.50 | | 205.50 | 245 |
| 2450 Maximum | | 3 | 2.667 | | 182 | 97.50 | | 279.50 | 360 |
| 2500 Roof to wall joint with extruded aluminum cover | 1 Shee | 115 | .070 | L.F. | 31.50 | 3.59 | | 35.09 | 40.50 |

# 07 71 Roof Specialties

## 07 71 29 – Manufactured Roof Expansion Joints

| 07 71 29.10 Expansion Joints | Crew | Daily Output | Labor-Hours | Unit | Material | 2011 Bare Costs Labor | Equipment | Total | Total Incl O&P |
|---|---|---|---|---|---|---|---|---|---|
| 2700 | Wall joint, closed cell foam on PVC cover, 9" wide | 1 Rofc | 125 | .064 | L.F. | 4.90 | 2.34 | | 7.24 | 9.30 |
| 2800 | 12" wide | " | 115 | .070 | | 5.50 | 2.54 | | 8.04 | 10.30 |

## 07 71 43 – Drip Edge

### 07 71 43.10 Drip Edge, Rake Edge, Ice Belts

| | | Crew | Daily Output | Labor-Hours | Unit | Material | Labor | Equipment | Total | Total Incl O&P |
|---|---|---|---|---|---|---|---|---|---|---|
| 0010 | **DRIP EDGE, RAKE EDGE, ICE BELTS** | | | | | | | | | |
| 0020 | Aluminum, .016" thick, 5" wide, mill finish | 1 Carp | 400 | .020 | L.F. | .38 | .86 | | 1.24 | 1.74 |
| 0100 | White finish | | 400 | .020 | | .38 | .86 | | 1.24 | 1.74 |
| 0200 | 8" wide, mill finish | | 400 | .020 | | .46 | .86 | | 1.32 | 1.83 |
| 0300 | Ice belt, 28" wide, mill finish | | 100 | .080 | | 3.85 | 3.44 | | 7.29 | 9.55 |
| 0310 | Vented, mill finish | | 400 | .020 | | 2.76 | .86 | | 3.62 | 4.36 |
| 0320 | Painted finish | | 400 | .020 | | 2.86 | .86 | | 3.72 | 4.47 |
| 0400 | Galvanized, 5" wide | | 400 | .020 | | .36 | .86 | | 1.22 | 1.72 |
| 0500 | 8" wide, mill finish | | 400 | .020 | | .69 | .86 | | 1.55 | 2.08 |
| 0510 | Rake edge, aluminum, 1-1/2" x 1-1/2" | | 400 | .020 | | .28 | .86 | | 1.14 | 1.63 |
| 0520 | 3-1/2" x 1-1/2" | | 400 | .020 | | .38 | .86 | | 1.24 | 1.74 |

# 07 72 Roof Accessories

## 07 72 23 – Relief Vents

### 07 72 23.10 Roof Vents

| | | Crew | Daily Output | Labor-Hours | Unit | Material | Labor | Equipment | Total | Total Incl O&P |
|---|---|---|---|---|---|---|---|---|---|---|
| 0010 | **ROOF VENTS** | | | | | | | | | |
| 0020 | Mushroom shape, for built-up roofs, aluminum | 1 Rofc | 30 | .267 | Ea. | 88.50 | 9.75 | | 98.25 | 113 |
| 0100 | PVC, 6" high | " | 30 | .267 | " | 28.50 | 9.75 | | 38.25 | 48 |

## 07 72 26 – Ridge Vents

### 07 72 26.10 Ridge Vents and Accessories

| | | Crew | Daily Output | Labor-Hours | Unit | Material | Labor | Equipment | Total | Total Incl O&P |
|---|---|---|---|---|---|---|---|---|---|---|
| 0010 | **RIDGE VENTS AND ACCESSORIES** | | | | | | | | | |
| 0100 | Aluminum strips, mill finish | 1 Rofc | 160 | .050 | L.F. | 2.62 | 1.83 | | 4.45 | 5.90 |
| 0150 | Painted finish | | 160 | .050 | " | 3.74 | 1.83 | | 5.57 | 7.15 |
| 0200 | Connectors | | 48 | .167 | Ea. | 3.68 | 6.10 | | 9.78 | 14.20 |
| 0300 | End caps | | 48 | .167 | " | 1.37 | 6.10 | | 7.47 | 11.65 |
| 0400 | Galvanized strips | | 160 | .050 | L.F. | 2 | 1.83 | | 3.83 | 5.25 |
| 0430 | Molded polyethylene, shingles not included | | 160 | .050 | " | 2.99 | 1.83 | | 4.82 | 6.35 |
| 0440 | End plugs | | 48 | .167 | Ea. | 1.37 | 6.10 | | 7.47 | 11.65 |
| 0450 | Flexible roll, shingles not included | | 160 | .050 | L.F. | 2.15 | 1.83 | | 3.98 | 5.40 |
| 2300 | Ridge vent strip, mill finish | 1 Shee | 155 | .052 | " | 2.73 | 2.67 | | 5.40 | 7.05 |

## 07 72 33 – Roof Hatches

### 07 72 33.10 Roof Hatch Options

| | | Crew | Daily Output | Labor-Hours | Unit | Material | Labor | Equipment | Total | Total Incl O&P |
|---|---|---|---|---|---|---|---|---|---|---|
| 0010 | **ROOF HATCH OPTIONS** | | | | | | | | | |
| 0500 | 2'-6" x 3', aluminum curb and cover | G-3 | 10 | 3.200 | Ea. | 985 | 138 | | 1,123 | 1,275 |
| 0520 | Galvanized steel curb and aluminum cover | | 10 | 3.200 | | 595 | 138 | | 733 | 865 |
| 0540 | Galvanized steel curb and cover | | 10 | 3.200 | | 550 | 138 | | 688 | 815 |
| 0600 | 2'-6" x 4'-6", aluminum curb and cover | | 9 | 3.556 | | 850 | 153 | | 1,003 | 1,175 |
| 0800 | Galvanized steel curb and aluminum cover | | 9 | 3.556 | | 600 | 153 | | 753 | 895 |
| 0900 | Galvanized steel curb and cover | | 9 | 3.556 | | 825 | 153 | | 978 | 1,150 |
| 1100 | 4' x 4' aluminum curb and cover | | 8 | 4 | | 1,675 | 172 | | 1,847 | 2,075 |
| 1120 | Galvanized steel curb and aluminum cover | | 8 | 4 | | 1,675 | 172 | | 1,847 | 2,075 |
| 1140 | Galvanized steel curb and cover | | 8 | 4 | | 1,500 | 172 | | 1,672 | 1,900 |
| 1200 | 2'-6" x 8'-0", aluminum curb and cover | | 6.60 | 4.848 | | 1,675 | 208 | | 1,883 | 2,150 |
| 1400 | Galvanized steel curb and aluminum cover | | 6.60 | 4.848 | | 1,600 | 208 | | 1,808 | 2,075 |
| 1500 | Galvanized steel curb and cover | | 6.60 | 4.848 | | 1,200 | 208 | | 1,408 | 1,650 |

# Looking for your next job to bid?

Get up-to-date construction project information, plans and specs with SmartProject News from Reed Construction Data.

- Find projects bidding in your area
- Browse full project details, plans and specs
- Identify key project participants
- Track projects with e-mail notifications

**Register for free access at www.smartprojectnews.com.**

**Smart**Project News
Reed Construction Data

# 07 72 Roof Accessories

## 07 72 33 – Roof Hatches

### 07 72 33.10 Roof Hatch Options

| | | Crew | Daily Output | Labor-Hours | Unit | Material | 2011 Bare Costs Labor | Equipment | Total | Total Incl O&P |
|---|---|---|---|---|---|---|---|---|---|---|
| 1800 | For plexiglass panels, 2'-6" x 3'-0", add to above | | | | Ea. | 400 | | | 400 | 440 |

## 07 72 36 – Smoke Vents

### 07 72 36.10 Smoke Hatches

| 0010 | **SMOKE HATCHES** | | | | | | | | | |
|---|---|---|---|---|---|---|---|---|---|---|
| 0200 | For 3'-0" long, add to roof hatches from Section 07 72 33.10 | | | | Ea. | 25% | 5% | | | |
| 0250 | For 4'-0" long, add to roof hatches from Section 07 72 33.10 | | | | | 20% | 5% | | | |
| 0300 | For 8'-0" long, add to roof hatches from Section 07 72 33.10 | | | | | 10% | 5% | | | |

### 07 72 36.20 Smoke Vent Options

| 0010 | **SMOKE VENT OPTIONS** | | | | | | | | | |
|---|---|---|---|---|---|---|---|---|---|---|
| 0100 | 4' x 4' aluminum cover and frame | G-3 | 13 | 2.462 | Ea. | 1,900 | 106 | | 2,006 | 2,250 |
| 0200 | Galvanized steel cover and frame | | 13 | 2.462 | | 1,675 | 106 | | 1,781 | 1,975 |
| 0300 | 4' x 8' aluminum cover and frame | | 8 | 4 | | 2,600 | 172 | | 2,772 | 3,100 |
| 0400 | Galvanized steel cover and frame | | 8 | 4 | | 2,200 | 172 | | 2,372 | 2,675 |

## 07 72 53 – Snow Guards

### 07 72 53.10 Snow Guard Options

| 0010 | **SNOW GUARD OPTIONS** | | | | | | | | | |
|---|---|---|---|---|---|---|---|---|---|---|
| 0100 | Slate & asphalt shingle roofs, fastened with nails | 1 Rofc | 160 | .050 | Ea. | 9.40 | 1.83 | | 11.23 | 13.40 |
| 0200 | Standing seam metal roofs, fastened with set screws | | 48 | .167 | | 14.60 | 6.10 | | 20.70 | 26 |
| 0300 | Surface mount for metal roofs, fastened with solder | | 48 | .167 | | 6.60 | 6.10 | | 12.70 | 17.40 |
| 0400 | Double rail pipe type, including pipe | | 130 | .062 | L.F. | 22.50 | 2.25 | | 24.75 | 28.50 |

## 07 72 73 – Pitch Pockets

### 07 72 73.10 Pitch Pockets, Variable Sizes

| 0010 | **PITCH POCKETS, VARIABLE SIZES** | | | | | | | | | |
|---|---|---|---|---|---|---|---|---|---|---|
| 0100 | Adjustable, 4" to 7", welded corners, 4" deep | 1 Rofc | 48 | .167 | Ea. | 12.90 | 6.10 | | 19 | 24.50 |
| 0200 | Side extenders, 6" | " | 240 | .033 | " | 1.97 | 1.22 | | 3.19 | 4.20 |

## 07 72 80 – Vents

### 07 72 80.30 Vent Options

| 0010 | **VENT OPTIONS** | | | | | | | | | |
|---|---|---|---|---|---|---|---|---|---|---|
| 0020 | Plastic, for insulated decks, 1 per M.S.F., minimum | 1 Rofc | 40 | .200 | Ea. | 19 | 7.30 | | 26.30 | 33 |
| 0100 | Maximum | | 20 | .400 | | 38 | 14.60 | | 52.60 | 66.50 |
| 0300 | Aluminum | | 30 | .267 | | 24 | 9.75 | | 33.75 | 43 |
| 0800 | Polystyrene baffles, 12" wide for 16" O.C. rafter spacing | 1 Carp | 90 | .089 | | .40 | 3.83 | | 4.23 | 6.30 |
| 0900 | For 24" O.C. rafter spacing | " | 110 | .073 | | .75 | 3.13 | | 3.88 | 5.65 |

# 07 81 Applied Fireproofing

## 07 81 16 – Cementitious Fireproofing

### 07 81 16.10 Sprayed Cementitious Fireproofing

| 0010 | **SPRAYED CEMENTITIOUS FIREPROOFING** | | | | | | | | | |
|---|---|---|---|---|---|---|---|---|---|---|
| 0050 | Not incl tamping or canvas protection | | | | | | | | | |
| 0100 | 1" thick, on flat plate steel | G-2 | 3000 | .008 | S.F. | .53 | .29 | .04 | .86 | 1.07 |
| 0200 | Flat decking | | 2400 | .010 | | .53 | .36 | .05 | .94 | 1.18 |
| 0400 | Beams | | 1500 | .016 | | .53 | .58 | .08 | 1.19 | 1.54 |
| 0500 | Corrugated or fluted decks | | 1250 | .019 | | .79 | .70 | .10 | 1.59 | 2.03 |
| 0700 | Columns, 1-1/8" thick | | 1100 | .022 | | .59 | .79 | .11 | 1.49 | 1.97 |
| 0800 | 2-3/16" thick | | 700 | .034 | | 1.13 | 1.24 | .18 | 2.55 | 3.31 |
| 0850 | For tamping, add | | | | | | 10% | | | |
| 0900 | For canvas protection, add | G-2 | 5000 | .005 | S.F. | .07 | .17 | .03 | .27 | .37 |
| 1500 | Intumescent epoxy fireproofing on wire mesh, 3/16" thick | | | | | | | | | |

# 07 81 Applied Fireproofing

## 07 81 16 – Cementitious Fireproofing

| 07 81 16.10 Sprayed Cementitious Fireproofing | Crew | Daily Output | Labor-Hours | Unit | Material | 2011 Bare Costs Labor | Equipment | Total | Total Incl O&P |
|---|---|---|---|---|---|---|---|---|---|
| 1550 | 1 hour rating, exterior use | G-2 | 136 | .176 | S.F. | 6.90 | 6.40 | .92 | 14.22 | 18.15 |
| 1600 | Magnesium oxychloride, 35# to 40# density, 1/4" thick | | 3000 | .008 | | 1.44 | .29 | .04 | 1.77 | 2.07 |
| 1650 | 1/2" thick | | 2000 | .012 | | 2.90 | .44 | .06 | 3.40 | 3.91 |
| 1700 | 60# to 70# density, 1/4" thick | | 3000 | .008 | | 1.92 | .29 | .04 | 2.25 | 2.60 |
| 1750 | 1/2" thick | | 2000 | .012 | | 3.85 | .44 | .06 | 4.35 | 4.96 |
| 2000 | Vermiculite cement, troweled or sprayed, 1/4" thick | | 3000 | .008 | | 1.31 | .29 | .04 | 1.64 | 1.93 |
| 2050 | 1/2" thick | | 2000 | .012 | | 2.60 | .44 | .06 | 3.10 | 3.58 |

# 07 84 Firestopping

## 07 84 13 – Penetration Firestopping

### 07 84 13.10 Firestopping

| | 07 84 13.10 Firestopping | Crew | Daily Output | Labor-Hours | Unit | Material | 2011 Bare Costs Labor | Equipment | Total | Total Incl O&P |
|---|---|---|---|---|---|---|---|---|---|---|
| 0010 | **FIRESTOPPING** R078413-30 | | | | | | | | | |
| 0100 | Metallic piping, non insulated | | | | | | | | | |
| 0110 | Through walls, 2" diameter | 1 Carp | 16 | .500 | Ea. | 13.75 | 21.50 | | 35.25 | 48 |
| 0120 | 4" diameter | | 14 | .571 | | 21 | 24.50 | | 45.50 | 60.50 |
| 0130 | 6" diameter | | 12 | .667 | | 28.50 | 28.50 | | 57 | 75 |
| 0140 | 12" diameter | | 10 | .800 | | 50 | 34.50 | | 84.50 | 108 |
| 0150 | Through floors, 2" diameter | | 32 | .250 | | 8.35 | 10.75 | | 19.10 | 25.50 |
| 0160 | 4" diameter | | 28 | .286 | | 12 | 12.30 | | 24.30 | 32 |
| 0170 | 6" diameter | | 24 | .333 | | 15.75 | 14.35 | | 30.10 | 39.50 |
| 0180 | 12" diameter | | 20 | .400 | | 26.50 | 17.20 | | 43.70 | 56 |
| 0190 | Metallic piping, insulated | | | | | | | | | |
| 0200 | Through walls, 2" diameter | 1 Carp | 16 | .500 | Ea. | 19.50 | 21.50 | | 41 | 54.50 |
| 0210 | 4" diameter | | 14 | .571 | | 26.50 | 24.50 | | 51 | 67 |
| 0220 | 6" diameter | | 12 | .667 | | 34 | 28.50 | | 62.50 | 81.50 |
| 0230 | 12" diameter | | 10 | .800 | | 55.50 | 34.50 | | 90 | 114 |
| 0240 | Through floors, 2" diameter | | 32 | .250 | | 14.10 | 10.75 | | 24.85 | 32 |
| 0250 | 4" diameter | | 28 | .286 | | 17.70 | 12.30 | | 30 | 38.50 |
| 0260 | 6" diameter | | 24 | .333 | | 21.50 | 14.35 | | 35.85 | 45.50 |
| 0270 | 12" diameter | | 20 | .400 | | 26.50 | 17.20 | | 43.70 | 56 |
| 0280 | Non metallic piping, non insulated | | | | | | | | | |
| 0290 | Through walls, 2" diameter | 1 Carp | 12 | .667 | Ea. | 56.50 | 28.50 | | 85 | 107 |
| 0300 | 4" diameter | | 10 | .800 | | 71 | 34.50 | | 105.50 | 131 |
| 0310 | 6" diameter | | 8 | 1 | | 99 | 43 | | 142 | 175 |
| 0330 | Through floors, 2" diameter | | 16 | .500 | | 44.50 | 21.50 | | 66 | 82 |
| 0340 | 4" diameter | | 6 | 1.333 | | 55 | 57.50 | | 112.50 | 149 |
| 0350 | 6" diameter | | 6 | 1.333 | | 66 | 57.50 | | 123.50 | 161 |
| 0370 | Ductwork, insulated & non insulated, round | | | | | | | | | |
| 0380 | Through walls, 6" diameter | 1 Carp | 12 | .667 | Ea. | 29 | 28.50 | | 57.50 | 76 |
| 0390 | 12" diameter | | 10 | .800 | | 57 | 34.50 | | 91.50 | 116 |
| 0400 | 18" diameter | | 8 | 1 | | 93 | 43 | | 136 | 169 |
| 0410 | Through floors, 6" diameter | | 16 | .500 | | 15.75 | 21.50 | | 37.25 | 50.50 |
| 0420 | 12" diameter | | 14 | .571 | | 29 | 24.50 | | 53.50 | 69.50 |
| 0430 | 18" diameter | | 12 | .667 | | 50 | 28.50 | | 78.50 | 99 |
| 0440 | Ductwork, insulated & non insulated, rectangular | | | | | | | | | |
| 0450 | With stiffener/closure angle, through walls, 6" x 12" | 1 Carp | 8 | 1 | Ea. | 24 | 43 | | 67 | 92 |
| 0460 | 12" x 24" | | 6 | 1.333 | | 31.50 | 57.50 | | 89 | 123 |
| 0470 | 24" x 48" | | 4 | 2 | | 90.50 | 86 | | 176.50 | 232 |
| 0480 | With stiffener/closure angle, through floors, 6" x 12" | | 10 | .800 | | 12.95 | 34.50 | | 47.45 | 67 |
| 0490 | 12" x 24" | | 8 | 1 | | 23.50 | 43 | | 66.50 | 91.50 |
| 0500 | 24" x 48" | | 6 | 1.333 | | 45.50 | 57.50 | | 103 | 138 |

# 07 84 Firestopping

## 07 84 13 – Penetration Firestopping

| 07 84 13.10 Firestopping | | Crew | Daily Output | Labor-Hours | Unit | Material | 2011 Bare Costs Labor | Equipment | Total | Total Incl O&P |
|---|---|---|---|---|---|---|---|---|---|---|
| 0510 | Multi trade openings | | | | | | | | | |
| 0520 | Through walls, 6" x 12" | 1 Carp | 2 | 4 | Ea. | 50 | 172 | | 222 | 320 |
| 0530 | 12" x 24" | " | 1 | 8 | | 202 | 345 | | 547 | 750 |
| 0540 | 24" x 48" | 2 Carp | 1 | 16 | | 810 | 690 | | 1,500 | 1,950 |
| 0550 | 48" x 96" | " | .75 | 21.333 | | 3,250 | 920 | | 4,170 | 4,975 |
| 0560 | Through floors, 6" x 12" | 1 Carp | 2 | 4 | | 50 | 172 | | 222 | 320 |
| 0570 | 12" x 24" | " | 1 | 8 | | 202 | 345 | | 547 | 750 |
| 0580 | 24" x 48" | 2 Carp | .75 | 21.333 | | 810 | 920 | | 1,730 | 2,300 |
| 0590 | 48" x 96" | " | .50 | 32 | ▼ | 3,250 | 1,375 | | 4,625 | 5,700 |
| 0600 | Structural penetrations, through walls | | | | | | | | | |
| 0610 | Steel beams, W8 x 10 | 1 Carp | 8 | 1 | Ea. | 31.50 | 43 | | 74.50 | 101 |
| 0620 | W12 x 14 | | 6 | 1.333 | | 50 | 57.50 | | 107.50 | 143 |
| 0630 | W21 x 44 | | 5 | 1.600 | | 100 | 69 | | 169 | 217 |
| 0640 | W36 x 135 | | 3 | 2.667 | | 244 | 115 | | 359 | 445 |
| 0650 | Bar joists, 18" deep | | 6 | 1.333 | | 46 | 57.50 | | 103.50 | 139 |
| 0660 | 24" deep | | 6 | 1.333 | | 57 | 57.50 | | 114.50 | 151 |
| 0670 | 36" deep | | 5 | 1.600 | | 86 | 69 | | 155 | 201 |
| 0680 | 48" deep | ▼ | 4 | 2 | ▼ | 100 | 86 | | 186 | 243 |
| 0690 | Construction joints, floor slab at exterior wall | | | | | | | | | |
| 0700 | Precast, brick, block or drywall exterior | | | | | | | | | |
| 0710 | 2" wide joint | 1 Carp | 125 | .064 | L.F. | 7.15 | 2.76 | | 9.91 | 12.15 |
| 0720 | 4" wide joint | " | 75 | .107 | " | 14.30 | 4.59 | | 18.89 | 23 |
| 0730 | Metal panel, glass or curtain wall exterior | | | | | | | | | |
| 0740 | 2" wide joint | 1 Carp | 40 | .200 | L.F. | 17 | 8.60 | | 25.60 | 32 |
| 0750 | 4" wide joint | " | 25 | .320 | " | 23.50 | 13.80 | | 37.30 | 46.50 |
| 0760 | Floor slab to drywall partition | | | | | | | | | |
| 0770 | Flat joint | 1 Carp | 100 | .080 | L.F. | 7.05 | 3.44 | | 10.49 | 13.05 |
| 0780 | Fluted joint | | 50 | .160 | | 14.30 | 6.90 | | 21.20 | 26.50 |
| 0790 | Etched fluted joint | ▼ | 75 | .107 | ▼ | 9.30 | 4.59 | | 13.89 | 17.30 |
| 0800 | Floor slab to concrete/masonry partition | | | | | | | | | |
| 0810 | Flat joint | 1 Carp | 75 | .107 | L.F. | 15.75 | 4.59 | | 20.34 | 24.50 |
| 0820 | Fluted joint | " | 50 | .160 | " | 18.65 | 6.90 | | 25.55 | 31 |
| 0830 | Concrete/CMU wall joints | | | | | | | | | |
| 0840 | 1" wide | 1 Carp | 100 | .080 | L.F. | 8.60 | 3.44 | | 12.04 | 14.75 |
| 0850 | 2" wide | | 75 | .107 | | 15.75 | 4.59 | | 20.34 | 24.50 |
| 0860 | 4" wide | ▼ | 50 | .160 | ▼ | 30 | 6.90 | | 36.90 | 43.50 |
| 0870 | Concrete/CMU floor joints | | | | | | | | | |
| 0880 | 1" wide | 1 Carp | 200 | .040 | L.F. | 4.31 | 1.72 | | 6.03 | 7.40 |
| 0890 | 2" wide | | 150 | .053 | | 7.90 | 2.30 | | 10.20 | 12.15 |
| 0900 | 4" wide | ▼ | 100 | .080 | ▼ | 15.05 | 3.44 | | 18.49 | 22 |

# 07 91 Preformed Joint Seals

## 07 91 13 – Compression Seals

| 07 91 13.10 Compression Seals | Crew | Daily Output | Labor-Hours | Unit | Material | 2011 Bare Costs Labor | Equipment | Total | Total Incl O&P |
|---|---|---|---|---|---|---|---|---|---|
| 0010 **COMPRESSION SEALS** | | | | | | | | | |
| 4900 O-ring type cord, 1/4" | 1 Bric | 472 | .017 | L.F. | .47 | .73 | | 1.20 | 1.62 |
| 4910 1/2" | | 440 | .018 | | 1.78 | .78 | | 2.56 | 3.14 |
| 4920 3/4" | | 424 | .019 | | 3.61 | .81 | | 4.42 | 5.20 |
| 4930 1" | | 408 | .020 | | 6.95 | .85 | | 7.80 | 8.90 |
| 4940 1-1/4" | | 384 | .021 | | 13.60 | .90 | | 14.50 | 16.30 |
| 4950 1-1/2" | | 368 | .022 | | 17.20 | .94 | | 18.14 | 20.50 |
| 4960 1-3/4" | | 352 | .023 | | 36 | .98 | | 36.98 | 41 |
| 4970 2" | | 344 | .023 | | 63.50 | 1 | | 64.50 | 71.50 |

## 07 91 16 – Joint Gaskets

### 07 91 16.10 Joint Gaskets

| | Crew | Daily Output | Labor-Hours | Unit | Material | Labor | Equipment | Total | Total Incl O&P |
|---|---|---|---|---|---|---|---|---|---|
| 0010 **JOINT GASKETS** | | | | | | | | | |
| 4400 Joint gaskets, neoprene, closed cell w/adh, 1/8" x 3/8" | 1 Bric | 240 | .033 | L.F. | .26 | 1.44 | | 1.70 | 2.45 |
| 4500 1/4" x 3/4" | | 215 | .037 | | .58 | 1.61 | | 2.19 | 3.05 |
| 4700 1/2" x 1" | | 200 | .040 | | 1.30 | 1.73 | | 3.03 | 4.02 |
| 4800 3/4" x 1-1/2" | | 165 | .048 | | 1.44 | 2.09 | | 3.53 | 4.72 |

## 07 91 23 – Backer Rods

### 07 91 23.10 Backer Rods

| | Crew | Daily Output | Labor-Hours | Unit | Material | Labor | Equipment | Total | Total Incl O&P |
|---|---|---|---|---|---|---|---|---|---|
| 0010 **BACKER RODS** | | | | | | | | | |
| 0030 Backer rod, polyethylene, 1/4" diameter | 1 Bric | 4.60 | 1.739 | C.L.F. | 2.83 | 75 | | 77.83 | 116 |
| 0050 1/2" diameter | | 4.60 | 1.739 | | 6.60 | 75 | | 81.60 | 120 |
| 0070 3/4" diameter | | 4.60 | 1.739 | | 10.10 | 75 | | 85.10 | 124 |
| 0090 1" diameter | | 4.60 | 1.739 | | 18.70 | 75 | | 93.70 | 134 |

## 07 91 26 – Joint Fillers

### 07 91 26.10 Joint Fillers

| | Crew | Daily Output | Labor-Hours | Unit | Material | Labor | Equipment | Total | Total Incl O&P |
|---|---|---|---|---|---|---|---|---|---|
| 0010 **JOINT FILLERS** | | | | | | | | | |
| 4360 Butyl rubber filler, 1/4" x 1/4" | 1 Bric | 290 | .028 | L.F. | .14 | 1.19 | | 1.33 | 1.95 |
| 4365 1/2" x 1/2" | | 250 | .032 | | .57 | 1.38 | | 1.95 | 2.69 |
| 4370 1/2" x 3/4" | | 210 | .038 | | .85 | 1.64 | | 2.49 | 3.41 |
| 4375 3/4" x 3/4" | | 230 | .035 | | 1.28 | 1.50 | | 2.78 | 3.66 |
| 4380 1" x 1" | | 180 | .044 | | 1.71 | 1.92 | | 3.63 | 4.76 |
| 4390 For coloring, add | | | | | 12% | | | | |
| 4980 Polyethylene joint backing, 1/4" x 2" | 1 Bric | 2.08 | 3.846 | C.L.F. | 12 | 166 | | 178 | 262 |
| 4990 1/4" x 6" | | 1.28 | 6.250 | " | 36 | 270 | | 306 | 445 |
| 5600 Silicone, room temp vulcanizing foam seal, 1/4" x 1/2" | | 1312 | .006 | L.F. | .29 | .26 | | .55 | .72 |
| 5610 1/2" x 1/2" | | 656 | .012 | | .57 | .53 | | 1.10 | 1.42 |
| 5620 1/2" x 3/4" | | 442 | .018 | | .86 | .78 | | 1.64 | 2.12 |
| 5630 3/4" x 3/4" | | 328 | .024 | | 1.29 | 1.05 | | 2.34 | 3 |
| 5640 1/8" x 1" | | 1312 | .006 | | .29 | .26 | | .55 | .72 |
| 5650 1/8" x 3" | | 442 | .018 | | .86 | .78 | | 1.64 | 2.12 |
| 5670 1/4" x 3" | | 295 | .027 | | 1.72 | 1.17 | | 2.89 | 3.65 |
| 5680 1/4" x 6" | | 148 | .054 | | 3.44 | 2.33 | | 5.77 | 7.30 |
| 5690 1/2" x 6" | | 82 | .098 | | 6.90 | 4.21 | | 11.11 | 13.85 |
| 5700 1/2" x 9" | | 52.50 | .152 | | 10.30 | 6.60 | | 16.90 | 21 |
| 5710 1/2" x 12" | | 33 | .242 | | 13.75 | 10.45 | | 24.20 | 31 |

# 07 92 Joint Sealants

## 07 92 13 – Elastomeric Joint Sealants

### 07 92 13.20 Caulking and Sealant Options

| 07 92 13.20 Caulking and Sealant Options | Crew | Daily Output | Labor-Hours | Unit | Material | 2011 Bare Costs Labor | Equipment | Total | Total Incl O&P |
|---|---|---|---|---|---|---|---|---|---|
| 0010 **CAULKING AND SEALANT OPTIONS** | | | | | | | | | |
| 0050 Latex acrylic based, bulk | | | | Gal. | 27 | | | 27 | 29.50 |
| 0055 Bulk in place 1/4" x 1/4" bead | 1 Bric | 300 | .027 | L.F. | .09 | 1.15 | | 1.24 | 1.82 |
| 0060 1/4" x 3/8" | | 294 | .027 | | .14 | 1.17 | | 1.31 | 1.92 |
| 0065 1/4" x 1/2" | | 288 | .028 | | .19 | 1.20 | | 1.39 | 2.01 |
| 0075 3/8" x 3/8" | | 284 | .028 | | .21 | 1.22 | | 1.43 | 2.05 |
| 0080 3/8" x 1/2" | | 280 | .029 | | .28 | 1.23 | | 1.51 | 2.16 |
| 0085 3/8" x 5/8" | | 276 | .029 | | .35 | 1.25 | | 1.60 | 2.27 |
| 0095 3/8" x 3/4" | | 272 | .029 | | .42 | 1.27 | | 1.69 | 2.37 |
| 0100 1/2" x 1/2" | | 275 | .029 | | .38 | 1.26 | | 1.64 | 2.30 |
| 0105 1/2" x 5/8" | | 269 | .030 | | .47 | 1.28 | | 1.75 | 2.45 |
| 0110 1/2" x 3/4" | | 263 | .030 | | .57 | 1.31 | | 1.88 | 2.60 |
| 0115 1/2" x 7/8" | | 256 | .031 | | .66 | 1.35 | | 2.01 | 2.75 |
| 0120 1/2" x 1" | | 250 | .032 | | .76 | 1.38 | | 2.14 | 2.90 |
| 0125 3/4" x 3/4" | | 244 | .033 | | .85 | 1.41 | | 2.26 | 3.06 |
| 0130 3/4" x 1" | | 225 | .036 | | 1.14 | 1.53 | | 2.67 | 3.55 |
| 0135 1" x 1" | ▼ | 200 | .040 | ▼ | 1.52 | 1.73 | | 3.25 | 4.26 |
| 0190 Cartridges | | | | Gal. | 22.50 | | | 22.50 | 24.50 |
| 0200 11 fl. oz cartridge | | | | Ea. | 1.93 | | | 1.93 | 2.12 |
| 0500 1/4" x 1/2" | 1 Bric | 288 | .028 | L.F. | .16 | 1.20 | | 1.36 | 1.97 |
| 0600 1/2" x 1/2" | | 275 | .029 | | .32 | 1.26 | | 1.58 | 2.23 |
| 0800 3/4" x 3/4" | | 244 | .033 | | .71 | 1.41 | | 2.12 | 2.90 |
| 0900 3/4" x 1" | | 225 | .036 | | .95 | 1.53 | | 2.48 | 3.34 |
| 1000 1" x 1" | ▼ | 200 | .040 | ▼ | 1.18 | 1.73 | | 2.91 | 3.89 |
| 1400 Butyl based, bulk | | | | Gal. | 26 | | | 26 | 28.50 |
| 1500 Cartridges | | | | " | 32 | | | 32 | 35 |
| 1700 1/4" x 1/2", 154 L.F./gal. | 1 Bric | 288 | .028 | L.F. | .17 | 1.20 | | 1.37 | 1.99 |
| 1800 1/2" x 1/2", 77 L.F./gal. | " | 275 | .029 | " | .34 | 1.26 | | 1.60 | 2.25 |
| 2300 Polysulfide compounds, 1 component, bulk | | | | Gal. | 56 | | | 56 | 61.50 |
| 2600 1 or 2 component, in place, 1/4" x 1/4", 308 L.F./gal. | 1 Bric | 300 | .027 | L.F. | .18 | 1.15 | | 1.33 | 1.93 |
| 2700 1/2" x 1/4", 154 L.F./gal. | | 288 | .028 | | .36 | 1.20 | | 1.56 | 2.20 |
| 2900 3/4" x 3/8", 68 L.F./gal. | | 272 | .029 | | .82 | 1.27 | | 2.09 | 2.81 |
| 3000 1" x 1/2", 38 L.F./gal. | ▼ | 250 | .032 | ▼ | 1.47 | 1.38 | | 2.85 | 3.69 |
| 3200 Polyurethane, 1 or 2 component | | | | Gal. | 50 | | | 50 | 55 |
| 3500 Bulk, in place, 1/4" x 1/4" | 1 Bric | 300 | .027 | L.F. | .16 | 1.15 | | 1.31 | 1.91 |
| 3655 1/2" x 1/4" | | 288 | .028 | | .32 | 1.20 | | 1.52 | 2.16 |
| 3800 3/4" x 3/8" | | 272 | .029 | | .74 | 1.27 | | 2.01 | 2.71 |
| 3900 1" x 1/2" | ▼ | 250 | .032 | ▼ | 1.30 | 1.38 | | 2.68 | 3.50 |
| 4100 Silicone rubber, bulk | | | | Gal. | 41 | | | 41 | 45 |
| 4200 Cartridges | | | | " | 41 | | | 41 | 45 |

## 07 92 16 – Rigid Joint Sealants

### 07 92 16.10 Rigid Joint Sealants

| | Crew | Daily Output | Labor-Hours | Unit | Material | Labor | Equipment | Total | Total Incl O&P |
|---|---|---|---|---|---|---|---|---|---|
| 0010 **RIGID JOINT SEALANTS** | | | | | | | | | |
| 5800 Tapes, sealant, P.V.C. foam adhesive, 1/16" x 1/4" | | | | C.L.F. | 5.65 | | | 5.65 | 6.20 |
| 5900 1/16" x 1/2" | | | | | 8.35 | | | 8.35 | 9.20 |
| 5950 1/16" x 1" | | | | | 13.85 | | | 13.85 | 15.25 |
| 6000 1/8" x 1/2" | | | | ▼ | 9.35 | | | 9.35 | 10.30 |

## 07 92 19 – Acoustical Joint Sealants

### 07 92 19.10 Acoustical Sealant

| | Crew | Daily Output | Labor-Hours | Unit | Material | Labor | Equipment | Total | Total Incl O&P |
|---|---|---|---|---|---|---|---|---|---|
| 0010 **ACOUSTICAL SEALANT** | | | | | | | | | |
| 0020 Acoustical sealant, elastomeric, cartridges | | | | Ea. | 2.30 | | | 2.30 | 2.53 |
| 0025 In place, 1/4" x 1/4" | 1 Bric | 300 | .027 | L.F. | .09 | 1.15 | | 1.24 | 1.83 |

# 07 92 Joint Sealants

## 07 92 19 – Acoustical Joint Sealants

| 07 92 19.10 Acoustical Sealant | | Crew | Daily Output | Labor-Hours | Unit | Material | 2011 Bare Costs Labor | Equipment | Total | Total Incl O&P |
|---|---|---|---|---|---|---|---|---|---|---|
| 0030 | 1/4" x 1/2" | 1 Bric | 288 | .028 | L.F. | .19 | 1.20 | | 1.39 | 2.01 |
| 0035 | 1/2" x 1/2" | | 275 | .029 | | .38 | 1.26 | | 1.64 | 2.29 |
| 0040 | 1/2" x 3/4" | | 263 | .030 | | .56 | 1.31 | | 1.87 | 2.59 |
| 0045 | 3/4" x 3/4" | | 244 | .033 | | .85 | 1.41 | | 2.26 | 3.05 |
| 0050 | 1" x 1" | | 200 | .040 | | 1.50 | 1.73 | | 3.23 | 4.24 |

# 07 95 Expansion Control

## 07 95 13 – Expansion Joint Cover Assemblies

| 07 95 13.50 Expansion Joint Assemblies | | Crew | Daily Output | Labor-Hours | Unit | Material | 2011 Bare Costs Labor | Equipment | Total | Total Incl O&P |
|---|---|---|---|---|---|---|---|---|---|---|
| 0010 | **EXPANSION JOINT ASSEMBLIES** | | | | | | | | | |
| 0200 | Floor cover assemblies, 1" space, aluminum | 1 Sswk | 38 | .211 | L.F. | 18.35 | 10.20 | | 28.55 | 38 |
| 0300 | Bronze | | 38 | .211 | | 43.50 | 10.20 | | 53.70 | 65.50 |
| 0500 | 2" space, aluminum | | 38 | .211 | | 22.50 | 10.20 | | 32.70 | 42.50 |
| 0600 | Bronze | | 38 | .211 | | 46.50 | 10.20 | | 56.70 | 69.50 |
| 0800 | Wall and ceiling assemblies, 1" space, aluminum | | 38 | .211 | | 10.95 | 10.20 | | 21.15 | 30 |
| 0900 | Bronze | | 38 | .211 | | 40.50 | 10.20 | | 50.70 | 62.50 |
| 1100 | 2" space, aluminum | | 38 | .211 | | 18.60 | 10.20 | | 28.80 | 38.50 |
| 1200 | Bronze | | 38 | .211 | | 50 | 10.20 | | 60.20 | 73 |
| 1400 | Floor to wall assemblies, 1" space, aluminum | | 38 | .211 | | 16.70 | 10.20 | | 26.90 | 36.50 |
| 1500 | Bronze or stainless | | 38 | .211 | | 49 | 10.20 | | 59.20 | 72 |
| 1700 | Gym floor angle covers, aluminum, 3" x 3" angle | | 46 | .174 | | 14.45 | 8.45 | | 22.90 | 31 |
| 1800 | 3" x 4" angle | | 46 | .174 | | 20 | 8.45 | | 28.45 | 37 |
| 2000 | Roof closures, aluminum, flat roof, low profile, 1" space | | 57 | .140 | | 33.50 | 6.80 | | 40.30 | 48.50 |
| 2100 | High profile | | 57 | .140 | | 48 | 6.80 | | 54.80 | 65 |
| 2300 | Roof to wall, low profile, 1" space | | 57 | .140 | | 18.45 | 6.80 | | 25.25 | 32.50 |
| 2400 | High profile | | 57 | .140 | | 24 | 6.80 | | 30.80 | 38.50 |

# Estimating Tips

## 08 10 00 Doors and Frames

All exterior doors should be addressed for their energy conservation.

- Most metal doors and frames look alike, but there may be significant differences among them. When estimating these items, be sure to choose the line item that most closely compares to the specification or door schedule requirements regarding:
  - type of metal
  - metal gauge
  - door core material
  - fire rating
  - finish
- Wood and plastic doors vary considerably in price. The primary determinant is the veneer material. Lauan, birch, and oak are the most common veneers. Other variables include the following:
  - hollow or solid core
  - fire rating
  - flush or raised panel
  - finish

## 08 30 00 Specialty Doors and Frames

- There are many varieties of special doors, and they are usually priced per each. Add frames, hardware, or operators required for a complete installation.

## 08 40 00 Entrances, Storefronts, and Curtain Walls

- Glazed curtain walls consist of the metal tube framing and the glazing material. The cost data in this subdivision is presented for the metal tube framing alone or the composite wall. If your estimate requires a detailed takeoff of the framing, be sure to add the glazing cost.

## 08 50 00 Windows

- Most metal windows are delivered preglazed. However, some metal windows are priced without glass. Refer to 08 80 00 Glazing for glass pricing. The grade C indicates commercial grade windows, usually ASTM C-35.
- All wood windows and vinyl are priced preglazed. The glazing is insulating glass. Some wood windows may have single pane float glass. Add the cost of screens and grills if required, and not already included.

## 08 70 00 Hardware

- Hardware costs add considerably to the cost of a door. The most efficient method to determine the hardware requirements for a project is to review the door schedule.

- Door hinges are priced by the pair, with most doors requiring 1-1/2 pairs per door. The hinge prices do not include installation labor because it is included in door installation. Hinges are classified according to the frequency of use.

## 08 80 00 Glazing

- Different openings require different types of glass. The three most common types are:
  - float
  - tempered
  - insulating
  - impact-resistant
  - ballistic-resistant
- Most exterior windows are glazed with insulating glass. Entrance doors and window walls, where the glass is less than 18" from the floor, are generally glazed with tempered glass. Interior windows and some residential windows are glazed with float glass.
- Coastal communities are requiring the use of impact-resistant glass.
- The insulation or 'u' value is a strong consideration, along with solar heat gain.

## Reference Numbers

Reference numbers are shown in shaded boxes at the beginning of some major classifications. These numbers refer to related items in the Reference Section. The reference information may be an estimating procedure, an alternate pricing method, or technical information.

*Note:* Not all subdivisions listed here necessarily appear in this publication.

## 08 05 05 – Selective Windows and Doors Demolition

### 08 05 05.10 Selective Demolition Doors

| | | Crew | Daily Output | Labor-Hours | Unit | Material | 2011 Bare Costs Labor | 2011 Bare Costs Equipment | Total | Total Incl O&P |
|---|---|---|---|---|---|---|---|---|---|---|
| 0010 | **SELECTIVE DEMOLITION DOORS**          R024119-10 | | | | | | | | | |
| 0200 | Doors, exterior, 1-3/4" thick, single, 3' x 7' high | 1 Clab | 16 | .500 | Ea. | | 17.20 | | 17.20 | 26.50 |
| 0220 | Double, 6' x 7' high | | 12 | .667 | | | 23 | | 23 | 35 |
| 0500 | Interior, 1-3/8" thick, single, 3' x 7' high | | 20 | .400 | | | 13.75 | | 13.75 | 21 |
| 0520 | Double, 6' x 7' high | | 16 | .500 | | | 17.20 | | 17.20 | 26.50 |
| 0700 | Bi-folding, 3' x 6'-8" high | | 20 | .400 | | | 13.75 | | 13.75 | 21 |
| 0720 | 6' x 6'-8" high | | 18 | .444 | | | 15.25 | | 15.25 | 23.50 |
| 0900 | Bi-passing, 3' x 6'-8" high | | 16 | .500 | | | 17.20 | | 17.20 | 26.50 |
| 0940 | 6' x 6'-8" high | ▼ | 14 | .571 | | | 19.65 | | 19.65 | 30 |
| 1500 | Remove and reset, minimum | 1 Carp | 8 | 1 | | | 43 | | 43 | 66 |
| 1520 | Maximum | | 6 | 1.333 | | | 57.50 | | 57.50 | 88 |
| 2000 | Frames, including trim, metal | ▼ | 8 | 1 | | | 43 | | 43 | 66 |
| 2200 | Wood | 2 Carp | 32 | .500 | | | 21.50 | | 21.50 | 33 |
| 3000 | Special doors, counter doors | | 6 | 2.667 | | | 115 | | 115 | 176 |
| 3100 | Double acting | | 10 | 1.600 | | | 69 | | 69 | 106 |
| 3200 | Floor door (trap type), or access type | | 8 | 2 | | | 86 | | 86 | 132 |
| 3300 | Glass, sliding, including frames | | 12 | 1.333 | | | 57.50 | | 57.50 | 88 |
| 3400 | Overhead, commercial, 12' x 12' high | | 4 | 4 | | | 172 | | 172 | 264 |
| 3440 | up to 20' x 16' high | | 3 | 5.333 | | | 230 | | 230 | 350 |
| 3445 | up to 35' x 30' high | | 1 | 16 | | | 690 | | 690 | 1,050 |
| 3500 | Residential, 9' x 7' high | | 8 | 2 | | | 86 | | 86 | 132 |
| 3540 | 16' x 7' high | | 7 | 2.286 | | | 98.50 | | 98.50 | 151 |
| 3600 | Remove and reset, minimum | | 4 | 4 | | | 172 | | 172 | 264 |
| 3620 | Maximum | | 2.50 | 6.400 | | | 276 | | 276 | 425 |
| 3700 | Roll-up grille | | 5 | 3.200 | | | 138 | | 138 | 211 |
| 3800 | Revolving door | | 2 | 8 | | | 345 | | 345 | 530 |
| 3900 | Storefront swing door | ▼ | 3 | 5.333 | ▼ | | 230 | | 230 | 350 |
| 6600 | Demo flexible transparent strip entrance | 3 Shee | 115 | .209 | SF Surf | | 10.80 | | 10.80 | 16.30 |
| 7100 | Remove double swing pneumatic doors, openers and sensors | 2 Skwk | .50 | 32 | Opng. | | 1,425 | | 1,425 | 2,175 |
| 7110 | Remove automatic operators, industrial, sliding doors, to 12' wide | " | .40 | 40 | " | | 1,775 | | 1,775 | 2,725 |
| 7550 | Hangar door demo | 2 Sswk | 330 | .048 | S.F. | | 2.35 | | 2.35 | 4.16 |
| 7570 | Remove shock absorbing door | " | 1.90 | 8.421 | Opng. | | 410 | | 410 | 720 |

### 08 05 05.20 Selective Demolition of Windows

| | | Crew | Daily Output | Labor-Hours | Unit | Material | 2011 Bare Costs Labor | 2011 Bare Costs Equipment | Total | Total Incl O&P |
|---|---|---|---|---|---|---|---|---|---|---|
| 0010 | **SELECTIVE DEMOLITION OF WINDOWS**          R024119-10 | | | | | | | | | |
| 0200 | Aluminum, including trim, to 12 S.F. | 1 Clab | 16 | .500 | Ea. | | 17.20 | | 17.20 | 26.50 |
| 0240 | To 25 S.F. | | 11 | .727 | | | 25 | | 25 | 38.50 |
| 0280 | To 50 S.F. | | 5 | 1.600 | | | 55 | | 55 | 84.50 |
| 0320 | Storm windows/screens, to 12 S.F. | | 27 | .296 | | | 10.20 | | 10.20 | 15.60 |
| 0360 | To 25 S.F. | | 21 | .381 | | | 13.10 | | 13.10 | 20 |
| 0400 | To 50 S.F. | | 16 | .500 | ▼ | | 17.20 | | 17.20 | 26.50 |
| 0600 | Glass, minimum | | 200 | .040 | S.F. | | 1.37 | | 1.37 | 2.11 |
| 0620 | Maximum | | 150 | .053 | " | | 1.83 | | 1.83 | 2.81 |
| 1000 | Steel, including trim, to 12 S.F. | | 13 | .615 | Ea. | | 21 | | 21 | 32.50 |
| 1020 | To 25 S.F. | | 9 | .889 | | | 30.50 | | 30.50 | 47 |
| 1040 | To 50 S.F. | | 4 | 2 | | | 68.50 | | 68.50 | 105 |
| 2000 | Wood, including trim, to 12 S.F. | | 22 | .364 | | | 12.50 | | 12.50 | 19.15 |
| 2020 | To 25 S.F. | | 18 | .444 | | | 15.25 | | 15.25 | 23.50 |
| 2060 | To 50 S.F. | | 13 | .615 | | | 21 | | 21 | 32.50 |
| 2065 | To 180 S.F. | ▼ | 8 | 1 | ▼ | | 34.50 | | 34.50 | 52.50 |
| 4300 | Remove bay/bow window | 2 Carp | 6 | 2.667 | ▼ | | 115 | | 115 | 176 |
| 4410 | Remove skylight, plstc domes, flush/curb mtd | G-3 | 395 | .081 | S.F. | | 3.48 | | 3.48 | 5.30 |
| 5020 | Remove and reset window, minimum | 1 Carp | 6 | 1.333 | Ea. | | 57.50 | | 57.50 | 88 |

# 08 05 Common Work Results for Openings

## 08 05 05 – Selective Windows and Doors Demolition

### 08 05 05.20 Selective Demolition of Windows

| | | Crew | Daily Output | Labor-Hours | Unit | Material | 2011 Bare Costs Labor | Equipment | Total | Total Incl O&P |
|---|---|---|---|---|---|---|---|---|---|---|
| 5040 | Average | 1 Carp | 4 | 2 | Ea. | | 86 | | 86 | 132 |
| 5080 | Maximum | ↓ | 2 | 4 | ↓ | | 172 | | 172 | 264 |

# 08 11 Metal Doors and Frames

## 08 11 16 – Aluminum Doors and Frames

### 08 11 16.10 Entrance Doors

| | | Crew | Daily Output | Labor-Hours | Unit | Material | 2011 Bare Costs Labor | Equipment | Total | Total Incl O&P |
|---|---|---|---|---|---|---|---|---|---|---|
| 0010 | **ENTRANCE DOORS** Aluminum, narrow stile | | | | | | | | | |
| 0011 | Including standard hardware, clear finish, no glass | | | | | | | | | |
| 0020 | 3'-0" x 7'-0" opening | 2 Sswk | 2 | 8 | Ea. | 760 | 390 | | 1,150 | 1,525 |
| 0030 | 3'-6" x 7'-0" opening | | 2 | 8 | | 740 | 390 | | 1,130 | 1,500 |
| 0100 | 3'-0" x 10'-0" opening, 3' high transom | | 1.80 | 8.889 | | 1,200 | 430 | | 1,630 | 2,050 |
| 0200 | 3'-6" x 10'-0" opening, 3' high transom | | 1.80 | 8.889 | | 1,250 | 430 | | 1,680 | 2,125 |
| 0280 | 5'-0" x 7'-0" opening | | 2 | 8 | | 1,350 | 390 | | 1,740 | 2,150 |
| 0300 | 6'-0" x 7'-0" opening | | 1.30 | 12.308 | ↓ | 1,150 | 600 | | 1,750 | 2,300 |
| 0400 | 6'-0" x 10'-0" opening, 3' high transom | | 1.10 | 14.545 | Pr. | 1,525 | 705 | | 2,230 | 2,925 |
| 0420 | 7'-0" x 7'-0" opening | | 1 | 16 | " | 1,275 | 775 | | 2,050 | 2,775 |
| 0520 | 3'-0" x 7'-0" opening, wide stile | | 2 | 8 | Ea. | 910 | 390 | | 1,300 | 1,675 |
| 0540 | 3'-6" x 7'-0" opening | | 2 | 8 | | 1,125 | 390 | | 1,515 | 1,925 |
| 0560 | 5'-0" x 7'-0" opening | | 2 | 8 | ↓ | 1,550 | 390 | | 1,940 | 2,375 |
| 0580 | 6'-0" x 7'-0" opening | | 1.30 | 12.308 | Pr. | 1,475 | 600 | | 2,075 | 2,675 |
| 0600 | 7'-0" x 7'-0" opening | ↓ | 1 | 16 | " | 1,625 | 775 | | 2,400 | 3,150 |
| 1100 | For full vision doors, with 1/2" glass, add | | | | Leaf | 55% | | | | |
| 1200 | For non-standard size, add | | | | | 67% | | | | |
| 1300 | Light bronze finish, add | | | | | 36% | | | | |
| 1400 | Dark bronze finish, add | | | | | 18% | | | | |
| 1500 | For black finish, add | | | | | 36% | | | | |
| 1600 | Concealed panic device, add | | | | ↓ | 910 | | | 910 | 1,000 |
| 1700 | Electric striker release, add | | | | Opng. | 252 | | | 252 | 277 |
| 1800 | Floor check, add | | | | Leaf | 630 | | | 630 | 695 |
| 1900 | Concealed closer, add | | | | " | 505 | | | 505 | 555 |
| 2000 | Flush 3' x 7' Insulated, 12" x 12" lite, clear finish | 2 Sswk | 2 | 8 | Ea. | 1,300 | 390 | | 1,690 | 2,100 |

## 08 11 63 – Metal Screen and Storm Doors and Frames

### 08 11 63.23 Aluminum Screen and Storm Doors and Frames

| | | Crew | Daily Output | Labor-Hours | Unit | Material | 2011 Bare Costs Labor | Equipment | Total | Total Incl O&P |
|---|---|---|---|---|---|---|---|---|---|---|
| 0010 | **ALUMINUM SCREEN AND STORM DOORS AND FRAMES** | | | | | | | | | |
| 0020 | Combination storm and screen | | | | | | | | | |
| 0420 | Clear anodic coating, 2'-8" wide | 2 Carp | 14 | 1.143 | Ea. | 158 | 49 | | 207 | 250 |
| 0440 | 3'-0" wide | " | 14 | 1.143 | " | 158 | 49 | | 207 | 250 |
| 0500 | For 7' door height, add | | | | | 5% | | | | |
| 1020 | Mill finish, 2'-8" wide | 2 Carp | 14 | 1.143 | Ea. | 225 | 49 | | 274 | 325 |
| 1040 | 3'-0" wide | " | 14 | 1.143 | | 246 | 49 | | 295 | 345 |
| 1100 | For 7'-0" door, add | | | | | 5% | | | | |
| 1520 | White painted, 2'-8" wide | 2 Carp | 14 | 1.143 | | 281 | 49 | | 330 | 385 |
| 1540 | 3'-0" wide | " | 14 | 1.143 | | 298 | 49 | | 347 | 400 |
| 1600 | For 7'-0" door, add | | | | ↓ | 5% | | | | |
| 2000 | Wood door & screen, see Section 08 14 33.20 | | | | | | | | | |

## 08 11 74 – Sliding Metal Grilles

### 08 11 74.10 Rolling Grille Supports

| | | Crew | Daily Output | Labor-Hours | Unit | Material | 2011 Bare Costs Labor | Equipment | Total | Total Incl O&P |
|---|---|---|---|---|---|---|---|---|---|---|
| 0010 | **ROLLING GRILLE SUPPORTS** | | | | | | | | | |
| 0020 | Rolling grille supports, overhead framed | E-4 | 36 | .889 | L.F. | 24.50 | 43.50 | 3.03 | 71.03 | 107 |

# 08 12 Metal Frames

## 08 12 13 – Hollow Metal Frames

| 08 12 13.13 Standard Hollow Metal Frames | | Crew | Daily Output | Labor-Hours | Unit | Material | 2011 Bare Costs Labor | Equipment | Total | Total Incl O&P |
|---|---|---|---|---|---|---|---|---|---|---|
| 0010 | **STANDARD HOLLOW METAL FRAMES** | | | | | | | | | |
| 0020 | 16 ga., up to 5-3/4" jamb depth | | | | | | | | | |
| 0025 | 6'-8" high, 3'-0" wide, single | G | 2 Carp | 16 | 1 | Ea. | 138 | 43 | | 181 | 218 |
| 0028 | 3'-6" wide, single | G | | 16 | 1 | | 146 | 43 | | 189 | 227 |
| 0030 | 4'-0" wide, single | G | | 16 | 1 | | 146 | 43 | | 189 | 227 |
| 0040 | 6'-0" wide, double | G | | 14 | 1.143 | | 195 | 49 | | 244 | 290 |
| 0045 | 8'-0" wide, double | G | | 14 | 1.143 | | 203 | 49 | | 252 | 299 |
| 0100 | 7'-0" high, 3'-0" wide, single | G | | 16 | 1 | | 143 | 43 | | 186 | 223 |
| 0110 | 3'-6" wide, single | G | | 16 | 1 | | 151 | 43 | | 194 | 232 |
| 0112 | 4'-0" wide, single | G | | 16 | 1 | | 151 | 43 | | 194 | 232 |
| 0140 | 6'-0" wide, double | G | | 14 | 1.143 | | 185 | 49 | | 234 | 280 |
| 0145 | 8'-0" wide, double | G | | 14 | 1.143 | | 219 | 49 | | 268 | 315 |
| 1000 | 16 ga., up to 4-7/8" deep, 7'-0" H, 3'-0" W, single | G | | 16 | 1 | | 157 | 43 | | 200 | 238 |
| 1140 | 6'-0" wide, double | G | | 14 | 1.143 | | 177 | 49 | | 226 | 271 |
| 1200 | 16 ga., 8-3/4" deep, 7'-0" H, 3'-0" W, single | G | | 16 | 1 | | 190 | 43 | | 233 | 275 |
| 1240 | 6'-0" wide, double | G | | 14 | 1.143 | | 223 | 49 | | 272 | 320 |
| 2800 | 14 ga., up to 3-7/8" deep, 7'-0" high, 3'-0" wide, single | G | | 16 | 1 | | 171 | 43 | | 214 | 254 |
| 2840 | 6'-0" wide, double | G | | 14 | 1.143 | | 205 | 49 | | 254 | 300 |
| 3000 | 14 ga., up to 5-3/4" deep, 6'-8" high, 3'-0" wide, single | G | | 16 | 1 | | 145 | 43 | | 188 | 225 |
| 3002 | 3'-6" wide, single | G | | 16 | 1 | | 188 | 43 | | 231 | 273 |
| 3005 | 4'-0" wide, single | G | | 16 | 1 | | 188 | 43 | | 231 | 273 |
| 3600 | up to 5-3/4" jamb depth, 7'-0" high, 4'-0" wide, single | G | | 15 | 1.067 | | 176 | 46 | | 222 | 265 |
| 3620 | 6'-0" wide, double | G | | 12 | 1.333 | | 224 | 57.50 | | 281.50 | 335 |
| 3640 | 8'-0" wide, double | G | | 12 | 1.333 | | 235 | 57.50 | | 292.50 | 345 |
| 3700 | 8'-0" high, 4'-0" wide, single | G | | 15 | 1.067 | | 224 | 46 | | 270 | 320 |
| 3740 | 8'-0" wide, double | G | | 12 | 1.333 | | 276 | 57.50 | | 333.50 | 395 |
| 4000 | 6-3/4" deep, 7'-0" high, 4'-0" wide, single | G | | 15 | 1.067 | | 206 | 46 | | 252 | 298 |
| 4020 | 6'-0" wide, double | G | | 12 | 1.333 | | 257 | 57.50 | | 314.50 | 370 |
| 4040 | 8'-0", wide double | G | | 12 | 1.333 | | 264 | 57.50 | | 321.50 | 380 |
| 4100 | 8'-0" high, 4'-0" wide, single | G | | 15 | 1.067 | | 263 | 46 | | 309 | 360 |
| 4140 | 8'-0" wide, double | G | | 12 | 1.333 | | 305 | 57.50 | | 362.50 | 430 |
| 4400 | 8-3/4" deep, 7'-0" high, 4'-0" wide, single | G | | 15 | 1.067 | | 241 | 46 | | 287 | 335 |
| 4440 | 8'-0" wide, double | G | | 12 | 1.333 | | 299 | 57.50 | | 356.50 | 420 |
| 4500 | 8'-0" high, 4'-0" wide, single | G | | 15 | 1.067 | | 270 | 46 | | 316 | 370 |
| 4540 | 8'-0" wide, double | G | | 12 | 1.333 | | 335 | 57.50 | | 392.50 | 460 |
| 4900 | For welded frames, add | | | | | | 60 | | | 60 | 66 |
| 5400 | 14 ga., "B" label, up to 5-3/4" deep, 7'-0" high, 4'-0" wide, single | G | 2 Carp | 15 | 1.067 | | 197 | 46 | | 243 | 288 |
| 5440 | 8'-0" wide, double | G | | 12 | 1.333 | | 258 | 57.50 | | 315.50 | 370 |
| 5800 | 6-3/4" deep, 7'-0" high, 4'-0" wide, single | G | | 15 | 1.067 | | 208 | 46 | | 254 | 300 |
| 5840 | 8'-0" wide, double | G | | 12 | 1.333 | | 365 | 57.50 | | 422.50 | 490 |
| 6200 | 8-3/4" deep, 7'-0" high, 4'-0" wide, single | G | | 15 | 1.067 | | 296 | 46 | | 342 | 395 |
| 6240 | 8'-0" wide, double | G | | 12 | 1.333 | | 340 | 57.50 | | 397.50 | 465 |
| 6300 | For "A" label use same price as "B" label | | | | | | | | | | |
| 6400 | For baked enamel finish, add | | | | | | 30% | 15% | | | |
| 6500 | For galvanizing, add | | | | | | 15% | | | | |
| 6600 | For hospital stop, add | | | | | Ea. | 282 | | | 282 | 310 |
| 6620 | For hospital stop, stainless steel add | | | | | " | 365 | | | 365 | 400 |
| 7900 | Transom lite frames, fixed, add | | 2 Carp | 155 | .103 | S.F. | 52 | 4.44 | | 56.44 | 64 |
| 8000 | Movable, add | | " | 130 | .123 | " | 65 | 5.30 | | 70.30 | 79.50 |

## 08 12 13.25 Channel Metal Frames

| 0010 | **CHANNEL METAL FRAMES** | | | | | | | | | |
|---|---|---|---|---|---|---|---|---|---|---|
| 0020 | Steel channels with anchors and bar stops | | | | | | | | | |

# 08 12 Metal Frames

## 08 12 13 – Hollow Metal Frames

| 08 12 13.25 Channel Metal Frames | | Crew | Daily Output | Labor-Hours | Unit | Material | 2011 Bare Costs Labor | Equipment | Total | Total Incl O&P |
|---|---|---|---|---|---|---|---|---|---|---|
| 0100 | 6" channel @ 8.2#/L.F., 3' x 7' door, weighs 150# | G | E-4 | 13 | 2.462 | Ea. | 203 | 121 | 8.40 | 332.40 | 445 |
| 0200 | 8" channel @ 11.5#/L.F., 6' x 8' door, weighs 275# | G | | 9 | 3.556 | | 370 | 174 | 12.10 | 556.10 | 735 |
| 0300 | 8' x 12' door, weighs 400# | G | | 6.50 | 4.923 | | 540 | 241 | 16.80 | 797.80 | 1,050 |
| 0400 | 10" channel @ 15.3#/L.F., 10' x 10' door, weighs 500# | G | | 6 | 5.333 | | 675 | 262 | 18.20 | 955.20 | 1,225 |
| 0500 | 12' x 12' door, weighs 600# | G | | 5.50 | 5.818 | | 810 | 285 | 19.85 | 1,114.85 | 1,425 |
| 0600 | 12" channel @ 20.7#/L.F., 12' x 12' door, weighs 825# | G | | 4.50 | 7.111 | | 1,125 | 350 | 24.50 | 1,499.50 | 1,875 |
| 0700 | 12' x 16' door, weighs 1000# | G | | 4 | 8 | | 1,350 | 390 | 27.50 | 1,767.50 | 2,200 |
| 0800 | For frames without bar stops, light sections, deduct | | | | | | 15% | | | | |
| 0900 | Heavy sections, deduct | | | | | | 10% | | | | |

# 08 13 Metal Doors

## 08 13 13 – Hollow Metal Doors

### 08 13 13.13 Standard Hollow Metal Doors

| 0010 | STANDARD HOLLOW METAL DOORS | R081313-20 | | Crew | Daily Output | Labor-Hours | Unit | Material | Labor | Equipment | Total | Total Incl O&P |
|---|---|---|---|---|---|---|---|---|---|---|---|---|
| 0015 | Flush, full panel, hollow core | | | | | | | | | | | |
| 0017 | When noted doors are prepared but do not include glass or louvers | | | | | | | | | | | |
| 0020 | 1-3/8" thick, 20 ga., 2'-0" x 6'-8" | G | | 2 Carp | 20 | .800 | Ea. | 320 | 34.50 | | 354.50 | 405 |
| 0040 | 2'-8" x 6'-8" | G | | | 18 | .889 | | 335 | 38.50 | | 373.50 | 430 |
| 0060 | 3'-0" x 6'-8" | G | | | 17 | .941 | | 335 | 40.50 | | 375.50 | 430 |
| 0100 | 3'-0" x 7'-0" | G | | | 17 | .941 | | 345 | 40.50 | | 385.50 | 440 |
| 0120 | For vision lite, add | | | | | | | 90 | | | 90 | 99 |
| 0140 | For narrow lite, add | | | | | | | 97.50 | | | 97.50 | 107 |
| 0320 | Half glass, 20 ga., 2'-0" x 6'-8" | G | | 2 Carp | 20 | .800 | | 470 | 34.50 | | 504.50 | 570 |
| 0340 | 2'-8" x 6'-8" | G | | | 18 | .889 | | 490 | 38.50 | | 528.50 | 600 |
| 0360 | 3'-0" x 6'-8" | G | | | 17 | .941 | | 485 | 40.50 | | 525.50 | 595 |
| 0400 | 3'-0" x 7'-0" | G | | | 17 | .941 | | 595 | 40.50 | | 635.50 | 715 |
| 0410 | 1-3/8" thick, 18 ga., 2'-0" x 6'-8" | G | | | 20 | .800 | | 380 | 34.50 | | 414.50 | 475 |
| 0420 | 3'-0" x 6'-8" | G | | | 17 | .941 | | 385 | 40.50 | | 425.50 | 485 |
| 0425 | 3'-0" x 7'-0" | G | | | 17 | .941 | | 400 | 40.50 | | 440.50 | 500 |
| 0450 | For vision lite, add | | | | | | | 90 | | | 90 | 99 |
| 0452 | For narrow lite, add | | | | | | | 97.50 | | | 97.50 | 107 |
| 0460 | Half glass, 18 ga., 2'-0" x 6'-8" | G | | 2 Carp | 20 | .800 | | 530 | 34.50 | | 564.50 | 640 |
| 0465 | 2'-8" x 6'-8" | G | | | 18 | .889 | | 550 | 38.50 | | 588.50 | 665 |
| 0470 | 3'-0" x 6'-8" | G | | | 17 | .941 | | 535 | 40.50 | | 575.50 | 650 |
| 0475 | 3'-0" x 7'-0" | G | | | 17 | .941 | | 550 | 40.50 | | 590.50 | 665 |
| 0500 | Hollow core, 1-3/4" thick, full panel, 20 ga., 2'-8" x 6'-8" | G | | | 18 | .889 | | 400 | 38.50 | | 438.50 | 500 |
| 0520 | 3'-0" x 6'-8" | G | | | 17 | .941 | | 400 | 40.50 | | 440.50 | 500 |
| 0640 | 3'-0" x 7'-0" | G | | | 17 | .941 | | 420 | 40.50 | | 460.50 | 520 |
| 0680 | 4'-0" x 7'-0" | G | | | 15 | 1.067 | | 605 | 46 | | 651 | 735 |
| 0700 | 4'-0" x 8'-0" | G | | | 13 | 1.231 | | 705 | 53 | | 758 | 855 |
| 1000 | 18 ga., 2'-8" x 6'-8" | G | | | 17 | .941 | | 465 | 40.50 | | 505.50 | 575 |
| 1020 | 3'-0" x 6'-8" | G | | | 16 | 1 | | 455 | 43 | | 498 | 565 |
| 1120 | 3'-0" x 7'-0" | G | | | 17 | .941 | | 475 | 40.50 | | 515.50 | 585 |
| 1180 | 4'-0" x 7'-0" | G | | | 14 | 1.143 | | 605 | 49 | | 654 | 740 |
| 1200 | 4'-0" x 8'-0" | G | | | 17 | .941 | | 705 | 40.50 | | 745.50 | 835 |
| 1212 | For vision lite, add | | | | | | | 90 | | | 90 | 99 |
| 1214 | For narrow lite, add | | | | | | | 97.50 | | | 97.50 | 107 |
| 1230 | Half glass, 20 ga., 2'-8" x 6'-8" | G | | 2 Carp | 20 | .800 | | 550 | 34.50 | | 584.50 | 660 |
| 1240 | 3'-0" x 6'-8" | G | | | 18 | .889 | | 550 | 38.50 | | 588.50 | 665 |
| 1260 | 3'-0" x 7'-0" | G | | | 18 | .889 | | 565 | 38.50 | | 603.50 | 680 |
| 1320 | 18 ga., 2'-8" x 6'-8" | G | | | 18 | .889 | | 610 | 38.50 | | 648.50 | 730 |

# 08 13 Metal Doors

## 08 13 13 – Hollow Metal Doors

### 08 13 13.13 Standard Hollow Metal Doors

| | | | Crew | Daily Output | Labor-Hours | Unit | Material | 2011 Bare Costs Labor | 2011 Bare Costs Equipment | Total | Total Incl O&P |
|---|---|---|---|---|---|---|---|---|---|---|---|
| 1340 | 3'-0" x 6'-8" | G | 2 Carp | 17 | .941 | Ea. | 605 | 40.50 | | 645.50 | 725 |
| 1360 | 3'-0" x 7'-0" | G | | 17 | .941 | | 615 | 40.50 | | 655.50 | 735 |
| 1380 | 4'-0" x 7'-0" | G | | 15 | 1.067 | | 750 | 46 | | 796 | 895 |
| 1400 | 4'-0" x 8'-0" | G | | 14 | 1.143 | | 850 | 49 | | 899 | 1,000 |
| 1720 | Insulated, 1-3/4" thick, full panel, 18 ga., 3'-0" x 6'-8" | G | | 15 | 1.067 | | 455 | 46 | | 501 | 570 |
| 1740 | 2'-8" x 7'-0" | G | | 16 | 1 | | 480 | 43 | | 523 | 595 |
| 1760 | 3'-0" x 7'-0" | G | | 15 | 1.067 | | 465 | 46 | | 511 | 585 |
| 1800 | 4'-0" x 8'-0" | G | | 13 | 1.231 | | 705 | 53 | | 758 | 855 |
| 1805 | For vision lite, add | | | | | | 90 | | | 90 | 99 |
| 1810 | For narrow lite, add | | | | | | 97.50 | | | 97.50 | 107 |
| 1820 | Half glass, 18 ga., 3'-0" x 6'-8" | G | 2 Carp | 16 | 1 | | 605 | 43 | | 648 | 730 |
| 1840 | 2'-8" x 7'-0" | G | | 17 | .941 | | 630 | 40.50 | | 670.50 | 750 |
| 1860 | 3'-0" x 7'-0" | G | | 16 | 1 | | 660 | 43 | | 703 | 790 |
| 1900 | 4'-0" x 8'-0" | G | | 14 | 1.143 | | 705 | 49 | | 754 | 850 |
| 2000 | For bottom louver, add | | | | | | 264 | | | 264 | 291 |
| 2020 | For baked enamel finish, add | | | | | | 30% | 15% | | | |
| 2040 | For galvanizing, add | | | | | | 15% | | | | |

### 08 13 13.15 Metal Fire Doors

| | | | Crew | Daily Output | Labor-Hours | Unit | Material | 2011 Bare Costs Labor | 2011 Bare Costs Equipment | Total | Total Incl O&P |
|---|---|---|---|---|---|---|---|---|---|---|---|
| 0010 | **METAL FIRE DOORS** | R081313-20 | | | | | | | | | |
| 0015 | Steel, flush, "B" label, 90 minute | | | | | | | | | | |
| 0020 | Full panel, 20 ga., 2'-0" x 6'-8" | | 2 Carp | 20 | .800 | Ea. | 400 | 34.50 | | 434.50 | 495 |
| 0040 | 2'-8" x 6'-8" | | | 18 | .889 | | 415 | 38.50 | | 453.50 | 520 |
| 0060 | 3'-0" x 6'-8" | | | 17 | .941 | | 415 | 40.50 | | 455.50 | 520 |
| 0080 | 3'-0" x 7'-0" | | | 17 | .941 | | 430 | 40.50 | | 470.50 | 535 |
| 0140 | 18 ga., 3'-0" x 6'-8" | | | 16 | 1 | | 470 | 43 | | 513 | 580 |
| 0160 | 2'-8" x 7'-0" | | | 17 | .941 | | 495 | 40.50 | | 535.50 | 605 |
| 0180 | 3'-0" x 7'-0" | | | 16 | 1 | | 480 | 43 | | 523 | 595 |
| 0200 | 4'-0" x 7'-0" | | | 15 | 1.067 | | 620 | 46 | | 666 | 750 |
| 0220 | For "A" label, 3 hour, 18 ga., use same price as "B" label | | | | | | | | | | |
| 0240 | For vision lite, add | | | | | Ea. | 149 | | | 149 | 164 |
| 0520 | Flush, "B" label 90 min., egress core, 20 ga., 2'-0" x 6'-8" | | 2 Carp | 18 | .889 | | 625 | 38.50 | | 663.50 | 745 |
| 0540 | 2'-8" x 6'-8" | | | 17 | .941 | | 635 | 40.50 | | 675.50 | 755 |
| 0560 | 3'-0" x 6'-8" | | | 16 | 1 | | 635 | 43 | | 678 | 760 |
| 0580 | 3'-0" x 7'-0" | | | 16 | 1 | | 655 | 43 | | 698 | 785 |
| 0640 | Flush, "A" label 3 hour, egress core, 18 ga., 3'-0" x 6'-8" | | | 15 | 1.067 | | 695 | 46 | | 741 | 830 |
| 0660 | 2'-8" x 7'-0" | | | 16 | 1 | | 715 | 43 | | 758 | 850 |
| 0680 | 3'-0" x 7'-0" | | | 15 | 1.067 | | 705 | 46 | | 751 | 845 |
| 0700 | 4'-0" x 7'-0" | | | 14 | 1.143 | | 840 | 49 | | 889 | 1,000 |

### 08 13 13.20 Residential Steel Doors

| | | | Crew | Daily Output | Labor-Hours | Unit | Material | 2011 Bare Costs Labor | 2011 Bare Costs Equipment | Total | Total Incl O&P |
|---|---|---|---|---|---|---|---|---|---|---|---|
| 0010 | **RESIDENTIAL STEEL DOORS** | | | | | | | | | | |
| 0020 | Prehung, insulated, exterior | | | | | | | | | | |
| 0030 | Embossed, full panel, 2'-8" x 6'-8" | G | 2 Carp | 17 | .941 | Ea. | 293 | 40.50 | | 333.50 | 385 |
| 0040 | 3'-0" x 6'-8" | G | | 15 | 1.067 | | 242 | 46 | | 288 | 335 |
| 0060 | 3'-0" x 7'-0" | G | | 15 | 1.067 | | 325 | 46 | | 371 | 430 |
| 0070 | 5'-4" x 6'-8", double | G | | 8 | 2 | | 625 | 86 | | 711 | 815 |
| 0220 | Half glass, 2'-8" x 6'-8" | | | 17 | .941 | | 300 | 40.50 | | 340.50 | 390 |
| 0240 | 3'-0" x 6'-8" | | | 16 | 1 | | 300 | 43 | | 343 | 395 |
| 0260 | 3'-0" x 7'-0" | | | 16 | 1 | | 345 | 43 | | 388 | 445 |
| 0270 | 5'-4" x 6'-8", double | | | 8 | 2 | | 640 | 86 | | 726 | 835 |
| 1320 | Flush face, full panel, 2'-8" x 6'-8" | | | 16 | 1 | | 230 | 43 | | 273 | 320 |
| 1340 | 3'-0" x 6'-8" | | | 15 | 1.067 | | 230 | 46 | | 276 | 325 |
| 1360 | 3'-0" x 7'-0" | | | 15 | 1.067 | | 290 | 46 | | 336 | 390 |

# 08 13 Metal Doors

## 08 13 13 – Hollow Metal Doors

### 08 13 13.20 Residential Steel Doors

| | | Crew | Daily Output | Labor-Hours | Unit | Material | 2011 Bare Costs Labor | Equipment | Total | Total Incl O&P |
|---|---|---|---|---|---|---|---|---|---|---|
| 1380 | 5'-4" x 6'-8", double | 2 Carp | 8 | 2 | Ea. | 470 | 86 | | 556 | 645 |
| 1420 | Half glass, 2'-8" x 6'-8" | | 17 | .941 | | 320 | 40.50 | | 360.50 | 410 |
| 1440 | 3'-0" x 6'-8" | | 16 | 1 | | 320 | 43 | | 363 | 415 |
| 1460 | 3'-0" x 7'-0" | | 16 | 1 | | 345 | 43 | | 388 | 445 |
| 1480 | 5'-4" x 6'-8", double | | 8 | 2 | | 635 | 86 | | 721 | 825 |
| 1500 | Sidelight, full lite, 1'-0" x 6'-8"with grille | | | | | 236 | | | 236 | 260 |
| 1510 | 1'-0" x 6'-8", low e | | | | | 257 | | | 257 | 282 |
| 1520 | 1'-0" x 6'-8", half lite | | | | | 254 | | | 254 | 280 |
| 1530 | 1'-0" x 6'-8", half lite, low e | | | | | 262 | | | 262 | 289 |
| 2300 | Interior, residential, closet, bi-fold, 6'-8" x 2'-0" wide | 2 Carp | 16 | 1 | | 161 | 43 | | 204 | 243 |
| 2330 | 3'-0" wide | | 16 | 1 | | 200 | 43 | | 243 | 286 |
| 2360 | 4'-0" wide | | 15 | 1.067 | | 260 | 46 | | 306 | 355 |
| 2400 | 5'-0" wide | | 14 | 1.143 | | 315 | 49 | | 364 | 425 |
| 2420 | 6'-0" wide | | 13 | 1.231 | | 330 | 53 | | 383 | 445 |

### 08 13 13.25 Doors Hollow Metal

| | | Crew | Daily Output | Labor-Hours | Unit | Material | 2011 Bare Costs Labor | Equipment | Total | Total Incl O&P |
|---|---|---|---|---|---|---|---|---|---|---|
| 0010 | **DOORS HOLLOW METAL** | | | | | | | | | |
| 0500 | Exterior, commercial, flush, 20 ga., 1-3/4" x 7'-0" x 2'-6" wide | 2 Carp | 15 | 1.067 | Ea. | 375 | 46 | | 421 | 485 |
| 0530 | 2'-8" wide | | 15 | 1.067 | | 415 | 46 | | 461 | 530 |
| 0560 | 3'-0" wide | | 14 | 1.143 | | 415 | 49 | | 464 | 535 |
| 1000 | 18 ga., 1-3/4" x 7'-0" x 2'-6" wide | | 15 | 1.067 | | 475 | 46 | | 521 | 595 |
| 1030 | 2'-8" wide | | 15 | 1.067 | | 480 | 46 | | 526 | 600 |
| 1060 | 3'-0" wide | | 14 | 1.143 | | 465 | 49 | | 514 | 590 |
| 1500 | 16 ga., 1-3/4 x 7'-0" x 2'-6" wide | | 15 | 1.067 | | 540 | 46 | | 586 | 665 |
| 1530 | 2'-8" wide | | 15 | 1.067 | | 545 | 46 | | 591 | 670 |
| 1560 | 3'-0" wide | | 14 | 1.143 | | 530 | 49 | | 579 | 660 |
| 1590 | 3'-6" wide | | 14 | 1.143 | | 625 | 49 | | 674 | 760 |
| 2900 | Fire door, "A" label, 1-3/4" x 7'-0" x 2'-6" wide | | 15 | 1.067 | | 630 | 46 | | 676 | 760 |
| 2930 | 2'-8" wide | | 15 | 1.067 | | 630 | 46 | | 676 | 765 |
| 2960 | 3'-0" wide | | 14 | 1.143 | | 620 | 49 | | 669 | 755 |
| 2990 | 3'-6" wide | | 14 | 1.143 | | 705 | 49 | | 754 | 850 |
| 3100 | "B" label, 2'-6" wide | | 15 | 1.067 | | 565 | 46 | | 611 | 690 |
| 3130 | 2'-8" wide | | 15 | 1.067 | | 565 | 46 | | 611 | 695 |
| 3160 | 3'-0" wide | | 14 | 1.143 | | 560 | 49 | | 609 | 690 |

## 08 13 16 – Aluminum Doors

### 08 13 16.10 Commercial Aluminum Doors

| | | Crew | Daily Output | Labor-Hours | Unit | Material | 2011 Bare Costs Labor | Equipment | Total | Total Incl O&P |
|---|---|---|---|---|---|---|---|---|---|---|
| 0010 | **COMMERCIAL ALUMINUM DOORS**, no glazing | | | | | | | | | |
| 0020 | Incl. hinges, push/pull, deadlock, cyl., threshold | | | | | | | | | |
| 1000 | Narrow stile, no glazing, standard hardware, 3'-0" x 7'-0", single | 2 Carp | 3 | 5.333 | Ea. | 535 | 230 | | 765 | 935 |
| 1200 | Pair of 3'-0" x 7'-0" | | 1.70 | 9.412 | Pr. | 1,200 | 405 | | 1,605 | 1,950 |
| 1500 | 3'-6" x 7'-0", single | | 3 | 5.333 | Ea. | 700 | 230 | | 930 | 1,125 |
| 2100 | Medium stile, 3'-0" x 7'-0", single | | 3 | 5.333 | " | 740 | 230 | | 970 | 1,175 |
| 2200 | Pair of 3'-0" x 7'-0" | | 1.70 | 9.412 | Pr. | 1,475 | 405 | | 1,880 | 2,250 |
| 2300 | 3'-6" x 7'-0", single | | 3 | 5.333 | Ea. | 955 | 230 | | 1,185 | 1,400 |
| 5000 | Flush panel doors, pair of 2'-6" x 7'-0" | 2 Sswk | 2 | 8 | Pr. | 1,375 | 390 | | 1,765 | 2,175 |
| 5050 | 3'-0" x 7'-0", single | | 2.50 | 6.400 | Ea. | 780 | 310 | | 1,090 | 1,400 |
| 5100 | Pair of 3'-0" x 7'-0" | | 2 | 8 | Pr. | 1,450 | 390 | | 1,840 | 2,275 |
| 5150 | 3'-6" x 7'-0", single | | 2.50 | 6.400 | Ea. | 930 | 310 | | 1,240 | 1,575 |

## 08 13 73 – Sliding Metal Doors

### 08 13 73.10 Steel Sliding Doors

| | | Crew | Daily Output | Labor-Hours | Unit | Material | 2011 Bare Costs Labor | Equipment | Total | Total Incl O&P |
|---|---|---|---|---|---|---|---|---|---|---|
| 0010 | **STEEL SLIDING DOORS** | | | | | | | | | |
| 0020 | Up to 50' x 18', electric, standard duty, minimum | L-5 | 360 | .156 | S.F. | 29 | 7.55 | 2.08 | 38.63 | 47 |

# 08 13 Metal Doors

## 08 13 73 – Sliding Metal Doors

| 08 13 73.10 Steel Sliding Doors | Crew | Daily Output | Labor-Hours | Unit | Material | 2011 Bare Costs Labor | Equipment | Total | Total Incl O&P |
|---|---|---|---|---|---|---|---|---|---|
| 0100 | Maximum | L-5 | 340 | .165 | S.F. | 45.50 | 8 | 2.20 | 55.70 | 66 |
| 0500 | Heavy duty, minimum | | 297 | .189 | | 37 | 9.15 | 2.52 | 48.67 | 59 |
| 0600 | Maximum | ↓ | 277 | .202 | ↓ | 92.50 | 9.80 | 2.70 | 105 | 122 |

# 08 14 Wood Doors

## 08 14 13 – Carved Wood Doors

### 08 14 13.10 Types of Wood Doors, Carved

| 08 14 13.10 | | Crew | Daily Output | Labor-Hours | Unit | Material | Labor | Equipment | Total | Total Incl O&P |
|---|---|---|---|---|---|---|---|---|---|---|
| 0010 | **TYPES OF WOOD DOORS, CARVED** | | | | | | | | | |
| 3000 | Solid wood, 1-3/4" thick stile and rail | | | | | | | | | |
| 3020 | Mahogany, 3'-0" x 7'-0", minimum | 2 Carp | 14 | 1.143 | Ea. | 945 | 49 | | 994 | 1,125 |
| 3030 | Maximum | | 10 | 1.600 | | 1,500 | 69 | | 1,569 | 1,750 |
| 3040 | 3'-6" x 8'-0", minimum | | 10 | 1.600 | | 1,250 | 69 | | 1,319 | 1,475 |
| 3050 | Maximum | | 8 | 2 | | 1,850 | 86 | | 1,936 | 2,175 |
| 3100 | Pine, 3'-0" x 7'-0", minimum | | 14 | 1.143 | | 490 | 49 | | 539 | 610 |
| 3110 | Maximum | | 10 | 1.600 | | 825 | 69 | | 894 | 1,025 |
| 3120 | 3'-6" x 8'-0", minimum | | 10 | 1.600 | | 885 | 69 | | 954 | 1,075 |
| 3130 | Maximum | | 8 | 2 | | 1,800 | 86 | | 1,886 | 2,100 |
| 3200 | Red oak, 3'-0" x 7'-0", minimum | | 14 | 1.143 | | 1,625 | 49 | | 1,674 | 1,875 |
| 3210 | Maximum | | 10 | 1.600 | | 2,025 | 69 | | 2,094 | 2,325 |
| 3220 | 3'-6" x 8'-0", minimum | | 10 | 1.600 | | 2,200 | 69 | | 2,269 | 2,525 |
| 3230 | Maximum | ↓ | 8 | 2 | ↓ | 3,200 | 86 | | 3,286 | 3,650 |
| 4000 | Hand carved door, mahogany | | | | | | | | | |
| 4020 | 3'-0" x 7'-0", minimum | 2 Carp | 14 | 1.143 | Ea. | 1,600 | 49 | | 1,649 | 1,825 |
| 4030 | Maximum | | 11 | 1.455 | | 3,600 | 62.50 | | 3,662.50 | 4,050 |
| 4040 | 3'-6" x 8'-0", minimum | | 10 | 1.600 | | 2,500 | 69 | | 2,569 | 2,850 |
| 4050 | Maximum | | 8 | 2 | | 3,500 | 86 | | 3,586 | 3,975 |
| 4200 | Rose wood, 3'-0" x 7'-0", minimum | | 14 | 1.143 | | 5,000 | 49 | | 5,049 | 5,575 |
| 4210 | Maximum | | 11 | 1.455 | | 13,600 | 62.50 | | 13,662.50 | 15,100 |
| 4220 | 3'-6" x 8'-0", minimum | | 10 | 1.600 | | 5,700 | 69 | | 5,769 | 6,375 |
| 4225 | maximum | ↓ | 10 | 1.600 | | 8,700 | 69 | | 8,769 | 9,675 |
| 4280 | For 6'-8" high door, deduct from 7'-0" door | | | | | 34 | | | 34 | 37.50 |
| 4400 | For custom finish, add | | | | | 400 | | | 400 | 440 |
| 4600 | Side light, mahogany, 7'-0" x 1'-6" wide, minimum | 2 Carp | 18 | .889 | | 900 | 38.50 | | 938.50 | 1,050 |
| 4610 | Maximum | | 14 | 1.143 | | 2,625 | 49 | | 2,674 | 2,975 |
| 4620 | 8'-0" x 1'-6" wide, minimum | | 14 | 1.143 | | 1,700 | 49 | | 1,749 | 1,950 |
| 4630 | Maximum | | 10 | 1.600 | | 1,950 | 69 | | 2,019 | 2,250 |
| 4640 | Side light, oak, 7'-0" x 1'-6" wide, minimum | | 18 | .889 | | 1,100 | 38.50 | | 1,138.50 | 1,250 |
| 4650 | Maximum | | 14 | 1.143 | | 1,900 | 49 | | 1,949 | 2,175 |
| 4660 | 8'-0" x 1'-6" wide, minimum | | 14 | 1.143 | | 1,000 | 49 | | 1,049 | 1,175 |
| 4670 | Maximum | ↓ | 10 | 1.600 | ↓ | 1,900 | 69 | | 1,969 | 2,200 |

## 08 14 16 – Flush Wood Doors

### 08 14 16.09 Smooth Wood Doors

| 08 14 16.09 | | Crew | Daily Output | Labor-Hours | Unit | Material | Labor | Equipment | Total | Total Incl O&P |
|---|---|---|---|---|---|---|---|---|---|---|
| 0010 | **SMOOTH WOOD DOORS** | | | | | | | | | |
| 0015 | Flush, interior, hollow core | | | | | | | | | |
| 0025 | Lauan face, 1-3/8", 3'-0" x 6'-8" | 2 Carp | 17 | .941 | Ea. | 52.50 | 40.50 | | 93 | 120 |
| 0030 | 4'-0" x 6'-8" | | 16 | 1 | | 126 | 43 | | 169 | 205 |
| 0080 | 1-3/4", 2'-0" x 6'-8" | | 17 | .941 | | 35 | 40.50 | | 75.50 | 101 |
| 0108 | 3'-0" x 7'-0" | | 16 | 1 | ↓ | 95.50 | 43 | | 138.50 | 171 |
| 0112 | Pair of 3'-0" x 7'-0" | | 9 | 1.778 | Pr. | 182 | 76.50 | | 258.50 | 320 |
| 0140 | Birch face, 1-3/8", 2'-6" x 6'-8" | | 17 | .941 | Ea. | 82.50 | 40.50 | | 123 | 153 |

| 08 14 16.09 Smooth Wood Doors | | Crew | Daily Output | Labor-Hours | Unit | Material | 2011 Bare Costs Labor | Equipment | Total | Total Incl O&P |
|---|---|---|---|---|---|---|---|---|---|---|
| 0180 | 3'-0" x 6'-8" | 2 Carp | 17 | .941 | Ea. | 94 | 40.50 | | 134.50 | 165 |
| 0200 | 4'-0" x 6'-8" | | 16 | 1 | | 158 | 43 | | 201 | 240 |
| 0202 | 1-3/4", 2'-0" x 6'-8" | | 17 | .941 | | 43.50 | 40.50 | | 84 | 110 |
| 0210 | 3'-0" x 7'-0" | | 16 | 1 | | 103 | 43 | | 146 | 179 |
| 0214 | Pair of 3'-0" x 7'-0" | | 9 | 1.778 | Pr. | 197 | 76.50 | | 273.50 | 335 |
| 0220 | Oak face, 1-3/8", 2'-0" x 6'-8" | | 17 | .941 | Ea. | 96 | 40.50 | | 136.50 | 168 |
| 0280 | 3'-0" x 6'-8" | | 17 | .941 | | 110 | 40.50 | | 150.50 | 183 |
| 0300 | 4'-0" x 6'-8" | | 16 | 1 | | 134 | 43 | | 177 | 214 |
| 0305 | 1-3/4", 2'-6" x 6'-8" | | 17 | .941 | | 106 | 40.50 | | 146.50 | 178 |
| 0320 | Walnut face, 1-3/8", 2'-0" x 6'-8" | | 17 | .941 | | 178 | 40.50 | | 218.50 | 257 |
| 0340 | 2'-6" x 6'-8" | | 17 | .941 | | 184 | 40.50 | | 224.50 | 265 |
| 0380 | 3'-0" x 6'-8" | | 17 | .941 | | 192 | 40.50 | | 232.50 | 273 |
| 0400 | 4'-0" x 6'-8" | ▼ | 16 | 1 | | 213 | 43 | | 256 | 300 |
| 0430 | For 7'-0" high, add | | | | | 16 | | | 16 | 17.60 |
| 0440 | For 8'-0" high, add | | | | | 30 | | | 30 | 33 |
| 0480 | For prefinishing, clear, add | | | | | 35 | | | 35 | 38.50 |
| 0500 | For prefinishing, stain, add | | | | | 45 | | | 45 | 49.50 |
| 1320 | M.D. overlay on hardboard, 1-3/8", 2'-0" x 6'-8" | 2 Carp | 17 | .941 | | 110 | 40.50 | | 150.50 | 183 |
| 1340 | 2'-6" x 6'-8" | | 17 | .941 | | 110 | 40.50 | | 150.50 | 183 |
| 1380 | 3'-0" x 6'-8" | | 17 | .941 | | 120 | 40.50 | | 160.50 | 194 |
| 1400 | 4'-0" x 6'-8" | ▼ | 16 | 1 | | 170 | 43 | | 213 | 253 |
| 1420 | For 7'-0" high, add | | | | | 12 | | | 12 | 13.20 |
| 1440 | For 8'-0" high, add | | | | | 25 | | | 25 | 27.50 |
| 1720 | H.P. plastic laminate, 1-3/8", 2'-0" x 6'-8" | 2 Carp | 16 | 1 | | 228 | 43 | | 271 | 315 |
| 1740 | 2'-6" x 6'-8" | | 16 | 1 | | 228 | 43 | | 271 | 315 |
| 1780 | 3'-0" x 6'-8" | | 15 | 1.067 | | 276 | 46 | | 322 | 375 |
| 1800 | 4'-0" x 6'-8" | ▼ | 14 | 1.143 | | 360 | 49 | | 409 | 470 |
| 1820 | For 7'-0" high, add | | | | | 11 | | | 11 | 12.10 |
| 1840 | For 8'-0" high, add | | | | | 26 | | | 26 | 28.50 |
| 2020 | Particle core, lauan face, 1-3/8", 2'-6" x 6'-8" | 2 Carp | 15 | 1.067 | | 90 | 46 | | 136 | 170 |
| 2040 | 3'-0" x 6'-8" | | 14 | 1.143 | | 92 | 49 | | 141 | 177 |
| 2080 | 3'-0" x 7'-0" | | 13 | 1.231 | | 93.50 | 53 | | 146.50 | 185 |
| 2085 | 4'-0" x 7'-0" | | 12 | 1.333 | | 120 | 57.50 | | 177.50 | 220 |
| 2120 | Birch face, 1-3/8", 2'-6" x 6'-8" | | 15 | 1.067 | | 100 | 46 | | 146 | 181 |
| 2140 | 3'-0" x 6'-8" | | 14 | 1.143 | | 110 | 49 | | 159 | 197 |
| 2180 | 3'-0" x 7'-0" | | 13 | 1.231 | | 120 | 53 | | 173 | 214 |
| 2200 | 4'-0" x 7'-0" | | 12 | 1.333 | | 135 | 57.50 | | 192.50 | 237 |
| 2220 | Oak face, 2'-6" x 6'-8" | | 15 | 1.067 | | 113 | 46 | | 159 | 195 |
| 2240 | 3'-0" x 6'-8" | | 14 | 1.143 | | 125 | 49 | | 174 | 214 |
| 2280 | 3'-0" x 7'-0" | | 13 | 1.231 | | 130 | 53 | | 183 | 225 |
| 2300 | 4'-0" x 7'-0" | | 12 | 1.333 | | 152 | 57.50 | | 209.50 | 255 |
| 2320 | Walnut face, 1-3/8", 2'-0" x 6'-8" | | 15 | 1.067 | | 120 | 46 | | 166 | 203 |
| 2340 | 2'-6" x 6'-8" | | 14 | 1.143 | | 135 | 49 | | 184 | 225 |
| 2380 | 3'-0" x 6'-8" | | 13 | 1.231 | | 152 | 53 | | 205 | 249 |
| 2400 | 4'-0" x 6'-8" | ▼ | 12 | 1.333 | | 200 | 57.50 | | 257.50 | 310 |
| 2440 | For 8'-0" high, add | | | | | 35 | | | 35 | 38.50 |
| 2460 | For 8'-0" high walnut, add | | | | | 20 | | | 20 | 22 |
| 2480 | For solid wood core, add | | | | | 39 | | | 39 | 43 |
| 2720 | For prefinishing, clear, add | | | | | 25 | | | 25 | 27.50 |
| 2740 | For prefinishing, stain, add | | | | | 52 | | | 52 | 57 |
| 3320 | M.D. overlay on hardboard, 1-3/8', 2'-6" x 6'-8" | 2 Carp | 14 | 1.143 | | 106 | 49 | | 155 | 193 |
| 3340 | 3'-0" x 6'-8" | | 13 | 1.231 | | 112 | 53 | | 165 | 205 |
| 3380 | 3'-0" x 7'-0" | ▼ | 12 | 1.333 | ▼ | 114 | 57.50 | | 171.50 | 213 |

# 08 14 Wood Doors

## 08 14 16 - Flush Wood Doors

| 08 14 16.09 **Smooth Wood Doors** | | Crew | Daily Output | Labor-Hours | Unit | Material | 2011 Bare Costs Labor | Equipment | Total | Total Incl O&P |
|---|---|---|---|---|---|---|---|---|---|---|
| 3400 | 4'-0" x 7'-0" | 2 Carp | 10 | 1.600 | Ea. | 150 | 69 | | 219 | 271 |
| 3440 | For 8'-0" height, add | | | | | 35 | | | 35 | 38.50 |
| 3460 | For solid wood core, add | | | | | 40 | | | 40 | 44 |
| 3720 | H.P. plastic laminate, 1-3/8", 2'-6" x 6'-8" | 2 Carp | 13 | 1.231 | | 155 | 53 | | 208 | 253 |
| 3740 | 3'-0" x 6'-8" | | 12 | 1.333 | | 180 | 57.50 | | 237.50 | 286 |
| 3780 | 3'-0" x 7'-0" | | 11 | 1.455 | | 185 | 62.50 | | 247.50 | 300 |
| 3800 | 4'-0" x 7'-0" | | 8 | 2 | | 222 | 86 | | 308 | 375 |
| 3840 | For 8'-0" height, add | | | | | 35 | | | 35 | 38.50 |
| 3860 | For solid wood core, add | | | | | 40 | | | 40 | 44 |
| 4000 | Exterior, flush, solid core, birch, 1-3/4" x 2'-6" x 7'-0" | 2 Carp | 15 | 1.067 | | 165 | 46 | | 211 | 253 |
| 4020 | 2'-8" wide | | 15 | 1.067 | | 153 | 46 | | 199 | 239 |
| 4040 | 3'-0" wide | | 14 | 1.143 | | 195 | 49 | | 244 | 291 |
| 4100 | Oak faced 1-3/4" x 2'-6" x 7'-0" | | 15 | 1.067 | | 185 | 46 | | 231 | 275 |
| 4120 | 2'-8" wide | | 15 | 1.067 | | 200 | 46 | | 246 | 291 |
| 4140 | 3'-0" wide | | 14 | 1.143 | | 210 | 49 | | 259 | 305 |
| 4200 | Walnut faced, 1-3/4" x 2'-6" x 7'-0" | | 15 | 1.067 | | 270 | 46 | | 316 | 370 |
| 4220 | 2'-8" wide | | 15 | 1.067 | | 285 | 46 | | 331 | 385 |
| 4240 | 3'-0" wide | | 14 | 1.143 | | 295 | 49 | | 344 | 400 |
| 4300 | For 6'-8" high door, deduct from 7'-0" door | | | | | 16 | | | 16 | 17.60 |
| 5000 | Wood doors, for vision lite, add | | | | | 90 | | | 90 | 99 |
| 5010 | Wood doors, for narrow lite, add | | | | | 97.50 | | | 97.50 | 107 |
| 5015 | Wood doors, for bottom (or top) louver, add | | | | | 264 | | | 264 | 291 |
| **08 14 16.10 Wood Doors Decorator** | | | | | | | | | | |
| 0010 | **WOOD DOORS DECORATOR** | | | | | | | | | |
| 0040 | 7 ply hollow core lauan face, 2'-6" x 6'-8" | 2 Carp | 17 | .941 | Ea. | 38.50 | 40.50 | | 79 | 104 |
| **08 14 16.20 Wood Fire Doors** | | | | | | | | | | |
| 0010 | **WOOD FIRE DOORS** | | | | | | | | | |
| 0020 | Particle core, 7 face plys, "B" label, | | | | | | | | | |
| 0040 | 1 hour, birch face, 1-3/4" x 2'-6" x 6'-8" | 2 Carp | 14 | 1.143 | Ea. | 385 | 49 | | 434 | 500 |
| 0080 | 3'-0" x 6'-8" | | 13 | 1.231 | | 390 | 53 | | 443 | 510 |
| 0090 | 3'-0" x 7'-0" | | 12 | 1.333 | | 435 | 57.50 | | 492.50 | 565 |
| 0100 | 4'-0" x 7'-0" | | 12 | 1.333 | | 510 | 57.50 | | 567.50 | 650 |
| 0140 | Oak face, 2'-6" x 6'-8" | | 14 | 1.143 | | 335 | 49 | | 384 | 445 |
| 0180 | 3'-0" x 6'-8" | | 13 | 1.231 | | 350 | 53 | | 403 | 465 |
| 0190 | 3'-0" x 7'-0" | | 12 | 1.333 | | 365 | 57.50 | | 422.50 | 490 |
| 0200 | 4'-0" x 7'-0" | | 12 | 1.333 | | 480 | 57.50 | | 537.50 | 620 |
| 0240 | Walnut face, 2'-6" x 6'-8" | | 14 | 1.143 | | 450 | 49 | | 499 | 570 |
| 0280 | 3'-0" x 6'-8" | | 13 | 1.231 | | 470 | 53 | | 523 | 595 |
| 0290 | 3'-0" x 7'-0" | | 12 | 1.333 | | 490 | 57.50 | | 547.50 | 630 |
| 0300 | 4'-0" x 7'-0" | | 12 | 1.333 | | 610 | 57.50 | | 667.50 | 760 |
| 0440 | M.D. overlay on hardboard, 2'-6" x 6'-8" | | 15 | 1.067 | | 300 | 46 | | 346 | 400 |
| 0480 | 3'-0" x 6'-8" | | 14 | 1.143 | | 315 | 49 | | 364 | 420 |
| 0490 | 3'-0" x 7'-0" | | 13 | 1.231 | | 330 | 53 | | 383 | 445 |
| 0500 | 4'-0" x 7'-0" | | 12 | 1.333 | | 400 | 57.50 | | 457.50 | 530 |
| 0740 | 90 minutes, birch face, 1-3/4" x 2'-6" x 6'-8" | | 14 | 1.143 | | 335 | 49 | | 384 | 440 |
| 0780 | 3'-0" x 6'-8" | | 13 | 1.231 | | 310 | 53 | | 363 | 420 |
| 0790 | 3'-0" x 7'-0" | | 12 | 1.333 | | 375 | 57.50 | | 432.50 | 505 |
| 0800 | 4'-0" x 7'-0" | | 12 | 1.333 | | 490 | 57.50 | | 547.50 | 625 |
| 0840 | Oak face, 2'-6" x 6'-8" | | 14 | 1.143 | | 425 | 49 | | 474 | 540 |
| 0880 | 3'-0" x 6'-8" | | 13 | 1.231 | | 435 | 53 | | 488 | 555 |
| 0890 | 3'-0" x 7'-0" | | 12 | 1.333 | | 450 | 57.50 | | 507.50 | 585 |
| 0900 | 4'-0" x 7'-0" | | 12 | 1.333 | | 580 | 57.50 | | 637.50 | 730 |

# 08 14 Wood Doors

## 08 14 16 – Flush Wood Doors

### 08 14 16.20 Wood Fire Doors

| | | Crew | Daily Output | Labor-Hours | Unit | Material | 2011 Bare Costs Labor | Equipment | Total | Total Incl O&P |
|---|---|---|---|---|---|---|---|---|---|---|
| 0940 | Walnut face, 2'-6" x 6'-8" | 2 Carp | 14 | 1.143 | Ea. | 390 | 49 | | 439 | 505 |
| 0980 | 3'-0" x 6'-8" | | 13 | 1.231 | | 400 | 53 | | 453 | 520 |
| 0990 | 3'-0" x 7'-0" | | 12 | 1.333 | | 460 | 57.50 | | 517.50 | 595 |
| 1000 | 4'-0" x 7'-0" | | 12 | 1.333 | | 610 | 57.50 | | 667.50 | 760 |
| 1140 | M.D. overlay on hardboard, 2'-6" x 6'-8" | | 15 | 1.067 | | 350 | 46 | | 396 | 455 |
| 1180 | 3'-0" x 6'-8" | | 14 | 1.143 | | 370 | 49 | | 419 | 485 |
| 1190 | 3'-0" x 7'-0" | | 13 | 1.231 | | 400 | 53 | | 453 | 520 |
| 1200 | 4'-0" x 7'-0" | | 12 | 1.333 | | 450 | 57.50 | | 507.50 | 585 |
| 1240 | For 8'-0" height, add | | | | | 65 | | | 65 | 71.50 |
| 1260 | For 8'-0" height walnut, add | | | | | 80 | | | 80 | 88 |
| 2200 | Custom architectural "B" label, flush, 1-3/4" thick, birch, | | | | | | | | | |
| 2210 | Solid core | | | | | | | | | |
| 2220 | 2'-6" x 7'-0" | 2 Carp | 15 | 1.067 | Ea. | 296 | 46 | | 342 | 395 |
| 2260 | 3'-0" x 7'-0" | | 14 | 1.143 | | 305 | 49 | | 354 | 410 |
| 2300 | 4'-0" x 7'-0" | | 13 | 1.231 | | 400 | 53 | | 453 | 515 |
| 2420 | 4'-0" x 8'-0" | | 11 | 1.455 | | 430 | 62.50 | | 492.50 | 565 |
| 2480 | For oak veneer, add | | | | | 50% | | | | |
| 2500 | For walnut veneer, add | | | | | 75% | | | | |

## 08 14 23 – Clad Wood Doors

### 08 14 23.13 Metal-Faced Wood Doors

| | | Crew | Daily Output | Labor-Hours | Unit | Material | 2011 Bare Costs Labor | Equipment | Total | Total Incl O&P |
|---|---|---|---|---|---|---|---|---|---|---|
| 0010 | **METAL-FACED WOOD DOORS** | | | | | | | | | |
| 0020 | Interior, flush type, 3' x 7' | 2 Carp | 4.30 | 3.721 | Opng. | 340 | 160 | | 500 | 620 |

### 08 14 23.20 Tin Clad Wood Doors

| | | Crew | Daily Output | Labor-Hours | Unit | Material | 2011 Bare Costs Labor | Equipment | Total | Total Incl O&P |
|---|---|---|---|---|---|---|---|---|---|---|
| 0010 | **TIN CLAD WOOD DOORS** | | | | | | | | | |
| 0020 | 3 ply, 6' x 7', double sliding, doors only | 2 Carp | 1 | 16 | Opng. | 1,900 | 690 | | 2,590 | 3,150 |
| 1000 | For electric operator, add | 1 Elec | 2 | 4 | " | 3,250 | 201 | | 3,451 | 3,875 |

## 08 14 33 – Stile and Rail Wood Doors

### 08 14 33.10 Wood Doors Paneled

| | | Crew | Daily Output | Labor-Hours | Unit | Material | 2011 Bare Costs Labor | Equipment | Total | Total Incl O&P |
|---|---|---|---|---|---|---|---|---|---|---|
| 0010 | **WOOD DOORS PANELED** | | | | | | | | | |
| 0020 | Interior, six panel, hollow core, 1-3/8" thick | | | | | | | | | |
| 0040 | Molded hardboard, 2'-0" x 6'-8" | 2 Carp | 17 | .941 | Ea. | 60 | 40.50 | | 100.50 | 128 |
| 0060 | 2'-6" x 6'-8" | | 17 | .941 | | 62 | 40.50 | | 102.50 | 130 |
| 0070 | 2'-8" x 6'-8" | | 17 | .941 | | 65 | 40.50 | | 105.50 | 134 |
| 0080 | 3'-0" x 6'-8" | | 17 | .941 | | 68 | 40.50 | | 108.50 | 137 |
| 0140 | Embossed print, molded hardboard, 2'-0" x 6'-8" | | 17 | .941 | | 62 | 40.50 | | 102.50 | 130 |
| 0160 | 2'-6" x 6'-8" | | 17 | .941 | | 62 | 40.50 | | 102.50 | 130 |
| 0180 | 3'-0" x 6'-8" | | 17 | .941 | | 68 | 40.50 | | 108.50 | 137 |
| 0540 | Six panel, solid, 1-3/8" thick, pine, 2'-0" x 6'-8" | | 15 | 1.067 | | 150 | 46 | | 196 | 236 |
| 0560 | 2'-6" x 6'-8" | | 14 | 1.143 | | 165 | 49 | | 214 | 258 |
| 0580 | 3'-0" x 6'-8" | | 13 | 1.231 | | 140 | 53 | | 193 | 236 |
| 1020 | Two panel, bored rail, solid, 1-3/8" thick, pine, 1'-6" x 6'-8" | | 16 | 1 | | 265 | 43 | | 308 | 360 |
| 1040 | 2'-0" x 6'-8" | | 15 | 1.067 | | 350 | 46 | | 396 | 455 |
| 1060 | 2'-6" x 6'-8" | | 14 | 1.143 | | 395 | 49 | | 444 | 510 |
| 1340 | Two panel, solid, 1-3/8" thick, fir, 2'-0" x 6'-8" | | 15 | 1.067 | | 155 | 46 | | 201 | 242 |
| 1360 | 2'-6" x 6'-8" | | 14 | 1.143 | | 205 | 49 | | 254 | 300 |
| 1380 | 3'-0" x 6'-8" | | 13 | 1.231 | | 410 | 53 | | 463 | 530 |
| 1740 | Five panel, solid, 1-3/8" thick, fir, 2'-0" x 6'-8" | | 15 | 1.067 | | 275 | 46 | | 321 | 375 |
| 1760 | 2'-6" x 6'-8" | | 14 | 1.143 | | 415 | 49 | | 464 | 530 |
| 1780 | 3'-0" x 6'-8" | | 13 | 1.231 | | 415 | 53 | | 468 | 535 |

# 08 14 Wood Doors

## 08 14 33 – Stile and Rail Wood Doors

| 08 14 33.20 Wood Doors Residential | Crew | Daily Output | Labor-Hours | Unit | Material | 2011 Bare Costs Labor | Equipment | Total | Total Incl O&P |
|---|---|---|---|---|---|---|---|---|---|
| 0010 **WOOD DOORS RESIDENTIAL** | | | | | | | | | |
| 0200 Exterior, combination storm & screen, pine | | | | | | | | | |
| 0260 2'-8" wide | 2 Carp | 10 | 1.600 | Ea. | 298 | 69 | | 367 | 435 |
| 0280 3'-0" wide | | 9 | 1.778 | | 305 | 76.50 | | 381.50 | 450 |
| 0300 7'-1" x 3'-0" wide | | 9 | 1.778 | | 325 | 76.50 | | 401.50 | 475 |
| 0400 Full lite, 6'-9" x 2'-6" wide | | 11 | 1.455 | | 310 | 62.50 | | 372.50 | 435 |
| 0420 2'-8" wide | | 10 | 1.600 | | 310 | 69 | | 379 | 445 |
| 0440 3'-0" wide | | 9 | 1.778 | | 320 | 76.50 | | 396.50 | 465 |
| 0500 7'-1" x 3'-0" wide | | 9 | 1.778 | | 345 | 76.50 | | 421.50 | 490 |
| 0700 Dutch door, pine, 1-3/4" x 6'-8" x 2'-8" wide, minimum | | 12 | 1.333 | | 675 | 57.50 | | 732.50 | 835 |
| 0720 Maximum | | 10 | 1.600 | | 825 | 69 | | 894 | 1,025 |
| 0800 3'-0" wide, minimum | | 12 | 1.333 | | 690 | 57.50 | | 747.50 | 845 |
| 0820 Maximum | | 10 | 1.600 | | 900 | 69 | | 969 | 1,100 |
| 1000 Entrance door, colonial, 1-3/4" x 6'-8" x 2'-8" wide | | 16 | 1 | | 465 | 43 | | 508 | 575 |
| 1020 6 panel pine, 3'-0" wide | | 15 | 1.067 | | 435 | 46 | | 481 | 550 |
| 1100 8 panel pine, 2'-8" wide | | 16 | 1 | | 570 | 43 | | 613 | 695 |
| 1120 3'-0" wide | ▼ | 15 | 1.067 | | 560 | 46 | | 606 | 685 |
| 1200 For tempered safety glass lites, (min of 2) add | | | | | 75.50 | | | 75.50 | 83 |
| 1300 Flush, birch, solid core, 1-3/4" x 6'-8" x 2'-8" wide | 2 Carp | 16 | 1 | | 105 | 43 | | 148 | 181 |
| 1320 3'-0" wide | | 15 | 1.067 | | 115 | 46 | | 161 | 197 |
| 1350 7'-0" x 2'-8" wide | | 16 | 1 | | 114 | 43 | | 157 | 192 |
| 1360 3'-0" wide | ▼ | 15 | 1.067 | | 123 | 46 | | 169 | 206 |
| 1380 For tempered safety glass lites, add | | | | ▼ | 103 | | | 103 | 113 |
| 2700 Interior, closet, bi-fold, w/hardware, no frame or trim incl. | | | | | | | | | |
| 2720 Flush, birch, 6'-6" or 6'-8" x 2'-6" wide | 2 Carp | 13 | 1.231 | Ea. | 52.50 | 53 | | 105.50 | 140 |
| 2740 3'-0" wide | | 13 | 1.231 | | 56.50 | 53 | | 109.50 | 144 |
| 2760 4'-0" wide | | 12 | 1.333 | | 101 | 57.50 | | 158.50 | 200 |
| 2780 5'-0" wide | | 11 | 1.455 | | 101 | 62.50 | | 163.50 | 207 |
| 2800 6'-0" wide | | 10 | 1.600 | | 127 | 69 | | 196 | 246 |
| 3000 Raised panel pine, 6'-6" or 6'-8" x 2'-6" wide | | 13 | 1.231 | | 187 | 53 | | 240 | 288 |
| 3020 3'-0" wide | | 13 | 1.231 | | 275 | 53 | | 328 | 385 |
| 3040 4'-0" wide | | 12 | 1.333 | | 305 | 57.50 | | 362.50 | 425 |
| 3060 5'-0" wide | | 11 | 1.455 | | 365 | 62.50 | | 427.50 | 495 |
| 3080 6'-0" wide | | 10 | 1.600 | | 400 | 69 | | 469 | 545 |
| 3200 Louvered, pine 6'-6" or 6'-8" x 2'-6" wide | | 13 | 1.231 | | 137 | 53 | | 190 | 233 |
| 3220 3'-0" wide | | 13 | 1.231 | | 197 | 53 | | 250 | 299 |
| 3240 4'-0" wide | | 12 | 1.333 | | 222 | 57.50 | | 279.50 | 330 |
| 3260 5'-0" wide | | 11 | 1.455 | | 256 | 62.50 | | 318.50 | 380 |
| 3280 6'-0" wide | ▼ | 10 | 1.600 | ▼ | 287 | 69 | | 356 | 420 |
| 4400 Bi-passing closet, incl. hardware and frame, no trim incl. | | | | | | | | | |
| 4420 Flush, lauan, 6'-8" x 4'-0" wide | 2 Carp | 12 | 1.333 | Opng. | 169 | 57.50 | | 226.50 | 273 |
| 4440 5'-0" wide | | 11 | 1.455 | | 186 | 62.50 | | 248.50 | 300 |
| 4460 6'-0" wide | | 10 | 1.600 | | 209 | 69 | | 278 | 335 |
| 4600 Flush, birch, 6'-8" x 4'-0" wide | | 12 | 1.333 | | 220 | 57.50 | | 277.50 | 330 |
| 4620 5'-0" wide | | 11 | 1.455 | | 212 | 62.50 | | 274.50 | 330 |
| 4640 6'-0" wide | | 10 | 1.600 | | 256 | 69 | | 325 | 390 |
| 4800 Louvered, pine, 6'-8" x 4'-0" wide | | 12 | 1.333 | | 400 | 57.50 | | 457.50 | 530 |
| 4820 5'-0" wide | | 11 | 1.455 | | 395 | 62.50 | | 457.50 | 530 |
| 4840 6'-0" wide | | 10 | 1.600 | | 515 | 69 | | 584 | 670 |
| 5000 Paneled, pine, 6'-8" x 4'-0" wide | | 12 | 1.333 | | 495 | 57.50 | | 552.50 | 635 |
| 5020 5'-0" wide | | 11 | 1.455 | | 410 | 62.50 | | 472.50 | 545 |
| 5040 6'-0" wide | ▼ | 10 | 1.600 | ▼ | 590 | 69 | | 659 | 750 |

# 08 14 Wood Doors

## 08 14 33 – Stile and Rail Wood Doors

### 08 14 33.20 Wood Doors Residential

| | | Crew | Daily Output | Labor-Hours | Unit | Material | 2011 Bare Costs Labor | Equipment | Total | Total Incl O&P |
|---|---|---|---|---|---|---|---|---|---|---|
| 5042 | 8'-0" wide | 2 Carp | 12 | 1.333 | Opng. | 985 | 57.50 | | 1,042.50 | 1,175 |
| 6100 | Folding accordion, closet, including track and frame | | | | | | | | | |
| 6120 | Vinyl, 2 layer, stock | 2 Carp | 10 | 1.600 | Ea. | 62 | 69 | | 131 | 174 |
| 6140 | Woven mahogany and vinyl, stock | | 10 | 1.600 | | 45.50 | 69 | | 114.50 | 156 |
| 6160 | Wood slats with vinyl overlay, stock | | 10 | 1.600 | | 145 | 69 | | 214 | 266 |
| 6180 | Economy vinyl, stock | | 10 | 1.600 | | 28 | 69 | | 97 | 137 |
| 6200 | Rigid PVC | | 10 | 1.600 | | 34.50 | 69 | | 103.50 | 144 |
| 6220 | For custom partition, add | | | | | 25% | 10% | | | |
| 7310 | Passage doors, flush, no frame included | | | | | | | | | |
| 7320 | Hardboard, hollow core, 1-3/8" x 6'-8" x 1'-6" wide | 2 Carp | 18 | .889 | Ea. | 42.50 | 38.50 | | 81 | 105 |
| 7330 | 2'-0" wide | | 18 | .889 | | 42 | 38.50 | | 80.50 | 105 |
| 7340 | 2'-6" wide | | 18 | .889 | | 46.50 | 38.50 | | 85 | 110 |
| 7350 | 2'-8" wide | | 18 | .889 | | 48 | 38.50 | | 86.50 | 112 |
| 7360 | 3'-0" wide | | 17 | .941 | | 51 | 40.50 | | 91.50 | 118 |
| 7420 | Lauan, hollow core, 1-3/8" x 6'-8" x 1'-6" wide | | 18 | .889 | | 31 | 38.50 | | 69.50 | 92.50 |
| 7440 | 2'-0" wide | | 18 | .889 | | 34 | 38.50 | | 72.50 | 95.50 |
| 7450 | 2'-4" wide | | 18 | .889 | | 37.50 | 38.50 | | 76 | 100 |
| 7460 | 2'-6" wide | | 18 | .889 | | 37.50 | 38.50 | | 76 | 100 |
| 7480 | 2'-8" wide | | 18 | .889 | | 39 | 38.50 | | 77.50 | 102 |
| 7500 | 3'-0" wide | | 17 | .941 | | 41.50 | 40.50 | | 82 | 108 |
| 7700 | Birch, hollow core, 1-3/8" x 6'-8" x 1'-6" wide | | 18 | .889 | | 38.50 | 38.50 | | 77 | 101 |
| 7720 | 2'-0" wide | | 18 | .889 | | 41 | 38.50 | | 79.50 | 104 |
| 7740 | 2'-6" wide | | 18 | .889 | | 49 | 38.50 | | 87.50 | 112 |
| 7760 | 2'-8" wide | | 18 | .889 | | 49.50 | 38.50 | | 88 | 113 |
| 7780 | 3'-0" wide | | 17 | .941 | | 51 | 40.50 | | 91.50 | 118 |
| 8000 | Pine louvered, 1-3/8" x 6'-8" x 1'-6" wide | | 19 | .842 | | 101 | 36.50 | | 137.50 | 167 |
| 8020 | 2'-0" wide | | 18 | .889 | | 128 | 38.50 | | 166.50 | 199 |
| 8040 | 2'-6" wide | | 18 | .889 | | 139 | 38.50 | | 177.50 | 212 |
| 8060 | 2'-8" wide | | 18 | .889 | | 146 | 38.50 | | 184.50 | 220 |
| 8080 | 3'-0" wide | | 17 | .941 | | 173 | 40.50 | | 213.50 | 252 |
| 8300 | Pine paneled, 1-3/8" x 6'-8" x 1'-6" wide | | 19 | .842 | | 111 | 36.50 | | 147.50 | 178 |
| 8320 | 2'-0" wide | | 18 | .889 | | 130 | 38.50 | | 168.50 | 202 |
| 8330 | 2'-4" wide | | 18 | .889 | | 150 | 38.50 | | 188.50 | 224 |
| 8340 | 2'-6" wide | | 18 | .889 | | 158 | 38.50 | | 196.50 | 232 |
| 8360 | 2'-8" wide | | 18 | .889 | | 161 | 38.50 | | 199.50 | 236 |
| 8380 | 3'-0" wide | | 17 | .941 | | 167 | 40.50 | | 207.50 | 245 |

## 08 14 40 – Interior Cafe Doors

### 08 14 40.10 Cafe Style Doors

| | | Crew | Daily Output | Labor-Hours | Unit | Material | 2011 Bare Costs Labor | Equipment | Total | Total Incl O&P |
|---|---|---|---|---|---|---|---|---|---|---|
| 0010 | **CAFE STYLE DOORS** | | | | | | | | | |
| 6520 | Interior cafe doors, 2'-6" opening, stock, panel pine | 2 Carp | 16 | 1 | Ea. | 191 | 43 | | 234 | 276 |
| 6540 | 3'-0" opening | " | 16 | 1 | " | 199 | 43 | | 242 | 285 |
| 6550 | Louvered pine | | | | | | | | | |
| 6560 | 2'-6" opening | 2 Carp | 16 | 1 | Ea. | 165 | 43 | | 208 | 248 |
| 8000 | 3'-0" opening | | 16 | 1 | | 176 | 43 | | 219 | 260 |
| 8010 | 2'-6" opening, hardwood | | 16 | 1 | | 242 | 43 | | 285 | 330 |
| 8020 | 3'-0" opening | | 16 | 1 | | 268 | 43 | | 311 | 360 |

# 08 16 Composite Doors

## 08 16 13 – Fiberglass Doors

### 08 16 13.10 Entrance Doors, Fiberous Glass

| | Crew | Daily Output | Labor-Hours | Unit | Material | 2011 Bare Costs Labor | Equipment | Total | Total Incl O&P |
|---|---|---|---|---|---|---|---|---|---|
| 0010 **ENTRANCE DOORS, FIBEROUS GLASS** | | | | | | | | | |
| 0020 Exterior, fiberglass, door, 2'-8" wide x 6'-8" high **G** | 2 Carp | 15 | 1.067 | Ea. | 252 | 46 | | 298 | 350 |
| 0040 3'-0" wide x 6'-8" high **G** | | 15 | 1.067 | | 252 | 46 | | 298 | 350 |
| 0060 3'-0" wide x 7'-0" high **G** | | 15 | 1.067 | | 455 | 46 | | 501 | 570 |
| 0080 3'-0" wide x 6'-8" high, with two lites **G** | | 15 | 1.067 | | 294 | 46 | | 340 | 395 |
| 0100 3'-0" wide x 8'-0" high, with two lites **G** | | 15 | 1.067 | | 490 | 46 | | 536 | 605 |
| 0110 Half glass, 3'-0" wide x 6'-8" high **G** | | 15 | 1.067 | | 294 | 46 | | 340 | 395 |
| 0120 3'-0" wide x 6'-8" high, low e **G** | | 15 | 1.067 | | 325 | 46 | | 371 | 425 |
| 0130 3'-0" wide x 8'-0" high **G** | | 15 | 1.067 | | 580 | 46 | | 626 | 705 |
| 0140 3'-0" wide x 8'-0" high, low e **G** | | 15 | 1.067 | | 640 | 46 | | 686 | 770 |
| 0150 Side lights, 1'-0" wide x 6'-8" high, **G** | | | | | 247 | | | 247 | 271 |
| 0160 1'-0" wide x 6'-8" high, low e **G** | | | | | 256 | | | 256 | 281 |
| 0180 1'-0" wide x 6'-8" high, full glass **G** | | | | | 275 | | | 275 | 305 |
| 0190 1'-0" wide x 6'-8" high, low e **G** | | | | | 284 | | | 284 | 315 |

## 08 16 14 – French Doors

### 08 16 14.10 Exterior Doors With Glass Lites

| | Crew | Daily Output | Labor-Hours | Unit | Material | 2011 Bare Costs Labor | Equipment | Total | Total Incl O&P |
|---|---|---|---|---|---|---|---|---|---|
| 0010 **EXTERIOR DOORS WITH GLASS LITES** | | | | | | | | | |
| 0020 French, Fir, 1-3/4", 3'-0"wide x 6'-8" high | 2 Carp | 12 | 1.333 | Ea. | 600 | 57.50 | | 657.50 | 750 |
| 0025 double | | 12 | 1.333 | | 1,200 | 57.50 | | 1,257.50 | 1,425 |
| 0030 Maple, 1-3/4", 3'-0"wide x 6'-8" high | | 12 | 1.333 | | 675 | 57.50 | | 732.50 | 835 |
| 0035 double | | 12 | 1.333 | | 1,350 | 57.50 | | 1,407.50 | 1,575 |
| 0040 Cherry, 1-3/4", 3'-0"wide x 6'-8" high | | 12 | 1.333 | | 790 | 57.50 | | 847.50 | 955 |
| 0045 double | | 12 | 1.333 | | 1,575 | 57.50 | | 1,632.50 | 1,825 |
| 0100 Mahogany, 1-3/4", 3'-0"wide x 8'-0" high | | 10 | 1.600 | | 700 | 69 | | 769 | 875 |
| 0105 double | | 10 | 1.600 | | 1,400 | 69 | | 1,469 | 1,650 |
| 0110 Fir, 1-3/4", 3'-0"wide x 8'-0" high | | 10 | 1.600 | | 1,100 | 69 | | 1,169 | 1,300 |
| 0115 double | | 10 | 1.600 | | 2,200 | 69 | | 2,269 | 2,525 |
| 0120 Oak, 1-3/4", 3'-0"wide x 8'-0" high | | 10 | 1.600 | | 1,825 | 69 | | 1,894 | 2,100 |
| 0125 double | | 10 | 1.600 | | 3,650 | 69 | | 3,719 | 4,100 |

# 08 17 Integrated Door Opening Assemblies

## 08 17 23 – Integrated Wood Door Opening Assemblies

### 08 17 23.10 Pre-Hung Doors

| | Crew | Daily Output | Labor-Hours | Unit | Material | 2011 Bare Costs Labor | Equipment | Total | Total Incl O&P |
|---|---|---|---|---|---|---|---|---|---|
| 0010 **PRE-HUNG DOORS** | | | | | | | | | |
| 0300 Exterior, wood, comb. storm & screen, 6'-9" x 2'-6" wide | 2 Carp | 15 | 1.067 | Ea. | 289 | 46 | | 335 | 390 |
| 0320 2'-8" wide | | 15 | 1.067 | | 289 | 46 | | 335 | 390 |
| 0340 3'-0" wide | | 15 | 1.067 | | 296 | 46 | | 342 | 395 |
| 0360 For 7'-0" high door, add | | | | | 24 | | | 24 | 26.50 |
| 1600 Entrance door, flush, birch, solid core | | | | | | | | | |
| 1620 4-5/8" solid jamb, 1-3/4" x 6'-8" x 2'-8" wide | 2 Carp | 16 | 1 | Ea. | 285 | 43 | | 328 | 380 |
| 1640 3'-0" wide | | 16 | 1 | | 325 | 43 | | 368 | 425 |
| 1642 5-5/8" jamb | | 16 | 1 | | 325 | 43 | | 368 | 420 |
| 1680 For 7'-0" high door, add | | | | | 21 | | | 21 | 23 |
| 2000 Entrance door, colonial, 6 panel pine | | | | | | | | | |
| 2020 4-5/8" solid jamb, 1-3/4" x 6'-8" x 2'-8" wide | 2 Carp | 16 | 1 | Ea. | 635 | 43 | | 678 | 765 |
| 2040 3'-0" wide | " | 16 | 1 | | 665 | 43 | | 708 | 800 |
| 2060 For 7'-0" high door, add | | | | | 52.50 | | | 52.50 | 58 |
| 2200 For 5-5/8" solid jamb, add | | | | | 40.50 | | | 40.50 | 44.50 |
| 4000 Interior, passage door, 4-5/8" solid jamb | | | | | | | | | |
| 4400 Lauan, flush, solid core, 1-3/8" x 6'-8" x 2'-6" wide | 2 Carp | 17 | .941 | Ea. | 184 | 40.50 | | 224.50 | 264 |

# 08 17 Integrated Door Opening Assemblies

## 08 17 23 – Integrated Wood Door Opening Assemblies

| 08 17 23.10 Pre-Hung Doors | Crew | Daily Output | Labor-Hours | Unit | Material | 2011 Bare Costs Labor | Equipment | Total | Total Incl O&P |
|---|---|---|---|---|---|---|---|---|---|
| 4420 | 2'-8" wide | 2 Carp | 17 | .941 | Ea. | 184 | 40.50 | | 224.50 | 264 |
| 4440 | 3'-0" wide | | 16 | 1 | | 199 | 43 | | 242 | 285 |
| 4600 | Hollow core, 1-3/8" x 6'-8" x 2'-6" wide | | 17 | .941 | | 124 | 40.50 | | 164.50 | 198 |
| 4620 | 2'-8" wide | | 17 | .941 | | 124 | 40.50 | | 164.50 | 198 |
| 4640 | 3'-0" wide | | 16 | 1 | | 139 | 43 | | 182 | 219 |
| 4700 | For 7'-0" high door, add | | | | | 30 | | | 30 | 33 |
| 5000 | Birch, flush, solid core, 1-3/8" x 6'-8" x 2'-6" wide | 2 Carp | 17 | .941 | | 238 | 40.50 | | 278.50 | 325 |
| 5020 | 2'-8" wide | | 17 | .941 | | 199 | 40.50 | | 239.50 | 281 |
| 5040 | 3'-0" wide | | 16 | 1 | | 270 | 43 | | 313 | 365 |
| 5200 | Hollow core, 1-3/8" x 6'-8" x 2'-6" wide | | 17 | .941 | | 182 | 40.50 | | 222.50 | 262 |
| 5220 | 2'-8" wide | | 17 | .941 | | 158 | 40.50 | | 198.50 | 235 |
| 5240 | 3'-0" wide | | 16 | 1 | | 195 | 43 | | 238 | 280 |
| 5280 | For 7'-0" high door, add | | | | | 25 | | | 25 | 27.50 |
| 5500 | Hardboard paneled, 1-3/8" x 6'-8" x 2'-6" wide | 2 Carp | 17 | .941 | | 143 | 40.50 | | 183.50 | 220 |
| 5520 | 2'-8" wide | | 17 | .941 | | 152 | 40.50 | | 192.50 | 229 |
| 5540 | 3'-0" wide | | 16 | 1 | | 149 | 43 | | 192 | 230 |
| 6000 | Pine paneled, 1-3/8" x 6'-8" x 2'-6" wide | | 17 | .941 | | 247 | 40.50 | | 287.50 | 335 |
| 6020 | 2'-8" wide | | 17 | .941 | | 267 | 40.50 | | 307.50 | 355 |
| 6040 | 3'-0" wide | | 16 | 1 | | 272 | 43 | | 315 | 365 |
| 6500 | For 5-5/8" solid jamb, add | | | | | 20 | | | 20 | 22 |
| 6520 | For split jamb, deduct | | | | | 21.50 | | | 21.50 | 23.50 |

# 08 31 Access Doors and Panels

## 08 31 13 – Access Doors and Frames

| 08 31 13.10 Types of Framed Access Doors | Crew | Daily Output | Labor-Hours | Unit | Material | 2011 Bare Costs Labor | Equipment | Total | Total Incl O&P |
|---|---|---|---|---|---|---|---|---|---|
| 0010 | **TYPES OF FRAMED ACCESS DOORS** | | | | | | | | | |
| 1000 | Fire rated door with lock | | | | | | | | | |
| 1100 | Metal, 12" x 12" | 1 Carp | 10 | .800 | Ea. | 158 | 34.50 | | 192.50 | 227 |
| 1150 | 18" x 18" | | 9 | .889 | | 205 | 38.50 | | 243.50 | 285 |
| 1200 | 24" x 24" | | 9 | .889 | | 325 | 38.50 | | 363.50 | 420 |
| 1250 | 24" x 36" | | 8 | 1 | | 335 | 43 | | 378 | 430 |
| 1300 | 24" x 48" | | 8 | 1 | | 410 | 43 | | 453 | 520 |
| 1350 | 36" x 36" | | 7.50 | 1.067 | | 495 | 46 | | 541 | 615 |
| 1400 | 48" x 48" | | 7.50 | 1.067 | | 635 | 46 | | 681 | 765 |
| 1600 | Stainless steel, 12" x 12" | | 10 | .800 | | 282 | 34.50 | | 316.50 | 365 |
| 1650 | 18" x 18" | | 9 | .889 | | 410 | 38.50 | | 448.50 | 510 |
| 1700 | 24" x 24" | | 9 | .889 | | 500 | 38.50 | | 538.50 | 610 |
| 1750 | 24" x 36" | | 8 | 1 | | 640 | 43 | | 683 | 770 |
| 2000 | Flush door for finishing | | | | | | | | | |
| 2100 | Metal 8" x 8" | 1 Carp | 10 | .800 | Ea. | 35 | 34.50 | | 69.50 | 91.50 |
| 2150 | 12" x 12" | " | 10 | .800 | " | 40 | 34.50 | | 74.50 | 97 |
| 3000 | Recessed door for acoustic tile | | | | | | | | | |
| 3100 | Metal, 12" x 12" | 1 Carp | 4.50 | 1.778 | Ea. | 72 | 76.50 | | 148.50 | 196 |
| 3150 | 12" x 24" | | 4.50 | 1.778 | | 90 | 76.50 | | 166.50 | 216 |
| 3200 | 24" x 24" | | 4 | 2 | | 121 | 86 | | 207 | 265 |
| 3250 | 24" x 36" | | 4 | 2 | | 170 | 86 | | 256 | 320 |
| 4000 | Recessed door for drywall | | | | | | | | | |
| 4100 | Metal 12" x 12" | 1 Carp | 6 | 1.333 | Ea. | 77 | 57.50 | | 134.50 | 173 |
| 4150 | 12" x 24" | | 5.50 | 1.455 | | 114 | 62.50 | | 176.50 | 221 |
| 4200 | 24" x 36" | | 5 | 1.600 | | 182 | 69 | | 251 | 305 |
| 6000 | Standard door | | | | | | | | | |

# 08 31 Access Doors and Panels

## 08 31 13 – Access Doors and Frames

| 08 31 13.10 Types of Framed Access Doors | Crew | Daily Output | Labor-Hours | Unit | Material | 2011 Bare Costs Labor | Equipment | Total | Total Incl O&P |
|---|---|---|---|---|---|---|---|---|---|
| 6100 | Metal, 8" x 8" | 1 Carp | 10 | .800 | Ea. | 35 | 34.50 | | 69.50 | 91.50 |
| 6150 | 12" x 12" | | 10 | .800 | | 40 | 34.50 | | 74.50 | 97 |
| 6200 | 18" x 18" | | 9 | .889 | | 62 | 38.50 | | 100.50 | 127 |
| 6250 | 24" x 24" | | 9 | .889 | | 75 | 38.50 | | 113.50 | 141 |
| 6300 | 24" x 36" | | 8 | 1 | | 115 | 43 | | 158 | 193 |
| 6350 | 36" x 36" | | 8 | 1 | | 140 | 43 | | 183 | 220 |
| 6500 | Stainless steel, 8" x 8" | | 10 | .800 | | 80 | 34.50 | | 114.50 | 141 |
| 6550 | 12" x 12" | | 10 | .800 | | 107 | 34.50 | | 141.50 | 171 |
| 6600 | 18" x 18" | | 9 | .889 | | 198 | 38.50 | | 236.50 | 277 |
| 6650 | 24" x 24" | | 9 | .889 | | 260 | 38.50 | | 298.50 | 345 |
| 7000 | Ceiling hatches, 2'-6" x 2'-6", swing up, single leaf, st fr & cover | G-3 | 11 | 2.909 | | 570 | 125 | | 695 | 815 |
| 7010 | Aluminum cover | " | 11 | 2.909 | | 550 | 125 | | 675 | 795 |

## 08 31 13.20 Bulkhead/Cellar Doors

| | | Crew | Daily Output | Labor-Hours | Unit | Material | Labor | Equipment | Total | Total Incl O&P |
|---|---|---|---|---|---|---|---|---|---|---|
| 0010 | **BULKHEAD/CELLAR DOORS** | | | | | | | | | |
| 0020 | Steel, not incl. sides, 44" x 62" | 1 Carp | 5.50 | 1.455 | Ea. | 490 | 62.50 | | 552.50 | 635 |
| 0100 | 52" x 73" | | 5.10 | 1.569 | | 670 | 67.50 | | 737.50 | 840 |
| 0500 | With sides and foundation plates, 57" x 45" x 24" | | 4.70 | 1.702 | | 700 | 73.50 | | 773.50 | 880 |
| 0600 | 42" x 49" x 51" | | 4.30 | 1.860 | | 790 | 80 | | 870 | 995 |

## 08 31 13.30 Commercial Floor Doors

| | | Crew | Daily Output | Labor-Hours | Unit | Material | Labor | Equipment | Total | Total Incl O&P |
|---|---|---|---|---|---|---|---|---|---|---|
| 0010 | **COMMERCIAL FLOOR DOORS** | | | | | | | | | |
| 0020 | Aluminum tile, steel frame, one leaf, 2' x 2' opng. | 2 Sswk | 3.50 | 4.571 | Opng. | 775 | 222 | | 997 | 1,250 |
| 0050 | 3'-6" x 3'-6" opening | | 3.50 | 4.571 | | 1,475 | 222 | | 1,697 | 2,025 |
| 0500 | Double leaf, 4' x 4' opening | | 3 | 5.333 | | 1,575 | 259 | | 1,834 | 2,175 |
| 0550 | 5' x 5' opening | | 3 | 5.333 | | 2,800 | 259 | | 3,059 | 3,525 |

## 08 31 13.35 Industrial Floor Doors

| | | Crew | Daily Output | Labor-Hours | Unit | Material | Labor | Equipment | Total | Total Incl O&P |
|---|---|---|---|---|---|---|---|---|---|---|
| 0010 | **INDUSTRIAL FLOOR DOORS** | | | | | | | | | |
| 0020 | Steel 300 psf L.L., single leaf, 2' x 2', 175# | 2 Sswk | 6 | 2.667 | Opng. | 735 | 129 | | 864 | 1,050 |
| 0050 | 3' x 3' opening, 300# | | 5.50 | 2.909 | | 1,000 | 141 | | 1,141 | 1,350 |
| 0300 | Double leaf, 4' x 4' opening, 455# | | 5 | 3.200 | | 1,625 | 155 | | 1,780 | 2,075 |
| 0350 | 5' x 5' opening, 645# | | 4.50 | 3.556 | | 2,450 | 173 | | 2,623 | 3,000 |
| 1000 | Aluminum, 300 psf L.L., single leaf, 2' x 2', 60# | | 6 | 2.667 | | 740 | 129 | | 869 | 1,050 |
| 1050 | 3' x 3' opening, 100# | | 5.50 | 2.909 | | 1,100 | 141 | | 1,241 | 1,450 |
| 1500 | Double leaf, 4' x 4' opening, 160# | | 5 | 3.200 | | 1,950 | 155 | | 2,105 | 2,425 |
| 1550 | 5' x 5' opening, 235# | | 4.50 | 3.556 | | 2,675 | 173 | | 2,848 | 3,250 |
| 2000 | Aluminum, 150 psf L.L., single leaf, 2' x 2', 60# | | 6 | 2.667 | | 610 | 129 | | 739 | 900 |
| 2050 | 3' x 3' opening, 95# | | 5.50 | 2.909 | | 920 | 141 | | 1,061 | 1,250 |
| 2500 | Double leaf, 4' x 4' opening, 150# | | 5 | 3.200 | | 1,350 | 155 | | 1,505 | 1,750 |
| 2550 | 5' x 5' opening, 230# | | 4.50 | 3.556 | | 1,825 | 173 | | 1,998 | 2,325 |

## 08 31 13.40 Kennel Doors

| | | Crew | Daily Output | Labor-Hours | Unit | Material | Labor | Equipment | Total | Total Incl O&P |
|---|---|---|---|---|---|---|---|---|---|---|
| 0010 | **KENNEL DOORS** | | | | | | | | | |
| 0020 | 2 way, swinging type, 13" x 19" opening | 2 Carp | 11 | 1.455 | Opng. | 90 | 62.50 | | 152.50 | 195 |
| 0100 | 17" x 29" opening | | 11 | 1.455 | | 110 | 62.50 | | 172.50 | 217 |
| 0200 | 9" x 9" opening, electronic with accessories | | 11 | 1.455 | | 144 | 62.50 | | 206.50 | 254 |

# 08 32 Sliding Glass Doors

## 08 32 13 – Sliding Aluminum-Framed Glass Doors

| 08 32 13.10 Sliding Aluminum Doors | | Crew | Daily Output | Labor-Hours | Unit | Material | 2011 Bare Costs Labor | Equipment | Total | Total Incl O&P |
|---|---|---|---|---|---|---|---|---|---|---|
| 0010 | SLIDING ALUMINUM DOORS | | | | | | | | | |
| 0350 | Aluminum, 5/8" tempered insulated glass, 6' wide | | | | | | | | | |
| 0400 | Premium | 2 Carp | 4 | 4 | Ea. | 1,550 | 172 | | 1,722 | 2,000 |
| 0450 | Economy | | 4 | 4 | | 885 | 172 | | 1,057 | 1,250 |
| 0500 | 8' wide, premium | | 3 | 5.333 | | 1,625 | 230 | | 1,855 | 2,125 |
| 0550 | Economy | | 3 | 5.333 | | 1,400 | 230 | | 1,630 | 1,875 |
| 0600 | 12' wide, premium | | 2.50 | 6.400 | | 2,950 | 276 | | 3,226 | 3,675 |
| 0650 | Economy | | 2.50 | 6.400 | | 1,375 | 276 | | 1,651 | 1,925 |
| 0700 | Non-insulated, 6' wide | | 4 | 4 | | 1,100 | 172 | | 1,272 | 1,475 |
| 4000 | Aluminum, baked on enamel, temp glass, 6'-8" x 10'-0" wide | | 4 | 4 | | 1,025 | 172 | | 1,197 | 1,400 |
| 4020 | Insulating glass, 6'-8" x 6'-0" wide | | 4 | 4 | | 895 | 172 | | 1,067 | 1,250 |
| 4040 | 8'-0" wide | | 3 | 5.333 | | 1,050 | 230 | | 1,280 | 1,500 |
| 4060 | 10'-0" wide | | 2 | 8 | | 1,300 | 345 | | 1,645 | 1,950 |
| 4080 | Anodized, temp glass, 6'-8" x 6'-0" wide | | 4 | 4 | | 450 | 172 | | 622 | 760 |
| 4100 | 8'-0" wide | | 3 | 5.333 | | 570 | 230 | | 800 | 980 |
| 4120 | 10'-0" wide | | 2 | 8 | | 645 | 345 | | 990 | 1,225 |
| 4200 | Vinyl clad, anodized, temp glass, 6'-8" x 6'-0" wide | | 4 | 4 | | 420 | 172 | | 592 | 725 |
| 4240 | 8'-0" wide | | 3 | 5.333 | | 580 | 230 | | 810 | 990 |
| 4260 | 10'-0" wide | | 2 | 8 | | 630 | 345 | | 975 | 1,225 |
| 4280 | Insulating glass, 6'-8" x 6'-0' wide | | 4 | 4 | | 1,100 | 172 | | 1,272 | 1,475 |
| 4300 | 8'-0" wide | | 3 | 5.333 | | 920 | 230 | | 1,150 | 1,350 |
| 4320 | 10'-0" wide | | 4 | 4 | | 1,625 | 172 | | 1,797 | 2,050 |

## 08 32 19 – Sliding Wood-Framed Glass Doors

| 08 32 19.15 Sliding Glass Vinyl-Clad Wood Doors | | | Crew | Daily Output | Labor-Hours | Unit | Material | 2011 Bare Costs Labor | Equipment | Total | Total Incl O&P |
|---|---|---|---|---|---|---|---|---|---|---|---|
| 0010 | SLIDING GLASS VINYL-CLAD WOOD DOORS | | | | | | | | | | |
| 0020 | Glass, sliding vinyl clad, insul. glass, 6'-0" x 6'-8" | G | 2 Carp | 4 | 4 | Ea. | 1,500 | 172 | | 1,672 | 1,925 |
| 0025 | 6'-0" x 6'-10" high | | | 4 | 4 | Opng. | 1,625 | 172 | | 1,797 | 2,050 |
| 0030 | 6'-0" x 8'-0" high | G | | 4 | 4 | Ea. | 1,925 | 172 | | 2,097 | 2,400 |
| 0050 | 5'-0" x 6'-8" high | G | | 4 | 4 | " | 1,475 | 172 | | 1,647 | 1,900 |
| 0100 | 8'-0" x 6'-10" high | G | | 4 | 4 | Opng. | 2,000 | 172 | | 2,172 | 2,475 |
| 0500 | 4 leaf, 9'-0" x 6'-10" high | G | | 3 | 5.333 | | 3,175 | 230 | | 3,405 | 3,850 |
| 0600 | 12'-0" x 6'-10" high | G | | 3 | 5.333 | | 3,800 | 230 | | 4,030 | 4,525 |

# 08 33 Coiling Doors and Grilles

## 08 33 13 – Coiling Counter Doors

| 08 33 13.10 Counter Doors, Coiling Type | | Crew | Daily Output | Labor-Hours | Unit | Material | 2011 Bare Costs Labor | Equipment | Total | Total Incl O&P |
|---|---|---|---|---|---|---|---|---|---|---|
| 0010 | COUNTER DOORS, COILING TYPE | | | | | | | | | |
| 0020 | Manual, incl. frm and hdwe, galv. stl., 4' roll-up, 6' long | 2 Carp | 2 | 8 | Opng. | 1,150 | 345 | | 1,495 | 1,800 |
| 0300 | Galvanized steel, UL label | | 1.80 | 8.889 | | 1,200 | 385 | | 1,585 | 1,900 |
| 0600 | Stainless steel, 4' high roll-up, 6' long | | 2 | 8 | | 2,100 | 345 | | 2,445 | 2,825 |
| 0700 | 10' long | | 1.80 | 8.889 | | 2,450 | 385 | | 2,835 | 3,275 |
| 2000 | Aluminum, 4' high, 4' long | | 2.20 | 7.273 | | 1,325 | 315 | | 1,640 | 1,950 |
| 2020 | 6' long | | 2 | 8 | | 1,475 | 345 | | 1,820 | 2,150 |
| 2040 | 8' long | | 1.90 | 8.421 | | 1,675 | 365 | | 2,040 | 2,400 |
| 2060 | 10' long | | 1.80 | 8.889 | | 1,775 | 385 | | 2,160 | 2,525 |
| 2080 | 14' long | | 1.40 | 11.429 | | 2,500 | 490 | | 2,990 | 3,500 |
| 2100 | 6' high, 4' long | | 2 | 8 | | 1,500 | 345 | | 1,845 | 2,175 |
| 2120 | 6' long | | 1.60 | 10 | | 1,550 | 430 | | 1,980 | 2,375 |
| 2140 | 10' long | | 1.40 | 11.429 | | 2,050 | 490 | | 2,540 | 3,000 |

# 08 33 Coiling Doors and Grilles

## 08 33 16 – Coiling Counter Grilles

| 08 33 16.10 Coiling Grilles | Crew | Daily Output | Labor-Hours | Unit | Material | 2011 Bare Costs Labor | Equipment | Total | Total Incl O&P |
|---|---|---|---|---|---|---|---|---|---|
| 0010 **COILING GRILLES** | | | | | | | | | |
| 0015 Aluminum, manual, incl. frame, mill finish | | | | | | | | | |
| 0020 Top coiling, 4' high, 4' long | 2 Sswk | 3.20 | 5 | Opng. | 1,475 | 243 | | 1,718 | 2,025 |
| 0030 6' long | | 3.20 | 5 | | 1,700 | 243 | | 1,943 | 2,300 |
| 0040 8' long | | 2.40 | 6.667 | | 1,900 | 325 | | 2,225 | 2,650 |
| 0050 12' long | | 2.40 | 6.667 | | 2,300 | 325 | | 2,625 | 3,100 |
| 0060 16' long | | 1.60 | 10 | | 2,475 | 485 | | 2,960 | 3,550 |
| 0070 6' high, 4' long | | 3.20 | 5 | | 1,700 | 243 | | 1,943 | 2,300 |
| 0080 6' long | | 3.20 | 5 | | 1,725 | 243 | | 1,968 | 2,325 |
| 0090 8' long | | 2.40 | 6.667 | | 1,725 | 325 | | 2,050 | 2,475 |
| 0100 12' long | | 1.60 | 10 | | 2,250 | 485 | | 2,735 | 3,325 |
| 0110 16' long | | 1.20 | 13.333 | | 2,700 | 645 | | 3,345 | 4,125 |
| 0200 Side coiling, 8' high, 12' long | | .60 | 26.667 | | 2,225 | 1,300 | | 3,525 | 4,725 |
| 0220 18' long | | .50 | 32 | | 3,925 | 1,550 | | 5,475 | 7,075 |
| 0240 24' long | | .40 | 40 | | 3,525 | 1,950 | | 5,475 | 7,300 |
| 0260 12' high, 12' long | | .50 | 32 | | 3,250 | 1,550 | | 4,800 | 6,325 |
| 0280 18' long | | .40 | 40 | | 4,800 | 1,950 | | 6,750 | 8,700 |
| 0300 24' long | | .28 | 57.143 | | 6,425 | 2,775 | | 9,200 | 12,000 |

## 08 33 23 – Overhead Coiling Doors

| 08 33 23.10 Rolling Service Doors | Crew | Daily Output | Labor-Hours | Unit | Material | 2011 Bare Costs Labor | Equipment | Total | Total Incl O&P |
|---|---|---|---|---|---|---|---|---|---|
| 0010 **ROLLING SERVICE DOORS** Steel, manual, 20 ga., incl. hardware | | | | | | | | | |
| 0050 8' x 8' high | 2 Sswk | 1.60 | 10 | Ea. | 1,075 | 485 | | 1,560 | 2,025 |
| 0100 10' x 10' high | | 1.40 | 11.429 | | 1,775 | 555 | | 2,330 | 2,925 |
| 0200 20' x 10' high | | 1 | 16 | | 3,000 | 775 | | 3,775 | 4,675 |
| 0300 12' x 12' high | | 1.20 | 13.333 | | 1,800 | 645 | | 2,445 | 3,150 |
| 0400 20' x 12' high | | .90 | 17.778 | | 2,050 | 865 | | 2,915 | 3,800 |
| 0500 14' x 14' high | | .80 | 20 | | 2,950 | 970 | | 3,920 | 4,950 |
| 0600 20' x 16' high | | .60 | 26.667 | | 3,325 | 1,300 | | 4,625 | 5,925 |
| 0700 10' x 20' high | | .50 | 32 | | 2,350 | 1,550 | | 3,900 | 5,325 |
| 1000 12' x 12', crank operated, crank on door side | | .80 | 20 | | 1,600 | 970 | | 2,570 | 3,475 |
| 1100 Crank thru wall | | .70 | 22.857 | | 1,850 | 1,100 | | 2,950 | 3,975 |
| 1300 For vision panel, add | | | | | 315 | | | 315 | 350 |
| 1400 For 22 ga., deduct | | | | S.F. | 1.10 | | | 1.10 | 1.21 |
| 1600 3' x 7' pass door within rolling steel door, new construction | | | | Ea. | 1,725 | | | 1,725 | 1,900 |
| 1700 Existing construction | 2 Sswk | 2 | 8 | | 1,825 | 390 | | 2,215 | 2,675 |
| 2000 Class A fire doors, manual, 20 ga., 8' x 8' high | | 1.40 | 11.429 | | 1,425 | 555 | | 1,980 | 2,550 |
| 2100 10' x 10' high | | 1.10 | 14.545 | | 1,950 | 705 | | 2,655 | 3,400 |
| 2200 20' x 10' high | | .80 | 20 | | 3,975 | 970 | | 4,945 | 6,100 |
| 2300 12' x 12' high | | 1 | 16 | | 3,050 | 775 | | 3,825 | 4,750 |
| 2400 20' x 12' high | | .80 | 20 | | 4,325 | 970 | | 5,295 | 6,475 |
| 2500 14' x 14' high | | .60 | 26.667 | | 3,350 | 1,300 | | 4,650 | 5,975 |
| 2600 20' x 16' high | | .50 | 32 | | 5,400 | 1,550 | | 6,950 | 8,700 |
| 2700 10' x 20' high | | .40 | 40 | | 4,175 | 1,950 | | 6,125 | 8,025 |
| 3000 For 18 ga. doors, add | | | | S.F. | 1.40 | | | 1.40 | 1.54 |
| 3300 For enamel finish, add | | | | " | 1.65 | | | 1.65 | 1.82 |
| 3600 For safety edge bottom bar, pneumatic, add | | | | L.F. | 20.50 | | | 20.50 | 22.50 |
| 3700 Electric, add | | | | | 38.50 | | | 38.50 | 42 |
| 4000 For weatherstripping, extruded rubber, jambs, add | | | | | 12.90 | | | 12.90 | 14.20 |
| 4100 Hood, add | | | | | 7.90 | | | 7.90 | 8.70 |
| 4200 Sill, add | | | | | 4.55 | | | 4.55 | 5 |
| 4500 Motor operators, to 14' x 14' opening | 2 Sswk | 5 | 3.200 | Ea. | 1,025 | 155 | | 1,180 | 1,425 |
| 4600 Over 14' x 14', jack shaft type | " | 5 | 3.200 | | 1,075 | 155 | | 1,230 | 1,475 |

# 08 33 Coiling Doors and Grilles

## 08 33 23 – Overhead Coiling Doors

| 08 33 23.10 Rolling Service Doors | Crew | Daily Output | Labor-Hours | Unit | Material | 2011 Bare Costs Labor | Equipment | Total | Total Incl O&P |
|---|---|---|---|---|---|---|---|---|---|
| 4700 | For fire door, additional fusible link, add | | | | Ea. | 25 | | | 25 | 27.50 |

# 08 34 Special Function Doors

## 08 34 13 – Cold Storage Doors

### 08 34 13.10 Doors for Cold Area Storage

| | | Crew | Daily Output | Labor-Hours | Unit | Material | 2011 Bare Costs Labor | Equipment | Total | Total Incl O&P |
|---|---|---|---|---|---|---|---|---|---|---|
| 0010 | **DOORS FOR COLD AREA STORAGE** | | | | | | | | | |
| 0020 | Single, 20 ga. galvanized steel | | | | | | | | | |
| 0300 | Horizontal sliding, 5' x 7', manual operation, 3.5" thick | 2 Carp | 2 | 8 | Ea. | 3,175 | 345 | | 3,520 | 4,000 |
| 0400 | 4" thick | | 2 | 8 | | 3,150 | 345 | | 3,495 | 3,975 |
| 0500 | 6" thick | | 2 | 8 | | 3,300 | 345 | | 3,645 | 4,150 |
| 0800 | 5' x 7', power operation, 2" thick | | 1.90 | 8.421 | | 5,375 | 365 | | 5,740 | 6,450 |
| 0900 | 4" thick | | 1.90 | 8.421 | | 5,500 | 365 | | 5,865 | 6,600 |
| 1000 | 6" thick | | 1.90 | 8.421 | | 6,275 | 365 | | 6,640 | 7,450 |
| 1300 | 9' x 10', manual operation, 2" insulation | | 1.70 | 9.412 | | 4,375 | 405 | | 4,780 | 5,425 |
| 1400 | 4" insulation | | 1.70 | 9.412 | | 4,475 | 405 | | 4,880 | 5,550 |
| 1500 | 6" insulation | | 1.70 | 9.412 | | 5,375 | 405 | | 5,780 | 6,550 |
| 1800 | Power operation, 2" insulation | | 1.60 | 10 | | 7,400 | 430 | | 7,830 | 8,800 |
| 1900 | 4" insulation | | 1.60 | 10 | | 7,650 | 430 | | 8,080 | 9,050 |
| 2000 | 6" insulation | ▼ | 1.70 | 9.412 | ▼ | 8,600 | 405 | | 9,005 | 10,100 |
| 2300 | For stainless steel face, add | | | | | 22% | | | | |
| 3000 | Hinged, lightweight, 3' x 7'-0", 2" thick | 2 Carp | 2 | 8 | Ea. | 1,375 | 345 | | 1,720 | 2,025 |
| 3050 | 4" thick | | 1.90 | 8.421 | | 1,700 | 365 | | 2,065 | 2,425 |
| 3300 | Aluminum doors, 3' x 7'-0", 4" thick | | 1.90 | 8.421 | | 1,300 | 365 | | 1,665 | 1,975 |
| 3350 | 6" thick | | 1.40 | 11.429 | | 2,275 | 490 | | 2,765 | 3,250 |
| 3600 | Stainless steel, 3' x 7'-0", 4" thick | | 1.90 | 8.421 | | 1,675 | 365 | | 2,040 | 2,375 |
| 3650 | 6" thick | | 1.40 | 11.429 | | 2,850 | 490 | | 3,340 | 3,875 |
| 3900 | Painted, 3' x 7'-0", 4" thick | | 1.90 | 8.421 | | 1,225 | 365 | | 1,590 | 1,900 |
| 3950 | 6" thick | ▼ | 1.40 | 11.429 | ▼ | 2,275 | 490 | | 2,765 | 3,250 |
| 5000 | Bi-parting, electric operated | | | | | | | | | |
| 5010 | 6' x 8' opening, galv. faces, 4" thick for cooler | 2 Carp | .80 | 20 | Opng. | 7,000 | 860 | | 7,860 | 9,025 |
| 5050 | For freezer, 4" thick | | .80 | 20 | | 7,725 | 860 | | 8,585 | 9,825 |
| 5300 | For door buck framing and door protection, add | | 2.50 | 6.400 | | 620 | 276 | | 896 | 1,100 |
| 6000 | Galvanized batten door, galvanized hinges, 4' x 7' | | 2 | 8 | | 1,750 | 345 | | 2,095 | 2,450 |
| 6050 | 6' x 8' | | 1.80 | 8.889 | | 2,400 | 385 | | 2,785 | 3,200 |
| 6500 | Fire door, 3 hr., 6' x 8', single slide | | .80 | 20 | | 8,025 | 860 | | 8,885 | 10,200 |
| 6550 | Double, bi-parting | ▼ | .70 | 22.857 | ▼ | 12,200 | 985 | | 13,185 | 14,900 |

## 08 34 16 – Hangar Doors

### 08 34 16.10 Aircraft Hangar Doors

| | | Crew | Daily Output | Labor-Hours | Unit | Material | 2011 Bare Costs Labor | Equipment | Total | Total Incl O&P |
|---|---|---|---|---|---|---|---|---|---|---|
| 0010 | **AIRCRAFT HANGAR DOORS** | | | | | | | | | |
| 0020 | Bi-fold, ovhd., 20 PSF wind load, incl. elec. oper. | | | | | | | | | |
| 0100 | 12' high x 40' | 2 Sswk | 240 | .067 | S.F. | 15.45 | 3.24 | | 18.69 | 22.50 |
| 0200 | 16' high x 60' | | 230 | .070 | | 18.50 | 3.38 | | 21.88 | 26.50 |
| 0300 | 20' high x 80' | ▼ | 220 | .073 | ▼ | 20 | 3.53 | | 23.53 | 28.50 |

## 08 34 36 – Darkroom Doors

### 08 34 36.10 Various Types of Darkroom Doors

| | | Crew | Daily Output | Labor-Hours | Unit | Material | 2011 Bare Costs Labor | Equipment | Total | Total Incl O&P |
|---|---|---|---|---|---|---|---|---|---|---|
| 0010 | **VARIOUS TYPES OF DARKROOM DOORS** | | | | | | | | | |
| 0015 | Revolving, standard, 2 way, 36" diameter | 2 Carp | 3.10 | 5.161 | Opng. | 2,250 | 222 | | 2,472 | 2,825 |
| 0020 | 41" diameter | | 3.10 | 5.161 | | 2,425 | 222 | | 2,647 | 3,025 |
| 0050 | 3 way, 51" diameter | | 1.40 | 11.429 | | 3,075 | 490 | | 3,565 | 4,125 |
| 1000 | 4 way, 49" diameter | | 1.40 | 11.429 | | 3,625 | 490 | | 4,115 | 4,750 |

## 08 34 36 – Darkroom Doors

| 08 34 36.10 Various Types of Darkroom Doors | Crew | Daily Output | Labor-Hours | Unit | Material | 2011 Bare Costs Labor | Equipment | Total | Total Incl O&P |
|---|---|---|---|---|---|---|---|---|---|
| 2000 | Hinged safety, 2 way, 41" diameter | 2 Carp | 2.30 | 6.957 | Opng. | 2,825 | 299 | | 3,124 | 3,575 |
| 2500 | 3 way, 51" diameter | | 1.40 | 11.429 | | 3,625 | 490 | | 4,115 | 4,750 |
| 3000 | Pop out safety, 2 way, 41" diameter | | 3.10 | 5.161 | | 3,600 | 222 | | 3,822 | 4,300 |
| 4000 | 3 way, 51" diameter | | 1.40 | 11.429 | | 3,550 | 490 | | 4,040 | 4,650 |
| 5000 | Wheelchair-type, pop out, 51" diameter | | 1.40 | 11.429 | | 3,525 | 490 | | 4,015 | 4,625 |
| 5020 | 72" diameter | | .90 | 17.778 | | 7,225 | 765 | | 7,990 | 9,125 |
| 9300 | For complete darkrooms, see Section 13 21 53.50 | | | | | | | | | |

## 08 34 53 – Security Doors and Frames

### 08 34 53.20 Steel Door

| | | Crew | Daily Output | Labor-Hours | Unit | Material | 2011 Bare Costs Labor | Equipment | Total | Total Incl O&P |
|---|---|---|---|---|---|---|---|---|---|---|
| 0010 | **STEEL DOOR** with ballistic core and welded frame both 14 gauge | | | | | | | | | |
| 0050 | Flush, UL 752 Level 3, 1-3/4", 3'-0" x 6'-8" | 2 Carp | 1.50 | 10.667 | Opng. | 2,150 | 460 | | 2,610 | 3,050 |
| 0055 | 1-3/4", 3'-6" x 6'-8" | | 1.50 | 10.667 | | 2,250 | 460 | | 2,710 | 3,175 |
| 0060 | 1-3/4", 4'-0" x 6'-8" | | 1.20 | 13.333 | | 2,425 | 575 | | 3,000 | 3,550 |
| 0100 | UL 752 Level 8, 1-3/4", 3'-0" x 6'-8" | | 1.50 | 10.667 | | 9,475 | 460 | | 9,935 | 11,100 |
| 0105 | 1-3/4", 3'-6" x 6'-8" | | 1.50 | 10.667 | | 9,575 | 460 | | 10,035 | 11,200 |
| 0110 | 1-3/4", 4'-0" x 6'-8" | | 1.20 | 13.333 | | 10,900 | 575 | | 11,475 | 12,900 |
| 0120 | UL 752 Level 3, 1-3/4", 3'-0" x 7'-0" | | 1.50 | 10.667 | | 2,175 | 460 | | 2,635 | 3,075 |
| 0125 | 1-3/4", 3'-6" x 7'-0" | | 1.50 | 10.667 | | 2,275 | 460 | | 2,735 | 3,200 |
| 0130 | 1-3/4", 4'-0" x 7'-0" | | 1.20 | 13.333 | | 2,475 | 575 | | 3,050 | 3,575 |
| 0150 | UL 752 Level 8, 1-3/4", 3'-0" x 7'-0" | | 1.50 | 10.667 | | 9,500 | 460 | | 9,960 | 11,200 |
| 0155 | 1-3/4", 3'-6" x 7'-0" | | 1.50 | 10.667 | | 9,600 | 460 | | 10,060 | 11,300 |
| 0160 | 1-3/4", 4'-0" x 7'-0" | | 1.20 | 13.333 | | 10,900 | 575 | | 11,475 | 12,900 |
| 1000 | Safe Room sliding door and hardware, 1-3/4", 3'-0" x 7'-0" UL 752 Level 3 | | .50 | 32 | | 22,700 | 1,375 | | 24,075 | 27,000 |
| 1050 | Safe Room swinging door and hardware, 1-3/4", 3'-0" x 7'-0" UL 752 Level 3 | | .50 | 32 | | 26,800 | 1,375 | | 28,175 | 31,600 |

### 08 34 53.30 Wood Ballistic Doors

| | | Crew | Daily Output | Labor-Hours | Unit | Material | 2011 Bare Costs Labor | Equipment | Total | Total Incl O&P |
|---|---|---|---|---|---|---|---|---|---|---|
| 0010 | **WOOD BALLISTIC DOORS** with frames and hardware | | | | | | | | | |
| 0050 | Wood, 1-3/4", 3'-0" x 7'-0" UL 752 Level 3 | 2 Carp | 1.50 | 10.667 | Opng. | 2,050 | 460 | | 2,510 | 2,975 |

## 08 34 56 – SECURITY GATES

### 08 34 56.10 Gates

| | | Crew | Daily Output | Labor-Hours | Unit | Material | 2011 Bare Costs Labor | Equipment | Total | Total Incl O&P |
|---|---|---|---|---|---|---|---|---|---|---|
| 0010 | **GATES** | | | | | | | | | |
| 0015 | Gates include mounting hardware | | | | | | | | | |
| 0500 | Wood, security gate, driveway, dual, 10' wide | H-4 | .80 | 25 | Opng. | 3,675 | 1,000 | | 4,675 | 5,575 |
| 0505 | 12' wide | | .80 | 25 | | 3,975 | 1,000 | | 4,975 | 5,900 |
| 0510 | 15' wide | | .80 | 25 | | 4,575 | 1,000 | | 5,575 | 6,550 |
| 0600 | Steel, security gate, driveway, single, 10' wide | | .80 | 25 | | 2,150 | 1,000 | | 3,150 | 3,875 |
| 0605 | 12' wide | | .80 | 25 | | 2,225 | 1,000 | | 3,225 | 3,950 |
| 0620 | Steel, security gate, driveway, steel, dual, 12' wide | | .80 | 25 | | 2,050 | 1,000 | | 3,050 | 3,800 |
| 0625 | 14' wide | | .80 | 25 | | 2,200 | 1,000 | | 3,200 | 3,950 |
| 0630 | 16' wide | | .80 | 25 | | 2,525 | 1,000 | | 3,525 | 4,300 |
| 0700 | Aluminum, security gate, driveway, dual, 10' wide | | .80 | 25 | | 2,075 | 1,000 | | 3,075 | 3,825 |
| 0705 | 12' wide | | .80 | 25 | | 2,475 | 1,000 | | 3,475 | 4,250 |
| 0710 | 16' wide | | .80 | 25 | | 3,300 | 1,000 | | 4,300 | 5,175 |
| 1000 | Security gate, driveway, opener 12 VDC | | | | Ea. | 650 | | | 650 | 715 |
| 1010 | Wireless | | | | | 1,600 | | | 1,600 | 1,750 |
| 1020 | Security gate, driveway, opener 24 VDC | | | | | 1,000 | | | 1,000 | 1,100 |
| 1030 | Wireless | | | | | 1,100 | | | 1,100 | 1,200 |
| 1040 | Security gate, driveway, opener 12 VDC, solar panel 10 watt | 1 Elec | 2 | 4 | | 160 | 201 | | 361 | 475 |
| 1050 | 20 watt | " | 2 | 4 | | 300 | 201 | | 501 | 630 |

# 08 34 Special Function Doors

## 08 34 59 – Vault Doors and Day Gates

### 08 34 59.10 Secure Storage Doors

| | | Crew | Daily Output | Labor-Hours | Unit | Material | 2011 Bare Costs Labor | Equipment | Total | Total Incl O&P |
|---|---|---|---|---|---|---|---|---|---|---|
| 0010 | **SECURE STORAGE DOORS** | | | | | | | | | |
| 0020 | Door and frame, 32" x 78", clear opening | | | | | | | | | |
| 0100 | 1 hour test, 32" door, weighs 750 lbs. | 2 Sswk | 1.50 | 10.667 | Opng. | 2,750 | 520 | | 3,270 | 3,950 |
| 0200 | 2 hour test, 32" door, weighs 950 lbs. | | 1.30 | 12.308 | | 2,950 | 600 | | 3,550 | 4,300 |
| 0250 | 40" door, weighs 1130 lbs. | | 1 | 16 | | 3,350 | 775 | | 4,125 | 5,050 |
| 0300 | 4 hour test, 32" door, weighs 1025 lbs. | | 1.20 | 13.333 | | 3,250 | 645 | | 3,895 | 4,725 |
| 0350 | 40" door, weighs 1140 lbs. | | .90 | 17.778 | | 3,775 | 865 | | 4,640 | 5,675 |
| 0500 | For stainless steel front, including frame, add to above | | | | | 1,900 | | | 1,900 | 2,100 |
| 0550 | Back, add | | | | | 1,900 | | | 1,900 | 2,100 |
| 0600 | For time lock, two movement, add | 1 Elec | 2 | 4 | Ea. | 1,525 | 201 | | 1,726 | 1,975 |
| 0650 | Three movement, add | " | 2 | 4 | | 2,200 | 201 | | 2,401 | 2,725 |
| 0800 | Day gate, painted, steel, 32" wide | 2 Sswk | 1.50 | 10.667 | | 1,850 | 520 | | 2,370 | 2,975 |
| 0850 | 40" wide | | 1.40 | 11.429 | | 2,050 | 555 | | 2,605 | 3,225 |
| 0900 | Aluminum, 32" wide | | 1.50 | 10.667 | | 2,950 | 520 | | 3,470 | 4,175 |
| 0950 | 40" wide | | 1.40 | 11.429 | | 3,275 | 555 | | 3,830 | 4,575 |
| 2050 | Security vault door, class I, 3' wide, 3 1/2" thick | E-24 | .19 | 166 | Opng. | 15,200 | 7,950 | 3,900 | 27,050 | 34,500 |
| 2100 | Class II, 3' wide, 7" thick | | .19 | 166 | | 17,700 | 7,950 | 3,900 | 29,550 | 37,200 |
| 2150 | Class III, 9R, 3' wide, 10" thick, minimum | | .13 | 250 | | 22,200 | 11,900 | 5,850 | 39,950 | 51,000 |
| 2160 | Class V, type 1, 40" door | | 2.48 | 12.903 | Ea. | 6,475 | 615 | 300 | 7,390 | 8,500 |
| 2170 | Class V, type 2, 40" door | | 2.48 | 12.903 | | 6,425 | 615 | 300 | 7,340 | 8,450 |
| 2180 | Day gate for class V vault | 2 Sswk | 2 | 8 | | 1,700 | 390 | | 2,090 | 2,550 |

## 08 34 63 – Detention Doors and Frames

### 08 34 63.13 Steel Detention Doors and Frames

| | | Crew | Daily Output | Labor-Hours | Unit | Material | 2011 Bare Costs Labor | Equipment | Total | Total Incl O&P |
|---|---|---|---|---|---|---|---|---|---|---|
| 0010 | **STEEL DETENTION DOORS AND FRAMES** | | | | | | | | | |
| 0500 | Rolling cell door, bar front, 7/8" bars, 4" O.C., 7' H, 5' W, with hardware ⬛G | E-4 | 2 | 16 | Ea. | 5,025 | 785 | 54.50 | 5,864.50 | 6,950 |
| 0550 | Actuator for rolling cell door, bar front | 2 Skwk | 2 | 8 | | 4,200 | 355 | | 4,555 | 5,175 |
| 1000 | Doors & frames, 3' x 7', complete, with hardware, single plate | E-4 | 4 | 8 | | 4,200 | 390 | 27.50 | 4,617.50 | 5,350 |
| 1650 | Double plate | " | 4 | 8 | | 5,175 | 390 | 27.50 | 5,592.50 | 6,425 |

## 08 34 73 – Sound Control Door Assemblies

### 08 34 73.10 Acoustical Doors

| | | Crew | Daily Output | Labor-Hours | Unit | Material | 2011 Bare Costs Labor | Equipment | Total | Total Incl O&P |
|---|---|---|---|---|---|---|---|---|---|---|
| 0010 | **ACOUSTICAL DOORS** | | | | | | | | | |
| 0020 | Including framed seals, 3' x 7', wood, 27 STC rating | 2 Carp | 1.50 | 10.667 | Ea. | 900 | 460 | | 1,360 | 1,700 |
| 0100 | Steel, 40 STC rating | | 1.50 | 10.667 | | 2,900 | 460 | | 3,360 | 3,875 |
| 0200 | 45 STC rating | | 1.50 | 10.667 | | 3,400 | 460 | | 3,860 | 4,450 |
| 0300 | 48 STC rating | | 1.50 | 10.667 | | 3,925 | 460 | | 4,385 | 5,000 |
| 0400 | 52 STC rating | | 1.50 | 10.667 | | 4,425 | 460 | | 4,885 | 5,575 |

# 08 36 Panel Doors

## 08 36 13 – Sectional Doors

### 08 36 13.10 Overhead Commercial Doors

| | | Crew | Daily Output | Labor-Hours | Unit | Material | 2011 Bare Costs Labor | Equipment | Total | Total Incl O&P |
|---|---|---|---|---|---|---|---|---|---|---|
| 0010 | **OVERHEAD COMMERCIAL DOORS** | | | | | | | | | |
| 1000 | Stock, sectional, heavy duty, wood, 1-3/4" thick, 8' x 8' high | 2 Carp | 2 | 8 | Ea. | 810 | 345 | | 1,155 | 1,425 |
| 1100 | 10' x 10' high | | 1.80 | 8.889 | | 1,175 | 385 | | 1,560 | 1,875 |
| 1200 | 12' x 12' high | | 1.50 | 10.667 | | 1,700 | 460 | | 2,160 | 2,550 |
| 1300 | Chain hoist, 14' x 14' high | | 1.30 | 12.308 | | 2,600 | 530 | | 3,130 | 3,700 |
| 1400 | 12' x 16' high | | 1 | 16 | | 2,575 | 690 | | 3,265 | 3,875 |
| 1500 | 20' x 8' high | | 1.30 | 12.270 | | 2,325 | 530 | | 2,855 | 3,375 |
| 1600 | 20' x 16' high | | .65 | 24.615 | | 4,700 | 1,050 | | 5,750 | 6,800 |
| 1800 | Center mullion openings, 8' high | | 4 | 4 | | 1,200 | 172 | | 1,372 | 1,600 |

# 08 36 Panel Doors

## 08 36 13 – Sectional Doors

| 08 36 13.10 Overhead Commercial Doors | | Crew | Daily Output | Labor-Hours | Unit | Material | 2011 Bare Costs Labor | 2011 Bare Costs Equipment | Total | Total Incl O&P |
|---|---|---|---|---|---|---|---|---|---|---|
| 1900 | 20' high | 2 Carp | 2 | 8 | Ea. | 2,000 | 345 | | 2,345 | 2,725 |
| 2100 | For medium duty custom door, deduct | | | | | 5% | 5% | | | |
| 2150 | For medium duty stock doors, deduct | | | | | 10% | 5% | | | |
| 2300 | Fiberglass and aluminum, heavy duty, sectional, 12' x 12' high | 2 Carp | 1.50 | 10.667 | Ea. | 2,625 | 460 | | 3,085 | 3,575 |
| 2450 | Chain hoist, 20' x 20' high | | .50 | 32 | | 6,550 | 1,375 | | 7,925 | 9,325 |
| 2600 | Steel, 24 ga. sectional, manual, 8' x 8' high | | 2 | 8 | | 805 | 345 | | 1,150 | 1,425 |
| 2650 | 10' x 10' high | | 1.80 | 8.889 | | 1,025 | 385 | | 1,410 | 1,700 |
| 2700 | 12' x 12' high | | 1.50 | 10.667 | | 1,250 | 460 | | 1,710 | 2,075 |
| 2800 | Chain hoist, 20' x 14' high | | .70 | 22.857 | | 3,450 | 985 | | 4,435 | 5,300 |
| 2850 | For 1-1/4" rigid insulation and 26 ga. galv. | | | | | | | | | |
| 2860 | back panel, add | | | | S.F. | 4.50 | | | 4.50 | 4.95 |
| 2900 | For electric trolley operator, 1/3 H.P., to 12' x 12', add | 1 Carp | 2 | 4 | Ea. | 905 | 172 | | 1,077 | 1,250 |
| 2950 | Over 12' x 12', 1/2 H.P., add | " | 1 | 8 | " | 1,050 | 345 | | 1,395 | 1,675 |

## 08 36 13.20 Residential Garage Doors

| | | Crew | Daily Output | Labor-Hours | Unit | Material | 2011 Bare Costs Labor | 2011 Bare Costs Equipment | Total | Total Incl O&P |
|---|---|---|---|---|---|---|---|---|---|---|
| 0010 | **RESIDENTIAL GARAGE DOORS** | | | | | | | | | |
| 0050 | Hinged, wood, custom, double door, 9' x 7' | 2 Carp | 4 | 4 | Ea. | 640 | 172 | | 812 | 970 |
| 0070 | 16' x 7' | | 3 | 5.333 | | 1,000 | 230 | | 1,230 | 1,450 |
| 0200 | Overhead, sectional, incl. hardware, fiberglass, 9' x 7', standard | | 5.28 | 3.030 | | 855 | 130 | | 985 | 1,150 |
| 0220 | Deluxe | | 5.28 | 3.030 | | 1,075 | 130 | | 1,205 | 1,400 |
| 0300 | 16' x 7', standard | | 6 | 2.667 | | 1,550 | 115 | | 1,665 | 1,875 |
| 0320 | Deluxe | | 6 | 2.667 | | 2,100 | 115 | | 2,215 | 2,475 |
| 0500 | Hardboard, 9' x 7', standard | | 8 | 2 | | 575 | 86 | | 661 | 765 |
| 0520 | Deluxe | | 8 | 2 | | 800 | 86 | | 886 | 1,000 |
| 0600 | 16' x 7', standard | | 6 | 2.667 | | 1,200 | 115 | | 1,315 | 1,500 |
| 0620 | Deluxe | | 6 | 2.667 | | 1,400 | 115 | | 1,515 | 1,725 |
| 0700 | Metal, 9' x 7', standard | | 5.28 | 3.030 | | 625 | 130 | | 755 | 890 |
| 0720 | Deluxe | | 8 | 2 | | 860 | 86 | | 946 | 1,075 |
| 0800 | 16' x 7', standard | | 3 | 5.333 | | 760 | 230 | | 990 | 1,175 |
| 0820 | Deluxe | | 6 | 2.667 | | 1,375 | 115 | | 1,490 | 1,675 |
| 0900 | Wood, 9' x 7', standard | | 8 | 2 | | 900 | 86 | | 986 | 1,125 |
| 0920 | Deluxe | | 8 | 2 | | 2,100 | 86 | | 2,186 | 2,425 |
| 1000 | 16' x 7', standard | | 6 | 2.667 | | 1,600 | 115 | | 1,715 | 1,925 |
| 1020 | Deluxe | | 6 | 2.667 | | 2,975 | 115 | | 3,090 | 3,450 |
| 1800 | Door hardware, sectional | 1 Carp | 4 | 2 | | 320 | 86 | | 406 | 480 |
| 1810 | Door tracks only | | 4 | 2 | | 139 | 86 | | 225 | 285 |
| 1820 | One side only | | 7 | 1.143 | | 110 | 49 | | 159 | 197 |
| 3000 | Swing-up, including hardware, fiberglass, 9' x 7', standard | 2 Carp | 8 | 2 | | 980 | 86 | | 1,066 | 1,200 |
| 3020 | Deluxe | | 8 | 2 | | 1,000 | 86 | | 1,086 | 1,225 |
| 3100 | 16' x 7', standard | | 6 | 2.667 | | 1,225 | 115 | | 1,340 | 1,525 |
| 3120 | Deluxe | | 6 | 2.667 | | 1,500 | 115 | | 1,615 | 1,825 |
| 3200 | Hardboard, 9' x 7', standard | | 8 | 2 | | 500 | 86 | | 586 | 680 |
| 3220 | Deluxe | | 8 | 2 | | 600 | 86 | | 686 | 790 |
| 3300 | 16' x 7', standard | | 6 | 2.667 | | 620 | 115 | | 735 | 855 |
| 3320 | Deluxe | | 6 | 2.667 | | 800 | 115 | | 915 | 1,050 |
| 3400 | Metal, 9' x 7', standard | | 8 | 2 | | 420 | 86 | | 506 | 595 |
| 3420 | Deluxe | | 8 | 2 | | 900 | 86 | | 986 | 1,125 |
| 3500 | 16' x 7', standard | | 6 | 2.667 | | 685 | 115 | | 800 | 930 |
| 3520 | Deluxe | | 6 | 2.667 | | 1,000 | 115 | | 1,115 | 1,275 |
| 3600 | Wood, 9' x 7', standard | | 8 | 2 | | 600 | 86 | | 686 | 790 |
| 3620 | Deluxe | | 8 | 2 | | 1,025 | 86 | | 1,111 | 1,250 |
| 3700 | 16' x 7', standard | | 6 | 2.667 | | 800 | 115 | | 915 | 1,050 |
| 3720 | Deluxe | | 6 | 2.667 | | 2,000 | 115 | | 2,115 | 2,375 |

# 08 36 Panel Doors

## 08 36 13 – Sectional Doors

### 08 36 13.20 Residential Garage Doors

| | | Crew | Daily Output | Labor-Hours | Unit | Material | 2011 Bare Costs Labor | Equipment | Total | Total Incl O&P |
|---|---|---|---|---|---|---|---|---|---|---|
| 3900 | Door hardware only, swing up | 1 Carp | 4 | 2 | Ea. | 150 | 86 | | 236 | 297 |
| 3920 | One side only | | 7 | 1.143 | | 80 | 49 | | 129 | 164 |
| 4000 | For electric operator, economy, add | | 8 | 1 | | 410 | 43 | | 453 | 515 |
| 4100 | Deluxe, including remote control | | 8 | 1 | | 600 | 43 | | 643 | 725 |
| 4500 | For transmitter/receiver control , add to operator | | | | Total | 110 | | | 110 | 121 |
| 4600 | Transmitters, additional | | | | " | 55 | | | 55 | 60.50 |

## 08 36 19 – Multi-Leaf Vertical Lift Doors

### 08 36 19.10 Sectional Vertical Lift Doors

| | | Crew | Daily Output | Labor-Hours | Unit | Material | 2011 Bare Costs Labor | Equipment | Total | Total Incl O&P |
|---|---|---|---|---|---|---|---|---|---|---|
| 0010 | **SECTIONAL VERTICAL LIFT DOORS** | | | | | | | | | |
| 0020 | Motorized, 14 ga. stl, incl., frm and ctrl pnl | | | | | | | | | |
| 0050 | 16' x 16' high | L-10 | .50 | 48 | Ea. | 21,500 | 2,325 | 1,300 | 25,125 | 29,100 |
| 0100 | 10' x 20' high | | 1.30 | 18.462 | | 31,500 | 895 | 505 | 32,900 | 36,800 |
| 0120 | 15' x 20' high | | 1.30 | 18.462 | | 39,000 | 895 | 505 | 40,400 | 45,000 |
| 0140 | 20' x 20' high | | 1 | 24 | | 45,400 | 1,175 | 655 | 47,230 | 52,500 |
| 0160 | 25' x 20' high | | 1 | 24 | | 52,000 | 1,175 | 655 | 53,830 | 59,500 |
| 0170 | 32' x 24' high | | .75 | 32 | | 46,400 | 1,550 | 875 | 48,825 | 54,500 |
| 0180 | 20' x 25' high | | 1 | 24 | | 53,000 | 1,175 | 655 | 54,830 | 61,000 |
| 0200 | 25' x 25' high | | .70 | 34.286 | | 62,000 | 1,675 | 935 | 64,610 | 72,500 |
| 0220 | 25' x 30' high | | .70 | 34.286 | | 67,000 | 1,675 | 935 | 69,610 | 78,000 |
| 0240 | 30' x 30' high | | .70 | 34.286 | | 78,000 | 1,675 | 935 | 80,610 | 90,000 |
| 0260 | 35' x 30' high | | .70 | 34.286 | | 88,000 | 1,675 | 935 | 90,610 | 101,000 |

## 08 36 23 – Telescoping Vertical Lift Doors

### 08 36 23.10 Telescoping Steel Doors

| | | Crew | Daily Output | Labor-Hours | Unit | Material | 2011 Bare Costs Labor | Equipment | Total | Total Incl O&P |
|---|---|---|---|---|---|---|---|---|---|---|
| 0010 | **TELESCOPING STEEL DOORS** | | | | | | | | | |
| 1000 | Overhead, .03" thick, electric operated, 10' x 10' | E-3 | .80 | 30 | Ea. | 14,400 | 1,475 | 136 | 16,011 | 18,600 |
| 2000 | 20' x 10' | | .60 | 40 | | 18,500 | 1,975 | 182 | 20,657 | 24,100 |
| 3000 | 20' x 16' | | .40 | 60 | | 19,000 | 2,950 | 272 | 22,222 | 26,400 |

# 08 38 Traffic Doors

## 08 38 13 – Flexible Strip Doors

### 08 38 13.10 Flexible Transparent Strip Doors

| | | Crew | Daily Output | Labor-Hours | Unit | Material | 2011 Bare Costs Labor | Equipment | Total | Total Incl O&P |
|---|---|---|---|---|---|---|---|---|---|---|
| 0010 | **FLEXIBLE TRANSPARENT STRIP DOORS** | | | | | | | | | |
| 0100 | 12" strip width, 2/3 overlap | 3 Shee | 135 | .178 | SF Surf | 7.50 | 9.20 | | 16.70 | 22 |
| 0200 | Full overlap | | 115 | .209 | | 9.40 | 10.80 | | 20.20 | 26.50 |
| 0220 | 8" strip width, 1/2 overlap | | 140 | .171 | | 6 | 8.85 | | 14.85 | 20 |
| 0240 | Full overlap | | 120 | .200 | | 7.60 | 10.35 | | 17.95 | 24 |
| 0300 | Add for suspension system, header mount | | | | L.F. | 8.90 | | | 8.90 | 9.80 |
| 0400 | Wall mount | | | | " | 9 | | | 9 | 9.90 |

## 08 38 19 – Rigid Traffic Doors

### 08 38 19.20 Double Acting Swing Doors

| | | Crew | Daily Output | Labor-Hours | Unit | Material | 2011 Bare Costs Labor | Equipment | Total | Total Incl O&P |
|---|---|---|---|---|---|---|---|---|---|---|
| 0010 | **DOUBLE ACTING SWING DOORS** | | | | | | | | | |
| 1000 | .063" aluminum, 7'-0" high, 4'-0" wide | 2 Carp | 4.20 | 3.810 | Pr. | 1,900 | 164 | | 2,064 | 2,350 |
| 1025 | 6'-0" wide | | 4 | 4 | | 2,100 | 172 | | 2,272 | 2,575 |
| 1050 | 6'-8" wide | | 4 | 4 | | 2,300 | 172 | | 2,472 | 2,800 |
| 2000 | Solid core wood, 3/4" thick, metal frame, stainless steel | | | | | | | | | |
| 2010 | base plate, 7' high opening, 4' wide | 2 Carp | 4 | 4 | Pr. | 2,200 | 172 | | 2,372 | 2,700 |
| 2050 | 7' wide | " | 3.80 | 4.211 | " | 2,500 | 181 | | 2,681 | 3,025 |

# 08 38 Traffic Doors

## 08 38 19 – Rigid Traffic Doors

| 08 38 19.30 Shock Absorbing Doors | | Crew | Daily Output | Labor-Hours | Unit | Material | 2011 Bare Costs Labor | Equipment | Total | Total Incl O&P |
|---|---|---|---|---|---|---|---|---|---|---|
| 0010 | **SHOCK ABSORBING DOORS** | | | | | | | | | |
| 0020 | Rigid, no frame, 1-1/2" thick, 5' x 7' | 2 Sswk | 1.90 | 8.421 | Opng. | 1,400 | 410 | | 1,810 | 2,275 |
| 0100 | 8' x 8' | | 1.80 | 8.889 | | 2,000 | 430 | | 2,430 | 2,950 |
| 0500 | Flexible, no frame, insulated, .16" thick, economy, 5' x 7' | | 2 | 8 | | 1,750 | 390 | | 2,140 | 2,600 |
| 0600 | Deluxe | | 1.90 | 8.421 | | 2,625 | 410 | | 3,035 | 3,600 |
| 1000 | 8' x 8' opening, economy | | 2 | 8 | | 2,750 | 390 | | 3,140 | 3,700 |
| 1100 | Deluxe | | 1.90 | 8.421 | | 3,500 | 410 | | 3,910 | 4,575 |

# 08 41 Entrances and Storefronts

## 08 41 13 – Aluminum-Framed Entrances and Storefronts

| 08 41 13.20 Tube Framing | | Crew | Daily Output | Labor-Hours | Unit | Material | 2011 Bare Costs Labor | Equipment | Total | Total Incl O&P |
|---|---|---|---|---|---|---|---|---|---|---|
| 0010 | **TUBE FRAMING**, For window walls and store fronts, aluminum stock | | | | | | | | | |
| 0050 | Plain tube frame, mill finish, 1-3/4" x 1-3/4" | 2 Glaz | 103 | .155 | L.F. | 9.05 | 6.55 | | 15.60 | 19.75 |
| 0150 | 1-3/4" x 4" | | 98 | .163 | | 12.50 | 6.85 | | 19.35 | 24 |
| 0200 | 1-3/4" x 4-1/2" | | 95 | .168 | | 14.35 | 7.10 | | 21.45 | 26.50 |
| 0250 | 2" x 6" | | 89 | .180 | | 21.50 | 7.55 | | 29.05 | 35 |
| 0350 | 4" x 4" | | 87 | .184 | | 24.50 | 7.75 | | 32.25 | 38 |
| 0400 | 4-1/2" x 4-1/2" | | 85 | .188 | | 25.50 | 7.90 | | 33.40 | 40 |
| 0450 | Glass bead | | 240 | .067 | | 2.73 | 2.80 | | 5.53 | 7.20 |
| 1000 | Flush tube frame, mill finish, 1/4" glass, 1-3/4" x 4", open header | | 80 | .200 | | 12.05 | 8.40 | | 20.45 | 26 |
| 1050 | Open sill | | 82 | .195 | | 9.75 | 8.20 | | 17.95 | 23 |
| 1100 | Closed back header | | 83 | .193 | | 17.20 | 8.10 | | 25.30 | 31 |
| 1150 | Closed back sill | | 85 | .188 | | 16.40 | 7.90 | | 24.30 | 30 |
| 1160 | Tube fmg, spandrel cover both sides, alum 1" wide | 1 Sswk | 85 | .094 | S.F. | 91.50 | 4.57 | | 96.07 | 108 |
| 1170 | Tube fmg, spandrel cover both sides, alum 2" wide | " | 85 | .094 | " | 35 | 4.57 | | 39.57 | 46.50 |
| 1200 | Vertical mullion, one piece | 2 Glaz | 75 | .213 | L.F. | 18 | 8.95 | | 26.95 | 33.50 |
| 1250 | Two piece | | 73 | .219 | | 19.15 | 9.20 | | 28.35 | 35 |
| 1300 | 90° or 180° vertical corner post | | 75 | .213 | | 29 | 8.95 | | 37.95 | 45.50 |
| 1400 | 1-3/4" x 4-1/2", open header | | 80 | .200 | | 14.60 | 8.40 | | 23 | 28.50 |
| 1450 | Open sill | | 82 | .195 | | 12.20 | 8.20 | | 20.40 | 25.50 |
| 1500 | Closed back header | | 83 | .193 | | 17.45 | 8.10 | | 25.55 | 31.50 |
| 1550 | Closed back sill | | 85 | .188 | | 17.15 | 7.90 | | 25.05 | 31 |
| 1600 | Vertical mullion, one piece | | 75 | .213 | | 19.20 | 8.95 | | 28.15 | 34.50 |
| 1650 | Two piece | | 73 | .219 | | 20 | 9.20 | | 29.20 | 36.50 |
| 1700 | 90° or 180° vertical corner post | | 75 | .213 | | 20.50 | 8.95 | | 29.45 | 36.50 |
| 2000 | Flush tube frame, mil fin.,ins. glass w/thml brk, 2" x 4-1/2", open header | | 75 | .213 | | 15 | 8.95 | | 23.95 | 30 |
| 2050 | Open sill | | 77 | .208 | | 12.60 | 8.75 | | 21.35 | 27 |
| 2100 | Closed back header | | 78 | .205 | | 14.15 | 8.65 | | 22.80 | 28.50 |
| 2150 | Closed back sill | | 80 | .200 | | 15.15 | 8.40 | | 23.55 | 29.50 |
| 2200 | Vertical mullion, one piece | | 70 | .229 | | 15.70 | 9.60 | | 25.30 | 32 |
| 2250 | Two piece | | 68 | .235 | | 16.95 | 9.90 | | 26.85 | 33.50 |
| 2300 | 90° or 180° vertical corner post | | 70 | .229 | | 16.20 | 9.60 | | 25.80 | 32.50 |
| 5000 | Flush tube frame, mill fin., thermal brk., 2-1/4" x 4-1/2", open header | | 74 | .216 | | 15.60 | 9.10 | | 24.70 | 31 |
| 5050 | Open sill | | 75 | .213 | | 13.90 | 8.95 | | 22.85 | 29 |
| 5100 | Vertical mullion, one piece | | 69 | .232 | | 17.05 | 9.75 | | 26.80 | 33.50 |
| 5150 | Two piece | | 67 | .239 | | 19.75 | 10.05 | | 29.80 | 36.50 |
| 5200 | 90° or 180° vertical corner post | | 69 | .232 | | 17.40 | 9.75 | | 27.15 | 34 |
| 6980 | Door stop (snap in) | | 380 | .042 | | 3 | 1.77 | | 4.77 | 5.95 |
| 7000 | For joints, 90°, clip type, add | | | | Ea. | 23.50 | | | 23.50 | 26 |
| 7050 | Screw spline joint, add | | | | | 22 | | | 22 | 24 |

# 08 41 Entrances and Storefronts

## 08 41 13 – Aluminum-Framed Entrances and Storefronts

| 08 41 13.20 Tube Framing | Crew | Daily Output | Labor-Hours | Unit | Material | 2011 Bare Costs Labor | Equipment | Total | Total Incl O&P |
|---|---|---|---|---|---|---|---|---|---|
| 7100      For joint other than 90°, add | | | | Ea. | 45.50 | | | 45.50 | 50 |
| 8000    For bronze anodized aluminum, add | | | | | 15% | | | | |
| 8020    For black finish, add | | | | | 27% | | | | |
| 8050    For stainless steel materials, add | | | | | 350% | | | | |
| 8100    For monumental grade, add | | | | | 52% | | | | |
| 8150    For steel stiffener, add | 2 Glaz | 200 | .080 | L.F. | 10.50 | 3.36 | | 13.86 | 16.60 |
| 8200    For 2 to 5 stories, add per story | | | | Story | | 6% | | | |

## 08 41 19 – Stainless-Steel-Framed Entrances and Storefronts

### 08 41 19.10 Stainless-Steel and Glass Entrance Unit

| | Crew | Daily Output | Labor-Hours | Unit | Material | 2011 Bare Costs Labor | Equipment | Total | Total Incl O&P |
|---|---|---|---|---|---|---|---|---|---|
| 0010 **STAINLESS-STEEL AND GLASS ENTRANCE UNIT**, narrow stiles | | | | | | | | | |
| 0020    3' x 7' opening, including hardware, minimum | 2 Sswk | 1.60 | 10 | Opng. | 6,400 | 485 | | 6,885 | 7,900 |
| 0050      Average | | 1.40 | 11.429 | | 6,800 | 555 | | 7,355 | 8,450 |
| 0100      Maximum | ↓ | 1.20 | 13.333 | | 7,350 | 645 | | 7,995 | 9,225 |
| 1000    For solid bronze entrance units, statuary finish, add | | | | | 62% | | | | |
| 1100      Without statuary finish, add | | | | ↓ | 45% | | | | |
| 2000    Balanced doors, 3' x 7', economy | 2 Sswk | .90 | 17.778 | Ea. | 8,600 | 865 | | 9,465 | 11,000 |
| 2100      Premium | " | .70 | 22.857 | " | 14,800 | 1,100 | | 15,900 | 18,300 |

## 08 41 26 – All-Glass Entrances and Storefronts

### 08 41 26.10 Window Walls Aluminum, Stock

| | Crew | Daily Output | Labor-Hours | Unit | Material | 2011 Bare Costs Labor | Equipment | Total | Total Incl O&P |
|---|---|---|---|---|---|---|---|---|---|
| 0010 **WINDOW WALLS ALUMINUM, STOCK**, including glazing | | | | | | | | | |
| 0020      Minimum | H-2 | 160 | .150 | S.F. | 42 | 5.90 | | 47.90 | 55.50 |
| 0050      Average | | 140 | .171 | | 58 | 6.75 | | 64.75 | 74 |
| 0100      Maximum | ↓ | 110 | .218 | ↓ | 162 | 8.60 | | 170.60 | 191 |
| 0500    For translucent sandwich wall systems, see Section 07 41 33.10 | | | | | | | | | |
| 0850      Cost of the above walls depends on material, | | | | | | | | | |
| 0860        finish, repetition, and size of units. | | | | | | | | | |
| 0870      The larger the opening, the lower the S.F. cost | | | | | | | | | |
| 1200    Double glazed acoustical window wall for airports, | | | | | | | | | |
| 1220      including 1" thick glass with 2" x 4-1/2" tube frame | H-2 | 40 | .600 | S.F. | 112 | 23.50 | | 135.50 | 159 |

# 08 42 Entrances

## 08 42 26 – All-Glass Entrances

### 08 42 26.10 Swinging Glass Doors

| | Crew | Daily Output | Labor-Hours | Unit | Material | 2011 Bare Costs Labor | Equipment | Total | Total Incl O&P |
|---|---|---|---|---|---|---|---|---|---|
| 0010 **SWINGING GLASS DOORS** | | | | | | | | | |
| 0020    Including hardware, 1/2" thick, tempered, 3' x 7' opening | 2 Glaz | 2 | 8 | Opng. | 2,100 | 335 | | 2,435 | 2,825 |
| 0100      6' x 7' opening | " | 1.40 | 11.429 | " | 4,025 | 480 | | 4,505 | 5,150 |

## 08 42 29 – Automatic Entrances

### 08 42 29.23 Sliding Automatic Entrances

| | Crew | Daily Output | Labor-Hours | Unit | Material | 2011 Bare Costs Labor | Equipment | Total | Total Incl O&P |
|---|---|---|---|---|---|---|---|---|---|
| 0010 **SLIDING AUTOMATIC ENTRANCES** 12' x 7'-6" opng., 5' x 7' door, 2 way traffic | | | | | | | | | |
| 0020    Mat or electronic activated, panic pushout, incl. operator & hardware, | | | | | | | | | |
| 0030      not including glass or glazing | 2 Glaz | .70 | 22.857 | Opng. | 8,150 | 960 | | 9,110 | 10,400 |

## 08 42 33 – Revolving Door Entrances

### 08 42 33.10 Circular Rotating Entrance Doors

| | Crew | Daily Output | Labor-Hours | Unit | Material | 2011 Bare Costs Labor | Equipment | Total | Total Incl O&P |
|---|---|---|---|---|---|---|---|---|---|
| 0010 **CIRCULAR ROTATING ENTRANCE DOORS**, Aluminum | | | | | | | | | |
| 0020    6'-10" to 7' high, stock units, minimum | 4 Sswk | .75 | 42.667 | Opng. | 19,100 | 2,075 | | 21,175 | 24,700 |
| 0050      Average | | .60 | 53.333 | | 23,100 | 2,600 | | 25,700 | 30,000 |
| 0100      Maximum | | .45 | 71.111 | | 39,700 | 3,450 | | 43,150 | 49,700 |
| 1000      Stainless steel | | .30 | 105 | | 39,000 | 5,150 | | 44,150 | 52,000 |
| 1100      Solid bronze | ↓ | .15 | 213 | | 46,000 | 10,400 | | 56,400 | 69,000 |

# 08 42 Entrances

## 08 42 33 – Revolving Door Entrances

| 08 42 33.10 Circular Rotating Entrance Doors | Crew | Daily Output | Labor-Hours | Unit | Material | 2011 Bare Costs Labor | Equipment | Total | Total Incl O&P |
|---|---|---|---|---|---|---|---|---|---|
| 1500 | For automatic controls, add | 2 Elec | 2 | 8 | Opng. | 14,500 | 400 | | 14,900 | 16,500 |

## 08 42 36 – Balanced Door Entrances

### 08 42 36.10 Balanced Entrance Doors

| | | Crew | Daily Output | Labor-Hours | Unit | Material | Labor | Equipment | Total | Total Incl O&P |
|---|---|---|---|---|---|---|---|---|---|---|
| 0010 | **BALANCED ENTRANCE DOORS** | | | | | | | | | |
| 0020 | Hardware & frame, alum. & glass, 3' x 7', econ. | 2 Sswk | .90 | 17.778 | Ea. | 6,050 | 865 | | 6,915 | 8,175 |
| 0150 | Premium | " | .70 | 22.857 | " | 7,400 | 1,100 | | 8,500 | 10,100 |

# 08 43 Storefronts

## 08 43 13 – Aluminum-Framed Storefronts

### 08 43 13.10 Aluminum-Framed Entrance Doors

| | | Crew | Daily Output | Labor-Hours | Unit | Material | Labor | Equipment | Total | Total Incl O&P |
|---|---|---|---|---|---|---|---|---|---|---|
| 0010 | **ALUMINUM-FRAMED ENTRANCE DOORS** | | | | | | | | | |
| 0015 | Standard hardware and glass stops but no glazing | | | | | | | | | |
| 0020 | Entrance door, 3' x 7' opening, clear anodized finish | 2 Sswk | 7 | 2.286 | Opng. | 355 | 111 | | 466 | 590 |
| 0040 | Bronze finish | | 7 | 2.286 | | 375 | 111 | | 486 | 610 |
| 0060 | Black finish | | 7 | 2.286 | | 420 | 111 | | 531 | 655 |
| 0500 | 6' x 7' opening, clear finish | | 6 | 2.667 | | 720 | 129 | | 849 | 1,025 |
| 0520 | Bronze finish | | 6 | 2.667 | | 755 | 129 | | 884 | 1,050 |
| 0540 | Black finish | | 6 | 2.667 | | 845 | 129 | | 974 | 1,150 |
| 1000 | With 3' high transom above, 3' x 7' opening, clear finish | | 5.50 | 2.909 | | 415 | 141 | | 556 | 710 |
| 1050 | Bronze finish | | 5.50 | 2.909 | | 440 | 141 | | 581 | 735 |
| 1100 | Black finish | | 5.50 | 2.909 | | 475 | 141 | | 616 | 770 |
| 1300 | 3'-6" x 7'-0" opening, clear finish | | 5.50 | 2.909 | | 320 | 141 | | 461 | 605 |
| 1320 | Bronze finish | | 5.50 | 2.909 | | 330 | 141 | | 471 | 615 |
| 1340 | Black finish | | 5.50 | 2.909 | | 350 | 141 | | 491 | 635 |
| 1500 | 6' x 7' opening, clear finish | | 5.50 | 2.909 | | 505 | 141 | | 646 | 805 |
| 1550 | Bronze finish | | 5.50 | 2.909 | | 530 | 141 | | 671 | 835 |
| 1600 | Black finish | | 5.50 | 2.909 | | 590 | 141 | | 731 | 895 |

### 08 43 13.20 Storefront Systems

| | | Crew | Daily Output | Labor-Hours | Unit | Material | Labor | Equipment | Total | Total Incl O&P |
|---|---|---|---|---|---|---|---|---|---|---|
| 0010 | **STOREFRONT SYSTEMS**, aluminum frame clear 3/8" plate glass | | | | | | | | | |
| 0020 | incl. 3' x 7' door with hardware (400 sq. ft. max. wall) | | | | | | | | | |
| 0500 | Wall height to 12' high, commercial grade | 2 Glaz | 150 | .107 | S.F. | 20 | 4.49 | | 24.49 | 29 |
| 0600 | Institutional grade | | 130 | .123 | | 25 | 5.20 | | 30.20 | 35.50 |
| 0700 | Monumental grade | | 115 | .139 | | 38 | 5.85 | | 43.85 | 50.50 |
| 1000 | 6' x 7' door with hardware, commercial grade | | 135 | .119 | | 20 | 4.98 | | 24.98 | 29.50 |
| 1100 | Institutional grade | | 115 | .139 | | 28 | 5.85 | | 33.85 | 39.50 |
| 1200 | Monumental grade | | 100 | .160 | | 51 | 6.75 | | 57.75 | 66.50 |
| 1500 | For bronze anodized finish, add | | | | | 15% | | | | |
| 1600 | For black anodized finish, add | | | | | 35% | | | | |
| 1700 | For stainless steel framing, add to monumental | | | | | 76% | | | | |

## 08 43 29 – Sliding Storefronts

### 08 43 29.10 Sliding Panels

| | | Crew | Daily Output | Labor-Hours | Unit | Material | Labor | Equipment | Total | Total Incl O&P |
|---|---|---|---|---|---|---|---|---|---|---|
| 0010 | **SLIDING PANELS** | | | | | | | | | |
| 0020 | Mall fronts, aluminum & glass, 15' x 9' high | 2 Glaz | 1.30 | 12.308 | Opng. | 3,250 | 520 | | 3,770 | 4,350 |
| 0100 | 24' x 9' high | | .70 | 22.857 | | 4,625 | 960 | | 5,585 | 6,525 |
| 0200 | 48' x 9' high, with fixed panels | | .90 | 17.778 | | 8,600 | 750 | | 9,350 | 10,600 |
| 0500 | For bronze finish, add | | | | | 17% | | | | |

# 08 44 Curtain Wall and Glazed Assemblies

## 08 44 13 – Glazed Aluminum Curtain Walls

| 08 44 13.10 Glazed Curtain Walls | Crew | Daily Output | Labor-Hours | Unit | Material | 2011 Bare Costs Labor | Equipment | Total | Total Incl O&P |
|---|---|---|---|---|---|---|---|---|---|
| 0010 **GLAZED CURTAIN WALLS**, aluminum, stock, including glazing | | | | | | | | | |
| 0020 Minimum | H-1 | 205 | .156 | S.F. | 33.50 | 7.05 | | 40.55 | 48 |
| 0050 Average, single glazed | | 195 | .164 | | 48 | 7.45 | | 55.45 | 65 |
| 0150 Average, double glazed | | 180 | .178 | | 62 | 8.05 | | 70.05 | 82 |
| 0200 Maximum | ↓ | 160 | .200 | ↓ | 169 | 9.05 | | 178.05 | 201 |

# 08 45 Translucent Wall and Roof Assemblies

## 08 45 10 – Translucent Roof Assemblies

### 08 45 10.10 Skyroofs

| | Crew | Daily Output | Labor-Hours | Unit | Material | 2011 Bare Costs Labor | Equipment | Total | Total Incl O&P |
|---|---|---|---|---|---|---|---|---|---|
| 0010 **SKYROOFS**, Translucent panels, 2-3/4" thick | | | | | | | | | |
| 0020 Under 5000 S.F. | G-3 | 395 | .081 | SF Hor. | 33 | 3.48 | | 36.48 | 42 |
| 0100 Over 5000 S.F. | | 465 | .069 | | 31 | 2.96 | | 33.96 | 38.50 |
| 0300 Continuous vaulted, semi-circular, to 8' wide, double glazed [G] | | 145 | .221 | | 51.50 | 9.50 | | 61 | 71 |
| 0400 Single glazed | | 160 | .200 | | 34.50 | 8.60 | | 43.10 | 51 |
| 0600 To 20' wide, single glazed | | 175 | .183 | | 45 | 7.85 | | 52.85 | 61.50 |
| 0700 Over 20' wide, single glazed | | 200 | .160 | | 44 | 6.90 | | 50.90 | 59 |
| 0900 Motorized opening type, single glazed, 1/3 opening | | 145 | .221 | | 45.50 | 9.50 | | 55 | 64.50 |
| 1000 Full opening | ↓ | 130 | .246 | ↓ | 51.50 | 10.60 | | 62.10 | 73 |
| 1200 Pyramid type units, self-supporting, to 30' clear opening, | | | | | | | | | |
| 1300 square or circular, single glazed, minimum | G-3 | 200 | .160 | SF Hor. | 26.50 | 6.90 | | 33.40 | 39.50 |
| 1310 Average | | 165 | .194 | | 36 | 8.35 | | 44.35 | 52.50 |
| 1400 Maximum | | 130 | .246 | | 52 | 10.60 | | 62.60 | 73 |
| 1500 Grid type, 4' to 10' modules, single glass glazed, minimum | | 200 | .160 | | 35.50 | 6.90 | | 42.40 | 49.50 |
| 1550 Maximum | | 128 | .250 | | 55 | 10.75 | | 65.75 | 77 |
| 1600 Preformed acrylic, minimum | | 300 | .107 | | 42.50 | 4.59 | | 47.09 | 53.50 |
| 1650 Maximum | ↓ | 175 | .183 | ↓ | 58 | 7.85 | | 65.85 | 76 |
| 1800 Skyroofs, dome type, self-supporting, to 100' clear opening, circular | | | | | | | | | |
| 1900 Rise to span ratio = 0.20 | | | | | | | | | |
| 1920 Minimum | G-3 | 197 | .162 | SF Hor. | 15.50 | 7 | | 22.50 | 27.50 |
| 1950 Maximum | | 113 | .283 | | 52 | 12.20 | | 64.20 | 75.50 |
| 2100 Rise to span ratio = 0.33, minimum | | 169 | .189 | | 31 | 8.15 | | 39.15 | 46.50 |
| 2150 Maximum | | 101 | .317 | | 61.50 | 13.60 | | 75.10 | 88 |
| 2200 Rise to span ratio = 0.50, minimum | | 148 | .216 | | 47 | 9.30 | | 56.30 | 65.50 |
| 2250 Maximum | | 87 | .368 | | 68 | 15.80 | | 83.80 | 99 |
| 2400 Ridge units, continuous, to 8' wide, double [G] | | 130 | .246 | | 119 | 10.60 | | 129.60 | 147 |
| 2500 Single [G] | | 200 | .160 | | 82 | 6.90 | | 88.90 | 100 |
| 2700 Ridge and furrow units, over 4' O.C., double, minimum [G] | | 200 | .160 | | 25.50 | 6.90 | | 32.40 | 38.50 |
| 2750 Maximum [G] | | 120 | .267 | | 50.50 | 11.45 | | 61.95 | 73 |
| 2800 Single, minimum | | 214 | .150 | | 22 | 6.45 | | 28.45 | 34 |
| 2850 Maximum | | 153 | .209 | ↓ | 46 | 9 | | 55 | 64.50 |
| 3000 Rolling roof, translucent panels, flat roof, minimum | | 253 | .126 | S.F. | 19.80 | 5.45 | | 25.25 | 30.50 |
| 3030 Maximum [G] | | 160 | .200 | | 39.50 | 8.60 | | 48.10 | 56.50 |
| 3100 Lean-to skyroof, long span, double, minimum [G] | | 197 | .162 | | 26.50 | 7 | | 33.50 | 39.50 |
| 3150 Maximum [G] | | 101 | .317 | | 53 | 13.60 | | 66.60 | 78.50 |
| 3300 Single, minimum | | 321 | .100 | | 20 | 4.29 | | 24.29 | 28.50 |
| 3350 Maximum | ↓ | 160 | .200 | ↓ | 33 | 8.60 | | 41.60 | 49.50 |

# 08 51 Metal Windows

## 08 51 13 – Aluminum Windows

### 08 51 13.10 Aluminum Sash

| | | Crew | Daily Output | Labor-Hours | Unit | Material | 2011 Bare Costs Labor | Equipment | Total | Total Incl O&P |
|---|---|---|---|---|---|---|---|---|---|---|
| 0010 | **ALUMINUM SASH** | | | | | | | | | |
| 0020 | Stock, grade C, glaze & trim not incl., casement | 2 Sswk | 200 | .080 | S.F. | 37.50 | 3.88 | | 41.38 | 48 |
| 0050 | Double hung | | 200 | .080 | | 38 | 3.88 | | 41.88 | 48.50 |
| 0100 | Fixed casement | | 200 | .080 | | 16.55 | 3.88 | | 20.43 | 25 |
| 0150 | Picture window | | 200 | .080 | | 17.55 | 3.88 | | 21.43 | 26 |
| 0200 | Projected window | | 200 | .080 | | 34 | 3.88 | | 37.88 | 44 |
| 0250 | Single hung | | 200 | .080 | | 15.80 | 3.88 | | 19.68 | 24.50 |
| 0300 | Sliding | | 200 | .080 | | 20.50 | 3.88 | | 24.38 | 29.50 |
| 1000 | Mullions for above, tubular | | 240 | .067 | L.F. | 5.90 | 3.24 | | 9.14 | 12.15 |
| 2000 | Custom aluminum sash, grade HC, glazing not included, minimum | | 200 | .080 | S.F. | 38 | 3.88 | | 41.88 | 49 |
| 2100 | Maximum | | 85 | .188 | " | 49.50 | 9.15 | | 58.65 | 70.50 |

### 08 51 13.20 Aluminum Windows

| | | Crew | Daily Output | Labor-Hours | Unit | Material | 2011 Bare Costs Labor | Equipment | Total | Total Incl O&P |
|---|---|---|---|---|---|---|---|---|---|---|
| 0010 | **ALUMINUM WINDOWS**, incl. frame and glazing, commercial grade | | | | | | | | | |
| 1000 | Stock units, casement, 3'-1" x 3'-2" opening | 2 Sswk | 10 | 1.600 | Ea. | 355 | 77.50 | | 432.50 | 525 |
| 1050 | Add for storms | | | | | 115 | | | 115 | 126 |
| 1600 | Projected, with screen, 3'-1" x 3'-2" opening | 2 Sswk | 10 | 1.600 | | 335 | 77.50 | | 412.50 | 505 |
| 1700 | Add for storms | | | | | 112 | | | 112 | 123 |
| 2000 | 4'-5" x 5'-3" opening | 2 Sswk | 8 | 2 | | 380 | 97 | | 477 | 590 |
| 2100 | Add for storms | | | | | 120 | | | 120 | 132 |
| 2500 | Enamel finish windows, 3'-1" x 3'-2" | 2 Sswk | 10 | 1.600 | | 340 | 77.50 | | 417.50 | 510 |
| 2600 | 4'-5" x 5'-3" | | 8 | 2 | | 385 | 97 | | 482 | 595 |
| 3000 | Single hung, 2' x 3' opening, enameled, standard glazed | | 10 | 1.600 | | 198 | 77.50 | | 275.50 | 355 |
| 3100 | Insulating glass | | 10 | 1.600 | | 240 | 77.50 | | 317.50 | 400 |
| 3300 | 2'-8" x 6'-8" opening, standard glazed | | 8 | 2 | | 350 | 97 | | 447 | 555 |
| 3400 | Insulating glass | | 8 | 2 | | 450 | 97 | | 547 | 665 |
| 3700 | 3'-4" x 5'-0" opening, standard glazed | | 9 | 1.778 | | 286 | 86.50 | | 372.50 | 465 |
| 3800 | Insulating glass | | 9 | 1.778 | | 315 | 86.50 | | 401.50 | 500 |
| 3890 | Awning type, 3' x 3' opening standard glass | | 14 | 1.143 | | 405 | 55.50 | | 460.50 | 545 |
| 3900 | Insulating glass | | 14 | 1.143 | | 430 | 55.50 | | 485.50 | 575 |
| 3910 | 3' x 4' opening, standard glass | | 10 | 1.600 | | 470 | 77.50 | | 547.50 | 655 |
| 3920 | Insulating glass | | 10 | 1.600 | | 545 | 77.50 | | 622.50 | 730 |
| 3930 | 3' x 5'-4" opening, standard glass | | 10 | 1.600 | | 570 | 77.50 | | 647.50 | 760 |
| 3940 | Insulating glass | | 10 | 1.600 | | 670 | 77.50 | | 747.50 | 870 |
| 3950 | 4' x 5'-4" opening, standard glass | | 9 | 1.778 | | 625 | 86.50 | | 711.50 | 835 |
| 3960 | Insulating glass | | 9 | 1.778 | | 750 | 86.50 | | 836.50 | 975 |
| 4000 | Sliding aluminum, 3' x 2' opening, standard glazed | | 10 | 1.600 | | 209 | 77.50 | | 286.50 | 365 |
| 4100 | Insulating glass | | 10 | 1.600 | | 224 | 77.50 | | 301.50 | 385 |
| 4300 | 5' x 3' opening, standard glazed | | 9 | 1.778 | | 320 | 86.50 | | 406.50 | 500 |
| 4400 | Insulating glass | | 9 | 1.778 | | 370 | 86.50 | | 456.50 | 560 |
| 4600 | 8' x 4' opening, standard glazed | | 6 | 2.667 | | 335 | 129 | | 464 | 600 |
| 4700 | Insulating glass | | 6 | 2.667 | | 545 | 129 | | 674 | 825 |
| 5000 | 9' x 5' opening, standard glazed | | 4 | 4 | | 510 | 194 | | 704 | 905 |
| 5100 | Insulating glass | | 4 | 4 | | 820 | 194 | | 1,014 | 1,250 |
| 5500 | Sliding, with thermal barrier and screen, 6' x 4', 2 track | | 8 | 2 | | 695 | 97 | | 792 | 935 |
| 5700 | 4 track | | 8 | 2 | | 880 | 97 | | 977 | 1,150 |
| 6000 | For above units with bronze finish, add | | | | | 12% | | | | |
| 6200 | For installation in concrete openings, add | | | | | 6% | | | | |

## 08 51 23 – Steel Windows

### 08 51 23.10 Steel Sash

| | | Crew | Daily Output | Labor-Hours | Unit | Material | 2011 Bare Costs Labor | Equipment | Total | Total Incl O&P |
|---|---|---|---|---|---|---|---|---|---|---|
| 0010 | **STEEL SASH** Custom units, glazing and trim not included | R085123-10 | | | | | | | | |
| 0100 | Casement, 100% vented | 2 Sswk | 200 | .080 | S.F. | 62 | 3.88 | | 65.88 | 75 |
| 0200 | 50% vented | | 200 | .080 | | 51 | 3.88 | | 54.88 | 63 |

# 08 51 Metal Windows

## 08 51 23 - Steel Windows

### 08 51 23.10 Steel Sash

| | | Crew | Daily Output | Labor-Hours | Unit | Material | 2011 Bare Costs Labor | Equipment | Total | Total Incl O&P |
|---|---|---|---|---|---|---|---|---|---|---|
| 0300 | Fixed | 2 Sswk | 200 | .080 | S.F. | 27.50 | 3.88 | | 31.38 | 37 |
| 1000 | Projected, commercial, 40% vented | | 200 | .080 | | 48.50 | 3.88 | | 52.38 | 60.50 |
| 1100 | Intermediate, 50% vented | | 200 | .080 | | 54.50 | 3.88 | | 58.38 | 67 |
| 1500 | Industrial, horizontally pivoted | | 200 | .080 | | 50.50 | 3.88 | | 54.38 | 62.50 |
| 1600 | Fixed | | 200 | .080 | | 29.50 | 3.88 | | 33.38 | 39 |
| 2000 | Industrial security sash, 50% vented | | 200 | .080 | | 54.50 | 3.88 | | 58.38 | 67 |
| 2100 | Fixed | | 200 | .080 | | 44.50 | 3.88 | | 48.38 | 56 |
| 2500 | Picture window | | 200 | .080 | | 28.50 | 3.88 | | 32.38 | 38.50 |
| 3000 | Double hung | | 200 | .080 | | 56 | 3.88 | | 59.88 | 68.50 |
| 5000 | Mullions for above, open interior face | | 240 | .067 | L.F. | 9.80 | 3.24 | | 13.04 | 16.45 |
| 5100 | With interior cover | | 240 | .067 | " | 16.35 | 3.24 | | 19.59 | 23.50 |

### 08 51 23.20 Steel Windows

| | | | Crew | Daily Output | Labor-Hours | Unit | Material | 2011 Bare Costs Labor | Equipment | Total | Total Incl O&P |
|---|---|---|---|---|---|---|---|---|---|---|---|
| 0010 | **STEEL WINDOWS** Stock, including frame, trim and insul. glass | | | | | | | | | | |
| 0020 | See Section 13 34 19.50 | | | | | | | | | | |
| 1000 | Custom units, double hung, 2'-8" x 4'-6" opening | R085123-10 | 2 Sswk | 12 | 1.333 | Ea. | 680 | 64.50 | | 744.50 | 865 |
| 1100 | 2'-4" x 3'-9" opening | | | 12 | 1.333 | | 565 | 64.50 | | 629.50 | 735 |
| 1500 | Commercial projected, 3'-9" x 5'-5" opening | | | 10 | 1.600 | | 1,200 | 77.50 | | 1,277.50 | 1,425 |
| 1600 | 6'-9" x 4'-1" opening | | | 7 | 2.286 | | 1,575 | 111 | | 1,686 | 1,925 |
| 2000 | Intermediate projected, 2'-9" x 4'-1" opening | | | 12 | 1.333 | | 670 | 64.50 | | 734.50 | 850 |
| 2100 | 4'-1" x 5'-5" opening | | | 10 | 1.600 | | 1,350 | 77.50 | | 1,427.50 | 1,600 |

## 08 51 66 - Metal Window Screens

### 08 51 66.10 Screens

| | | Crew | Daily Output | Labor-Hours | Unit | Material | 2011 Bare Costs Labor | Equipment | Total | Total Incl O&P |
|---|---|---|---|---|---|---|---|---|---|---|
| 0010 | **SCREENS** | | | | | | | | | |
| 0020 | For metal sash, aluminum or bronze mesh, flat screen | 2 Sswk | 1200 | .013 | S.F. | 4.10 | .65 | | 4.75 | 5.65 |
| 0500 | Wicket screen, inside window | | 1000 | .016 | | 6.30 | .78 | | 7.08 | 8.30 |
| 0800 | Security screen, aluminum frame with stainless steel cloth | | 1200 | .013 | | 22.50 | .65 | | 23.15 | 26 |
| 0900 | Steel grate, painted, on steel frame | | 1600 | .010 | | 12.50 | .49 | | 12.99 | 14.60 |
| 1000 | For solar louvers, add | | 160 | .100 | | 23.50 | 4.86 | | 28.36 | 34 |
| 4000 | See Section 05 58 27.90 | | | | | | | | | |

# 08 52 Wood Windows

## 08 52 10 - Plain Wood Windows

### 08 52 10.20 Awning Window

| | | Crew | Daily Output | Labor-Hours | Unit | Material | 2011 Bare Costs Labor | Equipment | Total | Total Incl O&P |
|---|---|---|---|---|---|---|---|---|---|---|
| 0010 | **AWNING WINDOW**, Including frame, screens and grilles | | | | | | | | | |
| 0100 | Average quality, 34" x 22", double insulated glass | 1 Carp | 10 | .800 | Ea. | 256 | 34.50 | | 290.50 | 335 |
| 0200 | Low E glass | | 10 | .800 | | 265 | 34.50 | | 299.50 | 345 |
| 0300 | 40" x 28", double insulated glass | | 9 | .889 | | 310 | 38.50 | | 348.50 | 400 |
| 0400 | Low E Glass | | 9 | .889 | | 335 | 38.50 | | 373.50 | 430 |
| 0500 | 48" x 36", double insulated glass | | 8 | 1 | | 460 | 43 | | 503 | 570 |
| 0600 | Low E glass | | 8 | 1 | | 485 | 43 | | 528 | 595 |
| 1000 | Vinyl clad, 34" x 22" | | 10 | .800 | | 260 | 34.50 | | 294.50 | 340 |
| 1100 | 40" x 22" | | 10 | .800 | | 284 | 34.50 | | 318.50 | 365 |
| 1200 | 36" x 28" | | 9 | .889 | | 300 | 38.50 | | 338.50 | 390 |
| 1300 | 36" x 36" | | 9 | .889 | | 335 | 38.50 | | 373.50 | 430 |
| 1400 | 48" x 28" | | 8 | 1 | | 360 | 43 | | 403 | 465 |
| 1500 | 60" x 36" | | 8 | 1 | | 510 | 43 | | 553 | 630 |
| 2200 | Metal clad, 36" x 25" | | 9 | .889 | | 269 | 38.50 | | 307.50 | 355 |
| 2300 | 40" x 30" | | 9 | .889 | | 335 | 38.50 | | 373.50 | 430 |
| 2400 | 48" x 28" | | 8 | 1 | | 345 | 43 | | 388 | 445 |
| 2500 | 60" x 36" | | 8 | 1 | | 365 | 43 | | 408 | 470 |

# 08 52 Wood Windows

## 08 52 10 – Plain Wood Windows

### 08 52 10.20 Awning Window

| | | Crew | Daily Output | Labor-Hours | Unit | Material | 2011 Bare Costs Labor | Equipment | Total | Total Incl O&P |
|---|---|---|---|---|---|---|---|---|---|---|
| 4000 | Impact windows, minimum, add | | | | Ea. | 60% | | | | |
| 4010 | Impact windows, maximum, add | | | | | 160% | | | | |

### 08 52 10.40 Casement Window

| | | | Crew | Daily Output | Labor-Hours | Unit | Material | Labor | Equipment | Total | Total Incl O&P |
|---|---|---|---|---|---|---|---|---|---|---|---|
| 0010 | **CASEMENT WINDOW**, including frame, screen and grilles | | | | | | | | | | |
| 0100 | Avg. quality, bldrs. model, 2'-0" x 3'-0" H, dbl. insulated glass | G | 1 Carp | 10 | .800 | Ea. | 276 | 34.50 | | 310.50 | 360 |
| 0150 | Low E glass | G | | 10 | .800 | | 225 | 34.50 | | 259.50 | 300 |
| 0200 | 2'-0" x 4'-6" high, double insulated glass | G | | 9 | .889 | | 345 | 38.50 | | 383.50 | 440 |
| 0250 | Low E glass | G | | 9 | .889 | | 263 | 38.50 | | 301.50 | 350 |
| 0260 | Casement 4'-2" x 4'-2" double insulated glass | G | | 11 | .727 | | 855 | 31.50 | | 886.50 | 990 |
| 0270 | 4'-0" x 4'-0" Low E glass | G | | 11 | .727 | | 465 | 31.50 | | 496.50 | 565 |
| 0290 | 6'-4" x 5'-7" Low E glass | G | | 9 | .889 | | 1,100 | 38.50 | | 1,138.50 | 1,250 |
| 0300 | 2'-4" x 6'-0" high, double insulated glass | G | | 8 | 1 | | 400 | 43 | | 443 | 505 |
| 0350 | Low E glass | G | | 8 | 1 | | 465 | 43 | | 508 | 580 |
| 0522 | Vinyl clad, premium, double insulated glass, 2'-0" x 3'-0" | G | | 10 | .800 | | 266 | 34.50 | | 300.50 | 345 |
| 0524 | 2'-0" x 4'-0" | G | | 9 | .889 | | 310 | 38.50 | | 348.50 | 400 |
| 0525 | 2'-0" x 5'-0" | G | | 8 | 1 | | 355 | 43 | | 398 | 455 |
| 0528 | 2'-0" x 6'-0" | G | | 8 | 1 | | 355 | 43 | | 398 | 455 |
| 0600 | 3'-0" x 5'-0" | | | 8 | 1 | | 650 | 43 | | 693 | 780 |
| 0700 | 4'-0" x 3'-0" | | | 8 | 1 | | 715 | 43 | | 758 | 855 |
| 0710 | 4'-0" x 4'-0" | G | | 8 | 1 | | 610 | 43 | | 653 | 740 |
| 0720 | 4'-8" x 4'-0" | | | 8 | 1 | | 675 | 43 | | 718 | 810 |
| 0730 | 4'-8" x 5'-0" | | | 6 | 1.333 | | 770 | 57.50 | | 827.50 | 940 |
| 0740 | 4'-8" x 6'-0" | | | 6 | 1.333 | | 870 | 57.50 | | 927.50 | 1,050 |
| 0750 | 6'-0" x 4'-0" | | | 6 | 1.333 | | 790 | 57.50 | | 847.50 | 955 |
| 0800 | 6'-0" x 5'-0" | | | 6 | 1.333 | | 885 | 57.50 | | 942.50 | 1,050 |
| 0900 | 5'-6" x 5'-6" | G | 2 Carp | 15 | 1.067 | | 1,400 | 46 | | 1,446 | 1,600 |
| 2000 | Bay, casement units, 8' x 5', w/screens, dbl. insul. glass | | | 2.50 | 6.400 | Opng. | 1,575 | 276 | | 1,851 | 2,150 |
| 2100 | Low E glass | | | 2.50 | 6.400 | " | 1,650 | 276 | | 1,926 | 2,225 |
| 8100 | Metal clad, deluxe, dbl. insul. glass, 2'-0" x 3'-0" high | G | 1 Carp | 10 | .800 | Ea. | 272 | 34.50 | | 306.50 | 350 |
| 8120 | 2'-0" x 4'-0" high | G | | 9 | .889 | | 305 | 38.50 | | 343.50 | 395 |
| 8140 | 2'-0" x 5'-0" high | G | | 8 | 1 | | 320 | 43 | | 363 | 415 |
| 8160 | 2'-0" x 6'-0" high | G | | 8 | 1 | | 355 | 43 | | 398 | 455 |
| 8190 | For installation, add per leaf | | | | | | | 15% | | | |
| 8200 | For multiple leaf units, deduct for stationary sash | | | | | | | | | | |
| 8220 | 2' high | | | | | Ea. | 22.50 | | | 22.50 | 24.50 |
| 8240 | 4'-6" high | | | | | | 25.50 | | | 25.50 | 28 |
| 8260 | 6' high | | | | | | 33.50 | | | 33.50 | 37 |
| 8300 | Impact windows, minimum, add | | | | | | 60% | | | | |
| 8310 | Impact windows, maximum, add | | | | | | 160% | | | | |

### 08 52 10.50 Double Hung

| | | | Crew | Daily Output | Labor-Hours | Unit | Material | Labor | Equipment | Total | Total Incl O&P |
|---|---|---|---|---|---|---|---|---|---|---|---|
| 0010 | **DOUBLE HUNG**, Including frame, screens and grilles  R085216-10 | | | | | | | | | | |
| 0100 | Avg. quality, bldrs. model, 2'-0" x 3'-0" high, low e insul. glass | G | 1 Carp | 10 | .800 | Ea. | 208 | 34.50 | | 242.50 | 282 |
| 0200 | 3'-0" x 4'-0" high, double insulated glass | G | | 9 | .889 | | 264 | 38.50 | | 302.50 | 350 |
| 0300 | 4'-0" x 4'-6" high, low e insulated glass | G | | 8 | 1 | | 315 | 43 | | 358 | 410 |
| 1000 | Vinyl clad, premium, double insulated glass, 2'-6" x 3'-0" | G | | 10 | .800 | | 300 | 34.50 | | 334.50 | 385 |
| 1005 | 2'-6" x 4'-0" | G | | 10 | .800 | | 355 | 34.50 | | 389.50 | 445 |
| 1100 | 3'-0" x 3'-6" | G | | 10 | .800 | | 330 | 34.50 | | 364.50 | 415 |
| 1200 | 3'-0" x 4'-0" | G | | 9 | .889 | | 375 | 38.50 | | 413.50 | 475 |
| 1300 | 3'-0" x 4'-6" | G | | 9 | .889 | | 395 | 38.50 | | 433.50 | 495 |
| 1400 | 3'-0" x 5'-0" | G | | 8 | 1 | | 435 | 43 | | 478 | 545 |
| 1490 | 3'-4" x 5'-0" | G | | 8 | 1 | | 445 | 43 | | 488 | 555 |
| 1500 | 3'-6" x 6'-0" | G | | 8 | 1 | | 485 | 43 | | 528 | 595 |

# 08 52 Wood Windows

## 08 52 10 – Plain Wood Windows

### 08 52 10.50 Double Hung

| | | | Crew | Daily Output | Labor-Hours | Unit | Material | 2011 Bare Costs Labor | Equipment | Total | Total Incl O&P |
|---|---|---|---|---|---|---|---|---|---|---|---|
| 1520 | 4'-0" x 5'-0" | G | 1 Carp | 7 | 1.143 | Ea. | 550 | 49 | | 599 | 680 |
| 1530 | 4'-0" x 6'-0" | G | | 7 | 1.143 | | 685 | 49 | | 734 | 830 |
| 2000 | Metal clad, deluxe, dbl. insul. glass, 2'-6" x 3'-0" high | G | | 10 | .800 | | 265 | 34.50 | | 299.50 | 345 |
| 2100 | 3'-0" x 3'-6" high | G | | 10 | .800 | | 305 | 34.50 | | 339.50 | 390 |
| 2200 | 3'-0" x 4'-0" high | G | | 9 | .889 | | 320 | 38.50 | | 358.50 | 410 |
| 2300 | 3'-0" x 4'-6" high | G | | 9 | .889 | | 335 | 38.50 | | 373.50 | 430 |
| 2400 | 3'-0" x 5'-0" high | G | | 8 | 1 | | 365 | 43 | | 408 | 465 |
| 2500 | 3'-6" x 6'-0" high | G | | 8 | 1 | | 440 | 43 | | 483 | 545 |
| 8000 | Impact windows, minimum, add | | | | | | 60% | | | | |
| 8010 | Impact windows, maximum, add | | | | | | 160% | | | | |

### 08 52 10.55 Picture Window

| | | | Crew | Daily Output | Labor-Hours | Unit | Material | 2011 Bare Costs Labor | Equipment | Total | Total Incl O&P |
|---|---|---|---|---|---|---|---|---|---|---|---|
| 0010 | **PICTURE WINDOW**, Including frame and grilles | | | | | | | | | | |
| 0100 | Average quality, bldrs. model, 3'-6" x 4'-0" high, dbl. insulated glass | | 2 Carp | 12 | 1.333 | Ea. | 395 | 57.50 | | 452.50 | 525 |
| 0150 | Low E glass | | | 12 | 1.333 | | 410 | 57.50 | | 467.50 | 540 |
| 0200 | 4'-0" x 4'-6" high, double insulated glass | | | 11 | 1.455 | | 490 | 62.50 | | 552.50 | 635 |
| 0250 | Low E glass | | | 11 | 1.455 | | 520 | 62.50 | | 582.50 | 665 |
| 0300 | 5'-0" x 4'-0" high, double insulated glass | | | 11 | 1.455 | | 570 | 62.50 | | 632.50 | 720 |
| 0350 | Low E glass | | | 11 | 1.455 | | 590 | 62.50 | | 652.50 | 745 |
| 0400 | 6'-0" x 4'-6" high, double insulated glass | | | 10 | 1.600 | | 610 | 69 | | 679 | 780 |
| 0450 | Low E glass | | | 10 | 1.600 | | 620 | 69 | | 689 | 790 |

### 08 52 10.65 Wood Sash

| | | | Crew | Daily Output | Labor-Hours | Unit | Material | 2011 Bare Costs Labor | Equipment | Total | Total Incl O&P |
|---|---|---|---|---|---|---|---|---|---|---|---|
| 0010 | **WOOD SASH**, Including glazing but not trim | | | | | | | | | | |
| 0050 | Custom, 5'-0" x 4'-0", 1" dbl. glazed, 3/16" thick lites | | 2 Carp | 3.20 | 5 | Ea. | 225 | 215 | | 440 | 580 |
| 0100 | 1/4" thick lites | | | 5 | 3.200 | | 240 | 138 | | 378 | 475 |
| 0200 | 1" thick, triple glazed | | | 5 | 3.200 | | 405 | 138 | | 543 | 655 |
| 0300 | 7'-0" x 4'-6" high, 1" double glazed, 3/16" thick lites | | | 4.30 | 3.721 | | 415 | 160 | | 575 | 700 |
| 0400 | 1/4" thick lites | | | 4.30 | 3.721 | | 465 | 160 | | 625 | 755 |
| 0500 | 1" thick, triple glazed | | | 4.30 | 3.721 | | 530 | 160 | | 690 | 830 |
| 0600 | 8'-6" x 5'-0" high, 1" double glazed, 3/16" thick lites | | | 3.50 | 4.571 | | 555 | 197 | | 752 | 910 |
| 0700 | 1/4" thick lites | | | 3.50 | 4.571 | | 605 | 197 | | 802 | 965 |
| 0800 | 1" thick, triple glazed | | | 3.50 | 4.571 | | 610 | 197 | | 807 | 975 |
| 0900 | Window frames only, based on perimeter length | | | | | L.F. | 3.94 | | | 3.94 | 4.33 |
| 1200 | Window sill, stock, per lineal foot | | | | | | 8.45 | | | 8.45 | 9.30 |
| 1250 | Casing, stock | | | | | | 3.23 | | | 3.23 | 3.55 |

### 08 52 10.70 Sliding Windows

| | | | Crew | Daily Output | Labor-Hours | Unit | Material | 2011 Bare Costs Labor | Equipment | Total | Total Incl O&P |
|---|---|---|---|---|---|---|---|---|---|---|---|
| 0010 | **SLIDING WINDOWS** | | | | | | | | | | |
| 0100 | Average quality, bldrs. model, 3'-0" x 3'-0" high, double insulated | G | 1 Carp | 10 | .800 | Ea. | 272 | 34.50 | | 306.50 | 350 |
| 0120 | Low E glass | G | | 10 | .800 | | 293 | 34.50 | | 327.50 | 375 |
| 0200 | 4'-0" x 3'-6" high, double insulated | G | | 9 | .889 | | 325 | 38.50 | | 363.50 | 420 |
| 0220 | Low E glass | G | | 9 | .889 | | 350 | 38.50 | | 388.50 | 445 |
| 0300 | 6'-0" x 5'-0" high, double insulated | G | | 8 | 1 | | 460 | 43 | | 503 | 575 |
| 0320 | Low E glass | G | | 8 | 1 | | 500 | 43 | | 543 | 615 |

## 08 52 13 – Metal-Clad Wood Windows

### 08 52 13.10 Awning Windows, Metal-Clad

| | | | Crew | Daily Output | Labor-Hours | Unit | Material | 2011 Bare Costs Labor | Equipment | Total | Total Incl O&P |
|---|---|---|---|---|---|---|---|---|---|---|---|
| 0010 | **AWNING WINDOWS, METAL-CLAD** | | | | | | | | | | |
| 2000 | Metal clad, awning deluxe, double insulated glass, 34" x 22" | | 1 Carp | 10 | .800 | Ea. | 243 | 34.50 | | 277.50 | 320 |
| 2100 | 40" x 22" | | " | 10 | .800 | " | 286 | 34.50 | | 320.50 | 370 |

### 08 52 13.35 Picture and Sliding Windows Metal-Clad

| | | | Crew | Daily Output | Labor-Hours | Unit | Material | 2011 Bare Costs Labor | Equipment | Total | Total Incl O&P |
|---|---|---|---|---|---|---|---|---|---|---|---|
| 0010 | **PICTURE AND SLIDING WINDOWS METAL-CLAD** | | | | | | | | | | |
| 2000 | Metal clad, dlx picture, dbl. insul. glass, 4'-0" x 4'-0" high | | 2 Carp | 12 | 1.333 | Ea. | 365 | 57.50 | | 422.50 | 495 |
| 2100 | 4'-0" x 6'-0" high | | | 11 | 1.455 | | 535 | 62.50 | | 597.50 | 685 |

## 08 52 13 – Metal-Clad Wood Windows

| 08 52 13.35 Picture and Sliding Windows Metal-Clad | | Crew | Daily Output | Labor-Hours | Unit | Material | 2011 Bare Costs Labor | Equipment | Total | Total Incl O&P |
|---|---|---|---|---|---|---|---|---|---|---|
| 2200 | 5'-0" x 6'-0" high | 2 Carp | 10 | 1.600 | Ea. | 595 | 69 | | 664 | 760 |
| 2300 | 6'-0" x 6'-0" high | ↓ | 10 | 1.600 | | 680 | 69 | | 749 | 855 |
| 2400 | Metal clad, dlx sliding, double insulated glass, 3'-0" x 3'-0" high  G | 1 Carp | 10 | .800 | | 330 | 34.50 | | 364.50 | 415 |
| 2420 | 4'-0" x 3'-6" high  G | | 9 | .889 | | 400 | 38.50 | | 438.50 | 500 |
| 2440 | 5'-0" x 4'-0" high  G | | 9 | .889 | | 480 | 38.50 | | 518.50 | 585 |
| 2460 | 6'-0" x 5'-0" high  G | ↓ | 8 | 1 | ↓ | 705 | 43 | | 748 | 840 |

## 08 52 16 – Plastic-Clad Wood Windows

### 08 52 16.10 Bow Window

| | | Crew | Daily Output | Labor-Hours | Unit | Material | Labor | Equipment | Total | Total Incl O&P |
|---|---|---|---|---|---|---|---|---|---|---|
| 0010 | **BOW WINDOW** Including frames, screens, and grilles | | | | | | | | | |
| 0020 | End panels operable | | | | | | | | | |
| 1000 | Bow type, casement, wood, bldrs mdl, 8' x 5' dbl. insltd glass, 4 panel | 2 Carp | 10 | 1.600 | Ea. | 1,500 | 69 | | 1,569 | 1,750 |
| 1050 | Low E glass | | 10 | 1.600 | | 1,325 | 69 | | 1,394 | 1,550 |
| 1100 | 10'-0" x 5'-0", double insulated glass, 6 panels | | 6 | 2.667 | | 1,350 | 115 | | 1,465 | 1,675 |
| 1200 | Low E glass, 6 panels | | 6 | 2.667 | | 1,450 | 115 | | 1,565 | 1,775 |
| 1300 | Vinyl clad, bldrs model, double insulated glass, 6'-0" x 4'-0", 3 panel | | 10 | 1.600 | | 1,025 | 69 | | 1,094 | 1,225 |
| 1340 | 9'-0" x 4'-0", 4 panel | | 8 | 2 | | 1,350 | 86 | | 1,436 | 1,625 |
| 1380 | 10'-0" x 6'-0", 5 panels | | 7 | 2.286 | | 2,250 | 98.50 | | 2,348.50 | 2,625 |
| 1420 | 12'-0" x 6'-0", 6 panels | | 6 | 2.667 | | 2,925 | 115 | | 3,040 | 3,400 |
| 1600 | Metal clad, casement, bldrs mdl, 6'-0" x 4'-0", dbl. insltd gls, 3 panels | | 10 | 1.600 | | 1,200 | 69 | | 1,269 | 1,425 |
| 1640 | 9'-0" x 4'-0", 4 panels | | 8 | 2 | | 1,550 | 86 | | 1,636 | 1,825 |
| 1680 | 10'-0" x 5'-0", 5 panels | | 7 | 2.286 | | 2,100 | 98.50 | | 2,198.50 | 2,475 |
| 1720 | 12'-0" x 6'-0", 6 panels | | 6 | 2.667 | | 2,700 | 115 | | 2,815 | 3,125 |
| 2000 | Bay window, 8' x 5', dbl. insul glass | | 10 | 1.600 | | 1,825 | 69 | | 1,894 | 2,125 |
| 2050 | Low E glass | | 10 | 1.600 | | 2,225 | 69 | | 2,294 | 2,550 |
| 2100 | 12'-0" x 6'-0", double insulated glass, 6 panels | | 6 | 2.667 | | 2,300 | 115 | | 2,415 | 2,700 |
| 2200 | Low E glass | | 6 | 2.667 | | 2,350 | 115 | | 2,465 | 2,775 |
| 2280 | 6'-0" x 4'-0" | | 11 | 1.455 | | 1,225 | 62.50 | | 1,287.50 | 1,450 |
| 2300 | Vinyl clad, premium, double insulated glass, 8'-0" x 5'-0" | | 10 | 1.600 | | 1,750 | 69 | | 1,819 | 2,025 |
| 2340 | 10'-0" x 5'-0" | | 8 | 2 | | 2,275 | 86 | | 2,361 | 2,625 |
| 2380 | 10'-0" x 6'-0" | | 7 | 2.286 | | 2,600 | 98.50 | | 2,698.50 | 3,000 |
| 2420 | 12'-0" x 6'-0" | | 6 | 2.667 | | 3,200 | 115 | | 3,315 | 3,700 |
| 2600 | Metal clad, deluxe, dbl. insul. glass, 8'-0" x 5'-0" high, 4 panels | | 10 | 1.600 | | 1,625 | 69 | | 1,694 | 1,875 |
| 2640 | 10'-0" x 5'-0" high, 5 panels | | 8 | 2 | | 1,750 | 86 | | 1,836 | 2,050 |
| 2680 | 10'-0" x 6'-0" high, 5 panels | | 7 | 2.286 | | 2,050 | 98.50 | | 2,148.50 | 2,425 |
| 2720 | 12'-0" x 6'-0" high, 6 panels | | 6 | 2.667 | | 2,850 | 115 | | 2,965 | 3,325 |
| 3000 | Double hung, bldrs. model, bay, 8' x 4' high, dbl. insulated glass | | 10 | 1.600 | | 1,300 | 69 | | 1,369 | 1,525 |
| 3050 | Low E glass | | 10 | 1.600 | | 1,375 | 69 | | 1,444 | 1,625 |
| 3100 | 9'-0" x 5'-0" high, double insulated glass | | 6 | 2.667 | | 1,400 | 115 | | 1,515 | 1,700 |
| 3200 | Low E glass | | 6 | 2.667 | | 1,475 | 115 | | 1,590 | 1,800 |
| 3300 | Vinyl clad, premium, double insulated glass, 7'-0" x 4'-6" | | 10 | 1.600 | | 1,325 | 69 | | 1,394 | 1,575 |
| 3340 | 8'-0" x 4'-6" | | 8 | 2 | | 1,375 | 86 | | 1,461 | 1,625 |
| 3380 | 8'-0" x 5'-0" | | 7 | 2.286 | | 1,425 | 98.50 | | 1,523.50 | 1,725 |
| 3420 | 9'-0" x 5'-0" | | 6 | 2.667 | | 1,475 | 115 | | 1,590 | 1,800 |
| 3600 | Metal clad, deluxe, dbl. insul. glass, 7'-0" x 4'-0" high | | 10 | 1.600 | | 1,225 | 69 | | 1,294 | 1,450 |
| 3640 | 8'-0" x 4'-0" high | | 8 | 2 | | 1,275 | 86 | | 1,361 | 1,525 |
| 3680 | 8'-0" x 5'-0" high | | 7 | 2.286 | | 1,325 | 98.50 | | 1,423.50 | 1,600 |
| 3720 | 9'-0" x 5'-0" high | ↓ | 6 | 2.667 | ↓ | 1,400 | 115 | | 1,515 | 1,725 |

### 08 52 16.30 Palladian Windows

| | | Crew | Daily Output | Labor-Hours | Unit | Material | Labor | Equipment | Total | Total Incl O&P |
|---|---|---|---|---|---|---|---|---|---|---|
| 0010 | **PALLADIAN WINDOWS** | | | | | | | | | |
| 0020 | Vinyl clad, double insulated glass, including frame and grilles | | | | | | | | | |
| 0040 | 3'-2" x 2'-6" high | 2 Carp | 11 | 1.455 | Ea. | 1,250 | 62.50 | | 1,312.50 | 1,475 |
| 0060 | 3'-2" x 4'-10" | ↓ | 11 | 1.455 | ↓ | 1,750 | 62.50 | | 1,812.50 | 2,000 |

# 08 52 Wood Windows

## 08 52 16 – Plastic-Clad Wood Windows

### 08 52 16.30 Palladian Windows

| | | Crew | Daily Output | Labor-Hours | Unit | Material | 2011 Bare Costs Labor | Equipment | Total | Total Incl O&P |
|---|---|---|---|---|---|---|---|---|---|---|
| 0080 | 3'-2" x 6'-4" | 2 Carp | 10 | 1.600 | Ea. | 1,700 | 69 | | 1,769 | 1,975 |
| 0100 | 4'-0" x 4'-0" | ↓ | 10 | 1.600 | | 1,525 | 69 | | 1,594 | 1,775 |
| 0120 | 4'-0" x 5'-4" | 3 Carp | 10 | 2.400 | | 1,875 | 103 | | 1,978 | 2,200 |
| 0140 | 4'-0" x 6'-0" | | 9 | 2.667 | | 1,950 | 115 | | 2,065 | 2,325 |
| 0160 | 4'-0" x 7'-4" | | 9 | 2.667 | | 2,125 | 115 | | 2,240 | 2,500 |
| 0180 | 5'-5" x 4'-10" | | 9 | 2.667 | | 2,275 | 115 | | 2,390 | 2,675 |
| 0200 | 5'-5" x 6'-10" | | 9 | 2.667 | | 2,575 | 115 | | 2,690 | 3,025 |
| 0220 | 5'-5" x 7'-9" | | 9 | 2.667 | | 2,800 | 115 | | 2,915 | 3,250 |
| 0240 | 6'-0" x 7'-11" | | 8 | 3 | | 3,475 | 129 | | 3,604 | 4,025 |
| 0260 | 8'-0" x 6'-0" | ↓ | 8 | 3 | ↓ | 3,075 | 129 | | 3,204 | 3,575 |

### 08 52 16.40 Transom Windows

| | | Crew | Daily Output | Labor-Hours | Unit | Material | Labor | Equipment | Total | Total Incl O&P |
|---|---|---|---|---|---|---|---|---|---|---|
| 0010 | **TRANSOM WINDOWS** | | | | | | | | | |
| 1000 | Vinyl clad, premium, dbl. insul. glass, 4'-0" x 4'-0" | 2 Carp | 12 | 1.333 | Ea. | 480 | 57.50 | | 537.50 | 620 |
| 1100 | 4'-0" x 6'-0" | | 11 | 1.455 | | 910 | 62.50 | | 972.50 | 1,100 |
| 1200 | 5'-0" x 6'-0" | | 10 | 1.600 | | 1,000 | 69 | | 1,069 | 1,225 |
| 1300 | 6'-0" x 6'-0" | ↓ | 10 | 1.600 | ↓ | 1,025 | 69 | | 1,094 | 1,225 |

### 08 52 16.70 Vinyl Clad, Premium, Dbl. Insulated Glass

| | | | Crew | Daily Output | Labor-Hours | Unit | Material | Labor | Equipment | Total | Total Incl O&P |
|---|---|---|---|---|---|---|---|---|---|---|---|
| 0010 | **VINYL CLAD, PREMIUM, DBL. INSULATED GLASS** | | | | | | | | | | |
| 1000 | Sliding, 3'-0" x 3'-0" | G | 1 Carp | 10 | .800 | Ea. | 575 | 34.50 | | 609.50 | 685 |
| 1050 | 4'-0" x 3'-6" | G | | 9 | .889 | | 670 | 38.50 | | 708.50 | 795 |
| 1100 | 5'-0" x 4'-0" | G | | 9 | .889 | | 860 | 38.50 | | 898.50 | 1,000 |
| 1150 | 6'-0" x 5'-0" | G | ↓ | 8 | 1 | ↓ | 1,075 | 43 | | 1,118 | 1,275 |

## 08 52 50 – Window Accessories

### 08 52 50.10 Window Grille or Muntin

| | | Crew | Daily Output | Labor-Hours | Unit | Material | Labor | Equipment | Total | Total Incl O&P |
|---|---|---|---|---|---|---|---|---|---|---|
| 0010 | **WINDOW GRILLE OR MUNTIN**, snap in type | | | | | | | | | |
| 0020 | Standard pattern interior grilles | | | | | | | | | |
| 2000 | Wood, awning window, glass size 28" x 16" high | 1 Carp | 30 | .267 | Ea. | 25.50 | 11.50 | | 37 | 45.50 |
| 2060 | 44" x 24" high | | 32 | .250 | | 37 | 10.75 | | 47.75 | 57 |
| 2100 | Casement, glass size, 20" x 36" high | | 30 | .267 | | 29.50 | 11.50 | | 41 | 50 |
| 2180 | 20" x 56" high | | 32 | .250 | ↓ | 41 | 10.75 | | 51.75 | 61.50 |
| 2200 | Double hung, glass size, 16" x 24" high | | 24 | .333 | Set | 49 | 14.35 | | 63.35 | 75.50 |
| 2280 | 32" x 32" high | | 34 | .235 | " | 128 | 10.15 | | 138.15 | 157 |
| 2500 | Picture, glass size, 48" x 48" high | | 30 | .267 | Ea. | 114 | 11.50 | | 125.50 | 143 |
| 2580 | 60" x 68" high | | 28 | .286 | " | 178 | 12.30 | | 190.30 | 214 |
| 2600 | Sliding, glass size, 14" x 36" high | | 24 | .333 | Set | 34 | 14.35 | | 48.35 | 59 |
| 2680 | 36" x 36" high | ↓ | 22 | .364 | " | 41.50 | 15.65 | | 57.15 | 69.50 |

## 08 52 66 – Wood Window Screens

### 08 52 66.10 Wood Screens

| | | Crew | Daily Output | Labor-Hours | Unit | Material | Labor | Equipment | Total | Total Incl O&P |
|---|---|---|---|---|---|---|---|---|---|---|
| 0010 | **WOOD SCREENS** | | | | | | | | | |
| 0020 | Over 3 S.F., 3/4" frames | 2 Carp | 375 | .043 | S.F. | 4.66 | 1.84 | | 6.50 | 7.95 |
| 0100 | 1-1/8" frames | " | 375 | .043 | " | 7.85 | 1.84 | | 9.69 | 11.40 |

## 08 52 69 – Wood Storm Windows

### 08 52 69.10 Storm Windows

| | | | Crew | Daily Output | Labor-Hours | Unit | Material | Labor | Equipment | Total | Total Incl O&P |
|---|---|---|---|---|---|---|---|---|---|---|---|
| 0010 | **STORM WINDOWS**, aluminum residential | | | | | | | | | | |
| 0300 | Basement, mill finish, incl. fiberglass screen | | | | | | | | | | |
| 0320 | 1'-10" x 1'-0" high | G | 2 Carp | 30 | .533 | Ea. | 35 | 23 | | 58 | 73.50 |
| 0340 | 2'-9" x 1'-6" high | G | " | 30 | .533 | " | 38 | 23 | | 61 | 77 |
| 1600 | Double-hung, combination, storm & screen | | | | | | | | | | |
| 2000 | Average quality, clear anodic coating, 2'-0" x 3'-5" high | G | 2 Carp | 30 | .533 | Ea. | 85 | 23 | | 108 | 129 |
| 2020 | 2'-6" x 5'-0" high | G | ↓ | 28 | .571 | ↓ | 107 | 24.50 | | 131.50 | 155 |

# 08 52 Wood Windows

## 08 52 69 – Wood Storm Windows

### 08 52 69.10 Storm Windows

| | | Crew | Daily Output | Labor-Hours | Unit | Material | 2011 Bare Costs Labor | Equipment | Total | Total Incl O&P |
|---|---|---|---|---|---|---|---|---|---|---|
| 2040 | 4'-0" x 6'-0" high | G | 25 | .640 | Ea. | 126 | 27.50 | | 153.50 | 182 |
| 2400 | White painted, 2'-0" x 3'-5" high | G | 30 | .533 | | 85 | 23 | | 108 | 129 |
| 2420 | 2'-6" x 5'-0" high | G | 28 | .571 | | 90 | 24.50 | | 114.50 | 137 |
| 2440 | 4'-0" x 6'-0" high | G | 25 | .640 | | 98 | 27.50 | | 125.50 | 151 |
| 2600 | Mill finish, 2'-0" x 3'-5" high | G | 30 | .533 | | 75 | 23 | | 98 | 118 |
| 2620 | 2'-6" x 5'-0" high | G | 28 | .571 | | 85 | 24.50 | | 109.50 | 131 |
| 2640 | 4'-0" x 6'-8" high | G | 25 | .640 | | 99 | 27.50 | | 126.50 | 152 |

# 08 53 Plastic Windows

## 08 53 13 – Vinyl Windows

### 08 53 13.20 Vinyl Single Hung Windows

| | | Crew | Daily Output | Labor-Hours | Unit | Material | 2011 Bare Costs Labor | Equipment | Total | Total Incl O&P |
|---|---|---|---|---|---|---|---|---|---|---|
| 0010 | **VINYL SINGLE HUNG WINDOWS**, insulated glass | | | | | | | | | |
| 0100 | Grids, low E, J fin, ext. jambs, 21" x 53" | G | 2 Carp | 18 | .889 | Ea. | 198 | 38.50 | 236.50 | 277 |
| 0110 | 21" x 57" | G | 17 | .941 | | 195 | 40.50 | | 235.50 | 277 |
| 0120 | 21" x 65" | G | 16 | 1 | | 200 | 43 | | 243 | 286 |
| 0130 | 25" x 41" | G | 20 | .800 | | 188 | 34.50 | | 222.50 | 260 |
| 0140 | 25" x 49" | G | 18 | .889 | | 198 | 38.50 | | 236.50 | 277 |
| 0150 | 25" x 57" | G | 17 | .941 | | 201 | 40.50 | | 241.50 | 283 |
| 0160 | 25" x 65" | G | 16 | 1 | | 207 | 43 | | 250 | 294 |
| 0170 | 29" x 41" | G | 18 | .889 | | 192 | 38.50 | | 230.50 | 270 |
| 0180 | 29" x 53" | G | 18 | .889 | | 202 | 38.50 | | 240.50 | 281 |
| 0190 | 29" x 57" | G | 17 | .941 | | 206 | 40.50 | | 246.50 | 289 |
| 0200 | 29" x 65" | G | 16 | 1 | | 212 | 43 | | 255 | 299 |
| 0210 | 33" x 41" | G | 20 | .800 | | 198 | 34.50 | | 232.50 | 271 |
| 0220 | 33" x 53" | G | 18 | .889 | | 209 | 38.50 | | 247.50 | 289 |
| 0230 | 33" x 57" | G | 17 | .941 | | 212 | 40.50 | | 252.50 | 295 |
| 0240 | 33" x 65" | G | 16 | 1 | | 218 | 43 | | 261 | 305 |
| 0250 | 37" x 41" | G | 20 | .800 | | 206 | 34.50 | | 240.50 | 280 |
| 0260 | 37" x 53" | G | 18 | .889 | | 218 | 38.50 | | 256.50 | 299 |
| 0270 | 37" x 57" | G | 17 | .941 | | 221 | 40.50 | | 261.50 | 305 |
| 0280 | 37" x 65" | G | 16 | 1 | | 232 | 43 | | 275 | 320 |

### 08 53 13.30 Vinyl Double Hung Windows

| | | Crew | Daily Output | Labor-Hours | Unit | Material | 2011 Bare Costs Labor | Equipment | Total | Total Incl O&P |
|---|---|---|---|---|---|---|---|---|---|---|
| 0010 | **VINYL DOUBLE HUNG WINDOWS**, insulated glass | | | | | | | | | |
| 0100 | Grids, low E, J fin, ext. jambs, 21" x 53" | G | 2 Carp | 18 | .889 | Ea. | 215 | 38.50 | 253.50 | 296 |
| 0102 | 21" x 37" | G | 18 | .889 | | 197 | 38.50 | | 235.50 | 276 |
| 0104 | 21" x 41" | G | 18 | .889 | | 201 | 38.50 | | 239.50 | 280 |
| 0106 | 21" x 49" | G | 18 | .889 | | 221 | 38.50 | | 259.50 | 300 |
| 0110 | 21" x 57" | G | 17 | .941 | | 218 | 40.50 | | 258.50 | 300 |
| 0120 | 21" x 65" | G | 16 | 1 | | 228 | 43 | | 271 | 315 |
| 0128 | 25" x 37" | G | 20 | .800 | | 206 | 34.50 | | 240.50 | 280 |
| 0130 | 25" x 41" | G | 20 | .800 | | 209 | 34.50 | | 243.50 | 283 |
| 0140 | 25" x 49" | G | 18 | .889 | | 213 | 38.50 | | 251.50 | 293 |
| 0145 | 25" x 53" | G | 18 | .889 | | 221 | 38.50 | | 259.50 | 300 |
| 0150 | 25" x 57" | G | 17 | .941 | | 221 | 40.50 | | 261.50 | 305 |
| 0160 | 25" x 65" | G | 16 | 1 | | 231 | 43 | | 274 | 320 |
| 0162 | 25" x 69" | G | 16 | 1 | | 238 | 43 | | 281 | 330 |
| 0164 | 25" x 77" | G | 16 | 1 | | 248 | 43 | | 291 | 340 |
| 0168 | 29" x 37" | G | 18 | .889 | | 211 | 38.50 | | 249.50 | 291 |
| 0170 | 29" x 41" | G | 18 | .889 | | 214 | 38.50 | | 252.50 | 294 |
| 0172 | 29" x 49" | G | 18 | .889 | | 221 | 38.50 | | 259.50 | 300 |
| 0180 | 29" x 53" | G | 18 | .889 | | 225 | 38.50 | | 263.50 | 305 |

# 08 53 Plastic Windows

## 08 53 13 – Vinyl Windows

### 08 53 13.30 Vinyl Double Hung Windows

| | | Crew | Daily Output | Labor-Hours | Unit | Material | 2011 Bare Costs Labor | Equipment | Total | Total Incl O&P |
|---|---|---|---|---|---|---|---|---|---|---|
| 0190 | 29" x 57" G | 2 Carp | 17 | .941 | Ea. | 229 | 40.50 | | 269.50 | 315 |
| 0200 | 29" x 65" G | | 16 | 1 | | 236 | 43 | | 279 | 325 |
| 0202 | 29" x 69" G | | 16 | 1 | | 243 | 43 | | 286 | 335 |
| 0205 | 29" x 77" G | | 16 | 1 | | 254 | 43 | | 297 | 345 |
| 0208 | 33" x 37" G | | 20 | .800 | | 215 | 34.50 | | 249.50 | 290 |
| 0210 | 33" x 41" G | | 20 | .800 | | 219 | 34.50 | | 253.50 | 294 |
| 0215 | 33" x 49" G | | 20 | .800 | | 227 | 34.50 | | 261.50 | 305 |
| 0220 | 33" x 53" G | | 18 | .889 | | 231 | 38.50 | | 269.50 | 315 |
| 0230 | 33" x 57" G | | 17 | .941 | | 235 | 40.50 | | 275.50 | 320 |
| 0240 | 33" x 65" G | | 16 | 1 | | 239 | 43 | | 282 | 330 |
| 0242 | 33" x 69" G | | 16 | 1 | | 245 | 43 | | 288 | 335 |
| 0246 | 33" x 77" G | | 16 | 1 | | 251 | 43 | | 294 | 340 |
| 0250 | 37" x 41" G | | 20 | .800 | | 223 | 34.50 | | 257.50 | 298 |
| 0255 | 37" x 49" G | | 20 | .800 | | 231 | 34.50 | | 265.50 | 305 |
| 0260 | 37" x 53" G | | 18 | .889 | | 239 | 38.50 | | 277.50 | 320 |
| 0270 | 37" x 57" G | | 17 | .941 | | 243 | 40.50 | | 283.50 | 330 |
| 0280 | 37" x 65" G | | 16 | 1 | | 255 | 43 | | 298 | 345 |
| 0282 | 37" x 69" G | | 16 | 1 | | 262 | 43 | | 305 | 355 |
| 0286 | 37" x 77" G | | 16 | 1 | | 270 | 43 | | 313 | 365 |
| 0300 | Solid vinyl, average quality, double insulated glass, 2'-0" x 3'-0" G | 1 Carp | 10 | .800 | | 287 | 34.50 | | 321.50 | 370 |
| 0310 | 3'-0" x 4'-0" G | | 9 | .889 | | 198 | 38.50 | | 236.50 | 276 |
| 0320 | 4'-0" x 4'-6" G | | 8 | 1 | | 298 | 43 | | 341 | 395 |
| 0330 | Premium, double insulated glass, 2'-6" x 3'-0" G | | 10 | .800 | | 230 | 34.50 | | 264.50 | 305 |
| 0340 | 3'-0" x 3'-6" G | | 9 | .889 | | 254 | 38.50 | | 292.50 | 340 |
| 0350 | 3'-0" x 4"-0" G | | 9 | .889 | | 274 | 38.50 | | 312.50 | 360 |
| 0360 | 3'-0" x 4'-6" G | | 9 | .889 | | 266 | 38.50 | | 304.50 | 350 |
| 0370 | 3'-0" x 5'-0" G | | 8 | 1 | | 286 | 43 | | 329 | 380 |
| 0380 | 3'-6" x 6"-0" G | | 8 | 1 | | 320 | 43 | | 363 | 415 |

### 08 53 13.40 Vinyl Casement Windows

| | | Crew | Daily Output | Labor-Hours | Unit | Material | 2011 Bare Costs Labor | Equipment | Total | Total Incl O&P |
|---|---|---|---|---|---|---|---|---|---|---|
| 0010 | **VINYL CASEMENT WINDOWS**, insulated glass | | | | | | | | | |
| 0015 | Grids, low E, J fin, extension jambs, screens | | | | | | | | | |
| 0100 | One lite, 21" x 41" G | 2 Carp | 20 | .800 | Ea. | 271 | 34.50 | | 305.50 | 350 |
| 0110 | 21" x 47" G | | 20 | .800 | | 294 | 34.50 | | 328.50 | 380 |
| 0120 | 21" x 53" G | | 20 | .800 | | 315 | 34.50 | | 349.50 | 405 |
| 0128 | 24" x 35" G | | 19 | .842 | | 263 | 36.50 | | 299.50 | 345 |
| 0130 | 24" x 41" G | | 19 | .842 | | 283 | 36.50 | | 319.50 | 365 |
| 0140 | 24" x 47" G | | 19 | .842 | | 305 | 36.50 | | 341.50 | 390 |
| 0150 | 24" x 53" G | | 19 | .842 | | 325 | 36.50 | | 361.50 | 415 |
| 0158 | 28" x 35" G | | 19 | .842 | | 278 | 36.50 | | 314.50 | 360 |
| 0160 | 28" x 41" G | | 19 | .842 | | 298 | 36.50 | | 334.50 | 385 |
| 0170 | 28" x 47" G | | 19 | .842 | | 320 | 36.50 | – | 356.50 | 405 |
| 0180 | 28" x 53" G | | 19 | .842 | | 350 | 36.50 | | 386.50 | 440 |
| 0184 | 28" x 59" G | | 19 | .842 | | 350 | 36.50 | | 386.50 | 440 |
| 0188 | Two lites, 33" x 35" G | | 18 | .889 | | 430 | 38.50 | | 468.50 | 530 |
| 0190 | 33" x 41" G | | 18 | .889 | | 455 | 38.50 | | 493.50 | 565 |
| 0200 | 33" x 47" G | | 18 | .889 | | 490 | 38.50 | | 528.50 | 595 |
| 0210 | 33" x 53" G | | 18 | .889 | | 520 | 38.50 | | 558.50 | 630 |
| 0212 | 33" x 59" G | | 18 | .889 | | 550 | 38.50 | | 588.50 | 665 |
| 0215 | 33" x 72" G | | 18 | .889 | | 570 | 38.50 | | 608.50 | 685 |
| 0220 | 41" x 41" G | | 18 | .889 | | 495 | 38.50 | | 533.50 | 605 |
| 0230 | 41" x 47" G | | 18 | .889 | | 530 | 38.50 | | 568.50 | 640 |
| 0240 | 41" x 53" G | | 17 | .941 | | 560 | 40.50 | | 600.50 | 675 |

# 08 53 Plastic Windows

## 08 53 13 – Vinyl Windows

### 08 53 13.40 Vinyl Casement Windows

| | | Crew | Daily Output | Labor-Hours | Unit | Material | 2011 Bare Costs Labor | Equipment | Total | Total Incl O&P |
|---|---|---|---|---|---|---|---|---|---|---|
| 0242 | 41" x 59" | **G** 2 Carp | 17 | .941 | Ea. | 590 | 40.50 | | 630.50 | 705 |
| 0246 | 41" x 72" | **G** | 17 | .941 | | 615 | 40.50 | | 655.50 | 735 |
| 0250 | 47" x 41" | **G** | 17 | .941 | | 500 | 40.50 | | 540.50 | 610 |
| 0260 | 47" x 47" | **G** | 17 | .941 | | 530 | 40.50 | | 570.50 | 645 |
| 0270 | 47" x 53" | **G** | 17 | .941 | | 560 | 40.50 | | 600.50 | 680 |
| 0272 | 47" x 59" | **G** | 17 | .941 | | 615 | 40.50 | | 655.50 | 735 |
| 0280 | 56" x 41" | **G** | 15 | 1.067 | | 535 | 46 | | 581 | 660 |
| 0290 | 56" x 47" | **G** | 15 | 1.067 | | 560 | 46 | | 606 | 690 |
| 0300 | 56" x 53" | **G** | 15 | 1.067 | | 610 | 46 | | 656 | 740 |
| 0302 | 56" x 59" | **G** | 15 | 1.067 | | 635 | 46 | | 681 | 770 |
| 0310 | 56" x 72" | **G** | 15 | 1.067 | | 690 | 46 | | 736 | 830 |
| 0340 | Solid vinyl, premium, double insulated glass, 2'-0" x 3'-0" high | **G** 1 Carp | 10 | .800 | | 259 | 34.50 | | 293.50 | 340 |
| 0360 | 2'-0" x 4'-0" high | **G** | 9 | .889 | | 289 | 38.50 | | 327.50 | 380 |
| 0380 | 2'-0" x 5'-0" high | **G** | 8 | 1 | | 295 | 43 | | 338 | 390 |

### 08 53 13.50 Vinyl Picture Windows

| | | Crew | Daily Output | Labor-Hours | Unit | Material | 2011 Bare Costs Labor | Equipment | Total | Total Incl O&P |
|---|---|---|---|---|---|---|---|---|---|---|
| 0010 | **VINYL PICTURE WINDOWS**, insulated glass | | | | | | | | | |
| 0100 | Grids, low E, J fin, ext. jambs, 33" x 47" | 2 Carp | 12 | 1.333 | Ea. | 238 | 57.50 | | 295.50 | 350 |
| 0110 | 35" x 71" | | 12 | 1.333 | | 310 | 57.50 | | 367.50 | 430 |
| 0120 | 41" x 47" | | 12 | 1.333 | | 252 | 57.50 | | 309.50 | 365 |
| 0130 | 41" x 71" | | 12 | 1.333 | | 320 | 57.50 | | 377.50 | 445 |
| 0140 | 47" x 47" | | 12 | 1.333 | | 277 | 57.50 | | 334.50 | 395 |
| 0150 | 47" x 71" | | 11 | 1.455 | | 350 | 62.50 | | 412.50 | 480 |
| 0160 | 53" x 47" | | 11 | 1.455 | | 300 | 62.50 | | 362.50 | 425 |
| 0170 | 53" x 71" | | 11 | 1.455 | | 315 | 62.50 | | 377.50 | 440 |
| 0180 | 59" x 47" | | 11 | 1.455 | | 340 | 62.50 | | 402.50 | 470 |
| 0190 | 59" x 71" | | 11 | 1.455 | | 365 | 62.50 | | 427.50 | 495 |
| 0200 | 71" x 47" | | 10 | 1.600 | | 375 | 69 | | 444 | 520 |
| 0210 | 71" x 71" | | 10 | 1.600 | | 395 | 69 | | 464 | 540 |

# 08 54 Composite Windows

## 08 54 13 – Fiberglass Windows

### 08 54 13.10 Fiberglass Single Hung Windows

| | | Crew | Daily Output | Labor-Hours | Unit | Material | 2011 Bare Costs Labor | Equipment | Total | Total Incl O&P |
|---|---|---|---|---|---|---|---|---|---|---|
| 0010 | **FIBERGLASS SINGLE HUNG WINDOWS** | | | | | | | | | |
| 0100 | Grids, low E, 18" x 24" | **G** 2 Carp | 18 | .889 | Ea. | 335 | 38.50 | | 373.50 | 430 |
| 0110 | 18" x 40" | **G** | 17 | .941 | | 300 | 40.50 | | 340.50 | 390 |
| 0130 | 24" x 40" | **G** | 20 | .800 | | 330 | 34.50 | | 364.50 | 420 |
| 0230 | 36" x 36" | **G** | 17 | .941 | | 370 | 40.50 | | 410.50 | 465 |
| 0250 | 36" x 48" | **G** | 20 | .800 | | 405 | 34.50 | | 439.50 | 500 |
| 0260 | 36" x 60" | **G** | 18 | .889 | | 445 | 38.50 | | 483.50 | 550 |
| 0280 | 36" x 72" | **G** | 16 | 1 | | 470 | 43 | | 513 | 580 |
| 0290 | 48" x 40" | **G** | 16 | 1 | | 470 | 43 | | 513 | 580 |

# 08 56 Special Function Windows

## 08 56 63 – Detention Windows

### 08 56 63.13 Visitor Cubicle Windows

| | | Crew | Daily Output | Labor-Hours | Unit | Material | 2011 Bare Costs Labor | 2011 Bare Costs Equipment | Total | Total Incl O&P |
|---|---|---|---|---|---|---|---|---|---|---|
| 0010 | **VISITOR CUBICLE WINDOWS** | | | | | | | | | |
| 4000 | Visitor cubicle, vision panel, no intercom | E-4 | 2 | 16 | Ea. | 2,925 | 785 | 54.50 | 3,764.50 | 4,650 |

# 08 62 Unit Skylights

## 08 62 13 – Domed Unit Skylights

### 08 62 13.20 Skylights

| | | | Crew | Daily Output | Labor-Hours | Unit | Material | Labor | Equipment | Total | Total Incl O&P |
|---|---|---|---|---|---|---|---|---|---|---|---|
| 0010 | **SKYLIGHTS**, Plastic domes, flush or curb mounted | | | | | | | | | | |
| 0300 | Nominal size under 10 S.F., double | G | G-3 | 130 | .246 | S.F. | 29 | 10.60 | | 39.60 | 48 |
| 0400 | Single | | | 160 | .200 | | 25.50 | 8.60 | | 34.10 | 41 |
| 0600 | 10 S.F. to 20 S.F., double | G | | 315 | .102 | | 25 | 4.37 | | 29.37 | 34 |
| 0700 | Single | | | 395 | .081 | | 24.50 | 3.48 | | 27.98 | 32.50 |
| 0900 | 20 S.F. to 30 S.F., double | G | | 395 | .081 | | 23 | 3.48 | | 26.48 | 31 |
| 1000 | Single | | | 465 | .069 | | 20.50 | 2.96 | | 23.46 | 27 |
| 1200 | 30 S.F. to 65 S.F., double | G | | 465 | .069 | | 23.50 | 2.96 | | 26.46 | 30 |
| 1300 | Single | | | 610 | .052 | | 17.40 | 2.26 | | 19.66 | 22.50 |
| 1500 | For insulated 4" curbs, double, add | | | | | | 27% | | | | |
| 1600 | Single, add | | | | | | 30% | | | | |
| 1800 | For integral insulated 9" curbs, double, add | | | | | | 30% | | | | |
| 1900 | Single, add | | | | | | 40% | | | | |
| 2120 | Ventilating insulated plexiglass dome with | | | | | | | | | | |
| 2130 | curb mounting, 36" x 36" | G | G-3 | 12 | 2.667 | Ea. | 460 | 115 | | 575 | 680 |
| 2150 | 52" x 52" | G | | 12 | 2.667 | | 640 | 115 | | 755 | 880 |
| 2160 | 28" x 52" | G | | 10 | 3.200 | | 470 | 138 | | 608 | 725 |
| 2170 | 36" x 52" | G | | 10 | 3.200 | | 520 | 138 | | 658 | 780 |
| 2180 | For electric opening system, add | G | | | | | 300 | | | 300 | 330 |
| 2300 | Insulated safety glass with aluminum frame | G | G-3 | 160 | .200 | S.F. | 90 | 8.60 | | 98.60 | 112 |
| 2400 | Sandwich panels, fiberglass, for walls, 1-9/16" thick, to 250 SF | G | | 200 | .160 | | 16.20 | 6.90 | | 23.10 | 28.50 |
| 2700 | As above, but for roofs, 2-3/4" thick, to 250 SF | G | | 295 | .108 | | 23.50 | 4.66 | | 28.16 | 32.50 |
| 2800 | 250 SF and up | G | | 330 | .097 | | 19 | 4.17 | | 23.17 | 27.50 |

# 08 63 Metal-Framed Skylights

## 08 63 23 – Ridge Metal-Framed Skylights

### 08 63 23.10 Prefabricated

| | | | Crew | Daily Output | Labor-Hours | Unit | Material | Labor | Equipment | Total | Total Incl O&P |
|---|---|---|---|---|---|---|---|---|---|---|---|
| 0010 | **PREFABRICATED** glass block with metal frame | | | | | | | | | | |
| 0020 | Minimum | G | G-3 | 265 | .121 | S.F. | 56 | 5.20 | | 61.20 | 69.50 |
| 0100 | Maximum | G | " | 160 | .200 | " | 108 | 8.60 | | 116.60 | 132 |

# 08 71 Door Hardware

## 08 71 13 – Automatic Door Operators

### 08 71 13.10 Automatic Openers Commercial

| | | Crew | Daily Output | Labor-Hours | Unit | Material | Labor | Equipment | Total | Total Incl O&P |
|---|---|---|---|---|---|---|---|---|---|---|
| 0010 | **AUTOMATIC OPENERS COMMERCIAL** | | | | | | | | | |
| 0020 | Pneumatic, incl opener, motion sens, control box, tubing, compressor | | | | | | | | | |
| 0050 | For single swing door, per opening | 2 Skwk | .80 | 20 | Ea. | 4,300 | 890 | | 5,190 | 6,075 |
| 0100 | Pair, per opening | | .50 | 32 | Opng. | 7,000 | 1,425 | | 8,425 | 9,875 |
| 1000 | For single sliding door, per opening | | .60 | 26.667 | | 4,700 | 1,175 | | 5,875 | 7,000 |
| 1300 | Bi-parting pair | | .50 | 32 | | 7,100 | 1,425 | | 8,525 | 9,975 |
| 1420 | Electronic door opener incl motion sens, 12 V control box, motor | | | | | | | | | |

# 08 71 Door Hardware

## 08 71 13 – Automatic Door Operators

| | | Crew | Daily Output | Labor-Hours | Unit | Material | 2011 Bare Costs Labor | 2011 Bare Costs Equipment | Total | Total Incl O&P |
|---|---|---|---|---|---|---|---|---|---|---|
| **08 71 13.10** | **Automatic Openers Commercial** | | | | | | | | | |
| 1450 | For single swing door, per opening | 2 Skwk | .80 | 20 | Opng. | 3,500 | 890 | | 4,390 | 5,200 |
| 1500 | Pair, per opening | | .50 | 32 | | 5,700 | 1,425 | | 7,125 | 8,450 |
| 1600 | For single sliding door, per opening | | .60 | 26.667 | | 3,800 | 1,175 | | 4,975 | 6,000 |
| 1700 | Bi-parting pair | | .50 | 32 | | 5,800 | 1,425 | | 7,225 | 8,550 |
| 1750 | Handicap actuator buttons, 2, including 12 V DC wiring, add | 1 Carp | 1.50 | 5.333 | Pr. | 430 | 230 | | 660 | 825 |
| **08 71 13.20** | **Automatic Openers Industrial** | | | | | | | | | |
| 0010 | **AUTOMATIC OPENERS INDUSTRIAL** | | | | | | | | | |
| 0015 | Sliding doors up to 6' wide | 2 Skwk | .60 | 26.667 | Opng. | 5,500 | 1,175 | | 6,675 | 7,875 |
| 0200 | To 12' wide | " | .40 | 40 | " | 6,600 | 1,775 | | 8,375 | 9,975 |
| 0400 | Over 12' wide, add per L.F. of excess | | | | L.F. | 750 | | | 750 | 825 |
| 1000 | Swing doors, to 5' wide | 2 Skwk | .80 | 20 | Ea. | 3,200 | 890 | | 4,090 | 4,875 |
| 1860 | Add for controls, wall pushbutton, 3 button | | 4 | 4 | | 215 | 178 | | 393 | 510 |
| 1870 | Ceiling pull cord | | 4.30 | 3.721 | | 175 | 165 | | 340 | 445 |

## 08 71 20 – Hardware

| | | Crew | Daily Output | Labor-Hours | Unit | Material | 2011 Bare Costs Labor | 2011 Bare Costs Equipment | Total | Total Incl O&P |
|---|---|---|---|---|---|---|---|---|---|---|
| **08 71 20.10** | **Bolts, Flush** | | | | | | | | | |
| 0010 | **BOLTS, FLUSH** | | | | | | | | | |
| 0020 | Standard, concealed | 1 Carp | 7 | 1.143 | Ea. | 23 | 49 | | 72 | 101 |
| 0800 | Automatic fire exit | " | 5 | 1.600 | | 299 | 69 | | 368 | 435 |
| 1600 | For electric release, add | 1 Elec | 3 | 2.667 | | 125 | 134 | | 259 | 335 |
| 3000 | Barrel, brass, 2" long | 1 Carp | 40 | .200 | | 5.20 | 8.60 | | 13.80 | 18.90 |
| 3020 | 4" long | | 40 | .200 | | 5.30 | 8.60 | | 13.90 | 19.05 |
| 3060 | 6" long | | 40 | .200 | | 10.40 | 8.60 | | 19 | 24.50 |
| **08 71 20.15** | **Hardware** | | | | | | | | | |
| 0009 | **HARDWARE** | | | | | | | | | |
| 0010 | Average percentage for hardware, total job cost | | | | | | | | | |
| 0025 | Minimum | | | | Job | | | | | 1% |
| 0050 | Maximum | | | | | | | | | 4% |
| 0500 | Total hardware for building, average distribution | | | | | 85% | 15% | | | |
| 1000 | Door hardware, apartment, interior | 1 Carp | 4 | 2 | Door | 510 | 86 | | 596 | 690 |
| 1300 | Average, door hardware, motel/hotel interior, with access card | | 4 | 2 | | 625 | 86 | | 711 | 820 |
| 1500 | Hospital bedroom, minimum | | 4 | 2 | | 595 | 86 | | 681 | 785 |
| 2000 | Maximum | | 3 | 2.667 | | 705 | 115 | | 820 | 950 |
| 2100 | Pocket door | | 6 | 1.333 | Ea. | 134 | 57.50 | | 191.50 | 235 |
| 2250 | School, single exterior, incl. lever, incl. panic device | | 3 | 2.667 | Door | 1,100 | 115 | | 1,215 | 1,400 |
| 2500 | Single interior, regular use, lever included | | 3 | 2.667 | | 595 | 115 | | 710 | 825 |
| 2550 | Average, door hdwe, school, classroom, ANSI F84, lever handle | | 3 | 2.667 | | 750 | 115 | | 865 | 1,000 |
| 2600 | Average, door hdwe, school, classroom, ANSI F88, incl. lever | | 3 | 2.667 | | 790 | 115 | | 905 | 1,050 |
| 2850 | Stairway, single interior | | 3 | 2.667 | | 525 | 115 | | 640 | 750 |
| 3100 | Double exterior, with panic device | | 2 | 4 | Pr. | 2,075 | 172 | | 2,247 | 2,550 |
| 3600 | Toilet, public, single interior | | | | Door | 210 | | | 210 | 231 |
| 6020 | Add for fire alarm door holder, electro-magnetic | 1 Elec | 4 | 2 | Ea. | 99 | 101 | | 200 | 258 |
| **08 71 20.20** | **Door Protectors** | | | | | | | | | |
| 0010 | **DOOR PROTECTORS** | | | | | | | | | |
| 0020 | 1-3/4" x 3/4" U channel | 2 Carp | 80 | .200 | L.F. | 25 | 8.60 | | 33.60 | 40.50 |
| 0021 | 1-3/4" x 1-1/4" U channel | | 80 | .200 | " | 15 | 8.60 | | 23.60 | 29.50 |
| 1000 | Tear drop, spring-stl, 8" high x 19" long | | 15 | 1.067 | Ea. | 92 | 46 | | 138 | 172 |
| 1010 | 8" high x 32" long | | 15 | 1.067 | | 114 | 46 | | 160 | 196 |
| 1100 | Tear drop, stainless stl, 8" high x 19" long | | 15 | 1.067 | | 240 | 46 | | 286 | 335 |
| 1200 | 8" high x 32" long | | 15 | 1.067 | | 300 | 46 | | 346 | 400 |

# 08 71 Door Hardware

## 08 71 20 – Hardware

| 08 71 20.30 Door Closers | Crew | Daily Output | Labor-Hours | Unit | Material | 2011 Bare Costs Labor | 2011 Bare Costs Equipment | Total | Total Incl O&P |
|---|---|---|---|---|---|---|---|---|---|
| 0010 | **DOOR CLOSERS** | | | | | | | | | |
| 0020 | Adjustable backcheck, 3 way mount, all sizes, regular arm | 1 Carp | 6 | 1.333 | Ea. | 182 | 57.50 | | 239.50 | 288 |
| 0040 | Hold open arm | | 6 | 1.333 | | 230 | 57.50 | | 287.50 | 340 |
| 0100 | Fusible link | | 6.50 | 1.231 | | 155 | 53 | | 208 | 253 |
| 0200 | Non sized, regular arm | | 6 | 1.333 | | 165 | 57.50 | | 222.50 | 270 |
| 0240 | Hold open arm | | 6 | 1.333 | | 200 | 57.50 | | 257.50 | 310 |
| 0400 | 4 way mount, non sized, regular arm | | 6 | 1.333 | | 220 | 57.50 | | 277.50 | 330 |
| 0440 | Hold open arm | | 6 | 1.333 | | 230 | 57.50 | | 287.50 | 340 |
| 2000 | Backcheck and adjustable power, hinge face mount | | | | | | | | | |
| 2010 | All sizes, regular arm | 1 Carp | 6.50 | 1.231 | Ea. | 210 | 53 | | 263 | 315 |
| 2040 | Hold open arm | | 6.50 | 1.231 | | 245 | 53 | | 298 | 350 |
| 2400 | Top jamb mount, all sizes, regular arm | | 6 | 1.333 | | 200 | 57.50 | | 257.50 | 310 |
| 2440 | Hold open arm | | 6 | 1.333 | | 210 | 57.50 | | 267.50 | 320 |
| 2800 | Top face mount, all sizes, regular arm | | 6.50 | 1.231 | | 195 | 53 | | 248 | 297 |
| 2840 | Hold open arm | | 6.50 | 1.231 | | 215 | 53 | | 268 | 320 |
| 4000 | Backcheck, overhead concealed, all sizes, regular arm | | 5.50 | 1.455 | | 216 | 62.50 | | 278.50 | 335 |
| 4040 | Concealed arm | | 5 | 1.600 | | 227 | 69 | | 296 | 355 |
| 4400 | Compact overhead, concealed, all sizes, regular arm | | 5.50 | 1.455 | | 345 | 62.50 | | 407.50 | 475 |
| 4440 | Concealed arm | | 5 | 1.600 | | 380 | 69 | | 449 | 525 |
| 4800 | Concealed in door, all sizes, regular arm | | 5.50 | 1.455 | | 355 | 62.50 | | 417.50 | 485 |
| 4840 | Concealed arm | | 5 | 1.600 | | 375 | 69 | | 444 | 520 |
| 4900 | Floor concealed, all sizes, single acting | | 2.20 | 3.636 | | 380 | 157 | | 537 | 660 |
| 4940 | Double acting | | 2.20 | 3.636 | | 320 | 157 | | 477 | 590 |
| 5000 | For cast aluminum cylinder, deduct | | | | | 35 | | | 35 | 38.50 |
| 5040 | For delayed action, add | | | | | 45 | | | 45 | 49.50 |
| 5080 | For fusible link arm, add | | | | | 30 | | | 30 | 33 |
| 5120 | For shock absorbing arm, add | | | | | 50 | | | 50 | 55 |
| 5160 | For spring power adjustment, add | | | | | 40 | | | 40 | 44 |
| 6000 | Closer-holder, hinge face mount, all sizes, exposed arm | 1 Carp | 6.50 | 1.231 | | 165 | 53 | | 218 | 264 |
| 7000 | Electronic closer-holder, hinge facemount, concealed arm | | 5 | 1.600 | | 270 | 69 | | 339 | 405 |
| 7400 | With built-in detector | | 5 | 1.600 | | 665 | 69 | | 734 | 835 |
| 8000 | Surface mounted, stand. duty, parallel arm, primed, traditional | | 6 | 1.333 | | 187 | 57.50 | | 244.50 | 294 |
| 8030 | Light duty | | 6 | 1.333 | | 123 | 57.50 | | 180.50 | 223 |
| 8050 | Heavy duty | | 6 | 1.333 | | 213 | 57.50 | | 270.50 | 320 |
| 8100 | Standard duty, parallel arm, modern | | 6 | 1.333 | | 213 | 57.50 | | 270.50 | 320 |
| 8150 | Heavy duty | | 6 | 1.333 | | 228 | 57.50 | | 285.50 | 340 |

## 08 71 20.31 Door Closers

| | | Crew | Daily Output | Labor-Hours | Unit | Material | Labor | Equipment | Total | Total Incl O&P |
|---|---|---|---|---|---|---|---|---|---|---|
| 0010 | **DOOR CLOSERS** | | | | | | | | | |
| 0015 | Door closer, rack and pinion | 1 Carp | 6.50 | 1.231 | Ea. | 168 | 53 | | 221 | 266 |
| 1520 | Door, single acting, standard arm | | 1 | 8 | | 199 | 345 | | 544 | 750 |
| 1522 | Hold open arm | | 1 | 8 | | 216 | 345 | | 561 | 770 |
| 1526 | Frame, single acting, standard arm | | 1 | 8 | | 335 | 345 | | 680 | 900 |
| 1530 | Hold open arm | | 1 | 8 | | 355 | 345 | | 700 | 920 |
| 1534 | Double acting, standard arm | | 1 | 8 | | 450 | 345 | | 795 | 1,025 |
| 1536 | Hold open arm | | 1 | 8 | | 460 | 345 | | 805 | 1,025 |
| 1554 | Floor, center hung, single acting, bottom arm | | 1 | 8 | | 320 | 345 | | 665 | 880 |
| 1558 | Double acting | | 1 | 8 | | 330 | 345 | | 675 | 895 |
| 1560 | Offset hung, single acting, bottom arm | | 1 | 8 | | 370 | 345 | | 715 | 940 |
| 6500 | Electro magnetic closer/holder | | | | | | | | | |
| 6510 | Single point, no detector | 1 Carp | 1 | 8 | Ea. | 350 | 345 | | 695 | 915 |
| 6515 | Including detector | | 1 | 8 | | 670 | 345 | | 1,015 | 1,275 |
| 6520 | Multi-point, no detector | | 1 | 8 | | 770 | 345 | | 1,115 | 1,375 |

273

## 08 71 20 – Hardware

| 08 71 20.31 Door Closers | | Crew | Daily Output | Labor-Hours | Unit | Material | 2011 Bare Costs Labor | Equipment | Total | Total Incl O&P |
|---|---|---|---|---|---|---|---|---|---|---|
| 6524 | Including detector | 1 Carp | 1 | 8 | Ea. | 780 | 345 | | 1,125 | 1,400 |
| 6550 | Electric automatic operators | | | | | | | | | |
| 6555 | Operator | 1 Carp | 1 | 8 | Ea. | 1,525 | 345 | | 1,870 | 2,200 |
| 6570 | Wall plate actuator | | 1 | 8 | | 195 | 345 | | 540 | 745 |
| 8010 | Light duty, regular arm | | 1 | 8 | | 111 | 345 | | 456 | 650 |
| 8032 | Parallel arm | | 1 | 8 | | 115 | 345 | | 460 | 655 |
| 8034 | Hold open arm | | 1 | 8 | | 121 | 345 | | 466 | 665 |
| 8036 | Fusible link arm | | 1 | 8 | | 176 | 345 | | 521 | 725 |
| 8040 | Medium duty, regular arm | | 1 | 8 | | 140 | 345 | | 485 | 685 |
| 8042 | Extra duty parallel arm | | 1 | 8 | | 150 | 345 | | 495 | 695 |
| 8044 | Hold open arm | | 1 | 8 | | 157 | 345 | | 502 | 700 |
| 8046 | Positive stop arm | | 1 | 8 | | 170 | 345 | | 515 | 715 |
| 8052 | Heavy duty, regular arm | | 1 | 8 | | 172 | 345 | | 517 | 720 |
| 8054 | Top jamb mount | | 1 | 8 | | 172 | 345 | | 517 | 720 |
| 8056 | Extra duty parallel arm | | 1 | 8 | | 172 | 345 | | 517 | 720 |
| 8058 | Hold open arm | | 1 | 8 | | 187 | 345 | | 532 | 735 |
| 8060 | Positive stop arm | | 1 | 8 | | 189 | 345 | | 534 | 740 |
| 8062 | Fusible link arm | | 1 | 8 | | 217 | 345 | | 562 | 770 |
| 8080 | Universal heavy duty, regular arm | | 1 | 8 | | 186 | 345 | | 531 | 735 |
| 8084 | Parallel arm | | 1 | 8 | | 186 | 345 | | 531 | 735 |
| 8088 | Extra duty, parallel arm | | 1 | 8 | | 191 | 345 | | 536 | 740 |
| 8090 | Hold open arm | | 1 | 8 | | 272 | 345 | | 617 | 830 |
| 8094 | Positive stop arm | | 1 | 8 | | 277 | 345 | | 622 | 835 |
| 8098 | For delayed action add | | 1 | 8 | | 30 | 345 | | 375 | 565 |

| 08 71 20.35 Panic Devices | | | | | | | | | | |
|---|---|---|---|---|---|---|---|---|---|---|
| 0010 | **PANIC DEVICES** | | | | | | | | | |
| 0015 | For rim locks, single door exit only | 1 Carp | 6 | 1.333 | Ea. | 530 | 57.50 | | 587.50 | 675 |
| 0020 | Outside key and pull | | 5 | 1.600 | | 575 | 69 | | 644 | 735 |
| 0200 | Bar and vertical rod, exit only | | 5 | 1.600 | | 735 | 69 | | 804 | 915 |
| 0210 | Outside key and pull | | 4 | 2 | | 810 | 86 | | 896 | 1,025 |
| 0400 | Bar and concealed rod | | 4 | 2 | | 595 | 86 | | 681 | 785 |
| 0600 | Touch bar, exit only | | 6 | 1.333 | | 505 | 57.50 | | 562.50 | 645 |
| 0610 | Outside key and pull | | 5 | 1.600 | | 600 | 69 | | 669 | 770 |
| 0700 | Touch bar and vertical rod, exit only | | 5 | 1.600 | | 670 | 69 | | 739 | 840 |
| 0710 | Outside key and pull | | 4 | 2 | | 780 | 86 | | 866 | 985 |
| 1000 | Mortise, bar, exit only | | 4 | 2 | | 550 | 86 | | 636 | 735 |
| 1600 | Touch bar, exit only | | 4 | 2 | | 620 | 86 | | 706 | 815 |
| 2000 | Narrow stile, rim mounted, bar, exit only | | 6 | 1.333 | | 650 | 57.50 | | 707.50 | 805 |
| 2010 | Outside key and pull | | 5 | 1.600 | | 700 | 69 | | 769 | 875 |
| 2200 | Bar and vertical rod, exit only | | 5 | 1.600 | | 680 | 69 | | 749 | 850 |
| 2210 | Outside key and pull | | 4 | 2 | | 680 | 86 | | 766 | 875 |
| 2400 | Bar and concealed rod, exit only | | 3 | 2.667 | | 765 | 115 | | 880 | 1,025 |
| 3000 | Mortise, bar, exit only | | 4 | 2 | | 610 | 86 | | 696 | 800 |
| 3600 | Touch bar, exit only | | 4 | 2 | | 880 | 86 | | 966 | 1,100 |

| 08 71 20.40 Lockset | | | | | | | | | | |
|---|---|---|---|---|---|---|---|---|---|---|
| 0010 | **LOCKSET**, Standard duty | | | | | | | | | |
| 0020 | Non-keyed, passage, w/sect.trim | 1 Carp | 12 | .667 | Ea. | 54.50 | 28.50 | | 83 | 104 |
| 0100 | Privacy | | 12 | .667 | | 62 | 28.50 | | 90.50 | 112 |
| 0400 | Keyed, single cylinder function | | 10 | .800 | | 92.50 | 34.50 | | 127 | 155 |
| 0420 | Hotel (see also Section 08 71 20.15) | | 8 | 1 | | 135 | 43 | | 178 | 215 |
| 0500 | Lever handled, keyed, single cylinder function | | 10 | .800 | | 159 | 34.50 | | 193.50 | 228 |
| 1000 | Heavy duty with sectional trim, non-keyed, passages | | 12 | .667 | | 145 | 28.50 | | 173.50 | 204 |

# 08 71 Door Hardware

## 08 71 20 – Hardware

### 08 71 20.40 Lockset

| | | Crew | Daily Output | Labor-Hours | Unit | Material | 2011 Bare Costs Labor | Equipment | Total | Total Incl O&P |
|---|---|---|---|---|---|---|---|---|---|---|
| 1100 | Privacy | 1 Carp | 12 | .667 | Ea. | 185 | 28.50 | | 213.50 | 248 |
| 1400 | Keyed, single cylinder function | | 10 | .800 | | 220 | 34.50 | | 254.50 | 295 |
| 1420 | Hotel | | 8 | 1 | | 325 | 43 | | 368 | 425 |
| 1600 | Communicating | | 10 | .800 | | 250 | 34.50 | | 284.50 | 330 |
| 1690 | For re-core cylinder, add | | | | | 40 | | | 40 | 44 |
| 1700 | Residential, interior door, minimum | 1 Carp | 16 | .500 | | 25 | 21.50 | | 46.50 | 60.50 |
| 1720 | Maximum | | 8 | 1 | | 70 | 43 | | 113 | 143 |
| 1800 | Exterior, minimum | | 14 | .571 | | 45 | 24.50 | | 69.50 | 87 |
| 1810 | Average | | 8 | 1 | | 72 | 43 | | 115 | 146 |
| 1820 | Maximum | | 8 | 1 | | 170 | 43 | | 213 | 253 |

### 08 71 20.41 Dead Locks

| | | Crew | Daily Output | Labor-Hours | Unit | Material | 2011 Bare Costs Labor | Equipment | Total | Total Incl O&P |
|---|---|---|---|---|---|---|---|---|---|---|
| 0010 | **DEAD LOCKS** | | | | | | | | | |
| 0011 | Mortise heavy duty outside key (security item) | 1 Carp | 9 | .889 | Ea. | 269 | 38.50 | | 307.50 | 355 |
| 0020 | Double cylinder | | 9 | .889 | | 265 | 38.50 | | 303.50 | 350 |
| 0100 | Medium duty, outside key | | 10 | .800 | | 140 | 34.50 | | 174.50 | 207 |
| 0110 | Double cylinder | | 10 | .800 | | 155 | 34.50 | | 189.50 | 224 |
| 1000 | Tubular, standard duty, outside key | | 10 | .800 | | 65 | 34.50 | | 99.50 | 125 |
| 1010 | Double cylinder | | 10 | .800 | | 82 | 34.50 | | 116.50 | 143 |
| 1200 | Night latch, outside key | | 10 | .800 | | 81 | 34.50 | | 115.50 | 142 |

### 08 71 20.42 Mortise Locksets

| | | Crew | Daily Output | Labor-Hours | Unit | Material | 2011 Bare Costs Labor | Equipment | Total | Total Incl O&P |
|---|---|---|---|---|---|---|---|---|---|---|
| 0010 | **MORTISE LOCKSETS**, Comm., wrought knobs & full escutcheon trim | | | | | | | | | |
| 0020 | Non-keyed, passage, minimum | 1 Carp | 9 | .889 | Ea. | 190 | 38.50 | | 228.50 | 268 |
| 0030 | Maximum | | 8 | 1 | | 300 | 43 | | 343 | 395 |
| 0040 | Privacy, minimum | | 9 | .889 | | 200 | 38.50 | | 238.50 | 279 |
| 0050 | Maximum | | 8 | 1 | | 325 | 43 | | 368 | 425 |
| 0100 | Keyed, office/entrance/apartment, minimum | | 8 | 1 | | 230 | 43 | | 273 | 320 |
| 0110 | Maximum | | 7 | 1.143 | | 385 | 49 | | 434 | 500 |
| 0120 | Single cylinder, typical, minimum | | 8 | 1 | | 198 | 43 | | 241 | 284 |
| 0130 | Maximum | | 7 | 1.143 | | 360 | 49 | | 409 | 470 |
| 0200 | Hotel, minimum (see also Section 08 71 20.15) | | 7 | 1.143 | | 230 | 49 | | 279 | 330 |
| 0210 | Maximum (see also Section 08 71 20.15) | | 6 | 1.333 | | 375 | 57.50 | | 432.50 | 505 |
| 0300 | Communication, double cylinder, minimum | | 8 | 1 | | 225 | 43 | | 268 | 315 |
| 0310 | Maximum | | 7 | 1.143 | | 300 | 49 | | 349 | 405 |
| 1000 | Wrought knobs and sectional trim, non-keyed, passage, minimum | | 10 | .800 | | 130 | 34.50 | | 164.50 | 196 |
| 1010 | Maximum | | 9 | .889 | | 240 | 38.50 | | 278.50 | 325 |
| 1040 | Privacy, minimum | | 10 | .800 | | 145 | 34.50 | | 179.50 | 213 |
| 1050 | Maximum | | 9 | .889 | | 260 | 38.50 | | 298.50 | 345 |
| 1100 | Keyed, entrance, office/apartment, minimum | | 9 | .889 | | 220 | 38.50 | | 258.50 | 300 |
| 1110 | Maximum | | 8 | 1 | | 310 | 43 | | 353 | 405 |
| 1120 | Single cylinder, typical, minimum | | 9 | .889 | | 215 | 38.50 | | 253.50 | 296 |
| 1130 | Maximum | | 8 | 1 | | 295 | 43 | | 338 | 390 |
| 2000 | Cast knobs and full escutcheon trim | | | | | | | | | |
| 2010 | Non-keyed, passage, minimum | 1 Carp | 9 | .889 | Ea. | 260 | 38.50 | | 298.50 | 345 |
| 2020 | Maximum | | 8 | 1 | | 410 | 43 | | 453 | 515 |
| 2040 | Privacy, minimum | | 9 | .889 | | 310 | 38.50 | | 348.50 | 400 |
| 2050 | Maximum | | 8 | 1 | | 445 | 43 | | 488 | 555 |
| 2120 | Keyed, single cylinder, typical, minimum | | 8 | 1 | | 320 | 43 | | 363 | 415 |
| 2130 | Maximum | | 7 | 1.143 | | 490 | 49 | | 539 | 615 |
| 3000 | Cast knob and sectional trim, non-keyed, passage, minimum | | 10 | .800 | | 210 | 34.50 | | 244.50 | 284 |
| 3010 | Maximum | | 10 | .800 | | 400 | 34.50 | | 434.50 | 495 |
| 3040 | Privacy, minimum | | 10 | .800 | | 225 | 34.50 | | 259.50 | 300 |
| 3050 | Maximum | | 10 | .800 | | 400 | 34.50 | | 434.50 | 495 |

## 08 71 20 – Hardware

### 08 71 20.42 Mortise Locksets

| | | Crew | Daily Output | Labor-Hours | Unit | Material | 2011 Bare Costs Labor | Equipment | Total | Total Incl O&P |
|---|---|---|---|---|---|---|---|---|---|---|
| 3100 | Keyed, office/entrance/apartment, minimum | 1 Carp | 9 | .889 | Ea. | 255 | 38.50 | | 293.50 | 340 |
| 3110 | Maximum | | 9 | .889 | | 400 | 38.50 | | 438.50 | 500 |
| 3120 | Single cylinder, typical, minimum | | 9 | .889 | | 255 | 38.50 | | 293.50 | 340 |
| 3130 | Maximum | | 9 | .889 | | 490 | 38.50 | | 528.50 | 600 |
| 3190 | For re-core cylinder, add | | | | | 35 | | | 35 | 38.50 |
| 3800 | Cipher lockset w/key pad (security item) | 1 Carp | 13 | .615 | | 830 | 26.50 | | 856.50 | 955 |
| 3900 | Cipher lockset, with dial for swinging doors (security item) | | 13 | .615 | | 1,725 | 26.50 | | 1,751.50 | 1,950 |
| 3920 | with dial for swinging doors & drill resistant plate (security item) | | 12 | .667 | | 2,100 | 28.50 | | 2,128.50 | 2,350 |
| 3950 | Cipher lockset with dial for safe/vault door (security item) | | 12 | .667 | | 1,400 | 28.50 | | 1,428.50 | 1,600 |
| 3980 | Keyless, pushbutton type | | | | | | | | | |
| 4000 | Residential/light commercial, deadbolt, standard | 1 Carp | 9 | .889 | Ea. | 130 | 38.50 | | 168.50 | 202 |
| 4010 | Heavy duty | | 9 | .889 | | 150 | 38.50 | | 188.50 | 224 |
| 4020 | Industrial, heavy duty, with deadbolt | | 9 | .889 | | 284 | 38.50 | | 322.50 | 375 |
| 4030 | Key override | | 9 | .889 | | 320 | 38.50 | | 358.50 | 410 |
| 4040 | Lever activated handle | | 9 | .889 | | 335 | 38.50 | | 373.50 | 430 |
| 4050 | Key override | | 9 | .889 | | 345 | 38.50 | | 383.50 | 440 |
| 4060 | Double sided pushbutton type | | 8 | 1 | | 695 | 43 | | 738 | 825 |
| 4070 | Key override | | 8 | 1 | | 690 | 43 | | 733 | 820 |

### 08 71 20.45 Peepholes

| | | Crew | Daily Output | Labor-Hours | Unit | Material | 2011 Bare Costs Labor | Equipment | Total | Total Incl O&P |
|---|---|---|---|---|---|---|---|---|---|---|
| 0010 | **PEEPHOLES** | | | | | | | | | |
| 2010 | Peephole | 1 Carp | 32 | .250 | Ea. | 15.80 | 10.75 | | 26.55 | 34 |

### 08 71 20.50 Door Stops

| | | Crew | Daily Output | Labor-Hours | Unit | Material | 2011 Bare Costs Labor | Equipment | Total | Total Incl O&P |
|---|---|---|---|---|---|---|---|---|---|---|
| 0010 | **DOOR STOPS** | | | | | | | | | |
| 0020 | Holder & bumper, floor or wall | 1 Carp | 32 | .250 | Ea. | 35.50 | 10.75 | | 46.25 | 55.50 |
| 1300 | Wall bumper, 4" diameter, with rubber pad, aluminum | | 32 | .250 | | 10.95 | 10.75 | | 21.70 | 28.50 |
| 1600 | Door bumper, floor type, aluminum | | 32 | .250 | | 6.70 | 10.75 | | 17.45 | 24 |
| 1900 | Plunger type, door mounted | | 32 | .250 | | 25.50 | 10.75 | | 36.25 | 44.50 |

### 08 71 20.55 Push-Pull Plates

| | | Crew | Daily Output | Labor-Hours | Unit | Material | 2011 Bare Costs Labor | Equipment | Total | Total Incl O&P |
|---|---|---|---|---|---|---|---|---|---|---|
| 0010 | **PUSH-PULL PLATES** | | | | | | | | | |
| 0100 | Push plate, .050 thick, 4" x 16", aluminum | 1 Carp | 12 | .667 | Ea. | 10.20 | 28.50 | | 38.70 | 55.50 |
| 0500 | Bronze | | 12 | .667 | | 21.50 | 28.50 | | 50 | 67.50 |
| 1500 | Pull handle and push bar, aluminum | | 11 | .727 | | 131 | 31.50 | | 162.50 | 192 |
| 2000 | Bronze | | 10 | .800 | | 169 | 34.50 | | 203.50 | 238 |
| 3000 | Push plate both sides, aluminum | | 14 | .571 | | 19.95 | 24.50 | | 44.45 | 59.50 |
| 3500 | Bronze | | 13 | .615 | | 44 | 26.50 | | 70.50 | 89 |
| 4000 | Door pull, designer style, cast aluminum, minimum | | 12 | .667 | | 71.50 | 28.50 | | 100 | 123 |
| 5000 | Maximum | | 8 | 1 | | 370 | 43 | | 413 | 470 |
| 6000 | Cast bronze, minimum | | 12 | .667 | | 87 | 28.50 | | 115.50 | 140 |
| 7000 | Maximum | | 8 | 1 | | 405 | 43 | | 448 | 510 |
| 8000 | Walnut, minimum | | 12 | .667 | | 67 | 28.50 | | 95.50 | 118 |
| 9000 | Maximum | | 8 | 1 | | 370 | 43 | | 413 | 470 |

### 08 71 20.60 Entrance Locks

| | | Crew | Daily Output | Labor-Hours | Unit | Material | 2011 Bare Costs Labor | Equipment | Total | Total Incl O&P |
|---|---|---|---|---|---|---|---|---|---|---|
| 0010 | **ENTRANCE LOCKS** | | | | | | | | | |
| 0015 | Cylinder, grip handle deadlocking latch | 1 Carp | 9 | .889 | Ea. | 132 | 38.50 | | 170.50 | 204 |
| 0020 | Deadbolt | | 8 | 1 | | 160 | 43 | | 203 | 242 |
| 0100 | Push and pull plate, dead bolt | | 8 | 1 | | 150 | 43 | | 193 | 231 |
| 0900 | For handicapped lever, add | | | | | 170 | | | 170 | 187 |

### 08 71 20.65 Thresholds

| | | Crew | Daily Output | Labor-Hours | Unit | Material | 2011 Bare Costs Labor | Equipment | Total | Total Incl O&P |
|---|---|---|---|---|---|---|---|---|---|---|
| 0010 | **THRESHOLDS** | | | | | | | | | |
| 0011 | Threshold 3' long saddles aluminum | 1 Carp | 48 | .167 | L.F. | 4.93 | 7.20 | | 12.13 | 16.40 |
| 0100 | Aluminum, 8" wide, 1/2" thick | | 12 | .667 | Ea. | 39 | 28.50 | | 67.50 | 87 |

# 08 71 Door Hardware

## 08 71 20 – Hardware

### 08 71 20.65 Thresholds

| | | Crew | Daily Output | Labor-Hours | Unit | Material | 2011 Bare Costs Labor | Equipment | Total | Total Incl O&P |
|---|---|---|---|---|---|---|---|---|---|---|
| 0500 | Bronze | 1 Carp | 60 | .133 | L.F. | 41.50 | 5.75 | | 47.25 | 54.50 |
| 0600 | Bronze, panic threshold, 5" wide, 1/2" thick | | 12 | .667 | Ea. | 67.50 | 28.50 | | 96 | 118 |
| 0700 | Rubber, 1/2" thick, 5-1/2" wide | | 20 | .400 | | 40 | 17.20 | | 57.20 | 70.50 |
| 0800 | 2-3/4" wide | | 20 | .400 | | 29 | 17.20 | | 46.20 | 58.50 |

### 08 71 20.70 Floor Checks

| | | | | | | | | | | |
|---|---|---|---|---|---|---|---|---|---|---|
| 0010 | **FLOOR CHECKS** | | | | | | | | | |
| 0020 | For over 3' wide doors single acting | 1 Carp | 2.50 | 3.200 | Ea. | 520 | 138 | | 658 | 780 |
| 0500 | Double acting | " | 2.50 | 3.200 | " | 670 | 138 | | 808 | 945 |

### 08 71 20.75 Door Hardware Accessories

| | | | | | | | | | | |
|---|---|---|---|---|---|---|---|---|---|---|
| 0010 | **DOOR HARDWARE ACCESSORIES** | | | | | | | | | |
| 0050 | Door closing coordinator, 36" (for paired openings up to 56") | 1 Carp | 8 | 1 | Ea. | 94 | 43 | | 137 | 169 |
| 0060 | 48" (for paired openings up to 84") | | 8 | 1 | | 101 | 43 | | 144 | 177 |
| 0070 | 56" (for paired openings up to 96") | | 8 | 1 | | 111 | 43 | | 154 | 188 |

### 08 71 20.80 Hasps

| | | | | | | | | | | |
|---|---|---|---|---|---|---|---|---|---|---|
| 0010 | **HASPS**, steel assembly | | | | | | | | | |
| 0015 | 3" | 1 Carp | 26 | .308 | Ea. | 4.71 | 13.25 | | 17.96 | 25.50 |
| 0020 | 4-1/2" | | 13 | .615 | | 6.50 | 26.50 | | 33 | 47.50 |
| 0040 | 6" | | 12.50 | .640 | | 9.25 | 27.50 | | 36.75 | 52.50 |

### 08 71 20.90 Hinges

| | | | | | | | | | | |
|---|---|---|---|---|---|---|---|---|---|---|
| 0010 | **HINGES**                    R087120-10 | | | | | | | | | |
| 0012 | Full mortise, avg. freq., steel base, USP, 4-1/2" x 4-1/2" | | | | Pr. | 27 | | | 27 | 29.50 |
| 0100 | 5" x 5", USP | | | | | 43 | | | 43 | 47.50 |
| 0200 | 6" x 6", USP | | | | | 91 | | | 91 | 100 |
| 0400 | Brass base, 4-1/2" x 4-1/2", US10 | | | | | 53.50 | | | 53.50 | 59 |
| 0500 | 5" x 5", US10 | | | | | 77 | | | 77 | 85 |
| 0600 | 6" x 6", US10 | | | | | 130 | | | 130 | 143 |
| 0800 | Stainless steel base, 4-1/2" x 4-1/2", US32 | | | | | 79.50 | | | 79.50 | 87.50 |
| 0900 | For non removable pin, add (security item) | | | | Ea. | 4.72 | | | 4.72 | 5.20 |
| 0910 | For floating pin, driven tips, add | | | | | 3.25 | | | 3.25 | 3.58 |
| 0930 | For hospital type tip on pin, add | | | | | 14.10 | | | 14.10 | 15.50 |
| 0940 | For steeple type tip on pin, add | | | | | 13 | | | 13 | 14.30 |
| 0950 | Full mortise, high frequency, steel base, 3-1/2" x 3-1/2", US26D | | | | Pr. | 27 | | | 27 | 29.50 |
| 1000 | 4-1/2" x 4-1/2", USP | | | | | 62 | | | 62 | 68.50 |
| 1100 | 5" x 5", USP | | | | | 58.50 | | | 58.50 | 64 |
| 1200 | 6" x 6", USP | | | | | 142 | | | 142 | 156 |
| 1400 | Brass base, 3-1/2" x 3-1/2", US4 | | | | | 48 | | | 48 | 52.50 |
| 1430 | 4-1/2" x 4-1/2", US10 | | | | | 84 | | | 84 | 92 |
| 1500 | 5" x 5", US10 | | | | | 123 | | | 123 | 135 |
| 1600 | 6" x 6", US10 | | | | | 178 | | | 178 | 196 |
| 1800 | Stainless steel base, 4-1/2" x 4-1/2", US32 | | | | | 127 | | | 127 | 140 |
| 1810 | 5" x 4-1/2", US32 | | | | | 182 | | | 182 | 200 |
| 1930 | For hospital type tip on pin, add | | | | Ea. | 12.60 | | | 12.60 | 13.85 |
| 1950 | Full mortise, low frequency, steel base, 3-1/2" x 3-1/2", US26D | | | | Pr. | 11.55 | | | 11.55 | 12.75 |
| 2000 | 4-1/2" x 4-1/2", USP | | | | | 12.80 | | | 12.80 | 14.05 |
| 2100 | 5" x 5", USP | | | | | 31.50 | | | 31.50 | 34.50 |
| 2200 | 6" x 6", USP | | | | | 63.50 | | | 63.50 | 69.50 |
| 2300 | 4-1/2" x 4-1/2", US3 | | | | | 18.40 | | | 18.40 | 20 |
| 2310 | 5" x 5", US3 | | | | | 44 | | | 44 | 48.50 |
| 2400 | Brass bass, 4-1/2" x 4-1/2", US10 | | | | | 44.50 | | | 44.50 | 49 |
| 2500 | 5" x 5", US10 | | | | | 67 | | | 67 | 74 |
| 2800 | Stainless steel base, 4-1/2" x 4-1/2", US32 | | | | | 75.50 | | | 75.50 | 83 |

### 08 71 20.91 Special Hinges

| | Crew | Daily Output | Labor-Hours | Unit | Material | 2011 Bare Costs Labor | Equipment | Total | Total Incl O&P |
|---|---|---|---|---|---|---|---|---|---|
| 0010 **SPECIAL HINGES** | | | | | | | | | |
| 0015 Paumelle, high frequency | | | | | | | | | |
| 0020 Steel base, 6" x 4-1/2", US10 | | | | Pr. | 145 | | | 145 | 160 |
| 0100 Brass base, 5" x 4-1/2", US10 | | | | Ea. | 205 | | | 205 | 226 |
| 0200 Paumelle, average frequency, steel base, 4-1/2" x 3-1/2", US10 | | | | Pr. | 98.50 | | | 98.50 | 108 |
| 0400 Olive knuckle, low frequency, brass base, 6" x 4-1/2", US10 | | | | Ea. | 152 | | | 152 | 167 |
| 1000 Electric hinge with concealed conductor, average frequency | | | | | | | | | |
| 1010 Steel base, 4-1/2" x 4-1/2", US26D | | | | Pr. | 315 | | | 315 | 345 |
| 1100 Bronze base, 4-1/2" x 4-1/2", US26D | | | | " | 320 | | | 320 | 350 |
| 1200 Electric hinge with concealed conductor, high frequency | | | | | | | | | |
| 1210 Steel base, 4-1/2" x 4-1/2", US26D | | | | Pr. | 241 | | | 241 | 265 |
| 1600 Double weight, 800 lb., steel base, removable pin, 5" x 6", USP | | | | | 132 | | | 132 | 145 |
| 1700 Steel base-welded pin, 5" x 6", USP | | | | | 163 | | | 163 | 179 |
| 1800 Triple weight, 2000 lb., steel base, welded pin, 5" x 6", USP | | | | | 151 | | | 151 | 166 |
| 2000 Pivot reinf., high frequency, steel base, 7-3/4" door plate, USP | | | | | 187 | | | 187 | 206 |
| 2200 Bronze base, 7-3/4" door plate, US10 | | | | | 232 | | | 232 | 255 |
| 3000 Swing clear, full mortise, full or half surface, high frequency, | | | | | | | | | |
| 3010 Steel base, 5" high, USP | | | | Pr. | 159 | | | 159 | 175 |
| 3200 Swing clear, full mortise, average frequency | | | | | | | | | |
| 3210 Steel base, 4-1/2" high, USP | | | | Pr. | 127 | | | 127 | 140 |
| 4000 Wide throw, average frequency, steel base, 4-1/2" x 6", USP | | | | | 97 | | | 97 | 107 |
| 4200 High frequency, steel base, 4-1/2" x 6", USP | | | | | 148 | | | 148 | 163 |
| 4600 Spring hinge, single acting, 6" flange, steel | | | | Ea. | 54.50 | | | 54.50 | 60 |
| 4700 Brass | | | | | 95.50 | | | 95.50 | 105 |
| 4900 Double acting, 6" flange, steel | | | | | 96.50 | | | 96.50 | 106 |
| 4950 Brass | | | | | 158 | | | 158 | 173 |
| 9000 Continuous hinge, steel, full mortise, heavy duty, 96 inch | 1 Carp | 2 | 4 | | 565 | 172 | | 737 | 890 |

### 08 71 20.95 Kick Plates

| | Crew | Daily Output | Labor-Hours | Unit | Material | 2011 Bare Costs Labor | Equipment | Total | Total Incl O&P |
|---|---|---|---|---|---|---|---|---|---|
| 0010 **KICK PLATES** | | | | | | | | | |
| 0020 Stainless steel, .050, 16 ga, 8" x 28", US32 | 1 Carp | 15 | .533 | Ea. | 35 | 23 | | 58 | 73.50 |
| 0030 8" x 30" | | 15 | .533 | | 36 | 23 | | 59 | 74.50 |
| 0040 8" x 34" | | 15 | .533 | | 40 | 23 | | 63 | 79 |
| 0050 10" x 28" | | 15 | .533 | | 67 | 23 | | 90 | 109 |
| 0060 10" x 30" | | 15 | .533 | | 72 | 23 | | 95 | 114 |
| 0070 10" x 34" | | 15 | .533 | | 81 | 23 | | 104 | 124 |
| 0080 Mop/Kick, 4" x 28" | | 15 | .533 | | 30 | 23 | | 53 | 68 |
| 0090 4" x 30" | | 15 | .533 | | 32 | 23 | | 55 | 70 |
| 0100 4" x 34" | | 15 | .533 | | 36 | 23 | | 59 | 74.50 |
| 0110 6" x 28" | | 15 | .533 | | 38 | 23 | | 61 | 77 |
| 0120 6" x 30" | | 15 | .533 | | 41 | 23 | | 64 | 80 |
| 0130 6" x 34" | | 15 | .533 | | 47 | 23 | | 70 | 86.50 |
| 0500 Bronze, .050", 8" x 28" | | 15 | .533 | | 47 | 23 | | 70 | 86.50 |
| 0510 8" x 30" | | 15 | .533 | | 63 | 23 | | 86 | 105 |
| 0520 8" x 34" | | 15 | .533 | | 71 | 23 | | 94 | 113 |
| 0530 10" x 28" | | 15 | .533 | | 73 | 23 | | 96 | 116 |
| 0540 10" x 30" | | 15 | .533 | | 78 | 23 | | 101 | 121 |
| 0550 10" x 34" | | 15 | .533 | | 89 | 23 | | 112 | 133 |
| 0560 Mop/Kick, 4" x 28" | | 15 | .533 | | 31 | 23 | | 54 | 69 |
| 0570 4" x 30" | | 15 | .533 | | 34 | 23 | | 57 | 72.50 |
| 0580 4" x 34" | | 15 | .533 | | 35 | 23 | | 58 | 73.50 |
| 0590 6" x 28" | | 15 | .533 | | 44 | 23 | | 67 | 83.50 |
| 0600 6" x 30" | | 15 | .533 | | 50 | 23 | | 73 | 90 |

# 08 71 Door Hardware

## 08 71 20 – Hardware

### 08 71 20.95 Kick Plates

| | | Crew | Daily Output | Labor- Hours | Unit | Material | 2011 Bare Costs Labor | Equipment | Total | Total Incl O&P |
|---|---|---|---|---|---|---|---|---|---|---|
| 0610 | 6" x 34" | 1 Carp | 15 | .533 | Ea. | 54 | 23 | | 77 | 94.50 |
| 1000 | Acrylic, .125", 8" x 26" | | 15 | .533 | | 28 | 23 | | 51 | 66 |
| 1010 | 8" x 36" | | 15 | .533 | | 38 | 23 | | 61 | 77 |
| 1020 | 8" x 42" | | 15 | .533 | | 45 | 23 | | 68 | 84.50 |
| 1030 | 10" x 26" | | 15 | .533 | | 35 | 23 | | 58 | 73.50 |
| 1040 | 10" x 36" | | 15 | .533 | | 48 | 23 | | 71 | 88 |
| 1050 | 10" x 42" | | 15 | .533 | | 68.50 | 23 | | 91.50 | 111 |
| 1060 | Mop/Kick, 4" x 26" | | 15 | .533 | | 17 | 23 | | 40 | 53.50 |
| 1070 | 4" x 36" | | 15 | .533 | | 24 | 23 | | 47 | 61.50 |
| 1080 | 4" x 42" | | 15 | .533 | | 27 | 23 | | 50 | 64.50 |
| 1090 | 6" x 26" | | 15 | .533 | | 23 | 23 | | 46 | 60.50 |
| 1100 | 6" x 36" | | 15 | .533 | | 34 | 23 | | 57 | 72.50 |
| 1110 | 6" x 42" | | 15 | .533 | | 39 | 23 | | 62 | 78 |
| 1220 | Brass, .050", 8" x 26" | | 15 | .533 | | 56 | 23 | | 79 | 96.50 |
| 1230 | 8" x 36" | | 15 | .533 | | 75 | 23 | | 98 | 118 |
| 1240 | 8" x 42" | | 15 | .533 | | 86 | 23 | | 109 | 130 |
| 1250 | 10" x 26" | | 15 | .533 | | 71 | 23 | | 94 | 113 |
| 1260 | 10" x 36" | | 15 | .533 | | 91 | 23 | | 114 | 135 |
| 1270 | 10" x 42" | | 15 | .533 | | 105 | 23 | | 128 | 151 |
| 1320 | Mop/Kick, 4" x 26" | | 15 | .533 | | 28 | 23 | | 51 | 66 |
| 1330 | 4" x 36" | | 15 | .533 | | 39 | 23 | | 62 | 78 |
| 1340 | 4" x 42" | | 15 | .533 | | 44 | 23 | | 67 | 83.50 |
| 1350 | 6" x 26" | | 15 | .533 | | 38 | 23 | | 61 | 77 |
| 1360 | 6" x 36" | | 15 | .533 | | 48 | 23 | | 71 | 88 |
| 1370 | 6" x 42" | | 15 | .533 | | 55 | 23 | | 78 | 95.50 |
| 1800 | Aluminum, .050", 8" x 26" | | 15 | .533 | | 22 | 23 | | 45 | 59 |
| 1810 | 8" x 36" | | 15 | .533 | | 30.50 | 23 | | 53.50 | 68.50 |
| 1820 | 8" x 42" | | 15 | .533 | | 35.50 | 23 | | 58.50 | 74 |
| 1830 | 10" x 26" | | 15 | .533 | | 27.50 | 23 | | 50.50 | 65.50 |
| 1840 | 10" x 36" | | 15 | .533 | | 38 | 23 | | 61 | 77 |
| 1850 | 10" x 42" | | 15 | .533 | | 44.50 | 23 | | 67.50 | 84 |
| 1860 | Mop/Kick, 4" x 26" | | 15 | .533 | | 11 | 23 | | 34 | 47 |
| 1870 | 4" x 36" | | 15 | .533 | | 15.50 | 23 | | 38.50 | 52 |
| 1880 | 4" x 42" | | 15 | .533 | | 17 | 23 | | 40 | 53.50 |
| 1890 | 6" x 26" | | 15 | .533 | | 16.50 | 23 | | 39.50 | 53 |
| 1900 | 6" x 36" | | 15 | .533 | | 23 | 23 | | 46 | 60.50 |
| 1910 | 6" x 42" | | 15 | .533 | | 27 | 23 | | 50 | 64.50 |

## 08 71 21 – Astragals

### 08 71 21.10 Exterior Mouldings, Astragals

| | | Crew | Daily Output | Labor- Hours | Unit | Material | 2011 Bare Costs Labor | Equipment | Total | Total Incl O&P |
|---|---|---|---|---|---|---|---|---|---|---|
| 0010 | **EXTERIOR MOULDINGS, ASTRAGALS** | | | | | | | | | |
| 0400 | One piece, overlapping cadmium plated steel, flat, 3/16" x 2" | 1 Carp | 90 | .089 | L.F. | 3.60 | 3.83 | | 7.43 | 9.80 |
| 0600 | Prime coated steel, flat, 1/8" x 3" | | 90 | .089 | | 4.75 | 3.83 | | 8.58 | 11.10 |
| 0800 | Stainless steel, flat, 3/32" x 1-5/8" | | 90 | .089 | | 17.95 | 3.83 | | 21.78 | 25.50 |
| 1000 | Aluminum, flat, 1/8" x 2" | | 90 | .089 | | 3.60 | 3.83 | | 7.43 | 9.80 |
| 1200 | Nail on, "T" extrusion | | 120 | .067 | | 1 | 2.87 | | 3.87 | 5.50 |
| 1300 | Vinyl bulb insert | | 105 | .076 | | 1.40 | 3.28 | | 4.68 | 6.60 |
| 1600 | Screw on, "T" extrusion | | 90 | .089 | | 5.20 | 3.83 | | 9.03 | 11.55 |
| 1700 | Vinyl insert | | 75 | .107 | | 3.50 | 4.59 | | 8.09 | 10.90 |
| 2000 | "L" extrusion, neoprene bulbs | | 75 | .107 | | 2.10 | 4.59 | | 6.69 | 9.35 |
| 2100 | Neoprene sponge insert | | 75 | .107 | | 6.05 | 4.59 | | 10.64 | 13.70 |
| 2200 | Magnetic | | 75 | .107 | | 10 | 4.59 | | 14.59 | 18.05 |
| 2400 | Spring hinged security seal, with cam | | 75 | .107 | | 6.45 | 4.59 | | 11.04 | 14.15 |

# 08 71 Door Hardware

## 08 71 21 – Astragals

### 08 71 21.10 Exterior Mouldings, Astragals

| | 08 71 21.10 Exterior Mouldings, Astragals | Crew | Daily Output | Labor-Hours | Unit | Material | 2011 Bare Costs Labor | Equipment | Total | Total Incl O&P |
|---|---|---|---|---|---|---|---|---|---|---|
| 2600 | Spring loaded locking bolt, vinyl insert | 1 Carp | 45 | .178 | L.F. | 8.70 | 7.65 | | 16.35 | 21.50 |
| 2800 | Neoprene sponge strip, "Z" shaped, aluminum | | 60 | .133 | | 4.25 | 5.75 | | 10 | 13.50 |
| 2900 | Solid neoprene strip, nail on aluminum strip | | 90 | .089 | | 3.50 | 3.83 | | 7.33 | 9.70 |
| 3000 | One piece stile protection | | | | | | | | | |
| 3020 | Neoprene fabric loop, nail on aluminum strips | 1 Carp | 60 | .133 | L.F. | .90 | 5.75 | | 6.65 | 9.80 |
| 3110 | Flush mounted aluminum extrusion, 1/2" x 1-1/4" | | 60 | .133 | | 3.35 | 5.75 | | 9.10 | 12.50 |
| 3140 | 3/4" x 1-3/8" | | 60 | .133 | | 4.25 | 5.75 | | 10 | 13.50 |
| 3160 | 1-1/8" x 1-3/4" | | 60 | .133 | | 6.70 | 5.75 | | 12.45 | 16.15 |
| 3300 | Mortise, 9/16" x 3/4" | | 60 | .133 | | 3.65 | 5.75 | | 9.40 | 12.80 |
| 3320 | 13/16" x 1-3/8" | | 60 | .133 | | 3.90 | 5.75 | | 9.65 | 13.10 |
| 3600 | Spring bronze strip, nail on type | | 105 | .076 | | 3.15 | 3.28 | | 6.43 | 8.50 |
| 3620 | Screw on, with retainer | | 75 | .107 | | 2.55 | 4.59 | | 7.14 | 9.85 |
| 3800 | Flexible stainless steel housing, pile insert, 1/2" door | | 105 | .076 | | 6.85 | 3.28 | | 10.13 | 12.60 |
| 3820 | 3/4" door | | 105 | .076 | | 7.70 | 3.28 | | 10.98 | 13.50 |
| 4000 | Extruded aluminum retainer, flush mount, pile insert | | 105 | .076 | | 2.45 | 3.28 | | 5.73 | 7.75 |
| 4080 | Mortise, felt insert | | 90 | .089 | | 4.30 | 3.83 | | 8.13 | 10.60 |
| 4160 | Mortise with spring, pile insert | | 90 | .089 | | 3.40 | 3.83 | | 7.23 | 9.60 |
| 4400 | Rigid vinyl retainer, mortise, pile insert | | 105 | .076 | | 2.45 | 3.28 | | 5.73 | 7.75 |
| 4600 | Wool pile filler strip, aluminum backing | | 105 | .076 | | 2.50 | 3.28 | | 5.78 | 7.80 |
| 5000 | Two piece overlapping astragal, extruded aluminum retainer | | | | | | | | | |
| 5010 | Pile insert | 1 Carp | 60 | .133 | L.F. | 3.15 | 5.75 | | 8.90 | 12.25 |
| 5020 | Vinyl bulb insert | | 60 | .133 | | 2.10 | 5.75 | | 7.85 | 11.10 |
| 5040 | Vinyl flap insert | | 60 | .133 | | 6.25 | 5.75 | | 12 | 15.70 |
| 5060 | Solid neoprene flap insert | | 60 | .133 | | 6.20 | 5.75 | | 11.95 | 15.60 |
| 5080 | Hypalon rubber flap insert | | 60 | .133 | | 6.30 | 5.75 | | 12.05 | 15.75 |
| 5090 | Snap on cover, pile insert | | 60 | .133 | | 7.20 | 5.75 | | 12.95 | 16.70 |
| 5400 | Magnetic aluminum, surface mounted | | 60 | .133 | | 24.50 | 5.75 | | 30.25 | 36 |
| 5500 | Interlocking aluminum, 5/8" x 1" neoprene bulb insert | | 45 | .178 | | 3.95 | 7.65 | | 11.60 | 16.10 |
| 5600 | Adjustable aluminum, 9/16" x 21/32", pile insert | | 45 | .178 | | 18.25 | 7.65 | | 25.90 | 32 |
| 5790 | For vinyl bulb, deduct | | | | | .65 | | | .65 | .72 |
| 5800 | Magnetic, adjustable, 9/16" x 21/32" | 1 Carp | 45 | .178 | | 23.50 | 7.65 | | 31.15 | 38 |
| 6000 | Two piece stile protection | | | | | | | | | |
| 6010 | Cloth backed rubber loop, 1" gap, nail on aluminum strips | 1 Carp | 45 | .178 | L.F. | 4.10 | 7.65 | | 11.75 | 16.25 |
| 6040 | Screw on aluminum strips | | 45 | .178 | | 6.20 | 7.65 | | 13.85 | 18.55 |
| 6100 | 1-1/2" gap, screw on aluminum extrusion | | 45 | .178 | | 5.55 | 7.65 | | 13.20 | 17.85 |
| 6240 | Vinyl fabric loop, slotted aluminum extrusion, 1" gap | | 45 | .178 | | 2.05 | 7.65 | | 9.70 | 14 |
| 6300 | 1-1/4" gap | | 45 | .178 | | 5.85 | 7.65 | | 13.50 | 18.20 |

## 08 71 25 – Weatherstripping

### 08 71 25.10 Mechanical Seals, Weatherstripping

| | 08 71 25.10 Mechanical Seals, Weatherstripping | Crew | Daily Output | Labor-Hours | Unit | Material | 2011 Bare Costs Labor | Equipment | Total | Total Incl O&P |
|---|---|---|---|---|---|---|---|---|---|---|
| 0010 | **MECHANICAL SEALS, WEATHERSTRIPPING** | | | | | | | | | |
| 1000 | Doors, wood frame, interlocking, for 3' x 7' door, zinc | 1 Carp | 3 | 2.667 | Opng. | 16 | 115 | | 131 | 194 |
| 1100 | Bronze | | 3 | 2.667 | | 25 | 115 | | 140 | 204 |
| 1300 | 6' x 7' opening, zinc | | 2 | 4 | | 18 | 172 | | 190 | 284 |
| 1400 | Bronze | | 2 | 4 | | 33 | 172 | | 205 | 300 |
| 1700 | Wood frame, spring type, bronze | | | | | | | | | |
| 1800 | 3' x 7' door | 1 Carp | 7.60 | 1.053 | Opng. | 19.35 | 45.50 | | 64.85 | 91 |
| 1900 | 6' x 7' door | " | 7 | 1.143 | " | 28 | 49 | | 77 | 106 |
| 2200 | Metal frame, spring type, bronze | | | | | | | | | |
| 2300 | 3' x 7' door | 1 Carp | 3 | 2.667 | Opng. | 37 | 115 | | 152 | 217 |
| 2400 | 6' x 7' door | " | 2.50 | 3.200 | " | 47 | 138 | | 185 | 263 |
| 2500 | For stainless steel, spring type, add | | | | | 133% | | | | |
| 2700 | Metal frame, extruded sections, 3' x 7' door, aluminum | 1 Carp | 3 | 2.667 | Opng. | 46 | 115 | | 161 | 227 |

# 08 71 Door Hardware

## 08 71 25 – Weatherstripping

| 08 71 25.10 Mechanical Seals, Weatherstripping | Crew | Daily Output | Labor-Hours | Unit | Material | 2011 Bare Costs Labor | Equipment | Total | Total Incl O&P |
|---|---|---|---|---|---|---|---|---|---|
| 2800 | Bronze | 1 Carp | 3 | 2.667 | Opng. | 120 | 115 | | 235 | 310 |
| 3100 | 6' x 7' door, aluminum | | 1.50 | 5.333 | | 58 | 230 | | 288 | 415 |
| 3200 | Bronze | | 1.50 | 5.333 | | 137 | 230 | | 367 | 500 |
| 3500 | Threshold weatherstripping | | | | | | | | | |
| 3650 | Door sweep, flush mounted, aluminum | 1 Carp | 25 | .320 | Ea. | 13.55 | 13.80 | | 27.35 | 36 |
| 3700 | Vinyl | | 25 | .320 | | 16 | 13.80 | | 29.80 | 38.50 |
| 5000 | Garage door bottom weatherstrip, 12' aluminum, clear | | 14 | .571 | | 21.50 | 24.50 | | 46 | 61.50 |
| 5010 | Bronze | | 14 | .571 | | 84 | 24.50 | | 108.50 | 130 |
| 5050 | Bottom protection, Rubber | | 14 | .571 | | 24.50 | 24.50 | | 49 | 64 |
| 5100 | Threshold | | 14 | .571 | | 91.50 | 24.50 | | 116 | 139 |

# 08 74 Access Control Hardware

## 08 74 13 – Card Key Access Control Hardware

| 08 74 13.50 Card Key Access | Crew | Daily Output | Labor-Hours | Unit | Material | 2011 Bare Costs Labor | Equipment | Total | Total Incl O&P |
|---|---|---|---|---|---|---|---|---|---|
| 0010 | **CARD KEY ACCESS**, does not include door hardware | | | | | | | | | |
| 0020 | Card type, 1 time zone, minimum | | | | Ea. | 665 | | | 665 | 730 |
| 0040 | Maximum | | | | | 1,775 | | | 1,775 | 1,950 |
| 0060 | 3 time zones, minimum | | | | | 2,200 | | | 2,200 | 2,425 |
| 0080 | Maximum | | | | | 2,450 | | | 2,450 | 2,700 |
| 0100 | System with printer, and control console, 3 zones | | | | Total | 9,150 | | | 9,150 | 10,100 |
| 0120 | 6 zones | | | | " | 12,000 | | | 12,000 | 13,200 |
| 0140 | For each door, minimum, add | | | | Ea. | 1,400 | | | 1,400 | 1,550 |
| 0160 | Maximum, add | | | | " | 2,000 | | | 2,000 | 2,200 |

## 08 74 19 – Biometric Identity Access Control Hardware

| 08 74 19.50 Biometric Identity Access | Crew | Daily Output | Labor-Hours | Unit | Material | 2011 Bare Costs Labor | Equipment | Total | Total Incl O&P |
|---|---|---|---|---|---|---|---|---|---|
| 0010 | **BIOMETRIC IDENTITY ACCESS** | | | | | | | | | |
| 0220 | Hand geometry scanner, mem of 512 users, excl striker/powr | 1 Elec | 3 | 2.667 | Ea. | 2,025 | 134 | | 2,159 | 2,425 |
| 0230 | Memory upgrade for, adds 9,700 user profiles | | 8 | 1 | | 264 | 50.50 | | 314.50 | 365 |
| 0240 | Adds 32,500 user profiles | | 8 | 1 | | 580 | 50.50 | | 630.50 | 715 |
| 0250 | Prison type, memory of 256 users, excl striker, power | | 3 | 2.667 | | 2,600 | 134 | | 2,734 | 3,050 |
| 0260 | Memory upgrade for, adds 3,300 user profiles | | 8 | 1 | | 211 | 50.50 | | 261.50 | 305 |
| 0270 | Adds 9,700 user profiles | | 8 | 1 | | 425 | 50.50 | | 475.50 | 540 |
| 0280 | Adds 27,900 user profiles | | 8 | 1 | | 580 | 50.50 | | 630.50 | 715 |
| 0290 | All weather, mem of 512 users, excl striker/pwr | | 3 | 2.667 | | 3,900 | 134 | | 4,034 | 4,475 |
| 0300 | Facial & fingerprint scanner, combination unit, excl striker/power | | 3 | 2.667 | | 4,200 | 134 | | 4,334 | 4,825 |
| 0310 | Access for, for initial setup, excl striker/power | | 3 | 2.667 | | 1,000 | 134 | | 1,134 | 1,300 |

# 08 75 Window Hardware

## 08 75 30 – Weatherstripping

| 08 75 30.10 Mechanical Weather Seals | Crew | Daily Output | Labor-Hours | Unit | Material | 2011 Bare Costs Labor | Equipment | Total | Total Incl O&P |
|---|---|---|---|---|---|---|---|---|---|
| 0010 | **MECHANICAL WEATHER SEALS**, Window, double hung, 3' X 5' | | | | | | | | | |
| 0020 | Zinc | 1 Carp | 7.20 | 1.111 | Opng. | 12.80 | 48 | | 60.80 | 87.50 |
| 0100 | Bronze | | 7.20 | 1.111 | | 28 | 48 | | 76 | 105 |
| 0500 | As above but heavy duty, zinc | | 4.60 | 1.739 | | 18 | 75 | | 93 | 135 |
| 0600 | Bronze | | 4.60 | 1.739 | | 32 | 75 | | 107 | 150 |

# 08 79 Hardware Accessories

## 08 79 13 – Key Storage Equipment

| 08 79 13.10 Key Cabinets | Crew | Daily Output | Labor-Hours | Unit | Material | 2011 Bare Costs Labor | Equipment | Total | Total Incl O&P |
|---|---|---|---|---|---|---|---|---|---|
| 0010 **KEY CABINETS** | | | | | | | | | |
| 0020 Wall mounted, 60 key capacity | 1 Carp | 20 | .400 | Ea. | 95.50 | 17.20 | | 112.70 | 132 |
| 0200 Drawer type, 600 key capacity | 1 Clab | 15 | .533 | | 725 | 18.30 | | 743.30 | 830 |
| 0300 2,400 key capacity | | 20 | .400 | | 4,200 | 13.75 | | 4,213.75 | 4,650 |
| 0400 Tray type, 20 key capacity | | 50 | .160 | | 41.50 | 5.50 | | 47 | 54 |
| 0500 50 key capacity | | 40 | .200 | | 92.50 | 6.85 | | 99.35 | 113 |

## 08 79 20 – Door Accessories

| 08 79 20.10 Door Hardware Accessories | Crew | Daily Output | Labor-Hours | Unit | Material | 2011 Bare Costs Labor | Equipment | Total | Total Incl O&P |
|---|---|---|---|---|---|---|---|---|---|
| 0010 **DOOR HARDWARE ACCESSORIES** | | | | | | | | | |
| 0140 Door bolt, surface, 4" | 1 Carp | 32 | .250 | Ea. | 9 | 10.75 | | 19.75 | 26.50 |
| 0160 Door latch | " | 12 | .667 | " | 8.45 | 28.50 | | 36.95 | 53.50 |
| 0200 Sliding closet door | | | | | | | | | |
| 0220 Track and hanger, single | 1 Carp | 10 | .800 | Ea. | 54 | 34.50 | | 88.50 | 113 |
| 0240 Double | | 8 | 1 | | 75 | 43 | | 118 | 149 |
| 0260 Door guide, single | | 48 | .167 | | 25 | 7.20 | | 32.20 | 38.50 |
| 0280 Double | | 48 | .167 | | 34 | 7.20 | | 41.20 | 48.50 |
| 0600 Deadbolt and lock cover plate, brass or stainless steel | | 30 | .267 | | 28 | 11.50 | | 39.50 | 48.50 |
| 0620 Hole cover plate, brass or chrome | | 35 | .229 | | 7.20 | 9.85 | | 17.05 | 23 |
| 2240 Mortise lockset, passage, lever handle | | 9 | .889 | | 198 | 38.50 | | 236.50 | 277 |
| 4000 Security chain, standard | | 18 | .444 | | 7.90 | 19.15 | | 27.05 | 38 |

# 08 81 Glass Glazing

## 08 81 10 – Float Glass

| 08 81 10.10 Various Types and Thickness of Float Glass | Crew | Daily Output | Labor-Hours | Unit | Material | 2011 Bare Costs Labor | Equipment | Total | Total Incl O&P |
|---|---|---|---|---|---|---|---|---|---|
| 0010 **VARIOUS TYPES AND THICKNESS OF FLOAT GLASS** R088110-10 | | | | | | | | | |
| 0020 3/16" Plain | 2 Glaz | 130 | .123 | S.F. | 4.63 | 5.20 | | 9.83 | 12.85 |
| 0200 Tempered, clear | | 130 | .123 | | 6.45 | 5.20 | | 11.65 | 14.85 |
| 0300 Tinted | | 130 | .123 | | 7.35 | 5.20 | | 12.55 | 15.80 |
| 0600 1/4" thick, clear, plain | | 120 | .133 | | 5.40 | 5.60 | | 11 | 14.35 |
| 0700 Tinted | | 120 | .133 | | 8 | 5.60 | | 13.60 | 17.20 |
| 0800 Tempered, clear | | 120 | .133 | | 7.85 | 5.60 | | 13.45 | 17.05 |
| 0900 Tinted | | 120 | .133 | | 10.05 | 5.60 | | 15.65 | 19.45 |
| 1600 3/8" thick, clear, plain | | 75 | .213 | | 9.45 | 8.95 | | 18.40 | 24 |
| 1700 Tinted | | 75 | .213 | | 14.80 | 8.95 | | 23.75 | 30 |
| 1800 Tempered, clear | | 75 | .213 | | 15.60 | 8.95 | | 24.55 | 30.50 |
| 1900 Tinted | | 75 | .213 | | 17.40 | 8.95 | | 26.35 | 32.50 |
| 2200 1/2" thick, clear, plain | | 55 | .291 | | 16 | 12.25 | | 28.25 | 36 |
| 2300 Tinted | | 55 | .291 | | 26 | 12.25 | | 38.25 | 47 |
| 2400 Tempered, clear | | 55 | .291 | | 23 | 12.25 | | 35.25 | 44 |
| 2500 Tinted | | 55 | .291 | | 24 | 12.25 | | 36.25 | 45 |
| 2800 5/8" thick, clear, plain | | 45 | .356 | | 26 | 14.95 | | 40.95 | 51 |
| 2900 Tempered, clear | | 45 | .356 | | 29.50 | 14.95 | | 44.45 | 55 |
| 3200 3/4" thick, clear, plain | | 35 | .457 | | 33.50 | 19.20 | | 52.70 | 65.50 |
| 3300 Tempered, clear | | 35 | .457 | | 39 | 19.20 | | 58.20 | 72 |
| 3600 1" thick, clear, plain | | 30 | .533 | | 55.50 | 22.50 | | 78 | 94.50 |
| 8900 For low emissivity coating for 3/16" & 1/4" only, add to above | | | | | 16% | | | | |

## 08 81 13 – Decorative Glass Glazing

| 08 81 13.10 Beveled Glass | Crew | Daily Output | Labor-Hours | Unit | Material | 2011 Bare Costs Labor | Equipment | Total | Total Incl O&P |
|---|---|---|---|---|---|---|---|---|---|
| 0010 **BEVELED GLASS**, with design patterns | | | | | | | | | |
| 0020 Minimum | 2 Glaz | 150 | .107 | S.F. | 58 | 4.49 | | 62.49 | 71 |

# 08 81 Glass Glazing

## 08 81 13 – Decorative Glass Glazing

### 08 81 13.10 Beveled Glass

| | | Crew | Daily Output | Labor-Hours | Unit | Material | 2011 Bare Costs Labor | 2011 Bare Costs Equipment | Total | Total Incl O&P |
|---|---|---|---|---|---|---|---|---|---|---|
| 0050 | Average | 2 Glaz | 125 | .128 | S.F. | 129 | 5.40 | | 134.40 | 150 |
| 0100 | Maximum | ↓ | 100 | .160 | ↓ | 229 | 6.75 | | 235.75 | 261 |
| 0102 | Maximum, per LF for 30" cabinet | 1 Glaz | 40 | .200 | L.F. | 570 | 8.40 | | 578.40 | 645 |

### 08 81 13.20 Faceted Glass

| | | Crew | Daily Output | Labor-Hours | Unit | Material | Labor | Equipment | Total | Total Incl O&P |
|---|---|---|---|---|---|---|---|---|---|---|
| 0010 | **FACETED GLASS**, Color tinted 3/4" thick | | | | | | | | | |
| 0020 | Minimum | 2 Glaz | 95 | .168 | S.F. | 46.50 | 7.10 | | 53.60 | 61.50 |
| 0100 | Maximum | " | 75 | .213 | " | 81.50 | 8.95 | | 90.45 | 103 |

### 08 81 13.30 Sandblasted Glass

| | | Crew | Daily Output | Labor-Hours | Unit | Material | Labor | Equipment | Total | Total Incl O&P |
|---|---|---|---|---|---|---|---|---|---|---|
| 0010 | **SANDBLASTED GLASS**, float glass | | | | | | | | | |
| 0020 | 1/8" thick | 2 Glaz | 160 | .100 | S.F. | 10.40 | 4.21 | | 14.61 | 17.75 |
| 0100 | 3/16" thick | | 130 | .123 | | 11.50 | 5.20 | | 16.70 | 20.50 |
| 0500 | 1/4" thick | | 120 | .133 | | 12 | 5.60 | | 17.60 | 21.50 |
| 0600 | 3/8" thick | ↓ | 75 | .213 | ↓ | 12.90 | 8.95 | | 21.85 | 27.50 |

## 08 81 20 – Vision Panels

### 08 81 20.10 Full Vision

| | | Crew | Daily Output | Labor-Hours | Unit | Material | Labor | Equipment | Total | Total Incl O&P |
|---|---|---|---|---|---|---|---|---|---|---|
| 0010 | **FULL VISION**, window system with 3/4" glass mullions | | | | | | | | | |
| 0020 | Up to 10' high | H-2 | 130 | .185 | S.F. | 59.50 | 7.30 | | 66.80 | 76.50 |
| 0100 | 10' to 20' high, minimum | | 110 | .218 | | 63 | 8.60 | | 71.60 | 82 |
| 0150 | Average | | 100 | .240 | | 68 | 9.50 | | 77.50 | 89.50 |
| 0200 | Maximum | ↓ | 80 | .300 | ↓ | 76.50 | 11.85 | | 88.35 | 102 |

## 08 81 25 – Glazing Variables

### 08 81 25.10 Applications of Glazing

| | | | Crew | Daily Output | Labor-Hours | Unit | Material | Labor | Equipment | Total | Total Incl O&P |
|---|---|---|---|---|---|---|---|---|---|---|---|
| 0010 | **APPLICATIONS OF GLAZING** | R088110-10 | | | | | | | | | |
| 0500 | For high rise glazing, exterior, add per S.F. per story | | | | | S.F. | | | | | .45 |
| 0600 | For glass replacement, add | | | | | " | | 100% | | | |
| 0700 | For gasket settings, add | | | | | L.F. | 5.55 | | | 5.55 | 6.10 |
| 0900 | For sloped glazing, add | | | | | S.F. | | 25% | | | |
| 2000 | Fabrication, polished edges, 1/4" thick | | | | | Inch | .50 | | | .50 | .55 |
| 2100 | 1/2" thick | | | | | | 1.23 | | | 1.23 | 1.35 |
| 2500 | Mitered edges, 1/4" thick | | | | | | 1.23 | | | 1.23 | 1.35 |
| 2600 | 1/2" thick | | | | | ↓ | 2 | | | 2 | 2.20 |

## 08 81 30 – Insulating Glass

### 08 81 30.10 Reduce Heat Transfer Glass

| | | | Crew | Daily Output | Labor-Hours | Unit | Material | Labor | Equipment | Total | Total Incl O&P |
|---|---|---|---|---|---|---|---|---|---|---|---|
| 0010 | **REDUCE HEAT TRANSFER GLASS** | R088110-10 | | | | | | | | | |
| 0015 | 2 lites 1/8" float, 1/2" thk under 15 S.F. | | | | | | | | | | |
| 0020 | Clear | G | 2 Glaz | 95 | .168 | S.F. | 9.40 | 7.10 | | 16.50 | 21 |
| 0100 | Tinted | G | | 95 | .168 | | 13.10 | 7.10 | | 20.20 | 25 |
| 0200 | 2 lites 3/16" float, for 5/8" thk unit, 15 to 30 S.F., clear | G | | 90 | .178 | | 12.95 | 7.50 | | 20.45 | 25.50 |
| 0300 | Tinted | G | | 90 | .178 | | 13.30 | 7.50 | | 20.80 | 26 |
| 0400 | 1" thk, dbl. glazed, 1/4" float, 30-70 S.F., clear | G | | 75 | .213 | | 15.65 | 8.95 | | 24.60 | 30.50 |
| 0500 | Tinted | G | | 75 | .213 | | 22 | 8.95 | | 30.95 | 38 |
| 0600 | 1" thick double glazed, 1/4" float, 1/4" wire | | | 75 | .213 | | 22 | 8.95 | | 30.95 | 37.50 |
| 0700 | 1/4" float, 1/4" tempered | | | 75 | .213 | | 18.55 | 8.95 | | 27.50 | 34 |
| 0800 | 1/4" wire, 1/4" tempered | | | 75 | .213 | | 27.50 | 8.95 | | 36.45 | 44 |
| 0900 | Both lites, 1/4" wire | | | 75 | .213 | | 35 | 8.95 | | 43.95 | 52 |
| 2000 | Both lites, light & heat reflective | G | | 85 | .188 | | 29.50 | 7.90 | | 37.40 | 44.50 |
| 2500 | Heat reflective, film inside, 1" thick unit, clear | G | | 85 | .188 | | 26 | 7.90 | | 33.90 | 40.50 |
| 2600 | Tinted | G | | 85 | .188 | | 27 | 7.90 | | 34.90 | 41.50 |
| 3000 | Film on weatherside, clear, 1/2" thick unit | G | | 95 | .168 | | 18.60 | 7.10 | | 25.70 | 31 |
| 3100 | 5/8" thick unit | G | ↓ | 90 | .178 | ↓ | 18.60 | 7.50 | | 26.10 | 32 |

# 08 81 Glass Glazing

## 08 81 30 – Insulating Glass

| 08 81 30.10 Reduce Heat Transfer Glass | | Crew | Daily Output | Labor-Hours | Unit | Material | 2011 Bare Costs Labor | Equipment | Total | Total Incl O&P |
|---|---|---|---|---|---|---|---|---|---|---|
| 3200 | 1" thick unit [G] | 2 Glaz | 85 | .188 | S.F. | 25.50 | 7.90 | | 33.40 | 40 |

## 08 81 35 – Translucent Glass

### 08 81 35.10 Obscure Glass

| | | | | | | | | | | |
|---|---|---|---|---|---|---|---|---|---|---|
| 0010 | **OBSCURE GLASS** | | | | | | | | | |
| 0020 | 1/8" thick, Minimum | 2 Glaz | 140 | .114 | S.F. | 11 | 4.81 | | 15.81 | 19.30 |
| 0100 | Maximum | | 125 | .128 | | 13 | 5.40 | | 18.40 | 22.50 |
| 0300 | 7/32" thick, minimum | | 120 | .133 | | 12 | 5.60 | | 17.60 | 21.50 |
| 0400 | Maximum | ↓ | 105 | .152 | ↓ | 15.10 | 6.40 | | 21.50 | 26 |

### 08 81 35.20 Patterned Glass

| | | | | | | | | | | |
|---|---|---|---|---|---|---|---|---|---|---|
| 0010 | **PATTERNED GLASS**, colored | | | | | | | | | |
| 0020 | 1/8" thick, Minimum | 2 Glaz | 140 | .114 | S.F. | 9 | 4.81 | | 13.81 | 17.10 |
| 0100 | Maximum | | 125 | .128 | | 15 | 5.40 | | 20.40 | 24.50 |
| 0300 | 7/32" thick, minimum | | 120 | .133 | | 11.05 | 5.60 | | 16.65 | 20.50 |
| 0400 | Maximum | ↓ | 105 | .152 | ↓ | 16 | 6.40 | | 22.40 | 27 |

## 08 81 45 – Sheet Glass

### 08 81 45.10 Window Glass, Sheet

| | | | | | | | | | | |
|---|---|---|---|---|---|---|---|---|---|---|
| 0010 | **WINDOW GLASS, SHEET** gray | | | | | | | | | |
| 0020 | 1/8" thick | 2 Glaz | 160 | .100 | S.F. | 5.40 | 4.21 | | 9.61 | 12.25 |
| 0200 | 1/4" thick | " | 130 | .123 | " | 6.60 | 5.20 | | 11.80 | 15 |

## 08 81 50 – Spandrel Glass

### 08 81 50.10 Glass for Non Vision Areas

| | | | | | | | | | | |
|---|---|---|---|---|---|---|---|---|---|---|
| 0010 | **GLASS FOR NON VISION AREAS**, 1/4" thick standard colors | | | | | | | | | |
| 0020 | Up to 1000 S.F. | 2 Glaz | 110 | .145 | S.F. | 15.55 | 6.10 | | 21.65 | 26.50 |
| 0200 | 1,000 to 2,000 S.F. | " | 120 | .133 | " | 14.45 | 5.60 | | 20.05 | 24.50 |
| 0300 | For custom colors, add | | | | Total | 10% | | | | |
| 0500 | For 3/8" thick, add | | | | S.F. | 10.70 | | | 10.70 | 11.75 |
| 1000 | For double coated, 1/4" thick, add | | | | | 4 | | | 4 | 4.40 |
| 1200 | For insulation on panels, add | | | | | 6.60 | | | 6.60 | 7.25 |
| 2000 | Panels, insulated, with aluminum backed fiberglass, 1" thick | 2 Glaz | 120 | .133 | | 15.90 | 5.60 | | 21.50 | 26 |
| 2100 | 2" thick | " | 120 | .133 | | 18.90 | 5.60 | | 24.50 | 29.50 |
| 2500 | With galvanized steel backing, add | | | | ↓ | 5.50 | | | 5.50 | 6.05 |

## 08 81 55 – Window Glass

### 08 81 55.10 Sheet Glass

| | | | | | | | | | | |
|---|---|---|---|---|---|---|---|---|---|---|
| 0010 | **SHEET GLASS**(window), clear float, stops, putty bed | | | | | | | | | |
| 0015 | 1/8" thick, clear float | 2 Glaz | 480 | .033 | S.F. | 3.40 | 1.40 | | 4.80 | 5.85 |
| 0500 | 3/16" thick, clear | | 480 | .033 | | 5.40 | 1.40 | | 6.80 | 8.05 |
| 0600 | Tinted | | 480 | .033 | | 7.15 | 1.40 | | 8.55 | 9.95 |
| 0700 | Tempered | ↓ | 480 | .033 | ↓ | 8.65 | 1.40 | | 10.05 | 11.60 |

## 08 81 65 – Wire Glass

### 08 81 65.10 Glass Reinforced With Wire

| | | | | | | | | | | |
|---|---|---|---|---|---|---|---|---|---|---|
| 0010 | **GLASS REINFORCED WITH WIRE** | | | | | | | | | |
| 0012 | 1/4" thick rough obscure | 2 Glaz | 135 | .119 | S.F. | 17.75 | 4.98 | | 22.73 | 27 |
| 1000 | Polished wire, 1/4" thick, diamond, clear | | 135 | .119 | | 20.50 | 4.98 | | 25.48 | 30 |
| 1500 | Pinstripe, obscure | ↓ | 135 | .119 | ↓ | 22 | 4.98 | | 26.98 | 31.50 |

# 08 83 Mirrors

## 08 83 13 – Mirrored Glass Glazing

| 08 83 13.10 Mirrors | | Crew | Daily Output | Labor-Hours | Unit | Material | 2011 Bare Costs Labor | Equipment | Total | Total Incl O&P |
|---|---|---|---|---|---|---|---|---|---|---|
| 0010 | **MIRRORS**, No frames, wall type, 1/4" plate glass, polished edge | | | | | | | | | |
| 0100 | Up to 5 S.F. | 2 Glaz | 125 | .128 | S.F. | 9.35 | 5.40 | | 14.75 | 18.40 |
| 0200 | Over 5 S.F. | | 160 | .100 | | 9.05 | 4.21 | | 13.26 | 16.25 |
| 0500 | Door type, 1/4" plate glass, up to 12 S.F. | | 160 | .100 | | 8.20 | 4.21 | | 12.41 | 15.30 |
| 1000 | Float glass, up to 10 S.F., 1/8" thick | | 160 | .100 | | 5.45 | 4.21 | | 9.66 | 12.25 |
| 1100 | 3/16" thick | | 150 | .107 | | 6.65 | 4.49 | | 11.14 | 14.05 |
| 1500 | 12" x 12" wall tiles, square edge, clear | | 195 | .082 | | 1.90 | 3.45 | | 5.35 | 7.30 |
| 1600 | Veined | | 195 | .082 | | 4.86 | 3.45 | | 8.31 | 10.55 |
| 2000 | 1/4" thick, stock sizes, one way transparent | | 125 | .128 | | 18.10 | 5.40 | | 23.50 | 28 |
| 2010 | Bathroom, unframed, laminated | | 160 | .100 | | 12.60 | 4.21 | | 16.81 | 20 |

## 08 83 13.15 Reflective Glass

| 0010 | **REFLECTIVE GLASS** | | | | | | | | | | |
|---|---|---|---|---|---|---|---|---|---|---|---|
| 0100 | 1/4" float with fused metallic oxide fixed | G | 2 Glaz | 115 | .139 | S.F. | 16 | 5.85 | | 21.85 | 26.50 |
| 0500 | 1/4" float glass with reflective applied coating | G | " | 115 | .139 | " | 13 | 5.85 | | 18.85 | 23 |

# 08 84 Plastic Glazing

## 08 84 10 – Plexiglass Glazing

### 08 84 10.10 Plexiglass Acrylic

| 0010 | **PLEXIGLASS ACRYLIC**, clear, masked, | | | | | | | | | |
|---|---|---|---|---|---|---|---|---|---|---|
| 0020 | 1/8" thick, cut sheets | 2 Glaz | 170 | .094 | S.F. | 5.30 | 3.96 | | 9.26 | 11.80 |
| 0200 | Full sheets | | 195 | .082 | | 2.75 | 3.45 | | 6.20 | 8.25 |
| 0500 | 1/4" thick, cut sheets | | 165 | .097 | | 9.20 | 4.08 | | 13.28 | 16.20 |
| 0600 | Full sheets | | 185 | .086 | | 5 | 3.64 | | 8.64 | 10.95 |
| 0900 | 3/8" thick, cut sheets | | 155 | .103 | | 16.85 | 4.34 | | 21.19 | 25 |
| 1000 | Full sheets | | 180 | .089 | | 9.10 | 3.74 | | 12.84 | 15.60 |
| 1300 | 1/2" thick, cut sheets | | 135 | .119 | | 19.40 | 4.98 | | 24.38 | 29 |
| 1400 | Full sheets | | 150 | .107 | | 18.80 | 4.49 | | 23.29 | 27.50 |
| 1700 | 3/4" thick, cut sheets | | 115 | .139 | | 69 | 5.85 | | 74.85 | 85 |
| 1800 | Full sheets | | 130 | .123 | | 40 | 5.20 | | 45.20 | 52 |
| 2100 | 1" thick, cut sheets | | 105 | .152 | | 77.50 | 6.40 | | 83.90 | 95 |
| 2200 | Full sheets | | 125 | .128 | | 48 | 5.40 | | 53.40 | 61 |
| 3000 | Colored, 1/8" thick, cut sheets | | 170 | .094 | | 16 | 3.96 | | 19.96 | 23.50 |
| 3200 | Full sheets | | 195 | .082 | | 10.30 | 3.45 | | 13.75 | 16.55 |
| 3500 | 1/4" thick, cut sheets | | 165 | .097 | | 18 | 4.08 | | 22.08 | 26 |
| 3600 | Full sheets | | 185 | .086 | | 12.20 | 3.64 | | 15.84 | 18.85 |
| 4000 | Mirrors, untinted, cut sheets, 1/8" thick | | 185 | .086 | | 7.50 | 3.64 | | 11.14 | 13.70 |
| 4200 | 1/4" thick | | 180 | .089 | | 12.10 | 3.74 | | 15.84 | 18.90 |

## 08 84 20 – Polycarbonate

### 08 84 20.10 Thermoplastic

| 0010 | **THERMOPLASTIC**, clear, masked, cut sheets | | | | | | | | | |
|---|---|---|---|---|---|---|---|---|---|---|
| 0020 | 1/8" thick | 2 Glaz | 170 | .094 | S.F. | 9.70 | 3.96 | | 13.66 | 16.60 |
| 0500 | 3/16" thick | | 165 | .097 | | 11.10 | 4.08 | | 15.18 | 18.30 |
| 1000 | 1/4" thick | | 155 | .103 | | 12.50 | 4.34 | | 16.84 | 20.50 |
| 1500 | 3/8" thick | | 150 | .107 | | 22 | 4.49 | | 26.49 | 31.50 |

## 08 84 30 – Vinyl Glass

### 08 84 30.10 Vinyl Sheets

| 0010 | **VINYL SHEETS**, Steel mesh reinforced, stock sizes | | | | | | | | | |
|---|---|---|---|---|---|---|---|---|---|---|
| 0020 | .090" thick | 2 Glaz | 170 | .094 | S.F. | 13.10 | 3.96 | | 17.06 | 20.50 |
| 0500 | .120" thick | | 170 | .094 | | 16 | 3.96 | | 19.96 | 23.50 |
| 1000 | .250" thick | | 155 | .103 | | 21.50 | 4.34 | | 25.84 | 30 |

# 08 84 Plastic Glazing

## 08 84 30 – Vinyl Glass

| 08 84 30.10 Vinyl Sheets | Crew | Daily Output | Labor-Hours | Unit | Material | 2011 Bare Costs Labor | Equipment | Total | Total Incl O&P |
|---|---|---|---|---|---|---|---|---|---|
| 1500 | For non-standard sizes, add | | | | S.F. | 18% | | | | |

# 08 87 Glazing Surface Films

## 08 87 13 – Solar Control Films

### 08 87 13.10 Solar Films On Glass

| | | Crew | Daily Output | Labor-Hours | Unit | Material | 2011 Bare Costs Labor | Equipment | Total | Total Incl O&P |
|---|---|---|---|---|---|---|---|---|---|---|
| 0010 | **SOLAR FILMS ON GLASS** (glass not included) | | | | | | | | | |
| 2000 | Minimum | G 2 Glaz | 180 | .089 | S.F. | 6.50 | 3.74 | | 10.24 | 12.75 |
| 2050 | Maximum | G " | 225 | .071 | " | 14.50 | 2.99 | | 17.49 | 20.50 |

## 08 87 23 – Safety and Security Films

### 08 87 23.13 Safety Films

| | | Crew | Daily Output | Labor-Hours | Unit | Material | 2011 Bare Costs Labor | Equipment | Total | Total Incl O&P |
|---|---|---|---|---|---|---|---|---|---|---|
| 0010 | **SAFETY FILMS** | | | | | | | | | |
| 2010 | Window protection film, controls blast damage | G 2 Glaz | 80 | .200 | S.F. | 6.50 | 8.40 | | 14.90 | 19.80 |

### 08 87 23.16 Security Films

| | | Crew | Daily Output | Labor-Hours | Unit | Material | 2011 Bare Costs Labor | Equipment | Total | Total Incl O&P |
|---|---|---|---|---|---|---|---|---|---|---|
| 0010 | **SECURITY FILMS**, clear, 32000 psi tensile strength, adhered to glass | | | | | | | | | |
| 0100 | .002" thick, daylight installation | H-2 | 950 | .025 | S.F. | 2.20 | 1 | | 3.20 | 3.93 |
| 0150 | .004" thick, daylight installation | R088110-10 | 800 | .030 | | 2.30 | 1.18 | | 3.48 | 4.32 |
| 0200 | .006" thick, daylight installation | | 700 | .034 | | 2.45 | 1.35 | | 3.80 | 4.75 |
| 0210 | Install for anchorage | | 600 | .040 | | 2.72 | 1.58 | | 4.30 | 5.40 |
| 0400 | .007" thick, daylight installation | | 600 | .040 | | 2.65 | 1.58 | | 4.23 | 5.30 |
| 0410 | Install for anchorage | | 500 | .048 | | 2.94 | 1.90 | | 4.84 | 6.10 |
| 0500 | .008" thick, daylight installation | | 500 | .048 | | 2.85 | 1.90 | | 4.75 | 6 |
| 0510 | Install for anchorage | | 500 | .048 | | 3.17 | 1.90 | | 5.07 | 6.35 |
| 0600 | .015" thick, daylight installation | | 400 | .060 | | 4.75 | 2.37 | | 7.12 | 8.85 |
| 0610 | Install for anchorage | | 400 | .060 | | 3.17 | 2.37 | | 5.54 | 7.05 |
| 0900 | Security film anchorage, mechanical attachment and cover plate | H-3 | 370 | .043 | L.F. | 9.05 | 1.62 | | 10.67 | 12.40 |
| 0950 | Security film anchorage, wet glaze structural caulking | 1 Glaz | 225 | .036 | " | .83 | 1.50 | | 2.33 | 3.16 |
| 1000 | Adhered security film removal | 1 Clab | 275 | .029 | S.F. | | 1 | | 1 | 1.53 |

# 08 88 Special Function Glazing

## 08 88 40 – Acoustical Glass Units

### 08 88 40.10 Sound Reduction Units

| | | Crew | Daily Output | Labor-Hours | Unit | Material | 2011 Bare Costs Labor | Equipment | Total | Total Incl O&P |
|---|---|---|---|---|---|---|---|---|---|---|
| 0010 | **SOUND REDUCTION UNITS**, 1 lite at 3/8", 1 lite at 3/16" | | | | | | | | | |
| 0020 | For 1" thick | 2 Glaz | 100 | .160 | S.F. | 32 | 6.75 | | 38.75 | 45.50 |
| 0100 | For 4" thick | " | 80 | .200 | " | 55 | 8.40 | | 63.40 | 73 |

## 08 88 56 – Ballistics-Resistant Glazing

### 08 88 56.10 Laminated Glass

| | | Crew | Daily Output | Labor-Hours | Unit | Material | 2011 Bare Costs Labor | Equipment | Total | Total Incl O&P |
|---|---|---|---|---|---|---|---|---|---|---|
| 0010 | **LAMINATED GLASS** | | | | | | | | | |
| 0020 | Clear float .03" vinyl 1/4" | 2 Glaz | 90 | .178 | S.F. | 11.55 | 7.50 | | 19.05 | 24 |
| 0100 | 3/8" thick | | 78 | .205 | | 21 | 8.65 | | 29.65 | 36 |
| 0200 | .06" vinyl, 1/2" thick | | 65 | .246 | | 24 | 10.35 | | 34.35 | 42 |
| 1000 | 5/8" thick | | 90 | .178 | | 28 | 7.50 | | 35.50 | 42.50 |
| 2000 | Bullet-resisting, 1-3/16" thick, to 15 S.F. | | 16 | 1 | | 79.50 | 42 | | 121.50 | 151 |
| 2100 | Over 15 S.F. | | 16 | 1 | | 96 | 42 | | 138 | 168 |
| 2500 | 2-1/4" thick, to 15 S.F. | | 12 | 1.333 | | 93 | 56 | | 149 | 186 |
| 2600 | Over 15 S.F. | | 12 | 1.333 | | 83 | 56 | | 139 | 176 |
| 2700 | Level 2 (.357 magnum) | | 12 | 1.333 | | 65 | 56 | | 121 | 156 |
| 2750 | Level 3A (.44 magnum) | | 12 | 1.333 | | 70 | 56 | | 126 | 161 |
| 2800 | Level 4 (AK-47) | | 12 | 1.333 | | 91.50 | 56 | | 147.50 | 185 |

# 08 88 Special Function Glazing

## 08 88 56 – Ballistics-Resistant Glazing

| 08 88 56.10 Laminated Glass | | Crew | Daily Output | Labor-Hours | Unit | Material | 2011 Bare Costs Labor | Equipment | Total | Total Incl O&P |
|---|---|---|---|---|---|---|---|---|---|---|
| 2850 | Level 5 (M-16) | 2 Glaz | 12 | 1.333 | S.F. | 102 | 56 | | 158 | 196 |
| 2900 | Level 3 (7.62 Armor Piercing) | ↓ | 12 | 1.333 | ↓ | 122 | 56 | | 178 | 218 |

# 08 91 Louvers

## 08 91 19 – Fixed Louvers

### 08 91 19.10 Aluminum Louvers

| | | Crew | Daily Output | Labor-Hours | Unit | Material | 2011 Bare Costs Labor | Equipment | Total | Total Incl O&P |
|---|---|---|---|---|---|---|---|---|---|---|
| 0010 | **ALUMINUM LOUVERS** | | | | | | | | | |
| 0020 | Aluminum with screen, residential, 8" x 8" | 1 Carp | 38 | .211 | Ea. | 10.55 | 9.05 | | 19.60 | 25.50 |
| 0100 | 12" x 12" | | 38 | .211 | | 15 | 9.05 | | 24.05 | 30.50 |
| 0200 | 12" x 18" | | 35 | .229 | | 18.85 | 9.85 | | 28.70 | 35.50 |
| 0250 | 14" x 24" | | 30 | .267 | | 26 | 11.50 | | 37.50 | 46 |
| 0300 | 18" x 24" | | 27 | .296 | | 29 | 12.75 | | 41.75 | 51.50 |
| 0500 | 24" x 30" | | 24 | .333 | | 32 | 14.35 | | 46.35 | 57 |
| 0700 | Triangle, adjustable, small | | 20 | .400 | | 32 | 17.20 | | 49.20 | 61.50 |
| 0800 | Large | ↓ | 15 | .533 | ↓ | 41 | 23 | | 64 | 80 |
| 1200 | Extruded aluminum, see Section 23 37 15.40 | | | | | | | | | |
| 2100 | Midget, aluminum, 3/4" deep, 1" diameter | 1 Carp | 85 | .094 | Ea. | .69 | 4.05 | | 4.74 | 6.95 |
| 2150 | 3" diameter | | 60 | .133 | | 1.87 | 5.75 | | 7.62 | 10.85 |
| 2200 | 4" diameter | | 50 | .160 | | 3.64 | 6.90 | | 10.54 | 14.55 |
| 2250 | 6" diameter | ↓ | 30 | .267 | ↓ | 3.74 | 11.50 | | 15.24 | 21.50 |

## 08 91 26 – Door Louvers

### 08 91 26.10 Steel Louvers, 18 Gauge, Fixed Blade

| | | Crew | Daily Output | Labor-Hours | Unit | Material | 2011 Bare Costs Labor | Equipment | Total | Total Incl O&P |
|---|---|---|---|---|---|---|---|---|---|---|
| 0010 | **STEEL LOUVERS, 18 GAUGE, FIXED BLADE** | | | | | | | | | |
| 0050 | 12" x 12", with enamel or powder coat | 1 Carp | 20 | .400 | Ea. | 67.50 | 17.20 | | 84.70 | 101 |
| 0055 | 18" x 12" | | 20 | .400 | | 76 | 17.20 | | 93.20 | 110 |
| 0060 | 18" x 18" | | 20 | .400 | | 83 | 17.20 | | 100.20 | 118 |
| 0065 | 24" x 12" | | 20 | .400 | | 91.50 | 17.20 | | 108.70 | 128 |
| 0070 | 24" x 18" | | 20 | .400 | | 126 | 17.20 | | 143.20 | 166 |
| 0075 | 24" x 24" | | 20 | .400 | | 130 | 17.20 | | 147.20 | 170 |
| 0100 | 12" x 12", galvanized | | 20 | .400 | | 58 | 17.20 | | 75.20 | 90.50 |
| 0105 | 18" x 12" | | 20 | .400 | | 65.50 | 17.20 | | 82.70 | 98.50 |
| 0115 | 24" x 12" | | 20 | .400 | | 82 | 17.20 | | 99.20 | 117 |
| 0125 | 24" x 24" | ↓ | 20 | .400 | ↓ | 119 | 17.20 | | 136.20 | 157 |

# 08 95 Vents

## 08 95 13 – Soffit Vents

### 08 95 13.10 Wall Louvers

| | | Crew | Daily Output | Labor-Hours | Unit | Material | 2011 Bare Costs Labor | Equipment | Total | Total Incl O&P |
|---|---|---|---|---|---|---|---|---|---|---|
| 0010 | **WALL LOUVERS** | | | | | | | | | |
| 2330 | Soffit vent, continuous, 3" wide, aluminum, mill finish | 1 Carp | 200 | .040 | L.F. | .48 | 1.72 | | 2.20 | 3.17 |
| 2340 | Baked enamel finish | | 200 | .040 | " | .52 | 1.72 | | 2.24 | 3.21 |
| 2400 | Under eaves vent, aluminum, mill finish, 16" x 4" | | 48 | .167 | Ea. | 1.37 | 7.20 | | 8.57 | 12.50 |
| 2500 | 16" x 8" | ↓ | 48 | .167 | " | 1.93 | 7.20 | | 9.13 | 13.10 |

## 08 95 16 – Wall Vents

### 08 95 16.10 Louvers

| | | Crew | Daily Output | Labor-Hours | Unit | Material | 2011 Bare Costs Labor | Equipment | Total | Total Incl O&P |
|---|---|---|---|---|---|---|---|---|---|---|
| 0010 | **LOUVERS** | | | | | | | | | |
| 0020 | Redwood, 2'-0" diameter, full circle | 1 Carp | 16 | .500 | Ea. | 169 | 21.50 | | 190.50 | 219 |
| 0100 | Half circle | | 16 | .500 | | 189 | 21.50 | | 210.50 | 241 |
| 0200 | Octagonal | ↓ | 16 | .500 | ↓ | 129 | 21.50 | | 150.50 | 175 |

# 08 95 Vents

## 08 95 16 — Wall Vents

| 08 95 16.10 Louvers | | Crew | Daily Output | Labor-Hours | Unit | Material | 2011 Bare Costs Labor | Equipment | Total | Total Incl O&P |
|---|---|---|---|---|---|---|---|---|---|---|
| 0300 | Triangular, 5/12 pitch, 5'-0" at base | 1 Carp | 16 | .500 | Ea. | 189 | 21.50 | | 210.50 | 241 |
| 7000 | Vinyl gable vent, 8" x 8" | | 38 | .211 | | 11.55 | 9.05 | | 20.60 | 26.50 |
| 7020 | 12" x 12" | | 38 | .211 | | 22.50 | 9.05 | | 31.55 | 39 |
| 7080 | 12" x 18" | | 35 | .229 | | 29.50 | 9.85 | | 39.35 | 47.50 |
| 7200 | 18" x 24" | | 30 | .267 | | 36.50 | 11.50 | | 48 | 57.50 |

# Estimating Tips

## General

- Room Finish Schedule: A complete set of plans should contain a room finish schedule. If one is not available, it would be well worth the time and effort to obtain one.

## 09 20 00 Plaster and Gypsum Board

- Lath is estimated by the square yard plus a 5% allowance for waste. Furring, channels, and accessories are measured by the linear foot. An extra foot should be allowed for each accessory miter or stop.

- Plaster is also estimated by the square yard. Deductions for openings vary by preference, from zero deduction to 50% of all openings over 2 feet in width. The estimator should allow one extra square foot for each linear foot of horizontal interior or exterior angle located below the ceiling level. Also, double the areas of small radius work.

- Drywall accessories, studs, track, and acoustical caulking are all measured by the linear foot. Drywall taping is figured by the square foot. Gypsum wallboard is estimated by the square foot. No material deductions should be made for door or window openings under 32 S.F.

## 09 60 00 Flooring

- Tile and terrazzo areas are taken off on a square foot basis. Trim and base materials are measured by the linear foot. Accent tiles are listed per each. Two basic methods of installation are used. Mud set is approximately 30% more expensive than thin set. In terrazzo work, be sure to include the linear footage of embedded decorative strips, grounds, machine rubbing, and power cleanup.

- Wood flooring is available in strip, parquet, or block configuration. The latter two types are set in adhesives with quantities estimated by the square foot. The laying pattern will influence labor costs and material waste. In addition to the material and labor for laying wood floors, the estimator must make allowances for sanding and finishing these areas unless the flooring is prefinished.

- Sheet flooring is measured by the square yard. Roll widths vary, so consideration should be given to use the most economical width, as waste must be figured into the total quantity. Consider also the installation methods available, direct glue down or stretched.

## 09 70 00 Wall Finishes

- Wall coverings are estimated by the square foot. The area to be covered is measured, length by height of wall above baseboards, to calculate the square footage of each wall. This figure is divided by the number of square feet in the single roll which is being used. Deduct, in full, the areas of openings such as doors and windows. Where a pattern match is required allow 25%–30% waste.

## 09 80 00 Acoustic Treatment

- Acoustical systems fall into several categories. The takeoff of these materials should be by the square foot of area with a 5% allowance for waste. Do not forget about scaffolding, if applicable, when estimating these systems.

## 09 90 00 Painting and Coating

- A major portion of the work in painting involves surface preparation. Be sure to include cleaning, sanding, filling, and masking costs in the estimate.

- Protection of adjacent surfaces is not included in painting costs. When considering the method of paint application, an important factor is the amount of protection and masking required. These must be estimated separately and may be the determining factor in choosing the method of application.

# Reference Numbers

Reference numbers are shown in shaded boxes at the beginning of some major classifications. These numbers refer to related items in the Reference Section. The reference information may be an estimating procedure, an alternate pricing method, or technical information.

*Note:* Not all subdivisions listed here necessarily appear in this publication.

# 09 01 Maintenance of Finishes

## 09 01 60 – Maintenance of Flooring

### 09 01 60.10 Carpet Maintenance

| | | Crew | Daily Output | Labor-Hours | Unit | Material | 2011 Bare Costs Labor | 2011 Bare Costs Equipment | Total | Total Incl O&P |
|---|---|---|---|---|---|---|---|---|---|---|
| 0010 | **CARPET MAINTENANCE** | | | | | | | | | |
| 0020 | Steam clean, per cleaning, minimum | 1 Clab | 3000 | .003 | S.F. | .07 | .09 | | .16 | .22 |
| 0500 | Maximum | " | 2000 | .004 | " | .12 | .14 | | .26 | .34 |

## 09 01 70 – Maintenance of Wall Finishes

### 09 01 70.10 Gypsum Wallboard Repairs

| | | Crew | Daily Output | Labor-Hours | Unit | Material | 2011 Bare Costs Labor | 2011 Bare Costs Equipment | Total | Total Incl O&P |
|---|---|---|---|---|---|---|---|---|---|---|
| 0010 | **GYPSUM WALLBOARD REPAIRS** | | | | | | | | | |
| 0100 | Fill and sand, pin/nail holes | 1 Carp | 960 | .008 | Ea. | | .36 | | .36 | .55 |
| 0110 | Screw head pops | | 480 | .017 | | | .72 | | .72 | 1.10 |
| 0120 | Dents, up to 2" square | | 48 | .167 | | .01 | 7.20 | | 7.21 | 11 |
| 0130 | 2" to 4" square | | 24 | .333 | | .03 | 14.35 | | 14.38 | 22 |
| 0140 | Cut square, patch, sand and finish, holes, up to 2" square | | 12 | .667 | | .04 | 28.50 | | 28.54 | 44 |
| 0150 | 2" to 4" square | | 11 | .727 | | .09 | 31.50 | | 31.59 | 48 |
| 0160 | 4" to 8" square | | 10 | .800 | | .22 | 34.50 | | 34.72 | 53 |
| 0170 | 8" to 12" square | | 8 | 1 | | .43 | 43 | | 43.43 | 66.50 |
| 0180 | 12" to 32" square | | 6 | 1.333 | | 1.39 | 57.50 | | 58.89 | 89.50 |
| 0210 | 16" by 48" | | 5 | 1.600 | | 2.33 | 69 | | 71.33 | 109 |
| 0220 | 32" by 48" | | 4 | 2 | | 3.30 | 86 | | 89.30 | 136 |
| 0230 | 48" square | | 3.50 | 2.286 | | 4.62 | 98.50 | | 103.12 | 156 |
| 0240 | 60" square | | 3.20 | 2.500 | | 8.30 | 108 | | 116.30 | 174 |
| 0500 | Skim coat surface with joint compound | | 1600 | .005 | S.F. | .03 | .22 | | .25 | .36 |
| 0510 | Prepare, retape and refinish joints | | 60 | .133 | L.F. | .58 | 5.75 | | 6.33 | 9.45 |

# 09 05 Common Work Results for Finishes

## 09 05 05 – Selective Finishes Demolition

### 09 05 05.10 Selective Demolition, Ceilings

| | | Crew | Daily Output | Labor-Hours | Unit | Material | 2011 Bare Costs Labor | 2011 Bare Costs Equipment | Total | Total Incl O&P |
|---|---|---|---|---|---|---|---|---|---|---|
| 0010 | **SELECTIVE DEMOLITION, CEILINGS** R024119-10 | | | | | | | | | |
| 0200 | Ceiling, drywall, furred and nailed or screwed | 2 Clab | 800 | .020 | S.F. | | .69 | | .69 | 1.05 |
| 0220 | On metal frame | | 760 | .021 | | | .72 | | .72 | 1.11 |
| 0240 | On suspension system, including system | | 720 | .022 | | | .76 | | .76 | 1.17 |
| 1000 | Plaster, lime and horse hair, on wood lath, incl. lath | | 700 | .023 | | | .79 | | .79 | 1.20 |
| 1020 | On metal lath | | 570 | .028 | | | .96 | | .96 | 1.48 |
| 1100 | Gypsum, on gypsum lath | | 720 | .022 | | | .76 | | .76 | 1.17 |
| 1120 | On metal lath | | 500 | .032 | | | 1.10 | | 1.10 | 1.69 |
| 1200 | Suspended ceiling, mineral fiber, 2' x 2' or 2' x 4' | | 1500 | .011 | | | .37 | | .37 | .56 |
| 1250 | On suspension system, incl. system | | 1200 | .013 | | | .46 | | .46 | .70 |
| 1500 | Tile, wood fiber, 12" x 12", glued | | 900 | .018 | | | .61 | | .61 | .94 |
| 1540 | Stapled | | 1500 | .011 | | | .37 | | .37 | .56 |
| 1580 | On suspension system, incl. system | | 760 | .021 | | | .72 | | .72 | 1.11 |
| 2000 | Wood, tongue and groove, 1" x 4" | | 1000 | .016 | | | .55 | | .55 | .84 |
| 2040 | 1" x 8" | | 1100 | .015 | | | .50 | | .50 | .77 |
| 2400 | Plywood or wood fiberboard, 4' x 8' sheets | | 1200 | .013 | | | .46 | | .46 | .70 |

### 09 05 05.20 Selective Demolition, Flooring

| | | Crew | Daily Output | Labor-Hours | Unit | Material | 2011 Bare Costs Labor | 2011 Bare Costs Equipment | Total | Total Incl O&P |
|---|---|---|---|---|---|---|---|---|---|---|
| 0010 | **SELECTIVE DEMOLITION, FLOORING** R024119-10 | | | | | | | | | |
| 0200 | Brick with mortar | 2 Clab | 475 | .034 | S.F. | | 1.16 | | 1.16 | 1.77 |
| 0400 | Carpet, bonded, including surface scraping | | 2000 | .008 | | | .27 | | .27 | .42 |
| 0440 | Scrim applied | | 8000 | .002 | | | .07 | | .07 | .11 |
| 0480 | Tackless | | 9000 | .002 | | | .06 | | .06 | .09 |
| 0550 | Carpet tile, releasable adhesive | | 5000 | .003 | | | .11 | | .11 | .17 |
| 0560 | Permanent adhesive | | 1850 | .009 | | | .30 | | .30 | .46 |
| 0600 | Composition, acrylic or epoxy | | 400 | .040 | | | 1.37 | | 1.37 | 2.11 |

## 09 05 05 – Selective Finishes Demolition

### 09 05 05.20 Selective Demolition, Flooring

| | | Crew | Daily Output | Labor-Hours | Unit | Material | 2011 Bare Costs Labor | 2011 Bare Costs Equipment | Total | Total Incl O&P |
|---|---|---|---|---|---|---|---|---|---|---|
| 0700 | Concrete, scarify skin | A-1A | 225 | .036 | S.F. | | 1.58 | .96 | 2.54 | 3.48 |
| 0800 | Resilient, sheet goods | 2 Clab | 1400 | .011 | | | .39 | | .39 | .60 |
| 0820 | For gym floors | " | 900 | .018 | | | .61 | | .61 | .94 |
| 0850 | Vinyl or rubber cove base | 1 Clab | 1000 | .008 | L.F. | | .27 | | .27 | .42 |
| 0860 | Vinyl or rubber cove base, molded corner | " | 1000 | .008 | Ea. | | .27 | | .27 | .42 |
| 0870 | For glued and caulked installation, add to labor | | | | | | 50% | | | |
| 0900 | Vinyl composition tile, 12" x 12" | 2 Clab | 1000 | .016 | S.F. | | .55 | | .55 | .84 |
| 2000 | Tile, ceramic, thin set | | 675 | .024 | | | .81 | | .81 | 1.25 |
| 2020 | Mud set | | 625 | .026 | | | .88 | | .88 | 1.35 |
| 2200 | Marble, slate, thin set | | 675 | .024 | | | .81 | | .81 | 1.25 |
| 2220 | Mud set | | 625 | .026 | | | .88 | | .88 | 1.35 |
| 2600 | Terrazzo, thin set | | 450 | .036 | | | 1.22 | | 1.22 | 1.87 |
| 2620 | Mud set | | 425 | .038 | | | 1.29 | | 1.29 | 1.98 |
| 2630 | Grout removal, carbide tipped, rotary grinder | 1 Clab | 240 | .033 | L.F. | | 1.14 | | 1.14 | 1.76 |
| 2640 | Terrazzo, cast in place | 2 Clab | 300 | .053 | S.F. | | 1.83 | | 1.83 | 2.81 |
| 3000 | Wood, block, on end | 1 Carp | 400 | .020 | | | .86 | | .86 | 1.32 |
| 3200 | Parquet | | 450 | .018 | | | .77 | | .77 | 1.17 |
| 3400 | Strip flooring, interior, 2-1/4" x 25/32" thick | | 325 | .025 | | | 1.06 | | 1.06 | 1.63 |
| 3500 | Exterior, porch flooring, 1" x 4" | | 220 | .036 | | | 1.57 | | 1.57 | 2.40 |
| 3800 | Subfloor, tongue and groove, 1" x 6" | | 325 | .025 | | | 1.06 | | 1.06 | 1.63 |
| 3820 | 1" x 8" | | 430 | .019 | | | .80 | | .80 | 1.23 |
| 3840 | 1" x 10" | | 520 | .015 | | | .66 | | .66 | 1.02 |
| 4000 | Plywood, nailed | | 600 | .013 | | | .57 | | .57 | .88 |
| 4100 | Glued and nailed | | 400 | .020 | | | .86 | | .86 | 1.32 |
| 4200 | Hardboard, 1/4" thick | | 760 | .011 | | | .45 | | .45 | .70 |
| 8000 | Remove flooring, bead blast, minimum | A-1A | 1000 | .008 | | | .36 | .22 | .58 | .78 |
| 8100 | Maximum | | 400 | .020 | | | .89 | .54 | 1.43 | 1.95 |
| 8150 | Mastic only | | 1500 | .005 | | | .24 | .14 | .38 | .52 |

### 09 05 05.30 Selective Demolition, Walls and Partitions

| | | Crew | Daily Output | Labor-Hours | Unit | Material | 2011 Bare Costs Labor | 2011 Bare Costs Equipment | Total | Total Incl O&P |
|---|---|---|---|---|---|---|---|---|---|---|
| 0010 | **SELECTIVE DEMOLITION, WALLS AND PARTITIONS** R024119-10 | | | | | | | | | |
| 0020 | Walls, concrete, reinforced | B-39 | 120 | .400 | C.F. | | 14.50 | 1.62 | 16.12 | 24 |
| 0025 | Plain | " | 160 | .300 | | | 10.85 | 1.22 | 12.07 | 17.95 |
| 0100 | Brick, 4" to 12" thick | B-9 | 220 | .182 | | | 6.30 | .89 | 7.19 | 10.65 |
| 0200 | Concrete block, 4" thick | | 1000 | .040 | S.F. | | 1.39 | .20 | 1.59 | 2.34 |
| 0280 | 8" thick | | 810 | .049 | | | 1.72 | .24 | 1.96 | 2.89 |
| 0300 | Exterior stucco 1" thick over mesh | | 3200 | .013 | | | .43 | .06 | .49 | .74 |
| 1000 | Drywall, nailed or screwed | 1 Clab | 1000 | .008 | | | .27 | | .27 | .42 |
| 1010 | 2 layers | | 400 | .020 | | | .69 | | .69 | 1.05 |
| 1020 | Glued and nailed | | 900 | .009 | | | .31 | | .31 | .47 |
| 1500 | Fiberboard, nailed | | 900 | .009 | | | .31 | | .31 | .47 |
| 1520 | Glued and nailed | | 800 | .010 | | | .34 | | .34 | .53 |
| 1568 | Plenum barrier, sheet lead | | 300 | .027 | | | .92 | | .92 | 1.41 |
| 2000 | Movable walls, metal, 5' high | | 300 | .027 | | | .92 | | .92 | 1.41 |
| 2020 | 8' high | | 400 | .020 | | | .69 | | .69 | 1.05 |
| 2200 | Metal or wood studs, finish 2 sides, fiberboard | B-1 | 520 | .046 | | | 1.62 | | 1.62 | 2.48 |
| 2250 | Lath and plaster | | 260 | .092 | | | 3.23 | | 3.23 | 4.96 |
| 2300 | Plasterboard (drywall) | | 520 | .046 | | | 1.62 | | 1.62 | 2.48 |
| 2350 | Plywood | | 450 | .053 | | | 1.87 | | 1.87 | 2.86 |
| 2800 | Paneling, 4' x 8' sheets | 1 Clab | 475 | .017 | | | .58 | | .58 | .89 |
| 3000 | Plaster, lime and horsehair, on wood lath | | 400 | .020 | | | .69 | | .69 | 1.05 |
| 3020 | On metal lath | | 335 | .024 | | | .82 | | .82 | 1.26 |
| 3400 | Gypsum or perlite, on gypsum lath | | 410 | .020 | | | .67 | | .67 | 1.03 |

# 09 05 Common Work Results for Finishes

## 09 05 05 – Selective Finishes Demolition

| 09 05 05.30 Selective Demolition, Walls and Partitions | Crew | Daily Output | Labor-Hours | Unit | Material | 2011 Bare Costs Labor | Equipment | Total | Total Incl O&P |
|---|---|---|---|---|---|---|---|---|---|
| 3420 | On metal lath | 1 Clab | 300 | .027 | S.F. | | .92 | | .92 | 1.41 |
| 3450 | Plaster, interior gypsum, acoustic, or cement | | 60 | .133 | S.Y. | | 4.58 | | 4.58 | 7.05 |
| 3500 | Stucco, on masonry | | 145 | .055 | | | 1.90 | | 1.90 | 2.91 |
| 3510 | Commercial 3-coat | | 80 | .100 | | | 3.44 | | 3.44 | 5.25 |
| 3520 | Interior stucco | | 25 | .320 | | | 11 | | 11 | 16.85 |
| 3600 | Plywood, one side | B-1 | 1500 | .016 | S.F. | | .56 | | .56 | .86 |
| 3750 | Terra cotta block and plaster, to 6" thick | " | 175 | .137 | | | 4.80 | | 4.80 | 7.35 |
| 3760 | Tile, ceramic, on walls, thin set | 1 Clab | 300 | .027 | | | .92 | | .92 | 1.41 |
| 3765 | Mud set | | 250 | .032 | | | 1.10 | | 1.10 | 1.69 |
| 3800 | Toilet partitions, slate or marble | | 5 | 1.600 | Ea. | | 55 | | 55 | 84.50 |
| 3820 | Metal or plastic | | 8 | 1 | " | | 34.50 | | 34.50 | 52.50 |
| 5000 | Wallcovering, vinyl | 1 Pape | 700 | .011 | S.F. | | .43 | | .43 | .64 |
| 5010 | With release agent | | 1500 | .005 | | | .20 | | .20 | .30 |
| 5025 | Wallpaper, 2 layers or less, by hand | | 250 | .032 | | | 1.21 | | 1.21 | 1.80 |
| 5035 | 3 layers or more | | 165 | .048 | | | 1.84 | | 1.84 | 2.72 |
| 5040 | Designer | | 480 | .017 | | | .63 | | .63 | .94 |

# 09 21 Plaster and Gypsum Board Assemblies

## 09 21 13 – Plaster Assemblies

### 09 21 13.10 Plaster Partition Wall

| | | Crew | Daily Output | Labor-Hours | Unit | Material | Labor | Equipment | Total | Total Incl O&P |
|---|---|---|---|---|---|---|---|---|---|---|
| 0010 | **PLASTER PARTITION WALL** | | | | | | | | | |
| 0400 | Stud walls, 3.4 lb. metal lath, 3 coat gypsum plaster, 2 sides | | | | | | | | | |
| 0600 | 2" x 4" wood studs, 16" O.C. | J-2 | 315 | .152 | S.F. | 3.31 | 5.75 | .39 | 9.45 | 12.60 |
| 0700 | 2-1/2" metal studs, 25 ga., 12" O.C. | | 325 | .148 | | 3.06 | 5.60 | .38 | 9.04 | 12.10 |
| 0800 | 3-5/8" metal studs, 25 ga., 16" O.C. | | 320 | .150 | | 3.08 | 5.65 | .39 | 9.12 | 12.20 |
| 0900 | Gypsum lath, 2 coat vermiculite plaster, 2 sides | | | | | | | | | |
| 1000 | 2" x 4" wood studs, 16" O.C. | J-2 | 355 | .135 | S.F. | 3.69 | 5.10 | .35 | 9.14 | 12.05 |
| 1200 | 2-1/2" metal studs, 25 ga., 12" O.C. | | 365 | .132 | | 3.27 | 4.97 | .34 | 8.58 | 11.30 |
| 1300 | 3-5/8" metal studs, 25 ga., 16" O.C. | | 360 | .133 | | 3.36 | 5.05 | .34 | 8.75 | 11.55 |

## 09 21 16 – Gypsum Board Assemblies

### 09 21 16.23 Gypsum Board Shaft Wall Assemblies

| | | Crew | Daily Output | Labor-Hours | Unit | Material | Labor | Equipment | Total | Total Incl O&P |
|---|---|---|---|---|---|---|---|---|---|---|
| 0010 | **GYPSUM BOARD SHAFT WALL ASSEMBLIES** | | | | | | | | | |
| 0020 | Cavity type on 25 ga. J-track & C-H studs, 24" O.C. | | | | | | | | | |
| 0030 | 1" thick coreboard wall liner on shaft side | | | | | | | | | |
| 0040 | 2-hour assembly with double layer | | | | | | | | | |
| 0060 | 5/8" fire rated gypsum board on room side | 2 Carp | 220 | .073 | S.F. | 1.68 | 3.13 | | 4.81 | 6.65 |
| 0100 | 3-hour assembly with triple layer | | | | | | | | | |
| 0300 | 5/8" fire rated gypsum board on room side | 2 Carp | 180 | .089 | S.F. | 1.40 | 3.83 | | 5.23 | 7.40 |
| 0400 | 4-hour assembly, 1" coreboard, 5/8" fire rated gypsum board | | | | | | | | | |
| 0600 | and 3/4" galv. metal furring channels, 24" O.C., with | | | | | | | | | |
| 0700 | Double layer 5/8" fire rated gypsum board on room side | 2 Carp | 110 | .145 | S.F. | 1.30 | 6.25 | | 7.55 | 11.05 |
| 0900 | For taping & finishing, add per side | 1 Carp | 1050 | .008 | " | .04 | .33 | | .37 | .55 |
| 1000 | For insulation, see Section 07 21 | | | | | | | | | |
| 5200 | For work over 8' high, add | 2 Carp | 3060 | .005 | S.F. | | .23 | | .23 | .35 |
| 5300 | For distribution cost over 3 stories high, add per story | " | 6100 | .003 | " | | .11 | | .11 | .17 |

### 09 21 16.33 Partition Wall

| | | Crew | Daily Output | Labor-Hours | Unit | Material | Labor | Equipment | Total | Total Incl O&P |
|---|---|---|---|---|---|---|---|---|---|---|
| 0010 | **PARTITION WALL** Stud wall, 8' to 12' high | | | | | | | | | |
| 0050 | 1/2", interior, gypsum board, std, tape & finish 2 sides | | | | | | | | | |
| 0500 | Installed on and incl., 2" x 4" wood studs, 16" O.C. | 2 Carp | 310 | .052 | S.F. | .92 | 2.22 | | 3.14 | 4.42 |
| 1000 | Metal studs, NLB, 25 ga., 16" O.C., 3-5/8" wide | | 350 | .046 | | .86 | 1.97 | | 2.83 | 3.96 |

## 09 21 16 – Gypsum Board Assemblies

| 09 21 16.33 **Partition Wall** | Crew | Daily Output | Labor-Hours | Unit | Material | 2011 Bare Costs Labor | Equipment | Total | Total Incl O&P |
|---|---|---|---|---|---|---|---|---|---|
| 1200 | 6" wide | 2 Carp | 330 | .048 | S.F. | .99 | 2.09 | | 3.08 | 4.29 |
| 1400 | Water resistant, on 2" x 4" wood studs, 16" O.C. | | 310 | .052 | | 1.18 | 2.22 | | 3.40 | 4.71 |
| 1600 | Metal studs, NLB, 25 ga., 16" O.C., 3-5/8" wide | | 350 | .046 | | 1.12 | 1.97 | | 3.09 | 4.25 |
| 1800 | 6" wide | | 330 | .048 | | 1.25 | 2.09 | | 3.34 | 4.57 |
| 2000 | Fire res., 2 layers, 1-1/2 hr., on 2" x 4" wood studs, 16" O.C. | | 210 | .076 | | 1.58 | 3.28 | | 4.86 | 6.80 |
| 2200 | Metal studs, NLB, 25 ga., 16" O.C., 3-5/8" wide | | 250 | .064 | | 1.52 | 2.76 | | 4.28 | 5.90 |
| 2400 | 6" wide | | 230 | .070 | | 1.65 | 3 | | 4.65 | 6.40 |
| 2600 | Fire & water res., 2 layers, 1-1/2 hr., 2" x 4" studs, 16" O.C. | | 210 | .076 | | 1.58 | 3.28 | | 4.86 | 6.80 |
| 2800 | Metal studs, NLB, 25 ga., 16" O.C., 3-5/8" wide | | 250 | .064 | | 1.52 | 2.76 | | 4.28 | 5.90 |
| 3000 | 6" wide | ▼ | 230 | .070 | ▼ | 1.65 | 3 | | 4.65 | 6.40 |
| 3200 | 5/8", interior, gypsum board, std, tape & finish 2 sides | | | | | | | | | |
| 3400 | Installed on and including 2" x 4" wood studs, 16" O.C. | 2 Carp | 300 | .053 | S.F. | 1.02 | 2.30 | | 3.32 | 4.64 |
| 3600 | 24" O.C. | | 330 | .048 | | .94 | 2.09 | | 3.03 | 4.23 |
| 3800 | Metal studs, NLB, 25 ga., 16" O.C., 3-5/8" wide | | 340 | .047 | | .96 | 2.03 | | 2.99 | 4.16 |
| 4000 | 6" wide | | 320 | .050 | | 1.09 | 2.15 | | 3.24 | 4.50 |
| 4200 | 24" O.C., 3-5/8" wide | | 360 | .044 | | .88 | 1.91 | | 2.79 | 3.90 |
| 4400 | 6" wide | | 340 | .047 | | .97 | 2.03 | | 3 | 4.18 |
| 4800 | Water resistant, on 2" x 4" wood studs, 16" O.C. | | 300 | .053 | | 1.26 | 2.30 | | 3.56 | 4.91 |
| 5000 | 24" O.C. | | 330 | .048 | | 1.18 | 2.09 | | 3.27 | 4.50 |
| 5200 | Metal studs, NLB, 25 ga. 16" O.C., 3-5/8" wide | | 340 | .047 | | 1.20 | 2.03 | | 3.23 | 4.43 |
| 5400 | 6" wide | | 320 | .050 | | 1.33 | 2.15 | | 3.48 | 4.76 |
| 5600 | 24" O.C., 3-5/8" wide | | 360 | .044 | | 1.12 | 1.91 | | 3.03 | 4.17 |
| 5800 | 6" wide | | 340 | .047 | | 1.21 | 2.03 | | 3.24 | 4.44 |
| 6000 | Fire resistant, 2 layers, 2 hr., on 2" x 4" wood studs, 16" O.C. | | 205 | .078 | | 1.50 | 3.36 | | 4.86 | 6.80 |
| 6200 | 24" O.C. | | 235 | .068 | | 1.50 | 2.93 | | 4.43 | 6.15 |
| 6400 | Metal studs, NLB, 25 ga., 16" O.C., 3-5/8" wide | | 245 | .065 | | 1.55 | 2.81 | | 4.36 | 6 |
| 6600 | 6" wide | | 225 | .071 | | 1.65 | 3.06 | | 4.71 | 6.50 |
| 6800 | 24" O.C., 3-5/8" wide | | 265 | .060 | | 1.44 | 2.60 | | 4.04 | 5.55 |
| 7000 | 6" wide | | 245 | .065 | | 1.53 | 2.81 | | 4.34 | 6 |
| 7200 | Fire & water resistant, 2 layers, 2 hr., 2" x 4" studs, 16" O.C. | | 205 | .078 | | 1.58 | 3.36 | | 4.94 | 6.90 |
| 7400 | 24" O.C. | | 235 | .068 | | 1.50 | 2.93 | | 4.43 | 6.15 |
| 7600 | Metal studs, NLB, 25 ga., 16" O.C., 3-5/8" wide | | 245 | .065 | | 1.52 | 2.81 | | 4.33 | 6 |
| 7800 | 6" wide | | 225 | .071 | | 1.65 | 3.06 | | 4.71 | 6.50 |
| 8000 | 24" O.C., 3-5/8" wide | | 265 | .060 | | 1.44 | 2.60 | | 4.04 | 5.55 |
| 8200 | 6" wide | ▼ | 245 | .065 | ▼ | 1.53 | 2.81 | | 4.34 | 6 |
| 8600 | 1/2" blueboard, mesh tape both sides | | | | | | | | | |
| 8620 | Installed on and including 2" x 4" wood studs, 16" O.C. | 2 Carp | 300 | .053 | S.F. | 1.02 | 2.30 | | 3.32 | 4.64 |
| 8640 | Metal studs, NLB, 25 ga., 16" O.C., 3-5/8" wide | | 340 | .047 | | .96 | 2.03 | | 2.99 | 4.16 |
| 8660 | 6" wide | ▼ | 320 | .050 | ▼ | 1.09 | 2.15 | | 3.24 | 4.50 |
| 9000 | Exterior, 1/2" gypsum sheathing, 1/2" gypsum finished, interior, | | | | | | | | | |
| 9100 | including foil faced insulation, metal studs, 20 ga. | | | | | | | | | |
| 9200 | 16" O.C., 3-5/8" wide | 2 Carp | 290 | .055 | S.F. | 1.62 | 2.38 | | 4 | 5.45 |
| 9400 | 6" wide | | 270 | .059 | | 1.79 | 2.55 | | 4.34 | 5.90 |
| 9600 | Partitions, for work over 8' high, add | ▼ | 1530 | .010 | ▼ | | .45 | | .45 | .69 |

# 09 22 Supports for Plaster and Gypsum Board

## 09 22 03 – Fastening Methods for Finishes

| 09 22 03.20 Drilling Plaster/Drywall | Crew | Daily Output | Labor-Hours | Unit | Material | 2011 Bare Costs Labor | Equipment | Total | Total Incl O&P |
|---|---|---|---|---|---|---|---|---|---|
| 0010 **DRILLING PLASTER/DRYWALL** | | | | | | | | | |
| 1100 Drilling & layout for drywall/plaster walls, up to 1" deep, no anchor | | | | | | | | | |
| 1200 Holes, 1/4" diameter | 1 Carp | 150 | .053 | Ea. | .01 | 2.30 | | 2.31 | 3.53 |
| 1300 3/8" diameter | | 140 | .057 | | .01 | 2.46 | | 2.47 | 3.78 |
| 1400 1/2" diameter | | 130 | .062 | | .01 | 2.65 | | 2.66 | 4.07 |
| 1500 3/4" diameter | | 120 | .067 | | .01 | 2.87 | | 2.88 | 4.41 |
| 1600 1" diameter | | 110 | .073 | | .02 | 3.13 | | 3.15 | 4.82 |
| 1700 1-1/4" diameter | | 100 | .080 | | .03 | 3.44 | | 3.47 | 5.35 |
| 1800 1-1/2" diameter | | 90 | .089 | | .04 | 3.83 | | 3.87 | 5.90 |
| 1900 For ceiling installations, add | | | | | | 40% | | | |

## 09 22 13 – Metal Furring

### 09 22 13.13 Metal Channel Furring

| 09 22 13.13 Metal Channel Furring | Crew | Daily Output | Labor-Hours | Unit | Material | 2011 Bare Costs Labor | Equipment | Total | Total Incl O&P |
|---|---|---|---|---|---|---|---|---|---|
| 0010 **METAL CHANNEL FURRING** | | | | | | | | | |
| 0030 Beams and columns, 7/8" channels, galvanized, 12" O.C. | 1 Lath | 155 | .052 | S.F. | .31 | 1.99 | | 2.30 | 3.24 |
| 0050 16" O.C. | | 170 | .047 | | .25 | 1.81 | | 2.06 | 2.93 |
| 0070 24" O.C. | | 185 | .043 | | .17 | 1.67 | | 1.84 | 2.62 |
| 0100 Ceilings, on steel, 7/8" channels, galvanized, 12" O.C. | | 210 | .038 | | .28 | 1.47 | | 1.75 | 2.45 |
| 0300 16" O.C. | | 290 | .028 | | .25 | 1.06 | | 1.31 | 1.83 |
| 0400 24" O.C. | | 420 | .019 | | .17 | .73 | | .90 | 1.26 |
| 0600 1-5/8" channels, galvanized, 12" O.C. | | 190 | .042 | | .38 | 1.62 | | 2 | 2.79 |
| 0700 16" O.C. | | 260 | .031 | | .34 | 1.19 | | 1.53 | 2.10 |
| 0900 24" O.C. | | 390 | .021 | | .23 | .79 | | 1.02 | 1.40 |
| 0930 7/8" channels with sound isolation clips, 12" O.C. | | 120 | .067 | | 1.35 | 2.57 | | 3.92 | 5.25 |
| 0940 16" O.C. | | 100 | .080 | | 1.88 | 3.08 | | 4.96 | 6.55 |
| 0950 24" O.C. | | 165 | .048 | | 1.23 | 1.87 | | 3.10 | 4.09 |
| 0960 1-5/8" channels, galvanized, 12" O.C. | | 110 | .073 | | 1.44 | 2.80 | | 4.24 | 5.70 |
| 0970 16" O.C. | | 100 | .080 | | 1.97 | 3.08 | | 5.05 | 6.65 |
| 0980 24" O.C. | | 155 | .052 | | 1.29 | 1.99 | | 3.28 | 4.32 |
| 1000 Walls, 7/8" channels, galvanized, 12" O.C. | | 235 | .034 | | .28 | 1.31 | | 1.59 | 2.22 |
| 1200 16" O.C. | | 265 | .030 | | .25 | 1.16 | | 1.41 | 1.98 |
| 1300 24" O.C. | | 350 | .023 | | .17 | .88 | | 1.05 | 1.48 |
| 1500 1-5/8" channels, galvanized, 12" O.C. | | 210 | .038 | | .38 | 1.47 | | 1.85 | 2.56 |
| 1600 16" O.C. | | 240 | .033 | | .34 | 1.28 | | 1.62 | 2.24 |
| 1800 24" O.C. | | 305 | .026 | | .23 | 1.01 | | 1.24 | 1.73 |
| 1920 7/8" channels with sound isolation clips, 12" O.C. | | 125 | .064 | | 1.35 | 2.47 | | 3.82 | 5.10 |
| 1940 16" O.C. | | 100 | .080 | | 1.88 | 3.08 | | 4.96 | 6.55 |
| 1950 24" O.C. | | 150 | .053 | | 1.23 | 2.06 | | 3.29 | 4.36 |
| 1960 1-5/8" channels, galvanized, 12" O.C. | | 115 | .070 | | 1.44 | 2.68 | | 4.12 | 5.50 |
| 1970 16" O.C. | | 95 | .084 | | 1.97 | 3.25 | | 5.22 | 6.90 |
| 1980 24" O.C. | | 140 | .057 | | 1.29 | 2.20 | | 3.49 | 4.63 |

## 09 22 16 – Non-Structural Metal Framing

### 09 22 16.13 Non-Structural Metal Stud Framing

| 09 22 16.13 Non-Structural Metal Stud Framing | Crew | Daily Output | Labor-Hours | Unit | Material | 2011 Bare Costs Labor | Equipment | Total | Total Incl O&P |
|---|---|---|---|---|---|---|---|---|---|
| 0010 **NON-STRUCTURAL METAL STUD FRAMING** | | | | | | | | | |
| 1600 Non-load bearing, galv, 8' high, 25 ga. 1-5/8" wide, 16" O.C. | 1 Carp | 619 | .013 | S.F. | .24 | .56 | | .80 | 1.12 |
| 1610 24" O.C. | | 950 | .008 | | .18 | .36 | | .54 | .76 |
| 1620 2-1/2" wide, 16" O.C. | | 613 | .013 | | .28 | .56 | | .84 | 1.17 |
| 1630 24" O.C. | | 938 | .009 | | .21 | .37 | | .58 | .79 |
| 1640 3-5/8" wide, 16" O.C. | | 600 | .013 | | .32 | .57 | | .89 | 1.24 |
| 1650 24" O.C. | | 925 | .009 | | .24 | .37 | | .61 | .84 |
| 1660 4" wide, 16" O.C. | | 594 | .013 | | .37 | .58 | | .95 | 1.29 |
| 1670 24" O.C. | | 925 | .009 | | .27 | .37 | | .64 | .87 |
| 1680 6" wide, 16" O.C. | | 588 | .014 | | .46 | .59 | | 1.05 | 1.41 |

## 09 22 16 – Non-Structural Metal Framing

### 09 22 16.13 Non-Structural Metal Stud Framing

| | | Crew | Daily Output | Labor-Hours | Unit | Material | 2011 Bare Costs Labor | Equipment | Total | Total Incl O&P |
|---|---|---|---|---|---|---|---|---|---|---|
| 1690 | 24" O.C. | 1 Carp | 906 | .009 | S.F. | .35 | .38 | | .73 | .96 |
| 1700 | 20 ga. studs, 1-5/8" wide, 16" O.C. | | 494 | .016 | | .37 | .70 | | 1.07 | 1.47 |
| 1710 | 24" O.C. | | 763 | .010 | | .27 | .45 | | .72 | .99 |
| 1720 | 2-1/2" wide, 16" O.C. | | 488 | .016 | | .42 | .71 | | 1.13 | 1.54 |
| 1730 | 24" O.C. | | 750 | .011 | | .32 | .46 | | .78 | 1.05 |
| 1740 | 3-5/8" wide, 16" O.C. | | 481 | .017 | | .45 | .72 | | 1.17 | 1.60 |
| 1750 | 24" O.C. | | 738 | .011 | | .34 | .47 | | .81 | 1.09 |
| 1760 | 4" wide, 16" O.C. | | 475 | .017 | | .54 | .73 | | 1.27 | 1.70 |
| 1770 | 24" O.C. | | 738 | .011 | | .40 | .47 | | .87 | 1.16 |
| 1780 | 6" wide, 16" O.C. | | 469 | .017 | | .63 | .73 | | 1.36 | 1.82 |
| 1790 | 24" O.C. | | 725 | .011 | | .47 | .47 | | .94 | 1.25 |
| 2000 | Non-load bearing, galv, 10' high, 25 ga. 1-5/8" wide, 16" O.C. | | 495 | .016 | | .23 | .70 | | .93 | 1.32 |
| 2100 | 24" O.C. | | 760 | .011 | | .17 | .45 | | .62 | .89 |
| 2200 | 2-1/2" wide, 16" O.C. | | 490 | .016 | | .27 | .70 | | .97 | 1.37 |
| 2250 | 24" O.C. | | 750 | .011 | | .20 | .46 | | .66 | .91 |
| 2300 | 3-5/8" wide, 16" O.C. | | 480 | .017 | | .31 | .72 | | 1.03 | 1.44 |
| 2350 | 24" O.C. | | 740 | .011 | | .23 | .47 | | .70 | .96 |
| 2400 | 4" wide, 16" O.C. | | 475 | .017 | | .35 | .73 | | 1.08 | 1.49 |
| 2450 | 24" O.C. | | 740 | .011 | | .25 | .47 | | .72 | .99 |
| 2500 | 6" wide, 16" O.C. | | 470 | .017 | | .44 | .73 | | 1.17 | 1.60 |
| 2550 | 24" O.C. | | 725 | .011 | | .32 | .47 | | .79 | 1.08 |
| 2600 | 20 ga. studs, 1-5/8" wide, 16" O.C. | | 395 | .020 | | .35 | .87 | | 1.22 | 1.72 |
| 2650 | 24" O.C. | | 610 | .013 | | .25 | .56 | | .81 | 1.15 |
| 2700 | 2-1/2" wide, 16" O.C. | | 390 | .021 | | .40 | .88 | | 1.28 | 1.79 |
| 2750 | 24" O.C. | | 600 | .013 | | .29 | .57 | | .86 | 1.20 |
| 2800 | 3-5/8" wide, 16" OC | | 385 | .021 | | .43 | .89 | | 1.32 | 1.84 |
| 2850 | 24" O.C. | | 590 | .014 | | .32 | .58 | | .90 | 1.25 |
| 2900 | 4" wide, 16" O.C. | | 380 | .021 | | .51 | .91 | | 1.42 | 1.95 |
| 2950 | 24" O.C. | | 590 | .014 | | .37 | .58 | | .95 | 1.31 |
| 3000 | 6" wide, 16" O.C. | | 375 | .021 | | .60 | .92 | | 1.52 | 2.07 |
| 3050 | 24" O.C. | | 580 | .014 | | .44 | .59 | | 1.03 | 1.39 |
| 3060 | Non-load bearing, galv, 12' high, 25 ga. 1-5/8" wide, 16" O.C. | | 413 | .019 | | .22 | .83 | | 1.05 | 1.52 |
| 3070 | 24" O.C. | | 633 | .013 | | .16 | .54 | | .70 | 1.01 |
| 3080 | 2-1/2" wide, 16" O.C. | | 408 | .020 | | .25 | .84 | | 1.09 | 1.58 |
| 3090 | 24" O.C. | | 625 | .013 | | .18 | .55 | | .73 | 1.05 |
| 3100 | 3-5/8" wide, 16" O.C. | | 400 | .020 | | .29 | .86 | | 1.15 | 1.64 |
| 3110 | 24" O.C. | | 617 | .013 | | .21 | .56 | | .77 | 1.10 |
| 3120 | 4" wide, 16" O.C. | | 396 | .020 | | .33 | .87 | | 1.20 | 1.70 |
| 3130 | 24" O.C. | | 617 | .013 | | .24 | .56 | | .80 | 1.13 |
| 3140 | 6" wide, 16" O.C. | | 392 | .020 | | .42 | .88 | | 1.30 | 1.81 |
| 3150 | 24" O.C. | | 604 | .013 | | .30 | .57 | | .87 | 1.21 |
| 3160 | 20 ga. studs, 1-5/8" wide, 16" O.C. | | 329 | .024 | | .33 | 1.05 | | 1.38 | 1.97 |
| 3170 | 24" O.C. | | 508 | .016 | | .24 | .68 | | .92 | 1.31 |
| 3180 | 2-1/2" wide, 16" O.C. | | 325 | .025 | | .38 | 1.06 | | 1.44 | 2.05 |
| 3190 | 24" O.C. | | 500 | .016 | | .28 | .69 | | .97 | 1.36 |
| 3200 | 3-5/8" wide, 16" O.C. | | 321 | .025 | | .41 | 1.07 | | 1.48 | 2.10 |
| 3210 | 24" O.C. | | 492 | .016 | | .30 | .70 | | 1 | 1.40 |
| 3220 | 4" wide, 16" O.C. | | 317 | .025 | | .49 | 1.09 | | 1.58 | 2.21 |
| 3230 | 24" O.C. | | 492 | .016 | | .35 | .70 | | 1.05 | 1.46 |
| 3240 | 6" wide, 16" O.C. | | 313 | .026 | | .57 | 1.10 | | 1.67 | 2.32 |
| 3250 | 24" O.C. | | 483 | .017 | | .42 | .71 | | 1.13 | 1.55 |
| 5000 | Load bearing studs, see Section 05 41 13.30 | | | | | | | | | |

## 09 22 26 – Suspension Systems

| 09 22 26.13 Ceiling Suspension Systems | Crew | Daily Output | Labor-Hours | Unit | Material | 2011 Bare Costs Labor | Equipment | Total | Total Incl O&P |
|---|---|---|---|---|---|---|---|---|---|
| 0010 **CEILING SUSPENSION SYSTEMS** For gypsum board or plaster | | | | | | | | | |
| 8000 Suspended ceilings, including carriers | | | | | | | | | |
| 8200 1-1/2" carriers, 24" O.C. with: | | | | | | | | | |
| 8300 7/8" channels, 16" O.C. | 1 Lath | 275 | .029 | S.F. | .43 | 1.12 | | 1.55 | 2.11 |
| 8320 24" O.C. | | 310 | .026 | | .34 | 1 | | 1.34 | 1.83 |
| 8400 1-5/8" channels, 16" O.C. | | 205 | .039 | | .51 | 1.50 | | 2.01 | 2.76 |
| 8420 24" O.C. | | 250 | .032 | | .40 | 1.23 | | 1.63 | 2.24 |
| 8600 2" carriers, 24" O.C. with: | | | | | | | | | |
| 8700 7/8" channels, 16" O.C. | 1 Lath | 250 | .032 | S.F. | .50 | 1.23 | | 1.73 | 2.36 |
| 8720 24" O.C. | | 285 | .028 | | .42 | 1.08 | | 1.50 | 2.04 |
| 8800 1-5/8" channels, 16" O.C. | | 190 | .042 | | .59 | 1.62 | | 2.21 | 3.02 |
| 8820 24" O.C. | | 225 | .036 | | .48 | 1.37 | | 1.85 | 2.52 |

## 09 22 36 – Lath

### 09 22 36.13 Gypsum Lath

| 09 22 36.13 Gypsum Lath | Crew | Daily Output | Labor-Hours | Unit | Material | 2011 Bare Costs Labor | Equipment | Total | Total Incl O&P |
|---|---|---|---|---|---|---|---|---|---|
| 0010 **GYPSUM LATH** R092000-50 | | | | | | | | | |
| 0020 Plain or perforated, nailed, 3/8" thick | 1 Lath | 85 | .094 | S.Y. | 4.95 | 3.63 | | 8.58 | 10.75 |
| 0100 1/2" thick | | 80 | .100 | | 4.86 | 3.86 | | 8.72 | 11 |
| 0300 Clipped to steel studs, 3/8" thick | | 75 | .107 | | 4.95 | 4.11 | | 9.06 | 11.45 |
| 0400 1/2" thick | | 70 | .114 | | 4.86 | 4.41 | | 9.27 | 11.80 |
| 1500 For ceiling installations, add | | 216 | .037 | | | 1.43 | | 1.43 | 2.08 |
| 1600 For columns and beams, add | | 170 | .047 | | | 1.81 | | 1.81 | 2.65 |

### 09 22 36.23 Metal Lath

| 09 22 36.23 Metal Lath | Crew | Daily Output | Labor-Hours | Unit | Material | 2011 Bare Costs Labor | Equipment | Total | Total Incl O&P |
|---|---|---|---|---|---|---|---|---|---|
| 0010 **METAL LATH** R092000-50 | | | | | | | | | |
| 0020 Diamond, expanded, 2.5 lb. per S.Y., painted | | | | S.Y. | 3.62 | | | 3.62 | 3.98 |
| 0100 Galvanized | | | | | 2.63 | | | 2.63 | 2.89 |
| 0300 3.4 lb. per S.Y., painted | | | | | 4.38 | | | 4.38 | 4.82 |
| 0400 Galvanized | | | | | 4 | | | 4 | 4.40 |
| 0600 For 15# asphalt sheathing paper, add | | | | | .51 | | | .51 | .56 |
| 0900 Flat rib, 1/8" high, 2.75 lb., painted | | | | | 3.14 | | | 3.14 | 3.45 |
| 1000 Foil backed | | | | | 2.80 | | | 2.80 | 3.08 |
| 1200 3.4 lb. per S.Y., painted | | | | | 4 | | | 4 | 4.40 |
| 1300 Galvanized | | | | | 3.48 | | | 3.48 | 3.83 |
| 1500 For 15# asphalt sheathing paper, add | | | | | .51 | | | .51 | .56 |
| 1800 High rib, 3/8" high, 3.4 lb. per S.Y., painted | | | | | 4.84 | | | 4.84 | 5.30 |
| 1900 Galvanized | | | | | 4.84 | | | 4.84 | 5.30 |
| 2400 3/4" high, painted, .60 lb. per S.F. | | | | S.F. | .50 | | | .50 | .55 |
| 2500 .75 lb. per S.F. | | | | " | 1.08 | | | 1.08 | 1.19 |
| 2800 Stucco mesh, painted, 3.6 lb. | | | | S.Y. | 3.03 | | | 3.03 | 3.33 |
| 3000 K-lath, perforated, absorbent paper, regular | | | | | 4.21 | | | 4.21 | 4.63 |
| 3100 Heavy duty | | | | | 4.97 | | | 4.97 | 5.45 |
| 3300 Waterproof, heavy duty, grade B backing | | | | | 4.86 | | | 4.86 | 5.35 |
| 3400 Fire resistant backing | | | | | 5.40 | | | 5.40 | 5.90 |
| 3600 2.5 lb. diamond painted, on wood framing, on walls | 1 Lath | 85 | .094 | | 3.62 | 3.63 | | 7.25 | 9.30 |
| 3700 On ceilings | | 75 | .107 | | 3.62 | 4.11 | | 7.73 | 10 |
| 3900 3.4 lb. diamond painted, on wood framing, on walls | | 80 | .100 | | 4 | 3.86 | | 7.86 | 10.05 |
| 4000 On ceilings | | 70 | .114 | | 4 | 4.41 | | 8.41 | 10.85 |
| 4200 3.4 lb. diamond painted, wired to steel framing | | 75 | .107 | | 4 | 4.11 | | 8.11 | 10.40 |
| 4300 On ceilings | | 60 | .133 | | 4 | 5.15 | | 9.15 | 11.90 |
| 4500 Columns and beams, wired to steel | | 40 | .200 | | 4 | 7.70 | | 11.70 | 15.65 |
| 4600 Cornices, wired to steel | | 35 | .229 | | 4 | 8.80 | | 12.80 | 17.25 |
| 4800 Screwed to steel studs, 2.5 lb. | | 80 | .100 | | 3.62 | 3.86 | | 7.48 | 9.65 |

## 09 22 36 – Lath

| 09 22 36.23 Metal Lath | | Crew | Daily Output | Labor-Hours | Unit | Material | 2011 Bare Costs Labor | Equipment | Total | Total Incl O&P |
|---|---|---|---|---|---|---|---|---|---|---|
| 4900 | 3.4 lb. | 1 Lath | 75 | .107 | S.Y. | 4.38 | 4.11 | | 8.49 | 10.80 |
| 5100 | Rib lath, painted, wired to steel, on walls, 2.5 lb. | | 75 | .107 | | 3.14 | 4.11 | | 7.25 | 9.45 |
| 5200 | 3.4 lb. | | 70 | .114 | | 4.84 | 4.41 | | 9.25 | 11.75 |
| 5400 | 4.0 lb. | | 65 | .123 | | 5.45 | 4.74 | | 10.19 | 12.85 |
| 5500 | For self-furring lath, add | | | | | .10 | | | .10 | .11 |
| 5700 | Suspended ceiling system, incl. 3.4 lb. diamond lath, painted | 1 Lath | 15 | .533 | | 3.35 | 20.50 | | 23.85 | 33.50 |
| 5800 | Galvanized | " | 15 | .533 | | 6.50 | 20.50 | | 27 | 37 |
| 6000 | Hollow metal stud partitions, 3.4 lb. painted lath both sides | | | | | | | | | |
| 6010 | Non-load bearing, 25 ga., w/rib lath 2-1/2" studs, 12" O.C. | 1 Lath | 20.30 | .394 | S.Y. | 12.70 | 15.20 | | 27.90 | 36 |
| 6300 | 16" O.C. | | 21.10 | .379 | | 12.05 | 14.60 | | 26.65 | 35 |
| 6350 | 24" O.C. | | 22.70 | .352 | | 11.45 | 13.60 | | 25.05 | 32.50 |
| 6400 | 3-5/8" studs, 16" O.C. | | 19.50 | .410 | | 12.45 | 15.80 | | 28.25 | 36.50 |
| 6600 | 24" O.C. | | 20.40 | .392 | | 11.70 | 15.10 | | 26.80 | 35 |
| 6700 | 4" studs, 16" O.C. | | 20.40 | .392 | | 12.80 | 15.10 | | 27.90 | 36 |
| 6900 | 24" O.C. | | 21.60 | .370 | | 11.95 | 14.30 | | 26.25 | 34 |
| 7000 | 6" studs, 16" O.C. | | 19.50 | .410 | | 13.60 | 15.80 | | 29.40 | 38 |
| 7100 | 24" O.C. | | 21.10 | .379 | | 12.60 | 14.60 | | 27.20 | 35.50 |
| 7200 | L.B. partitions, 16 ga., w/rib lath, 2-1/2" studs, 16" O.C. | | 20 | .400 | | 12.30 | 15.40 | | 27.70 | 36 |
| 7300 | 3-5/8" studs, 16 ga. | | 19.70 | .406 | | 13.70 | 15.65 | | 29.35 | 38 |
| 7500 | 4" studs, 16 ga. | | 19.50 | .410 | | 14.20 | 15.80 | | 30 | 38.50 |
| 7600 | 6" studs, 16 ga. | | 18.70 | .428 | | 16.50 | 16.50 | | 33 | 42 |

### 09 22 36.43 Security Mesh

| | | Crew | Daily Output | Labor-Hours | Unit | Material | Labor | Equipment | Total | Total Incl O&P |
|---|---|---|---|---|---|---|---|---|---|---|
| 0010 | **SECURITY MESH**, expanded metal, flat, screwed to framing | | | | | | | | | |
| 0100 | On walls, 3/4", 1.76 lb./S.F. | 2 Carp | 1500 | .011 | S.F. | 1.49 | .46 | | 1.95 | 2.34 |
| 0110 | 1-1/2", 1.14 lb./S.F. | | 1600 | .010 | | 1.21 | .43 | | 1.64 | 1.99 |
| 0200 | On ceilings, 3/4", 1.76 lb./S.F. | | 1350 | .012 | | 1.49 | .51 | | 2 | 2.42 |
| 0210 | 1-1/2", 1.14 lb./S.F. | | 1450 | .011 | | 1.21 | .47 | | 1.68 | 2.06 |

### 09 22 36.83 Accessories, Plaster

| | | Crew | Daily Output | Labor-Hours | Unit | Material | Labor | Equipment | Total | Total Incl O&P |
|---|---|---|---|---|---|---|---|---|---|---|
| 0010 | **ACCESSORIES, PLASTER** | | | | | | | | | |
| 0020 | Casing bead, expanded flange, galvanized | 1 Lath | 2.70 | 2.963 | C.L.F. | 43.50 | 114 | | 157.50 | 215 |
| 0200 | Foundation weep screed, galvanized | " | 2.70 | 2.963 | | 58 | 114 | | 172 | 231 |
| 0900 | Channels, cold rolled, 16 ga., 3/4" deep, galvanized | | | | | 28.50 | | | 28.50 | 31 |
| 1200 | 1-1/2" deep, 16 ga., galvanized | | | | | 38 | | | 38 | 41.50 |
| 1620 | Corner bead, expanded bullnose, 3/4" radius, #10, galvanized | 1 Lath | 2.60 | 3.077 | | 30.50 | 119 | | 149.50 | 207 |
| 1650 | #1, galvanized | | 2.55 | 3.137 | | 48 | 121 | | 169 | 229 |
| 1670 | Expanded wing, 2-3/4" wide, #1, galvanized | | 2.65 | 3.019 | | 33.50 | 116 | | 149.50 | 207 |
| 1700 | Inside corner (corner rite), 3" x 3", painted | | 2.60 | 3.077 | | 28.50 | 119 | | 147.50 | 205 |
| 1750 | Strip-ex, 4" wide, painted | | 2.55 | 3.137 | | 29 | 121 | | 150 | 208 |
| 1800 | Expansion joint, 3/4" grounds, limited expansion, galv., 1 piece | | 2.70 | 2.963 | | 98 | 114 | | 212 | 275 |
| 2100 | Extreme expansion, galvanized, 2 piece | | 2.60 | 3.077 | | 157 | 119 | | 276 | 345 |

# 09 23 Gypsum Plastering

## 09 23 13 - Acoustical Gypsum Plastering

### 09 23 13.10 Perlite or Vermiculite Plaster

| | Crew | Daily Output | Labor-Hours | Unit | Material | 2011 Bare Costs Labor | 2011 Bare Costs Equipment | Total | Total Incl O&P |
|---|---|---|---|---|---|---|---|---|---|
| 0010 **PERLITE OR VERMICULITE PLASTER** R092000-50 | | | | | | | | | |
| 0020 In 100 lb. bags, under 200 bags | | | | Bag | 14.40 | | | 14.40 | 15.80 |
| 0100 Over 200 bags | | | | " | 13.05 | | | 13.05 | 14.35 |
| 0300 2 coats, no lath included, on walls | J-1 | 92 | .435 | S.Y. | 3.78 | 16.35 | 1.34 | 21.47 | 30 |
| 0400 On ceilings | " | 79 | .506 | | 3.78 | 19.05 | 1.56 | 24.39 | 34.50 |
| 0600 On and incl. 3/8" gypsum lath, on metal studs | J-2 | 84 | .571 | | 8.60 | 21.50 | 1.47 | 31.57 | 43 |
| 0700 On ceilings | " | 70 | .686 | | 8.60 | 26 | 1.77 | 36.37 | 50 |
| 0900 3 coats, no lath included, on walls | J-1 | 74 | .541 | | 6.30 | 20.50 | 1.67 | 28.47 | 39.50 |
| 1000 On ceilings | " | 63 | .635 | | 6.30 | 24 | 1.96 | 32.26 | 44.50 |
| 1200 On and incl. painted metal lath, on metal studs | J-2 | 72 | .667 | | 11.15 | 25 | 1.72 | 37.87 | 51.50 |
| 1300 On ceilings | | 61 | .787 | | 11.15 | 30 | 2.03 | 43.18 | 58.50 |
| 1500 On and incl. suspended metal lath ceiling | ↓ | 37 | 1.297 | ↓ | 9.65 | 49 | 3.35 | 62 | 87 |
| 1700 For irregular or curved surfaces, add to above | | | | | | 30% | | | |
| 1800 For columns and beams, add to above | | | | | | 50% | | | |
| 1900 For soffits, add to ceiling prices | | | | | | 40% | | | |

## 09 23 20 - Gypsum Plaster

### 09 23 20.10 Gypsum Plaster On Walls and Ceilings

| | Crew | Daily Output | Labor-Hours | Unit | Material | 2011 Bare Costs Labor | 2011 Bare Costs Equipment | Total | Total Incl O&P |
|---|---|---|---|---|---|---|---|---|---|
| 0010 **GYPSUM PLASTER ON WALLS AND CEILINGS** R092000-50 | | | | | | | | | |
| 0020 80# bag, less than 1 ton | | | | Bag | 17.30 | | | 17.30 | 19 |
| 0100 Over 1 ton | | | | " | 15.40 | | | 15.40 | 16.95 |
| 0300 2 coats, no lath included, on walls | J-1 | 105 | .381 | S.Y. | 3.87 | 14.35 | 1.18 | 19.40 | 27 |
| 0400 On ceilings | " | 92 | .435 | | 3.87 | 16.35 | 1.34 | 21.56 | 30 |
| 0600 On and incl. 3/8" gypsum lath on steel, on walls | J-2 | 97 | .495 | | 8.80 | 18.70 | 1.28 | 28.78 | 38.50 |
| 0700 On ceilings | " | 83 | .578 | | 8.80 | 22 | 1.49 | 32.29 | 44 |
| 0900 3 coats, no lath included, on walls | J-1 | 87 | .460 | | 5.55 | 17.30 | 1.42 | 24.27 | 33.50 |
| 1000 On ceilings | " | 78 | .513 | | 5.55 | 19.30 | 1.58 | 26.43 | 36.50 |
| 1200 On and including painted metal lath, on wood studs | J-2 | 86 | .558 | | 10.50 | 21 | 1.44 | 32.94 | 44.50 |
| 1300 On ceilings | " | 76.50 | .627 | ↓ | 10.50 | 23.50 | 1.62 | 35.62 | 48.50 |
| 1600 For irregular or curved surfaces, add | | | | | | 30% | | | |
| 1800 For columns & beams, add | | | | | | 50% | | | |

### 09 23 20.20 Gauging Plaster

| | Crew | Daily Output | Labor-Hours | Unit | Material | 2011 Bare Costs Labor | 2011 Bare Costs Equipment | Total | Total Incl O&P |
|---|---|---|---|---|---|---|---|---|---|
| 0010 **GAUGING PLASTER** R092000-50 | | | | | | | | | |
| 0020 100 lb. bags, less than 1 ton | | | | Bag | 18.10 | | | 18.10 | 19.90 |
| 0100 Over 1 ton | | | | " | 15.85 | | | 15.85 | 17.45 |

### 09 23 20.30 Keenes Cement

| | Crew | Daily Output | Labor-Hours | Unit | Material | 2011 Bare Costs Labor | 2011 Bare Costs Equipment | Total | Total Incl O&P |
|---|---|---|---|---|---|---|---|---|---|
| 0010 **KEENES CEMENT** R092000-50 | | | | | | | | | |
| 0020 In 100 lb. bags, less than 1 ton | | | | Bag | 26 | | | 26 | 28.50 |
| 0100 Over 1 ton | | | | " | 24.50 | | | 24.50 | 27 |
| 0300 Finish only, add to plaster prices, standard | J-1 | 215 | .186 | S.Y. | 2.36 | 7 | .57 | 9.93 | 13.65 |
| 0400 High quality | " | 144 | .278 | " | 2.39 | 10.45 | .86 | 13.70 | 19.10 |

298

# 09 24 Cement Plastering

## 09 24 23 – Cement Stucco

### 09 24 23.40 Stucco

| | | Crew | Daily Output | Labor-Hours | Unit | Material | 2011 Bare Costs Labor | Equipment | Total | Total Incl O&P |
|---|---|---|---|---|---|---|---|---|---|---|
| 0010 | **STUCCO** R092000-50 | | | | | | | | | |
| 0015 | 3 coats 1" thick, float finish, with mesh, on wood frame | J-2 | 63 | .762 | S.Y. | 5.30 | 29 | 1.97 | 36.27 | 50.50 |
| 0100 | On masonry construction, no mesh incl. | J-1 | 67 | .597 | | 2.28 | 22.50 | 1.84 | 26.62 | 38 |
| 0300 | For trowel finish, add | 1 Plas | 170 | .047 | | | 1.85 | | 1.85 | 2.76 |
| 0400 | For 3/4" thick, on masonry, deduct | J-1 | 880 | .045 | | .61 | 1.71 | .14 | 2.46 | 3.37 |
| 0600 | For coloring and special finish, add, minimum | | 685 | .058 | | .39 | 2.20 | .18 | 2.77 | 3.90 |
| 0700 | Maximum | | 200 | .200 | | 1.36 | 7.55 | .62 | 9.53 | 13.40 |
| 0900 | For soffits, add | J-2 | 155 | .310 | | 2.11 | 11.70 | .80 | 14.61 | 20.50 |
| 1000 | Exterior stucco, with bonding agent, 3 coats, on walls, no mesh incl. | J-1 | 200 | .200 | | 3.54 | 7.55 | .62 | 11.71 | 15.75 |
| 1200 | Ceilings | | 180 | .222 | | 3.54 | 8.35 | .69 | 12.58 | 17.10 |
| 1300 | Beams | | 80 | .500 | | 3.54 | 18.85 | 1.55 | 23.94 | 33.50 |
| 1500 | Columns | | 100 | .400 | | 3.54 | 15.05 | 1.24 | 19.83 | 28 |
| 1600 | Mesh, painted, nailed to wood, 1.8 lb. | 1 Lath | 60 | .133 | | 5 | 5.15 | | 10.15 | 13 |
| 1800 | 3.6 lb. | | 55 | .145 | | 3.03 | 5.60 | | 8.63 | 11.55 |
| 1900 | Wired to steel, painted, 1.8 lb. | | 53 | .151 | | 5 | 5.80 | | 10.80 | 14 |
| 2100 | 3.6 lb. | | 50 | .160 | | 3.03 | 6.15 | | 9.18 | 12.35 |

# 09 26 Veneer Plastering

## 09 26 13 – Gypsum Veneer Plastering

### 09 26 13.20 Blueboard

| | | Crew | Daily Output | Labor-Hours | Unit | Material | 2011 Bare Costs Labor | Equipment | Total | Total Incl O&P |
|---|---|---|---|---|---|---|---|---|---|---|
| 0010 | **BLUEBOARD** For use with thin coat | | | | | | | | | |
| 0100 | plaster application (see Section 09 26 13.80) | | | | | | | | | |
| 1000 | 3/8" thick, on walls or ceilings, standard, no finish included | 2 Carp | 1900 | .008 | S.F. | .26 | .36 | | .62 | .85 |
| 1100 | With thin coat plaster finish | | 875 | .018 | | .36 | .79 | | 1.15 | 1.60 |
| 1400 | On beams, columns, or soffits, standard, no finish included | | 675 | .024 | | .30 | 1.02 | | 1.32 | 1.90 |
| 1450 | With thin coat plaster finish | | 475 | .034 | | .40 | 1.45 | | 1.85 | 2.65 |
| 3000 | 1/2" thick, on walls or ceilings, standard, no finish included | | 1900 | .008 | | .29 | .36 | | .65 | .88 |
| 3100 | With thin coat plaster finish | | 875 | .018 | | .39 | .79 | | 1.18 | 1.63 |
| 3300 | Fire resistant, no finish included | | 1900 | .008 | | .29 | .36 | | .65 | .88 |
| 3400 | With thin coat plaster finish | | 875 | .018 | | .39 | .79 | | 1.18 | 1.63 |
| 3450 | On beams, columns, or soffits, standard, no finish included | | 675 | .024 | | .33 | 1.02 | | 1.35 | 1.94 |
| 3500 | With thin coat plaster finish | | 475 | .034 | | .43 | 1.45 | | 1.88 | 2.69 |
| 3700 | Fire resistant, no finish included | | 675 | .024 | | .33 | 1.02 | | 1.35 | 1.94 |
| 3800 | With thin coat plaster finish | | 475 | .034 | | .43 | 1.45 | | 1.88 | 2.69 |
| 5000 | 5/8" thick, on walls or ceilings, fire resistant, no finish included | | 1900 | .008 | | .30 | .36 | | .66 | .89 |
| 5100 | With thin coat plaster finish | | 875 | .018 | | .40 | .79 | | 1.19 | 1.65 |
| 5500 | On beams, columns, or soffits, no finish included | | 675 | .024 | | .35 | 1.02 | | 1.37 | 1.95 |
| 5600 | With thin coat plaster finish | | 475 | .034 | | .44 | 1.45 | | 1.89 | 2.71 |
| 6000 | For high ceilings, over 8' high, add | | 3060 | .005 | | | .23 | | .23 | .35 |
| 6500 | For over 3 stories high, add per story | | 6100 | .003 | | | .11 | | .11 | .17 |

### 09 26 13.80 Thin Coat Plaster

| | | Crew | Daily Output | Labor-Hours | Unit | Material | 2011 Bare Costs Labor | Equipment | Total | Total Incl O&P |
|---|---|---|---|---|---|---|---|---|---|---|
| 0010 | **THIN COAT PLASTER** | | | | | | | | | |
| 0012 | 1 coat veneer, not incl. lath | J-1 | 3600 | .011 | S.F. | .10 | .42 | .03 | .55 | .77 |
| 1000 | In 50 lb. bags | | | | Bag | 12.95 | | | 12.95 | 14.25 |

# 09 28 Backing Boards and Underlayments

## 09 28 13 – Cementitious Backing Boards

| 09 28 13.10 Cementitious Backerboard | Crew | Daily Output | Labor-Hours | Unit | Material | 2011 Bare Costs Labor | Equipment | Total | Total Incl O&P |
|---|---|---|---|---|---|---|---|---|---|
| 0010 **CEMENTITIOUS BACKERBOARD** | | | | | | | | | |
| 0070 Cementitious backerboard, on floor, 3' x 4' x 1/2" sheets | 2 Carp | 525 | .030 | S.F. | .73 | 1.31 | | 2.04 | 2.81 |
| 0080     3' x 5' x 1/2" sheets | | 525 | .030 | | .71 | 1.31 | | 2.02 | 2.79 |
| 0090     3' x 6' x 1/2" sheets | | 525 | .030 | | .68 | 1.31 | | 1.99 | 2.76 |
| 0100     3' x 4' x 5/8" sheets | | 525 | .030 | | .96 | 1.31 | | 2.27 | 3.06 |
| 0110     3' x 5' x 5/8" sheets | | 525 | .030 | | .97 | 1.31 | | 2.28 | 3.08 |
| 0120     3' x 6' x 5/8" sheets | | 525 | .030 | | .82 | 1.31 | | 2.13 | 2.92 |
| 0150     On wall, 3' x 4' x 1/2" sheets | | 350 | .046 | | .73 | 1.97 | | 2.70 | 3.82 |
| 0160     3' x 5' x 1/2" sheets | | 350 | .046 | | .71 | 1.97 | | 2.68 | 3.80 |
| 0170     3' x 6' x 1/2" sheets | | 350 | .046 | | .68 | 1.97 | | 2.65 | 3.77 |
| 0180     3' x 4' x 5/8" sheets | | 350 | .046 | | .96 | 1.97 | | 2.93 | 4.07 |
| 0190     3' x 5' x 5/8" sheets | | 350 | .046 | | .97 | 1.97 | | 2.94 | 4.09 |
| 0200     3' x 6' x 5/8" sheets | | 350 | .046 | | .82 | 1.97 | | 2.79 | 3.93 |
| 0250     On counter, 3' x 4' x 1/2" sheets | | 180 | .089 | | .73 | 3.83 | | 4.56 | 6.65 |
| 0260     3' x 5' x 1/2" sheets | | 180 | .089 | | .71 | 3.83 | | 4.54 | 6.65 |
| 0270     3' x 6' x 1/2" sheets | | 180 | .089 | | .68 | 3.83 | | 4.51 | 6.60 |
| 0300     3' x 4' x 5/8" sheets | | 180 | .089 | | .96 | 3.83 | | 4.79 | 6.90 |
| 0310     3' x 5' x 5/8" sheets | | 180 | .089 | | .97 | 3.83 | | 4.80 | 6.90 |
| 0320     3' x 6' x 5/8" sheets | | 180 | .089 | | .82 | 3.83 | | 4.65 | 6.75 |

# 09 29 Gypsum Board

## 09 29 10 – Gypsum Board Panels

| 09 29 10.30 Gypsum Board | Crew | Daily Output | Labor-Hours | Unit | Material | 2011 Bare Costs Labor | Equipment | Total | Total Incl O&P |
|---|---|---|---|---|---|---|---|---|---|
| 0010 **GYPSUM BOARD** on walls & ceilings   R092910-10 | | | | | | | | | |
| 0100     Nailed or screwed to studs unless otherwise noted | | | | | | | | | |
| 0150     3/8" thick, on walls, standard, no finish included | 2 Carp | 2000 | .008 | S.F. | .24 | .34 | | .58 | .79 |
| 0200     On ceilings, standard, no finish included | | 1800 | .009 | | .24 | .38 | | .62 | .85 |
| 0250     On beams, columns, or soffits, no finish included | | 675 | .024 | | .24 | 1.02 | | 1.26 | 1.83 |
| 0300     1/2" thick, on walls, standard, no finish included | | 2000 | .008 | | .23 | .34 | | .57 | .78 |
| 0350     Taped and finished (level 4 finish) | | 965 | .017 | | .27 | .71 | | .98 | 1.40 |
| 0390     With compound skim coat (level 5 finish) | | 775 | .021 | | .32 | .89 | | 1.21 | 1.72 |
| 0400     Fire resistant, no finish included | | 2000 | .008 | | .28 | .34 | | .62 | .84 |
| 0450     Taped and finished (level 4 finish) | | 965 | .017 | | .32 | .71 | | 1.03 | 1.46 |
| 0490     With compound skim coat (level 5 finish) | | 775 | .021 | | .37 | .89 | | 1.26 | 1.77 |
| 0500     Water resistant, no finish included | | 2000 | .008 | | .36 | .34 | | .70 | .93 |
| 0550     Taped and finished (level 4 finish) | | 965 | .017 | | .40 | .71 | | 1.11 | 1.55 |
| 0590     With compound skim coat (level 5 finish) | | 775 | .021 | | .45 | .89 | | 1.34 | 1.86 |
| 0600     Prefinished, vinyl, clipped to studs | | 900 | .018 | | .46 | .77 | | 1.23 | 1.68 |
| 0700     Mold resistant, no finish included | | 2000 | .008 | | .37 | .34 | | .71 | .94 |
| 0710     Taped and finished (level 4 finish) | | 965 | .017 | | .41 | .71 | | 1.12 | 1.56 |
| 0720     With compound skim coat (level 5 finish) | | 775 | .021 | | .46 | .89 | | 1.35 | 1.87 |
| 1000     On ceilings, standard, no finish included | | 1800 | .009 | | .23 | .38 | | .61 | .84 |
| 1050     Taped and finished (level 4 finish) | | 765 | .021 | | .27 | .90 | | 1.17 | 1.68 |
| 1090     With compound skim coat (level 5 finish) | | 610 | .026 | | .32 | 1.13 | | 1.45 | 2.09 |
| 1100     Fire resistant, no finish included | | 1800 | .009 | | .28 | .38 | | .66 | .90 |
| 1150     Taped and finished (level 4 finish) | | 765 | .021 | | .32 | .90 | | 1.22 | 1.74 |
| 1195     With compound skim coat (level 5 finish) | | 610 | .026 | | .37 | 1.13 | | 1.50 | 2.14 |
| 1200     Water resistant, no finish included | | 1800 | .009 | | .36 | .38 | | .74 | .99 |
| 1250     Taped and finished (level 4 finish) | | 765 | .021 | | .40 | .90 | | 1.30 | 1.83 |
| 1290     With compound skim coat (level 5 finish) | | 610 | .026 | | .45 | 1.13 | | 1.58 | 2.23 |
| 1310     Mold resistant, no finish included | | 1800 | .009 | | .37 | .38 | | .75 | 1 |

| 09 29 10.30 Gypsum Board | | Crew | Daily Output | Labor-Hours | Unit | Material | 2011 Bare Costs Labor | Equipment | Total | Total Incl O&P |
|---|---|---|---|---|---|---|---|---|---|---|
| 1320 | Taped and finished (level 4 finish) | 2 Carp | 765 | .021 | S.F. | .41 | .90 | | 1.31 | 1.84 |
| 1330 | With compound skim coat (level 5 finish) | | 610 | .026 | | .46 | 1.13 | | 1.59 | 2.24 |
| 1500 | On beams, columns, or soffits, standard, no finish included | | 675 | .024 | | .26 | 1.02 | | 1.28 | 1.86 |
| 1550 | Taped and finished (level 4 finish) | | 475 | .034 | | .27 | 1.45 | | 1.72 | 2.52 |
| 1590 | With compound skim coat (level 5 finish) | | 540 | .030 | | .32 | 1.28 | | 1.60 | 2.32 |
| 1600 | Fire resistant, no finish included | | 675 | .024 | | .28 | 1.02 | | 1.30 | 1.88 |
| 1650 | Taped and finished (level 4 finish) | | 475 | .034 | | .32 | 1.45 | | 1.77 | 2.58 |
| 1690 | With compound skim coat (level 5 finish) | | 540 | .030 | | .37 | 1.28 | | 1.65 | 2.37 |
| 1700 | Water resistant, no finish included | | 675 | .024 | | .41 | 1.02 | | 1.43 | 2.03 |
| 1750 | Taped and finished (level 4 finish) | | 475 | .034 | | .40 | 1.45 | | 1.85 | 2.67 |
| 1790 | With compound skim coat (level 5 finish) | | 540 | .030 | | .45 | 1.28 | | 1.73 | 2.46 |
| 1800 | Mold resistant, no finish included | | 675 | .024 | | .43 | 1.02 | | 1.45 | 2.04 |
| 1810 | Taped and finished (level 4 finish) | | 475 | .034 | | .41 | 1.45 | | 1.86 | 2.68 |
| 1820 | With compound skim coat (level 5 finish) | | 540 | .030 | | .46 | 1.28 | | 1.74 | 2.47 |
| 2000 | 5/8" thick, on walls, standard, no finish included | | 2000 | .008 | | .28 | .34 | | .62 | .84 |
| 2050 | Taped and finished (level 4 finish) | | 965 | .017 | | .32 | .71 | | 1.03 | 1.46 |
| 2090 | With compound skim coat (level 5 finish) | | 775 | .021 | | .37 | .89 | | 1.26 | 1.77 |
| 2100 | Fire resistant, no finish included | | 2000 | .008 | | .28 | .34 | | .62 | .84 |
| 2150 | Taped and finished (level 4 finish) | | 965 | .017 | | .32 | .71 | | 1.03 | 1.46 |
| 2195 | With compound skim coat (level 5 finish) | | 775 | .021 | | .37 | .89 | | 1.26 | 1.77 |
| 2200 | Water resistant, no finish included | | 2000 | .008 | | .40 | .34 | | .74 | .97 |
| 2250 | Taped and finished (level 4 finish) | | 965 | .017 | | .44 | .71 | | 1.15 | 1.59 |
| 2290 | With compound skim coat (level 5 finish) | | 775 | .021 | | .49 | .89 | | 1.38 | 1.90 |
| 2300 | Prefinished, vinyl, clipped to studs | | 900 | .018 | | .72 | .77 | | 1.49 | 1.96 |
| 2510 | Mold resistant, no finish included | | 2000 | .008 | | .42 | .34 | | .76 | .99 |
| 2520 | Taped and finished (level 4 finish) | | 965 | .017 | | .46 | .71 | | 1.17 | 1.61 |
| 2530 | With compound skim coat (level 5 finish) | | 775 | .021 | | .51 | .89 | | 1.40 | 1.92 |
| 3000 | On ceilings, standard, no finish included | | 1800 | .009 | | .28 | .38 | | .66 | .90 |
| 3050 | Taped and finished (level 4 finish) | | 765 | .021 | | .32 | .90 | | 1.22 | 1.74 |
| 3090 | With compound skim coat (level 5 finish) | | 615 | .026 | | .37 | 1.12 | | 1.49 | 2.13 |
| 3100 | Fire resistant, no finish included | | 1800 | .009 | | .28 | .38 | | .66 | .90 |
| 3150 | Taped and finished (level 4 finish) | | 765 | .021 | | .32 | .90 | | 1.22 | 1.74 |
| 3190 | With compound skim coat (level 5 finish) | | 615 | .026 | | .37 | 1.12 | | 1.49 | 2.13 |
| 3200 | Water resistant, no finish included | | 1800 | .009 | | .40 | .38 | | .78 | 1.03 |
| 3250 | Taped and finished (level 4 finish) | | 765 | .021 | | .44 | .90 | | 1.34 | 1.87 |
| 3290 | With compound skim coat (level 5 finish) | | 615 | .026 | | .49 | 1.12 | | 1.61 | 2.26 |
| 3300 | Mold resistant, no finish included | | 1800 | .009 | | .42 | .38 | | .80 | 1.05 |
| 3310 | Taped and finished (level 4 finish) | | 765 | .021 | | .46 | .90 | | 1.36 | 1.89 |
| 3320 | With compound skim coat (level 5 finish) | | 615 | .026 | | .51 | 1.12 | | 1.63 | 2.28 |
| 3500 | On beams, columns, or soffits, no finish included | | 675 | .024 | | .32 | 1.02 | | 1.34 | 1.92 |
| 3550 | Taped and finished (level 4 finish) | | 475 | .034 | | .37 | 1.45 | | 1.82 | 2.63 |
| 3590 | With compound skim coat (level 5 finish) | | 380 | .042 | | .43 | 1.81 | | 2.24 | 3.25 |
| 3600 | Fire resistant, no finish included | | 675 | .024 | | .32 | 1.02 | | 1.34 | 1.92 |
| 3650 | Taped and finished (level 4 finish) | | 475 | .034 | | .37 | 1.45 | | 1.82 | 2.63 |
| 3690 | With compound skim coat (level 5 finish) | | 380 | .042 | | .37 | 1.81 | | 2.18 | 3.19 |
| 3700 | Water resistant, no finish included | | 675 | .024 | | .46 | 1.02 | | 1.48 | 2.08 |
| 3750 | Taped and finished (level 4 finish) | | 475 | .034 | | .51 | 1.45 | | 1.96 | 2.78 |
| 3790 | With compound skim coat (level 5 finish) | | 380 | .042 | | .49 | 1.81 | | 2.30 | 3.32 |
| 3800 | Mold resistant, no finish included | | 675 | .024 | | .48 | 1.02 | | 1.50 | 2.10 |
| 3810 | Taped and finished (level 4 finish) | | 475 | .034 | | .53 | 1.45 | | 1.98 | 2.81 |
| 3820 | With compound skim coat (level 5 finish) | | 380 | .042 | | .51 | 1.81 | | 2.32 | 3.34 |
| 4000 | Fireproofing, beams or columns, 2 layers, 1/2" thick, incl finish | | 330 | .048 | | .60 | 2.09 | | 2.69 | 3.87 |
| 4010 | Mold resistant | | 330 | .048 | | .78 | 2.09 | | 2.87 | 4.06 |

# 09 29 Gypsum Board

## 09 29 10 - Gypsum Board Panels

### 09 29 10.30 Gypsum Board

| | | Crew | Daily Output | Labor-Hours | Unit | Material | 2011 Bare Costs Labor | Equipment | Total | Total Incl O&P |
|---|---|---|---|---|---|---|---|---|---|---|
| 4050 | 5/8" thick | 2 Carp | 300 | .053 | S.F. | .65 | 2.30 | | 2.95 | 4.23 |
| 4060 | Mold resistant | | 300 | .053 | | .93 | 2.30 | | 3.23 | 4.54 |
| 4100 | 3 layers, 1/2" thick | | 225 | .071 | | .88 | 3.06 | | 3.94 | 5.65 |
| 4110 | Mold resistant | | 225 | .071 | | 1.15 | 3.06 | | 4.21 | 5.95 |
| 4150 | 5/8" thick | | 210 | .076 | | .97 | 3.28 | | 4.25 | 6.10 |
| 4160 | Mold resistant | | 210 | .076 | | 1.39 | 3.28 | | 4.67 | 6.60 |
| 5050 | For 1" thick coreboard on columns | ↓ | 480 | .033 | | .56 | 1.43 | | 1.99 | 2.82 |
| 5100 | For foil-backed board, add | | | | | .14 | | | .14 | .15 |
| 5200 | For work over 8' high, add | 2 Carp | 3060 | .005 | | | .23 | | .23 | .35 |
| 5270 | For textured spray, add | 2 Lath | 1600 | .010 | | .05 | .39 | | .44 | .62 |
| 5300 | For distribution cost over 3 stories high, add per story | 2 Carp | 6100 | .003 | ↓ | | .11 | | .11 | .17 |
| 5350 | For finishing inner corners, add | | 950 | .017 | L.F. | .10 | .73 | | .83 | 1.22 |
| 5355 | For finishing outer corners, add | ↓ | 1250 | .013 | | .21 | .55 | | .76 | 1.09 |
| 5500 | For acoustical sealant, add per bead | 1 Carp | 500 | .016 | ↓ | .04 | .69 | | .73 | 1.11 |
| 5550 | Sealant, 1 quart tube | | | | Ea. | 7.05 | | | 7.05 | 7.80 |
| 5600 | Sound deadening board, 1/4" gypsum | 2 Carp | 1800 | .009 | S.F. | .34 | .38 | | .72 | .96 |
| 5650 | 1/2" wood fiber | " | 1800 | .009 | " | .34 | .38 | | .72 | .96 |

### 09 29 10.50 High Abuse Gypsum Board

| | | Crew | Daily Output | Labor-Hours | Unit | Material | 2011 Bare Costs Labor | Equipment | Total | Total Incl O&P |
|---|---|---|---|---|---|---|---|---|---|---|
| 0010 | **HIGH ABUSE GYPSUM BOARD**, fiber reinforced, nailed or | | | | | | | | | |
| 0100 | screwed to studs unless otherwise noted | | | | | | | | | |
| 0110 | 1/2" thick, on walls, no finish included | 2 Carp | 1800 | .009 | S.F. | .86 | .38 | | 1.24 | 1.54 |
| 0120 | Taped and finished (level 4 finish) | | 870 | .018 | | .90 | .79 | | 1.69 | 2.21 |
| 0130 | With compound skim coat (level 5 finish) | | 700 | .023 | | .95 | .98 | | 1.93 | 2.56 |
| 0150 | On ceilings, no finish included | | 1620 | .010 | | .86 | .43 | | 1.29 | 1.60 |
| 0160 | Taped and finished (level 4 finish) | | 690 | .023 | | .90 | 1 | | 1.90 | 2.53 |
| 0170 | With compound skim coat (level 5 finish) | | 550 | .029 | | .95 | 1.25 | | 2.20 | 2.97 |
| 0210 | 5/8" thick, on walls, no finish included | | 1800 | .009 | | .90 | .38 | | 1.28 | 1.58 |
| 0220 | Taped and finished (level 4 finish) | | 870 | .018 | | .94 | .79 | | 1.73 | 2.25 |
| 0230 | With compound skim coat (level 5 finish) | | 700 | .023 | | .99 | .98 | | 1.97 | 2.60 |
| 0250 | On ceilings, no finish included | | 1620 | .010 | | .90 | .43 | | 1.33 | 1.64 |
| 0260 | Taped and finished (level 4 finish) | | 690 | .023 | | .94 | 1 | | 1.94 | 2.57 |
| 0270 | With compound skim coat (level 5 finish) | | 550 | .029 | | .99 | 1.25 | | 2.24 | 3.01 |
| 0310 | 5/8" thick, on walls, very high impact, no finish included | | 1800 | .009 | | .98 | .38 | | 1.36 | 1.67 |
| 0320 | Taped and finished (level 4 finish) | | 870 | .018 | | 1.02 | .79 | | 1.81 | 2.34 |
| 0330 | With compound skim coat (level 5 finish) | | 700 | .023 | | 1.07 | .98 | | 2.05 | 2.69 |
| 0350 | On ceilings, no finish included | | 1620 | .010 | | .98 | .43 | | 1.41 | 1.73 |
| 0360 | Taped and finished (level 4 finish) | | 690 | .023 | | 1.02 | 1 | | 2.02 | 2.66 |
| 0370 | With compound skim coat (level 5 finish) | | 550 | .029 | | 1.07 | 1.25 | | 2.32 | 3.10 |
| 0400 | High abuse, gypsum core, paper face | | | | | | | | | |
| 0410 | 1/2" thick, on walls, no finish included | 2 Carp | 1800 | .009 | S.F. | .79 | .38 | | 1.17 | 1.46 |
| 0420 | Taped and finished (level 4 finish) | | 870 | .018 | | .83 | .79 | | 1.62 | 2.13 |
| 0430 | With compound skim coat (level 5 finish) | | 700 | .023 | | .88 | .98 | | 1.86 | 2.48 |
| 0450 | On ceilings, no finish included | | 1620 | .010 | | .79 | .43 | | 1.22 | 1.52 |
| 0460 | Taped and finished (level 4 finish) | | 690 | .023 | | .83 | 1 | | 1.83 | 2.45 |
| 0470 | With compound skim coat (level 5 finish) | | 550 | .029 | | .88 | 1.25 | | 2.13 | 2.89 |
| 0510 | 5/8" thick, on walls, no finish included | | 1800 | .009 | | .79 | .38 | | 1.17 | 1.46 |
| 0520 | Taped and finished (level 4 finish) | | 870 | .018 | | .83 | .79 | | 1.62 | 2.13 |
| 0530 | With compound skim coat (level 5 finish) | | 700 | .023 | | .88 | .98 | | 1.86 | 2.48 |
| 0550 | On ceilings, no finish included | | 1620 | .010 | | .79 | .43 | | 1.22 | 1.52 |
| 0560 | Taped and finished (level 4 finish) | | 690 | .023 | | .83 | 1 | | 1.83 | 2.45 |
| 0570 | With compound skim coat (level 5 finish) | | 550 | .029 | | .88 | 1.25 | | 2.13 | 2.89 |
| 1000 | For high ceilings, over 8' high, add | ↓ | 2750 | .006 | ↓ | | .25 | | .25 | .38 |

# 09 29 Gypsum Board

## 09 29 10 – Gypsum Board Panels

| 09 29 10.50 High Abuse Gypsum Board | Crew | Daily Output | Labor-Hours | Unit | Material | 2011 Bare Costs Labor | Equipment | Total | Total Incl O&P |
|---|---|---|---|---|---|---|---|---|---|
| 1010 | For distribution cost over 3 stories high, add per story | 2 Carp | 5500 | .003 | S.F. | | .13 | | .13 | .19 |

## 09 29 15 – Gypsum Board Accessories

### 09 29 15.10 Accessories, Gypsum Board

| | | Crew | Daily Output | Labor-Hours | Unit | Material | 2011 Bare Costs Labor | Equipment | Total | Total Incl O&P |
|---|---|---|---|---|---|---|---|---|---|---|
| 0010 | **ACCESSORIES, GYPSUM BOARD** | | | | | | | | | |
| 0020 | Casing bead, galvanized steel | 1 Carp | 2.90 | 2.759 | C.L.F. | 27 | 119 | | 146 | 212 |
| 0100 | Vinyl | | 3 | 2.667 | | 23 | 115 | | 138 | 202 |
| 0300 | Corner bead, galvanized steel, 1" x 1" | | 4 | 2 | | 14.35 | 86 | | 100.35 | 148 |
| 0400 | 1-1/4" x 1-1/4" | | 3.50 | 2.286 | | 15.30 | 98.50 | | 113.80 | 168 |
| 0600 | Vinyl | | 4 | 2 | | 16.75 | 86 | | 102.75 | 150 |
| 0900 | Furring channel, galv. steel, 7/8" deep, standard | | 2.60 | 3.077 | | 24.50 | 132 | | 156.50 | 230 |
| 1000 | Resilient | | 2.55 | 3.137 | | 27 | 135 | | 162 | 237 |
| 1100 | J trim, galvanized steel, 1/2" wide | | 3 | 2.667 | | 29 | 115 | | 144 | 208 |
| 1120 | 5/8" wide | | 2.95 | 2.712 | | 28 | 117 | | 145 | 210 |
| 1140 | L trim, galvanized | | 3 | 2.667 | | 23 | 115 | | 138 | 201 |
| 1150 | U trim, galvanized | | 2.95 | 2.712 | | 22 | 117 | | 139 | 203 |
| 1160 | Screws #6 x 1" A | | | | M | 8.45 | | | 8.45 | 9.30 |
| 1170 | #6 x 1-5/8" A | | | | " | 12.40 | | | 12.40 | 13.65 |
| 1200 | For stud partitions, see Section 05 41 13.30 and 09 22 16.13 | | | | | | | | | |
| 1500 | Z stud, galvanized steel, 1-1/2" wide | 1 Carp | 2.60 | 3.077 | C.L.F. | 38.50 | 132 | | 170.50 | 246 |
| 1600 | 2" wide | " | 2.55 | 3.137 | " | 64.50 | 135 | | 199.50 | 278 |

# 09 30 Tiling

## 09 30 13 – Ceramic Tiling

### 09 30 13.10 Ceramic Tile

| | | Crew | Daily Output | Labor-Hours | Unit | Material | 2011 Bare Costs Labor | Equipment | Total | Total Incl O&P |
|---|---|---|---|---|---|---|---|---|---|---|
| 0010 | **CERAMIC TILE** | | | | | | | | | |
| 0020 | Backsplash, thinset, average grade tiles | 1 Tilf | 50 | .160 | S.F. | 2.32 | 6.45 | | 8.77 | 12 |
| 0022 | Custom grade tiles | | 50 | .160 | | 4.65 | 6.45 | | 11.10 | 14.55 |
| 0024 | Luxury grade tiles | | 50 | .160 | | 9.30 | 6.45 | | 15.75 | 19.70 |
| 0026 | Economy grade tiles | | 50 | .160 | | 2.12 | 6.45 | | 8.57 | 11.80 |
| 0050 | Base, using 1' x 4" high pc. with 1" x 1" tiles, mud set | D-7 | 82 | .195 | L.F. | 4.81 | 7.05 | | 11.86 | 15.60 |
| 0100 | Thin set | " | 128 | .125 | | 4.46 | 4.52 | | 8.98 | 11.50 |
| 0300 | For 6" high base, 1" x 1" tile face, add | | | | | .72 | | | .72 | .79 |
| 0400 | For 2" x 2" tile face, add to above | | | | | .38 | | | .38 | .42 |
| 0600 | Cove base, 4-1/4" x 4-1/4" high, mud set | D-7 | 91 | .176 | | 3.75 | 6.35 | | 10.10 | 13.40 |
| 0700 | Thin set | | 128 | .125 | | 3.81 | 4.52 | | 8.33 | 10.80 |
| 0900 | 6" x 4-1/4" high, mud set | | 100 | .160 | | 3.20 | 5.80 | | 9 | 11.95 |
| 1000 | Thin set | | 137 | .117 | | 3.07 | 4.22 | | 7.29 | 9.55 |
| 1200 | Sanitary cove base, 6" x 4-1/4" high, mud set | | 93 | .172 | | 3.85 | 6.20 | | 10.05 | 13.30 |
| 1300 | Thin set | | 124 | .129 | | 3.75 | 4.66 | | 8.41 | 10.95 |
| 1500 | 6" x 6" high, mud set | | 84 | .190 | | 4.34 | 6.90 | | 11.24 | 14.85 |
| 1600 | Thin set | | 117 | .137 | | 4.21 | 4.94 | | 9.15 | 11.85 |
| 1800 | Bathroom accessories, average (soap dish, tooth brush holder) | | 82 | .195 | Ea. | 11 | 7.05 | | 18.05 | 22.50 |
| 1900 | Bathtub, 5', rec. 4-1/4" x 4-1/4" tile wainscot, adhesive set 6' high | | 2.90 | 5.517 | | 156 | 199 | | 355 | 465 |
| 2100 | 7' high wainscot | | 2.50 | 6.400 | | 179 | 231 | | 410 | 530 |
| 2200 | 8' high wainscot | | 2.20 | 7.273 | | 190 | 263 | | 453 | 595 |
| 2400 | Bullnose trim, 4-1/4" x 4-1/4", mud set | | 82 | .195 | L.F. | 3.52 | 7.05 | | 10.57 | 14.15 |
| 2500 | Thin set | | 128 | .125 | | 3.37 | 4.52 | | 7.89 | 10.30 |
| 2700 | 6" x 4-1/4" bullnose trim, mud set | | 84 | .190 | | 2.70 | 6.90 | | 9.60 | 13 |
| 2800 | Thin set | | 124 | .129 | | 2.61 | 4.66 | | 7.27 | 9.65 |
| 3000 | Floors, natural clay, random or uniform, thin set, color group 1 | | 183 | .087 | S.F. | 3.83 | 3.16 | | 6.99 | 8.85 |

## 09 30 13 – Ceramic Tiling

| 09 30 13.10 Ceramic Tile | | Crew | Daily Output | Labor-Hours | Unit | Material | 2011 Bare Costs Labor | 2011 Bare Costs Equipment | Total | Total Incl O&P |
|---|---|---|---|---|---|---|---|---|---|---|
| 3100 | Color group 2 | D-7 | 183 | .087 | S.F. | 5.15 | 3.16 | | 8.31 | 10.25 |
| 3255 | Floors, glazed, thin set, 6" x 6", color group 1 | | 200 | .080 | | 4.43 | 2.89 | | 7.32 | 9.10 |
| 3260 | 8" x 8" tile | | 250 | .064 | | 4.43 | 2.31 | | 6.74 | 8.25 |
| 3270 | 12" x 12" tile | | 325 | .049 | | 5.25 | 1.78 | | 7.03 | 8.35 |
| 3280 | 16" x 16" tile | | 550 | .029 | | 6.05 | 1.05 | | 7.10 | 8.20 |
| 3281 | 18" x 18" tile | | 600 | .027 | | 6.05 | .96 | | 7.01 | 8.05 |
| 3283 | 24" x 24" tile | | 650 | .025 | | 6.05 | .89 | | 6.94 | 7.95 |
| 3285 | Border, 6" x 12" tile | | 275 | .058 | | 11.55 | 2.10 | | 13.65 | 15.80 |
| 3290 | 3" x 12" tile | | 200 | .080 | | 34 | 2.89 | | 36.89 | 41 |
| 3300 | Porcelain type, 1 color, color group 2, 1" x 1" | | 183 | .087 | | 4.39 | 3.16 | | 7.55 | 9.45 |
| 3310 | 2" x 2" or 2" x 1", thin set | ▼ | 190 | .084 | | 4.26 | 3.04 | | 7.30 | 9.15 |
| 3350 | For random blend, 2 colors, add | | | | | .88 | | | .88 | .97 |
| 3360 | 4 colors, add | | | | | 1.24 | | | 1.24 | 1.36 |
| 3370 | For color group 3, add | | | | | .50 | | | .50 | .55 |
| 3380 | For abrasive non-slip tile, add | | | | | .49 | | | .49 | .54 |
| 4300 | Specialty tile, 4-1/4" x 4-1/4" x 1/2", decorator finish | D-7 | 183 | .087 | | 9.90 | 3.16 | | 13.06 | 15.50 |
| 4500 | Add for epoxy grout, 1/16" joint, 1" x 1" tile | | 800 | .020 | | .62 | .72 | | 1.34 | 1.73 |
| 4600 | 2" x 2" tile | ▼ | 820 | .020 | ▼ | .59 | .71 | | 1.30 | 1.68 |
| 4800 | Pregrouted sheets, walls, 4-1/4" x 4-1/4", 6" x 4-1/4" | | | | | | | | | |
| 4810 | and 8-1/2" x 4-1/4", 4 S.F. sheets, silicone grout | D-7 | 240 | .067 | S.F. | 4.73 | 2.41 | | 7.14 | 8.70 |
| 5100 | Floors, unglazed, 2 S.F. sheets, | | | | | | | | | |
| 5110 | Urethane adhesive | D-7 | 180 | .089 | S.F. | 4.71 | 3.21 | | 7.92 | 9.90 |
| 5400 | Walls, interior, thin set, 4-1/4" x 4-1/4" tile | | 190 | .084 | | 2.14 | 3.04 | | 5.18 | 6.80 |
| 5500 | 6" x 4-1/4" tile | | 190 | .084 | | 2.73 | 3.04 | | 5.77 | 7.45 |
| 5700 | 8-1/2" x 4-1/4" tile | | 190 | .084 | | 3.80 | 3.04 | | 6.84 | 8.60 |
| 5800 | 6" x 6" tile | | 200 | .080 | | 2.38 | 2.89 | | 5.27 | 6.85 |
| 5810 | 8" x 8" tile | | 225 | .071 | | 4.08 | 2.57 | | 6.65 | 8.25 |
| 5820 | 12" x 12" tile | | 300 | .053 | | 3.53 | 1.93 | | 5.46 | 6.70 |
| 5830 | 16" x 16" tile | | 500 | .032 | | 4.13 | 1.16 | | 5.29 | 6.25 |
| 6000 | Decorated wall tile, 4-1/4" x 4-1/4", minimum | | 270 | .059 | | 2.64 | 2.14 | | 4.78 | 6.05 |
| 6100 | Maximum | | 180 | .089 | | 44 | 3.21 | | 47.21 | 52.50 |
| 6300 | Exterior walls, frostproof, mud set, 4-1/4" x 4-1/4" | | 102 | .157 | | 7 | 5.65 | | 12.65 | 15.95 |
| 6400 | 1-3/8" x 1-3/8" | | 93 | .172 | | 4.65 | 6.20 | | 10.85 | 14.15 |
| 6600 | Crystalline glazed, 4-1/4" x 4-1/4", mud set, plain | | 100 | .160 | | 3.78 | 5.80 | | 9.58 | 12.60 |
| 6700 | 4-1/4" x 4-1/4", scored tile | | 100 | .160 | | 4.95 | 5.80 | | 10.75 | 13.90 |
| 6900 | 6" x 6" plain | | 93 | .172 | | 5.15 | 6.20 | | 11.35 | 14.70 |
| 7000 | For epoxy grout, 1/16" joints, 4-1/4" tile, add | | 800 | .020 | | .39 | .72 | | 1.11 | 1.48 |
| 7200 | For tile set in dry mortar, add | | 1735 | .009 | | | .33 | | .33 | .49 |
| 7300 | For tile set in Portland cement mortar, add | ▼ | 290 | .055 | | .16 | 1.99 | | 2.15 | 3.09 |
| 9000 | Regrout tile 4-1/2 x 4-1/2, or larger, wall | 1 Tilf | 100 | .080 | | .14 | 3.24 | | 3.38 | 4.87 |
| 9220 | Floor | " | 125 | .064 | | .15 | 2.59 | | 2.74 | 3.95 |
| 9300 | Ceramic tiles, recycled glass, standard colors, 2" x 2" thru 6" x 6" G | D-7 | 190 | .084 | | 17 | 3.04 | | 20.04 | 23 |
| 9310 | 6" x 6" G | | 200 | .080 | | 17 | 2.89 | | 19.89 | 23 |
| 9320 | 8" x 8" G | | 225 | .071 | | 18.45 | 2.57 | | 21.02 | 24.50 |
| 9330 | 12" x 12" G | | 300 | .053 | | 18.45 | 1.93 | | 20.38 | 23.50 |
| 9340 | Earthtones, 2" x 2" to 4" x 8" G | | 190 | .084 | | 21.50 | 3.04 | | 24.54 | 28 |
| 9350 | 6" x 6" G | | 200 | .080 | | 21.50 | 2.89 | | 24.39 | 27.50 |
| 9360 | 8" x 8" G | | 225 | .071 | | 23 | 2.57 | | 25.57 | 29 |
| 9370 | 12" x 12" G | | 300 | .053 | | 23 | 1.93 | | 24.93 | 28 |
| 9380 | Deep colors, 2" x 2" to 4" x 8" G | | 190 | .084 | | 31 | 3.04 | | 34.04 | 38.50 |
| 9390 | 6" x 6" G | | 200 | .080 | | 31 | 2.89 | | 33.89 | 38 |
| 9400 | 8" x 8" G | | 225 | .071 | | 32 | 2.57 | | 34.57 | 39.50 |
| 9410 | 12" x 12" G | | 300 | .053 | ▼ | 32 | 1.93 | | 33.93 | 38.50 |

# 09 30 Tiling

## 09 30 13 – Ceramic Tiling

| 09 30 13.45 Ceramic Tile Accessories | Crew | Daily Output | Labor-Hours | Unit | Material | 2011 Bare Costs Labor | Equipment | Total | Total Incl O&P |
|---|---|---|---|---|---|---|---|---|---|
| 0010 **CERAMIC TILE ACCESSORIES** | | | | | | | | | |
| 0100 Spacers, 1/8" | | | | Ea. | 1.44 | | | 1.44 | 1.58 |
| 1310 Sealer, installed | 1 Tilf | 650 | .012 | S.F. | .38 | .50 | | .88 | 1.15 |

## 09 30 16 – Quarry Tiling

### 09 30 16.10 Quarry Tile

| | Crew | Daily Output | Labor-Hours | Unit | Material | Labor | Equipment | Total | Total Incl O&P |
|---|---|---|---|---|---|---|---|---|---|
| 0010 **QUARRY TILE** | | | | | | | | | |
| 0100 Base, cove or sanitary, mud set, to 5" high, 1/2" thick | D-7 | 110 | .145 | L.F. | 5.30 | 5.25 | | 10.55 | 13.45 |
| 0300 Bullnose trim, red, mud set, 6" x 6" x 1/2" thick | | 120 | .133 | | 4.34 | 4.82 | | 9.16 | 11.85 |
| 0400 4" x 4" x 1/2" thick | | 110 | .145 | | 4.82 | 5.25 | | 10.07 | 12.95 |
| 0600 4" x 8" x 1/2" thick, using 8" as edge | | 130 | .123 | | 4.26 | 4.45 | | 8.71 | 11.20 |
| 0700 Floors, mud set, 1,000 S.F. lots, red, 4" x 4" x 1/2" thick | | 120 | .133 | S.F. | 6.10 | 4.82 | | 10.92 | 13.75 |
| 0900 6" x 6" x 1/2" thick | | 140 | .114 | | 5.50 | 4.13 | | 9.63 | 12.10 |
| 1000 4" x 8" x 1/2" thick | | 130 | .123 | | 6.10 | 4.45 | | 10.55 | 13.20 |
| 1300 For waxed coating, add | | | | | .71 | | | .71 | .78 |
| 1500 For non-standard colors, add | | | | | .42 | | | .42 | .46 |
| 1600 For abrasive surface, add | | | | | .49 | | | .49 | .54 |
| 1800 Brown tile, imported, 6" x 6" x 3/4" | D-7 | 120 | .133 | | 7.30 | 4.82 | | 12.12 | 15.05 |
| 1900 8" x 8" x 1" | | 110 | .145 | | 7.95 | 5.25 | | 13.20 | 16.40 |
| 2100 For thin set mortar application, deduct | | 700 | .023 | | | .83 | | .83 | 1.21 |
| 2200 For epoxy grout & mortar, 6" x 6" x 1/2", add | | 350 | .046 | | 1.94 | 1.65 | | 3.59 | 4.54 |
| 2700 Stair tread, 6" x 6" x 3/4", plain | | 50 | .320 | | 5.15 | 11.55 | | 16.70 | 22.50 |
| 2800 Abrasive | | 47 | .340 | | 5.60 | 12.30 | | 17.90 | 24 |
| 3000 Wainscot, 6" x 6" x 1/2", thin set, red | | 105 | .152 | | 4.31 | 5.50 | | 9.81 | 12.80 |
| 3100 Non-standard colors | | 105 | .152 | | 4.78 | 5.50 | | 10.28 | 13.30 |
| 3300 Window sill, 6" wide, 3/4" thick | | 90 | .178 | L.F. | 5.20 | 6.45 | | 11.65 | 15.10 |
| 3400 Corners | | 80 | .200 | Ea. | 6 | 7.25 | | 13.25 | 17.15 |

## 09 30 23 – Glass Mosaic Tiling

### 09 30 23.10 Glass Mosaics

| | Crew | Daily Output | Labor-Hours | Unit | Material | Labor | Equipment | Total | Total Incl O&P |
|---|---|---|---|---|---|---|---|---|---|
| 0010 **GLASS MOSAICS** 3/4" tile on 12" sheets, standard grout | | | | | | | | | |
| 0300 Color group 1 & 2 | D-7 | 73 | .219 | S.F. | 19 | 7.90 | | 26.90 | 32.50 |
| 0350 Color group 3 | | 73 | .219 | | 19.30 | 7.90 | | 27.20 | 33 |
| 0400 Color group 4 | | 73 | .219 | | 25 | 7.90 | | 32.90 | 39 |
| 0450 Color group 5 | | 73 | .219 | | 28.50 | 7.90 | | 36.40 | 43 |
| 0500 Color group 6 | | 73 | .219 | | 40.50 | 7.90 | | 48.40 | 56 |
| 0600 Color group 7 | | 73 | .219 | | 43.50 | 7.90 | | 51.40 | 59.50 |
| 0700 Color group 8, golds, silvers & specialties | | 64 | .250 | | 62 | 9.05 | | 71.05 | 81.50 |

## 09 30 29 – Metal Tiling

### 09 30 29.10 Metal Tile

| | Crew | Daily Output | Labor-Hours | Unit | Material | Labor | Equipment | Total | Total Incl O&P |
|---|---|---|---|---|---|---|---|---|---|
| 0010 **METAL TILE** 4' x 4' sheet, 24 ga., tile pattern, nailed | | | | | | | | | |
| 0200 Stainless steel | 2 Carp | 512 | .031 | S.F. | 25.50 | 1.35 | | 26.85 | 30 |
| 0400 Aluminized steel | " | 512 | .031 | " | 13.80 | 1.35 | | 15.15 | 17.25 |

# 09 34 Waterproofing-Membrane Tiling

## 09 34 13 – Waterproofing-Membrane Ceramic Tiling

| 09 34 13.10 Ceramic Tile Waterproofing Membrane | Crew | Daily Output | Labor-Hours | Unit | Material | 2011 Bare Costs Labor | Equipment | Total | Total Incl O&P |
|---|---|---|---|---|---|---|---|---|---|
| 0010 **CERAMIC TILE WATERPROOFING MEMBRANE** | | | | | | | | | |
| 0020 On floors, including thinset | | | | | | | | | |
| 0030 Fleece laminated polyethylene grid, 1/8" thick | D-7 | 250 | .064 | S.F. | 2.24 | 2.31 | | 4.55 | 5.85 |
| 0040 5/16" thick | " | 250 | .064 | " | 2.60 | 2.31 | | 4.91 | 6.25 |
| 0050 On walls, including thinset | | | | | | | | | |
| 0060 Fleece laminated polyethylene sheet, 8 mil thick | D-7 | 480 | .033 | S.F. | 2.24 | 1.20 | | 3.44 | 4.22 |
| 0070 Accessories, including thinset | | | | | | | | | |
| 0080 Joint and corner sheet, 4 mils thick, 5" wide | 1 Tilf | 240 | .033 | L.F. | 1.45 | 1.35 | | 2.80 | 3.56 |
| 0090 7-1/4" wide | | 180 | .044 | | 1.84 | 1.80 | | 3.64 | 4.64 |
| 0100 10" wide | | 120 | .067 | ▼ | 2.24 | 2.70 | | 4.94 | 6.40 |
| 0110 Pre-formed corners, inside | | 32 | .250 | Ea. | 3.93 | 10.10 | | 14.03 | 19.05 |
| 0120 Outside | | 32 | .250 | | 7.40 | 10.10 | | 17.50 | 23 |
| 0130 2" flanged floor drain with 6" stainless steel grate | | 16 | .500 | ▼ | 355 | 20 | | 375 | 420 |
| 0140 EPS, sloped shower floor | | 480 | .017 | S.F. | 4.82 | .67 | | 5.49 | 6.30 |
| 0150 Curb | ▼ | 32 | .250 | L.F. | 13.55 | 10.10 | | 23.65 | 29.50 |

# 09 51 Acoustical Ceilings

## 09 51 23 – Acoustical Tile Ceilings

### 09 51 23.10 Suspended Acoustic Ceiling Tiles

| 09 51 23.10 Suspended Acoustic Ceiling Tiles | Crew | Daily Output | Labor-Hours | Unit | Material | 2011 Bare Costs Labor | Equipment | Total | Total Incl O&P |
|---|---|---|---|---|---|---|---|---|---|
| 0010 **SUSPENDED ACOUSTIC CEILING TILES**, not including | | | | | | | | | |
| 0100 suspension system | | | | | | | | | |
| 0300 Fiberglass boards, film faced, 2' x 2' or 2' x 4', 5/8" thick | 1 Carp | 625 | .013 | S.F. | .72 | .55 | | 1.27 | 1.64 |
| 0400 3/4" thick | | 600 | .013 | | 1.66 | .57 | | 2.23 | 2.71 |
| 0500 3" thick, thermal, R11 | | 450 | .018 | | 1.53 | .77 | | 2.30 | 2.85 |
| 0600 Glass cloth faced fiberglass, 3/4" thick | | 500 | .016 | | 1.49 | .69 | | 2.18 | 2.70 |
| 0700 1" thick | | 485 | .016 | | 2.18 | .71 | | 2.89 | 3.49 |
| 0820 1-1/2" thick, nubby face | | 475 | .017 | | 2.66 | .73 | | 3.39 | 4.04 |
| 1110 Mineral fiber tile, lay-in, 2' x 2' or 2' x 4', 5/8" thick, fine texture | | 625 | .013 | | .59 | .55 | | 1.14 | 1.50 |
| 1115 Rough textured | | 625 | .013 | | 1.44 | .55 | | 1.99 | 2.43 |
| 1125 3/4" thick, fine textured | | 600 | .013 | | 1.50 | .57 | | 2.07 | 2.53 |
| 1130 Rough textured | | 600 | .013 | | 1.79 | .57 | | 2.36 | 2.85 |
| 1135 Fissured | | 600 | .013 | | 2.45 | .57 | | 3.02 | 3.58 |
| 1150 Tegular, 5/8" thick, fine textured | | 470 | .017 | | 1.45 | .73 | | 2.18 | 2.72 |
| 1155 Rough textured | | 470 | .017 | | 1.90 | .73 | | 2.63 | 3.21 |
| 1165 3/4" thick, fine textured | | 450 | .018 | | 2.08 | .77 | | 2.85 | 3.46 |
| 1170 Rough textured | | 450 | .018 | | 2.35 | .77 | | 3.12 | 3.76 |
| 1175 Fissured | ▼ | 450 | .018 | | 3.63 | .77 | | 4.40 | 5.15 |
| 1180 For aluminum face, add | | | | | 6.05 | | | 6.05 | 6.65 |
| 1185 For plastic film face, add | | | | | .94 | | | .94 | 1.03 |
| 1190 For fire rating, add | | | | | .45 | | | .45 | .50 |
| 1300 Mirror tile, acrylic, 2' x 2' | 1 Carp | 500 | .016 | | 8.70 | .69 | | 9.39 | 10.60 |
| 1900 Eggcrate, acrylic, 1/2" x 1/2" x 1/2" cubes | | 500 | .016 | | 1.82 | .69 | | 2.51 | 3.06 |
| 2100 Polystyrene eggcrate, 3/8" x 3/8" x 1/2" cubes | | 510 | .016 | | 1.53 | .68 | | 2.21 | 2.72 |
| 2200 1/2" x 1/2" x 1/2" cubes | | 500 | .016 | | 1.75 | .69 | | 2.44 | 2.99 |
| 2400 Luminous panels, prismatic, acrylic | | 400 | .020 | | 2.22 | .86 | | 3.08 | 3.76 |
| 2500 Polystyrene | | 400 | .020 | | 1.14 | .86 | | 2 | 2.57 |
| 2700 Flat white acrylic | | 400 | .020 | | 3.86 | .86 | | 4.72 | 5.55 |
| 2800 Polystyrene | | 400 | .020 | | 2.65 | .86 | | 3.51 | 4.24 |
| 3000 Drop pan, white, acrylic | | 400 | .020 | | 5.65 | .86 | | 6.51 | 7.55 |
| 3100 Polystyrene | | 400 | .020 | | 4.73 | .86 | | 5.59 | 6.50 |
| 3600 Perforated aluminum sheets, .024" thick, corrugated, painted | ▼ | 490 | .016 | | 2.25 | .70 | | 2.95 | 3.56 |

# 09 51 Acoustical Ceilings

## 09 51 23 – Acoustical Tile Ceilings

### 09 51 23.10 Suspended Acoustic Ceiling Tiles

| | | Crew | Daily Output | Labor-Hours | Unit | Material | 2011 Bare Costs Labor | Equipment | Total | Total Incl O&P |
|---|---|---|---|---|---|---|---|---|---|---|
| 3700 | Plain | 1 Carp | 500 | .016 | S.F. | 3.76 | .69 | | 4.45 | 5.20 |
| 3720 | Mineral fiber, 24" x 24" or 48", reveal edge, painted, 5/8" thick | | 600 | .013 | | 1.38 | .57 | | 1.95 | 2.40 |
| 3740 | 3/4" thick | | 575 | .014 | | 2.24 | .60 | | 2.84 | 3.38 |
| 5020 | 66 – 78% recycled content, 3/4" thick  G | | 600 | .013 | | 1.82 | .57 | | 2.39 | 2.88 |
| 5040 | Mylar, 42% recycled content, 3/4" thick  G | | 600 | .013 | | 4.28 | .57 | | 4.85 | 5.60 |

### 09 51 23.30 Suspended Ceilings, Complete

| | | Crew | Daily Output | Labor-Hours | Unit | Material | 2011 Bare Costs Labor | Equipment | Total | Total Incl O&P |
|---|---|---|---|---|---|---|---|---|---|---|
| 0010 | **SUSPENDED CEILINGS, COMPLETE** Including standard | | | | | | | | | |
| 0100 | suspension system but not incl. 1-1/2" carrier channels | | | | | | | | | |
| 0600 | Fiberglass ceiling board, 2' x 4' x 5/8', plain faced | 1 Carp | 500 | .016 | S.F. | 1.37 | .69 | | 2.06 | 2.57 |
| 0700 | Offices, 2' x 4' x 3/4" | | 380 | .021 | | 2.31 | .91 | | 3.22 | 3.93 |
| 0800 | Mineral fiber, on 15/16" T bar susp. 2' x 2' x 3/4" lay-in board | | 345 | .023 | | 2.31 | 1 | | 3.31 | 4.07 |
| 0810 | 2' x 4' x 5/8" tile | | 380 | .021 | | 1.24 | .91 | | 2.15 | 2.75 |
| 0820 | Tegular, 2' x 2' x 5/8" tile on 9/16" grid | | 250 | .032 | | 2.42 | 1.38 | | 3.80 | 4.77 |
| 0830 | 2' x 4' x 3/4" tile | | 275 | .029 | | 2.89 | 1.25 | | 4.14 | 5.10 |
| 0900 | Luminous panels, prismatic, acrylic | | 255 | .031 | | 2.87 | 1.35 | | 4.22 | 5.25 |
| 1200 | Metal pan with acoustic pad, steel | | 75 | .107 | | 4.18 | 4.59 | | 8.77 | 11.65 |
| 1300 | Painted aluminum | | 75 | .107 | | 2.90 | 4.59 | | 7.49 | 10.25 |
| 1500 | Aluminum, degreased finish | | 75 | .107 | | 4.82 | 4.59 | | 9.41 | 12.35 |
| 1600 | Stainless steel | | 75 | .107 | | 9.05 | 4.59 | | 13.64 | 17 |
| 1800 | Tile, Z bar suspension, 5/8" mineral fiber tile | | 150 | .053 | | 2.16 | 2.30 | | 4.46 | 5.90 |
| 1900 | 3/4" mineral fiber tile | | 150 | .053 | | 2.31 | 2.30 | | 4.61 | 6.05 |
| 2400 | For strip lighting, see Section 26 51 13.50 | | | | | | | | | |
| 2500 | For rooms under 500 S.F., add | | | | S.F. | | 25% | | | |

## 09 51 53 – Direct-Applied Acoustical Ceilings

### 09 51 53.10 Ceiling Tile

| | | Crew | Daily Output | Labor-Hours | Unit | Material | 2011 Bare Costs Labor | Equipment | Total | Total Incl O&P |
|---|---|---|---|---|---|---|---|---|---|---|
| 0010 | **CEILING TILE**, Stapled or cemented | | | | | | | | | |
| 0100 | 12" x 12" or 12" x 24", not including furring | | | | | | | | | |
| 0600 | Mineral fiber, vinyl coated, 5/8" thick | 1 Carp | 300 | .027 | S.F. | 1.79 | 1.15 | | 2.94 | 3.73 |
| 0700 | 3/4" thick | | 300 | .027 | | 1.91 | 1.15 | | 3.06 | 3.86 |
| 0900 | Fire rated, 3/4" thick, plain faced | | 300 | .027 | | 1.40 | 1.15 | | 2.55 | 3.30 |
| 1000 | Plastic coated face | | 300 | .027 | | 1.18 | 1.15 | | 2.33 | 3.06 |
| 1200 | Aluminum faced, 5/8" thick, plain | | 300 | .027 | | 1.51 | 1.15 | | 2.66 | 3.42 |
| 3700 | Wall application of above, add | | 1000 | .008 | | | .34 | | .34 | .53 |
| 3900 | For ceiling primer, add | | | | | .13 | | | .13 | .14 |
| 4000 | For ceiling cement, add | | | | | .37 | | | .37 | .41 |

# 09 53 Acoustical Ceiling Suspension Assemblies

## 09 53 23 – Metal Acoustical Ceiling Suspension Assemblies

### 09 53 23.30 Ceiling Suspension Systems

| | | Crew | Daily Output | Labor-Hours | Unit | Material | 2011 Bare Costs Labor | Equipment | Total | Total Incl O&P |
|---|---|---|---|---|---|---|---|---|---|---|
| 0010 | **CEILING SUSPENSION SYSTEMS** for boards and tile | | | | | | | | | |
| 0050 | Class A suspension system, 15/16" T bar, 2' x 4' grid | 1 Carp | 800 | .010 | S.F. | .65 | .43 | | 1.08 | 1.38 |
| 0300 | 2' x 2' grid | | 650 | .012 | | .81 | .53 | | 1.34 | 1.70 |
| 0310 | 25% recycled steel, 2' x 4' grid  G | | 800 | .010 | | .74 | .43 | | 1.17 | 1.47 |
| 0320 | 2' x 2' grid  G | | 650 | .012 | | .93 | .53 | | 1.46 | 1.83 |
| 0350 | For 9/16" grid, add | | | | | .16 | | | .16 | .18 |
| 0360 | For fire rated grid, add | | | | | .09 | | | .09 | .10 |
| 0370 | For colored grid, add | | | | | .21 | | | .21 | .23 |
| 0400 | Concealed Z bar suspension system, 12" module | 1 Carp | 520 | .015 | | .66 | .66 | | 1.32 | 1.75 |
| 0600 | 1-1/2" carrier channels, 4' O.C., add | " | 470 | .017 | | .10 | .73 | | .83 | 1.22 |
| 0700 | Carrier channels for ceilings with | | | | | | | | | |

# 09 53 Acoustical Ceiling Suspension Assemblies

## 09 53 23 – Metal Acoustical Ceiling Suspension Assemblies

| 09 53 23.30 Ceiling Suspension Systems | Crew | Daily Output | Labor-Hours | Unit | Material | 2011 Bare Costs Labor | Equipment | Total | Total Incl O&P |
|---|---|---|---|---|---|---|---|---|---|
| 0900 | recessed lighting fixtures, add | 1 Carp | 460 | .017 | S.F. | .17 | .75 | | .92 | 1.34 |
| 1040 | Hanging wire, 12 ga., 4' long | | 65 | .123 | C.S.F. | 7.55 | 5.30 | | 12.85 | 16.45 |
| 1080 | 8' long | | 65 | .123 | " | 15.15 | 5.30 | | 20.45 | 25 |
| 1200 | Seismic bracing, 2-1/2" compression post with four bracing wires | | 50 | .160 | Ea. | 2.74 | 6.90 | | 9.64 | 13.55 |

# 09 54 Specialty Ceilings

## 09 54 33 – Decorative Panel Ceilings

### 09 54 33.20 Metal Panel Ceilings

| | | Crew | Daily Output | Labor-Hours | Unit | Material | Labor | Equipment | Total | Total Incl O&P |
|---|---|---|---|---|---|---|---|---|---|---|
| 0010 | **METAL PANEL CEILINGS** | | | | | | | | | |
| 0020 | Lay-in or screwed to furring, not including grid | | | | | | | | | |
| 0100 | Tin ceilings, 2' x 2' or 2' x 4', bare steel finish | 2 Carp | 300 | .053 | S.F. | 2.37 | 2.30 | | 4.67 | 6.15 |
| 0120 | Painted white finish | | 300 | .053 | " | 3.66 | 2.30 | | 5.96 | 7.55 |
| 0140 | Copper, chrome or brass finish | | 300 | .053 | L.F. | 6.40 | 2.30 | | 8.70 | 10.55 |
| 0200 | Cornice molding, 2-1/2" to 3-1/2" wide, 4' long, bare steel finish | | 200 | .080 | S.F. | 2.25 | 3.44 | | 5.69 | 7.75 |
| 0220 | Painted white finish | | 200 | .080 | | 2.94 | 3.44 | | 6.38 | 8.55 |
| 0240 | Copper, chrome or brass finish | | 200 | .080 | | 3.78 | 3.44 | | 7.22 | 9.45 |
| 0320 | 5" to 6-1/2" wide, 4' long, bare steel finish | | 150 | .107 | | 3.19 | 4.59 | | 7.78 | 10.55 |
| 0340 | Painted white finish | | 150 | .107 | | 4.20 | 4.59 | | 8.79 | 11.65 |
| 0360 | Copper, chrome or brass finish | | 150 | .107 | | 6.40 | 4.59 | | 10.99 | 14.10 |
| 0420 | Flat molding, 3-1/2" to 5" wide, 4' long, bare steel finish | | 250 | .064 | | 3.43 | 2.76 | | 6.19 | 8 |
| 0440 | Painted white finish | | 250 | .064 | | 4.09 | 2.76 | | 6.85 | 8.75 |
| 0460 | Copper, chrome or brass finish | | 250 | .064 | | 7.35 | 2.76 | | 10.11 | 12.35 |

# 09 62 Specialty Flooring

## 09 62 19 – Laminate Flooring

### 09 62 19.10 Floating Floor

| | | Crew | Daily Output | Labor-Hours | Unit | Material | Labor | Equipment | Total | Total Incl O&P |
|---|---|---|---|---|---|---|---|---|---|---|
| 0010 | **FLOATING FLOOR** | | | | | | | | | |
| 8300 | Floating floor, laminate, wood pattern strip, complete | 1 Clab | 133 | .060 | S.F. | 4.57 | 2.07 | | 6.64 | 8.20 |
| 8310 | Components, T & G wood composite strips | | | | | 3.66 | | | 3.66 | 4.03 |
| 8320 | Film | | | | | .15 | | | .15 | .16 |
| 8330 | Foam | | | | | .24 | | | .24 | .27 |
| 8340 | Adhesive | | | | | .50 | | | .50 | .55 |
| 8350 | Installation kit | | | | | .17 | | | .17 | .19 |
| 8360 | Trim, 2" wide x 3' long | | | | L.F. | 3.85 | | | 3.85 | 4.24 |
| 8370 | Reducer moulding | | | | " | 5.20 | | | 5.20 | 5.70 |

## 09 62 23 – Bamboo Flooring

### 09 62 23.10 Flooring, Bamboo

| | | | Crew | Daily Output | Labor-Hours | Unit | Material | Labor | Equipment | Total | Total Incl O&P |
|---|---|---|---|---|---|---|---|---|---|---|---|
| 0010 | **FLOORING, BAMBOO** | | | | | | | | | | |
| 8600 | Flooring, wood, bamboo strips, unfinished, 5/8" x 4" x 3' | G | 1 Carp | 255 | .031 | S.F. | 4.44 | 1.35 | | 5.79 | 6.95 |
| 8610 | 5/8" x 4" x 4' | G | | 275 | .029 | | 4.61 | 1.25 | | 5.86 | 6.95 |
| 8620 | 5/8" x 4" x 6' | G | | 295 | .027 | | 5.05 | 1.17 | | 6.22 | 7.35 |
| 8630 | Finished, 5/8" x 4" x 3' | G | | 255 | .031 | | 4.89 | 1.35 | | 6.24 | 7.45 |
| 8640 | 5/8" x 4" x 4' | G | | 275 | .029 | | 5.10 | 1.25 | | 6.35 | 7.55 |
| 8650 | 5/8" x 4" x 6' | G | | 295 | .027 | | 4.74 | 1.17 | | 5.91 | 7 |
| 8660 | Stair treads, unfinished, 1-1/16" x 11-1/2" x 4' | G | | 18 | .444 | Ea. | 42.50 | 19.15 | | 61.65 | 76.50 |
| 8670 | Finished, 1-1/16" x 11-1/2" x 4' | G | | 18 | .444 | | 71.50 | 19.15 | | 90.65 | 109 |
| 8680 | Stair risers, unfinished, 5/8" x 7-1/2" x 4' | G | | 18 | .444 | | 15.75 | 19.15 | | 34.90 | 47 |
| 8690 | Finished, 5/8" x 7-1/2" x 4' | G | | 18 | .444 | | 30 | 19.15 | | 49.15 | 62.50 |
| 8700 | Stair nosing, unfinished, 6' long | G | | 16 | .500 | | 35 | 21.50 | | 56.50 | 71.50 |

# 09 62 Specialty Flooring

## 09 62 23 – Bamboo Flooring

| 09 62 23.10 Flooring, Bamboo | | Crew | Daily Output | Labor-Hours | Unit | Material | 2011 Bare Costs Labor | Equipment | Total | Total Incl O&P |
|---|---|---|---|---|---|---|---|---|---|---|
| 8710 | Finished, 6' long | G | 1 Carp | 16 | .500 | Ea. | 39.50 | 21.50 | | 61 | 76.50 |

# 09 63 Masonry Flooring

## 09 63 13 – Brick Flooring

### 09 63 13.10 Miscellaneous Brick Flooring

| | 09 63 13.10 Miscellaneous Brick Flooring | | Crew | Daily Output | Labor-Hours | Unit | Material | 2011 Bare Costs Labor | Equipment | Total | Total Incl O&P |
|---|---|---|---|---|---|---|---|---|---|---|---|
| 0010 | **MISCELLANEOUS BRICK FLOORING** | | | | | | | | | | |
| 0020 | Acid-proof shales, red, 8" x 3-3/4" x 1-1/4" thick | | D-7 | .43 | 37.209 | M | 860 | 1,350 | | 2,210 | 2,900 |
| 0050 | 2-1/4" thick | | D-1 | .40 | 40 | | 1,000 | 1,550 | | 2,550 | 3,450 |
| 0200 | Acid-proof clay brick, 8" x 3-3/4" x 2-1/4" thick | G | | .40 | 40 | | 895 | 1,550 | | 2,445 | 3,325 |
| 0250 | 9" x 4-1/2" x 3" | G | | 95 | .168 | S.F. | 3.91 | 6.55 | | 10.46 | 14.15 |
| 0260 | Cast ceramic, pressed, 4" x 8" x 1/2", unglazed | | D-7 | 100 | .160 | | 6.25 | 5.80 | | 12.05 | 15.35 |
| 0270 | Glazed | | | 100 | .160 | | 8.35 | 5.80 | | 14.15 | 17.60 |
| 0280 | Hand molded flooring, 4" x 8" x 3/4", unglazed | | | 95 | .168 | | 8.25 | 6.10 | | 14.35 | 18 |
| 0290 | Glazed | | | 95 | .168 | | 10.35 | 6.10 | | 16.45 | 20.50 |
| 0300 | 8" hexagonal, 3/4" thick, unglazed | | | 85 | .188 | | 9.05 | 6.80 | | 15.85 | 19.90 |
| 0310 | Glazed | | | 85 | .188 | | 16.35 | 6.80 | | 23.15 | 28 |
| 0400 | Heavy duty industrial, cement mortar bed, 2" thick, not incl. brick | | D-1 | 80 | .200 | | .83 | 7.80 | | 8.63 | 12.60 |
| 0450 | Acid-proof joints, 1/4" wide | " | | 65 | .246 | | 1.44 | 9.60 | | 11.04 | 16 |
| 0500 | Pavers, 8" x 4", 1" to 1-1/4" thick, red | | D-7 | 95 | .168 | | 3.64 | 6.10 | | 9.74 | 12.90 |
| 0510 | Ironspot | " | | 95 | .168 | | 5.15 | 6.10 | | 11.25 | 14.55 |
| 0540 | 1-3/8" to 1-3/4" thick, red | | D-1 | 95 | .168 | | 3.51 | 6.55 | | 10.06 | 13.70 |
| 0560 | Ironspot | | | 95 | .168 | | 5.10 | 6.55 | | 11.65 | 15.45 |
| 0580 | 2-1/4" thick, red | | | 90 | .178 | | 3.57 | 6.95 | | 10.52 | 14.35 |
| 0590 | Ironspot | | | 90 | .178 | | 5.55 | 6.95 | | 12.50 | 16.50 |
| 0700 | Paver, adobe brick, 6" x 12", 1/2" joint | G | | 42 | .381 | | .84 | 14.85 | | 15.69 | 23.50 |
| 0710 | Mexican red, 12" x 12" | G | 1 Tilf | 48 | .167 | | 1.38 | 6.75 | | 8.13 | 11.35 |
| 0720 | Saltillo, 12" x 12" | G | " | 48 | .167 | | 1.38 | 6.75 | | 8.13 | 11.35 |
| 0800 | For sidewalks and patios with pavers, see Section 32 14 16.10 | | | | | | | | | | |
| 0870 | For epoxy joints, add | | D-1 | 600 | .027 | S.F. | 2.72 | 1.04 | | 3.76 | 4.55 |
| 0880 | For Furan underlayment, add | " | | 600 | .027 | | 2.25 | 1.04 | | 3.29 | 4.04 |
| 0890 | For waxed surface, steam cleaned, add | | A-1H | 1000 | .008 | | .20 | .27 | .06 | .53 | .71 |

## 09 63 40 – Stone Flooring

### 09 63 40.10 Marble

| | 09 63 40.10 Marble | | Crew | Daily Output | Labor-Hours | Unit | Material | 2011 Bare Costs Labor | Equipment | Total | Total Incl O&P |
|---|---|---|---|---|---|---|---|---|---|---|---|
| 0010 | **MARBLE** | | | | | | | | | | |
| 0020 | Thin gauge tile, 12" x 6", 3/8", white Carara | | D-7 | 60 | .267 | S.F. | 11.25 | 9.65 | | 20.90 | 26.50 |
| 0100 | Travertine | | | 60 | .267 | | 12.30 | 9.65 | | 21.95 | 27.50 |
| 0200 | 12" x 12" x 3/8", thin set, floors | | | 60 | .267 | | 6.05 | 9.65 | | 15.70 | 20.50 |
| 0300 | On walls | | | 52 | .308 | | 9.50 | 11.10 | | 20.60 | 26.50 |
| 1000 | Marble threshold, 4" wide x 36" long x 5/8" thick, white | | | 60 | .267 | Ea. | 6.05 | 9.65 | | 15.70 | 20.50 |

### 09 63 40.20 Slate Tile

| | 09 63 40.20 Slate Tile | | Crew | Daily Output | Labor-Hours | Unit | Material | 2011 Bare Costs Labor | Equipment | Total | Total Incl O&P |
|---|---|---|---|---|---|---|---|---|---|---|---|
| 0010 | **SLATE TILE** | | | | | | | | | | |
| 0020 | Vermont, 6" x 6" x 1/4" thick, thin set | | D-7 | 180 | .089 | S.F. | 6.80 | 3.21 | | 10.01 | 12.15 |
| 0200 | See also Section 32 14 40.10 | | | | | | | | | | |

# 09 64 Wood Flooring

## 09 64 16 – Wood Block Flooring

| 09 64 16.10 End Grain Block Flooring | Crew | Daily Output | Labor-Hours | Unit | Material | 2011 Bare Costs Labor | Equipment | Total | Total Incl O&P |
|---|---|---|---|---|---|---|---|---|---|
| 0010 **END GRAIN BLOCK FLOORING** | | | | | | | | | |
| 0020 End grain flooring, coated, 2" thick | 1 Carp | 295 | .027 | S.F. | 3.29 | 1.17 | | 4.46 | 5.40 |
| 0400 Natural finish, 1" thick, fir | | 125 | .064 | | 3.40 | 2.76 | | 6.16 | 7.95 |
| 0600 1-1/2" thick, pine | | 125 | .064 | | 3.34 | 2.76 | | 6.10 | 7.90 |
| 0700 2" thick, pine | | 125 | .064 | | 4.64 | 2.76 | | 7.40 | 9.35 |

## 09 64 19 – Wood Composition Flooring

| 09 64 19.10 Wood Composition | Crew | Daily Output | Labor-Hours | Unit | Material | 2011 Bare Costs Labor | Equipment | Total | Total Incl O&P |
|---|---|---|---|---|---|---|---|---|---|
| 0010 **WOOD COMPOSITION** Gym floors | | | | | | | | | |
| 0100 2-1/4" x 6-7/8" x 3/8", on 2" grout setting bed | D-7 | 150 | .107 | S.F. | 5.85 | 3.86 | | 9.71 | 12.05 |
| 0200 Thin set, on concrete | " | 250 | .064 | | 5.35 | 2.31 | | 7.66 | 9.25 |
| 0300 Sanding and finishing, add | 1 Carp | 200 | .040 | | .79 | 1.72 | | 2.51 | 3.51 |

## 09 64 23 – Wood Parquet Flooring

| 09 64 23.10 Wood Parquet | Crew | Daily Output | Labor-Hours | Unit | Material | 2011 Bare Costs Labor | Equipment | Total | Total Incl O&P |
|---|---|---|---|---|---|---|---|---|---|
| 0010 **WOOD PARQUET** flooring | | | | | | | | | |
| 5200 Parquetry, standard, 5/16" thick, not incl. finish, oak, minimum | 1 Carp | 160 | .050 | S.F. | 4.55 | 2.15 | | 6.70 | 8.30 |
| 5300 Maximum | | 100 | .080 | | 5.55 | 3.44 | | 8.99 | 11.40 |
| 5500 Teak, minimum | | 160 | .050 | | 4.86 | 2.15 | | 7.01 | 8.65 |
| 5600 Maximum | | 100 | .080 | | 8.50 | 3.44 | | 11.94 | 14.65 |
| 5650 13/16" thick, select grade oak, minimum | | 160 | .050 | | 9.35 | 2.15 | | 11.50 | 13.60 |
| 5700 Maximum | | 100 | .080 | | 14.20 | 3.44 | | 17.64 | 21 |
| 5800 Custom parquetry, including finish, minimum | | 100 | .080 | | 15.65 | 3.44 | | 19.09 | 22.50 |
| 5900 Maximum | | 50 | .160 | | 21 | 6.90 | | 27.90 | 33.50 |
| 6700 Parquetry, prefinished white oak, 5/16" thick, minimum | | 160 | .050 | | 3.71 | 2.15 | | 5.86 | 7.40 |
| 6800 Maximum | | 100 | .080 | | 5.50 | 3.44 | | 8.94 | 11.35 |
| 7000 Walnut or teak, parquetry, minimum | | 160 | .050 | | 5.35 | 2.15 | | 7.50 | 9.20 |
| 7100 Maximum | | 100 | .080 | | 9.35 | 3.44 | | 12.79 | 15.60 |
| 7200 Acrylic wood parquet blocks, 12" x 12" x 5/16", | | | | | | | | | |
| 7210 Irradiated, set in epoxy | 1 Carp | 160 | .050 | S.F. | 7.80 | 2.15 | | 9.95 | 11.90 |

## 09 64 29 – Wood Strip and Plank Flooring

| 09 64 29.10 Wood | Crew | Daily Output | Labor-Hours | Unit | Material | 2011 Bare Costs Labor | Equipment | Total | Total Incl O&P |
|---|---|---|---|---|---|---|---|---|---|
| 0010 **WOOD** | | | | | | | | | |
| 0020 Fir, vertical grain, 1" x 4", not incl. finish, grade B & better | 1 Carp | 255 | .031 | S.F. | 2.74 | 1.35 | | 4.09 | 5.10 |
| 0100 C grade & better | | 255 | .031 | | 2.58 | 1.35 | | 3.93 | 4.91 |
| 4000 Maple, strip, 25/32" x 2-1/4", not incl. finish, select | | 170 | .047 | | 4.37 | 2.03 | | 6.40 | 7.90 |
| 4100 #2 & better | | 170 | .047 | | 3.49 | 2.03 | | 5.52 | 6.95 |
| 4300 33/32" x 3-1/4", not incl. finish, #1 grade | | 170 | .047 | | 4.19 | 2.03 | | 6.22 | 7.70 |
| 4400 #2 & better | | 170 | .047 | | 3.73 | 2.03 | | 5.76 | 7.20 |
| 4600 Oak, white or red, 25/32" x 2-1/4", not incl. finish | | | | | | | | | |
| 4700 #1 common | 1 Carp | 170 | .047 | S.F. | 2.49 | 2.03 | | 4.52 | 5.85 |
| 4900 Select quartered, 2-1/4" wide | | 170 | .047 | | 3.09 | 2.03 | | 5.12 | 6.50 |
| 5000 Clear | | 170 | .047 | | 3.95 | 2.03 | | 5.98 | 7.45 |
| 6100 Prefinished, white oak, prime grade, 2-1/4" wide | | 170 | .047 | | 4.49 | 2.03 | | 6.52 | 8.05 |
| 6200 3-1/4" wide | | 185 | .043 | | 4.99 | 1.86 | | 6.85 | 8.35 |
| 6400 Ranch plank | | 145 | .055 | | 6.75 | 2.38 | | 9.13 | 11.05 |
| 6500 Hardwood blocks, 9" x 9", 25/32" thick | | 160 | .050 | | 5.75 | 2.15 | | 7.90 | 9.65 |
| 7400 Yellow pine, 3/4" x 3-1/8", T & G, C & better, not incl. finish | | 200 | .040 | | 1.49 | 1.72 | | 3.21 | 4.28 |
| 7500 Refinish wood floor, sand, 2 cts poly, wax, soft wood, min. | 1 Clab | 400 | .020 | | .78 | .69 | | 1.47 | 1.91 |
| 7600 Hard wood, max | | 130 | .062 | | 1.16 | 2.11 | | 3.27 | 4.52 |
| 7800 Sanding and finishing, 2 coats polyurethane | | 295 | .027 | | .78 | .93 | | 1.71 | 2.29 |
| 7900 Subfloor and underlayment, see Section 06 16 | | | | | | | | | |
| 8015 Transition molding, 2 1/4" wide, 5' long | 1 Carp | 19.20 | .417 | Ea. | 15 | 17.95 | | 32.95 | 44 |

# 09 64 Wood Flooring

## 09 64 66 - Wood Athletic Flooring

| 09 64 66.10 Gymnasium Flooring | Crew | Daily Output | Labor-Hours | Unit | Material | 2011 Bare Costs Labor | Equipment | Total | Total Incl O&P |
|---|---|---|---|---|---|---|---|---|---|
| 0010 **GYMNASIUM FLOORING** | | | | | | | | | |
| 0600 Gym floor, in mastic, over 2 ply felt, #2 & better | | | | | | | | | |
| 0700 25/32" thick maple | 1 Carp | 100 | .080 | S.F. | 4.14 | 3.44 | | 7.58 | 9.85 |
| 0900 33/32" thick maple | | 98 | .082 | | 4.99 | 3.51 | | 8.50 | 10.90 |
| 1000 For 1/2" corkboard underlayment, add | | 750 | .011 | | .90 | .46 | | 1.36 | 1.69 |
| 1300 For #1 grade maple, add | | | | | .47 | | | .47 | .52 |
| 1600 Maple flooring, over sleepers, #2 & better | | | | | | | | | |
| 1700 25/32" thick | 1 Carp | 85 | .094 | S.F. | 4.32 | 4.05 | | 8.37 | 10.95 |
| 1900 33/32" thick | " | 83 | .096 | | 5.05 | 4.15 | | 9.20 | 11.90 |
| 2000 For #1 grade, add | | | | | .50 | | | .50 | .55 |
| 2200 For 3/4" subfloor, add | 1 Carp | 350 | .023 | | 1.12 | .98 | | 2.10 | 2.74 |
| 2300 With two 1/2" subfloors, 25/32" thick | " | 69 | .116 | | 5.40 | 4.99 | | 10.39 | 13.60 |
| 2500 Maple, incl. finish, #2 & btr., 25/32" thick, on rubber | | | | | | | | | |
| 2600 Sleepers, with two 1/2" subfloors | 1 Carp | 76 | .105 | S.F. | 5.80 | 4.53 | | 10.33 | 13.35 |
| 2800 With steel spline, double connection to channels | " | 73 | .110 | | 6.20 | 4.72 | | 10.92 | 14.05 |
| 2900 For 33/32" maple, add | | | | | .67 | | | .67 | .74 |
| 3100 For #1 grade maple, add | | | | | .50 | | | .50 | .55 |
| 3500 For termite proofing all of the above, add | | | | | .27 | | | .27 | .30 |
| 3700 Portable hardwood, prefinished panels | 1 Carp | 83 | .096 | | 7.75 | 4.15 | | 11.90 | 14.85 |
| 3720 Insulated with polystyrene, 1" thick, add | | 165 | .048 | | .67 | 2.09 | | 2.76 | 3.94 |
| 3750 Running tracks, Sitka spruce surface, 25/32" x 2-1/4" | | 62 | .129 | | 14.15 | 5.55 | | 19.70 | 24 |
| 3770 3/4" plywood surface, finished | | 100 | .080 | | 3.36 | 3.44 | | 6.80 | 9 |

# 09 65 Resilient Flooring

## 09 65 10 - Resilient Tile Underlayment

### 09 65 10.10 Latex Underlayment

| | Crew | Daily Output | Labor-Hours | Unit | Material | 2011 Bare Costs Labor | Equipment | Total | Total Incl O&P |
|---|---|---|---|---|---|---|---|---|---|
| 0010 **LATEX UNDERLAYMENT** | | | | | | | | | |
| 3600 Latex underlayment, 1/8" thk., cementitious for resilient flooring | 1 Tilf | 160 | .050 | S.F. | .82 | 2.02 | | 2.84 | 3.85 |
| 4000 Liquid, fortified | | | | Gal. | 40.50 | | | 40.50 | 44.50 |

## 09 65 13 - Resilient Base and Accessories

### 09 65 13.13 Resilient Base

| | Crew | Daily Output | Labor-Hours | Unit | Material | 2011 Bare Costs Labor | Equipment | Total | Total Incl O&P |
|---|---|---|---|---|---|---|---|---|---|
| 0010 **RESILIENT BASE** | | | | | | | | | |
| 0800 Base, cove, rubber or vinyl | | | | | | | | | |
| 1100 Standard colors, 0.080" thick, 2-1/2" high | 1 Tilf | 315 | .025 | L.F. | 1 | 1.03 | | 2.03 | 2.60 |
| 1150 4" high | | 315 | .025 | | 1.08 | 1.03 | | 2.11 | 2.69 |
| 1200 6" high | | 315 | .025 | | 1.50 | 1.03 | | 2.53 | 3.15 |
| 1450 1/8" thick, 2-1/2" high | | 315 | .025 | | 1 | 1.03 | | 2.03 | 2.60 |
| 1500 4" high | | 315 | .025 | | .90 | 1.03 | | 1.93 | 2.49 |
| 1600 Corners, 2-1/2" high | | 315 | .025 | Ea. | 1.60 | 1.03 | | 2.63 | 3.26 |
| 1630 4" high | | 315 | .025 | | 3 | 1.03 | | 4.03 | 4.80 |
| 1660 6" high | | 315 | .025 | | 3 | 1.03 | | 4.03 | 4.80 |

### 09 65 13.23 Resilient Stair Treads and Risers

| | Crew | Daily Output | Labor-Hours | Unit | Material | 2011 Bare Costs Labor | Equipment | Total | Total Incl O&P |
|---|---|---|---|---|---|---|---|---|---|
| 0010 **RESILIENT STAIR TREADS AND RISERS** | | | | | | | | | |
| 0300 Rubber, molded tread, 12" wide, 5/16" thick, black | 1 Tilf | 115 | .070 | L.F. | 12.75 | 2.81 | | 15.56 | 18.15 |
| 0400 Colors | | 115 | .070 | | 13.35 | 2.81 | | 16.16 | 18.80 |
| 0600 1/4" thick, black | | 115 | .070 | | 12.50 | 2.81 | | 15.31 | 17.85 |
| 0700 Colors | | 115 | .070 | | 12.75 | 2.81 | | 15.56 | 18.15 |
| 0900 Grip strip safety tread, colors, 5/16" thick | | 115 | .070 | | 16.95 | 2.81 | | 19.76 | 22.50 |
| 1000 3/16" thick | | 120 | .067 | | 11.75 | 2.70 | | 14.45 | 16.85 |
| 1200 Landings, smooth sheet rubber, 1/8" thick | | 120 | .067 | S.F. | 6.30 | 2.70 | | 9 | 10.90 |

# 09 65 Resilient Flooring

## 09 65 13 – Resilient Base and Accessories

| 09 65 13.23 Resilient Stair Treads and Risers | | Crew | Daily Output | Labor-Hours | Unit | Material | 2011 Bare Costs Labor | Equipment | Total | Total Incl O&P |
|---|---|---|---|---|---|---|---|---|---|---|
| 1300 | 3/16" thick | 1 Tilf | 120 | .067 | S.F. | 7.95 | 2.70 | | 10.65 | 12.70 |
| 1500 | Nosings, 3" wide, 3/16" thick, black | | 140 | .057 | L.F. | 3.67 | 2.31 | | 5.98 | 7.40 |
| 1600 | Colors | | 140 | .057 | | 3.67 | 2.31 | | 5.98 | 7.40 |
| 1800 | Risers, 7" high, 1/8" thick, flat | | 250 | .032 | | 4.55 | 1.29 | | 5.84 | 6.90 |
| 1900 | Coved | | 250 | .032 | | 5.95 | 1.29 | | 7.24 | 8.45 |
| 2100 | Vinyl, molded tread, 12" wide, colors, 1/8" thick | | 115 | .070 | | 6.25 | 2.81 | | 9.06 | 11 |
| 2200 | 1/4" thick | | 115 | .070 | | 8 | 2.81 | | 10.81 | 12.90 |
| 2300 | Landing material, 1/8" thick | | 200 | .040 | S.F. | 6.25 | 1.62 | | 7.87 | 9.25 |
| 2400 | Riser, 7" high, 1/8" thick, coved | | 175 | .046 | L.F. | 3.50 | 1.85 | | 5.35 | 6.55 |
| 2500 | Tread and riser combined, 1/8" thick | | 80 | .100 | " | 9.50 | 4.05 | | 13.55 | 16.35 |

## 09 65 16 – Resilient Sheet Flooring

### 09 65 16.10 Rubber and Vinyl Sheet Flooring

| | | | Crew | Daily Output | Labor-Hours | Unit | Material | 2011 Bare Costs Labor | Equipment | Total | Total Incl O&P |
|---|---|---|---|---|---|---|---|---|---|---|---|
| 0010 | **RUBBER AND VINYL SHEET FLOORING** | | | | | | | | | | |
| 5500 | Linoleum, sheet goods | G | 1 Tilf | 360 | .022 | S.F. | 4.28 | .90 | | 5.18 | 6 |
| 5900 | Rubber, sheet goods, 36" wide, 1/8" thick | | | 120 | .067 | | 6.85 | 2.70 | | 9.55 | 11.50 |
| 5950 | 3/16" thick | | | 100 | .080 | | 9.25 | 3.24 | | 12.49 | 14.90 |
| 6000 | 1/4" thick | | | 90 | .089 | | 10.95 | 3.60 | | 14.55 | 17.30 |
| 8000 | Vinyl sheet goods, backed, .065" thick, minimum | | | 250 | .032 | | 3.75 | 1.29 | | 5.04 | 6 |
| 8050 | Maximum | | | 200 | .040 | | 3.69 | 1.62 | | 5.31 | 6.40 |
| 8100 | .080" thick, minimum | | | 230 | .035 | | 3.80 | 1.41 | | 5.21 | 6.25 |
| 8150 | Maximum | | | 200 | .040 | | 5.40 | 1.62 | | 7.02 | 8.30 |
| 8200 | .125" thick, minimum | | | 230 | .035 | | 4.35 | 1.41 | | 5.76 | 6.85 |
| 8250 | Maximum | | | 200 | .040 | | 6.50 | 1.62 | | 8.12 | 9.50 |
| 8700 | Adhesive cement, 1 gallon per 200 to 300 S.F. | | | | | Gal. | 25 | | | 25 | 27.50 |
| 8800 | Asphalt primer, 1 gallon per 300 S.F. | | | | | | 12.75 | | | 12.75 | 14.05 |
| 8900 | Emulsion, 1 gallon per 140 S.F. | | | | | | 16.25 | | | 16.25 | 17.90 |

## 09 65 19 – Resilient Tile Flooring

### 09 65 19.10 Miscellaneous Resilient Tile Flooring

| | | | Crew | Daily Output | Labor-Hours | Unit | Material | 2011 Bare Costs Labor | Equipment | Total | Total Incl O&P |
|---|---|---|---|---|---|---|---|---|---|---|---|
| 0010 | **MISCELLANEOUS RESILIENT TILE FLOORING** | | | | | | | | | | |
| 2200 | Cork tile, standard finish, 1/8" thick | G | 1 Tilf | 315 | .025 | S.F. | 6.25 | 1.03 | | 7.28 | 8.40 |
| 2250 | 3/16" thick | G | | 315 | .025 | | 6.30 | 1.03 | | 7.33 | 8.40 |
| 2300 | 5/16" thick | G | | 315 | .025 | | 7.75 | 1.03 | | 8.78 | 10.05 |
| 2350 | 1/2" thick | G | | 315 | .025 | | 9.20 | 1.03 | | 10.23 | 11.60 |
| 2500 | Urethane finish, 1/8" thick | G | | 315 | .025 | | 7.45 | 1.03 | | 8.48 | 9.65 |
| 2550 | 3/16" thick | G | | 315 | .025 | | 7.80 | 1.03 | | 8.83 | 10.10 |
| 2600 | 5/16" thick | G | | 315 | .025 | | 10 | 1.03 | | 11.03 | 12.50 |
| 2650 | 1/2" thick | G | | 315 | .025 | | 13.50 | 1.03 | | 14.53 | 16.35 |
| 6050 | Rubber tile, marbleized colors, 12" x 12", 1/8" thick | | | 400 | .020 | | 6.65 | .81 | | 7.46 | 8.55 |
| 6100 | 3/16" thick | | | 400 | .020 | | 10 | .81 | | 10.81 | 12.20 |
| 6300 | Special tile, plain colors, 1/8" thick | | | 400 | .020 | | 10 | .81 | | 10.81 | 12.20 |
| 6350 | 3/16" thick | | | 400 | .020 | | 10 | .81 | | 10.81 | 12.20 |
| 6410 | Raised, radial or square, minimum | | | 400 | .020 | | 9.25 | .81 | | 10.06 | 11.40 |
| 6430 | Maximum | | | 400 | .020 | | 12.50 | .81 | | 13.31 | 14.95 |
| 6450 | For golf course, skating rink, etc., 1/4" thick | | | 275 | .029 | | 12.50 | 1.18 | | 13.68 | 15.45 |
| 6700 | Synthetic turf, 3/8" thick | | | 90 | .089 | | 6.20 | 3.60 | | 9.80 | 12.05 |
| 6750 | Interlocking 2' x 2' squares, 1/2" thick, not | | | | | | | | | | |
| 6810 | cemented, for playgrounds, minimum | | 1 Tilf | 210 | .038 | S.F. | 4.75 | 1.54 | | 6.29 | 7.50 |
| 6850 | Maximum | | | 190 | .042 | | 12 | 1.70 | | 13.70 | 15.70 |
| 7000 | Vinyl composition tile, 12" x 12", 1/16" thick | | | 500 | .016 | | .93 | .65 | | 1.58 | 1.96 |
| 7050 | Embossed | | | 500 | .016 | | 2.17 | .65 | | 2.82 | 3.33 |
| 7100 | Marbleized | | | 500 | .016 | | 2.17 | .65 | | 2.82 | 3.33 |
| 7150 | Solid | | | 500 | .016 | | 2.79 | .65 | | 3.44 | 4.01 |

# 09 65 Resilient Flooring

## 09 65 19 – Resilient Tile Flooring

| 09 65 19.10 Miscellaneous Resilient Tile Flooring | | Crew | Daily Output | Labor-Hours | Unit | Material | 2011 Bare Costs Labor | Equipment | Total | Total Incl O&P |
|---|---|---|---|---|---|---|---|---|---|---|
| 7200 | 3/32" thick, embossed | 1 Tilf | 500 | .016 | S.F. | 1.35 | .65 | | 2 | 2.43 |
| 7250 | Marbleized | | 500 | .016 | | 2.49 | .65 | | 3.14 | 3.68 |
| 7300 | Solid | | 500 | .016 | | 2.31 | .65 | | 2.96 | 3.48 |
| 7350 | 1/8" thick, marbleized | | 500 | .016 | | 1.64 | .65 | | 2.29 | 2.74 |
| 7400 | Solid | | 500 | .016 | | 1.73 | .65 | | 2.38 | 2.84 |
| 7450 | Conductive | | 500 | .016 | | 5.95 | .65 | | 6.60 | 7.50 |
| 7500 | Vinyl tile, 12" x 12", .050" thick, minimum | | 500 | .016 | | 3 | .65 | | 3.65 | 4.24 |
| 7550 | Maximum | | 500 | .016 | | 6 | .65 | | 6.65 | 7.55 |
| 7600 | 1/8" thick, minimum | | 500 | .016 | | 4.50 | .65 | | 5.15 | 5.90 |
| 7650 | Solid colors | | 500 | .016 | | 6.20 | .65 | | 6.85 | 7.75 |
| 7700 | Marbleized or Travertine pattern | | 500 | .016 | | 5.65 | .65 | | 6.30 | 7.15 |
| 7750 | Florentine pattern | | 500 | .016 | | 6 | .65 | | 6.65 | 7.55 |
| 7800 | Maximum | | 500 | .016 | | 13.50 | .65 | | 14.15 | 15.80 |

## 09 65 33 – Conductive Resilient Flooring

### 09 65 33.10 Conductive Rubber and Vinyl Flooring

| | | Crew | Daily Output | Labor-Hours | Unit | Material | Labor | Equipment | Total | Total Incl O&P |
|---|---|---|---|---|---|---|---|---|---|---|
| 0010 | **CONDUCTIVE RUBBER AND VINYL FLOORING** | | | | | | | | | |
| 1700 | Conductive flooring, rubber tile, 1/8" thick | 1 Tilf | 315 | .025 | S.F. | 6 | 1.03 | | 7.03 | 8.10 |
| 1800 | Homogeneous vinyl tile, 1/8" thick | " | 315 | .025 | " | 8 | 1.03 | | 9.03 | 10.30 |

# 09 66 Terrazzo Flooring

## 09 66 13 – Portland Cement Terrazzo Flooring

### 09 66 13.10 Portland Cement Terrazzo

| | | Crew | Daily Output | Labor-Hours | Unit | Material | Labor | Equipment | Total | Total Incl O&P |
|---|---|---|---|---|---|---|---|---|---|---|
| 0010 | **PORTLAND CEMENT TERRAZZO**, cast-in-place    R096613-10 | | | | | | | | | |
| 0020 | Cove base, 6" high, 16 ga. zinc | 1 Mstz | 20 | .400 | L.F. | 2.98 | 16.05 | | 19.03 | 27 |
| 0100 | Curb, 6" high and 6" wide | | 6 | 1.333 | | 5.35 | 53.50 | | 58.85 | 84 |
| 0300 | Divider strip for floors, 14 ga., 1-1/4" deep, zinc | | 375 | .021 | | 1.39 | .86 | | 2.25 | 2.78 |
| 0400 | Brass | | 375 | .021 | | 2.43 | .86 | | 3.29 | 3.92 |
| 0600 | Heavy top strip 1/4" thick, 1-1/4" deep, zinc | | 300 | .027 | | 1.90 | 1.07 | | 2.97 | 3.65 |
| 0900 | Galv. bottoms, brass | | 300 | .027 | | 3.13 | 1.07 | | 4.20 | 5 |
| 1200 | For thin set floors, 16 ga., 1/2" x 1/2", zinc | | 350 | .023 | | .82 | .92 | | 1.74 | 2.24 |
| 1300 | Brass | | 350 | .023 | | 2.27 | .92 | | 3.19 | 3.84 |
| 1500 | Floor, bonded to concrete, 1-3/4" thick, gray cement | J-3 | 75 | .213 | S.F. | 2.92 | 7.85 | 3.22 | 13.99 | 18.20 |
| 1600 | White cement, mud set | | 75 | .213 | | 3.32 | 7.85 | 3.22 | 14.39 | 18.65 |
| 1800 | Not bonded, 3" total thickness, gray cement | | 70 | .229 | | 3.65 | 8.40 | 3.45 | 15.50 | 20 |
| 1900 | White cement, mud set | | 70 | .229 | | 3.98 | 8.40 | 3.45 | 15.83 | 20.50 |
| 2100 | For Venetian terrazzo, 1" topping, add | | | | | 50% | 50% | | | |
| 2200 | For heavy duty abrasive terrazzo, add | | | | | 50% | 50% | | | |
| 2700 | Monolithic terrazzo, 1/2" thick | | | | | | | | | |
| 2710 | 10' panels | J-3 | 125 | .128 | S.F. | 2.65 | 4.71 | 1.93 | 9.29 | 11.90 |
| 3000 | Stairs, cast in place, pan filled treads | | 30 | .533 | L.F. | 2.71 | 19.65 | 8.05 | 30.41 | 40.50 |
| 3100 | Treads and risers | | 14 | 1.143 | " | 5.55 | 42 | 17.25 | 64.80 | 86.50 |
| 3300 | For stair landings, add to floor prices | | | | | | 50% | | | |
| 3400 | Stair stringers and fascia | J-3 | 30 | .533 | S.F. | 4.44 | 19.65 | 8.05 | 32.14 | 42 |
| 3600 | For abrasive metal nosings on stairs, add | | 150 | .107 | L.F. | 8.60 | 3.93 | 1.61 | 14.14 | 16.90 |
| 3700 | For abrasive surface finish, add | | 600 | .027 | S.F. | 1.39 | .98 | .40 | 2.77 | 3.40 |
| 3900 | For raised abrasive strips, add | | 150 | .107 | L.F. | 1.13 | 3.93 | 1.61 | 6.67 | 8.70 |
| 4000 | Wainscot, bonded, 1-1/2" thick | | 30 | .533 | S.F. | 3.50 | 19.65 | 8.05 | 31.20 | 41 |
| 4200 | Epoxy terrazzo, 1/4" thick | | 40 | .400 | " | 4.96 | 14.70 | 6.05 | 25.71 | 33.50 |
| 4300 | Stone chips, onyx gemstone, per 50 lb. bag | | | | Bag | 16.30 | | | 16.30 | 17.95 |

## 09 66 13 – Portland Cement Terrazzo Flooring

### 09 66 13.30 Terrazzo, Precast

| | | Crew | Daily Output | Labor-Hours | Unit | Material | 2011 Bare Costs Labor | Equipment | Total | Total Incl O&P |
|---|---|---|---|---|---|---|---|---|---|---|
| 0010 | **TERRAZZO, PRECAST** | | | | | | | | | |
| 0020 | Base, 6" high, straight | 1 Mstz | 70 | .114 | L.F. | 10.55 | 4.58 | | 15.13 | 18.35 |
| 0100 | Cove | | 60 | .133 | | 12.15 | 5.35 | | 17.50 | 21 |
| 0300 | 8" high, straight | | 60 | .133 | | 10.85 | 5.35 | | 16.20 | 19.70 |
| 0400 | Cove | ↓ | 50 | .160 | | 15.95 | 6.40 | | 22.35 | 27 |
| 0600 | For white cement, add | | | | | .43 | | | .43 | .47 |
| 0700 | For 16 ga. zinc toe strip, add | | | | | 1.63 | | | 1.63 | 1.79 |
| 0900 | Curbs, 4" x 4" high | 1 Mstz | 40 | .200 | | 30.50 | 8 | | 38.50 | 45 |
| 1000 | 8" x 8" high | " | 30 | .267 | | 35.50 | 10.70 | | 46.20 | 54.50 |
| 1200 | Floor tiles, non-slip, 1" thick, 12" x 12" | D-1 | 60 | .267 | S.F. | 18.60 | 10.40 | | 29 | 36 |
| 1300 | 1-1/4" thick, 12" x 12" | | 60 | .267 | | 19.60 | 10.40 | | 30 | 37 |
| 1500 | 16" x 16" | | 50 | .320 | | 21.50 | 12.45 | | 33.95 | 42 |
| 1600 | 1-1/2" thick, 16" x 16" | ↓ | 45 | .356 | | 19.45 | 13.85 | | 33.30 | 42.50 |
| 1800 | For Venetian terrazzo, add | | | | | 5.85 | | | 5.85 | 6.40 |
| 1900 | For white cement, add | | | | | .55 | | | .55 | .61 |
| 2400 | Stair treads, 1-1/2" thick, non-slip, three line pattern | 2 Mstz | 70 | .229 | L.F. | 39.50 | 9.15 | | 48.65 | 57 |
| 2500 | Nosing and two lines | | 70 | .229 | | 39.50 | 9.15 | | 48.65 | 57 |
| 2700 | 2" thick treads, straight | | 60 | .267 | | 43 | 10.70 | | 53.70 | 62.50 |
| 2800 | Curved | | 50 | .320 | | 56 | 12.85 | | 68.85 | 80 |
| 3000 | Stair risers, 1" thick, to 6" high, straight sections | | 60 | .267 | | 9.80 | 10.70 | | 20.50 | 26.50 |
| 3100 | Cove | | 50 | .320 | | 14.35 | 12.85 | | 27.20 | 34.50 |
| 3300 | Curved, 1" thick, to 6" high, vertical | | 48 | .333 | | 19.55 | 13.35 | | 32.90 | 41 |
| 3400 | Cove | | 38 | .421 | | 37 | 16.90 | | 53.90 | 65.50 |
| 3600 | Stair tread and riser, single piece, straight, minimum | | 60 | .267 | | 50 | 10.70 | | 60.70 | 70.50 |
| 3700 | Maximum | | 40 | .400 | | 65 | 16.05 | | 81.05 | 94.50 |
| 3900 | Curved tread and riser, minimum | | 40 | .400 | | 70 | 16.05 | | 86.05 | 101 |
| 4000 | Maximum | | 32 | .500 | | 88.50 | 20 | | 108.50 | 126 |
| 4200 | Stair stringers, notched, 1" thick | | 25 | .640 | | 28.50 | 25.50 | | 54 | 69 |
| 4300 | 2" thick | | 22 | .727 | ↓ | 34 | 29 | | 63 | 79.50 |
| 4500 | Stair landings, structural, non-slip, 1-1/2" thick | | 85 | .188 | S.F. | 31 | 7.55 | | 38.55 | 45.50 |
| 4600 | 3" thick | ↓ | 75 | .213 | | 44.50 | 8.55 | | 53.05 | 61 |
| 4800 | Wainscot, 12" x 12" x 1" tiles | 1 Mstz | 12 | .667 | | 6.50 | 26.50 | | 33 | 46 |
| 4900 | 16" x 16" x 1-1/2" tiles | " | 8 | 1 | ↓ | 13.70 | 40 | | 53.70 | 73.50 |

## 09 66 16 – Terrazzo Floor Tile

### 09 66 16.10 Tile or Terrazzo Base

| | | Crew | Daily Output | Labor-Hours | Unit | Material | 2011 Bare Costs Labor | Equipment | Total | Total Incl O&P |
|---|---|---|---|---|---|---|---|---|---|---|
| 0010 | **TILE OR TERRAZZO BASE** | | | | | | | | | |
| 0020 | Scratch coat only | 1 Mstz | 150 | .053 | S.F. | .41 | 2.14 | | 2.55 | 3.57 |
| 0500 | Scratch and brown coat only | " | 75 | .107 | " | .78 | 4.28 | | 5.06 | 7.10 |

## 09 66 33 – Conductive Terrazzo Flooring

### 09 66 33.10 Conductive Terrazzo

| | | Crew | Daily Output | Labor-Hours | Unit | Material | 2011 Bare Costs Labor | Equipment | Total | Total Incl O&P |
|---|---|---|---|---|---|---|---|---|---|---|
| 0010 | **CONDUCTIVE TERRAZZO** | | | | | | | | | |
| 2400 | Bonded conductive floor for hospitals | J-3 | 90 | .178 | S.F. | 3.98 | 6.55 | 2.68 | 13.21 | 16.90 |
| 2500 | Epoxy terrazzo, 1/4" thick, minimum | | 100 | .160 | | 4.50 | 5.90 | 2.42 | 12.82 | 16.20 |
| 2550 | Average | | 75 | .213 | | 4.96 | 7.85 | 3.22 | 16.03 | 20.50 |
| 2600 | Maximum | ↓ | 60 | .267 | ↓ | 5.45 | 9.80 | 4.03 | 19.28 | 24.50 |

## 09 67 20 – Epoxy-Marble Chip Flooring

### 09 67 20.13 Elastomeric Liquid Flooring

| | | Daily Output | Labor-Hours | Unit | Material | 2011 Bare Costs Labor | Equipment | Total | Total Incl O&P |
|---|---|---|---|---|---|---|---|---|---|
| | | Crew | | | | | | | |
| 0010 | **ELASTOMERIC LIQUID FLOORING** | | | | | | | | |
| 0020 | Cementitious acrylic, 1/4" thick | C-6 | 520 | .092 | S.F. | 1.47 | 3.30 | .09 | 4.86 | 6.70 |
| 0100 | 3/8" thick | " | 450 | .107 | | 1.91 | 3.82 | .10 | 5.83 | 8 |
| 0200 | Methyl methacrylate, 1/4" thick | C-8A | 3000 | .016 | | 5.40 | .59 | | 5.99 | 6.85 |
| 0210 | 1/8" thick | " | 3000 | .016 | | 2.97 | .59 | | 3.56 | 4.16 |
| 0300 | Cupric oxychloride, on bond coat, minimum | C-6 | 480 | .100 | | 3.25 | 3.58 | .10 | 6.93 | 9.15 |
| 0400 | Maximum | | 420 | .114 | | 5.40 | 4.09 | .11 | 9.60 | 12.25 |
| 2400 | Mastic, hot laid, 2 coat, 1-1/2" thick, standard, minimum | | 690 | .070 | | 3.77 | 2.49 | .07 | 6.33 | 8 |
| 2500 | Maximum | | 520 | .092 | | 4.84 | 3.30 | .09 | 8.23 | 10.40 |
| 2700 | Acid-proof, minimum | | 605 | .079 | | 4.84 | 2.84 | .08 | 7.76 | 9.70 |
| 2800 | Maximum | | 350 | .137 | | 6.70 | 4.91 | .13 | 11.74 | 14.95 |
| 3000 | Neoprene, trowelled on, 1/4" thick, minimum | | 545 | .088 | | 3.71 | 3.15 | .09 | 6.95 | 8.95 |
| 3100 | Maximum | | 430 | .112 | | 5.10 | 3.99 | .11 | 9.20 | 11.75 |
| 4300 | Polyurethane, with suspended vinyl chips, minimum | | 1065 | .045 | | 7.20 | 1.61 | .04 | 8.85 | 10.45 |
| 4500 | Maximum | | 860 | .056 | | 10.40 | 2 | .05 | 12.45 | 14.55 |

### 09 67 20.16 Epoxy Terrazzo

| | | Crew | Daily Output | Labor-Hours | Unit | Material | Labor | Equipment | Total | Total Incl O&P |
|---|---|---|---|---|---|---|---|---|---|---|
| 0010 | **EPOXY TERRAZZO** | | | | | | | | | |
| 1800 | Epoxy terrazzo, 1/4" thick, chemical resistant, minimum | J-3 | 200 | .080 | S.F. | 5.15 | 2.94 | 1.21 | 9.30 | 11.30 |
| 1900 | Maximum | | 150 | .107 | | 7.85 | 3.93 | 1.61 | 13.39 | 16.10 |
| 2100 | Conductive, minimum | | 210 | .076 | | 7.05 | 2.80 | 1.15 | 11 | 13.15 |
| 2200 | Maximum | | 160 | .100 | | 9.10 | 3.68 | 1.51 | 14.29 | 17 |

### 09 67 20.19 Polyacrylate Terrazzo

| | | Crew | Daily Output | Labor-Hours | Unit | Material | Labor | Equipment | Total | Total Incl O&P |
|---|---|---|---|---|---|---|---|---|---|---|
| 0010 | **POLYACRYLATE TERRAZZO** | | | | | | | | | |
| 3150 | Polyacrylate, 1/4" thick, minimum | C-6 | 735 | .065 | S.F. | 3.07 | 2.34 | .06 | 5.47 | 7 |
| 3170 | Maximum | | 480 | .100 | | 3.91 | 3.58 | .10 | 7.59 | 9.85 |
| 3200 | 3/8" thick, minimum | | 620 | .077 | | 4 | 2.77 | .08 | 6.85 | 8.70 |
| 3220 | Maximum | | 480 | .100 | | 5.65 | 3.58 | .10 | 9.33 | 11.80 |
| 3300 | Conductive, 1/4" thick, minimum | | 450 | .107 | | 6.40 | 3.82 | .10 | 10.32 | 12.95 |
| 3330 | Maximum | | 305 | .157 | | 8.80 | 5.65 | .15 | 14.60 | 18.40 |
| 3350 | 3/8" thick, minimum | | 365 | .132 | | 8.40 | 4.70 | .13 | 13.23 | 16.55 |
| 3370 | Maximum | | 255 | .188 | | 11.25 | 6.75 | .18 | 18.18 | 23 |
| 3450 | Granite, conductive, 1/4" thick, minimum | | 695 | .069 | | 8.15 | 2.47 | .07 | 10.69 | 12.75 |
| 3470 | Maximum | | 420 | .114 | | 10.50 | 4.09 | .11 | 14.70 | 17.85 |
| 3500 | 3/8" thick, minimum | | 695 | .069 | | 11.90 | 2.47 | .07 | 14.44 | 16.90 |
| 3520 | Maximum | | 380 | .126 | | 14.40 | 4.52 | .12 | 19.04 | 23 |

### 09 67 20.26 Quartz Flooring

| | | Crew | Daily Output | Labor-Hours | Unit | Material | Labor | Equipment | Total | Total Incl O&P |
|---|---|---|---|---|---|---|---|---|---|---|
| 0010 | **QUARTZ FLOORING** | | | | | | | | | |
| 0600 | Epoxy, with colored quartz chips, broadcast, minimum | C-6 | 675 | .071 | S.F. | 2.57 | 2.54 | .07 | 5.18 | 6.75 |
| 0700 | Maximum | | 490 | .098 | | 3.12 | 3.50 | .10 | 6.72 | 8.85 |
| 0900 | Trowelled, minimum | | 560 | .086 | | 3.31 | 3.07 | .08 | 6.46 | 8.40 |
| 1000 | Maximum | | 480 | .100 | | 4.84 | 3.58 | .10 | 8.52 | 10.85 |
| 1200 | Heavy duty epoxy topping, 1/4" thick, | | | | | | | | | |
| 1300 | 500 to 1,000 S.F. | C-6 | 420 | .114 | S.F. | 4.89 | 4.09 | .11 | 9.09 | 11.70 |
| 1500 | 1,000 to 2,000 S.F. | | 450 | .107 | | 4.06 | 3.82 | .10 | 7.98 | 10.40 |
| 1600 | Over 10,000 S.F. | | 480 | .100 | | 3.83 | 3.58 | .10 | 7.51 | 9.75 |
| 3600 | Polyester, with colored quartz chips, 1/16" thick, minimum | | 1065 | .045 | | 2.74 | 1.61 | .04 | 4.39 | 5.50 |
| 3700 | Maximum | | 560 | .086 | | 3.73 | 3.07 | .08 | 6.88 | 8.85 |
| 3900 | 1/8" thick, minimum | | 810 | .059 | | 3.19 | 2.12 | .06 | 5.37 | 6.80 |
| 4000 | Maximum | | 675 | .071 | | 4.23 | 2.54 | .07 | 6.84 | 8.60 |
| 4200 | Polyester, heavy duty, compared to epoxy, add | | 2590 | .019 | | 1.31 | .66 | .02 | 1.99 | 2.47 |

# 09 67 Fluid-Applied Flooring

## 09 67 66 – Fluid-Applied Athletic Flooring

| 09 67 66.10 Polyurethane | Crew | Daily Output | Labor-Hours | Unit | Material | 2011 Bare Costs Labor | Equipment | Total | Total Incl O&P |
|---|---|---|---|---|---|---|---|---|---|
| 0010 **POLYURETHANE** | | | | | | | | | |
| 4400 Thermoset, prefabricated in place, indoor | | | | | | | | | |
| 4500 3/8" thick for basketball, gyms, etc. | 1 Tilf | 100 | .080 | S.F. | 4.95 | 3.24 | | 8.19 | 10.15 |
| 4600 1/2" thick for professional sports | | 95 | .084 | | 6.65 | 3.41 | | 10.06 | 12.25 |
| 4700 Outdoor, 1/4" thick, smooth, for tennis | | 100 | .080 | | 5.15 | 3.24 | | 8.39 | 10.35 |
| 4800 Rough, for track, 3/8" thick | | 95 | .084 | | 5.95 | 3.41 | | 9.36 | 11.50 |
| 5000 Poured in place, indoor, with finish, 1/4" thick | | 80 | .100 | | 3.66 | 4.05 | | 7.71 | 9.95 |
| 5050 3/8" thick | | 65 | .123 | | 4.45 | 4.98 | | 9.43 | 12.15 |
| 5100 1/2" thick | | 50 | .160 | | 5.85 | 6.45 | | 12.30 | 15.90 |

# 09 68 Carpeting

## 09 68 05 – Carpet Accessories

### 09 68 05.11 Flooring Transition Strip

| | Crew | Daily Output | Labor-Hours | Unit | Material | Labor | Equipment | Total | Total Incl O&P |
|---|---|---|---|---|---|---|---|---|---|
| 0010 **FLOORING TRANSITION STRIP** | | | | | | | | | |
| 0107 Clamp down brass divider, 12' strip, vinyl to carpet | 1 Tilf | 31.25 | .256 | Ea. | 10.45 | 10.35 | | 20.80 | 26.50 |
| 0117 Vinyl to hard surface | " | 31.25 | .256 | " | 10.45 | 10.35 | | 20.80 | 26.50 |

## 09 68 10 – Carpet Pad

### 09 68 10.10 Commercial Grade Carpet Pad

| | Crew | Daily Output | Labor-Hours | Unit | Material | Labor | Equipment | Total | Total Incl O&P |
|---|---|---|---|---|---|---|---|---|---|
| 0010 **COMMERCIAL GRADE CARPET PAD** | | | | | | | | | |
| 9000 Sponge rubber pad, minimum | 1 Tilf | 150 | .053 | S.Y. | 4.06 | 2.16 | | 6.22 | 7.60 |
| 9100 Maximum | | 150 | .053 | | 10 | 2.16 | | 12.16 | 14.15 |
| 9200 Felt pad, minimum | | 150 | .053 | | 4.14 | 2.16 | | 6.30 | 7.70 |
| 9300 Maximum | | 150 | .053 | | 7.10 | 2.16 | | 9.26 | 10.95 |
| 9400 Bonded urethane pad, minimum | | 150 | .053 | | 4.57 | 2.16 | | 6.73 | 8.20 |
| 9500 Maximum | | 150 | .053 | | 7.80 | 2.16 | | 9.96 | 11.75 |
| 9600 Prime urethane pad, minimum | | 150 | .053 | | 3 | 2.16 | | 5.16 | 6.45 |
| 9700 Maximum | | 150 | .053 | | 5 | 2.16 | | 7.16 | 8.65 |

## 09 68 13 – Tile Carpeting

### 09 68 13.10 Carpet Tile

| | Crew | Daily Output | Labor-Hours | Unit | Material | Labor | Equipment | Total | Total Incl O&P |
|---|---|---|---|---|---|---|---|---|---|
| 0010 **CARPET TILE** | | | | | | | | | |
| 0100 Tufted nylon, 18" x 18", hard back, 20 oz. | 1 Tilf | 80 | .100 | S.Y. | 22 | 4.05 | | 26.05 | 30.50 |
| 0110 26 oz. | | 80 | .100 | | 38 | 4.05 | | 42.05 | 47.50 |
| 0200 Cushion back, 20 oz. | | 80 | .100 | | 28 | 4.05 | | 32.05 | 36.50 |
| 0210 26 oz. | | 80 | .100 | | 43.50 | 4.05 | | 47.55 | 54 |
| 1100 Tufted, 24" x 24", hard back, 24 oz. nylon | | 80 | .100 | | 29 | 4.05 | | 33.05 | 37.50 |
| 1180 35 oz. | | 80 | .100 | | 34 | 4.05 | | 38.05 | 43.50 |
| 5060 42 oz. | | 80 | .100 | | 45.50 | 4.05 | | 49.55 | 56 |

## 09 68 16 – Sheet Carpeting

### 09 68 16.10 Sheet Carpet

| | Crew | Daily Output | Labor-Hours | Unit | Material | Labor | Equipment | Total | Total Incl O&P |
|---|---|---|---|---|---|---|---|---|---|
| 0010 **SHEET CARPET** | | | | | | | | | |
| 0700 Nylon, level loop, 26 oz., light to medium traffic | 1 Tilf | 75 | .107 | S.Y. | 28 | 4.31 | | 32.31 | 37.50 |
| 0720 28 oz., light to medium traffic | | 75 | .107 | | 29 | 4.31 | | 33.31 | 38.50 |
| 0900 32 oz., medium traffic | | 75 | .107 | | 35 | 4.31 | | 39.31 | 44.50 |
| 1100 40 oz., medium to heavy traffic | | 75 | .107 | | 51 | 4.31 | | 55.31 | 62.50 |
| 2920 Nylon plush, 30 oz., medium traffic | | 57 | .140 | | 27 | 5.70 | | 32.70 | 38 |
| 3000 36 oz., medium traffic | | 75 | .107 | | 35.50 | 4.31 | | 39.81 | 45.50 |
| 3100 42 oz., medium to heavy traffic | | 70 | .114 | | 40 | 4.62 | | 44.62 | 51 |
| 3200 46 oz., medium to heavy traffic | | 70 | .114 | | 46.50 | 4.62 | | 51.12 | 58.50 |
| 3300 54 oz., heavy traffic | | 70 | .114 | | 51 | 4.62 | | 55.62 | 63 |

# 09 68 Carpeting

## 09 68 16 – Sheet Carpeting

### 09 68 16.10 Sheet Carpet

| | | Crew | Daily Output | Labor-Hours | Unit | Material | 2011 Bare Costs Labor | 2011 Bare Costs Equipment | Total | Total Incl O&P |
|---|---|---|---|---|---|---|---|---|---|---|
| 3340 | 60 oz., heavy traffic | 1 Tilf | 70 | .114 | S.Y. | 70 | 4.62 | | 74.62 | 84 |
| 3665 | Olefin, 24 oz., light to medium traffic | | 75 | .107 | | 16 | 4.31 | | 20.31 | 24 |
| 3670 | 26 oz., medium traffic | | 75 | .107 | | 17.35 | 4.31 | | 21.66 | 25.50 |
| 3680 | 28 oz., medium to heavy traffic | | 75 | .107 | | 18.95 | 4.31 | | 23.26 | 27.50 |
| 3700 | 32 oz., medium to heavy traffic | | 75 | .107 | | 24 | 4.31 | | 28.31 | 33 |
| 3730 | 42 oz., heavy traffic | | 70 | .114 | | 25.50 | 4.62 | | 30.12 | 35 |
| 4110 | Wool, level loop, 40 oz., medium traffic | | 75 | .107 | | 119 | 4.31 | | 123.31 | 137 |
| 4500 | 50 oz., medium to heavy traffic | | 75 | .107 | | 120 | 4.31 | | 124.31 | 138 |
| 4700 | Patterned, 32 oz., medium to heavy traffic | | 70 | .114 | | 119 | 4.62 | | 123.62 | 138 |
| 4900 | 48 oz., heavy traffic | | 70 | .114 | | 122 | 4.62 | | 126.62 | 141 |
| 5000 | For less than full roll (approx. 1500 S.F.), add | | | | | 25% | | | | |
| 5100 | For small rooms, less than 12' wide, add | | | | | | 25% | | | |
| 5200 | For large open areas (no cuts), deduct | | | | | | 25% | | | |
| 5600 | For bound carpet baseboard, add | 1 Tilf | 300 | .027 | L.F. | 2.75 | 1.08 | | 3.83 | 4.60 |
| 5610 | For stairs, not incl. price of carpet, add | " | 30 | .267 | Riser | | 10.80 | | 10.80 | 15.75 |
| 5620 | For borders and patterns, add to labor | | | | | | 18% | | | |
| 8950 | For tackless, stretched installation, add padding from 09 68 10.10 to above | | | | | | | | | |
| 9850 | For brand-named specific fiber, add | | | | S.Y. | 25% | | | | |

## 09 68 20 – Athletic Carpet

### 09 68 20.10 Indoor Athletic Carpet

| | | Crew | Daily Output | Labor-Hours | Unit | Material | 2011 Bare Costs Labor | 2011 Bare Costs Equipment | Total | Total Incl O&P |
|---|---|---|---|---|---|---|---|---|---|---|
| 0010 | **INDOOR ATHLETIC CARPET** | | | | | | | | | |
| 3700 | Polyethylene, in rolls, no base incl., landscape surfaces | 1 Tilf | 275 | .029 | S.F. | 3.30 | 1.18 | | 4.48 | 5.35 |
| 3800 | Nylon action surface, 1/8" thick | | 275 | .029 | | 3.46 | 1.18 | | 4.64 | 5.55 |
| 3900 | 1/4" thick | | 275 | .029 | | 4.99 | 1.18 | | 6.17 | 7.20 |
| 4000 | 3/8" thick | | 275 | .029 | | 6.25 | 1.18 | | 7.43 | 8.60 |
| 4100 | Golf tee surface with foam back | | 235 | .034 | | 6.20 | 1.38 | | 7.58 | 8.80 |
| 4200 | Practice putting, knitted nylon surface | | 235 | .034 | | 5.25 | 1.38 | | 6.63 | 7.80 |
| 5500 | Polyvinyl chloride, sheet goods for gyms, 1/4" thick | | 80 | .100 | | 8.25 | 4.05 | | 12.30 | 15 |
| 5600 | 3/8" thick | | 60 | .133 | | 9.25 | 5.40 | | 14.65 | 18.05 |

# 09 69 Access Flooring

## 09 69 13 – Rigid-Grid Access Flooring

### 09 69 13.10 Access Floors

| | | Crew | Daily Output | Labor-Hours | Unit | Material | 2011 Bare Costs Labor | 2011 Bare Costs Equipment | Total | Total Incl O&P |
|---|---|---|---|---|---|---|---|---|---|---|
| 0010 | **ACCESS FLOORS** | | | | | | | | | |
| 0015 | Access floor package including panel, pedestal, stringers & laminate cover | | | | | | | | | |
| 0100 | Computer room, greater than 6,000 S.F. | 4 Carp | 750 | .043 | S.F. | 7 | 1.84 | | 8.84 | 10.50 |
| 0110 | Less than 6,000 S.F. | 2 Carp | 375 | .043 | | 7.65 | 1.84 | | 9.49 | 11.20 |
| 0120 | Office, greater than 6,000 S.F. | 4 Carp | 1050 | .030 | | 5 | 1.31 | | 6.31 | 7.50 |
| 0250 | Panels, particle board or steel, 1250# load, no covering, under 6,000 S.F. | 2 Carp | 600 | .027 | | 3.50 | 1.15 | | 4.65 | 5.60 |
| 0300 | Over 6,000 S.F. | | 640 | .025 | | 3 | 1.08 | | 4.08 | 4.95 |
| 0400 | Aluminum, 24" panels | | 500 | .032 | | 31.50 | 1.38 | | 32.88 | 37 |
| 0600 | For carpet covering, add | | | | | 8.55 | | | 8.55 | 9.40 |
| 0700 | For vinyl floor covering, add | | | | | 6.70 | | | 6.70 | 7.35 |
| 0900 | For high pressure laminate covering, add | | | | | 5.45 | | | 5.45 | 6 |
| 0910 | For snap on stringer system, add | 2 Carp | 1000 | .016 | | 1.44 | .69 | | 2.13 | 2.64 |
| 0950 | Office applications, steel or concrete panels, | | | | | | | | | |
| 0960 | no covering, over 6,000 S.F. | 2 Carp | 960 | .017 | S.F. | 10.10 | .72 | | 10.82 | 12.20 |
| 1000 | Machine cutouts after initial installation | 1 Carp | 50 | .160 | Ea. | 5.05 | 6.90 | | 11.95 | 16.10 |
| 1050 | Pedestals, 6" to 12" | 2 Carp | 85 | .188 | | 7.95 | 8.10 | | 16.05 | 21 |
| 1100 | Air conditioning grilles, 4" x 12" | 1 Carp | 17 | .471 | | 66 | 20.50 | | 86.50 | 104 |

# 09 69 Access Flooring

## 09 69 13 – Rigid-Grid Access Flooring

| 09 69 13.10 Access Floors | | Crew | Daily Output | Labor-Hours | Unit | Material | 2011 Bare Costs Labor | Equipment | Total | Total Incl O&P |
|---|---|---|---|---|---|---|---|---|---|---|
| 1150 | 4" x 18" | 1 Carp | 14 | .571 | Ea. | 90.50 | 24.50 | | 115 | 137 |
| 1200 | Approach ramps, minimum | 2 Carp | 60 | .267 | S.F. | 24.50 | 11.50 | | 36 | 44.50 |
| 1300 | Maximum | " | 40 | .400 | " | 33.50 | 17.20 | | 50.70 | 63 |
| 1500 | Handrail, 2 rail, aluminum | 1 Carp | 15 | .533 | L.F. | 94.50 | 23 | | 117.50 | 139 |

# 09 72 Wall Coverings

## 09 72 23 – Wallpapering

| 09 72 23.10 Wallpaper | | Crew | Daily Output | Labor-Hours | Unit | Material | 2011 Bare Costs Labor | Equipment | Total | Total Incl O&P |
|---|---|---|---|---|---|---|---|---|---|---|
| 0010 | **WALLPAPER** including sizing; add 10-30 percent waste @ takeoff  R097223-10 | | | | | | | | | |
| 0050 | Aluminum foil | 1 Pape | 275 | .029 | S.F. | 1.02 | 1.10 | | 2.12 | 2.75 |
| 0100 | Copper sheets, .025" thick, vinyl backing | | 240 | .033 | | 5.45 | 1.26 | | 6.71 | 7.85 |
| 0300 | Phenolic backing | | 240 | .033 | | 7.05 | 1.26 | | 8.31 | 9.65 |
| 0600 | Cork tiles, light or dark, 12" x 12" x 3/16" | | 240 | .033 | | 4.47 | 1.26 | | 5.73 | 6.80 |
| 0700 | 5/16" thick | | 235 | .034 | | 3.80 | 1.29 | | 5.09 | 6.10 |
| 0900 | 1/4" basketweave | | 240 | .033 | | 5.85 | 1.26 | | 7.11 | 8.30 |
| 1000 | 1/2" natural, non-directional pattern | | 240 | .033 | | 7.85 | 1.26 | | 9.11 | 10.50 |
| 1100 | 3/4" natural, non-directional pattern | | 240 | .033 | | 10.80 | 1.26 | | 12.06 | 13.70 |
| 1200 | Granular surface, 12" x 36", 1/2" thick | | 385 | .021 | | 1.26 | .79 | | 2.05 | 2.56 |
| 1300 | 1" thick | | 370 | .022 | | 1.63 | .82 | | 2.45 | 3 |
| 1500 | Polyurethane coated, 12" x 12" x 3/16" thick | | 240 | .033 | | 3.95 | 1.26 | | 5.21 | 6.20 |
| 1600 | 5/16" thick | | 235 | .034 | | 5.65 | 1.29 | | 6.94 | 8.10 |
| 1800 | Cork wallpaper, paperbacked, natural | | 480 | .017 | | 1.99 | .63 | | 2.62 | 3.13 |
| 1900 | Colors | | 480 | .017 | | 2.78 | .63 | | 3.41 | 4 |
| 2100 | Flexible wood veneer, 1/32" thick, plain woods | | 100 | .080 | | 2.36 | 3.03 | | 5.39 | 7.10 |
| 2200 | Exotic woods | ▼ | 95 | .084 | ▼ | 3.58 | 3.19 | | 6.77 | 8.65 |
| 2400 | Gypsum-based, fabric-backed, fire resistant | | | | | | | | | |
| 2500 | for masonry walls, minimum, 21 oz./S.Y. | 1 Pape | 800 | .010 | S.F. | .82 | .38 | | 1.20 | 1.46 |
| 2700 | Maximum (small quantities) | " | 640 | .013 | | 1.37 | .47 | | 1.84 | 2.21 |
| 2750 | Acrylic, modified, semi-rigid PVC, .028" thick | 2 Carp | 330 | .048 | | 1.13 | 2.09 | | 3.22 | 4.44 |
| 2800 | .040" thick | " | 320 | .050 | | 1.49 | 2.15 | | 3.64 | 4.94 |
| 3000 | Vinyl wall covering, fabric-backed, lightweight, type 1 (12-15 oz./S.Y.) | 1 Pape | 640 | .013 | | .69 | .47 | | 1.16 | 1.46 |
| 3300 | Medium weight, type 2 (20-24 oz./S.Y.) | | 480 | .017 | | .82 | .63 | | 1.45 | 1.84 |
| 3400 | Heavy weight, type 3 (28 oz./S.Y.) | ▼ | 435 | .018 | ▼ | 1.64 | .70 | | 2.34 | 2.83 |
| 3600 | Adhesive, 5 gal. lots (18 S.Y./gal.) | | | | Gal. | 10.40 | | | 10.40 | 11.45 |
| 3700 | Wallpaper, average workmanship, solid pattern, low cost paper | 1 Pape | 640 | .013 | S.F. | .37 | .47 | | .84 | 1.11 |
| 3900 | basic patterns (matching required), avg. cost paper | | 535 | .015 | | .78 | .57 | | 1.35 | 1.70 |
| 4000 | Paper at $85 per double roll, quality workmanship | ▼ | 435 | .018 | ▼ | 2.47 | .70 | | 3.17 | 3.75 |
| 4100 | Linen wall covering, paper backed | | | | | | | | | |
| 4150 | Flame treatment, minimum | | | | S.F. | 1.04 | | | 1.04 | 1.14 |
| 4180 | Maximum | | | | | 1.55 | | | 1.55 | 1.71 |
| 4190 | Grass cloth, natural fabric  [G] | 1 Pape | 400 | .020 | | 2.99 | .76 | | 3.75 | 4.41 |
| 4200 | Grass cloths with lining paper, minimum  [G] | | 400 | .020 | | .78 | .76 | | 1.54 | 1.98 |
| 4300 | Maximum  [G] | ▼ | 350 | .023 | ▼ | 2.49 | .87 | | 3.36 | 4.02 |

# 09 77 Special Wall Surfacing

## 09 77 33 – Fiberglass Reinforced Panels

| 09 77 33.10 Fiberglass Reinforced Plastic Panels | Crew | Daily Output | Labor-Hours | Unit | Material | 2011 Bare Costs Labor | Equipment | Total | Total Incl O&P |
|---|---|---|---|---|---|---|---|---|---|
| 0010 | **FIBERGLASS REINFORCED PLASTIC PANELS**, .090" thick | | | | | | | | | |
| 0020 | On walls, adhesive mounted, embossed surface | 2 Carp | 640 | .025 | S.F. | 1.23 | 1.08 | | 2.31 | 3 |
| 0030 | Smooth surface | | 640 | .025 | | 1.51 | 1.08 | | 2.59 | 3.31 |
| 0040 | Fire rated, embossed surface | | 640 | .025 | | 2.19 | 1.08 | | 3.27 | 4.06 |
| 0050 | Nylon rivet mounted, on drywall, embossed surface | | 480 | .033 | | 1.15 | 1.43 | | 2.58 | 3.47 |
| 0060 | Smooth surface | | 480 | .033 | | 1.43 | 1.43 | | 2.86 | 3.77 |
| 0070 | Fire rated, embossed surface | | 480 | .033 | | 2.11 | 1.43 | | 3.54 | 4.52 |
| 0080 | On masonry, embossed surface | | 320 | .050 | | 1.15 | 2.15 | | 3.30 | 4.57 |
| 0090 | Smooth surface | | 320 | .050 | | 1.43 | 2.15 | | 3.58 | 4.87 |
| 0100 | Fire rated, embossed surface | | 320 | .050 | | 2.11 | 2.15 | | 4.26 | 5.60 |
| 0110 | Nylon rivet and adhesive mounted, on drywall, embossed surface | | 240 | .067 | | 1.30 | 2.87 | | 4.17 | 5.85 |
| 0120 | Smooth surface | | 240 | .067 | | 1.58 | 2.87 | | 4.45 | 6.15 |
| 0130 | Fire rated, embossed surface | | 240 | .067 | | 2.26 | 2.87 | | 5.13 | 6.90 |
| 0140 | On masonry, embossed surface | | 190 | .084 | | 1.30 | 3.63 | | 4.93 | 7 |
| 0150 | Smooth surface | | 190 | .084 | | 1.58 | 3.63 | | 5.21 | 7.30 |
| 0160 | Fire rated, embossed surface | | 190 | .084 | | 2.26 | 3.63 | | 5.89 | 8.05 |
| 0170 | For moldings add | 1 Carp | 250 | .032 | L.F. | .28 | 1.38 | | 1.66 | 2.42 |
| 0180 | On ceilings, for lay in grid system, embossed surface | | 400 | .020 | S.F. | 1.23 | .86 | | 2.09 | 2.67 |
| 0190 | Smooth surface | | 400 | .020 | | 1.51 | .86 | | 2.37 | 2.98 |
| 0200 | Fire rated, embossed surface | | 400 | .020 | | 2.19 | .86 | | 3.05 | 3.73 |

## 09 77 43 – Panel Systems

| 09 77 43.20 Slatwall Panels and Accessories | Crew | Daily Output | Labor-Hours | Unit | Material | 2011 Bare Costs Labor | Equipment | Total | Total Incl O&P |
|---|---|---|---|---|---|---|---|---|---|
| 0010 | **SLATWALL PANELS AND ACCESSORIES** | | | | | | | | | |
| 0100 | Slatwall panel, 4' x 8' x 3/4" T, MDF, paint grade | 1 Carp | 500 | .016 | S.F. | 1.32 | .69 | | 2.01 | 2.51 |
| 0110 | Melamine finish | | 500 | .016 | | 1.87 | .69 | | 2.56 | 3.12 |
| 0120 | High pressure plastic laminate finish | | 500 | .016 | | 3.15 | .69 | | 3.84 | 4.53 |
| 0130 | Aluminum channel inserts, add | | | | | 2.94 | | | 2.94 | 3.23 |
| 0200 | Accessories, corner forms, 8' L | | | | L.F. | 4.28 | | | 4.28 | 4.71 |
| 0210 | T-connector, 8' L | | | | | 5.15 | | | 5.15 | 5.70 |
| 0220 | J-mold, 8' L | | | | | 1.33 | | | 1.33 | 1.46 |
| 0230 | Edge cap, 8' L | | | | | 2.05 | | | 2.05 | 2.26 |
| 0240 | Finish end cap, 8' L | | | | | 3.18 | | | 3.18 | 3.50 |
| 0300 | Display hook, metal, 4" L | | | | Ea. | .44 | | | .44 | .48 |
| 0310 | 6" L | | | | | .50 | | | .50 | .55 |
| 0320 | 8" L | | | | | .54 | | | .54 | .59 |
| 0330 | 10" L | | | | | .58 | | | .58 | .64 |
| 0340 | 12" L | | | | | .63 | | | .63 | .69 |
| 0350 | Acrylic, 4" L | | | | | .91 | | | .91 | 1 |
| 0360 | 6" L | | | | | 1.05 | | | 1.05 | 1.16 |
| 0370 | 8" L | | | | | 1.10 | | | 1.10 | 1.21 |
| 0380 | 10" L | | | | | 1.12 | | | 1.12 | 1.23 |
| 0400 | Waterfall hanger, metal, 12" - 16" | | | | | 4.52 | | | 4.52 | 4.97 |
| 0410 | Acrylic | | | | | 9.85 | | | 9.85 | 10.85 |
| 0500 | Shelf bracket, metal, 8" | | | | | 4.43 | | | 4.43 | 4.87 |
| 0510 | 10" | | | | | 4.60 | | | 4.60 | 5.05 |
| 0520 | 12" | | | | | 4.49 | | | 4.49 | 4.94 |
| 0530 | 14" | | | | | 4.79 | | | 4.79 | 5.25 |
| 0540 | 16" | | | | | 6.05 | | | 6.05 | 6.65 |
| 0550 | Acrylic, 8" | | | | | 3.04 | | | 3.04 | 3.34 |
| 0560 | 10" | | | | | 3.37 | | | 3.37 | 3.71 |
| 0570 | 12" | | | | | 3.72 | | | 3.72 | 4.09 |
| 0580 | 14" | | | | | 3.99 | | | 3.99 | 4.39 |

# 09 77 Special Wall Surfacing

## 09 77 43 – Panel Systems

| 09 77 43.20 Slatwall Panels and Accessories | Crew | Daily Output | Labor-Hours | Unit | Material | 2011 Bare Costs Labor | Equipment | Total | Total Incl O&P |
|---|---|---|---|---|---|---|---|---|---|
| 0600 | Shelf, acrylic, 12" x 16" x 1/4" | | | | Ea. | 19.90 | | | 19.90 | 22 |
| 0610 | 12" x 24" x 1/4" | | | | ↓ | 30 | | | 30 | 33 |

# 09 81 Acoustic Insulation

## 09 81 16 – Acoustic Blanket Insulation

### 09 81 16.10 Sound Attenuation Blanket

| | | Crew | Daily Output | Labor-Hours | Unit | Material | 2011 Bare Costs Labor | Equipment | Total | Total Incl O&P |
|---|---|---|---|---|---|---|---|---|---|---|
| 0010 | **SOUND ATTENUATION BLANKET** | | | | | | | | | |
| 0020 | Blanket, 1" thick | 1 Carp | 925 | .009 | S.F. | .25 | .37 | | .62 | .85 |
| 0500 | 1-1/2" thick | | 920 | .009 | | .25 | .37 | | .62 | .85 |
| 1000 | 2" thick | | 915 | .009 | | .34 | .38 | | .72 | .95 |
| 1500 | 3" thick | | 910 | .009 | | .50 | .38 | | .88 | 1.13 |
| 3000 | Thermal or acoustical batt above ceiling, 2" thick | | 900 | .009 | | .49 | .38 | | .87 | 1.13 |
| 3100 | 3" thick | | 900 | .009 | | .74 | .38 | | 1.12 | 1.40 |
| 3200 | 4" thick | | 900 | .009 | | .92 | .38 | | 1.30 | 1.60 |
| 3400 | Urethane plastic foam, open cell, on wall, 2" thick | 2 Carp | 2050 | .008 | | 3.05 | .34 | | 3.39 | 3.88 |
| 3500 | 3" thick | | 1550 | .010 | | 4.05 | .44 | | 4.49 | 5.15 |
| 3600 | 4" thick | | 1050 | .015 | | 5.70 | .66 | | 6.36 | 7.25 |
| 3700 | On ceiling, 2" thick | | 1700 | .009 | | 3.04 | .41 | | 3.45 | 3.96 |
| 3800 | 3" thick | | 1300 | .012 | | 4.05 | .53 | | 4.58 | 5.25 |
| 3900 | 4" thick | | 900 | .018 | ↓ | 5.70 | .77 | | 6.47 | 7.40 |
| 4000 | Nylon matting 0.4" thick, with carbon black spinerette | | | | | | | | | |
| 4010 | plus polyester fabric, on floor | D-7 | 1600 | .010 | S.F. | 1.10 | .36 | | 1.46 | 1.74 |
| 4200 | Fiberglass reinf. backer board underlayment, 7/16" thick, on floor | " | 1500 | .011 | " | 1.90 | .39 | | 2.29 | 2.65 |

# 09 84 Acoustic Room Components

## 09 84 13 – Fixed Sound-Absorptive Panels

### 09 84 13.10 Fixed Panels

| | | Crew | Daily Output | Labor-Hours | Unit | Material | 2011 Bare Costs Labor | Equipment | Total | Total Incl O&P |
|---|---|---|---|---|---|---|---|---|---|---|
| 0010 | **FIXED PANELS** Perforated steel facing, painted with | | | | | | | | | |
| 0100 | Fiberglass or mineral filler, no backs, 2-1/4" thick, modular | | | | | | | | | |
| 0200 | space units, ceiling or wall hung, white or colored | 1 Carp | 100 | .080 | S.F. | 7.35 | 3.44 | | 10.79 | 13.40 |
| 0300 | Fiberboard sound deadening panels, 1/2" thick | " | 600 | .013 | " | .32 | .57 | | .89 | 1.23 |
| 0500 | Fiberglass panels, 4' x 8' x 1" thick, with | | | | | | | | | |
| 0600 | glass cloth face for walls, cemented | 1 Carp | 155 | .052 | S.F. | 6.50 | 2.22 | | 8.72 | 10.55 |
| 0700 | 1-1/2" thick, dacron covered, inner aluminum frame, | | | | | | | | | |
| 0710 | wall mounted | 1 Carp | 300 | .027 | S.F. | 8.75 | 1.15 | | 9.90 | 11.35 |
| 0900 | Mineral fiberboard panels, fabric covered, 30" x 108", | | | | | | | | | |
| 1000 | 3/4" thick, concealed spline, wall mounted | 1 Carp | 150 | .053 | S.F. | 6.25 | 2.30 | | 8.55 | 10.40 |

## 09 84 36 – Sound-Absorbing Ceiling Units

### 09 84 36.10 Barriers

| | | Crew | Daily Output | Labor-Hours | Unit | Material | 2011 Bare Costs Labor | Equipment | Total | Total Incl O&P |
|---|---|---|---|---|---|---|---|---|---|---|
| 0010 | **BARRIERS** Plenum | | | | | | | | | |
| 0600 | Aluminum foil, fiberglass reinf., parallel with joists | 1 Carp | 275 | .029 | S.F. | .92 | 1.25 | | 2.17 | 2.93 |
| 0700 | Perpendicular to joists | | 180 | .044 | | .92 | 1.91 | | 2.83 | 3.95 |
| 0900 | Aluminum mesh, kraft paperbacked | | 275 | .029 | | .77 | 1.25 | | 2.02 | 2.77 |
| 0970 | Fiberglass batts, kraft faced, 3-1/2" thick | | 1400 | .006 | | .36 | .25 | | .61 | .78 |
| 0980 | 6" thick | | 1300 | .006 | | .65 | .26 | | .91 | 1.13 |
| 1000 | Sheet lead, 1 lb., 1/64" thick, perpendicular to joists | | 150 | .053 | | 3.47 | 2.30 | | 5.77 | 7.35 |
| 1100 | Vinyl foam reinforced, 1/8" thick, 1.0 lb. per S.F. | ↓ | 150 | .053 | ↓ | 4 | 2.30 | | 6.30 | 7.90 |

# 09 91 Painting

## 09 91 03 – Paint Restoration

### 09 91 03.20 Sanding

| | | Crew | Daily Output | Labor-Hours | Unit | Material | 2011 Bare Costs Labor | 2011 Bare Costs Equipment | Total | Total Incl O&P |
|---|---|---|---|---|---|---|---|---|---|---|
| 0010 | **SANDING** and puttying interior trim, compared to | R099100-10 | | | | | | | | |
| 0100 | Painting 1 coat, on quality work | | | | L.F. | | 100% | | | |
| 0300 | Medium work | | | | | | 50% | | | |
| 0400 | Industrial grade | | | | | | 25% | | | |
| 0500 | Surface protection, placement and removal | | | | | | | | | |
| 0510 | Basic drop cloths | 1 Pord | 6400 | .001 | S.F. | | .05 | | .05 | .07 |
| 0520 | Masking with paper | | 800 | .010 | | .05 | .38 | | .43 | .62 |
| 0530 | Volume cover up (using plastic sheathing, or building paper) | | 16000 | .001 | | | .02 | | .02 | .03 |

### 09 91 03.30 Exterior Surface Preparation

| | | Crew | Daily Output | Labor-Hours | Unit | Material | 2011 Bare Costs Labor | 2011 Bare Costs Equipment | Total | Total Incl O&P |
|---|---|---|---|---|---|---|---|---|---|---|
| 0010 | **EXTERIOR SURFACE PREPARATION** | R099100-10 | | | | | | | | |
| 0015 | Doors, per side, not incl. frames or trim | | | | | | | | | |
| 0020 | Scrape & sand | | | | | | | | | |
| 0030 | Wood, flush | 1 Pord | 616 | .013 | S.F. | | .49 | | .49 | .72 |
| 0040 | Wood, detail | | 496 | .016 | | | .61 | | .61 | .90 |
| 0050 | Wood, louvered | | 280 | .029 | | | 1.07 | | 1.07 | 1.59 |
| 0060 | Wood, overhead | | 616 | .013 | | | .49 | | .49 | .72 |
| 0070 | Wire brush | | | | | | | | | |
| 0080 | Metal, flush | 1 Pord | 640 | .013 | S.F. | | .47 | | .47 | .70 |
| 0090 | Metal, detail | | 520 | .015 | | | .58 | | .58 | .86 |
| 0100 | Metal, louvered | | 360 | .022 | | | .83 | | .83 | 1.24 |
| 0110 | Metal or fibr., overhead | | 640 | .013 | | | .47 | | .47 | .70 |
| 0120 | Metal, roll up | | 560 | .014 | | | .54 | | .54 | .80 |
| 0130 | Metal, bulkhead | | 640 | .013 | | | .47 | | .47 | .70 |
| 0140 | Power wash, based on 2500 lb. operating pressure | | | | | | | | | |
| 0150 | Metal, flush | A-1H | 2240 | .004 | S.F. | | .12 | .03 | .15 | .22 |
| 0160 | Metal, detail | | 2120 | .004 | | | .13 | .03 | .16 | .23 |
| 0170 | Metal, louvered | | 2000 | .004 | | | .14 | .03 | .17 | .24 |
| 0180 | Metal or fibr., overhead | | 2400 | .003 | | | .11 | .03 | .14 | .21 |
| 0190 | Metal, roll up | | 2400 | .003 | | | .11 | .03 | .14 | .21 |
| 0200 | Metal, bulkhead | | 2200 | .004 | | | .13 | .03 | .16 | .22 |
| 0400 | Windows, per side, not incl. trim | | | | | | | | | |
| 0410 | Scrape & sand | | | | | | | | | |
| 0420 | Wood, 1-2 lite | 1 Pord | 320 | .025 | S.F. | | .94 | | .94 | 1.39 |
| 0430 | Wood, 3-6 lite | | 280 | .029 | | | 1.07 | | 1.07 | 1.59 |
| 0440 | Wood, 7-10 lite | | 240 | .033 | | | 1.25 | | 1.25 | 1.86 |
| 0450 | Wood, 12 lite | | 200 | .040 | | | 1.50 | | 1.50 | 2.23 |
| 0460 | Wood, Bay/Bow | | 320 | .025 | | | .94 | | .94 | 1.39 |
| 0470 | Wire brush | | | | | | | | | |
| 0480 | Metal, 1-2 lite | 1 Pord | 480 | .017 | S.F. | | .63 | | .63 | .93 |
| 0490 | Metal, 3-6 lite | | 400 | .020 | | | .75 | | .75 | 1.11 |
| 0500 | Metal, Bay/Bow | | 480 | .017 | | | .63 | | .63 | .93 |
| 0510 | Power wash, based on 2500 lb. operating pressure | | | | | | | | | |
| 0520 | 1-2 lite | A-1H | 4400 | .002 | S.F. | | .06 | .01 | .07 | .12 |
| 0530 | 3-6 lite | | 4320 | .002 | | | .06 | .01 | .07 | .12 |
| 0540 | 7-10 lite | | 4240 | .002 | | | .06 | .01 | .07 | .12 |
| 0550 | 12 lite | | 4160 | .002 | | | .07 | .01 | .08 | .12 |
| 0560 | Bay/Bow | | 4400 | .002 | | | .06 | .01 | .07 | .12 |
| 0600 | Siding, scrape and sand, light=10-30%, med.=30-70% | | | | | | | | | |
| 0610 | Heavy=70-100% of surface to sand | | | | | | | | | |
| 0650 | Texture 1-11, light | 1 Pord | 480 | .017 | S.F. | | .63 | | .63 | .93 |
| 0660 | Med. | | 440 | .018 | | | .68 | | .68 | 1.01 |
| 0670 | Heavy | | 360 | .022 | | | .83 | | .83 | 1.24 |

| 09 91 03.30 Exterior Surface Preparation | | Crew | Daily Output | Labor-Hours | Unit | Material | 2011 Bare Costs Labor | 2011 Bare Costs Equipment | Total | Total Incl O&P |
|---|---|---|---|---|---|---|---|---|---|---|
| 0680 | Wood shingles, shakes, light | 1 Pord | 440 | .018 | S.F. | | .68 | | .68 | 1.01 |
| 0690 | Med. | | 360 | .022 | | | .83 | | .83 | 1.24 |
| 0700 | Heavy | | 280 | .029 | | | 1.07 | | 1.07 | 1.59 |
| 0710 | Clapboard, light | | 520 | .015 | | | .58 | | .58 | .86 |
| 0720 | Med. | | 480 | .017 | | | .63 | | .63 | .93 |
| 0730 | Heavy | | 400 | .020 | | | .75 | | .75 | 1.11 |
| 0740 | Wire brush | | | | | | | | | |
| 0750 | Aluminum, light | 1 Pord | 600 | .013 | S.F. | | .50 | | .50 | .74 |
| 0760 | Med. | | 520 | .015 | | | .58 | | .58 | .86 |
| 0770 | Heavy | | 440 | .018 | | | .68 | | .68 | 1.01 |
| 0780 | Pressure wash, based on 2500 lb. operating pressure | | | | | | | | | |
| 0790 | Stucco | A-1H | 3080 | .003 | S.F. | | .09 | .02 | .11 | .16 |
| 0800 | Aluminum or vinyl | | 3200 | .003 | | | .09 | .02 | .11 | .15 |
| 0810 | Siding, masonry, brick & block | | 2400 | .003 | | | .11 | .03 | .14 | .21 |
| 1300 | Miscellaneous, wire brush | | | | | | | | | |
| 1310 | Metal, pedestrian gate | 1 Pord | 100 | .080 | S.F. | | 3 | | 3 | 4.46 |
| 8000 | For chemical washing, see Section 04 01 30 | | | | | | | | | |
| 8010 | For steam cleaning, see Section 04 01 30.20 | | | | | | | | | |
| 8020 | For sand blasting, see Section 05 01 10.51 and 03 35 29.60 | | | | | | | | | |

| 09 91 03.40 Interior Surface Preparation | | Crew | Daily Output | Labor-Hours | Unit | Material | 2011 Bare Costs Labor | 2011 Bare Costs Equipment | Total | Total Incl O&P |
|---|---|---|---|---|---|---|---|---|---|---|
| 0010 | **INTERIOR SURFACE PREPARATION** R099100-10 | | | | | | | | | |
| 0020 | Doors, per side, not incl. frames or trim | | | | | | | | | |
| 0030 | Scrape & sand | | | | | | | | | |
| 0040 | Wood, flush | 1 Pord | 616 | .013 | S.F. | | .49 | | .49 | .72 |
| 0050 | Wood, detail | | 496 | .016 | | | .61 | | .61 | .90 |
| 0060 | Wood, louvered | | 280 | .029 | | | 1.07 | | 1.07 | 1.59 |
| 0070 | Wire brush | | | | | | | | | |
| 0080 | Metal, flush | 1 Pord | 640 | .013 | S.F. | | .47 | | .47 | .70 |
| 0090 | Metal, detail | | 520 | .015 | | | .58 | | .58 | .86 |
| 0100 | Metal, louvered | | 360 | .022 | | | .83 | | .83 | 1.24 |
| 0110 | Hand wash | | | | | | | | | |
| 0120 | Wood, flush | 1 Pord | 2160 | .004 | S.F. | | .14 | | .14 | .21 |
| 0130 | Wood, detailed | | 2000 | .004 | | | .15 | | .15 | .22 |
| 0140 | Wood, louvered | | 1360 | .006 | | | .22 | | .22 | .33 |
| 0150 | Metal, flush | | 2160 | .004 | | | .14 | | .14 | .21 |
| 0160 | Metal, detail | | 2000 | .004 | | | .15 | | .15 | .22 |
| 0170 | Metal, louvered | | 1360 | .006 | | | .22 | | .22 | .33 |
| 0400 | Windows, per side, not incl. trim | | | | | | | | | |
| 0410 | Scrape & sand | | | | | | | | | |
| 0420 | Wood, 1-2 lite | 1 Pord | 360 | .022 | S.F. | | .83 | | .83 | 1.24 |
| 0430 | Wood, 3-6 lite | | 320 | .025 | | | .94 | | .94 | 1.39 |
| 0440 | Wood, 7-10 lite | | 280 | .029 | | | 1.07 | | 1.07 | 1.59 |
| 0450 | Wood, 12 lite | | 240 | .033 | | | 1.25 | | 1.25 | 1.86 |
| 0460 | Wood, Bay/Bow | | 360 | .022 | | | .83 | | .83 | 1.24 |
| 0470 | Wire brush | | | | | | | | | |
| 0480 | Metal, 1-2 lite | 1 Pord | 520 | .015 | S.F. | | .58 | | .58 | .86 |
| 0490 | Metal, 3-6 lite | | 440 | .018 | | | .68 | | .68 | 1.01 |
| 0500 | Metal, Bay/Bow | | 520 | .015 | | | .58 | | .58 | .86 |
| 0600 | Walls, sanding, light=10-30%, medium - 30-70%, | | | | | | | | | |
| 0610 | heavy=70-100% of surface to sand | | | | | | | | | |
| 0650 | Walls, sand | | | | | | | | | |
| 0660 | Gypsum board or plaster, light | 1 Pord | 3077 | .003 | S.F. | | .10 | | .10 | .14 |

# 09 91 Painting

## 09 91 03 – Paint Restoration

### 09 91 03.40 Interior Surface Preparation

| | | Crew | Daily Output | Labor-Hours | Unit | Material | 2011 Bare Costs Labor | Equipment | Total | Total Incl O&P |
|---|---|---|---|---|---|---|---|---|---|---|
| 0670 | Gypsum board or plaster, medium | 1 Pord | 2160 | .004 | S.F. | | .14 | | .14 | .21 |
| 0680 | Gypsum board or plaster, heavy | | 923 | .009 | | | .33 | | .33 | .48 |
| 0690 | Wood, T&G, light | | 2400 | .003 | | | .13 | | .13 | .19 |
| 0700 | Wood, T&G, med. | | 1600 | .005 | | | .19 | | .19 | .28 |
| 0710 | Wood, T&G, heavy | ↓ | 800 | .010 | ↓ | | .38 | | .38 | .56 |
| 0720 | Walls, wash | | | | | | | | | |
| 0730 | Gypsum board or plaster | 1 Pord | 3200 | .003 | S.F. | | .09 | | .09 | .14 |
| 0740 | Wood, T&G | | 3200 | .003 | | | .09 | | .09 | .14 |
| 0750 | Masonry, brick & block, smooth | | 2800 | .003 | | | .11 | | .11 | .16 |
| 0760 | Masonry, brick & block, coarse | ↓ | 2000 | .004 | ↓ | | .15 | | .15 | .22 |
| 8000 | For chemical washing, see Section 04 01 30 | | | | | | | | | |
| 8010 | For steam cleaning, see Section 04 01 30.20 | | | | | | | | | |
| 8020 | For sand blasting, see Section 03 35 29.60 and 05 01 10.51 | | | | | | | | | |

## 09 91 13 – Exterior Painting

### 09 91 13.30 Fences

| | | | Crew | Daily Output | Labor-Hours | Unit | Material | 2011 Bare Costs Labor | Equipment | Total | Total Incl O&P |
|---|---|---|---|---|---|---|---|---|---|---|---|
| 0010 | **FENCES** | R099100-20 | | | | | | | | | |
| 0100 | Chain link or wire metal, one side, water base | | | | | | | | | | |
| 0110 | Roll & brush, first coat | | 1 Pord | 960 | .008 | S.F. | .06 | .31 | | .37 | .53 |
| 0120 | Second coat | | | 1280 | .006 | | .06 | .23 | | .29 | .41 |
| 0130 | Spray, first coat | | | 2275 | .004 | | .06 | .13 | | .19 | .27 |
| 0140 | Second coat | | ↓ | 2600 | .003 | ↓ | .06 | .12 | | .18 | .24 |
| 0150 | Picket, water base | | | | | | | | | | |
| 0160 | Roll & brush, first coat | | 1 Pord | 865 | .009 | S.F. | .06 | .35 | | .41 | .59 |
| 0170 | Second coat | | | 1050 | .008 | | .06 | .29 | | .35 | .49 |
| 0180 | Spray, first coat | | | 2275 | .004 | | .06 | .13 | | .19 | .27 |
| 0190 | Second coat | | ↓ | 2600 | .003 | ↓ | .06 | .12 | | .18 | .24 |
| 0200 | Stockade, water base | | | | | | | | | | |
| 0210 | Roll & brush, first coat | | 1 Pord | 1040 | .008 | S.F. | .06 | .29 | | .35 | .50 |
| 0220 | Second coat | | | 1200 | .007 | | .06 | .25 | | .31 | .44 |
| 0230 | Spray, first coat | | | 2275 | .004 | | .06 | .13 | | .19 | .27 |
| 0240 | Second coat | | ↓ | 2600 | .003 | ↓ | .06 | .12 | | .18 | .24 |

### 09 91 13.42 Miscellaneous, Exterior

| | | | Crew | Daily Output | Labor-Hours | Unit | Material | 2011 Bare Costs Labor | Equipment | Total | Total Incl O&P |
|---|---|---|---|---|---|---|---|---|---|---|---|
| 0010 | **MISCELLANEOUS, EXTERIOR** | R099100-20 | | | | | | | | | |
| 0015 | For painting metals, see Section 09 97 13.23 | | | | | | | | | | |
| 0100 | Railing, ext., decorative wood, incl. cap & baluster | | | | | | | | | | |
| 0110 | Newels & spindles @ 12" O.C. | | | | | | | | | | |
| 0120 | Brushwork, stain, sand, seal & varnish | | | | | | | | | | |
| 0130 | First coat | | 1 Pord | 90 | .089 | L.F. | .69 | 3.34 | | 4.03 | 5.70 |
| 0140 | Second coat | | " | 120 | .067 | " | .69 | 2.50 | | 3.19 | 4.47 |
| 0150 | Rough sawn wood, 42" high, 2" x 2" verticals, 6" O.C. | | | | | | | | | | |
| 0160 | Brushwork, stain, each coat | | 1 Pord | 90 | .089 | L.F. | .22 | 3.34 | | 3.56 | 5.20 |
| 0170 | Wrought iron, 1" rail, 1/2" sq. verticals | | | | | | | | | | |
| 0180 | Brushwork, zinc chromate, 60" high, bars 6" O.C. | | | | | | | | | | |
| 0190 | Primer | | 1 Pord | 130 | .062 | L.F. | .59 | 2.31 | | 2.90 | 4.08 |
| 0200 | Finish coat | | | 130 | .062 | | .18 | 2.31 | | 2.49 | 3.63 |
| 0210 | Additional coat | | ↓ | 190 | .042 | ↓ | .21 | 1.58 | | 1.79 | 2.58 |
| 0220 | Shutters or blinds, single panel, 2' x 4', paint all sides | | | | | | | | | | |
| 0230 | Brushwork, primer | | 1 Pord | 20 | .400 | Ea. | .58 | 15 | | 15.58 | 23 |
| 0240 | Finish coat, exterior latex | | | 20 | .400 | | .49 | 15 | | 15.49 | 23 |
| 0250 | Primer & 1 coat, exterior latex | | | 13 | .615 | | .94 | 23 | | 23.94 | 35.50 |
| 0260 | Spray, primer | | | 35 | .229 | | .85 | 8.60 | | 9.45 | 13.70 |
| 0270 | Finish coat, exterior latex | | ↓ | 35 | .229 | ↓ | 1.04 | 8.60 | | 9.64 | 13.90 |

# 09 91 Painting

## 09 91 13 – Exterior Painting

### 09 91 13.42 Miscellaneous, Exterior

| | | Crew | Daily Output | Labor-Hours | Unit | Material | 2011 Bare Costs Labor | Equipment | Total | Total Incl O&P |
|---|---|---|---|---|---|---|---|---|---|---|
| 0280 | Primer & 1 coat, exterior latex | 1 Pord | 20 | .400 | Ea. | .91 | 15 | | 15.91 | 23.50 |
| 0290 | For louvered shutters, add | | | | S.F. | 10% | | | | |
| 0300 | Stair stringers, exterior, metal | | | | | | | | | |
| 0310 | Roll & brush, zinc chromate, to 14", each coat | 1 Pord | 320 | .025 | L.F. | .06 | .94 | | 1 | 1.46 |
| 0320 | Rough sawn wood, 4" x 12" | | | | | | | | | |
| 0330 | Roll & brush, exterior latex, each coat | 1 Pord | 215 | .037 | L.F. | .07 | 1.40 | | 1.47 | 2.15 |
| 0340 | Trellis/lattice, 2" x 2" @ 3" O.C. with 2" x 8" supports | | | | | | | | | |
| 0350 | Spray, latex, per side, each coat | 1 Pord | 475 | .017 | S.F. | .07 | .63 | | .70 | 1.02 |
| 0450 | Decking, ext., sealer, alkyd, brushwork, sealer coat | | 1140 | .007 | | .09 | .26 | | .35 | .49 |
| 0460 | 1st coat | | 1140 | .007 | | .09 | .26 | | .35 | .48 |
| 0470 | 2nd coat | | 1300 | .006 | | .06 | .23 | | .29 | .41 |
| 0500 | Paint, alkyd, brushwork, primer coat | | 1140 | .007 | | .08 | .26 | | .34 | .47 |
| 0510 | 1st coat | | 1140 | .007 | | .16 | .26 | | .42 | .56 |
| 0520 | 2nd coat | | 1300 | .006 | | .12 | .23 | | .35 | .47 |
| 0600 | Sand paint, alkyd, brushwork, 1 coat | | 150 | .053 | | .12 | 2 | | 2.12 | 3.10 |

### 09 91 13.60 Siding Exterior

| | | Crew | Daily Output | Labor-Hours | Unit | Material | 2011 Bare Costs Labor | Equipment | Total | Total Incl O&P |
|---|---|---|---|---|---|---|---|---|---|---|
| 0010 | **SIDING EXTERIOR**, Alkyd (oil base)   R099100-10 | | | | | | | | | |
| 0450 | Steel siding, oil base, paint 1 coat, brushwork | 2 Pord | 2015 | .008 | S.F. | .13 | .30 | | .43 | .58 |
| 0500 | Spray   R099100-20 | | 4550 | .004 | | .19 | .13 | | .32 | .41 |
| 0800 | Paint 2 coats, brushwork | | 1300 | .012 | | .25 | .46 | | .71 | .97 |
| 1000 | Spray | | 2750 | .006 | | .17 | .22 | | .39 | .51 |
| 1200 | Stucco, rough, oil base, paint 2 coats, brushwork | | 1300 | .012 | | .25 | .46 | | .71 | .97 |
| 1400 | Roller | | 1625 | .010 | | .27 | .37 | | .64 | .84 |
| 1600 | Spray | | 2925 | .005 | | .28 | .21 | | .49 | .61 |
| 1800 | Texture 1-11 or clapboard, oil base, primer coat, brushwork | | 1300 | .012 | | .10 | .46 | | .56 | .80 |
| 2000 | Spray | | 4550 | .004 | | .10 | .13 | | .23 | .31 |
| 2400 | Paint 2 coats, brushwork | | 810 | .020 | | .37 | .74 | | 1.11 | 1.50 |
| 2600 | Spray | | 2600 | .006 | | .41 | .23 | | .64 | .79 |
| 3400 | Stain 2 coats, brushwork | | 950 | .017 | | .15 | .63 | | .78 | 1.10 |
| 4000 | Spray | | 3050 | .005 | | .17 | .20 | | .37 | .47 |
| 4200 | Wood shingles, oil base primer coat, brushwork | | 1300 | .012 | | .09 | .46 | | .55 | .79 |
| 4400 | Spray | | 3900 | .004 | | .09 | .15 | | .24 | .32 |
| 5000 | Paint 2 coats, brushwork | | 810 | .020 | | .31 | .74 | | 1.05 | 1.44 |
| 5200 | Spray | | 2275 | .007 | | .29 | .26 | | .55 | .71 |
| 6500 | Stain 2 coats, brushwork | | 950 | .017 | | .15 | .63 | | .78 | 1.10 |
| 7000 | Spray | | 2660 | .006 | | .21 | .23 | | .44 | .57 |
| 8000 | For latex paint, deduct | | | | | 10% | | | | |
| 8100 | For work over 12' H, from pipe scaffolding, add | | | | | | 15% | | | |
| 8200 | For work over 12' H, from extension ladder, add | | | | | | 25% | | | |
| 8300 | For work over 12' H, from swing staging, add | | | | | | 35% | | | |

### 09 91 13.62 Siding, Misc.

| | | Crew | Daily Output | Labor-Hours | Unit | Material | 2011 Bare Costs Labor | Equipment | Total | Total Incl O&P |
|---|---|---|---|---|---|---|---|---|---|---|
| 0010 | **SIDING, MISC.**, latex paint   R099100-10 | | | | | | | | | |
| 0100 | Aluminum siding | | | | | | | | | |
| 0110 | Brushwork, primer   R099100-20 | 2 Pord | 2275 | .007 | S.F. | .05 | .26 | | .31 | .45 |
| 0120 | Finish coat, exterior latex | | 2275 | .007 | | .04 | .26 | | .30 | .44 |
| 0130 | Primer & 1 coat exterior latex | | 1300 | .012 | | .10 | .46 | | .56 | .80 |
| 0140 | Primer & 2 coats exterior latex | | 975 | .016 | | .14 | .62 | | .76 | 1.07 |
| 0150 | Mineral fiber shingles | | | | | | | | | |
| 0160 | Brushwork, primer | 2 Pord | 1495 | .011 | S.F. | .10 | .40 | | .50 | .71 |
| 0170 | Finish coat, industrial enamel | | 1495 | .011 | | .15 | .40 | | .55 | .76 |
| 0180 | Primer & 1 coat enamel | | 810 | .020 | | .25 | .74 | | .99 | 1.37 |
| 0190 | Primer & 2 coats enamel | | 540 | .030 | | .40 | 1.11 | | 1.51 | 2.08 |

# 09 91 Painting

## 09 91 13 – Exterior Painting

### 09 91 13.62 Siding, Misc.

| | | Crew | Daily Output | Labor-Hours | Unit | Material | 2011 Bare Costs Labor | Equipment | Total | Total Incl O&P |
|---|---|---|---|---|---|---|---|---|---|---|
| 0200 | Roll, primer | 2 Pord | 1625 | .010 | S.F. | .11 | .37 | | .48 | .67 |
| 0210 | Finish coat, industrial enamel | | 1625 | .010 | | .16 | .37 | | .53 | .73 |
| 0220 | Primer & 1 coat enamel | | 975 | .016 | | .27 | .62 | | .89 | 1.21 |
| 0230 | Primer & 2 coats enamel | | 650 | .025 | | .43 | .92 | | 1.35 | 1.84 |
| 0240 | Spray, primer | | 3900 | .004 | | .09 | .15 | | .24 | .32 |
| 0250 | Finish coat, industrial enamel | | 3900 | .004 | | .13 | .15 | | .28 | .38 |
| 0260 | Primer & 1 coat enamel | | 2275 | .007 | | .22 | .26 | | .48 | .63 |
| 0270 | Primer & 2 coats enamel | | 1625 | .010 | | .35 | .37 | | .72 | .94 |
| 0280 | Waterproof sealer, first coat | | 4485 | .004 | | .07 | .13 | | .20 | .27 |
| 0290 | Second coat | ▼ | 5235 | .003 | ▼ | .06 | .11 | | .17 | .24 |
| 0300 | Rough wood incl. shingles, shakes or rough sawn siding | | | | | | | | | |
| 0310 | Brushwork, primer | 2 Pord | 1280 | .013 | S.F. | .12 | .47 | | .59 | .83 |
| 0320 | Finish coat, exterior latex | | 1280 | .013 | | .07 | .47 | | .54 | .78 |
| 0330 | Primer & 1 coat exterior latex | | 960 | .017 | | .19 | .63 | | .82 | 1.14 |
| 0340 | Primer & 2 coats exterior latex | | 700 | .023 | | .26 | .86 | | 1.12 | 1.56 |
| 0350 | Roll, primer | | 2925 | .005 | | .16 | .21 | | .37 | .47 |
| 0360 | Finish coat, exterior latex | | 2925 | .005 | | .09 | .21 | | .30 | .40 |
| 0370 | Primer & 1 coat exterior latex | | 1790 | .009 | | .24 | .34 | | .58 | .77 |
| 0380 | Primer & 2 coats exterior latex | | 1300 | .012 | | .33 | .46 | | .79 | 1.05 |
| 0390 | Spray, primer | | 3900 | .004 | | .13 | .15 | | .28 | .38 |
| 0400 | Finish coat, exterior latex | | 3900 | .004 | | .07 | .15 | | .22 | .30 |
| 0410 | Primer & 1 coat exterior latex | | 2600 | .006 | | .20 | .23 | | .43 | .56 |
| 0420 | Primer & 2 coats exterior latex | | 2080 | .008 | | .27 | .29 | | .56 | .72 |
| 0430 | Waterproof sealer, first coat | | 4485 | .004 | | .12 | .13 | | .25 | .33 |
| 0440 | Second coat | ▼ | 4485 | .004 | ▼ | .07 | .13 | | .20 | .27 |
| 0450 | Smooth wood incl. butt, T&G, beveled, drop or B&B siding | | | | | | | | | |
| 0460 | Brushwork, primer | 2 Pord | 2325 | .007 | S.F. | .08 | .26 | | .34 | .47 |
| 0470 | Finish coat, exterior latex | | 1280 | .013 | | .07 | .47 | | .54 | .78 |
| 0480 | Primer & 1 coat exterior latex | | 800 | .020 | | .16 | .75 | | .91 | 1.28 |
| 0490 | Primer & 2 coats exterior latex | | 630 | .025 | | .23 | .95 | | 1.18 | 1.66 |
| 0500 | Roll, primer | | 2275 | .007 | | .09 | .26 | | .35 | .49 |
| 0510 | Finish coat, exterior latex | | 2275 | .007 | | .08 | .26 | | .34 | .48 |
| 0520 | Primer & 1 coat exterior latex | | 1300 | .012 | | .17 | .46 | | .63 | .88 |
| 0530 | Primer & 2 coats exterior latex | | 975 | .016 | | .25 | .62 | | .87 | 1.19 |
| 0540 | Spray, primer | | 4550 | .004 | | .07 | .13 | | .20 | .28 |
| 0550 | Finish coat, exterior latex | | 4550 | .004 | | .07 | .13 | | .20 | .27 |
| 0560 | Primer & 1 coat exterior latex | | 2600 | .006 | | .14 | .23 | | .37 | .49 |
| 0570 | Primer & 2 coats exterior latex | | 1950 | .008 | | .21 | .31 | | .52 | .69 |
| 0580 | Waterproof sealer, first coat | | 5230 | .003 | | .07 | .11 | | .18 | .24 |
| 0590 | Second coat | ▼ | 5980 | .003 | ▼ | .07 | .10 | | .17 | .22 |
| 0600 | For oil base paint, add | | | | | 10% | | | | |

### 09 91 13.70 Doors and Windows, Exterior

| | | | Crew | Daily Output | Labor-Hours | Unit | Material | 2011 Bare Costs Labor | Equipment | Total | Total Incl O&P |
|---|---|---|---|---|---|---|---|---|---|---|---|
| 0010 | **DOORS AND WINDOWS, EXTERIOR** | R099100-10 | | | | | | | | | |
| 0100 | Door frames & trim, only | | | | | | | | | | |
| 0110 | Brushwork, primer | R099100-20 | 1 Pord | 512 | .016 | L.F. | .05 | .59 | | .64 | .93 |
| 0120 | Finish coat, exterior latex | | | 512 | .016 | | .06 | .59 | | .65 | .94 |
| 0130 | Primer & 1 coat, exterior latex | | | 300 | .027 | | .11 | 1 | | 1.11 | 1.62 |
| 0140 | Primer & 2 coats, exterior latex | | ▼ | 265 | .030 | ▼ | .18 | 1.13 | | 1.31 | 1.87 |
| 0150 | Doors, flush, both sides, incl. frame & trim | | | | | | | | | | |
| 0160 | Roll & brush, primer | | 1 Pord | 10 | .800 | Ea. | 4.03 | 30 | | 34.03 | 49 |
| 0170 | Finish coat, exterior latex | | | 10 | .800 | | 4.65 | 30 | | 34.65 | 49.50 |
| 0180 | Primer & 1 coat, exterior latex | | ▼ | 7 | 1.143 | ▼ | 8.70 | 43 | | 51.70 | 73 |

# 09 91 Painting

## 09 91 13 – Exterior Painting

### 09 91 13.70 Doors and Windows, Exterior

| | | Crew | Daily Output | Labor-Hours | Unit | Material | 2011 Bare Costs Labor | 2011 Bare Costs Equipment | Total | Total Incl O&P |
|---|---|---|---|---|---|---|---|---|---|---|
| 0190 | Primer & 2 coats, exterior latex | 1 Pord | 5 | 1.600 | Ea. | 13.30 | 60 | | 73.30 | 104 |
| 0200 | Brushwork, stain, sealer & 2 coats polyurethane | ↓ | 4 | 2 | ↓ | 24 | 75 | | 99 | 137 |
| 0210 | Doors, French, both sides, 10-15 lite, incl. frame & trim | | | | | | | | | |
| 0220 | Brushwork, primer | 1 Pord | 6 | 1.333 | Ea. | 2.02 | 50 | | 52.02 | 76.50 |
| 0230 | Finish coat, exterior latex | | 6 | 1.333 | | 2.32 | 50 | | 52.32 | 77 |
| 0240 | Primer & 1 coat, exterior latex | | 3 | 2.667 | | 4.34 | 100 | | 104.34 | 154 |
| 0250 | Primer & 2 coats, exterior latex | | 2 | 4 | | 6.55 | 150 | | 156.55 | 230 |
| 0260 | Brushwork, stain, sealer & 2 coats polyurethane | ↓ | 2.50 | 3.200 | ↓ | 8.70 | 120 | | 128.70 | 188 |
| 0270 | Doors, louvered, both sides, incl. frame & trim | | | | | | | | | |
| 0280 | Brushwork, primer | 1 Pord | 7 | 1.143 | Ea. | 4.03 | 43 | | 47.03 | 68 |
| 0290 | Finish coat, exterior latex | | 7 | 1.143 | | 4.65 | 43 | | 47.65 | 68.50 |
| 0300 | Primer & 1 coat, exterior latex | | 4 | 2 | | 8.70 | 75 | | 83.70 | 121 |
| 0310 | Primer & 2 coats, exterior latex | | 3 | 2.667 | | 13.05 | 100 | | 113.05 | 163 |
| 0320 | Brushwork, stain, sealer & 2 coats polyurethane | ↓ | 4.50 | 1.778 | ↓ | 24 | 67 | | 91 | 125 |
| 0330 | Doors, panel, both sides, incl. frame & trim | | | | | | | | | |
| 0340 | Roll & brush, primer | 1 Pord | 6 | 1.333 | Ea. | 4.03 | 50 | | 54.03 | 79 |
| 0350 | Finish coat, exterior latex | | 6 | 1.333 | | 4.65 | 50 | | 54.65 | 79.50 |
| 0360 | Primer & 1 coat, exterior latex | | 3 | 2.667 | | 8.70 | 100 | | 108.70 | 159 |
| 0370 | Primer & 2 coats, exterior latex | | 2.50 | 3.200 | | 13.05 | 120 | | 133.05 | 192 |
| 0380 | Brushwork, stain, sealer & 2 coats polyurethane | ↓ | 3 | 2.667 | ↓ | 24 | 100 | | 124 | 175 |
| 0400 | Windows, per ext. side, based on 15 SF | | | | | | | | | |
| 0410 | 1 to 6 lite | | | | | | | | | |
| 0420 | Brushwork, primer | 1 Pord | 13 | .615 | Ea. | .80 | 23 | | 23.80 | 35.50 |
| 0430 | Finish coat, exterior latex | | 13 | .615 | | .92 | 23 | | 23.92 | 35.50 |
| 0440 | Primer & 1 coat, exterior latex | | 8 | 1 | | 1.71 | 37.50 | | 39.21 | 57.50 |
| 0450 | Primer & 2 coats, exterior latex | | 6 | 1.333 | | 2.58 | 50 | | 52.58 | 77.50 |
| 0460 | Stain, sealer & 1 coat varnish | ↓ | 7 | 1.143 | ↓ | 3.44 | 43 | | 46.44 | 67.50 |
| 0470 | 7 to 10 lite | | | | | | | | | |
| 0480 | Brushwork, primer | 1 Pord | 11 | .727 | Ea. | .80 | 27.50 | | 28.30 | 41.50 |
| 0490 | Finish coat, exterior latex | | 11 | .727 | | .92 | 27.50 | | 28.42 | 41.50 |
| 0500 | Primer & 1 coat, exterior latex | | 7 | 1.143 | | 1.71 | 43 | | 44.71 | 65.50 |
| 0510 | Primer & 2 coats, exterior latex | | 5 | 1.600 | | 2.58 | 60 | | 62.58 | 92 |
| 0520 | Stain, sealer & 1 coat varnish | ↓ | 6 | 1.333 | ↓ | 3.44 | 50 | | 53.44 | 78.50 |
| 0530 | 12 lite | | | | | | | | | |
| 0540 | Brushwork, primer | 1 Pord | 10 | .800 | Ea. | .80 | 30 | | 30.80 | 45.50 |
| 0550 | Finish coat, exterior latex | | 10 | .800 | | .92 | 30 | | 30.92 | 45.50 |
| 0560 | Primer & 1 coat, exterior latex | | 6 | 1.333 | | 1.71 | 50 | | 51.71 | 76.50 |
| 0570 | Primer & 2 coats, exterior latex | | 5 | 1.600 | | 2.58 | 60 | | 62.58 | 92 |
| 0580 | Stain, sealer & 1 coat varnish | ↓ | 6 | 1.333 | ↓ | 3.37 | 50 | | 53.37 | 78 |
| 0590 | For oil base paint, add | | | | | 10% | | | | |

### 09 91 13.80 Trim, Exterior

| | | | Crew | Daily Output | Labor-Hours | Unit | Material | 2011 Bare Costs Labor | 2011 Bare Costs Equipment | Total | Total Incl O&P |
|---|---|---|---|---|---|---|---|---|---|---|---|
| 0010 | **TRIM, EXTERIOR** | R099100-10 | | | | | | | | | |
| 0100 | Door frames & trim (see Doors, interior or exterior) | | | | | | | | | | |
| 0110 | Fascia, latex paint, one coat coverage | R099100-20 | | | | | | | | | |
| 0120 | 1" x 4", brushwork | | 1 Pord | 640 | .013 | L.F. | .02 | .47 | | .49 | .72 |
| 0130 | Roll | | | 1280 | .006 | | .02 | .23 | | .25 | .37 |
| 0140 | Spray | | | 2080 | .004 | | .02 | .14 | | .16 | .23 |
| 0150 | 1" x 6" to 1" x 10", brushwork | | | 640 | .013 | | .06 | .47 | | .53 | .77 |
| 0160 | Roll | | | 1230 | .007 | | .07 | .24 | | .31 | .44 |
| 0170 | Spray | | | 2100 | .004 | | .05 | .14 | | .19 | .27 |
| 0180 | 1" x 12", brushwork | | | 640 | .013 | | .06 | .47 | | .53 | .77 |
| 0190 | Roll | | ↓ | 1050 | .008 | | .07 | .29 | | .36 | .50 |

# 09 91 Painting

## 09 91 13 – Exterior Painting

### 09 91 13.80 Trim, Exterior

| | Crew | Daily Output | Labor-Hours | Unit | Material | 2011 Bare Costs Labor | Equipment | Total | Total Incl O&P |
|---|---|---|---|---|---|---|---|---|---|
| 0200 Spray | 1 Pord | 2200 | .004 | L.F. | .05 | .14 | | .19 | .26 |
| 0210 Gutters & downspouts, metal, zinc chromate paint | | | | | | | | | |
| 0220 Brushwork, gutters, 5", first coat | 1 Pord | 640 | .013 | L.F. | .06 | .47 | | .53 | .77 |
| 0230 Second coat | | 960 | .008 | | .06 | .31 | | .37 | .53 |
| 0240 Third coat | | 1280 | .006 | | .05 | .23 | | .28 | .40 |
| 0250 Downspouts, 4", first coat | | 640 | .013 | | .06 | .47 | | .53 | .77 |
| 0260 Second coat | | 960 | .008 | | .06 | .31 | | .37 | .53 |
| 0270 Third coat | | 1280 | .006 | | .05 | .23 | | .28 | .40 |
| 0280 Gutters & downspouts, wood | | | | | | | | | |
| 0290 Brushwork, gutters, 5", primer | 1 Pord | 640 | .013 | L.F. | .05 | .47 | | .52 | .76 |
| 0300 Finish coat, exterior latex | | 640 | .013 | | .05 | .47 | | .52 | .76 |
| 0310 Primer & 1 coat exterior latex | | 400 | .020 | | .11 | .75 | | .86 | 1.24 |
| 0320 Primer & 2 coats exterior latex | | 325 | .025 | | .18 | .92 | | 1.10 | 1.56 |
| 0330 Downspouts, 4", primer | | 640 | .013 | | .05 | .47 | | .52 | .76 |
| 0340 Finish coat, exterior latex | | 640 | .013 | | .05 | .47 | | .52 | .76 |
| 0350 Primer & 1 coat exterior latex | | 400 | .020 | | .11 | .75 | | .86 | 1.24 |
| 0360 Primer & 2 coats exterior latex | | 325 | .025 | | .09 | .92 | | 1.01 | 1.47 |
| 0370 Molding, exterior, up to 14" wide | | | | | | | | | |
| 0380 Brushwork, primer | 1 Pord | 640 | .013 | L.F. | .06 | .47 | | .53 | .77 |
| 0390 Finish coat, exterior latex | | 640 | .013 | | .07 | .47 | | .54 | .77 |
| 0400 Primer & 1 coat exterior latex | | 400 | .020 | | .14 | .75 | | .89 | 1.26 |
| 0410 Primer & 2 coats exterior latex | | 315 | .025 | | .14 | .95 | | 1.09 | 1.56 |
| 0420 Stain & fill | | 1050 | .008 | | .09 | .29 | | .38 | .52 |
| 0430 Shellac | | 1850 | .004 | | .09 | .16 | | .25 | .34 |
| 0440 Varnish | | 1275 | .006 | | .10 | .24 | | .34 | .46 |

### 09 91 13.90 Walls, Masonry (CMU), Exterior

| | Crew | Daily Output | Labor-Hours | Unit | Material | 2011 Bare Costs Labor | Equipment | Total | Total Incl O&P |
|---|---|---|---|---|---|---|---|---|---|
| 0350 **WALLS, MASONRY (CMU), EXTERIOR** R099100-10 | | | | | | | | | |
| 0360 Concrete masonry units (CMU), smooth surface | | | | | | | | | |
| 0370 Brushwork, latex, first coat | 1 Pord | 640 | .013 | S.F. | .06 | .47 | | .53 | .76 |
| 0380 Second coat | | 960 | .008 | | .05 | .31 | | .36 | .51 |
| 0390 Waterproof sealer, first coat | | 736 | .011 | | .25 | .41 | | .66 | .88 |
| 0400 Second coat | | 1104 | .007 | | .25 | .27 | | .52 | .67 |
| 0410 Roll, latex, paint, first coat | | 1465 | .005 | | .07 | .21 | | .28 | .38 |
| 0420 Second coat | | 1790 | .004 | | .05 | .17 | | .22 | .31 |
| 0430 Waterproof sealer, first coat | | 1680 | .005 | | .25 | .18 | | .43 | .54 |
| 0440 Second coat | | 2060 | .004 | | .25 | .15 | | .40 | .49 |
| 0450 Spray, latex, paint, first coat | | 1950 | .004 | | .05 | .15 | | .20 | .29 |
| 0460 Second coat | | 2600 | .003 | | .04 | .12 | | .16 | .22 |
| 0470 Waterproof sealer, first coat | | 2245 | .004 | | .25 | .13 | | .38 | .47 |
| 0480 Second coat | | 2990 | .003 | | .25 | .10 | | .35 | .42 |
| 0490 Concrete masonry unit (CMU), porous | | | | | | | | | |
| 0500 Brushwork, latex, first coat | 1 Pord | 640 | .013 | S.F. | .12 | .47 | | .59 | .83 |
| 0510 Second coat | | 960 | .008 | | .06 | .31 | | .37 | .52 |
| 0520 Waterproof sealer, first coat | | 736 | .011 | | .25 | .41 | | .66 | .88 |
| 0530 Second coat | | 1104 | .007 | | .25 | .27 | | .52 | .67 |
| 0540 Roll latex, first coat | | 1465 | .005 | | .09 | .21 | | .30 | .40 |
| 0550 Second coat | | 1790 | .004 | | .06 | .17 | | .23 | .31 |
| 0560 Waterproof sealer, first coat | | 1680 | .005 | | .25 | .18 | | .43 | .54 |
| 0570 Second coat | | 2060 | .004 | | .25 | .15 | | .40 | .49 |
| 0580 Spray latex, first coat | | 1950 | .004 | | .06 | .15 | | .21 | .30 |
| 0590 Second coat | | 2600 | .003 | | .04 | .12 | | .16 | .22 |
| 0600 Waterproof sealer, first coat | | 2245 | .004 | | .25 | .13 | | .38 | .47 |

# 09 91 Painting

## 09 91 13 – Exterior Painting

| 09 91 13.90 Walls, Masonry (CMU), Exterior | Crew | Daily Output | Labor-Hours | Unit | Material | 2011 Bare Costs Labor | Equipment | Total | Total Incl O&P |
|---|---|---|---|---|---|---|---|---|---|
| 0610 | Second coat | 1 Pord | 2990 | .003 | S.F. | .25 | .10 | | .35 | .42 |

## 09 91 23 – Interior Painting

### 09 91 23.20 Cabinets and Casework

| | | Crew | Daily Output | Labor-Hours | Unit | Material | Labor | Equipment | Total | Total Incl O&P |
|---|---|---|---|---|---|---|---|---|---|---|
| 0010 | **CABINETS AND CASEWORK**  R099100-10 | | | | | | | | | |
| 1000 | Primer coat, oil base, brushwork | 1 Pord | 650 | .012 | S.F. | .06 | .46 | | .52 | .76 |
| 2000 | Paint, oil base, brushwork, 1 coat  R099100-20 | | 650 | .012 | | .08 | .46 | | .54 | .78 |
| 3000 | Stain, brushwork, wipe off | | 650 | .012 | | .07 | .46 | | .53 | .77 |
| 4000 | Shellac, 1 coat, brushwork | | 650 | .012 | | .07 | .46 | | .53 | .77 |
| 4500 | Varnish, 3 coats, brushwork, sand after 1st coat | | 325 | .025 | | .24 | .92 | | 1.16 | 1.63 |
| 5000 | For latex paint, deduct | | | | | 10% | | | | |

### 09 91 23.33 Doors and Windows, Interior Alkyd (Oil Base)

| | | Crew | Daily Output | Labor-Hours | Unit | Material | Labor | Equipment | Total | Total Incl O&P |
|---|---|---|---|---|---|---|---|---|---|---|
| 0010 | **DOORS AND WINDOWS, INTERIOR ALKYD (OIL BASE)**  R099100-10 | | | | | | | | | |
| 0500 | Flush door & frame, 3' x 7', oil, primer, brushwork | 1 Pord | 10 | .800 | Ea. | 3.04 | 30 | | 33.04 | 48 |
| 1000 | Paint, 1 coat  R099100-20 | | 10 | .800 | | 3.62 | 30 | | 33.62 | 48.50 |
| 1400 | Stain, brushwork, wipe off | | 18 | .444 | | 1.57 | 16.70 | | 18.27 | 26.50 |
| 1600 | Shellac, 1 coat, brushwork | | 25 | .320 | | 1.54 | 12 | | 13.54 | 19.50 |
| 1800 | Varnish, 3 coats, brushwork, sand after 1st coat | | 9 | .889 | | 5 | 33.50 | | 38.50 | 55 |
| 2000 | Panel door & frame, 3' x 7', oil, primer, brushwork | | 6 | 1.333 | | 2.31 | 50 | | 52.31 | 77 |
| 2200 | Paint, 1 coat | | 6 | 1.333 | | 3.62 | 50 | | 53.62 | 78.50 |
| 2600 | Stain, brushwork, panel door, 3' x 7', not incl. frame | | 16 | .500 | | 1.57 | 18.80 | | 20.37 | 29.50 |
| 2800 | Shellac, 1 coat, brushwork | | 22 | .364 | | 1.54 | 13.65 | | 15.19 | 22 |
| 3000 | Varnish, 3 coats, brushwork, sand after 1st coat | | 7.50 | 1.067 | | 5 | 40 | | 45 | 65 |
| 4400 | Windows, including frame and trim, per side | | | | | | | | | |
| 4600 | Colonial type, 6/6 lites, 2' x 3', oil, primer, brushwork | 1 Pord | 14 | .571 | Ea. | .36 | 21.50 | | 21.86 | 32.50 |
| 5800 | Paint, 1 coat | | 14 | .571 | | .57 | 21.50 | | 22.07 | 32.50 |
| 6200 | 3' x 5' opening, 6/6 lites, primer coat, brushwork | | 12 | .667 | | .91 | 25 | | 25.91 | 38 |
| 6400 | Paint, 1 coat | | 12 | .667 | | 1.43 | 25 | | 26.43 | 38.50 |
| 6800 | 4' x 8' opening, 6/6 lites, primer coat, brushwork | | 8 | 1 | | 1.95 | 37.50 | | 39.45 | 57.50 |
| 7000 | Paint, 1 coat | | 8 | 1 | | 3.05 | 37.50 | | 40.55 | 59 |
| 8000 | Single lite type, 2' x 3', oil base, primer coat, brushwork | | 33 | .242 | | .36 | 9.10 | | 9.46 | 13.90 |
| 8200 | Paint, 1 coat | | 33 | .242 | | .57 | 9.10 | | 9.67 | 14.15 |
| 8600 | 3' x 5' opening, primer coat, brushwork | | 20 | .400 | | .91 | 15 | | 15.91 | 23.50 |
| 8800 | Paint, 1 coat | | 20 | .400 | | 1.43 | 15 | | 16.43 | 24 |
| 9200 | 4' x 8' opening, primer coat, brushwork | | 14 | .571 | | 1.95 | 21.50 | | 23.45 | 34 |
| 9400 | Paint, 1 coat | | 14 | .571 | | 3.05 | 21.50 | | 24.55 | 35.50 |

### 09 91 23.35 Doors and Windows, Interior Latex

| | | Crew | Daily Output | Labor-Hours | Unit | Material | Labor | Equipment | Total | Total Incl O&P |
|---|---|---|---|---|---|---|---|---|---|---|
| 0010 | **DOORS & WINDOWS, INTERIOR LATEX**  R099100-10 | | | | | | | | | |
| 0100 | Doors, flush, both sides, incl. frame & trim | | | | | | | | | |
| 0110 | Roll & brush, primer  R099100-20 | 1 Pord | 10 | .800 | Ea. | 3.73 | 30 | | 33.73 | 48.50 |
| 0120 | Finish coat, latex | | 10 | .800 | | 4.13 | 30 | | 34.13 | 49 |
| 0130 | Primer & 1 coat latex | | 7 | 1.143 | | 7.85 | 43 | | 50.85 | 72 |
| 0140 | Primer & 2 coats latex | | 5 | 1.600 | | 11.75 | 60 | | 71.75 | 102 |
| 0160 | Spray, both sides, primer | | 20 | .400 | | 3.93 | 15 | | 18.93 | 27 |
| 0170 | Finish coat, latex | | 20 | .400 | | 4.33 | 15 | | 19.33 | 27.50 |
| 0180 | Primer & 1 coat latex | | 11 | .727 | | 8.30 | 27.50 | | 35.80 | 49.50 |
| 0190 | Primer & 2 coats latex | | 8 | 1 | | 12.45 | 37.50 | | 49.95 | 69 |
| 0200 | Doors, French, both sides, 10-15 lite, incl. frame & trim | | | | | | | | | |
| 0210 | Roll & brush, primer | 1 Pord | 6 | 1.333 | Ea. | 1.87 | 50 | | 51.87 | 76.50 |
| 0220 | Finish coat, latex | | 6 | 1.333 | | 2.06 | 50 | | 52.06 | 77 |
| 0230 | Primer & 1 coat latex | | 3 | 2.667 | | 3.93 | 100 | | 103.93 | 153 |
| 0240 | Primer & 2 coats latex | | 2 | 4 | | 5.90 | 150 | | 155.90 | 229 |
| 0260 | Doors, louvered, both sides, incl. frame & trim | | | | | | | | | |

## 09 91 23 – Interior Painting

### 09 91 23.35 Doors and Windows, Interior Latex

| | | Crew | Daily Output | Labor-Hours | Unit | Material | 2011 Bare Costs Labor | 2011 Bare Costs Equipment | Total | Total Incl O&P |
|---|---|---|---|---|---|---|---|---|---|---|
| 0270 | Roll & brush, primer | 1 Pord | 7 | 1.143 | Ea. | 3.73 | 43 | | 46.73 | 67.50 |
| 0280 | Finish coat, latex | | 7 | 1.143 | | 4.13 | 43 | | 47.13 | 68 |
| 0290 | Primer & 1 coat, latex | | 4 | 2 | | 7.65 | 75 | | 82.65 | 119 |
| 0300 | Primer & 2 coats, latex | | 3 | 2.667 | | 12 | 100 | | 112 | 162 |
| 0320 | Spray, both sides, primer | | 20 | .400 | | 3.93 | 15 | | 18.93 | 27 |
| 0330 | Finish coat, latex | | 20 | .400 | | 4.33 | 15 | | 19.33 | 27.50 |
| 0340 | Primer & 1 coat, latex | | 11 | .727 | | 8.30 | 27.50 | | 35.80 | 49.50 |
| 0350 | Primer & 2 coats, latex | | 8 | 1 | | 12.70 | 37.50 | | 50.20 | 69.50 |
| 0360 | Doors, panel, both sides, incl. frame & trim | | | | | | | | | |
| 0370 | Roll & brush, primer | 1 Pord | 6 | 1.333 | Ea. | 3.93 | 50 | | 53.93 | 79 |
| 0380 | Finish coat, latex | | 6 | 1.333 | | 4.13 | 50 | | 54.13 | 79 |
| 0390 | Primer & 1 coat, latex | | 3 | 2.667 | | 7.85 | 100 | | 107.85 | 158 |
| 0400 | Primer & 2 coats, latex | | 2.50 | 3.200 | | 12 | 120 | | 132 | 191 |
| 0420 | Spray, both sides, primer | | 10 | .800 | | 3.93 | 30 | | 33.93 | 49 |
| 0430 | Finish coat, latex | | 10 | .800 | | 4.33 | 30 | | 34.33 | 49.50 |
| 0440 | Primer & 1 coat, latex | | 5 | 1.600 | | 8.30 | 60 | | 68.30 | 98 |
| 0450 | Primer & 2 coats, latex | | 4 | 2 | | 12.70 | 75 | | 87.70 | 125 |
| 0460 | Windows, per interior side, based on 15 SF | | | | | | | | | |
| 0470 | 1 to 6 lite | | | | | | | | | |
| 0480 | Brushwork, primer | 1 Pord | 13 | .615 | Ea. | .74 | 23 | | 23.74 | 35.50 |
| 0490 | Finish coat, enamel | | 13 | .615 | | .81 | 23 | | 23.81 | 35.50 |
| 0500 | Primer & 1 coat enamel | | 8 | 1 | | 1.55 | 37.50 | | 39.05 | 57 |
| 0510 | Primer & 2 coats enamel | | 6 | 1.333 | | 2.37 | 50 | | 52.37 | 77 |
| 0530 | 7 to 10 lite | | | | | | | | | |
| 0540 | Brushwork, primer | 1 Pord | 11 | .727 | Ea. | .74 | 27.50 | | 28.24 | 41.50 |
| 0550 | Finish coat, enamel | | 11 | .727 | | .81 | 27.50 | | 28.31 | 41.50 |
| 0560 | Primer & 1 coat enamel | | 7 | 1.143 | | 1.55 | 43 | | 44.55 | 65 |
| 0570 | Primer & 2 coats enamel | | 5 | 1.600 | | 2.37 | 60 | | 62.37 | 91.50 |
| 0590 | 12 lite | | | | | | | | | |
| 0600 | Brushwork, primer | 1 Pord | 10 | .800 | Ea. | .74 | 30 | | 30.74 | 45.50 |
| 0610 | Finish coat, enamel | | 10 | .800 | | .81 | 30 | | 30.81 | 45.50 |
| 0620 | Primer & 1 coat enamel | | 6 | 1.333 | | 1.55 | 50 | | 51.55 | 76 |
| 0630 | Primer & 2 coats enamel | | 5 | 1.600 | | 2.37 | 60 | | 62.37 | 91.50 |
| 0650 | For oil base paint, add | | | | | 10% | | | | |

### 09 91 23.39 Doors and Windows, Interior Latex, Zero Voc

| | | | Crew | Daily Output | Labor-Hours | Unit | Material | 2011 Bare Costs Labor | 2011 Bare Costs Equipment | Total | Total Incl O&P |
|---|---|---|---|---|---|---|---|---|---|---|---|
| 0010 | **DOORS & WINDOWS, INTERIOR LATEX, ZERO VOC** | | | | | | | | | | |
| 0100 | Doors flush, both sides, incl. frame & trim | | | | | | | | | | |
| 0110 | Roll & brush, primer | G | 1 Pord | 10 | .800 | Ea. | 5 | 30 | | 35 | 50 |
| 0120 | Finish coat, latex | G | | 10 | .800 | | 5.25 | 30 | | 35.25 | 50.50 |
| 0130 | Primer & 1 coat latex | G | | 7 | 1.143 | | 10.25 | 43 | | 53.25 | 75 |
| 0140 | Primer & 2 coats latex | G | | 5 | 1.600 | | 15.20 | 60 | | 75.20 | 106 |
| 0160 | Spray, both sides, primer | G | | 20 | .400 | | 5.25 | 15 | | 20.25 | 28.50 |
| 0170 | Finish coat, latex | G | | 20 | .400 | | 5.50 | 15 | | 20.50 | 28.50 |
| 0180 | Primer & 1 coat latex | G | | 11 | .727 | | 10.85 | 27.50 | | 38.35 | 52.50 |
| 0190 | Primer & 2 coats latex | G | | 8 | 1 | | 16.10 | 37.50 | | 53.60 | 73 |
| 0200 | Doors, French, both sides, 10-15 lite, incl. frame & trim | | | | | | | | | | |
| 0210 | Roll & brush, primer | G | 1 Pord | 6 | 1.333 | Ea. | 2.50 | 50 | | 52.50 | 77.50 |
| 0220 | Finish coat, latex | G | | 6 | 1.333 | | 2.63 | 50 | | 52.63 | 77.50 |
| 0230 | Primer & 1 coat latex | G | | 3 | 2.667 | | 5.15 | 100 | | 105.15 | 155 |
| 0240 | Primer & 2 coats latex | G | | 2 | 4 | | 7.60 | 150 | | 157.60 | 231 |
| 0360 | Doors, panel, both sides, incl. frame & trim | | | | | | | | | | |
| 0370 | Roll & brush, primer | G | 1 Pord | 6 | 1.333 | Ea. | 5.25 | 50 | | 55.25 | 80.50 |

## 09 91 23 – Interior Painting

| 09 91 23.39 Doors and Windows, Interior Latex, Zero Voc | | Crew | Daily Output | Labor-Hours | Unit | Material | 2011 Bare Costs Labor | Equipment | Total | Total Incl O&P |
|---|---|---|---|---|---|---|---|---|---|---|
| 0380 | Finish coat, latex | G | 1 Pord | 6 | 1.333 | Ea. | 5.25 | 50 | | 55.25 | 80.50 |
| 0390 | Primer & 1 coat, latex | G | | 3 | 2.667 | | 10.25 | 100 | | 110.25 | 160 |
| 0400 | Primer & 2 coats, latex | G | | 2.50 | 3.200 | | 15.50 | 120 | | 135.50 | 195 |
| 0420 | Spray, both sides, primer | G | | 10 | .800 | | 5.25 | 30 | | 35.25 | 50.50 |
| 0430 | Finish coat, latex | G | | 10 | .800 | | 5.50 | 30 | | 35.50 | 50.50 |
| 0440 | Primer & 1 coat, latex | G | | 5 | 1.600 | | 10.85 | 60 | | 70.85 | 101 |
| 0450 | Primer & 2 coats, latex | G | | 4 | 2 | | 16.45 | 75 | | 91.45 | 129 |
| 0460 | Windows, per interior side, based on 15 SF | | | | | | | | | | |
| 0470 | 1 to 6 lite | | | | | | | | | | |
| 0480 | Brushwork, primer | G | 1 Pord | 13 | .615 | Ea. | .99 | 23 | | 23.99 | 35.50 |
| 0490 | Finish coat, enamel | G | | 13 | .615 | | 1.04 | 23 | | 24.04 | 35.50 |
| 0500 | Primer & 1 coat enamel | G | | 8 | 1 | | 2.02 | 37.50 | | 39.52 | 57.50 |
| 0510 | Primer & 2 coats enamel | G | | 6 | 1.333 | | 3.06 | 50 | | 53.06 | 78 |

### 09 91 23.40 Floors, Interior

| | | | Crew | Daily Output | Labor-Hours | Unit | Material | Labor | Equipment | Total | Total Incl O&P |
|---|---|---|---|---|---|---|---|---|---|---|---|
| 0010 | **FLOORS, INTERIOR** | R099100-10 | | | | | | | | | |
| 0100 | Concrete paint, latex | | | | | | | | | | |
| 0110 | Brushwork | | | | | | | | | | |
| 0120 | 1st coat | | 1 Pord | 975 | .008 | S.F. | .15 | .31 | | .46 | .63 |
| 0130 | 2nd coat | | | 1150 | .007 | | .10 | .26 | | .36 | .50 |
| 0140 | 3rd coat | | | 1300 | .006 | | .08 | .23 | | .31 | .43 |
| 0150 | Roll | | | | | | | | | | |
| 0160 | 1st coat | | 1 Pord | 2600 | .003 | S.F. | .20 | .12 | | .32 | .39 |
| 0170 | 2nd coat | | | 3250 | .002 | | .12 | .09 | | .21 | .27 |
| 0180 | 3rd coat | | | 3900 | .002 | | .09 | .08 | | .17 | .21 |
| 0190 | Spray | | | | | | | | | | |
| 0200 | 1st coat | | 1 Pord | 2600 | .003 | S.F. | .17 | .12 | | .29 | .36 |
| 0210 | 2nd coat | | | 3250 | .002 | | .09 | .09 | | .18 | .24 |
| 0220 | 3rd coat | | | 3900 | .002 | | .08 | .08 | | .16 | .19 |
| 0300 | Acid stain and sealer | | | | | | | | | | |
| 0310 | Stain, one coat | | 1 Pord | 650 | .012 | S.F. | .11 | .46 | | .57 | .81 |
| 0320 | Two coats | | | 570 | .014 | | .22 | .53 | | .75 | 1.02 |
| 0330 | Acrylic sealer, one coat | | | 2600 | .003 | | .18 | .12 | | .30 | .37 |
| 0340 | Two coats | | | 1400 | .006 | | .36 | .21 | | .57 | .71 |

### 09 91 23.52 Miscellaneous, Interior

| | | | Crew | Daily Output | Labor-Hours | Unit | Material | Labor | Equipment | Total | Total Incl O&P |
|---|---|---|---|---|---|---|---|---|---|---|---|
| 0010 | **MISCELLANEOUS, INTERIOR** | R099100-10 | | | | | | | | | |
| 2400 | Floors, conc./wood, oil base, primer/sealer coat, brushwork | | 2 Pord | 1950 | .008 | S.F. | .06 | .31 | | .37 | .53 |
| 2450 | Roller | | | 5200 | .003 | | .06 | .12 | | .18 | .24 |
| 2600 | Spray | | | 6000 | .003 | | .06 | .10 | | .16 | .22 |
| 2650 | Paint 1 coat, brushwork | | | 1950 | .008 | | .10 | .31 | | .41 | .57 |
| 2800 | Roller | | | 5200 | .003 | | .10 | .12 | | .22 | .28 |
| 2850 | Spray | | | 6000 | .003 | | .11 | .10 | | .21 | .27 |
| 3000 | Stain, wood floor, brushwork, 1 coat | | | 4550 | .004 | | .07 | .13 | | .20 | .28 |
| 3200 | Roller | | | 5200 | .003 | | .08 | .12 | | .20 | .26 |
| 3250 | Spray | | | 6000 | .003 | | .08 | .10 | | .18 | .24 |
| 3400 | Varnish, wood floor, brushwork | | | 4550 | .004 | | .08 | .13 | | .21 | .29 |
| 3450 | Roller | | | 5200 | .003 | | .08 | .12 | | .20 | .26 |
| 3600 | Spray | | | 6000 | .003 | | .09 | .10 | | .19 | .25 |
| 3650 | For dust proofing or anti skid, see Section 03 35 29.30 | | | | | | | | | | |
| 3800 | Grilles, per side, oil base, primer coat, brushwork | | 1 Pord | 520 | .015 | S.F. | .12 | .58 | | .70 | .99 |
| 3850 | Spray | | | 1140 | .007 | | .13 | .26 | | .39 | .53 |
| 3920 | Paint 2 coats, brushwork | | | 325 | .025 | | .37 | .92 | | 1.29 | 1.78 |
| 3940 | Spray | | | 650 | .012 | | .42 | .46 | | .88 | 1.16 |

## 09 91 23 – Interior Painting

### 09 91 23.52 Miscellaneous, Interior

| | | Crew | Daily Output | Labor-Hours | Unit | Material | 2011 Bare Costs Labor | Equipment | Total | Total Incl O&P |
|---|---|---|---|---|---|---|---|---|---|---|
| 5000 | Pipe, 1" - 4" diameter, primer or sealer coat, oil base, brushwork | 2 Pord | 1250 | .013 | L.F. | .06 | .48 | | .54 | .78 |
| 5100 | Spray | | 2165 | .007 | | .06 | .28 | | .34 | .48 |
| 5350 | Paint 2 coats, brushwork | | 775 | .021 | | .18 | .78 | | .96 | 1.35 |
| 5400 | Spray | | 1240 | .013 | | .20 | .48 | | .68 | .94 |
| 6300 | 13" - 16" diameter, primer or sealer coat, brushwork | | 310 | .052 | | .26 | 1.94 | | 2.20 | 3.16 |
| 6350 | Spray | | 540 | .030 | | .29 | 1.11 | | 1.40 | 1.97 |
| 6500 | Paint 2 coats, brushwork | | 195 | .082 | | .72 | 3.08 | | 3.80 | 5.35 |
| 6550 | Spray | | 310 | .052 | | .80 | 1.94 | | 2.74 | 3.75 |
| 7000 | Trim, wood, incl. puttying, under 6" wide | | | | | | | | | |
| 7200 | Primer coat, oil base, brushwork | 1 Pord | 650 | .012 | L.F. | .03 | .46 | | .49 | .72 |
| 7250 | Paint, 1 coat, brushwork | | 650 | .012 | | .05 | .46 | | .51 | .74 |
| 7450 | 3 coats | | 325 | .025 | | .14 | .92 | | 1.06 | 1.52 |
| 7500 | Over 6" wide, primer coat, brushwork | | 650 | .012 | | .06 | .46 | | .52 | .76 |
| 7550 | Paint, 1 coat, brushwork | | 650 | .012 | | .10 | .46 | | .56 | .79 |
| 7650 | 3 coats | | 325 | .025 | | .27 | .92 | | 1.19 | 1.67 |
| 8000 | Cornice, simple design, primer coat, oil base, brushwork | | 650 | .012 | S.F. | .06 | .46 | | .52 | .76 |
| 8250 | Paint, 1 coat | | 650 | .012 | | .10 | .46 | | .56 | .79 |
| 8350 | Ornate design, primer coat | | 350 | .023 | | .06 | .86 | | .92 | 1.34 |
| 8400 | Paint, 1 coat | | 350 | .023 | | .10 | .86 | | .96 | 1.37 |
| 8600 | Balustrades, primer coat, oil base, brushwork | | 520 | .015 | | .06 | .58 | | .64 | .93 |
| 8650 | Paint, 1 coat | | 520 | .015 | | .10 | .58 | | .68 | .96 |
| 8900 | Trusses and wood frames, primer coat, oil base, brushwork | | 800 | .010 | | .06 | .38 | | .44 | .63 |
| 8950 | Spray | | 1200 | .007 | | .06 | .25 | | .31 | .44 |
| 9220 | Paint 2 coats, brushwork | | 500 | .016 | | .19 | .60 | | .79 | 1.09 |
| 9240 | Spray | | 600 | .013 | | .21 | .50 | | .71 | .97 |
| 9260 | Stain, brushwork, wipe off | | 600 | .013 | | .07 | .50 | | .57 | .82 |
| 9280 | Varnish, 3 coats, brushwork | | 275 | .029 | | .24 | 1.09 | | 1.33 | 1.88 |
| 9350 | For latex paint, deduct | | | | | 10% | | | | |

### 09 91 23.62 Electrostatic Painting

| | | Crew | Daily Output | Labor-Hours | Unit | Material | 2011 Bare Costs Labor | Equipment | Total | Total Incl O&P |
|---|---|---|---|---|---|---|---|---|---|---|
| 0010 | **ELECTROSTATIC PAINTING** | | | | | | | | | |
| 0100 | In Shop | | | | | | | | | |
| 0200 | Flat Surfaces (lockers, casework, elev doors. etc) | | | | | | | | | |
| 0300 | One coat | 1 Pord | 200 | .040 | S.F. | .46 | 1.50 | | 1.96 | 2.74 |
| 0400 | Two coats | " | 120 | .067 | " | .67 | 2.50 | | 3.17 | 4.45 |
| 0500 | Irregular Surfaces (furniture, door frames, etc) | | | | | | | | | |
| 0600 | One coat | 1 Pord | 150 | .053 | S.F. | .46 | 2 | | 2.46 | 3.48 |
| 0700 | Two coats | " | 100 | .080 | " | .67 | 3 | | 3.67 | 5.20 |
| 0800 | On Site | | | | | | | | | |
| 0900 | Flat Surfaces (lockers, casework, elev doors. etc) | | | | | | | | | |
| 1000 | One coat | 1 Pord | 150 | .053 | S.F. | .46 | 2 | | 2.46 | 3.48 |
| 1100 | Two coats | " | 100 | .080 | " | .67 | 3 | | 3.67 | 5.20 |
| 1200 | Irregular Surfaces (furniture, door frames, etc) | | | | | | | | | |
| 1300 | One coat | 1 Pord | 115 | .070 | S.F. | .46 | 2.61 | | 3.07 | 4.39 |
| 1400 | Two coats | " | 70 | .114 | " | .67 | 4.29 | | 4.96 | 7.10 |

### 09 91 23.72 Walls and Ceilings, Interior

| | | | Crew | Daily Output | Labor-Hours | Unit | Material | 2011 Bare Costs Labor | Equipment | Total | Total Incl O&P |
|---|---|---|---|---|---|---|---|---|---|---|---|
| 0010 | **WALLS AND CEILINGS, INTERIOR** | R099100-10 | | | | | | | | | |
| 0100 | Concrete, drywall or plaster, latex , primer or sealer coat | R099100-20 | | | | | | | | | |
| 0200 | Smooth finish, brushwork | | 1 Pord | 1150 | .007 | S.F. | .06 | .26 | | .32 | .45 |
| 0240 | Roller | | | 1350 | .006 | | .06 | .22 | | .28 | .39 |
| 0280 | Spray | | | 2750 | .003 | | .05 | .11 | | .16 | .21 |
| 0300 | Sand finish, brushwork | | | 975 | .008 | | .06 | .31 | | .37 | .52 |
| 0340 | Roller | | | 1150 | .007 | | .06 | .26 | | .32 | .45 |

# 09 91 Painting

## 09 91 23.72 Walls and Ceilings, Interior

| | | Crew | Daily Output | Labor-Hours | Unit | Material | 2011 Bare Costs Labor | Equipment | Total | Total Incl O&P |
|---|---|---|---|---|---|---|---|---|---|---|
| 0380 | Spray | 1 Pord | 2275 | .004 | S.F. | .05 | .13 | | .18 | .25 |
| 0800 | Paint 2 coats, smooth finish, brushwork | | 680 | .012 | | .11 | .44 | | .55 | .78 |
| 0840 | Roller | | 800 | .010 | | .11 | .38 | | .49 | .68 |
| 0880 | Spray | | 1625 | .005 | | .10 | .18 | | .28 | .38 |
| 0900 | Sand finish, brushwork | | 605 | .013 | | .11 | .50 | | .61 | .86 |
| 0940 | Roller | | 1020 | .008 | | .11 | .29 | | .40 | .56 |
| 0980 | Spray | | 1700 | .005 | | .10 | .18 | | .28 | .37 |
| 1200 | Paint 3 coats, smooth finish, brushwork | | 510 | .016 | | .16 | .59 | | .75 | 1.05 |
| 1240 | Roller | | 650 | .012 | | .16 | .46 | | .62 | .87 |
| 1280 | Spray | | 1625 | .005 | | .15 | .18 | | .33 | .44 |
| 1600 | Glaze coating, 2 coats, spray, clear | | 1200 | .007 | | .49 | .25 | | .74 | .91 |
| 1640 | Multicolor | | 1200 | .007 | | .95 | .25 | | 1.20 | 1.42 |
| 1660 | Painting walls, complete, including surface prep, primer & | | | | | | | | | |
| 1670 | 2 coats finish, on drywall or plaster, with roller | 1 Pord | 325 | .025 | S.F. | .17 | .92 | | 1.09 | 1.56 |
| 1700 | For oil base paint, add | | | | | 10% | | | | |
| 1800 | For ceiling installations, add | | | | | | 25% | | | |
| 2000 | Masonry or concrete block, primer/sealer, latex paint | | | | | | | | | |
| 2100 | Primer, smooth finish, brushwork | 1 Pord | 1000 | .008 | S.F. | .10 | .30 | | .40 | .56 |
| 2110 | Roller | | 1150 | .007 | | .10 | .26 | | .36 | .50 |
| 2180 | Spray | | 2400 | .003 | | .09 | .13 | | .22 | .29 |
| 2200 | Sand finish, brushwork | | 850 | .009 | | .10 | .35 | | .45 | .63 |
| 2210 | Roller | | 975 | .008 | | .10 | .31 | | .41 | .57 |
| 2280 | Spray | | 2050 | .004 | | .09 | .15 | | .24 | .32 |
| 2400 | Finish coat, smooth finish, brush | | 1100 | .007 | | .08 | .27 | | .35 | .48 |
| 2410 | Roller | | 1300 | .006 | | .08 | .23 | | .31 | .42 |
| 2480 | Spray | | 2400 | .003 | | .07 | .13 | | .20 | .26 |
| 2500 | Sand finish, brushwork | | 950 | .008 | | .08 | .32 | | .40 | .55 |
| 2510 | Roller | | 1090 | .007 | | .08 | .28 | | .36 | .49 |
| 2580 | Spray | | 2040 | .004 | | .07 | .15 | | .22 | .29 |
| 2800 | Primer plus one finish coat, smooth brush | | 525 | .015 | | .27 | .57 | | .84 | 1.15 |
| 2810 | Roller | | 615 | .013 | | .17 | .49 | | .66 | .91 |
| 2880 | Spray | | 1200 | .007 | | .15 | .25 | | .40 | .54 |
| 2900 | Sand finish, brushwork | | 450 | .018 | | .17 | .67 | | .84 | 1.18 |
| 2910 | Roller | | 515 | .016 | | .17 | .58 | | .75 | 1.06 |
| 2980 | Spray | | 1025 | .008 | | .15 | .29 | | .44 | .60 |
| 3200 | Primer plus 2 finish coats, smooth, brush | | 355 | .023 | | .25 | .85 | | 1.10 | 1.54 |
| 3210 | Roller | | 415 | .019 | | .25 | .72 | | .97 | 1.35 |
| 3280 | Spray | | 800 | .010 | | .22 | .38 | | .60 | .80 |
| 3300 | Sand finish, brushwork | | 305 | .026 | | .25 | .98 | | 1.23 | 1.74 |
| 3310 | Roller | | 350 | .023 | | .25 | .86 | | 1.11 | 1.55 |
| 3380 | Spray | | 675 | .012 | | .22 | .45 | | .67 | .90 |
| 3600 | Glaze coating, 3 coats, spray, clear | | 900 | .009 | | .70 | .33 | | 1.03 | 1.27 |
| 3620 | Multicolor | | 900 | .009 | | 1.10 | .33 | | 1.43 | 1.71 |
| 4000 | Block filler, 1 coat, brushwork | | 425 | .019 | | .13 | .71 | | .84 | 1.20 |
| 4100 | Silicone, water repellent, 2 coats, spray | | 2000 | .004 | | .26 | .15 | | .41 | .50 |
| 4120 | For oil base paint, add | | | | | 10% | | | | |
| 8200 | For work 8' - 15' H, add | | | | | | 10% | | | |
| 8300 | For work over 15' H, add | | | | | | 20% | | | |
| 8400 | For light textured surfaces, add | | | | | | 10% | | | |
| 8410 | Heavy textured, add | | | | | | 25% | | | |

# 09 91 Painting

## 09 91 23 – Interior Painting

| 09 91 23.74 Walls and Ceilings, Interior, Zero VOC Latex | Crew | Daily Output | Labor-Hours | Unit | Material | 2011 Bare Costs Labor | 2011 Bare Costs Equipment | Total | Total Incl O&P |
|---|---|---|---|---|---|---|---|---|---|
| 0010 **WALLS AND CEILINGS, INTERIOR, ZERO VOC LATEX** | | | | | | | | | |
| 0100 Concrete, dry wall or plaster, latex, primer or sealer coat | | | | | | | | | |
| 0200 Smooth finish, brushwork [G] | 1 Pord | 1150 | .007 | S.F. | .07 | .26 | | .33 | .46 |
| 0240 Roller [G] | | 1350 | .006 | | .06 | .22 | | .28 | .40 |
| 0280 Spray [G] | | 2750 | .003 | | .05 | .11 | | .16 | .21 |
| 0300 Sand finish, brushwork [G] | | 975 | .008 | | .06 | .31 | | .37 | .53 |
| 0340 Roller [G] | | 1150 | .007 | | .07 | .26 | | .33 | .47 |
| 0380 Spray [G] | | 2275 | .004 | | .05 | .13 | | .18 | .26 |
| 0800 Paint 2 coats, smooth finish, brushwork [G] | | 680 | .012 | | .13 | .44 | | .57 | .80 |
| 0840 Roller [G] | | 800 | .010 | | .13 | .38 | | .51 | .70 |
| 0880 Spray [G] | | 1625 | .005 | | .11 | .18 | | .29 | .39 |
| 0900 Sand finish, brushwork [G] | | 605 | .013 | | .12 | .50 | | .62 | .88 |
| 0940 Roller [G] | | 1020 | .008 | | .13 | .29 | | .42 | .58 |
| 0980 Spray [G] | | 1700 | .005 | | .11 | .18 | | .29 | .38 |
| 1200 Paint 3 coats, smooth finish, brushwork [G] | | 510 | .016 | | .18 | .59 | | .77 | 1.07 |
| 1240 Roller [G] | | 650 | .012 | | .19 | .46 | | .65 | .90 |
| 1280 Spray [G] | | 1625 | .005 | | .17 | .18 | | .35 | .45 |
| 1800 For ceiling installations, add [G] | | | | | | 25% | | | |
| 8200 For work 8' - 15' H, add | | | | | | 10% | | | |
| 8300 For work over 15' H, add | | | | | | 20% | | | |

### 09 91 23.75 Dry Fall Painting

| | Crew | Daily Output | Labor-Hours | Unit | Material | 2011 Bare Costs Labor | 2011 Bare Costs Equipment | Total | Total Incl O&P |
|---|---|---|---|---|---|---|---|---|---|
| 0010 **DRY FALL PAINTING** R099100-10 | | | | | | | | | |
| 0100 Sprayed on walls, gypsum board or plaster | | | | | | | | | |
| 0220 One coat R099100-20 | 1 Pord | 2600 | .003 | S.F. | .05 | .12 | | .17 | .22 |
| 0250 Two coats | | 1560 | .005 | | .10 | .19 | | .29 | .40 |
| 0280 Concrete or textured plaster, one coat | | 1560 | .005 | | .05 | .19 | | .24 | .34 |
| 0310 Two coats | | 1300 | .006 | | .10 | .23 | | .33 | .45 |
| 0340 Concrete block, one coat | | 1560 | .005 | | .05 | .19 | | .24 | .34 |
| 0370 Two coats | | 1300 | .006 | | .10 | .23 | | .33 | .45 |
| 0400 Wood, one coat | | 877 | .009 | | .05 | .34 | | .39 | .56 |
| 0430 Two coats | | 650 | .012 | | .10 | .46 | | .56 | .80 |
| 0440 On ceilings, gypsum board or plaster | | | | | | | | | |
| 0470 One coat | 1 Pord | 1560 | .005 | S.F. | .05 | .19 | | .24 | .34 |
| 0500 Two coats | | 1300 | .006 | | .10 | .23 | | .33 | .45 |
| 0530 Concrete or textured plaster, one coat | | 1560 | .005 | | .05 | .19 | | .24 | .34 |
| 0560 Two coats | | 1300 | .006 | | .10 | .23 | | .33 | .45 |
| 0570 Structural steel, bar joists or metal deck, one coat | | 1560 | .005 | | .05 | .19 | | .24 | .34 |
| 0580 Two coats | | 1040 | .008 | | .10 | .29 | | .39 | .54 |

# 09 93 Staining and Transparent Finishing

## 09 93 23 – Interior Staining and Finishing

### 09 93 23.10 Varnish

| | Crew | Daily Output | Labor-Hours | Unit | Material | 2011 Bare Costs Labor | 2011 Bare Costs Equipment | Total | Total Incl O&P |
|---|---|---|---|---|---|---|---|---|---|
| 0010 **VARNISH** | | | | | | | | | |
| 0012 1 coat + sealer, on wood trim, brush, no sanding included | 1 Pord | 400 | .020 | S.F. | .07 | .75 | | .82 | 1.19 |
| 0100 Hardwood floors, 2 coats, no sanding included, roller | " | 1890 | .004 | " | .15 | .16 | | .31 | .41 |

# 09 96 High-Performance Coatings

## 09 96 23 – Graffiti-Resistant Coatings

| 09 96 23.10 Graffiti Resistant Treatments | Crew | Daily Output | Labor-Hours | Unit | Material | 2011 Bare Costs Labor | Equipment | Total | Total Incl O&P |
|---|---|---|---|---|---|---|---|---|---|
| 0010 **GRAFFITI RESISTANT TREATMENTS**, sprayed on walls | | | | | | | | | |
| 0100 Non-sacrificial, permanent non-stick coating, clear, on metals | 1 Pord | 2000 | .004 | S.F. | 2.06 | .15 | | 2.21 | 2.49 |
| 0200 Concrete | | 2000 | .004 | | 2.35 | .15 | | 2.50 | 2.80 |
| 0300 Concrete block | | 2000 | .004 | | 3.03 | .15 | | 3.18 | 3.56 |
| 0400 Brick | | 2000 | .004 | | 3.44 | .15 | | 3.59 | 4 |
| 0500 Stone | | 2000 | .004 | | 3.44 | .15 | | 3.59 | 4 |
| 0600 Unpainted wood | | 2000 | .004 | | 3.97 | .15 | | 4.12 | 4.59 |
| 2000 Semi-permanent cross linking polymer primer, on metals | | 2000 | .004 | | .40 | .15 | | .55 | .66 |
| 2100 Concrete | | 2000 | .004 | | .48 | .15 | | .63 | .75 |
| 2200 Concrete block | | 2000 | .004 | | .60 | .15 | | .75 | .88 |
| 2300 Brick | | 2000 | .004 | | .48 | .15 | | .63 | .75 |
| 2400 Stone | | 2000 | .004 | | .48 | .15 | | .63 | .75 |
| 2500 Unpainted wood | | 2000 | .004 | | .67 | .15 | | .82 | .95 |
| 3000 Top coat, on metals | | 2000 | .004 | | .43 | .15 | | .58 | .69 |
| 3100 Concrete | | 2000 | .004 | | .49 | .15 | | .64 | .75 |
| 3200 Concrete block | | 2000 | .004 | | .68 | .15 | | .83 | .97 |
| 3300 Brick | | 2000 | .004 | | .57 | .15 | | .72 | .84 |
| 3400 Stone | | 2000 | .004 | | .57 | .15 | | .72 | .84 |
| 3500 Unpainted wood | | 2000 | .004 | | .68 | .15 | | .83 | .97 |
| 5000 Sacrificial, water based, on metal | | 2000 | .004 | | .24 | .15 | | .39 | .49 |
| 5100 Concrete | | 2000 | .004 | | .24 | .15 | | .39 | .49 |
| 5200 Concrete block | | 2000 | .004 | | .24 | .15 | | .39 | .49 |
| 5300 Brick | | 2000 | .004 | | .24 | .15 | | .39 | .49 |
| 5400 Stone | | 2000 | .004 | | .24 | .15 | | .39 | .49 |
| 5500 Unpainted wood | | 2000 | .004 | | .24 | .15 | | .39 | .49 |
| 8000 Cleaner for use after treatment | | | | | | | | | |
| 8100 Towels or wipes, per package of 30 | | | | Ea. | .63 | | | .63 | .70 |
| 8200 Aerosol spray, 24 oz. can | | | | " | 18 | | | 18 | 19.80 |

## 09 96 56 – Epoxy Coatings

### 09 96 56.20 Wall Coatings

| 09 96 56.20 Wall Coatings | Crew | Daily Output | Labor-Hours | Unit | Material | 2011 Bare Costs Labor | Equipment | Total | Total Incl O&P |
|---|---|---|---|---|---|---|---|---|---|
| 0010 **WALL COATINGS** | | | | | | | | | |
| 0100 Acrylic glazed coatings, minimum | 1 Pord | 525 | .015 | S.F. | .29 | .57 | | .86 | 1.17 |
| 0200 Maximum | | 305 | .026 | | .62 | .98 | | 1.60 | 2.14 |
| 0300 Epoxy coatings, minimum | | 525 | .015 | | .38 | .57 | | .95 | 1.27 |
| 0400 Maximum | | 170 | .047 | | 1.15 | 1.77 | | 2.92 | 3.89 |
| 0600 Exposed aggregate, troweled on, 1/16" to 1/4", minimum | | 235 | .034 | | .58 | 1.28 | | 1.86 | 2.54 |
| 0700 Maximum (epoxy or polyacrylate) | | 130 | .062 | | 1.24 | 2.31 | | 3.55 | 4.79 |
| 0900 1/2" to 5/8" aggregate, minimum | | 130 | .062 | | 1.14 | 2.31 | | 3.45 | 4.68 |
| 1000 Maximum | | 80 | .100 | | 1.96 | 3.76 | | 5.72 | 7.70 |
| 1500 Exposed aggregate, sprayed on, 1/8" aggregate, minimum | | 295 | .027 | | .53 | 1.02 | | 1.55 | 2.09 |
| 1600 Maximum | | 145 | .055 | | .99 | 2.07 | | 3.06 | 4.16 |
| 1800 High build epoxy, 50 mil, minimum | | 390 | .021 | | .64 | .77 | | 1.41 | 1.84 |
| 1900 Maximum | | 95 | .084 | | 1.09 | 3.16 | | 4.25 | 5.90 |
| 2100 Laminated epoxy with fiberglass, minimum | | 295 | .027 | | .69 | 1.02 | | 1.71 | 2.27 |
| 2200 Maximum | | 145 | .055 | | 1.23 | 2.07 | | 3.30 | 4.42 |
| 2400 Sprayed perlite or vermiculite, 1/16" thick, minimum | | 2935 | .003 | | .25 | .10 | | .35 | .43 |
| 2500 Maximum | | 640 | .013 | | .70 | .47 | | 1.17 | 1.47 |
| 2700 Vinyl plastic wall coating, minimum | | 735 | .011 | | .31 | .41 | | .72 | .95 |
| 2800 Maximum | | 240 | .033 | | .77 | 1.25 | | 2.02 | 2.71 |
| 3000 Urethane on smooth surface, 2 coats, minimum | | 1135 | .007 | | .25 | .26 | | .51 | .67 |
| 3100 Maximum | | 665 | .012 | | .55 | .45 | | 1 | 1.28 |
| 3600 Ceramic-like glazed coating, cementitious, minimum | | 440 | .018 | | .45 | .68 | | 1.13 | 1.51 |

# 09 96 High-Performance Coatings

## 09 96 56 – Epoxy Coatings

| | 09 96 56.20 Wall Coatings | Crew | Daily Output | Labor-Hours | Unit | Material | 2011 Bare Costs Labor | Equipment | Total | Total Incl O&P |
|---|---|---|---|---|---|---|---|---|---|---|
| 3700 | Maximum | 1 Pord | 345 | .023 | S.F. | .76 | .87 | | 1.63 | 2.13 |
| 3900 | Resin base, minimum | | 640 | .013 | | .31 | .47 | | .78 | 1.04 |
| 4000 | Maximum | | 330 | .024 | | .50 | .91 | | 1.41 | 1.90 |

# 09 97 Special Coatings

## 09 97 13 – Steel Coatings

### 09 97 13.23 Exterior Steel Coatings

| | | Crew | Daily Output | Labor-Hours | Unit | Material | 2011 Bare Costs Labor | Equipment | Total | Total Incl O&P |
|---|---|---|---|---|---|---|---|---|---|---|
| 0010 | **EXTERIOR STEEL COATINGS** R050516-30 | | | | | | | | | |
| 6100 | Cold galvanizing, brush in field | 1 Psst | 1100 | .007 | S.F. | .07 | .28 | | .35 | .57 |
| 6510 | Paints & protective coatings, sprayed in field | | | | | | | | | |
| 6520 | Alkyds, primer | 2 Psst | 3600 | .004 | S.F. | .06 | .17 | | .23 | .38 |
| 6540 | Gloss topcoats | | 3200 | .005 | | .07 | .19 | | .26 | .43 |
| 6560 | Silicone alkyd | | 3200 | .005 | | .12 | .19 | | .31 | .48 |
| 6610 | Epoxy, primer | | 3000 | .005 | | .24 | .21 | | .45 | .63 |
| 6630 | Intermediate or topcoat | | 2800 | .006 | | .21 | .22 | | .43 | .63 |
| 6650 | Enamel coat | | 2800 | .006 | | .27 | .22 | | .49 | .68 |
| 6700 | Epoxy ester, primer | | 2800 | .006 | | .52 | .22 | | .74 | .96 |
| 6720 | Topcoats | | 2800 | .006 | | .18 | .22 | | .40 | .59 |
| 6810 | Latex primer | | 3600 | .004 | | .05 | .17 | | .22 | .37 |
| 6830 | Topcoats | | 3200 | .005 | | .06 | .19 | | .25 | .41 |
| 6910 | Universal primers, one part, phenolic, modified alkyd | | 2000 | .008 | | .11 | .31 | | .42 | .67 |
| 6940 | Two part, epoxy spray | | 2000 | .008 | | .28 | .31 | | .59 | .86 |
| 7000 | Zinc rich primers, self cure, spray, inorganic | | 1800 | .009 | | .59 | .34 | | .93 | 1.26 |
| 7010 | Epoxy, spray, organic | | 1800 | .009 | | .21 | .34 | | .55 | .85 |
| 7020 | Above one story, spray painting simple structures, add | | | | | | 25% | | | |
| 7030 | Intricate structures, add | | | | | | 50% | | | |

# Division Notes

| | CREW | DAILY OUTPUT | LABOR-HOURS | UNIT | MAT. | LABOR | EQUIP. | TOTAL | TOTAL INCL O&P |
|---|---|---|---|---|---|---|---|---|---|
| | | | | | | | | | |
| | | | | | | | | | |
| | | | | | | | | | |
| | | | | | | | | | |
| | | | | | | | | | |
| | | | | | | | | | |
| | | | | | | | | | |
| | | | | | | | | | |
| | | | | | | | | | |
| | | | | | | | | | |
| | | | | | | | | | |
| | | | | | | | | | |
| | | | | | | | | | |
| | | | | | | | | | |
| | | | | | | | | | |
| | | | | | | | | | |
| | | | | | | | | | |
| | | | | | | | | | |
| | | | | | | | | | |
| | | | | | | | | | |
| | | | | | | | | | |

The 2011 Bare Costs columns span MAT., LABOR, EQUIP., and TOTAL.

# Estimating Tips

## General

- The items in this division are usually priced per square foot or each.

- Many items in Division 10 require some type of support system or special anchors that are not usually furnished with the item. The required anchors must be added to the estimate in the appropriate division.

- Some items in Division 10, such as lockers, may require assembly before installation. Verify the amount of assembly required. Assembly can often exceed installation time.

## 10 20 00 Interior Specialties

- Support angles and blocking are not included in the installation of toilet compartments, shower/dressing compartments, or cubicles. Appropriate line items from Divisions 5 or 6 may need to be added to support the installations.

- Toilet partitions are priced by the stall. A stall consists of a side wall, pilaster, and door with hardware. Toilet tissue holders and grab bars are extra.

- The required acoustical rating of a folding partition can have a significant impact on costs. Verify the sound transmission coefficient rating of the panel priced to the specification requirements.

- Grab bar installation does not include supplemental blocking or backing to support the required load. When grab bars are installed at an existing facility, provisions must be made to attach the grab bars to solid structure.

## Reference Numbers

Reference numbers are shown in shaded boxes at the beginning of some major classifications. These numbers refer to related items in the Reference Section. The reference information may be an estimating procedure, an alternate pricing method, or technical information.

*Note:* Not all subdivisions listed here necessarily appear in this publication.

## 10 05 05 – Selective Specialties Demolition

### 10 05 05.10 Selective Demolition, Specialties

| | | Crew | Daily Output | Labor-Hours | Unit | Material | 2011 Bare Costs Labor | Equipment | Total | Total Incl O&P |
|---|---|---|---|---|---|---|---|---|---|---|
| 0010 | **SELECTIVE DEMOLITION, SPECIALTIES** | | | | | | | | | |
| 1100 | Boards and panels, wall mounted | 2 Clab | 15 | 1.067 | Ea. | | 36.50 | | 36.50 | 56 |
| 1200 | Cases, for directory and/or bulletin boards, incl doors | | 24 | .667 | | | 23 | | 23 | 35 |
| 1850 | Shower partitions, cabinet or stall, incl base and door | ↓ | 8 | 2 | | | 68.50 | | 68.50 | 105 |
| 1855 | Shower receptor, terrazzo or concrete | 1 Clab | 14 | .571 | ↓ | | 19.65 | | 19.65 | 30 |
| 1900 | Curtain track or rod, hospital type, ceiling mounted or suspended | " | 220 | .036 | L.F. | | 1.25 | | 1.25 | 1.92 |
| 1910 | Toilet Cubicles, remove | 2 Clab | 8 | 2 | Ea. | | 68.50 | | 68.50 | 105 |
| 1930 | Urinal Screen, remove | 1 Clab | 12 | .667 | " | | 23 | | 23 | 35 |
| 2650 | Wall guard, misc. wall or corner protection | " | 320 | .025 | L.F. | | .86 | | .86 | 1.32 |
| 2750 | Access floor, metal panel system, incl pedestals, covering | 2 Clab | 850 | .019 | S.F. | | .65 | | .65 | .99 |
| 3050 | Fireplace, prefab, freestanding or wall hung, incl hood and screen | 1 Clab | 2 | 4 | Ea. | | 137 | | 137 | 211 |
| 3054 | Chimney top, simulated brick, 4' high | " | 15 | .533 | | | 18.30 | | 18.30 | 28 |
| 3200 | Stove, woodburning, cast iron | 2 Clab | 2 | 8 | | | 275 | | 275 | 420 |
| 3440 | Weathervane, residential | 1 Clab | 12 | .667 | | | 23 | | 23 | 35 |
| 3500 | Flagpole, groundset, to 70' high, excl base/fndn | K-1 | 1 | 16 | | | 610 | 217 | 827 | 1,175 |
| 3555 | To 30' high | " | 2.50 | 6.400 | | | 245 | 87 | 332 | 470 |
| 4300 | Letter, signs or plaques, exterior on wall | 1 Clab | 20 | .400 | | | 13.75 | | 13.75 | 21 |
| 4310 | Signs, street, reflective aluminum, incl post and bracket | | 60 | .133 | | | 4.58 | | 4.58 | 7.05 |
| 4320 | Door signs interior on door 6" x 6", selective demolition | ↓ | 20 | .400 | | | 13.75 | | 13.75 | 21 |
| 4550 | Turnstiles, manual or electric | 2 Clab | 2 | 8 | ↓ | | 275 | | 275 | 420 |
| 5050 | Lockers | 1 Clab | 15 | .533 | Opng. | | 18.30 | | 18.30 | 28 |
| 5250 | Cabinets, recessed | Q-12 | 12 | 1.333 | Ea. | | 63.50 | | 63.50 | 95 |
| 5260 | Mail boxes, Horiz., Key Lock, front loading, Remove | 1 Carp | 34 | .235 | " | | 10.15 | | 10.15 | 15.55 |
| 5350 | Awning, fabric, including frame | 2 Clab | 100 | .160 | S.F. | | 5.50 | | 5.50 | 8.45 |
| 6050 | Partition, woven wire | | 1400 | .011 | | | .39 | | .39 | .60 |
| 6100 | Folding gate, security, door or window | | 500 | .032 | | | 1.10 | | 1.10 | 1.69 |
| 6580 | Acoustic air wall | | 650 | .025 | ↓ | | .85 | | .85 | 1.30 |
| 7550 | Telephone enclosure, exterior, post mounted | | 3 | 5.333 | Ea. | | 183 | | 183 | 281 |
| 8850 | Scale, platform, excludes foundation or pit | ↓ | .25 | 64 | " | | 2,200 | | 2,200 | 3,375 |

# 10 11 Visual Display Units

## 10 11 13 – Chalkboards

### 10 11 13.13 Fixed Chalkboards

| | | Crew | Daily Output | Labor-Hours | Unit | Material | 2011 Bare Costs Labor | Equipment | Total | Total Incl O&P |
|---|---|---|---|---|---|---|---|---|---|---|
| 0010 | **FIXED CHALKBOARDS** Porcelain enamel steel | | | | | | | | | |
| 3900 | Wall hung | | | | | | | | | |
| 4000 | Aluminum frame and chalktrough | | | | | | | | | |
| 4200 | 3' x 4' | 2 Carp | 16 | 1 | Ea. | 185 | 43 | | 228 | 270 |
| 4300 | 3' x 5' | | 15 | 1.067 | | 248 | 46 | | 294 | 345 |
| 4500 | 4' x 8' | | 14 | 1.143 | | 355 | 49 | | 404 | 465 |
| 4600 | 4' x 12' | ↓ | 13 | 1.231 | ↓ | 515 | 53 | | 568 | 645 |
| 4700 | Wood frame and chalktrough | | | | | | | | | |
| 4800 | 3' x 4' | 2 Carp | 16 | 1 | Ea. | 188 | 43 | | 231 | 273 |
| 5000 | 3' x 5' | | 15 | 1.067 | | 233 | 46 | | 279 | 325 |
| 5100 | 4' x 5' | | 14 | 1.143 | | 248 | 49 | | 297 | 350 |
| 5300 | 4' x 8' | ↓ | 13 | 1.231 | ↓ | 350 | 53 | | 403 | 465 |
| 5400 | Liquid chalk, white porcelain enamel, wall hung | | | | | | | | | |
| 5420 | Deluxe units, aluminum trim and chalktrough | | | | | | | | | |
| 5450 | 4' x 4' | 2 Carp | 16 | 1 | Ea. | 219 | 43 | | 262 | 305 |
| 5500 | 4' x 8' | | 14 | 1.143 | | 335 | 49 | | 384 | 445 |
| 5550 | 4' x 12' | ↓ | 12 | 1.333 | ↓ | 495 | 57.50 | | 552.50 | 635 |
| 5700 | Wood trim and chalktrough | | | | | | | | | |

# 10 11 Visual Display Units

## 10 11 13 – Chalkboards

### 10 11 13.13 Fixed Chalkboards

| | | Crew | Daily Output | Labor-Hours | Unit | Material | 2011 Bare Costs Labor | Equipment | Total | Total Incl O&P |
|---|---|---|---|---|---|---|---|---|---|---|
| 5900 | 4' x 4' | 2 Carp | 16 | 1 | Ea. | 550 | 43 | | 593 | 670 |
| 6000 | 4' x 6' | | 15 | 1.067 | | 655 | 46 | | 701 | 790 |
| 6200 | 4' x 8' | | 14 | 1.143 | | 760 | 49 | | 809 | 910 |
| 6300 | Liquid chalk, felt tip markers | | | | | 1.52 | | | 1.52 | 1.67 |
| 6500 | Erasers | | | | | 1.54 | | | 1.54 | 1.69 |
| 6600 | Board cleaner, 8 oz. bottle | | | | | 4.49 | | | 4.49 | 4.94 |

### 10 11 13.23 Modular-Support-Mounted Chalkboards

| | | Crew | Daily Output | Labor-Hours | Unit | Material | 2011 Bare Costs Labor | Equipment | Total | Total Incl O&P |
|---|---|---|---|---|---|---|---|---|---|---|
| 0010 | **MODULAR-SUPPORT-MOUNTED CHALKBOARDS** | | | | | | | | | |
| 0400 | Sliding chalkboards | | | | | | | | | |
| 0450 | Vertical, one sliding board with back panel, wall mounted | | | | | | | | | |
| 0500 | 8' x 4' | 2 Carp | 8 | 2 | Ea. | 2,050 | 86 | | 2,136 | 2,375 |
| 0520 | 8' x 8' | | 7.50 | 2.133 | | 2,875 | 92 | | 2,967 | 3,325 |
| 0540 | 8' x 12' | | 7 | 2.286 | | 3,500 | 98.50 | | 3,598.50 | 4,000 |
| 0600 | Two sliding boards, with back panel | | | | | | | | | |
| 0620 | 8' x 4' | 2 Carp | 8 | 2 | Ea. | 3,300 | 86 | | 3,386 | 3,750 |
| 0640 | 8' x 8' | | 7.50 | 2.133 | | 4,900 | 92 | | 4,992 | 5,525 |
| 0660 | 8' x 12' | | 7 | 2.286 | | 7,925 | 98.50 | | 8,023.50 | 8,875 |
| 0700 | Horizontal, two track | | | | | | | | | |
| 0800 | 4' x 8', 2 sliding panels | 2 Carp | 8 | 2 | Ea. | 1,825 | 86 | | 1,911 | 2,150 |
| 0820 | 4' x 12', 2 sliding panels | | 7.50 | 2.133 | | 2,400 | 92 | | 2,492 | 2,775 |
| 0840 | 4' x 16', 4 sliding panels | | 7 | 2.286 | | 3,225 | 98.50 | | 3,323.50 | 3,700 |
| 0900 | Four track, four sliding panels | | | | | | | | | |
| 0920 | 4' x 8' | 2 Carp | 8 | 2 | Ea. | 2,975 | 86 | | 3,061 | 3,375 |
| 0940 | 4' x 12' | | 7.50 | 2.133 | | 3,875 | 92 | | 3,967 | 4,400 |
| 0960 | 4' x 16' | | 7 | 2.286 | | 5,050 | 98.50 | | 5,148.50 | 5,700 |
| 1200 | Vertical, motor operated | | | | | | | | | |
| 1400 | One sliding panel with back panel | | | | | | | | | |
| 1450 | 10' x 4' | 2 Carp | 4 | 4 | Ea. | 5,325 | 172 | | 5,497 | 6,150 |
| 1500 | 10' x 10' | | 3.75 | 4.267 | | 6,425 | 184 | | 6,609 | 7,350 |
| 1550 | 10' x 16' | | 3.50 | 4.571 | | 7,575 | 197 | | 7,772 | 8,650 |
| 1700 | Two sliding panels with back panel | | | | | | | | | |
| 1750 | 10' x 4' | 2 Carp | 4 | 4 | Ea. | 9,475 | 172 | | 9,647 | 10,700 |
| 1800 | 10' x 10' | | 3.75 | 4.267 | | 11,000 | 184 | | 11,184 | 12,400 |
| 1850 | 10' x 16' | | 3.50 | 4.571 | | 12,600 | 197 | | 12,797 | 14,200 |
| 2000 | Three sliding panels with back panel | | | | | | | | | |
| 2100 | 10' x 4' | 2 Carp | 4 | 4 | Ea. | 13,200 | 172 | | 13,372 | 14,800 |
| 2150 | 10' x 10' | | 3.75 | 4.267 | | 14,500 | 184 | | 14,684 | 16,200 |
| 2200 | 10' x 16' | | 3.50 | 4.571 | | 17,500 | 197 | | 17,697 | 19,500 |
| 2400 | For projection screen, glass beaded, add | | | | S.F. | 7.65 | | | 7.65 | 8.40 |
| 2500 | For remote control, 1 panel control, add | | | | Ea. | 360 | | | 360 | 395 |
| 2600 | 2 panel control, add | | | | " | 620 | | | 620 | 680 |
| 2800 | For units without back panels, deduct | | | | S.F. | 4.84 | | | 4.84 | 5.30 |
| 2850 | For liquid chalk porcelain panels, add | | | | " | 5.15 | | | 5.15 | 5.70 |
| 3000 | Swing leaf, any comb. of chalkboard & cork, aluminum frame | | | | | | | | | |
| 3100 | Floor style, 6 panels | | | | | | | | | |
| 3150 | 30" x 40" panels | | | | Ea. | 1,575 | | | 1,575 | 1,750 |
| 3200 | 48" x 40" panels | | | | " | 2,650 | | | 2,650 | 2,925 |
| 3300 | Wall mounted, 6 panels | | | | | | | | | |
| 3400 | 30" x 40" panels | 2 Carp | 16 | 1 | Ea. | 1,500 | 43 | | 1,543 | 1,725 |
| 3450 | 48" x 40" panels | " | 16 | 1 | " | 1,850 | 43 | | 1,893 | 2,100 |
| 3600 | Extra panels for swing leaf units | | | | | | | | | |
| 3700 | 30" x 40" panels | | | | Ea. | 299 | | | 299 | 330 |

# 10 11 Visual Display Units

## 10 11 13 – Chalkboards

| 10 11 13.23 Modular-Support-Mounted Chalkboards | Crew | Daily Output | Labor-Hours | Unit | Material | 2011 Bare Costs Labor | Equipment | Total | Total Incl O&P |
|---|---|---|---|---|---|---|---|---|---|
| 3750 | 48" x 40" panels | | | | Ea. | 370 | | | 370 | 405 |

### 10 11 13.43 Portable Chalkboards

| | | Crew | Daily Output | Labor-Hours | Unit | Material | Labor | Equipment | Total | Total Incl O&P |
|---|---|---|---|---|---|---|---|---|---|---|
| 0010 | **PORTABLE CHALKBOARDS** | | | | | | | | | |
| 0100 | Freestanding, reversible | | | | | | | | | |
| 0120 | Economy, wood frame, 4' x 6' | | | | | | | | | |
| 0140 | Chalkboard both sides | | | | Ea. | 525 | | | 525 | 575 |
| 0160 | Chalkboard one side, cork other side | | | | " | 500 | | | 500 | 555 |
| 0200 | Standard, lightweight satin finished aluminum, 4' x 6' | | | | | | | | | |
| 0220 | Chalkboard both sides | | | | Ea. | 565 | | | 565 | 620 |
| 0240 | Chalkboard one side, cork other side | | | | " | 595 | | | 595 | 655 |
| 0300 | Deluxe, heavy duty extruded aluminum, 4' x 6' | | | | | | | | | |
| 0320 | Chalkboard both sides | | | | Ea. | 990 | | | 990 | 1,100 |
| 0340 | Chalkboard one side, cork other side | | | | " | 890 | | | 890 | 980 |

## 10 11 23 – Tackboards

### 10 11 23.10 Fixed Tackboards

| | | Crew | Daily Output | Labor-Hours | Unit | Material | Labor | Equipment | Total | Total Incl O&P |
|---|---|---|---|---|---|---|---|---|---|---|
| 0010 | **FIXED TACKBOARDS** | | | | | | | | | |
| 0020 | Cork sheets, unbacked, no frame, 1/4" thick | 2 Carp | 290 | .055 | S.F. | 1.94 | 2.38 | | 4.32 | 5.75 |
| 0100 | 1/2" thick | | 290 | .055 | | 3.85 | 2.38 | | 6.23 | 7.90 |
| 0300 | Fabric-face, no frame, on 7/32" cork underlay | | 290 | .055 | | 6 | 2.38 | | 8.38 | 10.25 |
| 0400 | On 1/4" cork on 1/4" hardboard | | 290 | .055 | | 7.85 | 2.38 | | 10.23 | 12.30 |
| 0600 | With edges wrapped | | 290 | .055 | | 8.45 | 2.38 | | 10.83 | 12.90 |
| 0700 | On 7/16" fire retardant core | | 290 | .055 | | 7.60 | 2.38 | | 9.98 | 12 |
| 0900 | With edges wrapped | | 290 | .055 | | 12.30 | 2.38 | | 14.68 | 17.15 |
| 1000 | Designer fabric only, cut to size | | | | | 2.07 | | | 2.07 | 2.28 |
| 1200 | 1/4" vinyl cork, on 1/4" hardboard, no frame | 2 Carp | 290 | .055 | | 8.40 | 2.38 | | 10.78 | 12.85 |
| 1300 | On 1/4" coreboard | | 290 | .055 | | 7.95 | 2.38 | | 10.33 | 12.40 |
| 2000 | For map and display rail, economy, add | | 385 | .042 | L.F. | 2.52 | 1.79 | | 4.31 | 5.50 |
| 2100 | Deluxe, add | | 350 | .046 | " | 4.44 | 1.97 | | 6.41 | 7.90 |
| 2120 | Prefabricated, 1/4" cork, 3' x 5' with aluminum frame | | 16 | 1 | Ea. | 113 | 43 | | 156 | 190 |
| 2140 | Wood frame | | 16 | 1 | | 143 | 43 | | 186 | 223 |
| 2160 | 4' x 4' with aluminum frame | | 16 | 1 | | 98.50 | 43 | | 141.50 | 174 |
| 2180 | Wood frame | | 16 | 1 | | 128 | 43 | | 171 | 207 |
| 2200 | 4' x 8' with aluminum frame | | 14 | 1.143 | | 177 | 49 | | 226 | 270 |
| 2210 | With wood frame | | 14 | 1.143 | | 177 | 49 | | 226 | 270 |
| 2220 | 4' x 12' with aluminum frame | | 12 | 1.333 | | 263 | 57.50 | | 320.50 | 375 |
| 2230 | Bulletin board case, single glass door, with lock | | | | | | | | | |
| 2240 | 36" x 24", economy | 2 Carp | 12 | 1.333 | Ea. | 297 | 57.50 | | 354.50 | 415 |
| 2250 | Deluxe | | 12 | 1.333 | | 520 | 57.50 | | 577.50 | 660 |
| 2260 | 42" x 30", economy | | 12 | 1.333 | | 340 | 57.50 | | 397.50 | 465 |
| 2270 | Deluxe | | 12 | 1.333 | | 300 | 57.50 | | 357.50 | 425 |
| 2300 | Glass enclosed cabinets, alum., cork panel, hinged doors | | | | | | | | | |
| 2400 | 3' x 3', 1 door | 2 Carp | 12 | 1.333 | Ea. | 570 | 57.50 | | 627.50 | 720 |
| 2500 | 4' x 4', 2 door | | 11 | 1.455 | | 955 | 62.50 | | 1,017.50 | 1,150 |
| 2600 | 4' x 7', 3 door | | 10 | 1.600 | | 1,650 | 69 | | 1,719 | 1,925 |
| 2800 | 4' x 10', 4 door | | 8 | 2 | | 2,200 | 86 | | 2,286 | 2,525 |
| 2900 | For lights, add per door opening | 1 Elec | 13 | .615 | | 131 | 31 | | 162 | 191 |
| 3100 | Horizontal sliding units, 4 doors, 4' x 8', 8' x 4' | 2 Carp | 9 | 1.778 | | 1,825 | 76.50 | | 1,901.50 | 2,150 |
| 3200 | 4' x 12' | | 7 | 2.286 | | 2,400 | 98.50 | | 2,498.50 | 2,775 |
| 3400 | 8 doors, 4' x 16' | | 5 | 3.200 | | 3,250 | 138 | | 3,388 | 3,775 |
| 3500 | 4' x 24' | | 4 | 4 | | 4,375 | 172 | | 4,547 | 5,100 |

# 10 11 Visual Display Units

## 10 11 23 – Tackboards

| 10 11 23.20 Control Boards | Crew | Daily Output | Labor-Hours | Unit | Material | 2011 Bare Costs Labor | Equipment | Total | Total Incl O&P |
|---|---|---|---|---|---|---|---|---|---|
| 0010 **CONTROL BOARDS** | | | | | | | | | |
| 0020 Magnetic, porcelain finish, 18" x 24", framed | 2 Carp | 8 | 2 | Ea. | 199 | 86 | | 285 | 350 |
| 0100 24" x 36" | | 7.50 | 2.133 | | 295 | 92 | | 387 | 465 |
| 0200 36" x 48" | | 7 | 2.286 | | 390 | 98.50 | | 488.50 | 580 |
| 0300 48" x 72" | | 6 | 2.667 | | 860 | 115 | | 975 | 1,125 |
| 0400 48" x 96" | | 5 | 3.200 | | 1,075 | 138 | | 1,213 | 1,375 |

# 10 13 Directories

## 10 13 10 – Building Directories

| 10 13 10.10 Directory Boards | Crew | Daily Output | Labor-Hours | Unit | Material | 2011 Bare Costs Labor | Equipment | Total | Total Incl O&P |
|---|---|---|---|---|---|---|---|---|---|
| 0010 **DIRECTORY BOARDS** | | | | | | | | | |
| 0050 Plastic, glass covered, 30" x 20" | 2 Carp | 3 | 5.333 | Ea. | 260 | 230 | | 490 | 635 |
| 0100 36" x 48" | | 2 | 8 | | 745 | 345 | | 1,090 | 1,350 |
| 0300 Grooved cork, 30" x 20" | | 3 | 5.333 | | 375 | 230 | | 605 | 760 |
| 0400 36" x 48" | | 2 | 8 | | 505 | 345 | | 850 | 1,075 |
| 0600 Black felt, 30" x 20" | | 3 | 5.333 | | 227 | 230 | | 457 | 600 |
| 0700 36" x 48" | | 2 | 8 | | 440 | 345 | | 785 | 1,025 |
| 0900 Outdoor, weatherproof, black plastic, 36" x 24" | | 2 | 8 | | 720 | 345 | | 1,065 | 1,325 |
| 1000 36" x 36" | | 1.50 | 10.667 | | 830 | 460 | | 1,290 | 1,625 |
| 1800 Indoor, economy, open face, 18" x 24" | | 7 | 2.286 | | 134 | 98.50 | | 232.50 | 299 |
| 1900 24" x 36" | | 7 | 2.286 | | 124 | 98.50 | | 222.50 | 287 |
| 2000 36" x 24" | | 6 | 2.667 | | 141 | 115 | | 256 | 330 |
| 2100 36" x 48" | | 6 | 2.667 | | 218 | 115 | | 333 | 415 |
| 2400 Building directory, alum., black felt panels, 1 door, 24" x 18" | | 4 | 4 | | 300 | 172 | | 472 | 595 |
| 2500 36" x 24" | | 3.50 | 4.571 | | 355 | 197 | | 552 | 690 |
| 2600 48" x 32" | | 3 | 5.333 | | 575 | 230 | | 805 | 985 |
| 2700 2 door, 36" x 48" | | 2.50 | 6.400 | | 630 | 276 | | 906 | 1,125 |
| 2800 36" x 60" | | 2 | 8 | | 825 | 345 | | 1,170 | 1,425 |
| 2900 48" x 60" | | 1 | 16 | | 915 | 690 | | 1,605 | 2,050 |
| 3100 For bronze enamel finish, add | | | | | 15% | | | | |
| 3200 For bronze anodized finish, add | | | | | 25% | | | | |
| 3400 For illuminated directory, single door unit, add | | | | | 110 | | | 110 | 120 |
| 3500 For 6" header panel, 6 letters per foot, add | | | | L.F. | 20.50 | | | 20.50 | 22.50 |
| 6050 Building directory, electronic display, alum. frame, wall mounted | 2 Carp | 32 | .500 | S.F. | 550 | 21.50 | | 571.50 | 640 |
| 6100 Free standing | " | 60 | .267 | " | 655 | 11.50 | | 666.50 | 740 |

# 10 14 Signage

## 10 14 19 – Dimensional Letter Signage

| 10 14 19.10 Exterior Signs | Crew | Daily Output | Labor-Hours | Unit | Material | 2011 Bare Costs Labor | Equipment | Total | Total Incl O&P |
|---|---|---|---|---|---|---|---|---|---|
| 0010 **EXTERIOR SIGNS** | | | | | | | | | |
| 0020 Letters, 2" high, 3/8" deep, cast bronze | 1 Carp | 24 | .333 | Ea. | 22.50 | 14.35 | | 36.85 | 47 |
| 0140 1/2" deep, cast aluminum | | 18 | .444 | | 22.50 | 19.15 | | 41.65 | 54.50 |
| 0160 Cast bronze | | 32 | .250 | | 36 | 10.75 | | 46.75 | 56 |
| 0300 6" high, 5/8" deep, cast aluminum | | 24 | .333 | | 28.50 | 14.35 | | 42.85 | 53 |
| 0400 Cast bronze | | 24 | .333 | | 59 | 14.35 | | 73.35 | 87 |
| 0600 8" high, 3/4" deep, cast aluminum | | 14 | .571 | | 34.50 | 24.50 | | 59 | 75.50 |
| 0700 Cast bronze | | 20 | .400 | | 84.50 | 17.20 | | 101.70 | 120 |
| 0900 10" high, 1" deep, cast aluminum | | 18 | .444 | | 52 | 19.15 | | 71.15 | 86.50 |
| 1000 Bronze | | 18 | .444 | | 114 | 19.15 | | 133.15 | 155 |

# 10 14 Signage

## 10 14 19 – Dimensional Letter Signage

### 10 14 19.10 Exterior Signs

| | | Crew | Daily Output | Labor-Hours | Unit | Material | 2011 Bare Costs Labor | Equipment | Total | Total Incl O&P |
|---|---|---|---|---|---|---|---|---|---|---|
| 1200 | 12" high, 1-1/4" deep, cast aluminum | 1 Carp | 12 | .667 | Ea. | 55 | 28.50 | | 83.50 | 105 |
| 1500 | Cast bronze | | 18 | .444 | | 129 | 19.15 | | 148.15 | 172 |
| 1600 | 14" high, 2-5/16" deep, cast aluminum | | 12 | .667 | | 79 | 28.50 | | 107.50 | 131 |
| 1800 | Fabricated stainless steel, 6" high, 2" deep | | 20 | .400 | | 53 | 17.20 | | 70.20 | 84.50 |
| 1900 | 12" high, 3" deep | | 18 | .444 | | 118 | 19.15 | | 137.15 | 159 |
| 2100 | 18" high, 3" deep | | 12 | .667 | | 228 | 28.50 | | 256.50 | 294 |
| 2200 | 24" high, 4" deep | | 10 | .800 | | 385 | 34.50 | | 419.50 | 475 |
| 2700 | Acrylic, on high density foam, 12" high, 2" deep | | 20 | .400 | | 22.50 | 17.20 | | 39.70 | 51 |
| 2800 | 18" high, 2" deep | ↓ | 18 | .444 | | 61 | 19.15 | | 80.15 | 96.50 |
| 3900 | Plaques, custom, 20" x 30", for up to 450 letters, cast aluminum | 2 Carp | 4 | 4 | | 1,150 | 172 | | 1,322 | 1,525 |
| 4000 | Cast bronze | | 4 | 4 | | 1,675 | 172 | | 1,847 | 2,125 |
| 4200 | 30" x 36", up to 900 letters cast aluminum | | 3 | 5.333 | | 1,925 | 230 | | 2,155 | 2,475 |
| 4300 | Cast bronze | | 3 | 5.333 | | 3,475 | 230 | | 3,705 | 4,175 |
| 4500 | 36" x 48", for up to 1300 letters, cast bronze | | 2 | 8 | | 4,250 | 345 | | 4,595 | 5,225 |
| 4800 | Signs, reflective alum. directional signs, dbl. face, 2-way, w/bracket | | 30 | .533 | | 92.50 | 23 | | 115.50 | 137 |
| 4900 | 4-way | ↓ | 30 | .533 | | 148 | 23 | | 171 | 198 |
| 5100 | Exit signs, 24 ga. alum., 14" x 12" surface mounted | 1 Carp | 30 | .267 | | 41.50 | 11.50 | | 53 | 63 |
| 5200 | 10" x 7" | | 20 | .400 | | 26.50 | 17.20 | | 43.70 | 56 |
| 5400 | Bracket mounted, double face, 12" x 10" | ↓ | 30 | .267 | | 42 | 11.50 | | 53.50 | 64 |
| 5500 | Sticky back, stock decals, 14" x 10" | 1 Clab | 50 | .160 | | 19.60 | 5.50 | | 25.10 | 30 |
| 6000 | Interior elec., wall mount, fiberglass panels, 2 lamps, 6" | 1 Elec | 8 | 1 | | 121 | 50.50 | | 171.50 | 208 |
| 6100 | 8" | " | 8 | 1 | | 95.50 | 50.50 | | 146 | 180 |
| 6400 | Replacement sign faces, 6" or 8" | 1 Clab | 50 | .160 | ↓ | 54.50 | 5.50 | | 60 | 68 |

## 10 14 23 – Panel Signage

### 10 14 23.13 Engraved Panel Signage

| | | Crew | Daily Output | Labor-Hours | Unit | Material | 2011 Bare Costs Labor | Equipment | Total | Total Incl O&P |
|---|---|---|---|---|---|---|---|---|---|---|
| 0010 | **ENGRAVED PANEL SIGNAGE**, Interior | | | | | | | | | |
| 1010 | Flexible door sign, adhesive back, w/Braille, 5/8" letters, 4" x 4" | 1 Clab | 32 | .250 | Ea. | 29 | 8.60 | | 37.60 | 45 |
| 1050 | 6" x 6" | | 32 | .250 | | 42.50 | 8.60 | | 51.10 | 60 |
| 1100 | 8" x 2" | | 32 | .250 | | 29.50 | 8.60 | | 38.10 | 45.50 |
| 1150 | 8" x 4" | | 32 | .250 | | 37.50 | 8.60 | | 46.10 | 54.50 |
| 1200 | 8" x 8" | | 32 | .250 | | 59.50 | 8.60 | | 68.10 | 78.50 |
| 1250 | 12" x 2" | | 32 | .250 | | 35 | 8.60 | | 43.60 | 51.50 |
| 1300 | 12" x 6" | | 32 | .250 | | 50 | 8.60 | | 58.60 | 68 |
| 1350 | 12" x 12" | | 32 | .250 | | 119 | 8.60 | | 127.60 | 144 |
| 1500 | Graphic symbols, 2" x 2" | | 32 | .250 | | 12 | 8.60 | | 20.60 | 26.50 |
| 1550 | 6" x 6" | | 32 | .250 | | 31 | 8.60 | | 39.60 | 47 |
| 1600 | 8" x 8" | ↓ | 32 | .250 | ↓ | 31 | 8.60 | | 39.60 | 47 |

## 10 14 53 – Traffic Signage

### 10 14 53.20 Traffic Signs

| | | Crew | Daily Output | Labor-Hours | Unit | Material | 2011 Bare Costs Labor | Equipment | Total | Total Incl O&P |
|---|---|---|---|---|---|---|---|---|---|---|
| 0010 | **TRAFFIC SIGNS** | | | | | | | | | |
| 0012 | Stock, 24" x 24", no posts, .080" alum. reflectorized | B-80 | 70 | .457 | Ea. | 78.50 | 16.90 | 8.85 | 104.25 | 122 |
| 0100 | High intensity | | 70 | .457 | | 78.50 | 16.90 | 8.85 | 104.25 | 122 |
| 0300 | 30" x 30", reflectorized | | 70 | .457 | | 161 | 16.90 | 8.85 | 186.75 | 212 |
| 0400 | High intensity | | 70 | .457 | | 161 | 16.90 | 8.85 | 186.75 | 212 |
| 0600 | Guide and directional signs, 12" x 18", reflectorized | | 70 | .457 | | 34.50 | 16.90 | 8.85 | 60.25 | 73 |
| 0700 | High intensity | | 70 | .457 | | 49.50 | 16.90 | 8.85 | 75.25 | 89.50 |
| 0900 | 18" x 24", stock signs, reflectorized | | 70 | .457 | | 52.50 | 16.90 | 8.85 | 78.25 | 92.50 |
| 1000 | High intensity | | 70 | .457 | | 52.50 | 16.90 | 8.85 | 78.25 | 92.50 |
| 1200 | 24" x 24", stock signs, reflectorized | | 70 | .457 | | 64.50 | 16.90 | 8.85 | 90.25 | 106 |
| 1300 | High intensity | | 70 | .457 | | 64.50 | 16.90 | 8.85 | 90.25 | 106 |
| 1500 | Add to above for steel posts, galvanized, 10'-0" upright, bolted | | 200 | .160 | | 35 | 5.90 | 3.09 | 43.99 | 50.50 |
| 1600 | 12'-0" upright, bolted | | 140 | .229 | ↓ | 33 | 8.45 | 4.42 | 45.87 | 54 |

# 10 14 Signage

## 10 14 53 – Traffic Signage

### 10 14 53.20 Traffic Signs

| | 10 14 53.20 Traffic Signs | Crew | Daily Output | Labor-Hours | Unit | Material | Labor | Equipment | Total | Total Incl O&P |
|---|---|---|---|---|---|---|---|---|---|---|
| | | | | | | | 2011 Bare Costs | | | |
| 1800 | Highway road signs, aluminum, over 20 S. F., reflectorized | B-80 | 350 | .091 | S.F. | 28 | 3.38 | 1.77 | 33.15 | 38 |
| 2000 | High intensity | | 350 | .091 | | 28 | 3.38 | 1.77 | 33.15 | 38 |
| 2200 | Highway, suspended over road, 80 S.F. min., reflectorized | | 165 | .194 | | 28 | 7.15 | 3.75 | 38.90 | 46 |
| 2300 | High intensity | | 165 | .194 | | 28 | 7.15 | 3.75 | 38.90 | 46 |

# 10 17 Telephone Specialties

## 10 17 16 – Telephone Enclosures

### 10 17 16.10 Commercial Telephone Enclosures

| | 10 17 16.10 Commercial Telephone Enclosures | Crew | Daily Output | Labor-Hours | Unit | Material | Labor | Equipment | Total | Total Incl O&P |
|---|---|---|---|---|---|---|---|---|---|---|
| 0010 | **COMMERCIAL TELEPHONE ENCLOSURES** | | | | | | | | | |
| 0300 | Shelf type, wall hung, minimum | 2 Carp | 5 | 3.200 | Ea. | 1,050 | 138 | | 1,188 | 1,350 |
| 0400 | Maximum | | 5 | 3.200 | | 2,350 | 138 | | 2,488 | 2,775 |
| 0600 | Booth type, painted steel, indoor or outdoor, minimum | | 1.50 | 10.667 | | 3,475 | 460 | | 3,935 | 4,525 |
| 0700 | Maximum (stainless steel) | | 1.50 | 10.667 | | 11,600 | 460 | | 12,060 | 13,400 |
| 1300 | Outdoor, acoustical, on post | | 3 | 5.333 | | 1,525 | 230 | | 1,755 | 2,025 |
| 1400 | Phone carousel, pedestal mounted with dividers | | .60 | 26.667 | | 5,800 | 1,150 | | 6,950 | 8,125 |
| 1900 | Outdoor, drive-up type, wall mounted | | 4 | 4 | | 930 | 172 | | 1,102 | 1,300 |
| 2000 | Post mounted, stainless steel posts | | 3 | 5.333 | | 1,425 | 230 | | 1,655 | 1,925 |
| 2200 | Directory shelf, wall mounted, stainless steel | | | | | | | | | |
| 2300 | 3 binders | 2 Carp | 8 | 2 | Ea. | 1,100 | 86 | | 1,186 | 1,350 |
| 2500 | 4 binders | | 7 | 2.286 | | 1,275 | 98.50 | | 1,373.50 | 1,550 |
| 2600 | 5 binders | | 6 | 2.667 | | 1,625 | 115 | | 1,740 | 1,975 |
| 2800 | Table type, stainless steel, 4 binders | | 8 | 2 | | 1,300 | 86 | | 1,386 | 1,575 |
| 2900 | 7 binders | | 7 | 2.286 | | 1,625 | 98.50 | | 1,723.50 | 1,950 |

# 10 21 Compartments and Cubicles

## 10 21 13 – Toilet Compartments

### 10 21 13.13 Metal Toilet Compartments

| | 10 21 13.13 Metal Toilet Compartments | Crew | Daily Output | Labor-Hours | Unit | Material | Labor | Equipment | Total | Total Incl O&P |
|---|---|---|---|---|---|---|---|---|---|---|
| 0010 | **METAL TOILET COMPARTMENTS** | | | | | | | | | |
| 0110 | Cubicles, ceiling hung | | | | | | | | | |
| 0200 | Powder coated steel | 2 Carp | 4 | 4 | Ea. | 480 | 172 | | 652 | 795 |
| 0500 | Stainless steel | " | 4 | 4 | | 755 | 172 | | 927 | 1,100 |
| 0600 | For handicap units, incl. 52" grab bars, add | | | | | 410 | | | 410 | 450 |
| 0900 | Floor and ceiling anchored | | | | | | | | | |
| 1000 | Powder coated steel | 2 Carp | 5 | 3.200 | Ea. | 590 | 138 | | 728 | 855 |
| 1300 | Stainless steel | " | 5 | 3.200 | | 1,050 | 138 | | 1,188 | 1,350 |
| 1400 | For handicap units, incl. 52" grab bars, add | | | | | 287 | | | 287 | 315 |
| 1610 | Floor anchored | | | | | | | | | |
| 1700 | Powder coated steel | 2 Carp | 7 | 2.286 | Ea. | 555 | 98.50 | | 653.50 | 760 |
| 2000 | Stainless steel | " | 7 | 2.286 | | 1,250 | 98.50 | | 1,348.50 | 1,525 |
| 2100 | For handicap units, incl. 52" grab bars, add | | | | | 285 | | | 285 | 315 |
| 2200 | For juvenile units, deduct | | | | | 41.50 | | | 41.50 | 45.50 |
| 2450 | Floor anchored, headrail braced | | | | | | | | | |
| 2500 | Powder coated steel | 2 Carp | 6 | 2.667 | Ea. | 485 | 115 | | 600 | 710 |
| 2804 | Stainless steel | " | 4.60 | 3.478 | | 1,175 | 150 | | 1,325 | 1,525 |
| 2900 | For handicap units, incl. 52" grab bars, add | | | | | 330 | | | 330 | 365 |
| 3000 | Wall hung partitions, powder coated steel | 2 Carp | 7 | 2.286 | | 625 | 98.50 | | 723.50 | 835 |
| 3300 | Stainless steel | " | 7 | 2.286 | | 1,650 | 98.50 | | 1,748.50 | 1,975 |
| 3400 | For handicap units, incl. 52" grab bars, add | | | | | 330 | | | 330 | 365 |
| 4000 | Screens, entrance, floor mounted, 58" high, 48" wide | | | | | | | | | |

# 10 21 Compartments and Cubicles

## 10 21 13 – Toilet Compartments

### 10 21 13.13 Metal Toilet Compartments

| | | Crew | Daily Output | Labor-Hours | Unit | Material | 2011 Bare Costs Labor | 2011 Bare Costs Equipment | Total | Total Incl O&P |
|---|---|---|---|---|---|---|---|---|---|---|
| 4200 | Powder coated steel | 2 Carp | 15 | 1.067 | Ea. | 238 | 46 | | 284 | 335 |
| 4500 | Stainless steel | " | 15 | 1.067 | " | 895 | 46 | | 941 | 1,050 |
| 4650 | Urinal screen, 18" wide | | | | | | | | | |
| 4704 | Powder coated steel | 2 Carp | 6.15 | 2.602 | Ea. | 220 | 112 | | 332 | 415 |
| 5004 | Stainless steel | " | 6.15 | 2.602 | " | 580 | 112 | | 692 | 810 |
| 5100 | Floor mounted, head rail braced | | | | | | | | | |
| 5300 | Powder coated steel | 2 Carp | 8 | 2 | Ea. | 244 | 86 | | 330 | 400 |
| 5600 | Stainless steel | " | 8 | 2 | " | 575 | 86 | | 661 | 760 |
| 5750 | Pilaster, flush | | | | | | | | | |
| 5800 | Powder coated steel | 2 Carp | 10 | 1.600 | Ea. | 276 | 69 | | 345 | 410 |
| 6100 | Stainless steel | | 10 | 1.600 | | 670 | 69 | | 739 | 840 |
| 6300 | Post braced, powder coated steel | | 10 | 1.600 | | 175 | 69 | | 244 | 299 |
| 6600 | Stainless steel | | 10 | 1.600 | | 450 | 69 | | 519 | 600 |
| 6700 | Wall hung, bracket supported | | | | | | | | | |
| 6800 | Powder coated steel | 2 Carp | 10 | 1.600 | Ea. | 305 | 69 | | 374 | 440 |
| 7100 | Stainless steel | | 10 | 1.600 | | 214 | 69 | | 283 | 340 |
| 7400 | Flange supported, powder coated steel | | 10 | 1.600 | | 106 | 69 | | 175 | 222 |
| 7700 | Stainless steel | | 10 | 1.600 | | 310 | 69 | | 379 | 445 |
| 7800 | Wedge type, powder coated steel | | 10 | 1.600 | | 134 | 69 | | 203 | 254 |
| 8100 | Stainless steel | | 10 | 1.600 | | 565 | 69 | | 634 | 725 |

### 10 21 13.14 Metal Toilet Compartment Components

| | | Crew | Daily Output | Labor-Hours | Unit | Material | 2011 Bare Costs Labor | 2011 Bare Costs Equipment | Total | Total Incl O&P |
|---|---|---|---|---|---|---|---|---|---|---|
| 0010 | **METAL TOILET COMPARTMENT COMPONENTS** | | | | | | | | | |
| 0100 | Pilasters | | | | | | | | | |
| 0110 | Overhead braced, powder coated steel, 7" wide x 82" high | 2 Carp | 22.20 | .721 | Ea. | 82.50 | 31 | | 113.50 | 138 |
| 0120 | Stainless steel | | 22.20 | .721 | | 168 | 31 | | 199 | 233 |
| 0130 | Floor braced, powder coated steel, 7" wide x 70" high | | 23.30 | .687 | | 122 | 29.50 | | 151.50 | 181 |
| 0140 | Stainless steel | | 23.30 | .687 | | 232 | 29.50 | | 261.50 | 300 |
| 0150 | Ceiling hung, powder coated steel, 7" wide x 83" high | | 13.30 | 1.203 | | 129 | 52 | | 181 | 222 |
| 0160 | Stainless steel | | 13.30 | 1.203 | | 258 | 52 | | 310 | 365 |
| 0170 | Wall hung, powder coated steel, 3" wide x 58" high | | 18.90 | .847 | | 123 | 36.50 | | 159.50 | 191 |
| 0180 | Stainless steel | | 18.90 | .847 | | 171 | 36.50 | | 207.50 | 244 |
| 0200 | Panels | | | | | | | | | |
| 0210 | Powder coated steel, 31" wide x 58" high | 2 Carp | 18.90 | .847 | Ea. | 136 | 36.50 | | 172.50 | 205 |
| 0220 | Stainless steel | | 18.90 | .847 | | 345 | 36.50 | | 381.50 | 430 |
| 0230 | Powder coated steel, 53" wide x 58" high | | 18.90 | .847 | | 244 | 36.50 | | 280.50 | 325 |
| 0240 | Stainless steel | | 18.90 | .847 | | 455 | 36.50 | | 491.50 | 555 |
| 0250 | Powder coated steel, 63" wide x 58" high | | 18.90 | .847 | | 211 | 36.50 | | 247.50 | 288 |
| 0260 | Stainless steel | | 18.90 | .847 | | 515 | 36.50 | | 551.50 | 620 |
| 0300 | Doors | | | | | | | | | |
| 0310 | Powder coated steel, 24" wide x 58" high | 2 Carp | 14.10 | 1.135 | Ea. | 149 | 49 | | 198 | 239 |
| 0320 | Stainless steel | | 14.10 | 1.135 | | 335 | 49 | | 384 | 445 |
| 0330 | Powder coated steel, 26" wide x 58" high | | 14.10 | 1.135 | | 148 | 49 | | 197 | 238 |
| 0340 | Stainless steel | | 14.10 | 1.135 | | 340 | 49 | | 389 | 450 |
| 0350 | Powder coated steel, 28" wide x 58" high | | 14.10 | 1.135 | | 157 | 49 | | 206 | 248 |
| 0360 | Stainless steel | | 14.10 | 1.135 | | 360 | 49 | | 409 | 470 |
| 0370 | Powder coated steel, 36" wide x 58" high | | 14.10 | 1.135 | | 182 | 49 | | 231 | 275 |
| 0380 | Stainless steel | | 14.10 | 1.135 | | 405 | 49 | | 454 | 520 |
| 0400 | Headrails | | | | | | | | | |
| 0410 | For powder coated steel, 62" long | 2 Carp | 65 | .246 | Ea. | 24 | 10.60 | | 34.60 | 43 |
| 0420 | Stainless steel | | 65 | .246 | | 24 | 10.60 | | 34.60 | 43 |
| 0430 | For powder coated steel, 84" long | | 50 | .320 | | 32.50 | 13.80 | | 46.30 | 57 |
| 0440 | Stainless steel | | 50 | .320 | | 31 | 13.80 | | 44.80 | 55 |

# 10 21 Compartments and Cubicles

## 10 21 13 – Toilet Compartments

### 10 21 13.14 Metal Toilet Compartment Components

| | | Crew | Daily Output | Labor-Hours | Unit | Material | 2011 Bare Costs Labor | Equipment | Total | Total Incl O&P |
|---|---|---|---|---|---|---|---|---|---|---|
| 0450 | For powder coated steel, 120" long | 2 Carp | 30 | .533 | Ea. | 44.50 | 23 | | 67.50 | 84 |
| 0460 | Stainless steel | ↓ | 30 | .533 | ↓ | 44.50 | 23 | | 67.50 | 84 |

### 10 21 13.16 Plastic-Laminate-Clad Toilet Compartments

| | | Crew | Daily Output | Labor-Hours | Unit | Material | 2011 Bare Costs Labor | Equipment | Total | Total Incl O&P |
|---|---|---|---|---|---|---|---|---|---|---|
| 0010 | **PLASTIC-LAMINATE-CLAD TOILET COMPARTMENTS** | | | | | | | | | |
| 0110 | Cubicles, ceiling hung | | | | | | | | | |
| 0300 | Plastic laminate on particle board | 2 Carp | 4 | 4 | Ea. | 575 | 172 | | 747 | 895 |
| 0600 | For handicap units, incl. 52" grab bars, add | | | | " | 410 | | | 410 | 450 |
| 0900 | Floor and ceiling anchored | | | | | | | | | |
| 1100 | Plastic laminate on particle board | 2 Carp | 5 | 3.200 | Ea. | 750 | 138 | | 888 | 1,025 |
| 1400 | For handicap units, incl. 52" grab bars, add | | | | " | 287 | | | 287 | 315 |
| 1610 | Floor mounted | | | | | | | | | |
| 1800 | Plastic laminate on particle board | 2 Carp | 7 | 2.286 | Ea. | 535 | 98.50 | | 633.50 | 735 |
| 2450 | Floor mounted, headrail braced | | | | | | | | | |
| 2600 | Plastic laminate on particle board | 2 Carp | 6 | 2.667 | Ea. | 800 | 115 | | 915 | 1,050 |
| 3400 | For handicap units, incl. 52" grab bars, add | | | | | 330 | | | 330 | 365 |
| 4300 | Entrance screen, floor mtd., plas. lam., 58" high, 48" wide | 2 Carp | 15 | 1.067 | | 580 | 46 | | 626 | 710 |
| 4800 | Urinal screen, 18" wide, ceiling braced, plastic laminate | | 8 | 2 | | 174 | 86 | | 260 | 325 |
| 5400 | Floor mounted, headrail braced | | 8 | 2 | | 180 | 86 | | 266 | 330 |
| 5900 | Pilaster, flush, plastic laminate | | 10 | 1.600 | | 485 | 69 | | 554 | 635 |
| 6400 | Post braced, plastic laminate | ↓ | 10 | 1.600 | ↓ | 285 | 69 | | 354 | 420 |
| 6700 | Wall hung, bracket supported | | | | | | | | | |
| 6900 | Plastic laminate on particle board | 2 Carp | 10 | 1.600 | Ea. | 89.50 | 69 | | 158.50 | 205 |
| 7450 | Flange supported | | | | | | | | | |
| 7500 | Plastic laminate on particle board | 2 Carp | 10 | 1.600 | Ea. | 221 | 69 | | 290 | 350 |

### 10 21 13.17 Plastic-Laminate Clad Toilet Compartment Components

| | | Crew | Daily Output | Labor-Hours | Unit | Material | 2011 Bare Costs Labor | Equipment | Total | Total Incl O&P |
|---|---|---|---|---|---|---|---|---|---|---|
| 0010 | **PLASTIC-LAMINATE CLAD TOILET COMPARTMENT COMPONENTS** | | | | | | | | | |
| 0100 | Pilasters | | | | | | | | | |
| 0110 | Overhead braced, 7" wide x 82" high | 2 Carp | 22.20 | .721 | Ea. | 98 | 31 | | 129 | 156 |
| 0130 | Floor anchored, 7" wide x 70" high | | 23.30 | .687 | | 93 | 29.50 | | 122.50 | 148 |
| 0150 | Ceiling hung, 7" wide x 83" high | | 13.30 | 1.203 | | 97.50 | 52 | | 149.50 | 187 |
| 0180 | Wall hung, 3" wide x 58" high | ↓ | 18.90 | .847 | ↓ | 87.50 | 36.50 | | 124 | 153 |
| 0200 | Panels | | | | | | | | | |
| 0210 | 31" wide x 58" high | 2 Carp | 18.90 | .847 | Ea. | 143 | 36.50 | | 179.50 | 214 |
| 0230 | 51" wide x 58" high | | 18.90 | .847 | | 193 | 36.50 | | 229.50 | 268 |
| 0250 | 63" wide x 58" high | ↓ | 18.90 | .847 | ↓ | 223 | 36.50 | | 259.50 | 300 |
| 0300 | Doors | | | | | | | | | |
| 0310 | 24" wide x 58" high | 2 Carp | 14.10 | 1.135 | Ea. | 133 | 49 | | 182 | 222 |
| 0330 | 26" wide x 58" high | | 14.10 | 1.135 | | 138 | 49 | | 187 | 226 |
| 0350 | 28" wide x 58" high | | 14.10 | 1.135 | | 143 | 49 | | 192 | 232 |
| 0370 | 36" wide x 58" high | ↓ | 14.10 | 1.135 | ↓ | 170 | 49 | | 219 | 262 |
| 0400 | Headrails | | | | | | | | | |
| 0410 | 62" long | 2 Carp | 65 | .246 | Ea. | 25.50 | 10.60 | | 36.10 | 44.50 |
| 0430 | 84" long | | 60 | .267 | | 31 | 11.50 | | 42.50 | 51.50 |
| 0450 | 120" long | ↓ | 30 | .533 | ↓ | 44.50 | 23 | | 67.50 | 84 |

### 10 21 13.19 Plastic Toilet Compartments

| | | Crew | Daily Output | Labor-Hours | Unit | Material | 2011 Bare Costs Labor | Equipment | Total | Total Incl O&P |
|---|---|---|---|---|---|---|---|---|---|---|
| 0010 | **PLASTIC TOILET COMPARTMENTS** | | | | | | | | | |
| 0110 | Cubicles, ceiling hung | | | | | | | | | |
| 0250 | Phenolic | 2 Carp | 4 | 4 | Ea. | 805 | 172 | | 977 | 1,150 |
| 0600 | For handicap units, incl. 52" grab bars, add | | | | " | 410 | | | 410 | 450 |
| 0900 | Floor and ceiling anchored | | | | | | | | | |
| 1050 | Phenolic | 2 Carp | 5 | 3.200 | Ea. | 805 | 138 | | 943 | 1,100 |
| 1400 | For handicap units, incl. 52" grab bars, add | | | | " | 287 | | | 287 | 315 |

## 10 21 13 – Toilet Compartments

| 10 21 13.19 Plastic Toilet Compartments | | Crew | Daily Output | Labor-Hours | Unit | Material | 2011 Bare Costs Labor | Equipment | Total | Total Incl O&P |
|---|---|---|---|---|---|---|---|---|---|---|
| 1610 | Floor mounted | | | | | | | | | |
| 1750 | Phenolic | 2 Carp | 7 | 2.286 | Ea. | 1,350 | 98.50 | | 1,448.50 | 1,625 |
| 2100 | For handicap units, incl. 52" grab bars, add | | | | | 285 | | | 285 | 315 |
| 2200 | For juvenile units, deduct | | | | | 41.50 | | | 41.50 | 45.50 |
| 2450 | Floor mounted, headrail braced | | | | | | | | | |
| 2550 | Phenolic | 2 Carp | 6 | 2.667 | Ea. | 895 | 115 | | 1,010 | 1,150 |

## 10 21 13.20 Plastic Toilet Compartment Components

| | | Crew | Daily Output | Labor-Hours | Unit | Material | 2011 Bare Costs Labor | Equipment | Total | Total Incl O&P |
|---|---|---|---|---|---|---|---|---|---|---|
| 0010 | **PLASTIC TOILET COMPARTMENT COMPONENTS** | | | | | | | | | |
| 0100 | Pilasters | | | | | | | | | |
| 0110 | Overhead braced, polymer plastic, 7" wide x 82" high | 2 Carp | 22.20 | .721 | Ea. | 128 | 31 | | 159 | 189 |
| 0120 | Phenolic | | 22.20 | .721 | | 136 | 31 | | 167 | 198 |
| 0130 | Floor braced, polymer plastic, 7" wide x 70" high | | 23.30 | .687 | | 170 | 29.50 | | 199.50 | 233 |
| 0140 | Phenolic | | 23.30 | .687 | | 128 | 29.50 | | 157.50 | 187 |
| 0150 | Ceiling hung, polymer plastic, 7" wide x 83" high | | 13.30 | 1.203 | | 151 | 52 | | 203 | 246 |
| 0160 | Phenolic | | 13.30 | 1.203 | | 145 | 52 | | 197 | 240 |
| 0180 | Wall hung, phenolic, 3" wide x 58" high | | 18.90 | .847 | | 90.50 | 36.50 | | 127 | 156 |
| 0200 | Panels | | | | | | | | | |
| 0210 | Polymer plastic, 31" high x 55" high | 2 Carp | 18.90 | .847 | Ea. | 289 | 36.50 | | 325.50 | 375 |
| 0220 | Phenolic, 31" wide x 58" high | | 18.90 | .847 | | 251 | 36.50 | | 287.50 | 330 |
| 0230 | Polymer plastic, 51" wide x 55" high | | 18.90 | .847 | | 435 | 36.50 | | 471.50 | 535 |
| 0240 | Phenolic, 51" wide x 58" high | | 18.90 | .847 | | 375 | 36.50 | | 411.50 | 465 |
| 0250 | Polymer plastic, 63" wide x 55" high | | 18.90 | .847 | | 520 | 36.50 | | 556.50 | 625 |
| 0260 | Phenolic, 63" wide x 58" high | | 18.90 | .847 | | 445 | 36.50 | | 481.50 | 545 |
| 0300 | Doors | | | | | | | | | |
| 0310 | Polymer plastic, 24" wide x 55" high | 2 Carp | 14.10 | 1.135 | Ea. | 200 | 49 | | 249 | 295 |
| 0320 | Phenolic, 24" wide x 58" high | | 14.10 | 1.135 | | 274 | 49 | | 323 | 375 |
| 0330 | Polymer plastic, 26" high x 55" high | | 14.10 | 1.135 | | 213 | 49 | | 262 | 310 |
| 0340 | Phenolic, 26" wide x 58" high | | 14.10 | 1.135 | | 291 | 49 | | 340 | 395 |
| 0350 | Polymer plastic, 28" wide x 55" high | | 14.10 | 1.135 | | 225 | 49 | | 274 | 325 |
| 0360 | Phenolic, 28" wide x 58" high | | 14.10 | 1.135 | | 310 | 49 | | 359 | 415 |
| 0370 | Polymer plastic, 36" wide x 55" high | | 14.10 | 1.135 | | 285 | 49 | | 334 | 390 |
| 0380 | Phenolic, 36" wide x 58" high | | 14.10 | 1.135 | | 380 | 49 | | 429 | 495 |
| 0400 | Headrails | | | | | | | | | |
| 0410 | For polymer plastic, 62" long | 2 Carp | 65 | .246 | Ea. | 32 | 10.60 | | 42.60 | 52 |
| 0420 | Phenolic | | 65 | .246 | | 23 | 10.60 | | 33.60 | 42 |
| 0430 | For polymer plastic, 84" long | | 50 | .320 | | 36 | 13.80 | | 49.80 | 61 |
| 0440 | Phenolic | | 50 | .320 | | 29 | 13.80 | | 42.80 | 53 |
| 0450 | For polymer plastic, 120" long | | 30 | .533 | | 51 | 23 | | 74 | 91 |
| 0460 | Phenolic | | 30 | .533 | | 39 | 23 | | 62 | 78 |

## 10 21 13.40 Stone Toilet Compartments

| | | Crew | Daily Output | Labor-Hours | Unit | Material | 2011 Bare Costs Labor | Equipment | Total | Total Incl O&P |
|---|---|---|---|---|---|---|---|---|---|---|
| 0010 | **STONE TOILET COMPARTMENTS** | | | | | | | | | |
| 0100 | Cubicles, ceiling hung, marble | 2 Marb | 2 | 8 | Ea. | 1,450 | 325 | | 1,775 | 2,100 |
| 0600 | For handicap units, incl. 52" grab bars, add | | | | | 410 | | | 410 | 450 |
| 0800 | Floor & ceiling anchored, marble | 2 Marb | 2.50 | 6.400 | | 1,825 | 261 | | 2,086 | 2,400 |
| 1400 | For handicap units, incl. 52" grab bars, add | | | | | 287 | | | 287 | 315 |
| 1600 | Floor mounted, marble | 2 Marb | 3 | 5.333 | | 1,075 | 217 | | 1,292 | 1,500 |
| 2400 | Floor mounted, headrail braced, marble | " | 3 | 5.333 | | 1,025 | 217 | | 1,242 | 1,450 |
| 2900 | For handicap units, incl. 52" grab bars, add | | | | | 330 | | | 330 | 365 |
| 4100 | Entrance screen, floor mounted marble, 58" high, 48" wide | 2 Marb | 9 | 1.778 | | 675 | 72.50 | | 747.50 | 855 |
| 4600 | Urinal screen, 18" wide, ceiling braced, marble | D-1 | 6 | 2.667 | | 675 | 104 | | 779 | 900 |
| 5100 | Floor mounted, head rail braced | | | | | | | | | |
| 5200 | Marble | D-1 | 6 | 2.667 | Ea. | 575 | 104 | | 679 | 785 |

# 10 21 Compartments and Cubicles

## 10 21 13 – Toilet Compartments

| 10 21 13.40 Stone Toilet Compartments | Crew | Daily Output | Labor-Hours | Unit | Material | 2011 Bare Costs Labor | Equipment | Total | Total Incl O&P |
|---|---|---|---|---|---|---|---|---|---|
| 5700 | Pilaster, flush, marble | D-1 | 9 | 1.778 | Ea. | 750 | 69.50 | | 819.50 | 930 |
| 6200 | Post braced, marble | ↓ | 9 | 1.778 | ↓ | 745 | 69.50 | | 814.50 | 925 |

## 10 21 16 – Shower and Dressing Compartments

### 10 21 16.10 Partitions, Shower

| | | Crew | Daily Output | Labor-Hours | Unit | Material | 2011 Bare Costs Labor | Equipment | Total | Total Incl O&P |
|---|---|---|---|---|---|---|---|---|---|---|
| 0010 | **PARTITIONS, SHOWER** Floor mounted, no plumbing | | | | | | | | | |
| 0400 | Cabinet, one piece, fiberglass, 32" x 32" | 2 Carp | 5 | 3.200 | Ea. | 475 | 138 | | 613 | 730 |
| 0420 | 36" x 36" | | 5 | 3.200 | | 530 | 138 | | 668 | 790 |
| 0440 | 36" x 48" | | 5 | 3.200 | | 1,375 | 138 | | 1,513 | 1,725 |
| 0460 | Acrylic, 32" x 32" | | 5 | 3.200 | | 325 | 138 | | 463 | 565 |
| 0480 | 36" x 36" | | 5 | 3.200 | | 980 | 138 | | 1,118 | 1,275 |
| 0500 | 36" x 48" | ↓ | 5 | 3.200 | | 2,100 | 138 | | 2,238 | 2,525 |
| 0520 | Shower door for above, clear plastic, 24" wide | 1 Carp | 8 | 1 | | 162 | 43 | | 205 | 244 |
| 0540 | 28" wide | | 8 | 1 | | 184 | 43 | | 227 | 268 |
| 0560 | Tempered glass, 24" wide | | 8 | 1 | | 175 | 43 | | 218 | 258 |
| 0580 | 28" wide | ↓ | 8 | 1 | | 197 | 43 | | 240 | 283 |
| 2400 | Glass stalls, with doors, no receptors, chrome on brass | 2 Shee | 3 | 5.333 | | 1,525 | 275 | | 1,800 | 2,100 |
| 2700 | Anodized aluminum | " | 4 | 4 | | 1,050 | 207 | | 1,257 | 1,450 |
| 2900 | Marble shower stall, stock design, with shower door | 2 Marb | 1.20 | 13.333 | | 2,275 | 545 | | 2,820 | 3,325 |
| 3000 | With curtain | | 1.30 | 12.308 | | 2,000 | 500 | | 2,500 | 2,950 |
| 3200 | Receptors, precast terrazzo, 32" x 32" | | 14 | 1.143 | | 147 | 46.50 | | 193.50 | 232 |
| 3300 | 48" x 34" | | 9.50 | 1.684 | | 203 | 68.50 | | 271.50 | 325 |
| 3500 | Plastic, simulated terrazzo receptor, 32" x 32" | | 14 | 1.143 | | 120 | 46.50 | | 166.50 | 202 |
| 3600 | 32" x 48" | | 12 | 1.333 | | 206 | 54.50 | | 260.50 | 310 |
| 3800 | Precast concrete, colors, 32" x 32" | | 14 | 1.143 | | 225 | 46.50 | | 271.50 | 320 |
| 3900 | 48" x 48" | ↓ | 8 | 2 | | 400 | 81.50 | | 481.50 | 560 |
| 4100 | Shower doors, economy plastic, 24" wide | 1 Shee | 9 | .889 | | 132 | 46 | | 178 | 215 |
| 4200 | Tempered glass door, economy | | 8 | 1 | | 218 | 51.50 | | 269.50 | 320 |
| 4400 | Folding, tempered glass, aluminum frame | | 6 | 1.333 | | 395 | 69 | | 464 | 540 |
| 4500 | Sliding, tempered glass, 48" opening | | 6 | 1.333 | | 263 | 69 | | 332 | 395 |
| 4700 | Deluxe, tempered glass, chrome on brass frame, minimum | | 8 | 1 | | 235 | 51.50 | | 286.50 | 335 |
| 4800 | Maximum | | 1 | 8 | | 910 | 415 | | 1,325 | 1,625 |
| 4850 | On anodized aluminum frame, minimum | | 2 | 4 | | 148 | 207 | | 355 | 475 |
| 4900 | Maximum | ↓ | 1 | 8 | ↓ | 535 | 415 | | 950 | 1,225 |
| 5100 | Shower enclosure, tempered glass, anodized alum. frame | | | | | | | | | |
| 5120 | 2 panel & door, corner unit, 32" x 32" | 1 Shee | 2 | 4 | Ea. | 520 | 207 | | 727 | 885 |
| 5140 | Neo-angle corner unit, 16" x 24" x 16" | " | 2 | 4 | | 960 | 207 | | 1,167 | 1,350 |
| 5200 | Shower surround, 3 wall, polypropylene, 32" x 32" | 1 Carp | 4 | 2 | | 385 | 86 | | 471 | 555 |
| 5220 | PVC, 32" x 32" | | 4 | 2 | | 350 | 86 | | 436 | 515 |
| 5240 | Fiberglass | | 4 | 2 | | 385 | 86 | | 471 | 550 |
| 5250 | 2 wall, polypropylene, 32" x 32" | | 4 | 2 | | 290 | 86 | | 376 | 450 |
| 5270 | PVC | | 4 | 2 | | 360 | 86 | | 446 | 525 |
| 5290 | Fiberglass | ↓ | 4 | 2 | | 365 | 86 | | 451 | 535 |
| 5300 | Tub doors, tempered glass & frame, minimum | 1 Shee | 8 | 1 | | 210 | 51.50 | | 261.50 | 310 |
| 5400 | Maximum | | 6 | 1.333 | | 490 | 69 | | 559 | 645 |
| 5600 | Chrome plated, brass frame, minimum | | 8 | 1 | | 278 | 51.50 | | 329.50 | 385 |
| 5700 | Maximum | | 6 | 1.333 | | 680 | 69 | | 749 | 855 |
| 5900 | Tub/shower enclosure, temp. glass, alum. frame, minimum | | 2 | 4 | | 375 | 207 | | 582 | 725 |
| 6200 | Maximum | | 1.50 | 5.333 | | 785 | 275 | | 1,060 | 1,275 |
| 6500 | On chrome-plated brass frame, minimum | | 2 | 4 | | 520 | 207 | | 727 | 880 |
| 6600 | Maximum | ↓ | 1.50 | 5.333 | | 1,100 | 275 | | 1,375 | 1,650 |
| 6800 | Tub surround, 3 wall, polypropylene | 1 Carp | 4 | 2 | | 235 | 86 | | 321 | 390 |
| 6900 | PVC | ↓ | 4 | 2 | ↓ | 355 | 86 | | 441 | 525 |

# 10 21 Compartments and Cubicles

## 10 21 16 – Shower and Dressing Compartments

| 10 21 16.10 Partitions, Shower | | Crew | Daily Output | Labor-Hours | Unit | Material | 2011 Bare Costs Labor | Equipment | Total | Total Incl O&P |
|---|---|---|---|---|---|---|---|---|---|---|
| 7000 | Fiberglass, minimum | 1 Carp | 4 | 2 | Ea. | 370 | 86 | | 456 | 535 |
| 7100 | Maximum | ↓ | 3 | 2.667 | ↓ | 625 | 115 | | 740 | 860 |

## 10 21 23 – Cubicle Curtains and Track

### 10 21 23.16 Cubicle Track and Hardware

| | | Crew | Daily Output | Labor-Hours | Unit | Material | 2011 Bare Costs Labor | Equipment | Total | Total Incl O&P |
|---|---|---|---|---|---|---|---|---|---|---|
| 0010 | **CUBICLE TRACK AND HARDWARE** | | | | | | | | | |
| 0020 | Curtain track, box channel, ceiling mounted | 1 Carp | 135 | .059 | L.F. | 6 | 2.55 | | 8.55 | 10.50 |
| 0100 | Suspended | " | 100 | .080 | " | 7.60 | 3.44 | | 11.04 | 13.65 |
| 0300 | Curtains, nylon mesh tops, fire resistant, 11 oz. per lineal yard | | | | | | | | | |
| 0310 | Polyester oxford cloth, 9' ceiling height | 1 Carp | 425 | .019 | L.F. | 12 | .81 | | 12.81 | 14.45 |
| 0500 | 8' ceiling height | | 425 | .019 | | 10.15 | .81 | | 10.96 | 12.40 |
| 0700 | Designer oxford cloth | ↓ | 425 | .019 | ↓ | 25.50 | .81 | | 26.31 | 29.50 |
| 0800 | I.V. track systems | | | | | | | | | |
| 0820 | I.V. track, oval | 1 Carp | 135 | .059 | L.F. | 5.25 | 2.55 | | 7.80 | 9.70 |
| 0830 | I.V. trolley | | 32 | .250 | Ea. | 36 | 10.75 | | 46.75 | 56.50 |
| 0840 | I.V. pendent, (tree, 5 hook) | ↓ | 32 | .250 | " | 129 | 10.75 | | 139.75 | 159 |

# 10 22 Partitions

## 10 22 13 – Wire Mesh Partitions

### 10 22 13.10 Partitions, Woven Wire

| | | Crew | Daily Output | Labor-Hours | Unit | Material | 2011 Bare Costs Labor | Equipment | Total | Total Incl O&P |
|---|---|---|---|---|---|---|---|---|---|---|
| 0010 | **PARTITIONS, WOVEN WIRE** For tool or stockroom enclosures | | | | | | | | | |
| 0100 | Channel frame, 1-1/2" diamond mesh, 10 ga. wire, painted | | | | | | | | | |
| 0300 | Wall panels, 4'-0" wide, 7' high | 2 Carp | 25 | .640 | Ea. | 126 | 27.50 | | 153.50 | 182 |
| 0400 | 8' high | | 23 | .696 | | 147 | 30 | | 177 | 208 |
| 0600 | 10' high | ↓ | 18 | .889 | | 173 | 38.50 | | 211.50 | 249 |
| 0700 | For 5' wide panels, add | | | | | 5% | | | | |
| 0900 | Ceiling panels, 10' long, 2' wide | 2 Carp | 25 | .640 | | 130 | 27.50 | | 157.50 | 186 |
| 1000 | 4' wide | | 15 | 1.067 | | 164 | 46 | | 210 | 251 |
| 1200 | Panel with service window & shelf, 5' wide, 7' high | | 20 | .800 | | 405 | 34.50 | | 439.50 | 500 |
| 1300 | 8' high | | 15 | 1.067 | | 400 | 46 | | 446 | 510 |
| 1500 | Sliding doors, full height, 3' wide, 7' high | | 6 | 2.667 | | 420 | 115 | | 535 | 635 |
| 1600 | 10' high | | 5 | 3.200 | | 375 | 138 | | 513 | 620 |
| 1800 | 6' wide sliding door, 7' full height | | 5 | 3.200 | | 575 | 138 | | 713 | 840 |
| 1900 | 10' high | | 4 | 4 | | 710 | 172 | | 882 | 1,050 |
| 2100 | Swinging doors, 3' wide, 7' high, no transom | | 6 | 2.667 | | 265 | 115 | | 380 | 470 |
| 2200 | 7' high, 3' transom | ↓ | 5 | 3.200 | ↓ | 320 | 138 | | 458 | 560 |

## 10 22 16 – Folding Gates

### 10 22 16.10 Security Gates

| | | Crew | Daily Output | Labor-Hours | Unit | Material | 2011 Bare Costs Labor | Equipment | Total | Total Incl O&P |
|---|---|---|---|---|---|---|---|---|---|---|
| 0010 | **SECURITY GATES** For roll up type, see Section 08 33 13.10 | | | | | | | | | |
| 0300 | Scissors type folding gate, ptd. steel, single, 6-1/2' high, 5-1/2' wide | 2 Sswk | 4 | 4 | Opng. | 196 | 194 | | 390 | 560 |
| 0350 | 6-1/2' wide | | 4 | 4 | | 218 | 194 | | 412 | 585 |
| 0400 | 7-1/2' wide | | 4 | 4 | | 200 | 194 | | 394 | 565 |
| 0600 | Double gate, 8' high, 8' wide | | 2.50 | 6.400 | | 345 | 310 | | 655 | 930 |
| 0650 | 10' wide | | 2.50 | 6.400 | | 375 | 310 | | 685 | 965 |
| 0700 | 12' wide | | 2 | 8 | | 550 | 390 | | 940 | 1,300 |
| 0750 | 14' wide | | 2 | 8 | | 630 | 390 | | 1,020 | 1,375 |
| 0900 | Door gate, folding steel, 4' wide, 61" high | | 4 | 4 | | 121 | 194 | | 315 | 480 |
| 1000 | 71" high | | 4 | 4 | | 109 | 194 | | 303 | 465 |
| 1200 | 81" high | | 4 | 4 | | 118 | 194 | | 312 | 475 |
| 1300 | Window gates, 2' to 4' wide, 31" high | | 4 | 4 | | 59.50 | 194 | | 253.50 | 410 |
| 1500 | 55" high | ↓ | 3.75 | 4.267 | ↓ | 93 | 207 | | 300 | 465 |

# 10 22 Partitions

## 10 22 16 – Folding Gates

| 10 22 16.10 Security Gates | Crew | Daily Output | Labor-Hours | Unit | Material | 2011 Bare Costs Labor | Equipment | Total | Total Incl O&P |
|---|---|---|---|---|---|---|---|---|---|
| 1600 | 79" high | 2 Sswk | 3.50 | 4.571 | Opng. | 108 | 222 | | 330 | 510 |

## 10 22 19 – Demountable Partitions

### 10 22 19.43 Demountable Composite Partitions

| | | Crew | Daily Output | Labor-Hours | Unit | Material | 2011 Bare Costs Labor | Equipment | Total | Total Incl O&P |
|---|---|---|---|---|---|---|---|---|---|---|
| 0010 | **DEMOUNTABLE COMPOSITE PARTITIONS**, add for doors | | | | | | | | | |
| 0100 | Do not deduct door openings from total L.F. | | | | | | | | | |
| 0900 | Demountable gypsum system on 2" to 2-1/2" | | | | | | | | | |
| 1000 | steel studs, 9' high, 3" to 3-3/4" thick | | | | | | | | | |
| 1200 | Vinyl clad gypsum | 2 Carp | 48 | .333 | L.F. | 58 | 14.35 | | 72.35 | 85.50 |
| 1300 | Fabric clad gypsum | | 44 | .364 | | 144 | 15.65 | | 159.65 | 182 |
| 1500 | Steel clad gypsum | | 40 | .400 | | 156 | 17.20 | | 173.20 | 199 |
| 1600 | 1.75 system, aluminum framing, vinyl clad hardboard, | | | | | | | | | |
| 1800 | paper honeycomb core panel, 1-3/4" to 2-1/2" thick | | | | | | | | | |
| 1900 | 9' high | 2 Carp | 48 | .333 | L.F. | 97 | 14.35 | | 111.35 | 129 |
| 2100 | 7' high | | 60 | .267 | | 87 | 11.50 | | 98.50 | 113 |
| 2200 | 5' high | | 80 | .200 | | 73.50 | 8.60 | | 82.10 | 94 |
| 2250 | Unitized gypsum system | | | | | | | | | |
| 2300 | Unitized panel, 9' high, 2" to 2-1/2" thick | | | | | | | | | |
| 2350 | Vinyl clad gypsum | 2 Carp | 48 | .333 | L.F. | 125 | 14.35 | | 139.35 | 159 |
| 2400 | Fabric clad gypsum | " | 44 | .364 | " | 206 | 15.65 | | 221.65 | 250 |
| 2500 | Unitized mineral fiber system | | | | | | | | | |
| 2510 | Unitized panel, 9' high, 2-1/4" thick, aluminum frame | | | | | | | | | |
| 2550 | Vinyl clad mineral fiber | 2 Carp | 48 | .333 | L.F. | 124 | 14.35 | | 138.35 | 159 |
| 2600 | Fabric clad mineral fiber | " | 44 | .364 | " | 185 | 15.65 | | 200.65 | 228 |
| 2800 | Movable steel walls, modular system | | | | | | | | | |
| 2900 | Unitized panels, 9' high, 48" wide | | | | | | | | | |
| 3100 | Baked enamel, pre-finished | 2 Carp | 60 | .267 | L.F. | 141 | 11.50 | | 152.50 | 173 |
| 3200 | Fabric clad steel | | 56 | .286 | " | 203 | 12.30 | | 215.30 | 243 |
| 5310 | Trackless wall, cork finish, semi-acoustic, 1-5/8" thick, minimum | | 325 | .049 | S.F. | 38.50 | 2.12 | | 40.62 | 46 |
| 5320 | Maximum | | 190 | .084 | | 36.50 | 3.63 | | 40.13 | 45.50 |
| 5330 | Acoustic, 2" thick, minimum | | 305 | .052 | | 31 | 2.26 | | 33.26 | 37.50 |
| 5340 | Maximum | | 225 | .071 | | 53.50 | 3.06 | | 56.56 | 63.50 |
| 5500 | For acoustical partitions, add, minimum | | | | | 2.22 | | | 2.22 | 2.44 |
| 5550 | Maximum | | | | | 10.35 | | | 10.35 | 11.40 |
| 5700 | For doors, see Sections 08 11 & 08 16 | | | | | | | | | |
| 5800 | For door hardware, see Section 08 71 | | | | | | | | | |
| 6100 | In-plant modular office system, w/prehung hollow core door | | | | | | | | | |
| 6200 | 3" thick polystyrene core panels | | | | | | | | | |
| 6250 | 12' x 12', 2 wall | 2 Clab | 3.80 | 4.211 | Ea. | 3,700 | 145 | | 3,845 | 4,300 |
| 6300 | 4 wall | | 1.90 | 8.421 | | 4,700 | 289 | | 4,989 | 5,625 |
| 6350 | 16' x 16', 2 wall | | 3.60 | 4.444 | | 4,925 | 153 | | 5,078 | 5,625 |
| 6400 | 4 wall | | 1.80 | 8.889 | | 6,625 | 305 | | 6,930 | 7,750 |

## 10 22 23 – Portable Partitions, Screens, and Panels

### 10 22 23.13 Wall Screens

| | | Crew | Daily Output | Labor-Hours | Unit | Material | 2011 Bare Costs Labor | Equipment | Total | Total Incl O&P |
|---|---|---|---|---|---|---|---|---|---|---|
| 0010 | **WALL SCREENS**, divider panels, free standing, fiber core | | | | | | | | | |
| 0020 | Fabric face straight | | | | | | | | | |
| 0100 | 3'-0" long, 4'-0" high | 2 Carp | 100 | .160 | L.F. | 146 | 6.90 | | 152.90 | 172 |
| 0200 | 5'-0" high | | 90 | .178 | | 103 | 7.65 | | 110.65 | 125 |
| 0500 | 6'-0" high | | 75 | .213 | | 120 | 9.20 | | 129.20 | 146 |
| 0900 | 5'-0" long, 4'-0" high | | 175 | .091 | | 90.50 | 3.94 | | 94.44 | 106 |
| 1000 | 5'-0" high | | 150 | .107 | | 98.50 | 4.59 | | 103.09 | 115 |
| 1500 | 6'-0" high | | 125 | .128 | | 114 | 5.50 | | 119.50 | 134 |
| 1600 | 6'-0" long, 5'-0" high | | 162 | .099 | | 95.50 | 4.25 | | 99.75 | 112 |

# 10 22 Partitions

## 10 22 23 – Portable Partitions, Screens, and Panels

### 10 22 23.13 Wall Screens

| | | Crew | Daily Output | Labor-Hours | Unit | Material | 2011 Bare Costs Labor | Equipment | Total | Total Incl O&P |
|---|---|---|---|---|---|---|---|---|---|---|
| 3200 | Economical panels, fabric face, 4'-0" long, 5'-0" high | 2 Carp | 132 | .121 | L.F. | 51.50 | 5.20 | | 56.70 | 64.50 |
| 3250 | 6'-0" high | | 112 | .143 | | 54 | 6.15 | | 60.15 | 68.50 |
| 3300 | 5'-0" long, 5'-0" high | | 150 | .107 | | 44 | 4.59 | | 48.59 | 55.50 |
| 3350 | 6'-0" high | | 125 | .128 | | 46 | 5.50 | | 51.50 | 59 |
| 3450 | Acoustical panels, 60 to 90 NRC, 3'-0" long, 5'-0" high | | 90 | .178 | | 89.50 | 7.65 | | 97.15 | 110 |
| 3550 | 6'-0" high | | 75 | .213 | | 99.50 | 9.20 | | 108.70 | 123 |
| 3600 | 5'-0" long, 5'-0" high | | 150 | .107 | | 55.50 | 4.59 | | 60.09 | 68.50 |
| 3650 | 6'-0" high | | 125 | .128 | | 68 | 5.50 | | 73.50 | 83 |
| 3700 | 6'-0" long, 5'-0" high | | 162 | .099 | | 51.50 | 4.25 | | 55.75 | 63 |
| 3750 | 6'-0" high | | 138 | .116 | | 56 | 4.99 | | 60.99 | 69 |
| 3800 | Economy acoustical panels, 40 N.R.C., 4'-0" long, 5'-0" high | | 132 | .121 | | 51.50 | 5.20 | | 56.70 | 64.50 |
| 3850 | 6'-0" high | | 112 | .143 | | 54 | 6.15 | | 60.15 | 68.50 |
| 3900 | 5'-0" long, 6'-0" high | | 125 | .128 | | 46 | 5.50 | | 51.50 | 59 |
| 3950 | 6'-0" long, 5'-0" high | | 162 | .099 | | 40.50 | 4.25 | | 44.75 | 51 |
| 4000 | Metal chalkboard, 6'-6" high, chalkboard, 1 side | | 125 | .128 | | 106 | 5.50 | | 111.50 | 124 |
| 4100 | Metal chalkboard, 2 sides | | 120 | .133 | | 121 | 5.75 | | 126.75 | 142 |
| 4300 | Tackboard, both sides | | 123 | .130 | | 95.50 | 5.60 | | 101.10 | 114 |

## 10 22 33 – Accordion Folding Partitions

### 10 22 33.10 Partitions, Accordian Folding

| | | Crew | Daily Output | Labor-Hours | Unit | Material | 2011 Bare Costs Labor | Equipment | Total | Total Incl O&P |
|---|---|---|---|---|---|---|---|---|---|---|
| 0010 | **PARTITIONS, ACCORDIAN FOLDING** | | | | | | | | | |
| 0100 | Vinyl covered, over 150 S.F., frame not included | | | | | | | | | |
| 0300 | Residential, 1.25 lb. per S.F., 8' maximum height | 2 Carp | 300 | .053 | S.F. | 19.75 | 2.30 | | 22.05 | 25 |
| 0400 | Commercial, 1.75 lb. per S.F., 8' maximum height | | 225 | .071 | | 22.50 | 3.06 | | 25.56 | 29.50 |
| 0600 | 2 lb. per S.F., 17' maximum height | | 150 | .107 | | 23.50 | 4.59 | | 28.09 | 32.50 |
| 0700 | Industrial, 4 lb. per S.F., 20' maximum height | | 75 | .213 | | 33.50 | 9.20 | | 42.70 | 50.50 |
| 0900 | Acoustical, 3 lb. per S.F., 17' maximum height | | 100 | .160 | | 26 | 6.90 | | 32.90 | 39.50 |
| 1200 | 5 lb. per S.F., 20' maximum height | | 95 | .168 | | 36 | 7.25 | | 43.25 | 51 |
| 1300 | 5.5 lb. per S.F., 17' maximum height | | 90 | .178 | | 42.50 | 7.65 | | 50.15 | 59 |
| 1400 | Fire rated, 4.5 psf, 20' maximum height | | 160 | .100 | | 42.50 | 4.31 | | 46.81 | 53.50 |
| 1500 | Vinyl clad wood or steel, electric operation, 5.0 psf | | 160 | .100 | | 49.50 | 4.31 | | 53.81 | 60.50 |
| 1900 | Wood, non-acoustic, birch or mahogany, to 10' high | | 300 | .053 | | 26.50 | 2.30 | | 28.80 | 32.50 |

## 10 22 39 – Folding Panel Partitions

### 10 22 39.10 Partitions, Folding Panel

| | | Crew | Daily Output | Labor-Hours | Unit | Material | 2011 Bare Costs Labor | Equipment | Total | Total Incl O&P |
|---|---|---|---|---|---|---|---|---|---|---|
| 0010 | **PARTITIONS, FOLDING PANEL,** acoustic, wood | | | | | | | | | |
| 0100 | Vinyl faced, to 18' high, 6 psf, minimum | 2 Carp | 60 | .267 | S.F. | 51.50 | 11.50 | | 63 | 74 |
| 0150 | Average | | 45 | .356 | | 61.50 | 15.30 | | 76.80 | 91 |
| 0200 | Maximum | | 30 | .533 | | 79.50 | 23 | | 102.50 | 123 |
| 0400 | Formica or hardwood finish, minimum | | 60 | .267 | | 53 | 11.50 | | 64.50 | 76 |
| 0500 | Maximum | | 30 | .533 | | 56.50 | 23 | | 79.50 | 97.50 |
| 0600 | Wood, low acoustical type, 4.5 psf, to 14' high | | 50 | .320 | | 38.50 | 13.80 | | 52.30 | 63.50 |
| 1100 | Steel, acoustical, 9 to 12 lb. per S.F., vinyl faced, minimum | | 60 | .267 | | 55 | 11.50 | | 66.50 | 78 |
| 1200 | Maximum | | 30 | .533 | | 67 | 23 | | 90 | 109 |
| 1700 | Aluminum framed, acoustical, to 12' high, 5.5 psf, minimum | | 60 | .267 | | 37 | 11.50 | | 48.50 | 58 |
| 1800 | Maximum | | 30 | .533 | | 44.50 | 23 | | 67.50 | 84 |
| 2000 | 6.5 lb. per S.F., minimum | | 60 | .267 | | 39 | 11.50 | | 50.50 | 60.50 |
| 2100 | Maximum | | 30 | .533 | | 48 | 23 | | 71 | 88 |

## 10 22 43 – Sliding Partitions

### 10 22 43.10 Partitions, Sliding

| | | Crew | Daily Output | Labor-Hours | Unit | Material | 2011 Bare Costs Labor | Equipment | Total | Total Incl O&P |
|---|---|---|---|---|---|---|---|---|---|---|
| 0010 | **PARTITIONS, SLIDING** | | | | | | | | | |
| 0020 | Acoustic air wall, 1-5/8" thick, minimum | 2 Carp | 375 | .043 | S.F. | 31.50 | 1.84 | | 33.34 | 37.50 |
| 0100 | Maximum | | 365 | .044 | | 54 | 1.89 | | 55.89 | 62.50 |

# 10 22 Partitions

## 10 22 43 – Sliding Partitions

| 10 22 43.10 Partitions, Sliding | Crew | Daily Output | Labor-Hours | Unit | Material | 2011 Bare Costs Labor | 2011 Bare Costs Equipment | Total | Total Incl O&P |
|---|---|---|---|---|---|---|---|---|---|
| 0300 | 2-1/4" thick, minimum | 2 Carp | 360 | .044 | S.F. | 35.50 | 1.91 | | 37.41 | 42 |
| 0400 | Maximum | ↓ | 330 | .048 | ↓ | 62 | 2.09 | | 64.09 | 71 |
| 0600 | For track type, add to above | | | | L.F. | 115 | | | 115 | 127 |
| 0700 | Overhead track type, acoustical, 3" thick, 11 psf, minimum | 2 Carp | 350 | .046 | S.F. | 79.50 | 1.97 | | 81.47 | 90.50 |
| 0800 | Maximum | " | 300 | .053 | " | 95.50 | 2.30 | | 97.80 | 109 |

# 10 26 Wall and Door Protection

## 10 26 13 – Corner Guards

### 10 26 13.10 Metal Corner Guards

| | | Crew | Daily Output | Labor-Hours | Unit | Material | Labor | Equipment | Total | Total Incl O&P |
|---|---|---|---|---|---|---|---|---|---|---|
| 0010 | **METAL CORNER GUARDS** | | | | | | | | | |
| 0020 | Steel angle w/anchors, 1" x 1" x 1/4", 1.5#/L.F. | 2 Carp | 160 | .100 | L.F. | 6.85 | 4.31 | | 11.16 | 14.15 |
| 0100 | 2" x 2" x 1/4" angles, 3.2#/L.F. | | 150 | .107 | | 9.60 | 4.59 | | 14.19 | 17.60 |
| 0200 | 3" x 3" x 5/16" angles, 6.1#/L.F. | | 140 | .114 | | 13.70 | 4.92 | | 18.62 | 22.50 |
| 0300 | 4" x 4" x 5/16" angles, 8.2#/L.F. | ↓ | 120 | .133 | | 14.85 | 5.75 | | 20.60 | 25 |
| 0350 | For angles drilled and anchored to masonry, add | | | | | 15% | 120% | | | |
| 0370 | Drilled and anchored to concrete, add | | | | | 20% | 170% | | | |
| 0400 | For galvanized angles, add | | | | | 35% | | | | |
| 0450 | For stainless steel angles, add | | | | ↓ | 100% | | | | |
| 0500 | Steel door track/wheel guards, 4'-0" high | E-4 | 22 | 1.455 | Ea. | 153 | 71.50 | 4.96 | 229.46 | 299 |
| 0800 | Pipe bumper for truck doors, 8' long, 6" diameter, filled | | 20 | 1.600 | | 665 | 78.50 | 5.45 | 748.95 | 875 |
| 0900 | 8" diameter | ↓ | 20 | 1.600 | ↓ | 980 | 78.50 | 5.45 | 1,063.95 | 1,225 |

### 10 26 13.20 Corner Protection

| | | Crew | Daily Output | Labor-Hours | Unit | Material | Labor | Equipment | Total | Total Incl O&P |
|---|---|---|---|---|---|---|---|---|---|---|
| 0010 | **CORNER PROTECTION** | | | | | | | | | |
| 0100 | Stainless steel, 16 ga., adhesive mount, 3-1/2" leg | 1 Sswk | 80 | .100 | L.F. | 18.60 | 4.86 | | 23.46 | 29 |
| 0200 | 12 ga. stainless, adhesive mount | " | 80 | .100 | | 21 | 4.86 | | 25.86 | 31.50 |
| 0300 | For screw mount, add | | | | | | 10% | | | |
| 0500 | Vinyl acrylic, adhesive mount, 3" leg | 1 Carp | 128 | .063 | | 7.75 | 2.69 | | 10.44 | 12.65 |
| 0550 | 1-1/2" leg | | 160 | .050 | | 4.38 | 2.15 | | 6.53 | 8.10 |
| 0600 | Screw mounted, 3" leg | | 80 | .100 | | 8.30 | 4.31 | | 12.61 | 15.75 |
| 0650 | 1-1/2" leg | | 100 | .080 | | 4.12 | 3.44 | | 7.56 | 9.85 |
| 0700 | Clear plastic, screw mounted, 2-1/2" | | 60 | .133 | | 4.55 | 5.75 | | 10.30 | 13.80 |
| 1000 | Vinyl cover, alum. retainer, surface mount, 3" x 3" | | 48 | .167 | | 9.75 | 7.20 | | 16.95 | 21.50 |
| 1050 | 2" x 2" | | 48 | .167 | | 9.30 | 7.20 | | 16.50 | 21.50 |
| 1100 | Flush mounted, 3" x 3" | | 32 | .250 | | 19.25 | 10.75 | | 30 | 37.50 |
| 1150 | 2" x 2" | ↓ | 32 | .250 | ↓ | 15.65 | 10.75 | | 26.40 | 34 |

## 10 26 16 – Bumper Guards

### 10 26 16.10 Wallguard

| | | Crew | Daily Output | Labor-Hours | Unit | Material | Labor | Equipment | Total | Total Incl O&P |
|---|---|---|---|---|---|---|---|---|---|---|
| 0010 | **WALLGUARD** | | | | | | | | | |
| 0400 | Rub rail, vinyl, adhesive mounted | 1 Carp | 185 | .043 | L.F. | 7.70 | 1.86 | | 9.56 | 11.30 |
| 0500 | Neoprene, aluminum backing, 1-1/2" x 2" | | 110 | .073 | | 8.65 | 3.13 | | 11.78 | 14.30 |
| 1000 | Trolley rail, PVC, clipped to wall, 5" high | | 185 | .043 | | 8.65 | 1.86 | | 10.51 | 12.35 |
| 1050 | 8" high | | 180 | .044 | ↓ | 10.30 | 1.91 | | 12.21 | 14.25 |
| 1200 | Bed bumper, vinyl acrylic, alum. retainer, 21" long | | 10 | .800 | Ea. | 39.50 | 34.50 | | 74 | 96.50 |
| 1300 | 53" long with aligner | | 9 | .889 | " | 100 | 38.50 | | 138.50 | 169 |
| 1400 | Bumper, vinyl cover, alum. retain., cush. mnt., 1-1/2" x 2-3/4" | | 80 | .100 | L.F. | 13.95 | 4.31 | | 18.26 | 22 |
| 1500 | 2" x 4-1/4" | | 80 | .100 | | 20.50 | 4.31 | | 24.81 | 29 |
| 1600 | Surface mounted, 1-3/4" x 3-5/8" | | 80 | .100 | | 11.70 | 4.31 | | 16.01 | 19.50 |
| 2000 | Crash rail, vinyl cover, alum. retainer, 1" x 4" | | 110 | .073 | | 10.80 | 3.13 | | 13.93 | 16.70 |
| 2100 | 1" x 8" | | 90 | .089 | | 16.90 | 3.83 | | 20.73 | 24.50 |
| 2150 | Vinyl inserts, aluminum plate, 1" x 2-1/2" | ↓ | 110 | .073 | ↓ | 14.60 | 3.13 | | 17.73 | 21 |

# 10 26 Wall and Door Protection

## 10 26 16 – Bumper Guards

| | 10 26 16.10 Wallguard | Crew | Daily Output | Labor-Hours | Unit | Material | Labor | 2011 Bare Costs Equipment | Total | Total Incl O&P |
|---|---|---|---|---|---|---|---|---|---|---|
| 2200 | 1" x 5" | 1 Carp | 90 | .089 | L.F. | 22.50 | 3.83 | | 26.33 | 31 |
| 3000 | Handrail/bumper, vinyl cover, alum. retainer | | | | | | | | | |
| 3010 | Bracket mounted, flat rail, 5-1/2" | 1 Carp | 80 | .100 | L.F. | 17.55 | 4.31 | | 21.86 | 26 |
| 3100 | 6-1/2" | | 80 | .100 | | 22 | 4.31 | | 26.31 | 30.50 |
| 3200 | Bronze bracket, 1-3/4" diam. rail | | 80 | .100 | | 15.85 | 4.31 | | 20.16 | 24 |

# 10 28 Toilet, Bath, and Laundry Accessories

## 10 28 13 – Toilet Accessories

### 10 28 13.13 Commercial Toilet Accessories

| | | Crew | Daily Output | Labor-Hours | Unit | Material | Labor | 2011 Bare Costs Equipment | Total | Total Incl O&P |
|---|---|---|---|---|---|---|---|---|---|---|
| 0010 | **COMMERCIAL TOILET ACCESSORIES** | | | | | | | | | |
| 0200 | Curtain rod, stainless steel, 5' long, 1" diameter | 1 Carp | 13 | .615 | Ea. | 34.50 | 26.50 | | 61 | 78.50 |
| 0300 | 1-1/4" diameter | | 13 | .615 | | 23.50 | 26.50 | | 50 | 66.50 |
| 0350 | Chrome, 1" diameter | | 13 | .615 | | 25 | 26.50 | | 51.50 | 68 |
| 0360 | For vinyl curtain, add | | 13 | .615 | S.F. | .76 | 26.50 | | 27.26 | 41.50 |
| 0400 | Diaper changing station, horizontal, wall mounted, plastic | | 10 | .800 | Ea. | 224 | 34.50 | | 258.50 | 299 |
| 0500 | Dispenser units, combined soap & towel dispensers, | | | | | | | | | |
| 0510 | mirror and shelf, flush mounted | 1 Carp | 10 | .800 | Ea. | 310 | 34.50 | | 344.50 | 395 |
| 0600 | Towel dispenser and waste receptacle, | | | | | | | | | |
| 0610 | 18 gallon capacity | 1 Carp | 10 | .800 | Ea. | 263 | 34.50 | | 297.50 | 340 |
| 0800 | Grab bar, straight, 1-1/4" diameter, stainless steel, 18" long | | 24 | .333 | | 29 | 14.35 | | 43.35 | 54 |
| 0900 | 24" long | | 23 | .348 | | 29.50 | 14.95 | | 44.45 | 55.50 |
| 1000 | 30" long | | 22 | .364 | | 31.50 | 15.65 | | 47.15 | 58.50 |
| 1100 | 36" long | | 20 | .400 | | 39 | 17.20 | | 56.20 | 69.50 |
| 1105 | 42" long | | 20 | .400 | | 42 | 17.20 | | 59.20 | 73 |
| 1200 | 1-1/2" diameter, 24" long | | 23 | .348 | | 29.50 | 14.95 | | 44.45 | 55.50 |
| 1300 | 36" long | | 20 | .400 | | 34 | 17.20 | | 51.20 | 64 |
| 1310 | 42" long | | 18 | .444 | | 43 | 19.15 | | 62.15 | 77 |
| 1500 | Tub bar, 1-1/4" diameter, 24" x 36" | | 14 | .571 | | 86 | 24.50 | | 110.50 | 132 |
| 1600 | Plus vertical arm | | 12 | .667 | | 85.50 | 28.50 | | 114 | 138 |
| 1900 | End tub bar, 1" diameter, 90° angle, 16" x 32" | | 12 | .667 | | 96 | 28.50 | | 124.50 | 149 |
| 2010 | Tub/shower/toilet, 2-wall, 36" x 24" | | 12 | .667 | | 91 | 28.50 | | 119.50 | 144 |
| 2300 | Hand dryer, surface mounted, electric, 115 volt, 20 amp | | 4 | 2 | | 490 | 86 | | 576 | 670 |
| 2400 | 230 volt, 10 amp | | 4 | 2 | | 655 | 86 | | 741 | 850 |
| 2600 | Hat and coat strip, stainless steel, 4 hook, 36" long | | 24 | .333 | | 71 | 14.35 | | 85.35 | 101 |
| 2700 | 6 hook, 60" long | | 20 | .400 | | 112 | 17.20 | | 129.20 | 150 |
| 3000 | Mirror, with stainless steel 3/4" square frame, 18" x 24" | | 20 | .400 | | 47.50 | 17.20 | | 64.70 | 78.50 |
| 3100 | 36" x 24" | | 15 | .533 | | 101 | 23 | | 124 | 146 |
| 3200 | 48" x 24" | | 10 | .800 | | 151 | 34.50 | | 185.50 | 219 |
| 3300 | 72" x 24" | | 6 | 1.333 | | 196 | 57.50 | | 253.50 | 305 |
| 3500 | With 5" stainless steel shelf, 18" x 24" | | 20 | .400 | | 188 | 17.20 | | 205.20 | 234 |
| 3600 | 36" x 24" | | 15 | .533 | | 195 | 23 | | 218 | 250 |
| 3700 | 48" x 24" | | 10 | .800 | | 214 | 34.50 | | 248.50 | 289 |
| 3800 | 72" x 24" | | 6 | 1.333 | | 264 | 57.50 | | 321.50 | 380 |
| 4100 | Mop holder strip, stainless steel, 5 holders, 48" long | | 20 | .400 | | 78 | 17.20 | | 95.20 | 113 |
| 4200 | Napkin/tampon dispenser, recessed | | 15 | .533 | | 525 | 23 | | 548 | 610 |
| 4250 | Napkin receptacle, recessed | | 6.50 | 1.231 | | 143 | 53 | | 196 | 239 |
| 4300 | Robe hook, single, regular | | 36 | .222 | | 11.30 | 9.55 | | 20.85 | 27 |
| 4400 | Heavy duty, concealed mounting | | 36 | .222 | | 14.75 | 9.55 | | 24.30 | 31 |
| 4600 | Soap dispenser, chrome, surface mounted, liquid | | 20 | .400 | | 42 | 17.20 | | 59.20 | 73 |
| 4700 | Powder | | 20 | .400 | | 50.50 | 17.20 | | 67.70 | 82 |
| 5000 | Recessed stainless steel, liquid | | 10 | .800 | | 123 | 34.50 | | 157.50 | 188 |

# 10 28 Toilet, Bath, and Laundry Accessories

## 10 28 13 – Toilet Accessories

### 10 28 13.13 Commercial Toilet Accessories

| | | Crew | Daily Output | Labor-Hours | Unit | Material | 2011 Bare Costs Labor | Equipment | Total | Total Incl O&P |
|---|---|---|---|---|---|---|---|---|---|---|
| 5100 | Powder | 1 Carp | 10 | .800 | Ea. | 200 | 34.50 | | 234.50 | 273 |
| 5300 | Soap tank, stainless steel, 1 gallon | | 10 | .800 | | 200 | 34.50 | | 234.50 | 273 |
| 5400 | 5 gallon | | 5 | 1.600 | | 275 | 69 | | 344 | 405 |
| 5600 | Shelf, stainless steel, 5" wide, 18 ga., 24" long | | 24 | .333 | | 53.50 | 14.35 | | 67.85 | 81 |
| 5700 | 48" long | | 16 | .500 | | 108 | 21.50 | | 129.50 | 152 |
| 5800 | 8" wide shelf, 18 ga., 24" long | | 22 | .364 | | 61.50 | 15.65 | | 77.15 | 91.50 |
| 5900 | 48" long | | 14 | .571 | | 119 | 24.50 | | 143.50 | 168 |
| 6000 | Toilet seat cover dispenser, stainless steel, recessed | | 20 | .400 | | 160 | 17.20 | | 177.20 | 203 |
| 6050 | Surface mounted | | 15 | .533 | | 30 | 23 | | 53 | 68 |
| 6100 | Toilet tissue dispenser, surface mounted, SS, single roll | | 30 | .267 | | 16.95 | 11.50 | | 28.45 | 36.50 |
| 6200 | Double roll | | 24 | .333 | | 22.50 | 14.35 | | 36.85 | 47 |
| 6240 | Plastic, twin/jumbo dbl. roll | | 24 | .333 | | 27.50 | 14.35 | | 41.85 | 52.50 |
| 6400 | Towel bar, stainless steel, 18" long | | 23 | .348 | | 39.50 | 14.95 | | 54.45 | 66.50 |
| 6500 | 30" long | | 21 | .381 | | 111 | 16.40 | | 127.40 | 148 |
| 6700 | Towel dispenser, stainless steel, surface mounted | | 16 | .500 | | 33.50 | 21.50 | | 55 | 69.50 |
| 6800 | Flush mounted, recessed | | 10 | .800 | | 233 | 34.50 | | 267.50 | 310 |
| 6900 | Plastic, touchless, battery operated | | 16 | .500 | | 76 | 21.50 | | 97.50 | 117 |
| 7000 | Towel holder, hotel type, 2 guest size | | 20 | .400 | | 43.50 | 17.20 | | 60.70 | 74.50 |
| 7200 | Towel shelf, stainless steel, 24" long, 8" wide | | 20 | .400 | | 56 | 17.20 | | 73.20 | 88 |
| 7400 | Tumbler holder, tumbler only | | 30 | .267 | | 41.50 | 11.50 | | 53 | 63 |
| 7500 | Soap, tumbler & toothbrush | | 30 | .267 | | 18.30 | 11.50 | | 29.80 | 37.50 |
| 7700 | Wall urn ash receiver, surface mount, 11" long | | 12 | .667 | | 72 | 28.50 | | 100.50 | 124 |
| 7800 | 7-1/2", long | | 18 | .444 | | 104 | 19.15 | | 123.15 | 144 |
| 8000 | Waste receptacles, stainless steel, with top, 13 gallon | | 10 | .800 | | 270 | 34.50 | | 304.50 | 350 |
| 8100 | 36 gallon | ▼ | 8 | 1 | ▼ | 325 | 43 | | 368 | 425 |

## 10 28 16 – Bath Accessories

### 10 28 16.20 Medicine Cabinets

| | | Crew | Daily Output | Labor-Hours | Unit | Material | 2011 Bare Costs Labor | Equipment | Total | Total Incl O&P |
|---|---|---|---|---|---|---|---|---|---|---|
| 0010 | **MEDICINE CABINETS** | | | | | | | | | |
| 0020 | With mirror, st. st. frame, 16" x 22", unlighted | 1 Carp | 14 | .571 | Ea. | 64.50 | 24.50 | | 89 | 109 |
| 0100 | Wood frame | | 14 | .571 | | 126 | 24.50 | | 150.50 | 177 |
| 0300 | Sliding mirror doors, 20" x 16" x 4-3/4", unlighted | | 7 | 1.143 | | 110 | 49 | | 159 | 197 |
| 0400 | 24" x 19" x 8-1/2", lighted | | 5 | 1.600 | | 159 | 69 | | 228 | 281 |
| 0600 | Triple door, 30" x 32", unlighted, plywood body | | 7 | 1.143 | | 280 | 49 | | 329 | 385 |
| 0700 | Steel body | | 7 | 1.143 | | 310 | 49 | | 359 | 415 |
| 0900 | Oak door, wood body, beveled mirror, single door | | 7 | 1.143 | | 180 | 49 | | 229 | 274 |
| 1000 | Double door | | 6 | 1.333 | | 375 | 57.50 | | 432.50 | 500 |
| 1200 | Hotel cabinets, stainless, with lower shelf, unlighted | | 10 | .800 | | 202 | 34.50 | | 236.50 | 275 |
| 1300 | Lighted | ▼ | 5 | 1.600 | ▼ | 300 | 69 | | 369 | 435 |

## 10 28 23 – Laundry Accessories

### 10 28 23.13 Built-In Ironing Boards

| | | Crew | Daily Output | Labor-Hours | Unit | Material | 2011 Bare Costs Labor | Equipment | Total | Total Incl O&P |
|---|---|---|---|---|---|---|---|---|---|---|
| 0010 | **BUILT-IN IRONING BOARDS** | | | | | | | | | |
| 0020 | Including cabinet, board & light, minimum | 1 Carp | 2 | 4 | Ea. | 315 | 172 | | 487 | 610 |
| 0100 | Maximum, see also Section 11 23 13.13 | " | 1.50 | 5.333 | " | 525 | 230 | | 755 | 930 |

# 10 31 Manufactured Fireplaces

## 10 31 13 – Manufactured Fireplace Chimneys

### 10 31 13.10 Fireplace Chimneys

| | | Crew | Daily Output | Labor-Hours | Unit | Material | 2011 Bare Costs Labor | 2011 Bare Costs Equipment | Total | Total Incl O&P |
|---|---|---|---|---|---|---|---|---|---|---|
| 0010 | **FIREPLACE CHIMNEYS** | | | | | | | | | |
| 0500 | Chimney dbl. wall, all stainless, over 8'-6", 7" diam., add to fireplace | 1 Carp | 33 | .242 | V.L.F. | 72 | 10.45 | | 82.45 | 95 |
| 0600 | 10" diameter, add to fireplace | | 32 | .250 | | 106 | 10.75 | | 116.75 | 134 |
| 0700 | 12" diameter, add to fireplace | | 31 | .258 | | 139 | 11.10 | | 150.10 | 169 |
| 0800 | 14" diameter, add to fireplace | | 30 | .267 | | 183 | 11.50 | | 194.50 | 219 |
| 1000 | Simulated brick chimney top, 4' high, 16" x 16" | | 10 | .800 | Ea. | 258 | 34.50 | | 292.50 | 335 |
| 1100 | 24" x 24" | | 7 | 1.143 | " | 480 | 49 | | 529 | 605 |

### 10 31 13.20 Chimney Accessories

| | | Crew | Daily Output | Labor-Hours | Unit | Material | 2011 Bare Costs Labor | 2011 Bare Costs Equipment | Total | Total Incl O&P |
|---|---|---|---|---|---|---|---|---|---|---|
| 0010 | **CHIMNEY ACCESSORIES** | | | | | | | | | |
| 0020 | Chimney screens, galv., 13" x 13" flue | 1 Bric | 8 | 1 | Ea. | 51 | 43 | | 94 | 121 |
| 0050 | 24" x 24" flue | | 5 | 1.600 | | 118 | 69 | | 187 | 234 |
| 0200 | Stainless steel, 13" x 13" flue | | 8 | 1 | | 315 | 43 | | 358 | 410 |
| 0250 | 20" x 20" flue | | 5 | 1.600 | | 425 | 69 | | 494 | 575 |
| 2400 | Squirrel and bird screens, galvanized, 8" x 8" flue | | 16 | .500 | | 44.50 | 21.50 | | 66 | 81.50 |
| 2450 | 13" x 13" flue | | 12 | .667 | | 48 | 29 | | 77 | 96 |

## 10 31 16 – Manufactured Fireplace Forms

### 10 31 16.10 Fireplace Forms

| | | Crew | Daily Output | Labor-Hours | Unit | Material | 2011 Bare Costs Labor | 2011 Bare Costs Equipment | Total | Total Incl O&P |
|---|---|---|---|---|---|---|---|---|---|---|
| 0010 | **FIREPLACE FORMS** | | | | | | | | | |
| 1800 | Fireplace forms, no accessories, 32" opening | 1 Bric | 3 | 2.667 | Ea. | 650 | 115 | | 765 | 890 |
| 1900 | 36" opening | | 2.50 | 3.200 | | 830 | 138 | | 968 | 1,125 |
| 2000 | 40" opening | | 2 | 4 | | 1,100 | 173 | | 1,273 | 1,450 |
| 2100 | 78" opening | | 1.50 | 5.333 | | 1,600 | 230 | | 1,830 | 2,100 |

## 10 31 23 – Prefabricated Fireplaces

### 10 31 23.10 Fireplace, Prefabricated

| | | Crew | Daily Output | Labor-Hours | Unit | Material | 2011 Bare Costs Labor | 2011 Bare Costs Equipment | Total | Total Incl O&P |
|---|---|---|---|---|---|---|---|---|---|---|
| 0010 | **FIREPLACE, PREFABRICATED**, free standing or wall hung | | | | | | | | | |
| 0100 | With hood & screen, minimum | 1 Carp | 1.30 | 6.154 | Ea. | 1,200 | 265 | | 1,465 | 1,700 |
| 0150 | Average | | 1 | 8 | | 1,675 | 345 | | 2,020 | 2,375 |
| 0200 | Maximum | | .90 | 8.889 | | 3,475 | 385 | | 3,860 | 4,400 |
| 1500 | Simulated logs, gas fired, 40,000 BTU, 2' long, minimum | | 7 | 1.143 | Set | 445 | 49 | | 494 | 565 |
| 1600 | Maximum | | 6 | 1.333 | | 1,025 | 57.50 | | 1,082.50 | 1,225 |
| 1700 | Electric, 1,500 BTU, 1'-6" long, minimum | | 7 | 1.143 | | 175 | 49 | | 224 | 269 |
| 1800 | 11,500 BTU, maximum | | 6 | 1.333 | | 395 | 57.50 | | 452.50 | 520 |
| 2000 | Fireplace, built-in, 36" hearth, radiant | | 1.30 | 6.154 | Ea. | 565 | 265 | | 830 | 1,025 |
| 2100 | Recirculating, small fan | | 1 | 8 | | 900 | 345 | | 1,245 | 1,525 |
| 2150 | Large fan | | .90 | 8.889 | | 1,775 | 385 | | 2,160 | 2,550 |
| 2200 | 42" hearth, radiant | | 1.20 | 6.667 | | 800 | 287 | | 1,087 | 1,325 |
| 2300 | Recirculating, small fan | | .90 | 8.889 | | 1,075 | 385 | | 1,460 | 1,775 |
| 2350 | Large fan | | .80 | 10 | | 1,775 | 430 | | 2,205 | 2,625 |
| 2400 | 48" hearth, radiant | | 1.10 | 7.273 | | 1,675 | 315 | | 1,990 | 2,300 |
| 2500 | Recirculating, small fan | | .80 | 10 | | 2,075 | 430 | | 2,505 | 2,925 |
| 2550 | Large fan | | .70 | 11.429 | | 3,200 | 490 | | 3,690 | 4,275 |
| 3000 | See through, including doors | | .80 | 10 | | 2,725 | 430 | | 3,155 | 3,650 |
| 3200 | Corner (2 wall) | | 1 | 8 | | 2,725 | 345 | | 3,070 | 3,525 |

# 10 32 Fireplace Specialties

## 10 32 13 – Fireplace Dampers

| 10 32 13.10 Dampers | | Crew | Daily Output | Labor-Hours | Unit | Material | Labor | Equipment | Total | Total Incl O&P |
|---|---|---|---|---|---|---|---|---|---|---|
| 0010 | **DAMPERS** | | | | | | | | | |
| 0800 | Damper, rotary control, steel, 30" opening | 1 Bric | 6 | 1.333 | Ea. | 75.50 | 57.50 | | 133 | 170 |
| 0850 | Cast iron, 30" opening | | 6 | 1.333 | | 84 | 57.50 | | 141.50 | 179 |
| 0880 | 36" opening | | 6 | 1.333 | | 91.50 | 57.50 | | 149 | 188 |
| 0900 | 48" opening | | 6 | 1.333 | | 131 | 57.50 | | 188.50 | 231 |
| 0920 | 60" opening | | 6 | 1.333 | | 292 | 57.50 | | 349.50 | 405 |
| 0950 | 72" opening | | 5 | 1.600 | | 350 | 69 | | 419 | 490 |
| 1000 | 84" opening, special order | | 5 | 1.600 | | 755 | 69 | | 824 | 935 |
| 1050 | 96" opening, special order | | 4 | 2 | | 765 | 86.50 | | 851.50 | 970 |
| 1200 | Steel plate, poker control, 60" opening | | 8 | 1 | | 267 | 43 | | 310 | 360 |
| 1250 | 84" opening, special order | | 5 | 1.600 | | 485 | 69 | | 554 | 640 |
| 1400 | "Universal" type, chain operated, 32" x 20" opening | | 8 | 1 | | 207 | 43 | | 250 | 293 |
| 1450 | 48" x 24" opening | | 5 | 1.600 | | 310 | 69 | | 379 | 445 |

## 10 32 23 – Fireplace Doors

| 10 32 23.10 Doors | | Crew | Daily Output | Labor-Hours | Unit | Material | Labor | Equipment | Total | Total Incl O&P |
|---|---|---|---|---|---|---|---|---|---|---|
| 0010 | **DOORS** | | | | | | | | | |
| 0400 | Cleanout doors and frames, cast iron, 8" x 8" | 1 Bric | 12 | .667 | Ea. | 35.50 | 29 | | 64.50 | 82 |
| 0450 | 12" x 12" | | 10 | .800 | | 43.50 | 34.50 | | 78 | 99.50 |
| 0500 | 18" x 24" | | 8 | 1 | | 124 | 43 | | 167 | 202 |
| 0550 | Cast iron frame, steel door, 24" x 30" | | 5 | 1.600 | | 269 | 69 | | 338 | 400 |
| 1600 | Dutch Oven door and frame, cast iron, 12" x 15" opening | | 13 | .615 | | 109 | 26.50 | | 135.50 | 160 |
| 1650 | Copper plated, 12" x 15" opening | | 13 | .615 | | 213 | 26.50 | | 239.50 | 274 |

# 10 35 Stoves

## 10 35 13 – Heating Stoves

| 10 35 13.10 Woodburning Stoves | | Crew | Daily Output | Labor-Hours | Unit | Material | Labor | Equipment | Total | Total Incl O&P |
|---|---|---|---|---|---|---|---|---|---|---|
| 0010 | **WOODBURNING STOVES** | | | | | | | | | |
| 0015 | Cast iron, minimum | 2 Carp | 1.30 | 12.308 | Ea. | 1,200 | 530 | | 1,730 | 2,150 |
| 0020 | Average | | 1 | 16 | | 1,600 | 690 | | 2,290 | 2,800 |
| 0030 | Maximum | | .80 | 20 | | 2,875 | 860 | | 3,735 | 4,475 |
| 0050 | For gas log lighter, add | | | | | 42 | | | 42 | 46.50 |

# 10 44 Fire Protection Specialties

## 10 44 13 – Fire Protection Cabinets

| 10 44 13.53 Fire Equipment Cabinets | | Crew | Daily Output | Labor-Hours | Unit | Material | Labor | Equipment | Total | Total Incl O&P |
|---|---|---|---|---|---|---|---|---|---|---|
| 0010 | **FIRE EQUIPMENT CABINETS**, not equipped, 20 ga. steel box | | | | | | | | | |
| 0040 | recessed, D.S. glass in door, box size given | | | | | | | | | |
| 1000 | Portable extinguisher, single, 8" x 12" x 27", alum. door & frame | Q-12 | 8 | 2 | Ea. | 124 | 95.50 | | 219.50 | 280 |
| 1100 | Steel door and frame | " | 8 | 2 | " | 93 | 95.50 | | 188.50 | 245 |
| 3000 | Hose rack assy., 1-1/2" valve & 100' hose, 24" x 40" x 5-1/2" | | | | | | | | | |
| 3100 | Aluminum door and frame | Q-12 | 6 | 2.667 | Ea. | 305 | 127 | | 432 | 525 |
| 3200 | Steel door and frame | | 6 | 2.667 | | 370 | 127 | | 497 | 600 |
| 3300 | Stainless steel door and frame | | 6 | 2.667 | | 400 | 127 | | 527 | 630 |
| 4000 | Hose rack assy., 2-1/2" x 1-1/2" valve, 100' hose, 24" x 40" x 8" | | | | | | | | | |
| 4100 | Aluminum door and frame | Q-12 | 6 | 2.667 | Ea. | 305 | 127 | | 432 | 525 |
| 4200 | Steel door and frame | | 6 | 2.667 | | 208 | 127 | | 335 | 420 |
| 4300 | Stainless steel door and frame | | 6 | 2.667 | | 410 | 127 | | 537 | 640 |
| 5000 | Hose rack assy., 2-1/2" x 1-1/2" valve, 100' hose | | | | | | | | | |
| 5010 | and extinguisher, 30" x 40" x 8" | | | | | | | | | |

## 10 44 13 – Fire Protection Cabinets

| 10 44 13.53 Fire Equipment Cabinets | Crew | Daily Output | Labor-Hours | Unit | Material | 2011 Bare Costs Labor | Equipment | Total | Total Incl O&P |
|---|---|---|---|---|---|---|---|---|---|
| 5100 | Aluminum door and frame | Q-12 | 5 | 3.200 | Ea. | 390 | 153 | | 543 | 660 |
| 5200 | Steel door and frame | | 5 | 3.200 | | 229 | 153 | | 382 | 480 |
| 5300 | Stainless steel door and frame | | 5 | 3.200 | | 465 | 153 | | 618 | 745 |
| 8000 | Valve cabinet for 2-1/2" FD angle valve, 18" x 18" x 8" | | | | | | | | | |
| 8100 | Aluminum door and frame | Q-12 | 12 | 1.333 | Ea. | 135 | 63.50 | | 198.50 | 244 |
| 8200 | Steel door and frame | | 12 | 1.333 | | 111 | 63.50 | | 174.50 | 218 |
| 8300 | Stainless steel door and frame | | 12 | 1.333 | | 182 | 63.50 | | 245.50 | 295 |

## 10 44 16 – Fire Extinguishers

### 10 44 16.13 Portable Fire Extinguishers

| | | Crew | Daily Output | Labor-Hours | Unit | Material | Labor | Equipment | Total | Total Incl O&P |
|---|---|---|---|---|---|---|---|---|---|---|
| 0010 | **PORTABLE FIRE EXTINGUISHERS** | | | | | | | | | |
| 0140 | CO₂, with hose and "H" horn, 10 lb. | | | | Ea. | 191 | | | 191 | 210 |
| 0160 | 15 lb. | | | | | 250 | | | 250 | 276 |
| 0180 | 20 lb. | | | | | 277 | | | 277 | 305 |
| 1000 | Dry chemical, pressurized | | | | | | | | | |
| 1040 | Standard type, portable, painted, 2-1/2 lb. | | | | Ea. | 31.50 | | | 31.50 | 34.50 |
| 1060 | 5 lb. | | | | | 44 | | | 44 | 48.50 |
| 1080 | 10 lb. | | | | | 69.50 | | | 69.50 | 76.50 |
| 1100 | 20 lb. | | | | | 113 | | | 113 | 124 |
| 1120 | 30 lb. | | | | | 375 | | | 375 | 415 |
| 1300 | Standard type, wheeled, 150 lb. | | | | | 1,875 | | | 1,875 | 2,075 |
| 2000 | ABC all purpose type, portable, 2-1/2 lb. | | | | | 21 | | | 21 | 23.50 |
| 2060 | 5 lb. | | | | | 26.50 | | | 26.50 | 29 |
| 2080 | 9-1/2 lb. | | | | | 44 | | | 44 | 48 |
| 2100 | 20 lb. | | | | | 79.50 | | | 79.50 | 87.50 |
| 5000 | Pressurized water, 2-1/2 gallon, stainless steel | | | | | 97.50 | | | 97.50 | 107 |
| 5060 | With anti-freeze | | | | | 106 | | | 106 | 116 |
| 9400 | Installation of extinguishers, 12 or more, on nailable surface | 1 Carp | 30 | .267 | | | 11.50 | | 11.50 | 17.60 |
| 9420 | On masonry or concrete | " | 15 | .533 | | | 23 | | 23 | 35 |

### 10 44 16.16 Wheeled Fire Extinguisher Units

| | | Crew | Daily Output | Labor-Hours | Unit | Material | Labor | Equipment | Total | Total Incl O&P |
|---|---|---|---|---|---|---|---|---|---|---|
| 0010 | **WHEELED FIRE EXTINGUISHER UNITS** | | | | | | | | | |
| 0350 | CO₂, portable, with swivel horn | | | | | | | | | |
| 0360 | Wheeled type, cart mounted, 50 lb. | | | | Ea. | 940 | | | 940 | 1,025 |
| 0400 | 100 lb. | | | | " | 2,750 | | | 2,750 | 3,025 |
| 2200 | ABC all purpose type | | | | | | | | | |
| 2300 | Wheeled, 45 lb. | | | | Ea. | 735 | | | 735 | 810 |
| 2360 | 150 lb. | | | | " | 1,850 | | | 1,850 | 2,025 |

# 10 51 Lockers

## 10 51 13 – Metal Lockers

### 10 51 13.10 Lockers

| | | Crew | Daily Output | Labor-Hours | Unit | Material | Labor | Equipment | Total | Total Incl O&P |
|---|---|---|---|---|---|---|---|---|---|---|
| 0011 | **LOCKERS** Steel, baked enamel, pre-assembled | | | | | | | | | |
| 0110 | Single tier box locker, 12" x 15" x 72" | 1 Shee | 20 | .400 | Ea. | 213 | 20.50 | | 233.50 | 266 |
| 0120 | 18" x 15" x 72" | | 20 | .400 | | 236 | 20.50 | | 256.50 | 291 |
| 0130 | 12" x 18" x 72" | | 20 | .400 | | 243 | 20.50 | | 263.50 | 298 |
| 0140 | 18" x 18" x 72" | | 20 | .400 | | 247 | 20.50 | | 267.50 | 300 |
| 0410 | Double tier, 12" x 15" x 36" | | 30 | .267 | | 227 | 13.75 | | 240.75 | 271 |
| 0420 | 18" x 15" x 36" | | 30 | .267 | | 263 | 13.75 | | 276.75 | 310 |
| 0430 | 12" x 18" x 36" | | 30 | .267 | | 259 | 13.75 | | 272.75 | 305 |
| 0440 | 18" x 18" x 36" | | 30 | .267 | | 241 | 13.75 | | 254.75 | 286 |
| 0500 | Two person, 18" x 15" x 72" | | 20 | .400 | | 264 | 20.50 | | 284.50 | 320 |

# 10 51 Lockers

## 10 51 13 – Metal Lockers

### 10 51 13.10 Lockers

| | | Crew | Daily Output | Labor-Hours | Unit | Material | 2011 Bare Costs Labor | Equipment | Total | Total Incl O&P |
|---|---|---|---|---|---|---|---|---|---|---|
| 0510 | 18" x 18" x 72" | 1 Shee | 20 | .400 | Ea. | 266 | 20.50 | | 286.50 | 325 |
| 0520 | Duplex, 15" x 15" x 72" | | 20 | .400 | | 291 | 20.50 | | 311.50 | 350 |
| 0530 | 15" x 21" x 72" | | 20 | .400 | | 315 | 20.50 | | 335.50 | 380 |
| 0600 | 5 tier box lockers, minimum | | 30 | .267 | Opng. | 43.50 | 13.75 | | 57.25 | 69 |
| 0700 | Maximum | | 24 | .333 | | 51.50 | 17.20 | | 68.70 | 82.50 |
| 0900 | 6 tier box lockers, minimum | | 36 | .222 | | 36 | 11.50 | | 47.50 | 57.50 |
| 1000 | Maximum | | 30 | .267 | | 47 | 13.75 | | 60.75 | 73 |
| 1100 | Wire meshed wardrobe, floor. mtd., open front varsity type | | 7.50 | 1.067 | Ea. | 236 | 55 | | 291 | 345 |
| 2400 | 16-person locker unit with clothing rack | | | | | | | | | |
| 2500 | 72 wide x 15" deep x 72" high | 1 Shee | 15 | .533 | Ea. | 425 | 27.50 | | 452.50 | 505 |
| 2550 | 18" deep | " | 15 | .533 | " | 595 | 27.50 | | 622.50 | 695 |
| 3000 | Wall mounted lockers, 4 person, with coat bar | | | | | | | | | |
| 3100 | 48" wide x 18" deep x 12" high | 1 Shee | 20 | .400 | Ea. | 300 | 20.50 | | 320.50 | 360 |
| 3250 | Rack w/24 wire mesh baskets | | 1.50 | 5.333 | Set | 365 | 275 | | 640 | 815 |
| 3260 | 30 baskets | | 1.25 | 6.400 | | 330 | 330 | | 660 | 865 |
| 3270 | 36 baskets | | .95 | 8.421 | | 360 | 435 | | 795 | 1,050 |
| 3280 | 42 baskets | | .80 | 10 | | 415 | 515 | | 930 | 1,225 |
| 3300 | For built-in lock with 2 keys, add | | | | Ea. | 12.90 | | | 12.90 | 14.20 |
| 3600 | For hanger rods, add | | | | | 1.86 | | | 1.86 | 2.05 |
| 3650 | For number plate kit, 100 plates #1 - #100, add | 1 Shee | 4 | 2 | | 61.50 | 103 | | 164.50 | 224 |
| 3700 | For locker base, closed front panel | | 90 | .089 | | 6.45 | 4.59 | | 11.04 | 14.05 |
| 3710 | End panel, bolted | | 36 | .222 | | 7.85 | 11.50 | | 19.35 | 26 |
| 3800 | For sloping top, 12" wide | | 24 | .333 | | 24.50 | 17.20 | | 41.70 | 53 |
| 3810 | 15" wide | | 24 | .333 | | 31.50 | 17.20 | | 48.70 | 61 |
| 3820 | 18" wide | | 24 | .333 | | 34 | 17.20 | | 51.20 | 63.50 |
| 3850 | Sloping top end panel, 12" deep | | 72 | .111 | | 10.35 | 5.75 | | 16.10 | 20 |
| 3860 | 15" deep | | 72 | .111 | | 10.50 | 5.75 | | 16.25 | 20.50 |
| 3870 | 18" deep | | 72 | .111 | | 10.95 | 5.75 | | 16.70 | 21 |
| 3900 | For finish end panels, steel, 60" high, 15" deep | | 12 | .667 | | 50 | 34.50 | | 84.50 | 107 |
| 3910 | 72" high, 12" deep | | 12 | .667 | | 50 | 34.50 | | 84.50 | 107 |
| 3920 | 18" deep | | 12 | .667 | | 58.50 | 34.50 | | 93 | 117 |
| 5000 | For 'ready to assemble' lockers, | | | | | | | | | |
| 5010 | Add to labor | | | | | | 75% | | | |
| 5020 | Deduct from material | | | | | 20% | | | | |

## 10 51 26 – Plastic Lockers

### 10 51 26.13 Recycled Plastic Lockers

| | | | Crew | Daily Output | Labor-Hours | Unit | Material | 2011 Bare Costs Labor | Equipment | Total | Total Incl O&P |
|---|---|---|---|---|---|---|---|---|---|---|---|
| 0011 | **RECYCLED PLASTIC LOCKERS**, 30% recycled | | | | | | | | | | |
| 0110 | Single tier box locker, 12" x 12" x 72" | G | 1 Shee | 8 | 1 | Ea. | 435 | 51.50 | | 486.50 | 560 |
| 0120 | 12" x 15" x 72" | G | | 8 | 1 | | 420 | 51.50 | | 471.50 | 540 |
| 0130 | 12" x 18" x 72" | G | | 8 | 1 | | 350 | 51.50 | | 401.50 | 465 |
| 0410 | Double tier, 12" x 12" x 72" | G | | 21 | .381 | | 445 | 19.70 | | 464.70 | 520 |
| 0420 | 12" x 15" x 72" | G | | 21 | .381 | | 400 | 19.70 | | 419.70 | 470 |
| 0430 | 12" x 18" x 72" | G | | 21 | .381 | | 400 | 19.70 | | 419.70 | 470 |

## 10 51 53 – Locker Room Benches

### 10 51 53.10 Benches

| | | Crew | Daily Output | Labor-Hours | Unit | Material | 2011 Bare Costs Labor | Equipment | Total | Total Incl O&P |
|---|---|---|---|---|---|---|---|---|---|---|
| 0010 | **BENCHES** | | | | | | | | | |
| 2100 | Locker bench, laminated maple, top only | 1 Shee | 100 | .080 | L.F. | 14.75 | 4.13 | | 18.88 | 22.50 |
| 2200 | Pedestals, steel pipe | | 25 | .320 | Ea. | 40 | 16.55 | | 56.55 | 69 |
| 2250 | Plastic, 9.5" top with PVC pedestals | | 80 | .100 | L.F. | 40.50 | 5.15 | | 45.65 | 52.50 |

# 10 55 Postal Specialties

## 10 55 23 – Mail Boxes

| 10 55 23.10 Commercial Mail Boxes | Crew | Daily Output | Labor-Hours | Unit | Material | 2011 Bare Costs Labor | Equipment | Total | Total Incl O&P |
|---|---|---|---|---|---|---|---|---|---|
| 0010 **COMMERCIAL MAIL BOXES** | | | | | | | | | |
| 0020 Horiz., key lock, 5"H x 6"W x 15"D, alum., rear load | 1 Carp | 34 | .235 | Ea. | 35 | 10.15 | | 45.15 | 54 |
| 0100 Front loading | | 34 | .235 | | 35 | 10.15 | | 45.15 | 54 |
| 0200 Double, 5"H x 12"W x 15"D, rear loading | | 26 | .308 | | 71.50 | 13.25 | | 84.75 | 99.50 |
| 0300 Front loading | | 26 | .308 | | 66 | 13.25 | | 79.25 | 93 |
| 0500 Quadruple, 10"H x 12"W x 15"D, rear loading | | 20 | .400 | | 91 | 17.20 | | 108.20 | 127 |
| 0600 Front loading | | 20 | .400 | | 95.50 | 17.20 | | 112.70 | 132 |
| 0800 Vertical, front load, 15"H x 5"W x 6"D, alum., per compartment | | 34 | .235 | | 30 | 10.15 | | 40.15 | 48.50 |
| 0900 Bronze, duranodic finish | | 34 | .235 | | 35 | 10.15 | | 45.15 | 54 |
| 1000 Steel, enameled | | 34 | .235 | | 30 | 10.15 | | 40.15 | 48.50 |
| 1700 Alphabetical directories, 120 names | | 10 | .800 | | 127 | 34.50 | | 161.50 | 193 |
| 1800 Letter collection box | | 6 | 1.333 | | 660 | 57.50 | | 717.50 | 820 |
| 1830 Lobby collection boxes, aluminum | 2 Shee | 5 | 3.200 | | 2,025 | 165 | | 2,190 | 2,475 |
| 1840 Bronze or stainless | " | 4.50 | 3.556 | | 2,500 | 184 | | 2,684 | 3,025 |
| 1900 Letter slot, residential | 1 Carp | 20 | .400 | | 89.50 | 17.20 | | 106.70 | 125 |
| 2000 Post office type | | 8 | 1 | | 105 | 43 | | 148 | 182 |
| 2200 Post office counter window, with grille | | 2 | 4 | | 550 | 172 | | 722 | 870 |
| 2250 Key keeper, single key, aluminum | | 26 | .308 | | 37 | 13.25 | | 50.25 | 61.50 |
| 2300 Steel, enameled | | 26 | .308 | | 107 | 13.25 | | 120.25 | 139 |

# 10 56 Storage Assemblies

## 10 56 13 – Metal Storage Shelving

| 10 56 13.10 Shelving | Crew | Daily Output | Labor-Hours | Unit | Material | 2011 Bare Costs Labor | Equipment | Total | Total Incl O&P |
|---|---|---|---|---|---|---|---|---|---|
| 0010 **SHELVING** | | | | | | | | | |
| 0020 Metal, industrial, cross-braced, 3' wide, 12" deep | 1 Sswk | 175 | .046 | SF Shlf | 6.65 | 2.22 | | 8.87 | 11.25 |
| 0100 24" deep | | 330 | .024 | | 4.56 | 1.18 | | 5.74 | 7.10 |
| 0300 4' wide, 12" deep | | 185 | .043 | | 6 | 2.10 | | 8.10 | 10.30 |
| 0400 24" deep | | 380 | .021 | | 4.44 | 1.02 | | 5.46 | 6.70 |
| 1200 Enclosed sides, cross-braced back, 3' wide, 12" deep | | 175 | .046 | | 11.85 | 2.22 | | 14.07 | 16.90 |
| 1300 24" deep | | 290 | .028 | | 9.60 | 1.34 | | 10.94 | 12.90 |
| 1500 Fully enclosed, sides and back, 3' wide, 12" deep | | 150 | .053 | | 12.95 | 2.59 | | 15.54 | 18.80 |
| 1600 24" deep | | 255 | .031 | | 8.60 | 1.52 | | 10.12 | 12.15 |
| 1800 4' wide, 12" deep | | 150 | .053 | | 9.95 | 2.59 | | 12.54 | 15.50 |
| 1900 24" deep | | 290 | .028 | | 7.20 | 1.34 | | 8.54 | 10.25 |
| 2200 Wide span, 1600 lb. capacity per shelf, 6' wide, 24" deep | | 380 | .021 | | 6.70 | 1.02 | | 7.72 | 9.20 |
| 2400 36" deep | | 440 | .018 | | 6.40 | .88 | | 7.28 | 8.60 |
| 2600 8' wide, 24" deep | | 440 | .018 | | 6.70 | .88 | | 7.58 | 8.90 |
| 2800 36" deep | | 520 | .015 | | 5.20 | .75 | | 5.95 | 7 |
| 4000 Pallet racks, steel frame 5,000 lb. capacity, 8' long, 36" deep | 2 Sswk | 450 | .036 | | 8.80 | 1.73 | | 10.53 | 12.75 |
| 4200 42" deep | | 500 | .032 | | 8 | 1.55 | | 9.55 | 11.55 |
| 4400 48" deep | | 520 | .031 | | 7.15 | 1.49 | | 8.64 | 10.50 |

| 10 56 13.20 Parts Bins | Crew | Daily Output | Labor-Hours | Unit | Material | 2011 Bare Costs Labor | Equipment | Total | Total Incl O&P |
|---|---|---|---|---|---|---|---|---|---|
| 0010 **PARTS BINS** Metal, gray baked enamel finish | | | | | | | | | |
| 0100 6'-3" high, 3' wide | | | | | | | | | |
| 0300 12 bins, 18" wide x 12" high, 12" deep | 2 Clab | 10 | 1.600 | Ea. | 300 | 55 | | 355 | 415 |
| 0400 24" deep | | 10 | 1.600 | | 415 | 55 | | 470 | 540 |
| 0600 72 bins, 6" wide x 6" high, 12" deep | | 8 | 2 | | 500 | 68.50 | | 568.50 | 655 |
| 0700 18" deep | | 8 | 2 | | 750 | 68.50 | | 818.50 | 930 |
| 1000 7'-3" high, 3' wide | | | | | | | | | |
| 1200 14 bins, 18" wide x 12" high, 12" deep | 2 Clab | 10 | 1.600 | Ea. | 310 | 55 | | 365 | 425 |
| 1300 24" deep | | 10 | 1.600 | | 405 | 55 | | 460 | 530 |

# 10 56 Storage Assemblies

## 10 56 13 – Metal Storage Shelving

| 10 56 13.20 Parts Bins | | Crew | Daily Output | Labor-Hours | Unit | Material | 2011 Bare Costs Labor | Equipment | Total | Total Incl O&P |
|---|---|---|---|---|---|---|---|---|---|---|
| 1500 | 84 bins, 6" wide x 6" high, 12" deep | 2 Clab | 8 | 2 | Ea. | 730 | 68.50 | | 798.50 | 910 |
| 1600 | 24" deep | ↓ | 8 | 2 | ↓ | 905 | 68.50 | | 973.50 | 1,100 |

# 10 57 Wardrobe and Closet Specialties

## 10 57 13 – Hat and Coat Racks

### 10 57 13.10 Coat Racks and Wardrobes

| | | Crew | Daily Output | Labor-Hours | Unit | Material | 2011 Bare Costs Labor | Equipment | Total | Total Incl O&P |
|---|---|---|---|---|---|---|---|---|---|---|
| 0010 | **COAT RACKS AND WARDROBES** | | | | | | | | | |
| 0020 | Hat & coat rack, floor model, 6 hangers | | | | | | | | | |
| 0050 | Standing, beech wood, 21" x 21" x 72", chrome | | | | Ea. | 237 | | | 237 | 260 |
| 0100 | 18 gauge tubular steel, 21" x 21" x 69", wood walnut | | | | " | 281 | | | 281 | 310 |
| 0500 | 16 gauge steel frame, 22 gauge steel shelves | | | | | | | | | |
| 0650 | Single pedestal, 30" x 18" x 63" | | | | Ea. | 253 | | | 253 | 278 |
| 0800 | Single face rack, 29" x 18-1/2" x 62" | | | | | 254 | | | 254 | 279 |
| 0900 | 51" x 18-1/2" x 70" | | | | | 385 | | | 385 | 420 |
| 0910 | Double face rack, 39" x 26" x 70" | | | | | 340 | | | 340 | 375 |
| 0920 | 63" x 26" x 70" | | | | ↓ | 410 | | | 410 | 455 |
| 0940 | For 2" ball casters, add | | | | Set | 88 | | | 88 | 97 |
| 1400 | Utility hook strips, 3/8" x 2-1/2" x 18", 6 hooks | 1 Carp | 48 | .167 | Ea. | 57 | 7.20 | | 64.20 | 74 |
| 1500 | 34" long, 12 hooks | " | 48 | .167 | " | 51 | 7.20 | | 58.20 | 67 |
| 1650 | Wall mounted racks, 16 gauge steel frame, 22 gauge steel shelves | | | | | | | | | |
| 1850 | 12" x 15" x 26", 6 hangers | 1 Carp | 32 | .250 | Ea. | 155 | 10.75 | | 165.75 | 187 |
| 2000 | 12" x 15" x 50", 12 hangers | " | 32 | .250 | " | 181 | 10.75 | | 191.75 | 216 |
| 2150 | Wardrobe cabinet, steel, baked enamel finish | | | | | | | | | |
| 2300 | 36" x 21" x 78", incl. top shelf & hanger rod | | | | Ea. | 282 | | | 282 | 310 |

## 10 57 23 – Closet and Utility Shelving

### 10 57 23.19 Wood Closet and Utility Shelving

| | | Crew | Daily Output | Labor-Hours | Unit | Material | 2011 Bare Costs Labor | Equipment | Total | Total Incl O&P |
|---|---|---|---|---|---|---|---|---|---|---|
| 0010 | **WOOD CLOSET AND UTILITY SHELVING** | | | | | | | | | |
| 0020 | Pine, clear grade, no edge band, 1" x 8" | 1 Carp | 115 | .070 | L.F. | 2.83 | 3 | | 5.83 | 7.70 |
| 0100 | 1" x 10" | | 110 | .073 | | 3.52 | 3.13 | | 6.65 | 8.65 |
| 0200 | 1" x 12" | | 105 | .076 | | 4.24 | 3.28 | | 7.52 | 9.70 |
| 0600 | Plywood, 3/4" thick with lumber edge, 12" wide | | 75 | .107 | | 1.58 | 4.59 | | 6.17 | 8.80 |
| 0700 | 24" wide | | 70 | .114 | ↓ | 2.81 | 4.92 | | 7.73 | 10.65 |
| 0900 | Bookcase, clear grade pine, shelves 12" O.C., 8" deep, per SF shelf | | 70 | .114 | S.F. | 9.20 | 4.92 | | 14.12 | 17.65 |
| 1000 | 12" deep shelves | | 65 | .123 | " | 13.80 | 5.30 | | 19.10 | 23.50 |
| 1200 | Adjustable closet rod and shelf, 12" wide, 3' long | | 20 | .400 | Ea. | 8.25 | 17.20 | | 25.45 | 35.50 |
| 1300 | 8' long | | 15 | .533 | " | 17.15 | 23 | | 40.15 | 54 |
| 1500 | Prefinished shelves with supports, stock, 8" wide | | 75 | .107 | L.F. | 3.14 | 4.59 | | 7.73 | 10.50 |
| 1600 | 10" wide | ↓ | 70 | .114 | " | 4.27 | 4.92 | | 9.19 | 12.25 |

# 10 71 Exterior Protection

## 10 71 13 – Exterior Sun Control Devices

| 10 71 13.19 Rolling Exterior Shutters | Crew | Daily Output | Labor-Hours | Unit | Material | 2011 Bare Costs Labor | Equipment | Total | Total Incl O&P |
|---|---|---|---|---|---|---|---|---|---|
| 0010 **ROLLING EXTERIOR SHUTTERS** | | | | | | | | | |
| 0020 Roll-up, manual operation, aluminum, 3' x 4', incl. frame | 2 Carp | 8 | 2 | Ea. | 750 | 86 | | 836 | 955 |
| 0030      6' x 7' | " | 8 | 2 | " | 1,550 | 86 | | 1,636 | 1,850 |

# 10 73 Protective Covers

## 10 73 13 – Awnings

### 10 73 13.10 Awnings, Fabric

| 10 73 13.10 Awnings, Fabric | Crew | Daily Output | Labor-Hours | Unit | Material | 2011 Bare Costs Labor | Equipment | Total | Total Incl O&P |
|---|---|---|---|---|---|---|---|---|---|
| 0010 **AWNINGS, FABRIC** | | | | | | | | | |
| 0020 Including acrylic canvas and frame, standard design | | | | | | | | | |
| 0100 Door and window, slope, 3' high, 4' wide | 1 Carp | 4.50 | 1.778 | Ea. | 690 | 76.50 | | 766.50 | 875 |
| 0110      6' wide | | 3.50 | 2.286 | | 890 | 98.50 | | 988.50 | 1,125 |
| 0120      8' wide | | 3 | 2.667 | | 1,100 | 115 | | 1,215 | 1,375 |
| 0200 Quarter round convex, 4' wide | | 3 | 2.667 | | 1,075 | 115 | | 1,190 | 1,350 |
| 0210      6' wide | | 2.25 | 3.556 | | 1,400 | 153 | | 1,553 | 1,750 |
| 0220      8' wide | | 1.80 | 4.444 | | 1,700 | 191 | | 1,891 | 2,175 |
| 0300 Dome, 4' wide | | 7.50 | 1.067 | | 415 | 46 | | 461 | 525 |
| 0310      6' wide | | 3.50 | 2.286 | | 935 | 98.50 | | 1,033.50 | 1,175 |
| 0320      8' wide | | 2 | 4 | | 1,650 | 172 | | 1,822 | 2,100 |
| 0350 Elongated dome, 4' wide | | 1.33 | 6.015 | | 1,550 | 259 | | 1,809 | 2,100 |
| 0360      6' wide | | 1.11 | 7.207 | | 1,850 | 310 | | 2,160 | 2,525 |
| 0370      8' wide | | 1 | 8 | | 2,175 | 345 | | 2,520 | 2,925 |
| 1000 Entry or walkway, peak, 12' long, 4' wide | 2 Carp | .90 | 17.778 | | 4,925 | 765 | | 5,690 | 6,575 |
| 1010      6' wide | | .60 | 26.667 | | 7,575 | 1,150 | | 8,725 | 10,100 |
| 1020      8' wide | | .40 | 40 | | 10,500 | 1,725 | | 12,225 | 14,200 |
| 1100 Radius with dome end, 4' wide | | 1.10 | 14.545 | | 3,725 | 625 | | 4,350 | 5,050 |
| 1110      6' wide | | .70 | 22.857 | | 6,000 | 985 | | 6,985 | 8,100 |
| 1120      8' wide | | .50 | 32 | | 8,525 | 1,375 | | 9,900 | 11,500 |
| 2000 Retractable lateral arm awning, manual | | | | | | | | | |
| 2010      To 12' wide, 8'-6" projection | 2 Carp | 1.70 | 9.412 | Ea. | 1,100 | 405 | | 1,505 | 1,850 |
| 2020      To 14' wide, 8'-6" projection | | 1.10 | 14.545 | | 1,300 | 625 | | 1,925 | 2,375 |
| 2030      To 19' wide, 8'-6" projection | | .85 | 18.824 | | 1,750 | 810 | | 2,560 | 3,175 |
| 2040      To 24' wide, 8'-6" projection | | .67 | 23.881 | | 2,225 | 1,025 | | 3,250 | 4,025 |
| 2050 Motor for above, add | 1 Carp | 2.67 | 3 | | 955 | 129 | | 1,084 | 1,250 |
| 3000 Patio/deck canopy with frame | | | | | | | | | |
| 3010      12' wide, 12' projection | 2 Carp | 2 | 8 | Ea. | 1,575 | 345 | | 1,920 | 2,250 |
| 3020      16' wide, 14' projection | " | 1.20 | 13.333 | | 2,450 | 575 | | 3,025 | 3,550 |
| 9000 For fire retardant canvas, add | | | | | 7% | | | | |
| 9010 For lettering or graphics, add | | | | | 35% | | | | |
| 9020 For painted or coated acrylic canvas, deduct | | | | | 8% | | | | |
| 9030 For translucent or opaque vinyl canvas, add | | | | | 10% | | | | |
| 9040 For 6 or more units, deduct | | | | | 20% | 15% | | | |

## 10 73 16 – Canopies

### 10 73 16.20 Metal Canopies

| 10 73 16.20 Metal Canopies | Crew | Daily Output | Labor-Hours | Unit | Material | 2011 Bare Costs Labor | Equipment | Total | Total Incl O&P |
|---|---|---|---|---|---|---|---|---|---|
| 0010 **METAL CANOPIES** | | | | | | | | | |
| 0020 Wall hung, .032", aluminum, prefinished, 8' x 10' | K-2 | 1.30 | 18.462 | Ea. | 2,150 | 815 | 167 | 3,132 | 3,925 |
| 0300      8' x 20' | | 1.10 | 21.818 | | 4,275 | 965 | 197 | 5,437 | 6,575 |
| 0500      10' x 10' | | 1.30 | 18.462 | | 2,400 | 815 | 167 | 3,382 | 4,200 |
| 0700      10' x 20' | | 1.10 | 21.818 | | 4,450 | 965 | 197 | 5,612 | 6,775 |
| 1000      12' x 20' | | 1 | 24 | | 5,475 | 1,050 | 217 | 6,742 | 8,075 |
| 1360      12' x 30' | | .80 | 30 | | 8,050 | 1,325 | 271 | 9,646 | 11,400 |

# 10 73 Protective Covers

## 10 73 16 – Canopies

### 10 73 16.20 Metal Canopies

| 10 73 16.20 Metal Canopies | | Crew | Daily Output | Labor-Hours | Unit | Material | 2011 Bare Costs Labor | Equipment | Total | Total Incl O&P |
|---|---|---|---|---|---|---|---|---|---|---|
| 1700 | 12' x 40' | K-2 | .60 | 40 | Ea. | 9,800 | 1,775 | 360 | 11,935 | 14,200 |
| 1900 | For free standing units, add | | | | | 20% | 10% | | | |
| 2300 | Aluminum entrance canopies, flat soffit, .032" | | | | | | | | | |
| 2500 | 3'-6" x 4'-0", clear anodized | 2 Carp | 4 | 4 | Ea. | 900 | 172 | | 1,072 | 1,250 |
| 2700 | Bronze anodized | | 4 | 4 | | 1,575 | 172 | | 1,747 | 2,025 |
| 3000 | Polyurethane painted | | 4 | 4 | | 1,275 | 172 | | 1,447 | 1,675 |
| 3300 | 4'-6" x 10'-0", clear anodized | | 2 | 8 | | 2,475 | 345 | | 2,820 | 3,250 |
| 3500 | Bronze anodized | | 2 | 8 | | 3,175 | 345 | | 3,520 | 4,000 |
| 3700 | Polyurethane painted | | 2 | 8 | | 2,650 | 345 | | 2,995 | 3,450 |
| 4000 | Wall downspout, 10 L.F., clear anodized | 1 Carp | 7 | 1.143 | | 154 | 49 | | 203 | 246 |
| 4300 | Bronze anodized | | 7 | 1.143 | | 269 | 49 | | 318 | 370 |
| 4500 | Polyurethane painted | | 7 | 1.143 | | 231 | 49 | | 280 | 330 |
| 7000 | Carport, baked vinyl finish, .032", 20' x 10', no foundations, min. | K-2 | 4 | 6 | Car | 3,850 | 265 | 54 | 4,169 | 4,725 |
| 7250 | Maximum | | 2 | 12 | " | 7,000 | 530 | 108 | 7,638 | 8,725 |
| 7500 | Walkway cover, to 12' wide, stl., vinyl finish, .032",no fndtns., min. | | 250 | .096 | S.F. | 22.50 | 4.24 | .87 | 27.61 | 33 |
| 7750 | Maximum | | 200 | .120 | " | 24.50 | 5.30 | 1.08 | 30.88 | 37 |

# 10 74 Manufactured Exterior Specialties

## 10 74 23 – Cupolas

### 10 74 23.10 Wood Cupolas

| 10 74 23.10 Wood Cupolas | | Crew | Daily Output | Labor-Hours | Unit | Material | 2011 Bare Costs Labor | Equipment | Total | Total Incl O&P |
|---|---|---|---|---|---|---|---|---|---|---|
| 0010 | **WOOD CUPOLAS** | | | | | | | | | |
| 0020 | Stock units, pine, painted, 18" sq., 28" high, alum. roof | 1 Carp | 4.10 | 1.951 | Ea. | 162 | 84 | | 246 | 305 |
| 0100 | Copper roof | | 3.80 | 2.105 | | 208 | 90.50 | | 298.50 | 370 |
| 0300 | 23" square, 33" high, aluminum roof | | 3.70 | 2.162 | | 385 | 93 | | 478 | 570 |
| 0400 | Copper roof | | 3.30 | 2.424 | | 465 | 104 | | 569 | 670 |
| 0600 | 30" square, 37" high, aluminum roof | | 3.70 | 2.162 | | 495 | 93 | | 588 | 690 |
| 0700 | Copper roof | | 3.30 | 2.424 | | 590 | 104 | | 694 | 810 |
| 0900 | Hexagonal, 31" wide, 46" high, copper roof | | 4 | 2 | | 730 | 86 | | 816 | 935 |
| 1000 | 36" wide, 50" high, copper roof | | 3.50 | 2.286 | | 1,475 | 98.50 | | 1,573.50 | 1,775 |
| 1200 | For deluxe stock units, add to above | | | | | 25% | | | | |
| 1400 | For custom built units, add to above | | | | | 50% | 50% | | | |

## 10 74 29 – Steeples

### 10 74 29.10 Prefabricated Steeples

| 10 74 29.10 Prefabricated Steeples | | Crew | Daily Output | Labor-Hours | Unit | Material | 2011 Bare Costs Labor | Equipment | Total | Total Incl O&P |
|---|---|---|---|---|---|---|---|---|---|---|
| 0010 | **PREFABRICATED STEEPLES** | | | | | | | | | |
| 4000 | Steeples, translucent fiberglass, 30" square, 15' high | F-3 | 2 | 20 | Ea. | 5,825 | 875 | 330 | 7,030 | 8,100 |
| 4150 | 25' high | | 1.80 | 22.222 | | 6,800 | 970 | 365 | 8,135 | 9,350 |
| 4350 | Opaque fiberglass, 24" square, 14' high | | 2 | 20 | | 4,850 | 875 | 330 | 6,055 | 7,025 |
| 4500 | 28' high | | 1.80 | 22.222 | | 5,000 | 970 | 365 | 6,335 | 7,375 |
| 4600 | Aluminum, baked finish, 16" square, 14' high | | | | | 4,400 | | | 4,400 | 4,850 |
| 4620 | 20' high, 3'-6" base | | | | | 7,400 | | | 7,400 | 8,150 |
| 4640 | 35' high, 8' base | | | | | 28,600 | | | 28,600 | 31,500 |
| 4660 | 60' high, 14' base | | | | | 61,500 | | | 61,500 | 67,500 |
| 4680 | 152' high, custom | | | | | 512,500 | | | 512,500 | 564,000 |
| 4700 | Porcelain enamel steeples, custom, 40' high | F-3 | .50 | 80 | | 14,100 | 3,500 | 1,300 | 18,900 | 22,400 |
| 4800 | 60' high | " | .30 | 133 | | 24,500 | 5,825 | 2,175 | 32,500 | 38,300 |

## 10 74 46 – Window Wells

### 10 74 46.10 Area Window Wells

| 10 74 46.10 Area Window Wells | | Crew | Daily Output | Labor-Hours | Unit | Material | 2011 Bare Costs Labor | Equipment | Total | Total Incl O&P |
|---|---|---|---|---|---|---|---|---|---|---|
| 0010 | **AREA WINDOW WELLS**, Galvanized steel | | | | | | | | | |
| 0020 | 20 ga., 3'-2" wide, 1' deep | 1 Sswk | 29 | .276 | Ea. | 14 | 13.40 | | 27.40 | 39 |
| 0100 | 2' deep | | 23 | .348 | | 25 | 16.90 | | 41.90 | 57.50 |

# 10 74 Manufactured Exterior Specialties

## 10 74 46 – Window Wells

| 10 74 46.10 Area Window Wells | Crew | Daily Output | Labor-Hours | Unit | Material | 2011 Bare Costs Labor | Equipment | Total | Total Incl O&P |
|---|---|---|---|---|---|---|---|---|---|
| 0300 | 16 ga., 3'-2" wide, 1' deep | 1 Sswk | 29 | .276 | Ea. | 20 | 13.40 | | 33.40 | 45.50 |
| 0400 | 3' deep | | 23 | .348 | | 39 | 16.90 | | 55.90 | 73 |
| 0600 | Welded grating for above, 15 lbs., painted | | 45 | .178 | | 85 | 8.65 | | 93.65 | 109 |
| 0700 | Galvanized | | 45 | .178 | | 114 | 8.65 | | 122.65 | 140 |
| 0900 | Translucent plastic cap for above | | 60 | .133 | | 16 | 6.45 | | 22.45 | 29 |

# 10 75 Flagpoles

## 10 75 16 – Ground-Set Flagpoles

| 10 75 16.10 Flagpoles | Crew | Daily Output | Labor-Hours | Unit | Material | 2011 Bare Costs Labor | Equipment | Total | Total Incl O&P |
|---|---|---|---|---|---|---|---|---|---|
| 0010 | **FLAGPOLES**, Ground set | | | | | | | | | |
| 0050 | Not including base or foundation | | | | | | | | | |
| 0100 | Aluminum, tapered, ground set 20' high | K-1 | 2 | 8 | Ea. | 880 | 305 | 108 | 1,293 | 1,550 |
| 0200 | 25' high | | 1.70 | 9.412 | | 1,275 | 360 | 128 | 1,763 | 2,100 |
| 0300 | 30' high | | 1.50 | 10.667 | | 1,225 | 410 | 145 | 1,780 | 2,125 |
| 0400 | 35' high | | 1.40 | 11.429 | | 1,650 | 435 | 155 | 2,240 | 2,650 |
| 0500 | 40' high | | 1.20 | 13.333 | | 2,350 | 510 | 181 | 3,041 | 3,550 |
| 0600 | 50' high | | 1 | 16 | | 2,850 | 610 | 217 | 3,677 | 4,325 |
| 0700 | 60' high | | .90 | 17.778 | | 5,225 | 680 | 241 | 6,146 | 7,050 |
| 0800 | 70' high | | .80 | 20 | | 6,600 | 765 | 271 | 7,636 | 8,725 |
| 1100 | Counterbalanced, internal halyard, 20' high | | 1.80 | 8.889 | | 2,350 | 340 | 121 | 2,811 | 3,225 |
| 1200 | 30' high | | 1.50 | 10.667 | | 2,600 | 410 | 145 | 3,155 | 3,625 |
| 1300 | 40' high | | 1.30 | 12.308 | | 5,675 | 470 | 167 | 6,312 | 7,150 |
| 1400 | 50' high | | 1 | 16 | | 7,900 | 610 | 217 | 8,727 | 9,850 |
| 2820 | Aluminum, electronically operated, 30' high | | 1.40 | 11.429 | | 4,300 | 435 | 155 | 4,890 | 5,550 |
| 2840 | 35' high | | 1.30 | 12.308 | | 4,625 | 470 | 167 | 5,262 | 6,000 |
| 2860 | 39' high | | 1.10 | 14.545 | | 5,050 | 555 | 197 | 5,802 | 6,625 |
| 2880 | 45' high | | 1 | 16 | | 6,450 | 610 | 217 | 7,277 | 8,275 |
| 2900 | 50' high | | .90 | 17.778 | | 6,875 | 680 | 241 | 7,796 | 8,875 |
| 3000 | Fiberglass, tapered, ground set, 23' high | | 2 | 8 | | 1,100 | 305 | 108 | 1,513 | 1,775 |
| 3100 | 29'-7" high | | 1.50 | 10.667 | | 1,450 | 410 | 145 | 2,005 | 2,350 |
| 3200 | 36'-1" high | | 1.40 | 11.429 | | 1,900 | 435 | 155 | 2,490 | 2,925 |
| 3300 | 39'-5" high | | 1.20 | 13.333 | | 2,225 | 510 | 181 | 2,916 | 3,425 |
| 3400 | 49'-2" high | | 1 | 16 | | 3,975 | 610 | 217 | 4,802 | 5,550 |
| 3500 | 59' high | | .90 | 17.778 | | 4,800 | 680 | 241 | 5,721 | 6,600 |
| 4300 | Steel, direct imbedded installation | | | | | | | | | |
| 4400 | Internal halyard, 20' high | K-1 | 2.50 | 6.400 | Ea. | 1,550 | 245 | 87 | 1,882 | 2,200 |
| 4500 | 25' high | | 2.50 | 6.400 | | 1,750 | 245 | 87 | 2,082 | 2,375 |
| 4600 | 30' high | | 2.30 | 6.957 | | 2,050 | 266 | 94.50 | 2,410.50 | 2,750 |
| 4700 | 40' high | | 2.10 | 7.619 | | 3,050 | 291 | 103 | 3,444 | 3,900 |
| 4800 | 50' high | | 1.90 | 8.421 | | 4,500 | 320 | 114 | 4,934 | 5,575 |
| 5000 | 60' high | | 1.80 | 8.889 | | 6,800 | 340 | 121 | 7,261 | 8,125 |
| 5100 | 70' high | | 1.60 | 10 | | 8,300 | 380 | 136 | 8,816 | 9,850 |
| 5200 | 80' high | | 1.40 | 11.429 | | 10,900 | 435 | 155 | 11,490 | 12,800 |
| 5300 | 90' high | | 1.20 | 13.333 | | 15,000 | 510 | 181 | 15,691 | 17,500 |
| 5500 | 100' high | | 1 | 16 | | 15,500 | 610 | 217 | 16,327 | 18,300 |
| 6400 | Wood poles, tapered, clear vertical grain fir with tilting | | | | | | | | | |
| 6410 | base, not incl. foundation, 4" butt, 25' high | K-1 | 1.90 | 8.421 | Ea. | 1,375 | 320 | 114 | 1,809 | 2,150 |
| 6800 | 6" butt, 30' high | " | 1.30 | 12.308 | " | 1,925 | 470 | 167 | 2,562 | 3,025 |
| 7300 | Foundations for flagpoles, including | | | | | | | | | |
| 7400 | excavation and concrete, to 35' high poles | C-1 | 10 | 3.200 | Ea. | 540 | 131 | | 671 | 795 |
| 7600 | 40' to 50' high | | 3.50 | 9.143 | | 985 | 375 | | 1,360 | 1,650 |

# 10 75 Flagpoles

## 10 75 16 – Ground-Set Flagpoles

### 10 75 16.10 Flagpoles

| | Crew | Daily Output | Labor-Hours | Unit | Material | 2011 Bare Costs Labor | Equipment | Total | Total Incl O&P |
|---|---|---|---|---|---|---|---|---|---|
| 7700 Over 60' high | C-1 | 2 | 16 | Ea. | 1,100 | 655 | | 1,755 | 2,200 |

## 10 75 23 – Wall-Mounted Flagpoles

### 10 75 23.10 Flagpoles

| | | Crew | Daily Output | Labor-Hours | Unit | Material | Labor | Equipment | Total | Total Incl O&P |
|---|---|---|---|---|---|---|---|---|---|---|
| 0010 | **FLAGPOLES**, Structure mounted | | | | | | | | | |
| 0100 | Fiberglass, vertical wall set, 19'-8" long | K-1 | 1.50 | 10.667 | Ea. | 1,300 | 410 | 145 | 1,855 | 2,200 |
| 0200 | 23' long | | 1.40 | 11.429 | | 1,375 | 435 | 155 | 1,965 | 2,325 |
| 0300 | 26'-3" long | | 1.30 | 12.308 | | 1,775 | 470 | 167 | 2,412 | 2,850 |
| 0800 | 19'-8" long outrigger | | 1.30 | 12.308 | | 1,325 | 470 | 167 | 1,962 | 2,350 |
| 1300 | Aluminum, vertical wall set, tapered, with base, 20' high | | 1.20 | 13.333 | | 1,050 | 510 | 181 | 1,741 | 2,125 |
| 1400 | 29'-6" high | | 1 | 16 | | 2,250 | 610 | 217 | 3,077 | 3,650 |
| 2400 | Outrigger poles with base, 12' long | | 1.30 | 12.308 | | 930 | 470 | 167 | 1,567 | 1,925 |
| 2500 | 14' long | | 1 | 16 | | 1,425 | 610 | 217 | 2,252 | 2,750 |

# 10 81 Pest Control Devices

## 10 81 13 – Bird Control Devices

### 10 81 13.10 Bird Control Netting

| | | Crew | Daily Output | Labor-Hours | Unit | Material | Labor | Equipment | Total | Total Incl O&P |
|---|---|---|---|---|---|---|---|---|---|---|
| 0010 | **BIRD CONTROL NETTING** | | | | | | | | | |
| 0020 | 1/8" square mesh | 4 Clab | 4000 | .008 | S.F. | .06 | .27 | | .33 | .49 |
| 0100 | 1/4" square mesh | | 4000 | .008 | | .07 | .27 | | .34 | .50 |
| 0120 | 1/2" square mesh | | 4000 | .008 | | .08 | .27 | | .35 | .51 |
| 0140 | 5/8" x 3/4" mesh | | 4000 | .008 | | .08 | .27 | | .35 | .51 |
| 0160 | 1-1/4" x 1-1/2" mesh | | 4000 | .008 | | .10 | .27 | | .37 | .53 |
| 0200 | 4" square mesh | | 4000 | .008 | | .11 | .27 | | .38 | .54 |
| 1000 | Poly clips | | | | Ea. | .02 | | | .02 | .02 |

# 10 88 Scales

## 10 88 05 – Commercial Scales

### 10 88 05.10 Scales

| | | Crew | Daily Output | Labor-Hours | Unit | Material | Labor | Equipment | Total | Total Incl O&P |
|---|---|---|---|---|---|---|---|---|---|---|
| 0010 | **SCALES** | | | | | | | | | |
| 0700 | Truck scales, incl. steel weigh bridge, | | | | | | | | | |
| 0800 | not including foundation, pits | | | | | | | | | |
| 1550 | Digital, electronic, 100 ton capacity, steel deck 12' x 10' platform | 3 Carp | .20 | 120 | Ea. | 12,500 | 5,175 | | 17,675 | 21,600 |
| 1600 | 40' x 10' platform | | .14 | 171 | | 25,300 | 7,375 | | 32,675 | 39,200 |
| 1640 | 60' x 10' platform | | .13 | 184 | | 31,400 | 7,950 | | 39,350 | 46,700 |
| 1680 | 70' x 10' platform | | .12 | 200 | | 35,000 | 8,600 | | 43,600 | 51,500 |
| 2000 | For standard automatic printing device, add | | | | | 2,525 | | | 2,525 | 2,775 |
| 2100 | For remote reading electronic system, add | | | | | 2,275 | | | 2,275 | 2,500 |
| 2300 | Concrete foundation pits for above, 8' x 6', 5 C.Y. required | C-1 | .50 | 64 | | 955 | 2,625 | | 3,580 | 5,075 |
| 2400 | 14' x 6' platform, 10 C.Y. required | | .35 | 91.429 | | 1,400 | 3,750 | | 5,150 | 7,250 |
| 2600 | 50' x 10' platform, 30 C.Y. required | | .25 | 128 | | 1,875 | 5,225 | | 7,100 | 10,100 |
| 2700 | 70' x 10' platform, 40 C.Y. required | | .15 | 213 | | 4,125 | 8,725 | | 12,850 | 17,900 |
| 2750 | Crane scales, dial, 1 ton capacity | | | | | 1,025 | | | 1,025 | 1,125 |
| 2780 | 5 ton capacity | | | | | 1,400 | | | 1,400 | 1,525 |
| 2800 | Digital, 1 ton capacity | | | | | 1,775 | | | 1,775 | 1,950 |
| 2850 | 10 ton capacity | | | | | 4,475 | | | 4,475 | 4,900 |
| 2900 | Low profile electronic warehouse scale, | | | | | | | | | |
| 3000 | not incl. printer, 4' x 4' platform, 10,000 lb. capacity | 2 Carp | .30 | 53.333 | Ea. | 1,425 | 2,300 | | 3,725 | 5,075 |
| 3300 | 5' x 7' platform, 10,000 lb. capacity | | .25 | 64 | | 4,125 | 2,750 | | 6,875 | 8,750 |
| 3400 | 20,000 lb. capacity | | .20 | 80 | | 5,300 | 3,450 | | 8,750 | 11,100 |

# 10 88 Scales

## 10 88 05 – Commercial Scales

| 10 88 05.10 Scales | | Crew | Daily Output | Labor-Hours | Unit | Material | 2011 Bare Costs Labor | Equipment | Total | Total Incl O&P |
|---|---|---|---|---|---|---|---|---|---|---|
| 3500 | For printers, incl. time, date & numbering, add | | | | Ea. | 1,325 | | | 1,325 | 1,450 |
| 3800 | Portable, beam type, capacity 1000#, platform 18" x 24" | | | | | 745 | | | 745 | 820 |
| 3900 | Dial type, capacity 2000#, platform 24" x 24" | | | | | 1,325 | | | 1,325 | 1,450 |
| 4000 | Digital type, capacity 1000#, platform 24" x 30" | | | | | 2,125 | | | 2,125 | 2,325 |
| 4100 | Portable contractor truck scales, 50 ton cap., 40' x 10' platform | | | | | 32,000 | | | 32,000 | 35,200 |
| 4200 | 60' x 10' platform | | | | | 29,500 | | | 29,500 | 32,500 |

# Estimating Tips

## General

- The items in this division are usually priced per square foot or each. Many of these items are purchased by the owner for installation by the contractor. Check the specifications for responsibilities and include time for receiving, storage, installation, and mechanical and electrical hookups in the appropriate divisions.

- Many items in Division 11 require some type of support system that is not usually furnished with the item. Examples of these systems include blocking for the attachment of casework and support angles for ceiling-hung projection screens. The required blocking or supports must be added to the estimate in the appropriate division.

- Some items in Division 11 may require assembly or electrical hookups. Verify the amount of assembly required or the need for a hard electrical connection and add the appropriate costs.

# Reference Numbers

Reference numbers are shown in shaded boxes at the beginning of some major classifications. These numbers refer to related items in the Reference Section. The reference information may be an estimating procedure, an alternate pricing method, or technical information.

*Note:* Not all subdivisions listed here necessarily appear in this publication.

| 11 05 05.10 Selective Demolition | Crew | Daily Output | Labor-Hours | Unit | Material | 2011 Bare Costs Labor | Equipment | Total | Total Incl O&P |
|---|---|---|---|---|---|---|---|---|---|
| **0010** | **SELECTIVE DEMOLITION** | | | | | | | | | |
| 0130 | Central vacuum, motor unit, residential or commercial | 1 Clab | 2 | 4 | Ea. | | 137 | | 137 | 211 |
| 0210 | Vault door and frame | 2 Skwk | 2 | 8 | | | 355 | | 355 | 545 |
| 0215 | Day gate, for vault | " | 3 | 5.333 | | | 237 | | 237 | 365 |
| 0380 | Bank equipment, teller window, bullet resistant | 1 Clab | 1.20 | 6.667 | | | 229 | | 229 | 350 |
| 0381 | Counter | 2 Clab | 1.50 | 10.667 | Station | | 365 | | 365 | 560 |
| 0382 | Drive-up window, including drawer and glass | | 1.50 | 10.667 | " | | 365 | | 365 | 560 |
| 0383 | Thru-wall boxes and chests, selective demolition | | 2.50 | 6.400 | Ea. | | 220 | | 220 | 335 |
| 0384 | Bullet resistant partitions | | 20 | .800 | L.F. | | 27.50 | | 27.50 | 42 |
| 0385 | Pneumatic tube system, 2 lane drive-up | L-3 | .45 | 35.556 | Ea. | | 1,675 | | 1,675 | 2,525 |
| 0386 | Safety deposit box | 1 Clab | 50 | .160 | Opng. | | 5.50 | | 5.50 | 8.45 |
| 0387 | Surveillance system, video, complete | 2 Elec | 2 | 8 | Ea. | | 400 | | 400 | 595 |
| 0410 | Church equipment, misc moveable fixtures | 2 Clab | 1 | 16 | | | 550 | | 550 | 845 |
| 0412 | Steeple, to 28' high | F-3 | 3 | 13.333 | | | 585 | 219 | 804 | 1,125 |
| 0414 | 40' to 60' high | " | .80 | 50 | | | 2,175 | 820 | 2,995 | 4,225 |
| 0510 | Library equipment, bookshelves, wood, to 90" high | 1 Clab | 20 | .400 | L.F. | | 13.75 | | 13.75 | 21 |
| 0515 | Carrels, hardwood, 36" x 24" | " | 9 | .889 | Ea. | | 30.50 | | 30.50 | 47 |
| 0630 | Stage equipment, light control panel | 1 Elec | 1 | 8 | " | | 400 | | 400 | 595 |
| 0632 | Border lights | | 40 | .200 | L.F. | | 10.05 | | 10.05 | 14.90 |
| 0634 | Spotlights | | 8 | 1 | Ea. | | 50.50 | | 50.50 | 74.50 |
| 0636 | Telescoping platforms and risers | 2 Clab | 175 | .091 | SF Stg. | | 3.14 | | 3.14 | 4.82 |
| 1020 | Barber equipment, hydraulic chair | 1 Clab | 40 | .200 | Ea. | | 6.85 | | 6.85 | 10.55 |
| 1030 | Checkout counter, supermarket or warehouse conveyor | 2 Clab | 18 | .889 | | | 30.50 | | 30.50 | 47 |
| 1040 | Food cases, refrigerated or frozen | Q-5 | 6 | 2.667 | | | 130 | | 130 | 194 |
| 1190 | Laundry equipment, commercial | L-6 | 3 | 4 | | | 210 | | 210 | 315 |
| 1360 | Movie equipment, lamphouse, to 4000 watt, incl rectifier | 1 Elec | 4 | 2 | | | 101 | | 101 | 149 |
| 1365 | Sound system, incl amplifier | " | 1.25 | 6.400 | | | 320 | | 320 | 475 |
| 1410 | Air compressor, to 5 H.P. | 2 Clab | 2.50 | 6.400 | | | 220 | | 220 | 335 |
| 1412 | Lubrication equipment, automotive, 3 reel type, incl pump, excl piping | L-4 | 1 | 24 | Set | | 970 | | 970 | 1,500 |
| 1414 | Booth, spray paint, complete, to 26' long | " | .80 | 30 | " | | 1,225 | | 1,225 | 1,850 |
| 1560 | Parking equipment, cashier booth | B-22 | 2 | 15 | Ea. | | 600 | 97 | 697 | 1,025 |
| 1600 | Loading dock equipment, dock bumpers, rubber | 1 Clab | 50 | .160 | " | | 5.50 | | 5.50 | 8.45 |
| 1610 | Door seal for door perimeter | " | 50 | .160 | L.F. | | 5.50 | | 5.50 | 8.45 |
| 1620 | Platform lifter, fixed, 6' x 8', 5000 lb capacity | E-16 | 1.50 | 10.667 | Ea. | | 530 | 72.50 | 602.50 | 1,025 |
| 1630 | Dock leveller | " | 2 | 8 | | | 395 | 54.50 | 449.50 | 760 |
| 1640 | Lights, single or double arm | 1 Elec | 8 | 1 | | | 50.50 | | 50.50 | 74.50 |
| 1650 | Shelter, fabric, truck or train | 1 Clab | 1.50 | 5.333 | | | 183 | | 183 | 281 |
| 1790 | Waste handling equipment, commercial compactor | L-4 | 2 | 12 | | | 485 | | 485 | 745 |
| 1792 | Commercial or municipal incinerator, gas | " | 2 | 12 | | | 485 | | 485 | 745 |
| 1795 | Crematory, excluding building | Q-3 | .25 | 128 | | | 6,525 | | 6,525 | 9,750 |
| 1910 | Detection equipment, cell bar front | E-4 | 4 | 8 | | | 390 | 27.50 | 417.50 | 725 |
| 1912 | Cell door and frame | | 8 | 4 | | | 196 | 13.65 | 209.65 | 360 |
| 1914 | Prefab cell, 4' to 5' wide, 7' to 8' high, 7' deep | | 8 | 4 | | | 196 | 13.65 | 209.65 | 360 |
| 1916 | Cot, bolted, single | | 40 | .800 | | | 39 | 2.73 | 41.73 | 72.50 |
| 1918 | Visitor cubicle | | 4 | 8 | | | 390 | 27.50 | 417.50 | 725 |
| 2850 | Hydraulic gates, canal, flap, knife, slide or sluice, to 18" diameter | L-5A | 8 | 4 | | | 194 | 70 | 264 | 405 |
| 2852 | 19" to 36" diameter | | 6 | 5.333 | | | 259 | 93.50 | 352.50 | 545 |
| 2854 | 37" to 48" diameter | | 2 | 16 | | | 775 | 280 | 1,055 | 1,625 |
| 2856 | 49" to 60" diameter | | 1 | 32 | | | 1,550 | 560 | 2,110 | 3,275 |
| 2858 | Over 60" diameter | | .30 | 106 | | | 5,175 | 1,875 | 7,050 | 10,900 |
| 3100 | Sewage pumping system, prefabricated, to 1000 GPM | C-17D | .20 | 420 | | | 18,800 | 4,225 | 23,025 | 33,600 |
| 3110 | Sewage treatment, holding tank for recirc chemical water closet | 1 Plum | 8 | 1 | | | 53.50 | | 53.50 | 80 |
| 3900 | Wastewater treatment system, to 1500 gallons | B-21 | 2 | 14 | | | 555 | 64.50 | 619.50 | 915 |

## 11 05 05 – Selective Equipment Demolition

| 11 05 05.10 Selective Demolition | | Crew | Daily Output | Labor-Hours | Unit | Material | 2011 Bare Costs Labor | 2011 Bare Costs Equipment | Total | Total Incl O&P |
|---|---|---|---|---|---|---|---|---|---|---|
| 4050 | Food storage equipment, walk-in refrigerator/freezer | 2 Clab | 64 | .250 | S.F. | | 8.60 | | 8.60 | 13.20 |
| 4052 | Shelving, stainless steel, 4 tier or dunnage rack | 1 Clab | 12 | .667 | Ea. | | 23 | | 23 | 35 |
| 4100 | Food preparation equipment, small countertop | | 18 | .444 | | | 15.25 | | 15.25 | 23.50 |
| 4150 | Food delivery carts, heated cabinets | | 18 | .444 | | | 15.25 | | 15.25 | 23.50 |
| 4200 | Cooking equipment, commercial range | Q-1 | 12 | 1.333 | | | 64.50 | | 64.50 | 96 |
| 4250 | Hood and ventilation equipment, kitchen exhaust hood, excl fire prot | 1 Clab | 3 | 2.667 | | | 91.50 | | 91.50 | 141 |
| 4255 | Fire protection system | Q-1 | 3 | 5.333 | | | 257 | | 257 | 385 |
| 4300 | Food dispensing equipment, countertop items | 1 Clab | 15 | .533 | | | 18.30 | | 18.30 | 28 |
| 4310 | Serving counter | " | 65 | .123 | L.F. | | 4.23 | | 4.23 | 6.50 |
| 4350 | Ice machine, ice cube maker, flakers and storage bins, to 2000 lb/day | Q-1 | 1.60 | 10 | Ea. | | 485 | | 485 | 720 |
| 4400 | Cleaning and disposal, commercial dishwasher, to 50 racks per hour | L-6 | 1 | 12 | | | 630 | | 630 | 940 |
| 4405 | To 275 racks per hour | L-4 | 1 | 24 | | | 970 | | 970 | 1,500 |
| 4410 | Dishwasher hood | 2 Clab | 5 | 3.200 | | | 110 | | 110 | 169 |
| 4420 | Garbage disposal, commercial, to 5 H.P. | L-1 | 8 | 2 | | | 104 | | 104 | 155 |
| 4540 | Water heater, residential, to 80 gal/day | " | 5 | 3.200 | | | 166 | | 166 | 247 |
| 4542 | Water softener, automatic | 2 Plum | 10 | 1.600 | | | 86 | | 86 | 128 |
| 4544 | Disappearing stairway, to 15' floor height | 2 Clab | 6 | 2.667 | | | 91.50 | | 91.50 | 141 |
| 4710 | Darkroom equipment, light | L-7 | 10 | 2.800 | | | 116 | | 116 | 178 |
| 4712 | Heavy | " | 1.50 | 18.667 | | | 775 | | 775 | 1,175 |
| 4720 | Doors | 2 Clab | 3.50 | 4.571 | Opng. | | 157 | | 157 | 241 |
| 4830 | Bowling alley, complete, incl pinsetter, scorer, counters, misc supplies | 4 Clab | .40 | 80 | Lane | | 2,750 | | 2,750 | 4,225 |
| 4840 | Health club equipment, circuit training apparatus | 2 Clab | 2 | 8 | Set | | 275 | | 275 | 420 |
| 4842 | Squat racks | " | 10 | 1.600 | Ea. | | 55 | | 55 | 84.50 |
| 4860 | School equipment, basketball backstop | L-2 | 2 | 8 | | | 305 | | 305 | 465 |
| 4862 | Table and benches, folding, in wall, 14' long | L-4 | 4 | 6 | | | 243 | | 243 | 370 |
| 4864 | Bleachers, telescoping, to 30 tier | F-5 | 120 | .267 | Seat | | 11.60 | | 11.60 | 17.80 |
| 4866 | Boxing ring, elevated | L-4 | .20 | 120 | Ea. | | 4,850 | | 4,850 | 7,450 |
| 4867 | Boxing ring, floor level | " | 2 | 12 | | | 485 | | 485 | 745 |
| 4868 | Exercise equipment | 1 Clab | 6 | 1.333 | | | 46 | | 46 | 70.50 |
| 4870 | Gym divider | L-4 | 1000 | .024 | S.F. | | .97 | | .97 | 1.49 |
| 4875 | Scoreboard | R-3 | 2 | 10 | Ea. | | 495 | 64.50 | 559.50 | 810 |
| 4880 | Shooting range, incl bullet traps, targets, excl structure | L-9 | 1 | 36 | Point | | 1,425 | | 1,425 | 2,250 |
| 5200 | Vocational shop equipment | 2 Clab | 8 | 2 | Ea. | | 68.50 | | 68.50 | 105 |
| 6200 | Fume hood, incl countertop, excl HVAC | " | 6 | 2.667 | L.F. | | 91.50 | | 91.50 | 141 |
| 7100 | Medical sterilizing, distiller, water, steam heated, 50 gal. capacity | 1 Plum | 2.80 | 2.857 | Ea. | | 153 | | 153 | 229 |
| 7200 | Medical equipment, surgery table, minor | 1 Clab | 1 | 8 | | | 275 | | 275 | 420 |
| 7210 | Surgical lights, doctors office, single or double arm | 2 Elec | 3 | 5.333 | | | 268 | | 268 | 395 |
| 7300 | Physical therapy, table | 2 Clab | 4 | 4 | | | 137 | | 137 | 211 |
| 7310 | Whirlpool bath, fixed, incl mixing valves | 1 Plum | 4 | 2 | | | 107 | | 107 | 160 |
| 7400 | Dental equipment, chair, electric or hydraulic | 1 Clab | .75 | 10.667 | | | 365 | | 365 | 560 |
| 7410 | Central suction system | 1 Plum | 2 | 4 | | | 214 | | 214 | 320 |
| 7420 | Drill console with accessories | 1 Clab | 3.20 | 2.500 | | | 86 | | 86 | 132 |
| 7430 | X-ray unit | " | 4 | 2 | | | 68.50 | | 68.50 | 105 |
| 7440 | X-ray developer | 1 Plum | 10 | .800 | | | 43 | | 43 | 64 |

## 11 05 10 – Equipment Installation

### 11 05 10.10 Industrial Equipment Installation

| 0010 | INDUSTRIAL EQUIPMENT INSTALLATION | | | | | | | | | |
|---|---|---|---|---|---|---|---|---|---|---|
| 0020 | Industrial equipment, minimum | E-2 | 12 | 4.667 | Ton | | 221 | 135 | 356 | 525 |
| 0200 | Maximum | " | 2 | 28 | " | | 1,325 | 810 | 2,135 | 3,150 |

# 11 11 Vehicle Service Equipment

## 11 11 13 – Compressed-Air Vehicle Service Equipment

| 11 11 13.10 Compressed Air Equipment | Crew | Daily Output | Labor-Hours | Unit | Material | 2011 Bare Costs Labor | Equipment | Total | Total Incl O&P |
|---|---|---|---|---|---|---|---|---|---|
| 0010 **COMPRESSED AIR EQUIPMENT** | | | | | | | | | |
| 0030 Compressors, electric, 1-1/2 H.P., standard controls | L-4 | 1.50 | 16 | Ea. | 410 | 650 | | 1,060 | 1,450 |
| 0550 Dual controls | | 1.50 | 16 | | 740 | 650 | | 1,390 | 1,800 |
| 0600 5 H.P., 115/230 volt, standard controls | | 1 | 24 | | 2,175 | 970 | | 3,145 | 3,900 |
| 0650 Dual controls | ↓ | 1 | 24 | ↓ | 2,350 | 970 | | 3,320 | 4,075 |

## 11 11 19 – Vehicle Lubrication Equipment

### 11 11 19.10 Lubrication Equipment

| | Crew | Daily Output | Labor-Hours | Unit | Material | Labor | Equipment | Total | Total Incl O&P |
|---|---|---|---|---|---|---|---|---|---|
| 0010 **LUBRICATION EQUIPMENT** | | | | | | | | | |
| 3000 Lube equipment, 3 reel type, with pumps, not including piping | L-4 | .50 | 48 | Set | 7,900 | 1,950 | | 9,850 | 11,700 |

## 11 11 33 – Vehicle Spray Painting Equipment

### 11 11 33.10 Spray Painting Equipment

| | Crew | Daily Output | Labor-Hours | Unit | Material | Labor | Equipment | Total | Total Incl O&P |
|---|---|---|---|---|---|---|---|---|---|
| 0010 **SPRAY PAINTING EQUIPMENT** | | | | | | | | | |
| 4000 Spray painting booth, 26' long, complete | L-4 | .40 | 60 | Ea. | 15,900 | 2,425 | | 18,325 | 21,100 |

# 11 12 Parking Control Equipment

## 11 12 13 – Parking Key and Card Control Units

### 11 12 13.10 Parking Control Units

| | Crew | Daily Output | Labor-Hours | Unit | Material | Labor | Equipment | Total | Total Incl O&P |
|---|---|---|---|---|---|---|---|---|---|
| 0010 **PARKING CONTROL UNITS** | | | | | | | | | |
| 5100 Card reader | 1 Elec | 2 | 4 | Ea. | 1,875 | 201 | | 2,076 | 2,350 |
| 5120 Proximity with customer display | 2 Elec | 1 | 16 | | 5,500 | 805 | | 6,305 | 7,250 |
| 6000 Parking control software, minimum | 1 Elec | .50 | 16 | | 22,600 | 805 | | 23,405 | 26,000 |
| 6020 Maximum | " | .20 | 40 | ↓ | 98,500 | 2,000 | | 100,500 | 111,500 |

## 11 12 16 – Parking Ticket Dispensers

### 11 12 16.10 Ticket Dispensers

| | Crew | Daily Output | Labor-Hours | Unit | Material | Labor | Equipment | Total | Total Incl O&P |
|---|---|---|---|---|---|---|---|---|---|
| 0010 **TICKET DISPENSERS** | | | | | | | | | |
| 5900 Ticket spitter with time/date stamp, standard | 2 Elec | 2 | 8 | Ea. | 6,150 | 400 | | 6,550 | 7,350 |
| 5920 Mag stripe encoding | " | 2 | 8 | " | 18,600 | 400 | | 19,000 | 21,000 |

## 11 12 26 – Parking Fee Collection Equipment

### 11 12 26.13 Parking Fee Coin Collection Equipment

| | Crew | Daily Output | Labor-Hours | Unit | Material | Labor | Equipment | Total | Total Incl O&P |
|---|---|---|---|---|---|---|---|---|---|
| 0010 **PARKING FEE COIN COLLECTION EQUIPMENT** | | | | | | | | | |
| 5200 Cashier booth, average | B-22 | 1 | 30 | Ea. | 9,875 | 1,200 | 194 | 11,269 | 12,900 |
| 5300 Collector station, pay on foot | 2 Elec | .20 | 80 | | 109,500 | 4,025 | | 113,525 | 126,500 |
| 5320 Credit card only | " | .50 | 32 | ↓ | 20,200 | 1,600 | | 21,800 | 24,700 |

### 11 12 26.23 Fee Equipment

| | Crew | Daily Output | Labor-Hours | Unit | Material | Labor | Equipment | Total | Total Incl O&P |
|---|---|---|---|---|---|---|---|---|---|
| 0010 **FEE EQUIPMENT** | | | | | | | | | |
| 5600 Fee computer | 1 Elec | 1.50 | 5.333 | Ea. | 13,600 | 268 | | 13,868 | 15,400 |

## 11 12 33 – Parking Gates

### 11 12 33.13 Parking Gate Equipment

| | Crew | Daily Output | Labor-Hours | Unit | Material | Labor | Equipment | Total | Total Incl O&P |
|---|---|---|---|---|---|---|---|---|---|
| 0010 **PARKING GATE EQUIPMENT** | | | | | | | | | |
| 5000 Barrier gate with programmable controller | 2 Elec | 3 | 5.333 | Ea. | 3,425 | 268 | | 3,693 | 4,175 |
| 5020 Industrial | | 3 | 5.333 | | 4,725 | 268 | | 4,993 | 5,600 |
| 5500 Exit verifier | ↓ | 1 | 16 | | 17,400 | 805 | | 18,205 | 20,400 |
| 5700 Full sign, 4" letters | 1 Elec | 2 | 4 | | 1,225 | 201 | | 1,426 | 1,625 |
| 5800 Inductive loop | 2 Elec | 4 | 4 | | 169 | 201 | | 370 | 485 |
| 5950 Vehicle detector, microprocessor based | 1 Elec | 3 | 2.667 | ↓ | 395 | 134 | | 529 | 635 |

# 11 13 Loading Dock Equipment

## 11 13 13 – Loading Dock Bumpers

### 11 13 13.10 Dock Bumpers

| | 11 13 13.10 Dock Bumpers | Crew | Daily Output | Labor-Hours | Unit | Material | 2011 Bare Costs Labor | Equipment | Total | Total Incl O&P |
|---|---|---|---|---|---|---|---|---|---|---|
| 0010 | **DOCK BUMPERS** Bolts not included | | | | | | | | | |
| 0020 | 2" x 6" to 4" x 8", average | 1 Carp | .30 | 26.667 | M.B.F. | 1,325 | 1,150 | | 2,475 | 3,225 |
| 0050 | Bumpers, rubber blocks 4-1/2" thick, 10" high, 14" long | | 26 | .308 | Ea. | 50.50 | 13.25 | | 63.75 | 76 |
| 0200 | 24" long | | 22 | .364 | | 95 | 15.65 | | 110.65 | 128 |
| 0300 | 36" long | | 17 | .471 | | 96 | 20.50 | | 116.50 | 137 |
| 0500 | 12" high, 14" long | | 25 | .320 | | 100 | 13.80 | | 113.80 | 131 |
| 0550 | 24" long | | 20 | .400 | | 111 | 17.20 | | 128.20 | 149 |
| 0600 | 36" long | | 15 | .533 | | 123 | 23 | | 146 | 171 |
| 0800 | Rubber blocks 6" thick, 10" high, 14" long | | 22 | .364 | | 93 | 15.65 | | 108.65 | 126 |
| 0850 | 24" long | | 18 | .444 | | 121 | 19.15 | | 140.15 | 163 |
| 0900 | 36" long | | 13 | .615 | | 147 | 26.50 | | 173.50 | 203 |
| 0910 | 20" high, 11" long | | 13 | .615 | | 147 | 26.50 | | 173.50 | 202 |
| 0920 | Extruded rubber bumpers, T section, 22" x 22" x 3" thick | | 41 | .195 | | 55 | 8.40 | | 63.40 | 73.50 |
| 0940 | Molded rubber bumpers, 24" x 12" x 3" thick | | 20 | .400 | | 54 | 17.20 | | 71.20 | 85.50 |
| 1000 | Welded installation of above bumpers | E-14 | 8 | 1 | | 3.74 | 50.50 | 13.65 | 67.89 | 109 |
| 1100 | For drilled anchors, add per anchor | 1 Carp | 36 | .222 | | 6.70 | 9.55 | | 16.25 | 22 |
| 1300 | Steel bumpers, see Section 10 26 13.10 | | | | | | | | | |

## 11 13 16 – Loading Dock Seals and Shelters

### 11 13 16.10 Dock Seals and Shelters

| | 11 13 16.10 Dock Seals and Shelters | Crew | Daily Output | Labor-Hours | Unit | Material | 2011 Bare Costs Labor | Equipment | Total | Total Incl O&P |
|---|---|---|---|---|---|---|---|---|---|---|
| 0010 | **DOCK SEALS AND SHELTERS** | | | | | | | | | |
| 3600 | Door seal for door perimeter, 12" x 12", vinyl covered | 1 Carp | 26 | .308 | L.F. | 30.50 | 13.25 | | 43.75 | 54 |
| 3900 | Folding gates, see Section 10 22 16.10 | | | | | | | | | |
| 6200 | Shelters, fabric, for truck or train, scissor arms, minimum | 1 Carp | 1 | 8 | Ea. | 1,575 | 345 | | 1,920 | 2,275 |
| 6300 | Maximum | " | .50 | 16 | " | 2,050 | 690 | | 2,740 | 3,300 |

## 11 13 19 – Stationary Loading Dock Equipment

### 11 13 19.10 Dock Equipment

| | 11 13 19.10 Dock Equipment | Crew | Daily Output | Labor-Hours | Unit | Material | 2011 Bare Costs Labor | Equipment | Total | Total Incl O&P |
|---|---|---|---|---|---|---|---|---|---|---|
| 0010 | **DOCK EQUIPMENT** | | | | | | | | | |
| 2200 | Dock boards, heavy duty, 60" x 60", aluminum, 5,000 lb. capacity | | | | Ea. | 1,200 | | | 1,200 | 1,325 |
| 2700 | 9,000 lb. capacity | | | | | 1,375 | | | 1,375 | 1,525 |
| 3200 | 15,000 lb. capacity | | | | | 1,625 | | | 1,625 | 1,775 |
| 4200 | Platform lifter, 6' x 6', portable, 3,000 lb. capacity | | | | | 9,150 | | | 9,150 | 10,100 |
| 4250 | 4,000 lb. capacity | | | | | 11,300 | | | 11,300 | 12,400 |
| 4400 | Fixed, 6' x 8', 5,000 lb. capacity | E-16 | .70 | 22.857 | | 9,850 | 1,125 | 156 | 11,131 | 13,000 |
| 4500 | Levelers, hinged for trucks, 10 ton capacity, 6' x 8' | | 1.08 | 14.815 | | 4,975 | 735 | 101 | 5,811 | 6,875 |
| 4650 | 7' x 8' | | 1.08 | 14.815 | | 5,350 | 735 | 101 | 6,186 | 7,275 |
| 4670 | Air bag power operated, 10 ton cap., 6' x 8' | | 1.08 | 14.815 | | 5,775 | 735 | 101 | 6,611 | 7,750 |
| 4680 | 7' x 8' | | 1.08 | 14.815 | | 5,800 | 735 | 101 | 6,636 | 7,775 |
| 4700 | Hydraulic, 10 ton capacity, 6' x 8' | | 1.08 | 14.815 | | 8,650 | 735 | 101 | 9,486 | 10,900 |
| 4800 | 7' x 8' | | 1.08 | 14.815 | | 9,325 | 735 | 101 | 10,161 | 11,600 |
| 5800 | Loading dock safety restraints, manual style | | 1.08 | 14.815 | | 3,175 | 735 | 101 | 4,011 | 4,875 |
| 5900 | Automatic style | | 1.08 | 14.815 | | 5,325 | 735 | 101 | 6,161 | 7,250 |

## 11 13 26 – Loading Dock Lights

### 11 13 26.10 Dock Lights

| | 11 13 26.10 Dock Lights | Crew | Daily Output | Labor-Hours | Unit | Material | 2011 Bare Costs Labor | Equipment | Total | Total Incl O&P |
|---|---|---|---|---|---|---|---|---|---|---|
| 0010 | **DOCK LIGHTS** | | | | | | | | | |
| 5000 | Lights for loading docks, single arm, 24" long | 1 Elec | 3.80 | 2.105 | Ea. | 139 | 106 | | 245 | 310 |
| 5700 | Double arm, 60" long | " | 3.80 | 2.105 | " | 184 | 106 | | 290 | 360 |

# 11 14 Pedestrian Control Equipment

## 11 14 13 – Pedestrian Gates

| 11 14 13.13 Portable Posts and Railings | Crew | Daily Output | Labor-Hours | Unit | Material | 2011 Bare Costs Labor | Equipment | Total | Total Incl O&P |
|---|---|---|---|---|---|---|---|---|---|
| 0010 **PORTABLE POSTS AND RAILINGS** | | | | | | | | | |
| 0020 Portable for pedestrian traffic control, standard, minimum | | | | Ea. | 122 | | | 122 | 135 |
| 0100 Maximum | | | | | 181 | | | 181 | 199 |
| 0300 Deluxe posts, minimum | | | | | 172 | | | 172 | 189 |
| 0400 Maximum | | | | | 325 | | | 325 | 355 |
| 0600 Ropes for above posts, plastic covered, 1-1/2" diameter | | | | L.F. | 15.55 | | | 15.55 | 17.10 |
| 0700 Chain core | | | | " | 10.90 | | | 10.90 | 12 |
| 1500 Portable security or safety barrier, black with 7' yellow strap | | | | Ea. | 250 | | | 250 | 275 |
| 1510 12' yellow strap | | | | | 290 | | | 290 | 320 |
| 1550 Sign holder, standard design | | | | | 135 | | | 135 | 148 |

### 11 14 13.19 Turnstiles

| | Crew | Daily Output | Labor-Hours | Unit | Material | 2011 Bare Costs Labor | Equipment | Total | Total Incl O&P |
|---|---|---|---|---|---|---|---|---|---|
| 0010 **TURNSTILES** | | | | | | | | | |
| 0020 One way, 4 arm, 46" diameter, economy, manual | 2 Carp | 5 | 3.200 | Ea. | 805 | 138 | | 943 | 1,100 |
| 0100 Electric | | 1.20 | 13.333 | | 1,300 | 575 | | 1,875 | 2,325 |
| 0300 High security, galv., 5'-5" diameter, 7' high, manual | | 1 | 16 | | 3,975 | 690 | | 4,665 | 5,425 |
| 0350 Electric | | .60 | 26.667 | | 4,500 | 1,150 | | 5,650 | 6,700 |
| 0420 Three arm, 24" opening, light duty, manual | | 2 | 8 | | 1,250 | 345 | | 1,595 | 1,925 |
| 0450 Heavy duty | | 1.50 | 10.667 | | 3,125 | 460 | | 3,585 | 4,150 |
| 0460 Manual, with registering & controls, light duty | | 2 | 8 | | 2,725 | 345 | | 3,070 | 3,525 |
| 0470 Heavy duty | | 1.50 | 10.667 | | 3,300 | 460 | | 3,760 | 4,325 |
| 0480 Electric, heavy duty | | 1.10 | 14.545 | | 3,825 | 625 | | 4,450 | 5,150 |
| 0500 For coin or token operating, add | | | | | 610 | | | 610 | 670 |
| 1200 One way gate with horizontal bars, 5'-5" diameter | | | | | | | | | |
| 1300 7' high, recreation or transit type | 2 Carp | .80 | 20 | Ea. | 4,425 | 860 | | 5,285 | 6,175 |
| 1500 For electronic counter, add | " | | | " | 210 | | | 210 | 231 |

# 11 16 Vault Equipment

## 11 16 16 – Safes

| 11 16 16.10 Commercial Safes | Crew | Daily Output | Labor-Hours | Unit | Material | 2011 Bare Costs Labor | Equipment | Total | Total Incl O&P |
|---|---|---|---|---|---|---|---|---|---|
| 0010 **COMMERCIAL SAFES** | | | | | | | | | |
| 0200 Office, 1 hr. rating, 30" x 18" x 18" | | | | Ea. | 2,175 | | | 2,175 | 2,400 |
| 0250 40" x 18" x 18" | | | | | 4,100 | | | 4,100 | 4,525 |
| 0300 60" x 36" x 18", double door | | | | | 8,150 | | | 8,150 | 8,975 |
| 0600 Data, 1 hr. rating, 27" x 19" x 16" | | | | | 7,600 | | | 7,600 | 8,350 |
| 0700 63" x 34" x 16" | | | | | 14,500 | | | 14,500 | 15,900 |
| 0750 Diskette, 1 hr., 14" x 12" x 11", inside | | | | | 4,200 | | | 4,200 | 4,625 |
| 0800 Money, "B" label, 9" x 14" x 14" | | | | | 525 | | | 525 | 580 |
| 0900 Tool resistive, 24" x 24" x 20" | | | | | 3,575 | | | 3,575 | 3,925 |
| 1050 Tool and torch resistive, 24" x 24" x 20" | | | | | 9,150 | | | 9,150 | 10,100 |
| 1150 Jewelers, 23" x 20" x 18" | | | | | 11,700 | | | 11,700 | 12,800 |
| 1200 63" x 25" x 18" | | | | | 22,400 | | | 22,400 | 24,700 |
| 1300 For handling into building, add, minimum | A-2 | 8.50 | 2.824 | | | 96 | 20 | 116 | 169 |
| 1400 Maximum | " | .78 | 30.769 | | | 1,050 | 220 | 1,270 | 1,850 |

# 11 17 Teller and Service Equipment

## 11 17 13 – Teller Equipment Systems

### 11 17 13.10 Bank Equipment

| 11 17 13.10 Bank Equipment | Crew | Daily Output | Labor-Hours | Unit | Material | 2011 Bare Costs Labor | Equipment | Total | Total Incl O&P |
|---|---|---|---|---|---|---|---|---|---|
| 0010 **BANK EQUIPMENT** | | | | | | | | | |
| 0020 Alarm system, police | 2 Elec | 1.60 | 10 | Ea. | 4,675 | 505 | | 5,180 | 5,900 |
| 0100 With vault alarm | " | .40 | 40 | | 18,600 | 2,000 | | 20,600 | 23,400 |
| 0400 Bullet resistant teller window, 44" x 60" | 1 Glaz | .60 | 13.333 | | 3,525 | 560 | | 4,085 | 4,725 |
| 0500 48" x 60" | " | .60 | 13.333 | | 4,700 | 560 | | 5,260 | 6,025 |
| 3000 Counters for banks, frontal only | 2 Carp | 1 | 16 | Station | 1,750 | 690 | | 2,440 | 2,975 |
| 3100 Complete with steel undercounter | " | .50 | 32 | " | 3,400 | 1,375 | | 4,775 | 5,875 |
| 4600 Door and frame, bullet-resistant, with vision panel, minimum | 2 Sswk | 1.10 | 14.545 | Ea. | 5,000 | 705 | | 5,705 | 6,750 |
| 4700 Maximum | | 1.10 | 14.545 | | 6,775 | 705 | | 7,480 | 8,725 |
| 4800 Drive-up window, drawer & mike, not incl. glass, minimum | | 1 | 16 | | 5,800 | 775 | | 6,575 | 7,750 |
| 4900 Maximum | | .50 | 32 | | 8,875 | 1,550 | | 10,425 | 12,500 |
| 5000 Night depository, with chest, minimum | | 1 | 16 | | 7,250 | 775 | | 8,025 | 9,350 |
| 5100 Maximum | | .50 | 32 | | 10,300 | 1,550 | | 11,850 | 14,100 |
| 5200 Package receiver, painted | | 3.20 | 5 | | 1,300 | 243 | | 1,543 | 1,850 |
| 5300 Stainless steel | | 3.20 | 5 | | 2,225 | 243 | | 2,468 | 2,850 |
| 5400 Partitions, bullet-resistant, 1-3/16" glass, 8' high | 2 Carp | 10 | 1.600 | L.F. | 189 | 69 | | 258 | 315 |
| 5450 Acrylic | " | 10 | 1.600 | " | 360 | 69 | | 429 | 500 |
| 5500 Pneumatic tube systems, 2 lane drive-up, complete | L-3 | .25 | 64 | Total | 24,600 | 3,000 | | 27,600 | 31,700 |
| 5550 With T.V. viewer | " | .20 | 80 | " | 47,600 | 3,750 | | 51,350 | 58,000 |
| 5570 Safety deposit boxes, minimum | 1 Sswk | 44 | .182 | Opng. | 54.50 | 8.85 | | 63.35 | 75.50 |
| 5580 Maximum, 10" x 15" opening | | 19 | .421 | | 116 | 20.50 | | 136.50 | 163 |
| 5590 Teller locker, average | | 15 | .533 | | 1,500 | 26 | | 1,526 | 1,700 |
| 5600 Pass thru, bullet-res. window, painted steel, 24" x 36" | 2 Sswk | 1.60 | 10 | Ea. | 2,375 | 485 | | 2,860 | 3,475 |
| 5700 48" x 48" | | 1.20 | 13.333 | | 2,550 | 645 | | 3,195 | 3,950 |
| 5800 72" x 40" | | .80 | 20 | | 3,725 | 970 | | 4,695 | 5,825 |
| 5900 For stainless steel frames, add | | | | | 20% | | | | |
| 6100 Surveillance system, video camera, complete | 2 Elec | 1 | 16 | Ea. | 9,700 | 805 | | 10,505 | 11,900 |
| 6110 For each additional camera, add | | | | " | 945 | | | 945 | 1,050 |
| 6120 CCTV system, see Section 27 41 19.10 | | | | | | | | | |
| 6200 Twenty-four hour teller, single unit, | | | | | | | | | |
| 6300 automated deposit, cash and memo | L-3 | .25 | 64 | Ea. | 42,800 | 3,000 | | 45,800 | 51,500 |
| 7000 Vault front, see Section 08 34 59.10 | | | | | | | | | |

# 11 19 Detention Equipment

## 11 19 30 – Detention Cell Equipment

### 11 19 30.10 Cell Equipment

| | Crew | Daily Output | Labor-Hours | Unit | Material | Labor | Equipment | Total | Total Incl O&P |
|---|---|---|---|---|---|---|---|---|---|
| 0010 **CELL EQUIPMENT** | | | | | | | | | |
| 3000 Toilet apparatus including wash basin, average | L-8 | 1.50 | 13.333 | Ea. | 3,350 | 600 | | 3,950 | 4,625 |

# 11 21 Mercantile and Service Equipment

## 11 21 13 – Cash Registers and Checking Equipment

### 11 21 13.10 Checkout Counter

| | Crew | Daily Output | Labor-Hours | Unit | Material | Labor | Equipment | Total | Total Incl O&P |
|---|---|---|---|---|---|---|---|---|---|
| 0010 **CHECKOUT COUNTER** | | | | | | | | | |
| 0020 Supermarket conveyor, single belt | 2 Clab | 10 | 1.600 | Ea. | 2,900 | 55 | | 2,955 | 3,275 |
| 0100 Double belt, power take-away | | 9 | 1.778 | | 4,175 | 61 | | 4,236 | 4,700 |
| 0400 Double belt, power take-away, incl. side scanning | | 7 | 2.286 | | 4,900 | 78.50 | | 4,978.50 | 5,525 |
| 0800 Warehouse or bulk type | | 6 | 2.667 | | 5,800 | 91.50 | | 5,891.50 | 6,525 |
| 1000 Scanning system, 2 lanes, w/registers, scan gun & memory | | | | System | 15,200 | | | 15,200 | 16,700 |
| 1100 10 lanes, single processor, full scan, with scales | | | | " | 144,500 | | | 144,500 | 158,500 |

# 11 21 Mercantile and Service Equipment

## 11 21 13 – Cash Registers and Checking Equipment

| 11 21 13.10 Checkout Counter | Crew | Daily Output | Labor-Hours | Unit | Material | 2011 Bare Costs Labor | Equipment | Total | Total Incl O&P |
|---|---|---|---|---|---|---|---|---|---|
| 2000 | Register, restaurant, minimum | | | | Ea. | 675 | | | 675 | 745 |
| 2100 | Maximum | | | | | 2,850 | | | 2,850 | 3,125 |
| 2150 | Store, minimum | | | | | 675 | | | 675 | 745 |
| 2200 | Maximum | | | | | 2,850 | | | 2,850 | 3,125 |

## 11 21 53 – Barber and Beauty Shop Equipment

### 11 21 53.10 Barber Equipment

| | | Crew | Daily Output | Labor-Hours | Unit | Material | 2011 Bare Costs Labor | Equipment | Total | Total Incl O&P |
|---|---|---|---|---|---|---|---|---|---|---|
| 0010 | **BARBER EQUIPMENT** | | | | | | | | | |
| 0020 | Chair, hydraulic, movable, minimum | 1 Carp | 24 | .333 | Ea. | 460 | 14.35 | | 474.35 | 525 |
| 0050 | Maximum | " | 16 | .500 | | 2,975 | 21.50 | | 2,996.50 | 3,300 |
| 0200 | Wall hung styling station with mirrors, minimum | L-2 | 8 | 2 | | 395 | 75.50 | | 470.50 | 550 |
| 0300 | Maximum | " | 4 | 4 | | 1,975 | 151 | | 2,126 | 2,400 |
| 0500 | Sink, hair washing basin, rough plumbing not incl. | 1 Plum | 8 | 1 | | 425 | 53.50 | | 478.50 | 545 |
| 1000 | Sterilizer, liquid solution for tools | | | | | 138 | | | 138 | 151 |
| 1100 | Total equipment, rule of thumb, per chair, minimum | L-8 | 1 | 20 | | 1,725 | 905 | | 2,630 | 3,275 |
| 1150 | Maximum | " | 1 | 20 | | 4,400 | 905 | | 5,305 | 6,225 |

# 11 23 Commercial Laundry and Dry Cleaning Equipment

## 11 23 13 – Dry Cleaning Equipment

### 11 23 13.13 Dry Cleaners

| | | Crew | Daily Output | Labor-Hours | Unit | Material | 2011 Bare Costs Labor | Equipment | Total | Total Incl O&P |
|---|---|---|---|---|---|---|---|---|---|---|
| 0010 | **DRY CLEANERS** Not incl. rough-in | | | | | | | | | |
| 2000 | Dry cleaners, electric, 20 lb. capacity | L-1 | .20 | 80 | Ea. | 33,300 | 4,150 | | 37,450 | 42,800 |
| 2050 | 25 lb. capacity | | .17 | 94.118 | | 45,800 | 4,900 | | 50,700 | 58,000 |
| 2100 | 30 lb. capacity | | .15 | 106 | | 48,300 | 5,550 | | 53,850 | 61,500 |
| 2150 | 60 lb. capacity | | .09 | 177 | | 73,000 | 9,225 | | 82,225 | 93,500 |

## 11 23 16 – Drying and Conditioning Equipment

### 11 23 16.13 Dryers

| | | Crew | Daily Output | Labor-Hours | Unit | Material | 2011 Bare Costs Labor | Equipment | Total | Total Incl O&P |
|---|---|---|---|---|---|---|---|---|---|---|
| 0010 | **DRYERS**, Not including rough-in | | | | | | | | | |
| 1500 | Industrial, 30 lb. capacity | 1 Plum | 2 | 4 | Ea. | 2,975 | 214 | | 3,189 | 3,600 |
| 1600 | 50 lb. capacity | " | 1.70 | 4.706 | | 3,200 | 252 | | 3,452 | 3,900 |
| 4700 | Lint collector, ductwork not included, 8,000 to 10,000 C.F.M. | Q-10 | .30 | 80 | | 8,650 | 3,850 | | 12,500 | 15,400 |

## 11 23 19 – Finishing Equipment

### 11 23 19.13 Folders and Spreaders

| | | Crew | Daily Output | Labor-Hours | Unit | Material | 2011 Bare Costs Labor | Equipment | Total | Total Incl O&P |
|---|---|---|---|---|---|---|---|---|---|---|
| 0010 | **FOLDERS AND SPREADERS** | | | | | | | | | |
| 3500 | Folders, blankets & sheets, minimum | 1 Elec | .17 | 47.059 | Ea. | 31,100 | 2,375 | | 33,475 | 37,700 |
| 3700 | King size with automatic stacker | | .10 | 80 | | 56,500 | 4,025 | | 60,525 | 68,000 |
| 3800 | For conveyor delivery, add | | .45 | 17.778 | | 6,000 | 895 | | 6,895 | 7,925 |
| 4900 | Spreader feeders, 240V, 2 station | L-6 | .70 | 17.143 | | 51,500 | 900 | | 52,400 | 58,000 |
| 4920 | 4 station | " | .35 | 34.286 | | 62,000 | 1,800 | | 63,800 | 71,000 |

## 11 23 23 – Commercial Ironing Equipment

### 11 23 23.13 Irons and Pressers

| | | Crew | Daily Output | Labor-Hours | Unit | Material | 2011 Bare Costs Labor | Equipment | Total | Total Incl O&P |
|---|---|---|---|---|---|---|---|---|---|---|
| 0010 | **IRONS AND PRESSERS** | | | | | | | | | |
| 4500 | Ironers, institutional, 110", single roll | 1 Elec | .20 | 40 | Ea. | 30,200 | 2,000 | | 32,200 | 36,200 |
| 4800 | Pressers, low capacity air operated | L-6 | 1.75 | 6.857 | | 8,775 | 360 | | 9,135 | 10,200 |
| 4820 | Hand operated | | 1.75 | 6.857 | | 8,125 | 360 | | 8,485 | 9,475 |
| 4840 | Ironer 48", 240V | | 3.50 | 3.429 | | 103,000 | 180 | | 103,180 | 114,000 |
| 6600 | Hand operated presser | | .70 | 17.143 | | 5,525 | 900 | | 6,425 | 7,425 |
| 6620 | Mushroom press 115 V | | .70 | 17.143 | | 6,925 | 900 | | 7,825 | 8,950 |

# 11 23 Commercial Laundry and Dry Cleaning Equipment

## 11 23 26 – Commercial Washers and Extractors

### 11 23 26.13 Washers and Extractors

| 11 23 26.13 Washers and Extractors | Crew | Daily Output | Labor-Hours | Unit | Material | 2011 Bare Costs Labor | 2011 Bare Costs Equipment | Total | Total Incl O&P |
|---|---|---|---|---|---|---|---|---|---|
| 0010 **WASHERS AND EXTRACTORS**, not including rough-in | | | | | | | | | |
| 6000 Combination washer/extractor, 20 lb. capacity | L-6 | 1.50 | 8 | Ea. | 5,400 | 420 | | 5,820 | 6,575 |
| 6100      30 lb. capacity | | .80 | 15 | | 8,550 | 790 | | 9,340 | 10,600 |
| 6200      50 lb. capacity | | .68 | 17.647 | | 10,000 | 925 | | 10,925 | 12,400 |
| 6300      75 lb. capacity | | .30 | 40 | | 18,700 | 2,100 | | 20,800 | 23,700 |
| 6350      125 lb. capacity | | .16 | 75 | | 24,900 | 3,950 | | 28,850 | 33,300 |
| 6380 Washer extractor/dryer, 110 lb., 240V | | 1 | 12 | | 8,150 | 630 | | 8,780 | 9,925 |
| 6400 Washer extractor, 135 lb, 240V | | 1 | 12 | | 23,800 | 630 | | 24,430 | 27,000 |
| 6450      Pass through | | 1 | 12 | | 64,000 | 630 | | 64,630 | 71,500 |
| 6500 200 lb. washer extractor | | 1 | 12 | | 66,500 | 630 | | 67,130 | 74,500 |
| 6550      Pass through | | 1 | 12 | | 69,500 | 630 | | 70,130 | 77,500 |
| 6600 Extractor, low capacity | | 1.75 | 6.857 | | 6,625 | 360 | | 6,985 | 7,800 |

## 11 23 33 – Coin-Operated Laundry Equipment

### 11 23 33.13 Coin Operated Washers and Dryers

| | Crew | Daily Output | Labor-Hours | Unit | Material | 2011 Bare Costs Labor | 2011 Bare Costs Equipment | Total | Total Incl O&P |
|---|---|---|---|---|---|---|---|---|---|
| 0010 **COIN OPERATED WASHERS AND DRYERS** | | | | | | | | | |
| 0990 Dryer, gas fired | | | | | | | | | |
| 1000 Commercial, 30 lb. capacity, coin operated, single | 1 Plum | 3 | 2.667 | Ea. | 3,100 | 143 | | 3,243 | 3,625 |
| 1100      Double stacked | " | 2 | 4 | | 6,275 | 214 | | 6,489 | 7,225 |
| 4860 Coin dry cleaner 20 lb. | L-6 | 1.75 | 6.857 | | 25,900 | 360 | | 26,260 | 29,000 |
| 5290 Clothes washer | | | | | | | | | |
| 5300      Commercial, coin operated, average | 1 Plum | 3 | 2.667 | Ea. | 1,125 | 143 | | 1,268 | 1,475 |

# 11 24 Maintenance Equipment

## 11 24 19 – Vacuum Cleaning Systems

### 11 24 19.10 Vacuum Cleaning

| | Crew | Daily Output | Labor-Hours | Unit | Material | 2011 Bare Costs Labor | 2011 Bare Costs Equipment | Total | Total Incl O&P |
|---|---|---|---|---|---|---|---|---|---|
| 0010 **VACUUM CLEANING** | | | | | | | | | |
| 0020 Central, 3 inlet, residential | 1 Skwk | .90 | 8.889 | Total | 1,025 | 395 | | 1,420 | 1,750 |
| 0200      Commercial | | .70 | 11.429 | | 1,225 | 505 | | 1,730 | 2,125 |
| 0400 5 inlet system, residential | | .50 | 16 | | 1,450 | 710 | | 2,160 | 2,700 |
| 0600 7 inlet system, commercial | | .40 | 20 | | 1,675 | 890 | | 2,565 | 3,200 |
| 0800 9 inlet system, residential | | .30 | 26.667 | | 3,600 | 1,175 | | 4,775 | 5,775 |
| 4010 Rule of thumb: First 1200 S.F., installed | | | | | | | | 1,325 | 1,450 |
| 4020      For each additional S.F., add | | | | S.F. | | | | | .24 |

# 11 26 Unit Kitchens

## 11 26 13 – Metal Unit Kitchens

### 11 26 13.10 Commercial Unit Kitchens

| | Crew | Daily Output | Labor-Hours | Unit | Material | 2011 Bare Costs Labor | 2011 Bare Costs Equipment | Total | Total Incl O&P |
|---|---|---|---|---|---|---|---|---|---|
| 0010 **COMMERCIAL UNIT KITCHENS** | | | | | | | | | |
| 1500 Combination range, refrigerator and sink, 30" wide, minimum | L-1 | 2 | 8 | Ea. | 925 | 415 | | 1,340 | 1,650 |
| 1550      Maximum | | 1 | 16 | | 3,675 | 830 | | 4,505 | 5,275 |
| 1570      60" wide, average | | 1.40 | 11.429 | | 3,225 | 595 | | 3,820 | 4,400 |
| 1590      72" wide, average | | 1.20 | 13.333 | | 4,200 | 695 | | 4,895 | 5,650 |
| 1600 Office model, 48" wide | | 2 | 8 | | 2,650 | 415 | | 3,065 | 3,525 |
| 1620      Refrigerator and sink only | | 2.40 | 6.667 | | 2,725 | 345 | | 3,070 | 3,500 |
| 1640 Combination range, refrigerator, sink, microwave | | | | | | | | | |
| 1660      Oven and ice maker | L-1 | .80 | 20 | Ea. | 4,450 | 1,050 | | 5,500 | 6,450 |

# 11 27 Photographic Processing Equipment

## 11 27 13 – Darkroom Processing Equipment

### 11 27 13.10 Darkroom Equipment

| | 11 27 13.10 Darkroom Equipment | Crew | Daily Output | Labor-Hours | Unit | Material | 2011 Bare Costs Labor | Equipment | Total | Total Incl O&P |
|---|---|---|---|---|---|---|---|---|---|---|
| 0010 | **DARKROOM EQUIPMENT** | | | | | | | | | |
| 0020 | Developing sink, 5" deep, 24" x 48" | Q-1 | 2 | 8 | Ea. | 5,625 | 385 | | 6,010 | 6,750 |
| 0050 | 48" x 52" | | 1.70 | 9.412 | | 5,675 | 455 | | 6,130 | 6,900 |
| 0200 | 10" deep, 24" x 48" | | 1.70 | 9.412 | | 7,775 | 455 | | 8,230 | 9,225 |
| 0250 | 24" x 108" | | 1.50 | 10.667 | | 10,200 | 515 | | 10,715 | 12,100 |
| 0500 | Dryers, dehumidified filtered air, 36" x 25" x 68" high | L-7 | 6 | 4.667 | | 5,075 | 194 | | 5,269 | 5,875 |
| 0550 | 48" x 25" x 68" high | | 5 | 5.600 | | 10,000 | 233 | | 10,233 | 11,400 |
| 2000 | Processors, automatic, color print, minimum | | 4 | 7 | | 15,700 | 291 | | 15,991 | 17,700 |
| 2050 | Maximum | | .60 | 46.667 | | 24,700 | 1,950 | | 26,650 | 30,200 |
| 2300 | Black and white print, minimum | | 2 | 14 | | 11,900 | 580 | | 12,480 | 14,000 |
| 2350 | Maximum | | .80 | 35 | | 61,500 | 1,450 | | 62,950 | 69,500 |
| 2600 | Manual processor, 16" x 20" maximum print size | | 2 | 14 | | 9,450 | 580 | | 10,030 | 11,300 |
| 2650 | 20" x 24" maximum print size | | 1 | 28 | | 8,975 | 1,175 | | 10,150 | 11,700 |
| 3000 | Viewing lights, 20" x 24" | | 6 | 4.667 | | 380 | 194 | | 574 | 710 |
| 3100 | 20" x 24" with color correction | | 6 | 4.667 | | 535 | 194 | | 729 | 885 |
| 3500 | Washers, round, minimum sheet 11" x 14" | Q-1 | 2 | 8 | | 3,400 | 385 | | 3,785 | 4,325 |
| 3550 | Maximum sheet 20" x 24" | | 1 | 16 | | 3,475 | 770 | | 4,245 | 4,975 |
| 3800 | Square, minimum sheet 20" x 24" | | 1 | 16 | | 3,125 | 770 | | 3,895 | 4,575 |
| 3900 | Maximum sheet 50" x 56" | | .80 | 20 | | 4,850 | 965 | | 5,815 | 6,775 |
| 4500 | Combination tank sink, tray sink, washers, with | | | | | | | | | |
| 4510 | Dry side tables, average | Q-1 | .45 | 35.556 | Ea. | 10,600 | 1,725 | | 12,325 | 14,300 |

# 11 31 Residential Appliances

## 11 31 13 – Residential Kitchen Appliances

### 11 31 13.13 Cooking Equipment

| | 11 31 13.13 Cooking Equipment | Crew | Daily Output | Labor-Hours | Unit | Material | 2011 Bare Costs Labor | Equipment | Total | Total Incl O&P |
|---|---|---|---|---|---|---|---|---|---|---|
| 0010 | **COOKING EQUIPMENT** | | | | | | | | | |
| 0020 | Cooking range, 30" free standing, 1 oven, minimum | 2 Clab | 10 | 1.600 | Ea. | 505 | 55 | | 560 | 640 |
| 0050 | Maximum | | 4 | 4 | | 1,700 | 137 | | 1,837 | 2,075 |
| 0150 | 2 oven, minimum | | 10 | 1.600 | | 1,050 | 55 | | 1,105 | 1,225 |
| 0200 | Maximum | | 10 | 1.600 | | 1,100 | 55 | | 1,155 | 1,275 |
| 0350 | Built-in, 30" wide, 1 oven, minimum | 1 Elec | 6 | 1.333 | | 605 | 67 | | 672 | 765 |
| 0400 | Maximum | 2 Carp | 2 | 8 | | 1,150 | 345 | | 1,495 | 1,775 |
| 0500 | 2 oven, conventional, minimum | | 4 | 4 | | 1,350 | 172 | | 1,522 | 1,750 |
| 0550 | 1 conventional, 1 microwave, maximum | | 2 | 8 | | 1,600 | 345 | | 1,945 | 2,300 |
| 0700 | Free-standing, 1 oven, 21" wide range, minimum | 2 Clab | 10 | 1.600 | | 330 | 55 | | 385 | 445 |
| 0750 | 21" wide, maximum | " | 4 | 4 | | 335 | 137 | | 472 | 580 |
| 0900 | Countertop cooktops, 4 burner, standard, minimum | 1 Elec | 6 | 1.333 | | 284 | 67 | | 351 | 410 |
| 0950 | Maximum | | 3 | 2.667 | | 1,100 | 134 | | 1,234 | 1,400 |
| 1050 | As above, but with grill and griddle attachment, minimum | | 6 | 1.333 | | 1,225 | 67 | | 1,292 | 1,450 |
| 1100 | Maximum | | 3 | 2.667 | | 3,125 | 134 | | 3,259 | 3,650 |
| 1200 | Induction cooktop, 30" wide | | 3 | 2.667 | | 1,125 | 134 | | 1,259 | 1,450 |
| 1250 | Microwave oven, minimum | | 4 | 2 | | 110 | 101 | | 211 | 270 |
| 1300 | Maximum | | 2 | 4 | | 495 | 201 | | 696 | 845 |

### 11 31 13.23 Refrigeration Equipment

| | | Crew | Daily Output | Labor-Hours | Unit | Material | 2011 Bare Costs Labor | Equipment | Total | Total Incl O&P |
|---|---|---|---|---|---|---|---|---|---|---|
| 0010 | **REFRIGERATION EQUIPMENT** | | | | | | | | | |
| 2000 | Deep freeze, 15 to 23 C.F., minimum | 2 Clab | 10 | 1.600 | Ea. | 550 | 55 | | 605 | 690 |
| 2050 | Maximum | | 5 | 3.200 | | 730 | 110 | | 840 | 975 |
| 2200 | 30 C.F., minimum | | 8 | 2 | | 800 | 68.50 | | 868.50 | 985 |
| 2250 | Maximum | | 3 | 5.333 | | 850 | 183 | | 1,033 | 1,225 |
| 5200 | Icemaker, automatic, 20 lb. per day | 1 Plum | 7 | 1.143 | | 880 | 61.50 | | 941.50 | 1,050 |
| 5350 | 51 lb. per day | " | 2 | 4 | | 1,250 | 214 | | 1,464 | 1,700 |

# 11 31 Residential Appliances

## 11 31 13 – Residential Kitchen Appliances

### 11 31 13.23 Refrigeration Equipment

| | | Crew | Daily Output | Labor-Hours | Unit | Material | 2011 Bare Costs Labor | 2011 Bare Costs Equipment | Total | Total Incl O&P |
|---|---|---|---|---|---|---|---|---|---|---|
| 5500 | Refrigerator, no frost, 10 C.F. to 12 C.F., minimum | 2 Clab | 10 | 1.600 | Ea. | 415 | 55 | | 470 | 540 |
| 5600 | Maximum | | 6 | 2.667 | | 505 | 91.50 | | 596.50 | 695 |
| 5750 | 14 C.F. to 16 C.F., minimum | | 9 | 1.778 | | 480 | 61 | | 541 | 625 |
| 5800 | Maximum | | 5 | 3.200 | | 635 | 110 | | 745 | 870 |
| 5950 | 18 C.F. to 20 C.F., minimum | | 8 | 2 | | 610 | 68.50 | | 678.50 | 775 |
| 6000 | Maximum | | 4 | 4 | | 1,175 | 137 | | 1,312 | 1,500 |
| 6150 | 21 C.F. to 29 C.F., minimum | | 7 | 2.286 | | 950 | 78.50 | | 1,028.50 | 1,175 |
| 6200 | Maximum | | 3 | 5.333 | | 3,175 | 183 | | 3,358 | 3,750 |
| 6790 | Energy-star qualified, 18 C.F., minimum  G | 2 Carp | 4 | 4 | | 600 | 172 | | 772 | 925 |
| 6795 | Maximum  G | | 2 | 8 | | 1,300 | 345 | | 1,645 | 1,950 |
| 6797 | 21.7 C.F., minimum  G | | 4 | 4 | | 920 | 172 | | 1,092 | 1,275 |
| 6799 | Maximum  G | | 4 | 4 | | 1,500 | 172 | | 1,672 | 1,925 |

### 11 31 13.33 Kitchen Cleaning Equipment

| | | Crew | Daily Output | Labor-Hours | Unit | Material | Labor | Equipment | Total | Total Incl O&P |
|---|---|---|---|---|---|---|---|---|---|---|
| 0010 | **KITCHEN CLEANING EQUIPMENT** | | | | | | | | | |
| 2750 | Dishwasher, built-in, 2 cycles, minimum | L-1 | 4 | 4 | Ea. | 221 | 208 | | 429 | 555 |
| 2800 | Maximum | | 2 | 8 | | 400 | 415 | | 815 | 1,050 |
| 2950 | 4 or more cycles, minimum | | 4 | 4 | | 325 | 208 | | 533 | 665 |
| 2960 | Average | | 4 | 4 | | 430 | 208 | | 638 | 785 |
| 3000 | Maximum | | 2 | 8 | | 1,075 | 415 | | 1,490 | 1,825 |
| 3100 | Energy-star qualified, minimum  G | | 4 | 4 | | 284 | 208 | | 492 | 620 |
| 3110 | Maximum  G | | 2 | 8 | | 1,300 | 415 | | 1,715 | 2,050 |

### 11 31 13.43 Waste Disposal Equipment

| | | Crew | Daily Output | Labor-Hours | Unit | Material | Labor | Equipment | Total | Total Incl O&P |
|---|---|---|---|---|---|---|---|---|---|---|
| 0010 | **WASTE DISPOSAL EQUIPMENT** | | | | | | | | | |
| 1750 | Compactor, residential size, 4 to 1 compaction, minimum | 1 Carp | 5 | 1.600 | Ea. | 535 | 69 | | 604 | 690 |
| 1800 | Maximum | " | 3 | 2.667 | | 850 | 115 | | 965 | 1,100 |
| 3300 | Garbage disposal, sink type, minimum | L-1 | 10 | 1.600 | | 72.50 | 83 | | 155.50 | 204 |
| 3350 | Maximum | " | 10 | 1.600 | | 191 | 83 | | 274 | 335 |

### 11 31 13.53 Kitchen Ventilation Equipment

| | | Crew | Daily Output | Labor-Hours | Unit | Material | Labor | Equipment | Total | Total Incl O&P |
|---|---|---|---|---|---|---|---|---|---|---|
| 0010 | **KITCHEN VENTILATION EQUIPMENT** | | | | | | | | | |
| 4150 | Hood for range, 2 speed, vented, 30" wide, minimum | L-3 | 5 | 3.200 | Ea. | 62.50 | 150 | | 212.50 | 297 |
| 4200 | Maximum | | 3 | 5.333 | | 1,000 | 251 | | 1,251 | 1,475 |
| 4300 | 42" wide, minimum | | 5 | 3.200 | | 188 | 150 | | 338 | 435 |
| 4330 | Custom | | 5 | 3.200 | | 1,650 | 150 | | 1,800 | 2,050 |
| 4350 | Maximum | | 3 | 5.333 | | 2,025 | 251 | | 2,276 | 2,600 |
| 4500 | For ventless hood, 2 speed, add | | | | | 17.05 | | | 17.05 | 18.75 |
| 4650 | For vented 1 speed, deduct from maximum | | | | | 45.50 | | | 45.50 | 50 |

## 11 31 23 – Residential Laundry Appliances

### 11 31 23.13 Washers

| | | Crew | Daily Output | Labor-Hours | Unit | Material | Labor | Equipment | Total | Total Incl O&P |
|---|---|---|---|---|---|---|---|---|---|---|
| 0010 | **WASHERS** | | | | | | | | | |
| 5000 | Residential, 4 cycle, average | 1 Plum | 3 | 2.667 | Ea. | 800 | 143 | | 943 | 1,100 |
| 6650 | Washing machine, automatic, minimum | | 3 | 2.667 | | 400 | 143 | | 543 | 655 |
| 6700 | Maximum | | 1 | 8 | | 1,300 | 430 | | 1,730 | 2,075 |
| 6750 | Energy star qualified, front loading, minimum  G | | 3 | 2.667 | | 620 | 143 | | 763 | 900 |
| 6760 | Maximum  G | | 1 | 8 | | 1,275 | 430 | | 1,705 | 2,050 |
| 6764 | Top loading, minimum  G | | 3 | 2.667 | | 370 | 143 | | 513 | 625 |
| 6766 | Maximum  G | | 3 | 2.667 | | 1,075 | 143 | | 1,218 | 1,400 |

### 11 31 23.23 Dryers

| | | Crew | Daily Output | Labor-Hours | Unit | Material | Labor | Equipment | Total | Total Incl O&P |
|---|---|---|---|---|---|---|---|---|---|---|
| 0010 | **DRYERS** | | | | | | | | | |
| 0500 | Gas fired residential, 16 lb. capacity, average | 1 Plum | 3 | 2.667 | Ea. | 635 | 143 | | 778 | 915 |
| 6770 | Electric, front loading, energy-star qualified, minimum  G | L-2 | 3 | 5.333 | | 500 | 202 | | 702 | 860 |
| 6780 | Maximum  G | " | 2 | 8 | | 1,600 | 305 | | 1,905 | 2,225 |

# 11 31 Residential Appliances

## 11 31 23 – Residential Laundry Appliances

### 11 31 23.23 Dryers

| | | Crew | Daily Output | Labor-Hours | Unit | Material | 2011 Bare Costs Labor | 2011 Bare Costs Equipment | Total | Total Incl O&P |
|---|---|---|---|---|---|---|---|---|---|---|
| 7450 | Vent kits for dryers | 1 Carp | 10 | .800 | Ea. | 37.50 | 34.50 | | 72 | 94 |

## 11 31 33 – Miscellaneous Residential Appliances

### 11 31 33.13 Sump Pumps

| | | Crew | Daily Output | Labor-Hours | Unit | Material | Labor | Equipment | Total | Total Incl O&P |
|---|---|---|---|---|---|---|---|---|---|---|
| 0010 | **SUMP PUMPS** | | | | | | | | | |
| 6400 | Cellar drainer, pedestal, 1/3 H.P., molded PVC base | 1 Plum | 3 | 2.667 | Ea. | 119 | 143 | | 262 | 345 |
| 6450 | Solid brass | " | 2 | 4 | " | 263 | 214 | | 477 | 610 |
| 6460 | Sump pump, see also Section 22 14 29.16 | | | | | | | | | |

### 11 31 33.23 Water Heaters

| | | Crew | Daily Output | Labor-Hours | Unit | Material | Labor | Equipment | Total | Total Incl O&P |
|---|---|---|---|---|---|---|---|---|---|---|
| 0010 | **WATER HEATERS** | | | | | | | | | |
| 6900 | Electric, glass lined, 30 gallon, minimum | L-1 | 5 | 3.200 | Ea. | 520 | 166 | | 686 | 815 |
| 6950 | Maximum | | 3 | 5.333 | | 720 | 277 | | 997 | 1,200 |
| 7100 | 80 gallon, minimum | | 2 | 8 | | 935 | 415 | | 1,350 | 1,650 |
| 7150 | Maximum | | 1 | 16 | | 1,300 | 830 | | 2,130 | 2,650 |
| 7180 | Gas, glass lined, 30 gallon, minimum | 2 Plum | 5 | 3.200 | | 805 | 172 | | 977 | 1,150 |
| 7220 | Maximum | | 3 | 5.333 | | 1,125 | 286 | | 1,411 | 1,650 |
| 7260 | 50 gallon, minimum | | 2.50 | 6.400 | | 845 | 345 | | 1,190 | 1,425 |
| 7300 | Maximum | | 1.50 | 10.667 | | 1,175 | 570 | | 1,745 | 2,125 |
| 7310 | Water heater, see also Section 22 33 30.13 | | | | | | | | | |

### 11 31 33.43 Air Quality

| | | Crew | Daily Output | Labor-Hours | Unit | Material | Labor | Equipment | Total | Total Incl O&P |
|---|---|---|---|---|---|---|---|---|---|---|
| 0010 | **AIR QUALITY** | | | | | | | | | |
| 2450 | Dehumidifier, portable, automatic, 15 pint | | | | Ea. | 154 | | | 154 | 170 |
| 2550 | 40 pint | | | | | 171 | | | 171 | 188 |
| 3550 | Heater, electric, built-in, 1250 watt, ceiling type, minimum | 1 Elec | 4 | 2 | | 98.50 | 101 | | 199.50 | 258 |
| 3600 | Maximum | | 3 | 2.667 | | 161 | 134 | | 295 | 375 |
| 3700 | Wall type, minimum | | 4 | 2 | | 159 | 101 | | 260 | 325 |
| 3750 | Maximum | | 3 | 2.667 | | 171 | 134 | | 305 | 385 |
| 3900 | 1500 watt wall type, with blower | | 4 | 2 | | 159 | 101 | | 260 | 325 |
| 3950 | 3000 watt | | 3 | 2.667 | | 325 | 134 | | 459 | 555 |
| 4850 | Humidifier, portable, 8 gallons per day | | | | | 165 | | | 165 | 182 |
| 5000 | 15 gallons per day | | | | | 198 | | | 198 | 218 |

# 11 33 Retractable Stairs

## 11 33 10 – Disappearing Stairs

### 11 33 10.10 Disappearing Stairway

| | | Crew | Daily Output | Labor-Hours | Unit | Material | Labor | Equipment | Total | Total Incl O&P |
|---|---|---|---|---|---|---|---|---|---|---|
| 0010 | **DISAPPEARING STAIRWAY** No trim included | | | | | | | | | |
| 0100 | Custom grade, pine, 8'-6" ceiling, minimum | 1 Carp | 4 | 2 | Ea. | 142 | 86 | | 228 | 288 |
| 0150 | Average | | 3.50 | 2.286 | | 217 | 98.50 | | 315.50 | 390 |
| 0200 | Maximum | | 3 | 2.667 | | 289 | 115 | | 404 | 495 |
| 0500 | Heavy duty, pivoted, from 7'-7" to 12'-10" floor to floor | | 3 | 2.667 | | 685 | 115 | | 800 | 925 |
| 0600 | 16'-0" ceiling | | 2 | 4 | | 1,400 | 172 | | 1,572 | 1,800 |
| 0800 | Economy folding, pine, 8'-6" ceiling | | 4 | 2 | | 160 | 86 | | 246 | 310 |
| 0900 | 9'-6" ceiling | | 4 | 2 | | 160 | 86 | | 246 | 310 |
| 1100 | Automatic electric, aluminum, floor to floor height, 8' to 9' | 2 Carp | 1 | 16 | | 8,675 | 690 | | 9,365 | 10,600 |
| 1400 | 11' to 12' | | .90 | 17.778 | | 9,325 | 765 | | 10,090 | 11,500 |
| 1700 | 14' to 15' | | .70 | 22.857 | | 9,950 | 985 | | 10,935 | 12,400 |

# 11 41 Foodservice Storage Equipment

## 11 41 13 – Refrigerated Food Storage Cases

### 11 41 13.10 Refrigerated Food Cases

| | | Crew | Daily Output | Labor-Hours | Unit | Material | 2011 Bare Costs Labor | Equipment | Total | Total Incl O&P |
|---|---|---|---|---|---|---|---|---|---|---|
| 0010 | **REFRIGERATED FOOD CASES** | | | | | | | | | |
| 0030 | Dairy, multi-deck, 12' long | Q-5 | 3 | 5.333 | Ea. | 10,700 | 260 | | 10,960 | 12,200 |
| 0100 | For rear sliding doors, add | | | | | 1,525 | | | 1,525 | 1,700 |
| 0200 | Delicatessen case, service deli, 12' long, single deck | Q-5 | 3.90 | 4.103 | | 7,225 | 200 | | 7,425 | 8,250 |
| 0300 | Multi-deck, 18 S.F. shelf display | | 3 | 5.333 | | 6,450 | 260 | | 6,710 | 7,475 |
| 0400 | Freezer, self-contained, chest-type, 30 C.F. | | 3.90 | 4.103 | | 7,400 | 200 | | 7,600 | 8,450 |
| 0500 | Glass door, upright, 78 C.F. | | 3.30 | 4.848 | | 10,000 | 236 | | 10,236 | 11,400 |
| 0600 | Frozen food, chest type, 12' long | | 3.30 | 4.848 | | 7,250 | 236 | | 7,486 | 8,325 |
| 0700 | Glass door, reach-in, 5 door | | 3 | 5.333 | | 13,900 | 260 | | 14,160 | 15,700 |
| 0800 | Island case, 12' long, single deck | | 3.30 | 4.848 | | 8,225 | 236 | | 8,461 | 9,375 |
| 0900 | Multi-deck | | 3 | 5.333 | | 17,400 | 260 | | 17,660 | 19,500 |
| 1000 | Meat case, 12' long, single deck | | 3.30 | 4.848 | | 5,975 | 236 | | 6,211 | 6,900 |
| 1050 | Multi-deck | | 3.10 | 5.161 | | 10,300 | 251 | | 10,551 | 11,700 |
| 1100 | Produce, 12' long, single deck | | 3.30 | 4.848 | | 7,900 | 236 | | 8,136 | 9,050 |
| 1200 | Multi-deck | | 3.10 | 5.161 | | 8,625 | 251 | | 8,876 | 9,850 |

### 11 41 13.20 Refrigerated Food Storage Equipment

| | | Crew | Daily Output | Labor-Hours | Unit | Material | 2011 Bare Costs Labor | Equipment | Total | Total Incl O&P |
|---|---|---|---|---|---|---|---|---|---|---|
| 0010 | **REFRIGERATED FOOD STORAGE EQUIPMENT** | | | | | | | | | |
| 2350 | Cooler, reach-in, beverage, 6' long | Q-1 | 6 | 2.667 | Ea. | 4,625 | 129 | | 4,754 | 5,275 |
| 4300 | Freezers, reach-in, 44 C.F. | | 4 | 4 | | 2,700 | 193 | | 2,893 | 3,275 |
| 4500 | 68 C.F. | | 3 | 5.333 | | 3,650 | 257 | | 3,907 | 4,375 |
| 4600 | Freezer, pre-fab, 8' x 8' w/refrigeration | 2 Carp | .45 | 35.556 | | 10,800 | 1,525 | | 12,325 | 14,300 |
| 4620 | 8' x 12' | | .35 | 45.714 | | 12,800 | 1,975 | | 14,775 | 17,100 |
| 4640 | 8' x 16' | | .25 | 64 | | 15,800 | 2,750 | | 18,550 | 21,600 |
| 4660 | 8' x 20' | | .17 | 94.118 | | 19,100 | 4,050 | | 23,150 | 27,200 |
| 4680 | Reach-in, 1 compartment | Q-1 | 4 | 4 | | 2,275 | 193 | | 2,468 | 2,800 |
| 4685 | Energy star rated   **G** | R-18 | 7.80 | 3.333 | | 2,225 | 132 | | 2,357 | 2,650 |
| 4700 | 2 compartment | Q-1 | 3 | 5.333 | | 3,600 | 257 | | 3,857 | 4,325 |
| 4705 | Energy star rated   **G** | R-18 | 6.20 | 4.194 | | 3,200 | 166 | | 3,366 | 3,775 |
| 4710 | 3 compartment | Q-1 | 3 | 5.333 | | 4,875 | 257 | | 5,132 | 5,750 |
| 4715 | Energy star rated   **G** | R-18 | 5.60 | 4.643 | | 4,025 | 183 | | 4,208 | 4,700 |
| 8320 | Refrigerator, reach-in, 1 compartment | | 7.80 | 3.333 | | 2,175 | 132 | | 2,307 | 2,600 |
| 8325 | Energy star rated   **G** | | 7.80 | 3.333 | | 2,225 | 132 | | 2,357 | 2,650 |
| 8330 | 2 compartment | | 6.20 | 4.194 | | 3,275 | 166 | | 3,441 | 3,850 |
| 8335 | Energy star rated   **G** | | 6.20 | 4.194 | | 3,200 | 166 | | 3,366 | 3,775 |
| 8340 | 3 compartment | | 5.60 | 4.643 | | 4,000 | 183 | | 4,183 | 4,675 |
| 8345 | Energy star rated   **G** | | 5.60 | 4.643 | | 4,025 | 183 | | 4,208 | 4,700 |
| 8350 | Pre-fab, with refrigeration, 8' x 8' | 2 Carp | .45 | 35.556 | | 6,350 | 1,525 | | 7,875 | 9,325 |
| 8360 | 8' x 12' | | .35 | 45.714 | | 7,525 | 1,975 | | 9,500 | 11,300 |
| 8370 | 8' x 16' | | .25 | 64 | | 11,200 | 2,750 | | 13,950 | 16,500 |
| 8380 | 8' x 20' | | .17 | 94.118 | | 14,200 | 4,050 | | 18,250 | 21,800 |
| 8390 | Pass-thru/roll-in, 1 compartment | R-18 | 7.80 | 3.333 | | 4,075 | 132 | | 4,207 | 4,675 |
| 8400 | 2 compartment | | 6.24 | 4.167 | | 5,800 | 165 | | 5,965 | 6,625 |
| 8410 | 3 compartment | | 5.60 | 4.643 | | 7,925 | 183 | | 8,108 | 8,975 |
| 8420 | Walk-in, alum, door & floor only, no refrig, 6' x 6' x 7'-6" | 2 Carp | 1.40 | 11.429 | | 8,725 | 490 | | 9,215 | 10,300 |
| 8430 | 10' x 6' x 7'-6" | | .55 | 29.091 | | 12,500 | 1,250 | | 13,750 | 15,600 |
| 8440 | 12' x 14' x 7'-6" | | .25 | 64 | | 17,500 | 2,750 | | 20,250 | 23,400 |
| 8450 | 12' x 20' x 7'-6" | | .17 | 94.118 | | 21,300 | 4,050 | | 25,350 | 29,600 |
| 8460 | Refrigerated cabinets, mobile | | | | | 3,650 | | | 3,650 | 4,025 |
| 8470 | Refrigerator/freezer, reach-in, 1 compartment | R-18 | 5.60 | 4.643 | | 4,825 | 183 | | 5,008 | 5,575 |
| 8480 | 2 compartment | " | 4.80 | 5.417 | | 6,775 | 214 | | 6,989 | 7,800 |

# 11 41 Foodservice Storage Equipment

## 11 41 13 – Refrigerated Food Storage Cases

| 11 41 13.30 Wine Cellar | Crew | Daily Output | Labor-Hours | Unit | Material | 2011 Bare Costs Labor | Equipment | Total | Total Incl O&P |
|---|---|---|---|---|---|---|---|---|---|
| 0010 **WINE CELLAR**, refrigerated, Redwood interior, carpeted, walk-in type | | | | | | | | | |
| 0020 6'-8" high, including racks | | | | | | | | | |
| 0200 80" W x 48" D for 900 bottles | 2 Carp | 1.50 | 10.667 | Ea. | 3,275 | 460 | | 3,735 | 4,300 |
| 0250 80" W x 72" D for 1300 bottles | | 1.33 | 12.030 | | 4,125 | 520 | | 4,645 | 5,350 |
| 0300 80" W x 94" D for 1900 bottles | | 1.17 | 13.675 | | 4,900 | 590 | | 5,490 | 6,275 |
| 0400 80" W x 124" D for 2500 bottles | ↓ | 1 | 16 | ↓ | 6,000 | 690 | | 6,690 | 7,650 |
| 0600 Portable cabinets, red oak, reach-in temp.& humidity controlled | | | | | | | | | |
| 0650 26-5/8"W x 26-1/2"D x 68"H for 235 bottles | | | | Ea. | 2,675 | | | 2,675 | 2,950 |
| 0660 32"W x 21-1/2"D x 73-1/2"H for 144 bottles | | | | | 2,625 | | | 2,625 | 2,900 |
| 0670 32"W x 29-1/2"D x 73-1/2"H for 288 bottles | | | | | 2,700 | | | 2,700 | 2,975 |
| 0680 39-1/2"W x 29-1/2"D x 86-1/2"H for 440 bottles | | | | | 4,025 | | | 4,025 | 4,425 |
| 0690 52-1/2"W x 29-1/2"D x 73-1/2"H for 468 bottles | | | | | 3,750 | | | 3,750 | 4,100 |
| 0700 52-1/2"W x 29-1/2"D x 86-1/2"H for 572 bottles | | | | ↓ | 3,850 | | | 3,850 | 4,250 |
| 0730 Portable, red oak, can be built-in with glass door | | | | | | | | | |
| 0750 23-7/8"W x 24"D x 34-1/2"H for 50 bottles | | | | Ea. | 965 | | | 965 | 1,050 |

## 11 41 33 – Foodservice Shelving

### 11 41 33.20 Metal Food Storage Shelving

| 11 41 33.20 Metal Food Storage Shelving | Crew | Daily Output | Labor-Hours | Unit | Material | 2011 Bare Costs Labor | Equipment | Total | Total Incl O&P |
|---|---|---|---|---|---|---|---|---|---|
| 0010 **METAL FOOD STORAGE SHELVING** | | | | | | | | | |
| 8600 Stainless steel shelving, louvered 4-tier, 20" x 3' | 1 Clab | 6 | 1.333 | Ea. | 1,250 | 46 | | 1,296 | 1,450 |
| 8605 20" x 4' | | 6 | 1.333 | | 1,800 | 46 | | 1,846 | 2,050 |
| 8610 20" x 6' | | 6 | 1.333 | | 2,600 | 46 | | 2,646 | 2,925 |
| 8615 24" x 3' | | 6 | 1.333 | | 1,650 | 46 | | 1,696 | 1,875 |
| 8620 24" x 4' | | 6 | 1.333 | | 1,925 | 46 | | 1,971 | 2,175 |
| 8625 24" x 6' | | 6 | 1.333 | | 2,775 | 46 | | 2,821 | 3,125 |
| 8630 Flat 4-tier, 20" x 3' | | 6 | 1.333 | | 1,125 | 46 | | 1,171 | 1,300 |
| 8635 20" x 4' | | 6 | 1.333 | | 1,350 | 46 | | 1,396 | 1,575 |
| 8640 20" x 5' | | 6 | 1.333 | | 1,575 | 46 | | 1,621 | 1,800 |
| 8645 24" x 3' | | 6 | 1.333 | | 1,275 | 46 | | 1,321 | 1,475 |
| 8650 24" x 4' | | 6 | 1.333 | | 2,100 | 46 | | 2,146 | 2,375 |
| 8655 24" x 6' | | 6 | 1.333 | | 2,500 | 46 | | 2,546 | 2,825 |
| 8700 Galvanized shelving, louvered 4-tier, 20" x 3' | | 6 | 1.333 | | 640 | 46 | | 686 | 775 |
| 8705 20" x 4' | | 6 | 1.333 | | 715 | 46 | | 761 | 855 |
| 8710 20" x 6' | | 6 | 1.333 | | 870 | 46 | | 916 | 1,025 |
| 8715 24" x 3' | | 6 | 1.333 | | 565 | 46 | | 611 | 695 |
| 8720 24" x 4' | | 6 | 1.333 | | 765 | 46 | | 811 | 910 |
| 8725 24" x 6' | | 6 | 1.333 | | 1,100 | 46 | | 1,146 | 1,275 |
| 8730 Flat 4-tier, 20" x 3' | | 6 | 1.333 | | 565 | 46 | | 611 | 690 |
| 8735 20" x 4' | | 6 | 1.333 | | 620 | 46 | | 666 | 755 |
| 8740 20" x 6' | | 6 | 1.333 | | 895 | 46 | | 941 | 1,050 |
| 8745 24" x 3' | | 6 | 1.333 | | 625 | 46 | | 671 | 760 |
| 8750 24" x 4' | | 6 | 1.333 | | 635 | 46 | | 681 | 770 |
| 8755 24" x 6' | | 6 | 1.333 | | 930 | 46 | | 976 | 1,100 |
| 8760 Stainless steel dunnage rack, 24" x 3' | | 8 | 1 | | 320 | 34.50 | | 354.50 | 405 |
| 8765 24" x 4' | | 8 | 1 | | 405 | 34.50 | | 439.50 | 500 |
| 8770 Galvanized dunnage rack, 24" x 3' | | 8 | 1 | | 162 | 34.50 | | 196.50 | 231 |
| 8775 24" x 4' | ↓ | 8 | 1 | ↓ | 182 | 34.50 | | 216.50 | 254 |

# 11 42 Food Preparation Equipment

## 11 42 10 – Commercial Food Preparation Equipment

### 11 42 10.10 Choppers, Mixers and Misc. Equipment

| 11 42 10.10 Choppers, Mixers and Misc. Equipment | | Crew | Daily Output | Labor-Hours | Unit | Material | 2011 Bare Costs | | Total | Total Incl O&P |
|---|---|---|---|---|---|---|---|---|---|---|
| | | | | | | | Labor | Equipment | | |
| 0010 | **CHOPPERS, MIXERS AND MISC. EQUIPMENT** | | | | | | | | | |
| 1700 | Choppers, 5 pounds | R-18 | 7 | 3.714 | Ea. | 1,825 | 147 | | 1,972 | 2,225 |
| 1720 | 16 pounds | | 5 | 5.200 | | 2,200 | 205 | | 2,405 | 2,725 |
| 1740 | 35 to 40 pounds | | 4 | 6.500 | | 3,300 | 257 | | 3,557 | 4,000 |
| 1840 | Coffee brewer, 5 burners | 1 Plum | 3 | 2.667 | | 1,250 | 143 | | 1,393 | 1,600 |
| 1850 | Coffee urn, twin 6 gallon urns | | 2 | 4 | | 2,725 | 214 | | 2,939 | 3,325 |
| 1860 | Single, 3 gallon | | 3 | 2.667 | | 1,975 | 143 | | 2,118 | 2,400 |
| 3000 | Fast food equipment, total package, minimum | 6 Skwk | .08 | 600 | | 192,500 | 26,600 | | 219,100 | 253,000 |
| 3100 | Maximum | " | .07 | 685 | | 262,500 | 30,400 | | 292,900 | 335,000 |
| 3800 | Food mixers, 20 quarts | L-7 | 7 | 4 | | 2,525 | 166 | | 2,691 | 3,025 |
| 3850 | 40 quarts | | 5.40 | 5.185 | | 6,525 | 216 | | 6,741 | 7,500 |
| 3900 | 60 quarts | | 5 | 5.600 | | 9,925 | 233 | | 10,158 | 11,300 |
| 4040 | 80 quarts | | 3.90 | 7.179 | | 11,700 | 299 | | 11,999 | 13,400 |
| 4080 | 130 quarts | | 2.20 | 12.727 | | 17,100 | 530 | | 17,630 | 19,600 |
| 4100 | Floor type, 20 quarts | | 15 | 1.867 | | 3,625 | 77.50 | | 3,702.50 | 4,125 |
| 4120 | 60 quarts | | 14 | 2 | | 9,050 | 83 | | 9,133 | 10,100 |
| 4140 | 80 quarts | | 12 | 2.333 | | 12,000 | 97 | | 12,097 | 13,200 |
| 4160 | 140 quarts | | 8.60 | 3.256 | | 18,700 | 135 | | 18,835 | 20,800 |
| 6700 | Peelers, small | R-18 | 8 | 3.250 | | 1,600 | 128 | | 1,728 | 1,950 |
| 6720 | Large | " | 6 | 4.333 | | 4,750 | 171 | | 4,921 | 5,475 |
| 6800 | Pulper/extractor, close coupled, 5 HP | 1 Plum | 1.90 | 4.211 | | 3,600 | 226 | | 3,826 | 4,275 |
| 8580 | Slicer with table | R-18 | 9 | 2.889 | | 4,925 | 114 | | 5,039 | 5,600 |

# 11 43 Food Delivery Carts and Conveyors

## 11 43 13 – Food Delivery Carts

### 11 43 13.10 Mobile Carts, Racks and Trays

| 11 43 13.10 Mobile Carts, Racks and Trays | | Crew | Daily Output | Labor-Hours | Unit | Material | 2011 Bare Costs | | Total | Total Incl O&P |
|---|---|---|---|---|---|---|---|---|---|---|
| | | | | | | | Labor | Equipment | | |
| 0010 | **MOBILE CARTS, RACKS AND TRAYS** | | | | | | | | | |
| 1650 | Cabinet, heated, 1 compartment, reach-in | R-18 | 5.60 | 4.643 | Ea. | 3,625 | 183 | | 3,808 | 4,275 |
| 1655 | Pass-thru roll-in | | 5.60 | 4.643 | | 4,950 | 183 | | 5,133 | 5,725 |
| 1660 | 2 compartment, reach-in | | 4.80 | 5.417 | | 7,625 | 214 | | 7,839 | 8,725 |
| 1670 | Mobile | | | | | 2,925 | | | 2,925 | 3,225 |
| 6850 | Mobile rack w/pan slide | | | | | 1,300 | | | 1,300 | 1,425 |
| 9180 | Tray and silver dispenser, mobile | 1 Clab | 16 | .500 | | 855 | 17.20 | | 872.20 | 965 |

# 11 44 Food Cooking Equipment

## 11 44 13 – Commercial Ranges

### 11 44 13.10 Cooking Equipment

| 11 44 13.10 Cooking Equipment | | Crew | Daily Output | Labor-Hours | Unit | Material | 2011 Bare Costs | | Total | Total Incl O&P |
|---|---|---|---|---|---|---|---|---|---|---|
| | | | | | | | Labor | Equipment | | |
| 0010 | **COOKING EQUIPMENT** | | | | | | | | | |
| 0020 | Bake oven, gas, one section | Q-1 | 8 | 2 | Ea. | 4,650 | 96.50 | | 4,746.50 | 5,275 |
| 0300 | Two sections | | 7 | 2.286 | | 9,300 | 110 | | 9,410 | 10,400 |
| 0600 | Three sections | | 6 | 2.667 | | 13,500 | 129 | | 13,629 | 15,000 |
| 0900 | Electric convection, single deck | L-7 | 4 | 7 | | 6,475 | 291 | | 6,766 | 7,575 |
| 1300 | Broiler, without oven, standard | Q-1 | 8 | 2 | | 3,550 | 96.50 | | 3,646.50 | 4,050 |
| 1550 | Infrared | L-7 | 4 | 7 | | 9,325 | 291 | | 9,616 | 10,700 |
| 4750 | Fryer, with twin baskets, modular model | Q-1 | 7 | 2.286 | | 1,575 | 110 | | 1,685 | 1,900 |
| 5000 | Floor model, on 6" legs | " | 5 | 3.200 | | 1,975 | 154 | | 2,129 | 2,400 |
| 5100 | Extra single basket, large | | | | | 102 | | | 102 | 112 |
| 5170 | Energy star rated, 50 lb. capacity Ⓖ | R-18 | 4 | 6.500 | | 3,950 | 257 | | 4,207 | 4,700 |
| 5175 | 85 lb. capacity Ⓖ | " | 4 | 6.500 | | 7,450 | 257 | | 7,707 | 8,550 |

# 11 44 Food Cooking Equipment

## 11 44 13 – Commercial Ranges

| 11 44 13.10 Cooking Equipment | Crew | Daily Output | Labor-Hours | Unit | Material | 2011 Bare Costs Labor | Equipment | Total | Total Incl O&P |
|---|---|---|---|---|---|---|---|---|---|
| 5300 | Griddle, SS, 24" plate, w/4" legs, elec, 208 V, 3 phase, 3' long | Q-1 | 7 | 2.286 | Ea. | 1,300 | 110 | | 1,410 | 1,600 |
| 5550 | 4' long | " | 6 | 2.667 | | 2,225 | 129 | | 2,354 | 2,650 |
| 6200 | Iced tea brewer | 1 Plum | 3.44 | 2.326 | | 700 | 125 | | 825 | 960 |
| 6350 | Kettle, w/steam jacket, tilting, w/positive lock, SS, 20 gallons | L-7 | 7 | 4 | | 7,400 | 166 | | 7,566 | 8,375 |
| 6600 | 60 gallons | " | 6 | 4.667 | | 9,800 | 194 | | 9,994 | 11,100 |
| 6900 | Range, restaurant type, 6 burners and 1 standard oven, 36" wide | Q-1 | 7 | 2.286 | | 2,825 | 110 | | 2,935 | 3,275 |
| 6950 | Convection | | 7 | 2.286 | | 5,325 | 110 | | 5,435 | 6,025 |
| 7150 | 2 standard ovens, 24" griddle, 60" wide | | 6 | 2.667 | | 4,000 | 129 | | 4,129 | 4,600 |
| 7200 | 1 standard, 1 convection oven | | 6 | 2.667 | | 8,150 | 129 | | 8,279 | 9,150 |
| 7450 | Heavy duty, single 34" standard oven, open top | | 5 | 3.200 | | 4,975 | 154 | | 5,129 | 5,700 |
| 7500 | Convection oven | | 5 | 3.200 | | 4,650 | 154 | | 4,804 | 5,325 |
| 7700 | Griddle top | | 6 | 2.667 | | 2,525 | 129 | | 2,654 | 3,000 |
| 7750 | Convection oven | ↓ | 6 | 2.667 | | 7,000 | 129 | | 7,129 | 7,900 |
| 8850 | Steamer, electric 27 KW | L-7 | 7 | 4 | | 8,125 | 166 | | 8,291 | 9,200 |
| 9100 | Electric, 10 KW or gas 100,000 BTU | " | 5 | 5.600 | | 5,975 | 233 | | 6,208 | 6,925 |
| 9150 | Toaster, conveyor type, 16-22 slices per minute | | | | | 975 | | | 975 | 1,075 |
| 9160 | Pop-up, 2 slot | | | | ↓ | 600 | | | 600 | 660 |
| 9200 | For deluxe models of above equipment, add | | | | | 75% | | | | |
| 9400 | Rule of thumb: Equipment cost based | | | | | | | | | |
| 9410 | on kitchen work area | | | | | | | | | |
| 9420 | Office buildings, minimum | L-7 | 77 | .364 | S.F. | 85.50 | 15.15 | | 100.65 | 118 |
| 9450 | Maximum | | 58 | .483 | | 145 | 20 | | 165 | 190 |
| 9550 | Public eating facilities, minimum | | 77 | .364 | | 112 | 15.15 | | 127.15 | 147 |
| 9600 | Maximum | | 46 | .609 | | 183 | 25.50 | | 208.50 | 240 |
| 9750 | Hospitals, minimum | | 58 | .483 | | 115 | 20 | | 135 | 158 |
| 9800 | Maximum | ↓ | 39 | .718 | ↓ | 213 | 30 | | 243 | 280 |

# 11 46 Food Dispensing Equipment

## 11 46 16 – Service Line Equipment

### 11 46 16.10 Commercial Food Dispensing Equipment

| | | Crew | Daily Output | Labor-Hours | Unit | Material | Labor | Equipment | Total | Total Incl O&P |
|---|---|---|---|---|---|---|---|---|---|---|
| 0010 | **COMMERCIAL FOOD DISPENSING EQUIPMENT** | | | | | | | | | |
| 1050 | Butter pat dispenser | 1 Clab | 13 | .615 | Ea. | 925 | 21 | | 946 | 1,050 |
| 1100 | Bread dispenser, counter top | | 13 | .615 | | 885 | 21 | | 906 | 1,000 |
| 1900 | Cup and glass dispenser, drop in | | 4 | 2 | | 960 | 68.50 | | 1,028.50 | 1,150 |
| 1920 | Disposable cup, drop in | | 16 | .500 | | 435 | 17.20 | | 452.20 | 505 |
| 2650 | Dish dispenser, drop in, 12" | | 11 | .727 | | 2,025 | 25 | | 2,050 | 2,275 |
| 2660 | Mobile | ↓ | 10 | .800 | | 2,450 | 27.50 | | 2,477.50 | 2,750 |
| 3300 | Food warmer, counter, 1.2 KW | | | | | 480 | | | 480 | 530 |
| 3550 | 1.6 KW | | | | | 1,975 | | | 1,975 | 2,175 |
| 3600 | Well, hot food, built-in, rectangular, 12" x 20" | R-30 | 10 | 2.600 | | 640 | 106 | | 746 | 860 |
| 3610 | Circular, 7 qt. | | 10 | 2.600 | | 360 | 106 | | 466 | 560 |
| 3620 | Refrigerated, 2 compartments | | 10 | 2.600 | | 2,725 | 106 | | 2,831 | 3,150 |
| 3630 | 3 compartments | | 9 | 2.889 | | 3,325 | 117 | | 3,442 | 3,850 |
| 3640 | 4 compartments | | 8 | 3.250 | | 3,900 | 132 | | 4,032 | 4,500 |
| 4720 | Frost cold plate | ↓ | 9 | 2.889 | | 17,400 | 117 | | 17,517 | 19,400 |
| 5700 | Hot chocolate dispenser | 1 Plum | 4 | 2 | | 1,075 | 107 | | 1,182 | 1,325 |
| 6250 | Jet spray dispenser | R-18 | 4.50 | 5.778 | | 3,100 | 228 | | 3,328 | 3,750 |
| 6300 | Juice dispenser, concentrate | " | 4.50 | 5.778 | | 2,150 | 228 | | 2,378 | 2,700 |
| 6690 | Milk dispenser, bulk, 2 flavor | R-30 | 8 | 3.250 | | 1,525 | 132 | | 1,657 | 1,875 |
| 6695 | 3 flavor | " | 8 | 3.250 | ↓ | 2,100 | 132 | | 2,232 | 2,500 |
| 8800 | Serving counter, straight | 1 Carp | 40 | .200 | L.F. | 805 | 8.60 | | 813.60 | 900 |

# 11 46 Food Dispensing Equipment

## 11 46 16 – Service Line Equipment

| 11 46 16.10 Commercial Food Dispensing Equipment | Crew | Daily Output | Labor-Hours | Unit | Material | 2011 Bare Costs Labor | Equipment | Total | Total Incl O&P |
|---|---|---|---|---|---|---|---|---|---|
| 8820 | Curved section | 1 Carp | 30 | .267 | L.F. | 1,000 | 11.50 | | 1,011.50 | 1,125 |
| 8825 | Solid surface, see Section 12 36 61.16 | | | | | | | | | |
| 8830 | Soft serve ice cream machine, medium | R-18 | 11 | 2.364 | Ea. | 11,700 | 93.50 | | 11,793.50 | 12,900 |
| 8840 | Large | " | 9 | 2.889 | " | 19,500 | 114 | | 19,614 | 21,700 |

# 11 47 Ice Machines

## 11 47 10 – Commercial Ice Machines

### 11 47 10.10 Commercial Ice Equipment

| | | Crew | Daily Output | Labor-Hours | Unit | Material | 2011 Bare Costs Labor | Equipment | Total | Total Incl O&P |
|---|---|---|---|---|---|---|---|---|---|---|
| 0010 | **COMMERCIAL ICE EQUIPMENT** | | | | | | | | | |
| 5800 | Ice cube maker, 50 pounds per day | Q-1 | 6 | 2.667 | Ea. | 1,475 | 129 | | 1,604 | 1,825 |
| 5900 | 250 pounds per day | | 1.20 | 13.333 | | 2,600 | 645 | | 3,245 | 3,800 |
| 6050 | 500 pounds per day | | 4 | 4 | | 2,825 | 193 | | 3,018 | 3,400 |
| 6060 | With bin | | 1.20 | 13.333 | | 2,900 | 645 | | 3,545 | 4,150 |
| 6090 | 1000 pounds per day, with bin | | 1 | 16 | | 4,725 | 770 | | 5,495 | 6,325 |
| 6100 | Ice flakers, 300 pounds per day | | 1.60 | 10 | | 2,700 | 485 | | 3,185 | 3,700 |
| 6120 | 600 pounds per day | | .95 | 16.842 | | 3,850 | 815 | | 4,665 | 5,450 |
| 6130 | 1000 pounds per day | | .75 | 21.333 | | 4,325 | 1,025 | | 5,350 | 6,275 |
| 6140 | 2000 pounds per day | | .65 | 24.615 | | 20,400 | 1,200 | | 21,600 | 24,200 |
| 6160 | Ice storage bin, 500 pound capacity | Q-5 | 1 | 16 | | 1,000 | 780 | | 1,780 | 2,275 |
| 6180 | 1000 pound | " | .56 | 28.571 | | 2,425 | 1,400 | | 3,825 | 4,750 |

# 11 48 Cleaning and Disposal Equipment

## 11 48 13 – Commercial Dishwashers

### 11 48 13.10 Dishwashers

| | | Crew | Daily Output | Labor-Hours | Unit | Material | 2011 Bare Costs Labor | Equipment | Total | Total Incl O&P |
|---|---|---|---|---|---|---|---|---|---|---|
| 0010 | **DISHWASHERS** | | | | | | | | | |
| 2700 | Dishwasher, commercial, rack type | | | | | | | | | |
| 2720 | 10 to 12 racks per hour | Q-1 | 3.20 | 5 | Ea. | 4,300 | 241 | | 4,541 | 5,075 |
| 2730 | Energy star rated, 35 to 40 racks/hour   G | | 1.30 | 12.308 | | 4,650 | 595 | | 5,245 | 5,975 |
| 2740 | 50 to 60 racks/hour   G | | 1.30 | 12.308 | | 10,100 | 595 | | 10,695 | 12,000 |
| 2800 | Automatic, 190 to 230 racks per hour | L-6 | .35 | 34.286 | | 14,300 | 1,800 | | 16,100 | 18,400 |
| 2820 | 235 to 275 racks per hour | | .25 | 48 | | 30,300 | 2,525 | | 32,825 | 37,100 |
| 2840 | 8,750 to 12,500 dishes per hour | | .10 | 120 | | 49,500 | 6,300 | | 55,800 | 64,000 |
| 2950 | Dishwasher hood, canopy type | L-3A | 10 | 1.200 | L.F. | 800 | 56.50 | | 856.50 | 965 |
| 2960 | Pant leg type | " | 2.50 | 4.800 | Ea. | 8,225 | 227 | | 8,452 | 9,400 |
| 5200 | Garbage disposal 1.5 HP, 100 GPH | L-1 | 4.80 | 3.333 | | 1,950 | 173 | | 2,123 | 2,375 |
| 5210 | 3 HP, 120 GPH | | 4.60 | 3.478 | | 2,075 | 181 | | 2,256 | 2,550 |
| 5220 | 5 HP, 250 GPH | | 4.50 | 3.556 | | 3,275 | 185 | | 3,460 | 3,875 |
| 6750 | Pot sink, 3 compartment | 1 Plum | 7.25 | 1.103 | L.F. | 1,000 | 59 | | 1,059 | 1,200 |
| 6760 | Pot washer, low temp wash/rinse | | 1.60 | 5 | Ea. | 3,825 | 268 | | 4,093 | 4,600 |
| 6770 | High pressure wash, high temperature rinse | | 1.20 | 6.667 | | 36,900 | 355 | | 37,255 | 41,100 |
| 9170 | Trash compactor, small, up to 125 lb. compacted weight | L-4 | 4 | 6 | | 22,800 | 243 | | 23,043 | 25,500 |
| 9175 | Large, up to 175 lb. compacted weight | " | 3 | 8 | | 27,900 | 325 | | 28,225 | 31,200 |

# 11 52 Audio-Visual Equipment

## 11 52 13 – Projection Screens

### 11 52 13.10 Projection Screens, Wall or Ceiling Hung

| | 11 52 13.10 Projection Screens, Wall or Ceiling Hung | Crew | Daily Output | Labor-Hours | Unit | Material | 2011 Bare Costs Labor | Equipment | Total | Total Incl O&P |
|---|---|---|---|---|---|---|---|---|---|---|
| 0010 | **PROJECTION SCREENS, WALL OR CEILING HUNG**, matte white | | | | | | | | | |
| 0100 | Manually operated, economy | 2 Carp | 500 | .032 | S.F. | 5.85 | 1.38 | | 7.23 | 8.55 |
| 0300 | Intermediate | | 450 | .036 | | 6.80 | 1.53 | | 8.33 | 9.85 |
| 0400 | Deluxe | | 400 | .040 | | 9.45 | 1.72 | | 11.17 | 13 |
| 0600 | Electric operated, matte white, 25 S.F., economy | | 5 | 3.200 | Ea. | 855 | 138 | | 993 | 1,150 |
| 0700 | Deluxe | | 4 | 4 | | 1,775 | 172 | | 1,947 | 2,225 |
| 0900 | 50 S.F., economy | | 3 | 5.333 | | 750 | 230 | | 980 | 1,175 |
| 1000 | Deluxe | | 2 | 8 | | 1,975 | 345 | | 2,320 | 2,700 |
| 1200 | Heavy duty, electric operated, 200 S.F. | | 1.50 | 10.667 | | 3,900 | 460 | | 4,360 | 5,000 |
| 1300 | 400 S.F. | | 1 | 16 | | 4,800 | 690 | | 5,490 | 6,350 |
| 1500 | Rigid acrylic in wall, for rear projection, 1/4" thick | 2 Glaz | 30 | .533 | S.F. | 47 | 22.50 | | 69.50 | 85.50 |
| 1600 | 1/2" thick (maximum size 10' x 20') | " | 25 | .640 | " | 83.50 | 27 | | 110.50 | 132 |

## 11 52 16 – Projectors

### 11 52 16.10 Movie Equipment

| | 11 52 16.10 Movie Equipment | Crew | Daily Output | Labor-Hours | Unit | Material | 2011 Bare Costs Labor | Equipment | Total | Total Incl O&P |
|---|---|---|---|---|---|---|---|---|---|---|
| 0010 | **MOVIE EQUIPMENT** | | | | | | | | | |
| 0020 | Changeover, minimum | | | | Ea. | 465 | | | 465 | 510 |
| 0100 | Maximum | | | | | 905 | | | 905 | 995 |
| 0400 | Film transport, incl. platters and autowind, minimum | | | | | 5,025 | | | 5,025 | 5,525 |
| 0500 | Maximum | | | | | 14,300 | | | 14,300 | 15,700 |
| 0800 | Lamphouses, incl. rectifiers, xenon, 1,000 watt | 1 Elec | 2 | 4 | | 6,650 | 201 | | 6,851 | 7,600 |
| 0900 | 1,600 watt | | 2 | 4 | | 7,100 | 201 | | 7,301 | 8,100 |
| 1000 | 2,000 watt | | 1.50 | 5.333 | | 7,625 | 268 | | 7,893 | 8,775 |
| 1100 | 4,000 watt | | 1.50 | 5.333 | | 9,425 | 268 | | 9,693 | 10,800 |
| 1400 | Lenses, anamorphic, minimum | | | | | 1,275 | | | 1,275 | 1,400 |
| 1500 | Maximum | | | | | 2,875 | | | 2,875 | 3,150 |
| 1800 | Flat 35 mm, minimum | | | | | 1,100 | | | 1,100 | 1,225 |
| 1900 | Maximum | | | | | 1,725 | | | 1,725 | 1,900 |
| 2200 | Pedestals, for projectors | | | | | 1,475 | | | 1,475 | 1,625 |
| 2300 | Console type | | | | | 10,700 | | | 10,700 | 11,800 |
| 2600 | Projector mechanisms, incl. soundhead, 35 mm, minimum | | | | | 11,100 | | | 11,100 | 12,200 |
| 2700 | Maximum | | | | | 15,200 | | | 15,200 | 16,700 |
| 3000 | Projection screens, rigid, in wall, acrylic, 1/4" thick | 2 Glaz | 195 | .082 | S.F. | 42 | 3.45 | | 45.45 | 51 |
| 3100 | 1/2" thick | " | 130 | .123 | " | 48.50 | 5.20 | | 53.70 | 61.50 |
| 3300 | Electric operated, heavy duty, 400 S.F. | 2 Carp | 1 | 16 | Ea. | 2,950 | 690 | | 3,640 | 4,300 |
| 3320 | Theater projection screens, matte white, including frames | " | 200 | .080 | S.F. | 6.65 | 3.44 | | 10.09 | 12.60 |
| 3400 | Also see Section 11 52 13.10 | | | | | | | | | |
| 3700 | Sound systems, incl. amplifier, mono, minimum | 1 Elec | .90 | 8.889 | Ea. | 3,300 | 445 | | 3,745 | 4,275 |
| 3800 | Dolby/Super Sound, maximum | | .40 | 20 | | 18,100 | 1,000 | | 19,100 | 21,400 |
| 4100 | Dual system, 2 channel, front surround, minimum | | .70 | 11.429 | | 4,625 | 575 | | 5,200 | 5,950 |
| 4200 | Dolby/Super Sound, 4 channel, maximum | | .40 | 20 | | 16,500 | 1,000 | | 17,500 | 19,700 |
| 4500 | Sound heads, 35 mm | | | | | 5,300 | | | 5,300 | 5,825 |
| 4900 | Splicer, tape system, minimum | | | | | 740 | | | 740 | 815 |
| 5000 | Tape type, maximum | | | | | 1,325 | | | 1,325 | 1,450 |
| 5300 | Speakers, recessed behind screen, minimum | 1 Elec | 2 | 4 | | 1,050 | 201 | | 1,251 | 1,475 |
| 5400 | Maximum | " | 1 | 8 | | 3,100 | 400 | | 3,500 | 4,000 |
| 5700 | Seating, painted steel, upholstered, minimum | 2 Carp | 35 | .457 | | 132 | 19.70 | | 151.70 | 175 |
| 5800 | Maximum | " | 28 | .571 | | 420 | 24.50 | | 444.50 | 505 |
| 6100 | Rewind tables, minimum | | | | | 2,650 | | | 2,650 | 2,900 |
| 6200 | Maximum | | | | | 4,725 | | | 4,725 | 5,175 |
| 7000 | For automation, varying sophistication, minimum | 1 Elec | 1 | 8 | System | 2,375 | 400 | | 2,775 | 3,225 |
| 7100 | Maximum | 2 Elec | .30 | 53.333 | " | 5,575 | 2,675 | | 8,250 | 10,100 |

# 11 52 Audio-Visual Equipment

## 11 52 16 – Projectors

| 11 52 16.20 Movie Equipment- Digital | Crew | Daily Output | Labor-Hours | Unit | Material | 2011 Bare Costs Labor | Equipment | Total | Total Incl O&P |
|---|---|---|---|---|---|---|---|---|---|
| 0010 **MOVIE EQUIPMENT- DIGITAL** | | | | | | | | | |
| 1000 Digital 2K projection system, 98" DMD | 1 Elec | 2 | 4 | Ea. | 35,800 | 201 | | 36,001 | 39,600 |
| 1100 OEM lens | | 2 | 4 | | 5,125 | 201 | | 5,326 | 5,925 |
| 2000 Pedestal with power distribution | | 2 | 4 | | 1,950 | 201 | | 2,151 | 2,450 |
| 3000 Software | ↓ | 2 | 4 | ↓ | 1,600 | 201 | | 1,801 | 2,050 |

# 11 53 Laboratory Equipment

## 11 53 03 – Laboratory Test Equipment

### 11 53 03.13 Test Equipment

| | Crew | Daily Output | Labor-Hours | Unit | Material | Labor | Equipment | Total | Total Incl O&P |
|---|---|---|---|---|---|---|---|---|---|
| 0010 **TEST EQUIPMENT** | | | | | | | | | |
| 1700 Thermometer, electric, portable | | | | Ea. | 365 | | | 365 | 400 |
| 1800 Titration unit, four 2000 ml reservoirs | | | | " | 5,500 | | | 5,500 | 6,050 |

## 11 53 13 – Laboratory Fume Hoods

### 11 53 13.13 Recirculating Laboratory Fume Hoods

| | Crew | Daily Output | Labor-Hours | Unit | Material | Labor | Equipment | Total | Total Incl O&P |
|---|---|---|---|---|---|---|---|---|---|
| 0010 **RECIRCULATING LABORATORY FUME HOODS** | | | | | | | | | |
| 0600 Fume hood, with countertop & base, not including HVAC | | | | | | | | | |
| 0610 Simple, minimum | 2 Carp | 5.40 | 2.963 | L.F. | 635 | 128 | | 763 | 895 |
| 0620 Complex, including fixtures | | 2.40 | 6.667 | | 1,800 | 287 | | 2,087 | 2,425 |
| 0630 Special, maximum | ↓ | 1.70 | 9.412 | ↓ | 1,825 | 405 | | 2,230 | 2,625 |
| 0670 Service fixtures, average | | | | Ea. | 287 | | | 287 | 315 |
| 0680 For sink assembly with hot and cold water, add | 1 Plum | 1.40 | 5.714 | | 705 | 305 | | 1,010 | 1,225 |
| 0750 Glove box, fiberglass, bacteriological | | | | | 16,900 | | | 16,900 | 18,600 |
| 0760 Controlled atmosphere | | | | | 18,900 | | | 18,900 | 20,700 |
| 0770 Radioisotope | | | | | 16,900 | | | 16,900 | 18,600 |
| 0780 Carcinogenic | | | | ↓ | 16,900 | | | 16,900 | 18,600 |

### 11 53 13.23 Exhaust Hoods

| | Crew | Daily Output | Labor-Hours | Unit | Material | Labor | Equipment | Total | Total Incl O&P |
|---|---|---|---|---|---|---|---|---|---|
| 0010 **EXHAUST HOODS** | | | | | | | | | |
| 0650 Ductwork, minimum | 2 Shee | 1 | 16 | Hood | 3,625 | 825 | | 4,450 | 5,250 |
| 0660 Maximum | " | .50 | 32 | " | 5,575 | 1,650 | | 7,225 | 8,650 |

## 11 53 16 – Laboratory Incubators

### 11 53 16.13 Incubators

| | Crew | Daily Output | Labor-Hours | Unit | Material | Labor | Equipment | Total | Total Incl O&P |
|---|---|---|---|---|---|---|---|---|---|
| 0010 **INCUBATORS** | | | | | | | | | |
| 1000 Incubators, minimum | | | | Ea. | 3,400 | | | 3,400 | 3,750 |
| 1010 Maximum | | | | " | 20,100 | | | 20,100 | 22,100 |

## 11 53 19 – Laboratory Sterilizers

### 11 53 19.13 Sterilizers

| | Crew | Daily Output | Labor-Hours | Unit | Material | Labor | Equipment | Total | Total Incl O&P |
|---|---|---|---|---|---|---|---|---|---|
| 0010 **STERILIZERS** | | | | | | | | | |
| 0700 Glassware washer, undercounter, minimum | L-1 | 1.80 | 8.889 | Ea. | 5,850 | 460 | | 6,310 | 7,125 |
| 0710 Maximum | " | 1 | 16 | | 11,700 | 830 | | 12,530 | 14,000 |
| 1850 Utensil washer-sanitizer | 1 Plum | 2 | 4 | ↓ | 10,900 | 214 | | 11,114 | 12,300 |

## 11 53 23 – Laboratory Refrigerators

### 11 53 23.13 Refrigerators

| | Crew | Daily Output | Labor-Hours | Unit | Material | Labor | Equipment | Total | Total Incl O&P |
|---|---|---|---|---|---|---|---|---|---|
| 0010 **REFRIGERATORS** | | | | | | | | | |
| 1200 Blood bank, 28.6 C.F. emergency signal | | | | Ea. | 8,675 | | | 8,675 | 9,550 |
| 1210 Reach-in, 16.9 C.F. | | | | " | 8,175 | | | 8,175 | 9,000 |

# 11 53 Laboratory Equipment

## 11 53 33 – Emergency Safety Appliances

| 11 53 33.13 Emergency Equipment | Crew | Daily Output | Labor-Hours | Unit | Material | 2011 Bare Costs Labor | 2011 Bare Costs Equipment | Total | Total Incl O&P |
|---|---|---|---|---|---|---|---|---|---|
| 0010 **EMERGENCY EQUIPMENT** | | | | | | | | | |
| 1400 Safety equipment, eye wash, hand held | | | | Ea. | 405 | | | 405 | 445 |
| 1450 Deluge shower | | | | " | 740 | | | 740 | 815 |

## 11 53 43 – Service Fittings and Accessories

### 11 53 43.13 Fittings

| | Crew | Daily Output | Labor-Hours | Unit | Material | 2011 Bare Costs Labor | 2011 Bare Costs Equipment | Total | Total Incl O&P |
|---|---|---|---|---|---|---|---|---|---|
| 0010 **FITTINGS** | | | | | | | | | |
| 1600 Sink, one piece plastic, flask wash, hose, free standing | 1 Plum | 1.60 | 5 | Ea. | 1,800 | 268 | | 2,068 | 2,375 |
| 1610 Epoxy resin sink, 25" x 16" x 10" | " | 2 | 4 | " | 198 | 214 | | 412 | 540 |
| 1950 Utility table, acid resistant top with drawers | 2 Carp | 30 | .533 | L.F. | 147 | 23 | | 170 | 196 |
| 8000 Alternate pricing method: as percent of lab furniture | | | | | | | | | |
| 8050 Installation, not incl. plumbing & duct work | | | | % Furn. | | | | 20% | 22% |
| 8100 Plumbing, final connections, simple system | | | | | | | | 9.09% | 10% |
| 8110 Moderately complex system | | | | | | | | 13.64% | 15% |
| 8120 Complex system | | | | | | | | 18.18% | 20% |
| 8150 Electrical, simple system | | | | | | | | 9.09% | 10% |
| 8160 Moderately complex system | | | | | | | | 18.18% | 20% |
| 8170 Complex system | | | | | | | | 31.80% | 35% |

# 11 57 Vocational Shop Equipment

## 11 57 10 – Shop Equipment

### 11 57 10.10 Vocational School Shop Equipment

| | Crew | Daily Output | Labor-Hours | Unit | Material | 2011 Bare Costs Labor | 2011 Bare Costs Equipment | Total | Total Incl O&P |
|---|---|---|---|---|---|---|---|---|---|
| 0010 **VOCATIONAL SCHOOL SHOP EQUIPMENT** | | | | | | | | | |
| 0020 Benches, work, wood, average | 2 Carp | 5 | 3.200 | Ea. | 525 | 138 | | 663 | 785 |
| 0100 Metal, average | | 5 | 3.200 | | 475 | 138 | | 613 | 730 |
| 0400 Combination belt & disc sander, 6" | | 4 | 4 | | 1,400 | 172 | | 1,572 | 1,800 |
| 0700 Drill press, floor mounted, 12", 1/2 H.P. | | 4 | 4 | | 425 | 172 | | 597 | 730 |
| 0800 Dust collector, not incl. ductwork, 6" diameter | 1 Shee | 1.10 | 7.273 | | 4,500 | 375 | | 4,875 | 5,525 |
| 0810 Dust collector bag, 20" diameter | " | 5 | 1.600 | | 310 | 82.50 | | 392.50 | 465 |
| 1000 Grinders, double wheel, 1/2 H.P. | 2 Carp | 5 | 3.200 | | 254 | 138 | | 392 | 490 |
| 1300 Jointer, 4", 3/4 H.P. | | 4 | 4 | | 1,325 | 172 | | 1,497 | 1,725 |
| 1600 Kilns, 16 C.F., to 2000° | | 4 | 4 | | 2,275 | 172 | | 2,447 | 2,775 |
| 1900 Lathe, woodworking, 10", 1/2 H.P. | | 4 | 4 | | 735 | 172 | | 907 | 1,075 |
| 2200 Planer, 13" x 6" | | 4 | 4 | | 1,000 | 172 | | 1,172 | 1,375 |
| 2500 Potter's wheel, motorized | | 4 | 4 | | 1,175 | 172 | | 1,347 | 1,575 |
| 2800 Saws, band, 14", 3/4 H.P. | | 4 | 4 | | 865 | 172 | | 1,037 | 1,225 |
| 3100 Metal cutting band saw, 14" | | 4 | 4 | | 2,200 | 172 | | 2,372 | 2,675 |
| 3400 Radial arm saw, 10", 2 H.P. | | 4 | 4 | | 1,325 | 172 | | 1,497 | 1,725 |
| 3700 Scroll saw, 24" | | 4 | 4 | | 590 | 172 | | 762 | 915 |
| 4000 Table saw, 10", 3 H.P. | | 4 | 4 | | 2,150 | 172 | | 2,322 | 2,625 |
| 4300 Welder AC arc, 30 amp capacity | | 4 | 4 | | 2,100 | 172 | | 2,272 | 2,600 |

# 11 61 Theater and Stage Equipment

## 11 61 23 – Folding and Portable Stages

### 11 61 23.10 Portable Stages

| | | Crew | Daily Output | Labor-Hours | Unit | Material | 2011 Bare Costs Labor | Equipment | Total | Total Incl O&P |
|---|---|---|---|---|---|---|---|---|---|---|
| 0010 | **PORTABLE STAGES** | | | | | | | | | |
| 1500 | Flooring, portable oak parquet, 3' x 3' sections | | | | S.F. | 13.40 | | | 13.40 | 14.70 |
| 1600 | Cart to carry 225 S.F. of flooring | | | | Ea. | 400 | | | 400 | 440 |
| 5000 | Stages, portable with steps, folding legs, stock, 8" high | | | | SF Stg. | 30.50 | | | 30.50 | 33.50 |
| 5100 | 16" high | | | | | 23.50 | | | 23.50 | 26 |
| 5200 | 32" high | | | | | 41.50 | | | 41.50 | 46 |
| 5300 | 40" high | | | | | 59.50 | | | 59.50 | 65 |
| 6000 | Telescoping platforms, extruded alum., straight, minimum | 4 Carp | 157 | .204 | | 33 | 8.75 | | 41.75 | 49.50 |
| 6100 | Maximum | | 77 | .416 | | 44.50 | 17.90 | | 62.40 | 76.50 |
| 6500 | Pie-shaped, minimum | | 150 | .213 | | 71 | 9.20 | | 80.20 | 92.50 |
| 6600 | Maximum | | 70 | .457 | | 79.50 | 19.70 | | 99.20 | 118 |
| 6800 | For 3/4" plywood covered deck, deduct | | | | | 4.05 | | | 4.05 | 4.46 |
| 7000 | Band risers, steel frame, plywood deck, minimum | 4 Carp | 275 | .116 | | 30.50 | 5 | | 35.50 | 41 |
| 7100 | Maximum | " | 138 | .232 | | 67 | 10 | | 77 | 89 |
| 7500 | Chairs for above, self-storing, minimum | 2 Carp | 43 | .372 | Ea. | 107 | 16 | | 123 | 143 |
| 7600 | Maximum | " | 40 | .400 | " | 189 | 17.20 | | 206.20 | 235 |

## 11 61 33 – Rigging Systems and Controls

### 11 61 33.10 Controls

| | | Crew | Daily Output | Labor-Hours | Unit | Material | 2011 Bare Costs Labor | Equipment | Total | Total Incl O&P |
|---|---|---|---|---|---|---|---|---|---|---|
| 0010 | **CONTROLS** | | | | | | | | | |
| 0050 | Control boards with dimmers and breakers, minimum | 1 Elec | 1 | 8 | Ea. | 12,300 | 400 | | 12,700 | 14,100 |
| 0100 | Average | | .50 | 16 | | 38,600 | 805 | | 39,405 | 43,600 |
| 0150 | Maximum | | .20 | 40 | | 126,000 | 2,000 | | 128,000 | 141,500 |
| 8000 | Rule of thumb: total stage equipment, minimum | 4 Carp | 100 | .320 | SF Stg. | 97 | 13.80 | | 110.80 | 128 |
| 8100 | Maximum | " | 25 | 1.280 | " | 545 | 55 | | 600 | 685 |

## 11 61 43 – Stage Curtains

### 11 61 43.10 Curtains

| | | Crew | Daily Output | Labor-Hours | Unit | Material | 2011 Bare Costs Labor | Equipment | Total | Total Incl O&P |
|---|---|---|---|---|---|---|---|---|---|---|
| 0010 | **CURTAINS** | | | | | | | | | |
| 0500 | Curtain track, straight, light duty | 2 Carp | 20 | .800 | L.F. | 27 | 34.50 | | 61.50 | 82.50 |
| 0600 | Heavy duty | | 18 | .889 | | 61 | 38.50 | | 99.50 | 126 |
| 0700 | Curved sections | | 12 | 1.333 | | 173 | 57.50 | | 230.50 | 278 |
| 1000 | Curtains, velour, medium weight | | 600 | .027 | S.F. | 7.90 | 1.15 | | 9.05 | 10.45 |
| 1150 | Silica based yarn, inherently fire retardant | | 50 | .320 | " | 15.20 | 13.80 | | 29 | 37.50 |

# 11 62 Musical Equipment

## 11 62 16 – Carillons

### 11 62 16.10 Bell Tower Equipment

| | | Crew | Daily Output | Labor-Hours | Unit | Material | 2011 Bare Costs Labor | Equipment | Total | Total Incl O&P |
|---|---|---|---|---|---|---|---|---|---|---|
| 0010 | **BELL TOWER EQUIPMENT** | | | | | | | | | |
| 0300 | Carillon, 4 octave (48 bells), with keyboard | | | | System | 937,000 | | | 937,000 | 1,031,000 |
| 0320 | 2 octave (24 bells) | | | | | 441,000 | | | 441,000 | 485,000 |
| 0340 | 3 to 4 bell peal, minimum | | | | | 110,500 | | | 110,500 | 121,500 |
| 0360 | Maximum | | | | | 661,500 | | | 661,500 | 727,500 |
| 0380 | Cast bronze bell, average | | | | Ea. | 99,000 | | | 99,000 | 109,000 |
| 0400 | Electronic, digital, minimum | | | | | 16,500 | | | 16,500 | 18,200 |
| 0410 | With keyboard, maximum | | | | | 82,500 | | | 82,500 | 91,000 |

## 11 66 13 – Exercise Equipment

| 11 66 13.10 Physical Training Equipment | Crew | Daily Output | Labor-Hours | Unit | Material | 2011 Bare Costs Labor | 2011 Bare Costs Equipment | Total | Total Incl O&P |
|---|---|---|---|---|---|---|---|---|---|
| 0010 **PHYSICAL TRAINING EQUIPMENT** | | | | | | | | | |
| 0020 Abdominal rack, 2 board capacity | | | | Ea. | 485 | | | 485 | 535 |
| 0050 Abdominal board, upholstered | | | | | 605 | | | 605 | 665 |
| 0200 Bicycle trainer, minimum | | | | | 400 | | | 400 | 440 |
| 0300 Deluxe, electric | | | | | 3,800 | | | 3,800 | 4,175 |
| 0400 Barbell set, chrome plated steel, 25 lbs. | | | | | 281 | | | 281 | 310 |
| 0420 100 lbs. | | | | | 420 | | | 420 | 460 |
| 0450 200 lbs. | | | | | 685 | | | 685 | 755 |
| 0500 Weight plates, cast iron, per lb. | | | | Lb. | 5.20 | | | 5.20 | 5.70 |
| 0520 Storage rack, 10 station | | | | Ea. | 900 | | | 900 | 990 |
| 0600 Circuit training apparatus, 12 machines minimum | 2 Clab | 1.25 | 12.800 | Set | 28,700 | 440 | | 29,140 | 32,200 |
| 0700 Average | | 1 | 16 | | 35,200 | 550 | | 35,750 | 39,500 |
| 0800 Maximum | | .75 | 21.333 | | 41,700 | 735 | | 42,435 | 47,000 |
| 0820 Dumbbell set, cast iron, with rack and 5 pair | | | | | 605 | | | 605 | 670 |
| 0900 Squat racks | 2 Clab | 5 | 3.200 | Ea. | 890 | 110 | | 1,000 | 1,150 |
| 1200 Multi-station gym machine, 5 station | | | | | 5,000 | | | 5,000 | 5,500 |
| 1250 9 station | | | | | 10,000 | | | 10,000 | 11,000 |
| 1280 Rowing machine, hydraulic | | | | | 1,700 | | | 1,700 | 1,875 |
| 1300 Treadmill, manual | | | | | 1,000 | | | 1,000 | 1,100 |
| 1320 Motorized | | | | | 3,625 | | | 3,625 | 4,000 |
| 1340 Electronic | | | | | 3,700 | | | 3,700 | 4,075 |
| 1360 Cardio-testing | | | | | 4,000 | | | 4,000 | 4,400 |
| 1400 Treatment/massage tables, minimum | | | | | 590 | | | 590 | 650 |
| 1420 Deluxe, with accessories | | | | | 715 | | | 715 | 785 |
| 4150 Exercise equipment, bicycle trainer | | | | | 595 | | | 595 | 655 |
| 4180 Chinning bar, adjustable, wall mounted | 1 Carp | 5 | 1.600 | | 300 | 69 | | 369 | 435 |
| 4200 Exercise ladder, 16' x 1'-7", suspended | L-2 | 3 | 5.333 | | 1,525 | 202 | | 1,727 | 2,000 |
| 4210 High bar, floor plate attached | 1 Carp | 4 | 2 | | 1,150 | 86 | | 1,236 | 1,400 |
| 4240 Parallel bars, adjustable | | 4 | 2 | | 2,125 | 86 | | 2,211 | 2,475 |
| 4270 Uneven parallel bars, adjustable | | 4 | 2 | | 2,775 | 86 | | 2,861 | 3,175 |
| 4280 Wall mounted, adjustable | L-2 | 1.50 | 10.667 | Set | 820 | 405 | | 1,225 | 1,525 |
| 4300 Rope, ceiling mounted, 18' long | 1 Carp | 3.66 | 2.186 | Ea. | 268 | 94 | | 362 | 440 |
| 4330 Side horse, vaulting | | 5 | 1.600 | | 1,775 | 69 | | 1,844 | 2,050 |
| 4360 Treadmill, motorized, deluxe, training type | | 5 | 1.600 | | 3,000 | 69 | | 3,069 | 3,400 |
| 4390 Weight lifting multi-station, minimum | 2 Clab | 1 | 16 | | 450 | 550 | | 1,000 | 1,350 |
| 4450 Maximum | " | .50 | 32 | | 14,700 | 1,100 | | 15,800 | 17,900 |

## 11 66 23 – Gymnasium Equipment

### 11 66 23.13 Basketball Equipment

| | Crew | Daily Output | Labor-Hours | Unit | Material | 2011 Bare Costs Labor | 2011 Bare Costs Equipment | Total | Total Incl O&P |
|---|---|---|---|---|---|---|---|---|---|
| 0010 **BASKETBALL EQUIPMENT** | | | | | | | | | |
| 1000 Backstops, wall mtd., 6' extended, fixed, minimum | L-2 | 1 | 16 | Ea. | 1,175 | 605 | | 1,780 | 2,225 |
| 1100 Maximum | | 1 | 16 | | 1,750 | 605 | | 2,355 | 2,850 |
| 1200 Swing up, minimum | | 1 | 16 | | 1,425 | 605 | | 2,030 | 2,500 |
| 1250 Maximum | | 1 | 16 | | 5,575 | 605 | | 6,180 | 7,050 |
| 1300 Portable, manual, heavy duty, spring operated | | 1.90 | 8.421 | | 13,800 | 320 | | 14,120 | 15,700 |
| 1400 Ceiling suspended, stationary, minimum | | .78 | 20.513 | | 3,550 | 775 | | 4,325 | 5,100 |
| 1450 Fold up, with accessories, maximum | | .40 | 40 | | 6,350 | 1,525 | | 7,875 | 9,300 |
| 1600 For electrically operated, add | 1 Elec | 1 | 8 | | 2,075 | 400 | | 2,475 | 2,900 |
| 5800 Wall pads, 1-1/2" thick, standard (not fire rated) | 2 Carp | 640 | .025 | S.F. | 6.95 | 1.08 | | 8.03 | 9.30 |

### 11 66 23.19 Boxing Ring

| | Crew | Daily Output | Labor-Hours | Unit | Material | 2011 Bare Costs Labor | 2011 Bare Costs Equipment | Total | Total Incl O&P |
|---|---|---|---|---|---|---|---|---|---|
| 0010 **BOXING RING** | | | | | | | | | |
| 4100 Elevated, 22' x 22' | L-4 | .10 | 240 | Ea. | 7,725 | 9,725 | | 17,450 | 23,400 |
| 4110 For cellular plastic foam padding, add | | .10 | 240 | | 900 | 9,725 | | 10,625 | 15,900 |

# 11 66 Athletic Equipment

## 11 66 23 – Gymnasium Equipment

### 11 66 23.19 Boxing Ring

| | | Daily Output | Labor-Hours | Unit | Material | 2011 Bare Costs Labor | Equipment | Total | Total Incl O&P |
|---|---|---|---|---|---|---|---|---|---|
| | | | | | | Crew | | | |
| 4120 | Floor level, including posts and ropes only, 20' x 20' | L-4 | .80 | 30 | Ea. | 3,750 | 1,225 | | 4,975 | 5,975 |
| 4130 | Canvas, 30' x 30' | ↓ | 5 | 4.800 | ↓ | 1,200 | 194 | | 1,394 | 1,625 |

*(Header note: Crew column appears between Labor-Hours description and Daily Output)*

### 11 66 23.47 Gym Mats

| | | Crew | Daily Output | Labor-Hours | Unit | Material | Labor | Equipment | Total | Total Incl O&P |
|---|---|---|---|---|---|---|---|---|---|---|
| 0010 | **GYM MATS** | | | | | | | | | |
| 5500 | 2" thick, naugahyde covered | | | | S.F. | 3.77 | | | 3.77 | 4.15 |
| 5600 | Vinyl/nylon covered | | | | | 6.35 | | | 6.35 | 7 |
| 6000 | Wrestling mats, 1" thick, heavy duty | | | | ↓ | 5.55 | | | 5.55 | 6.10 |

## 11 66 43 – Interior Scoreboards

### 11 66 43.10 Scoreboards

| | | Crew | Daily Output | Labor-Hours | Unit | Material | Labor | Equipment | Total | Total Incl O&P |
|---|---|---|---|---|---|---|---|---|---|---|
| 0010 | **SCOREBOARDS** | | | | | | | | | |
| 7000 | Baseball, minimum | R-3 | 1.30 | 15.385 | Ea. | 4,000 | 765 | 99 | 4,864 | 5,625 |
| 7200 | Maximum | | .05 | 400 | | 17,500 | 19,900 | 2,575 | 39,975 | 51,500 |
| 7300 | Football, minimum | | .86 | 23.256 | | 5,175 | 1,150 | 150 | 6,475 | 7,575 |
| 7400 | Maximum | | .20 | 100 | | 15,400 | 4,975 | 645 | 21,020 | 25,100 |
| 7500 | Basketball (one side), minimum | | 2.07 | 9.662 | | 2,325 | 480 | 62.50 | 2,867.50 | 3,325 |
| 7600 | Maximum | | .30 | 66.667 | | 3,400 | 3,325 | 430 | 7,155 | 9,150 |
| 7700 | Hockey-basketball (four sides), minimum | | .25 | 80 | | 5,500 | 3,975 | 515 | 9,990 | 12,500 |
| 7800 | Maximum | ↓ | .15 | 133 | ↓ | 5,575 | 6,625 | 860 | 13,060 | 16,900 |

## 11 66 53 – Gymnasium Dividers

### 11 66 53.10 Divider Curtains

| | | Crew | Daily Output | Labor-Hours | Unit | Material | Labor | Equipment | Total | Total Incl O&P |
|---|---|---|---|---|---|---|---|---|---|---|
| 0010 | **DIVIDER CURTAINS** | | | | | | | | | |
| 4500 | Gym divider curtain, mesh top, vinyl bottom, manual | L-4 | 500 | .048 | S.F. | 8.70 | 1.94 | | 10.64 | 12.60 |
| 4700 | Electric roll up | L-7 | 400 | .070 | " | 10.65 | 2.91 | | 13.56 | 16.15 |

# 11 67 Recreational Equipment

## 11 67 13 – Bowling Alley Equipment

### 11 67 13.10 Bowling Alleys

| | | Crew | Daily Output | Labor-Hours | Unit | Material | Labor | Equipment | Total | Total Incl O&P |
|---|---|---|---|---|---|---|---|---|---|---|
| 0010 | **BOWLING ALLEYS** Including alley, pinsetter, scorer, | | | | | | | | | |
| 0020 | Counters and misc. supplies, minimum | 4 Carp | .20 | 160 | Lane | 41,300 | 6,900 | | 48,200 | 56,000 |
| 0150 | Average | | .19 | 168 | | 45,400 | 7,250 | | 52,650 | 61,000 |
| 0300 | Maximum | ↓ | .18 | 177 | | 52,000 | 7,650 | | 59,650 | 69,000 |
| 0400 | Combo table ball rack, add | | | | | 1,150 | | | 1,150 | 1,250 |
| 0600 | For automatic scorer, add, minimum | | | | | 8,225 | | | 8,225 | 9,050 |
| 0700 | Maximum | | | | ↓ | 9,875 | | | 9,875 | 10,900 |

## 11 67 23 – Shooting Range Equipment

### 11 67 23.10 Shooting Range

| | | Crew | Daily Output | Labor-Hours | Unit | Material | Labor | Equipment | Total | Total Incl O&P |
|---|---|---|---|---|---|---|---|---|---|---|
| 0010 | **SHOOTING RANGE** Incl. bullet traps, target provisions, controls, | | | | | | | | | |
| 0100 | Separators, ceiling system, etc. Not incl. structural shell | | | | | | | | | |
| 0200 | Commercial | L-9 | .64 | 56.250 | Point | 17,500 | 2,225 | | 19,725 | 22,800 |
| 0300 | Law enforcement | | .28 | 128 | | 33,000 | 5,075 | | 38,075 | 44,400 |
| 0400 | National Guard armories | | .71 | 50.704 | | 21,000 | 2,000 | | 23,000 | 26,300 |
| 0500 | Reserve training centers | | .71 | 50.704 | | 15,000 | 2,000 | | 17,000 | 19,700 |
| 0600 | Schools and colleges | | .32 | 112 | | 31,000 | 4,425 | | 35,425 | 41,200 |
| 0700 | Major academies | ↓ | .19 | 189 | | 46,000 | 7,475 | | 53,475 | 62,500 |
| 0800 | For acoustical treatment, add | | | | | 10% | 10% | | | |
| 0900 | For lighting, add | | | | | 28% | 25% | | | |
| 1000 | For plumbing, add | | | | | 5% | 5% | | | |
| 1100 | For ventilating system, add, minimum | | | | | 40% | 40% | | | |
| 1200 | Add, average | | | | ↓ | 25% | 25% | | | |

# 11 67 Recreational Equipment

## 11 67 23 – Shooting Range Equipment

| 11 67 23.10 Shooting Range | | Crew | Daily Output | Labor-Hours | Unit | Material | 2011 Bare Costs Labor | Equipment | Total | Total Incl O&P |
|---|---|---|---|---|---|---|---|---|---|---|
| 1300 | Add, maximum | | | | Point | 35% | 35% | | | |

# 11 68 Play Field Equipment and Structures

## 11 68 13 – Playground Equipment

### 11 68 13.10 Free-Standing Playground Equipment

| | | | Crew | Daily Output | Labor-Hours | Unit | Material | Labor | Equipment | Total | Total Incl O&P |
|---|---|---|---|---|---|---|---|---|---|---|---|
| 0010 | **FREE-STANDING PLAYGROUND EQUIPMENT** See also individual items | | | | | | | | | | |
| 0200 | Bike rack, 10' long, permanent | G | B-1 | 12 | 2 | Ea. | 555 | 70 | | 625 | 715 |
| 0400 | Horizontal monkey ladder, 14' long, 6' high | | | 4 | 6 | | 940 | 210 | | 1,150 | 1,350 |
| 0590 | Parallel bars, 10' long | | | 4 | 6 | | 273 | 210 | | 483 | 620 |
| 0600 | Posts, tether ball set, 2-3/8" O.D. | | | 12 | 2 | | 300 | 70 | | 370 | 435 |
| 0800 | Poles, multiple purpose, 10'-6" long | | | 12 | 2 | Pr. | 160 | 70 | | 230 | 283 |
| 1000 | Ground socket for movable posts, 2-3/8" post | | | 10 | 2.400 | | 102 | 84 | | 186 | 241 |
| 1100 | 3-1/2" post | | | 10 | 2.400 | | 139 | 84 | | 223 | 282 |
| 1300 | See-saw, spring, steel, 2 units | | | 6 | 4 | Ea. | 760 | 140 | | 900 | 1,050 |
| 1400 | 4 units | | | 4 | 6 | | 1,375 | 210 | | 1,585 | 1,825 |
| 1500 | 6 units | | | 3 | 8 | | 1,675 | 280 | | 1,955 | 2,250 |
| 1700 | Shelter, fiberglass golf tee, 3 person | | | 4.60 | 5.217 | | 2,675 | 183 | | 2,858 | 3,225 |
| 1900 | Slides, stainless steel bed, 12' long, 6' high | | | 3 | 8 | | 3,525 | 280 | | 3,805 | 4,300 |
| 2000 | 20' long, 10' high | | | 2 | 12 | | 4,250 | 420 | | 4,670 | 5,325 |
| 2200 | Swings, plain seats, 8' high, 4 seats | | | 2 | 12 | | 1,025 | 420 | | 1,445 | 1,775 |
| 2300 | 8 seats | | | 1.30 | 18.462 | | 1,900 | 645 | | 2,545 | 3,100 |
| 2500 | 12' high, 4 seats | | | 2 | 12 | | 1,975 | 420 | | 2,395 | 2,825 |
| 2600 | 8 seats | | | 1.30 | 18.462 | | 3,450 | 645 | | 4,095 | 4,775 |
| 2800 | Whirlers, 8' diameter | | | 3 | 8 | | 2,900 | 280 | | 3,180 | 3,625 |
| 2900 | 10' diameter | | | 3 | 8 | | 6,050 | 280 | | 6,330 | 7,075 |

### 11 68 13.20 Modular Playground

| | | Crew | Daily Output | Labor-Hours | Unit | Material | Labor | Equipment | Total | Total Incl O&P |
|---|---|---|---|---|---|---|---|---|---|---|
| 0010 | **MODULAR PLAYGROUND** Basic components | | | | | | | | | |
| 0100 | Deck, square, steel, 48" x 48" | B-1 | 1 | 24 | Ea. | 525 | 840 | | 1,365 | 1,875 |
| 0110 | Recycled polyurethane | | 1 | 24 | | 525 | 840 | | 1,365 | 1,875 |
| 0120 | Triangular, steel, 48" side | | 1 | 24 | | 630 | 840 | | 1,470 | 2,000 |
| 0130 | Post, steel, 5" square | | 18 | 1.333 | L.F. | 35 | 46.50 | | 81.50 | 110 |
| 0140 | Aluminum, 2-3/8" square | | 20 | 1.200 | | 30 | 42 | | 72 | 97.50 |
| 0150 | 5" square | | 18 | 1.333 | | 38 | 46.50 | | 84.50 | 114 |
| 0160 | Roof, square poly, 54" side | | 18 | 1.333 | Ea. | 1,150 | 46.50 | | 1,196.50 | 1,350 |
| 0170 | Wheelchair transfer module, for 3' high deck | | 3 | 8 | " | 2,725 | 280 | | 3,005 | 3,425 |
| 0180 | Guardrail, pipe, 36" high | | 60 | .400 | L.F. | 202 | 14 | | 216 | 244 |
| 0190 | Steps, deck-to-deck, 3 – 8" steps | | 8 | 3 | Ea. | 990 | 105 | | 1,095 | 1,225 |
| 0200 | Activity panel, crawl through panel | | 2 | 12 | | 365 | 420 | | 785 | 1,050 |
| 0210 | Alphabet/spelling panel | | 2 | 12 | | 495 | 420 | | 915 | 1,200 |
| 0360 | With guardrails | | 3 | 8 | | 1,625 | 280 | | 1,905 | 2,200 |
| 0370 | Crawl tunnel, straight, 56" long | | 4 | 6 | | 925 | 210 | | 1,135 | 1,350 |
| 0380 | 90°, 4' long | | 4 | 6 | | 965 | 210 | | 1,175 | 1,400 |
| 1200 | Slide, tunnel, for 56" high deck | | 8 | 3 | | 1,275 | 105 | | 1,380 | 1,575 |
| 1210 | Straight, poly | | 8 | 3 | | 2,325 | 105 | | 2,430 | 2,700 |
| 1220 | Stainless steel, 54" high deck | | 6 | 4 | | 520 | 140 | | 660 | 790 |
| 1230 | Curved, poly, 40" high deck | | 6 | 4 | | 610 | 140 | | 750 | 890 |
| 1240 | Spiral slide, 56" - 72" high | | 5 | 4.800 | | 3,225 | 168 | | 3,393 | 3,800 |
| 1300 | Ladder, vertical, for 24" - 72" high deck | | 5 | 4.800 | | 600 | 168 | | 768 | 920 |
| 1310 | Horizontal, 8' long | | 5 | 4.800 | | 580 | 168 | | 748 | 895 |
| 1320 | Corkscrew climber, 6' high | | 3 | 8 | | 595 | 280 | | 875 | 1,075 |
| 1330 | Fire pole for 72" high deck | | 6 | 4 | | 550 | 140 | | 690 | 820 |

# 11 68 Play Field Equipment and Structures

## 11 68 13 – Playground Equipment

| 11 68 13.20 Modular Playground | Crew | Daily Output | Labor-Hours | Unit | Material | 2011 Bare Costs Labor | Equipment | Total | Total Incl O&P |
|---|---|---|---|---|---|---|---|---|---|
| 1340 | Bridge, ring climber, 8' long | B-1 | 4 | 6 | Ea. | 1,475 | 210 | | 1,685 | 1,950 |
| 1350 | Suspension | ↓ | 4 | 6 | L.F. | 1,025 | 210 | | 1,235 | 1,450 |

## 11 68 16 – Play Structures

### 11 68 16.10 Handball/Squash Court

| | | Crew | Daily Output | Labor-Hours | Unit | Material | Labor | Equipment | Total | Total Incl O&P |
|---|---|---|---|---|---|---|---|---|---|---|
| 0010 | **HANDBALL/SQUASH COURT**, outdoor | | | | | | | | | |
| 0900 | Handball or squash court, outdoor, wood | 2 Carp | .50 | 32 | Ea. | 5,250 | 1,375 | | 6,625 | 7,925 |
| 1000 | Masonry handball/squash court | D-1 | .30 | 53.333 | " | 24,000 | 2,075 | | 26,075 | 29,400 |

### 11 68 16.30 Platform/Paddle Tennis Court

| | | Crew | Daily Output | Labor-Hours | Unit | Material | Labor | Equipment | Total | Total Incl O&P |
|---|---|---|---|---|---|---|---|---|---|---|
| 0010 | **PLATFORM/PADDLE TENNIS COURT** Complete with lighting, etc. | | | | | | | | | |
| 0100 | Aluminum slat deck with aluminum frame | B-1 | .08 | 300 | Court | 49,500 | 10,500 | | 60,000 | 70,500 |
| 0500 | Aluminum slat deck with wood frame | C-1 | .12 | 266 | | 49,500 | 10,900 | | 60,400 | 71,000 |
| 0800 | Aluminum deck heater, add | B-1 | 1.18 | 20.339 | | 4,900 | 710 | | 5,610 | 6,500 |
| 0900 | Douglas fir planking with wood frame 2" x 6" x 30' | C-1 | .12 | 266 | | 46,500 | 10,900 | | 57,400 | 67,500 |
| 1000 | Plywood deck with steel frame | | .12 | 266 | | 46,500 | 10,900 | | 57,400 | 67,500 |
| 1100 | Steel slat deck with wood frame | ↓ | .12 | 266 | ↓ | 33,000 | 10,900 | | 43,900 | 53,000 |

## 11 68 33 – Athletic Field Equipment

### 11 68 33.13 Football Field Equipment

| | | Crew | Daily Output | Labor-Hours | Unit | Material | Labor | Equipment | Total | Total Incl O&P |
|---|---|---|---|---|---|---|---|---|---|---|
| 0010 | **FOOTBALL FIELD EQUIPMENT** | | | | | | | | | |
| 0020 | Goal posts, steel, football, double post | B-1 | 1.50 | 16 | Pr. | 3,375 | 560 | | 3,935 | 4,575 |
| 0100 | Deluxe, single post | | 1.50 | 16 | | 2,425 | 560 | | 2,985 | 3,500 |
| 0300 | Football, convertible to soccer | | 1.50 | 16 | | 2,025 | 560 | | 2,585 | 3,075 |
| 0500 | Soccer, regulation | ↓ | 2 | 12 | ↓ | 1,075 | 420 | | 1,495 | 1,825 |

# 11 71 Medical Sterilizing Equipment

## 11 71 10 – Medical Sterilizers & Distillers

### 11 71 10.10 Sterilizers and Distillers

| | | Crew | Daily Output | Labor-Hours | Unit | Material | Labor | Equipment | Total | Total Incl O&P |
|---|---|---|---|---|---|---|---|---|---|---|
| 0010 | **STERILIZERS AND DISTILLERS** | | | | | | | | | |
| 0700 | Distiller, water, steam heated, 50 gal. capacity | 1 Plum | 1.40 | 5.714 | Ea. | 18,500 | 305 | | 18,805 | 20,800 |
| 5600 | Sterilizers, floor loading, 26" x 62" x 42", single door, steam | | | | | 158,500 | | | 158,500 | 174,500 |
| 5650 | Double door, steam | | | | | 204,000 | | | 204,000 | 224,000 |
| 5800 | General purpose, 20" x 20" x 38", single door | | | | | 15,300 | | | 15,300 | 16,800 |
| 6000 | Portable, counter top, steam, minimum | | | | | 3,800 | | | 3,800 | 4,175 |
| 6020 | Maximum | | | | | 5,950 | | | 5,950 | 6,550 |
| 6050 | Portable, counter top, gas, 17" x 15" x 32-1/2" | | | | | 39,300 | | | 39,300 | 43,200 |
| 6150 | Manual washer/sterilizer, 16" x 16" x 26" | 1 Plum | 2 | 4 | ↓ | 54,000 | 214 | | 54,214 | 59,500 |
| 6200 | Steam generators, electric 10 kW to 180 kW, freestanding | | | | | | | | | |
| 6250 | Minimum | 1 Elec | 3 | 2.667 | Ea. | 8,000 | 134 | | 8,134 | 9,000 |
| 6300 | Maximum | " | .70 | 11.429 | | 28,800 | 575 | | 29,375 | 32,600 |
| 8200 | Bed pan washer-sanitizer | 1 Plum | 2 | 4 | ↓ | 7,125 | 214 | | 7,339 | 8,175 |

# 11 72 Examination and Treatment Equipment

## 11 72 13 – Examination Equipment

### 11 72 13.13 Examination Equipment

| | 11 72 13.13 Examination Equipment | Crew | Daily Output | Labor-Hours | Unit | Material | 2011 Bare Costs Labor | Equipment | Total | Total Incl O&P |
|---|---|---|---|---|---|---|---|---|---|---|
| 0010 | **EXAMINATION EQUIPMENT** | | | | | | | | | |
| 0300 | Blood pressure unit, mercurial, wall | | | | Ea. | 198 | | | 198 | 218 |
| 0400 | Diagnostic set, wall | | | | | 650 | | | 650 | 715 |
| 4400 | Scale, physician's, with height rod | | | | | 286 | | | 286 | 315 |

## 11 72 53 – Treatment Equipment

### 11 72 53.13 Medical Treatment Equipment

| | 11 72 53.13 Medical Treatment Equipment | Crew | Daily Output | Labor-Hours | Unit | Material | Labor | Equipment | Total | Total Incl O&P |
|---|---|---|---|---|---|---|---|---|---|---|
| 0010 | **MEDICAL TREATMENT EQUIPMENT** | | | | | | | | | |
| 6500 | Surgery table, minor minimum | 1 Sswk | .70 | 11.429 | Ea. | 13,900 | 555 | | 14,455 | 16,200 |
| 6520 | Maximum | " | .50 | 16 | | 26,100 | 775 | | 26,875 | 30,100 |
| 6700 | Surgical lights, doctor's office, single arm | 2 Elec | 2 | 8 | | 1,225 | 400 | | 1,625 | 1,950 |
| 6750 | Dual arm | " | 1 | 16 | | 5,800 | 805 | | 6,605 | 7,600 |

# 11 73 Patient Care Equipment

## 11 73 10 – Patient Treatment Equipment

### 11 73 10.10 Treatment Equipment

| | 11 73 10.10 Treatment Equipment | Crew | Daily Output | Labor-Hours | Unit | Material | Labor | Equipment | Total | Total Incl O&P |
|---|---|---|---|---|---|---|---|---|---|---|
| 0010 | **TREATMENT EQUIPMENT** | | | | | | | | | |
| 0750 | Exam room furnishings, average per room | | | | Ea. | 6,450 | | | 6,450 | 7,100 |
| 1800 | Heat therapy unit, humidified, 26" x 78" x 28" | | | | " | 3,400 | | | 3,400 | 3,725 |
| 2100 | Hubbard tank with accessories, stainless steel, | | | | | | | | | |
| 2110 | 125 GPM at 45 psi water pressure | | | | Ea. | 26,300 | | | 26,300 | 29,000 |
| 2150 | For electric overhead hoist, add | | | | | 2,875 | | | 2,875 | 3,150 |
| 2900 | K-Module for heat therapy, 20 oz. capacity, 75°F to 110°F | | | | | 405 | | | 405 | 445 |
| 3600 | Paraffin bath, 126°F, auto controlled | | | | | 1,475 | | | 1,475 | 1,600 |
| 3900 | Parallel bars for walking training, 12'-0" | | | | | 1,200 | | | 1,200 | 1,325 |
| 4600 | Station, dietary, medium, with ice | | | | | 16,000 | | | 16,000 | 17,600 |
| 4700 | Medicine | | | | | 7,250 | | | 7,250 | 7,975 |
| 7000 | Tables, physical therapy, walk off, electric | 2 Carp | 3 | 5.333 | | 3,200 | 230 | | 3,430 | 3,875 |
| 7150 | Standard, vinyl top with base cabinets, minimum | | 3 | 5.333 | | 1,300 | 230 | | 1,530 | 1,775 |
| 7200 | Maximum | | 2 | 8 | | 3,850 | 345 | | 4,195 | 4,775 |
| 8400 | Whirlpool bath, mobile, sst, 18" x 24" x 60" | | | | | 4,500 | | | 4,500 | 4,950 |
| 8450 | Fixed, incl. mixing valves | 1 Plum | 2 | 4 | | 9,200 | 214 | | 9,414 | 10,400 |

# 11 74 Dental Equipment

## 11 74 10 – Dental Office Equipment

### 11 74 10.10 Diagnostic and Treatment Equipment

| | 11 74 10.10 Diagnostic and Treatment Equipment | Crew | Daily Output | Labor-Hours | Unit | Material | Labor | Equipment | Total | Total Incl O&P |
|---|---|---|---|---|---|---|---|---|---|---|
| 0010 | **DIAGNOSTIC AND TREATMENT EQUIPMENT** | | | | | | | | | |
| 0020 | Central suction system, minimum | 1 Plum | 1.20 | 6.667 | Ea. | 1,550 | 355 | | 1,905 | 2,225 |
| 0100 | Maximum | " | .90 | 8.889 | | 4,250 | 475 | | 4,725 | 5,375 |
| 0300 | Air compressor, minimum | 1 Skwk | .80 | 10 | | 3,075 | 445 | | 3,520 | 4,050 |
| 0400 | Maximum | | .50 | 16 | | 7,200 | 710 | | 7,910 | 9,025 |
| 0600 | Chair, electric or hydraulic, minimum | | .50 | 16 | | 2,625 | 710 | | 3,335 | 3,975 |
| 0700 | Maximum | | .25 | 32 | | 14,400 | 1,425 | | 15,825 | 18,100 |
| 0800 | Doctor's/assistant's stool, minimum | | | | | 119 | | | 119 | 131 |
| 0850 | Maximum | | | | | 810 | | | 810 | 890 |
| 1000 | Drill console with accessories, minimum | 1 Skwk | 1.60 | 5 | | 1,575 | 222 | | 1,797 | 2,075 |
| 1100 | Maximum | | 1.60 | 5 | | 5,375 | 222 | | 5,597 | 6,275 |
| 2000 | Light, ceiling mounted, minimum | | 8 | 1 | | 1,225 | 44.50 | | 1,269.50 | 1,425 |
| 2100 | Maximum | | 8 | 1 | | 1,625 | 44.50 | | 1,669.50 | 1,875 |
| 2200 | Unit light, minimum | 2 Skwk | 5.33 | 3.002 | | 600 | 133 | | 733 | 865 |

# 11 74 Dental Equipment

## 11 74 10 – Dental Office Equipment

| 11 74 10.10 Diagnostic and Treatment Equipment | Crew | Daily Output | Labor-Hours | Unit | Material | 2011 Bare Costs Labor | Equipment | Total | Total Incl O&P |
|---|---|---|---|---|---|---|---|---|---|
| 2210 | Maximum | 2 Skwk | 5.33 | 3.002 | Ea. | 1,625 | 133 | | 1,758 | 2,000 |
| 2220 | Track light, minimum | | 3.20 | 5 | | 1,525 | 222 | | 1,747 | 2,025 |
| 2230 | Maximum | | 3.20 | 5 | | 3,725 | 222 | | 3,947 | 4,450 |
| 2300 | Sterilizers, steam portable, minimum | | | | | 1,850 | | | 1,850 | 2,050 |
| 2350 | Maximum | | | | | 10,400 | | | 10,400 | 11,400 |
| 2600 | Steam, institutional | | | | | 4,250 | | | 4,250 | 4,675 |
| 2650 | Dry heat, electric, portable, 3 trays | | | | | 1,225 | | | 1,225 | 1,325 |
| 2700 | Ultra-sonic cleaner, portable, minimum | | | | | 239 | | | 239 | 263 |
| 2750 | Maximum (institutional) | | | | | 1,525 | | | 1,525 | 1,675 |
| 3000 | X-ray unit, wall, minimum | 1 Skwk | 4 | 2 | | 2,300 | 89 | | 2,389 | 2,650 |
| 3010 | Maximum | | 4 | 2 | | 5,125 | 89 | | 5,214 | 5,775 |
| 3100 | Panoramic unit | | .60 | 13.333 | | 14,600 | 590 | | 15,190 | 16,900 |
| 3105 | Deluxe, minimum | 2 Skwk | 1.60 | 10 | | 16,800 | 445 | | 17,245 | 19,200 |
| 3110 | Maximum | " | 1.60 | 10 | | 54,000 | 445 | | 54,445 | 60,000 |
| 3500 | Developers, X-ray, average | 1 Plum | 5.33 | 1.501 | | 4,950 | 80.50 | | 5,030.50 | 5,575 |
| 3600 | Maximum | " | 5.33 | 1.501 | | 8,100 | 80.50 | | 8,180.50 | 9,025 |

# 11 76 Operating Room Equipment

## 11 76 10 – Operating Room Equipment

### 11 76 10.10 Surgical Equipment

| | | Crew | Daily Output | Labor-Hours | Unit | Material | 2011 Bare Costs Labor | Equipment | Total | Total Incl O&P |
|---|---|---|---|---|---|---|---|---|---|---|
| 0010 | **SURGICAL EQUIPMENT** | | | | | | | | | |
| 5000 | Scrub, surgical, stainless steel, single station, minimum | 1 Plum | 3 | 2.667 | Ea. | 4,200 | 143 | | 4,343 | 4,850 |
| 5100 | Maximum | | | | | 7,875 | | | 7,875 | 8,650 |
| 6550 | Major surgery table, minimum | 1 Sswk | .50 | 16 | | 31,400 | 775 | | 32,175 | 36,000 |
| 6570 | Maximum | " | .50 | 16 | | 103,500 | 775 | | 104,275 | 115,500 |
| 6800 | Surgical lights, major operating room, dual head, minimum | 2 Elec | 1 | 16 | | 17,500 | 805 | | 18,305 | 20,400 |
| 6850 | Maximum | " | 1 | 16 | | 29,100 | 805 | | 29,905 | 33,200 |

# 11 77 Radiology Equipment

## 11 77 10 – Radiology Equipment

### 11 77 10.10 X-Ray Equipment

| | | Crew | Daily Output | Labor-Hours | Unit | Material | 2011 Bare Costs Labor | Equipment | Total | Total Incl O&P |
|---|---|---|---|---|---|---|---|---|---|---|
| 0010 | **X-RAY EQUIPMENT** | | | | | | | | | |
| 8700 | X-ray, mobile, minimum | | | | Ea. | 13,400 | | | 13,400 | 14,700 |
| 8750 | Maximum | | | | | 75,500 | | | 75,500 | 83,000 |
| 8900 | Stationary, minimum | | | | | 41,800 | | | 41,800 | 46,000 |
| 8950 | Maximum | | | | | 217,500 | | | 217,500 | 239,500 |
| 9150 | Developing processors, minimum | | | | | 11,600 | | | 11,600 | 12,800 |
| 9200 | Maximum | | | | | 30,600 | | | 30,600 | 33,600 |

# 11 78 Mortuary Equipment

## 11 78 13 – Mortuary Refrigerators

| 11 78 13.10 Mortuary and Autopsy Equipment | Crew | Daily Output | Labor-Hours | Unit | Material | 2011 Bare Costs Labor | Equipment | Total | Total Incl O&P |
|---|---|---|---|---|---|---|---|---|---|
| 0010 **MORTUARY AND AUTOPSY EQUIPMENT** | | | | | | | | | |
| 0015 Autopsy table, standard | 1 Plum | 1 | 8 | Ea. | 8,850 | 430 | | 9,280 | 10,400 |
| 0020 Deluxe | " | .60 | 13.333 | | 15,600 | 715 | | 16,315 | 18,300 |
| 3200 Mortuary refrigerator, end operated, 2 capacity | | | | | 12,800 | | | 12,800 | 14,100 |
| 3300 6 capacity | | | | | 23,900 | | | 23,900 | 26,300 |

## 11 78 16 – Crematorium Equipment

### 11 78 16.10 Crematory

| | Crew | Daily Output | Labor-Hours | Unit | Material | 2011 Bare Costs Labor | Equipment | Total | Total Incl O&P |
|---|---|---|---|---|---|---|---|---|---|
| 0010 **CREMATORY** | | | | | | | | | |
| 1500 Crematory, not including building, 1 place | Q-3 | .20 | 160 | Ea. | 63,000 | 8,175 | | 71,175 | 81,500 |
| 1750 2 place | " | .10 | 320 | " | 90,000 | 16,300 | | 106,300 | 123,500 |

# 11 82 Solid Waste Handling Equipment

## 11 82 19 – Packaged Incinerators

### 11 82 19.10 Packaged Gas Fired Incinerators

| | Crew | Daily Output | Labor-Hours | Unit | Material | 2011 Bare Costs Labor | Equipment | Total | Total Incl O&P |
|---|---|---|---|---|---|---|---|---|---|
| 0010 **PACKAGED GAS FIRED INCINERATORS** | | | | | | | | | |
| 4400 Incinerator, gas, not incl. chimney, elec. or pipe, 50#/hr., minimum | Q-3 | .80 | 40 | Ea. | 11,500 | 2,050 | | 13,550 | 15,800 |
| 4420 Maximum | | .70 | 45.714 | | 37,100 | 2,325 | | 39,425 | 44,300 |
| 4440 200 lb. per hr., minimum (batch type) | | .60 | 53.333 | | 63,500 | 2,725 | | 66,225 | 74,000 |
| 4460 Maximum (with feeder) | | .50 | 64 | | 90,000 | 3,275 | | 93,275 | 104,000 |
| 4480 400 lb. per hr., minimum (batch type) | | .30 | 106 | | 70,500 | 5,450 | | 75,950 | 85,500 |
| 4500 Maximum (with feeder) | | .25 | 128 | | 95,500 | 6,525 | | 102,025 | 115,000 |
| 4520 800 lb. per hr., with feeder, minimum | | .20 | 160 | | 115,500 | 8,175 | | 123,675 | 139,000 |
| 4540 Maximum | | .17 | 188 | | 159,000 | 9,600 | | 168,600 | 189,500 |
| 4560 1,200 lb. per hr., with feeder, minimum | | .15 | 213 | | 155,000 | 10,900 | | 165,900 | 187,000 |
| 4580 Maximum | | .11 | 290 | | 185,000 | 14,900 | | 199,900 | 225,500 |
| 4600 2,000 lb. per hr., with feeder, minimum | | .10 | 320 | | 379,000 | 16,300 | | 395,300 | 441,500 |
| 4620 Maximum | | .05 | 640 | | 519,000 | 32,700 | | 551,700 | 620,000 |
| 4700 For heat recovery system, add, minimum | | .25 | 128 | | 73,000 | 6,525 | | 79,525 | 90,500 |
| 4710 Add, maximum | | .11 | 290 | | 234,000 | 14,900 | | 248,900 | 279,500 |
| 4720 For automatic ash conveyer, add | | .50 | 64 | | 30,700 | 3,275 | | 33,975 | 38,700 |
| 4750 Large municipal incinerators, incl. stack, minimum | | .25 | 128 | Ton/day | 18,700 | 6,525 | | 25,225 | 30,400 |
| 4850 Maximum | | .10 | 320 | " | 49,700 | 16,300 | | 66,000 | 79,000 |

## 11 82 26 – Facility Waste Compactors

### 11 82 26.10 Compactors

| | Crew | Daily Output | Labor-Hours | Unit | Material | 2011 Bare Costs Labor | Equipment | Total | Total Incl O&P |
|---|---|---|---|---|---|---|---|---|---|
| 0010 **COMPACTORS** | | | | | | | | | |
| 0020 Compactors, 115 volt, 250#/hr., chute fed | L-4 | 1 | 24 | Ea. | 11,600 | 970 | | 12,570 | 14,200 |
| 0100 Hand fed | | 2.40 | 10 | | 8,450 | 405 | | 8,855 | 9,925 |
| 0300 Multi-bag, 230 volt, 600#/hr, chute fed | | 1 | 24 | | 10,500 | 970 | | 11,470 | 13,100 |
| 0400 Hand fed | | 1 | 24 | | 9,025 | 970 | | 9,995 | 11,400 |
| 0500 Containerized, hand fed, 2 to 6 C.Y. containers, 250#/hr. | | 1 | 24 | | 11,100 | 970 | | 12,070 | 13,800 |
| 0550 For chute fed, add per floor | | 1 | 24 | | 1,275 | 970 | | 2,245 | 2,900 |
| 1000 Heavy duty industrial compactor, 0.5 C.Y. capacity | | 1 | 24 | | 7,525 | 970 | | 8,495 | 9,775 |
| 1050 1.0 C.Y. capacity | | 1 | 24 | | 11,500 | 970 | | 12,470 | 14,100 |
| 1100 3.0 C.Y. capacity | | .50 | 48 | | 15,700 | 1,950 | | 17,650 | 20,200 |
| 1150 5.0 C.Y. capacity | | .50 | 48 | | 19,400 | 1,950 | | 21,350 | 24,300 |
| 1200 Combination shredder/compactor (5,000 lb./hr.) | | .50 | 48 | | 38,300 | 1,950 | | 40,250 | 45,200 |
| 1400 For handling hazardous waste materials, 55 gallon drum packer, std. | | | | | 17,800 | | | 17,800 | 19,600 |
| 1410 55 gallon drum packer w/HEPA filter | | | | | 22,200 | | | 22,200 | 24,400 |
| 1420 55 gallon drum packer w/charcoal & HEPA filter | | | | | 29,700 | | | 29,700 | 32,600 |
| 1430 All of the above made explosion proof, add | | | | | 13,500 | | | 13,500 | 14,900 |

# 11 82 Solid Waste Handling Equipment

## 11 82 26 – Facility Waste Compactors

| 11 82 26.10 Compactors | Crew | Daily Output | Labor-Hours | Unit | Material | 2011 Bare Costs Labor | Equipment | Total | Total Incl O&P |
|---|---|---|---|---|---|---|---|---|---|
| 5500 | Shredder, municipal use, 35 ton per hour | | | | Ea. | 278,000 | | | 278,000 | 306,000 |
| 5600 | 60 ton per hour | | | | | 592,500 | | | 592,500 | 651,500 |
| 5750 | Shredder & baler, 50 ton per day | | | | | 555,500 | | | 555,500 | 611,000 |
| 5800 | Shredder, industrial, minimum | | | | | 21,900 | | | 21,900 | 24,100 |
| 5850 | Maximum | | | | | 117,500 | | | 117,500 | 129,000 |
| 5900 | Baler, industrial, minimum | | | | | 8,775 | | | 8,775 | 9,650 |
| 5950 | Maximum | | | | | 512,000 | | | 512,000 | 563,500 |
| 6000 | Transfer station compactor, with power unit | | | | | | | | | |
| 6050 | and pedestal, not including pit, 50 ton per hour | | | | Ea. | 175,500 | | | 175,500 | 193,000 |

# 11 91 Religious Equipment

## 11 91 13 – Baptisteries

### 11 91 13.10 Baptistry

| | | Crew | Daily Output | Labor-Hours | Unit | Material | Labor | Equipment | Total | Total Incl O&P |
|---|---|---|---|---|---|---|---|---|---|---|
| 0010 | **BAPTISTRY** | | | | | | | | | |
| 0150 | Fiberglass, 3'-6" deep, x 13'-7" long, | | | | | | | | | |
| 0160 | steps at both ends, incl. plumbing, minimum | L-8 | 1 | 20 | Ea. | 4,100 | 905 | | 5,005 | 5,875 |
| 0200 | Maximum | " | .70 | 28.571 | | 6,675 | 1,300 | | 7,975 | 9,325 |
| 0250 | Add for filter, heater and lights | | | | | 1,475 | | | 1,475 | 1,600 |

## 11 91 23 – Sanctuary Equipment

### 11 91 23.10 Sanctuary Furnishings

| | | Crew | Daily Output | Labor-Hours | Unit | Material | Labor | Equipment | Total | Total Incl O&P |
|---|---|---|---|---|---|---|---|---|---|---|
| 0010 | **SANCTUARY FURNISHINGS** | | | | | | | | | |
| 0020 | Altar, wood, custom design, plain | 1 Carp | 1.40 | 5.714 | Ea. | 2,000 | 246 | | 2,246 | 2,575 |
| 0050 | Deluxe | " | .20 | 40 | | 9,700 | 1,725 | | 11,425 | 13,400 |
| 0070 | Granite or marble, average | 2 Marb | .50 | 32 | | 11,200 | 1,300 | | 12,500 | 14,300 |
| 0090 | Deluxe | " | .20 | 80 | | 31,900 | 3,250 | | 35,150 | 40,000 |
| 0100 | Arks, prefabricated, plain | 2 Carp | .80 | 20 | | 8,425 | 860 | | 9,285 | 10,600 |
| 0130 | Deluxe, maximum | " | .20 | 80 | | 119,500 | 3,450 | | 122,950 | 137,000 |
| 0500 | Reconciliation room, wood, prefabricated, single, plain | 1 Carp | .60 | 13.333 | | 2,550 | 575 | | 3,125 | 3,675 |
| 0550 | Deluxe | | .40 | 20 | | 7,025 | 860 | | 7,885 | 9,050 |
| 0650 | Double, plain | | .40 | 20 | | 5,100 | 860 | | 5,960 | 6,950 |
| 0700 | Deluxe | | .20 | 40 | | 15,900 | 1,725 | | 17,625 | 20,200 |
| 1000 | Lecterns, wood, plain | | 5 | 1.600 | | 770 | 69 | | 839 | 955 |
| 1100 | Deluxe | | 2 | 4 | | 5,375 | 172 | | 5,547 | 6,200 |
| 2000 | Pulpits, hardwood, prefabricated, plain | | 2 | 4 | | 1,275 | 172 | | 1,447 | 1,675 |
| 2100 | Deluxe | | 1.60 | 5 | | 8,725 | 215 | | 8,940 | 9,925 |
| 2500 | Railing, hardwood, average | | 25 | .320 | L.F. | 175 | 13.80 | | 188.80 | 213 |
| 3000 | Seating, individual, oak, contour, laminated | | 21 | .381 | Person | 154 | 16.40 | | 170.40 | 194 |
| 3100 | Cushion seat | | 21 | .381 | | 139 | 16.40 | | 155.40 | 178 |
| 3200 | Fully upholstered | | 21 | .381 | | 136 | 16.40 | | 152.40 | 175 |
| 3300 | Combination, self-rising | | 21 | .381 | | 350 | 16.40 | | 366.40 | 410 |
| 3500 | For cherry, add | | | | | 30% | | | | |
| 5000 | Wall cross, aluminum, extruded, 2" x 2" section | 1 Carp | 34 | .235 | L.F. | 146 | 10.15 | | 156.15 | 176 |
| 5150 | 4" x 4" section | | 29 | .276 | | 209 | 11.90 | | 220.90 | 248 |
| 5300 | Bronze, extruded, 1" x 2" section | | 31 | .258 | | 287 | 11.10 | | 298.10 | 330 |
| 5350 | 2-1/2" x 2-1/2" section | | 34 | .235 | | 435 | 10.15 | | 445.15 | 495 |
| 5450 | Solid bar stock, 1/2" x 3" section | | 29 | .276 | | 570 | 11.90 | | 581.90 | 645 |
| 5600 | Fiberglass, stock | | 34 | .235 | | 118 | 10.15 | | 128.15 | 146 |
| 5700 | Stainless steel, 4" deep, channel section | | 29 | .276 | | 460 | 11.90 | | 471.90 | 525 |
| 5800 | 4" deep box section | | 29 | .276 | | 625 | 11.90 | | 636.90 | 710 |

# Division Notes

| | | CREW | DAILY OUTPUT | LABOR-HOURS | UNIT | 2011 BARE COSTS | | | | TOTAL INCL O&P |
|---|---|---|---|---|---|---|---|---|---|---|
| | | | | | | MAT. | LABOR | EQUIP. | TOTAL | |
| | | | | | | | | | | |
| | | | | | | | | | | |
| | | | | | | | | | | |
| | | | | | | | | | | |
| | | | | | | | | | | |
| | | | | | | | | | | |
| | | | | | | | | | | |
| | | | | | | | | | | |
| | | | | | | | | | | |
| | | | | | | | | | | |
| | | | | | | | | | | |
| | | | | | | | | | | |
| | | | | | | | | | | |
| | | | | | | | | | | |
| | | | | | | | | | | |
| | | | | | | | | | | |
| | | | | | | | | | | |
| | | | | | | | | | | |
| | | | | | | | | | | |
| | | | | | | | | | | |
| | | | | | | | | | | |

# Estimating Tips

## General

- The items in this division are usually priced per square foot or each. Most of these items are purchased by the owner and placed by the contractor. Do not assume the items in Division 12 will be purchased and installed by the contractor. Check the specifications for responsibilities and include receiving, storage, installation, and mechanical and electrical hookups in the appropriate divisions.

- Some items in this division require some type of support system that is not usually furnished with the item. Examples of these systems include blocking for the attachment of casework and heavy drapery rods. The required blocking must be added to the estimate in the appropriate division.

# Reference Numbers

Reference numbers are shown in shaded boxes at the beginning of some major classifications. These numbers refer to related items in the Reference Section. The reference information may be an estimating procedure, an alternate pricing method, or technical information.

*Note:* Not all subdivisions listed here necessarily appear in this publication.

# 12 21 Window Blinds

## 12 21 13 – Horizontal Louver Blinds

### 12 21 13.13 Metal Horizontal Louver Blinds

| | | Crew | Daily Output | Labor-Hours | Unit | Material | 2011 Bare Costs Labor | Equipment | Total | Total Incl O&P |
|---|---|---|---|---|---|---|---|---|---|---|
| 0010 | **METAL HORIZONTAL LOUVER BLINDS** | | | | | | | | | |
| 0020 | Horizontal, 1" aluminum slats, solid color, stock | 1 Carp | 590 | .014 | S.F. | 4.75 | .58 | | 5.33 | 6.15 |
| 0090 | Custom, minimum | | 590 | .014 | | 5.25 | .58 | | 5.83 | 6.70 |
| 0100 | Maximum | | 440 | .018 | | 7.15 | .78 | | 7.93 | 9.05 |
| 0250 | 2" aluminum slats, custom, minimum | | 590 | .014 | | 8.10 | .58 | | 8.68 | 9.85 |
| 0350 | Maximum | | 440 | .018 | | 13.15 | .78 | | 13.93 | 15.70 |
| 0450 | Stock, minimum | | 590 | .014 | | 4.86 | .58 | | 5.44 | 6.25 |
| 0500 | Maximum | | 440 | .018 | | 7.90 | .78 | | 8.68 | 9.90 |
| 0600 | 2" steel slats, stock, minimum | | 590 | .014 | | 1.84 | .58 | | 2.42 | 2.92 |
| 0630 | Maximum | | 440 | .018 | | 5.15 | .78 | | 5.93 | 6.90 |
| 0750 | Custom, minimum | | 590 | .014 | | 1.67 | .58 | | 2.25 | 2.74 |
| 0850 | Maximum | ▼ | 400 | .020 | ▼ | 8.25 | .86 | | 9.11 | 10.40 |

### 12 21 13.33 Vinyl Horizontal Louver Blinds

| | | Crew | Daily Output | Labor-Hours | Unit | Material | 2011 Bare Costs Labor | Equipment | Total | Total Incl O&P |
|---|---|---|---|---|---|---|---|---|---|---|
| 0010 | **VINYL HORIZONTAL LOUVER BLINDS** | | | | | | | | | |
| 0100 | 2" custom composite, 48" wide, 48" high | 1 Carp | 30 | .267 | Ea. | 89 | 11.50 | | 100.50 | 116 |
| 0120 | 72" high | | 29 | .276 | | 126 | 11.90 | | 137.90 | 157 |
| 0140 | 96" high | | 28 | .286 | | 161 | 12.30 | | 173.30 | 196 |
| 0200 | 60" wide, 60" high | | 27 | .296 | | 93.50 | 12.75 | | 106.25 | 123 |
| 0220 | 72" high | | 25 | .320 | | 108 | 13.80 | | 121.80 | 139 |
| 0240 | 96" high | | 24 | .333 | | 151 | 14.35 | | 165.35 | 188 |
| 0300 | 72" wide, 72" high | | 25 | .320 | | 169 | 13.80 | | 182.80 | 206 |
| 0320 | 96" high | | 23 | .348 | | 249 | 14.95 | | 263.95 | 297 |
| 0400 | 96" wide, 96" high | | 20 | .400 | | 295 | 17.20 | | 312.20 | 350 |
| 1000 | 2" custom faux wood, 48" wide, 48" high | | 30 | .267 | | 55.50 | 11.50 | | 67 | 78.50 |
| 1020 | 72" high | | 29 | .276 | | 74 | 11.90 | | 85.90 | 99.50 |
| 1040 | 96" high | | 28 | .286 | | 92.50 | 12.30 | | 104.80 | 121 |
| 1300 | 72" wide, 72" high | | 25 | .320 | | 115 | 13.80 | | 128.80 | 147 |
| 1320 | 96" high | | 23 | .348 | | 164 | 14.95 | | 178.95 | 204 |
| 1400 | 96" wide, 96" high | ▼ | 20 | .400 | ▼ | 180 | 17.20 | | 197.20 | 225 |

## 12 21 16 – Vertical Louver Blinds

### 12 21 16.13 Metal Vertical Louver Blinds

| | | Crew | Daily Output | Labor-Hours | Unit | Material | 2011 Bare Costs Labor | Equipment | Total | Total Incl O&P |
|---|---|---|---|---|---|---|---|---|---|---|
| 0010 | **METAL VERTICAL LOUVER BLINDS** | | | | | | | | | |
| 1500 | Vertical, 3" to 5" PVC or cloth strips, minimum | 1 Carp | 460 | .017 | S.F. | 5.95 | .75 | | 6.70 | 7.70 |
| 1600 | Maximum | | 400 | .020 | | 24 | .86 | | 24.86 | 28 |
| 1800 | 4" aluminum slats, minimum | | 460 | .017 | | 7.65 | .75 | | 8.40 | 9.55 |
| 1900 | Maximum | | 400 | .020 | | 9.10 | .86 | | 9.96 | 11.30 |
| 1950 | Mylar mirror-finish strips, to 8" wide, minimum | | 460 | .017 | | 4.62 | .75 | | 5.37 | 6.25 |
| 1970 | Maximum | ▼ | 400 | .020 | ▼ | 10.50 | .86 | | 11.36 | 12.85 |

# 12 22 Curtains and Drapes

## 12 22 16 – Drapery Track and Accessories

### 12 22 16.10 Drapery Hardware

| | | Crew | Daily Output | Labor-Hours | Unit | Material | 2011 Bare Costs Labor | Equipment | Total | Total Incl O&P |
|---|---|---|---|---|---|---|---|---|---|---|
| 0010 | **DRAPERY HARDWARE** | | | | | | | | | |
| 0030 | Standard traverse, per foot, minimum | 1 Carp | 59 | .136 | L.F. | 3.42 | 5.85 | | 9.27 | 12.70 |
| 0100 | Maximum | | 51 | .157 | " | 8.05 | 6.75 | | 14.80 | 19.20 |
| 4000 | Traverse rods, adjustable, 28" to 48" | | 22 | .364 | Ea. | 21 | 15.65 | | 36.65 | 47 |
| 4020 | 48" to 84" | | 20 | .400 | | 29.50 | 17.20 | | 46.70 | 59 |
| 4040 | 66" to 120" | | 18 | .444 | | 31.50 | 19.15 | | 50.65 | 64 |
| 4060 | 84" to 156" | | 16 | .500 | | 33 | 21.50 | | 54.50 | 69 |
| 4080 | 100" to 180" | ▼ | 14 | .571 | ▼ | 42.50 | 24.50 | | 67 | 84 |

# 12 22 Curtains and Drapes

## 12 22 16 – Drapery Track and Accessories

| 12 22 16.10 Drapery Hardware | Crew | Daily Output | Labor-Hours | Unit | Material | 2011 Bare Costs Labor | Equipment | Total | Total Incl O&P |
|---|---|---|---|---|---|---|---|---|---|
| 4090 | 156" to 228" | 1 Carp | 13 | .615 | Ea. | 47.50 | 26.50 | | 74 | 93 |
| 4100 | 228" to 312" | | 13 | .615 | | 59.50 | 26.50 | | 86 | 106 |
| 4200 | Double rods, adjustable, 30" to 48" | | 9 | .889 | | 38 | 38.50 | | 76.50 | 101 |
| 4220 | 48" to 86" | | 9 | .889 | | 52.50 | 38.50 | | 91 | 117 |
| 4240 | 86" to 150" | | 8 | 1 | | 60 | 43 | | 103 | 132 |
| 4260 | 100" to 180" | | 7 | 1.143 | | 63 | 49 | | 112 | 145 |
| 4300 | Tray and curtain rod, adjustable, 30" to 48" | | 9 | .889 | | 24.50 | 38.50 | | 63 | 85.50 |
| 4320 | 48" to 86" | | 9 | .889 | | 35 | 38.50 | | 73.50 | 97 |
| 4340 | 86" to 150" | | 8 | 1 | | 43.50 | 43 | | 86.50 | 114 |
| 4360 | 100" to 180" | | 7 | 1.143 | | 47.50 | 49 | | 96.50 | 128 |
| 4600 | Valance, pinch pleated fabric, 12" deep, up to 54" long, minimum | | | | | 36.50 | | | 36.50 | 40.50 |
| 4610 | Maximum | | | | | 91.50 | | | 91.50 | 101 |
| 4620 | Up to 77" long, minimum | | | | | 56.50 | | | 56.50 | 62 |
| 4630 | Maximum | | | | | 148 | | | 148 | 163 |
| 5000 | Stationary rods, first 2' | | | | | 9.35 | | | 9.35 | 10.25 |
| 5020 | Each additional foot, add | | | | L.F. | 3.64 | | | 3.64 | 4 |

## 12 22 16.20 Blast Curtains

| | | Crew | Daily Output | Labor-Hours | Unit | Material | Labor | Equipment | Total | Total Incl O&P |
|---|---|---|---|---|---|---|---|---|---|---|
| 0010 | **BLAST CURTAINS** per L.F. horizontal opening width, off-white or gray fabric | | | | | | | | | |
| 0100 | Blast curtains, drapery system, complete, including hardware, minimum | 1 Carp | 10.25 | .780 | L.F. | 170 | 33.50 | | 203.50 | 239 |
| 0120 | Average | | 10.25 | .780 | | 200 | 33.50 | | 233.50 | 272 |
| 0140 | Maximum | | 10.25 | .780 | | 230 | 33.50 | | 263.50 | 305 |

# 12 23 Interior Shutters

## 12 23 10 – Wood Interior Shutters

### 12 23 10.10 Wood Interior Shutters

| | | Crew | Daily Output | Labor-Hours | Unit | Material | Labor | Equipment | Total | Total Incl O&P |
|---|---|---|---|---|---|---|---|---|---|---|
| 0010 | **WOOD INTERIOR SHUTTERS**, louvered | | | | | | | | | |
| 0200 | Two panel, 27" wide, 36" high | 1 Carp | 5 | 1.600 | Set | 125 | 69 | | 194 | 244 |
| 0300 | 33" wide, 36" high | | 5 | 1.600 | | 164 | 69 | | 233 | 286 |
| 0500 | 47" wide, 36" high | | 5 | 1.600 | | 219 | 69 | | 288 | 345 |
| 1000 | Four panel, 27" wide, 36" high | | 5 | 1.600 | | 148 | 69 | | 217 | 269 |
| 1100 | 33" wide, 36" high | | 5 | 1.600 | | 181 | 69 | | 250 | 305 |
| 1300 | 47" wide, 36" high | | 5 | 1.600 | | 258 | 69 | | 327 | 390 |

### 12 23 10.13 Wood Panels

| | | Crew | Daily Output | Labor-Hours | Unit | Material | Labor | Equipment | Total | Total Incl O&P |
|---|---|---|---|---|---|---|---|---|---|---|
| 0010 | **WOOD PANELS** | | | | | | | | | |
| 3000 | Wood folding panels with movable louvers, 7" x 20" each | 1 Carp | 17 | .471 | Pr. | 50 | 20.50 | | 70.50 | 86 |
| 3300 | 8" x 28" each | | 17 | .471 | | 64 | 20.50 | | 84.50 | 102 |
| 3450 | 9" x 36" each | | 17 | .471 | | 84 | 20.50 | | 104.50 | 124 |
| 3600 | 10" x 40" each | | 17 | .471 | | 90 | 20.50 | | 110.50 | 130 |
| 4000 | Fixed louver type, stock units, 8" x 20" each | | 17 | .471 | | 58 | 20.50 | | 78.50 | 95 |
| 4150 | 10" x 28" each | | 17 | .471 | | 68 | 20.50 | | 88.50 | 106 |
| 4300 | 12" x 36" each | | 17 | .471 | | 86 | 20.50 | | 106.50 | 126 |
| 4450 | 18" x 40" each | | 17 | .471 | | 122 | 20.50 | | 142.50 | 165 |
| 5000 | Insert panel type, stock, 7" x 20" each | | 17 | .471 | | 19.65 | 20.50 | | 40.15 | 52.50 |
| 5150 | 8" x 28" each | | 17 | .471 | | 36 | 20.50 | | 56.50 | 70.50 |
| 5300 | 9" x 36" each | | 17 | .471 | | 46 | 20.50 | | 66.50 | 81.50 |
| 5450 | 10" x 40" each | | 17 | .471 | | 49 | 20.50 | | 69.50 | 85 |
| 5600 | Raised panel type, stock, 10" x 24" each | | 17 | .471 | | 80 | 20.50 | | 100.50 | 119 |
| 5650 | 12" x 26" each | | 17 | .471 | | 86 | 20.50 | | 106.50 | 126 |
| 5700 | 14" x 30" each | | 17 | .471 | | 258 | 20.50 | | 278.50 | 315 |
| 5750 | 16" x 36" each | | 17 | .471 | | 286 | 20.50 | | 306.50 | 345 |

# 12 23 Interior Shutters

## 12 23 10 – Wood Interior Shutters

| 12 23 10.13 Wood Panels | Crew | Daily Output | Labor-Hours | Unit | Material | 2011 Bare Costs Labor | Equipment | Total | Total Incl O&P |
|---|---|---|---|---|---|---|---|---|---|
| 6000     For custom built pine, add | | | | Pr. | 22% | | | | |
| 6500     For custom built hardwood blinds, add | | | | ↓ | 42% | | | | |

# 12 24 Window Shades

## 12 24 13 – Roller Window Shades

### 12 24 13.10 Shades

| | Crew | Daily Output | Labor-Hours | Unit | Material | 2011 Bare Costs Labor | Equipment | Total | Total Incl O&P |
|---|---|---|---|---|---|---|---|---|---|
| 0010 **SHADES** | | | | | | | | | |
| 0020     Basswood, roll-up, stain finish, 3/8" slats | 1 Carp | 300 | .027 | S.F. | 13.95 | 1.15 | | 15.10 | 17.10 |
| 0200        7/8" slats | | 300 | .027 | | 13.15 | 1.15 | | 14.30 | 16.25 |
| 0300        Vertical side slide, stain finish, 3/8" slats | | 300 | .027 | | 18.05 | 1.15 | | 19.20 | 21.50 |
| 0400        7/8" slats | ↓ | 300 | .027 | | 18.05 | 1.15 | | 19.20 | 21.50 |
| 0500     For fire retardant finishes, add | | | | | 16% | | | | |
| 0600     For "B" rated finishes, add | | | | | 20% | | | | |
| 0900     Mylar, single layer, non-heat reflective | 1 Carp | 685 | .012 | | 6.70 | .50 | | 7.20 | 8.10 |
| 1000        Double layered, heat reflective | | 685 | .012 | | 9.25 | .50 | | 9.75 | 10.95 |
| 1100        Triple layered, heat reflective | ↓ | 685 | .012 | | 10.80 | .50 | | 11.30 | 12.60 |
| 1200     For metal roller instead of wood, add per | | | | Shade | 4.20 | | | 4.20 | 4.62 |
| 1300     Vinyl coated cotton, standard | 1 Carp | 685 | .012 | S.F. | 3.43 | .50 | | 3.93 | 4.54 |
| 1400        Lightproof decorator shades | | 685 | .012 | | 2.15 | .50 | | 2.65 | 3.14 |
| 1500     Vinyl, lightweight, 4 gauge | | 685 | .012 | | .56 | .50 | | 1.06 | 1.39 |
| 1600        Heavyweight, 6 gauge | | 685 | .012 | | 1.70 | .50 | | 2.20 | 2.64 |
| 1700     Vinyl laminated fiberglass, 6 ga., translucent | | 685 | .012 | | 2.41 | .50 | | 2.91 | 3.42 |
| 1800        Lightproof | | 685 | .012 | | 3.91 | .50 | | 4.41 | 5.05 |
| 3000     Woven aluminum, 3/8" thick, lightproof and fireproof | ↓ | 350 | .023 | ↓ | 5.50 | .98 | | 6.48 | 7.55 |

# 12 32 Manufactured Wood Casework

## 12 32 16 – Manufactured Plastic-Laminate-Clad Casework

### 12 32 16.20 Plastic Laminate Casework Doors

| | Crew | Daily Output | Labor-Hours | Unit | Material | 2011 Bare Costs Labor | Equipment | Total | Total Incl O&P |
|---|---|---|---|---|---|---|---|---|---|
| 0010 **PLASTIC LAMINATE CASEWORK DOORS** | | | | | | | | | |
| 1000     For casework frames, see Section 12 32 23.15 | | | | | | | | | |
| 1100     For casework hardware, see Section 12 32 23.35 | | | | | | | | | |
| 6000     Plastic laminate on particle board | | | | | | | | | |
| 6100        12" wide, 18" high | 1 Carp | 25 | .320 | Ea. | 10.65 | 13.80 | | 24.45 | 32.50 |
| 6140        30" high | | 23 | .348 | | 17.70 | 14.95 | | 32.65 | 42.50 |
| 6500        18" wide, 18" high | | 24 | .333 | | 15.95 | 14.35 | | 30.30 | 39.50 |
| 6600        30" high | ↓ | 22 | .364 | ↓ | 26.50 | 15.65 | | 42.15 | 53 |

### 12 32 16.25 Plastic Laminate Drawer Fronts

| | Crew | Daily Output | Labor-Hours | Unit | Material | 2011 Bare Costs Labor | Equipment | Total | Total Incl O&P |
|---|---|---|---|---|---|---|---|---|---|
| 0010 **PLASTIC LAMINATE DRAWER FRONTS** | | | | | | | | | |
| 2800     Plastic laminate on particle board front | | | | | | | | | |
| 3000        4" high, 12" wide | 1 Carp | 17 | .471 | Ea. | 5.05 | 20.50 | | 25.55 | 36.50 |
| 3200        18" wide | " | 16 | .500 | " | 7.60 | 21.50 | | 29.10 | 41.50 |

## 12 32 23 – Hardwood Casework

### 12 32 23.10 Manufactured Wood Casework, Stock Units

| | Crew | Daily Output | Labor-Hours | Unit | Material | 2011 Bare Costs Labor | Equipment | Total | Total Incl O&P |
|---|---|---|---|---|---|---|---|---|---|
| 0010 **MANUFACTURED WOOD CASEWORK, STOCK UNITS** | | | | | | | | | |
| 0300     Built-in drawer units, pine, 18" deep, 32" high, unfinished | | | | | | | | | |
| 0400        Minimum | 2 Carp | 53 | .302 | L.F. | 145 | 13 | | 158 | 180 |
| 0500        Maximum | " | 40 | .400 | " | 180 | 17.20 | | 197.20 | 225 |
| 0700     Kitchen base cabinets, hardwood, not incl. counter tops, | | | | | | | | | |

# 12 32 Manufactured Wood Casework

## 12 32 23 – Hardwood Casework

### 12 32 23.10 Manufactured Wood Casework, Stock Units

| | | Crew | Daily Output | Labor-Hours | Unit | Material | 2011 Bare Costs Labor | Equipment | Total | Total Incl O&P |
|---|---|---|---|---|---|---|---|---|---|---|
| 0710 | 24" deep, 35" high, prefinished | | | | | | | | | |
| 0800 | One top drawer, one door below, 12" wide | 2 Carp | 24.80 | .645 | Ea. | 229 | 28 | | 257 | 295 |
| 0840 | 18" wide | | 23.30 | .687 | | 259 | 29.50 | | 288.50 | 330 |
| 0880 | 24" wide | | 22.30 | .717 | | 310 | 31 | | 341 | 395 |
| 1000 | Four drawers, 12" wide | | 24.80 | .645 | | 240 | 28 | | 268 | 305 |
| 1040 | 18" wide | | 23.30 | .687 | | 268 | 29.50 | | 297.50 | 340 |
| 1060 | 24" wide | | 22.30 | .717 | | 305 | 31 | | 336 | 385 |
| 1200 | Two top drawers, two doors below, 27" wide | | 22 | .727 | | 330 | 31.50 | | 361.50 | 415 |
| 1260 | 36" wide | | 20.30 | .788 | | 400 | 34 | | 434 | 490 |
| 1300 | 48" wide | | 18.90 | .847 | | 460 | 36.50 | | 496.50 | 560 |
| 1500 | Range or sink base, two doors below, 30" wide | | 21.40 | .748 | | 305 | 32 | | 337 | 385 |
| 1540 | 36" wide | | 20.30 | .788 | | 345 | 34 | | 379 | 430 |
| 1580 | 48" wide | ▼ | 18.90 | .847 | | 385 | 36.50 | | 421.50 | 480 |
| 1800 | For sink front units, deduct | | | | | 145 | | | 145 | 160 |
| 2000 | Corner base cabinets, 36" wide, standard | 2 Carp | 18 | .889 | | 540 | 38.50 | | 578.50 | 655 |
| 2100 | Lazy Susan with revolving door | " | 16.50 | .970 | ▼ | 725 | 42 | | 767 | 860 |
| 4000 | Kitchen wall cabinets, hardwood, 12" deep with two doors | | | | | | | | | |
| 4050 | 12" high, 30" wide | 2 Carp | 24.80 | .645 | Ea. | 205 | 28 | | 233 | 268 |
| 4100 | 36" wide | | 24 | .667 | | 240 | 28.50 | | 268.50 | 310 |
| 4400 | 15" high, 30" wide | | 24 | .667 | | 165 | 28.50 | | 193.50 | 225 |
| 4440 | 36" wide | | 22.70 | .705 | | 240 | 30.50 | | 270.50 | 310 |
| 4700 | 24" high, 30" wide | | 23.30 | .687 | | 275 | 29.50 | | 304.50 | 350 |
| 4720 | 36" wide | | 22.70 | .705 | | 305 | 30.50 | | 335.50 | 380 |
| 5000 | 30" high, one door, 12" wide | | 22 | .727 | | 185 | 31.50 | | 216.50 | 252 |
| 5040 | 18" wide | | 20.90 | .766 | | 229 | 33 | | 262 | 305 |
| 5060 | 24" wide | | 20.30 | .788 | | 267 | 34 | | 301 | 345 |
| 5300 | Two doors, 27" wide | | 19.80 | .808 | | 294 | 35 | | 329 | 380 |
| 5340 | 36" wide | | 18.80 | .851 | | 350 | 36.50 | | 386.50 | 440 |
| 5380 | 48" wide | | 18.40 | .870 | | 430 | 37.50 | | 467.50 | 535 |
| 6000 | Corner wall, 30" high, 24" wide | | 18 | .889 | | 233 | 38.50 | | 271.50 | 315 |
| 6050 | 30" wide | | 17.20 | .930 | | 264 | 40 | | 304 | 350 |
| 6100 | 36" wide | | 16.50 | .970 | | 295 | 42 | | 337 | 390 |
| 6500 | Revolving Lazy Susan | | 15.20 | 1.053 | | 415 | 45.50 | | 460.50 | 525 |
| 7000 | Broom cabinet, 84" high, 24" deep, 18" wide | | 10 | 1.600 | | 575 | 69 | | 644 | 740 |
| 7500 | Oven cabinets, 84" high, 24" deep, 27" wide | ▼ | 8 | 2 | ▼ | 860 | 86 | | 946 | 1,075 |
| 7750 | Valance board trim | ▼ | 396 | .040 | L.F. | 12.40 | 1.74 | | 14.14 | 16.30 |
| 9000 | For deluxe models of all cabinets, add | | | | | 40% | | | | |
| 9500 | For custom built in place, add | | | | | 25% | 10% | | | |
| 9558 | Rule of thumb, kitchen cabinets not including | | | | | | | | | |
| 9560 | appliances & counter top, minimum | 2 Carp | 30 | .533 | L.F. | 146 | 23 | | 169 | 196 |
| 9600 | Maximum | " | 25 | .640 | " | 310 | 27.50 | | 337.50 | 390 |
| 9610 | For metal cabinets, see Section 12 35 70.13 | | | | | | | | | |

### 12 32 23.15 Manufactured Wood Casework Frames

| | | Crew | Daily Output | Labor-Hours | Unit | Material | 2011 Bare Costs Labor | Equipment | Total | Total Incl O&P |
|---|---|---|---|---|---|---|---|---|---|---|
| 0010 | **MANUFACTURED WOOD CASEWORK FRAMES** | | | | | | | | | |
| 0050 | Base cabinets, counter storage, 36" high | | | | | | | | | |
| 0100 | One bay, 18" wide | 1 Carp | 2.70 | 2.963 | Ea. | 134 | 128 | | 262 | 345 |
| 0400 | Two bay, 36" wide | | 2.20 | 3.636 | | 204 | 157 | | 361 | 465 |
| 1100 | Three bay, 54" wide | | 1.50 | 5.333 | | 243 | 230 | | 473 | 615 |
| 2800 | Bookcases, one bay, 7' high, 18" wide | | 2.40 | 3.333 | | 157 | 144 | | 301 | 395 |
| 3500 | Two bay, 36" wide | | 1.60 | 5 | | 228 | 215 | | 443 | 580 |
| 4100 | Three bay, 54" wide | | 1.20 | 6.667 | | 375 | 287 | | 662 | 855 |
| 5100 | Coat racks, one bay, 7' high, 24" wide | | 4.50 | 1.778 | | 157 | 76.50 | | 233.50 | 290 |

# 12 32 Manufactured Wood Casework

## 12 32 23 - Hardwood Casework

| 12 32 23.15 Manufactured Wood Casework Frames | Crew | Daily Output | Labor-Hours | Unit | Material | 2011 Bare Costs Labor | Equipment | Total | Total Incl O&P |
|---|---|---|---|---|---|---|---|---|---|
| 5300 | Two bay, 48" wide | 1 Carp | 2.75 | 2.909 | Ea. | 219 | 125 | | 344 | 435 |
| 5800 | Three bay, 72" wide | | 2.10 | 3.810 | | 320 | 164 | | 484 | 605 |
| 6100 | Wall mounted cabinet, one bay, 24" high, 18" wide | | 3.60 | 2.222 | | 86.50 | 95.50 | | 182 | 242 |
| 6800 | Two bay, 36" wide | | 2.20 | 3.636 | | 126 | 157 | | 283 | 380 |
| 7400 | Three bay, 54" wide | | 1.70 | 4.706 | | 157 | 203 | | 360 | 485 |
| 8400 | 30" high, one bay, 18" wide | | 3.60 | 2.222 | | 94 | 95.50 | | 189.50 | 250 |
| 9000 | Two bay, 36" wide | | 2.15 | 3.721 | | 125 | 160 | | 285 | 385 |
| 9400 | Three bay, 54" wide | | 1.60 | 5 | | 156 | 215 | | 371 | 500 |
| 9800 | Wardrobe, 7' high, single, 24" wide | | 2.70 | 2.963 | | 173 | 128 | | 301 | 385 |
| 9880 | Partition & adjustable shelves, 48" wide | | 1.70 | 4.706 | | 220 | 203 | | 423 | 550 |
| 9950 | Partition, adjustable shelves & drawers, 48" wide | ▼ | 1.40 | 5.714 | ▼ | 330 | 246 | | 576 | 740 |

## 12 32 23.20 Manufactured Hardwood Casework Doors

| | | Crew | Daily Output | Labor-Hours | Unit | Material | 2011 Bare Costs Labor | Equipment | Total | Total Incl O&P |
|---|---|---|---|---|---|---|---|---|---|---|
| 0010 | **MANUFACTURED HARDWOOD CASEWORK DOORS** | | | | | | | | | |
| 2000 | Glass panel, hardwood frame | | | | | | | | | |
| 2200 | 12" wide, 18" high | 1 Carp | 34 | .235 | Ea. | 22 | 10.15 | | 32.15 | 39.50 |
| 2600 | 30" high | | 32 | .250 | | 36.50 | 10.75 | | 47.25 | 56.50 |
| 4450 | 18" wide, 18" high | | 32 | .250 | | 32.50 | 10.75 | | 43.25 | 52.50 |
| 4550 | 30" high | ▼ | 29 | .276 | ▼ | 54.50 | 11.90 | | 66.40 | 78 |
| 5000 | Hardwood, raised panel | | | | | | | | | |
| 5100 | 12" wide, 18" high | 1 Carp | 16 | .500 | Ea. | 26 | 21.50 | | 47.50 | 61.50 |
| 5200 | 30" high | | 15 | .533 | | 43.50 | 23 | | 66.50 | 82.50 |
| 5500 | 18" wide, 18" high | | 15 | .533 | | 39 | 23 | | 62 | 78 |
| 5600 | 30" high | ▼ | 14 | .571 | ▼ | 65 | 24.50 | | 89.50 | 109 |

## 12 32 23.25 Manufactured Wood Casework Drawer Fronts

| | | Crew | Daily Output | Labor-Hours | Unit | Material | 2011 Bare Costs Labor | Equipment | Total | Total Incl O&P |
|---|---|---|---|---|---|---|---|---|---|---|
| 0010 | **MANUFACTURED WOOD CASEWORK DRAWER FRONTS** | | | | | | | | | |
| 0100 | Solid hardwood front | | | | | | | | | |
| 1000 | 4" high, 12" wide | 1 Carp | 17 | .471 | Ea. | 3.33 | 20.50 | | 23.83 | 34.50 |
| 1200 | 18" wide | " | 16 | .500 | " | 4.95 | 21.50 | | 26.45 | 38.50 |

## 12 32 23.30 Manufactured Wood Casework Vanities

| | | Crew | Daily Output | Labor-Hours | Unit | Material | 2011 Bare Costs Labor | Equipment | Total | Total Incl O&P |
|---|---|---|---|---|---|---|---|---|---|---|
| 0010 | **MANUFACTURED WOOD CASEWORK VANITIES** | | | | | | | | | |
| 8000 | Vanity bases, 2 doors, 30" high, 21" deep, 24" wide | 2 Carp | 20 | .800 | Ea. | 227 | 34.50 | | 261.50 | 305 |
| 8050 | 30" wide | | 16 | 1 | | 262 | 43 | | 305 | 355 |
| 8100 | 36" wide | | 13.33 | 1.200 | | 350 | 51.50 | | 401.50 | 465 |
| 8150 | 48" wide | ▼ | 11.43 | 1.400 | | 415 | 60.50 | | 475.50 | 555 |
| 9000 | For deluxe models of all vanities, add to above | | | | | 40% | | | | |
| 9500 | For custom built in place, add to above | | | | ▼ | 25% | 10% | | | |

## 12 32 23.35 Manufactured Wood Casework Hardware

| | | Crew | Daily Output | Labor-Hours | Unit | Material | 2011 Bare Costs Labor | Equipment | Total | Total Incl O&P |
|---|---|---|---|---|---|---|---|---|---|---|
| 0010 | **MANUFACTURED WOOD CASEWORK HARDWARE** | | | | | | | | | |
| 1000 | Catches, minimum | 1 Carp | 235 | .034 | Ea. | 1 | 1.47 | | 2.47 | 3.35 |
| 1040 | Maximum | " | 80 | .100 | " | 6.25 | 4.31 | | 10.56 | 13.45 |
| 2000 | Door/drawer pulls, handles | | | | | | | | | |
| 2200 | Handles and pulls, projecting, metal, minimum | 1 Carp | 48 | .167 | Ea. | 4.52 | 7.20 | | 11.72 | 15.95 |
| 2240 | Maximum | | 24 | .333 | | 9.55 | 14.35 | | 23.90 | 32.50 |
| 2300 | Wood, minimum | | 48 | .167 | | 4.76 | 7.20 | | 11.96 | 16.25 |
| 2340 | Maximum | | 24 | .333 | | 8.75 | 14.35 | | 23.10 | 31.50 |
| 2600 | Flush, metal, minimum | | 48 | .167 | | 4.76 | 7.20 | | 11.96 | 16.25 |
| 2640 | Maximum | | 24 | .333 | ▼ | 8.75 | 14.35 | | 23.10 | 31.50 |
| 3000 | Drawer tracks/glides, minimum | | 48 | .167 | Pr. | 8.05 | 7.20 | | 15.25 | 19.90 |
| 3040 | Maximum | | 24 | .333 | | 23.50 | 14.35 | | 37.85 | 48 |
| 4000 | Cabinet hinges, minimum | | 160 | .050 | | 2.74 | 2.15 | | 4.89 | 6.30 |
| 4040 | Maximum | ▼ | 68 | .118 | ▼ | 10.35 | 5.05 | | 15.40 | 19.15 |

# 12 35 Specialty Casework

## 12 35 50 – Educational/Library Casework

### 12 35 50.13 Educational Casework

| | 12 35 50.13 Educational Casework | Crew | Daily Output | Labor-Hours | Unit | Material | 2011 Bare Costs Labor | Equipment | Total | Total Incl O&P |
|---|---|---|---|---|---|---|---|---|---|---|
| 0010 | **EDUCATIONAL CASEWORK** | | | | | | | | | |
| 5000 | School, 24" deep, metal, 84" high units | 2 Carp | 15 | 1.067 | L.F. | 385 | 46 | | 431 | 495 |
| 5150 | Counter height units | | 20 | .800 | | 258 | 34.50 | | 292.50 | 335 |
| 5450 | Wood, custom fabricated, 32" high counter | | 20 | .800 | | 215 | 34.50 | | 249.50 | 289 |
| 5600 | Add for counter top | | 56 | .286 | | 23 | 12.30 | | 35.30 | 44 |
| 5800 | 84" high wall units | ↓ | 15 | 1.067 | ↓ | 420 | 46 | | 466 | 530 |
| 6000 | Laminated plastic finish is same price as wood | | | | | | | | | |

## 12 35 53 – Laboratory Casework

### 12 35 53.13 Metal Laboratory Casework

| | 12 35 53.13 Metal Laboratory Casework | Crew | Daily Output | Labor-Hours | Unit | Material | Labor | Equipment | Total | Total Incl O&P |
|---|---|---|---|---|---|---|---|---|---|---|
| 0010 | **METAL LABORATORY CASEWORK** | | | | | | | | | |
| 0020 | Cabinets, base, door units, metal | 2 Carp | 18 | .889 | L.F. | 202 | 38.50 | | 240.50 | 281 |
| 0300 | Drawer units | | 18 | .889 | | 450 | 38.50 | | 488.50 | 555 |
| 0700 | Tall storage cabinets, open, 7' high | | 20 | .800 | | 430 | 34.50 | | 464.50 | 530 |
| 0900 | With glazed doors | | 20 | .800 | | 545 | 34.50 | | 579.50 | 655 |
| 1300 | Wall cabinets, metal, 12-1/2" deep, open | | 20 | .800 | | 142 | 34.50 | | 176.50 | 209 |
| 1500 | With doors | ↓ | 20 | .800 | ↓ | 295 | 34.50 | | 329.50 | 380 |
| 6300 | Rule of thumb: lab furniture including installation & connection | | | | | | | | | |
| 6320 | High school | | | | S.F. | | | | 33.50 | 37 |
| 6340 | College | | | | | | | | 49.50 | 54.50 |
| 6360 | Clinical, health care | | | | | | | | 43 | 47 |
| 6380 | Industrial | | | | ↓ | | | | 69 | 75.50 |

## 12 35 59 – Display Casework

### 12 35 59.10 Display Cases

| | 12 35 59.10 Display Cases | Crew | Daily Output | Labor-Hours | Unit | Material | Labor | Equipment | Total | Total Incl O&P |
|---|---|---|---|---|---|---|---|---|---|---|
| 0010 | **DISPLAY CASES** Free standing, all glass | | | | | | | | | |
| 0020 | Aluminum frame, 42" high x 36" x 12" deep | 2 Carp | 8 | 2 | Ea. | 1,250 | 86 | | 1,336 | 1,475 |
| 0100 | 70" high x 48" x 18" deep | " | 6 | 2.667 | | 3,075 | 115 | | 3,190 | 3,550 |
| 0500 | For wood bases, add | | | | | 9% | | | | |
| 0600 | For hardwood frames, deduct | | | | | 8% | | | | |
| 0700 | For bronze, baked enamel finish, add | | | | ↓ | 10% | | | | |
| 2000 | Wall mounted, glass front, aluminum frame | | | | | | | | | |
| 2010 | Non-illuminated, one section 3' x 4' x 1'-4" | 2 Carp | 5 | 3.200 | Ea. | 2,150 | 138 | | 2,288 | 2,550 |
| 2100 | 5' x 4' x 1'-4" | | 5 | 3.200 | | 2,475 | 138 | | 2,613 | 2,925 |
| 2200 | 6' x 4' x 1'-4" | | 4 | 4 | | 2,975 | 172 | | 3,147 | 3,550 |
| 2500 | Two sections, 8' x 4' x 1'-4" | | 2 | 8 | | 1,900 | 345 | | 2,245 | 2,625 |
| 2600 | 10' x 4' x 1'-4" | | 2 | 8 | | 2,250 | 345 | | 2,595 | 3,000 |
| 3000 | Three sections, 16' x 4' x 1'-4" | ↓ | 1.50 | 10.667 | ↓ | 3,900 | 460 | | 4,360 | 5,000 |
| 3500 | For fluorescent lights, add | | | | Section | 330 | | | 330 | 365 |
| 4000 | Table exhibit cases, 2' wide, 3' high, 4' long, flat top | 2 Carp | 5 | 3.200 | Ea. | 1,150 | 138 | | 1,288 | 1,475 |
| 4100 | 3' wide, 3' high, 4' long, sloping top | " | 3 | 5.333 | " | 1,850 | 230 | | 2,080 | 2,375 |

## 12 35 70 – Healthcare Casework

### 12 35 70.13 Hospital Casework

| | 12 35 70.13 Hospital Casework | Crew | Daily Output | Labor-Hours | Unit | Material | Labor | Equipment | Total | Total Incl O&P |
|---|---|---|---|---|---|---|---|---|---|---|
| 0010 | **HOSPITAL CASEWORK** | | | | | | | | | |
| 0500 | Base cabinets, laminated plastic | 2 Carp | 10 | 1.600 | L.F. | 246 | 69 | | 315 | 375 |
| 1000 | Stainless steel | " | 10 | 1.600 | | 455 | 69 | | 524 | 605 |
| 1200 | For all drawers, add | | | | | 26 | | | 26 | 28.50 |
| 1300 | Cabinet base trim, 4" high, enameled steel | 2 Carp | 200 | .080 | | 41.50 | 3.44 | | 44.94 | 51 |
| 1400 | Stainless steel | | 200 | .080 | | 82.50 | 3.44 | | 85.94 | 96.50 |
| 1450 | Countertop, laminated plastic, no backsplash | | 40 | .400 | | 42.50 | 17.20 | | 59.70 | 73.50 |
| 1650 | With backsplash | | 40 | .400 | ↓ | 53 | 17.20 | | 70.20 | 84.50 |
| 1800 | For sink cutout, add | | 12.20 | 1.311 | Ea. | | 56.50 | | 56.50 | 86.50 |
| 1900 | Stainless steel counter top | ↓ | 40 | .400 | L.F. | 138 | 17.20 | | 155.20 | 178 |

# 12 35 Specialty Casework

## 12 35 70 – Healthcare Casework

### 12 35 70.13 Hospital Casework

| | | Crew | Daily Output | Labor-Hours | Unit | Material | 2011 Bare Costs Labor | Equipment | Total | Total Incl O&P |
|---|---|---|---|---|---|---|---|---|---|---|
| 2000 | For drop-in stainless 43" x 21" sink, add | | | | Ea. | 905 | | | 905 | 995 |
| 2500 | Wall cabinets, laminated plastic | 2 Carp | 15 | 1.067 | L.F. | 184 | 46 | | 230 | 274 |
| 2600 | Enameled steel | | 15 | 1.067 | | 227 | 46 | | 273 | 320 |
| 2700 | Stainless steel | ↓ | 15 | 1.067 | | 450 | 46 | | 496 | 565 |
| 2800 | For glass doors, add | | | | ↓ | 31 | | | 31 | 34 |

### 12 35 70.16 Nurse Station Casework

| | | Crew | Daily Output | Labor-Hours | Unit | Material | 2011 Bare Costs Labor | Equipment | Total | Total Incl O&P |
|---|---|---|---|---|---|---|---|---|---|---|
| 0010 | **NURSE STATION CASEWORK** | | | | | | | | | |
| 2100 | Door type, laminated plastic | 2 Carp | 10 | 1.600 | L.F. | 286 | 69 | | 355 | 420 |
| 2200 | Enameled steel | | 10 | 1.600 | | 273 | 69 | | 342 | 405 |
| 2300 | Stainless steel | ↓ | 10 | 1.600 | | 545 | 69 | | 614 | 705 |
| 2400 | For drawer type, add | | | | | 232 | | | 232 | 255 |

## 12 35 80 – Commercial Kitchen Casework

### 12 35 80.13 Metal Kitchen Casework

| | | Crew | Daily Output | Labor-Hours | Unit | Material | 2011 Bare Costs Labor | Equipment | Total | Total Incl O&P |
|---|---|---|---|---|---|---|---|---|---|---|
| 0010 | **METAL KITCHEN CASEWORK** | | | | | | | | | |
| 3500 | Base cabinets, metal, minimum | 2 Carp | 30 | .533 | L.F. | 66.50 | 23 | | 89.50 | 108 |
| 3600 | Maximum | | 25 | .640 | | 169 | 27.50 | | 196.50 | 229 |
| 3700 | Wall cabinets, metal, minimum | | 30 | .533 | | 66.50 | 23 | | 89.50 | 108 |
| 3800 | Maximum | ↓ | 25 | .640 | ↓ | 153 | 27.50 | | 180.50 | 211 |

# 12 36 Countertops

## 12 36 16 – Metal Countertops

### 12 36 16.10 Stainless Steel Countertops

| | | Crew | Daily Output | Labor-Hours | Unit | Material | 2011 Bare Costs Labor | Equipment | Total | Total Incl O&P |
|---|---|---|---|---|---|---|---|---|---|---|
| 0010 | **STAINLESS STEEL COUNTERTOPS** | | | | | | | | | |
| 3200 | Stainless steel, custom | 1 Carp | 24 | .333 | S.F. | 145 | 14.35 | | 159.35 | 182 |

## 12 36 19 – Wood Countertops

### 12 36 19.10 Maple Countertops

| | | Crew | Daily Output | Labor-Hours | Unit | Material | 2011 Bare Costs Labor | Equipment | Total | Total Incl O&P |
|---|---|---|---|---|---|---|---|---|---|---|
| 0010 | **MAPLE COUNTERTOPS** | | | | | | | | | |
| 2900 | Solid, laminated, 1-1/2" thick, no splash | 1 Carp | 28 | .286 | L.F. | 74 | 12.30 | | 86.30 | 100 |
| 3000 | With square splash | | 28 | .286 | " | 88 | 12.30 | | 100.30 | 116 |
| 3400 | Recessed cutting block with trim, 16" x 20" x 1" | ↓ | 8 | 1 | Ea. | 74 | 43 | | 117 | 148 |

## 12 36 23 – Plastic Countertops

### 12 36 23.13 Plastic-Laminate-Clad Countertops

| | | Crew | Daily Output | Labor-Hours | Unit | Material | 2011 Bare Costs Labor | Equipment | Total | Total Incl O&P |
|---|---|---|---|---|---|---|---|---|---|---|
| 0010 | **PLASTIC-LAMINATE-CLAD COUNTERTOPS** | | | | | | | | | |
| 0020 | Stock, 24" wide w/backsplash, minimum | 1 Carp | 30 | .267 | L.F. | 9.85 | 11.50 | | 21.35 | 28.50 |
| 0100 | Maximum | | 25 | .320 | | 21 | 13.80 | | 34.80 | 44 |
| 0300 | Custom plastic, 7/8" thick, aluminum molding, no splash | | 30 | .267 | | 22.50 | 11.50 | | 34 | 42.50 |
| 0400 | Cove splash | | 30 | .267 | | 29 | 11.50 | | 40.50 | 49.50 |
| 0600 | 1-1/4" thick, no splash | | 28 | .286 | | 31.50 | 12.30 | | 43.80 | 53.50 |
| 0700 | Square splash | | 28 | .286 | | 32 | 12.30 | | 44.30 | 54 |
| 0900 | Square edge, plastic face, 7/8" thick, no splash | | 30 | .267 | | 28 | 11.50 | | 39.50 | 48.50 |
| 1000 | With splash | ↓ | 30 | .267 | | 36 | 11.50 | | 47.50 | 57 |
| 1200 | For stainless channel edge, 7/8" thick, add | | | | | 3 | | | 3 | 3.30 |
| 1300 | 1-1/4" thick, add | | | | | 3.50 | | | 3.50 | 3.85 |
| 1500 | For solid color suede finish, add | | | | ↓ | 3 | | | 3 | 3.30 |
| 1700 | For end splash, add | | | | Ea. | 18 | | | 18 | 19.80 |
| 1900 | For cut outs, standard, add, minimum | 1 Carp | 32 | .250 | | 4 | 10.75 | | 14.75 | 21 |
| 2000 | Maximum | | 8 | 1 | ↓ | 6 | 43 | | 49 | 72.50 |
| 2100 | Postformed, including backsplash and front edge | | 30 | .267 | L.F. | 11 | 11.50 | | 22.50 | 29.50 |
| 2110 | Mitred, add | ↓ | 12 | .667 | Ea. | | 28.50 | | 28.50 | 44 |

# 12 36 Countertops

## 12 36 23 – Plastic Countertops

### 12 36 23.13 Plastic-Laminate-Clad Countertops

| | | Crew | Daily Output | Labor-Hours | Unit | Material | 2011 Bare Costs Labor | Equipment | Total | Total Incl O&P |
|---|---|---|---|---|---|---|---|---|---|---|
| 2200 | Built-in place, 25" wide, plastic laminate | 1 Carp | 25 | .320 | L.F. | 30 | 13.80 | | 43.80 | 54 |

## 12 36 33 – Tile Countertops

### 12 36 33.10 Ceramic Tile Countertops

| 0010 | **CERAMIC TILE COUNTERTOPS** | | | | | | | | | |
|---|---|---|---|---|---|---|---|---|---|---|
| 2300 | Ceramic tile mosaic | 1 Carp | 25 | .320 | L.F. | 33 | 13.80 | | 46.80 | 57.50 |

## 12 36 40 – Stone Countertops

### 12 36 40.10 Natural Stone Countertops

| 0010 | **NATURAL STONE COUNTERTOPS** | | | | | | | | | |
|---|---|---|---|---|---|---|---|---|---|---|
| 2500 | Marble, stock, with splash, 1/2" thick, minimum | 1 Bric | 17 | .471 | L.F. | 41 | 20.50 | | 61.50 | 75.50 |
| 2700 | 3/4" thick, maximum | | 13 | .615 | | 103 | 26.50 | | 129.50 | 153 |
| 2800 | Granite, average, 1-1/4" thick, 24" wide, no splash | ↓ | 13.01 | .615 | ↓ | 135 | 26.50 | | 161.50 | 189 |

## 12 36 53 – Laboratory Countertops

### 12 36 53.10 Laboratory Countertops and Sinks

| 0010 | **LABORATORY COUNTERTOPS AND SINKS** | | | | | | | | | |
|---|---|---|---|---|---|---|---|---|---|---|
| 0020 | Countertops, not incl. base cabinets, acid-proof, minimum | 2 Carp | 82 | .195 | S.F. | 34.50 | 8.40 | | 42.90 | 51 |
| 0030 | Maximum | | 70 | .229 | | 44.50 | 9.85 | | 54.35 | 64 |
| 0040 | Stainless steel | ↓ | 82 | .195 | ↓ | 110 | 8.40 | | 118.40 | 133 |

## 12 36 61 – Simulated Stone Countertops

### 12 36 61.16 Solid Surface Countertops

| 0010 | **SOLID SURFACE COUNTERTOPS**, Acrylic polymer | | | | | | | | | |
|---|---|---|---|---|---|---|---|---|---|---|
| 0020 | Pricing for orders of 100 L.F. or greater | | | | | | | | | |
| 0100 | 25" wide, solid colors | 2 Carp | 28 | .571 | L.F. | 44.50 | 24.50 | | 69 | 86.50 |
| 0200 | Patterned colors | | 28 | .571 | | 56.50 | 24.50 | | 81 | 99.50 |
| 0300 | Premium patterned colors | | 28 | .571 | | 70.50 | 24.50 | | 95 | 116 |
| 0400 | With silicone attached 4" backsplash, solid colors | | 27 | .593 | | 49 | 25.50 | | 74.50 | 93 |
| 0500 | Patterned colors | | 27 | .593 | | 62 | 25.50 | | 87.50 | 107 |
| 0600 | Premium patterned colors | | 27 | .593 | | 77 | 25.50 | | 102.50 | 124 |
| 0700 | With hard seam attached 4" backsplash, solid colors | | 23 | .696 | | 49 | 30 | | 79 | 100 |
| 0800 | Patterned colors | | 23 | .696 | | 62 | 30 | | 92 | 114 |
| 0900 | Premium patterned colors | ↓ | 23 | .696 | ↓ | 77 | 30 | | 107 | 131 |
| 1000 | Pricing for order of 51 – 99 L.F. | | | | | | | | | |
| 1100 | 25" wide, solid colors | 2 Carp | 24 | .667 | L.F. | 51.50 | 28.50 | | 80 | 101 |
| 1200 | Patterned colors | | 24 | .667 | | 65 | 28.50 | | 93.50 | 116 |
| 1300 | Premium patterned colors | | 24 | .667 | | 81.50 | 28.50 | | 110 | 134 |
| 1400 | With silicone attached 4" backsplash, solid colors | | 23 | .696 | | 56.50 | 30 | | 86.50 | 108 |
| 1500 | Patterned colors | | 23 | .696 | | 71.50 | 30 | | 101.50 | 125 |
| 1600 | Premium patterned colors | | 23 | .696 | | 89 | 30 | | 119 | 144 |
| 1700 | With hard seam attached 4" backsplash, solid colors | | 20 | .800 | | 56.50 | 34.50 | | 91 | 115 |
| 1800 | Patterned colors | | 20 | .800 | | 71.50 | 34.50 | | 106 | 132 |
| 1900 | Premium patterned colors | ↓ | 20 | .800 | ↓ | 89 | 34.50 | | 123.50 | 151 |
| 2000 | Pricing for order of 1 – 50 L.F. | | | | | | | | | |
| 2100 | 25" wide, solid colors | 2 Carp | 20 | .800 | L.F. | 60 | 34.50 | | 94.50 | 119 |
| 2200 | Patterned colors | | 20 | .800 | | 76.50 | 34.50 | | 111 | 137 |
| 2300 | Premium patterned colors | | 20 | .800 | | 95.50 | 34.50 | | 130 | 158 |
| 2400 | With silicone attached 4" backsplash, solid colors | | 19 | .842 | | 66 | 36.50 | | 102.50 | 128 |
| 2500 | Patterned colors | | 19 | .842 | | 83.50 | 36.50 | | 120 | 148 |
| 2600 | Premium patterned colors | | 19 | .842 | | 104 | 36.50 | | 140.50 | 171 |
| 2700 | With hard seam attached 4" backsplash, solid colors | | 15 | 1.067 | | 66 | 46 | | 112 | 143 |
| 2800 | Patterned colors | | 15 | 1.067 | | 83.50 | 46 | | 129.50 | 163 |
| 2900 | Premium patterned colors | ↓ | 15 | 1.067 | ↓ | 104 | 46 | | 150 | 186 |

# 12 36 Countertops

## 12 36 61 – Simulated Stone Countertops

### 12 36 61.16 Solid Surface Countertops

| | | Crew | Daily Output | Labor-Hours | Unit | Material | 2011 Bare Costs Labor | 2011 Bare Costs Equipment | Total | Total Incl O&P |
|---|---|---|---|---|---|---|---|---|---|---|
| 3000 | Sinks, pricing for order of 100 or greater units | | | | | | | | | |
| 3100 | Single bowl, hard seamed, solid colors, 13" x 17" | 1 Carp | 3 | 2.667 | Ea. | 300 | 115 | | 415 | 505 |
| 3200 | 10" x 15" | | 7 | 1.143 | | 139 | 49 | | 188 | 229 |
| 3300 | Cutouts for sinks | | 8 | 1 | | | 43 | | 43 | 66 |
| 3400 | Sinks, pricing for order of 51 – 99 units | | | | | | | | | |
| 3500 | Single bowl, hard seamed, solid colors, 13" x 17" | 1 Carp | 2.55 | 3.137 | Ea. | 345 | 135 | | 480 | 585 |
| 3600 | 10" x 15" | | 6 | 1.333 | | 160 | 57.50 | | 217.50 | 264 |
| 3700 | Cutouts for sinks | | 7 | 1.143 | | | 49 | | 49 | 75.50 |
| 3800 | Sinks, pricing for order of 1 – 50 units | | | | | | | | | |
| 3900 | Single bowl, hard seamed, solid colors, 13" x 17" | 1 Carp | 2 | 4 | Ea. | 405 | 172 | | 577 | 710 |
| 4000 | 10" x 15" | | 4.55 | 1.758 | | 188 | 75.50 | | 263.50 | 325 |
| 4100 | Cutouts for sinks | | 5.25 | 1.524 | | | 65.50 | | 65.50 | 101 |
| 4200 | Cooktop cutouts, pricing for 100 or greater units | | 4 | 2 | | 22 | 86 | | 108 | 157 |
| 4300 | 51 – 99 units | | 3.40 | 2.353 | | 25.50 | 101 | | 126.50 | 183 |
| 4400 | 1 – 50 units | | 3 | 2.667 | | 30 | 115 | | 145 | 209 |

### 12 36 61.17 Solid Surface Vanity Tops

| | | Crew | Daily Output | Labor-Hours | Unit | Material | 2011 Bare Costs Labor | 2011 Bare Costs Equipment | Total | Total Incl O&P |
|---|---|---|---|---|---|---|---|---|---|---|
| 0010 | **SOLID SURFACE VANITY TOPS** | | | | | | | | | |
| 0015 | Solid surface, center bowl, 17" x 19" | 1 Carp | 12 | .667 | Ea. | 189 | 28.50 | | 217.50 | 252 |
| 0020 | 19" x 25" | | 12 | .667 | | 229 | 28.50 | | 257.50 | 296 |
| 0030 | 19" x 31" | | 12 | .667 | | 278 | 28.50 | | 306.50 | 350 |
| 0040 | 19" x 37" | | 12 | .667 | | 325 | 28.50 | | 353.50 | 400 |
| 0050 | 22" x 25" | | 10 | .800 | | 202 | 34.50 | | 236.50 | 275 |
| 0060 | 22" x 31" | | 10 | .800 | | 236 | 34.50 | | 270.50 | 310 |
| 0070 | 22" x 37" | | 10 | .800 | | 274 | 34.50 | | 308.50 | 355 |
| 0080 | 22" x 43" | | 10 | .800 | | 315 | 34.50 | | 349.50 | 400 |
| 0090 | 22" x 49" | | 10 | .800 | | 345 | 34.50 | | 379.50 | 435 |
| 0110 | 22" x 55" | | 8 | 1 | | 395 | 43 | | 438 | 500 |
| 0120 | 22" x 61" | | 8 | 1 | | 450 | 43 | | 493 | 560 |
| 0220 | Double bowl, 22" x 61" | | 8 | 1 | | 505 | 43 | | 548 | 625 |
| 0230 | Double bowl, 22" x 73" | | 8 | 1 | | 705 | 43 | | 748 | 840 |
| 0240 | For aggregate colors, add | | | | | 35% | | | | |
| 0250 | For faucets and fittings, see Section 22 41 39.10 | | | | | | | | | |

### 12 36 61.19 Engineered Stone Countertops

| | | Crew | Daily Output | Labor-Hours | Unit | Material | 2011 Bare Costs Labor | 2011 Bare Costs Equipment | Total | Total Incl O&P |
|---|---|---|---|---|---|---|---|---|---|---|
| 0010 | **ENGINEERED STONE COUNTERTOPS** | | | | | | | | | |
| 0100 | 25" wide, 4" backsplash, color group A, minimum | 2 Carp | 15 | 1.067 | L.F. | 54 | 46 | | 100 | 130 |
| 0110 | Maximum | | 15 | 1.067 | | 75 | 46 | | 121 | 153 |
| 0120 | Color group B, minimum | | 15 | 1.067 | | 58 | 46 | | 104 | 135 |
| 0130 | Maximum | | 15 | 1.067 | | 80 | 46 | | 126 | 159 |
| 0140 | Color group C, minimum | | 15 | 1.067 | | 65 | 46 | | 111 | 142 |
| 0150 | Maximum | | 15 | 1.067 | | 91 | 46 | | 137 | 171 |
| 0160 | Color group D, minimum | | 15 | 1.067 | | 71 | 46 | | 117 | 149 |
| 0170 | Maximum | | 15 | 1.067 | | 98 | 46 | | 144 | 179 |

# 12 46 Furnishing Accessories

## 12 46 13 – Ash Receptacles

### 12 46 13.10 Ash/Trash Receivers

| | | Crew | Daily Output | Labor-Hours | Unit | Material | 2011 Bare Costs Labor | Equipment | Total | Total Incl O&P |
|---|---|---|---|---|---|---|---|---|---|---|
| 0010 | **ASH/TRASH RECEIVERS** | | | | | | | | | |
| 1000 | Ash urn, cylindrical metal | | | | | | | | | |
| 1020 | 8" diameter, 20" high | 1 Clab | 60 | .133 | Ea. | 85 | 4.58 | | 89.58 | 101 |
| 1060 | 10" diameter, 26" high | " | 60 | .133 | " | 100 | 4.58 | | 104.58 | 117 |
| 2000 | Combination ash/trash urn, metal | | | | | | | | | |
| 2020 | 8" diameter, 20" high | 1 Clab | 60 | .133 | Ea. | 85 | 4.58 | | 89.58 | 101 |
| 2050 | 10" diameter, 26" high | " | 60 | .133 | " | 105 | 4.58 | | 109.58 | 122 |

## 12 46 19 – Clocks

### 12 46 19.50 Wall Clocks

| | | Crew | Daily Output | Labor-Hours | Unit | Material | 2011 Bare Costs Labor | Equipment | Total | Total Incl O&P |
|---|---|---|---|---|---|---|---|---|---|---|
| 0010 | **WALL CLOCKS** | | | | | | | | | |
| 0080 | 12" diameter, single face | 1 Elec | 8 | 1 | Ea. | 139 | 50.50 | | 189.50 | 227 |
| 0100 | Double face | " | 6.20 | 1.290 | " | 350 | 65 | | 415 | 480 |

## 12 46 33 – Waste Receptacles

### 12 46 33.13 Trash Receptacles

| | | Crew | Daily Output | Labor-Hours | Unit | Material | 2011 Bare Costs Labor | Equipment | Total | Total Incl O&P |
|---|---|---|---|---|---|---|---|---|---|---|
| 0010 | **TRASH RECEPTACLES** | | | | | | | | | |
| 4000 | Trash receptacle, metal | | | | | | | | | |
| 4020 | 8" diameter, 15" high | 1 Clab | 60 | .133 | Ea. | 67 | 4.58 | | 71.58 | 80.50 |
| 4040 | 10" diameter, 18" high | | 60 | .133 | | 135 | 4.58 | | 139.58 | 155 |
| 5040 | 16" x 8" x 14" high | ↓ | 60 | .133 | ↓ | 21 | 4.58 | | 25.58 | 30 |
| 5500 | Plastic, with lid | | | | | | | | | |
| 5520 | 35 gallon | 1 Clab | 60 | .133 | Ea. | 135 | 4.58 | | 139.58 | 155 |
| 5540 | 45 gallon | | 60 | .133 | | 220 | 4.58 | | 224.58 | 249 |
| 5550 | Plastic recycling barrel, w/lid & wheels, 32 gal. [G] | | 60 | .133 | | 75 | 4.58 | | 79.58 | 89.50 |
| 5560 | 65 gal. [G] | | 60 | .133 | | 135 | 4.58 | | 139.58 | 155 |
| 5570 | 95 gal. [G] | ↓ | 60 | .133 | ↓ | 220 | 4.58 | | 224.58 | 249 |

# 12 48 Rugs and Mats

## 12 48 13 – Entrance Floor Mats and Frames

### 12 48 13.13 Entrance Floor Mats

| | | Crew | Daily Output | Labor-Hours | Unit | Material | 2011 Bare Costs Labor | Equipment | Total | Total Incl O&P |
|---|---|---|---|---|---|---|---|---|---|---|
| 0010 | **ENTRANCE FLOOR MATS** | | | | | | | | | |
| 0020 | Recessed, in-laid black rubber, 3/8" thick, solid | 1 Clab | 155 | .052 | S.F. | 19.85 | 1.77 | | 21.62 | 24.50 |
| 0050 | Perforated | | 155 | .052 | | 13.80 | 1.77 | | 15.57 | 17.90 |
| 0100 | 1/2" thick, solid | | 155 | .052 | | 16.60 | 1.77 | | 18.37 | 21 |
| 0150 | Perforated | | 155 | .052 | | 22 | 1.77 | | 23.77 | 26.50 |
| 0200 | In colors, 3/8" thick, solid | | 155 | .052 | | 17.85 | 1.77 | | 19.62 | 22.50 |
| 0250 | Perforated | | 155 | .052 | | 18.45 | 1.77 | | 20.22 | 23 |
| 0300 | 1/2" thick, solid | | 155 | .052 | | 23 | 1.77 | | 24.77 | 28 |
| 0350 | Perforated | | 155 | .052 | | 23.50 | 1.77 | | 25.27 | 28.50 |
| 2000 | Recycled rubber tire tile, 12" x 12" x 3/8" thick [G] | | 125 | .064 | | 8.50 | 2.20 | | 10.70 | 12.70 |
| 2510 | Natural cocoa fiber, 1/2" thick [G] | | 125 | .064 | | 8.10 | 2.20 | | 10.30 | 12.30 |
| 2520 | 3/4" thick [G] | | 125 | .064 | | 6.25 | 2.20 | | 8.45 | 10.20 |
| 2530 | 1" thick [G] | ↓ | 125 | .064 | ↓ | 6.90 | 2.20 | | 9.10 | 10.95 |

# 12 51 Office Furniture

## 12 51 16 – Case Goods

### 12 51 16.13 Metal Case Goods

| | Crew | Daily Output | Labor-Hours | Unit | Material | 2011 Bare Costs Labor | Equipment | Total | Total Incl O&P |
|---|---|---|---|---|---|---|---|---|---|
| 0010 **METAL CASE GOODS** | | | | | | | | | |
| 0020 Desks, 29" high, double pedestal, 30" x 60", metal, minimum | | | | Ea. | 500 | | | 500 | 550 |
| 0030 Maximum | | | | | 1,275 | | | 1,275 | 1,400 |
| 0600 Desks, single pedestal, 30" x 60", metal, minimum | | | | | 475 | | | 475 | 520 |
| 0620 Maximum | | | | | 1,050 | | | 1,050 | 1,150 |
| 0720 Desks, secretarial, 30" x 60", metal, minimum | | | | | 435 | | | 435 | 480 |
| 0730 Maximum | | | | | 750 | | | 750 | 825 |
| 0740 Return, 20" x 42", minimum | | | | | 325 | | | 325 | 355 |
| 0750 Maximum | | | | | 490 | | | 490 | 540 |
| 0940 59" x 12" x 23" high, steel, minimum | | | | | 286 | | | 286 | 315 |
| 0960 Maximum | | | | | 350 | | | 350 | 385 |

### 12 51 16.16 Wood Case Goods

| | Crew | Daily Output | Labor-Hours | Unit | Material | 2011 Bare Costs Labor | Equipment | Total | Total Incl O&P |
|---|---|---|---|---|---|---|---|---|---|
| 0010 **WOOD CASE GOODS** | | | | | | | | | |
| 0150 Desk, 29" high, double pedestal, 30" x 60" | | | | | | | | | |
| 0160 Wood, minimum | | | | Ea. | 330 | | | 330 | 365 |
| 0180 Maximum | | | | " | 2,475 | | | 2,475 | 2,725 |
| 0630 Single pedestal, 30" x 60" | | | | | | | | | |
| 0640 Wood, minimum | | | | Ea. | 560 | | | 560 | 615 |
| 0650 Maximum | | | | | 780 | | | 780 | 855 |
| 0670 Executive return, 24" x 42", with box, file, wood, minimum | | | | | 395 | | | 395 | 430 |
| 0680 Maximum | | | | | 780 | | | 780 | 855 |
| 0790 Desk, 29" high, secretarial, 30" x 60" | | | | | | | | | |
| 0800 Wood, minimum | | | | Ea. | 495 | | | 495 | 545 |
| 0810 Maximum | | | | | 2,525 | | | 2,525 | 2,775 |
| 0820 Return, 20" x 42", minimum | | | | | 290 | | | 290 | 320 |
| 0830 Maximum | | | | | 995 | | | 995 | 1,100 |
| 0900 Desktop organizer, 72" x 14" x 36" high, wood, minimum | | | | | 169 | | | 169 | 186 |
| 0920 Maximum | | | | | 400 | | | 400 | 440 |
| 1110 Furniture, credenza, 29" high, 18" to 22" x 60" to 72" | | | | | | | | | |
| 1120 Wood, minimum | | | | Ea. | 645 | | | 645 | 710 |
| 1140 Maximum | | | | " | 3,025 | | | 3,025 | 3,325 |

## 12 51 23 – Office Tables

### 12 51 23.33 Conference Tables

| | Crew | Daily Output | Labor-Hours | Unit | Material | 2011 Bare Costs Labor | Equipment | Total | Total Incl O&P |
|---|---|---|---|---|---|---|---|---|---|
| 0010 **CONFERENCE TABLES** | | | | | | | | | |
| 6050 Boat, 96" x 42", minimum | | | | Ea. | 735 | | | 735 | 810 |
| 6150 Maximum | | | | | 3,075 | | | 3,075 | 3,375 |
| 6720 Rectangle, 96" x 42", minimum | | | | | 1,075 | | | 1,075 | 1,175 |
| 6740 Maximum | | | | | 3,075 | | | 3,075 | 3,375 |

# 12 52 Seating

## 12 52 23 – Office Seating

### 12 52 23.13 Office Chairs

| | Crew | Daily Output | Labor-Hours | Unit | Material | 2011 Bare Costs Labor | Equipment | Total | Total Incl O&P |
|---|---|---|---|---|---|---|---|---|---|
| 0010 **OFFICE CHAIRS** | | | | | | | | | |
| 2000 Standard office chair, executive, minimum | | | | Ea. | 330 | | | 330 | 360 |
| 2150 Maximum | | | | | 1,800 | | | 1,800 | 2,000 |
| 2200 Management, minimum | | | | | 206 | | | 206 | 226 |
| 2250 Maximum | | | | | 2,300 | | | 2,300 | 2,525 |
| 2280 Task, minimum | | | | | 136 | | | 136 | 150 |
| 2290 Maximum | | | | | 575 | | | 575 | 630 |
| 2300 Arm kit, minimum | | | | | 61 | | | 61 | 67 |

# 12 52 Seating

## 12 52 23 – Office Seating

| 12 52 23.13 Office Chairs | | Crew | Daily Output | Labor-Hours | Unit | Material | 2011 Bare Costs Labor | Equipment | Total | Total Incl O&P |
|---|---|---|---|---|---|---|---|---|---|---|
| 2320 | Maximum | | | | Ea. | 91 | | | 91 | 100 |

# 12 54 Hospitality Furniture

## 12 54 13 – Hotel and Motel Furniture

### 12 54 13.10 Hotel Furniture

| | | Crew | Daily Output | Labor-Hours | Unit | Material | Labor | Equipment | Total | Total Incl O&P |
|---|---|---|---|---|---|---|---|---|---|---|
| 0010 | **HOTEL FURNITURE** | | | | | | | | | |
| 0020 | Standard quality set, minimum | | | | Room | 2,325 | | | 2,325 | 2,550 |
| 0200 | Maximum | | | | " | 8,300 | | | 8,300 | 9,125 |

## 12 54 16 – Restaurant Furniture

### 12 54 16.10 Tables, Folding

| | | Crew | Daily Output | Labor-Hours | Unit | Material | Labor | Equipment | Total | Total Incl O&P |
|---|---|---|---|---|---|---|---|---|---|---|
| 0010 | **TABLES, FOLDING** Laminated plastic tops | | | | | | | | | |
| 1000 | Tubular steel legs with glides | | | | | | | | | |
| 1020 | 18" x 60", minimum | | | | Ea. | 236 | | | 236 | 260 |
| 1040 | Maximum | | | | | 1,300 | | | 1,300 | 1,425 |
| 1840 | 36" x 96", minimum | | | | | 272 | | | 272 | 299 |
| 1860 | Maximum | | | | | 2,600 | | | 2,600 | 2,875 |
| 2000 | Round, wood stained, plywood top, 60" diameter, minimum | | | | | 385 | | | 385 | 420 |
| 2020 | Maximum | | | | | 1,400 | | | 1,400 | 1,525 |

### 12 54 16.20 Furniture, Restaurant

| | | Crew | Daily Output | Labor-Hours | Unit | Material | Labor | Equipment | Total | Total Incl O&P |
|---|---|---|---|---|---|---|---|---|---|---|
| 0010 | **FURNITURE, RESTAURANT** | | | | | | | | | |
| 0020 | Bars, built-in, front bar | 1 Carp | 5 | 1.600 | L.F. | 250 | 69 | | 319 | 380 |
| 0200 | Back bar | " | 5 | 1.600 | " | 182 | 69 | | 251 | 305 |
| 0300 | Booth seating, see Section 12 54 16.70 | | | | | | | | | |
| 2000 | Chair, bentwood side chair, metal, minimum | | | | Ea. | 83.50 | | | 83.50 | 92 |
| 2020 | Maximum | | | | | 98 | | | 98 | 108 |
| 2600 | Upholstered seat & back, arms, minimum | | | | | 136 | | | 136 | 150 |
| 2620 | Maximum | | | | | 385 | | | 385 | 425 |

### 12 54 16.70 Booths

| | | Crew | Daily Output | Labor-Hours | Unit | Material | Labor | Equipment | Total | Total Incl O&P |
|---|---|---|---|---|---|---|---|---|---|---|
| 0010 | **BOOTHS** | | | | | | | | | |
| 1000 | Banquet, upholstered seat and back, custom | | | | | | | | | |
| 1500 | Straight, minimum | 2 Carp | 40 | .400 | L.F. | 169 | 17.20 | | 186.20 | 213 |
| 1520 | Maximum | | 36 | .444 | | 330 | 19.15 | | 349.15 | 390 |
| 1600 | "L" or "U" shape, minimum | | 35 | .457 | | 172 | 19.70 | | 191.70 | 220 |
| 1620 | Maximum | | 30 | .533 | | 305 | 23 | | 328 | 370 |
| 1800 | Upholstered outside finished backs for | | | | | | | | | |
| 1810 | single booths and custom banquets | | | | | | | | | |
| 1820 | Minimum | 2 Carp | 44 | .364 | L.F. | 19.50 | 15.65 | | 35.15 | 45.50 |
| 1840 | Maximum | " | 40 | .400 | " | 58.50 | 17.20 | | 75.70 | 91 |
| 3000 | Fixed seating, one piece plastic chair and | | | | | | | | | |
| 3010 | plastic laminate table top | | | | | | | | | |
| 3100 | Two seat, 24" x 24" table, minimum | F-7 | 30 | 1.067 | Ea. | 680 | 41.50 | | 721.50 | 815 |
| 3120 | Maximum | | 26 | 1.231 | | 970 | 47.50 | | 1,017.50 | 1,150 |
| 3200 | Four seat, 24" x 48" table, minimum | | 28 | 1.143 | | 675 | 44 | | 719 | 815 |
| 3220 | Maximum | | 24 | 1.333 | | 1,150 | 51.50 | | 1,201.50 | 1,350 |
| 5000 | Mount in floor, wood fiber core with | | | | | | | | | |
| 5010 | plastic laminate face, single booth | | | | | | | | | |
| 5050 | 24" wide | F-7 | 30 | 1.067 | Ea. | 261 | 41.50 | | 302.50 | 350 |
| 5100 | 48" wide | " | 28 | 1.143 | " | 330 | 44 | | 374 | 435 |

# 12 55 Detention Furniture

## 12 55 13 – Detention Bunks

| 12 55 13.13 Cots | Crew | Daily Output | Labor-Hours | Unit | Material | 2011 Bare Costs Labor | Equipment | Total | Total Incl O&P |
|---|---|---|---|---|---|---|---|---|---|
| 0010 **COTS** | | | | | | | | | |
| 2500 Bolted, single, painted steel | E-4 | 20 | 1.600 | Ea. | 330 | 78.50 | 5.45 | 413.95 | 510 |
| 2700 Stainless steel | " | 20 | 1.600 | " | 960 | 78.50 | 5.45 | 1,043.95 | 1,200 |

# 12 56 Institutional Furniture

## 12 56 33 – Classroom Furniture

| 12 56 33.10 Furniture, School | Crew | Daily Output | Labor-Hours | Unit | Material | 2011 Bare Costs Labor | Equipment | Total | Total Incl O&P |
|---|---|---|---|---|---|---|---|---|---|
| 0010 **FURNITURE, SCHOOL** | | | | | | | | | |
| 0500 Classroom, movable chair & desk type, minimum | | | | Set | | | | 67 | 74 |
| 0600 Maximum | | | | " | | | | 141 | 155 |
| 1000 Chair, molded plastic | | | | | | | | | |
| 1100 Integral tablet arm, minimum | | | | Ea. | 77.50 | | | 77.50 | 85 |
| 1150 Maximum | | | | | 151 | | | 151 | 166 |
| 2000 Desk, single pedestal, top book compartment, minimum | | | | | 55 | | | 55 | 60.50 |
| 2020 Maximum | | | | | 67.50 | | | 67.50 | 74.50 |
| 2200 Flip top, minimum | | | | | 95.50 | | | 95.50 | 105 |
| 2220 Maximum | | | | | 122 | | | 122 | 134 |

## 12 56 43 – Dormitory Furniture

| 12 56 43.10 Dormitory Furnishings | Crew | Daily Output | Labor-Hours | Unit | Material | 2011 Bare Costs Labor | Equipment | Total | Total Incl O&P |
|---|---|---|---|---|---|---|---|---|---|
| 0010 **DORMITORY FURNISHINGS** | | | | | | | | | |
| 0300 Bunkable bed, twin, minimum | | | | Ea. | 300 | | | 300 | 330 |
| 0320 Maximum | | | | | 685 | | | 685 | 750 |
| 1000 Chest, four drawer, minimum | | | | | 440 | | | 440 | 485 |
| 1020 Maximum | | | | | 580 | | | 580 | 635 |
| 1050 Built-in, minimum | 2 Carp | 13 | 1.231 | L.F. | 145 | 53 | | 198 | 242 |
| 1150 Maximum | | 10 | 1.600 | | 268 | 69 | | 337 | 400 |
| 1200 Desk top, built-in, laminated plastic, 24" deep, minimum | | 50 | .320 | | 39.50 | 13.80 | | 53.30 | 64.50 |
| 1300 Maximum | | 40 | .400 | | 118 | 17.20 | | 135.20 | 157 |
| 1450 30" deep, minimum | | 50 | .320 | | 50.50 | 13.80 | | 64.30 | 76.50 |
| 1550 Maximum | | 40 | .400 | | 221 | 17.20 | | 238.20 | 270 |
| 1750 Dressing unit, built-in, minimum | | 12 | 1.333 | | 218 | 57.50 | | 275.50 | 325 |
| 1850 Maximum | | 8 | 2 | | 655 | 86 | | 741 | 850 |
| 8000 Rule of thumb: total cost for furniture, minimum | | | | Student | | | | 2,300 | 2,525 |
| 8050 Maximum | | | | " | | | | 4,425 | 4,875 |

## 12 56 51 – Library Furniture

| 12 56 51.10 Library Furnishings | Crew | Daily Output | Labor-Hours | Unit | Material | 2011 Bare Costs Labor | Equipment | Total | Total Incl O&P |
|---|---|---|---|---|---|---|---|---|---|
| 0010 **LIBRARY FURNISHINGS** | | | | | | | | | |
| 0100 Attendant desk, 36" x 62" x 29" high | 1 Carp | 16 | .500 | Ea. | 2,700 | 21.50 | | 2,721.50 | 3,000 |
| 0200 Book display, "A" frame display, both sides, 42" x 42" x 60" high | | 16 | .500 | | 1,500 | 21.50 | | 1,521.50 | 1,675 |
| 0220 Table with bulletin board, 42" x 24" x 49" high | | 16 | .500 | | 1,325 | 21.50 | | 1,346.50 | 1,475 |
| 0800 Card catalogue, 30 tray unit | | 16 | .500 | | 2,850 | 21.50 | | 2,871.50 | 3,150 |
| 0840 60 tray unit | | 16 | .500 | | 5,225 | 21.50 | | 5,246.50 | 5,775 |
| 0880 72 tray unit | 2 Carp | 16 | 1 | | 6,400 | 43 | | 6,443 | 7,125 |
| 1000 Carrels, single face, initial unit | 1 Carp | 16 | .500 | | 720 | 21.50 | | 741.50 | 830 |
| 1500 Double face, initial unit | 2 Carp | 16 | 1 | | 1,150 | 43 | | 1,193 | 1,325 |
| 1710 Carrels, hardwood, 36" x 24", minimum | 1 Carp | 5 | 1.600 | | 520 | 69 | | 589 | 680 |
| 1720 Maximum | " | 4 | 2 | | 1,275 | 86 | | 1,361 | 1,525 |
| 2700 Card catalog file, 60 trays, complete | | | | | 5,600 | | | 5,600 | 6,150 |
| 2720 Alternate method: each tray | | | | | 93.50 | | | 93.50 | 103 |
| 3800 Charging desk, built-in, with counter, plastic laminated top | 1 Carp | 7 | 1.143 | L.F. | 410 | 49 | | 459 | 530 |

# 12 56 Institutional Furniture

## 12 56 51 – Library Furniture

### 12 56 51.10 Library Furnishings

| | | Crew | Daily Output | Labor-Hours | Unit | Material | 2011 Bare Costs Labor | Equipment | Total | Total Incl O&P |
|---|---|---|---|---|---|---|---|---|---|---|
| 4000 | Dictionary stand, stationary | 1 Carp | 16 | .500 | Ea. | 880 | 21.50 | | 901.50 | 1,000 |
| 4020 | Revolving | | 16 | .500 | | 345 | 21.50 | | 366.50 | 415 |
| 4200 | Exhibit case, table style, 60" x 28" x 36" | | 11 | .727 | | 2,800 | 31.50 | | 2,831.50 | 3,150 |
| 6010 | Bookshelf, metal, 90" high, 10" shelf, double face | | 11.50 | .696 | L.F. | 118 | 30 | | 148 | 176 |
| 6020 | Single face | | 12 | .667 | " | 113 | 28.50 | | 141.50 | 168 |
| 6050 | For 8" shelving, subtract from above | | | | | 10% | | | | |
| 6060 | For 12" shelving, add to above | | | | | 10% | | | | |
| 6070 | For 42" high with countertop, subtract from above | | | | | 20% | | | | |
| 6100 | Mobile compacted shelving, hand crank, 9'-0" high | | | | | | | | | |
| 6110 | Double face, including track, 3' section | | | | Ea. | 1,125 | | | 1,125 | 1,250 |
| 6150 | For electrical operation, add | | | | | 25% | | | | |
| 6200 | Magazine shelving, 82" high, 12" deep, single face | 1 Carp | 11.50 | .696 | L.F. | 92.50 | 30 | | 122.50 | 148 |
| 6210 | Double face | | 11.50 | .696 | " | 119 | 30 | | 149 | 176 |
| 7000 | Tables, card catalog reference, 24" x 60" x 42" | | 16 | .500 | Ea. | 975 | 21.50 | | 996.50 | 1,100 |
| 7200 | Reading table, laminated top, 60" x 36" | | | | " | 720 | | | 720 | 790 |

## 12 56 70 – Healthcare Furniture

### 12 56 70.10 Furniture, Hospital

| | | Crew | Daily Output | Labor-Hours | Unit | Material | Labor | Equipment | Total | Total Incl O&P |
|---|---|---|---|---|---|---|---|---|---|---|
| 0010 | **FURNITURE, HOSPITAL** | | | | | | | | | |
| 0020 | Beds, manual, minimum | | | | Ea. | 755 | | | 755 | 830 |
| 0100 | Maximum | | | | | 2,575 | | | 2,575 | 2,825 |
| 0300 | Manual and electric beds, minimum | | | | | 1,050 | | | 1,050 | 1,150 |
| 0400 | Maximum | | | | | 2,975 | | | 2,975 | 3,275 |
| 0600 | All electric hospital beds, minimum | | | | | 1,325 | | | 1,325 | 1,450 |
| 0700 | Maximum | | | | | 4,300 | | | 4,300 | 4,750 |
| 0900 | Manual, nursing home beds, minimum | | | | | 850 | | | 850 | 935 |
| 1000 | Maximum | | | | | 1,775 | | | 1,775 | 1,950 |
| 1020 | Overbed table, laminated top, minimum | | | | | 360 | | | 360 | 400 |
| 1040 | Maximum | | | | | 900 | | | 900 | 990 |
| 1100 | Patient wall systems, not incl. plumbing, minimum | | | | Room | 1,075 | | | 1,075 | 1,200 |
| 1200 | Maximum | | | | " | 2,000 | | | 2,000 | 2,200 |
| 2000 | Geriatric chairs, minimum | | | | Ea. | 420 | | | 420 | 465 |
| 2020 | Maximum | | | | " | 720 | | | 720 | 790 |

# 12 61 Fixed Audience Seating

## 12 61 13 – Upholstered Audience Seating

### 12 61 13.13 Auditorium Chairs

| | | Crew | Daily Output | Labor-Hours | Unit | Material | Labor | Equipment | Total | Total Incl O&P |
|---|---|---|---|---|---|---|---|---|---|---|
| 0010 | **AUDITORIUM CHAIRS** | | | | | | | | | |
| 2000 | All veneer construction | 2 Carp | 22 | .727 | Ea. | 175 | 31.50 | | 206.50 | 241 |
| 2200 | Veneer back, padded seat | | 22 | .727 | | 221 | 31.50 | | 252.50 | 291 |
| 2350 | Fully upholstered, spring seat | | 22 | .727 | | 211 | 31.50 | | 242.50 | 280 |
| 2450 | For tablet arms, add | | | | | 58 | | | 58 | 63.50 |
| 2500 | For fire retardancy, CATB-133, add | | | | | 25 | | | 25 | 27.50 |

### 12 61 13.23 Lecture Hall Seating

| | | Crew | Daily Output | Labor-Hours | Unit | Material | Labor | Equipment | Total | Total Incl O&P |
|---|---|---|---|---|---|---|---|---|---|---|
| 0010 | **LECTURE HALL SEATING** | | | | | | | | | |
| 1000 | Pedestal type, minimum | 2 Carp | 22 | .727 | Ea. | 165 | 31.50 | | 196.50 | 230 |
| 1200 | Maximum | " | 14.50 | 1.103 | " | 555 | 47.50 | | 602.50 | 685 |

# 12 63 Stadium and Arena Seating

## 12 63 13 – Stadium and Arena Bench Seating

| 12 63 13.13 Bleachers | Crew | Daily Output | Labor-Hours | Unit | Material | 2011 Bare Costs Labor | Equipment | Total | Total Incl O&P |
|---|---|---|---|---|---|---|---|---|---|
| 0010 **BLEACHERS** | | | | | | | | | |
| 3000 Telescoping, manual to 15 tier, minimum | F-5 | 65 | .492 | Seat | 86 | 21.50 | | 107.50 | 128 |
| 3100 Maximum | | 60 | .533 | | 129 | 23 | | 152 | 178 |
| 3300 16 to 20 tier, minimum | | 60 | .533 | | 207 | 23 | | 230 | 264 |
| 3400 Maximum | | 55 | .582 | | 259 | 25.50 | | 284.50 | 325 |
| 3600 21 to 30 tier, minimum | | 50 | .640 | | 215 | 28 | | 243 | 280 |
| 3700 Maximum | | 40 | .800 | | 259 | 35 | | 294 | 340 |
| 3900 For integral power operation, add, minimum | 2 Elec | 300 | .053 | | 43 | 2.68 | | 45.68 | 51.50 |
| 4000 Maximum | " | 250 | .064 | | 69 | 3.22 | | 72.22 | 81 |
| 5000 Benches, folding, in wall, 14' table, 2 benches | L-4 | 2 | 12 | Set | 730 | 485 | | 1,215 | 1,550 |

# 12 67 Pews and Benches

## 12 67 13 – Pews

| 12 67 13.13 Sanctuary Pews | Crew | Daily Output | Labor-Hours | Unit | Material | 2011 Bare Costs Labor | Equipment | Total | Total Incl O&P |
|---|---|---|---|---|---|---|---|---|---|
| 0010 **SANCTUARY PEWS** | | | | | | | | | |
| 1500 Bench type, hardwood, minimum | 1 Carp | 20 | .400 | L.F. | 77.50 | 17.20 | | 94.70 | 112 |
| 1550 Maximum | " | 15 | .533 | | 154 | 23 | | 177 | 204 |
| 1570 For kneeler, add | | | | | 19.05 | | | 19.05 | 21 |

# 12 92 Interior Planters and Artificial Plants

## 12 92 33 – Interior Planters

| 12 92 33.10 Planters | Crew | Daily Output | Labor-Hours | Unit | Material | 2011 Bare Costs Labor | Equipment | Total | Total Incl O&P |
|---|---|---|---|---|---|---|---|---|---|
| 0010 **PLANTERS** | | | | | | | | | |
| 1000 Fiberglass, hanging, 12" diameter, 7" high | | | | Ea. | 105 | | | 105 | 115 |
| 1500 Rectangular, 48" long, 16" high x 15" wide | | | | | 690 | | | 690 | 760 |
| 1650 60" long, 30" high, 28" wide | | | | | 1,050 | | | 1,050 | 1,150 |
| 2000 Round, 12" diameter, 13" high | | | | | 119 | | | 119 | 131 |
| 2050 25" high | | | | | 135 | | | 135 | 148 |
| 5000 Square, 10" side, 20" high | | | | | 158 | | | 158 | 173 |
| 5100 14" side, 15" high | | | | | 190 | | | 190 | 209 |
| 6000 Metal bowl, 32" diameter, 8" high, minimum | | | | | 480 | | | 480 | 525 |
| 6050 Maximum | | | | | 620 | | | 620 | 680 |
| 8750 Wood, fiberglass liner, square | | | | | | | | | |
| 8780 14" square, 15" high, minimum | | | | Ea. | 300 | | | 300 | 330 |
| 8800 Maximum | | | | | 370 | | | 370 | 410 |
| 9400 Plastic cylinder, molded, 10" diameter, 10" high | | | | | 12.25 | | | 12.25 | 13.45 |
| 9500 11" diameter, 11" high | | | | | 39 | | | 39 | 42.50 |

# 12 93 Site Furnishings

## 12 93 23 – Trash and Litter Receptors

### 12 93 23.10 Trash Receptacles

| | | Crew | Daily Output | Labor-Hours | Unit | Material | 2011 Bare Costs Labor | Equipment | Total | Total Incl O&P |
|---|---|---|---|---|---|---|---|---|---|---|
| 0010 | **TRASH RECEPTACLES** | | | | | | | | | |
| 0020 | Fiberglass, 2' square, 18" high | 2 Clab | 30 | .533 | Ea. | 525 | 18.30 | | 543.30 | 605 |
| 0100 | 2' square, 2'-6" high | | 30 | .533 | | 525 | 18.30 | | 543.30 | 605 |
| 0300 | Circular , 2' diameter, 18" high | | 30 | .533 | | 515 | 18.30 | | 533.30 | 595 |
| 0400 | 2' diameter, 2'-6" high | | 30 | .533 | | 595 | 18.30 | | 613.30 | 685 |
| 0500 | Recycled plastic, var colors, round, 32 gal., 28" x 38" H  G | | 5 | 3.200 | | 390 | 110 | | 500 | 595 |
| 0510 | 32 gal., 31" x 32" H  G | | 5 | 3.200 | | 430 | 110 | | 540 | 640 |

### 12 93 23.20 Trash Closure

| | | Crew | Daily Output | Labor-Hours | Unit | Material | 2011 Bare Costs Labor | Equipment | Total | Total Incl O&P |
|---|---|---|---|---|---|---|---|---|---|---|
| 0010 | **TRASH CLOSURE** | | | | | | | | | |
| 0020 | Steel with pullover cover, 2'-3" wide, 4'-7" high, 6'-2" long | 2 Clab | 5 | 3.200 | Ea. | 1,375 | 110 | | 1,485 | 1,700 |
| 0100 | 10'-1" long | | 4 | 4 | | 1,950 | 137 | | 2,087 | 2,350 |
| 0300 | Wood, 10' wide, 6' high, 10' long | | 1.20 | 13.333 | | 1,450 | 460 | | 1,910 | 2,300 |

## 12 93 33 – Manufactured Planters

### 12 93 33.10 Planters

| | | Crew | Daily Output | Labor-Hours | Unit | Material | 2011 Bare Costs Labor | Equipment | Total | Total Incl O&P |
|---|---|---|---|---|---|---|---|---|---|---|
| 0010 | **PLANTERS** | | | | | | | | | |
| 0012 | Concrete, sandblasted, precast, 48" diameter, 24" high | 2 Clab | 15 | 1.067 | Ea. | 630 | 36.50 | | 666.50 | 750 |
| 0100 | Fluted, precast, 7' diameter, 36" high | | 10 | 1.600 | | 1,300 | 55 | | 1,355 | 1,500 |
| 0300 | Fiberglass, circular, 36" diameter, 24" high | | 15 | 1.067 | | 600 | 36.50 | | 636.50 | 715 |
| 0320 | 36" diameter, 27" high | | 12 | 1.333 | | 710 | 46 | | 756 | 850 |
| 0330 | 33" high | | 15 | 1.067 | | 760 | 36.50 | | 796.50 | 890 |
| 0335 | 24" diameter, 36" high | | 15 | 1.067 | | 485 | 36.50 | | 521.50 | 590 |
| 0340 | 60" diameter, 39" high | | 8 | 2 | | 1,225 | 68.50 | | 1,293.50 | 1,450 |
| 0400 | 60" diameter, 24" high | | 10 | 1.600 | | 705 | 55 | | 760 | 860 |
| 0600 | Square, 24" side, 36" high | | 15 | 1.067 | | 615 | 36.50 | | 651.50 | 735 |
| 0610 | 24" side, 27" high | | 12 | 1.333 | | 615 | 46 | | 661 | 750 |
| 0620 | 24" side, 16" high | | 20 | .800 | | 330 | 27.50 | | 357.50 | 400 |
| 0700 | 48" side, 36" high | | 15 | 1.067 | | 900 | 36.50 | | 936.50 | 1,050 |
| 0900 | Planter/bench, 72" square, 36" high | | 5 | 3.200 | | 1,300 | 110 | | 1,410 | 1,600 |
| 1000 | 96" square, 27" high | | 5 | 3.200 | | 1,975 | 110 | | 2,085 | 2,350 |
| 1200 | Wood, square, 48" side, 24" high | | 15 | 1.067 | | 1,225 | 36.50 | | 1,261.50 | 1,375 |
| 1300 | Circular, 48" diameter, 30" high | | 10 | 1.600 | | 910 | 55 | | 965 | 1,075 |
| 1500 | 72" diameter, 30" high | | 10 | 1.600 | | 1,425 | 55 | | 1,480 | 1,625 |
| 1600 | Planter/bench, 72" | | 5 | 3.200 | | 3,150 | 110 | | 3,260 | 3,650 |

## 12 93 43 – Site Seating and Tables

### 12 93 43.13 Site Seating

| | | Crew | Daily Output | Labor-Hours | Unit | Material | 2011 Bare Costs Labor | Equipment | Total | Total Incl O&P |
|---|---|---|---|---|---|---|---|---|---|---|
| 0010 | **SITE SEATING** | | | | | | | | | |
| 0012 | Seating, benches, park, precast conc, w/backs, wood rails, 4' long | 2 Clab | 5 | 3.200 | Ea. | 470 | 110 | | 580 | 685 |
| 0100 | 8' long | | 4 | 4 | | 930 | 137 | | 1,067 | 1,225 |
| 0300 | Fiberglass, without back, one piece, 4' long | | 10 | 1.600 | | 605 | 55 | | 660 | 750 |
| 0400 | 8' long | | 7 | 2.286 | | 805 | 78.50 | | 883.50 | 1,000 |
| 0500 | Steel barstock pedestals w/backs, 2" x 3" wood rails, 4' long | | 10 | 1.600 | | 1,075 | 55 | | 1,130 | 1,250 |
| 0510 | 8' long | | 7 | 2.286 | | 1,300 | 78.50 | | 1,378.50 | 1,550 |
| 0520 | 3" x 8" wood plank, 4' long | | 10 | 1.600 | | 1,075 | 55 | | 1,130 | 1,250 |
| 0530 | 8' long | | 7 | 2.286 | | 1,300 | 78.50 | | 1,378.50 | 1,550 |
| 0540 | Backless, 4" x 4" wood plank, 4' square | | 10 | 1.600 | | 1,075 | 55 | | 1,130 | 1,275 |
| 0550 | 8' long | | 7 | 2.286 | | 835 | 78.50 | | 913.50 | 1,050 |
| 0600 | Aluminum pedestals, with backs, aluminum slats, 8' long | | 8 | 2 | | 335 | 68.50 | | 403.50 | 475 |
| 0610 | 15' long | | 5 | 3.200 | | 400 | 110 | | 510 | 610 |
| 0620 | Portable, aluminum slats, 8' long | | 8 | 2 | | 340 | 68.50 | | 408.50 | 475 |
| 0630 | 15' long | | 5 | 3.200 | | 435 | 110 | | 545 | 650 |
| 0800 | Cast iron pedestals, back & arms, wood slats, 4' long | | 8 | 2 | | 320 | 68.50 | | 388.50 | 460 |

# 12 93 Site Furnishings

## 12 93 43 – Site Seating and Tables

| 12 93 43.13 Site Seating | | Crew | Daily Output | Labor-Hours | Unit | Material | 2011 Bare Costs Labor | Equipment | Total | Total Incl O&P |
|---|---|---|---|---|---|---|---|---|---|---|
| 0820 | 8' long | 2 Clab | 5 | 3.200 | Ea. | 890 | 110 | | 1,000 | 1,150 |
| 0840 | Backless, wood slats, 4' long | | 8 | 2 | | 395 | 68.50 | | 463.50 | 540 |
| 0860 | 8' long | | 5 | 3.200 | | 925 | 110 | | 1,035 | 1,200 |
| 1700 | Steel frame, fir seat, 10' long | | 10 | 1.600 | | 330 | 55 | | 385 | 445 |

# Estimating Tips

## General

- The items and systems in this division are usually estimated, purchased, supplied, and installed as a unit by one or more subcontractors. The estimator must ensure that all parties are operating from the same set of specifications and assumptions, and that all necessary items are estimated and will be provided. Many times the complex items and systems are covered, but the more common ones, such as excavation or a crane, are overlooked for the very reason that everyone assumes nobody could miss them. The estimator should be the central focus and be able to ensure that all systems are complete.

- Another area where problems can develop in this division is at the interface between systems. The estimator must ensure, for instance, that anchor bolts, nuts, and washers are estimated and included for the air-supported structures and pre-engineered buildings to be bolted to their foundations. Utility supply is a common area where essential items or pieces of equipment can be missed or overlooked due to the fact that each subcontractor may feel it is another's responsibility. The estimator should also be aware of certain items which may be supplied as part of a package but installed by others, and ensure that the installing contractor's estimate includes the cost of installation. Conversely, the estimator must also ensure that items are not costed by two different subcontractors, resulting in an inflated overall estimate.

## 13 30 00 Special Structures

- The foundations and floor slab, as well as rough mechanical and electrical, should be estimated, as this work is required for the assembly and erection of the structure. Generally, as noted in the book, the pre-engineered building comes as a shell. Pricing is based on the size and structural design parameters stated in the reference section. Additional features, such as windows and doors with their related structural framing, must also be included by the estimator. Here again, the estimator must have a clear understanding of the scope of each portion of the work and all the necessary interfaces.

# Reference Numbers

Reference numbers are shown in shaded boxes at the beginning of some major classifications. These numbers refer to related items in the Reference Section. The reference information may be an estimating procedure, an alternate pricing method, or technical information.

*Note:* Not all subdivisions listed here necessarily appear in this publication.

# 13 05 Common Work Results for Special Construction

## 13 05 05 – Selective Special Construction Demolition

### 13 05 05.10 Selective Demolition, Air Supported Structures

| 13 05 05.10 Selective Demolition, Air Supported Structures | Crew | Daily Output | Labor-Hours | Unit | Material | 2011 Bare Costs Labor | 2011 Bare Costs Equipment | Total | Total Incl O&P |
|---|---|---|---|---|---|---|---|---|---|
| 0010 **SELECTIVE DEMOLITION, AIR SUPPORTED STRUCTURES** | | | | | | | | | |
| 0020 Tank covers, scrim, dbl. layer, vinyl poly w/hdw, blower & controls | | | | | | | | | |
| 0050 Round and rectangular R024119-10 | B-2 | 9000 | .004 | S.F. | | .15 | | .15 | .24 |
| 0100 Warehouse structures | | | | | | | | | |
| 0120 Poly/vinyl fabric, 28 oz, incl tension cables & inflation system | 4 Clab | 9000 | .004 | SF Flr. | | .12 | | .12 | .19 |
| 0150 Reinforced vinyl, 12 oz., 3000 S.F. | " | 5000 | .006 | | | .22 | | .22 | .34 |
| 0200 12,000 to 24,000 S.F. | 8 Clab | 20000 | .003 | | | .11 | | .11 | .17 |
| 0250 Tedlar vinyl fabric, 28 oz. w/liner, to 3000 S.F. | 4 Clab | 5000 | .006 | | | .22 | | .22 | .34 |
| 0300 12,000 to 24,000 S.F. | 8 Clab | 20000 | .003 | ↓ | | .11 | | .11 | .17 |
| 0350 Greenhouse/shelter, woven polyethylene with liner | | | | | | | | | |
| 0400 3000 S.F. | 4 Clab | 5000 | .006 | SF Flr. | | .22 | | .22 | .34 |
| 0450 12,000 to 24,000 S.F. | 8 Clab | 20000 | .003 | | | .11 | | .11 | .17 |
| 0500 Tennis/gymnasium, poly/vinyl fabric, 28 oz., incl thermal liner | 4 Clab | 9000 | .004 | | | .12 | | .12 | .19 |
| 0600 Stadium/convention center, teflon coated fiberglass, incl thermal liner | 9 Clab | 40000 | .002 | ↓ | | .06 | | .06 | .09 |
| 0700 Doors, air lock, 15' long, 10' x 10' | 2 Carp | 1.50 | 10.667 | Ea. | | 460 | | 460 | 705 |
| 0720 15' x 15' | | .80 | 20 | | | 860 | | 860 | 1,325 |
| 0750 Revolving personnel door, 6' dia. x 6'-6" high | ↓ | 1.50 | 10.667 | ↓ | | 460 | | 460 | 705 |

### 13 05 05.20 Selective Demolition, Garden Houses

| | Crew | Daily Output | Labor-Hours | Unit | Material | Labor | Equipment | Total | Total Incl O&P |
|---|---|---|---|---|---|---|---|---|---|
| 0010 **SELECTIVE DEMOLITION, GARDEN HOUSES** R024119-10 | | | | | | | | | |
| 0020 Prefab, wood, excl foundation, average | 2 Clab | 400 | .040 | SF Flr. | | 1.37 | | 1.37 | 2.11 |

### 13 05 05.25 Selective Demolition, Geodesic Domes

| | Crew | Daily Output | Labor-Hours | Unit | Material | Labor | Equipment | Total | Total Incl O&P |
|---|---|---|---|---|---|---|---|---|---|
| 0010 **SELECTIVE DEMOLITION, GEODESIC DOMES** | | | | | | | | | |
| 0050 Shell only, interlocking plywood panels, 30' diameter | F-5 | 3.20 | 10 | Ea. | | 435 | | 435 | 670 |
| 0060 34' diameter | | 2.30 | 13.913 | | | 605 | | 605 | 930 |
| 0070 39' diameter | ↓ | 2 | 16 | | | 695 | | 695 | 1,075 |
| 0080 45' diameter | F-3 | 2.20 | 18.182 | | | 795 | 298 | 1,093 | 1,550 |
| 0090 55' diameter | | 2 | 20 | | | 875 | 330 | 1,205 | 1,675 |
| 0100 60' diameter | | 2 | 20 | | | 875 | 330 | 1,205 | 1,675 |
| 0110 65' diameter | ↓ | 1.60 | 25 | ↓ | | 1,100 | 410 | 1,510 | 2,125 |

### 13 05 05.30 Selective Demolition, Greenhouses

| | Crew | Daily Output | Labor-Hours | Unit | Material | Labor | Equipment | Total | Total Incl O&P |
|---|---|---|---|---|---|---|---|---|---|
| 0010 **SELECTIVE DEMOLITION, GREENHOUSES** R024119-10 | | | | | | | | | |
| 0020 Resi-type, free standing, excl foundations, 9' long x 8' wide | 2 Clab | 160 | .100 | SF Flr. | | 3.44 | | 3.44 | 5.25 |
| 0030 9' long x 11' wide | | 170 | .094 | | | 3.23 | | 3.23 | 4.96 |
| 0040 9' long x 14' wide | | 220 | .073 | | | 2.50 | | 2.50 | 3.83 |
| 0050 9' long x 17' wide | | 320 | .050 | | | 1.72 | | 1.72 | 2.64 |
| 0060 Lean-to type, 4' wide | | 64 | .250 | | | 8.60 | | 8.60 | 13.20 |
| 0070 7' wide | | 120 | .133 | ↓ | | 4.58 | | 4.58 | 7.05 |
| 0080 Geodesic hemisphere, 1/8" plexiglass glazing, 8' diam. | | 4 | 4 | Ea. | | 137 | | 137 | 211 |
| 0090 24' dia | | .80 | 20 | | | 685 | | 685 | 1,050 |
| 0100 48' dia | ↓ | .40 | 40 | | | 1,375 | | 1,375 | 2,100 |

### 13 05 05.35 Selective Demolition, Hangars

| | Crew | Daily Output | Labor-Hours | Unit | Material | Labor | Equipment | Total | Total Incl O&P |
|---|---|---|---|---|---|---|---|---|---|
| 0010 **SELECTIVE DEMOLITION, HANGARS** | | | | | | | | | |
| 0020 T type hangars, prefab, steel , galv roof & walls, incl doors, excl fndtn | E-2 | 2550 | .022 | SF Flr. | | 1.04 | .64 | 1.68 | 2.46 |
| 0030 Circular type, prefab, steel frame, plastic skin, incl foundation, 80' diam | " | .50 | 112 | Total | | 5,300 | 3,250 | 8,550 | 12,600 |

### 13 05 05.45 Selective Demolition, Lightning Protection

| | Crew | Daily Output | Labor-Hours | Unit | Material | Labor | Equipment | Total | Total Incl O&P |
|---|---|---|---|---|---|---|---|---|---|
| 0010 **SELECTIVE DEMOLITION, LIGHTNING PROTECTION** | | | | | | | | | |
| 0020 Air terminal & base, copper, 3/8" diam. x 10", to 75' h | 1 Clab | 16 | .500 | Ea. | | 17.20 | | 17.20 | 26.50 |
| 0030 1/2" diam. x 12", over 75' h | | 16 | .500 | | | 17.20 | | 17.20 | 26.50 |
| 0050 Aluminum, 1/2" diam. x 12", to 75' h | | 16 | .500 | | | 17.20 | | 17.20 | 26.50 |
| 0060 5/8" diam. x 12", over 75' h | | 16 | .500 | ↓ | | 17.20 | | 17.20 | 26.50 |
| 0070 Cable, copper, 220 lb per thousand feet, to 75' high | ↓ | 640 | .013 | L.F. | | .43 | | .43 | .66 |

# 13 05 Common Work Results for Special Construction

## 13 05 05 – Selective Special Construction Demolition

### 13 05 05.45 Selective Demolition, Lightning Protection

| | | Crew | Daily Output | Labor-Hours | Unit | Material | 2011 Bare Costs Labor | Equipment | Total | Total Incl O&P |
|---|---|---|---|---|---|---|---|---|---|---|
| 0080 | 375 lb per thousand feet, over 75' high | 1 Clab | 460 | .017 | L.F. | | .60 | | .60 | .92 |
| 0090 | Aluminum, 101 lb per thousand feet, to 75' high | | 560 | .014 | | | .49 | | .49 | .75 |
| 0100 | 199 lb per thousand feet, over 75' high | | 480 | .017 | | | .57 | | .57 | .88 |
| 0110 | Arrester, 175 V AC, to ground | | 16 | .500 | Ea. | | 17.20 | | 17.20 | 26.50 |
| 0120 | 650 V AC, to ground | | 13 | .615 | " | | 21 | | 21 | 32.50 |

### 13 05 05.50 Selective Demolition, Pre-Engineered Steel Buildings

| | | Crew | Daily Output | Labor-Hours | Unit | Material | 2011 Bare Costs Labor | Equipment | Total | Total Incl O&P |
|---|---|---|---|---|---|---|---|---|---|---|
| 0010 | **SELECTIVE DEMOLITION, PRE-ENGINEERED STEEL BUILDINGS** | | | | | | | | | |
| 0500 | Pre-engd steel bldgs, rigid frame, clear span & multi post, excl salvage | | | | | | | | | |
| 0550 | 3,500 to 7,500 S.F. | L-10 | 1000 | .024 | SF Flr. | | 1.16 | .66 | 1.82 | 2.68 |
| 0600 | 7,501 to 12,500 S.F. | | 1500 | .016 | | | .78 | .44 | 1.22 | 1.78 |
| 0650 | 12,500 S.F. or greater | | 1650 | .015 | | | .71 | .40 | 1.11 | 1.63 |
| 0700 | Pre-engd steel building components | | | | | | | | | |
| 0710 | Entrance canopy, including frame 4' x 4' | E-24 | 8 | 4 | Ea. | | 191 | 93.50 | 284.50 | 430 |
| 0720 | 4' x 8' | " | 7 | 4.571 | | | 218 | 107 | 325 | 490 |
| 0730 | H.M. doors, self framing, single leaf | 2 Skwk | 8 | 2 | | | 89 | | 89 | 136 |
| 0740 | Double leaf | | 5 | 3.200 | | | 142 | | 142 | 218 |
| 0760 | Gutter, eave type | | 600 | .027 | L.F. | | 1.18 | | 1.18 | 1.82 |
| 0770 | Sash, single slide, double slide or fixed | | 24 | .667 | Ea. | | 29.50 | | 29.50 | 45.50 |
| 0780 | Skylight, fiberglass, to 30 S.F. | | 16 | 1 | | | 44.50 | | 44.50 | 68 |
| 0785 | Roof vents, circular, 12" to 24" diameter | | 12 | 1.333 | | | 59 | | 59 | 91 |
| 0790 | Continuous, 10' long | | 8 | 2 | | | 89 | | 89 | 136 |
| 0900 | Shelters, aluminum frame | | | | | | | | | |
| 0910 | Acrylic glazing, 3' x 9' x 8' high | 2 Skwk | 2 | 8 | Ea. | | 355 | | 355 | 545 |
| 0920 | 9' x 12' x 8' high | " | 1.50 | 10.667 | " | | 475 | | 475 | 725 |

### 13 05 05.60 Selective Demolition, Silos

| | | Crew | Daily Output | Labor-Hours | Unit | Material | 2011 Bare Costs Labor | Equipment | Total | Total Incl O&P |
|---|---|---|---|---|---|---|---|---|---|---|
| 0010 | **SELECTIVE DEMOLITION, SILOS** | | | | | | | | | |
| 0020 | Conc stave, indstrl, conical/sloping bott, excl fndtn, 12' diam., 35' h | E-24 | .18 | 177 | Ea. | | 8,500 | 4,150 | 12,650 | 19,000 |
| 0030 | 16' diam., 45' h | | .12 | 266 | | | 12,700 | 6,225 | 18,925 | 28,600 |
| 0040 | 25' dia, 75' h | | .08 | 400 | | | 19,100 | 9,350 | 28,450 | 42,800 |
| 0050 | Steel, factory fabricated, 30,000 gal. cap, painted or epoxy lined | L-5 | 2 | 28 | | | 1,350 | 375 | 1,725 | 2,750 |

### 13 05 05.65 Selective Demolition, Sound Control

| | | Crew | Daily Output | Labor-Hours | Unit | Material | 2011 Bare Costs Labor | Equipment | Total | Total Incl O&P |
|---|---|---|---|---|---|---|---|---|---|---|
| 0010 | **SELECTIVE DEMOLITION, SOUND CONTROL**   R024119-10 | | | | | | | | | |
| 0120 | Acoustical enclosure, 4" thick walls & ceiling panels, 8 lb./S.F. | 3 Carp | 144 | .167 | SF Surf | | 7.20 | | 7.20 | 11 |
| 0130 | 10.5 lb./S.F. | | 128 | .188 | | | 8.05 | | 8.05 | 12.40 |
| 0140 | Reverb chamber, parallel walls, 4" thick | | 120 | .200 | | | 8.60 | | 8.60 | 13.20 |
| 0150 | Skewed walls, parallel roof, 4" thick | | 110 | .218 | | | 9.40 | | 9.40 | 14.40 |
| 0160 | Skewed walls/roof, 4" layer/air space | | 96 | .250 | | | 10.75 | | 10.75 | 16.50 |
| 0170 | Sound-absorbing panels, painted metal, 2'-6" x 8', under 1,000 SF | | 430 | .056 | | | 2.40 | | 2.40 | 3.69 |
| 0180 | Over 1,000 SF | | 480 | .050 | | | 2.15 | | 2.15 | 3.30 |
| 0190 | Flexible transparent curtain, clear | 3 Shee | 430 | .056 | | | 2.88 | | 2.88 | 4.36 |
| 0192 | 50% clear, 50% foam | | 430 | .056 | | | 2.88 | | 2.88 | 4.36 |
| 0194 | 25% clear, 75% foam | | 430 | .056 | | | 2.88 | | 2.88 | 4.36 |
| 0196 | 100% foam | | 430 | .056 | | | 2.88 | | 2.88 | 4.36 |
| 0200 | Audio-masking sys, incl speakers, amplfr, signal gnrtr, clng mntd | | | | | | | | | |
| 0205 | Ceiling mounted, 5,000 SF | 2 Elec | 4800 | .003 | S.F. | | .17 | | .17 | .25 |
| 0210 | 10,000 SF | | 5600 | .003 | | | .14 | | .14 | .21 |
| 0220 | Plenum mounted, 5,000 SF | | 7600 | .002 | | | .11 | | .11 | .16 |
| 0230 | 10,000 SF | | 8800 | .002 | | | .09 | | .09 | .14 |

### 13 05 05.70 Selective Demolition, Special Purpose Rooms

| | | Crew | Daily Output | Labor-Hours | Unit | Material | 2011 Bare Costs Labor | Equipment | Total | Total Incl O&P |
|---|---|---|---|---|---|---|---|---|---|---|
| 0010 | **SELECTIVE DEMOLITION, SPECIAL PURPOSE ROOMS**   R024119-10 | | | | | | | | | |
| 0100 | Audiometric rooms, under 500 S.F. surface | 4 Carp | 200 | .160 | SF Surf | | 6.90 | | 6.90 | 10.55 |
| 0110 | Over 500 S.F. surface | " | 240 | .133 | " | | 5.75 | | 5.75 | 8.80 |

### 13 05 05.70 Selective Demolition, Special Purpose Rooms

| | | Crew | Daily Output | Labor-Hours | Unit | Material | 2011 Bare Costs Labor | 2011 Bare Costs Equipment | Total | Total Incl O&P |
|---|---|---|---|---|---|---|---|---|---|---|
| 0200 | Clean rooms, 12' x 12' soft wall, class 100 | 1 Carp | .30 | 26.667 | Ea. | | 1,150 | | 1,150 | 1,750 |
| 0210 | Class 1000 | | .30 | 26.667 | | | 1,150 | | 1,150 | 1,750 |
| 0220 | Class 10,000 | | .35 | 22.857 | | | 985 | | 985 | 1,500 |
| 0230 | Class 100,000 | | .35 | 22.857 | | | 985 | | 985 | 1,500 |
| 0300 | Darkrooms, shell complete, 8' high | 2 Carp | 220 | .073 | SF Flr. | | 3.13 | | 3.13 | 4.80 |
| 0310 | 12' high | | 110 | .145 | " | | 6.25 | | 6.25 | 9.60 |
| 0350 | Darkrooms doors, mini-cylindrical, revolving | | 4 | 4 | Ea. | | 172 | | 172 | 264 |
| 0400 | Music room, practice modular | | 140 | .114 | SF Surf | | 4.92 | | 4.92 | 7.55 |
| 0500 | Refrigeration structures and finishes | | | | | | | | | |
| 0510 | Wall finish, 2 coat portland cement plaster, 1/2" thick | 1 Clab | 200 | .040 | S.F. | | 1.37 | | 1.37 | 2.11 |
| 0520 | Fiberglass panels, 1/8" thick | | 400 | .020 | | | .69 | | .69 | 1.05 |
| 0530 | Ceiling finish, polystyrene plastic, 1" to 2" thick | | 500 | .016 | | | .55 | | .55 | .84 |
| 0540 | 4" thick | | 450 | .018 | | | .61 | | .61 | .94 |
| 0550 | Refrigerator, prefab aluminum walk-in, 7'-6" high, 6' x 6' OD | 2 Carp | 100 | .160 | SF Flr. | | 6.90 | | 6.90 | 10.55 |
| 0560 | 10' x 10' OD | | 160 | .100 | | | 4.31 | | 4.31 | 6.60 |
| 0570 | Over 150 S.F. | | 200 | .080 | | | 3.44 | | 3.44 | 5.30 |
| 0600 | Sauna, prefabricated, including heater & controls, 7' high, to 30 S.F. | | 120 | .133 | | | 5.75 | | 5.75 | 8.80 |
| 0610 | To 40 S.F. | | 140 | .114 | | | 4.92 | | 4.92 | 7.55 |
| 0620 | To 60 S.F. | | 175 | .091 | | | 3.94 | | 3.94 | 6.05 |
| 0630 | To 100 S.F. | | 220 | .073 | | | 3.13 | | 3.13 | 4.80 |
| 0640 | To 130 S.F. | | 250 | .064 | | | 2.76 | | 2.76 | 4.23 |
| 0650 | Steam bath, heater, timer, head, single, to 140 C.F. | 1 Plum | 2.20 | 3.636 | Ea. | | 195 | | 195 | 291 |
| 0660 | To 300 C.F. | | 2.20 | 3.636 | | | 195 | | 195 | 291 |
| 0670 | Steam bath, comm. size, w/blow-down assembly, to 800 C.F. | | 1.80 | 4.444 | | | 238 | | 238 | 355 |
| 0680 | To 2500 C.F. | | 1.60 | 5 | | | 268 | | 268 | 400 |
| 0690 | Steam bath, comm. size, multiple, for motels, apts, 500 C.F., 2 baths | | 2 | 4 | | | 214 | | 214 | 320 |
| 0700 | 1,000 C.F., 4 baths | | 1.40 | 5.714 | | | 305 | | 305 | 455 |

### 13 05 05.75 Selective Demolition, Storage Tanks

| | | Crew | Daily Output | Labor-Hours | Unit | Material | 2011 Bare Costs Labor | 2011 Bare Costs Equipment | Total | Total Incl O&P |
|---|---|---|---|---|---|---|---|---|---|---|
| 0010 | **SELECTIVE DEMOLITION, STORAGE TANKS** | | | | | | | | | |
| 0500 | Steel tank, single wall, above ground, not incl fdn, pumps or piping | | | | | | | | | |
| 0510 | Single wall, 275 gallon    R024119-10 | Q-1 | 3 | 5.333 | Ea. | | 257 | | 257 | 385 |
| 0520 | 550 thru 2,000 gallon | B-34P | 2 | 12 | | | 530 | 269 | 799 | 1,100 |
| 0530 | 5,000 thru 10,000 gallon | B-34Q | 2 | 12 | | | 535 | 590 | 1,125 | 1,450 |
| 0540 | 15,000 thru 30,000 gallon | B-34S | 2 | 16 | | | 755 | 1,775 | 2,530 | 3,075 |
| 0600 | Steel tank, double wall, above ground not incl fdn, pumps & piping | | | | | | | | | |
| 0620 | 500 thru 2,000 gallon | B-34P | 2 | 12 | Ea. | | 530 | 269 | 799 | 1,100 |

### 13 05 05.85 Selective Demolition, Swimming Pool Equip

| | | Crew | Daily Output | Labor-Hours | Unit | Material | 2011 Bare Costs Labor | 2011 Bare Costs Equipment | Total | Total Incl O&P |
|---|---|---|---|---|---|---|---|---|---|---|
| 0010 | **SELECTIVE DEMOLITION, SWIMMING POOL EQUIP** | | | | | | | | | |
| 0020 | Diving stand, stainless steel, 3 meter | 2 Clab | 3 | 5.333 | Ea. | | 183 | | 183 | 281 |
| 0030 | 1 meter | | 5 | 3.200 | | | 110 | | 110 | 169 |
| 0040 | Diving board, 16' long, aluminum | | 5.40 | 2.963 | | | 102 | | 102 | 156 |
| 0050 | Fiberglass | | 5.40 | 2.963 | | | 102 | | 102 | 156 |
| 0070 | Ladders, heavy duty, stainless steel, 2 tread | | 14 | 1.143 | | | 39.50 | | 39.50 | 60 |
| 0080 | 4 tread | | 12 | 1.333 | | | 46 | | 46 | 70.50 |
| 0090 | Lifeguard chair, stainless steel, fixed | | 5 | 3.200 | | | 110 | | 110 | 169 |
| 0100 | Slide, tubular, fiberglass, aluminum handrails & ladder, 5', straight | | 4 | 4 | | | 137 | | 137 | 211 |
| 0110 | 8', curved | | 6 | 2.667 | | | 91.50 | | 91.50 | 141 |
| 0120 | 10', curved | | 3 | 5.333 | | | 183 | | 183 | 281 |
| 0130 | 12' straight, with platform | | 2.50 | 6.400 | | | 220 | | 220 | 335 |
| 0140 | Removable access ramp, stainless steel | | 4 | 4 | | | 137 | | 137 | 211 |
| 0150 | Removable stairs, stainless steel, collapsible | | 4 | 4 | | | 137 | | 137 | 211 |

# 13 05 Common Work Results for Special Construction

## 13 05 05 – Selective Special Construction Demolition

### 13 05 05.90 Selective Demolition, Tension Structures

| | | Crew | Daily Output | Labor-Hours | Unit | Material | 2011 Bare Costs Labor | Equipment | Total | Total Incl O&P |
|---|---|---|---|---|---|---|---|---|---|---|
| 0010 | **SELECTIVE DEMOLITION, TENSION STRUCTURES** | | | | | | | | | |
| 0020 | Tension structure, steel/alum frame, fabric shell, 60' clear span, 6,000 SF | B-41 | 2000 | .022 | SF Flr. | | .78 | .14 | .92 | 1.36 |
| 0030 | 12,000 SF | | 2200 | .020 | | | .71 | .13 | .84 | 1.23 |
| 0040 | 80' clear span, 20,800 SF | ↓ | 2440 | .018 | | | .64 | .12 | .76 | 1.11 |
| 0050 | 100' clear span, 10,000 SF | L-5 | 4350 | .013 | | | .62 | .17 | .79 | 1.27 |
| 0060 | 26,000 SF | " | 4600 | .012 | ↓ | | .59 | .16 | .75 | 1.20 |

### 13 05 05.95 Selective Demo, X-Ray/Radio Freq Protection

| | | Crew | Daily Output | Labor-Hours | Unit | Material | 2011 Bare Costs Labor | Equipment | Total | Total Incl O&P |
|---|---|---|---|---|---|---|---|---|---|---|
| 0010 | **SELECTIVE DEMO, X-RAY/RADIO FREQ PROTECTION** | | | | | | | | | |
| 0020 | Shielding lead, lined door frame, excl hdwe, 1/16" thick | 1 Clab | 4.80 | 1.667 | Ea. | | 57.50 | | 57.50 | 88 |
| 0030 | Lead sheets, 1/16" thick                R024119-10 | 2 Clab | 270 | .059 | S.F. | | 2.04 | | 2.04 | 3.12 |
| 0040 | 1/8" thick | | 240 | .067 | | | 2.29 | | 2.29 | 3.51 |
| 0050 | Lead shielding, 1/4" thick | | 270 | .059 | | | 2.04 | | 2.04 | 3.12 |
| 0060 | 1/2" thick | ↓ | 240 | .067 | ↓ | | 2.29 | | 2.29 | 3.51 |
| 0070 | Lead glass, 1/4" thick, 2.0 mm LE, 12" x 16" | 2 Glaz | 16 | 1 | Ea. | | 42 | | 42 | 63 |
| 0080 | 24" x 36" | | 8 | 2 | | | 84 | | 84 | 126 |
| 0090 | 36" x 60" | | 4 | 4 | | | 168 | | 168 | 253 |
| 0100 | Lead glass window frame, with 1/16" lead & voice passage, 36" x 60" | | 4 | 4 | | | 168 | | 168 | 253 |
| 0110 | Lead glass window frame, 24" x 36" | ↓ | 8 | 2 | ↓ | | 84 | | 84 | 126 |
| 0120 | Lead gypsum board, 5/8" thick with 1/16" lead | 2 Clab | 320 | .050 | S.F. | | 1.72 | | 1.72 | 2.64 |
| 0130 | 1/8" lead | | 280 | .057 | | | 1.96 | | 1.96 | 3.01 |
| 0140 | 1/32" lead | | 400 | .040 | ↓ | | 1.37 | | 1.37 | 2.11 |
| 0150 | Butt joints, 1/8" lead or thicker, 2" x 7' long batten strip | | 480 | .033 | Ea. | | 1.14 | | 1.14 | 1.76 |
| 0160 | X-ray protection, average radiography room, up to 300 SF, 1/16" lead, min | | .50 | 32 | Total | | 1,100 | | 1,100 | 1,675 |
| 0170 | Maximum | | .30 | 53.333 | | | 1,825 | | 1,825 | 2,800 |
| 0180 | Deep therapy X-ray room, 250 kV cap, up to 300 SF, 1/4" lead, min | | .20 | 80 | | | 2,750 | | 2,750 | 4,225 |
| 0190 | Maximum | | .12 | 133 | ↓ | | 4,575 | | 4,575 | 7,025 |
| 0880 | Radio frequency shielding, prefab or screen-type copper or steel, minimum | | 360 | .044 | SF Surf | | 1.53 | | 1.53 | 2.34 |
| 0890 | Average | | 310 | .052 | | | 1.77 | | 1.77 | 2.72 |
| 0895 | Maximum | ↓ | 290 | .055 | ↓ | | 1.90 | | 1.90 | 2.91 |

# 13 11 Swimming Pools

## 13 11 13 – Below-Grade Swimming Pools

### 13 11 13.50 Swimming Pools

| | | Crew | Daily Output | Labor-Hours | Unit | Material | 2011 Bare Costs Labor | Equipment | Total | Total Incl O&P |
|---|---|---|---|---|---|---|---|---|---|---|
| 0010 | **SWIMMING POOLS** Residential in-ground, vinyl lined, concrete | | | | | | | | | |
| 0020 | Sides including equipment, sand bottom | B-52 | 300 | .187 | SF Surf | 15.95 | 7.45 | 1.68 | 25.08 | 31 |
| 0100 | Metal or polystyrene sides                R131113-20 | B-14 | 410 | .117 | | 13.35 | 4.24 | .78 | 18.37 | 22 |
| 0200 | Add for vermiculite bottom | | | | ↓ | 1.02 | | | 1.02 | 1.12 |
| 0500 | Gunite bottom and sides, white plaster finish | | | | | | | | | |
| 0600 | 12' x 30' pool | B-52 | 145 | .386 | SF Surf | 29.50 | 15.40 | 3.48 | 48.38 | 60 |
| 0720 | 16' x 32' pool | ↓ | 155 | .361 | ↓ | 27 | 14.40 | 3.25 | 44.65 | 55 |
| 0750 | 20' x 40' pool | ↓ | 250 | .224 | ↓ | 24 | 8.90 | 2.02 | 34.92 | 42.50 |
| 0810 | Concrete bottom and sides, tile finish | | | | | | | | | |
| 0820 | 12' x 30' pool | B-52 | 80 | .700 | SF Surf | 30 | 28 | 6.30 | 64.30 | 82.50 |
| 0830 | 16' x 32' pool | | 95 | .589 | | 25 | 23.50 | 5.30 | 53.80 | 69.50 |
| 0840 | 20' x 40' pool | ↓ | 130 | .431 | ↓ | 19.70 | 17.15 | 3.88 | 40.73 | 52 |
| 1100 | Motel, gunite with plaster finish, incl. medium | | | | | | | | | |
| 1150 | capacity filtration & chlorination | B-52 | 115 | .487 | SF Surf | 36.50 | 19.40 | 4.38 | 60.28 | 74.50 |
| 1200 | Municipal, gunite with plaster finish, incl. high | | | | | | | | | |
| 1250 | capacity filtration & chlorination | B-52 | 100 | .560 | SF Surf | 47.50 | 22.50 | 5.05 | 75.05 | 91.50 |
| 1350 | Add for formed gutters | | | | L.F. | 69.50 | | | 69.50 | 76.50 |

# 13 11 Swimming Pools

## 13 11 13 – Below-Grade Swimming Pools

### 13 11 13.50 Swimming Pools

| | | Crew | Daily Output | Labor-Hours | Unit | Material | 2011 Bare Costs Labor | 2011 Bare Costs Equipment | Total | Total Incl O&P |
|---|---|---|---|---|---|---|---|---|---|---|
| 1360 | Add for stainless steel gutters | | | | L.F. | 206 | | | 206 | 226 |
| 1600 | For water heating system, see Section 23 52 28.10 | | | | | | | | | |
| 1700 | Filtration and deck equipment only, as % of total | | | | Total | | | | 20% | 20% |
| 1800 | Deck equipment, rule of thumb, 20' x 40' pool | | | | SF Pool | | | | 1.18 | 1.30 |
| 1900 | 5000 S.F. pool | | | | " | | | | 1.73 | 1.90 |
| 3000 | Painting pools, preparation + 3 coats, 20' x 40' pool, epoxy | 2 Pord | .33 | 48.485 | Total | 1,700 | 1,825 | | 3,525 | 4,550 |
| 3100 | Rubber base paint, 18 gallons | " | .33 | 48.485 | | 1,100 | 1,825 | | 2,925 | 3,900 |
| 3500 | 42' x 82' pool, 75 gallons, epoxy paint | 3 Pord | .14 | 171 | | 7,150 | 6,425 | | 13,575 | 17,400 |
| 3600 | Rubber base paint | " | .14 | 171 | ↓ | 4,500 | 6,425 | | 10,925 | 14,500 |

## 13 11 46 – Swimming Pool Accessories

### 13 11 46.50 Swimming Pool Equipment

| | | Crew | Daily Output | Labor-Hours | Unit | Material | 2011 Bare Costs Labor | 2011 Bare Costs Equipment | Total | Total Incl O&P |
|---|---|---|---|---|---|---|---|---|---|---|
| 0010 | **SWIMMING POOL EQUIPMENT** | | | | | | | | | |
| 0020 | Diving stand, stainless steel, 3 meter | 2 Carp | .40 | 40 | Ea. | 11,000 | 1,725 | | 12,725 | 14,800 |
| 0300 | 1 meter | | 2.70 | 5.926 | | 6,375 | 255 | | 6,630 | 7,400 |
| 0600 | Diving boards, 16' long, aluminum | | 2.70 | 5.926 | | 3,550 | 255 | | 3,805 | 4,300 |
| 0700 | Fiberglass | ↓ | 2.70 | 5.926 | ↓ | 2,625 | 255 | | 2,880 | 3,300 |
| 1100 | Gutter system, stainless steel, with grating, stock, | | | | | | | | | |
| 1110 | contains supply and drainage system | E-1 | 20 | 1.200 | L.F. | 268 | 57 | 5.45 | 330.45 | 395 |
| 1120 | Integral gutter and 5' high wall system, stainless steel | " | 10 | 2.400 | " | 360 | 114 | 10.90 | 484.90 | 600 |
| 1200 | Ladders, heavy duty, stainless steel, 2 tread | 2 Carp | 7 | 2.286 | Ea. | 635 | 98.50 | | 733.50 | 845 |
| 1500 | 4 tread | | 6 | 2.667 | | 835 | 115 | | 950 | 1,100 |
| 1800 | Lifeguard chair, stainless steel, fixed | ↓ | 2.70 | 5.926 | | 3,050 | 255 | | 3,305 | 3,750 |
| 1900 | Portable | | | | | 2,750 | | | 2,750 | 3,025 |
| 2100 | Lights, underwater, 12 volt, with transformer, 300 watt | 1 Elec | 1 | 8 | | 263 | 400 | | 663 | 885 |
| 2200 | 110 volt, 500 watt, standard | | 1 | 8 | | 209 | 400 | | 609 | 825 |
| 2400 | Low water cutoff type | ↓ | 1 | 8 | ↓ | 254 | 400 | | 654 | 875 |
| 2800 | Heaters, see Section 23 52 28.10 | | | | | | | | | |
| 3000 | Pool covers, reinforced vinyl | 3 Clab | 1800 | .013 | S.F. | .35 | .46 | | .81 | 1.09 |
| 3050 | Automatic, electric | | | | | | | | 8.75 | 9.65 |
| 3100 | Vinyl water tube | 3 Clab | 3200 | .008 | | .50 | .26 | | .76 | .95 |
| 3200 | Maximum | " | 3000 | .008 | | .63 | .27 | | .90 | 1.11 |
| 3250 | Sealed air bubble polyethylene solar blanket, 16 mils | | | | ↓ | .36 | | | .36 | .40 |
| 3300 | Slides, tubular, fiberglass, aluminum handrails & ladder, 5'-0", straight | 2 Carp | 1.60 | 10 | Ea. | 3,075 | 430 | | 3,505 | 4,025 |
| 3320 | 8'-0", curved | | 3 | 5.333 | | 6,575 | 230 | | 6,805 | 7,575 |
| 3400 | 10'-0", curved | | 1 | 16 | | 28,700 | 690 | | 29,390 | 32,600 |
| 3420 | 12'-0", straight with platform | ↓ | 1.20 | 13.333 | ↓ | 13,000 | 575 | | 13,575 | 15,200 |
| 4500 | Hydraulic lift, movable pool bottom, single ram | | | | | | | | | |
| 4520 | Under 1,000 S.F. area | L-9 | 72 | .500 | S.F. | 130 | 19.70 | | 149.70 | 175 |
| 4600 | Four ram lift, over 1,000 S.F. | " | 109 | .330 | " | 105 | 13 | | 118 | 137 |
| 5000 | Removable access ramp, stainless steel | 2 Clab | 2 | 8 | Ea. | 10,500 | 275 | | 10,775 | 11,900 |
| 5500 | Removable stairs, stainless steel, collapsible | " | 2 | 8 | " | 4,450 | 275 | | 4,725 | 5,300 |

# 13 17 Tubs and Pools

## 13 17 33 – Whirlpool Tubs

| 13 17 33.10 Whirlpool Bath | Crew | Daily Output | Labor-Hours | Unit | Material | 2011 Bare Costs Labor | Equipment | Total | Total Incl O&P |
|---|---|---|---|---|---|---|---|---|---|
| 0010 **WHIRLPOOL BATH** | | | | | | | | | |
| 6000 Whirlpool, bath with vented overflow, molded fiberglass | | | | | | | | | |
| 6100 66" x 48" x 24" | Q-1 | 1 | 16 | Ea. | 3,525 | 770 | | 4,295 | 5,025 |

# 13 18 Ice Rinks

## 13 18 13 – Ice Rink Floor Systems

### 13 18 13.50 Ice Skating

| | Crew | Daily Output | Labor-Hours | Unit | Material | Labor | Equipment | Total | Total Incl O&P |
|---|---|---|---|---|---|---|---|---|---|
| 0010 **ICE SKATING** Equipment incl. refrigeration, plumbing & cooling | | | | | | | | | |
| 0020 coils & concrete slab, 85' x 200' rink | | | | | | | | | |
| 0300 55° system, 5 mos., 100 ton | | | | Total | 575,000 | | | 575,000 | 632,500 |
| 0700 90° system, 12 mos., 135 ton | | | | " | 600,000 | | | 600,000 | 660,000 |
| 1200 Subsoil heating system (recycled from compressor), 85' x 200' | Q-7 | .27 | 118 | Ea. | 25,000 | 6,100 | | 31,100 | 36,600 |
| 1300 Subsoil insulation, 2 lb. polystyrene with vapor barrier, 85' x 200' | 2 Carp | .14 | 114 | " | 30,000 | 4,925 | | 34,925 | 40,600 |

## 13 18 16 – Ice Rink Dasher Boards

### 13 18 16.50 Ice Rink Dasher Boards

| | Crew | Daily Output | Labor-Hours | Unit | Material | Labor | Equipment | Total | Total Incl O&P |
|---|---|---|---|---|---|---|---|---|---|
| 0010 **ICE RINK DASHER BOARDS** | | | | | | | | | |
| 1000 Dasher boards, 1/2" H.D. polyethylene faced steel frame, 3' acrylic | | | | | | | | | |
| 1020 screen at sides, 5' acrylic ends, 85' x 200' | F-5 | .06 | 533 | Ea. | 125,000 | 23,200 | | 148,200 | 173,000 |
| 1100 Fiberglass & aluminum construction, same sides and ends | " | .06 | 533 | " | 125,000 | 23,200 | | 148,200 | 173,000 |

# 13 21 Controlled Environment Rooms

## 13 21 13 – Clean Rooms

### 13 21 13.50 Clean Room Components

| | Crew | Daily Output | Labor-Hours | Unit | Material | Labor | Equipment | Total | Total Incl O&P |
|---|---|---|---|---|---|---|---|---|---|
| 0010 **CLEAN ROOM COMPONENTS** | | | | | | | | | |
| 1100 Clean room, soft wall, 12' x 12', Class 100 | 1 Carp | .18 | 44.444 | Ea. | 15,500 | 1,925 | | 17,425 | 20,000 |
| 1110 Class 1,000 | | .18 | 44.444 | | 12,600 | 1,925 | | 14,525 | 16,800 |
| 1120 Class 10,000 | | .21 | 38.095 | | 11,100 | 1,650 | | 12,750 | 14,700 |
| 1130 Class 100,000 | | .21 | 38.095 | | 10,400 | 1,650 | | 12,050 | 14,000 |
| 2800 Ceiling grid support, slotted channel struts 4'-0" O.C., ea. way | | | | S.F. | | | | 5.90 | 6.50 |
| 3000 Ceiling panel, vinyl coated foil on mineral substrate | | | | | | | | | |
| 3020 Sealed, non-perforated | | | | S.F. | | | | 1.27 | 1.40 |
| 4000 Ceiling panel seal, silicone sealant, 150 L.F./gal. | 1 Carp | 150 | .053 | L.F. | .30 | 2.30 | | 2.60 | 3.85 |
| 4100 Two sided adhesive tape | " | 240 | .033 | " | .13 | 1.43 | | 1.56 | 2.34 |
| 4200 Clips, one per panel | | | | Ea. | 1.04 | | | 1.04 | 1.14 |
| 6000 HEPA filter, 2' x 4', 99.97% eff., 3" dp beveled frame (silicone seal) | | | | | 365 | | | 365 | 400 |
| 6040 6" deep skirted frame (channel seal) | | | | | 390 | | | 390 | 430 |
| 6100 99.99% efficient, 3" deep beveled frame (silicone seal) | | | | | 380 | | | 380 | 420 |
| 6140 6" deep skirted frame (channel seal) | | | | | 405 | | | 405 | 445 |
| 6200 99.999% efficient, 3" deep beveled frame (silicone seal) | | | | | 395 | | | 395 | 435 |
| 6240 6" deep skirted frame (channel seal) | | | | | 420 | | | 420 | 465 |
| 7000 Wall panel systems, including channel strut framing | | | | | | | | | |
| 7020 Polyester coated aluminum, particle board | | | | S.F. | | | | 18.20 | 20 |
| 7100 Porcelain coated aluminum, particle board | | | | | | | | 32 | 35 |
| 7400 Wall panel support, slotted channel struts, to 12' high | | | | | | | | 16.35 | 18 |

## 13 21 26 – Cold Storage Rooms

### 13 21 26.50 Refrigeration

| | Crew | Daily Output | Labor-Hours | Unit | Material | Labor | Equipment | Total | Total Incl O&P |
|---|---|---|---|---|---|---|---|---|---|
| 0010 **REFRIGERATION** | | | | | | | | | |
| 0020 Curbs, 12" high, 4" thick, concrete | 2 Carp | 58 | .276 | L.F. | 4.37 | 11.90 | | 16.27 | 23 |

# 13 21 Controlled Environment Rooms

## 13 21 26 - Cold Storage Rooms

### 13 21 26.50 Refrigeration

| 13 21 26.50 Refrigeration | Crew | Daily Output | Labor-Hours | Unit | Material | 2011 Bare Costs Labor | Equipment | Total | Total Incl O&P |
|---|---|---|---|---|---|---|---|---|---|
| 1000 | Doors, see Section 08 34 13.10 | | | | | | | | |
| 2400 | Finishes, 2 coat portland cement plaster, 1/2" thick | 1 Plas | 48 | .167 | S.F. | 1.20 | 6.55 | | 7.75 | 11.05 |
| 2500 | For galvanized reinforcing mesh, add | 1 Lath | 335 | .024 | | .80 | .92 | | 1.72 | 2.22 |
| 2700 | 3/16" thick latex cement | 1 Plas | 88 | .091 | | 2.05 | 3.58 | | 5.63 | 7.60 |
| 2900 | For glass cloth reinforced ceilings, add | " | 450 | .018 | | .49 | .70 | | 1.19 | 1.58 |
| 3100 | Fiberglass panels, 1/8" thick | 1 Carp | 149.45 | .054 | | 2.71 | 2.30 | | 5.01 | 6.50 |
| 3200 | Polystyrene, plastic finish ceiling, 1" thick | | 274 | .029 | | 2.50 | 1.26 | | 3.76 | 4.68 |
| 3400 | 2" thick | | 274 | .029 | | 2.85 | 1.26 | | 4.11 | 5.05 |
| 3500 | 4" thick | | 219 | .037 | | 3.17 | 1.57 | | 4.74 | 5.90 |
| 3800 | Floors, concrete, 4" thick | 1 Cefi | 93 | .086 | | 1.29 | 3.51 | | 4.80 | 6.50 |
| 3900 | 6" thick | " | 85 | .094 | | 2.01 | 3.84 | | 5.85 | 7.80 |
| 4000 | Insulation, 1" to 6" thick, cork | | | | B.F. | 1.16 | | | 1.16 | 1.28 |
| 4100 | Urethane | | | | | .51 | | | .51 | .56 |
| 4300 | Polystyrene, regular | | | | | .50 | | | .50 | .55 |
| 4400 | Bead board | | | | | .29 | | | .29 | .32 |
| 4600 | Installation of above, add per layer | 2 Carp | 657.60 | .024 | S.F. | .38 | 1.05 | | 1.43 | 2.03 |
| 4700 | Wall and ceiling juncture | | 298.90 | .054 | L.F. | 1.87 | 2.30 | | 4.17 | 5.60 |
| 4900 | Partitions, galvanized sandwich panels, 4" thick, stock | | 219.20 | .073 | S.F. | 7.85 | 3.14 | | 10.99 | 13.40 |
| 5000 | Aluminum or fiberglass | | 219.20 | .073 | " | 8.55 | 3.14 | | 11.69 | 14.25 |
| 5200 | Prefab walk-in, 7'-6" high, aluminum, incl. door & floors, | | | | | | | | | |
| 5210 | not incl. partitions or refrigeration, 6' x 6' O.D. nominal | 2 Carp | 54.80 | .292 | SF Flr. | 140 | 12.55 | | 152.55 | 173 |
| 5500 | 10' x 10' O.D. nominal | | 82.20 | .195 | | 113 | 8.40 | | 121.40 | 137 |
| 5700 | 12' x 14' O.D. nominal | | 109.60 | .146 | | 101 | 6.30 | | 107.30 | 121 |
| 5800 | 12' x 20' O.D. nominal | | 109.60 | .146 | | 88 | 6.30 | | 94.30 | 107 |
| 6100 | For 8'-6" high, add | | | | | 5% | | | | |
| 6300 | Rule of thumb for complete units, w/o doors & refrigeration, cooler | 2 Carp | 146 | .110 | | 127 | 4.72 | | 131.72 | 147 |
| 6400 | Freezer | | 109.60 | .146 | | 150 | 6.30 | | 156.30 | 175 |
| 6600 | Shelving, plated or galvanized, steel wire type | | 360 | .044 | SF Hor. | 11.10 | 1.91 | | 13.01 | 15.20 |
| 6700 | Slat shelf type | | 375 | .043 | | 13.70 | 1.84 | | 15.54 | 17.90 |
| 6900 | For stainless steel shelving, add | | | | | 300% | | | | |
| 7000 | Vapor barrier, on wood walls | 2 Carp | 1644 | .010 | S.F. | .16 | .42 | | .58 | .82 |
| 7200 | On masonry walls | " | 1315 | .012 | " | .42 | .52 | | .94 | 1.26 |
| 7500 | For air curtain doors, see Section 23 34 33.10 | | | | | | | | | |

## 13 21 48 - Sound-Conditioned Rooms

### 13 21 48.10 Anechoic Chambers

| | | Crew | Daily Output | Labor-Hours | Unit | Material | Labor | Equipment | Total | Total Incl O&P |
|---|---|---|---|---|---|---|---|---|---|---|
| 0010 | **ANECHOIC CHAMBERS** Standard units, 7' ceiling heights | | | | | | | | | |
| 0100 | Area for pricing is net inside dimensions | | | | | | | | | |
| 0300 | 200 cycles per second cutoff, 25 S.F. floor area | | | | SF Flr. | 1,600 | | | 1,600 | 1,750 |
| 0400 | 50 S.F. | | | | | | | | 1,050 | 1,150 |
| 0600 | 75 S.F. | | | | | | | | 1,000 | 1,100 |
| 0700 | 100 S.F. | | | | | 1,200 | | | 1,200 | 1,325 |
| 0900 | For 150 cycles per second cutoff, add to 100 S.F. room | | | | | | | | | 30% |
| 1000 | For 100 cycles per second cutoff, add to 100 S.F. room | | | | | | | | | 45% |

### 13 21 48.15 Audiometric Rooms

| | | Crew | Daily Output | Labor-Hours | Unit | Material | Labor | Equipment | Total | Total Incl O&P |
|---|---|---|---|---|---|---|---|---|---|---|
| 0010 | **AUDIOMETRIC ROOMS** | | | | | | | | | |
| 0020 | Under 500 S.F. surface | 4 Carp | 98 | .327 | SF Surf | 51.50 | 14.05 | | 65.55 | 78 |
| 0100 | Over 500 S.F. surface | " | 120 | .267 | " | 49 | 11.50 | | 60.50 | 71 |

## 13 21 53 - Darkrooms

### 13 21 53.50 Darkrooms

| | | Crew | Daily Output | Labor-Hours | Unit | Material | Labor | Equipment | Total | Total Incl O&P |
|---|---|---|---|---|---|---|---|---|---|---|
| 0010 | **DARKROOMS** | | | | | | | | | |
| 0020 | Shell, complete except for door, 64 S.F., 8' high | 2 Carp | 128 | .125 | SF Flr. | 53.50 | 5.40 | | 58.90 | 67 |

# 13 21 Controlled Environment Rooms

## 13 21 53 – Darkrooms

### 13 21 53.50 Darkrooms

| 13 21 53.50 Darkrooms | | Crew | Daily Output | Labor-Hours | Unit | Material | 2011 Bare Costs Labor | Equipment | Total | Total Incl O&P |
|---|---|---|---|---|---|---|---|---|---|---|
| 0100 | 12' high | 2 Carp | 64 | .250 | SF Flr. | 68.50 | 10.75 | | 79.25 | 91.50 |
| 0500 | 120 S.F. floor, 8' high | | 120 | .133 | | 37.50 | 5.75 | | 43.25 | 50 |
| 0600 | 12' high | | 60 | .267 | | 51.50 | 11.50 | | 63 | 74 |
| 0800 | 240 S.F. floor, 8' high | | 120 | .133 | | 27 | 5.75 | | 32.75 | 38.50 |
| 0900 | 12' high | | 60 | .267 | | 38 | 11.50 | | 49.50 | 59 |
| 1200 | Mini-cylindrical, revolving, unlined, 4' diameter | | 3.50 | 4.571 | Ea. | 2,300 | 197 | | 2,497 | 2,825 |
| 1400 | 5'-6" diameter | ↓ | 2.50 | 6.400 | | 3,950 | 276 | | 4,226 | 4,775 |
| 1600 | Add for lead lining, inner cylinder, 1/32" thick | | | | | 1,475 | | | 1,475 | 1,625 |
| 1700 | 1/16" thick | | | | | 3,975 | | | 3,975 | 4,375 |
| 1800 | Add for lead lining, inner and outer cylinder, 1/32" thick | | | | | 2,725 | | | 2,725 | 3,000 |
| 1900 | 1/16" thick | | | | ↓ | 6,100 | | | 6,100 | 6,700 |
| 2000 | For darkroom door, see Section 08 34 36.10 | | | | | | | | | |

## 13 21 56 – Music Rooms

### 13 21 56.50 Music Rooms

| | | Crew | Daily Output | Labor-Hours | Unit | Material | 2011 Bare Costs Labor | Equipment | Total | Total Incl O&P |
|---|---|---|---|---|---|---|---|---|---|---|
| 0010 | **MUSIC ROOMS** | | | | | | | | | |
| 0020 | Practice room, modular, perforated steel, under 500 S.F. | 2 Carp | 70 | .229 | SF Surf | 31.50 | 9.85 | | 41.35 | 49.50 |
| 0100 | Over 500 S.F. | " | 80 | .200 | " | 26.50 | 8.60 | | 35.10 | 42.50 |

# 13 24 Special Activity Rooms

## 13 24 16 – Saunas

### 13 24 16.50 Saunas and Heaters

| | | Crew | Daily Output | Labor-Hours | Unit | Material | 2011 Bare Costs Labor | Equipment | Total | Total Incl O&P |
|---|---|---|---|---|---|---|---|---|---|---|
| 0010 | **SAUNAS AND HEATERS** | | | | | | | | | |
| 0020 | Prefabricated, incl. heater & controls, 7' high, 6' x 4', C/C | L-7 | 2.20 | 12.727 | Ea. | 5,125 | 530 | | 5,655 | 6,450 |
| 0050 | 6' x 4', C/P | | 2 | 14 | | 4,300 | 580 | | 4,880 | 5,625 |
| 0400 | 6' x 5', C/C | | 2 | 14 | | 5,225 | 580 | | 5,805 | 6,650 |
| 0450 | 6' x 5', C/P | | 2 | 14 | | 4,450 | 580 | | 5,030 | 5,800 |
| 0600 | 6' x 6', C/C | | 1.80 | 15.556 | | 5,550 | 645 | | 6,195 | 7,075 |
| 0650 | 6' x 6', C/P | | 1.80 | 15.556 | | 4,750 | 645 | | 5,395 | 6,200 |
| 0800 | 6' x 9', C/C | | 1.60 | 17.500 | | 6,875 | 730 | | 7,605 | 8,650 |
| 0850 | 6' x 9', C/P | | 1.60 | 17.500 | | 6,425 | 730 | | 7,155 | 8,175 |
| 1000 | 8' x 12', C/C | | 1.10 | 25.455 | | 11,200 | 1,050 | | 12,250 | 13,900 |
| 1050 | 8' x 12', C/P | | 1.10 | 25.455 | | 9,900 | 1,050 | | 10,950 | 12,500 |
| 1200 | 8' x 8', C/C | | 1.40 | 20 | | 8,525 | 830 | | 9,355 | 10,700 |
| 1250 | 8' x 8', C/P | | 1.40 | 20 | | 7,750 | 830 | | 8,580 | 9,800 |
| 1400 | 8' x 10', C/C | | 1.20 | 23.333 | | 9,450 | 970 | | 10,420 | 11,900 |
| 1450 | 8' x 10', C/P | | 1.20 | 23.333 | | 8,500 | 970 | | 9,470 | 10,800 |
| 1600 | 10' x 12', C/C | | 1 | 28 | | 11,800 | 1,175 | | 12,975 | 14,800 |
| 1650 | 10' x 12', C/P | ↓ | 1 | 28 | | 10,300 | 1,175 | | 11,475 | 13,200 |
| 1700 | Door only, cedar, 2' x 6', with tempered insulated glass window | 2 Carp | 3.40 | 4.706 | | 620 | 203 | | 823 | 990 |
| 1800 | Prehung, incl. jambs, pulls & hardware | " | 12 | 1.333 | | 620 | 57.50 | | 677.50 | 775 |
| 2500 | Heaters only (incl. above), wall mounted, to 200 C.F. | | | | | 615 | | | 615 | 675 |
| 2750 | To 300 C.F. | | | | | 795 | | | 795 | 875 |
| 3000 | Floor standing, to 720 C.F., 10,000 watts, w/controls | 1 Elec | 3 | 2.667 | | 2,100 | 134 | | 2,234 | 2,500 |
| 3250 | To 1,000 C.F., 16,000 watts | " | 3 | 2.667 | ↓ | 2,700 | 134 | | 2,834 | 3,150 |

## 13 24 26 – Steam Baths

### 13 24 26.50 Steam Baths and Components

| | | Crew | Daily Output | Labor-Hours | Unit | Material | 2011 Bare Costs Labor | Equipment | Total | Total Incl O&P |
|---|---|---|---|---|---|---|---|---|---|---|
| 0010 | **STEAM BATHS AND COMPONENTS** | | | | | | | | | |
| 0020 | Heater, timer & head, single, to 140 C.F. | 1 Plum | 1.20 | 6.667 | Ea. | 2,100 | 355 | | 2,455 | 2,825 |
| 0500 | To 300 C.F. | | 1.10 | 7.273 | | 2,425 | 390 | | 2,815 | 3,250 |
| 1000 | Commercial size, with blow-down assembly, to 800 C.F. | ↓ | .90 | 8.889 | | 7,800 | 475 | | 8,275 | 9,275 |

# 13 24 Special Activity Rooms

## 13 24 26 – Steam Baths

| 13 24 26.50 Steam Baths and Components | Crew | Daily Output | Labor-Hours | Unit | Material | 2011 Bare Costs Labor | Equipment | Total | Total Incl O&P |
|---|---|---|---|---|---|---|---|---|---|
| 1500 | To 2500 C.F. | 1 Plum | .80 | 10 | Ea. | 8,775 | 535 | | 9,310 | 10,500 |
| 2000 | Multiple, motels, apts., 2 baths, w/blow-down assm., 500 C.F. | Q-1 | 1.30 | 12.308 | | 5,900 | 595 | | 6,495 | 7,375 |
| 2500 | 4 baths | " | .70 | 22.857 | | 7,450 | 1,100 | | 8,550 | 9,850 |
| 2700 | Conversion unit for residential tub, including door | | | | | 4,100 | | | 4,100 | 4,500 |

# 13 28 Athletic and Recreational Special Construction

## 13 28 33 – Athletic and Recreational Court Walls

### 13 28 33.50 Sport Court

| | | Crew | Daily Output | Labor-Hours | Unit | Material | 2011 Bare Costs Labor | Equipment | Total | Total Incl O&P |
|---|---|---|---|---|---|---|---|---|---|---|
| 0010 | **SPORT COURT** | | | | | | | | | |
| 0020 | Floors, No. 2 & better maple, 25/32" thick | | | | SF Flr. | | | | 6.05 | 6.65 |
| 0100 | Walls, laminated plastic bonded to galv. steel studs | | | | SF Wall | | | | 7.70 | 8.45 |
| 0150 | Laminated fiberglass surfacing, minimum | | | | | | | | 1.91 | 2.10 |
| 0180 | Maximum | | | | | | | | 2.15 | 2.37 |
| 0300 | Squash, regulation court in existing building, minimum | | | | Court | 36,800 | | | 36,800 | 40,400 |
| 0400 | Maximum | | | | " | 41,000 | | | 41,000 | 45,000 |
| 0450 | Rule of thumb for components: | | | | | | | | | |
| 0470 | Walls | 3 Carp | .15 | 160 | Court | 11,000 | 6,900 | | 17,900 | 22,700 |
| 0500 | Floor | " | .25 | 96 | | 8,725 | 4,125 | | 12,850 | 15,900 |
| 0550 | Lighting | 2 Elec | .60 | 26.667 | | 2,100 | 1,350 | | 3,450 | 4,275 |
| 0600 | Handball, racquetball court in existing building, minimum | C-1 | .20 | 160 | | 39,800 | 6,550 | | 46,350 | 54,000 |
| 0800 | Maximum | " | .10 | 320 | | 43,100 | 13,100 | | 56,200 | 67,500 |
| 0900 | Rule of thumb for components: walls | 3 Carp | .12 | 200 | | 12,600 | 8,600 | | 21,200 | 27,100 |
| 1000 | Floor | | .25 | 96 | | 8,725 | 4,125 | | 12,850 | 15,900 |
| 1100 | Ceiling | | .33 | 72.727 | | 4,200 | 3,125 | | 7,325 | 9,425 |
| 1200 | Lighting | 2 Elec | .60 | 26.667 | | 2,200 | 1,350 | | 3,550 | 4,400 |

# 13 31 Fabric Structures

## 13 31 13 – Air-Supported Fabric Structures

### 13 31 13.09 Air Supported Tank Covers

| | | Crew | Daily Output | Labor-Hours | Unit | Material | 2011 Bare Costs Labor | Equipment | Total | Total Incl O&P |
|---|---|---|---|---|---|---|---|---|---|---|
| 0010 | **AIR SUPPORTED TANK COVERS**, vinyl polyester | | | | | | | | | |
| 0100 | Scrim, double layer, with hardware, blower, standby & controls | | | | | | | | | |
| 0200 | Round, 75' diameter | B-2 | 4500 | .009 | S.F. | 10 | .31 | | 10.31 | 11.45 |
| 0300 | 100' diameter | | 5000 | .008 | | 9.10 | .28 | | 9.38 | 10.45 |
| 0400 | 150' diameter | | 5000 | .008 | | 7.20 | .28 | | 7.48 | 8.40 |
| 0500 | Rectangular, 20' x 20' | | 4500 | .009 | | 23 | .31 | | 23.31 | 25.50 |
| 0600 | 30' x 40' | | 4500 | .009 | | 23 | .31 | | 23.31 | 25.50 |
| 0700 | 50' x 60' | | 4500 | .009 | | 23 | .31 | | 23.31 | 25.50 |
| 0800 | For single wall construction, deduct, minimum | | | | | .77 | | | .77 | .85 |
| 0900 | Maximum | | | | | 2.29 | | | 2.29 | 2.52 |
| 1000 | For maximum resistance to atmosphere or cold, add | | | | | 1.12 | | | 1.12 | 1.23 |
| 1100 | For average shipping charges, add | | | | Total | 1,925 | | | 1,925 | 2,125 |

### 13 31 13.13 Single-Walled Air-Supported Structures

| | | Crew | Daily Output | Labor-Hours | Unit | Material | 2011 Bare Costs Labor | Equipment | Total | Total Incl O&P |
|---|---|---|---|---|---|---|---|---|---|---|
| 0010 | **SINGLE-WALLED AIR-SUPPORTED STRUCTURES** R133113-10 | | | | | | | | | |
| 0020 | Site preparation, incl. anchor placement and utilities | B-11B | 1000 | .016 | SF Flr. | 1.14 | .62 | .25 | 2.01 | 2.46 |
| 0030 | For concrete, see Section 03 30 53.40 | | | | | | | | | |
| 0050 | Warehouse, polyester/vinyl fabric, 28 oz., over 10 yr. life, welded | | | | | | | | | |
| 0060 | Seams, tension cables, primary & auxiliary inflation system, | | | | | | | | | |
| 0070 | airlock, personnel doors and liner | | | | | | | | | |
| 0100 | 5,000 S.F. | 4 Clab | 5000 | .006 | SF Flr. | 24.50 | .22 | | 24.72 | 27.50 |

# 13 31 Fabric Structures

## 13 31 13 – Air-Supported Fabric Structures

| 13 31 13.13 Single-Walled Air-Supported Structures | Crew | Daily Output | Labor-Hours | Unit | Material | 2011 Bare Costs Labor | Equipment | Total | Total Incl O&P |
|---|---|---|---|---|---|---|---|---|---|
| 0250    12,000 S.F. | 4 Clab | 6000 | .005 | SF Flr. | 17.65 | .18 | | 17.83 | 19.70 |
| 0400    24,000 S.F. | 8 Clab | 12000 | .005 | | 12.40 | .18 | | 12.58 | 13.95 |
| 0500    50,000 S.F. | " | 12500 | .005 | | 11.50 | .18 | | 11.68 | 12.90 |
| 0700   12 oz. reinforced vinyl fabric, 5 yr. life, sewn seams, | | | | | | | | | |
| 0710     accordian door, including liner | | | | | | | | | |
| 0750    3000 S.F. | 4 Clab | 3000 | .011 | SF Flr. | 11.75 | .37 | | 12.12 | 13.45 |
| 0800    12,000 S.F. | " | 6000 | .005 | | 10 | .18 | | 10.18 | 11.30 |
| 0850    24,000 S.F. | 8 Clab | 12000 | .005 | | 8.45 | .18 | | 8.63 | 9.60 |
| 0950    Deduct for single layer | | | | | 1.01 | | | 1.01 | 1.11 |
| 1000    Add for welded seams | | | | | 1.01 | | | 1.01 | 1.11 |
| 1050    Add for double layer, welded seams included | | | | | 2.08 | | | 2.08 | 2.29 |
| 1250   Tedlar/vinyl fabric, 28 oz., with liner, over 10 yr. life, | | | | | | | | | |
| 1260    incl. overhead and personnel doors | | | | | | | | | |
| 1300    3000 S.F. | 4 Clab | 3000 | .011 | SF Flr. | 24 | .37 | | 24.37 | 27 |
| 1450    12,000 S.F. | " | 6000 | .005 | | 16.85 | .18 | | 17.03 | 18.85 |
| 1550    24,000 S.F. | 8 Clab | 12000 | .005 | | 13.05 | .18 | | 13.23 | 14.65 |
| 1700    Deduct for single layer | | | | | 1.44 | | | 1.44 | 1.58 |
| 2250   Greenhouse/shelter, woven polyethylene with liner, 2 yr. life, | | | | | | | | | |
| 2260    sewn seams, including doors | | | | | | | | | |
| 2300    3000 S.F. | 4 Clab | 3000 | .011 | SF Flr. | 6.60 | .37 | | 6.97 | 7.80 |
| 2350    12,000 S.F. | " | 6000 | .005 | | 6.70 | .18 | | 6.88 | 7.70 |
| 2450    24,000 S.F. | 8 Clab | 12000 | .005 | | 5.60 | .18 | | 5.78 | 6.45 |
| 2550    Deduct for single layer | | | | | .63 | | | .63 | .69 |
| 2600   Tennis/gymnasium, polyester/vinyl fabric, 28 oz., over 10 yr. life, | | | | | | | | | |
| 2610    including thermal liner, heat and lights | | | | | | | | | |
| 2650    7,200 S.F. | 4 Clab | 6000 | .005 | SF Flr. | 23 | .18 | | 23.18 | 26 |
| 2750    13,000 S.F. | " | 6500 | .005 | | 17.65 | .17 | | 17.82 | 19.65 |
| 2850    Over 24,000 S.F. | 8 Clab | 12000 | .005 | | 16.10 | .18 | | 16.28 | 18.05 |
| 2860   For low temperature conditions, add | | | | | 1.12 | | | 1.12 | 1.23 |
| 2870   For average shipping charges, add | | | | Total | 5,475 | | | 5,475 | 6,000 |
| 2900   Thermal liner, translucent reinforced vinyl | | | | SF Flr. | 1.12 | | | 1.12 | 1.23 |
| 2950    Metalized mylar fabric and mesh, double liner | | | | " | 2.29 | | | 2.29 | 2.52 |
| 3050   Stadium/convention center, teflon coated fiberglass, heavy weight, | | | | | | | | | |
| 3060    over 20 yr. life, incl. thermal liner and heating system | | | | | | | | | |
| 3100     Minimum | 9 Clab | 26000 | .003 | SF Flr. | 56 | .10 | | 56.10 | 62 |
| 3110     Maximum | " | 19000 | .004 | " | 67 | .13 | | 67.13 | 73.50 |
| 3400   Doors, air lock, 15' long, 10' x 10' | 2 Carp | .80 | 20 | Ea. | 20,100 | 860 | | 20,960 | 23,400 |
| 3600    15' x 15' | " | .50 | 32 | | 29,700 | 1,375 | | 31,075 | 34,800 |
| 3700    For each added 5' length, add | | | | | 5,300 | | | 5,300 | 5,850 |
| 3900   Revolving personnel door, 6' diameter, 6'-6" high | 2 Carp | .80 | 20 | | 14,900 | 860 | | 15,760 | 17,700 |
| 4200   Double wall, self supporting, shell only, minimum | | | | SF Flr. | | | | 19.10 | 21 |
| 4300    Maximum | | | | " | | | | 35.50 | 39 |

## 13 31 23 – Tensioned Fabric Structures

### 13 31 23.50 Tension Structures

| | Crew | Daily Output | Labor-Hours | Unit | Material | 2011 Bare Costs Labor | Equipment | Total | Total Incl O&P |
|---|---|---|---|---|---|---|---|---|---|
| 0010   **TENSION STRUCTURES** Rigid steel/alum. frame, vinyl coated poly | | | | | | | | | |
| 0100   Fabric shell, 60' clear span, not incl. foundations or floors | | | | | | | | | |
| 0200    6,000 S.F. | B-41 | 1000 | .044 | SF Flr. | 14.65 | 1.56 | .28 | 16.49 | 18.80 |
| 0300    12,000 S.F. | | 1100 | .040 | | 13.55 | 1.42 | .26 | 15.23 | 17.35 |
| 0400    80' clear span, 20,800 S.F. | | 1220 | .036 | | 13.45 | 1.28 | .23 | 14.96 | 16.95 |
| 0410    100' clear span, 10,000 S.F. | L-5 | 2175 | .026 | | 14.80 | 1.25 | .34 | 16.39 | 18.85 |
| 0430    26,000 S.F. | | 2300 | .024 | | 13.95 | 1.18 | .33 | 15.46 | 17.75 |
| 0450    36,000 S.F. | | 2500 | .022 | | 13.60 | 1.09 | .30 | 14.99 | 17.15 |

# 13 31 Fabric Structures

## 13 31 23 – Tensioned Fabric Structures

| 13 31 23.50 Tension Structures | Crew | Daily Output | Labor-Hours | Unit | Material | 2011 Bare Costs Labor | 2011 Bare Costs Equipment | Total | Total Incl O&P |
|---|---|---|---|---|---|---|---|---|---|
| 0460 | 120' clear span, 24,000 S.F. | L-5 | 3000 | .019 | SF Flr. | 15.55 | .91 | .25 | 16.71 | 18.95 |
| 0470 | 150' clear span, 30,000 S.F. | ↓ | 6000 | .009 | | 16.70 | .45 | .12 | 17.27 | 19.25 |
| 0480 | 200' clear span, 40,000 S.F. | E-6 | 8000 | .016 | ↓ | 19.05 | .77 | .24 | 20.06 | 22.50 |
| 0500 | For roll-up door, 12' x 14', add | L-2 | 1 | 16 | Ea. | 6,075 | 605 | | 6,680 | 7,600 |
| 0600 | For personnel doors, add, minimum | | | | SF Flr. | 5% | | | | |
| 0700 | Add, maximum | | | | | 15% | | | | |
| 0800 | For site work, simple foundation, etc., add, minimum | | | | | | | | 1.25 | 1.95 |
| 0900 | Add, maximum | | | | ↓ | | | | 2.75 | 3.05 |

# 13 34 Fabricated Engineered Structures

## 13 34 13 – Glazed Structures

### 13 34 13.13 Greenhouses

| | | Crew | Daily Output | Labor-Hours | Unit | Material | Labor | Equipment | Total | Total Incl O&P |
|---|---|---|---|---|---|---|---|---|---|---|
| 0010 | **GREENHOUSES**, Shell only, stock units, not incl. 2' stub walls, | | | | | | | | | |
| 0020 | foundation, floors, heat or compartments | | | | | | | | | |
| 0300 | Residential type, free standing, 8'-6" long x 7'-6" wide | 2 Carp | 59 | .271 | SF Flr. | 30.50 | 11.65 | | 42.15 | 51.50 |
| 0400 | 10'-6" wide | | 85 | .188 | | 26 | 8.10 | | 34.10 | 41 |
| 0600 | 13'-6" wide | | 108 | .148 | | 20.50 | 6.40 | | 26.90 | 33 |
| 0700 | 17'-0" wide | | 160 | .100 | | 25 | 4.31 | | 29.31 | 34 |
| 0900 | Lean-to type, 3'-10" wide | | 34 | .471 | | 33.50 | 20.50 | | 54 | 68 |
| 1000 | 6'-10" wide | ↓ | 58 | .276 | ↓ | 41 | 11.90 | | 52.90 | 63 |
| 1500 | Commercial, custom, truss frame, incl. equip., plumbing, elec., | | | | | | | | | |
| 1510 | benches and controls, under 2,000 S.F., minimum | | | | SF Flr. | | | | 27.50 | 30 |
| 1550 | Maximum | | | | | 14.45 | | | 14.45 | 15.90 |
| 1700 | Over 5,000 S.F., minimum | | | | | 14.45 | | | 14.45 | 15.90 |
| 1750 | Maximum | | | | ↓ | | | | 27.50 | 30 |
| 2000 | Institutional, custom, rigid frame, including compartments and | | | | | | | | | |
| 2010 | multi-controls, under 500 S.F., minimum | | | | SF Flr. | 45 | | | 45 | 49.50 |
| 2050 | Maximum | | | | | 32 | | | 32 | 35.50 |
| 2150 | Over 2,000 S.F., minimum | | | | | 11.70 | | | 11.70 | 12.90 |
| 2200 | Maximum | | | | | 59 | | | 59 | 65 |
| 2400 | Concealed rigid frame, under 500 S.F., minimum | | | | | | | | 82 | 90 |
| 2450 | Maximum | | | | | | | | 100 | 110 |
| 2550 | Over 2,000 S.F., minimum | | | | | | | | 62 | 68 |
| 2600 | Maximum | | | | | | | | 72 | 79 |
| 2800 | Lean-to type, under 500 S.F., minimum | | | | | | | | 72.50 | 80 |
| 2850 | Maximum | | | | | | | | 109 | 120 |
| 3000 | Over 2,000 S.F., minimum | | | | | | | | 40 | 44 |
| 3050 | Maximum | | | | ↓ | | | | 66.50 | 73 |
| 3600 | For 1/4" clear plate glass, add | | | | SF Surf | 1.86 | | | 1.86 | 2.05 |
| 3700 | For 1/4" tempered glass, add | | | | " | 4.21 | | | 4.21 | 4.63 |
| 3900 | For cooling, add, minimum | | | | SF Flr. | 2.81 | | | 2.81 | 3.09 |
| 4000 | Maximum | | | | | 6.95 | | | 6.95 | 7.65 |
| 4200 | For heaters, 13.6 MBH, add | | | | | 5.35 | | | 5.35 | 5.85 |
| 4300 | 60 MBH, add | | | | ↓ | 2 | | | 2 | 2.20 |
| 4500 | For benches, 2' x 3'-6", add | | | | SF Hor. | 24 | | | 24 | 26.50 |
| 4600 | 3' x 10', add | | | | S.F. | 12.95 | | | 12.95 | 14.25 |
| 4800 | For controls, add, minimum | | | | Total | 2,400 | | | 2,400 | 2,650 |
| 4900 | Maximum | | | | " | 14,300 | | | 14,300 | 15,700 |
| 5100 | For humidification equipment, add | | | | M.C.F. | 6.15 | | | 6.15 | 6.75 |
| 5200 | For vinyl shading, add | | | | S.F. | 1.29 | | | 1.29 | 1.42 |
| 6000 | Geodesic hemisphere, 1/8" plexiglass glazing | | | | | | | | | |

# 13 34 Fabricated Engineered Structures

## 13 34 13 – Glazed Structures

### 13 34 13.13 Greenhouses

| | | Crew | Daily Output | Labor-Hours | Unit | Material | 2011 Bare Costs Labor | Equipment | Total | Total Incl O&P |
|---|---|---|---|---|---|---|---|---|---|---|
| 6050 | 8' diameter | 2 Carp | 2 | 8 | Ea. | 2,700 | 345 | | 3,045 | 3,500 |
| 6150 | 24' diameter | | .35 | 45.714 | | 13,900 | 1,975 | | 15,875 | 18,300 |
| 6250 | 48' diameter | | .20 | 80 | | 37,600 | 3,450 | | 41,050 | 46,700 |

### 13 34 13.19 Swimming Pool Enclosures

| | | Crew | Daily Output | Labor-Hours | Unit | Material | 2011 Bare Costs Labor | Equipment | Total | Total Incl O&P |
|---|---|---|---|---|---|---|---|---|---|---|
| 0010 | **SWIMMING POOL ENCLOSURES** Translucent, free standing | | | | | | | | | |
| 0020 | not including foundations, heat or light | | | | | | | | | |
| 0200 | Economy, minimum | 2 Carp | 200 | .080 | SF Hor. | 14.80 | 3.44 | | 18.24 | 21.50 |
| 0300 | Maximum | | 100 | .160 | | 45.50 | 6.90 | | 52.40 | 60.50 |
| 0400 | Deluxe, minimum | | 100 | .160 | | 50.50 | 6.90 | | 57.40 | 66 |
| 0600 | Maximum | | 70 | .229 | | 227 | 9.85 | | 236.85 | 265 |
| 0700 | For motorized roof, 40% opening, solid roof, add | | | | | 11.70 | | | 11.70 | 12.85 |
| 0800 | Skylight type roof, add | | | | | 9.35 | | | 9.35 | 10.30 |
| 0900 | Air-inflated, including blowers and heaters, minimum | | | | | | | | 3.18 | 3.50 |
| 1000 | Maximum | | | | | | | | 6.10 | 6.70 |

## 13 34 16 – Grandstands and Bleachers

### 13 34 16.13 Grandstands

| | | Crew | Daily Output | Labor-Hours | Unit | Material | 2011 Bare Costs Labor | Equipment | Total | Total Incl O&P |
|---|---|---|---|---|---|---|---|---|---|---|
| 0010 | **GRANDSTANDS** Permanent, municipal, including foundation | | | | | | | | | |
| 0050 | Steel understructure w/aluminum closed deck, minimum | | | | Seat | | | | 118 | 130 |
| 0100 | Maximum | | | | | | | | 205 | 225 |
| 0300 | Steel, minimum | | | | | 55 | | | 55 | 60.50 |
| 0400 | Maximum | | | | | 110 | | | 110 | 121 |
| 0600 | Aluminum, extruded, stock design, minimum | | | | | | | | 63.50 | 70 |
| 0700 | Maximum | | | | | | | | 105 | 115 |
| 0900 | Composite, steel, wood and plastic, stock design, minimum | | | | | 130 | | | 130 | 143 |
| 1000 | Maximum | | | | | 260 | | | 260 | 286 |

### 13 34 16.53 Bleachers

| | | Crew | Daily Output | Labor-Hours | Unit | Material | 2011 Bare Costs Labor | Equipment | Total | Total Incl O&P |
|---|---|---|---|---|---|---|---|---|---|---|
| 0010 | **BLEACHERS** | | | | | | | | | |
| 0020 | Bleachers, outdoor, portable, 3 to 5 tiers, to 300' long, min | 2 Sswk | 120 | .133 | Seat | 48 | 6.45 | | 54.45 | 64 |
| 0100 | Maximum, less than 15' long, prefabricated | | 80 | .200 | | 78 | 9.70 | | 87.70 | 103 |
| 0200 | 6 to 20 tiers, minimum, up to 300' long | | 120 | .133 | | 66.50 | 6.45 | | 72.95 | 85 |
| 0300 | Max., under 15', (highly prefabricated, on wheels) | | 80 | .200 | | 85.50 | 9.70 | | 95.20 | 111 |
| 0500 | Permanent grandstands, wood seat, steel frame, 24" row | | | | | | | | | |
| 0600 | 3 to 15 tiers, minimum | 2 Sswk | 60 | .267 | Seat | 103 | 12.95 | | 115.95 | 136 |
| 0700 | Maximum | | 48 | .333 | | 123 | 16.20 | | 139.20 | 164 |
| 0900 | 16 to 30 tiers, minimum | | 60 | .267 | | 189 | 12.95 | | 201.95 | 231 |
| 0950 | Average | | 55 | .291 | | 211 | 14.10 | | 225.10 | 257 |
| 1000 | Maximum | | 48 | .333 | | 240 | 16.20 | | 256.20 | 293 |
| 1200 | Seat backs only, 30" row, fiberglass | | 160 | .100 | | 40 | 4.86 | | 44.86 | 52.50 |
| 1300 | Steel and wood | | 160 | .100 | | 62 | 4.86 | | 66.86 | 76.50 |
| 1400 | NOTE: average seating is 1.5' in width | | | | | | | | | |

## 13 34 19 – Metal Building Systems

### 13 34 19.50 Pre-Engineered Steel Buildings

| | | Crew | Daily Output | Labor-Hours | Unit | Material | 2011 Bare Costs Labor | Equipment | Total | Total Incl O&P |
|---|---|---|---|---|---|---|---|---|---|---|
| 0010 | **PRE-ENGINEERED STEEL BUILDINGS** R133419-10 | | | | | | | | | |
| 0100 | Clear span rigid frame, 26 ga. colored roofing and siding | | | | | | | | | |
| 0150 | 20' to 29' wide, 10' eave height | E-2 | 425 | .132 | SF Flr. | 8.85 | 6.25 | 3.82 | 18.92 | 24.50 |
| 0160 | 14' eave height | | 350 | .160 | | 9.35 | 7.60 | 4.63 | 21.58 | 28.50 |
| 0170 | 16' eave height | | 320 | .175 | | 9.80 | 8.30 | 5.05 | 23.15 | 30.50 |
| 0180 | 20' eave height | | 275 | .204 | | 10.85 | 9.65 | 5.90 | 26.40 | 35 |
| 0190 | 24' eave height | | 240 | .233 | | 12.40 | 11.05 | 6.75 | 30.20 | 40 |
| 0200 | 30' to 40' wide, 10' eave height | | 535 | .105 | | 7.25 | 4.96 | 3.03 | 15.24 | 19.70 |
| 0300 | 14' eave height | | 450 | .124 | | 7.60 | 5.90 | 3.60 | 17.10 | 22.50 |

| 13 34 19.50 Pre-Engineered Steel Buildings | | Crew | Daily Output | Labor-Hours | Unit | Material | 2011 Bare Costs Labor | Equipment | Total | Total Incl O&P |
|---|---|---|---|---|---|---|---|---|---|---|
| 0400 | 16' eave height | E-2 | 415 | .135 | SF Flr. | 8 | 6.40 | 3.91 | 18.31 | 24 |
| 0500 | 20' eave height | | 360 | .156 | | 8.75 | 7.35 | 4.51 | 20.61 | 27 |
| 0600 | 24' eave height | | 320 | .175 | | 9.90 | 8.30 | 5.05 | 23.25 | 30.50 |
| 0700 | 50' to 100' wide, 10' eave height | | 770 | .073 | | 6.20 | 3.44 | 2.11 | 11.75 | 15 |
| 0800 | 14' eave height | | 650 | .086 | | 6.50 | 4.08 | 2.49 | 13.07 | 16.80 |
| 0900 | 16' eave height | | 600 | .093 | | 7 | 4.42 | 2.70 | 14.12 | 18.15 |
| 1000 | 20' eave height | | 490 | .114 | | 7.50 | 5.40 | 3.31 | 16.21 | 21 |
| 1100 | 24' eave height | | 435 | .129 | | 8.25 | 6.10 | 3.73 | 18.08 | 23.50 |
| 1200 | Clear span tapered beam frame, 26 ga. colored roofing/siding | | | | | | | | | |
| 1300 | 30' to 39' wide, 10' eave height | E-2 | 535 | .105 | SF Flr. | 7.95 | 4.96 | 3.03 | 15.94 | 20.50 |
| 1400 | 14' eave height | | 450 | .124 | | 8.75 | 5.90 | 3.60 | 18.25 | 23.50 |
| 1500 | 16' eave height | | 415 | .135 | | 9.40 | 6.40 | 3.91 | 19.71 | 25.50 |
| 1600 | 20' eave height | | 360 | .156 | | 10.35 | 7.35 | 4.51 | 22.21 | 29 |
| 1700 | 40' wide, 10' eave height | | 600 | .093 | | 7.30 | 4.42 | 2.70 | 14.42 | 18.50 |
| 1800 | 14' eave height | | 510 | .110 | | 7.95 | 5.20 | 3.18 | 16.33 | 21 |
| 1900 | 16' eave height | | 475 | .118 | | 8.30 | 5.60 | 3.41 | 17.31 | 22.50 |
| 2000 | 20' eave height | | 415 | .135 | | 9.10 | 6.40 | 3.91 | 19.41 | 25 |
| 2100 | 50' to 79' wide, 10' eave height | | 770 | .073 | | 6.75 | 3.44 | 2.11 | 12.30 | 15.60 |
| 2200 | 14' eave height | | 675 | .083 | | 7.35 | 3.93 | 2.40 | 13.68 | 17.40 |
| 2300 | 16' eave height | | 635 | .088 | | 7.65 | 4.18 | 2.55 | 14.38 | 18.35 |
| 2400 | 20' eave height | | 490 | .114 | | 8.35 | 5.40 | 3.31 | 17.06 | 22.00 |
| 2410 | 80' to 100' wide, 10' eave height | | 935 | .060 | | 6.00 | 2.84 | 1.73 | 10.57 | 13.30 |
| 2420 | 14' eave height | | 750 | .075 | | 6.60 | 3.54 | 2.16 | 12.30 | 15.65 |
| 2430 | 16' eave height | | 685 | .082 | | 6.90 | 3.87 | 2.37 | 13.14 | 16.75 |
| 2440 | 20' eave height | | 560 | .100 | | 7.40 | 4.74 | 2.90 | 15.04 | 19.35 |
| 2460 | 101' to 120' wide, 10' eave height | | 950 | .059 | | 5.50 | 2.79 | 1.71 | 10.00 | 12.65 |
| 2470 | 14' eave height | | 770 | .073 | | 6.10 | 3.44 | 2.11 | 11.65 | 14.85 |
| 2480 | 16' eave height | | 675 | .083 | | 6.50 | 3.93 | 2.40 | 12.83 | 16.45 |
| 2490 | 20' eave height | | 560 | .100 | | 6.95 | 4.74 | 2.90 | 14.59 | 18.90 |
| 2500 | Single post 2-span frame, 26 ga. colored roofing and siding | | | | | | | | | |
| 2600 | 80' wide, 14' eave height | E-2 | 740 | .076 | SF Flr. | 5.95 | 3.58 | 2.19 | 11.72 | 15 |
| 2700 | 16' eave height | | 695 | .081 | | 6.30 | 3.82 | 2.33 | 12.45 | 15.95 |
| 2800 | 20' eave height | | 625 | .090 | | 6.90 | 4.24 | 2.59 | 13.73 | 17.65 |
| 2900 | 24' eave height | | 570 | .098 | | 7.45 | 4.65 | 2.85 | 14.95 | 19.25 |
| 3000 | 100' wide, 14' eave height | | 835 | .067 | | 5.75 | 3.18 | 1.94 | 10.87 | 13.90 |
| 3100 | 16' eave height | | 795 | .070 | | 5.70 | 3.34 | 2.04 | 11.08 | 14.15 |
| 3200 | 20' eave height | | 730 | .077 | | 6.65 | 3.63 | 2.22 | 12.50 | 15.90 |
| 3300 | 24' eave height | | 670 | .084 | | 7.15 | 3.96 | 2.42 | 13.53 | 17.25 |
| 3400 | 120' wide, 14' eave height | | 870 | .064 | | 5.70 | 3.05 | 1.86 | 10.61 | 13.50 |
| 3500 | 16' eave height | | 830 | .067 | | 5.95 | 3.20 | 1.95 | 11.10 | 14.10 |
| 3600 | 20' eave height | | 765 | .073 | | 6.45 | 3.47 | 2.12 | 12.04 | 15.30 |
| 3700 | 24' eave height | | 705 | .079 | | 7 | 3.76 | 2.30 | 13.06 | 16.65 |
| 3800 | Double post 3-span frame, 26 ga. colored roofing and siding | | | | | | | | | |
| 3900 | 150' wide, 14' eave height | E-2 | 925 | .061 | SF Flr. | 4.92 | 2.87 | 1.75 | 9.54 | 12.20 |
| 4000 | 16' eave height | | 890 | .063 | | 5.05 | 2.98 | 1.82 | 9.85 | 12.65 |
| 4100 | 20' eave height | | 820 | .068 | | 5.35 | 3.23 | 1.98 | 10.56 | 13.60 |
| 4200 | 24' eave height | | 765 | .073 | | 6.40 | 3.47 | 2.12 | 11.99 | 15.25 |
| 4300 | Triple post 4-span frame, 26 ga. colored roofing and siding | | | | | | | | | |
| 4400 | 160' wide, 14' eave height | E-2 | 970 | .058 | SF Flr. | 4.77 | 2.73 | 1.67 | 9.17 | 11.70 |
| 4500 | 16' eave height | | 930 | .060 | | 4.99 | 2.85 | 1.74 | 9.58 | 12.25 |
| 4600 | 20' eave height | | 870 | .064 | | 4.99 | 3.05 | 1.86 | 9.90 | 12.70 |
| 4700 | 24' eave height | | 815 | .069 | | 5.95 | 3.25 | 1.99 | 11.19 | 14.25 |
| 4800 | 200' wide, 14' eave height | | 1030 | .054 | | 4.58 | 2.58 | 1.57 | 8.73 | 11.15 |

## 13 34 19 – Metal Building Systems

| 13 34 19.50 Pre-Engineered Steel Buildings | Crew | Daily Output | Labor-Hours | Unit | Material | 2011 Bare Costs Labor | Equipment | Total | Total Incl O&P |
|---|---|---|---|---|---|---|---|---|---|
| 4900 | 16' eave height | E-2 | 995 | .056 | SF Flr. | 4.70 | 2.67 | 1.63 | 9 | 11.45 |
| 5000 | 20' eave height | | 935 | .060 | | 5.10 | 2.84 | 1.73 | 9.67 | 12.30 |
| 5100 | 24' eave height | ↓ | 885 | .063 | ↓ | 6 | 3 | 1.83 | 10.83 | 13.70 |
| 5200 | Accessory items: add to the basic building cost above | | | | | | | | | |
| 5250 | Eave overhang, 2' wide, 26 ga., with soffit | E-2 | 360 | .156 | L.F. | 25 | 7.35 | 4.51 | 36.86 | 45 |
| 5300 | 4' wide, without soffit | | 300 | .187 | | 24.50 | 8.85 | 5.40 | 38.75 | 48 |
| 5350 | With soffit | | 250 | .224 | | 34.50 | 10.60 | 6.50 | 51.60 | 63 |
| 5400 | 6' wide, without soffit | | 250 | .224 | | 33.50 | 10.60 | 6.50 | 50.60 | 62 |
| 5450 | With soffit | | 200 | .280 | ↓ | 43.50 | 13.25 | 8.10 | 64.85 | 79.50 |
| 5500 | Entrance canopy, incl. frame, 4' x 4' | | 25 | 2.240 | Ea. | 350 | 106 | 65 | 521 | 635 |
| 5550 | 4' x 8' | | 19 | 2.947 | " | 480 | 140 | 85.50 | 705.50 | 860 |
| 5600 | End wall roof overhang, 4' wide, without soffit | | 850 | .066 | L.F. | 16.40 | 3.12 | 1.91 | 21.43 | 25.50 |
| 5650 | With soffit | ↓ | 500 | .112 | " | 24 | 5.30 | 3.24 | 32.54 | 39 |
| 5700 | Doors, H.M. self-framing, incl. butts, lockset and trim | | | | | | | | | |
| 5750 | Single leaf, 3070 (3' x 7'), economy | 2 Sswk | 5 | 3.200 | Opng. | 560 | 155 | | 715 | 890 |
| 5800 | Deluxe | | 4 | 4 | | 610 | 194 | | 804 | 1,025 |
| 5825 | Glazed | | 4 | 4 | | 710 | 194 | | 904 | 1,125 |
| 5850 | 3670 (3'-6" x 7') | | 4 | 4 | | 780 | 194 | | 974 | 1,200 |
| 5900 | 4070 (4' x 7') | | 3 | 5.333 | | 845 | 259 | | 1,104 | 1,375 |
| 5950 | Double leaf, 6070 (6' x 7') | | 2 | 8 | | 1,050 | 390 | | 1,440 | 1,825 |
| 6000 | Glazed | | 2 | 8 | | 1,325 | 390 | | 1,715 | 2,150 |
| 6050 | Framing only, for openings, 3' x 7' | | 4 | 4 | | 157 | 194 | | 351 | 520 |
| 6100 | 10' x 10' | | 3 | 5.333 | | 515 | 259 | | 774 | 1,025 |
| 6150 | For windows below, 2020 (2' x 2') | | 6 | 2.667 | | 166 | 129 | | 295 | 410 |
| 6200 | 4030 (4' x 3') | | 5 | 3.200 | ↓ | 203 | 155 | | 358 | 495 |
| 6250 | Flashings, 26 ga., corner or eave, painted | | 240 | .067 | L.F. | 4.08 | 3.24 | | 7.32 | 10.20 |
| 6300 | Galvanized | | 240 | .067 | | 3.50 | 3.24 | | 6.74 | 9.55 |
| 6350 | Rake flashing, painted | | 240 | .067 | | 4.37 | 3.24 | | 7.61 | 10.50 |
| 6400 | Galvanized | | 240 | .067 | | 3.84 | 3.24 | | 7.08 | 9.90 |
| 6450 | Ridge flashing, 18" wide, painted | | 240 | .067 | | 5.35 | 3.24 | | 8.59 | 11.60 |
| 6500 | Galvanized | | 240 | .067 | | 4.95 | 3.24 | | 8.19 | 11.15 |
| 6550 | Gutter, eave type, 26 ga., painted | | 320 | .050 | | 5.60 | 2.43 | | 8.03 | 10.45 |
| 6600 | Galvanized | | 320 | .050 | | 3.54 | 2.43 | | 5.97 | 8.20 |
| 6650 | Valley type, between buildings, painted | ↓ | 120 | .133 | ↓ | 9.35 | 6.45 | | 15.80 | 22 |
| 6710 | Insulation, rated .6 lb density, unfaced 4" thick, R13 | 2 Carp | 2300 | .007 | S.F. | .30 | .30 | | .60 | .79 |
| 6720 | 6" thick, R-19 | | 2300 | .007 | | .42 | .30 | | .72 | .92 |
| 6730 | 10" thick, R-30 | ↓ | 2300 | .007 | ↓ | .72 | .30 | | 1.02 | 1.25 |
| 6750 | Insulation, rated .6 lb density, poly/scrim/foil (PSF) faced | | | | | | | | | |
| 6760 | 4" thick R-13 | 2 Carp | 2300 | .007 | S.F. | .44 | .30 | | .74 | .94 |
| 6770 | 6" thick, R-19 | | 2300 | .007 | | .58 | .30 | | .88 | 1.10 |
| 6780 | 10" thick, R-30 | | 2300 | .007 | | .78 | .30 | | 1.08 | 1.32 |
| 6800 | Insulation, rated .6 lb density, vinyl faced 1-1/2" thick, R5 | | 2300 | .007 | | .27 | .30 | | .57 | .76 |
| 6850 | 3" thick, R10 | | 2300 | .007 | | .29 | .30 | | .59 | .78 |
| 6900 | 4" thick, R13 | | 2300 | .007 | | .42 | .30 | | .72 | .92 |
| 6920 | 6" thick, R19 | | 2300 | .007 | | .59 | .30 | | .89 | 1.11 |
| 6930 | 10" thick, R30 | | 2300 | .007 | | .92 | .30 | | 1.22 | 1.47 |
| 6950 | Foil/scrim/kraft (FSK) faced, 1-1/2" thick, R5 | | 2300 | .007 | | .29 | .30 | | .59 | .78 |
| 7000 | 2" thick, R6 | | 2300 | .007 | | .37 | .30 | | .67 | .87 |
| 7050 | 3" thick, R10 | | 2300 | .007 | | .38 | .30 | | .68 | .88 |
| 7100 | 4" thick, R13 | | 2300 | .007 | | .47 | .30 | | .77 | .98 |
| 7110 | 6" thick, R19 | | 2300 | .007 | | .53 | .30 | | .83 | 1.04 |
| 7120 | 10" thick, R30 | | 2300 | .007 | | .78 | .30 | | 1.08 | 1.32 |
| 7150 | Met polyester/scrim/kraft (PSK) facing,1-1/2" thk, R5 | ↓ | 2300 | .007 | ↓ | .47 | .30 | | .77 | .98 |

## 13 34 19 – Metal Building Systems

| 13 34 19.50 Pre-Engineered Steel Buildings | Crew | Daily Output | Labor-Hours | Unit | Material | 2011 Bare Costs Labor | Equipment | Total | Total Incl O&P |
|---|---|---|---|---|---|---|---|---|---|
| 7200 | 2" thick, R6 | 2 Carp | 2300 | .007 | S.F. | .55 | .30 | | .85 | 1.07 |
| 7250 | 3" thick, R11 | | 2300 | .007 | | .52 | .30 | | .82 | 1.03 |
| 7300 | 4" thick, R13 | | 2300 | .007 | | .63 | .30 | | .93 | 1.15 |
| 7310 | 6" thick, R19 | | 2300 | .007 | | .77 | .30 | | 1.07 | 1.31 |
| 7320 | 10" thick, R30 | | 2300 | .007 | | .83 | .30 | | 1.13 | 1.37 |
| 7350 | Vinyl/scrim/foil (VSF), 1-1/2" thick, R5 | | 2300 | .007 | | .41 | .30 | | .71 | .91 |
| 7400 | 2" thick, R6 | | 2300 | .007 | | .52 | .30 | | .82 | 1.03 |
| 7450 | 3" thick, R10 | | 2300 | .007 | | .58 | .30 | | .88 | 1.10 |
| 7500 | 4" thick, R13 | | 2300 | .007 | | .67 | .30 | | .97 | 1.20 |
| 7510 | Vinyl/scrim/vinyl (VSV) 4" thick, R13 | | 2300 | .007 | | .47 | .30 | | .77 | .98 |
| 7520 | 6" thick, R19 | | 2300 | .007 | | .62 | .30 | | .92 | 1.14 |
| 7530 | 10" thick, R30 | | 2300 | .007 | | .83 | .30 | | 1.13 | 1.37 |
| 7540 | Polyprop/scrim/polyester (PSP), 4", R13 | | 2300 | .007 | | .40 | .30 | | .70 | .90 |
| 7550 | 6" thick, R19 | | 2300 | .007 | | .54 | .30 | | .84 | 1.05 |
| 7555 | 10" thick, R30 | | 2300 | .007 | | .99 | .30 | | 1.29 | 1.55 |
| 7560 | Polyprop/scrim/kraft/polyester (PSKP), 4" thick, R13 | | 2300 | .007 | | .42 | .30 | | .72 | .92 |
| 7570 | 6" thick, R19 | | 2300 | .007 | | .55 | .30 | | .85 | 1.07 |
| 7580 | 10" thick, R19 | | 2300 | .007 | | 1.05 | .30 | | 1.35 | 1.62 |
| 7585 | Vinyl/scrim/polyester (VSP), 4" thick, R13 | | 2300 | .007 | | .43 | .30 | | .73 | .93 |
| 7590 | 6" thick, R19 | | 2300 | .007 | | .57 | .30 | | .87 | 1.09 |
| 7600 | 10" thick, R30 | | 2300 | .007 | | 1.03 | .30 | | 1.33 | 1.59 |
| 7635 | Insulation installation, over the purlin, second layer, up to 4" thick, add | | | | | | 90% | | | |
| 7640 | Insulation installation, between the purlins, up to 4" thick, add | | | | | | 100% | | | |
| 7650 | Sash, single slide, glazed, with screens, 2020 (2' x 2') | E-1 | 22 | 1.091 | Opng. | 118 | 52 | 4.95 | 174.95 | 223 |
| 7700 | 3030 (3' x 3') | | 14 | 1.714 | | 266 | 81.50 | 7.80 | 355.30 | 440 |
| 7750 | 4030 (4' x 3') | | 13 | 1.846 | | 355 | 88 | 8.40 | 451.40 | 545 |
| 7800 | 6040 (6' x 4') | | 12 | 2 | | 710 | 95 | 9.10 | 814.10 | 950 |
| 7850 | Double slide sash, 3030 (3' x 3') | | 14 | 1.714 | | 207 | 81.50 | 7.80 | 296.30 | 375 |
| 7900 | 6040 (6' x 4') | | 12 | 2 | | 550 | 95 | 9.10 | 654.10 | 775 |
| 7950 | Fixed glass, no screens, 3030 (3' x 3') | | 14 | 1.714 | | 220 | 81.50 | 7.80 | 309.30 | 390 |
| 8000 | 6040 (6' x 4') | | 12 | 2 | | 585 | 95 | 9.10 | 689.10 | 815 |
| 8050 | Prefinished storm sash, 3030 (3' x 3') | | 70 | .343 | | 72 | 16.30 | 1.56 | 89.86 | 108 |
| 8100 | Siding and roofing, see Sections 07 41 13.00 & 07 42 13.00 | | | | | | | | | |
| 8200 | Skylight, fiberglass panels, to 30 S.F. | E-1 | 10 | 2.400 | Ea. | 103 | 114 | 10.90 | 227.90 | 320 |
| 8250 | Larger sizes, add for excess over 30 S.F. | " | 300 | .080 | S.F. | 3.44 | 3.81 | .36 | 7.61 | 10.60 |
| 8300 | Roof vents, circular with damper, birdscreen | | | | | | | | | |
| 8350 | and operator hardware, painted | | | | | | | | | |
| 8400 | 26 ga., 12" diameter | 1 Sswk | 4 | 2 | Ea. | 69 | 97 | | 166 | 248 |
| 8450 | 20" diameter | | 3 | 2.667 | | 142 | 129 | | 271 | 385 |
| 8500 | 24 ga., 24" diameter | | 2 | 4 | | 282 | 194 | | 476 | 655 |
| 8550 | Galvanized | | 2 | 4 | | 236 | 194 | | 430 | 605 |
| 8600 | Continuous, 26 ga., 10' long, 9" wide | 2 Sswk | 4 | 4 | | 20 | 194 | | 214 | 365 |
| 8650 | 12" wide | " | 4 | 4 | | 24 | 194 | | 218 | 370 |

## 13 34 23 – Fabricated Structures

### 13 34 23.10 Comfort Stations

| | | Crew | Daily Output | Labor-Hours | Unit | Material | Labor | Equipment | Total | Total Incl O&P |
|---|---|---|---|---|---|---|---|---|---|---|
| 0010 | **COMFORT STATIONS** Prefab., stock, w/doors, windows & fixt. | | | | | | | | | |
| 0100 | Not incl. interior finish or electrical | | | | | | | | | |
| 0300 | Mobile, on steel frame, minimum | | | | S.F. | 46.50 | | | 46.50 | 51 |
| 0350 | Maximum | | | | | 75 | | | 75 | 82.50 |
| 0400 | Permanent, including concrete slab, minimum | B-12J | 50 | .320 | | 199 | 12.95 | 17.15 | 229.10 | 257 |
| 0500 | Maximum | " | 43 | .372 | | 288 | 15.05 | 19.95 | 323 | 360 |
| 0600 | Alternate pricing method, mobile, minimum | | | | Fixture | 2,150 | | | 2,150 | 2,375 |

# 13 34 Fabricated Engineered Structures

## 13 34 23 – Fabricated Structures

| 13 34 23.10 Comfort Stations | Crew | Daily Output | Labor-Hours | Unit | Material | 2011 Bare Costs Labor | Equipment | Total | Total Incl O&P |
|---|---|---|---|---|---|---|---|---|---|
| 0650     Maximum | | | | Fixture | 3,175 | | | 3,175 | 3,500 |
| 0700     Permanent, minimum | B-12J | .70 | 22.857 | | 12,200 | 925 | 1,225 | 14,350 | 16,300 |
| 0750     Maximum | " | .50 | 32 | ↓ | 20,400 | 1,300 | 1,725 | 23,425 | 26,200 |

### 13 34 23.15 Domes

| | Crew | Daily Output | Labor-Hours | Unit | Material | Labor | Equipment | Total | Total Incl O&P |
|---|---|---|---|---|---|---|---|---|---|
| 0010 **DOMES** | | | | | | | | | |
| 0020   Domes, rev alum, elec drive, for astronomy obsv shell only, stock units | | | | | | | | | |
| 0600     10'-0" diameter, 800#, dome | 2 Carp | .25 | 64 | Ea. | 9,875 | 2,750 | | 12,625 | 15,100 |
| 0700       Base | | .67 | 23.881 | | 3,425 | 1,025 | | 4,450 | 5,325 |
| 0900     18'-0" diameter, 2,500#, dome | | .17 | 94.118 | | 28,900 | 4,050 | | 32,950 | 38,000 |
| 1000       Base | | .33 | 48.485 | | 9,575 | 2,075 | | 11,650 | 13,700 |
| 1200     24'-0" diameter, 4,500#, dome | | .08 | 200 | | 53,500 | 8,600 | | 62,100 | 72,000 |
| 1300       Base | ↓ | .25 | 64 | ↓ | 17,200 | 2,750 | | 19,950 | 23,200 |
| 1500   Domes, bulk storage, shell only, dual radius hemisphere, arch, steel | | | | | | | | | |
| 1600     framing, corrugated steel covering, 150' diameter | E-2 | 550 | .102 | SF Flr. | 28 | 4.82 | 2.95 | 35.77 | 42 |
| 1700       400' diameter | " | 720 | .078 | | 22.50 | 3.68 | 2.25 | 28.43 | 33.50 |
| 1800     Wood framing, wood decking, to 400' diameter | F-4 | 400 | .120 | ↓ | 20.50 | 5.20 | 3.02 | 28.72 | 33.50 |
| 1900   Radial framed wood (2" x 6"), 1/2" thick | | | | | | | | | |
| 2000     plywood, asphalt shingles, 50' diameter | F-3 | 2000 | .020 | SF Flr. | 40.50 | .87 | .33 | 41.70 | 46 |
| 2100       60' diameter | | 1900 | .021 | | 31 | .92 | .35 | 32.27 | 36 |
| 2200       72' diameter | | 1800 | .022 | | 27.50 | .97 | .36 | 28.83 | 32.50 |
| 2300       116' diameter | | 1730 | .023 | | 24.50 | 1.01 | .38 | 25.89 | 29 |
| 2400       150' diameter | ↓ | 1500 | .027 | ↓ | 21 | 1.17 | .44 | 22.61 | 25.50 |

### 13 34 23.16 Fabricated Control Booths

| | Crew | Daily Output | Labor-Hours | Unit | Material | Labor | Equipment | Total | Total Incl O&P |
|---|---|---|---|---|---|---|---|---|---|
| 0010 **FABRICATED CONTROL BOOTHS** | | | | | | | | | |
| 0100   Guard House, prefab conc w/bullet resistant doors & windows, roof & wiring | | | | | | | | | |
| 0110     8' x 8', Level III | L-10 | 1 | 24 | Ea. | 24,200 | 1,175 | 655 | 26,030 | 29,300 |
| 0120     8' x 8', Level IV | " | 1 | 24 | " | 30,300 | 1,175 | 655 | 32,130 | 36,000 |

### 13 34 23.25 Garage Costs

| | Crew | Daily Output | Labor-Hours | Unit | Material | Labor | Equipment | Total | Total Incl O&P |
|---|---|---|---|---|---|---|---|---|---|
| 0010 **GARAGE COSTS** | | | | | | | | | |
| 0020   Public parking, average | | | | Car | | | | 16,600 | 18,300 |
| 0100     See also Square Foot Costs in Reference Section | | | | | | | | | |
| 0300   Residential, prefab shell, stock, wood, single car, minimum | 2 Carp | 1 | 16 | Total | 5,100 | 690 | | 5,790 | 6,650 |
| 0350     Maximum | | .67 | 23.881 | | 5,875 | 1,025 | | 6,900 | 8,050 |
| 0400   Two car, minimum | | .67 | 23.881 | | 10,200 | 1,025 | | 11,225 | 12,800 |
| 0450     Maximum | ↓ | .50 | 32 | ↓ | 11,800 | 1,375 | | 13,175 | 15,000 |

### 13 34 23.30 Garden House

| | Crew | Daily Output | Labor-Hours | Unit | Material | Labor | Equipment | Total | Total Incl O&P |
|---|---|---|---|---|---|---|---|---|---|
| 0010 **GARDEN HOUSE** Prefab wood, no floors or foundations | | | | | | | | | |
| 0100   32 to 200 S.F., minimum | 2 Carp | 200 | .080 | SF Flr. | 27.50 | 3.44 | | 30.94 | 35.50 |
| 0300     Maximum | " | 48 | .333 | " | 47 | 14.35 | | 61.35 | 73.50 |

### 13 34 23.35 Geodesic Domes

| | Crew | Daily Output | Labor-Hours | Unit | Material | Labor | Equipment | Total | Total Incl O&P |
|---|---|---|---|---|---|---|---|---|---|
| 0010 **GEODESIC DOMES** Shell only, interlocking plywood panels   R133423-30 | | | | | | | | | |
| 0400     30' diameter | F-5 | 1.60 | 20 | Ea. | 17,700 | 870 | | 18,570 | 20,800 |
| 0500     34' diameter | | 1.14 | 28.070 | | 21,800 | 1,225 | | 23,025 | 25,800 |
| 0600     39' diameter | ↓ | 1 | 32 | | 25,300 | 1,400 | | 26,700 | 30,100 |
| 0700     45' diameter | F-3 | 1.13 | 35.556 | | 30,500 | 1,550 | 585 | 32,635 | 36,600 |
| 0750     55' diameter | | 1 | 40 | | 45,400 | 1,750 | 655 | 47,805 | 53,500 |
| 0800     60' diameter | | 1 | 40 | | 59,000 | 1,750 | 655 | 61,405 | 68,500 |
| 0850     65' diameter | ↓ | .80 | 50 | ↓ | 70,000 | 2,175 | 820 | 72,995 | 81,000 |
| 1100   Aluminum panel, with 6" insulation | | | | | | | | | |
| 1200     100' diameter | | | | SF Flr. | 37.50 | | | 37.50 | 41 |
| 1300     500' diameter | | | | " | 32 | | | 32 | 35.50 |

# 13 34 Fabricated Engineered Structures

## 13 34 23 – Fabricated Structures

| 13 34 23.35 Geodesic Domes | | Crew | Daily Output | Labor-Hours | Unit | Material | 2011 Bare Costs Labor | Equipment | Total | Total Incl O&P |
|---|---|---|---|---|---|---|---|---|---|---|
| 1600 | Aluminum framed, plexiglass closure panels | | | | | 95 | | | 95 | 104 |
| 1700 | 40' diameter | | | | SF Flr. | 95 | | | 95 | 104 |
| 1800 | 200' diameter | | | | " | 75.50 | | | 75.50 | 83.50 |
| 2100 | Aluminum framed, aluminum closure panels | | | | | | | | | |
| 2200 | 40' diameter | | | | SF Flr. | 30 | | | 30 | 33 |
| 2300 | 100' diameter | | | | | 22.50 | | | 22.50 | 24.50 |
| 2400 | 200' diameter | | | | | 21.50 | | | 21.50 | 23.50 |
| 2500 | For VRP faced bonded fiberglass insulation, add | | | | | | | | | 10 |
| 2700 | Aluminum framed, fiberglass sandwich panel closure | | | | | | | | | |
| 2800 | 6' diameter | 2 Carp | 150 | .107 | SF Flr. | 33.50 | 4.59 | | 38.09 | 43.50 |
| 2900 | 28' diameter | " | 350 | .046 | " | 30.50 | 1.97 | | 32.47 | 36.50 |

## 13 34 23.45 Kiosks

| | 13 34 23.45 Kiosks | Crew | Daily Output | Labor-Hours | Unit | Material | Labor | Equipment | Total | Total Incl O&P |
|---|---|---|---|---|---|---|---|---|---|---|
| 0010 | **KIOSKS** | | | | | | | | | |
| 0020 | Round, 5' diameter, 8' high, 1/4" fiberglass wall | | | | Ea. | 5,800 | | | 5,800 | 6,375 |
| 0100 | 1" insulated double wall, fiberglass | | | | | 6,600 | | | 6,600 | 7,250 |
| 0500 | Rectangular, 5' x 9', 7'-6" high, 1/4" fiberglass wall | | | | | 8,450 | | | 8,450 | 9,300 |
| 0600 | 1" insulated double wall, fiberglass | | | | | 10,000 | | | 10,000 | 11,000 |

## 13 34 23.60 Portable Booths

| | 13 34 23.60 Portable Booths | Crew | Daily Output | Labor-Hours | Unit | Material | Labor | Equipment | Total | Total Incl O&P |
|---|---|---|---|---|---|---|---|---|---|---|
| 0010 | **PORTABLE BOOTHS** Prefab aluminum with doors, windows, ext. roof | | | | | | | | | |
| 0100 | lights wiring & insulation, 15 S.F. building, O.D., painted, minimum | | | | S.F. | 300 | | | 300 | 330 |
| 0300 | 30 S.F. building, minimum | | | | | 254 | | | 254 | 279 |
| 0400 | 50 S.F. building, minimum | | | | | 189 | | | 189 | 208 |
| 0600 | 80 S.F. building, minimum | | | | | 154 | | | 154 | 169 |
| 0700 | 100 S.F. building, minimum | | | | | 144 | | | 144 | 158 |
| 0900 | Acoustical booth, 27 Db @ 1,000 Hz, 15 S.F. floor | | | | Ea. | 4,350 | | | 4,350 | 4,775 |
| 1000 | 7' x 7'-6", including light & ventilation | | | | | 8,900 | | | 8,900 | 9,800 |
| 1200 | Ticket booth, galv. steel, not incl. foundations., 4' x 4' | | | | | 5,625 | | | 5,625 | 6,200 |
| 1300 | 4' x 6' | | | | | 7,225 | | | 7,225 | 7,950 |

## 13 34 23.70 Shelters

| | 13 34 23.70 Shelters | Crew | Daily Output | Labor-Hours | Unit | Material | Labor | Equipment | Total | Total Incl O&P |
|---|---|---|---|---|---|---|---|---|---|---|
| 0010 | **SHELTERS** | | | | | | | | | |
| 0020 | Aluminum frame, acrylic glazing, 3' x 9' x 8' high | 2 Sswk | 1.14 | 14.035 | Ea. | 2,250 | 680 | | 2,930 | 3,675 |
| 0100 | 9' x 12' x 8' high | " | .73 | 21.918 | " | 4,775 | 1,075 | | 5,850 | 7,150 |

## 13 34 43 – Aircraft Hangars

### 13 34 43.50 Hangars

| | 13 34 43.50 Hangars | Crew | Daily Output | Labor-Hours | Unit | Material | Labor | Equipment | Total | Total Incl O&P |
|---|---|---|---|---|---|---|---|---|---|---|
| 0010 | **HANGARS** Prefabricated steel T hangars, Galv. steel roof & | | | | | | | | | |
| 0100 | walls, incl. electric bi-folding doors, 4 or more units, | | | | | | | | | |
| 0110 | not including floors or foundations, minimum | E-2 | 1275 | .044 | SF Flr. | 13.20 | 2.08 | 1.27 | 16.55 | 19.50 |
| 0130 | Maximum | | 1063 | .053 | | 14.65 | 2.49 | 1.53 | 18.67 | 22 |
| 0900 | With bottom rolling doors, minimum | | 1386 | .040 | | 12.85 | 1.91 | 1.17 | 15.93 | 18.70 |
| 1000 | Maximum | | 966 | .058 | | 14.25 | 2.75 | 1.68 | 18.68 | 22 |
| 1200 | Alternate pricing method: | | | | | | | | | |
| 1300 | Galv. roof and walls, electric bi-folding doors, minimum | E-2 | 1.06 | 52.830 | Plane | 17,100 | 2,500 | 1,525 | 21,125 | 24,700 |
| 1500 | Maximum | | .91 | 61.538 | | 20,400 | 2,925 | 1,775 | 25,100 | 29,400 |
| 1600 | With bottom rolling doors, minimum | | 1.25 | 44.800 | | 12,900 | 2,125 | 1,300 | 16,325 | 19,200 |
| 1800 | Maximum | | .97 | 57.732 | | 15,000 | 2,725 | 1,675 | 19,400 | 23,000 |
| 2000 | Circular type, prefab., steel frame, plastic skin, electric | | | | | | | | | |
| 2010 | door, including foundations, 80' diameter, | | | | | | | | | |
| 2020 | for up to 5 light planes, minimum | E-2 | .50 | 112 | Total | 72,500 | 5,300 | 3,250 | 81,050 | 92,500 |
| 2200 | Maximum | " | .25 | 224 | " | 82,000 | 10,600 | 6,475 | 99,075 | 115,000 |

# 13 34 Fabricated Engineered Structures

## 13 34 53 – Agricultural Structures

### 13 34 53.50 Silos

| | | Crew | Daily Output | Labor-Hours | Unit | Material | 2011 Bare Costs Labor | Equipment | Total | Total Incl O&P |
|---|---|---|---|---|---|---|---|---|---|---|
| 0010 | **SILOS** Concrete stave industrial, conical or sloping bottom | | | | | | | | | |
| 0100 | excluding foundation, 12' diameter, 35' high | E-24 | .09 | 355 | Ea. | 42,400 | 17,000 | 8,325 | 67,725 | 84,500 |
| 0200 | 16' diameter, 45' high | | .06 | 533 | | 69,000 | 25,500 | 12,500 | 107,000 | 133,000 |
| 0400 | 25' diameter, 75' high | ↓ | .04 | 800 | | 159,000 | 38,200 | 18,700 | 215,900 | 260,500 |
| 0500 | Steel, factory fab., 30,000 gallon cap., painted, minimum | L-5 | 1 | 56 | | 21,000 | 2,725 | 750 | 24,475 | 28,600 |
| 0700 | Maximum | | .50 | 112 | | 33,400 | 5,425 | 1,500 | 40,325 | 47,900 |
| 0800 | Epoxy lined, minimum | | 1 | 56 | | 34,300 | 2,725 | 750 | 37,775 | 43,300 |
| 1000 | Maximum | ↓ | .50 | 112 | ↓ | 43,600 | 5,425 | 1,500 | 50,525 | 59,000 |

## 13 34 63 – Natural Fiber Construction

### 13 34 63.50 Straw Bale Construction

| | | | Crew | Daily Output | Labor-Hours | Unit | Material | 2011 Bare Costs Labor | Equipment | Total | Total Incl O&P |
|---|---|---|---|---|---|---|---|---|---|---|---|
| 0010 | **STRAW BALE CONSTRUCTION** | | | | | | | | | | |
| 2000 | Straw bales wall, incl labor & material complete, minimum | G | | | | SF Flr. | | | | | 130 |
| 2010 | Maximum | G | | | | " | | | | | 160 |
| 2020 | Straw bales in walls w/modified post and beam frame | G | 2 Carp | 320 | .050 | S.F. | 4.10 | 2.15 | | 6.25 | 7.80 |

# 13 36 Towers

## 13 36 13 – Metal Towers

### 13 36 13.50 Control Towers

| | | Crew | Daily Output | Labor-Hours | Unit | Material | 2011 Bare Costs Labor | Equipment | Total | Total Incl O&P |
|---|---|---|---|---|---|---|---|---|---|---|
| 0010 | **CONTROL TOWERS** | | | | | | | | | |
| 0020 | Modular 12' x 10', incl. instruments, min. | | | | Ea. | 670,000 | | | 670,000 | 737,000 |
| 0100 | Maximum | | | | | | | | 927,500 | 1,020,000 |
| 0500 | With standard 40' tower, average | | | | ↓ | 935,000 | | | 935,000 | 1,028,500 |
| 1000 | Temporary portable control towers, 8' x 12', | | | | | | | | | |
| 1010 | complete with one position communications, minimum | | | | Ea. | | | | 266,000 | 293,000 |
| 2000 | For fixed facilities, depending on height, minimum | | | | | | | | 57,000 | 63,000 |
| 2010 | Maximum | | | | ↓ | | | | 115,000 | 127,000 |

# 13 42 Building Modules

## 13 42 63 – Detention Cell Modules

### 13 42 63.16 Steel Detention Cell Modules

| | | Crew | Daily Output | Labor-Hours | Unit | Material | 2011 Bare Costs Labor | Equipment | Total | Total Incl O&P |
|---|---|---|---|---|---|---|---|---|---|---|
| 0010 | **STEEL DETENTION CELL MODULES** | | | | | | | | | |
| 2000 | Cells, prefab., 5' to 6' wide, 7' to 8' high, 7' to 8' deep, | | | | | | | | | |
| 2010 | bar front, cot, not incl. plumbing | E-4 | 1.50 | 21.333 | Ea. | 9,650 | 1,050 | 73 | 10,773 | 12,500 |

# 13 48 Sound, Vibration, and Seismic Control

## 13 48 13 – Manufactured Sound and Vibration Control Components

### 13 48 13.50 Audio Masking

| | | Crew | Daily Output | Labor-Hours | Unit | Material | 2011 Bare Costs Labor | Equipment | Total | Total Incl O&P |
|---|---|---|---|---|---|---|---|---|---|---|
| 0010 | **AUDIO MASKING**, acoustical enclosure, 4" thick wall and ceiling | | | | | | | | | |
| 0020 | 8# per S.F., up to 12' span | 3 Carp | 72 | .333 | SF Surf | 32 | 14.35 | | 46.35 | 57.50 |
| 0300 | Better quality panels, 10.5# per S.F. | | 64 | .375 | | 36.50 | 16.15 | | 52.65 | 65 |
| 0400 | Reverb-chamber, 4" thick, parallel walls | | 60 | .400 | | 45.50 | 17.20 | | 62.70 | 76.50 |
| 0600 | Skewed wall, parallel roof, 4" thick panels | | 55 | .436 | | 52 | 18.80 | | 70.80 | 86 |
| 0700 | Skewed walls, skewed roof, 4" layers, 4" air space | | 48 | .500 | | 58.50 | 21.50 | | 80 | 97.50 |
| 0900 | Sound-absorbing panels, pntd mtl, 2'-6" x 8', under 1,000 S.F. | | 215 | .112 | | 12.10 | 4.81 | | 16.91 | 20.50 |
| 1100 | Over 1000 S.F. | | 240 | .100 | | 11.65 | 4.31 | | 15.96 | 19.40 |
| 1200 | Fabric faced | ↓ | 240 | .100 | | 9.45 | 4.31 | | 13.76 | 16.95 |

# 13 48 Sound, Vibration, and Seismic Control

## 13 48 13 – Manufactured Sound and Vibration Control Components

| 13 48 13.50 Audio Masking | Crew | Daily Output | Labor-Hours | Unit | Material | 2011 Bare Costs Labor | Equipment | Total | Total Incl O&P |
|---|---|---|---|---|---|---|---|---|---|
| 1500 Flexible transparent curtain, clear | 3 Shee | 215 | .112 | SF Surf | 7.30 | 5.75 | | 13.05 | 16.70 |
| 1600    50% foam | | 215 | .112 | | 10.20 | 5.75 | | 15.95 | 19.90 |
| 1700    75% foam | | 215 | .112 | | 10.20 | 5.75 | | 15.95 | 19.90 |
| 1800    100% foam | ↓ | 215 | .112 | ↓ | 10.20 | 5.75 | | 15.95 | 19.90 |
| 3100 Audio masking system, including speakers, amplification | | | | | | | | | |
| 3110    and signal generator | | | | | | | | | |
| 3200    Ceiling mounted, 5,000 S.F. | 2 Elec | 2400 | .007 | S.F. | 1.25 | .34 | | 1.59 | 1.88 |
| 3300    10,000 S.F. | | 2800 | .006 | | 1.01 | .29 | | 1.30 | 1.54 |
| 3400    Plenum mounted, 5,000 S.F. | | 3800 | .004 | | 1.06 | .21 | | 1.27 | 1.48 |
| 3500    10,000 S.F. | ↓ | 4400 | .004 | ↓ | .72 | .18 | | .90 | 1.06 |

# 13 49 Radiation Protection

## 13 49 13 – Integrated X-Ray Shielding Assemblies

### 13 49 13.50 Lead Sheets

| | | Crew | Daily Output | Labor-Hours | Unit | Material | Labor | Equipment | Total | Total Incl O&P |
|---|---|---|---|---|---|---|---|---|---|---|
| 0010 | **LEAD SHEETS** | | | | | | | | | |
| 0300 | Lead sheets, 1/16" thick | 2 Lath | 135 | .119 | S.F. | 7.15 | 4.57 | | 11.72 | 14.55 |
| 0400 |   1/8" thick | | 120 | .133 | | 14.45 | 5.15 | | 19.60 | 23.50 |
| 0500 | Lead shielding, 1/4" thick | | 135 | .119 | | 34 | 4.57 | | 38.57 | 44 |
| 0550 |   1/2" thick | ↓ | 120 | .133 | ↓ | 59.50 | 5.15 | | 64.65 | 73 |
| 0950 | Lead headed nails (average 1 lb. per sheet) | | | | Lb. | 7.15 | | | 7.15 | 7.90 |
| 1000 | Butt joints in 1/8" lead or thicker, 2" batten strip x 7' long | 2 Lath | 240 | .067 | Ea. | 16.95 | 2.57 | | 19.52 | 22.50 |
| 1200 | X-ray protection, average radiography or fluoroscopy | | | | | | | | | |
| 1210 |   room, up to 300 S.F. floor, 1/16" lead, minimum | 2 Lath | .25 | 64 | Total | 8,075 | 2,475 | | 10,550 | 12,500 |
| 1500 |     Maximum, 7'-0" walls | " | .15 | 106 | " | 9,700 | 4,100 | | 13,800 | 16,700 |
| 1600 | Deep therapy X-ray room, 250 kV capacity, | | | | | | | | | |
| 1800 |   up to 300 S.F. floor, 1/4" lead, minimum | 2 Lath | .08 | 200 | Total | 22,600 | 7,700 | | 30,300 | 36,100 |
| 1900 |     Maximum, 7'-0" walls | " | .06 | 266 | " | 27,800 | 10,300 | | 38,100 | 45,600 |

## 13 49 19 – Lead-Lined Materials

### 13 49 19.50 Shielding Lead

| | | Crew | Daily Output | Labor-Hours | Unit | Material | Labor | Equipment | Total | Total Incl O&P |
|---|---|---|---|---|---|---|---|---|---|---|
| 0010 | **SHIELDING LEAD** | | | | | | | | | |
| 0100 | Laminated lead in wood doors, 1/16" thick, no hardware | | | | S.F. | 50 | | | 50 | 55 |
| 0200 | Lead lined door frame, not incl. hardware, | | | | | | | | | |
| 0210 |   1/16" thick lead, butt prepared for hardware | 1 Lath | 2.40 | 3.333 | Ea. | 575 | 129 | | 704 | 820 |
| 0850 |   Window frame with 1/16" lead and voice passage, 36" x 60" | 2 Glaz | 2 | 8 | | 1,275 | 335 | | 1,610 | 1,900 |
| 0870 |     24" x 36" frame | | 4 | 4 | ↓ | 1,025 | 168 | | 1,193 | 1,375 |
| 0900 | Lead gypsum board, 5/8" thick with 1/16" lead | | 160 | .100 | S.F. | 7.10 | 4.21 | | 11.31 | 14.15 |
| 0910 |   1/8" lead | | 140 | .114 | | 14.45 | 4.81 | | 19.26 | 23 |
| 0930 |   1/32" lead | 2 Lath | 200 | .080 | ↓ | 4.43 | 3.08 | | 7.51 | 9.35 |

## 13 49 21 – Lead Glazing

### 13 49 21.50 Lead Glazing

| | | Crew | Daily Output | Labor-Hours | Unit | Material | Labor | Equipment | Total | Total Incl O&P |
|---|---|---|---|---|---|---|---|---|---|---|
| 0010 | **LEAD GLAZING** | | | | | | | | | |
| 0600 | Lead glass, 1/4" thick, 2.0 mm LE, 12" x 16" | 2 Glaz | 13 | 1.231 | Ea. | 265 | 52 | | 317 | 370 |
| 0700 |   24" x 36" | | 8 | 2 | | 980 | 84 | | 1,064 | 1,200 |
| 0800 |   36" x 60" | ↓ | 2 | 8 | ↓ | 4,125 | 335 | | 4,460 | 5,025 |
| 2000 | X-ray viewing panels, clear lead plastic | | | | | | | | | |
| 2010 |   7 mm thick, 0.3 mm LE, 2.3 lb./S.F. | H-3 | 139 | .115 | S.F. | 179 | 4.30 | | 183.30 | 204 |
| 2020 |   12 mm thick, 0.5 mm LE, 3.9 lb./S.F. | | 82 | .195 | | 242 | 7.30 | | 249.30 | 277 |
| 2030 |   18 mm thick, 0.8 mm LE, 5.9 lb./S.F. | | 54 | .296 | | 263 | 11.05 | | 274.05 | 305 |
| 2040 |   22 mm thick, 1.0 mm LE, 7.2 lb./S.F. | | 44 | .364 | | 269 | 13.60 | | 282.60 | 315 |
| 2050 |   35 mm thick, 1.5 mm LE, 11.5 lb./S.F. | ↓ | 28 | .571 | ↓ | 300 | 21.50 | | 321.50 | 365 |

432

# 13 49 Radiation Protection

## 13 49 21 – Lead Glazing

| 13 49 21.50 Lead Glazing | Crew | Daily Output | Labor-Hours | Unit | Material | 2011 Bare Costs Labor | Equipment | Total | Total Incl O&P |
|---|---|---|---|---|---|---|---|---|---|
| 2060 | 46 mm thick, 2.0 mm LE, 15.0 lb./S.F. | H-3 | 21 | .762 | S.F. | 395 | 28.50 | | 423.50 | 480 |
| 2090 | For panels 12 S.F. to 48 S.F., add crating charge | | | | Ea. | | | | | 50 |

## 13 49 23 – Integrated RFI/EMI Shielding Assemblies

### 13 49 23.50 Modular Shielding Partitions

| | 13 49 23.50 Modular Shielding Partitions | Crew | Daily Output | Labor-Hours | Unit | Material | Labor | Equipment | Total | Total Incl O&P |
|---|---|---|---|---|---|---|---|---|---|---|
| 0010 | **MODULAR SHIELDING PARTITIONS** | | | | | | | | | |
| 4000 | X-ray barriers, modular, panels mounted within framework for | | | | | | | | | |
| 4002 | attaching to floor, wall or ceiling, upper portion is clear lead | | | | | | | | | |
| 4005 | plastic window panels 48"H, lower portion is opaque leaded | | | | | | | | | |
| 4008 | steel panels 36"H, structural supports not incl. | | | | | | | | | |
| 4010 | 1-section barrier, 36"W x 84"H overall | | | | | | | | | |
| 4020 | 0.5 mm LE panels | H-3 | 6.40 | 2.500 | Ea. | 5,650 | 93.50 | | 5,743.50 | 6,375 |
| 4030 | 0.8 mm LE panels | | 6.40 | 2.500 | | 6,025 | 93.50 | | 6,118.50 | 6,775 |
| 4040 | 1.0 mm LE panels | | 5.33 | 3.002 | | 6,150 | 112 | | 6,262 | 6,925 |
| 4050 | 1.5 mm LE panels | | 5.33 | 3.002 | | 6,550 | 112 | | 6,662 | 7,375 |
| 4060 | 2-section barrier, 72"W x 84"H overall | | | | | | | | | |
| 4070 | 0.5 mm LE panels | H-3 | 4 | 4 | Ea. | 11,600 | 149 | | 11,749 | 12,900 |
| 4080 | 0.8 mm LE panels | | 4 | 4 | | 12,200 | 149 | | 12,349 | 13,600 |
| 4090 | 1.0 mm LE panels | | 3.56 | 4.494 | | 12,500 | 168 | | 12,668 | 14,000 |
| 5000 | 1.5 mm LE panels | | 3.20 | 5 | | 14,200 | 187 | | 14,387 | 15,900 |
| 5010 | 3-section barrier, 108"W x 84"H overall | | | | | | | | | |
| 5020 | 0.5 mm LE panels | H-3 | 3.20 | 5 | Ea. | 17,400 | 187 | | 17,587 | 19,400 |
| 5030 | 0.8 mm LE panels | | 3.20 | 5 | | 18,300 | 187 | | 18,487 | 20,400 |
| 5040 | 1.0 mm LE panels | | 2.67 | 5.993 | | 18,600 | 224 | | 18,824 | 20,800 |
| 5050 | 1.5 mm LE panels | | 2.46 | 6.504 | | 20,200 | 243 | | 20,443 | 22,700 |
| 7000 | X-ray barriers, mobile, mounted within framework w/casters on | | | | | | | | | |
| 7005 | bottom, clear lead plastic window panels on upper portion, | | | | | | | | | |
| 7010 | opaque on lower, 30"W x 75"H overall, incl. framework | | | | | | | | | |
| 7020 | 24"H upper w/0.5 mm LE, 48"H lower w/0.8 mm LE | 1 Carp | 16 | .500 | Ea. | 3,025 | 21.50 | | 3,046.50 | 3,350 |
| 7030 | 48"W x 75"H overall, incl. framework | | | | | | | | | |
| 7040 | 36"H upper w/0.5 mm LE, 36"H lower w/0.8 mm LE | 1 Carp | 16 | .500 | Ea. | 5,525 | 21.50 | | 5,546.50 | 6,100 |
| 7050 | 36"H upper w/1.0 mm LE, 36"H lower w/1.5 mm LE | " | 16 | .500 | " | 6,550 | 21.50 | | 6,571.50 | 7,250 |
| 7060 | 72"W x 75"H overall, incl. framework | | | | | | | | | |
| 7070 | 36"H upper w/0.5 mm LE, 36"H lower w/0.8 mm LE | 1 Carp | 16 | .500 | Ea. | 6,550 | 21.50 | | 6,571.50 | 7,225 |
| 7080 | 36"H upper w/1.0 mm LE, 36"H lower w/1.5 mm LE | " | 16 | .500 | " | 8,175 | 21.50 | | 8,196.50 | 9,025 |

## 13 49 33 – Radio Frequency Shielding

### 13 49 33.50 Shielding, Radio Frequency

| | 13 49 33.50 Shielding, Radio Frequency | Crew | Daily Output | Labor-Hours | Unit | Material | Labor | Equipment | Total | Total Incl O&P |
|---|---|---|---|---|---|---|---|---|---|---|
| 0010 | **SHIELDING, RADIO FREQUENCY** | | | | | | | | | |
| 0020 | Prefabricated or screen-type copper or steel, minimum | 2 Carp | 180 | .089 | SF Surf | 30 | 3.83 | | 33.83 | 39.50 |
| 0100 | Average | | 155 | .103 | | 32 | 4.44 | | 36.44 | 42.50 |
| 0150 | Maximum | | 145 | .110 | | 34 | 4.75 | | 38.75 | 45 |

# 13 53 Meteorological Instrumentation

## 13 53 09 – Weather Instrumentation

| 13 53 09.50 Weather Station | Crew | Daily Output | Labor-Hours | Unit | Material | 2011 Bare Costs Labor | Equipment | Total | Total Incl O&P |
|---|---|---|---|---|---|---|---|---|---|
| 0010 **WEATHER STATION** | | | | | | | | | |
| 0020     Remote recording, solar powered, with rain gauge & display, 400 ft range | | | | Ea. | 1,375 | | | 1,375 | 1,500 |
| 0100         1 mile range | | | | " | 1,650 | | | 1,650 | 1,800 |

# Estimating Tips

## General

- Many products in Division 14 will require some type of support or blocking for installation not included with the item itself. Examples are supports for conveyors or tube systems, attachment points for lifts, and footings for hoists or cranes. Add these supports in the appropriate division.

## 14 10 00 Dumbwaiters
## 14 20 00 Elevators

- Dumbwaiters and elevators are estimated and purchased in a method similar to buying a car. The manufacturer has a base unit with standard features. Added to this base unit price will be whatever options the owner or specifications require. Increased load capacity, additional vertical travel, additional stops, higher speed, and cab finish options are items to be considered. When developing an estimate for dumbwaiters and elevators, remember that some items needed by the installers may have to be included as part of the general contract.

Examples are:
— shaftway
— rail support brackets
— machine room
— electrical supply
— sill angles
— electrical connections
— pits
— roof penthouses
— pit ladders

Check the job specifications and drawings before pricing.

- Installation of elevators and handicapped lifts in historic structures can require significant additional costs. The associated structural requirements may involve cutting into and repairing finishes, mouldings, flooring, etc. The estimator must account for these special conditions.

## 14 30 00 Escalators and Moving Walks

- Escalators and moving walks are specialty items installed by specialty contractors. There are numerous options associated with these items. For specific options, contact a manufacturer or contractor. In a method similar to estimating dumbwaiters and elevators, you should verify the extent of general contract work and add items as necessary.

## 14 40 00 Lifts
## 14 90 00 Other Conveying Equipment

- Products such as correspondence lifts, chutes, and pneumatic tube systems, as well as other items specified in this subdivision, may require trained installers. The general contractor might not have any choice as to who will perform the installation or when it will be performed. Long lead times are often required for these products, making early decisions in scheduling necessary.

## Reference Numbers

Reference numbers are shown in shaded boxes at the beginning of some major classifications. These numbers refer to related items in the Reference Section. The reference information may be an estimating procedure, an alternate pricing method, or technical information.

*Note:* Not all subdivisions listed here necessarily appear in this publication.

# 14 11 Manual Dumbwaiters

## 14 11 10 - Hand Operated Dumbwaiters

| 14 11 10.20 Manual Dumbwaiters | Crew | Daily Output | Labor-Hours | Unit | Material | 2011 Bare Costs Labor | 2011 Bare Costs Equipment | Total | Total Incl O&P |
|---|---|---|---|---|---|---|---|---|---|
| 0010 **MANUAL DUMBWAITERS** | | | | | | | | | |
| 0020   2 stop, minimum | 2 Elev | .75 | 21.333 | Ea. | 2,925 | 1,450 | | 4,375 | 5,350 |
| 0100     Maximum | | .50 | 32 | " | 6,650 | 2,150 | | 8,800 | 10,500 |
| 0300   For each additional stop, add | | .75 | 21.333 | Stop | 1,050 | 1,450 | | 2,500 | 3,300 |

# 14 12 Electric Dumbwaiters

## 14 12 10 - Dumbwaiters

### 14 12 10.10 Electric Dumbwaiters

| | Crew | Daily Output | Labor-Hours | Unit | Material | 2011 Bare Costs Labor | 2011 Bare Costs Equipment | Total | Total Incl O&P |
|---|---|---|---|---|---|---|---|---|---|
| 0010 **ELECTRIC DUMBWAITERS** | | | | | | | | | |
| 0020   2 stop, minimum | 2 Elev | .13 | 123 | Ea. | 7,375 | 8,300 | | 15,675 | 20,400 |
| 0100     Maximum | | .11 | 145 | " | 22,200 | 9,800 | | 32,000 | 38,900 |
| 0600   For each additional stop, add | | .54 | 29.630 | Stop | 3,250 | 2,000 | | 5,250 | 6,550 |
| 0750   Correspondence lift, 1 floor, 2 stop, 45 lb capacity | 2 Elec | .20 | 80 | Ea. | 8,350 | 4,025 | | 12,375 | 15,200 |

# 14 21 Electric Traction Elevators

## 14 21 13 - Electric Traction Freight Elevators

### 14 21 13.10 Electric Traction Freight Elevators and Options

| | Crew | Daily Output | Labor-Hours | Unit | Material | 2011 Bare Costs Labor | 2011 Bare Costs Equipment | Total | Total Incl O&P |
|---|---|---|---|---|---|---|---|---|---|
| 0010 **ELECTRIC TRACTION FREIGHT ELEVATORS AND OPTIONS** R142000-10 | | | | | | | | | |
| 0425   Electric freight, base unit, 4000 lb, 200 fpm, 4 stop, std. fin. | 2 Elev | .05 | 320 | Ea. | 96,500 | 21,600 | | 118,100 | 138,000 |
| 0450   For 5000 lb capacity, add | | | | | 5,450 | | | 5,450 | 6,000 |
| 0500   For 6000 lb capacity, add | | | | | 16,900 | | | 16,900 | 18,600 |
| 0525   For 7000 lb capacity, add | | | | | 20,700 | | | 20,700 | 22,800 |
| 0550   For 8000 lb capacity, add | | | | | 28,000 | | | 28,000 | 30,800 |
| 0575   For 10000 lb capacity, add | | | | | 31,500 | | | 31,500 | 34,700 |
| 0600   For 12000 lb capacity, add | | | | | 38,200 | | | 38,200 | 42,000 |
| 0625   For 16000 lb capacity, add | | | | | 48,700 | | | 48,700 | 53,500 |
| 0650   For 20000 lb capacity, add | | | | | 58,000 | | | 58,000 | 63,500 |
| 0675   For increased speed, 250 fpm, add | | | | | 21,800 | | | 21,800 | 23,900 |
| 0700     300 fpm, geared electric, add | | | | | 27,600 | | | 27,600 | 30,300 |
| 0725     350 fpm, geared electric, add | | | | | 33,300 | | | 33,300 | 36,700 |
| 0750     400 fpm, geared electric, add | | | | | 38,800 | | | 38,800 | 42,600 |
| 0775     500 fpm, gearless electric, add | | | | | 46,400 | | | 46,400 | 51,000 |
| 0800     600 fpm, gearless electric, add | | | | | 56,000 | | | 56,000 | 62,000 |
| 0825     700 fpm, gearless electric, add | | | | | 67,000 | | | 67,000 | 74,000 |
| 0850     800 fpm, gearless electric, add | | | | | 77,500 | | | 77,500 | 85,500 |
| 0875   For class "B" loading, add | | | | | 6,375 | | | 6,375 | 7,000 |
| 0900   For class "C-1" loading, add | | | | | 8,800 | | | 8,800 | 9,700 |
| 0925   For class "C-2" loading, add | | | | | 10,000 | | | 10,000 | 11,000 |
| 0950   For class "C-3" loading, add | | | | | 12,700 | | | 12,700 | 13,900 |
| 0975   For travel over 40 V.L.F., add | 2 Elev | 7.25 | 2.207 | V.L.F. | 540 | 149 | | 689 | 815 |
| 1000   For number of stops over 4, add | " | .27 | 59.259 | Stop | 3,850 | 4,000 | | 7,850 | 10,200 |

## 14 21 23 - Electric Traction Passenger Elevators

### 14 21 23.10 Electric Traction Passenger Elevators and Options

| | Crew | Daily Output | Labor-Hours | Unit | Material | 2011 Bare Costs Labor | 2011 Bare Costs Equipment | Total | Total Incl O&P |
|---|---|---|---|---|---|---|---|---|---|
| 0010 **ELECTRIC TRACTION PASSENGER ELEVATORS AND OPTIONS** | | | | | | | | | |
| 1625   Electric pass., base unit, 2000 lb, 200 fpm, 4 stop, std. fin. | 2 Elev | .05 | 320 | Ea. | 83,500 | 21,600 | | 105,100 | 124,000 |
| 1650   For 2500 lb capacity, add | | | | | 3,475 | | | 3,475 | 3,825 |
| 1675   For 3000 lb capacity, add | | | | | 4,150 | | | 4,150 | 4,575 |
| 1700   For 3500 lb capacity, add | | | | | 6,950 | | | 6,950 | 7,625 |
| 1725   For 4000 lb capacity, add | | | | | 6,525 | | | 6,525 | 7,175 |

# 14 21 Electric Traction Elevators

## 14 21 23 – Electric Traction Passenger Elevators

### 14 21 23.10 Electric Traction Passenger Elevators and Options

| | | Crew | Daily Output | Labor-Hours | Unit | Material | 2011 Bare Costs Labor | 2011 Bare Costs Equipment | Total | Total Incl O&P |
|---|---|---|---|---|---|---|---|---|---|---|
| 1750 | For 4500 lb capacity, add | | | | Ea. | 8,250 | | | 8,250 | 9,075 |
| 1775 | For 5000 lb capacity, add | | | | | 9,925 | | | 9,925 | 10,900 |
| 1800 | For increased speed, 250 fpm, geared electric, add | | | | | 6,600 | | | 6,600 | 7,250 |
| 1825 | 300 fpm, geared electric, add | | | | | 9,975 | | | 9,975 | 11,000 |
| 1850 | 350 fpm, geared electric, add | | | | | 11,300 | | | 11,300 | 12,400 |
| 1875 | 400 fpm, geared electric, add | | | | | 14,400 | | | 14,400 | 15,800 |
| 1900 | 500 fpm, gearless electric, add | | | | | 29,500 | | | 29,500 | 32,500 |
| 1925 | 600 fpm, gearless electric, add | | | | | 32,700 | | | 32,700 | 36,000 |
| 1950 | 700 fpm, gearless electric, add | | | | | 37,500 | | | 37,500 | 41,300 |
| 1975 | 800 fpm, gearless electric, add | | | | | 42,800 | | | 42,800 | 47,100 |
| 2000 | For travel over 40 V.L.F., add | 2 Elev | 7.25 | 2.207 | V.L.F. | 650 | 149 | | 799 | 935 |
| 2025 | For number of stops over 4, add | | .27 | 59.259 | Stop | 7,400 | 4,000 | | 11,400 | 14,100 |
| 2400 | Electric hospital, base unit, 4000 lb, 200 fpm, 4 stop, std fin. | | .05 | 320 | Ea. | 72,000 | 21,600 | | 93,600 | 111,000 |
| 2425 | For 4500 lb capacity, add | | | | | 5,125 | | | 5,125 | 5,650 |
| 2450 | For 5000 lb capacity, add | | | | | 6,175 | | | 6,175 | 6,775 |
| 2475 | For increased speed, 250 fpm, geared electric, add | | | | | 6,625 | | | 6,625 | 7,300 |
| 2500 | 300 fpm, geared electric, add | | | | | 9,825 | | | 9,825 | 10,800 |
| 2525 | 350 fpm, geared electric, add | | | | | 11,200 | | | 11,200 | 12,300 |
| 2550 | 400 fpm, geared electric, add | | | | | 14,200 | | | 14,200 | 15,600 |
| 2575 | 500 fpm, gearless electric, add | | | | | 29,200 | | | 29,200 | 32,100 |
| 2600 | 600 fpm, gearless electric, add | | | | | 33,400 | | | 33,400 | 36,800 |
| 2625 | 700 fpm, gearless electric, add | | | | | 37,900 | | | 37,900 | 41,700 |
| 2650 | 800 fpm, gearless electric, add | | | | | 43,700 | | | 43,700 | 48,000 |
| 2675 | For travel over 40 V.L.F., add | 2 Elev | 7.25 | 2.207 | V.L.F. | 635 | 149 | | 784 | 915 |
| 2700 | For number of stops over 4, add | " | .27 | 59.259 | Stop | 3,825 | 4,000 | | 7,825 | 10,200 |

## 14 21 33 – Electric Traction Residential Elevators

### 14 21 33.20 Residential Elevators

| | | Crew | Daily Output | Labor-Hours | Unit | Material | Labor | Equipment | Total | Total Incl O&P |
|---|---|---|---|---|---|---|---|---|---|---|
| 0010 | **RESIDENTIAL ELEVATORS** | | | | | | | | | |
| 7000 | Residential, cab type, 1 floor, 2 stop, minimum | 2 Elev | .20 | 80 | Ea. | 11,000 | 5,400 | | 16,400 | 20,100 |
| 7100 | Maximum | | .10 | 160 | | 18,600 | 10,800 | | 29,400 | 36,500 |
| 7200 | 2 floor, 3 stop, minimum | | .12 | 133 | | 16,400 | 8,975 | | 25,375 | 31,300 |
| 7300 | Maximum | | .06 | 266 | | 26,700 | 18,000 | | 44,700 | 56,000 |

# 14 24 Hydraulic Elevators

## 14 24 13 – Hydraulic Freight Elevators

### 14 24 13.10 Hydraulic Freight Elevators and Options

| | | Crew | Daily Output | Labor-Hours | Unit | Material | Labor | Equipment | Total | Total Incl O&P |
|---|---|---|---|---|---|---|---|---|---|---|
| 0010 | **HYDRAULIC FREIGHT ELEVATORS AND OPTIONS** | | | | | | | | | |
| 1025 | Hydraulic freight, base unit, 2000 lb, 50 fpm, 2 stop, std. fin. | 2 Elev | .10 | 160 | Ea. | 75,000 | 10,800 | | 85,800 | 98,500 |
| 1050 | For 2500 lb capacity, add | | | | | 4,725 | | | 4,725 | 5,200 |
| 1075 | For 3000 lb capacity, add | | | | | 8,375 | | | 8,375 | 9,225 |
| 1100 | For 3500 lb capacity, add | | | | | 15,200 | | | 15,200 | 16,700 |
| 1125 | For 4000 lb capacity, add | | | | | 16,200 | | | 16,200 | 17,800 |
| 1150 | For 4500 lb capacity, add | | | | | 19,000 | | | 19,000 | 20,900 |
| 1175 | For 5000 lb capacity, add | | | | | 23,900 | | | 23,900 | 26,300 |
| 1200 | For 6000 lb capacity, add | | | | | 29,900 | | | 29,900 | 32,900 |
| 1225 | For 7000 lb capacity, add | | | | | 35,900 | | | 35,900 | 39,500 |
| 1250 | For 8000 lb capacity, add | | | | | 39,500 | | | 39,500 | 43,500 |
| 1275 | For 10000 lb capacity, add | | | | | 50,000 | | | 50,000 | 55,000 |
| 1300 | For 12000 lb capacity, add | | | | | 65,500 | | | 65,500 | 72,000 |
| 1325 | For 16000 lb capacity, add | | | | | 90,000 | | | 90,000 | 99,000 |

# 14 24 Hydraulic Elevators

## 14 24 13 – Hydraulic Freight Elevators

### 14 24 13.10 Hydraulic Freight Elevators and Options

| | 14 24 13.10 Hydraulic Freight Elevators and Options | Crew | Daily Output | Labor-Hours | Unit | Material | 2011 Bare Costs Labor | Equipment | Total | Total Incl O&P |
|---|---|---|---|---|---|---|---|---|---|---|
| 1350 | For 20000 lb capacity, add | | | | Ea. | 109,500 | | | 109,500 | 120,500 |
| 1375 | For increased speed, 100 fpm, add | | | | | 2,100 | | | 2,100 | 2,325 |
| 1400 | 125 fpm, add | | | | | 4,525 | | | 4,525 | 4,975 |
| 1425 | 150 fpm, add | | | | | 5,400 | | | 5,400 | 5,925 |
| 1450 | 175 fpm, add | | | | | 9,200 | | | 9,200 | 10,100 |
| 1475 | For class "B" loading, add | | | | | 6,350 | | | 6,350 | 6,975 |
| 1500 | For class "C-1" loading, add | | | | | 8,775 | | | 8,775 | 9,650 |
| 1525 | For class "C-2" loading, add | | | | | 10,000 | | | 10,000 | 11,000 |
| 1550 | For class "C-3" loading, add | | | | | 12,600 | | | 12,600 | 13,900 |
| 1575 | For travel over 20 V.L.F., add | 2 Elev | 7.25 | 2.207 | V.L.F. | 1,650 | 149 | | 1,799 | 2,050 |
| 1600 | For number of stops over 2, add | " | .27 | 59.259 | Stop | 4,450 | 4,000 | | 8,450 | 10,800 |

## 14 24 23 – Hydraulic Passenger Elevators

### 14 24 23.10 Hydraulic Passenger Elevators and Options

| | 14 24 23.10 Hydraulic Passenger Elevators and Options | Crew | Daily Output | Labor-Hours | Unit | Material | Labor | Equipment | Total | Total Incl O&P |
|---|---|---|---|---|---|---|---|---|---|---|
| 0010 | **HYDRAULIC PASSENGER ELEVATORS AND OPTIONS** | | | | | | | | | |
| 2050 | Hyd. pass., base unit, 1500 lb, 100 fpm, 2 stop, std. fin. | 2 Elev | .10 | 160 | Ea. | 34,700 | 10,800 | | 45,500 | 54,000 |
| 2075 | For 2000 lb capacity, add | | | | | 1,900 | | | 1,900 | 2,100 |
| 2100 | For 2500 lb. capacity, add | | | | | 2,975 | | | 2,975 | 3,250 |
| 2125 | For 3000 lb capacity, add | | | | | 3,900 | | | 3,900 | 4,300 |
| 2150 | For 3500 lb capacity, add | | | | | 5,650 | | | 5,650 | 6,225 |
| 2175 | For 4000 lb capacity, add | | | | | 6,950 | | | 6,950 | 7,650 |
| 2200 | For 4500 lb capacity, add | | | | | 8,175 | | | 8,175 | 9,000 |
| 2225 | For 5000 lb capacity, add | | | | | 10,800 | | | 10,800 | 11,900 |
| 2250 | For increased speed, 125 fpm, add | | | | | 2,800 | | | 2,800 | 3,075 |
| 2275 | 150 fpm, add | | | | | 3,525 | | | 3,525 | 3,875 |
| 2300 | 175 fpm, add | | | | | 5,275 | | | 5,275 | 5,800 |
| 2325 | 200 fpm, add | | | | | 7,300 | | | 7,300 | 8,025 |
| 2350 | For travel over 12 V.L.F., add | 2 Elev | 7.25 | 2.207 | V.L.F. | 865 | 149 | | 1,014 | 1,175 |
| 2375 | For number of stops over 2, add | | .27 | 59.259 | Stop | 2,450 | 4,000 | | 6,450 | 8,600 |
| 2725 | Hydraulic hospital, base unit, 4000 lb, 100 fpm, 2 stop, std. fin. | | .10 | 160 | Ea. | 56,500 | 10,800 | | 67,300 | 78,000 |
| 2775 | For 4500 lb capacity, add | | | | | 6,400 | | | 6,400 | 7,050 |
| 2800 | For 5000 lb capacity, add | | | | | 8,475 | | | 8,475 | 9,325 |
| 2825 | For increased speed, 125 fpm, add | | | | | 3,225 | | | 3,225 | 3,550 |
| 2850 | 150 fpm, add | | | | | 3,875 | | | 3,875 | 4,275 |
| 2875 | 175 fpm, add | | | | | 5,625 | | | 5,625 | 6,200 |
| 2900 | 200 fpm, add | | | | | 6,800 | | | 6,800 | 7,475 |
| 2925 | For travel over 12 V.L.F., add | 2 Elev | 7.25 | 2.207 | V.L.F. | 715 | 149 | | 864 | 1,000 |
| 2950 | For number of stops over 2, add | " | .27 | 59.259 | Stop | 3,850 | 4,000 | | 7,850 | 10,200 |

# 14 27 Custom Elevator Cabs and Doors

## 14 27 13 – Custom Elevator Cab Finishes

### 14 27 13.10 Cab Finishes

| | 14 27 13.10 Cab Finishes | Crew | Daily Output | Labor-Hours | Unit | Material | Labor | Equipment | Total | Total Incl O&P |
|---|---|---|---|---|---|---|---|---|---|---|
| 0010 | **CAB FINISHES** | | | | | | | | | |
| 3325 | Passenger elevator cab finishes (based on 3500 lb cab size) | | | | | | | | | |
| 3350 | Acrylic panel ceiling | | | | Ea. | 705 | | | 705 | 775 |
| 3375 | Aluminum eggcrate ceiling | | | | | 1,150 | | | 1,150 | 1,275 |
| 3400 | Stainless steel doors | | | | | 2,850 | | | 2,850 | 3,125 |
| 3425 | Carpet flooring | | | | | 570 | | | 570 | 630 |
| 3450 | Epoxy flooring | | | | | 435 | | | 435 | 480 |
| 3475 | Quarry tile flooring | | | | | 810 | | | 810 | 890 |
| 3500 | Slate flooring | | | | | 1,425 | | | 1,425 | 1,550 |

# 14 27 Custom Elevator Cabs and Doors

## 14 27 13 – Custom Elevator Cab Finishes

| 14 27 13.10 Cab Finishes | | Crew | Daily Output | Labor-Hours | Unit | Material | 2011 Bare Costs Labor | Equipment | Total | Total Incl O&P |
|---|---|---|---|---|---|---|---|---|---|---|
| 3525 | Textured rubber flooring | | | | Ea. | 188 | | | 188 | 207 |
| 3550 | Stainless steel walls | | | | | 4,975 | | | 4,975 | 5,475 |
| 3575 | Stainless steel returns at door | | | | | 1,150 | | | 1,150 | 1,275 |
| 4450 | Hospital elevator cab finishes (based on 3500 lb cab size) | | | | | | | | | |
| 4475 | Aluminum eggcrate ceiling | | | | Ea. | 1,150 | | | 1,150 | 1,250 |
| 4500 | Stainless steel doors | | | | | 2,575 | | | 2,575 | 2,825 |
| 4525 | Epoxy flooring | | | | | 350 | | | 350 | 385 |
| 4550 | Quarry tile flooring | | | | | 820 | | | 820 | 900 |
| 4575 | Textured rubber flooring | | | | | 625 | | | 625 | 685 |
| 4600 | Stainless steel walls | | | | | 4,975 | | | 4,975 | 5,475 |
| 4625 | Stainless steel returns at door | | | | | 1,150 | | | 1,150 | 1,275 |

# 14 28 Elevator Equipment and Controls

## 14 28 10 – Elevator Equipment and Control Options

### 14 28 10.10 Elevator Controls and Doors

| | | Crew | Daily Output | Labor-Hours | Unit | Material | 2011 Bare Costs Labor | Equipment | Total | Total Incl O&P |
|---|---|---|---|---|---|---|---|---|---|---|
| 0010 | **ELEVATOR CONTROLS AND DOORS** | | | | | | | | | |
| 2975 | Passenger elevator options | | | | | | | | | |
| 3000 | 2 car group automatic controls | 2 Elev | .66 | 24.242 | Ea. | 6,275 | 1,625 | | 7,900 | 9,325 |
| 3025 | 3 car group automatic controls | | .44 | 36.364 | | 12,400 | 2,450 | | 14,850 | 17,200 |
| 3050 | 4 car group automatic controls | | .33 | 48.485 | | 25,200 | 3,275 | | 28,475 | 32,600 |
| 3075 | 5 car group automatic controls | | .26 | 61.538 | | 46,200 | 4,150 | | 50,350 | 57,000 |
| 3100 | 6 car group automatic controls | | .22 | 72.727 | | 95,000 | 4,900 | | 99,900 | 112,000 |
| 3125 | Intercom service | | 3 | 5.333 | | 1,325 | 360 | | 1,685 | 2,000 |
| 3150 | Duplex car selective collective | | .66 | 24.242 | | 11,700 | 1,625 | | 13,325 | 15,200 |
| 3175 | Center opening 1 speed doors | | 2 | 8 | | 1,725 | 540 | | 2,265 | 2,700 |
| 3200 | Center opening 2 speed doors | | 2 | 8 | | 2,425 | 540 | | 2,965 | 3,475 |
| 3225 | Rear opening doors (opposite front) | | 2 | 8 | | 2,875 | 540 | | 3,415 | 3,975 |
| 3250 | Side opening 2 speed doors | | 2 | 8 | | 4,525 | 540 | | 5,065 | 5,775 |
| 3275 | Automatic emergency power switching | | .66 | 24.242 | | 1,100 | 1,625 | | 2,725 | 3,625 |
| 3300 | Manual emergency power switching | | 8 | 2 | | 480 | 135 | | 615 | 730 |
| 3625 | Hall finishes, stainless steel doors | | | | | 1,925 | | | 1,925 | 2,125 |
| 3650 | Stainless steel frames | | | | | 1,400 | | | 1,400 | 1,550 |
| 3675 | 12 month maintenance contract | | | | | | | | 3,600 | 3,950 |
| 3700 | Signal devices, hall lanterns | 2 Elev | 8 | 2 | | 525 | 135 | | 660 | 780 |
| 3725 | Position indicators, up to 3 | | 9.40 | 1.702 | | 615 | 115 | | 730 | 845 |
| 3750 | Position indicators, per each over 3 | | 32 | .500 | | 595 | 33.50 | | 628.50 | 705 |
| 3775 | High speed heavy duty door opener | | | | | 4,050 | | | 4,050 | 4,450 |
| 3800 | Variable voltage, O.H. gearless machine, min. | 2 Elev | .16 | 100 | | 66,500 | 6,750 | | 73,250 | 83,000 |
| 3815 | Maximum | | .07 | 228 | | 98,000 | 15,400 | | 113,400 | 131,000 |
| 3825 | Basement installed geared machine | | .33 | 48.485 | | 69,000 | 3,275 | | 72,275 | 81,000 |
| 3850 | Freight elevator options | | | | | | | | | |
| 3875 | Doors, bi-parting | 2 Elev | .66 | 24.242 | Ea. | 9,525 | 1,625 | | 11,150 | 12,900 |
| 3900 | Power operated door and gate | " | .66 | 24.242 | | 22,100 | 1,625 | | 23,725 | 26,700 |
| 3925 | Finishes, steel plate floor | | | | | 2,375 | | | 2,375 | 2,625 |
| 3950 | 14 ga. 1/4" x 4' steel plate walls | | | | | 1,800 | | | 1,800 | 1,975 |
| 3975 | 12 month maintenance contract | | | | | | | | 3,600 | 3,950 |
| 4000 | Signal devices, hall lanterns | 2 Elev | 8 | 2 | | 520 | 135 | | 655 | 770 |
| 4025 | Position indicators, up to 3 | | 9.40 | 1.702 | | 625 | 115 | | 740 | 855 |
| 4050 | Position indicators, per each over 3 | | 32 | .500 | | 600 | 33.50 | | 633.50 | 710 |
| 4075 | Variable voltage basement installed geared machine | | .66 | 24.242 | | 18,200 | 1,625 | | 19,825 | 22,500 |
| 4100 | Hospital elevator options | | | | | | | | | |

## 14 28 10 – Elevator Equipment and Control Options

| 14 28 10.10 Elevator Controls and Doors | Crew | Daily Output | Labor-Hours | Unit | Material | 2011 Bare Costs Labor | Equipment | Total | Total Incl O&P |
|---|---|---|---|---|---|---|---|---|---|
| 4125 | 2 car group automatic controls | 2 Elev | .66 | 24.242 | Ea. | 6,275 | 1,625 | | 7,900 | 9,325 |
| 4150 | 3 car group automatic controls | | .44 | 36.364 | | 12,400 | 2,450 | | 14,850 | 17,200 |
| 4175 | 4 car group automatic controls | | .33 | 48.485 | | 25,200 | 3,275 | | 28,475 | 32,600 |
| 4200 | 5 car group automatic controls | | .26 | 61.538 | | 46,000 | 4,150 | | 50,150 | 56,500 |
| 4225 | 6 car group automatic controls | | .22 | 72.727 | | 94,500 | 4,900 | | 99,400 | 111,500 |
| 4250 | Intercom service | | 3 | 5.333 | | 1,325 | 360 | | 1,685 | 1,975 |
| 4275 | Duplex car selective collective | | .66 | 24.242 | | 11,600 | 1,625 | | 13,225 | 15,200 |
| 4300 | Center opening 1 speed doors | | 2 | 8 | | 1,725 | 540 | | 2,265 | 2,700 |
| 4325 | Center opening 2 speed doors | | 2 | 8 | | 2,350 | 540 | | 2,890 | 3,400 |
| 4350 | Rear opening doors (opposite front) | | 2 | 8 | | 2,875 | 540 | | 3,415 | 3,975 |
| 4375 | Side opening 2 speed doors | | 2 | 8 | | 3,600 | 540 | | 4,140 | 4,750 |
| 4400 | Automatic emergency power switching | | .66 | 24.242 | | 1,075 | 1,625 | | 2,700 | 3,600 |
| 4425 | Manual emergency power switching | | 8 | 2 | | 470 | 135 | | 605 | 715 |
| 4675 | Hall finishes, stainless steel doors | | | | | 1,925 | | | 1,925 | 2,125 |
| 4700 | Stainless steel frames | | | | | 1,400 | | | 1,400 | 1,550 |
| 4725 | 12 month maintenance contract | | | | | | | | 3,600 | 3,950 |
| 4750 | Signal devices, hall lanterns | 2 Elev | 8 | 2 | | 515 | 135 | | 650 | 765 |
| 4775 | Position indicators, up to 3 | | 9.40 | 1.702 | | 615 | 115 | | 730 | 845 |
| 4800 | Position indicators, per each over 3 | | 32 | .500 | | 595 | 33.50 | | 628.50 | 705 |
| 4825 | High speed heavy duty door opener | | | | | 4,050 | | | 4,050 | 4,450 |
| 4850 | Variable voltage, O.H. gearless machine, min. | 2 Elev | .16 | 100 | | 66,000 | 6,750 | | 72,750 | 82,500 |
| 4865 | Maximum | | .07 | 228 | | 97,500 | 15,400 | | 112,900 | 130,500 |
| 4875 | Basement installed geared machine | | .33 | 48.485 | | 25,000 | 3,275 | | 28,275 | 32,400 |
| 5000 | Drilling for piston, casing included, 18" diameter | B-48 | 80 | .700 | V.L.F. | 47 | 27 | 35 | 109 | 131 |

# 14 31 Escalators

## 14 31 10 – Glass and Steel Escalators

### 14 31 10.10 Escalators

| | | Crew | Daily Output | Labor-Hours | Unit | Material | Labor | Equipment | Total | Total Incl O&P |
|---|---|---|---|---|---|---|---|---|---|---|
| 0010 | **ESCALATORS** R143110-10 | | | | | | | | | |
| 1000 | Glass, 32" wide x 10' floor to floor height | M-1 | .07 | 457 | Ea. | 80,000 | 29,300 | 700 | 110,000 | 132,000 |
| 1010 | 48" wide x 10' floor to floor height | | .07 | 457 | | 86,000 | 29,300 | 700 | 116,000 | 139,000 |
| 1020 | 32" wide x 15' floor to floor height | | .06 | 533 | | 84,000 | 34,100 | 815 | 118,915 | 144,000 |
| 1030 | 48" wide x 15' floor to floor height | | .06 | 533 | | 89,000 | 34,100 | 815 | 123,915 | 149,500 |
| 1040 | 32" wide x 20' floor to floor height | | .05 | 653 | | 88,500 | 41,800 | 1,000 | 131,300 | 160,000 |
| 1050 | 48" wide x 20' floor to floor height | | .05 | 653 | | 96,000 | 41,800 | 1,000 | 138,800 | 168,500 |
| 1060 | 32" wide x 25' floor to floor height | | .04 | 800 | | 96,000 | 51,000 | 1,225 | 148,225 | 183,000 |
| 1070 | 48" wide x 25' floor to floor height | | .04 | 800 | | 111,500 | 51,000 | 1,225 | 163,725 | 200,000 |
| 1080 | Enameled steel, 32" wide x 10' floor to floor height | | .07 | 457 | | 85,500 | 29,300 | 700 | 115,500 | 138,000 |
| 1090 | 48" wide x 10' floor to floor height | | .07 | 457 | | 92,500 | 29,300 | 700 | 122,500 | 146,000 |
| 1110 | 32" wide x 15' floor to floor height | | .06 | 533 | | 92,500 | 34,100 | 815 | 127,415 | 153,500 |
| 1120 | 48" wide x 15' floor to floor height | | .06 | 533 | | 92,500 | 34,100 | 815 | 127,415 | 153,500 |
| 1130 | 32" wide x 20' floor to floor height | | .05 | 653 | | 94,500 | 41,800 | 1,000 | 137,300 | 167,000 |
| 1140 | 48" wide x 20' floor to floor height | | .05 | 653 | | 101,500 | 41,800 | 1,000 | 144,300 | 174,500 |
| 1150 | 32" wide x 25' floor to floor height | | .04 | 800 | | 101,500 | 51,000 | 1,225 | 153,725 | 189,000 |
| 1160 | 48" wide x 25' floor to floor height | | .04 | 800 | | 101,500 | 51,000 | 1,225 | 153,725 | 189,000 |
| 1170 | Stainless steel, 32" wide x 10' floor to floor height | | .07 | 457 | | 101,500 | 29,300 | 700 | 131,500 | 155,500 |
| 1180 | 48" wide x 10' floor to floor height | | .07 | 457 | | 101,500 | 29,300 | 700 | 131,500 | 155,500 |
| 1500 | 32" wide x 15' floor to floor height | | .06 | 533 | | 93,000 | 34,100 | 815 | 127,915 | 153,500 |
| 1700 | 48" wide x 15' floor to floor height | | .06 | 533 | | 94,000 | 34,100 | 815 | 128,915 | 155,000 |
| 2300 | 32" wide x 25' floor to floor height | | .04 | 800 | | 101,500 | 51,000 | 1,225 | 153,725 | 189,500 |
| 2500 | 48" wide x 25' floor to floor height | | .04 | 800 | | 123,000 | 51,000 | 1,225 | 175,225 | 212,500 |

# 14 32 Moving Walks

## 14 32 10 – Moving Walkways

| 14 32 10.10 Moving Walks | | Crew | Daily Output | Labor-Hours | Unit | Material | 2011 Bare Costs Labor | Equipment | Total | Total Incl O&P |
|---|---|---|---|---|---|---|---|---|---|---|
| 0010 | **MOVING WALKS** | R143210-20 | | | | | | | | |
| 0020 | Walk, 27" tread width, minimum | M-1 | 6.50 | 4.923 | L.F. | 850 | 315 | 7.55 | 1,172.55 | 1,425 |
| 0100 | 300' to 500', maximum | | 4.43 | 7.223 | | 1,175 | 465 | 11.05 | 1,651.05 | 2,000 |
| 0300 | 48" tread width walk, minimum | | 4.43 | 7.223 | | 1,925 | 465 | 11.05 | 2,401.05 | 2,825 |
| 0400 | 100' to 350', maximum | | 3.82 | 8.377 | | 2,250 | 535 | 12.80 | 2,797.80 | 3,275 |
| 0600 | Ramp, 12° incline, 36" tread width, minimum | | 5.27 | 6.072 | | 1,575 | 390 | 9.30 | 1,974.30 | 2,300 |
| 0700 | 70' to 90' maximum | | 3.82 | 8.377 | | 2,250 | 535 | 12.80 | 2,797.80 | 3,275 |
| 0900 | 48" tread width, minimum | | 3.57 | 8.964 | | 2,300 | 575 | 13.70 | 2,888.70 | 3,400 |
| 1000 | 40' to 70', maximum | ↓ | 2.91 | 10.997 | ↓ | 2,900 | 705 | 16.80 | 3,621.80 | 4,275 |

# 14 42 Wheelchair Lifts

## 14 42 13 – Inclined Wheelchair Lifts

### 14 42 13.10 Inclined Wheelchair Lifts and Stairclimbers

| 14 42 13.10 | | Crew | Daily Output | Labor-Hours | Unit | Material | Labor | Equipment | Total | Total Incl O&P |
|---|---|---|---|---|---|---|---|---|---|---|
| 0010 | **INCLINED WHEELCHAIR LIFTS AND STAIRCLIMBERS** | | | | | | | | | |
| 7700 | Stair climber (chair lift), single seat, minimum | 2 Elev | 1 | 16 | Ea. | 5,325 | 1,075 | | 6,400 | 7,475 |
| 7800 | Maximum | | .20 | 80 | | 7,350 | 5,400 | | 12,750 | 16,100 |
| 8700 | Stair lift, minimum | | 1 | 16 | | 14,500 | 1,075 | | 15,575 | 17,500 |
| 8900 | Maximum | ↓ | .20 | 80 | ↓ | 22,900 | 5,400 | | 28,300 | 33,200 |

## 14 42 16 – Vertical Wheelchair Lifts

### 14 42 16.10 Wheelchair Lifts

| 14 42 16.10 | | Crew | Daily Output | Labor-Hours | Unit | Material | Labor | Equipment | Total | Total Incl O&P |
|---|---|---|---|---|---|---|---|---|---|---|
| 0010 | **WHEELCHAIR LIFTS** | | | | | | | | | |
| 8000 | Wheelchair lift, minimum | 2 Elev | 1 | 16 | Ea. | 7,325 | 1,075 | | 8,400 | 9,650 |
| 8500 | Maximum | " | .50 | 32 | " | 17,300 | 2,150 | | 19,450 | 22,200 |

# 14 45 Vehicle Lifts

## 14 45 10 – Hydraulic Vehicle Lifts

### 14 45 10.10 Hydraulic Lifts

| 14 45 10.10 | | Crew | Daily Output | Labor-Hours | Unit | Material | Labor | Equipment | Total | Total Incl O&P |
|---|---|---|---|---|---|---|---|---|---|---|
| 0010 | **HYDRAULIC LIFTS** | | | | | | | | | |
| 0200 | Double post, 9000 lb frame | L-4 | 1.15 | 20.870 | Ea. | 5,525 | 845 | | 6,370 | 7,375 |
| 0400 | 26,000 lb frame | | .22 | 109 | | 25,400 | 4,425 | | 29,825 | 34,800 |
| 2200 | Hoists, single post, 8000 lb capacity, swivel arms | | .40 | 60 | | 4,975 | 2,425 | | 7,400 | 9,200 |
| 2400 | Two posts, adjustable frames, 11,000 lb capacity | | .25 | 96 | | 6,400 | 3,875 | | 10,275 | 13,000 |
| 2500 | 24,000 lb capacity | | .15 | 160 | | 8,550 | 6,475 | | 15,025 | 19,300 |
| 2700 | 7500 lb capacity, frame supports | | .50 | 48 | | 7,125 | 1,950 | | 9,075 | 10,800 |
| 2800 | Four post, roll on ramp | | .50 | 48 | | 6,400 | 1,950 | | 8,350 | 10,000 |
| 2810 | Hydraulic lifts, above ground, 2 post, clear floor, 6000 lb cap | | 2.67 | 8.989 | | 7,025 | 365 | | 7,390 | 8,275 |
| 2815 | 9000 lb capacity | | 2.29 | 10.480 | | 16,700 | 425 | | 17,125 | 19,000 |
| 2820 | 15,000 lb capacity | | 2 | 12 | | 18,800 | 485 | | 19,285 | 21,400 |
| 2825 | 30,000 lb capacity | | 1.60 | 15 | | 41,600 | 605 | | 42,205 | 46,700 |
| 2830 | 4 post, ramp style, 25,000 lb capacity | | 2 | 12 | | 16,300 | 485 | | 16,785 | 18,600 |
| 2835 | 35,000 lb capacity | | 1 | 24 | | 76,000 | 970 | | 76,970 | 85,000 |
| 2840 | 50,000 lb capacity | | 1 | 24 | | 85,000 | 970 | | 85,970 | 95,000 |
| 2845 | 75,000 lb capacity | ↓ | 1 | 24 | | 98,500 | 970 | | 99,470 | 110,000 |
| 2850 | For drive thru tracks, add, minimum | | | | | 1,050 | | | 1,050 | 1,150 |
| 2855 | Maximum | | | | | 1,800 | | | 1,800 | 1,975 |
| 2860 | Ramp extensions, 3' (set of 2) | | | | | 860 | | | 860 | 945 |
| 2865 | Rolling jack platform | | | | | 3,000 | | | 3,000 | 3,300 |
| 2870 | Elec/hyd jacking beam | | | | | 8,000 | | | 8,000 | 8,800 |
| 2880 | Scissor lift, portable, 6000 lb capacity | | | | ↓ | 7,850 | | | 7,850 | 8,625 |

# 14 51 Correspondence and Parcel Lifts

## 14 51 10 – Electric Correspondence and Parcel Lifts

| 14 51 10.10 Correspondence Lifts | Crew | Daily Output | Labor-Hours | Unit | Material | 2011 Bare Costs Labor | Equipment | Total | Total Incl O&P |
|---|---|---|---|---|---|---|---|---|---|
| 0010 **CORRESPONDENCE LIFTS** | | | | | | | | | |
| 0020 1 floor, 2 stop, 25 lb capacity, electric | 2 Elev | .20 | 80 | Ea. | 5,975 | 5,400 | | 11,375 | 14,600 |
| 0100 Hand, 5 lb capacity | " | .20 | 80 | " | 2,250 | 5,400 | | 7,650 | 10,500 |

### 14 51 10.20 Parcel Lifts

| | Crew | Daily Output | Labor-Hours | Unit | Material | Labor | Equipment | Total | Total Incl O&P |
|---|---|---|---|---|---|---|---|---|---|
| 0010 **PARCEL LIFTS** | | | | | | | | | |
| 0020 20" x 20", 100 lb capacity, electric, per floor | 2 Mill | .25 | 64 | Ea. | 9,350 | 2,900 | | 12,250 | 14,500 |

# 14 91 Facility Chutes

## 14 91 33 – Laundry and Linen Chutes

### 14 91 33.10 Chutes

| | Crew | Daily Output | Labor-Hours | Unit | Material | Labor | Equipment | Total | Total Incl O&P |
|---|---|---|---|---|---|---|---|---|---|
| 0011 **CHUTES**, linen, trash or refuse | | | | | | | | | |
| 0050 Aluminized steel, 16 ga., 18" diameter | 2 Shee | 3.50 | 4.571 | Floor | 1,625 | 236 | | 1,861 | 2,150 |
| 0100 24" diameter | | 3.20 | 5 | | 1,725 | 258 | | 1,983 | 2,275 |
| 0200 30" diameter | | 3 | 5.333 | | 2,050 | 275 | | 2,325 | 2,675 |
| 0300 36" diameter | | 2.80 | 5.714 | | 2,475 | 295 | | 2,770 | 3,175 |
| 0400 Galvanized steel, 16 ga., 18" diameter | | 3.50 | 4.571 | | 1,000 | 236 | | 1,236 | 1,450 |
| 0500 24" diameter | | 3.20 | 5 | | 1,125 | 258 | | 1,383 | 1,650 |
| 0600 30" diameter | | 3 | 5.333 | | 1,275 | 275 | | 1,550 | 1,825 |
| 0700 36" diameter | | 2.80 | 5.714 | | 1,500 | 295 | | 1,795 | 2,100 |
| 0800 Stainless steel, 18" diameter | | 3.50 | 4.571 | | 2,725 | 236 | | 2,961 | 3,350 |
| 0900 24" diameter | | 3.20 | 5 | | 2,950 | 258 | | 3,208 | 3,650 |
| 1000 30" diameter | | 3 | 5.333 | | 3,525 | 275 | | 3,800 | 4,300 |
| 1005 36" diameter | | 2.80 | 5.714 | | 3,825 | 295 | | 4,120 | 4,650 |
| 1200 Linen chute bottom collector, aluminized steel | | 4 | 4 | Ea. | 1,400 | 207 | | 1,607 | 1,850 |
| 1300 Stainless steel | | 4 | 4 | | 1,800 | 207 | | 2,007 | 2,275 |
| 1500 Refuse, bottom hopper, aluminized steel, 18" diameter | | 3 | 5.333 | | 1,025 | 275 | | 1,300 | 1,550 |
| 1600 24" diameter | | 3 | 5.333 | | 1,275 | 275 | | 1,550 | 1,825 |
| 1800 36" diameter | | 3 | 5.333 | | 2,375 | 275 | | 2,650 | 3,025 |

## 14 91 82 – Trash Chutes

### 14 91 82.10 Trash Chutes and Accessories

| | Crew | Daily Output | Labor-Hours | Unit | Material | Labor | Equipment | Total | Total Incl O&P |
|---|---|---|---|---|---|---|---|---|---|
| 0010 **TRASH CHUTES AND ACCESSORIES** | | | | | | | | | |
| 2900 Package chutes, spiral type, minimum | 2 Shee | 4.50 | 3.556 | Floor | 2,425 | 184 | | 2,609 | 2,950 |
| 3000 Maximum | " | 1.50 | 10.667 | " | 6,325 | 550 | | 6,875 | 7,800 |

# 14 92 Pneumatic Tube Systems

## 14 92 10 – Conventional, Automatic and Computer Controlled Pneumatic Tube Systems

### 14 92 10.10 Pneumatic Tube Systems

| | Crew | Daily Output | Labor-Hours | Unit | Material | Labor | Equipment | Total | Total Incl O&P |
|---|---|---|---|---|---|---|---|---|---|
| 0010 **PNEUMATIC TUBE SYSTEMS** | | | | | | | | | |
| 0020 100' long, single tube, 2 stations, stock | | | | | | | | | |
| 0100 3" diameter | 2 Stpi | .12 | 133 | Total | 6,600 | 7,225 | | 13,825 | 18,100 |
| 0300 4" diameter | " | .09 | 177 | " | 7,450 | 9,625 | | 17,075 | 22,600 |
| 0400 Twin tube, two stations or more, conventional system | | | | | | | | | |
| 0600 2-1/2" round | 2 Stpi | 62.50 | .256 | L.F. | 13.30 | 13.85 | | 27.15 | 35 |
| 0700 3" round | | 46 | .348 | | 15.25 | 18.80 | | 34.05 | 45 |
| 0900 4" round | | 49.60 | .323 | | 16.90 | 17.45 | | 34.35 | 44.50 |
| 1000 4" x 7" oval | | 37.60 | .426 | | 24.50 | 23 | | 47.50 | 61.50 |
| 1050 Add for blower | | 2 | 8 | System | 5,200 | 435 | | 5,635 | 6,375 |
| 1110 Plus for each round station, add | | 7.50 | 2.133 | Ea. | 585 | 115 | | 700 | 815 |

# 14 92 Pneumatic Tube Systems

## 14 92 10 – Conventional, Automatic and Computer Controlled Pneumatic Tube Systems

| 14 92 10.10 Pneumatic Tube Systems | | Crew | Daily Output | Labor-Hours | Unit | Material | 2011 Bare Costs | | | Total Incl O&P |
|---|---|---|---|---|---|---|---|---|---|---|
| | | | | | | | Labor | Equipment | Total | |
| 1150 | Plus for each oval station, add | 2 Stpi | 7.50 | 2.133 | Ea. | 585 | 115 | | 700 | 815 |
| 1200 | Alternate pricing method: base cost, minimum | | .75 | 21.333 | Total | 6,375 | 1,150 | | 7,525 | 8,750 |
| 1300 | Maximum | | .25 | 64 | " | 12,700 | 3,450 | | 16,150 | 19,100 |
| 1500 | Plus total system length, add, minimum | | 93.40 | .171 | L.F. | 8.20 | 9.25 | | 17.45 | 23 |
| 1600 | Maximum | | 37.60 | .426 | " | 24.50 | 23 | | 47.50 | 61 |
| 1800 | Completely automatic system, 4" round, 15 to 50 stations | | .29 | 55.172 | Station | 20,100 | 2,975 | | 23,075 | 26,600 |
| 2200 | 51 to 144 stations | | .32 | 50 | | 15,600 | 2,700 | | 18,300 | 21,300 |
| 2400 | 6" round or 4" x 7" oval, 15 to 50 stations | | .24 | 66.667 | | 25,200 | 3,600 | | 28,800 | 33,100 |
| 2800 | 51 to 144 stations | | .23 | 69.565 | | 21,100 | 3,775 | | 24,875 | 28,800 |

# Division Notes

| | | CREW | DAILY OUTPUT | LABOR-HOURS | UNIT | 2011 BARE COSTS | | | | TOTAL INCL O&P |
|---|---|---|---|---|---|---|---|---|---|---|
| | | | | | | MAT. | LABOR | EQUIP. | TOTAL | |
| | | | | | | | | | | |
| | | | | | | | | | | |
| | | | | | | | | | | |
| | | | | | | | | | | |
| | | | | | | | | | | |
| | | | | | | | | | | |
| | | | | | | | | | | |
| | | | | | | | | | | |
| | | | | | | | | | | |
| | | | | | | | | | | |
| | | | | | | | | | | |
| | | | | | | | | | | |
| | | | | | | | | | | |
| | | | | | | | | | | |
| | | | | | | | | | | |
| | | | | | | | | | | |
| | | | | | | | | | | |
| | | | | | | | | | | |
| | | | | | | | | | | |
| | | | | | | | | | | |
| | | | | | | | | | | |

## Estimating Tips

Pipe for fire protection and all uses is located in Subdivision 22 11 13.

The labor adjustment factors listed in Subdivision 22 01 02.20 also apply to Division 21.

Many, but not all, areas in the U.S. require backflow protection in the fire system. It is advisable to check local building codes for specific requirements.

For your reference, the following is a list of the most applicable Fire Codes and Standards which may be purchased from the NFPA, 1 Batterymarch Park, Quincy, MA 02169-7471.

NFPA 1: Uniform Fire Code

NFPA 10: Portable Fire Extinguishers

NFPA 11: Low-, Medium-, and High-Expansion Foam

NFPA 12: Carbon Dioxide Extinguishing Systems (Also companion 12A)

NFPA 13: Installation of Sprinkler Systems (Also companion 13D, 13E, and 13R)

NFPA 14: Installation of Standpipe and Hose Systems

NFPA 15: Water Spray Fixed Systems for Fire Protection

NFPA 16: Installation of Foam-Water Sprinkler and Foam-Water Spray Systems

NFPA 17: Dry Chemical Extinguishing Systems (Also companion 17A)

NFPA 18: Wetting Agents

NFPA 20: Installation of Stationary Pumps for Fire Protection

NFPA 22: Water Tanks for Private Fire Protection

NFPA 24: Installation of Private Fire Service Mains and their Appurtenances

NFPA 25: Inspection, Testing and Maintenance of Water-Based Fire Protection

## Reference Numbers

Reference numbers are shown in shaded boxes at the beginning of some major classifications. These numbers refer to related items in the Reference Section. The reference information may be an estimating procedure, an alternate pricing method, or technical information.

*Note:* Not all subdivisions listed here necessarily appear in this publication.

*Note:* **Trade Service,** *in part, has been used as a reference source for some of the material prices used in Division 21.*

# 21 05 Common Work Results for Fire Suppression

## 21 05 23 – General-Duty Valves for Water-Based Fire-Suppression Piping

| 21 05 23.50 General-Duty Valves for Water-Based Fire Supp. | Crew | Daily Output | Labor-Hours | Unit | Material | 2011 Bare Costs Labor | Equipment | Total | Total Incl O&P |
|---|---|---|---|---|---|---|---|---|---|
| 0010 **GENERAL-DUTY VALVES FOR WATER-BASED FIRE SUPPRESSION PIPING** | | | | | | | | | |
| 6200 Valves and components | | | | | | | | | |
| 6500 Check, swing, C.I. body, brass fittings, auto. ball drip | | | | | | | | | |
| 6520 4" size | Q-12 | 3 | 5.333 | Ea. | 270 | 255 | | 525 | 675 |
| 6800 Check, wafer, butterfly type, C.I. body, bronze fittings | | | | | | | | | |
| 6820 4" size | Q-12 | 4 | 4 | Ea. | 298 | 191 | | 489 | 615 |

# 21 11 Facility Fire-Suppression Water-Service Piping

## 21 11 16 – Facility Fire Hydrants

### 21 11 16.50 Fire Hydrants for Buildings

| | Crew | Daily Output | Labor-Hours | Unit | Material | 2011 Bare Costs Labor | Equipment | Total | Total Incl O&P |
|---|---|---|---|---|---|---|---|---|---|
| 0010 **FIRE HYDRANTS FOR BUILDINGS** | | | | | | | | | |
| 3750 Hydrants, wall, w/caps, single, flush, polished brass | | | | | | | | | |
| 3800 2-1/2" x 2-1/2" | Q-12 | 5 | 3.200 | Ea. | 175 | 153 | | 328 | 420 |
| 3840 2-1/2" x 3" | " | 5 | 3.200 | " | 355 | 153 | | 508 | 620 |
| 3950 Double, flush, polished brass | | | | | | | | | |
| 4000 2-1/2" x 2-1/2" x 4" | Q-12 | 5 | 3.200 | Ea. | 470 | 153 | | 623 | 750 |
| 4040 2-1/2" x 2-1/2" x 6" | " | 4.60 | 3.478 | ↓ | 675 | 166 | | 841 | 995 |
| 4200 For polished chrome, add | | | | | 10% | | | | |
| 4350 Double, projecting, polished brass | | | | | | | | | |
| 4400 2-1/2" x 2-1/2" x 4" | Q-12 | 5 | 3.200 | Ea. | 209 | 153 | | 362 | 460 |
| 4450 2-1/2" x 2-1/2" x 6" | " | 4.60 | 3.478 | " | 425 | 166 | | 591 | 720 |
| 4460 Valve control, dbl. flush/projecting hydrant, cap & | | | | | | | | | |
| 4470 chain, extension rod & cplg., escutcheon, polished brass | Q-12 | 8 | 2 | Ea. | 249 | 95.50 | | 344.50 | 415 |

## 21 11 19 – Fire-Department Connections

### 21 11 19.50 Connections for the Fire-Department

| | Crew | Daily Output | Labor-Hours | Unit | Material | 2011 Bare Costs Labor | Equipment | Total | Total Incl O&P |
|---|---|---|---|---|---|---|---|---|---|
| 0010 **CONNECTIONS FOR THE FIRE-DEPARTMENT** | | | | | | | | | |
| 7140 Standpipe connections, wall, w/plugs & chains | | | | | | | | | |
| 7160 Single, flush, brass, 2-1/2" x 2-1/2" | Q-12 | 5 | 3.200 | Ea. | 123 | 153 | | 276 | 365 |
| 7180 2-1/2" x 3" | " | 5 | 3.200 | " | 135 | 153 | | 288 | 380 |
| 7240 For polished chrome, add | | | | | 15% | | | | |
| 7280 Double, flush, polished brass | | | | | | | | | |
| 7300 2-1/2" x 2-1/2" x 4" | Q-12 | 5 | 3.200 | Ea. | 425 | 153 | | 578 | 700 |
| 7330 2-1/2" x 2-1/2" x 6" | " | 4.60 | 3.478 | " | 595 | 166 | | 761 | 905 |
| 7400 For polished chrome, add | | | | | 15% | | | | |
| 7440 For sill cock combination, add | | | | Ea. | 50.50 | | | 50.50 | 56 |
| 7900 Three way, flush, polished brass | | | | | | | | | |
| 7920 2-1/2" (3) x 4" | Q-12 | 4.80 | 3.333 | Ea. | 960 | 159 | | 1,119 | 1,300 |
| 7930 2-1/2" (3) x 6" | " | 4.80 | 3.333 | ↓ | 1,150 | 159 | | 1,309 | 1,525 |
| 8000 For polished chrome, add | | | | | 9% | | | | |
| 8020 Three way, projecting, polished brass | | | | | | | | | |
| 8040 2-1/2" (3) x 4" | Q-12 | 4.80 | 3.333 | Ea. | 985 | 159 | | 1,144 | 1,325 |

# 21 12 Fire-Suppression Standpipes

## 21 12 13 – Fire-Suppression Hoses and Nozzles

| 21 12 13.50 Fire Hoses and Nozzles | | Crew | Daily Output | Labor-Hours | Unit | Material | 2011 Bare Costs Labor | Equipment | Total | Total Incl O&P |
|---|---|---|---|---|---|---|---|---|---|---|
| 0010 | **FIRE HOSES AND NOZZLES** | | | | | | | | | |
| 0200 | Adapters, rough brass, straight hose threads | | | | | | | | | |
| 0220 | One piece, female to male, rocker lugs | | | | | | | | | |
| 0240 | 1" x 1" | | | | Ea. | 39.50 | | | 39.50 | 43.50 |
| 2200 | Hose, less couplings | | | | | | | | | |
| 2260 | Synthetic jacket, lined, 300 lb. test, 1-1/2" diameter | Q-12 | 2600 | .006 | L.F. | 2.39 | .29 | | 2.68 | 3.07 |
| 2280 | 2-1/2" diameter | | 2200 | .007 | | 3.98 | .35 | | 4.33 | 4.90 |
| 2360 | High strength, 500 lb. test, 1-1/2" diameter | | 2600 | .006 | | 2.47 | .29 | | 2.76 | 3.16 |
| 2380 | 2-1/2" diameter | ▼ | 2200 | .007 | ▼ | 4.29 | .35 | | 4.64 | 5.25 |
| 5600 | Nozzles, brass | | | | | | | | | |
| 5620 | Adjustable fog, 3/4" booster line | | | | Ea. | 91 | | | 91 | 100 |
| 5630 | 1" booster line | | | | | 109 | | | 109 | 119 |
| 5640 | 1-1/2" leader line | | | | | 66 | | | 66 | 72.50 |
| 5660 | 2-1/2" direct connection | | | | ▼ | 172 | | | 172 | 189 |
| 5780 | For chrome plated, add | | | | | 8% | | | | |
| 5850 | Electrical fire, adjustable fog, no shock | | | | | | | | | |
| 5900 | 1-1/2" | | | | Ea. | 340 | | | 340 | 375 |
| 5920 | 2-1/2" | | | | | 455 | | | 455 | 505 |
| 5980 | For polished chrome, add | | | | ▼ | 6% | | | | |
| 6200 | Heavy duty, comb. adj. fog and str. stream, with handle | | | | | | | | | |
| 6210 | 1" booster line | | | | Ea. | 325 | | | 325 | 360 |

## 21 12 19 – Fire-Suppression Hose Racks

### 21 12 19.50 Fire Hose Racks

| 0010 | **FIRE HOSE RACKS** | | | | | | | | | |
|---|---|---|---|---|---|---|---|---|---|---|
| 2600 | Hose rack, swinging, for 1-1/2" diameter hose, | | | | | | | | | |
| 2620 | Enameled steel, 50' & 75' lengths of hose | Q-12 | 20 | .800 | Ea. | 48.50 | 38 | | 86.50 | 110 |
| 2640 | 100' and 125' lengths of hose | " | 20 | .800 | " | 49 | 38 | | 87 | 111 |

## 21 12 23 – Fire-Suppression Hose Valves

### 21 12 23.70 Fire Hose Valves

| 0010 | **FIRE HOSE VALVES** | | | | | | | | | |
|---|---|---|---|---|---|---|---|---|---|---|
| 0080 | Wheel handle, 300 lb., 1-1/2" | 1 Spri | 12 | .667 | Ea. | 74.50 | 35.50 | | 110 | 135 |
| 0090 | 2-1/2" | " | 7 | 1.143 | " | 60 | 60.50 | | 120.50 | 157 |
| 0100 | For polished brass, add | | | | | 35% | | | | |
| 0110 | For polished chrome, add | | | | | 50% | | | | |

# 21 13 Fire-Suppression Sprinkler Systems

## 21 13 13 – Wet-Pipe Sprinkler Systems

### 21 13 13.50 Wet-Pipe Sprinkler System Components

| 0010 | **WET-PIPE SPRINKLER SYSTEM COMPONENTS** | | | | | | | | | |
|---|---|---|---|---|---|---|---|---|---|---|
| 2600 | Sprinkler heads, not including supply piping | | | | | | | | | |
| 3700 | Standard spray, pendent or upright, brass, 135°F to 286°F | | | | | | | | | |
| 3730 | 1/2" NPT, 7/16" orifice | 1 Spri | 16 | .500 | Ea. | 12 | 26.50 | | 38.50 | 52.50 |
| 3740 | 1/2" NPT, 1/2" orifice | " | 16 | .500 | | 7.80 | 26.50 | | 34.30 | 48 |
| 3860 | For wax and lead coating, add | | | | | 25.50 | | | 25.50 | 28 |
| 3880 | For wax coating, add | | | | | 15.25 | | | 15.25 | 16.80 |
| 3900 | For lead coating, add | | | | ▼ | 16.50 | | | 16.50 | 18.15 |
| 3920 | For 360°F, same cost | | | | | | | | | |
| 3930 | For 400°F, add | | | | Ea. | 58.50 | | | 58.50 | 64 |
| 3940 | For 500°F, add | | | | " | 58.50 | | | 58.50 | 64 |
| 4500 | Sidewall, horizontal, brass, 135°F to 286°F | | | | | | | | | |

# 21 13 Fire-Suppression Sprinkler Systems

## 21 13 13 – Wet-Pipe Sprinkler Systems

### 21 13 13.50 Wet-Pipe Sprinkler System Components

| | | Crew | Daily Output | Labor-Hours | Unit | Material | 2011 Bare Costs Labor | Equipment | Total | Total Incl O&P |
|---|---|---|---|---|---|---|---|---|---|---|
| 4520 | 1/2" NPT, 1/2" orifice | 1 Spri | 16 | .500 | Ea. | 15 | 26.50 | | 41.50 | 56 |
| 4540 | For 360°F, same cost | | | | | | | | | |
| 4800 | Recessed pendent, brass, 135°F to 286°F | | | | | | | | | |
| 4820 | 1/2" NPT, 3/8" orifice | 1 Spri | 10 | .800 | Ea. | 35 | 42.50 | | 77.50 | 102 |
| 4830 | 1/2" NPT, 7/16" orifice | | 10 | .800 | | 19.70 | 42.50 | | 62.20 | 85 |
| 4840 | 1/2" NPT, 1/2" orifice | | 10 | .800 | | 14.30 | 42.50 | | 56.80 | 79.50 |

## 21 13 16 – Dry-Pipe Sprinkler Systems

### 21 13 16.50 Dry-Pipe Sprinkler System Components

| | | Crew | Daily Output | Labor-Hours | Unit | Material | 2011 Bare Costs Labor | Equipment | Total | Total Incl O&P |
|---|---|---|---|---|---|---|---|---|---|---|
| 0010 | **DRY-PIPE SPRINKLER SYSTEM COMPONENTS** | | | | | | | | | |
| 0600 | Accelerator | 1 Spri | 8 | 1 | Ea. | 595 | 53 | | 648 | 735 |
| 2600 | Sprinkler heads, not including supply piping | | | | | | | | | |
| 2640 | Dry, pendent, 1/2" orifice, 3/4" or 1" NPT | | | | | | | | | |
| 2700 | 15-1/4" to 18" length | 1 Spri | 14 | .571 | Ea. | 112 | 30.50 | | 142.50 | 169 |
| 2710 | 18-1/4" to 21" length | | 13 | .615 | | 116 | 32.50 | | 148.50 | 176 |
| 2720 | 21-1/4" to 24" length | | 13 | .615 | | 120 | 32.50 | | 152.50 | 181 |
| 2730 | 24-1/4" to 27" length | | 13 | .615 | | 125 | 32.50 | | 157.50 | 186 |

## 21 13 26 – Deluge Fire-Suppression Sprinkler Systems

### 21 13 26.50 Deluge Fire-Suppression Sprinkler Sys. Comp.

| | | Crew | Daily Output | Labor-Hours | Unit | Material | 2011 Bare Costs Labor | Equipment | Total | Total Incl O&P |
|---|---|---|---|---|---|---|---|---|---|---|
| 0010 | **DELUGE FIRE-SUPPRESSION SPRINKLER SYSTEM COMPONENTS** | | | | | | | | | |
| 6200 | Valves and components | | | | | | | | | |
| 7000 | Deluge, assembly, incl. trim, pressure | | | | | | | | | |
| 7020 | operated relief, emergency release, gauges | | | | | | | | | |
| 7040 | 2" size | Q-12 | 2 | 8 | Ea. | 2,900 | 380 | | 3,280 | 3,775 |
| 7060 | 3" size | " | 1.50 | 10.667 | " | 3,250 | 510 | | 3,760 | 4,325 |

## 21 13 39 – Foam-Water Systems

### 21 13 39.50 Foam-Water System Components

| | | Crew | Daily Output | Labor-Hours | Unit | Material | 2011 Bare Costs Labor | Equipment | Total | Total Incl O&P |
|---|---|---|---|---|---|---|---|---|---|---|
| 0010 | **FOAM-WATER SYSTEM COMPONENTS** | | | | | | | | | |
| 2600 | Sprinkler heads, not including supply piping | | | | | | | | | |
| 3600 | Foam-water, pendent or upright, 1/2" NPT | 1 Spri | 12 | .667 | Ea. | 136 | 35.50 | | 171.50 | 203 |

# 21 21 Carbon-Dioxide Fire-Extinguishing Systems

## 21 21 16 – Carbon-Dioxide Fire-Extinguishing Equipment

### 21 21 16.50 CO2 Fire Extinguishing System

| | | Crew | Daily Output | Labor-Hours | Unit | Material | 2011 Bare Costs Labor | Equipment | Total | Total Incl O&P |
|---|---|---|---|---|---|---|---|---|---|---|
| 0010 | **$CO_2$ FIRE EXTINGUISHING SYSTEM** | | | | | | | | | |
| 0042 | For detectors and control stations, see Section 28 31 23.50 | | | | | | | | | |
| 1000 | Dispersion nozzle, $CO_2$, 3" x 5" | 1 Plum | 18 | .444 | Ea. | 56 | 24 | | 80 | 97.50 |
| 2000 | Extinguisher, $CO_2$ system, high pressure, 75 lb. cylinder | Q-1 | 6 | 2.667 | | 1,075 | 129 | | 1,204 | 1,375 |
| 3000 | Electro/mechanical release | L-1 | 4 | 4 | | 140 | 208 | | 348 | 465 |
| 3400 | Manual pull station | 1 Plum | 6 | 1.333 | | 50.50 | 71.50 | | 122 | 163 |

# 21 22 Clean-Agent Fire-Extinguishing Systems

## 21 22 16 – Clean-Agent Fire-Extinguishing Equipment

### 21 22 16.50 FM200 Fire Extinguishing System

| | 21 22 16.50 FM200 Fire Extinguishing System | Crew | Daily Output | Labor-Hours | Unit | Material | 2011 Bare Costs Labor | Equipment | Total | Total Incl O&P |
|---|---|---|---|---|---|---|---|---|---|---|
| 0010 | **FM200 FIRE EXTINGUISHING SYSTEM** | | | | | | | | | |
| 1100 | Dispersion nozzle FM200, 1-1/2" | 1 Plum | 14 | .571 | Ea. | 56 | 30.50 | | 86.50 | 108 |
| 2400 | Extinguisher, FM 200 system, filled, with mounting bracket | | | | | | | | | |
| 2540 | 196 lb. container | Q-1 | 4 | 4 | Ea. | 6,525 | 193 | | 6,718 | 7,475 |
| 6000 | Average FM200 system, minimum | | | | C.F. | 1.48 | | | 1.48 | 1.63 |
| 6020 | Maximum | | | | " | 2.94 | | | 2.94 | 3.23 |

# 21 31 Centrifugal Fire Pumps

## 21 31 13 – Electric-Drive, Centrifugal Fire Pumps

### 21 31 13.50 Electric-Drive Fire Pumps

| | | Crew | Daily Output | Labor-Hours | Unit | Material | 2011 Bare Costs Labor | Equipment | Total | Total Incl O&P |
|---|---|---|---|---|---|---|---|---|---|---|
| 0010 | **ELECTRIC-DRIVE FIRE PUMPS** Including controller, fittings and relief valve | | | | | | | | | |
| 3100 | 250 GPM, 55 psi, 15 HP, 3,550 RPM, 2" pump | Q-13 | .70 | 45.714 | Ea. | 14,000 | 2,300 | | 16,300 | 18,900 |
| 3200 | 500 GPM, 50 psi, 27 HP, 1,770 RPM, 4" pump | | .68 | 47.059 | | 14,700 | 2,375 | | 17,075 | 19,700 |
| 3350 | 750 GPM, 50 psi, 44 HP, 1,770 RPM, 5" pump | | .64 | 50 | | 15,200 | 2,525 | | 17,725 | 20,500 |
| 3400 | 750 GPM, 100 psi, 66 HP, 3,550 RPM, 4" pump | | .58 | 55.172 | | 17,300 | 2,800 | | 20,100 | 23,300 |
| 5000 | For jockey pump 1", 3 HP, with control, add | Q-12 | 2 | 8 | | 2,125 | 380 | | 2,505 | 2,925 |

## 21 31 16 – Diesel-Drive, Centrifugal Fire Pumps

### 21 31 16.50 Diesel-Drive Fire Pumps

| | | Crew | Daily Output | Labor-Hours | Unit | Material | 2011 Bare Costs Labor | Equipment | Total | Total Incl O&P |
|---|---|---|---|---|---|---|---|---|---|---|
| 0010 | **DIESEL-DRIVE FIRE PUMPS** Including controller, fittings and relief valve | | | | | | | | | |
| 0050 | 500 GPM, 50 psi, 27 HP, 4" pump | Q-13 | .64 | 50 | Ea. | 26,400 | 2,525 | | 28,925 | 32,800 |
| 0200 | 750 GPM, 50 psi, 44 HP, 5" pump | | .60 | 53.333 | | 27,300 | 2,700 | | 30,000 | 34,000 |
| 0400 | 1,000 GPM, 100 psi, 89 HP, 4" pump | | .56 | 57.143 | | 31,100 | 2,900 | | 34,000 | 38,500 |
| 0700 | 2,000 GPM, 100 psi, 167 HP, 6" pump | | .34 | 94.118 | | 34,200 | 4,750 | | 38,950 | 44,700 |
| 0950 | 3,500 GPM, 100 psi, 300 HP, 10" pump | | .24 | 133 | | 63,500 | 6,750 | | 70,250 | 80,000 |

# Division Notes

| | CREW | DAILY OUTPUT | LABOR-HOURS | UNIT | 2011 BARE COSTS | | | | TOTAL INCL O&P |
|---|---|---|---|---|---|---|---|---|---|
| | | | | | MAT. | LABOR | EQUIP. | TOTAL | |
| | | | | | | | | | |
| | | | | | | | | | |
| | | | | | | | | | |
| | | | | | | | | | |
| | | | | | | | | | |
| | | | | | | | | | |
| | | | | | | | | | |
| | | | | | | | | | |
| | | | | | | | | | |
| | | | | | | | | | |
| | | | | | | | | | |
| | | | | | | | | | |
| | | | | | | | | | |
| | | | | | | | | | |
| | | | | | | | | | |
| | | | | | | | | | |
| | | | | | | | | | |
| | | | | | | | | | |
| | | | | | | | | | |
| | | | | | | | | | |
| | | | | | | | | | |

# Estimating Tips

## 22 10 00 Plumbing Piping and Pumps

This subdivision is primarily basic pipe and related materials. The pipe may be used by any of the mechanical disciplines, i.e., plumbing, fire protection, heating, and air conditioning.

- The labor adjustment factors listed in Subdivision 22 01 02.20 apply throughout Divisions 21, 22, and 23. CAUTION: the correct percentage may vary for the same items. For example, the percentage add for the basic pipe installation should be based on the maximum height that the craftsman must install for that particular section. If the pipe is to be located 14' above the floor but it is suspended on threaded rod from beams, the bottom flange of which is 18' high (4' rods), then the height is actually 18' and the add is 20%. The pipe coverer, however, does not have to go above the 14', and so his or her add should be 10%.

- Most pipe is priced first as straight pipe with a joint (coupling, weld, etc.) every 10' and a hanger usually every 10'.

There are exceptions with hanger spacing such as for cast iron pipe (5') and plastic pipe (3 per 10'). Following each type of pipe there are several lines listing sizes and the amount to be subtracted to delete couplings and hangers. This is for pipe that is to be buried or supported together on trapeze hangers. The reason that the couplings are deleted is that these runs are usually long, and frequently longer lengths of pipe are used. By deleting the couplings, the estimator is expected to look up and add back the correct reduced number of couplings.

- When preparing an estimate, it may be necessary to approximate the fittings. Fittings usually run between 25% and 50% of the cost of the pipe. The lower percentage is for simpler runs, and the higher number is for complex areas, such as mechanical rooms.

- For historic restoration projects, the systems must be as invisible as possible, and pathways must be sought for pipes, conduit, and ductwork. While installations in accessible spaces (such as basements and attics) are relatively straightforward to estimate, labor costs may be more difficult to determine when delivery systems must be concealed.

## 22 40 00 Plumbing Fixtures

- Plumbing fixture costs usually require two lines: the fixture itself and its "rough-in, supply, and waste."

- In the Assemblies Section (Plumbing D2010) for the desired fixture, the System Components Group at the center of the page shows the fixture on the first line. The rest of the list (fittings, pipe, tubing, etc.) will total up to what we refer to in the Unit Price section as "Rough-in, supply, waste, and vent." Note that for most fixtures we allow a nominal 5' of tubing to reach from the fixture to a main or riser.

- Remember that gas- and oil-fired units need venting.

# Reference Numbers

Reference numbers are shown in shaded boxes at the beginning of some major classifications. These numbers refer to related items in the Reference Section. The reference information may be an estimating procedure, an alternate pricing method, or technical information.

*Note:* Not all subdivisions listed here necessarily appear in this publication.

*Note:* **Trade Service,** *in part, has been used as a reference source for some of the material prices used in Division 22.*

## 22 01 02 – Labor Adjustments

### 22 01 02.10 Boilers, General

| | | Crew | Daily Output | Labor-Hours | Unit | Material | 2011 Bare Costs Labor | 2011 Bare Costs Equipment | Total | Total Incl O&P |
|---|---|---|---|---|---|---|---|---|---|---|
| 0010 | **BOILERS, GENERAL**, Prices do not include flue piping, elec. wiring, | | | | | | | | | |
| 0020 | gas or oil piping, boiler base, pad, or tankless unless noted | | | | | | | | | |
| 0100 | Boiler H.P.: 10 KW = 34 lbs/steam/hr = 33,475 BTU/hr.    R235000-50 | | | | | | | | | |
| 0150 | To convert SFR to BTU rating: Hot water, 150 x SFR; | | | | | | | | | |
| 0160 | Forced hot water, 180 x SFR; steam, 240 x SFR | | | | | | | | | |

### 22 01 02.20 Labor Adjustment Factors

| | | Crew | Daily Output | Labor-Hours | Unit | Material | 2011 Bare Costs Labor | 2011 Bare Costs Equipment | Total | Total Incl O&P |
|---|---|---|---|---|---|---|---|---|---|---|
| 0010 | **LABOR ADJUSTMENT FACTORS**, (For Div. 21, 22 and 23) | | | | | | | | | |
| 1000 | Add to labor for elevated installation (Above floor level) | | | | | | | | | |
| 1080 | 10' to 14.5' high | | | | | | 10% | | | |
| 1100 | 15' to 19.5' high | | | | | | 20% | | | |
| 1120 | 20' to 24.5' high | | | | | | 25% | | | |
| 1140 | 25' to 29.5' high | | | | | | 35% | | | |
| 1160 | 30' to 34.5' high | | | | | | 40% | | | |
| 1180 | 35' to 39.5' high | | | | | | 50% | | | |
| 1200 | 40' and higher | | | | | | 55% | | | |
| 2000 | Add to labor for crawl space | | | | | | | | | |
| 2100 | 3' high | | | | | | 40% | | | |
| 2140 | 4' high | | | | | | 30% | | | |
| 3000 | Add to labor for multi-story building | | | | | | | | | |
| 3100 | Add per floor for floors 3 thru 19 | | | | | | 2% | | | |
| 3140 | Add per floor for floors 20 and up | | | | | | 4% | | | |
| 4000 | Add to labor for working in existing occupied buildings | | | | | | | | | |
| 4100 | Hospital | | | | | | 35% | | | |
| 4140 | Office building | | | | | | 25% | | | |
| 4180 | School | | | | | | 20% | | | |
| 4220 | Factory or warehouse | | | | | | 15% | | | |
| 4260 | Multi dwelling | | | | | | 15% | | | |
| 5000 | Add to labor, miscellaneous | | | | | | | | | |
| 5100 | Cramped shaft | | | | | | 35% | | | |
| 5140 | Congested area | | | | | | 15% | | | |
| 5180 | Excessive heat or cold | | | | | | 30% | | | |
| 9000 | Labor factors, The above are reasonable suggestions, however | | | | | | | | | |
| 9010 | each project should be evaluated for its own peculiarities. | | | | | | | | | |
| 9100 | Other factors to be considered are: | | | | | | | | | |
| 9140 | Movement of material and equipment through finished areas | | | | | | | | | |
| 9180 | Equipment room | | | | | | | | | |
| 9220 | Attic space | | | | | | | | | |
| 9260 | No service road | | | | | | | | | |
| 9300 | Poor unloading/storage area | | | | | | | | | |
| 9340 | Congested site area/heavy traffic | | | | | | | | | |

# 22 05 Common Work Results for Plumbing

## 22 05 05 – Selective Plumbing Demolition

### 22 05 05.10 Plumbing Demolition

| 22 05 05.10 Plumbing Demolition | Crew | Daily Output | Labor-Hours | Unit | Material | 2011 Bare Costs Labor | Equipment | Total | Total Incl O&P |
|---|---|---|---|---|---|---|---|---|---|
| 0010 **PLUMBING DEMOLITION** | | | | | | | | | |
| 1020 Fixtures, including 10' piping | | | | | | | | | |
| 1100 Bathtubs, cast iron | 1 Plum | 4 | 2 | Ea. | | 107 | | 107 | 160 |
| 1120 Fiberglass | | 6 | 1.333 | | | 71.50 | | 71.50 | 107 |
| 1140 Steel | | 5 | 1.600 | | | 86 | | 86 | 128 |
| 1200 Lavatory, wall hung | | 10 | .800 | | | 43 | | 43 | 64 |
| 1220 Counter top | | 8 | 1 | | | 53.50 | | 53.50 | 80 |
| 1300 Sink, single compartment | | 8 | 1 | | | 53.50 | | 53.50 | 80 |
| 1320 Double compartment | | 7 | 1.143 | | | 61.50 | | 61.50 | 91.50 |
| 1400 Water closet, floor mounted | | 8 | 1 | | | 53.50 | | 53.50 | 80 |
| 1420 Wall mounted | | 7 | 1.143 | | | 61.50 | | 61.50 | 91.50 |
| 1500 Urinal, floor mounted | | 4 | 2 | | | 107 | | 107 | 160 |
| 1520 Wall mounted | | 7 | 1.143 | | | 61.50 | | 61.50 | 91.50 |
| 1600 Water fountains, free standing | | 8 | 1 | | | 53.50 | | 53.50 | 80 |
| 1620 Wall or deck mounted | | 6 | 1.333 | | | 71.50 | | 71.50 | 107 |
| 2000 Piping, metal, up thru 1-1/2" diameter | | 200 | .040 | L.F. | | 2.14 | | 2.14 | 3.20 |
| 2050 2" thru 3-1/2" diameter | | 150 | .053 | | | 2.86 | | 2.86 | 4.27 |
| 2100 4" thru 6" diameter | 2 Plum | 100 | .160 | | | 8.60 | | 8.60 | 12.80 |
| 2150 8" thru 14" diameter | " | 60 | .267 | | | 14.30 | | 14.30 | 21.50 |
| 2153 16" thru 20" diameter | Q-18 | 70 | .343 | | | 17.30 | .46 | 17.76 | 26.50 |
| 2155 24" thru 26" diameter | | 55 | .436 | | | 22 | .59 | 22.59 | 33.50 |
| 2156 30" thru 36" diameter | | 40 | .600 | | | 30.50 | .81 | 31.31 | 46 |
| 2212 Deduct for salvage, aluminum scrap | | | | Ton | | | | 700 | 770 |
| 2214 Brass scrap | | | | | | | | 2,450 | 2,675 |
| 2216 Copper scrap | | | | | | | | 3,200 | 3,525 |
| 2218 Lead scrap | | | | | | | | 520 | 570 |
| 2220 Steel scrap | | | | | | | | 180 | 200 |
| 2250 Water heater, 40 gal. | 1 Plum | 6 | 1.333 | Ea. | | 71.50 | | 71.50 | 107 |
| 9470 Water softener | Q-1 | 2 | 8 | " | | 385 | | 385 | 575 |

## 22 05 23 – General-Duty Valves for Plumbing Piping

### 22 05 23.10 Valves, Brass

| 22 05 23.10 Valves, Brass | Crew | Daily Output | Labor-Hours | Unit | Material | 2011 Bare Costs Labor | Equipment | Total | Total Incl O&P |
|---|---|---|---|---|---|---|---|---|---|
| 0010 **VALVES, BRASS** | | | | | | | | | |
| 0500 Gas cocks, threaded | | | | | | | | | |
| 0530 1/2" | 1 Plum | 24 | .333 | Ea. | 8.90 | 17.85 | | 26.75 | 36.50 |
| 0540 3/4" | | 22 | .364 | | 11.20 | 19.50 | | 30.70 | 41.50 |
| 0550 1" | | 19 | .421 | | 17.65 | 22.50 | | 40.15 | 53 |
| 0560 1-1/4" | | 15 | .533 | | 38 | 28.50 | | 66.50 | 84.50 |

### 22 05 23.20 Valves, Bronze

| 22 05 23.20 Valves, Bronze | Crew | Daily Output | Labor-Hours | Unit | Material | 2011 Bare Costs Labor | Equipment | Total | Total Incl O&P |
|---|---|---|---|---|---|---|---|---|---|
| 0010 **VALVES, BRONZE** | | | | | | | | | |
| 1020 Angle, 150 lb., rising stem, threaded | | | | | | | | | |
| 1030 1/8" | 1 Plum | 24 | .333 | Ea. | 102 | 17.85 | | 119.85 | 139 |
| 1040 1/4" | | 24 | .333 | | 102 | 17.85 | | 119.85 | 139 |
| 1050 3/8" | | 24 | .333 | | 102 | 17.85 | | 119.85 | 139 |
| 1060 1/2" | | 22 | .364 | | 102 | 19.50 | | 121.50 | 141 |
| 1070 3/4" | | 20 | .400 | | 139 | 21.50 | | 160.50 | 185 |
| 1080 1" | | 19 | .421 | | 200 | 22.50 | | 222.50 | 254 |
| 1100 1-1/2" | | 13 | .615 | | 335 | 33 | | 368 | 420 |
| 1110 2" | | 11 | .727 | | 540 | 39 | | 579 | 655 |
| 1380 Ball, 150 psi, threaded | | | | | | | | | |
| 1400 1/4" | 1 Plum | 24 | .333 | Ea. | 11.15 | 17.85 | | 29 | 39 |
| 1430 3/8" | | 24 | .333 | | 11.15 | 17.85 | | 29 | 39 |
| 1450 1/2" | | 22 | .364 | | 11.30 | 19.50 | | 30.80 | 41.50 |

| 22 05 23.20 Valves, Bronze | | Crew | Daily Output | Labor-Hours | Unit | Material | 2011 Bare Costs Labor | Equipment | Total | Total Incl O&P |
|---|---|---|---|---|---|---|---|---|---|---|
| 1460 | 3/4" | 1 Plum | 20 | .400 | Ea. | 18.60 | 21.50 | | 40.10 | 52.50 |
| 1470 | 1" | | 19 | .421 | | 23.50 | 22.50 | | 46 | 59.50 |
| 1480 | 1-1/4" | | 15 | .533 | | 39.50 | 28.50 | | 68 | 86 |
| 1490 | 1-1/2" | | 13 | .615 | | 55 | 33 | | 88 | 110 |
| 1500 | 2" | | 11 | .727 | | 66.50 | 39 | | 105.50 | 131 |
| 1750 | Check, swing, class 150, regrinding disc, threaded | | | | | | | | | |
| 1800 | 1/8" | 1 Plum | 24 | .333 | Ea. | 47.50 | 17.85 | | 65.35 | 78.50 |
| 1830 | 1/4" | | 24 | .333 | | 47.50 | 17.85 | | 65.35 | 78.50 |
| 1840 | 3/8" | | 24 | .333 | | 47.50 | 17.85 | | 65.35 | 78.50 |
| 1850 | 1/2" | | 24 | .333 | | 49.50 | 17.85 | | 67.35 | 81 |
| 1860 | 3/4" | | 20 | .400 | | 72 | 21.50 | | 93.50 | 111 |
| 1870 | 1" | | 19 | .421 | | 104 | 22.50 | | 126.50 | 148 |
| 1880 | 1-1/4" | | 15 | .533 | | 149 | 28.50 | | 177.50 | 207 |
| 1890 | 1-1/2" | | 13 | .615 | | 174 | 33 | | 207 | 240 |
| 1900 | 2" | | 11 | .727 | | 256 | 39 | | 295 | 340 |
| 1910 | 2-1/2" | Q-1 | 15 | 1.067 | | 575 | 51.50 | | 626.50 | 705 |
| 2000 | For 200 lb, add | | | | | 5% | 10% | | | |
| 2040 | For 300 lb, add | | | | | 15% | 15% | | | |
| 2850 | Gate, N.R.S., soldered, 125 psi | | | | | | | | | |
| 2900 | 3/8" | 1 Plum | 24 | .333 | Ea. | 42 | 17.85 | | 59.85 | 72.50 |
| 2920 | 1/2" | | 24 | .333 | | 35 | 17.85 | | 52.85 | 65 |
| 2940 | 3/4" | | 20 | .400 | | 40.50 | 21.50 | | 62 | 76.50 |
| 2950 | 1" | | 19 | .421 | | 57 | 22.50 | | 79.50 | 96 |
| 2960 | 1-1/4" | | 15 | .533 | | 91 | 28.50 | | 119.50 | 143 |
| 2970 | 1-1/2" | | 13 | .615 | | 97 | 33 | | 130 | 156 |
| 2980 | 2" | | 11 | .727 | | 138 | 39 | | 177 | 210 |
| 2990 | 2-1/2" | Q-1 | 15 | 1.067 | | 345 | 51.50 | | 396.50 | 455 |
| 3000 | 3" | " | 13 | 1.231 | | 450 | 59.50 | | 509.50 | 585 |
| 3850 | Rising stem, soldered, 300 psi | | | | | | | | | |
| 3950 | 1" | 1 Plum | 19 | .421 | Ea. | 136 | 22.50 | | 158.50 | 184 |
| 3980 | 2" | " | 11 | .727 | | 360 | 39 | | 399 | 455 |
| 4000 | 3" | Q-1 | 13 | 1.231 | | 1,150 | 59.50 | | 1,209.50 | 1,350 |
| 4250 | Threaded, class 150 | | | | | | | | | |
| 4310 | 1/4" | 1 Plum | 24 | .333 | Ea. | 53.50 | 17.85 | | 71.35 | 85 |
| 4320 | 3/8" | | 24 | .333 | | 53.50 | 17.85 | | 71.35 | 85 |
| 4330 | 1/2" | | 24 | .333 | | 49.50 | 17.85 | | 67.35 | 81 |
| 4340 | 3/4" | | 20 | .400 | | 57.50 | 21.50 | | 79 | 95.50 |
| 4350 | 1" | | 19 | .421 | | 77 | 22.50 | | 99.50 | 118 |
| 4360 | 1-1/4" | | 15 | .533 | | 104 | 28.50 | | 132.50 | 158 |
| 4370 | 1-1/2" | | 13 | .615 | | 132 | 33 | | 165 | 194 |
| 4380 | 2" | | 11 | .727 | | 174 | 39 | | 213 | 250 |
| 4390 | 2-1/2" | Q-1 | 15 | 1.067 | | 415 | 51.50 | | 466.50 | 530 |
| 4400 | 3" | " | 13 | 1.231 | | 580 | 59.50 | | 639.50 | 725 |
| 4500 | For 300 psi, threaded, add | | | | | 100% | 15% | | | |
| 4540 | For chain operated type, add | | | | | 15% | | | | |
| 4850 | Globe, class 150, rising stem, threaded | | | | | | | | | |
| 4920 | 1/4" | 1 Plum | 24 | .333 | Ea. | 76 | 17.85 | | 93.85 | 110 |
| 4940 | 3/8" size | | 24 | .333 | | 76 | 17.85 | | 93.85 | 110 |
| 4950 | 1/2" | | 24 | .333 | | 76 | 17.85 | | 93.85 | 110 |
| 4960 | 3/4" | | 20 | .400 | | 102 | 21.50 | | 123.50 | 144 |
| 4970 | 1" | | 19 | .421 | | 160 | 22.50 | | 182.50 | 210 |
| 4980 | 1-1/4" | | 15 | .533 | | 254 | 28.50 | | 282.50 | 325 |
| 4990 | 1-1/2" | | 13 | .615 | | 310 | 33 | | 343 | 390 |

## 22 05 23 – General-Duty Valves for Plumbing Piping

### 22 05 23.20 Valves, Bronze

| | | Crew | Daily Output | Labor-Hours | Unit | Material | 2011 Bare Costs Labor | Equipment | Total | Total Incl O&P |
|---|---|---|---|---|---|---|---|---|---|---|
| 5000 | 2" | 1 Plum | 11 | .727 | Ea. | 465 | 39 | | 504 | 570 |
| 5010 | 2-1/2" | Q-1 | 15 | 1.067 | | 935 | 51.50 | | 986.50 | 1,100 |
| 5020 | 3" | " | 13 | 1.231 | ↓ | 1,325 | 59.50 | | 1,384.50 | 1,550 |
| 5120 | For 300 lb threaded, add | | | | | 50% | 15% | | | |
| 5600 | Relief, pressure & temperature, self-closing, ASME, threaded | | | | | | | | | |
| 5640 | 3/4" | 1 Plum | 28 | .286 | Ea. | 132 | 15.30 | | 147.30 | 169 |
| 5650 | 1" | | 24 | .333 | | 202 | 17.85 | | 219.85 | 249 |
| 5660 | 1-1/4" | | 20 | .400 | | 390 | 21.50 | | 411.50 | 455 |
| 5670 | 1-1/2" | | 18 | .444 | | 745 | 24 | | 769 | 850 |
| 5680 | 2" | ↓ | 16 | .500 | ↓ | 790 | 27 | | 817 | 910 |
| 5950 | Pressure, poppet type, threaded | | | | | | | | | |
| 6000 | 1/2" | 1 Plum | 30 | .267 | Ea. | 35 | 14.30 | | 49.30 | 60 |
| 6040 | 3/4" | " | 28 | .286 | " | 37.50 | 15.30 | | 52.80 | 64 |
| 6400 | Pressure, water, ASME, threaded | | | | | | | | | |
| 6440 | 3/4" | 1 Plum | 28 | .286 | Ea. | 74.50 | 15.30 | | 89.80 | 105 |
| 6450 | 1" | | 24 | .333 | | 158 | 17.85 | | 175.85 | 201 |
| 6460 | 1-1/4" | | 20 | .400 | | 251 | 21.50 | | 272.50 | 310 |
| 6470 | 1-1/2" | | 18 | .444 | | 345 | 24 | | 369 | 415 |
| 6480 | 2" | | 16 | .500 | | 500 | 27 | | 527 | 590 |
| 6490 | 2-1/2" | ↓ | 15 | .533 | ↓ | 1,900 | 28.50 | | 1,928.50 | 2,150 |
| 6900 | Reducing, water pressure | | | | | | | | | |
| 6920 | 300 psi to 25-75 psi, threaded or sweat | | | | | | | | | |
| 6940 | 1/2" | 1 Plum | 24 | .333 | Ea. | 239 | 17.85 | | 256.85 | 290 |
| 6950 | 3/4" | | 20 | .400 | | 239 | 21.50 | | 260.50 | 295 |
| 6960 | 1" | | 19 | .421 | | 370 | 22.50 | | 392.50 | 445 |
| 6970 | 1-1/4" | | 15 | .533 | | 665 | 28.50 | | 693.50 | 775 |
| 6980 | 1-1/2" | ↓ | 13 | .615 | ↓ | 1,000 | 33 | | 1,033 | 1,150 |
| 8350 | Tempering, water, sweat connections | | | | | | | | | |
| 8400 | 1/2" | 1 Plum | 24 | .333 | Ea. | 98 | 17.85 | | 115.85 | 135 |
| 8440 | 3/4" | " | 20 | .400 | " | 114 | 21.50 | | 135.50 | 157 |
| 8650 | Threaded connections | | | | | | | | | |
| 8700 | 1/2" | 1 Plum | 24 | .333 | Ea. | 141 | 17.85 | | 158.85 | 182 |
| 8740 | 3/4" | | 20 | .400 | | 445 | 21.50 | | 466.50 | 515 |
| 8750 | 1" | | 19 | .421 | | 500 | 22.50 | | 522.50 | 585 |
| 8760 | 1-1/4" | | 15 | .533 | | 770 | 28.50 | | 798.50 | 890 |
| 8770 | 1-1/2" | | 13 | .615 | | 845 | 33 | | 878 | 975 |
| 8780 | 2" | ↓ | 11 | .727 | ↓ | 1,275 | 39 | | 1,314 | 1,450 |

### 22 05 23.60 Valves, Plastic

| | | Crew | Daily Output | Labor-Hours | Unit | Material | 2011 Bare Costs Labor | Equipment | Total | Total Incl O&P |
|---|---|---|---|---|---|---|---|---|---|---|
| 0010 | **VALVES, PLASTIC** | | | | | | | | | |
| 1100 | Angle, PVC, threaded | | | | | | | | | |
| 1110 | 1/4" | 1 Plum | 26 | .308 | Ea. | 69.50 | 16.50 | | 86 | 101 |
| 1120 | 1/2" | | 26 | .308 | | 69.50 | 16.50 | | 86 | 101 |
| 1130 | 3/4" | | 25 | .320 | | 82.50 | 17.15 | | 99.65 | 117 |
| 1140 | 1" | ↓ | 23 | .348 | ↓ | 100 | 18.65 | | 118.65 | 138 |
| 1150 | Ball, PVC, socket or threaded, single union | | | | | | | | | |
| 1230 | 1/2" | 1 Plum | 26 | .308 | Ea. | 26.50 | 16.50 | | 43 | 53.50 |
| 1240 | 3/4" | | 25 | .320 | | 30 | 17.15 | | 47.15 | 58.50 |
| 1250 | 1" | | 23 | .348 | | 38 | 18.65 | | 56.65 | 69.50 |
| 1260 | 1-1/4" | | 21 | .381 | | 51 | 20.50 | | 71.50 | 86.50 |
| 1270 | 1-1/2" | | 20 | .400 | | 60.50 | 21.50 | | 82 | 98.50 |
| 1280 | 2" | ↓ | 17 | .471 | | 87.50 | 25 | | 112.50 | 134 |
| 1360 | For PVC, flanged, add | | | | ↓ | 100% | 15% | | | |

## 22 05 23 – General-Duty Valves for Plumbing Piping

| 22 05 23.60 Valves, Plastic | | Crew | Daily Output | Labor-Hours | Unit | Material | 2011 Bare Costs Labor | 2011 Bare Costs Equipment | Total | Total Incl O&P |
|---|---|---|---|---|---|---|---|---|---|---|
| 1650 | CPVC, socket or threaded, single union | | | | | | | | | |
| 1700 | 1/2" | 1 Plum | 26 | .308 | Ea. | 50 | 16.50 | | 66.50 | 79.50 |
| 1720 | 3/4" | | 25 | .320 | | 62.50 | 17.15 | | 79.65 | 94.50 |
| 1730 | 1" | | 23 | .348 | | 75 | 18.65 | | 93.65 | 111 |
| 1750 | 1-1/4" | | 21 | .381 | | 120 | 20.50 | | 140.50 | 162 |
| 1760 | 1-1/2" | | 20 | .400 | | 120 | 21.50 | | 141.50 | 163 |
| 1840 | For CPVC, flanged, add | | | | | 65% | 15% | | | |
| 1880 | For true union, socket or threaded, add | | | | | 50% | 5% | | | |
| 2050 | Polypropylene, threaded | | | | | | | | | |
| 2100 | 1/4" | 1 Plum | 26 | .308 | Ea. | 38.50 | 16.50 | | 55 | 66.50 |
| 2120 | 3/8" | | 26 | .308 | | 38.50 | 16.50 | | 55 | 66.50 |
| 2130 | 1/2" | | 26 | .308 | | 38.50 | 16.50 | | 55 | 66.50 |
| 2140 | 3/4" | | 25 | .320 | | 46.50 | 17.15 | | 63.65 | 76.50 |
| 2150 | 1" | | 23 | .348 | | 53.50 | 18.65 | | 72.15 | 87 |
| 2160 | 1-1/4" | | 21 | .381 | | 72 | 20.50 | | 92.50 | 110 |
| 2170 | 1-1/2" | | 20 | .400 | | 88 | 21.50 | | 109.50 | 129 |
| 2180 | 2" | | 17 | .471 | | 119 | 25 | | 144 | 169 |
| 4850 | Foot valve, PVC, socket or threaded | | | | | | | | | |
| 4900 | 1/2" | 1 Plum | 34 | .235 | Ea. | 66.50 | 12.60 | | 79.10 | 92 |
| 4930 | 3/4" | | 32 | .250 | | 75 | 13.40 | | 88.40 | 103 |
| 4940 | 1" | | 28 | .286 | | 97.50 | 15.30 | | 112.80 | 130 |
| 4950 | 1-1/4" | | 27 | .296 | | 187 | 15.90 | | 202.90 | 230 |
| 4960 | 1-1/2" | | 26 | .308 | | 187 | 16.50 | | 203.50 | 231 |
| 6350 | Y sediment strainer, PVC, socket or threaded | | | | | | | | | |
| 6400 | 1/2" | 1 Plum | 26 | .308 | Ea. | 44 | 16.50 | | 60.50 | 73 |
| 6440 | 3/4" | | 24 | .333 | | 47 | 17.85 | | 64.85 | 78.50 |
| 6450 | 1" | | 23 | .348 | | 57 | 18.65 | | 75.65 | 90.50 |
| 6460 | 1-1/4" | | 21 | .381 | | 94.50 | 20.50 | | 115 | 135 |
| 6470 | 1-1/2" | | 20 | .400 | | 94.50 | 21.50 | | 116 | 136 |

## 22 05 48 – Vibration and Seismic Controls for Plumbing Piping and Equipment

### 22 05 48.10 Seismic Bracing Supports

| 22 05 48.10 Seismic Bracing Supports | | Crew | Daily Output | Labor-Hours | Unit | Material | 2011 Bare Costs Labor | 2011 Bare Costs Equipment | Total | Total Incl O&P |
|---|---|---|---|---|---|---|---|---|---|---|
| 0010 | **SEISMIC BRACING SUPPORTS** | | | | | | | | | |
| 0020 | Clamps | | | | | | | | | |
| 0030 | C-clamp, for mounting on steel beam | | | | | | | | | |
| 0040 | 3/8" threaded rod | 1 Skwk | 160 | .050 | Ea. | 3.15 | 2.22 | | 5.37 | 6.90 |
| 0050 | 1/2" threaded rod | | 160 | .050 | | 3.91 | 2.22 | | 6.13 | 7.70 |
| 0060 | 5/8" threaded rod | | 160 | .050 | | 6 | 2.22 | | 8.22 | 10 |
| 0070 | 3/4" threaded rod | | 160 | .050 | | 8.05 | 2.22 | | 10.27 | 12.30 |
| 0100 | Brackets | | | | | | | | | |
| 0110 | Beam side or wall malleable iron | | | | | | | | | |
| 0120 | 3/8" threaded rod | 1 Skwk | 48 | .167 | Ea. | 3.34 | 7.40 | | 10.74 | 15 |
| 0130 | 1/2" threaded rod | | 48 | .167 | | 5.80 | 7.40 | | 13.20 | 17.75 |
| 0140 | 5/8" threaded rod | | 48 | .167 | | 9.10 | 7.40 | | 16.50 | 21.50 |
| 0150 | 3/4" threaded rod | | 48 | .167 | | 10.95 | 7.40 | | 18.35 | 23.50 |
| 0160 | 7/8" threaded rod | | 48 | .167 | | 12.85 | 7.40 | | 20.25 | 25.50 |
| 0170 | For concrete installation, add | | | | | | 30% | | | |
| 0180 | Wall, welded steel | | | | | | | | | |
| 0190 | 0 size 12" wide 18" deep | 1 Skwk | 34 | .235 | Ea. | 234 | 10.45 | | 244.45 | 273 |
| 0200 | 1 size 18" wide 24" deep | | 34 | .235 | | 278 | 10.45 | | 288.45 | 320 |
| 0210 | 2 size 24" wide 30" deep | | 34 | .235 | | 365 | 10.45 | | 375.45 | 420 |
| 0300 | Rod, carbon steel | | | | | | | | | |
| 0310 | Continuous thread | | | | | | | | | |

## 22 05 48 – Vibration and Seismic Controls for Plumbing Piping and Equipment

| 22 05 48.10 Seismic Bracing Supports | | Crew | Daily Output | Labor-Hours | Unit | Material | 2011 Bare Costs Labor | Equipment | Total | Total Incl O&P |
|---|---|---|---|---|---|---|---|---|---|---|
| 0320 | 1/4" thread | 1 Skwk | 144 | .056 | L.F. | 1.35 | 2.47 | | 3.82 | 5.25 |
| 0330 | 3/8" thread | | 144 | .056 | | 1.44 | 2.47 | | 3.91 | 5.35 |
| 0340 | 1/2" thread | | 144 | .056 | | 2.26 | 2.47 | | 4.73 | 6.25 |
| 0350 | 5/8" thread | | 144 | .056 | | 3.21 | 2.47 | | 5.68 | 7.30 |
| 0360 | 3/4" thread | | 144 | .056 | | 5.65 | 2.47 | | 8.12 | 10.05 |
| 0370 | 7/8" thread | | 144 | .056 | | 7.10 | 2.47 | | 9.57 | 11.60 |
| 0380 | For galvanized, add | | | | | 30% | | | | |
| 0400 | Channel, steel | | | | | | | | | |
| 0410 | 3/4" x 1-1/2" | 1 Skwk | 80 | .100 | L.F. | 6.10 | 4.44 | | 10.54 | 13.50 |
| 0420 | 1-1/2" x 1-1/2" | | 70 | .114 | | 8.05 | 5.05 | | 13.10 | 16.70 |
| 0430 | 1-7/8" x 1-1/2" | | 60 | .133 | | 26 | 5.90 | | 31.90 | 37.50 |
| 0440 | 3" x 1-1/2" | | 50 | .160 | | 30 | 7.10 | | 37.10 | 44 |
| 0450 | Spring nuts | | | | | | | | | |
| 0460 | 3/8" | 1 Skwk | 100 | .080 | Ea. | 1.85 | 3.55 | | 5.40 | 7.50 |
| 0470 | 1/2" | " | 80 | .100 | " | 2 | 4.44 | | 6.44 | 9 |
| 0500 | Welding, field | | | | | | | | | |
| 0510 | Cleaning and welding plates, bars, or rods | | | | | | | | | |
| 0520 | To existing beams, columns, or trusses | | | | | | | | | |
| 0530 | 1" weld | 1 Skwk | 144 | .056 | Ea. | .20 | 2.47 | | 2.67 | 4 |
| 0540 | 2" weld | | 72 | .111 | | .35 | 4.93 | | 5.28 | 7.95 |
| 0550 | 3" weld | | 54 | .148 | | .54 | 6.60 | | 7.14 | 10.70 |
| 0560 | 4" weld | | 36 | .222 | | .75 | 9.85 | | 10.60 | 16 |
| 0570 | 5" weld | | 30 | .267 | | .94 | 11.85 | | 12.79 | 19.20 |
| 0580 | 6" weld | | 24 | .333 | | 1.07 | 14.80 | | 15.87 | 23.50 |
| 0600 | Vibration absorbers | | | | | | | | | |
| 0610 | Hangers, neoprene flex | | | | | | | | | |
| 0620 | 10-120 lb. capacity | 1 Skwk | 8 | 1 | Ea. | 22.50 | 44.50 | | 67 | 93 |
| 0630 | 75-550 lb. capacity | | 8 | 1 | | 30.50 | 44.50 | | 75 | 102 |
| 0640 | 250-1100 lb. capacity | | 6 | 1.333 | | 63 | 59 | | 122 | 161 |
| 0650 | 1000-4000 lb. capacity | | 6 | 1.333 | | 100 | 59 | | 159 | 201 |
| 0660 | Spring flex | | | | | | | | | |
| 0670 | 60-450 lb. capacity | 1 Skwk | 8 | 1 | Ea. | 62.50 | 44.50 | | 107 | 137 |
| 0680 | 85-450 lb. capacity | | 8 | 1 | | 83.50 | 44.50 | | 128 | 160 |
| 0690 | 600-900 lb. capacity | | 8 | 1 | | 86.50 | 44.50 | | 131 | 163 |
| 0700 | 1100-1300 lb. capacity | | 6 | 1.333 | | 92 | 59 | | 151 | 192 |
| 0710 | Mounts, neoprene | | | | | | | | | |
| 0720 | 135-380 lb. capacity | 1 Skwk | 7 | 1.143 | Ea. | 15.35 | 50.50 | | 65.85 | 95 |
| 0730 | 250-1100 lb. capacity | | 7 | 1.143 | | 40.50 | 50.50 | | 91 | 123 |
| 0740 | 1000-4000 lb. capacity | | 5 | 1.600 | | 90.50 | 71 | | 161.50 | 209 |
| 0900 | Wire support | | | | | | | | | |
| 0910 | Light fixture, recessed (pair) | 1 Skwk | 96 | .083 | Ea. | 3.45 | 3.70 | | 7.15 | 9.45 |

## 22 05 76 – Facility Drainage Piping Cleanouts

### 22 05 76.10 Cleanouts

| | | Crew | Daily Output | Labor-Hours | Unit | Material | Labor | Equipment | Total | Total Incl O&P |
|---|---|---|---|---|---|---|---|---|---|---|
| 0010 | **CLEANOUTS** | | | | | | | | | |
| 0060 | Floor type | | | | | | | | | |
| 0080 | Round or square, scoriated nickel bronze top | | | | | | | | | |
| 0100 | 2" pipe size | 1 Plum | 10 | .800 | Ea. | 155 | 43 | | 198 | 234 |
| 0120 | 3" pipe size | | 8 | 1 | | 232 | 53.50 | | 285.50 | 335 |
| 0140 | 4" pipe size | | 6 | 1.333 | | 232 | 71.50 | | 303.50 | 365 |
| 0980 | Round top, recessed for terrazzo | | | | | | | | | |
| 1000 | 2" pipe size | 1 Plum | 9 | .889 | Ea. | 155 | 47.50 | | 202.50 | 241 |
| 1080 | 3" pipe size | | 6 | 1.333 | | 232 | 71.50 | | 303.50 | 365 |

## 22 05 76 – Facility Drainage Piping Cleanouts

### 22 05 76.10 Cleanouts

| | | Crew | Daily Output | Labor-Hours | Unit | Material | 2011 Bare Costs Labor | Equipment | Total | Total Incl O&P |
|---|---|---|---|---|---|---|---|---|---|---|
| 1100 | 4" pipe size | 1 Plum | 4 | 2 | Ea. | 232 | 107 | | 339 | 415 |
| 1120 | 5" pipe size | Q-1 | 6 | 2.667 | | 295 | 129 | | 424 | 515 |

### 22 05 76.20 Cleanout Tees

| | | Crew | Daily Output | Labor-Hours | Unit | Material | 2011 Bare Costs Labor | Equipment | Total | Total Incl O&P |
|---|---|---|---|---|---|---|---|---|---|---|
| 0010 | **CLEANOUT TEES** | | | | | | | | | |
| 0100 | Cast iron, B&S, with countersunk plug | | | | | | | | | |
| 0200 | 2" pipe size | 1 Plum | 4 | 2 | Ea. | 211 | 107 | | 318 | 390 |
| 0220 | 3" pipe size | | 3.60 | 2.222 | | 230 | 119 | | 349 | 430 |
| 0240 | 4" pipe size | | 3.30 | 2.424 | | 286 | 130 | | 416 | 510 |
| 0280 | 6" pipe size | Q-1 | 5 | 3.200 | | 770 | 154 | | 924 | 1,075 |
| 0500 | For round smooth access cover, same price | | | | | | | | | |
| 4000 | Plastic, tees and adapters. Add plugs | | | | | | | | | |
| 4010 | ABS, DWV | | | | | | | | | |
| 4020 | Cleanout tee, 1-1/2" pipe size | 1 Plum | 15 | .533 | Ea. | 7.25 | 28.50 | | 35.75 | 50.50 |
| 4030 | 2" pipe size | Q-1 | 27 | .593 | | 7.95 | 28.50 | | 36.45 | 51.50 |
| 4040 | 3" pipe size | | 21 | .762 | | 14.90 | 37 | | 51.90 | 71.50 |
| 4050 | 4" pipe size | | 16 | 1 | | 32 | 48.50 | | 80.50 | 107 |
| 4100 | Cleanout plug, 1-1/2" pipe size | 1 Plum | 32 | .250 | | 1.26 | 13.40 | | 14.66 | 21.50 |
| 4110 | 2" pipe size | Q-1 | 56 | .286 | | 1.68 | 13.80 | | 15.48 | 22.50 |
| 4120 | 3" pipe size | | 36 | .444 | | 2.66 | 21.50 | | 24.16 | 35 |
| 4130 | 4" pipe size | | 30 | .533 | | 4.65 | 25.50 | | 30.15 | 43.50 |
| 4180 | Cleanout adapter fitting, 1-1/2" pipe size | 1 Plum | 32 | .250 | | 1.98 | 13.40 | | 15.38 | 22 |
| 4190 | 2" pipe size | Q-1 | 56 | .286 | | 3.04 | 13.80 | | 16.84 | 24 |
| 4200 | 3" pipe size | | 36 | .444 | | 7.80 | 21.50 | | 29.30 | 40.50 |
| 4210 | 4" pipe size | | 30 | .533 | | 14.35 | 25.50 | | 39.85 | 54.50 |
| 5000 | PVC, DWV | | | | | | | | | |
| 5010 | Cleanout tee, 1-1/2" pipe size | 1 Plum | 15 | .533 | Ea. | 5.35 | 28.50 | | 33.85 | 48.50 |
| 5020 | 2" pipe size | Q-1 | 27 | .593 | | 6.20 | 28.50 | | 34.70 | 49.50 |
| 5030 | 3" pipe size | | 21 | .762 | | 12.05 | 37 | | 49.05 | 68.50 |
| 5040 | 4" pipe size | | 16 | 1 | | 21.50 | 48.50 | | 70 | 96 |
| 5090 | Cleanout plug, 1-1/2" pipe size | 1 Plum | 32 | .250 | | 1.27 | 13.40 | | 14.67 | 21.50 |
| 5100 | 2" pipe size | Q-1 | 56 | .286 | | 1.41 | 13.80 | | 15.21 | 22 |
| 5110 | 3" pipe size | | 36 | .444 | | 2.52 | 21.50 | | 24.02 | 35 |
| 5120 | 4" pipe size | | 30 | .533 | | 3.72 | 25.50 | | 29.22 | 42.50 |
| 5130 | 6" pipe size | | 24 | .667 | | 12.35 | 32 | | 44.35 | 61.50 |
| 5170 | Cleanout adapter fitting, 1-1/2" pipe size | 1 Plum | 32 | .250 | | 1.69 | 13.40 | | 15.09 | 22 |
| 5180 | 2" pipe size | Q-1 | 56 | .286 | | 2.18 | 13.80 | | 15.98 | 23 |
| 5190 | 3" pipe size | | 36 | .444 | | 6.15 | 21.50 | | 27.65 | 39 |
| 5200 | 4" pipe size | | 30 | .533 | | 10.10 | 25.50 | | 35.60 | 49.50 |
| 5210 | 6" pipe size | | 24 | .667 | | 29.50 | 32 | | 61.50 | 80.50 |

# 22 07 Plumbing Insulation

## 22 07 19 – Plumbing Piping Insulation

### 22 07 19.10 Piping Insulation

| | | | Crew | Daily Output | Labor-Hours | Unit | Material | 2011 Bare Costs Labor | Equipment | Total | Total Incl O&P |
|---|---|---|---|---|---|---|---|---|---|---|---|
| 0010 | **PIPING INSULATION** | | | | | | | | | | |
| 0100 | Rule of thumb, as a percentage of total mechanical costs | | | | | Job | | | | 10% | |
| 0110 | Insulation req'd is based on the surface size/area to be covered | | | | | | | | | | |
| 4000 | Pipe covering (price copper tube one size less than IPS) | | | | | | | | | | |
| 6600 | Fiberglass, with all service jacket | | | | | | | | | | |
| 6840 | 1" wall, 1/2" iron pipe size | G | Q-14 | 240 | .067 | L.F. | .95 | 2.81 | | 3.76 | 5.40 |
| 6870 | 1" iron pipe size | G | | 220 | .073 | | 1.12 | 3.07 | | 4.19 | 6 |
| 6900 | 2" iron pipe size | G | | 200 | .080 | | 1.41 | 3.38 | | 4.79 | 6.75 |

## 22 07 19 – Plumbing Piping Insulation

| 22 07 19.10 Piping Insulation | | Crew | Daily Output | Labor-Hours | Unit | Material | 2011 Bare Costs Labor | Equipment | Total | Total Incl O&P |
|---|---|---|---|---|---|---|---|---|---|---|
| 6920 | 3" iron pipe size | G | Q-14 | 180 | .089 | L.F. | 1.72 | 3.75 | | 5.47 | 7.70 |
| 6940 | 4" iron pipe size | G | | 150 | .107 | | 2.27 | 4.50 | | 6.77 | 9.45 |
| 7320 | 2" wall, 1/2" iron pipe size | G | | 220 | .073 | | 2.81 | 3.07 | | 5.88 | 7.85 |
| 7440 | 6" iron pipe size | G | | 100 | .160 | | 5.65 | 6.75 | | 12.40 | 16.65 |
| 7460 | 8" iron pipe size | G | | 80 | .200 | | 6.85 | 8.45 | | 15.30 | 20.50 |
| 7480 | 10" iron pipe size | G | | 70 | .229 | | 8.20 | 9.65 | | 17.85 | 24 |
| 7490 | 12" iron pipe size | G | | 65 | .246 | | 9.20 | 10.40 | | 19.60 | 26 |
| 7800 | For fiberglass with standard canvas jacket, deduct | | | | | | 5% | | | | |
| 7802 | For fittings, add 3 L.F. for each fitting | | | | | | | | | | |
| 7804 | plus 4 L.F. for each flange of the fitting | | | | | | | | | | |
| 7810 | Finishes | | | | | | | | | | |
| 7811 | Finishes, for .010" aluminum jacket, add | G | Q-14 | 200 | .080 | S.F. | .77 | 3.38 | | 4.15 | 6.05 |
| 7812 | For .016" aluminum jacket, add | G | | 200 | .080 | | 1.12 | 3.38 | | 4.50 | 6.45 |
| 7813 | For .010" stainless steel, add | G | | 160 | .100 | | 3.11 | 4.22 | | 7.33 | 9.95 |
| 7814 | For single layer of felt, add | | | | | | 10% | 10% | | | |
| 7816 | For roofing paper, 45 lb to 55 lb, add | | | | | | 25% | 10% | | | |
| 7879 | Rubber tubing, flexible closed cell foam | | | | | | | | | | |
| 7880 | 3/8" wall, 1/4" iron pipe size | G | 1 Asbe | 120 | .067 | L.F. | .55 | 3.13 | | 3.68 | 5.45 |
| 7910 | 1/2" iron pipe size | G | | 115 | .070 | | .70 | 3.26 | | 3.96 | 5.80 |
| 7920 | 3/4" iron pipe size | G | | 115 | .070 | | .85 | 3.26 | | 4.11 | 6 |
| 7930 | 1" iron pipe size | G | | 110 | .073 | | .98 | 3.41 | | 4.39 | 6.35 |
| 7950 | 1-1/2" iron pipe size | G | | 110 | .073 | | 1.30 | 3.41 | | 4.71 | 6.70 |
| 8100 | 1/2" wall, 1/4" iron pipe size | G | | 90 | .089 | | .50 | 4.17 | | 4.67 | 7 |
| 8130 | 1/2" iron pipe size | G | | 89 | .090 | | .61 | 4.22 | | 4.83 | 7.15 |
| 8140 | 3/4" iron pipe size | G | | 89 | .090 | | .68 | 4.22 | | 4.90 | 7.25 |
| 8150 | 1" iron pipe size | G | | 88 | .091 | | .75 | 4.26 | | 5.01 | 7.45 |
| 8170 | 1-1/2" iron pipe size | G | | 87 | .092 | | 1.06 | 4.31 | | 5.37 | 7.80 |
| 8180 | 2" iron pipe size | G | | 86 | .093 | | 1.38 | 4.36 | | 5.74 | 8.25 |
| 8200 | 3" iron pipe size | G | | 85 | .094 | | 1.95 | 4.41 | | 6.36 | 8.95 |
| 8300 | 3/4" wall, 1/4" iron pipe size | G | | 90 | .089 | | .77 | 4.17 | | 4.94 | 7.30 |
| 8330 | 1/2" iron pipe size | G | | 89 | .090 | | 1.01 | 4.22 | | 5.23 | 7.60 |
| 8340 | 3/4" iron pipe size | G | | 89 | .090 | | 1.23 | 4.22 | | 5.45 | 7.85 |
| 8350 | 1" iron pipe size | G | | 88 | .091 | | 1.41 | 4.26 | | 5.67 | 8.15 |
| 8370 | 1-1/2" iron pipe size | G | | 87 | .092 | | 2.15 | 4.31 | | 6.46 | 9 |
| 8380 | 2" iron pipe size | G | | 86 | .093 | | 2.52 | 4.36 | | 6.88 | 9.50 |
| 8400 | 3" iron pipe size | G | | 85 | .094 | | 3.85 | 4.41 | | 8.26 | 11.05 |
| 8444 | 1" wall, 1/2" iron pipe size | G | | 86 | .093 | | 1.93 | 4.36 | | 6.29 | 8.85 |
| 8445 | 3/4" iron pipe size | G | | 84 | .095 | | 2.34 | 4.47 | | 6.81 | 9.45 |
| 8446 | 1" iron pipe size | G | | 84 | .095 | | 2.72 | 4.47 | | 7.19 | 9.90 |
| 8447 | 1-1/4" iron pipe size | G | | 82 | .098 | | 3.08 | 4.58 | | 7.66 | 10.45 |
| 8448 | 1-1/2" iron pipe size | G | | 82 | .098 | | 3.58 | 4.58 | | 8.16 | 11 |
| 8449 | 2" iron pipe size | G | | 80 | .100 | | 4.78 | 4.69 | | 9.47 | 12.50 |
| 8450 | 2-1/2" iron pipe size | G | | 80 | .100 | | 6.25 | 4.69 | | 10.94 | 14.10 |
| 8456 | Rubber insulation tape, 1/8" x 2" x 30' | G | | | | Ea. | 11.40 | | | 11.40 | 12.50 |

## 22 11 13 – Facility Water Distribution Piping

### 22 11 13.14 Pipe, Brass

| | | Crew | Daily Output | Labor-Hours | Unit | Material | 2011 Bare Costs Labor | Equipment | Total | Total Incl O&P |
|---|---|---|---|---|---|---|---|---|---|---|
| 0010 | **PIPE, BRASS**, Plain end | | | | | | | | | |
| 0900 | Field threaded, coupling & clevis hanger assembly 10' O.C. | | | | | | | | | |
| 0920 | Regular weight | | | | | | | | | |
| 1120 | 1/2" diameter | 1 Plum | 48 | .167 | L.F. | 6.50 | 8.95 | | 15.45 | 20.50 |
| 1140 | 3/4" diameter | | 46 | .174 | | 8.60 | 9.30 | | 17.90 | 23.50 |
| 1160 | 1" diameter | | 43 | .186 | | 12.50 | 9.95 | | 22.45 | 28.50 |
| 1180 | 1-1/4" diameter | Q-1 | 72 | .222 | | 18.90 | 10.70 | | 29.60 | 37 |
| 1200 | 1-1/2" diameter | | 65 | .246 | | 22.50 | 11.90 | | 34.40 | 42.50 |
| 1220 | 2" diameter | | 53 | .302 | | 31 | 14.55 | | 45.55 | 56 |

### 22 11 13.23 Pipe/Tube, Copper

| | | Crew | Daily Output | Labor-Hours | Unit | Material | 2011 Bare Costs Labor | Equipment | Total | Total Incl O&P |
|---|---|---|---|---|---|---|---|---|---|---|
| 0010 | **PIPE/TUBE, COPPER**, Solder joints | | | | | | | | | |
| 1000 | Type K tubing, couplings & clevis hanger assemblies 10' O.C. | | | | | | | | | |
| 1100 | 1/4" diameter | 1 Plum | 84 | .095 | L.F. | 4.41 | 5.10 | | 9.51 | 12.45 |
| 1200 | 1" diameter | | 66 | .121 | | 11.80 | 6.50 | | 18.30 | 22.50 |
| 1260 | 2" diameter | | 40 | .200 | | 28.50 | 10.70 | | 39.20 | 47.50 |
| 2000 | Type L tubing, couplings & clevis hanger assemblies 10' O.C. | | | | | | | | | |
| 2100 | 1/4" diameter | 1 Plum | 88 | .091 | L.F. | 2.71 | 4.87 | | 7.58 | 10.25 |
| 2120 | 3/8" diameter | | 84 | .095 | | 3.65 | 5.10 | | 8.75 | 11.60 |
| 2140 | 1/2" diameter | | 81 | .099 | | 3.97 | 5.30 | | 9.27 | 12.25 |
| 2160 | 5/8" diameter | | 79 | .101 | | 5.85 | 5.45 | | 11.30 | 14.55 |
| 2180 | 3/4" diameter | | 76 | .105 | | 5.95 | 5.65 | | 11.60 | 14.95 |
| 2200 | 1" diameter | | 68 | .118 | | 9 | 6.30 | | 15.30 | 19.35 |
| 2220 | 1-1/4" diameter | | 58 | .138 | | 12.45 | 7.40 | | 19.85 | 24.50 |
| 2240 | 1-1/2" diameter | | 52 | .154 | | 15.70 | 8.25 | | 23.95 | 29.50 |
| 2260 | 2" diameter | | 42 | .190 | | 24 | 10.20 | | 34.20 | 42 |
| 2280 | 2-1/2" diameter | Q-1 | 62 | .258 | | 37 | 12.45 | | 49.45 | 59.50 |
| 2300 | 3" diameter | | 56 | .286 | | 50 | 13.80 | | 63.80 | 75.50 |
| 2320 | 3-1/2" diameter | | 43 | .372 | | 69.50 | 17.95 | | 87.45 | 104 |
| 2340 | 4" diameter | | 39 | .410 | | 85.50 | 19.80 | | 105.30 | 124 |
| 2360 | 5" diameter | | 34 | .471 | | 183 | 22.50 | | 205.50 | 235 |
| 2380 | 6" diameter | Q-2 | 40 | .600 | | 258 | 30 | | 288 | 330 |
| 2400 | 8" diameter | " | 36 | .667 | | 430 | 33.50 | | 463.50 | 520 |
| 2410 | For other than full hard temper, add | | | | | 21% | | | | |
| 2590 | For silver solder, add | | | | | | 15% | | | |
| 4000 | Type DWV tubing, couplings & clevis hanger assemblies 10' O.C. | | | | | | | | | |
| 4100 | 1-1/4" diameter | 1 Plum | 60 | .133 | L.F. | 11.05 | 7.15 | | 18.20 | 23 |
| 4120 | 1-1/2" diameter | | 54 | .148 | | 13.65 | 7.95 | | 21.60 | 27 |
| 4140 | 2" diameter | | 44 | .182 | | 18.30 | 9.75 | | 28.05 | 34.50 |
| 4160 | 3" diameter | Q-1 | 58 | .276 | | 33.50 | 13.30 | | 46.80 | 57 |
| 4180 | 4" diameter | | 40 | .400 | | 59.50 | 19.30 | | 78.80 | 94.50 |
| 4200 | 5" diameter | | 36 | .444 | | 167 | 21.50 | | 188.50 | 215 |
| 4220 | 6" diameter | Q-2 | 42 | .571 | | 239 | 28.50 | | 267.50 | 305 |

### 22 11 13.44 Pipe, Steel

| | | Crew | Daily Output | Labor-Hours | Unit | Material | 2011 Bare Costs Labor | Equipment | Total | Total Incl O&P |
|---|---|---|---|---|---|---|---|---|---|---|
| 0010 | **PIPE, STEEL** | | | | | | | | | |
| 0012 | The steel pipe in this section does not include fittings such as ells, tees | | | | | | | | | |
| 0014 | For fittings either add a % (usually 25 to 35%) or see | | | | | | | | | |
| 0015 | the Mechanical or Plumbing Cost Data | | | | | | | | | |
| 0020 | All pipe sizes are to Spec. A-53 unless noted otherwise   R221113-50 | | | | | | | | | |
| 0050 | Schedule 40, threaded, with couplings, and clevis hanger | | | | | | | | | |
| 0060 | assemblies sized for covering, 10' O.C. | | | | | | | | | |
| 0540 | Black, 1/4" diameter | 1 Plum | 66 | .121 | L.F. | 6.50 | 6.50 | | 13 | 16.85 |
| 0550 | 3/8" diameter | | 65 | .123 | | 7.25 | 6.60 | | 13.85 | 17.85 |

## 22 11 13 – Facility Water Distribution Piping

### 22 11 13.44 Pipe, Steel

| | | Crew | Daily Output | Labor-Hours | Unit | Material | 2011 Bare Costs Labor | Equipment | Total | Total Incl O&P |
|---|---|---|---|---|---|---|---|---|---|---|
| 0560 | 1/2" diameter | 1 Plum | 63 | .127 | L.F. | 3.80 | 6.80 | | 10.60 | 14.35 |
| 0570 | 3/4" diameter | | 61 | .131 | | 4.37 | 7.05 | | 11.42 | 15.30 |
| 0580 | 1" diameter | | 53 | .151 | | 6.15 | 8.10 | | 14.25 | 18.85 |
| 0590 | 1-1/4" diameter | Q-1 | 89 | .180 | | 7.55 | 8.65 | | 16.20 | 21.50 |
| 0600 | 1-1/2" diameter | | 80 | .200 | | 8.70 | 9.65 | | 18.35 | 24 |
| 0610 | 2" diameter | | 64 | .250 | | 11.45 | 12.05 | | 23.50 | 30.50 |
| 0620 | 2-1/2" diameter | | 50 | .320 | | 17.30 | 15.45 | | 32.75 | 42 |
| 0630 | 3" diameter | | 43 | .372 | | 22 | 17.95 | | 39.95 | 51 |
| 0640 | 3-1/2" diameter | | 40 | .400 | | 30 | 19.30 | | 49.30 | 61.50 |
| 0650 | 4" diameter | | 36 | .444 | | 32.50 | 21.50 | | 54 | 67.50 |
| 1280 | All pipe sizes are to Spec. A-53 unless noted otherwise | | | | | | | | | |
| 1281 | Schedule 40, threaded, with couplings and clevis hanger | | | | | | | | | |
| 1282 | assemblies sized for covering, 10' O.C. | | | | | | | | | |
| 1290 | Galvanized, 1/4" diameter | 1 Plum | 66 | .121 | L.F. | 8.60 | 6.50 | | 15.10 | 19.20 |
| 1300 | 3/8" diameter | | 65 | .123 | | 9.45 | 6.60 | | 16.05 | 20.50 |
| 1310 | 1/2" diameter | | 63 | .127 | | 5.15 | 6.80 | | 11.95 | 15.80 |
| 1320 | 3/4" diameter | | 61 | .131 | | 10.25 | 7.05 | | 17.30 | 22 |
| 1330 | 1" diameter | | 53 | .151 | | 8.10 | 8.10 | | 16.20 | 21 |
| 1340 | 1-1/4" diameter | Q-1 | 89 | .180 | | 10.15 | 8.65 | | 18.80 | 24 |
| 1350 | 1-1/2" diameter | | 80 | .200 | | 11.80 | 9.65 | | 21.45 | 27.50 |
| 1360 | 2" diameter | | 64 | .250 | | 15.60 | 12.05 | | 27.65 | 35 |
| 1370 | 2-1/2" diameter | | 50 | .320 | | 25 | 15.45 | | 40.45 | 51 |
| 1380 | 3" diameter | | 43 | .372 | | 32 | 17.95 | | 49.95 | 62 |
| 1390 | 3-1/2" diameter | | 40 | .400 | | 40.50 | 19.30 | | 59.80 | 73.50 |
| 1400 | 4" diameter | | 36 | .444 | | 46 | 21.50 | | 67.50 | 82.50 |
| 2000 | Welded, sch. 40, on yoke & roll hanger assy's, sized for covering, 10' O.C. | | | | | | | | | |
| 2040 | Black, 1" diameter | Q-15 | 93 | .172 | L.F. | 4.64 | 8.30 | .35 | 13.29 | 17.90 |
| 2070 | 2" diameter | | 61 | .262 | | 7.20 | 12.65 | .53 | 20.38 | 27.50 |
| 2090 | 3" diameter | | 43 | .372 | | 11.40 | 17.95 | .76 | 30.11 | 40.50 |
| 2110 | 4" diameter | | 37 | .432 | | 15.80 | 21 | .88 | 37.68 | 49.50 |
| 2120 | 5" diameter | | 32 | .500 | | 21 | 24 | 1.02 | 46.02 | 60 |
| 2130 | 6" diameter | Q-16 | 36 | .667 | | 25 | 33.50 | .90 | 59.40 | 78.50 |
| 2140 | 8" diameter | | 29 | .828 | | 36.50 | 41.50 | 1.12 | 79.12 | 103 |
| 2150 | 10" diameter | | 24 | 1 | | 51 | 50 | 1.35 | 102.35 | 132 |
| 2160 | 12" diameter | | 19 | 1.263 | | 87.50 | 63 | 1.71 | 152.21 | 192 |

### 22 11 13.48 Pipe, Fittings and Valves, Steel, Grooved-Joint

| | | Crew | Daily Output | Labor-Hours | Unit | Material | 2011 Bare Costs Labor | Equipment | Total | Total Incl O&P |
|---|---|---|---|---|---|---|---|---|---|---|
| 0010 | **PIPE, FITTINGS AND VALVES, STEEL, GROOVED-JOINT** | | | | | | | | | |
| 0012 | Fittings are ductile iron. Steel fittings noted. | | | | | | | | | |
| 0020 | Pipe includes coupling & clevis type hanger assemblies, 10' O.C. | | | | | | | | | |
| 1000 | Schedule 40, black | | | | | | | | | |
| 1040 | 3/4" diameter | 1 Plum | 71 | .113 | L.F. | 5.40 | 6.05 | | 11.45 | 14.95 |
| 1050 | 1" diameter | | 63 | .127 | | 4.09 | 6.80 | | 10.89 | 14.65 |
| 1060 | 1-1/4" diameter | | 58 | .138 | | 5.10 | 7.40 | | 12.50 | 16.70 |
| 1070 | 1-1/2" diameter | | 51 | .157 | | 5.75 | 8.40 | | 14.15 | 18.85 |
| 1080 | 2" diameter | | 40 | .200 | | 6.60 | 10.70 | | 17.30 | 23.50 |
| 1090 | 2-1/2" diameter | Q-1 | 57 | .281 | | 9.25 | 13.55 | | 22.80 | 30 |
| 1100 | 3" diameter | | 50 | .320 | | 11.30 | 15.45 | | 26.75 | 35.50 |
| 1110 | 4" diameter | | 45 | .356 | | 15.75 | 17.15 | | 32.90 | 43 |
| 1120 | 5" diameter | | 37 | .432 | | 21.50 | 21 | | 42.50 | 54.50 |
| 1130 | 6" diameter | Q-2 | 42 | .571 | | 27.50 | 28.50 | | 56 | 72.50 |
| 1800 | Galvanized | | | | | | | | | |
| 1840 | 3/4" diameter | 1 Plum | 71 | .113 | L.F. | 11.25 | 6.05 | | 17.30 | 21.50 |

## 22 11 13 – Facility Water Distribution Piping

### 22 11 13.48 Pipe, Fittings and Valves, Steel, Grooved-Joint

| | | Crew | Daily Output | Labor-Hours | Unit | Material | 2011 Bare Costs Labor | Equipment | Total | Total Incl O&P |
|---|---|---|---|---|---|---|---|---|---|---|
| 1850 | 1" diameter | 1 Plum | 63 | .127 | L.F. | 8.05 | 6.80 | | 14.85 | 19 |
| 1860 | 1-1/4" diameter | | 58 | .138 | | 10.15 | 7.40 | | 17.55 | 22 |
| 1870 | 1-1/2" diameter | | 51 | .157 | | 11.70 | 8.40 | | 20.10 | 25.50 |
| 1880 | 2" diameter | | 40 | .200 | | 14.85 | 10.70 | | 25.55 | 32.50 |
| 1890 | 2-1/2" diameter | Q-1 | 57 | .281 | | 22 | 13.55 | | 35.55 | 44.50 |
| 1900 | 3" diameter | | 50 | .320 | | 28 | 15.45 | | 43.45 | 54 |
| 1910 | 4" diameter | | 45 | .356 | | 39 | 17.15 | | 56.15 | 68.50 |
| 1920 | 5" diameter | | 37 | .432 | | 46 | 21 | | 67 | 81.50 |
| 1930 | 6" diameter | Q-2 | 42 | .571 | | 51.50 | 28.50 | | 80 | 99 |
| 3990 | Fittings: coupling material required at joints not incl. in fitting price. | | | | | | | | | |
| 3994 | Add 1 selected coupling, material only, per joint for installed price. | | | | | | | | | |
| 4000 | Elbow, 90° or 45°, painted | | | | | | | | | |
| 4030 | 3/4" diameter | 1 Plum | 50 | .160 | Ea. | 43 | 8.60 | | 51.60 | 60.50 |
| 4040 | 1" diameter | | 50 | .160 | | 23 | 8.60 | | 31.60 | 38 |
| 4050 | 1-1/4" diameter | | 40 | .200 | | 23 | 10.70 | | 33.70 | 41 |
| 4060 | 1-1/2" diameter | | 33 | .242 | | 23 | 13 | | 36 | 44.50 |
| 4070 | 2" diameter | | 25 | .320 | | 23 | 17.15 | | 40.15 | 50.50 |
| 4080 | 2-1/2" diameter | Q-1 | 40 | .400 | | 23 | 19.30 | | 42.30 | 54 |
| 4090 | 3" diameter | | 33 | .485 | | 40.50 | 23.50 | | 64 | 79.50 |
| 4100 | 4" diameter | | 25 | .640 | | 44 | 31 | | 75 | 94.50 |
| 4110 | 5" diameter | | 20 | .800 | | 105 | 38.50 | | 143.50 | 174 |
| 4120 | 6" diameter | Q-2 | 25 | .960 | | 124 | 48 | | 172 | 208 |
| 4250 | For galvanized elbows, add | | | | | 26% | | | | |
| 4690 | Tee, painted | | | | | | | | | |
| 4700 | 3/4" diameter | 1 Plum | 38 | .211 | Ea. | 46 | 11.30 | | 57.30 | 68 |
| 4740 | 1" diameter | | 33 | .242 | | 35.50 | 13 | | 48.50 | 58.50 |
| 4750 | 1-1/4" diameter | | 27 | .296 | | 35.50 | 15.90 | | 51.40 | 62.50 |
| 4760 | 1-1/2" diameter | | 22 | .364 | | 35.50 | 19.50 | | 55 | 68 |
| 4770 | 2" diameter | | 17 | .471 | | 35.50 | 25 | | 60.50 | 76.50 |
| 4780 | 2-1/2" diameter | Q-1 | 27 | .593 | | 35.50 | 28.50 | | 64 | 81.50 |
| 4790 | 3" diameter | | 22 | .727 | | 49 | 35 | | 84 | 106 |
| 4800 | 4" diameter | | 17 | .941 | | 74.50 | 45.50 | | 120 | 150 |
| 4810 | 5" diameter | | 13 | 1.231 | | 174 | 59.50 | | 233.50 | 280 |
| 4820 | 6" diameter | Q-2 | 17 | 1.412 | | 201 | 70.50 | | 271.50 | 325 |
| 4900 | For galvanized tees, add | | | | | 24% | | | | |
| 4906 | Couplings, rigid style, painted | | | | | | | | | |
| 4908 | 1" diameter | 1 Plum | 100 | .080 | Ea. | 17.20 | 4.29 | | 21.49 | 25.50 |
| 4909 | 1-1/4" diameter | | 100 | .080 | | 17.20 | 4.29 | | 21.49 | 25.50 |
| 4910 | 1-1/2" diameter | | 67 | .119 | | 17.20 | 6.40 | | 23.60 | 28.50 |
| 4912 | 2" diameter | | 50 | .160 | | 18.80 | 8.60 | | 27.40 | 33.50 |
| 4914 | 2-1/2" diameter | Q-1 | 80 | .200 | | 22 | 9.65 | | 31.65 | 38.50 |
| 4916 | 3" diameter | | 67 | .239 | | 25.50 | 11.50 | | 37 | 45 |
| 4918 | 4" diameter | | 50 | .320 | | 35.50 | 15.45 | | 50.95 | 62 |
| 4920 | 5" diameter | | 40 | .400 | | 46 | 19.30 | | 65.30 | 80 |
| 4922 | 6" diameter | Q-2 | 50 | .480 | | 61 | 24 | | 85 | 103 |
| 4940 | Flexible, standard, painted | | | | | | | | | |
| 4950 | 3/4" diameter | 1 Plum | 100 | .080 | Ea. | 12.40 | 4.29 | | 16.69 | 20 |
| 4960 | 1" diameter | | 100 | .080 | | 12.40 | 4.29 | | 16.69 | 20 |
| 4970 | 1-1/4" diameter | | 80 | .100 | | 16.40 | 5.35 | | 21.75 | 26 |
| 4980 | 1-1/2" diameter | | 67 | .119 | | 18 | 6.40 | | 24.40 | 29.50 |
| 4990 | 2" diameter | | 50 | .160 | | 19 | 8.60 | | 27.60 | 34 |
| 5000 | 2-1/2" diameter | Q-1 | 80 | .200 | | 22 | 9.65 | | 31.65 | 39 |
| 5010 | 3" diameter | | 67 | .239 | | 25 | 11.50 | | 36.50 | 44.50 |

## 22 11 13 – Facility Water Distribution Piping

| 22 11 13.48 Pipe, Fittings and Valves, Steel, Grooved-Joint | Crew | Daily Output | Labor-Hours | Unit | Material | 2011 Bare Costs Labor | Equipment | Total | Total Incl O&P |
|---|---|---|---|---|---|---|---|---|---|
| 5020    3-1/2" diameter | Q-1 | 57 | .281 | Ea. | 36 | 13.55 | | 49.55 | 59.50 |
| 5030    4" diameter | | 50 | .320 | | 36 | 15.45 | | 51.45 | 62.50 |
| 5040    5" diameter | | 40 | .400 | | 54.50 | 19.30 | | 73.80 | 89 |
| 5050    6" diameter | Q-2 | 50 | .480 | | 64 | 24 | | 88 | 107 |

### 22 11 13.64 Pipe, Stainless Steel

| | Crew | Daily Output | Labor-Hours | Unit | Material | 2011 Bare Costs Labor | Equipment | Total | Total Incl O&P |
|---|---|---|---|---|---|---|---|---|---|
| 0010  **PIPE, STAINLESS STEEL** | | | | | | | | | |
| 3500    Threaded, couplings and clevis hanger assemblies, 10' O.C. | | | | | | | | | |
| 3520       Schedule 40, type 304 | | | | | | | | | |
| 3540          1/4" diameter | 1 Plum | 54 | .148 | L.F. | 6.45 | 7.95 | | 14.40 | 18.95 |
| 3550          3/8" diameter | | 53 | .151 | | 8.15 | 8.10 | | 16.25 | 21 |
| 3560          1/2" diameter | | 52 | .154 | | 9.60 | 8.25 | | 17.85 | 23 |
| 3580          1" diameter | | 45 | .178 | | 15.65 | 9.55 | | 25.20 | 31.50 |
| 3610          2" diameter | Q-1 | 57 | .281 | | 34 | 13.55 | | 47.55 | 57.50 |
| 3640          4" diameter | Q-2 | 51 | .471 | | 97.50 | 23.50 | | 121 | 142 |
| 3740       For small quantities, add | | | | | 10% | | | | |
| 4250       Schedule 40, type 316 | | | | | | | | | |
| 4290          1/4" diameter | 1 Plum | 54 | .148 | L.F. | 7.75 | 7.95 | | 15.70 | 20.50 |
| 4300          3/8" diameter | | 53 | .151 | | 8.75 | 8.10 | | 16.85 | 21.50 |
| 4310          1/2" diameter | | 52 | .154 | | 11.65 | 8.25 | | 19.90 | 25 |
| 4320          3/4" diameter | | 51 | .157 | | 14.20 | 8.40 | | 22.60 | 28 |
| 4330          1" diameter | | 45 | .178 | | 20 | 9.55 | | 29.55 | 36 |
| 4360          2" diameter | Q-1 | 57 | .281 | | 44.50 | 13.55 | | 58.05 | 68.50 |
| 4390          4" diameter | Q-2 | 51 | .471 | | 145 | 23.50 | | 168.50 | 194 |
| 4490       For small quantities, add | | | | | 10% | | | | |

### 22 11 13.74 Pipe, Plastic

| | Crew | Daily Output | Labor-Hours | Unit | Material | 2011 Bare Costs Labor | Equipment | Total | Total Incl O&P |
|---|---|---|---|---|---|---|---|---|---|
| 0010  **PIPE, PLASTIC** | | | | | | | | | |
| 1800    PVC, couplings 10' O.C., clevis hanger assemblies, 3 per 10' | | | | | | | | | |
| 1820       Schedule 40 | | | | | | | | | |
| 1860          1/2" diameter | 1 Plum | 54 | .148 | L.F. | 3.53 | 7.95 | | 11.48 | 15.75 |
| 1870          3/4" diameter | | 51 | .157 | | 3.81 | 8.40 | | 12.21 | 16.75 |
| 1880          1" diameter | | 46 | .174 | | 4.89 | 9.30 | | 14.19 | 19.30 |
| 1890          1-1/4" diameter | | 42 | .190 | | 5.70 | 10.20 | | 15.90 | 21.50 |
| 1900          1-1/2" diameter | | 36 | .222 | | 5.75 | 11.90 | | 17.65 | 24 |
| 1910          2" diameter | Q-1 | 59 | .271 | | 6.75 | 13.10 | | 19.85 | 27 |
| 1920          2-1/2" diameter | | 56 | .286 | | 9.05 | 13.80 | | 22.85 | 30.50 |
| 1930          3" diameter | | 53 | .302 | | 11.10 | 14.55 | | 25.65 | 34 |
| 1940          4" diameter | | 48 | .333 | | 14.20 | 16.10 | | 30.30 | 39.50 |
| 1950          5" diameter | | 43 | .372 | | 19.20 | 17.95 | | 37.15 | 48 |
| 1960          6" diameter | | 39 | .410 | | 24.50 | 19.80 | | 44.30 | 56.50 |
| 4100    DWV type, schedule 40, couplings 10' O.C., clevis hanger assy's, 3 per 10' | | | | | | | | | |
| 4210       ABS, schedule 40, foam core type | | | | | | | | | |
| 4212          Plain end black | | | | | | | | | |
| 4214             1-1/2" diameter | 1 Plum | 39 | .205 | L.F. | 4.48 | 11 | | 15.48 | 21.50 |
| 4216             2" diameter | Q-1 | 62 | .258 | | 4.93 | 12.45 | | 17.38 | 24 |
| 4218             3" diameter | | 56 | .286 | | 7.45 | 13.80 | | 21.25 | 28.50 |
| 4220             4" diameter | | 51 | .314 | | 9.70 | 15.15 | | 24.85 | 33 |
| 4222             6" diameter | | 42 | .381 | | 17.85 | 18.40 | | 36.25 | 47 |
| 4240          To delete coupling & hangers, subtract | | | | | | | | | |
| 4244             1-1/2" diam. to 6" diam. | | | | | 43% | 48% | | | |
| 4400       PVC | | | | | | | | | |
| 4410          1-1/4" diameter | 1 Plum | 42 | .190 | L.F. | 4.65 | 10.20 | | 14.85 | 20.50 |
| 4420          1-1/2" diameter | " | 36 | .222 | | 4.53 | 11.90 | | 16.43 | 23 |

| 22 11 13.74 Pipe, Plastic | Crew | Daily Output | Labor-Hours | Unit | Material | 2011 Bare Costs Labor | Equipment | Total | Total Incl O&P |
|---|---|---|---|---|---|---|---|---|---|
| 4460 | 2" diameter | Q-1 | 59 | .271 | L.F. | 4.97 | 13.10 | | 18.07 | 25 |
| 4470 | 3" diameter | | 53 | .302 | | 7.50 | 14.55 | | 22.05 | 30.50 |
| 4480 | 4" diameter | | 48 | .333 | | 9.50 | 16.10 | | 25.60 | 34.50 |
| 4490 | 6" diameter | | 39 | .410 | | 16.10 | 19.80 | | 35.90 | 47 |
| 5300 | CPVC, socket joint, couplings 10' O.C., clevis hanger assemblies, 3 per 10' | | | | | | | | | |
| 5302 | Schedule 40 | | | | | | | | | |
| 5304 | 1/2" diameter | 1 Plum | 54 | .148 | L.F. | 4.22 | 7.95 | | 12.17 | 16.50 |
| 5305 | 3/4" diameter | | 51 | .157 | | 4.75 | 8.40 | | 13.15 | 17.80 |
| 5306 | 1" diameter | | 46 | .174 | | 6.45 | 9.30 | | 15.75 | 21 |
| 5307 | 1-1/4" diameter | | 42 | .190 | | 7.65 | 10.20 | | 17.85 | 23.50 |
| 5308 | 1-1/2" diameter | | 36 | .222 | | 8.55 | 11.90 | | 20.45 | 27 |
| 5309 | 2" diameter | Q-1 | 59 | .271 | | 10.05 | 13.10 | | 23.15 | 30.50 |
| 5310 | 2-1/2" diameter | | 56 | .286 | | 15.85 | 13.80 | | 29.65 | 38 |
| 5311 | 3" diameter | | 53 | .302 | | 18.65 | 14.55 | | 33.20 | 42.50 |
| 5360 | CPVC, threaded, couplings 10' O.C., clevis hanger assemblies, 3 per 10' | | | | | | | | | |
| 5380 | Schedule 40 | | | | | | | | | |
| 5460 | 1/2" diameter | 1 Plum | 54 | .148 | L.F. | 4.93 | 7.95 | | 12.88 | 17.30 |
| 5470 | 3/4" diameter | | 51 | .157 | | 6 | 8.40 | | 14.40 | 19.15 |
| 5480 | 1" diameter | | 46 | .174 | | 7.70 | 9.30 | | 17 | 22.50 |
| 5490 | 1-1/4" diameter | | 42 | .190 | | 8.65 | 10.20 | | 18.85 | 25 |
| 5500 | 1-1/2" diameter | | 36 | .222 | | 9.40 | 11.90 | | 21.30 | 28 |
| 5510 | 2" diameter | Q-1 | 59 | .271 | | 11.15 | 13.10 | | 24.25 | 32 |
| 5520 | 2-1/2" diameter | | 56 | .286 | | 16.95 | 13.80 | | 30.75 | 39 |
| 5530 | 3" diameter | | 53 | .302 | | 20.50 | 14.55 | | 35.05 | 44.50 |
| 7280 | PEX, flexible, no couplings or hangers | | | | | | | | | |
| 7282 | Note: For labor costs add 25% to the couplings and fittings labor total. | | | | | | | | | |
| 7300 | Non-barrier type, hot/cold tubing rolls | | | | | | | | | |
| 7310 | 1/4" diameter x 100' | | | | L.F. | .46 | | | .46 | .51 |
| 7350 | 3/8" diameter x 100' | | | | | .51 | | | .51 | .56 |
| 7360 | 1/2" diameter x 100' | | | | | .57 | | | .57 | .63 |
| 7370 | 1/2" diameter x 500' | | | | | .57 | | | .57 | .63 |
| 7380 | 1/2" diameter x 1000' | | | | | .57 | | | .57 | .63 |
| 7400 | 3/4" diameter x 100' | | | | | 1.05 | | | 1.05 | 1.16 |
| 7410 | 3/4" diameter x 500' | | | | | 1.05 | | | 1.05 | 1.16 |
| 7420 | 3/4" diameter x 1000' | | | | | 1.05 | | | 1.05 | 1.16 |
| 7460 | 1" diameter x 100' | | | | | 1.80 | | | 1.80 | 1.98 |
| 7470 | 1" diameter x 300' | | | | | 1.80 | | | 1.80 | 1.98 |
| 7480 | 1" diameter x 500' | | | | | 1.78 | | | 1.78 | 1.96 |
| 7500 | 1-1/4" diameter x 100' | | | | | 3.04 | | | 3.04 | 3.34 |
| 7510 | 1-1/4" diameter x 300' | | | | | 3.04 | | | 3.04 | 3.34 |
| 7540 | 1-1/2" diameter x 100' | | | | | 3.82 | | | 3.82 | 4.20 |
| 7550 | 1-1/2" diameter x 300' | | | | | 3.81 | | | 3.81 | 4.19 |
| 7596 | Most sizes available in red or blue | | | | | | | | | |
| 7700 | Non-barrier type, hot/cold tubing straight lengths | | | | | | | | | |
| 7710 | 1/2" diameter x 20' | | | | L.F. | .57 | | | .57 | .63 |
| 7750 | 3/4" diameter x 20' | | | | | 1.05 | | | 1.05 | 1.16 |
| 7760 | 1" diameter x 20' | | | | | 1.80 | | | 1.80 | 1.98 |
| 7770 | 1-1/4" diameter x 20' | | | | | 3.04 | | | 3.04 | 3.34 |
| 7780 | 1-1/2" diameter x 20' | | | | | 3.92 | | | 3.92 | 4.31 |
| 7790 | 2" diameter | | | | | 5.30 | | | 5.30 | 5.85 |
| 7796 | Most sizes available in red or blue | | | | | | | | | |

# 22 11 Facility Water Distribution

## 22 11 19 – Domestic Water Piping Specialties

| 22 11 19.10 Flexible Connectors | Crew | Daily Output | Labor-Hours | Unit | Material | 2011 Bare Costs Labor | Equipment | Total | Total Incl O&P |
|---|---|---|---|---|---|---|---|---|---|
| 0010 **FLEXIBLE CONNECTORS**, Corrugated, 7/8" O.D., 1/2" I.D. | | | | | | | | | |
| 0050 Gas, seamless brass, steel fittings | | | | | | | | | |
| 0200 12" long | 1 Plum | 36 | .222 | Ea. | 15.10 | 11.90 | | 27 | 34.50 |
| 0220 18" long | | 36 | .222 | | 18.70 | 11.90 | | 30.60 | 38.50 |
| 0240 24" long | | 34 | .235 | | 22 | 12.60 | | 34.60 | 43.50 |
| 0280 36" long | | 32 | .250 | | 26.50 | 13.40 | | 39.90 | 49 |
| 0340 60" long | | 30 | .267 | | 40 | 14.30 | | 54.30 | 65.50 |
| 2000 Water, copper tubing, dielectric separators | | | | | | | | | |
| 2100 12" long | 1 Plum | 36 | .222 | Ea. | 14 | 11.90 | | 25.90 | 33 |
| 2260 24" long | " | 34 | .235 | " | 20.50 | 12.60 | | 33.10 | 41.50 |

### 22 11 19.14 Flexible Metal Hose

| | Crew | Daily Output | Labor-Hours | Unit | Material | Labor | Equipment | Total | Total Incl O&P |
|---|---|---|---|---|---|---|---|---|---|
| 0010 **FLEXIBLE METAL HOSE**, Connectors, standard lengths | | | | | | | | | |
| 0100 Bronze braided, bronze ends | | | | | | | | | |
| 0120 3/8" diameter x 12" | 1 Stpi | 26 | .308 | Ea. | 18.85 | 16.65 | | 35.50 | 46 |
| 0160 3/4" diameter x 12" | | 20 | .400 | | 26.50 | 21.50 | | 48 | 61.50 |
| 0180 1" diameter x 18" | | 19 | .421 | | 34 | 23 | | 57 | 71.50 |
| 0200 1-1/2" diameter x 18" | | 13 | .615 | | 53.50 | 33.50 | | 87 | 109 |
| 0220 2" diameter x 18" | | 11 | .727 | | 64.50 | 39.50 | | 104 | 130 |

### 22 11 19.26 Pressure Regulators

| | Crew | Daily Output | Labor-Hours | Unit | Material | Labor | Equipment | Total | Total Incl O&P |
|---|---|---|---|---|---|---|---|---|---|
| 0010 **PRESSURE REGULATORS** | | | | | | | | | |
| 3000 Steam, high capacity, bronze body, stainless steel trim | | | | | | | | | |
| 3020 Threaded, 1/2" diameter | 1 Stpi | 24 | .333 | Ea. | 1,450 | 18.05 | | 1,468.05 | 1,625 |
| 3030 3/4" diameter | | 24 | .333 | | 1,450 | 18.05 | | 1,468.05 | 1,625 |
| 3040 1" diameter | | 19 | .421 | | 1,625 | 23 | | 1,648 | 1,800 |
| 3060 1-1/4" diameter | | 15 | .533 | | 1,800 | 29 | | 1,829 | 2,025 |
| 3080 1-1/2" diameter | | 13 | .615 | | 2,050 | 33.50 | | 2,083.50 | 2,300 |
| 3100 2" diameter | | 11 | .727 | | 2,500 | 39.50 | | 2,539.50 | 2,800 |
| 3120 2-1/2" diameter | Q-5 | 12 | 1.333 | | 3,125 | 65 | | 3,190 | 3,550 |
| 3140 3" diameter | " | 11 | 1.455 | | 3,575 | 71 | | 3,646 | 4,050 |
| 3500 Flanged connection, iron body, 125 lb. W.S.P. | | | | | | | | | |
| 3520 3" diameter | Q-5 | 11 | 1.455 | Ea. | 3,925 | 71 | | 3,996 | 4,425 |
| 3540 4" diameter | " | 5 | 3.200 | " | 4,950 | 156 | | 5,106 | 5,675 |

### 22 11 19.38 Water Supply Meters

| | Crew | Daily Output | Labor-Hours | Unit | Material | Labor | Equipment | Total | Total Incl O&P |
|---|---|---|---|---|---|---|---|---|---|
| 0010 **WATER SUPPLY METERS** | | | | | | | | | |
| 2000 Domestic/commercial, bronze | | | | | | | | | |
| 2020 Threaded | | | | | | | | | |
| 2060 5/8" diameter, to 20 GPM | 1 Plum | 16 | .500 | Ea. | 43.50 | 27 | | 70.50 | 88 |
| 2080 3/4" diameter, to 30 GPM | | 14 | .571 | | 79.50 | 30.50 | | 110 | 133 |
| 2100 1" diameter, to 50 GPM | | 12 | .667 | | 121 | 35.50 | | 156.50 | 187 |
| 2300 Threaded/flanged | | | | | | | | | |
| 2340 1-1/2" diameter, to 100 GPM | 1 Plum | 8 | 1 | Ea. | 295 | 53.50 | | 348.50 | 405 |
| 2360 2" diameter, to 160 GPM | " | 6 | 1.333 | " | 400 | 71.50 | | 471.50 | 545 |
| 2600 Flanged, compound | | | | | | | | | |
| 2640 3" diameter, 320 GPM | Q-1 | 3 | 5.333 | Ea. | 2,725 | 257 | | 2,982 | 3,375 |
| 2660 4" diameter, to 500 GPM | | 1.50 | 10.667 | | 4,350 | 515 | | 4,865 | 5,575 |
| 2680 6" diameter, to 1,000 GPM | | 1 | 16 | | 6,950 | 770 | | 7,720 | 8,800 |
| 2700 8" diameter, to 1,800 GPM | | .80 | 20 | | 10,900 | 965 | | 11,865 | 13,500 |

### 22 11 19.42 Backflow Preventers

| | | | | | | | | | |
|---|---|---|---|---|---|---|---|---|---|
| 0010 **BACKFLOW PREVENTERS**, Includes valves | | | | | | | | | |
| 0020 and four test cocks, corrosion resistant, automatic operation | | | | | | | | | |
| 4000 Reduced pressure principle | | | | | | | | | |

# 22 11 Facility Water Distribution

## 22 11 19 – Domestic Water Piping Specialties

### 22 11 19.42 Backflow Preventers

| | | Crew | Daily Output | Labor-Hours | Unit | Material | 2011 Bare Costs Labor | 2011 Bare Costs Equipment | Total | Total Incl O&P |
|---|---|---|---|---|---|---|---|---|---|---|
| 4100 | Threaded, bronze, valves are ball | | | | | | | | | |
| 4120 | 3/4" pipe size | 1 Plum | 16 | .500 | Ea. | 345 | 27 | | 372 | 420 |
| 4140 | 1" pipe size | | 14 | .571 | | 375 | 30.50 | | 405.50 | 455 |
| 4150 | 1-1/4" pipe size | | 12 | .667 | | 640 | 35.50 | | 675.50 | 760 |
| 4160 | 1-1/2" pipe size | | 10 | .800 | | 700 | 43 | | 743 | 840 |
| 4180 | 2" pipe size | | 7 | 1.143 | | 615 | 61.50 | | 676.50 | 765 |
| 5000 | Flanged, bronze, valves are OS&Y | | | | | | | | | |
| 5060 | 2-1/2" pipe size | Q-1 | 5 | 3.200 | Ea. | 5,600 | 154 | | 5,754 | 6,375 |
| 5080 | 3" pipe size | | 4.50 | 3.556 | | 6,525 | 172 | | 6,697 | 7,425 |
| 5100 | 4" pipe size | | 3 | 5.333 | | 7,650 | 257 | | 7,907 | 8,800 |
| 5120 | 6" pipe size | Q-2 | 3 | 8 | | 12,200 | 400 | | 12,600 | 14,000 |
| 5600 | Flanged, iron, valves are OS&Y | | | | | | | | | |
| 5660 | 2-1/2" pipe size | Q-1 | 5 | 3.200 | Ea. | 2,625 | 154 | | 2,779 | 3,125 |
| 5680 | 3" pipe size | | 4.50 | 3.556 | | 2,775 | 172 | | 2,947 | 3,300 |
| 5700 | 4" pipe size | | 3 | 5.333 | | 3,475 | 257 | | 3,732 | 4,200 |
| 5720 | 6" pipe size | Q-2 | 3 | 8 | | 5,025 | 400 | | 5,425 | 6,125 |
| 5740 | 8" pipe size | | 2 | 12 | | 10,300 | 600 | | 10,900 | 12,200 |
| 5760 | 10" pipe size | | 1 | 24 | | 11,800 | 1,200 | | 13,000 | 14,800 |

### 22 11 19.50 Vacuum Breakers

| | | Crew | Daily Output | Labor-Hours | Unit | Material | 2011 Bare Costs Labor | 2011 Bare Costs Equipment | Total | Total Incl O&P |
|---|---|---|---|---|---|---|---|---|---|---|
| 0010 | **VACUUM BREAKERS** | | | | | | | | | |
| 0013 | See also backflow preventers Section 22 11 19.42 | | | | | | | | | |
| 1000 | Anti-siphon continuous pressure type | | | | | | | | | |
| 1010 | Max. 150 PSI - 210°F | | | | | | | | | |
| 1020 | Bronze body | | | | | | | | | |
| 1030 | 1/2" size | 1 Stpi | 24 | .333 | Ea. | 160 | 18.05 | | 178.05 | 203 |
| 1040 | 3/4" size | | 20 | .400 | | 160 | 21.50 | | 181.50 | 209 |
| 1050 | 1" size | | 19 | .421 | | 165 | 23 | | 188 | 216 |
| 1060 | 1-1/4" size | | 15 | .533 | | 325 | 29 | | 354 | 405 |
| 1070 | 1-1/2" size | | 13 | .615 | | 400 | 33.50 | | 433.50 | 490 |
| 1080 | 2" size | | 11 | .727 | | 415 | 39.50 | | 454.50 | 515 |
| 1200 | Max. 125 PSI with atmospheric vent | | | | | | | | | |
| 1210 | Brass, in-line construction | | | | | | | | | |
| 1220 | 1/4" size | 1 Stpi | 24 | .333 | Ea. | 63 | 18.05 | | 81.05 | 96.50 |
| 1230 | 3/8" size | " | 24 | .333 | | 63 | 18.05 | | 81.05 | 96.50 |
| 1260 | For polished chrome finish, add | | | | | 13% | | | | |
| 2000 | Anti-siphon, non-continuous pressure type | | | | | | | | | |
| 2010 | Hot or cold water 125 PSI - 210°F | | | | | | | | | |
| 2020 | Bronze body | | | | | | | | | |
| 2030 | 1/4" size | 1 Stpi | 24 | .333 | Ea. | 40.50 | 18.05 | | 58.55 | 71.50 |
| 2040 | 3/8" size | | 24 | .333 | | 40.50 | 18.05 | | 58.55 | 71.50 |
| 2050 | 1/2" size | | 24 | .333 | | 46 | 18.05 | | 64.05 | 77.50 |
| 2060 | 3/4" size | | 20 | .400 | | 54 | 21.50 | | 75.50 | 92 |
| 2070 | 1" size | | 19 | .421 | | 84.50 | 23 | | 107.50 | 127 |
| 2080 | 1-1/4" size | | 15 | .533 | | 148 | 29 | | 177 | 206 |
| 2090 | 1-1/2" size | | 13 | .615 | | 174 | 33.50 | | 207.50 | 241 |
| 2100 | 2" size | | 11 | .727 | | 271 | 39.50 | | 310.50 | 355 |
| 2110 | 2-1/2" size | | 8 | 1 | | 780 | 54 | | 834 | 935 |
| 2120 | 3" size | | 6 | 1.333 | | 1,025 | 72 | | 1,097 | 1,250 |
| 2150 | For polished chrome finish, add | | | | | 50% | | | | |

### 22 11 19.54 Water Hammer Arresters/Shock Absorbers

| | | Crew | Daily Output | Labor-Hours | Unit | Material | 2011 Bare Costs Labor | 2011 Bare Costs Equipment | Total | Total Incl O&P |
|---|---|---|---|---|---|---|---|---|---|---|
| 0010 | **WATER HAMMER ARRESTERS/SHOCK ABSORBERS** | | | | | | | | | |
| 0490 | Copper | | | | | | | | | |

# 22 11 Facility Water Distribution

## 22 11 19 – Domestic Water Piping Specialties

| 22 11 19.54 Water Hammer Arresters/Shock Absorbers | Crew | Daily Output | Labor-Hours | Unit | Material | 2011 Bare Costs Labor | Equipment | Total | Total Incl O&P |
|---|---|---|---|---|---|---|---|---|---|
| 0500 | 3/4" male I.P.S. For 1 to 11 fixtures | 1 Plum | 12 | .667 | Ea. | 20.50 | 35.50 | | 56 | 76 |
| 0600 | 1" male I.P.S. For 12 to 32 fixtures | | 8 | 1 | | 41.50 | 53.50 | | 95 | 126 |
| 0700 | 1-1/4" male I.P.S. For 33 to 60 fixtures | | 8 | 1 | | 44 | 53.50 | | 97.50 | 129 |
| 0800 | 1-1/2" male I.P.S. For 61 to 113 fixtures | | 8 | 1 | | 63 | 53.50 | | 116.50 | 150 |
| 0900 | 2" male I.P.S. For 114 to 154 fixtures | | 8 | 1 | | 92 | 53.50 | | 145.50 | 181 |
| 1000 | 2-1/2" male I.P.S. For 155 to 330 fixtures | ↓ | 4 | 2 | ↓ | 285 | 107 | | 392 | 475 |

### 22 11 19.64 Hydrants

| | | Crew | Daily Output | Labor-Hours | Unit | Material | 2011 Bare Costs Labor | Equipment | Total | Total Incl O&P |
|---|---|---|---|---|---|---|---|---|---|---|
| 0010 | **HYDRANTS** | | | | | | | | | |
| 0050 | Wall type, moderate climate, bronze, encased | | | | | | | | | |
| 0200 | 3/4" IPS connection | 1 Plum | 16 | .500 | Ea. | 600 | 27 | | 627 | 700 |
| 0300 | 1" IPS connection | " | 14 | .571 | | 690 | 30.50 | | 720.50 | 805 |
| 0500 | Anti-siphon type, 3/4" connection | | | | ↓ | 520 | | | 520 | 570 |
| 1000 | Non-freeze, bronze, exposed | | | | | | | | | |
| 1100 | 3/4" IPS connection, 4" to 9" thick wall | 1 Plum | 14 | .571 | Ea. | 405 | 30.50 | | 435.50 | 490 |
| 1120 | 10" to 14" thick wall | | 12 | .667 | | 520 | 35.50 | | 555.50 | 625 |
| 1140 | 15" to 19" thick wall | | 12 | .667 | | 520 | 35.50 | | 555.50 | 625 |
| 1160 | 20" to 24" thick wall | ↓ | 10 | .800 | ↓ | 530 | 43 | | 573 | 650 |
| 1200 | For 1" IPS connection, add | | | | | 15% | 10% | | | |
| 1240 | For 3/4" adapter type vacuum breaker, add | | | | Ea. | 51 | | | 51 | 56 |
| 1280 | For anti-siphon type, add | | | | " | 106 | | | 106 | 117 |
| 2000 | Non-freeze bronze, encased, anti-siphon type | | | | | | | | | |
| 2100 | 3/4" IPS connection, 5" to 9" thick wall | 1 Plum | 14 | .571 | Ea. | 1,025 | 30.50 | | 1,055.50 | 1,175 |
| 2140 | 15" to 19" thick wall | " | 12 | .667 | " | 1,100 | 35.50 | | 1,135.50 | 1,275 |
| 3000 | Ground box type, bronze frame, 3/4" IPS connection | | | | | | | | | |
| 3080 | Non-freeze, all bronze, polished face, set flush | | | | | | | | | |
| 3100 | 2 feet depth of bury | 1 Plum | 8 | 1 | Ea. | 765 | 53.50 | | 818.50 | 925 |
| 3180 | 6 feet depth of bury | | 7 | 1.143 | | 1,000 | 61.50 | | 1,061.50 | 1,200 |
| 3220 | 8 feet depth of bury | ↓ | 5 | 1.600 | | 1,125 | 86 | | 1,211 | 1,350 |
| 3400 | For 1" IPS connection, add | | | | | 15% | 10% | | | |
| 3550 | For 2" connection, add | | | | | 445% | 24% | | | |
| 3600 | For tapped drain port in box, add | | | | ↓ | 69.50 | | | 69.50 | 76 |
| 5000 | Moderate climate, all bronze, polished face | | | | | | | | | |
| 5020 | and scoriated cover, set flush | | | | | | | | | |
| 5100 | 3/4" IPS connection | 1 Plum | 16 | .500 | Ea. | 530 | 27 | | 557 | 625 |
| 5120 | 1" IPS connection | " | 14 | .571 | | 655 | 30.50 | | 685.50 | 765 |
| 5200 | For tapped drain port in box, add | | | | ↓ | 69.50 | | | 69.50 | 76 |

## 22 11 23 – Domestic Water Pumps

### 22 11 23.10 General Utility Pumps

| | | Crew | Daily Output | Labor-Hours | Unit | Material | 2011 Bare Costs Labor | Equipment | Total | Total Incl O&P |
|---|---|---|---|---|---|---|---|---|---|---|
| 0010 | **GENERAL UTILITY PUMPS** | | | | | | | | | |
| 2000 | Single stage | | | | | | | | | |
| 3000 | Double suction, | | | | | | | | | |
| 3190 | 75 HP, to 2500 GPM | Q-3 | .28 | 114 | Ea. | 16,800 | 5,825 | | 22,625 | 27,200 |
| 3220 | 100 HP, to 3000 GPM | | .26 | 123 | | 21,200 | 6,275 | | 27,475 | 32,800 |
| 3240 | 150 HP, to 4000 GPM | ↓ | .24 | 133 | ↓ | 32,800 | 6,800 | | 39,600 | 46,300 |

## 22 13 16.20 Pipe, Cast Iron

| 22 13 16.20 Pipe, Cast Iron | Crew | Daily Output | Labor-Hours | Unit | Material | 2011 Bare Costs Labor | Equipment | Total | Total Incl O&P |
|---|---|---|---|---|---|---|---|---|---|
| 0010 **PIPE, CAST IRON**, Soil, on clevis hanger assemblies, 5' O.C. | | | | | | | | | |
| 0020 Single hub, service wt., lead & oakum joints 10' O.C. | | | | | | | | | |
| 2120 2" diameter | Q-1 | 63 | .254 | L.F. | 9 | 12.25 | | 21.25 | 28 |
| 2140 3" diameter | | 60 | .267 | | 12.20 | 12.85 | | 25.05 | 32.50 |
| 2160 4" diameter | ↓ | 55 | .291 | | 15.50 | 14.05 | | 29.55 | 38 |
| 2180 5" diameter | Q-2 | 76 | .316 | | 20.50 | 15.80 | | 36.30 | 46.50 |
| 2200 6" diameter | " | 73 | .329 | | 25.50 | 16.45 | | 41.95 | 52.50 |
| 2220 8" diameter | Q-3 | 59 | .542 | | 39 | 27.50 | | 66.50 | 84.50 |
| 2240 10" diameter | | 54 | .593 | | 62.50 | 30.50 | | 93 | 114 |
| 2260 12" diameter | ↓ | 48 | .667 | | 88 | 34 | | 122 | 148 |
| 2320 For service weight, double hub, add | | | | | 10% | | | | |
| 2340 For extra heavy, single hub, add | | | | | 48% | 4% | | | |
| 2360 For extra heavy, double hub, add | | | | ↓ | 71% | 4% | | | |
| 2400 Lead for caulking, (1#/dia in) | Q-1 | 160 | .100 | Lb. | 1.04 | 4.83 | | 5.87 | 8.35 |
| 2420 Oakum for caulking, (1/8#/dia in) | " | 40 | .400 | " | 4.88 | 19.30 | | 24.18 | 34.50 |
| 4000 No hub, couplings 10' O.C. | | | | | | | | | |
| 4100 1-1/2" diameter | Q-1 | 71 | .225 | L.F. | 9 | 10.85 | | 19.85 | 26 |
| 4120 2" diameter | | 67 | .239 | | 9.25 | 11.50 | | 20.75 | 27.50 |
| 4160 4" diameter | ↓ | 58 | .276 | | 15.80 | 13.30 | | 29.10 | 37 |

## 22 13 16.50 Shower Drains

| 22 13 16.50 Shower Drains | Crew | Daily Output | Labor-Hours | Unit | Material | 2011 Bare Costs Labor | Equipment | Total | Total Incl O&P |
|---|---|---|---|---|---|---|---|---|---|
| 0010 **SHOWER DRAINS** | | | | | | | | | |
| 2780 Shower, with strainer, uniform diam. trap, bronze top | | | | | | | | | |
| 2800 2" and 3" pipe size | Q-1 | 8 | 2 | Ea. | 335 | 96.50 | | 431.50 | 510 |
| 2820 4" pipe size | " | 7 | 2.286 | | 370 | 110 | | 480 | 570 |
| 2840 For galvanized body, add | | | | ↓ | 148 | | | 148 | 163 |

## 22 13 16.60 Traps

| 22 13 16.60 Traps | Crew | Daily Output | Labor-Hours | Unit | Material | 2011 Bare Costs Labor | Equipment | Total | Total Incl O&P |
|---|---|---|---|---|---|---|---|---|---|
| 0010 **TRAPS** | | | | | | | | | |
| 0030 Cast iron, service weight | | | | | | | | | |
| 0050 Running P trap, without vent | | | | | | | | | |
| 1100 2" | Q-1 | 16 | 1 | Ea. | 113 | 48.50 | | 161.50 | 196 |
| 1140 3" | | 14 | 1.143 | | 113 | 55 | | 168 | 207 |
| 1150 4" | ↓ | 13 | 1.231 | | 113 | 59.50 | | 172.50 | 213 |
| 1160 6" | Q-2 | 17 | 1.412 | ↓ | 490 | 70.50 | | 560.50 | 645 |
| 1180 Running trap, single hub, with vent | | | | | | | | | |
| 2080 3" pipe size, 3" vent | Q-1 | 14 | 1.143 | Ea. | 89.50 | 55 | | 144.50 | 181 |
| 2120 4" pipe size, 4" vent | " | 13 | 1.231 | | 121 | 59.50 | | 180.50 | 223 |
| 2300 For double hub, vent, add | | | | | 10% | 20% | | | |
| 3000 P trap, B&S, 2" pipe size | Q-1 | 16 | 1 | | 26.50 | 48.50 | | 75 | 102 |
| 3040 3" pipe size | " | 14 | 1.143 | ↓ | 40 | 55 | | 95 | 127 |
| 3350 Deep seal trap, B&S | | | | | | | | | |
| 3400 1-1/4" pipe size | Q-1 | 14 | 1.143 | Ea. | 42.50 | 55 | | 97.50 | 130 |
| 3410 1-1/2" pipe size | | 14 | 1.143 | | 42.50 | 55 | | 97.50 | 130 |
| 3420 2" pipe size | | 14 | 1.143 | | 36 | 55 | | 91 | 122 |
| 3440 3" pipe size | ↓ | 12 | 1.333 | ↓ | 49 | 64.50 | | 113.50 | 150 |
| 4700 Copper, drainage, drum trap | | | | | | | | | |
| 4800 3" x 5" solid, 1-1/2" pipe size | 1 Plum | 16 | .500 | Ea. | 177 | 27 | | 204 | 234 |
| 4840 3" x 6" swivel, 1-1/2" pipe size | " | 16 | .500 | " | 280 | 27 | | 307 | 350 |
| 5100 P trap, standard pattern | | | | | | | | | |
| 5200 1-1/4" pipe size | 1 Plum | 18 | .444 | Ea. | 130 | 24 | | 154 | 179 |
| 5240 1-1/2" pipe size | | 17 | .471 | | 126 | 25 | | 151 | 177 |
| 5260 2" pipe size | | 15 | .533 | | 195 | 28.50 | | 223.50 | 257 |
| 5280 3" pipe size | ↓ | 11 | .727 | ↓ | 465 | 39 | | 504 | 575 |

# 22 13 Facility Sanitary Sewerage

## 22 13 16 − Sanitary Waste and Vent Piping

### 22 13 16.60 Traps

| 22 13 16.60 Traps | | Crew | Daily Output | Labor-Hours | Unit | Material | 2011 Bare Costs Labor | Equipment | Total | Total Incl O&P |
|---|---|---|---|---|---|---|---|---|---|---|
| 5340 | With cleanout, swivel joint and slip joint | | | | | | | | | |
| 5360 | 1-1/4" pipe size | 1 Plum | 18 | .444 | Ea. | 92 | 24 | | 116 | 137 |
| 5400 | 1-1/2" pipe size | " | 17 | .471 | " | 192 | 25 | | 217 | 249 |

### 22 13 16.80 Vent Flashing and Caps

| 22 13 16.80 Vent Flashing and Caps | | Crew | Daily Output | Labor-Hours | Unit | Material | 2011 Bare Costs Labor | Equipment | Total | Total Incl O&P |
|---|---|---|---|---|---|---|---|---|---|---|
| 0010 | **VENT FLASHING AND CAPS** | | | | | | | | | |
| 0120 | Vent caps | | | | | | | | | |
| 0140 | Cast iron | | | | | | | | | |
| 0180 | 2-1/2" - 3-5/8" pipe | 1 Plum | 21 | .381 | Ea. | 40 | 20.50 | | 60.50 | 74.50 |
| 0190 | 4" - 4-1/8" pipe | " | 19 | .421 | " | 48 | 22.50 | | 70.50 | 86.50 |
| 0900 | Vent flashing | | | | | | | | | |
| 1000 | Aluminum with lead ring | | | | | | | | | |
| 1020 | 1-1/4" pipe | 1 Plum | 20 | .400 | Ea. | 10.70 | 21.50 | | 32.20 | 44 |
| 1030 | 1-1/2" pipe | | 20 | .400 | | 10.95 | 21.50 | | 32.45 | 44 |
| 1040 | 2" pipe | | 18 | .444 | | 11.45 | 24 | | 35.45 | 48 |
| 1050 | 3" pipe | | 17 | .471 | | 12.70 | 25 | | 37.70 | 51.50 |
| 1060 | 4" pipe | | 16 | .500 | | 15.30 | 27 | | 42.30 | 57 |
| 1350 | Copper with neoprene ring | | | | | | | | | |
| 1400 | 1-1/4" pipe | 1 Plum | 20 | .400 | Ea. | 19.95 | 21.50 | | 41.45 | 54 |
| 1430 | 1-1/2" pipe | | 20 | .400 | | 19.95 | 21.50 | | 41.45 | 54 |
| 1440 | 2" pipe | | 18 | .444 | | 21 | 24 | | 45 | 58.50 |
| 1450 | 3" pipe | | 17 | .471 | | 24.50 | 25 | | 49.50 | 64.50 |
| 1460 | 4" pipe | | 16 | .500 | | 27 | 27 | | 54 | 70 |

## 22 13 19 − Sanitary Waste Piping Specialties

### 22 13 19.13 Sanitary Drains

| 22 13 19.13 Sanitary Drains | | Crew | Daily Output | Labor-Hours | Unit | Material | 2011 Bare Costs Labor | Equipment | Total | Total Incl O&P |
|---|---|---|---|---|---|---|---|---|---|---|
| 0010 | **SANITARY DRAINS** | | | | | | | | | |
| 0400 | Deck, auto park, C.I., 13" top | | | | | | | | | |
| 0440 | 3", 4", 5", and 6" pipe size | Q-1 | 8 | 2 | Ea. | 1,125 | 96.50 | | 1,221.50 | 1,400 |
| 0480 | For galvanized body, add | | | | " | 610 | | | 610 | 675 |
| 2000 | Floor, medium duty, C.I., deep flange, 7" dia top | | | | | | | | | |
| 2040 | 2" and 3" pipe size | Q-1 | 12 | 1.333 | Ea. | 158 | 64.50 | | 222.50 | 270 |
| 2080 | For galvanized body, add | | | | | 75.50 | | | 75.50 | 83 |
| 2120 | For polished bronze top, add | | | | | 89 | | | 89 | 98 |
| 2400 | Heavy duty, with sediment bucket, C.I., 12" dia. loose grate | | | | | | | | | |
| 2420 | 2", 3", 4", 5", and 6" pipe size | Q-1 | 9 | 1.778 | Ea. | 545 | 86 | | 631 | 730 |
| 2460 | For polished bronze top, add | | | | " | 225 | | | 225 | 248 |
| 2500 | Heavy duty, cleanout & trap w/bucket, C.I., 15" top | | | | | | | | | |
| 2540 | 2", 3", and 4" pipe size | Q-1 | 6 | 2.667 | Ea. | 5,175 | 129 | | 5,304 | 5,875 |
| 2560 | For galvanized body, add | | | | | 1,325 | | | 1,325 | 1,450 |
| 2580 | For polished bronze top, add | | | | | 570 | | | 570 | 630 |

## 22 13 23 − Sanitary Waste Interceptors

### 22 13 23.10 Interceptors

| 22 13 23.10 Interceptors | | Crew | Daily Output | Labor-Hours | Unit | Material | 2011 Bare Costs Labor | Equipment | Total | Total Incl O&P |
|---|---|---|---|---|---|---|---|---|---|---|
| 0010 | **INTERCEPTORS** | | | | | | | | | |
| 0150 | Grease, fabricated steel, 4 GPM, 8 lb. fat capacity | 1 Plum | 4 | 2 | Ea. | 935 | 107 | | 1,042 | 1,175 |
| 0200 | 7 GPM, 14 lb. fat capacity | | 4 | 2 | | 1,300 | 107 | | 1,407 | 1,575 |
| 1000 | 10 GPM, 20 lb. fat capacity | | 4 | 2 | | 1,525 | 107 | | 1,632 | 1,825 |
| 1040 | 15 GPM, 30 lb. fat capacity | | 4 | 2 | | 2,250 | 107 | | 2,357 | 2,625 |
| 1060 | 20 GPM, 40 lb. fat capacity | | 3 | 2.667 | | 2,775 | 143 | | 2,918 | 3,275 |
| 1120 | 50 GPM, 100 lb. fat capacity | Q-1 | 2 | 8 | | 8,150 | 385 | | 8,535 | 9,525 |
| 1160 | 100 GPM, 200 lb. fat capacity | " | 2 | 8 | | 11,500 | 385 | | 11,885 | 13,300 |
| 1580 | For seepage pan, add | | | | | 7% | | | | |
| 3000 | Hair, cast iron, 1-1/4" and 1-1/2" pipe connection | 1 Plum | 8 | 1 | Ea. | 350 | 53.50 | | 403.50 | 465 |

# 22 13 Facility Sanitary Sewerage

## 22 13 23 – Sanitary Waste Interceptors

| 22 13 23.10 Interceptors | | Crew | Daily Output | Labor-Hours | Unit | Material | 2011 Bare Costs Labor | Equipment | Total | Total Incl O&P |
|---|---|---|---|---|---|---|---|---|---|---|
| 3100 | For chrome-plated cast iron, add | | | | Ea. | 215 | | | 215 | 236 |
| 4000 | Oil, fabricated steel, 10 GPM, 2" pipe size | 1 Plum | 4 | 2 | | 2,075 | 107 | | 2,182 | 2,450 |
| 4100 | 15 GPM, 2" or 3" pipe size | | 4 | 2 | | 2,850 | 107 | | 2,957 | 3,300 |
| 4120 | 20 GPM, 2" or 3" pipe size | | 3 | 2.667 | | 3,450 | 143 | | 3,593 | 4,025 |
| 4220 | 100 GPM, 3" pipe size | Q-1 | 2 | 8 | | 11,500 | 385 | | 11,885 | 13,300 |
| 6000 | Solids, precious metals recovery, C.I., 1-1/4" to 2" pipe | 1 Plum | 4 | 2 | | 530 | 107 | | 637 | 740 |
| 6100 | Dental Lab., large, C.I., 1-1/2" to 2" pipe | " | 3 | 2.667 | | 1,850 | 143 | | 1,993 | 2,250 |

## 22 13 29 – Sanitary Sewerage Pumps

### 22 13 29.13 Wet-Pit-Mounted, Vertical Sewerage Pumps

| | | Crew | Daily Output | Labor-Hours | Unit | Material | Labor | Equipment | Total | Total Incl O&P |
|---|---|---|---|---|---|---|---|---|---|---|
| 0010 | **WET-PIT-MOUNTED, VERTICAL SEWERAGE PUMPS** | | | | | | | | | |
| 0020 | Controls incl. alarm/disconnect panel w/wire. Excavation not included | | | | | | | | | |
| 0260 | Simplex, 9 GPM at 60 PSIG, 91 gal. tank | | | | Ea. | 2,975 | | | 2,975 | 3,275 |
| 0300 | Unit with manway, 26" I.D., 18" high | | | | | 3,375 | | | 3,375 | 3,700 |
| 0340 | 26" I.D., 36" high | | | | | 3,500 | | | 3,500 | 3,850 |
| 0380 | 43" I.D., 4' high | | | | | 3,575 | | | 3,575 | 3,925 |
| 3000 | Indoor residential type installation | | | | | | | | | |
| 3020 | Simplex, 9 GPM at 60 PSIG, 91 gal. HDPE tank | | | | Ea. | 2,950 | | | 2,950 | 3,225 |

### 22 13 29.14 Sewage Ejector Pumps

| | | Crew | Daily Output | Labor-Hours | Unit | Material | Labor | Equipment | Total | Total Incl O&P |
|---|---|---|---|---|---|---|---|---|---|---|
| 0010 | **SEWAGE EJECTOR PUMPS**, With operating and level controls | | | | | | | | | |
| 0100 | Simplex system incl. tank, cover, pump 15' head | | | | | | | | | |
| 0500 | 37 gal. PE tank, 12 GPM, 1/2 HP, 2" discharge | Q-1 | 3.20 | 5 | Ea. | 440 | 241 | | 681 | 845 |
| 0510 | 3" discharge | | 3.10 | 5.161 | | 480 | 249 | | 729 | 895 |
| 0530 | 87 GPM, .7 HP, 2" discharge | | 3.20 | 5 | | 670 | 241 | | 911 | 1,100 |
| 0540 | 3" discharge | | 3.10 | 5.161 | | 725 | 249 | | 974 | 1,175 |
| 0600 | 45 gal. coated stl tank, 12 GPM, 1/2 HP, 2" discharge | | 3 | 5.333 | | 780 | 257 | | 1,037 | 1,250 |
| 0610 | 3" discharge | | 2.90 | 5.517 | | 810 | 266 | | 1,076 | 1,275 |
| 0630 | 87 GPM, .7 HP, 2" discharge | | 3 | 5.333 | | 1,000 | 257 | | 1,257 | 1,475 |
| 0640 | 3" discharge | | 2.90 | 5.517 | | 1,050 | 266 | | 1,316 | 1,550 |
| 0660 | 134 GPM, 1 HP, 2" discharge | | 2.80 | 5.714 | | 1,075 | 276 | | 1,351 | 1,600 |
| 0680 | 3" discharge | | 2.70 | 5.926 | | 1,150 | 286 | | 1,436 | 1,675 |
| 0700 | 70 gal. PE tank, 12 GPM, 1/2 HP, 2" discharge | | 2.60 | 6.154 | | 855 | 297 | | 1,152 | 1,375 |
| 0710 | 3" discharge | | 2.40 | 6.667 | | 910 | 320 | | 1,230 | 1,475 |
| 0730 | 87 GPM, 0.7 HP, 2" discharge | | 2.50 | 6.400 | | 1,100 | 310 | | 1,410 | 1,650 |
| 0740 | 3" discharge | | 2.30 | 6.957 | | 1,150 | 335 | | 1,485 | 1,775 |
| 0760 | 134 GPM, 1 HP, 2" discharge | | 2.20 | 7.273 | | 1,200 | 350 | | 1,550 | 1,825 |
| 0770 | 3" discharge | | 2 | 8 | | 1,275 | 385 | | 1,660 | 1,975 |

# 22 14 Facility Storm Drainage

## 22 14 23 – Storm Drainage Piping Specialties

### 22 14 23.33 Backwater Valves

| | | Crew | Daily Output | Labor-Hours | Unit | Material | Labor | Equipment | Total | Total Incl O&P |
|---|---|---|---|---|---|---|---|---|---|---|
| 0010 | **BACKWATER VALVES**, C.I. Body | | | | | | | | | |
| 6980 | Bronze gate and automatic flapper valves | | | | | | | | | |
| 7000 | 3" and 4" pipe size | Q-1 | 13 | 1.231 | Ea. | 1,600 | 59.50 | | 1,659.50 | 1,875 |
| 7100 | 5" and 6" pipe size | " | 13 | 1.231 | " | 2,475 | 59.50 | | 2,534.50 | 2,800 |
| 7240 | Bronze flapper valve, bolted cover | | | | | | | | | |
| 7260 | 2" pipe size | Q-1 | 16 | 1 | Ea. | 470 | 48.50 | | 518.50 | 585 |
| 7300 | 4" pipe size | " | 13 | 1.231 | | 905 | 59.50 | | 964.50 | 1,075 |
| 7340 | 6" pipe size | Q-2 | 17 | 1.412 | | 1,300 | 70.50 | | 1,370.50 | 1,525 |

## 22 14 26 – Facility Storm Drains

| 22 14 26.13 Roof Drains | | Crew | Daily Output | Labor-Hours | Unit | Material | 2011 Bare Costs Labor | Equipment | Total | Total Incl O&P |
|---|---|---|---|---|---|---|---|---|---|---|
| 0010 | **ROOF DRAINS** | | | | | | | | | |
| 0140 | Cornice, C.I., 45° or 90° outlet | | | | | | | | | |
| 0200 | 3" and 4" pipe size | Q-1 | 12 | 1.333 | Ea. | 259 | 64.50 | | 323.50 | 380 |
| 0260 | For galvanized body, add | | | | | 59.50 | | | 59.50 | 65.50 |
| 0280 | For polished bronze dome, add | | | | ↓ | 46 | | | 46 | 51 |
| 3860 | Roof, flat metal deck, C.I. body, 12" C.I. dome | | | | | | | | | |
| 3890 | 3" pipe size | Q-1 | 14 | 1.143 | Ea. | 310 | 55 | | 365 | 425 |
| 3920 | 6" pipe size | " | 10 | 1.600 | " | 540 | 77 | | 617 | 705 |
| 4620 | Main, all aluminum, 12" low profile dome | | | | | | | | | |
| 4640 | 2", 3" and 4" pipe size | Q-1 | 14 | 1.143 | Ea. | 345 | 55 | | 400 | 465 |

## 22 14 26.16 Facility Area Drains

| 22 14 26.16 Facility Area Drains | | Crew | Daily Output | Labor-Hours | Unit | Material | 2011 Bare Costs Labor | Equipment | Total | Total Incl O&P |
|---|---|---|---|---|---|---|---|---|---|---|
| 0010 | **FACILITY AREA DRAINS** | | | | | | | | | |
| 4980 | Scupper floor, oblique strainer, C.I. | | | | | | | | | |
| 5000 | 6" x 7" top, 2", 3" and 4" pipe size | Q-1 | 16 | 1 | Ea. | 223 | 48.50 | | 271.50 | 315 |
| 5100 | 8" x 12" top, 5" and 6" pipe size | " | 14 | 1.143 | | 435 | 55 | | 490 | 560 |
| 5160 | For galvanized body, add | | | | | 40% | | | | |
| 5200 | For polished bronze strainer, add | | | | ↓ | 85% | | | | |

## 22 14 26.19 Facility Trench Drains

| 22 14 26.19 Facility Trench Drains | | Crew | Daily Output | Labor-Hours | Unit | Material | 2011 Bare Costs Labor | Equipment | Total | Total Incl O&P |
|---|---|---|---|---|---|---|---|---|---|---|
| 0010 | **FACILITY TRENCH DRAINS** | | | | | | | | | |
| 5980 | Trench, floor, hvy duty, modular, C.I., 12" x 12" top | | | | | | | | | |
| 6000 | 2", 3", 4", 5", & 6" pipe size | Q-1 | 8 | 2 | Ea. | 705 | 96.50 | | 801.50 | 920 |
| 6100 | For unit with polished bronze top | " | 8 | 2 | " | 1,050 | 96.50 | | 1,146.50 | 1,300 |
| 6600 | Trench, floor, for cement concrete encasement | | | | | | | | | |
| 6610 | Not including trenching or concrete | | | | | | | | | |
| 6640 | Polyester polymer concrete | | | | | | | | | |
| 6650 | 4" internal width, with grate | | | | | | | | | |
| 6660 | Light duty steel grate | Q-1 | 120 | .133 | L.F. | 26.50 | 6.45 | | 32.95 | 38.50 |
| 6670 | Medium duty steel grate | | 115 | .139 | | 33 | 6.70 | | 39.70 | 46.50 |
| 6680 | Heavy duty iron grate | ↓ | 110 | .145 | ↓ | 51 | 7 | | 58 | 66.50 |
| 6700 | 12" internal width, with grate | | | | | | | | | |
| 6770 | Heavy duty galvanized grate | Q-1 | 80 | .200 | L.F. | 128 | 9.65 | | 137.65 | 155 |
| 6800 | Fiberglass | | | | | | | | | |
| 6810 | 8" internal width, with grate | | | | | | | | | |
| 6820 | Medium duty galvanized grate | Q-1 | 115 | .139 | L.F. | 89 | 6.70 | | 95.70 | 108 |
| 6830 | Heavy duty iron grate | " | 110 | .145 | " | 86 | 7 | | 93 | 105 |

## 22 14 29 – Sump Pumps

### 22 14 29.13 Wet-Pit-Mounted, Vertical Sump Pumps

| 22 14 29.13 Wet-Pit-Mounted, Vertical Sump Pumps | | Crew | Daily Output | Labor-Hours | Unit | Material | 2011 Bare Costs Labor | Equipment | Total | Total Incl O&P |
|---|---|---|---|---|---|---|---|---|---|---|
| 0010 | **WET-PIT-MOUNTED, VERTICAL SUMP PUMPS** | | | | | | | | | |
| 0400 | Molded PVC base, 21 GPM at 15' head, 1/3 HP | 1 Plum | 5 | 1.600 | Ea. | 119 | 86 | | 205 | 259 |
| 0800 | Iron base, 21 GPM at 15' head, 1/3 HP | | 5 | 1.600 | | 150 | 86 | | 236 | 293 |
| 1200 | Solid brass, 21 GPM at 15' head, 1/3 HP | ↓ | 5 | 1.600 | ↓ | 263 | 86 | | 349 | 420 |

### 22 14 29.16 Submersible Sump Pumps

| 22 14 29.16 Submersible Sump Pumps | | Crew | Daily Output | Labor-Hours | Unit | Material | 2011 Bare Costs Labor | Equipment | Total | Total Incl O&P |
|---|---|---|---|---|---|---|---|---|---|---|
| 0010 | **SUBMERSIBLE SUMP PUMPS** | | | | | | | | | |
| 7000 | Sump pump, automatic | | | | | | | | | |
| 7100 | Plastic, 1-1/4" discharge, 1/4 HP | 1 Plum | 6 | 1.333 | Ea. | 148 | 71.50 | | 219.50 | 269 |
| 7140 | 1/3 HP | | 5 | 1.600 | | 181 | 86 | | 267 | 325 |
| 7160 | 1/2 HP | | 5 | 1.600 | | 221 | 86 | | 307 | 370 |
| 7180 | 1-1/2" discharge, 1/2 HP | | 4 | 2 | | 246 | 107 | | 353 | 430 |
| 7500 | Cast iron, 1-1/4" discharge, 1/4 HP | | 6 | 1.333 | | 170 | 71.50 | | 241.50 | 294 |
| 7540 | 1/3 HP | | 6 | 1.333 | | 200 | 71.50 | | 271.50 | 325 |
| 7560 | 1/2 HP | ↓ | 5 | 1.600 | ↓ | 242 | 86 | | 328 | 395 |

# 22 31 Domestic Water Softeners

## 22 31 13 – Residential Domestic Water Softeners

### 22 31 13.10 Residential Water Softeners

| | | Crew | Daily Output | Labor-Hours | Unit | Material | 2011 Bare Costs Labor | Equipment | Total | Total Incl O&P |
|---|---|---|---|---|---|---|---|---|---|---|
| 0010 | **RESIDENTIAL WATER SOFTENERS** | | | | | | | | | |
| 7350 | Water softener, automatic, to 30 grains per gallon | 2 Plum | 5 | 3.200 | Ea. | 380 | 172 | | 552 | 675 |
| 7400 | To 100 grains per gallon | " | 4 | 4 | " | 750 | 214 | | 964 | 1,150 |

## 22 31 16 – Commercial Domestic Water Softeners

### 22 31 16.10 Water Softeners

| | | Crew | Daily Output | Labor-Hours | Unit | Material | 2011 Bare Costs Labor | Equipment | Total | Total Incl O&P |
|---|---|---|---|---|---|---|---|---|---|---|
| 0010 | **WATER SOFTENERS** | | | | | | | | | |
| 5800 | Softener systems, automatic, intermediate sizes | | | | | | | | | |
| 5820 | available, may be used in multiples. | | | | | | | | | |
| 6000 | Hardness capacity between regenerations and flow | | | | | | | | | |
| 6100 | 150,000 grains, 37 GPM cont., 51 GPM peak | Q-1 | 1.20 | 13.333 | Ea. | 3,900 | 645 | | 4,545 | 5,225 |
| 6200 | 300,000 grains, 81 GPM cont., 113 GPM peak | | 1 | 16 | | 8,675 | 770 | | 9,445 | 10,700 |
| 6300 | 750,000 grains, 160 GPM cont., 230 GPM peak | | .80 | 20 | | 11,300 | 965 | | 12,265 | 14,000 |
| 6400 | 900,000 grains, 185 GPM cont., 270 GPM peak | | .70 | 22.857 | | 18,200 | 1,100 | | 19,300 | 21,700 |

# 22 33 Electric Domestic Water Heaters

## 22 33 13 – Instantaneous Electric Domestic Water Heaters

### 22 33 13.10 Hot Water Dispensers

| | | Crew | Daily Output | Labor-Hours | Unit | Material | 2011 Bare Costs Labor | Equipment | Total | Total Incl O&P |
|---|---|---|---|---|---|---|---|---|---|---|
| 0010 | **HOT WATER DISPENSERS** | | | | | | | | | |
| 0160 | Commercial, 100 cup, 11.3 amp | 1 Plum | 14 | .571 | Ea. | 450 | 30.50 | | 480.50 | 540 |
| 3180 | Household, 60 cup | " | 14 | .571 | " | 238 | 30.50 | | 268.50 | 310 |

## 22 33 30 – Residential, Electric Domestic Water Heaters

### 22 33 30.13 Residential, Small-Capacity Elec. Water Heaters

| | | Crew | Daily Output | Labor-Hours | Unit | Material | 2011 Bare Costs Labor | Equipment | Total | Total Incl O&P |
|---|---|---|---|---|---|---|---|---|---|---|
| 0010 | **RESIDENTIAL, SMALL-CAPACITY ELECTRIC DOMESTIC WATER HEATERS** | | | | | | | | | |
| 1000 | Residential, electric, glass lined tank, 5 yr., 10 gal., single element | 1 Plum | 2.30 | 3.478 | Ea. | 385 | 186 | | 571 | 705 |
| 1040 | 20 gallon, single element | | 2.20 | 3.636 | | 440 | 195 | | 635 | 775 |
| 1060 | 30 gallon, double element | | 2.20 | 3.636 | | 575 | 195 | | 770 | 925 |
| 1080 | 40 gallon, double element | | 2 | 4 | | 615 | 214 | | 829 | 995 |
| 1100 | 52 gallon, double element | | 2 | 4 | | 690 | 214 | | 904 | 1,075 |
| 1180 | 120 gallon, double element | | 1.40 | 5.714 | | 1,450 | 305 | | 1,755 | 2,050 |

## 22 33 33 – Light-Commercial Electric Domestic Water Heaters

### 22 33 33.10 Commercial Electric Water Heaters

| | | Crew | Daily Output | Labor-Hours | Unit | Material | 2011 Bare Costs Labor | Equipment | Total | Total Incl O&P |
|---|---|---|---|---|---|---|---|---|---|---|
| 0010 | **COMMERCIAL ELECTRIC WATER HEATERS** | | | | | | | | | |
| 4000 | Commercial, 100° rise. NOTE: for each size tank, a range of | | | | | | | | | |
| 4010 | heaters between the ones shown are available | | | | | | | | | |
| 4020 | Electric | | | | | | | | | |
| 4100 | 5 gal., 3 kW, 12 GPH, 208 volt | 1 Plum | 2 | 4 | Ea. | 2,350 | 214 | | 2,564 | 2,900 |
| 4120 | 10 gal., 6 kW, 25 GPH, 208 volt | | 2 | 4 | | 2,600 | 214 | | 2,814 | 3,200 |
| 4140 | 50 gal., 9 kW, 37 GPH, 208 volt | | 1.80 | 4.444 | | 3,275 | 238 | | 3,513 | 3,950 |
| 4160 | 50 gal., 36 kW, 148 GPH, 208 volt | | 1.80 | 4.444 | | 5,475 | 238 | | 5,713 | 6,375 |
| 4300 | 200 gal., 15 kW, 61 GPH, 480 volt | Q-1 | 1.70 | 9.412 | | 16,900 | 455 | | 17,355 | 19,200 |
| 4320 | 200 gal., 120 kW , 490 GPH, 480 volt | | 1.70 | 9.412 | | 23,000 | 455 | | 23,455 | 26,000 |
| 4460 | 400 gal., 30 kW, 123 GPH, 480 volt | | 1 | 16 | | 22,900 | 770 | | 23,670 | 26,400 |
| 5400 | Modulating step control, 2-5 steps | 1 Elec | 5.30 | 1.509 | | 810 | 76 | | 886 | 1,000 |
| 5440 | 6-10 steps | | 3.20 | 2.500 | | 1,025 | 126 | | 1,151 | 1,300 |
| 5460 | 11-15 steps | | 2.70 | 2.963 | | 1,475 | 149 | | 1,624 | 1,850 |
| 5480 | 16-20 steps | | 1.60 | 5 | | 2,175 | 252 | | 2,427 | 2,750 |

# 22 34 Fuel-Fired Domestic Water Heaters

## 22 34 13 – Instantaneous, Tankless, Gas Domestic Water Heaters

### 22 34 13.10 Instantaneous, Tankless, Gas Water Heaters

| | | Crew | Daily Output | Labor-Hours | Unit | Material | 2011 Bare Costs Labor | 2011 Bare Costs Equipment | Total | Total Incl O&P |
|---|---|---|---|---|---|---|---|---|---|---|
| 0010 | INSTANTANEOUS, TANKLESS, GAS WATER HEATERS | | | | | | | | | |
| 9410 | Natural gas/propane, incl vent, 3.2 GPM  **G** | 1 Plum | 2 | 4 | Ea. | 535 | 214 | | 749 | 910 |

## 22 34 30 – Residential Gas Domestic Water Heaters

### 22 34 30.13 Residential, Atmos, Gas Domestic Wtr Heaters

| | | Crew | Daily Output | Labor-Hours | Unit | Material | Labor | Equipment | Total | Total Incl O&P |
|---|---|---|---|---|---|---|---|---|---|---|
| 0010 | RESIDENTIAL, ATMOSPHERIC, GAS DOMESTIC WATER HEATERS | | | | | | | | | |
| 2000 | Gas fired, foam lined tank, 10 yr., vent not incl. | | | | | | | | | |
| 2040 | 30 gallon | 1 Plum | 2 | 4 | Ea. | 895 | 214 | | 1,109 | 1,300 |
| 2100 | 75 gallon | | 1.50 | 5.333 | | 1,225 | 286 | | 1,511 | 1,775 |
| 2120 | 100 gallon | ↓ | 1.30 | 6.154 | ↓ | 1,525 | 330 | | 1,855 | 2,175 |

## 22 34 36 – Commercial Gas Domestic Water Heaters

### 22 34 36.13 Commercial, Atmos., Gas Domestic Water Htrs.

| | | Crew | Daily Output | Labor-Hours | Unit | Material | Labor | Equipment | Total | Total Incl O&P |
|---|---|---|---|---|---|---|---|---|---|---|
| 0010 | COMMERCIAL, ATMOSPHERIC, GAS DOMESTIC WATER HEATERS | | | | | | | | | |
| 6000 | Gas fired, flush jacket, std. controls, vent not incl. | | | | | | | | | |
| 6040 | 75 MBH input, 73 GPH | 1 Plum | 1.40 | 5.714 | Ea. | 1,575 | 305 | | 1,880 | 2,175 |
| 6060 | 98 MBH input, 95 GPH | | 1.40 | 5.714 | | 4,575 | 305 | | 4,880 | 5,500 |
| 6080 | 120 MBH input, 110 GPH | | 1.20 | 6.667 | | 4,975 | 355 | | 5,330 | 5,975 |
| 6180 | 200 MBH input, 192 GPH | | .60 | 13.333 | | 7,650 | 715 | | 8,365 | 9,475 |
| 6200 | 250 MBH input, 245 GPH | | .50 | 16 | | 7,975 | 860 | | 8,835 | 10,100 |
| 6900 | For low water cutoff, add | | 8 | 1 | | 275 | 53.50 | | 328.50 | 380 |
| 6960 | For bronze body hot water circulator, add | ↓ | 4 | 2 | ↓ | 1,750 | 107 | | 1,857 | 2,075 |

## 22 34 46 – Oil-Fired Domestic Water Heaters

### 22 34 46.10 Residential Oil-Fired Water Heaters

| | | Crew | Daily Output | Labor-Hours | Unit | Material | Labor | Equipment | Total | Total Incl O&P |
|---|---|---|---|---|---|---|---|---|---|---|
| 0010 | RESIDENTIAL OIL-FIRED WATER HEATERS | | | | | | | | | |
| 3000 | Oil fired, glass lined tank, 5 yr., vent not included, 30 gallon | 1 Plum | 2 | 4 | Ea. | 665 | 214 | | 879 | 1,050 |
| 3040 | 50 gallon | | 1.80 | 4.444 | | 1,050 | 238 | | 1,288 | 1,500 |
| 3060 | 70 gallon | | 1.50 | 5.333 | | 1,275 | 286 | | 1,561 | 1,825 |
| 3080 | 85 gallon | ↓ | 1.40 | 5.714 | ↓ | 7,125 | 305 | | 7,430 | 8,275 |

### 22 34 46.20 Commercial Oil-Fired Water Heaters

| | | Crew | Daily Output | Labor-Hours | Unit | Material | Labor | Equipment | Total | Total Incl O&P |
|---|---|---|---|---|---|---|---|---|---|---|
| 0010 | COMMERCIAL OIL-FIRED WATER HEATERS | | | | | | | | | |
| 8000 | Oil fired, glass lined, UL listed, std. controls, vent not incl. | | | | | | | | | |
| 8060 | 140 gal., 140 MBH input, 134 GPH | Q-1 | 2.13 | 7.512 | Ea. | 14,900 | 360 | | 15,260 | 16,900 |
| 8080 | 140 gal., 199 MBH input, 191 GPH | | 2 | 8 | | 15,400 | 385 | | 15,785 | 17,500 |
| 8100 | 140 gal., 255 MBH input, 247 GPH | | 1.60 | 10 | | 15,900 | 485 | | 16,385 | 18,100 |
| 8160 | 140 gal., 540 MBH input, 519 GPH | | .96 | 16.667 | | 21,000 | 805 | | 21,805 | 24,300 |
| 8180 | 140 gal., 720 MBH input, 691 GPH | ↓ | .92 | 17.391 | | 21,400 | 840 | | 22,240 | 24,800 |
| 8280 | 201 gal., 1250 MBH input, 1200 GPH | Q-2 | 1.22 | 19.672 | | 32,900 | 985 | | 33,885 | 37,600 |
| 8300 | 201 gal., 1500 MBH input, 1441 GPH | " | 1.16 | 20.690 | | 35,700 | 1,025 | | 36,725 | 40,900 |
| 8900 | For low water cutoff, add | 1 Plum | 8 | 1 | | 300 | 53.50 | | 353.50 | 410 |
| 8960 | For bronze body hot water circulator, add | " | 4 | 2 | ↓ | 710 | 107 | | 817 | 940 |

# 22 35 Domestic Water Heat Exchangers

## 22 35 30 – Water Heating by Steam

| 22 35 30.10 Water Heating Transfer Package | Crew | Daily Output | Labor-Hours | Unit | Material | 2011 Bare Costs Labor | Equipment | Total | Total Incl O&P |
|---|---|---|---|---|---|---|---|---|---|
| 0010 **WATER HEATING TRANSFER PACKAGE**, Complete controls, | | | | | | | | | |
| 0020     expansion tank, converter, air separator | | | | | | | | | |
| 1000   Hot water, 180°F enter, 200°F leaving, 15# steam | | | | | | | | | |
| 1010     One pump system, 28 GPM | Q-6 | .75 | 32 | Ea. | 16,700 | 1,625 | | 18,325 | 20,700 |
| 1020     35 GPM | | .70 | 34.286 | | 18,100 | 1,725 | | 19,825 | 22,500 |
| 1040     55 GPM | | .65 | 36.923 | | 21,300 | 1,875 | | 23,175 | 26,300 |
| 1060     130 GPM | | .55 | 43.636 | | 27,000 | 2,200 | | 29,200 | 33,000 |
| 1080     255 GPM | | .40 | 60 | | 35,300 | 3,025 | | 38,325 | 43,400 |
| 1100     550 GPM | ▼ | .30 | 80 | ▼ | 44,800 | 4,050 | | 48,850 | 55,000 |

# 22 41 Residential Plumbing Fixtures

## 22 41 06 – Plumbing Fixtures General

### 22 41 06.10 Plumbing Fixture Notes

| | | | | | | | | | |
|---|---|---|---|---|---|---|---|---|---|
| 0010 **PLUMBING FIXTURE NOTES**, Incl. trim fittings unless otherwise noted R224000-40 | | | | | | | | | |
| 0080   For rough-in, supply, waste, and vent, see add for each type | | | | | | | | | |
| 0122   For electric water coolers, see Section 22 47 16.10 | | | | | | | | | |
| 0160   For color, unless otherwise noted, add | | | | Ea. | 20% | | | | |

## 22 41 13 – Residential Water Closets, Urinals, and Bidets

### 22 41 13.40 Water Closets

| | Crew | Daily Output | Labor-Hours | Unit | Material | Labor | Equipment | Total | Total Incl O&P |
|---|---|---|---|---|---|---|---|---|---|
| 0010 **WATER CLOSETS** R224000-40 | | | | | | | | | |
| 0032   For automatic flush, see Line 22 42 39.10 0972 | | | | | | | | | |
| 0150   Tank type, vitreous china, incl. seat, supply pipe w/stop, 1.6 gpf or noted | | | | | | | | | |
| 0200     Wall hung | | | | | | | | | |
| 0400       Two piece, close coupled | Q-1 | 5.30 | 3.019 | Ea. | 590 | 146 | | 736 | 865 |
| 0960       For rough-in, supply, waste, vent and carrier | " | 2.73 | 5.861 | " | 910 | 283 | | 1,193 | 1,425 |
| 0999     Floor mounted | | | | | | | | | |
| 1100       Two piece, close coupled | Q-1 | 5.30 | 3.019 | Ea. | 240 | 146 | | 386 | 480 |
| 1102         Economy | | 5.30 | 3.019 | | 134 | 146 | | 280 | 365 |
| 1110       Two piece, close coupled, dual flush | | 5.30 | 3.019 | | 256 | 146 | | 402 | 500 |
| 1140       Two piece, close coupled, 1.28 gpf, ADA  G | ▼ | 5.30 | 3.019 | ▼ | 295 | 146 | | 441 | 540 |
| 1960     For color, add | | | | | 30% | | | | |
| 1980     For rough-in, supply, waste and vent | Q-1 | 3.05 | 5.246 | Ea. | 305 | 253 | | 558 | 715 |

## 22 41 16 – Residential Lavatories and Sinks

### 22 41 16.10 Lavatories

| | Crew | Daily Output | Labor-Hours | Unit | Material | Labor | Equipment | Total | Total Incl O&P |
|---|---|---|---|---|---|---|---|---|---|
| 0010 **LAVATORIES**, With trim, white unless noted otherwise R224000-40 | | | | | | | | | |
| 0500   Vanity top, porcelain enamel on cast iron | | | | | | | | | |
| 0600     20" x 18" | Q-1 | 6.40 | 2.500 | Ea. | 296 | 121 | | 417 | 505 |
| 0640     33" x 19" oval | | 6.40 | 2.500 | | 605 | 121 | | 726 | 845 |
| 0720     19" round | ▼ | 6.40 | 2.500 | ▼ | 228 | 121 | | 349 | 430 |
| 0860     For color, add | | | | | 25% | | | | |
| 1000   Cultured marble, 19" x 17", single bowl | Q-1 | 6.40 | 2.500 | Ea. | 211 | 121 | | 332 | 410 |
| 1040     25" x 19", single bowl | " | 6.40 | 2.500 | " | 242 | 121 | | 363 | 445 |
| 1580     For color, same price | | | | | | | | | |
| 1900   Stainless steel, self-rimming, 25" x 22", single bowl, ledge | Q-1 | 6.40 | 2.500 | Ea. | 360 | 121 | | 481 | 575 |
| 1960     17" x 22", single bowl | | 6.40 | 2.500 | | 350 | 121 | | 471 | 565 |
| 2600   Steel, enameled, 20" x 17", single bowl | | 5.80 | 2.759 | | 186 | 133 | | 319 | 405 |
| 2660     19" round | | 5.80 | 2.759 | | 161 | 133 | | 294 | 375 |
| 2900   Vitreous china, 20" x 16", single bowl | | 5.40 | 2.963 | | 260 | 143 | | 403 | 500 |
| 2960     20" x 17", single bowl | | 5.40 | 2.963 | | 176 | 143 | | 319 | 405 |
| 3020     19" round, single bowl | ▼ | 5.40 | 2.963 | ▼ | 174 | 143 | | 317 | 405 |

## 22 41 16 – Residential Lavatories and Sinks

| 22 41 16.10 Lavatories | | Crew | Daily Output | Labor-Hours | Unit | Material | 2011 Bare Costs Labor | 2011 Bare Costs Equipment | Total | Total Incl O&P |
|---|---|---|---|---|---|---|---|---|---|---|
| 3200 | 22" x 13", single bowl | Q-1 | 5.40 | 2.963 | Ea. | 267 | 143 | | 410 | 505 |
| 3560 | For color, add | | | | | 50% | | | | |
| 3580 | Rough-in, supply, waste and vent for all above lavatories | Q-1 | 2.30 | 6.957 | Ea. | 320 | 335 | | 655 | 850 |
| 4000 | Wall hung | | | | | | | | | |
| 4040 | Porcelain enamel on cast iron, 16" x 14", single bowl | Q-1 | 8 | 2 | Ea. | 445 | 96.50 | | 541.50 | 635 |
| 4180 | 20" x 18", single bowl | | 8 | 2 | | 345 | 96.50 | | 441.50 | 525 |
| 4240 | 22" x 19", single bowl | ↓ | 8 | 2 | ↓ | 595 | 96.50 | | 691.50 | 800 |
| 4580 | For color, add | | | | | 30% | | | | |
| 6000 | Vitreous china, 18" x 15", single bowl with backsplash | Q-1 | 7 | 2.286 | Ea. | 231 | 110 | | 341 | 420 |
| 6500 | For color, add | | | | | 30% | | | | |
| 6960 | Rough-in, supply, waste and vent for above lavatories | Q-1 | 1.66 | 9.639 | Ea. | 420 | 465 | | 885 | 1,150 |
| 7000 | Pedestal type | | | | | | | | | |
| 7600 | Vitreous china, 27" x 21", white | Q-1 | 6.60 | 2.424 | Ea. | 545 | 117 | | 662 | 775 |
| 7610 | 27" x 21", colored | | 6.60 | 2.424 | | 655 | 117 | | 772 | 895 |
| 7620 | 27" x 21", premium color | | 6.60 | 2.424 | | 720 | 117 | | 837 | 965 |
| 7660 | 26" x 20", white | | 6.60 | 2.424 | | 485 | 117 | | 602 | 710 |
| 7670 | 26" x 20", colored | | 6.60 | 2.424 | | 580 | 117 | | 697 | 810 |
| 7680 | 26" x 20", premium color | | 6.60 | 2.424 | | 635 | 117 | | 752 | 875 |
| 7700 | 24" x 18", white | | 6.60 | 2.424 | | 470 | 117 | | 587 | 695 |
| 7710 | 24" x 18", colored | | 6.60 | 2.424 | | 560 | 117 | | 677 | 795 |
| 7720 | 24" x 18", premium color | | 6.60 | 2.424 | | 655 | 117 | | 772 | 895 |
| 7760 | 21" x 18", white | | 6.60 | 2.424 | | 215 | 117 | | 332 | 410 |
| 7770 | 21" x 18", colored | | 6.60 | 2.424 | | 232 | 117 | | 349 | 430 |
| 7990 | Rough-in, supply, waste and vent for pedestal lavatories | ↓ | 1.66 | 9.639 | ↓ | 420 | 465 | | 885 | 1,150 |

## 22 41 16.30 Sinks

| | | Crew | Daily Output | Labor-Hours | Unit | Material | 2011 Bare Costs Labor | 2011 Bare Costs Equipment | Total | Total Incl O&P |
|---|---|---|---|---|---|---|---|---|---|---|
| 0010 | **SINKS**, With faucets and drain      R224000-40 | | | | | | | | | |
| 2000 | Kitchen, counter top style, P.E. on C.I., 24" x 21" single bowl | Q-1 | 5.60 | 2.857 | Ea. | 257 | 138 | | 395 | 490 |
| 2100 | 31" x 22" single bowl | | 5.60 | 2.857 | | 500 | 138 | | 638 | 755 |
| 2200 | 32" x 21" double bowl | | 4.80 | 3.333 | | 277 | 161 | | 438 | 545 |
| 2300 | 42" x 21" double bowl | | 4.80 | 3.333 | | 1,025 | 161 | | 1,186 | 1,375 |
| 3000 | Stainless steel, self rimming, 19" x 18" single bowl | | 5.60 | 2.857 | | 580 | 138 | | 718 | 840 |
| 3100 | 25" x 22" single bowl | | 5.60 | 2.857 | | 645 | 138 | | 783 | 915 |
| 4000 | Steel, enameled, with ledge, 24" x 21" single bowl | | 5.60 | 2.857 | | 420 | 138 | | 558 | 665 |
| 4100 | 32" x 21" double bowl | ↓ | 4.80 | 3.333 | | 400 | 161 | | 561 | 680 |
| 4960 | For color sinks except stainless steel, add | | | | | 10% | | | | |
| 4980 | For rough-in, supply, waste and vent, counter top sinks | Q-1 | 2.14 | 7.477 | ↓ | 365 | 360 | | 725 | 940 |
| 5000 | Kitchen, raised deck, P.E. on C.I. | | | | | | | | | |
| 5100 | 32" x 21", dual level, double bowl | Q-1 | 2.60 | 6.154 | Ea. | 330 | 297 | | 627 | 810 |
| 5700 | For color, add | | | | | 20% | | | | |
| 5790 | For rough-in, supply, waste & vent, sinks | Q-1 | 1.85 | 8.649 | ↓ | 365 | 415 | | 780 | 1,025 |

## 22 41 19 – Residential Bathtubs

### 22 41 19.10 Baths

| | | Crew | Daily Output | Labor-Hours | Unit | Material | 2011 Bare Costs Labor | 2011 Bare Costs Equipment | Total | Total Incl O&P |
|---|---|---|---|---|---|---|---|---|---|---|
| 0010 | **BATHS** | | | | | | | | | |
| 0100 | Tubs, recessed porcelain enamel on cast iron, with trim | | | | | | | | | |
| 0180 | 48" x 42" | Q-1 | 4 | 4 | Ea. | 2,225 | 193 | | 2,418 | 2,725 |
| 0220 | 72" x 36" | | 3 | 5.333 | | 2,275 | 257 | | 2,532 | 2,900 |
| 2000 | Enameled formed steel, 4'-6" long | | 5.80 | 2.759 | | 415 | 133 | | 548 | 655 |
| 4000 | Soaking, acrylic with pop-up drain, 60" x 32" x 21" deep | | 5.50 | 2.909 | | 1,025 | 140 | | 1,165 | 1,325 |
| 4100 | 60" x 48" x 18-1/2" deep | | 5 | 3.200 | | 960 | 154 | | 1,114 | 1,275 |
| 9600 | Rough-in, supply, waste and vent, for all above tubs, add | ↓ | 2.07 | 7.729 | ↓ | 395 | 375 | | 770 | 985 |

# 22 41 Residential Plumbing Fixtures

## 22 41 23 – Residential Showers

### 22 41 23.20 Showers

| 22 41 23.20 Showers | Crew | Daily Output | Labor-Hours | Unit | Material | 2011 Bare Costs Labor | Equipment | Total | Total Incl O&P |
|---|---|---|---|---|---|---|---|---|---|
| 0010 **SHOWERS** | | | | | | | | | |
| 1500 Stall, with drain only. Add for valve and door/curtain | | | | | | | | | |
| 3000 Fiberglass, one piece, with 3 walls, 32" x 32" square | Q-1 | 5.50 | 2.909 | Ea. | 545 | 140 | | 685 | 805 |
| 3100 36" x 36" square | | 5.50 | 2.909 | | 555 | 140 | | 695 | 825 |
| 3250 64" x 65-3/4" x 81-1/2" fold. seat, whlchr. | | 3.80 | 4.211 | | 2,050 | 203 | | 2,253 | 2,575 |
| 4000 Polypropylene, stall only, w/molded-stone floor, 30" x 30" | | 2 | 8 | | 500 | 385 | | 885 | 1,125 |
| 4200 Rough-in, supply, waste and vent for above showers | | 2.05 | 7.805 | | 485 | 375 | | 860 | 1,100 |

### 22 41 23.40 Shower System Components

| 22 41 23.40 Shower System Components | Crew | Daily Output | Labor-Hours | Unit | Material | 2011 Bare Costs Labor | Equipment | Total | Total Incl O&P |
|---|---|---|---|---|---|---|---|---|---|
| 0010 **SHOWER SYSTEM COMPONENTS** | | | | | | | | | |
| 4500 Receptor only | | | | | | | | | |
| 4510 For tile, 36" x 36" | 1 Plum | 4 | 2 | Ea. | 325 | 107 | | 432 | 520 |
| 4520 Fiberglass receptor only, 32" x 32" | | 8 | 1 | | 107 | 53.50 | | 160.50 | 198 |
| 4530 34" x 34" | | 7.80 | 1.026 | | 126 | 55 | | 181 | 221 |
| 4540 36" x 36" | | 7.60 | 1.053 | | 125 | 56.50 | | 181.50 | 222 |
| 4600 Rectangular | | | | | | | | | |
| 4620 32" x 48" | 1 Plum | 7.40 | 1.081 | Ea. | 149 | 58 | | 207 | 250 |
| 4630 32" x 54" | | 7.20 | 1.111 | | 196 | 59.50 | | 255.50 | 305 |
| 4640 32" x 60" | | 7 | 1.143 | | 208 | 61.50 | | 269.50 | 320 |
| 5000 Built-in, head, arm, 2.5 GPM valve | | 4 | 2 | | 80 | 107 | | 187 | 248 |
| 5200 Head, arm, by-pass, integral stops, handles | | 3.60 | 2.222 | | 216 | 119 | | 335 | 415 |

## 22 41 36 – Residential Laundry Trays

### 22 41 36.10 Laundry Sinks

| 22 41 36.10 Laundry Sinks | Crew | Daily Output | Labor-Hours | Unit | Material | 2011 Bare Costs Labor | Equipment | Total | Total Incl O&P |
|---|---|---|---|---|---|---|---|---|---|
| 0010 **LAUNDRY SINKS**, With trim | | | | | | | | | |
| 0020 Porcelain enamel on cast iron, black iron frame | | | | | | | | | |
| 0050 24" x 21", single compartment | Q-1 | 6 | 2.667 | Ea. | 465 | 129 | | 594 | 700 |
| 0100 26" x 21", single compartment | " | 6 | 2.667 | " | 450 | 129 | | 579 | 685 |
| 2000 Molded stone, on wall hanger or legs | | | | | | | | | |
| 2020 22" x 23", single compartment | Q-1 | 6 | 2.667 | Ea. | 138 | 129 | | 267 | 345 |
| 2100 45" x 21", double compartment | " | 5 | 3.200 | " | 259 | 154 | | 413 | 515 |
| 3000 Plastic, on wall hanger or legs | | | | | | | | | |
| 3020 18" x 23", single compartment | Q-1 | 6.50 | 2.462 | Ea. | 145 | 119 | | 264 | 335 |
| 3300 40" x 24", double compartment | | 5.50 | 2.909 | | 257 | 140 | | 397 | 495 |
| 5000 Stainless steel, counter top, 22" x 17" single compartment | | 6 | 2.667 | | 61.50 | 129 | | 190.50 | 260 |
| 5200 33" x 22", double compartment | | 5 | 3.200 | | 73.50 | 154 | | 227.50 | 310 |
| 9600 Rough-in, supply, waste and vent, for all laundry sinks | | 2.14 | 7.477 | | 365 | 360 | | 725 | 940 |

## 22 41 39 – Residential Faucets, Supplies and Trim

### 22 41 39.10 Faucets and Fittings

| 22 41 39.10 Faucets and Fittings | Crew | Daily Output | Labor-Hours | Unit | Material | 2011 Bare Costs Labor | Equipment | Total | Total Incl O&P |
|---|---|---|---|---|---|---|---|---|---|
| 0010 **FAUCETS AND FITTINGS** | | | | | | | | | |
| 0150 Bath, faucets, diverter spout combination, sweat | 1 Plum | 8 | 1 | Ea. | 86.50 | 53.50 | | 140 | 175 |
| 0200 For integral stops, IPS unions, add | | | | | 109 | | | 109 | 120 |
| 0420 Bath, press-bal mix valve w/diverter, spout, shower hd, arm/flange | 1 Plum | 8 | 1 | | 155 | 53.50 | | 208.50 | 251 |
| 0810 Bidet | | | | | | | | | |
| 0812 Fitting, over the rim, swivel spray/pop-up drain | 1 Plum | 8 | 1 | Ea. | 172 | 53.50 | | 225.50 | 269 |
| 1000 Kitchen sink faucets, top mount, cast spout | | 10 | .800 | | 61.50 | 43 | | 104.50 | 132 |
| 1100 For spray, add | | 24 | .333 | | 16.15 | 17.85 | | 34 | 44.50 |
| 1300 Single control lever handle | | | | | | | | | |
| 1310 With pull out spray | | | | | | | | | |
| 1320 Polished chrome | 1 Plum | 10 | .800 | Ea. | 171 | 43 | | 214 | 253 |
| 2000 Laundry faucets, shelf type, IPS or copper unions | | 12 | .667 | | 49.50 | 35.50 | | 85 | 108 |
| 2100 Lavatory faucet, centerset, without drain | | 10 | .800 | | 44.50 | 43 | | 87.50 | 113 |
| 2210 Porcelain cross handles and pop-up drain | | | | | | | | | |

# 22 41 Residential Plumbing Fixtures

## 22 41 39 – Residential Faucets, Supplies and Trim

| 22 41 39.10 Faucets and Fittings | | Crew | Daily Output | Labor-Hours | Unit | Material | 2011 Bare Costs Labor | Equipment | Total | Total Incl O&P |
|---|---|---|---|---|---|---|---|---|---|---|
| 2220 | Polished chrome | 1 Plum | 6.66 | 1.201 | Ea. | 128 | 64.50 | | 192.50 | 237 |
| 2230 | Polished brass | " | 6.66 | 1.201 | " | 193 | 64.50 | | 257.50 | 310 |
| 2260 | Single lever handle and pop-up drain | | | | | | | | | |
| 2270 | Black nickel | 1 Plum | 6.66 | 1.201 | Ea. | 220 | 64.50 | | 284.50 | 340 |
| 2280 | Polished brass | | 6.66 | 1.201 | | 220 | 64.50 | | 284.50 | 340 |
| 2290 | Polished chrome | | 6.66 | 1.201 | | 172 | 64.50 | | 236.50 | 285 |
| 2810 | Automatic sensor and operator, with faucet head [G] | | 6.15 | 1.301 | | 370 | 69.50 | | 439.50 | 515 |
| 4000 | Shower by-pass valve with union | | 18 | .444 | | 68.50 | 24 | | 92.50 | 111 |
| 4200 | Shower thermostatic mixing valve, concealed, with shower head trim kit | | 8 | 1 | | 305 | 53.50 | | 358.50 | 415 |
| 4220 | Shower pressure balancing mixing valve, | | | | | | | | | |
| 4230 | With shower head, arm, flange and diverter tub spout | | | | | | | | | |
| 4240 | Chrome | 1 Plum | 6.14 | 1.303 | Ea. | 222 | 70 | | 292 | 350 |
| 4250 | Polished brass | | 6.14 | 1.303 | | 279 | 70 | | 349 | 410 |
| 4260 | Satin | | 6.14 | 1.303 | | 279 | 70 | | 349 | 410 |
| 4270 | Polished chrome/brass | | 6.14 | 1.303 | | 229 | 70 | | 299 | 355 |
| 5000 | Sillcock, compact, brass, IPS or copper to hose | | 24 | .333 | | 10.60 | 17.85 | | 28.45 | 38 |

# 22 42 Commercial Plumbing Fixtures

## 22 42 13 – Commercial Water Closets, Urinals, and Bidets

### 22 42 13.30 Urinals

| | | Crew | Daily Output | Labor-Hours | Unit | Material | 2011 Bare Costs Labor | Equipment | Total | Total Incl O&P |
|---|---|---|---|---|---|---|---|---|---|---|
| 0010 | **URINALS** | | | | | | | | | |
| 3000 | Wall hung, vitreous china, with hanger & self-closing valve | | | | | | | | | |
| 3100 | Siphon jet type | Q-1 | 3 | 5.333 | Ea. | 289 | 257 | | 546 | 700 |
| 3120 | Blowout type | | 3 | 5.333 | | 405 | 257 | | 662 | 830 |
| 3140 | Water saving .5 gpf [G] | | 3 | 5.333 | | 645 | 257 | | 902 | 1,100 |
| 3300 | Rough-in, supply, waste & vent | | 2.83 | 5.654 | | 525 | 273 | | 798 | 985 |
| 5000 | Stall type, vitreous china, includes valve | | 2.50 | 6.400 | | 640 | 310 | | 950 | 1,175 |
| 6980 | Rough-in, supply, waste and vent | | 1.99 | 8.040 | | 465 | 390 | | 855 | 1,100 |
| 8000 | Waterless (no flush) urinal | | | | | | | | | |
| 8010 | Wall hung | | | | | | | | | |
| 8014 | Fiberglass reinforced polyester | | | | | | | | | |
| 8020 | Standard unit [G] | Q-1 | 21.30 | .751 | Ea. | 385 | 36 | | 421 | 475 |
| 8030 | ADA compliant unit [G] | " | 21.30 | .751 | | 400 | 36 | | 436 | 495 |
| 8070 | For solid color, add [G] | | | | | 60 | | | 60 | 66 |
| 8080 | For 2" brass flange, (new const.), add [G] | Q-1 | 96 | .167 | | 19.20 | 8.05 | | 27.25 | 33 |
| 8200 | Vitreous china | | | | | | | | | |
| 8220 | Standard unit [G] | Q-1 | 21.30 | .751 | Ea. | 280 | 36 | | 316 | 365 |
| 8240 | ADA compliant unit [G] | | 21.30 | .751 | | 320 | 36 | | 356 | 405 |
| 8250 | ADA compliant unit, compact | | 21.30 | .751 | | 272 | 36 | | 308 | 355 |
| 8270 | For solid color, add [G] | | | | | 60 | | | 60 | 66 |
| 8290 | Rough-in, supply, waste & vent [G] | Q-1 | 2.92 | 5.479 | | 480 | 264 | | 744 | 925 |
| 8400 | Trap liquid | | | | | | | | | |
| 8410 | 1 quart [G] | | | | Ea. | 14.35 | | | 14.35 | 15.80 |
| 8420 | 1 gallon [G] | | | | " | 56 | | | 56 | 61.50 |

### 22 42 13.40 Water Closets

| | | Crew | Daily Output | Labor-Hours | Unit | Material | 2011 Bare Costs Labor | Equipment | Total | Total Incl O&P |
|---|---|---|---|---|---|---|---|---|---|---|
| 0010 | **WATER CLOSETS** | | | | | | | | | |
| 3000 | Bowl only, with flush valve, seat, 1.6 gpf unless noted | | | | | | | | | |
| 3100 | Wall hung | Q-1 | 5.80 | 2.759 | Ea. | 445 | 133 | | 578 | 690 |
| 3200 | For rough-in, supply, waste and vent, single WC | | 2.56 | 6.250 | | 975 | 300 | | 1,275 | 1,525 |
| 3300 | Floor mounted | | 5.80 | 2.759 | | 315 | 133 | | 448 | 545 |
| 3350 | With wall outlet | | 5.80 | 2.759 | | 445 | 133 | | 578 | 690 |

# 22 42 Commercial Plumbing Fixtures

## 22 42 13 – Commercial Water Closets, Urinals, and Bidets

### 22 42 13.40 Water Closets

| | | Crew | Daily Output | Labor-Hours | Unit | Material | 2011 Bare Costs Labor | 2011 Bare Costs Equipment | Total | Total Incl O&P |
|---|---|---|---|---|---|---|---|---|---|---|
| 3360 | With floor outlet, 1.28 gpf  [G] | Q-1 | 5.80 | 2.759 | Ea. | 555 | 133 | | 688 | 810 |
| 3362 | With floor outlet, 1.28 gpf, ADA  [G] | | 5.80 | 2.759 | | 595 | 133 | | 728 | 855 |
| 3400 | For rough-in, supply, waste and vent, single WC | | 2.84 | 5.634 | | 365 | 272 | | 637 | 810 |

## 22 42 16 – Commercial Lavatories and Sinks

### 22 42 16.10 Handwasher-Dryer Module

| | | Crew | Daily Output | Labor-Hours | Unit | Material | Labor | Equipment | Total | Total Incl O&P |
|---|---|---|---|---|---|---|---|---|---|---|
| 0010 | **HANDWASHER-DRYER MODULE** | | | | | | | | | |
| 0030 | Wall mounted | | | | | | | | | |
| 0040 | With electric dryer | | | | | | | | | |
| 0050 | Sensor operated | Q-1 | 8 | 2 | Ea. | 3,650 | 96.50 | | 3,746.50 | 4,150 |
| 0110 | Sensor operated (ADA) | | 8 | 2 | | 3,650 | 96.50 | | 3,746.50 | 4,175 |
| 0140 | Sensor operated (ADA), surface mounted | | 8 | 2 | | 4,400 | 96.50 | | 4,496.50 | 5,000 |
| 0150 | With paper towels | | | | | | | | | |
| 0180 | Sensor operated | Q-1 | 8 | 2 | Ea. | 3,550 | 96.50 | | 3,646.50 | 4,050 |

### 22 42 16.14 Lavatories

| | | | | | | | | | | |
|---|---|---|---|---|---|---|---|---|---|---|
| 0010 | **LAVATORIES**, With trim, white unless noted otherwise | | | | | | | | | |
| 0020 | Commercial lavatories same as residential. See Section 22 41 16.10 | | | | | | | | | |

### 22 42 16.40 Service Sinks

| | | Crew | Daily Output | Labor-Hours | Unit | Material | Labor | Equipment | Total | Total Incl O&P |
|---|---|---|---|---|---|---|---|---|---|---|
| 0010 | **SERVICE SINKS** | | | | | | | | | |
| 6650 | Service, floor, corner, P.E. on C.I., 28" x 28" | Q-1 | 4.40 | 3.636 | Ea. | 845 | 175 | | 1,020 | 1,200 |
| 6790 | For rough-in, supply, waste & vent, floor service sinks | | 1.64 | 9.756 | | 1,025 | 470 | | 1,495 | 1,850 |
| 7000 | Service, wall, P.E. on C.I., roll rim, 22" x 18" | | 4 | 4 | | 645 | 193 | | 838 | 1,000 |
| 7100 | 24" x 20" | | 4 | 4 | | 715 | 193 | | 908 | 1,075 |
| 8600 | Vitreous china, 22" x 20" | | 4 | 4 | | 530 | 193 | | 723 | 875 |
| 8960 | For stainless steel rim guard, front or side, add | | | | | 38.50 | | | 38.50 | 42 |
| 8980 | For rough-in, supply, waste & vent, wall service sinks | Q-1 | 1.30 | 12.308 | | 1,300 | 595 | | 1,895 | 2,300 |

## 22 42 23 – Commercial Showers

### 22 42 23.30 Group Showers

| | | Crew | Daily Output | Labor-Hours | Unit | Material | Labor | Equipment | Total | Total Incl O&P |
|---|---|---|---|---|---|---|---|---|---|---|
| 0010 | **GROUP SHOWERS** | | | | | | | | | |
| 6000 | Group, w/pressure balancing valve, rough-in and rigging not included | | | | | | | | | |
| 6800 | Column, 6 heads, no receptors, less partitions | Q-1 | 3 | 5.333 | Ea. | 3,000 | 257 | | 3,257 | 3,675 |
| 6900 | With stainless steel partitions | | 1 | 16 | | 9,250 | 770 | | 10,020 | 11,400 |
| 7600 | 5 heads, no receptors, less partitions | | 3 | 5.333 | | 2,475 | 257 | | 2,732 | 3,100 |
| 7620 | 4 heads (1 handicap) no receptors, less partitions | | 3 | 5.333 | | 3,975 | 257 | | 4,232 | 4,750 |
| 7700 | With stainless steel partitions | | 1 | 16 | | 7,600 | 770 | | 8,370 | 9,500 |
| 8000 | Wall, 2 heads, no receptors, less partitions | | 4 | 4 | | 1,375 | 193 | | 1,568 | 1,825 |
| 8100 | With stainless steel partitions | | 2 | 8 | | 3,225 | 385 | | 3,610 | 4,125 |

## 22 42 33 – Wash Fountains

### 22 42 33.20 Commercial Wash Fountains

| | | Crew | Daily Output | Labor-Hours | Unit | Material | Labor | Equipment | Total | Total Incl O&P |
|---|---|---|---|---|---|---|---|---|---|---|
| 0010 | **COMMERCIAL WASH FOUNTAINS** | | | | | | | | | |
| 1900 | Group, foot control | | | | | | | | | |
| 2000 | Precast terrazzo, circular, 36" diam., 5 or 6 persons | Q-2 | 3 | 8 | Ea. | 4,200 | 400 | | 4,600 | 5,200 |
| 2100 | 54" diameter for 8 or 10 persons | | 2.50 | 9.600 | | 5,225 | 480 | | 5,705 | 6,450 |
| 2400 | Semi-circular, 36" diam. for 3 persons | | 3 | 8 | | 3,850 | 400 | | 4,250 | 4,850 |
| 2500 | 54" diam. for 4 or 5 persons | | 2.50 | 9.600 | | 4,650 | 480 | | 5,130 | 5,850 |
| 2700 | Quarter circle (corner), 54" for 3 persons | | 3.50 | 6.857 | | 4,525 | 345 | | 4,870 | 5,475 |
| 3000 | Stainless steel, circular, 36" diameter | | 3.50 | 6.857 | | 4,100 | 345 | | 4,445 | 5,025 |
| 3100 | 54" diameter | | 2.80 | 8.571 | | 5,300 | 430 | | 5,730 | 6,475 |
| 3400 | Semi-circular, 36" diameter | | 3.50 | 6.857 | | 3,575 | 345 | | 3,920 | 4,425 |
| 3500 | 54" diameter | | 2.80 | 8.571 | | 4,575 | 430 | | 5,005 | 5,675 |
| 5610 | Group, infrared control, barrier free | | | | | | | | | |

# 22 42 Commercial Plumbing Fixtures

## 22 42 33 – Wash Fountains

| 22 42 33.20 Commercial Wash Fountains | Crew | Daily Output | Labor-Hours | Unit | Material | 2011 Bare Costs Labor | Equipment | Total | Total Incl O&P |
|---|---|---|---|---|---|---|---|---|---|
| 5614 | Precast terrazzo | | | | | | | | |
| 5620 | Semi-circular 36" diam. for 3 persons | Q-2 | 3 | 8 | Ea. | 5,875 | 400 | | 6,275 | 7,075 |
| 5630 | 46" diam. for 4 persons | | 2.80 | 8.571 | | 6,425 | 430 | | 6,855 | 7,700 |
| 5640 | Circular, 54" diam. for 8 persons, button control | | 2.50 | 9.600 | | 7,700 | 480 | | 8,180 | 9,200 |
| 5700 | Rough-in, supply, waste and vent for above wash fountains | Q-1 | 1.82 | 8.791 | | 525 | 425 | | 950 | 1,225 |
| 6200 | Duo for small washrooms, stainless steel | | 2 | 8 | | 2,725 | 385 | | 3,110 | 3,550 |
| 6500 | Rough-in, supply, waste & vent for duo fountains | | 2.02 | 7.921 | | 310 | 380 | | 690 | 910 |

## 22 42 39 – Commercial Faucets, Supplies, and Trim

### 22 42 39.10 Faucets and Fittings

| | | Crew | Daily Output | Labor-Hours | Unit | Material | Labor | Equipment | Total | Total Incl O&P |
|---|---|---|---|---|---|---|---|---|---|---|
| 0010 | **FAUCETS AND FITTINGS** | | | | | | | | | |
| 0840 | Flush valves, with vacuum breaker | | | | | | | | | |
| 0850 | Water closet | | | | | | | | | |
| 0860 | Exposed, rear spud | 1 Plum | 8 | 1 | Ea. | 137 | 53.50 | | 190.50 | 231 |
| 0870 | Top spud | | 8 | 1 | | 126 | 53.50 | | 179.50 | 219 |
| 0880 | Concealed, rear spud | | 8 | 1 | | 180 | 53.50 | | 233.50 | 278 |
| 0890 | Top spud | | 8 | 1 | | 149 | 53.50 | | 202.50 | 244 |
| 0900 | Wall hung | | 8 | 1 | | 159 | 53.50 | | 212.50 | 255 |
| 0920 | Urinal | | | | | | | | | |
| 0930 | Exposed, stall | 1 Plum | 8 | 1 | Ea. | 126 | 53.50 | | 179.50 | 219 |
| 0940 | Wall, (washout) | | 8 | 1 | | 126 | 53.50 | | 179.50 | 219 |
| 0950 | Pedestal, top spud | | 8 | 1 | | 129 | 53.50 | | 182.50 | 222 |
| 0960 | Concealed, stall | | 8 | 1 | | 145 | 53.50 | | 198.50 | 239 |
| 0970 | Wall (washout) | | 8 | 1 | | 156 | 53.50 | | 209.50 | 252 |
| 0971 | Automatic flush sensor and operator for &#9855; G | | | | | | | | | |
| 0972 | urinals or water closets, standard G | 1 Plum | 8 | 1 | Ea. | 415 | 53.50 | | 468.50 | 540 |
| 0980 | High efficiency water saving | | | | | | | | | |
| 0984 | Water closets, 1.28 gpf G | 1 Plum | 8 | 1 | Ea. | 395 | 53.50 | | 448.50 | 515 |
| 0988 | Urinals, .5 gpf G | " | 8 | 1 | " | 395 | 53.50 | | 448.50 | 515 |
| 2790 | Faucets for lavatories | | | | | | | | | |
| 2800 | Self-closing, center set | 1 Plum | 10 | .800 | Ea. | 131 | 43 | | 174 | 208 |
| 2810 | Automatic sensor and operator, with faucet head | | 6.15 | 1.301 | | 370 | 69.50 | | 439.50 | 515 |
| 3000 | Service sink faucet, cast spout, pail hook, hose end | | 14 | .571 | | 80 | 30.50 | | 110.50 | 134 |

### 22 42 39.30 Carriers and Supports

| | | Crew | Daily Output | Labor-Hours | Unit | Material | Labor | Equipment | Total | Total Incl O&P |
|---|---|---|---|---|---|---|---|---|---|---|
| 0010 | **CARRIERS AND SUPPORTS**, For plumbing fixtures | | | | | | | | | |
| 0500 | Drinking fountain, wall mounted | | | | | | | | | |
| 0600 | Plate type with studs, top back plate | 1 Plum | 7 | 1.143 | Ea. | 70.50 | 61.50 | | 132 | 170 |
| 0700 | Top front and back plate | | 7 | 1.143 | | 87 | 61.50 | | 148.50 | 187 |
| 0800 | Top & bottom, front & back plates, w/bearing jacks | | 7 | 1.143 | | 126 | 61.50 | | 187.50 | 231 |
| 3000 | Lavatory, concealed arm | | | | | | | | | |
| 3050 | Floor mounted, single | | | | | | | | | |
| 3100 | High back fixture | 1 Plum | 6 | 1.333 | Ea. | 395 | 71.50 | | 466.50 | 535 |
| 3200 | Flat slab fixture | | 6 | 1.333 | | 340 | 71.50 | | 411.50 | 480 |
| 3220 | Paraplegic &#9855; | | 6 | 1.333 | | 440 | 71.50 | | 511.50 | 590 |
| 3250 | Floor mounted, back to back | | | | | | | | | |
| 3300 | High back fixtures | 1 Plum | 5 | 1.600 | Ea. | 560 | 86 | | 646 | 745 |
| 3400 | Flat slab fixtures | | 5 | 1.600 | | 690 | 86 | | 776 | 890 |
| 3430 | Paraplegic &#9855; | | 5 | 1.600 | | 640 | 86 | | 726 | 835 |
| 3500 | Wall mounted, in stud or masonry | | | | | | | | | |
| 3600 | High back fixture | 1 Plum | 6 | 1.333 | Ea. | 232 | 71.50 | | 303.50 | 365 |
| 3700 | Flat slab fixture | " | 6 | 1.333 | " | 201 | 71.50 | | 272.50 | 330 |
| 4600 | Sink, floor mounted | | | | | | | | | |
| 4650 | Exposed arm system | | | | | | | | | |

## 22 42 39 – Commercial Faucets, Supplies, and Trim

| 22 42 39.30 Carriers and Supports | | Crew | Daily Output | Labor-Hours | Unit | Material | 2011 Bare Costs Labor | Equipment | Total | Total Incl O&P |
|---|---|---|---|---|---|---|---|---|---|---|
| 4700 | Single heavy fixture | 1 Plum | 5 | 1.600 | Ea. | 655 | 86 | | 741 | 855 |
| 4750 | Single heavy sink with slab | | 5 | 1.600 | | 810 | 86 | | 896 | 1,025 |
| 4800 | Back to back, standard fixtures | | 5 | 1.600 | | 480 | 86 | | 566 | 660 |
| 4850 | Back to back, heavy fixtures | | 5 | 1.600 | | 690 | 86 | | 776 | 890 |
| 4900 | Back to back, heavy sink with slab | | 5 | 1.600 | | 690 | 86 | | 776 | 890 |
| 4950 | Exposed offset arm system | | | | | | | | | |
| 5000 | Single heavy deep fixture | 1 Plum | 5 | 1.600 | Ea. | 625 | 86 | | 711 | 820 |
| 5100 | Plate type system | | | | | | | | | |
| 5200 | With bearing jacks, single fixture | 1 Plum | 5 | 1.600 | Ea. | 720 | 86 | | 806 | 920 |
| 5300 | With exposed arms, single heavy fixture | | 5 | 1.600 | | 920 | 86 | | 1,006 | 1,150 |
| 5400 | Wall mounted, exposed arms, single heavy fixture | | 5 | 1.600 | | 335 | 86 | | 421 | 500 |
| 6000 | Urinal, floor mounted, 2" or 3" coupling, blowout type | | 6 | 1.333 | | 415 | 71.50 | | 486.50 | 565 |
| 6100 | With fixture or hanger bolts, blowout or washout | | 6 | 1.333 | | 298 | 71.50 | | 369.50 | 435 |
| 6200 | With bearing plate | | 6 | 1.333 | | 330 | 71.50 | | 401.50 | 470 |
| 6300 | Wall mounted, plate type system | | 6 | 1.333 | | 257 | 71.50 | | 328.50 | 390 |
| 6980 | Water closet, siphon jet | | | | | | | | | |
| 7000 | Horizontal, adjustable, caulk | | | | | | | | | |
| 7040 | Single, 4" pipe size | 1 Plum | 5.33 | 1.501 | Ea. | 755 | 80.50 | | 835.50 | 950 |
| 7050 | 4" pipe size, paraplegic | | 5.33 | 1.501 | | 755 | 80.50 | | 835.50 | 950 |
| 7060 | 5" pipe size | | 5.33 | 1.501 | | 955 | 80.50 | | 1,035.50 | 1,175 |
| 7100 | Double, 4" pipe size | | 5 | 1.600 | | 1,100 | 86 | | 1,186 | 1,350 |
| 7110 | 4" pipe size, paraplegic | | 5 | 1.600 | | 1,100 | 86 | | 1,186 | 1,350 |
| 7120 | 5" pipe size | | 5 | 1.600 | | 1,600 | 86 | | 1,686 | 1,875 |
| 7160 | Horizontal, adjustable, extended, caulk | | | | | | | | | |
| 7180 | Single, 4" pipe size | 1 Plum | 5.33 | 1.501 | Ea. | 755 | 80.50 | | 835.50 | 950 |
| 7200 | 5" pipe size | | 5.33 | 1.501 | | 955 | 80.50 | | 1,035.50 | 1,175 |
| 7240 | Double, 4" pipe size | | 5 | 1.600 | | 1,375 | 86 | | 1,461 | 1,625 |
| 7260 | 5" pipe size | | 5 | 1.600 | | 1,600 | 86 | | 1,686 | 1,875 |
| 7400 | Vertical, adjustable, caulk or thread | | | | | | | | | |
| 7440 | Single, 4" pipe size | 1 Plum | 5.33 | 1.501 | Ea. | 765 | 80.50 | | 845.50 | 965 |
| 7460 | 5" pipe size | | 5.33 | 1.501 | | 945 | 80.50 | | 1,025.50 | 1,175 |
| 7480 | 6" pipe size | | 5 | 1.600 | | 1,075 | 86 | | 1,161 | 1,325 |
| 7520 | Double, 4" pipe size | | 5 | 1.600 | | 1,150 | 86 | | 1,236 | 1,375 |
| 7540 | 5" pipe size | | 5 | 1.600 | | 1,325 | 86 | | 1,411 | 1,575 |
| 7560 | 6" pipe size | | 4 | 2 | | 1,450 | 107 | | 1,557 | 1,750 |
| 7600 | Vertical, adjustable, extended, caulk | | | | | | | | | |
| 7620 | Single, 4" pipe size | 1 Plum | 5.33 | 1.501 | Ea. | 765 | 80.50 | | 845.50 | 965 |
| 7640 | 5" pipe size | | 5.33 | 1.501 | | 945 | 80.50 | | 1,025.50 | 1,175 |
| 7680 | 6" pipe size | | 5 | 1.600 | | 1,075 | 86 | | 1,161 | 1,325 |
| 7720 | Double, 4" pipe size | | 5 | 1.600 | | 1,150 | 86 | | 1,236 | 1,375 |
| 7740 | 5" pipe size | | 5 | 1.600 | | 1,325 | 86 | | 1,411 | 1,575 |
| 7760 | 6" pipe size | | 4 | 2 | | 1,450 | 107 | | 1,557 | 1,750 |
| 7780 | Water closet, blow out | | | | | | | | | |
| 7800 | Vertical offset, caulk or thread | | | | | | | | | |
| 7820 | Single, 4" pipe size | 1 Plum | 5.33 | 1.501 | Ea. | 565 | 80.50 | | 645.50 | 745 |
| 7840 | Double, 4" pipe size | " | 5 | 1.600 | " | 970 | 86 | | 1,056 | 1,200 |
| 7880 | Vertical offset, extended, caulk | | | | | | | | | |
| 7900 | Single, 4" pipe size | 1 Plum | 5.33 | 1.501 | Ea. | 710 | 80.50 | | 790.50 | 900 |
| 7920 | Double, 4" pipe size | " | 5 | 1.600 | " | 1,125 | 86 | | 1,211 | 1,350 |
| 7960 | Vertical, for floor mounted back-outlet | | | | | | | | | |
| 7980 | Single, 4" thread, 2" vent | 1 Plum | 5.33 | 1.501 | Ea. | 500 | 80.50 | | 580.50 | 670 |
| 8000 | Double, 4" thread, 2" vent | " | 6 | 1.333 | " | 1,475 | 71.50 | | 1,546.50 | 1,700 |
| 8040 | Vertical, for floor mounted back-outlet, extended | | | | | | | | | |

# 22 42 Commercial Plumbing Fixtures

## 22 42 39 – Commercial Faucets, Supplies, and Trim

| 22 42 39.30 Carriers and Supports | | Crew | Daily Output | Labor-Hours | Unit | Material | 2011 Bare Costs Labor | 2011 Bare Costs Equipment | Total | Total Incl O&P |
|---|---|---|---|---|---|---|---|---|---|---|
| 8060 | Single, 4" caulk, 2" vent | 1 Plum | 6 | 1.333 | Ea. | 500 | 71.50 | | 571.50 | 655 |
| 8080 | Double, 4" caulk, 2" vent | " | 6 | 1.333 | " | 1,475 | 71.50 | | 1,546.50 | 1,700 |
| 8200 | Water closet, residential | | | | | | | | | |
| 8220 | Vertical centerline, floor mount | | | | | | | | | |
| 8240 | Single, 3" caulk, 2" or 3" vent | 1 Plum | 6 | 1.333 | Ea. | 455 | 71.50 | | 526.50 | 605 |
| 8260 | 4" caulk, 2" or 4" vent | | 6 | 1.333 | | 575 | 71.50 | | 646.50 | 740 |
| 8280 | 3" copper sweat, 3" vent | | 6 | 1.333 | | 405 | 71.50 | | 476.50 | 550 |
| 8300 | 4" copper sweat, 4" vent | | 6 | 1.333 | | 490 | 71.50 | | 561.50 | 645 |
| 8400 | Vertical offset, floor mount | | | | | | | | | |
| 8420 | Single, 3" or 4" caulk, vent | 1 Plum | 4 | 2 | Ea. | 565 | 107 | | 672 | 785 |
| 8440 | 3" or 4" copper sweat, vent | | 5 | 1.600 | | 565 | 86 | | 651 | 755 |
| 8460 | Double, 3" or 4" caulk, vent | | 4 | 2 | | 970 | 107 | | 1,077 | 1,225 |
| 8480 | 3" or 4" copper sweat, vent | | 5 | 1.600 | | 970 | 86 | | 1,056 | 1,200 |
| 9000 | Water cooler (electric), floor mounted | | | | | | | | | |
| 9100 | Plate type with bearing plate, single | 1 Plum | 6 | 1.333 | Ea. | 286 | 71.50 | | 357.50 | 420 |

# 22 45 Emergency Plumbing Fixtures

## 22 45 13 – Emergency Showers

### 22 45 13.10 Emergency Showers

| | | Crew | Daily Output | Labor-Hours | Unit | Material | 2011 Bare Costs Labor | 2011 Bare Costs Equipment | Total | Total Incl O&P |
|---|---|---|---|---|---|---|---|---|---|---|
| 0010 | **EMERGENCY SHOWERS**, Rough-in not included | | | | | | | | | |
| 5000 | Shower, single head, drench, ball valve, pull, freestanding | Q-1 | 4 | 4 | Ea. | 340 | 193 | | 533 | 665 |
| 5200 | Horizontal or vertical supply | | 4 | 4 | | 490 | 193 | | 683 | 830 |
| 6000 | Multi-nozzle, eye/face wash combination | | 4 | 4 | | 595 | 193 | | 788 | 945 |
| 6400 | Multi-nozzle, 12 spray, shower only | | 4 | 4 | | 1,600 | 193 | | 1,793 | 2,050 |
| 6600 | For freeze-proof, add | | 6 | 2.667 | | 365 | 129 | | 494 | 595 |

## 22 45 16 – Eyewash Equipment

### 22 45 16.10 Eyewash Safety Equipment

| | | Crew | Daily Output | Labor-Hours | Unit | Material | 2011 Bare Costs Labor | 2011 Bare Costs Equipment | Total | Total Incl O&P |
|---|---|---|---|---|---|---|---|---|---|---|
| 0010 | **EYEWASH SAFETY EQUIPMENT**, Rough-in not included | | | | | | | | | |
| 1000 | Eye wash fountain | | | | | | | | | |
| 1400 | Plastic bowl, pedestal mounted | Q-1 | 4 | 4 | Ea. | 250 | 193 | | 443 | 565 |
| 1600 | Unmounted | | 4 | 4 | | 186 | 193 | | 379 | 490 |
| 1800 | Wall mounted | | 4 | 4 | | 380 | 193 | | 573 | 705 |
| 2000 | Stainless steel, pedestal mounted | | 4 | 4 | | 330 | 193 | | 523 | 655 |
| 2200 | Unmounted | | 4 | 4 | | 248 | 193 | | 441 | 560 |
| 2400 | Wall mounted | | 4 | 4 | | 244 | 193 | | 437 | 555 |

## 22 45 19 – Self-Contained Eyewash Equipment

### 22 45 19.10 Self-Contained Eyewash Safety Equipment

| | | Crew | Daily Output | Labor-Hours | Unit | Material | 2011 Bare Costs Labor | 2011 Bare Costs Equipment | Total | Total Incl O&P |
|---|---|---|---|---|---|---|---|---|---|---|
| 0010 | **SELF-CONTAINED EYEWASH SAFETY EQUIPMENT** | | | | | | | | | |
| 3000 | Eye wash, portable, self-contained | | | | Ea. | 865 | | | 865 | 950 |

## 22 45 26 – Eye/Face Wash Equipment

### 22 45 26.10 Eye/Face Wash Safety Equipment

| | | Crew | Daily Output | Labor-Hours | Unit | Material | 2011 Bare Costs Labor | 2011 Bare Costs Equipment | Total | Total Incl O&P |
|---|---|---|---|---|---|---|---|---|---|---|
| 0010 | **EYE/FACE WASH SAFETY EQUIPMENT**, Rough-in not included | | | | | | | | | |
| 4000 | Eye and face wash, combination fountain | | | | | | | | | |
| 4200 | Stainless steel, pedestal mounted | Q-1 | 4 | 4 | Ea. | 490 | 193 | | 683 | 830 |
| 4400 | Unmounted | | 4 | 4 | | 345 | 193 | | 538 | 670 |
| 4600 | Wall mounted | | 4 | 4 | | 224 | 193 | | 417 | 535 |

# 22 47 Drinking Fountains and Water Coolers

## 22 47 13 – Drinking Fountains

| 22 47 13.10 Drinking Water Fountains | Crew | Daily Output | Labor-Hours | Unit | Material | 2011 Bare Costs Labor | 2011 Bare Costs Equipment | Total | Total Incl O&P |
|---|---|---|---|---|---|---|---|---|---|
| 0010 **DRINKING WATER FOUNTAINS,** For connection to cold water supply  R224000-40 | | | | | | | | | |
| 1000    Wall mounted, non-recessed | | | | | | | | | |
| 1400       Bronze, with no back | 1 Plum | 4 | 2 | Ea. | 930 | 107 | | 1,037 | 1,175 |
| 1800       Cast aluminum, enameled, for correctional institutions | | 4 | 2 | | 1,375 | 107 | | 1,482 | 1,650 |
| 2000       Fiberglass, 12" back, single bubbler unit | | 4 | 2 | | 1,750 | 107 | | 1,857 | 2,075 |
| 2040          Dual bubbler | | 3.20 | 2.500 | | 1,950 | 134 | | 2,084 | 2,350 |
| 2400       Precast stone, no back | | 4 | 2 | | 760 | 107 | | 867 | 995 |
| 2700       Stainless steel, single bubbler, no back | | 4 | 2 | | 920 | 107 | | 1,027 | 1,150 |
| 2740          With back | | 4 | 2 | | 635 | 107 | | 742 | 860 |
| 2780          Dual handle & wheelchair projection type | | 4 | 2 | | 655 | 107 | | 762 | 880 |
| 2820          Dual level for handicapped type | | 3.20 | 2.500 | | 1,325 | 134 | | 1,459 | 1,675 |
| 3300       Vitreous china | | | | | | | | | |
| 3340          7" back | 1 Plum | 4 | 2 | Ea. | 575 | 107 | | 682 | 795 |
| 3940       For vandal-resistant bottom plate, add | | | | | 59.50 | | | 59.50 | 65.50 |
| 3960       For freeze-proof valve system, add | 1 Plum | 2 | 4 | | 550 | 214 | | 764 | 925 |
| 3980       For rough-in, supply and waste, add | " | 2.21 | 3.620 | | 254 | 194 | | 448 | 570 |
| 4000    Wall mounted, semi-recessed | | | | | | | | | |
| 4200       Poly-marble, single bubbler | 1 Plum | 4 | 2 | Ea. | 925 | 107 | | 1,032 | 1,175 |
| 4600       Stainless steel, satin finish, single bubbler | | 4 | 2 | | 995 | 107 | | 1,102 | 1,250 |
| 4900       Vitreous china, single bubbler | | 4 | 2 | | 830 | 107 | | 937 | 1,075 |
| 5980       For rough-in, supply and waste, add | | 1.83 | 4.372 | | 254 | 234 | | 488 | 630 |
| 6000    Wall mounted, fully recessed | | | | | | | | | |
| 6400       Poly-marble, single bubbler | 1 Plum | 4 | 2 | Ea. | 1,525 | 107 | | 1,632 | 1,825 |
| 6800       Stainless steel, single bubbler | | 4 | 2 | | 1,350 | 107 | | 1,457 | 1,625 |
| 7560       For freeze-proof valve system, add | | 2 | 4 | | 700 | 214 | | 914 | 1,100 |
| 7580       For rough-in, supply and waste, add | | 1.83 | 4.372 | | 254 | 234 | | 488 | 630 |
| 7600    Floor mounted, pedestal type | | | | | | | | | |
| 7700       Aluminum, architectural style, C.I. base | 1 Plum | 2 | 4 | Ea. | 2,200 | 214 | | 2,414 | 2,725 |
| 7780          Wheelchair handicap unit | | 2 | 4 | | 1,500 | 214 | | 1,714 | 1,975 |
| 8400       Stainless steel, architectural style | | 2 | 4 | | 2,650 | 214 | | 2,864 | 3,250 |
| 8600       Enameled iron, heavy duty service, 2 bubblers | | 2 | 4 | | 2,500 | 214 | | 2,714 | 3,100 |
| 8660          4 bubblers | | 2 | 4 | | 3,825 | 214 | | 4,039 | 4,525 |
| 8880       For freeze-proof valve system, add | | 2 | 4 | | 415 | 214 | | 629 | 775 |
| 8900       For rough-in, supply and waste, add | | 1.83 | 4.372 | | 254 | 234 | | 488 | 630 |
| 9100    Deck mounted | | | | | | | | | |
| 9500       Stainless steel, circular receptor | 1 Plum | 4 | 2 | Ea. | 370 | 107 | | 477 | 570 |
| 9760       White enameled steel, 14" x 9" receptor | | 4 | 2 | | 320 | 107 | | 427 | 515 |
| 9860       White enameled cast iron, 24" x 16" receptor | | 3 | 2.667 | | 315 | 143 | | 458 | 565 |
| 9980       For rough-in, supply and waste, add | | 1.83 | 4.372 | | 254 | 234 | | 488 | 630 |

## 22 47 16 – Pressure Water Coolers

### 22 47 16.10 Electric Water Coolers

| | Crew | Daily Output | Labor-Hours | Unit | Material | 2011 Bare Costs Labor | 2011 Bare Costs Equipment | Total | Total Incl O&P |
|---|---|---|---|---|---|---|---|---|---|
| 0010 **ELECTRIC WATER COOLERS**  R224000-50 | | | | | | | | | |
| 0100    Wall mounted, non-recessed | | | | | | | | | |
| 0140       4 GPH | Q-1 | 4 | 4 | Ea. | 615 | 193 | | 808 | 965 |
| 0160       8 GPH, barrier free, sensor operated | | 4 | 4 | | 890 | 193 | | 1,083 | 1,275 |
| 0180       8.2 GPH | | 4 | 4 | | 680 | 193 | | 873 | 1,050 |
| 0600       8 GPH hot and cold water | | 4 | 4 | | 1,175 | 193 | | 1,368 | 1,600 |
| 0640       For stainless steel cabinet, add | | | | | 84 | | | 84 | 92 |
| 1000       Dual height, 8.2 GPH | Q-1 | 3.80 | 4.211 | | 1,975 | 203 | | 2,178 | 2,475 |
| 1040          14.3 GPH | " | 3.80 | 4.211 | | 1,125 | 203 | | 1,328 | 1,550 |
| 1240       For stainless steel cabinet, add | | | | | 150 | | | 150 | 165 |
| 2600       Wheelchair type, 8 GPH | Q-1 | 4 | 4 | | 975 | 193 | | 1,168 | 1,375 |

# 22 47 Drinking Fountains and Water Coolers

## 22 47 16 – Pressure Water Coolers

| 22 47 16.10 Electric Water Coolers | Crew | Daily Output | Labor-Hours | Unit | Material | 2011 Bare Costs Labor | 2011 Bare Costs Equipment | Total | Total Incl O&P |
|---|---|---|---|---|---|---|---|---|---|
| 3300 | Semi-recessed, 8.1 GPH | Q-1 | 4 | 4 | Ea. | 630 | 193 | | 823 | 980 |
| 3320 | 12 GPH | ↓ | 4 | 4 | ↓ | 725 | 193 | | 918 | 1,075 |
| 4600 | Floor mounted, flush-to-wall | | | | | | | | | |
| 4640 | 4 GPH | 1 Plum | 3 | 2.667 | Ea. | 620 | 143 | | 763 | 900 |
| 4680 | 8.2 GPH | | 3 | 2.667 | | 655 | 143 | | 798 | 935 |
| 4720 | 14.3 GPH | | 3 | 2.667 | | 750 | 143 | | 893 | 1,050 |
| 4960 | 14 GPH hot and cold water | ↓ | 3 | 2.667 | | 1,050 | 143 | | 1,193 | 1,375 |
| 4980 | For stainless steel cabinet, add | | | | | 128 | | | 128 | 141 |
| 5000 | Dual height, 8.2 GPH | 1 Plum | 2 | 4 | | 1,000 | 214 | | 1,214 | 1,425 |
| 5040 | 14.3 GPH | " | 2 | 4 | | 1,050 | 214 | | 1,264 | 1,475 |
| 5120 | For stainless steel cabinet, add | | | | | 168 | | | 168 | 185 |
| 9800 | For supply, waste & vent, all coolers | 1 Plum | 2.21 | 3.620 | ↓ | 254 | 194 | | 448 | 570 |

# 22 51 Swimming Pool Plumbing Systems

## 22 51 19 – Swimming Pool Water Treatment Equipment

### 22 51 19.50 Swimming Pool Filtration Equipment

| | | Crew | Daily Output | Labor-Hours | Unit | Material | 2011 Bare Costs Labor | 2011 Bare Costs Equipment | Total | Total Incl O&P |
|---|---|---|---|---|---|---|---|---|---|---|
| 0010 | **SWIMMING POOL FILTRATION EQUIPMENT** | | | | | | | | | |
| 0900 | Filter system, sand or diatomite type, incl. pump, 6,000 gal./hr. | 2 Plum | 1.80 | 8.889 | Total | 1,750 | 475 | | 2,225 | 2,625 |
| 1020 | Add for chlorination system, 800 S.F. pool | | 3 | 5.333 | Ea. | 103 | 286 | | 389 | 540 |
| 1040 | 5,000 S.F. pool | ↓ | 3 | 5.333 | " | 1,875 | 286 | | 2,161 | 2,500 |

# 22 52 Fountain Plumbing Systems

## 22 52 16 – Fountain Pumps

### 22 52 16.10 Fountain Water Pumps

| | | Crew | Daily Output | Labor-Hours | Unit | Material | 2011 Bare Costs Labor | 2011 Bare Costs Equipment | Total | Total Incl O&P |
|---|---|---|---|---|---|---|---|---|---|---|
| 0010 | **FOUNTAIN WATER PUMPS** | | | | | | | | | |
| 0100 | Pump w/controls | | | | | | | | | |
| 0200 | Single phase, 100' cord, 1/2 H.P. pump | 2 Skwk | 4.40 | 3.636 | Ea. | 1,200 | 161 | | 1,361 | 1,575 |
| 0300 | 3/4 H.P. pump | | 4.30 | 3.721 | | 2,050 | 165 | | 2,215 | 2,500 |
| 0400 | 1 H.P. pump | | 4.20 | 3.810 | | 2,225 | 169 | | 2,394 | 2,700 |
| 0500 | 1-1/2 H.P. pump | | 4.10 | 3.902 | | 2,650 | 173 | | 2,823 | 3,175 |
| 0600 | 2 H.P. pump | | 4 | 4 | | 3,575 | 178 | | 3,753 | 4,200 |
| 0700 | Three phase, 200' cord, 5 H.P. pump | | 3.90 | 4.103 | | 4,900 | 182 | | 5,082 | 5,650 |
| 0800 | 7-1/2 H.P. pump | | 3.80 | 4.211 | | 8,500 | 187 | | 8,687 | 9,625 |
| 0900 | 10 H.P. pump | | 3.70 | 4.324 | | 9,275 | 192 | | 9,467 | 10,500 |
| 1000 | 15 H.P. pump | ↓ | 3.60 | 4.444 | ↓ | 11,200 | 197 | | 11,397 | 12,600 |
| 2000 | DESIGN NOTE: Use two horsepower per surface acre. | | | | | | | | | |

## 22 52 33 – Fountain Ancillary

### 22 52 33.10 Fountain Miscellaneous

| | | Crew | Daily Output | Labor-Hours | Unit | Material | 2011 Bare Costs Labor | 2011 Bare Costs Equipment | Total | Total Incl O&P |
|---|---|---|---|---|---|---|---|---|---|---|
| 0010 | **FOUNTAIN MISCELLANEOUS** | | | | | | | | | |
| 1100 | Nozzles, minimum | 2 Skwk | 8 | 2 | Ea. | 145 | 89 | | 234 | 296 |
| 1200 | Maximum | | 8 | 2 | | 310 | 89 | | 399 | 475 |
| 1300 | Lights w/mounting kits, 200 watt | | 18 | .889 | | 1,025 | 39.50 | | 1,064.50 | 1,175 |
| 1400 | 300 watt | | 18 | .889 | | 1,225 | 39.50 | | 1,264.50 | 1,400 |
| 1500 | 500 watt | | 18 | .889 | | 1,400 | 39.50 | | 1,439.50 | 1,575 |
| 1600 | Color blender | ↓ | 12 | 1.333 | ↓ | 540 | 59 | | 599 | 685 |

## 22 66 53 – Laboratory Chemical-Waste and Vent Piping

| 22 66 53.30 Glass Pipe | Crew | Daily Output | Labor-Hours | Unit | Material | 2011 Bare Costs Labor | Equipment | Total | Total Incl O&P |
|---|---|---|---|---|---|---|---|---|---|
| 0010 **GLASS PIPE**, Borosilicate, couplings & clevis hanger assemblies, 10' O.C. | | | | | | | | | |
| 0020 Drainage | | | | | | | | | |
| 1100 1-1/2" diameter | Q-1 | 52 | .308 | L.F. | 10.20 | 14.85 | | 25.05 | 33.50 |
| 1120 2" diameter | | 44 | .364 | | 13.20 | 17.55 | | 30.75 | 40.50 |
| 1140 3" diameter | | 39 | .410 | | 17.65 | 19.80 | | 37.45 | 49 |
| 1160 4" diameter | | 30 | .533 | | 31 | 25.50 | | 56.50 | 72.50 |
| 1180 6" diameter | | 26 | .615 | | 57 | 29.50 | | 86.50 | 107 |

| 22 66 53.60 Corrosion Resistant Pipe | Crew | Daily Output | Labor-Hours | Unit | Material | 2011 Bare Costs Labor | Equipment | Total | Total Incl O&P |
|---|---|---|---|---|---|---|---|---|---|
| 0010 **CORROSION RESISTANT PIPE**, No couplings or hangers | | | | | | | | | |
| 0020 Iron alloy, drain, mechanical joint | | | | | | | | | |
| 1000 1-1/2" diameter | Q-1 | 70 | .229 | L.F. | 46 | 11.05 | | 57.05 | 67 |
| 1100 2" diameter | | 66 | .242 | | 47 | 11.70 | | 58.70 | 69 |
| 1120 3" diameter | | 60 | .267 | | 60.50 | 12.85 | | 73.35 | 85.50 |
| 1140 4" diameter | | 52 | .308 | | 77.50 | 14.85 | | 92.35 | 108 |
| 2980 Plastic, epoxy, fiberglass filament wound, B&S joint | | | | | | | | | |
| 3000 2" diameter | Q-1 | 62 | .258 | L.F. | 11.40 | 12.45 | | 23.85 | 31 |
| 3100 3" diameter | | 51 | .314 | | 13.30 | 15.15 | | 28.45 | 37 |
| 3120 4" diameter | | 45 | .356 | | 19 | 17.15 | | 36.15 | 46.50 |
| 3140 6" diameter | | 32 | .500 | | 26.50 | 24 | | 50.50 | 65.50 |
| 3980 Polyester, fiberglass filament wound, B&S joint | | | | | | | | | |
| 4000 2" diameter | Q-1 | 62 | .258 | L.F. | 12.35 | 12.45 | | 24.80 | 32 |
| 4100 3" diameter | | 51 | .314 | | 16.15 | 15.15 | | 31.30 | 40.50 |
| 4120 4" diameter | | 45 | .356 | | 24 | 17.15 | | 41.15 | 51.50 |
| 4140 6" diameter | | 32 | .500 | | 34 | 24 | | 58 | 73.50 |
| 4980 Polypropylene, acid resistant, fire retardant, schedule 40 | | | | | | | | | |
| 5000 1-1/2" diameter | Q-1 | 68 | .235 | L.F. | 6.05 | 11.35 | | 17.40 | 23.50 |
| 5100 2" diameter | | 62 | .258 | | 8.20 | 12.45 | | 20.65 | 27.50 |
| 5120 3" diameter | | 51 | .314 | | 16.80 | 15.15 | | 31.95 | 41 |
| 5140 4" diameter | | 45 | .356 | | 21.50 | 17.15 | | 38.65 | 49 |
| 5980 Proxylene, fire retardant, Schedule 40 | | | | | | | | | |
| 6000 1-1/2" diameter | Q-1 | 68 | .235 | L.F. | 9.40 | 11.35 | | 20.75 | 27.50 |
| 6100 2" diameter | | 62 | .258 | | 12.90 | 12.45 | | 25.35 | 33 |
| 6120 3" diameter | | 51 | .314 | | 23.50 | 15.15 | | 38.65 | 48.50 |
| 6140 4" diameter | | 45 | .356 | | 33 | 17.15 | | 50.15 | 62 |

# Estimating Tips

The labor adjustment factors listed in Subdivision 22 01 02.20 also apply to Division 23.

## 23 10 00 Facility Fuel Systems

- The prices in this subdivision for above- and below-ground storage tanks do not include foundations or hold-down slabs, unless noted. The estimator should refer to Divisions 3 and 31 for foundation system pricing. In addition to the foundations, required tank accessories, such as tank gauges, leak detection devices, and additional manholes and piping, must be added to the tank prices.

## 23 50 00 Central Heating Equipment

- When estimating the cost of an HVAC system, check to see who is responsible for providing and installing the temperature control system. It is possible to overlook controls, assuming that they would be included in the electrical estimate.
- When looking up a boiler, be careful on specified capacity. Some manufacturers rate their products on output while others use input.
- Include HVAC insulation for pipe, boiler, and duct (wrap and liner).
- Be careful when looking up mechanical items to get the correct pressure rating and connection type (thread, weld, flange).

## 23 70 00 Central HVAC Equipment

- Combination heating and cooling units are sized by the air conditioning requirements. (See Reference No. R236000-20 for preliminary sizing guide.)
- A ton of air conditioning is nominally 400 CFM.
- Rectangular duct is taken off by the linear foot for each size, but its cost is usually estimated by the pound. Remember that SMACNA standards now base duct on internal pressure.
- Prefabricated duct is estimated and purchased like pipe: straight sections and fittings.
- Note that cranes or other lifting equipment are not included on any lines in Division 23. For example, if a crane is required to lift a heavy piece of pipe into place high above a gym floor, or to put a rooftop unit on the roof of a four-story building, etc., it must be added. Due to the potential for extreme variation—from nothing additional required to a major crane or helicopter—we feel that including a nominal amount for "lifting contingency" would be useless and detract from the accuracy of the estimate. When using equipment rental cost data from RSMeans, do not forget to include the cost of the operator(s).

# Reference Numbers

Reference numbers are shown in shaded boxes at the beginning of some major classifications. These numbers refer to related items in the Reference Section. The reference information may be an estimating procedure, an alternate pricing method, or technical information.

*Note:* Not all subdivisions listed here necessarily appear in this publication.

*Note:* **Trade Service,** *in part, has been used as a reference source for some of the material prices used in Division 23.*

## 23 05 02 – HVAC General

| 23 05 02.10 Air Conditioning, General | Crew | Daily Output | Labor-Hours | Unit | Material | 2011 Bare Costs Labor | 2011 Bare Costs Equipment | Total | Total Incl O&P |
|---|---|---|---|---|---|---|---|---|---|
| 0010 **AIR CONDITIONING, GENERAL** Prices are for standard efficiencies (SEER 13) | | | | | | | | | |
| 0020     for upgrade to SEER 14 add | | | | | 20% | | | | |

## 23 05 05 – Selective HVAC Demolition

### 23 05 05.10 HVAC Demolition

| | | Crew | Daily Output | Labor-Hours | Unit | Material | Labor | Equipment | Total | Total Incl O&P |
|---|---|---|---|---|---|---|---|---|---|---|
| 0010 | **HVAC DEMOLITION** | | | | | | | | | |
| 0100 | Air conditioner, split unit, 3 ton | Q-5 | 2 | 8 | Ea. | | 390 | | 390 | 580 |
| 0150 | Package unit, 3 ton | Q-6 | 3 | 8 | " | | 405 | | 405 | 605 |
| 0298 | Boilers | | | | | | | | | |
| 0300 | Electric, up thru 148 kW | Q-19 | 2 | 12 | Ea. | | 590 | | 590 | 880 |
| 0310 | 150 thru 518 kW | " | 1 | 24 | | | 1,175 | | 1,175 | 1,750 |
| 0320 | 550 thru 2000 kW | Q-21 | .40 | 80 | | | 4,025 | | 4,025 | 6,025 |
| 0330 | 2070 kW and up | " | .30 | 106 | | | 5,375 | | 5,375 | 8,025 |
| 0340 | Gas and/or oil, up thru 150 MBH | Q-7 | 2.20 | 14.545 | | | 750 | | 750 | 1,125 |
| 0350 | 160 thru 2000 MBH | | .80 | 40 | | | 2,050 | | 2,050 | 3,075 |
| 0360 | 2100 thru 4500 MBH | | .50 | 64 | | | 3,300 | | 3,300 | 4,925 |
| 0370 | 4600 thru 7000 MBH | | .30 | 106 | | | 5,500 | | 5,500 | 8,200 |
| 0380 | 7100 thru 12,000 MBH | | .16 | 200 | | | 10,300 | | 10,300 | 15,400 |
| 0390 | 12,200 thru 25,000 MBH | | .12 | 266 | | | 13,700 | | 13,700 | 20,500 |
| 1000 | Ductwork, 4" high, 8" wide | 1 Clab | 200 | .040 | L.F. | | 1.37 | | 1.37 | 2.11 |
| 1100 | 6" high, 8" wide | | 165 | .048 | | | 1.67 | | 1.67 | 2.55 |
| 1200 | 10" high, 12" wide | | 125 | .064 | | | 2.20 | | 2.20 | 3.37 |
| 1300 | 12"-14" high, 16"-18" wide | | 85 | .094 | | | 3.23 | | 3.23 | 4.96 |
| 1400 | 18" high, 24" wide | | 67 | .119 | | | 4.10 | | 4.10 | 6.30 |
| 1500 | 30" high, 36" wide | | 56 | .143 | | | 4.91 | | 4.91 | 7.55 |
| 1540 | 72" wide | | 50 | .160 | | | 5.50 | | 5.50 | 8.45 |
| 3000 | Mechanical equipment, light items. Unit is weight, not cooling. | Q-5 | .90 | 17.778 | Ton | | 865 | | 865 | 1,300 |
| 3600 | Heavy items | " | 1.10 | 14.545 | " | | 710 | | 710 | 1,050 |
| 3700 | Deduct for salvage (when applicable), minimum | | | | Job | | | | 73 | 80 |
| 3710 | Maximum | | | | " | | | | 455 | 500 |

## 23 05 23 – General-Duty Valves for HVAC Piping

### 23 05 23.30 Valves, Iron Body

| | | Crew | Daily Output | Labor-Hours | Unit | Material | Labor | Equipment | Total | Total Incl O&P |
|---|---|---|---|---|---|---|---|---|---|---|
| 0010 | **VALVES, IRON BODY** | | | | | | | | | |
| 1020 | Butterfly, wafer type, gear actuator, 200 lb. | | | | | | | | | |
| 1030 | 2" | 1 Plum | 14 | .571 | Ea. | 167 | 30.50 | | 197.50 | 229 |
| 1040 | 2-1/2" | Q-1 | 9 | 1.778 | | 172 | 86 | | 258 | 315 |
| 1050 | 3" | | 8 | 2 | | 178 | 96.50 | | 274.50 | 340 |
| 1060 | 4" | | 5 | 3.200 | | 202 | 154 | | 356 | 455 |
| 1070 | 5" | Q-2 | 5 | 4.800 | | 269 | 240 | | 509 | 655 |
| 1080 | 6" | " | 5 | 4.800 | | 305 | 240 | | 545 | 695 |
| 1650 | Gate, 125 lb., N.R.S. | | | | | | | | | |
| 2150 | Flanged | | | | | | | | | |
| 2200 | 2" | 1 Plum | 5 | 1.600 | Ea. | 550 | 86 | | 636 | 735 |
| 2240 | 2-1/2" | Q-1 | 5 | 3.200 | | 565 | 154 | | 719 | 850 |
| 2260 | 3" | | 4.50 | 3.556 | | 630 | 172 | | 802 | 950 |
| 2280 | 4" | | 3 | 5.333 | | 905 | 257 | | 1,162 | 1,375 |
| 2300 | 6" | Q-2 | 3 | 8 | | 1,550 | 400 | | 1,950 | 2,300 |
| 3550 | OS&Y, 125 lb., flanged | | | | | | | | | |
| 3600 | 2" | 1 Plum | 5 | 1.600 | Ea. | 350 | 86 | | 436 | 515 |
| 3660 | 3" | Q-1 | 4.50 | 3.556 | | 395 | 172 | | 567 | 690 |
| 3680 | 4" | " | 3 | 5.333 | | 565 | 257 | | 822 | 1,000 |
| 3700 | 6" | Q-2 | 3 | 8 | | 930 | 400 | | 1,330 | 1,625 |
| 3900 | For 175 lb, flanged, add | | | | | 200% | 10% | | | |

## 23 05 23 – General-Duty Valves for HVAC Piping

### 23 05 23.30 Valves, Iron Body

| | 23 05 23.30 Valves, Iron Body | Crew | Daily Output | Labor-Hours | Unit | Material | 2011 Bare Costs Labor | Equipment | Total | Total Incl O&P |
|---|---|---|---|---|---|---|---|---|---|---|
| 5450 | Swing check, 125 lb., threaded | | | | | | | | | |
| 5470 | 1" | 1 Plum | 13 | .615 | Ea. | 575 | 33 | | 608 | 685 |
| 5500 | 2" | " | 11 | .727 | | 705 | 39 | | 744 | 835 |
| 5540 | 2-1/2" | Q-1 | 15 | 1.067 | | 675 | 51.50 | | 726.50 | 820 |
| 5550 | 3" | | 13 | 1.231 | | 750 | 59.50 | | 809.50 | 915 |
| 5560 | 4" | | 10 | 1.600 | | 1,300 | 77 | | 1,377 | 1,550 |
| 5950 | Flanged | | | | | | | | | |
| 6000 | 2" | 1 Plum | 5 | 1.600 | Ea. | 284 | 86 | | 370 | 440 |
| 6040 | 2-1/2" | Q-1 | 5 | 3.200 | | 286 | 154 | | 440 | 545 |
| 6050 | 3" | | 4.50 | 3.556 | | 305 | 172 | | 477 | 590 |
| 6060 | 4" | | 3 | 5.333 | | 480 | 257 | | 737 | 915 |
| 6070 | 6" | Q-2 | 3 | 8 | | 820 | 400 | | 1,220 | 1,500 |

### 23 05 23.80 Valves, Steel

| | 23 05 23.80 Valves, Steel | Crew | Daily Output | Labor-Hours | Unit | Material | 2011 Bare Costs Labor | Equipment | Total | Total Incl O&P |
|---|---|---|---|---|---|---|---|---|---|---|
| 0010 | **VALVES, STEEL** | | | | | | | | | |
| 0800 | Cast | | | | | | | | | |
| 1350 | Check valve, swing type, 150 lb., flanged | | | | | | | | | |
| 1370 | 1" | 1 Plum | 10 | .800 | Ea. | 545 | 43 | | 588 | 665 |
| 1400 | 2" | " | 8 | 1 | | 660 | 53.50 | | 713.50 | 805 |
| 1440 | 2-1/2" | Q-1 | 5 | 3.200 | | 765 | 154 | | 919 | 1,075 |
| 1450 | 3" | | 4.50 | 3.556 | | 780 | 172 | | 952 | 1,100 |
| 1460 | 4" | | 3 | 5.333 | | 1,125 | 257 | | 1,382 | 1,600 |
| 1540 | For 300 lb., flanged, add | | | | | 50% | 15% | | | |
| 1548 | For 600 lb., flanged, add | | | | | 110% | 20% | | | |
| 1950 | Gate valve, 150 lb., flanged | | | | | | | | | |
| 2000 | 2" | 1 Plum | 8 | 1 | Ea. | 710 | 53.50 | | 763.50 | 860 |
| 2040 | 2-1/2" | Q-1 | 5 | 3.200 | | 1,000 | 154 | | 1,154 | 1,325 |
| 2050 | 3" | | 4.50 | 3.556 | | 1,000 | 172 | | 1,172 | 1,350 |
| 2060 | 4" | | 3 | 5.333 | | 1,250 | 257 | | 1,507 | 1,750 |
| 2070 | 6" | Q-2 | 3 | 8 | | 1,850 | 400 | | 2,250 | 2,625 |
| 3650 | Globe valve, 150 lb., flanged | | | | | | | | | |
| 3700 | 2" | 1 Plum | 8 | 1 | Ea. | 890 | 53.50 | | 943.50 | 1,050 |
| 3740 | 2-1/2" | Q-1 | 5 | 3.200 | | 1,125 | 154 | | 1,279 | 1,475 |
| 3750 | 3" | | 4.50 | 3.556 | | 1,125 | 172 | | 1,297 | 1,500 |
| 3760 | 4" | | 3 | 5.333 | | 1,650 | 257 | | 1,907 | 2,200 |
| 3770 | 6" | Q-2 | 3 | 8 | | 2,600 | 400 | | 3,000 | 3,450 |
| 5150 | Forged | | | | | | | | | |
| 5650 | Check valve, class 800, horizontal, socket | | | | | | | | | |
| 5698 | Threaded | | | | | | | | | |
| 5700 | 1/4" | 1 Plum | 24 | .333 | Ea. | 88 | 17.85 | | 105.85 | 123 |
| 5720 | 3/8" | | 24 | .333 | | 88 | 17.85 | | 105.85 | 123 |
| 5730 | 1/2" | | 24 | .333 | | 88 | 17.85 | | 105.85 | 123 |
| 5740 | 3/4" | | 20 | .400 | | 94.50 | 21.50 | | 116 | 136 |
| 5750 | 1" | | 19 | .421 | | 111 | 22.50 | | 133.50 | 156 |
| 5760 | 1-1/4" | | 15 | .533 | | 217 | 28.50 | | 245.50 | 282 |

## 23 05 93 – Testing, Adjusting, and Balancing for HVAC

### 23 05 93.10 Balancing, Air

| | 23 05 93.10 Balancing, Air | Crew | Daily Output | Labor-Hours | Unit | Material | 2011 Bare Costs Labor | Equipment | Total | Total Incl O&P |
|---|---|---|---|---|---|---|---|---|---|---|
| 0010 | **BALANCING, AIR** (Subcontractor's quote incl. material and labor) | | | | | | | | | |
| 0900 | Heating and ventilating equipment | | | | | | | | | |
| 1000 | Centrifugal fans, utility sets | | | | Ea. | | | | 340 | 375 |
| 1100 | Heating and ventilating unit | | | | | | | | 510 | 560 |
| 1200 | In-line fan | | | | | | | | 510 | 560 |
| 1300 | Propeller and wall fan | | | | | | | | 96.50 | 106 |

## 23 05 93 – Testing, Adjusting, and Balancing for HVAC

| 23 05 93.10 Balancing, Air | | Crew | Daily Output | Labor-Hours | Unit | Material | 2011 Bare Costs Labor | Equipment | Total | Total Incl O&P |
|---|---|---|---|---|---|---|---|---|---|---|
| 1400 | Roof exhaust fan | | | | Ea. | | | | 227 | 250 |
| 2000 | Air conditioning equipment, central station | | | | | | | | 740 | 810 |
| 2100 | Built-up low pressure unit | | | | | | | | 680 | 750 |
| 2200 | Built-up high pressure unit | | | | | | | | 795 | 875 |
| 2500 | Multi-zone A.C. and heating unit | | | | | | | | 510 | 560 |
| 2600 | For each zone over one, add | | | | | | | | 114 | 125 |
| 2700 | Package A.C. unit | | | | | | | | 285 | 310 |
| 2800 | Rooftop heating and cooling unit | | | | | | | | 400 | 440 |
| 3000 | Supply, return, exhaust, registers & diffusers, avg. height ceiling | | | | | | | | 68 | 75 |
| 3100 | High ceiling | | | | | | | | 102 | 112 |
| 3200 | Floor height | | | | | | | | 57 | 62 |

### 23 05 93.20 Balancing, Water

| | | Crew | Daily Output | Labor-Hours | Unit | Material | Labor | Equipment | Total | Total Incl O&P |
|---|---|---|---|---|---|---|---|---|---|---|
| 0010 | **BALANCING, WATER** (Subcontractor's quote incl. material and labor) | | | | | | | | | |
| 0050 | Air cooled condenser | | | | Ea. | | | | 206 | 226 |
| 0080 | Boiler | | | | | | | | 415 | 455 |
| 0100 | Cabinet unit heater | | | | | | | | 70.50 | 77.50 |
| 0200 | Chiller | | | | | | | | 500 | 550 |
| 0300 | Convector | | | | | | | | 59 | 64.50 |
| 0500 | Cooling tower | | | | | | | | 380 | 420 |
| 0600 | Fan coil unit, unit ventilator | | | | | | | | 106 | 116 |
| 0700 | Fin tube and radiant panels | | | | | | | | 118 | 129 |
| 0800 | Main and duct re-heat coils | | | | | | | | 109 | 120 |
| 0810 | Heat exchanger | | | | | | | | 109 | 120 |
| 1000 | Pumps | | | | | | | | 259 | 284 |
| 1100 | Unit heater | | | | | | | | 82.50 | 90.50 |

# 23 07 HVAC Insulation

## 23 07 13 – Duct Insulation

### 23 07 13.10 Duct Thermal Insulation

| | | | Crew | Daily Output | Labor-Hours | Unit | Material | Labor | Equipment | Total | Total Incl O&P |
|---|---|---|---|---|---|---|---|---|---|---|---|
| 0010 | **DUCT THERMAL INSULATION** | | | | | | | | | | |
| 0100 | Rule of thumb, as a percentage of total mechanical costs | | | | | Job | | | | 10% | |
| 0110 | Insulation req'd is based on the surface size/area to be covered | | | | | | | | | | |
| 3000 | Ductwork | | | | | | | | | | |
| 3020 | Blanket type, fiberglass, flexible | | | | | | | | | | |
| 3140 | FSK vapor barrier wrap, .75 lb. density | | | | | | | | | | |
| 3160 | 1" thick | G | Q-14 | 350 | .046 | S.F. | .20 | 1.93 | | 2.13 | 3.20 |
| 3170 | 1-1/2" thick | G | | 320 | .050 | | .23 | 2.11 | | 2.34 | 3.51 |
| 3180 | 2" thick | G | | 300 | .053 | | .28 | 2.25 | | 2.53 | 3.79 |
| 3190 | 3" thick | G | | 260 | .062 | | .40 | 2.60 | | 3 | 4.46 |
| 3200 | 4" thick | G | | 242 | .066 | | .60 | 2.79 | | 3.39 | 4.97 |
| 3212 | Vinyl Jacket, .75 lb. density, 1-1/2" thick | G | | 320 | .050 | | .23 | 2.11 | | 2.34 | 3.51 |
| 3280 | Unfaced, 1 lb. density | | | | | | | | | | |
| 3310 | 1" thick | G | Q-14 | 360 | .044 | S.F. | .19 | 1.88 | | 2.07 | 3.11 |
| 3320 | 1-1/2" thick | G | | 330 | .048 | | .23 | 2.05 | | 2.28 | 3.41 |
| 3330 | 2" thick | G | | 310 | .052 | | .31 | 2.18 | | 2.49 | 3.71 |
| 3400 | FSK facing, 1 lb. density | | | | | | | | | | |
| 3420 | 1-1/2" thick | G | Q-14 | 310 | .052 | S.F. | .18 | 2.18 | | 2.36 | 3.57 |
| 3430 | 2" thick | G | " | 300 | .053 | " | .20 | 2.25 | | 2.45 | 3.70 |
| 3450 | FSK facing, 1.5 lb. density | | | | | | | | | | |
| 3470 | 1-1/2" thick | G | Q-14 | 300 | .053 | S.F. | .27 | 2.25 | | 2.52 | 3.78 |
| 3480 | 2" thick | G | " | 290 | .055 | " | .33 | 2.33 | | 2.66 | 3.96 |

# 23 07 HVAC Insulation

## 23 07 13 – Duct Insulation

### 23 07 13.10 Duct Thermal Insulation

| | | Crew | Daily Output | Labor-Hours | Unit | Material | 2011 Bare Costs Labor | 2011 Bare Costs Equipment | Total | Total Incl O&P |
|---|---|---|---|---|---|---|---|---|---|---|
| 3795 | Finishes | | | | | | | | | |
| 3800 | Stainless steel woven mesh | Q-14 | 100 | .160 | S.F. | .84 | 6.75 | | 7.59 | 11.35 |
| 3820 | 18 oz. fiberglass cloth, pasted on | | 170 | .094 | | .77 | 3.97 | | 4.74 | 7 |
| 3900 | 8 oz. canvas, pasted on | | 180 | .089 | | .49 | 3.75 | | 4.24 | 6.35 |

## 23 07 16 – HVAC Equipment Insulation

### 23 07 16.10 HVAC Equipment Thermal Insulation

| | | | Crew | Daily Output | Labor-Hours | Unit | Material | Labor | Equipment | Total | Total Incl O&P |
|---|---|---|---|---|---|---|---|---|---|---|---|
| 0010 | **HVAC EQUIPMENT THERMAL INSULATION** | | | | | | | | | | |
| 0100 | Rule of thumb, as a percentage of total mechanical costs | | | | | Job | | | | 10% | |
| 0110 | Insulation req'd is based on the surface size/area to be covered | | | | | | | | | | |
| 1000 | Boiler, 1-1/2" calcium silicate only | G | Q-14 | 110 | .145 | S.F. | 2.92 | 6.15 | | 9.07 | 12.70 |
| 1020 | Plus 2" fiberglass | G | " | 80 | .200 | " | 3.82 | 8.45 | | 12.27 | 17.25 |
| 2000 | Breeching, 2" calcium silicate | | | | | | | | | | |
| 2020 | Rectangular | G | Q-14 | 42 | .381 | S.F. | 5.70 | 16.10 | | 21.80 | 31.50 |
| 2040 | Round | G | " | 38.70 | .413 | " | 5.95 | 17.45 | | 23.40 | 33.50 |

# 23 09 Instrumentation and Control for HVAC

## 23 09 33 – Electric and Electronic Control System for HVAC

### 23 09 33.10 Electronic Control Systems

| | | | | | Unit | | | | | Total Incl O&P |
|---|---|---|---|---|---|---|---|---|---|---|
| 0010 | **ELECTRONIC CONTROL SYSTEMS** | | | | | | | | | |
| 0020 | For electronic costs, add to Section 23 09 43.10 | | | | Ea. | | | | | 15% |

## 23 09 43 – Pneumatic Control System for HVAC

### 23 09 43.10 Pneumatic Control Systems

| | | | Crew | Daily Output | Labor-Hours | Unit | Material | Labor | Equipment | Total | Total Incl O&P |
|---|---|---|---|---|---|---|---|---|---|---|---|
| 0010 | **PNEUMATIC CONTROL SYSTEMS** | | | | | | | | | | |
| 0011 | Including a nominal 50 ft. of tubing. Add control panelboard if req'd. | | | | | | | | | | |
| 0100 | Heating and ventilating, split system | | | | | | | | | | |
| 0200 | Mixed air control, economizer cycle, panel readout, tubing | | | | | | | | | | |
| 0220 | Up to 10 tons | G | Q-19 | .68 | 35.294 | Ea. | 3,675 | 1,750 | | 5,425 | 6,625 |
| 0240 | For 10 to 20 tons | G | | .63 | 37.915 | | 3,925 | 1,875 | | 5,800 | 7,100 |
| 0260 | For over 20 tons | G | | .58 | 41.096 | | 4,275 | 2,025 | | 6,300 | 7,700 |
| 0300 | Heating coil, hot water, 3 way valve, | | | | | | | | | | |
| 0320 | Freezestat, limit control on discharge, readout | | Q-5 | .69 | 23.088 | Ea. | 2,725 | 1,125 | | 3,850 | 4,675 |
| 0500 | Cooling coil, chilled water, room | | | | | | | | | | |
| 0520 | Thermostat, 3 way valve | | Q-5 | 2 | 8 | Ea. | 1,225 | 390 | | 1,615 | 1,925 |
| 0600 | Cooling tower, fan cycle, damper control, | | | | | | | | | | |
| 0620 | Control system including water readout in/out at panel | | Q-19 | .67 | 35.821 | Ea. | 4,850 | 1,775 | | 6,625 | 7,950 |
| 1000 | Unit ventilator, day/night operation, | | | | | | | | | | |
| 1100 | freezestat, ASHRAE, cycle 2 | | Q-19 | .91 | 26.374 | Ea. | 2,675 | 1,300 | | 3,975 | 4,875 |
| 2000 | Compensated hot water from boiler, valve control, | | | | | | | | | | |
| 2100 | readout and reset at panel, up to 60 GPM | | Q-19 | .55 | 43.956 | Ea. | 5,000 | 2,175 | | 7,175 | 8,725 |
| 2120 | For 120 GPM | | | .51 | 47.059 | | 5,350 | 2,325 | | 7,675 | 9,350 |
| 2140 | For 240 GPM | | | .49 | 49.180 | | 5,600 | 2,425 | | 8,025 | 9,775 |
| 3000 | Boiler room combustion air, damper to 5 SF, controls | | | 1.37 | 17.582 | | 2,425 | 865 | | 3,290 | 3,950 |
| 3500 | Fan coil, heating and cooling valves, 4 pipe control system | | | 3 | 8 | | 1,100 | 395 | | 1,495 | 1,775 |
| 3600 | Heat exchanger system controls | | | .86 | 27.907 | | 2,350 | 1,375 | | 3,725 | 4,625 |
| 4000 | Pneumatic thermostat, including controlling room radiator valve | | Q-5 | 2.43 | 6.593 | | 730 | 320 | | 1,050 | 1,275 |
| 4060 | Pump control system | | Q-19 | 3 | 8 | | 1,125 | 395 | | 1,520 | 1,800 |
| 4500 | Air supply for pneumatic control system | | | | | | | | | | |
| 4600 | Tank mounted duplex compressor, starter, alternator, | | | | | | | | | | |
| 4620 | piping, dryer, PRV station and filter | | | | | | | | | | |
| 4630 | 1/2 HP | | Q-19 | .68 | 35.139 | Ea. | 8,975 | 1,725 | | 10,700 | 12,500 |

# 23 09 Instrumentation and Control for HVAC

## 23 09 43 – Pneumatic Control System for HVAC

| 23 09 43.10 Pneumatic Control Systems | Crew | Daily Output | Labor-Hours | Unit | Material | 2011 Bare Costs Labor | Equipment | Total | Total Incl O&P |
|---|---|---|---|---|---|---|---|---|---|
| 4660 | 1-1/2 HP | Q-19 | .58 | 41.739 | Ea. | 11,000 | 2,050 | | 13,050 | 15,200 |
| 4690 | 5 HP | ↓ | .42 | 57.143 | ↓ | 26,100 | 2,825 | | 28,925 | 32,900 |

# 23 13 Facility Fuel-Storage Tanks

## 23 13 13 – Facility Underground Fuel-Oil, Storage Tanks

### 23 13 13.09 Single-Wall Steel Fuel-Oil Tanks

| | | Crew | Daily Output | Labor-Hours | Unit | Material | 2011 Bare Costs Labor | Equipment | Total | Total Incl O&P |
|---|---|---|---|---|---|---|---|---|---|---|
| 0010 | **SINGLE-WALL STEEL FUEL-OIL TANKS** | | | | | | | | | |
| 5000 | Steel underground, sti-P3, set in place, not incl. hold-down bars. | | | | | | | | | |
| 5500 | Excavation, pad, pumps and piping not included | | | | | | | | | |
| 5510 | Single wall, 500 gallon capacity, 7 gauge shell | Q-5 | 2.70 | 5.926 | Ea. | 2,050 | 289 | | 2,339 | 2,675 |
| 5520 | 1,000 gallon capacity, 7 gauge shell | " | 2.50 | 6.400 | | 3,150 | 310 | | 3,460 | 3,950 |
| 5530 | 2,000 gallon capacity, 1/4" thick shell | Q-7 | 4.60 | 6.957 | | 5,125 | 360 | | 5,485 | 6,150 |
| 5535 | 2,500 gallon capacity, 7 gauge shell | Q-5 | 3 | 5.333 | | 5,650 | 260 | | 5,910 | 6,625 |
| 5540 | 5,000 gallon capacity, 1/4" thick shell | Q-7 | 3.20 | 10 | | 7,300 | 515 | | 7,815 | 8,825 |
| 5580 | 15,000 gallon capacity, 5/16" thick shell | | 1.70 | 18.824 | | 17,400 | 970 | | 18,370 | 20,600 |
| 5600 | 20,000 gallon capacity, 5/16" thick shell | | 1.50 | 21.333 | | 23,000 | 1,100 | | 24,100 | 27,000 |
| 5610 | 25,000 gallon capacity, 3/8" thick shell | | 1.30 | 24.615 | | 31,100 | 1,275 | | 32,375 | 36,100 |
| 5620 | 30,000 gallon capacity, 3/8" thick shell | | 1.10 | 29.091 | | 31,700 | 1,500 | | 33,200 | 37,100 |
| 5630 | 40,000 gallon capacity, 3/8" thick shell | | .90 | 35.556 | | 35,600 | 1,825 | | 37,425 | 41,900 |
| 5640 | 50,000 gallon capacity, 3/8" thick shell | ↓ | .80 | 40 | ↓ | 39,500 | 2,050 | | 41,550 | 46,600 |

### 23 13 13.23 Glass-Fiber-Reinfcd-Plastic, Fuel-Oil, Storage

| | | Crew | Daily Output | Labor-Hours | Unit | Material | 2011 Bare Costs Labor | Equipment | Total | Total Incl O&P |
|---|---|---|---|---|---|---|---|---|---|---|
| 0010 | **GLASS-FIBER-REINFCD-PLASTIC, UNDERGRND FUEL-OIL, STORAGE** | | | | | | | | | |
| 0210 | Fiberglass, underground, single wall, U.L. listed, not including | | | | | | | | | |
| 0220 | manway or hold-down strap | | | | | | | | | |
| 0240 | 2,000 gallon capacity | Q-7 | 4.57 | 7.002 | Ea. | 6,075 | 360 | | 6,435 | 7,250 |
| 0250 | 4,000 gallon capacity | | 3.55 | 9.014 | | 8,650 | 465 | | 9,115 | 10,200 |
| 0260 | 6,000 gallon capacity | | 2.67 | 11.985 | | 9,600 | 620 | | 10,220 | 11,500 |
| 0280 | 10,000 gallon capacity | | 2 | 16 | | 12,900 | 825 | | 13,725 | 15,400 |
| 0284 | 15,000 gallon capacity | | 1.68 | 19.048 | | 23,900 | 980 | | 24,880 | 27,800 |
| 0290 | 20,000 gallon capacity | ↓ | 1.45 | 22.069 | | 25,800 | 1,125 | | 26,925 | 30,100 |
| 0500 | For manway, fittings and hold-downs, add | | | | ↓ | 20% | 15% | | | |
| 1020 | Fiberglass, underground, double wall, U.L. listed | | | | | | | | | |
| 1030 | includes manways, not incl. hold-down straps | | | | | | | | | |
| 1040 | 600 gallon capacity | Q-5 | 2.42 | 6.612 | Ea. | 6,925 | 320 | | 7,245 | 8,100 |
| 1050 | 1,000 gallon capacity | " | 2.25 | 7.111 | | 8,650 | 345 | | 8,995 | 10,000 |
| 1060 | 2,500 gallon capacity | Q-7 | 4.16 | 7.692 | | 13,400 | 395 | | 13,795 | 15,400 |
| 1070 | 3,000 gallon capacity | | 3.90 | 8.205 | | 14,500 | 425 | | 14,925 | 16,600 |
| 1080 | 4,000 gallon capacity | | 3.64 | 8.791 | | 16,000 | 455 | | 16,455 | 18,400 |
| 1090 | 6,000 gallon capacity | | 2.42 | 13.223 | | 18,000 | 680 | | 18,680 | 20,800 |
| 1100 | 8,000 gallon capacity | | 2.08 | 15.385 | | 21,300 | 795 | | 22,095 | 24,700 |
| 1110 | 10,000 gallon capacity | | 1.82 | 17.582 | | 24,000 | 905 | | 24,905 | 27,800 |
| 1120 | 12,000 gallon capacity | ↓ | 1.70 | 18.824 | ↓ | 29,600 | 970 | | 30,570 | 34,000 |
| 2210 | Fiberglass, underground, single wall, U.L. listed, including | | | | | | | | | |
| 2220 | hold-down straps, no manways | | | | | | | | | |
| 2240 | 2,000 gallon capacity | Q-7 | 3.55 | 9.014 | Ea. | 6,525 | 465 | | 6,990 | 7,875 |
| 2250 | 4,000 gallon capacity | | 2.90 | 11.034 | | 9,100 | 570 | | 9,670 | 10,900 |
| 2260 | 6,000 gallon capacity | | 2 | 16 | | 10,500 | 825 | | 11,325 | 12,700 |
| 2280 | 10,000 gallon capacity | | 1.60 | 20 | | 13,800 | 1,025 | | 14,825 | 16,800 |
| 2284 | 15,000 gallon capacity | | 1.39 | 23.022 | | 24,800 | 1,175 | | 25,975 | 29,100 |
| 2290 | 20,000 gallon capacity | ↓ | 1.14 | 28.070 | ↓ | 27,100 | 1,450 | | 28,550 | 32,100 |
| 3020 | Fiberglass, underground, double wall, U.L. listed | | | | | | | | | |

# 23 13 Facility Fuel-Storage Tanks

## 23 13 13 – Facility Underground Fuel-Oil, Storage Tanks

### 23 13 13.23 Glass-Fiber-Reinfcd-Plastic, Fuel-Oil, Storage

| | 23 13 13.23 Glass-Fiber-Reinfcd-Plastic, Fuel-Oil, Storage | Crew | Daily Output | Labor-Hours | Unit | Material | 2011 Bare Costs Labor | Equipment | Total | Total Incl O&P |
|---|---|---|---|---|---|---|---|---|---|---|
| 3030 | includes manways and hold-down straps | | | | | | | | | |
| 3040 | 600 gallon capacity | Q-5 | 1.86 | 8.602 | Ea. | 7,375 | 420 | | 7,795 | 8,725 |
| 3050 | 1,000 gallon capacity | " | 1.70 | 9.412 | | 9,100 | 460 | | 9,560 | 10,700 |
| 3060 | 2,500 gallon capacity | Q-7 | 3.29 | 9.726 | | 13,900 | 500 | | 14,400 | 16,100 |
| 3070 | 3,000 gallon capacity | | 3.13 | 10.224 | | 15,000 | 525 | | 15,525 | 17,300 |
| 3080 | 4,000 gallon capacity | | 2.93 | 10.922 | | 16,500 | 565 | | 17,065 | 18,900 |
| 3090 | 6,000 gallon capacity | | 1.86 | 17.204 | | 18,900 | 885 | | 19,785 | 22,100 |
| 3100 | 8,000 gallon capacity | | 1.65 | 19.394 | | 22,200 | 1,000 | | 23,200 | 25,900 |
| 3110 | 10,000 gallon capacity | | 1.48 | 21.622 | | 24,900 | 1,125 | | 26,025 | 29,100 |
| 3120 | 12,000 gallon capacity | | 1.40 | 22.857 | | 30,500 | 1,175 | | 31,675 | 35,300 |

## 23 13 23 – Facility Aboveground Fuel-Oil, Storage Tanks

### 23 13 23.13 Vertical, Steel, Abvground Fuel-Oil, Stor. Tanks

| | | Crew | Daily Output | Labor-Hours | Unit | Material | Labor | Equipment | Total | Total Incl O&P |
|---|---|---|---|---|---|---|---|---|---|---|
| 0010 | **VERTICAL, STEEL, ABOVEGROUND FUEL-OIL, STORAGE TANKS** | | | | | | | | | |
| 4000 | Fixed roof oil storage tanks, steel, (1 BBL=42 gal. w/foundation 3'D x 1'W) | | | | | | | | | |
| 4200 | 5,000 barrels | | | | Ea. | | | | 194,000 | 213,500 |
| 4300 | 24,000 barrels | | | | | | | | 333,500 | 367,000 |
| 4500 | 56,000 barrels | | | | | | | | 729,000 | 802,000 |
| 4600 | 110,000 barrels | | | | | | | | 1,060,000 | 1,166,000 |
| 4800 | 143,000 barrels | | | | | | | | 1,250,000 | 1,375,000 |
| 4900 | 224,000 barrels | | | | | | | | 1,360,000 | 1,496,000 |
| 5100 | Floating roof gasoline tanks, steel, 5,000 barrels (w/foundation 3'D x 1'W) | | | | | | | | 204,000 | 225,000 |
| 5200 | 25,000 barrels | | | | | | | | 381,000 | 419,000 |
| 5400 | 55,000 barrels | | | | | | | | 839,000 | 923,000 |
| 5500 | 100,000 barrels | | | | | | | | 1,253,000 | 1,379,000 |
| 5700 | 150,000 barrels | | | | | | | | 1,532,000 | 1,685,000 |
| 5800 | 225,000 barrels | | | | | | | | 2,300,000 | 2,783,000 |

### 23 13 23.16 Horizontal, Stl, Abvgrd Fuel-Oil, Storage Tanks

| | | Crew | Daily Output | Labor-Hours | Unit | Material | Labor | Equipment | Total | Total Incl O&P |
|---|---|---|---|---|---|---|---|---|---|---|
| 0010 | **HORIZONTAL, STEEL, ABOVEGROUND FUEL-OIL, STORAGE TANKS** | | | | | | | | | |
| 3000 | Steel, storage, above ground, including cradles, coating, | | | | | | | | | |
| 3020 | fittings, not including foundation, pumps or piping | | | | | | | | | |
| 3040 | Single wall, 275 gallon | Q-5 | 5 | 3.200 | Ea. | 405 | 156 | | 561 | 680 |
| 3060 | 550 gallon | " | 2.70 | 5.926 | | 2,800 | 289 | | 3,089 | 3,525 |
| 3080 | 1,000 gallon | Q-7 | 5 | 6.400 | | 3,175 | 330 | | 3,505 | 4,000 |
| 3100 | 1,500 gallon | | 4.75 | 6.737 | | 5,025 | 345 | | 5,370 | 6,075 |
| 3120 | 2,000 gallon | | 4.60 | 6.957 | | 5,775 | 360 | | 6,135 | 6,875 |
| 3140 | 5,000 gallon | | 3.20 | 10 | | 6,075 | 515 | | 6,590 | 7,475 |
| 3150 | 10,000 gallon | | 2 | 16 | | 26,400 | 825 | | 27,225 | 30,200 |
| 3160 | 15,000 gallon | | 1.70 | 18.824 | | 24,800 | 970 | | 25,770 | 28,800 |
| 3170 | 20,000 gallon | | 1.45 | 22.069 | | 31,700 | 1,125 | | 32,825 | 36,500 |
| 3180 | 25,000 gallon | | 1.30 | 24.615 | | 32,500 | 1,275 | | 33,775 | 37,600 |
| 3190 | 30,000 gallon | | 1.10 | 29.091 | | 42,300 | 1,500 | | 43,800 | 48,700 |
| 3320 | Double wall, 500 gallon capacity | Q-5 | 2.40 | 6.667 | | 2,200 | 325 | | 2,525 | 2,900 |
| 3330 | 2000 gallon capacity | Q-7 | 4.15 | 7.711 | | 8,250 | 395 | | 8,645 | 9,675 |
| 3340 | 4000 gallon capacity | | 3.60 | 8.889 | | 14,700 | 460 | | 15,160 | 16,900 |
| 3350 | 6000 gallon capacity | | 2.40 | 13.333 | | 17,400 | 685 | | 18,085 | 20,100 |
| 3360 | 8000 gallon capacity | | 2 | 16 | | 22,300 | 825 | | 23,125 | 25,700 |
| 3370 | 10000 gallon capacity | | 1.80 | 17.778 | | 25,000 | 915 | | 25,915 | 28,900 |
| 3380 | 15000 gallon capacity | | 1.50 | 21.333 | | 37,900 | 1,100 | | 39,000 | 43,400 |
| 3390 | 20000 gallon capacity | | 1.30 | 24.615 | | 43,200 | 1,275 | | 44,475 | 49,400 |
| 3400 | 25000 gallon capacity | | 1.15 | 27.826 | | 52,500 | 1,425 | | 53,925 | 59,500 |
| 3410 | 30000 gallon capacity | | 1 | 32 | | 57,500 | 1,650 | | 59,150 | 66,000 |

# 23 21 Hydronic Piping and Pumps

## 23 21 20 – Hydronic HVAC Piping Specialties

| 23 21 20.10 Air Control | | Crew | Daily Output | Labor-Hours | Unit | Material | 2011 Bare Costs Labor | 2011 Bare Costs Equipment | Total | Total Incl O&P |
|---|---|---|---|---|---|---|---|---|---|---|
| 0010 | **AIR CONTROL** | | | | | | | | | |
| 0030 | Air separator, with strainer | | | | | | | | | |
| 0040 | 2" diameter | Q-5 | 6 | 2.667 | Ea. | 865 | 130 | | 995 | 1,150 |
| 0080 | 2-1/2" diameter | | 5 | 3.200 | | 980 | 156 | | 1,136 | 1,300 |
| 0100 | 3" diameter | | 4 | 4 | | 1,500 | 195 | | 1,695 | 1,925 |
| 0120 | 4" diameter | | 3 | 5.333 | | 2,150 | 260 | | 2,410 | 2,775 |
| 0130 | 5" diameter | Q-6 | 3.60 | 6.667 | | 2,750 | 335 | | 3,085 | 3,525 |
| 0140 | 6" diameter | " | 3.40 | 7.059 | | 3,275 | 355 | | 3,630 | 4,125 |

| 23 21 20.18 Automatic Air Vent | | Crew | Daily Output | Labor-Hours | Unit | Material | Labor | Equipment | Total | Total Incl O&P |
|---|---|---|---|---|---|---|---|---|---|---|
| 0010 | **AUTOMATIC AIR VENT** | | | | | | | | | |
| 0020 | Cast iron body, stainless steel internals, float type | | | | | | | | | |
| 0060 | 1/2" NPT inlet, 300 psi | 1 Stpi | 12 | .667 | Ea. | 90.50 | 36 | | 126.50 | 154 |
| 0220 | 3/4" NPT inlet, 250 psi | " | 10 | .800 | | 289 | 43.50 | | 332.50 | 380 |
| 0340 | 1-1/2" NPT inlet, 250 psi | Q-5 | 12 | 1.333 | | 895 | 65 | | 960 | 1,075 |

| 23 21 20.42 Expansion Joints | | Crew | Daily Output | Labor-Hours | Unit | Material | Labor | Equipment | Total | Total Incl O&P |
|---|---|---|---|---|---|---|---|---|---|---|
| 0010 | **EXPANSION JOINTS** | | | | | | | | | |
| 0100 | Bellows type, neoprene cover, flanged spool | | | | | | | | | |
| 0140 | 6" face to face, 1-1/4" diameter | 1 Stpi | 11 | .727 | Ea. | 248 | 39.50 | | 287.50 | 330 |
| 0160 | 1-1/2" diameter | " | 10.60 | .755 | | 248 | 41 | | 289 | 335 |
| 0180 | 2" diameter | Q-5 | 13.30 | 1.203 | | 251 | 58.50 | | 309.50 | 365 |
| 0190 | 2-1/2" diameter | | 12.40 | 1.290 | | 260 | 63 | | 323 | 380 |
| 0200 | 3" diameter | | 11.40 | 1.404 | | 291 | 68.50 | | 359.50 | 420 |
| 0480 | 10" face to face, 2" diameter | | 13 | 1.231 | | 360 | 60 | | 420 | 485 |
| 0500 | 2-1/2" diameter | | 12 | 1.333 | | 380 | 65 | | 445 | 515 |
| 0520 | 3" diameter | | 11 | 1.455 | | 390 | 71 | | 461 | 530 |
| 0540 | 4" diameter | | 8 | 2 | | 440 | 97.50 | | 537.50 | 630 |
| 0560 | 5" diameter | | 7 | 2.286 | | 525 | 111 | | 636 | 745 |
| 0580 | 6" diameter | | 6 | 2.667 | | 545 | 130 | | 675 | 795 |

| 23 21 20.46 Expansion Tanks | | Crew | Daily Output | Labor-Hours | Unit | Material | Labor | Equipment | Total | Total Incl O&P |
|---|---|---|---|---|---|---|---|---|---|---|
| 0010 | **EXPANSION TANKS** | | | | | | | | | |
| 1507 | Underground fuel-oil storage tanks, see Section 23 13 13 | | | | | | | | | |
| 1512 | Tank leak detection systems, see Section 28 33 33.50 | | | | | | | | | |
| 2000 | Steel, liquid expansion, ASME, painted, 15 gallon capacity | Q-5 | 17 | .941 | Ea. | 555 | 46 | | 601 | 685 |
| 2020 | 24 gallon capacity | | 14 | 1.143 | | 595 | 55.50 | | 650.50 | 735 |
| 2040 | 30 gallon capacity | | 12 | 1.333 | | 625 | 65 | | 690 | 780 |
| 2060 | 40 gallon capacity | | 10 | 1.600 | | 730 | 78 | | 808 | 915 |
| 2080 | 60 gallon capacity | | 8 | 2 | | 875 | 97.50 | | 972.50 | 1,100 |
| 2100 | 80 gallon capacity | | 7 | 2.286 | | 935 | 111 | | 1,046 | 1,200 |
| 2120 | 100 gallon capacity | | 6 | 2.667 | | 1,275 | 130 | | 1,405 | 1,600 |
| 3000 | Steel ASME expansion, rubber diaphragm, 19 gal. cap. accept. | | 12 | 1.333 | | 2,225 | 65 | | 2,290 | 2,550 |
| 3020 | 31 gallon capacity | | 8 | 2 | | 2,475 | 97.50 | | 2,572.50 | 2,875 |
| 3040 | 61 gallon capacity | | 6 | 2.667 | | 3,475 | 130 | | 3,605 | 4,025 |
| 3080 | 119 gallon capacity | | 4 | 4 | | 3,750 | 195 | | 3,945 | 4,425 |
| 3100 | 158 gallon capacity | | 3.80 | 4.211 | | 5,200 | 205 | | 5,405 | 6,025 |
| 3140 | 317 gallon capacity | | 2.80 | 5.714 | | 7,850 | 278 | | 8,128 | 9,075 |
| 3180 | 528 gallon capacity | | 2.40 | 6.667 | | 12,800 | 325 | | 13,125 | 14,500 |

| 23 21 20.58 Hydronic Heating Control Valves | | Crew | Daily Output | Labor-Hours | Unit | Material | Labor | Equipment | Total | Total Incl O&P |
|---|---|---|---|---|---|---|---|---|---|---|
| 0010 | **HYDRONIC HEATING CONTROL VALVES** | | | | | | | | | |
| 0100 | Radiator supply, 1/2" diameter | 1 Stpi | 24 | .333 | Ea. | 58 | 18.05 | | 76.05 | 90.50 |
| 0120 | 3/4" diameter | | 20 | .400 | | 59.50 | 21.50 | | 81 | 97.50 |
| 0140 | 1" diameter | | 19 | .421 | | 74.50 | 23 | | 97.50 | 116 |
| 0160 | 1-1/4" diameter | | 15 | .533 | | 107 | 29 | | 136 | 160 |

# 23 21 Hydronic Piping and Pumps

## 23 21 20 – Hydronic HVAC Piping Specialties

| 23 21 20.58 Hydronic Heating Control Valves | Crew | Daily Output | Labor-Hours | Unit | Material | 2011 Bare Costs Labor | Equipment | Total | Total Incl O&P |
|---|---|---|---|---|---|---|---|---|---|
| 0500 | For low pressure steam, add | | | | | 25% | | | | |

### 23 21 20.70 Steam Traps

| | | Crew | Daily Output | Labor-Hours | Unit | Material | Labor | Equipment | Total | Total Incl O&P |
|---|---|---|---|---|---|---|---|---|---|---|
| 0010 | **STEAM TRAPS** | | | | | | | | | |
| 0030 | Cast iron body, threaded | | | | | | | | | |
| 0040 | Inverted bucket | | | | | | | | | |
| 0050 | 1/2" pipe size | 1 Stpi | 12 | .667 | Ea. | 131 | 36 | | 167 | 198 |
| 0070 | 3/4" pipe size | | 10 | .800 | | 228 | 43.50 | | 271.50 | 315 |
| 0100 | 1" pipe size | | 9 | .889 | | 350 | 48 | | 398 | 455 |
| 0120 | 1-1/4" pipe size | | 8 | 1 | | 530 | 54 | | 584 | 665 |
| 1000 | Float & thermostatic, 15 psi | | | | | | | | | |
| 1010 | 3/4" pipe size | 1 Stpi | 16 | .500 | Ea. | 105 | 27 | | 132 | 157 |
| 1020 | 1" pipe size | | 15 | .533 | | 127 | 29 | | 156 | 182 |
| 1040 | 1-1/2" pipe size | | 9 | .889 | | 224 | 48 | | 272 | 320 |
| 1060 | 2" pipe size | | 6 | 1.333 | | 410 | 72 | | 482 | 565 |

### 23 21 20.76 Strainers, Y Type, Bronze Body

| | | Crew | Daily Output | Labor-Hours | Unit | Material | Labor | Equipment | Total | Total Incl O&P |
|---|---|---|---|---|---|---|---|---|---|---|
| 0010 | **STRAINERS, Y TYPE, BRONZE BODY** | | | | | | | | | |
| 0050 | Screwed, 150 lb., 1/4" pipe size | 1 Stpi | 24 | .333 | Ea. | 17.30 | 18.05 | | 35.35 | 46 |
| 0070 | 3/8" pipe size | | 24 | .333 | | 23 | 18.05 | | 41.05 | 52 |
| 0100 | 1/2" pipe size | | 20 | .400 | | 23 | 21.50 | | 44.50 | 57.50 |
| 0140 | 1" pipe size | | 17 | .471 | | 30.50 | 25.50 | | 56 | 72 |
| 0160 | 1-1/2" pipe size | | 14 | .571 | | 66 | 31 | | 97 | 119 |
| 0180 | 2" pipe size | | 13 | .615 | | 88.50 | 33.50 | | 122 | 147 |
| 0182 | 3" pipe size | | 12 | .667 | | 670 | 36 | | 706 | 790 |
| 0200 | 300 lb., 2-1/2" pipe size | Q-5 | 17 | .941 | | 420 | 46 | | 466 | 535 |
| 0220 | 3" pipe size | | 16 | 1 | | 835 | 48.50 | | 883.50 | 995 |
| 0240 | 4" pipe size | | 15 | 1.067 | | 1,900 | 52 | | 1,952 | 2,175 |
| 0500 | For 300 lb rating 1/4" thru 2", add | | | | | 15% | | | | |
| 1000 | Flanged, 150 lb., 1-1/2" pipe size | 1 Stpi | 11 | .727 | Ea. | 435 | 39.50 | | 474.50 | 540 |
| 1020 | 2" pipe size | " | 8 | 1 | | 590 | 54 | | 644 | 730 |
| 1030 | 2-1/2" pipe size | Q-5 | 5 | 3.200 | | 875 | 156 | | 1,031 | 1,200 |
| 1040 | 3" pipe size | | 4.50 | 3.556 | | 1,075 | 173 | | 1,248 | 1,425 |
| 1060 | 4" pipe size | | 3 | 5.333 | | 1,625 | 260 | | 1,885 | 2,200 |
| 1100 | 6" pipe size | Q-6 | 3 | 8 | | 3,125 | 405 | | 3,530 | 4,025 |
| 1106 | 8" pipe size | " | 2.60 | 9.231 | | 3,425 | 465 | | 3,890 | 4,475 |
| 1500 | For 300 lb rating, add | | | | | 40% | | | | |

### 23 21 20.78 Strainers, Y Type, Iron Body

| | | Crew | Daily Output | Labor-Hours | Unit | Material | Labor | Equipment | Total | Total Incl O&P |
|---|---|---|---|---|---|---|---|---|---|---|
| 0010 | **STRAINERS, Y TYPE, IRON BODY** | | | | | | | | | |
| 0050 | Screwed, 250 lb., 1/4" pipe size | 1 Stpi | 20 | .400 | Ea. | 9.60 | 21.50 | | 31.10 | 43 |
| 0070 | 3/8" pipe size | | 20 | .400 | | 9.60 | 21.50 | | 31.10 | 43 |
| 0100 | 1/2" pipe size | | 20 | .400 | | 9.60 | 21.50 | | 31.10 | 43 |
| 0140 | 1" pipe size | | 16 | .500 | | 15.85 | 27 | | 42.85 | 58 |
| 0160 | 1-1/2" pipe size | | 12 | .667 | | 26 | 36 | | 62 | 82.50 |
| 0180 | 2" pipe size | | 8 | 1 | | 39 | 54 | | 93 | 124 |
| 0220 | 3" pipe size | Q-5 | 11 | 1.455 | | 251 | 71 | | 322 | 380 |
| 0240 | 4" pipe size | " | 5 | 3.200 | | 425 | 156 | | 581 | 700 |
| 0500 | For galvanized body, add | | | | | 50% | | | | |
| 1000 | Flanged, 125 lb., 1-1/2" pipe size | 1 Stpi | 11 | .727 | Ea. | 135 | 39.50 | | 174.50 | 207 |
| 1020 | 2" pipe size | " | 8 | 1 | | 142 | 54 | | 196 | 237 |
| 1040 | 3" pipe size | Q-5 | 4.50 | 3.556 | | 133 | 173 | | 306 | 405 |
| 1060 | 4" pipe size | " | 3 | 5.333 | | 242 | 260 | | 502 | 655 |
| 1080 | 5" pipe size | Q-6 | 3.40 | 7.059 | | 380 | 355 | | 735 | 945 |
| 1100 | 6" pipe size | " | 3 | 8 | | 460 | 405 | | 865 | 1,125 |

# 23 21 Hydronic Piping and Pumps

## 23 21 20 – Hydronic HVAC Piping Specialties

### 23 21 20.78 Strainers, Y Type, Iron Body

| | | Crew | Daily Output | Labor-Hours | Unit | Material | 2011 Bare Costs Labor | Equipment | Total | Total Incl O&P |
|---|---|---|---|---|---|---|---|---|---|---|
| 1500 | For 250 lb rating, add | | | | | 20% | | | | |
| 2000 | For galvanized body, add | | | | | 50% | | | | |
| 2500 | For steel body, add | | | | | 40% | | | | |

### 23 21 20.88 Venturi Flow

| | | Crew | Daily Output | Labor-Hours | Unit | Material | 2011 Bare Costs Labor | Equipment | Total | Total Incl O&P |
|---|---|---|---|---|---|---|---|---|---|---|
| 0010 | **VENTURI FLOW**, Measuring device | | | | | | | | | |
| 0050 | 1/2" diameter | 1 Stpi | 24 | .333 | Ea. | 281 | 18.05 | | 299.05 | 335 |
| 0120 | 1" diameter | | 19 | .421 | | 276 | 23 | | 299 | 340 |
| 0140 | 1-1/4" diameter | | 15 | .533 | | 340 | 29 | | 369 | 420 |
| 0160 | 1-1/2" diameter | | 13 | .615 | | 355 | 33.50 | | 388.50 | 440 |
| 0180 | 2" diameter | ↓ | 11 | .727 | | 365 | 39.50 | | 404.50 | 465 |
| 0220 | 3" diameter | Q-5 | 14 | 1.143 | | 515 | 55.50 | | 570.50 | 655 |
| 0240 | 4" diameter | " | 11 | 1.455 | | 775 | 71 | | 846 | 955 |
| 0280 | 6" diameter | Q-6 | 3.50 | 6.857 | | 1,125 | 345 | | 1,470 | 1,775 |
| 0500 | For meter, add | | | | ↓ | 2,125 | | | 2,125 | 2,350 |

## 23 21 23 – Hydronic Pumps

### 23 21 23.13 In-Line Centrifugal Hydronic Pumps

| | | Crew | Daily Output | Labor-Hours | Unit | Material | 2011 Bare Costs Labor | Equipment | Total | Total Incl O&P |
|---|---|---|---|---|---|---|---|---|---|---|
| 0010 | **IN-LINE CENTRIFUGAL HYDRONIC PUMPS** | | | | | | | | | |
| 0600 | Bronze, sweat connections, 1/40 HP, in line | | | | | | | | | |
| 0640 | 3/4" size | Q-1 | 16 | 1 | Ea. | 192 | 48.50 | | 240.50 | 283 |
| 1000 | Flange connection, 3/4" to 1-1/2" size | | | | | | | | | |
| 1040 | 1/12 HP | Q-1 | 6 | 2.667 | Ea. | 495 | 129 | | 624 | 735 |
| 1060 | 1/8 HP | | 6 | 2.667 | | 855 | 129 | | 984 | 1,125 |
| 1100 | 1/3 HP | | 6 | 2.667 | | 955 | 129 | | 1,084 | 1,250 |
| 1140 | 2" size, 1/6 HP | | 5 | 3.200 | | 1,025 | 154 | | 1,179 | 1,350 |
| 1180 | 2-1/2" size, 1/4 HP | ↓ | 5 | 3.200 | ↓ | 1,525 | 154 | | 1,679 | 1,925 |
| 2000 | Cast iron, flange connection | | | | | | | | | |
| 2040 | 3/4" to 1-1/2" size, in line, 1/12 HP | Q-1 | 6 | 2.667 | Ea. | 320 | 129 | | 449 | 540 |
| 2100 | 1/3 HP | | 6 | 2.667 | | 595 | 129 | | 724 | 845 |
| 2140 | 2" size, 1/6 HP | | 5 | 3.200 | | 655 | 154 | | 809 | 950 |
| 2180 | 2-1/2" size, 1/4 HP | | 5 | 3.200 | | 830 | 154 | | 984 | 1,150 |
| 2220 | 3" size, 1/4 HP | ↓ | 4 | 4 | ↓ | 840 | 193 | | 1,033 | 1,225 |
| 2600 | For non-ferrous impeller, add | | | | | 3% | | | | |

## 23 21 29 – Automatic Condensate Pump Units

### 23 21 29.10 Condensate Removal Pump System

| | | Crew | Daily Output | Labor-Hours | Unit | Material | 2011 Bare Costs Labor | Equipment | Total | Total Incl O&P |
|---|---|---|---|---|---|---|---|---|---|---|
| 0010 | **CONDENSATE REMOVAL PUMP SYSTEM** | | | | | | | | | |
| 0020 | Pump with 1 gal. ABS tank | G | | | | | | | | |
| 0100 | 115 V | G | | | | | | | | |
| 0120 | 1/50 HP, 200 GPH | G | 1 Stpi | 12 | .667 | Ea. | 176 | 36 | | 212 | 248 |
| 0140 | 1/18 HP, 270 GPH | G | | 10 | .800 | | 183 | 43.50 | | 226.50 | 266 |
| 0160 | 1/5 HP, 450 GPH | G | ↓ | 8 | 1 | ↓ | 420 | 54 | | 474 | 545 |
| 0200 | 230 V | G | | | | | | | | |
| 0240 | 1/18 HP, 270 GPH | G | 1 Stpi | 10 | .800 | Ea. | 195 | 43.50 | | 238.50 | 280 |
| 0260 | 1/5 HP, 450 GPH | G | " | 8 | 1 | " | 465 | 54 | | 519 | 595 |

# 23 22 Steam and Condensate Piping and Pumps

## 23 22 13 – Steam and Condensate Heating Piping

| 23 22 13.23 Aboveground Steam and Condensate Piping | Crew | Daily Output | Labor-Hours | Unit | Material | 2011 Bare Costs Labor | Equipment | Total | Total Incl O&P |
|---|---|---|---|---|---|---|---|---|---|
| 0010 **ABOVEGROUND STEAM AND CONDENSATE HEATING PIPING** | | | | | | | | | |
| 0020 Condensate meter | | | | | | | | | |
| 0100 500 lb. per hour | 1 Stpi | 14 | .571 | Ea. | 2,350 | 31 | | 2,381 | 2,650 |
| 0140 1500 lb. per hour | " | 7 | 1.143 | " | 2,550 | 62 | | 2,612 | 2,900 |

## 23 22 23 – Steam Condensate Pumps

### 23 22 23.10 Condensate Return System

| | Crew | Daily Output | Labor-Hours | Unit | Material | 2011 Bare Costs Labor | Equipment | Total | Total Incl O&P |
|---|---|---|---|---|---|---|---|---|---|
| 0010 **CONDENSATE RETURN SYSTEM** | | | | | | | | | |
| 2000 Simplex | | | | | | | | | |
| 2010 With pump, motor, CI receiver, float switch | | | | | | | | | |
| 2020 3/4 HP, 15 GPM | Q-1 | 1.80 | 8.889 | Ea. | 5,550 | 430 | | 5,980 | 6,750 |
| 2100 Duplex | | | | | | | | | |
| 2110 With 2 pumps and motors, CI receiver, float switch, alternator | | | | | | | | | |
| 2120 3/4 HP, 15 GPM, 15 gal. CI rcvr | Q-1 | 1.40 | 11.429 | Ea. | 6,050 | 550 | | 6,600 | 7,475 |
| 2130 1 HP, 25 GPM | | 1.20 | 13.333 | | 7,250 | 645 | | 7,895 | 8,925 |
| 2140 1-1/2 HP, 45 GPM | | 1 | 16 | | 8,425 | 770 | | 9,195 | 10,400 |
| 2150 1-1/2 HP, 60 GPM | | 1 | 16 | | 9,450 | 770 | | 10,220 | 11,600 |

# 23 31 HVAC Ducts and Casings

## 23 31 13 – Metal Ducts

### 23 31 13.13 Rectangular Metal Ducts

| | Crew | Daily Output | Labor-Hours | Unit | Material | 2011 Bare Costs Labor | Equipment | Total | Total Incl O&P |
|---|---|---|---|---|---|---|---|---|---|
| 0010 **RECTANGULAR METAL DUCTS**      R233100-20 | | | | | | | | | |
| 0020 Fabricated rectangular, includes fittings, joints, supports, | | | | | | | | | |
| 0030 allowance for flexible connections, no insulation | | | | | | | | | |
| 0031 NOTE: Fabrication and installation are combined | | | | | | | | | |
| 0040 as LABOR cost. Approx. 25% fittings assumed. | | | | | | | | | |
| 0050 Add to labor for elevated installation | | | | | | | | | |
| 0051 of fabricated ductwork | | | | | | | | | |
| 0052 10' to 15' high | | | | | | 6% | | | |
| 0053 15' to 20' high | | | | | | 12% | | | |
| 0054 20' to 25' high | | | | | | 15% | | | |
| 0055 25' to 30' high | | | | | | 21% | | | |
| 0056 30' to 35' high | | | | | | 24% | | | |
| 0057 35' to 40' high | | | | | | 30% | | | |
| 0058 Over 40' high | | | | | | 33% | | | |
| 0072 For duct insulation see Line 23 07 13.10 3000 | | | | | | | | | |
| 0100 Aluminum, alloy 3003-H14, under 100 lb. | Q-10 | 75 | .320 | Lb. | 3.51 | 15.40 | | 18.91 | 27.50 |
| 0110 100 to 500 lb. | | 80 | .300 | | 2.31 | 14.45 | | 16.76 | 24.50 |
| 0120 500 to 1,000 lb. | | 95 | .253 | | 2.19 | 12.20 | | 14.39 | 21 |
| 0140 1,000 to 2,000 lb. | | 120 | .200 | | 2.11 | 9.65 | | 11.76 | 16.90 |
| 0150 2,000 to 5,000 lb. | | 130 | .185 | | 2.11 | 8.90 | | 11.01 | 15.75 |
| 0160 Over 5,000 lb. | | 145 | .166 | | 2.11 | 8 | | 10.11 | 14.35 |
| 0500 Galvanized steel, under 200 lb. | | 235 | .102 | | .66 | 4.92 | | 5.58 | 8.20 |
| 0520 200 to 500 lb. | | 245 | .098 | | .65 | 4.72 | | 5.37 | 7.85 |
| 0540 500 to 1,000 lb. | | 255 | .094 | | .64 | 4.54 | | 5.18 | 7.55 |
| 0560 1,000 to 2,000 lb. | | 265 | .091 | | .63 | 4.37 | | 5 | 7.30 |
| 0570 2,000 to 5,000 lb. | | 275 | .087 | | .62 | 4.21 | | 4.83 | 7.05 |
| 0580 Over 5,000 lb. | | 285 | .084 | | .62 | 4.06 | | 4.68 | 6.85 |
| 1000 Stainless steel, type 304, under 100 lb. | | 165 | .145 | | 2.40 | 7 | | 9.40 | 13.25 |
| 1020 100 to 500 lb. | | 175 | .137 | | 4.02 | 6.60 | | 10.62 | 14.40 |
| 1030 500 to 1,000 lb. | | 190 | .126 | | 3.68 | 6.10 | | 9.78 | 13.25 |

# 23 31 HVAC Ducts and Casings

## 23 31 13 – Metal Ducts

### 23 31 13.13 Rectangular Metal Ducts

| | | Crew | Daily Output | Labor-Hours | Unit | Material | 2011 Bare Costs Labor | Equipment | Total | Total Incl O&P |
|---|---|---|---|---|---|---|---|---|---|---|
| 1040 | 1,000 to 2,000 lb. | Q-10 | 200 | .120 | Lb. | 3.65 | 5.80 | | 9.45 | 12.75 |
| 1050 | 2,000 to 5,000 lb. | | 225 | .107 | | 3.31 | 5.15 | | 8.46 | 11.40 |
| 1060 | Over 5,000 lb. | ↓ | 235 | .102 | | 3.26 | 4.92 | | 8.18 | 11.05 |
| 1100 | For medium pressure ductwork, add | | | | | | 15% | | | |
| 1200 | For high pressure ductwork, add | | | | | | 40% | | | |
| 1210 | For welded ductwork, add | | | | | | 85% | | | |
| 1220 | For 30% fittings, add | | | | | | 11% | | | |
| 1224 | For 40% fittings, add | | | | | | 34% | | | |
| 1228 | For 50% fittings, add | | | | | | 56% | | | |
| 1232 | For 60% fittings, add | | | | | | 79% | | | |
| 1236 | For 70% fittings, add | | | | | | 101% | | | |
| 1240 | For 80% fittings, add | | | | | | 124% | | | |
| 1244 | For 90% fittings, add | | | | | | 147% | | | |
| 1248 | For 100% fittings, add | | | | | | 169% | | | |
| 1252 | Note: Fittings add includes time for detailing and installation. | | | | | | | | | |

### 23 31 13.19 Metal Duct Fittings

| | | Crew | Daily Output | Labor-Hours | Unit | Material | 2011 Bare Costs Labor | Equipment | Total | Total Incl O&P |
|---|---|---|---|---|---|---|---|---|---|---|
| 0010 | **METAL DUCT FITTINGS** | | | | | | | | | |
| 0050 | Air extractors, 12" x 4" | 1 Shee | 24 | .333 | Ea. | 16.10 | 17.20 | | 33.30 | 44 |
| 0100 | 8" x 6" | | 22 | .364 | | 16.10 | 18.80 | | 34.90 | 46.50 |
| 0200 | 20" x 8" | | 16 | .500 | | 36.50 | 26 | | 62.50 | 79 |
| 0280 | 24" x 12" | | 10 | .800 | ↓ | 49.50 | 41.50 | | 91 | 117 |
| 2000 | Fabrics for flexible connections, with metal edge | | 100 | .080 | L.F. | 2.13 | 4.13 | | 6.26 | 8.60 |
| 2100 | Without metal edge | ↓ | 160 | .050 | " | 1.53 | 2.58 | | 4.11 | 5.60 |

## 23 31 16 – Nonmetal Ducts

### 23 31 16.13 Fibrous-Glass Ducts

| | | Crew | Daily Output | Labor-Hours | Unit | Material | 2011 Bare Costs Labor | Equipment | Total | Total Incl O&P |
|---|---|---|---|---|---|---|---|---|---|---|
| 0010 | **FIBROUS-GLASS DUCTS** | R233100-20 | | | | | | | | |
| 3490 | Rigid fiberglass duct board, foil reinf. kraft facing | | | | | | | | | |
| 3500 | Rectangular, 1" thick, alum. faced, (FRK), std. weight | Q-10 | 350 | .069 | SF Surf | .77 | 3.31 | | 4.08 | 5.85 |

# 23 33 Air Duct Accessories

## 23 33 13 – Dampers

### 23 33 13.13 Volume-Control Dampers

| | | Crew | Daily Output | Labor-Hours | Unit | Material | 2011 Bare Costs Labor | Equipment | Total | Total Incl O&P |
|---|---|---|---|---|---|---|---|---|---|---|
| 0010 | **VOLUME-CONTROL DAMPERS** | | | | | | | | | |
| 5990 | Multi-blade dampers, opposed blade, 8" x 6" | 1 Shee | 24 | .333 | Ea. | 20 | 17.20 | | 37.20 | 48 |
| 5994 | 8" x 8" | | 22 | .364 | | 21 | 18.80 | | 39.80 | 51.50 |
| 5996 | 10" x 10" | | 21 | .381 | | 24.50 | 19.70 | | 44.20 | 57 |
| 6000 | 12" x 12" | | 21 | .381 | | 27.50 | 19.70 | | 47.20 | 60 |
| 6020 | 12" x 18" | | 18 | .444 | | 36.50 | 23 | | 59.50 | 75 |
| 6030 | 14" x 10" | | 20 | .400 | | 26.50 | 20.50 | | 47 | 60.50 |
| 6031 | 14" x 14" | | 17 | .471 | | 32.50 | 24.50 | | 57 | 72.50 |
| 6033 | 16" x 12" | | 17 | .471 | | 32.50 | 24.50 | | 57 | 72.50 |
| 6035 | 16" x 16" | | 16 | .500 | | 40.50 | 26 | | 66.50 | 83.50 |
| 6037 | 18" x 16" | | 15 | .533 | | 44.50 | 27.50 | | 72 | 90.50 |
| 6038 | 18" x 18" | | 15 | .533 | | 48 | 27.50 | | 75.50 | 94.50 |
| 6070 | 20" x 16" | | 14 | .571 | | 48 | 29.50 | | 77.50 | 97.50 |
| 6072 | 20" x 20" | | 13 | .615 | | 57.50 | 32 | | 89.50 | 112 |
| 6074 | 22" x 18" | | 14 | .571 | | 57.50 | 29.50 | | 87 | 108 |
| 6076 | 24" x 16" | | 11 | .727 | | 57 | 37.50 | | 94.50 | 120 |
| 6078 | 24" x 20" | | 8 | 1 | | 67 | 51.50 | | 118.50 | 152 |
| 6080 | 24" x 24" | ↓ | 8 | 1 | ↓ | 78.50 | 51.50 | | 130 | 165 |

# 23 33 Air Duct Accessories

## 23 33 13 – Dampers

### 23 33 13.13 Volume-Control Dampers

| | | Crew | Daily Output | Labor-Hours | Unit | Material | 2011 Bare Costs Labor | Equipment | Total | Total Incl O&P |
|---|---|---|---|---|---|---|---|---|---|---|
| 6110 | 26" x 26" | 1 Shee | 6 | 1.333 | Ea. | 86.50 | 69 | | 155.50 | 199 |
| 6133 | 30" x 30" | Q-9 | 6.60 | 2.424 | | 124 | 113 | | 237 | 305 |
| 6135 | 32" x 32" | | 6.40 | 2.500 | | 144 | 116 | | 260 | 335 |
| 6180 | 48" x 36" | ↓ | 5.60 | 2.857 | ↓ | 237 | 133 | | 370 | 460 |
| 8000 | Multi-blade dampers, parallel blade | | | | | | | | | |
| 8100 | 8" x 8" | 1 Shee | 24 | .333 | Ea. | 73 | 17.20 | | 90.20 | 107 |
| 8140 | 16" x 10" | | 20 | .400 | | 94 | 20.50 | | 114.50 | 134 |
| 8200 | 24" x 16" | | 11 | .727 | | 120 | 37.50 | | 157.50 | 188 |
| 8260 | 30" x 18" | ↓ | 7 | 1.143 | ↓ | 164 | 59 | | 223 | 270 |

### 23 33 13.16 Fire Dampers

| | | Crew | Daily Output | Labor-Hours | Unit | Material | 2011 Bare Costs Labor | Equipment | Total | Total Incl O&P |
|---|---|---|---|---|---|---|---|---|---|---|
| 0010 | **FIRE DAMPERS** | | | | | | | | | |
| 3000 | Fire damper, curtain type, 1-1/2 hr rated, vertical, 6" x 6" | 1 Shee | 24 | .333 | Ea. | 21.50 | 17.20 | | 38.70 | 50 |
| 3020 | 8" x 6" | | 22 | .364 | | 21.50 | 18.80 | | 40.30 | 52.50 |
| 3240 | 16" x 14" | | 18 | .444 | | 29.50 | 23 | | 52.50 | 67 |
| 3400 | 24" x 20" | ↓ | 8 | 1 | ↓ | 42 | 51.50 | | 93.50 | 124 |

### 23 33 13.28 Splitter Damper Assembly

| | | Crew | Daily Output | Labor-Hours | Unit | Material | 2011 Bare Costs Labor | Equipment | Total | Total Incl O&P |
|---|---|---|---|---|---|---|---|---|---|---|
| 0009 | **SPLITTER DAMPER ASSEMBLY** | | | | | | | | | |
| 0010 | Self locking, 1' rod | 1 Shee | 24 | .333 | Ea. | 21.50 | 17.20 | | 38.70 | 50 |
| 7020 | 3' rod | | 22 | .364 | | 28 | 18.80 | | 46.80 | 59.50 |
| 7040 | 4' rod | | 20 | .400 | | 31.50 | 20.50 | | 52 | 65.50 |
| 7060 | 6' rod | ↓ | 18 | .444 | ↓ | 38 | 23 | | 61 | 76.50 |

## 23 33 19 – Duct Silencers

### 23 33 19.10 Duct Silencers

| | | Crew | Daily Output | Labor-Hours | Unit | Material | 2011 Bare Costs Labor | Equipment | Total | Total Incl O&P |
|---|---|---|---|---|---|---|---|---|---|---|
| 0009 | **DUCT SILENCERS** | | | | | | | | | |
| 0010 | Silencers, noise control for air flow, duct | | | | MCFM | 55.50 | | | 55.50 | 61 |

## 23 33 33 – Duct-Mounting Access Doors

### 23 33 33.13 Duct Access Doors

| | | Crew | Daily Output | Labor-Hours | Unit | Material | 2011 Bare Costs Labor | Equipment | Total | Total Incl O&P |
|---|---|---|---|---|---|---|---|---|---|---|
| 0010 | **DUCT ACCESS DOORS** | | | | | | | | | |
| 1000 | Duct access door, insulated, 6" x 6" | 1 Shee | 14 | .571 | Ea. | 15 | 29.50 | | 44.50 | 61 |
| 1020 | 10" x 10" | | 11 | .727 | | 18.15 | 37.50 | | 55.65 | 77 |
| 1040 | 12" x 12" | | 10 | .800 | | 19.40 | 41.50 | | 60.90 | 84 |
| 1050 | 12" x 18" | | 9 | .889 | | 34.50 | 46 | | 80.50 | 108 |
| 1070 | 18" x 18" | | 8 | 1 | | 34.50 | 51.50 | | 86 | 116 |
| 1074 | 24" x 18" | ↓ | 8 | 1 | ↓ | 41 | 51.50 | | 92.50 | 124 |

## 23 33 46 – Flexible Ducts

### 23 33 46.10 Flexible Air Ducts

| | | | Crew | Daily Output | Labor-Hours | Unit | Material | 2011 Bare Costs Labor | Equipment | Total | Total Incl O&P |
|---|---|---|---|---|---|---|---|---|---|---|---|
| 0010 | **FLEXIBLE AIR DUCTS** | R233100-20 | | | | | | | | | |
| 1280 | Add to labor for elevated installation | | | | | | | | | | |
| 1282 | of prefabricated (purchased) ductwork | | | | | | | | | | |
| 1283 | 10' to 15' high | | | | | | | 10% | | | |
| 1284 | 15' to 20' high | | | | | | | 20% | | | |
| 1285 | 20' to 25' high | | | | | | | 25% | | | |
| 1286 | 25' to 30' high | | | | | | | 35% | | | |
| 1287 | 30' to 35' high | | | | | | | 40% | | | |
| 1288 | 35' to 40' high | | | | | | | 50% | | | |
| 1289 | Over 40' high | | | | | | | 55% | | | |
| 1300 | Flexible, coated fiberglass fabric on corr. resist. metal helix | | | | | | | | | | |
| 1400 | pressure to 12" (WG) UL-181 | | | | | | | | | | |
| 1500 | Non-insulated, 3" diameter | | Q-9 | 400 | .040 | L.F. | 1.02 | 1.86 | | 2.88 | 3.93 |
| 1540 | 5" diameter | | ↓ | 320 | .050 | ↓ | 1.25 | 2.32 | | 3.57 | 4.89 |

# 23 33 Air Duct Accessories

## 23 33 46 – Flexible Ducts

### 23 33 46.10 Flexible Air Ducts

| | | Crew | Daily Output | Labor-Hours | Unit | Material | 2011 Bare Costs Labor | Equipment | Total | Total Incl O&P |
|---|---|---|---|---|---|---|---|---|---|---|
| 1560 | 6" diameter | Q-9 | 280 | .057 | L.F. | 1.53 | 2.66 | | 4.19 | 5.70 |
| 1580 | 7" diameter | | 240 | .067 | | 1.82 | 3.10 | | 4.92 | 6.70 |
| 1600 | 8" diameter | | 200 | .080 | | 2.10 | 3.72 | | 5.82 | 7.90 |
| 1640 | 10" diameter | | 160 | .100 | | 2.61 | 4.65 | | 7.26 | 9.90 |
| 1660 | 12" diameter | | 120 | .133 | | 3.17 | 6.20 | | 9.37 | 12.85 |
| 1900 | Insulated, 1" thick, PE jacket, 3" diameter [G] | | 380 | .042 | | 2.15 | 1.96 | | 4.11 | 5.35 |
| 1910 | 4" diameter [G] | | 340 | .047 | | 2.15 | 2.19 | | 4.34 | 5.70 |
| 1920 | 5" diameter [G] | | 300 | .053 | | 2.15 | 2.48 | | 4.63 | 6.10 |
| 1940 | 6" diameter [G] | | 260 | .062 | | 2.38 | 2.86 | | 5.24 | 6.95 |
| 1960 | 7" diameter [G] | | 220 | .073 | | 2.78 | 3.38 | | 6.16 | 8.15 |
| 1980 | 8" diameter [G] | | 180 | .089 | | 3 | 4.13 | | 7.13 | 9.55 |
| 2020 | 10" diameter [G] | | 140 | .114 | | 3.51 | 5.30 | | 8.81 | 11.90 |
| 2040 | 12" diameter [G] | | 100 | .160 | | 4.36 | 7.45 | | 11.81 | 16.05 |

## 23 33 53 – Duct Liners

### 23 33 53.10 Duct Liner Board

| | | | Crew | Daily Output | Labor-Hours | Unit | Material | 2011 Bare Costs Labor | Equipment | Total | Total Incl O&P |
|---|---|---|---|---|---|---|---|---|---|---|---|
| 0010 | **DUCT LINER BOARD** | | | | | | | | | | |
| 3490 | Board type, fiberglass liner, 3 lb. density | | | | | | | | | | |
| 3500 | Fire resistant, black pigmented, 1 side | | | | | | | | | | |
| 3520 | 1" thick | [G] | Q-14 | 150 | .107 | S.F. | .77 | 4.50 | | 5.27 | 7.80 |
| 3540 | 1-1/2" thick | [G] | | 130 | .123 | | .97 | 5.20 | | 6.17 | 9.10 |
| 3560 | 2" thick | [G] | | 120 | .133 | | 1.18 | 5.65 | | 6.83 | 10 |
| 3600 | FSK vapor barrier | | | | | | | | | | |
| 3620 | 1" thick | [G] | Q-14 | 150 | .107 | S.F. | .77 | 4.50 | | 5.27 | 7.80 |
| 3630 | 1-1/2" thick | [G] | | 130 | .123 | | .97 | 5.20 | | 6.17 | 9.10 |
| 3640 | 2" thick | [G] | | 120 | .133 | | 1.18 | 5.65 | | 6.83 | 10 |
| 3680 | No finish | | | | | | | | | | |
| 3700 | 1" thick | [G] | Q-14 | 170 | .094 | S.F. | .40 | 3.97 | | 4.37 | 6.60 |
| 3710 | 1-1/2" thick | [G] | | 140 | .114 | | .60 | 4.82 | | 5.42 | 8.10 |
| 3720 | 2" thick | [G] | | 130 | .123 | | .81 | 5.20 | | 6.01 | 8.95 |
| 3940 | Board type, non-fibrous foam | | | | | | | | | | |
| 3950 | Temperature, bacteria and fungi resistant | | | | | | | | | | |
| 3960 | 1" thick | [G] | Q-14 | 150 | .107 | S.F. | 2.25 | 4.50 | | 6.75 | 9.45 |
| 3970 | 1-1/2" thick | [G] | | 130 | .123 | | 3 | 5.20 | | 8.20 | 11.35 |
| 3980 | 2" thick | [G] | | 120 | .133 | | 3.60 | 5.65 | | 9.25 | 12.65 |

# 23 34 HVAC Fans

## 23 34 13 – Axial HVAC Fans

### 23 34 13.10 Axial Flow HVAC Fans

| | | | Crew | Daily Output | Labor-Hours | Unit | Material | 2011 Bare Costs Labor | Equipment | Total | Total Incl O&P |
|---|---|---|---|---|---|---|---|---|---|---|---|
| 0010 | **AXIAL FLOW HVAC FANS** | R233400-10 | | | | | | | | | |
| 0020 | Air conditioning and process air handling | | | | | | | | | | |
| 0030 | Axial flow, compact, low sound, 2.5" S.P. | | | | | | | | | | |
| 0050 | 3,800 CFM, 5 HP | | Q-20 | 3.40 | 5.882 | Ea. | 4,625 | 278 | | 4,903 | 5,525 |
| 0080 | 6,400 CFM, 5 HP | | | 2.80 | 7.143 | | 5,175 | 335 | | 5,510 | 6,200 |
| 0100 | 10,500 CFM, 7-1/2 HP | | | 2.40 | 8.333 | | 6,450 | 395 | | 6,845 | 7,700 |
| 0120 | 15,600 CFM, 10 HP | | | 1.60 | 12.500 | | 8,125 | 590 | | 8,715 | 9,825 |
| 1500 | Vaneaxial, low pressure, 2000 CFM, 1/2 HP | | | 3.60 | 5.556 | | 1,975 | 262 | | 2,237 | 2,575 |
| 1520 | 4,000 CFM, 1 HP | | | 3.20 | 6.250 | | 2,275 | 295 | | 2,570 | 2,950 |
| 1540 | 8,000 CFM, 2 HP | | | 2.80 | 7.143 | | 2,900 | 335 | | 3,235 | 3,700 |

# 23 34 HVAC Fans

## 23 34 14 – Blower HVAC Fans

### 23 34 14.10 Blower Type HVAC Fans

| 23 34 14.10 Blower Type HVAC Fans | Crew | Daily Output | Labor-Hours | Unit | Material | 2011 Bare Costs Labor | Equipment | Total | Total Incl O&P |
|---|---|---|---|---|---|---|---|---|---|
| 0010 **BLOWER TYPE HVAC FANS** | | | | | | | | | |
| 2500 Ceiling fan, right angle, extra quiet, 0.10" S.P. | | | | | | | | | |
| 2520     95 CFM | Q-20 | 20 | 1 | Ea. | 252 | 47 | | 299 | 350 |
| 2540     210 CFM | | 19 | 1.053 | | 298 | 49.50 | | 347.50 | 400 |
| 2560     385 CFM | | 18 | 1.111 | | 380 | 52.50 | | 432.50 | 495 |
| 2580     885 CFM | | 16 | 1.250 | | 745 | 59 | | 804 | 910 |
| 2600     1,650 CFM | | 13 | 1.538 | | 1,025 | 72.50 | | 1,097.50 | 1,225 |
| 2620     2,960 CFM | | 11 | 1.818 | | 1,375 | 86 | | 1,461 | 1,625 |
| 2640     For wall or roof cap, add | 1 Shee | 16 | .500 | | 252 | 26 | | 278 | 315 |
| 2660     For straight thru fan, add | | | | | 10% | | | | |
| 2680     For speed control switch, add | 1 Elec | 16 | .500 | | 137 | 25 | | 162 | 189 |
| 7500 Utility set, steel construction, pedestal, 1/4" S.P. | | | | | | | | | |
| 7520     Direct drive, 150 CFM, 1/8 HP | Q-20 | 6.40 | 3.125 | Ea. | 775 | 148 | | 923 | 1,075 |
| 7540     485 CFM, 1/6 HP | | 5.80 | 3.448 | | 980 | 163 | | 1,143 | 1,325 |
| 7560     1950 CFM, 1/2 HP | | 4.80 | 4.167 | | 1,150 | 197 | | 1,347 | 1,550 |
| 7580     2410 CFM, 3/4 HP | | 4.40 | 4.545 | | 2,125 | 215 | | 2,340 | 2,650 |
| 7600     3328 CFM, 1-1/2 HP | | 3 | 6.667 | | 2,350 | 315 | | 2,665 | 3,075 |
| 7680     V-belt drive, drive cover, 3 phase | | | | | | | | | |
| 7700     800 CFM, 1/4 HP | Q-20 | 6 | 3.333 | Ea. | 855 | 157 | | 1,012 | 1,175 |
| 7720     1,300 CFM, 1/3 HP | | 5 | 4 | | 895 | 189 | | 1,084 | 1,275 |
| 7740     2,000 CFM, 1 HP | | 4.60 | 4.348 | | 1,050 | 205 | | 1,255 | 1,475 |
| 7760     2,900 CFM, 3/4 HP | | 4.20 | 4.762 | | 1,425 | 225 | | 1,650 | 1,925 |

## 23 34 16 – Centrifugal HVAC Fans

### 23 34 16.10 Centrifugal Type HVAC Fans

| 23 34 16.10 Centrifugal Type HVAC Fans | Crew | Daily Output | Labor-Hours | Unit | Material | 2011 Bare Costs Labor | Equipment | Total | Total Incl O&P |
|---|---|---|---|---|---|---|---|---|---|
| 0010 **CENTRIFUGAL TYPE HVAC FANS** | | | | | | | | | |
| 0200 In-line centrifugal, supply/exhaust booster | | | | | | | | | |
| 0220     aluminum wheel/hub, disconnect switch, 1/4" S.P. | | | | | | | | | |
| 0240     500 CFM, 10" diameter connection | Q-20 | 3 | 6.667 | Ea. | 1,150 | 315 | | 1,465 | 1,750 |
| 0260     1,380 CFM, 12" diameter connection | | 2 | 10 | | 1,225 | 470 | | 1,695 | 2,050 |
| 0280     1,520 CFM, 16" diameter connection | | 2 | 10 | | 1,325 | 470 | | 1,795 | 2,150 |
| 0300     2,560 CFM, 18" diameter connection | | 1 | 20 | | 1,450 | 945 | | 2,395 | 3,025 |
| 0320     3,480 CFM, 20" diameter connection | | .80 | 25 | | 1,725 | 1,175 | | 2,900 | 3,650 |
| 0326     5,080 CFM, 20" diameter connection | | .75 | 26.667 | | 1,875 | 1,250 | | 3,125 | 3,950 |
| 3500 Centrifugal, airfoil, motor and drive, complete | | | | | | | | | |
| 3520     1000 CFM, 1/2 HP | Q-20 | 2.50 | 8 | Ea. | 1,675 | 380 | | 2,055 | 2,425 |
| 3540     2,000 CFM, 1 HP | | 2 | 10 | | 1,875 | 470 | | 2,345 | 2,750 |
| 3560     4,000 CFM, 3 HP | | 1.80 | 11.111 | | 2,375 | 525 | | 2,900 | 3,400 |
| 3580     8,000 CFM, 7-1/2 HP | | 1.40 | 14.286 | | 3,675 | 675 | | 4,350 | 5,075 |
| 3600     12,000 CFM, 10 HP | | 1 | 20 | | 4,425 | 945 | | 5,370 | 6,300 |
| 4500 Corrosive fume resistant, plastic | | | | | | | | | |
| 4600     roof ventilators, centrifugal, V belt drive, motor | | | | | | | | | |
| 4620     1/4" S.P., 250 CFM, 1/4 HP | Q-20 | 6 | 3.333 | Ea. | 3,975 | 157 | | 4,132 | 4,600 |
| 4640     895 CFM, 1/3 HP | | 5 | 4 | | 4,325 | 189 | | 4,514 | 5,025 |
| 4660     1630 CFM, 1/2 HP | | 4 | 5 | | 5,125 | 236 | | 5,361 | 5,975 |
| 4680     2240 CFM, 1 HP | | 3 | 6.667 | | 5,325 | 315 | | 5,640 | 6,350 |
| 5000 Utility set, centrifugal, V belt drive, motor | | | | | | | | | |
| 5020     1/4" S.P., 1200 CFM, 1/4 HP | Q-20 | 6 | 3.333 | Ea. | 1,550 | 157 | | 1,707 | 1,950 |
| 5040     1520 CFM, 1/3 HP | | 5 | 4 | | 1,725 | 189 | | 1,914 | 2,175 |
| 5060     1850 CFM, 1/2 HP | | 4 | 5 | | 1,725 | 236 | | 1,961 | 2,250 |
| 5080     2180 CFM, 3/4 HP | | 3 | 6.667 | | 2,050 | 315 | | 2,365 | 2,725 |
| 5100     1/2" S.P., 3600 CFM, 1 HP | | 2 | 10 | | 2,350 | 470 | | 2,820 | 3,300 |
| 5120     4250 CFM, 1-1/2 HP | | 1.60 | 12.500 | | 2,575 | 590 | | 3,165 | 3,750 |

# 23 34 HVAC Fans

## 23 34 16 – Centrifugal HVAC Fans

### 23 34 16.10 Centrifugal Type HVAC Fans

| | | Crew | Daily Output | Labor-Hours | Unit | Material | 2011 Bare Costs Labor | Equipment | Total | Total Incl O&P |
|---|---|---|---|---|---|---|---|---|---|---|
| 5140 | 4800 CFM, 2 HP | Q-20 | 1.40 | 14.286 | Ea. | 2,900 | 675 | | 3,575 | 4,225 |
| 7000 | Roof exhauster, centrifugal, aluminum housing, 12" galvanized | | | | | | | | | |
| 7020 | curb, bird screen, back draft damper, 1/4" S.P. | | | | | | | | | |
| 7100 | Direct drive, 320 CFM, 11" sq. damper | Q-20 | 7 | 2.857 | Ea. | 610 | 135 | | 745 | 875 |
| 7120 | 600 CFM, 11" sq. damper | | 6 | 3.333 | | 760 | 157 | | 917 | 1,075 |
| 7140 | 815 CFM, 13" sq. damper | | 5 | 4 | | 760 | 189 | | 949 | 1,125 |
| 7160 | 1450 CFM, 13" sq. damper | | 4.20 | 4.762 | | 1,100 | 225 | | 1,325 | 1,550 |
| 7180 | 2050 CFM, 16" sq. damper | | 4 | 5 | | 1,450 | 236 | | 1,686 | 1,950 |
| 7200 | V-belt drive, 1650 CFM, 12" sq. damper | | 6 | 3.333 | | 1,100 | 157 | | 1,257 | 1,425 |
| 7220 | 2750 CFM, 21" sq. damper | | 5 | 4 | | 1,300 | 189 | | 1,489 | 1,700 |
| 7230 | 3500 CFM, 21" sq. damper | | 4.50 | 4.444 | | 1,475 | 210 | | 1,685 | 1,950 |
| 7240 | 4910 CFM, 23" sq. damper | | 4 | 5 | | 1,775 | 236 | | 2,011 | 2,325 |
| 7260 | 8525 CFM, 28" sq. damper | | 3 | 6.667 | | 2,400 | 315 | | 2,715 | 3,125 |
| 7280 | 13,760 CFM, 35" sq. damper | | 2 | 10 | | 3,225 | 470 | | 3,695 | 4,250 |
| 7300 | 20,558 CFM, 43" sq. damper | ↓ | 1 | 20 | | 6,550 | 945 | | 7,495 | 8,650 |
| 7320 | For 2 speed winding, add | | | | | 15% | | | | |
| 7340 | For explosionproof motor, add | | | | | 495 | | | 495 | 545 |
| 7360 | For belt driven, top discharge, add | | | | ↓ | 15% | | | | |
| 8500 | Wall exhausters, centrifugal, auto damper, 1/8" S.P. | | | | | | | | | |
| 8520 | Direct drive, 610 CFM, 1/20 HP | Q-20 | 14 | 1.429 | Ea. | 365 | 67.50 | | 432.50 | 500 |
| 8540 | 796 CFM, 1/12 HP | | 13 | 1.538 | | 735 | 72.50 | | 807.50 | 915 |
| 8560 | 822 CFM, 1/6 HP | | 12 | 1.667 | | 925 | 78.50 | | 1,003.50 | 1,150 |
| 8580 | 1,320 CFM, 1/4 HP | ↓ | 12 | 1.667 | ↓ | 1,100 | 78.50 | | 1,178.50 | 1,325 |
| 9500 | V-belt drive, 3 phase | | | | | | | | | |
| 9520 | 2,800 CFM, 1/4 HP | Q-20 | 9 | 2.222 | Ea. | 1,650 | 105 | | 1,755 | 1,975 |
| 9540 | 3,740 CFM, 1/2 HP | " | 8 | 2.500 | " | 1,725 | 118 | | 1,843 | 2,075 |

## 23 34 23 – HVAC Power Ventilators

### 23 34 23.10 HVAC Power Circulators and Ventilators

| | | | Crew | Daily Output | Labor-Hours | Unit | Material | 2011 Bare Costs Labor | Equipment | Total | Total Incl O&P |
|---|---|---|---|---|---|---|---|---|---|---|---|
| 0010 | **HVAC POWER CIRCULATORS AND VENTILATORS** | | | | | | | | | | |
| 3000 | Paddle blade air circulator, 3 speed switch | | | | | | | | | | |
| 3020 | 42", 5,000 CFM high, 3000 CFM low | G | 1 Elec | 2.40 | 3.333 | Ea. | 154 | 168 | | 322 | 415 |
| 3040 | 52", 6,500 CFM high, 4000 CFM low | G | " | 2.20 | 3.636 | " | 154 | 183 | | 337 | 440 |
| 3100 | For antique white motor, same cost | | | | | | | | | | |
| 3200 | For brass plated motor, same cost | | | | | | | | | | |
| 3300 | For light adaptor kit, add | G | | | | Ea. | 38.50 | | | 38.50 | 42.50 |
| 6000 | Propeller exhaust, wall shutter, 1/4" S.P. | | | | | | | | | | |
| 6020 | Direct drive, two speed | | | | | | | | | | |
| 6100 | 375 CFM, 1/10 HP | | Q-20 | 10 | 2 | Ea. | 430 | 94.50 | | 524.50 | 615 |
| 6120 | 730 CFM, 1/7 HP | | | 9 | 2.222 | | 480 | 105 | | 585 | 685 |
| 6140 | 1000 CFM, 1/8 HP | | | 8 | 2.500 | | 665 | 118 | | 783 | 915 |
| 6200 | 4720 CFM, 1 HP | | ↓ | 5 | 4 | ↓ | 1,075 | 189 | | 1,264 | 1,450 |
| 6300 | V-belt drive, 3 phase | | | | | | | | | | |
| 6320 | 6175 CFM, 3/4 HP | | Q-20 | 5 | 4 | Ea. | 870 | 189 | | 1,059 | 1,250 |
| 6340 | 7500 CFM, 3/4 HP | | | 5 | 4 | | 960 | 189 | | 1,149 | 1,325 |
| 6360 | 10,100 CFM, 1 HP | | | 4.50 | 4.444 | | 1,050 | 210 | | 1,260 | 1,475 |
| 6380 | 14,300 CFM, 1-1/2 HP | | ↓ | 4 | 5 | ↓ | 1,200 | 236 | | 1,436 | 1,675 |
| 6650 | Residential, bath exhaust, grille, back draft damper | | | | | | | | | | |
| 6660 | 50 CFM | | Q-20 | 24 | .833 | Ea. | 47.50 | 39.50 | | 87 | 112 |
| 6670 | 110 CFM | | | 22 | .909 | | 79 | 43 | | 122 | 151 |
| 6680 | Light combination, squirrel cage, 100 watt, 70 CFM | | ↓ | 24 | .833 | ↓ | 85.50 | 39.50 | | 125 | 154 |
| 6700 | Light/heater combination, ceiling mounted | | | | | | | | | | |
| 6710 | 70 CFM, 1450 watt | | Q-20 | 24 | .833 | Ea. | 127 | 39.50 | | 166.50 | 200 |

# 23 34 HVAC Fans

## 23 34 23 – HVAC Power Ventilators

| 23 34 23.10 HVAC Power Circulators and Ventilators | Crew | Daily Output | Labor-Hours | Unit | Material | 2011 Bare Costs Labor | Equipment | Total | Total Incl O&P |
|---|---|---|---|---|---|---|---|---|---|
| 6800 | Heater combination, recessed, 70 CFM | Q-20 | 24 | .833 | Ea. | 51 | 39.50 | | 90.50 | 116 |
| 6820 | With 2 infrared bulbs | | 23 | .870 | | 79.50 | 41 | | 120.50 | 150 |
| 6900 | Kitchen exhaust, grille, complete, 160 CFM | | 22 | .909 | | 87.50 | 43 | | 130.50 | 161 |
| 6910 | 180 CFM | | 20 | 1 | | 74.50 | 47 | | 121.50 | 153 |
| 6920 | 270 CFM | | 18 | 1.111 | | 134 | 52.50 | | 186.50 | 227 |
| 6930 | 350 CFM | ▼ | 16 | 1.250 | ▼ | 105 | 59 | | 164 | 204 |
| 6940 | Residential roof jacks and wall caps | | | | | | | | | |
| 6944 | Wall cap with back draft damper | | | | | | | | | |
| 6946 | 3" & 4" dia. round duct | 1 Shee | 11 | .727 | Ea. | 20.50 | 37.50 | | 58 | 80 |
| 6948 | 6" dia. round duct | " | 11 | .727 | " | 52.50 | 37.50 | | 90 | 115 |
| 6958 | Roof jack with bird screen and back draft damper | | | | | | | | | |
| 6960 | 3" & 4" dia. round duct | 1 Shee | 11 | .727 | Ea. | 21.50 | 37.50 | | 59 | 80.50 |
| 6962 | 3-1/4" x 10" rectangular duct | " | 10 | .800 | " | 39 | 41.50 | | 80.50 | 105 |
| 6980 | Transition | | | | | | | | | |
| 6982 | 3-1/4" x 10" to 6" dia. round | 1 Shee | 20 | .400 | Ea. | 23 | 20.50 | | 43.50 | 56.50 |

## 23 34 33 – Air Curtains

| 23 34 33.10 Air Barrier Curtains | Crew | Daily Output | Labor-Hours | Unit | Material | 2011 Bare Costs Labor | Equipment | Total | Total Incl O&P |
|---|---|---|---|---|---|---|---|---|---|
| 0010 | **AIR BARRIER CURTAINS**, Incl. motor starters, transformers, | | | | | | | | | |
| 0050 | door switches & temperature controls | | | | | | | | | |
| 0100 | Shipping and receiving doors, unheated, minimal wind stoppage | | | | | | | | | |
| 0150 | 8' high, multiples of 3' wide | 2 Shee | 6 | 2.667 | L.F. | 244 | 138 | | 382 | 475 |
| 0160 | 5' wide | | 10 | 1.600 | | 235 | 82.50 | | 317.50 | 385 |
| 0210 | 10' high, multiples of 4' wide | | 8 | 2 | | 194 | 103 | | 297 | 370 |
| 0250 | 12' high, 3'-6" wide | | 7 | 2.286 | | 210 | 118 | | 328 | 410 |
| 0260 | 12' wide | | 6 | 2.667 | | 225 | 138 | | 363 | 455 |
| 0350 | 16' high, 3'-6" wide | | 7 | 2.286 | | 210 | 118 | | 328 | 410 |
| 0360 | 12' wide | ▼ | 6 | 2.667 | ▼ | 225 | 138 | | 363 | 455 |
| 0500 | Maximum wind stoppage | | | | | | | | | |
| 0550 | 10' high, multiples of 4' wide | 2 Shee | 8 | 2 | L.F. | 194 | 103 | | 297 | 370 |
| 0650 | 14' high, multiples of 4' wide | | 8 | 2 | | 585 | 103 | | 688 | 800 |
| 0750 | 20' high, multiples of 8' wide | ▼ | 6 | 2.667 | ▼ | 600 | 138 | | 738 | 865 |
| 1100 | Heated, maximum wind stoppage, steam heat | | | | | | | | | |
| 1150 | 10' high, multiples of 4' wide | 2 Shee | 6 | 2.667 | L.F. | 930 | 138 | | 1,068 | 1,225 |
| 1250 | 14' high, multiples of 4' wide | | 6 | 2.667 | | 930 | 138 | | 1,068 | 1,225 |
| 1350 | 20' high, multiples of 8' wide | ▼ | 4 | 4 | ▼ | 790 | 207 | | 997 | 1,175 |
| 1500 | Customer entrance doors, unheated, minimal wind stoppage | | | | | | | | | |
| 1550 | 10' high, multiples of 3' wide | 2 Shee | 6 | 2.667 | L.F. | 220 | 138 | | 358 | 450 |
| 1560 | 5' wide | | 10 | 1.600 | | 235 | 82.50 | | 317.50 | 385 |
| 1650 | Maximum wind stoppage, 12' high, multiples of 4' wide | ▼ | 8 | 2 | ▼ | 350 | 103 | | 453 | 540 |
| 1700 | Heated, minimal wind stoppage, electric heat | | | | | | | | | |
| 1750 | 8' high, multiples of 3' wide | 2 Shee | 6 | 2.667 | L.F. | 244 | 138 | | 382 | 475 |
| 1760 | Multiples of 5' wide | | 10 | 1.600 | | 235 | 82.50 | | 317.50 | 385 |
| 1850 | 10' high, multiples of 3' wide | | 6 | 2.667 | | 220 | 138 | | 358 | 450 |
| 1860 | Multiples of 5' wide | ▼ | 10 | 1.600 | ▼ | 510 | 82.50 | | 592.50 | 685 |
| 1950 | Maximum wind stoppage, steam heat | | | | | | | | | |
| 1960 | 12' high, multiples of 4' wide | 2 Shee | 8 | 2 | L.F. | 930 | 103 | | 1,033 | 1,175 |
| 2000 | Walk-in coolers and freezers, ambient air, minimal wind stoppage | | | | | | | | | |
| 2050 | 8' high, multiples of 3' wide | 2 Shee | 6 | 2.667 | L.F. | 220 | 138 | | 358 | 450 |
| 2060 | Multiples of 5' wide | | 10 | 1.600 | | 510 | 82.50 | | 592.50 | 685 |
| 2250 | Maximum wind stoppage, 12' high, multiples of 3' wide | | 6 | 2.667 | | 220 | 138 | | 358 | 450 |
| 2450 | Conveyor openings or service windows, unheated, 5' high | | 5 | 3.200 | | 172 | 165 | | 337 | 440 |
| 2460 | Heated, electric, 5' high, 2'-6" wide | ▼ | 5 | 3.200 | ▼ | 460 | 165 | | 625 | 755 |

# 23 37 Air Outlets and Inlets

## 23 37 13 – Diffusers, Registers, and Grilles

### 23 37 13.10 Diffusers

| | | Crew | Daily Output | Labor-Hours | Unit | Material | 2011 Bare Costs Labor | Equipment | Total | Total Incl O&P |
|---|---|---|---|---|---|---|---|---|---|---|
| 0010 | **DIFFUSERS**, Aluminum, opposed blade damper unless noted | | | | | | | | | |
| 0100 | Ceiling, linear, also for sidewall | | | | | | | | | |
| 0500 | Perforated, 24" x 24" lay-in panel size, 6" x 6" | 1 Shee | 16 | .500 | Ea. | 57.50 | 26 | | 83.50 | 103 |
| 0520 | 8" x 8" | | 15 | .533 | | 59 | 27.50 | | 86.50 | 107 |
| 0530 | 9" x 9" | | 14 | .571 | | 63 | 29.50 | | 92.50 | 114 |
| 0540 | 10" x 10" | | 14 | .571 | | 65 | 29.50 | | 94.50 | 116 |
| 0560 | 12" x 12" | | 12 | .667 | | 66 | 34.50 | | 100.50 | 125 |
| 0590 | 16" x 16" | | 11 | .727 | | 109 | 37.50 | | 146.50 | 177 |
| 0600 | 18" x 18" | | 10 | .800 | | 115 | 41.50 | | 156.50 | 190 |
| 0610 | 20" x 20" | | 10 | .800 | | 133 | 41.50 | | 174.50 | 209 |
| 0620 | 24" x 24" | | 9 | .889 | | 165 | 46 | | 211 | 252 |
| 1000 | Rectangular, 1 to 4 way blow, 6" x 6" | | 16 | .500 | | 46.50 | 26 | | 72.50 | 90 |
| 1010 | 8" x 8" | | 15 | .533 | | 55.50 | 27.50 | | 83 | 103 |
| 1014 | 9" x 9" | | 15 | .533 | | 61 | 27.50 | | 88.50 | 109 |
| 1016 | 10" x 10" | | 15 | .533 | | 75.50 | 27.50 | | 103 | 125 |
| 1020 | 12" x 6" | | 15 | .533 | | 67.50 | 27.50 | | 95 | 116 |
| 1040 | 12" x 9" | | 14 | .571 | | 73.50 | 29.50 | | 103 | 126 |
| 1060 | 12" x 12" | | 12 | .667 | | 78.50 | 34.50 | | 113 | 139 |
| 1070 | 14" x 6" | | 13 | .615 | | 73 | 32 | | 105 | 129 |
| 1074 | 14" x 14" | | 12 | .667 | | 121 | 34.50 | | 155.50 | 185 |
| 1150 | 18" x 18" | | 9 | .889 | | 132 | 46 | | 178 | 215 |
| 1160 | 21" x 21" | | 8 | 1 | | 210 | 51.50 | | 261.50 | 310 |
| 1170 | 24" x 12" | | 10 | .800 | | 160 | 41.50 | | 201.50 | 239 |
| 1500 | Round, butterfly damper, steel, diffuser size, 6" diameter | | 18 | .444 | | 24 | 23 | | 47 | 61 |
| 1520 | 8" diameter | | 16 | .500 | | 25.50 | 26 | | 51.50 | 67 |
| 1540 | 10" diameter | | 14 | .571 | | 31.50 | 29.50 | | 61 | 79.50 |
| 1560 | 12" diameter | | 12 | .667 | | 41.50 | 34.50 | | 76 | 97.50 |
| 1580 | 14" diameter | | 10 | .800 | | 51.50 | 41.50 | | 93 | 119 |
| 2000 | T bar mounting, 24" x 24" lay-in frame, 6" x 6" | | 16 | .500 | | 44 | 26 | | 70 | 87.50 |
| 2020 | 8" x 8" | | 14 | .571 | | 46 | 29.50 | | 75.50 | 95 |
| 2040 | 12" x 12" | | 12 | .667 | | 54.50 | 34.50 | | 89 | 112 |
| 2060 | 16" x 16" | | 11 | .727 | | 73 | 37.50 | | 110.50 | 138 |
| 2080 | 18" x 18" | | 10 | .800 | | 81 | 41.50 | | 122.50 | 152 |
| 6000 | For steel diffusers instead of aluminum, deduct | | | | | 10% | | | | |

### 23 37 13.30 Grilles

| | | Crew | Daily Output | Labor-Hours | Unit | Material | 2011 Bare Costs Labor | Equipment | Total | Total Incl O&P |
|---|---|---|---|---|---|---|---|---|---|---|
| 0010 | **GRILLES** | | | | | | | | | |
| 0020 | Aluminum | | | | | | | | | |
| 1000 | Air return, 6" x 6" | 1 Shee | 26 | .308 | Ea. | 17.55 | 15.90 | | 33.45 | 43.50 |
| 1020 | 10" x 6" | | 24 | .333 | | 17.55 | 17.20 | | 34.75 | 45.50 |
| 1080 | 16" x 8" | | 22 | .364 | | 22 | 18.80 | | 40.80 | 53 |
| 1100 | 12" x 12" | | 22 | .364 | | 24 | 18.80 | | 42.80 | 54.50 |
| 1120 | 24" x 12" | | 18 | .444 | | 33 | 23 | | 56 | 70.50 |
| 1220 | 24" x 18" | | 16 | .500 | | 41.50 | 26 | | 67.50 | 84.50 |
| 1280 | 36" x 24" | | 14 | .571 | | 71 | 29.50 | | 100.50 | 123 |
| 3000 | Filter grille with filter, 12" x 12" | | 24 | .333 | | 47 | 17.20 | | 64.20 | 77.50 |
| 3020 | 18" x 12" | | 20 | .400 | | 65.50 | 20.50 | | 86 | 104 |
| 3040 | 24" x 18" | | 18 | .444 | | 77 | 23 | | 100 | 119 |
| 3060 | 24" x 24" | | 16 | .500 | | 89 | 26 | | 115 | 137 |
| 6000 | For steel grilles instead of aluminum in above, deduct | | | | | 10% | | | | |

### 23 37 13.60 Registers

| | | Crew | Daily Output | Labor-Hours | Unit | Material | 2011 Bare Costs Labor | Equipment | Total | Total Incl O&P |
|---|---|---|---|---|---|---|---|---|---|---|
| 0010 | **REGISTERS** | | | | | | | | | |
| 0980 | Air supply | | | | | | | | | |

## 23 37 13 – Diffusers, Registers, and Grilles

| 23 37 13.60 Registers | | Crew | Daily Output | Labor-Hours | Unit | Material | 2011 Bare Costs Labor | Equipment | Total | Total Incl O&P |
|---|---|---|---|---|---|---|---|---|---|---|
| 1000 | Ceiling/wall, O.B. damper, anodized aluminum | | | | | | | | | |
| 1010 | One or two way deflection, adj. curved face bars | | | | | | | | | |
| 1020 | 8" x 4" | 1 Shee | 26 | .308 | Ea. | 34 | 15.90 | | 49.90 | 61 |
| 1120 | 12" x 12" | | 18 | .444 | | 62 | 23 | | 85 | 103 |
| 1240 | 20" x 6" | | 18 | .444 | | 55 | 23 | | 78 | 95 |
| 1340 | 24" x 8" | | 13 | .615 | | 74.50 | 32 | | 106.50 | 130 |
| 1350 | 24" x 18" | | 12 | .667 | | 139 | 34.50 | | 173.50 | 204 |
| 2700 | Above registers in steel instead of aluminum, deduct | | | | | 10% | | | | |
| 4000 | Floor, toe operated damper, enameled steel | | | | | | | | | |
| 4020 | 4" x 8" | 1 Shee | 32 | .250 | Ea. | 26.50 | 12.90 | | 39.40 | 49 |
| 4100 | 8" x 10" | | 22 | .364 | | 32.50 | 18.80 | | 51.30 | 64.50 |
| 4140 | 10" x 10" | | 20 | .400 | | 39 | 20.50 | | 59.50 | 74 |
| 4220 | 14" x 14" | | 16 | .500 | | 137 | 26 | | 163 | 190 |
| 4240 | 14" x 20" | | 15 | .533 | | 183 | 27.50 | | 210.50 | 243 |
| 4980 | Air return | | | | | | | | | |
| 5000 | Ceiling or wall, fixed 45° face blades | | | | | | | | | |
| 5010 | Adjustable O.B. damper, anodized aluminum | | | | | | | | | |
| 5020 | 4" x 8" | 1 Shee | 26 | .308 | Ea. | 24 | 15.90 | | 39.90 | 50 |
| 5060 | 6" x 10" | | 19 | .421 | | 27 | 22 | | 49 | 63 |
| 5280 | 24" x 24" | | 11 | .727 | | 92 | 37.50 | | 129.50 | 158 |
| 5300 | 24" x 36" | | 8 | 1 | | 144 | 51.50 | | 195.50 | 237 |
| 6000 | For steel construction instead of aluminum, deduct | | | | | 10% | | | | |

## 23 37 15 – Louvers

### 23 37 15.40 HVAC Louvers

| 23 37 15.40 HVAC Louvers | | Crew | Daily Output | Labor-Hours | Unit | Material | 2011 Bare Costs Labor | Equipment | Total | Total Incl O&P |
|---|---|---|---|---|---|---|---|---|---|---|
| 0010 | **HVAC LOUVERS** | | | | | | | | | |
| 0100 | Aluminum, extruded, with screen, mill finish | | | | | | | | | |
| 1002 | Brick vent, see also Section 04 05 23.19 | | | | | | | | | |
| 1100 | Standard, 4" deep, 8" wide, 5" high | 1 Shee | 24 | .333 | Ea. | 34.50 | 17.20 | | 51.70 | 63.50 |
| 1200 | Modular, 4" deep, 7-3/4" wide, 5" high | | 24 | .333 | | 36 | 17.20 | | 53.20 | 65.50 |
| 1300 | Speed brick, 4" deep, 11-5/8" wide, 3-7/8" high | | 24 | .333 | | 36 | 17.20 | | 53.20 | 65.50 |
| 1400 | Fuel oil brick, 4" deep, 8" wide, 5" high | | 24 | .333 | | 61.50 | 17.20 | | 78.70 | 94 |
| 2000 | Cooling tower and mechanical equip., screens, light weight | | 40 | .200 | S.F. | 15.40 | 10.35 | | 25.75 | 32.50 |
| 2020 | Standard weight | | 35 | .229 | | 41 | 11.80 | | 52.80 | 63 |
| 2500 | Dual combination, automatic, intake or exhaust | | 20 | .400 | | 56 | 20.50 | | 76.50 | 92.50 |
| 2520 | Manual operation | | 20 | .400 | | 41.50 | 20.50 | | 62 | 77 |
| 2540 | Electric or pneumatic operation | | 20 | .400 | | 41.50 | 20.50 | | 62 | 77 |
| 2560 | Motor, for electric or pneumatic | | 14 | .571 | Ea. | 480 | 29.50 | | 509.50 | 575 |
| 3000 | Fixed blade, continuous line | | | | | | | | | |
| 3100 | Mullion type, stormproof | 1 Shee | 28 | .286 | S.F. | 41.50 | 14.75 | | 56.25 | 68.50 |
| 3200 | Stormproof | | 28 | .286 | | 41.50 | 14.75 | | 56.25 | 68.50 |
| 3300 | Vertical line | | 28 | .286 | | 49.50 | 14.75 | | 64.25 | 76.50 |
| 3500 | For damper to use with above, add | | | | | 50% | 30% | | | |
| 3520 | Motor, for damper, electric or pneumatic | 1 Shee | 14 | .571 | Ea. | 480 | 29.50 | | 509.50 | 575 |
| 4000 | Operating, 45°, manual, electric or pneumatic | | 24 | .333 | S.F. | 50 | 17.20 | | 67.20 | 81 |
| 4100 | Motor, for electric or pneumatic | | 14 | .571 | Ea. | 480 | 29.50 | | 509.50 | 575 |
| 4200 | Penthouse, roof | | 56 | .143 | S.F. | 24.50 | 7.40 | | 31.90 | 38 |
| 4300 | Walls | | 40 | .200 | | 58 | 10.35 | | 68.35 | 79.50 |
| 5000 | Thinline, under 4" thick, fixed blade | | 40 | .200 | | 24 | 10.35 | | 34.35 | 42 |
| 5010 | Finishes, applied by mfr. at additional cost, available in colors | | | | | | | | | |
| 5020 | Prime coat only, add | | | | S.F. | 3.30 | | | 3.30 | 3.63 |
| 5040 | Baked enamel finish coating, add | | | | | 6.10 | | | 6.10 | 6.70 |
| 5060 | Anodized finish, add | | | | | 6.60 | | | 6.60 | 7.25 |

# 23 37 Air Outlets and Inlets

## 23 37 15 – Louvers

| | | Daily | Labor- | | | 2011 Bare Costs | | | Total |
|---|---|---|---|---|---|---|---|---|---|
| **23 37 15.40 HVAC Louvers** | Crew | Output | Hours | Unit | Material | Labor | Equipment | Total | Incl O&P |
| 5080 | Duranodic finish, add | | | | S.F. | 12 | | | 12 | 13.20 |
| 5100 | Fluoropolymer finish coating, add | | | | | 18.90 | | | 18.90 | 21 |
| 9980 | For small orders (under 10 pieces), add | | | | ↓ | 25% | | | | |

## 23 37 23 – HVAC Gravity Ventilators

### 23 37 23.10 HVAC Gravity Air Ventilators

| | | Crew | Daily Output | Labor-Hours | Unit | Material | Labor | Equipment | Total | Total Incl O&P |
|---|---|---|---|---|---|---|---|---|---|---|
| 0010 | **HVAC GRAVITY AIR VENTILATORS**, Incl. base and damper | | | | | | | | | |
| 1280 | Rotary ventilators, wind driven, galvanized | | | | | | | | | |
| 1300 | 4" neck diameter | Q-9 | 20 | .800 | Ea. | 50 | 37 | | 87 | 111 |
| 1340 | 6" neck diameter | | 16 | 1 | | 56 | 46.50 | | 102.50 | 132 |
| 1400 | 12" neck diameter | | 10 | 1.600 | | 81 | 74.50 | | 155.50 | 202 |
| 1500 | 24" neck diameter, 3,100 CFM | | 8 | 2 | | 236 | 93 | | 329 | 400 |
| 1540 | 36" neck diameter, 5,500 CFM | ↓ | 6 | 2.667 | ↓ | 695 | 124 | | 819 | 950 |
| 2000 | Stationary, gravity, syphon, galvanized | | | | | | | | | |
| 2160 | 6" neck diameter, 66 CFM | Q-9 | 16 | 1 | Ea. | 53 | 46.50 | | 99.50 | 129 |
| 2240 | 12" neck diameter, 160 CFM | | 10 | 1.600 | | 91 | 74.50 | | 165.50 | 212 |
| 2340 | 24" neck diameter, 900 CFM | | 8 | 2 | | 248 | 93 | | 341 | 415 |
| 2380 | 36" neck diameter, 2,000 CFM | | 6 | 2.667 | | 775 | 124 | | 899 | 1,050 |
| 4200 | Stationary mushroom, aluminum, 16" orifice diameter | | 10 | 1.600 | | 535 | 74.50 | | 609.50 | 700 |
| 4220 | 26" orifice diameter | | 6.15 | 2.602 | | 790 | 121 | | 911 | 1,050 |
| 4230 | 30" orifice diameter | | 5.71 | 2.802 | | 1,150 | 130 | | 1,280 | 1,475 |
| 4240 | 38" orifice diameter | | 5 | 3.200 | | 1,650 | 149 | | 1,799 | 2,050 |
| 4250 | 42" orifice diameter | | 4.70 | 3.404 | | 2,200 | 158 | | 2,358 | 2,675 |
| 4260 | 50" orifice diameter | | 4.44 | 3.604 | ↓ | 2,625 | 168 | | 2,793 | 3,125 |
| 5000 | Relief vent | | | | | | | | | |
| 5500 | Rectangular, aluminum, galvanized curb | | | | | | | | | |
| 5510 | intake/exhaust, 0.033" SP | | | | | | | | | |
| 5580 | 500 CFM, 12" x 12" | Q-9 | 8.60 | 1.860 | Ea. | 625 | 86.50 | | 711.50 | 820 |
| 5600 | 600 CFM, 12" x 16" | | 8 | 2 | | 705 | 93 | | 798 | 915 |
| 5640 | 1000 CFM, 12" x 24" | | 6.60 | 2.424 | | 795 | 113 | | 908 | 1,050 |
| 5680 | 3000 CFM, 20" x 42" | ↓ | 4 | 4 | ↓ | 1,400 | 186 | | 1,586 | 1,800 |
| 5880 | Size is throat area, volume is at 500 fpm | | | | | | | | | |

# 23 38 Ventilation Hoods

## 23 38 13 – Commercial-Kitchen Hoods

### 23 38 13.10 Hood and Ventilation Equipment

| | | Crew | Daily Output | Labor-Hours | Unit | Material | Labor | Equipment | Total | Total Incl O&P |
|---|---|---|---|---|---|---|---|---|---|---|
| 0010 | **HOOD AND VENTILATION EQUIPMENT** | | | | | | | | | |
| 2970 | Exhaust hood, sst, gutter on all sides, 4' x 4' x 2' | 1 Carp | 1.80 | 4.444 | Ea. | 4,525 | 191 | | 4,716 | 5,275 |
| 2980 | 4' x 4' x 7' | " | 1.60 | 5 | | 7,100 | 215 | | 7,315 | 8,150 |
| 7950 | Hood fire protection system, minimum | Q-1 | 3 | 5.333 | | 4,400 | 257 | | 4,657 | 5,225 |
| 8050 | Maximum | " | 1 | 16 | ↓ | 35,300 | 770 | | 36,070 | 40,100 |

# 23 41 Particulate Air Filtration

## 23 41 13 – Panel Air Filters

### 23 41 13.10 Panel Type Air Filters

| | | Crew | Daily Output | Labor-Hours | Unit | Material | 2011 Bare Costs Labor | Equipment | Total | Total Incl O&P |
|---|---|---|---|---|---|---|---|---|---|---|
| 0010 | **PANEL TYPE AIR FILTERS** | | | | | | | | | |
| 2950 | Mechanical media filtration units | | | | | | | | | |
| 3000 | High efficiency type, with frame, non-supported [G] | | | | MCFM | 45 | | | 45 | 49.50 |
| 3100 | Supported type [G] | | | | " | 60 | | | 60 | 66 |
| 5500 | Throwaway glass or paper media type | | | | Ea. | 3.35 | | | 3.35 | 3.69 |

## 23 41 16 – Renewable-Media Air Filters

### 23 41 16.10 Disposable Media Air Filters

| | | Crew | Daily Output | Labor-Hours | Unit | Material | Labor | Equipment | Total | Total Incl O&P |
|---|---|---|---|---|---|---|---|---|---|---|
| 0010 | **DISPOSABLE MEDIA AIR FILTERS** | | | | | | | | | |
| 5000 | Renewable disposable roll | | | | MCFM | 19.35 | | | 19.35 | 21.50 |

## 23 41 19 – Washable Air Filters

### 23 41 19.10 Permanent Air Filters

| | | Crew | Daily Output | Labor-Hours | Unit | Material | Labor | Equipment | Total | Total Incl O&P |
|---|---|---|---|---|---|---|---|---|---|---|
| 0010 | **PERMANENT AIR FILTERS** | | | | | | | | | |
| 4500 | Permanent washable [G] | | | | MCFM | 20 | | | 20 | 22 |

## 23 41 23 – Extended Surface Filters

### 23 41 23.10 Expanded Surface Filters

| | | Crew | Daily Output | Labor-Hours | Unit | Material | Labor | Equipment | Total | Total Incl O&P |
|---|---|---|---|---|---|---|---|---|---|---|
| 0010 | **EXPANDED SURFACE FILTERS** | | | | | | | | | |
| 4000 | Medium efficiency, extended surface [G] | | | | MCFM | 5.50 | | | 5.50 | 6.05 |

# 23 42 Gas-Phase Air Filtration

## 23 42 13 – Activated-Carbon Air Filtration

### 23 42 13.10 Charcoal Type Air Filtration

| | | Crew | Daily Output | Labor-Hours | Unit | Material | Labor | Equipment | Total | Total Incl O&P |
|---|---|---|---|---|---|---|---|---|---|---|
| 0010 | **CHARCOAL TYPE AIR FILTRATION** | | | | | | | | | |
| 0050 | Activated charcoal type, full flow | | | | MCFM | 600 | | | 600 | 660 |
| 0060 | Full flow, impregnated media 12" deep | | | | | 225 | | | 225 | 248 |
| 0070 | HEPA filter & frame for field erection | | | | | 300 | | | 300 | 330 |
| 0080 | HEPA filter-diffuser, ceiling install. | | | | ↓ | 300 | | | 300 | 330 |

# 23 43 Electronic Air Cleaners

## 23 43 13 – Washable Electronic Air Cleaners

### 23 43 13.10 Electronic Air Cleaners

| | | Crew | Daily Output | Labor-Hours | Unit | Material | Labor | Equipment | Total | Total Incl O&P |
|---|---|---|---|---|---|---|---|---|---|---|
| 0010 | **ELECTRONIC AIR CLEANERS** | | | | | | | | | |
| 2000 | Electronic air cleaner, duct mounted | | | | | | | | | |
| 2150 | 400 – 1000 CFM | 1 Shee | 2.30 | 3.478 | Ea. | 178 | 180 | | 358 | 470 |
| 2200 | 1000 – 1400 CFM | ↓ | 2.20 | 3.636 | | 1,600 | 188 | | 1,788 | 2,050 |
| 2250 | 1400 – 2000 CFM | ↓ | 2.10 | 3.810 | ↓ | 1,625 | 197 | | 1,822 | 2,075 |

# 23 51 Breechings, Chimneys, and Stacks

## 23 51 13 - Draft Control Devices

### 23 51 13.13 Draft-Induction Fans

| | | Crew | Daily Output | Labor-Hours | Unit | Material | 2011 Bare Costs Labor | 2011 Bare Costs Equipment | Total | Total Incl O&P |
|---|---|---|---|---|---|---|---|---|---|---|
| 0010 | **DRAFT-INDUCTION FANS** | | | | | | | | | |
| 1000 | Breeching installation | | | | | | | | | |
| 1800 | Hot gas, 600°F, variable pitch pulley and motor | | | | | | | | | |
| 1860 | 8" diam. inlet, 1/4 H.P., 1phase, 1120 CFM | Q-9 | 4 | 4 | Ea. | 2,675 | 186 | | 2,861 | 3,225 |
| 1900 | 12" diam. inlet, 3/4 H.P., 3 phase, 2960 CFM | ↓ | 3 | 5.333 | | 3,650 | 248 | | 3,898 | 4,400 |
| 1980 | 24" diam. inlet, 7-1/2 H.P., 3 phase, 17,760 CFM | ↓ | .80 | 20 | | 10,500 | 930 | | 11,430 | 12,900 |
| 2300 | For multi-blade damper at fan inlet, add | | | | ↓ | 20% | | | | |

## 23 51 23 - Gas Vents

### 23 51 23.10 Gas Chimney Vents

| | | Crew | Daily Output | Labor-Hours | Unit | Material | 2011 Bare Costs Labor | 2011 Bare Costs Equipment | Total | Total Incl O&P |
|---|---|---|---|---|---|---|---|---|---|---|
| 0010 | **GAS CHIMNEY VENTS,** Prefab metal, U.L. listed | | | | | | | | | |
| 0020 | Gas, double wall, galvanized steel | | | | | | | | | |
| 0080 | 3" diameter | Q-9 | 72 | .222 | V.L.F. | 5.55 | 10.35 | | 15.90 | 22 |
| 0100 | 4" diameter | | 68 | .235 | | 6.95 | 10.95 | | 17.90 | 24 |
| 0120 | 5" diameter | | 64 | .250 | | 8.10 | 11.60 | | 19.70 | 26.50 |
| 0140 | 6" diameter | | 60 | .267 | | 9.45 | 12.40 | | 21.85 | 29 |
| 0160 | 7" diameter | | 56 | .286 | | 13.85 | 13.30 | | 27.15 | 35.50 |
| 0180 | 8" diameter | | 52 | .308 | | 15.45 | 14.30 | | 29.75 | 38.50 |
| 0200 | 10" diameter | | 48 | .333 | | 35 | 15.50 | | 50.50 | 62 |
| 0220 | 12" diameter | | 44 | .364 | | 47 | 16.90 | | 63.90 | 77 |
| 0260 | 16" diameter | ↓ | 40 | .400 | | 107 | 18.60 | | 125.60 | 146 |
| 0300 | 20" diameter | Q-10 | 36 | .667 | | 145 | 32 | | 177 | 208 |
| 0480 | 38" diameter | " | 24 | 1 | ↓ | 520 | 48 | | 568 | 645 |

## 23 51 26 - All-Fuel Vent Chimneys

### 23 51 26.30 All-Fuel Vent Chimneys, Double Wall, St. Stl.

| | | Crew | Daily Output | Labor-Hours | Unit | Material | 2011 Bare Costs Labor | 2011 Bare Costs Equipment | Total | Total Incl O&P |
|---|---|---|---|---|---|---|---|---|---|---|
| 0010 | **ALL-FUEL VENT CHIMNEYS, DOUBLE WALL, STAINLESS STEEL** | | | | | | | | | |
| 7800 | All fuel, double wall, stainless steel, 6" diameter | Q-9 | 60 | .267 | V.L.F. | 58.50 | 12.40 | | 70.90 | 83.50 |
| 7802 | 7" diameter | | 56 | .286 | | 76 | 13.30 | | 89.30 | 104 |
| 7804 | 8" diameter | | 52 | .308 | | 90 | 14.30 | | 104.30 | 121 |
| 7806 | 10" diameter | | 48 | .333 | | 130 | 15.50 | | 145.50 | 167 |
| 7808 | 12" diameter | | 44 | .364 | | 174 | 16.90 | | 190.90 | 217 |
| 7810 | 14" diameter | ↓ | 42 | .381 | ↓ | 228 | 17.70 | | 245.70 | 278 |

## 23 51 33 - Insulated Sectional Chimneys

### 23 51 33.10 Prefabricated Insulated Sectional Chimneys

| | | Crew | Daily Output | Labor-Hours | Unit | Material | 2011 Bare Costs Labor | 2011 Bare Costs Equipment | Total | Total Incl O&P |
|---|---|---|---|---|---|---|---|---|---|---|
| 0010 | **PREFABRICATED INSULATED SECTIONAL CHIMNEYS** | | | | | | | | | |
| 9000 | High temp. (2000°F), steel jacket, acid resistant refractory lining | | | | | | | | | |
| 9010 | 11 ga. galvanized jacket, U.L. listed | | | | | | | | | |
| 9020 | Straight section, 48" long, 10" diameter | Q-10 | 13.30 | 1.805 | Ea. | 400 | 87 | | 487 | 570 |
| 9040 | 18" diameter | ↓ | 7.40 | 3.243 | | 615 | 156 | | 771 | 910 |
| 9070 | 36" diameter | ↓ | 2.70 | 8.889 | | 1,325 | 430 | | 1,755 | 2,100 |
| 9090 | 48" diameter | Q-11 | 2.70 | 11.852 | | 1,925 | 585 | | 2,510 | 2,975 |
| 9120 | Tee section, 10" diameter | Q-10 | 4.40 | 5.455 | | 855 | 263 | | 1,118 | 1,350 |
| 9140 | 18" diameter | ↓ | 2.40 | 10 | | 1,225 | 480 | | 1,705 | 2,075 |
| 9170 | 36" diameter | ↓ | .80 | 30 | | 3,975 | 1,450 | | 5,425 | 6,550 |
| 9190 | 48" diameter | Q-11 | .90 | 35.556 | | 6,775 | 1,750 | | 8,525 | 10,100 |
| 9220 | Cleanout pier section, 10" diameter | Q-10 | 3.50 | 6.857 | | 655 | 330 | | 985 | 1,225 |
| 9240 | 18" diameter | ↓ | 1.90 | 12.632 | | 885 | 610 | | 1,495 | 1,900 |
| 9270 | 36" diameter | ↓ | .75 | 32 | | 1,625 | 1,550 | | 3,175 | 4,100 |
| 9290 | 48" diameter | Q-11 | .75 | 42.667 | | 2,475 | 2,100 | | 4,575 | 5,900 |
| 9320 | For drain, add | | | | | 53% | | | | |
| 9330 | Elbow, 30° and 45°, 10" diameter | Q-10 | 6.60 | 3.636 | | 555 | 175 | | 730 | 875 |
| 9350 | 18" diameter | ↓ | 3.70 | 6.486 | ↓ | 900 | 315 | | 1,215 | 1,475 |

# 23 51 Breechings, Chimneys, and Stacks

## 23 51 33 – Insulated Sectional Chimneys

| 23 51 33.10 Prefabricated Insulated Sectional Chimneys | | Crew | Daily Output | Labor-Hours | Unit | Material | 2011 Bare Costs Labor | Equipment | Total | Total Incl O&P |
|---|---|---|---|---|---|---|---|---|---|---|
| 9380 | 36" diameter | Q-10 | 1.30 | 18.462 | Ea. | 2,225 | 890 | | 3,115 | 3,800 |
| 9400 | 48" diameter | Q-11 | 1.35 | 23.704 | | 3,900 | 1,175 | | 5,075 | 6,050 |
| 9430 | For 60° and 90° elbow, add | | | | | 112% | | | | |
| 9440 | End cap, 10" diameter | Q-10 | 26 | .923 | | 955 | 44.50 | | 999.50 | 1,125 |
| 9460 | 18" diameter | | 15 | 1.600 | | 1,350 | 77 | | 1,427 | 1,600 |
| 9490 | 36" diameter | ↓ | 5 | 4.800 | | 2,350 | 231 | | 2,581 | 2,950 |
| 9510 | 48" diameter | Q-11 | 5.30 | 6.038 | | 3,475 | 297 | | 3,772 | 4,250 |
| 9540 | Increaser (1 diameter), 10" diameter | Q-10 | 6.60 | 3.636 | | 475 | 175 | | 650 | 785 |
| 9560 | 18" diameter | | 3.70 | 6.486 | | 740 | 315 | | 1,055 | 1,300 |
| 9590 | 36" diameter | ↓ | 1.30 | 18.462 | | 1,450 | 890 | | 2,340 | 2,950 |
| 9592 | 42" diameter | Q-11 | 1.60 | 20 | | 1,600 | 985 | | 2,585 | 3,250 |
| 9594 | 48" diameter | | 1.35 | 23.704 | | 1,925 | 1,175 | | 3,100 | 3,875 |
| 9596 | 54" diameter | ↓ | 1.10 | 29.091 | | 2,325 | 1,425 | | 3,750 | 4,725 |
| 9600 | For expansion joints, add to straight section | | | | | 7% | | | | |
| 9610 | For 1/4" hot rolled steel jacket, add | | | | | 157% | | | | |
| 9620 | For 2950°F very high temperature, add | | | | ↓ | 89% | | | | |
| 9630 | 26 ga. aluminized jacket, straight section, 48" long | | | | | | | | | |
| 9640 | 10" diameter | Q-10 | 15.30 | 1.569 | V.L.F. | 237 | 75.50 | | 312.50 | 375 |
| 9660 | 18" diameter | | 8.50 | 2.824 | | 365 | 136 | | 501 | 610 |
| 9690 | 36" diameter | ↓ | 3.10 | 7.742 | ↓ | 845 | 375 | | 1,220 | 1,500 |
| 9810 | Draw band, galv. stl., 11 gauge, 10" diameter | | 32 | .750 | Ea. | 65 | 36 | | 101 | 126 |
| 9820 | 12" diameter | | 30 | .800 | | 69 | 38.50 | | 107.50 | 134 |
| 9830 | 18" diameter | | 26 | .923 | | 86 | 44.50 | | 130.50 | 162 |
| 9860 | 36" diameter | ↓ | 18 | 1.333 | | 146 | 64.50 | | 210.50 | 257 |
| 9880 | 48" diameter | Q-11 | 20 | 1.600 | ↓ | 251 | 78.50 | | 329.50 | 395 |

# 23 52 Heating Boilers

## 23 52 13 – Electric Boilers

### 23 52 13.10 Electric Boilers, ASME

| | | Crew | Daily Output | Labor-Hours | Unit | Material | 2011 Bare Costs Labor | Equipment | Total | Total Incl O&P |
|---|---|---|---|---|---|---|---|---|---|---|
| 0010 | **ELECTRIC BOILERS, ASME**, Standard controls and trim | | | | | | | | | |
| 1000 | Steam, 6 KW, 20.5 MBH | Q-19 | 1.20 | 20 | Ea. | 3,625 | 985 | | 4,610 | 5,475 |
| 1160 | 60 KW, 205 MBH | | 1 | 24 | | 6,250 | 1,175 | | 7,425 | 8,625 |
| 1220 | 112 KW, 382 MBH | | .75 | 32 | | 8,550 | 1,575 | | 10,125 | 11,800 |
| 1280 | 222 KW, 758 MBH | ↓ | .55 | 43.636 | | 13,500 | 2,150 | | 15,650 | 18,100 |
| 1380 | 518 KW, 1768 MBH | Q-21 | .36 | 88.889 | | 22,800 | 4,475 | | 27,275 | 31,800 |
| 1480 | 814 KW, 2778 MBH | | .25 | 128 | | 31,900 | 6,450 | | 38,350 | 44,700 |
| 1600 | 2,340 KW, 7984 MBH | ↓ | .16 | 200 | | 65,500 | 10,100 | | 75,600 | 87,000 |
| 2000 | Hot water, 7.5 KW, 25.6 MBH | Q-19 | 1.30 | 18.462 | | 3,825 | 910 | | 4,735 | 5,550 |
| 2100 | 90 KW, 307 MBH | | 1.10 | 21.818 | | 5,575 | 1,075 | | 6,650 | 7,725 |
| 2220 | 296 KW, 1010 MBH | ↓ | .55 | 43.636 | | 11,900 | 2,150 | | 14,050 | 16,300 |
| 2500 | 1036 KW, 3536 MBH | Q-21 | .34 | 94.118 | | 27,700 | 4,750 | | 32,450 | 37,600 |
| 2680 | 2400 KW, 8191 MBH | | .25 | 128 | | 58,000 | 6,450 | | 64,450 | 73,000 |
| 2820 | 3600 KW, 12,283 MBH | ↓ | .16 | 200 | ↓ | 81,500 | 10,100 | | 91,600 | 104,500 |

## 23 52 23 – Cast-Iron Boilers

### 23 52 23.20 Gas-Fired Boilers

| | | Crew | Daily Output | Labor-Hours | Unit | Material | 2011 Bare Costs Labor | Equipment | Total | Total Incl O&P |
|---|---|---|---|---|---|---|---|---|---|---|
| 0010 | **GAS-FIRED BOILERS,** Natural or propane, standard controls, packaged | | | | | | | | | |
| 1000 | Cast iron, with insulated jacket | | | | | | | | | |
| 2000 | Steam, gross output, 81 MBH | Q-7 | 1.40 | 22.857 | Ea. | 2,025 | 1,175 | | 3,200 | 3,975 |
| 2080 | 203 MBH | | .90 | 35.556 | | 3,225 | 1,825 | | 5,050 | 6,275 |
| 2180 | 400 MBH | ↓ | .56 | 56.838 | ↓ | 5,400 | 2,925 | | 8,325 | 10,300 |

# 23 52 Heating Boilers

## 23 52 23 – Cast-Iron Boilers

### 23 52 23.20 Gas-Fired Boilers

| | | Crew | Daily Output | Labor-Hours | Unit | Material | 2011 Bare Costs Labor | Equipment | Total | Total Incl O&P |
|---|---|---|---|---|---|---|---|---|---|---|
| 2240 | 765 MBH | Q-7 | .43 | 74.419 | Ea. | 10,600 | 3,825 | | 14,425 | 17,400 |
| 2320 | 1,875 MBH | | .30 | 106 | | 19,000 | 5,500 | | 24,500 | 29,100 |
| 2440 | 4,720 MBH | | .15 | 207 | | 70,000 | 10,700 | | 80,700 | 93,000 |
| 2480 | 6,100 MBH | | .13 | 246 | | 77,500 | 12,700 | | 90,200 | 104,000 |
| 2540 | 6,970 MBH | | .10 | 320 | | 84,500 | 16,500 | | 101,000 | 117,500 |
| 3000 | Hot water, gross output, 80 MBH | | 1.46 | 21.918 | | 1,850 | 1,125 | | 2,975 | 3,725 |
| 3140 | 320 MBH | | .80 | 40 | | 4,375 | 2,050 | | 6,425 | 7,900 |
| 3260 | 1,088 MBH | | .40 | 80 | | 12,900 | 4,125 | | 17,025 | 20,400 |
| 3360 | 2,856 MBH | | .20 | 160 | | 22,400 | 8,250 | | 30,650 | 36,900 |
| 3380 | 3,264 MBH | | .18 | 179 | | 23,900 | 9,275 | | 33,175 | 40,100 |
| 3480 | 6,100 MBH | | .13 | 250 | | 77,500 | 12,900 | | 90,400 | 104,000 |
| 3540 | 6,970 MBH | | .09 | 359 | | 84,500 | 18,500 | | 103,000 | 120,500 |
| 7000 | For tankless water heater, add | | | | | 10% | | | | |

### 23 52 23.30 Gas/Oil Fired Boilers

| | | Crew | Daily Output | Labor-Hours | Unit | Material | 2011 Bare Costs Labor | Equipment | Total | Total Incl O&P |
|---|---|---|---|---|---|---|---|---|---|---|
| 0010 | **GAS/OIL FIRED BOILERS,** Combination with burners and controls, packaged | | | | | | | | | |
| 1000 | Cast iron with insulated jacket | | | | | | | | | |
| 2000 | Steam, gross output, 720 MBH | Q-7 | .43 | 74.074 | Ea. | 14,400 | 3,825 | | 18,225 | 21,600 |
| 2080 | 1,600 MBH | | .30 | 107 | | 17,700 | 5,525 | | 23,225 | 27,600 |
| 2140 | 2,700 MBH | | .19 | 165 | | 28,600 | 8,550 | | 37,150 | 44,300 |
| 2280 | 5,520 MBH | | .14 | 235 | | 87,000 | 12,100 | | 99,100 | 114,000 |
| 2340 | 6,390 MBH | | .11 | 296 | | 93,500 | 15,300 | | 108,800 | 125,500 |
| 2380 | 6,970 MBH | | .09 | 372 | | 99,000 | 19,200 | | 118,200 | 137,500 |
| 2900 | Hot water, gross output | | | | | | | | | |
| 2910 | 200 MBH | Q-6 | .62 | 39.024 | Ea. | 8,275 | 1,975 | | 10,250 | 12,100 |
| 2920 | 300 MBH | | .49 | 49.080 | | 8,275 | 2,475 | | 10,750 | 12,800 |
| 2930 | 400 MBH | | .41 | 57.971 | | 9,650 | 2,925 | | 12,575 | 15,000 |
| 2940 | 500 MBH | | .36 | 67.039 | | 10,400 | 3,375 | | 13,775 | 16,600 |
| 3000 | 584 MBH | Q-7 | .44 | 72.072 | | 11,500 | 3,725 | | 15,225 | 18,300 |
| 3060 | 1,460 MBH | | .28 | 113 | | 23,900 | 5,825 | | 29,725 | 35,000 |
| 3160 | 4,088 MBH | | .16 | 195 | | 48,300 | 10,100 | | 58,400 | 68,000 |
| 3300 | 13,500 MBH, 403.3 BHP | | .04 | 727 | | 149,500 | 37,500 | | 187,000 | 220,500 |

### 23 52 23.40 Oil-Fired Boilers

| | | Crew | Daily Output | Labor-Hours | Unit | Material | 2011 Bare Costs Labor | Equipment | Total | Total Incl O&P |
|---|---|---|---|---|---|---|---|---|---|---|
| 0010 | **OIL-FIRED BOILERS,** Standard controls, flame retention burner, packaged | | | | | | | | | |
| 1000 | Cast iron, with insulated flush jacket | | | | | | | | | |
| 2000 | Steam, gross output, 109 MBH | Q-7 | 1.20 | 26.667 | Ea. | 2,050 | 1,375 | | 3,425 | 4,300 |
| 2060 | 207 MBH | | .90 | 35.556 | | 2,800 | 1,825 | | 4,625 | 5,825 |
| 2180 | 1,084 MBH | | .38 | 85.106 | | 9,400 | 4,375 | | 13,775 | 16,900 |
| 2280 | 3,000 MBH | | .19 | 170 | | 21,600 | 8,775 | | 30,375 | 36,800 |
| 2380 | 5,520 MBH | | .14 | 235 | | 68,500 | 12,100 | | 80,600 | 93,500 |
| 2460 | 6,970 MBH | | .09 | 363 | | 88,000 | 18,700 | | 106,700 | 125,000 |
| 3000 | Hot water, same price as steam | | | | | | | | | |
| 4000 | For tankless coil in smaller sizes, add | | | | Ea. | 15% | | | | |

## 23 52 26 – Steel Boilers

### 23 52 26.40 Oil-Fired Boilers

| | | Crew | Daily Output | Labor-Hours | Unit | Material | 2011 Bare Costs Labor | Equipment | Total | Total Incl O&P |
|---|---|---|---|---|---|---|---|---|---|---|
| 0010 | **OIL-FIRED BOILERS,** Standard controls, flame retention burner | | | | | | | | | |
| 5000 | Steel, with insulated flush jacket | | | | | | | | | |
| 7000 | Hot water, gross output, 103 MBH | Q-6 | 1.60 | 15 | Ea. | 1,725 | 760 | | 2,485 | 3,025 |
| 7120 | 420 MBH | | .70 | 34.483 | | 5,625 | 1,750 | | 7,375 | 8,775 |
| 7320 | 3,150 MBH | | .13 | 184 | | 39,300 | 9,325 | | 48,625 | 57,000 |
| 7340 | For tankless coil in steam or hot water, add | | | | | 7% | | | | |

# 23 52 Heating Boilers

## 23 52 28 – Swimming Pool Boilers

| 23 52 28.10 Swimming Pool Heaters | Crew | Daily Output | Labor-Hours | Unit | Material | 2011 Bare Costs Labor | Equipment | Total | Total Incl O&P |
|---|---|---|---|---|---|---|---|---|---|
| 0010 **SWIMMING POOL HEATERS**, Not including wiring, external | | | | | | | | | |
| 0020 piping, base or pad, | | | | | | | | | |
| 0160 Gas fired, input, 155 MBH | Q-6 | 1.50 | 16 | Ea. | 1,625 | 810 | | 2,435 | 2,975 |
| 0200 199 MBH | | 1 | 24 | | 1,725 | 1,200 | | 2,925 | 3,700 |
| 0280 500 MBH | | .40 | 60 | | 7,325 | 3,025 | | 10,350 | 12,600 |
| 0400 1,800 MBH | | .14 | 171 | | 15,900 | 8,650 | | 24,550 | 30,400 |
| 2000 Electric, 12 KW, 4,800 gallon pool | Q-19 | 3 | 8 | | 2,025 | 395 | | 2,420 | 2,800 |
| 2020 15 KW, 7,200 gallon pool | | 2.80 | 8.571 | | 2,175 | 420 | | 2,595 | 3,000 |
| 2040 24 KW, 9,600 gallon pool | | 2.40 | 10 | | 2,375 | 490 | | 2,865 | 3,325 |
| 2100 57 KW, 24,000 gallon pool | | 1.20 | 20 | | 3,475 | 985 | | 4,460 | 5,300 |

# 23 54 Furnaces

## 23 54 13 – Electric-Resistance Furnaces

### 23 54 13.10 Electric Furnaces

| | Crew | Daily Output | Labor-Hours | Unit | Material | 2011 Bare Costs Labor | Equipment | Total | Total Incl O&P |
|---|---|---|---|---|---|---|---|---|---|
| 0010 **ELECTRIC FURNACES**, Hot air, blowers, std. controls | | | | | | | | | |
| 0011 not including gas, oil or flue piping | | | | | | | | | |
| 1000 Electric, UL listed | | | | | | | | | |
| 1020 10.2 MBH | Q-20 | 5 | 4 | Ea. | 475 | 189 | | 664 | 805 |
| 1040 17.1 MBH | | 4.80 | 4.167 | | 495 | 197 | | 692 | 840 |
| 1060 27.3 MBH | | 4.60 | 4.348 | | 600 | 205 | | 805 | 970 |
| 1100 34.1 MBH | | 4.40 | 4.545 | | 635 | 215 | | 850 | 1,025 |

## 23 54 16 – Fuel-Fired Furnaces

### 23 54 16.13 Gas-Fired Furnaces

| | Crew | Daily Output | Labor-Hours | Unit | Material | 2011 Bare Costs Labor | Equipment | Total | Total Incl O&P |
|---|---|---|---|---|---|---|---|---|---|
| 0010 **GAS-FIRED FURNACES** | | | | | | | | | |
| 3000 Gas, AGA certified, upflow, direct drive models | | | | | | | | | |
| 3020 45 MBH input | Q-9 | 4 | 4 | Ea. | 575 | 186 | | 761 | 915 |
| 3040 60 MBH input | | 3.80 | 4.211 | | 615 | 196 | | 811 | 970 |
| 3060 75 MBH input | | 3.60 | 4.444 | | 685 | 207 | | 892 | 1,075 |
| 3100 100 MBH input | | 3.20 | 5 | | 690 | 232 | | 922 | 1,100 |

### 23 54 16.16 Oil-Fired Furnaces

| | Crew | Daily Output | Labor-Hours | Unit | Material | 2011 Bare Costs Labor | Equipment | Total | Total Incl O&P |
|---|---|---|---|---|---|---|---|---|---|
| 0010 **OIL-FIRED FURNACES** | | | | | | | | | |
| 6000 Oil, UL listed, atomizing gun type burner | | | | | | | | | |
| 6020 56 MBH output | Q-9 | 3.60 | 4.444 | Ea. | 1,475 | 207 | | 1,682 | 1,925 |
| 6030 84 MBH output | | 3.50 | 4.571 | | 1,850 | 212 | | 2,062 | 2,375 |
| 6040 95 MBH output | | 3.40 | 4.706 | | 1,875 | 219 | | 2,094 | 2,400 |
| 6060 134 MBH output | | 3.20 | 5 | | 2,175 | 232 | | 2,407 | 2,750 |
| 6080 151 MBH output | | 3 | 5.333 | | 2,250 | 248 | | 2,498 | 2,850 |

# 23 55 Fuel-Fired Heaters

## 23 55 13 – Fuel-Fired Duct Heaters

### 23 55 13.16 Gas-Fired Duct Heaters

| | | Crew | Daily Output | Labor-Hours | Unit | Material | 2011 Bare Costs Labor | Equipment | Total | Total Incl O&P |
|---|---|---|---|---|---|---|---|---|---|---|
| 0010 | **GAS-FIRED DUCT HEATERS**, Includes burner, controls, stainless steel | | | | | | | | | |
| 0020 | heat exchanger. Gas fired, electric ignition | | | | | | | | | |
| 0030 | Indoor installation | | | | | | | | | |
| 0100 | 120 MBH output | Q-5 | 4 | 4 | Ea. | 2,125 | 195 | | 2,320 | 2,650 |
| 0130 | 200 MBH output | | 2.70 | 5.926 | | 2,850 | 289 | | 3,139 | 3,550 |
| 0140 | 240 MBH output | | 2.30 | 6.957 | | 2,975 | 340 | | 3,315 | 3,750 |
| 0180 | 320 MBH output | | 1.60 | 10 | | 3,550 | 485 | | 4,035 | 4,650 |
| 0300 | For powered venter and adapter, add | | | | | 460 | | | 460 | 510 |
| 0502 | For required flue pipe, see Section 23 51 23.10 | | | | | | | | | |
| 1000 | Outdoor installation, with power venter | | | | | | | | | |
| 1020 | 75 MBH output | Q-5 | 4 | 4 | Ea. | 2,775 | 195 | | 2,970 | 3,350 |
| 1060 | 120 MBH output | | 4 | 4 | | 3,525 | 195 | | 3,720 | 4,200 |
| 1100 | 187 MBH output | | 3 | 5.333 | | 5,050 | 260 | | 5,310 | 5,950 |
| 1140 | 300 MBH output | | 1.80 | 8.889 | | 6,100 | 435 | | 6,535 | 7,350 |
| 1180 | 450 MBH output | | 1.40 | 11.429 | | 7,450 | 555 | | 8,005 | 9,025 |

## 23 55 33 – Fuel-Fired Unit Heaters

### 23 55 33.13 Oil-Fired Unit Heaters

| | | Crew | Daily Output | Labor-Hours | Unit | Material | Labor | Equipment | Total | Total Incl O&P |
|---|---|---|---|---|---|---|---|---|---|---|
| 0010 | **OIL-FIRED UNIT HEATERS**, Cabinet, grilles, fan, ctrl, burner, no piping | | | | | | | | | |
| 6000 | Oil fired, suspension mounted, 94 MBH output | Q-5 | 4 | 4 | Ea. | 3,400 | 195 | | 3,595 | 4,025 |
| 6040 | 140 MBH output | | 3 | 5.333 | | 3,625 | 260 | | 3,885 | 4,375 |
| 6060 | 184 MBH output | | 3 | 5.333 | | 3,900 | 260 | | 4,160 | 4,700 |

### 23 55 33.16 Gas-Fired Unit Heaters

| | | Crew | Daily Output | Labor-Hours | Unit | Material | Labor | Equipment | Total | Total Incl O&P |
|---|---|---|---|---|---|---|---|---|---|---|
| 0010 | **GAS-FIRED UNIT HEATERS**, Cabinet, grilles, fan, ctrls., burner, no piping | | | | | | | | | |
| 0022 | thermostat, no piping. For flue see Section 23 51 23.10 | | | | | | | | | |
| 1000 | Gas fired, floor mounted | | | | | | | | | |
| 1100 | 60 MBH output | Q-5 | 10 | 1.600 | Ea. | 805 | 78 | | 883 | 1,000 |
| 1140 | 100 MBH output | | 8 | 2 | | 890 | 97.50 | | 987.50 | 1,125 |
| 1180 | 180 MBH output | | 6 | 2.667 | | 1,275 | 130 | | 1,405 | 1,600 |
| 2000 | Suspension mounted, propeller fan, 20 MBH output | | 8.50 | 1.882 | | 985 | 91.50 | | 1,076.50 | 1,200 |
| 2040 | 60 MBH output | | 7 | 2.286 | | 1,200 | 111 | | 1,311 | 1,500 |
| 2060 | 80 MBH output | | 6 | 2.667 | | 1,225 | 130 | | 1,355 | 1,550 |
| 2100 | 130 MBH output | | 5 | 3.200 | | 1,500 | 156 | | 1,656 | 1,875 |
| 2240 | 320 MBH output | | 2 | 8 | | 2,650 | 390 | | 3,040 | 3,475 |
| 2500 | For powered venter and adapter, add | | | | | 355 | | | 355 | 390 |
| 5000 | Wall furnace, 17.5 MBH output | Q-5 | 6 | 2.667 | | 715 | 130 | | 845 | 980 |
| 5020 | 24 MBH output | | 5 | 3.200 | | 740 | 156 | | 896 | 1,050 |
| 5040 | 35 MBH output | | 4 | 4 | | 895 | 195 | | 1,090 | 1,275 |

# 23 56 Solar Energy Heating Equipment

## 23 56 16 – Packaged Solar Heating Equipment

### 23 56 16.40 Solar Heating Systems

| | | | Crew | Daily Output | Labor-Hours | Unit | Material | Labor | Equipment | Total | Total Incl O&P |
|---|---|---|---|---|---|---|---|---|---|---|---|
| 0010 | **SOLAR HEATING SYSTEMS** | R235616-60 | | | | | | | | | |
| 0020 | System/Package prices, not including connecting | | | | | | | | | | |
| 0030 | pipe, insulation, or special heating/plumbing fixtures | | | | | | | | | | |
| 0500 | Hot water, standard package, low temperature | | | | | | | | | | |
| 0540 | 1 collector, circulator, fittings, 65 gal. tank | G | Q-1 | .50 | 32 | Ea. | 1,550 | 1,550 | | 3,100 | 4,000 |
| 0580 | 2 collectors, circulator, fittings, 120 gal. tank | G | | .40 | 40 | | 2,250 | 1,925 | | 4,175 | 5,350 |
| 0620 | 3 collectors, circulator, fittings, 120 gal. tank | G | | .34 | 47.059 | | 3,075 | 2,275 | | 5,350 | 6,775 |
| 0700 | Medium temperature package | | | | | | | | | | |
| 0720 | 1 collector, circulator, fittings, 80 gal. tank | G | Q-1 | .50 | 32 | Ea. | 2,250 | 1,550 | | 3,800 | 4,775 |

# 23 56 Solar Energy Heating Equipment

## 23 56 16 – Packaged Solar Heating Equipment

| 23 56 16.40 Solar Heating Systems | | Crew | Daily Output | Labor-Hours | Unit | Material | 2011 Bare Costs Labor | Equipment | Total | Total Incl O&P |
|---|---|---|---|---|---|---|---|---|---|---|
| 0740 | 2 collectors, circulator, fittings, 120 gal. tank | G | Q-1 | .40 | 40 | Ea. | 3,275 | 1,925 | | 5,200 | 6,475 |
| 0780 | 3 collectors, circulator, fittings, 120 gal. tank | G | ↓ | .30 | 53.333 | | 4,250 | 2,575 | | 6,825 | 8,525 |
| 0980 | For each additional 120 gal. tank, add | G | | | | ↓ | 1,475 | | | 1,475 | 1,625 |

## 23 56 19 – Solar Heating Components

### 23 56 19.50 Solar Heating Ancillary

| | | Crew | Daily Output | Labor-Hours | Unit | Material | 2011 Bare Costs Labor | Equipment | Total | Total Incl O&P |
|---|---|---|---|---|---|---|---|---|---|---|
| 0010 | **SOLAR HEATING ANCILLARY** | | | | | | | | | | |
| 2300 | Circulators, air | | | | | | | | | | |
| 2310 | Blowers | | | | | | | | | | |
| 2400 | Reversible fan, 20" diameter, 2 speed | G | Q-9 | 18 | .889 | Ea. | 108 | 41.50 | | 149.50 | 182 |
| 2870 | 1/12 HP, 30 GPM | G | Q-1 | 10 | 1.600 | " | 330 | 77 | | 407 | 480 |
| 3000 | Collector panels, air with aluminum absorber plate | | | | | | | | | | |
| 3010 | Wall or roof mount | | | | | | | | | | |
| 3040 | Flat black, plastic glazing | | | | | | | | | | |
| 3080 | 4' x 8' | G | Q-9 | 6 | 2.667 | Ea. | 635 | 124 | | 759 | 880 |
| 3200 | Flush roof mount, 10' to 16' x 22" wide | G | " | 96 | .167 | L.F. | 450 | 7.75 | | 457.75 | 505 |
| 3300 | Collector panels, liquid with copper absorber plate | | | | | | | | | | |
| 3330 | Alum. frame, 4' x 8', 5/32" single glazing | G | Q-1 | 9.50 | 1.684 | Ea. | 945 | 81.50 | | 1,026.50 | 1,175 |
| 3390 | Alum. frame, 4' x 10', 5/32" single glazing | G | | 6 | 2.667 | | 1,075 | 129 | | 1,204 | 1,400 |
| 3450 | Flat black, alum. frame, 3.5' x 7.5' | G | | 9 | 1.778 | | 690 | 86 | | 776 | 890 |
| 3500 | 4' x 8' | G | | 5.50 | 2.909 | | 865 | 140 | | 1,005 | 1,150 |
| 3520 | 4' x 10' | G | | 10 | 1.600 | | 970 | 77 | | 1,047 | 1,200 |
| 3540 | 4' x 12.5' | G | | 5 | 3.200 | | 1,175 | 154 | | 1,329 | 1,500 |
| 3600 | Liquid, full wetted, plastic, alum. frame, 3' x 10' | G | | 5 | 3.200 | | 234 | 154 | | 388 | 490 |
| 3650 | Collector panel mounting, flat roof or ground rack | G | | 7 | 2.286 | ↓ | 217 | 110 | | 327 | 405 |
| 3670 | Roof clamps | G | ↓ | 70 | .229 | Set | 2.35 | 11.05 | | 13.40 | 19.05 |
| 3700 | Roof strap, teflon | G | 1 Plum | 205 | .039 | L.F. | 19.95 | 2.09 | | 22.04 | 25 |
| 3900 | Differential controller with two sensors | | | | | | | | | | |
| 3930 | Thermostat, hard wired | G | 1 Plum | 8 | 1 | Ea. | 79.50 | 53.50 | | 133 | 167 |
| 4100 | Five station with digital read-out | G | " | 3 | 2.667 | " | 218 | 143 | | 361 | 450 |
| 4300 | Heat exchanger | | | | | | | | | | |
| 4580 | Fluid to fluid package includes two circulating pumps | | | | | | | | | | |
| 4590 | expansion tank, check valve, relief valve | | | | | | | | | | |
| 4600 | controller, high temperature cutoff and sensors | G | Q-1 | 2.50 | 6.400 | Ea. | 695 | 310 | | 1,005 | 1,225 |
| 4650 | Heat transfer fluid | | | | | | | | | | |
| 4700 | Propylene glycol, inhibited anti-freeze | G | 1 Plum | 28 | .286 | Gal. | 12.05 | 15.30 | | 27.35 | 36.50 |
| 8250 | Water storage tank with heat exchanger and electric element | | | | | | | | | | |
| 8300 | 80 gal. with 2" x 2 lb. density insulation | G | 1 Plum | 1.60 | 5 | Ea. | 1,100 | 268 | | 1,368 | 1,600 |
| 8380 | 120 gal. with 2" x 2 lb. density insulation | G | | 1.40 | 5.714 | | 1,225 | 305 | | 1,530 | 1,800 |
| 8400 | 120 gal. with 2" x 2 lb. density insul., 40 S.F. heat coil | G | ↓ | 1.40 | 5.714 | ↓ | 1,575 | 305 | | 1,880 | 2,175 |

# 23 57 Heat Exchangers for HVAC

## 23 57 16 – Steam-to-Water Heat Exchangers

### 23 57 16.10 Shell/Tube Type Steam-to-Water Heat Exch.

| | | Crew | Daily Output | Labor-Hours | Unit | Material | 2011 Bare Costs Labor | Equipment | Total | Total Incl O&P |
|---|---|---|---|---|---|---|---|---|---|---|
| 0010 | **SHELL AND TUBE TYPE STEAM-TO-WATER HEAT EXCHANGERS** | | | | | | | | | | |
| 0016 | Shell & tube type, 2 or 4 pass, 3/4" O.D. copper tubes, | | | | | | | | | | |
| 0020 | C.I. heads, C.I. tube sheet, steel shell | | | | | | | | | | |
| 0100 | Hot water 40°F to 180°F, by steam at 10 PSI | | | | | | | | | | |
| 0120 | 8 GPM | | Q-5 | 6 | 2.667 | Ea. | 1,825 | 130 | | 1,955 | 2,200 |
| 0140 | 10 GPM | | | 5 | 3.200 | | 2,750 | 156 | | 2,906 | 3,250 |
| 0160 | 40 GPM | | ↓ | 4 | 4 | ↓ | 4,250 | 195 | | 4,445 | 4,975 |

# 23 57 Heat Exchangers for HVAC

## 23 57 16 – Steam-to-Water Heat Exchangers

| 23 57 16.10 Shell/Tube Type Steam-to-Water Heat Exch. | Crew | Daily Output | Labor-Hours | Unit | Material | 2011 Bare Costs Labor | Equipment | Total | Total Incl O&P |
|---|---|---|---|---|---|---|---|---|---|
| 0180 | 64 GPM | Q-5 | 2 | 8 | Ea. | 6,525 | 390 | | 6,915 | 7,750 |
| 0200 | 96 GPM | ↓ | 1 | 16 | | 8,725 | 780 | | 9,505 | 10,800 |
| 0220 | 120 GPM | Q-6 | 1.50 | 16 | ↓ | 11,400 | 810 | | 12,210 | 13,800 |

## 23 57 19 – Liquid-to-Liquid Heat Exchangers

### 23 57 19.13 Plate-Type, Liquid-to-Liquid Heat Exchangers

| | | Crew | Daily Output | Labor-Hours | Unit | Material | Labor | Equipment | Total | Total Incl O&P |
|---|---|---|---|---|---|---|---|---|---|---|
| 0010 | PLATE-TYPE, LIQUID-TO-LIQUID HEAT EXCHANGERS | | | | | | | | | |
| 3000 | Plate type, | | | | | | | | | |
| 3100 | 400 GPM | Q-6 | .80 | 30 | Ea. | 32,300 | 1,525 | | 33,825 | 37,800 |
| 3120 | 800 GPM | " | .50 | 48 | | 55,500 | 2,425 | | 57,925 | 64,500 |
| 3140 | 1200 GPM | Q-7 | .34 | 94.118 | | 82,500 | 4,850 | | 87,350 | 98,500 |
| 3160 | 1800 GPM | " | .24 | 133 | ↓ | 110,000 | 6,875 | | 116,875 | 131,500 |

### 23 57 19.16 Shell-Type, Liquid-to-Liquid Heat Exchangers

| | | Crew | Daily Output | Labor-Hours | Unit | Material | Labor | Equipment | Total | Total Incl O&P |
|---|---|---|---|---|---|---|---|---|---|---|
| 0010 | SHELL-TYPE, LIQUID-TO-LIQUID HEAT EXCHANGERS | | | | | | | | | |
| 1000 | Hot water 40°F to 140°F, by water at 200°F | | | | | | | | | |
| 1020 | 7 GPM | Q-5 | 6 | 2.667 | Ea. | 2,225 | 130 | | 2,355 | 2,650 |
| 1040 | 16 GPM | | 5 | 3.200 | | 3,175 | 156 | | 3,331 | 3,725 |
| 1060 | 34 GPM | | 4 | 4 | | 4,800 | 195 | | 4,995 | 5,575 |
| 1100 | 74 GPM | ↓ | 1.50 | 10.667 | ↓ | 8,700 | 520 | | 9,220 | 10,400 |

# 23 62 Packaged Compressor and Condenser Units

## 23 62 13 – Packaged Air-Cooled Refrigerant Compressor and Condenser Units

### 23 62 13.10 Packaged Air-Cooled Refrig. Condensing Units

| | | Crew | Daily Output | Labor-Hours | Unit | Material | Labor | Equipment | Total | Total Incl O&P |
|---|---|---|---|---|---|---|---|---|---|---|
| 0010 | PACKAGED AIR-COOLED REFRIGERANT CONDENSING UNITS | | | | | | | | | |
| 0020 | Condensing unit | | | | | | | | | |
| 0030 | Air cooled, compressor, standard controls | | | | | | | | | |
| 0050 | 1.5 ton | Q-5 | 2.50 | 6.400 | Ea. | 1,125 | 310 | | 1,435 | 1,700 |
| 0500 | 5 ton | ↓ | .60 | 26.667 | | 2,225 | 1,300 | | 3,525 | 4,400 |
| 0600 | 10 ton | ↓ | .50 | 32 | | 4,950 | 1,550 | | 6,500 | 7,775 |
| 0700 | 20 ton | Q-6 | .40 | 60 | ↓ | 11,000 | 3,025 | | 14,025 | 16,600 |

# 23 63 Refrigerant Condensers

## 23 63 13 – Air-Cooled Refrigerant Condensers

### 23 63 13.10 Air-Cooled Refrig. Condensers

| | | Crew | Daily Output | Labor-Hours | Unit | Material | Labor | Equipment | Total | Total Incl O&P |
|---|---|---|---|---|---|---|---|---|---|---|
| 0010 | AIR-COOLED REFRIG. CONDENSERS, Ratings for 30°F TD, R-22. | | | | | | | | | |
| 0080 | Air cooled, belt drive, propeller fan | | | | | | | | | |
| 0240 | 50 ton | Q-6 | .69 | 34.985 | Ea. | 9,975 | 1,775 | | 11,750 | 13,600 |
| 0280 | 59 ton | | .58 | 41.308 | | 12,000 | 2,075 | | 14,075 | 16,300 |
| 0320 | 73 ton | | .47 | 51.173 | | 15,600 | 2,575 | | 18,175 | 21,100 |
| 0360 | 86 ton | | .40 | 60.302 | | 18,100 | 3,050 | | 21,150 | 24,500 |
| 0380 | 88 ton | ↓ | .39 | 61.697 | ↓ | 19,400 | 3,125 | | 22,525 | 26,100 |
| 1550 | Air cooled, direct drive, propeller fan | | | | | | | | | |
| 1590 | 1 ton | Q-5 | 3.80 | 4.211 | Ea. | 685 | 205 | | 890 | 1,050 |
| 1600 | 1-1/2 ton | | 3.60 | 4.444 | | 840 | 216 | | 1,056 | 1,250 |
| 1620 | 2 ton | | 3.20 | 5 | | 900 | 244 | | 1,144 | 1,350 |
| 1640 | 5 ton | | 2 | 8 | | 1,825 | 390 | | 2,215 | 2,600 |
| 1660 | 10 ton | | 1.40 | 11.429 | | 3,025 | 555 | | 3,580 | 4,150 |
| 1690 | 16 ton | | 1.10 | 14.545 | | 4,150 | 710 | | 4,860 | 5,625 |
| 1720 | 26 ton | ↓ | .84 | 19.002 | | 5,400 | 925 | | 6,325 | 7,325 |

# 23 63 Refrigerant Condensers

## 23 63 13 – Air-Cooled Refrigerant Condensers

| 23 63 13.10 Air-Cooled Refrig. Condensers | | Crew | Daily Output | Labor-Hours | Unit | Material | 2011 Bare Costs Labor | Equipment | Total | Total Incl O&P |
|---|---|---|---|---|---|---|---|---|---|---|
| 1760 | 41 ton | Q-6 | .77 | 31.008 | Ea. | 9,700 | 1,575 | | 11,275 | 13,000 |
| 1800 | 63 ton | " | .55 | 44.037 | ↓ | 15,300 | 2,225 | | 17,525 | 20,100 |

# 23 64 Packaged Water Chillers

## 23 64 13 – Absorption Water Chillers

### 23 64 13.16 Indirect-Fired Absorption Water Chillers

| | | Crew | Daily Output | Labor-Hours | Unit | Material | 2011 Bare Costs Labor | Equipment | Total | Total Incl O&P |
|---|---|---|---|---|---|---|---|---|---|---|
| 0010 | **INDIRECT-FIRED ABSORPTION WATER CHILLERS** | | | | | | | | | |
| 0020 | Steam or hot water, water cooled | | | | | | | | | |
| 0050 | 100 ton | Q-7 | .13 | 240 | Ea. | 125,000 | 12,400 | | 137,400 | 156,000 |
| 0400 | 420 ton | " | .10 | 323 | " | 335,000 | 16,700 | | 351,700 | 394,000 |

## 23 64 16 – Centrifugal Water Chillers

### 23 64 16.10 Centrifugal Type Water Chillers

| | | Crew | Daily Output | Labor-Hours | Unit | Material | 2011 Bare Costs Labor | Equipment | Total | Total Incl O&P |
|---|---|---|---|---|---|---|---|---|---|---|
| 0010 | **CENTRIFUGAL TYPE WATER CHILLERS**, With standard controls | | | | | | | | | |
| 0020 | Centrifugal liquid chiller, water cooled | | | | | | | | | |
| 0030 | not including water tower | | | | | | | | | |
| 0100 | 2000 ton (twin 1000 ton units) | Q-7 | .07 | 477 | Ea. | 847,500 | 24,600 | | 872,100 | 969,000 |

## 23 64 19 – Reciprocating Water Chillers

### 23 64 19.10 Reciprocating Type Water Chillers

| | | Crew | Daily Output | Labor-Hours | Unit | Material | 2011 Bare Costs Labor | Equipment | Total | Total Incl O&P |
|---|---|---|---|---|---|---|---|---|---|---|
| 0010 | **RECIPROCATING TYPE WATER CHILLERS**, With standard controls | | | | | | | | | |
| 0494 | Water chillers, integral air cooled condenser | | | | | | | | | |
| 0600 | 100 ton cooling | Q-7 | .25 | 129 | Ea. | 71,500 | 6,650 | | 78,150 | 88,500 |
| 0980 | Water cooled, multiple compressor, semi-hermetic, tower not incl. | | | | | | | | | |
| 1000 | 15 ton cooling | Q-6 | .36 | 65.934 | Ea. | 17,500 | 3,325 | | 20,825 | 24,300 |
| 1020 | 25 ton cooling | Q-7 | .41 | 78.049 | | 18,900 | 4,025 | | 22,925 | 26,800 |
| 1060 | 35 ton cooling | | .31 | 101 | | 22,600 | 5,250 | | 27,850 | 32,700 |
| 1100 | 50 ton cooling | | .28 | 113 | | 32,300 | 5,875 | | 38,175 | 44,300 |
| 1160 | 100 ton cooling | | .18 | 179 | | 58,500 | 9,275 | | 67,775 | 78,500 |
| 1180 | 125 ton cooling | | .16 | 196 | | 66,500 | 10,100 | | 76,600 | 88,000 |
| 1200 | 145 ton cooling | ↓ | .16 | 202 | ↓ | 77,500 | 10,400 | | 87,900 | 101,000 |
| 1451 | Water cooled, dual compressors, semi-hermetic, tower not incl. | | | | | | | | | |
| 1500 | 80 ton cooling | Q-7 | .14 | 222 | Ea. | 27,400 | 11,500 | | 38,900 | 47,200 |
| 1520 | 100 ton cooling | | .14 | 228 | | 37,000 | 11,800 | | 48,800 | 58,000 |
| 1540 | 120 ton cooling | ↓ | .14 | 231 | ↓ | 44,800 | 11,900 | | 56,700 | 67,000 |

## 23 64 23 – Scroll Water Chillers

### 23 64 23.10 Scroll Water Chillers

| | | Crew | Daily Output | Labor-Hours | Unit | Material | 2011 Bare Costs Labor | Equipment | Total | Total Incl O&P |
|---|---|---|---|---|---|---|---|---|---|---|
| 0010 | **SCROLL WATER CHILLERS**, With standard controls | | | | | | | | | |
| 0490 | Packaged w/integral air cooled condenser, 15 ton cool | Q-7 | .37 | 86.486 | Ea. | 20,400 | 4,450 | | 24,850 | 29,200 |
| 0500 | 20 ton cooling | | .34 | 94.118 | | 26,400 | 4,850 | | 31,250 | 36,300 |
| 0520 | 40 ton cooling | ↓ | .30 | 108 | ↓ | 34,100 | 5,575 | | 39,675 | 45,800 |
| 0680 | Scroll water cooled, single compressor, hermetic, tower not incl. | | | | | | | | | |
| 0700 | 2 ton cooling | Q-5 | .57 | 28.070 | Ea. | 3,225 | 1,375 | | 4,600 | 5,600 |
| 0710 | 5 ton cooling | | .57 | 28.070 | | 3,850 | 1,375 | | 5,225 | 6,300 |
| 0740 | 8 ton cooling | ↓ | .31 | 52.117 | | 5,325 | 2,550 | | 7,875 | 9,650 |
| 0760 | 10 ton cooling | Q-6 | .36 | 67.039 | | 6,125 | 3,375 | | 9,500 | 11,800 |
| 0800 | 20 ton cooling | Q-7 | .38 | 83.990 | | 10,900 | 4,325 | | 15,225 | 18,500 |
| 0820 | 30 ton cooling | " | .33 | 96.096 | ↓ | 12,200 | 4,950 | | 17,150 | 20,900 |

# 23 64 Packaged Water Chillers

## 23 64 26 – Rotary-Screw Water Chillers

### 23 64 26.10 Rotary-Screw Type Water Chillers

| | | Crew | Daily Output | Labor-Hours | Unit | Material | 2011 Bare Costs Labor | 2011 Bare Costs Equipment | Total | Total Incl O&P |
|---|---|---|---|---|---|---|---|---|---|---|
| 0010 | **ROTARY-SCREW TYPE WATER CHILLERS**, With standard controls | | | | | | | | | |
| 0110 | Screw, liquid chiller, air cooled, insulated evaporator | | | | | | | | | |
| 0120 | 130 ton | Q-7 | .14 | 228 | Ea. | 83,500 | 11,800 | | 95,300 | 109,500 |
| 0124 | 160 ton | | .13 | 246 | | 103,000 | 12,700 | | 115,700 | 132,000 |
| 0128 | 180 ton | | .13 | 250 | | 115,500 | 12,900 | | 128,400 | 146,000 |
| 0132 | 210 ton | | .12 | 258 | | 127,500 | 13,300 | | 140,800 | 160,000 |
| 0136 | 270 ton | | .12 | 266 | | 145,500 | 13,700 | | 159,200 | 180,500 |
| 0140 | 320 ton | | .12 | 275 | | 183,000 | 14,200 | | 197,200 | 222,000 |
| 0200 | Packaged unit, water cooled, not incl. tower | | | | | | | | | |
| 0210 | 80 ton | Q-7 | .14 | 223 | Ea. | 42,700 | 11,500 | | 54,200 | 64,000 |
| 0240 | 200 ton | | .13 | 251 | | 82,000 | 13,000 | | 95,000 | 109,500 |
| 0270 | 350 ton | | .12 | 275 | | 158,500 | 14,200 | | 172,700 | 195,500 |
| 1450 | Water cooled, tower not included | | | | | | | | | |
| 1580 | 150 ton cooling, screw compressors | Q-7 | .13 | 240 | Ea. | 63,500 | 12,400 | | 75,900 | 88,500 |
| 1620 | 200 ton cooling, screw compressors | | .13 | 250 | | 87,500 | 12,900 | | 100,400 | 115,500 |
| 1660 | 291 ton cooling, screw compressors | | .12 | 260 | | 91,500 | 13,400 | | 104,900 | 121,000 |

# 23 65 Cooling Towers

## 23 65 13 – Forced-Draft Cooling Towers

### 23 65 13.10 Forced-Draft Type Cooling Towers

| | | Crew | Daily Output | Labor-Hours | Unit | Material | 2011 Bare Costs Labor | 2011 Bare Costs Equipment | Total | Total Incl O&P |
|---|---|---|---|---|---|---|---|---|---|---|
| 0010 | **FORCED-DRAFT TYPE COOLING TOWERS**, Packaged units | | | | | | | | | |
| 0070 | Galvanized steel | | | | | | | | | |
| 0080 | Induced draft, crossflow | | | | | | | | | |
| 0100 | Vertical, belt drive, 61 tons | Q-6 | 90 | .267 | TonAC | 171 | 13.45 | | 184.45 | 208 |
| 0150 | 100 ton | | 100 | .240 | | 163 | 12.10 | | 175.10 | 198 |
| 0200 | 115 ton | | 109 | .220 | | 153 | 11.10 | | 164.10 | 185 |
| 0250 | 131 ton | | 120 | .200 | | 151 | 10.10 | | 161.10 | 181 |
| 0260 | 162 ton | | 132 | .182 | | 149 | 9.20 | | 158.20 | 178 |
| 1500 | Induced air, double flow | | | | | | | | | |
| 1900 | Vertical, gear drive, 167 ton | Q-6 | 126 | .190 | TonAC | 145 | 9.60 | | 154.60 | 173 |
| 2000 | 297 ton | | 129 | .186 | | 90.50 | 9.40 | | 99.90 | 114 |
| 2100 | 582 ton | | 132 | .182 | | 73 | 9.20 | | 82.20 | 94 |
| 2150 | 849 ton | | 142 | .169 | | 70.50 | 8.55 | | 79.05 | 91 |
| 2200 | 1016 ton | | 150 | .160 | | 65.50 | 8.10 | | 73.60 | 84 |
| 3000 | For higher capacities, use multiples | | | | | | | | | |
| 3500 | For pumps and piping, add | Q-6 | 38 | .632 | TonAC | 85.50 | 32 | | 117.50 | 142 |
| 4000 | For absorption systems, add | | | | " | 75% | 75% | | | |
| 5000 | Fiberglass tower on galvanized steel support structure | | | | | | | | | |
| 5010 | Draw thru | | | | | | | | | |
| 5100 | 100 ton | Q-6 | 1.40 | 17.143 | Ea. | 10,100 | 865 | | 10,965 | 12,400 |
| 5120 | 120 ton | | 1.20 | 20 | | 11,700 | 1,000 | | 12,700 | 14,400 |
| 5140 | 140 ton | | 1 | 24 | | 12,700 | 1,200 | | 13,900 | 15,800 |
| 5160 | 160 ton | | .80 | 30 | | 14,100 | 1,525 | | 15,625 | 17,800 |
| 5180 | 180 ton | | .65 | 36.923 | | 16,000 | 1,875 | | 17,875 | 20,400 |
| 5200 | 200 ton | | .48 | 50 | | 18,100 | 2,525 | | 20,625 | 23,700 |
| 5300 | For stainless steel support structure, add | | | | | 30% | | | | |
| 5360 | For higher capacities, use multiples of each size | | | | | | | | | |
| 6000 | Stainless steel | | | | | | | | | |
| 6010 | Induced draft, crossflow, horizontal, belt drive | | | | | | | | | |
| 6100 | 57 ton | Q-6 | 1.50 | 16 | Ea. | 24,100 | 810 | | 24,910 | 27,700 |

514

# 23 65 Cooling Towers

## 23 65 13 – Forced-Draft Cooling Towers

| 23 65 13.10 Forced-Draft Type Cooling Towers | | Crew | Daily Output | Labor-Hours | Unit | Material | 2011 Bare Costs Labor | 2011 Bare Costs Equipment | Total | Total Incl O&P |
|---|---|---|---|---|---|---|---|---|---|---|
| 6120 | 91 ton | Q-6 | .99 | 24.242 | Ea. | 29,000 | 1,225 | | 30,225 | 33,800 |
| 6140 | 111 ton | | .43 | 55.814 | | 37,000 | 2,825 | | 39,825 | 44,900 |
| 6160 | 126 ton | ↓ | .22 | 109 | ↓ | 37,000 | 5,500 | | 42,500 | 48,900 |

# 23 73 Indoor Central-Station Air-Handling Units

## 23 73 13 – Modular Indoor Central-Station Air-Handling Units

### 23 73 13.10 Air Handling Units

| | | Crew | Daily Output | Labor-Hours | Unit | Material | 2011 Bare Costs Labor | 2011 Bare Costs Equipment | Total | Total Incl O&P |
|---|---|---|---|---|---|---|---|---|---|---|
| 0010 | **AIR HANDLING UNITS**, Built-Up | | | | | | | | | |
| 0100 | With cooling/heating coil section, filters, mixing box | | | | | | | | | |
| 0880 | Single zone, horizontal/vertical | | | | | | | | | |
| 0890 | Constant volume | | | | | | | | | |
| 0900 | 1600 CFM | Q-5 | 1.20 | 13.333 | Ea. | 3,325 | 650 | | 3,975 | 4,625 |
| 0920 | 5000 CFM | Q-6 | 1.40 | 17.143 | | 8,625 | 865 | | 9,490 | 10,800 |
| 0940 | 11,500 CFM | | 1 | 24 | | 17,500 | 1,200 | | 18,700 | 21,000 |
| 0970 | 22,000 CFM | | .60 | 40 | | 33,400 | 2,025 | | 35,425 | 39,800 |
| 1000 | 40,000 CFM | ↓ | .30 | 80 | | 51,000 | 4,050 | | 55,050 | 62,500 |
| 1140 | 60,000 CFM | Q-7 | .18 | 177 | ↓ | 75,000 | 9,150 | | 84,150 | 95,500 |

## 23 73 39 – Indoor, Direct Gas-Fired Heating and Ventilating Units

### 23 73 39.10 Make-Up Air Unit

| | | Crew | Daily Output | Labor-Hours | Unit | Material | 2011 Bare Costs Labor | 2011 Bare Costs Equipment | Total | Total Incl O&P |
|---|---|---|---|---|---|---|---|---|---|---|
| 0010 | **MAKE-UP AIR UNIT** | | | | | | | | | |
| 0020 | Indoor suspension, natural/LP gas, direct fired, | | | | | | | | | |
| 0032 | standard control. For flue see Section 23 51 23.10 | | | | | | | | | |
| 0040 | 70°F temperature rise, MBH is input | | | | | | | | | |
| 0100 | 75 MBH input | Q-6 | 3.60 | 6.667 | Ea. | 2,400 | 335 | | 2,735 | 3,150 |
| 0160 | 150 MBH input | | 3 | 8 | | 2,850 | 405 | | 3,255 | 3,725 |
| 0220 | 225 MBH input | | 2.40 | 10 | | 3,300 | 505 | | 3,805 | 4,375 |
| 0300 | 400 MBH input | ↓ | 1.60 | 15 | | 7,450 | 760 | | 8,210 | 9,300 |
| 0600 | For discharge louver assembly, add | | | | | 5% | | | | |
| 0700 | For filters, add | | | | | 10% | | | | |
| 0800 | For air shut-off damper section, add | | | | ↓ | 30% | | | | |

# 23 74 Packaged Outdoor HVAC Equipment

## 23 74 33 – Dedicated Outdoor-Air Units

### 23 74 33.10 Roof Top Air Conditioners

| | | | Crew | Daily Output | Labor-Hours | Unit | Material | 2011 Bare Costs Labor | 2011 Bare Costs Equipment | Total | Total Incl O&P |
|---|---|---|---|---|---|---|---|---|---|---|---|
| 0010 | **ROOF TOP AIR CONDITIONERS**, Standard controls, curb, economizer | | | | | | | | | | |
| 1000 | Single zone, electric cool, gas heat | | | | | | | | | | |
| 1100 | 3 ton cooling, 60 MBH heating | R236000-20 | Q-5 | .70 | 22.857 | Ea. | 3,600 | 1,125 | | 4,725 | 5,600 |
| 1120 | 4 ton cooling, 95 MBH heating | | | .61 | 26.403 | | 4,250 | 1,275 | | 5,525 | 6,600 |
| 1140 | 5 ton cooling, 112 MBH heating | | | .56 | 28.521 | | 4,975 | 1,400 | | 6,375 | 7,550 |
| 1145 | 6 ton cooling, 140 MBH heating | | | .52 | 30.769 | | 5,825 | 1,500 | | 7,325 | 8,650 |
| 1150 | 7.5 ton cooling, 170 MBH heating | | ↓ | .50 | 32.258 | | 7,650 | 1,575 | | 9,225 | 10,800 |
| 1160 | 10 ton cooling, 200 MBH heating | | Q-6 | .67 | 35.982 | | 10,600 | 1,825 | | 12,425 | 14,300 |
| 1170 | 12.5 ton cooling, 230 MBH heating | | | .63 | 37.975 | | 13,400 | 1,925 | | 15,325 | 17,600 |
| 1190 | 17.5 ton cooling, 330 MBH heating | | ↓ | .52 | 45.889 | | 18,800 | 2,325 | | 21,125 | 24,200 |
| 1200 | 20 ton cooling, 360 MBH heating | | Q-7 | .67 | 47.976 | | 21,900 | 2,475 | | 24,375 | 27,800 |
| 1210 | 25 ton cooling, 450 MBH heating | | | .56 | 57.554 | | 25,000 | 2,975 | | 27,975 | 31,900 |
| 1220 | 30 ton cooling, 540 MBH heating | | | .47 | 68.376 | | 25,500 | 3,525 | | 29,025 | 33,400 |
| 1240 | 40 ton cooling, 675 MBH heating | | ↓ | .35 | 91.168 | ↓ | 33,400 | 4,700 | | 38,100 | 43,700 |
| 2000 | Multizone, electric cool, gas heat, economizer | | | | | | | | | | |

# 23 74 Packaged Outdoor HVAC Equipment

## 23 74 33 – Dedicated Outdoor-Air Units

### 23 74 33.10 Roof Top Air Conditioners

| | | Crew | Daily Output | Labor-Hours | Unit | Material | 2011 Bare Costs Labor | 2011 Bare Costs Equipment | Total | Total Incl O&P |
|---|---|---|---|---|---|---|---|---|---|---|
| 2100 | 15 ton cooling, 360 MBH heating | Q-7 | .61 | 52.545 | Ea. | 57,000 | 2,700 | | 59,700 | 66,500 |
| 2120 | 20 ton cooling, 360 MBH heating | | .53 | 60.038 | | 61,000 | 3,100 | | 64,100 | 72,000 |
| 2200 | 40 ton cooling, 540 MBH heating | | .28 | 113 | | 107,000 | 5,875 | | 112,875 | 127,000 |
| 2210 | 50 ton cooling, 540 MBH heating | | .23 | 142 | | 133,500 | 7,325 | | 140,825 | 157,500 |
| 2220 | 70 ton cooling, 1500 MBH heating | | .16 | 198 | | 144,500 | 10,200 | | 154,700 | 174,500 |
| 2240 | 80 ton cooling, 1500 MBH heating | | .14 | 228 | | 165,000 | 11,800 | | 176,800 | 199,000 |
| 2260 | 90 ton cooling, 1500 MBH heating | | .13 | 256 | | 173,000 | 13,200 | | 186,200 | 209,500 |
| 2280 | 105 ton cooling, 1500 MBH heating | ↓ | .11 | 290 | | 190,500 | 15,000 | | 205,500 | 232,000 |
| 2400 | For hot water heat coil, deduct | | | | | 5% | | | | |
| 2500 | For steam heat coil, deduct | | | | | 2% | | | | |
| 2600 | For electric heat, deduct | | | | ↓ | 3% | 5% | | | |

# 23 81 Decentralized Unitary HVAC Equipment

## 23 81 13 – Packaged Terminal Air-Conditioners

### 23 81 13.10 Packaged Cabinet Type Air-Conditioners

| | | Crew | Daily Output | Labor-Hours | Unit | Material | Labor | Equipment | Total | Total Incl O&P |
|---|---|---|---|---|---|---|---|---|---|---|
| 0010 | **PACKAGED CABINET TYPE AIR-CONDITIONERS**, Cabinet, wall sleeve, | | | | | | | | | |
| 0100 | louver, electric heat, thermostat, manual changeover, 208 V | | | | | | | | | |
| 0200 | 6,000 BTUH cooling, 8800 BTU heat | Q-5 | 6 | 2.667 | Ea. | 715 | 130 | | 845 | 980 |
| 0220 | 9,000 BTUH cooling, 13,900 BTU heat | | 5 | 3.200 | | 810 | 156 | | 966 | 1,125 |
| 0240 | 12,000 BTUH cooling, 13,900 BTU heat | | 4 | 4 | | 895 | 195 | | 1,090 | 1,275 |
| 0260 | 15,000 BTUH cooling, 13,900 BTU heat | ↓ | 3 | 5.333 | | 985 | 260 | | 1,245 | 1,475 |
| 0500 | For hot water coil, increase heat by 10%, add | | | | | 5% | 10% | | | |
| 1000 | For steam, increase heat output by 30%, add | | | | ↓ | 8% | 10% | | | |

## 23 81 19 – Self-Contained Air-Conditioners

### 23 81 19.20 Self-Contained Single Package

| | | Crew | Daily Output | Labor-Hours | Unit | Material | Labor | Equipment | Total | Total Incl O&P |
|---|---|---|---|---|---|---|---|---|---|---|
| 0010 | **SELF-CONTAINED SINGLE PACKAGE** | | | | | | | | | |
| 0100 | Air cooled, for free blow or duct, not incl. remote condenser | | | | | | | | | |
| 0200 | 3 ton cooling | Q-5 | 1 | 16 | Ea. | 3,475 | 780 | | 4,255 | 5,000 |
| 0220 | 5 ton cooling | Q-6 | 1.20 | 20 | | 4,150 | 1,000 | | 5,150 | 6,075 |
| 0240 | 10 ton cooling | Q-7 | 1 | 32 | | 7,000 | 1,650 | | 8,650 | 10,200 |
| 0260 | 20 ton cooling | | .90 | 35.556 | | 15,600 | 1,825 | | 17,425 | 19,900 |
| 0280 | 30 ton cooling | ↓ | .80 | 40 | ↓ | 23,600 | 2,050 | | 25,650 | 29,100 |
| 0340 | 60 ton cooling | Q-8 | .40 | 80 | | 53,500 | 4,125 | 81 | 57,706 | 65,000 |
| 0490 | For duct mounting no price change | | | | | | | | | |
| 0500 | For steam heating coils, add | | | | Ea. | 10% | 10% | | | |
| 1000 | Water cooled for free blow or duct, not including tower | | | | | | | | | |
| 1010 | Constant volume | | | | | | | | | |
| 1100 | 3 ton cooling | Q-6 | 1 | 24 | Ea. | 3,475 | 1,200 | | 4,675 | 5,625 |
| 1120 | 5 ton cooling | " | 1 | 24 | | 4,550 | 1,200 | | 5,750 | 6,825 |
| 1140 | 10 ton cooling | Q-7 | .90 | 35.556 | | 8,550 | 1,825 | | 10,375 | 12,100 |
| 1160 | 20 ton cooling | | .80 | 40 | | 26,800 | 2,050 | | 28,850 | 32,600 |
| 1180 | 30 ton cooling | ↓ | .70 | 45.714 | ↓ | 35,500 | 2,350 | | 37,850 | 42,600 |

## 23 81 23 – Computer-Room Air-Conditioners

### 23 81 23.10 Computer Room Units

| | | Crew | Daily Output | Labor-Hours | Unit | Material | Labor | Equipment | Total | Total Incl O&P |
|---|---|---|---|---|---|---|---|---|---|---|
| 0010 | **COMPUTER ROOM UNITS** | | | | | | | | | |
| 1000 | Air cooled, includes remote condenser but not | | | | | | | | | |
| 1020 | interconnecting tubing or refrigerant | | | | | | | | | |
| 1080 | 3 ton | Q-5 | .50 | 32 | Ea. | 17,000 | 1,550 | | 18,550 | 21,000 |
| 1120 | 5 ton | | .45 | 35.556 | | 18,100 | 1,725 | | 19,825 | 22,600 |
| 1160 | 6 ton | ↓ | .30 | 53.333 | ↓ | 32,000 | 2,600 | | 34,600 | 39,100 |

# 23 81 Decentralized Unitary HVAC Equipment

## 23 81 23 – Computer-Room Air-Conditioners

| 23 81 23.10 Computer Room Units | | Crew | Daily Output | Labor-Hours | Unit | Material | 2011 Bare Costs Labor | Equipment | Total | Total Incl O&P |
|---|---|---|---|---|---|---|---|---|---|---|
| 1200 | 8 ton | Q-5 | .27 | 59.259 | Ea. | 33,900 | 2,875 | | 36,775 | 41,600 |
| 1240 | 10 ton | | .25 | 64 | | 35,400 | 3,125 | | 38,525 | 43,700 |
| 1260 | 12 ton | | .24 | 66.667 | | 36,700 | 3,250 | | 39,950 | 45,300 |
| 1280 | 15 ton | | .22 | 72.727 | | 39,000 | 3,550 | | 42,550 | 48,200 |
| 1290 | 18 ton | | .20 | 80 | | 44,400 | 3,900 | | 48,300 | 54,500 |
| 1300 | 20 ton | Q-6 | .26 | 92.308 | | 46,700 | 4,650 | | 51,350 | 58,500 |
| 1320 | 22 ton | | .24 | 100 | | 47,300 | 5,050 | | 52,350 | 59,500 |
| 1360 | 30 ton | | .21 | 114 | | 58,000 | 5,775 | | 63,775 | 72,500 |
| 2200 | Chilled water, for connection to | | | | | | | | | |
| 2220 | existing chiller system of adequate capacity | | | | | | | | | |
| 2260 | 5 ton | Q-5 | .74 | 21.622 | Ea. | 13,000 | 1,050 | | 14,050 | 15,900 |

## 23 81 43 – Air-Source Unitary Heat Pumps

### 23 81 43.10 Air-Source Heat Pumps

| | | Crew | Daily Output | Labor-Hours | Unit | Material | 2011 Bare Costs Labor | Equipment | Total | Total Incl O&P |
|---|---|---|---|---|---|---|---|---|---|---|
| 0010 | **AIR-SOURCE HEAT PUMPS**, Not including interconnecting tubing | | | | | | | | | |
| 1000 | Air to air, split system, not including curbs, pads, fan coil and ductwork | | | | | | | | | |
| 1012 | Outside condensing unit only, for fan coil see Section 23 82 19.10 | | | | | | | | | |
| 1020 | 2 ton cooling, 8.5 MBH heat @ 0°F | Q-5 | 2 | 8 | Ea. | 2,125 | 390 | | 2,515 | 2,900 |
| 1060 | 5 ton cooling, 27 MBH heat @ 0°F | | .50 | 32 | | 3,425 | 1,550 | | 4,975 | 6,100 |
| 1080 | 7.5 ton cooling, 33 MBH heat @ 0°F | | .45 | 35.556 | | 5,925 | 1,725 | | 7,650 | 9,075 |
| 1100 | 10 ton cooling, 50 MBH heat @ 0°F | Q-6 | .64 | 37.500 | | 8,250 | 1,900 | | 10,150 | 11,900 |
| 1120 | 15 ton cooling, 64 MBH heat @ 0°F | | .50 | 48 | | 12,200 | 2,425 | | 14,625 | 17,000 |
| 1130 | 20 ton cooling, 85 MBH heat @ 0°F | | .35 | 68.571 | | 16,000 | 3,475 | | 19,475 | 22,800 |
| 1140 | 25 ton cooling, 119 MBH heat @ 0°F | | .25 | 96 | | 19,300 | 4,850 | | 24,150 | 28,400 |
| 1500 | Single package, not including curbs, pads, or plenums | | | | | | | | | |
| 1520 | 2 ton cooling, 6.5 MBH heat @ 0°F | Q-5 | 1.50 | 10.667 | Ea. | 2,750 | 520 | | 3,270 | 3,800 |
| 1580 | 4 ton cooling, 13 MBH heat @ 0°F | | .96 | 16.667 | | 3,850 | 810 | | 4,660 | 5,450 |
| 1640 | 7.5 ton cooling, 35 MBH heat @ 0°F | | .40 | 40 | | 6,850 | 1,950 | | 8,800 | 10,400 |

## 23 81 46 – Water-Source Unitary Heat Pumps

### 23 81 46.10 Water Source Heat Pumps

| | | Crew | Daily Output | Labor-Hours | Unit | Material | 2011 Bare Costs Labor | Equipment | Total | Total Incl O&P |
|---|---|---|---|---|---|---|---|---|---|---|
| 0010 | **WATER SOURCE HEAT PUMPS**, Not incl. connecting tubing or water source | | | | | | | | | |
| 2000 | Water source to air, single package | | | | | | | | | |
| 2100 | 1 ton cooling, 13 MBH heat @ 75°F | Q-5 | 2 | 8 | Ea. | 1,700 | 390 | | 2,090 | 2,425 |
| 2140 | 2 ton cooling, 19 MBH heat @ 75°F | | 1.70 | 9.412 | | 1,875 | 460 | | 2,335 | 2,750 |
| 2220 | 5 ton cooling, 29 MBH heat @ 75°F | | .90 | 17.778 | | 2,825 | 865 | | 3,690 | 4,400 |
| 3960 | For supplementary heat coil, add | | | | | 10% | | | | |
| 4000 | For increase in capacity thru use | | | | | | | | | |
| 4020 | of solar collector, size boiler at 60% | | | | | | | | | |

# 23 82 Convection Heating and Cooling Units

## 23 82 16 – Air Coils

### 23 82 16.10 Flanged Coils

| | | Crew | Daily Output | Labor-Hours | Unit | Material | 2011 Bare Costs Labor | Equipment | Total | Total Incl O&P |
|---|---|---|---|---|---|---|---|---|---|---|
| 0010 | **FLANGED COILS** | | | | | | | | | |
| 0500 | Chilled water cooling, 6 rows, 24" x 48" | Q-5 | 3.20 | 5 | Ea. | 3,725 | 244 | | 3,969 | 4,475 |
| 1000 | Direct expansion cooling, 6 rows, 24" x 48" | | 2.80 | 5.714 | | 4,025 | 278 | | 4,303 | 4,850 |
| 1500 | Hot water heating, 1 row, 24" x 48" | | 4 | 4 | | 1,475 | 195 | | 1,670 | 1,925 |
| 2000 | Steam heating, 1 row, 24" x 48" | | 3.06 | 5.229 | | 2,100 | 255 | | 2,355 | 2,675 |

### 23 82 16.20 Duct Heaters

| | | Crew | Daily Output | Labor-Hours | Unit | Material | 2011 Bare Costs Labor | Equipment | Total | Total Incl O&P |
|---|---|---|---|---|---|---|---|---|---|---|
| 0010 | **DUCT HEATERS**, Electric, 480 V, 3 Ph. | | | | | | | | | |
| 0020 | Finned tubular insert, 500°F | | | | | | | | | |
| 0100 | 8" wide x 6" high, 4.0 kW | Q-20 | 16 | 1.250 | Ea. | 705 | 59 | | 764 | 865 |

## 23 82 16 – Air Coils

| 23 82 16.20 Duct Heaters | | Crew | Daily Output | Labor-Hours | Unit | Material | 2011 Bare Costs Labor | 2011 Bare Costs Equipment | Total | Total Incl O&P |
|---|---|---|---|---|---|---|---|---|---|---|
| 0120 | 12" high, 8.0 kW | Q-20 | 15 | 1.333 | Ea. | 1,175 | 63 | | 1,238 | 1,375 |
| 0140 | 18" high, 12.0 kW | | 14 | 1.429 | | 1,650 | 67.50 | | 1,717.50 | 1,900 |
| 0160 | 24" high, 16.0 kW | | 13 | 1.538 | | 2,100 | 72.50 | | 2,172.50 | 2,425 |
| 0180 | 30" high, 20.0 kW | | 12 | 1.667 | | 2,575 | 78.50 | | 2,653.50 | 2,950 |
| 0300 | 12" wide x 6" high, 6.7 kW | | 15 | 1.333 | | 750 | 63 | | 813 | 920 |
| 0360 | 24" high, 26.7 kW | | 12 | 1.667 | | 2,175 | 78.50 | | 2,253.50 | 2,525 |
| 0700 | 24" wide x 6" high, 17.8 kW | | 13 | 1.538 | | 890 | 72.50 | | 962.50 | 1,100 |
| 0760 | 24" high, 71.1 kW | | 10 | 2 | | 2,725 | 94.50 | | 2,819.50 | 3,150 |
| 8000 | To obtain BTU multiply kW by 3413 | | | | | | | | | |

## 23 82 19 – Fan Coil Units

### 23 82 19.10 Fan Coil Air Conditioning

| | | Crew | Daily Output | Labor-Hours | Unit | Material | 2011 Bare Costs Labor | 2011 Bare Costs Equipment | Total | Total Incl O&P |
|---|---|---|---|---|---|---|---|---|---|---|
| 0010 | **FAN COIL AIR CONDITIONING** Cabinet mounted, filters, controls | | | | | | | | | |
| 0100 | Chilled water, 1/2 ton cooling | Q-5 | 8 | 2 | Ea. | 710 | 97.50 | | 807.50 | 925 |
| 0120 | 1 ton cooling | | 6 | 2.667 | | 815 | 130 | | 945 | 1,100 |
| 0140 | 1.5 ton cooling | | 5.50 | 2.909 | | 920 | 142 | | 1,062 | 1,200 |
| 0150 | 2 ton cooling | | 5.25 | 3.048 | | 1,150 | 148 | | 1,298 | 1,500 |
| 0180 | 3 ton cooling | | 4 | 4 | | 1,725 | 195 | | 1,920 | 2,200 |
| 0262 | For hot water coil, add | | | | | 40% | 10% | | | |
| 0940 | Direct expansion, for use w/air cooled condensing unit, 1.5 ton cooling | Q-5 | 5 | 3.200 | Ea. | 690 | 156 | | 846 | 995 |
| 1000 | 5 ton cooling | " | 3 | 5.333 | | 1,325 | 260 | | 1,585 | 1,850 |
| 1040 | 10 ton cooling | Q-6 | 2.60 | 9.231 | | 2,975 | 465 | | 3,440 | 3,975 |
| 1060 | 20 ton cooling | " | .70 | 34.286 | | 5,575 | 1,725 | | 7,300 | 8,725 |

### 23 82 19.20 Heating and Ventilating Units

| | | Crew | Daily Output | Labor-Hours | Unit | Material | 2011 Bare Costs Labor | 2011 Bare Costs Equipment | Total | Total Incl O&P |
|---|---|---|---|---|---|---|---|---|---|---|
| 0010 | **HEATING AND VENTILATING UNITS**, Classroom units | | | | | | | | | |
| 0020 | Includes filter, heating/cooling coils, standard controls | | | | | | | | | |
| 0080 | 750 CFM, 2 tons cooling | Q-6 | 2 | 12 | Ea. | 3,775 | 605 | | 4,380 | 5,050 |
| 0120 | 1250 CFM, 3 tons cooling | | 1.40 | 17.143 | | 4,600 | 865 | | 5,465 | 6,350 |
| 0140 | 1500 CFM, 4 tons cooling | | .80 | 30 | | 4,925 | 1,525 | | 6,450 | 7,650 |
| 0500 | For electric heat, add | | | | | 35% | | | | |
| 1000 | For no cooling, deduct | | | | | 25% | 10% | | | |

## 23 82 27 – Infrared Units

### 23 82 27.10 Infrared Type Heating Units

| | | Crew | Daily Output | Labor-Hours | Unit | Material | 2011 Bare Costs Labor | 2011 Bare Costs Equipment | Total | Total Incl O&P |
|---|---|---|---|---|---|---|---|---|---|---|
| 0010 | **INFRARED TYPE HEATING UNITS** | | | | | | | | | |
| 0020 | Gas fired, unvented, electric ignition, 100% shutoff. | | | | | | | | | |
| 0030 | Piping and wiring not included | | | | | | | | | |
| 0060 | Input, 15 MBH | Q-5 | 7 | 2.286 | Ea. | 505 | 111 | | 616 | 720 |
| 0120 | 45 MBH | | 5 | 3.200 | | 1,500 | 156 | | 1,656 | 1,875 |
| 0160 | 60 MBH | | 4 | 4 | | 1,600 | 195 | | 1,795 | 2,050 |
| 0240 | 120 MBH | | 2 | 8 | | 1,975 | 390 | | 2,365 | 2,725 |

## 23 82 29 – Radiators

### 23 82 29.10 Hydronic Heating

| | | Crew | Daily Output | Labor-Hours | Unit | Material | 2011 Bare Costs Labor | 2011 Bare Costs Equipment | Total | Total Incl O&P |
|---|---|---|---|---|---|---|---|---|---|---|
| 0010 | **HYDRONIC HEATING**, Terminal units, not incl. main supply pipe | | | | | | | | | |
| 1000 | Radiation | | | | | | | | | |
| 1100 | Panel, baseboard, C.I., including supports, no covers | Q-5 | 46 | .348 | L.F. | 32.50 | 16.95 | | 49.45 | 61 |
| 3000 | Radiators, cast iron | | | | | | | | | |
| 3100 | Free standing or wall hung, 6 tube, 25" high | Q-5 | 96 | .167 | Section | 36 | 8.10 | | 44.10 | 51.50 |
| 3200 | 4 tube, 19" high | " | 96 | .167 | " | 25 | 8.10 | | 33.10 | 39.50 |
| 3250 | Adj. brackets, 2 per wall radiator up to 30 sections | 1 Stpi | 32 | .250 | Ea. | 28 | 13.55 | | 41.55 | 50.50 |
| 9500 | To convert SFR to BTU rating: Hot water, 150 x SFR | | | | | | | | | |
| 9510 | Forced hot water, 180 x SFR; steam, 240 x SFR | | | | | | | | | |

# 23 82 Convection Heating and Cooling Units

## 23 82 33 – Convectors

### 23 82 33.10 Convector Units

| | | Crew | Daily Output | Labor-Hours | Unit | Material | 2011 Bare Costs Labor | Equipment | Total | Total Incl O&P |
|---|---|---|---|---|---|---|---|---|---|---|
| 0010 | **CONVECTOR UNITS,** Terminal units, not incl. main supply pipe | | | | | | | | | |
| 2204 | Convector, multifin, 2 pipe w/cabinet | | | | | | | | | |
| 2210 | 17" H x 24" L | Q-5 | 10 | 1.600 | Ea. | 98.50 | 78 | | 176.50 | 224 |
| 2214 | 17" H x 36" L | | 8.60 | 1.860 | | 148 | 90.50 | | 238.50 | 297 |
| 2218 | 17" H x 48" L | | 7.40 | 2.162 | | 197 | 105 | | 302 | 375 |
| 2222 | 21" H x 24" L | | 9 | 1.778 | | 96.50 | 86.50 | | 183 | 235 |
| 2226 | 21" H x 36" L | | 8.20 | 1.951 | | 145 | 95 | | 240 | 300 |
| 2228 | 21" H x 48" L | | 6.80 | 2.353 | | 193 | 115 | | 308 | 385 |
| 2240 | For knob operated damper, add | | | | | 140% | | | | |
| 2241 | For metal trim strips, add | Q-5 | 64 | .250 | Ea. | 19.25 | 12.20 | | 31.45 | 39 |
| 2243 | For snap-on inlet grille, add | | | | | 10% | 10% | | | |
| 2245 | For hinged access door, add | Q-5 | 64 | .250 | Ea. | 35.50 | 12.20 | | 47.70 | 57 |
| 2246 | For air chamber, auto-venting, add | " | 58 | .276 | " | 6.35 | 13.45 | | 19.80 | 27 |

## 23 82 36 – Finned-Tube Radiation Heaters

### 23 82 36.10 Finned Tube Radiation

| | | Crew | Daily Output | Labor-Hours | Unit | Material | 2011 Bare Costs Labor | Equipment | Total | Total Incl O&P |
|---|---|---|---|---|---|---|---|---|---|---|
| 0010 | **FINNED TUBE RADIATION,** Terminal units, not incl. main supply pipe | | | | | | | | | |
| 1150 | Fin tube, wall hung, 14" slope top cover, with damper | | | | | | | | | |
| 1200 | 1-1/4" copper tube, 4-1/4" alum. fin | Q-5 | 38 | .421 | L.F. | 40.50 | 20.50 | | 61 | 75.50 |
| 1250 | 1-1/4" steel tube, 4-1/4" steel fin | " | 36 | .444 | " | 38.50 | 21.50 | | 60 | 75 |
| 1500 | Note: fin tube may also require corners, caps, etc. | | | | | | | | | |

## 23 82 39 – Unit Heaters

### 23 82 39.16 Propeller Unit Heaters

| | | Crew | Daily Output | Labor-Hours | Unit | Material | 2011 Bare Costs Labor | Equipment | Total | Total Incl O&P |
|---|---|---|---|---|---|---|---|---|---|---|
| 0010 | **PROPELLER UNIT HEATERS** | | | | | | | | | |
| 3950 | Unit heaters, propeller, 115 V 2 psi steam, 60°F entering air | | | | | | | | | |
| 4000 | Horizontal, 12 MBH | Q-5 | 12 | 1.333 | Ea. | 335 | 65 | | 400 | 465 |
| 4060 | 43.9 MBH | | 8 | 2 | | 505 | 97.50 | | 602.50 | 700 |
| 4140 | 96.8 MBH | | 6 | 2.667 | | 735 | 130 | | 865 | 1,000 |
| 4180 | 157.6 MBH | | 4 | 4 | | 980 | 195 | | 1,175 | 1,375 |
| 4240 | 286.9 MBH | | 2 | 8 | | 1,600 | 390 | | 1,990 | 2,325 |
| 4250 | 326.0 MBH | | 1.90 | 8.421 | | 1,825 | 410 | | 2,235 | 2,600 |
| 4260 | 364 MBH | | 1.80 | 8.889 | | 1,900 | 435 | | 2,335 | 2,750 |
| 4270 | 404 MBH | | 1.60 | 10 | | 2,500 | 485 | | 2,985 | 3,475 |
| 4300 | For vertical diffuser, add | | | | | 174 | | | 174 | 191 |
| 4310 | Vertical flow, 40 MBH | Q-5 | 11 | 1.455 | | 500 | 71 | | 571 | 655 |
| 4314 | 58.5 MBH | | 8 | 2 | | 520 | 97.50 | | 617.50 | 715 |
| 4326 | 131.0 MBH | | 4 | 4 | | 780 | 195 | | 975 | 1,150 |
| 4346 | 297.0 MBH | | 1.80 | 8.889 | | 1,500 | 435 | | 1,935 | 2,300 |
| 4354 | 420 MBH, (460 V) | Q-6 | 1.80 | 13.333 | | 2,000 | 675 | | 2,675 | 3,200 |
| 4358 | 500 MBH, (460 V) | | 1.71 | 14.035 | | 2,650 | 710 | | 3,360 | 3,950 |
| 4362 | 570 MBH, (460 V) | | 1.40 | 17.143 | | 3,725 | 865 | | 4,590 | 5,400 |
| 4366 | 620 MBH, (460 V) | | 1.30 | 18.462 | | 4,275 | 930 | | 5,205 | 6,100 |
| 4370 | 960 MBH, (460 V) | | 1.10 | 21.818 | | 7,200 | 1,100 | | 8,300 | 9,550 |

# 23 83 Radiant Heating Units

## 23 83 16 – Radiant-Heating Hydronic Piping

| 23 83 16.10 Radiant Floor Heating | Crew | Daily Output | Labor-Hours | Unit | Material | 2011 Bare Costs Labor | Equipment | Total | Total Incl O&P |
|---|---|---|---|---|---|---|---|---|---|
| 0010 **RADIANT FLOOR HEATING** | | | | | | | | | |
| 0100 Tubing, PEX (cross-linked polyethylene) | | | | | | | | | |
| 0110 Oxygen barrier type for systems with ferrous materials | | | | | | | | | |
| 0120 1/2" | Q-5 | 800 | .020 | L.F. | .99 | .97 | | 1.96 | 2.54 |
| 0130 3/4" | | 535 | .030 | | 1.40 | 1.46 | | 2.86 | 3.71 |
| 0140 1" | ↓ | 400 | .040 | ↓ | 2.19 | 1.95 | | 4.14 | 5.30 |
| 0200 Non barrier type for ferrous free systems | | | | | | | | | |
| 0210 1/2" | Q-5 | 800 | .020 | L.F. | .57 | .97 | | 1.54 | 2.08 |
| 0220 3/4" | | 535 | .030 | | 1.05 | 1.46 | | 2.51 | 3.33 |
| 0230 1" | ↓ | 400 | .040 | ↓ | 1.80 | 1.95 | | 3.75 | 4.89 |
| 1000 Manifolds | | | | | | | | | |
| 1110 Brass | | | | | | | | | |
| 1120 With supply and return valves, flow meter, thermometer, | | | | | | | | | |
| 1122 auto air vent and drain/fill valve. | | | | | | | | | |
| 1130 1", 2 circuit | Q-5 | 14 | 1.143 | Ea. | 236 | 55.50 | | 291.50 | 340 |
| 1140 1", 3 circuit | | 13.50 | 1.185 | | 269 | 57.50 | | 326.50 | 380 |
| 1150 1", 4 circuit | | 13 | 1.231 | | 292 | 60 | | 352 | 410 |
| 1154 1", 5 circuit | | 12.50 | 1.280 | | 350 | 62.50 | | 412.50 | 480 |
| 1158 1", 6 circuit | | 12 | 1.333 | | 380 | 65 | | 445 | 510 |
| 1162 1", 7 circuit | | 11.50 | 1.391 | | 415 | 68 | | 483 | 555 |
| 1166 1", 8 circuit | | 11 | 1.455 | | 460 | 71 | | 531 | 610 |
| 1172 1", 9 circuit | | 10.50 | 1.524 | | 495 | 74 | | 569 | 655 |
| 1174 1", 10 circuit | | 10 | 1.600 | | 530 | 78 | | 608 | 700 |
| 1178 1", 11 circuit | | 9.50 | 1.684 | | 555 | 82 | | 637 | 730 |
| 1182 1", 12 circuit | ↓ | 9 | 1.778 | ↓ | 610 | 86.50 | | 696.50 | 805 |
| 1610 Copper manifold header, (cut to size) | | | | | | | | | |
| 1620 1" header, 12 – 1/2" sweat outlets | Q-5 | 3.33 | 4.805 | Ea. | 76 | 234 | | 310 | 435 |
| 1630 1-1/4" header, 12 – 1/2" sweat outlets | | 3.20 | 5 | | 88 | 244 | | 332 | 460 |
| 1640 1-1/4" header, 12 – 3/4" sweat outlets | | 3 | 5.333 | | 95 | 260 | | 355 | 495 |
| 1650 1-1/2" header, 12 – 3/4" sweat outlets | | 3.10 | 5.161 | | 114 | 251 | | 365 | 500 |
| 1660 2" header, 12 – 3/4" sweat outlets | ↓ | 2.90 | 5.517 | ↓ | 168 | 269 | | 437 | 585 |
| 3000 Valves | | | | | | | | | |
| 3110 Thermostatic zone valve actuator with end switch | Q-5 | 40 | .400 | Ea. | 36.50 | 19.50 | | 56 | 69 |
| 3114 Thermostatic zone valve actuator | " | 36 | .444 | " | 81 | 21.50 | | 102.50 | 122 |
| 3120 Motorized straight zone valve with operator complete | | | | | | | | | |
| 3130 3/4" | Q-5 | 35 | .457 | Ea. | 125 | 22.50 | | 147.50 | 171 |
| 3140 1" | | 32 | .500 | | 137 | 24.50 | | 161.50 | 188 |
| 3150 1-1/4" | ↓ | 29.60 | .541 | ↓ | 165 | 26.50 | | 191.50 | 222 |
| 3500 4 Way mixing valve, manual, brass | | | | | | | | | |
| 3530 1" | Q-5 | 13.30 | 1.203 | Ea. | 126 | 58.50 | | 184.50 | 227 |
| 3540 1-1/4" | | 11.40 | 1.404 | | 137 | 68.50 | | 205.50 | 253 |
| 3550 1-1/2" | | 11 | 1.455 | | 203 | 71 | | 274 | 330 |
| 3560 2" | | 10.60 | 1.509 | | 288 | 73.50 | | 361.50 | 425 |
| 3800 Mixing valve motor, 4 way for valves, 1" and 1-1/4" | | 34 | .471 | | 310 | 23 | | 333 | 375 |
| 3810 Mixing valve motor, 4 way for valves, 1-1/2" and 2" | ↓ | 30 | .533 | ↓ | 355 | 26 | | 381 | 430 |
| 5000 Radiant floor heating, zone control panel | | | | | | | | | |
| 5120 4 Zone actuator valve control, expandable | Q-5 | 20 | .800 | Ea. | 163 | 39 | | 202 | 237 |
| 5130 6 Zone actuator valve control, expandable | | 18 | .889 | | 199 | 43.50 | | 242.50 | 284 |
| 6070 Thermal track, straight panel for long continuous runs, 5.333 S.F. | | 40 | .400 | | 16.60 | 19.50 | | 36.10 | 47.50 |
| 6080 Thermal track, utility panel, for direction reverse at run end, 5.333 S.F. | | 40 | .400 | | 16.60 | 19.50 | | 36.10 | 47.50 |
| 6090 Combination panel, for direction reverse plus straight run, 5.333 S.F. | ↓ | 40 | .400 | ↓ | 16.60 | 19.50 | | 36.10 | 47.50 |
| 7000 PEX tubing fittings | | | | | | | | | |
| 7100 Compression type | | | | | | | | | |

# 23 83 Radiant Heating Units

## 23 83 16 – Radiant-Heating Hydronic Piping

### 23 83 16.10 Radiant Floor Heating

| | | Crew | Daily Output | Labor-Hours | Unit | Material | 2011 Bare Costs Labor | Equipment | Total | Total Incl O&P |
|---|---|---|---|---|---|---|---|---|---|---|
| 7116 | Coupling | | | | | | | | | |
| 7120 | 1/2" x 1/2" | 1 Stpi | 27 | .296 | Ea. | 6.40 | 16.05 | | 22.45 | 31 |
| 7124 | 3/4" x 3/4" | " | 23 | .348 | " | 10.10 | 18.80 | | 28.90 | 39 |
| 7130 | Adapter | | | | | | | | | |
| 7132 | 1/2" x female sweat 1/2" | 1 Stpi | 27 | .296 | Ea. | 4.15 | 16.05 | | 20.20 | 28.50 |
| 7134 | 1/2" x female sweat 3/4" | | 26 | .308 | | 4.88 | 16.65 | | 21.53 | 30.50 |
| 7136 | 5/8" x female sweat 3/4" | ↓ | 24 | .333 | ↓ | 7 | 18.05 | | 25.05 | 34.50 |
| 7140 | Elbow | | | | | | | | | |
| 7142 | 1/2" x female sweat 1/2" | 1 Stpi | 27 | .296 | Ea. | 6.65 | 16.05 | | 22.70 | 31.50 |
| 7144 | 1/2" x female sweat 3/4" | | 26 | .308 | | 7.80 | 16.65 | | 24.45 | 33.50 |
| 7146 | 5/8" x female sweat 3/4" | ↓ | 24 | .333 | ↓ | 8.75 | 18.05 | | 26.80 | 36.50 |
| 7200 | Insert type | | | | | | | | | |
| 7206 | PEX x male NPT | | | | | | | | | |
| 7210 | 1/2" x 1/2" | 1 Stpi | 29 | .276 | Ea. | 3.80 | 14.90 | | 18.70 | 26.50 |
| 7220 | 3/4" x 3/4" | | 27 | .296 | | 4.61 | 16.05 | | 20.66 | 29 |
| 7230 | 1" x 1" | ↓ | 26 | .308 | | 6.20 | 16.65 | | 22.85 | 32 |
| 7300 | PEX coupling | | | | | | | | | |
| 7310 | 1/2" x 1/2" | 1 Stpi | 30 | .267 | Ea. | 1.07 | 14.45 | | 15.52 | 22.50 |
| 7320 | 3/4" x 3/4" | | 29 | .276 | | 1.62 | 14.90 | | 16.52 | 24.50 |
| 7330 | 1" x 1" | ↓ | 28 | .286 | ↓ | 3.01 | 15.45 | | 18.46 | 26.50 |
| 7400 | PEX stainless crimp ring | | | | | | | | | |
| 7410 | 1/2" x 1/2" | 1 Stpi | 86 | .093 | Ea. | .26 | 5.05 | | 5.31 | 7.80 |
| 7420 | 3/4" x 3/4" | | 84 | .095 | | .36 | 5.15 | | 5.51 | 8.10 |
| 7430 | 1" x 1" | ↓ | 82 | .098 | ↓ | .42 | 5.30 | | 5.72 | 8.35 |

## 23 83 33 – Electric Radiant Heaters

### 23 83 33.10 Electric Heating

| | | Crew | Daily Output | Labor-Hours | Unit | Material | 2011 Bare Costs Labor | Equipment | Total | Total Incl O&P |
|---|---|---|---|---|---|---|---|---|---|---|
| 0010 | **ELECTRIC HEATING**, not incl. conduit or feed wiring | | | | | | | | | |
| 1100 | Rule of thumb: Baseboard units, including control | 1 Elec | 4.40 | 1.818 | kW | 95.50 | 91.50 | | 187 | 240 |
| 1300 | Baseboard heaters, 2' long, 375 watt | | 8 | 1 | Ea. | 40 | 50.50 | | 90.50 | 119 |
| 1400 | 3' long, 500 watt | | 8 | 1 | | 46 | 50.50 | | 96.50 | 125 |
| 1600 | 4' long, 750 watt | | 6.70 | 1.194 | | 54 | 60 | | 114 | 149 |
| 1800 | 5' long, 935 watt | | 5.70 | 1.404 | | 68 | 70.50 | | 138.50 | 180 |
| 2000 | 6' long, 1125 watt | | 5 | 1.600 | | 74 | 80.50 | | 154.50 | 201 |
| 2400 | 8' long, 1500 watt | ↓ | 4 | 2 | ↓ | 92 | 101 | | 193 | 250 |
| 2950 | Wall heaters with fan, 120 to 277 volt | | | | | | | | | |
| 3600 | Thermostats, integral | 1 Elec | 16 | .500 | Ea. | 30 | 25 | | 55 | 70.50 |
| 3800 | Line voltage, 1 pole | " | 8 | 1 | " | 31 | 50.50 | | 81.50 | 109 |

# 23 84 Humidity Control Equipment

## 23 84 13 – Humidifiers

### 23 84 13.10 Humidifier Units

| | | Crew | Daily Output | Labor-Hours | Unit | Material | 2011 Bare Costs Labor | Equipment | Total | Total Incl O&P |
|---|---|---|---|---|---|---|---|---|---|---|
| 0010 | **HUMIDIFIER UNITS** | | | | | | | | | |
| 0520 | Steam, room or duct, filter, regulators, auto. controls, 220 V | | | | | | | | | |
| 0540 | 11 lb. per hour | Q-5 | 6 | 2.667 | Ea. | 2,450 | 130 | | 2,580 | 2,900 |
| 0560 | 22 lb. per hour | | 5 | 3.200 | | 2,700 | 156 | | 2,856 | 3,200 |
| 0580 | 33 lb. per hour | | 4 | 4 | | 2,775 | 195 | | 2,970 | 3,350 |
| 0600 | 50 lb. per hour | | 4 | 4 | | 3,400 | 195 | | 3,595 | 4,050 |
| 0620 | 100 lb. per hour | ↓ | 3 | 5.333 | ↓ | 4,075 | 260 | | 4,335 | 4,875 |

# Division Notes

| | CREW | DAILY OUTPUT | LABOR-HOURS | UNIT | 2011 BARE COSTS | | | | TOTAL INCL O&P |
|---|---|---|---|---|---|---|---|---|---|
| | | | | | MAT. | LABOR | EQUIP. | TOTAL | |
| | | | | | | | | | |
| | | | | | | | | | |
| | | | | | | | | | |
| | | | | | | | | | |
| | | | | | | | | | |
| | | | | | | | | | |
| | | | | | | | | | |
| | | | | | | | | | |
| | | | | | | | | | |
| | | | | | | | | | |
| | | | | | | | | | |
| | | | | | | | | | |
| | | | | | | | | | |
| | | | | | | | | | |
| | | | | | | | | | |
| | | | | | | | | | |
| | | | | | | | | | |
| | | | | | | | | | |
| | | | | | | | | | |
| | | | | | | | | | |
| | | | | | | | | | |
| | | | | | | | | | |
| | | | | | | | | | |
| | | | | | | | | | |
| | | | | | | | | | |

# Estimating Tips

## 26 05 00 Common Work Results for Electrical

- Conduit should be taken off in three main categories—power distribution, branch power, and branch lighting—so the estimator can concentrate on systems and components, therefore making it easier to ensure all items have been accounted for.

- For cost modifications for elevated conduit installation, add the percentages to labor according to the height of installation, and only to the quantities exceeding the different height levels, not to the total conduit quantities.

- Remember that aluminum wiring of equal ampacity is larger in diameter than copper and may require larger conduit.

- If more than three wires at a time are being pulled, deduct percentages from the labor hours of that grouping of wires.

- When taking off grounding systems, identify separately the type and size of wire, and list each unique type of ground connection.

- The estimator should take the weights of materials into consideration when completing a takeoff. Topics to consider include: How will the materials be supported? What methods of support are available? How high will the support structure have to reach? Will the final support structure be able to withstand the total burden? Is the support material included or separate from the fixture, equipment, and material specified?

- Do not overlook the costs for equipment used in the installation. If scaffolding or highlifts are available in the field, contractors may use them in lieu of the proposed ladders and rolling staging.

## 26 20 00 Low-Voltage Electrical Transmission

- Supports and concrete pads may be shown on drawings for the larger equipment, or the support system may be only a piece of plywood for the back of a panelboard. In either case, it must be included in the costs.

## 26 40 00 Electrical and Cathodic Protection

- When taking off cathodic protections systems, identify the type and size of cable, and list each unique type of anode connection.

## 26 50 00 Lighting

- Fixtures should be taken off room by room, using the fixture schedule, specifications, and the ceiling plan. For large concentrations of lighting fixtures in the same area, deduct the percentages from labor hours.

# Reference Numbers

Reference numbers are shown in shaded boxes at the beginning of some major classifications. These numbers refer to related items in the Reference Section. The reference information may be an estimating procedure, an alternate pricing method, or technical information.

*Note:* Not all subdivisions listed here necessarily appear in this publication.

*Note:* **Trade Service,** *in part, has been used as a reference source for some of the material prices used in Division 26.*

| 26 05 05.10 Electrical Demolition | Crew | Daily Output | Labor-Hours | Unit | Material | 2011 Bare Costs Labor | Equipment | Total | Total Incl O&P |
|---|---|---|---|---|---|---|---|---|---|
| **0010** | **ELECTRICAL DEMOLITION** | | | | | | | | | |
| 0020 | Conduit to 15' high, including fittings & hangers | | | | | | | | |
| 0100 | Rigid galvanized steel, 1/2" to 1" diameter | 1 Elec | 242 | .033 | L.F. | | 1.66 | | 1.66 | 2.46 |
| 0120 | 1-1/4" to 2" | " | 200 | .040 | | | 2.01 | | 2.01 | 2.98 |
| 0140 | 2-1/2" to 3-1/2" | 2 Elec | 302 | .053 | | | 2.66 | | 2.66 | 3.95 |
| 0160 | 4" to 6" | " | 160 | .100 | | | 5.05 | | 5.05 | 7.45 |
| 0200 | Electric metallic tubing (EMT), 1/2" to 1" | 1 Elec | 394 | .020 | | | 1.02 | | 1.02 | 1.51 |
| 0220 | 1-1/4" to 1-1/2" | " | 326 | .025 | | | 1.23 | | 1.23 | 1.83 |
| 0260 | 3-1/2" to 4" | 2 Elec | 310 | .052 | | | 2.60 | | 2.60 | 3.84 |
| 0270 | Armored cable, (BX) avg. 50' runs | | | | | | | | |
| 0280 | #14, 2 wire | 1 Elec | 690 | .012 | L.F. | | .58 | | .58 | .86 |
| 0290 | #14, 3 wire | | 571 | .014 | | | .70 | | .70 | 1.04 |
| 0300 | #12, 2 wire | | 605 | .013 | | | .67 | | .67 | .98 |
| 0310 | #12, 3 wire | | 514 | .016 | | | .78 | | .78 | 1.16 |
| 0320 | #10, 2 wire | | 514 | .016 | | | .78 | | .78 | 1.16 |
| 0330 | #10, 3 wire | | 425 | .019 | | | .95 | | .95 | 1.40 |
| 0340 | #8, 3 wire | | 342 | .023 | | | 1.18 | | 1.18 | 1.74 |
| 0350 | Non metallic sheathed cable (Romex) | | | | | | | | |
| 0360 | #14, 2 wire | 1 Elec | 720 | .011 | L.F. | | .56 | | .56 | .83 |
| 0370 | #14, 3 wire | | 657 | .012 | | | .61 | | .61 | .91 |
| 0380 | #12, 2 wire | | 629 | .013 | | | .64 | | .64 | .95 |
| 0390 | #10, 3 wire | | 450 | .018 | | | .89 | | .89 | 1.32 |
| 0400 | Wiremold raceway, including fittings & hangers | | | | | | | | |
| 0420 | No. 3000 | 1 Elec | 250 | .032 | L.F. | | 1.61 | | 1.61 | 2.38 |
| 0440 | No. 4000 | | 217 | .037 | | | 1.85 | | 1.85 | 2.75 |
| 0460 | No. 6000 | | 166 | .048 | | | 2.42 | | 2.42 | 3.59 |
| 0465 | Telephone/power pole | | 12 | .667 | Ea. | | 33.50 | | 33.50 | 49.50 |
| 0470 | Non-metallic, straight section | | 480 | .017 | L.F. | | .84 | | .84 | 1.24 |
| 0500 | Channels, steel, including fittings & hangers | | | | | | | | |
| 0520 | 3/4" x 1-1/2" | 1 Elec | 308 | .026 | L.F. | | 1.31 | | 1.31 | 1.93 |
| 0540 | 1-1/2" x 1-1/2" | | 269 | .030 | | | 1.50 | | 1.50 | 2.22 |
| 0560 | 1-1/2" x 1-7/8" | | 229 | .035 | | | 1.76 | | 1.76 | 2.60 |
| 0600 | Copper bus duct, indoor, 3 phase | | | | | | | | |
| 0610 | Including hangers & supports | | | | | | | | |
| 0620 | 225 amp | 2 Elec | 135 | .119 | L.F. | | 5.95 | | 5.95 | 8.85 |
| 0640 | 400 amp | | 106 | .151 | | | 7.60 | | 7.60 | 11.25 |
| 0660 | 600 amp | | 86 | .186 | | | 9.35 | | 9.35 | 13.85 |
| 0680 | 1000 amp | | 60 | .267 | | | 13.40 | | 13.40 | 19.85 |
| 0700 | 1600 amp | | 40 | .400 | | | 20 | | 20 | 30 |
| 0720 | 3000 amp | | 10 | 1.600 | | | 80.50 | | 80.50 | 119 |
| 1300 | Transformer, dry type, 1 phase, incl. removal of | | | | | | | | |
| 1320 | supports, wire & conduit terminations | | | | | | | | |
| 1340 | 1 kVA | 1 Elec | 7.70 | 1.039 | Ea. | | 52.50 | | 52.50 | 77.50 |
| 1420 | 75 kVA | 2 Elec | 2.50 | 6.400 | " | | 320 | | 320 | 475 |
| 1440 | 3 phase to 600V, primary | | | | | | | | |
| 1460 | 3 kVA | 1 Elec | 3.85 | 2.078 | Ea. | | 105 | | 105 | 155 |
| 1520 | 75 kVA | 2 Elec | 2.70 | 5.926 | | | 298 | | 298 | 440 |
| 1550 | 300 kVA | R-3 | 1.80 | 11.111 | | | 555 | 71.50 | 626.50 | 900 |
| 1570 | 750 kVA | " | 1.10 | 18.182 | | | 905 | 117 | 1,022 | 1,475 |
| 1800 | Wire, THW-THWN-THHN, removed from | | | | | | | | |
| 1810 | in place conduit, to 15' high | | | | | | | | |
| 1830 | #14 | 1 Elec | 65 | .123 | C.L.F. | | 6.20 | | 6.20 | 9.15 |
| 1840 | #12 | | 55 | .145 | | | 7.30 | | 7.30 | 10.85 |

## 26 05 05 – Selective Electrical Demolition

### 26 05 05.10 Electrical Demolition

| | | Crew | Daily Output | Labor-Hours | Unit | Material | 2011 Bare Costs Labor | 2011 Bare Costs Equipment | Total | Total Incl O&P |
|---|---|---|---|---|---|---|---|---|---|---|
| 1850 | #10 | 1 Elec | 45.50 | .176 | C.L.F. | | 8.85 | | 8.85 | 13.10 |
| 1860 | #8 | | 40.40 | .198 | | | 9.95 | | 9.95 | 14.75 |
| 1870 | #6 | | 32.60 | .245 | | | 12.35 | | 12.35 | 18.30 |
| 1880 | #4 | 2 Elec | 53 | .302 | | | 15.20 | | 15.20 | 22.50 |
| 1890 | #3 | | 50 | .320 | | | 16.10 | | 16.10 | 24 |
| 1900 | #2 | | 44.60 | .359 | | | 18.05 | | 18.05 | 26.50 |
| 1910 | 1/0 | | 33.20 | .482 | | | 24 | | 24 | 36 |
| 1920 | 2/0 | | 29.20 | .548 | | | 27.50 | | 27.50 | 41 |
| 1930 | 3/0 | | 25 | .640 | | | 32 | | 32 | 47.50 |
| 1940 | 4/0 | | 22 | .727 | | | 36.50 | | 36.50 | 54 |
| 1950 | 250 kcmil | | 20 | .800 | | | 40 | | 40 | 59.50 |
| 1960 | 300 kcmil | | 19 | .842 | | | 42.50 | | 42.50 | 62.50 |
| 1970 | 350 kcmil | | 18 | .889 | | | 44.50 | | 44.50 | 66 |
| 1980 | 400 kcmil | | 17 | .941 | | | 47.50 | | 47.50 | 70 |
| 1990 | 500 kcmil | | 16.20 | .988 | | | 49.50 | | 49.50 | 73.50 |
| 2000 | Interior fluorescent fixtures, incl. supports | | | | | | | | | |
| 2010 | & whips, to 15' high | | | | | | | | | |
| 2100 | Recessed drop-in 2' x 2', 2 lamp | 2 Elec | 35 | .457 | Ea. | | 23 | | 23 | 34 |
| 2120 | 2' x 4', 2 lamp | | 33 | .485 | | | 24.50 | | 24.50 | 36 |
| 2140 | 2' x 4', 4 lamp | | 30 | .533 | | | 27 | | 27 | 39.50 |
| 2160 | 4' x 4', 4 lamp | | 20 | .800 | | | 40 | | 40 | 59.50 |
| 2180 | Surface mount, acrylic lens & hinged frame | | | | | | | | | |
| 2200 | 1' x 4', 2 lamp | 2 Elec | 44 | .364 | Ea. | | 18.30 | | 18.30 | 27 |
| 2220 | 2' x 2', 2 lamp | | 44 | .364 | | | 18.30 | | 18.30 | 27 |
| 2260 | 2' x 4', 4 lamp | | 33 | .485 | | | 24.50 | | 24.50 | 36 |
| 2280 | 4' x 4', 4 lamp | | 23 | .696 | | | 35 | | 35 | 52 |
| 2300 | Strip fixtures, surface mount | | | | | | | | | |
| 2320 | 4' long, 1 lamp | 2 Elec | 53 | .302 | Ea. | | 15.20 | | 15.20 | 22.50 |
| 2340 | 4' long, 2 lamp | | 50 | .320 | | | 16.10 | | 16.10 | 24 |
| 2360 | 8' long, 1 lamp | | 42 | .381 | | | 19.15 | | 19.15 | 28.50 |
| 2380 | 8' long, 2 lamp | | 40 | .400 | | | 20 | | 20 | 30 |
| 2400 | Pendant mount, industrial, incl. removal | | | | | | | | | |
| 2410 | of chain or rod hangers, to 15' high | | | | | | | | | |
| 2420 | 4' long, 2 lamp | 2 Elec | 35 | .457 | Ea. | | 23 | | 23 | 34 |
| 2440 | 8' long, 2 lamp | " | 27 | .593 | " | | 30 | | 30 | 44 |

## 26 05 13 – Medium-Voltage Cables

### 26 05 13.16 Medium-Voltage, Single Cable

| | | Crew | Daily Output | Labor-Hours | Unit | Material | 2011 Bare Costs Labor | 2011 Bare Costs Equipment | Total | Total Incl O&P |
|---|---|---|---|---|---|---|---|---|---|---|
| 0010 | **MEDIUM-VOLTAGE, SINGLE CABLE** Splicing & terminations not included | | | | | | | | | |
| 0040 | Copper, XLP shielding, 5 kV, #6 | 2 Elec | 4.40 | 3.636 | C.L.F. | 139 | 183 | | 322 | 425 |
| 0050 | #4 | | 4.40 | 3.636 | | 181 | 183 | | 364 | 470 |
| 0100 | #2 | | 4 | 4 | | 214 | 201 | | 415 | 535 |
| 0200 | #1 | | 4 | 4 | | 246 | 201 | | 447 | 570 |
| 0400 | 1/0 | | 3.80 | 4.211 | | 271 | 212 | | 483 | 615 |
| 0600 | 2/0 | | 3.60 | 4.444 | | 340 | 224 | | 564 | 705 |
| 0800 | 4/0 | | 3.20 | 5 | | 445 | 252 | | 697 | 865 |
| 1000 | 250 kcmil | 3 Elec | 4.50 | 5.333 | | 550 | 268 | | 818 | 1,000 |
| 1200 | 350 kcmil | | 3.90 | 6.154 | | 705 | 310 | | 1,015 | 1,225 |
| 1400 | 500 kcmil | | 3.60 | 6.667 | | 865 | 335 | | 1,200 | 1,450 |
| 1600 | 15 kV, ungrounded neutral, #1 | 2 Elec | 4 | 4 | | 299 | 201 | | 500 | 630 |
| 1800 | 1/0 | | 3.80 | 4.211 | | 355 | 212 | | 567 | 710 |
| 2000 | 2/0 | | 3.60 | 4.444 | | 405 | 224 | | 629 | 775 |
| 2200 | 4/0 | | 3.20 | 5 | | 540 | 252 | | 792 | 970 |

# 26 05 Common Work Results for Electrical

## 26 05 13 – Medium-Voltage Cables

| 26 05 13.16 Medium-Voltage, Single Cable | | Crew | Daily Output | Labor-Hours | Unit | Material | 2011 Bare Costs Labor | Equipment | Total | Total Incl O&P |
|---|---|---|---|---|---|---|---|---|---|---|
| 2400 | 250 kcmil | 3 Elec | 4.50 | 5.333 | C.L.F. | 600 | 268 | | 868 | 1,050 |
| 2600 | 350 kcmil | | 3.90 | 6.154 | | 755 | 310 | | 1,065 | 1,300 |
| 2800 | 500 kcmil | ▼ | 3.60 | 6.667 | ▼ | 930 | 335 | | 1,265 | 1,525 |

## 26 05 19 – Low-Voltage Electrical Power Conductors and Cables

### 26 05 19.20 Armored Cable

| | | Crew | Daily Output | Labor-Hours | Unit | Material | Labor | Equipment | Total | Total Incl O&P |
|---|---|---|---|---|---|---|---|---|---|---|
| 0010 | **ARMORED CABLE** | | | | | | | | | |
| 0050 | 600 volt, copper (BX), #14, 2 conductor, solid | 1 Elec | 2.40 | 3.333 | C.L.F. | 71.50 | 168 | | 239.50 | 325 |
| 0100 | 3 conductor, solid | | 2.20 | 3.636 | | 111 | 183 | | 294 | 395 |
| 0150 | #12, 2 conductor, solid | | 2.30 | 3.478 | | 72.50 | 175 | | 247.50 | 340 |
| 0200 | 3 conductor, solid | | 2 | 4 | | 117 | 201 | | 318 | 425 |
| 0250 | #10, 2 conductor, solid | | 2 | 4 | | 133 | 201 | | 334 | 445 |
| 0300 | 3 conductor, solid | | 1.60 | 5 | | 184 | 252 | | 436 | 575 |
| 0350 | #8, 3 conductor, solid | | 1.30 | 6.154 | | 345 | 310 | | 655 | 840 |
| 0400 | 3 conductor with PVC jacket, in cable tray, #6 | ▼ | 3.10 | 2.581 | | 385 | 130 | | 515 | 615 |
| 0450 | #4 | 2 Elec | 5.40 | 2.963 | | 495 | 149 | | 644 | 765 |
| 0500 | #2 | | 4.60 | 3.478 | | 660 | 175 | | 835 | 985 |
| 0550 | #1 | | 4 | 4 | | 895 | 201 | | 1,096 | 1,275 |
| 0600 | 1/0 | | 3.60 | 4.444 | | 1,050 | 224 | | 1,274 | 1,475 |
| 0650 | 2/0 | | 3.40 | 4.706 | | 1,275 | 237 | | 1,512 | 1,750 |
| 0700 | 3/0 | | 3.20 | 5 | | 1,500 | 252 | | 1,752 | 2,025 |
| 0750 | 4/0 | ▼ | 3 | 5.333 | | 1,700 | 268 | | 1,968 | 2,275 |
| 0800 | 250 kcmil | 3 Elec | 3.60 | 6.667 | | 1,925 | 335 | | 2,260 | 2,625 |
| 0850 | 350 kcmil | | 3.30 | 7.273 | | 2,575 | 365 | | 2,940 | 3,375 |
| 0900 | 500 kcmil | ▼ | 3 | 8 | ▼ | 3,350 | 400 | | 3,750 | 4,275 |
| 1050 | 5 kV, copper, 3 conductor with PVC jacket, | | | | | | | | | |
| 1060 | non-shielded, in cable tray, #4 | 2 Elec | 380 | .042 | L.F. | 7.60 | 2.12 | | 9.72 | 11.50 |
| 1100 | #2 | | 360 | .044 | | 9.85 | 2.24 | | 12.09 | 14.15 |
| 1200 | #1 | | 300 | .053 | | 12.55 | 2.68 | | 15.23 | 17.75 |
| 1400 | 1/0 | | 290 | .055 | | 14.50 | 2.78 | | 17.28 | 20 |
| 1600 | 2/0 | | 260 | .062 | | 16.75 | 3.10 | | 19.85 | 23 |
| 2000 | 4/0 | ▼ | 240 | .067 | | 22.50 | 3.35 | | 25.85 | 29.50 |
| 2100 | 250 kcmil | 3 Elec | 330 | .073 | | 30.50 | 3.66 | | 34.16 | 39 |
| 2150 | 350 kcmil | | 315 | .076 | | 37.50 | 3.83 | | 41.33 | 47 |
| 2200 | 500 kcmil | ▼ | 270 | .089 | ▼ | 50 | 4.47 | | 54.47 | 61.50 |
| 2400 | 15 kV, copper, 3 conductor with PVC jacket galv steel armored | | | | | | | | | |
| 2500 | grounded neutral, in cable tray, #2 | 2 Elec | 300 | .053 | L.F. | 15.95 | 2.68 | | 18.63 | 21.50 |
| 2600 | #1 | | 280 | .057 | | 16.95 | 2.87 | | 19.82 | 23 |
| 2800 | 1/0 | | 260 | .062 | | 19.50 | 3.10 | | 22.60 | 26 |
| 2900 | 2/0 | | 220 | .073 | | 26 | 3.66 | | 29.66 | 34 |
| 3000 | 4/0 | ▼ | 190 | .084 | | 29.50 | 4.24 | | 33.74 | 38.50 |
| 3100 | 250 kcmil | 3 Elec | 270 | .089 | | 32.50 | 4.47 | | 36.97 | 42.50 |
| 3150 | 350 kcmil | | 240 | .100 | | 38.50 | 5.05 | | 43.55 | 49.50 |
| 3200 | 500 kcmil | ▼ | 210 | .114 | ▼ | 51.50 | 5.75 | | 57.25 | 65.50 |
| 3400 | 15 kV, copper, 3 conductor with PVC jacket, | | | | | | | | | |
| 3450 | ungrounded neutral, in cable tray, #2 | 2 Elec | 260 | .062 | L.F. | 17.20 | 3.10 | | 20.30 | 23.50 |
| 3500 | #1 | | 230 | .070 | | 19.05 | 3.50 | | 22.55 | 26 |
| 3600 | 1/0 | | 200 | .080 | | 22 | 4.02 | | 26.02 | 30 |
| 3700 | 2/0 | | 190 | .084 | | 27 | 4.24 | | 31.24 | 36 |
| 3800 | 4/0 | ▼ | 160 | .100 | | 32.50 | 5.05 | | 37.55 | 43 |
| 4000 | 250 kcmil | 3 Elec | 210 | .114 | | 37.50 | 5.75 | | 43.25 | 50 |
| 4050 | 350 kcmil | | 195 | .123 | | 49.50 | 6.20 | | 55.70 | 63.50 |
| 4100 | 500 kcmil | ▼ | 180 | .133 | ▼ | 60.50 | 6.70 | | 67.20 | 76.50 |

# 26 05 Common Work Results for Electrical

## 26 05 19 – Low-Voltage Electrical Power Conductors and Cables

### 26 05 19.20 Armored Cable

| | | Crew | Daily Output | Labor-Hours | Unit | Material | 2011 Bare Costs Labor | 2011 Bare Costs Equipment | Total | Total Incl O&P |
|---|---|---|---|---|---|---|---|---|---|---|
| 9010 | 600 volt, copper (MC) steel clad, #14, 2 wire | 1 Elec | 2.40 | 3.333 | C.L.F. | 72.50 | 168 | | 240.50 | 330 |
| 9020 | 3 wire | | 2.20 | 3.636 | | 111 | 183 | | 294 | 395 |
| 9040 | #12, 2 wire | | 2.30 | 3.478 | | 73 | 175 | | 248 | 340 |
| 9050 | 3 wire | | 2 | 4 | | 122 | 201 | | 323 | 430 |
| 9070 | #10, 2 wire | | 2 | 4 | | 152 | 201 | | 353 | 465 |
| 9080 | 3 wire | | 1.60 | 5 | | 211 | 252 | | 463 | 605 |
| 9100 | #8, 2 wire, stranded | | 1.80 | 4.444 | | 278 | 224 | | 502 | 635 |
| 9110 | 3 wire, stranded | | 1.30 | 6.154 | | 430 | 310 | | 740 | 935 |
| 9200 | 600 volt, copper (MC) aluminum clad, #14, 2 wire | | 2.65 | 3.019 | | 72.50 | 152 | | 224.50 | 305 |
| 9210 | 3 wire | | 2.45 | 3.265 | | 111 | 164 | | 275 | 365 |
| 9220 | 4 wire | | 2.20 | 3.636 | | 151 | 183 | | 334 | 435 |
| 9230 | #12, 2 wire | | 2.55 | 3.137 | | 73 | 158 | | 231 | 315 |
| 9240 | 3 wire | | 2.20 | 3.636 | | 122 | 183 | | 305 | 405 |
| 9250 | 4 wire | | 2 | 4 | | 162 | 201 | | 363 | 475 |
| 9260 | #10, 2 wire | | 2.20 | 3.636 | | 152 | 183 | | 335 | 440 |
| 9270 | 3 wire | | 1.80 | 4.444 | | 211 | 224 | | 435 | 560 |
| 9280 | 4 wire | | 1.55 | 5.161 | | 325 | 260 | | 585 | 745 |

### 26 05 19.35 Cable Terminations

| | | Crew | Daily Output | Labor-Hours | Unit | Material | 2011 Bare Costs Labor | 2011 Bare Costs Equipment | Total | Total Incl O&P |
|---|---|---|---|---|---|---|---|---|---|---|
| 0010 | **CABLE TERMINATIONS** | | | | | | | | | |
| 0015 | Wire connectors, screw type, #22 to #14 | 1 Elec | 260 | .031 | Ea. | .08 | 1.55 | | 1.63 | 2.38 |
| 0020 | #18 to #12 | | 240 | .033 | | .10 | 1.68 | | 1.78 | 2.59 |
| 0025 | #18 to #10 | | 240 | .033 | | .16 | 1.68 | | 1.84 | 2.66 |
| 0030 | Screw-on connectors, insulated, #18 to #12 | | 240 | .033 | | .12 | 1.68 | | 1.80 | 2.61 |
| 0035 | #16 to #10 | | 230 | .035 | | .17 | 1.75 | | 1.92 | 2.78 |
| 0040 | #14 to #8 | | 210 | .038 | | .38 | 1.92 | | 2.30 | 3.26 |
| 0045 | #12 to #6 | | 180 | .044 | | .63 | 2.24 | | 2.87 | 4 |
| 0050 | Terminal lugs, solderless, #16 to #10 | | 50 | .160 | | .60 | 8.05 | | 8.65 | 12.55 |
| 0100 | #8 to #4 | | 30 | .267 | | .93 | 13.40 | | 14.33 | 21 |
| 0150 | #2 to #1 | | 22 | .364 | | 1.25 | 18.30 | | 19.55 | 28.50 |
| 0200 | 1/0 to 2/0 | | 16 | .500 | | 2.22 | 25 | | 27.22 | 40 |
| 0250 | 3/0 | | 12 | .667 | | 4.30 | 33.50 | | 37.80 | 54 |
| 0300 | 4/0 | | 11 | .727 | | 4.30 | 36.50 | | 40.80 | 58.50 |
| 0350 | 250 kcmil | | 9 | .889 | | 4.30 | 44.50 | | 48.80 | 70.50 |
| 0400 | 350 kcmil | | 7 | 1.143 | | 5.50 | 57.50 | | 63 | 91 |
| 0450 | 500 kcmil | | 6 | 1.333 | | 10.70 | 67 | | 77.70 | 111 |
| 1600 | Crimp 1 hole lugs, copper or aluminum, 600 volt | | | | | | | | | |
| 1620 | #14 | 1 Elec | 60 | .133 | Ea. | .43 | 6.70 | | 7.13 | 10.40 |
| 1630 | #12 | | 50 | .160 | | .58 | 8.05 | | 8.63 | 12.55 |
| 1640 | #10 | | 45 | .178 | | .58 | 8.95 | | 9.53 | 13.90 |
| 1780 | #8 | | 36 | .222 | | 2.44 | 11.20 | | 13.64 | 19.25 |
| 1800 | #6 | | 30 | .267 | | 3.20 | 13.40 | | 16.60 | 23.50 |
| 2000 | #4 | | 27 | .296 | | 4.36 | 14.90 | | 19.26 | 27 |
| 2200 | #2 | | 24 | .333 | | 6.80 | 16.75 | | 23.55 | 32.50 |
| 2400 | #1 | | 20 | .400 | | 7.05 | 20 | | 27.05 | 38 |
| 2600 | 2/0 | | 15 | .533 | | 8.70 | 27 | | 35.70 | 49 |
| 2800 | 3/0 | | 12 | .667 | | 10.25 | 33.50 | | 43.75 | 61 |
| 3000 | 4/0 | | 11 | .727 | | 11.45 | 36.50 | | 47.95 | 66.50 |
| 3200 | 250 kcmil | | 9 | .889 | | 13.45 | 44.50 | | 57.95 | 81 |
| 3400 | 300 kcmil | | 8 | 1 | | 15.50 | 50.50 | | 66 | 91.50 |
| 3500 | 350 kcmil | | 7 | 1.143 | | 15.95 | 57.50 | | 73.45 | 103 |
| 3600 | 400 kcmil | | 6.50 | 1.231 | | 19.45 | 62 | | 81.45 | 113 |
| 3800 | 500 kcmil | | 6 | 1.333 | | 22.50 | 67 | | 89.50 | 125 |

## 26 05 19 – Low-Voltage Electrical Power Conductors and Cables

| 26 05 19.50 Mineral Insulated Cable | | Crew | Daily Output | Labor-Hours | Unit | Material | 2011 Bare Costs Labor | Equipment | Total | Total Incl O&P |
|---|---|---|---|---|---|---|---|---|---|---|
| 0010 | **MINERAL INSULATED CABLE** 600 volt | | | | | | | | | |
| 0100 | 1 conductor, #12 | 1 Elec | 1.60 | 5 | C.L.F. | 370 | 252 | | 622 | 785 |
| 0200 | #10 | | 1.60 | 5 | | 480 | 252 | | 732 | 900 |
| 0400 | #8 | | 1.50 | 5.333 | | 570 | 268 | | 838 | 1,025 |
| 0500 | #6 | | 1.40 | 5.714 | | 645 | 287 | | 932 | 1,125 |
| 0600 | #4 | 2 Elec | 2.40 | 6.667 | | 890 | 335 | | 1,225 | 1,475 |
| 0800 | #2 | | 2.20 | 7.273 | | 1,200 | 365 | | 1,565 | 1,850 |
| 0900 | #1 | | 2.10 | 7.619 | | 1,350 | 385 | | 1,735 | 2,075 |
| 1000 | 1/0 | | 2 | 8 | | 1,550 | 400 | | 1,950 | 2,325 |
| 1100 | 2/0 | | 1.90 | 8.421 | | 1,850 | 425 | | 2,275 | 2,675 |
| 1200 | 3/0 | | 1.80 | 8.889 | | 2,200 | 445 | | 2,645 | 3,075 |
| 1400 | 4/0 | | 1.60 | 10 | | 2,525 | 505 | | 3,030 | 3,525 |
| 1410 | 250 kcmil | 3 Elec | 2.40 | 10 | | 2,875 | 505 | | 3,380 | 3,900 |
| 1420 | 350 kcmil | | 1.95 | 12.308 | | 3,225 | 620 | | 3,845 | 4,450 |
| 1430 | 500 kcmil | | 1.95 | 12.308 | | 4,025 | 620 | | 4,645 | 5,350 |

### 26 05 19.55 Non-Metallic Sheathed Cable

| | | Crew | Daily Output | Labor-Hours | Unit | Material | Labor | Equipment | Total | Total Incl O&P |
|---|---|---|---|---|---|---|---|---|---|---|
| 0010 | **NON-METALLIC SHEATHED CABLE** 600 volt | | | | | | | | | |
| 0100 | Copper with ground wire, (Romex) | | | | | | | | | |
| 0150 | #14, 2 conductor | 1 Elec | 2.70 | 2.963 | C.L.F. | 25.50 | 149 | | 174.50 | 249 |
| 0200 | 3 conductor | | 2.40 | 3.333 | | 35.50 | 168 | | 203.50 | 287 |
| 0250 | #12, 2 conductor | | 2.50 | 3.200 | | 39 | 161 | | 200 | 281 |
| 0300 | 3 conductor | | 2.20 | 3.636 | | 55 | 183 | | 238 | 330 |
| 0350 | #10, 2 conductor | | 2.20 | 3.636 | | 61 | 183 | | 244 | 340 |
| 0400 | 3 conductor | | 1.80 | 4.444 | | 86.50 | 224 | | 310.50 | 425 |
| 0450 | #8, 3 conductor | | 1.50 | 5.333 | | 138 | 268 | | 406 | 545 |
| 0500 | #6, 3 conductor | | 1.40 | 5.714 | | 223 | 287 | | 510 | 670 |
| 0550 | SE type SER aluminum cable, 3 RHW and | | | | | | | | | |
| 0600 | 1 bare neutral, 3 #8 & 1 #8 | 1 Elec | 1.60 | 5 | C.L.F. | 119 | 252 | | 371 | 505 |
| 0650 | 3 #6 & 1 #6 | " | 1.40 | 5.714 | | 134 | 287 | | 421 | 575 |
| 0700 | 3 #4 & 1 #6 | 2 Elec | 2.40 | 6.667 | | 151 | 335 | | 486 | 660 |
| 0750 | 3 #2 & 1 #4 | | 2.20 | 7.273 | | 222 | 365 | | 587 | 785 |
| 0800 | 3 #1/0 & 1 #2 | | 2 | 8 | | 350 | 400 | | 750 | 980 |
| 0850 | 3 #2/0 & 1 #1 | | 1.80 | 8.889 | | 395 | 445 | | 840 | 1,100 |
| 0900 | 3 #4/0 & 1 #2/0 | | 1.60 | 10 | | 565 | 505 | | 1,070 | 1,375 |

### 26 05 19.90 Wire

| | | | Crew | Daily Output | Labor-Hours | Unit | Material | Labor | Equipment | Total | Total Incl O&P |
|---|---|---|---|---|---|---|---|---|---|---|---|
| 0010 | **WIRE** | R260519-92 | | | | | | | | | |
| 0020 | 600 volt, copper type THW, solid, #14 | | 1 Elec | 13 | .615 | C.L.F. | 7.85 | 31 | | 38.85 | 54.50 |
| 0030 | #12 | | | 11 | .727 | | 11.95 | 36.50 | | 48.45 | 67 |
| 0040 | #10 | | | 10 | .800 | | 18.70 | 40 | | 58.70 | 80 |
| 0050 | Stranded, #14 | R260533-22 | | 13 | .615 | | 8.95 | 31 | | 39.95 | 56 |
| 0100 | #12 | | | 11 | .727 | | 12.90 | 36.50 | | 49.40 | 68 |
| 0120 | #10 | | | 10 | .800 | | 19.90 | 40 | | 59.90 | 81.50 |
| 0140 | #8 | | | 8 | 1 | | 32.50 | 50.50 | | 83 | 110 |
| 0160 | #6 | | | 6.50 | 1.231 | | 55 | 62 | | 117 | 152 |
| 0180 | #4 | | 2 Elec | 10.60 | 1.509 | | 86.50 | 76 | | 162.50 | 207 |
| 0200 | #3 | | | 10 | 1.600 | | 109 | 80.50 | | 189.50 | 239 |
| 0220 | #2 | | | 9 | 1.778 | | 137 | 89.50 | | 226.50 | 282 |
| 0240 | #1 | | | 8 | 2 | | 173 | 101 | | 274 | 340 |
| 0260 | 1/0 | | | 6.60 | 2.424 | | 216 | 122 | | 338 | 420 |
| 0280 | 2/0 | | | 5.80 | 2.759 | | 271 | 139 | | 410 | 505 |
| 0300 | 3/0 | | | 5 | 3.200 | | 340 | 161 | | 501 | 615 |
| 0350 | 4/0 | | | 4.40 | 3.636 | | 425 | 183 | | 608 | 740 |

## 26 05 19 – Low-Voltage Electrical Power Conductors and Cables

### 26 05 19.90 Wire

| | | Crew | Daily Output | Labor-Hours | Unit | Material | 2011 Bare Costs Labor | Equipment | Total | Total Incl O&P |
|---|---|---|---|---|---|---|---|---|---|---|
| 0400 | 250 kcmil | 3 Elec | 6 | 4 | C.L.F. | 505 | 201 | | 706 | 855 |
| 0420 | 300 kcmil | | 5.70 | 4.211 | | 605 | 212 | | 817 | 980 |
| 0450 | 350 kcmil | | 5.40 | 4.444 | | 705 | 224 | | 929 | 1,100 |
| 0480 | 400 kcmil | | 5.10 | 4.706 | | 805 | 237 | | 1,042 | 1,225 |
| 0490 | 500 kcmil | | 4.80 | 5 | | 1,000 | 252 | | 1,252 | 1,475 |
| 0540 | 600 volt, aluminum type THHN, stranded, #6 | 1 Elec | 8 | 1 | | 27 | 50.50 | | 77.50 | 105 |
| 0560 | #4 | 2 Elec | 13 | 1.231 | | 33.50 | 62 | | 95.50 | 129 |
| 0580 | #2 | | 10.60 | 1.509 | | 46 | 76 | | 122 | 163 |
| 0600 | #1 | | 9 | 1.778 | | 67 | 89.50 | | 156.50 | 206 |
| 0620 | 1/0 | | 8 | 2 | | 80.50 | 101 | | 181.50 | 238 |
| 0640 | 2/0 | | 7.20 | 2.222 | | 95 | 112 | | 207 | 270 |
| 0680 | 3/0 | | 6.60 | 2.424 | | 118 | 122 | | 240 | 310 |
| 0700 | 4/0 | | 6.20 | 2.581 | | 131 | 130 | | 261 | 335 |
| 0720 | 250 kcmil | 3 Elec | 8.70 | 2.759 | | 160 | 139 | | 299 | 380 |
| 0740 | 300 kcmil | | 8.10 | 2.963 | | 221 | 149 | | 370 | 465 |
| 0760 | 350 kcmil | | 7.50 | 3.200 | | 225 | 161 | | 386 | 485 |
| 0780 | 400 kcmil | | 6.90 | 3.478 | | 263 | 175 | | 438 | 550 |
| 0800 | 500 kcmil | | 6 | 4 | | 290 | 201 | | 491 | 620 |
| 0850 | 600 kcmil | | 5.70 | 4.211 | | 365 | 212 | | 577 | 720 |
| 0880 | 700 kcmil | | 5.10 | 4.706 | | 425 | 237 | | 662 | 815 |
| 0900 | 750 kcmil | | 4.80 | 5 | | 430 | 252 | | 682 | 845 |
| 0920 | 600 volt, copper type THWN-THHN, solid, #14 | 1 Elec | 13 | .615 | | 7.85 | 31 | | 38.85 | 54.50 |
| 0940 | #12 | | 11 | .727 | | 11.95 | 36.50 | | 48.45 | 67 |
| 0960 | #10 | | 10 | .800 | | 18.70 | 40 | | 58.70 | 80 |
| 1000 | Stranded, #14 | | 13 | .615 | | 8.95 | 31 | | 39.95 | 56 |
| 1200 | #12 | | 11 | .727 | | 12.90 | 36.50 | | 49.40 | 68 |
| 1250 | #10 | | 10 | .800 | | 19.90 | 40 | | 59.90 | 81.50 |
| 1300 | #8 | | 8 | 1 | | 32.50 | 50.50 | | 83 | 110 |
| 1350 | #6 | | 6.50 | 1.231 | | 55 | 62 | | 117 | 152 |
| 1400 | #4 | 2 Elec | 10.60 | 1.509 | | 86.50 | 76 | | 162.50 | 207 |

## 26 05 23 – Control-Voltage Electrical Power Cables

### 26 05 23.10 Control Cable

| | | Crew | Daily Output | Labor-Hours | Unit | Material | 2011 Bare Costs Labor | Equipment | Total | Total Incl O&P |
|---|---|---|---|---|---|---|---|---|---|---|
| 0010 | **CONTROL CABLE** | | | | | | | | | |
| 0020 | 600 volt, copper, #14 THWN wire with PVC jacket, 2 wires | 1 Elec | 9 | .889 | C.L.F. | 19.60 | 44.50 | | 64.10 | 87.50 |
| 0100 | 4 wires | | 7 | 1.143 | | 33.50 | 57.50 | | 91 | 122 |
| 0200 | 6 wires | | 6 | 1.333 | | 63 | 67 | | 130 | 169 |
| 0300 | 8 wires | | 5.30 | 1.509 | | 77.50 | 76 | | 153.50 | 198 |
| 0400 | 10 wires | | 4.80 | 1.667 | | 91 | 84 | | 175 | 224 |
| 0500 | 12 wires | | 4.30 | 1.860 | | 106 | 93.50 | | 199.50 | 256 |
| 0600 | 14 wires | | 3.80 | 2.105 | | 127 | 106 | | 233 | 297 |
| 0700 | 16 wires | | 3.50 | 2.286 | | 130 | 115 | | 245 | 315 |
| 0800 | 18 wires | | 3.30 | 2.424 | | 144 | 122 | | 266 | 340 |
| 0900 | 20 wires | | 3 | 2.667 | | 175 | 134 | | 309 | 390 |
| 1000 | 22 wires | | 2.80 | 2.857 | | 180 | 144 | | 324 | 410 |

## 26 05 26 – Grounding and Bonding for Electrical Systems

### 26 05 26.80 Grounding

| | | Crew | Daily Output | Labor-Hours | Unit | Material | 2011 Bare Costs Labor | Equipment | Total | Total Incl O&P |
|---|---|---|---|---|---|---|---|---|---|---|
| 0010 | **GROUNDING** | | | | | | | | | |
| 0030 | Rod, copper clad, 8' long, 1/2" diameter | 1 Elec | 5.50 | 1.455 | Ea. | 18.65 | 73 | | 91.65 | 129 |
| 0050 | 3/4" diameter | | 5.30 | 1.509 | | 32.50 | 76 | | 108.50 | 148 |
| 0080 | 10' long, 1/2" diameter | | 4.80 | 1.667 | | 21.50 | 84 | | 105.50 | 148 |
| 0100 | 3/4" diameter | | 4.40 | 1.818 | | 38.50 | 91.50 | | 130 | 178 |
| 0130 | 15' long, 3/4" diameter | | 4 | 2 | | 83.50 | 101 | | 184.50 | 241 |

## 26 05 26 – Grounding and Bonding for Electrical Systems

| 26 05 26.80 Grounding | | Crew | Daily Output | Labor-Hours | Unit | Material | 2011 Bare Costs Labor | Equipment | Total | Total Incl O&P |
|---|---|---|---|---|---|---|---|---|---|---|
| 0390 | Bare copper wire, stranded, #8 | 1 Elec | 11 | .727 | C.L.F. | 30.50 | 36.50 | | 67 | 87.50 |
| 0400 | #6 | | 10 | .800 | | 53.50 | 40 | | 93.50 | 119 |
| 0600 | #2 | 2 Elec | 10 | 1.600 | | 135 | 80.50 | | 215.50 | 268 |
| 0800 | 3/0 | | 6.60 | 2.424 | | 320 | 122 | | 442 | 530 |
| 1000 | 4/0 | | 5.70 | 2.807 | | 400 | 141 | | 541 | 650 |
| 1200 | 250 kcmil | 3 Elec | 7.20 | 3.333 | | 475 | 168 | | 643 | 775 |
| 1800 | Water pipe ground clamps, heavy duty | | | | | | | | | |
| 2000 | Bronze, 1/2" to 1" diameter | 1 Elec | 8 | 1 | Ea. | 26 | 50.50 | | 76.50 | 103 |
| 2100 | 1-1/4" to 2" diameter | | 8 | 1 | | 34 | 50.50 | | 84.50 | 112 |
| 2200 | 2-1/2" to 3" diameter | | 6 | 1.333 | | 59.50 | 67 | | 126.50 | 165 |
| 2800 | Brazed connections, #6 wire | | 12 | .667 | | 13.65 | 33.50 | | 47.15 | 64.50 |
| 3000 | #2 wire | | 10 | .800 | | 18.30 | 40 | | 58.30 | 79.50 |
| 3100 | 3/0 wire | | 8 | 1 | | 27.50 | 50.50 | | 78 | 105 |
| 3200 | 4/0 wire | | 7 | 1.143 | | 31.50 | 57.50 | | 89 | 120 |
| 3400 | 250 kcmil wire | | 5 | 1.600 | | 36.50 | 80.50 | | 117 | 159 |
| 3600 | 500 kcmil wire | | 4 | 2 | | 45 | 101 | | 146 | 199 |

## 26 05 33 – Raceway and Boxes for Electrical Systems

### 26 05 33.13 Conduit

| | | Crew | Daily Output | Labor-Hours | Unit | Material | 2011 Bare Costs Labor | Equipment | Total | Total Incl O&P |
|---|---|---|---|---|---|---|---|---|---|---|
| 0010 | **CONDUIT** To 15' high, includes 2 terminations, 2 elbows, | R260533-22 | | | | | | | | |
| 0020 | 11 beam clamps, and 11 couplings per 100 L.F. | | | | | | | | | |
| 0300 | Aluminum, 1/2" diameter | 1 Elec | 100 | .080 | L.F. | 2.26 | 4.02 | | 6.28 | 8.45 |
| 0500 | 3/4" diameter | | 90 | .089 | | 3.06 | 4.47 | | 7.53 | 9.95 |
| 0700 | 1" diameter | | 80 | .100 | | 4.03 | 5.05 | | 9.08 | 11.90 |
| 1000 | 1-1/4" diameter | | 70 | .114 | | 5.40 | 5.75 | | 11.15 | 14.45 |
| 1030 | 1-1/2" diameter | | 65 | .123 | | 6.50 | 6.20 | | 12.70 | 16.30 |
| 1050 | 2" diameter | | 60 | .133 | | 8.85 | 6.70 | | 15.55 | 19.70 |
| 1070 | 2-1/2" diameter | | 50 | .160 | | 14.45 | 8.05 | | 22.50 | 28 |
| 1100 | 3" diameter | 2 Elec | 90 | .178 | | 20 | 8.95 | | 28.95 | 35.50 |
| 1130 | 3-1/2" diameter | | 80 | .200 | | 26.50 | 10.05 | | 36.55 | 44 |
| 1140 | 4" diameter | | 70 | .229 | | 32 | 11.50 | | 43.50 | 52 |
| 1750 | Rigid galvanized steel, 1/2" diameter | 1 Elec | 90 | .089 | | 2.55 | 4.47 | | 7.02 | 9.40 |
| 1770 | 3/4" diameter | | 80 | .100 | | 2.70 | 5.05 | | 7.75 | 10.40 |
| 1800 | 1" diameter | | 65 | .123 | | 3.62 | 6.20 | | 9.82 | 13.15 |
| 1830 | 1-1/4" diameter | | 60 | .133 | | 5.05 | 6.70 | | 11.75 | 15.50 |
| 1850 | 1-1/2" diameter | | 55 | .145 | | 5.80 | 7.30 | | 13.10 | 17.25 |
| 1870 | 2" diameter | | 45 | .178 | | 7.55 | 8.95 | | 16.50 | 21.50 |
| 1900 | 2-1/2" diameter | | 35 | .229 | | 14.40 | 11.50 | | 25.90 | 33 |
| 1930 | 3" diameter | 2 Elec | 50 | .320 | | 17.65 | 16.10 | | 33.75 | 43.50 |
| 1950 | 3-1/2" diameter | | 44 | .364 | | 21.50 | 18.30 | | 39.80 | 50.50 |
| 1970 | 4" diameter | | 40 | .400 | | 25.50 | 20 | | 45.50 | 58.50 |
| 2500 | Steel, intermediate conduit (IMC), 1/2" diameter | 1 Elec | 100 | .080 | | 1.73 | 4.02 | | 5.75 | 7.85 |
| 2530 | 3/4" diameter | | 90 | .089 | | 2.07 | 4.47 | | 6.54 | 8.90 |
| 2550 | 1" diameter | | 70 | .114 | | 2.84 | 5.75 | | 8.59 | 11.60 |
| 2570 | 1-1/4" diameter | | 65 | .123 | | 3.72 | 6.20 | | 9.92 | 13.25 |
| 2600 | 1-1/2" diameter | | 60 | .133 | | 4.47 | 6.70 | | 11.17 | 14.85 |
| 2630 | 2" diameter | | 50 | .160 | | 5.65 | 8.05 | | 13.70 | 18.15 |
| 2650 | 2-1/2" diameter | | 40 | .200 | | 10.85 | 10.05 | | 20.90 | 27 |
| 2670 | 3" diameter | 2 Elec | 60 | .267 | | 14.75 | 13.40 | | 28.15 | 36 |
| 2700 | 3-1/2" diameter | | 54 | .296 | | 18.80 | 14.90 | | 33.70 | 42.50 |
| 2730 | 4" diameter | | 50 | .320 | | 21 | 16.10 | | 37.10 | 47 |
| 5000 | Electric metallic tubing (EMT), 1/2" diameter | 1 Elec | 170 | .047 | | .64 | 2.37 | | 3.01 | 4.21 |
| 5020 | 3/4" diameter | | 130 | .062 | | 1 | 3.10 | | 4.10 | 5.70 |

## 26 05 33 – Raceway and Boxes for Electrical Systems

| 26 05 33.13 Conduit | | Crew | Daily Output | Labor-Hours | Unit | Material | 2011 Bare Costs Labor | 2011 Bare Costs Equipment | Total | Total Incl O&P |
|---|---|---|---|---|---|---|---|---|---|---|
| 5040 | 1" diameter | 1 Elec | 115 | .070 | L.F. | 1.70 | 3.50 | | 5.20 | 7.05 |
| 5060 | 1-1/4" diameter | | 100 | .080 | | 2.81 | 4.02 | | 6.83 | 9.05 |
| 5080 | 1-1/2" diameter | | 90 | .089 | | 3.66 | 4.47 | | 8.13 | 10.60 |
| 5100 | 2" diameter | | 80 | .100 | | 4.74 | 5.05 | | 9.79 | 12.65 |
| 5120 | 2-1/2" diameter | | 60 | .133 | | 11.35 | 6.70 | | 18.05 | 22.50 |
| 5140 | 3" diameter | 2 Elec | 100 | .160 | | 13.40 | 8.05 | | 21.45 | 26.50 |
| 5160 | 3-1/2" diameter | | 90 | .178 | | 16.90 | 8.95 | | 25.85 | 32 |
| 5180 | 4" diameter | | 80 | .200 | | 18.75 | 10.05 | | 28.80 | 35.50 |
| 9900 | Add to labor for higher elevated installation | | | | | | | | | |
| 9910 | 15' to 20' high, add | | | | | | 10% | | | |
| 9920 | 20' to 25' high, add | | | | | | 20% | | | |
| 9930 | 25' to 30' high, add | | | | | | 25% | | | |
| 9940 | 30' to 35' high, add | | | | | | 30% | | | |
| 9950 | 35' to 40' high, add | | | | | | 35% | | | |
| 9960 | Over 40' high, add | | | | | | 40% | | | |
| 9980 | Allow. for cond. ftngs., 5% min.-20% max. | | | | | | | | | |

## 26 05 33.16 Outlet Boxes

| 26 05 33.16 Outlet Boxes | | Crew | Daily Output | Labor-Hours | Unit | Material | 2011 Bare Costs Labor | 2011 Bare Costs Equipment | Total | Total Incl O&P |
|---|---|---|---|---|---|---|---|---|---|---|
| 0010 | **OUTLET BOXES** | | | | | | | | | |
| 0020 | Pressed steel, octagon, 4" | 1 Elec | 20 | .400 | Ea. | 3.63 | 20 | | 23.63 | 34 |
| 0060 | Covers, blank | | 64 | .125 | | 1.50 | 6.30 | | 7.80 | 10.95 |
| 0100 | Extension rings | | 40 | .200 | | 6 | 10.05 | | 16.05 | 21.50 |
| 0150 | Square, 4" | | 20 | .400 | | 2.48 | 20 | | 22.48 | 32.50 |
| 0200 | Extension rings | | 40 | .200 | | 6.10 | 10.05 | | 16.15 | 21.50 |
| 0250 | Covers, blank | | 64 | .125 | | 1.65 | 6.30 | | 7.95 | 11.10 |
| 0300 | Plaster rings | | 64 | .125 | | 3.36 | 6.30 | | 9.66 | 13 |
| 0650 | Switchbox | | 27 | .296 | | 5.80 | 14.90 | | 20.70 | 28.50 |
| 1100 | Concrete, floor, 1 gang | | 5.30 | 1.509 | | 89.50 | 76 | | 165.50 | 211 |
| 2000 | Poke-thru fitting, fire rated, for 3-3/4" floor | | 6.80 | 1.176 | | 119 | 59 | | 178 | 219 |
| 2040 | For 7" floor | | 6.80 | 1.176 | | 119 | 59 | | 178 | 219 |
| 2100 | Pedestal, 15 amp, duplex receptacle & blank plate | | 5.25 | 1.524 | | 131 | 76.50 | | 207.50 | 259 |
| 2120 | Duplex receptacle and telephone plate | | 5.25 | 1.524 | | 131 | 76.50 | | 207.50 | 259 |
| 2140 | Pedestal, 20 amp, duplex recept. & phone plate | | 5 | 1.600 | | 132 | 80.50 | | 212.50 | 265 |
| 2200 | Abandonment plate | | 32 | .250 | | 37.50 | 12.60 | | 50.10 | 59.50 |

## 26 05 33.18 Pull Boxes

| 26 05 33.18 Pull Boxes | | Crew | Daily Output | Labor-Hours | Unit | Material | 2011 Bare Costs Labor | 2011 Bare Costs Equipment | Total | Total Incl O&P |
|---|---|---|---|---|---|---|---|---|---|---|
| 0010 | **PULL BOXES** | | | | | | | | | |
| 0100 | Sheet metal, pull box, NEMA 1, type SC, 6" W x 6" H x 4" D | 1 Elec | 8 | 1 | Ea. | 14.90 | 50.50 | | 65.40 | 91 |
| 0200 | 8" W x 8" H x 4" D | | 8 | 1 | | 18.70 | 50.50 | | 69.20 | 95 |
| 0300 | 10" W x 12" H x 6" D | | 5.30 | 1.509 | | 38 | 76 | | 114 | 154 |
| 0400 | 16" W x 20" H x 8" D | | 4 | 2 | | 130 | 101 | | 231 | 292 |
| 0500 | 20" W x 24" H x 8" D | | 3.20 | 2.500 | | 141 | 126 | | 267 | 340 |
| 0600 | 24" W x 36" H x 8" D | | 2.70 | 2.963 | | 217 | 149 | | 366 | 460 |
| 0650 | Pull box, hinged , NEMA 1, 6" W x 6" H x 4" D | | 8 | 1 | | 14.90 | 50.50 | | 65.40 | 91 |
| 0800 | 12" W x 16" H x 6" D | | 4.70 | 1.702 | | 48.50 | 85.50 | | 134 | 181 |
| 1000 | 20" W x 20" H x 6" D | | 3.60 | 2.222 | | 97 | 112 | | 209 | 273 |
| 1200 | 20" W x 20" H x 8" D | | 3.20 | 2.500 | | 194 | 126 | | 320 | 400 |
| 1400 | 24" W x 36" H x 8" D | | 2.70 | 2.963 | | 310 | 149 | | 459 | 560 |
| 1600 | 24" W x 42" H x 8" D | | 2 | 4 | | 470 | 201 | | 671 | 820 |
| 2100 | Pull box, NEMA 3R, type SC, raintight & weatherproof | | | | | | | | | |
| 2150 | 6" L x 6" W x 6" D | 1 Elec | 10 | .800 | Ea. | 25 | 40 | | 65 | 87 |
| 2200 | 8" L x 6" W x 6" D | | 8 | 1 | | 31.50 | 50.50 | | 82 | 109 |
| 2250 | 10" L x 6" W x 6" D | | 7 | 1.143 | | 41.50 | 57.50 | | 99 | 131 |
| 2300 | 12" L x 12" W x 6" D | | 5 | 1.600 | | 59 | 80.50 | | 139.50 | 184 |

## 26 05 33 – Raceway and Boxes for Electrical Systems

### 26 05 33.18 Pull Boxes

| | | Crew | Daily Output | Labor-Hours | Unit | Material | 2011 Bare Costs Labor | Equipment | Total | Total Incl O&P |
|---|---|---|---|---|---|---|---|---|---|---|
| 2350 | 16" L x 16" W x 6" D | 1 Elec | 4.50 | 1.778 | Ea. | 117 | 89.50 | | 206.50 | 261 |
| 2400 | 20" L x 20" W x 6" D | | 4 | 2 | | 160 | 101 | | 261 | 325 |
| 2450 | 24" L x 18" W x 8" D | | 3 | 2.667 | | 166 | 134 | | 300 | 380 |
| 2500 | 24" L x 24" W x 10" D | | 2.50 | 3.200 | | 213 | 161 | | 374 | 470 |
| 2550 | 30" L x 24" W x 12" D | | 2 | 4 | | 430 | 201 | | 631 | 770 |
| 2600 | 36" L x 36" W x 12" D | | 1.50 | 5.333 | | 565 | 268 | | 833 | 1,025 |
| 2800 | Cast iron, pull boxes for surface mounting | | | | | | | | | |
| 3000 | NEMA 4, watertight & dust tight | | | | | | | | | |
| 3050 | 6" L x 6" W x 6" D | 1 Elec | 4 | 2 | Ea. | 233 | 101 | | 334 | 405 |
| 3100 | 8" L x 6" W x 6" D | | 3.20 | 2.500 | | 253 | 126 | | 379 | 465 |
| 3150 | 10" L x 6" W x 6" D | | 2.50 | 3.200 | | 315 | 161 | | 476 | 585 |
| 3200 | 12" L x 12" W x 6" D | | 2.30 | 3.478 | | 555 | 175 | | 730 | 875 |
| 3250 | 16" L x 16" W x 6" D | | 1.30 | 6.154 | | 1,450 | 310 | | 1,760 | 2,050 |
| 3300 | 20" L x 20" W x 6" D | | .80 | 10 | | 2,700 | 505 | | 3,205 | 3,725 |
| 3350 | 24" L x 18" W x 8" D | | .70 | 11.429 | | 2,825 | 575 | | 3,400 | 3,975 |
| 3400 | 24" L x 24" W x 10" D | | .50 | 16 | | 4,350 | 805 | | 5,155 | 5,975 |
| 3450 | 30" L x 24" W x 12" D | | .40 | 20 | | 7,275 | 1,000 | | 8,275 | 9,500 |
| 3500 | 36" L x 36" W x 12" D | | .20 | 40 | | 9,000 | 2,000 | | 11,000 | 12,900 |
| 6000 | J.I.C. wiring boxes, NEMA 12, dust tight & drip tight | | | | | | | | | |
| 6050 | 6" L x 8" W x 4" D | 1 Elec | 10 | .800 | Ea. | 59 | 40 | | 99 | 125 |
| 6100 | 8" L x 10" W x 4" D | | 8 | 1 | | 74 | 50.50 | | 124.50 | 156 |
| 6150 | 12" L x 14" W x 6" D | | 5.30 | 1.509 | | 124 | 76 | | 200 | 248 |
| 6200 | 14" L x 16" W x 6" D | | 4.70 | 1.702 | | 155 | 85.50 | | 240.50 | 298 |
| 6250 | 16" L x 20" W x 6" D | | 4.40 | 1.818 | | 305 | 91.50 | | 396.50 | 470 |
| 6300 | 24" L x 30" W x 6" D | | 3.20 | 2.500 | | 450 | 126 | | 576 | 680 |
| 6350 | 24" L x 30" W x 8" D | | 2.90 | 2.759 | | 480 | 139 | | 619 | 735 |
| 6400 | 24" L x 36" W x 8" D | | 2.70 | 2.963 | | 535 | 149 | | 684 | 805 |
| 6450 | 24" L x 42" W x 8" D | | 2.30 | 3.478 | | 585 | 175 | | 760 | 900 |
| 6500 | 24" L x 48" W x 8" D | | 2 | 4 | | 635 | 201 | | 836 | 1,000 |

### 26 05 33.23 Wireway

| | | Crew | Daily Output | Labor-Hours | Unit | Material | 2011 Bare Costs Labor | Equipment | Total | Total Incl O&P |
|---|---|---|---|---|---|---|---|---|---|---|
| 0010 | **WIREWAY** to 15' high | | | | | | | | | |
| 0100 | Screw cover, NEMA 1 w/fittings and supports, 2-1/2" x 2-1/2" | 1 Elec | 45 | .178 | L.F. | 11.45 | 8.95 | | 20.40 | 26 |
| 0200 | 4" x 4" | " | 40 | .200 | | 12.65 | 10.05 | | 22.70 | 29 |
| 0400 | 6" x 6" | 2 Elec | 60 | .267 | | 19.70 | 13.40 | | 33.10 | 41.50 |
| 0600 | 8" x 8" | " | 40 | .400 | | 33 | 20 | | 53 | 66.50 |
| 4475 | Screw cover, NEMA 3R w/fittings and supports, 4" x 4" | 1 Elec | 36 | .222 | | 27 | 11.20 | | 38.20 | 46 |
| 4480 | 6" x 6" | 2 Elec | 55 | .291 | | 34 | 14.65 | | 48.65 | 59 |
| 4485 | 8" x 8" | | 36 | .444 | | 55 | 22.50 | | 77.50 | 93.50 |
| 4490 | 12" x 12" | | 18 | .889 | | 77 | 44.50 | | 121.50 | 151 |
| 9980 | Allow. for wireway ftngs., 5% min.-20% max. | | | | | | | | | |

## 26 05 36 – Cable Trays for Electrical Systems

### 26 05 36.10 Cable Tray Ladder Type

| | | Crew | Daily Output | Labor-Hours | Unit | Material | 2011 Bare Costs Labor | Equipment | Total | Total Incl O&P |
|---|---|---|---|---|---|---|---|---|---|---|
| 0010 | **CABLE TRAY LADDER TYPE** w/ftngs. & supports, 4" dp., to 15' elev. | | | | | | | | | |
| 0160 | Galvanized steel tray | | | | | | | | | |
| 0170 | 4" rung spacing, 6" wide | 2 Elec | 98 | .163 | L.F. | 14.10 | 8.20 | | 22.30 | 27.50 |
| 0200 | 12" wide | | 86 | .186 | | 16.90 | 9.35 | | 26.25 | 32.50 |
| 0400 | 18" wide | | 82 | .195 | | 19.65 | 9.80 | | 29.45 | 36 |
| 0600 | 24" wide | | 78 | .205 | | 22.50 | 10.30 | | 32.80 | 40.50 |
| 3200 | Aluminum tray, 4" deep, 6" rung spacing, 6" wide | | 134 | .119 | | 17.70 | 6 | | 23.70 | 28.50 |
| 3220 | 12" wide | | 124 | .129 | | 19.80 | 6.50 | | 26.30 | 31.50 |
| 3230 | 18" wide | | 114 | .140 | | 22 | 7.05 | | 29.05 | 35 |
| 3240 | 24" wide | | 106 | .151 | | 25.50 | 7.60 | | 33.10 | 39.50 |

# 26 05 Common Work Results for Electrical

## 26 05 36 – Cable Trays for Electrical Systems

| 26 05 36.10 Cable Tray Ladder Type | Crew | Daily Output | Labor-Hours | Unit | Material | 2011 Bare Costs Labor | Equipment | Total | Total Incl O&P |
|---|---|---|---|---|---|---|---|---|---|
| 9980 | Allow. for tray ftngs., 5% min.-20% max. | | | | | | | | | |

## 26 05 39 – Underfloor Raceways for Electrical Systems

### 26 05 39.30 Conduit In Concrete Slab

| | 26 05 39.30 Conduit In Concrete Slab | Crew | Daily Output | Labor-Hours | Unit | Material | Labor | Equipment | Total | Total Incl O&P |
|---|---|---|---|---|---|---|---|---|---|---|
| 0010 | **CONDUIT IN CONCRETE SLAB** Including terminations, | | | | | | | | | |
| 0020 | fittings and supports | | | | | | | | | |
| 3230 | PVC, schedule 40, 1/2" diameter | 1 Elec | 270 | .030 | L.F. | .72 | 1.49 | | 2.21 | 3 |
| 3250 | 3/4" diameter | | 230 | .035 | | .86 | 1.75 | | 2.61 | 3.54 |
| 3270 | 1" diameter | | 200 | .040 | | 1.14 | 2.01 | | 3.15 | 4.24 |
| 3300 | 1-1/4" diameter | | 170 | .047 | | 1.61 | 2.37 | | 3.98 | 5.30 |
| 3330 | 1-1/2" diameter | | 140 | .057 | | 1.91 | 2.87 | | 4.78 | 6.35 |
| 3350 | 2" diameter | | 120 | .067 | | 2.45 | 3.35 | | 5.80 | 7.65 |
| 4350 | Rigid galvanized steel, 1/2" diameter | | 200 | .040 | | 2.27 | 2.01 | | 4.28 | 5.50 |
| 4400 | 3/4" diameter | | 170 | .047 | | 2.42 | 2.37 | | 4.79 | 6.15 |
| 4450 | 1" diameter | | 130 | .062 | | 3.34 | 3.10 | | 6.44 | 8.25 |
| 4500 | 1-1/4" diameter | | 110 | .073 | | 4.48 | 3.66 | | 8.14 | 10.30 |
| 4600 | 1-1/2" diameter | | 100 | .080 | | 5.25 | 4.02 | | 9.27 | 11.75 |
| 4800 | 2" diameter | | 90 | .089 | | 6.75 | 4.47 | | 11.22 | 14.05 |

### 26 05 39.40 Conduit In Trench

| | 26 05 39.40 Conduit In Trench | Crew | Daily Output | Labor-Hours | Unit | Material | Labor | Equipment | Total | Total Incl O&P |
|---|---|---|---|---|---|---|---|---|---|---|
| 0010 | **CONDUIT IN TRENCH** Includes terminations and fittings | | | | | | | | | |
| 0020 | Does not include excavation or backfill, see Section 31 23 16.00 | | | | | | | | | |
| 0200 | Rigid galvanized steel, 2" diameter | 1 Elec | 150 | .053 | L.F. | 6.45 | 2.68 | | 9.13 | 11 |
| 0400 | 2-1/2" diameter | " | 100 | .080 | | 12.30 | 4.02 | | 16.32 | 19.45 |
| 0600 | 3" diameter | 2 Elec | 160 | .100 | | 15.45 | 5.05 | | 20.50 | 24.50 |
| 0800 | 3-1/2" diameter | | 140 | .114 | | 19.95 | 5.75 | | 25.70 | 30.50 |
| 1000 | 4" diameter | | 100 | .160 | | 22.50 | 8.05 | | 30.55 | 36.50 |
| 1200 | 5" diameter | | 80 | .200 | | 46 | 10.05 | | 56.05 | 66 |
| 1400 | 6" diameter | | 60 | .267 | | 67.50 | 13.40 | | 80.90 | 94.50 |

## 26 05 43 – Underground Ducts and Raceways for Electrical Systems

### 26 05 43.10 Trench Duct

| | 26 05 43.10 Trench Duct | Crew | Daily Output | Labor-Hours | Unit | Material | Labor | Equipment | Total | Total Incl O&P |
|---|---|---|---|---|---|---|---|---|---|---|
| 0010 | **TRENCH DUCT** Steel with cover | | | | | | | | | |
| 0020 | Standard adjustable, depths to 4" | | | | | | | | | |
| 0100 | Straight, single compartment, 9" wide | 2 Elec | 40 | .400 | L.F. | 111 | 20 | | 131 | 152 |
| 0200 | 12" wide | | 32 | .500 | | 126 | 25 | | 151 | 176 |
| 0400 | 18" wide | | 26 | .615 | | 167 | 31 | | 198 | 230 |
| 0600 | 24" wide | | 22 | .727 | | 215 | 36.50 | | 251.50 | 290 |
| 0800 | 30" wide | | 20 | .800 | | 259 | 40 | | 299 | 345 |
| 1000 | 36" wide | | 16 | 1 | | 310 | 50.50 | | 360.50 | 420 |
| 1200 | Horizontal elbow, 9" wide | | 5.40 | 2.963 | Ea. | 410 | 149 | | 559 | 670 |
| 1400 | 12" wide | | 4.60 | 3.478 | | 475 | 175 | | 650 | 780 |
| 1600 | 18" wide | | 4 | 4 | | 605 | 201 | | 806 | 965 |
| 1800 | 24" wide | | 3.20 | 5 | | 850 | 252 | | 1,102 | 1,300 |
| 2000 | 30" wide | | 2.60 | 6.154 | | 1,125 | 310 | | 1,435 | 1,700 |
| 2200 | 36" wide | | 2.40 | 6.667 | | 1,500 | 335 | | 1,835 | 2,125 |
| 2400 | Vertical elbow, 9" wide | | 5.40 | 2.963 | | 143 | 149 | | 292 | 380 |
| 2600 | 12" wide | | 4.60 | 3.478 | | 155 | 175 | | 330 | 430 |
| 2800 | 18" wide | | 4 | 4 | | 178 | 201 | | 379 | 495 |
| 3000 | 24" wide | | 3.20 | 5 | | 221 | 252 | | 473 | 620 |
| 3200 | 30" wide | | 2.60 | 6.154 | | 244 | 310 | | 554 | 730 |
| 3400 | 36" wide | | 2.40 | 6.667 | | 269 | 335 | | 604 | 790 |
| 3600 | Cross, 9" wide | | 4 | 4 | | 670 | 201 | | 871 | 1,050 |
| 3800 | 12" wide | | 3.20 | 5 | | 710 | 252 | | 962 | 1,150 |

## 26 05 43 – Underground Ducts and Raceways for Electrical Systems

### 26 05 43.10 Trench Duct

| | | Crew | Daily Output | Labor-Hours | Unit | Material | 2011 Bare Costs Labor | Equipment | Total | Total Incl O&P |
|---|---|---|---|---|---|---|---|---|---|---|
| 4000 | 18" wide | 2 Elec | 2.60 | 6.154 | Ea. | 850 | 310 | | 1,160 | 1,400 |
| 4200 | 24" wide | | 2.20 | 7.273 | | 1,075 | 365 | | 1,440 | 1,725 |
| 4400 | 30" wide | | 2 | 8 | | 1,400 | 400 | | 1,800 | 2,150 |
| 4600 | 36" wide | | 1.80 | 8.889 | | 1,750 | 445 | | 2,195 | 2,575 |
| 4800 | End closure, 9" wide | | 14.40 | 1.111 | | 42 | 56 | | 98 | 129 |
| 5000 | 12" wide | | 12 | 1.333 | | 48 | 67 | | 115 | 153 |
| 5200 | 18" wide | | 10 | 1.600 | | 73.50 | 80.50 | | 154 | 200 |
| 5400 | 24" wide | | 8 | 2 | | 97 | 101 | | 198 | 256 |
| 5600 | 30" wide | | 6.60 | 2.424 | | 122 | 122 | | 244 | 315 |
| 5800 | 36" wide | | 5.80 | 2.759 | | 144 | 139 | | 283 | 365 |
| 6000 | Tees, 9" wide | | 4 | 4 | | 410 | 201 | | 611 | 750 |
| 6200 | 12" wide | | 3.60 | 4.444 | | 475 | 224 | | 699 | 850 |
| 6400 | 18" wide | | 3.20 | 5 | | 605 | 252 | | 857 | 1,050 |
| 6600 | 24" wide | | 3 | 5.333 | | 870 | 268 | | 1,138 | 1,350 |
| 6800 | 30" wide | | 2.60 | 6.154 | | 1,125 | 310 | | 1,435 | 1,700 |
| 7000 | 36" wide | | 2 | 8 | | 1,500 | 400 | | 1,900 | 2,225 |
| 7200 | Riser, and cabinet connector, 9" wide | | 5.40 | 2.963 | | 178 | 149 | | 327 | 415 |
| 7400 | 12" wide | | 4.60 | 3.478 | | 208 | 175 | | 383 | 490 |
| 7600 | 18" wide | | 4 | 4 | | 255 | 201 | | 456 | 580 |
| 7800 | 24" wide | | 3.20 | 5 | | 310 | 252 | | 562 | 715 |
| 8000 | 30" wide | | 2.60 | 6.154 | | 355 | 310 | | 665 | 850 |
| 8200 | 36" wide | | 2 | 8 | | 415 | 400 | | 815 | 1,050 |
| 8400 | Insert assembly, cell to conduit adapter, 1-1/4" | 1 Elec | 16 | .500 | | 70.50 | 25 | | 95.50 | 115 |

### 26 05 43.20 Underfloor Duct

| | | Crew | Daily Output | Labor-Hours | Unit | Material | 2011 Bare Costs Labor | Equipment | Total | Total Incl O&P |
|---|---|---|---|---|---|---|---|---|---|---|
| 0010 | **UNDERFLOOR DUCT** | | | | | | | | | |
| 0100 | Duct, 1-3/8" x 3-1/8" blank, standard | 2 Elec | 160 | .100 | L.F. | 13.50 | 5.05 | | 18.55 | 22.50 |
| 0200 | 1-3/8" x 7-1/4" blank, super duct | | 120 | .133 | | 27 | 6.70 | | 33.70 | 39.50 |
| 0400 | 7/8" or 1-3/8" insert type, 24" O.C., 1-3/8" x 3-1/8", std. | | 140 | .114 | | 18.10 | 5.75 | | 23.85 | 28.50 |
| 0600 | 1-3/8" x 7-1/4", super duct | | 100 | .160 | | 31.50 | 8.05 | | 39.55 | 47 |
| 0800 | Junction box, single duct, 1 level, 3-1/8" | 1 Elec | 4 | 2 | Ea. | 405 | 101 | | 506 | 595 |
| 1000 | Junction box, single duct, 1 level, 7-1/4" | | 2.70 | 2.963 | | 475 | 149 | | 624 | 740 |
| 1200 | 1 level, 2 duct, 3-1/8" | | 3.20 | 2.500 | | 540 | 126 | | 666 | 780 |
| 1400 | Junction box, 1 level, 2 duct, 7-1/4" | | 2.30 | 3.478 | | 1,400 | 175 | | 1,575 | 1,775 |
| 1600 | Triple duct, 3-1/8" | | 2.30 | 3.478 | | 920 | 175 | | 1,095 | 1,275 |
| 1800 | Insert to conduit adapter, 3/4" & 1" | | 32 | .250 | | 33 | 12.60 | | 45.60 | 55 |
| 2000 | Support, single cell | | 27 | .296 | | 50 | 14.90 | | 64.90 | 77 |
| 2200 | Super duct | | 16 | .500 | | 50 | 25 | | 75 | 92.50 |
| 2400 | Double cell | | 16 | .500 | | 50 | 25 | | 75 | 92.50 |
| 2600 | Triple cell | | 11 | .727 | | 50 | 36.50 | | 86.50 | 109 |
| 2800 | Vertical elbow, standard duct | | 10 | .800 | | 89.50 | 40 | | 129.50 | 158 |
| 3000 | Super duct | | 8 | 1 | | 89.50 | 50.50 | | 140 | 173 |
| 3200 | Cabinet connector, standard duct | | 32 | .250 | | 67.50 | 12.60 | | 80.10 | 93 |
| 3400 | Super duct | | 27 | .296 | | 67.50 | 14.90 | | 82.40 | 96.50 |
| 3600 | Conduit adapter, 1" to 1-1/4" | | 32 | .250 | | 67.50 | 12.60 | | 80.10 | 93 |
| 3800 | 2" to 1-1/4" | | 27 | .296 | | 81 | 14.90 | | 95.90 | 111 |
| 4000 | Outlet, low tension (tele, computer, etc.) | | 8 | 1 | | 94.50 | 50.50 | | 145 | 179 |
| 4200 | High tension, receptacle (120 volt) | | 8 | 1 | | 94.50 | 50.50 | | 145 | 179 |

## 26 05 80 – Wiring Connections

### 26 05 80.10 Motor Connections

| | | Crew | Daily Output | Labor-Hours | Unit | Material | 2011 Bare Costs Labor | Equipment | Total | Total Incl O&P |
|---|---|---|---|---|---|---|---|---|---|---|
| 0010 | **MOTOR CONNECTIONS** | | | | | | | | | |
| 0020 | Flexible conduit and fittings, 115 volt, 1 phase, up to 1 HP motor | 1 Elec | 8 | 1 | Ea. | 9.70 | 50.50 | | 60.20 | 85 |
| 0120 | 230 volt, 10 HP motor, 3 phase | | 4.20 | 1.905 | | 16.40 | 96 | | 112.40 | 160 |

## 26 05 80 – Wiring Connections

### 26 05 80.10 Motor Connections

| | | Crew | Daily Output | Labor-Hours | Unit | Material | 2011 Bare Costs Labor | Equipment | Total | Total Incl O&P |
|---|---|---|---|---|---|---|---|---|---|---|
| 0200 | 25 HP motor | 1 Elec | 2.70 | 2.963 | Ea. | 34 | 149 | | 183 | 258 |
| 0400 | 50 HP motor | | 2.20 | 3.636 | | 83 | 183 | | 266 | 365 |
| 0600 | 100 HP motor | ↓ | 1.50 | 5.333 | ↓ | 201 | 268 | | 469 | 615 |

## 26 05 90 – Residential Applications

### 26 05 90.10 Residential Wiring

| | | Crew | Daily Output | Labor-Hours | Unit | Material | 2011 Bare Costs Labor | Equipment | Total | Total Incl O&P |
|---|---|---|---|---|---|---|---|---|---|---|
| 0010 | **RESIDENTIAL WIRING** | | | | | | | | | |
| 0020 | 20' avg. runs and #14/2 wiring incl. unless otherwise noted | | | | | | | | | |
| 1000 | Service & panel, includes 24' SE-AL cable, service eye, meter, | | | | | | | | | |
| 1010 | Socket, panel board, main bkr., ground rod, 15 or 20 amp | | | | | | | | | |
| 1020 | 1-pole circuit breakers, and misc. hardware | | | | | | | | | |
| 1100 | 100 amp, with 10 branch breakers | 1 Elec | 1.19 | 6.723 | Ea. | 560 | 340 | | 900 | 1,125 |
| 1110 | With PVC conduit and wire | | .92 | 8.696 | | 630 | 435 | | 1,065 | 1,350 |
| 1120 | With RGS conduit and wire | | .73 | 10.959 | | 790 | 550 | | 1,340 | 1,675 |
| 1150 | 150 amp, with 14 branch breakers | | 1.03 | 7.767 | | 865 | 390 | | 1,255 | 1,525 |
| 1170 | With PVC conduit and wire | | .82 | 9.756 | | 1,000 | 490 | | 1,490 | 1,825 |
| 1180 | With RGS conduit and wire | ↓ | .67 | 11.940 | | 1,300 | 600 | | 1,900 | 2,350 |
| 1200 | 200 amp, with 18 branch breakers | 2 Elec | 1.80 | 8.889 | | 1,150 | 445 | | 1,595 | 1,900 |
| 1220 | With PVC conduit and wire | | 1.46 | 10.959 | | 1,275 | 550 | | 1,825 | 2,225 |
| 1230 | With RGS conduit and wire | ↓ | 1.24 | 12.903 | | 1,700 | 650 | | 2,350 | 2,825 |
| 1800 | Lightning surge suppressor for above services, add | 1 Elec | 32 | .250 | ↓ | 55 | 12.60 | | 67.60 | 79 |
| 2000 | Switch devices | | | | | | | | | |
| 2100 | Single pole, 15 amp, Ivory, with a 1-gang box, cover plate, | | | | | | | | | |
| 2110 | Type NM (Romex) cable | 1 Elec | 17.10 | .468 | Ea. | 10.70 | 23.50 | | 34.20 | 47 |
| 2120 | Type MC (BX) cable | | 14.30 | .559 | | 28.50 | 28 | | 56.50 | 73 |
| 2130 | EMT & wire | | 5.71 | 1.401 | | 34 | 70.50 | | 104.50 | 141 |
| 2150 | 3-way, #14/3, type NM cable | | 14.55 | .550 | | 14.85 | 27.50 | | 42.35 | 57.50 |
| 2170 | Type MC cable | | 12.31 | .650 | | 38.50 | 32.50 | | 71 | 91 |
| 2180 | EMT & wire | | 5 | 1.600 | | 37.50 | 80.50 | | 118 | 160 |
| 2200 | 4-way, #14/3, type NM cable | | 14.55 | .550 | | 31.50 | 27.50 | | 59 | 76 |
| 2220 | Type MC cable | | 12.31 | .650 | | 55.50 | 32.50 | | 88 | 110 |
| 2230 | EMT & wire | | 5 | 1.600 | | 54.50 | 80.50 | | 135 | 179 |
| 2250 | S.P., 20 amp, #12/2, type NM cable | | 13.33 | .600 | | 18.05 | 30 | | 48.05 | 64.50 |
| 2270 | Type MC cable | | 11.43 | .700 | | 33.50 | 35 | | 68.50 | 88.50 |
| 2280 | EMT & wire | | 4.85 | 1.649 | | 42.50 | 83 | | 125.50 | 170 |
| 2290 | S.P. rotary dimmer, 600W, no wiring | | 17 | .471 | | 21 | 23.50 | | 44.50 | 58 |
| 2300 | S.P. rotary dimmer, 600W, type NM cable | | 14.55 | .550 | | 26 | 27.50 | | 53.50 | 70 |
| 2320 | Type MC cable | | 12.31 | .650 | | 44 | 32.50 | | 76.50 | 97 |
| 2330 | EMT & wire | | 5 | 1.600 | | 51 | 80.50 | | 131.50 | 175 |
| 2350 | 3-way rotary dimmer, type NM cable | | 13.33 | .600 | | 33.50 | 30 | | 63.50 | 81 |
| 2370 | Type MC cable | | 11.43 | .700 | | 51 | 35 | | 86 | 109 |
| 2380 | EMT & wire | ↓ | 4.85 | 1.649 | ↓ | 58 | 83 | | 141 | 187 |
| 2400 | Interval timer wall switch, 20 amp, 1-30 min., #12/2 | | | | | | | | | |
| 2410 | Type NM cable | 1 Elec | 14.55 | .550 | Ea. | 53 | 27.50 | | 80.50 | 99.50 |
| 2420 | Type MC cable | | 12.31 | .650 | | 61 | 32.50 | | 93.50 | 116 |
| 2430 | EMT & wire | ↓ | 5 | 1.600 | ↓ | 77.50 | 80.50 | | 158 | 204 |
| 2500 | Decorator style | | | | | | | | | |
| 2510 | S.P., 15 amp, type NM cable | 1 Elec | 17.10 | .468 | Ea. | 15.55 | 23.50 | | 39.05 | 52 |
| 2520 | Type MC cable | | 14.30 | .559 | | 33.50 | 28 | | 61.50 | 78 |
| 2530 | EMT & wire | | 5.71 | 1.401 | | 38.50 | 70.50 | | 109 | 147 |
| 2550 | 3-way, #14/3, type NM cable | | 14.55 | .550 | | 19.70 | 27.50 | | 47.20 | 62.50 |
| 2570 | Type MC cable | | 12.31 | .650 | | 43.50 | 32.50 | | 76 | 96 |
| 2580 | EMT & wire | ↓ | 5 | 1.600 | ↓ | 42.50 | 80.50 | | 123 | 166 |

| 26 05 90.10 Residential Wiring | Crew | Daily Output | Labor-Hours | Unit | Material | 2011 Bare Costs Labor | Equipment | Total | Total Incl O&P |
|---|---|---|---|---|---|---|---|---|---|
| 2600  4-way, #14/3, type NM cable | 1 Elec | 14.55 | .550 | Ea. | 36.50 | 27.50 | | 64 | 81.50 |
| 2620  Type MC cable | | 12.31 | .650 | | 60.50 | 32.50 | | 93 | 115 |
| 2630  EMT & wire | | 5 | 1.600 | | 59.50 | 80.50 | | 140 | 184 |
| 2650  S.P., 20 amp, #12/2, type NM cable | | 13.33 | .600 | | 23 | 30 | | 53 | 69.50 |
| 2670  Type MC cable | | 11.43 | .700 | | 38 | 35 | | 73 | 94 |
| 2680  EMT & wire | | 4.85 | 1.649 | | 47.50 | 83 | | 130.50 | 175 |
| 2700  S.P., slide dimmer, type NM cable | | 17.10 | .468 | | 33.50 | 23.50 | | 57 | 72 |
| 2720  Type MC cable | | 14.30 | .559 | | 51.50 | 28 | | 79.50 | 98 |
| 2730  EMT & wire | | 5.71 | 1.401 | | 58.50 | 70.50 | | 129 | 168 |
| 2750  S.P., touch dimmer, type NM cable | | 17.10 | .468 | | 28.50 | 23.50 | | 52 | 66.50 |
| 2770  Type MC cable | | 14.30 | .559 | | 46.50 | 28 | | 74.50 | 92.50 |
| 2780  EMT & wire | | 5.71 | 1.401 | | 53 | 70.50 | | 123.50 | 163 |
| 2800  3-way touch dimmer, type NM cable | | 13.33 | .600 | | 48 | 30 | | 78 | 97 |
| 2820  Type MC cable | | 11.43 | .700 | | 65.50 | 35 | | 100.50 | 125 |
| 2830  EMT & wire | ▼ | 4.85 | 1.649 | ▼ | 72.50 | 83 | | 155.50 | 203 |
| 3000  Combination devices | | | | | | | | | |
| 3100  S.P. switch/15 amp recpt., Ivory, 1-gang box, plate | | | | | | | | | |
| 3110  Type NM cable | 1 Elec | 11.43 | .700 | Ea. | 22.50 | 35 | | 57.50 | 77 |
| 3120  Type MC cable | | 10 | .800 | | 40.50 | 40 | | 80.50 | 104 |
| 3130  EMT & wire | | 4.40 | 1.818 | | 47 | 91.50 | | 138.50 | 187 |
| 3150  S.P. switch/pilot light, type NM cable | | 11.43 | .700 | | 23.50 | 35 | | 58.50 | 78 |
| 3170  Type MC cable | | 10 | .800 | | 41.50 | 40 | | 81.50 | 105 |
| 3180  EMT & wire | | 4.43 | 1.806 | | 48 | 91 | | 139 | 188 |
| 3190  2-S.P. switches, 2-#14/2, no wiring | | 14 | .571 | | 8.40 | 28.50 | | 36.90 | 51.50 |
| 3200  2-S.P. switches, 2-#14/2, type NM cables | | 10 | .800 | | 26.50 | 40 | | 66.50 | 89 |
| 3220  Type MC cable | | 8.89 | .900 | | 53.50 | 45.50 | | 99 | 126 |
| 3230  EMT & wire | | 4.10 | 1.951 | | 51 | 98 | | 149 | 201 |
| 3250  3-way switch/15 amp recpt., #14/3, type NM cable | | 10 | .800 | | 30.50 | 40 | | 70.50 | 93 |
| 3270  Type MC cable | | 8.89 | .900 | | 54.50 | 45.50 | | 100 | 127 |
| 3280  EMT & wire | | 4.10 | 1.951 | | 53.50 | 98 | | 151.50 | 204 |
| 3300  2-3 way switches, 2-#14/3, type NM cables | | 8.89 | .900 | | 40.50 | 45.50 | | 86 | 112 |
| 3320  Type MC cable | | 8 | 1 | | 79.50 | 50.50 | | 130 | 162 |
| 3330  EMT & wire | | 4 | 2 | | 61 | 101 | | 162 | 216 |
| 3350  S.P. switch/20 amp recpt., #12/2, type NM cable | | 10 | .800 | | 34.50 | 40 | | 74.50 | 97.50 |
| 3370  Type MC cable | | 8.89 | .900 | | 42.50 | 45.50 | | 88 | 114 |
| 3380  EMT & wire | ▼ | 4.10 | 1.951 | ▼ | 59 | 98 | | 157 | 210 |
| 3400  Decorator style | | | | | | | | | |
| 3410  S.P. switch/15 amp recpt., type NM cable | 1 Elec | 11.43 | .700 | Ea. | 27.50 | 35 | | 62.50 | 82 |
| 3420  Type MC cable | | 10 | .800 | | 45.50 | 40 | | 85.50 | 110 |
| 3430  EMT & wire | | 4.40 | 1.818 | | 52 | 91.50 | | 143.50 | 193 |
| 3450  S.P. switch/pilot light, type NM cable | | 11.43 | .700 | | 28.50 | 35 | | 63.50 | 83 |
| 3470  Type MC cable | | 10 | .800 | | 46 | 40 | | 86 | 111 |
| 3480  EMT & wire | | 4.40 | 1.818 | | 53 | 91.50 | | 144.50 | 193 |
| 3500  2-S.P. switches, 2-#14/2, type NM cables | | 10 | .800 | | 31.50 | 40 | | 71.50 | 94 |
| 3520  Type MC cable | | 8.89 | .900 | | 58.50 | 45.50 | | 104 | 132 |
| 3530  EMT & wire | | 4.10 | 1.951 | | 56 | 98 | | 154 | 207 |
| 3550  3-way/15 amp recpt., #14/3, type NM cable | | 10 | .800 | | 35.50 | 40 | | 75.50 | 98.50 |
| 3570  Type MC cable | | 8.89 | .900 | | 59 | 45.50 | | 104.50 | 132 |
| 3580  EMT & wire | | 4.10 | 1.951 | | 58 | 98 | | 156 | 209 |
| 3650  2-3 way switches, 2-#14/3, type NM cables | | 8.89 | .900 | | 45.50 | 45.50 | | 91 | 117 |
| 3670  Type MC cable | | 8 | 1 | | 84 | 50.50 | | 134.50 | 167 |
| 3680  EMT & wire | | 4 | 2 | | 65.50 | 101 | | 166.50 | 222 |
| 3700  S.P. switch/20 amp recpt., #12/2, type NM cable | ▼ | 10 | .800 | ▼ | 39.50 | 40 | | 79.50 | 103 |

| 26 05 90.10 Residential Wiring | | Crew | Daily Output | Labor-Hours | Unit | Material | 2011 Bare Costs Labor | Equipment | Total | Total Incl O&P |
|---|---|---|---|---|---|---|---|---|---|---|
| 3720 | Type MC cable | 1 Elec | 8.89 | .900 | Ea. | 47.50 | 45.50 | | 93 | 119 |
| 3730 | EMT & wire | ↓ | 4.10 | 1.951 | ↓ | 64 | 98 | | 162 | 216 |
| 4000 | Receptacle devices | | | | | | | | | |
| 4010 | Duplex outlet, 15 amp recpt., Ivory, 1-gang box, plate | | | | | | | | | |
| 4015 | Type NM cable | 1 Elec | 14.55 | .550 | Ea. | 9.10 | 27.50 | | 36.60 | 51 |
| 4020 | Type MC cable | | 12.31 | .650 | | 27 | 32.50 | | 59.50 | 78 |
| 4030 | EMT & wire | | 5.33 | 1.501 | | 32 | 75.50 | | 107.50 | 148 |
| 4050 | With #12/2, type NM cable | | 12.31 | .650 | | 11.80 | 32.50 | | 44.30 | 61.50 |
| 4070 | Type MC cable | | 10.67 | .750 | | 27 | 37.50 | | 64.50 | 86 |
| 4080 | EMT & wire | | 4.71 | 1.699 | | 36 | 85.50 | | 121.50 | 167 |
| 4100 | 20 amp recpt., #12/2, type NM cable | | 12.31 | .650 | | 21 | 32.50 | | 53.50 | 71.50 |
| 4120 | Type MC cable | | 10.67 | .750 | | 36.50 | 37.50 | | 74 | 96 |
| 4130 | EMT & wire | ↓ | 4.71 | 1.699 | ↓ | 45.50 | 85.50 | | 131 | 177 |
| 4140 | For GFI see Section 26 05 90.10 line 4300 below | | | | | | | | | |
| 4150 | Decorator style, 15 amp recpt., type NM cable | 1 Elec | 14.55 | .550 | Ea. | 13.95 | 27.50 | | 41.45 | 56.50 |
| 4170 | Type MC cable | | 12.31 | .650 | | 32 | 32.50 | | 64.50 | 83.50 |
| 4180 | EMT & wire | | 5.33 | 1.501 | | 37 | 75.50 | | 112.50 | 153 |
| 4200 | With #12/2, type NM cable | | 12.31 | .650 | | 16.65 | 32.50 | | 49.15 | 67 |
| 4220 | Type MC cable | | 10.67 | .750 | | 32 | 37.50 | | 69.50 | 91 |
| 4230 | EMT & wire | | 4.71 | 1.699 | | 41 | 85.50 | | 126.50 | 172 |
| 4250 | 20 amp recpt. #12/2, type NM cable | | 12.31 | .650 | | 26 | 32.50 | | 58.50 | 77 |
| 4270 | Type MC cable | | 10.67 | .750 | | 41 | 37.50 | | 78.50 | 102 |
| 4280 | EMT & wire | | 4.71 | 1.699 | | 50.50 | 85.50 | | 136 | 183 |
| 4300 | GFI, 15 amp recpt., type NM cable | | 12.31 | .650 | | 43.50 | 32.50 | | 76 | 96.50 |
| 4320 | Type MC cable | | 10.67 | .750 | | 61.50 | 37.50 | | 99 | 124 |
| 4330 | EMT & wire | | 4.71 | 1.699 | | 66.50 | 85.50 | | 152 | 201 |
| 4350 | GFI with #12/2, type NM cable | | 10.67 | .750 | | 46.50 | 37.50 | | 84 | 107 |
| 4370 | Type MC cable | | 9.20 | .870 | | 61.50 | 43.50 | | 105 | 133 |
| 4380 | EMT & wire | | 4.21 | 1.900 | | 71 | 95.50 | | 166.50 | 220 |
| 4400 | 20 amp recpt., #12/2 type NM cable | | 10.67 | .750 | | 48.50 | 37.50 | | 86 | 109 |
| 4420 | Type MC cable | | 9.20 | .870 | | 63.50 | 43.50 | | 107 | 135 |
| 4430 | EMT & wire | | 4.21 | 1.900 | | 73 | 95.50 | | 168.50 | 222 |
| 4500 | Weather-proof cover for above receptacles, add | ↓ | 32 | .250 | ↓ | 5.25 | 12.60 | | 17.85 | 24.50 |
| 4550 | Air conditioner outlet, 20 amp-240 volt recpt. | | | | | | | | | |
| 4560 | 30' of #12/2, 2 pole circuit breaker | | | | | | | | | |
| 4570 | Type NM cable | 1 Elec | 10 | .800 | Ea. | 61 | 40 | | 101 | 127 |
| 4580 | Type MC cable | | 9 | .889 | | 79.50 | 44.50 | | 124 | 154 |
| 4590 | EMT & wire | | 4 | 2 | | 85 | 101 | | 186 | 243 |
| 4600 | Decorator style, type NM cable | | 10 | .800 | | 65.50 | 40 | | 105.50 | 132 |
| 4620 | Type MC cable | | 9 | .889 | | 84 | 44.50 | | 128.50 | 159 |
| 4630 | EMT & wire | ↓ | 4 | 2 | ↓ | 89.50 | 101 | | 190.50 | 248 |
| 4650 | Dryer outlet, 30 amp-240 volt recpt., 20' of #10/3 | | | | | | | | | |
| 4660 | 2 pole circuit breaker | | | | | | | | | |
| 4670 | Type NM cable | 1 Elec | 6.41 | 1.248 | Ea. | 67.50 | 63 | | 130.50 | 168 |
| 4680 | Type MC cable | | 5.71 | 1.401 | | 80 | 70.50 | | 150.50 | 192 |
| 4690 | EMT & wire | ↓ | 3.48 | 2.299 | ↓ | 80.50 | 116 | | 196.50 | 260 |
| 4700 | Range outlet, 50 amp-240 volt recpt., 30' of #8/3 | | | | | | | | | |
| 4710 | Type NM cable | 1 Elec | 4.21 | 1.900 | Ea. | 96.50 | 95.50 | | 192 | 248 |
| 4720 | Type MC cable | | 4 | 2 | | 165 | 101 | | 266 | 330 |
| 4730 | EMT & wire | | 2.96 | 2.703 | | 110 | 136 | | 246 | 320 |
| 4750 | Central vacuum outlet, Type NM cable | | 6.40 | 1.250 | | 65 | 63 | | 128 | 165 |
| 4770 | Type MC cable | | 5.71 | 1.401 | | 91 | 70.50 | | 161.50 | 204 |
| 4780 | EMT & wire | ↓ | 3.48 | 2.299 | ↓ | 90 | 116 | | 206 | 270 |

| 26 05 90.10 Residential Wiring | Crew | Daily Output | Labor-Hours | Unit | Material | 2011 Bare Costs Labor | Equipment | Total | Total Incl O&P |
|---|---|---|---|---|---|---|---|---|---|
| 4800 | 30 amp-110 volt locking recpt., #10/2 circ. bkr. | | | | | | | | |
| 4810 | Type NM cable | 1 Elec | 6.20 | 1.290 | Ea. | 75.50 | 65 | | 140.50 | 179 |
| 4820 | Type MC cable | | 5.40 | 1.481 | | 105 | 74.50 | | 179.50 | 226 |
| 4830 | EMT & wire | ↓ | 3.20 | 2.500 | ↓ | 100 | 126 | | 226 | 296 |
| 4900 | Low voltage outlets | | | | | | | | | |
| 4910 | Telephone recpt., 20' of 4/C phone wire | 1 Elec | 26 | .308 | Ea. | 11 | 15.50 | | 26.50 | 35 |
| 4920 | TV recpt., 20' of RG59U coax wire, F type connector | " | 16 | .500 | " | 19.25 | 25 | | 44.25 | 58.50 |
| 4950 | Door bell chime, transformer, 2 buttons, 60' of bellwire | | | | | | | | | |
| 4970 | Economy model | 1 Elec | 11.50 | .696 | Ea. | 77.50 | 35 | | 112.50 | 138 |
| 4980 | Custom model | | 11.50 | .696 | | 121 | 35 | | 156 | 185 |
| 4990 | Luxury model, 3 buttons | ↓ | 9.50 | .842 | ↓ | 325 | 42.50 | | 367.50 | 420 |
| 6000 | Lighting outlets | | | | | | | | | |
| 6050 | Wire only (for fixture), type NM cable | 1 Elec | 32 | .250 | Ea. | 7.15 | 12.60 | | 19.75 | 26.50 |
| 6070 | Type MC cable | | 24 | .333 | | 17.25 | 16.75 | | 34 | 44 |
| 6080 | EMT & wire | | 10 | .800 | | 21 | 40 | | 61 | 82.50 |
| 6100 | Box (4"), and wire (for fixture), type NM cable | | 25 | .320 | | 17.70 | 16.10 | | 33.80 | 43.50 |
| 6120 | Type MC cable | | 20 | .400 | | 28 | 20 | | 48 | 60.50 |
| 6130 | EMT & wire | ↓ | 11 | .727 | ↓ | 31.50 | 36.50 | | 68 | 89 |
| 6200 | Fixtures (use with lines 6050 or 6100 above) | | | | | | | | | |
| 6210 | Canopy style, economy grade | 1 Elec | 40 | .200 | Ea. | 34.50 | 10.05 | | 44.55 | 53 |
| 6220 | Custom grade | | 40 | .200 | | 53 | 10.05 | | 63.05 | 73.50 |
| 6250 | Dining room chandelier, economy grade | | 19 | .421 | | 86 | 21 | | 107 | 126 |
| 6260 | Custom grade | | 19 | .421 | | 253 | 21 | | 274 | 310 |
| 6270 | Luxury grade | | 15 | .533 | | 570 | 27 | | 597 | 665 |
| 6310 | Kitchen fixture (fluorescent), economy grade | | 30 | .267 | | 64 | 13.40 | | 77.40 | 90.50 |
| 6320 | Custom grade | | 25 | .320 | | 175 | 16.10 | | 191.10 | 217 |
| 6350 | Outdoor, wall mounted, economy grade | | 30 | .267 | | 32.50 | 13.40 | | 45.90 | 56 |
| 6360 | Custom grade | | 30 | .267 | | 112 | 13.40 | | 125.40 | 143 |
| 6370 | Luxury grade | | 25 | .320 | | 253 | 16.10 | | 269.10 | 300 |
| 6410 | Outdoor PAR floodlights, 1 lamp, 150 watt | | 20 | .400 | | 35 | 20 | | 55 | 68.50 |
| 6420 | 2 lamp, 150 watt each | | 20 | .400 | | 58 | 20 | | 78 | 94 |
| 6425 | Motion sensing, 2 lamp, 150 watt each | | 20 | .400 | | 70 | 20 | | 90 | 107 |
| 6430 | For infrared security sensor, add | | 32 | .250 | | 117 | 12.60 | | 129.60 | 148 |
| 6450 | Outdoor, quartz-halogen, 300 watt flood | | 20 | .400 | | 42 | 20 | | 62 | 76.50 |
| 6600 | Recessed downlight, round, pre-wired, 50 or 75 watt trim | | 30 | .267 | | 43.50 | 13.40 | | 56.90 | 67.50 |
| 6610 | With shower light trim | | 30 | .267 | | 53.50 | 13.40 | | 66.90 | 79 |
| 6620 | With wall washer trim | | 28 | .286 | | 64.50 | 14.35 | | 78.85 | 92.50 |
| 6630 | With eye-ball trim | ↓ | 28 | .286 | | 64.50 | 14.35 | | 78.85 | 92.50 |
| 6640 | For direct contact with insulation, add | | | | | 2.30 | | | 2.30 | 2.53 |
| 6700 | Porcelain lamp holder | 1 Elec | 40 | .200 | | 4.50 | 10.05 | | 14.55 | 19.85 |
| 6710 | With pull switch | | 40 | .200 | | 5.25 | 10.05 | | 15.30 | 20.50 |
| 6750 | Fluorescent strip, 1-20 watt tube, wrap around diffuser, 24" | | 24 | .333 | | 54 | 16.75 | | 70.75 | 84.50 |
| 6760 | 1-34 watt tube, 48" | | 24 | .333 | | 69.50 | 16.75 | | 86.25 | 102 |
| 6770 | 2-34 watt tubes, 48" | | 20 | .400 | | 82.50 | 20 | | 102.50 | 121 |
| 6780 | With residential ballast | | 20 | .400 | | 92.50 | 20 | | 112.50 | 132 |
| 6800 | Bathroom heat lamp, 1-250 watt | | 28 | .286 | | 47.50 | 14.35 | | 61.85 | 74 |
| 6810 | 2-250 watt lamps | ↓ | 28 | .286 | ↓ | 73.50 | 14.35 | | 87.85 | 103 |
| 6820 | For timer switch, see Section 26 05 90.10 line 2400 | | | | | | | | | |
| 6900 | Outdoor post lamp, incl. post, fixture, 35' of #14/2 | | | | | | | | | |
| 6910 | Type NMC cable | 1 Elec | 3.50 | 2.286 | Ea. | 203 | 115 | | 318 | 395 |
| 6920 | Photo-eye, add | | 27 | .296 | | 35 | 14.90 | | 49.90 | 60.50 |
| 6950 | Clock dial time switch, 24 hr., w/enclosure, type NM cable | | 11.43 | .700 | | 67 | 35 | | 102 | 126 |
| 6970 | Type MC cable | ↓ | 11 | .727 | ↓ | 84.50 | 36.50 | | 121 | 147 |

| 26 05 90.10 Residential Wiring | Crew | Daily Output | Labor-Hours | Unit | Material | 2011 Bare Costs Labor | Equipment | Total | Total Incl O&P |
|---|---|---|---|---|---|---|---|---|---|
| 6980　　　EMT & wire | 1 Elec | 4.85 | 1.649 | Ea. | 90 | 83 | | 173 | 222 |
| 7000　Alarm systems | | | | | | | | | |
| 7050　　Smoke detectors, box, #14/3, type NM cable | 1 Elec | 14.55 | .550 | Ea. | 35 | 27.50 | | 62.50 | 79.50 |
| 7070　　　Type MC cable | ↓ | 12.31 | .650 | | 51.50 | 32.50 | | 84 | 105 |
| 7080　　　EMT & wire | ↓ | 5 | 1.600 | ↓ | 50.50 | 80.50 | | 131 | 175 |
| 7090　　For relay output to security system, add | | | | ↓ | 13 | | | 13 | 14.30 |
| 8000　Residential equipment | | | | | | | | | |
| 8050　　Disposal hook-up, incl. switch, outlet box, 3' of flex | | | | | | | | | |
| 8060　　　20 amp-1 pole circ. bkr., and 25' of #12/2 | | | | | | | | | |
| 8070　　　Type NM cable | 1 Elec | 10 | .800 | Ea. | 31.50 | 40 | | 71.50 | 94 |
| 8080　　　Type MC cable | ↓ | 8 | 1 | | 48.50 | 50.50 | | 99 | 128 |
| 8090　　　EMT & wire | ↓ | 5 | 1.600 | ↓ | 58 | 80.50 | | 138.50 | 183 |
| 8100　　Trash compactor or dishwasher hook-up, incl. outlet box, | | | | | | | | | |
| 8110　　　3' of flex, 15 amp-1 pole circ. bkr., and 25' of #14/2 | | | | | | | | | |
| 8120　　　Type NM cable | 1 Elec | 10 | .800 | Ea. | 23 | 40 | | 63 | 84.50 |
| 8130　　　Type MC cable | ↓ | 8 | 1 | | 43 | 50.50 | | 93.50 | 122 |
| 8140　　　EMT & wire | ↓ | 5 | 1.600 | ↓ | 50 | 80.50 | | 130.50 | 174 |
| 8150　　Hot water sink dispensor hook-up, use line 8100 | | | | | | | | | |
| 8200　　Vent/exhaust fan hook-up, type NM cable | 1 Elec | 32 | .250 | Ea. | 7.15 | 12.60 | | 19.75 | 26.50 |
| 8220　　　Type MC cable | ↓ | 24 | .333 | | 17.25 | 16.75 | | 34 | 44 |
| 8230　　　EMT & wire | ↓ | 10 | .800 | ↓ | 21 | 40 | | 61 | 82.50 |
| 8250　　Bathroom vent fan, 50 CFM (use with above hook-up) | | | | | | | | | |
| 8260　　　Economy model | 1 Elec | 15 | .533 | Ea. | 25 | 27 | | 52 | 67 |
| 8270　　　Low noise model | | 15 | .533 | | 34 | 27 | | 61 | 77 |
| 8280　　　Custom model | ↓ | 12 | .667 | ↓ | 117 | 33.50 | | 150.50 | 179 |
| 8300　　Bathroom or kitchen vent fan, 110 CFM | | | | | | | | | |
| 8310　　　Economy model | 1 Elec | 15 | .533 | Ea. | 69.50 | 27 | | 96.50 | 116 |
| 8320　　　Low noise model | " | 15 | .533 | " | 86 | 27 | | 113 | 134 |
| 8350　　Paddle fan, variable speed (w/o lights) | | | | | | | | | |
| 8360　　　Economy model (AC motor) | 1 Elec | 10 | .800 | Ea. | 115 | 40 | | 155 | 187 |
| 8362　　　With light kit | | 10 | .800 | | 153 | 40 | | 193 | 229 |
| 8370　　　Custom model (AC motor) | | 10 | .800 | | 182 | 40 | | 222 | 260 |
| 8372　　　With light kit | | 10 | .800 | | 220 | 40 | | 260 | 305 |
| 8380　　　Luxury model (DC motor) | | 8 | 1 | | 360 | 50.50 | | 410.50 | 470 |
| 8382　　　With light kit | | 8 | 1 | | 400 | 50.50 | | 450.50 | 515 |
| 8390　　　Remote speed switch for above, add | ↓ | 12 | .667 | ↓ | 30 | 33.50 | | 63.50 | 82.50 |
| 8500　　Whole house exhaust fan, ceiling mount, 36", variable speed | | | | | | | | | |
| 8510　　　Remote switch, incl. shutters, 20 amp-1 pole circ. bkr. | | | | | | | | | |
| 8520　　　30' of #12/2, type NM cable | 1 Elec | 4 | 2 | Ea. | 1,125 | 101 | | 1,226 | 1,400 |
| 8530　　　Type MC cable | ↓ | 3.50 | 2.286 | | 1,150 | 115 | | 1,265 | 1,450 |
| 8540　　　EMT & wire | ↓ | 3 | 2.667 | ↓ | 1,150 | 134 | | 1,284 | 1,475 |
| 8600　　Whirlpool tub hook-up, incl. timer switch, outlet box | | | | | | | | | |
| 8610　　　3' of flex, 20 amp-1 pole GFI circ. bkr. | | | | | | | | | |
| 8620　　　30' of #12/2, type NM cable | 1 Elec | 5 | 1.600 | Ea. | 128 | 80.50 | | 208.50 | 259 |
| 8630　　　Type MC cable | ↓ | 4.20 | 1.905 | | 139 | 96 | | 235 | 295 |
| 8640　　　EMT & wire | ↓ | 3.40 | 2.353 | ↓ | 148 | 118 | | 266 | 340 |
| 8650　　Hot water heater hook-up, incl. 1-2 pole circ. bkr., box; | | | | | | | | | |
| 8660　　　3' of flex, 20' of #10/2, type NM cable | 1 Elec | 5 | 1.600 | Ea. | 33 | 80.50 | | 113.50 | 155 |
| 8670　　　Type MC cable | ↓ | 4.20 | 1.905 | | 56 | 96 | | 152 | 204 |
| 8680　　　EMT & wire | ↓ | 3.40 | 2.353 | ↓ | 53 | 118 | | 171 | 233 |
| 9000　Heating/air conditioning | | | | | | | | | |
| 9050　　Furnace/boiler hook-up, incl. firestat, local on-off switch | | | | | | | | | |
| 9060　　　Emergency switch, and 40' of type NM cable | 1 Elec | 4 | 2 | Ea. | 56.50 | 101 | | 157.50 | 211 |

# 26 05 Common Work Results for Electrical

## 26 05 90 – Residential Applications

| 26 05 90.10 Residential Wiring | Crew | Daily Output | Labor-Hours | Unit | Material | 2011 Bare Costs Labor | Equipment | Total | Total Incl O&P |
|---|---|---|---|---|---|---|---|---|---|
| 9070 | Type MC cable | 1 Elec | 3.50 | 2.286 | Ea. | 83.50 | 115 | | 198.50 | 262 |
| 9080 | EMT & wire | ↓ | 1.50 | 5.333 | ↓ | 92.50 | 268 | | 360.50 | 495 |
| 9100 | Air conditioner hook-up, incl. local 60 amp disc. switch | | | | | | | | | |
| 9110 | 3' sealtite, 40 amp, 2 pole circuit breaker | | | | | | | | | |
| 9130 | 40' of #8/2, type NM cable | 1 Elec | 3.50 | 2.286 | Ea. | 225 | 115 | | 340 | 415 |
| 9140 | Type MC cable | | 3 | 2.667 | | 325 | 134 | | 459 | 560 |
| 9150 | EMT & wire | ↓ | 1.30 | 6.154 | ↓ | 259 | 310 | | 569 | 745 |
| 9200 | Heat pump hook-up, 1-40 & 1-100 amp 2 pole circ. bkr. | | | | | | | | | |
| 9210 | Local disconnect switch, 3' sealtite | | | | | | | | | |
| 9220 | 40' of #8/2 & 30' of #3/2 | | | | | | | | | |
| 9230 | Type NM cable | 1 Elec | 1.30 | 6.154 | Ea. | 570 | 310 | | 880 | 1,100 |
| 9240 | Type MC cable | | 1.08 | 7.407 | | 755 | 375 | | 1,130 | 1,375 |
| 9250 | EMT & wire | ↓ | .94 | 8.511 | ↓ | 620 | 430 | | 1,050 | 1,325 |
| 9500 | Thermostat hook-up, using low voltage wire | | | | | | | | | |
| 9520 | Heating only, 25' of #18-3 | 1 Elec | 24 | .333 | Ea. | 9.70 | 16.75 | | 26.45 | 35.50 |
| 9530 | Heating/cooling, 25' of #18-4 | " | 20 | .400 | " | 12.25 | 20 | | 32.25 | 43.50 |

# 26 09 Instrumentation and Control for Electrical Systems

## 26 09 13 – Electrical Power Monitoring

### 26 09 13.10 Switchboard Instruments

| | | Crew | Daily Output | Labor-Hours | Unit | Material | 2011 Bare Costs Labor | Equipment | Total | Total Incl O&P |
|---|---|---|---|---|---|---|---|---|---|---|
| 0010 | **SWITCHBOARD INSTRUMENTS** 3 phase, 4 wire | | | | | | | | | |
| 0100 | AC indicating, ammeter & switch | 1 Elec | 8 | 1 | Ea. | 2,025 | 50.50 | | 2,075.50 | 2,300 |
| 0200 | Voltmeter & switch | | 8 | 1 | | 2,025 | 50.50 | | 2,075.50 | 2,300 |
| 0300 | Wattmeter | | 8 | 1 | | 4,000 | 50.50 | | 4,050.50 | 4,475 |
| 0400 | AC recording, ammeter | | 4 | 2 | | 7,150 | 101 | | 7,251 | 8,000 |
| 0500 | Voltmeter | | 4 | 2 | | 7,150 | 101 | | 7,251 | 8,000 |
| 0600 | Ground fault protection, zero sequence | | 2.70 | 2.963 | | 6,300 | 149 | | 6,449 | 7,150 |
| 0700 | Ground return path | | 2.70 | 2.963 | | 6,300 | 149 | | 6,449 | 7,150 |
| 0800 | 3 current transformers, 5 to 800 amp | | 2 | 4 | | 2,925 | 201 | | 3,126 | 3,525 |
| 0900 | 1000 to 1500 amp | | 1.30 | 6.154 | | 4,225 | 310 | | 4,535 | 5,100 |
| 1200 | 2000 to 4000 amp | | 1 | 8 | | 4,975 | 400 | | 5,375 | 6,075 |
| 1300 | Fused potential transformer, maximum 600 volt | ↓ | 8 | 1 | ↓ | 1,100 | 50.50 | | 1,150.50 | 1,300 |

# 26 12 Medium-Voltage Transformers

## 26 12 19 – Pad-Mounted, Liquid-Filled, Medium-Voltage Transformers

### 26 12 19.10 Transformer, Oil-Filled

| | | Crew | Daily Output | Labor-Hours | Unit | Material | 2011 Bare Costs Labor | Equipment | Total | Total Incl O&P |
|---|---|---|---|---|---|---|---|---|---|---|
| 0010 | **TRANSFORMER, OIL-FILLED** primary delta or Y, | | | | | | | | | |
| 0050 | Pad mounted 5 kV or 15 kV, with taps, 277/480 V secondary, 3 phase | | | | | | | | | |
| 0100 | 150 kVA | R-3 | .65 | 30.769 | Ea. | 9,850 | 1,525 | 198 | 11,573 | 13,300 |
| 0200 | 300 kVA | | .45 | 44.444 | | 12,700 | 2,200 | 287 | 15,187 | 17,600 |
| 0300 | 500 kVA | | .40 | 50 | | 19,700 | 2,475 | 325 | 22,500 | 25,800 |
| 0400 | 750 kVA | | .38 | 52.632 | | 22,900 | 2,625 | 340 | 25,865 | 29,500 |
| 0500 | 1000 kVA | | .26 | 76.923 | | 27,200 | 3,825 | 495 | 31,520 | 36,100 |
| 0600 | 1500 kVA | | .23 | 86.957 | | 32,200 | 4,325 | 560 | 37,085 | 42,500 |
| 0700 | 2000 kVA | | .20 | 100 | | 40,700 | 4,975 | 645 | 46,320 | 53,000 |
| 0800 | 3750 kVA | ↓ | .16 | 125 | ↓ | 76,500 | 6,225 | 805 | 83,530 | 94,000 |

# 26 22 Low-Voltage Transformers

## 26 22 13 – Low-Voltage Distribution Transformers

### 26 22 13.10 Transformer, Dry-Type

| 26 22 13.10 Transformer, Dry-Type | Crew | Daily Output | Labor-Hours | Unit | Material | 2011 Bare Costs Labor | Equipment | Total | Total Incl O&P |
|---|---|---|---|---|---|---|---|---|---|
| 0010 **TRANSFORMER, DRY-TYPE** | | | | | | | | | |
| 0050 Single phase, 240/480 volt primary, 120/240 volt secondary | | | | | | | | | |
| 0100 1 kVA | 1 Elec | 2 | 4 | Ea. | 282 | 201 | | 483 | 610 |
| 0300 2 kVA | | 1.60 | 5 | | 425 | 252 | | 677 | 840 |
| 0500 3 kVA | | 1.40 | 5.714 | | 525 | 287 | | 812 | 1,000 |
| 0700 5 kVA | | 1.20 | 6.667 | | 720 | 335 | | 1,055 | 1,300 |
| 0900 7.5 kVA | 2 Elec | 2.20 | 7.273 | | 1,000 | 365 | | 1,365 | 1,650 |
| 1100 10 kVA | | 1.60 | 10 | | 1,275 | 505 | | 1,780 | 2,150 |
| 1300 15 kVA | | 1.20 | 13.333 | | 1,750 | 670 | | 2,420 | 2,925 |
| 1500 25 kVA | | 1 | 16 | | 2,400 | 805 | | 3,205 | 3,825 |
| 1700 37.5 kVA | | .80 | 20 | | 3,100 | 1,000 | | 4,100 | 4,900 |
| 1900 50 kVA | | .70 | 22.857 | | 3,675 | 1,150 | | 4,825 | 5,750 |
| 2100 75 kVA | | .65 | 24.615 | | 4,875 | 1,250 | | 6,125 | 7,175 |
| 2190 480 V primary 120/240 V secondary, nonvent., 15 kVA | | 1.20 | 13.333 | | 1,775 | 670 | | 2,445 | 2,950 |
| 2200 25 kVA | | .90 | 17.778 | | 2,600 | 895 | | 3,495 | 4,200 |
| 2210 37 kVA | | .75 | 21.333 | | 3,100 | 1,075 | | 4,175 | 5,000 |
| 2220 50 kVA | | .65 | 24.615 | | 3,675 | 1,250 | | 4,925 | 5,875 |
| 2300 3 phase, 480 volt primary 120/208 volt secondary | | | | | | | | | |
| 2310 Ventilated, 3 kVA | 1 Elec | 1 | 8 | Ea. | 915 | 400 | | 1,315 | 1,600 |
| 2700 6 kVA | | .80 | 10 | | 1,250 | 505 | | 1,755 | 2,125 |
| 2900 9 kVA | | .70 | 11.429 | | 1,425 | 575 | | 2,000 | 2,425 |
| 3100 15 kVA | 2 Elec | 1.10 | 14.545 | | 1,900 | 730 | | 2,630 | 3,175 |
| 3300 30 kVA | | .90 | 17.778 | | 2,225 | 895 | | 3,120 | 3,775 |
| 3500 45 kVA | | .80 | 20 | | 2,675 | 1,000 | | 3,675 | 4,450 |
| 3700 75 kVA | | .70 | 22.857 | | 4,025 | 1,150 | | 5,175 | 6,125 |
| 3900 112.5 kVA | R-3 | .90 | 22.222 | | 5,350 | 1,100 | 143 | 6,593 | 7,675 |
| 4100 150 kVA | | .85 | 23.529 | | 6,975 | 1,175 | 152 | 8,302 | 9,575 |
| 4300 225 kVA | | .65 | 30.769 | | 9,475 | 1,525 | 198 | 11,198 | 12,900 |
| 4500 300 kVA | | .55 | 36.364 | | 12,000 | 1,800 | 235 | 14,035 | 16,100 |
| 4700 500 kVA | | .45 | 44.444 | | 19,800 | 2,200 | 287 | 22,287 | 25,400 |
| 4800 750 kVA | | .35 | 57.143 | | 34,700 | 2,850 | 370 | 37,920 | 42,800 |

# 26 24 Switchboards and Panelboards

## 26 24 13 – Switchboards

### 26 24 13.10 Incoming Switchboards

| 26 24 13.10 Incoming Switchboards | Crew | Daily Output | Labor-Hours | Unit | Material | 2011 Bare Costs Labor | Equipment | Total | Total Incl O&P |
|---|---|---|---|---|---|---|---|---|---|
| 0010 **INCOMING SWITCHBOARDS** main service section | | | | | | | | | |
| 0100 Aluminum bus bars, not including CT's or PT's | | | | | | | | | |
| 0200 No main disconnect, includes CT compartment | | | | | | | | | |
| 0300 120/208 volt, 4 wire, 600 amp | 2 Elec | 1 | 16 | Ea. | 4,025 | 805 | | 4,830 | 5,625 |
| 0400 800 amp | | .88 | 18.182 | | 4,025 | 915 | | 4,940 | 5,775 |
| 0500 1000 amp | | .80 | 20 | | 4,850 | 1,000 | | 5,850 | 6,825 |
| 0600 1200 amp | | .72 | 22.222 | | 4,850 | 1,125 | | 5,975 | 6,975 |
| 0700 1600 amp | | .66 | 24.242 | | 4,850 | 1,225 | | 6,075 | 7,125 |
| 0800 2000 amp | | .62 | 25.806 | | 5,225 | 1,300 | | 6,525 | 7,650 |
| 1000 3000 amp | | .56 | 28.571 | | 6,875 | 1,425 | | 8,300 | 9,700 |
| 2000 Fused switch & CT compartment | | | | | | | | | |
| 2100 120/208 volt, 4 wire, 400 amp | 2 Elec | 1.12 | 14.286 | Ea. | 3,875 | 720 | | 4,595 | 5,325 |
| 2200 600 amp | | .94 | 17.021 | | 4,575 | 855 | | 5,430 | 6,325 |
| 2300 800 amp | | .84 | 19.048 | | 10,900 | 960 | | 11,860 | 13,400 |
| 2400 1200 amp | | .68 | 23.529 | | 14,200 | 1,175 | | 15,375 | 17,400 |
| 2900 Pressure switch & CT compartment | | | | | | | | | |

## 26 24 13 – Switchboards

### 26 24 13.10 Incoming Switchboards

| | | Crew | Daily Output | Labor-Hours | Unit | Material | 2011 Bare Costs Labor | Equipment | Total | Total Incl O&P |
|---|---|---|---|---|---|---|---|---|---|---|
| 3000 | 120/208 volt, 4 wire, 800 amp | 2 Elec | .80 | 20 | Ea. | 9,800 | 1,000 | | 10,800 | 12,300 |
| 3100 | 1200 amp | | .66 | 24.242 | | 19,000 | 1,225 | | 20,225 | 22,600 |
| 3200 | 1600 amp | | .62 | 25.806 | | 20,100 | 1,300 | | 21,400 | 24,100 |
| 3300 | 2000 amp | | .56 | 28.571 | | 21,400 | 1,425 | | 22,825 | 25,700 |
| 4400 | Circuit breaker, molded case & CT compartment | | | | | | | | | |
| 4600 | 3 pole, 4 wire, 600 amp | 2 Elec | .94 | 17.021 | Ea. | 8,425 | 855 | | 9,280 | 10,600 |
| 4800 | 800 amp | | .84 | 19.048 | | 10,100 | 960 | | 11,060 | 12,500 |
| 5000 | 1200 amp | | .68 | 23.529 | | 13,800 | 1,175 | | 14,975 | 16,900 |
| 5100 | Copper bus bars, not incl. CT's or PT's, add, minimum | | | | | 15% | | | | |

### 26 24 13.30 Distribution Switchboards Section

| | | Crew | Daily Output | Labor-Hours | Unit | Material | 2011 Bare Costs Labor | Equipment | Total | Total Incl O&P |
|---|---|---|---|---|---|---|---|---|---|---|
| 0010 | **DISTRIBUTION SWITCHBOARDS SECTION** | | | | | | | | | |
| 0100 | Aluminum bus bars, not including breakers | | | | | | | | | |
| 0200 | 120/208 or 277/480 volt, 4 wire, 600 amp | 2 Elec | 1 | 16 | Ea. | 2,300 | 805 | | 3,105 | 3,725 |
| 0300 | 800 amp | | .88 | 18.182 | | 3,050 | 915 | | 3,965 | 4,700 |
| 0400 | 1000 amp | | .80 | 20 | | 3,050 | 1,000 | | 4,050 | 4,850 |
| 0500 | 1200 amp | | .72 | 22.222 | | 3,775 | 1,125 | | 4,900 | 5,800 |
| 0600 | 1600 amp | | .66 | 24.242 | | 4,325 | 1,225 | | 5,550 | 6,550 |
| 0700 | 2000 amp | | .62 | 25.806 | | 5,075 | 1,300 | | 6,375 | 7,500 |
| 0800 | 2500 amp | | .60 | 26.667 | | 5,725 | 1,350 | | 7,075 | 8,250 |
| 0900 | 3000 amp | | .56 | 28.571 | | 6,925 | 1,425 | | 8,350 | 9,750 |
| 0950 | 4000 amp | | .52 | 30.769 | | 10,100 | 1,550 | | 11,650 | 13,400 |

### 26 24 13.40 Switchboards Feeder Section

| | | Crew | Daily Output | Labor-Hours | Unit | Material | 2011 Bare Costs Labor | Equipment | Total | Total Incl O&P |
|---|---|---|---|---|---|---|---|---|---|---|
| 0010 | **SWITCHBOARDS FEEDER SECTION** group mounted devices | | | | | | | | | |
| 0030 | Circuit breakers | | | | | | | | | |
| 0160 | FA frame, 15 to 60 amp, 240 volt, 1 pole | 1 Elec | 8 | 1 | Ea. | 115 | 50.50 | | 165.50 | 201 |
| 0280 | FA frame, 70 to 100 amp, 240 volt, 1 pole | | 7 | 1.143 | | 151 | 57.50 | | 208.50 | 252 |
| 0420 | KA frame, 70 to 225 amp | | 3.20 | 2.500 | | 1,275 | 126 | | 1,401 | 1,575 |
| 0430 | LA frame, 125 to 400 amp | | 2.30 | 3.478 | | 2,875 | 175 | | 3,050 | 3,400 |
| 0460 | MA frame, 450 to 600 amp | | 1.60 | 5 | | 4,725 | 252 | | 4,977 | 5,575 |
| 0470 | 700 to 800 amp | | 1.30 | 6.154 | | 6,150 | 310 | | 6,460 | 7,225 |
| 0480 | MAL frame, 1000 amp | | 1 | 8 | | 6,375 | 400 | | 6,775 | 7,625 |
| 0490 | PA frame, 1200 amp | | .80 | 10 | | 13,000 | 505 | | 13,505 | 15,000 |
| 0500 | Branch circuit, fusible switch, 600 volt, double 30/30 amp | | 4 | 2 | | 1,200 | 101 | | 1,301 | 1,475 |
| 0550 | 60/60 amp | | 3.20 | 2.500 | | 1,225 | 126 | | 1,351 | 1,525 |
| 0600 | 100/100 amp | | 2.70 | 2.963 | | 1,725 | 149 | | 1,874 | 2,125 |
| 0650 | Single, 30 amp | | 5.30 | 1.509 | | 740 | 76 | | 816 | 925 |
| 0700 | 60 amp | | 4.70 | 1.702 | | 740 | 85.50 | | 825.50 | 940 |
| 0750 | 100 amp | | 4 | 2 | | 1,175 | 101 | | 1,276 | 1,425 |
| 0800 | 200 amp | | 2.70 | 2.963 | | 1,700 | 149 | | 1,849 | 2,100 |
| 0850 | 400 amp | | 2.30 | 3.478 | | 3,200 | 175 | | 3,375 | 3,775 |
| 0900 | 600 amp | | 1.80 | 4.444 | | 3,800 | 224 | | 4,024 | 4,525 |
| 0950 | 800 amp | | 1.30 | 6.154 | | 6,225 | 310 | | 6,535 | 7,300 |
| 1000 | 1200 amp | | .80 | 10 | | 7,425 | 505 | | 7,930 | 8,925 |

## 26 24 16 – Panelboards

### 26 24 16.20 Panelboard and Load Center Circuit Breakers

| | | Crew | Daily Output | Labor-Hours | Unit | Material | 2011 Bare Costs Labor | Equipment | Total | Total Incl O&P |
|---|---|---|---|---|---|---|---|---|---|---|
| 0010 | **PANELBOARD AND LOAD CENTER CIRCUIT BREAKERS** | | | | | | | | | |
| 0050 | Bolt-on, 10,000 amp IC, 120 volt, 1 pole | | | | | | | | | |
| 0100 | 15 to 50 amp | 1 Elec | 10 | .800 | Ea. | 14.80 | 40 | | 54.80 | 76 |
| 0200 | 60 amp | | 8 | 1 | | 14.80 | 50.50 | | 65.30 | 91 |
| 0300 | 70 amp | | 8 | 1 | | 28 | 50.50 | | 78.50 | 106 |
| 0350 | 240 volt, 2 pole | | | | | | | | | |
| 0400 | 15 to 50 amp | 1 Elec | 8 | 1 | Ea. | 32.50 | 50.50 | | 83 | 111 |

# 26 24 Switchboards and Panelboards

## 26 24 16 – Panelboards

### 26 24 16.20 Panelboard and Load Center Circuit Breakers

| | | Crew | Daily Output | Labor-Hours | Unit | Material | 2011 Bare Costs Labor | 2011 Bare Costs Equipment | Total | Total Incl O&P |
|---|---|---|---|---|---|---|---|---|---|---|
| 0500 | 60 amp | 1 Elec | 7.50 | 1.067 | Ea. | 32.50 | 53.50 | | 86 | 116 |
| 0600 | 80 to 100 amp | | 5 | 1.600 | | 83.50 | 80.50 | | 164 | 211 |
| 0700 | 3 pole, 15 to 60 amp | | 6.20 | 1.290 | | 103 | 65 | | 168 | 209 |
| 0800 | 70 amp | | 5 | 1.600 | | 130 | 80.50 | | 210.50 | 261 |
| 0900 | 80 to 100 amp | | 3.60 | 2.222 | | 147 | 112 | | 259 | 330 |
| 1000 | 22,000 amp I.C., 240 volt, 2 pole, 70 – 225 amp | | 2.70 | 2.963 | | 635 | 149 | | 784 | 915 |
| 1100 | 3 pole, 70 – 225 amp | | 2.30 | 3.478 | | 705 | 175 | | 880 | 1,025 |
| 1200 | 14,000 amp I.C., 277 volts, 1 pole, 15 – 30 amp | | 8 | 1 | | 39 | 50.50 | | 89.50 | 118 |
| 1300 | 22,000 amp I.C., 480 volts, 2 pole, 70 – 225 amp | | 2.70 | 2.963 | | 635 | 149 | | 784 | 915 |
| 1400 | 3 pole, 70 – 225 amp | | 2.30 | 3.478 | | 780 | 175 | | 955 | 1,125 |

### 26 24 16.30 Panelboards Commercial Applications

| | | Crew | Daily Output | Labor-Hours | Unit | Material | 2011 Bare Costs Labor | 2011 Bare Costs Equipment | Total | Total Incl O&P |
|---|---|---|---|---|---|---|---|---|---|---|
| 0010 | **PANELBOARDS COMMERCIAL APPLICATIONS** | | | | | | | | | |
| 0050 | NQOD, w/20 amp 1 pole bolt-on circuit breakers | | | | | | | | | |
| 0100 | 3 wire, 120/240 volts, 100 amp main lugs | | | | | | | | | |
| 0150 | 10 circuits | 1 Elec | 1 | 8 | Ea. | 545 | 400 | | 945 | 1,200 |
| 0200 | 14 circuits | | .88 | 9.091 | | 630 | 455 | | 1,085 | 1,375 |
| 0250 | 18 circuits | | .75 | 10.667 | | 685 | 535 | | 1,220 | 1,550 |
| 0300 | 20 circuits | | .65 | 12.308 | | 770 | 620 | | 1,390 | 1,750 |
| 0350 | 225 amp main lugs, 24 circuits | 2 Elec | 1.20 | 13.333 | | 870 | 670 | | 1,540 | 1,950 |
| 0400 | 30 circuits | | .90 | 17.778 | | 1,000 | 895 | | 1,895 | 2,425 |
| 0450 | 36 circuits | | .80 | 20 | | 1,150 | 1,000 | | 2,150 | 2,750 |
| 0500 | 38 circuits | | .72 | 22.222 | | 1,225 | 1,125 | | 2,350 | 3,000 |
| 0550 | 42 circuits | | .66 | 24.242 | | 1,275 | 1,225 | | 2,500 | 3,200 |
| 0600 | 4 wire, 120/208 volts, 100 amp main lugs, 12 circuits | 1 Elec | 1 | 8 | | 610 | 400 | | 1,010 | 1,275 |
| 0650 | 16 circuits | | .75 | 10.667 | | 700 | 535 | | 1,235 | 1,575 |
| 0700 | 20 circuits | | .65 | 12.308 | | 810 | 620 | | 1,430 | 1,800 |
| 0750 | 24 circuits | | .60 | 13.333 | | 875 | 670 | | 1,545 | 1,950 |
| 0800 | 30 circuits | | .53 | 15.094 | | 1,000 | 760 | | 1,760 | 2,225 |
| 0850 | 225 amp main lugs, 32 circuits | 2 Elec | .90 | 17.778 | | 1,125 | 895 | | 2,020 | 2,550 |
| 0900 | 34 circuits | | .84 | 19.048 | | 1,150 | 960 | | 2,110 | 2,700 |
| 0950 | 36 circuits | | .80 | 20 | | 1,175 | 1,000 | | 2,175 | 2,800 |
| 1000 | 42 circuits | | .68 | 23.529 | | 1,325 | 1,175 | | 2,500 | 3,200 |
| 1200 | NEHB, w/20 amp, 1 pole bolt-on circuit breakers | | | | | | | | | |
| 1250 | 4 wire, 277/480 volts, 100 amp main lugs, 12 circuits | 1 Elec | .88 | 9.091 | Ea. | 1,125 | 455 | | 1,580 | 1,925 |
| 1300 | 20 circuits | " | .60 | 13.333 | | 1,675 | 670 | | 2,345 | 2,850 |
| 1350 | 225 amp main lugs, 24 circuits | 2 Elec | .90 | 17.778 | | 1,900 | 895 | | 2,795 | 3,425 |
| 1400 | 30 circuits | | .80 | 20 | | 2,275 | 1,000 | | 3,275 | 4,025 |
| 1450 | 36 circuits | | .72 | 22.222 | | 2,650 | 1,125 | | 3,775 | 4,575 |
| 1600 | NQOD panel, w/20 amp, 1 pole, circuit breakers | | | | | | | | | |
| 1650 | 3 wire, 120/240 volt with main circuit breaker | | | | | | | | | |
| 1700 | 100 amp main, 12 circuits | 1 Elec | .80 | 10 | Ea. | 740 | 505 | | 1,245 | 1,550 |
| 1750 | 20 circuits | " | .60 | 13.333 | | 940 | 670 | | 1,610 | 2,025 |
| 1800 | 225 amp main, 30 circuits | 2 Elec | .68 | 23.529 | | 1,775 | 1,175 | | 2,950 | 3,700 |
| 1850 | 42 circuits | | .52 | 30.769 | | 2,050 | 1,550 | | 3,600 | 4,550 |
| 1900 | 400 amp main, 30 circuits | | .54 | 29.630 | | 2,425 | 1,500 | | 3,925 | 4,875 |
| 1950 | 42 circuits | | .50 | 32 | | 2,700 | 1,600 | | 4,300 | 5,350 |
| 2000 | 4 wire, 120/208 volts with main circuit breaker | | | | | | | | | |
| 2050 | 100 amp main, 24 circuits | 1 Elec | .47 | 17.021 | Ea. | 1,100 | 855 | | 1,955 | 2,475 |
| 2100 | 30 circuits | " | .40 | 20 | | 1,225 | 1,000 | | 2,225 | 2,850 |
| 2200 | 225 amp main, 32 circuits | 2 Elec | .72 | 22.222 | | 2,050 | 1,125 | | 3,175 | 3,900 |
| 2250 | 42 circuits | | .56 | 28.571 | | 2,250 | 1,425 | | 3,675 | 4,600 |
| 2300 | 400 amp main, 42 circuits | | .48 | 33.333 | | 3,000 | 1,675 | | 4,675 | 5,775 |

## 26 24 16 – Panelboards

| 26 24 16.30 Panelboards Commercial Applications | Crew | Daily Output | Labor-Hours | Unit | Material | 2011 Bare Costs Labor | Equipment | Total | Total Incl O&P |
|---|---|---|---|---|---|---|---|---|---|
| 2350 | 600 amp main, 42 circuits | 2 Elec | .40 | 40 | Ea. | 4,450 | 2,000 | | 6,450 | 7,875 |
| 2400 | NEHB, with 20 amp, 1 pole circuit breaker | | | | | | | | | |
| 2450 | 4 wire, 277/480 volts with main circuit breaker | | | | | | | | | |
| 2500 | 100 amp main, 24 circuits | 1 Elec | .42 | 19.048 | Ea. | 2,200 | 960 | | 3,160 | 3,850 |
| 2550 | 30 circuits | " | .38 | 21.053 | | 2,575 | 1,050 | | 3,625 | 4,400 |
| 2600 | 225 amp main, 30 circuits | 2 Elec | .72 | 22.222 | | 3,225 | 1,125 | | 4,350 | 5,200 |
| 2650 | 42 circuits | " | .56 | 28.571 | | 3,975 | 1,425 | | 5,400 | 6,500 |

## 26 24 19 – Motor-Control Centers

### 26 24 19.40 Motor Starters and Controls

| | 26 24 19.40 Motor Starters and Controls | Crew | Daily Output | Labor-Hours | Unit | Material | 2011 Bare Costs Labor | Equipment | Total | Total Incl O&P |
|---|---|---|---|---|---|---|---|---|---|---|
| 0010 | **MOTOR STARTERS AND CONTROLS** | | | | | | | | | |
| 0050 | Magnetic, FVNR, with enclosure and heaters, 480 volt | | | | | | | | | |
| 0100 | 5 HP, size 0 | 1 Elec | 2.30 | 3.478 | Ea. | 255 | 175 | | 430 | 540 |
| 0200 | 10 HP, size 1 | " | 1.60 | 5 | | 287 | 252 | | 539 | 690 |
| 0300 | 25 HP, size 2 | 2 Elec | 2.20 | 7.273 | | 540 | 365 | | 905 | 1,125 |
| 0400 | 50 HP, size 3 | | 1.80 | 8.889 | | 880 | 445 | | 1,325 | 1,625 |
| 0500 | 100 HP, size 4 | | 1.20 | 13.333 | | 1,950 | 670 | | 2,620 | 3,150 |
| 0600 | 200 HP, size 5 | | .90 | 17.778 | | 4,550 | 895 | | 5,445 | 6,350 |
| 0700 | Combination, with motor circuit protectors, 5 HP, size 0 | 1 Elec | 1.80 | 4.444 | | 830 | 224 | | 1,054 | 1,250 |
| 0800 | 10 HP, size 1 | " | 1.30 | 6.154 | | 860 | 310 | | 1,170 | 1,400 |
| 0900 | 25 HP, size 2 | 2 Elec | 2 | 8 | | 1,200 | 400 | | 1,600 | 1,925 |
| 1000 | 50 HP, size 3 | | 1.32 | 12.121 | | 1,750 | 610 | | 2,360 | 2,825 |
| 1200 | 100 HP, size 4 | | .80 | 20 | | 3,800 | 1,000 | | 4,800 | 5,675 |
| 1400 | Combination, with fused switch, 5 HP, size 0 | 1 Elec | 1.80 | 4.444 | | 635 | 224 | | 859 | 1,025 |
| 1600 | 10 HP, size 1 | " | 1.30 | 6.154 | | 675 | 310 | | 985 | 1,200 |
| 1800 | 25 HP, size 2 | 2 Elec | 2 | 8 | | 1,100 | 400 | | 1,500 | 1,800 |
| 2000 | 50 HP, size 3 | | 1.32 | 12.121 | | 1,850 | 610 | | 2,460 | 2,950 |
| 2200 | 100 HP, size 4 | | .80 | 20 | | 3,250 | 1,000 | | 4,250 | 5,075 |

## 26 25 13 – Bus Duct/Busway and Fittings

### 26 25 13.40 Copper Bus Duct

| | 26 25 13.40 Copper Bus Duct | Crew | Daily Output | Labor-Hours | Unit | Material | 2011 Bare Costs Labor | Equipment | Total | Total Incl O&P |
|---|---|---|---|---|---|---|---|---|---|---|
| 0010 | **COPPER BUS DUCT** 10 ft. long | | | | | | | | | |
| 0050 | Indoor 3 pole 4 wire, plug-in, straight section, 225 amp | 2 Elec | 40 | .400 | L.F. | 190 | 20 | | 210 | 239 |
| 1000 | 400 amp | | 32 | .500 | | 190 | 25 | | 215 | 247 |
| 1500 | 600 amp | | 26 | .615 | | 190 | 31 | | 221 | 255 |
| 2400 | 800 amp | | 20 | .800 | | 226 | 40 | | 266 | 310 |
| 2450 | 1000 amp | | 18 | .889 | | 250 | 44.50 | | 294.50 | 340 |
| 2500 | 1350 amp | | 16 | 1 | | 355 | 50.50 | | 405.50 | 470 |
| 2510 | 1600 amp | | 12 | 1.333 | | 405 | 67 | | 472 | 545 |
| 2520 | 2000 amp | | 10 | 1.600 | | 510 | 80.50 | | 590.50 | 685 |
| 2550 | Feeder, 600 amp | | 28 | .571 | | 167 | 28.50 | | 195.50 | 226 |
| 2600 | 800 amp | | 22 | .727 | | 202 | 36.50 | | 238.50 | 277 |
| 2700 | 1000 amp | | 20 | .800 | | 226 | 40 | | 266 | 310 |
| 2800 | 1350 amp | | 18 | .889 | | 335 | 44.50 | | 379.50 | 430 |
| 2900 | 1600 amp | | 14 | 1.143 | | 380 | 57.50 | | 437.50 | 505 |
| 3000 | 2000 amp | | 12 | 1.333 | | 490 | 67 | | 557 | 635 |
| 3100 | Elbows, 225 amp | | 4 | 4 | Ea. | 1,125 | 201 | | 1,326 | 1,550 |
| 3200 | 400 amp | | 3.60 | 4.444 | | 1,125 | 224 | | 1,349 | 1,575 |
| 3300 | 600 amp | | 3.20 | 5 | | 1,125 | 252 | | 1,377 | 1,625 |
| 3400 | 800 amp | | 2.80 | 5.714 | | 1,225 | 287 | | 1,512 | 1,775 |

## 26 25 13 – Bus Duct/Busway and Fittings

| 26 25 13.40 Copper Bus Duct | | Crew | Daily Output | Labor-Hours | Unit | Material | 2011 Bare Costs Labor | Equipment | Total | Total Incl O&P |
|---|---|---|---|---|---|---|---|---|---|---|
| 3500 | 1000 amp | 2 Elec | 2.60 | 6.154 | Ea. | 1,375 | 310 | | 1,685 | 1,950 |
| 3600 | 1350 amp | | 2.40 | 6.667 | | 1,600 | 335 | | 1,935 | 2,275 |
| 3700 | 1600 amp | | 2.20 | 7.273 | | 1,750 | 365 | | 2,115 | 2,475 |
| 3800 | 2000 amp | | 1.80 | 8.889 | | 2,150 | 445 | | 2,595 | 3,025 |
| 4000 | End box, 225 amp | | 34 | .471 | | 152 | 23.50 | | 175.50 | 203 |
| 4100 | 400 amp | | 32 | .500 | | 152 | 25 | | 177 | 206 |
| 4200 | 600 amp | | 28 | .571 | | 152 | 28.50 | | 180.50 | 211 |
| 4300 | 800 amp | | 26 | .615 | | 152 | 31 | | 183 | 214 |
| 4400 | 1000 amp | | 24 | .667 | | 152 | 33.50 | | 185.50 | 218 |
| 4500 | 1350 amp | | 22 | .727 | | 152 | 36.50 | | 188.50 | 222 |
| 4600 | 1600 amp | | 20 | .800 | | 152 | 40 | | 192 | 228 |
| 4700 | 2000 amp | | 18 | .889 | | 187 | 44.50 | | 231.50 | 272 |
| 4800 | Cable tap box end, 225 amp | | 3.20 | 5 | | 1,250 | 252 | | 1,502 | 1,725 |
| 5000 | 400 amp | | 2.60 | 6.154 | | 1,250 | 310 | | 1,560 | 1,800 |
| 5100 | 600 amp | | 2.20 | 7.273 | | 1,250 | 365 | | 1,615 | 1,900 |
| 5200 | 800 amp | | 2 | 8 | | 1,425 | 400 | | 1,825 | 2,175 |
| 5300 | 1000 amp | | 1.60 | 10 | | 1,525 | 505 | | 2,030 | 2,425 |
| 5400 | 1350 amp | | 1.40 | 11.429 | | 1,900 | 575 | | 2,475 | 2,950 |
| 5500 | 1600 amp | | 1.20 | 13.333 | | 2,125 | 670 | | 2,795 | 3,350 |
| 5600 | 2000 amp | | 1 | 16 | | 2,375 | 805 | | 3,180 | 3,825 |
| 5700 | Switchboard stub, 225 amp | | 5.40 | 2.963 | | 1,150 | 149 | | 1,299 | 1,475 |
| 5800 | 400 amp | | 4.60 | 3.478 | | 1,150 | 175 | | 1,325 | 1,500 |
| 5900 | 600 amp | | 4 | 4 | | 1,150 | 201 | | 1,351 | 1,550 |
| 6000 | 800 amp | | 3.20 | 5 | | 1,375 | 252 | | 1,627 | 1,900 |
| 6100 | 1000 amp | | 3 | 5.333 | | 1,600 | 268 | | 1,868 | 2,150 |
| 6200 | 1350 amp | | 2.60 | 6.154 | | 2,050 | 310 | | 2,360 | 2,725 |
| 6300 | 1600 amp | | 2.40 | 6.667 | | 2,325 | 335 | | 2,660 | 3,050 |
| 6400 | 2000 amp | | 2 | 8 | | 2,825 | 400 | | 3,225 | 3,700 |
| 6490 | Tee fittings, 225 amp | | 2.40 | 6.667 | | 1,550 | 335 | | 1,885 | 2,200 |
| 6500 | 400 amp | | 2 | 8 | | 1,550 | 400 | | 1,950 | 2,300 |
| 6600 | 600 amp | | 1.80 | 8.889 | | 1,550 | 445 | | 1,995 | 2,350 |
| 6700 | 800 amp | | 1.60 | 10 | | 1,800 | 505 | | 2,305 | 2,725 |
| 6800 | 1350 amp | | 1.20 | 13.333 | | 2,575 | 670 | | 3,245 | 3,825 |
| 7000 | 1600 amp | | 1 | 16 | | 2,925 | 805 | | 3,730 | 4,425 |
| 7100 | 2000 amp | | .80 | 20 | | 3,475 | 1,000 | | 4,475 | 5,325 |
| 7200 | Plug-in fusible switches w/3 fuses, 600 volt, 3 pole, 30 amp | 1 Elec | 4 | 2 | | 695 | 101 | | 796 | 915 |
| 7300 | 60 amp | | 3.60 | 2.222 | | 780 | 112 | | 892 | 1,025 |
| 7400 | 100 amp | | 2.70 | 2.963 | | 1,200 | 149 | | 1,349 | 1,525 |
| 7500 | 200 amp | 2 Elec | 3.20 | 5 | | 2,125 | 252 | | 2,377 | 2,725 |
| 7600 | 400 amp | | 1.40 | 11.429 | | 6,250 | 575 | | 6,825 | 7,700 |
| 7700 | 600 amp | | .90 | 17.778 | | 7,075 | 895 | | 7,970 | 9,100 |
| 7800 | 800 amp | | .66 | 24.242 | | 9,925 | 1,225 | | 11,150 | 12,700 |
| 7900 | 1200 amp | | .50 | 32 | | 18,600 | 1,600 | | 20,200 | 22,900 |
| 8000 | Plug-in circuit breakers, molded case, 15 to 50 amp | 1 Elec | 4.40 | 1.818 | | 660 | 91.50 | | 751.50 | 860 |
| 8100 | 70 to 100 amp | " | 3.10 | 2.581 | | 730 | 130 | | 860 | 995 |
| 8200 | 150 to 225 amp | 2 Elec | 3.40 | 4.706 | | 1,975 | 237 | | 2,212 | 2,525 |
| 8300 | 250 to 400 amp | | 1.40 | 11.429 | | 3,475 | 575 | | 4,050 | 4,675 |
| 8400 | 500 to 600 amp | | 1 | 16 | | 4,700 | 805 | | 5,505 | 6,375 |
| 8500 | 700 to 800 amp | | .64 | 25 | | 5,800 | 1,250 | | 7,050 | 8,250 |
| 8600 | 900 to 1000 amp | | .56 | 28.571 | | 8,275 | 1,425 | | 9,700 | 11,300 |
| 8700 | 1200 amp | | .44 | 36.364 | | 9,950 | 1,825 | | 11,775 | 13,700 |

# 26 27 Low-Voltage Distribution Equipment

## 26 27 16 – Electrical Cabinets and Enclosures

### 26 27 16.10 Cabinets

| | | Crew | Daily Output | Labor-Hours | Unit | Material | 2011 Bare Costs Labor | 2011 Bare Costs Equipment | Total | Total Incl O&P |
|---|---|---|---|---|---|---|---|---|---|---|
| 0010 | **CABINETS** | | | | | | | | | |
| 7000 | Cabinets, current transformer | | | | | | | | | |
| 7050 | Single door, 24" H x 24" W x 10" D | 1 Elec | 1.60 | 5 | Ea. | 206 | 252 | | 458 | 600 |
| 7100 | 30" H x 24" W x 10" D | | 1.30 | 6.154 | | 210 | 310 | | 520 | 690 |
| 7150 | 36" H x 24" W x 10" D | | 1.10 | 7.273 | | 215 | 365 | | 580 | 775 |
| 7200 | 30" H x 30" W x 10" D | | 1 | 8 | | 225 | 400 | | 625 | 845 |
| 7250 | 36" H x 30" W x 10" D | | .90 | 8.889 | | 305 | 445 | | 750 | 995 |
| 7300 | 36" H x 36" W x 10" D | | .80 | 10 | | 325 | 505 | | 830 | 1,100 |
| 7500 | Double door, 48" H x 36" W x 10" D | | .60 | 13.333 | | 670 | 670 | | 1,340 | 1,725 |
| 7550 | 24" H x 24" W x 12" D | | 1 | 8 | | 274 | 400 | | 674 | 895 |

## 26 27 23 – Indoor Service Poles

### 26 27 23.40 Surface Raceway

| | | Crew | Daily Output | Labor-Hours | Unit | Material | 2011 Bare Costs Labor | 2011 Bare Costs Equipment | Total | Total Incl O&P |
|---|---|---|---|---|---|---|---|---|---|---|
| 0010 | **SURFACE RACEWAY** | | | | | | | | | |
| 0090 | Metal, straight section | | | | | | | | | |
| 0100 | No. 500 | 1 Elec | 100 | .080 | L.F. | 1.03 | 4.02 | | 5.05 | 7.10 |
| 0110 | No. 700 | | 100 | .080 | | 1.16 | 4.02 | | 5.18 | 7.25 |
| 0400 | No. 1500, small pancake | | 90 | .089 | | 2.10 | 4.47 | | 6.57 | 8.90 |
| 0600 | No. 2000, base & cover, blank | | 90 | .089 | | 2.15 | 4.47 | | 6.62 | 8.95 |
| 0800 | No. 3000, base & cover, blank | | 75 | .107 | | 4.10 | 5.35 | | 9.45 | 12.45 |
| 1000 | No. 4000, base & cover, blank | | 65 | .123 | | 6.70 | 6.20 | | 12.90 | 16.50 |
| 1200 | No. 6000, base & cover, blank | | 50 | .160 | | 11.10 | 8.05 | | 19.15 | 24 |
| 2400 | Fittings, elbows, No. 500 | | 40 | .200 | Ea. | 1.85 | 10.05 | | 11.90 | 16.95 |
| 2800 | Elbow cover, No. 2000 | | 40 | .200 | | 3.50 | 10.05 | | 13.55 | 18.75 |
| 2880 | Tee, No. 500 | | 42 | .190 | | 3.60 | 9.60 | | 13.20 | 18.15 |
| 2900 | No. 2000 | | 27 | .296 | | 11.70 | 14.90 | | 26.60 | 35 |
| 3000 | Switch box, No. 500 | | 16 | .500 | | 12.40 | 25 | | 37.40 | 51 |
| 3400 | Telephone outlet, No. 1500 | | 16 | .500 | | 14.50 | 25 | | 39.50 | 53.50 |
| 3600 | Junction box, No. 1500 | | 16 | .500 | | 9.90 | 25 | | 34.90 | 48.50 |
| 3800 | Plugmold wired sections, No. 2000 | | | | | | | | | |
| 4000 | 1 circuit, 6 outlets, 3 ft. long | 1 Elec | 8 | 1 | Ea. | 36 | 50.50 | | 86.50 | 114 |
| 4100 | 2 circuits, 8 outlets, 6 ft. long | " | 5.30 | 1.509 | " | 59.50 | 76 | | 135.50 | 178 |

## 26 27 26 – Wiring Devices

### 26 27 26.10 Low Voltage Switching

| | | Crew | Daily Output | Labor-Hours | Unit | Material | 2011 Bare Costs Labor | 2011 Bare Costs Equipment | Total | Total Incl O&P |
|---|---|---|---|---|---|---|---|---|---|---|
| 0010 | **LOW VOLTAGE SWITCHING** | | | | | | | | | |
| 3600 | Relays, 120 V or 277 V standard | 1 Elec | 12 | .667 | Ea. | 38.50 | 33.50 | | 72 | 92 |
| 3800 | Flush switch, standard | | 40 | .200 | | 11.30 | 10.05 | | 21.35 | 27.50 |
| 4000 | Interchangeable | | 40 | .200 | | 14.75 | 10.05 | | 24.80 | 31 |
| 4100 | Surface switch, standard | | 40 | .200 | | 8 | 10.05 | | 18.05 | 23.50 |
| 4200 | Transformer 115 V to 25 V | | 12 | .667 | | 128 | 33.50 | | 161.50 | 191 |
| 4400 | Master control, 12 circuit, manual | | 4 | 2 | | 118 | 101 | | 219 | 279 |
| 4500 | 25 circuit, motorized | | 4 | 2 | | 131 | 101 | | 232 | 293 |
| 4600 | Rectifier, silicon | | 12 | .667 | | 42.50 | 33.50 | | 76 | 96.50 |
| 4800 | Switchplates, 1 gang, 1, 2 or 3 switch, plastic | | 80 | .100 | | 4.67 | 5.05 | | 9.72 | 12.60 |
| 5000 | Stainless steel | | 80 | .100 | | 10.60 | 5.05 | | 15.65 | 19.10 |
| 5400 | 2 gang, 3 switch, stainless steel | | 53 | .151 | | 20 | 7.60 | | 27.60 | 33.50 |
| 5500 | 4 switch, plastic | | 53 | .151 | | 9.65 | 7.60 | | 17.25 | 22 |
| 5800 | 3 gang, 9 switch, stainless steel | | 32 | .250 | | 64 | 12.60 | | 76.60 | 89 |

### 26 27 26.20 Wiring Devices Elements

| | | Crew | Daily Output | Labor-Hours | Unit | Material | 2011 Bare Costs Labor | 2011 Bare Costs Equipment | Total | Total Incl O&P |
|---|---|---|---|---|---|---|---|---|---|---|
| 0010 | **WIRING DEVICES ELEMENTS** | | | | | | | | | |
| 0200 | Toggle switch, quiet type, single pole, 15 amp | 1 Elec | 40 | .200 | Ea. | 5.40 | 10.05 | | 15.45 | 21 |
| 0600 | 3 way, 15 amp | | 23 | .348 | | 7.90 | 17.50 | | 25.40 | 34.50 |

# 26 27 Low-Voltage Distribution Equipment

## 26 27 26 – Wiring Devices

### 26 27 26.20 Wiring Devices Elements

| | | Crew | Daily Output | Labor-Hours | Unit | Material | 2011 Bare Costs Labor | 2011 Bare Costs Equipment | Total | Total Incl O&P |
|---|---|---|---|---|---|---|---|---|---|---|
| 0900 | 4 way, 15 amp | 1 Elec | 15 | .533 | Ea. | 26 | 27 | | 53 | 68 |
| 1650 | Dimmer switch, 120 volt, incandescent, 600 watt, 1 pole [G] | | 16 | .500 | | 13.60 | 25 | | 38.60 | 52.50 |
| 2460 | Receptacle, duplex, 120 volt, grounded, 15 amp | | 40 | .200 | | 1.37 | 10.05 | | 11.42 | 16.40 |
| 2470 | 20 amp | | 27 | .296 | | 10.55 | 14.90 | | 25.45 | 33.50 |
| 2490 | Dryer, 30 amp | | 15 | .533 | | 6.55 | 27 | | 33.55 | 46.50 |
| 2500 | Range, 50 amp | | 11 | .727 | | 14.60 | 36.50 | | 51.10 | 70 |
| 2600 | Wall plates, stainless steel, 1 gang | | 80 | .100 | | 2.50 | 5.05 | | 7.55 | 10.20 |
| 2800 | 2 gang | | 53 | .151 | | 4.25 | 7.60 | | 11.85 | 15.95 |
| 3200 | Lampholder, keyless | | 26 | .308 | | 12.80 | 15.50 | | 28.30 | 37 |
| 3400 | Pullchain with receptacle | ▼ | 22 | .364 | ▼ | 13.60 | 18.30 | | 31.90 | 42 |

## 26 27 73 – Door Chimes

### 26 27 73.10 Doorbell System

| | | Crew | Daily Output | Labor-Hours | Unit | Material | 2011 Bare Costs Labor | 2011 Bare Costs Equipment | Total | Total Incl O&P |
|---|---|---|---|---|---|---|---|---|---|---|
| 0010 | **DOORBELL SYSTEM**, incl. transformer, button & signal | | | | | | | | | |
| 0100 | 6" bell | 1 Elec | 4 | 2 | Ea. | 114 | 101 | | 215 | 275 |
| 0200 | Buzzer | " | 4 | 2 | " | 90.50 | 101 | | 191.50 | 249 |

# 26 28 Low-Voltage Circuit Protective Devices

## 26 28 16 – Enclosed Switches and Circuit Breakers

### 26 28 16.10 Circuit Breakers

| | | Crew | Daily Output | Labor-Hours | Unit | Material | 2011 Bare Costs Labor | 2011 Bare Costs Equipment | Total | Total Incl O&P |
|---|---|---|---|---|---|---|---|---|---|---|
| 0010 | **CIRCUIT BREAKERS** (in enclosure) | | | | | | | | | |
| 0100 | Enclosed (NEMA 1), 600 volt, 3 pole, 30 amp | 1 Elec | 3.20 | 2.500 | Ea. | 600 | 126 | | 726 | 840 |
| 0200 | 60 amp | | 2.80 | 2.857 | | 600 | 144 | | 744 | 870 |
| 0400 | 100 amp | | 2.30 | 3.478 | | 685 | 175 | | 860 | 1,000 |
| 0600 | 225 amp | ▼ | 1.50 | 5.333 | | 1,575 | 268 | | 1,843 | 2,150 |
| 0700 | 400 amp | 2 Elec | 1.60 | 10 | | 2,700 | 505 | | 3,205 | 3,725 |
| 0800 | 600 amp | | 1.20 | 13.333 | | 3,900 | 670 | | 4,570 | 5,300 |
| 1000 | 800 amp | ▼ | .94 | 17.021 | ▼ | 5,100 | 855 | | 5,955 | 6,875 |

### 26 28 16.20 Safety Switches

| | | Crew | Daily Output | Labor-Hours | Unit | Material | 2011 Bare Costs Labor | 2011 Bare Costs Equipment | Total | Total Incl O&P |
|---|---|---|---|---|---|---|---|---|---|---|
| 0010 | **SAFETY SWITCHES** | | | | | | | | | |
| 0100 | General duty 240 volt, 3 pole NEMA 1, fusible, 30 amp | 1 Elec | 3.20 | 2.500 | Ea. | 119 | 126 | | 245 | 315 |
| 0200 | 60 amp | | 2.30 | 3.478 | | 202 | 175 | | 377 | 480 |
| 0300 | 100 amp | | 1.90 | 4.211 | | 345 | 212 | | 557 | 695 |
| 0400 | 200 amp | ▼ | 1.30 | 6.154 | | 745 | 310 | | 1,055 | 1,275 |
| 0500 | 400 amp | 2 Elec | 1.80 | 8.889 | | 1,875 | 445 | | 2,320 | 2,725 |
| 0600 | 600 amp | " | 1.20 | 13.333 | ▼ | 3,500 | 670 | | 4,170 | 4,850 |
| 2900 | Heavy duty, 240 volt, 3 pole NEMA 1 fusible | | | | | | | | | |
| 2910 | 30 amp | 1 Elec | 3.20 | 2.500 | Ea. | 192 | 126 | | 318 | 395 |
| 3000 | 60 amp | | 2.30 | 3.478 | | 325 | 175 | | 500 | 620 |
| 3300 | 100 amp | | 1.90 | 4.211 | | 515 | 212 | | 727 | 880 |
| 3500 | 200 amp | ▼ | 1.30 | 6.154 | | 880 | 310 | | 1,190 | 1,425 |
| 3700 | 400 amp | 2 Elec | 1.80 | 8.889 | | 2,250 | 445 | | 2,695 | 3,150 |
| 3900 | 600 amp | " | 1.20 | 13.333 | ▼ | 3,900 | 670 | | 4,570 | 5,275 |

# 26 29 Low-Voltage Controllers

## 26 29 13 – Enclosed Controllers

| 26 29 13.20 Control Stations | Crew | Daily Output | Labor-Hours | Unit | Material | 2011 Bare Costs Labor | Equipment | Total | Total Incl O&P |
|---|---|---|---|---|---|---|---|---|---|
| 0010 **CONTROL STATIONS** | | | | | | | | | |
| 0050 NEMA 1, heavy duty, stop/start | 1 Elec | 8 | 1 | Ea. | 137 | 50.50 | | 187.50 | 225 |
| 0100 Stop/start, pilot light | | 6.20 | 1.290 | | 186 | 65 | | 251 | 300 |
| 0200 Hand/off/automatic | | 6.20 | 1.290 | | 101 | 65 | | 166 | 207 |
| 0400 Stop/start/reverse | | 5.30 | 1.509 | | 184 | 76 | | 260 | 315 |

# 26 32 Packaged Generator Assemblies

## 26 32 13 – Engine Generators

### 26 32 13.13 Diesel-Engine-Driven Generator Sets

| | Crew | Daily Output | Labor-Hours | Unit | Material | 2011 Bare Costs Labor | Equipment | Total | Total Incl O&P |
|---|---|---|---|---|---|---|---|---|---|
| 0010 **DIESEL-ENGINE-DRIVEN GENERATOR SETS** | | | | | | | | | |
| 2000 Diesel engine, including battery, charger, | | | | | | | | | |
| 2010 muffler, automatic transfer switch & day tank, 30 kW | R-3 | .55 | 36.364 | Ea. | 21,000 | 1,800 | 235 | 23,035 | 26,000 |
| 2100 50 kW | | .42 | 47.619 | | 24,900 | 2,375 | 305 | 27,580 | 31,300 |
| 2200 75 kW | | .35 | 57.143 | | 32,300 | 2,850 | 370 | 35,520 | 40,200 |
| 2300 100 kW | | .31 | 64.516 | | 36,000 | 3,200 | 415 | 39,615 | 44,700 |
| 2400 125 kW | | .29 | 68.966 | | 37,900 | 3,425 | 445 | 41,770 | 47,300 |
| 2500 150 kW | | .26 | 76.923 | | 43,500 | 3,825 | 495 | 47,820 | 54,000 |
| 2600 175 kW | | .25 | 80 | | 47,300 | 3,975 | 515 | 51,790 | 58,500 |
| 2700 200 kW | | .24 | 83.333 | | 48,800 | 4,150 | 540 | 53,490 | 60,000 |
| 2800 250 kW | | .23 | 86.957 | | 57,500 | 4,325 | 560 | 62,385 | 70,000 |
| 2900 300 kW | | .22 | 90.909 | | 62,000 | 4,525 | 585 | 67,110 | 76,000 |
| 3000 350 kW | | .20 | 100 | | 70,000 | 4,975 | 645 | 75,620 | 85,000 |
| 3100 400 kW | | .19 | 105 | | 86,500 | 5,225 | 680 | 92,405 | 104,000 |
| 3200 500 kW | | .18 | 111 | | 109,000 | 5,525 | 715 | 115,240 | 129,000 |

### 26 32 13.16 Gas-Engine-Driven Generator Sets

| | Crew | Daily Output | Labor-Hours | Unit | Material | 2011 Bare Costs Labor | Equipment | Total | Total Incl O&P |
|---|---|---|---|---|---|---|---|---|---|
| 0010 **GAS-ENGINE-DRIVEN GENERATOR SETS** | | | | | | | | | |
| 0020 Gas or gasoline operated, includes battery, | | | | | | | | | |
| 0050 charger, muffler & transfer switch | | | | | | | | | |
| 0200 3 phase 4 wire, 277/480 volt, 7.5 kW | R-3 | .83 | 24.096 | Ea. | 7,925 | 1,200 | 155 | 9,280 | 10,700 |
| 0300 11.5 kW | | .71 | 28.169 | | 11,200 | 1,400 | 182 | 12,782 | 14,700 |
| 0400 20 kW | | .63 | 31.746 | | 13,200 | 1,575 | 205 | 14,980 | 17,200 |
| 0500 35 kW | | .55 | 36.364 | | 15,800 | 1,800 | 235 | 17,835 | 20,200 |
| 0600 80 kW | | .40 | 50 | | 25,900 | 2,475 | 325 | 28,700 | 32,500 |
| 0700 100 kW | | .33 | 60.606 | | 28,300 | 3,025 | 390 | 31,715 | 36,100 |
| 0800 125 kW | | .28 | 71.429 | | 58,000 | 3,550 | 460 | 62,010 | 70,000 |
| 0900 185 kW | | .25 | 80 | | 76,500 | 3,975 | 515 | 80,990 | 91,000 |

# 26 35 Power Filters and Conditioners

## 26 35 13 – Capacitors

### 26 35 13.10 Capacitors Indoor

| | Crew | Daily Output | Labor-Hours | Unit | Material | 2011 Bare Costs Labor | Equipment | Total | Total Incl O&P |
|---|---|---|---|---|---|---|---|---|---|
| 0010 **CAPACITORS INDOOR** | | | | | | | | | |
| 0020 240 volts, single & 3 phase, 0.5 kVAR | 1 Elec | 2.70 | 2.963 | Ea. | 395 | 149 | | 544 | 655 |
| 0100 1.0 kVAR | | 2.70 | 2.963 | | 475 | 149 | | 624 | 745 |
| 0150 2.5 kVAR | | 2 | 4 | | 535 | 201 | | 736 | 890 |
| 0200 5.0 kVAR | | 1.80 | 4.444 | | 640 | 224 | | 864 | 1,025 |
| 0250 7.5 kVAR | | 1.60 | 5 | | 745 | 252 | | 997 | 1,200 |
| 0300 10 kVAR | | 1.50 | 5.333 | | 900 | 268 | | 1,168 | 1,375 |
| 0350 15 kVAR | | 1.30 | 6.154 | | 1,225 | 310 | | 1,535 | 1,800 |
| 0400 20 kVAR | | 1.10 | 7.273 | | 1,550 | 365 | | 1,915 | 2,250 |

# 26 35 Power Filters and Conditioners

## 26 35 13 – Capacitors

### 26 35 13.10 Capacitors Indoor

| | | Crew | Daily Output | Labor-Hours | Unit | Material | 2011 Bare Costs Labor | Equipment | Total | Total Incl O&P |
|---|---|---|---|---|---|---|---|---|---|---|
| 0450 | 25 kVAR | 1 Elec | 1 | 8 | Ea. | 1,800 | 400 | | 2,200 | 2,600 |
| 1000 | 480 volts, single & 3 phase, 1 kVAR | | 2.70 | 2.963 | | 360 | 149 | | 509 | 615 |
| 1050 | 2 kVAR | | 2.70 | 2.963 | | 415 | 149 | | 564 | 675 |
| 1100 | 5 kVAR | | 2 | 4 | | 520 | 201 | | 721 | 875 |
| 1150 | 7.5 kVAR | | 2 | 4 | | 560 | 201 | | 761 | 920 |
| 1200 | 10 kVAR | | 2 | 4 | | 630 | 201 | | 831 | 990 |
| 1250 | 15 kVAR | | 2 | 4 | | 775 | 201 | | 976 | 1,150 |
| 1300 | 20 kVAR | | 1.60 | 5 | | 865 | 252 | | 1,117 | 1,325 |
| 1350 | 30 kVAR | | 1.50 | 5.333 | | 1,075 | 268 | | 1,343 | 1,600 |
| 1400 | 40 kVAR | | 1.20 | 6.667 | | 1,375 | 335 | | 1,710 | 2,025 |
| 1450 | 50 kVAR | | 1.10 | 7.273 | | 1,575 | 365 | | 1,940 | 2,275 |

# 26 42 Cathodic Protection

## 26 42 16 – Passive Cathodic Protection for Underground Storage Tank

### 26 42 16.50 Cathodic Protection Wiring Methods

| | | Crew | Daily Output | Labor-Hours | Unit | Material | 2011 Bare Costs Labor | Equipment | Total | Total Incl O&P |
|---|---|---|---|---|---|---|---|---|---|---|
| 0010 | **CATHODIC PROTECTION WIRING METHODS** | | | | | | | | | |
| 1000 | Anodes, magnesium type, 9 # | R-15 | 18.50 | 2.595 | Ea. | 38.50 | 128 | 17.20 | 183.70 | 251 |
| 1010 | 17 # | | 13 | 3.692 | | 69.50 | 182 | 24.50 | 276 | 375 |
| 1020 | 32 # | | 10 | 4.800 | | 122 | 237 | 32 | 391 | 520 |
| 1030 | 48 # | | 7.20 | 6.667 | | 161 | 330 | 44 | 535 | 710 |
| 1100 | Graphite type w/epoxy cap, 3" x 60" (32 #) | R-22 | 8.40 | 4.438 | | 123 | 190 | | 313 | 420 |
| 1110 | 4" x 80" (68 #) | | 6 | 6.213 | | 235 | 266 | | 501 | 660 |
| 1120 | 6" x 72" (80 #) | | 5.20 | 7.169 | | 1,475 | 305 | | 1,780 | 2,075 |
| 1130 | 6" x 36" (45 #) | | 9.60 | 3.883 | | 745 | 166 | | 911 | 1,075 |
| 2000 | Rectifiers, silicon type, air cooled, 28 V/10 A | R-19 | 3.50 | 5.714 | | 2,100 | 288 | | 2,388 | 2,750 |
| 2010 | 20 V/20 A | | 3.50 | 5.714 | | 2,200 | 288 | | 2,488 | 2,850 |
| 2100 | Oil immersed, 28 V/10 A | | 3 | 6.667 | | 2,975 | 335 | | 3,310 | 3,775 |
| 2110 | 20 V/20 A | | 3 | 6.667 | | 3,100 | 335 | | 3,435 | 3,900 |
| 3000 | Anode backfill, coke breeze | R-22 | 3850 | .010 | Lb. | .18 | .41 | | .59 | .82 |
| 4000 | Cable, HMWPE, No. 8 | | 2.40 | 15.533 | M.L.F. | 480 | 665 | | 1,145 | 1,525 |
| 4010 | No. 6 | | 2.40 | 15.533 | | 655 | 665 | | 1,320 | 1,725 |
| 4020 | No. 4 | | 2.40 | 15.533 | | 980 | 665 | | 1,645 | 2,075 |
| 4030 | No. 2 | | 2.40 | 15.533 | | 1,525 | 665 | | 2,190 | 2,675 |
| 4040 | No. 1 | | 2.20 | 16.945 | | 2,050 | 725 | | 2,775 | 3,350 |
| 4050 | No. 1/0 | | 2.20 | 16.945 | | 2,575 | 725 | | 3,300 | 3,900 |
| 4060 | No. 2/0 | | 2.20 | 16.945 | | 4,100 | 725 | | 4,825 | 5,575 |
| 4070 | No. 4/0 | | 2 | 18.640 | | 4,475 | 800 | | 5,275 | 6,125 |
| 5000 | Test station, 7 terminal box, flush curb type w/lockable cover | R-19 | 12 | 1.667 | Ea. | 70.50 | 84 | | 154.50 | 202 |
| 5010 | Reference cell, 2" dia PVC conduit, cplg., plug, set flush | " | 4.80 | 4.167 | " | 148 | 210 | | 358 | 475 |

## 26 51 13 – Interior Lighting Fixtures, Lamps, and Ballasts

| 26 51 13.50 Interior Lighting Fixtures | | Crew | Daily Output | Labor-Hours | Unit | Material | 2011 Bare Costs Labor | Equipment | Total | Total Incl O&P |
|---|---|---|---|---|---|---|---|---|---|---|
| 0010 | **INTERIOR LIGHTING FIXTURES** Including lamps, mounting | | | | | | | | | |
| 0030 | hardware and connections | | | | | | | | | |
| 0100 | Fluorescent, C.W. lamps, troffer, recess mounted in grid, RS | | | | | | | | | |
| 0130 | Grid ceiling mount | | | | | | | | | |
| 0200 | Acrylic lens, 1'W x 4'L, two 40 watt | 1 Elec | 5.70 | 1.404 | Ea. | 54 | 70.50 | | 124.50 | 165 |
| 0210 | 1'W x 4'L, three 40 watt | | 5.40 | 1.481 | | 62.50 | 74.50 | | 137 | 179 |
| 0300 | 2'W x 2'L, two U40 watt | | 5.70 | 1.404 | | 59.50 | 70.50 | | 130 | 171 |
| 0400 | 2'W x 4'L, two 40 watt | | 5.30 | 1.509 | | 58 | 76 | | 134 | 176 |
| 0500 | 2'W x 4'L, three 40 watt | | 5 | 1.600 | | 64 | 80.50 | | 144.50 | 190 |
| 0600 | 2'W x 4'L, four 40 watt | | 4.70 | 1.702 | | 67 | 85.50 | | 152.50 | 201 |
| 0700 | 4'W x 4'L, four 40 watt | 2 Elec | 6.40 | 2.500 | | 340 | 126 | | 466 | 560 |
| 0800 | 4'W x 4'L, six 40 watt | | 6.20 | 2.581 | | 350 | 130 | | 480 | 575 |
| 0900 | 4'W x 4'L, eight 40 watt | | 5.80 | 2.759 | | 365 | 139 | | 504 | 605 |
| 0910 | Acrylic lens, 1'W x 4'L, two 32 watt T8 [G] | 1 Elec | 5.70 | 1.404 | | 67 | 70.50 | | 137.50 | 179 |
| 0930 | 2'W x 2'L, two U32 watt T8 [G] | | 5.70 | 1.404 | | 80 | 70.50 | | 150.50 | 193 |
| 0940 | 2'W x 4'L, two 32 watt T8 [G] | | 5.30 | 1.509 | | 72.50 | 76 | | 148.50 | 192 |
| 0950 | 2'W x 4'L, three 32 watt T8 [G] | | 5 | 1.600 | | 76.50 | 80.50 | | 157 | 203 |
| 0960 | 2'W x 4'L, four 32 watt T8 [G] | | 4.70 | 1.702 | | 78.50 | 85.50 | | 164 | 213 |
| 1000 | Surface mounted, RS | | | | | | | | | |
| 1030 | Acrylic lens with hinged & latched door frame | | | | | | | | | |
| 1100 | 1'W x 4'L, two 40 watt | 1 Elec | 7 | 1.143 | Ea. | 74 | 57.50 | | 131.50 | 167 |
| 1110 | 1'W x 4'L, three 40 watt | | 6.70 | 1.194 | | 76.50 | 60 | | 136.50 | 173 |
| 1200 | 2'W x 2'L, two U40 watt | | 7 | 1.143 | | 79.50 | 57.50 | | 137 | 172 |
| 1300 | 2'W x 4'L, two 40 watt | | 6.20 | 1.290 | | 91 | 65 | | 156 | 196 |
| 1400 | 2'W x 4'L, three 40 watt | | 5.70 | 1.404 | | 92 | 70.50 | | 162.50 | 206 |
| 1500 | 2'W x 4'L, four 40 watt | | 5.30 | 1.509 | | 94.50 | 76 | | 170.50 | 216 |
| 1600 | 4'W x 4'L, four 40 watt | 2 Elec | 7.20 | 2.222 | | 465 | 112 | | 577 | 675 |
| 1700 | 4'W x 4'L, six 40 watt | | 6.60 | 2.424 | | 505 | 122 | | 627 | 735 |
| 1800 | 4'W x 4'L, eight 40 watt | | 6.20 | 2.581 | | 525 | 130 | | 655 | 770 |
| 1900 | 2'W x 8'L, four 40 watt | | 6.40 | 2.500 | | 185 | 126 | | 311 | 390 |
| 2000 | 2'W x 8'L, eight 40 watt | | 6.20 | 2.581 | | 199 | 130 | | 329 | 410 |
| 2100 | Strip fixture | | | | | | | | | |
| 2130 | Surface mounted | | | | | | | | | |
| 2200 | 4' long, one 40 watt, RS | 1 Elec | 8.50 | .941 | Ea. | 31.50 | 47.50 | | 79 | 105 |
| 2300 | 4' long, two 40 watt, RS | | 8 | 1 | | 38 | 50.50 | | 88.50 | 116 |
| 2400 | 4' long, one 40 watt, SL | | 8 | 1 | | 46 | 50.50 | | 96.50 | 125 |
| 2500 | 4' long, two 40 watt, SL | | 7 | 1.143 | | 62.50 | 57.50 | | 120 | 154 |
| 2600 | 8' long, one 75 watt, SL | 2 Elec | 13.40 | 1.194 | | 47 | 60 | | 107 | 141 |
| 2700 | 8' long, two 75 watt, SL | " | 12.40 | 1.290 | | 57 | 65 | | 122 | 159 |
| 2800 | 4' long, two 60 watt, HO | 1 Elec | 6.70 | 1.194 | | 91.50 | 60 | | 151.50 | 190 |
| 2900 | 8' long, two 110 watt, HO | 2 Elec | 10.60 | 1.509 | | 96.50 | 76 | | 172.50 | 218 |
| 2950 | High bay pendent mounted, 16" W x 4' L, four 54 watt, T5HO [G] | | 8.90 | 1.798 | | 224 | 90.50 | | 314.50 | 380 |
| 2952 | 2' W x 4' L, six 54 watt, T5HO [G] | | 8.50 | 1.882 | | 300 | 94.50 | | 394.50 | 470 |
| 2954 | 2' W x 4' L, six 32 watt, T8 [G] | | 8.50 | 1.882 | | 175 | 94.50 | | 269.50 | 335 |
| 3000 | Strip, pendent mounted, industrial, white porcelain enamel | | | | | | | | | |
| 3100 | 4' long, two 40 watt, RS | 1 Elec | 5.70 | 1.404 | Ea. | 52 | 70.50 | | 122.50 | 162 |
| 3200 | 4' long, two 60 watt, HO | " | 5 | 1.600 | | 81 | 80.50 | | 161.50 | 208 |
| 3300 | 8' long, two 75 watt, SL | 2 Elec | 8.80 | 1.818 | | 96 | 91.50 | | 187.50 | 241 |
| 3400 | 8' long, two 110 watt, HO | " | 8 | 2 | | 123 | 101 | | 224 | 284 |
| 3470 | Troffer, air handling, 2'W x 4'L with four 32 watt T8 [G] | 1 Elec | 4 | 2 | | 110 | 101 | | 211 | 270 |
| 3480 | 2'W x 2'L with two U32 watt T8 [G] | | 5.50 | 1.455 | | 106 | 73 | | 179 | 225 |
| 3490 | Air connector insulated, 5" diameter | | 20 | .400 | | 66 | 20 | | 86 | 103 |
| 3500 | 6" diameter | | 20 | .400 | | 67.50 | 20 | | 87.50 | 104 |

# 26 51 Interior Lighting

## 26 51 13 – Interior Lighting Fixtures, Lamps, and Ballasts

### 26 51 13.50 Interior Lighting Fixtures

| | 26 51 13.50 Interior Lighting Fixtures | | Crew | Daily Output | Labor-Hours | Unit | Material | 2011 Bare Costs Labor | 2011 Bare Costs Equipment | Total | Total Incl O&P |
|---|---|---|---|---|---|---|---|---|---|---|---|
| 3510 | Troffer parabolic lay-in, 1'W x 4'L with one 32 W T8 | G | 1 Elec | 5.70 | 1.404 | Ea. | 112 | 70.50 | | 182.50 | 228 |
| 3520 | 1'W x 4'L with two 32 W T8 | G | | 5.30 | 1.509 | | 134 | 76 | | 210 | 259 |
| 3525 | 2'W x 2'L with two U32 W T8 | G | | 5.70 | 1.404 | | 113 | 70.50 | | 183.50 | 229 |
| 3530 | 2'W x 4'L with three 32 W T8 | G | | 5 | 1.600 | | 125 | 80.50 | | 205.50 | 257 |
| 4450 | Incandescent, high hat can, round alzak reflector, prewired | | | | | | | | | | |
| 4470 | 100 watt | | 1 Elec | 8 | 1 | Ea. | 70 | 50.50 | | 120.50 | 152 |
| 4480 | 150 watt | | | 8 | 1 | | 102 | 50.50 | | 152.50 | 187 |
| 4500 | 300 watt | | | 6.70 | 1.194 | | 236 | 60 | | 296 | 350 |
| 4600 | Square glass lens with metal trim, prewired | | | | | | | | | | |
| 4630 | 100 watt | | 1 Elec | 6.70 | 1.194 | Ea. | 54 | 60 | | 114 | 149 |
| 4700 | 200 watt | | | 6.70 | 1.194 | | 95 | 60 | | 155 | 194 |
| 4800 | 300 watt | | | 5.70 | 1.404 | | 142 | 70.50 | | 212.50 | 261 |
| 4900 | Ceiling/wall, surface mounted, metal cylinder, 75 watt | | | 10 | .800 | | 55 | 40 | | 95 | 120 |
| 4920 | 150 watt | | | 10 | .800 | | 79 | 40 | | 119 | 147 |
| 5200 | Ceiling, surface mounted, opal glass drum | | | | | | | | | | |
| 5300 | 8", one 60 watt lamp | | 1 Elec | 10 | .800 | Ea. | 43.50 | 40 | | 83.50 | 108 |
| 5400 | 10", two 60 watt lamps | | | 8 | 1 | | 49 | 50.50 | | 99.50 | 129 |
| 5500 | 12", four 60 watt lamps | | | 6.70 | 1.194 | | 69 | 60 | | 129 | 165 |
| 6010 | Vapor tight, incandescent, ceiling mounted, 200 watt | | | 6.20 | 1.290 | | 81 | 65 | | 146 | 185 |
| 6100 | Fluorescent, surface mounted, 2 lamps, 4'L, RS, 40 watt | | | 3.20 | 2.500 | | 113 | 126 | | 239 | 310 |
| 6850 | Vandalproof, surface mounted, fluorescent, two 32 watt T8 | G | | 3.20 | 2.500 | | 248 | 126 | | 374 | 460 |
| 6860 | Incandescent, one 150 watt | | | 8 | 1 | | 84 | 50.50 | | 134.50 | 167 |
| 7500 | Ballast replacement, by weight of ballast, to 15' high | | | | | | | | | | |
| 7520 | Indoor fluorescent, less than 2 lbs. | | 1 Elec | 10 | .800 | Ea. | 25.50 | 40 | | 65.50 | 87.50 |
| 7540 | Two 40W, watt reducer, 2 to 5 lbs. | | | 9.40 | .851 | | 40.50 | 43 | | 83.50 | 108 |
| 7560 | Two F96 slimline, over 5 lbs. | | | 8 | 1 | | 75.50 | 50.50 | | 126 | 158 |
| 7580 | Vaportite ballast, less than 2 lbs. | | | 9.40 | .851 | | 25.50 | 43 | | 68.50 | 91.50 |
| 7600 | 2 lbs. to 5 lbs. | | | 8.90 | .899 | | 40.50 | 45 | | 85.50 | 112 |
| 7620 | Over 5 lbs. | | | 7.60 | 1.053 | | 75.50 | 53 | | 128.50 | 162 |
| 7630 | Electronic ballast for two tubes | | | 8 | 1 | | 36.50 | 50.50 | | 87 | 115 |
| 7640 | Dimmable ballast one lamp | G | | 8 | 1 | | 104 | 50.50 | | 154.50 | 189 |
| 7650 | Dimmable ballast two-lamp | G | | 7.60 | 1.053 | | 102 | 53 | | 155 | 191 |

# 26 52 Emergency Lighting

## 26 52 13 – Emergency Lighting Equipments

### 26 52 13.10 Emergency Lighting and Battery Units

| | 26 52 13.10 Emergency Lighting and Battery Units | Crew | Daily Output | Labor-Hours | Unit | Material | 2011 Bare Costs Labor | 2011 Bare Costs Equipment | Total | Total Incl O&P |
|---|---|---|---|---|---|---|---|---|---|---|
| 0010 | **EMERGENCY LIGHTING AND BATTERY UNITS** | | | | | | | | | |
| 0300 | Emergency light units, battery operated | | | | | | | | | |
| 0350 | Twin sealed beam light, 25 watt, 6 volt each | | | | | | | | | |
| 0500 | Lead battery operated | 1 Elec | 4 | 2 | Ea. | 137 | 101 | | 238 | 299 |
| 0700 | Nickel cadmium battery operated | | 4 | 2 | | 575 | 101 | | 676 | 785 |
| 0900 | Self-contained fluorescent lamp pack | | 10 | .800 | | 157 | 40 | | 197 | 233 |

# 26 53 Exit Signs

## 26 53 13 – Exit Lighting

| 26 53 13.10 Exit Lighting Fixtures | | Crew | Daily Output | Labor-Hours | Unit | Material | 2011 Bare Costs Labor | Equipment | Total | Total Incl O&P |
|---|---|---|---|---|---|---|---|---|---|---|
| 0010 | **EXIT LIGHTING FIXTURES** | | | | | | | | | |
| 0080 | Exit light ceiling or wall mount, incandescent, single face | 1 Elec | 8 | 1 | Ea. | 42.50 | 50.50 | | 93 | 122 |
| 0100 | Double face | | 6.70 | 1.194 | | 49 | 60 | | 109 | 143 |
| 0200 | L.E.D. standard, single face | G | 8 | 1 | | 74.50 | 50.50 | | 125 | 157 |
| 0220 | Double face | G | 6.70 | 1.194 | | 74.50 | 60 | | 134.50 | 171 |
| 0230 | L.E.D. vandal-resistant, single face | G | 7.27 | 1.100 | | 223 | 55.50 | | 278.50 | 325 |
| 0240 | L.E.D. w/battery unit, single face | G | 4.40 | 1.818 | | 120 | 91.50 | | 211.50 | 267 |
| 0260 | Double face | G | 4 | 2 | | 137 | 101 | | 238 | 300 |
| 0262 | L.E.D. w/battery unit, vandal-resistant, single face | G | 4.40 | 1.818 | | 370 | 91.50 | | 461.50 | 545 |
| 0270 | Combination emergency light units and exit sign | | 4 | 2 | | 184 | 101 | | 285 | 350 |

# 26 54 Classified Location Lighting

## 26 54 13 – Classified Lighting

| 26 54 13.20 Explosionproof | | Crew | Daily Output | Labor-Hours | Unit | Material | 2011 Bare Costs Labor | Equipment | Total | Total Incl O&P |
|---|---|---|---|---|---|---|---|---|---|---|
| 0010 | **EXPLOSIONPROOF**, incl lamps, mounting hardware and connections | | | | | | | | | |
| 6510 | Incandescent, ceiling mounted, 200 watt | 1 Elec | 4 | 2 | Ea. | 960 | 101 | | 1,061 | 1,200 |
| 6600 | Fluorescent, RS, 4' long, ceiling mounted, two 40 watt | " | 2.70 | 2.963 | " | 2,550 | 149 | | 2,699 | 3,025 |

# 26 55 Special Purpose Lighting

## 26 55 61 – Theatrical Lighting

| 26 55 61.10 Lights | | Crew | Daily Output | Labor-Hours | Unit | Material | 2011 Bare Costs Labor | Equipment | Total | Total Incl O&P |
|---|---|---|---|---|---|---|---|---|---|---|
| 0010 | **LIGHTS** | | | | | | | | | |
| 2000 | Lights, border, quartz, reflector, vented, | | | | | | | | | |
| 2100 | colored or white | 1 Elec | 20 | .400 | L.F. | 173 | 20 | | 193 | 220 |
| 2500 | Spotlight, follow spot, with transformer, 2,100 watt | " | 4 | 2 | Ea. | 3,075 | 101 | | 3,176 | 3,525 |
| 2600 | For no transformer, deduct | | | | | 910 | | | 910 | 1,000 |
| 3000 | Stationary spot, fresnel quartz, 6" lens | 1 Elec | 4 | 2 | | 131 | 101 | | 232 | 293 |
| 3100 | 8" lens | | 4 | 2 | | 231 | 101 | | 332 | 405 |
| 3500 | Ellipsoidal quartz, 1,000W, 6" lens | | 4 | 2 | | 335 | 101 | | 436 | 520 |
| 3600 | 12" lens | | 4 | 2 | | 600 | 101 | | 701 | 810 |
| 4000 | Strobe light, 1 to 15 flashes per second, quartz | | 3 | 2.667 | | 750 | 134 | | 884 | 1,025 |
| 4500 | Color wheel, portable, five hole, motorized | | 4 | 2 | | 194 | 101 | | 295 | 365 |

# 26 56 Exterior Lighting

## 26 56 13 – Lighting Poles and Standards

| 26 56 13.10 Lighting Poles | | Crew | Daily Output | Labor-Hours | Unit | Material | 2011 Bare Costs Labor | Equipment | Total | Total Incl O&P |
|---|---|---|---|---|---|---|---|---|---|---|
| 0010 | **LIGHTING POLES** | | | | | | | | | |
| 2800 | Light poles, anchor base | | | | | | | | | |
| 2820 | not including concrete bases | | | | | | | | | |
| 2840 | Aluminum pole, 8' high | 1 Elec | 4 | 2 | Ea. | 700 | 101 | | 801 | 920 |
| 3000 | 20' high | R-3 | 2.90 | 6.897 | | 930 | 345 | 44.50 | 1,319.50 | 1,575 |
| 3200 | 30' high | | 2.60 | 7.692 | | 1,675 | 385 | 49.50 | 2,109.50 | 2,475 |
| 3400 | 35' high | | 2.30 | 8.696 | | 1,900 | 435 | 56 | 2,391 | 2,800 |
| 3600 | 40' high | | 2 | 10 | | 2,175 | 495 | 64.50 | 2,734.50 | 3,200 |
| 3800 | Bracket arms, 1 arm | 1 Elec | 8 | 1 | | 125 | 50.50 | | 175.50 | 212 |
| 4000 | 2 arms | | 8 | 1 | | 247 | 50.50 | | 297.50 | 345 |
| 4200 | 3 arms | | 5.30 | 1.509 | | 380 | 76 | | 456 | 530 |

# 26 56 Exterior Lighting

## 26 56 13 – Lighting Poles and Standards

| 26 56 13.10 Lighting Poles | | Crew | Daily Output | Labor-Hours | Unit | Material | 2011 Bare Costs Labor | 2011 Bare Costs Equipment | Total | Total Incl O&P |
|---|---|---|---|---|---|---|---|---|---|---|
| 4400 | 4 arms | 1 Elec | 5.30 | 1.509 | Ea. | 510 | 76 | | 586 | 670 |
| 4500 | Steel pole, galvanized, 8' high | ↓ | 3.80 | 2.105 | | 625 | 106 | | 731 | 840 |
| 4600 | 20' high | R-3 | 2.60 | 7.692 | | 1,100 | 385 | 49.50 | 1,534.50 | 1,850 |
| 4800 | 30' high | | 2.30 | 8.696 | | 1,200 | 435 | 56 | 1,691 | 2,000 |
| 5000 | 35' high | | 2.20 | 9.091 | | 1,400 | 450 | 58.50 | 1,908.50 | 2,275 |
| 5200 | 40' high | ↓ | 1.70 | 11.765 | | 1,750 | 585 | 76 | 2,411 | 2,875 |
| 5400 | Bracket arms, 1 arm | 1 Elec | 8 | 1 | | 176 | 50.50 | | 226.50 | 269 |
| 5600 | 2 arms | | 8 | 1 | | 273 | 50.50 | | 323.50 | 375 |
| 5800 | 3 arms | | 5.30 | 1.509 | | 295 | 76 | | 371 | 435 |
| 6000 | 4 arms | ↓ | 5.30 | 1.509 | ↓ | 410 | 76 | | 486 | 565 |

## 26 56 19 – Roadway Lighting

### 26 56 19.20 Roadway Area Luminaire

| | | Crew | Daily Output | Labor-Hours | Unit | Material | Labor | Equipment | Total | Total Incl O&P |
|---|---|---|---|---|---|---|---|---|---|---|
| 0010 | **ROADWAY AREA LUMINAIRE** | | | | | | | | | |
| 2650 | Roadway area luminaire, low pressure sodium, 135 watt | 1 Elec | 2 | 4 | Ea. | 645 | 201 | | 846 | 1,000 |
| 2700 | 180 watt | " | 2 | 4 | | 675 | 201 | | 876 | 1,050 |
| 2750 | Metal halide, 400 watt | 2 Elec | 4.40 | 3.636 | | 530 | 183 | | 713 | 855 |
| 2760 | 1000 watt | | 4 | 4 | | 600 | 201 | | 801 | 960 |
| 2780 | High pressure sodium, 400 watt | | 4.40 | 3.636 | | 550 | 183 | | 733 | 875 |
| 2790 | 1000 watt | ↓ | 4 | 4 | ↓ | 630 | 201 | | 831 | 990 |

## 26 56 23 – Area Lighting

### 26 56 23.10 Exterior Fixtures

| | | Crew | Daily Output | Labor-Hours | Unit | Material | Labor | Equipment | Total | Total Incl O&P |
|---|---|---|---|---|---|---|---|---|---|---|
| 0010 | **EXTERIOR FIXTURES** With lamps | | | | | | | | | |
| 0200 | Wall mounted, incandescent, 100 watt | 1 Elec | 8 | 1 | Ea. | 33.50 | 50.50 | | 84 | 112 |
| 0400 | Quartz, 500 watt | | 5.30 | 1.509 | | 56 | 76 | | 132 | 174 |
| 1100 | Wall pack, low pressure sodium, 35 watt | | 4 | 2 | | 234 | 101 | | 335 | 405 |
| 1150 | 55 watt | ↓ | 4 | 2 | | 278 | 101 | | 379 | 455 |

## 26 56 36 – Flood Lighting

### 26 56 36.20 Floodlights

| | | Crew | Daily Output | Labor-Hours | Unit | Material | Labor | Equipment | Total | Total Incl O&P |
|---|---|---|---|---|---|---|---|---|---|---|
| 0010 | **FLOODLIGHTS** with ballast and lamp, | | | | | | | | | |
| 1400 | pole mounted, pole not included | | | | | | | | | |
| 1950 | Metal halide, 175 watt | 1 Elec | 2.70 | 2.963 | Ea. | 325 | 149 | | 474 | 580 |
| 2000 | 400 watt | 2 Elec | 4.40 | 3.636 | | 405 | 183 | | 588 | 715 |
| 2200 | 1000 watt | " | 4 | 4 | | 555 | 201 | | 756 | 910 |
| 2340 | High pressure sodium, 70 watt | 1 Elec | 2.70 | 2.963 | | 245 | 149 | | 394 | 490 |
| 2400 | 400 watt | 2 Elec | 4.40 | 3.636 | | 365 | 183 | | 548 | 670 |
| 2600 | 1000 watt | " | 4 | 4 | ↓ | 625 | 201 | | 826 | 990 |

# 26 61 Lighting Systems and Accessories

## 26 61 23 – Lamps Applications

### 26 61 23.10 Lamps

| | | | Crew | Daily Output | Labor-Hours | Unit | Material | Labor | Equipment | Total | Total Incl O&P |
|---|---|---|---|---|---|---|---|---|---|---|---|
| 0010 | **LAMPS** | | | | | | | | | | |
| 0080 | Fluorescent, rapid start, cool white, 2' long, 20 watt | | 1 Elec | 1 | 8 | C | 355 | 400 | | 755 | 985 |
| 0100 | 4' long, 40 watt | | | .90 | 8.889 | | 300 | 445 | | 745 | 990 |
| 0200 | Slimline, 4' long, 40 watt | | | .90 | 8.889 | | 895 | 445 | | 1,340 | 1,650 |
| 0210 | 4' long, 30 watt energy saver | G | | .90 | 8.889 | | 895 | 445 | | 1,340 | 1,650 |
| 0400 | High output, 4' long, 60 watt | | | .90 | 8.889 | | 1,025 | 445 | | 1,470 | 1,775 |
| 0410 | 8' long, 95 watt energy saver | G | | .80 | 10 | | 1,025 | 505 | | 1,530 | 1,900 |
| 0500 | 8' long, 110 watt | | | .80 | 10 | | 1,025 | 505 | | 1,530 | 1,900 |
| 0512 | 2' long, T5, 14 watt energy saver | G | ↓ | 1 | 8 | | 1,400 | 400 | | 1,800 | 2,150 |

# 26 61 Lighting Systems and Accessories

## 26 61 23 – Lamps Applications

### 26 61 23.10 Lamps

| 26 61 23.10 Lamps | | Crew | Daily Output | Labor-Hours | Unit | Material | 2011 Bare Costs Labor | Equipment | Total | Total Incl O&P |
|---|---|---|---|---|---|---|---|---|---|---|
| 0514 | 3' long, T5, 21 watt energy saver [G] | 1 Elec | .90 | 8.889 | C | 1,400 | 445 | | 1,845 | 2,200 |
| 0516 | 4' long, T5, 28 watt energy saver [G] | | .90 | 8.889 | | 1,225 | 445 | | 1,670 | 1,975 |
| 0517 | 4' long, T5, 54 watt energy saver [G] | | .90 | 8.889 | | 1,775 | 445 | | 2,220 | 2,600 |
| 0560 | Twin tube compact lamp [G] | | .90 | 8.889 | | 585 | 445 | | 1,030 | 1,300 |
| 0570 | Double twin tube compact lamp [G] | | .80 | 10 | | 1,300 | 505 | | 1,805 | 2,200 |
| 0600 | Mercury vapor, mogul base, deluxe white, 100 watt | | .30 | 26.667 | | 3,100 | 1,350 | | 4,450 | 5,400 |
| 0700 | 250 watt | | .30 | 26.667 | | 4,800 | 1,350 | | 6,150 | 7,275 |
| 0800 | 400 watt | | .30 | 26.667 | | 3,875 | 1,350 | | 5,225 | 6,225 |
| 0900 | 1000 watt | | .20 | 40 | | 9,025 | 2,000 | | 11,025 | 12,900 |
| 1000 | Metal halide, mogul base, 175 watt | | .30 | 26.667 | | 4,125 | 1,350 | | 5,475 | 6,525 |
| 1200 | 400 watt | | .30 | 26.667 | | 4,425 | 1,350 | | 5,775 | 6,850 |
| 1300 | 1000 watt | | .20 | 40 | | 10,700 | 2,000 | | 12,700 | 14,800 |
| 1350 | High pressure sodium, 70 watt | | .30 | 26.667 | | 4,800 | 1,350 | | 6,150 | 7,250 |
| 1380 | 250 watt | | .30 | 26.667 | | 5,475 | 1,350 | | 6,825 | 8,000 |
| 1400 | 400 watt | | .30 | 26.667 | | 5,625 | 1,350 | | 6,975 | 8,150 |
| 1450 | 1000 watt | | .20 | 40 | | 15,500 | 2,000 | | 17,500 | 20,100 |
| 3000 | Guards, fluorescent lamp, 4' long | | 1 | 8 | | 1,175 | 400 | | 1,575 | 1,875 |
| 3200 | 8' long | | .90 | 8.889 | | 2,325 | 445 | | 2,770 | 3,225 |

# 26 71 Electrical Machines

## 26 71 13 – Motors Applications

### 26 71 13.20 Motors

| 26 71 13.20 Motors | | Crew | Daily Output | Labor-Hours | Unit | Material | 2011 Bare Costs Labor | Equipment | Total | Total Incl O&P |
|---|---|---|---|---|---|---|---|---|---|---|
| 0010 | **MOTORS** 230/460 volts, 60 HZ | | | | | | | | | |
| 0050 | Dripproof, premium efficiency, 1.15 service factor | | | | | | | | | |
| 0060 | 1800 RPM, 1/4 HP | 1 Elec | 5.33 | 1.501 | Ea. | 179 | 75.50 | | 254.50 | 310 |
| 0070 | 1/3 HP | | 5.33 | 1.501 | | 183 | 75.50 | | 258.50 | 315 |
| 0080 | 1/2 HP | | 5.33 | 1.501 | | 274 | 75.50 | | 349.50 | 410 |
| 0090 | 3/4 HP | | 5.33 | 1.501 | | 300 | 75.50 | | 375.50 | 440 |
| 0100 | 1 HP | | 4.50 | 1.778 | | 315 | 89.50 | | 404.50 | 480 |
| 0250 | 5 HP | | 4.50 | 1.778 | | 460 | 89.50 | | 549.50 | 635 |
| 0350 | 10 HP | | 4 | 2 | | 685 | 101 | | 786 | 900 |
| 0450 | 20 HP | 2 Elec | 5.20 | 3.077 | | 1,100 | 155 | | 1,255 | 1,425 |

## Estimating Tips

### 27 20 00 Data Communications
### 27 30 00 Voice Communications
### 27 40 00 Audio-Video Communications

When estimating material costs for special systems, it is always prudent to obtain manufacturers' quotations for equipment prices and special installation requirements which will affect the total costs.

## Reference Numbers

Reference numbers are shown in shaded boxes at the beginning of some major classifications. These numbers refer to related items in the Reference Section. The reference information may be an estimating procedure, an alternate pricing method, or technical information.

*Note:* Not all subdivisions listed here necessarily appear in this publication.

*Note:* **Trade Service,** *in part, has been used as a reference source for some of the material prices used in Division 27.*

# 27 13 Communications Backbone Cabling

## 27 13 23 – Communications Optical Fiber Backbone Cabling

| 27 13 23.13 Communications Optical Fiber | Crew | Daily Output | Labor-Hours | Unit | Material | 2011 Bare Costs Labor | Equipment | Total | Total Incl O&P |
|---|---|---|---|---|---|---|---|---|---|
| 0010 **COMMUNICATIONS OPTICAL FIBER** | | | | | | | | | |
| 0040 Specialized tools & techniques cause installation costs to vary. | | | | | | | | | |
| 0070 Fiber optic, cable, bulk simplex, minimum | 1 Elec | 8 | 1 | C.L.F. | 23 | 50.50 | | 73.50 | 100 |
| 0080 Bulk plenum quad, maximum | " | 2.29 | 3.493 | " | 68 | 176 | | 244 | 335 |
| 0150 Jumper | | | | Ea. | 34.50 | | | 34.50 | 38 |
| 0200 Pigtail | | | | | 32 | | | 32 | 35 |
| 0300 Connector | 1 Elec | 24 | .333 | | 20 | 16.75 | | 36.75 | 47 |
| 0350 Finger splice | | 32 | .250 | | 32 | 12.60 | | 44.60 | 53.50 |
| 0400 Transceiver (low cost bi-directional) | | 8 | 1 | | 410 | 50.50 | | 460.50 | 530 |
| 0450 Rack housing, 4 rack spaces, 12 panels (144 fibers) | | 2 | 4 | | 375 | 201 | | 576 | 715 |
| 0500 Patch panel, 12 ports | | 6 | 1.333 | | 242 | 67 | | 309 | 365 |

# 27 41 Audio-Video Systems

## 27 41 19 – Portable Audio-Video Equipment

### 27 41 19.10 T.V. Systems

| | Crew | Daily Output | Labor-Hours | Unit | Material | 2011 Bare Costs Labor | Equipment | Total | Total Incl O&P |
|---|---|---|---|---|---|---|---|---|---|
| 0010 **T.V. SYSTEMS**, not including rough-in wires, cables & conduits | | | | | | | | | |
| 0100 Master TV antenna system | | | | | | | | | |
| 0200 VHF reception & distribution, 12 outlets | 1 Elec | 6 | 1.333 | Outlet | 205 | 67 | | 272 | 325 |
| 0400 30 outlets | | 10 | .800 | | 135 | 40 | | 175 | 208 |
| 0600 100 outlets | | 13 | .615 | | 140 | 31 | | 171 | 200 |
| 0800 VHF & UHF reception & distribution, 12 outlets | | 6 | 1.333 | | 203 | 67 | | 270 | 325 |
| 1000 30 outlets | | 10 | .800 | | 135 | 40 | | 175 | 208 |
| 1200 100 outlets | | 13 | .615 | | 137 | 31 | | 168 | 197 |
| 1400 School and deluxe systems, 12 outlets | | 2.40 | 3.333 | | 268 | 168 | | 436 | 545 |
| 1600 30 outlets | | 4 | 2 | | 235 | 101 | | 336 | 405 |
| 1800 80 outlets | | 5.30 | 1.509 | | 226 | 76 | | 302 | 360 |

# 27 51 Distributed Audio-Video Communications Systems

## 27 51 16 – Public Address and Mass Notification Systems

### 27 51 16.10 Public Address System

| | Crew | Daily Output | Labor-Hours | Unit | Material | 2011 Bare Costs Labor | Equipment | Total | Total Incl O&P |
|---|---|---|---|---|---|---|---|---|---|
| 0010 **PUBLIC ADDRESS SYSTEM** | | | | | | | | | |
| 0100 Conventional, office | 1 Elec | 5.33 | 1.501 | Speaker | 129 | 75.50 | | 204.50 | 253 |
| 0200 Industrial | " | 2.70 | 2.963 | " | 248 | 149 | | 397 | 495 |

## 27 51 19 – Sound Masking Systems

### 27 51 19.10 Sound System

| | Crew | Daily Output | Labor-Hours | Unit | Material | 2011 Bare Costs Labor | Equipment | Total | Total Incl O&P |
|---|---|---|---|---|---|---|---|---|---|
| 0010 **SOUND SYSTEM**, not including rough-in wires, cables & conduits | | | | | | | | | |
| 0100 Components, projector outlet | 1 Elec | 8 | 1 | Ea. | 55 | 50.50 | | 105.50 | 135 |
| 0200 Microphone | | 4 | 2 | | 73 | 101 | | 174 | 230 |
| 0400 Speakers, ceiling or wall | | 8 | 1 | | 112 | 50.50 | | 162.50 | 198 |
| 0600 Trumpets | | 4 | 2 | | 209 | 101 | | 310 | 380 |
| 0800 Privacy switch | | 8 | 1 | | 83 | 50.50 | | 133.50 | 166 |
| 1000 Monitor panel | | 4 | 2 | | 370 | 101 | | 471 | 560 |
| 1200 Antenna, AM/FM | | 4 | 2 | | 132 | 101 | | 233 | 294 |
| 1400 Volume control | | 8 | 1 | | 80 | 50.50 | | 130.50 | 163 |
| 1600 Amplifier, 250 watts | | 1 | 8 | | 1,175 | 400 | | 1,575 | 1,875 |
| 1800 Cabinets | | 1 | 8 | | 810 | 400 | | 1,210 | 1,475 |
| 2000 Intercom, 30 station capacity, master station | 2 Elec | 2 | 8 | | 2,300 | 400 | | 2,700 | 3,125 |
| 2200 Remote station | 1 Elec | 8 | 1 | | 156 | 50.50 | | 206.50 | 246 |
| 2400 Intercom outlets | | 8 | 1 | | 91.50 | 50.50 | | 142 | 176 |

# 27 51 Distributed Audio-Video Communications Systems

## 27 51 19 – Sound Masking Systems

| 27 51 19.10 Sound System | Crew | Daily Output | Labor-Hours | Unit | Material | 2011 Bare Costs Labor | Equipment | Total | Total Incl O&P |
|---|---|---|---|---|---|---|---|---|---|
| 2600 | Handset | 1 Elec | 4 | 2 | Ea. | 305 | 101 | | 406 | 485 |
| 2800 | Emergency call system, 12 zones, annunciator | | 1.30 | 6.154 | | 915 | 310 | | 1,225 | 1,450 |
| 3000 | Bell | | 5.30 | 1.509 | | 94.50 | 76 | | 170.50 | 216 |
| 3200 | Light or relay | | 8 | 1 | | 47 | 50.50 | | 97.50 | 127 |
| 3400 | Transformer | | 4 | 2 | | 207 | 101 | | 308 | 375 |
| 3600 | House telephone, talking station | | 1.60 | 5 | | 445 | 252 | | 697 | 865 |
| 3800 | Press to talk, release to listen | ↓ | 5.30 | 1.509 | | 104 | 76 | | 180 | 226 |
| 4000 | System-on button | | | | | 62 | | | 62 | 68 |
| 4200 | Door release | 1 Elec | 4 | 2 | | 111 | 101 | | 212 | 271 |
| 4400 | Combination speaker and microphone | | 8 | 1 | | 189 | 50.50 | | 239.50 | 283 |
| 4600 | Termination box | | 3.20 | 2.500 | | 59.50 | 126 | | 185.50 | 252 |
| 4800 | Amplifier or power supply | | 5.30 | 1.509 | ↓ | 680 | 76 | | 756 | 860 |
| 5000 | Vestibule door unit | | 16 | .500 | Name | 125 | 25 | | 150 | 176 |
| 5200 | Strip cabinet | | 27 | .296 | Ea. | 237 | 14.90 | | 251.90 | 283 |
| 5400 | Directory | ↓ | 16 | .500 | " | 112 | 25 | | 137 | 161 |

# 27 52 Healthcare Communications and Monitoring Systems

## 27 52 23 – Nurse Call/Code Blue Systems

### 27 52 23.10 Nurse Call Systems

| | | Crew | Daily Output | Labor-Hours | Unit | Material | Labor | Equipment | Total | Total Incl O&P |
|---|---|---|---|---|---|---|---|---|---|---|
| 0010 | **NURSE CALL SYSTEMS** | | | | | | | | | |
| 0100 | Single bedside call station | 1 Elec | 8 | 1 | Ea. | 227 | 50.50 | | 277.50 | 325 |
| 0200 | Ceiling speaker station | | 8 | 1 | | 65.50 | 50.50 | | 116 | 147 |
| 0400 | Emergency call station | | 8 | 1 | | 114 | 50.50 | | 164.50 | 200 |
| 0600 | Pillow speaker | | 8 | 1 | | 172 | 50.50 | | 222.50 | 265 |
| 0800 | Double bedside call station | | 4 | 2 | | 204 | 101 | | 305 | 375 |
| 1000 | Duty station | | 4 | 2 | | 138 | 101 | | 239 | 300 |
| 1200 | Standard call button | | 8 | 1 | | 106 | 50.50 | | 156.50 | 192 |
| 1400 | Lights, corridor, dome or zone indicator | ↓ | 8 | 1 | ↓ | 61 | 50.50 | | 111.50 | 142 |
| 1600 | Master control station for 20 stations | 2 Elec | .65 | 24.615 | Total | 3,675 | 1,250 | | 4,925 | 5,875 |

# 27 53 Distributed Systems

## 27 53 13 – Clock Systems

### 27 53 13.50 Clock Equipments

| | | Crew | Daily Output | Labor-Hours | Unit | Material | Labor | Equipment | Total | Total Incl O&P |
|---|---|---|---|---|---|---|---|---|---|---|
| 0010 | **CLOCK EQUIPMENTS**, not including wires & conduits | | | | | | | | | |
| 0100 | Time system components, master controller | 1 Elec | .33 | 24.242 | Ea. | 1,825 | 1,225 | | 3,050 | 3,800 |
| 0200 | Program bell | | 8 | 1 | | 85 | 50.50 | | 135.50 | 168 |
| 0400 | Combination clock & speaker | | 3.20 | 2.500 | | 210 | 126 | | 336 | 415 |
| 0600 | Frequency generator | | 2 | 4 | | 2,350 | 201 | | 2,551 | 2,900 |
| 0800 | Job time automatic stamp recorder | ↓ | 4 | 2 | | 460 | 101 | | 561 | 655 |
| 1600 | Master time clock system, clocks & bells, 20 room | 4 Elec | .20 | 160 | | 6,300 | 8,050 | | 14,350 | 18,800 |
| 1800 | 50 room | " | .08 | 400 | | 13,000 | 20,100 | | 33,100 | 44,100 |
| 2000 | Time clock, 100 cards in & out, 1 color | 1 Elec | 3.20 | 2.500 | | 1,325 | 126 | | 1,451 | 1,625 |
| 2200 | 2 colors | | 3.20 | 2.500 | | 1,250 | 126 | | 1,376 | 1,575 |
| 2800 | Metal rack for 25 cards | ↓ | 7 | 1.143 | ↓ | 38.50 | 57.50 | | 96 | 128 |

# Division Notes

| | CREW | DAILY OUTPUT | LABOR-HOURS | UNIT | MAT. | LABOR | EQUIP. | TOTAL | TOTAL INCL O&P |
|---|---|---|---|---|---|---|---|---|---|
| | | | | | | | | | |
| | | | | | | | | | |
| | | | | | | | | | |
| | | | | | | | | | |
| | | | | | | | | | |
| | | | | | | | | | |
| | | | | | | | | | |
| | | | | | | | | | |
| | | | | | | | | | |
| | | | | | | | | | |
| | | | | | | | | | |
| | | | | | | | | | |
| | | | | | | | | | |
| | | | | | | | | | |
| | | | | | | | | | |
| | | | | | | | | | |
| | | | | | | | | | |
| | | | | | | | | | |
| | | | | | | | | | |
| | | | | | | | | | |
| | | | | | | | | | |
| | | | | | | | | | |
| | | | | | | | | | |
| | | | | | | | | | |

The "2011 BARE COSTS" header spans the MAT., LABOR, EQUIP., and TOTAL columns.

## Estimating Tips

- When estimating material costs for electronic safety and security systems, it is always prudent to obtain manufacturers' quotations for equipment prices and special installation requirements that affect the total cost.

- Fire alarm systems consist of control panels, annunciator panels, battery with rack, charger, and fire alarm actuating and indicating devices. Some fire alarm systems include speakers, telephone lines, door closer controls, and other components. Be careful not to overlook the costs related to installation for these items.

Also be aware of costs for integrated automation instrumentation and terminal devices, control equipment, control wiring, and programming.

- Security equipment includes items such as CCTV, access control, and other detection and identification systems to perform alert and alarm functions. Be sure to consider the costs related to installation for this security equipment, such as for integrated automation instrumentation and terminal devices, control equipment, control wiring, and programming.

## Reference Numbers

Reference numbers are shown in shaded boxes at the beginning of some major classifications. These numbers refer to related items in the Reference Section. The reference information may be an estimating procedure, an alternate pricing method, or technical information.

*Note:* Not all subdivisions listed here necessarily appear in this publication.

**Division 28 - Electronic Safety and Security**

# 28 13 Access Control

## 28 13 53 – Security Access Detection

### 28 13 53.13 Security Access Metal Detectors

| | | Crew | Daily Output | Labor-Hours | Unit | Material | 2011 Bare Costs Labor | Equipment | Total | Total Incl O&P |
|---|---|---|---|---|---|---|---|---|---|---|
| 0010 | **SECURITY ACCESS METAL DETECTORS** | | | | Ea. | 89 | | | 89 | 98 |
| 0240 | Metal detector, hand-held, wand type, unit only | | | | | | | | | |
| 0250 | Metal detector, walk through portal type, single zone | 1 Elec | 2 | 4 | | 2,750 | 201 | | 2,951 | 3,325 |
| 0260 | Multi-zone | " | 2 | 4 | ↓ | 3,500 | 201 | | 3,701 | 4,150 |

### 28 13 53.16 Security Access X-Ray Equipment

| | | Crew | Daily Output | Labor-Hours | Unit | Material | 2011 Bare Costs Labor | Equipment | Total | Total Incl O&P |
|---|---|---|---|---|---|---|---|---|---|---|
| 0010 | **SECURITY ACCESS X-RAY EQUIPMENT** | | | | | | | | | |
| 0290 | X-ray machine, desk top, for mail/small packages/letters | 1 Elec | 4 | 2 | Ea. | 3,000 | 101 | | 3,101 | 3,450 |
| 0300 | Conveyor type, incl monitor, minimum | | 2 | 4 | | 14,000 | 201 | | 14,201 | 15,700 |
| 0310 | Maximum | ↓ | 2 | 4 | | 25,000 | 201 | | 25,201 | 27,800 |
| 0320 | X-ray machine, large unit, for airports, incl monitor, min | 2 Elec | 1 | 16 | | 35,000 | 805 | | 35,805 | 39,700 |
| 0330 | Maximum | " | .50 | 32 | ↓ | 60,000 | 1,600 | | 61,600 | 68,500 |

### 28 13 53.23 Security Access Explosive Detection Equipment

| | | Crew | Daily Output | Labor-Hours | Unit | Material | 2011 Bare Costs Labor | Equipment | Total | Total Incl O&P |
|---|---|---|---|---|---|---|---|---|---|---|
| 0010 | **SECURITY ACCESS EXPLOSIVE DETECTION EQUIPMENT** | | | | | | | | | |
| 0270 | Explosives detector, walk through portal type | 1 Elec | 2 | 4 | Ea. | 3,500 | 201 | | 3,701 | 4,150 |
| 0280 | Hand-held, battery operated | | | | " | | | | 25,500 | 28,100 |

# 28 16 Intrusion Detection

## 28 16 16 – Intrusion Detection Systems Infrastructure

### 28 16 16.50 Intrusion Detection

| | | Crew | Daily Output | Labor-Hours | Unit | Material | 2011 Bare Costs Labor | Equipment | Total | Total Incl O&P |
|---|---|---|---|---|---|---|---|---|---|---|
| 0010 | **INTRUSION DETECTION**, not including wires & conduits | | | | | | | | | |
| 0100 | Burglar alarm, battery operated, mechanical trigger | 1 Elec | 4 | 2 | Ea. | 272 | 101 | | 373 | 450 |
| 0200 | Electrical trigger | | 4 | 2 | | 325 | 101 | | 426 | 505 |
| 0400 | For outside key control, add | | 8 | 1 | | 82 | 50.50 | | 132.50 | 165 |
| 0600 | For remote signaling circuitry, add | | 8 | 1 | | 122 | 50.50 | | 172.50 | 210 |
| 0800 | Card reader, flush type, standard | | 2.70 | 2.963 | | 910 | 149 | | 1,059 | 1,225 |
| 1000 | Multi-code | | 2.70 | 2.963 | | 1,175 | 149 | | 1,324 | 1,525 |
| 1200 | Door switches, hinge switch | | 5.30 | 1.509 | | 57.50 | 76 | | 133.50 | 175 |
| 1400 | Magnetic switch | | 5.30 | 1.509 | | 67.50 | 76 | | 143.50 | 187 |
| 1600 | Exit control locks, horn alarm | | 4 | 2 | | 266 | 101 | | 367 | 440 |
| 1800 | Flashing light alarm | | 4 | 2 | | 297 | 101 | | 398 | 475 |
| 2000 | Indicating panels, 1 channel | ↓ | 2.70 | 2.963 | | 360 | 149 | | 509 | 615 |
| 2200 | 10 channel | 2 Elec | 3.20 | 5 | | 1,225 | 252 | | 1,477 | 1,725 |
| 2400 | 20 channel | | 2 | 8 | | 2,400 | 400 | | 2,800 | 3,250 |
| 2600 | 40 channel | ↓ | 1.14 | 14.035 | | 4,375 | 705 | | 5,080 | 5,850 |
| 2800 | Ultrasonic motion detector, 12 volt | 1 Elec | 2.30 | 3.478 | | 225 | 175 | | 400 | 505 |
| 3000 | Infrared photoelectric detector | " | 4 | 2 | ↓ | 186 | 101 | | 287 | 355 |

# 28 23 Video Surveillance

## 28 23 13 – Video Surveillance Control and Management Systems

### 28 23 13.10 Closed Circuit Television System

| | | Crew | Daily Output | Labor-Hours | Unit | Material | 2011 Bare Costs Labor | Equipment | Total | Total Incl O&P |
|---|---|---|---|---|---|---|---|---|---|---|
| 0010 | **CLOSED CIRCUIT TELEVISION SYSTEM** | | | | | | | | | |
| 2000 | Surveillance, one station (camera & monitor) | 2 Elec | 2.60 | 6.154 | Total | 1,300 | 310 | | 1,610 | 1,875 |
| 2200 | For additional camera stations, add | 1 Elec | 2.70 | 2.963 | Ea. | 725 | 149 | | 874 | 1,025 |
| 2400 | Industrial quality, one station (camera & monitor) | 2 Elec | 2.60 | 6.154 | Total | 2,700 | 310 | | 3,010 | 3,400 |
| 2600 | For additional camera stations, add | 1 Elec | 2.70 | 2.963 | Ea. | 1,650 | 149 | | 1,799 | 2,050 |
| 2610 | For low light, add | | 2.70 | 2.963 | | 1,325 | 149 | | 1,474 | 1,675 |
| 2620 | For very low light, add | | 2.70 | 2.963 | | 9,750 | 149 | | 9,899 | 10,900 |
| 2800 | For weatherproof camera station, add | | 1.30 | 6.154 | | 1,025 | 310 | | 1,335 | 1,575 |
| 3000 | For pan and tilt, add | ↓ | 1.30 | 6.154 | | 2,625 | 310 | | 2,935 | 3,350 |

# 28 23 Video Surveillance

## 28 23 13 – Video Surveillance Control and Management Systems

### 28 23 13.10 Closed Circuit Television System

| 28 23 13.10 Closed Circuit Television System | Crew | Daily Output | Labor-Hours | Unit | Material | 2011 Bare Costs Labor | 2011 Bare Costs Equipment | Total | Total Incl O&P |
|---|---|---|---|---|---|---|---|---|---|
| 3200 | For zoom lens - remote control, add, minimum | 1 Elec | 2 | 4 | Ea. | 2,425 | 201 | | 2,626 | 2,975 |
| 3400 | Maximum | | 2 | 4 | | 8,850 | 201 | | 9,051 | 10,000 |
| 3410 | For automatic iris for low light, add | | 2 | 4 | | 2,125 | 201 | | 2,326 | 2,625 |
| 3600 | Educational T.V. studio, basic 3 camera system, black & white, | | | | | | | | | |
| 3800 | electrical & electronic equip. only, minimum | 4 Elec | .80 | 40 | Total | 12,600 | 2,000 | | 14,600 | 16,800 |
| 4000 | Maximum (full console) | | .28 | 114 | | 53,500 | 5,750 | | 59,250 | 67,500 |
| 4100 | As above, but color system, minimum | | .28 | 114 | | 71,000 | 5,750 | | 76,750 | 86,500 |
| 4120 | Maximum | | .12 | 266 | | 308,500 | 13,400 | | 321,900 | 359,000 |
| 4200 | For film chain, black & white, add | 1 Elec | 1 | 8 | Ea. | 14,400 | 400 | | 14,800 | 16,400 |
| 4250 | Color, add | | .25 | 32 | | 17,500 | 1,600 | | 19,100 | 21,700 |
| 4400 | For video recorders, add, minimum | | 1 | 8 | | 3,025 | 400 | | 3,425 | 3,925 |
| 4600 | Maximum | 4 Elec | .40 | 80 | | 25,200 | 4,025 | | 29,225 | 33,700 |

## 28 23 23 – Video Surveillance Systems Infrastructure

### 28 23 23.50 Video Surveillance Equipments

| | | Crew | Daily Output | Labor-Hours | Unit | Material | 2011 Bare Costs Labor | 2011 Bare Costs Equipment | Total | Total Incl O&P |
|---|---|---|---|---|---|---|---|---|---|---|
| 0010 | **VIDEO SURVEILLANCE EQUIPMENTS** | | | | | | | | | |
| 0200 | Video cameras, wireless, hidden in exit signs, clocks, etc, incl receiver | 1 Elec | 3 | 2.667 | Ea. | 143 | 134 | | 277 | 355 |
| 0210 | Accessories for video recorder, single camera | | 3 | 2.667 | | 260 | 134 | | 394 | 485 |
| 0220 | For multiple cameras | | 3 | 2.667 | | 1,750 | 134 | | 1,884 | 2,125 |
| 0230 | Video cameras, wireless, for under vehicle searching, complete | | 2 | 4 | | 9,975 | 201 | | 10,176 | 11,300 |

# 28 31 Fire Detection and Alarm

## 28 31 23 – Fire Detection and Alarm Annunciation Panels and Fire Stations

### 28 31 23.50 Alarm Panels and Devices

| | | Crew | Daily Output | Labor-Hours | Unit | Material | 2011 Bare Costs Labor | 2011 Bare Costs Equipment | Total | Total Incl O&P |
|---|---|---|---|---|---|---|---|---|---|---|
| 0010 | **ALARM PANELS AND DEVICES**, not including wires & conduits | | | | | | | | | |
| 3594 | Fire, alarm control panel | | | | | | | | | |
| 3600 | 4 zone | 2 Elec | 2 | 8 | Ea. | 390 | 400 | | 790 | 1,025 |
| 3800 | 8 zone | | 1 | 16 | | 760 | 805 | | 1,565 | 2,025 |
| 4000 | 12 zone | | .67 | 23.988 | | 2,300 | 1,200 | | 3,500 | 4,300 |
| 4020 | Alarm device | 1 Elec | 8 | 1 | | 229 | 50.50 | | 279.50 | 325 |
| 4050 | Actuating device | | 8 | 1 | | 325 | 50.50 | | 375.50 | 430 |
| 4200 | Battery and rack | | 4 | 2 | | 395 | 101 | | 496 | 585 |
| 4400 | Automatic charger | | 8 | 1 | | 560 | 50.50 | | 610.50 | 695 |
| 4600 | Signal bell | | 8 | 1 | | 70.50 | 50.50 | | 121 | 152 |
| 4800 | Trouble buzzer or manual station | | 8 | 1 | | 80.50 | 50.50 | | 131 | 163 |
| 5600 | Strobe and horn | | 5.30 | 1.509 | | 147 | 76 | | 223 | 273 |
| 5800 | Fire alarm horn | | 6.70 | 1.194 | | 58.50 | 60 | | 118.50 | 154 |
| 6000 | Door holder, electro-magnetic | | 4 | 2 | | 99 | 101 | | 200 | 258 |
| 6200 | Combination holder and closer | | 3.20 | 2.500 | | 119 | 126 | | 245 | 315 |
| 6600 | Drill switch | | 8 | 1 | | 360 | 50.50 | | 410.50 | 470 |
| 6800 | Master box | | 2.70 | 2.963 | | 6,175 | 149 | | 6,324 | 7,025 |
| 7000 | Break glass station | | 8 | 1 | | 65 | 50.50 | | 115.50 | 146 |
| 7800 | Remote annunciator, 8 zone lamp | | 1.80 | 4.444 | | 201 | 224 | | 425 | 550 |
| 8000 | 12 zone lamp | 2 Elec | 2.60 | 6.154 | | 325 | 310 | | 635 | 815 |
| 8200 | 16 zone lamp | " | 2.20 | 7.273 | | 405 | 365 | | 770 | 985 |

## 28 31 43 – Fire Detection Sensors

### 28 31 43.50 Fire and Heat Detectors

| | | Crew | Daily Output | Labor-Hours | Unit | Material | 2011 Bare Costs Labor | 2011 Bare Costs Equipment | Total | Total Incl O&P |
|---|---|---|---|---|---|---|---|---|---|---|
| 0010 | **FIRE & HEAT DETECTORS** | | | | | | | | | |
| 5000 | Detector, rate of rise | 1 Elec | 8 | 1 | Ea. | 49 | 50.50 | | 99.50 | 129 |

# 28 31 Fire Detection and Alarm

## 28 31 46 – Smoke Detection Sensors

### 28 31 46.50 Smoke Detectors

| | 28 31 46.50 Smoke Detectors | Crew | Daily Output | Labor-Hours | Unit | Material | 2011 Bare Costs Labor | Equipment | Total | Total Incl O&P |
|---|---|---|---|---|---|---|---|---|---|---|
| 0010 | **SMOKE DETECTORS** | | | | | | | | | |
| 5200 | Smoke detector, ceiling type | 1 Elec | 6.20 | 1.290 | Ea. | 126 | 65 | | 191 | 235 |
| 5400 | Duct type | " | 3.20 | 2.500 | " | 315 | 126 | | 441 | 530 |

# 28 33 Fuel-Gas Detection and Alarm

## 28 33 33 – Fuel-Gas Detection Sensors

### 28 33 33.50 Tank Leak Detection Systems

| | | Crew | Daily Output | Labor-Hours | Unit | Material | 2011 Bare Costs Labor | Equipment | Total | Total Incl O&P |
|---|---|---|---|---|---|---|---|---|---|---|
| 0010 | **TANK LEAK DETECTION SYSTEMS** Liquid and vapor | | | | | | | | | |
| 0100 | For hydrocarbons and hazardous liquids/vapors | | | | | | | | | |
| 0120 | Controller, data acquisition, incl. printer, modem, RS232 port | | | | | | | | | |
| 0140 | 24 channel, for use with all probes | | | | Ea. | 3,500 | | | 3,500 | 3,850 |
| 0160 | 9 channel, for external monitoring | | | | " | 770 | | | 770 | 845 |
| 0200 | Probes | | | | | | | | | |
| 0210 | Well monitoring | | | | | | | | | |
| 0220 | Liquid phase detection | | | | Ea. | 400 | | | 400 | 440 |
| 0230 | Hydrocarbon vapor, fixed position | | | | | 400 | | | 400 | 440 |
| 0240 | Hydrocarbon vapor, float mounted | | | | | 400 | | | 400 | 440 |
| 0250 | Both liquid and vapor hydrocarbon | | | | | 400 | | | 400 | 440 |
| 0300 | Secondary containment, liquid phase | | | | | | | | | |
| 0310 | Pipe trench/manway sump | | | | Ea. | 690 | | | 690 | 760 |
| 0320 | Double wall pipe and manual sump | | | | | 690 | | | 690 | 760 |
| 0330 | Double wall fiberglass annular space | | | | | 272 | | | 272 | 299 |
| 0340 | Double wall steel tank annular space | | | | | 282 | | | 282 | 310 |
| 0500 | Accessories | | | | | | | | | |
| 0510 | Modem, non-dedicated phone line | | | | Ea. | 255 | | | 255 | 280 |
| 0600 | Monitoring, internal | | | | | | | | | |
| 0610 | Automatic tank gauge, incl. overfill | | | | Ea. | 965 | | | 965 | 1,075 |
| 0620 | Product line | | | | " | 965 | | | 965 | 1,075 |
| 0700 | Monitoring, special | | | | | | | | | |
| 0710 | Cathodic protection | | | | Ea. | 605 | | | 605 | 670 |
| 0720 | Annular space chemical monitor | | | | " | 825 | | | 825 | 910 |

# 28 39 Mass Notification Systems

## 28 39 10 – Notification Systems

### 28 39 10.10 Mass Notification System

| | | Crew | Daily Output | Labor-Hours | Unit | Material | 2011 Bare Costs Labor | Equipment | Total | Total Incl O&P |
|---|---|---|---|---|---|---|---|---|---|---|
| 0010 | **MASS NOTIFICATION SYSTEM** | | | | | | | | | |
| 0100 | Wireless command center, 10,000 devices | 2 Elec | 1.33 | 12.030 | Ea. | 3,000 | 605 | | 3,605 | 4,200 |
| 0200 | Option, email notification | | | | | 1,300 | | | 1,300 | 1,425 |
| 0210 | Remote device supervision & monitor | | | | | 1,800 | | | 1,800 | 1,975 |
| 0300 | Antenna VHF or UHF, for medium range | 1 Elec | 4 | 2 | | 95 | 101 | | 196 | 254 |
| 0310 | For high-power transmitter | | 2 | 4 | | 495 | 201 | | 696 | 845 |
| 0400 | Transmitter, 25 watt | | 4 | 2 | | 1,500 | 101 | | 1,601 | 1,800 |
| 0410 | 40 watt | | 2.66 | 3.008 | | 2,000 | 151 | | 2,151 | 2,425 |
| 0420 | 100 watt | | 1.33 | 6.015 | | 5,000 | 305 | | 5,305 | 5,950 |
| 0500 | Wireless receiver/control module for speaker | | 8 | 1 | | 195 | 50.50 | | 245.50 | 290 |
| 0600 | Desktop paging controller, stand alone | | 4 | 2 | | 245 | 101 | | 346 | 420 |

# Estimating Tips

## 31 05 00 Common Work Results for Earthwork

- Estimating the actual cost of performing earthwork requires careful consideration of the variables involved. This includes items such as type of soil, whether water will be encountered, dewatering, whether banks need bracing, disposal of excavated earth, and length of haul to fill or spoil sites, etc. If the project has large quantities of cut or fill, consider raising or lowering the site to reduce costs, while paying close attention to the effect on site drainage and utilities.

- If the project has large quantities of fill, creating a borrow pit on the site can significantly lower the costs.

- It is very important to consider what time of year the project is scheduled for completion. Bad weather can create large cost overruns from dewatering, site repair, and lost productivity from cold weather.

# Reference Numbers

Reference numbers are shown in shaded boxes at the beginning of some major classifications. These numbers refer to related items in the Reference Section. The reference information may be an estimating procedure, an alternate pricing method, or technical information.

*Note:* Not all subdivisions listed here necessarily appear in this publication.

# 31 05 Common Work Results for Earthwork

## 31 05 13 – Soils for Earthwork

### 31 05 13.10 Borrow

| | 31 05 13.10 Borrow | Crew | Daily Output | Labor-Hours | Unit | Material | 2011 Bare Costs Labor | 2011 Bare Costs Equipment | Total | Total Incl O&P |
|---|---|---|---|---|---|---|---|---|---|---|
| 0010 | **BORROW** R312316-40 | | | | | | | | | |
| 0020 | Spread, 200 H.P. dozer, no compaction, 2 mi. RT haul | | | | | | | | | |
| 0200 | Common borrow | B-15 | 600 | .047 | C.Y. | 15 | 1.75 | 3.44 | 20.19 | 23 |
| 0700 | Screened loam | | 600 | .047 | | 33 | 1.75 | 3.44 | 38.19 | 43 |
| 0800 | Topsoil, weed free | ↓ | 600 | .047 | | 24 | 1.75 | 3.44 | 29.19 | 33 |
| 0900 | For 5 mile haul, add | B-34B | 200 | .040 | ↓ | | 1.38 | 2.53 | 3.91 | 4.87 |

## 31 05 16 – Aggregates for Earthwork

### 31 05 16.10 Borrow

| | 31 05 16.10 Borrow | Crew | Daily Output | Labor-Hours | Unit | Material | 2011 Bare Costs Labor | 2011 Bare Costs Equipment | Total | Total Incl O&P |
|---|---|---|---|---|---|---|---|---|---|---|
| 0010 | **BORROW** R312316-40 | | | | | | | | | |
| 0020 | Spread, with 200 H.P. dozer, no compaction, 2 mi. RT haul | | | | | | | | | |
| 0100 | Bank run gravel | B-15 | 600 | .047 | C.Y. | 22 | 1.75 | 3.44 | 27.19 | 30.50 |
| 0300 | Crushed stone (1.40 tons per CY) , 1-1/2" | | 600 | .047 | | 38 | 1.75 | 3.44 | 43.19 | 48.50 |
| 0320 | 3/4" | | 600 | .047 | | 38 | 1.75 | 3.44 | 43.19 | 48.50 |
| 0340 | 1/2" | | 600 | .047 | | 47 | 1.75 | 3.44 | 52.19 | 58.50 |
| 0360 | 3/8" | | 600 | .047 | | 32 | 1.75 | 3.44 | 37.19 | 42 |
| 0400 | Sand, washed, concrete | | 600 | .047 | | 41 | 1.75 | 3.44 | 46.19 | 51.50 |
| 0500 | Dead or bank sand | | 600 | .047 | | 15 | 1.75 | 3.44 | 20.19 | 23 |
| 0600 | Select structural fill | ↓ | 600 | .047 | | 20 | 1.75 | 3.44 | 25.19 | 28.50 |
| 0900 | For 5 mile haul, add | B-34B | 200 | .040 | ↓ | | 1.38 | 2.53 | 3.91 | 4.87 |

## 31 05 23 – Cement and Concrete for Earthwork

### 31 05 23.30 Plant Mixed Bituminous Concrete

| | 31 05 23.30 Plant Mixed Bituminous Concrete | Crew | Daily Output | Labor-Hours | Unit | Material | 2011 Bare Costs Labor | 2011 Bare Costs Equipment | Total | Total Incl O&P |
|---|---|---|---|---|---|---|---|---|---|---|
| 0010 | **PLANT MIXED BITUMINOUS CONCRETE** | | | | | | | | | |
| 0020 | Asphaltic concrete plant mix (145 lb. per C.F.) | | | | Ton | 64 | | | 64 | 70.50 |
| 0040 | Asphaltic concrete less than 300 tons add trucking costs | | | | | | | | | |
| 0050 | See Section 31 23 23.20 for hauling costs | | | | | | | | | |
| 0200 | All weather patching mix, hot | | | | Ton | 65 | | | 65 | 71.50 |
| 0250 | Cold patch | | | | ↓ | 65 | | | 65 | 71.50 |
| 0300 | Berm mix | | | | ↓ | 65 | | | 65 | 71.50 |

# 31 06 Schedules for Earthwork

## 31 06 60 – Schedules for Special Foundations and Load Bearing Elements

### 31 06 60.14 Piling Special Costs

| | 31 06 60.14 Piling Special Costs | Crew | Daily Output | Labor-Hours | Unit | Material | 2011 Bare Costs Labor | 2011 Bare Costs Equipment | Total | Total Incl O&P |
|---|---|---|---|---|---|---|---|---|---|---|
| 0010 | **PILING SPECIAL COSTS** | | | | | | | | | |
| 0011 | Piling special costs, pile caps, see Section 03 30 53.40 | | | | | | | | | |
| 0500 | Cutoffs, concrete piles, plain | 1 Pile | 5.50 | 1.455 | Ea. | | 61 | | 61 | 98 |
| 0600 | With steel thin shell, add | | 38 | .211 | | | 8.85 | | 8.85 | 14.20 |
| 0700 | Steel pile or "H" piles | | 19 | .421 | | | 17.70 | | 17.70 | 28.50 |
| 0800 | Wood piles | ↓ | 38 | .211 | ↓ | | 8.85 | | 8.85 | 14.20 |
| 0900 | Pre-augering up to 30' deep, average soil, 24" diameter | B-43 | 180 | .267 | L.F. | | 10.05 | 13.80 | 23.85 | 30.50 |
| 0920 | 36" diameter | | 115 | .417 | | | 15.75 | 21.50 | 37.25 | 48 |
| 0960 | 48" diameter | | 70 | .686 | | | 26 | 35.50 | 61.50 | 78.50 |
| 0980 | 60" diameter | ↓ | 50 | .960 | ↓ | | 36 | 49.50 | 85.50 | 110 |
| 1000 | Testing, any type piles, test load is twice the design load | | | | | | | | | |
| 1050 | 50 ton design load, 100 ton test | | | | Ea. | | | | 14,000 | 15,500 |
| 1100 | 100 ton design load, 200 ton test | | | | | | | | 20,000 | 22,000 |
| 1150 | 150 ton design load, 300 ton test | | | | | | | | 26,000 | 28,500 |
| 1200 | 200 ton design load, 400 ton test | | | | | | | | 28,000 | 31,000 |
| 1250 | 400 ton design load, 800 ton test | | | | ↓ | | | | 32,000 | 35,000 |
| 1500 | Wet conditions, soft damp ground | | | | | | | | | |
| 1600 | Requiring mats for crane, add | | | | | | | | 40% | 40% |

# 31 06 Schedules for Earthwork

## 31 06 60 – Schedules for Special Foundations and Load Bearing Elements

### 31 06 60.14 Piling Special Costs

| | | Crew | Daily Output | Labor-Hours | Unit | Material | 2011 Bare Costs Labor | Equipment | Total | Total Incl O&P |
|---|---|---|---|---|---|---|---|---|---|---|
| 1700 | Barge mounted driving rig, add | | | | | | | | 30% | 30% |

### 31 06 60.15 Mobilization

| | | Crew | Daily Output | Labor-Hours | Unit | Material | Labor | Equipment | Total | Total Incl O&P |
|---|---|---|---|---|---|---|---|---|---|---|
| 0010 | **MOBILIZATION** | | | | | | | | | |
| 0020 | Set up & remove, air compressor, 600 C.F.M. | A-5 | 3.30 | 5.455 | Ea. | | 187 | 13 | 200 | 300 |
| 0100 | 1200 C.F.M. | " | 2.20 | 8.182 | | | 280 | 19.45 | 299.45 | 450 |
| 0200 | Crane, with pile leads and pile hammer, 75 ton | B-19 | .60 | 106 | | | 4,600 | 3,075 | 7,675 | 10,600 |
| 0300 | 150 ton | " | .36 | 177 | | | 7,675 | 5,125 | 12,800 | 17,700 |
| 0500 | Drill rig, for caissons, to 36", minimum | B-43 | 2 | 24 | | | 905 | 1,250 | 2,155 | 2,750 |
| 0600 | Up to 84" | " | 1 | 48 | | | 1,800 | 2,475 | 4,275 | 5,475 |
| 0800 | Auxiliary boiler, for steam small | A-5 | 1.66 | 10.843 | | | 370 | 26 | 396 | 600 |
| 0900 | Large | " | .83 | 21.687 | | | 745 | 51.50 | 796.50 | 1,200 |
| 1100 | Rule of thumb: complete pile driving set up, small | B-19 | .45 | 142 | | | 6,150 | 4,100 | 10,250 | 14,100 |
| 1200 | Large | " | .27 | 237 | | | 10,200 | 6,850 | 17,050 | 23,500 |
| 1500 | Mobilization, barge, by tug boat | B-83 | 25 | .640 | Mile | | 25.50 | 25.50 | 51 | 66.50 |

# 31 11 Clearing and Grubbing

## 31 11 10 – Clearing and Grubbing Land

### 31 11 10.10 Clear and Grub Site

| | | Crew | Daily Output | Labor-Hours | Unit | Material | Labor | Equipment | Total | Total Incl O&P |
|---|---|---|---|---|---|---|---|---|---|---|
| 0010 | **CLEAR AND GRUB SITE** | | | | | | | | | |
| 0020 | Cut & chip light trees to 6" diam. | B-7 | 1 | 48 | Acre | | 1,750 | 1,300 | 3,050 | 4,125 |
| 0150 | Grub stumps and remove | B-30 | 2 | 12 | | | 455 | 980 | 1,435 | 1,775 |
| 0200 | Cut & chip medium, trees to 12" diam. | B-7 | .70 | 68.571 | | | 2,500 | 1,875 | 4,375 | 5,875 |
| 0250 | Grub stumps and remove | B-30 | 1 | 24 | | | 915 | 1,950 | 2,865 | 3,525 |
| 0300 | Cut & chip heavy, trees to 24" diam. | B-7 | .30 | 160 | | | 5,850 | 4,350 | 10,200 | 13,700 |
| 0350 | Grub stumps and remove | B-30 | .50 | 48 | | | 1,825 | 3,925 | 5,750 | 7,050 |
| 0400 | If burning is allowed, reduce cut & chip | | | | | | | | | 40% |
| 3000 | Chipping stumps, to 18" deep, 12" diam. | B-86 | 20 | .400 | Ea. | | 18.15 | 8.45 | 26.60 | 36.50 |
| 3040 | 18" diameter | | 16 | .500 | | | 22.50 | 10.60 | 33.10 | 45.50 |
| 3080 | 24" diameter | | 14 | .571 | | | 26 | 12.10 | 38.10 | 52 |
| 3100 | 30" diameter | | 12 | .667 | | | 30 | 14.10 | 44.10 | 60.50 |
| 3120 | 36" diameter | | 10 | .800 | | | 36.50 | 16.95 | 53.45 | 72.50 |
| 3160 | 48" diameter | | 8 | 1 | | | 45.50 | 21 | 66.50 | 91.50 |
| 5000 | Tree thinning, feller buncher, conifer | | | | | | | | | |
| 5080 | Up to 8" diameter | B-93 | 240 | .033 | Ea. | | 1.51 | 2.35 | 3.86 | 4.84 |
| 5120 | 12" diameter | | 160 | .050 | | | 2.27 | 3.53 | 5.80 | 7.25 |
| 5240 | Hardwood, up to 4" diameter | | 240 | .033 | | | 1.51 | 2.35 | 3.86 | 4.84 |
| 5280 | 8" diameter | | 180 | .044 | | | 2.02 | 3.13 | 5.15 | 6.45 |
| 5320 | 12" diameter | | 120 | .067 | | | 3.02 | 4.70 | 7.72 | 9.65 |
| 7000 | Tree removal, congested area, aerial lift truck | | | | | | | | | |
| 7040 | 8" diameter | B-85 | 7 | 5.714 | Ea. | | 209 | 123 | 332 | 455 |
| 7080 | 12" diameter | | 6 | 6.667 | | | 244 | 143 | 387 | 530 |
| 7120 | 18" diameter | | 5 | 8 | | | 292 | 172 | 464 | 635 |
| 7160 | 24" diameter | | 4 | 10 | | | 365 | 215 | 580 | 790 |
| 7240 | 36" diameter | | 3 | 13.333 | | | 485 | 287 | 772 | 1,050 |
| 7280 | 48" diameter | | 2 | 20 | | | 730 | 430 | 1,160 | 1,575 |

# 31 13 Selective Tree and Shrub Removal and Trimming

## 31 13 13 – Selective Tree and Shrub Removal

| 31 13 13.10 Selective Clearing | Crew | Daily Output | Labor-Hours | Unit | Material | 2011 Bare Costs Labor | Equipment | Total | Total Incl O&P |
|---|---|---|---|---|---|---|---|---|---|
| **0010 SELECTIVE CLEARING** | | | | | | | | | |
| 0020 Clearing brush with brush saw | A-1C | .25 | 32 | Acre | | 1,100 | 103 | 1,203 | 1,800 |
| 0100 By hand | 1 Clab | .12 | 66.667 | | | 2,300 | | 2,300 | 3,525 |
| 0300 With dozer, ball and chain, light clearing | B-11A | 2 | 8 | | | 320 | 525 | 845 | 1,050 |
| 0400 Medium clearing | | 1.50 | 10.667 | | | 425 | 700 | 1,125 | 1,400 |
| 0500 With dozer and brush rake, light | | 10 | 1.600 | | | 64 | 105 | 169 | 213 |
| 0550 Medium brush to 4" diameter | | 8 | 2 | | | 79.50 | 131 | 210.50 | 265 |
| 0600 Heavy brush to 4" diameter | | 6.40 | 2.500 | | | 99.50 | 164 | 263.50 | 330 |
| 1000 Brush mowing, tractor w/rotary mower, no removal | | | | | | | | | |
| 1020 Light density | B-84 | 2 | 4 | Acre | | 181 | 147 | 328 | 435 |
| 1040 Medium density | | 1.50 | 5.333 | | | 242 | 196 | 438 | 575 |
| 1080 Heavy density | | 1 | 8 | | | 365 | 294 | 659 | 865 |

## 31 13 13.20 Selective Tree Removal

| 31 13 13.20 Selective Tree Removal | Crew | Daily Output | Labor-Hours | Unit | Material | 2011 Bare Costs Labor | Equipment | Total | Total Incl O&P |
|---|---|---|---|---|---|---|---|---|---|
| **0010 SELECTIVE TREE REMOVAL** | | | | | | | | | |
| 0011 With tractor, large tract, firm | | | | | | | | | |
| 0020 level terrain, no boulders, less than 12" diam. trees | | | | | | | | | |
| 0300 300 HP dozer, up to 400 trees/acre, 0 to 25% hardwoods | B-10M | .75 | 16 | Acre | | 665 | 2,150 | 2,815 | 3,350 |
| 0340 25% to 50% hardwoods | | .60 | 20 | | | 835 | 2,675 | 3,510 | 4,200 |
| 0370 75% to 100% hardwoods | | .45 | 26.667 | | | 1,100 | 3,575 | 4,675 | 5,600 |
| 0400 500 trees/acre, 0% to 25% hardwoods | | .60 | 20 | | | 835 | 2,675 | 3,510 | 4,200 |
| 0440 25% to 50% hardwoods | | .48 | 25 | | | 1,050 | 3,350 | 4,400 | 5,250 |
| 0470 75% to 100% hardwoods | | .36 | 33.333 | | | 1,400 | 4,475 | 5,875 | 7,000 |
| 0500 More than 600 trees/acre, 0 to 25% hardwoods | | .52 | 23.077 | | | 960 | 3,100 | 4,060 | 4,850 |
| 0540 25% to 50% hardwoods | | .42 | 28.571 | | | 1,200 | 3,825 | 5,025 | 6,000 |
| 0570 75% to 100% hardwoods | | .31 | 38.710 | | | 1,625 | 5,175 | 6,800 | 8,125 |
| 0900 Large tract clearing per tree | | | | | | | | | |
| 1500 300 HP dozer, to 12" diameter, softwood | B-10M | 320 | .038 | Ea. | | 1.56 | 5 | 6.56 | 7.85 |
| 1550 Hardwood | | 100 | .120 | | | 5 | 16.05 | 21.05 | 25.50 |
| 1600 12" to 24" diameter, softwood | | 200 | .060 | | | 2.50 | 8.05 | 10.55 | 12.60 |
| 1650 Hardwood | | 80 | .150 | | | 6.25 | 20 | 26.25 | 31.50 |
| 1700 24" to 36" diameter, softwood | | 100 | .120 | | | 5 | 16.05 | 21.05 | 25.50 |
| 1750 Hardwood | | 50 | .240 | | | 10 | 32 | 42 | 50.50 |
| 1800 36" to 48" diameter, softwood | | 70 | .171 | | | 7.15 | 23 | 30.15 | 36.50 |
| 1850 Hardwood | | 35 | .343 | | | 14.30 | 46 | 60.30 | 72 |
| 2000 Stump removal on site by hydraulic backhoe, 1-1/2 C.Y. | | | | | | | | | |
| 2040 4" to 6" diameter | B-17 | 60 | .533 | Ea. | | 19.55 | 10.60 | 30.15 | 41 |
| 2050 8" to 12" diameter | B-30 | 33 | .727 | | | 27.50 | 59.50 | 87 | 108 |
| 2100 14" to 24" diameter | | 25 | .960 | | | 36.50 | 78.50 | 115 | 141 |
| 2150 26" to 36" diameter | | 16 | 1.500 | | | 57 | 122 | 179 | 221 |
| 3000 Remove selective trees, on site using chain saws and chipper, | | | | | | | | | |
| 3050 not incl. stumps, up to 6" diameter | B-7 | 18 | 2.667 | Ea. | | 97.50 | 72.50 | 170 | 229 |
| 3100 8" to 12" diameter | | 12 | 4 | | | 146 | 109 | 255 | 345 |
| 3150 14" to 24" diameter | | 10 | 4.800 | | | 175 | 131 | 306 | 410 |
| 3200 26" to 36" diameter | | 8 | 6 | | | 219 | 164 | 383 | 515 |
| 3300 Machine load, 2 mile haul to dump, 12" diam. tree | A-3B | 8 | 2 | | | 80 | 113 | 193 | 244 |

# 31 14 Earth Stripping and Stockpiling

## 31 14 13 – Soil Stripping and Stockpiling

| 31 14 13.23 Topsoil Stripping and Stockpiling | | Crew | Daily Output | Labor-Hours | Unit | Material | 2011 Bare Costs | | Total | Total Incl O&P |
|---|---|---|---|---|---|---|---|---|---|---|
| | | | | | | | Labor | Equipment | | |
| 0010 | **TOPSOIL STRIPPING AND STOCKPILING** | | | | | | | | | |
| 0020 | 200 H.P. dozer, ideal conditions | B-10B | 2300 | .005 | C.Y. | | .22 | .46 | .68 | .83 |
| 0100 | Adverse conditions | " | 1150 | .010 | | | .43 | .91 | 1.34 | 1.65 |
| 0200 | 300 H.P. dozer, ideal conditions | B-10M | 3000 | .004 | | | .17 | .54 | .71 | .84 |
| 0300 | Adverse conditions | " | 1650 | .007 | | | .30 | .97 | 1.27 | 1.53 |
| 0400 | 400 H.P. dozer, ideal conditions | B-10X | 3900 | .003 | | | .13 | .54 | .67 | .78 |
| 0500 | Adverse conditions | " | 2000 | .006 | | | .25 | 1.05 | 1.30 | 1.54 |
| 0600 | Clay, dry and soft, 200 H.P. dozer, ideal conditions | B-10B | 1600 | .008 | | | .31 | .66 | .97 | 1.19 |
| 0700 | Adverse conditions | " | 800 | .015 | | | .63 | 1.31 | 1.94 | 2.39 |
| 1000 | Medium hard, 300 H.P. dozer, ideal conditions | B-10M | 2000 | .006 | | | .25 | .80 | 1.05 | 1.26 |
| 1100 | Adverse conditions | " | 1100 | .011 | | | .45 | 1.46 | 1.91 | 2.29 |
| 1200 | Very hard, 400 H.P. dozer, ideal conditions | B-10X | 2600 | .005 | | | .19 | .81 | 1 | 1.18 |
| 1300 | Adverse conditions | " | 1340 | .009 | ▼ | | .37 | 1.57 | 1.94 | 2.29 |
| 1400 | Loam or topsoil, remove and stockpile on site | | | | | | | | | |
| 1420 | 6" deep, 200' haul | B-10B | 865 | .014 | C.Y. | | .58 | 1.21 | 1.79 | 2.21 |
| 1430 | 300' haul | | 520 | .023 | | | .96 | 2.02 | 2.98 | 3.67 |
| 1440 | 500' haul | | 225 | .053 | ▼ | | 2.22 | 4.67 | 6.89 | 8.50 |
| 1450 | Alternate method: 6" deep, 200' haul | | 5090 | .002 | S.Y. | | .10 | .21 | .31 | .38 |
| 1460 | 500' haul | ▼ | 1325 | .009 | " | | .38 | .79 | 1.17 | 1.44 |

# 31 22 Grading

## 31 22 16 – Fine Grading

### 31 22 16.10 Finish Grading

| | | Crew | Daily Output | Labor-Hours | Unit | Material | 2011 Bare Costs | | Total | Total Incl O&P |
|---|---|---|---|---|---|---|---|---|---|---|
| | | | | | | | Labor | Equipment | | |
| 0010 | **FINISH GRADING** | | | | | | | | | |
| 0012 | Finish grading area to be paved with grader, small area | B-11L | 400 | .040 | S.Y. | | 1.59 | 1.36 | 2.95 | 3.90 |
| 0100 | Large area | | 2000 | .008 | | | .32 | .27 | .59 | .78 |
| 1100 | Fine grade for slab on grade, machine | ▼ | 1040 | .015 | | | .61 | .52 | 1.13 | 1.50 |
| 1150 | Hand grading | B-18 | 700 | .034 | ▼ | | 1.20 | .06 | 1.26 | 1.90 |
| 3500 | Finish grading lagoon bottoms | B-11L | 4 | 4 | M.S.F. | | 159 | 136 | 295 | 390 |

# 31 23 Excavation and Fill

## 31 23 16 – Excavation

### 31 23 16.13 Excavating, Trench

| | | Crew | Daily Output | Labor-Hours | Unit | Material | 2011 Bare Costs | | Total | Total Incl O&P |
|---|---|---|---|---|---|---|---|---|---|---|
| | | | | | | | Labor | Equipment | | |
| 0010 | **EXCAVATING, TRENCH** | | | | | | | | | |
| 0011 | Or continuous footing | | | | | | | | | |
| 0020 | Common earth with no sheeting or dewatering included | | | | | | | | | |
| 0050 | 1' to 4' deep, 3/8 C.Y. excavator | B-11C | 150 | .107 | B.C.Y. | | 4.25 | 2.14 | 6.39 | 8.75 |
| 0060 | 1/2 C.Y. excavator | B-11M | 200 | .080 | | | 3.19 | 1.75 | 4.94 | 6.75 |
| 0090 | 4' to 6' deep, 1/2 C.Y. excavator | " | 200 | .080 | | | 3.19 | 1.75 | 4.94 | 6.75 |
| 0100 | 5/8 C.Y. excavator | B-12Q | 250 | .064 | | | 2.59 | 2.12 | 4.71 | 6.25 |
| 0110 | 3/4 C.Y. excavator | B-12F | 300 | .053 | | | 2.16 | 2.13 | 4.29 | 5.60 |
| 0300 | 1/2 C.Y. excavator, truck mounted | B-12J | 200 | .080 | | | 3.23 | 4.29 | 7.52 | 9.60 |
| 0500 | 6' to 10' deep, 3/4 C.Y. excavator | B-12F | 225 | .071 | | | 2.88 | 2.85 | 5.73 | 7.45 |
| 0510 | 1 C.Y. excavator | B-12A | 400 | .040 | | | 1.62 | 1.82 | 3.44 | 4.45 |
| 0600 | 1 C.Y. excavator, truck mounted | B-12K | 400 | .040 | | | 1.62 | 2.50 | 4.12 | 5.20 |
| 0610 | 1-1/2 C.Y. excavator | B-12B | 600 | .027 | | | 1.08 | 1.58 | 2.66 | 3.37 |
| 0900 | 10' to 14' deep, 3/4 C.Y. excavator | B-12F | 200 | .080 | | | 3.23 | 3.20 | 6.43 | 8.40 |
| 0910 | 1 C.Y. excavator | B-12A | 360 | .044 | | | 1.80 | 2.03 | 3.83 | 4.94 |
| 1000 | 1-1/2 C.Y. excavator | B-12B | 540 | .030 | ▼ | | 1.20 | 1.75 | 2.95 | 3.74 |

## 31 23 16 – Excavation

| 31 23 16.13 Excavating, Trench | | Crew | Daily Output | Labor-Hours | Unit | Material | 2011 Bare Costs Labor | 2011 Bare Costs Equipment | Total | Total Incl O&P |
|---|---|---|---|---|---|---|---|---|---|---|
| 1300 | 14' to 20' deep, 1 C.Y. excavator | B-12A | 320 | .050 | B.C.Y. | | 2.02 | 2.28 | 4.30 | 5.55 |
| 1310 | 1-1/2 C.Y. excavator | B-12B | 480 | .033 | | | 1.35 | 1.97 | 3.32 | 4.21 |
| 1320 | 2-1/2 C.Y. excavator | B-12S | 765 | .021 | | | .85 | 2.11 | 2.96 | 3.60 |
| 1340 | 20' to 24' deep, 1 C.Y. excavator | B-12A | 288 | .056 | | | 2.25 | 2.53 | 4.78 | 6.20 |
| 1342 | 1-1/2 C.Y. excavator | B-12B | 432 | .037 | | | 1.50 | 2.19 | 3.69 | 4.67 |
| 1344 | 2-1/2 C.Y. excavator | B-12S | 685 | .023 | | | .94 | 2.35 | 3.29 | 4.02 |
| 1352 | 4' to 6' deep, 1/2 C.Y. excavator w/trench box | B-13H | 188 | .085 | | | 3.44 | 5.15 | 8.59 | 10.90 |
| 1354 | 5/8 C.Y. excavator | " | 235 | .068 | | | 2.75 | 4.13 | 6.88 | 8.70 |
| 1356 | 3/4 C.Y. excavator | B-13G | 282 | .057 | | | 2.29 | 2.67 | 4.96 | 6.40 |
| 1362 | 6' to 10' deep, 3/4 C.Y. excavator w/trench box | " | 212 | .075 | | | 3.05 | 3.56 | 6.61 | 8.50 |
| 1370 | 1 C.Y. excavator | B-13D | 376 | .043 | | | 1.72 | 2.24 | 3.96 | 5.05 |
| 1371 | 1-1/2 C.Y. excavator | B-13E | 564 | .028 | | | 1.15 | 1.88 | 3.03 | 3.80 |
| 1374 | 10' to 14' deep, 3/4 C.Y. excavator w/trench box | B-13G | 188 | .085 | | | 3.44 | 4.01 | 7.45 | 9.60 |
| 1375 | 1 C.Y. excavator | B-13D | 338 | .047 | | | 1.91 | 2.49 | 4.40 | 5.65 |
| 1376 | 1-1/2 C.Y. excavator | B-13E | 508 | .032 | | | 1.27 | 2.09 | 3.36 | 4.22 |
| 1381 | 14' to 20' deep, 1 C.Y. excavator w/trench box | B-13D | 301 | .053 | | | 2.15 | 2.80 | 4.95 | 6.35 |
| 1382 | 1-1/2 C.Y. excavator | B-13E | 451 | .035 | | | 1.43 | 2.35 | 3.78 | 4.76 |
| 1383 | 2-1/2 C.Y. excavator | B-13J | 720 | .022 | | | .90 | 2.40 | 3.30 | 4 |
| 1386 | 20' to 24' deep, 1 C.Y. excavator w/trench box | B-13D | 271 | .059 | | | 2.39 | 3.11 | 5.50 | 7.05 |
| 1387 | 1-1/2 C.Y. excavator | B-13E | 406 | .039 | | | 1.59 | 2.61 | 4.20 | 5.30 |
| 1388 | 2-1/2 C.Y. excavator | B-13J | 645 | .025 | ▽ | | 1 | 2.68 | 3.68 | 4.46 |
| 1391 | Shoring by S.F./day trench wall protected loose mat., 4' W | B-6 | 3200 | .008 | SF Wall | .39 | .28 | .10 | .77 | .97 |
| 1392 | Rent shoring per week per S.F. wall protected, loose mat., 4' W | | | | | 1.07 | | | 1.07 | 1.18 |
| 1395 | Hydraulic shoring, S.F. trench wall protected stable mat., 4' W | 2 Clab | 2700 | .006 | | .14 | .20 | | .34 | .47 |
| 1397 | semi-stable material, 4' W | " | 2400 | .007 | | .21 | .23 | | .44 | .58 |
| 1398 | Rent hydraulic shoring per day/S.F. wall, stable mat., 4' W | | | | | .23 | | | .23 | .26 |
| 1399 | semi-stable material | | | | ▽ | .30 | | | .30 | .33 |
| 1400 | By hand with pick and shovel 2' to 6' deep, light soil | 1 Clab | 8 | 1 | B.C.Y. | | 34.50 | | 34.50 | 52.50 |
| 1500 | Heavy soil | " | 4 | 2 | " | | 68.50 | | 68.50 | 105 |
| 1700 | For tamping backfilled trenches, air tamp, add | A-1G | 100 | .080 | E.C.Y. | | 2.75 | .50 | 3.25 | 4.77 |
| 1900 | Vibrating plate, add | B-18 | 180 | .133 | " | | 4.67 | .22 | 4.89 | 7.40 |
| 2100 | Trim sides and bottom for concrete pours, common earth | | 1500 | .016 | S.F. | | .56 | .03 | .59 | .89 |
| 2300 | Hardpan | ▽ | 600 | .040 | " | | 1.40 | .07 | 1.47 | 2.22 |
| 2400 | Pier and spread footing excavation, add to above | | | | B.C.Y. | | | | 30% | 30% |
| 3000 | Backfill trench, F.E. loader, wheel mtd., 1 C.Y. bucket | | | | | | | | | |
| 3020 | Minimal haul | B-10R | 400 | .030 | L.C.Y. | | 1.25 | .59 | 1.84 | 2.53 |
| 3040 | 100' haul | " | 200 | .060 | | | 2.50 | 1.18 | 3.68 | 5.05 |
| 3080 | 2-1/4 C.Y. bucket, minimum haul | B-10T | 600 | .020 | | | .83 | .66 | 1.49 | 1.98 |
| 3090 | 100' haul | " | 300 | .040 | ▽ | | 1.67 | 1.33 | 3 | 3.97 |
| 5020 | Loam & Sandy clay with no sheeting or dewatering included | | | | | | | | | |
| 5050 | 1' to 4' deep, 3/8 C.Y. tractor loader/backhoe | B-11C | 162 | .099 | B.C.Y. | | 3.94 | 1.98 | 5.92 | 8.15 |
| 5060 | 1/2 C.Y. excavator | B-11M | 216 | .074 | | | 2.95 | 1.62 | 4.57 | 6.25 |
| 5080 | 4' to 6' deep, 1/2 C.Y. excavator | " | 216 | .074 | | | 2.95 | 1.62 | 4.57 | 6.25 |
| 5090 | 5/8 C.Y. excavator | B-12Q | 276 | .058 | | | 2.34 | 1.92 | 4.26 | 5.65 |
| 5100 | 3/4 C.Y. excavator | B-12F | 324 | .049 | | | 2 | 1.98 | 3.98 | 5.20 |
| 5130 | 1/2 C.Y. excavator, truck mounted | B-12J | 216 | .074 | | | 2.99 | 3.97 | 6.96 | 8.90 |
| 5140 | 6' to 10' deep, 3/4 C.Y. excavator | B-12F | 243 | .066 | | | 2.66 | 2.64 | 5.30 | 6.90 |
| 5150 | 1 C.Y. excavator | B-12A | 432 | .037 | | | 1.50 | 1.69 | 3.19 | 4.12 |
| 5160 | 1 C.Y. excavator, truck mounted | B-12K | 432 | .037 | | | 1.50 | 2.32 | 3.82 | 4.81 |
| 5170 | 1-1/2 C.Y. excavator | B-12B | 648 | .025 | | | 1 | 1.46 | 2.46 | 3.12 |
| 5190 | 10' to 14' deep, 3/4 C.Y. excavator | B-12F | 216 | .074 | | | 2.99 | 2.97 | 5.96 | 7.80 |
| 5200 | 1 C.Y. excavator | B-12A | 389 | .041 | | | 1.66 | 1.88 | 3.54 | 4.57 |
| 5210 | 1-1/2 C.Y. excavator | B-12B | 583 | .027 | ▽ | | 1.11 | 1.62 | 2.73 | 3.47 |

## 31 23 16 – Excavation

### 31 23 16.13 Excavating, Trench

| | | Crew | Daily Output | Labor-Hours | Unit | Material | 2011 Bare Costs Labor | Equipment | Total | Total Incl O&P |
|---|---|---|---|---|---|---|---|---|---|---|
| 5250 | 14' to 20' deep, 1 C.Y. excavator | B-12A | 346 | .046 | B.C.Y. | | 1.87 | 2.11 | 3.98 | 5.15 |
| 5260 | 1-1/2 C.Y. excavator | B-12B | 518 | .031 | | | 1.25 | 1.83 | 3.08 | 3.90 |
| 5270 | 2-1/2 C.Y. excavator | B-12S | 826 | .019 | | | .78 | 1.95 | 2.73 | 3.33 |
| 5300 | 20' to 24' deep, 1 C.Y. excavator | B-12A | 311 | .051 | | | 2.08 | 2.35 | 4.43 | 5.70 |
| 5310 | 1-1/2 C.Y. excavator | B-12B | 467 | .034 | | | 1.39 | 2.03 | 3.42 | 4.32 |
| 5320 | 2-1/2 C.Y. excavator | B-12S | 740 | .022 | | | .87 | 2.18 | 3.05 | 3.72 |
| 5352 | 4' to 6' deep, 1/2 C.Y. excavator w/trench box | B-13H | 205 | .078 | | | 3.16 | 4.74 | 7.90 | 9.95 |
| 5354 | 5/8 C.Y. excavator | " | 257 | .062 | | | 2.52 | 3.78 | 6.30 | 7.95 |
| 5356 | 3/4 C.Y. excavator | B-13G | 308 | .052 | | | 2.10 | 2.45 | 4.55 | 5.85 |
| 5362 | 6' to 10' deep, 3/4 C.Y. excavator w/trench box | " | 231 | .069 | | | 2.80 | 3.26 | 6.06 | 7.80 |
| 5364 | 1 C.Y. excavator | B-13D | 410 | .039 | | | 1.58 | 2.06 | 3.64 | 4.64 |
| 5366 | 1-1/2 C.Y. excavator | B-13E | 616 | .026 | | | 1.05 | 1.72 | 2.77 | 3.48 |
| 5370 | 10' to 14' deep, 3/4 C.Y. excavator w/trench box | B-13G | 205 | .078 | | | 3.16 | 3.68 | 6.84 | 8.80 |
| 5372 | 1 C.Y. excavator | B-13D | 370 | .043 | | | 1.75 | 2.28 | 4.03 | 5.15 |
| 5374 | 1-1/2 C.Y. excavator | B-13E | 554 | .029 | | | 1.17 | 1.91 | 3.08 | 3.87 |
| 5382 | 14' to 20' deep, 1 C.Y. excavator w/trench box | B-13D | 329 | .049 | | | 1.97 | 2.56 | 4.53 | 5.80 |
| 5384 | 1-1/2 C.Y. excavator | B-13E | 492 | .033 | | | 1.31 | 2.16 | 3.47 | 4.36 |
| 5386 | 2-1/2 C.Y. excavator | B-13J | 780 | .021 | | | .83 | 2.21 | 3.04 | 3.68 |
| 5392 | 20' to 24' deep, 1 C.Y. excavator w/trench box | B-13D | 295 | .054 | | | 2.19 | 2.86 | 5.05 | 6.45 |
| 5394 | 1-1/2 C.Y. excavator | B-13E | 444 | .036 | | | 1.46 | 2.39 | 3.85 | 4.83 |
| 5396 | 2-1/2 C.Y. excavator | B-13J | 695 | .023 | | | .93 | 2.48 | 3.41 | 4.14 |
| 6020 | Sand & gravel with no sheeting or dewatering included | | | | | | | | | |
| 6050 | 1' to 4' deep, 3/8 C.Y. excavator | B-11C | 165 | .097 | B.C.Y. | | 3.86 | 1.94 | 5.80 | 8 |
| 6060 | 1/2 C.Y. excavator | B-11M | 220 | .073 | | | 2.90 | 1.59 | 4.49 | 6.15 |
| 6080 | 4' to 6' deep, 1/2 C.Y. excavator | " | 220 | .073 | | | 2.90 | 1.59 | 4.49 | 6.15 |
| 6090 | 5/8 C.Y. excavator | B-12Q | 275 | .058 | | | 2.35 | 1.93 | 4.28 | 5.65 |
| 6100 | 3/4 C.Y. excavator | B-12F | 330 | .048 | | | 1.96 | 1.94 | 3.90 | 5.10 |
| 6130 | 1/2 C.Y. excavator, truck mounted | B-12J | 220 | .073 | | | 2.94 | 3.90 | 6.84 | 8.75 |
| 6140 | 6' to 10' deep, 3/4 C.Y. excavator | B-12F | 248 | .065 | | | 2.61 | 2.58 | 5.19 | 6.80 |
| 6150 | 1 C.Y. excavator | B-12A | 440 | .036 | | | 1.47 | 1.66 | 3.13 | 4.04 |
| 6160 | 1 C.Y. excavator, truck mounted | B-12K | 440 | .036 | | | 1.47 | 2.27 | 3.74 | 4.72 |
| 6170 | 1-1/2 C.Y. excavator | B-12B | 660 | .024 | | | .98 | 1.44 | 2.42 | 3.06 |
| 6190 | 10' to 14' deep, 3/4 C.Y. excavator | B-12F | 220 | .073 | | | 2.94 | 2.91 | 5.85 | 7.65 |
| 6200 | 1 C.Y. excavator | B-12A | 396 | .040 | | | 1.63 | 1.84 | 3.47 | 4.50 |
| 6210 | 1-1/2 C.Y. excavator | B-12B | 594 | .027 | | | 1.09 | 1.60 | 2.69 | 3.40 |
| 6250 | 14' to 20' deep, 1 C.Y. excavator | B-12A | 352 | .045 | | | 1.84 | 2.07 | 3.91 | 5.05 |
| 6260 | 1-1/2 C.Y. excavator | B-12B | 528 | .030 | | | 1.23 | 1.79 | 3.02 | 3.82 |
| 6270 | 2-1/2 C.Y. excavator | B-12S | 840 | .019 | | | .77 | 1.92 | 2.69 | 3.27 |
| 6300 | 20' to 24' deep, 1 C.Y. excavator | B-12A | 317 | .050 | | | 2.04 | 2.30 | 4.34 | 5.60 |
| 6310 | 1-1/2 C.Y. excavator | B-12B | 475 | .034 | | | 1.36 | 1.99 | 3.35 | 4.25 |
| 6320 | 2-1/2 C.Y. excavator | B-12S | 755 | .021 | | | .86 | 2.14 | 3 | 3.64 |
| 6352 | 4' to 6' deep, 1/2 C.Y. excavator w/trench box | B-13H | 209 | .077 | | | 3.10 | 4.65 | 7.75 | 9.80 |
| 6354 | 5/8 C.Y. excavator | " | 261 | .061 | | | 2.48 | 3.72 | 6.20 | 7.85 |
| 6356 | 3/4 C.Y. excavator | B-13G | 314 | .051 | | | 2.06 | 2.40 | 4.46 | 5.75 |
| 6362 | 6' to 10' deep, 3/4 C.Y. excavator w/trench box | " | 236 | .068 | | | 2.74 | 3.19 | 5.93 | 7.65 |
| 6364 | 1 C.Y. excavator | B-13D | 418 | .038 | | | 1.55 | 2.02 | 3.57 | 4.56 |
| 6366 | 1-1/2 C.Y. excavator | B-13E | 627 | .026 | | | 1.03 | 1.69 | 2.72 | 3.42 |
| 6370 | 10' to 14' deep, 3/4 C.Y. excavator w/trench box | B-13G | 209 | .077 | | | 3.10 | 3.61 | 6.71 | 8.65 |
| 6372 | 1 C.Y. excavator | B-13D | 376 | .043 | | | 1.72 | 2.24 | 3.96 | 5.05 |
| 6374 | 1-1/2 C.Y. excavator | B-13E | 564 | .028 | | | 1.15 | 1.88 | 3.03 | 3.80 |
| 6382 | 14' to 20' deep, 1 C.Y. excavator w/trench box | B-13D | 334 | .048 | | | 1.94 | 2.52 | 4.46 | 5.70 |
| 6384 | 1-1/2 C.Y. excavator | B-13E | 502 | .032 | | | 1.29 | 2.11 | 3.40 | 4.27 |
| 6386 | 2-1/2 C.Y. excavator | B-13J | 790 | .020 | | | .82 | 2.19 | 3.01 | 3.64 |

## 31 23 16 – Excavation

### 31 23 16.13 Excavating, Trench

| | | Crew | Daily Output | Labor-Hours | Unit | Material | 2011 Bare Costs Labor | 2011 Bare Costs Equipment | Total | Total Incl O&P |
|---|---|---|---|---|---|---|---|---|---|---|
| 6392 | 20' to 24' deep, 1 C.Y. excavator w/trench box | B-13D | 301 | .053 | B.C.Y. | | 2.15 | 2.80 | 4.95 | 6.35 |
| 6394 | 1-1/2 C.Y. excavator | B-13E | 452 | .035 | | | 1.43 | 2.35 | 3.78 | 4.74 |
| 6396 | 2-1/2 C.Y. excavator | B-13J | 710 | .023 | ↓ | | .91 | 2.43 | 3.34 | 4.06 |
| 7020 | Dense hard clay with no sheeting or dewatering included | | | | | | | | | |
| 7050 | 1' to 4' deep, 3/8 C.Y. excavator | B-11C | 132 | .121 | B.C.Y. | | 4.83 | 2.43 | 7.26 | 9.95 |
| 7060 | 1/2 C.Y. excavator | B-11M | 176 | .091 | | | 3.62 | 1.99 | 5.61 | 7.70 |
| 7080 | 4' to 6' deep, 1/2 C.Y. excavator | " | 176 | .091 | | | 3.62 | 1.99 | 5.61 | 7.70 |
| 7090 | 5/8 C.Y. excavator | B-12Q | 220 | .073 | | | 2.94 | 2.41 | 5.35 | 7.10 |
| 7100 | 3/4 C.Y. excavator | B-12F | 264 | .061 | | | 2.45 | 2.43 | 4.88 | 6.35 |
| 7130 | 1/2 C.Y. excavator, truck mounted | B-12J | 176 | .091 | | | 3.68 | 4.87 | 8.55 | 10.90 |
| 7140 | 6' to 10' deep, 3/4 C.Y. excavator | B-12F | 198 | .081 | | | 3.27 | 3.23 | 6.50 | 8.50 |
| 7150 | 1 C.Y. excavator | B-12A | 352 | .045 | | | 1.84 | 2.07 | 3.91 | 5.05 |
| 7160 | 1 C.Y. excavator, truck mounted | B-12K | 352 | .045 | | | 1.84 | 2.84 | 4.68 | 5.90 |
| 7170 | 1-1/2 C.Y. excavator | B-12B | 528 | .030 | | | 1.23 | 1.79 | 3.02 | 3.82 |
| 7190 | 10' to 14' deep, 3/4 C.Y. excavator | B-12F | 176 | .091 | | | 3.68 | 3.64 | 7.32 | 9.55 |
| 7200 | 1 C.Y. excavator | B-12A | 317 | .050 | | | 2.04 | 2.30 | 4.34 | 5.60 |
| 7210 | 1-1/2 C.Y. excavator | B-12B | 475 | .034 | | | 1.36 | 1.99 | 3.35 | 4.25 |
| 7250 | 14' to 20' deep, 1 C.Y. excavator | B-12A | 282 | .057 | | | 2.29 | 2.59 | 4.88 | 6.30 |
| 7260 | 1-1/2 C.Y. excavator | B-12B | 422 | .038 | | | 1.53 | 2.24 | 3.77 | 4.79 |
| 7270 | 2-1/2 C.Y. excavator | B-12S | 675 | .024 | | | .96 | 2.39 | 3.35 | 4.08 |
| 7300 | 20' to 24' deep, 1 C.Y. excavator | B-12A | 254 | .063 | | | 2.55 | 2.87 | 5.42 | 7 |
| 7310 | 1-1/2 C.Y. excavator | B-12B | 380 | .042 | | | 1.70 | 2.49 | 4.19 | 5.30 |
| 7320 | 2-1/2 C.Y. excavator | B-12S | 605 | .026 | ↓ | | 1.07 | 2.67 | 3.74 | 4.55 |

### 31 23 16.14 Excavating, Utility Trench

| | | Crew | Daily Output | Labor-Hours | Unit | Material | 2011 Bare Costs Labor | 2011 Bare Costs Equipment | Total | Total Incl O&P |
|---|---|---|---|---|---|---|---|---|---|---|
| 0010 | **EXCAVATING, UTILITY TRENCH** | | | | | | | | | |
| 0011 | Common earth | | | | | | | | | |
| 0050 | Trenching with chain trencher, 12 H.P., operator walking | | | | | | | | | |
| 0100 | 4" wide trench, 12" deep | B-53 | 800 | .010 | L.F. | | .44 | .07 | .51 | .73 |
| 0150 | 18" deep | | 750 | .011 | | | .47 | .08 | .55 | .78 |
| 0200 | 24" deep | | 700 | .011 | | | .50 | .08 | .58 | .83 |
| 0300 | 6" wide trench, 12" deep | | 650 | .012 | | | .54 | .09 | .63 | .90 |
| 0350 | 18" deep | | 600 | .013 | | | .58 | .10 | .68 | .98 |
| 0400 | 24" deep | | 550 | .015 | | | .63 | .10 | .73 | 1.06 |
| 0450 | 36" deep | | 450 | .018 | | | .78 | .13 | .91 | 1.30 |
| 0600 | 8" wide trench, 12" deep | | 475 | .017 | | | .73 | .12 | .85 | 1.23 |
| 0650 | 18" deep | | 400 | .020 | | | .87 | .14 | 1.01 | 1.46 |
| 0700 | 24" deep | | 350 | .023 | | | 1 | .16 | 1.16 | 1.67 |
| 0750 | 36" deep | ↓ | 300 | .027 | ↓ | | 1.16 | .19 | 1.35 | 1.95 |
| 1000 | Backfill by hand including compaction, add | | | | | | | | | |
| 1050 | 4" wide trench, 12" deep | A-1G | 800 | .010 | L.F. | | .34 | .06 | .40 | .60 |
| 1100 | 18" deep | | 530 | .015 | | | .52 | .09 | .61 | .90 |
| 1150 | 24" deep | | 400 | .020 | | | .69 | .12 | .81 | 1.19 |
| 1300 | 6" wide trench, 12" deep | | 540 | .015 | | | .51 | .09 | .60 | .88 |
| 1350 | 18" deep | | 405 | .020 | | | .68 | .12 | .80 | 1.17 |
| 1400 | 24" deep | | 270 | .030 | | | 1.02 | .18 | 1.20 | 1.76 |
| 1450 | 36" deep | | 180 | .044 | | | 1.53 | .28 | 1.81 | 2.64 |
| 1600 | 8" wide trench, 12" deep | | 400 | .020 | | | .69 | .12 | .81 | 1.19 |
| 1650 | 18" deep | | 265 | .030 | | | 1.04 | .19 | 1.23 | 1.80 |
| 1700 | 24" deep | | 200 | .040 | | | 1.37 | .25 | 1.62 | 2.38 |
| 1750 | 36" deep | ↓ | 135 | .059 | ↓ | | 2.04 | .37 | 2.41 | 3.52 |
| 2000 | Chain trencher, 40 H.P. operator riding | | | | | | | | | |
| 2050 | 6" wide trench and backfill, 12" deep | B-54 | 1200 | .007 | L.F. | | .29 | .25 | .54 | .70 |

# 31 23 Excavation and Fill

## 31 23 16 – Excavation

### 31 23 16.14 Excavating, Utility Trench

| | | Crew | Daily Output | Labor-Hours | Unit | Material | 2011 Bare Costs Labor | Equipment | Total | Total Incl O&P |
|---|---|---|---|---|---|---|---|---|---|---|
| 2100 | 18" deep | B-54 | 1000 | .008 | L.F. | | .35 | .30 | .65 | .85 |
| 2150 | 24" deep | | 975 | .008 | | | .36 | .30 | .66 | .87 |
| 2200 | 36" deep | | 900 | .009 | | | .39 | .33 | .72 | .94 |
| 2250 | 48" deep | | 750 | .011 | | | .47 | .40 | .87 | 1.14 |
| 2300 | 60" deep | | 650 | .012 | | | .54 | .46 | 1 | 1.30 |
| 2400 | 8" wide trench and backfill, 12" deep | | 1000 | .008 | | | .35 | .30 | .65 | .85 |
| 2450 | 18" deep | | 950 | .008 | | | .37 | .31 | .68 | .89 |
| 2500 | 24" deep | | 900 | .009 | | | .39 | .33 | .72 | .94 |
| 2550 | 36" deep | | 800 | .010 | | | .44 | .37 | .81 | 1.06 |
| 2600 | 48" deep | | 650 | .012 | | | .54 | .46 | 1 | 1.30 |
| 2700 | 12" wide trench and backfill, 12" deep | | 975 | .008 | | | .36 | .30 | .66 | .87 |
| 2750 | 18" deep | | 860 | .009 | | | .41 | .35 | .76 | .99 |
| 2800 | 24" deep | | 800 | .010 | | | .44 | .37 | .81 | 1.06 |
| 2850 | 36" deep | | 725 | .011 | | | .48 | .41 | .89 | 1.17 |
| 3000 | 16" wide trench and backfill, 12" deep | | 835 | .010 | | | .42 | .36 | .78 | 1.01 |
| 3050 | 18" deep | | 750 | .011 | | | .47 | .40 | .87 | 1.14 |
| 3100 | 24" deep | ▼ | 700 | .011 | ▼ | | .50 | .42 | .92 | 1.21 |
| 3200 | Compaction with vibratory plate, add | | | | | | | | 35% | 35% |
| 5100 | Hand excavate and trim for pipe bells after trench excavation | | | | | | | | | |
| 5200 | 8" pipe | 1 Clab | 155 | .052 | L.F. | | 1.77 | | 1.77 | 2.72 |
| 5300 | 18" pipe | " | 130 | .062 | " | | 2.11 | | 2.11 | 3.24 |

### 31 23 16.16 Structural Excavation for Minor Structures

| | | Crew | Daily Output | Labor-Hours | Unit | Material | 2011 Bare Costs Labor | Equipment | Total | Total Incl O&P |
|---|---|---|---|---|---|---|---|---|---|---|
| 0010 | **STRUCTURAL EXCAVATION FOR MINOR STRUCTURES** R312316-40 | | | | | | | | | |
| 0015 | Hand, pits to 6' deep, sandy soil | 1 Clab | 8 | 1 | B.C.Y. | | 34.50 | | 34.50 | 52.50 |
| 0100 | Heavy soil or clay | | 4 | 2 | | | 68.50 | | 68.50 | 105 |
| 0300 | Pits 6' to 12' deep, sandy soil | | 5 | 1.600 | | | 55 | | 55 | 84.50 |
| 0500 | Heavy soil or clay | | 3 | 2.667 | | | 91.50 | | 91.50 | 141 |
| 0700 | Pits 12' to 18' deep, sandy soil | | 4 | 2 | | | 68.50 | | 68.50 | 105 |
| 0900 | Heavy soil or clay | | 2 | 4 | | | 137 | | 137 | 211 |
| 1100 | Hand loading trucks from stock pile, sandy soil | | 12 | .667 | | | 23 | | 23 | 35 |
| 1300 | Heavy soil or clay | ▼ | 8 | 1 | ▼ | | 34.50 | | 34.50 | 52.50 |
| 1500 | For wet or muck hand excavation, add to above | | | | % | | | | 50% | 50% |
| 6000 | Machine excavation, for spread and mat footings, elevator pits, | | | | | | | | | |
| 6001 | and small building foundations | | | | | | | | | |
| 6030 | Common earth, hydraulic backhoe, 1/2 C.Y. bucket | B-12E | 55 | .291 | B.C.Y. | | 11.75 | 6.65 | 18.40 | 25 |
| 6035 | 3/4 C.Y. bucket | B-12F | 90 | .178 | | | 7.20 | 7.10 | 14.30 | 18.70 |
| 6040 | 1 C.Y. bucket | B-12A | 108 | .148 | | | 6 | 6.75 | 12.75 | 16.50 |
| 6050 | 1-1/2 C.Y. bucket | B-12B | 144 | .111 | | | 4.49 | 6.60 | 11.09 | 14.05 |
| 6060 | 2 C.Y. bucket | B-12C | 200 | .080 | | | 3.23 | 6.10 | 9.33 | 11.65 |
| 6070 | Sand and gravel, 3/4 C.Y. bucket | B-12F | 100 | .160 | | | 6.45 | 6.40 | 12.85 | 16.80 |
| 6080 | 1 C.Y. bucket | B-12A | 120 | .133 | | | 5.40 | 6.10 | 11.50 | 14.85 |
| 6090 | 1-1/2 C.Y. bucket | B-12B | 160 | .100 | | | 4.04 | 5.90 | 9.94 | 12.60 |
| 6100 | 2 C.Y. bucket | B-12C | 220 | .073 | | | 2.94 | 5.55 | 8.49 | 10.55 |
| 6110 | Clay, till, or blasted rock, 3/4 C.Y. bucket | B-12F | 80 | .200 | | | 8.10 | 8 | 16.10 | 21 |
| 6120 | 1 C.Y. bucket | B-12A | 95 | .168 | | | 6.80 | 7.70 | 14.50 | 18.75 |
| 6130 | 1-1/2 C.Y. bucket | B-12B | 130 | .123 | | | 4.98 | 7.30 | 12.28 | 15.50 |
| 6140 | 2 C.Y. bucket | B-12C | 175 | .091 | | | 3.70 | 7 | 10.70 | 13.30 |
| 6230 | Sandy clay & loam, hydraulic backhoe, 1/2 C.Y. bucket | B-12E | 60 | .267 | | | 10.80 | 6.10 | 16.90 | 23 |
| 6235 | 3/4 C.Y. bucket | B-12F | 98 | .163 | | | 6.60 | 6.55 | 13.15 | 17.15 |
| 6240 | 1 C.Y. bucket | B-12A | 116 | .138 | | | 5.60 | 6.30 | 11.90 | 15.30 |
| 6250 | 1-1/2 C.Y. bucket | B-12B | 156 | .103 | ▼ | | 4.15 | 6.05 | 10.20 | 12.95 |
| 9010 | For mobilization or demobilization, see Section 01 54 36.50 | | | | | | | | | |

# 31 23 Excavation and Fill

## 31 23 16 – Excavation

### 31 23 16.16 Structural Excavation for Minor Structures

| | | Crew | Daily Output | Labor-Hours | Unit | Material | 2011 Bare Costs Labor | Equipment | Total | Total Incl O&P |
|---|---|---|---|---|---|---|---|---|---|---|
| 9020 | For dewatering, see Section 31 23 19.20 | | | | | | | | | |
| 9022 | For larger structures, see Bulk Excavation, Section 31 23 16.42 | | | | | | | | | |
| 9024 | For loading onto trucks, add | | | | | | | | 15% | 15% |
| 9026 | For hauling, see Section 31 23 23.20 | | | | | | | | | |
| 9030 | For sheeting or soldier bms/lagging, see Section 31 52 16.10 | | | | | | | | | |
| 9040 | For trench excavation of strip ftgs, see Section 31 23 16.13 | | | | | | | | | |

### 31 23 16.26 Rock Removal

| | | Crew | Daily Output | Labor-Hours | Unit | Material | 2011 Bare Costs Labor | Equipment | Total | Total Incl O&P |
|---|---|---|---|---|---|---|---|---|---|---|
| 0010 | **ROCK REMOVAL** R312316-40 | | | | | | | | | |
| 0015 | Drilling only rock, 2" hole for rock bolts | B-47 | 316 | .076 | L.F. | | 2.89 | 4.13 | 7.02 | 8.95 |
| 0800 | 2-1/2" hole for pre-splitting | | 600 | .040 | | | 1.52 | 2.17 | 3.69 | 4.70 |
| 4600 | Quarry operations, 2-1/2" to 3-1/2" diameter | ↓ | 715 | .034 | ↓ | | 1.28 | 1.82 | 3.10 | 3.95 |

### 31 23 16.30 Drilling and Blasting Rock

| | | Crew | Daily Output | Labor-Hours | Unit | Material | 2011 Bare Costs Labor | Equipment | Total | Total Incl O&P |
|---|---|---|---|---|---|---|---|---|---|---|
| 0010 | **DRILLING AND BLASTING ROCK** | | | | | | | | | |
| 0020 | Rock, open face, under 1500 C.Y. | B-47 | 225 | .107 | B.C.Y. | 2.82 | 4.06 | 5.80 | 12.68 | 15.65 |
| 0100 | Over 1500 C.Y. | | 300 | .080 | | 2.82 | 3.05 | 4.35 | 10.22 | 12.50 |
| 0200 | Areas where blasting mats are required, under 1500 C.Y. | | 175 | .137 | | 2.82 | 5.25 | 7.45 | 15.52 | 19.25 |
| 0250 | Over 1500 C.Y. | ↓ | 250 | .096 | | 2.82 | 3.66 | 5.20 | 11.68 | 14.40 |
| 0300 | Bulk drilling and blasting, can vary greatly, average | | | | | | | | 9.50 | 12 |
| 0500 | Pits, average | | | | | | | | 25 | 31 |
| 1300 | Deep hole method, up to 1500 C.Y. | B-47 | 50 | .480 | | 2.82 | 18.30 | 26 | 47.12 | 59.50 |
| 1400 | Over 1500 C.Y. | | 66 | .364 | | 2.82 | 13.85 | 19.75 | 36.42 | 45.50 |
| 1900 | Restricted areas, up to 1500 C.Y. | | 13 | 1.846 | | 2.82 | 70.50 | 100 | 173.32 | 220 |
| 2000 | Over 1500 C.Y. | | 20 | 1.200 | | 2.82 | 45.50 | 65 | 113.32 | 144 |
| 2200 | Trenches, up to 1500 C.Y. | | 22 | 1.091 | | 8.20 | 41.50 | 59.50 | 109.20 | 137 |
| 2300 | Over 1500 C.Y. | | 26 | .923 | | 8.20 | 35 | 50 | 93.20 | 118 |
| 2500 | Pier holes, up to 1500 C.Y. | | 22 | 1.091 | | 2.82 | 41.50 | 59.50 | 103.82 | 131 |
| 2600 | Over 1500 C.Y. | ↓ | 31 | .774 | | 2.82 | 29.50 | 42 | 74.32 | 94.50 |
| 2800 | Boulders under 1/2 C.Y., loaded on truck, no hauling | B-100 | 80 | .150 | | | 6.25 | 9.65 | 15.90 | 20 |
| 2900 | Boulders, drilled, blasted | B-47 | 100 | .240 | ↓ | 2.82 | 9.15 | 13.05 | 25.02 | 31.50 |
| 3100 | Jackhammer operators with foreman compressor, air tools | B-9 | 1 | 40 | Day | | 1,400 | 195 | 1,595 | 2,350 |
| 3300 | Track drill, compressor, operator and foreman | B-47 | 1 | 24 | " | | 915 | 1,300 | 2,215 | 2,825 |
| 3500 | Blasting caps | | | | Ea. | 5.25 | | | 5.25 | 5.75 |
| 3700 | Explosives | | | | | .38 | | | .38 | .41 |
| 3900 | Blasting mats, rent, for first day | | | | | 123 | | | 123 | 135 |
| 4000 | Per added day | | | | | 41 | | | 41 | 45 |
| 4200 | Preblast survey for 6 room house, individual lot, minimum | A-6 | 2.40 | 6.667 | | | 288 | 29 | 317 | 470 |
| 4300 | Maximum | " | 1.35 | 11.852 | ↓ | | 510 | 51.50 | 561.50 | 835 |
| 4500 | City block within zone of influence, minimum | A-8 | 25200 | .001 | S.F. | | .06 | | .06 | .09 |
| 4600 | Maximum | " | 15100 | .002 | " | | .10 | | .10 | .16 |

### 31 23 16.42 Excavating, Bulk Bank Measure

| | | Crew | Daily Output | Labor-Hours | Unit | Material | 2011 Bare Costs Labor | Equipment | Total | Total Incl O&P |
|---|---|---|---|---|---|---|---|---|---|---|
| 0010 | **EXCAVATING, BULK BANK MEASURE** R312316-40 | | | | | | | | | |
| 0011 | Common earth piled | | | | | | | | | |
| 0020 | For loading onto trucks, add | | | | | | | | 15% | 15% |
| 0050 | For mobilization and demobilization, see Section 01 54 36.50 R312316-45 | | | | | | | | | |
| 0100 | For hauling, see Section 31 23 23.20 | | | | | | | | | |
| 0200 | Excavator, hydraulic, crawler mtd., 1 C.Y. cap. = 100 C.Y./hr. | B-12A | 800 | .020 | B.C.Y. | | .81 | .91 | 1.72 | 2.22 |
| 0250 | 1-1/2 C.Y. cap. = 125 C.Y./hr. | B-12B | 1000 | .016 | | | .65 | .95 | 1.60 | 2.02 |
| 0260 | 2 C.Y. cap. = 165 C.Y./hr. | B-12C | 1320 | .012 | | | .49 | .93 | 1.42 | 1.76 |
| 0300 | 3 C.Y. cap. = 260 C.Y./hr. | B-12D | 2080 | .008 | | | .31 | 1.04 | 1.35 | 1.62 |
| 0305 | 3-1/2 C.Y. cap. = 300 C.Y./hr. | " | 2400 | .007 | | | .27 | .91 | 1.18 | 1.41 |
| 0310 | Wheel mounted, 1/2 C.Y. cap. = 40 C.Y./hr. | B-12E | 320 | .050 | | | 2.02 | 1.14 | 3.16 | 4.30 |
| 0360 | 3/4 C.Y. cap. = 60 C.Y./hr. | B-12F | 480 | .033 | ↓ | | 1.35 | 1.33 | 2.68 | 3.51 |

# 31 23 Excavation and Fill

## 31 23 16 – Excavation

### 31 23 16.42 Excavating, Bulk Bank Measure

| | | Crew | Daily Output | Labor-Hours | Unit | Material | 2011 Bare Costs Labor | Equipment | Total | Total Incl O&P |
|---|---|---|---|---|---|---|---|---|---|---|
| 0500 | Clamshell, 1/2 C.Y. cap. = 20 C.Y./hr. | B-12G | 160 | .100 | B.C.Y. | | 4.04 | 4.35 | 8.39 | 10.90 |
| 0550 | 1 C.Y. cap. = 35 C.Y./hr. | B-12H | 280 | .057 | | | 2.31 | 4.18 | 6.49 | 8.10 |
| 0950 | Dragline, 1/2 C.Y. cap. = 30 C.Y./hr. | B-12I | 240 | .067 | | | 2.70 | 3.60 | 6.30 | 8 |
| 1000 | 3/4 C.Y. cap. = 35 C.Y./hr. | " | 280 | .057 | | | 2.31 | 3.08 | 5.39 | 6.90 |
| 1050 | 1-1/2 C.Y. cap. = 65 C.Y./hr. | B-12P | 520 | .031 | | | 1.24 | 2.24 | 3.48 | 4.34 |
| 1200 | Front end loader, track mtd., 1-1/2 C.Y. cap. = 70 C.Y./hr. | B-10N | 560 | .021 | | | .89 | .83 | 1.72 | 2.26 |
| 1250 | 2-1/2 C.Y. cap. = 95 C.Y./hr. | B-100 | 760 | .016 | | | .66 | 1.02 | 1.68 | 2.11 |
| 1300 | 3 C.Y. cap. = 130 C.Y./hr. | B-10P | 1040 | .012 | | | .48 | .97 | 1.45 | 1.79 |
| 1350 | 5 C.Y. cap. = 160 C.Y./hr. | B-10Q | 1280 | .009 | | | .39 | 1.12 | 1.51 | 1.82 |
| 1500 | Wheel mounted, 3/4 C.Y. cap. = 45 C.Y./hr. | B-10R | 360 | .033 | | | 1.39 | .65 | 2.04 | 2.81 |
| 1550 | 1-1/2 C.Y. cap. = 80 C.Y./hr. | B-10S | 640 | .019 | | | .78 | .50 | 1.28 | 1.73 |
| 1600 | 2-1/4 C.Y. cap. = 100 C.Y./hr. | B-10T | 800 | .015 | | | .63 | .50 | 1.13 | 1.49 |
| 1650 | 5 C.Y. cap. = 185 C.Y./hr. | B-10U | 1480 | .008 | | | .34 | .63 | .97 | 1.20 |
| 1800 | Hydraulic excavator, truck mtd. 1/2 C.Y. = 30 C.Y./hr. | B-12J | 240 | .067 | | | 2.70 | 3.57 | 6.27 | 8 |
| 1850 | 48 inch bucket, 1 C.Y. = 45 C.Y./hr. | B-12K | 360 | .044 | | | 1.80 | 2.78 | 4.58 | 5.75 |
| 3700 | Shovel, 1/2 C.Y. capacity = 55 C.Y./hr. | B-12L | 440 | .036 | | | 1.47 | 1.62 | 3.09 | 4 |
| 3750 | 3/4 C.Y. capacity = 85 C.Y./hr. | B-12M | 680 | .024 | | | .95 | 1.33 | 2.28 | 2.90 |
| 3800 | 1 C.Y. capacity = 120 C.Y./hr. | B-12N | 960 | .017 | | | .67 | 1.24 | 1.91 | 2.38 |
| 3850 | 1-1/2 C.Y. capacity = 160 C.Y./hr. | B-120 | 1280 | .013 | | | .51 | .94 | 1.45 | 1.80 |
| 3900 | 3 C.Y. cap. = 250 C.Y./hr. | B-12T | 2000 | .008 | | | .32 | .77 | 1.09 | 1.34 |
| 4000 | For soft soil or sand, deduct | | | | | | | | 15% | 15% |
| 4100 | For heavy soil or stiff clay, add | | | | | | | | 60% | 60% |
| 4200 | For wet excavation with clamshell or dragline, add | | | | | | | | 100% | 100% |
| 4250 | All other equipment, add | | | | | | | | 50% | 50% |
| 4400 | Clamshell in sheeting or cofferdam, minimum | B-12H | 160 | .100 | | | 4.04 | 7.30 | 11.34 | 14.15 |
| 4450 | Maximum | " | 60 | .267 | | | 10.80 | 19.50 | 30.30 | 38 |
| 5000 | Excavating, bulk bank measure, sandy clay & loam piled | | | | | | | | | |
| 5020 | For loading onto trucks, add | | | | | | | | 15% | 15% |
| 5100 | Excavator, hydraulic, crawler mtd., 1 C.Y. cap. = 120 C.Y./hr. | B-12A | 960 | .017 | B.C.Y. | | .67 | .76 | 1.43 | 1.86 |
| 5150 | 1-1/2 C.Y. cap. = 150 C.Y./hr. | B-12B | 1200 | .013 | | | .54 | .79 | 1.33 | 1.68 |
| 5300 | 2 C.Y. cap. = 195 C.Y./hr. | B-12C | 1560 | .010 | | | .41 | .78 | 1.19 | 1.49 |
| 5400 | 3 C.Y. cap. = 300 C.Y./hr. | B-12D | 2400 | .007 | | | .27 | .91 | 1.18 | 1.41 |
| 5500 | 3.5 C.Y. cap. = 350 C.Y./hr. | " | 2800 | .006 | | | .23 | .77 | 1 | 1.20 |
| 5610 | Wheel mounted, 1/2 C.Y. cap. = 44 C.Y./hr. | B-12E | 352 | .045 | | | 1.84 | 1.04 | 2.88 | 3.92 |
| 5660 | 3/4 C.Y. cap. = 66 C.Y./hr. | B-12F | 528 | .030 | | | 1.23 | 1.21 | 2.44 | 3.18 |
| 8000 | For hauling excavated material, see Section 31 23 23.20 | | | | | | | | | |

### 31 23 16.46 Excavating, Bulk, Dozer

| | | Crew | Daily Output | Labor-Hours | Unit | Material | 2011 Bare Costs Labor | Equipment | Total | Total Incl O&P |
|---|---|---|---|---|---|---|---|---|---|---|
| 0010 | **EXCAVATING, BULK, DOZER** | | | | | | | | | |
| 0011 | Open site | | | | | | | | | |
| 2000 | 80 H.P., 50' haul, sand & gravel | B-10L | 460 | .026 | B.C.Y. | | 1.09 | .90 | 1.99 | 2.63 |
| 2010 | Sandy clay & loam | | 440 | .027 | | | 1.14 | .94 | 2.08 | 2.74 |
| 2020 | Common earth | | 400 | .030 | | | 1.25 | 1.03 | 2.28 | 3.01 |
| 2040 | Clay | | 250 | .048 | | | 2 | 1.65 | 3.65 | 4.83 |
| 2200 | 150' haul, sand & gravel | | 230 | .052 | | | 2.17 | 1.79 | 3.96 | 5.25 |
| 2210 | Sandy clay & loam | | 220 | .055 | | | 2.27 | 1.88 | 4.15 | 5.50 |
| 2220 | Common earth | | 200 | .060 | | | 2.50 | 2.06 | 4.56 | 6.05 |
| 2240 | Clay | | 125 | .096 | | | 4 | 3.30 | 7.30 | 9.65 |
| 2400 | 300' haul, sand & gravel | | 120 | .100 | | | 4.17 | 3.44 | 7.61 | 10.05 |
| 2410 | Sandy clay & loam | | 115 | .104 | | | 4.35 | 3.59 | 7.94 | 10.50 |
| 2420 | Common earth | | 100 | .120 | | | 5 | 4.13 | 9.13 | 12.10 |
| 2440 | Clay | | 65 | .185 | | | 7.70 | 6.35 | 14.05 | 18.60 |
| 3000 | 105 H.P., 50' haul, sand & gravel | B-10W | 700 | .017 | | | .71 | .72 | 1.43 | 1.87 |

## 31 23 16 – Excavation

| 31 23 16.46 Excavating, Bulk, Dozer | Crew | Daily Output | Labor-Hours | Unit | Material | 2011 Bare Costs Labor | Equipment | Total | Total Incl O&P |
|---|---|---|---|---|---|---|---|---|---|
| 3010 | Sandy clay & loam | B-10W | 680 | .018 | B.C.Y. | | .74 | .74 | 1.48 | 1.93 |
| 3020 | Common earth | | 610 | .020 | | | .82 | .83 | 1.65 | 2.14 |
| 3040 | Clay | | 385 | .031 | | | 1.30 | 1.31 | 2.61 | 3.40 |
| 3200 | 150' haul, sand & gravel | | 310 | .039 | | | 1.61 | 1.63 | 3.24 | 4.22 |
| 3210 | Sandy clay & loam | | 300 | .040 | | | 1.67 | 1.68 | 3.35 | 4.36 |
| 3220 | Common earth | | 270 | .044 | | | 1.85 | 1.87 | 3.72 | 4.85 |
| 3240 | Clay | | 170 | .071 | | | 2.94 | 2.97 | 5.91 | 7.70 |
| 3300 | 300' haul, sand & gravel | | 140 | .086 | | | 3.57 | 3.61 | 7.18 | 9.35 |
| 3310 | Sandy clay & loam | | 135 | .089 | | | 3.70 | 3.74 | 7.44 | 9.70 |
| 3320 | Common earth | | 120 | .100 | | | 4.17 | 4.21 | 8.38 | 10.90 |
| 3340 | Clay | | 100 | .120 | | | 5 | 5.05 | 10.05 | 13.10 |
| 4000 | 200 H.P., 50' haul, sand & gravel | B-10B | 1400 | .009 | | | .36 | .75 | 1.11 | 1.37 |
| 4010 | Sandy clay & loam | | 1360 | .009 | | | .37 | .77 | 1.14 | 1.40 |
| 4020 | Common earth | | 1230 | .010 | | | .41 | .85 | 1.26 | 1.55 |
| 4040 | Clay | | 770 | .016 | | | .65 | 1.36 | 2.01 | 2.48 |
| 4200 | 150' haul, sand & gravel | | 595 | .020 | | | .84 | 1.77 | 2.61 | 3.21 |
| 4210 | Sandy clay & loam | | 580 | .021 | | | .86 | 1.81 | 2.67 | 3.29 |
| 4220 | Common earth | | 516 | .023 | | | .97 | 2.04 | 3.01 | 3.70 |
| 4240 | Clay | | 325 | .037 | | | 1.54 | 3.23 | 4.77 | 5.90 |
| 4400 | 300' haul, sand & gravel | | 310 | .039 | | | 1.61 | 3.39 | 5 | 6.15 |
| 4410 | Sandy clay & loam | | 300 | .040 | | | 1.67 | 3.50 | 5.17 | 6.35 |
| 4420 | Common earth | | 270 | .044 | | | 1.85 | 3.89 | 5.74 | 7.05 |
| 4440 | Clay | | 170 | .071 | | | 2.94 | 6.20 | 9.14 | 11.25 |
| 5000 | 300 H.P., 50' haul, sand & gravel | B-10M | 1900 | .006 | | | .26 | .85 | 1.11 | 1.33 |
| 5010 | Sandy clay & loam | | 1850 | .006 | | | .27 | .87 | 1.14 | 1.37 |
| 5020 | Common earth | | 1650 | .007 | | | .30 | .97 | 1.27 | 1.53 |
| 5040 | Clay | | 1025 | .012 | | | .49 | 1.57 | 2.06 | 2.46 |
| 5200 | 150' haul, sand & gravel | | 920 | .013 | | | .54 | 1.75 | 2.29 | 2.74 |
| 5210 | Sandy clay & loam | | 895 | .013 | | | .56 | 1.80 | 2.36 | 2.82 |
| 5220 | Common earth | | 800 | .015 | | | .63 | 2.01 | 2.64 | 3.15 |
| 5240 | Clay | | 500 | .024 | | | 1 | 3.21 | 4.21 | 5.05 |
| 5400 | 300' haul, sand & gravel | | 470 | .026 | | | 1.06 | 3.42 | 4.48 | 5.35 |
| 5410 | Sandy clay & loam | | 455 | .026 | | | 1.10 | 3.53 | 4.63 | 5.55 |
| 5420 | Common earth | | 410 | .029 | | | 1.22 | 3.92 | 5.14 | 6.15 |
| 5440 | Clay | | 250 | .048 | | | 2 | 6.45 | 8.45 | 10.05 |
| 5500 | 460 H.P., 50' haul, sand & gravel | B-10X | 1930 | .006 | | | .26 | 1.09 | 1.35 | 1.59 |
| 5506 | Sandy clay & loam | | 1880 | .006 | | | .27 | 1.12 | 1.39 | 1.63 |
| 5510 | Common earth | | 1680 | .007 | | | .30 | 1.25 | 1.55 | 1.83 |
| 5520 | Clay | | 1050 | .011 | | | .48 | 2.01 | 2.49 | 2.93 |
| 5530 | 150' haul, sand & gravel | | 1290 | .009 | | | .39 | 1.63 | 2.02 | 2.37 |
| 5535 | Sandy clay & loam | | 1250 | .010 | | | .40 | 1.68 | 2.08 | 2.45 |
| 5540 | Common earth | | 1120 | .011 | | | .45 | 1.88 | 2.33 | 2.74 |
| 5550 | Clay | | 700 | .017 | | | .71 | 3.01 | 3.72 | 4.39 |
| 5560 | 300' haul, sand & gravel | | 660 | .018 | | | .76 | 3.19 | 3.95 | 4.65 |
| 5565 | Sandy clay & loam | | 640 | .019 | | | .78 | 3.29 | 4.07 | 4.80 |
| 5570 | Common earth | | 575 | .021 | | | .87 | 3.66 | 4.53 | 5.35 |
| 5580 | Clay | | 350 | .034 | | | 1.43 | 6 | 7.43 | 8.75 |
| 6000 | 700 H.P., 50' haul, sand & gravel | B-10V | 3500 | .003 | | | .14 | 1.19 | 1.33 | 1.52 |
| 6006 | Sandy clay & loam | | 3400 | .004 | | | .15 | 1.22 | 1.37 | 1.56 |
| 6010 | Common earth | | 3035 | .004 | | | .16 | 1.37 | 1.53 | 1.75 |
| 6020 | Clay | | 1925 | .006 | | | .26 | 2.15 | 2.41 | 2.76 |
| 6030 | 150' haul, sand & gravel | | 2025 | .006 | | | .25 | 2.05 | 2.30 | 2.63 |
| 6035 | Sandy clay & loam | | 1960 | .006 | | | .26 | 2.12 | 2.38 | 2.71 |

# 31 23 Excavation and Fill

## 31 23 16 - Excavation

### 31 23 16.46 Excavating, Bulk, Dozer

| | | Crew | Daily Output | Labor-Hours | Unit | Material | 2011 Bare Costs Labor | 2011 Bare Costs Equipment | Total | Total Incl O&P |
|---|---|---|---|---|---|---|---|---|---|---|
| 6040 | Common earth | B-10V | 1750 | .007 | B.C.Y. | | .29 | 2.37 | 2.66 | 3.04 |
| 6050 | Clay | | 1100 | .011 | | | .45 | 3.77 | 4.22 | 4.83 |
| 6060 | 300' haul, sand & gravel | | 1030 | .012 | | | .49 | 4.03 | 4.52 | 5.15 |
| 6065 | Sandy clay & loam | | 1005 | .012 | | | .50 | 4.13 | 4.63 | 5.30 |
| 6070 | Common earth | | 900 | .013 | | | .56 | 4.61 | 5.17 | 5.90 |
| 6080 | Clay | | 550 | .022 | | | .91 | 7.55 | 8.46 | 9.65 |

### 31 23 16.50 Excavation, Bulk, Scrapers

| | | Crew | Daily Output | Labor-Hours | Unit | Material | 2011 Bare Costs Labor | 2011 Bare Costs Equipment | Total | Total Incl O&P |
|---|---|---|---|---|---|---|---|---|---|---|
| 0010 | **EXCAVATION, BULK, SCRAPERS** R312316-40 | | | | | | | | | |
| 0100 | Elev. scraper 11 C.Y., sand & gravel 1500' haul, 1/4 dozer | B-33F | 690 | .020 | B.C.Y. | | .86 | 2.16 | 3.02 | 3.67 |
| 0150 | 3000' haul | | 610 | .023 | | | .97 | 2.44 | 3.41 | 4.15 |
| 0200 | 5000' haul | | 505 | .028 | | | 1.17 | 2.95 | 4.12 | 5 |
| 0300 | Common earth, 1500' haul | | 600 | .023 | | | .98 | 2.48 | 3.46 | 4.21 |
| 0350 | 3000' haul | | 530 | .026 | | | 1.12 | 2.81 | 3.93 | 4.77 |
| 0400 | 5000' haul | | 440 | .032 | | | 1.34 | 3.39 | 4.73 | 5.75 |
| 0410 | Sandy clay & loam, 1500' haul | | 648 | .022 | | | .91 | 2.30 | 3.21 | 3.90 |
| 0420 | 3000' haul | | 572 | .024 | | | 1.03 | 2.61 | 3.64 | 4.42 |
| 0430 | 5000' haul | | 475 | .029 | | | 1.24 | 3.14 | 4.38 | 5.30 |
| 0500 | Clay, 1500' haul | | 375 | .037 | | | 1.58 | 3.97 | 5.55 | 6.75 |
| 0550 | 3000' haul | | 330 | .042 | | | 1.79 | 4.52 | 6.31 | 7.65 |
| 0600 | 5000' haul | | 275 | .051 | | | 2.15 | 5.40 | 7.55 | 9.20 |
| 1000 | Self propelled scraper, 14 C.Y. 1/4 push dozer, sand | | | | | | | | | |
| 1050 | Sand and gravel, 1500' haul | B-33D | 920 | .015 | B.C.Y. | | .64 | 2.08 | 2.72 | 3.26 |
| 1100 | 3000' haul | | 805 | .017 | | | .73 | 2.38 | 3.11 | 3.71 |
| 1200 | 5000' haul | | 645 | .022 | | | .92 | 2.97 | 3.89 | 4.64 |
| 1300 | Common earth, 1500' haul | | 800 | .018 | | | .74 | 2.39 | 3.13 | 3.74 |
| 1350 | 3000' haul | | 700 | .020 | | | .84 | 2.73 | 3.57 | 4.28 |
| 1400 | 5000' haul | | 560 | .025 | | | 1.06 | 3.42 | 4.48 | 5.35 |
| 1420 | Sandy clay & loam, 1500' haul | | 864 | .016 | | | .68 | 2.21 | 2.89 | 3.47 |
| 1430 | 3000' haul | | 786 | .018 | | | .75 | 2.43 | 3.18 | 3.81 |
| 1440 | 5000' haul | | 605 | .023 | | | .98 | 3.16 | 4.14 | 4.95 |
| 1500 | Clay, 1500' haul | | 500 | .028 | | | 1.18 | 3.83 | 5.01 | 6 |
| 1550 | 3000' haul | | 440 | .032 | | | 1.34 | 4.35 | 5.69 | 6.80 |
| 1600 | 5000' haul | | 350 | .040 | | | 1.69 | 5.45 | 7.14 | 8.55 |
| 2000 | 21 C.Y., 1/4 push dozer, sand & gravel, 1500' haul | B-33E | 1180 | .012 | | | .50 | 1.93 | 2.43 | 2.87 |
| 2100 | 3000' haul | | 910 | .015 | | | .65 | 2.50 | 3.15 | 3.73 |
| 2200 | 5000' haul | | 750 | .019 | | | .79 | 3.03 | 3.82 | 4.52 |
| 2300 | Common earth, 1500' haul | | 1030 | .014 | | | .57 | 2.21 | 2.78 | 3.29 |
| 2350 | 3000' haul | | 790 | .018 | | | .75 | 2.88 | 3.63 | 4.29 |
| 2400 | 5000' haul | | 650 | .022 | | | .91 | 3.50 | 4.41 | 5.20 |
| 2420 | Sandy clay & loam, 1500' haul | | 1112 | .013 | | | .53 | 2.04 | 2.57 | 3.05 |
| 2430 | 3000' haul | | 854 | .016 | | | .69 | 2.66 | 3.35 | 3.97 |
| 2440 | 5000' haul | | 702 | .020 | | | .84 | 3.24 | 4.08 | 4.83 |
| 2500 | Clay, 1500' haul | | 645 | .022 | | | .92 | 3.53 | 4.45 | 5.25 |
| 2550 | 3000' haul | | 495 | .028 | | | 1.19 | 4.59 | 5.78 | 6.85 |
| 2600 | 5000' haul | | 405 | .035 | | | 1.46 | 5.60 | 7.06 | 8.40 |
| 2700 | Towed, 10 C.Y., 1/4 push dozer, sand & gravel, 1500' haul | B-33B | 560 | .025 | | | 1.06 | 3.84 | 4.90 | 5.80 |
| 2720 | 3000' haul | | 450 | .031 | | | 1.31 | 4.77 | 6.08 | 7.20 |
| 2730 | 5000' haul | | 365 | .038 | | | 1.62 | 5.90 | 7.52 | 8.95 |
| 2750 | Common earth, 1500' haul | | 420 | .033 | | | 1.41 | 5.10 | 6.51 | 7.75 |
| 2770 | 3000' haul | | 400 | .035 | | | 1.48 | 5.35 | 6.83 | 8.10 |
| 2780 | 5000' haul | | 310 | .045 | | | 1.91 | 6.95 | 8.86 | 10.45 |
| 2785 | Sandy clay & Loam, 1500' haul | | 454 | .031 | | | 1.30 | 4.73 | 6.03 | 7.15 |

# 31 23 Excavation and Fill

## 31 23 16 – Excavation

### 31 23 16.50 Excavation, Bulk, Scrapers

| | | Crew | Daily Output | Labor-Hours | Unit | Material | 2011 Bare Costs Labor | Equipment | Total | Total Incl O&P |
|---|---|---|---|---|---|---|---|---|---|---|
| 2790 | 3000' haul | B-33B | 432 | .032 | B.C.Y. | | 1.37 | 4.97 | 6.34 | 7.50 |
| 2795 | 5000' haul | | 340 | .041 | | | 1.74 | 6.30 | 8.04 | 9.55 |
| 2800 | Clay, 1500' haul | | 315 | .044 | | | 1.88 | 6.80 | 8.68 | 10.30 |
| 2820 | 3000' haul | | 300 | .047 | | | 1.97 | 7.15 | 9.12 | 10.85 |
| 2840 | 5000' haul | | 225 | .062 | | | 2.63 | 9.55 | 12.18 | 14.45 |
| 2900 | 15 C.Y., 1/4 push dozer, sand & gravel, 1500' haul | B-33C | 800 | .018 | | | .74 | 2.71 | 3.45 | 4.09 |
| 2920 | 3000' haul | | 640 | .022 | | | .92 | 3.39 | 4.31 | 5.10 |
| 2940 | 5000' haul | | 520 | .027 | | | 1.14 | 4.17 | 5.31 | 6.30 |
| 2960 | Common earth, 1500' haul | | 600 | .023 | | | .98 | 3.61 | 4.59 | 5.45 |
| 2980 | 3000' haul | | 560 | .025 | | | 1.06 | 3.87 | 4.93 | 5.85 |
| 3000 | 5000' haul | | 440 | .032 | | | 1.34 | 4.93 | 6.27 | 7.40 |
| 3005 | Sandy clay & Loam, 1500' haul | | 648 | .022 | | | .91 | 3.34 | 4.25 | 5.05 |
| 3010 | 3000' haul | | 605 | .023 | | | .98 | 3.58 | 4.56 | 5.40 |
| 3015 | 5000' haul | | 475 | .029 | | | 1.24 | 4.56 | 5.80 | 6.85 |
| 3020 | Clay, 1500' haul | | 450 | .031 | | | 1.31 | 4.82 | 6.13 | 7.25 |
| 3040 | 3000' haul | | 420 | .033 | | | 1.41 | 5.15 | 6.56 | 7.80 |
| 3060 | 5000' haul | | 320 | .044 | | | 1.85 | 6.75 | 8.60 | 10.25 |

## 31 23 19 – Dewatering

### 31 23 19.20 Dewatering Systems

| | | Crew | Daily Output | Labor-Hours | Unit | Material | 2011 Bare Costs Labor | Equipment | Total | Total Incl O&P |
|---|---|---|---|---|---|---|---|---|---|---|
| 0010 | **DEWATERING SYSTEMS** | | | | | | | | | |
| 0020 | Excavate drainage trench, 2' wide, 2' deep | B-11C | 90 | .178 | C.Y. | | 7.10 | 3.56 | 10.66 | 14.60 |
| 0100 | 2' wide, 3' deep, with backhoe loader | " | 135 | .119 | | | 4.72 | 2.37 | 7.09 | 9.75 |
| 0200 | Excavate sump pits by hand, light soil | 1 Clab | 7.10 | 1.127 | | | 38.50 | | 38.50 | 59.50 |
| 0300 | Heavy soil | " | 3.50 | 2.286 | | | 78.50 | | 78.50 | 120 |
| 0500 | Pumping 8 hr., attended 2 hrs. per day, including 20 L.F. | | | | | | | | | |
| 0550 | of suction hose & 100 L.F. discharge hose | | | | | | | | | |
| 0600 | 2" diaphragm pump used for 8 hours | B-10H | 4 | 3 | Day | | 125 | 17.60 | 142.60 | 207 |
| 0650 | 4" diaphragm pump used for 8 hours | B-10I | 4 | 3 | | | 125 | 29.50 | 154.50 | 221 |
| 0800 | 8 hrs. attended, 2" diaphragm pump | B-10H | 1 | 12 | | | 500 | 70.50 | 570.50 | 835 |
| 0900 | 3" centrifugal pump | B-10J | 1 | 12 | | | 500 | 77.50 | 577.50 | 840 |
| 1000 | 4" diaphragm pump | B-10I | 1 | 12 | | | 500 | 117 | 617 | 885 |
| 1100 | 6" centrifugal pump | B-10K | 1 | 12 | | | 500 | 315 | 815 | 1,100 |
| 1300 | CMP, incl. excavation 3' deep, 12" diameter | B-6 | 115 | .209 | L.F. | 8.65 | 7.80 | 2.79 | 19.24 | 24.50 |
| 1400 | 18" diameter | | 100 | .240 | " | 14.35 | 9 | 3.20 | 26.55 | 33 |
| 1600 | Sump hole construction, incl. excavation and gravel, pit | | 1250 | .019 | C.F. | .99 | .72 | .26 | 1.97 | 2.46 |
| 1700 | With 12" gravel collar, 12" pipe, corrugated, 16 ga. | | 70 | .343 | L.F. | 23 | 12.85 | 4.58 | 40.43 | 50 |
| 1800 | 15" pipe, corrugated, 16 ga. | | 55 | .436 | | 29.50 | 16.35 | 5.85 | 51.70 | 64 |
| 1900 | 18" pipe, corrugated, 16 ga. | | 50 | .480 | | 34.50 | 17.95 | 6.40 | 58.85 | 72.50 |
| 2000 | 24" pipe, corrugated, 14 ga. | | 40 | .600 | | 41.50 | 22.50 | 8 | 72 | 88.50 |
| 2200 | Wood lining, up to 4' x 4', add | | 300 | .080 | SFCA | 17 | 2.99 | 1.07 | 21.06 | 24.50 |
| 9950 | See Section 31 23 19.40 for wellpoints | | | | | | | | | |
| 9960 | See Section 31 23 19.30 for deep well systems | | | | | | | | | |

### 31 23 19.30 Wells

| | | Crew | Daily Output | Labor-Hours | Unit | Material | 2011 Bare Costs Labor | Equipment | Total | Total Incl O&P |
|---|---|---|---|---|---|---|---|---|---|---|
| 0010 | **WELLS** | | | | | | | | | |
| 0011 | For dewatering 10' to 20' deep, 2' diameter | | | | | | | | | |
| 0020 | with steel casing, minimum | B-6 | 165 | .145 | V.L.F. | 35 | 5.45 | 1.94 | 42.39 | 49 |
| 0050 | Average | | 98 | .245 | | 40 | 9.15 | 3.27 | 52.42 | 61.50 |
| 0100 | Maximum | | 49 | .490 | | 45 | 18.35 | 6.55 | 69.90 | 84.50 |
| 0300 | For dewatering pumps see 01 54 33 in Reference Section | | | | | | | | | |
| 0500 | For domestic water wells, see Section 33 21 13.10 | | | | | | | | | |

## 31 23 19 – Dewatering

### 31 23 19.40 Wellpoints

| | | Crew | Daily Output | Labor-Hours | Unit | Material | 2011 Bare Costs Labor | Equipment | Total | Total Incl O&P |
|---|---|---|---|---|---|---|---|---|---|---|
| 0010 | **WELLPOINTS** R312319-90 | | | | | | | | | |
| 0011 | For equipment rental, see 01 54 33 in Reference Section | | | | | | | | | |
| 0100 | Installation and removal of single stage system | | | | | | | | | |
| 0110 | Labor only, .75 labor-hours per L.F., minimum | 1 Clab | 10.70 | .748 | LF Hdr | | 25.50 | | 25.50 | 39.50 |
| 0200 | 2.0 labor-hours per L.F., maximum | " | 4 | 2 | " | | 68.50 | | 68.50 | 105 |
| 0400 | Pump operation, 4 @ 6 hr. shifts | | | | | | | | | |
| 0410 | Per 24 hour day | 4 Eqlt | 1.27 | 25.197 | Day | | 1,100 | | 1,100 | 1,650 |
| 0500 | Per 168 hour week, 160 hr. straight, 8 hr. double time | | .18 | 177 | Week | | 7,750 | | 7,750 | 11,600 |
| 0550 | Per 4.3 week month | | .04 | 800 | Month | | 34,900 | | 34,900 | 52,000 |
| 0600 | Complete installation, operation, equipment rental, fuel & | | | | | | | | | |
| 0610 | removal of system with 2" wellpoints 5' O.C. | | | | | | | | | |
| 0700 | 100' long header, 6" diameter, first month | 4 Eqlt | 3.23 | 9.907 | LF Hdr | 157 | 430 | | 587 | 815 |
| 0800 | Thereafter, per month | | 4.13 | 7.748 | | 125 | 340 | | 465 | 645 |
| 1000 | 200' long header, 8" diameter, first month | | 6 | 5.333 | | 142 | 233 | | 375 | 500 |
| 1100 | Thereafter, per month | | 8.39 | 3.814 | | 70.50 | 166 | | 236.50 | 325 |
| 1300 | 500' long header, 8" diameter, first month | | 10.63 | 3.010 | | 55 | 131 | | 186 | 257 |
| 1400 | Thereafter, per month | | 20.91 | 1.530 | | 39 | 66.50 | | 105.50 | 143 |
| 1600 | 1,000' long header, 10" diameter, first month | | 11.62 | 2.754 | | 47 | 120 | | 167 | 231 |
| 1700 | Thereafter, per month | | 41.81 | .765 | | 23.50 | 33.50 | | 57 | 76 |
| 1900 | Note: above figures include pumping 168 hrs. per week | | | | | | | | | |
| 1910 | and include the pump operator and one stand-by pump. | | | | | | | | | |

## 31 23 23 – Fill

### 31 23 23.13 Backfill

| | | Crew | Daily Output | Labor-Hours | Unit | Material | 2011 Bare Costs Labor | Equipment | Total | Total Incl O&P |
|---|---|---|---|---|---|---|---|---|---|---|
| 0010 | **BACKFILL** R312323-30 | | | | | | | | | |
| 0015 | By hand, no compaction, light soil | 1 Clab | 14 | .571 | L.C.Y. | | 19.65 | | 19.65 | 30 |
| 0100 | Heavy soil | | 11 | .727 | " | | 25 | | 25 | 38.50 |
| 0300 | Compaction in 6" layers, hand tamp, add to above | | 20.60 | .388 | E.C.Y. | | 13.35 | | 13.35 | 20.50 |
| 0400 | Roller compaction operator walking, add | B-10A | 100 | .120 | | | 5 | 1.47 | 6.47 | 9.15 |
| 0500 | Air tamp, add | B-9D | 190 | .211 | | | 7.30 | 1.20 | 8.50 | 12.50 |
| 0600 | Vibrating plate, add | A-1D | 60 | .133 | | | 4.58 | .51 | 5.09 | 7.60 |
| 0800 | Compaction in 12" layers, hand tamp, add to above | 1 Clab | 34 | .235 | | | 8.10 | | 8.10 | 12.40 |
| 0900 | Roller compaction operator walking, add | B-10A | 150 | .080 | | | 3.33 | .98 | 4.31 | 6.10 |
| 1000 | Air tamp, add | B-9 | 285 | .140 | | | 4.88 | .68 | 5.56 | 8.25 |
| 1100 | Vibrating plate, add | A-1E | 90 | .089 | | | 3.05 | .43 | 3.48 | 5.15 |
| 1300 | Dozer backfilling, bulk, up to 300' haul, no compaction | B-10B | 1200 | .010 | L.C.Y. | | .42 | .88 | 1.30 | 1.59 |
| 1400 | Air tamped, add | B-11B | 80 | .200 | E.C.Y. | | 7.80 | 3.11 | 10.91 | 15.20 |
| 1600 | Compacting backfill, 6" to 12" lifts, vibrating roller | B-10C | 800 | .015 | | | .63 | 1.81 | 2.44 | 2.93 |
| 1700 | Sheepsfoot roller | B-10D | 750 | .016 | | | .67 | 2 | 2.67 | 3.20 |
| 1900 | Dozer backfilling, trench, up to 300' haul, no compaction | B-10B | 900 | .013 | L.C.Y. | | .56 | 1.17 | 1.73 | 2.12 |
| 2000 | Air tamped, add | B-11B | 80 | .200 | E.C.Y. | | 7.80 | 3.11 | 10.91 | 15.20 |
| 2200 | Compacting backfill, 6" to 12" lifts, vibrating roller | B-10C | 700 | .017 | | | .71 | 2.06 | 2.77 | 3.35 |
| 2300 | Sheepsfoot roller | B-10D | 650 | .018 | | | .77 | 2.31 | 3.08 | 3.70 |
| 2350 | Spreading in 8" layers, small dozer | B-10B | 1060 | .011 | L.C.Y. | | .47 | .99 | 1.46 | 1.80 |

### 31 23 23.14 Backfill, Structural

| | | Crew | Daily Output | Labor-Hours | Unit | Material | 2011 Bare Costs Labor | Equipment | Total | Total Incl O&P |
|---|---|---|---|---|---|---|---|---|---|---|
| 0010 | **BACKFILL, STRUCTURAL** | | | | | | | | | |
| 0011 | Dozer or F.E. loader | | | | | | | | | |
| 0020 | From existing stockpile, no compaction | | | | | | | | | |
| 2000 | 80 H.P., 50' haul, sand & gravel | B-10L | 1100 | .011 | L.C.Y. | | .45 | .38 | .83 | 1.09 |
| 2010 | Sandy clay & loam | | 1070 | .011 | | | .47 | .39 | .86 | 1.12 |
| 2020 | Common earth | | 975 | .012 | | | .51 | .42 | .93 | 1.24 |
| 2040 | Clay | | 850 | .014 | | | .59 | .49 | 1.08 | 1.42 |

### 31 23 23.14 Backfill, Structural

| | | Crew | Daily Output | Labor-Hours | Unit | Material | 2011 Bare Costs Labor | 2011 Bare Costs Equipment | Total | Total Incl O&P |
|---|---|---|---|---|---|---|---|---|---|---|
| 2400 | 300' haul, sand & gravel | B-10L | 370 | .032 | L.C.Y. | | 1.35 | 1.11 | 2.46 | 3.26 |
| 2410 | Sandy clay & loam | | 360 | .033 | | | 1.39 | 1.15 | 2.54 | 3.35 |
| 2420 | Common earth | | 330 | .036 | | | 1.52 | 1.25 | 2.77 | 3.66 |
| 2440 | Clay | | 290 | .041 | | | 1.72 | 1.42 | 3.14 | 4.17 |
| 3000 | 105 H.P., 50' haul, sand & gravel | B-10W | 1350 | .009 | | | .37 | .37 | .74 | .97 |
| 3010 | Sandy clay & loam | | 1325 | .009 | | | .38 | .38 | .76 | .99 |
| 3020 | Common earth | | 1225 | .010 | | | .41 | .41 | .82 | 1.06 |
| 3040 | Clay | | 1100 | .011 | | | .45 | .46 | .91 | 1.19 |
| 3300 | 300' haul, sand & gravel | | 465 | .026 | | | 1.08 | 1.09 | 2.17 | 2.82 |
| 3310 | Sandy clay & loam | | 455 | .026 | | | 1.10 | 1.11 | 2.21 | 2.87 |
| 3320 | Common earth | | 415 | .029 | | | 1.21 | 1.22 | 2.43 | 3.15 |
| 3340 | Clay | | 370 | .032 | | | 1.35 | 1.37 | 2.72 | 3.53 |
| 4000 | 200 H.P., 50' haul, sand & gravel | B-10B | 2500 | .005 | | | .20 | .42 | .62 | .76 |
| 4010 | Sandy clay & loam | | 2435 | .005 | | | .21 | .43 | .64 | .79 |
| 4020 | Common earth | | 2200 | .005 | | | .23 | .48 | .71 | .87 |
| 4040 | Clay | | 1950 | .006 | | | .26 | .54 | .80 | .98 |
| 4400 | 300' haul, sand & gravel | | 805 | .015 | | | .62 | 1.31 | 1.93 | 2.38 |
| 4410 | Sandy clay & loam | | 790 | .015 | | | .63 | 1.33 | 1.96 | 2.41 |
| 4420 | Common earth | | 735 | .016 | | | .68 | 1.43 | 2.11 | 2.59 |
| 4440 | Clay | | 660 | .018 | | | .76 | 1.59 | 2.35 | 2.89 |
| 5000 | 300 H.P., 50' haul, sand & gravel | B-10M | 3170 | .004 | | | .16 | .51 | .67 | .80 |
| 5010 | Sandy clay & loam | | 3110 | .004 | | | .16 | .52 | .68 | .81 |
| 5020 | Common earth | | 2900 | .004 | | | .17 | .55 | .72 | .87 |
| 5040 | Clay | | 2700 | .004 | | | .19 | .59 | .78 | .93 |
| 5400 | 300' haul, sand & gravel | | 1500 | .008 | | | .33 | 1.07 | 1.40 | 1.68 |
| 5410 | Sandy clay & loam | | 1470 | .008 | | | .34 | 1.09 | 1.43 | 1.71 |
| 5420 | Common earth | | 1350 | .009 | | | .37 | 1.19 | 1.56 | 1.87 |
| 5440 | Clay | | 1225 | .010 | | | .41 | 1.31 | 1.72 | 2.05 |
| 6010 | For trench backfill, see Section 31 23 16.13 and 31 23 16.14 | | | | | | | | | |
| 6100 | For compaction, see Section 31 23 23.24 | | | | | | | | | |

### 31 23 23.16 Fill By Borrow and Utility Bedding

| | | Crew | Daily Output | Labor-Hours | Unit | Material | 2011 Bare Costs Labor | 2011 Bare Costs Equipment | Total | Total Incl O&P |
|---|---|---|---|---|---|---|---|---|---|---|
| 0010 | **FILL BY BORROW AND UTILITY BEDDING** | | | | | | | | | |
| 0015 | Fill by borrow, load, 1 mile haul, spread with dozer | | | | | | | | | |
| 0020 | for embankments | B-15 | 1200 | .023 | L.C.Y. | 15 | .88 | 1.72 | 17.60 | 19.70 |
| 0035 | Select fill for shoulders & embankments | " | 1200 | .023 | " | 20 | .88 | 1.72 | 22.60 | 25 |
| 0040 | Fill, for hauling over 1 mile, add to above per C.Y., see Section 31 23 23.20 | | | | Mile | | | | 1.29 | 1.55 |
| 0049 | Utility bedding, for pipe & conduit, not incl. compaction | | | | | | | | | |
| 0050 | Crushed or screened bank run gravel | B-6 | 150 | .160 | L.C.Y. | 25 | 6 | 2.14 | 33.14 | 39 |
| 0100 | Crushed stone 3/4" to 1/2" | | 150 | .160 | | 38 | 6 | 2.14 | 46.14 | 53.50 |
| 0200 | Sand, dead or bank | | 150 | .160 | | 15 | 6 | 2.14 | 23.14 | 28 |
| 0500 | Compacting bedding in trench | A-1D | 90 | .089 | E.C.Y. | | 3.05 | .34 | 3.39 | 5.05 |
| 0600 | If material source exceeds 2 miles, add for extra mileage. | | | | | | | | | |
| 0610 | See Section 31 23 23.20 for hauling mileage add. | | | | | | | | | |

### 31 23 23.17 General Fill

| | | Crew | Daily Output | Labor-Hours | Unit | Material | 2011 Bare Costs Labor | 2011 Bare Costs Equipment | Total | Total Incl O&P |
|---|---|---|---|---|---|---|---|---|---|---|
| 0010 | **GENERAL FILL** | | | | | | | | | |
| 0011 | Spread dumped material, no compaction | | | | | | | | | |
| 0020 | By dozer, no compaction | B-10B | 1000 | .012 | L.C.Y. | | .50 | 1.05 | 1.55 | 1.91 |
| 0100 | By hand | 1 Clab | 12 | .667 | " | | 23 | | 23 | 35 |
| 0500 | Gravel fill, compacted, under floor slabs, 4" deep | B-37 | 10000 | .005 | S.F. | .40 | .17 | .01 | .58 | .73 |
| 0600 | 6" deep | | 8600 | .006 | | .60 | .20 | .02 | .82 | .99 |
| 0700 | 9" deep | | 7200 | .007 | | 1 | .24 | .02 | 1.26 | 1.49 |
| 0800 | 12" deep | | 6000 | .008 | | 1.40 | .29 | .02 | 1.71 | 2.01 |

## 31 23 23 – Fill

### 31 23 23.17 General Fill

| | | Crew | Daily Output | Labor-Hours | Unit | Material | 2011 Bare Costs Labor | 2011 Bare Costs Equipment | Total | Total Incl O&P |
|---|---|---|---|---|---|---|---|---|---|---|
| 1000 | Alternate pricing method, 4" deep | B-37 | 120 | .400 | E.C.Y. | 30 | 14.50 | 1.21 | 45.71 | 56.50 |
| 1100 | 6" deep | | 160 | .300 | | 30 | 10.85 | .91 | 41.76 | 50.50 |
| 1200 | 9" deep | | 200 | .240 | | 30 | 8.70 | .73 | 39.43 | 47 |
| 1300 | 12" deep | | 220 | .218 | | 30 | 7.90 | .66 | 38.56 | 46 |
| 1500 | For fill under exterior paving, see Section 32 11 23.23 | | | | | | | | | |
| 1600 | For flowable fill, see Section 03 31 05.35 | | | | | | | | | |

### 31 23 23.20 Hauling

| | | Crew | Daily Output | Labor-Hours | Unit | Material | 2011 Bare Costs Labor | 2011 Bare Costs Equipment | Total | Total Incl O&P |
|---|---|---|---|---|---|---|---|---|---|---|
| 0010 | **HAULING** | | | | | | | | | |
| 0011 | Excavated or borrow, loose cubic yards | | | | | | | | | |
| 0012 | no loading equipment, including hauling, waiting, loading/dumping | | | | | | | | | |
| 0013 | time per cycle (wait, load, travel, unload or dump & return) | | | | | | | | | |
| 0014 | 8 CY truck, 15 MPH ave, cycle 0.5 miles, 10 min. wait/Ld./Uld. | B-34A | 320 | .025 | L.C.Y. | | .86 | .99 | 1.85 | 2.39 |
| 0016 | cycle 1 mile | | 272 | .029 | | | 1.01 | 1.16 | 2.17 | 2.82 |
| 0018 | cycle 2 miles | | 208 | .038 | | | 1.32 | 1.52 | 2.84 | 3.68 |
| 0020 | cycle 4 miles | | 144 | .056 | | | 1.91 | 2.19 | 4.10 | 5.30 |
| 0022 | cycle 6 miles | | 112 | .071 | | | 2.46 | 2.82 | 5.28 | 6.85 |
| 0024 | cycle 8 miles | | 88 | .091 | | | 3.13 | 3.58 | 6.71 | 8.70 |
| 0026 | 20 MPH ave, cycle 0.5 mile | | 336 | .024 | | | .82 | .94 | 1.76 | 2.27 |
| 0028 | cycle 1 mile | | 296 | .027 | | | .93 | 1.07 | 2 | 2.58 |
| 0030 | cycle 2 miles | | 240 | .033 | | | 1.15 | 1.31 | 2.46 | 3.19 |
| 0032 | cycle 4 miles | | 176 | .045 | | | 1.56 | 1.79 | 3.35 | 4.34 |
| 0034 | cycle 6 miles | | 136 | .059 | | | 2.02 | 2.32 | 4.34 | 5.60 |
| 0036 | cycle 8 miles | | 112 | .071 | | | 2.46 | 2.82 | 5.28 | 6.85 |
| 0044 | 25 MPH ave, cycle 4 miles | | 192 | .042 | | | 1.43 | 1.64 | 3.07 | 3.99 |
| 0046 | cycle 6 miles | | 160 | .050 | | | 1.72 | 1.97 | 3.69 | 4.78 |
| 0048 | cycle 8 miles | | 128 | .063 | | | 2.15 | 2.46 | 4.61 | 6 |
| 0050 | 30 MPH ave, cycle 4 miles | | 216 | .037 | | | 1.27 | 1.46 | 2.73 | 3.55 |
| 0052 | cycle 6 miles | | 176 | .045 | | | 1.56 | 1.79 | 3.35 | 4.34 |
| 0054 | cycle 8 miles | | 144 | .056 | | | 1.91 | 2.19 | 4.10 | 5.30 |
| 0114 | 15 MPH ave, cycle 0.5 mile, 15 min. wait/Ld./Uld. | | 224 | .036 | | | 1.23 | 1.41 | 2.64 | 3.42 |
| 0116 | cycle 1 mile | | 200 | .040 | | | 1.38 | 1.58 | 2.96 | 3.82 |
| 0118 | cycle 2 miles | | 168 | .048 | | | 1.64 | 1.88 | 3.52 | 4.56 |
| 0120 | cycle 4 miles | | 120 | .067 | | | 2.29 | 2.63 | 4.92 | 6.35 |
| 0122 | cycle 6 miles | | 96 | .083 | | | 2.87 | 3.29 | 6.16 | 7.95 |
| 0124 | cycle 8 miles | | 80 | .100 | | | 3.44 | 3.94 | 7.38 | 9.60 |
| 0126 | 20 MPH ave, cycle 0.5 mile | | 232 | .034 | | | 1.19 | 1.36 | 2.55 | 3.30 |
| 0128 | cycle 1 mile | | 208 | .038 | | | 1.32 | 1.52 | 2.84 | 3.68 |
| 0130 | cycle 2 miles | | 184 | .043 | | | 1.50 | 1.71 | 3.21 | 4.16 |
| 0132 | cycle 4 miles | | 144 | .056 | | | 1.91 | 2.19 | 4.10 | 5.30 |
| 0134 | cycle 6 miles | | 112 | .071 | | | 2.46 | 2.82 | 5.28 | 6.85 |
| 0136 | cycle 8 miles | | 96 | .083 | | | 2.87 | 3.29 | 6.16 | 7.95 |
| 0144 | 25 MPH ave, cycle 4 miles | | 152 | .053 | | | 1.81 | 2.08 | 3.89 | 5.05 |
| 0146 | cycle 6 miles | | 128 | .063 | | | 2.15 | 2.46 | 4.61 | 6 |
| 0148 | cycle 8 miles | | 112 | .071 | | | 2.46 | 2.82 | 5.28 | 6.85 |
| 0150 | 30 MPH ave, cycle 4 miles | | 168 | .048 | | | 1.64 | 1.88 | 3.52 | 4.56 |
| 0152 | cycle 6 miles | | 144 | .056 | | | 1.91 | 2.19 | 4.10 | 5.30 |
| 0154 | cycle 8 miles | | 120 | .067 | | | 2.29 | 2.63 | 4.92 | 6.35 |
| 0214 | 15 MPH ave, cycle 0.5 mile, 20 min wait/Ld./Uld. | | 176 | .045 | | | 1.56 | 1.79 | 3.35 | 4.34 |
| 0216 | cycle 1 mile | | 160 | .050 | | | 1.72 | 1.97 | 3.69 | 4.78 |
| 0218 | cycle 2 miles | | 136 | .059 | | | 2.02 | 2.32 | 4.34 | 5.60 |
| 0220 | cycle 4 miles | | 104 | .077 | | | 2.65 | 3.03 | 5.68 | 7.35 |
| 0222 | cycle 6 miles | | 88 | .091 | | | 3.13 | 3.58 | 6.71 | 8.70 |

| 31 23 23.20 Hauling | Crew | Daily Output | Labor-Hours | Unit | Material | 2011 Bare Costs Labor | 2011 Bare Costs Equipment | 2011 Bare Costs Total | Total Incl O&P |
|---|---|---|---|---|---|---|---|---|---|
| 0224  cycle 8 miles | B-34A | 72 | .111 | L.C.Y. | | 3.82 | 4.38 | 8.20 | 10.60 |
| 0226  20 MPH ave, cycle 0.5 mile | | 176 | .045 | | | 1.56 | 1.79 | 3.35 | 4.34 |
| 0228  cycle 1 mile | | 168 | .048 | | | 1.64 | 1.88 | 3.52 | 4.56 |
| 0230  cycle 2 miles | | 144 | .056 | | | 1.91 | 2.19 | 4.10 | 5.30 |
| 0232  cycle 4 miles | | 120 | .067 | | | 2.29 | 2.63 | 4.92 | 6.35 |
| 0234  cycle 6 miles | | 96 | .083 | | | 2.87 | 3.29 | 6.16 | 7.95 |
| 0236  cycle 8 miles | | 88 | .091 | | | 3.13 | 3.58 | 6.71 | 8.70 |
| 0244  25 MPH ave, cycle 4 miles | | 128 | .063 | | | 2.15 | 2.46 | 4.61 | 6 |
| 0246  cycle 6 miles | | 112 | .071 | | | 2.46 | 2.82 | 5.28 | 6.85 |
| 0248  cycle 8 miles | | 96 | .083 | | | 2.87 | 3.29 | 6.16 | 7.95 |
| 0250  30 MPH ave, cycle 4 miles | | 136 | .059 | | | 2.02 | 2.32 | 4.34 | 5.60 |
| 0252  cycle 6 miles | | 120 | .067 | | | 2.29 | 2.63 | 4.92 | 6.35 |
| 0254  cycle 8 miles | | 104 | .077 | | | 2.65 | 3.03 | 5.68 | 7.35 |
| 0314  15 MPH ave, cycle 0.5 mile, 25 min wait/Ld./Uld. | | 144 | .056 | | | 1.91 | 2.19 | 4.10 | 5.30 |
| 0316  cycle 1 mile | | 128 | .063 | | | 2.15 | 2.46 | 4.61 | 6 |
| 0318  cycle 2 miles | | 112 | .071 | | | 2.46 | 2.82 | 5.28 | 6.85 |
| 0320  cycle 4 miles | | 96 | .083 | | | 2.87 | 3.29 | 6.16 | 7.95 |
| 0322  cycle 6 miles | | 80 | .100 | | | 3.44 | 3.94 | 7.38 | 9.60 |
| 0324  cycle 8 miles | | 64 | .125 | | | 4.30 | 4.93 | 9.23 | 11.95 |
| 0326  20 MPH ave, cycle 0.5 mile | | 144 | .056 | | | 1.91 | 2.19 | 4.10 | 5.30 |
| 0328  cycle 1 mile | | 136 | .059 | | | 2.02 | 2.32 | 4.34 | 5.60 |
| 0330  cycle 2 miles | | 120 | .067 | | | 2.29 | 2.63 | 4.92 | 6.35 |
| 0332  cycle 4 miles | | 104 | .077 | | | 2.65 | 3.03 | 5.68 | 7.35 |
| 0334  cycle 6 miles | | 88 | .091 | | | 3.13 | 3.58 | 6.71 | 8.70 |
| 0336  cycle 8 miles | | 80 | .100 | | | 3.44 | 3.94 | 7.38 | 9.60 |
| 0344  25 MPH ave, cycle 4 miles | | 112 | .071 | | | 2.46 | 2.82 | 5.28 | 6.85 |
| 0346  cycle 6 miles | | 96 | .083 | | | 2.87 | 3.29 | 6.16 | 7.95 |
| 0348  cycle 8 miles | | 88 | .091 | | | 3.13 | 3.58 | 6.71 | 8.70 |
| 0350  30 MPH ave, cycle 4 miles | | 112 | .071 | | | 2.46 | 2.82 | 5.28 | 6.85 |
| 0352  cycle 6 miles | | 104 | .077 | | | 2.65 | 3.03 | 5.68 | 7.35 |
| 0354  cycle 8 miles | | 96 | .083 | | | 2.87 | 3.29 | 6.16 | 7.95 |
| 0414  15 MPH ave, cycle 0.5 mile, 30 min wait/Ld./Uld. | | 120 | .067 | | | 2.29 | 2.63 | 4.92 | 6.35 |
| 0416  cycle 1 mile | | 112 | .071 | | | 2.46 | 2.82 | 5.28 | 6.85 |
| 0418  cycle 2 miles | | 96 | .083 | | | 2.87 | 3.29 | 6.16 | 7.95 |
| 0420  cycle 4 miles | | 80 | .100 | | | 3.44 | 3.94 | 7.38 | 9.60 |
| 0422  cycle 6 miles | | 72 | .111 | | | 3.82 | 4.38 | 8.20 | 10.60 |
| 0424  cycle 8 miles | | 64 | .125 | | | 4.30 | 4.93 | 9.23 | 11.95 |
| 0426  20 MPH ave, cycle 0.5 mile | | 120 | .067 | | | 2.29 | 2.63 | 4.92 | 6.35 |
| 0428  cycle 1 mile | | 112 | .071 | | | 2.46 | 2.82 | 5.28 | 6.85 |
| 0430  cycle 2 miles | | 104 | .077 | | | 2.65 | 3.03 | 5.68 | 7.35 |
| 0432  cycle 4 miles | | 88 | .091 | | | 3.13 | 3.58 | 6.71 | 8.70 |
| 0434  cycle 6 miles | | 80 | .100 | | | 3.44 | 3.94 | 7.38 | 9.60 |
| 0436  cycle 8 miles | | 72 | .111 | | | 3.82 | 4.38 | 8.20 | 10.60 |
| 0444  25 MPH ave, cycle 4 miles | | 96 | .083 | | | 2.87 | 3.29 | 6.16 | 7.95 |
| 0446  cycle 6 miles | | 88 | .091 | | | 3.13 | 3.58 | 6.71 | 8.70 |
| 0448  cycle 8 miles | | 80 | .100 | | | 3.44 | 3.94 | 7.38 | 9.60 |
| 0450  30 MPH ave, cycle 4 miles | | 96 | .083 | | | 2.87 | 3.29 | 6.16 | 7.95 |
| 0452  cycle 6 miles | | 88 | .091 | | | 3.13 | 3.58 | 6.71 | 8.70 |
| 0454  cycle 8 miles | | 80 | .100 | | | 3.44 | 3.94 | 7.38 | 9.60 |
| 0514  15 MPH ave, cycle 0.5 mile, 35 min wait/Ld./Uld. | | 104 | .077 | | | 2.65 | 3.03 | 5.68 | 7.35 |
| 0516  cycle 1 mile | | 96 | .083 | | | 2.87 | 3.29 | 6.16 | 7.95 |
| 0518  cycle 2 miles | | 88 | .091 | | | 3.13 | 3.58 | 6.71 | 8.70 |
| 0520  cycle 4 miles | | 72 | .111 | | | 3.82 | 4.38 | 8.20 | 10.60 |

## 31 23 23 – Fill

| 31 23 23.20 Hauling | Crew | Daily Output | Labor-Hours | Unit | Material | 2011 Bare Costs Labor | Equipment | Total | Total Incl O&P |
|---|---|---|---|---|---|---|---|---|---|
| 0522 | cycle 6 miles | B-34A | 64 | .125 | L.C.Y. | | 4.30 | 4.93 | 9.23 | 11.95 |
| 0524 | cycle 8 miles | | 56 | .143 | | | 4.91 | 5.65 | 10.56 | 13.65 |
| 0526 | 20 MPH ave, cycle 0.5 mile | | 104 | .077 | | | 2.65 | 3.03 | 5.68 | 7.35 |
| 0528 | cycle 1 mile | | 96 | .083 | | | 2.87 | 3.29 | 6.16 | 7.95 |
| 0530 | cycle 2 miles | | 96 | .083 | | | 2.87 | 3.29 | 6.16 | 7.95 |
| 0532 | cycle 4 miles | | 80 | .100 | | | 3.44 | 3.94 | 7.38 | 9.60 |
| 0534 | cycle 6 miles | | 72 | .111 | | | 3.82 | 4.38 | 8.20 | 10.60 |
| 0536 | cycle 8 miles | | 64 | .125 | | | 4.30 | 4.93 | 9.23 | 11.95 |
| 0544 | 25 MPH ave, cycle 4 miles | | 88 | .091 | | | 3.13 | 3.58 | 6.71 | 8.70 |
| 0546 | cycle 6 miles | | 80 | .100 | | | 3.44 | 3.94 | 7.38 | 9.60 |
| 0548 | cycle 8 miles | | 72 | .111 | | | 3.82 | 4.38 | 8.20 | 10.60 |
| 0550 | 30 MPH ave, cycle 4 miles | | 88 | .091 | | | 3.13 | 3.58 | 6.71 | 8.70 |
| 0552 | cycle 6 miles | | 80 | .100 | | | 3.44 | 3.94 | 7.38 | 9.60 |
| 0554 | cycle 8 miles | | 72 | .111 | | | 3.82 | 4.38 | 8.20 | 10.60 |
| 1014 | 12 CY truck, cycle 0.5 mile, 15 MPH ave, 15 min. wait/Ld./Uld. | B-34B | 336 | .024 | | | .82 | 1.51 | 2.33 | 2.90 |
| 1016 | cycle 1 mile | | 300 | .027 | | | .92 | 1.69 | 2.61 | 3.25 |
| 1018 | cycle 2 miles | | 252 | .032 | | | 1.09 | 2.01 | 3.10 | 3.87 |
| 1020 | cycle 4 miles | | 180 | .044 | | | 1.53 | 2.81 | 4.34 | 5.40 |
| 1022 | cycle 6 miles | | 144 | .056 | | | 1.91 | 3.51 | 5.42 | 6.75 |
| 1024 | cycle 8 miles | | 120 | .067 | | | 2.29 | 4.22 | 6.51 | 8.10 |
| 1025 | cycle 10 miles | | 96 | .083 | | | 2.87 | 5.25 | 8.12 | 10.15 |
| 1026 | 20 MPH ave, cycle 0.5 mile | | 348 | .023 | | | .79 | 1.45 | 2.24 | 2.80 |
| 1028 | cycle 1 mile | | 312 | .026 | | | .88 | 1.62 | 2.50 | 3.12 |
| 1030 | cycle 2 miles | | 276 | .029 | | | 1 | 1.83 | 2.83 | 3.53 |
| 1032 | cycle 4 miles | | 216 | .037 | | | 1.27 | 2.34 | 3.61 | 4.52 |
| 1034 | cycle 6 miles | | 168 | .048 | | | 1.64 | 3.01 | 4.65 | 5.80 |
| 1036 | cycle 8 miles | | 144 | .056 | | | 1.91 | 3.51 | 5.42 | 6.75 |
| 1038 | cycle 10 miles | | 120 | .067 | | | 2.29 | 4.22 | 6.51 | 8.10 |
| 1040 | 25 MPH ave, cycle 4 miles | | 228 | .035 | | | 1.21 | 2.22 | 3.43 | 4.27 |
| 1042 | cycle 6 miles | | 192 | .042 | | | 1.43 | 2.64 | 4.07 | 5.10 |
| 1044 | cycle 8 miles | | 168 | .048 | | | 1.64 | 3.01 | 4.65 | 5.80 |
| 1046 | cycle 10 miles | | 144 | .056 | | | 1.91 | 3.51 | 5.42 | 6.75 |
| 1050 | 30 MPH ave, cycle 4 miles | | 252 | .032 | | | 1.09 | 2.01 | 3.10 | 3.87 |
| 1052 | cycle 6 miles | | 216 | .037 | | | 1.27 | 2.34 | 3.61 | 4.52 |
| 1054 | cycle 8 miles | | 180 | .044 | | | 1.53 | 2.81 | 4.34 | 5.40 |
| 1056 | cycle 10 miles | | 156 | .051 | | | 1.76 | 3.24 | 5 | 6.25 |
| 1060 | 35 MPH ave, cycle 4 miles | | 264 | .030 | | | 1.04 | 1.92 | 2.96 | 3.69 |
| 1062 | cycle 6 miles | | 228 | .035 | | | 1.21 | 2.22 | 3.43 | 4.27 |
| 1064 | cycle 8 miles | | 204 | .039 | | | 1.35 | 2.48 | 3.83 | 4.78 |
| 1066 | cycle 10 miles | | 180 | .044 | | | 1.53 | 2.81 | 4.34 | 5.40 |
| 1068 | cycle 20 miles | | 120 | .067 | | | 2.29 | 4.22 | 6.51 | 8.10 |
| 1069 | cycle 30 miles | | 84 | .095 | | | 3.28 | 6 | 9.28 | 11.65 |
| 1070 | cycle 40 miles | | 72 | .111 | | | 3.82 | 7.05 | 10.87 | 13.55 |
| 1072 | 40 MPH ave, cycle 6 miles | | 240 | .033 | | | 1.15 | 2.11 | 3.26 | 4.06 |
| 1074 | cycle 8 miles | | 216 | .037 | | | 1.27 | 2.34 | 3.61 | 4.52 |
| 1076 | cycle 10 miles | | 192 | .042 | | | 1.43 | 2.64 | 4.07 | 5.10 |
| 1078 | cycle 20 miles | | 120 | .067 | | | 2.29 | 4.22 | 6.51 | 8.10 |
| 1080 | cycle 30 miles | | 96 | .083 | | | 2.87 | 5.25 | 8.12 | 10.15 |
| 1082 | cycle 40 miles | | 72 | .111 | | | 3.82 | 7.05 | 10.87 | 13.55 |
| 1084 | cycle 50 miles | | 60 | .133 | | | 4.59 | 8.45 | 13.04 | 16.25 |
| 1094 | 45 MPH ave, cycle 8 miles | | 216 | .037 | | | 1.27 | 2.34 | 3.61 | 4.52 |
| 1096 | cycle 10 miles | | 204 | .039 | | | 1.35 | 2.48 | 3.83 | 4.78 |
| 1098 | cycle 20 miles | | 132 | .061 | | | 2.09 | 3.83 | 5.92 | 7.40 |

| 31 23 23.20 Hauling | Crew | Daily Output | Labor-Hours | Unit | Material | 2011 Bare Costs Labor | Equipment | Total | Total Incl O&P |
|---|---|---|---|---|---|---|---|---|---|
| 1100 | cycle 30 miles | B-34B | 108 | .074 | L.C.Y. | | 2.55 | 4.68 | 7.23 | 9 |
| 1102 | cycle 40 miles | | 84 | .095 | | | 3.28 | 6 | 9.28 | 11.65 |
| 1104 | cycle 50 miles | | 72 | .111 | | | 3.82 | 7.05 | 10.87 | 13.55 |
| 1106 | 50 MPH ave, cycle 10 miles | | 216 | .037 | | | 1.27 | 2.34 | 3.61 | 4.52 |
| 1108 | cycle 20 miles | | 144 | .056 | | | 1.91 | 3.51 | 5.42 | 6.75 |
| 1110 | cycle 30 miles | | 108 | .074 | | | 2.55 | 4.68 | 7.23 | 9 |
| 1112 | cycle 40 miles | | 84 | .095 | | | 3.28 | 6 | 9.28 | 11.65 |
| 1114 | cycle 50 miles | | 72 | .111 | | | 3.82 | 7.05 | 10.87 | 13.55 |
| 1214 | 15 MPH ave, cycle 0.5 mile, 20 min. wait/Ld./Uld. | | 264 | .030 | | | 1.04 | 1.92 | 2.96 | 3.69 |
| 1216 | cycle 1 mile | | 240 | .033 | | | 1.15 | 2.11 | 3.26 | 4.06 |
| 1218 | cycle 2 miles | | 204 | .039 | | | 1.35 | 2.48 | 3.83 | 4.78 |
| 1220 | cycle 4 miles | | 156 | .051 | | | 1.76 | 3.24 | 5 | 6.25 |
| 1222 | cycle 6 miles | | 132 | .061 | | | 2.09 | 3.83 | 5.92 | 7.40 |
| 1224 | cycle 8 miles | | 108 | .074 | | | 2.55 | 4.68 | 7.23 | 9 |
| 1225 | cycle 10 miles | | 96 | .083 | | | 2.87 | 5.25 | 8.12 | 10.15 |
| 1226 | 20 MPH ave, cycle 0.5 mile | | 264 | .030 | | | 1.04 | 1.92 | 2.96 | 3.69 |
| 1228 | cycle 1 mile | | 252 | .032 | | | 1.09 | 2.01 | 3.10 | 3.87 |
| 1230 | cycle 2 miles | | 216 | .037 | | | 1.27 | 2.34 | 3.61 | 4.52 |
| 1232 | cycle 4 miles | | 180 | .044 | | | 1.53 | 2.81 | 4.34 | 5.40 |
| 1234 | cycle 6 miles | | 144 | .056 | | | 1.91 | 3.51 | 5.42 | 6.75 |
| 1236 | cycle 8 miles | | 132 | .061 | | | 2.09 | 3.83 | 5.92 | 7.40 |
| 1238 | cycle 10 miles | | 108 | .074 | | | 2.55 | 4.68 | 7.23 | 9 |
| 1240 | 25 MPH ave, cycle 4 miles | | 192 | .042 | | | 1.43 | 2.64 | 4.07 | 5.10 |
| 1242 | cycle 6 miles | | 168 | .048 | | | 1.64 | 3.01 | 4.65 | 5.80 |
| 1244 | cycle 8 miles | | 144 | .056 | | | 1.91 | 3.51 | 5.42 | 6.75 |
| 1246 | cycle 10 miles | | 132 | .061 | | | 2.09 | 3.83 | 5.92 | 7.40 |
| 1250 | 30 MPH ave, cycle 4 miles | | 204 | .039 | | | 1.35 | 2.48 | 3.83 | 4.78 |
| 1252 | cycle 6 miles | | 180 | .044 | | | 1.53 | 2.81 | 4.34 | 5.40 |
| 1254 | cycle 8 miles | | 156 | .051 | | | 1.76 | 3.24 | 5 | 6.25 |
| 1256 | cycle 10 miles | | 144 | .056 | | | 1.91 | 3.51 | 5.42 | 6.75 |
| 1260 | 35 MPH ave, cycle 4 miles | | 216 | .037 | | | 1.27 | 2.34 | 3.61 | 4.52 |
| 1262 | cycle 6 miles | | 192 | .042 | | | 1.43 | 2.64 | 4.07 | 5.10 |
| 1264 | cycle 8 miles | | 168 | .048 | | | 1.64 | 3.01 | 4.65 | 5.80 |
| 1266 | cycle 10 miles | | 156 | .051 | | | 1.76 | 3.24 | 5 | 6.25 |
| 1268 | cycle 20 miles | | 108 | .074 | | | 2.55 | 4.68 | 7.23 | 9 |
| 1269 | cycle 30 miles | | 72 | .111 | | | 3.82 | 7.05 | 10.87 | 13.55 |
| 1270 | cycle 40 miles | | 60 | .133 | | | 4.59 | 8.45 | 13.04 | 16.25 |
| 1272 | 40 MPH ave, cycle 6 miles | | 192 | .042 | | | 1.43 | 2.64 | 4.07 | 5.10 |
| 1274 | cycle 8 miles | | 180 | .044 | | | 1.53 | 2.81 | 4.34 | 5.40 |
| 1276 | cycle 10 miles | | 156 | .051 | | | 1.76 | 3.24 | 5 | 6.25 |
| 1278 | cycle 20 miles | | 108 | .074 | | | 2.55 | 4.68 | 7.23 | 9 |
| 1280 | cycle 30 miles | | 84 | .095 | | | 3.28 | 6 | 9.28 | 11.65 |
| 1282 | cycle 40 miles | | 72 | .111 | | | 3.82 | 7.05 | 10.87 | 13.55 |
| 1284 | cycle 50 miles | | 60 | .133 | | | 4.59 | 8.45 | 13.04 | 16.25 |
| 1294 | 45 MPH ave, cycle 8 miles | | 180 | .044 | | | 1.53 | 2.81 | 4.34 | 5.40 |
| 1296 | cycle 10 miles | | 168 | .048 | | | 1.64 | 3.01 | 4.65 | 5.80 |
| 1298 | cycle 20 miles | | 120 | .067 | | | 2.29 | 4.22 | 6.51 | 8.10 |
| 1300 | cycle 30 miles | | 96 | .083 | | | 2.87 | 5.25 | 8.12 | 10.15 |
| 1302 | cycle 40 miles | | 72 | .111 | | | 3.82 | 7.05 | 10.87 | 13.55 |
| 1304 | cycle 50 miles | | 60 | .133 | | | 4.59 | 8.45 | 13.04 | 16.25 |
| 1306 | 50 MPH ave, cycle 10 miles | | 180 | .044 | | | 1.53 | 2.81 | 4.34 | 5.40 |
| 1308 | cycle 20 miles | | 132 | .061 | | | 2.09 | 3.83 | 5.92 | 7.40 |
| 1310 | cycle 30 miles | | 96 | .083 | | | 2.87 | 5.25 | 8.12 | 10.15 |

| 31 23 23.20 **Hauling** | | Crew | Daily Output | Labor-Hours | Unit | Material | 2011 Bare Costs | | Total | Total Incl O&P |
|---|---|---|---|---|---|---|---|---|---|---|
| | | | | | | | Labor | Equipment | | |
| 1312 | cycle 40 miles | B-34B | 84 | .095 | L.C.Y. | | 3.28 | 6 | 9.28 | 11.65 |
| 1314 | cycle 50 miles | | 72 | .111 | | | 3.82 | 7.05 | 10.87 | 13.55 |
| 1414 | 15 MPH ave, cycle 0.5 mile, 25 min. wait/Ld./Uld. | | 204 | .039 | | | 1.35 | 2.48 | 3.83 | 4.78 |
| 1416 | cycle 1 mile | | 192 | .042 | | | 1.43 | 2.64 | 4.07 | 5.10 |
| 1418 | cycle 2 miles | | 168 | .048 | | | 1.64 | 3.01 | 4.65 | 5.80 |
| 1420 | cycle 4 miles | | 132 | .061 | | | 2.09 | 3.83 | 5.92 | 7.40 |
| 1422 | cycle 6 miles | | 120 | .067 | | | 2.29 | 4.22 | 6.51 | 8.10 |
| 1424 | cycle 8 miles | | 96 | .083 | | | 2.87 | 5.25 | 8.12 | 10.15 |
| 1425 | cycle 10 miles | | 84 | .095 | | | 3.28 | 6 | 9.28 | 11.65 |
| 1426 | 20 MPH ave, cycle 0.5 mile | | 216 | .037 | | | 1.27 | 2.34 | 3.61 | 4.52 |
| 1428 | cycle 1 mile | | 204 | .039 | | | 1.35 | 2.48 | 3.83 | 4.78 |
| 1430 | cycle 2 miles | | 180 | .044 | | | 1.53 | 2.81 | 4.34 | 5.40 |
| 1432 | cycle 4 miles | | 156 | .051 | | | 1.76 | 3.24 | 5 | 6.25 |
| 1434 | cycle 6 miles | | 132 | .061 | | | 2.09 | 3.83 | 5.92 | 7.40 |
| 1436 | cycle 8 miles | | 120 | .067 | | | 2.29 | 4.22 | 6.51 | 8.10 |
| 1438 | cycle 10 miles | | 96 | .083 | | | 2.87 | 5.25 | 8.12 | 10.15 |
| 1440 | 25 MPH ave, cycle 4 miles | | 168 | .048 | | | 1.64 | 3.01 | 4.65 | 5.80 |
| 1442 | cycle 6 miles | | 144 | .056 | | | 1.91 | 3.51 | 5.42 | 6.75 |
| 1444 | cycle 8 miles | | 132 | .061 | | | 2.09 | 3.83 | 5.92 | 7.40 |
| 1446 | cycle 10 miles | | 108 | .074 | | | 2.55 | 4.68 | 7.23 | 9 |
| 1450 | 30 MPH ave, cycle 4 miles | | 168 | .048 | | | 1.64 | 3.01 | 4.65 | 5.80 |
| 1452 | cycle 6 miles | | 156 | .051 | | | 1.76 | 3.24 | 5 | 6.25 |
| 1454 | cycle 8 miles | | 132 | .061 | | | 2.09 | 3.83 | 5.92 | 7.40 |
| 1456 | cycle 10 miles | | 120 | .067 | | | 2.29 | 4.22 | 6.51 | 8.10 |
| 1460 | 35 MPH ave, cycle 4 miles | | 180 | .044 | | | 1.53 | 2.81 | 4.34 | 5.40 |
| 1462 | cycle 6 miles | | 156 | .051 | | | 1.76 | 3.24 | 5 | 6.25 |
| 1464 | cycle 8 miles | | 144 | .056 | | | 1.91 | 3.51 | 5.42 | 6.75 |
| 1466 | cycle 10 miles | | 132 | .061 | | | 2.09 | 3.83 | 5.92 | 7.40 |
| 1468 | cycle 20 miles | | 96 | .083 | | | 2.87 | 5.25 | 8.12 | 10.15 |
| 1469 | cycle 30 miles | | 72 | .111 | | | 3.82 | 7.05 | 10.87 | 13.55 |
| 1470 | cycle 40 miles | | 60 | .133 | | | 4.59 | 8.45 | 13.04 | 16.25 |
| 1472 | 40 MPH ave, cycle 6 miles | | 168 | .048 | | | 1.64 | 3.01 | 4.65 | 5.80 |
| 1474 | cycle 8 miles | | 156 | .051 | | | 1.76 | 3.24 | 5 | 6.25 |
| 1476 | cycle 10 miles | | 144 | .056 | | | 1.91 | 3.51 | 5.42 | 6.75 |
| 1478 | cycle 20 miles | | 96 | .083 | | | 2.87 | 5.25 | 8.12 | 10.15 |
| 1480 | cycle 30 miles | | 84 | .095 | | | 3.28 | 6 | 9.28 | 11.65 |
| 1482 | cycle 40 miles | | 60 | .133 | | | 4.59 | 8.45 | 13.04 | 16.25 |
| 1484 | cycle 50 miles | | 60 | .133 | | | 4.59 | 8.45 | 13.04 | 16.25 |
| 1494 | 45 MPH ave, cycle 8 miles | | 156 | .051 | | | 1.76 | 3.24 | 5 | 6.25 |
| 1496 | cycle 10 miles | | 144 | .056 | | | 1.91 | 3.51 | 5.42 | 6.75 |
| 1498 | cycle 20 miles | | 108 | .074 | | | 2.55 | 4.68 | 7.23 | 9 |
| 1500 | cycle 30 miles | | 84 | .095 | | | 3.28 | 6 | 9.28 | 11.65 |
| 1502 | cycle 40 miles | | 72 | .111 | | | 3.82 | 7.05 | 10.87 | 13.55 |
| 1504 | cycle 50 miles | | 60 | .133 | | | 4.59 | 8.45 | 13.04 | 16.25 |
| 1506 | 50 MPH ave, cycle 10 miles | | 156 | .051 | | | 1.76 | 3.24 | 5 | 6.25 |
| 1508 | cycle 20 miles | | 120 | .067 | | | 2.29 | 4.22 | 6.51 | 8.10 |
| 1510 | cycle 30 miles | | 96 | .083 | | | 2.87 | 5.25 | 8.12 | 10.15 |
| 1512 | cycle 40 miles | | 72 | .111 | | | 3.82 | 7.05 | 10.87 | 13.55 |
| 1514 | cycle 50 miles | | 60 | .133 | | | 4.59 | 8.45 | 13.04 | 16.25 |
| 1614 | 15 MPH, cycle 0.5 mile, 30 min. wait/Ld./Uld. | | 180 | .044 | | | 1.53 | 2.81 | 4.34 | 5.40 |
| 1616 | cycle 1 mile | | 168 | .048 | | | 1.64 | 3.01 | 4.65 | 5.80 |
| 1618 | cycle 2 miles | | 144 | .056 | | | 1.91 | 3.51 | 5.42 | 6.75 |
| 1620 | cycle 4 miles | | 120 | .067 | | | 2.29 | 4.22 | 6.51 | 8.10 |

## 31 23 23 — Fill

| 31 23 23.20 Hauling | | Crew | Daily Output | Labor-Hours | Unit | Material | 2011 Bare Costs | | Total | Total Incl O&P |
|---|---|---|---|---|---|---|---|---|---|---|
| | | | | | | | Labor | Equipment | | |
| 1622 | cycle 6 miles | B-34B | 108 | .074 | L.C.Y. | | 2.55 | 4.68 | 7.23 | 9 |
| 1624 | cycle 8 miles | | 84 | .095 | | | 3.28 | 6 | 9.28 | 11.65 |
| 1625 | cycle 10 miles | | 84 | .095 | | | 3.28 | 6 | 9.28 | 11.65 |
| 1626 | 20 MPH ave, cycle 0.5 mile | | 180 | .044 | | | 1.53 | 2.81 | 4.34 | 5.40 |
| 1628 | cycle 1 mile | | 168 | .048 | | | 1.64 | 3.01 | 4.65 | 5.80 |
| 1630 | cycle 2 miles | | 156 | .051 | | | 1.76 | 3.24 | 5 | 6.25 |
| 1632 | cycle 4 miles | | 132 | .061 | | | 2.09 | 3.83 | 5.92 | 7.40 |
| 1634 | cycle 6 miles | | 120 | .067 | | | 2.29 | 4.22 | 6.51 | 8.10 |
| 1636 | cycle 8 miles | | 108 | .074 | | | 2.55 | 4.68 | 7.23 | 9 |
| 1638 | cycle 10 miles | | 96 | .083 | | | 2.87 | 5.25 | 8.12 | 10.15 |
| 1640 | 25 MPH ave, cycle 4 miles | | 144 | .056 | | | 1.91 | 3.51 | 5.42 | 6.75 |
| 1642 | cycle 6 miles | | 132 | .061 | | | 2.09 | 3.83 | 5.92 | 7.40 |
| 1644 | cycle 8 miles | | 108 | .074 | | | 2.55 | 4.68 | 7.23 | 9 |
| 1646 | cycle 10 miles | | 108 | .074 | | | 2.55 | 4.68 | 7.23 | 9 |
| 1650 | 30 MPH ave, cycle 4 miles | | 144 | .056 | | | 1.91 | 3.51 | 5.42 | 6.75 |
| 1652 | cycle 6 miles | | 132 | .061 | | | 2.09 | 3.83 | 5.92 | 7.40 |
| 1654 | cycle 8 miles | | 120 | .067 | | | 2.29 | 4.22 | 6.51 | 8.10 |
| 1656 | cycle 10 miles | | 108 | .074 | | | 2.55 | 4.68 | 7.23 | 9 |
| 1660 | 35 MPH ave, cycle 4 miles | | 156 | .051 | | | 1.76 | 3.24 | 5 | 6.25 |
| 1662 | cycle 6 miles | | 144 | .056 | | | 1.91 | 3.51 | 5.42 | 6.75 |
| 1664 | cycle 8 miles | | 132 | .061 | | | 2.09 | 3.83 | 5.92 | 7.40 |
| 1666 | cycle 10 miles | | 120 | .067 | | | 2.29 | 4.22 | 6.51 | 8.10 |
| 1668 | cycle 20 miles | | 84 | .095 | | | 3.28 | 6 | 9.28 | 11.65 |
| 1669 | cycle 30 miles | | 72 | .111 | | | 3.82 | 7.05 | 10.87 | 13.55 |
| 1670 | cycle 40 miles | | 60 | .133 | | | 4.59 | 8.45 | 13.04 | 16.25 |
| 1672 | 40 MPH, cycle 6 miles | | 144 | .056 | | | 1.91 | 3.51 | 5.42 | 6.75 |
| 1674 | cycle 8 miles | | 132 | .061 | | | 2.09 | 3.83 | 5.92 | 7.40 |
| 1676 | cycle 10 miles | | 120 | .067 | | | 2.29 | 4.22 | 6.51 | 8.10 |
| 1678 | cycle 20 miles | | 96 | .083 | | | 2.87 | 5.25 | 8.12 | 10.15 |
| 1680 | cycle 30 miles | | 72 | .111 | | | 3.82 | 7.05 | 10.87 | 13.55 |
| 1682 | cycle 40 miles | | 60 | .133 | | | 4.59 | 8.45 | 13.04 | 16.25 |
| 1684 | cycle 50 miles | | 48 | .167 | | | 5.75 | 10.55 | 16.30 | 20.50 |
| 1694 | 45 MPH ave, cycle 8 miles | | 144 | .056 | | | 1.91 | 3.51 | 5.42 | 6.75 |
| 1696 | cycle 10 miles | | 132 | .061 | | | 2.09 | 3.83 | 5.92 | 7.40 |
| 1698 | cycle 20 miles | | 96 | .083 | | | 2.87 | 5.25 | 8.12 | 10.15 |
| 1700 | cycle 30 miles | | 84 | .095 | | | 3.28 | 6 | 9.28 | 11.65 |
| 1702 | cycle 40 miles | | 60 | .133 | | | 4.59 | 8.45 | 13.04 | 16.25 |
| 1704 | cycle 50 miles | | 60 | .133 | | | 4.59 | 8.45 | 13.04 | 16.25 |
| 1706 | 50 MPH ave, cycle 10 miles | | 132 | .061 | | | 2.09 | 3.83 | 5.92 | 7.40 |
| 1708 | cycle 20 miles | | 108 | .074 | | | 2.55 | 4.68 | 7.23 | 9 |
| 1710 | cycle 30 miles | | 84 | .095 | | | 3.28 | 6 | 9.28 | 11.65 |
| 1712 | cycle 40 miles | | 72 | .111 | | | 3.82 | 7.05 | 10.87 | 13.55 |
| 1714 | cycle 50 miles | | 60 | .133 | | | 4.59 | 8.45 | 13.04 | 16.25 |
| 2000 | Hauling, 8 CY truck, small project cost per hour | B-34A | 8 | 1 | Hr. | | 34.50 | 39.50 | 74 | 96 |
| 2100 | 12 CY Truck | B-34B | 8 | 1 | | | 34.50 | 63.50 | 98 | 122 |
| 2150 | 16.5 CY Truck | B-34C | 8 | 1 | | | 34.50 | 73 | 107.50 | 133 |
| 2175 | 18 CY 8 wheel Truck | B-34I | 8 | 1 | | | 34.50 | 82.50 | 117 | 144 |
| 2200 | 20 CY Truck | B-34D | 8 | 1 | | | 34.50 | 75 | 109.50 | 135 |
| 2300 | Grading at dump, or embankment if required, by dozer | B-10B | 1000 | .012 | L.C.Y. | | .50 | 1.05 | 1.55 | 1.91 |
| 2310 | Spotter at fill or cut, if required | 1 Clab | 8 | 1 | Hr. | | 34.50 | | 34.50 | 52.50 |
| 9014 | 18 CY truck, 8 wheels,15 min. wait/Ld./Uld.,15 MPH, cycle 0.5 mi. | B-34I | 504 | .016 | L.C.Y. | | .55 | 1.31 | 1.86 | 2.27 |
| 9016 | cycle 1 mile | | 450 | .018 | | | .61 | 1.47 | 2.08 | 2.55 |
| 9018 | cycle 2 miles | | 378 | .021 | | | .73 | 1.75 | 2.48 | 3.03 |

| 31 23 23.20 Hauling | | Crew | Daily Output | Labor-Hours | Unit | Material | 2011 Bare Costs Labor | Equipment | Total | Total Incl O&P |
|---|---|---|---|---|---|---|---|---|---|---|
| 9020 | cycle 4 miles | B-34I | 270 | .030 | L.C.Y. | | 1.02 | 2.45 | 3.47 | 4.24 |
| 9022 | cycle 6 miles | | 216 | .037 | | | 1.27 | 3.06 | 4.33 | 5.30 |
| 9024 | cycle 8 miles | | 180 | .044 | | | 1.53 | 3.67 | 5.20 | 6.35 |
| 9025 | cycle 10 miles | | 144 | .056 | | | 1.91 | 4.59 | 6.50 | 7.95 |
| 9026 | 20 MPH ave, cycle 0.5 mile | | 522 | .015 | | | .53 | 1.27 | 1.80 | 2.19 |
| 9028 | cycle 1 mile | | 468 | .017 | | | .59 | 1.41 | 2 | 2.44 |
| 9030 | cycle 2 miles | | 414 | .019 | | | .66 | 1.60 | 2.26 | 2.77 |
| 9032 | cycle 4 miles | | 324 | .025 | | | .85 | 2.04 | 2.89 | 3.53 |
| 9034 | cycle 6 miles | | 252 | .032 | | | 1.09 | 2.62 | 3.71 | 4.54 |
| 9036 | cycle 8 miles | | 216 | .037 | | | 1.27 | 3.06 | 4.33 | 5.30 |
| 9038 | cycle 10 miles | | 180 | .044 | | | 1.53 | 3.67 | 5.20 | 6.35 |
| 9040 | 25 MPH ave, cycle 4 miles | | 342 | .023 | | | .80 | 1.93 | 2.73 | 3.35 |
| 9042 | cycle 6 miles | | 288 | .028 | | | .96 | 2.29 | 3.25 | 3.97 |
| 9044 | cycle 8 miles | | 252 | .032 | | | 1.09 | 2.62 | 3.71 | 4.54 |
| 9046 | cycle 10 miles | | 216 | .037 | | | 1.27 | 3.06 | 4.33 | 5.30 |
| 9050 | 30 MPH ave, cycle 4 miles | | 378 | .021 | | | .73 | 1.75 | 2.48 | 3.03 |
| 9052 | cycle 6 miles | | 324 | .025 | | | .85 | 2.04 | 2.89 | 3.53 |
| 9054 | cycle 8 miles | | 270 | .030 | | | 1.02 | 2.45 | 3.47 | 4.24 |
| 9056 | cycle 10 miles | | 234 | .034 | | | 1.18 | 2.82 | 4 | 4.90 |
| 9060 | 35 MPH ave, cycle 4 miles | | 396 | .020 | | | .69 | 1.67 | 2.36 | 2.90 |
| 9062 | cycle 6 miles | | 342 | .023 | | | .80 | 1.93 | 2.73 | 3.35 |
| 9064 | cycle 8 miles | | 288 | .028 | | | .96 | 2.29 | 3.25 | 3.97 |
| 9066 | cycle 10 miles | | 270 | .030 | | | 1.02 | 2.45 | 3.47 | 4.24 |
| 9068 | cycle 20 miles | | 162 | .049 | | | 1.70 | 4.08 | 5.78 | 7.05 |
| 9070 | cycle 30 miles | | 126 | .063 | | | 2.18 | 5.25 | 7.43 | 9.05 |
| 9072 | cycle 40 miles | | 90 | .089 | | | 3.06 | 7.35 | 10.41 | 12.75 |
| 9074 | 40 MPH ave, cycle 6 miles | | 360 | .022 | | | .76 | 1.84 | 2.60 | 3.18 |
| 9076 | cycle 8 miles | | 324 | .025 | | | .85 | 2.04 | 2.89 | 3.53 |
| 9078 | cycle 10 miles | | 288 | .028 | | | .96 | 2.29 | 3.25 | 3.97 |
| 9080 | cycle 20 miles | | 180 | .044 | | | 1.53 | 3.67 | 5.20 | 6.35 |
| 9082 | cycle 30 miles | | 144 | .056 | | | 1.91 | 4.59 | 6.50 | 7.95 |
| 9084 | cycle 40 miles | | 108 | .074 | | | 2.55 | 6.10 | 8.65 | 10.60 |
| 9086 | cycle 50 miles | | 90 | .089 | | | 3.06 | 7.35 | 10.41 | 12.75 |
| 9094 | 45 MPH ave, cycle 8 miles | | 324 | .025 | | | .85 | 2.04 | 2.89 | 3.53 |
| 9096 | cycle 10 miles | | 306 | .026 | | | .90 | 2.16 | 3.06 | 3.75 |
| 9098 | cycle 20 miles | | 198 | .040 | | | 1.39 | 3.34 | 4.73 | 5.80 |
| 9100 | cycle 30 miles | | 144 | .056 | | | 1.91 | 4.59 | 6.50 | 7.95 |
| 9102 | cycle 40 miles | | 126 | .063 | | | 2.18 | 5.25 | 7.43 | 9.05 |
| 9104 | cycle 50 miles | | 108 | .074 | | | 2.55 | 6.10 | 8.65 | 10.60 |
| 9106 | 50 MPH ave, cycle 10 miles | | 324 | .025 | | | .85 | 2.04 | 2.89 | 3.53 |
| 9108 | cycle 20 miles | | 216 | .037 | | | 1.27 | 3.06 | 4.33 | 5.30 |
| 9110 | cycle 30 miles | | 162 | .049 | | | 1.70 | 4.08 | 5.78 | 7.05 |
| 9112 | cycle 40 miles | | 126 | .063 | | | 2.18 | 5.25 | 7.43 | 9.05 |
| 9114 | cycle 50 miles | | 108 | .074 | | | 2.55 | 6.10 | 8.65 | 10.60 |
| 9214 | 20 min. wait/Ld./Uld.,15 MPH, cycle 0.5 mi. | | 396 | .020 | | | .69 | 1.67 | 2.36 | 2.90 |
| 9216 | cycle 1 mile | | 360 | .022 | | | .76 | 1.84 | 2.60 | 3.18 |
| 9218 | cycle 2 miles | | 306 | .026 | | | .90 | 2.16 | 3.06 | 3.75 |
| 9220 | cycle 4 miles | | 234 | .034 | | | 1.18 | 2.82 | 4 | 4.90 |
| 9222 | cycle 6 miles | | 198 | .040 | | | 1.39 | 3.34 | 4.73 | 5.80 |
| 9224 | cycle 8 miles | | 162 | .049 | | | 1.70 | 4.08 | 5.78 | 7.05 |
| 9225 | cycle 10 miles | | 144 | .056 | | | 1.91 | 4.59 | 6.50 | 7.95 |
| 9226 | 20 MPH ave, cycle 0.5 mile | | 396 | .020 | | | .69 | 1.67 | 2.36 | 2.90 |
| 9228 | cycle 1 mile | | 378 | .021 | | | .73 | 1.75 | 2.48 | 3.03 |

| 31 23 23.20 Hauling | | Crew | Daily Output | Labor-Hours | Unit | Material | 2011 Bare Costs Labor | Equipment | Total | Total Incl O&P |
|---|---|---|---|---|---|---|---|---|---|---|
| 9230 | cycle 2 miles | B-34I | 324 | .025 | L.C.Y. | | .85 | 2.04 | 2.89 | 3.53 |
| 9232 | cycle 4 miles | | 270 | .030 | | | 1.02 | 2.45 | 3.47 | 4.24 |
| 9234 | cycle 6 miles | | 216 | .037 | | | 1.27 | 3.06 | 4.33 | 5.30 |
| 9236 | cycle 8 miles | | 198 | .040 | | | 1.39 | 3.34 | 4.73 | 5.80 |
| 9238 | cycle 10 miles | | 162 | .049 | | | 1.70 | 4.08 | 5.78 | 7.05 |
| 9240 | 25 MPH ave, cycle 4 miles | | 288 | .028 | | | .96 | 2.29 | 3.25 | 3.97 |
| 9242 | cycle 6 miles | | 252 | .032 | | | 1.09 | 2.62 | 3.71 | 4.54 |
| 9244 | cycle 8 miles | | 216 | .037 | | | 1.27 | 3.06 | 4.33 | 5.30 |
| 9246 | cycle 10 miles | | 198 | .040 | | | 1.39 | 3.34 | 4.73 | 5.80 |
| 9250 | 30 MPH ave, cycle 4 miles | | 306 | .026 | | | .90 | 2.16 | 3.06 | 3.75 |
| 9252 | cycle 6 miles | | 270 | .030 | | | 1.02 | 2.45 | 3.47 | 4.24 |
| 9254 | cycle 8 miles | | 234 | .034 | | | 1.18 | 2.82 | 4 | 4.90 |
| 9256 | cycle 10 miles | | 216 | .037 | | | 1.27 | 3.06 | 4.33 | 5.30 |
| 9260 | 35 MPH ave, cycle 4 miles | | 324 | .025 | | | .85 | 2.04 | 2.89 | 3.53 |
| 9262 | cycle 6 miles | | 288 | .028 | | | .96 | 2.29 | 3.25 | 3.97 |
| 9264 | cycle 8 miles | | 252 | .032 | | | 1.09 | 2.62 | 3.71 | 4.54 |
| 9266 | cycle 10 miles | | 234 | .034 | | | 1.18 | 2.82 | 4 | 4.90 |
| 9268 | cycle 20 miles | | 162 | .049 | | | 1.70 | 4.08 | 5.78 | 7.05 |
| 9270 | cycle 30 miles | | 108 | .074 | | | 2.55 | 6.10 | 8.65 | 10.60 |
| 9272 | cycle 40 miles | | 90 | .089 | | | 3.06 | 7.35 | 10.41 | 12.75 |
| 9274 | 40 MPH ave, cycle 6 miles | | 288 | .028 | | | .96 | 2.29 | 3.25 | 3.97 |
| 9276 | cycle 8 miles | | 270 | .030 | | | 1.02 | 2.45 | 3.47 | 4.24 |
| 9278 | cycle 10 miles | | 234 | .034 | | | 1.18 | 2.82 | 4 | 4.90 |
| 9280 | cycle 20 miles | | 162 | .049 | | | 1.70 | 4.08 | 5.78 | 7.05 |
| 9282 | cycle 30 miles | | 126 | .063 | | | 2.18 | 5.25 | 7.43 | 9.05 |
| 9284 | cycle 40 miles | | 108 | .074 | | | 2.55 | 6.10 | 8.65 | 10.60 |
| 9286 | cycle 50 miles | | 90 | .089 | | | 3.06 | 7.35 | 10.41 | 12.75 |
| 9294 | 45 MPH ave, cycle 8 miles | | 270 | .030 | | | 1.02 | 2.45 | 3.47 | 4.24 |
| 9296 | cycle 10 miles | | 252 | .032 | | | 1.09 | 2.62 | 3.71 | 4.54 |
| 9298 | cycle 20 miles | | 180 | .044 | | | 1.53 | 3.67 | 5.20 | 6.35 |
| 9300 | cycle 30 miles | | 144 | .056 | | | 1.91 | 4.59 | 6.50 | 7.95 |
| 9302 | cycle 40 miles | | 108 | .074 | | | 2.55 | 6.10 | 8.65 | 10.60 |
| 9304 | cycle 50 miles | | 90 | .089 | | | 3.06 | 7.35 | 10.41 | 12.75 |
| 9306 | 50 MPH ave, cycle 10 miles | | 270 | .030 | | | 1.02 | 2.45 | 3.47 | 4.24 |
| 9308 | cycle 20 miles | | 198 | .040 | | | 1.39 | 3.34 | 4.73 | 5.80 |
| 9310 | cycle 30 miles | | 144 | .056 | | | 1.91 | 4.59 | 6.50 | 7.95 |
| 9312 | cycle 40 miles | | 126 | .063 | | | 2.18 | 5.25 | 7.43 | 9.05 |
| 9314 | cycle 50 miles | | 108 | .074 | | | 2.55 | 6.10 | 8.65 | 10.60 |
| 9414 | 25 min. wait/Ld./Uld.,15 MPH, cycle 0.5 mi. | | 306 | .026 | | | .90 | 2.16 | 3.06 | 3.75 |
| 9416 | cycle 1 mile | | 288 | .028 | | | .96 | 2.29 | 3.25 | 3.97 |
| 9418 | cycle 2 miles | | 252 | .032 | | | 1.09 | 2.62 | 3.71 | 4.54 |
| 9420 | cycle 4 miles | | 198 | .040 | | | 1.39 | 3.34 | 4.73 | 5.80 |
| 9422 | cycle 6 miles | | 180 | .044 | | | 1.53 | 3.67 | 5.20 | 6.35 |
| 9424 | cycle 8 miles | | 144 | .056 | | | 1.91 | 4.59 | 6.50 | 7.95 |
| 9425 | cycle 10 miles | | 126 | .063 | | | 2.18 | 5.25 | 7.43 | 9.05 |
| 9426 | 20 MPH ave, cycle 0.5 mile | | 324 | .025 | | | .85 | 2.04 | 2.89 | 3.53 |
| 9428 | cycle 1 mile | | 306 | .026 | | | .90 | 2.16 | 3.06 | 3.75 |
| 9430 | cycle 2 miles | | 270 | .030 | | | 1.02 | 2.45 | 3.47 | 4.24 |
| 9432 | cycle 4 miles | | 234 | .034 | | | 1.18 | 2.82 | 4 | 4.90 |
| 9434 | cycle 6 miles | | 198 | .040 | | | 1.39 | 3.34 | 4.73 | 5.80 |
| 9436 | cycle 8 miles | | 180 | .044 | | | 1.53 | 3.67 | 5.20 | 6.35 |
| 9438 | cycle 10 miles | | 144 | .056 | | | 1.91 | 4.59 | 6.50 | 7.95 |
| 9440 | 25 MPH ave, cycle 4 miles | | 252 | .032 | | | 1.09 | 2.62 | 3.71 | 4.54 |

| 31 23 23.20 Hauling | | Crew | Daily Output | Labor-Hours | Unit | Material | 2011 Bare Costs Labor | 2011 Bare Costs Equipment | 2011 Bare Costs Total | Total Incl O&P |
|---|---|---|---|---|---|---|---|---|---|---|
| 9442 | cycle 6 miles | B-34I | 216 | .037 | L.C.Y. | | 1.27 | 3.06 | 4.33 | 5.30 |
| 9444 | cycle 8 miles | | 198 | .040 | | | 1.39 | 3.34 | 4.73 | 5.80 |
| 9446 | cycle 10 miles | | 180 | .044 | | | 1.53 | 3.67 | 5.20 | 6.35 |
| 9450 | 30 MPH ave, cycle 4 miles | | 252 | .032 | | | 1.09 | 2.62 | 3.71 | 4.54 |
| 9452 | cycle 6 miles | | 234 | .034 | | | 1.18 | 2.82 | 4 | 4.90 |
| 9454 | cycle 8 miles | | 198 | .040 | | | 1.39 | 3.34 | 4.73 | 5.80 |
| 9456 | cycle 10 miles | | 180 | .044 | | | 1.53 | 3.67 | 5.20 | 6.35 |
| 9460 | 35 MPH ave, cycle 4 miles | | 270 | .030 | | | 1.02 | 2.45 | 3.47 | 4.24 |
| 9462 | cycle 6 miles | | 234 | .034 | | | 1.18 | 2.82 | 4 | 4.90 |
| 9464 | cycle 8 miles | | 216 | .037 | | | 1.27 | 3.06 | 4.33 | 5.30 |
| 9466 | cycle 10 miles | | 198 | .040 | | | 1.39 | 3.34 | 4.73 | 5.80 |
| 9468 | cycle 20 miles | | 144 | .056 | | | 1.91 | 4.59 | 6.50 | 7.95 |
| 9470 | cycle 30 miles | | 108 | .074 | | | 2.55 | 6.10 | 8.65 | 10.60 |
| 9472 | cycle 40 miles | | 90 | .089 | | | 3.06 | 7.35 | 10.41 | 12.75 |
| 9474 | 40 MPH ave, cycle 6 miles | | 252 | .032 | | | 1.09 | 2.62 | 3.71 | 4.54 |
| 9476 | cycle 8 miles | | 234 | .034 | | | 1.18 | 2.82 | 4 | 4.90 |
| 9478 | cycle 10 miles | | 216 | .037 | | | 1.27 | 3.06 | 4.33 | 5.30 |
| 9480 | cycle 20 miles | | 144 | .056 | | | 1.91 | 4.59 | 6.50 | 7.95 |
| 9482 | cycle 30 miles | | 126 | .063 | | | 2.18 | 5.25 | 7.43 | 9.05 |
| 9484 | cycle 40 miles | | 90 | .089 | | | 3.06 | 7.35 | 10.41 | 12.75 |
| 9486 | cycle 50 miles | | 90 | .089 | | | 3.06 | 7.35 | 10.41 | 12.75 |
| 9494 | 45 MPH ave, cycle 8 miles | | 234 | .034 | | | 1.18 | 2.82 | 4 | 4.90 |
| 9496 | cycle 10 miles | | 216 | .037 | | | 1.27 | 3.06 | 4.33 | 5.30 |
| 9498 | cycle 20 miles | | 162 | .049 | | | 1.70 | 4.08 | 5.78 | 7.05 |
| 9500 | cycle 30 miles | | 126 | .063 | | | 2.18 | 5.25 | 7.43 | 9.05 |
| 9502 | cycle 40 miles | | 108 | .074 | | | 2.55 | 6.10 | 8.65 | 10.60 |
| 9504 | cycle 50 miles | | 90 | .089 | | | 3.06 | 7.35 | 10.41 | 12.75 |
| 9506 | 50 MPH ave, cycle 10 miles | | 234 | .034 | | | 1.18 | 2.82 | 4 | 4.90 |
| 9508 | cycle 20 miles | | 180 | .044 | | | 1.53 | 3.67 | 5.20 | 6.35 |
| 9510 | cycle 30 miles | | 144 | .056 | | | 1.91 | 4.59 | 6.50 | 7.95 |
| 9512 | cycle 40 miles | | 108 | .074 | | | 2.55 | 6.10 | 8.65 | 10.60 |
| 9514 | cycle 50 miles | | 90 | .089 | | | 3.06 | 7.35 | 10.41 | 12.75 |
| 9614 | 30 min. wait/Ld./Uld.,15 MPH, cycle 0.5 mi. | | 270 | .030 | | | 1.02 | 2.45 | 3.47 | 4.24 |
| 9616 | cycle 1 mile | | 252 | .032 | | | 1.09 | 2.62 | 3.71 | 4.54 |
| 9618 | cycle 2 miles | | 216 | .037 | | | 1.27 | 3.06 | 4.33 | 5.30 |
| 9620 | cycle 4 miles | | 180 | .044 | | | 1.53 | 3.67 | 5.20 | 6.35 |
| 9622 | cycle 6 miles | | 162 | .049 | | | 1.70 | 4.08 | 5.78 | 7.05 |
| 9624 | cycle 8 miles | | 126 | .063 | | | 2.18 | 5.25 | 7.43 | 9.05 |
| 9625 | cycle 10 miles | | 126 | .063 | | | 2.18 | 5.25 | 7.43 | 9.05 |
| 9626 | 20 MPH ave, cycle 0.5 mile | | 270 | .030 | | | 1.02 | 2.45 | 3.47 | 4.24 |
| 9628 | cycle 1 mile | | 252 | .032 | | | 1.09 | 2.62 | 3.71 | 4.54 |
| 9630 | cycle 2 miles | | 234 | .034 | | | 1.18 | 2.82 | 4 | 4.90 |
| 9632 | cycle 4 miles | | 198 | .040 | | | 1.39 | 3.34 | 4.73 | 5.80 |
| 9634 | cycle 6 miles | | 180 | .044 | | | 1.53 | 3.67 | 5.20 | 6.35 |
| 9636 | cycle 8 miles | | 162 | .049 | | | 1.70 | 4.08 | 5.78 | 7.05 |
| 9638 | cycle 10 miles | | 144 | .056 | | | 1.91 | 4.59 | 6.50 | 7.95 |
| 9640 | 25 MPH ave, cycle 4 miles | | 216 | .037 | | | 1.27 | 3.06 | 4.33 | 5.30 |
| 9642 | cycle 6 miles | | 198 | .040 | | | 1.39 | 3.34 | 4.73 | 5.80 |
| 9644 | cycle 8 miles | | 180 | .044 | | | 1.53 | 3.67 | 5.20 | 6.35 |
| 9646 | cycle 10 miles | | 162 | .049 | | | 1.70 | 4.08 | 5.78 | 7.05 |
| 9650 | 30 MPH ave, cycle 4 miles | | 216 | .037 | | | 1.27 | 3.06 | 4.33 | 5.30 |
| 9652 | cycle 6 miles | | 198 | .040 | | | 1.39 | 3.34 | 4.73 | 5.80 |
| 9654 | cycle 8 miles | | 180 | .044 | | | 1.53 | 3.67 | 5.20 | 6.35 |

# 31 23 Excavation and Fill

## 31 23 23 – Fill

### 31 23 23.20 Hauling

| | | Crew | Daily Output | Labor-Hours | Unit | Material | 2011 Bare Costs Labor | 2011 Bare Costs Equipment | Total | Total Incl O&P |
|---|---|---|---|---|---|---|---|---|---|---|
| 9656 | cycle 10 miles | B-34I | 162 | .049 | L.C.Y. | | 1.70 | 4.08 | 5.78 | 7.05 |
| 9660 | 35 MPH ave, cycle 4 miles | | 234 | .034 | | | 1.18 | 2.82 | 4 | 4.90 |
| 9662 | cycle 6 miles | | 216 | .037 | | | 1.27 | 3.06 | 4.33 | 5.30 |
| 9664 | cycle 8 miles | | 198 | .040 | | | 1.39 | 3.34 | 4.73 | 5.80 |
| 9666 | cycle 10 miles | | 180 | .044 | | | 1.53 | 3.67 | 5.20 | 6.35 |
| 9668 | cycle 20 miles | | 126 | .063 | | | 2.18 | 5.25 | 7.43 | 9.05 |
| 9670 | cycle 30 miles | | 108 | .074 | | | 2.55 | 6.10 | 8.65 | 10.60 |
| 9672 | cycle 40 miles | | 90 | .089 | | | 3.06 | 7.35 | 10.41 | 12.75 |
| 9674 | 40 MPH ave, cycle 6 miles | | 216 | .037 | | | 1.27 | 3.06 | 4.33 | 5.30 |
| 9676 | cycle 8 miles | | 198 | .040 | | | 1.39 | 3.34 | 4.73 | 5.80 |
| 9678 | cycle 10 miles | | 180 | .044 | | | 1.53 | 3.67 | 5.20 | 6.35 |
| 9680 | cycle 20 miles | | 144 | .056 | | | 1.91 | 4.59 | 6.50 | 7.95 |
| 9682 | cycle 30 miles | | 108 | .074 | | | 2.55 | 6.10 | 8.65 | 10.60 |
| 9684 | cycle 40 miles | | 90 | .089 | | | 3.06 | 7.35 | 10.41 | 12.75 |
| 9686 | cycle 50 miles | | 72 | .111 | | | 3.82 | 9.20 | 13.02 | 15.90 |
| 9694 | 45 MPH ave, cycle 8 miles | | 216 | .037 | | | 1.27 | 3.06 | 4.33 | 5.30 |
| 9696 | cycle 10 miles | | 198 | .040 | | | 1.39 | 3.34 | 4.73 | 5.80 |
| 9698 | cycle 20 miles | | 144 | .056 | | | 1.91 | 4.59 | 6.50 | 7.95 |
| 9700 | cycle 30 miles | | 126 | .063 | | | 2.18 | 5.25 | 7.43 | 9.05 |
| 9702 | cycle 40 miles | | 108 | .074 | | | 2.55 | 6.10 | 8.65 | 10.60 |
| 9704 | cycle 50 miles | | 90 | .089 | | | 3.06 | 7.35 | 10.41 | 12.75 |
| 9706 | 50 MPH ave, cycle 10 miles | | 198 | .040 | | | 1.39 | 3.34 | 4.73 | 5.80 |
| 9708 | cycle 20 miles | | 162 | .049 | | | 1.70 | 4.08 | 5.78 | 7.05 |
| 9710 | cycle 30 miles | | 126 | .063 | | | 2.18 | 5.25 | 7.43 | 9.05 |
| 9712 | cycle 40 miles | | 108 | .074 | | | 2.55 | 6.10 | 8.65 | 10.60 |
| 9714 | cycle 50 miles | | 90 | .089 | | | 3.06 | 7.35 | 10.41 | 12.75 |

### 31 23 23.24 Compaction, Structural

| | | | Crew | Daily Output | Labor-Hours | Unit | Material | Labor | Equipment | Total | Total Incl O&P |
|---|---|---|---|---|---|---|---|---|---|---|---|
| 0010 | **COMPACTION, STRUCTURAL** | R312323-30 | | | | | | | | | |
| 0020 | Steel wheel tandem roller, 5 tons | | B-10E | 8 | 1.500 | Hr. | | 62.50 | 18.15 | 80.65 | 114 |
| 0100 | 10 tons | | B-10F | 8 | 1.500 | " | | 62.50 | 28 | 90.50 | 125 |
| 0300 | Sheepsfoot or wobbly wheel roller, 8" lifts, common fill | | B-10G | 1300 | .009 | E.C.Y. | | .38 | .84 | 1.22 | 1.51 |
| 0400 | Select fill | | " | 1500 | .008 | | | .33 | .73 | 1.06 | 1.30 |
| 0600 | Vibratory plate, 8" lifts, common fill | | A-1D | 200 | .040 | | | 1.37 | .15 | 1.52 | 2.28 |
| 0700 | Select fill | | " | 216 | .037 | | | 1.27 | .14 | 1.41 | 2.11 |

# 31 25 Erosion and Sedimentation Controls

## 31 25 14 – Stabilization Measures for Erosion and Sedimentation Control

### 31 25 14.16 Rolled Erosion Control Mats and Blankets

| | | | Crew | Daily Output | Labor-Hours | Unit | Material | Labor | Equipment | Total | Total Incl O&P |
|---|---|---|---|---|---|---|---|---|---|---|---|
| 0010 | **ROLLED EROSION CONTROL MATS AND BLANKETS** | | | | | | | | | | |
| 0020 | Jute mesh, 100 S.Y. per roll, 4' wide, stapled | G | B-80A | 2400 | .010 | S.Y. | .76 | .34 | .09 | 1.19 | 1.47 |
| 0070 | Paper biodegradable mesh | G | B-1 | 2500 | .010 | | .12 | .34 | | .46 | .65 |
| 0080 | Paper mulch | G | B-64 | 20000 | .001 | | .08 | .03 | .02 | .13 | .15 |
| 0100 | Plastic netting, stapled, 2" x 1" mesh, 20 mil | G | B-1 | 2500 | .010 | | .36 | .34 | | .70 | .92 |
| 0200 | Polypropylene mesh, stapled, 6.5 oz./S.Y. | G | | 2500 | .010 | | 1.34 | .34 | | 1.68 | 1.99 |
| 0300 | Tobacco netting, or jute mesh #2, stapled | G | | 2500 | .010 | | .11 | .34 | | .45 | .64 |
| 1000 | Silt fence, polypropylene, 3' high, ideal conditions | G | 2 Clab | 1600 | .010 | L.F. | .41 | .34 | | .75 | .98 |
| 1100 | Adverse conditions | G | " | 950 | .017 | " | .41 | .58 | | .99 | 1.34 |
| 1200 | Place and remove hay bales | G | A-2 | 3 | 8 | Ton | 164 | 272 | 57 | 493 | 660 |
| 1250 | Hay bales, staked | G | " | 2500 | .010 | L.F. | 6.55 | .33 | .07 | 6.95 | 7.80 |

# 31 31 Soil Treatment

## 31 31 16 – Termite Control

| 31 31 16.13 Chemical Termite Control | Crew | Daily Output | Labor-Hours | Unit | Material | 2011 Bare Costs Labor | Equipment | Total | Total Incl O&P |
|---|---|---|---|---|---|---|---|---|---|
| 0010 **CHEMICAL TERMITE CONTROL** | | | | | | | | | |
| 0020 Slab and walls, residential | 1 Skwk | 1200 | .007 | SF Flr. | .32 | .30 | | .62 | .80 |
| 0100 Commercial, minimum | | 2496 | .003 | | .33 | .14 | | .47 | .58 |
| 0200 Maximum | | 1645 | .005 | ↓ | .50 | .22 | | .72 | .87 |
| 0400 Insecticides for termite control, minimum | | 14.20 | .563 | Gal. | 12.75 | 25 | | 37.75 | 52.50 |
| 0500 Maximum | ↓ | 11 | .727 | " | 22 | 32.50 | | 54.50 | 73.50 |

# 31 32 Soil Stabilization

## 31 32 13 – Soil Mixing Stabilization

| 31 32 13.30 Calcium Chloride | Crew | Daily Output | Labor-Hours | Unit | Material | 2011 Bare Costs Labor | Equipment | Total | Total Incl O&P |
|---|---|---|---|---|---|---|---|---|---|
| 0010 **CALCIUM CHLORIDE** | | | | | | | | | |
| 0020 Calcium chloride delivered, 100 lb. bags, truckload lots | | | | Ton | 650 | | | 650 | 715 |
| 0030 Solution, 4 lb. flake per gallon, tank truck delivery | | | | Gal. | 1.22 | | | 1.22 | 1.34 |

# 31 33 Rock Stabilization

## 31 33 13 – Rock Bolting and Grouting

| 31 33 13.10 Rock Bolting | Crew | Daily Output | Labor-Hours | Unit | Material | 2011 Bare Costs Labor | Equipment | Total | Total Incl O&P |
|---|---|---|---|---|---|---|---|---|---|
| 0010 **ROCK BOLTING** | | | | | | | | | |
| 2020 Hollow core, prestressable anchor, 1" diameter, 5' long | 2 Skwk | 32 | .500 | Ea. | 89.50 | 22 | | 111.50 | 133 |
| 2025 10' long | | 24 | .667 | | 179 | 29.50 | | 208.50 | 243 |
| 2060 2" diameter, 5' long | | 32 | .500 | | 340 | 22 | | 362 | 410 |
| 2065 10' long | | 24 | .667 | | 670 | 29.50 | | 699.50 | 780 |
| 2100 Super high-tensile, 3/4" diameter, 5' long | | 32 | .500 | | 20.50 | 22 | | 42.50 | 56.50 |
| 2105 10' long | | 24 | .667 | | 40.50 | 29.50 | | 70 | 90 |
| 2160 2" diameter, 5' long | | 32 | .500 | | 181 | 22 | | 203 | 233 |
| 2165 10' long | ↓ | 24 | .667 | | 360 | 29.50 | | 389.50 | 445 |
| 4400 Drill hole for rock bolt, 1-3/4" diam., 5' long (for 3/4" bolt) | B-56 | 17 | .941 | | | 36.50 | 76 | 112.50 | 139 |
| 4405 10' long | | 9 | 1.778 | | | 69.50 | 143 | 212.50 | 263 |
| 4420 2" diameter, 5' long (for 1" bolt) | | 13 | 1.231 | | | 48 | 99 | 147 | 182 |
| 4425 10' long | | 7 | 2.286 | | | 89 | 184 | 273 | 340 |
| 4460 3-1/2" diameter, 5' long (for 2" bolt) | | 10 | 1.600 | | | 62.50 | 129 | 191.50 | 237 |
| 4465 10' long | ↓ | 5 | 3.200 | ↓ | | 125 | 258 | 383 | 475 |

# 31 36 Gabions

## 31 36 13 – Gabion Boxes

| 31 36 13.10 Gabion Box Systems | Crew | Daily Output | Labor-Hours | Unit | Material | 2011 Bare Costs Labor | Equipment | Total | Total Incl O&P |
|---|---|---|---|---|---|---|---|---|---|
| 0010 **GABION BOX SYSTEMS** | | | | | | | | | |
| 0400 Gabions, galvanized steel mesh mats or boxes, stone filled, 6" deep | B-13 | 200 | .280 | S.Y. | 26.50 | 10.40 | 3.74 | 40.64 | 49 |
| 0500 9" deep | | 163 | .344 | | 33 | 12.80 | 4.59 | 50.39 | 61 |
| 0600 12" deep | | 153 | .366 | | 48 | 13.60 | 4.89 | 66.49 | 79 |
| 0700 18" deep | | 102 | .549 | | 63.50 | 20.50 | 7.35 | 91.35 | 109 |
| 0800 36" deep | ↓ | 60 | .933 | ↓ | 102 | 34.50 | 12.50 | 149 | 179 |

# 31 37 Riprap

## 31 37 13 – Machined Riprap

### 31 37 13.10 Riprap and Rock Lining

| | | Crew | Daily Output | Labor-Hours | Unit | Material | 2011 Bare Costs Labor | Equipment | Total | Total Incl O&P |
|---|---|---|---|---|---|---|---|---|---|---|
| 0010 | **RIPRAP AND ROCK LINING** | | | | | | | | | |
| 0011 | Random, broken stone | | | | | | | | | |
| 0100 | Machine placed for slope protection | B-12G | 62 | .258 | L.C.Y. | 31.50 | 10.45 | 11.20 | 53.15 | 62.50 |
| 0110 | 3/8 to 1/4 C.Y. pieces, grouted | B-13 | 80 | .700 | S.Y. | 65.50 | 26 | 9.35 | 100.85 | 122 |
| 0200 | 18" minimum thickness, not grouted | " | 53 | 1.057 | " | 19.70 | 39.50 | 14.15 | 73.35 | 97 |
| 0300 | Dumped, 50 lb. average | B-11A | 800 | .020 | Ton | 30.50 | .80 | 1.31 | 32.61 | 36 |
| 0350 | 100 lb. average | | 700 | .023 | | 30.50 | .91 | 1.50 | 32.91 | 36.50 |
| 0370 | 300 lb. average | | 600 | .027 | | 30.50 | 1.06 | 1.75 | 33.31 | 37 |

# 31 41 Shoring

## 31 41 13 – Timber Shoring

### 31 41 13.10 Building Shoring

| | | Crew | Daily Output | Labor-Hours | Unit | Material | 2011 Bare Costs Labor | Equipment | Total | Total Incl O&P |
|---|---|---|---|---|---|---|---|---|---|---|
| 0010 | **BUILDING SHORING** | | | | | | | | | |
| 0020 | Shoring, existing building, with timber, no salvage allowance | B-51 | 2.20 | 21.818 | M.B.F. | 680 | 755 | 78 | 1,513 | 1,975 |
| 1000 | On cribbing with 35 ton screw jacks, per box and jack | " | 3.60 | 13.333 | Jack | 65 | 460 | 47.50 | 572.50 | 830 |

## 31 41 16 – Sheet Piling

### 31 41 16.10 Sheet Piling Systems

| | | | Crew | Daily Output | Labor-Hours | Unit | Material | 2011 Bare Costs Labor | Equipment | Total | Total Incl O&P |
|---|---|---|---|---|---|---|---|---|---|---|---|
| 0010 | **SHEET PILING SYSTEMS** | | | | | | | | | | |
| 0020 | Sheet piling steel, not incl. wales, 22 psf, 15' excav., left in place | | B-40 | 10.81 | 5.920 | Ton | 1,550 | 256 | 300 | 2,106 | 2,425 |
| 0100 | Drive, extract & salvage | R314116-40 | | 6 | 10.667 | | 545 | 460 | 545 | 1,550 | 1,925 |
| 0300 | 20' deep excavation, 27 psf, left in place | | | 12.95 | 4.942 | | 1,550 | 213 | 252 | 2,015 | 2,325 |
| 0400 | Drive, extract & salvage | R314116-45 | | 6.55 | 9.771 | | 545 | 420 | 500 | 1,465 | 1,800 |
| 0600 | 25' deep excavation, 38 psf, left in place | | | 19 | 3.368 | | 1,550 | 145 | 172 | 1,867 | 2,125 |
| 0700 | Drive, extract & salvage | | | 10.50 | 6.095 | | 545 | 263 | 310 | 1,118 | 1,350 |
| 0900 | 40' deep excavation, 38 psf, left in place | | | 21.20 | 3.019 | | 1,550 | 130 | 154 | 1,834 | 2,075 |
| 1000 | Drive, extract & salvage | | | 12.25 | 5.224 | | 545 | 225 | 267 | 1,037 | 1,250 |
| 1200 | 15' deep excavation, 22 psf, left in place | | | 983 | .065 | S.F. | 18 | 2.81 | 3.32 | 24.13 | 28 |
| 1300 | Drive, extract & salvage | | | 545 | .117 | | 6.10 | 5.05 | 6 | 17.15 | 21 |
| 1500 | 20' deep excavation, 27 psf, left in place | | | 960 | .067 | | 22.50 | 2.88 | 3.40 | 28.78 | 33 |
| 1600 | Drive, extract & salvage | | | 485 | .132 | | 7.90 | 5.70 | 6.75 | 20.35 | 25 |
| 1800 | 25' deep excavation, 38 psf, left in place | | | 1000 | .064 | | 33.50 | 2.76 | 3.27 | 39.53 | 44.50 |
| 1900 | Drive, extract & salvage | | | 553 | .116 | | 10.85 | 4.99 | 5.90 | 21.74 | 26.50 |
| 2100 | Rent steel sheet piling and wales, first month | | | | | Ton | 310 | | | 310 | 340 |
| 2200 | Per added month | | | | | | 31 | | | 31 | 34 |
| 2300 | Rental piling left in place, add to rental | | | | | | 1,250 | | | 1,250 | 1,375 |
| 2500 | Wales, connections & struts, 2/3 salvage | | | | | | 495 | | | 495 | 545 |
| 2700 | High strength piling, 50,000 psi, add | | | | | | 71.50 | | | 71.50 | 78.50 |
| 2800 | 55,000 psi, add | | | | | | 81 | | | 81 | 89 |
| 3000 | Tie rod, not upset, 1-1/2" to 4" diameter with turnbuckle | | | | | | 2,250 | | | 2,250 | 2,475 |
| 3100 | No turnbuckle | | | | | | 1,750 | | | 1,750 | 1,925 |
| 3300 | Upset, 1-3/4" to 4" diameter with turnbuckle | | | | | | 2,550 | | | 2,550 | 2,825 |
| 3400 | No turnbuckle | | | | | | 2,200 | | | 2,200 | 2,425 |
| 3600 | Lightweight, 18" to 28" wide, 7 ga., 9.22 psf, and | | | | | | | | | | |
| 3610 | 9 ga., 8.6 psf, minimum | | | | | Lb. | .93 | | | .93 | 1.02 |
| 3700 | Average | | | | | | 1.01 | | | 1.01 | 1.11 |
| 3750 | Maximum | | | | | | 1.15 | | | 1.15 | 1.27 |
| 3900 | Wood, solid sheeting, incl. wales, braces and spacers, | R314116-40 | | | | | | | | | |
| 3910 | drive, extract & salvage, 8' deep excavation | | B-31 | 330 | .121 | S.F. | 1.51 | 4.42 | .55 | 6.48 | 9.05 |
| 4000 | 10' deep, 50 S.F./hr. in & 150 S.F./hr. out | | | 300 | .133 | | 1.55 | 4.87 | .60 | 7.02 | 9.80 |
| 4100 | 12' deep, 45 S.F./hr. in & 135 S.F./hr. out | | | 270 | .148 | | 1.60 | 5.40 | .67 | 7.67 | 10.80 |

# 31 41 Shoring

## 31 41 16 – Sheet Piling

### 31 41 16.10 Sheet Piling Systems

| | | Crew | Daily Output | Labor-Hours | Unit | Material | 2011 Bare Costs Labor | Equipment | Total | Total Incl O&P |
|---|---|---|---|---|---|---|---|---|---|---|
| 4200 | 14' deep, 42 S.F./hr. in & 126 S.F./hr. out | B-31 | 250 | .160 | S.F. | 1.65 | 5.85 | .72 | 8.22 | 11.55 |
| 4300 | 16' deep, 40 S.F./hr. in & 120 S.F./hr. out | | 240 | .167 | | 1.70 | 6.10 | .76 | 8.56 | 12.05 |
| 4400 | 18' deep, 38 S.F./hr. in & 114 S.F./hr. out | | 230 | .174 | | 1.75 | 6.35 | .79 | 8.89 | 12.55 |
| 4500 | 20' deep, 35 S.F./hr. in & 105 S.F./hr. out | | 210 | .190 | | 1.81 | 6.95 | .86 | 9.62 | 13.60 |
| 4520 | Left in place, 8' deep, 55 S.F./hr. | | 440 | .091 | | 2.72 | 3.32 | .41 | 6.45 | 8.55 |
| 4540 | 10' deep, 50 S.F./hr. | | 400 | .100 | | 2.86 | 3.65 | .45 | 6.96 | 9.25 |
| 4560 | 12' deep, 45 S.F./hr. | | 360 | .111 | | 3.02 | 4.05 | .50 | 7.57 | 10.05 |
| 4565 | 14' deep, 42 S.F./hr. | | 335 | .119 | | 3.20 | 4.36 | .54 | 8.10 | 10.80 |
| 4570 | 16' deep, 40 S.F./hr. | | 320 | .125 | | 3.40 | 4.56 | .57 | 8.53 | 11.35 |
| 4580 | 18' deep, 38 S.F./hr. | | 305 | .131 | | 3.63 | 4.79 | .59 | 9.01 | 12 |
| 4590 | 20' deep, 35 S.F./hr. | | 280 | .143 | | 3.89 | 5.20 | .65 | 9.74 | 13 |
| 4700 | Alternate pricing, left in place, 8' deep | | 1.76 | 22.727 | M.B.F. | 610 | 830 | 103 | 1,543 | 2,050 |
| 4800 | Drive, extract and salvage, 8' deep | | 1.32 | 30.303 | " | 545 | 1,100 | 137 | 1,782 | 2,450 |
| 5000 | For treated lumber add cost of treatment to lumber | | | | | | | | | |

# 31 43 Concrete Raising

## 31 43 13 – Pressure Grouting

### 31 43 13.13 Concrete Pressure Grouting

| | | Crew | Daily Output | Labor-Hours | Unit | Material | 2011 Bare Costs Labor | Equipment | Total | Total Incl O&P |
|---|---|---|---|---|---|---|---|---|---|---|
| 0010 | **CONCRETE PRESSURE GROUTING** | | | | | | | | | |
| 0020 | Grouting, pressure, cement & sand, 1:1 mix, minimum | B-61 | 124 | .323 | Bag | 12.40 | 11.80 | 2.58 | 26.78 | 34.50 |
| 0100 | Maximum | | 51 | .784 | " | 12.40 | 28.50 | 6.25 | 47.15 | 64.50 |
| 0200 | Cement and sand, 1:1 mix, minimum | | 250 | .160 | C.F. | 25 | 5.85 | 1.28 | 32.13 | 38 |
| 0300 | Maximum | | 100 | .400 | | 37 | 14.65 | 3.20 | 54.85 | 67 |
| 0400 | Epoxy cement grout, minimum | | 137 | .292 | | 525 | 10.70 | 2.33 | 538.03 | 600 |
| 0500 | Maximum | | 57 | .702 | | 525 | 25.50 | 5.60 | 556.10 | 625 |
| 0700 | Alternate pricing method: (Add for materials) | | | | | | | | | |
| 0710 | 5 person crew and equipment | B-61 | 1 | 40 | Day | | 1,475 | 320 | 1,795 | 2,575 |

# 31 45 Vibroflotation and Densification

## 31 45 13 – Vibroflotation

### 31 45 13.10 Vibroflotation Densification

| | | Crew | Daily Output | Labor-Hours | Unit | Material | 2011 Bare Costs Labor | Equipment | Total | Total Incl O&P |
|---|---|---|---|---|---|---|---|---|---|---|
| 0010 | **VIBROFLOTATION DENSIFICATION** | R314513-90 | | | | | | | | |
| 0900 | Vibroflotation compacted sand cylinder, minimum | B-60 | 750 | .075 | V.L.F. | | 2.98 | 2.85 | 5.83 | 7.60 |
| 0950 | Maximum | | 325 | .172 | | | 6.85 | 6.55 | 13.40 | 17.60 |
| 1100 | Vibro replacement compacted stone cylinder, minimum | | 500 | .112 | | | 4.46 | 4.27 | 8.73 | 11.45 |
| 1150 | Maximum | | 250 | .224 | | | 8.95 | 8.55 | 17.50 | 23 |
| 1300 | Mobilization and demobilization, minimum | | .47 | 119 | Total | | 4,750 | 4,550 | 9,300 | 12,200 |
| 1400 | Maximum | | .14 | 400 | " | | 15,900 | 15,300 | 31,200 | 40,900 |

# 31 46 Needle Beams

## 31 46 13 – Cantilever Needle Beams

### 31 46 13.10 Needle Beams

| | 31 46 13.10 Needle Beams | Crew | Daily Output | Labor-Hours | Unit | Material | 2011 Bare Costs Labor | Equipment | Total | Total Incl O&P |
|---|---|---|---|---|---|---|---|---|---|---|
| 0010 | **NEEDLE BEAMS** | | | | | | | | | |
| 0011 | Incl. wood shoring 10' x 10' opening | | | | | | | | | |
| 0400 | Block, concrete, 8" thick | B-9 | 7.10 | 5.634 | Ea. | 43 | 196 | 27.50 | 266.50 | 380 |
| 0420 | 12" thick | | 6.70 | 5.970 | | 51.50 | 207 | 29 | 287.50 | 410 |
| 0800 | Brick, 4" thick with 8" backup block | | 5.70 | 7.018 | | 51.50 | 244 | 34.50 | 330 | 470 |
| 1000 | Brick, solid, 8" thick | | 6.20 | 6.452 | | 43 | 224 | 31.50 | 298.50 | 425 |
| 1040 | 12" thick | | 4.90 | 8.163 | | 51.50 | 284 | 40 | 375.50 | 535 |
| 1080 | 16" thick | | 4.50 | 8.889 | | 67.50 | 310 | 43.50 | 421 | 595 |
| 2000 | Add for additional floors of shoring | B-1 | 6 | 4 | | 43 | 140 | | 183 | 263 |

# 31 48 Underpinning

## 31 48 13 – Underpinning Piers

### 31 48 13.10 Underpinning Foundations

| | 31 48 13.10 Underpinning Foundations | Crew | Daily Output | Labor-Hours | Unit | Material | 2011 Bare Costs Labor | Equipment | Total | Total Incl O&P |
|---|---|---|---|---|---|---|---|---|---|---|
| 0010 | **UNDERPINNING FOUNDATIONS** | | | | | | | | | |
| 0011 | Including excavation, | | | | | | | | | |
| 0020 | forming, reinforcing, concrete and equipment | | | | | | | | | |
| 0100 | 5' to 16' below grade, 100 to 500 C.Y. | B-52 | 2.30 | 24.348 | C.Y. | 278 | 970 | 219 | 1,467 | 2,025 |
| 0200 | Over 500 C.Y. | | 2.50 | 22.400 | | 250 | 890 | 202 | 1,342 | 1,850 |
| 0400 | 16' to 25' below grade, 100 to 500 C.Y. | | 2 | 28 | | 305 | 1,125 | 252 | 1,682 | 2,300 |
| 0500 | Over 500 C.Y. | | 2.10 | 26.667 | | 289 | 1,050 | 240 | 1,579 | 2,200 |
| 0700 | 26' to 40' below grade, 100 to 500 C.Y. | | 1.60 | 35 | | 335 | 1,400 | 315 | 2,050 | 2,825 |
| 0800 | Over 500 C.Y. | | 1.80 | 31.111 | | 305 | 1,250 | 280 | 1,835 | 2,550 |
| 0900 | For under 50 C.Y., add | | | | | 10% | 40% | | | |
| 1000 | For 50 C.Y. to 100 C.Y., add | | | | | 5% | 20% | | | |

# 31 52 Cofferdams

## 31 52 16 – Timber Cofferdams

### 31 52 16.10 Cofferdams

| | 31 52 16.10 Cofferdams | Crew | Daily Output | Labor-Hours | Unit | Material | 2011 Bare Costs Labor | Equipment | Total | Total Incl O&P |
|---|---|---|---|---|---|---|---|---|---|---|
| 0010 | **COFFERDAMS** | | | | | | | | | |
| 0011 | Incl. mobilization and temporary sheeting | | | | | | | | | |
| 0080 | Soldier beams & lagging H piles with 3" wood sheeting | | | | | | | | | |
| 0090 | horizontal between piles, including removal of wales & braces | | | | | | | | | |
| 0100 | No hydrostatic head, 15' deep, 1 line of braces, minimum | B-50 | 545 | .206 | S.F. | 6.80 | 8.45 | 4.24 | 19.49 | 25.50 |
| 0200 | Maximum | | 495 | .226 | | 7.55 | 9.30 | 4.67 | 21.52 | 28 |
| 0400 | 15' to 22' deep with 2 lines of braces, 10" H, minimum | | 360 | .311 | | 8 | 12.80 | 6.40 | 27.20 | 36 |
| 0500 | Maximum | | 330 | .339 | | 9.05 | 13.95 | 7 | 30 | 39.50 |
| 0700 | 23' to 35' deep with 3 lines of braces, 12" H, minimum | | 325 | .345 | | 10.45 | 14.20 | 7.10 | 31.75 | 41.50 |
| 0800 | Maximum | | 295 | .380 | | 11.35 | 15.65 | 7.85 | 34.85 | 45.50 |
| 1000 | 36' to 45' deep with 4 lines of braces, 14" H, minimum | | 290 | .386 | | 11.70 | 15.90 | 7.95 | 35.55 | 46.50 |
| 1100 | Maximum | | 265 | .423 | | 12.35 | 17.40 | 8.70 | 38.45 | 50 |
| 1300 | No hydrostatic head, left in place, 15' dp., 1 line of braces, min. | | 635 | .176 | | 9.05 | 7.25 | 3.64 | 19.94 | 25.50 |
| 1400 | Maximum | | 575 | .195 | | 9.70 | 8 | 4.02 | 21.72 | 27.50 |
| 1600 | 15' to 22' deep with 2 lines of braces, minimum | | 455 | .246 | | 13.60 | 10.15 | 5.10 | 28.85 | 36.50 |
| 1700 | Maximum | | 415 | .270 | | 15.10 | 11.10 | 5.55 | 31.75 | 40 |
| 1900 | 23' to 35' deep with 3 lines of braces, minimum | | 420 | .267 | | 16.20 | 11 | 5.50 | 32.70 | 41 |
| 2000 | Maximum | | 380 | .295 | | 17.90 | 12.15 | 6.10 | 36.15 | 45.50 |
| 2200 | 36' to 45' deep with 4 lines of braces, minimum | | 385 | .291 | | 19.45 | 12 | 6 | 37.45 | 47 |
| 2300 | Maximum | | 350 | .320 | | 22.50 | 13.15 | 6.60 | 42.25 | 53 |
| 2350 | Lagging only, 3" thick wood between piles 8' O.C., minimum | B-46 | 400 | .120 | | 1.51 | 4.62 | .10 | 6.23 | 9 |

# 31 52 Cofferdams

## 31 52 16 – Timber Cofferdams

### 31 52 16.10 Cofferdams

| | 31 52 16.10 Cofferdams | Crew | Daily Output | Labor-Hours | Unit | Material | 2011 Bare Costs Labor | Equipment | Total | Total Incl O&P |
|---|---|---|---|---|---|---|---|---|---|---|
| 2370 | Maximum | B-46 | 250 | .192 | S.F. | 2.27 | 7.40 | .17 | 9.84 | 14.30 |
| 2400 | Open sheeting no bracing, for trenches to 10' deep, min. | | 1736 | .028 | | .68 | 1.06 | .02 | 1.76 | 2.45 |
| 2450 | Maximum | ↓ | 1510 | .032 | | .76 | 1.22 | .03 | 2.01 | 2.79 |
| 2500 | Tie-back method, add to open sheeting, add, minimum | | | | ↓ | | | | 20% | 20% |
| 2550 | Maximum | | | | | | | | 60% | 60% |
| 2700 | Tie-backs only, based on tie-backs total length, minimum | B-46 | 86.80 | .553 | L.F. | 12.55 | 21.50 | .48 | 34.53 | 48 |
| 2750 | Maximum | | 38.50 | 1.247 | " | 22 | 48 | 1.08 | 71.08 | 101 |
| 3500 | Tie-backs only, typical average, 25' long | | 2 | 24 | Ea. | 555 | 925 | 21 | 1,501 | 2,075 |
| 3600 | 35' long | ↓ | 1.58 | 30.380 | " | 735 | 1,175 | 26.50 | 1,936.50 | 2,700 |

# 31 56 Slurry Walls

## 31 56 23 – Lean Concrete Slurry Walls

### 31 56 23.20 Slurry Trench

| | | Crew | Daily Output | Labor-Hours | Unit | Material | 2011 Bare Costs Labor | Equipment | Total | Total Incl O&P |
|---|---|---|---|---|---|---|---|---|---|---|
| 0010 | **SLURRY TRENCH** | | | | | | | | | |
| 0011 | Excavated slurry trench in wet soils | | | | | | | | | |
| 0020 | backfilled with 3000 PSI concrete, no reinforcing steel | | | | | | | | | |
| 0050 | Minimum | C-7 | 333 | .216 | C.F. | 7.35 | 8.05 | 3.84 | 19.24 | 24.50 |
| 0100 | Maximum | | 200 | .360 | " | 12.30 | 13.40 | 6.40 | 32.10 | 41 |
| 0200 | Alternate pricing method, minimum | | 150 | .480 | S.F. | 14.65 | 17.85 | 8.50 | 41 | 52.50 |
| 0300 | Maximum | ↓ | 120 | .600 | | 22 | 22.50 | 10.65 | 55.15 | 69.50 |
| 0500 | Reinforced slurry trench, minimum | B-48 | 177 | .316 | | 11 | 12.20 | 15.80 | 39 | 48 |
| 0600 | Maximum | " | 69 | .812 | ↓ | 36.50 | 31.50 | 40.50 | 108.50 | 133 |
| 0800 | Haul for disposal, 2 mile haul, excavated material, add | B-34B | 99 | .081 | C.Y. | | 2.78 | 5.10 | 7.88 | 9.80 |
| 0900 | Haul bentonite castings for disposal, add | " | 40 | .200 | " | | 6.90 | 12.65 | 19.55 | 24.50 |

# 31 62 Driven Piles

## 31 62 13 – Concrete Piles

### 31 62 13.23 Prestressed Concrete Piles

| | | Crew | Daily Output | Labor-Hours | Unit | Material | 2011 Bare Costs Labor | Equipment | Total | Total Incl O&P |
|---|---|---|---|---|---|---|---|---|---|---|
| 0010 | **PRESTRESSED CONCRETE PILES**, 200 piles | | | | | | | | | |
| 0020 | Unless specified otherwise, not incl. pile caps or mobilization | | | | | | | | | |
| 2200 | Precast, prestressed, 50' long, 12" diam., 2-3/8" wall | B-19 | 720 | .089 | V.L.F. | 22.50 | 3.84 | 2.57 | 28.91 | 33.50 |
| 2300 | 14" diameter, 2-1/2" wall | | 680 | .094 | | 30 | 4.06 | 2.72 | 36.78 | 42.50 |
| 2500 | 16" diameter, 3" wall | ↓ | 640 | .100 | | 42.50 | 4.32 | 2.89 | 49.71 | 57 |
| 2600 | 18" diameter, 3" wall | B-19A | 600 | .107 | | 53.50 | 4.60 | 3.75 | 61.85 | 70.50 |
| 2800 | 20" diameter, 3-1/2" wall | | 560 | .114 | | 64.50 | 4.93 | 4.02 | 73.45 | 83 |
| 2900 | 24" diameter, 3-1/2" wall | ↓ | 520 | .123 | | 73 | 5.30 | 4.33 | 82.63 | 93.50 |
| 3100 | Precast, prestressed, 40' long, 10" thick, square | B-19 | 700 | .091 | | 16.35 | 3.95 | 2.64 | 22.94 | 27 |
| 3200 | 12" thick, square | | 680 | .094 | | 20.50 | 4.06 | 2.72 | 27.28 | 32 |
| 3400 | 14" thick, square | | 600 | .107 | | 24 | 4.60 | 3.08 | 31.68 | 37 |
| 3500 | Octagonal | | 640 | .100 | | 33 | 4.32 | 2.89 | 40.21 | 46.50 |
| 3700 | 16" thick, square | | 560 | .114 | | 38 | 4.93 | 3.30 | 46.23 | 53.50 |
| 3800 | Octagonal | ↓ | 600 | .107 | | 39.50 | 4.60 | 3.08 | 47.18 | 54 |
| 4000 | 18" thick, square | B-19A | 520 | .123 | | 46 | 5.30 | 4.33 | 55.63 | 63.50 |
| 4100 | Octagonal | B-19 | 560 | .114 | | 47 | 4.93 | 3.30 | 55.23 | 63 |
| 4300 | 20" thick, square | B-19A | 480 | .133 | | 57 | 5.75 | 4.69 | 67.44 | 76.50 |
| 4400 | Octagonal | B-19 | 520 | .123 | | 52 | 5.30 | 3.55 | 60.85 | 69 |
| 4600 | 24" thick, square | B-19A | 440 | .145 | | 81 | 6.30 | 5.10 | 92.40 | 104 |
| 4700 | Octagonal | B-19 | 480 | .133 | | 74.50 | 5.75 | 3.85 | 84.10 | 95 |
| 4730 | Precast, prestressed, 60' long, 10" thick, square | ↓ | 700 | .091 | ↓ | 17.15 | 3.95 | 2.64 | 23.74 | 28 |

# 31 62 Driven Piles

## 31 62 13 – Concrete Piles

| 31 62 13.23 Prestressed Concrete Piles | Crew | Daily Output | Labor-Hours | Unit | Material | 2011 Bare Costs Labor | Equipment | Total | Total Incl O&P |
|---|---|---|---|---|---|---|---|---|---|
| 4740 | 12" thick, square ( 60' long) | B-19 | 680 | .094 | V.L.F. | 21.50 | 4.06 | 2.72 | 28.28 | 33 |
| 4750 | Mobilization for 10,000 L.F. pile job, add | | 3300 | .019 | | | .84 | .56 | 1.40 | 1.93 |
| 4800 | 25,000 L.F. pile job, add | | 8500 | .008 | | | .33 | .22 | .55 | .75 |

## 31 62 16 – Steel Piles

### 31 62 16.13 Steel Piles

| | | Crew | Daily Output | Labor-Hours | Unit | Material | 2011 Bare Costs Labor | Equipment | Total | Total Incl O&P |
|---|---|---|---|---|---|---|---|---|---|---|
| 0010 | **STEEL PILES** | | | | | | | | | |
| 0100 | Step tapered, round, concrete filled | | | | | | | | | |
| 0110 | 8" tip, 60 ton capacity, 30' depth | B-19 | 760 | .084 | V.L.F. | 10 | 3.63 | 2.43 | 16.06 | 19.30 |
| 0120 | 60' depth | | 740 | .086 | | 11.30 | 3.73 | 2.50 | 17.53 | 21 |
| 0130 | 80' depth | | 700 | .091 | | 11.70 | 3.95 | 2.64 | 18.29 | 22 |
| 0150 | 10" tip, 90 ton capacity, 30' depth | | 700 | .091 | | 12 | 3.95 | 2.64 | 18.59 | 22.50 |
| 0160 | 60' depth | | 690 | .093 | | 12.35 | 4 | 2.68 | 19.03 | 23 |
| 0170 | 80' depth | | 670 | .096 | | 13.30 | 4.12 | 2.76 | 20.18 | 24 |
| 0190 | 12" tip, 120 ton capacity, 30' depth | | 660 | .097 | | 16.30 | 4.19 | 2.80 | 23.29 | 27.50 |
| 0200 | 60' depth, 12" diameter | | 630 | .102 | | 16.20 | 4.38 | 2.93 | 23.51 | 28 |
| 0210 | 80' depth | | 590 | .108 | | 14.95 | 4.68 | 3.13 | 22.76 | 27 |
| 0250 | "H" Sections, 50' long, HP8 x 36 | | 640 | .100 | | 17.85 | 4.32 | 2.89 | 25.06 | 29.50 |
| 0400 | HP10 X 42 | | 610 | .105 | | 21 | 4.53 | 3.03 | 28.56 | 33.50 |
| 0500 | HP10 X 57 | | 610 | .105 | | 28 | 4.53 | 3.03 | 35.56 | 41.50 |
| 0700 | HP12 X 53 | | 590 | .108 | | 26.50 | 4.68 | 3.13 | 34.31 | 40.50 |
| 0800 | HP12 X 74 | B-19A | 590 | .108 | | 37.50 | 4.68 | 3.81 | 45.99 | 52.50 |
| 1000 | HP14 X 73 | | 540 | .119 | | 36.50 | 5.10 | 4.17 | 45.77 | 53 |
| 1100 | HP14 X 89 | | 540 | .119 | | 45.50 | 5.10 | 4.17 | 54.77 | 62.50 |
| 1300 | HP14 X 102 | | 510 | .125 | | 51.50 | 5.40 | 4.41 | 61.31 | 70.50 |
| 1400 | HP14 X 117 | | 510 | .125 | | 59.50 | 5.40 | 4.41 | 69.31 | 79 |
| 1600 | Splice on standard points, not in leads, 8" or 10" | 1 Sswl | 5 | 1.600 | Ea. | 117 | 77.50 | | 194.50 | 266 |
| 1700 | 12" or 14" | | 4 | 2 | | 171 | 97 | | 268 | 360 |
| 1900 | Heavy duty points, not in leads, 10" wide | | 4 | 2 | | 181 | 97 | | 278 | 370 |
| 2100 | 14" wide | | 3.50 | 2.286 | | 235 | 111 | | 346 | 455 |

## 31 62 19 – Timber Piles

### 31 62 19.10 Wood Piles

| | | Crew | Daily Output | Labor-Hours | Unit | Material | 2011 Bare Costs Labor | Equipment | Total | Total Incl O&P |
|---|---|---|---|---|---|---|---|---|---|---|
| 0010 | **WOOD PILES** | | | | | | | | | |
| 0011 | Friction or end bearing, not including | | | | | | | | | |
| 0050 | mobilization or demobilization | | | | | | | | | |
| 0100 | Untreated piles, up to 30' long, 12" butts, 8" points | B-19 | 625 | .102 | V.L.F. | 8 | 4.42 | 2.96 | 15.38 | 18.95 |
| 0200 | 30' to 39' long, 12" butts, 8" points | | 700 | .091 | | 8 | 3.95 | 2.64 | 14.59 | 17.85 |
| 0300 | 40' to 49' long, 12" butts, 7" points | | 720 | .089 | | 8 | 3.84 | 2.57 | 14.41 | 17.60 |
| 0400 | 50' to 59' long, 13"butts, 7" points | | 800 | .080 | | 8 | 3.45 | 2.31 | 13.76 | 16.75 |
| 0500 | 60' to 69' long, 13" butts, 7" points | | 840 | .076 | | 9 | 3.29 | 2.20 | 14.49 | 17.45 |
| 0600 | 70' to 80' long, 13" butts, 6" points | | 840 | .076 | | 10 | 3.29 | 2.20 | 15.49 | 18.55 |
| 0800 | Treated piles, 12 lb. per C.F., | | | | | | | | | |
| 0810 | friction or end bearing, ASTM class B | | | | | | | | | |
| 1000 | Up to 30' long, 12" butts, 8" points | B-19 | 625 | .102 | V.L.F. | 10.90 | 4.42 | 2.96 | 18.28 | 22 |
| 1100 | 30' to 39' long, 12" butts, 8" points | | 700 | .091 | | 10.65 | 3.95 | 2.64 | 17.24 | 21 |
| 1200 | 40' to 49' long, 12" butts, 7" points | | 720 | .089 | | 11.40 | 3.84 | 2.57 | 17.81 | 21.50 |
| 1300 | 50' to 59' long, 13" butts, 7" points | | 800 | .080 | | 12.60 | 3.45 | 2.31 | 18.36 | 22 |
| 1400 | 60' to 69' long, 13" butts, 6" points | B-19A | 840 | .076 | | 17.20 | 3.29 | 2.68 | 23.17 | 27 |
| 1500 | 70' to 80' long, 13" butts, 6" points | " | 840 | .076 | | 21.50 | 3.29 | 2.68 | 27.47 | 32 |
| 1600 | Treated piles, C.C.A. 2.5# per C.F. | | | | | | | | | |
| 1610 | 8" butts, 10' long | B-19 | 400 | .160 | V.L.F. | 7.95 | 6.90 | 4.62 | 19.47 | 24.50 |
| 1620 | 11' to 16' long | | 500 | .128 | | 7.95 | 5.50 | 3.70 | 17.15 | 21.50 |
| 1630 | 17' to 20' long | | 575 | .111 | | 7.95 | 4.80 | 3.21 | 15.96 | 19.80 |

# 31 62 Driven Piles

## 31 62 19 – Timber Piles

### 31 62 19.10 Wood Piles

| | | Crew | Daily Output | Labor-Hours | Unit | Material | 2011 Bare Costs Labor | Equipment | Total | Total Incl O&P |
|---|---|---|---|---|---|---|---|---|---|---|
| 1640 | 10" butts, 10' to 16' long | B-19 | 500 | .128 | V.L.F. | 9.05 | 5.50 | 3.70 | 18.25 | 22.50 |
| 1650 | 17' to 20' long | | 575 | .111 | | 9.05 | 4.80 | 3.21 | 17.06 | 21 |
| 1660 | 21' to 40' long | | 700 | .091 | | 9.05 | 3.95 | 2.64 | 15.64 | 19 |
| 1670 | 12" butts, 10' to 20' long | | 575 | .111 | | 9.80 | 4.80 | 3.21 | 17.81 | 22 |
| 1680 | 21' to 35' long | | 650 | .098 | | 9.80 | 4.25 | 2.84 | 16.89 | 20.50 |
| 1690 | 36' to 40' long | | 700 | .091 | | 9.80 | 3.95 | 2.64 | 16.39 | 19.80 |
| 1695 | 14" butts. to 40' long | | 700 | .091 | | 14 | 3.95 | 2.64 | 20.59 | 24.50 |
| 1700 | Boot for pile tip, minimum | 1 Pile | 27 | .296 | Ea. | 22.50 | 12.45 | | 34.95 | 44.50 |
| 1800 | Maximum | | 21 | .381 | | 67 | 16 | | 83 | 99 |
| 2000 | Point for pile tip, minimum | | 20 | .400 | | 22.50 | 16.80 | | 39.30 | 51.50 |
| 2100 | Maximum | | 15 | .533 | | 80.50 | 22.50 | | 103 | 125 |
| 2300 | Splice for piles over 50' long, minimum | B-46 | 35 | 1.371 | | 56 | 53 | 1.19 | 110.19 | 146 |
| 2400 | Maximum | | 20 | 2.400 | | 67 | 92.50 | 2.09 | 161.59 | 221 |
| 2600 | Concrete encasement with wire mesh and tube | | 331 | .145 | V.L.F. | 10.50 | 5.60 | .13 | 16.23 | 20.50 |
| 2700 | Mobilization for 10,000 L.F. pile job, add | B-19 | 3300 | .019 | | | .84 | .56 | 1.40 | 1.93 |
| 2800 | 25,000 L.F. pile job, add | " | 8500 | .008 | | | .33 | .22 | .55 | .75 |

## 31 62 23 – Composite Piles

### 31 62 23.13 Concrete-Filled Steel Piles

| | | Crew | Daily Output | Labor-Hours | Unit | Material | 2011 Bare Costs Labor | Equipment | Total | Total Incl O&P |
|---|---|---|---|---|---|---|---|---|---|---|
| 0010 | **CONCRETE-FILLED STEEL PILES** no mobilization or demobilization | | | | | | | | | |
| 2600 | Pipe piles, 50' lg. 8" diam., 29 lb. per L.F., no concrete | B-19 | 500 | .128 | V.L.F. | 19.50 | 5.50 | 3.70 | 28.70 | 34 |
| 2700 | Concrete filled | | 460 | .139 | | 20.50 | 6 | 4.02 | 30.52 | 36.50 |
| 2900 | 10" diameter, 34 lb. per L.F., no concrete | | 500 | .128 | | 24.50 | 5.50 | 3.70 | 33.70 | 39.50 |
| 3000 | Concrete filled | | 450 | .142 | | 26.50 | 6.15 | 4.11 | 36.76 | 43 |
| 3200 | 12" diameter, 44 lb. per L.F., no concrete | | 475 | .135 | | 29.50 | 5.80 | 3.89 | 39.19 | 46 |
| 3300 | Concrete filled | | 415 | .154 | | 30.50 | 6.65 | 4.45 | 41.60 | 49 |
| 3500 | 14" diameter, 46 lb. per L.F., no concrete | | 430 | .149 | | 32 | 6.40 | 4.30 | 42.70 | 50 |
| 3600 | Concrete filled | | 355 | .180 | | 33.50 | 7.80 | 5.20 | 46.50 | 55 |
| 3800 | 16" diameter, 52 lb. per L.F., no concrete | | 385 | .166 | | 35.50 | 7.15 | 4.80 | 47.45 | 55.50 |
| 3900 | Concrete filled | | 335 | .191 | | 38 | 8.25 | 5.50 | 51.75 | 61 |
| 4100 | 18" diameter, 59 lb. per L.F., no concrete | | 355 | .180 | | 46.50 | 7.80 | 5.20 | 59.50 | 69 |
| 4200 | Concrete filled | | 310 | .206 | | 52.50 | 8.90 | 5.95 | 67.35 | 78 |
| 4400 | Splices for pipe piles, not in leads, 8" diameter | 1 Sswl | 4.67 | 1.713 | Ea. | 86.50 | 83 | | 169.50 | 243 |
| 4500 | 14" diameter | | 3.79 | 2.111 | | 114 | 102 | | 216 | 305 |
| 4600 | 16" diameter | | 3.03 | 2.640 | | 140 | 128 | | 268 | 380 |
| 4800 | Points, standard, 8" diameter | | 4.61 | 1.735 | | 142 | 84.50 | | 226.50 | 305 |
| 4900 | 14" diameter | | 4.05 | 1.975 | | 197 | 96 | | 293 | 385 |
| 5000 | 16" diameter | | 3.37 | 2.374 | | 240 | 115 | | 355 | 470 |
| 5200 | Points, heavy duty, 10" diameter | | 2.89 | 2.768 | | 281 | 134 | | 415 | 545 |
| 5300 | 14" or 16" diameter | | 2.02 | 3.960 | | 445 | 192 | | 637 | 830 |
| 5500 | For reinforcing steel, add | | 1150 | .007 | Lb. | .78 | .34 | | 1.12 | 1.46 |
| 5700 | For thick wall sections, add | | | | " | .89 | | | .89 | .98 |

## 31 63 26 – Drilled Caissons

| 31 63 26.13 Fixed End Caisson Piles | Crew | Daily Output | Labor-Hours | Unit | Material | 2011 Bare Costs Labor | Equipment | Total | Total Incl O&P |
|---|---|---|---|---|---|---|---|---|---|
| 0010 **FIXED END CAISSON PILES** R316326-60 | | | | | | | | | |
| 0015    Including excavation, concrete, 50 lbs reinforcing | | | | | | | | | |
| 0020    per C.Y., not incl. mobilization, boulder removal, disposal | | | | | | | | | |
| 0100    Open style, machine drilled, to 50' deep, in stable ground, no | | | | | | | | | |
| 0110      casings or ground water, 18" diam., 0.065 C.Y./L.F. | B-43 | 200 | .240 | V.L.F. | 7.90 | 9.05 | 12.45 | 29.40 | 36 |
| 0200      24" diameter, 0.116 C.Y./L.F. | | 190 | .253 | | 14.10 | 9.50 | 13.10 | 36.70 | 44.50 |
| 0300      30" diameter, 0.182 C.Y./L.F. | | 150 | .320 | | 22 | 12.05 | 16.55 | 50.60 | 61 |
| 0400      36" diameter, 0.262 C.Y./L.F. | | 125 | .384 | | 32 | 14.50 | 19.90 | 66.40 | 79 |
| 0500      48" diameter, 0.465 C.Y./L.F. | | 100 | .480 | | 56.50 | 18.10 | 25 | 99.60 | 117 |
| 0600      60" diameter, 0.727 C.Y./L.F. | | 90 | .533 | | 88.50 | 20 | 27.50 | 136 | 158 |
| 0700      72" diameter, 1.05 C.Y./L.F. | | 80 | .600 | | 128 | 22.50 | 31 | 181.50 | 209 |
| 0800      84" diameter, 1.43 C.Y./L.F. | | 75 | .640 | | 174 | 24 | 33 | 231 | 264 |
| 1000    For bell excavation and concrete, add | | | | | | | | | |
| 1020      4' bell diameter, 24" shaft, 0.444 C.Y. | B-43 | 20 | 2.400 | Ea. | 44 | 90.50 | 124 | 258.50 | 325 |
| 1040      6' bell diameter, 30" shaft, 1.57 C.Y. | | 5.70 | 8.421 | | 155 | 315 | 435 | 905 | 1,125 |
| 1060      8' bell diameter, 36" shaft, 3.72 C.Y. | | 2.40 | 20 | | 370 | 755 | 1,025 | 2,150 | 2,700 |
| 1080      9' bell diameter, 48" shaft, 4.48 C.Y. | | 2 | 24 | | 445 | 905 | 1,250 | 2,600 | 3,250 |
| 1100      10' bell diameter, 60" shaft, 5.24 C.Y. | | 1.70 | 28.235 | | 520 | 1,075 | 1,450 | 3,045 | 3,800 |
| 1120      12' bell diameter, 72" shaft, 8.74 C.Y. | | 1 | 48 | | 865 | 1,800 | 2,475 | 5,140 | 6,425 |
| 1140      14' bell diameter, 84" shaft, 13.6 C.Y. | | .70 | 68.571 | | 1,350 | 2,575 | 3,550 | 7,475 | 9,300 |
| 1200    Open style, machine drilled, to 50' deep, in wet ground, pulled | | | | | | | | | |
| 1300      casing and pumping, 18" diameter, 0.065 C.Y./L.F. | B-48 | 160 | .350 | V.L.F. | 7.90 | 13.50 | 17.50 | 38.90 | 48.50 |
| 1400      24" diameter, 0.116 C.Y./L.F. | | 125 | .448 | | 14.10 | 17.25 | 22.50 | 53.85 | 66 |
| 1500      30" diameter, 0.182 C.Y./L.F. | | 85 | .659 | | 22 | 25.50 | 33 | 80.50 | 99 |
| 1600      36" diameter, 0.262 C.Y./L.F. | | 60 | .933 | | 32 | 36 | 46.50 | 114.50 | 141 |
| 1700      48" diameter, 0.465 C.Y./L.F. | B-49 | 55 | 1.600 | | 56.50 | 64 | 64.50 | 185 | 231 |
| 1800      60" diameter, 0.727 C.Y./L.F. | | 35 | 2.514 | | 88.50 | 101 | 101 | 290.50 | 360 |
| 1900      72" diameter, 1.05 C.Y./L.F. | | 30 | 2.933 | | 128 | 117 | 118 | 363 | 450 |
| 2000      84" diameter, 1.43 C.Y./L.F. | | 25 | 3.520 | | 174 | 141 | 142 | 457 | 560 |
| 2100    For bell excavation and concrete, add | | | | | | | | | |
| 2120      4' bell diameter, 24" shaft, 0.444 C.Y. | B-48 | 19.80 | 2.828 | Ea. | 44 | 109 | 141 | 294 | 370 |
| 2140      6' bell diameter, 30" shaft, 1.57 C.Y. | | 5.70 | 9.825 | | 155 | 380 | 490 | 1,025 | 1,275 |
| 2160      8' bell diameter, 36" shaft, 3.72 C.Y. | | 2.40 | 23.333 | | 370 | 900 | 1,175 | 2,445 | 3,025 |
| 2180      9' bell diameter, 48" shaft, 4.48 C.Y. | B-49 | 3.30 | 26.667 | | 445 | 1,075 | 1,075 | 2,595 | 3,300 |
| 2200      10' bell diameter, 60" shaft, 5.24 C.Y. | | 2.80 | 31.429 | | 520 | 1,250 | 1,275 | 3,045 | 3,900 |
| 2220      12' bell diameter, 72" shaft, 8.74 C.Y. | | 1.60 | 55 | | 865 | 2,200 | 2,225 | 5,290 | 6,750 |
| 2240      14' bell diameter, 84" shaft, 13.6 C.Y. | | 1 | 88 | | 1,350 | 3,525 | 3,550 | 8,425 | 10,800 |
| 2300    Open style, machine drilled, to 50' deep, in soft rocks and | | | | | | | | | |
| 2400      medium hard shales, 18" diameter, 0.065 C.Y./L.F. | B-49 | 50 | 1.760 | V.L.F. | 7.90 | 70.50 | 71 | 149.40 | 195 |
| 2500      24" diameter, 0.116 C.Y./L.F. | | 30 | 2.933 | | 14.10 | 117 | 118 | 249.10 | 325 |
| 2600      30" diameter, 0.182 C.Y./L.F. | | 20 | 4.400 | | 22 | 176 | 177 | 375 | 490 |
| 2700      36" diameter, 0.262 C.Y./L.F. | | 15 | 5.867 | | 32 | 235 | 236 | 503 | 655 |
| 2800      48" diameter, 0.465 C.Y./L.F. | | 10 | 8.800 | | 56.50 | 350 | 355 | 761.50 | 990 |
| 2900      60" diameter, 0.727 C.Y./L.F. | | 7 | 12.571 | | 88.50 | 505 | 505 | 1,098.50 | 1,425 |
| 3000      72" diameter, 1.05 C.Y./L.F. | | 6 | 14.667 | | 128 | 585 | 590 | 1,303 | 1,675 |
| 3100      84" diameter, 1.43 C.Y./L.F. | | 5 | 17.600 | | 174 | 705 | 710 | 1,589 | 2,050 |
| 3200    For bell excavation and concrete, add | | | | | | | | | |
| 3220      4' bell diameter, 24" shaft, 0.444 C.Y. | B-49 | 10.90 | 8.073 | Ea. | 44 | 325 | 325 | 694 | 905 |
| 3240      6' bell diameter, 30" shaft, 1.57 C.Y. | | 3.10 | 28.387 | | 155 | 1,125 | 1,150 | 2,430 | 3,150 |
| 3260      8' bell diameter, 36" shaft, 3.72 C.Y. | | 1.30 | 67.692 | | 370 | 2,700 | 2,725 | 5,795 | 7,550 |
| 3280      9' bell diameter, 48" shaft, 4.48 C.Y. | | 1.10 | 80 | | 445 | 3,200 | 3,225 | 6,870 | 8,950 |
| 3300      10' bell diameter, 60" shaft, 5.24 C.Y. | | .90 | 97.778 | | 520 | 3,925 | 3,950 | 8,395 | 10,900 |
| 3320      12' bell diameter, 72" shaft, 8.74 C.Y. | | .60 | 146 | | 865 | 5,875 | 5,900 | 12,640 | 16,400 |

# 31 63 Bored Piles

## 31 63 26 – Drilled Caissons

### 31 63 26.13 Fixed End Caisson Piles

| | | Crew | Daily Output | Labor-Hours | Unit | Material | 2011 Bare Costs Labor | 2011 Bare Costs Equipment | Total | Total Incl O&P |
|---|---|---|---|---|---|---|---|---|---|---|
| 3340 | 14' bell diameter, 84" shaft, 13.6 C.Y. | B-49 | .40 | 220 | Ea. | 1,350 | 8,800 | 8,850 | 19,000 | 24,700 |
| 3600 | For rock excavation, sockets, add, minimum | | 120 | .733 | C.F. | | 29.50 | 29.50 | 59 | 77.50 |
| 3650 | Average | | 95 | .926 | | | 37 | 37.50 | 74.50 | 97.50 |
| 3700 | Maximum | | 48 | 1.833 | | | 73.50 | 74 | 147.50 | 193 |
| 3900 | For 50' to 100' deep, add | | | | V.L.F. | | | | 7% | 7% |
| 4000 | For 100' to 150' deep, add | | | | | | | | 25% | 25% |
| 4100 | For 150' to 200' deep, add | | | | | | | | 30% | 30% |
| 4200 | For casings left in place, add | | | | Lb. | .99 | | | .99 | 1.09 |
| 4300 | For other than 50 lb. reinf. per C.Y., add or deduct | | | | " | 1.05 | | | 1.05 | 1.16 |
| 4400 | For steel "I" beam cores, add | B-49 | 8.30 | 10.602 | Ton | 1,800 | 425 | 425 | 2,650 | 3,100 |
| 4500 | Load and haul excess excavation, 2 miles | B-34B | 178 | .045 | L.C.Y. | | 1.55 | 2.84 | 4.39 | 5.50 |
| 4600 | For mobilization, 50 mile radius, rig to 36" | B-43 | 2 | 24 | Ea. | | 905 | 1,250 | 2,155 | 2,750 |
| 4650 | Rig to 84" | B-48 | 1.75 | 32 | | | 1,225 | 1,600 | 2,825 | 3,625 |
| 4700 | For low headroom, add | | | | | | | | 50% | |
| 5000 | Bottom inspection | 1 Skwk | 1.20 | 6.667 | | | 296 | | 296 | 455 |

### 31 63 26.16 Concrete Caissons for Marine Construction

| | | Crew | Daily Output | Labor-Hours | Unit | Material | 2011 Bare Costs Labor | 2011 Bare Costs Equipment | Total | Total Incl O&P |
|---|---|---|---|---|---|---|---|---|---|---|
| 0010 | **CONCRETE CAISSONS FOR MARINE CONSTRUCTION** | | | | | | | | | |
| 0100 | Caissons, incl. mobilization and demobilization, up to 50 miles | | | | | | | | | |
| 0200 | Uncased shafts, 30 to 80 tons cap., 17" diam., 10' depth | B-44 | 88 | .727 | V.L.F. | 19.80 | 31 | 20.50 | 71.30 | 93 |
| 0300 | 25' depth | | 165 | .388 | | 14.15 | 16.45 | 11 | 41.60 | 53.50 |
| 0400 | 80-150 ton capacity, 22" diameter, 10' depth | | 80 | .800 | | 25 | 34 | 22.50 | 81.50 | 105 |
| 0500 | 20' depth | | 130 | .492 | | 19.80 | 21 | 13.95 | 54.75 | 70 |
| 0700 | Cased shafts, 10 to 30 ton capacity, 10-5/8" diam., 20' depth | | 175 | .366 | | 14.15 | 15.50 | 10.40 | 40.05 | 51.50 |
| 0800 | 30' depth | | 240 | .267 | | 13.20 | 11.30 | 7.55 | 32.05 | 40.50 |
| 0850 | 30 to 60 ton capacity, 12" diameter, 20' depth | | 160 | .400 | | 19.80 | 16.95 | 11.35 | 48.10 | 61 |
| 0900 | 40' depth | | 230 | .278 | | 15.25 | 11.80 | 7.90 | 34.95 | 44 |
| 1000 | 80 to 100 ton capacity, 16" diameter, 20' depth | | 160 | .400 | | 28.50 | 16.95 | 11.35 | 56.80 | 70 |
| 1100 | 40' depth | | 230 | .278 | | 26.50 | 11.80 | 7.90 | 46.20 | 56 |
| 1200 | 110 to 140 ton capacity, 17-5/8" diameter, 20' depth | | 160 | .400 | | 30.50 | 16.95 | 11.35 | 58.80 | 72.50 |
| 1300 | 40' depth | | 230 | .278 | | 28.50 | 11.80 | 7.90 | 48.20 | 58 |
| 1400 | 140 to 175 ton capacity, 19" diameter, 20' depth | | 130 | .492 | | 33 | 21 | 13.95 | 67.95 | 84.50 |
| 1500 | 40' depth | | 210 | .305 | | 30.50 | 12.95 | 8.65 | 52.10 | 63.50 |
| 1700 | Over 30' long, L.F. cost tends to be lower | | | | | | | | | |
| 1900 | Maximum depth is about 90' | | | | | | | | | |

## 31 63 29 – Drilled Concrete Piers and Shafts

### 31 63 29.13 Uncased Drilled Concrete Piers

| | | Crew | Daily Output | Labor-Hours | Unit | Material | 2011 Bare Costs Labor | 2011 Bare Costs Equipment | Total | Total Incl O&P |
|---|---|---|---|---|---|---|---|---|---|---|
| 0010 | **UNCASED DRILLED CONCRETE PIERS** | | | | | | | | | |
| 0020 | Unless specified otherwise, not incl. pile caps or mobilization | | | | | | | | | |
| 0050 | Cast in place augered piles, no casing or reinforcing | | | | | | | | | |
| 0060 | 8" diameter | B-43 | 540 | .089 | V.L.F. | 3.95 | 3.35 | 4.60 | 11.90 | 14.50 |
| 0065 | 10" diameter | | 480 | .100 | | 6.30 | 3.77 | 5.20 | 15.27 | 18.35 |
| 0070 | 12" diameter | | 420 | .114 | | 8.85 | 4.31 | 5.90 | 19.06 | 23 |
| 0075 | 14" diameter | | 360 | .133 | | 11.95 | 5.05 | 6.90 | 23.90 | 28.50 |
| 0080 | 16" diameter | | 300 | .160 | | 16.10 | 6.05 | 8.30 | 30.45 | 36 |
| 0085 | 18" diameter | | 240 | .200 | | 19.90 | 7.55 | 10.35 | 37.80 | 45 |
| 0100 | Cast in place, thin wall shell pile, straight sided, | | | | | | | | | |
| 0110 | not incl. reinforcing, 8" diam., 16 ga., 5.8 lb./L.F. | B-19 | 700 | .091 | V.L.F. | 9.15 | 3.95 | 2.64 | 15.74 | 19.10 |
| 0200 | 10" diameter, 16 ga. corrugated, 7.3 lb./L.F. | | 650 | .098 | | 12 | 4.25 | 2.84 | 19.09 | 23 |
| 0300 | 12" diameter, 16 ga. corrugated, 8.7 lb./L.F. | | 600 | .107 | | 15.55 | 4.60 | 3.08 | 23.23 | 27.50 |
| 0400 | 14" diameter, 16 ga. corrugated, 10.0 lb./L.F. | | 550 | .116 | | 18.30 | 5 | 3.36 | 26.66 | 31.50 |
| 0500 | 16" diameter, 16 ga. corrugated, 11.6 lb./L.F. | | 500 | .128 | | 22.50 | 5.50 | 3.70 | 31.70 | 37 |
| 0800 | Cast in place friction pile, 50' long, fluted, | | | | | | | | | |

# 31 63 Bored Piles

## 31 63 29 – Drilled Concrete Piers and Shafts

| 31 63 29.13 Uncased Drilled Concrete Piers | | Crew | Daily Output | Labor-Hours | Unit | Material | 2011 Bare Costs Labor | Equipment | Total | Total Incl O&P |
|---|---|---|---|---|---|---|---|---|---|---|
| 0810 | tapered steel, 4000 psi concrete, no reinforcing | | | | | | | | | |
| 0900 | 12" diameter, 7 ga. | B-19 | 600 | .107 | V.L.F. | 27.50 | 4.60 | 3.08 | 35.18 | 40.50 |
| 1000 | 14" diameter, 7 ga. | | 560 | .114 | | 30 | 4.93 | 3.30 | 38.23 | 44 |
| 1100 | 16" diameter, 7 ga. | | 520 | .123 | | 35 | 5.30 | 3.55 | 43.85 | 50.50 |
| 1200 | 18" diameter, 7 ga. | | 480 | .133 | | 41 | 5.75 | 3.85 | 50.60 | 58 |
| 1300 | End bearing, fluted, constant diameter, | | | | | | | | | |
| 1320 | 4000 psi concrete, no reinforcing | | | | | | | | | |
| 1340 | 12" diameter, 7 ga. | B-19 | 600 | .107 | V.L.F. | 28.50 | 4.60 | 3.08 | 36.18 | 42 |
| 1360 | 14" diameter, 7 ga. | | 560 | .114 | | 35.50 | 4.93 | 3.30 | 43.73 | 50.50 |
| 1380 | 16" diameter, 7 ga. | | 520 | .123 | | 41.50 | 5.30 | 3.55 | 50.35 | 57.50 |
| 1400 | 18" diameter, 7 ga. | | 480 | .133 | | 45.50 | 5.75 | 3.85 | 55.10 | 63 |

## 31 63 29.20 Cast In Place Piles, Adds

| | | Crew | Daily Output | Labor-Hours | Unit | Material | 2011 Bare Costs Labor | Equipment | Total | Total Incl O&P |
|---|---|---|---|---|---|---|---|---|---|---|
| 0010 | **CAST IN PLACE PILES, ADDS** | | | | | | | | | |
| 1500 | For reinforcing steel, add | | | | Lb. | .90 | | | .90 | .99 |
| 1700 | For ball or pedestal end, add | B-19 | 11 | 5.818 | C.Y. | 147 | 251 | 168 | 566 | 735 |
| 1900 | For lengths above 60', concrete, add | " | 11 | 5.818 | " | 152 | 251 | 168 | 571 | 745 |
| 2000 | For steel thin shell, pipe only | | | | Lb. | 1.11 | | | 1.11 | 1.22 |

# Estimating Tips

## 32 01 00 Operations and Maintenance of Exterior Improvements

- Recycling of asphalt pavement is becoming very popular and is an alternative to removal and replacement. It can be a good value engineering proposal if removed pavement can be recycled, either at the project site or at another site that is reasonably close to the project site. Two new sections on repair of flexible and rigid pavement have been added.

## 32 10 00 Bases, Ballasts, and Paving

- When estimating paving, keep in mind the project schedule. If an asphaltic paving project is in a colder climate and runs through to the spring, consider placing the base course in the autumn and then topping it in the spring, just prior to completion. This could save considerable costs in spring repair. Keep in mind that prices for asphalt and concrete are generally higher in the cold seasons. New lines have been added for paving intersections and bridge approaches.

## 32 90 00 Planting

- The timing of planting and guarantee specifications often dictate the costs for establishing tree and shrub growth and a stand of grass or ground cover. Establish the work performance schedule to coincide with the local planting season. Maintenance and growth guarantees can add from 20%–100% to the total landscaping cost. The cost to replace trees and shrubs can be as high as 5% of the total cost, depending on the planting zone, soil conditions, and time of year.

# Reference Numbers

Reference numbers are shown in shaded boxes at the beginning of some major classifications. These numbers refer to related items in the Reference Section. The reference information may be an estimating procedure, an alternate pricing method, or technical information.

*Note:* Not all subdivisions listed here necessarily appear in this publication.

**Division 32 - Exterior Improvements**

# 32 01 Operation and Maintenance of Exterior Improvements

## 32 01 13 – Flexible Paving Surface Treatment

### 32 01 13.61 Slurry Seal (Latex Modified)

| | | Crew | Daily Output | Labor-Hours | Unit | Material | 2011 Bare Costs Labor | 2011 Bare Costs Equipment | Total | Total Incl O&P |
|---|---|---|---|---|---|---|---|---|---|---|
| 0010 | **SLURRY SEAL (LATEX MODIFIED)** | | | | | | | | | |
| 3780 | Rubberized asphalt (latex) seal | B-45 | 5000 | .003 | S.Y. | 2.35 | .13 | .15 | 2.63 | 2.95 |

### 32 01 13.64 Sand Seal

| | | Crew | Daily Output | Labor-Hours | Unit | Material | Labor | Equipment | Total | Total Incl O&P |
|---|---|---|---|---|---|---|---|---|---|---|
| 0010 | **SAND SEAL** | | | | | | | | | |
| 2080 | Sand sealing, sharp sand, asphalt emulsion, small area | B-91 | 10000 | .006 | S.Y. | 1.35 | .26 | .21 | 1.82 | 2.11 |
| 2120 | Roadway or large area | " | 18000 | .004 | " | 1.16 | .14 | .11 | 1.41 | 1.63 |

### 32 01 13.66 Fog Seal

| | | Crew | Daily Output | Labor-Hours | Unit | Material | Labor | Equipment | Total | Total Incl O&P |
|---|---|---|---|---|---|---|---|---|---|---|
| 0010 | **FOG SEAL** | | | | | | | | | |
| 0012 | Sealcoating, 2 coat coal tar pitch emulsion over 10,000 S.Y. | B-45 | 5000 | .003 | S.Y. | 1.20 | .13 | .15 | 1.48 | 1.68 |
| 0030 | 1000 to 10,000 S.Y. | " | 3000 | .005 | | 1.20 | .21 | .26 | 1.67 | 1.92 |
| 0100 | Under 1000 S.Y. | B-1 | 1050 | .023 | | 1.20 | .80 | | 2 | 2.55 |
| 0300 | Petroleum resistant, over 10,000 S.Y. | B-45 | 5000 | .003 | | 1.40 | .13 | .15 | 1.68 | 1.90 |
| 0320 | 1000 to 10,000 S.Y. | " | 3000 | .005 | | 1.40 | .21 | .26 | 1.87 | 2.14 |
| 0400 | Under 1000 S.Y. | B-1 | 1050 | .023 | | 1.40 | .80 | | 2.20 | 2.77 |
| 0600 | Non-skid pavement renewal, over 10,000 S.Y. | B-45 | 5000 | .003 | | 1.55 | .13 | .15 | 1.83 | 2.07 |
| 0620 | 1000 to 10,000 S.Y. | " | 3000 | .005 | | 1.55 | .21 | .26 | 2.02 | 2.31 |
| 0700 | Under 1000 S.Y. | B-1 | 1050 | .023 | | 1.55 | .80 | | 2.35 | 2.94 |
| 0800 | Prepare and clean surface for above | A-2 | 8545 | .003 | ↓ | | .10 | .02 | .12 | .17 |
| 1000 | Hand seal asphalt curbing | B-1 | 4420 | .005 | L.F. | .85 | .19 | | 1.04 | 1.23 |
| 1900 | Asphalt surface treatment, single course, small area | | | | | | | | | |
| 1901 | 0.30 gal/S.Y. asphalt material, 20#/S.Y. aggregate | B-91 | 5000 | .013 | S.Y. | 1.78 | .51 | .41 | 2.70 | 3.19 |
| 1910 | Roadway or large area | | 10000 | .006 | | 1.64 | .26 | .21 | 2.11 | 2.42 |
| 1950 | Asphalt surface treatment, dbl. course for small area | | 3000 | .021 | | 3.56 | .86 | .68 | 5.10 | 5.95 |
| 1960 | Roadway or large area | | 6000 | .011 | | 3.20 | .43 | .34 | 3.97 | 4.55 |
| 1980 | Asphalt surface treatment, single course, for shoulders | ↓ | 7500 | .009 | ↓ | 1.88 | .34 | .27 | 2.49 | 2.89 |

# 32 06 Schedules for Exterior Improvements

## 32 06 10 – Schedules for Bases, Ballasts, and Paving

### 32 06 10.10 Sidewalks, Driveways and Patios

| | | Crew | Daily Output | Labor-Hours | Unit | Material | Labor | Equipment | Total | Total Incl O&P |
|---|---|---|---|---|---|---|---|---|---|---|
| 0010 | **SIDEWALKS, DRIVEWAYS AND PATIOS** No base | | | | | | | | | |
| 0020 | Asphaltic concrete, 2" thick | B-37 | 720 | .067 | S.Y. | 6.80 | 2.42 | .20 | 9.42 | 11.35 |
| 0100 | 2-1/2" thick | " | 660 | .073 | " | 8.60 | 2.64 | .22 | 11.46 | 13.70 |
| 0300 | Concrete, 3000 psi, CIP, 6 x 6 - W1.4 x W1.4 mesh, | | | | | | | | | |
| 0310 | broomed finish, no base, 4" thick | B-24 | 600 | .040 | S.F. | 1.65 | 1.58 | | 3.23 | 4.20 |
| 0350 | 5" thick | | 545 | .044 | | 2.20 | 1.74 | | 3.94 | 5.05 |
| 0400 | 6" thick | ↓ | 510 | .047 | | 2.57 | 1.86 | | 4.43 | 5.60 |
| 0450 | For bank run gravel base, 4" thick, add | B-18 | 2500 | .010 | | .51 | .34 | .02 | .87 | 1.10 |
| 0520 | 8" thick, add | " | 1600 | .015 | | 1.03 | .53 | .02 | 1.58 | 1.98 |
| 0550 | Exposed aggregate finish, add to above, minimum | B-24 | 1875 | .013 | | .13 | .50 | | .63 | .90 |
| 0600 | Maximum | " | 455 | .053 | | .43 | 2.08 | | 2.51 | 3.60 |
| 1000 | Crushed stone, 1" thick, white marble | 2 Clab | 1700 | .009 | | .25 | .32 | | .57 | .78 |
| 1050 | Bluestone | " | 1700 | .009 | | .18 | .32 | | .50 | .70 |
| 1700 | Redwood, prefabricated, 4' x 4' sections | 2 Carp | 316 | .051 | | 4.74 | 2.18 | | 6.92 | 8.55 |
| 1750 | Redwood planks, 1" thick, on sleepers | " | 240 | .067 | ↓ | 4.74 | 2.87 | | 7.61 | 9.60 |
| 2250 | Stone dust, 4" thick | B-62 | 900 | .027 | S.Y. | 3.33 | 1 | .17 | 4.50 | 5.40 |

### 32 06 10.20 Steps

| | | Crew | Daily Output | Labor-Hours | Unit | Material | Labor | Equipment | Total | Total Incl O&P |
|---|---|---|---|---|---|---|---|---|---|---|
| 0010 | **STEPS** | | | | | | | | | |
| 0011 | Incl. excav., borrow & concrete base as required | | | | | | | | | |
| 0100 | Brick steps | B-24 | 35 | .686 | LF Riser | 12.85 | 27 | | 39.85 | 54.50 |
| 0200 | Railroad ties | 2 Clab | 25 | .640 | ↓ | 3.50 | 22 | | 25.50 | 37.50 |

# 32 06 Schedules for Exterior Improvements

## 32 06 10 – Schedules for Bases, Ballasts, and Paving

| 32 06 10.20 Steps | | Crew | Daily Output | Labor-Hours | Unit | Material | 2011 Bare Costs Labor | Equipment | Total | Total Incl O&P |
|---|---|---|---|---|---|---|---|---|---|---|
| 0300 | Bluestone treads, 12" x 2" or 12" x 1-1/2" | B-24 | 30 | .800 | LF Riser | 28 | 31.50 | | 59.50 | 78.50 |
| 0500 | Concrete, cast in place, see Section 03 30 53.40 | | | | | | | | | |
| 0600 | Precast concrete, see Section 03 41 23.50 | | | | | | | | | |
| 4025 | Steel edge strips, incl. stakes, 1/4" x 5" | B-1 | 390 | .062 | L.F. | 3.35 | 2.16 | | 5.51 | 7 |
| 4050 | Edging, landscape timber or railroad ties, 6" x 8" | 2 Carp | 170 | .094 | " | 3.50 | 4.05 | | 7.55 | 10.05 |

# 32 11 Base Courses

## 32 11 23 – Aggregate Base Courses

### 32 11 23.23 Base Course Drainage Layers

| 32 11 23.23 | | Crew | Daily Output | Labor-Hours | Unit | Material | 2011 Bare Costs Labor | Equipment | Total | Total Incl O&P |
|---|---|---|---|---|---|---|---|---|---|---|
| 0010 | **BASE COURSE DRAINAGE LAYERS** | | | | | | | | | |
| 0011 | For roadways and large areas | | | | | | | | | |
| 0050 | Crushed 3/4" stone base, compacted, 3" deep | B-36C | 5200 | .008 | S.Y. | 3.71 | .32 | .66 | 4.69 | 5.30 |
| 0100 | 6" deep | | 5000 | .008 | | 7.40 | .33 | .69 | 8.42 | 9.40 |
| 0200 | 9" deep | | 4600 | .009 | | 11.15 | .36 | .75 | 12.26 | 13.60 |
| 0300 | 12" deep | | 4200 | .010 | | 14.85 | .39 | .82 | 16.06 | 17.85 |
| 0301 | Crushed 1-1/2" stone base, compacted to 4" deep | B-36B | 6000 | .011 | | 5.10 | .43 | .67 | 6.20 | 7 |
| 0302 | 6" deep | | 5400 | .012 | | 7.65 | .48 | .74 | 8.87 | 9.95 |
| 0303 | 8" deep | | 4500 | .014 | | 10.20 | .57 | .89 | 11.66 | 13.10 |
| 0304 | 12" deep | | 3800 | .017 | | 15.30 | .68 | 1.06 | 17.04 | 19.05 |
| 0350 | Bank run gravel, spread and compacted | | | | | | | | | |
| 0370 | 6" deep | B-32 | 6000 | .005 | S.Y. | 4.29 | .23 | .30 | 4.82 | 5.40 |
| 0390 | 9" deep | | 4900 | .007 | | 6.45 | .28 | .37 | 7.10 | 7.90 |
| 0400 | 12" deep | | 4200 | .008 | | 8.55 | .32 | .43 | 9.30 | 10.40 |
| 6000 | Stabilization fabric, polypropylene, 6 oz./S.Y. | B-6 | 10000 | .002 | | 1.48 | .09 | .03 | 1.60 | 1.81 |
| 6900 | For small and irregular areas, add | | | | | | 50% | 50% | | |
| 7000 | Prepare and roll sub-base, small areas to 2500 S.Y. | B-32A | 1500 | .016 | S.Y. | | .67 | .76 | 1.43 | 1.84 |
| 8000 | Large areas over 2500 S.Y. | " | 3500 | .007 | | | .29 | .33 | .62 | .79 |
| 8050 | For roadways | B-32 | 4000 | .008 | | | .34 | .45 | .79 | 1.01 |

## 32 11 26 – Asphaltic Base Courses

### 32 11 26.19 Bituminous-Stabilized Base Courses

| 32 11 26.19 | | Crew | Daily Output | Labor-Hours | Unit | Material | 2011 Bare Costs Labor | Equipment | Total | Total Incl O&P |
|---|---|---|---|---|---|---|---|---|---|---|
| 0010 | **BITUMINOUS-STABILIZED BASE COURSES** | | | | | | | | | |
| 0020 | And large paved areas | | | | | | | | | |
| 0700 | Liquid application to gravel base, asphalt emulsion | B-45 | 6000 | .003 | Gal. | 5.90 | .11 | .13 | 6.14 | 6.80 |
| 0800 | Prime and seal, cut back asphalt | | 6000 | .003 | " | 6.95 | .11 | .13 | 7.19 | 7.95 |
| 1000 | Macadam penetration crushed stone, 2 gal. per S.Y., 4" thick | | 6000 | .003 | S.Y. | 11.80 | .11 | .13 | 12.04 | 13.30 |
| 1100 | 6" thick, 3 gal. per S.Y. | | 4000 | .004 | | 17.70 | .16 | .19 | 18.05 | 19.90 |
| 1200 | 8" thick, 4 gal. per S.Y. | | 3000 | .005 | | 23.50 | .21 | .26 | 23.97 | 26.50 |
| 8900 | For small and irregular areas, add | | | | | | 50% | 50% | | |

# 32 12 Flexible Paving

## 32 12 16 – Asphalt Paving

### 32 12 16.13 Plant-Mix Asphalt Paving

| | | Crew | Daily Output | Labor-Hours | Unit | Material | 2011 Bare Costs Labor | Equipment | Total | Total Incl O&P |
|---|---|---|---|---|---|---|---|---|---|---|
| 0010 | **PLANT-MIX ASPHALT PAVING** | | | | | | | | | |
| 0020 | And large paved areas with no hauling included | | | | | | | | | |
| 0025 | See Section 31 23 23.20 for hauling costs | | | | | | | | | |
| 0080 | Binder course, 1-1/2" thick | B-25 | 7725 | .011 | S.Y. | 5.25 | .43 | .32 | 6 | 6.75 |
| 0120 | 2" thick | | 6345 | .014 | | 6.95 | .52 | .39 | 7.86 | 8.85 |
| 0130 | 2-1/2" thick | | 5620 | .016 | | 8.70 | .59 | .44 | 9.73 | 10.95 |
| 0160 | 3" thick | | 4905 | .018 | | 10.45 | .67 | .51 | 11.63 | 13.10 |
| 0170 | 3-1/2" thick | | 4520 | .019 | | 12.20 | .73 | .55 | 13.48 | 15.10 |
| 0200 | 4" thick | | 4140 | .021 | | 13.90 | .80 | .60 | 15.30 | 17.15 |
| 0300 | Wearing course, 1" thick | B-25B | 10575 | .009 | | 3.45 | .35 | .26 | 4.06 | 4.61 |
| 0340 | 1-1/2" thick | | 7725 | .012 | | 5.80 | .47 | .35 | 6.62 | 7.50 |
| 0380 | 2" thick | | 6345 | .015 | | 7.80 | .58 | .43 | 8.81 | 9.90 |
| 0420 | 2-1/2" thick | | 5480 | .018 | | 9.60 | .67 | .49 | 10.76 | 12.15 |
| 0460 | 3" thick | | 4900 | .020 | | 11.45 | .75 | .55 | 12.75 | 14.35 |
| 0470 | 3-1/2" thick | | 4520 | .021 | | 13.45 | .81 | .60 | 14.86 | 16.70 |
| 0480 | 4" thick | | 4140 | .023 | | 15.35 | .89 | .65 | 16.89 | 18.95 |
| 0500 | Open graded friction course | B-25C | 5000 | .010 | | 3.76 | .37 | .43 | 4.56 | 5.20 |
| 0800 | Alternate method of figuring paving costs | | | | | | | | | |
| 0810 | Binder course, 1-1/2" thick | B-25 | 630 | .140 | Ton | 64 | 5.25 | 3.94 | 73.19 | 83 |
| 0811 | 2" thick | | 690 | .128 | | 64 | 4.79 | 3.60 | 72.39 | 82 |
| 0812 | 3" thick | | 800 | .110 | | 64 | 4.13 | 3.10 | 71.23 | 80 |
| 0813 | 4" thick | | 900 | .098 | | 64 | 3.67 | 2.76 | 70.43 | 79 |
| 0850 | Wearing course, 1" thick | B-25B | 575 | .167 | | 65 | 6.35 | 4.71 | 76.06 | 86.50 |
| 0851 | 1-1/2" thick | | 630 | .152 | | 65 | 5.80 | 4.30 | 75.10 | 85 |
| 0852 | 2" thick | | 690 | .139 | | 65 | 5.30 | 3.92 | 74.22 | 84 |
| 0853 | 2-1/2" thick | | 765 | .125 | | 65 | 4.79 | 3.54 | 73.33 | 82.50 |
| 0854 | 3" thick | | 800 | .120 | | 65 | 4.58 | 3.38 | 72.96 | 82 |
| 1000 | Pavement replacement over trench, 2" thick | B-37 | 90 | .533 | S.Y. | 7.20 | 19.30 | 1.62 | 28.12 | 39 |
| 1050 | 4" thick | | 70 | .686 | | 14.20 | 25 | 2.08 | 41.28 | 56 |
| 1080 | 6" thick | | 55 | .873 | | 22.50 | 31.50 | 2.64 | 56.64 | 76 |

### 32 12 16.14 Asphaltic Concrete Paving

| | | Crew | Daily Output | Labor-Hours | Unit | Material | 2011 Bare Costs Labor | Equipment | Total | Total Incl O&P |
|---|---|---|---|---|---|---|---|---|---|---|
| 0011 | **ASPHALTIC CONCRETE PAVING**, parking lots & driveways | | | | | | | | | |
| 0015 | No asphalt hauling included | | | | | | | | | |
| 0018 | Use 6.05 C. Y. per inch per M.S.F. for hauling | | | | | | | | | |
| 0020 | 6" stone base, 2" binder course, 1" topping | B-25C | 9000 | .005 | S.F. | 1.85 | .20 | .24 | 2.29 | 2.62 |
| 0025 | 2" binder course, 2" topping | | 9000 | .005 | | 2.25 | .20 | .24 | 2.69 | 3.05 |
| 0030 | 3" binder course, 2" topping | | 9000 | .005 | | 2.64 | .20 | .24 | 3.08 | 3.48 |
| 0035 | 4" binder course, 2" topping | | 9000 | .005 | | 3.02 | .20 | .24 | 3.46 | 3.90 |
| 0040 | 1.5" binder course, 1" topping | | 9000 | .005 | | 1.66 | .20 | .24 | 2.10 | 2.41 |
| 0042 | 3" binder course, 1" topping | | 9000 | .005 | | 2.24 | .20 | .24 | 2.68 | 3.05 |
| 0045 | 3" binder course, 3" topping | | 9000 | .005 | | 3.03 | .20 | .24 | 3.47 | 3.92 |
| 0050 | 4" binder course, 3" topping | | 9000 | .005 | | 3.42 | .20 | .24 | 3.86 | 4.34 |
| 0055 | 4" binder course, 4" topping | | 9000 | .005 | | 3.81 | .20 | .24 | 4.25 | 4.77 |
| 0300 | Binder course, 1-1/2" thick | | 35000 | .001 | | .58 | .05 | .06 | .69 | .79 |
| 0400 | 2" thick | | 25000 | .002 | | .75 | .07 | .09 | .91 | 1.04 |
| 0500 | 3" thick | | 15000 | .003 | | 1.16 | .12 | .14 | 1.42 | 1.63 |
| 0600 | 4" thick | | 10800 | .004 | | 1.52 | .17 | .20 | 1.89 | 2.16 |
| 0800 | Sand finish course, 3/4" thick | | 41000 | .001 | | .31 | .04 | .05 | .40 | .47 |
| 0900 | 1" thick | | 34000 | .001 | | .38 | .05 | .06 | .49 | .57 |
| 1000 | Fill pot holes, hot mix, 2" thick | B-16 | 4200 | .008 | | .80 | .27 | .12 | 1.19 | 1.42 |
| 1100 | 4" thick | | 3500 | .009 | | 1.17 | .32 | .14 | 1.63 | 1.94 |
| 1120 | 6" thick | | 3100 | .010 | | 1.58 | .36 | .16 | 2.10 | 2.46 |

# 32 12 Flexible Paving

## 32 12 16 – Asphalt Paving

| 32 12 16.14 Asphaltic Concrete Paving | | Crew | Daily Output | Labor-Hours | Unit | Material | 2011 Bare Costs Labor | Equipment | Total | Total Incl O&P |
|---|---|---|---|---|---|---|---|---|---|---|
| 1140 | Cold patch, 2" thick | B-51 | 3000 | .016 | S.F. | .81 | .55 | .06 | 1.42 | 1.80 |
| 1160 | 4" thick | | 2700 | .018 | | 1.55 | .61 | .06 | 2.22 | 2.71 |
| 1180 | 6" thick | | 1900 | .025 | | 2.41 | .87 | .09 | 3.37 | 4.09 |

# 32 13 Rigid Paving

## 32 13 13 – Concrete Paving

### 32 13 13.23 Concrete Paving Surface Treatment

| | | Crew | Daily Output | Labor-Hours | Unit | Material | 2011 Bare Costs Labor | Equipment | Total | Total Incl O&P |
|---|---|---|---|---|---|---|---|---|---|---|
| 0010 | **CONCRETE PAVING SURFACE TREATMENT** | | | | | | | | | |
| 0015 | Including joints, finishing and curing | | | | | | | | | |
| 0020 | Fixed form, 12' pass, unreinforced, 6" thick | B-26 | 3000 | .029 | S.Y. | 21 | 1.13 | 1.01 | 23.14 | 26 |
| 0100 | 8" thick | | 2750 | .032 | | 31.50 | 1.23 | 1.10 | 33.83 | 38 |
| 0110 | 8" thick, small area | | 1375 | .064 | | 31.50 | 2.46 | 2.20 | 36.16 | 41 |
| 0200 | 9" thick | | 2500 | .035 | | 35.50 | 1.35 | 1.21 | 38.06 | 42.50 |
| 0300 | 10" thick | | 2100 | .042 | | 39 | 1.61 | 1.44 | 42.05 | 46.50 |
| 0310 | 10" thick, small area | | 1050 | .084 | | 39 | 3.22 | 2.88 | 45.10 | 50.50 |
| 0400 | 12" thick | | 1800 | .049 | | 44 | 1.88 | 1.68 | 47.56 | 53 |
| 0410 | Conc pavement, w/jt,fnsh&curing,fix form,24' pass,unreinforced,6"T | | 6000 | .015 | | 19.35 | .56 | .50 | 20.41 | 23 |
| 0430 | 8" thick | | 5500 | .016 | | 28.50 | .61 | .55 | 29.66 | 32.50 |
| 0440 | 9" thick | | 5000 | .018 | | 32.50 | .68 | .61 | 33.79 | 37 |
| 0450 | 10" thick | | 4200 | .021 | | 35.50 | .80 | .72 | 37.02 | 41 |
| 0460 | 12" thick | | 3600 | .024 | | 41 | .94 | .84 | 42.78 | 47.50 |
| 0470 | 15" thick | | 3000 | .029 | | 53.50 | 1.13 | 1.01 | 55.64 | 62 |
| 0500 | Fixed form 12' pass 15" thick | | 1500 | .059 | | 57 | 2.25 | 2.02 | 61.27 | 68 |
| 0510 | For small irregular areas, add | | | | % | 10% | 100% | 100% | | |
| 0520 | Welded wire fabric, sheets for rigid paving 2.33 lb./S.Y. | 2 Rodm | 389 | .041 | S.Y. | 1.13 | 1.99 | | 3.12 | 4.42 |
| 0530 | Reinforcing steel for rigid paving 12 lb./S.Y. | | 666 | .024 | | 5.70 | 1.16 | | 6.86 | 8.10 |
| 0540 | Reinforcing steel for rigid paving 18 lb./S.Y. | | 444 | .036 | | 8.50 | 1.74 | | 10.24 | 12.15 |
| 0620 | Slip form, 12' pass, unreinforced, 6" thick | B-26A | 5600 | .016 | | 20.50 | .60 | .56 | 21.66 | 24 |
| 0624 | 8" thick | | 5300 | .017 | | 31 | .64 | .60 | 32.24 | 35.50 |
| 0626 | 9" thick | | 4820 | .018 | | 34.50 | .70 | .66 | 35.86 | 40 |
| 0628 | 10" thick | | 4050 | .022 | | 38 | .83 | .78 | 39.61 | 43.50 |
| 0630 | 12" thick | | 3470 | .025 | | 43 | .97 | .91 | 44.88 | 50 |
| 0632 | 15" thick | | 2890 | .030 | | 55.50 | 1.17 | 1.09 | 57.76 | 64 |
| 0640 | Slip form, 24' pass, unreinforced, 6" thick | | 11200 | .008 | | 19.65 | .30 | .28 | 20.23 | 22.50 |
| 0644 | 8" thick | | 10600 | .008 | | 28 | .32 | .30 | 28.62 | 31.50 |
| 0646 | 9" thick | | 9640 | .009 | | 32 | .35 | .33 | 32.68 | 36 |
| 0648 | 10" thick | | 8100 | .011 | | 35 | .42 | .39 | 35.81 | 39.50 |
| 0650 | 12" thick | | 6940 | .013 | | 40.50 | .49 | .46 | 41.45 | 45.50 |
| 0652 | 15" thick | | 5780 | .015 | | 51 | .58 | .55 | 52.13 | 58 |
| 0700 | Finishing, broom finish small areas | 2 Cefi | 120 | .133 | | | 5.45 | | 5.45 | 7.95 |
| 1000 | Curing, with sprayed membrane by hand | 2 Clab | 1500 | .011 | | .50 | .37 | | .87 | 1.11 |
| 1650 | For integral coloring, see Section 03 05 13.20 | | | | | | | | | |

# 32 14 Unit Paving

## 32 14 13 – Precast Concrete Unit Paving

### 32 14 13.13 Interlocking Precast Concrete Unit Paving

| | Crew | Daily Output | Labor-Hours | Unit | Material | 2011 Bare Costs Labor | Equipment | Total | Total Incl O&P |
|---|---|---|---|---|---|---|---|---|---|
| 0010 **INTERLOCKING PRECAST CONCRETE UNIT PAVING** | | | | | | | | | |
| 0020 "V" blocks for retaining soil | D-1 | 205 | .078 | S.F. | 9.90 | 3.04 | | 12.94 | 15.45 |

### 32 14 13.16 Precast Concrete Unit Paving Slabs

| | Crew | Daily Output | Labor-Hours | Unit | Material | 2011 Bare Costs Labor | Equipment | Total | Total Incl O&P |
|---|---|---|---|---|---|---|---|---|---|
| 0010 **PRECAST CONCRETE UNIT PAVING SLABS** | | | | | | | | | |
| 0710 Precast concrete patio blocks, 2-3/8" thick, colors, 8" x 16" | D-1 | 265 | .060 | S.F. | 1.80 | 2.35 | | 4.15 | 5.50 |
| 0750 Exposed local aggregate, natural | 2 Bric | 250 | .064 | | 7.25 | 2.76 | | 10.01 | 12.15 |
| 0800 Colors | | 250 | .064 | | 7.25 | 2.76 | | 10.01 | 12.15 |
| 0850 Exposed granite or limestone aggregate | | 250 | .064 | | 7.50 | 2.76 | | 10.26 | 12.40 |
| 0900 Exposed white tumblestone aggregate | | 250 | .064 | | 5.55 | 2.76 | | 8.31 | 10.25 |

### 32 14 13.18 Precast Concrete Plantable Pavers

| | Crew | Daily Output | Labor-Hours | Unit | Material | 2011 Bare Costs Labor | Equipment | Total | Total Incl O&P |
|---|---|---|---|---|---|---|---|---|---|
| 0010 **PRECAST CONCRETE PLANTABLE PAVERS** (50% grass) | | | | | | | | | |
| 0015 Subgrade preparation and grass planting not included | | | | | | | | | |
| 0100 Precast concrete plantable pavers with topsoil, 24" x 16" | B-63 | 800 | .050 | S.F. | 2.90 | 1.81 | .19 | 4.90 | 6.15 |
| 0200 Less than 600 Square Feet or irregular area | " | 500 | .080 | " | 2.90 | 2.90 | .30 | 6.10 | 7.95 |
| 0300 3/4" crushed stone base for plantable pavers, 6 inch depth | B-62 | 1000 | .024 | S.Y. | 7.40 | .90 | .15 | 8.45 | 9.70 |
| 0400 8 inch depth | | 900 | .027 | | 9.95 | 1 | .17 | 11.12 | 12.65 |
| 0500 10 inch depth | | 800 | .030 | | 12.35 | 1.12 | .19 | 13.66 | 15.50 |
| 0600 12 inch depth | | 700 | .034 | | 14.85 | 1.28 | .22 | 16.35 | 18.55 |
| 0700 Hydro seeding plantable pavers | B-81A | 20 | .800 | M.S.F. | 14.85 | 27 | 16.50 | 58.35 | 76 |
| 0800 Apply fertilizer and seed to plantable pavers | 1 Clab | 8 | 1 | " | 13.40 | 34.50 | | 47.90 | 67.50 |

## 32 14 16 – Brick Unit Paving

### 32 14 16.10 Brick Paving

| | Crew | Daily Output | Labor-Hours | Unit | Material | 2011 Bare Costs Labor | Equipment | Total | Total Incl O&P |
|---|---|---|---|---|---|---|---|---|---|
| 0010 **BRICK PAVING** | | | | | | | | | |
| 0012 4" x 8" x 1-1/2", without joints (4.5 brick/S.F.) | D-1 | 110 | .145 | S.F. | 3.27 | 5.65 | | 8.92 | 12.10 |
| 0100 Grouted, 3/8" joint (3.9 brick/S.F.) | | 90 | .178 | | 3.38 | 6.95 | | 10.33 | 14.10 |
| 0200 4" x 8" x 2-1/4", without joints (4.5 bricks/S.F.) | | 110 | .145 | | 3.74 | 5.65 | | 9.39 | 12.60 |
| 0300 Grouted, 3/8" joint (3.9 brick/S.F.) | | 90 | .178 | | 3.45 | 6.95 | | 10.40 | 14.20 |
| 0500 Bedding, asphalt, 3/4" thick | B-25 | 5130 | .017 | | .64 | .64 | .48 | 1.76 | 2.21 |
| 0540 Course washed sand bed, 1" thick | B-18 | 5000 | .005 | | .30 | .17 | .01 | .48 | .60 |
| 0580 Mortar, 1" thick | D-1 | 300 | .053 | | .66 | 2.08 | | 2.74 | 3.84 |
| 0620 2" thick | | 200 | .080 | | 1.31 | 3.12 | | 4.43 | 6.10 |
| 1500 Brick on 1" thick sand bed laid flat, 4.5 per S.F. | | 100 | .160 | | 3.38 | 6.25 | | 9.63 | 13.05 |
| 2000 Brick pavers, laid on edge, 7.2 per S.F. | | 70 | .229 | | 3.16 | 8.90 | | 12.06 | 16.80 |
| 2500 For 4" thick concrete bed and joints, add | | 595 | .027 | | 1.12 | 1.05 | | 2.17 | 2.81 |
| 2800 For steam cleaning, add | A-1H | 950 | .008 | | .07 | .29 | .06 | .42 | .59 |

## 32 14 23 – Asphalt Unit Paving

### 32 14 23.10 Asphalt Blocks

| | Crew | Daily Output | Labor-Hours | Unit | Material | 2011 Bare Costs Labor | Equipment | Total | Total Incl O&P |
|---|---|---|---|---|---|---|---|---|---|
| 0010 **ASPHALT BLOCKS** | | | | | | | | | |
| 0020 Rectangular, 6" x 12" x 1-1/4", w/bed & neopr. adhesive | D-1 | 135 | .119 | S.F. | 7.85 | 4.62 | | 12.47 | 15.60 |
| 0100 3" thick | | 130 | .123 | | 11 | 4.80 | | 15.80 | 19.30 |
| 0300 Hexagonal tile, 8" wide, 1-1/4" thick | | 135 | .119 | | 7.85 | 4.62 | | 12.47 | 15.60 |
| 0400 2" thick | | 130 | .123 | | 11 | 4.80 | | 15.80 | 19.30 |
| 0500 Square, 8" x 8", 1-1/4" thick | | 135 | .119 | | 7.85 | 4.62 | | 12.47 | 15.60 |
| 0600 2" thick | | 130 | .123 | | 11 | 4.80 | | 15.80 | 19.30 |
| 0900 For exposed aggregate (ground finish) add | | | | | .55 | | | .55 | .61 |
| 0910 For colors, add | | | | | .42 | | | .42 | .46 |

## 32 14 40 – Stone Paving

### 32 14 40.10 Stone Pavers

| | Crew | Daily Output | Labor-Hours | Unit | Material | 2011 Bare Costs Labor | Equipment | Total | Total Incl O&P |
|---|---|---|---|---|---|---|---|---|---|
| 0010 **STONE PAVERS** | | | | | | | | | |
| 1100 Flagging, bluestone, irregular, 1" thick, | D-1 | 81 | .198 | S.F. | 5.50 | 7.70 | | 13.20 | 17.60 |

# 32 14 Unit Paving

## 32 14 40 – Stone Paving

### 32 14 40.10 Stone Pavers

| 32 14 40.10 Stone Pavers | | Crew | Daily Output | Labor-Hours | Unit | Material | 2011 Bare Costs Labor | Equipment | Total | Total Incl O&P |
|---|---|---|---|---|---|---|---|---|---|---|
| 1150 | Snapped random rectangular, 1" thick | D-1 | 92 | .174 | S.F. | 8.35 | 6.80 | | 15.15 | 19.30 |
| 1200 | 1-1/2" thick | | 85 | .188 | | 10 | 7.35 | | 17.35 | 22 |
| 1250 | 2" thick | | 83 | .193 | | 11.65 | 7.50 | | 19.15 | 24 |
| 1300 | Slate, natural cleft, irregular, 3/4" thick | | 92 | .174 | | 7 | 6.80 | | 13.80 | 17.85 |
| 1350 | Random rectangular, gauged, 1/2" thick | | 105 | .152 | | 15.15 | 5.95 | | 21.10 | 25.50 |
| 1400 | Random rectangular, butt joint, gauged, 1/4" thick | ↓ | 150 | .107 | ↓ | 16.30 | 4.16 | | 20.46 | 24 |
| 1500 | For interior setting, add | | | | | | | | 25% | 25% |
| 1550 | Granite blocks, 3-1/2" x 3-1/2" x 3-1/2" | D-1 | 92 | .174 | S.F. | 10.10 | 6.80 | | 16.90 | 21.50 |
| 1600 | 4" to 12" long, 3" to 5" wide, 3" to 5" thick | | 98 | .163 | | 8.40 | 6.35 | | 14.75 | 18.80 |
| 1650 | 6" to 15" long, 3" to 6" wide, 3" to 5" thick | ↓ | 105 | .152 | ↓ | 4.49 | 5.95 | | 10.44 | 13.85 |

# 32 16 Curbs, Gutters, Sidewalks, and Driveways

## 32 16 13 – Curbs and Gutters

### 32 16 13.13 Cast-in-Place Concrete Curbs and Gutters

| | | Crew | Daily Output | Labor-Hours | Unit | Material | 2011 Bare Costs Labor | Equipment | Total | Total Incl O&P |
|---|---|---|---|---|---|---|---|---|---|---|
| 0010 | **CAST-IN-PLACE CONCRETE CURBS AND GUTTERS** | | | | | | | | | |
| 0290 | Forms only, no concrete | | | | | | | | | |
| 0300 | Concrete, wood forms, 6" x 18", straight | C-2 | 500 | .096 | L.F. | 2.03 | 4.03 | | 6.06 | 8.45 |
| 0400 | 6" x 18", radius | " | 200 | .240 | " | 2.13 | 10.05 | | 12.18 | 17.80 |
| 0402 | Forms and concrete complete | | | | | | | | | |
| 0404 | Concrete, wood forms, 6" x 18", straight & concrete | C-2A | 500 | .096 | L.F. | 4.80 | 3.99 | | 8.79 | 11.35 |
| 0406 | 6" x 18", radius | | 200 | .240 | | 4.90 | 10 | | 14.90 | 20.50 |
| 0410 | Steel forms, 6" x 18", straight | | 700 | .069 | | 4.21 | 2.85 | | 7.06 | 8.95 |
| 0411 | 6" x 18", radius | ↓ | 400 | .120 | | 6.55 | 4.99 | | 11.54 | 14.80 |
| 0415 | Machine formed, 6" x 18", straight | B-69A | 2000 | .024 | | 6.10 | .90 | .37 | 7.37 | 8.45 |
| 0416 | 6" x 18", radius | " | 900 | .053 | ↓ | 6.25 | 2.01 | .83 | 9.09 | 10.80 |
| 0421 | Curb and gutter, straight | | | | | | | | | |
| 0422 | with 6" high curb and 6" thick gutter, wood forms | | | | | | | | | |
| 0430 | 24" wide, .055 C.Y. per L.F. | C-2A | 375 | .128 | L.F. | 12.90 | 5.30 | | 18.20 | 22.50 |
| 0435 | 30" wide, .066 C.Y. per L.F. | | 340 | .141 | | 14.35 | 5.85 | | 20.20 | 25 |
| 0440 | Steel forms, 24" wide, straight | | 700 | .069 | | 10.95 | 2.85 | | 13.80 | 16.40 |
| 0441 | Radius | | 300 | .160 | | 10.95 | 6.65 | | 17.60 | 22 |
| 0442 | 30" wide, straight | | 700 | .069 | | 12.90 | 2.85 | | 15.75 | 18.55 |
| 0443 | Radius | ↓ | 300 | .160 | | 12.90 | 6.65 | | 19.55 | 24.50 |
| 0445 | Machine formed, 24" wide, straight | B-69A | 2000 | .024 | | 5.55 | .90 | .37 | 6.82 | 7.85 |
| 0446 | Radius | | 900 | .053 | | 5.55 | 2.01 | .83 | 8.39 | 10.05 |
| 0447 | 30" wide, straight | | 2000 | .024 | | 6.45 | .90 | .37 | 7.72 | 8.85 |
| 0448 | Radius | ↓ | 900 | .053 | ↓ | 6.45 | 2.01 | .83 | 9.29 | 11.05 |

### 32 16 13.23 Precast Concrete Curbs and Gutters

| | | Crew | Daily Output | Labor-Hours | Unit | Material | 2011 Bare Costs Labor | Equipment | Total | Total Incl O&P |
|---|---|---|---|---|---|---|---|---|---|---|
| 0010 | **PRECAST CONCRETE CURBS AND GUTTERS** | | | | | | | | | |
| 0550 | Precast, 6" x 18", straight | B-29 | 700 | .080 | L.F. | 8 | 2.98 | 1.22 | 12.20 | 14.70 |
| 0600 | 6" x 18", radius | " | 325 | .172 | " | 20 | 6.40 | 2.64 | 29.04 | 34.50 |

### 32 16 13.33 Asphalt Curbs

| | | Crew | Daily Output | Labor-Hours | Unit | Material | 2011 Bare Costs Labor | Equipment | Total | Total Incl O&P |
|---|---|---|---|---|---|---|---|---|---|---|
| 0010 | **ASPHALT CURBS** | | | | | | | | | |
| 0012 | Curbs, asphaltic, machine formed, 8" wide, 6" high, 40 L.F./ton | B-27 | 1000 | .032 | L.F. | 1.63 | 1.12 | .28 | 3.03 | 3.80 |
| 0100 | 8" wide, 8" high, 30 L.F. per ton | | 900 | .036 | | 2.17 | 1.24 | .31 | 3.72 | 4.63 |
| 0150 | Asphaltic berm, 12" W, 3"-6" H, 35 L.F./ton, before pavement | ↓ | 700 | .046 | | .04 | 1.59 | .40 | 2.03 | 2.92 |
| 0200 | 12" W, 1-1/2" to 4" H, 60 L.F. per ton, laid with pavement | B-2 | 1050 | .038 | ↓ | .02 | 1.32 | | 1.34 | 2.05 |

### 32 16 13.43 Stone Curbs

| | | Crew | Daily Output | Labor-Hours | Unit | Material | 2011 Bare Costs Labor | Equipment | Total | Total Incl O&P |
|---|---|---|---|---|---|---|---|---|---|---|
| 0010 | **STONE CURBS** | | | | | | | | | |
| 1000 | Granite, split face, straight, 5" x 16" | D-13 | 500 | .096 | L.F. | 13.50 | 3.96 | .87 | 18.33 | 22 |

# 32 16 Curbs, Gutters, Sidewalks, and Driveways

## 32 16 13 – Curbs and Gutters

### 32 16 13.43 Stone Curbs

| | | Crew | Daily Output | Labor-Hours | Unit | Material | 2011 Bare Costs Labor | Equipment | Total | Total Incl O&P |
|---|---|---|---|---|---|---|---|---|---|---|
| 1100 | 6" x 18" | D-13 | 450 | .107 | L.F. | 17.75 | 4.40 | .97 | 23.12 | 27 |
| 1300 | Radius curbing, 6" x 18", over 10' radius | B-29 | 260 | .215 | ↓ | 21.50 | 8 | 3.30 | 32.80 | 40 |
| 1400 | Corners, 2' radius | | 80 | .700 | Ea. | 73 | 26 | 10.70 | 109.70 | 131 |
| 1600 | Edging, 4-1/2" x 12", straight | | 300 | .187 | L.F. | 6.75 | 6.95 | 2.86 | 16.56 | 21 |
| 1800 | Curb inlets, (guttermouth) straight | ↓ | 41 | 1.366 | Ea. | 162 | 51 | 21 | 234 | 279 |
| 2000 | Indian granite (belgian block) | | | | | | | | | |
| 2100 | Jumbo, 10-1/2" x 7-1/2" x 4", grey | D-1 | 150 | .107 | L.F. | 5.70 | 4.16 | | 9.86 | 12.50 |
| 2150 | Pink | | 150 | .107 | | 5.80 | 4.16 | | 9.96 | 12.65 |
| 2200 | Regular, 9" x 4-1/2" x 4-1/2", grey | | 160 | .100 | | 5.45 | 3.90 | | 9.35 | 11.85 |
| 2250 | Pink | | 160 | .100 | | 5.80 | 3.90 | | 9.70 | 12.20 |
| 2300 | Cubes, 4" x 4" x 4", grey | | 175 | .091 | | 3.18 | 3.56 | | 6.74 | 8.85 |
| 2350 | Pink | | 175 | .091 | | 3.15 | 3.56 | | 6.71 | 8.80 |
| 2400 | 6" x 6" x 6", pink | ↓ | 155 | .103 | ↓ | 10.40 | 4.02 | | 14.42 | 17.50 |
| 2500 | Alternate pricing method for indian granite | | | | | | | | | |
| 2550 | Jumbo, 10-1/2" x 7-1/2" x 4" (30 lb), grey | | | | Ton | 325 | | | 325 | 360 |
| 2600 | Pink | | | | | 340 | | | 340 | 370 |
| 2650 | Regular, 9" x 4-1/2" x 4-1/2" (20 lb), grey | | | | | 385 | | | 385 | 425 |
| 2700 | Pink | | | | | 405 | | | 405 | 445 |
| 2750 | Cubes, 4" x 4" x 4" (5 lb), grey | | | | | 385 | | | 385 | 425 |
| 2800 | Pink | | | | | 405 | | | 405 | 445 |
| 2850 | 6" x 6" x 6" (25 lb), pink | | | | | 405 | | | 405 | 445 |
| 2900 | For pallets, add | | | | ↓ | 20 | | | 20 | 22 |

# 32 17 Paving Specialties

## 32 17 13 – Parking Bumpers

### 32 17 13.13 Metal Parking Bumpers

| | | Crew | Daily Output | Labor-Hours | Unit | Material | 2011 Bare Costs Labor | Equipment | Total | Total Incl O&P |
|---|---|---|---|---|---|---|---|---|---|---|
| 0010 | **METAL PARKING BUMPERS** | | | | | | | | | |
| 0015 | Bumper rails for garages, 12 Ga. rail, 6" wide, with steel | | | | | | | | | |
| 0020 | posts 12'-6" O.C., minimum | E-4 | 190 | .168 | L.F. | 17.75 | 8.25 | .57 | 26.57 | 34.50 |
| 0030 | Average | | 165 | .194 | | 22 | 9.50 | .66 | 32.16 | 42 |
| 0100 | Maximum | | 140 | .229 | | 26.50 | 11.20 | .78 | 38.48 | 50 |
| 0300 | 12" channel rail, minimum | | 160 | .200 | | 22 | 9.80 | .68 | 32.48 | 42.50 |
| 0400 | Maximum | ↓ | 120 | .267 | ↓ | 33.50 | 13.10 | .91 | 47.51 | 60.50 |
| 1300 | Pipe bollards, conc filled/paint, 8' L x 4' D hole, 6" diam. | B-6 | 20 | 1.200 | Ea. | 640 | 45 | 16 | 701 | 790 |
| 1400 | 8" diam. | | 15 | 1.600 | | 940 | 60 | 21.50 | 1,021.50 | 1,150 |
| 1500 | 12" diam. | | 12 | 2 | | 1,050 | 75 | 26.50 | 1,151.50 | 1,300 |
| 1592 | Bollards, steel, 3' H, retractable, incl hydraulic controls, min | | 4 | 6 | | 40,000 | 225 | 80 | 40,305 | 44,400 |
| 1594 | Max | ↓ | 2 | 12 | | 45,000 | 450 | 160 | 45,610 | 50,500 |
| 2030 | Folding with individual padlocks | B-2 | 50 | .800 | ↓ | 630 | 28 | | 658 | 735 |
| 8000 | Parking lot control, see Section 11 12 13.10 | | | | | | | | | |

### 32 17 13.16 Plastic Parking Bumpers

| | | Crew | Daily Output | Labor-Hours | Unit | Material | 2011 Bare Costs Labor | Equipment | Total | Total Incl O&P |
|---|---|---|---|---|---|---|---|---|---|---|
| 0010 | **PLASTIC PARKING BUMPERS** | | | | | | | | | |
| 1200 | Thermoplastic, 6" x 10" x 6'-0" | B-2 | 120 | .333 | Ea. | 159 | 11.60 | | 170.60 | 193 |

### 32 17 13.19 Precast Concrete Parking Bumpers

| | | Crew | Daily Output | Labor-Hours | Unit | Material | 2011 Bare Costs Labor | Equipment | Total | Total Incl O&P |
|---|---|---|---|---|---|---|---|---|---|---|
| 0010 | **PRECAST CONCRETE PARKING BUMPERS** | | | | | | | | | |
| 1000 | Wheel stops, precast concrete incl. dowels, 6" x 10" x 6'-0" | B-2 | 120 | .333 | Ea. | 60 | 11.60 | | 71.60 | 84 |
| 1100 | 8" x 13" x 6'-0" | " | 120 | .333 | " | 86 | 11.60 | | 97.60 | 112 |

### 32 17 13.26 Wood Parking Bumpers

| | | Crew | Daily Output | Labor-Hours | Unit | Material | 2011 Bare Costs Labor | Equipment | Total | Total Incl O&P |
|---|---|---|---|---|---|---|---|---|---|---|
| 0010 | **WOOD PARKING BUMPERS** | | | | | | | | | |
| 0020 | Parking barriers, timber w/saddles, treated type | | | | | | | | | |

# 32 17 Paving Specialties

## 32 17 13 – Parking Bumpers

### 32 17 13.26 Wood Parking Bumpers

| 32 17 13.26 Wood Parking Bumpers | Crew | Daily Output | Labor-Hours | Unit | Material | 2011 Bare Costs Labor | Equipment | Total | Total Incl O&P |
|---|---|---|---|---|---|---|---|---|---|
| 0100 | 4" x 4" for cars | B-2 | 520 | .077 | L.F. | 2.21 | 2.67 | | 4.88 | 6.55 |
| 0200 | 6" x 6" for trucks | | 520 | .077 | " | 4.53 | 2.67 | | 7.20 | 9.10 |
| 0600 | Flexible fixed stanchion, 2' high, 3" diameter | | 100 | .400 | Ea. | 40 | 13.90 | | 53.90 | 65.50 |

## 32 17 23 – Pavement Markings

### 32 17 23.13 Painted Pavement Markings

| | | Crew | Daily Output | Labor-Hours | Unit | Material | Labor | Equipment | Total | Total Incl O&P |
|---|---|---|---|---|---|---|---|---|---|---|
| 0010 | **PAINTED PAVEMENT MARKINGS** | | | | | | | | | |
| 0020 | Acrylic waterborne, white or yellow, 4" wide | B-78 | 20000 | .002 | L.F. | .19 | .08 | .02 | .29 | .37 |
| 0200 | 6" wide | | 11000 | .004 | | .25 | .15 | .04 | .44 | .56 |
| 0500 | 8" wide | | 10000 | .005 | | .30 | .17 | .05 | .52 | .63 |
| 0600 | 12" wide | | 4000 | .012 | | .46 | .41 | .12 | .99 | 1.26 |
| 0620 | Arrows or gore lines | | 2300 | .021 | S.F. | .15 | .72 | .20 | 1.07 | 1.49 |
| 0640 | Temporary paint, white or yellow | | 15000 | .003 | L.F. | .05 | .11 | .03 | .19 | .26 |
| 0660 | Removal | 1 Clab | 300 | .027 | | | .92 | | .92 | 1.41 |
| 0680 | Temporary tape | 2 Clab | 1500 | .011 | | .40 | .37 | | .77 | 1 |
| 0710 | Thermoplastic, white or yellow, 4" wide | B-79 | 15000 | .003 | | .20 | .09 | .07 | .36 | .44 |
| 0730 | 6" wide | | 14000 | .003 | | .29 | .10 | .08 | .47 | .55 |
| 0740 | 8" wide | | 12000 | .003 | | .39 | .12 | .09 | .60 | .71 |
| 0750 | 12" wide | | 6000 | .007 | | .58 | .23 | .18 | .99 | 1.19 |
| 0760 | Arrows | | 660 | .061 | S.F. | .40 | 2.09 | 1.64 | 4.13 | 5.45 |
| 0770 | Gore lines | | 2500 | .016 | | .40 | .55 | .43 | 1.38 | 1.77 |
| 0780 | Letters | | 660 | .061 | | .40 | 2.09 | 1.64 | 4.13 | 5.45 |
| 1000 | Airport painted markings | | | | | | | | | |
| 1100 | Painting, white or yellow, taxiway markings | B-79 | 4000 | .010 | S.F. | .19 | .35 | .27 | .81 | 1.04 |
| 1200 | Runway markings | | 3500 | .011 | | .19 | .40 | .31 | .90 | 1.15 |
| 1300 | Pavement location or direction signs | | 2500 | .016 | | .19 | .55 | .43 | 1.17 | 1.54 |
| 1350 | Mobilization airport pavement painting | | 4 | 10 | Ea. | | 345 | 270 | 615 | 825 |
| 1400 | Paint markings or pavement signs removal daytime | B-78B | 400 | .045 | S.F. | | 1.59 | .78 | 2.37 | 3.29 |
| 1500 | Removal nighttime | | 335 | .054 | " | | 1.90 | .93 | 2.83 | 3.94 |
| 1600 | Mobilization pavement paint removal | | 4 | 4.500 | Ea. | | 159 | 78.50 | 237.50 | 330 |

### 32 17 23.14 Pavement Parking Markings

| | | Crew | Daily Output | Labor-Hours | Unit | Material | Labor | Equipment | Total | Total Incl O&P |
|---|---|---|---|---|---|---|---|---|---|---|
| 0010 | **PAVEMENT PARKING MARKINGS** | | | | | | | | | |
| 0790 | Layout of pavement marking | A-2 | 25000 | .001 | L.F. | | .03 | .01 | .04 | .06 |
| 0800 | Lines on pvmt, parking stall, paint, white, 4" wide | B-78 | 440 | .109 | Stall | 4.50 | 3.77 | 1.07 | 9.34 | 11.85 |
| 1000 | Street letters and numbers | " | 1600 | .030 | S.F. | .65 | 1.04 | .29 | 1.98 | 2.63 |

# 32 18 Athletic and Recreational Surfacing

## 32 18 13 – Synthetic Grass Surfacing

### 32 18 13.10 Artificial Grass Surfacing

| | | Crew | Daily Output | Labor-Hours | Unit | Material | Labor | Equipment | Total | Total Incl O&P |
|---|---|---|---|---|---|---|---|---|---|---|
| 0010 | **ARTIFICIAL GRASS SURFACING** | | | | | | | | | |
| 0015 | Not including asphalt base or drainage, | | | | | | | | | |
| 0020 | but including cushion pad, over 50,000 S.F. | | | | | | | | | |
| 0200 | 1/2" pile and 5/16" cushion pad, standard | C-17 | 3200 | .025 | S.F. | 9.50 | 1.12 | | 10.62 | 12.15 |
| 0300 | Deluxe | | 2560 | .031 | | 11.25 | 1.40 | | 12.65 | 14.55 |
| 0500 | 1/2" pile and 5/8" cushion pad, standard | | 2844 | .028 | | 13.75 | 1.26 | | 15.01 | 17.10 |
| 0600 | Deluxe | | 2327 | .034 | | 15.15 | 1.54 | | 16.69 | 19 |
| 0800 | For asphaltic concrete base, 2-1/2" thick, | | | | | | | | | |
| 0900 | with 6" crushed stone sub-base, add | B-25 | 12000 | .007 | S.F. | 1.77 | .28 | .21 | 2.26 | 2.60 |

# 32 18 Athletic and Recreational Surfacing

## 32 18 23 – Athletic Surfacing

### 32 18 23.33 Running Track Surfacing

| | | Crew | Daily Output | Labor-Hours | Unit | Material | 2011 Bare Costs Labor | Equipment | Total | Total Incl O&P |
|---|---|---|---|---|---|---|---|---|---|---|
| 0010 | **RUNNING TRACK SURFACING** | | | | | | | | | |
| 0020 | Running track, asphalt, incl base, 3" thick | B-37 | 300 | .160 | S.Y. | 21 | 5.80 | .48 | 27.28 | 32.50 |
| 0100 | Surface, latex rubber system, 3/8" thick, black | B-20 | 125 | .192 | | 9.35 | 7.35 | | 16.70 | 21.50 |
| 0150 | Colors | | 125 | .192 | | 16.65 | 7.35 | | 24 | 29.50 |
| 0300 | Urethane rubber system, 3/8" thick, black | | 120 | .200 | | 25 | 7.65 | | 32.65 | 39.50 |
| 0400 | Color coating | | 115 | .209 | | 30.50 | 8 | | 38.50 | 46 |

### 32 18 23.53 Tennis Court Surfacing

| | | Crew | Daily Output | Labor-Hours | Unit | Material | 2011 Bare Costs Labor | Equipment | Total | Total Incl O&P |
|---|---|---|---|---|---|---|---|---|---|---|
| 0010 | **TENNIS COURT SURFACING** | | | | | | | | | |
| 0020 | Tennis court, asphalt, incl. base, 2-1/2" thick, one court | B-37 | 450 | .107 | S.Y. | 33.50 | 3.86 | .32 | 37.68 | 43.50 |
| 0200 | Two courts | | 675 | .071 | | 15.10 | 2.58 | .22 | 17.90 | 21 |
| 0300 | Clay courts | | 360 | .133 | | 37.50 | 4.83 | .40 | 42.73 | 49.50 |
| 0400 | Pulverized natural greenstone with 4" base, fast dry | | 250 | .192 | | 36 | 6.95 | .58 | 43.53 | 50.50 |
| 0800 | Rubber-acrylic base resilient pavement | | 600 | .080 | | 51 | 2.90 | .24 | 54.14 | 60.50 |
| 1000 | Colored sealer, acrylic emulsion, 3 coats | 2 Clab | 800 | .020 | | 5.25 | .69 | | 5.94 | 6.85 |
| 1100 | 3 coat, 2 colors | " | 900 | .018 | | 7.40 | .61 | | 8.01 | 9.05 |
| 1200 | For preparing old courts, add | 1 Clab | 825 | .010 | | | .33 | | .33 | .51 |
| 1400 | Posts for nets, 3-1/2" diameter with eye bolts | B-1 | 3.40 | 7.059 | Pr. | 212 | 247 | | 459 | 615 |
| 1500 | With pulley & reel | | 3.40 | 7.059 | " | 300 | 247 | | 547 | 710 |
| 1700 | Net, 42' long, nylon thread with binder | | 50 | .480 | Ea. | 279 | 16.80 | | 295.80 | 330 |
| 1800 | All metal | | 6.50 | 3.692 | " | 490 | 129 | | 619 | 740 |
| 2000 | Paint markings on asphalt, 2 coats | 1 Pord | 1.78 | 4.494 | Court | 131 | 169 | | 300 | 395 |
| 2200 | Complete court with fence, etc., asphaltic conc., minimum | B-37 | .20 | 240 | | 25,000 | 8,700 | 725 | 34,425 | 41,600 |
| 2300 | Maximum | | .16 | 300 | | 50,000 | 10,900 | 910 | 61,810 | 72,500 |
| 2800 | Clay courts, minimum | | .20 | 240 | | 25,000 | 8,700 | 725 | 34,425 | 41,600 |
| 2900 | Maximum | | .16 | 300 | | 40,000 | 10,900 | 910 | 51,810 | 61,500 |

# 32 31 Fences and Gates

## 32 31 13 – Chain Link Fences and Gates

### 32 31 13.20 Fence, Chain Link Industrial

| | | Crew | Daily Output | Labor-Hours | Unit | Material | 2011 Bare Costs Labor | Equipment | Total | Total Incl O&P |
|---|---|---|---|---|---|---|---|---|---|---|
| 0010 | **FENCE, CHAIN LINK INDUSTRIAL** | | | | | | | | | |
| 0011 | Schedule 40, including concrete | | | | | | | | | |
| 0020 | 3 strands barb wire, 2" post @ 10' O.C., set in concrete, 6' H | | | | | | | | | |
| 0200 | 9 ga. wire, galv. steel, in concrete | B-80C | 240 | .100 | L.F. | 16 | 3.40 | .74 | 20.14 | 23.50 |
| 0248 | Fence, add for vinyl coated fabric | | | | S.F. | .65 | | | .65 | .72 |
| 0300 | Aluminized steel | B-80C | 240 | .100 | L.F. | 17 | 3.40 | .74 | 21.14 | 24.50 |
| 0500 | 6 ga. wire, galv. steel | | 240 | .100 | | 18 | 3.40 | .74 | 22.14 | 26 |
| 0600 | Aluminized steel | | 240 | .100 | | 21.50 | 3.40 | .74 | 25.64 | 29.50 |
| 0800 | 6 ga. wire, 6' high but omit barbed wire, galv. steel | | 250 | .096 | | 17.20 | 3.27 | .71 | 21.18 | 24.50 |
| 0900 | Aluminized steel, in concrete | | 250 | .096 | | 20.50 | 3.27 | .71 | 24.48 | 28.50 |
| 0920 | 8' H, 6 ga. wire, 2-1/2" line post, galv. steel, in concrete | | 180 | .133 | | 27 | 4.54 | .99 | 32.53 | 38 |
| 0940 | Aluminized steel, in concrete | | 180 | .133 | | 33 | 4.54 | .99 | 38.53 | 44.50 |
| 1100 | Add for corner posts, 3" diam., galv. steel, in concrete | | 40 | .600 | Ea. | 80 | 20.50 | 4.46 | 104.96 | 124 |
| 1200 | Aluminized steel, in concrete | | 40 | .600 | | 92.50 | 20.50 | 4.46 | 117.46 | 138 |
| 1300 | Add for braces, galv. steel | | 80 | .300 | | 27.50 | 10.20 | 2.23 | 39.93 | 48 |
| 1350 | Aluminized steel | | 80 | .300 | | 36.50 | 10.20 | 2.23 | 48.93 | 58 |
| 1400 | Gate for 6' high fence, 1-5/8" frame, 3' wide, galv. steel | | 10 | 2.400 | | 147 | 81.50 | 17.85 | 246.35 | 305 |
| 1500 | Aluminized steel, in concrete | | 10 | 2.400 | | 248 | 81.50 | 17.85 | 347.35 | 420 |
| 2000 | 5'-0" high fence, 9 ga., no barbed wire, 2" line post, in concrete | | | | | | | | | |
| 2010 | 10' O.C., 1-5/8" top rail, in concrete | | | | | | | | | |
| 2100 | Galvanized steel, in concrete | B-80C | 300 | .080 | L.F. | 16.30 | 2.72 | .60 | 19.62 | 23 |

# 32 31 Fences and Gates

## 32 31 13 – Chain Link Fences and Gates

### 32 31 13.20 Fence, Chain Link Industrial

| | | Crew | Daily Output | Labor-Hours | Unit | Material | 2011 Bare Costs Labor | Equipment | Total | Total Incl O&P |
|---|---|---|---|---|---|---|---|---|---|---|
| 2200 | Aluminized steel, in concrete | B-80C | 300 | .080 | L.F. | 17 | 2.72 | .60 | 20.32 | 23.50 |
| 2400 | Gate, 4' wide, 5' high, 2" frame, galv. steel, in concrete | | 10 | 2.400 | Ea. | 143 | 81.50 | 17.85 | 242.35 | 300 |
| 2500 | Aluminized steel, in concrete | | 10 | 2.400 | " | 153 | 81.50 | 17.85 | 252.35 | 315 |
| 3100 | Overhead slide gate, chain link, 6' high, to 18' wide, in concrete | | 38 | .632 | L.F. | 159 | 21.50 | 4.70 | 185.20 | 213 |
| 3110 | Cantilever type, in concrete | B-80 | 48 | .667 | | 137 | 24.50 | 12.90 | 174.40 | 202 |
| 3120 | 8' high, in concrete | | 24 | 1.333 | | 141 | 49 | 26 | 216 | 260 |
| 3130 | 10' high, in concrete | | 18 | 1.778 | | 156 | 65.50 | 34.50 | 256 | 310 |
| 5000 | Double swing gates, incl. posts & hardware, in concrete | | | | | | | | | |
| 5010 | 5' high, 12' opening, in concrete | B-80C | 3.40 | 7.059 | Opng. | 550 | 240 | 52.50 | 842.50 | 1,025 |
| 5020 | 20' opening, in concrete | | 2.80 | 8.571 | | 865 | 292 | 64 | 1,221 | 1,475 |
| 5060 | 6' high, 12' opening, in concrete | | 3.20 | 7.500 | | 485 | 255 | 56 | 796 | 985 |
| 5070 | 20' opening, in concrete | | 2.60 | 9.231 | | 725 | 315 | 68.50 | 1,108.50 | 1,350 |
| 5080 | 8' high, 12' opening, in concrete | B-80 | 2.13 | 15.002 | | 515 | 555 | 290 | 1,360 | 1,725 |
| 5090 | 20' opening, in concrete | | 1.45 | 22.069 | | 790 | 815 | 425 | 2,030 | 2,575 |
| 5100 | 10' high, 12' opening, in concrete | | 1.31 | 24.427 | | 995 | 900 | 470 | 2,365 | 3,000 |
| 5110 | 20' opening, in concrete | | 1.03 | 31.068 | | 1,375 | 1,150 | 600 | 3,125 | 3,900 |
| 5120 | 12' high, 12' opening, in concrete | | 1.05 | 30.476 | | 1,350 | 1,125 | 590 | 3,065 | 3,825 |
| 5130 | 20' opening, in concrete | | .85 | 37.647 | | 1,900 | 1,400 | 730 | 4,030 | 5,000 |
| 5190 | For aluminized steel add | | | | | 20% | | | | |
| 7075 | Fence, for small jobs 100 L.F. or less fence w/or wo gate, add | | | | S.F. | 20% | | | 20% | |

### 32 31 13.25 Fence, Chain Link Residential

| | | Crew | Daily Output | Labor-Hours | Unit | Material | 2011 Bare Costs Labor | Equipment | Total | Total Incl O&P |
|---|---|---|---|---|---|---|---|---|---|---|
| 0010 | **FENCE, CHAIN LINK RESIDENTIAL** | | | | | | | | | |
| 0011 | Schedule 20, 11 gauge wire, 1-5/8" post | | | | | | | | | |
| 0020 | 10' O.C., 1-3/8" top rail, 2" corner post, galv. stl. 3' high | B-80C | 500 | .048 | L.F. | 2.01 | 1.63 | .36 | 4 | 5.10 |
| 0050 | 4' high | | 400 | .060 | | 4.72 | 2.04 | .45 | 7.21 | 8.80 |
| 0100 | 6' high | | 200 | .120 | | 5.95 | 4.08 | .89 | 10.92 | 13.80 |
| 0150 | Add for gate 3' wide, 1-3/8" frame, 3' high | | 12 | 2 | Ea. | 82 | 68 | 14.90 | 164.90 | 210 |
| 0170 | 4' high | | 10 | 2.400 | | 88 | 81.50 | 17.85 | 187.35 | 242 |
| 0190 | 6' high | | 10 | 2.400 | | 94 | 81.50 | 17.85 | 193.35 | 249 |
| 0200 | Add for gate 4' wide, 1-3/8" frame, 3' high | | 9 | 2.667 | | 88 | 91 | 19.85 | 198.85 | 258 |
| 0220 | 4' high | | 9 | 2.667 | | 92.50 | 91 | 19.85 | 203.35 | 263 |
| 0240 | 6' high | | 8 | 3 | | 101 | 102 | 22.50 | 225.50 | 293 |
| 0350 | Aluminized steel, 11 ga. wire, 3' high | | 500 | .048 | L.F. | 5.50 | 1.63 | .36 | 7.49 | 8.95 |
| 0380 | 4' high | | 400 | .060 | | 7.20 | 2.04 | .45 | 9.69 | 11.50 |
| 0400 | 6' high | | 200 | .120 | | 9.85 | 4.08 | .89 | 14.82 | 18.10 |
| 0450 | Add for gate 3' wide, 1-3/8" frame, 3' high | | 12 | 2 | Ea. | 120 | 68 | 14.90 | 202.90 | 252 |
| 0470 | 4' high | | 10 | 2.400 | | 126 | 81.50 | 17.85 | 225.35 | 283 |
| 0490 | 6' high | | 10 | 2.400 | | 137 | 81.50 | 17.85 | 236.35 | 296 |
| 0500 | Add for gate 4' wide, 1-3/8" frame, 3' high | | 10 | 2.400 | | 129 | 81.50 | 17.85 | 228.35 | 287 |
| 0520 | 4' high | | 9 | 2.667 | | 135 | 91 | 19.85 | 245.85 | 310 |
| 0540 | 6' high | | 8 | 3 | | 144 | 102 | 22.50 | 268.50 | 340 |
| 0620 | Vinyl covered, 9 ga. wire, 3' high | | 500 | .048 | L.F. | 4.58 | 1.63 | .36 | 6.57 | 7.95 |
| 0640 | 4' high | | 400 | .060 | | 5.10 | 2.04 | .45 | 7.59 | 9.20 |
| 0660 | 6' high | | 200 | .120 | | 6.65 | 4.08 | .89 | 11.62 | 14.55 |
| 0720 | Add for gate 3' wide, 1-3/8" frame, 3' high | | 12 | 2 | Ea. | 120 | 68 | 14.90 | 202.90 | 252 |
| 0740 | 4' high | | 10 | 2.400 | | 126 | 81.50 | 17.85 | 225.35 | 284 |
| 0760 | 6' high | | 10 | 2.400 | | 127 | 81.50 | 17.85 | 226.35 | 285 |
| 0780 | Add for gate 4' wide, 1-3/8" frame, 3' high | | 10 | 2.400 | | 129 | 81.50 | 17.85 | 228.35 | 287 |
| 0800 | 4' high | | 9 | 2.667 | | 135 | 91 | 19.85 | 245.85 | 310 |
| 0820 | 6' high | | 8 | 3 | | 144 | 102 | 22.50 | 268.50 | 340 |
| 7076 | Fence, for small jobs 100 L.F. fence or less w/or wo gate, add | | | | S.F. | 20% | | | 20% | |

# 32 31 Fences and Gates

## 32 31 13 – Chain Link Fences and Gates

### 32 31 13.26 Tennis Court Fences and Gates

| | 32 31 13.26 Tennis Court Fences and Gates | Crew | Daily Output | Labor-Hours | Unit | Material | 2011 Bare Costs Labor | Equipment | Total | Total Incl O&P |
|---|---|---|---|---|---|---|---|---|---|---|
| 0010 | **TENNIS COURT FENCES AND GATES** | | | | | | | | | |
| 0860 | Tennis courts, 11 ga. wire, 2-1/2" post set | | | | | | | | | |
| 0870 | in concrete, 10' O.C., 1-5/8" top rail | | | | | | | | | |
| 0900 | 10' high | B-80 | 190 | .168 | L.F. | 9.35 | 6.20 | 3.26 | 18.81 | 23.50 |
| 0920 | 12' high | | 170 | .188 | " | 11.70 | 6.95 | 3.64 | 22.29 | 27.50 |
| 1000 | Add for gate 4' wide, 1-5/8" frame 7' high | | 10 | 3.200 | Ea. | 140 | 118 | 62 | 320 | 400 |
| 1040 | Aluminized steel, 11 ga. wire 10' high | | 190 | .168 | L.F. | 17.30 | 6.20 | 3.26 | 26.76 | 32 |
| 1100 | 12' high | | 170 | .188 | " | 21.50 | 6.95 | 3.64 | 32.09 | 38 |
| 1140 | Add for gate 4' wide, 1-5/8" frame, 7' high | | 10 | 3.200 | Ea. | 161 | 118 | 62 | 341 | 425 |
| 1250 | Vinyl covered, 9 ga. wire, 10' high | | 190 | .168 | L.F. | 10.25 | 6.20 | 3.26 | 19.71 | 24.50 |
| 1300 | 12' high | | 170 | .188 | " | 12.40 | 6.95 | 3.64 | 22.99 | 28 |
| 1310 | Fence, CL, tennis ct, transom gate, single, galv, 4' x 7' x 3' | B-80A | 8.72 | 2.752 | Ea. | 435 | 94.50 | 25 | 554.50 | 655 |
| 1400 | Add for gate 4' wide, 1-5/8" frame, 7' high | B-80 | 10 | 3.200 | " | 262 | 118 | 62 | 442 | 535 |

### 32 31 13.33 Chain Link Backstops

| | 32 31 13.33 Chain Link Backstops | Crew | Daily Output | Labor-Hours | Unit | Material | 2011 Bare Costs Labor | Equipment | Total | Total Incl O&P |
|---|---|---|---|---|---|---|---|---|---|---|
| 0010 | **CHAIN LINK BACKSTOPS** | | | | | | | | | |
| 0015 | Backstops, baseball, prefabricated, 30' wide, 12' high & 1 overhang | B-1 | 1 | 24 | Ea. | 2,400 | 840 | | 3,240 | 3,950 |
| 0100 | 40' wide, 12' high & 2 overhangs | " | .75 | 32 | | 5,225 | 1,125 | | 6,350 | 7,475 |
| 0300 | Basketball, steel, single goal | B-13 | 3.04 | 18.421 | | 1,375 | 685 | 246 | 2,306 | 2,850 |
| 0400 | Double goal | " | 1.92 | 29.167 | | 990 | 1,075 | 390 | 2,455 | 3,175 |
| 0600 | Tennis, wire mesh with pair of ends | B-1 | 2.48 | 9.677 | Set | 2,250 | 340 | | 2,590 | 3,000 |
| 0700 | Enclosed court | " | 1.30 | 18.462 | Ea. | 9,700 | 645 | | 10,345 | 11,700 |

### 32 31 13.53 High-Security Chain Link Fences, Gates and Systems

| | 32 31 13.53 High-Security Chain Link Fences, Gates and Systems | Crew | Daily Output | Labor-Hours | Unit | Material | 2011 Bare Costs Labor | Equipment | Total | Total Incl O&P |
|---|---|---|---|---|---|---|---|---|---|---|
| 0010 | **HIGH-SECURITY CHAIN LINK FENCES, GATES AND SYSTEMS** | | | | | | | | | |
| 0100 | Fence, chain link, security, 7' H, standard FE-7, incl excavation & posts | B-80C | 480 | .050 | L.F. | 40 | 1.70 | .37 | 42.07 | 47 |
| 0200 | Fence, barbed wire, security, 7' high, with 3 wire barbed wire arm | " | 400 | .060 | " | 7.90 | 2.04 | .45 | 10.39 | 12.30 |
| 0300 | Complete systems, including material and installation | | | | | | | | | |
| 0310 | Taunt wire fence detection system | | | | M.L.F. | | | | 22,800 | 24,800 |
| 0410 | Microwave fence detection system | | | | | | | | 37,500 | 40,900 |
| 0510 | Passive magnetic fence detection system | | | | | | | | 17,700 | 19,300 |
| 0610 | Infrared fence detection system | | | | | | | | 11,800 | 12,900 |
| 0710 | Strain relief fence detection system | | | | | | | | 22,800 | 24,800 |
| 0810 | Electro-shock fence detection system | | | | | | | | 32,600 | 35,500 |
| 0910 | Photo-electric fence detection system | | | | | | | | 14,800 | 16,100 |

## 32 31 19 – Decorative Metal Fences and Gates

### 32 31 19.10 Decorative Fence

| | 32 31 19.10 Decorative Fence | Crew | Daily Output | Labor-Hours | Unit | Material | 2011 Bare Costs Labor | Equipment | Total | Total Incl O&P |
|---|---|---|---|---|---|---|---|---|---|---|
| 0010 | **DECORATIVE FENCE** | | | | | | | | | |
| 5300 | Tubular picket, steel, 6' sections, 1-9/16" posts, 4' high | B-80C | 300 | .080 | L.F. | 27.50 | 2.72 | .60 | 30.82 | 35 |
| 5400 | 2" posts, 5' high | | 240 | .100 | | 41 | 3.40 | .74 | 45.14 | 51.50 |
| 5600 | 2" posts, 6' high | | 200 | .120 | | 47.50 | 4.08 | .89 | 52.47 | 59.50 |
| 5700 | Staggered picket 1-9/16" posts, 4' high | | 300 | .080 | | 27.50 | 2.72 | .60 | 30.82 | 35 |
| 5800 | 2" posts, 5' high | | 240 | .100 | | 41 | 3.40 | .74 | 45.14 | 51.50 |
| 5900 | 2" posts, 6' high | | 200 | .120 | | 47.50 | 4.08 | .89 | 52.47 | 59.50 |
| 6200 | Gates, 4' high, 3' wide | B-1 | 10 | 2.400 | Ea. | 235 | 84 | | 319 | 390 |
| 6300 | 5' high, 3' wide | | 10 | 2.400 | | 248 | 84 | | 332 | 400 |
| 6400 | 6' high, 3' wide | | 10 | 2.400 | | 298 | 84 | | 382 | 460 |
| 6500 | 4' wide | | 10 | 2.400 | | 325 | 84 | | 409 | 485 |

## 32 31 26 – Wire Fences and Gates

### 32 31 26.10 Fences, Misc. Metal

| | 32 31 26.10 Fences, Misc. Metal | Crew | Daily Output | Labor-Hours | Unit | Material | 2011 Bare Costs Labor | Equipment | Total | Total Incl O&P |
|---|---|---|---|---|---|---|---|---|---|---|
| 0010 | **FENCES, MISC. METAL** | | | | | | | | | |
| 0012 | Chicken wire, posts @ 4', 1" mesh, 4' high | B-80C | 410 | .059 | L.F. | 2.46 | 1.99 | .44 | 4.89 | 6.25 |

# 32 31 Fences and Gates

## 32 31 26 – Wire Fences and Gates

### 32 31 26.10 Fences, Misc. Metal

| | 32 31 26.10 Fences, Misc. Metal | Crew | Daily Output | Labor-Hours | Unit | Material | 2011 Bare Costs Labor | Equipment | Total | Total Incl O&P |
|---|---|---|---|---|---|---|---|---|---|---|
| 0100 | 2" mesh, 6' high | B-80C | 350 | .069 | L.F. | 2.38 | 2.33 | .51 | 5.22 | 6.75 |
| 0200 | Galv. steel, 12 ga., 2" x 4" mesh, posts 5' O.C., 3' high | | 300 | .080 | | 2.44 | 2.72 | .60 | 5.76 | 7.50 |
| 0300 | 5' high | | 300 | .080 | | 3.74 | 2.72 | .60 | 7.06 | 8.95 |
| 0400 | 14 ga., 1" x 2" mesh, 3' high | | 300 | .080 | | 2.76 | 2.72 | .60 | 6.08 | 7.85 |
| 0500 | 5' high | | 300 | .080 | | 4.34 | 2.72 | .60 | 7.66 | 9.60 |
| 1000 | Kennel fencing, 1-1/2" mesh, 6' long, 3'-6" wide, 6'-2" high | 2 Clab | 4 | 4 | Ea. | 460 | 137 | | 597 | 715 |
| 1050 | 12' long | | 4 | 4 | | 600 | 137 | | 737 | 870 |
| 1200 | Top covers, 1-1/2" mesh, 6' long | | 15 | 1.067 | | 153 | 36.50 | | 189.50 | 225 |
| 1250 | 12' long | | 12 | 1.333 | | 200 | 46 | | 246 | 291 |
| 1300 | For kennel doors, see Section 08 31 13.40 | | | | | | | | | |
| 4500 | Security fence, prison grade, set in concrete, 12' high | B-80 | 25 | 1.280 | L.F. | 50 | 47.50 | 25 | 122.50 | 154 |
| 4600 | 16' high | " | 20 | 1.600 | " | 68 | 59 | 31 | 158 | 199 |

### 32 31 26.20 Wire Fencing, General

| | 32 31 26.20 Wire Fencing, General | Crew | Daily Output | Labor-Hours | Unit | Material | 2011 Bare Costs Labor | Equipment | Total | Total Incl O&P |
|---|---|---|---|---|---|---|---|---|---|---|
| 0010 | **WIRE FENCING, GENERAL** | | | | | | | | | |
| 0015 | Barbed wire, galvanized, domestic steel, hi-tensile 15-1/2 ga. | | | | M.L.F. | 117 | | | 117 | 128 |
| 0020 | Standard, 12-3/4 ga. | | | | | 133 | | | 133 | 147 |
| 0210 | Barbless wire, 2-strand galvanized, 12-1/2 ga. | | | | | 133 | | | 133 | 147 |
| 0500 | Helical razor ribbon, stainless steel, 18" dia x 18" spacing | | | | C.L.F. | 370 | | | 370 | 405 |
| 0600 | Hardware cloth galv., 1/4" mesh, 23 ga., 2' wide | | | | C.S.F. | 74.50 | | | 74.50 | 82 |
| 0700 | 3' wide | | | | | 74.50 | | | 74.50 | 82 |
| 0900 | 1/2" mesh, 19 ga., 2' wide | | | | | 74.50 | | | 74.50 | 82 |
| 1000 | 4' wide | | | | | 56 | | | 56 | 61.50 |
| 1200 | Chain link fabric, steel, 2" mesh, 6 ga, galvanized | | | | | 200 | | | 200 | 220 |
| 1300 | 9 ga, galvanized | | | | | 76.50 | | | 76.50 | 84 |
| 1350 | Vinyl coated | | | | | 85 | | | 85 | 93.50 |
| 1360 | Aluminized | | | | | 88 | | | 88 | 96.50 |
| 1400 | 2-1/4" mesh, 11.5 ga, galvanized | | | | | 48 | | | 48 | 52.50 |
| 1600 | 1-3/4" mesh (tennis courts), 11.5 ga (core), vinyl coated | | | | | 48 | | | 48 | 52.50 |
| 1700 | 9 ga, galvanized | | | | | 73 | | | 73 | 80.50 |
| 2100 | Welded wire fabric, galvanized, 1" x 2", 14 ga. | | | | | 54.50 | | | 54.50 | 60 |
| 2200 | 2" x 4", 12-1/2 ga. | | | | | 56 | | | 56 | 61.50 |

## 32 31 29 – Wood Fences and Gates

### 32 31 29.20 Fence, Wood Rail

| | 32 31 29.20 Fence, Wood Rail | Crew | Daily Output | Labor-Hours | Unit | Material | 2011 Bare Costs Labor | Equipment | Total | Total Incl O&P |
|---|---|---|---|---|---|---|---|---|---|---|
| 0010 | **FENCE, WOOD RAIL** | | | | | | | | | |
| 0012 | Picket, No. 2 cedar, Gothic, 2 rail, 3' high | B-1 | 160 | .150 | L.F. | 6.10 | 5.25 | | 11.35 | 14.75 |
| 0050 | Gate, 3'-6" wide | B-80C | 9 | 2.667 | Ea. | 52 | 91 | 19.85 | 162.85 | 218 |
| 0400 | 3 rail, 4' high | | 150 | .160 | L.F. | 7 | 5.45 | 1.19 | 13.64 | 17.35 |
| 0500 | Gate, 3'-6" wide | | 9 | 2.667 | Ea. | 63 | 91 | 19.85 | 173.85 | 231 |
| 0600 | Open rail, rustic, No. 1 cedar, 2 rail, 3' high | | 160 | .150 | L.F. | 5.40 | 5.10 | 1.12 | 11.62 | 15 |
| 0650 | Gate, 3' wide | | 9 | 2.667 | Ea. | 60.50 | 91 | 19.85 | 171.35 | 228 |
| 0700 | 3 rail, 4' high | | 150 | .160 | L.F. | 6.20 | 5.45 | 1.19 | 12.84 | 16.45 |
| 0900 | Gate, 3' wide | | 9 | 2.667 | Ea. | 78 | 91 | 19.85 | 188.85 | 247 |
| 1200 | Stockade, No. 2 cedar, treated wood rails, 6' high | | 160 | .150 | L.F. | 7.55 | 5.10 | 1.12 | 13.77 | 17.35 |
| 1250 | Gate, 3' wide | | 9 | 2.667 | Ea. | 62 | 91 | 19.85 | 172.85 | 229 |
| 1300 | No. 1 cedar, 3-1/4" cedar rails, 6' high | | 160 | .150 | L.F. | 18.55 | 5.10 | 1.12 | 24.77 | 29.50 |
| 1500 | Gate, 3' wide | | 9 | 2.667 | Ea. | 151 | 91 | 19.85 | 261.85 | 325 |
| 1520 | Open rail, split, No. 1 cedar, 2 rail, 3' high | | 160 | .150 | L.F. | 5.45 | 5.10 | 1.12 | 11.67 | 15.05 |
| 1540 | 3 rail, 4'-0" high | | 150 | .160 | | 6.60 | 5.45 | 1.19 | 13.24 | 16.90 |
| 3300 | Board, shadow box, 1" x 6", treated pine, 6' high | | 160 | .150 | | 10.25 | 5.10 | 1.12 | 16.47 | 20.50 |
| 3400 | No. 1 cedar, 6' high | | 150 | .160 | | 20 | 5.45 | 1.19 | 26.64 | 31.50 |
| 3900 | Basket weave, No. 1 cedar, 6' high | | 160 | .150 | | 28 | 5.10 | 1.12 | 34.22 | 39.50 |
| 3950 | Gate, 3'-6" wide | B-1 | 8 | 3 | Ea. | 159 | 105 | | 264 | 335 |

# 32 31 Fences and Gates

## 32 31 29 – Wood Fences and Gates

| 32 31 29.20 Fence, Wood Rail | Crew | Daily Output | Labor-Hours | Unit | Material | 2011 Bare Costs Labor | Equipment | Total | Total Incl O&P |
|---|---|---|---|---|---|---|---|---|---|
| 4000 | Treated pine, 6' high | B-1 | 150 | .160 | L.F. | 14.90 | 5.60 | | 20.50 | 25 |
| 4200 | Gate, 3'-6" wide | | 9 | 2.667 | Ea. | 82 | 93.50 | | 175.50 | 233 |
| 5000 | Fence rail, redwood, 2" x 4", merch grade 8' | | 2400 | .010 | L.F. | 1.80 | .35 | | 2.15 | 2.52 |
| 5050 | Select grade, 8' | | 2400 | .010 | " | 4 | .35 | | 4.35 | 4.94 |
| 6000 | Fence post, select redwood, earthpacked & treated, 4" x 4" x 6' | | 96 | .250 | Ea. | 11.30 | 8.75 | | 20.05 | 26 |
| 6010 | 4" x 4" x 8' | | 96 | .250 | | 13.40 | 8.75 | | 22.15 | 28 |
| 6020 | Set in concrete, 4" x 4" x 6' | | 50 | .480 | | 14.50 | 16.80 | | 31.30 | 42 |
| 6030 | 4" x 4" x 8' | | 50 | .480 | | 17.30 | 16.80 | | 34.10 | 45 |
| 6040 | Wood post, 4' high, set in concrete, incl. concrete | | 50 | .480 | | 8.10 | 16.80 | | 24.90 | 35 |
| 6050 | Earth packed | | 96 | .250 | | 5.55 | 8.75 | | 14.30 | 19.55 |
| 6060 | 6' high, set in concrete, incl. concrete | | 50 | .480 | | 10.35 | 16.80 | | 27.15 | 37.50 |
| 6070 | Earth packed | | 96 | .250 | | 7.55 | 8.75 | | 16.30 | 22 |

# 32 32 Retaining Walls

## 32 32 13 – Cast-in-Place Concrete Retaining Walls

### 32 32 13.10 Retaining Walls, Cast Concrete

| | | Crew | Daily Output | Labor-Hours | Unit | Material | 2011 Bare Costs Labor | Equipment | Total | Total Incl O&P |
|---|---|---|---|---|---|---|---|---|---|---|
| 0010 | **RETAINING WALLS, CAST CONCRETE** | | | | | | | | | |
| 1800 | Concrete gravity wall with vertical face including excavation & backfill | | | | | | | | | |
| 1850 | No reinforcing | | | | | | | | | |
| 1900 | 6' high, level embankment | C-17C | 36 | 2.306 | L.F. | 70.50 | 103 | 17.65 | 191.15 | 256 |
| 2000 | 33° slope embankment | | 32 | 2.594 | | 66 | 116 | 19.85 | 201.85 | 273 |
| 2200 | 8' high, no surcharge | | 27 | 3.074 | | 90 | 138 | 23.50 | 251.50 | 335 |
| 2300 | 33° slope embankment | | 24 | 3.458 | | 109 | 155 | 26.50 | 290.50 | 385 |
| 2500 | 10' high, level embankment | | 19 | 4.368 | | 129 | 196 | 33.50 | 358.50 | 480 |
| 2600 | 33° slope embankment | | 18 | 4.611 | | 178 | 207 | 35.50 | 420.50 | 550 |
| 2800 | Reinforced concrete cantilever, incl. excavation, backfill & reinf. | | | | | | | | | |
| 2900 | 6' high, 33° slope embankment | C-17C | 35 | 2.371 | L.F. | 66 | 106 | 18.15 | 190.15 | 255 |
| 3000 | 8' high, 33° slope embankment | | 29 | 2.862 | | 76 | 128 | 22 | 226 | 305 |
| 3100 | 10' high, 33° slope embankment | | 20 | 4.150 | | 99 | 186 | 32 | 317 | 430 |
| 3200 | 20' high, 500 lb. per L.F. surcharge | | 7.50 | 11.067 | | 297 | 495 | 84.50 | 876.50 | 1,175 |
| 3500 | Concrete cribbing, incl. excavation and backfill | | | | | | | | | |
| 3700 | 12' high, open face | B-13 | 210 | .267 | S.F. | 33 | 9.95 | 3.57 | 46.52 | 55.50 |
| 3900 | Closed face | " | 210 | .267 | " | 31 | 9.95 | 3.57 | 44.52 | 53 |
| 4100 | Concrete filled slurry trench, see Section 31 56 23.20 | | | | | | | | | |

## 32 32 23 – Segmental Retaining Walls

### 32 32 23.13 Segmental Conc. Unit Masonry Retaining Walls

| | | Crew | Daily Output | Labor-Hours | Unit | Material | 2011 Bare Costs Labor | Equipment | Total | Total Incl O&P |
|---|---|---|---|---|---|---|---|---|---|---|
| 0010 | **SEGMENTAL CONC. UNIT MASONRY RETAINING WALLS** | | | | | | | | | |
| 7100 | Segmental Retaining Wall system, incl pins, and void fill | | | | | | | | | |
| 7120 | base not included | | | | | | | | | |
| 7140 | Large unit, 8" high x 18" wide x 20" deep, 3 plane split | B-62 | 300 | .080 | S.F. | 10.20 | 2.99 | .51 | 13.70 | 16.30 |
| 7150 | Straight split | | 300 | .080 | | 10.10 | 2.99 | .51 | 13.60 | 16.20 |
| 7160 | Medium, ltwt, 8" high x 18" wide x 12" deep, 3 plane split | | 400 | .060 | | 8 | 2.25 | .38 | 10.63 | 12.65 |
| 7170 | Straight split | | 400 | .060 | | 7.90 | 2.25 | .38 | 10.53 | 12.55 |
| 7180 | Small unit, 4" x 18" x 10" deep, 3 plane split | | 400 | .060 | | 12 | 2.25 | .38 | 14.63 | 17.05 |
| 7190 | Straight split | | 400 | .060 | | 12.20 | 2.25 | .38 | 14.83 | 17.25 |
| 7200 | Cap unit, 3 plane split | | 300 | .080 | | 11.80 | 2.99 | .51 | 15.30 | 18.10 |
| 7210 | Cap unit, straight split | | 300 | .080 | | 11.80 | 2.99 | .51 | 15.30 | 18.10 |
| 7260 | For reinforcing, add | | | | | | | | 4.28 | 4.71 |
| 8000 | For higher walls, add components as necessary | | | | | | | | | |

# 32 32 Retaining Walls

## 32 32 26 – Metal Crib Retaining Walls

### 32 32 26.10 Metal Bin Retaining Walls

| 32 32 26.10 Metal Bin Retaining Walls | Crew | Daily Output | Labor-Hours | Unit | Material | 2011 Bare Costs Labor | Equipment | Total | Total Incl O&P |
|---|---|---|---|---|---|---|---|---|---|
| 0010 **METAL BIN RETAINING WALLS** | | | | | | | | | |
| 0011 Aluminized steel bin, excavation | | | | | | | | | |
| 0020 and backfill not included, 10' wide | | | | | | | | | |
| 0100 4' high, 5.5' deep | B-13 | 650 | .086 | S.F. | 23 | 3.21 | 1.15 | 27.36 | 31.50 |
| 0200 8' high, 5.5' deep | | 615 | .091 | | 26.50 | 3.39 | 1.22 | 31.11 | 35.50 |
| 0300 10' high, 7.7' deep | | 580 | .097 | | 28 | 3.59 | 1.29 | 32.88 | 38 |
| 0400 12' high, 7.7' deep | | 530 | .106 | | 30 | 3.93 | 1.41 | 35.34 | 41 |
| 0500 16' high, 7.7' deep | | 515 | .109 | | 32 | 4.05 | 1.45 | 37.50 | 43 |
| 0600 16' high, 9.9' deep | | 500 | .112 | | 33.50 | 4.17 | 1.50 | 39.17 | 45 |
| 0700 20' high, 9.9' deep | | 470 | .119 | | 38 | 4.43 | 1.59 | 44.02 | 50 |
| 0800 20' high, 12.1' deep | | 460 | .122 | | 40.50 | 4.53 | 1.63 | 46.66 | 53 |
| 0900 24' high, 12.1' deep | | 455 | .123 | | 43 | 4.58 | 1.65 | 49.23 | 56.50 |
| 1000 24' high, 14.3' deep | | 450 | .124 | | 44.50 | 4.63 | 1.66 | 50.79 | 58 |
| 1100 28' high, 14.3' deep | | 440 | .127 | | 46.50 | 4.74 | 1.70 | 52.94 | 60 |
| 1300 For plain galvanized bin type walls, deduct | | | | | 10% | | | | |

## 32 32 36 – Gabion Retaining Walls

### 32 32 36.10 Stone Gabion Retaining Walls

| | Crew | Daily Output | Labor-Hours | Unit | Material | 2011 Bare Costs Labor | Equipment | Total | Total Incl O&P |
|---|---|---|---|---|---|---|---|---|---|
| 0010 **STONE GABION RETAINING WALLS** | | | | | | | | | |
| 4300 Stone filled gabions, not incl. excavation, | | | | | | | | | |
| 4310 Stone, delivered, 3' wide | | | | | | | | | |
| 4350 Galvanized, 6' high, 33° slope embankment | B-13 | 49 | 1.143 | L.F. | 22 | 42.50 | 15.30 | 79.80 | 106 |
| 4500 Highway surcharge | | 27 | 2.074 | | 80 | 77 | 27.50 | 184.50 | 236 |
| 4600 9' high, up to 33° slope embankment | | 24 | 2.333 | | 50 | 87 | 31 | 168 | 222 |
| 4700 Highway surcharge | | 16 | 3.500 | | 140 | 130 | 47 | 317 | 405 |
| 4900 12' high, up to 33° slope embankment | | 14 | 4 | | 77.50 | 149 | 53.50 | 280 | 370 |
| 5000 Highway surcharge | | 11 | 5.091 | | 200 | 189 | 68 | 457 | 585 |
| 5950 For PVC coating, add | | | | | 20% | | | | |

## 32 32 60 – Stone Retaining Walls

### 32 32 60.10 Retaining Walls, Stone

| | Crew | Daily Output | Labor-Hours | Unit | Material | 2011 Bare Costs Labor | Equipment | Total | Total Incl O&P |
|---|---|---|---|---|---|---|---|---|---|
| 0010 **RETAINING WALLS, STONE** | | | | | | | | | |
| 0015 Including excavation, concrete footing and | | | | | | | | | |
| 0020 stone 3' below grade. Price is exposed face area. | | | | | | | | | |
| 0200 Decorative random stone, to 6' high, 1'-6" thick, dry set | D-1 | 35 | .457 | S.F. | 50 | 17.80 | | 67.80 | 81.50 |
| 0300 Mortar set | | 40 | .400 | | 55 | 15.60 | | 70.60 | 84 |
| 0500 Cut stone, to 6' high, 1'-6" thick, dry set | | 35 | .457 | | 50 | 17.80 | | 67.80 | 81.50 |
| 0600 Mortar set | | 40 | .400 | | 55 | 15.60 | | 70.60 | 84 |
| 0800 Random stone, 6' to 10' high, 2' thick, dry set | | 45 | .356 | | 55 | 13.85 | | 68.85 | 81.50 |
| 0900 Mortar set | | 50 | .320 | | 60 | 12.45 | | 72.45 | 84.50 |
| 1100 Cut stone, 6' to 10' high, 2' thick, dry set | | 45 | .356 | | 55 | 13.85 | | 68.85 | 81.50 |
| 1200 Mortar set | | 50 | .320 | | 60 | 12.45 | | 72.45 | 84.50 |

# 32 34 Fabricated Bridges

## 32 34 20 – Fabricated Pedestrian Bridges

| 32 34 20.10 Bridges, Pedestrian | Crew | Daily Output | Labor-Hours | Unit | Material | 2011 Bare Costs Labor | Equipment | Total | Total Incl O&P |
|---|---|---|---|---|---|---|---|---|---|
| 0010 **BRIDGES, PEDESTRIAN** | | | | | | | | | |
| 0011 Spans over streams, roadways, etc. | | | | | | | | | |
| 0020 including erection, not including foundations | | | | | | | | | |
| 0050 Precast concrete, complete in place, 8' wide, 60' span | E-2 | 215 | .260 | S.F. | 105 | 12.35 | 7.55 | 124.90 | 144 |
| 0100 100' span | | 185 | .303 | | 114 | 14.35 | 8.75 | 137.10 | 160 |
| 0150 120' span | | 160 | .350 | | 124 | 16.60 | 10.15 | 150.75 | 176 |
| 0200 150' span | | 145 | .386 | | 129 | 18.30 | 11.20 | 158.50 | 185 |
| 0300 Steel, trussed or arch spans, compl. in place, 8' wide, 40' span | | 320 | .175 | | 99 | 8.30 | 5.05 | 112.35 | 129 |
| 0400 50' span | | 395 | .142 | | 89 | 6.70 | 4.11 | 99.81 | 114 |
| 0500 60' span | | 465 | .120 | | 89 | 5.70 | 3.49 | 98.19 | 111 |
| 0600 80' span | | 570 | .098 | | 106 | 4.65 | 2.85 | 113.50 | 128 |
| 0700 100' span | | 465 | .120 | | 149 | 5.70 | 3.49 | 158.19 | 177 |
| 0800 120' span | | 365 | .153 | | 188 | 7.25 | 4.44 | 199.69 | 224 |
| 0900 150' span | | 310 | .181 | | 200 | 8.55 | 5.25 | 213.80 | 240 |
| 1000 160' span | | 255 | .220 | | 200 | 10.40 | 6.35 | 216.75 | 245 |
| 1100 10' wide, 80' span | | 640 | .088 | | 110 | 4.14 | 2.53 | 116.67 | 131 |
| 1200 120' span | | 415 | .135 | | 143 | 6.40 | 3.91 | 153.31 | 172 |
| 1300 150' span | | 445 | .126 | | 160 | 5.95 | 3.64 | 169.59 | 190 |
| 1400 200' span | | 205 | .273 | | 171 | 12.95 | 7.90 | 191.85 | 219 |
| 1600 Wood, laminated type, complete in place, 80' span | C-12 | 203 | .236 | | 65.50 | 10.05 | 3.23 | 78.78 | 91 |
| 1700 130' span | " | 153 | .314 | | 68 | 13.35 | 4.29 | 85.64 | 100 |

# 32 35 Screening Devices

## 32 35 16 – Sound Barriers

### 32 35 16.10 Traffic Barriers, Highway Sound Barriers

| | Crew | Daily Output | Labor-Hours | Unit | Material | 2011 Bare Costs Labor | Equipment | Total | Total Incl O&P |
|---|---|---|---|---|---|---|---|---|---|
| 0010 **TRAFFIC BARRIERS, HIGHWAY SOUND BARRIERS** | | | | | | | | | |
| 0020 Highway sound barriers, not including footing | | | | | | | | | |
| 0100 Precast concrete, concrete columns @ 30' OC, 8" T, 8' H | C-12 | 400 | .120 | L.F. | 128 | 5.10 | 1.64 | 134.74 | 151 |
| 0110 12' H | | 265 | .181 | | 192 | 7.70 | 2.47 | 202.17 | 225 |
| 0120 16' H | | 200 | .240 | | 256 | 10.20 | 3.28 | 269.48 | 300 |
| 0130 20' H | | 160 | .300 | | 320 | 12.75 | 4.10 | 336.85 | 375 |
| 0400 Lt. Wt. composite panel, cementitious face, St. posts @ 12' OC, 8' H | B-80B | 190 | .168 | | 143 | 6.15 | 1.19 | 150.34 | 169 |
| 0410 12' H | | 125 | .256 | | 215 | 9.40 | 1.80 | 226.20 | 253 |
| 0420 16' H | | 95 | .337 | | 287 | 12.35 | 2.37 | 301.72 | 335 |
| 0430 20' H | | 75 | .427 | | 360 | 15.65 | 3.01 | 378.66 | 420 |

# 32 84 Planting Irrigation

## 32 84 23 – Underground Sprinklers

### 32 84 23.10 Sprinkler Irrigation System

| | Crew | Daily Output | Labor-Hours | Unit | Material | 2011 Bare Costs Labor | Equipment | Total | Total Incl O&P |
|---|---|---|---|---|---|---|---|---|---|
| 0010 **SPRINKLER IRRIGATION SYSTEM** | | | | | | | | | |
| 0011 For lawns | | | | | | | | | |
| 0100 Golf course with fully automatic system | C-17 | .05 | 1600 | 9 holes | 100,000 | 71,500 | | 171,500 | 220,000 |
| 0200 24' diam. head at 15' O.C incl. piping, auto oper., minimum | B-20 | 70 | .343 | Head | 21 | 13.15 | | 34.15 | 43 |
| 0300 Maximum | | 40 | .600 | | 52 | 23 | | 75 | 92.50 |
| 0500 60' diam. head at 40' O.C. incl. piping, auto oper., minimum | | 28 | .857 | | 66 | 33 | | 99 | 123 |
| 0600 Maximum | | 23 | 1.043 | | 183 | 40 | | 223 | 263 |
| 0800 Residential system, custom, 1" supply | | 2000 | .012 | S.F. | .32 | .46 | | .78 | 1.06 |
| 0900 1-1/2" supply | | 1800 | .013 | " | .37 | .51 | | .88 | 1.19 |
| 1020 Pop up spray head w/risers, hi-pop, full circle pattern, 4" | 2 Skwk | 76 | .211 | Ea. | 4.24 | 9.35 | | 13.59 | 19 |

| 32 84 23.10 Sprinkler Irrigation System | Crew | Daily Output | Labor-Hours | Unit | Material | 2011 Bare Costs Labor | Equipment | Total | Total Incl O&P |
|---|---|---|---|---|---|---|---|---|---|
| 1030 | 1/2 circle pattern, 4" | 2 Skwk | 76 | .211 | Ea. | 5.15 | 9.35 | | 14.50 | 20 |
| 1040 | 6", full circle pattern | | 76 | .211 | | 10.20 | 9.35 | | 19.55 | 25.50 |
| 1050 | 1/2 circle pattern, 6" | | 76 | .211 | | 11.05 | 9.35 | | 20.40 | 26.50 |
| 1060 | 12", full circle pattern | | 76 | .211 | | 15.70 | 9.35 | | 25.05 | 31.50 |
| 1070 | 1/2 circle pattern, 12" | | 76 | .211 | | 17.35 | 9.35 | | 26.70 | 33.50 |
| 1080 | Pop up bubbler head w/risers, hi-pop bubbler head, 4" | | 76 | .211 | | 5.15 | 9.35 | | 14.50 | 20 |
| 1090 | 6" | | 76 | .211 | | 11.45 | 9.35 | | 20.80 | 27 |
| 1100 | 12" | | 76 | .211 | | 15.35 | 9.35 | | 24.70 | 31.50 |
| 1110 | Impact full/part circle sprinklers, 28'-54' 25-60 PSI | | 37 | .432 | | 15.15 | 19.20 | | 34.35 | 46 |
| 1120 | Spaced 37'-49' @ 25-50 PSI | | 37 | .432 | | 29 | 19.20 | | 48.20 | 61 |
| 1130 | Spaced 43'-61' @ 30-60 PSI | | 37 | .432 | | 56.50 | 19.20 | | 75.70 | 91.50 |
| 1140 | Spaced 54'-78' @ 40-80 PSI | | 37 | .432 | | 105 | 19.20 | | 124.20 | 145 |
| 1145 | Impact rotor pop-up full/part commercial circle sprinklers | | | | | | | | | |
| 1150 | Spaced 42'-65' 35-80 PSI | 2 Skwk | 25 | .640 | Ea. | 13 | 28.50 | | 41.50 | 58 |
| 1160 | Spaced 48'-76' 45-85 PSI | " | 25 | .640 | " | 14.05 | 28.50 | | 42.55 | 59 |
| 1165 | Impact rotor pop-up part. circle comm., 53'-75', 55-100 PSI, w/accessories | | | | | | | | | |
| 1170 | Plastic case, metal cover | 2 Skwk | 25 | .640 | Ea. | 120 | 28.50 | | 148.50 | 175 |
| 1180 | Rubber cover | | 25 | .640 | | 132 | 28.50 | | 160.50 | 189 |
| 1190 | Iron case, metal cover | | 22 | .727 | | 144 | 32.50 | | 176.50 | 208 |
| 1200 | Rubber cover | | 22 | .727 | | 166 | 32.50 | | 198.50 | 232 |
| 1250 | Plastic case, 2 nozzle, metal cover | | 25 | .640 | | 143 | 28.50 | | 171.50 | 201 |
| 1260 | Rubber cover | | 25 | .640 | | 150 | 28.50 | | 178.50 | 208 |
| 1270 | Iron case, 2 nozzle, metal cover | | 22 | .727 | | 213 | 32.50 | | 245.50 | 284 |
| 1280 | Rubber cover | | 22 | .727 | | 210 | 32.50 | | 242.50 | 281 |
| 1282 | Impact rotor pop-up full circle commercial, 39'-99', 30-100 PSI | | | | | | | | | |
| 1284 | Plastic case, metal cover | 2 Skwk | 25 | .640 | Ea. | 150 | 28.50 | | 178.50 | 209 |
| 1286 | Rubber cover | | 25 | .640 | | 169 | 28.50 | | 197.50 | 230 |
| 1288 | Iron case, metal cover | | 22 | .727 | | 218 | 32.50 | | 250.50 | 289 |
| 1290 | Rubber cover | | 22 | .727 | | 226 | 32.50 | | 258.50 | 298 |
| 1292 | Plastic case, 2 nozzle, metal cover | | 22 | .727 | | 164 | 32.50 | | 196.50 | 230 |
| 1294 | Rubber cover | | 22 | .727 | | 166 | 32.50 | | 198.50 | 233 |
| 1296 | Iron case, 2 nozzle, metal cover | | 20 | .800 | | 208 | 35.50 | | 243.50 | 283 |
| 1298 | Rubber cover | | 20 | .800 | | 221 | 35.50 | | 256.50 | 298 |
| 1305 | Electric remote control valve, plastic, 3/4" | | 18 | .889 | | 19 | 39.50 | | 58.50 | 81.50 |
| 1310 | 1" | | 18 | .889 | | 22.50 | 39.50 | | 62 | 85.50 |
| 1320 | 1-1/2" | | 18 | .889 | | 117 | 39.50 | | 156.50 | 190 |
| 1330 | 2" | | 18 | .889 | | 128 | 39.50 | | 167.50 | 201 |
| 1335 | Quick coupling valves, brass, locking cover | | | | | | | | | |
| 1340 | Inlet coupling valve, 3/4" | 2 Skwk | 18.75 | .853 | Ea. | 44 | 38 | | 82 | 107 |
| 1350 | 1" | | 18.75 | .853 | | 43 | 38 | | 81 | 106 |
| 1360 | Controller valve boxes, 6" round boxes | | 18.75 | .853 | | 4.98 | 38 | | 42.98 | 63.50 |
| 1370 | 10" round boxes | | 14.25 | 1.123 | | 12.90 | 50 | | 62.90 | 90.50 |
| 1380 | 12" square box | | 9.75 | 1.641 | | 26 | 73 | | 99 | 141 |
| 1388 | Electromech. control, 14 day 3-60 min, auto start to 23/day | | | | | | | | | |
| 1390 | 4 station | 2 Skwk | 1.04 | 15.385 | Ea. | 159 | 685 | | 844 | 1,225 |
| 1400 | 7 station | | .64 | 25 | | 315 | 1,100 | | 1,415 | 2,050 |
| 1410 | 12 station | | .40 | 40 | | 330 | 1,775 | | 2,105 | 3,100 |
| 1420 | Dual programs, 18 station | | .24 | 66.667 | | 1,100 | 2,950 | | 4,050 | 5,750 |
| 1430 | 23 station | | .16 | 100 | | 1,850 | 4,450 | | 6,300 | 8,825 |
| 1435 | Backflow preventer, bronze, 0-175 PSI, w/valves, test cocks | | | | | | | | | |
| 1440 | 3/4" | 2 Skwk | 2 | 8 | Ea. | 136 | 355 | | 491 | 695 |
| 1450 | 1" | | 2 | 8 | | 159 | 355 | | 514 | 720 |
| 1460 | 1-1/2" | | 2 | 8 | | 242 | 355 | | 597 | 810 |

# 32 84 Planting Irrigation

## 32 84 23 – Underground Sprinklers

| 32 84 23.10 Sprinkler Irrigation System | | Crew | Daily Output | Labor-Hours | Unit | Material | 2011 Bare Costs Labor | Equipment | Total | Total Incl O&P |
|---|---|---|---|---|---|---|---|---|---|---|
| 1470 | 2" | 2 Skwk | 2 | 8 | Ea. | 266 | 355 | | 621 | 840 |
| 1475 | Pressure vacuum breaker, brass, 15-150 PSI | 2 Skwk | 2 | 8 | Ea. | 81 | 355 | | 436 | 635 |
| 1480 | 3/4" | | 2 | 8 | | 104 | 355 | | 459 | 660 |
| 1490 | 1" | | 2 | 8 | | 360 | 355 | | 715 | 940 |
| 1500 | 1-1/2" | | 2 | 8 | | 410 | 355 | | 765 | 995 |
| 1510 | 2" | | 2 | 8 | | | | | | |

# 32 91 Planting Preparation

## 32 91 13 – Soil Preparation

### 32 91 13.16 Mulching

| 32 91 13.16 Mulching | | Crew | Daily Output | Labor-Hours | Unit | Material | 2011 Bare Costs Labor | Equipment | Total | Total Incl O&P |
|---|---|---|---|---|---|---|---|---|---|---|
| 0010 | **MULCHING** | | | | | | | | | |
| 0100 | Aged barks, 3" deep, hand spread | 1 Clab | 100 | .080 | S.Y. | 3.02 | 2.75 | | 5.77 | 7.55 |
| 0150 | Skid steer loader | B-63 | 13.50 | 2.963 | M.S.F. | 335 | 107 | 11.25 | 453.25 | 545 |
| 0200 | Hay, 1" deep, hand spread | 1 Clab | 475 | .017 | S.Y. | .50 | .58 | | 1.08 | 1.44 |
| 0250 | Power mulcher, small | B-64 | 180 | .089 | M.S.F. | 55.50 | 3.01 | 1.68 | 60.19 | 67.50 |
| 0350 | Large | B-65 | 530 | .030 | " | 55.50 | 1.02 | .81 | 57.33 | 63.50 |
| 0400 | Humus peat, 1" deep, hand spread | 1 Clab | 700 | .011 | S.Y. | 3.99 | .39 | | 4.38 | 4.99 |
| 0450 | Push spreader | " | 2500 | .003 | " | 3.99 | .11 | | 4.10 | 4.56 |
| 0550 | Tractor spreader | B-66 | 700 | .011 | M.S.F. | 445 | .50 | .32 | 445.82 | 490 |
| 0600 | Oat straw, 1" deep, hand spread | 1 Clab | 475 | .017 | S.Y. | .50 | .58 | | 1.08 | 1.44 |
| 0650 | Power mulcher, small | B-64 | 180 | .089 | M.S.F. | 55.50 | 3.01 | 1.68 | 60.19 | 67.50 |
| 0700 | Large | B-65 | 530 | .030 | " | 55.50 | 1.02 | .81 | 57.33 | 63.50 |
| 0750 | Add for asphaltic emulsion | B-45 | 1770 | .009 | Gal. | 2.14 | .36 | .43 | 2.93 | 3.37 |
| 0800 | Peat moss, 1" deep, hand spread | 1 Clab | 900 | .009 | S.Y. | 2.98 | .31 | | 3.29 | 3.75 |
| 0850 | Push spreader | " | 2500 | .003 | " | 2.98 | .11 | | 3.09 | 3.45 |
| 0950 | Tractor spreader | B-66 | 700 | .011 | M.S.F. | 330 | .50 | .32 | 330.82 | 365 |
| 1000 | Polyethylene film, 6 mil. | 2 Clab | 2000 | .008 | S.Y. | .23 | .27 | | .50 | .67 |
| 1100 | Redwood nuggets, 3" deep, hand spread | 1 Clab | 150 | .053 | " | 3 | 1.83 | | 4.83 | 6.10 |
| 1150 | Skid steer loader | B-63 | 13.50 | 2.963 | M.S.F. | 335 | 107 | 11.25 | 453.25 | 540 |
| 1200 | Stone mulch, hand spread, ceramic chips, economy | 1 Clab | 125 | .064 | S.Y. | 6.95 | 2.20 | | 9.15 | 11 |
| 1250 | Deluxe | " | 95 | .084 | " | 10.75 | 2.89 | | 13.64 | 16.30 |
| 1300 | Granite chips | B-1 | 10 | 2.400 | C.Y. | 35 | 84 | | 119 | 168 |
| 1400 | Marble chips | | 10 | 2.400 | | 130 | 84 | | 214 | 272 |
| 1600 | Pea gravel | | 28 | .857 | | 69 | 30 | | 99 | 122 |
| 1700 | Quartz | | 10 | 2.400 | | 170 | 84 | | 254 | 315 |
| 1800 | Tar paper, 15 lb. felt | 1 Clab | 800 | .010 | S.Y. | .51 | .34 | | .85 | 1.09 |
| 1900 | Wood chips, 2" deep, hand spread | " | 220 | .036 | " | 1.95 | 1.25 | | 3.20 | 4.07 |
| 1950 | Skid steer loader | B-63 | 20.30 | 1.970 | M.S.F. | 217 | 71.50 | 7.50 | 296 | 355 |

### 32 91 13.26 Planting Beds

| 32 91 13.26 Planting Beds | | Crew | Daily Output | Labor-Hours | Unit | Material | 2011 Bare Costs Labor | Equipment | Total | Total Incl O&P |
|---|---|---|---|---|---|---|---|---|---|---|
| 0010 | **PLANTING BEDS** | | | | | | | | | |
| 0100 | Backfill planting pit, by hand, on site topsoil | 2 Clab | 18 | .889 | C.Y. | | 30.50 | | 30.50 | 47 |
| 0200 | Prepared planting mix, by hand | " | 24 | .667 | | | 23 | | 23 | 35 |
| 0300 | Skid steer loader, on site topsoil | B-62 | 340 | .071 | | | 2.64 | .45 | 3.09 | 4.50 |
| 0400 | Prepared planting mix | " | 410 | .059 | | | 2.19 | .37 | 2.56 | 3.74 |
| 1000 | Excavate planting pit, by hand, sandy soil | 2 Clab | 16 | 1 | | | 34.50 | | 34.50 | 52.50 |
| 1100 | Heavy soil or clay | " | 8 | 2 | | | 68.50 | | 68.50 | 105 |
| 1200 | 1/2 C.Y. backhoe, sandy soil | B-11C | 150 | .107 | | | 4.25 | 2.14 | 6.39 | 8.75 |
| 1300 | Heavy soil or clay | " | 115 | .139 | | | 5.55 | 2.79 | 8.34 | 11.45 |
| 2000 | Mix planting soil, incl. loam, manure, peat, by hand | 2 Clab | 60 | .267 | | 42 | 9.15 | | 51.15 | 60 |
| 2100 | Skid steer loader | B-62 | 150 | .160 | | 42 | 6 | 1.01 | 49.01 | 56 |
| 3000 | Pile sod, skid steer loader | " | 2800 | .009 | S.Y. | | .32 | .05 | .37 | .55 |

# 32 91 Planting Preparation

## 32 91 13 – Soil Preparation

### 32 91 13.26 Planting Beds

| | | Crew | Daily Output | Labor-Hours | Unit | Material | 2011 Bare Costs Labor | Equipment | Total | Total Incl O&P |
|---|---|---|---|---|---|---|---|---|---|---|
| 3100 | By hand | 2 Clab | 400 | .040 | S.Y. | | 1.37 | | 1.37 | 2.11 |
| 4000 | Remove sod, F.E. loader | B-10S | 2000 | .006 | | | .25 | .16 | .41 | .56 |
| 4100 | Sod cutter | B-12K | 3200 | .005 | | | .20 | .31 | .51 | .65 |
| 4200 | By hand | 2 Clab | 240 | .067 | ↓ | | 2.29 | | 2.29 | 3.51 |

## 32 91 19 – Landscape Grading

### 32 91 19.13 Topsoil Placement and Grading

| | | Crew | Daily Output | Labor-Hours | Unit | Material | 2011 Bare Costs Labor | Equipment | Total | Total Incl O&P |
|---|---|---|---|---|---|---|---|---|---|---|
| 0010 | **TOPSOIL PLACEMENT AND GRADING** | | | | | | | | | |
| 0400 | Spread from pile to rough finish grade, F.E. loader, 1.5 CY | B-10S | 200 | .060 | C.Y. | | 2.50 | 1.60 | 4.10 | 5.50 |
| 0500 | Up to 200' radius, by hand | 1 Clab | 14 | .571 | | | 19.65 | | 19.65 | 30 |
| 0600 | Top dress by hand, 1 C.Y. for 600 S.F. | " | 11.50 | .696 | ↓ | 33 | 24 | | 57 | 73 |
| 0700 | Furnish and place, truck dumped, screened, 4" deep | B-10S | 1300 | .009 | S.Y. | 4.13 | .38 | .25 | 4.76 | 5.40 |
| 0800 | 6" deep | " | 820 | .015 | " | 5.30 | .61 | .39 | 6.30 | 7.15 |

# 32 92 Turf and Grasses

## 32 92 19 – Seeding

### 32 92 19.13 Mechanical Seeding

| | | Crew | Daily Output | Labor-Hours | Unit | Material | 2011 Bare Costs Labor | Equipment | Total | Total Incl O&P |
|---|---|---|---|---|---|---|---|---|---|---|
| 0010 | **MECHANICAL SEEDING** R329219-50 | | | | | | | | | |
| 0020 | Mechanical seeding, 215 lb./acre | B-66 | 1.50 | 5.333 | Acre | 570 | 233 | 149 | 952 | 1,125 |
| 0100 | 44 lb./M.S.Y. | " | 2500 | .003 | S.Y. | .20 | .14 | .09 | .43 | .53 |
| 0300 | Fine grading and seeding incl. lime, fertilizer & seed, | | | | | | | | | |
| 0310 | with equipment | B-14 | 1000 | .048 | S.Y. | .48 | 1.74 | .32 | 2.54 | 3.53 |
| 0600 | Limestone hand push spreader, 50 lbs. per M.S.F. | 1 Clab | 180 | .044 | M.S.F. | 3.95 | 1.53 | | 5.48 | 6.70 |
| 0800 | Grass seed hand push spreader, 4.5 lbs. per M.S.F. | " | 180 | .044 | " | 22 | 1.53 | | 23.53 | 27 |
| 1000 | Hydro or air seeding for large areas, incl. seed and fertilizer | B-81 | 8900 | .003 | S.Y. | .20 | .10 | .06 | .36 | .45 |
| 1100 | With wood fiber mulch added | " | 8900 | .003 | " | .34 | .10 | .06 | .50 | .60 |
| 1300 | Seed only, over 100 lbs., field seed, minimum | | | | Lb. | 1.15 | | | 1.15 | 1.27 |
| 1400 | Maximum | | | | | 2.75 | | | 2.75 | 3.03 |
| 1500 | Lawn seed, minimum | | | | | .80 | | | .80 | .88 |
| 1600 | Maximum | | | | ↓ | 1.40 | | | 1.40 | 1.54 |
| 1800 | Aerial operations, seeding only, field seed | B-58 | 50 | .480 | Acre | 375 | 17.95 | 58 | 450.95 | 500 |
| 1900 | Lawn seed | | 50 | .480 | | 260 | 17.95 | 58 | 335.95 | 380 |
| 2100 | Seed and liquid fertilizer, field seed | | 50 | .480 | | 465 | 17.95 | 58 | 540.95 | 600 |
| 2200 | Lawn seed | ↓ | 50 | .480 | ↓ | 350 | 17.95 | 58 | 425.95 | 475 |

## 32 92 23 – Sodding

### 32 92 23.10 Sodding Systems

| | | Crew | Daily Output | Labor-Hours | Unit | Material | 2011 Bare Costs Labor | Equipment | Total | Total Incl O&P |
|---|---|---|---|---|---|---|---|---|---|---|
| 0010 | **SODDING SYSTEMS** | | | | | | | | | |
| 0020 | Sodding, 1" deep, bluegrass sod, on level ground, over 8 M.S.F. | B-63 | 22 | 1.818 | M.S.F. | 215 | 66 | 6.90 | 287.90 | 345 |
| 0200 | 4 M.S.F. | | 17 | 2.353 | | 275 | 85 | 8.95 | 368.95 | 445 |
| 0300 | 1000 S.F. | | 13.50 | 2.963 | | 300 | 107 | 11.25 | 418.25 | 505 |
| 0500 | Sloped ground, over 8 M.S.F. | | 6 | 6.667 | | 215 | 241 | 25.50 | 481.50 | 635 |
| 0600 | 4 M.S.F. | | 5 | 8 | | 275 | 290 | 30.50 | 595.50 | 780 |
| 0700 | 1000 S.F. | | 4 | 10 | | 300 | 360 | 38 | 698 | 920 |
| 1000 | Bent grass sod, on level ground, over 6 M.S.F. | | 20 | 2 | | 530 | 72.50 | 7.60 | 610.10 | 700 |
| 1100 | 3 M.S.F. | | 18 | 2.222 | | 610 | 80.50 | 8.45 | 698.95 | 800 |
| 1200 | Sodding 1000 S.F. or less | | 14 | 2.857 | | 695 | 103 | 10.85 | 808.85 | 935 |
| 1500 | Sloped ground, over 6 M.S.F. | | 15 | 2.667 | | 530 | 96.50 | 10.15 | 636.65 | 740 |
| 1600 | 3 M.S.F. | | 13.50 | 2.963 | | 610 | 107 | 11.25 | 728.25 | 845 |
| 1700 | 1000 S.F. | ↓ | 12 | 3.333 | ↓ | 695 | 121 | 12.65 | 828.65 | 965 |

# 32 93 Plants

## 32 93 10 – General Planting Costs

| | 32 93 10.12 Travel | Crew | Daily Output | Labor-Hours | Unit | Material | 2011 Bare Costs Labor | Equipment | Total | Total Incl O&P |
|---|---|---|---|---|---|---|---|---|---|---|
| 0010 | **TRAVEL** to all nursery items, for 10 to 20 miles, add | | | | | | | | | |
| 0015 | 10 to 20 miles one way, add | | | | All | | | | | 5% |
| 0100 | 30 to 50 miles one way, add | | | | " | | | | | 10% |

## 32 93 13 – Ground Covers

### 32 93 13.10 Ground Cover Plants

| | | Crew | Daily Output | Labor-Hours | Unit | Material | Labor | Equipment | Total | Total Incl O&P |
|---|---|---|---|---|---|---|---|---|---|---|
| 0010 | **GROUND COVER PLANTS** | | | | | | | | | |
| 0012 | Plants, pachysandra, in prepared beds | B-1 | 15 | 1.600 | C | 30 | 56 | | 86 | 119 |
| 0200 | Vinca minor, 1 yr., bare root, in prepared beds | | 12 | 2 | " | 43 | 70 | | 113 | 155 |
| 0600 | Stone chips, in 50 lb. bags, Georgia marble | | 520 | .046 | Bag | 6 | 1.62 | | 7.62 | 9.10 |
| 0700 | Onyx gemstone | | 260 | .092 | | 23 | 3.23 | | 26.23 | 30 |
| 0800 | Quartz | | 260 | .092 | ↓ | 8.60 | 3.23 | | 11.83 | 14.40 |
| 0900 | Pea gravel, truckload lots | ↓ | 28 | .857 | Ton | 30 | 30 | | 60 | 79 |

## 32 93 33 – Shrubs

### 32 93 33.10 Shrubs and Trees

| | | Crew | Daily Output | Labor-Hours | Unit | Material | Labor | Equipment | Total | Total Incl O&P |
|---|---|---|---|---|---|---|---|---|---|---|
| 0010 | **SHRUBS AND TREES** | | | | | | | | | |
| 0011 | Evergreen, in prepared beds, B & B | | | | | | | | | |
| 0100 | Arborvitae pyramidal, 4'-5' | B-17 | 30 | 1.067 | Ea. | 49 | 39 | 21 | 109 | 137 |
| 0150 | Globe, 12"-15" | B-1 | 96 | .250 | | 20.50 | 8.75 | | 29.25 | 36 |
| 0300 | Cedar, blue, 8'-10' | B-17 | 18 | 1.778 | | 210 | 65 | 35.50 | 310.50 | 370 |
| 0500 | Hemlock, canadian, 2-1/2'-3' | B-1 | 36 | .667 | | 32 | 23.50 | | 55.50 | 71 |
| 0550 | Holly, Savannah, 8' - 10' H | | 9.68 | 2.479 | | 375 | 87 | | 462 | 545 |
| 0600 | Juniper, andorra, 18"-24" | | 80 | .300 | | 16.80 | 10.50 | | 27.30 | 34.50 |
| 0620 | Wiltoni, 15"-18" | ↓ | 80 | .300 | | 17.25 | 10.50 | | 27.75 | 35 |
| 0640 | Skyrocket, 4-1/2'-5' | B-17 | 55 | .582 | | 68.50 | 21.50 | 11.55 | 101.55 | 121 |
| 0660 | Blue pfitzer, 2'-2-1/2' | B-1 | 44 | .545 | | 32.50 | 19.10 | | 51.60 | 65 |
| 0680 | Ketleerie, 2-1/2'-3' | | 50 | .480 | | 24 | 16.80 | | 40.80 | 52 |
| 0700 | Pine, black, 2-1/2'-3' | | 50 | .480 | | 50.50 | 16.80 | | 67.30 | 81.50 |
| 0720 | Mugo, 18"-24" | ↓ | 60 | .400 | | 40 | 14 | | 54 | 65.50 |
| 0740 | White, 4'-5' | B-17 | 75 | .427 | | 57 | 15.65 | 8.50 | 81.15 | 96 |
| 0800 | Spruce, blue, 18"-24" | B-1 | 60 | .400 | | 41 | 14 | | 55 | 66.50 |
| 0840 | Norway, 4'-5' | B-17 | 75 | .427 | | 65 | 15.65 | 8.50 | 89.15 | 105 |
| 0900 | Yew, denisforma, 12"-15" | B-1 | 60 | .400 | | 30.50 | 14 | | 44.50 | 55 |
| 1000 | Capitata, 18"-24" | | 30 | .800 | | 25 | 28 | | 53 | 70 |
| 1100 | Hicksi, 2'-2-1/2' | ↓ | 30 | .800 | ↓ | 31.50 | 28 | | 59.50 | 78 |

### 32 93 33.20 Shrubs

| | | Crew | Daily Output | Labor-Hours | Unit | Material | Labor | Equipment | Total | Total Incl O&P |
|---|---|---|---|---|---|---|---|---|---|---|
| 0010 | **SHRUBS** | | | | | | | | | |
| 0011 | Broadleaf Evergreen, planted in prepared beds | | | | | | | | | |
| 0100 | Andromeda, 15"-18", container | B-1 | 96 | .250 | Ea. | 28.50 | 8.75 | | 37.25 | 44.50 |
| 0200 | Azalea, 15" - 18", container | | 96 | .250 | | 21 | 8.75 | | 29.75 | 36.50 |
| 0300 | Barberry, 9"-12", container | | 130 | .185 | | 12.30 | 6.45 | | 18.75 | 23.50 |
| 0400 | Boxwood, 15"-18", B&B | | 96 | .250 | | 32 | 8.75 | | 40.75 | 48.50 |
| 0500 | Euonymus, emerald gaiety, 12" to 15", container | | 115 | .209 | | 22 | 7.30 | | 29.30 | 35 |
| 0600 | Holly, 15"-18", B & B | | 96 | .250 | | 17 | 8.75 | | 25.75 | 32 |
| 0900 | Mount laurel, 18" - 24", B & B | | 80 | .300 | | 60 | 10.50 | | 70.50 | 82 |
| 1000 | Paxistema, 9 – 12" high | | 130 | .185 | | 19 | 6.45 | | 25.45 | 31 |
| 1100 | Rhododendron, 18"-24", container | | 48 | .500 | | 28 | 17.50 | | 45.50 | 58 |
| 1200 | Rosemary, 1 gal. container | | 600 | .040 | | 63 | 1.40 | | 64.40 | 71.50 |
| 2000 | Deciduous, planted in prepared beds, amelanchier, 2'-3', B & B | | 57 | .421 | | 89 | 14.75 | | 103.75 | 121 |
| 2100 | Azalea, 15"-18", B & B | | 96 | .250 | | 26.50 | 8.75 | | 35.25 | 42.50 |
| 2300 | Bayberry, 2'-3', B & B | | 57 | .421 | | 24.50 | 14.75 | | 39.25 | 49.50 |
| 2600 | Cotoneaster, 15"-18", B & B | ↓ | 80 | .300 | ↓ | 22.50 | 10.50 | | 33 | 41 |

# 32 93 Plants

## 32 93 33 – Shrubs

### 32 93 33.20 Shrubs

| | | Crew | Daily Output | Labor-Hours | Unit | Material | Labor | Equipment | Total | Total Incl O&P |
|---|---|---|---|---|---|---|---|---|---|---|
| | | | | | | | 2011 Bare Costs | | | |
| 2800 | Dogwood, 3'-4', B & B | B-17 | 40 | .800 | Ea. | 27.50 | 29.50 | 15.90 | 72.90 | 92.50 |
| 2900 | Euonymus, alatus compacta, 15" to 18", container | B-1 | 80 | .300 | | 22.50 | 10.50 | | 33 | 41 |
| 3200 | Forsythia, 2'-3', container | " | 60 | .400 | | 19.25 | 14 | | 33.25 | 42.50 |
| 3300 | Hibiscus, 3'-4', B & B | B-17 | 75 | .427 | | 39.50 | 15.65 | 8.50 | 63.65 | 77 |
| 3400 | Honeysuckle, 3'-4', B & B | B-1 | 60 | .400 | | 21.50 | 14 | | 35.50 | 45 |
| 3500 | Hydrangea, 2'-3', B & B | " | 57 | .421 | | 29 | 14.75 | | 43.75 | 54.50 |
| 3600 | Lilac, 3'-4', B & B | B-17 | 40 | .800 | | 23.50 | 29.50 | 15.90 | 68.90 | 88 |
| 3900 | Privet, bare root, 18"-24" | B-1 | 80 | .300 | | 12.55 | 10.50 | | 23.05 | 30 |
| 4100 | Quince, 2'-3', B & B | " | 57 | .421 | | 25 | 14.75 | | 39.75 | 50 |
| 4200 | Russian olive, 3'-4', B & B | B-17 | 75 | .427 | | 25.50 | 15.65 | 8.50 | 49.65 | 61.50 |
| 4400 | Spirea, 3'-4', B & B | B-1 | 70 | .343 | | 17.25 | 12 | | 29.25 | 37.50 |
| 4500 | Viburnum, 3'-4', B & B | B-17 | 40 | .800 | | 28 | 29.50 | 15.90 | 73.40 | 93 |

## 32 93 43 – Trees

### 32 93 43.20 Trees

| | | | Crew | Daily Output | Labor-Hours | Unit | Material | Labor | Equipment | Total | Total Incl O&P |
|---|---|---|---|---|---|---|---|---|---|---|---|
| 0010 | **TREES** | | | | | | | | | | |
| 0011 | Deciduous, in prep. beds, balled & burlapped (B&B) | | | | | | | | | | |
| 0100 | Ash, 2" caliper | G | B-17 | 8 | 4 | Ea. | 79.50 | 147 | 79.50 | 306 | 400 |
| 0200 | Beech, 5'-6' | G | | 50 | .640 | | 95 | 23.50 | 12.70 | 131.20 | 155 |
| 0300 | Birch, 6'-8', 3 stems | G | | 20 | 1.600 | | 112 | 58.50 | 32 | 202.50 | 247 |
| 0500 | Crabapple, 6'-8' | G | | 20 | 1.600 | | 125 | 58.50 | 32 | 215.50 | 262 |
| 0600 | Dogwood, 4'-5' | G | | 40 | .800 | | 122 | 29.50 | 15.90 | 167.40 | 196 |
| 0700 | Eastern redbud 4'-5' | G | | 40 | .800 | | 146 | 29.50 | 15.90 | 191.40 | 222 |
| 0800 | Elm, 8'-10' | G | | 20 | 1.600 | | 119 | 58.50 | 32 | 209.50 | 255 |
| 0900 | Ginkgo, 6'-7' | G | | 24 | 1.333 | | 180 | 49 | 26.50 | 255.50 | 300 |
| 1000 | Hawthorn, 8'-10', 1" caliper | G | | 20 | 1.600 | | 141 | 58.50 | 32 | 231.50 | 279 |
| 1100 | Honeylocust, 10'-12', 1-1/2" caliper | G | | 10 | 3.200 | | 153 | 117 | 63.50 | 333.50 | 415 |
| 1300 | Larch, 8' | G | | 32 | 1 | | 108 | 36.50 | 19.85 | 164.35 | 196 |
| 1400 | Linden, 8'-10', 1" caliper | G | | 20 | 1.600 | | 113 | 58.50 | 32 | 203.50 | 248 |
| 1500 | Magnolia, 4'-5' | G | | 20 | 1.600 | | 81.50 | 58.50 | 32 | 172 | 214 |
| 1600 | Maple, red, 8'-10', 1-1/2" caliper | G | | 10 | 3.200 | | 255 | 117 | 63.50 | 435.50 | 530 |
| 1700 | Mountain ash, 8'-10', 1" caliper | G | | 16 | 2 | | 228 | 73.50 | 39.50 | 341 | 405 |
| 1800 | Oak, 2-1/2"-3" caliper | G | | 6 | 5.333 | | 315 | 196 | 106 | 617 | 760 |
| 2100 | Planetree, 9'-11', 1-1/4" caliper | G | | 10 | 3.200 | | 125 | 117 | 63.50 | 305.50 | 385 |
| 2200 | Plum, 6'-8', 1" caliper | G | | 20 | 1.600 | | 95.50 | 58.50 | 32 | 186 | 229 |
| 2300 | Poplar, 9'-11', 1-1/4" caliper | G | | 10 | 3.200 | | 95 | 117 | 63.50 | 275.50 | 355 |
| 2500 | Sumac, 2'-3' | G | | 75 | .427 | | 21 | 15.65 | 8.50 | 45.15 | 56.50 |
| 2700 | Tulip, 5'-6' | G | | 40 | .800 | | 53.50 | 29.50 | 15.90 | 98.90 | 121 |
| 2800 | Willow, 6'-8', 1" caliper | G | | 20 | 1.600 | | 57.50 | 58.50 | 32 | 148 | 188 |

# 32 94 Planting Accessories

## 32 94 13 – Landscape Edging

### 32 94 13.20 Edging

| | | Crew | Daily Output | Labor-Hours | Unit | Material | Labor | Equipment | Total | Total Incl O&P |
|---|---|---|---|---|---|---|---|---|---|---|---|
| 0010 | **EDGING** | | | | | | | | | |
| 0050 | Aluminum alloy, including stakes, 1/8" x 4", mill finish | B-1 | 390 | .062 | L.F. | 2.08 | 2.16 | | 4.24 | 5.60 |
| 0051 | Black paint | | 390 | .062 | | 2.41 | 2.16 | | 4.57 | 5.95 |
| 0052 | Black anodized | | 390 | .062 | | 2.79 | 2.16 | | 4.95 | 6.40 |
| 0100 | Brick, set horizontally, 1-1/2 bricks per L.F. | D-1 | 370 | .043 | | 1.37 | 1.69 | | 3.06 | 4.03 |
| 0150 | Set vertically, 3 bricks per L.F. | " | 135 | .119 | | 4.78 | 4.62 | | 9.40 | 12.20 |
| 0200 | Corrugated aluminum, roll, 4" wide | 1 Carp | 650 | .012 | | .85 | .53 | | 1.38 | 1.75 |
| 0250 | 6" wide | " | 550 | .015 | | 1.06 | .63 | | 1.69 | 2.13 |

# 32 94 Planting Accessories

## 32 94 13 – Landscape Edging

### 32 94 13.20 Edging

| | | Crew | Daily Output | Labor-Hours | Unit | Material | 2011 Bare Costs Labor | Equipment | Total | Total Incl O&P |
|---|---|---|---|---|---|---|---|---|---|---|
| 0600 | Railroad ties, 6" x 8" | 2 Carp | 170 | .094 | L.F. | 3.50 | 4.05 | | 7.55 | 10.05 |
| 0650 | 7" x 9" | | 136 | .118 | | 3.89 | 5.05 | | 8.94 | 12 |
| 0750 | Redwood 2" x 4" | | 330 | .048 | | 2.26 | 2.09 | | 4.35 | 5.70 |
| 0800 | Steel edge strips, incl. stakes, 1/4" x 5" | B-1 | 390 | .062 | | 3.35 | 2.16 | | 5.51 | 7 |
| 0850 | 3/16" x 4" | " | 390 | .062 | | 2.65 | 2.16 | | 4.81 | 6.20 |

## 32 94 50 – Tree Guying

### 32 94 50.10 Tree Guying Systems

| | | Crew | Daily Output | Labor-Hours | Unit | Material | 2011 Bare Costs Labor | Equipment | Total | Total Incl O&P |
|---|---|---|---|---|---|---|---|---|---|---|
| 0010 | **TREE GUYING SYSTEMS** | | | | | | | | | |
| 0015 | Tree guying Including stakes, guy wire and wrap | | | | | | | | | |
| 0100 | Less than 3" caliper, 2 stakes | 2 Clab | 35 | .457 | Ea. | 18 | 15.70 | | 33.70 | 44 |
| 0200 | 3" to 4" caliper, 3 stakes | " | 21 | .762 | " | 20.50 | 26 | | 46.50 | 62.50 |
| 1000 | Including arrowhead anchor, cable, turnbuckles and wrap | | | | | | | | | |
| 1100 | Less than 3" caliper, 3" anchors | 2 Clab | 20 | .800 | Ea. | 51 | 27.50 | | 78.50 | 98 |
| 1200 | 3" to 6" caliper, 4" anchors | | 15 | 1.067 | | 51.50 | 36.50 | | 88 | 113 |
| 1300 | 6" caliper, 6" anchors | | 12 | 1.333 | | 96 | 46 | | 142 | 177 |
| 1400 | 8" caliper, 8" anchors | | 9 | 1.778 | | 99 | 61 | | 160 | 203 |

# 32 96 Transplanting

## 32 96 23 – Plant and Bulb Transplanting

### 32 96 23.23 Planting

| | | Crew | Daily Output | Labor-Hours | Unit | Material | 2011 Bare Costs Labor | Equipment | Total | Total Incl O&P |
|---|---|---|---|---|---|---|---|---|---|---|
| 0010 | **PLANTING** | | | | | | | | | |
| 0012 | Moving shrubs on site, 12" ball | B-62 | 28 | .857 | Ea. | | 32 | 5.45 | 37.45 | 54.50 |
| 0100 | 24" ball | " | 22 | 1.091 | " | | 41 | 6.90 | 47.90 | 69.50 |

### 32 96 23.43 Moving Trees

| | | Crew | Daily Output | Labor-Hours | Unit | Material | 2011 Bare Costs Labor | Equipment | Total | Total Incl O&P |
|---|---|---|---|---|---|---|---|---|---|---|
| 0290 | **MOVING TREES**, On site | | | | | | | | | |
| 0300 | Moving trees on site, 36" ball | B-6 | 3.75 | 6.400 | Ea. | | 240 | 85.50 | 325.50 | 460 |
| 0400 | 60" ball | " | 1 | 24 | " | | 900 | 320 | 1,220 | 1,725 |

# Estimating Tips

## 33 10 00 Water Utilities
## 33 30 00 Sanitary Sewerage Utilities
## 33 40 00 Storm Drainage Utilities

- Never assume that the water, sewer, and drainage lines will go in at the early stages of the project. Consider the site access needs before dividing the site in half with open trenches, loose pipe, and machinery obstructions. Always inspect the site to establish that the site drawings are complete. Check off all existing utilities on your drawings as you locate them. Be especially careful with underground utilities because appurtenances are sometimes buried during regrading or repaving operations. If you find any discrepancies, mark up the site plan for further research. Differing site conditions can be very costly if discovered later in the project.

- See also Section 33 01 00 for restoration of pipe where removal/replacement may be undesirable. Use of new types of piping materials can reduce the overall project cost. Owners/design engineers should consider the installing contractor as a valuable source of current information on utility products and local conditions that could lead to significant cost savings.

# Reference Numbers

Reference numbers are shown in shaded boxes at the beginning of some major classifications. These numbers refer to related items in the Reference Section. The reference information may be an estimating procedure, an alternate pricing method, or technical information.

*Note:* Not all subdivisions listed here necessarily appear in this publication.

*Note:* **Trade Service,** *in part, has been used as a reference source for some of the material prices used in Division 33.*

# 33 01 Operation and Maintenance of Utilities

## 33 01 10 – Operation and Maintenance of Water Utilities

### 33 01 10.10 Corrosion Resistance

| 33 01 10.10 Corrosion Resistance | Crew | Daily Output | Labor-Hours | Unit | Material | 2011 Bare Costs Labor | Equipment | Total | Total Incl O&P |
|---|---|---|---|---|---|---|---|---|---|
| 0010 **CORROSION RESISTANCE** | | | | | | | | | |
| 0012 Wrap & coat, add to pipe, 4" diameter | | | | L.F. | 2.15 | | | 2.15 | 2.37 |
| 0040 6" diameter | | | | | 3.23 | | | 3.23 | 3.55 |
| 0060 8" diameter | | | | | 4.19 | | | 4.19 | 4.61 |
| 0100 12" diameter | | | | | 6.35 | | | 6.35 | 6.95 |
| 0200 24" diameter | | | | | 12.20 | | | 12.20 | 13.45 |
| 0500 Coating, bituminous, per diameter inch, 1 coat, add | | | | | .46 | | | .46 | .51 |
| 0540 3 coat | | | | | 1.37 | | | 1.37 | 1.51 |
| 0560 Coal tar epoxy, per diameter inch, 1 coat, add | | | | | .27 | | | .27 | .30 |
| 0600 3 coat | | | | | .55 | | | .55 | .61 |

## 33 01 30 – Operation and Maintenance of Sewer Utilities

### 33 01 30.72 Relining Sewers

| 33 01 30.72 Relining Sewers | Crew | Daily Output | Labor-Hours | Unit | Material | Labor | Equipment | Total | Total Incl O&P |
|---|---|---|---|---|---|---|---|---|---|
| 0010 **RELINING SEWERS** | | | | | | | | | |
| 0011 With cement incl. bypass & cleaning | | | | | | | | | |
| 0020 Less than 10,000 L.F., urban, 6" to 10" | C-17E | 130 | .615 | L.F. | 9 | 27.50 | .68 | 37.18 | 53 |
| 0200 24" to 36" | | 90 | .889 | | 14.35 | 40 | .99 | 55.34 | 78 |
| 0300 48" to 72" | | 80 | 1 | | 23 | 45 | 1.11 | 69.11 | 94.50 |

# 33 05 Common Work Results for Utilities

## 33 05 16 – Utility Structures

### 33 05 16.13 Precast Concrete Utility Boxes

| 33 05 16.13 Precast Concrete Utility Boxes | Crew | Daily Output | Labor-Hours | Unit | Material | Labor | Equipment | Total | Total Incl O&P |
|---|---|---|---|---|---|---|---|---|---|
| 0010 **PRECAST CONCRETE UTILITY BOXES**, 6" thick | | | | | | | | | |
| 0050 5' x 10' x 6' high, I.D. | B-13 | 2 | 28 | Ea. | 2,425 | 1,050 | 375 | 3,850 | 4,650 |
| 0100 6' x 10' x 6' high, I.D. | | 2 | 28 | | 2,500 | 1,050 | 375 | 3,925 | 4,725 |
| 0150 5' x 12' x 6' high, I.D. | | 2 | 28 | | 2,675 | 1,050 | 375 | 4,100 | 4,900 |
| 0200 6' x 12' x 6' high, I.D. | | 1.80 | 31.111 | | 2,975 | 1,150 | 415 | 4,540 | 5,475 |
| 0250 6' x 13' x 6' high, I.D. | | 1.50 | 37.333 | | 3,925 | 1,400 | 500 | 5,825 | 6,975 |
| 0300 8' x 14' x 7' high, I.D. | | 1 | 56 | | 4,225 | 2,075 | 750 | 7,050 | 8,650 |
| 0350 Hand hole, precast concrete, 1-1/2" thick | | | | | | | | | |
| 0400 1'-0" x 2'-0" x 1'-9", I.D., light duty | B-1 | 4 | 6 | Ea. | 375 | 210 | | 585 | 735 |
| 0450 4'-6" x 3'-2" x 2'-0", O.D., heavy duty | B-6 | 3 | 8 | " | 1,100 | 299 | 107 | 1,506 | 1,775 |

## 33 05 23 – Trenchless Utility Installation

### 33 05 23.19 Microtunneling

| 33 05 23.19 Microtunneling | Crew | Daily Output | Labor-Hours | Unit | Material | Labor | Equipment | Total | Total Incl O&P |
|---|---|---|---|---|---|---|---|---|---|
| 0010 **MICROTUNNELING** | | | | | | | | | |
| 0011 Not including excavation, backfill, shoring, | | | | | | | | | |
| 0020 or dewatering, average 50'/day, slurry method | | | | | | | | | |
| 0100 24" to 48" outside diameter, minimum | | | | L.F. | | | | 790 | 870 |
| 0110 Adverse conditions, add | | | | % | | | | 50% | 50% |
| 1000 Rent microtunneling machine, average monthly lease | | | | Month | | | | 88,000 | 97,000 |
| 1010 Operating technician | | | | Day | | | | 585 | 650 |
| 1100 Mobilization and demobilization, minimum | | | | Job | | | | 38,000 | 42,000 |
| 1110 Maximum | | | | " | | | | 400,000 | 440,000 |

### 33 05 23.20 Horizontal Boring

| 33 05 23.20 Horizontal Boring | Crew | Daily Output | Labor-Hours | Unit | Material | Labor | Equipment | Total | Total Incl O&P |
|---|---|---|---|---|---|---|---|---|---|
| 0010 **HORIZONTAL BORING** | | | | | | | | | |
| 0011 Casing only, 100' minimum, | | | | | | | | | |
| 0020 not incl. jacking pits or dewatering | | | | | | | | | |
| 0100 Roadwork, 1/2" thick wall, 24" diameter casing | B-42 | 20 | 3.200 | L.F. | 142 | 124 | 65 | 331 | 420 |
| 0200 36" diameter | | 16 | 4 | | 214 | 155 | 81 | 450 | 565 |
| 0300 48" diameter | | 15 | 4.267 | | 285 | 165 | 86.50 | 536.50 | 670 |
| 0500 Railroad work, 24" diameter | | 15 | 4.267 | | 142 | 165 | 86.50 | 393.50 | 510 |

# 33 05 Common Work Results for Utilities

## 33 05 23 – Trenchless Utility Installation

### 33 05 23.20 Horizontal Boring

| | | Crew | Daily Output | Labor-Hours | Unit | Material | 2011 Bare Costs Labor | Equipment | Total | Total Incl O&P |
|---|---|---|---|---|---|---|---|---|---|---|
| 0600 | 36" diameter | B-42 | 14 | 4.571 | L.F. | 214 | 177 | 93 | 484 | 610 |
| 0700 | 48" diameter | ↓ | 12 | 5.333 | | 285 | 206 | 108 | 599 | 755 |
| 0900 | For ledge, add | | | | ↓ | | | | | 20% |

## 33 05 26 – Utility Identification

### 33 05 26.10 Utility Accessories

| | | Crew | Daily Output | Labor-Hours | Unit | Material | 2011 Bare Costs Labor | Equipment | Total | Total Incl O&P |
|---|---|---|---|---|---|---|---|---|---|---|
| 0010 | **UTILITY ACCESSORIES** | | | | | | | | | |
| 0400 | Underground tape, detectable, reinforced, alum. foil core, 2" | 1 Clab | 150 | .053 | C.L.F. | 2.25 | 1.83 | | 4.08 | 5.30 |
| 0500 | 6" | " | 140 | .057 | " | 6.55 | 1.96 | | 8.51 | 10.20 |

# 33 11 Water Utility Distribution Piping

## 33 11 13 – Public Water Utility Distribution Piping

### 33 11 13.15 Water Supply, Ductile Iron Pipe

| | | Crew | Daily Output | Labor-Hours | Unit | Material | 2011 Bare Costs Labor | Equipment | Total | Total Incl O&P |
|---|---|---|---|---|---|---|---|---|---|---|
| 0010 | **WATER SUPPLY, DUCTILE IRON PIPE**    R331113-80 | | | | | | | | | |
| 0020 | Not including excavation or backfill | | | | | | | | | |
| 2000 | Pipe, class 50 water piping, 18' lengths | | | | | | | | | |
| 2020 | Mechanical joint, 4" diameter | B-21A | 200 | .200 | L.F. | 12.50 | 8.55 | 2.17 | 23.22 | 29 |
| 2040 | 6" diameter | | 160 | .250 | | 14.60 | 10.70 | 2.72 | 28.02 | 35 |
| 2060 | 8" diameter | | 133.33 | .300 | | 21 | 12.80 | 3.26 | 37.06 | 46 |
| 2080 | 10" diameter | | 114.29 | .350 | | 26.50 | 14.95 | 3.80 | 45.25 | 55.50 |
| 2100 | 12" diameter | | 105.26 | .380 | | 32 | 16.25 | 4.13 | 52.38 | 64 |
| 2120 | 14" diameter | | 100 | .400 | | 48.50 | 17.10 | 4.34 | 69.94 | 84.50 |
| 2140 | 16" diameter | | 72.73 | .550 | | 50 | 23.50 | 5.95 | 79.45 | 97 |
| 2160 | 18" diameter | | 68.97 | .580 | | 53 | 25 | 6.30 | 84.30 | 103 |
| 2170 | 20" diameter | | 57.14 | .700 | | 56 | 30 | 7.60 | 93.60 | 115 |
| 2180 | 24" diameter | | 47.06 | .850 | | 61 | 36.50 | 9.25 | 106.75 | 132 |
| 3000 | Tyton, push-on joint, 4" diameter | | 400 | .100 | | 10.80 | 4.27 | 1.09 | 16.16 | 19.55 |
| 3020 | 6" diameter | | 333.33 | .120 | | 15.20 | 5.15 | 1.30 | 21.65 | 26 |
| 3040 | 8" diameter | | 200 | .200 | | 21.50 | 8.55 | 2.17 | 32.22 | 39.50 |
| 3060 | 10" diameter | | 181.82 | .220 | | 29.50 | 9.40 | 2.39 | 41.29 | 49.50 |
| 3080 | 12" diameter | | 160 | .250 | | 31 | 10.70 | 2.72 | 44.42 | 53 |
| 3100 | 14" diameter | | 133.33 | .300 | | 34 | 12.80 | 3.26 | 50.06 | 60.50 |
| 3120 | 16" diameter | | 114.29 | .350 | | 37 | 14.95 | 3.80 | 55.75 | 67 |
| 3140 | 18" diameter | | 100 | .400 | | 41 | 17.10 | 4.34 | 62.44 | 76 |
| 3160 | 20" diameter | | 88.89 | .450 | | 42.50 | 19.25 | 4.89 | 66.64 | 81 |
| 3180 | 24" diameter | ↓ | 76.92 | .520 | ↓ | 44.50 | 22 | 5.65 | 72.15 | 88.50 |
| 8000 | Fittings, mechanical joint | | | | | | | | | |
| 8006 | 90° bend, 4" diameter | B-20A | 16 | 2 | Ea. | 248 | 83.50 | | 331.50 | 400 |
| 8020 | 6" diameter | | 12.80 | 2.500 | | 370 | 105 | | 475 | 565 |
| 8040 | 8" diameter | ↓ | 10.67 | 2.999 | | 530 | 125 | | 655 | 775 |
| 8060 | 10" diameter | B-21A | 11.43 | 3.500 | | 810 | 150 | 38 | 998 | 1,150 |
| 8080 | 12" diameter | | 10.53 | 3.799 | | 1,075 | 162 | 41.50 | 1,278.50 | 1,500 |
| 8100 | 14" diameter | | 10 | 4 | | 1,525 | 171 | 43.50 | 1,739.50 | 1,975 |
| 8120 | 16" diameter | | 7.27 | 5.502 | | 1,925 | 235 | 60 | 2,220 | 2,550 |
| 8140 | 18" diameter | | 6.90 | 5.797 | | 2,725 | 248 | 63 | 3,036 | 3,450 |
| 8160 | 20" diameter | | 5.71 | 7.005 | | 3,400 | 299 | 76 | 3,775 | 4,275 |
| 8180 | 24" diameter | ↓ | 4.70 | 8.511 | | 5,975 | 365 | 92.50 | 6,432.50 | 7,200 |
| 8200 | Wye or tee, 4" diameter | B-20A | 10.67 | 2.999 | | 360 | 125 | | 485 | 585 |
| 8220 | 6" diameter | | 8.53 | 3.751 | | 530 | 157 | | 687 | 820 |
| 8240 | 8" diameter | ↓ | 7.11 | 4.501 | | 785 | 188 | | 973 | 1,150 |
| 8260 | 10" diameter | B-21A | 7.62 | 5.249 | ↓ | 1,325 | 224 | 57 | 1,606 | 1,850 |

# 33 11 Water Utility Distribution Piping

## 33 11 13 – Public Water Utility Distribution Piping

### 33 11 13.15 Water Supply, Ductile Iron Pipe

| | | Crew | Daily Output | Labor-Hours | Unit | Material | Labor | Equipment | Total | Total Incl O&P |
|---|---|---|---|---|---|---|---|---|---|---|
| | | | | | | | 2011 Bare Costs | | | |
| 8280 | 12" diameter | B-21A | 7.02 | 5.698 | Ea. | 1,750 | 244 | 62 | 2,056 | 2,350 |
| 8300 | 14" diameter | | 6.67 | 5.997 | | 1,575 | 256 | 65 | 1,896 | 2,200 |
| 8320 | 16" diameter | | 4.85 | 8.247 | | 2,925 | 350 | 89.50 | 3,364.50 | 3,850 |
| 8340 | 18" diameter | | 4.60 | 8.696 | | 4,100 | 370 | 94.50 | 4,564.50 | 5,175 |
| 8360 | 20" diameter | | 3.81 | 10.499 | | 5,325 | 450 | 114 | 5,889 | 6,650 |
| 8380 | 24" diameter | | 3.14 | 12.739 | | 9,525 | 545 | 138 | 10,208 | 11,500 |
| 8450 | Decreaser, 6" x 4" diameter | B-20A | 14.22 | 2.250 | | 260 | 94 | | 354 | 430 |
| 8460 | 8" x 6" diameter | " | 11.64 | 2.749 | | 380 | 115 | | 495 | 595 |
| 8470 | 10" x 6" diameter | B-21A | 13.33 | 3.001 | | 455 | 128 | 32.50 | 615.50 | 730 |
| 8480 | 12" x 6" diameter | | 12.70 | 3.150 | | 595 | 135 | 34 | 764 | 895 |
| 8490 | 16" x 6" diameter | | 10 | 4 | | 1,175 | 171 | 43.50 | 1,389.50 | 1,600 |
| 8500 | 20" x 6" diameter | | 8.42 | 4.751 | | 2,600 | 203 | 51.50 | 2,854.50 | 3,200 |
| 8550 | Piping, butterfly valves, cast iron | | | | | | | | | |
| 8560 | 4" diameter | B-20 | 6 | 4 | Ea. | 795 | 153 | | 948 | 1,100 |
| 8570 | 6" diameter | " | 5 | 4.800 | | 845 | 184 | | 1,029 | 1,200 |
| 8580 | 8" diameter | B-21 | 4 | 7 | | 1,125 | 277 | 32 | 1,434 | 1,675 |
| 8590 | 10" diameter | | 3.50 | 8 | | 1,675 | 315 | 37 | 2,027 | 2,375 |
| 8600 | 12" diameter | | 3 | 9.333 | | 2,425 | 370 | 43 | 2,838 | 3,250 |
| 8610 | 14" diameter | | 2 | 14 | | 4,075 | 555 | 64.50 | 4,694.50 | 5,400 |
| 8620 | 16" diameter | | 2 | 14 | | 4,775 | 555 | 64.50 | 5,394.50 | 6,175 |

### 33 11 13.25 Water Supply, Polyvinyl Chloride Pipe

| | | Crew | Daily Output | Labor-Hours | Unit | Material | Labor | Equipment | Total | Total Incl O&P |
|---|---|---|---|---|---|---|---|---|---|---|
| 0010 | **WATER SUPPLY, POLYVINYL CHLORIDE PIPE** | | | | | | | | | |
| 0020 | Not including excavation or backfill, unless specified | | | | | | | | | |
| 2100 | PVC pipe, Class 150, 1-1/2" diameter | Q-1A | 750 | .013 | L.F. | .32 | .72 | | 1.04 | 1.42 |
| 2120 | 2" diameter | | 686 | .015 | | .49 | .79 | | 1.28 | 1.72 |
| 2140 | 2-1/2" diameter | | 500 | .020 | | .73 | 1.08 | | 1.81 | 2.41 |
| 2160 | 3" diameter | B-20 | 430 | .056 | | 1.11 | 2.14 | | 3.25 | 4.50 |
| 3010 | AWWA C905, PR 100, DR 25 | | | | | | | | | |
| 3030 | 14" diameter | B-20A | 213 | .150 | L.F. | 10.70 | 6.30 | | 17 | 21.50 |
| 3040 | 16" diameter | | 200 | .160 | | 13.90 | 6.70 | | 20.60 | 25.50 |
| 3050 | 18" diameter | | 160 | .200 | | 21 | 8.35 | | 29.35 | 35.50 |
| 3060 | 20" diameter | | 133 | .241 | | 21 | 10.05 | | 31.05 | 38 |
| 3070 | 24" diameter | | 107 | .299 | | 29.50 | 12.50 | | 42 | 51.50 |
| 3080 | 30" diameter | | 80 | .400 | | 59 | 16.70 | | 75.70 | 90.50 |
| 3090 | 36" diameter | | 80 | .400 | | 91 | 16.70 | | 107.70 | 126 |
| 3100 | 42" diameter | | 60 | .533 | | 123 | 22.50 | | 145.50 | 169 |
| 3200 | 48" diameter | | 60 | .533 | | 161 | 22.50 | | 183.50 | 211 |
| 4520 | Pressure pipe Class 150, SDR 18, AWWA C900, 4" diameter | | 380 | .084 | | 1.76 | 3.52 | | 5.28 | 7.25 |
| 4530 | 6" diameter | | 316 | .101 | | 5.15 | 4.23 | | 9.38 | 12.05 |
| 4540 | 8" diameter | | 264 | .121 | | 8 | 5.05 | | 13.05 | 16.45 |
| 4550 | 10" diameter | | 220 | .145 | | 12.60 | 6.10 | | 18.70 | 23 |
| 4560 | 12" diameter | | 186 | .172 | | 17.20 | 7.20 | | 24.40 | 30 |
| 8000 | Fittings with rubber gasket | | | | | | | | | |
| 8003 | Class 150, D.R. 18 | | | | | | | | | |
| 8006 | 90° Bend , 4" diameter | B-20 | 100 | .240 | Ea. | 54 | 9.20 | | 63.20 | 73.50 |
| 8020 | 6" diameter | | 90 | .267 | | 97 | 10.25 | | 107.25 | 123 |
| 8040 | 8" diameter | | 80 | .300 | | 185 | 11.50 | | 196.50 | 221 |
| 8060 | 10" diameter | | 50 | .480 | | 282 | 18.40 | | 300.40 | 340 |
| 8080 | 12" diameter | | 30 | .800 | | 430 | 30.50 | | 460.50 | 520 |
| 8100 | Tee, 4" diameter | | 90 | .267 | | 57 | 10.25 | | 67.25 | 78.50 |
| 8120 | 6" diameter | | 80 | .300 | | 154 | 11.50 | | 165.50 | 187 |
| 8140 | 8" diameter | | 70 | .343 | | 215 | 13.15 | | 228.15 | 256 |

# 33 11 Water Utility Distribution Piping

## 33 11 13 – Public Water Utility Distribution Piping

### 33 11 13.25 Water Supply, Polyvinyl Chloride Pipe

| | | Crew | Daily Output | Labor-Hours | Unit | Material | 2011 Bare Costs Labor | 2011 Bare Costs Equipment | Total | Total Incl O&P |
|---|---|---|---|---|---|---|---|---|---|---|
| 8160 | 10" diameter | B-20 | 40 | .600 | Ea. | 455 | 23 | | 478 | 535 |
| 8180 | 12" diameter | | 20 | 1.200 | | 635 | 46 | | 681 | 770 |
| 8200 | 45° Bend, 4" diameter | | 100 | .240 | | 53.50 | 9.20 | | 62.70 | 73 |
| 8220 | 6" diameter | | 90 | .267 | | 96.50 | 10.25 | | 106.75 | 122 |
| 8240 | 8" diameter | | 50 | .480 | | 182 | 18.40 | | 200.40 | 229 |
| 8260 | 10" diameter | | 50 | .480 | | 286 | 18.40 | | 304.40 | 345 |
| 8280 | 12" diameter | | 30 | .800 | | 380 | 30.50 | | 410.50 | 465 |
| 8300 | Reducing tee 6" x 4" | | 100 | .240 | | 131 | 9.20 | | 140.20 | 158 |
| 8320 | 8" x 6" | | 90 | .267 | | 266 | 10.25 | | 276.25 | 310 |
| 8330 | 10" x 6" | | 90 | .267 | | 310 | 10.25 | | 320.25 | 355 |
| 8340 | 10" x 8" | | 90 | .267 | | 330 | 10.25 | | 340.25 | 375 |
| 8350 | 12" x 6" | | 90 | .267 | | 415 | 10.25 | | 425.25 | 470 |
| 8360 | 12" x 8" | | 90 | .267 | | 460 | 10.25 | | 470.25 | 520 |
| 8400 | Tapped service tee (threaded type) 6" x 6" x 3/4" | | 100 | .240 | | 86.50 | 9.20 | | 95.70 | 109 |
| 8430 | 6" x 6" x 1" | | 90 | .267 | | 86.50 | 10.25 | | 96.75 | 111 |
| 8440 | 6" x 6" x 1-1/2" | | 90 | .267 | | 86.50 | 10.25 | | 96.75 | 111 |
| 8450 | 6" x 6" x 2" | | 90 | .267 | | 86.50 | 10.25 | | 96.75 | 111 |
| 8460 | 8" x 8" x 3/4" | | 90 | .267 | | 128 | 10.25 | | 138.25 | 156 |
| 8470 | 8" x 8" x 1" | | 90 | .267 | | 128 | 10.25 | | 138.25 | 156 |
| 8480 | 8" x 8" x 1-1/2" | | 90 | .267 | | 128 | 10.25 | | 138.25 | 156 |
| 8490 | 8" x 8" x 2" | | 90 | .267 | | 128 | 10.25 | | 138.25 | 156 |
| 8500 | Repair coupling 4" | | 100 | .240 | | 91 | 9.20 | | 100.20 | 114 |
| 8520 | 6" diameter | | 90 | .267 | | 114 | 10.25 | | 124.25 | 142 |
| 8540 | 8" diameter | | 50 | .480 | | 133 | 18.40 | | 151.40 | 176 |
| 8560 | 10" diameter | | 50 | .480 | | 185 | 18.40 | | 203.40 | 233 |
| 8580 | 12" diameter | | 50 | .480 | | 223 | 18.40 | | 241.40 | 275 |
| 8600 | Plug end 4" | | 100 | .240 | | 34 | 9.20 | | 43.20 | 51 |
| 8620 | 6" diameter | | 90 | .267 | | 50 | 10.25 | | 60.25 | 70.50 |
| 8640 | 8" diameter | | 50 | .480 | | 76 | 18.40 | | 94.40 | 112 |
| 8660 | 10" diameter | | 50 | .480 | | 105 | 18.40 | | 123.40 | 145 |
| 8680 | 12" diameter | | 50 | .480 | | 129 | 18.40 | | 147.40 | 171 |

# 33 12 Water Utility Distribution Equipment

## 33 12 19 – Water Utility Distribution Fire Hydrants

### 33 12 19.10 Fire Hydrants

| | | Crew | Daily Output | Labor-Hours | Unit | Material | 2011 Bare Costs Labor | 2011 Bare Costs Equipment | Total | Total Incl O&P |
|---|---|---|---|---|---|---|---|---|---|---|
| 0010 | **FIRE HYDRANTS** | | | | | | | | | |
| 0020 | Mechanical joints unless otherwise noted | | | | | | | | | |
| 1000 | Fire hydrants, two way; excavation and backfill not incl. | | | | | | | | | |
| 1100 | 4-1/2" valve size, depth 2'-0" | B-21 | 10 | 2.800 | Ea. | 1,350 | 111 | 12.90 | 1,473.90 | 1,675 |
| 1120 | 2'-6" | | 10 | 2.800 | | 1,450 | 111 | 12.90 | 1,573.90 | 1,775 |
| 1140 | 3'-0" | | 10 | 2.800 | | 1,550 | 111 | 12.90 | 1,673.90 | 1,875 |
| 1300 | 7'-0" | | 6 | 4.667 | | 1,775 | 184 | 21.50 | 1,980.50 | 2,250 |
| 2400 | Lower barrel extensions with stems, 1'-0" | B-20 | 14 | 1.714 | | 375 | 66 | | 441 | 515 |
| 2480 | 3'-0" | " | 12 | 2 | | 730 | 76.50 | | 806.50 | 925 |

# 33 16 Water Utility Storage Tanks

## 33 16 13 – Aboveground Water Utility Storage Tanks

### 33 16 13.13 Steel Water Storage Tanks

| 33 16 13.13 Steel Water Storage Tanks | Crew | Daily Output | Labor-Hours | Unit | Material | 2011 Bare Costs Labor | Equipment | Total | Total Incl O&P |
|---|---|---|---|---|---|---|---|---|---|
| 0010 **STEEL WATER STORAGE TANKS** | | | | | | | | | |
| 0910 Steel, ground level, ht/diam. less than 1, not incl. fdn., 100,000 gallons | | | | Ea. | | | | 178,000 | 196,000 |
| 1000 250,000 gallons | | | | | | | | 237,000 | 260,000 |
| 1200 500,000 gallons | | | | | | | | 335,000 | 368,000 |
| 1250 750,000 gallons | | | | | | | | 432,000 | 475,000 |
| 1300 1,000,000 gallons | | | | | | | | 530,000 | 583,000 |
| 1500 2,000,000 gallons | | | | | | | | 921,000 | 1,013,000 |
| 1600 4,000,000 gallons | | | | | | | | 1,703,000 | 1,873,000 |
| 1800 6,000,000 gallons | | | | | | | | 2,485,000 | 2,733,000 |
| 1850 8,000,000 gallons | | | | | | | | 3,266,000 | 3,593,000 |
| 1910 10,000,000 gallons | | | | | | | | 4,055,000 | 4,460,000 |
| 2100 Steel standpipes, hgt/diam. more than 1, 100' to overflow, no fdn | | | | | | | | | |
| 2200 500,000 gallons | | | | Ea. | | | | 439,000 | 482,000 |
| 2400 750,000 gallons | | | | | | | | 580,000 | 638,000 |
| 2500 1,000,000 gallons | | | | | | | | 568,000 | 625,000 |
| 2700 1,500,000 gallons | | | | | | | | 1,404,000 | 1,544,000 |
| 2800 2,000,000 gallons | | | | | | | | 1,049,000 | 1,154,000 |

### 33 16 13.16 Prestressed Conc. Aboveground Water Utility Storage Tanks

| 33 16 13.16 Prestressed Conc. Aboveground Water Utility Storage Tanks | Crew | Daily Output | Labor-Hours | Unit | Material | 2011 Bare Costs Labor | Equipment | Total | Total Incl O&P |
|---|---|---|---|---|---|---|---|---|---|
| 0010 **PRESTRESSED CONC. ABOVEGROUND WATER UTILITY STORAGE TANKS** | | | | | | | | | |
| 0020 Not including fdn., pipe or pumps, 250,000 gallons | | | | Ea. | | | | 270,000 | 297,000 |
| 0100 500,000 gallons | | | | | | | | 441,000 | 485,000 |
| 0300 1,000,000 gallons | | | | | | | | 639,000 | 703,000 |
| 0400 2,000,000 gallons | | | | | | | | 970,000 | 1,067,000 |
| 0600 4,000,000 gallons | | | | | | | | 1,543,000 | 1,697,000 |
| 0700 6,000,000 gallons | | | | | | | | 2,050,000 | 2,255,000 |
| 0750 8,000,000 gallons | | | | | | | | 2,645,000 | 2,909,000 |
| 0800 10,000,000 gallons | | | | | | | | 3,196,000 | 3,519,000 |

### 33 16 13.23 Plastic-Coated Fabric Pillow Water Tanks

| 33 16 13.23 Plastic-Coated Fabric Pillow Water Tanks | Crew | Daily Output | Labor-Hours | Unit | Material | 2011 Bare Costs Labor | Equipment | Total | Total Incl O&P |
|---|---|---|---|---|---|---|---|---|---|
| 0010 **PLASTIC-COATED FABRIC PILLOW WATER TANKS** | | | | | | | | | |
| 7000 Water tanks, vinyl coated fabric pillow tanks, freestanding, 5,000 gallons | 4 Clab | 4 | 8 | Ea. | 9,500 | 275 | | 9,775 | 10,900 |
| 7100 Supporting embankment not included, 25,000 gallons | 6 Clab | 2 | 24 | | 14,900 | 825 | | 15,725 | 17,700 |
| 7200 50,000 gallons | 8 Clab | 1.50 | 42.667 | | 27,600 | 1,475 | | 29,075 | 32,700 |
| 7300 100,000 gallons | 9 Clab | .90 | 80 | | 60,000 | 2,750 | | 62,750 | 70,000 |
| 7400 150,000 gallons | | .50 | 144 | | 82,000 | 4,950 | | 86,950 | 97,500 |
| 7500 200,000 gallons | | .40 | 180 | | 100,500 | 6,175 | | 106,675 | 120,000 |
| 7600 250,000 gallons | | .30 | 240 | | 143,000 | 8,250 | | 151,250 | 170,000 |

## 33 16 19 – Elevated Water Utility Storage Tanks

### 33 16 19.50 Elevated Water Storage Tanks

| 33 16 19.50 Elevated Water Storage Tanks | Crew | Daily Output | Labor-Hours | Unit | Material | 2011 Bare Costs Labor | Equipment | Total | Total Incl O&P |
|---|---|---|---|---|---|---|---|---|---|
| 0010 **ELEVATED WATER STORAGE TANKS** | | | | | | | | | |
| 0011 Not incl. pipe, pumps or foundation | | | | | | | | | |
| 3000 Elevated water tanks, 100' to bottom capacity line, incl. painting | | | | | | | | | |
| 3010 50,000 gallons | | | | Ea. | | | | 182,000 | 200,000 |
| 3300 100,000 gallons | | | | | | | | 273,000 | 300,000 |
| 3400 250,000 gallons | | | | | | | | 767,000 | 844,000 |
| 3600 500,000 gallons | | | | | | | | 1,364,000 | 1,500,000 |
| 3700 750,000 gallons | | | | | | | | 1,655,000 | 1,820,000 |
| 3900 1,000,000 gallons | | | | | | | | 2,364,000 | 2,600,000 |

# 33 21 Water Supply Wells

## 33 21 13 – Public Water Supply Wells

### 33 21 13.10 Wells and Accessories

| | 33 21 13.10 Wells and Accessories | Crew | Daily Output | Labor-Hours | Unit | Material | 2011 Bare Costs Labor | Equipment | Total | Total Incl O&P |
|---|---|---|---|---|---|---|---|---|---|---|
| 0010 | **WELLS & ACCESSORIES** | | | | | | | | | |
| 0011 | Domestic | | | | | | | | | |
| 0100 | Drilled, 4" to 6" diameter | B-23 | 120 | .333 | L.F. | | 11.60 | 22.50 | 34.10 | 43 |
| 0200 | 8" diameter | " | 95.20 | .420 | " | | 14.60 | 28.50 | 43.10 | 53.50 |
| 0400 | Gravel pack well, 40' deep, incl. gravel & casing, complete | | | | | | | | | |
| 0500 | 24" diameter casing x 18" diameter screen | B-23 | .13 | 307 | Total | 33,800 | 10,700 | 20,800 | 65,300 | 76,500 |
| 0600 | 36" diameter casing x 18" diameter screen | | .12 | 333 | " | 35,900 | 11,600 | 22,500 | 70,000 | 82,000 |
| 0800 | Observation wells, 1-1/4" riser pipe | ↓ | 163 | .245 | V.L.F. | 17.70 | 8.55 | 16.60 | 42.85 | 51 |
| 0900 | For flush Buffalo roadway box, add | 1 Skwk | 16.60 | .482 | Ea. | 48.50 | 21.50 | | 70 | 86 |
| 1200 | Test well, 2-1/2" diameter, up to 50' deep (15 to 50 GPM) | B-23 | 1.51 | 26.490 | " | 725 | 920 | 1,800 | 3,445 | 4,175 |
| 1300 | Over 50' deep, add | " | 121.80 | .328 | L.F. | 19.25 | 11.40 | 22 | 52.65 | 63 |
| 1500 | Pumps, installed in wells to 100' deep, 4" submersible | | | | | | | | | |
| 1510 | 1/2 H.P. | Q-1 | 3.22 | 4.969 | Ea. | 395 | 240 | | 635 | 795 |
| 1520 | 3/4 H.P. | | 2.66 | 6.015 | | 485 | 290 | | 775 | 970 |
| 1600 | 1 H.P. | ↓ | 2.29 | 6.987 | | 570 | 335 | | 905 | 1,125 |
| 1700 | 1-1/2 H.P. | Q-22 | 1.60 | 10 | | 810 | 485 | 410 | 1,705 | 2,050 |
| 1800 | 2 H.P. | | 1.33 | 12.030 | | 1,125 | 580 | 495 | 2,200 | 2,650 |
| 1900 | 3 H.P. | | 1.14 | 14.035 | | 1,325 | 675 | 575 | 2,575 | 3,075 |
| 2000 | 5 H.P. | | 1.14 | 14.035 | | 2,275 | 675 | 575 | 3,525 | 4,125 |
| 3000 | Pump, 6" submersible, 25' to 150' deep, 25 H.P., 249 to 297 GPM | | .89 | 17.978 | | 6,850 | 865 | 735 | 8,450 | 9,650 |
| 3100 | 25' to 500' deep, 30 H.P., 100 to 300 GPM | ↓ | .73 | 21.918 | ↓ | 8,425 | 1,050 | 900 | 10,375 | 11,800 |
| 8000 | Steel well casing | B-23A | 3020 | .008 | Lb. | .78 | .31 | .86 | 1.95 | 2.27 |
| 9950 | See Section 31 23 19.40 for wellpoints | | | | | | | | | |
| 9960 | See Section 31 23 19.30 for drainage wells | | | | | | | | | |

### 33 21 13.20 Water Supply Wells, Pumps

| | 33 21 13.20 Water Supply Wells, Pumps | Crew | Daily Output | Labor-Hours | Unit | Material | Labor | Equipment | Total | Total Incl O&P |
|---|---|---|---|---|---|---|---|---|---|---|
| 0010 | **WATER SUPPLY WELLS, PUMPS** | | | | | | | | | |
| 0011 | With pressure control | | | | | | | | | |
| 1000 | Deep well, jet, 42 gal. galvanized tank | | | | | | | | | |
| 1040 | 3/4 HP | 1 Plum | .80 | 10 | Ea. | 1,450 | 535 | | 1,985 | 2,400 |
| 3000 | Shallow well, jet, 30 gal. galvanized tank | | | | | | | | | |
| 3040 | 1/2 HP | 1 Plum | 2 | 4 | Ea. | 1,200 | 214 | | 1,414 | 1,650 |

# 33 31 Sanitary Utility Sewerage Piping

## 33 31 13 – Public Sanitary Utility Sewerage Piping

### 33 31 13.15 Sewage Collection, Concrete Pipe

| | 33 31 13.15 Sewage Collection, Concrete Pipe | Crew | Daily Output | Labor-Hours | Unit | Material | Labor | Equipment | Total | Total Incl O&P |
|---|---|---|---|---|---|---|---|---|---|---|
| 0010 | **SEWAGE COLLECTION, CONCRETE PIPE** | | | | | | | | | |
| 0020 | See Section 33 41 13.60 for sewage/drainage collection, concrete pipe | | | | | | | | | |

### 33 31 13.25 Sewage Collection, Polyvinyl Chloride Pipe

| | 33 31 13.25 Sewage Collection, Polyvinyl Chloride Pipe | Crew | Daily Output | Labor-Hours | Unit | Material | Labor | Equipment | Total | Total Incl O&P |
|---|---|---|---|---|---|---|---|---|---|---|
| 0010 | **SEWAGE COLLECTION, POLYVINYL CHLORIDE PIPE** | | | | | | | | | |
| 0020 | Not including excavation or backfill | | | | | | | | | |
| 2000 | 20' lengths, S.D.R. 35, B&S, 4" diameter | B-20 | 375 | .064 | L.F. | 1.85 | 2.46 | | 4.31 | 5.80 |
| 2040 | 6" diameter | | 350 | .069 | | 3.68 | 2.63 | | 6.31 | 8.10 |
| 2080 | 13' lengths, S.D.R. 35, B&S, 8" diameter | ↓ | 335 | .072 | | 7.60 | 2.75 | | 10.35 | 12.55 |
| 2120 | 10" diameter | B-21 | 330 | .085 | | 11.30 | 3.35 | .39 | 15.04 | 18 |
| 2160 | 12" diameter | | 320 | .088 | | 13.05 | 3.46 | .40 | 16.91 | 20 |
| 2200 | 15" diameter | ↓ | 240 | .117 | | 14.35 | 4.61 | .54 | 19.50 | 23.50 |
| 4000 | Piping, DWV PVC, no exc/bkfill, 10' L, Sch 40, 4" diameter | B-20 | 375 | .064 | | 3.85 | 2.46 | | 6.31 | 8 |
| 4010 | 6" diameter | | 350 | .069 | | 7.65 | 2.63 | | 10.28 | 12.50 |
| 4020 | 8" diameter | ↓ | 335 | .072 | ↓ | 13.40 | 2.75 | | 16.15 | 18.90 |

# 33 36 Utility Septic Tanks

## 33 36 13 – Utility Septic Tank and Effluent Wet Wells

### 33 36 13.10 Septic Tanks

| | | Crew | Daily Output | Labor-Hours | Unit | Material | 2011 Bare Costs Labor | Equipment | Total | Total Incl O&P |
|---|---|---|---|---|---|---|---|---|---|---|
| 0010 | **SEPTIC TANKS** | | | | | | | | | |
| 0011 | Not including excavation or piping | | | | | | | | | |
| 0015 | Septic tanks, precast, 1,000 gallon | B-21 | 8 | 3.500 | Ea. | 870 | 138 | 16.10 | 1,024.10 | 1,200 |
| 0060 | 1,500 gallon | | 7 | 4 | | 1,475 | 158 | 18.40 | 1,651.40 | 1,875 |
| 0100 | 2,000 gallon | ▼ | 5 | 5.600 | | 1,825 | 221 | 26 | 2,072 | 2,375 |
| 0200 | 5,000 gallon | B-13 | 3.50 | 16 | | 9,675 | 595 | 214 | 10,484 | 11,700 |
| 0300 | 15,000 gallon, 4 piece | B-13B | 1.70 | 32.941 | | 21,500 | 1,225 | 710 | 23,435 | 26,300 |
| 0400 | 25,000 gallon, 4 piece | | 1.10 | 50.909 | | 45,300 | 1,900 | 1,100 | 48,300 | 54,000 |
| 0500 | 40,000 gallon, 4 piece | ▼ | .80 | 70 | | 53,500 | 2,600 | 1,500 | 57,600 | 64,500 |
| 0520 | 50,000 gallon, 5 piece | B-13C | .60 | 93.333 | | 61,500 | 3,475 | 2,750 | 67,725 | 76,000 |
| 0540 | 75,000 gallon, cast in place | C-14C | .25 | 448 | | 75,000 | 18,500 | 94 | 93,594 | 111,000 |
| 0560 | 100,000 gallon | " | .15 | 746 | | 92,500 | 30,800 | 157 | 123,457 | 149,500 |
| 0600 | High density polyethylene, 1,000 gallon | B-21 | 6 | 4.667 | | 1,050 | 184 | 21.50 | 1,255.50 | 1,450 |
| 0700 | 1,500 gallon | ▼ | 4 | 7 | | 1,375 | 277 | 32 | 1,684 | 1,975 |
| 0900 | Galley, 4' x 4' x 4' | ▼ | 16 | 1.750 | | 280 | 69 | 8.05 | 357.05 | 425 |
| 1000 | Distribution boxes, concrete, 7 outlets | 2 Clab | 16 | 1 | | 81 | 34.50 | | 115.50 | 142 |
| 1100 | 9 outlets | " | 8 | 2 | | 455 | 68.50 | | 523.50 | 605 |
| 1150 | Leaching field chambers, 13' x 3'-7" x 1'-4", standard | B-13 | 16 | 3.500 | | 505 | 130 | 47 | 682 | 805 |
| 1200 | Heavy duty, 8' x 4' x 1'-6" | | 14 | 4 | | 280 | 149 | 53.50 | 482.50 | 595 |
| 1300 | 13' x 3'-9" x 1'-6" | | 12 | 4.667 | | 1,025 | 174 | 62.50 | 1,261.50 | 1,450 |
| 1350 | 20' x 4' x 1'-6" | ▼ | 5 | 11.200 | | 1,075 | 415 | 150 | 1,640 | 2,000 |
| 1400 | Leaching pit, precast concrete, 3' diameter, 3' deep | B-21 | 8 | 3.500 | | 640 | 138 | 16.10 | 794.10 | 935 |
| 1500 | 6' diameter, 3' section | | 4.70 | 5.957 | | 870 | 236 | 27.50 | 1,133.50 | 1,350 |
| 2000 | Velocity reducing pit, precast conc., 6' diameter, 3' deep | ▼ | 4.70 | 5.957 | ▼ | 1,525 | 236 | 27.50 | 1,788.50 | 2,075 |
| 2200 | Excavation for septic tank, 3/4 C.Y. backhoe | B-12F | 145 | .110 | C.Y. | | 4.46 | 4.42 | 8.88 | 11.60 |
| 2400 | 4' trench for disposal field, 3/4 C.Y. backhoe | " | 335 | .048 | L.F. | | 1.93 | 1.91 | 3.84 | 5 |
| 2600 | Gravel fill, run of bank | B-6 | 150 | .160 | C.Y. | 20 | 6 | 2.14 | 28.14 | 33.50 |
| 2800 | Crushed stone, 3/4" | " | 150 | .160 | " | 38 | 6 | 2.14 | 46.14 | 53 |
| 3000 | Effluent filter, 4" dia. | 1 Skwk | 8 | 1 | Ea. | 47.50 | 44.50 | | 92 | 121 |

# 33 41 Storm Utility Drainage Piping

## 33 41 13 – Public Storm Utility Drainage Piping

### 33 41 13.40 Piping, Storm Drainage, Corrugated Metal

| | | Crew | Daily Output | Labor-Hours | Unit | Material | 2011 Bare Costs Labor | Equipment | Total | Total Incl O&P |
|---|---|---|---|---|---|---|---|---|---|---|
| 0010 | **PIPING, STORM DRAINAGE, CORRUGATED METAL** | | | | | | | | | |
| 0020 | Not including excavation or backfill | | | | | | | | | |
| 2000 | Corrugated metal pipe, galvanized | | | | | | | | | |
| 2020 | Bituminous coated with paved invert, 20' lengths | | | | | | | | | |
| 2040 | 8" diameter, 16 ga. | B-14 | 330 | .145 | L.F. | 10.70 | 5.25 | .97 | 16.92 | 21 |
| 2060 | 10" diameter, 16 ga. | | 260 | .185 | | 11.15 | 6.70 | 1.23 | 19.08 | 24 |
| 2080 | 12" diameter, 16 ga. | | 210 | .229 | | 11.50 | 8.30 | 1.53 | 21.33 | 27 |
| 2100 | 15" diameter, 16 ga. | | 200 | .240 | | 18.75 | 8.70 | 1.60 | 29.05 | 35.50 |
| 2120 | 18" diameter, 16 ga. | | 190 | .253 | | 19.10 | 9.15 | 1.69 | 29.94 | 37 |
| 2140 | 24" diameter, 14 ga. | ▼ | 160 | .300 | | 22 | 10.85 | 2 | 34.85 | 43 |
| 2160 | 30" diameter, 14 ga. | B-13 | 120 | .467 | | 26 | 17.35 | 6.25 | 49.60 | 62 |
| 2180 | 36" diameter, 12 ga. | | 120 | .467 | | 31.50 | 17.35 | 6.25 | 55.10 | 68 |
| 2200 | 48" diameter, 12 ga. | ▼ | 100 | .560 | | 52.50 | 21 | 7.50 | 81 | 97.50 |
| 2220 | 60" diameter, 10 ga. | B-13B | 75 | .747 | | 79 | 28 | 16.10 | 123.10 | 147 |
| 2240 | 72" diameter, 8 ga. | " | 45 | 1.244 | ▼ | 90.50 | 46.50 | 27 | 164 | 200 |
| 2500 | Galvanized, uncoated, 20' lengths | | | | | | | | | |
| 2520 | 8" diameter, 16 ga. | B-14 | 355 | .135 | L.F. | 8.25 | 4.90 | .90 | 14.05 | 17.55 |
| 2540 | 10" diameter, 16 ga. | ▼ | 280 | .171 | ▼ | 8.75 | 6.20 | 1.15 | 16.10 | 20.50 |

## 33 41 13 – Public Storm Utility Drainage Piping

### 33 41 13.40 Piping, Storm Drainage, Corrugated Metal

| | | Crew | Daily Output | Labor-Hours | Unit | Material | 2011 Bare Costs Labor | 2011 Bare Costs Equipment | Total | Total Incl O&P |
|---|---|---|---|---|---|---|---|---|---|---|
| 2560 | 12" diameter, 16 ga. | B-14 | 220 | .218 | L.F. | 9 | 7.90 | 1.46 | 18.36 | 23.50 |
| 2580 | 15" diameter, 16 ga. | | 220 | .218 | | 13 | 7.90 | 1.46 | 22.36 | 28 |
| 2600 | 18" diameter, 16 ga. | | 205 | .234 | | 14.75 | 8.50 | 1.56 | 24.81 | 31 |
| 2620 | 24" diameter, 14 ga. | | 175 | .274 | | 18.75 | 9.95 | 1.83 | 30.53 | 37.50 |
| 2640 | 30" diameter, 14 ga. | B-13 | 130 | .431 | | 23 | 16.05 | 5.75 | 44.80 | 56 |
| 2660 | 36" diameter, 12 ga. | | 130 | .431 | | 27.50 | 16.05 | 5.75 | 49.30 | 61.50 |
| 2680 | 48" diameter, 12 ga. | | 110 | .509 | | 45.50 | 18.95 | 6.80 | 71.25 | 86.50 |
| 2690 | 60" diameter, 10 ga. | B-13B | 78 | .718 | | 69 | 26.50 | 15.50 | 111 | 134 |
| 2780 | End sections, 8" diameter | B-14 | 24 | 2 | Ea. | 34 | 72.50 | 13.35 | 119.85 | 163 |
| 2785 | 10" diameter | | 22 | 2.182 | | 44.50 | 79 | 14.55 | 138.05 | 186 |
| 2790 | 12" diameter | | 35 | 1.371 | | 78 | 49.50 | 9.15 | 136.65 | 172 |
| 2800 | 18" diameter | | 30 | 1.600 | | 108 | 58 | 10.70 | 176.70 | 218 |
| 2810 | 24" diameter | B-13 | 25 | 2.240 | | 143 | 83.50 | 30 | 256.50 | 315 |
| 2820 | 30" diameter | | 25 | 2.240 | | 246 | 83.50 | 30 | 359.50 | 430 |
| 2825 | 36" diameter | | 20 | 2.800 | | 365 | 104 | 37.50 | 506.50 | 605 |
| 2830 | 48" diameter | | 10 | 5.600 | | 930 | 208 | 75 | 1,213 | 1,425 |
| 2835 | 60" diameter | B-13B | 5 | 11.200 | | 985 | 415 | 242 | 1,642 | 1,975 |
| 2840 | 72" diameter | " | 4 | 14 | | 1,575 | 520 | 300 | 2,395 | 2,850 |

### 33 41 13.60 Sewage/Drainage Collection, Concrete Pipe

| | | Crew | Daily Output | Labor-Hours | Unit | Material | 2011 Bare Costs Labor | 2011 Bare Costs Equipment | Total | Total Incl O&P |
|---|---|---|---|---|---|---|---|---|---|---|
| 0010 | **SEWAGE/DRAINAGE COLLECTION, CONCRETE PIPE** | | | | | | | | | |
| 0020 | Not including excavation or backfill | | | | | | | | | |
| 1000 | Non-reinforced pipe, extra strength, B&S or T&G joints | | | | | | | | | |
| 1010 | 6" diameter | B-14 | 265.04 | .181 | L.F. | 6.25 | 6.55 | 1.21 | 14.01 | 18.25 |
| 1020 | 8" diameter | | 224 | .214 | | 6.90 | 7.75 | 1.43 | 16.08 | 21 |
| 1030 | 10" diameter | | 216 | .222 | | 7.65 | 8.05 | 1.48 | 17.18 | 22.50 |
| 1040 | 12" diameter | | 200 | .240 | | 9.35 | 8.70 | 1.60 | 19.65 | 25.50 |
| 1050 | 15" diameter | | 180 | .267 | | 12.80 | 9.65 | 1.78 | 24.23 | 31 |
| 1060 | 18" diameter | | 144 | .333 | | 15.10 | 12.10 | 2.23 | 29.43 | 37.50 |
| 1070 | 21" diameter | | 112 | .429 | | 18.60 | 15.55 | 2.86 | 37.01 | 47 |
| 1080 | 24" diameter | | 100 | .480 | | 25 | 17.40 | 3.21 | 45.61 | 57.50 |
| 2000 | Reinforced culvert, class 3, no gaskets | | | | | | | | | |
| 2010 | 12" diameter | B-14 | 150 | .320 | L.F. | 11.50 | 11.60 | 2.14 | 25.24 | 32.50 |
| 2020 | 15" diameter | | 150 | .320 | | 15.70 | 11.60 | 2.14 | 29.44 | 37.50 |
| 2030 | 18" diameter | | 132 | .364 | | 19.70 | 13.15 | 2.43 | 35.28 | 44 |
| 2035 | 21" diameter | | 120 | .400 | | 23.50 | 14.50 | 2.67 | 40.67 | 51 |
| 2040 | 24" diameter | | 100 | .480 | | 27.50 | 17.40 | 3.21 | 48.11 | 60.50 |
| 2045 | 27" diameter | B-13 | 92 | .609 | | 40 | 22.50 | 8.15 | 70.65 | 87.50 |
| 2050 | 30" diameter | | 88 | .636 | | 48 | 23.50 | 8.50 | 80 | 98.50 |
| 2060 | 36" diameter | | 72 | .778 | | 58 | 29 | 10.40 | 97.40 | 119 |
| 2070 | 42" diameter | B-13B | 72 | .778 | | 76.50 | 29 | 16.80 | 122.30 | 147 |
| 2080 | 48" diameter | | 64 | .875 | | 95.50 | 32.50 | 18.85 | 146.85 | 176 |
| 2090 | 60" diameter | | 48 | 1.167 | | 123 | 43.50 | 25 | 191.50 | 229 |
| 2100 | 72" diameter | | 40 | 1.400 | | 200 | 52 | 30 | 282 | 330 |
| 2120 | 84" diameter | | 32 | 1.750 | | 288 | 65 | 38 | 391 | 455 |
| 2140 | 96" diameter | | 24 | 2.333 | | 345 | 87 | 50.50 | 482.50 | 570 |
| 2200 | With gaskets, class 3, 12" diameter | B-21 | 168 | .167 | | 12.65 | 6.60 | .77 | 20.02 | 25 |
| 2220 | 15" diameter | | 160 | .175 | | 17.25 | 6.90 | .81 | 24.96 | 30.50 |
| 2230 | 18" diameter | | 152 | .184 | | 21.50 | 7.30 | .85 | 29.65 | 36 |
| 2240 | 24" diameter | | 136 | .206 | | 33.50 | 8.15 | .95 | 42.60 | 50.50 |
| 2260 | 30" diameter | B-13 | 88 | .636 | | 55 | 23.50 | 8.50 | 87 | 106 |
| 2270 | 36" diameter | " | 72 | .778 | | 66.50 | 29 | 10.40 | 105.90 | 128 |
| 2290 | 48" diameter | B-13B | 64 | .875 | | 107 | 32.50 | 18.85 | 158.35 | 189 |

## 33 41 13 – Public Storm Utility Drainage Piping

| 33 41 13.60 Sewage/Drainage Collection, Concrete Pipe | Crew | Daily Output | Labor-Hours | Unit | Material | 2011 Bare Costs Labor | Equipment | Total | Total Incl O&P |
|---|---|---|---|---|---|---|---|---|---|
| 2310 | 72" diameter | B-13B | 40 | 1.400 | L.F. | 217 | 52 | 30 | 299 | 350 |
| 2330 | Flared ends, 6'-1" long, 12" diameter | B-21 | 190 | .147 | | 36.50 | 5.85 | .68 | 43.03 | 50 |
| 2340 | 15" diameter | | 155 | .181 | | 40 | 7.15 | .83 | 47.98 | 55.50 |
| 2400 | 6'-2" long, 18" diameter | | 122 | .230 | | 51 | 9.05 | 1.06 | 61.11 | 71.50 |
| 2420 | 24" diameter | | 88 | .318 | | 55 | 12.60 | 1.46 | 69.06 | 81.50 |
| 2440 | 36" diameter | B-13 | 60 | .933 | | 127 | 34.50 | 12.50 | 174 | 207 |
| 3080 | Radius pipe, add to pipe prices, 12" to 60" diameter | | | | | 50% | | | | |
| 3090 | Over 60" diameter, add | | | | | 20% | | | | |
| 3500 | Reinforced elliptical, 8' lengths, C507 class 3 | | | | | | | | | |
| 3520 | 14" x 23" inside, round equivalent 18" diameter | B-21 | 82 | .341 | L.F. | 54.50 | 13.50 | 1.57 | 69.57 | 82 |
| 3530 | 24" x 38" inside, round equivalent 30" diameter | B-13 | 58 | .966 | | 81 | 36 | 12.90 | 129.90 | 158 |
| 3540 | 29" x 45" inside, round equivalent 36" diameter | | 52 | 1.077 | | 97 | 40 | 14.40 | 151.40 | 184 |
| 3550 | 38" x 60" inside, round equivalent 48" diameter | | 38 | 1.474 | | 161 | 55 | 19.70 | 235.70 | 282 |
| 3560 | 48" x 76" inside, round equivalent 60" diameter | | 26 | 2.154 | | 204 | 80 | 29 | 313 | 380 |
| 3570 | 58" x 91" inside, round equivalent 72" diameter | | 22 | 2.545 | | 305 | 94.50 | 34 | 433.50 | 520 |
| 3780 | Concrete slotted pipe, class 4 mortar joint | | | | | | | | | |
| 3800 | 12" diameter | B-21 | 168 | .167 | L.F. | 35 | 6.60 | .77 | 42.37 | 49.50 |
| 3840 | 18" diameter | " | 152 | .184 | " | 40 | 7.30 | .85 | 48.15 | 56 |
| 3900 | Concrete slotted pipe, Class 4 O-ring joint | | | | | | | | | |
| 3940 | 12" diameter | B-21 | 168 | .167 | L.F. | 35 | 6.60 | .77 | 42.37 | 49.50 |
| 3960 | 18" diameter | " | 152 | .184 | " | 40 | 7.30 | .85 | 48.15 | 56 |

# 33 42 Culverts

## 33 42 16 – Concrete Culverts

### 33 42 16.15 Oval Arch Culverts

| | | Crew | Daily Output | Labor-Hours | Unit | Material | 2011 Bare Costs Labor | Equipment | Total | Total Incl O&P |
|---|---|---|---|---|---|---|---|---|---|---|
| 0010 | **OVAL ARCH CULVERTS** | | | | | | | | | |
| 3000 | Corrugated galvanized or aluminum, coated & paved | | | | | | | | | |
| 3020 | 17" x 13", 16 ga., 15" equivalent | B-14 | 200 | .240 | L.F. | 17.45 | 8.70 | 1.60 | 27.75 | 34 |
| 3040 | 21" x 15", 16 ga., 18" equivalent | | 150 | .320 | | 24.50 | 11.60 | 2.14 | 38.24 | 47 |
| 3060 | 28" x 20", 14 ga., 24" equivalent | | 125 | .384 | | 28 | 13.90 | 2.57 | 44.47 | 55 |
| 3080 | 35" x 24", 14 ga., 30" equivalent | | 100 | .480 | | 37 | 17.40 | 3.21 | 57.61 | 70.50 |
| 3100 | 42" x 29", 12 ga., 36" equivalent | B-13 | 100 | .560 | | 46 | 21 | 7.50 | 74.50 | 91 |
| 3120 | 49" x 33", 12 ga., 42" equivalent | | 90 | .622 | | 62.50 | 23 | 8.30 | 93.80 | 113 |
| 3140 | 57" x 38", 12 ga., 48" equivalent | | 75 | .747 | | 75.50 | 28 | 10 | 113.50 | 137 |
| 3160 | Steel, plain oval arch culverts, plain | | | | | | | | | |
| 3180 | 17" x 13", 16 ga., 15" equivalent | B-14 | 225 | .213 | L.F. | 10.25 | 7.75 | 1.43 | 19.43 | 24.50 |
| 3200 | 21" x 15", 16 ga., 18" equivalent | | 175 | .274 | | 12.40 | 9.95 | 1.83 | 24.18 | 31 |
| 3220 | 28" x 20", 14 ga., 24" equivalent | | 150 | .320 | | 16.60 | 11.60 | 2.14 | 30.34 | 38.50 |
| 3240 | 35" x 24", 14 ga., 30" equivalent | B-13 | 108 | .519 | | 21.50 | 19.30 | 6.95 | 47.75 | 61 |
| 3260 | 42" x 29", 12 ga., 36" equivalent | | 108 | .519 | | 27 | 19.30 | 6.95 | 53.25 | 67 |
| 3280 | 49" x 33", 12 ga., 42" equivalent | | 92 | .609 | | 31 | 22.50 | 8.15 | 61.65 | 77.50 |
| 3300 | 57" x 38", 12 ga., 48" equivalent | | 75 | .747 | | 47.50 | 28 | 10 | 85.50 | 106 |
| 3320 | End sections, 17" x 13" | | 22 | 2.545 | Ea. | 105 | 94.50 | 34 | 233.50 | 298 |
| 3340 | 42" x 29" | | 17 | 3.294 | " | 292 | 123 | 44 | 459 | 555 |
| 3360 | Multi-plate arch, steel | B-20 | 1690 | .014 | Lb. | 1.20 | .54 | | 1.74 | 2.16 |

# 33 44 Storm Utility Water Drains

## 33 44 13 – Utility Area Drains

### 33 44 13.13 Catchbasins

| 33 44 13.13 Catchbasins | Crew | Daily Output | Labor-Hours | Unit | Material | 2011 Bare Costs Labor | Equipment | Total | Total Incl O&P |
|---|---|---|---|---|---|---|---|---|---|
| 0010 **CATCHBASINS** | | | | | | | | | |
| 0011   Not including footing & excavation | | | | | | | | | |
| 1600   Frames & covers, C.I., 24" square, 500 lb. | B-6 | 7.80 | 3.077 | Ea. | 345 | 115 | 41 | 501 | 600 |
| 1700     26" D shape, 600 lb. | | 7 | 3.429 | | 510 | 128 | 46 | 684 | 805 |
| 1800   Light traffic, 18" diameter, 100 lb. | | 10 | 2.400 | | 154 | 90 | 32 | 276 | 340 |
| 1900     24" diameter, 300 lb. | | 8.70 | 2.759 | | 236 | 103 | 37 | 376 | 460 |
| 2000     36" diameter, 900 lb. | | 5.80 | 4.138 | | 510 | 155 | 55 | 720 | 855 |
| 2100   Heavy traffic, 24" diameter, 400 lb. | | 7.80 | 3.077 | | 220 | 115 | 41 | 376 | 460 |
| 2200     36" diameter, 1150 lb. | | 3 | 8 | | 830 | 299 | 107 | 1,236 | 1,500 |
| 2300   Mass. State standard, 26" diameter, 475 lb. | | 7 | 3.429 | | 250 | 128 | 46 | 424 | 520 |
| 2400     30" diameter, 620 lb. | | 7 | 3.429 | | 340 | 128 | 46 | 514 | 620 |
| 2500   Watertight, 24" diameter, 350 lb. | | 7.80 | 3.077 | | 235 | 115 | 41 | 391 | 480 |
| 2600     26" diameter, 500 lb. | | 7 | 3.429 | | 395 | 128 | 46 | 569 | 680 |
| 2700     32" diameter, 575 lb. | ↓ | 6 | 4 | ↓ | 800 | 150 | 53.50 | 1,003.50 | 1,175 |
| 2800   3 piece cover & frame, 10" deep, | | | | | | | | | |
| 2900     1200 lbs., for heavy equipment | B-6 | 3 | 8 | Ea. | 1,100 | 299 | 107 | 1,506 | 1,800 |
| 3000   Raised for paving 1-1/4" to 2" high | | | | | | | | | |
| 3100     4 piece expansion ring | | | | | | | | | |
| 3200     20" to 26" diameter | 1 Clab | 3 | 2.667 | Ea. | 143 | 91.50 | | 234.50 | 298 |
| 3300     30" to 36" diameter | " | 3 | 2.667 | " | 199 | 91.50 | | 290.50 | 360 |
| 3320   Frames and covers, existing, raised for paving, 2", including | | | | | | | | | |
| 3340     row of brick, concrete collar, up to 12" wide frame | B-6 | 18 | 1.333 | Ea. | 45 | 50 | 17.80 | 112.80 | 145 |
| 3360     20" to 26" wide frame | | 11 | 2.182 | | 67.50 | 81.50 | 29 | 178 | 230 |
| 3380     30" to 36" wide frame | ↓ | 9 | 2.667 | | 83.50 | 100 | 35.50 | 219 | 283 |
| 3400   Inverts, single channel brick | D-1 | 3 | 5.333 | | 97 | 208 | | 305 | 415 |
| 3500     Concrete | | 5 | 3.200 | | 75.50 | 125 | | 200.50 | 270 |
| 3600     Triple channel, brick | | 2 | 8 | | 148 | 310 | | 458 | 635 |
| 3700     Concrete | ↓ | 3 | 5.333 | ↓ | 128 | 208 | | 336 | 450 |

# 33 46 Subdrainage

## 33 46 16 – Subdrainage Piping

### 33 46 16.25 Piping, Subdrainage, Corrugated Metal

| 33 46 16.25 Piping, Subdrainage, Corrugated Metal | Crew | Daily Output | Labor-Hours | Unit | Material | 2011 Bare Costs Labor | Equipment | Total | Total Incl O&P |
|---|---|---|---|---|---|---|---|---|---|
| 0010 **PIPING, SUBDRAINAGE, CORRUGATED METAL** | | | | | | | | | |
| 0021   Not including excavation and backfill | | | | | | | | | |
| 2010   Aluminum, perforated | | | | | | | | | |
| 2020     6" diameter, 18 ga. | B-14 | 380 | .126 | L.F. | 7.55 | 4.58 | .84 | 12.97 | 16.25 |
| 2200     8" diameter, 16 ga. | | 370 | .130 | | 10.90 | 4.70 | .87 | 16.47 | 20 |
| 2220     10" diameter, 16 ga. | | 360 | .133 | | 13.65 | 4.83 | .89 | 19.37 | 23.50 |
| 2240     12" diameter, 16 ga. | | 285 | .168 | | 15.30 | 6.10 | 1.13 | 22.53 | 27.50 |
| 2260     18" diameter, 16 ga. | ↓ | 205 | .234 | ↓ | 23 | 8.50 | 1.56 | 33.06 | 39.50 |
| 3000   Uncoated galvanized, perforated | | | | | | | | | |
| 3020     6" diameter, 18 ga. | B-20 | 380 | .063 | L.F. | 7.55 | 2.42 | | 9.97 | 12 |
| 3200     8" diameter, 16 ga. | " | 370 | .065 | | 10.90 | 2.49 | | 13.39 | 15.80 |
| 3220     10" diameter, 16 ga. | B-21 | 360 | .078 | | 16.40 | 3.07 | .36 | 19.83 | 23 |
| 3240     12" diameter, 16 ga. | | 285 | .098 | | 17.15 | 3.88 | .45 | 21.48 | 25.50 |
| 3260     18" diameter, 16 ga. | ↓ | 205 | .137 | ↓ | 26 | 5.40 | .63 | 32.03 | 38 |
| 4000   Steel, perforated, asphalt coated | | | | | | | | | |
| 4020     6" diameter 18 ga. | B-20 | 380 | .063 | L.F. | 7.55 | 2.42 | | 9.97 | 12 |
| 4030     8" diameter 18 ga. | " | 370 | .065 | | 10.90 | 2.49 | | 13.39 | 15.80 |
| 4040     10" diameter 16 ga. | B-21 | 360 | .078 | | 13.25 | 3.07 | .36 | 16.68 | 19.70 |
| 4050     12" diameter 16 ga. | ↓ | 285 | .098 | ↓ | 15.10 | 3.88 | .45 | 19.43 | 23 |

# 33 46 Subdrainage

## 33 46 16 – Subdrainage Piping

| 33 46 16.25 Piping, Subdrainage, Corrugated Metal | Crew | Daily Output | Labor-Hours | Unit | Material | 2011 Bare Costs Labor | Equipment | Total | Total Incl O&P |
|---|---|---|---|---|---|---|---|---|---|
| 4060 | 18" diameter 16 ga. | B-21 | 205 | .137 | L.F. | 21 | 5.40 | .63 | 27.03 | 32 |

## 33 46 16.40 Piping, Subdrainage, Polyvinyl Chloride

| | | | | | | | | | | |
|---|---|---|---|---|---|---|---|---|---|---|
| 0010 | **PIPING, SUBDRAINAGE, POLYVINYL CHLORIDE** | | | | | | | | | |
| 0020 | Perforated, price as solid pipe, Section 33 31 13.25 | | | | | | | | | |

# 33 49 Storm Drainage Structures

## 33 49 13 – Storm Drainage Manholes, Frames, and Covers

### 33 49 13.10 Storm Drainage Manholes, Frames and Covers

| | 33 49 13.10 Storm Drainage Manholes, Frames and Covers | Crew | Daily Output | Labor-Hours | Unit | Material | 2011 Bare Costs Labor | Equipment | Total | Total Incl O&P |
|---|---|---|---|---|---|---|---|---|---|---|
| 0010 | **STORM DRAINAGE MANHOLES, FRAMES & COVERS** | | | | | | | | | |
| 0020 | Excludes footing, excavation, backfill (See line items for frame & cover) | | | | | | | | | |
| 0050 | Brick, 4' inside diameter, 4' deep | D-1 | 1 | 16 | Ea. | 445 | 625 | | 1,070 | 1,425 |
| 0100 | 6' deep | | .70 | 22.857 | | 625 | 890 | | 1,515 | 2,000 |
| 0150 | 8' deep | | .50 | 32 | | 805 | 1,250 | | 2,055 | 2,750 |
| 0200 | For depths over 8', add | | 4 | 4 | V.L.F. | 155 | 156 | | 311 | 405 |
| 0400 | Concrete blocks (radial), 4' I.D., 4' deep | | 1.50 | 10.667 | Ea. | 370 | 415 | | 785 | 1,025 |
| 0500 | 6' deep | | 1 | 16 | | 495 | 625 | | 1,120 | 1,475 |
| 0600 | 8' deep | | .70 | 22.857 | | 625 | 890 | | 1,515 | 2,000 |
| 0700 | For depths over 8', add | | 5.50 | 2.909 | V.L.F. | 65.50 | 113 | | 178.50 | 242 |
| 0800 | Concrete, cast in place, 4' x 4', 8" thick, 4' deep | C-14H | 2 | 24 | Ea. | 430 | 1,025 | 11.50 | 1,466.50 | 2,075 |
| 0900 | 6' deep | | 1.50 | 32 | | 615 | 1,350 | 15.35 | 1,980.35 | 2,775 |
| 1000 | 8' deep | | 1 | 48 | | 895 | 2,050 | 23 | 2,968 | 4,125 |
| 1100 | For depths over 8', add | | 8 | 6 | V.L.F. | 98 | 255 | 2.88 | 355.88 | 500 |
| 1110 | Precast, 4' I.D., 4' deep | B-22 | 4.10 | 7.317 | Ea. | 690 | 293 | 47 | 1,030 | 1,250 |
| 1120 | 6' deep | | 3 | 10 | | 795 | 400 | 64.50 | 1,259.50 | 1,550 |
| 1130 | 8' deep | | 2 | 15 | | 990 | 600 | 97 | 1,687 | 2,125 |
| 1140 | For depths over 8', add | | 16 | 1.875 | V.L.F. | 101 | 75 | 12.10 | 188.10 | 238 |
| 1150 | 5' I.D., 4' deep | B-6 | 3 | 8 | Ea. | 1,400 | 299 | 107 | 1,806 | 2,125 |
| 1160 | 6' deep | | 2 | 12 | | 1,700 | 450 | 160 | 2,310 | 2,725 |
| 1170 | 8' deep | | 1.50 | 16 | | 1,950 | 600 | 214 | 2,764 | 3,300 |
| 1180 | For depths over 8', add | | 12 | 2 | V.L.F. | 180 | 75 | 26.50 | 281.50 | 340 |
| 1190 | 6' I.D., 4' deep | | 2 | 12 | Ea. | 2,000 | 450 | 160 | 2,610 | 3,050 |
| 1200 | 6' deep | | 1.50 | 16 | | 2,700 | 600 | 214 | 3,514 | 4,125 |
| 1210 | 8' deep | | 1 | 24 | | 3,250 | 900 | 320 | 4,470 | 5,325 |
| 1220 | For depths over 8', add | | 8 | 3 | V.L.F. | 252 | 112 | 40 | 404 | 490 |
| 1250 | Slab tops, precast, 8" thick | | | | | | | | | |
| 1300 | 4' diameter manhole | B-6 | 8 | 3 | Ea. | 234 | 112 | 40 | 386 | 470 |
| 1400 | 5' diameter manhole | | 7.50 | 3.200 | | 400 | 120 | 42.50 | 562.50 | 670 |
| 1500 | 6' diameter manhole | | 7 | 3.429 | | 615 | 128 | 46 | 789 | 920 |
| 3800 | Steps, heavyweight cast iron, 7" x 9" | 1 Bric | 40 | .200 | | 16.70 | 8.65 | | 25.35 | 31.50 |
| 3900 | 8" x 9" | | 40 | .200 | | 25 | 8.65 | | 33.65 | 40.50 |
| 3928 | 12" x 10-1/2" | | 40 | .200 | | 23.50 | 8.65 | | 32.15 | 38.50 |
| 4000 | Standard sizes, galvanized steel | | 40 | .200 | | 22 | 8.65 | | 30.65 | 37.50 |
| 4100 | Aluminum | | 40 | .200 | | 22 | 8.65 | | 30.65 | 37.50 |

# 33 51 Natural-Gas Distribution

## 33 51 13 – Natural-Gas Piping

| 33 51 13.10 Piping, Gas Service and Distribution, P.E. | Crew | Daily Output | Labor-Hours | Unit | Material | 2011 Bare Costs Labor | Equipment | Total | Total Incl O&P |
|---|---|---|---|---|---|---|---|---|---|
| 0010 **PIPING, GAS SERVICE AND DISTRIBUTION, POLYETHYLENE** | | | | | | | | | |
| 0020 Not including excavation or backfill | | | | | | | | | |
| 1000 60 psi coils, compression coupling @ 100', 1/2" diameter, SDR 11 | B-20A | 608 | .053 | L.F. | .75 | 2.20 | | 2.95 | 4.15 |
| 1010 1" diameter, SDR 11 | | 544 | .059 | | 1.95 | 2.46 | | 4.41 | 5.85 |
| 1040 1-1/4" diameter, SDR 11 | | 544 | .059 | | 2.39 | 2.46 | | 4.85 | 6.35 |
| 1100 2" diameter, SDR 11 | | 488 | .066 | | 3.26 | 2.74 | | 6 | 7.70 |
| 1160 3" diameter, SDR 11 | | 408 | .078 | | 4.31 | 3.28 | | 7.59 | 9.70 |
| 1500 60 PSI 40' joints with coupling, 3" diameter, SDR 11 | B-21A | 408 | .098 | | 5.35 | 4.19 | 1.06 | 10.60 | 13.35 |
| 1540 4" diameter, SDR 11 | | 352 | .114 | | 7.75 | 4.86 | 1.23 | 13.84 | 17.15 |
| 1600 6" diameter, SDR 11 | | 328 | .122 | | 27 | 5.20 | 1.32 | 33.52 | 39 |
| 1640 8" diameter, SDR 11 | | 272 | .147 | | 32.50 | 6.30 | 1.60 | 40.40 | 46.50 |

# 33 52 Liquid Fuel Distribution

## 33 52 16 – Gasoline Distribution

### 33 52 16.13 Gasoline Piping

| | Crew | Daily Output | Labor-Hours | Unit | Material | 2011 Bare Costs Labor | Equipment | Total | Total Incl O&P |
|---|---|---|---|---|---|---|---|---|---|
| 0010 **GASOLINE PIPING** | | | | | | | | | |
| 0020 Primary containment pipe, fiberglass-reinforced | | | | | | | | | |
| 0030 Plastic pipe 15' & 30' lengths | | | | | | | | | |
| 0040 2" diameter | Q-6 | 425 | .056 | L.F. | 6.25 | 2.85 | | 9.10 | 11.10 |
| 0050 3" diameter | | 400 | .060 | | 6.65 | 3.03 | | 9.68 | 11.85 |
| 0060 4" diameter | | 375 | .064 | | 8.60 | 3.23 | | 11.83 | 14.25 |
| 0100 Fittings | | | | | | | | | |
| 0110 Elbows, 90° & 45°, bell-ends, 2" | Q-6 | 24 | 1 | Ea. | 51.50 | 50.50 | | 102 | 132 |
| 0120 3" diameter | | 22 | 1.091 | | 54 | 55 | | 109 | 142 |
| 0130 4" diameter | | 20 | 1.200 | | 69 | 60.50 | | 129.50 | 166 |
| 0200 Tees, bell ends, 2" | | 21 | 1.143 | | 51 | 57.50 | | 108.50 | 143 |
| 0210 3" diameter | | 18 | 1.333 | | 61 | 67.50 | | 128.50 | 168 |
| 0220 4" diameter | | 15 | 1.600 | | 153 | 81 | | 234 | 289 |
| 0230 Flanges bell ends, 2" | | 24 | 1 | | 36 | 50.50 | | 86.50 | 115 |
| 0240 3" diameter | | 22 | 1.091 | | 47 | 55 | | 102 | 134 |
| 0250 4" diameter | | 20 | 1.200 | | 57.50 | 60.50 | | 118 | 154 |
| 0260 Sleeve couplings, 2" | | 21 | 1.143 | | 21.50 | 57.50 | | 79 | 110 |
| 0270 3" diameter | | 18 | 1.333 | | 25 | 67.50 | | 92.50 | 128 |
| 0280 4" diameter | | 15 | 1.600 | | 29.50 | 81 | | 110.50 | 154 |
| 0290 Threaded adapters 2" | | 21 | 1.143 | | 18 | 57.50 | | 75.50 | 106 |
| 0300 3" diameter | | 18 | 1.333 | | 31.50 | 67.50 | | 99 | 136 |
| 0310 4" diameter | | 15 | 1.600 | | 42.50 | 81 | | 123.50 | 168 |
| 0320 Reducers, 2" | | 27 | .889 | | 38.50 | 45 | | 83.50 | 110 |
| 0330 3" diameter | | 22 | 1.091 | | 50.50 | 55 | | 105.50 | 138 |
| 0340 4" diameter | | 20 | 1.200 | | 68.50 | 60.50 | | 129 | 166 |
| 1010 Gas station product line for secondary containment (double wall) | | | | | | | | | |
| 1100 Fiberglass reinforced plastic pipe 25' lengths | | | | | | | | | |
| 1120 Pipe, plain end, 3" diameter | Q-6 | 375 | .064 | L.F. | 13.40 | 3.23 | | 16.63 | 19.50 |
| 1130 4" diameter | | 350 | .069 | | 15.30 | 3.46 | | 18.76 | 22 |
| 1140 5" diameter | | 325 | .074 | | 20 | 3.73 | | 23.73 | 27.50 |
| 1150 6" diameter | | 300 | .080 | | 20.50 | 4.04 | | 24.54 | 29 |
| 1200 Fittings | | | | | | | | | |
| 1230 Elbows, 90° & 45°, 3" diameter | Q-6 | 18 | 1.333 | Ea. | 117 | 67.50 | | 184.50 | 229 |
| 1240 4" diameter | | 16 | 1.500 | | 123 | 76 | | 199 | 248 |
| 1250 5" diameter | | 14 | 1.714 | | 282 | 86.50 | | 368.50 | 440 |
| 1260 6" diameter | | 12 | 2 | | 285 | 101 | | 386 | 465 |

# 33 52 Liquid Fuel Distribution

## 33 52 16 – Gasoline Distribution

### 33 52 16.13 Gasoline Piping

| | | Crew | Daily Output | Labor-Hours | Unit | Material | 2011 Bare Costs Labor | 2011 Bare Costs Equipment | Total | Total Incl O&P |
|---|---|---|---|---|---|---|---|---|---|---|
| 1270 | Tees, 3" diameter | Q-6 | 15 | 1.600 | Ea. | 121 | 81 | | 202 | 254 |
| 1280 | 4" diameter | | 12 | 2 | | 142 | 101 | | 243 | 305 |
| 1290 | 5" diameter | | 9 | 2.667 | | 290 | 135 | | 425 | 520 |
| 1300 | 6" diameter | | 6 | 4 | | 305 | 202 | | 507 | 635 |
| 1310 | Couplings, 3" diameter | | 18 | 1.333 | | 95 | 67.50 | | 162.50 | 206 |
| 1320 | 4" diameter | | 16 | 1.500 | | 125 | 76 | | 201 | 251 |
| 1330 | 5" diameter | | 14 | 1.714 | | 260 | 86.50 | | 346.50 | 415 |
| 1340 | 6" diameter | | 12 | 2 | | 270 | 101 | | 371 | 450 |
| 1350 | Cross-over nipples, 3" diameter | | 18 | 1.333 | | 16.50 | 67.50 | | 84 | 119 |
| 1360 | 4" diameter | | 16 | 1.500 | | 19.50 | 76 | | 95.50 | 135 |
| 1370 | 5" diameter | | 14 | 1.714 | | 25 | 86.50 | | 111.50 | 157 |
| 1380 | 6" diameter | | 12 | 2 | | 30.50 | 101 | | 131.50 | 185 |
| 1400 | Telescoping, reducers, concentric 4" x 3" | | 18 | 1.333 | | 168 | 67.50 | | 235.50 | 286 |
| 1410 | 5" x 4" | | 17 | 1.412 | | 168 | 71.50 | | 239.50 | 291 |
| 1420 | 6" x 5" | | 16 | 1.500 | | 168 | 76 | | 244 | 298 |

# 33 71 Electrical Utility Transmission and Distribution

## 33 71 16 – Electrical Utility Poles

### 33 71 16.33 Wood Electrical Utility Poles

| | | Crew | Daily Output | Labor-Hours | Unit | Material | 2011 Bare Costs Labor | 2011 Bare Costs Equipment | Total | Total Incl O&P |
|---|---|---|---|---|---|---|---|---|---|---|
| 0010 | **WOOD ELECTRICAL UTILITY POLES** | | | | | | | | | |
| 0011 | Excludes excavation, backfill and cast-in-place concrete | | | | | | | | | |
| 6200 | Electric & tel sitework, 20' high, treated wd, see Section 26 56 13.10 | R-3 | 3.10 | 6.452 | Ea. | 243 | 320 | 41.50 | 604.50 | 790 |
| 6400 | 25' high | | 2.90 | 6.897 | | 297 | 345 | 44.50 | 686.50 | 885 |
| 6600 | 30' high | | 2.60 | 7.692 | | 385 | 385 | 49.50 | 819.50 | 1,050 |
| 6800 | 35' high | | 2.40 | 8.333 | | 475 | 415 | 54 | 944 | 1,200 |
| 7000 | 40' high | | 2.30 | 8.696 | | 685 | 435 | 56 | 1,176 | 1,450 |
| 7200 | 45' high | | 1.70 | 11.765 | | 825 | 585 | 76 | 1,486 | 1,875 |
| 7400 | Cross arms with hardware & insulators | | | | | | | | | |
| 7600 | 4' long | 1 Elec | 2.50 | 3.200 | Ea. | 130 | 161 | | 291 | 380 |
| 7800 | 5' long | | 2.40 | 3.333 | | 151 | 168 | | 319 | 415 |
| 8000 | 6' long | | 2.20 | 3.636 | | 174 | 183 | | 357 | 460 |

## 33 71 19 – Electrical Underground Ducts and Manholes

### 33 71 19.17 Electric and Telephone Underground

| | | Crew | Daily Output | Labor-Hours | Unit | Material | 2011 Bare Costs Labor | 2011 Bare Costs Equipment | Total | Total Incl O&P |
|---|---|---|---|---|---|---|---|---|---|---|
| 0010 | **ELECTRIC AND TELEPHONE UNDERGROUND** | | | | | | | | | |
| 0011 | Not including excavation | | | | | | | | | |
| 0200 | backfill and cast in place concrete | | | | | | | | | |
| 0400 | Hand holes, precast concrete, with concrete cover | | | | | | | | | |
| 0600 | 2' x 2' x 3' deep | R-3 | 2.40 | 8.333 | Ea. | 380 | 415 | 54 | 849 | 1,100 |
| 0800 | 3' x 3' x 3' deep | | 1.90 | 10.526 | | 490 | 525 | 68 | 1,083 | 1,400 |
| 1000 | 4' x 4' x 4' deep | | 1.40 | 14.286 | | 1,200 | 710 | 92 | 2,002 | 2,475 |
| 1200 | Manholes, precast with iron racks & pulling irons, C.I. frame | | | | | | | | | |
| 1400 | and cover, 4' x 6' x 7' deep | B-13 | 2 | 28 | Ea. | 2,075 | 1,050 | 375 | 3,500 | 4,250 |
| 1600 | 6' x 8' x 7' deep | | 1.90 | 29.474 | | 2,550 | 1,100 | 395 | 4,045 | 4,900 |
| 1800 | 6' x 10' x 7' deep | | 1.80 | 31.111 | | 2,850 | 1,150 | 415 | 4,415 | 5,350 |
| 4200 | Underground duct, banks ready for concrete fill, min. of 7.5" | | | | | | | | | |
| 4400 | between conduits, center to center | | | | | | | | | |
| 4580 | PVC, type EB, 1 @ 2" diameter | 2 Elec | 480 | .033 | L.F. | .36 | 1.68 | | 2.04 | 2.87 |
| 4600 | 2 @ 2" diameter | | 240 | .067 | | .72 | 3.35 | | 4.07 | 5.75 |
| 4800 | 4 @ 2" diameter | | 120 | .133 | | 1.43 | 6.70 | | 8.13 | 11.55 |
| 5000 | 2 @ 3" diameter | | 200 | .080 | | 1 | 4.02 | | 5.02 | 7.05 |

# 33 71 Electrical Utility Transmission and Distribution

## 33 71 19 – Electrical Underground Ducts and Manholes

| 33 71 19.17 Electric and Telephone Underground | Crew | Daily Output | Labor- Hours | Unit | Material | 2011 Bare Costs Labor | Equipment | Total | Total Incl O&P |
|---|---|---|---|---|---|---|---|---|---|
| 5200 | 4 @ 3" diameter | 2 Elec | 100 | .160 | L.F. | 2 | 8.05 | | 10.05 | 14.10 |
| 5400 | 2 @ 4" diameter | | 160 | .100 | | 1.47 | 5.05 | | 6.52 | 9.05 |
| 5600 | 4 @ 4" diameter | | 80 | .200 | | 2.93 | 10.05 | | 12.98 | 18.15 |
| 5800 | 6 @ 4" diameter | | 54 | .296 | | 4.40 | 14.90 | | 19.30 | 27 |
| 6200 | Rigid galvanized steel, 2 @ 2" diameter | | 180 | .089 | | 12.75 | 4.47 | | 17.22 | 20.50 |
| 6400 | 4 @ 2" diameter | | 90 | .178 | | 25.50 | 8.95 | | 34.45 | 41.50 |
| 6800 | 2 @ 3" diameter | | 100 | .160 | | 30 | 8.05 | | 38.05 | 45 |
| 7000 | 4 @ 3" diameter | | 50 | .320 | | 60.50 | 16.10 | | 76.60 | 90.50 |
| 7200 | 2 @ 4" diameter | | 70 | .229 | | 43.50 | 11.50 | | 55 | 64.50 |
| 7400 | 4 @ 4" diameter | | 34 | .471 | | 87 | 23.50 | | 110.50 | 131 |
| 7600 | 6 @ 4" diameter | | 22 | .727 | | 130 | 36.50 | | 166.50 | 197 |

# 33 81 Communications Structures

## 33 81 13 – Communications Transmission Towers

### 33 81 13.10 Radio Towers

| | | Crew | Daily Output | Labor- Hours | Unit | Material | 2011 Bare Costs Labor | Equipment | Total | Total Incl O&P |
|---|---|---|---|---|---|---|---|---|---|---|
| 0010 | **RADIO TOWERS** | | | | | | | | | |
| 0020 | Guyed, 50' H, 40 lb. sec., 70 MPH basic wind spd. | 2 Sswk | 1 | 16 | Ea. | 2,600 | 775 | | 3,375 | 4,225 |
| 0100 | Wind load 90 MPH basic wind speed | " | 1 | 16 | | 3,225 | 775 | | 4,000 | 4,925 |
| 0300 | 190' high, 40 lb. section, wind load 70 MPH basic wind speed | K-2 | .33 | 72.727 | | 10,000 | 3,200 | 655 | 13,855 | 17,200 |
| 0400 | 200' high, 70 lb. section, wind load 90 MPH basic wind speed | | .33 | 72.727 | | 17,800 | 3,200 | 655 | 21,655 | 25,800 |
| 0600 | 300' high, 70 lb. section, wind load 70 MPH basic wind speed | | .20 | 120 | | 25,000 | 5,300 | 1,075 | 31,375 | 37,700 |
| 0700 | 270' high, 90 lb. section, wind load 90 MPH basic wind speed | | .20 | 120 | | 27,000 | 5,300 | 1,075 | 33,375 | 39,900 |
| 0800 | 400' high, 100 lb. section, wind load 70 MPH basic wind speed | | .14 | 171 | | 37,000 | 7,575 | 1,550 | 46,125 | 55,500 |
| 0900 | Self-supporting, 60' high, wind load 70 MPH basic wind speed | | .80 | 30 | | 5,500 | 1,325 | 271 | 7,096 | 8,600 |
| 0910 | 60' high, wind load 90 MPH basic wind speed | | .45 | 53.333 | | 6,600 | 2,350 | 480 | 9,430 | 11,800 |
| 1000 | 120' high, wind load 70 MPH basic wind speed | | .40 | 60 | | 12,000 | 2,650 | 540 | 15,190 | 18,300 |
| 1200 | 190' high, wind load 90 MPH basic wind speed | | .20 | 120 | | 31,000 | 5,300 | 1,075 | 37,375 | 44,300 |
| 2000 | For states west of Rocky Mountains, add for shipping | | | | | 10% | | | | |

# Division Notes

| | CREW | DAILY OUTPUT | LABOR-HOURS | UNIT | 2011 BARE COSTS | | | | TOTAL INCL O&P |
|---|---|---|---|---|---|---|---|---|---|
| | | | | | MAT. | LABOR | EQUIP. | TOTAL | |
| | | | | | | | | | |
| | | | | | | | | | |
| | | | | | | | | | |
| | | | | | | | | | |
| | | | | | | | | | |
| | | | | | | | | | |
| | | | | | | | | | |
| | | | | | | | | | |
| | | | | | | | | | |
| | | | | | | | | | |
| | | | | | | | | | |
| | | | | | | | | | |
| | | | | | | | | | |
| | | | | | | | | | |
| | | | | | | | | | |
| | | | | | | | | | |
| | | | | | | | | | |
| | | | | | | | | | |
| | | | | | | | | | |

# Estimating Tips

## 34 11 00 Rail Tracks

This subdivision includes items that may involve either repair of existing, or construction of new, railroad tracks. Additional preparation work, such as the roadbed earthwork, would be found in Division 31. Additional new construction siding and turnouts are found in Subdivision 34 72. Maintenance of railroads is found under 34 01 23 Operation and Maintenance of Railways.

## 34 41 13 Traffic Signals

This subdivision includes traffic signal systems. Other traffic control devices such as traffic signs are found in Subdivision 10 14 53 Traffic Signage.

## 34 71 13 Vehicle Barriers

This subdivision includes security vehicle barriers, guide and guard rails, crash barriers, and delineators. The actual maintenance and construction of concrete and asphalt pavement is found in Division 32.

## Reference Numbers

Reference numbers are shown in shaded boxes at the beginning of some major classifications. These numbers refer to related items in the Reference Section. The reference information may be an estimating procedure, an alternate pricing method, or technical information.

*Note:* Not all subdivisions listed here necessarily appear in this publication.

**Division 34 - Transportation**

# 34 01 Operation and Maintenance of Transportation

## 34 01 23 - Operation and Maintenance of Railways

### 34 01 23.51 Maintenance of Railroads

| | | Daily Output | Labor-Hours | Unit | Material | 2011 Bare Costs | | Total | Total Incl O&P |
|---|---|---|---|---|---|---|---|---|---|
| | | | | | | Labor | Equipment | | |
| 0010 | **MAINTENANCE OF RAILROADS** | | | | | | | | |
| 0400 | Resurface and realign existing track | B-14 | 200 | .240 | L.F. | | 8.70 | 1.60 | 10.30 | 15 |
| 0600 | For crushed stone ballast, add | " | 500 | .096 | " | 15.35 | 3.48 | .64 | 19.47 | 23 |

*(Crew column: B-14 for 0400)*

# 34 11 Rail Tracks

## 34 11 13 - Track Rails

### 34 11 13.23 Heavy Rail Track

| | | Crew | Daily Output | Labor-Hours | Unit | Material | Labor | Equipment | Total | Total Incl O&P |
|---|---|---|---|---|---|---|---|---|---|---|
| 0010 | **HEAVY RAIL TRACK** | R347216-10 | | | | | | | | |
| 1000 | Rail, 100 lb. prime grade | | | | L.F. | 29.50 | | | 29.50 | 32.50 |
| 1500 | Relay rail | | | | " | 14.70 | | | 14.70 | 16.15 |

## 34 11 33 - Track Cross Ties

### 34 11 33.13 Concrete Track Cross Ties

| | | Crew | Daily Output | Labor-Hours | Unit | Material | Labor | Equipment | Total | Total Incl O&P |
|---|---|---|---|---|---|---|---|---|---|---|
| 0010 | **CONCRETE TRACK CROSS TIES** | | | | | | | | | |
| 1400 | Ties, concrete, 8'-6" long, 30" O.C. | B-14 | 80 | .600 | Ea. | 145 | 21.50 | 4.01 | 170.51 | 197 |

### 34 11 33.16 Timber Track Cross Ties

| | | Crew | Daily Output | Labor-Hours | Unit | Material | Labor | Equipment | Total | Total Incl O&P |
|---|---|---|---|---|---|---|---|---|---|---|
| 0010 | **TIMBER TRACK CROSS TIES** | | | | | | | | | |
| 1600 | Wood, pressure treated, 6" x 8" x 8'-6", C.L. lots | B-14 | 90 | .533 | Ea. | 43.50 | 19.30 | 3.56 | 66.36 | 81.50 |
| 1700 | L.C.L. lots | | 90 | .533 | | 46 | 19.30 | 3.56 | 68.86 | 84 |
| 1900 | Heavy duty, 7" x 9" x 8'-6", C.L. lots | | 70 | .686 | | 48 | 25 | 4.58 | 77.58 | 95.50 |
| 2000 | L.C.L. lots | | 70 | .686 | | 48 | 25 | 4.58 | 77.58 | 95.50 |

### 34 11 33.17 Timber Switch Ties

| | | Crew | Daily Output | Labor-Hours | Unit | Material | Labor | Equipment | Total | Total Incl O&P |
|---|---|---|---|---|---|---|---|---|---|---|
| 0010 | **TIMBER SWITCH TIES** | | | | | | | | | |
| 1200 | Switch timber, for a #8 switch, pressure treated | B-14 | 3.70 | 12.973 | M.B.F. | 2,625 | 470 | 86.50 | 3,181.50 | 3,675 |
| 1300 | Complete set of timbers, 3.7 M.B.F. for #8 switch | " | 1 | 48 | Total | 10,100 | 1,750 | 320 | 12,170 | 14,100 |

## 34 11 93 - Track Appurtenances and Accessories

### 34 11 93.50 Track Accessories

| | | Crew | Daily Output | Labor-Hours | Unit | Material | Labor | Equipment | Total | Total Incl O&P |
|---|---|---|---|---|---|---|---|---|---|---|
| 0010 | **TRACK ACCESSORIES** | | | | | | | | | |
| 0020 | Car bumpers, test | B-14 | 2 | 24 | Ea. | 3,275 | 870 | 160 | 4,305 | 5,100 |
| 0100 | Heavy duty | R347216-20 | 2 | 24 | | 6,200 | 870 | 160 | 7,230 | 8,325 |
| 0200 | Derails hand throw (sliding) | | 10 | 4.800 | | 1,025 | 174 | 32 | 1,231 | 1,425 |
| 0300 | Hand throw with standard timbers, open stand & target | | 8 | 6 | | 1,125 | 217 | 40 | 1,382 | 1,600 |
| 2400 | Wheel stops, fixed | | 18 | 2.667 | Pr. | 810 | 96.50 | 17.80 | 924.30 | 1,050 |
| 2450 | Hinged | | 14 | 3.429 | " | 1,100 | 124 | 23 | 1,247 | 1,425 |

### 34 11 93.60 Track Material

| | | Crew | Daily Output | Labor-Hours | Unit | Material | Labor | Equipment | Total | Total Incl O&P |
|---|---|---|---|---|---|---|---|---|---|---|
| 0010 | **TRACK MATERIAL** | | | | | | | | | |
| 0020 | Track bolts | | | | Ea. | 3.36 | | | 3.36 | 3.70 |
| 0100 | Joint bars | | | | Pr. | 78 | | | 78 | 85.50 |
| 0200 | Spikes | | | | Ea. | 1.11 | | | 1.11 | 1.22 |
| 0300 | Tie plates | | | | " | 13.05 | | | 13.05 | 14.35 |

# 34 41 Roadway Signaling and Control Equipment

## 34 41 13 – Traffic Signals

### 34 41 13.10 Traffic Signals Systems

| | | Crew | Daily Output | Labor-Hours | Unit | Material | 2011 Bare Costs Labor | 2011 Bare Costs Equipment | Total | Total Incl O&P |
|---|---|---|---|---|---|---|---|---|---|---|
| 0010 | **TRAFFIC SIGNALS SYSTEMS** | | | | | | | | | |
| 0020 | Mid block pedestrian crosswalk with pushbutton and mast arms | R-11 | .30 | 186 | Total | 86,000 | 8,875 | 2,700 | 97,575 | 110,500 |
| 0600 | Traffic signals, school flashing system, solar powered, remote controlled | " | 1 | 56 | Signal | 19,800 | 2,675 | 810 | 23,285 | 26,700 |
| 1000 | Intersection traffic signals, LED, mast, programmable, no lane control | | | | | | | | | |
| 1010 | Includes all labor, material and equip. for complete installation | R-11 | 1 | 56 | Signal | 200,000 | 2,675 | 810 | 203,485 | 225,000 |
| 1100 | Intersection traffic signals, LED, mast, programmable, R/L lane control | | | | | | | | | |
| 1110 | Includes all labor, material and equip. for complete installation | R-11 | 1 | 56 | Signal | 262,000 | 2,675 | 810 | 265,485 | 293,500 |
| 1200 | Add protective/permissive left turns to existing traffic light | | | | | | | | | |
| 1210 | Includes all labor, material and equip. for complete installation | R-11 | 1 | 56 | Signal | 80,000 | 2,675 | 810 | 83,485 | 93,000 |
| 1300 | Replace existing light heads with LED heads wire hung | | | | | | | | | |
| 1310 | Includes all labor, material and equip. for complete installation | R-11 | 1 | 56 | Signal | 54,000 | 2,675 | 810 | 57,485 | 64,500 |
| 1400 | Replace existing light heads with LED heads mast arm hung | | | | | | | | | |
| 1410 | Includes all labor, material and equip. for complete installation | R-11 | 1 | 56 | Signal | 100,000 | 2,675 | 810 | 103,485 | 115,000 |

# 34 71 Roadway Construction

## 34 71 13 – Vehicle Barriers

### 34 71 13.17 Security Vehicle Barriers

| | | Crew | Daily Output | Labor-Hours | Unit | Material | 2011 Bare Costs Labor | 2011 Bare Costs Equipment | Total | Total Incl O&P |
|---|---|---|---|---|---|---|---|---|---|---|
| 0010 | **SECURITY VEHICLE BARRIERS** | | | | | | | | | |
| 0020 | Security planters excludes filling material | | | | | | | | | |
| 0100 | Concrete security planter, exposed aggregate 36" dia x 30" high | B-11M | 8 | 2 | Ea. | 465 | 79.50 | 43.50 | 588 | 680 |
| 0200 | 48" dia x 36" high | | 8 | 2 | | 570 | 79.50 | 43.50 | 693 | 800 |
| 0300 | 53" dia x 18" high | | 8 | 2 | | 500 | 79.50 | 43.50 | 623 | 720 |
| 0400 | 72" dia x 18" high with seats | | 8 | 2 | | 1,075 | 79.50 | 43.50 | 1,198 | 1,350 |
| 0450 | 84" dia x 36" high | | 8 | 2 | | 1,300 | 79.50 | 43.50 | 1,423 | 1,600 |
| 0500 | 36" x 36" x 24" high square | | 8 | 2 | | 395 | 79.50 | 43.50 | 518 | 605 |
| 0600 | 36" x 36" x 30" high square | | 8 | 2 | | 465 | 79.50 | 43.50 | 588 | 680 |
| 0700 | 48" L x 24" W x 30" H rectangle | | 8 | 2 | | 455 | 79.50 | 43.50 | 578 | 670 |
| 0800 | 72" L x 24" W x 30" H rectangle | | 8 | 2 | | 565 | 79.50 | 43.50 | 688 | 790 |
| 0900 | 96" L x 24" W x 30" H rectangle | | 8 | 2 | | 715 | 79.50 | 43.50 | 838 | 955 |
| 0950 | Decorative geometric concrete barrier, 96" L x 24" W x 36" H | | 8 | 2 | | 655 | 79.50 | 43.50 | 778 | 890 |
| 1000 | Concrete security planter, filling with washed sand or gravel <1 CY | | 8 | 2 | | 39 | 79.50 | 43.50 | 162 | 211 |
| 1050 | <2 CY | | 6 | 2.667 | | 78 | 106 | 58 | 242 | 310 |
| 1200 | Jersey concrete barrier, 10' L x 2' by 0.5' W x 30" H | B-21B | 16 | 2.500 | | 395 | 93 | 41 | 529 | 620 |
| 1300 | 10 or more same site | | 24 | 1.667 | | 395 | 62 | 27.50 | 484.50 | 560 |
| 1400 | 10' L x 2' by 0.5' W x 32" H | | 16 | 2.500 | | 425 | 93 | 41 | 559 | 655 |
| 1500 | 10 or more same site | | 24 | 1.667 | | 425 | 62 | 27.50 | 514.50 | 595 |
| 1600 | 20' L x 2' by 0.5' W x 30" H | | 12 | 3.333 | | 600 | 124 | 54.50 | 778.50 | 910 |
| 1700 | 10 or more same site | | 18 | 2.222 | | 600 | 82.50 | 36.50 | 719 | 825 |
| 1800 | 20' L x 2' by 0.5' W x 32" H | | 12 | 3.333 | | 725 | 124 | 54.50 | 903.50 | 1,050 |
| 1900 | 10 or more same site | | 18 | 2.222 | | 725 | 82.50 | 36.50 | 844 | 965 |
| 2000 | GFRC decorative security barrier per 10 feet section including concrete | | 4 | 10 | | 2,225 | 370 | 164 | 2,759 | 3,200 |
| 2100 | Per 12 feet section including concrete | | 4 | 10 | | 2,675 | 370 | 164 | 3,209 | 3,700 |
| 2210 | GFRC decorative security barrier will stop 30 MPH, 4000 lb vehicle | | | | | | | | | |
| 2300 | High security barrier base prep per 12 feet section on bare ground | B-11C | 4 | 4 | Ea. | 34.50 | 159 | 80 | 273.50 | 365 |
| 2310 | GFRC barrier base prep does not include haul away of excavated matl | | | | | | | | | |
| 2400 | GFRC decorative high security barrier per 12 feet section w/concrete | B-21B | 3 | 13.333 | Ea. | 5,575 | 495 | 219 | 6,289 | 7,125 |
| 2410 | GFRC decorative high security barrier will stop 50 MPH, 15000 lb vehl | | | | | | | | | |
| 2500 | GFRC decorative impaler security barrier per 10 feet section w/prep | B-6 | 4 | 6 | Ea. | 2,175 | 225 | 80 | 2,480 | 2,825 |
| 2600 | Per 12 feet section w/prep | " | 4 | 6 | " | 2,625 | 225 | 80 | 2,930 | 3,300 |
| 2610 | Impaler barrier should stop 50 MPH, 15000 lb vehl w/some penetr | | | | | | | | | |
| 2700 | Pipe bollards, steel, concrete filled/painted, 8' L x 4' D hole, 8" diam. | B-6 | 10 | 2.400 | Ea. | 870 | 90 | 32 | 992 | 1,125 |

# 34 71 Roadway Construction

## 34 71 13 – Vehicle Barriers

### 34 71 13.17 Security Vehicle Barriers

| | | Crew | Daily Output | Labor-Hours | Unit | Material | 2011 Bare Costs Labor | Equipment | Total | Total Incl O&P |
|---|---|---|---|---|---|---|---|---|---|---|
| 2710 | Schedule 80 concrete bollards will stop 4000 lb vehicle @ 30 MPH | | | | | | | | | |
| 2800 | GFRC decorative jersey barrier cover per 10 feet section excludes soil | B-6 | 8 | 3 | Ea. | 910 | 112 | 40 | 1,062 | 1,225 |
| 2900 | Per 12 feet section excludes soil | | 8 | 3 | | 1,100 | 112 | 40 | 1,252 | 1,425 |
| 3000 | GFRC decorative 8″ diameter bollard cover | | 12 | 2 | | 480 | 75 | 26.50 | 581.50 | 670 |
| 3100 | GFRC decorative barrier face 10 foot section excludes earth backing | | 10 | 2.400 | | 910 | 90 | 32 | 1,032 | 1,175 |
| 3200 | Drop arm crash barrier, 15000 lb vehl @ 50 MPH | | | | | | | | | |
| 3205 | Includes all material, labor for complete installation | | | | | | | | | |
| 3210 | 12′ width | | | | Ea. | | | | 62,000 | 68,000 |
| 3310 | 24′ width | | | | | | | | 83,000 | 91,000 |
| 3410 | Wedge crash barrier, 10′ width | | | | | | | | 98,000 | 108,000 |
| 3510 | 12.5′ width | | | | | | | | 109,000 | 120,000 |
| 3520 | 15′ width | | | | | | | | 121,000 | 133,000 |
| 3610 | Sliding crash barrier, 12′ width | | | | | | | | 175,000 | 192,000 |
| 3710 | Sliding roller crash barrier, 20′ width | | | | | | | | 194,000 | 213,000 |
| 3810 | Sliding cantilever crash barrier, 20′ width | | | | | | | | 194,000 | 213,000 |
| 3890 | Note: Raised bollard crash barriers should be used w/tire shredders | | | | | | | | | |
| 3910 | Raised bollard crash barrier, 10′ width | | | | Ea. | | | | 38,000 | 42,000 |
| 4010 | 12′ width | | | | | | | | 44,000 | 48,000 |
| 4110 | Raised bollard crash barrier, 10′ width, solar powered | | | | | | | | 46,000 | 51,000 |
| 4210 | 12′ width | | | | | | | | 51,000 | 56,000 |
| 4310 | In ground tire shredder, 16′ width | | | | | | | | 43,000 | 47,000 |

### 34 71 13.26 Vehicle Guide Rails

| | | Crew | Daily Output | Labor-Hours | Unit | Material | 2011 Bare Costs Labor | Equipment | Total | Total Incl O&P |
|---|---|---|---|---|---|---|---|---|---|---|
| 0010 | **VEHICLE GUIDE RAILS** | | | | | | | | | |
| 0012 | Corrugated stl, galv. stl posts, 6′-3″ O.C. | B-80 | 850 | .038 | L.F. | 22 | 1.39 | .73 | 24.12 | 27.50 |
| 0200 | End sections, galvanized, flared | | 50 | .640 | Ea. | 90 | 23.50 | 12.40 | 125.90 | 149 |
| 0300 | Wrap around end | | 50 | .640 | ″ | 136 | 23.50 | 12.40 | 171.90 | 200 |
| 0400 | Timber guide rail, 4″ x 8″ with 6″ x 8″ wood posts, treated | | 960 | .033 | L.F. | 25.50 | 1.23 | .64 | 27.37 | 30.50 |
| 0600 | Cable guide rail, 3 at 3/4″ cables, steel posts, single face | | 900 | .036 | | 10.35 | 1.31 | .69 | 12.35 | 14.10 |
| 0700 | Wood posts | | 950 | .034 | | 12.35 | 1.24 | .65 | 14.24 | 16.15 |
| 0900 | Guide rail, steel box beam, 6″ x 6″ | | 120 | .267 | | 32 | 9.85 | 5.15 | 47 | 55.50 |
| 1100 | Median barrier, steel box beam, 6″ x 8″ | | 215 | .149 | | 42 | 5.50 | 2.88 | 50.38 | 57.50 |
| 1400 | Resilient guide fence and light shield, 6′ high | B-2 | 130 | .308 | | 35 | 10.70 | | 45.70 | 55 |
| 1500 | Concrete posts, individual, 6′-5″, triangular | B-80 | 110 | .291 | Ea. | 65 | 10.75 | 5.65 | 81.40 | 94 |
| 1550 | Square | ″ | 110 | .291 | ″ | 69.50 | 10.75 | 5.65 | 85.90 | 99 |
| 2000 | Median, precast concrete, 3′-6″ high, 2′ wide, single face | B-29 | 380 | .147 | L.F. | 50.50 | 5.50 | 2.26 | 58.26 | 66.50 |
| 2200 | Double face | ″ | 340 | .165 | ″ | 58 | 6.15 | 2.52 | 66.67 | 76 |
| 2400 | Speed bumps, thermoplastic, 10-1/2″ x 2-1/4″ x 48″ long | B-2 | 120 | .333 | Ea. | 140 | 11.60 | | 151.60 | 172 |
| 3030 | Impact barrier, UTMCD, barrel type | B-16 | 30 | 1.067 | ″ | 470 | 37 | 16.85 | 523.85 | 590 |

## 34 71 19 – Vehicle Delineators

### 34 71 19.13 Fixed Vehicle Delineators

| | | Crew | Daily Output | Labor-Hours | Unit | Material | 2011 Bare Costs Labor | Equipment | Total | Total Incl O&P |
|---|---|---|---|---|---|---|---|---|---|---|
| 0010 | **FIXED VEHICLE DELINEATORS** | | | | | | | | | |
| 0020 | Crash barriers | | | | | | | | | |
| 0100 | Traffic channelizing pavement markers, layout only | A-7 | 2000 | .012 | Ea. | | .56 | .03 | .59 | .90 |
| 0110 | 13″ x 7-1/2″ x 2-1/2″ high, non-plowable install | 2 Clab | 96 | .167 | | 26.50 | 5.75 | | 32.25 | 38.50 |
| 0200 | 8″ x 8″ x 3-1/4″ high, non-plowable, install | | 96 | .167 | | 23.50 | 5.75 | | 29.25 | 35 |
| 0230 | 4″ x 4″ x 3/4″ high, non-plowable, install | | 120 | .133 | | 1.41 | 4.58 | | 5.99 | 8.60 |
| 0240 | 9-1/4″ x 5-7/8″ x 1/4″ high, plowable, concrete pav't | A-2A | 70 | .343 | | 20 | 11.65 | 4.36 | 36.01 | 44.50 |
| 0250 | 9-1/4″ x 5-7/8″ x 1/4″ high, plowable, asphalt pav't | ″ | 120 | .200 | | 4.25 | 6.80 | 2.55 | 13.60 | 17.90 |
| 0300 | Barrier and curb delineators, reflectorized, 2″ x 4″ | 2 Clab | 150 | .107 | | 2.30 | 3.66 | | 5.96 | 8.15 |
| 0310 | 3″ x 5″ | ″ | 150 | .107 | | 4.50 | 3.66 | | 8.16 | 10.55 |
| 0500 | Rumble strip, polycarbonate | | | | | | | | | |
| 0510 | 24″ x 3-1/2″ x 1/2″ high | 2 Clab | 50 | .320 | Ea. | 8.10 | 11 | | 19.10 | 26 |

# 34 72 Railway Construction

## 34 72 16 – Railway Siding

### 34 72 16.50 Railroad Sidings

| | | Crew | Daily Output | Labor-Hours | Unit | Material | 2011 Bare Costs Labor | Equipment | Total | Total Incl O&P |
|---|---|---|---|---|---|---|---|---|---|---|
| 0010 | **RAILROAD SIDINGS** R347216-10 | | | | | | | | | |
| 0800 | Siding, yard spur, level grade | | | | | | | | | |
| 0820 | 100 lb. new rail | B-14 | 57 | .842 | L.F. | 114 | 30.50 | 5.65 | 150.15 | 179 |
| 1002 | Steel ties in concrete, incl. fasteners & plates | | | | | | | | | |
| 1020 | 100 lb. new rail | B-14 | 22 | 2.182 | L.F. | 156 | 79 | 14.55 | 249.55 | 310 |

### 34 72 16.60 Railroad Turnouts

| | | Crew | Daily Output | Labor-Hours | Unit | Material | 2011 Bare Costs Labor | Equipment | Total | Total Incl O&P |
|---|---|---|---|---|---|---|---|---|---|---|
| 0010 | **RAILROAD TURNOUTS** | | | | | | | | | |
| 2200 | Turnout, #8 complete, w/rails, plates, bars, frog, switch point, | | | | | | | | | |
| 2250 | timbers, and ballast to 6" below bottom of ties | | | | | | | | | |
| 2280 | 90 lb. rails | B-13 | .25 | 224 | Ea. | 33,700 | 8,325 | 3,000 | 45,025 | 53,000 |
| 2290 | 90 lb. relay rails | | .25 | 224 | | 22,600 | 8,325 | 3,000 | 33,925 | 40,900 |
| 2300 | 100 lb. rails | | .25 | 224 | | 37,500 | 8,325 | 3,000 | 48,825 | 57,500 |
| 2310 | 100 lb. relay rails | | .25 | 224 | | 25,200 | 8,325 | 3,000 | 36,525 | 43,700 |
| 2320 | 110 lb. rails | | .25 | 224 | | 41,300 | 8,325 | 3,000 | 52,625 | 61,500 |
| 2330 | 110 lb. relay rails | | .25 | 224 | | 27,800 | 8,325 | 3,000 | 39,125 | 46,600 |
| 2340 | 115 lb. rails | | .25 | 224 | | 45,300 | 8,325 | 3,000 | 56,625 | 66,000 |
| 2350 | 115 lb. relay rails | | .25 | 224 | | 30,400 | 8,325 | 3,000 | 41,725 | 49,500 |
| 2360 | 132 lb. rails | | .25 | 224 | | 52,000 | 8,325 | 3,000 | 63,325 | 73,000 |
| 2370 | 132 lb. relay rails | | .25 | 224 | | 34,300 | 8,325 | 3,000 | 45,625 | 53,500 |

# Division Notes

| | CREW | DAILY OUTPUT | LABOR-HOURS | UNIT | MAT. | LABOR | EQUIP. | TOTAL | TOTAL INCL O&P |
|---|---|---|---|---|---|---|---|---|---|
| | | | | | | | | | |
| | | | | | | | | | |
| | | | | | | | | | |
| | | | | | | | | | |
| | | | | | | | | | |
| | | | | | | | | | |
| | | | | | | | | | |
| | | | | | | | | | |
| | | | | | | | | | |
| | | | | | | | | | |
| | | | | | | | | | |
| | | | | | | | | | |
| | | | | | | | | | |
| | | | | | | | | | |
| | | | | | | | | | |
| | | | | | | | | | |
| | | | | | | | | | |
| | | | | | | | | | |
| | | | | | | | | | |
| | | | | | | | | | |
| | | | | | | | | | |
| | | | | | | | | | |

# Estimating Tips

## 35 20 16 Hydraulic Gates

This subdivision includes various types of gates that are commonly used in waterway and canal construction. Various earthwork items and structural support is found in Division 31, and concrete work in Division 3.

## 35 20 23 Dredging

This subdivision includes barge and shore dredging systems for rivers, canals, and channels.

## 35 31 00 Shoreline Protection

This subdivision includes breakwaters, bulkheads, and revetments for ocean and river inlets. Additional earthwork may be required from Division 31, and concrete work from Division 3.

## 35 41 00 Levees

Information on levee construction, including estimated cost of clay cone material.

## 35 49 00 Waterway Structures

This subdivision includes breakwaters and bulkheads for canals.

## 35 51 00 Floating Construction

This section includes floating piers, docks, and dock accessories. Fixed Pier Timber Construction is found in 06 13 33. Driven piles are found in Division 31, as well as sheet piling, cofferdams, and riprap.

# Reference Numbers

Reference numbers are shown in shaded boxes at the beginning of some major classifications. These numbers refer to related items in the Reference Section. The reference information may be an estimating procedure, an alternate pricing method, or technical information.

*Note:* Not all subdivisions listed here necessarily appear in this publication.

**Division 35 - Waterway and Marine**

# 35 20 Waterway and Marine Construction and Equipment

## 35 20 23 – Dredging

### 35 20 23.13 Mechanical Dredging

| 35 20 23.13 Mechanical Dredging | Crew | Daily Output | Labor-Hours | Unit | Material | 2011 Bare Costs Labor | Equipment | Total | Total Incl O&P |
|---|---|---|---|---|---|---|---|---|---|
| 0010 **MECHANICAL DREDGING** | | | | | | | | | |
| 0020  Dredging mobilization and demobilization, add to below, minimum | B-8 | .53 | 120 | Total | | 4,600 | 5,225 | 9,825 | 12,700 |
| 0100  Maximum | " | .10 | 640 | " | | 24,400 | 27,700 | 52,100 | 67,500 |
| 0300  Barge mounted clamshell excavation into scows | | | | | | | | | |
| 0310   Dumped 20 miles at sea, minimum | B-57 | 310 | .155 | B.C.Y. | | 6.10 | 5.15 | 11.25 | 14.90 |
| 0400    Maximum | " | 213 | .225 | " | | 8.85 | 7.50 | 16.35 | 21.50 |
| 0500  Barge mounted dragline or clamshell, hopper dumped, | | | | | | | | | |
| 0510   pumped 1000' to shore dump, minimum | B-57 | 340 | .141 | B.C.Y. | | 5.55 | 4.71 | 10.26 | 13.60 |
| 0525  All pumping uses 2000 gallons of water per cubic yard | | | | | | | | | |
| 0600    Maximum | B-57 | 243 | .198 | B.C.Y. | | 7.75 | 6.60 | 14.35 | 18.95 |

### 35 20 23.23 Hydraulic Dredging

| 35 20 23.23 Hydraulic Dredging | Crew | Daily Output | Labor-Hours | Unit | Material | 2011 Bare Costs Labor | Equipment | Total | Total Incl O&P |
|---|---|---|---|---|---|---|---|---|---|
| 0010 **HYDRAULIC DREDGING** | | | | | | | | | |
| 1000  Hydraulic method, pumped 1000' to shore dump, minimum | B-57 | 460 | .104 | B.C.Y. | | 4.09 | 3.48 | 7.57 | 10.05 |
| 1100   Maximum | | 310 | .155 | | | 6.10 | 5.15 | 11.25 | 14.90 |
| 1400  Into scows dumped 20 miles, minimum | | 425 | .113 | | | 4.43 | 3.77 | 8.20 | 10.85 |
| 1500   Maximum | | 243 | .198 | | | 7.75 | 6.60 | 14.35 | 18.95 |
| 1600   For inland rivers and canals in South, deduct | | | | | | | | 30% | 30% |

# 35 51 Floating Construction

## 35 51 13 – Floating Piers

### 35 51 13.23 Floating Wood Piers

| 35 51 13.23 Floating Wood Piers | Crew | Daily Output | Labor-Hours | Unit | Material | 2011 Bare Costs Labor | Equipment | Total | Total Incl O&P |
|---|---|---|---|---|---|---|---|---|---|
| 0010 **FLOATING WOOD PIERS** | | | | | | | | | |
| 0020  Polyethylene encased polystyrene, no pilings included | F-3 | 330 | .121 | S.F. | 40 | 5.30 | 1.99 | 47.29 | 54.50 |
| 0200  Pile supported, shore constructed, bare, 3" decking | | 130 | .308 | | 26.50 | 13.45 | 5.05 | 45 | 55 |
| 0250   4" decking | | 120 | .333 | | 25.50 | 14.60 | 5.45 | 45.55 | 56 |
| 0400  Floating, small boat, prefab, no shore facilities, minimum | | 250 | .160 | | 13.35 | 7 | 2.62 | 22.97 | 28.50 |
| 0500   Maximum | | 150 | .267 | | 45 | 11.65 | 4.37 | 61.02 | 72 |
| 0700  Per slip, minimum (180 S.F. each) | | 1.59 | 25.157 | Ea. | 2,900 | 1,100 | 410 | 4,410 | 5,325 |
| 0800   Maximum | | 1.40 | 28.571 | " | 8,550 | 1,250 | 470 | 10,270 | 11,800 |

## Estimating Tips

Products such as conveyors, material handling cranes and hoists, as well as other items specified in this division, require trained installers. The general contractor may not have any choice as to who will perform the installation or when it will be performed. Long lead times are often required for these products, making early decisions in purchasing and scheduling necessary.

The installation of this type of equipment may require the embedment of mounting hardware during construction of floors, structural walls, or interior walls/partitions. Electrical connections will require coordination with the electrical contractor.

## Reference Numbers

Reference numbers are shown in shaded boxes at the beginning of some major classifications. These numbers refer to related items in the Reference Section. The reference information may be an estimating procedure, an alternate pricing method, or technical information.

*Note:* Not all subdivisions listed here necessarily appear in this publication.

# 41 21 Conveyors

## 41 21 23 – Piece Material Conveyors

| 41 21 23.16 Container Piece Material Conveyors | Crew | Daily Output | Labor-Hours | Unit | Material | 2011 Bare Costs Labor | Equipment | Total | Total Incl O&P |
|---|---|---|---|---|---|---|---|---|---|
| 0010 **CONTAINER PIECE MATERIAL CONVEYORS** | | | | | | | | | |
| 0020 Gravity fed, 2" rollers, 3" O.C. | | | | | | | | | |
| 0050     10' sections with 2 supports, 600 lb. capacity, 18" wide | | | | Ea. | 450 | | | 450 | 495 |
| 0100      24" wide | | | | | 515 | | | 515 | 570 |
| 0150     1400 lb. capacity, 18" wide | | | | | 415 | | | 415 | 455 |
| 0200      24" wide | | | | | 460 | | | 460 | 505 |
| 0350 Horizontal belt, center drive and takeup, 60 fpm | | | | | | | | | |
| 0400     16" belt, 26.5' length | 2 Mill | .50 | 32 | Ea. | 3,150 | 1,450 | | 4,600 | 5,600 |
| 0450     24" belt, 41.5' length | | .40 | 40 | | 4,900 | 1,825 | | 6,725 | 8,025 |
| 0500      61.5' length | | .30 | 53.333 | | 6,525 | 2,425 | | 8,950 | 10,700 |
| 0600 Inclined belt, 10' rise with horizontal loader and | | | | | | | | | |
| 0620     End idler assembly, 27.5' length, 18" belt | 2 Mill | .30 | 53.333 | Ea. | 9,200 | 2,425 | | 11,625 | 13,600 |
| 0700     24" belt | " | .15 | 106 | " | 10,300 | 4,825 | | 15,125 | 18,500 |
| 3600 Monorail, overhead, manual, channel type | | | | | | | | | |
| 3700     125 lb. per L.F. | 1 Mill | 26 | .308 | L.F. | 10.25 | 13.95 | | 24.20 | 32 |
| 3900     500 lb. per L.F. | " | 21 | .381 | " | 9.25 | 17.30 | | 26.55 | 35 |
| 4000     Trolleys for above, 2 wheel, 125 lb. capacity | | | | Ea. | 77 | | | 77 | 84.50 |
| 4200      4 wheel, 250 lb. capacity | | | | | 207 | | | 207 | 227 |
| 4300      8 wheel, 1,000 lb. capacity | | | | | 380 | | | 380 | 420 |

# 41 22 Cranes and Hoists

## 41 22 13 – Cranes

### 41 22 13.10 Crane Rail

| | | Crew | Daily Output | Labor-Hours | Unit | Material | 2011 Bare Costs Labor | Equipment | Total | Total Incl O&P |
|---|---|---|---|---|---|---|---|---|---|---|
| 0010 | **CRANE RAIL** | | | | | | | | | |
| 0020 | Box beam bridge, no equipment included | E-4 | 3400 | .009 | Lb. | 1.19 | .46 | .03 | 1.68 | 2.17 |
| 0200 | Running track only, 104 lb per yard | " | 5600 | .006 | " | .62 | .28 | .02 | .92 | 1.19 |

### 41 22 13.13 Bridge Cranes

| | | Crew | Daily Output | Labor-Hours | Unit | Material | 2011 Bare Costs Labor | Equipment | Total | Total Incl O&P |
|---|---|---|---|---|---|---|---|---|---|---|
| 0010 | **BRIDGE CRANES** | | | | | | | | | |
| 0100 | 1 girder, 20' span, 3 ton | M-3 | 1 | 34 | Ea. | 23,800 | 1,750 | 128 | 25,678 | 29,000 |
| 0125 |   5 ton | | 1 | 34 | | 26,000 | 1,750 | 128 | 27,878 | 31,400 |
| 0150 |   7.5 ton | | 1 | 34 | | 31,000 | 1,750 | 128 | 32,878 | 36,900 |
| 0175 |   10 ton | | .80 | 42.500 | | 41,000 | 2,200 | 160 | 43,360 | 48,500 |
| 0200 |   15 ton | | .80 | 42.500 | | 52,500 | 2,200 | 160 | 54,860 | 61,500 |
| 0225 | 30' span, 3 ton | | 1 | 34 | | 24,700 | 1,750 | 128 | 26,578 | 30,000 |
| 0250 |   5 ton | | 1 | 34 | | 27,200 | 1,750 | 128 | 29,078 | 32,700 |
| 0275 |   7.5 ton | | 1 | 34 | | 32,600 | 1,750 | 128 | 34,478 | 38,600 |
| 0300 |   10 ton | | .80 | 42.500 | | 42,400 | 2,200 | 160 | 44,760 | 50,000 |
| 0325 |   15 ton | | .80 | 42.500 | | 55,000 | 2,200 | 160 | 57,360 | 64,000 |
| 0350 | 2 girder, 40' span, 3 ton | M-4 | .50 | 72 | | 40,900 | 3,675 | 330 | 44,905 | 50,500 |
| 0375 |   5 ton | | .50 | 72 | | 42,800 | 3,675 | 330 | 46,805 | 53,000 |
| 0400 |   7.5 ton | | .50 | 72 | | 46,900 | 3,675 | 330 | 50,905 | 57,500 |
| 0425 |   10 ton | | .40 | 90 | | 55,000 | 4,600 | 410 | 60,010 | 68,000 |
| 0450 |   15 ton | | .40 | 90 | | 74,500 | 4,600 | 410 | 79,510 | 89,500 |
| 0475 |   25 ton | | .30 | 120 | | 88,000 | 6,125 | 550 | 94,675 | 106,500 |
| 0500 | 50' span, 3 ton | | .50 | 72 | | 46,500 | 3,675 | 330 | 50,505 | 57,000 |
| 0525 |   5 ton | | .50 | 72 | | 48,400 | 3,675 | 330 | 52,405 | 59,000 |
| 0550 |   7.5 ton | | .50 | 72 | | 52,000 | 3,675 | 330 | 56,005 | 63,000 |
| 0575 |   10 ton | | .40 | 90 | | 59,500 | 4,600 | 410 | 64,510 | 73,000 |
| 0600 |   15 ton | | .40 | 90 | | 78,000 | 4,600 | 410 | 83,010 | 93,500 |
| 0625 |   25 ton | | .30 | 120 | | 92,500 | 6,125 | 550 | 99,175 | 111,000 |

# 41 22 Cranes and Hoists

## 41 22 23 – Hoists

| 41 22 23.10 Material Handling | Crew | Daily Output | Labor-Hours | Unit | Material | 2011 Bare Costs Labor | Equipment | Total | Total Incl O&P |
|---|---|---|---|---|---|---|---|---|---|
| 0010 | **MATERIAL HANDLING**, cranes, hoists and lifts | | | | | | | | | |
| 1500 | Cranes, portable hydraulic, floor type, 2,000 lb capacity | | | | Ea. | 2,475 | | | 2,475 | 2,700 |
| 1600 | 4,000 lb capacity | | | | | 4,150 | | | 4,150 | 4,575 |
| 1800 | Movable gantry type, 12' to 15' range, 2,000 lb capacity | | | | | 3,600 | | | 3,600 | 3,975 |
| 1900 | 6,000 lb capacity | | | | | 5,225 | | | 5,225 | 5,750 |
| 2100 | Hoists, electric overhead, chain, hook hung, 15' lift, 1 ton cap. | | | | | 2,250 | | | 2,250 | 2,475 |
| 2200 | 3 ton capacity | | | | | 2,625 | | | 2,625 | 2,900 |
| 2500 | 5 ton capacity | | | | | 6,625 | | | 6,625 | 7,275 |
| 2600 | For hand-pushed trolley, add | | | | | 15% | | | | |
| 2700 | For geared trolley, add | | | | | 30% | | | | |
| 2800 | For motor trolley, add | | | | | 75% | | | | |
| 3000 | For lifts over 15', 1 ton, add | | | | L.F. | 23.50 | | | 23.50 | 25.50 |
| 3100 | 5 ton, add | | | | " | 52 | | | 52 | 57 |
| 3300 | Lifts, scissor type, portable, electric, 36" high, 2,000 lb | | | | Ea. | 4,275 | | | 4,275 | 4,700 |
| 3400 | 48" high, 4,000 lb | | | | " | 4,675 | | | 4,675 | 5,150 |

# Division Notes

| | CREW | DAILY OUTPUT | LABOR-HOURS | UNIT | 2011 BARE COSTS | | | | TOTAL INCL O&P |
|---|---|---|---|---|---|---|---|---|---|
| | | | | | MAT. | LABOR | EQUIP. | TOTAL | |
| | | | | | | | | | |
| | | | | | | | | | |
| | | | | | | | | | |
| | | | | | | | | | |
| | | | | | | | | | |
| | | | | | | | | | |
| | | | | | | | | | |
| | | | | | | | | | |
| | | | | | | | | | |
| | | | | | | | | | |
| | | | | | | | | | |
| | | | | | | | | | |
| | | | | | | | | | |
| | | | | | | | | | |
| | | | | | | | | | |
| | | | | | | | | | |
| | | | | | | | | | |
| | | | | | | | | | |
| | | | | | | | | | |
| | | | | | | | | | |
| | | | | | | | | | |
| | | | | | | | | | |

## Estimating Tips

There are four basic types of pollution: Air, Noise, Water, and Solid.

These systems may be interrelated and care must be taken that the complete systems are estimated. For example, Air Pollution Equipment may include dust and air-entrained particles that have to be collected. The vacuum systems could be noisy, requiring silencers to reduce noise pollution, and the collected solids have to be disposed of to prevent Solid Pollution.

Water Treatment may be accomplished as comparatively small packaged plants providing Chemical, Biological or Thermal Treatment. These are usually priced by GPH or GPD.

Large plants and municipal treatment facilities are sized MGD. These facilities are usually priced in two steps. First, the large concrete structures used as settling ponds, etc.; and second, the equipment required. Storm water does not need as much treatment as sanitary wastewater. The total cost of the system is greatly affected by the level of purification required by specifications and codes before release.

An additional cost may be encountered if there is a possibility that the water being treated may contain petroleum products, which have to be isolated and disposed of in an environmentally acceptable manner.

## Reference Numbers

Reference numbers are shown in shaded boxes at the beginning of some major classifications. These numbers refer to related items in the Reference Section. The reference information may be an estimating procedure, an alternate pricing method, or technical information.

*Note:* Not all subdivisions listed here necessarily appear in this publication.

| 44 11 16.10 Dust Collection Systems | Crew | Daily Output | Labor-Hours | Unit | Material | 2011 Bare Costs Labor | Equipment | Total | Total Incl O&P |
|---|---|---|---|---|---|---|---|---|---|
| 0010 **DUST COLLECTION SYSTEMS** Commercial/Industrial | | | | | | | | | |
| 0120 Central vacuum units | | | | | | | | | |
| 0130 Includes stand, filters and motorized shaker | | | | | | | | | |
| 0200 500 CFM, 10" inlet, 2 HP | Q-20 | 2.40 | 8.333 | Ea. | 3,900 | 395 | | 4,295 | 4,900 |
| 0220 1000 CFM, 10" inlet, 3 HP | | 2.20 | 9.091 | | 4,100 | 430 | | 4,530 | 5,175 |
| 0240 1500 CFM, 10" inlet, 5 HP | | 2 | 10 | | 4,325 | 470 | | 4,795 | 5,450 |
| 0260 3000 CFM, 13" inlet, 10 HP | | 1.50 | 13.333 | | 12,200 | 630 | | 12,830 | 14,400 |
| 0280 5000 CFM, 16" inlet, 2 @ 10 HP | | 1 | 20 | | 12,900 | 945 | | 13,845 | 15,600 |
| 1000 Vacuum tubing, galvanized | | | | | | | | | |
| 1100 2-1/8" OD, 16 ga. | Q-9 | 440 | .036 | L.F. | 2.55 | 1.69 | | 4.24 | 5.35 |
| 1110 2-1/2" OD, 16 ga. | | 420 | .038 | | 2.93 | 1.77 | | 4.70 | 5.90 |
| 1120 3" OD, 16 ga. | | 400 | .040 | | 3.35 | 1.86 | | 5.21 | 6.50 |
| 1130 3-1/2" OD, 16 ga. | | 380 | .042 | | 4.86 | 1.96 | | 6.82 | 8.30 |
| 1140 4" OD, 16 ga. | | 360 | .044 | | 5.45 | 2.07 | | 7.52 | 9.10 |
| 1150 5" OD, 14 ga. | | 320 | .050 | | 9.80 | 2.32 | | 12.12 | 14.25 |
| 1160 6" OD, 14 ga. | | 280 | .057 | | 11.30 | 2.66 | | 13.96 | 16.45 |
| 1170 8" OD, 14 ga. | | 200 | .080 | | 17.05 | 3.72 | | 20.77 | 24.50 |
| 1180 10" OD, 12 ga. | | 160 | .100 | | 27 | 4.65 | | 31.65 | 36.50 |
| 1190 12" OD, 12 ga. | | 120 | .133 | | 29.50 | 6.20 | | 35.70 | 42 |
| 1200 14" OD, 12 ga. | | 80 | .200 | | 36 | 9.30 | | 45.30 | 53.50 |
| 1940 Hose, flexible wire reinforced rubber | | | | | | | | | |
| 1956 3" dia. | Q-9 | 400 | .040 | L.F. | 7.60 | 1.86 | | 9.46 | 11.15 |
| 1960 4" dia. | | 360 | .044 | | 8.80 | 2.07 | | 10.87 | 12.75 |
| 1970 5" dia. | | 320 | .050 | | 11.05 | 2.32 | | 13.37 | 15.65 |
| 1980 6" dia. | | 280 | .057 | | 11.95 | 2.66 | | 14.61 | 17.15 |
| 2000 90° Elbow, slip fit | | | | | | | | | |
| 2110 2-1/8" dia. | Q-9 | 70 | .229 | Ea. | 10.65 | 10.60 | | 21.25 | 28 |
| 2120 2-1/2" dia. | | 65 | .246 | | 14.90 | 11.45 | | 26.35 | 33.50 |
| 2130 3" dia. | | 60 | .267 | | 20 | 12.40 | | 32.40 | 41 |
| 2140 3-1/2" dia. | | 55 | .291 | | 25 | 13.50 | | 38.50 | 48 |
| 2150 4" dia. | | 50 | .320 | | 31 | 14.85 | | 45.85 | 56.50 |
| 2160 5" dia. | | 45 | .356 | | 58.50 | 16.55 | | 75.05 | 89.50 |
| 2170 6" dia. | | 40 | .400 | | 81 | 18.60 | | 99.60 | 117 |
| 2180 8" dia. | | 30 | .533 | | 154 | 25 | | 179 | 207 |
| 2400 45° Elbow, slip fit | | | | | | | | | |
| 2410 2-1/8" dia. | Q-9 | 70 | .229 | Ea. | 9.30 | 10.60 | | 19.90 | 26.50 |
| 2420 2-1/2" dia. | | 65 | .246 | | 13.70 | 11.45 | | 25.15 | 32.50 |
| 2430 3" dia. | | 60 | .267 | | 16.55 | 12.40 | | 28.95 | 37 |
| 2440 3-1/2" dia. | | 55 | .291 | | 21 | 13.50 | | 34.50 | 43.50 |
| 2450 4" dia. | | 50 | .320 | | 27 | 14.85 | | 41.85 | 52 |
| 2460 5" dia. | | 45 | .356 | | 45.50 | 16.55 | | 62.05 | 75 |
| 2470 6" dia. | | 40 | .400 | | 61.50 | 18.60 | | 80.10 | 96 |
| 2480 8" dia. | | 35 | .457 | | 121 | 21.50 | | 142.50 | 165 |
| 2800 90° TY, slip fit thru 6" dia. | | | | | | | | | |
| 2810 2-1/8" dia. | Q-9 | 42 | .381 | Ea. | 21 | 17.70 | | 38.70 | 50 |
| 2820 2-1/2" dia. | | 39 | .410 | | 27 | 19.05 | | 46.05 | 59 |
| 2830 3" dia. | | 36 | .444 | | 35.50 | 20.50 | | 56 | 70.50 |
| 2840 3-1/2" dia. | | 33 | .485 | | 45.50 | 22.50 | | 68 | 84 |
| 2850 4" dia. | | 30 | .533 | | 65.50 | 25 | | 90.50 | 110 |
| 2860 5" dia. | | 27 | .593 | | 124 | 27.50 | | 151.50 | 179 |
| 2870 6" dia. | | 24 | .667 | | 169 | 31 | | 200 | 233 |
| 2880 8" dia., butt end | | | | | 360 | | | 360 | 395 |
| 2890 10" dia., butt end | | | | | 650 | | | 650 | 715 |

| 44 11 16.10 Dust Collection Systems | Crew | Daily Output | Labor-Hours | Unit | Material | 2011 Bare Costs Labor | Equipment | Total | Total Incl O&P |
|---|---|---|---|---|---|---|---|---|---|
| 2900 | 12" dia., butt end | | | | Ea. | 650 | | | 650 | 715 |
| 2910 | 14" dia., butt end | | | | | 1,175 | | | 1,175 | 1,300 |
| 2920 | 6" x 4" dia., butt end | | | | | 103 | | | 103 | 114 |
| 2930 | 8" x 4" dia., butt end | | | | | 196 | | | 196 | 215 |
| 2940 | 10" x 4" dia., butt end | | | | | 256 | | | 256 | 282 |
| 2950 | 12" x 4" dia., butt end | | | | | 294 | | | 294 | 325 |
| 3100 | 90° Elbow, butt end segmented | | | | | | | | | |
| 3110 | 8" dia., butt end, segmented | | | | Ea. | 154 | | | 154 | 169 |
| 3120 | 10" dia., butt end, segmented | | | | | 380 | | | 380 | 420 |
| 3130 | 12" dia., butt end, segmented | | | | | 485 | | | 485 | 530 |
| 3140 | 14" dia., butt end, segmented | | | | | 605 | | | 605 | 665 |
| 3200 | 45° Elbow, butt end segmented | | | | | | | | | |
| 3210 | 8" dia., butt end, segmented | | | | Ea. | 121 | | | 121 | 133 |
| 3220 | 10" dia., butt end, segmented | | | | | 276 | | | 276 | 305 |
| 3230 | 12" dia., butt end, segmented | | | | | 282 | | | 282 | 310 |
| 3240 | 14" dia., butt end, segmented | | | | | 345 | | | 345 | 380 |
| 3400 | All butt end fittings require one coupling per joint. | | | | | | | | | |
| 3410 | Labor for fitting included with couplings. | | | | | | | | | |
| 3460 | Compression coupling, galvanized, neoprene gasket | | | | | | | | | |
| 3470 | 2-1/8" dia. | Q-9 | 44 | .364 | Ea. | 12.35 | 16.90 | | 29.25 | 39 |
| 3480 | 2-1/2" dia. | | 44 | .364 | | 12.35 | 16.90 | | 29.25 | 39 |
| 3490 | 3" dia. | | 38 | .421 | | 17.80 | 19.55 | | 37.35 | 49 |
| 3500 | 3-1/2" dia. | | 35 | .457 | | 20 | 21.50 | | 41.50 | 54 |
| 3510 | 4" dia. | | 33 | .485 | | 22 | 22.50 | | 44.50 | 58 |
| 3520 | 5" dia. | | 29 | .552 | | 25 | 25.50 | | 50.50 | 66.50 |
| 3530 | 6" dia. | | 26 | .615 | | 25 | 28.50 | | 53.50 | 71 |
| 3540 | 8" dia. | | 22 | .727 | | 52 | 34 | | 86 | 109 |
| 3550 | 10" dia. | | 20 | .800 | | 70.50 | 37 | | 107.50 | 134 |
| 3560 | 12" dia. | | 18 | .889 | | 87 | 41.50 | | 128.50 | 159 |
| 3570 | 14" dia. | | 16 | 1 | | 111 | 46.50 | | 157.50 | 193 |
| 3800 | Air gate valves, galvanized | | | | | | | | | |
| 3810 | 2-1/8" dia. | Q-9 | 30 | .533 | Ea. | 180 | 25 | | 205 | 236 |
| 3820 | 2-1/2" dia. | | 28 | .571 | | 199 | 26.50 | | 225.50 | 259 |
| 3830 | 3" dia. | | 26 | .615 | | 217 | 28.50 | | 245.50 | 283 |
| 3840 | 4" dia. | | 23 | .696 | | 273 | 32.50 | | 305.50 | 350 |
| 3850 | 6" dia. | | 18 | .889 | | 370 | 41.50 | | 411.50 | 470 |

# Division Notes

| | | CREW | DAILY OUTPUT | LABOR-HOURS | UNIT | 2011 BARE COSTS | | | | TOTAL INCL O&P |
|---|---|---|---|---|---|---|---|---|---|---|
| | | | | | | MAT. | LABOR | EQUIP. | TOTAL | |
| | | | | | | | | | | |
| | | | | | | | | | | |
| | | | | | | | | | | |
| | | | | | | | | | | |
| | | | | | | | | | | |
| | | | | | | | | | | |
| | | | | | | | | | | |
| | | | | | | | | | | |
| | | | | | | | | | | |
| | | | | | | | | | | |
| | | | | | | | | | | |
| | | | | | | | | | | |
| | | | | | | | | | | |
| | | | | | | | | | | |
| | | | | | | | | | | |
| | | | | | | | | | | |
| | | | | | | | | | | |
| | | | | | | | | | | |
| | | | | | | | | | | |
| | | | | | | | | | | |
| | | | | | | | | | | |
| | | | | | | | | | | |
| | | | | | | | | | | |

## Estimating Tips

This new division contains information about water and wastewater equipment and systems, which was formerly located in Division 44.

The main areas of focus are total wastewater treatment plants and components of wastewater treatment plants. In addition, there are assemblies such as sewage treatment lagoons that can be found in some publications under G30 Site Mechnical Utilities.

Also included in this section are oil/water separators for wastewater treatment.

## Reference Numbers

Reference numbers are shown in shaded boxes at the beginning of some major classifications. These numbers refer to related items in the Reference Section. The reference information may be an estimating procedure, an alternate pricing method, or technical information.

*Note:* Not all subdivisions listed here necessarily appear in this publication.

## Division 46 - Waste and Wastewater Equip.

## 46 07 53 – Packaged Wastewater Treatment Equipment

| 46 07 53.10 Biological Packaged Water Treatment Plants | Crew | Daily Output | Labor-Hours | Unit | Material | 2011 Bare Costs Labor | Equipment | Total | Total Incl O&P |
|---|---|---|---|---|---|---|---|---|---|
| 0010 **BIOLOGICAL PACKAGED WATER TREATMENT PLANTS** | | | | | | | | | |
| 0011 Not including fencing or external piping | | | | | | | | | |
| 0020 Steel packaged, blown air aeration plants | | | | | | | | | |
| 0100 1,000 GPD | | | | Gal. | | | | 95 | 109 |
| 0200 5,000 GPD | | | | | | | | 57 | 66 |
| 0300 15,000 GPD | | | | | | | | 48 | 55 |
| 0400 30,000 GPD | | | | | | | | 43 | 50 |
| 0500 50,000 GPD | | | | | | | | 38 | 44 |
| 0600 100,000 GPD | | | | | | | | 33 | 38 |
| 0700 200,000 GPD | | | | | | | | 28 | 32 |
| 0800 500,000 GPD | | | | ▼ | | | | 24 | 28 |
| 1000 Concrete, extended aeration, primary and secondary treatment | | | | | | | | | |
| 1010 10,000 GPD | | | | Gal. | | | | 48 | 55 |
| 1100 30,000 GPD | | | | | | | | 43 | 50 |
| 1200 50,000 GPD | | | | | | | | 38 | 44 |
| 1400 100,000 GPD | | | | | | | | 33 | 38 |
| 1500 500,000 GPD | | | | ▼ | | | | 24 | 28 |
| 1700 Municipal wastewater treatment facility | | | | | | | | | |
| 1720 1.0 MGD | | | | Gal. | | | | 9.70 | 11 |
| 1740 1.5 MGD | | | | | | | | 9.20 | 10.60 |
| 1760 2.0 MGD | | | | | | | | 8.75 | 10 |
| 1780 3.0 MGD | | | | | | | | 6.80 | 7.80 |
| 1800 5.0 MGD | | | | ▼ | | | | 5.80 | 6.70 |
| 2000 Holding tank system, not incl. excavation or backfill | | | | | | | | | |
| 2010 Recirculating chemical water closet | 2 Plum | 4 | 4 | Ea. | 460 | 214 | | 674 | 825 |
| 2100 For voltage converter, add | " | 16 | 1 | | 250 | 53.50 | | 303.50 | 355 |
| 2200 For high level alarm, add | 1 Plum | 7.80 | 1.026 | ▼ | 143 | 55 | | 198 | 239 |

## 46 07 53.20 Wastewater Treatment System

| | Crew | Daily Output | Labor-Hours | Unit | Material | 2011 Bare Costs Labor | Equipment | Total | Total Incl O&P |
|---|---|---|---|---|---|---|---|---|---|
| 0010 **WASTEWATER TREATMENT SYSTEM** | | | | | | | | | |
| 0020 Fiberglass, 1,000 gallon | B-21 | 1.29 | 21.705 | Ea. | 3,450 | 860 | 100 | 4,410 | 5,200 |
| 0100 1,500 gallon | B-21 | 1.03 | 27.184 | Ea. | 7,550 | 1,075 | 125 | 8,750 | 10,100 |

# Reference Section

All the reference information is in one section, making it easy to find what you need to know . . . and easy to use the book on a daily basis. This section is visually identified by a vertical gray bar on the page edges.

In this Reference Section, we've included Equipment Rental Costs, a listing of rental and operating costs; Crew Listings, a full listing of all crews and equipment, and their costs; Historical Cost Indexes for cost comparisons over time; City Cost Indexes and Location Factors for adjusting costs to the region you are in; Reference Tables, where you will find explanations, estimating information and procedures, or technical data; Change Orders, information on pricing changes to contract documents; Square Foot Costs that allow you to make a rough estimate for the overall cost of a project; and an explanation of all the Abbreviations in the book.

## Table of Contents

# Estimating Tips

- This section contains the average costs to rent and operate hundreds of pieces of construction equipment. This is useful information when estimating the time and material requirements of any particular operation in order to establish a unit or total cost. Equipment costs include not only rental, but also operating costs for equipment under normal use.

# Rental Costs

- Equipment rental rates are obtained from industry sources throughout North America–contractors, suppliers, dealers, manufacturers, and distributors.
- Rental rates vary throughout the country, with larger cities generally having lower rates. Lease plans for new equipment are available for periods in excess of six months, with a percentage of payments applying toward purchase.
- Monthly rental rates vary from 2% to 5% of the purchase price of the equipment depending on the anticipated life of the equipment and its wearing parts.
- Weekly rental rates are about 1/3 the monthly rates, and daily rental rates are about 1/3 the weekly rate.
- Rental rates can also be treated as reimbursement costs for contractor-owned equipment. Owned equipment costs include depreciation, loan payments, interest, taxes, insurance, storage, and major repairs.

# Operating Costs

- The operating costs include parts and labor for routine servicing, such as repair and replacement of pumps, filters and worn lines. Normal operating expendables, such as fuel, lubricants, tires and electricity (where applicable), are also included.
- Extraordinary operating expendables with highly variable wear patterns, such as diamond bits and blades, are excluded. These costs can be found as material costs in the Unit Price section.
- The hourly operating costs listed do not include the operator's wages.

# Equipment Cost/Day

- Any power equipment required by a crew is shown in the Crew Listings with a daily cost.
- The daily cost of equipment needed by a crew is based on dividing the weekly rental rate by 5 (number of working days in the week), and then adding the hourly operating cost times 8 (the number of hours in a day). This "Equipment Cost/Day" is shown in the far right column of the Equipment Rental pages.
- If equipment is needed for only one or two days, it is best to develop your own cost by including components for daily rent and hourly operating cost. This is important when the listed Crew for a task does not contain the equipment needed, such as a crane for lifting mechanical heating/cooling equipment up onto a roof.

- If the quantity of work is less than the crew's Daily Output shown for a Unit Price line item that includes a bare unit equipment cost, it is recommended to estimate one day's rental cost and operating cost for equipment shown in the Crew Listing for that line item.

# Mobilization/ Demobilization

- The cost to move construction equipment from an equipment yard or rental company to the jobsite and back again is not included in equipment rental costs listed in the Reference section, nor in the bare equipment cost of any Unit Price line item, nor in any equipment costs shown in the Crew listings.
- Mobilization (to the site) and demobilization (from the site) costs can be found in the Unit Price section.
- If a piece of equipment is already at the jobsite, it is not appropriate to utilize mobil./ demob. costs again in an estimate.

## 01 54 33 | Equipment Rental

| | | UNIT | HOURLY OPER. COST | RENT PER DAY | RENT PER WEEK | RENT PER MONTH | EQUIPMENT COST/DAY | |
|---|---|---|---|---|---|---|---|---|
| **10** | 0010 | **CONCRETE EQUIPMENT RENTAL** without operators  R015433 -10 | | | | | | | **10** |
| | 0200 | Bucket, concrete lightweight, 1/2 C.Y. | Ea. | .75 | 23 | 69 | 207 | 19.80 | |
| | 0300 | 1 C.Y. | | .80 | 28 | 84 | 252 | 23.20 | |
| | 0400 | 1-1/2 C.Y. | | 1.00 | 38.50 | 115 | 345 | 31 | |
| | 0500 | 2 C.Y. | | 1.10 | 45 | 135 | 405 | 35.80 | |
| | 0580 | 8 C.Y. | | 5.70 | 265 | 795 | 2,375 | 204.60 | |
| | 0600 | Cart, concrete, self-propelled, operator walking, 10 C.F. | | 2.50 | 58.50 | 175 | 525 | 55 | |
| | 0700 | Operator riding, 18 C.F. | | 4.20 | 96.50 | 290 | 870 | 91.60 | |
| | 0800 | Conveyer for concrete, portable, gas, 16" wide, 26' long | | 9.20 | 127 | 380 | 1,150 | 149.60 | |
| | 0900 | 46' long | | 9.55 | 153 | 460 | 1,375 | 168.40 | |
| | 1000 | 56' long | | 9.70 | 162 | 485 | 1,450 | 174.60 | |
| | 1100 | Core drill, electric, 2-1/2 H.P., 1" to 8" bit diameter | | 1.40 | 59.50 | 179 | 535 | 47 | |
| | 1150 | 11 H.P., 8" to 18" cores | | 5.10 | 115 | 345 | 1,025 | 109.80 | |
| | 1200 | Finisher, concrete floor, gas, riding trowel, 96" wide | | 8.55 | 148 | 445 | 1,325 | 157.40 | |
| | 1300 | Gas, walk-behind, 3 blade, 36" trowel | | 1.40 | 20.50 | 62 | 186 | 23.60 | |
| | 1400 | 4 blade, 48" trowel | | 2.70 | 28 | 84 | 252 | 38.40 | |
| | 1500 | Float, hand-operated (Bull float) 48" wide | | .08 | 13.65 | 41 | 123 | 8.85 | |
| | 1570 | Curb builder, 14 H.P., gas, single screw | | 11.95 | 253 | 760 | 2,275 | 247.60 | |
| | 1590 | Double screw | | 12.60 | 293 | 880 | 2,650 | 276.80 | |
| | 1600 | Floor grinder, concrete and terrazzo, electric, 22" path | | 1.62 | 102 | 305 | 915 | 73.95 | |
| | 1700 | Edger, concrete, electric, 7" path | | .90 | 51.50 | 155 | 465 | 38.20 | |
| | 1750 | Vacuum pick-up system for floor grinders, wet/dry | | 1.35 | 81.50 | 245 | 735 | 59.80 | |
| | 1800 | Mixer, powered, mortar and concrete, gas, 6 C.F., 18 H.P. | | 6.45 | 120 | 360 | 1,075 | 123.60 | |
| | 1900 | 10 C.F., 25 H.P. | | 7.90 | 147 | 440 | 1,325 | 151.20 | |
| | 2000 | 16 C.F. | | 8.20 | 170 | 510 | 1,525 | 167.60 | |
| | 2100 | Concrete, stationary, tilt drum, 2 C.Y. | | 6.45 | 238 | 715 | 2,150 | 194.60 | |
| | 2120 | Pump, concrete, truck mounted 4" line 80' boom | | 23.25 | 925 | 2,775 | 8,325 | 741 | |
| | 2140 | 5" line, 110' boom | | 30.70 | 1,250 | 3,770 | 11,300 | 999.60 | |
| | 2160 | Mud jack, 50 C.F. per hr. | | 6.10 | 128 | 385 | 1,150 | 125.80 | |
| | 2180 | 225 C.F. per hr | | 8.15 | 147 | 440 | 1,325 | 153.20 | |
| | 2190 | Shotcrete pump rig, 12 C.Y./hr. | | 14.15 | 227 | 680 | 2,050 | 249.20 | |
| | 2200 | 35 C.Y./hr. | | 16.75 | 243 | 730 | 2,200 | 280 | |
| | 2600 | Saw, concrete, manual, gas, 18 H.P. | | 4.65 | 43.50 | 130 | 390 | 63.20 | |
| | 2650 | Self-propelled, gas, 30 H.P. | | 8.90 | 105 | 315 | 945 | 134.20 | |
| | 2700 | Vibrators, concrete, electric, 60 cycle, 2 H.P. | | .31 | 9 | 27 | 81 | 7.90 | |
| | 2800 | 3 H.P. | | .43 | 12 | 36 | 108 | 10.65 | |
| | 2900 | Gas engine, 5 H.P. | | 1.30 | 16.35 | 49 | 147 | 20.20 | |
| | 3000 | 8 H.P. | | 1.75 | 15.35 | 46 | 138 | 23.20 | |
| | 3050 | Vibrating screed, gas engine, 8 H.P. | | 2.81 | 71.50 | 215 | 645 | 65.50 | |
| | 3120 | Concrete transit mixer, 6 x 4, 250 H.P., 8 C.Y., rear discharge | | 43.95 | 590 | 1,765 | 5,300 | 704.60 | |
| | 3200 | Front discharge | | 51.35 | 725 | 2,175 | 6,525 | 845.80 | |
| | 3300 | 6 x 6, 285 H.P., 12 C.Y., rear discharge | | 50.45 | 685 | 2,050 | 6,150 | 813.60 | |
| | 3400 | Front discharge | | 52.65 | 735 | 2,200 | 6,600 | 861.20 | |
| **20** | 0010 | **EARTHWORK EQUIPMENT RENTAL** without operators  R015433 -10 | | | | | | | **20** |
| | 0040 | Aggregate spreader, push type 8' to 12' wide | Ea. | 2.25 | 26 | 78 | 234 | 33.60 | |
| | 0045 | Tailgate type, 8' wide | | 2.20 | 33.50 | 100 | 300 | 37.60 | |
| | 0055 | Earth auger, truck-mounted, for fence & sign posts, utility poles | | 11.60 | 515 | 1,545 | 4,625 | 401.80 | |
| | 0060 | For borings and monitoring wells | | 38.45 | 665 | 1,995 | 5,975 | 706.60 | |
| | 0070 | Portable, trailer mounted | | 2.45 | 32.50 | 98 | 294 | 39.20 | |
| | 0075 | Truck-mounted, for caissons, water wells | | 83.90 | 3,025 | 9,075 | 27,200 | 2,486 | |
| | 0080 | Horizontal boring machine, 12" to 36" diameter, 45 H.P. | | 20.85 | 200 | 600 | 1,800 | 286.80 | |
| | 0090 | 12" to 48" diameter, 65 H.P. | | 28.90 | 350 | 1,055 | 3,175 | 442.20 | |
| | 0095 | Auger, for fence posts, gas engine, hand held | | .45 | 6 | 18 | 54 | 7.20 | |
| | 0100 | Excavator, diesel hydraulic, crawler mounted, 1/2 C.Y. cap. | | 16.70 | 385 | 1,155 | 3,475 | 364.60 | |
| | 0120 | 5/8 C.Y. capacity | | 24.50 | 555 | 1,670 | 5,000 | 530 | |
| | 0140 | 3/4 C.Y. capacity | | 30.55 | 660 | 1,980 | 5,950 | 640.40 | |
| | 0150 | 1 C.Y. capacity | | 36.35 | 730 | 2,195 | 6,575 | 729.80 | |

## 01 54 33 | Equipment Rental

| | UNIT | HOURLY OPER. COST | RENT PER DAY | RENT PER WEEK | RENT PER MONTH | EQUIPMENT COST/DAY | |
|---|---|---|---|---|---|---|---|
| **20** | | | | | | | **20** |
| 0200 1-1/2 C.Y. capacity | Ea. | 44.55 | 985 | 2,955 | 8,875 | 947.40 | |
| 0300 2 C.Y. capacity | | 58.25 | 1,275 | 3,790 | 11,400 | 1,224 | |
| 0320 2-1/2 C.Y. capacity | | 77.80 | 1,650 | 4,955 | 14,900 | 1,613 | |
| 0325 3-1/2 C.Y. capacity | | 104.95 | 2,225 | 6,655 | 20,000 | 2,171 | |
| 0330 4-1/2 C.Y. capacity | | 128.05 | 2,750 | 8,225 | 24,700 | 2,669 | |
| 0335 6 C.Y. capacity | | 158.65 | 3,050 | 9,160 | 27,500 | 3,101 | |
| 0340 7 C.Y. capacity | | 160.90 | 3,150 | 9,460 | 28,400 | 3,179 | |
| 0342 Excavator attachments, bucket thumbs | | 3.00 | 245 | 735 | 2,200 | 171 | |
| 0345 Grapples | | 2.70 | 213 | 640 | 1,925 | 149.60 | |
| 0347 Hydraulic hammer for boom mounting, 5000 ft lbs | | 12.65 | 420 | 1,255 | 3,775 | 352.20 | |
| 0349 11,000 ft lbs | | 21.50 | 750 | 2,250 | 6,750 | 622 | |
| 0350 Gradall type, truck mounted, 3 ton @ 15' radius, 5/8 C.Y. | | 36.80 | 940 | 2,815 | 8,450 | 857.40 | |
| 0370 1 C.Y. capacity | | 41.00 | 1,125 | 3,365 | 10,100 | 1,001 | |
| 0400 Backhoe-loader, 40 to 45 H.P., 5/8 C.Y. capacity | | 11.20 | 223 | 670 | 2,000 | 223.60 | |
| 0450 45 H.P. to 60 H.P., 3/4 C.Y. capacity | | 16.30 | 315 | 950 | 2,850 | 320.40 | |
| 0460 80 H.P., 1-1/4 C.Y. capacity | | 17.80 | 345 | 1,035 | 3,100 | 349.40 | |
| 0470 112 H.P., 1-1/2 C.Y. capacity | | 30.25 | 610 | 1,830 | 5,500 | 608 | |
| 0482 Backhoe-loader attachment, compactor, 20,000 lb. | | 5.40 | 145 | 435 | 1,300 | 130.20 | |
| 0485 Hydraulic hammer, 750 ft lbs | | 3.10 | 103 | 310 | 930 | 86.80 | |
| 0486 Hydraulic hammer, 1200 ft lbs | | 5.50 | 190 | 570 | 1,700 | 158 | |
| 0500 Brush chipper, gas engine, 6" cutter head, 35 H.P. | | 8.10 | 110 | 330 | 990 | 130.80 | |
| 0550 12" cutter head, 130 H.P. | | 13.25 | 185 | 555 | 1,675 | 217 | |
| 0600 15" cutter head, 165 H.P. | | 19.45 | 222 | 665 | 2,000 | 288.60 | |
| 0750 Bucket, clamshell, general purpose, 3/8 C.Y. | | 1.20 | 38.50 | 115 | 345 | 32.60 | |
| 0800 1/2 C.Y. | | 1.30 | 45 | 135 | 405 | 37.40 | |
| 0850 3/4 C.Y. | | 1.45 | 56.50 | 170 | 510 | 45.60 | |
| 0900 1 C.Y. | | 1.50 | 60 | 180 | 540 | 48 | |
| 0950 1-1/2 C.Y. | | 2.40 | 81.50 | 245 | 735 | 68.20 | |
| 1000 2 C.Y. | | 2.50 | 91.50 | 275 | 825 | 75 | |
| 1010 Bucket, dragline, medium duty, 1/2 C.Y. | | .70 | 24.50 | 73 | 219 | 20.20 | |
| 1020 3/4 C.Y. | | .70 | 25.50 | 77 | 231 | 21 | |
| 1030 1 C.Y. | | .75 | 27 | 81 | 243 | 22.20 | |
| 1040 1-1/2 C.Y. | | 1.15 | 41.50 | 125 | 375 | 34.20 | |
| 1050 2 C.Y. | | 1.20 | 46.50 | 140 | 420 | 37.60 | |
| 1070 3 C.Y. | | 1.85 | 63.50 | 190 | 570 | 52.80 | |
| 1200 Compactor, manually guided 2-drum vibratory smooth roller, 7.5 H.P. | | 5.85 | 167 | 500 | 1,500 | 146.80 | |
| 1250 Rammer/tamper, gas, 8" | | 2.20 | 43.50 | 130 | 390 | 43.60 | |
| 1260 15" | | 2.45 | 50 | 150 | 450 | 49.60 | |
| 1300 Vibratory plate, gas, 18" plate, 3000 lb. blow | | 2.10 | 23.50 | 70 | 210 | 30.80 | |
| 1350 21" plate, 5000 lb. blow | | 2.60 | 30.50 | 91 | 273 | 39 | |
| 1370 Curb builder/extruder, 14 H.P., gas, single screw | | 11.95 | 253 | 760 | 2,275 | 247.60 | |
| 1390 Double screw | | 12.60 | 293 | 880 | 2,650 | 276.80 | |
| 1500 Disc harrow attachment, for tractor | | .41 | 68.50 | 205 | 615 | 44.30 | |
| 1810 Feller buncher, shearing & accumulating trees, 100 H.P. | | 27.75 | 570 | 1,710 | 5,125 | 564 | |
| 1860 Grader, self-propelled, 25,000 lb. | | 22.90 | 470 | 1,405 | 4,225 | 464.20 | |
| 1910 30,000 lb. | | 26.35 | 555 | 1,660 | 4,975 | 542.80 | |
| 1920 40,000 lb. | | 43.05 | 1,050 | 3,155 | 9,475 | 975.40 | |
| 1930 55,000 lb. | | 55.95 | 1,425 | 4,275 | 12,800 | 1,303 | |
| 1950 Hammer, pavement breaker, self-propelled, diesel, 1000 to 1250 lb. | | 22.90 | 360 | 1,080 | 3,250 | 399.20 | |
| 2000 1300 to 1500 lb. | | 34.35 | 720 | 2,160 | 6,475 | 706.80 | |
| 2050 Pile driving hammer, steam or air, 4150 ft lbs @ 225 bpm | | 9.80 | 500 | 1,495 | 4,475 | 377.40 | |
| 2100 8750 ft lbs @ 145 bpm | | 11.80 | 700 | 2,095 | 6,275 | 513.40 | |
| 2150 15,000 ft lbs @ 60 bpm | | 13.40 | 845 | 2,535 | 7,600 | 614.20 | |
| 2200 24,450 ft lbs @ 111 bpm | | 14.40 | 940 | 2,820 | 8,450 | 679.20 | |
| 2250 Leads, 60' high for pile driving hammers up to 20,000 ft lbs | | 3.10 | 82 | 246 | 740 | 74 | |
| 2300 90' high for hammers over 20,000 ft lbs | | 4.70 | 149 | 448 | 1,350 | 127.20 | |
| 2350 Diesel type hammer, 22,400 ft lbs | | 24.80 | 655 | 1,970 | 5,900 | 592.40 | |
| 2400 41,300 ft lbs | | 31.25 | 710 | 2,135 | 6,400 | 677 | |

## 01 54 33 | Equipment Rental

| | | UNIT | HOURLY OPER. COST | RENT PER DAY | RENT PER WEEK | RENT PER MONTH | EQUIPMENT COST/DAY | |
|---|---|---|---|---|---|---|---|---|
| 20 | 2450 | 141,000 ft lbs | Ea. | 47.65 | 1,225 | 3,660 | 11,000 | 1,113 | 20 |
| | 2500 | Vib. elec. hammer/extractor, 200 kW diesel generator, 34 H.P. | | 36.95 | 680 | 2,045 | 6,125 | 704.60 | |
| | 2550 | 80 H.P. | | 64.95 | 1,000 | 3,025 | 9,075 | 1,125 | |
| | 2600 | 150 H.P. | | 120.90 | 1,950 | 5,865 | 17,600 | 2,140 | |
| | 2700 | Extractor, steam or air, 700 ft lbs | | 17.75 | 520 | 1,560 | 4,675 | 454 | |
| | 2750 | 1000 ft lbs | | 20.20 | 635 | 1,910 | 5,725 | 543.60 | |
| | 2800 | Log chipper, up to 22" diameter, 600 H.P. | | 35.35 | 455 | 1,360 | 4,075 | 554.80 | |
| | 2850 | Logger, for skidding & stacking logs, 150 H.P. | | 41.75 | 850 | 2,550 | 7,650 | 844 | |
| | 2860 | Mulcher, diesel powered, trailer mounted | | 15.60 | 223 | 670 | 2,000 | 258.80 | |
| | 2900 | Rake, spring tooth, with tractor | | 12.66 | 294 | 882 | 2,650 | 277.70 | |
| | 3000 | Roller, vibratory, tandem, smooth drum, 20 H.P. | | 7.40 | 143 | 430 | 1,300 | 145.20 | |
| | 3050 | 35 H.P. | | 9.40 | 248 | 745 | 2,225 | 224.20 | |
| | 3100 | Towed type vibratory compactor, smooth drum, 50 H.P. | | 22.45 | 360 | 1,075 | 3,225 | 394.60 | |
| | 3150 | Sheepsfoot, 50 H.P. | | 24.75 | 420 | 1,255 | 3,775 | 449 | |
| | 3170 | Landfill compactor, 220 H.P. | | 68.75 | 1,425 | 4,290 | 12,900 | 1,408 | |
| | 3200 | Pneumatic tire roller, 80 H.P. | | 12.65 | 345 | 1,040 | 3,125 | 309.20 | |
| | 3250 | 120 H.P. | | 19.15 | 600 | 1,800 | 5,400 | 513.20 | |
| | 3300 | Sheepsfoot vibratory roller, 240 H.P. | | 55.10 | 1,100 | 3,265 | 9,800 | 1,094 | |
| | 3320 | 340 H.P. | | 73.05 | 1,675 | 5,025 | 15,100 | 1,589 | |
| | 3350 | Smooth drum vibratory roller, 75 H.P. | | 20.60 | 555 | 1,670 | 5,000 | 498.80 | |
| | 3400 | 125 H.P. | | 25.60 | 655 | 1,960 | 5,875 | 596.80 | |
| | 3410 | Rotary mower, brush, 60", with tractor | | 16.60 | 268 | 805 | 2,425 | 293.80 | |
| | 3420 | Rototiller, walk-behind, gas, 5 H.P. | | 2.23 | 67.50 | 203 | 610 | 58.45 | |
| | 3422 | 8 H.P. | | 3.47 | 103 | 310 | 930 | 89.75 | |
| | 3440 | Scrapers, towed type, 7 C.Y. capacity | | 5.25 | 110 | 330 | 990 | 108 | |
| | 3450 | 10 C.Y. capacity | | 6.00 | 153 | 460 | 1,375 | 140 | |
| | 3500 | 15 C.Y. capacity | | 6.45 | 178 | 535 | 1,600 | 158.60 | |
| | 3525 | Self-propelled, single engine, 14 C.Y. capacity | | 77.50 | 1,475 | 4,460 | 13,400 | 1,512 | |
| | 3550 | Dual engine, 21 C.Y. capacity | | 107.80 | 1,675 | 5,050 | 15,200 | 1,872 | |
| | 3600 | 31 C.Y. capacity | | 152.05 | 2,600 | 7,805 | 23,400 | 2,777 | |
| | 3640 | 44 C.Y. capacity | | 193.60 | 3,600 | 10,780 | 32,300 | 3,705 | |
| | 3650 | Elevating type, single engine, 11 C.Y. capacity | | 54.35 | 1,100 | 3,270 | 9,800 | 1,089 | |
| | 3700 | 22 C.Y. capacity | | 99.85 | 2,150 | 6,440 | 19,300 | 2,087 | |
| | 3710 | Screening plant 110 H.P. w/5' x 10' screen | | 30.50 | 420 | 1,260 | 3,775 | 496 | |
| | 3720 | 5' x 16' screen | | 32.65 | 520 | 1,560 | 4,675 | 573.20 | |
| | 3850 | Shovel, crawler-mounted, front-loading, 7 C.Y. capacity | | 168.30 | 2,950 | 8,815 | 26,400 | 3,109 | |
| | 3855 | 12 C.Y. capacity | | 225.05 | 3,750 | 11,230 | 33,700 | 4,046 | |
| | 3860 | Shovel/backhoe bucket, 1/2 C.Y. | | 2.25 | 63.50 | 190 | 570 | 56 | |
| | 3870 | 3/4 C.Y. | | 2.30 | 71.50 | 215 | 645 | 61.40 | |
| | 3880 | 1 C.Y. | | 2.40 | 80 | 240 | 720 | 67.20 | |
| | 3890 | 1-1/2 C.Y. | | 2.55 | 95 | 285 | 855 | 77.40 | |
| | 3910 | 3 C.Y. | | 2.85 | 130 | 390 | 1,175 | 100.80 | |
| | 3950 | Stump chipper, 18" deep, 30 H.P. | | 7.17 | 187 | 560 | 1,675 | 169.35 | |
| | 4110 | Dozer, crawler, torque converter, diesel 80 H.P. | | 21.70 | 400 | 1,195 | 3,575 | 412.60 | |
| | 4150 | 105 H.P. | | 25.80 | 500 | 1,495 | 4,475 | 505.40 | |
| | 4200 | 140 H.P. | | 37.35 | 800 | 2,400 | 7,200 | 778.80 | |
| | 4260 | 200 H.P. | | 50.75 | 1,075 | 3,225 | 9,675 | 1,051 | |
| | 4310 | 300 H.P. | | 71.80 | 1,725 | 5,165 | 15,500 | 1,607 | |
| | 4360 | 410 H.P. | | 97.20 | 2,200 | 6,635 | 19,900 | 2,105 | |
| | 4370 | 500 H.P. | | 130.15 | 3,000 | 9,020 | 27,100 | 2,845 | |
| | 4380 | 700 H.P. | | 191.35 | 4,375 | 13,095 | 39,300 | 4,150 | |
| | 4400 | Loader, crawler, torque conv., diesel, 1-1/2 C.Y., 80 H.P. | | 22.70 | 475 | 1,425 | 4,275 | 466.60 | |
| | 4450 | 1-1/2 to 1-3/4 C.Y., 95 H.P. | | 25.75 | 600 | 1,795 | 5,375 | 565 | |
| | 4510 | 1-3/4 to 2-1/4 C.Y., 130 H.P. | | 33.70 | 840 | 2,520 | 7,550 | 773.60 | |
| | 4530 | 2-1/2 to 3-1/4 C.Y., 190 H.P. | | 46.40 | 1,050 | 3,185 | 9,550 | 1,008 | |
| | 4560 | 3-1/2 to 5 C.Y., 275 H.P. | | 64.35 | 1,525 | 4,585 | 13,800 | 1,432 | |
| | 4610 | Front end loader, 4WD, articulated frame, 1 to 1-1/4 C.Y., 70 H.P. | | 13.35 | 215 | 645 | 1,925 | 235.80 | |
| | 4620 | 1-1/2 to 1-3/4 C.Y., 95 H.P. | | 17.30 | 300 | 905 | 2,725 | 319.40 | |

## 01 54 33 | Equipment Rental

| | | UNIT | HOURLY OPER. COST | RENT PER DAY | RENT PER WEEK | RENT PER MONTH | EQUIPMENT COST/DAY | |
|---|---|---|---|---|---|---|---|---|
| **20** | 4650 | 1-3/4 to 2 C.Y., 130 H.P. | Ea. | 19.90 | 370 | 1,105 | 3,325 | 380.20 | **20** |
| | 4710 | 2-1/2 to 3-1/2 C.Y., 145 H.P. | | 21.20 | 380 | 1,140 | 3,425 | 397.60 | |
| | 4730 | 3 to 4-1/2 C.Y., 185 H.P. | | 28.95 | 610 | 1,835 | 5,500 | 598.60 | |
| | 4760 | 5-1/4 to 5-3/4 C.Y., 270 H.P. | | 47.05 | 930 | 2,785 | 8,350 | 933.40 | |
| | 4810 | 7 to 9 C.Y., 475 H.P. | | 83.30 | 1,900 | 5,725 | 17,200 | 1,811 | |
| | 4870 | 9 - 11 C.Y., 620 H.P. | | 108.70 | 2,675 | 8,030 | 24,100 | 2,476 | |
| | 4880 | Skid steer loader, wheeled, 10 C.F., 30 H.P. gas | | 7.85 | 148 | 445 | 1,325 | 151.80 | |
| | 4890 | 1 C.Y., 78 H.P., diesel | | 13.90 | 250 | 750 | 2,250 | 261.20 | |
| | 4892 | Skid-steer attachment, auger | | .53 | 88 | 264 | 790 | 57.05 | |
| | 4893 | Backhoe | | .71 | 119 | 356 | 1,075 | 76.90 | |
| | 4894 | Broom | | .74 | 123 | 370 | 1,100 | 79.90 | |
| | 4895 | Forks | | .24 | 40.50 | 122 | 365 | 26.30 | |
| | 4896 | Grapple | | .55 | 91.50 | 274 | 820 | 59.20 | |
| | 4897 | Concrete hammer | | 1.07 | 178 | 534 | 1,600 | 115.35 | |
| | 4898 | Tree spade | | .78 | 129 | 388 | 1,175 | 83.85 | |
| | 4899 | Trencher | | .76 | 127 | 380 | 1,150 | 82.10 | |
| | 4900 | Trencher, chain, boom type, gas, operator walking, 12 H.P. | | 3.55 | 48.50 | 145 | 435 | 57.40 | |
| | 4910 | Operator riding, 40 H.P. | | 14.25 | 305 | 915 | 2,750 | 297 | |
| | 5000 | Wheel type, diesel, 4' deep, 12" wide | | 61.50 | 850 | 2,545 | 7,625 | 1,001 | |
| | 5100 | 6' deep, 20" wide | | 77.45 | 1,975 | 5,910 | 17,700 | 1,802 | |
| | 5150 | Chain type, diesel, 5' deep, 8" wide | | 27.10 | 600 | 1,805 | 5,425 | 577.80 | |
| | 5200 | Diesel, 8' deep, 16" wide | | 114.45 | 2,650 | 7,950 | 23,900 | 2,506 | |
| | 5202 | Rock trencher, wheel type, 6" wide x 18" deep | | 17.35 | 325 | 970 | 2,900 | 332.80 | |
| | 5206 | Chain type, 18" wide x 7' deep | | 95.95 | 2,950 | 8,850 | 26,600 | 2,538 | |
| | 5210 | Tree spade, self-propelled | | 13.73 | 267 | 800 | 2,400 | 269.85 | |
| | 5250 | Truck, dump, 2-axle, 12 ton, 8 C.Y. payload, 220 H.P. | | 21.80 | 235 | 705 | 2,125 | 315.40 | |
| | 5300 | Three axle dump, 16 ton, 12 CY payload, 400 H.P. | | 38.00 | 335 | 1,010 | 3,025 | 506 | |
| | 5310 | Four axle dump, 25 ton, 18 C.Y. payload, 450 H.P. | | 45.60 | 495 | 1,480 | 4,450 | 660.80 | |
| | 5350 | Dump trailer only, rear dump, 16-1/2 C.Y. | | 5.10 | 142 | 425 | 1,275 | 125.80 | |
| | 5400 | 20 C.Y. | | 5.50 | 160 | 480 | 1,450 | 140 | |
| | 5450 | Flatbed, single axle, 1-1/2 ton rating | | 16.05 | 71.50 | 215 | 645 | 171.40 | |
| | 5500 | 3 ton rating | | 19.50 | 102 | 305 | 915 | 217 | |
| | 5550 | Off highway rear dump, 25 ton capacity | | 57.25 | 1,350 | 4,030 | 12,100 | 1,264 | |
| | 5600 | 35 ton capacity | | 52.80 | 1,200 | 3,565 | 10,700 | 1,135 | |
| | 5610 | 50 ton capacity | | 71.20 | 1,625 | 4,895 | 14,700 | 1,549 | |
| | 5620 | 65 ton capacity | | 72.90 | 1,625 | 4,840 | 14,500 | 1,551 | |
| | 5630 | 100 ton capacity | | 93.85 | 2,100 | 6,290 | 18,900 | 2,009 | |
| | 6000 | Vibratory plow, 25 H.P., walking | | 5.95 | 63.50 | 190 | 570 | 85.60 | |
| **40** | 0010 | **GENERAL EQUIPMENT RENTAL** without operators | | | | | | | **40** |
| | 0150 | Aerial lift, scissor type, to 15' high, 1000 lb. cap., electric [R015433 -10] | Ea. | 2.90 | 56.50 | 170 | 510 | 57.20 | |
| | 0160 | To 25' high, 2000 lb. capacity | | 3.30 | 71.50 | 215 | 645 | 69.40 | |
| | 0170 | Telescoping boom to 40' high, 500 lb. capacity, gas | | 15.85 | 320 | 955 | 2,875 | 317.80 | |
| | 0180 | To 45' high, 500 lb. capacity | | 16.95 | 370 | 1,105 | 3,325 | 356.60 | |
| | 0190 | To 60' high, 600 lb. capacity | | 19.20 | 480 | 1,445 | 4,325 | 442.60 | |
| | 0195 | Air compressor, portable, 6.5 CFM, electric | | .33 | 14 | 42 | 126 | 11.05 | |
| | 0196 | Gasoline | | .80 | 21 | 63 | 189 | 19 | |
| | 0200 | Towed type, gas engine, 60 CFM | | 10.15 | 48.50 | 145 | 435 | 110.20 | |
| | 0300 | 160 CFM | | 11.75 | 51.50 | 155 | 465 | 125 | |
| | 0400 | Diesel engine, rotary screw, 250 CFM | | 12.35 | 107 | 320 | 960 | 162.80 | |
| | 0500 | 365 CFM | | 16.40 | 132 | 395 | 1,175 | 210.20 | |
| | 0550 | 450 CFM | | 20.70 | 165 | 495 | 1,475 | 264.60 | |
| | 0600 | 600 CFM | | 35.95 | 228 | 685 | 2,050 | 424.60 | |
| | 0700 | 750 CFM | | 36.10 | 238 | 715 | 2,150 | 431.80 | |
| | 0800 | For silenced models, small sizes, add to rent | | 3% | 5% | 5% | 5% | | |
| | 0900 | Large sizes, add to rent | | 5% | 7% | 7% | 7% | | |
| | 0930 | Air tools, breaker, pavement, 60 lb. | Ea. | .50 | 9.65 | 29 | 87 | 9.80 | |
| | 0940 | 80 lb. | | .50 | 10.35 | 31 | 93 | 10.20 | |
| | 0950 | Drills, hand (jackhammer) 65 lb. | | .55 | 17.35 | 52 | 156 | 14.80 | |

## 01 54 33 | Equipment Rental

| | | UNIT | HOURLY OPER. COST | RENT PER DAY | RENT PER WEEK | RENT PER MONTH | EQUIPMENT COST/DAY | |
|---|---|---|---|---|---|---|---|---|
| 40 | 0960 | Track or wagon, swing boom, 4" drifter | Ea. | 49.45 | 755 | 2,270 | 6,800 | 849.60 | 40 |
| | 0970 | 5" drifter | | 59.40 | 815 | 2,450 | 7,350 | 965.20 | |
| | 0975 | Track mounted quarry drill, 6" diameter drill | | 82.30 | 1,250 | 3,725 | 11,200 | 1,403 | |
| | 0980 | Dust control per drill | | .58 | 18.65 | 56 | 168 | 15.85 | |
| | 0990 | Hammer, chipping, 12 lb. | | .50 | 24.50 | 73 | 219 | 18.60 | |
| | 1000 | Hose, air with couplings, 50' long, 3/4" diameter | | .03 | 4.67 | 14 | 42 | 3.05 | |
| | 1100 | 1" diameter | | .04 | 6.35 | 19 | 57 | 4.10 | |
| | 1200 | 1-1/2" diameter | | .06 | 9.65 | 29 | 87 | 6.30 | |
| | 1300 | 2" diameter | | .07 | 12 | 36 | 108 | 7.75 | |
| | 1400 | 2-1/2" diameter | | .11 | 19 | 57 | 171 | 12.30 | |
| | 1410 | 3" diameter | | .14 | 23 | 69 | 207 | 14.90 | |
| | 1450 | Drill, steel, 7/8" x 2' | | .05 | 8.65 | 26 | 78 | 5.60 | |
| | 1460 | 7/8" x 6' | | .05 | 9 | 27 | 81 | 5.80 | |
| | 1520 | Moil points | | .02 | 4 | 12 | 36 | 2.55 | |
| | 1525 | Pneumatic nailer w/accessories | | .49 | 32.50 | 98 | 294 | 23.50 | |
| | 1530 | Sheeting driver for 60 lb. breaker | | .04 | 6 | 18 | 54 | 3.90 | |
| | 1540 | For 90 lb. breaker | | .12 | 8 | 24 | 72 | 5.75 | |
| | 1550 | Spade, 25 lb. | | .45 | 7.35 | 22 | 66 | 8 | |
| | 1560 | Tamper, single, 35 lb. | | .55 | 36.50 | 109 | 325 | 26.20 | |
| | 1570 | Triple, 140 lb. | | .82 | 54.50 | 164 | 490 | 39.35 | |
| | 1580 | Wrenches, impact, air powered, up to 3/4" bolt | | .40 | 10 | 30 | 90 | 9.20 | |
| | 1590 | Up to 1-1/4" bolt | | .50 | 23.50 | 71 | 213 | 18.20 | |
| | 1600 | Barricades, barrels, reflectorized, 1 to 50 barrels | | .03 | 4.60 | 13.80 | 41.50 | 3 | |
| | 1610 | 100 to 200 barrels | | .02 | 3.53 | 10.60 | 32 | 2.30 | |
| | 1620 | Barrels with flashers, 1 to 50 barrels | | .03 | 5.25 | 15.80 | 47.50 | 3.40 | |
| | 1630 | 100 to 200 barrels | | .03 | 4.20 | 12.60 | 38 | 2.75 | |
| | 1640 | Barrels with steady burn type C lights | | .04 | 7 | 21 | 63 | 4.50 | |
| | 1650 | Illuminated board, trailer mounted, with generator | | .75 | 122 | 365 | 1,100 | 79 | |
| | 1670 | Portable barricade, stock, with flashers, 1 to 6 units | | .03 | 5.25 | 15.80 | 47.50 | 3.40 | |
| | 1680 | 25 to 50 units | | .03 | 4.90 | 14.70 | 44 | 3.20 | |
| | 1685 | Butt fusion machine, wheeled, 1.5 HP electric, 2" - 8" diam. pipe | | 2.56 | 167 | 500 | 1,500 | 120.50 | |
| | 1690 | Tracked, 20 HP diesel, 4"-12" diam. pipe | | 10.43 | 490 | 1,465 | 4,400 | 376.45 | |
| | 1695 | 83 HP diesel, 8" - 24" diam. pipe | | 27.52 | 975 | 2,930 | 8,800 | 806.15 | |
| | 1700 | Carts, brick, hand powered, 1000 lb. capacity | | .44 | 72.50 | 218 | 655 | 47.10 | |
| | 1800 | Gas engine, 1500 lb., 7-1/2' lift | | 4.04 | 115 | 345 | 1,025 | 101.30 | |
| | 1822 | Dehumidifier, medium, 6 lb/hr, 150 CFM | | .93 | 60 | 180 | 540 | 43.45 | |
| | 1824 | Large, 18 lb/hr, 600 CFM | | 1.89 | 122 | 366 | 1,100 | 88.30 | |
| | 1830 | Distributor, asphalt, trailer mounted, 2000 gal., 38 H.P. diesel | | 9.60 | 340 | 1,020 | 3,050 | 280.80 | |
| | 1840 | 3000 gal., 38 H.P. diesel | | 11.00 | 370 | 1,105 | 3,325 | 309 | |
| | 1850 | Drill, rotary hammer, electric | | .82 | 26.50 | 80 | 240 | 22.55 | |
| | 1860 | Carbide bit, 1-1/2" diameter, add to electric rotary hammer | | .02 | 6.25 | 18.80 | 56.50 | 3.90 | |
| | 1865 | Rotary, crawler, 250 H.P. | | 122.20 | 2,050 | 6,125 | 18,400 | 2,203 | |
| | 1870 | Emulsion sprayer, 65 gal., 5 H.P. gas engine | | 2.89 | 99 | 297 | 890 | 82.50 | |
| | 1880 | 200 gal., 5 H.P. engine | | 6.85 | 165 | 495 | 1,475 | 153.80 | |
| | 1900 | Floor auto-scrubbing machine, walk-behind, 28" path | | 3.26 | 213 | 640 | 1,925 | 154.10 | |
| | 1930 | Floodlight, mercury vapor, or quartz, on tripod, 1000 watt | | .27 | 14 | 42 | 126 | 10.55 | |
| | 1940 | 2000 watt | | .50 | 27 | 81 | 243 | 20.20 | |
| | 1950 | Floodlights, trailer mounted with generator, 1 - 300 watt light | | 3.10 | 73.50 | 220 | 660 | 68.80 | |
| | 1960 | 2 - 1000 watt lights | | 3.95 | 98.50 | 295 | 885 | 90.60 | |
| | 2000 | 4 - 300 watt lights | | 3.75 | 93.50 | 280 | 840 | 86 | |
| | 2005 | Foam spray rig, incl. box trailer, compressor, generator, proportioner | | 23.78 | 495 | 1,490 | 4,475 | 488.25 | |
| | 2020 | Forklift, straight mast, 12' lift, 5000 lb., 2 wheel drive, gas | | 19.80 | 202 | 605 | 1,825 | 279.40 | |
| | 2040 | 21' lift, 5000 lb., 4 wheel drive, diesel | | 16.20 | 248 | 745 | 2,225 | 278.60 | |
| | 2050 | For rough terrain, 42' lift, 35' reach, 9000 lb., 110 H.P. | | 22.30 | 450 | 1,345 | 4,025 | 447.40 | |
| | 2060 | For plant, 4 ton capacity, 80 H.P., 2 wheel drive, gas | | 12.05 | 102 | 305 | 915 | 157.40 | |
| | 2080 | 10 ton capacity, 120 H.P., 2 wheel drive, diesel | | 17.70 | 177 | 530 | 1,600 | 247.60 | |
| | 2100 | Generator, electric, gas engine, 1.5 kW to 3 kW | | 2.80 | 11.35 | 34 | 102 | 29.20 | |
| | 2200 | 5 kW | | 3.55 | 15.35 | 46 | 138 | 37.60 | |

## 01 54 33 | Equipment Rental

| | | UNIT | HOURLY OPER. COST | RENT PER DAY | RENT PER WEEK | RENT PER MONTH | EQUIPMENT COST/DAY |
|---|---|---|---|---|---|---|---|
| 2300 | 10 kW | Ea. | 6.65 | 38.50 | 115 | 345 | 76.20 |
| 2400 | 25 kW | | 7.80 | 88.50 | 265 | 795 | 115.40 |
| 2500 | Diesel engine, 20 kW | | 9.00 | 75 | 225 | 675 | 117 |
| 2600 | 50 kW | | 15.85 | 105 | 315 | 945 | 189.80 |
| 2700 | 100 kW | | 30.25 | 145 | 435 | 1,300 | 329 |
| 2800 | 250 kW | | 59.65 | 253 | 760 | 2,275 | 629.20 |
| 2850 | Hammer, hydraulic, for mounting on boom, to 500 ft lbs | | 2.40 | 76.50 | 230 | 690 | 65.20 |
| 2860 | 1000 ft lbs | | 4.05 | 127 | 380 | 1,150 | 108.40 |
| 2900 | Heaters, space, oil or electric, 50 MBH | | 1.84 | 8 | 24 | 72 | 19.50 |
| 3000 | 100 MBH | | 3.31 | 10.65 | 32 | 96 | 32.90 |
| 3100 | 300 MBH | | 10.62 | 36.50 | 110 | 330 | 106.95 |
| 3150 | 500 MBH | | 21.42 | 48.50 | 145 | 435 | 200.35 |
| 3200 | Hose, water, suction with coupling, 20' long, 2" diameter | | .02 | 3 | 9 | 27 | 1.95 |
| 3210 | 3" diameter | | .03 | 4.67 | 14 | 42 | 3.05 |
| 3220 | 4" diameter | | .03 | 5 | 15 | 45 | 3.25 |
| 3230 | 6" diameter | | .11 | 18.35 | 55 | 165 | 11.90 |
| 3240 | 8" diameter | | .20 | 33.50 | 100 | 300 | 21.60 |
| 3250 | Discharge hose with coupling, 50' long, 2" diameter | | .01 | 1.33 | 4 | 12 | .90 |
| 3260 | 3" diameter | | .02 | 2.67 | 8 | 24 | 1.75 |
| 3270 | 4" diameter | | .02 | 3.67 | 11 | 33 | 2.35 |
| 3280 | 6" diameter | | .06 | 9.65 | 29 | 87 | 6.30 |
| 3290 | 8" diameter | | .20 | 33.50 | 100 | 300 | 21.60 |
| 3295 | Insulation blower | | .20 | 6 | 18 | 54 | 5.20 |
| 3300 | Ladders, extension type, 16' to 36' long | | .15 | 25.50 | 76 | 228 | 16.40 |
| 3400 | 40' to 60' long | | .22 | 36 | 108 | 325 | 23.35 |
| 3405 | Lance for cutting concrete | | 2.17 | 104 | 313 | 940 | 79.95 |
| 3407 | Lawn mower, rotary, 22", 5 H.P. | | 2.14 | 63.50 | 190 | 570 | 55.10 |
| 3408 | 48" self propelled | | 2.88 | 94.50 | 283 | 850 | 79.65 |
| 3410 | Level, electronic, automatic, with tripod and leveling rod | | 1.45 | 96.50 | 290 | 870 | 69.60 |
| 3430 | Laser type, for pipe and sewer line and grade | | .76 | 50.50 | 152 | 455 | 36.50 |
| 3440 | Rotating beam for interior control | | 1.16 | 77.50 | 232 | 695 | 55.70 |
| 3460 | Builder's optical transit, with tripod and rod | | .09 | 14.65 | 44 | 132 | 9.50 |
| 3500 | Light towers, towable, with diesel generator, 2000 watt | | 3.75 | 93.50 | 280 | 840 | 86 |
| 3600 | 4000 watt | | 3.95 | 98.50 | 295 | 885 | 90.60 |
| 3700 | Mixer, powered, plaster and mortar, 6 C.F., 7 H.P. | | 1.85 | 30 | 90 | 270 | 32.80 |
| 3800 | 10 C.F., 9 H.P. | | 2.00 | 31.50 | 95 | 285 | 35 |
| 3850 | Nailer, pneumatic | | .49 | 32.50 | 98 | 294 | 23.50 |
| 3900 | Paint sprayers complete, 8 CFM | | .73 | 48.50 | 145 | 435 | 34.85 |
| 4000 | 17 CFM | | 1.35 | 89.50 | 269 | 805 | 64.60 |
| 4020 | Pavers, bituminous, rubber tires, 8' wide, 50 H.P., diesel | | 38.00 | 1,050 | 3,165 | 9,500 | 937 |
| 4030 | 10' wide, 150 H.P. | | 80.45 | 1,700 | 5,120 | 15,400 | 1,668 |
| 4050 | Crawler, 8' wide, 100 H.P., diesel | | 78.80 | 1,800 | 5,415 | 16,200 | 1,713 |
| 4060 | 10' wide, 150 H.P. | | 90.60 | 2,125 | 6,405 | 19,200 | 2,006 |
| 4070 | Concrete paver, 12' to 24' wide, 250 H.P. | | 78.20 | 1,625 | 4,875 | 14,600 | 1,601 |
| 4080 | Placer-spreader-trimmer, 24' wide, 300 H.P. | | 111.00 | 2,650 | 7,975 | 23,900 | 2,483 |
| 4100 | Pump, centrifugal gas pump, 1-1/2" diam., 65 GPM | | 3.50 | 48.50 | 145 | 435 | 57 |
| 4200 | 2" diameter, 130 GPM | | 4.50 | 53.50 | 160 | 480 | 68 |
| 4300 | 3" diameter, 250 GPM | | 4.75 | 55 | 165 | 495 | 71 |
| 4400 | 6" diameter, 1500 GPM | | 22.95 | 180 | 540 | 1,625 | 291.60 |
| 4500 | Submersible electric pump, 1-1/4" diameter, 55 GPM | | .31 | 16.35 | 49 | 147 | 12.30 |
| 4600 | 1-1/2" diameter, 83 GPM | | .35 | 18.65 | 56 | 168 | 14 |
| 4700 | 2" diameter, 120 GPM | | 1.30 | 23.50 | 70 | 210 | 24.40 |
| 4800 | 3" diameter, 300 GPM | | 2.20 | 41.50 | 125 | 375 | 42.60 |
| 4900 | 4" diameter, 560 GPM | | 9.50 | 163 | 490 | 1,475 | 174 |
| 5000 | 6" diameter, 1590 GPM | | 13.95 | 220 | 660 | 1,975 | 243.60 |
| 5100 | Diaphragm pump, gas, single, 1-1/2" diameter | | 1.15 | 49.50 | 148 | 445 | 38.80 |
| 5200 | 2" diameter | | 3.70 | 61.50 | 185 | 555 | 66.60 |
| 5300 | 3" diameter | | 3.70 | 61.50 | 185 | 555 | 66.60 |

## 01 54 33 | Equipment Rental

| | | UNIT | HOURLY OPER. COST | RENT PER DAY | RENT PER WEEK | RENT PER MONTH | EQUIPMENT COST/DAY |
|---|---|---|---|---|---|---|---|
| 5400 | Double, 4" diameter | Ea. | 5.55 | 108 | 325 | 975 | 109.40 |
| 5450 | Pressure washer 5 GPM, 3000 psi | | 3.85 | 46.50 | 140 | 420 | 58.80 |
| 5500 | Trash pump, self-priming, gas, 2" diameter | | 3.70 | 21 | 63 | 189 | 42.20 |
| 5600 | Diesel, 4" diameter | | 9.10 | 58.50 | 175 | 525 | 107.80 |
| 5650 | Diesel, 6" diameter | | 29.10 | 130 | 390 | 1,175 | 310.80 |
| 5655 | Grout Pump | | 14.75 | 100 | 300 | 900 | 178 |
| 5700 | Salamanders, L.P. gas fired, 100,000 Btu | | 3.35 | 16.35 | 49 | 147 | 36.60 |
| 5705 | 50,000 Btu | | 2.50 | 10.35 | 31 | 93 | 26.20 |
| 5720 | Sandblaster, portable, open top, 3 C.F. capacity | | .55 | 27 | 81 | 243 | 20.60 |
| 5730 | 6 C.F. capacity | | .90 | 40 | 120 | 360 | 31.20 |
| 5740 | Accessories for above | | .12 | 20.50 | 61 | 183 | 13.15 |
| 5750 | Sander, floor | | .60 | 20 | 60 | 180 | 16.80 |
| 5760 | Edger | | .61 | 24.50 | 74 | 222 | 19.70 |
| 5800 | Saw, chain, gas engine, 18" long | | 1.75 | 19.65 | 59 | 177 | 25.80 |
| 5900 | Hydraulic powered, 36" long | | .70 | 60 | 180 | 540 | 41.60 |
| 5950 | 60" long | | .70 | 61.50 | 185 | 555 | 42.60 |
| 6000 | Masonry, table mounted, 14" diameter, 5 H.P. | | 1.25 | 56.50 | 170 | 510 | 44 |
| 6050 | Portable cut-off, 8 H.P. | | 1.90 | 33 | 99 | 297 | 35 |
| 6100 | Circular, hand held, electric, 7-1/4" diameter | | .14 | 4.67 | 14 | 42 | 3.90 |
| 6200 | 12" diameter | | .21 | 7.65 | 23 | 69 | 6.30 |
| 6250 | Wall saw, w/hydraulic power, 10 H.P. | | 6.50 | 61.50 | 185 | 555 | 89 |
| 6275 | Shot blaster, walk-behind, 20" wide | | 4.65 | 298 | 895 | 2,675 | 216.20 |
| 6280 | Sidewalk broom, walk-behind | | 2.09 | 60.50 | 182 | 545 | 53.10 |
| 6300 | Steam cleaner, 100 gallons per hour | | 3.15 | 60.50 | 182 | 545 | 61.60 |
| 6310 | 200 gallons per hour | | 4.25 | 73.50 | 220 | 660 | 78 |
| 6340 | Tar Kettle/Pot, 400 gallons | | 4.97 | 81.50 | 245 | 735 | 88.75 |
| 6350 | Torch, cutting, acetylene-oxygen, 150' hose | | .50 | 23.50 | 70 | 210 | 18 |
| 6360 | Hourly operating cost includes tips and gas | | 10.35 | | | | 82.80 |
| 6410 | Toilet, portable chemical | | .12 | 19.65 | 59 | 177 | 12.75 |
| 6420 | Recycle flush type | | .14 | 23.50 | 71 | 213 | 15.30 |
| 6430 | Toilet, fresh water flush, garden hose, | | .17 | 27.50 | 83 | 249 | 17.95 |
| 6440 | Hoisted, non-flush, for high rise | | .14 | 23.50 | 70 | 210 | 15.10 |
| 6450 | Toilet, trailers, minimum | | .24 | 40.50 | 122 | 365 | 26.30 |
| 6460 | Maximum | | .74 | 124 | 372 | 1,125 | 80.30 |
| 6465 | Tractor, farm with attachment | | 15.50 | 305 | 910 | 2,725 | 306 |
| 6500 | Trailers, platform, flush deck, 2 axle, 25 ton capacity | | 5.15 | 117 | 350 | 1,050 | 111.20 |
| 6600 | 40 ton capacity | | 6.65 | 162 | 485 | 1,450 | 150.20 |
| 6700 | 3 axle, 50 ton capacity | | 7.20 | 180 | 540 | 1,625 | 165.60 |
| 6800 | 75 ton capacity | | 8.95 | 237 | 710 | 2,125 | 213.60 |
| 6810 | Trailer mounted cable reel for high voltage line work | | 5.04 | 240 | 720 | 2,150 | 184.30 |
| 6820 | Trailer mounted cable tensioning rig | | 10.01 | 475 | 1,430 | 4,300 | 366.10 |
| 6830 | Cable pulling rig | | 67.52 | 2,675 | 8,020 | 24,100 | 2,144 |
| 6900 | Water tank trailer, engine driven discharge, 5000 gallons | | 6.65 | 155 | 465 | 1,400 | 146.20 |
| 6925 | 10,000 gallons | | 9.15 | 218 | 655 | 1,975 | 204.20 |
| 6950 | Water truck, off highway, 6000 gallons | | 64.15 | 860 | 2,585 | 7,750 | 1,030 |
| 7010 | Tram car for high voltage line work, powered, 2 conductor | | 4.39 | 130 | 391 | 1,175 | 113.30 |
| 7020 | Transit (builder's level) with tripod | | .09 | 14.65 | 44 | 132 | 9.50 |
| 7030 | Trench box, 3000 lbs. 6' x 8' | | .56 | 93 | 279 | 835 | 60.30 |
| 7040 | 7200 lbs. 6' x 20' | | 1.05 | 175 | 525 | 1,575 | 113.40 |
| 7050 | 8000 lbs., 8' x 16' | | 1.05 | 175 | 525 | 1,575 | 113.40 |
| 7060 | 9500 lbs., 8' x 20' | | 1.19 | 199 | 596 | 1,800 | 128.70 |
| 7065 | 11,000 lbs., 8' x 24' | | 1.25 | 209 | 627 | 1,875 | 135.40 |
| 7070 | 12,000 lbs., 10' x 20' | | 1.36 | 227 | 680 | 2,050 | 146.90 |
| 7100 | Truck, pickup, 3/4 ton, 2 wheel drive | | 8.70 | 58.50 | 175 | 525 | 104.60 |
| 7200 | 4 wheel drive | | 8.95 | 75 | 225 | 675 | 116.60 |
| 7250 | Crew carrier, 9 passenger | | 12.30 | 90 | 270 | 810 | 152.40 |
| 7290 | Flat bed truck, 20,000 lbs. GVW | | 13.75 | 133 | 400 | 1,200 | 190 |
| 7300 | Tractor, 4 x 2, 220 H.P. | | 19.40 | 207 | 620 | 1,850 | 279.20 |

40

## 01 54 33 | Equipment Rental

| | | UNIT | HOURLY OPER. COST | RENT PER DAY | RENT PER WEEK | RENT PER MONTH | EQUIPMENT COST/DAY | |
|---|---|---|---|---|---|---|---|---|
| **40** | 7410 | 330 H.P. | Ea. | 28.45 | 283 | 850 | 2,550 | 397.60 | **40** |
| | 7500 | 6 x 4, 380 H.P. | | 32.55 | 330 | 995 | 2,975 | 459.40 | |
| | 7600 | 450 H.P. | | 39.85 | 400 | 1,200 | 3,600 | 558.80 | |
| | 7610 | Tractor, with A frame, boom and winch, 225 H.P. | | 22.10 | 283 | 850 | 2,550 | 346.80 | |
| | 7620 | Vacuum truck, hazardous material, 2500 gallons | | 11.35 | 315 | 945 | 2,825 | 279.80 | |
| | 7625 | 5,000 gallons | | 13.89 | 440 | 1,320 | 3,950 | 375.10 | |
| | 7650 | Vacuum, HEPA, 16 gallon, wet/dry | | .36 | 18 | 54 | 162 | 13.70 | |
| | 7655 | 55 gallon, wet/dry | | .59 | 27 | 81 | 243 | 20.90 | |
| | 7660 | Water tank, portable | | .17 | 28.50 | 85.50 | 257 | 18.45 | |
| | 7690 | Sewer/catch basin vacuum, 14 C.Y., 1500 gallons | | 18.51 | 660 | 1,980 | 5,950 | 544.10 | |
| | 7700 | Welder, electric, 200 amp | | 1.70 | 17.35 | 52 | 156 | 24 | |
| | 7800 | 300 amp | | 2.48 | 21 | 63 | 189 | 32.45 | |
| | 7900 | Gas engine, 200 amp | | 10.15 | 25.50 | 76 | 228 | 96.40 | |
| | 8000 | 300 amp | | 11.60 | 27 | 81 | 243 | 109 | |
| | 8100 | Wheelbarrow, any size | | .08 | 12.65 | 38 | 114 | 8.25 | |
| | 8200 | Wrecking ball, 4000 lbs. | | 2.20 | 73.50 | 220 | 660 | 61.60 | |
| **50** | 0010 | **HIGHWAY EQUIPMENT RENTAL** without operators | R015433 -10 | | | | | | **50** |
| | 0050 | Asphalt batch plant, portable drum mixer, 100 ton/hr | Ea. | 68.35 | 1,475 | 4,400 | 13,200 | 1,427 | |
| | 0060 | 200 ton/hr | | 76.45 | 1,550 | 4,655 | 14,000 | 1,543 | |
| | 0070 | 300 ton/hr | | 88.65 | 1,825 | 5,485 | 16,500 | 1,806 | |
| | 0100 | Backhoe attachment, long stick, up to 185 H.P., 10.5' long | | .33 | 22 | 66 | 198 | 15.85 | |
| | 0140 | Up to 250 H.P., 12' long | | .36 | 23.50 | 71 | 213 | 17.10 | |
| | 0180 | Over 250 H.P., 15' long | | .49 | 32.50 | 97 | 291 | 23.30 | |
| | 0200 | Special dipper arm, up to 100 H.P., 32' long | | 1.00 | 66.50 | 199 | 595 | 47.80 | |
| | 0240 | Over 100 H.P., 33' long | | 1.24 | 82.50 | 248 | 745 | 59.50 | |
| | 0280 | Catch basin/sewer cleaning truck, 3 ton, 9 C.Y., 1000 gal | | 30.85 | 415 | 1,250 | 3,750 | 496.80 | |
| | 0300 | Concrete batch plant, portable, electric, 200 C.Y./hr. | | 18.20 | 530 | 1,585 | 4,750 | 462.60 | |
| | 0520 | Grader/dozer attachment, ripper/scarifier, rear mounted, up to 135 H.P. | | 3.30 | 68.50 | 205 | 615 | 67.40 | |
| | 0540 | Up to 180 H.P. | | 3.90 | 86.50 | 260 | 780 | 83.20 | |
| | 0580 | Up to 250 H.P. | | 4.30 | 100 | 300 | 900 | 94.40 | |
| | 0700 | Pvmt. removal bucket, for hyd. excavator, up to 90 H.P. | | 1.75 | 51.50 | 155 | 465 | 45 | |
| | 0740 | Up to 200 H.P. | | 1.95 | 73.50 | 220 | 660 | 59.60 | |
| | 0780 | Over 200 H.P. | | 2.10 | 85 | 255 | 765 | 67.80 | |
| | 0900 | Aggregate spreader, self-propelled, 187 H.P. | | 40.05 | 715 | 2,140 | 6,425 | 748.40 | |
| | 1000 | Chemical spreader, 3 C.Y. | | 2.75 | 45 | 135 | 405 | 49 | |
| | 1900 | Hammermill, traveling, 250 H.P. | | 67.55 | 1,925 | 5,760 | 17,300 | 1,692 | |
| | 2000 | Horizontal borer, 3" diameter, 13 H.P. gas driven | | 5.25 | 56.50 | 170 | 510 | 76 | |
| | 2150 | Horizontal directional drill, 20,000 lb. thrust, 78 H.P. diesel | | 26.20 | 690 | 2,065 | 6,200 | 622.60 | |
| | 2160 | 30,000 lb. thrust, 115 H.P. | | 32.30 | 1,050 | 3,150 | 9,450 | 888.40 | |
| | 2170 | 50,000 lb. thrust, 170 H.P. | | 46.00 | 1,350 | 4,030 | 12,100 | 1,174 | |
| | 2190 | Mud trailer for HDD, 1500 gallons, 175 H.P., gas | | 21.10 | 162 | 485 | 1,450 | 265.80 | |
| | 2200 | Hydromulcher, diesel, 3000 gallon, for truck mounting | | 15.00 | 263 | 790 | 2,375 | 278 | |
| | 2300 | Gas, 600 gallon | | 6.40 | 103 | 310 | 930 | 113.20 | |
| | 2400 | Joint & crack cleaner, walk behind, 25 H.P. | | 2.75 | 51.50 | 155 | 465 | 53 | |
| | 2500 | Filler, trailer mounted, 400 gallons, 20 H.P. | | 7.70 | 222 | 665 | 2,000 | 194.60 | |
| | 3000 | Paint striper, self-propelled, double line, 30 H.P. | | 6.10 | 165 | 495 | 1,475 | 147.80 | |
| | 3200 | Post drivers, 6" I-Beam frame, for truck mounting | | 11.45 | 440 | 1,325 | 3,975 | 356.60 | |
| | 3400 | Road sweeper, self-propelled, 8' wide, 90 H.P. | | 31.10 | 595 | 1,780 | 5,350 | 604.80 | |
| | 3450 | Road sweeper, vacuum assisted, 4 C.Y., 220 gallons | | 51.45 | 650 | 1,945 | 5,825 | 800.60 | |
| | 4000 | Road mixer, self-propelled, 130 H.P. | | 41.45 | 785 | 2,350 | 7,050 | 801.60 | |
| | 4100 | 310 H.P. | | 71.70 | 2,200 | 6,580 | 19,700 | 1,890 | |
| | 4220 | Cold mix paver, incl. pug mill and bitumen tank, 165 H.P. | | 90.60 | 2,150 | 6,455 | 19,400 | 2,016 | |
| | 4250 | Paver, asphalt, wheel or crawler, 130 H.P., diesel | | 89.10 | 2,050 | 6,180 | 18,500 | 1,949 | |
| | 4300 | Paver, road widener, gas 1' to 6', 67 H.P. | | 42.15 | 885 | 2,650 | 7,950 | 867.20 | |
| | 4400 | Diesel, 2' to 14', 88 H.P. | | 53.00 | 1,075 | 3,215 | 9,650 | 1,067 | |
| | 4600 | Slipform pavers, curb and gutter, 2 track, 75 H.P. | | 35.60 | 765 | 2,300 | 6,900 | 744.80 | |
| | 4700 | 4 track, 165 H.P. | | 44.65 | 820 | 2,455 | 7,375 | 848.20 | |
| | 4800 | Median barrier, 215 H.P. | | 45.40 | 850 | 2,550 | 7,650 | 873.20 | |

## 01 54 33 | Equipment Rental

| | | | UNIT | HOURLY OPER. COST | RENT PER DAY | RENT PER WEEK | RENT PER MONTH | EQUIPMENT COST/DAY | |
|---|---|---|---|---|---|---|---|---|---|
| 50 | 4901 | Trailer, low bed, 75 ton capacity | Ea. | 9.65 | 237 | 710 | 2,125 | 219.20 | 50 |
| | 5000 | Road planer, walk behind, 10" cutting width, 10 H.P. | | 2.70 | 33.50 | 100 | 300 | 41.60 | |
| | 5100 | Self-propelled, 12" cutting width, 64 H.P. | | 7.00 | 115 | 345 | 1,025 | 125 | |
| | 5120 | Traffic line remover, metal ball blaster, truck mounted, 115 H.P. | | 47.15 | 765 | 2,300 | 6,900 | 837.20 | |
| | 5140 | Grinder, truck mounted, 115 H.P. | | 50.10 | 835 | 2,505 | 7,525 | 901.80 | |
| | 5160 | Walk-behind, 11 H.P. | | 3.05 | 55 | 165 | 495 | 57.40 | |
| | 5200 | Pavement profiler, 4' to 6' wide, 450 H.P. | | 210.55 | 3,425 | 10,285 | 30,900 | 3,741 | |
| | 5300 | 8' to 10' wide, 750 H.P. | | 332.90 | 4,950 | 14,815 | 44,400 | 5,626 | |
| | 5400 | Roadway plate, steel, 1" x 8' x 20' | | .07 | 12 | 36 | 108 | 7.75 | |
| | 5600 | Stabilizer, self-propelled, 150 H.P. | | 39.35 | 615 | 1,850 | 5,550 | 684.80 | |
| | 5700 | 310 H.P. | | 66.85 | 1,325 | 3,955 | 11,900 | 1,326 | |
| | 5800 | Striper, thermal, truck mounted 120 gallon paint, 150 H.P. | | 42.85 | 520 | 1,555 | 4,675 | 653.80 | |
| | 6000 | Tar kettle, 330 gallon, trailer mounted | | 4.58 | 63.50 | 190 | 570 | 74.65 | |
| | 7000 | Tunnel locomotive, diesel, 8 to 12 ton | | 27.45 | 585 | 1,755 | 5,275 | 570.60 | |
| | 7005 | Electric, 10 ton | | 24.75 | 670 | 2,005 | 6,025 | 599 | |
| | 7010 | Muck cars, 1/2 C.Y. capacity | | 1.90 | 23.50 | 71 | 213 | 29.40 | |
| | 7020 | 1 C.Y. capacity | | 2.15 | 32.50 | 98 | 294 | 36.80 | |
| | 7030 | 2 C.Y. capacity | | 2.25 | 36.50 | 110 | 330 | 40 | |
| | 7040 | Side dump, 2 C.Y. capacity | | 2.45 | 45 | 135 | 405 | 46.60 | |
| | 7050 | 3 C.Y. capacity | | 3.30 | 51.50 | 155 | 465 | 57.40 | |
| | 7060 | 5 C.Y. capacity | | 4.70 | 65 | 195 | 585 | 76.60 | |
| | 7100 | Ventilating blower for tunnel, 7-1/2 H.P. | | 1.02 | 51.50 | 155 | 465 | 39.15 | |
| | 7110 | 10 H.P. | | 1.11 | 51.50 | 155 | 465 | 39.90 | |
| | 7120 | 20 H.P. | | 1.67 | 67.50 | 202 | 605 | 53.75 | |
| | 7140 | 40 H.P. | | 2.77 | 96.50 | 290 | 870 | 80.15 | |
| | 7160 | 60 H.P. | | 4.26 | 152 | 455 | 1,375 | 125.10 | |
| | 7175 | 75 H.P. | | 5.51 | 202 | 607 | 1,825 | 165.50 | |
| | 7180 | 200 H.P. | | 11.15 | 305 | 910 | 2,725 | 271.20 | |
| | 7800 | Windrow loader, elevating | | 49.40 | 1,350 | 4,075 | 12,200 | 1,210 | |
| 60 | 0010 | **LIFTING AND HOISTING EQUIPMENT RENTAL** without operators [R015433-10] | | | | | | | 60 |
| | 0120 | Aerial lift truck, 2 person, to 80' | Ea. | 24.90 | 730 | 2,185 | 6,550 | 636.20 | |
| | 0140 | Boom work platform, 40' snorkel [R015433-15] | | 14.25 | 278 | 835 | 2,500 | 281 | |
| | 0150 | Crane, flatbed mounted, 3 ton capacity | | 13.45 | 197 | 590 | 1,775 | 225.60 | |
| | 0200 | Crane, climbing, 106' jib, 6000 lb. capacity, 410 fpm [R312316-45] | | 29.50 | 1,525 | 4,580 | 13,700 | 1,152 | |
| | 0300 | 101' jib, 10,250 lb. capacity, 270 fpm | | 35.60 | 1,925 | 5,800 | 17,400 | 1,445 | |
| | 0500 | Tower, static, 130' high, 106' jib, 6200 lb. capacity at 400 fpm | | 33.05 | 1,775 | 5,290 | 15,900 | 1,322 | |
| | 0600 | Crawler mounted, lattice boom, 1/2 C.Y., 15 tons at 12' radius | | 30.74 | 685 | 2,060 | 6,175 | 657.90 | |
| | 0700 | 3/4 C.Y., 20 tons at 12' radius | | 40.99 | 855 | 2,570 | 7,700 | 841.90 | |
| | 0800 | 1 C.Y., 25 tons at 12' radius | | 54.65 | 1,150 | 3,430 | 10,300 | 1,123 | |
| | 0900 | 1-1/2 C.Y., 40 tons at 12' radius | | 54.65 | 1,150 | 3,455 | 10,400 | 1,128 | |
| | 1000 | 2 C.Y., 50 tons at 12' radius | | 58.55 | 1,350 | 4,055 | 12,200 | 1,279 | |
| | 1100 | 3 C.Y., 75 tons at 12' radius | | 63.05 | 1,575 | 4,710 | 14,100 | 1,446 | |
| | 1200 | 100 ton capacity, 60' boom | | 71.50 | 1,800 | 5,405 | 16,200 | 1,653 | |
| | 1300 | 165 ton capacity, 60' boom | | 89.20 | 2,100 | 6,325 | 19,000 | 1,979 | |
| | 1400 | 200 ton capacity, 70' boom | | 111.15 | 2,675 | 7,995 | 24,000 | 2,488 | |
| | 1500 | 350 ton capacity, 80' boom | | 157.20 | 4,050 | 12,160 | 36,500 | 3,690 | |
| | 1600 | Truck mounted, lattice boom, 6 x 4, 20 tons at 10' radius | | 37.39 | 1,225 | 3,680 | 11,000 | 1,035 | |
| | 1700 | 25 tons at 10' radius | | 40.40 | 1,325 | 4,000 | 12,000 | 1,123 | |
| | 1800 | 8 x 4, 30 tons at 10' radius | | 43.77 | 1,425 | 4,260 | 12,800 | 1,202 | |
| | 1900 | 40 tons at 12' radius | | 46.65 | 1,475 | 4,450 | 13,400 | 1,263 | |
| | 2000 | 60 tons at 15' radius | | 52.35 | 1,575 | 4,710 | 14,100 | 1,361 | |
| | 2050 | 82 tons at 15' radius | | 58.46 | 1,675 | 5,030 | 15,100 | 1,474 | |
| | 2100 | 90 tons at 15' radius | | 65.56 | 1,825 | 5,490 | 16,500 | 1,622 | |
| | 2200 | 115 tons at 15' radius | | 73.91 | 2,050 | 6,130 | 18,400 | 1,817 | |
| | 2300 | 150 tons at 18' radius | | 76.90 | 2,150 | 6,455 | 19,400 | 1,906 | |
| | 2350 | 165 tons at 18' radius | | 86.63 | 2,275 | 6,840 | 20,500 | 2,061 | |
| | 2400 | Truck mounted, hydraulic, 12 ton capacity | | 41.45 | 540 | 1,620 | 4,850 | 655.60 | |

## 01 54 33 | Equipment Rental

| | | UNIT | HOURLY OPER. COST | RENT PER DAY | RENT PER WEEK | RENT PER MONTH | EQUIPMENT COST/DAY | |
|---|---|---|---|---|---|---|---|---|
| 60 | 2500 | 25 ton capacity | Ea. | 43.70 | 665 | 1,995 | 5,975 | 748.60 | 60 |
| | 2550 | 33 ton capacity | | 44.25 | 680 | 2,045 | 6,125 | 763 | |
| | 2560 | 40 ton capacity | | 57.20 | 800 | 2,405 | 7,225 | 938.60 | |
| | 2600 | 55 ton capacity | | 75.80 | 1,000 | 3,010 | 9,025 | 1,208 | |
| | 2700 | 80 ton capacity | | 97.75 | 1,525 | 4,555 | 13,700 | 1,693 | |
| | 2720 | 100 ton capacity | | 91.65 | 1,525 | 4,585 | 13,800 | 1,650 | |
| | 2740 | 120 ton capacity | | 106.15 | 1,625 | 4,910 | 14,700 | 1,831 | |
| | 2760 | 150 ton capacity | | 124.30 | 2,225 | 6,710 | 20,100 | 2,336 | |
| | 2800 | Self-propelled, 4 x 4, with telescoping boom, 5 ton | | 14.60 | 235 | 705 | 2,125 | 257.80 | |
| | 2900 | 12-1/2 ton capacity | | 25.80 | 380 | 1,140 | 3,425 | 434.40 | |
| | 3000 | 15 ton capacity | | 26.45 | 405 | 1,210 | 3,625 | 453.60 | |
| | 3050 | 20 ton capacity | | 28.25 | 475 | 1,425 | 4,275 | 511 | |
| | 3100 | 25 ton capacity | | 29.80 | 535 | 1,610 | 4,825 | 560.40 | |
| | 3150 | 40 ton capacity | | 36.95 | 605 | 1,815 | 5,450 | 658.60 | |
| | 3200 | Derricks, guy, 20 ton capacity, 60' boom, 75' mast | | 23.35 | 375 | 1,122 | 3,375 | 411.20 | |
| | 3300 | 100' boom, 115' mast | | 36.76 | 645 | 1,930 | 5,800 | 680.10 | |
| | 3400 | Stiffleg, 20 ton capacity, 70' boom, 37' mast | | 25.72 | 485 | 1,460 | 4,375 | 497.75 | |
| | 3500 | 100' boom, 47' mast | | 39.56 | 775 | 2,330 | 7,000 | 782.50 | |
| | 3550 | Helicopter, small, lift to 1250 lbs. maximum, w/pilot | | 95.92 | 3,025 | 9,060 | 27,200 | 2,579 | |
| | 3600 | Hoists, chain type, overhead, manual, 3/4 ton | | .15 | 1 | 3 | 9 | 1.80 | |
| | 3900 | 10 ton | | .70 | 7 | 21 | 63 | 9.80 | |
| | 4000 | Hoist and tower, 5000 lb. cap., portable electric, 40' high | | 3.89 | 215 | 646 | 1,950 | 160.30 | |
| | 4100 | For each added 10' section, add | | .10 | 17 | 51 | 153 | 11 | |
| | 4200 | Hoist and single tubular tower, 5000 lb. electric, 100' high | | 5.34 | 300 | 902 | 2,700 | 223.10 | |
| | 4300 | For each added 6'-6" section, add | | .17 | 29 | 87 | 261 | 18.75 | |
| | 4400 | Hoist and double tubular tower, 5000 lb., 100' high | | 5.79 | 330 | 993 | 2,975 | 244.90 | |
| | 4500 | For each added 6'-6" section, add | | .19 | 32.50 | 97 | 291 | 20.90 | |
| | 4550 | Hoist and tower, mast type, 6000 lb., 100' high | | 6.14 | 345 | 1,030 | 3,100 | 255.10 | |
| | 4570 | For each added 10' section, add | | .12 | 20.50 | 61 | 183 | 13.15 | |
| | 4600 | Hoist and tower, personnel, electric, 2000 lb., 100' @ 125 fpm | | 14.36 | 915 | 2,740 | 8,225 | 662.90 | |
| | 4700 | 3000 lb., 100' @ 200 fpm | | 16.38 | 1,025 | 3,110 | 9,325 | 753.05 | |
| | 4800 | 3000 lb., 150' @ 300 fpm | | 18.23 | 1,150 | 3,480 | 10,400 | 841.85 | |
| | 4900 | 4000 lb., 100' @ 300 fpm | | 18.74 | 1,175 | 3,550 | 10,700 | 859.90 | |
| | 5000 | 6000 lb., 100' @ 275 fpm | | 19.92 | 1,250 | 3,720 | 11,200 | 903.35 | |
| | 5100 | For added heights up to 500', add | L.F. | .01 | 1.67 | 5 | 15 | 1.10 | |
| | 5200 | Jacks, hydraulic, 20 ton | Ea. | .05 | 2 | 6 | 18 | 1.60 | |
| | 5500 | 100 ton | | .35 | 11.65 | 35 | 105 | 9.80 | |
| | 6100 | Jacks, hydraulic, climbing w/50' jackrods, control console, 30 ton cap. | | 1.86 | 124 | 371 | 1,125 | 89.10 | |
| | 6150 | For each added 10' jackrod section, add | | .05 | 3.33 | 10 | 30 | 2.40 | |
| | 6300 | 50 ton capacity | | 2.99 | 199 | 597 | 1,800 | 143.30 | |
| | 6350 | For each added 10' jackrod section, add | | .06 | 4 | 12 | 36 | 2.90 | |
| | 6500 | 125 ton capacity | | 7.85 | 525 | 1,570 | 4,700 | 376.80 | |
| | 6550 | For each added 10' jackrod section, add | | .54 | 35.50 | 107 | 320 | 25.70 | |
| | 6600 | Cable jack, 10 ton capacity with 200' cable | | 1.56 | 104 | 311 | 935 | 74.70 | |
| | 6650 | For each added 50' of cable, add | | .18 | 12 | 36 | 108 | 8.65 | |
| 70 | 0010 | **WELLPOINT EQUIPMENT RENTAL** without operators | | | | | | | 70 |
| | 0020 | Based on 2 months rental | | | | | | | |
| | 0100 | Combination jetting & wellpoint pump, 60 H.P. diesel | Ea. | 15.74 | 305 | 920 | 2,750 | 309.90 | |
| | 0200 | High pressure gas jet pump, 200 H.P., 300 psi | " | 36.50 | 262 | 786 | 2,350 | 449.20 | |
| | 0300 | Discharge pipe, 8" diameter | L.F. | .01 | .50 | 1.50 | 4.50 | .40 | |
| | 0350 | 12" diameter | | .01 | .73 | 2.20 | 6.60 | .50 | |
| | 0400 | Header pipe, flows up to 150 GPM, 4" diameter | | .01 | .45 | 1.35 | 4.05 | .35 | |
| | 0500 | 400 GPM, 6" diameter | | .01 | .53 | 1.60 | 4.80 | .40 | |
| | 0600 | 800 GPM, 8" diameter | | .01 | .73 | 2.20 | 6.60 | .50 | |
| | 0700 | 1500 GPM, 10" diameter | | .01 | .77 | 2.31 | 6.95 | .55 | |
| | 0800 | 2500 GPM, 12" diameter | | .02 | 1.46 | 4.38 | 13.15 | 1.05 | |
| | 0900 | 4500 GPM, 16" diameter | | .03 | 1.87 | 5.61 | 16.85 | 1.35 | |

R015433 -10

## 01 54 33 | Equipment Rental

| | | | UNIT | HOURLY OPER. COST | RENT PER DAY | RENT PER WEEK | RENT PER MONTH | EQUIPMENT COST/DAY | |
|---|---|---|---|---|---|---|---|---|---|
| 70 | 0950 | For quick coupling aluminum and plastic pipe, add | L.F. | .03 | 1.93 | 5.80 | 17.40 | 1.40 | 70 |
| | 1100 | Wellpoint, 25' long, with fittings & riser pipe, 1-1/2" or 2" diameter | Ea. | .06 | 3.86 | 11.58 | 34.50 | 2.80 | |
| | 1200 | Wellpoint pump, diesel powered, 4" suction, 20 H.P. | | 6.81 | 177 | 530 | 1,600 | 160.50 | |
| | 1300 | 6" suction, 30 H.P. | | 9.26 | 219 | 658 | 1,975 | 205.70 | |
| | 1400 | 8" suction, 40 H.P. | | 12.51 | 300 | 902 | 2,700 | 280.50 | |
| | 1500 | 10" suction, 75 H.P. | | 19.00 | 350 | 1,054 | 3,150 | 362.80 | |
| | 1600 | 12" suction, 100 H.P. | | 27.26 | 560 | 1,680 | 5,050 | 554.10 | |
| | 1700 | 12" suction, 175 H.P. | | 40.08 | 615 | 1,850 | 5,550 | 690.65 | |
| 80 | 0010 | **MARINE EQUIPMENT RENTAL** without operators  R015433 -10 | | | | | | | 80 |
| | 0200 | Barge, 400 Ton, 30' wide x 90' long | Ea. | 16.05 | 1,150 | 3,460 | 10,400 | 820.40 | |
| | 0240 | 800 Ton, 45' wide x 90' long | | 19.50 | 1,400 | 4,190 | 12,600 | 994 | |
| | 2000 | Tugboat, diesel, 100 H.P. | | 25.55 | 215 | 645 | 1,925 | 333.40 | |
| | 2040 | 250 H.P. | | 49.40 | 395 | 1,185 | 3,550 | 632.20 | |
| | 2080 | 380 H.P. | | 108.20 | 1,175 | 3,505 | 10,500 | 1,567 | |

| Crew No. | Bare Costs | | Incl. Subs O&P | | Cost Per Labor-Hour | |
|---|---|---|---|---|---|---|

### Crew A-1

| | Hr. | Daily | Hr. | Daily | Bare Costs | Incl. O&P |
|---|---|---|---|---|---|---|
| 1 Building Laborer | $34.35 | $274.80 | $52.70 | $421.60 | $34.35 | $52.70 |
| 1 Concrete saw, gas manual | | 63.20 | | 69.52 | 7.90 | 8.69 |
| 8 L.H., Daily Totals | | $338.00 | | $491.12 | $42.25 | $61.39 |

### Crew A-1A

| | Hr. | Daily | Hr. | Daily | Bare Costs | Incl. O&P |
|---|---|---|---|---|---|---|
| 1 Skilled Worker | $44.40 | $355.20 | $68.10 | $544.80 | $44.40 | $68.10 |
| 1 Shot Blaster, 20" | | 216.20 | | 237.82 | 27.02 | 29.73 |
| 8 L.H., Daily Totals | | $571.40 | | $782.62 | $71.42 | $97.83 |

### Crew A-1B

| | Hr. | Daily | Hr. | Daily | Bare Costs | Incl. O&P |
|---|---|---|---|---|---|---|
| 1 Building Laborer | $34.35 | $274.80 | $52.70 | $421.60 | $34.35 | $52.70 |
| 1 Concrete Saw | | 134.20 | | 147.62 | 16.77 | 18.45 |
| 8 L.H., Daily Totals | | $409.00 | | $569.22 | $51.13 | $71.15 |

### Crew A-1C

| | Hr. | Daily | Hr. | Daily | Bare Costs | Incl. O&P |
|---|---|---|---|---|---|---|
| 1 Building Laborer | $34.35 | $274.80 | $52.70 | $421.60 | $34.35 | $52.70 |
| 1 Chain Saw, gas, 18" | | 25.80 | | 28.38 | 3.23 | 3.55 |
| 8 L.H., Daily Totals | | $300.60 | | $449.98 | $37.58 | $56.25 |

### Crew A-1D

| | Hr. | Daily | Hr. | Daily | Bare Costs | Incl. O&P |
|---|---|---|---|---|---|---|
| 1 Building Laborer | $34.35 | $274.80 | $52.70 | $421.60 | $34.35 | $52.70 |
| 1 Vibrating plate, gas, 18" | | 30.80 | | 33.88 | 3.85 | 4.24 |
| 8 L.H., Daily Totals | | $305.60 | | $455.48 | $38.20 | $56.94 |

### Crew A-1E

| | Hr. | Daily | Hr. | Daily | Bare Costs | Incl. O&P |
|---|---|---|---|---|---|---|
| 1 Building Laborer | $34.35 | $274.80 | $52.70 | $421.60 | $34.35 | $52.70 |
| 1 Vibratory Plate, gas, 21" | | 39.00 | | 42.90 | 4.88 | 5.36 |
| 8 L.H., Daily Totals | | $313.80 | | $464.50 | $39.23 | $58.06 |

### Crew A-1F

| | Hr. | Daily | Hr. | Daily | Bare Costs | Incl. O&P |
|---|---|---|---|---|---|---|
| 1 Building Laborer | $34.35 | $274.80 | $52.70 | $421.60 | $34.35 | $52.70 |
| 1 Rammer/tamper, gas, 8" | | 43.60 | | 47.96 | 5.45 | 6.00 |
| 8 L.H., Daily Totals | | $318.40 | | $469.56 | $39.80 | $58.70 |

### Crew A-1G

| | Hr. | Daily | Hr. | Daily | Bare Costs | Incl. O&P |
|---|---|---|---|---|---|---|
| 1 Building Laborer | $34.35 | $274.80 | $52.70 | $421.60 | $34.35 | $52.70 |
| 1 Rammer/tamper, gas, 15" | | 49.60 | | 54.56 | 6.20 | 6.82 |
| 8 L.H., Daily Totals | | $324.40 | | $476.16 | $40.55 | $59.52 |

### Crew A-1H

| | Hr. | Daily | Hr. | Daily | Bare Costs | Incl. O&P |
|---|---|---|---|---|---|---|
| 1 Building Laborer | $34.35 | $274.80 | $52.70 | $421.60 | $34.35 | $52.70 |
| 1 Exterior Steam Cleaner | | 61.60 | | 67.76 | 7.70 | 8.47 |
| 8 L.H., Daily Totals | | $336.40 | | $489.36 | $42.05 | $61.17 |

### Crew A-1J

| | Hr. | Daily | Hr. | Daily | Bare Costs | Incl. O&P |
|---|---|---|---|---|---|---|
| 1 Building Laborer | $34.35 | $274.80 | $52.70 | $421.60 | $34.35 | $52.70 |
| 1 Cultivator, Walk-Behind, 5 H.P. | | 58.45 | | 64.30 | 7.31 | 8.04 |
| 8 L.H., Daily Totals | | $333.25 | | $485.89 | $41.66 | $60.74 |

### Crew A-1K

| | Hr. | Daily | Hr. | Daily | Bare Costs | Incl. O&P |
|---|---|---|---|---|---|---|
| 1 Building Laborer | $34.35 | $274.80 | $52.70 | $421.60 | $34.35 | $52.70 |
| 1 Cultivator, Walk-Behind, 8 H.P. | | 89.75 | | 98.72 | 11.22 | 12.34 |
| 8 L.H., Daily Totals | | $364.55 | | $520.33 | $45.57 | $65.04 |

### Crew A-1M

| | Hr. | Daily | Hr. | Daily | Bare Costs | Incl. O&P |
|---|---|---|---|---|---|---|
| 1 Building Laborer | $34.35 | $274.80 | $52.70 | $421.60 | $34.35 | $52.70 |
| 1 Snow Blower, Walk-Behind | | 53.10 | | 58.41 | 6.64 | 7.30 |
| 8 L.H., Daily Totals | | $327.90 | | $480.01 | $40.99 | $60.00 |

| Crew No. | Bare Costs | | Incl. Subs O&P | | Cost Per Labor-Hour | |
|---|---|---|---|---|---|---|

### Crew A-2

| | Hr. | Daily | Hr. | Daily | Bare Costs | Incl. O&P |
|---|---|---|---|---|---|---|
| 2 Laborers | $34.35 | $549.60 | $52.70 | $843.20 | $34.03 | $52.05 |
| 1 Truck Driver (light) | 33.40 | 267.20 | 50.75 | 406.00 | | |
| 1 Flatbed Truck, Gas, 1.5 Ton | | 171.40 | | 188.54 | 7.14 | 7.86 |
| 24 L.H., Daily Totals | | $988.20 | | $1437.74 | $41.17 | $59.91 |

### Crew A-2A

| | Hr. | Daily | Hr. | Daily | Bare Costs | Incl. O&P |
|---|---|---|---|---|---|---|
| 2 Laborers | $34.35 | $549.60 | $52.70 | $843.20 | $34.03 | $52.05 |
| 1 Truck Driver (light) | 33.40 | 267.20 | 50.75 | 406.00 | | |
| 1 Flatbed Truck, Gas, 1.5 Ton | | 171.40 | | 188.54 | | |
| 1 Concrete Saw | | 134.20 | | 147.62 | 12.73 | 14.01 |
| 24 L.H., Daily Totals | | $1122.40 | | $1585.36 | $46.77 | $66.06 |

### Crew A-2B

| | Hr. | Daily | Hr. | Daily | Bare Costs | Incl. O&P |
|---|---|---|---|---|---|---|
| 1 Truck Driver (light) | $33.40 | $267.20 | $50.75 | $406.00 | $33.40 | $50.75 |
| 1 Flatbed Truck, Gas, 1.5 Ton | | 171.40 | | 188.54 | 21.43 | 23.57 |
| 8 L.H., Daily Totals | | $438.60 | | $594.54 | $54.83 | $74.32 |

### Crew A-3A

| | Hr. | Daily | Hr. | Daily | Bare Costs | Incl. O&P |
|---|---|---|---|---|---|---|
| 1 Truck Driver (light) | $33.40 | $267.20 | $50.75 | $406.00 | $33.40 | $50.75 |
| 1 Pickup truck, 4 x 4, 3/4 ton | | 116.60 | | 128.26 | 14.57 | 16.03 |
| 8 L.H., Daily Totals | | $383.80 | | $534.26 | $47.98 | $66.78 |

### Crew A-3B

| | Hr. | Daily | Hr. | Daily | Bare Costs | Incl. O&P |
|---|---|---|---|---|---|---|
| 1 Equip. Oper. (med.) | $45.35 | $362.80 | $67.75 | $542.00 | $39.88 | $60.00 |
| 1 Truck Driver (heav.) | 34.40 | 275.20 | 52.25 | 418.00 | | |
| 1 Dump Truck, 12 C.Y., 400 H.P. | | 506.00 | | 556.60 | | |
| 1 F.E. Loader, W.M.,2.5 C.Y. | | 397.60 | | 437.36 | 56.48 | 62.12 |
| 16 L.H., Daily Totals | | $1541.60 | | $1953.96 | $96.35 | $122.12 |

### Crew A-3C

| | Hr. | Daily | Hr. | Daily | Bare Costs | Incl. O&P |
|---|---|---|---|---|---|---|
| 1 Equip. Oper. (light) | $43.60 | $348.80 | $65.15 | $521.20 | $43.60 | $65.15 |
| 1 Loader, Skid Steer, 78 H.P. | | 261.20 | | 287.32 | 32.65 | 35.91 |
| 8 L.H., Daily Totals | | $610.00 | | $808.52 | $76.25 | $101.07 |

### Crew A-3D

| | Hr. | Daily | Hr. | Daily | Bare Costs | Incl. O&P |
|---|---|---|---|---|---|---|
| 1 Truck Driver, Light | $33.40 | $267.20 | $50.75 | $406.00 | $33.40 | $50.75 |
| 1 Pickup truck, 4 x 4, 3/4 ton | | 116.60 | | 128.26 | | |
| 1 Flatbed Trailer, 25 Ton | | 111.20 | | 122.32 | 28.48 | 31.32 |
| 8 L.H., Daily Totals | | $495.00 | | $656.58 | $61.88 | $82.07 |

### Crew A-3E

| | Hr. | Daily | Hr. | Daily | Bare Costs | Incl. O&P |
|---|---|---|---|---|---|---|
| 1 Equip. Oper. (crane) | $46.50 | $372.00 | $69.45 | $555.60 | $40.45 | $60.85 |
| 1 Truck Driver (heavy) | 34.40 | 275.20 | 52.25 | 418.00 | | |
| 1 Pickup truck, 4 x 4, 3/4 ton | | 116.60 | | 128.26 | 7.29 | 8.02 |
| 16 L.H., Daily Totals | | $763.80 | | $1101.86 | $47.74 | $68.87 |

### Crew A-3F

| | Hr. | Daily | Hr. | Daily | Bare Costs | Incl. O&P |
|---|---|---|---|---|---|---|
| 1 Equip. Oper. (crane) | $46.50 | $372.00 | $69.45 | $555.60 | $40.45 | $60.85 |
| 1 Truck Driver (heavy) | 34.40 | 275.20 | 52.25 | 418.00 | | |
| 1 Pickup truck, 4 x 4, 3/4 ton | | 116.60 | | 128.26 | | |
| 1 Truck Tractor, 6x4, 380 H.P. | | 459.40 | | 505.34 | | |
| 1 Lowbed Trailer, 75 Ton | | 219.20 | | 241.12 | 49.70 | 54.67 |
| 16 L.H., Daily Totals | | $1442.40 | | $1848.32 | $90.15 | $115.52 |

| Crew No. | Bare Costs | | Incl. Subs O&P | | Cost Per Labor-Hour | |
|---|---|---|---|---|---|---|

| Crew A-3G | Hr. | Daily | Hr. | Daily | Bare Costs | Incl. O&P |
|---|---|---|---|---|---|---|
| 1 Equip. Oper. (crane) | $46.50 | $372.00 | $69.45 | $555.60 | $40.45 | $60.85 |
| 1 Truck Driver (heavy) | 34.40 | 275.20 | 52.25 | 418.00 | | |
| 1 Pickup truck, 4 x 4, 3/4 ton | | 116.60 | | 128.26 | | |
| 1 Truck Tractor, 6x4, 450 H.P. | | 558.80 | | 614.68 | | |
| 1 Lowbed Trailer, 75 Ton | | 219.20 | | 241.12 | 55.91 | 61.50 |
| 16 L.H., Daily Totals | | $1541.80 | | $1957.66 | $96.36 | $122.35 |

| Crew A-3H | Hr. | Daily | Hr. | Daily | Bare Costs | Incl. O&P |
|---|---|---|---|---|---|---|
| 1 Equip. Oper. (crane) | $46.50 | $372.00 | $69.45 | $555.60 | $46.50 | $69.45 |
| 1 Hyd. crane, 12 Ton (daily) | | 871.60 | | 958.76 | 108.95 | 119.85 |
| 8 L.H., Daily Totals | | $1243.60 | | $1514.36 | $155.45 | $189.29 |

| Crew A-3I | Hr. | Daily | Hr. | Daily | Bare Costs | Incl. O&P |
|---|---|---|---|---|---|---|
| 1 Equip. Oper. (crane) | $46.50 | $372.00 | $69.45 | $555.60 | $46.50 | $69.45 |
| 1 Hyd. crane, 25 Ton (daily) | | 1015.00 | | 1116.50 | 126.88 | 139.56 |
| 8 L.H., Daily Totals | | $1387.00 | | $1672.10 | $173.38 | $209.01 |

| Crew A-3J | Hr. | Daily | Hr. | Daily | Bare Costs | Incl. O&P |
|---|---|---|---|---|---|---|
| 1 Equip. Oper. (crane) | $46.50 | $372.00 | $69.45 | $555.60 | $46.50 | $69.45 |
| 1 Hyd. crane, 40 Ton (daily) | | 1258.00 | | 1383.80 | 157.25 | 172.97 |
| 8 L.H., Daily Totals | | $1630.00 | | $1939.40 | $203.75 | $242.43 |

| Crew A-3K | Hr. | Daily | Hr. | Daily | Bare Costs | Incl. O&P |
|---|---|---|---|---|---|---|
| 1 Equip. Oper. (crane) | $46.50 | $372.00 | $69.45 | $555.60 | $43.40 | $64.83 |
| 1 Equip. Oper. Oiler | 40.30 | 322.40 | 60.20 | 481.60 | | |
| 1 Hyd. crane, 55 Ton (daily) | | 1611.00 | | 1772.10 | | |
| 1 P/U Truck, 3/4 Ton (daily) | | 129.60 | | 142.56 | 108.79 | 119.67 |
| 16 L.H., Daily Totals | | $2435.00 | | $2951.86 | $152.19 | $184.49 |

| Crew A-3L | Hr. | Daily | Hr. | Daily | Bare Costs | Incl. O&P |
|---|---|---|---|---|---|---|
| 1 Equip. Oper. (crane) | $46.50 | $372.00 | $69.45 | $555.60 | $43.40 | $64.83 |
| 1 Equip. Oper. Oiler | 40.30 | 322.40 | 60.20 | 481.60 | | |
| 1 Hyd. crane, 80 Ton (daily) | | 2302.00 | | 2532.20 | | |
| 1 P/U Truck, 3/4 Ton (daily) | | 129.60 | | 142.56 | 151.97 | 167.17 |
| 16 L.H., Daily Totals | | $3126.00 | | $3711.96 | $195.38 | $232.00 |

| Crew A-3M | Hr. | Daily | Hr. | Daily | Bare Costs | Incl. O&P |
|---|---|---|---|---|---|---|
| 1 Equip. Oper. (crane) | $46.50 | $372.00 | $69.45 | $555.60 | $43.40 | $64.83 |
| 1 Equip. Oper. Oiler | 40.30 | 322.40 | 60.20 | 481.60 | | |
| 1 Hyd. crane, 100 Ton (daily) | | 2263.00 | | 2489.30 | | |
| 1 P/U Truck, 3/4 Ton (daily) | | 129.60 | | 142.56 | 149.54 | 164.49 |
| 16 L.H., Daily Totals | | $3087.00 | | $3669.06 | $192.94 | $229.32 |

| Crew A-3N | Hr. | Daily | Hr. | Daily | Bare Costs | Incl. O&P |
|---|---|---|---|---|---|---|
| 1 Equip. Oper. (crane) | $46.50 | $372.00 | $69.45 | $555.60 | $46.50 | $69.45 |
| 1 Tower crane (monthly) | | 987.20 | | 1085.92 | 123.40 | 135.74 |
| 8 L.H., Daily Totals | | $1359.20 | | $1641.52 | $169.90 | $205.19 |

| Crew A-3P | Hr. | Daily | Hr. | Daily | Bare Costs | Incl. O&P |
|---|---|---|---|---|---|---|
| 1 Equip. Oper., Light | $43.60 | $348.80 | $65.15 | $521.20 | $43.60 | $65.15 |
| 1 A.T. Forklift, 42' lift | | 447.40 | | 492.14 | 55.92 | 61.52 |
| 8 L.H., Daily Totals | | $796.20 | | $1013.34 | $99.53 | $126.67 |

| Crew A-4 | Hr. | Daily | Hr. | Daily | Bare Costs | Incl. O&P |
|---|---|---|---|---|---|---|
| 2 Carpenters | $43.05 | $688.80 | $66.05 | $1056.80 | $41.22 | $62.60 |
| 1 Painter, Ordinary | 37.55 | 300.40 | 55.70 | 445.60 | | |
| 24 L.H., Daily Totals | | $989.20 | | $1502.40 | $41.22 | $62.60 |

| Crew A-5 | Hr. | Daily | Hr. | Daily | Bare Costs | Incl. O&P |
|---|---|---|---|---|---|---|
| 2 Laborers | $34.35 | $549.60 | $52.70 | $843.20 | $34.24 | $52.48 |
| .25 Truck Driver (light) | 33.40 | 66.80 | 50.75 | 101.50 | | |
| .25 Flatbed Truck, Gas, 1.5 Ton | | 42.85 | | 47.13 | 2.38 | 2.62 |
| 18 L.H., Daily Totals | | $659.25 | | $991.84 | $36.63 | $55.10 |

| Crew A-6 | Hr. | Daily | Hr. | Daily | Bare Costs | Incl. O&P |
|---|---|---|---|---|---|---|
| 1 Instrument Man | $44.40 | $355.20 | $68.10 | $544.80 | $43.23 | $65.63 |
| 1 Rodman/Chainman | 42.05 | 336.40 | 63.15 | 505.20 | | |
| 1 Level, electronic | | 69.60 | | 76.56 | 4.35 | 4.79 |
| 16 L.H., Daily Totals | | $761.20 | | $1126.56 | $47.58 | $70.41 |

| Crew A-7 | Hr. | Daily | Hr. | Daily | Bare Costs | Incl. O&P |
|---|---|---|---|---|---|---|
| 1 Chief Of Party | $54.45 | $435.60 | $83.10 | $664.80 | $46.97 | $71.45 |
| 1 Instrument Man | 44.40 | 355.20 | 68.10 | 544.80 | | |
| 1 Rodman/Chainman | 42.05 | 336.40 | 63.15 | 505.20 | | |
| 1 Level, electronic | | 69.60 | | 76.56 | 2.90 | 3.19 |
| 24 L.H., Daily Totals | | $1196.80 | | $1791.36 | $49.87 | $74.64 |

| Crew A-8 | Hr. | Daily | Hr. | Daily | Bare Costs | Incl. O&P |
|---|---|---|---|---|---|---|
| 1 Chief of Party | $54.45 | $435.60 | $83.10 | $664.80 | $45.74 | $69.38 |
| 1 Instrument Man | 44.40 | 355.20 | 68.10 | 544.80 | | |
| 2 Rodmen/Chainmen | 42.05 | 672.80 | 63.15 | 1010.40 | | |
| 1 Level, electronic | | 69.60 | | 76.56 | 2.17 | 2.39 |
| 32 L.H., Daily Totals | | $1533.20 | | $2296.56 | $47.91 | $71.77 |

| Crew A-9 | Hr. | Daily | Hr. | Daily | Bare Costs | Incl. O&P |
|---|---|---|---|---|---|---|
| 1 Asbestos Foreman | $47.40 | $379.20 | $73.30 | $586.40 | $46.96 | $72.60 |
| 7 Asbestos Workers | 46.90 | 2626.40 | 72.50 | 4060.00 | | |
| 64 L.H., Daily Totals | | $3005.60 | | $4646.40 | $46.96 | $72.60 |

| Crew A-10A | Hr. | Daily | Hr. | Daily | Bare Costs | Incl. O&P |
|---|---|---|---|---|---|---|
| 1 Asbestos Foreman | $47.40 | $379.20 | $73.30 | $586.40 | $47.07 | $72.77 |
| 2 Asbestos Workers | 46.90 | 750.40 | 72.50 | 1160.00 | | |
| 24 L.H., Daily Totals | | $1129.60 | | $1746.40 | $47.07 | $72.77 |

| Crew A-10B | Hr. | Daily | Hr. | Daily | Bare Costs | Incl. O&P |
|---|---|---|---|---|---|---|
| 1 Asbestos Foreman | $47.40 | $379.20 | $73.30 | $586.40 | $47.02 | $72.70 |
| 3 Asbestos Workers | 46.90 | 1125.60 | 72.50 | 1740.00 | | |
| 32 L.H., Daily Totals | | $1504.80 | | $2326.40 | $47.02 | $72.70 |

| Crew A-10C | Hr. | Daily | Hr. | Daily | Bare Costs | Incl. O&P |
|---|---|---|---|---|---|---|
| 3 Asbestos Workers | $46.90 | $1125.60 | $72.50 | $1740.00 | $46.90 | $72.50 |
| 1 Flatbed Truck, Gas, 1.5 Ton | | 171.40 | | 188.54 | 7.14 | 7.86 |
| 24 L.H., Daily Totals | | $1297.00 | | $1928.54 | $54.04 | $80.36 |

| Crew A-10D | Hr. | Daily | Hr. | Daily | Bare Costs | Incl. O&P |
|---|---|---|---|---|---|---|
| 2 Asbestos Workers | $46.90 | $750.40 | $72.50 | $1160.00 | $45.15 | $68.66 |
| 1 Equip. Oper. (crane) | 46.50 | 372.00 | 69.45 | 555.60 | | |
| 1 Equip. Oper. Oiler | 40.30 | 322.40 | 60.20 | 481.60 | | |
| 1 Hydraulic Crane, 33 Ton | | 763.00 | | 839.30 | 23.84 | 26.23 |
| 32 L.H., Daily Totals | | $2207.80 | | $3036.50 | $68.99 | $94.89 |

| Crew A-11 | Hr. | Daily | Hr. | Daily | Bare Costs | Incl. O&P |
|---|---|---|---|---|---|---|
| 1 Asbestos Foreman | $47.40 | $379.20 | $73.30 | $586.40 | $46.96 | $72.60 |
| 7 Asbestos Workers | 46.90 | 2626.40 | 72.50 | 4060.00 | | |
| 2 Chip. Hammers, 12 Lb., Elec. | | 37.20 | | 40.92 | .58 | .64 |
| 64 L.H., Daily Totals | | $3042.80 | | $4687.32 | $47.54 | $73.24 |

| Crew No. | Bare Costs | | Incl. Subs O&P | | Cost Per Labor-Hour | |
|---|---|---|---|---|---|---|

| **Crew A-12** | Hr. | Daily | Hr. | Daily | Bare Costs | Incl. O&P |
|---|---|---|---|---|---|---|
| 1 Asbestos Foreman | $47.40 | $379.20 | $73.30 | $586.40 | $46.96 | $72.60 |
| 7 Asbestos Workers | 46.90 | 2626.40 | 72.50 | 4060.00 | | |
| 1 Trk-mtd vac, 14 CY, 1500 Gal. | | 544.10 | | 598.51 | | |
| 1 Flatbed Truck, 20,000 GVW | | 190.00 | | 209.00 | 11.47 | 12.62 |
| 64 L.H., Daily Totals | | $3739.70 | | $5453.91 | $58.43 | $85.22 |

| **Crew A-13** | Hr. | Daily | Hr. | Daily | Bare Costs | Incl. O&P |
|---|---|---|---|---|---|---|
| 1 Equip. Oper. (light) | $43.60 | $348.80 | $65.15 | $521.20 | $43.60 | $65.15 |
| 1 Trk-mtd vac, 14 CY, 1500 Gal. | | 544.10 | | 598.51 | | |
| 1 Flatbed Truck, 20,000 GVW | | 190.00 | | 209.00 | 91.76 | 100.94 |
| 8 L.H., Daily Totals | | $1082.90 | | $1328.71 | $135.36 | $166.09 |

| **Crew B-1** | Hr. | Daily | Hr. | Daily | Bare Costs | Incl. O&P |
|---|---|---|---|---|---|---|
| 1 Labor Foreman (outside) | $36.35 | $290.80 | $55.75 | $446.00 | $35.02 | $53.72 |
| 2 Laborers | 34.35 | 549.60 | 52.70 | 843.20 | | |
| 24 L.H., Daily Totals | | $840.40 | | $1289.20 | $35.02 | $53.72 |

| **Crew B-1A** | Hr. | Daily | Hr. | Daily | Bare Costs | Incl. O&P |
|---|---|---|---|---|---|---|
| 1 Laborer Foreman | $36.35 | $290.80 | $55.75 | $446.00 | $35.02 | $53.72 |
| 2 Laborers | 34.35 | 549.60 | 52.70 | 843.20 | | |
| 2 Cutting Torches | | 36.00 | | 39.60 | | |
| 2 Set of Gases | | 165.60 | | 182.16 | 8.40 | 9.24 |
| 24 L.H., Daily Totals | | $1042.00 | | $1510.96 | $43.42 | $62.96 |

| **Crew B-1B** | Hr. | Daily | Hr. | Daily | Bare Costs | Incl. O&P |
|---|---|---|---|---|---|---|
| 1 Laborer Foreman | $36.35 | $290.80 | $55.75 | $446.00 | $37.89 | $57.65 |
| 2 Laborers | 34.35 | 549.60 | 52.70 | 843.20 | | |
| 1 Equip. Oper. (crane) | 46.50 | 372.00 | 69.45 | 555.60 | | |
| 2 Cutting Torches | | 36.00 | | 39.60 | | |
| 2 Set of Gases | | 165.60 | | 182.16 | | |
| 1 Hyd. Crane, 12 Ton | | 655.60 | | 721.16 | 26.79 | 29.47 |
| 32 L.H., Daily Totals | | $2069.60 | | $2787.72 | $64.67 | $87.12 |

| **Crew B-2** | Hr. | Daily | Hr. | Daily | Bare Costs | Incl. O&P |
|---|---|---|---|---|---|---|
| 1 Labor Foreman (outside) | $36.35 | $290.80 | $55.75 | $446.00 | $34.75 | $53.31 |
| 4 Laborers | 34.35 | 1099.20 | 52.70 | 1686.40 | | |
| 40 L.H., Daily Totals | | $1390.00 | | $2132.40 | $34.75 | $53.31 |

| **Crew B-3** | Hr. | Daily | Hr. | Daily | Bare Costs | Incl. O&P |
|---|---|---|---|---|---|---|
| 1 Labor Foreman (outside) | $36.35 | $290.80 | $55.75 | $446.00 | $36.53 | $55.57 |
| 2 Laborers | 34.35 | 549.60 | 52.70 | 843.20 | | |
| 1 Equip. Oper. (med.) | 45.35 | 362.80 | 67.75 | 542.00 | | |
| 2 Truck Drivers (heavy) | 34.40 | 550.40 | 52.25 | 836.00 | | |
| 1 Crawler Loader, 3 C.Y. | | 1008.00 | | 1108.80 | | |
| 2 Dump Trucks 12 C.Y., 400 H.P. | | 1012.00 | | 1113.20 | 42.08 | 46.29 |
| 48 L.H., Daily Totals | | $3773.60 | | $4889.20 | $78.62 | $101.86 |

| **Crew B-3A** | Hr. | Daily | Hr. | Daily | Bare Costs | Incl. O&P |
|---|---|---|---|---|---|---|
| 4 Laborers | $34.35 | $1099.20 | $52.70 | $1686.40 | $36.55 | $55.71 |
| 1 Equip. Oper. (med.) | 45.35 | 362.80 | 67.75 | 542.00 | | |
| 1 Hyd. Excavator, 1.5 C.Y. | | 947.40 | | 1042.14 | 23.68 | 26.05 |
| 40 L.H., Daily Totals | | $2409.40 | | $3270.54 | $60.23 | $81.76 |

| **Crew B-3B** | Hr. | Daily | Hr. | Daily | Bare Costs | Incl. O&P |
|---|---|---|---|---|---|---|
| 2 Laborers | $34.35 | $549.60 | $52.70 | $843.20 | $37.11 | $56.35 |
| 1 Equip. Oper. (med.) | 45.35 | 362.80 | 67.75 | 542.00 | | |
| 1 Truck Driver (heavy) | 34.40 | 275.20 | 52.25 | 418.00 | | |
| 1 Backhoe Loader, 80 H.P. | | 349.40 | | 384.34 | | |
| 1 Dump Truck, 12 C.Y., 400 H.P. | | 506.00 | | 556.60 | 26.73 | 29.40 |
| 32 L.H., Daily Totals | | $2043.00 | | $2744.14 | $63.84 | $85.75 |

| **Crew B-3C** | Hr. | Daily | Hr. | Daily | Bare Costs | Incl. O&P |
|---|---|---|---|---|---|---|
| 3 Laborers | $34.35 | $824.40 | $52.70 | $1264.80 | $37.10 | $56.46 |
| 1 Equip. Oper. (med.) | 45.35 | 362.80 | 67.75 | 542.00 | | |
| 1 Crawler Loader, 4 C.Y. | | 1432.00 | | 1575.20 | 44.75 | 49.23 |
| 32 L.H., Daily Totals | | $2619.20 | | $3382.00 | $81.85 | $105.69 |

| **Crew B-4** | Hr. | Daily | Hr. | Daily | Bare Costs | Incl. O&P |
|---|---|---|---|---|---|---|
| 1 Labor Foreman (outside) | $36.35 | $290.80 | $55.75 | $446.00 | $34.69 | $53.13 |
| 4 Laborers | 34.35 | 1099.20 | 52.70 | 1686.40 | | |
| 1 Truck Driver (heavy) | 34.40 | 275.20 | 52.25 | 418.00 | | |
| 1 Truck Tractor, 220 H.P. | | 279.20 | | 307.12 | | |
| 1 Flatbed Trailer, 40 Ton | | 150.20 | | 165.22 | 8.95 | 9.84 |
| 48 L.H., Daily Totals | | $2094.60 | | $3022.74 | $43.64 | $62.97 |

| **Crew B-5** | Hr. | Daily | Hr. | Daily | Bare Costs | Incl. O&P |
|---|---|---|---|---|---|---|
| 1 Labor Foreman (outside) | $36.35 | $290.80 | $55.75 | $446.00 | $37.78 | $57.44 |
| 4 Laborers | 34.35 | 1099.20 | 52.70 | 1686.40 | | |
| 2 Equip. Oper. (med.) | 45.35 | 725.60 | 67.75 | 1084.00 | | |
| 1 Air Compressor, 250 cfm | | 162.80 | | 179.08 | | |
| 2 Breakers, Pavement, 60 lb. | | 19.60 | | 21.56 | | |
| 2 -50' Air Hoses, 1.5" | | 12.60 | | 13.86 | | |
| 1 Crawler Loader, 3 C.Y. | | 1008.00 | | 1108.80 | 21.48 | 23.63 |
| 56 L.H., Daily Totals | | $3318.60 | | $4539.70 | $59.26 | $81.07 |

| **Crew B-5A** | Hr. | Daily | Hr. | Daily | Bare Costs | Incl. O&P |
|---|---|---|---|---|---|---|
| 1 Foreman | $36.35 | $290.80 | $55.75 | $446.00 | $37.13 | $56.42 |
| 6 Laborers | 34.35 | 1648.80 | 52.70 | 2529.60 | | |
| 2 Equip. Oper. (med.) | 45.35 | 725.60 | 67.75 | 1084.00 | | |
| 1 Equip. Oper. (light) | 43.60 | 348.80 | 65.15 | 521.20 | | |
| 2 Truck Drivers (heavy) | 34.40 | 550.40 | 52.25 | 836.00 | | |
| 1 Air Compressor, 365 cfm | | 210.20 | | 231.22 | | |
| 2 Breakers, Pavement, 60 lb. | | 19.60 | | 21.56 | | |
| 8 -50' Air Hoses, 1" | | 32.80 | | 36.08 | | |
| 2 Dump Trucks, 8 C.Y., 220 H.P. | | 630.80 | | 693.88 | 9.31 | 10.24 |
| 96 L.H., Daily Totals | | $4457.80 | | $6399.54 | $46.44 | $66.66 |

| **Crew B-5B** | Hr. | Daily | Hr. | Daily | Bare Costs | Incl. O&P |
|---|---|---|---|---|---|---|
| 1 Powderman | $44.40 | $355.20 | $68.10 | $544.80 | $39.72 | $60.06 |
| 2 Equip. Oper. (med.) | 45.35 | 725.60 | 67.75 | 1084.00 | | |
| 3 Truck Drivers (heavy) | 34.40 | 825.60 | 52.25 | 1254.00 | | |
| 1 F.E. Loader, W.M.,2.5 C.Y. | | 397.60 | | 437.36 | | |
| 3 Dump Trucks, 12 C.Y., 400 H.P. | | 1518.00 | | 1669.80 | | |
| 1 Air Compressor, 365 cfm | | 210.20 | | 231.22 | 44.29 | 48.72 |
| 48 L.H., Daily Totals | | $4032.20 | | $5221.18 | $84.00 | $108.77 |

| **Crew B-5C** | Hr. | Daily | Hr. | Daily | Bare Costs | Incl. O&P |
|---|---|---|---|---|---|---|
| 3 Laborers | $34.35 | $824.40 | $52.70 | $1264.80 | $38.00 | $57.50 |
| 1 Equip. Oper. (med.) | 45.35 | 362.80 | 67.75 | 542.00 | | |
| 2 Truck Drivers (heav.) | 34.40 | 550.40 | 52.25 | 836.00 | | |
| 1 Equip. Oper. (crane) | 46.50 | 372.00 | 69.45 | 555.60 | | |
| 1 Equip. Oper. Oiler | 40.30 | 322.40 | 60.20 | 481.60 | | |
| 2 Dump Trucks, 12 C.Y., 400 H.P. | | 1012.00 | | 1113.20 | | |
| 1 Crawler Loader, 4 C.Y. | | 1432.00 | | 1575.20 | | |
| 1 S.P. Crane, 4x4, 25 Ton | | 560.40 | | 616.44 | 46.94 | 51.64 |
| 64 L.H., Daily Totals | | $5436.40 | | $6984.84 | $84.94 | $109.14 |

| **Crew B-6** | Hr. | Daily | Hr. | Daily | Bare Costs | Incl. O&P |
|---|---|---|---|---|---|---|
| 2 Laborers | $34.35 | $549.60 | $52.70 | $843.20 | $37.43 | $56.85 |
| 1 Equip. Oper. (light) | 43.60 | 348.80 | 65.15 | 521.20 | | |
| 1 Backhoe Loader, 48 H.P. | | 320.40 | | 352.44 | 13.35 | 14.69 |
| 24 L.H., Daily Totals | | $1218.80 | | $1716.84 | $50.78 | $71.53 |

## Crew B-6A

| Crew No. | Bare Costs Hr. | Bare Costs Daily | Incl. Subs O&P Hr. | Incl. Subs O&P Daily | Cost Per Labor-Hour Bare Costs | Cost Per Labor-Hour Incl. O&P |
|---|---|---|---|---|---|---|
| .5 Labor Foreman (outside) | $36.35 | $145.40 | $55.75 | $223.00 | $39.15 | $59.33 |
| 1 Laborer | 34.35 | 274.80 | 52.70 | 421.60 | | |
| 1 Equip. Oper. (med.) | 45.35 | 362.80 | 67.75 | 542.00 | | |
| 1 Vacuum Trk.,5000 Gal. | | 375.10 | | 412.61 | 18.75 | 20.63 |
| 20 L.H., Daily Totals | | $1158.10 | | $1599.21 | $57.91 | $79.96 |

## Crew B-6B

| Crew No. | Bare Costs Hr. | Bare Costs Daily | Incl. Subs O&P Hr. | Incl. Subs O&P Daily | Cost Per Labor-Hour Bare Costs | Cost Per Labor-Hour Incl. O&P |
|---|---|---|---|---|---|---|
| 2 Labor Foremen (out) | $36.35 | $581.60 | $55.75 | $892.00 | $35.02 | $53.72 |
| 4 Laborers | 34.35 | 1099.20 | 52.70 | 1686.40 | | |
| 1 S.P. Crane, 4x4, 5 Ton | | 257.80 | | 283.58 | | |
| 1 Flatbed Truck, Gas, 1.5 Ton | | 171.40 | | 188.54 | | |
| 1 Butt Fusion Mach., 4"-12" diam. | | 376.45 | | 414.10 | 16.78 | 18.46 |
| 48 L.H., Daily Totals | | $2486.45 | | $3464.61 | $51.80 | $72.18 |

## Crew B-6C

| Crew No. | Bare Costs Hr. | Bare Costs Daily | Incl. Subs O&P Hr. | Incl. Subs O&P Daily | Cost Per Labor-Hour Bare Costs | Cost Per Labor-Hour Incl. O&P |
|---|---|---|---|---|---|---|
| 2 Labor Foremen (out) | $36.35 | $581.60 | $55.75 | $892.00 | $35.02 | $53.72 |
| 4 Laborers | 34.35 | 1099.20 | 52.70 | 1686.40 | | |
| 1 S.P. Crane, 4x4, 12 Ton | | 434.40 | | 477.84 | | |
| 1 Flatbed Truck, Gas, 3 Ton | | 217.00 | | 238.70 | | |
| 1 Butt Fusion Mach., 8"-24" diam. | | 806.15 | | 886.76 | 30.37 | 33.40 |
| 48 L.H., Daily Totals | | $3138.35 | | $4181.70 | $65.38 | $87.12 |

## Crew B-7

| Crew No. | Bare Costs Hr. | Bare Costs Daily | Incl. Subs O&P Hr. | Incl. Subs O&P Daily | Cost Per Labor-Hour Bare Costs | Cost Per Labor-Hour Incl. O&P |
|---|---|---|---|---|---|---|
| 1 Labor Foreman (outside) | $36.35 | $290.80 | $55.75 | $446.00 | $36.52 | $55.72 |
| 4 Laborers | 34.35 | 1099.20 | 52.70 | 1686.40 | | |
| 1 Equip. Oper. (med.) | 45.35 | 362.80 | 67.75 | 542.00 | | |
| 1 Brush Chipper, 12", 130 H.P. | | 217.00 | | 238.70 | | |
| 1 Crawler Loader, 3 C.Y. | | 1008.00 | | 1108.80 | | |
| 2 Chain Saws, gas, 36" Long | | 83.20 | | 91.52 | 27.25 | 29.98 |
| 48 L.H., Daily Totals | | $3061.00 | | $4113.42 | $63.77 | $85.70 |

## Crew B-7A

| Crew No. | Bare Costs Hr. | Bare Costs Daily | Incl. Subs O&P Hr. | Incl. Subs O&P Daily | Cost Per Labor-Hour Bare Costs | Cost Per Labor-Hour Incl. O&P |
|---|---|---|---|---|---|---|
| 2 Laborers | $34.35 | $549.60 | $52.70 | $843.20 | $37.43 | $56.85 |
| 1 Equip. Oper. (light) | 43.60 | 348.80 | 65.15 | 521.20 | | |
| 1 Rake w/Tractor | | 277.70 | | 305.47 | | |
| 2 Chain Saws, gas, 18" | | 51.60 | | 56.76 | 13.72 | 15.09 |
| 24 L.H., Daily Totals | | $1227.70 | | $1726.63 | $51.15 | $71.94 |

## Crew B-8

| Crew No. | Bare Costs Hr. | Bare Costs Daily | Incl. Subs O&P Hr. | Incl. Subs O&P Daily | Cost Per Labor-Hour Bare Costs | Cost Per Labor-Hour Incl. O&P |
|---|---|---|---|---|---|---|
| 1 Labor Foreman (outside) | $36.35 | $290.80 | $55.75 | $446.00 | $38.11 | $57.67 |
| 2 Laborers | 34.35 | 549.60 | 52.70 | 843.20 | | |
| 2 Equip. Oper. (med.) | 45.35 | 725.60 | 67.75 | 1084.00 | | |
| 1 Equip. Oper. Oiler | 40.30 | 322.40 | 60.20 | 481.60 | | |
| 2 Truck Drivers (heavy) | 34.40 | 550.40 | 52.25 | 836.00 | | |
| 1 Hyd. Crane, 25 Ton | | 748.60 | | 823.46 | | |
| 1 Crawler Loader, 3 C.Y. | | 1008.00 | | 1108.80 | | |
| 2 Dump Trucks, 12 C.Y., 400 H.P. | | 1012.00 | | 1113.20 | 43.26 | 47.59 |
| 64 L.H., Daily Totals | | $5207.40 | | $6736.26 | $81.37 | $105.25 |

## Crew B-9

| Crew No. | Bare Costs Hr. | Bare Costs Daily | Incl. Subs O&P Hr. | Incl. Subs O&P Daily | Cost Per Labor-Hour Bare Costs | Cost Per Labor-Hour Incl. O&P |
|---|---|---|---|---|---|---|
| 1 Labor Foreman (outside) | $36.35 | $290.80 | $55.75 | $446.00 | $34.75 | $53.31 |
| 4 Laborers | 34.35 | 1099.20 | 52.70 | 1686.40 | | |
| 1 Air Compressor, 250 cfm | | 162.80 | | 179.08 | | |
| 2 Breakers, Pavement, 60 lb. | | 19.60 | | 21.56 | | |
| 2 -50' Air Hoses, 1.5" | | 12.60 | | 13.86 | 4.88 | 5.36 |
| 40 L.H., Daily Totals | | $1585.00 | | $2346.90 | $39.63 | $58.67 |

## Crew B-9A

| Crew No. | Bare Costs Hr. | Bare Costs Daily | Incl. Subs O&P Hr. | Incl. Subs O&P Daily | Cost Per Labor-Hour Bare Costs | Cost Per Labor-Hour Incl. O&P |
|---|---|---|---|---|---|---|
| 2 Laborers | $34.35 | $549.60 | $52.70 | $843.20 | $34.37 | $52.55 |
| 1 Truck Driver (heavy) | 34.40 | 275.20 | 52.25 | 418.00 | | |
| 1 Water Tank Trailer, 5000 Gal. | | 146.20 | | 160.82 | | |
| 1 Truck Tractor, 220 H.P. | | 279.20 | | 307.12 | | |
| 2 -50' Discharge Hoses, 3" | | 3.50 | | 3.85 | 17.87 | 19.66 |
| 24 L.H., Daily Totals | | $1253.70 | | $1732.99 | $52.24 | $72.21 |

## Crew B-9B

| Crew No. | Bare Costs Hr. | Bare Costs Daily | Incl. Subs O&P Hr. | Incl. Subs O&P Daily | Cost Per Labor-Hour Bare Costs | Cost Per Labor-Hour Incl. O&P |
|---|---|---|---|---|---|---|
| 2 Laborers | $34.35 | $549.60 | $52.70 | $843.20 | $34.37 | $52.55 |
| 1 Truck Driver (heavy) | 34.40 | 275.20 | 52.25 | 418.00 | | |
| 2 -50' Discharge Hoses, 3" | | 3.50 | | 3.85 | | |
| 1 Water Tank Trailer, 5000 Gal. | | 146.20 | | 160.82 | | |
| 1 Truck Tractor, 220 H.P. | | 279.20 | | 307.12 | | |
| 1 Pressure Washer | | 58.80 | | 64.68 | 20.32 | 22.35 |
| 24 L.H., Daily Totals | | $1312.50 | | $1797.67 | $54.69 | $74.90 |

## Crew B-9D

| Crew No. | Bare Costs Hr. | Bare Costs Daily | Incl. Subs O&P Hr. | Incl. Subs O&P Daily | Cost Per Labor-Hour Bare Costs | Cost Per Labor-Hour Incl. O&P |
|---|---|---|---|---|---|---|
| 1 Labor Foreman (Outside) | $36.35 | $290.80 | $55.75 | $446.00 | $34.75 | $53.31 |
| 4 Common Laborers | 34.35 | 1099.20 | 52.70 | 1686.40 | | |
| 1 Air Compressor, 250 cfm | | 162.80 | | 179.08 | | |
| 2 -50' Air Hoses, 1.5" | | 12.60 | | 13.86 | | |
| 2 Air Powered Tampers | | 52.40 | | 57.64 | 5.70 | 6.26 |
| 40 L.H., Daily Totals | | $1617.80 | | $2382.98 | $40.45 | $59.57 |

## Crew B-10

| Crew No. | Bare Costs Hr. | Bare Costs Daily | Incl. Subs O&P Hr. | Incl. Subs O&P Daily | Cost Per Labor-Hour Bare Costs | Cost Per Labor-Hour Incl. O&P |
|---|---|---|---|---|---|---|
| 1 Equip. Oper. (med.) | $45.35 | $362.80 | $67.75 | $542.00 | $41.68 | $62.73 |
| .5 Laborer | 34.35 | 137.40 | 52.70 | 210.80 | | |
| 12 L.H., Daily Totals | | $500.20 | | $752.80 | $41.68 | $62.73 |

## Crew B-10A

| Crew No. | Bare Costs Hr. | Bare Costs Daily | Incl. Subs O&P Hr. | Incl. Subs O&P Daily | Cost Per Labor-Hour Bare Costs | Cost Per Labor-Hour Incl. O&P |
|---|---|---|---|---|---|---|
| 1 Equip. Oper. (med.) | $45.35 | $362.80 | $67.75 | $542.00 | $41.68 | $62.73 |
| .5 Laborer | 34.35 | 137.40 | 52.70 | 210.80 | | |
| 1 Roller, 2-Drum, W.B., 7.5 H.P. | | 146.80 | | 161.48 | 12.23 | 13.46 |
| 12 L.H., Daily Totals | | $647.00 | | $914.28 | $53.92 | $76.19 |

## Crew B-10B

| Crew No. | Bare Costs Hr. | Bare Costs Daily | Incl. Subs O&P Hr. | Incl. Subs O&P Daily | Cost Per Labor-Hour Bare Costs | Cost Per Labor-Hour Incl. O&P |
|---|---|---|---|---|---|---|
| 1 Equip. Oper. (med.) | $45.35 | $362.80 | $67.75 | $542.00 | $41.68 | $62.73 |
| .5 Laborer | 34.35 | 137.40 | 52.70 | 210.80 | | |
| 1 Dozer, 200 H.P. | | 1051.00 | | 1156.10 | 87.58 | 96.34 |
| 12 L.H., Daily Totals | | $1551.20 | | $1908.90 | $129.27 | $159.07 |

## Crew B-10C

| Crew No. | Bare Costs Hr. | Bare Costs Daily | Incl. Subs O&P Hr. | Incl. Subs O&P Daily | Cost Per Labor-Hour Bare Costs | Cost Per Labor-Hour Incl. O&P |
|---|---|---|---|---|---|---|
| 1 Equip. Oper. (med.) | $45.35 | $362.80 | $67.75 | $542.00 | $41.68 | $62.73 |
| .5 Laborer | 34.35 | 137.40 | 52.70 | 210.80 | | |
| 1 Dozer, 200 H.P. | | 1051.00 | | 1156.10 | | |
| 1 Vibratory Roller, Towed, 23 Ton | | 394.60 | | 434.06 | 120.47 | 132.51 |
| 12 L.H., Daily Totals | | $1945.80 | | $2342.96 | $162.15 | $195.25 |

## Crew B-10D

| Crew No. | Bare Costs Hr. | Bare Costs Daily | Incl. Subs O&P Hr. | Incl. Subs O&P Daily | Cost Per Labor-Hour Bare Costs | Cost Per Labor-Hour Incl. O&P |
|---|---|---|---|---|---|---|
| 1 Equip. Oper. (med.) | $45.35 | $362.80 | $67.75 | $542.00 | $41.68 | $62.73 |
| .5 Laborer | 34.35 | 137.40 | 52.70 | 210.80 | | |
| 1 Dozer, 200 H.P. | | 1051.00 | | 1156.10 | | |
| 1 Sheepsft. Roller, Towed | | 449.00 | | 493.90 | 125.00 | 137.50 |
| 12 L.H., Daily Totals | | $2000.20 | | $2402.80 | $166.68 | $200.23 |

## Crew B-10E

| Crew No. | Bare Costs Hr. | Bare Costs Daily | Incl. Subs O&P Hr. | Incl. Subs O&P Daily | Cost Per Labor-Hour Bare Costs | Cost Per Labor-Hour Incl. O&P |
|---|---|---|---|---|---|---|
| 1 Equip. Oper. (med.) | $45.35 | $362.80 | $67.75 | $542.00 | $41.68 | $62.73 |
| .5 Laborer | 34.35 | 137.40 | 52.70 | 210.80 | | |
| 1 Tandem Roller, 5 Ton | | 145.20 | | 159.72 | 12.10 | 13.31 |
| 12 L.H., Daily Totals | | $645.40 | | $912.52 | $53.78 | $76.04 |

| Crew No. | Bare Costs Hr. | Daily | Incl. Subs O&P Hr. | Daily | Cost Per Labor-Hour Bare Costs | Incl. O&P |
|---|---|---|---|---|---|---|
| **Crew B-10F** | Hr. | Daily | Hr. | Daily | Bare Costs | Incl. O&P |
| 1 Equip. Oper. (med.) | $45.35 | $362.80 | $67.75 | $542.00 | $41.68 | $62.73 |
| .5 Laborer | 34.35 | 137.40 | 52.70 | 210.80 | | |
| 1 Tandem Roller, 10 Ton | | 224.20 | | 246.62 | 18.68 | 20.55 |
| 12 L.H., Daily Totals | | $724.40 | | $999.42 | $60.37 | $83.28 |

| Crew No. | Bare Costs Hr. | Daily | Incl. Subs O&P Hr. | Daily | Cost Per Labor-Hour Bare Costs | Incl. O&P |
|---|---|---|---|---|---|---|
| **Crew B-10G** | Hr. | Daily | Hr. | Daily | Bare Costs | Incl. O&P |
| 1 Equip. Oper. (med.) | $45.35 | $362.80 | $67.75 | $542.00 | $41.68 | $62.73 |
| .5 Laborer | 34.35 | 137.40 | 52.70 | 210.80 | | |
| 1 Sheepsft. Roll., 240 H.P. | | 1094.00 | | 1203.40 | 91.17 | 100.28 |
| 12 L.H., Daily Totals | | $1594.20 | | $1956.20 | $132.85 | $163.02 |

| Crew No. | Bare Costs Hr. | Daily | Incl. Subs O&P Hr. | Daily | Cost Per Labor-Hour Bare Costs | Incl. O&P |
|---|---|---|---|---|---|---|
| **Crew B-10H** | Hr. | Daily | Hr. | Daily | Bare Costs | Incl. O&P |
| 1 Equip. Oper. (med.) | $45.35 | $362.80 | $67.75 | $542.00 | $41.68 | $62.73 |
| .5 Laborer | 34.35 | 137.40 | 52.70 | 210.80 | | |
| 1 Diaphragm Water Pump, 2" | | 66.60 | | 73.26 | | |
| 1 -20' Suction Hose, 2" | | 1.95 | | 2.15 | | |
| 2 -50' Discharge Hoses, 2" | | 1.80 | | 1.98 | 5.86 | 6.45 |
| 12 L.H., Daily Totals | | $570.55 | | $830.18 | $47.55 | $69.18 |

| Crew No. | Bare Costs Hr. | Daily | Incl. Subs O&P Hr. | Daily | Cost Per Labor-Hour Bare Costs | Incl. O&P |
|---|---|---|---|---|---|---|
| **Crew B-10I** | Hr. | Daily | Hr. | Daily | Bare Costs | Incl. O&P |
| 1 Equip. Oper. (med.) | $45.35 | $362.80 | $67.75 | $542.00 | $41.68 | $62.73 |
| .5 Laborer | 34.35 | 137.40 | 52.70 | 210.80 | | |
| 1 Diaphragm Water Pump, 4" | | 109.60 | | 120.34 | | |
| 1 -20' Suction Hose, 4" | | 3.25 | | 3.58 | | |
| 2 -50' Discharge Hoses, 4" | | 4.70 | | 5.17 | 9.78 | 10.76 |
| 12 L.H., Daily Totals | | $617.55 | | $881.88 | $51.46 | $73.49 |

| Crew No. | Bare Costs Hr. | Daily | Incl. Subs O&P Hr. | Daily | Cost Per Labor-Hour Bare Costs | Incl. O&P |
|---|---|---|---|---|---|---|
| **Crew B-10J** | Hr. | Daily | Hr. | Daily | Bare Costs | Incl. O&P |
| 1 Equip. Oper. (med.) | $45.35 | $362.80 | $67.75 | $542.00 | $41.68 | $62.73 |
| .5 Laborer | 34.35 | 137.40 | 52.70 | 210.80 | | |
| 1 Centrifugal Water Pump, 3" | | 71.00 | | 78.10 | | |
| 1 -20' Suction Hose, 3" | | 3.05 | | 3.36 | | |
| 2 -50' Discharge Hoses, 3" | | 3.50 | | 3.85 | 6.46 | 7.11 |
| 12 L.H., Daily Totals | | $577.75 | | $838.11 | $48.15 | $69.84 |

| Crew No. | Bare Costs Hr. | Daily | Incl. Subs O&P Hr. | Daily | Cost Per Labor-Hour Bare Costs | Incl. O&P |
|---|---|---|---|---|---|---|
| **Crew B-10K** | Hr. | Daily | Hr. | Daily | Bare Costs | Incl. O&P |
| 1 Equip. Oper. (med.) | $45.35 | $362.80 | $67.75 | $542.00 | $41.68 | $62.73 |
| .5 Laborer | 34.35 | 137.40 | 52.70 | 210.80 | | |
| 1 Centr. Water Pump, 6" | | 291.60 | | 320.76 | | |
| 1 -20' Suction Hose, 6" | | 11.90 | | 13.09 | | |
| 2 -50' Discharge Hoses, 6" | | 12.60 | | 13.86 | 26.34 | 28.98 |
| 12 L.H., Daily Totals | | $816.30 | | $1100.51 | $68.03 | $91.71 |

| Crew No. | Bare Costs Hr. | Daily | Incl. Subs O&P Hr. | Daily | Cost Per Labor-Hour Bare Costs | Incl. O&P |
|---|---|---|---|---|---|---|
| **Crew B-10L** | Hr. | Daily | Hr. | Daily | Bare Costs | Incl. O&P |
| 1 Equip. Oper. (med.) | $45.35 | $362.80 | $67.75 | $542.00 | $41.68 | $62.73 |
| .5 Laborer | 34.35 | 137.40 | 52.70 | 210.80 | | |
| 1 Dozer, 80 H.P. | | 412.60 | | 453.86 | 34.38 | 37.82 |
| 12 L.H., Daily Totals | | $912.80 | | $1206.66 | $76.07 | $100.56 |

| Crew No. | Bare Costs Hr. | Daily | Incl. Subs O&P Hr. | Daily | Cost Per Labor-Hour Bare Costs | Incl. O&P |
|---|---|---|---|---|---|---|
| **Crew B-10M** | Hr. | Daily | Hr. | Daily | Bare Costs | Incl. O&P |
| 1 Equip. Oper. (med.) | $45.35 | $362.80 | $67.75 | $542.00 | $41.68 | $62.73 |
| .5 Laborer | 34.35 | 137.40 | 52.70 | 210.80 | | |
| 1 Dozer, 300 H.P. | | 1607.00 | | 1767.70 | 133.92 | 147.31 |
| 12 L.H., Daily Totals | | $2107.20 | | $2520.50 | $175.60 | $210.04 |

| Crew No. | Bare Costs Hr. | Daily | Incl. Subs O&P Hr. | Daily | Cost Per Labor-Hour Bare Costs | Incl. O&P |
|---|---|---|---|---|---|---|
| **Crew B-10N** | Hr. | Daily | Hr. | Daily | Bare Costs | Incl. O&P |
| 1 Equip. Oper. (med.) | $45.35 | $362.80 | $67.75 | $542.00 | $41.68 | $62.73 |
| .5 Laborer | 34.35 | 137.40 | 52.70 | 210.80 | | |
| 1 F.E. Loader, T.M., 1.5 C.Y | | 466.60 | | 513.26 | 38.88 | 42.77 |
| 12 L.H., Daily Totals | | $966.80 | | $1266.06 | $80.57 | $105.51 |

| Crew No. | Bare Costs Hr. | Daily | Incl. Subs O&P Hr. | Daily | Cost Per Labor-Hour Bare Costs | Incl. O&P |
|---|---|---|---|---|---|---|
| **Crew B-100** | Hr. | Daily | Hr. | Daily | Bare Costs | Incl. O&P |
| 1 Equip. Oper. (med.) | $45.35 | $362.80 | $67.75 | $542.00 | $41.68 | $62.73 |
| .5 Laborer | 34.35 | 137.40 | 52.70 | 210.80 | | |
| 1 F.E. Loader, T.M., 2.25 C.Y. | | 773.60 | | 850.96 | 64.47 | 70.91 |
| 12 L.H., Daily Totals | | $1273.80 | | $1603.76 | $106.15 | $133.65 |

| Crew No. | Bare Costs Hr. | Daily | Incl. Subs O&P Hr. | Daily | Cost Per Labor-Hour Bare Costs | Incl. O&P |
|---|---|---|---|---|---|---|
| **Crew B-10P** | Hr. | Daily | Hr. | Daily | Bare Costs | Incl. O&P |
| 1 Equip. Oper. (med.) | $45.35 | $362.80 | $67.75 | $542.00 | $41.68 | $62.73 |
| .5 Laborer | 34.35 | 137.40 | 52.70 | 210.80 | | |
| 1 Crawler Loader, 3 C.Y. | | 1008.00 | | 1108.80 | 84.00 | 92.40 |
| 12 L.H., Daily Totals | | $1508.20 | | $1861.60 | $125.68 | $155.13 |

| Crew No. | Bare Costs Hr. | Daily | Incl. Subs O&P Hr. | Daily | Cost Per Labor-Hour Bare Costs | Incl. O&P |
|---|---|---|---|---|---|---|
| **Crew B-10Q** | Hr. | Daily | Hr. | Daily | Bare Costs | Incl. O&P |
| 1 Equip. Oper. (med.) | $45.35 | $362.80 | $67.75 | $542.00 | $41.68 | $62.73 |
| .5 Laborer | 34.35 | 137.40 | 52.70 | 210.80 | | |
| 1 Crawler Loader, 4 C.Y. | | 1432.00 | | 1575.20 | 119.33 | 131.27 |
| 12 L.H., Daily Totals | | $1932.20 | | $2328.00 | $161.02 | $194.00 |

| Crew No. | Bare Costs Hr. | Daily | Incl. Subs O&P Hr. | Daily | Cost Per Labor-Hour Bare Costs | Incl. O&P |
|---|---|---|---|---|---|---|
| **Crew B-10R** | Hr. | Daily | Hr. | Daily | Bare Costs | Incl. O&P |
| 1 Equip. Oper. (med.) | $45.35 | $362.80 | $67.75 | $542.00 | $41.68 | $62.73 |
| .5 Laborer | 34.35 | 137.40 | 52.70 | 210.80 | | |
| 1 F.E. Loader, W.M., 1 C.Y. | | 235.80 | | 259.38 | 19.65 | 21.61 |
| 12 L.H., Daily Totals | | $736.00 | | $1012.18 | $61.33 | $84.35 |

| Crew No. | Bare Costs Hr. | Daily | Incl. Subs O&P Hr. | Daily | Cost Per Labor-Hour Bare Costs | Incl. O&P |
|---|---|---|---|---|---|---|
| **Crew B-10S** | Hr. | Daily | Hr. | Daily | Bare Costs | Incl. O&P |
| 1 Equip. Oper. (med.) | $45.35 | $362.80 | $67.75 | $542.00 | $41.68 | $62.73 |
| .5 Laborer | 34.35 | 137.40 | 52.70 | 210.80 | | |
| 1 F.E. Loader, W.M., 1.5 C.Y. | | 319.40 | | 351.34 | 26.62 | 29.28 |
| 12 L.H., Daily Totals | | $819.60 | | $1104.14 | $68.30 | $92.01 |

| Crew No. | Bare Costs Hr. | Daily | Incl. Subs O&P Hr. | Daily | Cost Per Labor-Hour Bare Costs | Incl. O&P |
|---|---|---|---|---|---|---|
| **Crew B-10T** | Hr. | Daily | Hr. | Daily | Bare Costs | Incl. O&P |
| 1 Equip. Oper. (med.) | $45.35 | $362.80 | $67.75 | $542.00 | $41.68 | $62.73 |
| .5 Laborer | 34.35 | 137.40 | 52.70 | 210.80 | | |
| 1 F.E. Loader, W.M.,2.5 C.Y. | | 397.60 | | 437.36 | 33.13 | 36.45 |
| 12 L.H., Daily Totals | | $897.80 | | $1190.16 | $74.82 | $99.18 |

| Crew No. | Bare Costs Hr. | Daily | Incl. Subs O&P Hr. | Daily | Cost Per Labor-Hour Bare Costs | Incl. O&P |
|---|---|---|---|---|---|---|
| **Crew B-10U** | Hr. | Daily | Hr. | Daily | Bare Costs | Incl. O&P |
| 1 Equip. Oper. (med.) | $45.35 | $362.80 | $67.75 | $542.00 | $41.68 | $62.73 |
| .5 Laborer | 34.35 | 137.40 | 52.70 | 210.80 | | |
| 1 F.E. Loader, W.M., 5.5 C.Y. | | 933.40 | | 1026.74 | 77.78 | 85.56 |
| 12 L.H., Daily Totals | | $1433.60 | | $1779.54 | $119.47 | $148.29 |

| Crew No. | Bare Costs Hr. | Daily | Incl. Subs O&P Hr. | Daily | Cost Per Labor-Hour Bare Costs | Incl. O&P |
|---|---|---|---|---|---|---|
| **Crew B-10V** | Hr. | Daily | Hr. | Daily | Bare Costs | Incl. O&P |
| 1 Equip. Oper. (med.) | $45.35 | $362.80 | $67.75 | $542.00 | $41.68 | $62.73 |
| .5 Laborer | 34.35 | 137.40 | 52.70 | 210.80 | | |
| 1 Dozer, 700 H.P. | | 4150.00 | | 4565.00 | 345.83 | 380.42 |
| 12 L.H., Daily Totals | | $4650.20 | | $5317.80 | $387.52 | $443.15 |

| Crew No. | Bare Costs Hr. | Daily | Incl. Subs O&P Hr. | Daily | Cost Per Labor-Hour Bare Costs | Incl. O&P |
|---|---|---|---|---|---|---|
| **Crew B-10W** | Hr. | Daily | Hr. | Daily | Bare Costs | Incl. O&P |
| 1 Equip. Oper. (med.) | $45.35 | $362.80 | $67.75 | $542.00 | $41.68 | $62.73 |
| .5 Laborer | 34.35 | 137.40 | 52.70 | 210.80 | | |
| 1 Dozer, 105 H.P. | | 505.40 | | 555.94 | 42.12 | 46.33 |
| 12 L.H., Daily Totals | | $1005.60 | | $1308.74 | $83.80 | $109.06 |

| Crew No. | Bare Costs Hr. | Daily | Incl. Subs O&P Hr. | Daily | Cost Per Labor-Hour Bare Costs | Incl. O&P |
|---|---|---|---|---|---|---|
| **Crew B-10X** | Hr. | Daily | Hr. | Daily | Bare Costs | Incl. O&P |
| 1 Equip. Oper. (med.) | $45.35 | $362.80 | $67.75 | $542.00 | $41.68 | $62.73 |
| .5 Laborer | 34.35 | 137.40 | 52.70 | 210.80 | | |
| 1 Dozer, 410 H.P. | | 2105.00 | | 2315.50 | 175.42 | 192.96 |
| 12 L.H., Daily Totals | | $2605.20 | | $3068.30 | $217.10 | $255.69 |

| Crew No. | Bare Costs | | Incl. Subs O&P | | Cost Per Labor-Hour | |
|---|---|---|---|---|---|---|

### Crew B-10Y

| | Hr. | Daily | Hr. | Daily | Bare Costs | Incl. O&P |
|---|---|---|---|---|---|---|
| 1 Equip. Oper. (med.) | $45.35 | $362.80 | $67.75 | $542.00 | $41.68 | $62.73 |
| .5 Laborer | 34.35 | 137.40 | 52.70 | 210.80 | | |
| 1 Vibr. Roller, Towed, 12 Ton | | 498.80 | | 548.68 | 41.57 | 45.72 |
| 12 L.H., Daily Totals | | $999.00 | | $1301.48 | $83.25 | $108.46 |

### Crew B-11A

| | Hr. | Daily | Hr. | Daily | Bare Costs | Incl. O&P |
|---|---|---|---|---|---|---|
| 1 Equipment Oper. (med.) | $45.35 | $362.80 | $67.75 | $542.00 | $39.85 | $60.23 |
| 1 Laborer | 34.35 | 274.80 | 52.70 | 421.60 | | |
| 1 Dozer, 200 H.P. | | 1051.00 | | 1156.10 | 65.69 | 72.26 |
| 16 L.H., Daily Totals | | $1688.60 | | $2119.70 | $105.54 | $132.48 |

### Crew B-11B

| | Hr. | Daily | Hr. | Daily | Bare Costs | Incl. O&P |
|---|---|---|---|---|---|---|
| 1 Equipment Oper. (light) | $43.60 | $348.80 | $65.15 | $521.20 | $38.98 | $58.92 |
| 1 Laborer | 34.35 | 274.80 | 52.70 | 421.60 | | |
| 1 Air Powered Tamper | | 26.20 | | 28.82 | | |
| 1 Air Compressor, 365 cfm | | 210.20 | | 231.22 | | |
| 2 -50' Air Hoses, 1.5" | | 12.60 | | 13.86 | 15.56 | 17.12 |
| 16 L.H., Daily Totals | | $872.60 | | $1216.70 | $54.54 | $76.04 |

### Crew B-11C

| | Hr. | Daily | Hr. | Daily | Bare Costs | Incl. O&P |
|---|---|---|---|---|---|---|
| 1 Equipment Oper. (med.) | $45.35 | $362.80 | $67.75 | $542.00 | $39.85 | $60.23 |
| 1 Laborer | 34.35 | 274.80 | 52.70 | 421.60 | | |
| 1 Backhoe Loader, 48 H.P. | | 320.40 | | 352.44 | 20.02 | 22.03 |
| 16 L.H., Daily Totals | | $958.00 | | $1316.04 | $59.88 | $82.25 |

### Crew B-11J

| | Hr. | Daily | Hr. | Daily | Bare Costs | Incl. O&P |
|---|---|---|---|---|---|---|
| 1 Equipment Oper. (med.) | $45.35 | $362.80 | $67.75 | $542.00 | $39.85 | $60.23 |
| 1 Laborer | 34.35 | 274.80 | 52.70 | 421.60 | | |
| 1 Grader, 30,000 Lbs. | | 542.80 | | 597.08 | | |
| 1 Ripper, beam & 1 shank | | 83.20 | | 91.52 | 39.13 | 43.04 |
| 16 L.H., Daily Totals | | $1263.60 | | $1652.20 | $78.97 | $103.26 |

### Crew B-11K

| | Hr. | Daily | Hr. | Daily | Bare Costs | Incl. O&P |
|---|---|---|---|---|---|---|
| 1 Equipment Oper. (med.) | $45.35 | $362.80 | $67.75 | $542.00 | $39.85 | $60.23 |
| 1 Laborer | 34.35 | 274.80 | 52.70 | 421.60 | | |
| 1 Trencher, Chain Type, 8' D | | 2506.00 | | 2756.60 | 156.63 | 172.29 |
| 16 L.H., Daily Totals | | $3143.60 | | $3720.20 | $196.47 | $232.51 |

### Crew B-11L

| | Hr. | Daily | Hr. | Daily | Bare Costs | Incl. O&P |
|---|---|---|---|---|---|---|
| 1 Equipment Oper. (med.) | $45.35 | $362.80 | $67.75 | $542.00 | $39.85 | $60.23 |
| 1 Laborer | 34.35 | 274.80 | 52.70 | 421.60 | | |
| 1 Grader, 30,000 Lbs. | | 542.80 | | 597.08 | 33.92 | 37.32 |
| 16 L.H., Daily Totals | | $1180.40 | | $1560.68 | $73.78 | $97.54 |

### Crew B-11M

| | Hr. | Daily | Hr. | Daily | Bare Costs | Incl. O&P |
|---|---|---|---|---|---|---|
| 1 Equipment Oper. (med.) | $45.35 | $362.80 | $67.75 | $542.00 | $39.85 | $60.23 |
| 1 Laborer | 34.35 | 274.80 | 52.70 | 421.60 | | |
| 1 Backhoe Loader, 80 H.P. | | 349.40 | | 384.34 | 21.84 | 24.02 |
| 16 L.H., Daily Totals | | $987.00 | | $1347.94 | $61.69 | $84.25 |

### Crew B-11N

| | Hr. | Daily | Hr. | Daily | Bare Costs | Incl. O&P |
|---|---|---|---|---|---|---|
| 1 Labor Foreman | $36.35 | $290.80 | $55.75 | $446.00 | $37.05 | $56.08 |
| 2 Equipment Operators (med.) | 45.35 | 725.60 | 67.75 | 1084.00 | | |
| 6 Truck Drivers (hvy.) | 34.40 | 1651.20 | 52.25 | 2508.00 | | |
| 1 F.E. Loader, W.M., 5.5 C.Y. | | 933.40 | | 1026.74 | | |
| 1 Dozer, 410 H.P. | | 2105.00 | | 2315.50 | | |
| 6 Dump Trucks, Off Hwy., 50 Ton | | 9294.00 | | 10223.40 | 171.28 | 188.41 |
| 72 L.H., Daily Totals | | $15000.00 | | $17603.64 | $208.33 | $244.50 |

### Crew B-11Q

| | Hr. | Daily | Hr. | Daily | Bare Costs | Incl. O&P |
|---|---|---|---|---|---|---|
| 1 Equipment Operator (med.) | $45.35 | $362.80 | $67.75 | $542.00 | $41.68 | $62.73 |
| .5 Laborer | 34.35 | 137.40 | 52.70 | 210.80 | | |
| 1 Dozer, 140 H.P. | | 778.80 | | 856.68 | 64.90 | 71.39 |
| 12 L.H., Daily Totals | | $1279.00 | | $1609.48 | $106.58 | $134.12 |

### Crew B-11R

| | Hr. | Daily | Hr. | Daily | Bare Costs | Incl. O&P |
|---|---|---|---|---|---|---|
| 1 Equipment Operator (med.) | $45.35 | $362.80 | $67.75 | $542.00 | $41.68 | $62.73 |
| .5 Laborer | 34.35 | 137.40 | 52.70 | 210.80 | | |
| 1 Dozer, 200 H.P. | | 1051.00 | | 1156.10 | 87.58 | 96.34 |
| 12 L.H., Daily Totals | | $1551.20 | | $1908.90 | $129.27 | $159.07 |

### Crew B-11S

| | Hr. | Daily | Hr. | Daily | Bare Costs | Incl. O&P |
|---|---|---|---|---|---|---|
| 1 Equipment Operator (med.) | $45.35 | $362.80 | $67.75 | $542.00 | $41.68 | $62.73 |
| .5 Laborer | 34.35 | 137.40 | 52.70 | 210.80 | | |
| 1 Dozer, 300 H.P. | | 1607.00 | | 1767.70 | | |
| 1 Ripper, beam & 1 shank | | 83.20 | | 91.52 | 140.85 | 154.94 |
| 12 L.H., Daily Totals | | $2190.40 | | $2612.02 | $182.53 | $217.67 |

### Crew B-11T

| | Hr. | Daily | Hr. | Daily | Bare Costs | Incl. O&P |
|---|---|---|---|---|---|---|
| 1 Equipment Operator (med.) | $45.35 | $362.80 | $67.75 | $542.00 | $41.68 | $62.73 |
| .5 Laborer | 34.35 | 137.40 | 52.70 | 210.80 | | |
| 1 Dozer, 410 H.P. | | 2105.00 | | 2315.50 | | |
| 1 Ripper, beam & 2 shanks | | 94.40 | | 103.84 | 183.28 | 201.61 |
| 12 L.H., Daily Totals | | $2699.60 | | $3172.14 | $224.97 | $264.35 |

### Crew B-11U

| | Hr. | Daily | Hr. | Daily | Bare Costs | Incl. O&P |
|---|---|---|---|---|---|---|
| 1 Equipment Operator (med.) | $45.35 | $362.80 | $67.75 | $542.00 | $41.68 | $62.73 |
| .5 Laborer | 34.35 | 137.40 | 52.70 | 210.80 | | |
| 1 Dozer, 520 H.P. | | 2845.00 | | 3129.50 | 237.08 | 260.79 |
| 12 L.H., Daily Totals | | $3345.20 | | $3882.30 | $278.77 | $323.52 |

### Crew B-11V

| | Hr. | Daily | Hr. | Daily | Bare Costs | Incl. O&P |
|---|---|---|---|---|---|---|
| 3 Laborers | $34.35 | $824.40 | $52.70 | $1264.80 | $34.35 | $52.70 |
| 1 Roller, 2-Drum, W.B., 7.5 H.P. | | 146.80 | | 161.48 | 6.12 | 6.73 |
| 24 L.H., Daily Totals | | $971.20 | | $1426.28 | $40.47 | $59.43 |

### Crew B-11W

| | Hr. | Daily | Hr. | Daily | Bare Costs | Incl. O&P |
|---|---|---|---|---|---|---|
| 1 Equipment Operator (med.) | $45.35 | $362.80 | $67.75 | $542.00 | $35.31 | $53.58 |
| 1 Common Laborer | 34.35 | 274.80 | 52.70 | 421.60 | | |
| 10 Truck Drivers (hvy.) | 34.40 | 2752.00 | 52.25 | 4180.00 | | |
| 1 Dozer, 200 H.P. | | 1051.00 | | 1156.10 | | |
| 1 Vibratory Roller, Towed, 23 Ton | | 394.60 | | 434.06 | | |
| 10 Dump Trucks, 8 C.Y., 220 H.P. | | 3154.00 | | 3469.40 | 47.91 | 52.70 |
| 96 L.H., Daily Totals | | $7989.20 | | $10203.16 | $83.22 | $106.28 |

### Crew B-11Y

| | Hr. | Daily | Hr. | Daily | Bare Costs | Incl. O&P |
|---|---|---|---|---|---|---|
| 1 Labor Foreman (Outside) | $36.35 | $290.80 | $55.75 | $446.00 | $38.24 | $58.06 |
| 5 Common Laborers | 34.35 | 1374.00 | 52.70 | 2108.00 | | |
| 3 Equipment Operators (med.) | 45.35 | 1088.40 | 67.75 | 1626.00 | | |
| 1 Dozer, 80 H.P. | | 412.60 | | 453.86 | | |
| 2 Roller, 2-Drum, W.B., 7.5 H.P. | | 293.60 | | 322.96 | | |
| 4 Vibratory Plates, gas, 21" | | 156.00 | | 171.60 | 11.98 | 13.17 |
| 72 L.H., Daily Totals | | $3615.40 | | $5128.42 | $50.21 | $71.23 |

### Crew B-12A

| | Hr. | Daily | Hr. | Daily | Bare Costs | Incl. O&P |
|---|---|---|---|---|---|---|
| 1 Equip. Oper. (crane) | $46.50 | $372.00 | $69.45 | $555.60 | $40.42 | $61.08 |
| 1 Laborer | 34.35 | 274.80 | 52.70 | 421.60 | | |
| 1 Hyd. Excavator, 1 C.Y. | | 729.80 | | 802.78 | 45.61 | 50.17 |
| 16 L.H., Daily Totals | | $1376.60 | | $1779.98 | $86.04 | $111.25 |

| Crew No. | Bare Costs | | Incl. Subs O&P | | Cost Per Labor-Hour | |
|---|---|---|---|---|---|---|

| Crew B-12B | Hr. | Daily | Hr. | Daily | Bare Costs | Incl. O&P |
|---|---|---|---|---|---|---|
| 1 Equip. Oper. (crane) | $46.50 | $372.00 | $69.45 | $555.60 | $40.42 | $61.08 |
| 1 Laborer | 34.35 | 274.80 | 52.70 | 421.60 | | |
| 1 Hyd. Excavator, 1.5 C.Y. | | 947.40 | | 1042.14 | 59.21 | 65.13 |
| 16 L.H., Daily Totals | | $1594.20 | | $2019.34 | $99.64 | $126.21 |

| Crew B-12C | Hr. | Daily | Hr. | Daily | Bare Costs | Incl. O&P |
|---|---|---|---|---|---|---|
| 1 Equip. Oper. (crane) | $46.50 | $372.00 | $69.45 | $555.60 | $40.42 | $61.08 |
| 1 Laborer | 34.35 | 274.80 | 52.70 | 421.60 | | |
| 1 Hyd. Excavator, 2 C.Y. | | 1224.00 | | 1346.40 | 76.50 | 84.15 |
| 16 L.H., Daily Totals | | $1870.80 | | $2323.60 | $116.93 | $145.22 |

| Crew B-12D | Hr. | Daily | Hr. | Daily | Bare Costs | Incl. O&P |
|---|---|---|---|---|---|---|
| 1 Equip. Oper. (crane) | $46.50 | $372.00 | $69.45 | $555.60 | $40.42 | $61.08 |
| 1 Laborer | 34.35 | 274.80 | 52.70 | 421.60 | | |
| 1 Hyd. Excavator, 3.5 C.Y. | | 2171.00 | | 2388.10 | 135.69 | 149.26 |
| 16 L.H., Daily Totals | | $2817.80 | | $3365.30 | $176.11 | $210.33 |

| Crew B-12E | Hr. | Daily | Hr. | Daily | Bare Costs | Incl. O&P |
|---|---|---|---|---|---|---|
| 1 Equip. Oper. (crane) | $46.50 | $372.00 | $69.45 | $555.60 | $40.42 | $61.08 |
| 1 Laborer | 34.35 | 274.80 | 52.70 | 421.60 | | |
| 1 Hyd. Excavator, .5 C.Y. | | 364.60 | | 401.06 | 22.79 | 25.07 |
| 16 L.H., Daily Totals | | $1011.40 | | $1378.26 | $63.21 | $86.14 |

| Crew B-12F | Hr. | Daily | Hr. | Daily | Bare Costs | Incl. O&P |
|---|---|---|---|---|---|---|
| 1 Equip. Oper. (crane) | $46.50 | $372.00 | $69.45 | $555.60 | $40.42 | $61.08 |
| 1 Laborer | 34.35 | 274.80 | 52.70 | 421.60 | | |
| 1 Hyd. Excavator, .75 C.Y. | | 640.40 | | 704.44 | 40.02 | 44.03 |
| 16 L.H., Daily Totals | | $1287.20 | | $1681.64 | $80.45 | $105.10 |

| Crew B-12G | Hr. | Daily | Hr. | Daily | Bare Costs | Incl. O&P |
|---|---|---|---|---|---|---|
| 1 Equip. Oper. (crane) | $46.50 | $372.00 | $69.45 | $555.60 | $40.42 | $61.08 |
| 1 Laborer | 34.35 | 274.80 | 52.70 | 421.60 | | |
| 1 Crawler Crane, 15 Ton | | 657.90 | | 723.69 | | |
| 1 Clamshell Bucket, .5 C.Y. | | 37.40 | | 41.14 | 43.46 | 47.80 |
| 16 L.H., Daily Totals | | $1342.10 | | $1742.03 | $83.88 | $108.88 |

| Crew B-12H | Hr. | Daily | Hr. | Daily | Bare Costs | Incl. O&P |
|---|---|---|---|---|---|---|
| 1 Equip. Oper. (crane) | $46.50 | $372.00 | $69.45 | $555.60 | $40.42 | $61.08 |
| 1 Laborer | 34.35 | 274.80 | 52.70 | 421.60 | | |
| 1 Crawler Crane, 25 Ton | | 1123.00 | | 1235.30 | | |
| 1 Clamshell Bucket, 1 C.Y. | | 48.00 | | 52.80 | 73.19 | 80.51 |
| 16 L.H., Daily Totals | | $1817.80 | | $2265.30 | $113.61 | $141.58 |

| Crew B-12I | Hr. | Daily | Hr. | Daily | Bare Costs | Incl. O&P |
|---|---|---|---|---|---|---|
| 1 Equip. Oper. (crane) | $46.50 | $372.00 | $69.45 | $555.60 | $40.42 | $61.08 |
| 1 Laborer | 34.35 | 274.80 | 52.70 | 421.60 | | |
| 1 Crawler Crane, 20 Ton | | 841.90 | | 926.09 | | |
| 1 Dragline Bucket, .75 C.Y. | | 21.00 | | 23.10 | 53.93 | 59.32 |
| 16 L.H., Daily Totals | | $1509.70 | | $1926.39 | $94.36 | $120.40 |

| Crew B-12J | Hr. | Daily | Hr. | Daily | Bare Costs | Incl. O&P |
|---|---|---|---|---|---|---|
| 1 Equip. Oper. (crane) | $46.50 | $372.00 | $69.45 | $555.60 | $40.42 | $61.08 |
| 1 Laborer | 34.35 | 274.80 | 52.70 | 421.60 | | |
| 1 Gradall, 5/8 C.Y. | | 857.40 | | 943.14 | 53.59 | 58.95 |
| 16 L.H., Daily Totals | | $1504.20 | | $1920.34 | $94.01 | $120.02 |

| Crew B-12K | Hr. | Daily | Hr. | Daily | Bare Costs | Incl. O&P |
|---|---|---|---|---|---|---|
| 1 Equip. Oper. (crane) | $46.50 | $372.00 | $69.45 | $555.60 | $40.42 | $61.08 |
| 1 Laborer | 34.35 | 274.80 | 52.70 | 421.60 | | |
| 1 Gradall, 3 Ton, 1 C.Y. | | 1001.00 | | 1101.10 | 62.56 | 68.82 |
| 16 L.H., Daily Totals | | $1647.80 | | $2078.30 | $102.99 | $129.89 |

| Crew B-12L | Hr. | Daily | Hr. | Daily | Bare Costs | Incl. O&P |
|---|---|---|---|---|---|---|
| 1 Equip. Oper. (crane) | $46.50 | $372.00 | $69.45 | $555.60 | $40.42 | $61.08 |
| 1 Laborer | 34.35 | 274.80 | 52.70 | 421.60 | | |
| 1 Crawler Crane, 15 Ton | | 657.90 | | 723.69 | | |
| 1 F.E. Attachment, .5 C.Y. | | 56.00 | | 61.60 | 44.62 | 49.08 |
| 16 L.H., Daily Totals | | $1360.70 | | $1762.49 | $85.04 | $110.16 |

| Crew B-12M | Hr. | Daily | Hr. | Daily | Bare Costs | Incl. O&P |
|---|---|---|---|---|---|---|
| 1 Equip. Oper. (crane) | $46.50 | $372.00 | $69.45 | $555.60 | $40.42 | $61.08 |
| 1 Laborer | 34.35 | 274.80 | 52.70 | 421.60 | | |
| 1 Crawler Crane, 20 Ton | | 841.90 | | 926.09 | | |
| 1 F.E. Attachment, .75 C.Y. | | 61.40 | | 67.54 | 56.46 | 62.10 |
| 16 L.H., Daily Totals | | $1550.10 | | $1970.83 | $96.88 | $123.18 |

| Crew B-12N | Hr. | Daily | Hr. | Daily | Bare Costs | Incl. O&P |
|---|---|---|---|---|---|---|
| 1 Equip. Oper. (crane) | $46.50 | $372.00 | $69.45 | $555.60 | $40.42 | $61.08 |
| 1 Laborer | 34.35 | 274.80 | 52.70 | 421.60 | | |
| 1 Crawler Crane, 25 Ton | | 1123.00 | | 1235.30 | | |
| 1 F.E. Attachment, 1 C.Y. | | 67.20 | | 73.92 | 74.39 | 81.83 |
| 16 L.H., Daily Totals | | $1837.00 | | $2286.42 | $114.81 | $142.90 |

| Crew B-12O | Hr. | Daily | Hr. | Daily | Bare Costs | Incl. O&P |
|---|---|---|---|---|---|---|
| 1 Equip. Oper. (crane) | $46.50 | $372.00 | $69.45 | $555.60 | $40.42 | $61.08 |
| 1 Laborer | 34.35 | 274.80 | 52.70 | 421.60 | | |
| 1 Crawler Crane, 40 Ton | | 1128.00 | | 1240.80 | | |
| 1 F.E. Attachment, 1.5 C.Y. | | 77.40 | | 85.14 | 75.34 | 82.87 |
| 16 L.H., Daily Totals | | $1852.20 | | $2303.14 | $115.76 | $143.95 |

| Crew B-12P | Hr. | Daily | Hr. | Daily | Bare Costs | Incl. O&P |
|---|---|---|---|---|---|---|
| 1 Equip. Oper. (crane) | $46.50 | $372.00 | $69.45 | $555.60 | $40.42 | $61.08 |
| 1 Laborer | 34.35 | 274.80 | 52.70 | 421.60 | | |
| 1 Crawler Crane, 40 Ton | | 1128.00 | | 1240.80 | | |
| 1 Dragline Bucket, 1.5 C.Y. | | 34.20 | | 37.62 | 72.64 | 79.90 |
| 16 L.H., Daily Totals | | $1809.00 | | $2255.62 | $113.06 | $140.98 |

| Crew B-12Q | Hr. | Daily | Hr. | Daily | Bare Costs | Incl. O&P |
|---|---|---|---|---|---|---|
| 1 Equip. Oper. (crane) | $46.50 | $372.00 | $69.45 | $555.60 | $40.42 | $61.08 |
| 1 Laborer | 34.35 | 274.80 | 52.70 | 421.60 | | |
| 1 Hyd. Excavator, 5/8 C.Y. | | 530.00 | | 583.00 | 33.13 | 36.44 |
| 16 L.H., Daily Totals | | $1176.80 | | $1560.20 | $73.55 | $97.51 |

| Crew B-12S | Hr. | Daily | Hr. | Daily | Bare Costs | Incl. O&P |
|---|---|---|---|---|---|---|
| 1 Equip. Oper. (crane) | $46.50 | $372.00 | $69.45 | $555.60 | $40.42 | $61.08 |
| 1 Laborer | 34.35 | 274.80 | 52.70 | 421.60 | | |
| 1 Hyd. Excavator, 2.5 C.Y. | | 1613.00 | | 1774.30 | 100.81 | 110.89 |
| 16 L.H., Daily Totals | | $2259.80 | | $2751.50 | $141.24 | $171.97 |

| Crew B-12T | Hr. | Daily | Hr. | Daily | Bare Costs | Incl. O&P |
|---|---|---|---|---|---|---|
| 1 Equip. Oper. (crane) | $46.50 | $372.00 | $69.45 | $555.60 | $40.42 | $61.08 |
| 1 Laborer | 34.35 | 274.80 | 52.70 | 421.60 | | |
| 1 Crawler Crane, 75 Ton | | 1446.00 | | 1590.60 | | |
| 1 F.E. Attachment, 3 C.Y. | | 100.80 | | 110.88 | 96.67 | 106.34 |
| 16 L.H., Daily Totals | | $2193.60 | | $2678.68 | $137.10 | $167.42 |

# Crews

## Crew B-12V

| Crew B-12V | Bare Costs Hr. | Daily | Incl. Subs O&P Hr. | Daily | Bare Costs | Incl. O&P |
|---|---|---|---|---|---|---|
| 1 Equip. Oper. (crane) | $46.50 | $372.00 | $69.45 | $555.60 | $40.42 | $61.08 |
| 1 Laborer | 34.35 | 274.80 | 52.70 | 421.60 | | |
| 1 Crawler Crane, 75 Ton | | 1446.00 | | 1590.60 | | |
| 1 Dragline Bucket, 3 C.Y. | | 52.80 | | 58.08 | 93.67 | 103.04 |
| 16 L.H., Daily Totals | | $2145.60 | | $2625.88 | $134.10 | $164.12 |

## Crew B-12Y

| Crew B-12Y | Bare Costs Hr. | Daily | Incl. Subs O&P Hr. | Daily | Bare Costs | Incl. O&P |
|---|---|---|---|---|---|---|
| 1 Equip. Oper. (crane) | $46.50 | $372.00 | $69.45 | $555.60 | $38.40 | $58.28 |
| 2 Laborers | 34.35 | 549.60 | 52.70 | 843.20 | | |
| 1 Hyd. Excavator, 3.5 C.Y. | | 2171.00 | | 2388.10 | 90.46 | 99.50 |
| 24 L.H., Daily Totals | | $3092.60 | | $3786.90 | $128.86 | $157.79 |

## Crew B-12Z

| Crew B-12Z | Bare Costs Hr. | Daily | Incl. Subs O&P Hr. | Daily | Bare Costs | Incl. O&P |
|---|---|---|---|---|---|---|
| 1 Equip. Oper. (crane) | $46.50 | $372.00 | $69.45 | $555.60 | $38.40 | $58.28 |
| 2 Laborers | 34.35 | 549.60 | 52.70 | 843.20 | | |
| 1 Hyd. Excavator, 2.5 C.Y. | | 1613.00 | | 1774.30 | 67.21 | 73.93 |
| 24 L.H., Daily Totals | | $2534.60 | | $3173.10 | $105.61 | $132.21 |

## Crew B-13

| Crew B-13 | Bare Costs Hr. | Daily | Incl. Subs O&P Hr. | Daily | Bare Costs | Incl. O&P |
|---|---|---|---|---|---|---|
| 1 Labor Foreman (outside) | $36.35 | $290.80 | $55.75 | $446.00 | $37.22 | $56.60 |
| 4 Laborers | 34.35 | 1099.20 | 52.70 | 1686.40 | | |
| 1 Equip. Oper. (crane) | 46.50 | 372.00 | 69.45 | 555.60 | | |
| 1 Equip. Oper. Oiler | 40.30 | 322.40 | 60.20 | 481.60 | | |
| 1 Hyd. Crane, 25 Ton | | 748.60 | | 823.46 | 13.37 | 14.70 |
| 56 L.H., Daily Totals | | $2833.00 | | $3993.06 | $50.59 | $71.30 |

## Crew B-13A

| Crew B-13A | Bare Costs Hr. | Daily | Incl. Subs O&P Hr. | Daily | Bare Costs | Incl. O&P |
|---|---|---|---|---|---|---|
| 1 Foreman | $36.35 | $290.80 | $55.75 | $446.00 | $37.79 | $57.31 |
| 2 Laborers | 34.35 | 549.60 | 52.70 | 843.20 | | |
| 2 Equipment Operators (med.) | 45.35 | 725.60 | 67.75 | 1084.00 | | |
| 2 Truck Drivers (heavy) | 34.40 | 550.40 | 52.25 | 836.00 | | |
| 1 Crawler Crane, 75 Ton | | 1446.00 | | 1590.60 | | |
| 1 Crawler Loader, 4 C.Y. | | 1432.00 | | 1575.20 | | |
| 2 Dump Trucks, 8 C.Y., 220 H.P. | | 630.80 | | 693.88 | 62.66 | 68.92 |
| 56 L.H., Daily Totals | | $5625.20 | | $7068.88 | $100.45 | $126.23 |

## Crew B-13B

| Crew B-13B | Bare Costs Hr. | Daily | Incl. Subs O&P Hr. | Daily | Bare Costs | Incl. O&P |
|---|---|---|---|---|---|---|
| 1 Labor Foreman (outside) | $36.35 | $290.80 | $55.75 | $446.00 | $37.22 | $56.60 |
| 4 Laborers | 34.35 | 1099.20 | 52.70 | 1686.40 | | |
| 1 Equip. Oper. (crane) | 46.50 | 372.00 | 69.45 | 555.60 | | |
| 1 Equip. Oper. Oiler | 40.30 | 322.40 | 60.20 | 481.60 | | |
| 1 Hyd. Crane, 55 Ton | | 1208.00 | | 1328.80 | 21.57 | 23.73 |
| 56 L.H., Daily Totals | | $3292.40 | | $4498.40 | $58.79 | $80.33 |

## Crew B-13C

| Crew B-13C | Bare Costs Hr. | Daily | Incl. Subs O&P Hr. | Daily | Bare Costs | Incl. O&P |
|---|---|---|---|---|---|---|
| 1 Labor Foreman (outside) | $36.35 | $290.80 | $55.75 | $446.00 | $37.22 | $56.60 |
| 4 Laborers | 34.35 | 1099.20 | 52.70 | 1686.40 | | |
| 1 Equip. Oper. (crane) | 46.50 | 372.00 | 69.45 | 555.60 | | |
| 1 Equip. Oper. Oiler | 40.30 | 322.40 | 60.20 | 481.60 | | |
| 1 Crawler Crane, 100 Ton | | 1653.00 | | 1818.30 | 29.52 | 32.47 |
| 56 L.H., Daily Totals | | $3737.40 | | $4987.90 | $66.74 | $89.07 |

## Crew B-13D

| Crew B-13D | Bare Costs Hr. | Daily | Incl. Subs O&P Hr. | Daily | Bare Costs | Incl. O&P |
|---|---|---|---|---|---|---|
| 1 Laborer | $34.35 | $274.80 | $52.70 | $421.60 | $40.42 | $61.08 |
| 1 Equip. Oper. (crane) | 46.50 | 372.00 | 69.45 | 555.60 | | |
| 1 Hyd. Excavator, 1 C.Y. | | 729.80 | | 802.78 | | |
| 1 Trench Box | | 113.40 | | 124.74 | 52.70 | 57.97 |
| 16 L.H., Daily Totals | | $1490.00 | | $1904.72 | $93.13 | $119.05 |

## Crew B-13E

| Crew B-13E | Bare Costs Hr. | Daily | Incl. Subs O&P Hr. | Daily | Bare Costs | Incl. O&P |
|---|---|---|---|---|---|---|
| 1 Laborer | $34.35 | $274.80 | $52.70 | $421.60 | $40.42 | $61.08 |
| 1 Equip. Oper. (crane) | 46.50 | 372.00 | 69.45 | 555.60 | | |
| 1 Hyd. Excavator, 1.5 C.Y. | | 947.40 | | 1042.14 | | |
| 1 Trench Box | | 113.40 | | 124.74 | 66.30 | 72.93 |
| 16 L.H., Daily Totals | | $1707.60 | | $2144.08 | $106.72 | $134.01 |

## Crew B-13F

| Crew B-13F | Bare Costs Hr. | Daily | Incl. Subs O&P Hr. | Daily | Bare Costs | Incl. O&P |
|---|---|---|---|---|---|---|
| 1 Laborer | $34.35 | $274.80 | $52.70 | $421.60 | $40.42 | $61.08 |
| 1 Equip. Oper. (crane) | 46.50 | 372.00 | 69.45 | 555.60 | | |
| 1 Hyd. Excavator, 3.5 C.Y. | | 2171.00 | | 2388.10 | | |
| 1 Trench Box | | 113.40 | | 124.74 | 142.78 | 157.05 |
| 16 L.H., Daily Totals | | $2931.20 | | $3490.04 | $183.20 | $218.13 |

## Crew B-13G

| Crew B-13G | Bare Costs Hr. | Daily | Incl. Subs O&P Hr. | Daily | Bare Costs | Incl. O&P |
|---|---|---|---|---|---|---|
| 1 Laborer | $34.35 | $274.80 | $52.70 | $421.60 | $40.42 | $61.08 |
| 1 Equip. Oper. (crane) | 46.50 | 372.00 | 69.45 | 555.60 | | |
| 1 Hyd. Excavator, .75 C.Y. | | 640.40 | | 704.44 | | |
| 1 Trench Box | | 113.40 | | 124.74 | 47.11 | 51.82 |
| 16 L.H., Daily Totals | | $1400.60 | | $1806.38 | $87.54 | $112.90 |

## Crew B-13H

| Crew B-13H | Bare Costs Hr. | Daily | Incl. Subs O&P Hr. | Daily | Bare Costs | Incl. O&P |
|---|---|---|---|---|---|---|
| 1 Laborer | $34.35 | $274.80 | $52.70 | $421.60 | $40.42 | $61.08 |
| 1 Equip. Oper. (crane) | 46.50 | 372.00 | 69.45 | 555.60 | | |
| 1 Gradall, 5/8 C.Y. | | 857.40 | | 943.14 | | |
| 1 Trench Box | | 113.40 | | 124.74 | 60.67 | 66.74 |
| 16 L.H., Daily Totals | | $1617.60 | | $2045.08 | $101.10 | $127.82 |

## Crew B-13I

| Crew B-13I | Bare Costs Hr. | Daily | Incl. Subs O&P Hr. | Daily | Bare Costs | Incl. O&P |
|---|---|---|---|---|---|---|
| 1 Laborer | $34.35 | $274.80 | $52.70 | $421.60 | $40.42 | $61.08 |
| 1 Equip. Oper. (crane) | 46.50 | 372.00 | 69.45 | 555.60 | | |
| 1 Gradall, 3 Ton, 1 C.Y. | | 1001.00 | | 1101.10 | | |
| 1 Trench Box | | 113.40 | | 124.74 | 69.65 | 76.61 |
| 16 L.H., Daily Totals | | $1761.20 | | $2203.04 | $110.08 | $137.69 |

## Crew B-13J

| Crew B-13J | Bare Costs Hr. | Daily | Incl. Subs O&P Hr. | Daily | Bare Costs | Incl. O&P |
|---|---|---|---|---|---|---|
| 1 Laborer | $34.35 | $274.80 | $52.70 | $421.60 | $40.42 | $61.08 |
| 1 Equip. Oper. (crane) | 46.50 | 372.00 | 69.45 | 555.60 | | |
| 1 Hyd. Excavator, 2.5 C.Y. | | 1613.00 | | 1774.30 | | |
| 1 Trench Box | | 113.40 | | 124.74 | 107.90 | 118.69 |
| 16 L.H., Daily Totals | | $2373.20 | | $2876.24 | $148.32 | $179.76 |

## Crew B-14

| Crew B-14 | Bare Costs Hr. | Daily | Incl. Subs O&P Hr. | Daily | Bare Costs | Incl. O&P |
|---|---|---|---|---|---|---|
| 1 Labor Foreman (outside) | $36.35 | $290.80 | $55.75 | $446.00 | $36.23 | $55.28 |
| 4 Laborers | 34.35 | 1099.20 | 52.70 | 1686.40 | | |
| 1 Equip. Oper. (light) | 43.60 | 348.80 | 65.15 | 521.20 | | |
| 1 Backhoe Loader, 48 H.P. | | 320.40 | | 352.44 | 6.67 | 7.34 |
| 48 L.H., Daily Totals | | $2059.20 | | $3006.04 | $42.90 | $62.63 |

## Crew B-14A

| Crew B-14A | Bare Costs Hr. | Daily | Incl. Subs O&P Hr. | Daily | Bare Costs | Incl. O&P |
|---|---|---|---|---|---|---|
| 1 Equip. Oper. (crane) | $46.50 | $372.00 | $69.45 | $555.60 | $42.45 | $63.87 |
| .5 Laborer | 34.35 | 137.40 | 52.70 | 210.80 | | |
| 1 Hyd. Excavator, 4.5 C.Y. | | 2669.00 | | 2935.90 | 222.42 | 244.66 |
| 12 L.H., Daily Totals | | $3178.40 | | $3702.30 | $264.87 | $308.52 |

## Crew B-14B

| Crew B-14B | Bare Costs Hr. | Daily | Incl. Subs O&P Hr. | Daily | Bare Costs | Incl. O&P |
|---|---|---|---|---|---|---|
| 1 Equip. Oper. (crane) | $46.50 | $372.00 | $69.45 | $555.60 | $42.45 | $63.87 |
| .5 Laborer | 34.35 | 137.40 | 52.70 | 210.80 | | |
| 1 Hyd. Excavator, 6 C.Y. | | 3101.00 | | 3411.10 | 258.42 | 284.26 |
| 12 L.H., Daily Totals | | $3610.40 | | $4177.50 | $300.87 | $348.13 |

| Crew No. | Bare Costs | | Incl. Subs O&P | | Cost Per Labor-Hour | |
|---|---|---|---|---|---|---|

| **Crew B-14C** | Hr. | Daily | Hr. | Daily | Bare Costs | Incl. O&P |
|---|---|---|---|---|---|---|
| 1 Equip. Oper. (crane) | $46.50 | $372.00 | $69.45 | $555.60 | $42.45 | $63.87 |
| .5 Laborer | 34.35 | 137.40 | 52.70 | 210.80 | | |
| 1 Hyd. Excavator, 7 C.Y. | | 3179.00 | | 3496.90 | 264.92 | 291.41 |
| 12 L.H., Daily Totals | | $3688.40 | | $4263.30 | $307.37 | $355.27 |

| **Crew B-14F** | Hr. | Daily | Hr. | Daily | Bare Costs | Incl. O&P |
|---|---|---|---|---|---|---|
| 1 Equip. Oper. (crane) | $46.50 | $372.00 | $69.45 | $555.60 | $42.45 | $63.87 |
| .5 Laborer | 34.35 | 137.40 | 52.70 | 210.80 | | |
| 1 Hyd. Shovel, 7 C.Y. | | 3109.00 | | 3419.90 | 259.08 | 284.99 |
| 12 L.H., Daily Totals | | $3618.40 | | $4186.30 | $301.53 | $348.86 |

| **Crew B-14G** | Hr. | Daily | Hr. | Daily | Bare Costs | Incl. O&P |
|---|---|---|---|---|---|---|
| 1 Equip. Oper. (crane) | $46.50 | $372.00 | $69.45 | $555.60 | $42.45 | $63.87 |
| .5 Laborer | 34.35 | 137.40 | 52.70 | 210.80 | | |
| 1 Hyd. Shovel, 12 C.Y. | | 4046.00 | | 4450.60 | 337.17 | 370.88 |
| 12 L.H., Daily Totals | | $4555.40 | | $5217.00 | $379.62 | $434.75 |

| **Crew B-14J** | Hr. | Daily | Hr. | Daily | Bare Costs | Incl. O&P |
|---|---|---|---|---|---|---|
| 1 Equip. Oper. (med.) | $45.35 | $362.80 | $67.75 | $542.00 | $41.68 | $62.73 |
| .5 Laborer | 34.35 | 137.40 | 52.70 | 210.80 | | |
| 1 F.E. Loader, 8 C.Y. | | 1811.00 | | 1992.10 | 150.92 | 166.01 |
| 12 L.H., Daily Totals | | $2311.20 | | $2744.90 | $192.60 | $228.74 |

| **Crew B-14K** | Hr. | Daily | Hr. | Daily | Bare Costs | Incl. O&P |
|---|---|---|---|---|---|---|
| 1 Equip. Oper. (med.) | $45.35 | $362.80 | $67.75 | $542.00 | $41.68 | $62.73 |
| .5 Laborer | 34.35 | 137.40 | 52.70 | 210.80 | | |
| 1 F.E. Loader, 10 C.Y. | | 2476.00 | | 2723.60 | 206.33 | 226.97 |
| 12 L.H., Daily Totals | | $2976.20 | | $3476.40 | $248.02 | $289.70 |

| **Crew B-15** | Hr. | Daily | Hr. | Daily | Bare Costs | Incl. O&P |
|---|---|---|---|---|---|---|
| 1 Equipment Oper. (med.) | $45.35 | $362.80 | $67.75 | $542.00 | $37.52 | $56.74 |
| .5 Laborer | 34.35 | 137.40 | 52.70 | 210.80 | | |
| 2 Truck Drivers (heavy) | 34.40 | 550.40 | 52.25 | 836.00 | | |
| 2 Dump Trucks, 12 C.Y., 400 H.P. | | 1012.00 | | 1113.20 | | |
| 1 Dozer, 200 H.P. | | 1051.00 | | 1156.10 | 73.68 | 81.05 |
| 28 L.H., Daily Totals | | $3113.60 | | $3858.10 | $111.20 | $137.79 |

| **Crew B-16** | Hr. | Daily | Hr. | Daily | Bare Costs | Incl. O&P |
|---|---|---|---|---|---|---|
| 1 Labor Foreman (outside) | $36.35 | $290.80 | $55.75 | $446.00 | $34.86 | $53.35 |
| 2 Laborers | 34.35 | 549.60 | 52.70 | 843.20 | | |
| 1 Truck Driver (heavy) | 34.40 | 275.20 | 52.25 | 418.00 | | |
| 1 Dump Truck, 12 C.Y., 400 H.P. | | 506.00 | | 556.60 | 15.81 | 17.39 |
| 32 L.H., Daily Totals | | $1621.60 | | $2263.80 | $50.67 | $70.74 |

| **Crew B-17** | Hr. | Daily | Hr. | Daily | Bare Costs | Incl. O&P |
|---|---|---|---|---|---|---|
| 2 Laborers | $34.35 | $549.60 | $52.70 | $843.20 | $36.67 | $55.70 |
| 1 Equip. Oper. (light) | 43.60 | 348.80 | 65.15 | 521.20 | | |
| 1 Truck Driver (heavy) | 34.40 | 275.20 | 52.25 | 418.00 | | |
| 1 Backhoe Loader, 48 H.P. | | 320.40 | | 352.44 | | |
| 1 Dump Truck, 8 C.Y., 220 H.P. | | 315.40 | | 346.94 | 19.87 | 21.86 |
| 32 L.H., Daily Totals | | $1809.40 | | $2481.78 | $56.54 | $77.56 |

| **Crew B-17A** | Hr. | Daily | Hr. | Daily | Bare Costs | Incl. O&P |
|---|---|---|---|---|---|---|
| 2 Laborer Foremen | $36.35 | $581.60 | $55.75 | $892.00 | $36.96 | $56.70 |
| 6 Laborers | 34.35 | 1648.80 | 52.70 | 2529.60 | | |
| 1 Skilled Worker Foreman | 46.40 | 371.20 | 71.20 | 569.60 | | |
| 1 Skilled Worker | 44.40 | 355.20 | 68.10 | 544.80 | | |
| 80 L.H., Daily Totals | | $2956.80 | | $4536.00 | $36.96 | $56.70 |

| **Crew B-18** | Hr. | Daily | Hr. | Daily | Bare Costs | Incl. O&P |
|---|---|---|---|---|---|---|
| 1 Labor Foreman (outside) | $36.35 | $290.80 | $55.75 | $446.00 | $35.02 | $53.72 |
| 2 Laborers | 34.35 | 549.60 | 52.70 | 843.20 | | |
| 1 Vibratory Plate, gas, 21" | | 39.00 | | 42.90 | 1.63 | 1.79 |
| 24 L.H., Daily Totals | | $879.40 | | $1332.10 | $36.64 | $55.50 |

| **Crew B-19** | Hr. | Daily | Hr. | Daily | Bare Costs | Incl. O&P |
|---|---|---|---|---|---|---|
| 1 Pile Driver Foreman | $44.00 | $352.00 | $70.55 | $564.40 | $43.16 | $67.38 |
| 4 Pile Drivers | 42.00 | 1344.00 | 67.35 | 2155.20 | | |
| 2 Equip. Oper. (crane) | 46.50 | 744.00 | 69.45 | 1111.20 | | |
| 1 Equip. Oper. Oiler | 40.30 | 322.40 | 60.20 | 481.60 | | |
| 1 Crawler Crane, 40 Ton | | 1128.00 | | 1240.80 | | |
| 1 Lead, 90' high | | 127.20 | | 139.92 | | |
| 1 Hammer, Diesel, 22k ft-lb | | 592.40 | | 651.64 | 28.87 | 31.76 |
| 64 L.H., Daily Totals | | $4610.00 | | $6344.76 | $72.03 | $99.14 |

| **Crew B-19A** | Hr. | Daily | Hr. | Daily | Bare Costs | Incl. O&P |
|---|---|---|---|---|---|---|
| 1 Pile Driver Foreman | $44.00 | $352.00 | $70.55 | $564.40 | $43.16 | $67.38 |
| 4 Pile Drivers | 42.00 | 1344.00 | 67.35 | 2155.20 | | |
| 2 Equip. Oper. (crane) | 46.50 | 744.00 | 69.45 | 1111.20 | | |
| 1 Equip. Oper. Oiler | 40.30 | 322.40 | 60.20 | 481.60 | | |
| 1 Crawler Crane, 75 Ton | | 1446.00 | | 1590.60 | | |
| 1 Lead, 90' high | | 127.20 | | 139.92 | | |
| 1 Hammer, Diesel, 41k ft-lb | | 677.00 | | 744.70 | 35.16 | 38.68 |
| 64 L.H., Daily Totals | | $5012.60 | | $6787.62 | $78.32 | $106.06 |

| **Crew B-20** | Hr. | Daily | Hr. | Daily | Bare Costs | Incl. O&P |
|---|---|---|---|---|---|---|
| 1 Labor Foreman (out) | $36.35 | $290.80 | $55.75 | $446.00 | $38.37 | $58.85 |
| 1 Skilled Worker | 44.40 | 355.20 | 68.10 | 544.80 | | |
| 1 Laborer | 34.35 | 274.80 | 52.70 | 421.60 | | |
| 24 L.H., Daily Totals | | $920.80 | | $1412.40 | $38.37 | $58.85 |

| **Crew B-20A** | Hr. | Daily | Hr. | Daily | Bare Costs | Incl. O&P |
|---|---|---|---|---|---|---|
| 1 Labor Foreman | $36.35 | $290.80 | $55.75 | $446.00 | $41.80 | $63.13 |
| 1 Laborer | 34.35 | 274.80 | 52.70 | 421.60 | | |
| 1 Plumber | 53.60 | 428.80 | 80.00 | 640.00 | | |
| 1 Plumber Apprentice | 42.90 | 343.20 | 64.05 | 512.40 | | |
| 32 L.H., Daily Totals | | $1337.60 | | $2020.00 | $41.80 | $63.13 |

| **Crew B-21** | Hr. | Daily | Hr. | Daily | Bare Costs | Incl. O&P |
|---|---|---|---|---|---|---|
| 1 Labor Foreman (out) | $36.35 | $290.80 | $55.75 | $446.00 | $39.53 | $60.36 |
| 1 Skilled Worker | 44.40 | 355.20 | 68.10 | 544.80 | | |
| 1 Laborer | 34.35 | 274.80 | 52.70 | 421.60 | | |
| .5 Equip. Oper. (crane) | 46.50 | 186.00 | 69.45 | 277.80 | | |
| .5 S.P. Crane, 4x4, 5 Ton | | 128.90 | | 141.79 | 4.60 | 5.06 |
| 28 L.H., Daily Totals | | $1235.70 | | $1831.99 | $44.13 | $65.43 |

| **Crew B-21A** | Hr. | Daily | Hr. | Daily | Bare Costs | Incl. O&P |
|---|---|---|---|---|---|---|
| 1 Labor Foreman | $36.35 | $290.80 | $55.75 | $446.00 | $42.74 | $64.39 |
| 1 Laborer | 34.35 | 274.80 | 52.70 | 421.60 | | |
| 1 Plumber | 53.60 | 428.80 | 80.00 | 640.00 | | |
| 1 Plumber Apprentice | 42.90 | 343.20 | 64.05 | 512.40 | | |
| 1 Equip. Oper. (crane) | 46.50 | 372.00 | 69.45 | 555.60 | | |
| 1 S.P. Crane, 4x4, 12 Ton | | 434.40 | | 477.84 | 10.86 | 11.95 |
| 40 L.H., Daily Totals | | $2144.00 | | $3053.44 | $53.60 | $76.34 |

| **Crew B-21B** | Hr. | Daily | Hr. | Daily | Bare Costs | Incl. O&P |
|---|---|---|---|---|---|---|
| 1 Laborer Foreman | $36.35 | $290.80 | $55.75 | $446.00 | $37.18 | $56.66 |
| 3 Laborers | 34.35 | 824.40 | 52.70 | 1264.80 | | |
| 1 Equip. Oper. (crane) | 46.50 | 372.00 | 69.45 | 555.60 | | |
| 1 Hyd. Crane, 12 Ton | | 655.60 | | 721.16 | 16.39 | 18.03 |
| 40 L.H., Daily Totals | | $2142.80 | | $2987.56 | $53.57 | $74.69 |

## Crew B-21C

| | Bare Costs Hr. | Bare Costs Daily | Incl. Subs O&P Hr. | Incl. Subs O&P Daily | Cost Per L.H. Bare Costs | Cost Per L.H. Incl. O&P |
|---|---|---|---|---|---|---|
| 1 Laborer Foreman | $36.35 | $290.80 | $55.75 | $446.00 | $37.22 | $56.60 |
| 4 Laborers | 34.35 | 1099.20 | 52.70 | 1686.40 | | |
| 1 Equip. Oper. (crane) | 46.50 | 372.00 | 69.45 | 555.60 | | |
| 1 Equip. Oper. Oiler | 40.30 | 322.40 | 60.20 | 481.60 | | |
| 2 Cutting Torches | | 36.00 | | 39.60 | | |
| 2 Set of Gases | | 165.60 | | 182.16 | | |
| 1 Lattice Boom Crane, 90 Ton | | 1622.00 | | 1784.20 | 32.56 | 35.82 |
| 56 L.H., Daily Totals | | $3908.00 | | $5175.56 | $69.79 | $92.42 |

## Crew B-22

| | Bare Costs Hr. | Bare Costs Daily | Incl. Subs O&P Hr. | Incl. Subs O&P Daily | Cost Per L.H. Bare Costs | Cost Per L.H. Incl. O&P |
|---|---|---|---|---|---|---|
| 1 Labor Foreman (outside) | $36.35 | $290.80 | $55.75 | $446.00 | $39.99 | $60.97 |
| 1 Skilled Worker | 44.40 | 355.20 | 68.10 | 544.80 | | |
| 1 Laborer | 34.35 | 274.80 | 52.70 | 421.60 | | |
| .75 Equip. Oper. (crane) | 46.50 | 279.00 | 69.45 | 416.70 | | |
| .75 S.P. Crane, 4x4, 5 Ton | | 193.35 | | 212.69 | 6.45 | 7.09 |
| 30 L.H., Daily Totals | | $1393.15 | | $2041.79 | $46.44 | $68.06 |

## Crew B-22A

| | Bare Costs Hr. | Bare Costs Daily | Incl. Subs O&P Hr. | Incl. Subs O&P Daily | Cost Per L.H. Bare Costs | Cost Per L.H. Incl. O&P |
|---|---|---|---|---|---|---|
| 1 Labor Foreman (outside) | $36.35 | $290.80 | $55.75 | $446.00 | $39.19 | $59.74 |
| 1 Skilled Worker | 44.40 | 355.20 | 68.10 | 544.80 | | |
| 2 Laborers | 34.35 | 549.60 | 52.70 | 843.20 | | |
| 1 Equipment Oper. (crane) | 46.50 | 372.00 | 69.45 | 555.60 | | |
| 1 S.P. Crane, 4x4, 5 Ton | | 257.80 | | 283.58 | | |
| 1 Butt Fusion Mach., 4"-12" diam. | | 376.45 | | 414.10 | 15.86 | 17.44 |
| 40 L.H., Daily Totals | | $2201.85 | | $3087.28 | $55.05 | $77.18 |

## Crew B-22B

| | Bare Costs Hr. | Bare Costs Daily | Incl. Subs O&P Hr. | Incl. Subs O&P Daily | Cost Per L.H. Bare Costs | Cost Per L.H. Incl. O&P |
|---|---|---|---|---|---|---|
| 1 Labor Foreman (outside) | $36.35 | $290.80 | $55.75 | $446.00 | $39.19 | $59.74 |
| 1 Skilled Worker | 44.40 | 355.20 | 68.10 | 544.80 | | |
| 2 Laborers | 34.35 | 549.60 | 52.70 | 843.20 | | |
| 1 Equip. Oper. (crane) | 46.50 | 372.00 | 69.45 | 555.60 | | |
| 1 S.P. Crane, 4x4, 5 Ton | | 257.80 | | 283.58 | | |
| 1 Butt Fusion Mach., 8"-24" diam. | | 806.15 | | 886.76 | 26.60 | 29.26 |
| 40 L.H., Daily Totals | | $2631.55 | | $3559.95 | $65.79 | $89.00 |

## Crew B-22C

| | Bare Costs Hr. | Bare Costs Daily | Incl. Subs O&P Hr. | Incl. Subs O&P Daily | Cost Per L.H. Bare Costs | Cost Per L.H. Incl. O&P |
|---|---|---|---|---|---|---|
| 1 Skilled Worker | $44.40 | $355.20 | $68.10 | $544.80 | $39.38 | $60.40 |
| 1 Laborer | 34.35 | 274.80 | 52.70 | 421.60 | | |
| 1 Butt Fusion Mach., 2"-8" diam. | | 120.50 | | 132.55 | 7.53 | 8.28 |
| 16 L.H., Daily Totals | | $750.50 | | $1098.95 | $46.91 | $68.68 |

## Crew B-23

| | Bare Costs Hr. | Bare Costs Daily | Incl. Subs O&P Hr. | Incl. Subs O&P Daily | Cost Per L.H. Bare Costs | Cost Per L.H. Incl. O&P |
|---|---|---|---|---|---|---|
| 1 Labor Foreman (outside) | $36.35 | $290.80 | $55.75 | $446.00 | $34.75 | $53.31 |
| 4 Laborers | 34.35 | 1099.20 | 52.70 | 1686.40 | | |
| 1 Drill Rig, Truck-Mounted | | 2486.00 | | 2734.60 | | |
| 1 Flatbed Truck, Gas, 3 Ton | | 217.00 | | 238.70 | 67.58 | 74.33 |
| 40 L.H., Daily Totals | | $4093.00 | | $5105.70 | $102.33 | $127.64 |

## Crew B-23A

| | Bare Costs Hr. | Bare Costs Daily | Incl. Subs O&P Hr. | Incl. Subs O&P Daily | Cost Per L.H. Bare Costs | Cost Per L.H. Incl. O&P |
|---|---|---|---|---|---|---|
| 1 Labor Foreman (outside) | $36.35 | $290.80 | $55.75 | $446.00 | $38.68 | $58.73 |
| 1 Laborer | 34.35 | 274.80 | 52.70 | 421.60 | | |
| 1 Equip. Operator (med.) | 45.35 | 362.80 | 67.75 | 542.00 | | |
| 1 Drill Rig, Truck-Mounted | | 2486.00 | | 2734.60 | | |
| 1 Pickup Truck, 3/4 Ton | | 104.60 | | 115.06 | 107.94 | 118.74 |
| 24 L.H., Daily Totals | | $3519.00 | | $4259.26 | $146.63 | $177.47 |

## Crew B-23B

| | Bare Costs Hr. | Bare Costs Daily | Incl. Subs O&P Hr. | Incl. Subs O&P Daily | Cost Per L.H. Bare Costs | Cost Per L.H. Incl. O&P |
|---|---|---|---|---|---|---|
| 1 Labor Foreman (outside) | $36.35 | $290.80 | $55.75 | $446.00 | $38.68 | $58.73 |
| 1 Laborer | 34.35 | 274.80 | 52.70 | 421.60 | | |
| 1 Equip. Operator (med.) | 45.35 | 362.80 | 67.75 | 542.00 | | |
| 1 Drill Rig, Truck-Mounted | | 2486.00 | | 2734.60 | | |
| 1 Pickup Truck, 3/4 Ton | | 104.60 | | 115.06 | | |
| 1 Centr. Water Pump, 6" | | 291.60 | | 320.76 | 120.09 | 132.10 |
| 24 L.H., Daily Totals | | $3810.60 | | $4580.02 | $158.78 | $190.83 |

## Crew B-24

| | Bare Costs Hr. | Bare Costs Daily | Incl. Subs O&P Hr. | Incl. Subs O&P Daily | Cost Per L.H. Bare Costs | Cost Per L.H. Incl. O&P |
|---|---|---|---|---|---|---|
| 1 Cement Finisher | $40.85 | $326.80 | $59.50 | $476.00 | $39.42 | $59.42 |
| 1 Laborer | 34.35 | 274.80 | 52.70 | 421.60 | | |
| 1 Carpenter | 43.05 | 344.40 | 66.05 | 528.40 | | |
| 24 L.H., Daily Totals | | $946.00 | | $1426.00 | $39.42 | $59.42 |

## Crew B-25

| | Bare Costs Hr. | Bare Costs Daily | Incl. Subs O&P Hr. | Incl. Subs O&P Daily | Cost Per L.H. Bare Costs | Cost Per L.H. Incl. O&P |
|---|---|---|---|---|---|---|
| 1 Labor Foreman | $36.35 | $290.80 | $55.75 | $446.00 | $37.53 | $57.08 |
| 7 Laborers | 34.35 | 1923.60 | 52.70 | 2951.20 | | |
| 3 Equip. Oper. (med.) | 45.35 | 1088.40 | 67.75 | 1626.00 | | |
| 1 Asphalt Paver, 130 H.P. | | 1949.00 | | 2143.90 | | |
| 1 Tandem Roller, 10 Ton | | 224.20 | | 246.62 | | |
| 1 Roller, Pneum. Whl, 12 Ton | | 309.20 | | 340.12 | 28.21 | 31.03 |
| 88 L.H., Daily Totals | | $5785.20 | | $7753.84 | $65.74 | $88.11 |

## Crew B-25B

| | Bare Costs Hr. | Bare Costs Daily | Incl. Subs O&P Hr. | Incl. Subs O&P Daily | Cost Per L.H. Bare Costs | Cost Per L.H. Incl. O&P |
|---|---|---|---|---|---|---|
| 1 Labor Foreman | $36.35 | $290.80 | $55.75 | $446.00 | $38.18 | $57.97 |
| 7 Laborers | 34.35 | 1923.60 | 52.70 | 2951.20 | | |
| 4 Equip. Oper. (med.) | 45.35 | 1451.20 | 67.75 | 2168.00 | | |
| 1 Asphalt Paver, 130 H.P. | | 1949.00 | | 2143.90 | | |
| 2 Tandem Rollers, 10 Ton | | 448.40 | | 493.24 | | |
| 1 Roller, Pneum. Whl, 12 Ton | | 309.20 | | 340.12 | 28.19 | 31.01 |
| 96 L.H., Daily Totals | | $6372.20 | | $8542.46 | $66.38 | $88.98 |

## Crew B-25C

| | Bare Costs Hr. | Bare Costs Daily | Incl. Subs O&P Hr. | Incl. Subs O&P Daily | Cost Per L.H. Bare Costs | Cost Per L.H. Incl. O&P |
|---|---|---|---|---|---|---|
| 1 Labor Foreman | $36.35 | $290.80 | $55.75 | $446.00 | $38.35 | $58.23 |
| 3 Laborers | 34.35 | 824.40 | 52.70 | 1264.80 | | |
| 2 Equip. Oper. (med.) | 45.35 | 725.60 | 67.75 | 1084.00 | | |
| 1 Asphalt Paver, 130 H.P. | | 1949.00 | | 2143.90 | | |
| 1 Tandem Roller, 10 Ton | | 224.20 | | 246.62 | 45.27 | 49.80 |
| 48 L.H., Daily Totals | | $4014.00 | | $5185.32 | $83.63 | $108.03 |

## Crew B-26

| | Bare Costs Hr. | Bare Costs Daily | Incl. Subs O&P Hr. | Incl. Subs O&P Daily | Cost Per L.H. Bare Costs | Cost Per L.H. Incl. O&P |
|---|---|---|---|---|---|---|
| 1 Labor Foreman (outside) | $36.35 | $290.80 | $55.75 | $446.00 | $38.40 | $58.57 |
| 6 Laborers | 34.35 | 1648.80 | 52.70 | 2529.60 | | |
| 2 Equip. Oper. (med.) | 45.35 | 725.60 | 67.75 | 1084.00 | | |
| 1 Rodman (reinf.) | 48.40 | 387.20 | 77.35 | 618.80 | | |
| 1 Cement Finisher | 40.85 | 326.80 | 59.50 | 476.00 | | |
| 1 Grader, 30,000 Lbs. | | 542.80 | | 597.08 | | |
| 1 Paving Mach. & Equip. | | 2483.00 | | 2731.30 | 34.38 | 37.82 |
| 88 L.H., Daily Totals | | $6405.00 | | $8482.78 | $72.78 | $96.40 |

## Crew B-26A

| | Bare Costs Hr. | Bare Costs Daily | Incl. Subs O&P Hr. | Incl. Subs O&P Daily | Cost Per L.H. Bare Costs | Cost Per L.H. Incl. O&P |
|---|---|---|---|---|---|---|
| 1 Labor Foreman (outside) | $36.35 | $290.80 | $55.75 | $446.00 | $38.40 | $58.57 |
| 6 Laborers | 34.35 | 1648.80 | 52.70 | 2529.60 | | |
| 2 Equip. Oper. (med.) | 45.35 | 725.60 | 67.75 | 1084.00 | | |
| 1 Rodman (reinf.) | 48.40 | 387.20 | 77.35 | 618.80 | | |
| 1 Cement Finisher | 40.85 | 326.80 | 59.50 | 476.00 | | |
| 1 Grader, 30,000 Lbs. | | 542.80 | | 597.08 | | |
| 1 Paving Mach. & Equip. | | 2483.00 | | 2731.30 | | |
| 1 Concrete Saw | | 134.20 | | 147.62 | 35.91 | 39.50 |
| 88 L.H., Daily Totals | | $6539.20 | | $8630.40 | $74.31 | $98.07 |

# Crews

| Crew No. | Bare Costs Hr. | Daily | Incl. Subs O&P Hr. | Daily | Cost Per Labor-Hour Bare Costs | Incl. O&P |
|---|---|---|---|---|---|---|
| **Crew B-26B** | Hr. | Daily | Hr. | Daily | Bare Costs | Incl. O&P |
| 1 Labor Foreman (outside) | $36.35 | $290.80 | $55.75 | $446.00 | $38.98 | $59.34 |
| 6 Laborers | 34.35 | 1648.80 | 52.70 | 2529.60 | | |
| 3 Equip. Oper. (med.) | 45.35 | 1088.40 | 67.75 | 1626.00 | | |
| 1 Rodman (reinf.) | 48.40 | 387.20 | 77.35 | 618.80 | | |
| 1 Cement Finisher | 40.85 | 326.80 | 59.50 | 476.00 | | |
| 1 Grader, 30,000 Lbs. | | 542.80 | | 597.08 | | |
| 1 Paving Mach. & Equip. | | 2483.00 | | 2731.30 | | |
| 1 Concrete Pump, 110' Boom | | 999.60 | | 1099.56 | 41.93 | 46.12 |
| 96 L.H., Daily Totals | | $7767.40 | | $10124.34 | $80.91 | $105.46 |
| **Crew B-27** | Hr. | Daily | Hr. | Daily | Bare Costs | Incl. O&P |
| 1 Labor Foreman (outside) | $36.35 | $290.80 | $55.75 | $446.00 | $34.85 | $53.46 |
| 3 Laborers | 34.35 | 824.40 | 52.70 | 1264.80 | | |
| 1 Berm Machine | | 276.80 | | 304.48 | 8.65 | 9.52 |
| 32 L.H., Daily Totals | | $1392.00 | | $2015.28 | $43.50 | $62.98 |
| **Crew B-28** | Hr. | Daily | Hr. | Daily | Bare Costs | Incl. O&P |
| 2 Carpenters | $43.05 | $688.80 | $66.05 | $1056.80 | $40.15 | $61.60 |
| 1 Laborer | 34.35 | 274.80 | 52.70 | 421.60 | | |
| 24 L.H., Daily Totals | | $963.60 | | $1478.40 | $40.15 | $61.60 |
| **Crew B-29** | Hr. | Daily | Hr. | Daily | Bare Costs | Incl. O&P |
| 1 Labor Foreman (outside) | $36.35 | $290.80 | $55.75 | $446.00 | $37.22 | $56.60 |
| 4 Laborers | 34.35 | 1099.20 | 52.70 | 1686.40 | | |
| 1 Equip. Oper. (crane) | 46.50 | 372.00 | 69.45 | 555.60 | | |
| 1 Equip. Oper. Oiler | 40.30 | 322.40 | 60.20 | 481.60 | | |
| 1 Gradall, 5/8 C.Y. | | 857.40 | | 943.14 | 15.31 | 16.84 |
| 56 L.H., Daily Totals | | $2941.80 | | $4112.74 | $52.53 | $73.44 |
| **Crew B-30** | Hr. | Daily | Hr. | Daily | Bare Costs | Incl. O&P |
| 1 Equip. Oper. (med.) | $45.35 | $362.80 | $67.75 | $542.00 | $38.05 | $57.42 |
| 2 Truck Drivers (heavy) | 34.40 | 550.40 | 52.25 | 836.00 | | |
| 1 Hyd. Excavator, 1.5 C.Y. | | 947.40 | | 1042.14 | | |
| 2 Dump Trucks, 12 C.Y., 400 H.P. | | 1012.00 | | 1113.20 | 81.64 | 89.81 |
| 24 L.H., Daily Totals | | $2872.60 | | $3533.34 | $119.69 | $147.22 |
| **Crew B-31** | Hr. | Daily | Hr. | Daily | Bare Costs | Incl. O&P |
| 1 Labor Foreman (outside) | $36.35 | $290.80 | $55.75 | $446.00 | $36.49 | $55.98 |
| 3 Laborers | 34.35 | 824.40 | 52.70 | 1264.80 | | |
| 1 Carpenter | 43.05 | 344.40 | 66.05 | 528.40 | | |
| 1 Air Compressor, 250 cfm | | 162.80 | | 179.08 | | |
| 1 Sheeting Driver | | 5.75 | | 6.33 | | |
| 2 -50' Air Hoses, 1.5" | | 12.60 | | 13.86 | 4.53 | 4.98 |
| 40 L.H., Daily Totals | | $1640.75 | | $2438.47 | $41.02 | $60.96 |
| **Crew B-32** | Hr. | Daily | Hr. | Daily | Bare Costs | Incl. O&P |
| 1 Laborer | $34.35 | $274.80 | $52.70 | $421.60 | $42.60 | $63.99 |
| 3 Equip. Oper. (med.) | 45.35 | 1088.40 | 67.75 | 1626.00 | | |
| 1 Grader, 30,000 Lbs. | | 542.80 | | 597.08 | | |
| 1 Tandem Roller, 10 Ton | | 224.20 | | 246.62 | | |
| 1 Dozer, 200 H.P. | | 1051.00 | | 1156.10 | 56.81 | 62.49 |
| 32 L.H., Daily Totals | | $3181.20 | | $4047.40 | $99.41 | $126.48 |
| **Crew B-32A** | Hr. | Daily | Hr. | Daily | Bare Costs | Incl. O&P |
| 1 Laborer | $34.35 | $274.80 | $52.70 | $421.60 | $41.68 | $62.73 |
| 2 Equip. Oper. (med.) | 45.35 | 725.60 | 67.75 | 1084.00 | | |
| 1 Grader, 30,000 Lbs. | | 542.80 | | 597.08 | | |
| 1 Roller, Vibratory, 25 Ton | | 596.80 | | 656.48 | 47.48 | 52.23 |
| 24 L.H., Daily Totals | | $2140.00 | | $2759.16 | $89.17 | $114.97 |

| Crew No. | Bare Costs Hr. | Daily | Incl. Subs O&P Hr. | Daily | Cost Per Labor-Hour Bare Costs | Incl. O&P |
|---|---|---|---|---|---|---|
| **Crew B-32B** | Hr. | Daily | Hr. | Daily | Bare Costs | Incl. O&P |
| 1 Laborer | $34.35 | $274.80 | $52.70 | $421.60 | $41.68 | $62.73 |
| 2 Equip. Oper. (med.) | 45.35 | 725.60 | 67.75 | 1084.00 | | |
| 1 Dozer, 200 H.P. | | 1051.00 | | 1156.10 | | |
| 1 Roller, Vibratory, 25 Ton | | 596.80 | | 656.48 | 68.66 | 75.52 |
| 24 L.H., Daily Totals | | $2648.20 | | $3318.18 | $110.34 | $138.26 |
| **Crew B-32C** | Hr. | Daily | Hr. | Daily | Bare Costs | Incl. O&P |
| 1 Labor Foreman | $36.35 | $290.80 | $55.75 | $446.00 | $40.18 | $60.73 |
| 2 Laborers | 34.35 | 549.60 | 52.70 | 843.20 | | |
| 3 Equip. Oper. (med.) | 45.35 | 1088.40 | 67.75 | 1626.00 | | |
| 1 Grader, 30,000 Lbs. | | 542.80 | | 597.08 | | |
| 1 Tandem Roller, 10 Ton | | 224.20 | | 246.62 | | |
| 1 Dozer, 200 H.P. | | 1051.00 | | 1156.10 | 37.88 | 41.66 |
| 48 L.H., Daily Totals | | $3746.80 | | $4915.00 | $78.06 | $102.40 |
| **Crew B-33A** | Hr. | Daily | Hr. | Daily | Bare Costs | Incl. O&P |
| 1 Equip. Oper. (med.) | $45.35 | $362.80 | $67.75 | $542.00 | $42.21 | $63.45 |
| .5 Laborer | 34.35 | 137.40 | 52.70 | 210.80 | | |
| .25 Equip. Oper. (med.) | 45.35 | 90.70 | 67.75 | 135.50 | | |
| 1 Scraper, Towed, 7 C.Y. | | 108.00 | | 118.80 | | |
| 1.25 Dozers, 300 H.P. | | 2008.75 | | 2209.63 | 151.20 | 166.32 |
| 14 L.H., Daily Totals | | $2707.65 | | $3216.72 | $193.40 | $229.77 |
| **Crew B-33B** | Hr. | Daily | Hr. | Daily | Bare Costs | Incl. O&P |
| 1 Equip. Oper. (med.) | $45.35 | $362.80 | $67.75 | $542.00 | $42.21 | $63.45 |
| .5 Laborer | 34.35 | 137.40 | 52.70 | 210.80 | | |
| .25 Equip. Oper. (med.) | 45.35 | 90.70 | 67.75 | 135.50 | | |
| 1 Scraper, Towed, 10 C.Y. | | 140.00 | | 154.00 | | |
| 1.25 Dozers, 300 H.P. | | 2008.75 | | 2209.63 | 153.48 | 168.83 |
| 14 L.H., Daily Totals | | $2739.65 | | $3251.93 | $195.69 | $232.28 |
| **Crew B-33C** | Hr. | Daily | Hr. | Daily | Bare Costs | Incl. O&P |
| 1 Equip. Oper. (med.) | $45.35 | $362.80 | $67.75 | $542.00 | $42.21 | $63.45 |
| .5 Laborer | 34.35 | 137.40 | 52.70 | 210.80 | | |
| .25 Equip. Oper. (med.) | 45.35 | 90.70 | 67.75 | 135.50 | | |
| 1 Scraper, Towed, 15 C.Y. | | 158.60 | | 174.46 | | |
| 1.25 Dozers, 300 H.P. | | 2008.75 | | 2209.63 | 154.81 | 170.29 |
| 14 L.H., Daily Totals | | $2758.25 | | $3272.39 | $197.02 | $233.74 |
| **Crew B-33D** | Hr. | Daily | Hr. | Daily | Bare Costs | Incl. O&P |
| 1 Equip. Oper. (med.) | $45.35 | $362.80 | $67.75 | $542.00 | $42.21 | $63.45 |
| .5 Laborer | 34.35 | 137.40 | 52.70 | 210.80 | | |
| .25 Equip. Oper. (med.) | 45.35 | 90.70 | 67.75 | 135.50 | | |
| 1 S.P. Scraper, 14 C.Y. | | 1512.00 | | 1663.20 | | |
| .25 Dozer, 300 H.P. | | 401.75 | | 441.93 | 136.70 | 150.37 |
| 14 L.H., Daily Totals | | $2504.65 | | $2993.43 | $178.90 | $213.82 |
| **Crew B-33E** | Hr. | Daily | Hr. | Daily | Bare Costs | Incl. O&P |
| 1 Equip. Oper. (med.) | $45.35 | $362.80 | $67.75 | $542.00 | $42.21 | $63.45 |
| .5 Laborer | 34.35 | 137.40 | 52.70 | 210.80 | | |
| .25 Equip. Oper. (med.) | 45.35 | 90.70 | 67.75 | 135.50 | | |
| 1 S.P. Scraper, 21 C.Y. | | 1872.00 | | 2059.20 | | |
| .25 Dozer, 300 H.P. | | 401.75 | | 441.93 | 162.41 | 178.65 |
| 14 L.H., Daily Totals | | $2864.65 | | $3389.43 | $204.62 | $242.10 |

| Crew No. | Bare Costs | | Incl. Subs O&P | | Cost Per Labor-Hour | |
|---|---|---|---|---|---|---|

| Crew B-33F | Hr. | Daily | Hr. | Daily | Bare Costs | Incl. O&P |
|---|---|---|---|---|---|---|
| 1 Equip. Oper. (med.) | $45.35 | $362.80 | $67.75 | $542.00 | $42.21 | $63.45 |
| .5 Laborer | 34.35 | 137.40 | 52.70 | 210.80 | | |
| .25 Equip. Oper. (med.) | 45.35 | 90.70 | 67.75 | 135.50 | | |
| 1 Elev. Scraper, 11 C.Y. | | 1089.00 | | 1197.90 | | |
| .25 Dozer, 300 H.P. | | 401.75 | | 441.93 | 106.48 | 117.13 |
| 14 L.H., Daily Totals | | $2081.65 | | $2528.13 | $148.69 | $180.58 |

| Crew B-33G | Hr. | Daily | Hr. | Daily | Bare Costs | Incl. O&P |
|---|---|---|---|---|---|---|
| 1 Equip. Oper. (med.) | $45.35 | $362.80 | $67.75 | $542.00 | $42.21 | $63.45 |
| .5 Laborer | 34.35 | 137.40 | 52.70 | 210.80 | | |
| .25 Equip. Oper. (med.) | 45.35 | 90.70 | 67.75 | 135.50 | | |
| 1 Elev. Scraper, 22 C.Y. | | 2087.00 | | 2295.70 | | |
| .25 Dozer, 300 H.P. | | 401.75 | | 441.93 | 177.77 | 195.54 |
| 14 L.H., Daily Totals | | $3079.65 | | $3625.93 | $219.97 | $258.99 |

| Crew B-33H | Hr. | Daily | Hr. | Daily | Bare Costs | Incl. O&P |
|---|---|---|---|---|---|---|
| .5 Laborer | $34.35 | $137.40 | $52.70 | $210.80 | $42.21 | $63.45 |
| 1 Equipment Operator (med.) | 45.35 | 362.80 | 67.75 | 542.00 | | |
| .25 Equipment Operator (med.) | 45.35 | 90.70 | 67.75 | 135.50 | | |
| 1 S.P. Scraper, 44 C.Y. | | 3705.00 | | 4075.50 | | |
| .25 Dozer, 410 H.P. | | 526.25 | | 578.88 | 302.23 | 332.46 |
| 14 L.H., Daily Totals | | $4822.15 | | $5542.68 | $344.44 | $395.91 |

| Crew B-33J | Hr. | Daily | Hr. | Daily | Bare Costs | Incl. O&P |
|---|---|---|---|---|---|---|
| 1 Equipment Operator (med.) | $45.35 | $362.80 | $67.75 | $542.00 | $45.35 | $67.75 |
| 1 S.P. Scraper, 14 C.Y. | | 1512.00 | | 1663.20 | 189.00 | 207.90 |
| 8 L.H., Daily Totals | | $1874.80 | | $2205.20 | $234.35 | $275.65 |

| Crew B-33K | Hr. | Daily | Hr. | Daily | Bare Costs | Incl. O&P |
|---|---|---|---|---|---|---|
| 1 Equipment Operator (med.) | $45.35 | $362.80 | $67.75 | $542.00 | $42.21 | $63.45 |
| .25 Equipment Operator (med.) | 45.35 | 90.70 | 67.75 | 135.50 | | |
| .5 Laborer | 34.35 | 137.40 | 52.70 | 210.80 | | |
| 1 S.P. Scraper, 31 C.Y. | | 2777.00 | | 3054.70 | | |
| .25 Dozer, 410 H.P. | | 526.25 | | 578.88 | 235.95 | 259.54 |
| 14 L.H., Daily Totals | | $3894.15 | | $4521.88 | $278.15 | $322.99 |

| Crew B-34A | Hr. | Daily | Hr. | Daily | Bare Costs | Incl. O&P |
|---|---|---|---|---|---|---|
| 1 Truck Driver (heavy) | $34.40 | $275.20 | $52.25 | $418.00 | $34.40 | $52.25 |
| 1 Dump Truck, 8 C.Y., 220 H.P. | | 315.40 | | 346.94 | 39.42 | 43.37 |
| 8 L.H., Daily Totals | | $590.60 | | $764.94 | $73.83 | $95.62 |

| Crew B-34B | Hr. | Daily | Hr. | Daily | Bare Costs | Incl. O&P |
|---|---|---|---|---|---|---|
| 1 Truck Driver (heavy) | $34.40 | $275.20 | $52.25 | $418.00 | $34.40 | $52.25 |
| 1 Dump Truck, 12 C.Y., 400 H.P. | | 506.00 | | 556.60 | 63.25 | 69.58 |
| 8 L.H., Daily Totals | | $781.20 | | $974.60 | $97.65 | $121.83 |

| Crew B-34C | Hr. | Daily | Hr. | Daily | Bare Costs | Incl. O&P |
|---|---|---|---|---|---|---|
| 1 Truck Driver (heavy) | $34.40 | $275.20 | $52.25 | $418.00 | $34.40 | $52.25 |
| 1 Truck Tractor, 6x4, 380 H.P. | | 459.40 | | 505.34 | | |
| 1 Dump Trailer, 16.5 C.Y. | | 125.80 | | 138.38 | 73.15 | 80.47 |
| 8 L.H., Daily Totals | | $860.40 | | $1061.72 | $107.55 | $132.72 |

| Crew B-34D | Hr. | Daily | Hr. | Daily | Bare Costs | Incl. O&P |
|---|---|---|---|---|---|---|
| 1 Truck Driver (heavy) | $34.40 | $275.20 | $52.25 | $418.00 | $34.40 | $52.25 |
| 1 Truck Tractor, 6x4, 380 H.P. | | 459.40 | | 505.34 | | |
| 1 Dump Trailer, 20 C.Y. | | 140.00 | | 154.00 | 74.92 | 82.42 |
| 8 L.H., Daily Totals | | $874.60 | | $1077.34 | $109.33 | $134.67 |

| Crew B-34E | Hr. | Daily | Hr. | Daily | Bare Costs | Incl. O&P |
|---|---|---|---|---|---|---|
| 1 Truck Driver (heavy) | $34.40 | $275.20 | $52.25 | $418.00 | $34.40 | $52.25 |
| 1 Dump Truck, Off Hwy., 25 Ton | | 1264.00 | | 1390.40 | 158.00 | 173.80 |
| 8 L.H., Daily Totals | | $1539.20 | | $1808.40 | $192.40 | $226.05 |

| Crew B-34F | Hr. | Daily | Hr. | Daily | Bare Costs | Incl. O&P |
|---|---|---|---|---|---|---|
| 1 Truck Driver (heavy) | $34.40 | $275.20 | $52.25 | $418.00 | $34.40 | $52.25 |
| 1 Dump Truck, Off Hwy., 35 Ton | | 1135.00 | | 1248.50 | 141.88 | 156.06 |
| 8 L.H., Daily Totals | | $1410.20 | | $1666.50 | $176.28 | $208.31 |

| Crew B-34G | Hr. | Daily | Hr. | Daily | Bare Costs | Incl. O&P |
|---|---|---|---|---|---|---|
| 1 Truck Driver (heavy) | $34.40 | $275.20 | $52.25 | $418.00 | $34.40 | $52.25 |
| 1 Dump Truck, Off Hwy., 50 Ton | | 1549.00 | | 1703.90 | 193.63 | 212.99 |
| 8 L.H., Daily Totals | | $1824.20 | | $2121.90 | $228.03 | $265.24 |

| Crew B-34H | Hr. | Daily | Hr. | Daily | Bare Costs | Incl. O&P |
|---|---|---|---|---|---|---|
| 1 Truck Driver (heavy) | $34.40 | $275.20 | $52.25 | $418.00 | $34.40 | $52.25 |
| 1 Dump Truck, Off Hwy., 65 Ton | | 1551.00 | | 1706.10 | 193.88 | 213.26 |
| 8 L.H., Daily Totals | | $1826.20 | | $2124.10 | $228.28 | $265.51 |

| Crew B-34I | Hr. | Daily | Hr. | Daily | Bare Costs | Incl. O&P |
|---|---|---|---|---|---|---|
| 1 Truck Driver (heavy) | $34.40 | $275.20 | $52.25 | $418.00 | $34.40 | $52.25 |
| 1 Dump Truck, 18 C.Y., 450 H.P. | | 660.80 | | 726.88 | 82.60 | 90.86 |
| 8 L.H., Daily Totals | | $936.00 | | $1144.88 | $117.00 | $143.11 |

| Crew B-34J | Hr. | Daily | Hr. | Daily | Bare Costs | Incl. O&P |
|---|---|---|---|---|---|---|
| 1 Truck Driver (heavy) | $34.40 | $275.20 | $52.25 | $418.00 | $34.40 | $52.25 |
| 1 Dump Truck, Off Hwy., 100 Ton | | 2009.00 | | 2209.90 | 251.13 | 276.24 |
| 8 L.H., Daily Totals | | $2284.20 | | $2627.90 | $285.52 | $328.49 |

| Crew B-34K | Hr. | Daily | Hr. | Daily | Bare Costs | Incl. O&P |
|---|---|---|---|---|---|---|
| 1 Truck Driver (heavy) | $34.40 | $275.20 | $52.25 | $418.00 | $34.40 | $52.25 |
| 1 Truck Tractor, 6x4, 450 H.P. | | 558.80 | | 614.68 | | |
| 1 Lowbed Trailer, 75 Ton | | 219.20 | | 241.12 | 97.25 | 106.97 |
| 8 L.H., Daily Totals | | $1053.20 | | $1273.80 | $131.65 | $159.22 |

| Crew B-34L | Hr. | Daily | Hr. | Daily | Bare Costs | Incl. O&P |
|---|---|---|---|---|---|---|
| 1 Equip. Oper. (light) | $43.60 | $348.80 | $65.15 | $521.20 | $43.60 | $65.15 |
| 1 Flatbed Truck, Gas, 1.5 Ton | | 171.40 | | 188.54 | 21.43 | 23.57 |
| 8 L.H., Daily Totals | | $520.20 | | $709.74 | $65.03 | $88.72 |

| Crew B-34M | Hr. | Daily | Hr. | Daily | Bare Costs | Incl. O&P |
|---|---|---|---|---|---|---|
| 1 Equip. Oper. (light) | $43.60 | $348.80 | $65.15 | $521.20 | $43.60 | $65.15 |
| 1 Flatbed Truck, Gas, 3 Ton | | 217.00 | | 238.70 | 27.13 | 29.84 |
| 8 L.H., Daily Totals | | $565.80 | | $759.90 | $70.72 | $94.99 |

| Crew B-34N | Hr. | Daily | Hr. | Daily | Bare Costs | Incl. O&P |
|---|---|---|---|---|---|---|
| 1 Truck Driver (heavy) | $34.40 | $275.20 | $52.25 | $418.00 | $34.40 | $52.25 |
| 1 Dump Truck, 8 C.Y., 220 H.P. | | 315.40 | | 346.94 | | |
| 1 Flatbed Trailer, 40 Ton | | 150.20 | | 165.22 | 58.20 | 64.02 |
| 8 L.H., Daily Totals | | $740.80 | | $930.16 | $92.60 | $116.27 |

| Crew B-34P | Hr. | Daily | Hr. | Daily | Bare Costs | Incl. O&P |
|---|---|---|---|---|---|---|
| 1 Pipe Fitter | $54.10 | $432.80 | $80.75 | $646.00 | $44.28 | $66.42 |
| 1 Truck Driver (light) | 33.40 | 267.20 | 50.75 | 406.00 | | |
| 1 Equip. Oper. (med.) | 45.35 | 362.80 | 67.75 | 542.00 | | |
| 1 Flatbed Truck, Gas, 3 Ton | | 217.00 | | 238.70 | | |
| 1 Backhoe Loader, 48 H.P. | | 320.40 | | 352.44 | 22.39 | 24.63 |
| 24 L.H., Daily Totals | | $1600.20 | | $2185.14 | $66.67 | $91.05 |

| Crew No. | Bare Costs | | Incl. Subs O&P | | Cost Per Labor-Hour | |

**Crew B-34Q**

| Crew B-34Q | Hr. | Daily | Hr. | Daily | Bare Costs | Incl. O&P |
|---|---|---|---|---|---|---|
| 1 Pipe Fitter | $54.10 | $432.80 | $80.75 | $646.00 | $44.67 | $66.98 |
| 1 Truck Driver (light) | 33.40 | 267.20 | 50.75 | 406.00 | | |
| 1 Eqip. Oper. (crane) | 46.50 | 372.00 | 69.45 | 555.60 | | |
| 1 Flatbed Trailer, 25 Ton | | 111.20 | | 122.32 | | |
| 1 Dump Truck, 8 C.Y., 220 H.P. | | 315.40 | | 346.94 | | |
| 1 Hyd. Crane, 25 Ton | | 748.60 | | 823.46 | 48.97 | 53.86 |
| 24 L.H., Daily Totals | | $2247.20 | | $2900.32 | $93.63 | $120.85 |

**Crew B-34R**

| Crew B-34R | Hr. | Daily | Hr. | Daily | Bare Costs | Incl. O&P |
|---|---|---|---|---|---|---|
| 1 Pipe Fitter | $54.10 | $432.80 | $80.75 | $646.00 | $44.67 | $66.98 |
| 1 Truck Driver (light) | 33.40 | 267.20 | 50.75 | 406.00 | | |
| 1 Eqip. Oper. (crane) | 46.50 | 372.00 | 69.45 | 555.60 | | |
| 1 Flatbed Trailer, 25 Ton | | 111.20 | | 122.32 | | |
| 1 Dump Truck, 8 C.Y., 220 H.P. | | 315.40 | | 346.94 | | |
| 1 Hyd. Crane, 25 Ton | | 748.60 | | 823.46 | | |
| 1 Hyd. Excavator, 1 C.Y. | | 729.80 | | 802.78 | 79.38 | 87.31 |
| 24 L.H., Daily Totals | | $2977.00 | | $3703.10 | $124.04 | $154.30 |

**Crew B-34S**

| Crew B-34S | Hr. | Daily | Hr. | Daily | Bare Costs | Incl. O&P |
|---|---|---|---|---|---|---|
| 2 Pipe Fitters | $54.10 | $865.60 | $80.75 | $1292.00 | $47.27 | $70.80 |
| 1 Truck Driver (heavy) | 34.40 | 275.20 | 52.25 | 418.00 | | |
| 1 Eqip. Oper. (crane) | 46.50 | 372.00 | 69.45 | 555.60 | | |
| 1 Flatbed Trailer, 40 Ton | | 150.20 | | 165.22 | | |
| 1 Truck Tractor, 6x4, 380 H.P. | | 459.40 | | 505.34 | | |
| 1 Hyd. Crane, 80 Ton | | 1693.00 | | 1862.30 | | |
| 1 Hyd. Excavator, 2 C.Y. | | 1224.00 | | 1346.40 | 110.21 | 121.23 |
| 32 L.H., Daily Totals | | $5039.40 | | $6144.86 | $157.48 | $192.03 |

**Crew B-34T**

| Crew B-34T | Hr. | Daily | Hr. | Daily | Bare Costs | Incl. O&P |
|---|---|---|---|---|---|---|
| 2 Pipe Fitters | $54.10 | $865.60 | $80.75 | $1292.00 | $47.27 | $70.80 |
| 1 Truck Driver (heavy) | 34.40 | 275.20 | 52.25 | 418.00 | | |
| 1 Eqip. Oper. (crane) | 46.50 | 372.00 | 69.45 | 555.60 | | |
| 1 Flatbed Trailer, 40 Ton | | 150.20 | | 165.22 | | |
| 1 Truck Tractor, 6x4, 380 H.P. | | 459.40 | | 505.34 | | |
| 1 Hyd. Crane, 80 Ton | | 1693.00 | | 1862.30 | 71.96 | 79.15 |
| 32 L.H., Daily Totals | | $3815.40 | | $4798.46 | $119.23 | $149.95 |

**Crew B-35**

| Crew B-35 | Hr. | Daily | Hr. | Daily | Bare Costs | Incl. O&P |
|---|---|---|---|---|---|---|
| 1 Laborer Foreman (out) | $36.35 | $290.80 | $55.75 | $446.00 | $42.58 | $64.37 |
| 1 Skilled Worker | 44.40 | 355.20 | 68.10 | 544.80 | | |
| 1 Welder (plumber) | 53.60 | 428.80 | 80.00 | 640.00 | | |
| 1 Laborer | 34.35 | 274.80 | 52.70 | 421.60 | | |
| 1 Equip. Oper. (crane) | 46.50 | 372.00 | 69.45 | 555.60 | | |
| 1 Equip. Oper. Oiler | 40.30 | 322.40 | 60.20 | 481.60 | | |
| 1 Welder, electric, 300 amp | | 32.45 | | 35.70 | | |
| 1 Hyd. Excavator, .75 C.Y. | | 640.40 | | 704.44 | 14.02 | 15.42 |
| 48 L.H., Daily Totals | | $2716.85 | | $3829.74 | $56.60 | $79.79 |

**Crew B-35A**

| Crew B-35A | Hr. | Daily | Hr. | Daily | Bare Costs | Incl. O&P |
|---|---|---|---|---|---|---|
| 1 Laborer Foreman (out) | $36.35 | $290.80 | $55.75 | $446.00 | $41.41 | $62.70 |
| 2 Laborers | 34.35 | 549.60 | 52.70 | 843.20 | | |
| 1 Skilled Worker | 44.40 | 355.20 | 68.10 | 544.80 | | |
| 1 Welder (plumber) | 53.60 | 428.80 | 80.00 | 640.00 | | |
| 1 Equip. Oper. (crane) | 46.50 | 372.00 | 69.45 | 555.60 | | |
| 1 Equip. Oper. Oiler | 40.30 | 322.40 | 60.20 | 481.60 | | |
| 1 Welder, gas engine, 300 amp | | 109.00 | | 119.90 | | |
| 1 Crawler Crane, 75 Ton | | 1446.00 | | 1590.60 | 27.77 | 30.54 |
| 56 L.H., Daily Totals | | $3873.80 | | $5221.70 | $69.17 | $93.24 |

**Crew B-36**

| Crew B-36 | Hr. | Daily | Hr. | Daily | Bare Costs | Incl. O&P |
|---|---|---|---|---|---|---|
| 1 Labor Foreman (outside) | $36.35 | $290.80 | $55.75 | $446.00 | $39.15 | $59.33 |
| 2 Laborers | 34.35 | 549.60 | 52.70 | 843.20 | | |
| 2 Equip. Oper. (med.) | 45.35 | 725.60 | 67.75 | 1084.00 | | |
| 1 Dozer, 200 H.P. | | 1051.00 | | 1156.10 | | |
| 1 Aggregate Spreader | | 33.60 | | 36.96 | | |
| 1 Tandem Roller, 10 Ton | | 224.20 | | 246.62 | 32.72 | 35.99 |
| 40 L.H., Daily Totals | | $2874.80 | | $3812.88 | $71.87 | $95.32 |

**Crew B-36A**

| Crew B-36A | Hr. | Daily | Hr. | Daily | Bare Costs | Incl. O&P |
|---|---|---|---|---|---|---|
| 1 Labor Foreman (outside) | $36.35 | $290.80 | $55.75 | $446.00 | $40.92 | $61.74 |
| 2 Laborers | 34.35 | 549.60 | 52.70 | 843.20 | | |
| 4 Equip. Oper. (med.) | 45.35 | 1451.20 | 67.75 | 2168.00 | | |
| 1 Dozer, 200 H.P. | | 1051.00 | | 1156.10 | | |
| 1 Aggregate Spreader | | 33.60 | | 36.96 | | |
| 1 Tandem Roller, 10 Ton | | 224.20 | | 246.62 | | |
| 1 Roller, Pneum. Whl, 12 Ton | | 309.20 | | 340.12 | 28.89 | 31.78 |
| 56 L.H., Daily Totals | | $3909.60 | | $5237.00 | $69.81 | $93.52 |

**Crew B-36B**

| Crew B-36B | Hr. | Daily | Hr. | Daily | Bare Costs | Incl. O&P |
|---|---|---|---|---|---|---|
| 1 Labor Foreman (outside) | $36.35 | $290.80 | $55.75 | $446.00 | $40.11 | $60.55 |
| 2 Laborers | 34.35 | 549.60 | 52.70 | 843.20 | | |
| 4 Equip. Oper. (med.) | 45.35 | 1451.20 | 67.75 | 2168.00 | | |
| 1 Truck Driver, Heavy | 34.40 | 275.20 | 52.25 | 418.00 | | |
| 1 Grader, 30,000 Lbs. | | 542.80 | | 597.08 | | |
| 1 F.E. Loader, crl, 1.5 C.Y. | | 565.00 | | 621.50 | | |
| 1 Dozer, 300 H.P. | | 1607.00 | | 1767.70 | | |
| 1 Roller, Vibratory, 25 Ton | | 596.80 | | 656.48 | | |
| 1 Truck Tractor, 6x4, 450 H.P. | | 558.80 | | 614.68 | | |
| 1 Water Tank Trailer, 5000 Gal. | | 146.20 | | 160.82 | 62.76 | 69.04 |
| 64 L.H., Daily Totals | | $6583.40 | | $8293.46 | $102.87 | $129.59 |

**Crew B-36C**

| Crew B-36C | Hr. | Daily | Hr. | Daily | Bare Costs | Incl. O&P |
|---|---|---|---|---|---|---|
| 1 Labor Foreman (outside) | $36.35 | $290.80 | $55.75 | $446.00 | $41.36 | $62.25 |
| 3 Equip. Oper. (med.) | 45.35 | 1088.40 | 67.75 | 1626.00 | | |
| 1 Truck Driver, Heavy | 34.40 | 275.20 | 52.25 | 418.00 | | |
| 1 Grader, 30,000 Lbs. | | 542.80 | | 597.08 | | |
| 1 Dozer, 300 H.P. | | 1607.00 | | 1767.70 | | |
| 1 Roller, Vibratory, 25 Ton | | 596.80 | | 656.48 | | |
| 1 Truck Tractor, 6x4, 450 H.P. | | 558.80 | | 614.68 | | |
| 1 Water Tank Trailer, 5000 Gal. | | 146.20 | | 160.82 | 86.29 | 94.92 |
| 40 L.H., Daily Totals | | $5106.00 | | $6286.76 | $127.65 | $157.17 |

**Crew B-36D**

| Crew B-36D | Hr. | Daily | Hr. | Daily | Bare Costs | Incl. O&P |
|---|---|---|---|---|---|---|
| 1 Labor Foreman (outside) | $36.35 | $290.80 | $55.75 | $446.00 | $43.10 | $64.75 |
| 3 Equip. Oper. (medium) | 45.35 | 1088.40 | 67.75 | 1626.00 | | |
| 1 Grader, 30,000 Lbs. | | 542.80 | | 597.08 | | |
| 1 Dozer, 300 H.P. | | 1607.00 | | 1767.70 | | |
| 1 Roller, Vibratory, 25 Ton | | 596.80 | | 656.48 | 85.83 | 94.41 |
| 32 L.H., Daily Totals | | $4125.80 | | $5093.26 | $128.93 | $159.16 |

**Crew B-37**

| Crew B-37 | Hr. | Daily | Hr. | Daily | Bare Costs | Incl. O&P |
|---|---|---|---|---|---|---|
| 1 Labor Foreman (outside) | $36.35 | $290.80 | $55.75 | $446.00 | $36.23 | $55.28 |
| 4 Laborers | 34.35 | 1099.20 | 52.70 | 1686.40 | | |
| 1 Equip. Oper. (light) | 43.60 | 348.80 | 65.15 | 521.20 | | |
| 1 Tandem Roller, 5 Ton | | 145.20 | | 159.72 | 3.02 | 3.33 |
| 48 L.H., Daily Totals | | $1884.00 | | $2813.32 | $39.25 | $58.61 |

| Crew No. | Bare Costs | | Incl. Subs O&P | | Cost Per Labor-Hour | |
|---|---|---|---|---|---|---|

| Crew B-37A | Hr. | Daily | Hr. | Daily | Bare Costs | Incl. O&P |
|---|---|---|---|---|---|---|
| 2 Laborers | $34.35 | $549.60 | $52.70 | $843.20 | $34.03 | $52.05 |
| 1 Truck Driver (light) | 33.40 | 267.20 | 50.75 | 406.00 | | |
| 1 Flatbed Truck, Gas, 1.5 Ton | | 171.40 | | 188.54 | | |
| 1 Tar Kettle, T.M. | | 74.65 | | 82.11 | 10.25 | 11.28 |
| 24 L.H., Daily Totals | | $1062.85 | | $1519.86 | $44.29 | $63.33 |

| Crew B-37B | Hr. | Daily | Hr. | Daily | Bare Costs | Incl. O&P |
|---|---|---|---|---|---|---|
| 3 Laborers | $34.35 | $824.40 | $52.70 | $1264.80 | $34.11 | $52.21 |
| 1 Truck Driver (light) | 33.40 | 267.20 | 50.75 | 406.00 | | |
| 1 Flatbed Truck, Gas, 1.5 Ton | | 171.40 | | 188.54 | | |
| 1 Tar Kettle, T.M. | | 74.65 | | 82.11 | 7.69 | 8.46 |
| 32 L.H., Daily Totals | | $1337.65 | | $1941.45 | $41.80 | $60.67 |

| Crew B-37C | Hr. | Daily | Hr. | Daily | Bare Costs | Incl. O&P |
|---|---|---|---|---|---|---|
| 2 Laborers | $34.35 | $549.60 | $52.70 | $843.20 | $33.88 | $51.73 |
| 2 Truck Drivers (light) | 33.40 | 534.40 | 50.75 | 812.00 | | |
| 2 Flatbed Truck, Gas, 1.5 Ton | | 342.80 | | 377.08 | | |
| 1 Tar Kettle, T.M. | | 74.65 | | 82.11 | 13.05 | 14.35 |
| 32 L.H., Daily Totals | | $1501.45 | | $2114.40 | $46.92 | $66.07 |

| Crew B-37D | Hr. | Daily | Hr. | Daily | Bare Costs | Incl. O&P |
|---|---|---|---|---|---|---|
| 1 Laborer | $34.35 | $274.80 | $52.70 | $421.60 | $33.88 | $51.73 |
| 1 Truck Driver (light) | 33.40 | 267.20 | 50.75 | 406.00 | | |
| 1 Pickup Truck, 3/4 Ton | | 104.60 | | 115.06 | 6.54 | 7.19 |
| 16 L.H., Daily Totals | | $646.60 | | $942.66 | $40.41 | $58.92 |

| Crew B-37E | Hr. | Daily | Hr. | Daily | Bare Costs | Incl. O&P |
|---|---|---|---|---|---|---|
| 3 Laborers | $34.35 | $824.40 | $52.70 | $1264.80 | $36.97 | $56.07 |
| 1 Equip. Oper. (light) | 43.60 | 348.80 | 65.15 | 521.20 | | |
| 1 Equip. Oper. (medium) | 45.35 | 362.80 | 67.75 | 542.00 | | |
| 2 Truck Drivers (light) | 33.40 | 534.40 | 50.75 | 812.00 | | |
| 4 Barrel w/ Flasher | | 13.60 | | 14.96 | | |
| 1 Concrete Saw | | 134.20 | | 147.62 | | |
| 1 Rotary Hammer Drill | | 22.55 | | 24.81 | | |
| 1 Hammer Drill Bit | | 3.90 | | 4.29 | | |
| 1 Loader, Skid Steer, 30 H.P., gas | | 151.80 | | 166.98 | | |
| 1 Conc. Hammer Attach. | | 115.35 | | 126.89 | | |
| 1 Vibrating plate, gas, 18" | | 30.80 | | 33.88 | | |
| 2 Flatbed Truck, Gas, 1.5 Ton | | 342.80 | | 377.08 | 14.55 | 16.01 |
| 56 L.H., Daily Totals | | $2885.40 | | $4036.50 | $51.52 | $72.08 |

| Crew B-37F | Hr. | Daily | Hr. | Daily | Bare Costs | Incl. O&P |
|---|---|---|---|---|---|---|
| 3 Laborers | $34.35 | $824.40 | $52.70 | $1264.80 | $34.11 | $52.21 |
| 1 Truck Driver (light) | 33.40 | 267.20 | 50.75 | 406.00 | | |
| 4 Barrel w/ Flasher | | 13.60 | | 14.96 | | |
| 1 Concrete Mixer, 10 C.F. | | 151.20 | | 166.32 | | |
| 1 Air Compressor, 60 cfm | | 110.20 | | 121.22 | | |
| 1 -50' Air Hose, 3/4" | | 3.05 | | 3.36 | | |
| 1 Spade (chipper) | | 8.00 | | 8.80 | | |
| 1 Flatbed Truck, Gas, 1.5 Ton | | 171.40 | | 188.54 | 14.30 | 15.72 |
| 32 L.H., Daily Totals | | $1549.05 | | $2173.99 | $48.41 | $67.94 |

| Crew B-38 | Hr. | Daily | Hr. | Daily | Bare Costs | Incl. O&P |
|---|---|---|---|---|---|---|
| 1 Labor Foreman (outside) | $36.35 | $290.80 | $55.75 | $446.00 | $38.80 | $58.81 |
| 2 Laborers | 34.35 | 549.60 | 52.70 | 843.20 | | |
| 1 Equip. Oper. (light) | 43.60 | 348.80 | 65.15 | 521.20 | | |
| 1 Equip. Oper. (med.) | 45.35 | 362.80 | 67.75 | 542.00 | | |
| 1 Backhoe Loader, 48 H.P. | | 320.40 | | 352.44 | | |
| 1 Hyd.Hammer, (1200 lb.) | | 158.00 | | 173.80 | | |
| 1 F.E. Loader, W.M., 4 C.Y. | | 598.60 | | 658.46 | | |
| 1 Pvmt. Rem. Bucket | | 59.60 | | 65.56 | 28.41 | 31.26 |
| 40 L.H., Daily Totals | | $2688.60 | | $3602.66 | $67.22 | $90.07 |

| Crew B-39 | Hr. | Daily | Hr. | Daily | Bare Costs | Incl. O&P |
|---|---|---|---|---|---|---|
| 1 Labor Foreman (outside) | $36.35 | $290.80 | $55.75 | $446.00 | $36.23 | $55.28 |
| 4 Laborers | 34.35 | 1099.20 | 52.70 | 1686.40 | | |
| 1 Equip. Oper. (light) | 43.60 | 348.80 | 65.15 | 521.20 | | |
| 1 Air Compressor, 250 cfm | | 162.80 | | 179.08 | | |
| 2 Breakers, Pavement, 60 lb. | | 19.60 | | 21.56 | | |
| 2 -50' Air Hoses, 1.5" | | 12.60 | | 13.86 | 4.06 | 4.47 |
| 48 L.H., Daily Totals | | $1933.80 | | $2868.10 | $40.29 | $59.75 |

| Crew B-40 | Hr. | Daily | Hr. | Daily | Bare Costs | Incl. O&P |
|---|---|---|---|---|---|---|
| 1 Pile Driver Foreman (out) | $44.00 | $352.00 | $70.55 | $564.40 | $43.16 | $67.38 |
| 4 Pile Drivers | 42.00 | 1344.00 | 67.35 | 2155.20 | | |
| 2 Equip. Oper. (crane) | 46.50 | 744.00 | 69.45 | 1111.20 | | |
| 1 Equip. Oper. Oiler | 40.30 | 322.40 | 60.20 | 481.60 | | |
| 1 Crawler Crane, 40 Ton | | 1128.00 | | 1240.80 | | |
| 1 Vibratory Hammer & Gen. | | 2140.00 | | 2354.00 | 51.06 | 56.17 |
| 64 L.H., Daily Totals | | $6030.40 | | $7907.20 | $94.22 | $123.55 |

| Crew B-40B | Hr. | Daily | Hr. | Daily | Bare Costs | Incl. O&P |
|---|---|---|---|---|---|---|
| 1 Laborer Foreman | $36.35 | $290.80 | $55.75 | $446.00 | $37.70 | $57.25 |
| 3 Laborers | 34.35 | 824.40 | 52.70 | 1264.80 | | |
| 1 Equip. Oper. (crane) | 46.50 | 372.00 | 69.45 | 555.60 | | |
| 1 Equip. Oper. Oiler | 40.30 | 322.40 | 60.20 | 481.60 | | |
| 1 Lattice Boom Crane, 40 Ton | | 1263.00 | | 1389.30 | 26.31 | 28.94 |
| 48 L.H., Daily Totals | | $3072.60 | | $4137.30 | $64.01 | $86.19 |

| Crew B-41 | Hr. | Daily | Hr. | Daily | Bare Costs | Incl. O&P |
|---|---|---|---|---|---|---|
| 1 Labor Foreman (outside) | $36.35 | $290.80 | $55.75 | $446.00 | $35.54 | $54.36 |
| 4 Laborers | 34.35 | 1099.20 | 52.70 | 1686.40 | | |
| .25 Equip. Oper. (crane) | 46.50 | 93.00 | 69.45 | 138.90 | | |
| .25 Equip. Oper. Oiler | 40.30 | 80.60 | 60.20 | 120.40 | | |
| .25 Crawler Crane, 40 Ton | | 282.00 | | 310.20 | 6.41 | 7.05 |
| 44 L.H., Daily Totals | | $1845.60 | | $2701.90 | $41.95 | $61.41 |

| Crew B-42 | Hr. | Daily | Hr. | Daily | Bare Costs | Incl. O&P |
|---|---|---|---|---|---|---|
| 1 Labor Foreman (outside) | $36.35 | $290.80 | $55.75 | $446.00 | $38.64 | $60.24 |
| 4 Laborers | 34.35 | 1099.20 | 52.70 | 1686.40 | | |
| 1 Equip. Oper. (crane) | 46.50 | 372.00 | 69.45 | 555.60 | | |
| 1 Equip. Oper. Oiler | 40.30 | 322.40 | 60.20 | 481.60 | | |
| 1 Welder | 48.55 | 388.40 | 85.75 | 686.00 | | |
| 1 Hyd. Crane, 25 Ton | | 748.60 | | 823.46 | | |
| 1 Welder, gas engine, 300 amp | | 109.00 | | 119.90 | | |
| 1 Horz. Boring Csg. Mch. | | 442.20 | | 486.42 | 20.31 | 22.34 |
| 64 L.H., Daily Totals | | $3772.60 | | $5285.38 | $58.95 | $82.58 |

# Crews

| Crew No. | Bare Costs Hr. | Daily | Incl. Subs O&P Hr. | Daily | Cost Per Labor-Hour Bare Costs | Incl. O&P |
|---|---|---|---|---|---|---|
| **Crew B-43** | | | | | Bare Costs | Incl. O&P |
| 1 Labor Foreman (outside) | $36.35 | $290.80 | $55.75 | $446.00 | $37.70 | $57.25 |
| 3 Laborers | 34.35 | 824.40 | 52.70 | 1264.80 | | |
| 1 Equip. Oper. (crane) | 46.50 | 372.00 | 69.45 | 555.60 | | |
| 1 Equip. Oper. Oiler | 40.30 | 322.40 | 60.20 | 481.60 | | |
| 1 Drill Rig, Truck-Mounted | | 2486.00 | | 2734.60 | 51.79 | 56.97 |
| 48 L.H., Daily Totals | | $4295.60 | | $5482.60 | $89.49 | $114.22 |
| **Crew B-44** | | | | | Bare Costs | Incl. O&P |
| 1 Pile Driver Foreman | $44.00 | $352.00 | $70.55 | $564.40 | $42.42 | $66.44 |
| 4 Pile Drivers | 42.00 | 1344.00 | 67.35 | 2155.20 | | |
| 2 Equip. Oper. (crane) | 46.50 | 744.00 | 69.45 | 1111.20 | | |
| 1 Laborer | 34.35 | 274.80 | 52.70 | 421.60 | | |
| 1 Crawler Crane, 40 Ton | | 1128.00 | | 1240.80 | | |
| 1 Lead, 60' high | | 74.00 | | 81.40 | | |
| 1 Hammer, diesel, 15K ft.-lbs. | | 614.20 | | 675.62 | 28.38 | 31.22 |
| 64 L.H., Daily Totals | | $4531.00 | | $6250.22 | $70.80 | $97.66 |
| **Crew B-45** | | | | | Bare Costs | Incl. O&P |
| 1 Equip. Oper. (med.) | $45.35 | $362.80 | $67.75 | $542.00 | $39.88 | $60.00 |
| 1 Truck Driver (heavy) | 34.40 | 275.20 | 52.25 | 418.00 | | |
| 1 Dist. Tanker, 3000 Gallon | | 309.00 | | 339.90 | | |
| 1 Truck Tractor, 6x4, 380 H.P. | | 459.40 | | 505.34 | 48.02 | 52.83 |
| 16 L.H., Daily Totals | | $1406.40 | | $1805.24 | $87.90 | $112.83 |
| **Crew B-46** | | | | | Bare Costs | Incl. O&P |
| 1 Pile Driver Foreman | $44.00 | $352.00 | $70.55 | $564.40 | $38.51 | $60.56 |
| 2 Pile Drivers | 42.00 | 672.00 | 67.35 | 1077.60 | | |
| 3 Laborers | 34.35 | 824.40 | 52.70 | 1264.80 | | |
| 1 Chain Saw, gas, 36" Long | | 41.60 | | 45.76 | .87 | .95 |
| 48 L.H., Daily Totals | | $1890.00 | | $2952.56 | $39.38 | $61.51 |
| **Crew B-47** | | | | | Bare Costs | Incl. O&P |
| 1 Blast Foreman | $36.35 | $290.80 | $55.75 | $446.00 | $38.10 | $57.87 |
| 1 Driller | 34.35 | 274.80 | 52.70 | 421.60 | | |
| 1 Equip. Oper. (light) | 43.60 | 348.80 | 65.15 | 521.20 | | |
| 1 Air Track Drill, 4" | | 849.60 | | 934.56 | | |
| 1 Air Compressor, 600 cfm | | 424.60 | | 467.06 | | |
| 2 -50' Air Hoses, 3" | | 29.80 | | 32.78 | 54.33 | 59.77 |
| 24 L.H., Daily Totals | | $2218.40 | | $2823.20 | $92.43 | $117.63 |
| **Crew B-47A** | | | | | Bare Costs | Incl. O&P |
| 1 Drilling Foreman | $36.35 | $290.80 | $55.75 | $446.00 | $41.05 | $61.80 |
| 1 Equip. Oper. (heavy) | 46.50 | 372.00 | 69.45 | 555.60 | | |
| 1 Oiler | 40.30 | 322.40 | 60.20 | 481.60 | | |
| 1 Air Track Drill, 5" | | 965.20 | | 1061.72 | 40.22 | 44.24 |
| 24 L.H., Daily Totals | | $1950.40 | | $2544.92 | $81.27 | $106.04 |
| **Crew B-47C** | | | | | Bare Costs | Incl. O&P |
| 1 Laborer | $34.35 | $274.80 | $52.70 | $421.60 | $38.98 | $58.92 |
| 1 Equip. Oper. (light) | 43.60 | 348.80 | 65.15 | 521.20 | | |
| 1 Air Compressor, 750 cfm | | 431.80 | | 474.98 | | |
| 2 -50' Air Hoses, 3" | | 29.80 | | 32.78 | | |
| 1 Air Track Drill, 4" | | 849.60 | | 934.56 | 81.95 | 90.14 |
| 16 L.H., Daily Totals | | $1934.80 | | $2385.12 | $120.93 | $149.07 |
| **Crew B-47E** | | | | | Bare Costs | Incl. O&P |
| 1 Laborer Foreman | $36.35 | $290.80 | $55.75 | $446.00 | $34.85 | $53.46 |
| 3 Laborers | 34.35 | 824.40 | 52.70 | 1264.80 | | |
| 1 Flatbed Truck, Gas, 3 Ton | | 217.00 | | 238.70 | 6.78 | 7.46 |
| 32 L.H., Daily Totals | | $1332.20 | | $1949.50 | $41.63 | $60.92 |
| **Crew B-47G** | | | | | Bare Costs | Incl. O&P |
| 1 Laborer Foreman | $36.35 | $290.80 | $55.75 | $446.00 | $37.16 | $56.58 |
| 2 Laborers | 34.35 | 549.60 | 52.70 | 843.20 | | |
| 1 Equip. Oper. (light) | 43.60 | 348.80 | 65.15 | 521.20 | | |
| 1 Air Track Drill, 4" | | 849.60 | | 934.56 | | |
| 1 Air Compressor, 600 cfm | | 424.60 | | 467.06 | | |
| 2 -50' Air Hoses, 3" | | 29.80 | | 32.78 | | |
| 1 Gunnite Pump Rig | | 178.00 | | 195.80 | 46.31 | 50.94 |
| 32 L.H., Daily Totals | | $2671.20 | | $3440.60 | $83.47 | $107.52 |
| **Crew B-47H** | | | | | Bare Costs | Incl. O&P |
| 1 Skilled Worker Foreman | $46.40 | $371.20 | $71.20 | $569.60 | $44.90 | $68.88 |
| 3 Skilled Workers | 44.40 | 1065.60 | 68.10 | 1634.40 | | |
| 1 Flatbed Truck, Gas, 3 Ton | | 217.00 | | 238.70 | 6.78 | 7.46 |
| 32 L.H., Daily Totals | | $1653.80 | | $2442.70 | $51.68 | $76.33 |
| **Crew B-48** | | | | | Bare Costs | Incl. O&P |
| 1 Labor Foreman (outside) | $36.35 | $290.80 | $55.75 | $446.00 | $38.54 | $58.38 |
| 3 Laborers | 34.35 | 824.40 | 52.70 | 1264.80 | | |
| 1 Equip. Oper. (crane) | 46.50 | 372.00 | 69.45 | 555.60 | | |
| 1 Equip. Oper. Oiler | 40.30 | 322.40 | 60.20 | 481.60 | | |
| 1 Equip. Oper. (light) | 43.60 | 348.80 | 65.15 | 521.20 | | |
| 1 Centr. Water Pump, 6" | | 291.60 | | 320.76 | | |
| 1 -20' Suction Hose, 6" | | 11.90 | | 13.09 | | |
| 1 -50' Discharge Hose, 6" | | 6.30 | | 6.93 | | |
| 1 Drill Rig, Truck-Mounted | | 2486.00 | | 2734.60 | 49.92 | 54.92 |
| 56 L.H., Daily Totals | | $4954.20 | | $6344.58 | $88.47 | $113.30 |
| **Crew B-49** | | | | | Bare Costs | Incl. O&P |
| 1 Labor Foreman (outside) | $36.35 | $290.80 | $55.75 | $446.00 | $40.05 | $61.18 |
| 3 Laborers | 34.35 | 824.40 | 52.70 | 1264.80 | | |
| 2 Equip. Oper. (crane) | 46.50 | 744.00 | 69.45 | 1111.20 | | |
| 2 Equip. Oper. Oilers | 40.30 | 644.80 | 60.20 | 963.20 | | |
| 1 Equip. Oper. (light) | 43.60 | 348.80 | 65.15 | 521.20 | | |
| 2 Pile Drivers | 42.00 | 672.00 | 67.35 | 1077.60 | | |
| 1 Hyd. Crane, 25 Ton | | 748.60 | | 823.46 | | |
| 1 Centr. Water Pump, 6" | | 291.60 | | 320.76 | | |
| 1 -20' Suction Hose, 6" | | 11.90 | | 13.09 | | |
| 1 -50' Discharge Hose, 6" | | 6.30 | | 6.93 | | |
| 1 Drill Rig, Truck-Mounted | | 2486.00 | | 2734.60 | 40.28 | 44.31 |
| 88 L.H., Daily Totals | | $7069.20 | | $9282.84 | $80.33 | $105.49 |
| **Crew B-50** | | | | | Bare Costs | Incl. O&P |
| 2 Pile Driver Foremen | $44.00 | $704.00 | $70.55 | $1128.80 | $41.17 | $64.46 |
| 6 Pile Drivers | 42.00 | 2016.00 | 67.35 | 3232.80 | | |
| 2 Equip. Oper. (crane) | 46.50 | 744.00 | 69.45 | 1111.20 | | |
| 1 Equip. Oper. Oiler | 40.30 | 322.40 | 60.20 | 481.60 | | |
| 3 Laborers | 34.35 | 824.40 | 52.70 | 1264.80 | | |
| 1 Crawler Crane, 40 Ton | | 1128.00 | | 1240.80 | | |
| 1 Lead, 60' high | | 74.00 | | 81.40 | | |
| 1 Hammer, diesel, 15K ft.-lbs. | | 614.20 | | 675.62 | | |
| 1 Air Compressor, 600 cfm | | 424.60 | | 467.06 | | |
| 2 -50' Air Hoses, 3" | | 29.80 | | 32.78 | | |
| 1 Chain Saw, gas, 36" Long | | 41.60 | | 45.76 | 20.64 | 22.71 |
| 112 L.H., Daily Totals | | $6923.00 | | $9762.62 | $61.81 | $87.17 |
| **Crew B-51** | | | | | Bare Costs | Incl. O&P |
| 1 Labor Foreman (outside) | $36.35 | $290.80 | $55.75 | $446.00 | $34.52 | $52.88 |
| 4 Laborers | 34.35 | 1099.20 | 52.70 | 1686.40 | | |
| 1 Truck Driver (light) | 33.40 | 267.20 | 50.75 | 406.00 | | |
| 1 Flatbed Truck, Gas, 1.5 Ton | | 171.40 | | 188.54 | 3.57 | 3.93 |
| 48 L.H., Daily Totals | | $1828.60 | | $2726.94 | $38.10 | $56.81 |

| Crew No. | Bare Costs | | Incl. Subs O&P | | Cost Per Labor-Hour | |

| Crew B-52 | Hr. | Daily | Hr. | Daily | Bare Costs | Incl. O&P |
|---|---|---|---|---|---|---|
| 1 Carpenter Foreman | $45.05 | $360.40 | $69.10 | $552.80 | $39.84 | $60.76 |
| 1 Carpenter | 43.05 | 344.40 | 66.05 | 528.40 | | |
| 3 Laborers | 34.35 | 824.40 | 52.70 | 1264.80 | | |
| 1 Cement Finisher | 40.85 | 326.80 | 59.50 | 476.00 | | |
| .5 Rodman (reinf.) | 48.40 | 193.60 | 77.35 | 309.40 | | |
| .5 Equip. Oper. (med.) | 45.35 | 181.40 | 67.75 | 271.00 | | |
| .5 Crawler Loader, 3 C.Y. | | 504.00 | | 554.40 | 9.00 | 9.90 |
| 56 L.H., Daily Totals | | $2735.00 | | $3956.80 | $48.84 | $70.66 |

| Crew B-53 | Hr. | Daily | Hr. | Daily | Bare Costs | Incl. O&P |
|---|---|---|---|---|---|---|
| 1 Equip. Oper. (light) | $43.60 | $348.80 | $65.15 | $521.20 | $43.60 | $65.15 |
| 1 Trencher, Chain, 12 H.P. | | 57.40 | | 63.14 | 7.17 | 7.89 |
| 8 L.H., Daily Totals | | $406.20 | | $584.34 | $50.77 | $73.04 |

| Crew B-54 | Hr. | Daily | Hr. | Daily | Bare Costs | Incl. O&P |
|---|---|---|---|---|---|---|
| 1 Equip. Oper. (light) | $43.60 | $348.80 | $65.15 | $521.20 | $43.60 | $65.15 |
| 1 Trencher, Chain, 40 H.P. | | 297.00 | | 326.70 | 37.13 | 40.84 |
| 8 L.H., Daily Totals | | $645.80 | | $847.90 | $80.72 | $105.99 |

| Crew B-54A | Hr. | Daily | Hr. | Daily | Bare Costs | Incl. O&P |
|---|---|---|---|---|---|---|
| .17 Labor Foreman (outside) | $36.35 | $49.44 | $55.75 | $75.82 | $44.04 | $66.01 |
| 1 Equipment Operator (med.) | 45.35 | 362.80 | 67.75 | 542.00 | | |
| 1 Wheel Trencher, 67 H.P. | | 1001.00 | | 1101.10 | 106.94 | 117.64 |
| 9.36 L.H., Daily Totals | | $1413.24 | | $1718.92 | $150.99 | $183.65 |

| Crew B-54B | Hr. | Daily | Hr. | Daily | Bare Costs | Incl. O&P |
|---|---|---|---|---|---|---|
| .25 Labor Foreman (outside) | $36.35 | $72.70 | $55.75 | $111.50 | $43.55 | $65.35 |
| 1 Equipment Operator (med.) | 45.35 | 362.80 | 67.75 | 542.00 | | |
| 1 Wheel Trencher, 150 H.P. | | 1802.00 | | 1982.20 | 180.20 | 198.22 |
| 10 L.H., Daily Totals | | $2237.50 | | $2635.70 | $223.75 | $263.57 |

| Crew B-54C | Hr. | Daily | Hr. | Daily | Bare Costs | Incl. O&P |
|---|---|---|---|---|---|---|
| 1 Laborer | $34.35 | $274.80 | $52.70 | $421.60 | $39.85 | $60.23 |
| 1 Equipment Operator (med.) | 45.35 | 362.80 | 67.75 | 542.00 | | |
| 1 Wheel Trencher, 67 H.P. | | 1001.00 | | 1101.10 | 62.56 | 68.82 |
| 16 L.H., Daily Totals | | $1638.60 | | $2064.70 | $102.41 | $129.04 |

| Crew B-54D | Hr. | Daily | Hr. | Daily | Bare Costs | Incl. O&P |
|---|---|---|---|---|---|---|
| 1 Laborer | $34.35 | $274.80 | $52.70 | $421.60 | $39.85 | $60.23 |
| 1 Equipment Operator (med.) | 45.35 | 362.80 | 67.75 | 542.00 | | |
| 1 Rock trencher, 6" width | | 332.80 | | 366.08 | 20.80 | 22.88 |
| 16 L.H., Daily Totals | | $970.40 | | $1329.68 | $60.65 | $83.11 |

| Crew B-54E | Hr. | Daily | Hr. | Daily | Bare Costs | Incl. O&P |
|---|---|---|---|---|---|---|
| 1 Laborer | $34.35 | $274.80 | $52.70 | $421.60 | $39.85 | $60.23 |
| 1 Equipment Operator (med.) | 45.35 | 362.80 | 67.75 | 542.00 | | |
| 1 Rock trencher, 18" width | | 2538.00 | | 2791.80 | 158.63 | 174.49 |
| 16 L.H., Daily Totals | | $3175.60 | | $3755.40 | $198.47 | $234.71 |

| Crew B-55 | Hr. | Daily | Hr. | Daily | Bare Costs | Incl. O&P |
|---|---|---|---|---|---|---|
| 2 Laborers | $34.35 | $549.60 | $52.70 | $843.20 | $34.03 | $52.05 |
| 1 Truck Driver (light) | 33.40 | 267.20 | 50.75 | 406.00 | | |
| 1 Truck-mounted earth auger | | 706.60 | | 777.26 | | |
| 1 Flatbed Truck, Gas, 3 Ton | | 217.00 | | 238.70 | 38.48 | 42.33 |
| 24 L.H., Daily Totals | | $1740.40 | | $2265.16 | $72.52 | $94.38 |

| Crew B-56 | Hr. | Daily | Hr. | Daily | Bare Costs | Incl. O&P |
|---|---|---|---|---|---|---|
| 1 Laborer | $34.35 | $274.80 | $52.70 | $421.60 | $38.98 | $58.92 |
| 1 Equip. Oper. (light) | 43.60 | 348.80 | 65.15 | 521.20 | | |
| 1 Air Track Drill, 4" | | 849.60 | | 934.56 | | |
| 1 Air Compressor, 600 cfm | | 424.60 | | 467.06 | | |
| 1 -50' Air Hose, 3" | | 14.90 | | 16.39 | 80.57 | 88.63 |
| 16 L.H., Daily Totals | | $1912.70 | | $2360.81 | $119.54 | $147.55 |

| Crew B-57 | Hr. | Daily | Hr. | Daily | Bare Costs | Incl. O&P |
|---|---|---|---|---|---|---|
| 1 Labor Foreman (outside) | $36.35 | $290.80 | $55.75 | $446.00 | $39.24 | $59.33 |
| 2 Laborers | 34.35 | 549.60 | 52.70 | 843.20 | | |
| 1 Equip. Oper. (crane) | 46.50 | 372.00 | 69.45 | 555.60 | | |
| 1 Equip. Oper. (light) | 43.60 | 348.80 | 65.15 | 521.20 | | |
| 1 Equip. Oper. Oiler | 40.30 | 322.40 | 60.20 | 481.60 | | |
| 1 Crawler Crane, 25 Ton | | 1123.00 | | 1235.30 | | |
| 1 Clamshell Bucket, 1 C.Y. | | 48.00 | | 52.80 | | |
| 1 Centr. Water Pump, 6" | | 291.60 | | 320.76 | | |
| 1 -20' Suction Hose, 6" | | 11.90 | | 13.09 | | |
| 20 -50' Discharge Hoses, 6" | | 126.00 | | 138.60 | 33.34 | 36.68 |
| 48 L.H., Daily Totals | | $3484.10 | | $4608.15 | $72.59 | $96.00 |

| Crew B-58 | Hr. | Daily | Hr. | Daily | Bare Costs | Incl. O&P |
|---|---|---|---|---|---|---|
| 2 Laborers | $34.35 | $549.60 | $52.70 | $843.20 | $37.43 | $56.85 |
| 1 Equip. Oper. (light) | 43.60 | 348.80 | 65.15 | 521.20 | | |
| 1 Backhoe Loader, 48 H.P. | | 320.40 | | 352.44 | | |
| 1 Small Helicopter, w/pilot | | 2579.00 | | 2836.90 | 120.81 | 132.89 |
| 24 L.H., Daily Totals | | $3797.80 | | $4553.74 | $158.24 | $189.74 |

| Crew B-59 | Hr. | Daily | Hr. | Daily | Bare Costs | Incl. O&P |
|---|---|---|---|---|---|---|
| 1 Truck Driver (heavy) | $34.40 | $275.20 | $52.25 | $418.00 | $34.40 | $52.25 |
| 1 Truck Tractor, 220 H.P. | | 279.20 | | 307.12 | | |
| 1 Water Tank Trailer, 5000 Gal. | | 146.20 | | 160.82 | 53.17 | 58.49 |
| 8 L.H., Daily Totals | | $700.60 | | $885.94 | $87.58 | $110.74 |

| Crew B-59A | Hr. | Daily | Hr. | Daily | Bare Costs | Incl. O&P |
|---|---|---|---|---|---|---|
| 2 Laborers | $34.35 | $549.60 | $52.70 | $843.20 | $34.37 | $52.55 |
| 1 Truck Driver (heavy) | 34.40 | 275.20 | 52.25 | 418.00 | | |
| 1 Water Tank Trailer, 5000 Gal. | | 146.20 | | 160.82 | | |
| 1 Truck Tractor, 220 H.P. | | 279.20 | | 307.12 | 17.73 | 19.50 |
| 24 L.H., Daily Totals | | $1250.20 | | $1729.14 | $52.09 | $72.05 |

| Crew B-60 | Hr. | Daily | Hr. | Daily | Bare Costs | Incl. O&P |
|---|---|---|---|---|---|---|
| 1 Labor Foreman (outside) | $36.35 | $290.80 | $55.75 | $446.00 | $39.86 | $60.16 |
| 2 Laborers | 34.35 | 549.60 | 52.70 | 843.20 | | |
| 1 Equip. Oper. (crane) | 46.50 | 372.00 | 69.45 | 555.60 | | |
| 2 Equip. Oper. (light) | 43.60 | 697.60 | 65.15 | 1042.40 | | |
| 1 Equip. Oper. Oiler | 40.30 | 322.40 | 60.20 | 481.60 | | |
| 1 Crawler Crane, 40 Ton | | 1128.00 | | 1240.80 | | |
| 1 Lead, 60' high | | 74.00 | | 81.40 | | |
| 1 Hammer, diesel, 15K ft.-lbs. | | 614.20 | | 675.62 | | |
| 1 Backhoe Loader, 48 H.P. | | 320.40 | | 352.44 | 38.15 | 41.97 |
| 56 L.H., Daily Totals | | $4369.00 | | $5719.06 | $78.02 | $102.13 |

| Crew B-61 | Hr. | Daily | Hr. | Daily | Bare Costs | Incl. O&P |
|---|---|---|---|---|---|---|
| 1 Labor Foreman (outside) | $36.35 | $290.80 | $55.75 | $446.00 | $36.60 | $55.80 |
| 3 Laborers | 34.35 | 824.40 | 52.70 | 1264.80 | | |
| 1 Equip. Oper. (light) | 43.60 | 348.80 | 65.15 | 521.20 | | |
| 1 Cement Mixer, 2 C.Y. | | 194.60 | | 214.06 | | |
| 1 Air Compressor, 160 cfm | | 125.00 | | 137.50 | 7.99 | 8.79 |
| 40 L.H., Daily Totals | | $1783.60 | | $2583.56 | $44.59 | $64.59 |

### Crew B-62

| Crew No. | Bare Costs Hr. | Daily | Incl. Subs O&P Hr. | Daily | Cost Per Labor-Hour Bare Costs | Incl. O&P |
|---|---|---|---|---|---|---|
| 2 Laborers | $34.35 | $549.60 | $52.70 | $843.20 | $37.43 | $56.85 |
| 1 Equip. Oper. (light) | 43.60 | 348.80 | 65.15 | 521.20 | | |
| 1 Loader, Skid Steer, 30 H.P., gas | | 151.80 | | 166.98 | 6.33 | 6.96 |
| 24 L.H., Daily Totals | | $1050.20 | | $1531.38 | $43.76 | $63.81 |

### Crew B-63

| Crew No. | Bare Costs Hr. | Daily | Incl. Subs O&P Hr. | Daily | Cost Per Labor-Hour Bare Costs | Incl. O&P |
|---|---|---|---|---|---|---|
| 4 Laborers | $34.35 | $1099.20 | $52.70 | $1686.40 | $36.20 | $55.19 |
| 1 Equip. Oper. (light) | 43.60 | 348.80 | 65.15 | 521.20 | | |
| 1 Loader, Skid Steer, 30 H.P., gas | | 151.80 | | 166.98 | 3.79 | 4.17 |
| 40 L.H., Daily Totals | | $1599.80 | | $2374.58 | $39.99 | $59.36 |

### Crew B-64

| Crew No. | Bare Costs Hr. | Daily | Incl. Subs O&P Hr. | Daily | Cost Per Labor-Hour Bare Costs | Incl. O&P |
|---|---|---|---|---|---|---|
| 1 Laborer | $34.35 | $274.80 | $52.70 | $421.60 | $33.88 | $51.73 |
| 1 Truck Driver (light) | 33.40 | 267.20 | 50.75 | 406.00 | | |
| 1 Power Mulcher (small) | | 130.80 | | 143.88 | | |
| 1 Flatbed Truck, Gas, 1.5 Ton | | 171.40 | | 188.54 | 18.89 | 20.78 |
| 16 L.H., Daily Totals | | $844.20 | | $1160.02 | $52.76 | $72.50 |

### Crew B-65

| Crew No. | Bare Costs Hr. | Daily | Incl. Subs O&P Hr. | Daily | Cost Per Labor-Hour Bare Costs | Incl. O&P |
|---|---|---|---|---|---|---|
| 1 Laborer | $34.35 | $274.80 | $52.70 | $421.60 | $33.88 | $51.73 |
| 1 Truck Driver (light) | 33.40 | 267.20 | 50.75 | 406.00 | | |
| 1 Power Mulcher (large) | | 258.80 | | 284.68 | | |
| 1 Flatbed Truck, Gas, 1.5 Ton | | 171.40 | | 188.54 | 26.89 | 29.58 |
| 16 L.H., Daily Totals | | $972.20 | | $1300.82 | $60.76 | $81.30 |

### Crew B-66

| Crew No. | Bare Costs Hr. | Daily | Incl. Subs O&P Hr. | Daily | Cost Per Labor-Hour Bare Costs | Incl. O&P |
|---|---|---|---|---|---|---|
| 1 Equip. Oper. (light) | $43.60 | $348.80 | $65.15 | $521.20 | $43.60 | $65.15 |
| 1 Loader-Backhoe | | 223.60 | | 245.96 | 27.95 | 30.75 |
| 8 L.H., Daily Totals | | $572.40 | | $767.16 | $71.55 | $95.89 |

### Crew B-67

| Crew No. | Bare Costs Hr. | Daily | Incl. Subs O&P Hr. | Daily | Cost Per Labor-Hour Bare Costs | Incl. O&P |
|---|---|---|---|---|---|---|
| 1 Millwright | $45.35 | $362.80 | $66.05 | $528.40 | $44.48 | $65.60 |
| 1 Equip. Oper. (light) | 43.60 | 348.80 | 65.15 | 521.20 | | |
| 1 Forklift, R/T, 4,000 Lb. | | 278.60 | | 306.46 | 17.41 | 19.15 |
| 16 L.H., Daily Totals | | $990.20 | | $1356.06 | $61.89 | $84.75 |

### Crew B-68

| Crew No. | Bare Costs Hr. | Daily | Incl. Subs O&P Hr. | Daily | Cost Per Labor-Hour Bare Costs | Incl. O&P |
|---|---|---|---|---|---|---|
| 2 Millwrights | $45.35 | $725.60 | $66.05 | $1056.80 | $44.77 | $65.75 |
| 1 Equip. Oper. (light) | 43.60 | 348.80 | 65.15 | 521.20 | | |
| 1 Forklift, R/T, 4,000 Lb. | | 278.60 | | 306.46 | 11.61 | 12.77 |
| 24 L.H., Daily Totals | | $1353.00 | | $1884.46 | $56.38 | $78.52 |

### Crew B-69

| Crew No. | Bare Costs Hr. | Daily | Incl. Subs O&P Hr. | Daily | Cost Per Labor-Hour Bare Costs | Incl. O&P |
|---|---|---|---|---|---|---|
| 1 Labor Foreman (outside) | $36.35 | $290.80 | $55.75 | $446.00 | $37.70 | $57.25 |
| 3 Laborers | 34.35 | 824.40 | 52.70 | 1264.80 | | |
| 1 Equip Oper. (crane) | 46.50 | 372.00 | 69.45 | 555.60 | | |
| 1 Equip Oper. Oiler | 40.30 | 322.40 | 60.20 | 481.60 | | |
| 1 Hyd. Crane, 80 Ton | | 1693.00 | | 1862.30 | 35.27 | 38.80 |
| 48 L.H., Daily Totals | | $3502.60 | | $4610.30 | $72.97 | $96.05 |

### Crew B-69A

| Crew No. | Bare Costs Hr. | Daily | Incl. Subs O&P Hr. | Daily | Cost Per Labor-Hour Bare Costs | Incl. O&P |
|---|---|---|---|---|---|---|
| 1 Labor Foreman | $36.35 | $290.80 | $55.75 | $446.00 | $37.60 | $56.85 |
| 3 Laborers | 34.35 | 824.40 | 52.70 | 1264.80 | | |
| 1 Equip. Oper. (med.) | 45.35 | 362.80 | 67.75 | 542.00 | | |
| 1 Concrete Finisher | 40.85 | 326.80 | 59.50 | 476.00 | | |
| 1 Curb/Gutter Paver, 2-Track | | 744.80 | | 819.28 | 15.52 | 17.07 |
| 48 L.H., Daily Totals | | $2549.60 | | $3548.08 | $53.12 | $73.92 |

### Crew B-69B

| Crew No. | Bare Costs Hr. | Daily | Incl. Subs O&P Hr. | Daily | Cost Per Labor-Hour Bare Costs | Incl. O&P |
|---|---|---|---|---|---|---|
| 1 Labor Foreman | $36.35 | $290.80 | $55.75 | $446.00 | $37.60 | $56.85 |
| 3 Laborers | 34.35 | 824.40 | 52.70 | 1264.80 | | |
| 1 Equip. Oper. (med.) | 45.35 | 362.80 | 67.75 | 542.00 | | |
| 1 Cement Finisher | 40.85 | 326.80 | 59.50 | 476.00 | | |
| 1 Curb/Gutter Paver, 4-Track | | 848.20 | | 933.02 | 17.67 | 19.44 |
| 48 L.H., Daily Totals | | $2653.00 | | $3661.82 | $55.27 | $76.29 |

### Crew B-70

| Crew No. | Bare Costs Hr. | Daily | Incl. Subs O&P Hr. | Daily | Cost Per Labor-Hour Bare Costs | Incl. O&P |
|---|---|---|---|---|---|---|
| 1 Labor Foreman (outside) | $36.35 | $290.80 | $55.75 | $446.00 | $39.35 | $59.59 |
| 3 Laborers | 34.35 | 824.40 | 52.70 | 1264.80 | | |
| 3 Equip. Oper. (med.) | 45.35 | 1088.40 | 67.75 | 1626.00 | | |
| 1 Grader, 30,000 Lbs. | | 542.80 | | 597.08 | | |
| 1 Ripper, beam & 1 shank | | 83.20 | | 91.52 | | |
| 1 Road Sweeper, S.P., 8' wide | | 604.80 | | 665.28 | | |
| 1 F.E. Loader, W.M., 1.5 C.Y. | | 319.40 | | 351.34 | 27.68 | 30.45 |
| 56 L.H., Daily Totals | | $3753.80 | | $5042.02 | $67.03 | $90.04 |

### Crew B-70A

| Crew No. | Bare Costs Hr. | Daily | Incl. Subs O&P Hr. | Daily | Cost Per Labor-Hour Bare Costs | Incl. O&P |
|---|---|---|---|---|---|---|
| 1 Laborer | $34.35 | $274.80 | $52.70 | $421.60 | $43.15 | $64.74 |
| 4 Equip. Oper. (med.) | 45.35 | 1451.20 | 67.75 | 2168.00 | | |
| 1 Grader, 40,000 Lbs. | | 975.40 | | 1072.94 | | |
| 1 F.E. Loader, W.M., 2.5 C.Y. | | 397.60 | | 437.36 | | |
| 1 Dozer, 80 H.P. | | 412.60 | | 453.86 | | |
| 1 Roller, Pneum. Whl, 12 Ton | | 309.20 | | 340.12 | 52.37 | 57.61 |
| 40 L.H., Daily Totals | | $3820.80 | | $4893.88 | $95.52 | $122.35 |

### Crew B-71

| Crew No. | Bare Costs Hr. | Daily | Incl. Subs O&P Hr. | Daily | Cost Per Labor-Hour Bare Costs | Incl. O&P |
|---|---|---|---|---|---|---|
| 1 Labor Foreman (outside) | $36.35 | $290.80 | $55.75 | $446.00 | $39.35 | $59.59 |
| 3 Laborers | 34.35 | 824.40 | 52.70 | 1264.80 | | |
| 3 Equip. Oper. (med.) | 45.35 | 1088.40 | 67.75 | 1626.00 | | |
| 1 Pvmt. Profiler, 750 H.P. | | 5626.00 | | 6188.60 | | |
| 1 Road Sweeper, S.P., 8' wide | | 604.80 | | 665.28 | | |
| 1 F.E. Loader, W.M., 1.5 C.Y. | | 319.40 | | 351.34 | 116.97 | 128.66 |
| 56 L.H., Daily Totals | | $8753.80 | | $10542.02 | $156.32 | $188.25 |

### Crew B-72

| Crew No. | Bare Costs Hr. | Daily | Incl. Subs O&P Hr. | Daily | Cost Per Labor-Hour Bare Costs | Incl. O&P |
|---|---|---|---|---|---|---|
| 1 Labor Foreman (outside) | $36.35 | $290.80 | $55.75 | $446.00 | $40.10 | $60.61 |
| 3 Laborers | 34.35 | 824.40 | 52.70 | 1264.80 | | |
| 4 Equip. Oper. (med.) | 45.35 | 1451.20 | 67.75 | 2168.00 | | |
| 1 Pvmt. Profiler, 750 H.P. | | 5626.00 | | 6188.60 | | |
| 1 Hammermill, 250 H.P. | | 1692.00 | | 1861.20 | | |
| 1 Windrow Loader | | 1210.00 | | 1331.00 | | |
| 1 Mix Paver 165 H.P. | | 2016.00 | | 2217.60 | | |
| 1 Roller, Pneum. Whl, 12 Ton | | 309.20 | | 340.12 | 169.58 | 186.54 |
| 64 L.H., Daily Totals | | $13419.60 | | $15817.32 | $209.68 | $247.15 |

### Crew B-73

| Crew No. | Bare Costs Hr. | Daily | Incl. Subs O&P Hr. | Daily | Cost Per Labor-Hour Bare Costs | Incl. O&P |
|---|---|---|---|---|---|---|
| 1 Labor Foreman (outside) | $36.35 | $290.80 | $55.75 | $446.00 | $41.48 | $62.49 |
| 2 Laborers | 34.35 | 549.60 | 52.70 | 843.20 | | |
| 5 Equip. Oper. (med.) | 45.35 | 1814.00 | 67.75 | 2710.00 | | |
| 1 Road Mixer, 310 H.P. | | 1890.00 | | 2079.00 | | |
| 1 Tandem Roller, 10 Ton | | 224.20 | | 246.62 | | |
| 1 Hammermill, 250 H.P. | | 1692.00 | | 1861.20 | | |
| 1 Grader, 30,000 Lbs. | | 542.80 | | 597.08 | | |
| .5 F.E. Loader, W.M., 1.5 C.Y. | | 159.70 | | 175.67 | | |
| .5 Truck Tractor, 220 H.P. | | 139.60 | | 153.56 | | |
| .5 Water Tank Trailer, 5000 Gal. | | 73.10 | | 80.41 | 73.77 | 81.15 |
| 64 L.H., Daily Totals | | $7375.80 | | $9192.74 | $115.25 | $143.64 |

## Crew B-74

| Crew No. | Bare Costs Hr. | Bare Costs Daily | Incl. Subs O&P Hr. | Incl. Subs O&P Daily | Cost Per Labor-Hour Bare Costs | Cost Per Labor-Hour Incl. O&P |
|---|---|---|---|---|---|---|
| 1 Labor Foreman (outside) | $36.35 | $290.80 | $55.75 | $446.00 | $40.11 | $60.49 |
| 1 Laborer | 34.35 | 274.80 | 52.70 | 421.60 | | |
| 4 Equip. Oper. (med.) | 45.35 | 1451.20 | 67.75 | 2168.00 | | |
| 2 Truck Drivers (heavy) | 34.40 | 550.40 | 52.25 | 836.00 | | |
| 1 Grader, 30,000 Lbs. | | 542.80 | | 597.08 | | |
| 1 Ripper, beam & 1 shank | | 83.20 | | 91.52 | | |
| 2 Stabilizers, 310 H.P. | | 2652.00 | | 2917.20 | | |
| 1 Flatbed Truck, Gas, 3 Ton | | 217.00 | | 238.70 | | |
| 1 Chem. Spreader, Towed | | 49.00 | | 53.90 | | |
| 1 Roller, Vibratory, 25 Ton | | 596.80 | | 656.48 | | |
| 1 Water Tank Trailer, 5000 Gal. | | 146.20 | | 160.82 | | |
| 1 Truck Tractor, 220 H.P. | | 279.20 | | 307.12 | 71.35 | 78.48 |
| 64 L.H., Daily Totals | | $7133.40 | | $8894.42 | $111.46 | $138.98 |

## Crew B-75

| Crew No. | Bare Costs Hr. | Bare Costs Daily | Incl. Subs O&P Hr. | Incl. Subs O&P Daily | Cost Per Labor-Hour Bare Costs | Cost Per Labor-Hour Incl. O&P |
|---|---|---|---|---|---|---|
| 1 Labor Foreman (outside) | $36.35 | $290.80 | $55.75 | $446.00 | $40.93 | $61.67 |
| 1 Laborer | 34.35 | 274.80 | 52.70 | 421.60 | | |
| 4 Equip. Oper. (med.) | 45.35 | 1451.20 | 67.75 | 2168.00 | | |
| 1 Truck Driver (heavy) | 34.40 | 275.20 | 52.25 | 418.00 | | |
| 1 Grader, 30,000 Lbs. | | 542.80 | | 597.08 | | |
| 1 Ripper, beam & 1 shank | | 83.20 | | 91.52 | | |
| 2 Stabilizers, 310 H.P. | | 2652.00 | | 2917.20 | | |
| 1 Dist. Tanker, 3000 Gallon | | 309.00 | | 339.90 | | |
| 1 Truck Tractor, 6x4, 380 H.P. | | 459.40 | | 505.34 | | |
| 1 Roller, Vibratory, 25 Ton | | 596.80 | | 656.48 | 82.91 | 91.21 |
| 56 L.H., Daily Totals | | $6935.20 | | $8561.12 | $123.84 | $152.88 |

## Crew B-76

| Crew No. | Bare Costs Hr. | Bare Costs Daily | Incl. Subs O&P Hr. | Incl. Subs O&P Daily | Cost Per Labor-Hour Bare Costs | Cost Per Labor-Hour Incl. O&P |
|---|---|---|---|---|---|---|
| 1 Dock Builder Foreman | $44.00 | $352.00 | $70.55 | $564.40 | $43.03 | $67.38 |
| 5 Dock Builders | 42.00 | 1680.00 | 67.35 | 2694.00 | | |
| 2 Equip. Oper. (crane) | 46.50 | 744.00 | 69.45 | 1111.20 | | |
| 1 Equip. Oper. Oiler | 40.30 | 322.40 | 60.20 | 481.60 | | |
| 1 Crawler Crane, 50 Ton | | 1279.00 | | 1406.90 | | |
| 1 Barge, 400 Ton | | 820.40 | | 902.44 | | |
| 1 Hammer, diesel, 15K ft.-lbs. | | 614.20 | | 675.62 | | |
| 1 Lead, 60' high | | 74.00 | | 81.40 | | |
| 1 Air Compressor, 600 cfm | | 424.60 | | 467.06 | | |
| 2 -50' Air Hoses, 3" | | 29.80 | | 32.78 | 45.03 | 49.53 |
| 72 L.H., Daily Totals | | $6340.40 | | $8417.40 | $88.06 | $116.91 |

## Crew B-76A

| Crew No. | Bare Costs Hr. | Bare Costs Daily | Incl. Subs O&P Hr. | Incl. Subs O&P Daily | Cost Per Labor-Hour Bare Costs | Cost Per Labor-Hour Incl. O&P |
|---|---|---|---|---|---|---|
| 1 Laborer Foreman | $36.35 | $290.80 | $55.75 | $446.00 | $36.86 | $56.11 |
| 5 Laborers | 34.35 | 1374.00 | 52.70 | 2108.00 | | |
| 1 Equip. Oper. (crane) | 46.50 | 372.00 | 69.45 | 555.60 | | |
| 1 Equip. Oper. Oiler | 40.30 | 322.40 | 60.20 | 481.60 | | |
| 1 Crawler Crane, 50 Ton | | 1279.00 | | 1406.90 | | |
| 1 Barge, 400 Ton | | 820.40 | | 902.44 | 32.80 | 36.08 |
| 64 L.H., Daily Totals | | $4458.60 | | $5900.54 | $69.67 | $92.20 |

## Crew B-77

| Crew No. | Bare Costs Hr. | Bare Costs Daily | Incl. Subs O&P Hr. | Incl. Subs O&P Daily | Cost Per Labor-Hour Bare Costs | Cost Per Labor-Hour Incl. O&P |
|---|---|---|---|---|---|---|
| 1 Labor Foreman | $36.35 | $290.80 | $55.75 | $446.00 | $34.56 | $52.92 |
| 3 Laborers | 34.35 | 824.40 | 52.70 | 1264.80 | | |
| 1 Truck Driver (light) | 33.40 | 267.20 | 50.75 | 406.00 | | |
| 1 Crack Cleaner, 25 H.P. | | 53.00 | | 58.30 | | |
| 1 Crack Filler, Trailer Mtd. | | 194.60 | | 214.06 | | |
| 1 Flatbed Truck, Gas, 3 Ton | | 217.00 | | 238.70 | 11.62 | 12.78 |
| 40 L.H., Daily Totals | | $1847.00 | | $2627.86 | $46.17 | $65.70 |

## Crew B-78

| Crew No. | Bare Costs Hr. | Bare Costs Daily | Incl. Subs O&P Hr. | Incl. Subs O&P Daily | Cost Per Labor-Hour Bare Costs | Cost Per Labor-Hour Incl. O&P |
|---|---|---|---|---|---|---|
| 1 Labor Foreman | $36.35 | $290.80 | $55.75 | $446.00 | $34.52 | $52.88 |
| 4 Laborers | 34.35 | 1099.20 | 52.70 | 1686.40 | | |
| 1 Truck Driver (light) | 33.40 | 267.20 | 50.75 | 406.00 | | |
| 1 Paint Striper, S.P. | | 147.80 | | 162.58 | | |
| 1 Flatbed Truck, Gas, 3 Ton | | 217.00 | | 238.70 | | |
| 1 Pickup Truck, 3/4 Ton | | 104.60 | | 115.06 | 9.78 | 10.76 |
| 48 L.H., Daily Totals | | $2126.60 | | $3054.74 | $44.30 | $63.64 |

## Crew B-78A

| Crew No. | Bare Costs Hr. | Bare Costs Daily | Incl. Subs O&P Hr. | Incl. Subs O&P Daily | Cost Per Labor-Hour Bare Costs | Cost Per Labor-Hour Incl. O&P |
|---|---|---|---|---|---|---|
| 1 Equip. Oper. (light) | $43.60 | $348.80 | $65.15 | $521.20 | $43.60 | $65.15 |
| 1 Line Rem. (metal balls) 115 H.P. | | 837.20 | | 920.92 | 104.65 | 115.11 |
| 8 L.H., Daily Totals | | $1186.00 | | $1442.12 | $148.25 | $180.26 |

## Crew B-78B

| Crew No. | Bare Costs Hr. | Bare Costs Daily | Incl. Subs O&P Hr. | Incl. Subs O&P Daily | Cost Per Labor-Hour Bare Costs | Cost Per Labor-Hour Incl. O&P |
|---|---|---|---|---|---|---|
| 2 Laborers | $34.35 | $549.60 | $52.70 | $843.20 | $35.38 | $54.08 |
| .25 Equip. Oper. (light) | 43.60 | 87.20 | 65.15 | 130.30 | | |
| 1 Pickup Truck, 3/4 Ton | | 104.60 | | 115.06 | | |
| 1 Line Rem., 11 H.P., walk behind | | 57.40 | | 63.14 | | |
| .25 Road Sweeper, S.P., 8' wide | | 151.20 | | 166.32 | 17.40 | 19.14 |
| 18 L.H., Daily Totals | | $950.00 | | $1318.02 | $52.78 | $73.22 |

## Crew B-79

| Crew No. | Bare Costs Hr. | Bare Costs Daily | Incl. Subs O&P Hr. | Incl. Subs O&P Daily | Cost Per Labor-Hour Bare Costs | Cost Per Labor-Hour Incl. O&P |
|---|---|---|---|---|---|---|
| 1 Labor Foreman | $36.35 | $290.80 | $55.75 | $446.00 | $34.56 | $52.92 |
| 3 Laborers | 34.35 | 824.40 | 52.70 | 1264.80 | | |
| 1 Truck Driver (light) | 33.40 | 267.20 | 50.75 | 406.00 | | |
| 1 Thermo. Striper, T.M. | | 653.80 | | 719.18 | | |
| 1 Flatbed Truck, Gas, 3 Ton | | 217.00 | | 238.70 | | |
| 2 Pickup Trucks, 3/4 Ton | | 209.20 | | 230.12 | 27.00 | 29.70 |
| 40 L.H., Daily Totals | | $2462.40 | | $3304.80 | $61.56 | $82.62 |

## Crew B-79A

| Crew No. | Bare Costs Hr. | Bare Costs Daily | Incl. Subs O&P Hr. | Incl. Subs O&P Daily | Cost Per Labor-Hour Bare Costs | Cost Per Labor-Hour Incl. O&P |
|---|---|---|---|---|---|---|
| 1.5 Equip. Oper. (light) | $43.60 | $523.20 | $65.15 | $781.80 | $43.60 | $65.15 |
| .5 Line Remov. (grinder) 115 H.P. | | 450.90 | | 495.99 | | |
| 1 Line Rem. (metal balls) 115 H.P. | | 837.20 | | 920.92 | 107.34 | 118.08 |
| 12 L.H., Daily Totals | | $1811.30 | | $2198.71 | $150.94 | $183.23 |

## Crew B-80

| Crew No. | Bare Costs Hr. | Bare Costs Daily | Incl. Subs O&P Hr. | Incl. Subs O&P Daily | Cost Per Labor-Hour Bare Costs | Cost Per Labor-Hour Incl. O&P |
|---|---|---|---|---|---|---|
| 1 Labor Foreman | $36.35 | $290.80 | $55.75 | $446.00 | $36.92 | $56.09 |
| 1 Laborer | 34.35 | 274.80 | 52.70 | 421.60 | | |
| 1 Truck Driver (light) | 33.40 | 267.20 | 50.75 | 406.00 | | |
| 1 Equip. Oper. (light) | 43.60 | 348.80 | 65.15 | 521.20 | | |
| 1 Flatbed Truck, Gas, 3 Ton | | 217.00 | | 238.70 | | |
| 1 Earth Auger, Truck-Mtd. | | 401.80 | | 441.98 | 19.34 | 21.27 |
| 32 L.H., Daily Totals | | $1800.40 | | $2475.48 | $56.26 | $77.36 |

## Crew B-80A

| Crew No. | Bare Costs Hr. | Bare Costs Daily | Incl. Subs O&P Hr. | Incl. Subs O&P Daily | Cost Per Labor-Hour Bare Costs | Cost Per Labor-Hour Incl. O&P |
|---|---|---|---|---|---|---|
| 3 Laborers | $34.35 | $824.40 | $52.70 | $1264.80 | $34.35 | $52.70 |
| 1 Flatbed Truck, Gas, 3 Ton | | 217.00 | | 238.70 | 9.04 | 9.95 |
| 24 L.H., Daily Totals | | $1041.40 | | $1503.50 | $43.39 | $62.65 |

## Crew B-80B

| Crew No. | Bare Costs Hr. | Bare Costs Daily | Incl. Subs O&P Hr. | Incl. Subs O&P Daily | Cost Per Labor-Hour Bare Costs | Cost Per Labor-Hour Incl. O&P |
|---|---|---|---|---|---|---|
| 3 Laborers | $34.35 | $824.40 | $52.70 | $1264.80 | $36.66 | $55.81 |
| 1 Equip. Oper. (light) | 43.60 | 348.80 | 65.15 | 521.20 | | |
| 1 Crane, Flatbed Mounted, 3 Ton | | 225.60 | | 248.16 | 7.05 | 7.75 |
| 32 L.H., Daily Totals | | $1398.80 | | $2034.16 | $43.71 | $63.57 |

| Crew No. | Bare Costs | | Incl. Subs O&P | | Cost Per Labor-Hour | |
|---|---|---|---|---|---|---|
| **Crew B-80C** | Hr. | Daily | Hr. | Daily | Bare Costs | Incl. O&P |
| 2 Laborers | $34.35 | $549.60 | $52.70 | $843.20 | $34.03 | $52.05 |
| 1 Truck Driver (light) | 33.40 | 267.20 | 50.75 | 406.00 | | |
| 1 Flatbed Truck, Gas, 1.5 Ton | | 171.40 | | 188.54 | | |
| 1 Manual fence post auger, gas | | 7.20 | | 7.92 | 7.44 | 8.19 |
| 24 L.H., Daily Totals | | $995.40 | | $1445.66 | $41.48 | $60.24 |
| **Crew B-81** | Hr. | Daily | Hr. | Daily | Bare Costs | Incl. O&P |
| 1 Laborer | $34.35 | $274.80 | $52.70 | $421.60 | $38.03 | $57.57 |
| 1 Equip. Oper. (med.) | 45.35 | 362.80 | 67.75 | 542.00 | | |
| 1 Truck Driver (heavy) | 34.40 | 275.20 | 52.25 | 418.00 | | |
| 1 Hydromulcher, T.M., 3000 Gal. | | 278.00 | | 305.80 | | |
| 1 Truck Tractor, 220 H.P. | | 279.20 | | 307.12 | 23.22 | 25.54 |
| 24 L.H., Daily Totals | | $1470.00 | | $1994.52 | $61.25 | $83.11 |
| **Crew B-81A** | Hr. | Daily | Hr. | Daily | Bare Costs | Incl. O&P |
| 1 Laborer | $34.35 | $274.80 | $52.70 | $421.60 | $33.88 | $51.73 |
| 1 Truck Driver (light) | 33.40 | 267.20 | 50.75 | 406.00 | | |
| 1 Hydromulcher, T.M., 600 Gal. | | 113.20 | | 124.52 | | |
| 1 Flatbed Truck, Gas, 3 Ton | | 217.00 | | 238.70 | 20.64 | 22.70 |
| 16 L.H., Daily Totals | | $872.20 | | $1190.82 | $54.51 | $74.43 |
| **Crew B-82** | Hr. | Daily | Hr. | Daily | Bare Costs | Incl. O&P |
| 1 Laborer | $34.35 | $274.80 | $52.70 | $421.60 | $38.98 | $58.92 |
| 1 Equip. Oper. (light) | 43.60 | 348.80 | 65.15 | 521.20 | | |
| 1 Horiz. Borer, 6 H.P. | | 76.00 | | 83.60 | 4.75 | 5.22 |
| 16 L.H., Daily Totals | | $699.60 | | $1026.40 | $43.73 | $64.15 |
| **Crew B-82A** | Hr. | Daily | Hr. | Daily | Bare Costs | Incl. O&P |
| 1 Laborer | $34.35 | $274.80 | $52.70 | $421.60 | $38.98 | $58.92 |
| 1 Equip. Oper. (light) | 43.60 | 348.80 | 65.15 | 521.20 | | |
| 1 Flatbed Truck, Gas, 3 Ton | | 217.00 | | 238.70 | | |
| 1 Flatbed Trailer, 25 Ton | | 111.20 | | 122.32 | | |
| 1 Horiz. Dir. Drill, 20k lb. thrust | | 622.60 | | 684.86 | 59.42 | 65.37 |
| 16 L.H., Daily Totals | | $1574.40 | | $1988.68 | $98.40 | $124.29 |
| **Crew B-82B** | Hr. | Daily | Hr. | Daily | Bare Costs | Incl. O&P |
| 2 Laborers | $34.35 | $549.60 | $52.70 | $843.20 | $37.43 | $56.85 |
| 1 Equip. Oper. (light) | 43.60 | 348.80 | 65.15 | 521.20 | | |
| 1 Flatbed Truck, Gas, 3 Ton | | 217.00 | | 238.70 | | |
| 1 Flatbed Trailer, 25 Ton | | 111.20 | | 122.32 | | |
| 1 Horiz. Dir. Drill, 30k lb. thrust | | 888.40 | | 977.24 | 50.69 | 55.76 |
| 24 L.H., Daily Totals | | $2115.00 | | $2702.66 | $88.13 | $112.61 |
| **Crew B-82C** | Hr. | Daily | Hr. | Daily | Bare Costs | Incl. O&P |
| 2 Laborers | $34.35 | $549.60 | $52.70 | $843.20 | $37.43 | $56.85 |
| 1 Equip. Oper. (light) | 43.60 | 348.80 | 65.15 | 521.20 | | |
| 1 Flatbed Truck, Gas, 3 Ton | | 217.00 | | 238.70 | | |
| 1 Flatbed Trailer, 25 Ton | | 111.20 | | 122.32 | | |
| 1 Horiz. Dir. Drill, 50k lb. thrust | | 1174.00 | | 1291.40 | 62.59 | 68.85 |
| 24 L.H., Daily Totals | | $2400.60 | | $3016.82 | $100.03 | $125.70 |
| **Crew B-82D** | Hr. | Daily | Hr. | Daily | Bare Costs | Incl. O&P |
| 1 Equip. Oper. (light) | $43.60 | $348.80 | $65.15 | $521.20 | $43.60 | $65.15 |
| 1 Mud Trailer for HDD, 1500 gallon | | 265.80 | | 292.38 | 33.23 | 36.55 |
| 8 L.H., Daily Totals | | $614.60 | | $813.58 | $76.83 | $101.70 |

| Crew No. | Bare Costs | | Incl. Subs O&P | | Cost Per Labor-Hour | |
|---|---|---|---|---|---|---|
| **Crew B-83** | Hr. | Daily | Hr. | Daily | Bare Costs | Incl. O&P |
| 1 Tugboat Captain | $45.35 | $362.80 | $67.75 | $542.00 | $39.85 | $60.23 |
| 1 Tugboat Hand | 34.35 | 274.80 | 52.70 | 421.60 | | |
| 1 Tugboat, 250 H.P. | | 632.20 | | 695.42 | 39.51 | 43.46 |
| 16 L.H., Daily Totals | | $1269.80 | | $1659.02 | $79.36 | $103.69 |
| **Crew B-84** | Hr. | Daily | Hr. | Daily | Bare Costs | Incl. O&P |
| 1 Equip. Oper. (med.) | $45.35 | $362.80 | $67.75 | $542.00 | $45.35 | $67.75 |
| 1 Rotary Mower/Tractor | | 293.80 | | 323.18 | 36.73 | 40.40 |
| 8 L.H., Daily Totals | | $656.60 | | $865.18 | $82.08 | $108.15 |
| **Crew B-85** | Hr. | Daily | Hr. | Daily | Bare Costs | Incl. O&P |
| 3 Laborers | $34.35 | $824.40 | $52.70 | $1264.80 | $36.56 | $55.62 |
| 1 Equip. Oper. (med.) | 45.35 | 362.80 | 67.75 | 542.00 | | |
| 1 Truck Driver (heavy) | 34.40 | 275.20 | 52.25 | 418.00 | | |
| 1 Aerial Lift Truck, 80' | | 636.20 | | 699.82 | | |
| 1 Brush Chipper, 12", 130 H.P. | | 217.00 | | 238.70 | | |
| 1 Pruning Saw, Rotary | | 6.30 | | 6.93 | 21.49 | 23.64 |
| 40 L.H., Daily Totals | | $2321.90 | | $3170.25 | $58.05 | $79.26 |
| **Crew B-86** | Hr. | Daily | Hr. | Daily | Bare Costs | Incl. O&P |
| 1 Equip. Oper. (med.) | $45.35 | $362.80 | $67.75 | $542.00 | $45.35 | $67.75 |
| 1 Stump Chipper, S.P. | | 169.35 | | 186.29 | 21.17 | 23.29 |
| 8 L.H., Daily Totals | | $532.15 | | $728.28 | $66.52 | $91.04 |
| **Crew B-86A** | Hr. | Daily | Hr. | Daily | Bare Costs | Incl. O&P |
| 1 Equip. Oper. (med.) | $45.35 | $362.80 | $67.75 | $542.00 | $45.35 | $67.75 |
| 1 Grader, 30,000 Lbs. | | 542.80 | | 597.08 | 67.85 | 74.64 |
| 8 L.H., Daily Totals | | $905.60 | | $1139.08 | $113.20 | $142.38 |
| **Crew B-86B** | Hr. | Daily | Hr. | Daily | Bare Costs | Incl. O&P |
| 1 Equip. Oper. (med.) | $45.35 | $362.80 | $67.75 | $542.00 | $45.35 | $67.75 |
| 1 Dozer, 200 H.P. | | 1051.00 | | 1156.10 | 131.38 | 144.51 |
| 8 L.H., Daily Totals | | $1413.80 | | $1698.10 | $176.72 | $212.26 |
| **Crew B-87** | Hr. | Daily | Hr. | Daily | Bare Costs | Incl. O&P |
| 1 Laborer | $34.35 | $274.80 | $52.70 | $421.60 | $43.15 | $64.74 |
| 4 Equip. Oper. (med.) | 45.35 | 1451.20 | 67.75 | 2168.00 | | |
| 2 Feller Bunchers, 100 H.P. | | 1128.00 | | 1240.80 | | |
| 1 Log Chipper, 22" Tree | | 554.80 | | 610.28 | | |
| 1 Dozer, 105 H.P. | | 505.40 | | 555.94 | | |
| 1 Chain Saw, gas, 36" Long | | 41.60 | | 45.76 | 55.74 | 61.32 |
| 40 L.H., Daily Totals | | $3955.80 | | $5042.38 | $98.89 | $126.06 |
| **Crew B-88** | Hr. | Daily | Hr. | Daily | Bare Costs | Incl. O&P |
| 1 Laborer | $34.35 | $274.80 | $52.70 | $421.60 | $43.78 | $65.60 |
| 6 Equip. Oper. (med.) | 45.35 | 2176.80 | 67.75 | 3252.00 | | |
| 2 Feller Bunchers, 100 H.P. | | 1128.00 | | 1240.80 | | |
| 1 Log Chipper, 22" Tree | | 554.80 | | 610.28 | | |
| 2 Log Skidders, 50 H.P. | | 1688.00 | | 1856.80 | | |
| 1 Dozer, 105 H.P. | | 505.40 | | 555.94 | | |
| 1 Chain Saw, gas, 36" Long | | 41.60 | | 45.76 | 69.96 | 76.96 |
| 56 L.H., Daily Totals | | $6369.40 | | $7983.18 | $113.74 | $142.56 |
| **Crew B-89** | Hr. | Daily | Hr. | Daily | Bare Costs | Incl. O&P |
| 1 Equip. Oper. (light) | $43.60 | $348.80 | $65.15 | $521.20 | $38.50 | $57.95 |
| 1 Truck Driver (light) | 33.40 | 267.20 | 50.75 | 406.00 | | |
| 1 Flatbed Truck, Gas, 3 Ton | | 217.00 | | 238.70 | | |
| 1 Concrete Saw | | 134.20 | | 147.62 | | |
| 1 Water Tank, 65 Gal. | | 18.45 | | 20.30 | 23.10 | 25.41 |
| 16 L.H., Daily Totals | | $985.65 | | $1333.82 | $61.60 | $83.36 |

| Crew No. | Bare Costs | | Incl. Subs O&P | | Cost Per Labor-Hour | |
|---|---|---|---|---|---|---|
| **Crew B-89A** | Hr. | Daily | Hr. | Daily | Bare Costs | Incl. O&P |
| 1 Skilled Worker | $44.40 | $355.20 | $68.10 | $544.80 | $39.38 | $60.40 |
| 1 Laborer | 34.35 | 274.80 | 52.70 | 421.60 | | |
| 1 Core Drill (large) | | 109.80 | | 120.78 | 6.86 | 7.55 |
| 16 L.H., Daily Totals | | $739.80 | | $1087.18 | $46.24 | $67.95 |

| | Bare Costs | | Incl. Subs O&P | | Cost Per Labor-Hour | |
|---|---|---|---|---|---|---|
| **Crew B-89B** | Hr. | Daily | Hr. | Daily | Bare Costs | Incl. O&P |
| 1 Equip. Oper. (light) | $43.60 | $348.80 | $65.15 | $521.20 | $38.50 | $57.95 |
| 1 Truck Driver, Light | 33.40 | 267.20 | 50.75 | 406.00 | | |
| 1 Wall Saw, Hydraulic, 10 H.P. | | 89.00 | | 97.90 | | |
| 1 Generator, Diesel, 100 kW | | 329.00 | | 361.90 | | |
| 1 Water Tank, 65 Gal. | | 18.45 | | 20.30 | | |
| 1 Flatbed Truck, Gas, 3 Ton | | 217.00 | | 238.70 | 40.84 | 44.92 |
| 16 L.H., Daily Totals | | $1269.45 | | $1645.99 | $79.34 | $102.87 |

| | Bare Costs | | Incl. Subs O&P | | Cost Per Labor-Hour | |
|---|---|---|---|---|---|---|
| **Crew B-90** | Hr. | Daily | Hr. | Daily | Bare Costs | Incl. O&P |
| 1 Labor Foreman (outside) | $36.35 | $290.80 | $55.75 | $446.00 | $36.92 | $56.08 |
| 3 Laborers | 34.35 | 824.40 | 52.70 | 1264.80 | | |
| 2 Equip. Oper. (light) | 43.60 | 697.60 | 65.15 | 1042.40 | | |
| 2 Truck Drivers (heavy) | 34.40 | 550.40 | 52.25 | 836.00 | | |
| 1 Road Mixer, 310 H.P. | | 1890.00 | | 2079.00 | | |
| 1 Dist. Truck, 2000 Gal. | | 280.80 | | 308.88 | 33.92 | 37.31 |
| 64 L.H., Daily Totals | | $4534.00 | | $5977.08 | $70.84 | $93.39 |

| | Bare Costs | | Incl. Subs O&P | | Cost Per Labor-Hour | |
|---|---|---|---|---|---|---|
| **Crew B-90A** | Hr. | Daily | Hr. | Daily | Bare Costs | Incl. O&P |
| 1 Labor Foreman | $36.35 | $290.80 | $55.75 | $446.00 | $40.92 | $61.74 |
| 2 Laborers | 34.35 | 549.60 | 52.70 | 843.20 | | |
| 4 Equip. Oper. (med.) | 45.35 | 1451.20 | 67.75 | 2168.00 | | |
| 2 Graders, 30,000 Lbs. | | 1085.60 | | 1194.16 | | |
| 1 Tandem Roller, 10 Ton | | 224.20 | | 246.62 | | |
| 1 Roller, Pneum. Whl, 12 Ton | | 309.20 | | 340.12 | 28.91 | 31.80 |
| 56 L.H., Daily Totals | | $3910.60 | | $5238.10 | $69.83 | $93.54 |

| | Bare Costs | | Incl. Subs O&P | | Cost Per Labor-Hour | |
|---|---|---|---|---|---|---|
| **Crew B-90B** | Hr. | Daily | Hr. | Daily | Bare Costs | Incl. O&P |
| 1 Labor Foreman | $36.35 | $290.80 | $55.75 | $446.00 | $40.18 | $60.73 |
| 2 Laborers | 34.35 | 549.60 | 52.70 | 843.20 | | |
| 3 Equip. Oper. (med.) | 45.35 | 1088.40 | 67.75 | 1626.00 | | |
| 1 Tandem Roller, 10 Ton | | 224.20 | | 246.62 | | |
| 1 Roller, Pneum. Whl, 12 Ton | | 309.20 | | 340.12 | | |
| 1 Road Mixer, 310 H.P. | | 1890.00 | | 2079.00 | 50.49 | 55.54 |
| 48 L.H., Daily Totals | | $4352.20 | | $5580.94 | $90.67 | $116.27 |

| | Bare Costs | | Incl. Subs O&P | | Cost Per Labor-Hour | |
|---|---|---|---|---|---|---|
| **Crew B-91** | Hr. | Daily | Hr. | Daily | Bare Costs | Incl. O&P |
| 1 Labor Foreman (outside) | $36.35 | $290.80 | $55.75 | $446.00 | $40.11 | $60.55 |
| 2 Laborers | 34.35 | 549.60 | 52.70 | 843.20 | | |
| 4 Equip. Oper. (med.) | 45.35 | 1451.20 | 67.75 | 2168.00 | | |
| 1 Truck Driver (heavy) | 34.40 | 275.20 | 52.25 | 418.00 | | |
| 1 Dist. Tanker, 3000 Gallon | | 309.00 | | 339.90 | | |
| 1 Truck Tractor, 6x4, 380 H.P. | | 459.40 | | 505.34 | | |
| 1 Aggreg. Spreader, S.P. | | 748.40 | | 823.24 | | |
| 1 Roller, Pneum. Whl, 12 Ton | | 309.20 | | 340.12 | | |
| 1 Tandem Roller, 10 Ton | | 224.20 | | 246.62 | 32.03 | 35.24 |
| 64 L.H., Daily Totals | | $4617.00 | | $6130.42 | $72.14 | $95.79 |

| | Bare Costs | | Incl. Subs O&P | | Cost Per Labor-Hour | |
|---|---|---|---|---|---|---|
| **Crew B-92** | Hr. | Daily | Hr. | Daily | Bare Costs | Incl. O&P |
| 1 Labor Foreman (outside) | $36.35 | $290.80 | $55.75 | $446.00 | $34.85 | $53.46 |
| 3 Laborers | 34.35 | 824.40 | 52.70 | 1264.80 | | |
| 1 Crack Cleaner, 25 H.P. | | 53.00 | | 58.30 | | |
| 1 Air Compressor, 60 cfm | | 110.20 | | 121.22 | | |
| 1 Tar Kettle, T.M. | | 74.65 | | 82.11 | | |
| 1 Flatbed Truck, Gas, 3 Ton | | 217.00 | | 238.70 | 14.21 | 15.64 |
| 32 L.H., Daily Totals | | $1570.05 | | $2211.14 | $49.06 | $69.10 |

| | Bare Costs | | Incl. Subs O&P | | Cost Per Labor-Hour | |
|---|---|---|---|---|---|---|
| **Crew B-93** | Hr. | Daily | Hr. | Daily | Bare Costs | Incl. O&P |
| 1 Equip. Oper. (med.) | $45.35 | $362.80 | $67.75 | $542.00 | $45.35 | $67.75 |
| 1 Feller Buncher, 100 H.P. | | 564.00 | | 620.40 | 70.50 | 77.55 |
| 8 L.H., Daily Totals | | $926.80 | | $1162.40 | $115.85 | $145.30 |

| | Bare Costs | | Incl. Subs O&P | | Cost Per Labor-Hour | |
|---|---|---|---|---|---|---|
| **Crew B-94A** | Hr. | Daily | Hr. | Daily | Bare Costs | Incl. O&P |
| 1 Laborer | $34.35 | $274.80 | $52.70 | $421.60 | $34.35 | $52.70 |
| 1 Diaphragm Water Pump, 2" | | 66.60 | | 73.26 | | |
| 1 -20' Suction Hose, 2" | | 1.95 | | 2.15 | | |
| 2 -50' Discharge Hoses, 2" | | 1.80 | | 1.98 | 8.79 | 9.67 |
| 8 L.H., Daily Totals | | $345.15 | | $498.99 | $43.14 | $62.37 |

| | Bare Costs | | Incl. Subs O&P | | Cost Per Labor-Hour | |
|---|---|---|---|---|---|---|
| **Crew B-94B** | Hr. | Daily | Hr. | Daily | Bare Costs | Incl. O&P |
| 1 Laborer | $34.35 | $274.80 | $52.70 | $421.60 | $34.35 | $52.70 |
| 1 Diaphragm Water Pump, 4" | | 109.40 | | 120.34 | | |
| 1 -20' Suction Hose, 4" | | 3.25 | | 3.58 | | |
| 2 -50' Discharge Hoses, 4" | | 4.70 | | 5.17 | 14.67 | 16.14 |
| 8 L.H., Daily Totals | | $392.15 | | $550.68 | $49.02 | $68.84 |

| | Bare Costs | | Incl. Subs O&P | | Cost Per Labor-Hour | |
|---|---|---|---|---|---|---|
| **Crew B-94C** | Hr. | Daily | Hr. | Daily | Bare Costs | Incl. O&P |
| 1 Laborer | $34.35 | $274.80 | $52.70 | $421.60 | $34.35 | $52.70 |
| 1 Centrifugal Water Pump, 3" | | 71.00 | | 78.10 | | |
| 1 -20' Suction Hose, 3" | | 3.05 | | 3.36 | | |
| 2 -50' Discharge Hoses, 3" | | 3.50 | | 3.85 | 9.69 | 10.66 |
| 8 L.H., Daily Totals | | $352.35 | | $506.90 | $44.04 | $63.36 |

| | Bare Costs | | Incl. Subs O&P | | Cost Per Labor-Hour | |
|---|---|---|---|---|---|---|
| **Crew B-94D** | Hr. | Daily | Hr. | Daily | Bare Costs | Incl. O&P |
| 1 Laborer | $34.35 | $274.80 | $52.70 | $421.60 | $34.35 | $52.70 |
| 1 Centr. Water Pump, 6" | | 291.60 | | 320.76 | | |
| 1 -20' Suction Hose, 6" | | 11.90 | | 13.09 | | |
| 2 -50' Discharge Hoses, 6" | | 12.60 | | 13.86 | 39.51 | 43.46 |
| 8 L.H., Daily Totals | | $590.90 | | $769.31 | $73.86 | $96.16 |

| | Bare Costs | | Incl. Subs O&P | | Cost Per Labor-Hour | |
|---|---|---|---|---|---|---|
| **Crew C-1** | Hr. | Daily | Hr. | Daily | Bare Costs | Incl. O&P |
| 3 Carpenters | $43.05 | $1033.20 | $66.05 | $1585.20 | $40.88 | $62.71 |
| 1 Laborer | 34.35 | 274.80 | 52.70 | 421.60 | | |
| 32 L.H., Daily Totals | | $1308.00 | | $2006.80 | $40.88 | $62.71 |

| | Bare Costs | | Incl. Subs O&P | | Cost Per Labor-Hour | |
|---|---|---|---|---|---|---|
| **Crew C-2** | Hr. | Daily | Hr. | Daily | Bare Costs | Incl. O&P |
| 1 Carpenter Foreman (out) | $45.05 | $360.40 | $69.10 | $552.80 | $41.93 | $64.33 |
| 4 Carpenters | 43.05 | 1377.60 | 66.05 | 2113.60 | | |
| 1 Laborer | 34.35 | 274.80 | 52.70 | 421.60 | | |
| 48 L.H., Daily Totals | | $2012.80 | | $3088.00 | $41.93 | $64.33 |

| | Bare Costs | | Incl. Subs O&P | | Cost Per Labor-Hour | |
|---|---|---|---|---|---|---|
| **Crew C-2A** | Hr. | Daily | Hr. | Daily | Bare Costs | Incl. O&P |
| 1 Carpenter Foreman (out) | $45.05 | $360.40 | $69.10 | $552.80 | $41.57 | $63.24 |
| 3 Carpenters | 43.05 | 1033.20 | 66.05 | 1585.20 | | |
| 1 Cement Finisher | 40.85 | 326.80 | 59.50 | 476.00 | | |
| 1 Laborer | 34.35 | 274.80 | 52.70 | 421.60 | | |
| 48 L.H., Daily Totals | | $1995.20 | | $3035.60 | $41.57 | $63.24 |

| | Bare Costs | | Incl. Subs O&P | | Cost Per Labor-Hour | |
|---|---|---|---|---|---|---|
| **Crew C-3** | Hr. | Daily | Hr. | Daily | Bare Costs | Incl. O&P |
| 1 Rodman Foreman | $50.40 | $403.20 | $80.55 | $644.40 | $44.54 | $70.06 |
| 4 Rodmen (reinf.) | 48.40 | 1548.80 | 77.35 | 2475.20 | | |
| 1 Equip. Oper. (light) | 43.60 | 348.80 | 65.15 | 521.20 | | |
| 2 Laborers | 34.35 | 549.60 | 52.70 | 843.20 | | |
| 3 Stressing Equipment | | 29.40 | | 32.34 | | |
| .5 Grouting Equipment | | 76.60 | | 84.26 | 1.66 | 1.82 |
| 64 L.H., Daily Totals | | $2956.40 | | $4600.60 | $46.19 | $71.88 |

### Crew C-4

| Crew No. | Bare Costs Hr. | Daily | Incl. Subs O&P Hr. | Daily | Cost Per Labor-Hour Bare Costs | Incl. O&P |
|---|---|---|---|---|---|---|
| 1 Rodman Foreman | $50.40 | $403.20 | $80.55 | $644.40 | $48.90 | $78.15 |
| 3 Rodmen (reinf.) | 48.40 | 1161.60 | 77.35 | 1856.40 | | |
| 3 Stressing Equipment | | 29.40 | | 32.34 | .92 | 1.01 |
| 32 L.H., Daily Totals | | $1594.20 | | $2533.14 | $49.82 | $79.16 |

### Crew C-4A

| Crew No. | Bare Costs Hr. | Daily | Incl. Subs O&P Hr. | Daily | Cost Per Labor-Hour Bare Costs | Incl. O&P |
|---|---|---|---|---|---|---|
| 2 Rodmen (reinf.) | $48.40 | $774.40 | $77.35 | $1237.60 | $48.40 | $77.35 |
| 4 Stressing Equipment | | 39.20 | | 43.12 | 2.45 | 2.69 |
| 16 L.H., Daily Totals | | $813.60 | | $1280.72 | $50.85 | $80.05 |

### Crew C-5

| Crew No. | Bare Costs Hr. | Daily | Incl. Subs O&P Hr. | Daily | Cost Per Labor-Hour Bare Costs | Incl. O&P |
|---|---|---|---|---|---|---|
| 1 Rodman Foreman | $50.40 | $403.20 | $80.55 | $644.40 | $47.26 | $74.23 |
| 4 Rodmen (reinf.) | 48.40 | 1548.80 | 77.35 | 2475.20 | | |
| 1 Equip. Oper. (crane) | 46.50 | 372.00 | 69.45 | 555.60 | | |
| 1 Equip. Oper. Oiler | 40.30 | 322.40 | 60.20 | 481.60 | | |
| 1 Hyd. Crane, 25 Ton | | 748.60 | | 823.46 | 13.37 | 14.70 |
| 56 L.H., Daily Totals | | $3395.00 | | $4980.26 | $60.63 | $88.93 |

### Crew C-6

| Crew No. | Bare Costs Hr. | Daily | Incl. Subs O&P Hr. | Daily | Cost Per Labor-Hour Bare Costs | Incl. O&P |
|---|---|---|---|---|---|---|
| 1 Labor Foreman (outside) | $36.35 | $290.80 | $55.75 | $446.00 | $35.77 | $54.34 |
| 4 Laborers | 34.35 | 1099.20 | 52.70 | 1686.40 | | |
| 1 Cement Finisher | 40.85 | 326.80 | 59.50 | 476.00 | | |
| 2 Gas Engine Vibrators | | 46.40 | | 51.04 | .97 | 1.06 |
| 48 L.H., Daily Totals | | $1763.20 | | $2659.44 | $36.73 | $55.41 |

### Crew C-7

| Crew No. | Bare Costs Hr. | Daily | Incl. Subs O&P Hr. | Daily | Cost Per Labor-Hour Bare Costs | Incl. O&P |
|---|---|---|---|---|---|---|
| 1 Labor Foreman (outside) | $36.35 | $290.80 | $55.75 | $446.00 | $37.18 | $56.30 |
| 5 Laborers | 34.35 | 1374.00 | 52.70 | 2108.00 | | |
| 1 Cement Finisher | 40.85 | 326.80 | 59.50 | 476.00 | | |
| 1 Equip. Oper. (med.) | 45.35 | 362.80 | 67.75 | 542.00 | | |
| 1 Equip. Oper. (oiler) | 40.30 | 322.40 | 60.20 | 481.60 | | |
| 2 Gas Engine Vibrators | | 46.40 | | 51.04 | | |
| 1 Concrete Bucket, 1 C.Y. | | 23.20 | | 25.52 | | |
| 1 Hyd. Crane, 55 Ton | | 1208.00 | | 1328.80 | 17.74 | 19.52 |
| 72 L.H., Daily Totals | | $3954.40 | | $5458.96 | $54.92 | $75.82 |

### Crew C-7A

| Crew No. | Bare Costs Hr. | Daily | Incl. Subs O&P Hr. | Daily | Cost Per Labor-Hour Bare Costs | Incl. O&P |
|---|---|---|---|---|---|---|
| 1 Labor Foreman (outside) | $36.35 | $290.80 | $55.75 | $446.00 | $34.61 | $52.97 |
| 5 Laborers | 34.35 | 1374.00 | 52.70 | 2108.00 | | |
| 2 Truck Drivers (Heavy) | 34.40 | 550.40 | 52.25 | 836.00 | | |
| 2 Conc. Transit Mixers | | 1722.40 | | 1894.64 | 26.91 | 29.60 |
| 64 L.H., Daily Totals | | $3937.60 | | $5284.64 | $61.52 | $82.57 |

### Crew C-7B

| Crew No. | Bare Costs Hr. | Daily | Incl. Subs O&P Hr. | Daily | Cost Per Labor-Hour Bare Costs | Incl. O&P |
|---|---|---|---|---|---|---|
| 1 Labor Foreman (outside) | $36.35 | $290.80 | $55.75 | $446.00 | $36.86 | $56.11 |
| 5 Laborers | 34.35 | 1374.00 | 52.70 | 2108.00 | | |
| 1 Equipment Oper. (crane) | 46.50 | 372.00 | 69.45 | 555.60 | | |
| 1 Equipment Oiler | 40.30 | 322.40 | 60.20 | 481.60 | | |
| 1 Conc. Bucket, 2 C.Y. | | 35.80 | | 39.38 | | |
| 1 Lattice Boom Crane, 165 Ton | | 2061.00 | | 2267.10 | 32.76 | 36.04 |
| 64 L.H., Daily Totals | | $4456.00 | | $5897.68 | $69.63 | $92.15 |

### Crew C-7C

| Crew No. | Bare Costs Hr. | Daily | Incl. Subs O&P Hr. | Daily | Cost Per Labor-Hour Bare Costs | Incl. O&P |
|---|---|---|---|---|---|---|
| 1 Labor Foreman (outside) | $36.35 | $290.80 | $55.75 | $446.00 | $37.35 | $56.84 |
| 5 Laborers | 34.35 | 1374.00 | 52.70 | 2108.00 | | |
| 2 Equipment Operators (med.) | 45.35 | 725.60 | 67.75 | 1084.00 | | |
| 2 F.E. Loaders, W.M., 4 C.Y. | | 1197.20 | | 1316.92 | 18.71 | 20.58 |
| 64 L.H., Daily Totals | | $3587.60 | | $4954.92 | $56.06 | $77.42 |

### Crew C-7D

| Crew No. | Bare Costs Hr. | Daily | Incl. Subs O&P Hr. | Daily | Cost Per Labor-Hour Bare Costs | Incl. O&P |
|---|---|---|---|---|---|---|
| 1 Labor Foreman (outside) | $36.35 | $290.80 | $55.75 | $446.00 | $36.21 | $55.29 |
| 5 Laborers | 34.35 | 1374.00 | 52.70 | 2108.00 | | |
| 1 Equip. Oper. (med.) | 45.35 | 362.80 | 67.75 | 542.00 | | |
| 1 Concrete Conveyer | | 174.60 | | 192.06 | 3.12 | 3.43 |
| 56 L.H., Daily Totals | | $2202.20 | | $3288.06 | $39.33 | $58.72 |

### Crew C-8

| Crew No. | Bare Costs Hr. | Daily | Incl. Subs O&P Hr. | Daily | Cost Per Labor-Hour Bare Costs | Incl. O&P |
|---|---|---|---|---|---|---|
| 1 Labor Foreman (outside) | $36.35 | $290.80 | $55.75 | $446.00 | $38.06 | $57.23 |
| 3 Laborers | 34.35 | 824.40 | 52.70 | 1264.80 | | |
| 2 Cement Finishers | 40.85 | 653.60 | 59.50 | 952.00 | | |
| 1 Equip. Oper. (med.) | 45.35 | 362.80 | 67.75 | 542.00 | | |
| 1 Concrete Pump (small) | | 741.00 | | 815.10 | 13.23 | 14.56 |
| 56 L.H., Daily Totals | | $2872.60 | | $4019.90 | $51.30 | $71.78 |

### Crew C-8A

| Crew No. | Bare Costs Hr. | Daily | Incl. Subs O&P Hr. | Daily | Cost Per Labor-Hour Bare Costs | Incl. O&P |
|---|---|---|---|---|---|---|
| 1 Labor Foreman (outside) | $36.35 | $290.80 | $55.75 | $446.00 | $36.85 | $55.48 |
| 3 Laborers | 34.35 | 824.40 | 52.70 | 1264.80 | | |
| 2 Cement Finishers | 40.85 | 653.60 | 59.50 | 952.00 | | |
| 48 L.H., Daily Totals | | $1768.80 | | $2662.80 | $36.85 | $55.48 |

### Crew C-8B

| Crew No. | Bare Costs Hr. | Daily | Incl. Subs O&P Hr. | Daily | Cost Per Labor-Hour Bare Costs | Incl. O&P |
|---|---|---|---|---|---|---|
| 1 Labor Foreman (outside) | $36.35 | $290.80 | $55.75 | $446.00 | $36.95 | $56.32 |
| 3 Laborers | 34.35 | 824.40 | 52.70 | 1264.80 | | |
| 1 Equip. Oper. (med.) | 45.35 | 362.80 | 67.75 | 542.00 | | |
| 1 Vibrating Power Screed | | 65.50 | | 72.05 | | |
| 1 Roller, Vibratory, 25 Ton | | 596.80 | | 656.48 | | |
| 1 Dozer, 200 H.P. | | 1051.00 | | 1156.10 | 42.83 | 47.12 |
| 40 L.H., Daily Totals | | $3191.30 | | $4137.43 | $79.78 | $103.44 |

### Crew C-8C

| Crew No. | Bare Costs Hr. | Daily | Incl. Subs O&P Hr. | Daily | Cost Per Labor-Hour Bare Costs | Incl. O&P |
|---|---|---|---|---|---|---|
| 1 Labor Foreman (outside) | $36.35 | $290.80 | $55.75 | $446.00 | $37.60 | $56.85 |
| 3 Laborers | 34.35 | 824.40 | 52.70 | 1264.80 | | |
| 1 Cement Finisher | 40.85 | 326.80 | 59.50 | 476.00 | | |
| 1 Equip. Oper. (med.) | 45.35 | 362.80 | 67.75 | 542.00 | | |
| 1 Shotcrete Rig, 12 C.Y./hr | | 249.20 | | 274.12 | | |
| 1 Air Compressor, 160 cfm | | 125.00 | | 137.50 | | |
| 4 -50' Air Hoses, 1" | | 16.40 | | 18.04 | | |
| 4 -50' Air Hoses, 2" | | 31.00 | | 34.10 | 8.78 | 9.66 |
| 48 L.H., Daily Totals | | $2226.40 | | $3192.56 | $46.38 | $66.51 |

### Crew C-8D

| Crew No. | Bare Costs Hr. | Daily | Incl. Subs O&P Hr. | Daily | Cost Per Labor-Hour Bare Costs | Incl. O&P |
|---|---|---|---|---|---|---|
| 1 Labor Foreman (outside) | $36.35 | $290.80 | $55.75 | $446.00 | $38.79 | $58.27 |
| 1 Laborer | 34.35 | 274.80 | 52.70 | 421.60 | | |
| 1 Cement Finisher | 40.85 | 326.80 | 59.50 | 476.00 | | |
| 1 Equipment Oper. (light) | 43.60 | 348.80 | 65.15 | 521.20 | | |
| 1 Air Compressor, 250 cfm | | 162.80 | | 179.08 | | |
| 2 -50' Air Hoses, 1" | | 8.20 | | 9.02 | 5.34 | 5.88 |
| 32 L.H., Daily Totals | | $1412.20 | | $2052.90 | $44.13 | $64.15 |

### Crew C-8E

| Crew No. | Bare Costs Hr. | Daily | Incl. Subs O&P Hr. | Daily | Cost Per Labor-Hour Bare Costs | Incl. O&P |
|---|---|---|---|---|---|---|
| 1 Labor Foreman (outside) | $36.35 | $290.80 | $55.75 | $446.00 | $37.31 | $56.42 |
| 3 Laborers | 34.35 | 824.40 | 52.70 | 1264.80 | | |
| 1 Cement Finisher | 40.85 | 326.80 | 59.50 | 476.00 | | |
| 1 Equipment Oper. (light) | 43.60 | 348.80 | 65.15 | 521.20 | | |
| 1 Shotcrete Rig, 35 C.Y./hr | | 280.00 | | 308.00 | | |
| 1 Air Compressor, 250 cfm | | 162.80 | | 179.08 | | |
| 4 -50' Air Hoses, 1" | | 16.40 | | 18.04 | | |
| 4 -50' Air Hoses, 2" | | 31.00 | | 34.10 | 10.21 | 11.23 |
| 48 L.H., Daily Totals | | $2281.00 | | $3247.22 | $47.52 | $67.65 |

## Crew C-10

| Crew No. | Bare Costs Hr. | Daily | Incl. Subs O&P Hr. | Daily | Cost Per Labor-Hour Bare Costs | Incl. O&P |
|---|---|---|---|---|---|---|
| 1 Laborer | $34.35 | $274.80 | $52.70 | $421.60 | $38.68 | $57.23 |
| 2 Cement Finishers | 40.85 | 653.60 | 59.50 | 952.00 | | |
| 24 L.H., Daily Totals | | $928.40 | | $1373.60 | $38.68 | $57.23 |

## Crew C-10B

| Crew No. | Bare Costs Hr. | Daily | Incl. Subs O&P Hr. | Daily | Cost Per Labor-Hour Bare Costs | Incl. O&P |
|---|---|---|---|---|---|---|
| 3 Laborers | $34.35 | $824.40 | $52.70 | $1264.80 | $36.95 | $55.42 |
| 2 Cement Finishers | 40.85 | 653.60 | 59.50 | 952.00 | | |
| 1 Concrete Mixer, 10 C.F. | | 151.20 | | 166.32 | | |
| 2 Trowels, 48" Walk-Behind | | 76.80 | | 84.48 | 5.70 | 6.27 |
| 40 L.H., Daily Totals | | $1706.00 | | $2467.60 | $42.65 | $61.69 |

## Crew C-10C

| Crew No. | Bare Costs Hr. | Daily | Incl. Subs O&P Hr. | Daily | Cost Per Labor-Hour Bare Costs | Incl. O&P |
|---|---|---|---|---|---|---|
| 1 Laborer | $34.35 | $274.80 | $52.70 | $421.60 | $38.68 | $57.23 |
| 2 Cement Finishers | 40.85 | 653.60 | 59.50 | 952.00 | | |
| 1 Trowel, 48" Walk-Behind | | 38.40 | | 42.24 | 1.60 | 1.76 |
| 24 L.H., Daily Totals | | $966.80 | | $1415.84 | $40.28 | $58.99 |

## Crew C-10D

| Crew No. | Bare Costs Hr. | Daily | Incl. Subs O&P Hr. | Daily | Cost Per Labor-Hour Bare Costs | Incl. O&P |
|---|---|---|---|---|---|---|
| 1 Laborer | $34.35 | $274.80 | $52.70 | $421.60 | $38.68 | $57.23 |
| 2 Cement Finishers | 40.85 | 653.60 | 59.50 | 952.00 | | |
| 1 Vibrating Power Screed | | 65.50 | | 72.05 | | |
| 1 Trowel, 48" Walk-Behind | | 38.40 | | 42.24 | 4.33 | 4.76 |
| 24 L.H., Daily Totals | | $1032.30 | | $1487.89 | $43.01 | $62.00 |

## Crew C-10E

| Crew No. | Bare Costs Hr. | Daily | Incl. Subs O&P Hr. | Daily | Cost Per Labor-Hour Bare Costs | Incl. O&P |
|---|---|---|---|---|---|---|
| 1 Laborer | $34.35 | $274.80 | $52.70 | $421.60 | $38.68 | $57.23 |
| 2 Cement Finishers | 40.85 | 653.60 | 59.50 | 952.00 | | |
| 1 Vibrating Power Screed | | 65.50 | | 72.05 | | |
| 1 Cement Trowel, 96" Ride-On | | 157.40 | | 173.14 | 9.29 | 10.22 |
| 24 L.H., Daily Totals | | $1151.30 | | $1618.79 | $47.97 | $67.45 |

## Crew C-11

| Crew No. | Bare Costs Hr. | Daily | Incl. Subs O&P Hr. | Daily | Cost Per Labor-Hour Bare Costs | Incl. O&P |
|---|---|---|---|---|---|---|
| 1 Struc. Steel Foreman | $50.55 | $404.40 | $89.25 | $714.00 | $47.63 | $81.49 |
| 6 Struc. Steel Workers | 48.55 | 2330.40 | 85.75 | 4116.00 | | |
| 1 Equip. Oper. (crane) | 46.50 | 372.00 | 69.45 | 555.60 | | |
| 1 Equip. Oper. Oiler | 40.30 | 322.40 | 60.20 | 481.60 | | |
| 1 Lattice Boom Crane, 150 Ton | | 1906.00 | | 2096.60 | 26.47 | 29.12 |
| 72 L.H., Daily Totals | | $5335.20 | | $7963.80 | $74.10 | $110.61 |

## Crew C-12

| Crew No. | Bare Costs Hr. | Daily | Incl. Subs O&P Hr. | Daily | Cost Per Labor-Hour Bare Costs | Incl. O&P |
|---|---|---|---|---|---|---|
| 1 Carpenter Foreman (out) | $45.05 | $360.40 | $69.10 | $552.80 | $42.51 | $64.90 |
| 3 Carpenters | 43.05 | 1033.20 | 66.05 | 1585.20 | | |
| 1 Laborer | 34.35 | 274.80 | 52.70 | 421.60 | | |
| 1 Equip. Oper. (crane) | 46.50 | 372.00 | 69.45 | 555.60 | | |
| 1 Hyd. Crane, 12 Ton | | 655.60 | | 721.16 | 13.66 | 15.02 |
| 48 L.H., Daily Totals | | $2696.00 | | $3836.36 | $56.17 | $79.92 |

## Crew C-13

| Crew No. | Bare Costs Hr. | Daily | Incl. Subs O&P Hr. | Daily | Cost Per Labor-Hour Bare Costs | Incl. O&P |
|---|---|---|---|---|---|---|
| 1 Struc. Steel Worker | $48.55 | $388.40 | $85.75 | $686.00 | $46.72 | $79.18 |
| 1 Welder | 48.55 | 388.40 | 85.75 | 686.00 | | |
| 1 Carpenter | 43.05 | 344.40 | 66.05 | 528.40 | | |
| 1 Welder, gas engine, 300 amp | | 109.00 | | 119.90 | 4.54 | 5.00 |
| 24 L.H., Daily Totals | | $1230.20 | | $2020.30 | $51.26 | $84.18 |

## Crew C-14

| Crew No. | Bare Costs Hr. | Daily | Incl. Subs O&P Hr. | Daily | Cost Per Labor-Hour Bare Costs | Incl. O&P |
|---|---|---|---|---|---|---|
| 1 Carpenter Foreman (out) | $45.05 | $360.40 | $69.10 | $552.80 | $42.21 | $64.90 |
| 5 Carpenters | 43.05 | 1722.00 | 66.05 | 2642.00 | | |
| 4 Laborers | 34.35 | 1099.20 | 52.70 | 1686.40 | | |
| 4 Rodmen (reinf.) | 48.40 | 1548.80 | 77.35 | 2475.20 | | |
| 2 Cement Finishers | 40.85 | 653.60 | 59.50 | 952.00 | | |
| 1 Equip. Oper. (crane) | 46.50 | 372.00 | 69.45 | 555.60 | | |
| 1 Equip. Oper. Oiler | 40.30 | 322.40 | 60.20 | 481.60 | | |
| 1 Hyd. Crane, 80 Ton | | 1693.00 | | 1862.30 | 11.76 | 12.93 |
| 144 L.H., Daily Totals | | $7771.40 | | $11207.90 | $53.97 | $77.83 |

## Crew C-14A

| Crew No. | Bare Costs Hr. | Daily | Incl. Subs O&P Hr. | Daily | Cost Per Labor-Hour Bare Costs | Incl. O&P |
|---|---|---|---|---|---|---|
| 1 Carpenter Foreman (out) | $45.05 | $360.40 | $69.10 | $552.80 | $43.29 | $66.72 |
| 16 Carpenters | 43.05 | 5510.40 | 66.05 | 8454.40 | | |
| 4 Rodmen (reinf.) | 48.40 | 1548.80 | 77.35 | 2475.20 | | |
| 2 Laborers | 34.35 | 549.60 | 52.70 | 843.20 | | |
| 1 Cement Finisher | 40.85 | 326.80 | 59.50 | 476.00 | | |
| 1 Equip. Oper. (med.) | 45.35 | 362.80 | 67.75 | 542.00 | | |
| 1 Gas Engine Vibrator | | 23.20 | | 25.52 | | |
| 1 Concrete Pump (small) | | 741.00 | | 815.10 | 3.82 | 4.20 |
| 200 L.H., Daily Totals | | $9423.00 | | $14184.22 | $47.12 | $70.92 |

## Crew C-14B

| Crew No. | Bare Costs Hr. | Daily | Incl. Subs O&P Hr. | Daily | Cost Per Labor-Hour Bare Costs | Incl. O&P |
|---|---|---|---|---|---|---|
| 1 Carpenter Foreman (out) | $45.05 | $360.40 | $69.10 | $552.80 | $43.20 | $66.44 |
| 16 Carpenters | 43.05 | 5510.40 | 66.05 | 8454.40 | | |
| 4 Rodmen (reinf.) | 48.40 | 1548.80 | 77.35 | 2475.20 | | |
| 2 Laborers | 34.35 | 549.60 | 52.70 | 843.20 | | |
| 2 Cement Finishers | 40.85 | 653.60 | 59.50 | 952.00 | | |
| 1 Equip. Oper. (med.) | 45.35 | 362.80 | 67.75 | 542.00 | | |
| 1 Gas Engine Vibrator | | 23.20 | | 25.52 | | |
| 1 Concrete Pump (small) | | 741.00 | | 815.10 | 3.67 | 4.04 |
| 208 L.H., Daily Totals | | $9749.80 | | $14660.22 | $46.87 | $70.48 |

## Crew C-14C

| Crew No. | Bare Costs Hr. | Daily | Incl. Subs O&P Hr. | Daily | Cost Per Labor-Hour Bare Costs | Incl. O&P |
|---|---|---|---|---|---|---|
| 1 Carpenter Foreman (out) | $45.05 | $360.40 | $69.10 | $552.80 | $41.31 | $63.60 |
| 6 Carpenters | 43.05 | 2066.40 | 66.05 | 3170.40 | | |
| 2 Rodmen (reinf.) | 48.40 | 774.40 | 77.35 | 1237.60 | | |
| 4 Laborers | 34.35 | 1099.20 | 52.70 | 1686.40 | | |
| 1 Cement Finisher | 40.85 | 326.80 | 59.50 | 476.00 | | |
| 1 Gas Engine Vibrator | | 23.20 | | 25.52 | .21 | .23 |
| 112 L.H., Daily Totals | | $4650.40 | | $7148.72 | $41.52 | $63.83 |

## Crew C-14D

| Crew No. | Bare Costs Hr. | Daily | Incl. Subs O&P Hr. | Daily | Cost Per Labor-Hour Bare Costs | Incl. O&P |
|---|---|---|---|---|---|---|
| 1 Carpenter Foreman (out) | $45.05 | $360.40 | $69.10 | $552.80 | $42.87 | $65.81 |
| 18 Carpenters | 43.05 | 6199.20 | 66.05 | 9511.20 | | |
| 2 Rodmen (reinf.) | 48.40 | 774.40 | 77.35 | 1237.60 | | |
| 2 Laborers | 34.35 | 549.60 | 52.70 | 843.20 | | |
| 1 Cement Finisher | 40.85 | 326.80 | 59.50 | 476.00 | | |
| 1 Equip. Oper. (med.) | 45.35 | 362.80 | 67.75 | 542.00 | | |
| 1 Gas Engine Vibrator | | 23.20 | | 25.52 | | |
| 1 Concrete Pump (small) | | 741.00 | | 815.10 | 3.82 | 4.20 |
| 200 L.H., Daily Totals | | $9337.40 | | $14003.42 | $46.69 | $70.02 |

## Crew C-14E

| Crew No. | Bare Costs Hr. | Daily | Incl. Subs O&P Hr. | Daily | Cost Per Labor-Hour Bare Costs | Incl. O&P |
|---|---|---|---|---|---|---|
| 1 Carpenter Foreman (out) | $45.05 | $360.40 | $69.10 | $552.80 | $42.60 | $66.20 |
| 2 Carpenters | 43.05 | 688.80 | 66.05 | 1056.80 | | |
| 4 Rodmen (reinf.) | 48.40 | 1548.80 | 77.35 | 2475.20 | | |
| 3 Laborers | 34.35 | 824.40 | 52.70 | 1264.80 | | |
| 1 Cement Finisher | 40.85 | 326.80 | 59.50 | 476.00 | | |
| 1 Gas Engine Vibrator | | 23.20 | | 25.52 | .26 | .29 |
| 88 L.H., Daily Totals | | $3772.40 | | $5851.12 | $42.87 | $66.49 |

| Crew No. | Bare Costs | | Incl. Subs O&P | | Cost Per Labor-Hour | |
|---|---|---|---|---|---|---|

### Crew C-14F

| | Hr. | Daily | Hr. | Daily | Bare Costs | Incl. O&P |
|---|---|---|---|---|---|---|
| 1 Laborer Foreman (out) | $36.35 | $290.80 | $55.75 | $446.00 | $38.91 | $57.57 |
| 2 Laborers | 34.35 | 549.60 | 52.70 | 843.20 | | |
| 6 Cement Finishers | 40.85 | 1960.80 | 59.50 | 2856.00 | | |
| 1 Gas Engine Vibrator | | 23.20 | | 25.52 | .32 | .35 |
| 72 L.H., Daily Totals | | $2824.40 | | $4170.72 | $39.23 | $57.93 |

### Crew C-14G

| | Hr. | Daily | Hr. | Daily | Bare Costs | Incl. O&P |
|---|---|---|---|---|---|---|
| 1 Laborer Foreman (out) | $36.35 | $290.80 | $55.75 | $446.00 | $38.35 | $57.02 |
| 2 Laborers | 34.35 | 549.60 | 52.70 | 843.20 | | |
| 4 Cement Finishers | 40.85 | 1307.20 | 59.50 | 1904.00 | | |
| 1 Gas Engine Vibrator | | 23.20 | | 25.52 | .41 | .46 |
| 56 L.H., Daily Totals | | $2170.80 | | $3218.72 | $38.76 | $57.48 |

### Crew C-14H

| | Hr. | Daily | Hr. | Daily | Bare Costs | Incl. O&P |
|---|---|---|---|---|---|---|
| 1 Carpenter Foreman (out) | $45.05 | $360.40 | $69.10 | $552.80 | $42.46 | $65.13 |
| 2 Carpenters | 43.05 | 688.80 | 66.05 | 1056.80 | | |
| 1 Rodman (reinf.) | 48.40 | 387.20 | 77.35 | 618.80 | | |
| 1 Laborer | 34.35 | 274.80 | 52.70 | 421.60 | | |
| 1 Cement Finisher | 40.85 | 326.80 | 59.50 | 476.00 | | |
| 1 Gas Engine Vibrator | | 23.20 | | 25.52 | .48 | .53 |
| 48 L.H., Daily Totals | | $2061.20 | | $3151.52 | $42.94 | $65.66 |

### Crew C-14L

| | Hr. | Daily | Hr. | Daily | Bare Costs | Incl. O&P |
|---|---|---|---|---|---|---|
| 1 Carpenter Foreman (out) | $45.05 | $360.40 | $69.10 | $552.80 | $40.13 | $61.31 |
| 6 Carpenters | 43.05 | 2066.40 | 66.05 | 3170.40 | | |
| 4 Laborers | 34.35 | 1099.20 | 52.70 | 1686.40 | | |
| 1 Cement Finisher | 40.85 | 326.80 | 59.50 | 476.00 | | |
| 1 Gas Engine Vibrator | | 23.20 | | 25.52 | .24 | .27 |
| 96 L.H., Daily Totals | | $3876.00 | | $5911.12 | $40.38 | $61.57 |

### Crew C-15

| | Hr. | Daily | Hr. | Daily | Bare Costs | Incl. O&P |
|---|---|---|---|---|---|---|
| 1 Carpenter Foreman (out) | $45.05 | $360.40 | $69.10 | $552.80 | $40.48 | $61.74 |
| 2 Carpenters | 43.05 | 688.80 | 66.05 | 1056.80 | | |
| 3 Laborers | 34.35 | 824.40 | 52.70 | 1264.80 | | |
| 2 Cement Finishers | 40.85 | 653.60 | 59.50 | 952.00 | | |
| 1 Rodman (reinf.) | 48.40 | 387.20 | 77.35 | 618.80 | | |
| 72 L.H., Daily Totals | | $2914.40 | | $4445.20 | $40.48 | $61.74 |

### Crew C-16

| | Hr. | Daily | Hr. | Daily | Bare Costs | Incl. O&P |
|---|---|---|---|---|---|---|
| 1 Labor Foreman (outside) | $36.35 | $290.80 | $55.75 | $446.00 | $38.06 | $57.23 |
| 3 Laborers | 34.35 | 824.40 | 52.70 | 1264.80 | | |
| 2 Cement Finishers | 40.85 | 653.60 | 59.50 | 952.00 | | |
| 1 Equip. Oper. (med.) | 45.35 | 362.80 | 67.75 | 542.00 | | |
| 1 Gunite Pump Rig | | 178.00 | | 195.80 | | |
| 2 -50' Air Hoses, 3/4" | | 6.10 | | 6.71 | | |
| 2 -50' Air Hoses, 2" | | 15.50 | | 17.05 | 3.56 | 3.92 |
| 56 L.H., Daily Totals | | $2331.20 | | $3424.36 | $41.63 | $61.15 |

### Crew C-17

| | Hr. | Daily | Hr. | Daily | Bare Costs | Incl. O&P |
|---|---|---|---|---|---|---|
| 2 Skilled Worker Foremen | $46.40 | $742.40 | $71.20 | $1139.20 | $44.80 | $68.72 |
| 8 Skilled Workers | 44.40 | 2841.60 | 68.10 | 4358.40 | | |
| 80 L.H., Daily Totals | | $3584.00 | | $5497.60 | $44.80 | $68.72 |

### Crew C-17A

| | Hr. | Daily | Hr. | Daily | Bare Costs | Incl. O&P |
|---|---|---|---|---|---|---|
| 2 Skilled Worker Foremen | $46.40 | $742.40 | $71.20 | $1139.20 | $44.82 | $68.73 |
| 8 Skilled Workers | 44.40 | 2841.60 | 68.10 | 4358.40 | | |
| .125 Equip. Oper. (crane) | 46.50 | 46.50 | 69.45 | 69.45 | | |
| .125 Hyd. Crane, 80 Ton | | 211.63 | | 232.79 | 2.61 | 2.87 |
| 81 L.H., Daily Totals | | $3842.13 | | $5799.84 | $47.43 | $71.60 |

### Crew C-17B

| | Hr. | Daily | Hr. | Daily | Bare Costs | Incl. O&P |
|---|---|---|---|---|---|---|
| 2 Skilled Worker Foremen | $46.40 | $742.40 | $71.20 | $1139.20 | $44.84 | $68.74 |
| 8 Skilled Workers | 44.40 | 2841.60 | 68.10 | 4358.40 | | |
| .25 Equip. Oper. (crane) | 46.50 | 93.00 | 69.45 | 138.90 | | |
| .25 Hyd. Crane, 80 Ton | | 423.25 | | 465.57 | | |
| .25 Trowel, 48" Walk-Behind | | 9.60 | | 10.56 | 5.28 | 5.81 |
| 82 L.H., Daily Totals | | $4109.85 | | $6112.64 | $50.12 | $74.54 |

### Crew C-17C

| | Hr. | Daily | Hr. | Daily | Bare Costs | Incl. O&P |
|---|---|---|---|---|---|---|
| 2 Skilled Worker Foremen | $46.40 | $742.40 | $71.20 | $1139.20 | $44.86 | $68.75 |
| 8 Skilled Workers | 44.40 | 2841.60 | 68.10 | 4358.40 | | |
| .375 Equip. Oper. (crane) | 46.50 | 139.50 | 69.45 | 208.35 | | |
| .375 Hyd. Crane, 80 Ton | | 634.88 | | 698.36 | 7.65 | 8.41 |
| 83 L.H., Daily Totals | | $4358.38 | | $6404.31 | $52.51 | $77.16 |

### Crew C-17D

| | Hr. | Daily | Hr. | Daily | Bare Costs | Incl. O&P |
|---|---|---|---|---|---|---|
| 2 Skilled Worker Foremen | $46.40 | $742.40 | $71.20 | $1139.20 | $44.88 | $68.75 |
| 8 Skilled Workers | 44.40 | 2841.60 | 68.10 | 4358.40 | | |
| .5 Equip. Oper. (crane) | 46.50 | 186.00 | 69.45 | 277.80 | | |
| .5 Hyd. Crane, 80 Ton | | 846.50 | | 931.15 | 10.08 | 11.09 |
| 84 L.H., Daily Totals | | $4616.50 | | $6706.55 | $54.96 | $79.84 |

### Crew C-17E

| | Hr. | Daily | Hr. | Daily | Bare Costs | Incl. O&P |
|---|---|---|---|---|---|---|
| 2 Skilled Worker Foremen | $46.40 | $742.40 | $71.20 | $1139.20 | $44.80 | $68.72 |
| 8 Skilled Workers | 44.40 | 2841.60 | 68.10 | 4358.40 | | |
| 1 Hyd. Jack with Rods | | 89.10 | | 98.01 | 1.11 | 1.23 |
| 80 L.H., Daily Totals | | $3673.10 | | $5595.61 | $45.91 | $69.95 |

### Crew C-18

| | Hr. | Daily | Hr. | Daily | Bare Costs | Incl. O&P |
|---|---|---|---|---|---|---|
| .125 Labor Foreman (out) | $36.35 | $36.35 | $55.75 | $55.75 | $34.57 | $53.04 |
| 1 Laborer | 34.35 | 274.80 | 52.70 | 421.60 | | |
| 1 Concrete Cart, 10 C.F. | | 55.00 | | 60.50 | 6.11 | 6.72 |
| 9 L.H., Daily Totals | | $366.15 | | $537.85 | $40.68 | $59.76 |

### Crew C-19

| | Hr. | Daily | Hr. | Daily | Bare Costs | Incl. O&P |
|---|---|---|---|---|---|---|
| .125 Labor Foreman (out) | $36.35 | $36.35 | $55.75 | $55.75 | $34.57 | $53.04 |
| 1 Laborer | 34.35 | 274.80 | 52.70 | 421.60 | | |
| 1 Concrete Cart, 18 C.F. | | 91.60 | | 100.76 | 10.18 | 11.20 |
| 9 L.H., Daily Totals | | $402.75 | | $578.11 | $44.75 | $64.23 |

### Crew C-20

| | Hr. | Daily | Hr. | Daily | Bare Costs | Incl. O&P |
|---|---|---|---|---|---|---|
| 1 Labor Foreman (outside) | $36.35 | $290.80 | $55.75 | $446.00 | $36.79 | $55.81 |
| 5 Laborers | 34.35 | 1374.00 | 52.70 | 2108.00 | | |
| 1 Cement Finisher | 40.85 | 326.80 | 59.50 | 476.00 | | |
| 1 Equip. Oper. (med.) | 45.35 | 362.80 | 67.75 | 542.00 | | |
| 2 Gas Engine Vibrators | | 46.40 | | 51.04 | | |
| 1 Concrete Pump (small) | | 741.00 | | 815.10 | 12.30 | 13.53 |
| 64 L.H., Daily Totals | | $3141.80 | | $4438.14 | $49.09 | $69.35 |

### Crew C-21

| | Hr. | Daily | Hr. | Daily | Bare Costs | Incl. O&P |
|---|---|---|---|---|---|---|
| 1 Labor Foreman (outside) | $36.35 | $290.80 | $55.75 | $446.00 | $36.79 | $55.81 |
| 5 Laborers | 34.35 | 1374.00 | 52.70 | 2108.00 | | |
| 1 Cement Finisher | 40.85 | 326.80 | 59.50 | 476.00 | | |
| 1 Equip. Oper. (med.) | 45.35 | 362.80 | 67.75 | 542.00 | | |
| 2 Gas Engine Vibrators | | 46.40 | | 51.04 | | |
| 1 Concrete Conveyer | | 174.60 | | 192.06 | 3.45 | 3.80 |
| 64 L.H., Daily Totals | | $2575.40 | | $3815.10 | $40.24 | $59.61 |

| Crew No. | Bare Costs | | Incl. Subs O&P | | Cost Per Labor-Hour | |
|---|---|---|---|---|---|---|

| Crew C-22 | Hr. | Daily | Hr. | Daily | Bare Costs | Incl. O&P |
|---|---|---|---|---|---|---|
| 1 Rodman Foreman | $50.40 | $403.20 | $80.55 | $644.40 | $48.54 | $77.36 |
| 4 Rodmen (reinf.) | 48.40 | 1548.80 | 77.35 | 2475.20 | | |
| .125 Equip. Oper. (crane) | 46.50 | 46.50 | 69.45 | 69.45 | | |
| .125 Equip. Oper. Oiler | 40.30 | 40.30 | 60.20 | 60.20 | | |
| .125 Hyd. Crane, 25 Ton | | 93.58 | | 102.93 | 2.23 | 2.45 |
| 42 L.H., Daily Totals | | $2132.38 | | $3352.18 | $50.77 | $79.81 |

| Crew C-23 | Hr. | Daily | Hr. | Daily | Bare Costs | Incl. O&P |
|---|---|---|---|---|---|---|
| 2 Skilled Worker Foremen | $46.40 | $742.40 | $71.20 | $1139.20 | $44.60 | $68.06 |
| 6 Skilled Workers | 44.40 | 2131.20 | 68.10 | 3268.80 | | |
| 1 Equip. Oper. (crane) | 46.50 | 372.00 | 69.45 | 555.60 | | |
| 1 Equip. Oper. Oiler | 40.30 | 322.40 | 60.20 | 481.60 | | |
| 1 Lattice Boom Crane, 90 Ton | | 1622.00 | | 1784.20 | 20.27 | 22.30 |
| 80 L.H., Daily Totals | | $5190.00 | | $7229.40 | $64.88 | $90.37 |

| Crew C-23A | Hr. | Daily | Hr. | Daily | Bare Costs | Incl. O&P |
|---|---|---|---|---|---|---|
| 1 Labor Foreman (outside) | $36.35 | $290.80 | $55.75 | $446.00 | $38.37 | $58.16 |
| 2 Laborers | 34.35 | 549.60 | 52.70 | 843.20 | | |
| 1 Equip. Oper. (crane) | 46.50 | 372.00 | 69.45 | 555.60 | | |
| 1 Equip. Oper. Oiler | 40.30 | 322.40 | 60.20 | 481.60 | | |
| 1 Crawler Crane, 100 Ton | | 1653.00 | | 1818.30 | | |
| 3 Conc. Buckets, 8 C.Y. | | 613.80 | | 675.18 | 56.67 | 62.34 |
| 40 L.H., Daily Totals | | $3801.60 | | $4819.88 | $95.04 | $120.50 |

| Crew C-24 | Hr. | Daily | Hr. | Daily | Bare Costs | Incl. O&P |
|---|---|---|---|---|---|---|
| 2 Skilled Worker Foremen | $46.40 | $742.40 | $71.20 | $1139.20 | $44.60 | $68.06 |
| 6 Skilled Workers | 44.40 | 2131.20 | 68.10 | 3268.80 | | |
| 1 Equip. Oper. (crane) | 46.50 | 372.00 | 69.45 | 555.60 | | |
| 1 Equip. Oper. Oiler | 40.30 | 322.40 | 60.20 | 481.60 | | |
| 1 Lattice Boom Crane, 150 Ton | | 1906.00 | | 2096.60 | 23.82 | 26.21 |
| 80 L.H., Daily Totals | | $5474.00 | | $7541.80 | $68.42 | $94.27 |

| Crew C-25 | Hr. | Daily | Hr. | Daily | Bare Costs | Incl. O&P |
|---|---|---|---|---|---|---|
| 2 Rodmen (reinf.) | $48.40 | $774.40 | $77.35 | $1237.60 | $37.75 | $61.23 |
| 2 Rodmen Helpers | 27.10 | 433.60 | 45.10 | 721.60 | | |
| 32 L.H., Daily Totals | | $1208.00 | | $1959.20 | $37.75 | $61.23 |

| Crew C-27 | Hr. | Daily | Hr. | Daily | Bare Costs | Incl. O&P |
|---|---|---|---|---|---|---|
| 2 Cement Finishers | $40.85 | $653.60 | $59.50 | $952.00 | $40.85 | $59.50 |
| 1 Concrete Saw | | 134.20 | | 147.62 | 8.39 | 9.23 |
| 16 L.H., Daily Totals | | $787.80 | | $1099.62 | $49.24 | $68.73 |

| Crew C-28 | Hr. | Daily | Hr. | Daily | Bare Costs | Incl. O&P |
|---|---|---|---|---|---|---|
| 1 Cement Finisher | $40.85 | $326.80 | $59.50 | $476.00 | $40.85 | $59.50 |
| 1 Portable Air Compressor, Gas | | 19.00 | | 20.90 | 2.38 | 2.61 |
| 8 L.H., Daily Totals | | $345.80 | | $496.90 | $43.23 | $62.11 |

| Crew C-29 | Hr. | Daily | Hr. | Daily | Bare Costs | Incl. O&P |
|---|---|---|---|---|---|---|
| 1 Laborer | $34.35 | $274.80 | $52.70 | $421.60 | $34.35 | $52.70 |
| 1 Pressure Washer | | 58.80 | | 64.68 | 7.35 | 8.09 |
| 8 L.H., Daily Totals | | $333.60 | | $486.28 | $41.70 | $60.78 |

| Crew D-1 | Hr. | Daily | Hr. | Daily | Bare Costs | Incl. O&P |
|---|---|---|---|---|---|---|
| 1 Bricklayer | $43.15 | $345.20 | $64.75 | $518.00 | $38.98 | $58.48 |
| 1 Bricklayer Helper | 34.80 | 278.40 | 52.20 | 417.60 | | |
| 16 L.H., Daily Totals | | $623.60 | | $935.60 | $38.98 | $58.48 |

| Crew D-2 | Hr. | Daily | Hr. | Daily | Bare Costs | Incl. O&P |
|---|---|---|---|---|---|---|
| 3 Bricklayers | $43.15 | $1035.60 | $64.75 | $1554.00 | $40.10 | $60.30 |
| 2 Bricklayer Helpers | 34.80 | 556.80 | 52.20 | 835.20 | | |
| .5 Carpenter | 43.05 | 172.20 | 66.05 | 264.20 | | |
| 44 L.H., Daily Totals | | $1764.60 | | $2653.40 | $40.10 | $60.30 |

| Crew D-3 | Hr. | Daily | Hr. | Daily | Bare Costs | Incl. O&P |
|---|---|---|---|---|---|---|
| 3 Bricklayers | $43.15 | $1035.60 | $64.75 | $1554.00 | $39.96 | $60.03 |
| 2 Bricklayer Helpers | 34.80 | 556.80 | 52.20 | 835.20 | | |
| .25 Carpenter | 43.05 | 86.10 | 66.05 | 132.10 | | |
| 42 L.H., Daily Totals | | $1678.50 | | $2521.30 | $39.96 | $60.03 |

| Crew D-4 | Hr. | Daily | Hr. | Daily | Bare Costs | Incl. O&P |
|---|---|---|---|---|---|---|
| 1 Bricklayer | $43.15 | $345.20 | $64.75 | $518.00 | $39.09 | $58.58 |
| 2 Bricklayer Helpers | 34.80 | 556.80 | 52.20 | 835.20 | | |
| 1 Equip. Oper. (light) | 43.60 | 348.80 | 65.15 | 521.20 | | |
| 1 Grout Pump, 50 C.F./hr. | | 125.80 | | 138.38 | 3.93 | 4.32 |
| 32 L.H., Daily Totals | | $1376.60 | | $2012.78 | $43.02 | $62.90 |

| Crew D-5 | Hr. | Daily | Hr. | Daily | Bare Costs | Incl. O&P |
|---|---|---|---|---|---|---|
| 1 Bricklayer | $43.15 | $345.20 | $64.75 | $518.00 | $43.15 | $64.75 |
| 8 L.H., Daily Totals | | $345.20 | | $518.00 | $43.15 | $64.75 |

| Crew D-6 | Hr. | Daily | Hr. | Daily | Bare Costs | Incl. O&P |
|---|---|---|---|---|---|---|
| 3 Bricklayers | $43.15 | $1035.60 | $64.75 | $1554.00 | $39.14 | $58.78 |
| 3 Bricklayer Helpers | 34.80 | 835.20 | 52.20 | 1252.80 | | |
| .25 Carpenter | 43.05 | 86.10 | 66.05 | 132.10 | | |
| 50 L.H., Daily Totals | | $1956.90 | | $2938.90 | $39.14 | $58.78 |

| Crew D-7 | Hr. | Daily | Hr. | Daily | Bare Costs | Incl. O&P |
|---|---|---|---|---|---|---|
| 1 Tile Layer | $40.45 | $323.60 | $59.00 | $472.00 | $36.15 | $52.73 |
| 1 Tile Layer Helper | 31.85 | 254.80 | 46.45 | 371.60 | | |
| 16 L.H., Daily Totals | | $578.40 | | $843.60 | $36.15 | $52.73 |

| Crew D-8 | Hr. | Daily | Hr. | Daily | Bare Costs | Incl. O&P |
|---|---|---|---|---|---|---|
| 3 Bricklayers | $43.15 | $1035.60 | $64.75 | $1554.00 | $39.81 | $59.73 |
| 2 Bricklayer Helpers | 34.80 | 556.80 | 52.20 | 835.20 | | |
| 40 L.H., Daily Totals | | $1592.40 | | $2389.20 | $39.81 | $59.73 |

| Crew D-9 | Hr. | Daily | Hr. | Daily | Bare Costs | Incl. O&P |
|---|---|---|---|---|---|---|
| 3 Bricklayers | $43.15 | $1035.60 | $64.75 | $1554.00 | $38.98 | $58.48 |
| 3 Bricklayer Helpers | 34.80 | 835.20 | 52.20 | 1252.80 | | |
| 48 L.H., Daily Totals | | $1870.80 | | $2806.80 | $38.98 | $58.48 |

| Crew D-10 | Hr. | Daily | Hr. | Daily | Bare Costs | Incl. O&P |
|---|---|---|---|---|---|---|
| 1 Bricklayer Foreman | $45.15 | $361.20 | $67.75 | $542.00 | $42.40 | $63.54 |
| 1 Bricklayer | 43.15 | 345.20 | 64.75 | 518.00 | | |
| 1 Bricklayer Helper | 34.80 | 278.40 | 52.20 | 417.60 | | |
| 1 Equip. Oper. (crane) | 46.50 | 372.00 | 69.45 | 555.60 | | |
| 1 S.P. Crane, 4x4, 12 Ton | | 434.40 | | 477.84 | 13.57 | 14.93 |
| 32 L.H., Daily Totals | | $1791.20 | | $2511.04 | $55.98 | $78.47 |

| Crew D-11 | Hr. | Daily | Hr. | Daily | Bare Costs | Incl. O&P |
|---|---|---|---|---|---|---|
| 1 Bricklayer Foreman | $45.15 | $361.20 | $67.75 | $542.00 | $41.03 | $61.57 |
| 1 Bricklayer | 43.15 | 345.20 | 64.75 | 518.00 | | |
| 1 Bricklayer Helper | 34.80 | 278.40 | 52.20 | 417.60 | | |
| 24 L.H., Daily Totals | | $984.80 | | $1477.60 | $41.03 | $61.57 |

| Crew No. | Bare Costs | | Incl. Subs O&P | | Cost Per Labor-Hour | |
|---|---|---|---|---|---|---|
| | Hr. | Daily | Hr. | Daily | Bare Costs | Incl. O&P |
| **Crew D-12** | | | | | | |
| 1 Bricklayer Foreman | $45.15 | $361.20 | $67.75 | $542.00 | $39.48 | $59.23 |
| 1 Bricklayer | 43.15 | 345.20 | 64.75 | 518.00 | | |
| 2 Bricklayer Helpers | 34.80 | 556.80 | 52.20 | 835.20 | | |
| 32 L.H., Daily Totals | | $1263.20 | | $1895.20 | $39.48 | $59.23 |
| **Crew D-13** | Hr. | Daily | Hr. | Daily | Bare Costs | Incl. O&P |
| 1 Bricklayer Foreman | $45.15 | $361.20 | $67.75 | $542.00 | $41.24 | $62.07 |
| 1 Bricklayer | 43.15 | 345.20 | 64.75 | 518.00 | | |
| 2 Bricklayer Helpers | 34.80 | 556.80 | 52.20 | 835.20 | | |
| 1 Carpenter | 43.05 | 344.40 | 66.05 | 528.40 | | |
| 1 Equip. Oper. (crane) | 46.50 | 372.00 | 69.45 | 555.60 | | |
| 1 S.P. Crane, 4x4, 12 Ton | | 434.40 | | 477.84 | 9.05 | 9.96 |
| 48 L.H., Daily Totals | | $2414.00 | | $3457.04 | $50.29 | $72.02 |
| **Crew E-1** | Hr. | Daily | Hr. | Daily | Bare Costs | Incl. O&P |
| 1 Welder Foreman | $50.55 | $404.40 | $89.25 | $714.00 | $47.57 | $80.05 |
| 1 Welder | 48.55 | 388.40 | 85.75 | 686.00 | | |
| 1 Equip. Oper. (light) | 43.60 | 348.80 | 65.15 | 521.20 | | |
| 1 Welder, gas engine, 300 amp | | 109.00 | | 119.90 | 4.54 | 5.00 |
| 24 L.H., Daily Totals | | $1250.60 | | $2041.10 | $52.11 | $85.05 |
| **Crew E-2** | Hr. | Daily | Hr. | Daily | Bare Costs | Incl. O&P |
| 1 Struc. Steel Foreman | $50.55 | $404.40 | $89.25 | $714.00 | $47.36 | $80.27 |
| 4 Struc. Steel Workers | 48.55 | 1553.60 | 85.75 | 2744.00 | | |
| 1 Equip. Oper. (crane) | 46.50 | 372.00 | 69.45 | 555.60 | | |
| 1 Equip. Oper. Oiler | 40.30 | 322.40 | 60.20 | 481.60 | | |
| 1 Lattice Boom Crane, 90 Ton | | 1622.00 | | 1784.20 | 28.96 | 31.86 |
| 56 L.H., Daily Totals | | $4274.40 | | $6279.40 | $76.33 | $112.13 |
| **Crew E-3** | Hr. | Daily | Hr. | Daily | Bare Costs | Incl. O&P |
| 1 Struc. Steel Foreman | $50.55 | $404.40 | $89.25 | $714.00 | $49.22 | $86.92 |
| 1 Struc. Steel Worker | 48.55 | 388.40 | 85.75 | 686.00 | | |
| 1 Welder | 48.55 | 388.40 | 85.75 | 686.00 | | |
| 1 Welder, gas engine, 300 amp | | 109.00 | | 119.90 | 4.54 | 5.00 |
| 24 L.H., Daily Totals | | $1290.20 | | $2205.90 | $53.76 | $91.91 |
| **Crew E-4** | Hr. | Daily | Hr. | Daily | Bare Costs | Incl. O&P |
| 1 Struc. Steel Foreman | $50.55 | $404.40 | $89.25 | $714.00 | $49.05 | $86.63 |
| 3 Struc. Steel Workers | 48.55 | 1165.20 | 85.75 | 2058.00 | | |
| 1 Welder, gas engine, 300 amp | | 109.00 | | 119.90 | 3.41 | 3.75 |
| 32 L.H., Daily Totals | | $1678.60 | | $2891.90 | $52.46 | $90.37 |
| **Crew E-5** | Hr. | Daily | Hr. | Daily | Bare Costs | Incl. O&P |
| 2 Struc. Steel Foremen | $50.55 | $808.80 | $89.25 | $1428.00 | $47.92 | $82.27 |
| 5 Struc. Steel Workers | 48.55 | 1942.00 | 85.75 | 3430.00 | | |
| 1 Equip. Oper. (crane) | 46.50 | 372.00 | 69.45 | 555.60 | | |
| 1 Welder | 48.55 | 388.40 | 85.75 | 686.00 | | |
| 1 Equip. Oper. Oiler | 40.30 | 322.40 | 60.20 | 481.60 | | |
| 1 Lattice Boom Crane, 90 Ton | | 1622.00 | | 1784.20 | | |
| 1 Welder, gas engine, 300 amp | | 109.00 | | 119.90 | 21.64 | 23.80 |
| 80 L.H., Daily Totals | | $5564.60 | | $8485.30 | $69.56 | $106.07 |

| Crew No. | Bare Costs | | Incl. Subs O&P | | Cost Per Labor-Hour | |
|---|---|---|---|---|---|---|
| | Hr. | Daily | Hr. | Daily | Bare Costs | Incl. O&P |
| **Crew E-6** | | | | | | |
| 3 Struc. Steel Foremen | $50.55 | $1213.20 | $89.25 | $2142.00 | $47.97 | $82.50 |
| 9 Struc. Steel Workers | 48.55 | 3495.60 | 85.75 | 6174.00 | | |
| 1 Equip. Oper. (crane) | 46.50 | 372.00 | 69.45 | 555.60 | | |
| 1 Welder | 48.55 | 388.40 | 85.75 | 686.00 | | |
| 1 Equip. Oper. Oiler | 40.30 | 322.40 | 60.20 | 481.60 | | |
| 1 Equip. Oper. (light) | 43.60 | 348.80 | 65.15 | 521.20 | | |
| 1 Lattice Boom Crane, 90 Ton | | 1622.00 | | 1784.20 | | |
| 1 Welder, gas engine, 300 amp | | 109.00 | | 119.90 | | |
| 1 Air Compressor, 160 cfm | | 125.00 | | 137.50 | | |
| 2 Impact Wrenches | | 36.40 | | 40.04 | 14.78 | 16.26 |
| 128 L.H., Daily Totals | | $8032.80 | | $12642.04 | $62.76 | $98.77 |
| **Crew E-7** | Hr. | Daily | Hr. | Daily | Bare Costs | Incl. O&P |
| 1 Struc. Steel Foreman | $50.55 | $404.40 | $89.25 | $714.00 | $47.92 | $82.27 |
| 4 Struc. Steel Workers | 48.55 | 1553.60 | 85.75 | 2744.00 | | |
| 1 Equip. Oper. (crane) | 46.50 | 372.00 | 69.45 | 555.60 | | |
| 1 Equip. Oper. Oiler | 40.30 | 322.40 | 60.20 | 481.60 | | |
| 1 Welder Foreman | 50.55 | 404.40 | 89.25 | 714.00 | | |
| 2 Welders | 48.55 | 776.80 | 85.75 | 1372.00 | | |
| 1 Lattice Boom Crane, 90 Ton | | 1622.00 | | 1784.20 | | |
| 2 Welders, gas engine, 300 amp | | 218.00 | | 239.80 | 23.00 | 25.30 |
| 80 L.H., Daily Totals | | $5673.60 | | $8605.20 | $70.92 | $107.57 |
| **Crew E-8** | Hr. | Daily | Hr. | Daily | Bare Costs | Incl. O&P |
| 1 Struc. Steel Foreman | $50.55 | $404.40 | $89.25 | $714.00 | $47.68 | $81.48 |
| 4 Struc. Steel Workers | 48.55 | 1553.60 | 85.75 | 2744.00 | | |
| 1 Welder Foreman | 50.55 | 404.40 | 89.25 | 714.00 | | |
| 4 Welders | 48.55 | 1553.60 | 85.75 | 2744.00 | | |
| 1 Equip. Oper. (crane) | 46.50 | 372.00 | 69.45 | 555.60 | | |
| 1 Equip. Oper. Oiler | 40.30 | 322.40 | 60.20 | 481.60 | | |
| 1 Equip. Oper. (light) | 43.60 | 348.80 | 65.15 | 521.20 | | |
| 1 Lattice Boom Crane, 90 Ton | | 1622.00 | | 1784.20 | | |
| 4 Welders, gas engine, 300 amp | | 436.00 | | 479.60 | 19.79 | 21.77 |
| 104 L.H., Daily Totals | | $7017.20 | | $10738.20 | $67.47 | $103.25 |
| **Crew E-9** | Hr. | Daily | Hr. | Daily | Bare Costs | Incl. O&P |
| 2 Struc. Steel Foremen | $50.55 | $808.80 | $89.25 | $1428.00 | $47.97 | $82.50 |
| 5 Struc. Steel Workers | 48.55 | 1942.00 | 85.75 | 3430.00 | | |
| 1 Welder Foreman | 50.55 | 404.40 | 89.25 | 714.00 | | |
| 5 Welders | 48.55 | 1942.00 | 85.75 | 3430.00 | | |
| 1 Equip. Oper. (crane) | 46.50 | 372.00 | 69.45 | 555.60 | | |
| 1 Equip. Oper. Oiler | 40.30 | 322.40 | 60.20 | 481.60 | | |
| 1 Equip. Oper. (light) | 43.60 | 348.80 | 65.15 | 521.20 | | |
| 1 Lattice Boom Crane, 90 Ton | | 1622.00 | | 1784.20 | | |
| 5 Welders, gas engine, 300 amp | | 545.00 | | 599.50 | 16.93 | 18.62 |
| 128 L.H., Daily Totals | | $8307.40 | | $12944.10 | $64.90 | $101.13 |
| **Crew E-10** | Hr. | Daily | Hr. | Daily | Bare Costs | Incl. O&P |
| 1 Welder Foreman | $50.55 | $404.40 | $89.25 | $714.00 | $49.55 | $87.50 |
| 1 Welder | 48.55 | 388.40 | 85.75 | 686.00 | | |
| 1 Welder, gas engine, 300 amp | | 109.00 | | 119.90 | | |
| 1 Flatbed Truck, Gas, 3 Ton | | 217.00 | | 238.70 | 20.38 | 22.41 |
| 16 L.H., Daily Totals | | $1118.80 | | $1758.60 | $69.92 | $109.91 |
| **Crew E-11** | Hr. | Daily | Hr. | Daily | Bare Costs | Incl. O&P |
| 2 Painters, Struc. Steel | $38.65 | $618.40 | $69.15 | $1106.40 | $38.81 | $64.04 |
| 1 Building Laborer | 34.35 | 274.80 | 52.70 | 421.60 | | |
| 1 Equip. Oper. (light) | 43.60 | 348.80 | 65.15 | 521.20 | | |
| 1 Air Compressor, 250 cfm | | 162.80 | | 179.08 | | |
| 1 Sandblaster, portable, 3 C.F. | | 20.60 | | 22.66 | | |
| 1 Set Sand Blasting Accessories | | 13.15 | | 14.47 | 6.14 | 6.76 |
| 32 L.H., Daily Totals | | $1438.55 | | $2265.41 | $44.95 | $70.79 |

### Crew E-12

| Crew No. | Bare Costs Hr. | Daily | Incl. Subs O&P Hr. | Daily | Cost Per Labor-Hour Bare Costs | Incl. O&P |
|---|---|---|---|---|---|---|
| 1 Welder Foreman | $50.55 | $404.40 | $89.25 | $714.00 | $47.08 | $77.20 |
| 1 Equip. Oper. (light) | 43.60 | 348.80 | 65.15 | 521.20 | | |
| 1 Welder, gas engine, 300 amp | | 109.00 | | 119.90 | 6.81 | 7.49 |
| 16 L.H., Daily Totals | | $862.20 | | $1355.10 | $53.89 | $84.69 |

### Crew E-13

| Crew No. | Bare Costs Hr. | Daily | Incl. Subs O&P Hr. | Daily | Cost Per Labor-Hour Bare Costs | Incl. O&P |
|---|---|---|---|---|---|---|
| 1 Welder Foreman | $50.55 | $404.40 | $89.25 | $714.00 | $48.23 | $81.22 |
| .5 Equip. Oper. (light) | 43.60 | 174.40 | 65.15 | 260.60 | | |
| 1 Welder, gas engine, 300 amp | | 109.00 | | 119.90 | 9.08 | 9.99 |
| 12 L.H., Daily Totals | | $687.80 | | $1094.50 | $57.32 | $91.21 |

### Crew E-14

| Crew No. | Bare Costs Hr. | Daily | Incl. Subs O&P Hr. | Daily | Cost Per Labor-Hour Bare Costs | Incl. O&P |
|---|---|---|---|---|---|---|
| 1 Welder Foreman | $50.55 | $404.40 | $89.25 | $714.00 | $50.55 | $89.25 |
| 1 Welder, gas engine, 300 amp | | 109.00 | | 119.90 | 13.63 | 14.99 |
| 8 L.H., Daily Totals | | $513.40 | | $833.90 | $64.17 | $104.24 |

### Crew E-16

| Crew No. | Bare Costs Hr. | Daily | Incl. Subs O&P Hr. | Daily | Cost Per Labor-Hour Bare Costs | Incl. O&P |
|---|---|---|---|---|---|---|
| 1 Welder Foreman | $50.55 | $404.40 | $89.25 | $714.00 | $49.55 | $87.50 |
| 1 Welder | 48.55 | 388.40 | 85.75 | 686.00 | | |
| 1 Welder, gas engine, 300 amp | | 109.00 | | 119.90 | 6.81 | 7.49 |
| 16 L.H., Daily Totals | | $901.80 | | $1519.90 | $56.36 | $94.99 |

### Crew E-17

| Crew No. | Bare Costs Hr. | Daily | Incl. Subs O&P Hr. | Daily | Cost Per Labor-Hour Bare Costs | Incl. O&P |
|---|---|---|---|---|---|---|
| 1 Structural Steel Foreman | $50.55 | $404.40 | $89.25 | $714.00 | $49.55 | $87.50 |
| 1 Structural Steel Worker | 48.55 | 388.40 | 85.75 | 686.00 | | |
| 16 L.H., Daily Totals | | $792.80 | | $1400.00 | $49.55 | $87.50 |

### Crew E-18

| Crew No. | Bare Costs Hr. | Daily | Incl. Subs O&P Hr. | Daily | Cost Per Labor-Hour Bare Costs | Incl. O&P |
|---|---|---|---|---|---|---|
| 1 Structural Steel Foreman | $50.55 | $404.40 | $89.25 | $714.00 | $48.31 | $82.85 |
| 3 Structural Steel Workers | 48.55 | 1165.20 | 85.75 | 2058.00 | | |
| 1 Equipment Operator (med.) | 45.35 | 362.80 | 67.75 | 542.00 | | |
| 1 Lattice Boom Crane, 20 Ton | | 1035.00 | | 1138.50 | 25.88 | 28.46 |
| 40 L.H., Daily Totals | | $2967.40 | | $4452.50 | $74.19 | $111.31 |

### Crew E-19

| Crew No. | Bare Costs Hr. | Daily | Incl. Subs O&P Hr. | Daily | Cost Per Labor-Hour Bare Costs | Incl. O&P |
|---|---|---|---|---|---|---|
| 1 Structural Steel Worker | $48.55 | $388.40 | $85.75 | $686.00 | $47.57 | $80.05 |
| 1 Structural Steel Foreman | 50.55 | 404.40 | 89.25 | 714.00 | | |
| 1 Equip. Oper. (light) | 43.60 | 348.80 | 65.15 | 521.20 | | |
| 1 Lattice Boom Crane, 20 Ton | | 1035.00 | | 1138.50 | 43.13 | 47.44 |
| 24 L.H., Daily Totals | | $2176.60 | | $3059.70 | $90.69 | $127.49 |

### Crew E-20

| Crew No. | Bare Costs Hr. | Daily | Incl. Subs O&P Hr. | Daily | Cost Per Labor-Hour Bare Costs | Incl. O&P |
|---|---|---|---|---|---|---|
| 1 Structural Steel Foreman | $50.55 | $404.40 | $89.25 | $714.00 | $47.51 | $80.96 |
| 5 Structural Steel Workers | 48.55 | 1942.00 | 85.75 | 3430.00 | | |
| 1 Equip. Oper. (crane) | 46.50 | 372.00 | 69.45 | 555.60 | | |
| 1 Oiler | 40.30 | 322.40 | 60.20 | 481.60 | | |
| 1 Lattice Boom Crane, 40 Ton | | 1263.00 | | 1389.30 | 19.73 | 21.71 |
| 64 L.H., Daily Totals | | $4303.80 | | $6570.50 | $67.25 | $102.66 |

### Crew E-22

| Crew No. | Bare Costs Hr. | Daily | Incl. Subs O&P Hr. | Daily | Cost Per Labor-Hour Bare Costs | Incl. O&P |
|---|---|---|---|---|---|---|
| 1 Skilled Worker Foreman | $46.40 | $371.20 | $71.20 | $569.60 | $45.07 | $69.13 |
| 2 Skilled Workers | 44.40 | 710.40 | 68.10 | 1089.60 | | |
| 24 L.H., Daily Totals | | $1081.60 | | $1659.20 | $45.07 | $69.13 |

### Crew E-24

| Crew No. | Bare Costs Hr. | Daily | Incl. Subs O&P Hr. | Daily | Cost Per Labor-Hour Bare Costs | Incl. O&P |
|---|---|---|---|---|---|---|
| 3 Structural Steel Workers | $48.55 | $1165.20 | $85.75 | $2058.00 | $47.75 | $81.25 |
| 1 Equipment Operator (med.) | 45.35 | 362.80 | 67.75 | 542.00 | | |
| 1 Hyd. Crane, 25 Ton | | 748.60 | | 823.46 | 23.39 | 25.73 |
| 32 L.H., Daily Totals | | $2276.60 | | $3423.46 | $71.14 | $106.98 |

### Crew E-25

| Crew No. | Bare Costs Hr. | Daily | Incl. Subs O&P Hr. | Daily | Cost Per Labor-Hour Bare Costs | Incl. O&P |
|---|---|---|---|---|---|---|
| 1 Welder Foreman | $50.55 | $404.40 | $89.25 | $714.00 | $50.55 | $89.25 |
| 1 Cutting Torch | | 18.00 | | 19.80 | | |
| 1 Set of Gases | | 82.80 | | 91.08 | 12.60 | 13.86 |
| 8 L.H., Daily Totals | | $505.20 | | $824.88 | $63.15 | $103.11 |

### Crew F-3

| Crew No. | Bare Costs Hr. | Daily | Incl. Subs O&P Hr. | Daily | Cost Per Labor-Hour Bare Costs | Incl. O&P |
|---|---|---|---|---|---|---|
| 4 Carpenters | $43.05 | $1377.60 | $66.05 | $2113.60 | $43.74 | $66.73 |
| 1 Equip. Oper. (crane) | 46.50 | 372.00 | 69.45 | 555.60 | | |
| 1 Hyd. Crane, 12 Ton | | 655.60 | | 721.16 | 16.39 | 18.03 |
| 40 L.H., Daily Totals | | $2405.20 | | $3390.36 | $60.13 | $84.76 |

### Crew F-4

| Crew No. | Bare Costs Hr. | Daily | Incl. Subs O&P Hr. | Daily | Cost Per Labor-Hour Bare Costs | Incl. O&P |
|---|---|---|---|---|---|---|
| 4 Carpenters | $43.05 | $1377.60 | $66.05 | $2113.60 | $43.17 | $65.64 |
| 1 Equip. Oper. (crane) | 46.50 | 372.00 | 69.45 | 555.60 | | |
| 1 Equip. Oper. Oiler | 40.30 | 322.40 | 60.20 | 481.60 | | |
| 1 Hyd. Crane, 55 Ton | | 1208.00 | | 1328.80 | 25.17 | 27.68 |
| 48 L.H., Daily Totals | | $3280.00 | | $4479.60 | $68.33 | $93.33 |

### Crew F-5

| Crew No. | Bare Costs Hr. | Daily | Incl. Subs O&P Hr. | Daily | Cost Per Labor-Hour Bare Costs | Incl. O&P |
|---|---|---|---|---|---|---|
| 1 Carpenter Foreman | $45.05 | $360.40 | $69.10 | $552.80 | $43.55 | $66.81 |
| 3 Carpenters | 43.05 | 1033.20 | 66.05 | 1585.20 | | |
| 32 L.H., Daily Totals | | $1393.60 | | $2138.00 | $43.55 | $66.81 |

### Crew F-6

| Crew No. | Bare Costs Hr. | Daily | Incl. Subs O&P Hr. | Daily | Cost Per Labor-Hour Bare Costs | Incl. O&P |
|---|---|---|---|---|---|---|
| 2 Carpenters | $43.05 | $688.80 | $66.05 | $1056.80 | $40.26 | $61.39 |
| 2 Building Laborers | 34.35 | 549.60 | 52.70 | 843.20 | | |
| 1 Equip. Oper. (crane) | 46.50 | 372.00 | 69.45 | 555.60 | | |
| 1 Hyd. Crane, 12 Ton | | 655.60 | | 721.16 | 16.39 | 18.03 |
| 40 L.H., Daily Totals | | $2266.00 | | $3176.76 | $56.65 | $79.42 |

### Crew F-7

| Crew No. | Bare Costs Hr. | Daily | Incl. Subs O&P Hr. | Daily | Cost Per Labor-Hour Bare Costs | Incl. O&P |
|---|---|---|---|---|---|---|
| 2 Carpenters | $43.05 | $688.80 | $66.05 | $1056.80 | $38.70 | $59.38 |
| 2 Building Laborers | 34.35 | 549.60 | 52.70 | 843.20 | | |
| 32 L.H., Daily Totals | | $1238.40 | | $1900.00 | $38.70 | $59.38 |

### Crew G-1

| Crew No. | Bare Costs Hr. | Daily | Incl. Subs O&P Hr. | Daily | Cost Per Labor-Hour Bare Costs | Incl. O&P |
|---|---|---|---|---|---|---|
| 1 Roofer Foreman | $38.55 | $308.40 | $64.20 | $513.60 | $34.14 | $56.83 |
| 4 Roofers, Composition | 36.55 | 1169.60 | 60.85 | 1947.20 | | |
| 2 Roofer Helpers | 27.10 | 433.60 | 45.10 | 721.60 | | |
| 1 Application Equipment | | 168.40 | | 185.24 | | |
| 1 Tar Kettle/Pot | | 88.75 | | 97.63 | | |
| 1 Crew Truck | | 152.40 | | 167.64 | 7.31 | 8.04 |
| 56 L.H., Daily Totals | | $2321.15 | | $3632.91 | $41.45 | $64.87 |

### Crew G-2

| Crew No. | Bare Costs Hr. | Daily | Incl. Subs O&P Hr. | Daily | Cost Per Labor-Hour Bare Costs | Incl. O&P |
|---|---|---|---|---|---|---|
| 1 Plasterer | $39.40 | $315.20 | $58.60 | $468.80 | $36.27 | $54.47 |
| 1 Plasterer Helper | 35.05 | 280.40 | 52.10 | 416.80 | | |
| 1 Building Laborer | 34.35 | 274.80 | 52.70 | 421.60 | | |
| 1 Grout Pump, 50 C.F./hr | | 125.80 | | 138.38 | 5.24 | 5.77 |
| 24 L.H., Daily Totals | | $996.20 | | $1445.58 | $41.51 | $60.23 |

### Crew G-2A

| Crew No. | Bare Costs Hr. | Daily | Incl. Subs O&P Hr. | Daily | Cost Per Labor-Hour Bare Costs | Incl. O&P |
|---|---|---|---|---|---|---|
| 1 Roofer, composition | $36.55 | $292.40 | $60.85 | $486.80 | $32.67 | $52.88 |
| 1 Roofer Helper | 27.10 | 216.80 | 45.10 | 360.80 | | |
| 1 Building Laborer | 34.35 | 274.80 | 52.70 | 421.60 | | |
| 1 Foam spray rig, trailer-mtd. | | 488.25 | | 537.08 | | |
| 1 Pickup Truck, 3/4 Ton | | 104.60 | | 115.06 | 24.70 | 27.17 |
| 24 L.H., Daily Totals | | $1376.85 | | $1921.34 | $57.37 | $80.06 |

| Crew No. | Bare Costs | | Incl. Subs O&P | | Cost Per Labor-Hour | |
|---|---|---|---|---|---|---|

**Crew G-3**

| | Hr. | Daily | Hr. | Daily | Bare Costs | Incl. O&P |
|---|---|---|---|---|---|---|
| 2 Sheet Metal Workers | $51.65 | $826.40 | $78.10 | $1249.60 | $43.00 | $65.40 |
| 2 Building Laborers | 34.35 | 549.60 | 52.70 | 843.20 | | |
| 32 L.H., Daily Totals | | $1376.00 | | $2092.80 | $43.00 | $65.40 |

**Crew G-4**

| | Hr. | Daily | Hr. | Daily | Bare Costs | Incl. O&P |
|---|---|---|---|---|---|---|
| 1 Labor Foreman (outside) | $36.35 | $290.80 | $55.75 | $446.00 | $35.02 | $53.72 |
| 2 Building Laborers | 34.35 | 549.60 | 52.70 | 843.20 | | |
| 1 Flatbed Truck, Gas, 1.5 Ton | | 171.40 | | 188.54 | | |
| 1 Air Compressor, 160 cfm | | 125.00 | | 137.50 | 12.35 | 13.59 |
| 24 L.H., Daily Totals | | $1136.80 | | $1615.24 | $47.37 | $67.30 |

**Crew G-5**

| | Hr. | Daily | Hr. | Daily | Bare Costs | Incl. O&P |
|---|---|---|---|---|---|---|
| 1 Roofer Foreman | $38.55 | $308.40 | $64.20 | $513.60 | $33.17 | $55.22 |
| 2 Roofers, Composition | 36.55 | 584.80 | 60.85 | 973.60 | | |
| 2 Roofer Helpers | 27.10 | 433.60 | 45.10 | 721.60 | | |
| 1 Application Equipment | | 168.40 | | 185.24 | 4.21 | 4.63 |
| 40 L.H., Daily Totals | | $1495.20 | | $2394.04 | $37.38 | $59.85 |

**Crew G-6A**

| | Hr. | Daily | Hr. | Daily | Bare Costs | Incl. O&P |
|---|---|---|---|---|---|---|
| 2 Roofers Composition | $36.55 | $584.80 | $60.85 | $973.60 | $36.55 | $60.85 |
| 1 Small Compressor, Electric | | 11.05 | | 12.15 | | |
| 2 Pneumatic Nailers | | 47.00 | | 51.70 | 3.63 | 3.99 |
| 16 L.H., Daily Totals | | $642.85 | | $1037.45 | $40.18 | $64.84 |

**Crew G-7**

| | Hr. | Daily | Hr. | Daily | Bare Costs | Incl. O&P |
|---|---|---|---|---|---|---|
| 1 Carpenter | $43.05 | $344.40 | $66.05 | $528.40 | $43.05 | $66.05 |
| 1 Small Compressor, Electric | | 11.05 | | 12.15 | | |
| 1 Pneumatic Nailer | | 23.50 | | 25.85 | 4.32 | 4.75 |
| 8 L.H., Daily Totals | | $378.95 | | $566.40 | $47.37 | $70.80 |

**Crew H-1**

| | Hr. | Daily | Hr. | Daily | Bare Costs | Incl. O&P |
|---|---|---|---|---|---|---|
| 2 Glaziers | $42.05 | $672.80 | $63.15 | $1010.40 | $45.30 | $74.45 |
| 2 Struc. Steel Workers | 48.55 | 776.80 | 85.75 | 1372.00 | | |
| 32 L.H., Daily Totals | | $1449.60 | | $2382.40 | $45.30 | $74.45 |

**Crew H-2**

| | Hr. | Daily | Hr. | Daily | Bare Costs | Incl. O&P |
|---|---|---|---|---|---|---|
| 2 Glaziers | $42.05 | $672.80 | $63.15 | $1010.40 | $39.48 | $59.67 |
| 1 Building Laborer | 34.35 | 274.80 | 52.70 | 421.60 | | |
| 24 L.H., Daily Totals | | $947.60 | | $1432.00 | $39.48 | $59.67 |

**Crew H-3**

| | Hr. | Daily | Hr. | Daily | Bare Costs | Incl. O&P |
|---|---|---|---|---|---|---|
| 1 Glazier | $42.05 | $336.40 | $63.15 | $505.20 | $37.35 | $56.52 |
| 1 Helper | 32.65 | 261.20 | 49.90 | 399.20 | | |
| 16 L.H., Daily Totals | | $597.60 | | $904.40 | $37.35 | $56.52 |

**Crew H-4**

| | Hr. | Daily | Hr. | Daily | Bare Costs | Incl. O&P |
|---|---|---|---|---|---|---|
| 1 Carpenter | $43.05 | $344.40 | $66.05 | $528.40 | $40.34 | $61.28 |
| 1 Carpenter Helper | 32.65 | 261.20 | 49.90 | 399.20 | | |
| .5 Electrician | 50.30 | 201.20 | 74.50 | 298.00 | | |
| 20 L.H., Daily Totals | | $806.80 | | $1225.60 | $40.34 | $61.28 |

**Crew J-1**

| | Hr. | Daily | Hr. | Daily | Bare Costs | Incl. O&P |
|---|---|---|---|---|---|---|
| 3 Plasterers | $39.40 | $945.60 | $58.60 | $1406.40 | $37.66 | $56.00 |
| 2 Plasterer Helpers | 35.05 | 560.80 | 52.10 | 833.60 | | |
| 1 Mixing Machine, 6 C.F. | | 123.60 | | 135.96 | 3.09 | 3.40 |
| 40 L.H., Daily Totals | | $1630.00 | | $2375.96 | $40.75 | $59.40 |

**Crew J-2**

| | Hr. | Daily | Hr. | Daily | Bare Costs | Incl. O&P |
|---|---|---|---|---|---|---|
| 3 Plasterers | $39.40 | $945.60 | $58.60 | $1406.40 | $37.81 | $56.04 |
| 2 Plasterer Helpers | 35.05 | 560.80 | 52.10 | 833.60 | | |
| 1 Lather | 38.55 | 308.40 | 56.25 | 450.00 | | |
| 1 Mixing Machine, 6 C.F. | | 123.60 | | 135.96 | 2.58 | 2.83 |
| 48 L.H., Daily Totals | | $1938.40 | | $2825.96 | $40.38 | $58.87 |

**Crew J-3**

| | Hr. | Daily | Hr. | Daily | Bare Costs | Incl. O&P |
|---|---|---|---|---|---|---|
| 1 Terrazzo Worker | $40.10 | $320.80 | $58.45 | $467.60 | $36.80 | $53.65 |
| 1 Terrazzo Helper | 33.50 | 268.00 | 48.85 | 390.80 | | |
| 1 Floor Grinder, 22" path | | 73.95 | | 81.34 | | |
| 1 Terrazzo Mixer | | 167.60 | | 184.36 | 15.10 | 16.61 |
| 16 L.H., Daily Totals | | $830.35 | | $1124.11 | $51.90 | $70.26 |

**Crew J-4**

| | Hr. | Daily | Hr. | Daily | Bare Costs | Incl. O&P |
|---|---|---|---|---|---|---|
| 2 Cement Finishers | $40.85 | $653.60 | $59.50 | $952.00 | $38.68 | $57.23 |
| 1 Laborer | 34.35 | 274.80 | 52.70 | 421.60 | | |
| 1 Floor Grinder, 22" path | | 73.95 | | 81.34 | | |
| 1 Floor Edger, 7" path | | 38.20 | | 42.02 | | |
| 1 Vacuum Pick-Up System | | 59.80 | | 65.78 | 7.16 | 7.88 |
| 24 L.H., Daily Totals | | $1100.35 | | $1562.74 | $45.85 | $65.11 |

**Crew J-4A**

| | Hr. | Daily | Hr. | Daily | Bare Costs | Incl. O&P |
|---|---|---|---|---|---|---|
| 2 Cement Finishers | $40.85 | $653.60 | $59.50 | $952.00 | $37.60 | $56.10 |
| 2 Laborers | 34.35 | 549.60 | 52.70 | 843.20 | | |
| 1 Floor Grinder, 22" path | | 73.95 | | 81.34 | | |
| 1 Floor Edger, 7" path | | 38.20 | | 42.02 | | |
| 1 Vacuum Pick-Up System | | 59.80 | | 65.78 | | |
| 1 Floor Auto Scrubber | | 154.10 | | 169.51 | 10.19 | 11.21 |
| 32 L.H., Daily Totals | | $1529.25 | | $2153.86 | $47.79 | $67.31 |

**Crew J-4B**

| | Hr. | Daily | Hr. | Daily | Bare Costs | Incl. O&P |
|---|---|---|---|---|---|---|
| 1 Laborer | $34.35 | $274.80 | $52.70 | $421.60 | $34.35 | $52.70 |
| 1 Floor Auto Scrubber | | 154.10 | | 169.51 | 19.26 | 21.19 |
| 8 L.H., Daily Totals | | $428.90 | | $591.11 | $53.61 | $73.89 |

**Crew K-1**

| | Hr. | Daily | Hr. | Daily | Bare Costs | Incl. O&P |
|---|---|---|---|---|---|---|
| 1 Carpenter | $43.05 | $344.40 | $66.05 | $528.40 | $38.23 | $58.40 |
| 1 Truck Driver (light) | 33.40 | 267.20 | 50.75 | 406.00 | | |
| 1 Flatbed Truck, Gas, 3 Ton | | 217.00 | | 238.70 | 13.56 | 14.92 |
| 16 L.H., Daily Totals | | $828.60 | | $1173.10 | $51.79 | $73.32 |

**Crew K-2**

| | Hr. | Daily | Hr. | Daily | Bare Costs | Incl. O&P |
|---|---|---|---|---|---|---|
| 1 Struc. Steel Foreman | $50.55 | $404.40 | $89.25 | $714.00 | $44.17 | $75.25 |
| 1 Struc. Steel Worker | 48.55 | 388.40 | 85.75 | 686.00 | | |
| 1 Truck Driver (light) | 33.40 | 267.20 | 50.75 | 406.00 | | |
| 1 Flatbed Truck, Gas, 3 Ton | | 217.00 | | 238.70 | 9.04 | 9.95 |
| 24 L.H., Daily Totals | | $1277.00 | | $2044.70 | $53.21 | $85.20 |

**Crew L-1**

| | Hr. | Daily | Hr. | Daily | Bare Costs | Incl. O&P |
|---|---|---|---|---|---|---|
| 1 Electrician | $50.30 | $402.40 | $74.50 | $596.00 | $51.95 | $77.25 |
| 1 Plumber | 53.60 | 428.80 | 80.00 | 640.00 | | |
| 16 L.H., Daily Totals | | $831.20 | | $1236.00 | $51.95 | $77.25 |

**Crew L-2**

| | Hr. | Daily | Hr. | Daily | Bare Costs | Incl. O&P |
|---|---|---|---|---|---|---|
| 1 Carpenter | $43.05 | $344.40 | $66.05 | $528.40 | $37.85 | $57.98 |
| 1 Carpenter Helper | 32.65 | 261.20 | 49.90 | 399.20 | | |
| 16 L.H., Daily Totals | | $605.60 | | $927.60 | $37.85 | $57.98 |

| Crew No. | Bare Costs | | Incl. Subs O&P | | Cost Per Labor-Hour | |
|---|---|---|---|---|---|---|

| Crew L-3 | Hr. | Daily | Hr. | Daily | Bare Costs | Incl. O&P |
|---|---|---|---|---|---|---|
| 1 Carpenter | $43.05 | $344.40 | $66.05 | $528.40 | $47.01 | $71.17 |
| .5 Electrician | 50.30 | 201.20 | 74.50 | 298.00 | | |
| .5 Sheet Metal Worker | 51.65 | 206.60 | 78.10 | 312.40 | | |
| 16 L.H., Daily Totals | | $752.20 | | $1138.80 | $47.01 | $71.17 |

| Crew L-3A | Hr. | Daily | Hr. | Daily | Bare Costs | Incl. O&P |
|---|---|---|---|---|---|---|
| 1 Carpenter Foreman (outside) | $45.05 | $360.40 | $69.10 | $552.80 | $47.25 | $72.10 |
| .5 Sheet Metal Worker | 51.65 | 206.60 | 78.10 | 312.40 | | |
| 12 L.H., Daily Totals | | $567.00 | | $865.20 | $47.25 | $72.10 |

| Crew L-4 | Hr. | Daily | Hr. | Daily | Bare Costs | Incl. O&P |
|---|---|---|---|---|---|---|
| 2 Skilled Workers | $44.40 | $710.40 | $68.10 | $1089.60 | $40.48 | $62.03 |
| 1 Helper | 32.65 | 261.20 | 49.90 | 399.20 | | |
| 24 L.H., Daily Totals | | $971.60 | | $1488.80 | $40.48 | $62.03 |

| Crew L-5 | Hr. | Daily | Hr. | Daily | Bare Costs | Incl. O&P |
|---|---|---|---|---|---|---|
| 1 Struc. Steel Foreman | $50.55 | $404.40 | $89.25 | $714.00 | $48.54 | $83.92 |
| 5 Struc. Steel Workers | 48.55 | 1942.00 | 85.75 | 3430.00 | | |
| 1 Equip. Oper. (crane) | 46.50 | 372.00 | 69.45 | 555.60 | | |
| 1 Hyd. Crane, 25 Ton | | 748.60 | | 823.46 | 13.37 | 14.70 |
| 56 L.H., Daily Totals | | $3467.00 | | $5523.06 | $61.91 | $98.63 |

| Crew L-5A | Hr. | Daily | Hr. | Daily | Bare Costs | Incl. O&P |
|---|---|---|---|---|---|---|
| 1 Structural Steel Foreman | $50.55 | $404.40 | $89.25 | $714.00 | $48.54 | $82.55 |
| 2 Structural Steel Workers | 48.55 | 776.80 | 85.75 | 1372.00 | | |
| 1 Equip. Oper. (crane) | 46.50 | 372.00 | 69.45 | 555.60 | | |
| 1 S.P. Crane, 4x4, 25 Ton | | 560.40 | | 616.44 | 17.51 | 19.26 |
| 32 L.H., Daily Totals | | $2113.60 | | $3258.04 | $66.05 | $101.81 |

| Crew L-5B | Hr. | Daily | Hr. | Daily | Bare Costs | Incl. O&P |
|---|---|---|---|---|---|---|
| 1 Structural Steel Foreman | $50.55 | $404.40 | $89.25 | $714.00 | $49.25 | $77.88 |
| 2 Structural Steel Workers | 48.55 | 776.80 | 85.75 | 1372.00 | | |
| 2 Electricians | 50.30 | 804.80 | 74.50 | 1192.00 | | |
| 2 Steamfitters/Pipefitters | 54.10 | 865.60 | 80.75 | 1292.00 | | |
| 1 Equip. Oper. (crane) | 46.50 | 372.00 | 69.45 | 555.60 | | |
| 1 Equip. Oper. Oiler | 40.30 | 322.40 | 60.20 | 481.60 | | |
| 1 Hyd. Crane, 80 Ton | | 1693.00 | | 1862.30 | 23.51 | 25.87 |
| 72 L.H., Daily Totals | | $5239.00 | | $7469.50 | $72.76 | $103.74 |

| Crew L-6 | Hr. | Daily | Hr. | Daily | Bare Costs | Incl. O&P |
|---|---|---|---|---|---|---|
| 1 Plumber | $53.60 | $428.80 | $80.00 | $640.00 | $52.50 | $78.17 |
| .5 Electrician | 50.30 | 201.20 | 74.50 | 298.00 | | |
| 12 L.H., Daily Totals | | $630.00 | | $938.00 | $52.50 | $78.17 |

| Crew L-7 | Hr. | Daily | Hr. | Daily | Bare Costs | Incl. O&P |
|---|---|---|---|---|---|---|
| 2 Carpenters | $43.05 | $688.80 | $66.05 | $1056.80 | $41.60 | $63.44 |
| 1 Building Laborer | 34.35 | 274.80 | 52.70 | 421.60 | | |
| .5 Electrician | 50.30 | 201.20 | 74.50 | 298.00 | | |
| 28 L.H., Daily Totals | | $1164.80 | | $1776.40 | $41.60 | $63.44 |

| Crew L-8 | Hr. | Daily | Hr. | Daily | Bare Costs | Incl. O&P |
|---|---|---|---|---|---|---|
| 2 Carpenters | $43.05 | $688.80 | $66.05 | $1056.80 | $45.16 | $68.84 |
| .5 Plumber | 53.60 | 214.40 | 80.00 | 320.00 | | |
| 20 L.H., Daily Totals | | $903.20 | | $1376.80 | $45.16 | $68.84 |

| Crew L-9 | Hr. | Daily | Hr. | Daily | Bare Costs | Incl. O&P |
|---|---|---|---|---|---|---|
| 1 Labor Foreman (inside) | $34.85 | $278.80 | $53.45 | $427.60 | $39.39 | $62.63 |
| 2 Building Laborers | 34.35 | 549.60 | 52.70 | 843.20 | | |
| 1 Struc. Steel Worker | 48.55 | 388.40 | 85.75 | 686.00 | | |
| .5 Electrician | 50.30 | 201.20 | 74.50 | 298.00 | | |
| 36 L.H., Daily Totals | | $1418.00 | | $2254.80 | $39.39 | $62.63 |

| Crew L-10 | Hr. | Daily | Hr. | Daily | Bare Costs | Incl. O&P |
|---|---|---|---|---|---|---|
| 1 Structural Steel Foreman | $50.55 | $404.40 | $89.25 | $714.00 | $48.53 | $81.48 |
| 1 Structural Steel Worker | 48.55 | 388.40 | 85.75 | 686.00 | | |
| 1 Equip. Oper. (crane) | 46.50 | 372.00 | 69.45 | 555.60 | | |
| 1 Hyd. Crane, 12 Ton | | 655.60 | | 721.16 | 27.32 | 30.05 |
| 24 L.H., Daily Totals | | $1820.40 | | $2676.76 | $75.85 | $111.53 |

| Crew L-11 | Hr. | Daily | Hr. | Daily | Bare Costs | Incl. O&P |
|---|---|---|---|---|---|---|
| 2 Wreckers | $34.35 | $549.60 | $57.80 | $924.80 | $39.70 | $62.55 |
| 1 Equip. Oper. (crane) | 46.50 | 372.00 | 69.45 | 555.60 | | |
| 1 Equip. Oper. (light) | 43.60 | 348.80 | 65.15 | 521.20 | | |
| 1 Hyd. Excavator, 2.5 C.Y. | | 1613.00 | | 1774.30 | | |
| 1 Loader, Skid Steer, 78 H.P. | | 261.20 | | 287.32 | 58.57 | 64.43 |
| 32 L.H., Daily Totals | | $3144.60 | | $4063.22 | $98.27 | $126.98 |

| Crew M-1 | Hr. | Daily | Hr. | Daily | Bare Costs | Incl. O&P |
|---|---|---|---|---|---|---|
| 3 Elevator Constructors | $67.40 | $1617.60 | $100.00 | $2400.00 | $64.03 | $95.00 |
| 1 Elevator Apprentice | 53.90 | 431.20 | 80.00 | 640.00 | | |
| 5 Hand Tools | | 49.00 | | 53.90 | 1.53 | 1.68 |
| 32 L.H., Daily Totals | | $2097.80 | | $3093.90 | $65.56 | $96.68 |

| Crew M-3 | Hr. | Daily | Hr. | Daily | Bare Costs | Incl. O&P |
|---|---|---|---|---|---|---|
| 1 Electrician Foreman (out) | $52.30 | $418.40 | $77.45 | $619.60 | $51.60 | $76.96 |
| 1 Common Laborer | 34.35 | 274.80 | 52.70 | 421.60 | | |
| .25 Equipment Operator, Med. | 45.35 | 90.70 | 67.75 | 135.50 | | |
| 1 Elevator Constructor | 67.40 | 539.20 | 100.00 | 800.00 | | |
| 1 Elevator Apprentice | 53.90 | 431.20 | 80.00 | 640.00 | | |
| .25 S.P. Crane, 4x4, 20 Ton | | 127.75 | | 140.53 | 3.76 | 4.13 |
| 34 L.H., Daily Totals | | $1882.05 | | $2757.22 | $55.35 | $81.09 |

| Crew M-4 | Hr. | Daily | Hr. | Daily | Bare Costs | Incl. O&P |
|---|---|---|---|---|---|---|
| 1 Electrician Foreman (out) | $52.30 | $418.40 | $77.45 | $619.60 | $51.03 | $76.13 |
| 1 Common Laborer | 34.35 | 274.80 | 52.70 | 421.60 | | |
| .25 Equipment Operator, Crane | 46.50 | 93.00 | 69.45 | 138.90 | | |
| .25 Equipment Operator, Oiler | 40.30 | 80.60 | 60.20 | 120.40 | | |
| 1 Elevator Constructor | 67.40 | 539.20 | 100.00 | 800.00 | | |
| 1 Elevator Apprentice | 53.90 | 431.20 | 80.00 | 640.00 | | |
| .25 S.P. Crane, 4x4, 40 Ton | | 164.65 | | 181.12 | 4.57 | 5.03 |
| 36 L.H., Daily Totals | | $2001.85 | | $2921.61 | $55.61 | $81.16 |

| Crew Q-1 | Hr. | Daily | Hr. | Daily | Bare Costs | Incl. O&P |
|---|---|---|---|---|---|---|
| 1 Plumber | $53.60 | $428.80 | $80.00 | $640.00 | $48.25 | $72.03 |
| 1 Plumber Apprentice | 42.90 | 343.20 | 64.05 | 512.40 | | |
| 16 L.H., Daily Totals | | $772.00 | | $1152.40 | $48.25 | $72.03 |

| Crew Q-1A | Hr. | Daily | Hr. | Daily | Bare Costs | Incl. O&P |
|---|---|---|---|---|---|---|
| .25 Plumber Foreman (out) | $55.60 | $111.20 | $83.00 | $166.00 | $54.00 | $80.60 |
| 1 Plumber | 53.60 | 428.80 | 80.00 | 640.00 | | |
| 10 L.H., Daily Totals | | $540.00 | | $806.00 | $54.00 | $80.60 |

| Crew Q-1C | Hr. | Daily | Hr. | Daily | Bare Costs | Incl. O&P |
|---|---|---|---|---|---|---|
| 1 Plumber | $53.60 | $428.80 | $80.00 | $640.00 | $47.28 | $70.60 |
| 1 Plumber Apprentice | 42.90 | 343.20 | 64.05 | 512.40 | | |
| 1 Equip. Oper. (medium) | 45.35 | 362.80 | 67.75 | 542.00 | | |
| 1 Trencher, Chain Type, 8' D | | 2506.00 | | 2756.60 | 104.42 | 114.86 |
| 24 L.H., Daily Totals | | $3640.80 | | $4451.00 | $151.70 | $185.46 |

| Crew Q-2 | Hr. | Daily | Hr. | Daily | Bare Costs | Incl. O&P |
|---|---|---|---|---|---|---|
| 2 Plumbers | $53.60 | $857.60 | $80.00 | $1280.00 | $50.03 | $74.68 |
| 1 Plumber Apprentice | 42.90 | 343.20 | 64.05 | 512.40 | | |
| 24 L.H., Daily Totals | | $1200.80 | | $1792.40 | $50.03 | $74.68 |

| Crew Q-3 | Hr. | Daily | Hr. | Daily | Bare Costs | Incl. O&P |
|---|---|---|---|---|---|---|
| 1 Plumber Foreman (inside) | $54.10 | $432.80 | $80.75 | $646.00 | $51.05 | $76.20 |
| 2 Plumbers | 53.60 | 857.60 | 80.00 | 1280.00 | | |
| 1 Plumber Apprentice | 42.90 | 343.20 | 64.05 | 512.40 | | |
| 32 L.H., Daily Totals | | $1633.60 | | $2438.40 | $51.05 | $76.20 |

| Crew Q-4 | Hr. | Daily | Hr. | Daily | Bare Costs | Incl. O&P |
|---|---|---|---|---|---|---|
| 1 Plumber Foreman (inside) | $54.10 | $432.80 | $80.75 | $646.00 | $51.05 | $76.20 |
| 1 Plumber | 53.60 | 428.80 | 80.00 | 640.00 | | |
| 1 Welder (plumber) | 53.60 | 428.80 | 80.00 | 640.00 | | |
| 1 Plumber Apprentice | 42.90 | 343.20 | 64.05 | 512.40 | | |
| 1 Welder, electric, 300 amp | | 32.45 | | 35.70 | 1.01 | 1.12 |
| 32 L.H., Daily Totals | | $1666.05 | | $2474.09 | $52.06 | $77.32 |

| Crew Q-5 | Hr. | Daily | Hr. | Daily | Bare Costs | Incl. O&P |
|---|---|---|---|---|---|---|
| 1 Steamfitter | $54.10 | $432.80 | $80.75 | $646.00 | $48.70 | $72.70 |
| 1 Steamfitter Apprentice | 43.30 | 346.40 | 64.65 | 517.20 | | |
| 16 L.H., Daily Totals | | $779.20 | | $1163.20 | $48.70 | $72.70 |

| Crew Q-6 | Hr. | Daily | Hr. | Daily | Bare Costs | Incl. O&P |
|---|---|---|---|---|---|---|
| 2 Steamfitters | $54.10 | $865.60 | $80.75 | $1292.00 | $50.50 | $75.38 |
| 1 Steamfitter Apprentice | 43.30 | 346.40 | 64.65 | 517.20 | | |
| 24 L.H., Daily Totals | | $1212.00 | | $1809.20 | $50.50 | $75.38 |

| Crew Q-7 | Hr. | Daily | Hr. | Daily | Bare Costs | Incl. O&P |
|---|---|---|---|---|---|---|
| 1 Steamfitter Foreman (inside) | $54.60 | $436.80 | $81.50 | $652.00 | $51.52 | $76.91 |
| 2 Steamfitters | 54.10 | 865.60 | 80.75 | 1292.00 | | |
| 1 Steamfitter Apprentice | 43.30 | 346.40 | 64.65 | 517.20 | | |
| 32 L.H., Daily Totals | | $1648.80 | | $2461.20 | $51.52 | $76.91 |

| Crew Q-8 | Hr. | Daily | Hr. | Daily | Bare Costs | Incl. O&P |
|---|---|---|---|---|---|---|
| 1 Steamfitter Foreman (inside) | $54.60 | $436.80 | $81.50 | $652.00 | $51.52 | $76.91 |
| 1 Steamfitter | 54.10 | 432.80 | 80.75 | 646.00 | | |
| 1 Welder (steamfitter) | 54.10 | 432.80 | 80.75 | 646.00 | | |
| 1 Steamfitter Apprentice | 43.30 | 346.40 | 64.65 | 517.20 | | |
| 1 Welder, electric, 300 amp | | 32.45 | | 35.70 | 1.01 | 1.12 |
| 32 L.H., Daily Totals | | $1681.25 | | $2496.90 | $52.54 | $78.03 |

| Crew Q-9 | Hr. | Daily | Hr. | Daily | Bare Costs | Incl. O&P |
|---|---|---|---|---|---|---|
| 1 Sheet Metal Worker | $51.65 | $413.20 | $78.10 | $624.80 | $46.48 | $70.28 |
| 1 Sheet Metal Apprentice | 41.30 | 330.40 | 62.45 | 499.60 | | |
| 16 L.H., Daily Totals | | $743.60 | | $1124.40 | $46.48 | $70.28 |

| Crew Q-10 | Hr. | Daily | Hr. | Daily | Bare Costs | Incl. O&P |
|---|---|---|---|---|---|---|
| 2 Sheet Metal Workers | $51.65 | $826.40 | $78.10 | $1249.60 | $48.20 | $72.88 |
| 1 Sheet Metal Apprentice | 41.30 | 330.40 | 62.45 | 499.60 | | |
| 24 L.H., Daily Totals | | $1156.80 | | $1749.20 | $48.20 | $72.88 |

| Crew Q-11 | Hr. | Daily | Hr. | Daily | Bare Costs | Incl. O&P |
|---|---|---|---|---|---|---|
| 1 Sheet Metal Foreman (inside) | $52.15 | $417.20 | $78.85 | $630.80 | $49.19 | $74.38 |
| 2 Sheet Metal Workers | 51.65 | 826.40 | 78.10 | 1249.60 | | |
| 1 Sheet Metal Apprentice | 41.30 | 330.40 | 62.45 | 499.60 | | |
| 32 L.H., Daily Totals | | $1574.00 | | $2380.00 | $49.19 | $74.38 |

| Crew Q-12 | Hr. | Daily | Hr. | Daily | Bare Costs | Incl. O&P |
|---|---|---|---|---|---|---|
| 1 Sprinkler Installer | $53.10 | $424.80 | $79.35 | $634.80 | $47.80 | $71.42 |
| 1 Sprinkler Apprentice | 42.50 | 340.00 | 63.50 | 508.00 | | |
| 16 L.H., Daily Totals | | $764.80 | | $1142.80 | $47.80 | $71.42 |

| Crew Q-13 | Hr. | Daily | Hr. | Daily | Bare Costs | Incl. O&P |
|---|---|---|---|---|---|---|
| 1 Sprinkler Foreman (inside) | $53.60 | $428.80 | $80.10 | $640.80 | $50.58 | $75.58 |
| 2 Sprinkler Installers | 53.10 | 849.60 | 79.35 | 1269.60 | | |
| 1 Sprinkler Apprentice | 42.50 | 340.00 | 63.50 | 508.00 | | |
| 32 L.H., Daily Totals | | $1618.40 | | $2418.40 | $50.58 | $75.58 |

| Crew Q-14 | Hr. | Daily | Hr. | Daily | Bare Costs | Incl. O&P |
|---|---|---|---|---|---|---|
| 1 Asbestos Worker | $46.90 | $375.20 | $72.50 | $580.00 | $42.20 | $65.25 |
| 1 Asbestos Apprentice | 37.50 | 300.00 | 58.00 | 464.00 | | |
| 16 L.H., Daily Totals | | $675.20 | | $1044.00 | $42.20 | $65.25 |

| Crew Q-15 | Hr. | Daily | Hr. | Daily | Bare Costs | Incl. O&P |
|---|---|---|---|---|---|---|
| 1 Plumber | $53.60 | $428.80 | $80.00 | $640.00 | $48.25 | $72.03 |
| 1 Plumber Apprentice | 42.90 | 343.20 | 64.05 | 512.40 | | |
| 1 Welder, electric, 300 amp | | 32.45 | | 35.70 | 2.03 | 2.23 |
| 16 L.H., Daily Totals | | $804.45 | | $1188.10 | $50.28 | $74.26 |

| Crew Q-16 | Hr. | Daily | Hr. | Daily | Bare Costs | Incl. O&P |
|---|---|---|---|---|---|---|
| 2 Plumbers | $53.60 | $857.60 | $80.00 | $1280.00 | $50.03 | $74.68 |
| 1 Plumber Apprentice | 42.90 | 343.20 | 64.05 | 512.40 | | |
| 1 Welder, electric, 300 amp | | 32.45 | | 35.70 | 1.35 | 1.49 |
| 24 L.H., Daily Totals | | $1233.25 | | $1828.10 | $51.39 | $76.17 |

| Crew Q-17 | Hr. | Daily | Hr. | Daily | Bare Costs | Incl. O&P |
|---|---|---|---|---|---|---|
| 1 Steamfitter | $54.10 | $432.80 | $80.75 | $646.00 | $48.70 | $72.70 |
| 1 Steamfitter Apprentice | 43.30 | 346.40 | 64.65 | 517.20 | | |
| 1 Welder, electric, 300 amp | | 32.45 | | 35.70 | 2.03 | 2.23 |
| 16 L.H., Daily Totals | | $811.65 | | $1198.90 | $50.73 | $74.93 |

| Crew Q-17A | Hr. | Daily | Hr. | Daily | Bare Costs | Incl. O&P |
|---|---|---|---|---|---|---|
| 1 Steamfitter | $54.10 | $432.80 | $80.75 | $646.00 | $47.97 | $71.62 |
| 1 Steamfitter Apprentice | 43.30 | 346.40 | 64.65 | 517.20 | | |
| 1 Equip. Oper. (crane) | 46.50 | 372.00 | 69.45 | 555.60 | | |
| 1 Hyd. Crane, 12 Ton | | 655.60 | | 721.16 | | |
| 1 Welder, electric, 300 amp | | 32.45 | | 35.70 | 28.67 | 31.54 |
| 24 L.H., Daily Totals | | $1839.25 | | $2475.66 | $76.64 | $103.15 |

| Crew Q-18 | Hr. | Daily | Hr. | Daily | Bare Costs | Incl. O&P |
|---|---|---|---|---|---|---|
| 2 Steamfitters | $54.10 | $865.60 | $80.75 | $1292.00 | $50.50 | $75.38 |
| 1 Steamfitter Apprentice | 43.30 | 346.40 | 64.65 | 517.20 | | |
| 1 Welder, electric, 300 amp | | 32.45 | | 35.70 | 1.35 | 1.49 |
| 24 L.H., Daily Totals | | $1244.45 | | $1844.90 | $51.85 | $76.87 |

| Crew Q-19 | Hr. | Daily | Hr. | Daily | Bare Costs | Incl. O&P |
|---|---|---|---|---|---|---|
| 1 Steamfitter | $54.10 | $432.80 | $80.75 | $646.00 | $49.23 | $73.30 |
| 1 Steamfitter Apprentice | 43.30 | 346.40 | 64.65 | 517.20 | | |
| 1 Electrician | 50.30 | 402.40 | 74.50 | 596.00 | | |
| 24 L.H., Daily Totals | | $1181.60 | | $1759.20 | $49.23 | $73.30 |

| Crew No. | Bare Costs | | Incl. Subs O&P | | Cost Per Labor-Hour | |
|---|---|---|---|---|---|---|
| **Crew Q-20** | Hr. | Daily | Hr. | Daily | Bare Costs | Incl. O&P |
| 1 Sheet Metal Worker | $51.65 | $413.20 | $78.10 | $624.80 | $47.24 | $71.12 |
| 1 Sheet Metal Apprentice | 41.30 | 330.40 | 62.45 | 499.60 | | |
| .5 Electrician | 50.30 | 201.20 | 74.50 | 298.00 | | |
| 20 L.H., Daily Totals | | $944.80 | | $1422.40 | $47.24 | $71.12 |

| Crew No. | Bare Costs | | Incl. Subs O&P | | Cost Per Labor-Hour | |
|---|---|---|---|---|---|---|
| **Crew Q-21** | Hr. | Daily | Hr. | Daily | Bare Costs | Incl. O&P |
| 2 Steamfitters | $54.10 | $865.60 | $80.75 | $1292.00 | $50.45 | $75.16 |
| 1 Steamfitter Apprentice | 43.30 | 346.40 | 64.65 | 517.20 | | |
| 1 Electrician | 50.30 | 402.40 | 74.50 | 596.00 | | |
| 32 L.H., Daily Totals | | $1614.40 | | $2405.20 | $50.45 | $75.16 |

| Crew No. | Bare Costs | | Incl. Subs O&P | | Cost Per Labor-Hour | |
|---|---|---|---|---|---|---|
| **Crew Q-22** | Hr. | Daily | Hr. | Daily | Bare Costs | Incl. O&P |
| 1 Plumber | $53.60 | $428.80 | $80.00 | $640.00 | $48.25 | $72.03 |
| 1 Plumber Apprentice | 42.90 | 343.20 | 64.05 | 512.40 | | |
| 1 Hyd. Crane, 12 Ton | | 655.60 | | 721.16 | 40.98 | 45.07 |
| 16 L.H., Daily Totals | | $1427.60 | | $1873.56 | $89.22 | $117.10 |

| Crew No. | Bare Costs | | Incl. Subs O&P | | Cost Per Labor-Hour | |
|---|---|---|---|---|---|---|
| **Crew Q-22A** | Hr. | Daily | Hr. | Daily | Bare Costs | Incl. O&P |
| 1 Plumber | $53.60 | $428.80 | $80.00 | $640.00 | $44.34 | $66.55 |
| 1 Plumber Apprentice | 42.90 | 343.20 | 64.05 | 512.40 | | |
| 1 Laborer | 34.35 | 274.80 | 52.70 | 421.60 | | |
| 1 Equip. Oper. (crane) | 46.50 | 372.00 | 69.45 | 555.60 | | |
| 1 Hyd. Crane, 12 Ton | | 655.60 | | 721.16 | 20.49 | 22.54 |
| 32 L.H., Daily Totals | | $2074.40 | | $2850.76 | $64.83 | $89.09 |

| Crew No. | Bare Costs | | Incl. Subs O&P | | Cost Per Labor-Hour | |
|---|---|---|---|---|---|---|
| **Crew Q-23** | Hr. | Daily | Hr. | Daily | Bare Costs | Incl. O&P |
| 1 Plumber Foreman | $55.60 | $444.80 | $83.00 | $664.00 | $51.52 | $76.92 |
| 1 Plumber | 53.60 | 428.80 | 80.00 | 640.00 | | |
| 1 Equip. Oper. (med.) | 45.35 | 362.80 | 67.75 | 542.00 | | |
| 1 Lattice Boom Crane, 20 Ton | | 1035.00 | | 1138.50 | 43.13 | 47.44 |
| 24 L.H., Daily Totals | | $2271.40 | | $2984.50 | $94.64 | $124.35 |

| Crew No. | Bare Costs | | Incl. Subs O&P | | Cost Per Labor-Hour | |
|---|---|---|---|---|---|---|
| **Crew R-1** | Hr. | Daily | Hr. | Daily | Bare Costs | Incl. O&P |
| 1 Electrician Foreman | $50.80 | $406.40 | $75.25 | $602.00 | $44.50 | $66.42 |
| 3 Electricians | 50.30 | 1207.20 | 74.50 | 1788.00 | | |
| 2 Helpers | 32.65 | 522.40 | 49.90 | 798.40 | | |
| 48 L.H., Daily Totals | | $2136.00 | | $3188.40 | $44.50 | $66.42 |

| Crew No. | Bare Costs | | Incl. Subs O&P | | Cost Per Labor-Hour | |
|---|---|---|---|---|---|---|
| **Crew R-1A** | Hr. | Daily | Hr. | Daily | Bare Costs | Incl. O&P |
| 1 Electrician | $50.30 | $402.40 | $74.50 | $596.00 | $41.48 | $62.20 |
| 1 Helper | 32.65 | 261.20 | 49.90 | 399.20 | | |
| 16 L.H., Daily Totals | | $663.60 | | $995.20 | $41.48 | $62.20 |

| Crew No. | Bare Costs | | Incl. Subs O&P | | Cost Per Labor-Hour | |
|---|---|---|---|---|---|---|
| **Crew R-2** | Hr. | Daily | Hr. | Daily | Bare Costs | Incl. O&P |
| 1 Electrician Foreman | $50.80 | $406.40 | $75.25 | $602.00 | $44.79 | $66.86 |
| 3 Electricians | 50.30 | 1207.20 | 74.50 | 1788.00 | | |
| 2 Helpers | 32.65 | 522.40 | 49.90 | 798.40 | | |
| 1 Equip. Oper. (crane) | 46.50 | 372.00 | 69.45 | 555.60 | | |
| 1 S.P. Crane, 4x4, 5 Ton | | 257.80 | | 283.58 | 4.60 | 5.06 |
| 56 L.H., Daily Totals | | $2765.80 | | $4027.58 | $49.39 | $71.92 |

| Crew No. | Bare Costs | | Incl. Subs O&P | | Cost Per Labor-Hour | |
|---|---|---|---|---|---|---|
| **Crew R-3** | Hr. | Daily | Hr. | Daily | Bare Costs | Incl. O&P |
| 1 Electrician Foreman | $50.80 | $406.40 | $75.25 | $602.00 | $49.74 | $73.79 |
| 1 Electrician | 50.30 | 402.40 | 74.50 | 596.00 | | |
| .5 Equip. Oper. (crane) | 46.50 | 186.00 | 69.45 | 277.80 | | |
| .5 S.P. Crane, 4x4, 5 Ton | | 128.90 | | 141.79 | 6.45 | 7.09 |
| 20 L.H., Daily Totals | | $1123.70 | | $1617.59 | $56.19 | $80.88 |

| Crew No. | Bare Costs | | Incl. Subs O&P | | Cost Per Labor-Hour | |
|---|---|---|---|---|---|---|
| **Crew R-4** | Hr. | Daily | Hr. | Daily | Bare Costs | Incl. O&P |
| 1 Struc. Steel Foreman | $50.55 | $404.40 | $89.25 | $714.00 | $49.30 | $84.20 |
| 3 Struc. Steel Workers | 48.55 | 1165.20 | 85.75 | 2058.00 | | |
| 1 Electrician | 50.30 | 402.40 | 74.50 | 596.00 | | |
| 1 Welder, gas engine, 300 amp | | 109.00 | | 119.90 | 2.73 | 3.00 |
| 40 L.H., Daily Totals | | $2081.00 | | $3487.90 | $52.02 | $87.20 |

| Crew No. | Bare Costs | | Incl. Subs O&P | | Cost Per Labor-Hour | |
|---|---|---|---|---|---|---|
| **Crew R-5** | Hr. | Daily | Hr. | Daily | Bare Costs | Incl. O&P |
| 1 Electrician Foreman | $50.80 | $406.40 | $75.25 | $602.00 | $43.93 | $65.62 |
| 4 Electrician Linemen | 50.30 | 1609.60 | 74.50 | 2384.00 | | |
| 2 Electrician Operators | 50.30 | 804.80 | 74.50 | 1192.00 | | |
| 4 Electrician Groundmen | 32.65 | 1044.80 | 49.90 | 1596.80 | | |
| 1 Crew Truck | | 152.40 | | 167.64 | | |
| 1 Flatbed Truck, 20,000 GVW | | 190.00 | | 209.00 | | |
| 1 Pickup Truck, 3/4 Ton | | 104.60 | | 115.06 | | |
| .2 Hyd. Crane, 55 Ton | | 241.60 | | 265.76 | | |
| .2 Hyd. Crane, 12 Ton | | 131.12 | | 144.23 | | |
| .2 Earth Auger, Truck-Mtd. | | 80.36 | | 88.40 | | |
| 1 Tractor w/Winch | | 346.80 | | 381.48 | 14.17 | 15.59 |
| 88 L.H., Daily Totals | | $5112.48 | | $7146.37 | $58.10 | $81.21 |

| Crew No. | Bare Costs | | Incl. Subs O&P | | Cost Per Labor-Hour | |
|---|---|---|---|---|---|---|
| **Crew R-6** | Hr. | Daily | Hr. | Daily | Bare Costs | Incl. O&P |
| 1 Electrician Foreman | $50.80 | $406.40 | $75.25 | $602.00 | $43.93 | $65.62 |
| 4 Electrician Linemen | 50.30 | 1609.60 | 74.50 | 2384.00 | | |
| 2 Electrician Operators | 50.30 | 804.80 | 74.50 | 1192.00 | | |
| 4 Electrician Groundmen | 32.65 | 1044.80 | 49.90 | 1596.80 | | |
| 1 Crew Truck | | 152.40 | | 167.64 | | |
| 1 Flatbed Truck, 20,000 GVW | | 190.00 | | 209.00 | | |
| 1 Pickup Truck, 3/4 Ton | | 104.60 | | 115.06 | | |
| .2 Hyd. Crane, 55 Ton | | 241.60 | | 265.76 | | |
| .2 Hyd. Crane, 12 Ton | | 131.12 | | 144.23 | | |
| .2 Earth Auger, Truck-Mtd. | | 80.36 | | 88.40 | | |
| 1 Tractor w/Winch | | 346.80 | | 381.48 | | |
| 3 Cable Trailers | | 552.90 | | 608.19 | | |
| .5 Tensioning Rig | | 183.05 | | 201.35 | | |
| .5 Cable Pulling Rig | | 1072.00 | | 1179.20 | 34.71 | 38.19 |
| 88 L.H., Daily Totals | | $6920.43 | | $9135.11 | $78.64 | $103.81 |

| Crew No. | Bare Costs | | Incl. Subs O&P | | Cost Per Labor-Hour | |
|---|---|---|---|---|---|---|
| **Crew R-7** | Hr. | Daily | Hr. | Daily | Bare Costs | Incl. O&P |
| 1 Electrician Foreman | $50.80 | $406.40 | $75.25 | $602.00 | $35.67 | $54.13 |
| 5 Electrician Groundmen | 32.65 | 1306.00 | 49.90 | 1996.00 | | |
| 1 Crew Truck | | 152.40 | | 167.64 | 3.17 | 3.49 |
| 48 L.H., Daily Totals | | $1864.80 | | $2765.64 | $38.85 | $57.62 |

| Crew No. | Bare Costs | | Incl. Subs O&P | | Cost Per Labor-Hour | |
|---|---|---|---|---|---|---|
| **Crew R-8** | Hr. | Daily | Hr. | Daily | Bare Costs | Incl. O&P |
| 1 Electrician Foreman | $50.80 | $406.40 | $75.25 | $602.00 | $44.50 | $66.42 |
| 3 Electrician Linemen | 50.30 | 1207.20 | 74.50 | 1788.00 | | |
| 2 Electrician Groundmen | 32.65 | 522.40 | 49.90 | 798.40 | | |
| 1 Pickup Truck, 3/4 Ton | | 104.60 | | 115.06 | | |
| 1 Crew Truck | | 152.40 | | 167.64 | 5.35 | 5.89 |
| 48 L.H., Daily Totals | | $2393.00 | | $3471.10 | $49.85 | $72.31 |

| Crew No. | Bare Costs | | Incl. Subs O&P | | Cost Per Labor-Hour | |
|---|---|---|---|---|---|---|
| **Crew R-9** | Hr. | Daily | Hr. | Daily | Bare Costs | Incl. O&P |
| 1 Electrician Foreman | $50.80 | $406.40 | $75.25 | $602.00 | $41.54 | $62.29 |
| 1 Electrician Lineman | 50.30 | 402.40 | 74.50 | 596.00 | | |
| 2 Electrician Operators | 50.30 | 804.80 | 74.50 | 1192.00 | | |
| 4 Electrician Groundmen | 32.65 | 1044.80 | 49.90 | 1596.80 | | |
| 1 Pickup Truck, 3/4 Ton | | 104.60 | | 115.06 | | |
| 1 Crew Truck | | 152.40 | | 167.64 | 4.02 | 4.42 |
| 64 L.H., Daily Totals | | $2915.40 | | $4269.50 | $45.55 | $66.71 |

| Crew No. | Bare Costs | | Incl.<br>Subs O&P | | Cost<br>Per Labor-Hour | |
|---|---|---|---|---|---|---|

| Crew R-10 | Hr. | Daily | Hr. | Daily | Bare<br>Costs | Incl.<br>O&P |
|---|---|---|---|---|---|---|
| 1 Electrician Foreman | $50.80 | $406.40 | $75.25 | $602.00 | $47.44 | $70.53 |
| 4 Electrician Linemen | 50.30 | 1609.60 | 74.50 | 2384.00 | | |
| 1 Electrician Groundman | 32.65 | 261.20 | 49.90 | 399.20 | | |
| 1 Crew Truck | | 152.40 | | 167.64 | | |
| 3 Tram Cars | | 339.90 | | 373.89 | 10.26 | 11.28 |
| 48 L.H., Daily Totals | | $2769.50 | | $3926.73 | $57.70 | $81.81 |

| Crew R-11 | Hr. | Daily | Hr. | Daily | Bare<br>Costs | Incl.<br>O&P |
|---|---|---|---|---|---|---|
| 1 Electrician Foreman | $50.80 | $406.40 | $75.25 | $602.00 | $47.55 | $70.77 |
| 4 Electricians | 50.30 | 1609.60 | 74.50 | 2384.00 | | |
| 1 Equip. Oper. (crane) | 46.50 | 372.00 | 69.45 | 555.60 | | |
| 1 Common Laborer | 34.35 | 274.80 | 52.70 | 421.60 | | |
| 1 Crew Truck | | 152.40 | | 167.64 | | |
| 1 Hyd. Crane, 12 Ton | | 655.60 | | 721.16 | 14.43 | 15.87 |
| 56 L.H., Daily Totals | | $3470.80 | | $4852.00 | $61.98 | $86.64 |

| Crew R-12 | Hr. | Daily | Hr. | Daily | Bare<br>Costs | Incl.<br>O&P |
|---|---|---|---|---|---|---|
| 1 Carpenter Foreman | $43.55 | $348.40 | $66.80 | $534.40 | $40.64 | $63.21 |
| 4 Carpenters | 43.05 | 1377.60 | 66.05 | 2113.60 | | |
| 4 Common Laborers | 34.35 | 1099.20 | 52.70 | 1686.40 | | |
| 1 Equip. Oper. (med.) | 45.35 | 362.80 | 67.75 | 542.00 | | |
| 1 Steel Worker | 48.55 | 388.40 | 85.75 | 686.00 | | |
| 1 Dozer, 200 H.P. | | 1051.00 | | 1156.10 | | |
| 1 Pickup Truck, 3/4 Ton | | 104.60 | | 115.06 | 13.13 | 14.45 |
| 88 L.H., Daily Totals | | $4732.00 | | $6833.56 | $53.77 | $77.65 |

| Crew R-13 | Hr. | Daily | Hr. | Daily | Bare<br>Costs | Incl.<br>O&P |
|---|---|---|---|---|---|---|
| 1 Electrician Foreman | $50.80 | $406.40 | $75.25 | $602.00 | $48.31 | $71.68 |
| 3 Electricians | 50.30 | 1207.20 | 74.50 | 1788.00 | | |
| .25 Equip. Oper. (crane) | 46.50 | 93.00 | 69.45 | 138.90 | | |
| 1 Equipment Oiler | 40.30 | 322.40 | 60.20 | 481.60 | | |
| .25 Hydraulic Crane, 33 Ton | | 190.75 | | 209.82 | 4.54 | 5.00 |
| 42 L.H., Daily Totals | | $2219.75 | | $3220.32 | $52.85 | $76.67 |

| Crew R-15 | Hr. | Daily | Hr. | Daily | Bare<br>Costs | Incl.<br>O&P |
|---|---|---|---|---|---|---|
| 1 Electrician Foreman | $50.80 | $406.40 | $75.25 | $602.00 | $49.27 | $73.07 |
| 4 Electricians | 50.30 | 1609.60 | 74.50 | 2384.00 | | |
| 1 Equipment Oper. (light) | 43.60 | 348.80 | 65.15 | 521.20 | | |
| 1 Aerial Lift Truck | | 317.80 | | 349.58 | 6.62 | 7.28 |
| 48 L.H., Daily Totals | | $2682.60 | | $3856.78 | $55.89 | $80.35 |

| Crew R-18 | Hr. | Daily | Hr. | Daily | Bare<br>Costs | Incl.<br>O&P |
|---|---|---|---|---|---|---|
| .25 Electrician Foreman | $50.80 | $101.60 | $75.25 | $150.50 | $39.48 | $59.42 |
| 1 Electrician | 50.30 | 402.40 | 74.50 | 596.00 | | |
| 2 Helpers | 32.65 | 522.40 | 49.90 | 798.40 | | |
| 26 L.H., Daily Totals | | $1026.40 | | $1544.90 | $39.48 | $59.42 |

| Crew R-19 | Hr. | Daily | Hr. | Daily | Bare<br>Costs | Incl.<br>O&P |
|---|---|---|---|---|---|---|
| .5 Electrician Foreman | $50.80 | $203.20 | $75.25 | $301.00 | $50.40 | $74.65 |
| 2 Electricians | 50.30 | 804.80 | 74.50 | 1192.00 | | |
| 20 L.H., Daily Totals | | $1008.00 | | $1493.00 | $50.40 | $74.65 |

| Crew R-21 | Hr. | Daily | Hr. | Daily | Bare<br>Costs | Incl.<br>O&P |
|---|---|---|---|---|---|---|
| 1 Electrician Foreman | $50.80 | $406.40 | $75.25 | $602.00 | $50.30 | $74.52 |
| 3 Electricians | 50.30 | 1207.20 | 74.50 | 1788.00 | | |
| .1 Equip. Oper. (med.) | 45.35 | 36.28 | 67.75 | 54.20 | | |
| .1 S.P. Crane, 4x4, 25 Ton | | 56.04 | | 61.64 | 1.71 | 1.88 |
| 32.8 L.H., Daily Totals | | $1705.92 | | $2505.84 | $52.01 | $76.40 |

| Crew R-22 | Hr. | Daily | Hr. | Daily | Bare<br>Costs | Incl.<br>O&P |
|---|---|---|---|---|---|---|
| .66 Electrician Foreman | $50.80 | $268.22 | $75.25 | $397.32 | $42.80 | $64.05 |
| 2 Helpers | 32.65 | 522.40 | 49.90 | 798.40 | | |
| 2 Electricians | 50.30 | 804.80 | 74.50 | 1192.00 | | |
| 37.28 L.H., Daily Totals | | $1595.42 | | $2387.72 | $42.80 | $64.05 |

| Crew R-30 | Hr. | Daily | Hr. | Daily | Bare<br>Costs | Incl.<br>O&P |
|---|---|---|---|---|---|---|
| .25 Electrician Foreman (out) | $52.30 | $104.60 | $77.45 | $154.90 | $40.64 | $61.31 |
| 1 Electrician | 50.30 | 402.40 | 74.50 | 596.00 | | |
| 2 Laborers, (Semi-Skilled) | 34.35 | 549.60 | 52.70 | 843.20 | | |
| 26 L.H., Daily Totals | | $1056.60 | | $1594.10 | $40.64 | $61.31 |

| Crew R-31 | Hr. | Daily | Hr. | Daily | Bare<br>Costs | Incl.<br>O&P |
|---|---|---|---|---|---|---|
| 1 Electrician | $50.30 | $402.40 | $74.50 | $596.00 | $50.30 | $74.50 |
| 1 Core Drill, Electric, 2.5 H.P. | | 47.00 | | 51.70 | 5.88 | 6.46 |
| 8 L.H., Daily Totals | | $449.40 | | $647.70 | $56.17 | $80.96 |

| Crew W-41E | Hr. | Daily | Hr. | Daily | Bare<br>Costs | Incl.<br>O&P |
|---|---|---|---|---|---|---|
| .5 Plumber Foreman (out) | $55.60 | $222.40 | $83.00 | $332.00 | $46.30 | $69.68 |
| 1 Plumber | 53.60 | 428.80 | 80.00 | 640.00 | | |
| 1 Laborer | 34.35 | 274.80 | 52.70 | 421.60 | | |
| 20 L.H., Daily Totals | | $926.00 | | $1393.60 | $46.30 | $69.68 |

# Historical Cost Indexes

The table below lists both the RSMeans Historical Cost Index based on Jan. 1, 1993 = 100 as well as the computed value of an index based on Jan. 1, 2011 costs. Since the Jan. 1, 2011 figure is estimated, space is left to write in the actual index figures as they become available through either the quarterly *Means Construction Cost Indexes* or as printed in the *Engineering News-Record*. To compute the actual index based on Jan. 1, 2011 = 100, divide the Historical Cost Index for a particular year by the actual Jan. 1, 2011 Construction Cost Index. Space has been left to advance the index figures as the year progresses.

| Year | Historical Cost Index Jan. 1, 1993 = 100 | | Current Index Based on Jan. 1, 2011 = 100 | | Year | Historical Cost Index Jan. 1, 1993 = 100 | Current Index Based on Jan. 1, 2011 = 100 | | Year | Historical Cost Index Jan. 1, 1993 = 100 | Current Index Based on Jan. 1, 2011 = 100 | |
|---|---|---|---|---|---|---|---|---|---|---|---|---|
| | Est. | Actual | Est. | Actual | | Actual | Est. | Actual | | Actual | Est. | Actual |
| Oct 2011* | | | | | July 1996 | 110.2 | 59.6 | | July 1978 | 53.5 | 28.9 | |
| July 2011* | | | | | 1995 | 107.6 | 58.2 | | 1977 | 49.5 | 26.8 | |
| April 2011* | | | | | 1994 | 104.4 | 56.4 | | 1976 | 46.9 | 25.4 | |
| Jan 2011* | 185.0 | | 100.0 | 100.0 | 1993 | 101.7 | 55.0 | | 1975 | 44.8 | 24.2 | |
| July 2010 | | 183.5 | 99.2 | | 1992 | 99.4 | 53.7 | | 1974 | 41.4 | 22.4 | |
| 2009 | | 180.1 | 97.4 | | 1991 | 96.8 | 52.3 | | 1973 | 37.7 | 20.4 | |
| 2008 | | 180.4 | 97.5 | | 1990 | 94.3 | 51.0 | | 1972 | 34.8 | 18.8 | |
| 2007 | | 169.4 | 91.6 | | 1989 | 92.1 | 49.8 | | 1971 | 32.1 | 17.4 | |
| 2006 | | 162.0 | 87.6 | | 1988 | 89.9 | 48.6 | | 1970 | 28.7 | 15.5 | |
| 2005 | | 151.6 | 81.9 | | 1987 | 87.7 | 47.4 | | 1969 | 26.9 | 14.5 | |
| 2004 | | 143.7 | 77.7 | | 1986 | 84.2 | 45.5 | | 1968 | 24.9 | 13.5 | |
| 2003 | | 132.0 | 71.4 | | 1985 | 82.6 | 44.7 | | 1967 | 23.5 | 12.7 | |
| 2002 | | 128.7 | 69.6 | | 1984 | 82.0 | 44.3 | | 1966 | 22.7 | 12.3 | |
| 2001 | | 125.1 | 67.6 | | 1983 | 80.2 | 43.3 | | 1965 | 21.7 | 11.7 | |
| 2000 | | 120.9 | 65.4 | | 1982 | 76.1 | 41.2 | | 1964 | 21.2 | 11.5 | |
| 1999 | | 117.6 | 63.6 | | 1981 | 70.0 | 37.8 | | 1963 | 20.7 | 11.2 | |
| 1998 | | 115.1 | 62.2 | | 1980 | 62.9 | 34.0 | | 1962 | 20.2 | 10.9 | |
| 1997 | | 112.8 | 61.0 | | 1979 | 57.8 | 31.2 | | 1961 | 19.8 | 10.7 | |

## Adjustments to Costs

The Historical Cost Index can be used to convert National Average building costs at a particular time to the approximate building costs for some other time.

## Example:

Estimate and compare construction costs for different years in the same city.

To estimate the National Average construction cost of a building in 1970, knowing that it cost $900,000 in 2011:

INDEX in 1970 = 28.7

INDEX in 2011 = 185.0

**Note:** The City Cost Indexes for Canada can be used to convert U.S. National averages to local costs in Canadian dollars.

*Historical Cost Index updates and other resources are provided on the following website.
**http://www.reedconstructiondata.com/rsmeans/chgnotice/842339**

**Time Adjustment using the Historical Cost Indexes:**

$$\frac{\text{Index for Year A}}{\text{Index for Year B}} \times \text{Cost in Year B} = \text{Cost in Year A}$$

$$\frac{\text{INDEX 1970}}{\text{INDEX 2011}} \times \text{Cost 2011} = \text{Cost 1970}$$

$$\frac{28.7}{185.0} \times \$900,000 = .155 \times \$900,000 = \$139,500$$

The construction cost of the building in 1970 is $139,500.

# How to Use the City Cost Indexes

## What you should know before you begin

RSMeans City Cost Indexes (CCI) are an extremely useful tool to use when you want to compare costs from city to city and region to region.

This publication contains average construction cost indexes for 731 U.S. and Canadian cities covering over 930 three-digit zip code locations, as listed directly under each city.

Keep in mind that a City Cost Index number is a percentage ratio of a specific city's cost to the national average cost of the same item at a stated time period.

In other words, these index figures represent relative construction factors (or, if you prefer, multipliers) for Material and Installation costs, as well as the weighted average for Total In Place costs for each CSI MasterFormat division. Installation costs include both labor and equipment rental costs. When estimating equipment rental rates only, for a specific location, use 015433 CONTRACTOR EQUIPMENT index.

The 30 City Average Index is the average of 30 major U.S. cities and serves as a National Average.

Index figures for both material and installation are based on the 30 major city average of 100 and represent the cost relationship as of July 1, 2010. The index for each division is computed from representative material and labor quantities for that division. The weighted average for each city is a weighted total of the components listed above it, but does not include relative productivity between trades or cities.

As changes occur in local material prices, labor rates, and equipment rental rates, (including fuel costs) the impact of these changes should be accurately measured by the change in the City Cost Index for each particular city (as compared to the 30 City Average).

Therefore, if you know (or have estimated) building costs in one city today, you can easily convert those costs to expected building costs in another city.

In addition, by using the Historical Cost Index, you can easily convert National Average building costs at a particular time to the approximate building costs for some other time. The City Cost Indexes can then be applied to calculate the costs for a particular city.

## Quick Calculations

**Location Adjustment Using the City Cost Indexes:**

$$\frac{\text{Index for City A}}{\text{Index for City B}} \times \text{Cost in City B} = \text{Cost in City A}$$

**Time Adjustment for the National Average Using the Historical Cost Index:**

$$\frac{\text{Index for Year A}}{\text{Index for Year B}} \times \text{Cost in Year B} = \text{Cost in Year A}$$

**Adjustment from the National Average:**

$$\frac{\text{Index for City A}}{100} \times \text{National Average Cost} = \text{Cost in City A}$$

Since each of the other RSMeans publications contains many different items, any *one* item multiplied by the particular city index may give incorrect results. However, the larger the number of items compiled, the closer the results should be to actual costs for that particular city.

The City Cost Indexes for Canadian cities are calculated using Canadian material and equipment prices and labor rates, in Canadian dollars. Therefore, indexes for Canadian cities can be used to convert U.S. National Average prices to local costs in Canadian dollars.

## How to use this section

**1. Compare costs from city to city.**

In using the RSMeans Indexes, remember that an index number is not a fixed number but a ratio: It's a percentage ratio of a building component's cost at any stated time to the National Average cost of that same component at the same time period. Put in the form of an equation:

$$\frac{\text{Specific City Cost}}{\text{National Average Cost}} \times 100 = \text{City Index Number}$$

Therefore, when making cost comparisons between cities, do not subtract one city's index number from the index number of another city and read the result as a percentage difference. Instead, divide one city's index number by that of the other city. The resulting number may then be used as a multiplier to calculate cost differences from city to city.

The formula used to find cost differences between cities for the purpose of comparison is as follows:

$$\frac{\text{City A Index}}{\text{City B Index}} \times \text{City B Cost (Known)} = \text{City A Cost (Unknown)}$$

In addition, you can use RSMeans CCI to calculate and compare costs division by division between cities using the same basic formula. (Just be sure that you're comparing similar divisions.)

**2. Compare a specific city's construction costs with the National Average.**

When you're studying construction location feasibility, it's advisable to compare a prospective project's cost index with an index of the National Average cost.

For example, divide the weighted average index of construction costs of a specific city by that of the 30 City Average, which = 100.

$$\frac{\text{City Index}}{100} = \% \text{ of National Average}$$

**As a result, you get a ratio that indicates the relative cost of construction in that city in comparison with the National Average.**

**3. Convert U.S. National Average to actual costs in Canadian City.**

$$\frac{\text{Index for Canadian City}}{100} \times \text{National Average Cost} = \text{Cost in Canadian City in \$ CAN}$$

**4. Adjust construction cost data based on a National Average.**

When you use a source of construction cost data which is based on a National Average (such as RSMeans cost data publications), it is necessary to adjust those costs to a specific location.

$$\frac{\text{City Index}}{100} \quad \text{x} \quad \frac{\text{"Book" Cost Based on}}{\text{National Average Costs}} \quad = \quad \frac{\text{City Cost}}{\text{(Unknown)}}$$

**5. When applying the City Cost Indexes to demolition projects, use the appropriate division installation index.** For example, for removal of existing doors and windows, use Division 8 (Openings) index.

# What you might like to know about how we developed the Indexes

The information presented in the CCI is organized according to the Construction Specifications Institute (CSI) MasterFormat 2010.

To create a reliable index, RSMeans researched the building type most often constructed in the United States and Canada. Because it was concluded that no one type of building completely represented the building construction industry, nine different types of buildings were combined to create a composite model.

The exact material, labor, and equipment quantities are based on detailed analysis of these nine building types, then each quantity is weighted in proportion to expected usage. These various material items, labor hours, and equipment rental rates are thus combined to form a composite building representing as closely as possible the actual usage of materials, labor and equipment used in the North American Building Construction Industry.

The following structures were chosen to make up that composite model:

1. Factory, 1 story
2. Office, 2–4 story
3. Store, Retail
4. Town Hall, 2–3 story
5. High School, 2–3 story
6. Hospital, 4–8 story
7. Garage, Parking
8. Apartment, 1–3 story
9. Hotel/Motel, 2–3 story

For the purposes of ensuring the timeliness of the data, the components of the index for the composite model have been streamlined. They currently consist of:

- specific quantities of 66 commonly used construction materials;
- specific labor-hours for 21 building construction trades; and
- specific days of equipment rental for 6 types of construction equipment (normally used to install the 66 material items by the 21 trades.) Fuel costs and routine maintenance costs are included in the equipment cost.

A sophisticated computer program handles the updating of all costs for each city on a quarterly basis. Material and equipment price quotations are gathered quarterly from 731 cities in the United States and Canada. These prices and the latest negotiated labor wage rates for 21 different building trades are used to compile the quarterly update of the City Cost Index.

The 30 major U.S. cities used to calculate the National Average are:

| | |
|---|---|
| Atlanta, GA | Memphis, TN |
| Baltimore, MD | Milwaukee, WI |
| Boston, MA | Minneapolis, MN |
| Buffalo, NY | Nashville, TN |
| Chicago, IL | New Orleans, LA |
| Cincinnati, OH | New York, NY |
| Cleveland, OH | Philadelphia, PA |
| Columbus, OH | Phoenix, AZ |
| Dallas, TX | Pittsburgh, PA |
| Denver, CO | St. Louis, MO |
| Detroit, MI | San Antonio, TX |
| Houston, TX | San Diego, CA |
| Indianapolis, IN | San Francisco, CA |
| Kansas City, MO | Seattle, WA |
| Los Angeles, CA | Washington, DC |

# What the CCI does not indicate

The weighted average for each city is a total of the divisional components weighted to reflect typical usage, but it does not include the productivity variations between trades or cities.

In addition, the CCI does not take into consideration factors such as the following:

- managerial efficiency
- competitive conditions
- automation
- restrictive union practices
- unique local requirements
- regional variations due to specific building codes

# City Cost Indexes

## ALABAMA (Part 1)

| DIVISION | | UNITED STATES 30 CITY AVERAGE | | | ANNISTON 362 | | | BIRMINGHAM 350-352 | | | BUTLER 369 | | | DECATUR 356 | | | DOTHAN 363 | | |
|---|---|---|---|---|---|---|---|---|---|---|---|---|---|---|---|---|---|---|---|
| | | MAT. | INST. | TOTAL | MAT. | INST. | TOTAL | MAT. | INST. | TOTAL | MAT. | INST. | TOTAL | MAT. | INST. | TOTAL | MAT. | INST. | TOTAL |
| 015433 | CONTRACTOR EQUIPMENT | | 100.0 | 100.0 | | 100.8 | 100.8 | | 100.9 | 100.9 | | 97.8 | 97.8 | | 100.8 | 100.8 | | 97.8 | 97.8 |
| 0241, 31-34 | SITE & INFRASTRUCTURE, DEMOLITION | 100.0 | 100.0 | 100.0 | 87.3 | 93.5 | 91.6 | 85.2 | 94.1 | 91.4 | 101.5 | 87.2 | 91.6 | 80.7 | 92.1 | 88.5 | 99.0 | 87.1 | 90.8 |
| 0310 | Concrete Forming & Accessories | 100.0 | 100.0 | 100.0 | 92.6 | 50.9 | 56.4 | 92.8 | 73.9 | 76.4 | 89.1 | 46.1 | 51.7 | 96.8 | 49.0 | 55.3 | 97.9 | 45.5 | 52.4 |
| 0320 | Concrete Reinforcing | 100.0 | 100.0 | 100.0 | 88.9 | 84.6 | 86.7 | 88.9 | 84.9 | 86.9 | 93.9 | 49.5 | 71.1 | 88.9 | 79.3 | 84.0 | 93.9 | 48.9 | 70.8 |
| 0330 | Cast-in-Place Concrete | 100.0 | 100.0 | 100.0 | 92.1 | 52.5 | 77.0 | 98.3 | 67.9 | 86.7 | 89.9 | 58.2 | 77.7 | 91.6 | 66.1 | 81.9 | 89.9 | 48.8 | 74.1 |
| 03 | CONCRETE | 100.0 | 100.0 | 100.0 | 91.5 | 59.4 | 76.2 | 90.8 | 74.9 | 83.2 | 87.6 | 52.7 | 73.3 | 87.6 | 62.2 | 75.5 | 91.6 | 49.1 | 71.3 |
| 04 | MASONRY | 100.0 | 100.0 | 100.0 | 85.7 | 72.1 | 77.5 | 88.1 | 80.9 | 83.7 | 89.5 | 52.4 | 67.0 | 86.6 | 55.4 | 67.7 | 90.7 | 42.9 | 61.7 |
| 05 | METALS | 100.0 | 100.0 | 100.0 | 93.6 | 92.1 | 93.1 | 94.5 | 93.9 | 94.3 | 92.6 | 79.2 | 88.0 | 95.8 | 88.5 | 93.3 | 92.7 | 77.9 | 87.6 |
| 06 | WOOD, PLASTICS & COMPOSITES | 100.0 | 100.0 | 100.0 | 94.5 | 46.2 | 66.8 | 96.4 | 72.5 | 82.7 | 89.6 | 45.9 | 64.5 | 99.5 | 46.5 | 69.1 | 101.1 | 45.9 | 69.4 |
| 07 | THERMAL & MOISTURE PROTECTION | 100.0 | 100.0 | 100.0 | 91.9 | 56.8 | 77.9 | 93.9 | 83.4 | 89.8 | 91.9 | 60.3 | 79.4 | 91.7 | 62.1 | 79.9 | 91.9 | 52.6 | 76.3 |
| 08 | OPENINGS | 100.0 | 100.0 | 100.0 | 93.9 | 52.1 | 83.3 | 97.6 | 74.8 | 91.8 | 94.0 | 47.7 | 82.3 | 97.9 | 53.6 | 86.7 | 94.0 | 47.7 | 82.3 |
| 0920 | Plaster & Gypsum Board | 100.0 | 100.0 | 100.0 | 89.7 | 45.1 | 58.3 | 98.4 | 72.2 | 79.9 | 87.5 | 44.8 | 57.4 | 94.6 | 45.4 | 60.0 | 97.7 | 44.8 | 60.4 |
| 0950, 0980 | Ceilings & Acoustic Treatment | 100.0 | 100.0 | 100.0 | 89.7 | 45.1 | 60.2 | 95.6 | 72.2 | 80.0 | 89.7 | 44.8 | 59.9 | 94.4 | 45.4 | 61.9 | 89.7 | 44.8 | 59.9 |
| 0960 | Flooring | 100.0 | 100.0 | 100.0 | 103.2 | 42.5 | 85.8 | 105.9 | 63.7 | 93.8 | 109.1 | 59.2 | 94.8 | 105.9 | 53.5 | 90.8 | 116.1 | 29.1 | 91.1 |
| 0970, 0990 | Wall Finishes & Painting/Coating | 100.0 | 100.0 | 100.0 | 96.5 | 33.9 | 58.5 | 96.5 | 59.8 | 74.3 | 96.5 | 57.4 | 72.8 | 96.5 | 59.9 | 74.3 | 96.5 | 57.4 | 72.8 |
| 09 | FINISHES | 100.0 | 100.0 | 100.0 | 96.4 | 46.3 | 68.8 | 99.5 | 70.0 | 83.2 | 100.0 | 49.2 | 72.0 | 98.2 | 50.1 | 71.7 | 103.1 | 43.2 | 70.0 |
| COVERS | DIVS. 10-14, 25, 28, 41, 43, 44, 46 | 100.0 | 100.0 | 100.0 | 100.0 | 76.6 | 95.1 | 100.0 | 88.5 | 97.6 | 100.0 | 46.1 | 88.8 | 100.0 | 48.2 | 89.2 | 100.0 | 46.2 | 88.8 |
| 21, 22, 23 | FIRE SUPPRESSION, PLUMBING & HVAC | 100.0 | 100.0 | 100.0 | 99.0 | 60.8 | 83.1 | 99.8 | 67.4 | 86.3 | 96.9 | 36.5 | 71.7 | 99.8 | 44.7 | 76.9 | 96.9 | 36.4 | 71.7 |
| 26, 27, 3370 | ELECTRICAL, COMMUNICATIONS & UTIL. | 100.0 | 100.0 | 100.0 | 96.4 | 62.2 | 78.2 | 105.6 | 62.7 | 82.8 | 98.6 | 43.6 | 69.4 | 98.2 | 66.2 | 81.2 | 97.3 | 62.3 | 78.7 |
| MF2010 | WEIGHTED AVERAGE | 100.0 | 100.0 | 100.0 | 95.1 | 65.5 | 81.9 | 97.0 | 75.6 | 87.4 | 95.7 | 52.6 | 76.4 | 95.8 | 60.8 | 80.1 | 95.8 | 52.6 | 76.5 |

## ALABAMA (Part 2)

| DIVISION | | EVERGREEN 364 | | | GADSDEN 359 | | | HUNTSVILLE 357-358 | | | JASPER 355 | | | MOBILE 365-366 | | | MONTGOMERY 360-361 | | |
|---|---|---|---|---|---|---|---|---|---|---|---|---|---|---|---|---|---|---|---|
| | | MAT. | INST. | TOTAL | MAT. | INST. | TOTAL | MAT. | INST. | TOTAL | MAT. | INST. | TOTAL | MAT. | INST. | TOTAL | MAT. | INST. | TOTAL |
| 015433 | CONTRACTOR EQUIPMENT | | 97.8 | 97.8 | | 100.8 | 100.8 | | 100.8 | 100.8 | | 100.8 | 100.8 | | 97.8 | 97.8 | | 97.8 | 97.8 |
| 0241, 31-34 | SITE & INFRASTRUCTURE, DEMOLITION | 101.9 | 87.4 | 91.9 | 86.0 | 93.0 | 90.8 | 80.5 | 93.1 | 89.2 | 85.5 | 92.8 | 90.5 | 93.6 | 88.2 | 89.9 | 92.3 | 88.1 | 89.4 |
| 0310 | Concrete Forming & Accessories | 85.6 | 47.9 | 52.8 | 89.1 | 44.6 | 50.5 | 96.8 | 65.6 | 69.7 | 94.3 | 37.1 | 44.6 | 96.9 | 58.0 | 63.1 | 94.7 | 46.7 | 53.1 |
| 0320 | Concrete Reinforcing | 94.0 | 49.5 | 71.1 | 94.1 | 83.7 | 88.8 | 88.9 | 79.8 | 84.2 | 88.8 | 83.6 | 86.2 | 91.8 | 83.0 | 87.2 | 91.8 | 83.6 | 87.5 |
| 0330 | Cast-in-Place Concrete | 89.9 | 61.0 | 78.8 | 91.6 | 51.6 | 76.3 | 89.2 | 64.4 | 79.7 | 101.5 | 46.3 | 80.4 | 94.2 | 70.7 | 85.2 | 96.2 | 49.2 | 78.2 |
| 03 | CONCRETE | 92.3 | 54.4 | 74.2 | 91.3 | 56.2 | 74.5 | 86.5 | 69.1 | 78.1 | 94.9 | 51.0 | 73.9 | 89.4 | 68.5 | 79.4 | 90.2 | 56.3 | 74.0 |
| 04 | MASONRY | 89.5 | 57.4 | 70.0 | 84.8 | 46.4 | 61.5 | 87.7 | 61.6 | 71.9 | 81.2 | 44.5 | 59.0 | 88.4 | 56.9 | 69.3 | 86.8 | 42.5 | 59.9 |
| 05 | METALS | 92.6 | 79.2 | 88.0 | 93.7 | 91.2 | 92.8 | 95.8 | 90.1 | 93.8 | 93.6 | 90.5 | 92.5 | 94.7 | 92.2 | 93.8 | 94.5 | 90.8 | 93.2 |
| 06 | WOOD, PLASTICS & COMPOSITES | 85.7 | 45.9 | 62.9 | 90.2 | 41.9 | 62.5 | 99.5 | 66.5 | 80.6 | 96.4 | 34.1 | 60.7 | 99.6 | 57.3 | 75.4 | 96.0 | 45.9 | 67.3 |
| 07 | THERMAL & MOISTURE PROTECTION | 91.9 | 61.2 | 79.7 | 91.8 | 67.0 | 82.0 | 91.6 | 75.9 | 85.3 | 91.8 | 48.6 | 74.6 | 91.6 | 74.5 | 84.8 | 91.1 | 65.8 | 81.1 |
| 08 | OPENINGS | 94.0 | 47.7 | 82.3 | 93.9 | 50.9 | 83.0 | 97.9 | 64.4 | 89.4 | 93.9 | 49.8 | 82.7 | 93.9 | 61.9 | 88.8 | 97.9 | 55.4 | 87.1 |
| 0920 | Plaster & Gypsum Board | 85.8 | 44.8 | 56.9 | 87.1 | 40.7 | 54.4 | 94.6 | 66.0 | 74.5 | 90.2 | 32.7 | 49.7 | 94.6 | 56.5 | 67.8 | 98.4 | 44.8 | 60.6 |
| 0950, 0980 | Ceilings & Acoustic Treatment | 89.7 | 44.8 | 59.9 | 89.7 | 40.7 | 57.2 | 96.5 | 66.0 | 76.3 | 89.7 | 32.7 | 51.9 | 96.5 | 56.5 | 70.0 | 95.6 | 44.8 | 61.9 |
| 0960 | Flooring | 106.5 | 59.2 | 92.9 | 101.4 | 47.5 | 85.9 | 105.9 | 57.1 | 91.9 | 104.1 | 42.5 | 86.4 | 115.1 | 59.2 | 99.0 | 113.5 | 31.3 | 89.9 |
| 0970, 0990 | Wall Finishes & Painting/Coating | 96.5 | 57.4 | 72.8 | 96.5 | 65.4 | 77.6 | 96.5 | 62.9 | 76.1 | 96.5 | 41.5 | 63.2 | 100.0 | 58.4 | 74.8 | 96.5 | 57.4 | 72.8 |
| 09 | FINISHES | 99.1 | 50.4 | 72.3 | 95.4 | 45.7 | 68.0 | 98.7 | 63.4 | 79.2 | 96.5 | 37.1 | 63.7 | 103.4 | 57.2 | 77.9 | 102.7 | 43.5 | 70.0 |
| COVERS | DIVS. 10-14, 25, 28, 41, 43, 44, 46 | 100.0 | 47.8 | 89.2 | 100.0 | 79.9 | 95.8 | 100.0 | 84.8 | 96.8 | 100.0 | 44.2 | 88.4 | 100.0 | 81.1 | 96.1 | 100.0 | 79.8 | 95.8 |
| 21, 22, 23 | FIRE SUPPRESSION, PLUMBING & HVAC | 96.9 | 39.0 | 72.8 | 101.4 | 39.4 | 75.6 | 99.8 | 56.8 | 81.9 | 101.4 | 57.7 | 83.2 | 99.8 | 63.6 | 84.7 | 99.9 | 37.2 | 73.8 |
| 26, 27, 3370 | ELECTRICAL, COMMUNICATIONS & UTIL. | 95.7 | 43.6 | 68.0 | 98.2 | 62.7 | 79.3 | 99.4 | 66.2 | 81.8 | 97.7 | 62.7 | 79.1 | 99.4 | 62.0 | 79.5 | 100.5 | 66.1 | 82.2 |
| MF2010 | WEIGHTED AVERAGE | 95.3 | 54.1 | 76.9 | 95.6 | 58.2 | 78.9 | 95.8 | 69.0 | 83.8 | 96.0 | 58.2 | 79.1 | 96.8 | 67.9 | 83.9 | 96.8 | 57.2 | 79.1 |

## ALABAMA (Part 3) / ALASKA

| DIVISION | | PHENIX CITY 368 | | | SELMA 367 | | | TUSCALOOSA 354 | | | ANCHORAGE 995-996 | | | FAIRBANKS 997 | | | JUNEAU 998 | | |
|---|---|---|---|---|---|---|---|---|---|---|---|---|---|---|---|---|---|---|---|
| | | MAT. | INST. | TOTAL | MAT. | INST. | TOTAL | MAT. | INST. | TOTAL | MAT. | INST. | TOTAL | MAT. | INST. | TOTAL | MAT. | INST. | TOTAL |
| 015433 | CONTRACTOR EQUIPMENT | | 97.8 | 97.8 | | 97.8 | 97.8 | | 100.8 | 100.8 | | 115.9 | 115.9 | | 115.9 | 115.9 | | 115.9 | 115.9 |
| 0241, 31-34 | SITE & INFRASTRUCTURE, DEMOLITION | 105.5 | 88.1 | 93.6 | 98.8 | 88.0 | 91.4 | 80.9 | 93.2 | 89.4 | 129.3 | 131.2 | 130.6 | 121.2 | 131.2 | 128.1 | 129.0 | 131.2 | 130.5 |
| 0310 | Concrete Forming & Accessories | 92.5 | 41.4 | 48.2 | 90.3 | 45.9 | 51.8 | 96.7 | 50.5 | 56.6 | 137.3 | 118.9 | 121.3 | 140.5 | 118.9 | 121.8 | 136.5 | 118.7 | 121.1 |
| 0320 | Concrete Reinforcing | 93.8 | 66.1 | 79.6 | 93.9 | 83.4 | 88.5 | 88.9 | 83.7 | 86.2 | 153.8 | 107.1 | 129.8 | 148.1 | 107.1 | 127.0 | 115.7 | 107.1 | 111.3 |
| 0330 | Cast-in-Place Concrete | 89.9 | 49.8 | 74.5 | 89.9 | 49.0 | 74.2 | 92.9 | 51.6 | 77.1 | 144.8 | 117.2 | 134.2 | 136.9 | 117.5 | 129.5 | 144.7 | 117.1 | 134.1 |
| 03 | CONCRETE | 94.6 | 50.8 | 73.7 | 91.1 | 55.8 | 74.2 | 88.3 | 58.8 | 74.2 | 132.9 | 115.4 | 124.5 | 120.4 | 115.5 | 118.1 | 123.5 | 115.3 | 119.6 |
| 04 | MASONRY | 89.4 | 44.3 | 62.1 | 92.8 | 41.7 | 61.8 | 86.7 | 49.6 | 64.2 | 200.7 | 120.0 | 151.8 | 203.5 | 120.0 | 152.9 | 185.9 | 120.0 | 146.0 |
| 05 | METALS | 92.6 | 84.2 | 89.7 | 92.6 | 89.7 | 91.6 | 94.9 | 91.1 | 93.6 | 122.1 | 101.5 | 115.0 | 119.1 | 101.6 | 113.1 | 119.1 | 101.4 | 113.0 |
| 06 | WOOD, PLASTICS & COMPOSITES | 94.2 | 38.3 | 62.1 | 91.2 | 45.9 | 65.2 | 99.5 | 48.1 | 70.0 | 112.6 | 118.7 | 116.1 | 117.6 | 118.7 | 118.2 | 112.6 | 118.7 | 116.1 |
| 07 | THERMAL & MOISTURE PROTECTION | 92.2 | 65.7 | 81.7 | 91.8 | 55.3 | 77.3 | 91.7 | 69.3 | 82.8 | 192.8 | 116.0 | 162.3 | 181.6 | 114.8 | 155.0 | 183.0 | 116.0 | 156.4 |
| 08 | OPENINGS | 93.9 | 47.0 | 82.1 | 93.9 | 55.4 | 84.2 | 97.9 | 61.5 | 88.7 | 120.1 | 114.0 | 118.5 | 116.6 | 114.0 | 116.0 | 115.9 | 114.0 | 115.4 |
| 0920 | Plaster & Gypsum Board | 91.1 | 37.0 | 53.0 | 88.9 | 44.8 | 57.8 | 94.6 | 47.0 | 61.1 | 146.7 | 119.0 | 127.2 | 153.1 | 119.0 | 129.1 | 134.6 | 119.0 | 123.6 |
| 0950, 0980 | Ceilings & Acoustic Treatment | 89.7 | 37.0 | 54.7 | 89.7 | 44.8 | 59.9 | 96.5 | 47.0 | 63.7 | 140.9 | 119.0 | 126.4 | 126.0 | 119.0 | 121.4 | 120.3 | 119.0 | 119.5 |
| 0960 | Flooring | 111.7 | 31.3 | 88.6 | 109.8 | 30.1 | 86.9 | 105.9 | 39.6 | 86.9 | 162.0 | 125.2 | 151.4 | 152.3 | 125.2 | 144.5 | 152.7 | 125.2 | 144.8 |
| 0970, 0990 | Wall Finishes & Painting/Coating | 96.5 | 57.4 | 72.8 | 96.5 | 49.4 | 68.0 | 96.5 | 49.4 | 68.0 | 150.8 | 117.3 | 130.5 | 149.8 | 122.3 | 133.1 | 150.8 | 117.3 | 130.5 |
| 09 | FINISHES | 101.7 | 39.3 | 67.3 | 100.1 | 43.3 | 68.8 | 98.7 | 47.1 | 70.2 | 152.6 | 120.4 | 134.8 | 145.2 | 120.9 | 131.8 | 142.4 | 120.4 | 130.2 |
| COVERS | DIVS. 10-14, 25, 28, 41, 43, 44, 46 | 100.0 | 79.3 | 95.7 | 100.0 | 46.1 | 88.8 | 100.0 | 81.8 | 96.2 | 100.0 | 112.3 | 102.6 | 100.0 | 112.3 | 102.6 | 100.0 | 112.3 | 102.6 |
| 21, 22, 23 | FIRE SUPPRESSION, PLUMBING & HVAC | 96.9 | 37.6 | 72.2 | 96.9 | 36.7 | 71.8 | 99.8 | 37.0 | 73.7 | 100.4 | 105.9 | 102.7 | 100.3 | 108.0 | 103.5 | 100.4 | 96.9 | 98.9 |
| 26, 27, 3370 | ELECTRICAL, COMMUNICATIONS & UTIL. | 98.0 | 74.2 | 85.4 | 97.0 | 43.6 | 68.6 | 98.8 | 62.7 | 79.6 | 141.8 | 116.3 | 128.2 | 147.2 | 116.3 | 130.8 | 137.5 | 116.3 | 126.3 |
| MF2010 | WEIGHTED AVERAGE | 96.2 | 56.3 | 78.4 | 95.5 | 52.5 | 76.3 | 95.8 | 59.2 | 79.4 | 127.2 | 114.5 | 121.5 | 124.4 | 115.0 | 120.2 | 122.8 | 112.6 | 118.3 |

| | | ALASKA | | | ARIZONA | | | | | | | | | | | | | |
|---|---|---|---|---|---|---|---|---|---|---|---|---|---|---|---|---|---|---|
| | DIVISION | KETCHIKAN | | | CHAMBERS | | | FLAGSTAFF | | | GLOBE | | | KINGMAN | | | MESA/TEMPE | | |
| | | 999 | | | 865 | | | 860 | | | 855 | | | 864 | | | 852 | | |
| | | MAT. | INST. | TOTAL | MAT. | INST. | TOTAL | MAT. | INST. | TOTAL | MAT. | INST. | TOTAL | MAT. | INST. | TOTAL | MAT. | INST. | TOTAL |
| 015433 | CONTRACTOR EQUIPMENT | | 115.9 | 115.9 | | 94.8 | 94.8 | | 94.8 | 94.8 | | 97.8 | 97.8 | | 94.8 | 94.8 | | 97.8 | 97.8 |
| 0241, 31 - 34 | SITE & INFRASTRUCTURE, DEMOLITION | 180.9 | 131.2 | 146.6 | 61.0 | 97.5 | 86.2 | 77.3 | 99.3 | 92.5 | 88.8 | 100.9 | 97.2 | 60.9 | 99.5 | 87.5 | 80.3 | 102.5 | 95.6 |
| 0310 | Concrete Forming & Accessories | 132.2 | 118.7 | 120.5 | 96.6 | 46.4 | 53.0 | 101.7 | 69.1 | 73.4 | 98.1 | 46.2 | 53.1 | 94.9 | 68.6 | 72.1 | 100.8 | 68.5 | 72.8 |
| 0320 | Concrete Reinforcing | 116.3 | 107.1 | 111.5 | 96.5 | 79.2 | 87.6 | 96.4 | 81.3 | 88.6 | 102.4 | 79.1 | 90.4 | 96.7 | 81.3 | 88.8 | 103.1 | 81.3 | 91.8 |
| 0330 | Cast-in-Place Concrete | 277.0 | 117.1 | 215.8 | 92.7 | 55.2 | 78.3 | 92.8 | 75.7 | 86.3 | 91.2 | 55.5 | 77.5 | 92.4 | 73.6 | 85.2 | 91.9 | 73.9 | 85.0 |
| 03 | CONCRETE | 205.7 | 115.3 | 162.4 | 96.9 | 56.0 | 77.3 | 116.0 | 73.6 | 95.7 | 109.6 | 56.0 | 84.0 | 96.5 | 72.7 | 85.1 | 101.0 | 72.8 | 87.5 |
| 04 | MASONRY | 212.0 | 120.0 | 156.3 | 93.1 | 42.1 | 62.2 | 93.1 | 63.7 | 75.3 | 107.8 | 42.1 | 67.9 | 93.1 | 62.6 | 74.6 | 107.9 | 62.6 | 80.5 |
| 05 | METALS | 119.3 | 101.4 | 113.2 | 94.6 | 69.0 | 85.8 | 95.1 | 75.0 | 88.2 | 92.4 | 69.3 | 84.5 | 95.2 | 74.9 | 88.2 | 92.7 | 75.3 | 86.7 |
| 06 | WOOD, PLASTICS & COMPOSITES | 108.8 | 118.7 | 114.5 | 97.1 | 44.3 | 66.8 | 102.9 | 69.8 | 83.9 | 91.7 | 44.4 | 64.5 | 93.1 | 69.8 | 79.7 | 94.9 | 69.9 | 80.6 |
| 07 | THERMAL & MOISTURE PROTECTION | 184.5 | 116.0 | 157.3 | 96.6 | 51.9 | 78.8 | 98.5 | 66.7 | 85.8 | 104.5 | 49.7 | 82.7 | 96.6 | 62.8 | 83.2 | 103.6 | 63.8 | 87.8 |
| 08 | OPENINGS | 118.3 | 114.0 | 117.2 | 102.7 | 51.8 | 89.8 | 102.8 | 72.1 | 95.1 | 97.4 | 51.9 | 85.9 | 102.9 | 72.1 | 95.1 | 97.4 | 72.2 | 91.0 |
| 0920 | Plaster & Gypsum Board | 141.1 | 119.0 | 125.6 | 94.9 | 42.6 | 58.1 | 99.2 | 68.9 | 77.8 | 101.8 | 42.6 | 60.1 | 89.0 | 68.9 | 74.8 | 104.4 | 68.9 | 79.4 |
| 0950, 0980 | Ceilings & Acoustic Treatment | 120.7 | 119.0 | 119.6 | 122.0 | 42.6 | 69.4 | 122.9 | 68.9 | 87.1 | 113.0 | 42.6 | 66.3 | 122.9 | 68.9 | 87.1 | 113.0 | 68.9 | 83.7 |
| 0960 | Flooring | 152.3 | 125.2 | 144.5 | 92.8 | 41.8 | 78.1 | 95.1 | 42.0 | 79.8 | 98.8 | 41.8 | 82.4 | 91.9 | 56.7 | 81.8 | 100.2 | 52.3 | 86.4 |
| 0970, 0990 | Wall Finishes & Painting/Coating | 149.8 | 117.3 | 130.1 | 91.5 | 39.7 | 60.1 | 91.5 | 58.9 | 71.8 | 96.5 | 39.7 | 62.1 | 91.5 | 58.9 | 71.8 | 96.5 | 58.9 | 73.7 |
| 09 | FINISHES | 147.6 | 120.4 | 132.6 | 99.5 | 43.6 | 68.7 | 102.7 | 62.4 | 80.5 | 100.2 | 43.7 | 69.1 | 98.7 | 64.7 | 80.0 | 100.0 | 63.9 | 80.1 |
| COVERS | DIVS. 10 - 14, 25, 28, 41, 43, 44, 46 | 100.0 | 112.3 | 102.6 | 100.0 | 77.5 | 95.3 | 100.0 | 82.7 | 96.4 | 100.0 | 77.8 | 95.4 | 100.0 | 82.4 | 96.3 | 100.0 | 82.7 | 96.4 |
| 21, 22, 23 | FIRE SUPPRESSION, PLUMBING & HVAC | 98.9 | 96.9 | 98.1 | 96.8 | 73.1 | 86.9 | 100.2 | 84.5 | 93.7 | 94.9 | 68.0 | 83.7 | 96.8 | 83.9 | 91.4 | 100.3 | 76.0 | 90.2 |
| 26, 27, 3370 | ELECTRICAL, COMMUNICATIONS & UTIL. | 147.1 | 116.3 | 130.8 | 101.4 | 70.0 | 84.7 | 100.2 | 63.6 | 80.7 | 97.0 | 65.1 | 80.1 | 101.4 | 63.6 | 81.3 | 93.5 | 63.6 | 77.6 |
| MF2010 | WEIGHTED AVERAGE | 136.9 | 112.6 | 126.0 | 96.8 | 63.5 | 81.9 | 100.7 | 74.4 | 88.9 | 98.5 | 62.0 | 82.2 | 96.7 | 74.2 | 86.7 | 98.2 | 72.8 | 86.8 |

| | | ARIZONA | | | | | | | | | | | | ARKANSAS | | | | | |
|---|---|---|---|---|---|---|---|---|---|---|---|---|---|---|---|---|---|---|---|
| | DIVISION | PHOENIX | | | PRESCOTT | | | SHOW LOW | | | TUCSON | | | BATESVILLE | | | CAMDEN | | |
| | | 850,853 | | | 863 | | | 859 | | | 856 - 857 | | | 725 | | | 717 | | |
| | | MAT. | INST. | TOTAL | MAT. | INST. | TOTAL | MAT. | INST. | TOTAL | MAT. | INST. | TOTAL | MAT. | INST. | TOTAL | MAT. | INST. | TOTAL |
| 015433 | CONTRACTOR EQUIPMENT | | 98.4 | 98.4 | | 94.8 | 94.8 | | 97.8 | 97.8 | | 97.8 | 97.8 | | 87.8 | 87.8 | | 87.8 | 87.8 |
| 0241, 31 - 34 | SITE & INFRASTRUCTURE, DEMOLITION | 80.7 | 103.0 | 96.1 | 66.6 | 97.5 | 87.9 | 90.8 | 100.9 | 97.8 | 76.3 | 102.7 | 94.5 | 73.8 | 84.1 | 80.9 | 74.6 | 83.7 | 80.9 |
| 0310 | Concrete Forming & Accessories | 101.7 | 71.5 | 75.5 | 98.1 | 46.2 | 53.0 | 104.5 | 46.5 | 54.2 | 101.1 | 71.1 | 75.0 | 83.6 | 48.6 | 53.2 | 83.4 | 32.9 | 39.6 |
| 0320 | Concrete Reinforcing | 101.5 | 81.4 | 91.1 | 96.4 | 79.4 | 87.6 | 103.1 | 79.3 | 90.8 | 85.1 | 81.3 | 83.1 | 99.3 | 65.5 | 81.9 | 100.6 | 65.3 | 82.4 |
| 0330 | Cast-in-Place Concrete | 92.0 | 76.1 | 85.9 | 92.7 | 55.1 | 78.3 | 91.2 | 55.6 | 77.6 | 94.6 | 76.0 | 87.5 | 81.3 | 47.8 | 68.5 | 83.4 | 40.1 | 66.8 |
| 03 | CONCRETE | 100.7 | 74.9 | 88.3 | 102.1 | 55.9 | 80.0 | 111.8 | 56.3 | 85.2 | 99.6 | 74.6 | 87.6 | 84.0 | 52.2 | 68.8 | 85.7 | 42.6 | 65.1 |
| 04 | MASONRY | 95.0 | 64.3 | 76.4 | 93.1 | 47.4 | 65.4 | 107.4 | 43.6 | 68.9 | 93.0 | 63.6 | 75.2 | 105.1 | 44.0 | 68.1 | 113.4 | 33.9 | 65.2 |
| 05 | METALS | 94.3 | 76.3 | 88.1 | 95.1 | 69.2 | 86.2 | 92.2 | 69.8 | 84.5 | 93.5 | 75.3 | 87.3 | 97.5 | 67.0 | 87.0 | 97.6 | 66.4 | 86.9 |
| 06 | WOOD, PLASTICS & COMPOSITES | 95.8 | 72.8 | 82.6 | 98.4 | 44.3 | 67.3 | 99.2 | 44.4 | 67.8 | 95.2 | 72.8 | 82.3 | 86.7 | 49.3 | 65.2 | 86.4 | 32.1 | 55.3 |
| 07 | THERMAL & MOISTURE PROTECTION | 103.5 | 67.0 | 88.8 | 97.1 | 51.2 | 78.9 | 104.7 | 52.1 | 83.8 | 104.8 | 64.1 | 88.6 | 95.8 | 46.8 | 76.3 | 95.6 | 37.9 | 72.7 |
| 08 | OPENINGS | 98.6 | 73.7 | 92.3 | 102.8 | 51.8 | 89.9 | 96.4 | 51.9 | 85.1 | 93.7 | 73.8 | 88.7 | 95.9 | 47.8 | 83.7 | 92.0 | 42.1 | 79.3 |
| 0920 | Plaster & Gypsum Board | 107.0 | 71.8 | 82.2 | 95.1 | 42.6 | 58.2 | 107.1 | 42.6 | 61.7 | 111.0 | 71.8 | 83.4 | 75.4 | 48.3 | 56.3 | 75.4 | 30.6 | 43.8 |
| 0950, 0980 | Ceilings & Acoustic Treatment | 120.5 | 71.8 | 88.2 | 121.0 | 42.6 | 69.1 | 113.0 | 42.6 | 66.3 | 113.9 | 71.8 | 86.0 | 86.6 | 48.3 | 61.2 | 86.6 | 30.6 | 49.5 |
| 0960 | Flooring | 100.4 | 54.6 | 87.3 | 93.7 | 41.8 | 78.8 | 101.9 | 47.6 | 86.3 | 91.0 | 42.0 | 76.9 | 101.4 | 62.4 | 90.2 | 101.3 | 41.8 | 84.2 |
| 0970, 0990 | Wall Finishes & Painting/Coating | 96.5 | 59.1 | 73.8 | 91.5 | 39.7 | 60.1 | 96.5 | 39.7 | 62.1 | 97.0 | 58.9 | 73.9 | 97.1 | 44.3 | 65.1 | 97.1 | 50.9 | 69.1 |
| 09 | FINISHES | 102.0 | 66.6 | 82.5 | 100.1 | 43.6 | 68.9 | 102.0 | 44.7 | 70.4 | 98.1 | 64.2 | 79.4 | 88.1 | 50.4 | 67.3 | 88.1 | 35.7 | 59.2 |
| COVERS | DIVS. 10 - 14, 25, 28, 41, 43, 44, 46 | 100.0 | 83.4 | 96.5 | 100.0 | 77.5 | 95.3 | 100.0 | 77.8 | 95.4 | 100.0 | 83.3 | 96.5 | 100.0 | 44.0 | 88.4 | 100.0 | 39.5 | 87.4 |
| 21, 22, 23 | FIRE SUPPRESSION, PLUMBING & HVAC | 100.2 | 84.6 | 93.7 | 100.2 | 72.8 | 88.8 | 94.9 | 73.2 | 85.8 | 100.2 | 72.0 | 88.5 | 94.7 | 54.9 | 78.1 | 94.7 | 52.1 | 77.0 |
| 26, 27, 3370 | ELECTRICAL, COMMUNICATIONS & UTIL. | 100.4 | 64.9 | 81.5 | 99.9 | 63.5 | 80.6 | 93.9 | 65.0 | 78.5 | 95.9 | 60.2 | 76.9 | 100.2 | 64.7 | 81.3 | 96.1 | 61.7 | 77.8 |
| MF2010 | WEIGHTED AVERAGE | 98.6 | 75.9 | 88.5 | 98.4 | 63.0 | 82.6 | 98.6 | 63.5 | 82.9 | 96.9 | 72.0 | 85.8 | 94.3 | 56.9 | 77.6 | 94.2 | 50.7 | 74.8 |

| | | ARKANSAS | | | | | | | | | | | | | | | | | |
|---|---|---|---|---|---|---|---|---|---|---|---|---|---|---|---|---|---|---|---|
| | DIVISION | FAYETTEVILLE | | | FORT SMITH | | | HARRISON | | | HOT SPRINGS | | | JONESBORO | | | LITTLE ROCK | | |
| | | 727 | | | 729 | | | 726 | | | 719 | | | 724 | | | 720 - 722 | | |
| | | MAT. | INST. | TOTAL | MAT. | INST. | TOTAL | MAT. | INST. | TOTAL | MAT. | INST. | TOTAL | MAT. | INST. | TOTAL | MAT. | INST. | TOTAL |
| 015433 | CONTRACTOR EQUIPMENT | | 87.8 | 87.8 | | 87.8 | 87.8 | | 87.8 | 87.8 | | 87.8 | 87.8 | | 106.7 | 106.7 | | 87.8 | 87.8 |
| 0241, 31 - 34 | SITE & INFRASTRUCTURE, DEMOLITION | 73.2 | 85.7 | 81.8 | 77.9 | 85.6 | 83.2 | 78.5 | 84.1 | 82.4 | 78.0 | 85.1 | 82.9 | 98.8 | 99.4 | 99.2 | 85.3 | 85.6 | 85.5 |
| 0310 | Concrete Forming & Accessories | 79.5 | 41.6 | 46.6 | 97.2 | 59.4 | 64.4 | 87.7 | 48.5 | 53.7 | 81.0 | 38.5 | 44.1 | 87.3 | 52.1 | 56.8 | 89.6 | 63.9 | 67.3 |
| 0320 | Concrete Reinforcing | 99.3 | 70.4 | 84.4 | 100.4 | 71.2 | 85.4 | 98.8 | 65.4 | 81.6 | 98.7 | 65.5 | 81.6 | 96.1 | 67.8 | 81.5 | 100.6 | 66.2 | 82.9 |
| 0330 | Cast-in-Place Concrete | 81.3 | 52.3 | 70.2 | 93.0 | 67.7 | 83.3 | 90.2 | 45.7 | 73.2 | 85.3 | 40.6 | 68.2 | 88.6 | 58.3 | 77.0 | 93.4 | 67.8 | 83.6 |
| 03 | CONCRETE | 83.8 | 51.6 | 68.4 | 92.0 | 64.9 | 79.0 | 91.2 | 51.5 | 72.2 | 89.0 | 45.3 | 68.1 | 88.7 | 58.6 | 74.3 | 91.7 | 66.0 | 79.4 |
| 04 | MASONRY | 94.5 | 44.7 | 64.3 | 99.8 | 53.1 | 71.5 | 105.5 | 42.1 | 67.1 | 84.5 | 36.2 | 55.2 | 95.9 | 45.3 | 65.2 | 96.6 | 53.1 | 70.2 |
| 05 | METALS | 97.5 | 69.1 | 87.8 | 99.9 | 70.9 | 89.9 | 98.7 | 66.7 | 87.6 | 97.5 | 67.0 | 87.0 | 94.4 | 80.5 | 89.6 | 96.1 | 69.3 | 86.8 |
| 06 | WOOD, PLASTICS & COMPOSITES | 82.3 | 38.4 | 57.1 | 102.6 | 61.9 | 79.3 | 91.9 | 49.3 | 67.5 | 83.6 | 38.6 | 57.8 | 90.1 | 52.3 | 68.5 | 97.2 | 67.8 | 80.4 |
| 07 | THERMAL & MOISTURE PROTECTION | 96.6 | 48.4 | 77.4 | 96.9 | 56.7 | 81.0 | 96.1 | 45.4 | 76.0 | 95.8 | 40.2 | 73.7 | 100.4 | 52.5 | 81.4 | 96.0 | 57.4 | 80.7 |
| 08 | OPENINGS | 95.9 | 47.3 | 83.5 | 96.7 | 58.1 | 87.0 | 96.7 | 48.5 | 84.5 | 91.9 | 44.0 | 79.8 | 98.8 | 56.5 | 88.1 | 96.7 | 61.4 | 87.8 |
| 0920 | Plaster & Gypsum Board | 72.3 | 37.1 | 47.5 | 81.4 | 61.3 | 67.2 | 80.0 | 48.3 | 57.7 | 73.7 | 37.3 | 48.1 | 91.9 | 51.1 | 63.2 | 86.8 | 67.4 | 73.2 |
| 0950, 0980 | Ceilings & Acoustic Treatment | 86.6 | 37.1 | 53.8 | 90.3 | 61.3 | 71.1 | 90.3 | 48.3 | 62.5 | 86.6 | 37.3 | 53.9 | 90.9 | 51.1 | 64.5 | 92.2 | 67.4 | 75.8 |
| 0960 | Flooring | 98.4 | 62.4 | 88.0 | 109.0 | 64.2 | 96.1 | 104.1 | 62.4 | 92.1 | 100.0 | 62.4 | 89.2 | 72.6 | 56.4 | 68.0 | 110.3 | 64.2 | 97.1 |
| 0970, 0990 | Wall Finishes & Painting/Coating | 97.1 | 31.6 | 57.4 | 97.1 | 57.9 | 73.4 | 97.1 | 44.3 | 65.1 | 97.1 | 50.9 | 69.1 | 85.9 | 52.1 | 65.4 | 97.1 | 59.5 | 74.3 |
| 09 | FINISHES | 86.9 | 43.1 | 62.7 | 92.1 | 60.4 | 74.6 | 90.6 | 50.4 | 68.4 | 87.9 | 43.2 | 63.3 | 85.2 | 52.3 | 67.1 | 93.8 | 64.1 | 77.4 |
| COVERS | DIVS. 10 - 14, 25, 28, 41, 43, 44, 46 | 100.0 | 54.4 | 90.5 | 100.0 | 77.3 | 95.3 | 100.0 | 54.9 | 90.6 | 100.0 | 40.6 | 87.7 | 100.0 | 50.3 | 89.7 | 100.0 | 78.0 | 95.4 |
| 21, 22, 23 | FIRE SUPPRESSION, PLUMBING & HVAC | 94.8 | 53.4 | 77.5 | 100.2 | 52.7 | 80.4 | 94.7 | 52.5 | 77.2 | 94.7 | 51.6 | 76.8 | 100.4 | 55.8 | 81.8 | 100.1 | 60.6 | 83.7 |
| 26, 27, 3370 | ELECTRICAL, COMMUNICATIONS & UTIL. | 93.1 | 55.6 | 73.2 | 97.1 | 70.1 | 82.7 | 98.5 | 40.3 | 67.5 | 98.6 | 68.6 | 82.7 | 103.9 | 64.8 | 83.1 | 101.6 | 71.2 | 85.5 |
| MF2010 | WEIGHTED AVERAGE | 92.9 | 54.9 | 75.9 | 97.0 | 63.4 | 82.0 | 95.7 | 53.0 | 76.6 | 93.4 | 53.6 | 75.6 | 96.7 | 61.6 | 81.0 | 97.0 | 65.9 | 83.1 |

## ARKANSAS / CALIFORNIA

| DIVISION | | PINE BLUFF 716 | | | RUSSELLVILLE 728 | | | TEXARKANA 718 | | | WEST MEMPHIS 723 | | | ALHAMBRA 917 - 918 | | | ANAHEIM 928 | | |
|---|---|---|---|---|---|---|---|---|---|---|---|---|---|---|---|---|---|---|---|
| | | MAT. | INST. | TOTAL | MAT. | INST. | TOTAL | MAT. | INST. | TOTAL | MAT. | INST. | TOTAL | MAT. | INST. | TOTAL | MAT. | INST. | TOTAL |
| 015433 | CONTRACTOR EQUIPMENT | | 87.8 | 87.8 | | 87.8 | 87.8 | | 88.5 | 88.5 | | 106.7 | 106.7 | | 99.1 | 99.1 | | 101.6 | 101.6 |
| 0241, 31 - 34 | SITE & INFRASTRUCTURE, DEMOLITION | 79.7 | 85.6 | 83.8 | 75.1 | 84.0 | 81.2 | 93.3 | 86.1 | 88.3 | 106.1 | 99.4 | 101.5 | 99.6 | 109.4 | 106.4 | 98.7 | 109.2 | 105.9 |
| 0310 | Concrete Forming & Accessories | 80.7 | 63.8 | 66.0 | 84.2 | 57.7 | 61.2 | 86.7 | 43.4 | 49.1 | 92.2 | 52.3 | 57.6 | 121.7 | 119.9 | 120.2 | 105.6 | 123.1 | 120.7 |
| 0320 | Concrete Reinforcing | 100.5 | 66.2 | 82.9 | 100.0 | 65.3 | 82.1 | 100.1 | 65.5 | 82.3 | 96.1 | 67.8 | 81.5 | 120.7 | 119.9 | 120.3 | 99.1 | 119.9 | 109.8 |
| 0330 | Cast-in-Place Concrete | 85.3 | 67.8 | 78.6 | 85.2 | 47.8 | 70.9 | 93.0 | 45.6 | 74.9 | 92.9 | 58.4 | 79.7 | 90.3 | 119.0 | 101.3 | 93.7 | 121.6 | 104.3 |
| 03 | CONCRETE | 89.7 | 65.9 | 78.4 | 87.2 | 56.2 | 72.4 | 88.6 | 49.2 | 69.8 | 95.5 | 58.7 | 77.9 | 103.7 | 118.6 | 110.8 | 101.9 | 120.9 | 111.0 |
| 04 | MASONRY | 122.1 | 53.1 | 80.3 | 99.1 | 40.3 | 63.5 | 99.8 | 35.8 | 61.0 | 81.4 | 45.3 | 59.5 | 113.5 | 116.9 | 115.1 | 82.6 | 115.1 | 102.3 |
| 05 | METALS | 98.4 | 69.1 | 88.3 | 97.5 | 66.5 | 86.9 | 91.7 | 67.0 | 83.2 | 93.5 | 80.7 | 89.1 | 86.6 | 102.1 | 91.9 | 102.7 | 103.1 | 102.8 |
| 06 | WOOD, PLASTICS & COMPOSITES | 83.1 | 67.8 | 74.4 | 88.0 | 61.9 | 73.0 | 91.4 | 45.5 | 65.1 | 95.9 | 52.3 | 70.9 | 93.3 | 119.2 | 108.1 | 93.6 | 123.1 | 110.5 |
| 07 | THERMAL & MOISTURE PROTECTION | 95.9 | 57.4 | 80.6 | 96.8 | 46.6 | 76.9 | 96.6 | 47.1 | 77.0 | 100.8 | 52.5 | 81.6 | 102.5 | 115.3 | 107.6 | 102.6 | 119.4 | 109.3 |
| 08 | OPENINGS | 91.9 | 61.4 | 84.2 | 95.9 | 57.4 | 86.1 | 97.2 | 48.5 | 84.9 | 98.8 | 56.5 | 88.1 | 91.0 | 119.0 | 98.1 | 103.2 | 123.3 | 108.3 |
| 0920 | Plaster & Gypsum Board | 73.2 | 67.4 | 69.1 | 75.4 | 61.3 | 65.5 | 78.3 | 44.4 | 54.4 | 94.6 | 51.1 | 64.0 | 108.3 | 119.8 | 116.4 | 107.5 | 123.7 | 118.9 |
| 0950, 0980 | Ceilings & Acoustic Treatment | 86.6 | 67.4 | 73.9 | 86.6 | 61.3 | 69.8 | 92.2 | 44.4 | 60.5 | 88.8 | 51.1 | 63.8 | 98.6 | 119.8 | 112.7 | 111.4 | 123.7 | 119.5 |
| 0960 | Flooring | 99.8 | 64.2 | 89.6 | 101.1 | 62.4 | 90.0 | 102.1 | 53.9 | 88.3 | 75.1 | 56.4 | 69.7 | 98.7 | 105.9 | 100.7 | 110.3 | 112.4 | 110.9 |
| 0970, 0990 | Wall Finishes & Painting/Coating | 97.1 | 59.5 | 74.3 | 97.1 | 37.7 | 61.1 | 97.1 | 32.6 | 58.0 | 85.9 | 52.1 | 65.4 | 105.9 | 109.0 | 107.8 | 102.5 | 109.0 | 106.4 |
| 09 | FINISHES | 87.9 | 64.1 | 74.7 | 88.2 | 57.1 | 71.0 | 91.0 | 43.5 | 64.8 | 86.5 | 52.3 | 67.7 | 99.4 | 116.3 | 108.7 | 105.9 | 119.7 | 113.5 |
| COVERS | DIVS. 10 - 14, 25, 28, 41, 43, 44, 46 | 100.0 | 78.0 | 95.4 | 100.0 | 45.4 | 88.7 | 100.0 | 36.2 | 86.7 | 100.0 | 50.3 | 89.7 | 100.0 | 109.5 | 102.0 | 100.0 | 110.3 | 102.1 |
| 21, 22, 23 | FIRE SUPPRESSION, PLUMBING & HVAC | 100.2 | 55.2 | 81.4 | 94.8 | 54.7 | 78.1 | 100.2 | 58.6 | 82.9 | 95.0 | 67.8 | 83.7 | 94.8 | 107.8 | 100.2 | 100.1 | 112.6 | 105.3 |
| 26, 27, 3370 | ELECTRICAL, COMMUNICATIONS & UTIL. | 96.3 | 71.2 | 83.0 | 97.0 | 48.5 | 71.2 | 98.5 | 39.1 | 66.9 | 105.8 | 61.6 | 82.3 | 116.1 | 114.5 | 115.3 | 93.8 | 107.7 | 101.2 |
| MF2010 | WEIGHTED AVERAGE | 96.7 | 64.8 | 82.4 | 94.2 | 56.1 | 77.2 | 95.9 | 51.9 | 76.5 | 95.9 | 63.7 | 81.5 | 98.6 | 112.7 | 104.9 | 99.9 | 113.7 | 106.1 |

## CALIFORNIA

| DIVISION | | BAKERSFIELD 932 - 933 | | | BERKELEY 947 | | | EUREKA 955 | | | FRESNO 936 - 938 | | | INGLEWOOD 903 - 905 | | | LONG BEACH 906 - 908 | | |
|---|---|---|---|---|---|---|---|---|---|---|---|---|---|---|---|---|---|---|---|
| | | MAT. | INST. | TOTAL | MAT. | INST. | TOTAL | MAT. | INST. | TOTAL | MAT. | INST. | TOTAL | MAT. | INST. | TOTAL | MAT. | INST. | TOTAL |
| 015433 | CONTRACTOR EQUIPMENT | | 99.4 | 99.4 | | 98.4 | 98.4 | | 99.1 | 99.1 | | 99.4 | 99.4 | | 98.8 | 98.8 | | 98.8 | 98.8 |
| 0241, 31 - 34 | SITE & INFRASTRUCTURE, DEMOLITION | 101.1 | 106.6 | 104.9 | 127.7 | 104.4 | 111.6 | 112.6 | 103.7 | 106.5 | 104.1 | 106.5 | 105.7 | 91.0 | 106.2 | 101.5 | 98.4 | 106.2 | 103.8 |
| 0310 | Concrete Forming & Accessories | 103.7 | 122.6 | 120.1 | 119.8 | 146.4 | 142.9 | 115.6 | 129.3 | 127.5 | 101.6 | 130.2 | 126.4 | 109.9 | 119.9 | 118.6 | 105.1 | 119.9 | 118.0 |
| 0320 | Concrete Reinforcing | 111.6 | 119.8 | 115.8 | 93.0 | 121.1 | 107.5 | 103.8 | 120.6 | 114.7 | 94.5 | 120.3 | 107.8 | 114.3 | 119.9 | 117.2 | 113.3 | 119.9 | 116.7 |
| 0330 | Cast-in-Place Concrete | 95.4 | 120.6 | 105.1 | 126.7 | 143.6 | 133.2 | 101.6 | 116.6 | 107.3 | 98.7 | 117.3 | 105.8 | 89.7 | 119.8 | 101.2 | 102.1 | 119.8 | 108.9 |
| 03 | CONCRETE | 102.2 | 120.4 | 110.9 | 115.4 | 138.8 | 126.6 | 112.8 | 122.0 | 117.2 | 103.5 | 122.7 | 112.7 | 99.3 | 119.0 | 108.8 | 109.4 | 119.0 | 114.0 |
| 04 | MASONRY | 102.5 | 114.5 | 109.8 | 139.6 | 131.5 | 134.7 | 108.3 | 130.1 | 121.5 | 105.8 | 121.8 | 115.5 | 75.0 | 117.0 | 100.5 | 83.6 | 117.0 | 103.9 |
| 05 | METALS | 98.0 | 102.3 | 99.5 | 105.9 | 103.7 | 105.1 | 102.6 | 101.7 | 102.3 | 102.6 | 103.0 | 102.7 | 93.4 | 103.5 | 96.9 | 93.3 | 103.6 | 96.8 |
| 06 | WOOD, PLASTICS & COMPOSITES | 87.8 | 123.2 | 108.1 | 115.3 | 151.1 | 135.8 | 106.6 | 132.6 | 121.5 | 99.0 | 132.6 | 118.3 | 92.1 | 119.1 | 107.6 | 85.8 | 119.1 | 104.9 |
| 07 | THERMAL & MOISTURE PROTECTION | 101.3 | 112.0 | 105.9 | 115.6 | 136.1 | 123.7 | 106.7 | 117.0 | 110.8 | 96.1 | 115.6 | 103.8 | 101.4 | 116.7 | 107.5 | 101.7 | 117.0 | 107.7 |
| 08 | OPENINGS | 100.7 | 118.3 | 105.2 | 101.2 | 140.0 | 111.0 | 102.0 | 112.8 | 104.7 | 102.5 | 123.4 | 107.8 | 92.8 | 119.0 | 99.4 | 92.8 | 119.0 | 99.4 |
| 0920 | Plaster & Gypsum Board | 99.3 | 123.7 | 116.5 | 110.5 | 152.1 | 139.8 | 112.6 | 133.4 | 127.3 | 101.9 | 133.4 | 124.1 | 104.9 | 119.8 | 115.4 | 100.5 | 119.8 | 114.1 |
| 0950, 0980 | Ceilings & Acoustic Treatment | 112.9 | 123.7 | 120.0 | 116.4 | 152.1 | 140.0 | 117.0 | 133.4 | 127.9 | 114.5 | 133.4 | 127.0 | 98.9 | 119.8 | 112.7 | 98.9 | 119.8 | 112.7 |
| 0960 | Flooring | 110.6 | 98.3 | 107.0 | 119.8 | 128.0 | 122.2 | 115.0 | 123.1 | 117.3 | 117.1 | 141.0 | 124.0 | 100.6 | 105.9 | 102.1 | 97.7 | 105.9 | 100.1 |
| 0970, 0990 | Wall Finishes & Painting/Coating | 104.4 | 95.2 | 98.8 | 106.2 | 143.2 | 128.7 | 103.7 | 49.8 | 71.0 | 120.9 | 107.8 | 113.0 | 99.3 | 109.0 | 105.2 | 99.3 | 109.0 | 105.2 |
| 09 | FINISHES | 107.2 | 116.0 | 112.0 | 115.3 | 144.1 | 131.2 | 112.3 | 121.7 | 117.5 | 111.5 | 131.2 | 122.4 | 101.1 | 116.3 | 109.5 | 100.2 | 116.3 | 109.1 |
| COVERS | DIVS. 10 - 14, 25, 28, 41, 43, 44, 46 | 100.0 | 110.5 | 102.2 | 100.0 | 125.5 | 105.3 | 100.0 | 122.1 | 104.6 | 100.0 | 122.0 | 104.6 | 100.0 | 109.4 | 102.0 | 100.0 | 109.4 | 102.0 |
| 21, 22, 23 | FIRE SUPPRESSION, PLUMBING & HVAC | 100.1 | 106.0 | 102.6 | 94.9 | 145.0 | 115.7 | 94.7 | 105.8 | 99.4 | 100.2 | 113.7 | 105.8 | 94.8 | 107.8 | 100.2 | 94.8 | 107.8 | 100.2 |
| 26, 27, 3370 | ELECTRICAL, COMMUNICATIONS & UTIL. | 105.8 | 103.4 | 104.5 | 104.0 | 138.6 | 122.4 | 101.7 | 119.4 | 111.1 | 93.2 | 99.4 | 96.5 | 106.2 | 114.5 | 110.6 | 105.9 | 114.5 | 110.5 |
| MF2010 | WEIGHTED AVERAGE | 101.4 | 110.4 | 105.4 | 106.8 | 133.6 | 118.8 | 103.1 | 115.2 | 108.5 | 101.7 | 115.3 | 107.8 | 96.1 | 112.6 | 103.5 | 97.8 | 112.6 | 104.5 |

## CALIFORNIA

| DIVISION | | LOS ANGELES 900 - 902 | | | MARYSVILLE 959 | | | MODESTO 953 | | | MOJAVE 935 | | | OAKLAND 946 | | | OXNARD 930 | | |
|---|---|---|---|---|---|---|---|---|---|---|---|---|---|---|---|---|---|---|---|
| | | MAT. | INST. | TOTAL | MAT. | INST. | TOTAL | MAT. | INST. | TOTAL | MAT. | INST. | TOTAL | MAT. | INST. | TOTAL | MAT. | INST. | TOTAL |
| 015433 | CONTRACTOR EQUIPMENT | | 100.5 | 100.5 | | 99.1 | 99.1 | | 99.1 | 99.1 | | 99.4 | 99.4 | | 98.4 | 98.4 | | 98.2 | 98.2 |
| 0241, 31 - 34 | SITE & INFRASTRUCTURE, DEMOLITION | 97.5 | 108.6 | 105.2 | 108.5 | 105.9 | 106.7 | 102.5 | 106.0 | 104.9 | 97.5 | 106.6 | 103.8 | 134.5 | 104.3 | 113.7 | 104.3 | 104.6 | 104.5 |
| 0310 | Concrete Forming & Accessories | 107.2 | 123.0 | 120.9 | 105.9 | 130.1 | 126.9 | 102.5 | 130.3 | 126.6 | 114.8 | 117.3 | 116.9 | 109.0 | 146.2 | 141.3 | 106.4 | 123.0 | 120.8 |
| 0320 | Concrete Reinforcing | 114.3 | 120.0 | 117.2 | 108.3 | 120.1 | 114.4 | 112.3 | 120.4 | 116.4 | 113.7 | 119.7 | 116.8 | 95.1 | 121.1 | 108.5 | 111.6 | 119.8 | 115.8 |
| 0330 | Cast-in-Place Concrete | 97.7 | 120.2 | 106.3 | 113.5 | 117.4 | 115.0 | 101.6 | 117.5 | 107.7 | 85.4 | 120.5 | 98.9 | 120.1 | 121.3 | 120.5 | 99.4 | 120.9 | 107.6 |
| 03 | CONCRETE | 105.9 | 120.5 | 112.9 | 114.6 | 122.6 | 118.5 | 105.3 | 122.8 | 113.7 | 95.6 | 117.9 | 106.3 | 115.2 | 131.1 | 122.8 | 104.4 | 120.6 | 112.2 |
| 04 | MASONRY | 88.9 | 117.8 | 106.4 | 109.3 | 118.1 | 114.6 | 107.1 | 122.6 | 116.5 | 103.9 | 113.8 | 109.9 | 147.9 | 131.5 | 138.0 | 107.6 | 113.2 | 111.0 |
| 05 | METALS | 100.7 | 103.9 | 101.8 | 102.0 | 103.0 | 102.4 | 99.2 | 103.6 | 100.7 | 99.6 | 102.0 | 100.5 | 100.1 | 103.6 | 101.3 | 97.8 | 102.6 | 99.5 |
| 06 | WOOD, PLASTICS & COMPOSITES | 88.6 | 122.9 | 108.3 | 94.1 | 132.6 | 116.2 | 90.2 | 132.6 | 114.5 | 99.2 | 116.2 | 109.0 | 102.6 | 151.1 | 130.4 | 94.0 | 123.2 | 110.7 |
| 07 | THERMAL & MOISTURE PROTECTION | 101.4 | 119.0 | 108.4 | 106.3 | 118.1 | 111.0 | 105.9 | 117.1 | 110.4 | 102.1 | 110.9 | 105.6 | 113.1 | 136.1 | 122.2 | 105.8 | 114.9 | 109.4 |
| 08 | OPENINGS | 99.1 | 123.2 | 105.2 | 101.2 | 124.3 | 107.0 | 100.0 | 124.5 | 106.2 | 96.6 | 113.2 | 100.8 | 101.3 | 140.0 | 111.1 | 99.6 | 123.3 | 105.6 |
| 0920 | Plaster & Gypsum Board | 104.9 | 123.7 | 118.1 | 104.8 | 133.4 | 125.0 | 108.0 | 133.4 | 125.9 | 115.5 | 116.5 | 116.2 | 103.2 | 152.1 | 137.6 | 107.5 | 123.7 | 118.9 |
| 0950, 0980 | Ceilings & Acoustic Treatment | 108.2 | 123.7 | 118.5 | 116.0 | 133.4 | 127.6 | 111.4 | 133.4 | 126.0 | 116.0 | 116.5 | 116.3 | 119.5 | 152.1 | 141.1 | 116.6 | 123.7 | 121.3 |
| 0960 | Flooring | 99.1 | 112.4 | 102.9 | 110.0 | 111.5 | 110.4 | 110.6 | 117.7 | 112.6 | 117.3 | 98.3 | 111.8 | 114.3 | 128.0 | 118.2 | 109.0 | 112.4 | 109.9 |
| 0970, 0990 | Wall Finishes & Painting/Coating | 99.3 | 109.0 | 105.2 | 103.7 | 117.7 | 112.2 | 103.7 | 118.5 | 112.7 | 103.7 | 94.6 | 98.2 | 106.2 | 143.2 | 128.7 | 103.7 | 100.7 | 101.9 |
| 09 | FINISHES | 102.9 | 119.6 | 112.1 | 109.2 | 127.2 | 119.1 | 108.1 | 128.4 | 119.3 | 111.8 | 111.7 | 111.7 | 114.0 | 144.1 | 130.6 | 108.7 | 118.9 | 114.3 |
| COVERS | DIVS. 10 - 14, 25, 28, 41, 43, 44, 46 | 100.0 | 109.9 | 102.1 | 100.0 | 122.0 | 104.6 | 100.0 | 122.1 | 104.6 | 100.0 | 109.7 | 102.0 | 100.0 | 125.5 | 105.3 | 100.0 | 110.5 | 102.2 |
| 21, 22, 23 | FIRE SUPPRESSION, PLUMBING & HVAC | 100.1 | 112.6 | 105.3 | 94.7 | 107.5 | 100.0 | 100.1 | 113.7 | 105.8 | 94.7 | 101.3 | 97.5 | 100.3 | 144.9 | 118.9 | 100.1 | 112.6 | 105.3 |
| 26, 27, 3370 | ELECTRICAL, COMMUNICATIONS & UTIL. | 102.8 | 118.2 | 111.0 | 97.8 | 103.9 | 101.0 | 100.6 | 109.9 | 105.6 | 92.0 | 100.4 | 96.5 | 103.1 | 146.7 | 126.3 | 97.7 | 109.8 | 104.1 |
| MF2010 | WEIGHTED AVERAGE | 100.6 | 115.4 | 107.2 | 102.4 | 113.8 | 107.5 | 101.9 | 116.6 | 108.5 | 98.2 | 107.7 | 102.4 | 107.5 | 133.7 | 119.2 | 101.4 | 113.1 | 106.6 |

## CALIFORNIA

| DIVISION | | PALM SPRINGS 922 | | | PALO ALTO 943 | | | PASADENA 910 - 912 | | | REDDING 960 | | | RICHMOND 948 | | | RIVERSIDE 925 | | |
|---|---|---|---|---|---|---|---|---|---|---|---|---|---|---|---|---|---|---|---|
| | | MAT. | INST. | TOTAL | MAT. | INST. | TOTAL | MAT. | INST. | TOTAL | MAT. | INST. | TOTAL | MAT. | INST. | TOTAL | MAT. | INST. | TOTAL |
| 015433 | CONTRACTOR EQUIPMENT | | 100.3 | 100.3 | | 98.4 | 98.4 | | 99.1 | 99.1 | | 99.1 | 99.1 | | 98.4 | 98.4 | | 100.3 | 100.3 |
| 0241, 31 - 34 | SITE & INFRASTRUCTURE, DEMOLITION | 90.2 | 107.2 | 101.9 | 123.1 | 104.3 | 110.1 | 96.4 | 109.4 | 105.4 | 108.3 | 105.9 | 106.7 | 134.2 | 104.3 | 113.6 | 96.9 | 107.2 | 104.0 |
| 0310 | Concrete Forming & Accessories | 102.2 | 120.1 | 117.7 | 107.1 | 139.7 | 135.4 | 111.1 | 119.9 | 118.8 | 105.7 | 130.1 | 126.9 | 122.2 | 145.8 | 142.6 | 105.9 | 123.0 | 120.7 |
| 0320 | Concrete Reinforcing | 113.9 | 119.8 | 116.9 | 93.0 | 120.8 | 107.3 | 121.7 | 119.9 | 120.8 | 108.3 | 120.1 | 114.4 | 93.0 | 120.8 | 107.3 | 110.6 | 119.8 | 115.4 |
| 0330 | Cast-in-Place Concrete | 89.3 | 121.5 | 101.6 | 107.1 | 121.0 | 112.4 | 85.7 | 119.0 | 98.5 | 112.6 | 117.4 | 114.4 | 123.2 | 121.1 | 122.4 | 97.1 | 121.6 | 106.5 |
| 03 | CONCRETE | 96.8 | 119.5 | 107.7 | 103.7 | 128.1 | 115.4 | 99.2 | 118.6 | 108.5 | 113.6 | 122.6 | 117.9 | 117.2 | 130.8 | 123.7 | 103.1 | 120.9 | 111.6 |
| 04 | MASONRY | 80.4 | 114.0 | 100.7 | 116.6 | 127.6 | 123.3 | 97.8 | 116.9 | 109.4 | 109.3 | 118.1 | 114.6 | 139.3 | 127.6 | 132.2 | 81.3 | 114.0 | 101.1 |
| 05 | METALS | 103.3 | 102.7 | 103.1 | 97.4 | 102.8 | 99.3 | 86.6 | 102.1 | 91.9 | 102.4 | 103.0 | 102.6 | 97.5 | 102.8 | 99.3 | 102.9 | 102.9 | 102.9 |
| 06 | WOOD, PLASTICS & COMPOSITES | 88.3 | 119.3 | 106.1 | 100.0 | 142.9 | 124.6 | 80.5 | 119.2 | 102.7 | 93.8 | 132.6 | 116.1 | 119.0 | 151.1 | 137.4 | 93.6 | 123.1 | 110.5 |
| 07 | THERMAL & MOISTURE PROTECTION | 102.4 | 116.5 | 108.0 | 112.7 | 129.5 | 119.4 | 102.2 | 115.3 | 107.4 | 106.3 | 118.1 | 111.0 | 113.2 | 132.5 | 120.9 | 102.9 | 118.3 | 109.0 |
| 08 | OPENINGS | 98.7 | 119.1 | 103.8 | 101.3 | 134.6 | 109.7 | 91.0 | 119.0 | 98.1 | 102.2 | 124.3 | 107.8 | 101.3 | 139.0 | 110.8 | 101.7 | 123.3 | 107.2 |
| 0920 | Plaster & Gypsum Board | 102.1 | 119.8 | 114.6 | 101.0 | 143.6 | 131.0 | 101.3 | 119.8 | 114.3 | 105.1 | 133.4 | 125.0 | 112.5 | 152.1 | 140.4 | 107.0 | 123.7 | 118.7 |
| 0950, 0980 | Ceilings & Acoustic Treatment | 109.5 | 119.8 | 116.3 | 117.4 | 143.6 | 134.8 | 98.6 | 119.8 | 112.7 | 122.2 | 133.4 | 129.6 | 117.4 | 152.1 | 140.4 | 114.7 | 123.7 | 120.7 |
| 0960 | Flooring | 112.5 | 105.9 | 110.6 | 113.3 | 121.9 | 115.8 | 93.2 | 105.9 | 96.8 | 110.0 | 111.5 | 110.4 | 121.8 | 121.9 | 121.8 | 114.4 | 112.4 | 113.8 |
| 0970, 0990 | Wall Finishes & Painting/Coating | 100.6 | 119.8 | 112.2 | 106.2 | 138.0 | 125.5 | 105.9 | 109.0 | 107.8 | 103.7 | 117.7 | 112.2 | 106.2 | 138.0 | 125.5 | 100.6 | 109.0 | 105.7 |
| 09 | FINISHES | 104.7 | 117.6 | 111.8 | 112.1 | 137.7 | 126.2 | 96.7 | 116.3 | 107.5 | 110.4 | 127.2 | 119.7 | 117.0 | 142.5 | 131.0 | 107.4 | 119.7 | 114.2 |
| COVERS | DIVS. 10 - 14, 25, 28, 41, 43, 44, 46 | 100.0 | 109.8 | 102.0 | 100.0 | 124.6 | 105.1 | 100.0 | 109.5 | 102.0 | 100.0 | 122.0 | 104.6 | 100.0 | 125.5 | 105.3 | 100.0 | 110.3 | 102.1 |
| 21, 22, 23 | FIRE SUPPRESSION, PLUMBING & HVAC | 94.7 | 107.8 | 100.2 | 94.9 | 138.9 | 113.2 | 94.8 | 107.8 | 100.2 | 100.1 | 107.5 | 103.2 | 94.9 | 135.4 | 111.8 | 100.1 | 112.6 | 105.3 |
| 26, 27, 3370 | ELECTRICAL, COMMUNICATIONS & UTIL. | 97.8 | 106.7 | 102.5 | 103.0 | 140.8 | 123.1 | 112.6 | 114.5 | 113.6 | 101.2 | 103.9 | 102.6 | 103.6 | 133.0 | 119.2 | 93.9 | 106.7 | 100.7 |
| MF2010 | WEIGHTED AVERAGE | 97.5 | 111.5 | 103.8 | 102.3 | 129.4 | 114.4 | 96.5 | 112.7 | 103.7 | 104.1 | 113.8 | 108.4 | 106.1 | 128.9 | 116.3 | 100.0 | 113.3 | 105.9 |

## CALIFORNIA

| DIVISION | | SACRAMENTO 942,956 - 958 | | | SALINAS 939 | | | SAN BERNARDINO 923 - 924 | | | SAN DIEGO 919 - 921 | | | SAN FRANCISCO 940 - 941 | | | SAN JOSE 951 | | |
|---|---|---|---|---|---|---|---|---|---|---|---|---|---|---|---|---|---|---|---|
| | | MAT. | INST. | TOTAL | MAT. | INST. | TOTAL | MAT. | INST. | TOTAL | MAT. | INST. | TOTAL | MAT. | INST. | TOTAL | MAT. | INST. | TOTAL |
| 015433 | CONTRACTOR EQUIPMENT | | 98.0 | 98.0 | | 99.4 | 99.4 | | 100.3 | 100.3 | | 99.1 | 99.1 | | 107.7 | 107.7 | | 99.8 | 99.8 |
| 0241, 31 - 34 | SITE & INFRASTRUCTURE, DEMOLITION | 105.8 | 109.9 | 108.6 | 119.0 | 106.6 | 110.5 | 75.9 | 107.1 | 97.5 | 102.2 | 103.1 | 102.8 | 136.9 | 110.1 | 118.4 | 138.8 | 100.8 | 112.6 |
| 0310 | Concrete Forming & Accessories | 107.1 | 132.2 | 128.9 | 109.1 | 133.6 | 130.3 | 109.3 | 120.0 | 118.6 | 108.5 | 111.9 | 111.5 | 108.5 | 146.9 | 141.8 | 98.7 | 121.1 | 110.2 |
| 0320 | Concrete Reinforcing | 88.0 | 120.4 | 104.7 | 112.3 | 120.8 | 116.7 | 110.6 | 119.8 | 115.3 | 117.9 | 119.7 | 118.8 | 108.3 | 121.4 | 115.0 | 98.7 | 121.0 | 110.2 |
| 0330 | Cast-in-Place Concrete | 101.6 | 118.1 | 107.9 | 98.2 | 117.8 | 105.7 | 67.1 | 121.4 | 87.9 | 93.9 | 107.6 | 99.2 | 123.2 | 122.6 | 122.9 | 118.1 | 120.8 | 119.1 |
| 03 | CONCRETE | 104.7 | 123.5 | 113.7 | 114.2 | 124.5 | 119.1 | 76.8 | 119.5 | 97.2 | 104.3 | 111.3 | 107.6 | 118.6 | 132.4 | 125.2 | 112.0 | 131.3 | 121.2 |
| 04 | MASONRY | 118.6 | 122.6 | 121.0 | 103.5 | 128.0 | 118.4 | 88.2 | 114.0 | 103.8 | 89.7 | 111.1 | 102.7 | 148.4 | 134.7 | 140.1 | 137.9 | 131.5 | 134.1 |
| 05 | METALS | 95.7 | 98.3 | 96.6 | 102.2 | 104.7 | 103.1 | 102.8 | 102.6 | 102.8 | 102.2 | 103.5 | 102.6 | 106.5 | 111.1 | 108.1 | 97.7 | 109.3 | 101.7 |
| 06 | WOOD, PLASTICS & COMPOSITES | 96.2 | 135.0 | 118.4 | 98.3 | 135.9 | 119.9 | 97.2 | 119.3 | 109.9 | 93.4 | 109.6 | 102.7 | 102.6 | 151.2 | 130.5 | 101.6 | 150.9 | 129.9 |
| 07 | THERMAL & MOISTURE PROTECTION | 119.3 | 120.7 | 119.9 | 102.7 | 125.2 | 111.7 | 101.6 | 116.5 | 107.5 | 108.2 | 104.6 | 106.8 | 115.3 | 137.5 | 124.1 | 102.2 | 137.3 | 116.2 |
| 08 | OPENINGS | 113.7 | 125.8 | 116.8 | 101.0 | 131.8 | 108.8 | 98.7 | 119.1 | 103.9 | 98.8 | 112.8 | 102.3 | 105.7 | 140.1 | 114.4 | 92.9 | 139.9 | 104.8 |
| 0920 | Plaster & Gypsum Board | 97.4 | 135.7 | 124.3 | 108.8 | 136.8 | 128.5 | 108.4 | 119.8 | 116.4 | 108.7 | 109.7 | 109.4 | 105.6 | 152.1 | 138.3 | 102.5 | 152.1 | 137.4 |
| 0950, 0980 | Ceilings & Acoustic Treatment | 117.4 | 135.7 | 129.5 | 116.0 | 136.8 | 129.8 | 111.4 | 119.8 | 117.0 | 106.1 | 109.7 | 108.5 | 125.8 | 152.1 | 143.2 | 108.2 | 152.1 | 137.3 |
| 0960 | Flooring | 113.5 | 111.7 | 113.0 | 111.3 | 128.0 | 116.1 | 116.6 | 105.9 | 113.5 | 101.5 | 112.4 | 104.6 | 114.3 | 128.0 | 118.2 | 102.8 | 128.0 | 110.0 |
| 0970, 0990 | Wall Finishes & Painting/Coating | 103.6 | 118.5 | 112.7 | 104.8 | 143.2 | 128.1 | 100.6 | 106.1 | 103.9 | 102.6 | 109.0 | 106.5 | 106.2 | 153.2 | 134.7 | 104.2 | 143.2 | 127.9 |
| 09 | FINISHES | 110.4 | 128.8 | 120.5 | 111.3 | 134.8 | 124.3 | 105.8 | 116.1 | 111.4 | 102.4 | 111.4 | 107.4 | 115.8 | 145.4 | 132.1 | 106.6 | 144.0 | 127.2 |
| COVERS | DIVS. 10 - 14, 25, 28, 41, 43, 44, 46 | 100.0 | 122.8 | 104.7 | 100.0 | 122.5 | 104.7 | 100.0 | 109.8 | 102.0 | 100.0 | 108.4 | 101.8 | 100.0 | 126.0 | 105.4 | 100.0 | 125.1 | 105.2 |
| 21, 22, 23 | FIRE SUPPRESSION, PLUMBING & HVAC | 100.1 | 114.9 | 106.3 | 94.7 | 120.6 | 105.5 | 94.7 | 107.8 | 100.2 | 100.1 | 111.7 | 105.0 | 100.2 | 165.7 | 127.5 | 100.1 | 144.4 | 118.6 |
| 26, 27, 3370 | ELECTRICAL, COMMUNICATIONS & UTIL. | 98.5 | 111.4 | 105.3 | 93.2 | 128.6 | 112.0 | 97.9 | 105.8 | 102.1 | 99.1 | 100.3 | 99.8 | 103.2 | 152.8 | 129.6 | 104.0 | 149.7 | 128.3 |
| MF2010 | WEIGHTED AVERAGE | 103.9 | 117.2 | 109.8 | 102.0 | 123.0 | 111.4 | 95.2 | 111.2 | 102.3 | 100.6 | 108.2 | 104.0 | 109.6 | 140.8 | 123.6 | 104.6 | 134.3 | 117.9 |

## CALIFORNIA

| DIVISION | | SAN LUIS OBISPO 934 | | | SAN MATEO 944 | | | SAN RAFAEL 949 | | | SANTA ANA 926 - 927 | | | SANTA BARBARA 931 | | | SANTA CRUZ 950 | | |
|---|---|---|---|---|---|---|---|---|---|---|---|---|---|---|---|---|---|---|---|
| | | MAT. | INST. | TOTAL | MAT. | INST. | TOTAL | MAT. | INST. | TOTAL | MAT. | INST. | TOTAL | MAT. | INST. | TOTAL | MAT. | INST. | TOTAL |
| 015433 | CONTRACTOR EQUIPMENT | | 99.4 | 99.4 | | 98.4 | 98.4 | | 98.5 | 98.5 | | 100.3 | 100.3 | | 99.4 | 99.4 | | 99.8 | 99.8 |
| 0241, 31 - 34 | SITE & INFRASTRUCTURE, DEMOLITION | 110.4 | 106.6 | 107.8 | 131.3 | 104.3 | 112.7 | 120.4 | 110.0 | 113.2 | 88.5 | 107.1 | 101.4 | 104.3 | 106.6 | 105.9 | 137.9 | 100.7 | 112.2 |
| 0310 | Concrete Forming & Accessories | 116.4 | 119.8 | 119.3 | 112.8 | 145.8 | 141.5 | 115.5 | 145.6 | 141.7 | 109.6 | 120.0 | 118.6 | 107.4 | 122.9 | 120.9 | 108.4 | 133.7 | 130.4 |
| 0320 | Concrete Reinforcing | 113.7 | 119.7 | 116.8 | 93.0 | 121.0 | 107.4 | 93.6 | 121.0 | 107.7 | 114.4 | 119.9 | 117.2 | 111.6 | 119.9 | 115.8 | 122.3 | 120.8 | 121.5 |
| 0330 | Cast-in-Place Concrete | 105.5 | 120.6 | 111.3 | 119.3 | 121.1 | 120.0 | 138.9 | 120.1 | 131.7 | 85.7 | 121.5 | 99.4 | 99.1 | 120.7 | 107.3 | 117.3 | 119.7 | 118.2 |
| 03 | CONCRETE | 112.4 | 119.1 | 115.6 | 113.4 | 130.9 | 121.8 | 136.7 | 130.3 | 133.6 | 94.4 | 119.5 | 106.4 | 104.3 | 120.5 | 112.1 | 115.1 | 125.4 | 120.0 |
| 04 | MASONRY | 105.5 | 113.8 | 110.5 | 139.0 | 130.8 | 134.0 | 113.9 | 130.8 | 124.1 | 77.4 | 114.3 | 99.8 | 104.1 | 113.8 | 110.0 | 104.6 | 128.1 | 133.7 |
| 05 | METALS | 100.3 | 102.2 | 100.9 | 97.3 | 103.1 | 99.3 | 98.9 | 100.2 | 99.3 | 102.9 | 102.7 | 102.8 | 98.2 | 102.5 | 99.7 | 104.6 | 108.2 | 105.8 |
| 06 | WOOD, PLASTICS & COMPOSITES | 101.5 | 119.4 | 111.8 | 107.7 | 151.1 | 132.6 | 104.5 | 150.8 | 131.1 | 98.8 | 119.3 | 110.6 | 94.0 | 123.2 | 110.7 | 101.6 | 136.0 | 121.3 |
| 07 | THERMAL & MOISTURE PROTECTION | 102.9 | 113.6 | 107.1 | 113.0 | 132.5 | 120.8 | 117.6 | 132.2 | 123.4 | 102.7 | 116.6 | 108.2 | 102.3 | 114.0 | 106.9 | 102.1 | 127.9 | 112.4 |
| 08 | OPENINGS | 98.8 | 115.0 | 102.9 | 101.2 | 139.0 | 110.8 | 111.0 | 138.9 | 118.0 | 97.9 | 119.1 | 103.3 | 100.7 | 123.3 | 106.4 | 94.1 | 131.8 | 103.7 |
| 0920 | Plaster & Gypsum Board | 117.2 | 119.8 | 119.0 | 106.8 | 152.1 | 138.6 | 109.2 | 152.1 | 139.4 | 109.3 | 119.8 | 116.7 | 107.5 | 123.7 | 118.9 | 108.3 | 136.8 | 128.3 |
| 0950, 0980 | Ceilings & Acoustic Treatment | 116.0 | 119.8 | 118.5 | 117.4 | 152.1 | 140.4 | 123.9 | 152.1 | 142.6 | 111.4 | 119.8 | 117.0 | 116.6 | 123.7 | 121.3 | 113.2 | 136.8 | 128.8 |
| 0960 | Flooring | 118.3 | 98.3 | 112.6 | 116.3 | 121.9 | 117.9 | 127.4 | 121.9 | 125.8 | 117.2 | 105.9 | 113.9 | 110.6 | 98.3 | 107.0 | 107.3 | 128.0 | 113.2 |
| 0970, 0990 | Wall Finishes & Painting/Coating | 103.7 | 100.7 | 101.9 | 106.2 | 138.0 | 125.5 | 102.3 | 138.0 | 124.0 | 100.6 | 106.1 | 103.9 | 103.7 | 100.7 | 101.9 | 104.5 | 143.2 | 128.0 |
| 09 | FINISHES | 113.4 | 113.4 | 113.4 | 114.4 | 142.1 | 129.7 | 116.4 | 142.3 | 130.7 | 107.3 | 116.1 | 112.2 | 109.4 | 116.5 | 113.3 | 109.7 | 134.9 | 123.6 |
| COVERS | DIVS. 10 - 14, 25, 28, 41, 43, 44, 46 | 100.0 | 120.7 | 104.3 | 100.0 | 125.5 | 105.3 | 100.0 | 124.8 | 105.2 | 100.0 | 109.8 | 102.0 | 100.0 | 110.5 | 102.2 | 100.0 | 122.9 | 104.7 |
| 21, 22, 23 | FIRE SUPPRESSION, PLUMBING & HVAC | 94.7 | 107.8 | 100.2 | 94.9 | 138.5 | 113.0 | 94.8 | 149.0 | 117.4 | 94.7 | 107.8 | 100.2 | 100.1 | 112.6 | 105.3 | 100.1 | 120.7 | 108.7 |
| 26, 27, 3370 | ELECTRICAL, COMMUNICATIONS & UTIL. | 92.0 | 98.7 | 95.6 | 103.0 | 144.2 | 124.9 | 99.9 | 118.8 | 109.9 | 97.9 | 107.2 | 102.8 | 90.9 | 110.5 | 101.3 | 102.8 | 128.6 | 116.5 |
| MF2010 | WEIGHTED AVERAGE | 101.2 | 109.8 | 105.0 | 105.1 | 131.4 | 116.9 | 107.4 | 130.3 | 117.6 | 97.2 | 111.4 | 103.6 | 100.6 | 113.1 | 106.2 | 106.4 | 123.1 | 113.9 |

706

# City Cost Indexes

## CALIFORNIA / COLORADO

| DIVISION | | Santa Rosa 954 MAT. | INST. | TOTAL | Stockton 952 MAT. | INST. | TOTAL | Susanville 961 MAT. | INST. | TOTAL | Vallejo 945 MAT. | INST. | TOTAL | Van Nuys 913-916 MAT. | INST. | TOTAL | Alamosa 811 MAT. | INST. | TOTAL |
|---|---|---|---|---|---|---|---|---|---|---|---|---|---|---|---|---|---|---|---|
| 015433 | CONTRACTOR EQUIPMENT | | 99.6 | 99.6 | | 99.1 | 99.1 | | 99.1 | 99.1 | | 98.5 | 98.5 | | 99.1 | 99.1 | | 96.4 | 96.4 |
| 0241, 31 - 34 | SITE & INFRASTRUCTURE, DEMOLITION | 103.1 | 106.1 | 105.1 | 102.2 | 106.0 | 104.8 | 115.8 | 105.9 | 109.0 | 105.0 | 110.0 | 108.4 | 114.3 | 109.4 | 111.0 | 122.3 | 92.7 | 101.9 |
| 0310 | Concrete Forming & Accessories | 103.1 | 145.5 | 139.9 | 106.3 | 130.3 | 127.1 | 106.9 | 130.1 | 127.0 | 106.2 | 145.1 | 139.9 | 117.2 | 119.9 | 119.6 | 106.0 | 72.8 | 77.2 |
| 0320 | Concrete Reinforcing | 109.3 | 121.2 | 115.4 | 112.3 | 120.4 | 116.4 | 108.3 | 120.1 | 114.4 | 94.7 | 121.2 | 108.3 | 121.7 | 119.9 | 120.8 | 104.4 | 79.6 | 91.6 |
| 0330 | Cast-in-Place Concrete | 111.5 | 119.1 | 114.4 | 99.0 | 117.5 | 106.1 | 102.4 | 117.4 | 108.2 | 110.6 | 119.3 | 113.9 | 90.4 | 119.0 | 101.3 | 103.4 | 80.6 | 94.7 |
| 03 | CONCRETE | 114.2 | 130.3 | 121.9 | 104.3 | 122.8 | 113.2 | 114.9 | 122.6 | 118.6 | 110.1 | 129.8 | 119.5 | 113.2 | 118.6 | 115.8 | 116.0 | 77.0 | 97.4 |
| 04 | MASONRY | 106.8 | 133.4 | 122.9 | 107.2 | 122.6 | 116.6 | 108.2 | 118.1 | 114.2 | 84.5 | 133.3 | 114.1 | 113.5 | 116.9 | 115.6 | 126.1 | 77.2 | 96.5 |
| 05 | METALS | 103.3 | 106.4 | 104.4 | 99.4 | 103.6 | 100.8 | 101.5 | 103.0 | 102.0 | 98.8 | 100.4 | 99.4 | 85.6 | 102.1 | 91.3 | 96.5 | 82.3 | 91.6 |
| 06 | WOOD, PLASTICS & COMPOSITES | 89.7 | 150.6 | 124.6 | 95.6 | 132.6 | 116.8 | 95.7 | 132.6 | 116.9 | 93.4 | 150.8 | 126.3 | 88.4 | 119.2 | 106.0 | 95.7 | 73.4 | 82.9 |
| 07 | THERMAL & MOISTURE PROTECTION | 103.1 | 134.9 | 115.8 | 106.0 | 118.5 | 110.9 | 106.8 | 118.1 | 111.3 | 115.2 | 134.6 | 122.9 | 103.1 | 115.3 | 108.0 | 106.3 | 81.2 | 96.3 |
| 08 | OPENINGS | 99.6 | 139.7 | 109.8 | 100.0 | 124.5 | 106.2 | 102.1 | 124.3 | 107.7 | 113.3 | 139.8 | 120.1 | 90.8 | 119.0 | 98.0 | 93.9 | 78.9 | 90.1 |
| 0920 | Plaster & Gypsum Board | 103.5 | 152.1 | 137.7 | 108.0 | 133.4 | 125.9 | 106.2 | 133.4 | 125.3 | 102.6 | 152.1 | 137.4 | 106.0 | 119.8 | 115.7 | 85.5 | 72.3 | 76.2 |
| 0950, 0980 | Ceilings & Acoustic Treatment | 111.4 | 152.1 | 138.3 | 116.6 | 133.4 | 127.7 | 116.0 | 133.4 | 127.6 | 126.0 | 152.1 | 143.3 | 96.8 | 119.8 | 112.0 | 118.2 | 72.3 | 87.8 |
| 0960 | Flooring | 113.1 | 121.4 | 115.5 | 110.6 | 117.7 | 112.6 | 110.6 | 111.5 | 110.8 | 122.3 | 128.0 | 123.9 | 96.4 | 105.9 | 99.1 | 112.3 | 57.9 | 96.6 |
| 0970, 0990 | Wall Finishes & Painting/Coating | 100.6 | 143.2 | 126.5 | 103.7 | 118.5 | 112.7 | 103.7 | 117.7 | 112.2 | 103.3 | 143.2 | 127.5 | 105.9 | 109.0 | 107.8 | 107.9 | 25.4 | 57.8 |
| 09 | FINISHES | 106.5 | 142.4 | 126.3 | 109.2 | 128.4 | 119.8 | 110.4 | 127.2 | 119.7 | 112.8 | 143.6 | 129.8 | 99.2 | 116.3 | 108.6 | 108.6 | 64.7 | 84.4 |
| COVERS | DIVS. 10 - 14, 25, 28, 41, 43, 44, 46 | 100.0 | 124.0 | 105.0 | 100.0 | 122.1 | 104.6 | 100.0 | 122.0 | 104.6 | 100.0 | 124.4 | 105.1 | 100.0 | 109.5 | 102.0 | 100.0 | 89.7 | 97.9 |
| 21, 22, 23 | FIRE SUPPRESSION, PLUMBING & HVAC | 94.7 | 164.9 | 123.9 | 100.1 | 113.7 | 105.8 | 94.7 | 107.5 | 100.0 | 100.2 | 129.0 | 112.2 | 94.8 | 107.8 | 100.2 | 94.6 | 69.2 | 84.1 |
| 26, 27, 3370 | ELECTRICAL, COMMUNICATIONS & UTIL. | 98.2 | 123.5 | 111.7 | 100.6 | 114.7 | 108.1 | 101.6 | 103.9 | 102.8 | 95.6 | 128.3 | 113.0 | 112.6 | 114.5 | 113.6 | 96.0 | 76.7 | 85.7 |
| MF2010 | WEIGHTED AVERAGE | 101.7 | 134.9 | 116.5 | 101.9 | 117.3 | 108.8 | 103.1 | 113.8 | 107.9 | 102.9 | 127.8 | 114.0 | 99.7 | 112.7 | 105.5 | 102.1 | 76.1 | 90.5 |

## COLORADO

| DIVISION | | Boulder 803 MAT. | INST. | TOTAL | Colorado Springs 808-809 MAT. | INST. | TOTAL | Denver 800-802 MAT. | INST. | TOTAL | Durango 813 MAT. | INST. | TOTAL | Fort Collins 805 MAT. | INST. | TOTAL | Fort Morgan 807 MAT. | INST. | TOTAL |
|---|---|---|---|---|---|---|---|---|---|---|---|---|---|---|---|---|---|---|---|
| 015433 | CONTRACTOR EQUIPMENT | | 96.2 | 96.2 | | 94.8 | 94.8 | | 99.2 | 99.2 | | 96.4 | 96.4 | | 96.2 | 96.2 | | 96.2 | 96.2 |
| 0241, 31 - 34 | SITE & INFRASTRUCTURE, DEMOLITION | 88.7 | 97.2 | 94.6 | 90.0 | 94.8 | 93.3 | 89.1 | 102.9 | 98.6 | 116.5 | 92.7 | 100.1 | 99.6 | 96.4 | 97.4 | 90.9 | 96.8 | 94.9 |
| 0310 | Concrete Forming & Accessories | 105.5 | 81.6 | 84.7 | 96.1 | 81.8 | 83.7 | 101.6 | 82.0 | 84.6 | 111.6 | 72.8 | 77.9 | 102.7 | 78.0 | 81.2 | 105.8 | 81.0 | 84.3 |
| 0320 | Concrete Reinforcing | 96.3 | 79.7 | 87.7 | 95.5 | 84.8 | 90.0 | 95.5 | 84.8 | 90.0 | 104.4 | 79.5 | 91.6 | 96.3 | 78.4 | 87.1 | 96.5 | 79.6 | 87.8 |
| 0330 | Cast-in-Place Concrete | 105.4 | 80.8 | 96.0 | 108.3 | 81.4 | 98.0 | 102.5 | 81.1 | 94.3 | 118.9 | 80.6 | 104.2 | 119.0 | 75.5 | 102.3 | 103.4 | 79.6 | 94.3 |
| 03 | CONCRETE | 103.3 | 81.2 | 92.7 | 106.6 | 82.5 | 95.0 | 100.9 | 82.4 | 92.1 | 119.1 | 77.0 | 98.9 | 114.1 | 77.4 | 96.5 | 101.8 | 80.4 | 91.5 |
| 04 | MASONRY | 91.6 | 77.8 | 83.2 | 93.4 | 77.8 | 83.9 | 95.3 | 77.8 | 84.7 | 113.6 | 77.2 | 91.6 | 108.0 | 49.1 | 72.3 | 106.3 | 77.4 | 88.8 |
| 05 | METALS | 98.9 | 85.7 | 94.3 | 102.0 | 88.3 | 97.3 | 104.5 | 88.3 | 98.9 | 96.5 | 82.3 | 91.6 | 100.1 | 80.6 | 93.4 | 98.6 | 82.4 | 93.0 |
| 06 | WOOD, PLASTICS & COMPOSITES | 99.8 | 83.9 | 90.7 | 90.3 | 84.1 | 86.7 | 97.7 | 84.0 | 89.8 | 103.9 | 73.4 | 86.4 | 97.5 | 83.9 | 89.7 | 100.1 | 83.9 | 90.8 |
| 07 | THERMAL & MOISTURE PROTECTION | 103.5 | 79.4 | 94.0 | 104.3 | 83.2 | 95.9 | 102.7 | 77.5 | 92.7 | 106.2 | 81.2 | 96.2 | 104.0 | 66.0 | 88.9 | 103.5 | 82.5 | 95.1 |
| 08 | OPENINGS | 98.0 | 84.6 | 94.6 | 102.5 | 86.2 | 98.4 | 102.6 | 86.1 | 98.4 | 101.7 | 78.9 | 95.9 | 98.0 | 84.6 | 94.6 | 98.0 | 84.6 | 94.6 |
| 0920 | Plaster & Gypsum Board | 108.6 | 83.7 | 91.0 | 88.7 | 83.7 | 85.1 | 104.7 | 83.7 | 89.9 | 97.8 | 72.3 | 79.9 | 102.9 | 83.7 | 89.3 | 109.5 | 83.7 | 91.3 |
| 0950, 0980 | Ceilings & Acoustic Treatment | 106.7 | 83.7 | 91.4 | 114.2 | 83.7 | 93.9 | 113.2 | 83.7 | 93.7 | 118.2 | 72.3 | 87.8 | 106.7 | 83.7 | 91.4 | 106.7 | 83.7 | 91.4 |
| 0960 | Flooring | 109.9 | 86.6 | 103.2 | 101.3 | 72.1 | 92.9 | 106.4 | 89.6 | 101.6 | 117.0 | 57.9 | 100.0 | 106.3 | 57.9 | 92.4 | 110.4 | 57.9 | 95.3 |
| 0970, 0990 | Wall Finishes & Painting/Coating | 103.6 | 71.7 | 84.2 | 103.3 | 43.6 | 67.1 | 103.6 | 71.7 | 84.2 | 107.9 | 25.4 | 57.8 | 103.6 | 43.3 | 67.0 | 103.6 | 57.7 | 75.7 |
| 09 | FINISHES | 106.6 | 81.8 | 92.9 | 103.3 | 76.1 | 88.3 | 106.4 | 82.4 | 93.2 | 110.7 | 64.7 | 85.3 | 105.4 | 71.0 | 86.4 | 106.8 | 74.4 | 88.9 |
| COVERS | DIVS. 10 - 14, 25, 28, 41, 43, 44, 46 | 100.0 | 89.9 | 97.9 | 100.0 | 90.3 | 98.0 | 100.0 | 90.0 | 97.9 | 100.0 | 89.7 | 97.9 | 100.0 | 87.4 | 97.4 | 100.0 | 90.0 | 97.9 |
| 21, 22, 23 | FIRE SUPPRESSION, PLUMBING & HVAC | 94.6 | 81.1 | 89.0 | 100.1 | 75.9 | 90.0 | 99.9 | 82.1 | 92.5 | 94.6 | 66.9 | 83.1 | 100.0 | 77.3 | 90.5 | 94.6 | 81.0 | 89.0 |
| 26, 27, 3370 | ELECTRICAL, COMMUNICATIONS & UTIL. | 98.3 | 85.8 | 91.7 | 102.2 | 86.5 | 93.9 | 103.9 | 85.8 | 94.3 | 95.3 | 74.5 | 84.3 | 98.3 | 85.8 | 91.6 | 98.7 | 85.8 | 91.8 |
| MF2010 | WEIGHTED AVERAGE | 98.5 | 83.7 | 91.8 | 101.3 | 82.4 | 92.8 | 101.4 | 84.9 | 94.0 | 102.6 | 75.3 | 90.4 | 102.3 | 77.1 | 91.0 | 99.1 | 82.3 | 91.6 |

## COLORADO

| DIVISION | | Glenwood Springs 816 MAT. | INST. | TOTAL | Golden 804 MAT. | INST. | TOTAL | Grand Junction 815 MAT. | INST. | TOTAL | Greeley 806 MAT. | INST. | TOTAL | Montrose 814 MAT. | INST. | TOTAL | Pueblo 810 MAT. | INST. | TOTAL |
|---|---|---|---|---|---|---|---|---|---|---|---|---|---|---|---|---|---|---|---|
| 015433 | CONTRACTOR EQUIPMENT | | 99.1 | 99.1 | | 96.2 | 96.2 | | 99.1 | 99.1 | | 96.2 | 96.2 | | 97.7 | 97.7 | | 96.4 | 96.4 |
| 0241, 31 - 34 | SITE & INFRASTRUCTURE, DEMOLITION | 130.5 | 99.6 | 109.2 | 100.7 | 97.1 | 98.2 | 115.8 | 99.3 | 104.4 | 87.3 | 95.2 | 92.8 | 124.6 | 95.8 | 104.7 | 108.1 | 94.5 | 98.7 |
| 0310 | Concrete Forming & Accessories | 103.2 | 80.8 | 83.7 | 98.6 | 81.1 | 83.4 | 110.3 | 80.3 | 84.3 | 100.9 | 78.8 | 81.7 | 102.8 | 80.5 | 83.4 | 107.9 | 82.0 | 85.5 |
| 0320 | Concrete Reinforcing | 103.3 | 79.5 | 91.1 | 96.5 | 79.6 | 87.8 | 103.6 | 79.5 | 91.2 | 96.3 | 78.2 | 86.9 | 103.1 | 79.5 | 91.0 | 99.9 | 84.8 | 92.1 |
| 0330 | Cast-in-Place Concrete | 103.4 | 79.9 | 94.4 | 103.5 | 79.7 | 94.4 | 114.5 | 79.2 | 101.0 | 99.4 | 52.4 | 81.4 | 103.4 | 79.9 | 94.4 | 102.6 | 82.5 | 94.9 |
| 03 | CONCRETE | 121.1 | 80.4 | 101.6 | 111.3 | 80.4 | 96.5 | 115.3 | 79.9 | 98.3 | 98.4 | 69.9 | 84.7 | 112.3 | 80.2 | 96.9 | 105.1 | 83.0 | 94.5 |
| 04 | MASONRY | 100.9 | 77.4 | 86.7 | 109.0 | 77.5 | 89.9 | 133.6 | 77.2 | 99.4 | 102.1 | 36.8 | 62.5 | 106.7 | 77.2 | 88.9 | 97.3 | 77.8 | 85.5 |
| 05 | METALS | 96.2 | 82.8 | 91.6 | 98.8 | 82.6 | 93.2 | 97.9 | 81.5 | 92.2 | 100.1 | 79.5 | 93.0 | 95.5 | 81.8 | 90.8 | 99.4 | 89.1 | 95.9 |
| 06 | WOOD, PLASTICS & COMPOSITES | 90.8 | 84.1 | 86.9 | 92.7 | 83.9 | 87.7 | 101.4 | 84.1 | 91.4 | 94.9 | 90.2 | 92.2 | 92.1 | 84.2 | 87.6 | 98.1 | 84.4 | 90.2 |
| 07 | THERMAL & MOISTURE PROTECTION | 106.1 | 82.3 | 96.7 | 104.4 | 77.3 | 93.7 | 105.0 | 71.9 | 91.8 | 103.4 | 57.5 | 85.2 | 106.3 | 82.3 | 96.8 | 104.6 | 83.1 | 96.0 |
| 08 | OPENINGS | 100.9 | 84.7 | 96.8 | 98.0 | 84.6 | 94.6 | 101.7 | 84.7 | 97.4 | 98.0 | 88.0 | 95.5 | 101.8 | 84.8 | 97.5 | 95.7 | 86.3 | 93.3 |
| 0920 | Plaster & Gypsum Board | 112.5 | 83.7 | 92.2 | 100.2 | 83.7 | 88.5 | 124.9 | 83.7 | 95.9 | 101.1 | 90.1 | 93.4 | 83.7 | 83.7 | 83.7 | 90.5 | 83.7 | 85.7 |
| 0950, 0980 | Ceilings & Acoustic Treatment | 117.3 | 83.7 | 95.0 | 106.7 | 83.7 | 91.4 | 117.3 | 83.7 | 95.0 | 106.7 | 90.1 | 95.7 | 118.2 | 83.7 | 95.3 | 128.5 | 83.7 | 98.8 |
| 0960 | Flooring | 111.4 | 53.3 | 94.7 | 104.1 | 57.9 | 90.8 | 116.4 | 57.9 | 99.6 | 105.2 | 57.9 | 91.6 | 114.5 | 47.9 | 95.4 | 113.2 | 89.6 | 106.4 |
| 0970, 0990 | Wall Finishes & Painting/Coating | 107.8 | 57.7 | 77.4 | 103.6 | 71.7 | 84.2 | 107.8 | 71.7 | 85.9 | 103.6 | 27.0 | 57.1 | 107.9 | 25.4 | 57.8 | 107.9 | 40.4 | 66.9 |
| 09 | FINISHES | 111.7 | 73.7 | 90.8 | 105.2 | 76.7 | 88.3 | 112.9 | 76.1 | 92.6 | 104.1 | 70.7 | 85.7 | 109.0 | 69.0 | 87.0 | 109.9 | 79.2 | 93.0 |
| COVERS | DIVS. 10 - 14, 25, 28, 41, 43, 44, 46 | 100.0 | 90.2 | 98.0 | 100.0 | 90.0 | 97.9 | 100.0 | 90.1 | 98.0 | 100.0 | 85.1 | 96.9 | 100.0 | 90.5 | 98.0 | 100.0 | 91.0 | 98.1 |
| 21, 22, 23 | FIRE SUPPRESSION, PLUMBING & HVAC | 94.6 | 66.7 | 83.0 | 94.6 | 80.7 | 88.8 | 99.9 | 65.5 | 85.6 | 100.0 | 72.8 | 88.7 | 94.6 | 66.7 | 83.0 | 99.9 | 69.4 | 87.2 |
| 26, 27, 3370 | ELECTRICAL, COMMUNICATIONS & UTIL. | 92.5 | 74.5 | 83.0 | 98.7 | 85.8 | 91.8 | 95.0 | 58.5 | 75.6 | 98.3 | 85.8 | 91.6 | 95.0 | 58.5 | 75.6 | 96.0 | 76.8 | 85.8 |
| MF2010 | WEIGHTED AVERAGE | 102.1 | 78.0 | 91.3 | 100.6 | 82.5 | 92.5 | 104.7 | 75.2 | 91.5 | 99.6 | 73.6 | 88.0 | 101.2 | 74.7 | 89.4 | 100.8 | 80.2 | 91.6 |

| DIVISION | | COLORADO SALIDA 812 | | | CONNECTICUT BRIDGEPORT 066 | | | BRISTOL 060 | | | HARTFORD 061 | | | MERIDEN 064 | | | NEW BRITAIN 060 | | |
|---|---|---|---|---|---|---|---|---|---|---|---|---|---|---|---|---|---|---|---|
| | | MAT. | INST. | TOTAL | MAT. | INST. | TOTAL | MAT. | INST. | TOTAL | MAT. | INST. | TOTAL | MAT. | INST. | TOTAL | MAT. | INST. | TOTAL |
| 015433 | CONTRACTOR EQUIPMENT | | 97.7 | 97.7 | | 100.3 | 100.3 | | 100.3 | 100.3 | | 100.3 | 100.3 | | 100.8 | 100.8 | | 100.3 | 100.3 |
| 0241, 31 - 34 | SITE & INFRASTRUCTURE, DEMOLITION | 116.2 | 96.1 | 102.3 | 103.5 | 103.8 | 103.7 | 102.7 | 103.8 | 103.5 | 96.7 | 103.8 | 101.6 | 101.0 | 104.6 | 103.5 | 102.9 | 103.8 | 103.5 |
| 0310 | Concrete Forming & Accessories | 110.2 | 80.7 | 84.6 | 96.5 | 118.3 | 115.4 | 96.5 | 121.8 | 118.5 | 94.5 | 121.8 | 118.2 | 96.2 | 121.8 | 118.4 | 96.8 | 121.8 | 118.5 |
| 0320 | Concrete Reinforcing | 102.9 | 79.5 | 90.9 | 108.1 | 127.0 | 117.8 | 108.1 | 127.0 | 117.8 | 108.1 | 127.0 | 117.8 | 108.1 | 127.0 | 117.8 | 108.1 | 127.0 | 117.8 |
| 0330 | Cast-in-Place Concrete | 118.4 | 79.9 | 103.7 | 101.4 | 124.5 | 110.3 | 95.0 | 124.5 | 106.3 | 100.7 | 124.5 | 109.8 | 91.4 | 124.5 | 104.1 | 96.6 | 124.5 | 107.3 |
| 03 | CONCRETE | 114.2 | 80.3 | 98.0 | 109.2 | 121.9 | 115.3 | 106.0 | 123.5 | 114.4 | 108.7 | 123.5 | 115.8 | 104.2 | 123.5 | 113.4 | 106.8 | 123.5 | 114.8 |
| 04 | MASONRY | 135.2 | 77.2 | 100.1 | 108.3 | 127.9 | 120.2 | 100.2 | 127.9 | 117.0 | 100.7 | 127.9 | 117.2 | 99.9 | 127.9 | 116.9 | 102.2 | 127.9 | 117.8 |
| 05 | METALS | 95.1 | 82.0 | 90.6 | 99.8 | 124.3 | 108.3 | 99.8 | 124.3 | 108.2 | 104.5 | 124.3 | 111.3 | 97.1 | 124.3 | 106.4 | 96.2 | 124.3 | 105.9 |
| 06 | WOOD, PLASTICS & COMPOSITES | 99.1 | 84.2 | 90.6 | 98.1 | 116.2 | 108.5 | 98.1 | 121.0 | 111.3 | 96.6 | 121.0 | 110.6 | 98.1 | 121.0 | 111.3 | 98.1 | 121.0 | 111.3 |
| 07 | THERMAL & MOISTURE PROTECTION | 105.1 | 82.3 | 96.1 | 96.4 | 125.4 | 107.9 | 96.5 | 123.6 | 107.3 | 97.7 | 123.6 | 108.0 | 96.5 | 123.6 | 107.2 | 96.5 | 123.6 | 107.3 |
| 08 | OPENINGS | 94.0 | 84.8 | 91.6 | 105.6 | 127.2 | 111.1 | 105.6 | 129.9 | 111.8 | 106.3 | 129.9 | 112.3 | 108.5 | 129.9 | 113.9 | 105.6 | 129.9 | 111.8 |
| 0920 | Plaster & Gypsum Board | 85.0 | 83.7 | 84.1 | 99.4 | 116.0 | 111.1 | 99.4 | 121.0 | 114.6 | 97.5 | 121.0 | 114.0 | 100.9 | 121.0 | 115.1 | 99.4 | 121.0 | 114.6 |
| 0950, 0980 | Ceilings & Acoustic Treatment | 118.2 | 83.7 | 95.3 | 99.4 | 116.0 | 110.4 | 99.4 | 121.0 | 113.7 | 97.5 | 121.0 | 113.1 | 102.9 | 121.0 | 114.9 | 99.4 | 121.0 | 113.7 |
| 0960 | Flooring | 119.5 | 47.9 | 99.0 | 95.8 | 120.8 | 103.0 | 95.8 | 120.8 | 103.0 | 95.8 | 120.8 | 103.0 | 95.8 | 120.8 | 103.0 | 95.8 | 120.8 | 103.0 |
| 0970, 0990 | Wall Finishes & Painting/Coating | 107.9 | 25.4 | 57.8 | 92.3 | 114.5 | 105.8 | 92.3 | 116.9 | 107.2 | 92.3 | 116.9 | 107.2 | 92.3 | 116.9 | 107.2 | 92.3 | 116.9 | 107.2 |
| 09 | FINISHES | 109.7 | 69.0 | 87.3 | 100.1 | 118.0 | 110.0 | 100.1 | 121.1 | 111.7 | 98.6 | 121.1 | 111.0 | 101.1 | 121.1 | 112.1 | 100.1 | 121.1 | 111.7 |
| COVERS | DIVS. 10 - 14, 25, 28, 41, 43, 44, 46 | 100.0 | 90.6 | 98.1 | 100.0 | 108.0 | 101.7 | 100.0 | 108.6 | 101.8 | 100.0 | 108.6 | 101.8 | 100.0 | 108.6 | 101.8 | 100.0 | 108.6 | 101.8 |
| 21, 22, 23 | FIRE SUPPRESSION, PLUMBING & HVAC | 94.6 | 69.2 | 84.0 | 100.2 | 112.8 | 105.4 | 100.2 | 112.8 | 105.4 | 100.2 | 112.8 | 105.5 | 94.8 | 112.8 | 102.3 | 100.2 | 112.8 | 105.4 |
| 26, 27, 3370 | ELECTRICAL, COMMUNICATIONS & UTIL. | 95.3 | 76.7 | 85.4 | 100.8 | 105.9 | 103.5 | 100.8 | 110.4 | 105.9 | 99.7 | 110.4 | 105.4 | 100.8 | 109.5 | 105.4 | 100.9 | 110.4 | 105.9 |
| MF2010 | WEIGHTED AVERAGE | 102.0 | 77.8 | 91.2 | 102.2 | 116.4 | 108.6 | 101.4 | 117.8 | 108.7 | 102.1 | 117.8 | 109.1 | 99.7 | 117.7 | 107.8 | 101.0 | 117.8 | 108.5 |

| DIVISION | | CONNECTICUT NEW HAVEN 065 | | | NEW LONDON 063 | | | NORWALK 068 | | | STAMFORD 069 | | | WATERBURY 067 | | | WILLIMANTIC 062 | | |
|---|---|---|---|---|---|---|---|---|---|---|---|---|---|---|---|---|---|---|---|
| | | MAT. | INST. | TOTAL | MAT. | INST. | TOTAL | MAT. | INST. | TOTAL | MAT. | INST. | TOTAL | MAT. | INST. | TOTAL | MAT. | INST. | TOTAL |
| 015433 | CONTRACTOR EQUIPMENT | | 100.8 | 100.8 | | 100.8 | 100.8 | | 100.3 | 100.3 | | 100.3 | 100.3 | | 100.3 | 100.3 | | 100.3 | 100.3 |
| 0241, 31 - 34 | SITE & INFRASTRUCTURE, DEMOLITION | 102.7 | 104.6 | 104.0 | 95.4 | 104.6 | 101.7 | 103.3 | 103.8 | 103.7 | 103.9 | 103.8 | 103.9 | 103.2 | 103.8 | 103.6 | 103.4 | 103.8 | 103.7 |
| 0310 | Concrete Forming & Accessories | 96.2 | 121.8 | 118.4 | 96.2 | 121.8 | 118.4 | 96.5 | 118.3 | 115.4 | 96.4 | 118.6 | 115.7 | 96.5 | 121.8 | 118.5 | 96.5 | 121.8 | 118.5 |
| 0320 | Concrete Reinforcing | 108.1 | 127.0 | 117.8 | 84.8 | 127.0 | 106.5 | 108.1 | 127.1 | 117.9 | 108.1 | 127.1 | 117.9 | 108.1 | 127.0 | 117.8 | 108.1 | 127.0 | 117.8 |
| 0330 | Cast-in-Place Concrete | 98.3 | 124.5 | 108.3 | 83.7 | 124.5 | 99.3 | 99.8 | 125.9 | 109.8 | 101.4 | 126.0 | 110.9 | 101.4 | 124.5 | 110.3 | 94.7 | 124.5 | 106.1 |
| 03 | CONCRETE | 123.0 | 123.5 | 123.2 | 93.7 | 123.5 | 108.0 | 108.4 | 122.4 | 115.1 | 109.2 | 122.6 | 115.6 | 109.2 | 123.5 | 116.0 | 105.9 | 123.5 | 114.3 |
| 04 | MASONRY | 100.5 | 127.9 | 117.1 | 98.9 | 127.9 | 116.5 | 99.9 | 129.5 | 117.8 | 100.7 | 129.5 | 118.2 | 100.7 | 127.9 | 117.2 | 100.2 | 127.9 | 117.0 |
| 05 | METALS | 96.4 | 124.3 | 106.0 | 96.1 | 124.3 | 105.8 | 99.8 | 124.4 | 108.3 | 99.8 | 124.8 | 108.4 | 99.8 | 124.3 | 108.2 | 99.6 | 124.3 | 108.1 |
| 06 | WOOD, PLASTICS & COMPOSITES | 98.1 | 121.0 | 111.3 | 98.1 | 121.0 | 111.3 | 98.1 | 116.2 | 108.5 | 98.1 | 116.2 | 108.5 | 98.1 | 121.0 | 111.3 | 98.1 | 121.0 | 111.3 |
| 07 | THERMAL & MOISTURE PROTECTION | 96.6 | 123.6 | 107.3 | 96.4 | 123.6 | 107.2 | 96.5 | 126.1 | 108.3 | 96.5 | 126.1 | 108.2 | 96.5 | 123.6 | 107.3 | 96.7 | 123.6 | 107.4 |
| 08 | OPENINGS | 105.6 | 129.9 | 111.8 | 108.9 | 129.9 | 114.2 | 105.6 | 127.2 | 111.1 | 105.6 | 127.2 | 111.1 | 105.6 | 129.9 | 111.8 | 108.9 | 129.9 | 114.2 |
| 0920 | Plaster & Gypsum Board | 99.4 | 121.0 | 114.6 | 99.4 | 121.0 | 114.6 | 99.4 | 116.0 | 111.1 | 99.4 | 116.0 | 111.1 | 99.4 | 121.0 | 114.6 | 99.4 | 121.0 | 114.6 |
| 0950, 0980 | Ceilings & Acoustic Treatment | 99.4 | 121.0 | 113.7 | 97.3 | 121.0 | 113.0 | 99.4 | 116.0 | 110.4 | 99.4 | 116.0 | 110.4 | 99.4 | 121.0 | 113.7 | 97.3 | 121.0 | 113.0 |
| 0960 | Flooring | 95.8 | 120.8 | 103.0 | 95.8 | 120.8 | 103.0 | 95.8 | 120.8 | 103.0 | 95.8 | 120.8 | 103.0 | 95.8 | 120.8 | 103.0 | 95.8 | 120.8 | 103.0 |
| 0970, 0990 | Wall Finishes & Painting/Coating | 92.3 | 114.5 | 105.8 | 92.3 | 116.9 | 107.2 | 92.3 | 114.5 | 105.8 | 92.3 | 114.5 | 105.8 | 92.3 | 116.9 | 107.2 | 92.3 | 116.9 | 107.2 |
| 09 | FINISHES | 100.2 | 120.9 | 111.6 | 99.1 | 121.1 | 111.3 | 100.1 | 118.0 | 110.0 | 100.2 | 118.0 | 110.0 | 100.0 | 121.1 | 111.6 | 99.8 | 121.1 | 111.6 |
| COVERS | DIVS. 10 - 14, 25, 28, 41, 43, 44, 46 | 100.0 | 108.6 | 101.8 | 100.0 | 108.6 | 101.8 | 100.0 | 108.0 | 101.7 | 100.0 | 108.2 | 101.7 | 100.0 | 108.6 | 101.8 | 100.0 | 108.6 | 101.8 |
| 21, 22, 23 | FIRE SUPPRESSION, PLUMBING & HVAC | 100.2 | 112.8 | 105.4 | 94.8 | 112.8 | 102.3 | 100.2 | 112.8 | 105.4 | 100.2 | 112.9 | 105.5 | 100.2 | 112.8 | 105.4 | 100.2 | 112.8 | 105.4 |
| 26, 27, 3370 | ELECTRICAL, COMMUNICATIONS & UTIL. | 100.8 | 109.5 | 105.4 | 97.2 | 109.5 | 103.7 | 100.8 | 104.4 | 102.7 | 100.8 | 151.0 | 127.5 | 100.2 | 109.8 | 105.3 | 101.6 | 110.4 | 105.9 |
| MF2010 | WEIGHTED AVERAGE | 102.9 | 117.7 | 109.5 | 97.7 | 117.7 | 106.6 | 101.6 | 116.5 | 108.3 | 101.8 | 123.0 | 111.3 | 101.7 | 117.7 | 108.9 | 101.6 | 117.8 | 108.9 |

| DIVISION | | D.C. WASHINGTON 200 - 205 | | | DELAWARE DOVER 199 | | | NEWARK 197 | | | WILMINGTON 198 | | | FLORIDA DAYTONA BEACH 321 | | | FORT LAUDERDALE 333 | | |
|---|---|---|---|---|---|---|---|---|---|---|---|---|---|---|---|---|---|---|---|
| | | MAT. | INST. | TOTAL | MAT. | INST. | TOTAL | MAT. | INST. | TOTAL | MAT. | INST. | TOTAL | MAT. | INST. | TOTAL | MAT. | INST. | TOTAL |
| 015433 | CONTRACTOR EQUIPMENT | | 103.6 | 103.6 | | 117.7 | 117.7 | | 117.7 | 117.7 | | 117.8 | 117.8 | | 97.8 | 97.8 | | 90.5 | 90.5 |
| 0241, 31 - 34 | SITE & INFRASTRUCTURE, DEMOLITION | 113.8 | 92.3 | 99.0 | 94.0 | 112.2 | 106.6 | 94.2 | 112.2 | 106.6 | 85.3 | 112.5 | 104.0 | 113.2 | 89.3 | 96.7 | 98.7 | 77.3 | 83.9 |
| 0310 | Concrete Forming & Accessories | 98.7 | 80.3 | 82.7 | 100.2 | 99.8 | 99.9 | 102.4 | 99.8 | 100.1 | 103.3 | 99.8 | 100.3 | 97.5 | 73.6 | 76.8 | 96.2 | 74.3 | 77.2 |
| 0320 | Concrete Reinforcing | 94.6 | 89.6 | 92.0 | 92.7 | 103.6 | 98.2 | 93.5 | 103.6 | 98.7 | 93.5 | 103.6 | 98.7 | 91.8 | 78.2 | 84.8 | 91.8 | 77.8 | 84.6 |
| 0330 | Cast-in-Place Concrete | 129.7 | 87.2 | 113.5 | 92.1 | 102.1 | 95.9 | 82.7 | 102.1 | 90.1 | 89.2 | 102.1 | 94.2 | 91.4 | 75.5 | 85.3 | 95.9 | 82.9 | 90.9 |
| 03 | CONCRETE | 116.2 | 85.7 | 101.6 | 97.1 | 102.3 | 99.6 | 92.7 | 102.3 | 97.3 | 96.0 | 102.3 | 99.0 | 88.1 | 76.2 | 82.4 | 90.3 | 79.0 | 84.9 |
| 04 | MASONRY | 102.2 | 81.9 | 89.9 | 104.4 | 97.6 | 100.3 | 102.7 | 97.6 | 99.6 | 108.7 | 97.6 | 102.0 | 87.1 | 70.9 | 77.3 | 87.4 | 73.8 | 79.1 |
| 05 | METALS | 93.5 | 107.2 | 98.2 | 104.6 | 118.0 | 109.2 | 104.6 | 118.0 | 109.2 | 101.6 | 118.0 | 107.2 | 96.1 | 93.0 | 95.1 | 96.0 | 93.9 | 95.3 |
| 06 | WOOD, PLASTICS & COMPOSITES | 96.8 | 79.2 | 86.7 | 100.4 | 98.1 | 99.1 | 103.4 | 98.1 | 100.3 | 105.8 | 98.1 | 101.4 | 100.5 | 75.1 | 86.0 | 96.3 | 72.9 | 82.9 |
| 07 | THERMAL & MOISTURE PROTECTION | 99.1 | 84.9 | 93.4 | 97.5 | 113.9 | 104.0 | 97.4 | 113.9 | 104.0 | 97.9 | 113.9 | 104.3 | 93.8 | 79.4 | 88.1 | 93.8 | 87.3 | 91.2 |
| 08 | OPENINGS | 104.2 | 88.8 | 100.3 | 95.8 | 109.0 | 99.2 | 95.8 | 109.0 | 99.2 | 95.6 | 109.0 | 99.0 | 100.5 | 71.9 | 93.2 | 97.9 | 70.9 | 91.1 |
| 0920 | Plaster & Gypsum Board | 104.3 | 78.6 | 86.2 | 105.2 | 97.9 | 100.1 | 106.6 | 97.9 | 100.5 | 107.1 | 97.9 | 100.6 | 94.6 | 74.9 | 80.7 | 93.7 | 72.6 | 78.8 |
| 0950, 0980 | Ceilings & Acoustic Treatment | 107.6 | 78.6 | 88.4 | 109.0 | 97.9 | 101.6 | 109.0 | 97.9 | 101.6 | 104.2 | 97.9 | 100.0 | 96.5 | 74.9 | 82.2 | 96.5 | 72.6 | 80.7 |
| 0960 | Flooring | 120.1 | 96.6 | 113.4 | 90.3 | 106.2 | 94.8 | 90.5 | 106.2 | 95.0 | 90.1 | 106.2 | 94.7 | 119.4 | 77.8 | 107.4 | 119.4 | 73.2 | 106.1 |
| 0970, 0990 | Wall Finishes & Painting/Coating | 121.0 | 88.5 | 101.3 | 94.7 | 101.6 | 98.9 | 94.7 | 101.6 | 98.9 | 94.7 | 101.6 | 98.9 | 111.3 | 79.8 | 92.2 | 105.9 | 75.6 | 87.5 |
| 09 | FINISHES | 108.4 | 83.5 | 94.7 | 99.8 | 100.3 | 100.1 | 100.3 | 100.3 | 100.3 | 99.1 | 100.3 | 99.8 | 107.6 | 74.8 | 89.5 | 104.9 | 73.5 | 87.6 |
| COVERS | DIVS. 10 - 14, 25, 28, 41, 43, 44, 46 | 100.0 | 98.1 | 99.6 | 100.0 | 96.5 | 99.3 | 100.0 | 96.5 | 99.3 | 100.0 | 96.5 | 99.3 | 100.0 | 81.2 | 96.1 | 100.0 | 88.0 | 97.5 |
| 21, 22, 23 | FIRE SUPPRESSION, PLUMBING & HVAC | 100.2 | 94.9 | 98.0 | 99.9 | 116.1 | 106.6 | 100.1 | 116.1 | 106.8 | 100.1 | 116.1 | 106.8 | 99.8 | 78.8 | 91.1 | 99.8 | 69.4 | 87.1 |
| 26, 27, 3370 | ELECTRICAL, COMMUNICATIONS & UTIL. | 101.4 | 100.2 | 100.7 | 101.2 | 112.5 | 107.2 | 99.6 | 112.5 | 106.4 | 103.5 | 112.5 | 108.3 | 98.1 | 59.6 | 77.6 | 98.1 | 75.9 | 86.3 |
| MF2010 | WEIGHTED AVERAGE | 102.8 | 91.9 | 97.9 | 99.9 | 108.6 | 103.8 | 99.3 | 108.6 | 103.4 | 99.6 | 108.6 | 103.6 | 98.0 | 76.4 | 88.3 | 97.3 | 76.7 | 88.1 |

# City Cost Indexes

## FLORIDA

| DIVISION | | FORT MYERS 339,341 | | | GAINESVILLE 326,344 | | | JACKSONVILLE 320,322 | | | LAKELAND 338 | | | MELBOURNE 329 | | | MIAMI 330 - 332,340 | | |
|---|---|---|---|---|---|---|---|---|---|---|---|---|---|---|---|---|---|---|---|
| | | MAT. | INST. | TOTAL | MAT. | INST. | TOTAL | MAT. | INST. | TOTAL | MAT. | INST. | TOTAL | MAT. | INST. | TOTAL | MAT. | INST. | TOTAL |
| 015433 | CONTRACTOR EQUIPMENT | | 97.8 | 97.8 | | 97.8 | 97.8 | | 97.8 | 97.8 | | 97.8 | 97.8 | | 97.8 | 97.8 | | 90.5 | 90.5 |
| 0241, 31 - 34 | SITE & INFRASTRUCTURE, DEMOLITION | 110.5 | 88.6 | 95.4 | 123.5 | 88.3 | 99.2 | 113.2 | 88.6 | 96.3 | 112.6 | 88.8 | 96.2 | 121.3 | 88.6 | 98.7 | 100.5 | 77.1 | 84.3 |
| 0310 | Concrete Forming & Accessories | 91.5 | 81.1 | 82.5 | 92.4 | 58.6 | 63.1 | 97.3 | 59.0 | 64.0 | 87.9 | 81.8 | 82.6 | 93.7 | 75.4 | 77.8 | 100.7 | 74.8 | 78.2 |
| 0320 | Concrete Reinforcing | 92.8 | 100.1 | 96.6 | 97.4 | 68.6 | 82.6 | 91.8 | 68.7 | 79.9 | 95.1 | 101.0 | 98.1 | 92.8 | 78.2 | 85.3 | 91.8 | 77.9 | 84.6 |
| 0330 | Cast-in-Place Concrete | 100.0 | 73.1 | 89.7 | 105.0 | 66.5 | 90.2 | 92.3 | 72.8 | 84.9 | 102.3 | 74.5 | 91.6 | 110.2 | 78.1 | 97.9 | 97.8 | 84.0 | 92.5 |
| 03 | CONCRETE | 91.2 | 82.9 | 87.2 | 98.8 | 64.7 | 82.5 | 88.5 | 67.0 | 78.2 | 93.1 | 83.8 | 88.6 | 99.6 | 77.9 | 89.2 | 91.5 | 79.5 | 85.8 |
| 04 | MASONRY | 80.7 | 66.9 | 72.3 | 100.2 | 66.0 | 79.5 | 87.0 | 66.1 | 74.3 | 95.7 | 81.9 | 87.3 | 84.0 | 75.3 | 78.8 | 85.9 | 77.7 | 80.9 |
| 05 | METALS | 98.1 | 101.5 | 99.3 | 95.1 | 87.7 | 92.5 | 94.7 | 88.0 | 92.4 | 98.0 | 102.3 | 99.5 | 104.0 | 93.1 | 100.3 | 100.7 | 92.7 | 98.0 |
| 06 | WOOD, PLASTICS & COMPOSITES | 93.2 | 84.1 | 88.0 | 94.4 | 56.1 | 72.5 | 100.5 | 56.1 | 75.1 | 88.9 | 84.1 | 86.1 | 96.0 | 75.1 | 84.0 | 102.3 | 72.9 | 85.4 |
| 07 | THERMAL & MOISTURE PROTECTION | 93.7 | 86.3 | 90.7 | 94.1 | 65.4 | 82.7 | 94.0 | 66.1 | 82.9 | 93.6 | 91.3 | 92.7 | 94.2 | 82.0 | 89.4 | 102.4 | 81.1 | 93.9 |
| 08 | OPENINGS | 99.0 | 80.6 | 94.4 | 98.4 | 56.7 | 87.9 | 100.5 | 60.1 | 90.2 | 99.0 | 81.3 | 94.5 | 99.6 | 75.4 | 93.4 | 97.9 | 70.9 | 91.1 |
| 0920 | Plaster & Gypsum Board | 88.9 | 84.1 | 85.5 | 89.7 | 55.3 | 65.5 | 94.6 | 55.3 | 66.9 | 86.6 | 84.1 | 84.9 | 89.7 | 74.9 | 79.3 | 92.0 | 72.6 | 78.4 |
| 0950, 0980 | Ceilings & Acoustic Treatment | 89.7 | 84.1 | 86.0 | 89.7 | 55.3 | 66.9 | 96.5 | 55.3 | 69.2 | 89.7 | 84.1 | 86.0 | 91.8 | 74.9 | 80.6 | 93.7 | 72.6 | 79.7 |
| 0960 | Flooring | 115.7 | 58.2 | 99.2 | 116.2 | 45.6 | 95.9 | 119.4 | 68.2 | 104.7 | 112.9 | 59.7 | 97.6 | 116.4 | 77.8 | 105.3 | 127.0 | 77.1 | 112.7 |
| 0970, 0990 | Wall Finishes & Painting/Coating | 111.3 | 65.4 | 83.5 | 111.3 | 65.4 | 83.5 | 111.3 | 66.7 | 84.3 | 111.3 | 65.4 | 83.5 | 111.3 | 99.0 | 103.8 | 105.9 | 75.6 | 87.5 |
| 09 | FINISHES | 104.2 | 75.5 | 88.3 | 105.7 | 55.8 | 78.2 | 107.7 | 60.5 | 81.6 | 103.2 | 75.8 | 88.1 | 105.9 | 78.1 | 90.5 | 106.5 | 74.7 | 89.0 |
| COVERS | DIVS. 10 - 14, 25, 28, 41, 43, 44, 46 | 100.0 | 81.5 | 96.2 | 100.0 | 79.4 | 95.7 | 100.0 | 77.4 | 95.3 | 100.0 | 81.5 | 96.2 | 100.0 | 82.8 | 96.4 | 100.0 | 88.6 | 97.6 |
| 21, 22, 23 | FIRE SUPPRESSION, PLUMBING & HVAC | 96.9 | 59.7 | 81.4 | 98.9 | 69.1 | 86.5 | 99.8 | 66.7 | 86.0 | 96.9 | 88.7 | 93.5 | 99.8 | 81.0 | 92.0 | 99.8 | 73.0 | 88.6 |
| 26, 27, 3370 | ELECTRICAL, COMMUNICATIONS & UTIL. | 100.6 | 66.9 | 82.7 | 98.4 | 77.9 | 87.5 | 97.8 | 67.6 | 81.7 | 98.4 | 66.8 | 81.6 | 99.4 | 69.5 | 83.5 | 103.9 | 78.5 | 90.4 |
| MF2010 | WEIGHTED AVERAGE | 97.3 | 75.4 | 87.5 | 99.5 | 70.6 | 86.6 | 97.8 | 69.7 | 85.2 | 98.0 | 83.5 | 91.5 | 100.4 | 79.6 | 91.1 | 99.1 | 78.1 | 89.7 |

## FLORIDA

| DIVISION | | ORLANDO 327 - 328,347 | | | PANAMA CITY 324 | | | PENSACOLA 325 | | | SARASOTA 342 | | | ST. PETERSBURG 337 | | | TALLAHASSEE 323 | | |
|---|---|---|---|---|---|---|---|---|---|---|---|---|---|---|---|---|---|---|---|
| | | MAT. | INST. | TOTAL | MAT. | INST. | TOTAL | MAT. | INST. | TOTAL | MAT. | INST. | TOTAL | MAT. | INST. | TOTAL | MAT. | INST. | TOTAL |
| 015433 | CONTRACTOR EQUIPMENT | | 97.8 | 97.8 | | 97.8 | 97.8 | | 97.8 | 97.8 | | 97.8 | 97.8 | | 97.8 | 97.8 | | 97.8 | 97.8 |
| 0241, 31 - 34 | SITE & INFRASTRUCTURE, DEMOLITION | 113.8 | 88.5 | 96.3 | 127.5 | 87.5 | 99.9 | 127.0 | 88.0 | 100.1 | 114.7 | 88.6 | 96.6 | 114.0 | 88.1 | 96.1 | 110.0 | 87.7 | 94.6 |
| 0310 | Concrete Forming & Accessories | 97.1 | 78.0 | 80.8 | 96.6 | 47.1 | 53.7 | 94.2 | 56.1 | 61.1 | 97.5 | 81.3 | 83.4 | 95.5 | 55.1 | 60.4 | 99.4 | 47.7 | 54.5 |
| 0320 | Concrete Reinforcing | 91.8 | 77.5 | 84.4 | 95.8 | 78.2 | 86.8 | 98.3 | 78.8 | 88.3 | 91.8 | 100.9 | 96.5 | 95.1 | 86.4 | 90.6 | 91.8 | 68.5 | 79.8 |
| 0330 | Cast-in-Place Concrete | 114.6 | 75.4 | 99.6 | 97.0 | 60.2 | 82.9 | 119.9 | 70.3 | 100.9 | 104.6 | 74.2 | 93.0 | 103.4 | 68.5 | 90.1 | 99.8 | 59.8 | 84.5 |
| 03 | CONCRETE | 100.9 | 78.0 | 89.9 | 96.2 | 59.2 | 78.5 | 107.1 | 66.8 | 87.8 | 94.7 | 83.4 | 89.3 | 94.5 | 67.1 | 81.4 | 92.4 | 57.5 | 75.7 |
| 04 | MASONRY | 90.4 | 70.9 | 78.6 | 91.4 | 51.0 | 66.9 | 109.7 | 58.7 | 78.8 | 88.1 | 81.9 | 84.3 | 133.0 | 52.8 | 84.4 | 89.8 | 58.1 | 70.6 |
| 05 | METALS | 101.6 | 92.3 | 98.4 | 95.9 | 89.7 | 93.8 | 97.1 | 91.6 | 95.2 | 98.7 | 101.7 | 99.8 | 99.0 | 94.1 | 97.3 | 89.9 | 87.4 | 89.1 |
| 06 | WOOD, PLASTICS & COMPOSITES | 100.1 | 81.7 | 89.5 | 99.3 | 45.2 | 68.3 | 97.4 | 55.9 | 73.6 | 100.5 | 84.1 | 91.1 | 97.8 | 54.3 | 72.9 | 97.9 | 44.7 | 67.4 |
| 07 | THERMAL & MOISTURE PROTECTION | 95.0 | 80.0 | 89.0 | 94.3 | 60.8 | 81.0 | 94.2 | 67.0 | 83.4 | 93.8 | 91.3 | 92.8 | 93.8 | 62.1 | 81.2 | 99.7 | 77.0 | 90.7 |
| 08 | OPENINGS | 100.5 | 74.4 | 93.8 | 98.0 | 50.3 | 85.9 | 97.9 | 61.6 | 88.7 | 100.4 | 80.3 | 95.3 | 99.0 | 64.2 | 90.2 | 100.0 | 50.9 | 88.3 |
| 0920 | Plaster & Gypsum Board | 105.4 | 81.6 | 88.7 | 92.4 | 44.1 | 58.4 | 92.9 | 55.0 | 66.3 | 94.6 | 84.1 | 87.2 | 91.5 | 53.4 | 64.7 | 110.8 | 43.6 | 63.5 |
| 0950, 0980 | Ceilings & Acoustic Treatment | 98.4 | 81.6 | 87.3 | 91.8 | 44.1 | 60.2 | 91.8 | 55.0 | 67.4 | 94.4 | 84.1 | 87.6 | 91.8 | 53.4 | 66.4 | 98.4 | 43.6 | 62.0 |
| 0960 | Flooring | 119.6 | 77.8 | 107.6 | 118.9 | 45.4 | 97.8 | 113.6 | 66.5 | 100.1 | 119.4 | 61.1 | 102.6 | 118.0 | 58.3 | 100.8 | 119.4 | 65.5 | 103.9 |
| 0970, 0990 | Wall Finishes & Painting/Coating | 111.3 | 75.6 | 89.6 | 111.3 | 67.7 | 84.9 | 111.3 | 67.7 | 84.9 | 111.3 | 65.4 | 83.5 | 111.3 | 65.4 | 83.5 | 111.3 | 67.7 | 84.9 |
| 09 | FINISHES | 109.1 | 78.2 | 92.1 | 107.7 | 48.1 | 74.8 | 105.8 | 58.5 | 79.7 | 107.2 | 76.0 | 90.0 | 105.9 | 55.8 | 78.3 | 109.4 | 51.8 | 77.6 |
| COVERS | DIVS. 10 - 14, 25, 28, 41, 43, 44, 46 | 100.0 | 81.9 | 96.2 | 100.0 | 51.5 | 89.9 | 100.0 | 51.6 | 89.9 | 100.0 | 81.5 | 96.2 | 100.0 | 62.4 | 92.2 | 100.0 | 73.8 | 94.6 |
| 21, 22, 23 | FIRE SUPPRESSION, PLUMBING & HVAC | 99.7 | 62.0 | 84.0 | 99.8 | 41.3 | 75.4 | 99.8 | 57.5 | 82.2 | 99.7 | 60.2 | 83.3 | 99.8 | 58.6 | 82.6 | 99.8 | 42.7 | 76.0 |
| 26, 27, 3370 | ELECTRICAL, COMMUNICATIONS & UTIL. | 104.1 | 63.6 | 82.6 | 96.7 | 58.5 | 76.4 | 100.9 | 58.4 | 78.3 | 98.1 | 66.8 | 81.5 | 98.4 | 66.8 | 81.6 | 104.1 | 58.6 | 79.9 |
| MF2010 | WEIGHTED AVERAGE | 101.2 | 74.2 | 89.1 | 99.2 | 57.5 | 80.6 | 101.9 | 65.1 | 85.4 | 99.2 | 77.3 | 89.4 | 101.3 | 66.1 | 85.6 | 98.6 | 59.7 | 81.2 |

## FLORIDA / GEORGIA

| DIVISION | | TAMPA 335 - 336,346 | | | WEST PALM BEACH 334,349 | | | ALBANY 317,398 | | | ATHENS 306 | | | ATLANTA 300 - 303,399 | | | AUGUSTA 308 - 309 | | |
|---|---|---|---|---|---|---|---|---|---|---|---|---|---|---|---|---|---|---|---|
| | | MAT. | INST. | TOTAL | MAT. | INST. | TOTAL | MAT. | INST. | TOTAL | MAT. | INST. | TOTAL | MAT. | INST. | TOTAL | MAT. | INST. | TOTAL |
| 015433 | CONTRACTOR EQUIPMENT | | 97.8 | 97.8 | | 90.5 | 90.5 | | 91.3 | 91.3 | | 92.7 | 92.7 | | 93.3 | 93.3 | | 92.7 | 92.7 |
| 0241, 31 - 34 | SITE & INFRASTRUCTURE, DEMOLITION | 114.4 | 88.7 | 96.7 | 95.6 | 77.4 | 83.0 | 98.9 | 78.5 | 84.8 | 101.2 | 92.3 | 95.0 | 97.2 | 94.1 | 95.1 | 93.8 | 92.0 | 92.6 |
| 0310 | Concrete Forming & Accessories | 98.3 | 81.9 | 84.1 | 99.5 | 73.9 | 77.3 | 96.8 | 46.6 | 53.2 | 94.8 | 44.3 | 51.0 | 98.2 | 74.1 | 77.3 | 95.7 | 68.4 | 72.0 |
| 0320 | Concrete Reinforcing | 91.8 | 101.0 | 96.5 | 94.4 | 77.6 | 85.7 | 91.4 | 84.5 | 87.8 | 99.4 | 84.0 | 91.4 | 98.8 | 85.9 | 92.2 | 99.7 | 76.6 | 87.8 |
| 0330 | Cast-in-Place Concrete | 101.1 | 74.5 | 90.9 | 91.2 | 79.7 | 86.8 | 97.6 | 48.4 | 78.8 | 110.0 | 53.6 | 88.4 | 110.0 | 72.3 | 95.5 | 103.9 | 50.4 | 83.4 |
| 03 | CONCRETE | 93.0 | 83.8 | 88.6 | 87.3 | 77.7 | 82.7 | 91.0 | 55.9 | 74.2 | 101.0 | 55.8 | 79.3 | 98.9 | 75.8 | 87.9 | 94.5 | 64.1 | 79.9 |
| 04 | MASONRY | 88.3 | 81.9 | 84.4 | 86.9 | 71.9 | 77.9 | 89.7 | 45.6 | 63.0 | 74.6 | 58.1 | 64.6 | 88.0 | 69.2 | 76.6 | 88.2 | 42.2 | 60.3 |
| 05 | METALS | 98.0 | 102.4 | 99.5 | 95.2 | 93.4 | 94.6 | 95.4 | 87.3 | 92.6 | 91.1 | 75.5 | 85.8 | 92.0 | 79.6 | 87.8 | 90.8 | 73.0 | 84.7 |
| 06 | WOOD, PLASTICS & COMPOSITES | 101.8 | 84.1 | 91.7 | 101.1 | 72.9 | 84.9 | 99.5 | 41.1 | 66.0 | 89.3 | 38.0 | 59.9 | 93.3 | 76.1 | 83.4 | 90.7 | 74.3 | 81.3 |
| 07 | THERMAL & MOISTURE PROTECTION | 94.0 | 91.3 | 92.9 | 93.6 | 75.6 | 86.4 | 91.5 | 61.7 | 79.7 | 94.5 | 54.3 | 78.6 | 94.4 | 74.7 | 86.6 | 94.1 | 58.3 | 79.9 |
| 08 | OPENINGS | 100.4 | 83.8 | 96.2 | 97.0 | 70.9 | 90.4 | 97.9 | 47.4 | 85.1 | 91.9 | 46.1 | 80.3 | 97.8 | 72.4 | 91.3 | 91.9 | 65.7 | 85.3 |
| 0920 | Plaster & Gypsum Board | 94.6 | 84.1 | 87.2 | 97.3 | 72.6 | 79.9 | 94.3 | 39.8 | 56.0 | 100.5 | 36.5 | 55.5 | 102.7 | 75.8 | 83.7 | 101.4 | 73.9 | 82.0 |
| 0950, 0980 | Ceilings & Acoustic Treatment | 96.5 | 84.1 | 88.3 | 89.7 | 72.6 | 78.4 | 95.6 | 39.8 | 58.6 | 100.6 | 36.5 | 58.1 | 100.6 | 75.8 | 84.1 | 101.6 | 73.9 | 83.2 |
| 0960 | Flooring | 119.4 | 59.7 | 102.2 | 121.8 | 70.2 | 107.0 | 119.4 | 37.9 | 96.0 | 89.5 | 56.9 | 80.1 | 90.9 | 68.9 | 84.6 | 89.9 | 41.5 | 76.0 |
| 0970, 0990 | Wall Finishes & Painting/Coating | 111.3 | 65.4 | 83.5 | 105.9 | 75.6 | 87.5 | 105.9 | 55.5 | 75.3 | 99.0 | 37.7 | 61.8 | 99.0 | 86.4 | 91.3 | 99.0 | 39.4 | 62.8 |
| 09 | FINISHES | 107.7 | 75.8 | 90.1 | 104.3 | 72.9 | 87.0 | 104.7 | 43.9 | 71.2 | 96.1 | 44.3 | 67.5 | 96.3 | 74.4 | 84.2 | 95.8 | 61.0 | 76.6 |
| COVERS | DIVS. 10 - 14, 25, 28, 41, 43, 44, 46 | 100.0 | 82.1 | 96.3 | 100.0 | 88.0 | 97.5 | 100.0 | 80.2 | 95.9 | 100.0 | 43.5 | 88.3 | 100.0 | 84.7 | 96.8 | 100.0 | 55.4 | 90.7 |
| 21, 22, 23 | FIRE SUPPRESSION, PLUMBING & HVAC | 99.8 | 88.7 | 95.2 | 96.9 | 68.8 | 85.2 | 99.8 | 69.8 | 87.3 | 94.6 | 74.5 | 86.2 | 99.9 | 74.4 | 89.3 | 99.9 | 67.2 | 86.3 |
| 26, 27, 3370 | ELECTRICAL, COMMUNICATIONS & UTIL. | 98.2 | 66.8 | 81.5 | 99.4 | 75.9 | 86.9 | 93.5 | 59.8 | 75.6 | 100.2 | 74.2 | 86.4 | 99.5 | 77.1 | 87.6 | 100.8 | 67.1 | 82.9 |
| MF2010 | WEIGHTED AVERAGE | 99.0 | 83.6 | 92.1 | 96.0 | 75.7 | 86.9 | 96.9 | 62.0 | 81.3 | 94.9 | 64.9 | 81.5 | 97.2 | 76.8 | 88.1 | 95.9 | 65.5 | 82.3 |

## GEORGIA

| DIVISION | | COLUMBUS 318 - 319 | | | DALTON 307 | | | GAINESVILLE 305 | | | MACON 310 - 312 | | | SAVANNAH 313 - 314 | | | STATESBORO 304 | | |
|---|---|---|---|---|---|---|---|---|---|---|---|---|---|---|---|---|---|---|---|
| | | MAT. | INST. | TOTAL | MAT. | INST. | TOTAL | MAT. | INST. | TOTAL | MAT. | INST. | TOTAL | MAT. | INST. | TOTAL | MAT. | INST. | TOTAL |
| 015433 | CONTRACTOR EQUIPMENT | | 91.3 | 91.3 | | 104.7 | 104.7 | | 92.7 | 92.7 | | 102.0 | 102.0 | | 92.3 | 92.3 | | 92.2 | 92.2 |
| 0241, 31 - 34 | SITE & INFRASTRUCTURE, DEMOLITION | 98.8 | 78.7 | 84.9 | 102.0 | 96.2 | 98.0 | 100.8 | 92.3 | 94.9 | 99.4 | 93.9 | 95.6 | 99.8 | 79.4 | 85.7 | 102.5 | 76.4 | 84.5 |
| 0310 | Concrete Forming & Accessories | 96.7 | 58.2 | 63.3 | 88.2 | 49.9 | 54.9 | 97.9 | 46.1 | 53.0 | 96.0 | 55.3 | 60.7 | 96.8 | 51.6 | 57.6 | 82.8 | 43.2 | 48.4 |
| 0320 | Concrete Reinforcing | 91.8 | 85.3 | 88.4 | 98.9 | 78.9 | 88.6 | 99.2 | 84.0 | 91.3 | 92.9 | 84.6 | 88.6 | 92.7 | 76.8 | 84.5 | 98.6 | 44.6 | 70.8 |
| 0330 | Cast-in-Place Concrete | 97.2 | 57.1 | 81.9 | 106.8 | 47.3 | 84.0 | 115.6 | 53.4 | 91.8 | 96.0 | 50.7 | 78.6 | 107.0 | 49.8 | 85.1 | 109.8 | 62.7 | 91.7 |
| 03 | CONCRETE | 90.9 | 64.3 | 78.1 | 99.8 | 56.1 | 78.9 | 103.1 | 56.6 | 80.8 | 90.4 | 60.7 | 76.1 | 95.9 | 57.2 | 77.4 | 100.4 | 51.9 | 77.2 |
| 04 | MASONRY | 89.2 | 59.7 | 71.3 | 75.5 | 34.6 | 50.7 | 83.6 | 58.5 | 68.4 | 101.6 | 40.3 | 64.4 | 88.6 | 52.7 | 66.8 | 77.4 | 42.5 | 56.2 |
| 05 | METALS | 94.9 | 89.1 | 92.9 | 91.9 | 85.3 | 89.6 | 90.4 | 75.2 | 85.1 | 91.2 | 88.1 | 90.2 | 91.6 | 84.6 | 89.2 | 95.4 | 72.4 | 87.5 |
| 06 | WOOD, PLASTICS & COMPOSITES | 99.5 | 56.5 | 74.8 | 73.3 | 51.4 | 60.7 | 93.1 | 41.0 | 63.2 | 109.0 | 55.5 | 78.3 | 115.1 | 48.5 | 76.9 | 66.9 | 42.6 | 53.0 |
| 07 | THERMAL & MOISTURE PROTECTION | 91.5 | 67.0 | 81.8 | 96.4 | 52.5 | 78.9 | 94.5 | 57.4 | 79.7 | 90.1 | 63.4 | 79.5 | 93.7 | 58.4 | 79.7 | 95.1 | 50.8 | 77.5 |
| 08 | OPENINGS | 97.9 | 57.7 | 87.7 | 93.6 | 47.9 | 82.0 | 91.9 | 43.2 | 79.5 | 96.1 | 56.2 | 86.0 | 99.1 | 49.4 | 86.5 | 94.3 | 34.1 | 79.1 |
| 0920 | Plaster & Gypsum Board | 94.3 | 55.7 | 67.1 | 85.0 | 50.4 | 60.6 | 102.7 | 39.6 | 58.3 | 102.9 | 54.6 | 68.9 | 93.3 | 47.4 | 61.0 | 85.7 | 41.3 | 54.5 |
| 0950, 0980 | Ceilings & Acoustic Treatment | 95.6 | 55.7 | 69.1 | 115.0 | 50.4 | 72.2 | 100.6 | 39.6 | 60.2 | 90.2 | 54.6 | 66.6 | 98.4 | 47.4 | 64.6 | 110.1 | 41.3 | 64.5 |
| 0960 | Flooring | 119.4 | 53.3 | 100.4 | 90.3 | 37.9 | 75.2 | 90.9 | 37.9 | 75.6 | 92.9 | 38.3 | 77.2 | 119.9 | 49.0 | 99.5 | 106.9 | 47.4 | 89.8 |
| 0970, 0990 | Wall Finishes & Painting/Coating | 105.9 | 68.9 | 83.4 | 89.7 | 57.5 | 70.2 | 99.0 | 37.7 | 61.8 | 108.2 | 55.5 | 76.2 | 107.3 | 58.4 | 77.6 | 97.0 | 35.3 | 59.6 |
| 09 | FINISHES | 104.6 | 57.4 | 78.6 | 105.9 | 47.8 | 73.9 | 96.5 | 41.6 | 66.2 | 91.8 | 51.0 | 69.3 | 105.5 | 50.9 | 75.4 | 108.4 | 42.4 | 72.0 |
| COVERS | DIVS. 10 - 14, 25, 28, 41, 43, 44, 46 | 100.0 | 82.0 | 96.3 | 100.0 | 24.7 | 84.4 | 100.0 | 43.8 | 88.3 | 100.0 | 79.9 | 95.8 | 100.0 | 54.3 | 90.5 | 100.0 | 40.4 | 87.6 |
| 21, 22, 23 | FIRE SUPPRESSION, PLUMBING & HVAC | 99.8 | 51.0 | 79.5 | 94.6 | 59.0 | 79.8 | 94.6 | 74.5 | 86.2 | 99.8 | 67.8 | 86.5 | 99.8 | 60.5 | 83.4 | 95.0 | 55.5 | 78.5 |
| 26, 27, 3370 | ELECTRICAL, COMMUNICATIONS & UTIL. | 93.7 | 74.3 | 83.4 | 109.9 | 73.5 | 90.6 | 100.2 | 74.2 | 86.4 | 92.7 | 63.4 | 77.1 | 97.7 | 58.7 | 77.0 | 100.3 | 58.7 | 78.2 |
| MF2010 | WEIGHTED AVERAGE | 96.8 | 65.3 | 82.7 | 96.8 | 60.5 | 80.6 | 95.5 | 64.7 | 81.7 | 95.6 | 65.0 | 81.9 | 97.4 | 60.8 | 81.2 | 96.8 | 54.1 | 77.7 |

| DIVISION | | GEORGIA | | | | | | HAWAII | | | | | | | | | IDAHO | | |
|---|---|---|---|---|---|---|---|---|---|---|---|---|---|---|---|---|---|---|---|
| | | VALDOSTA 316 | | | WAYCROSS 315 | | | HILO 967 | | | HONOLULU 968 | | | STATES & POSS., GUAM 969 | | | BOISE 836 - 837 | | |
| | | MAT. | INST. | TOTAL | MAT. | INST. | TOTAL | MAT. | INST. | TOTAL | MAT. | INST. | TOTAL | MAT. | INST. | TOTAL | MAT. | INST. | TOTAL |
| 015433 | CONTRACTOR EQUIPMENT | | 91.3 | 91.3 | | 91.3 | 91.3 | | 99.9 | 99.9 | | 99.9 | 99.9 | | 165.5 | 165.5 | | 100.6 | 100.6 |
| 0241, 31 - 34 | SITE & INFRASTRUCTURE, DEMOLITION | 108.8 | 78.5 | 87.9 | 105.7 | 77.2 | 86.1 | 127.4 | 108.0 | 114.0 | 129.9 | 108.0 | 114.8 | 169.4 | 105.6 | 125.4 | 74.1 | 99.4 | 91.6 |
| 0310 | Concrete Forming & Accessories | 86.3 | 45.7 | 51.1 | 88.2 | 67.8 | 70.5 | 108.6 | 134.5 | 131.1 | 115.7 | 134.5 | 132.0 | 110.8 | 68.4 | 74.0 | 100.7 | 75.0 | 78.4 |
| 0320 | Concrete Reinforcing | 93.6 | 72.5 | 82.8 | 93.6 | 78.4 | 85.8 | 112.7 | 120.1 | 116.5 | 111.6 | 120.1 | 116.0 | 196.9 | 32.5 | 112.2 | 98.5 | 76.0 | 86.9 |
| 0330 | Cast-in-Place Concrete | 95.6 | 52.4 | 79.0 | 108.0 | 51.0 | 86.2 | 189.4 | 125.4 | 164.9 | 159.8 | 125.4 | 146.6 | 164.2 | 110.7 | 143.7 | 93.6 | 76.4 | 87.0 |
| 03 | CONCRETE | 94.4 | 54.7 | 75.4 | 98.8 | 64.9 | 82.6 | 149.9 | 127.5 | 139.2 | 135.3 | 127.5 | 131.6 | 149.7 | 76.5 | 114.7 | 101.1 | 75.8 | 89.0 |
| 04 | MASONRY | 94.8 | 52.2 | 69.0 | 95.7 | 42.4 | 63.4 | 126.2 | 125.4 | 125.7 | 118.0 | 125.4 | 122.5 | 179.4 | 46.9 | 99.1 | 122.6 | 63.9 | 87.0 |
| 05 | METALS | 94.6 | 83.5 | 90.7 | 93.6 | 80.2 | 89.0 | 109.4 | 107.8 | 108.8 | 130.1 | 107.8 | 122.4 | 140.9 | 78.5 | 119.4 | 100.4 | 75.5 | 91.8 |
| 06 | WOOD, PLASTICS & COMPOSITES | 86.8 | 40.1 | 60.0 | 89.1 | 75.7 | 81.4 | 97.4 | 137.6 | 120.4 | 108.2 | 137.6 | 125.0 | 109.3 | 72.5 | 88.2 | 90.4 | 77.3 | 82.9 |
| 07 | THERMAL & MOISTURE PROTECTION | 91.7 | 64.8 | 81.0 | 91.5 | 52.9 | 76.1 | 105.8 | 123.9 | 113.0 | 115.7 | 123.9 | 118.9 | 121.8 | 69.3 | 100.9 | 96.2 | 68.0 | 85.0 |
| 08 | OPENINGS | 93.3 | 43.5 | 80.7 | 93.3 | 60.4 | 84.9 | 99.5 | 131.3 | 107.6 | 104.7 | 131.3 | 111.5 | 104.3 | 58.1 | 92.6 | 94.4 | 71.9 | 88.7 |
| 0920 | Plaster & Gypsum Board | 85.8 | 38.8 | 52.7 | 87.1 | 75.5 | 78.9 | 108.0 | 138.4 | 129.5 | 139.1 | 138.4 | 138.6 | 207.0 | 60.5 | 103.9 | 90.8 | 76.5 | 80.7 |
| 0950, 0980 | Ceilings & Acoustic Treatment | 91.8 | 38.8 | 56.7 | 89.7 | 75.5 | 80.3 | 113.2 | 138.4 | 130.0 | 118.5 | 138.4 | 131.7 | 224.9 | 60.5 | 115.9 | 122.2 | 76.5 | 91.9 |
| 0960 | Flooring | 112.0 | 49.1 | 94.0 | 113.4 | 31.6 | 89.9 | 134.9 | 131.0 | 133.8 | 158.6 | 131.0 | 150.7 | 158.3 | 49.0 | 126.9 | 97.3 | 67.1 | 88.6 |
| 0970, 0990 | Wall Finishes & Painting/Coating | 105.9 | 55.5 | 75.3 | 105.9 | 42.9 | 67.7 | 98.4 | 144.0 | 126.1 | 102.7 | 144.0 | 127.8 | 103.7 | 38.7 | 64.2 | 98.5 | 42.5 | 64.5 |
| 09 | FINISHES | 101.6 | 45.7 | 70.8 | 101.3 | 59.5 | 78.2 | 117.4 | 135.9 | 127.6 | 130.2 | 135.9 | 133.4 | 191.5 | 63.0 | 120.7 | 101.9 | 71.1 | 84.9 |
| COVERS | DIVS. 10 - 14, 25, 28, 41, 43, 44, 46 | 100.0 | 53.4 | 90.3 | 100.0 | 44.0 | 88.4 | 100.0 | 116.5 | 103.4 | 100.0 | 116.5 | 103.4 | 100.0 | 82.7 | 96.4 | 100.0 | 64.5 | 92.6 |
| 21, 22, 23 | FIRE SUPPRESSION, PLUMBING & HVAC | 99.8 | 69.7 | 87.3 | 96.3 | 59.4 | 80.9 | 100.1 | 108.6 | 103.7 | 100.2 | 108.6 | 103.7 | 102.3 | 40.9 | 76.7 | 100.0 | 67.2 | 86.3 |
| 26, 27, 3370 | ELECTRICAL, COMMUNICATIONS & UTIL. | 91.6 | 58.2 | 73.8 | 96.2 | 61.8 | 77.9 | 108.4 | 126.1 | 117.9 | 111.1 | 126.1 | 119.1 | 160.4 | 44.7 | 98.9 | 95.5 | 74.6 | 84.4 |
| MF2010 | WEIGHTED AVERAGE | 96.7 | 61.2 | 80.9 | 96.7 | 61.8 | 81.1 | 111.9 | 120.6 | 115.8 | 115.1 | 120.6 | 117.5 | 133.3 | 61.7 | 101.3 | 99.5 | 73.3 | 87.8 |

| DIVISION | | IDAHO | | | | | | | | | | | | | | | ILLINOIS | | |
|---|---|---|---|---|---|---|---|---|---|---|---|---|---|---|---|---|---|---|---|
| | | COEUR D'ALENE 838 | | | IDAHO FALLS 834 | | | LEWISTON 835 | | | POCATELLO 832 | | | TWIN FALLS 833 | | | BLOOMINGTON 617 | | |
| | | MAT. | INST. | TOTAL | MAT. | INST. | TOTAL | MAT. | INST. | TOTAL | MAT. | INST. | TOTAL | MAT. | INST. | TOTAL | MAT. | INST. | TOTAL |
| 015433 | CONTRACTOR EQUIPMENT | | 94.6 | 94.6 | | 100.6 | 100.6 | | 94.6 | 94.6 | | 100.6 | 100.6 | | 100.6 | 100.6 | | 101.0 | 101.0 |
| 0241, 31 - 34 | SITE & INFRASTRUCTURE, DEMOLITION | 74.0 | 94.7 | 88.3 | 72.6 | 98.5 | 90.5 | 79.9 | 95.2 | 90.5 | 75.6 | 99.4 | 92.0 | 81.6 | 97.6 | 92.7 | 98.8 | 96.8 | 97.4 |
| 0310 | Concrete Forming & Accessories | 117.8 | 58.5 | 66.3 | 95.2 | 42.1 | 49.1 | 122.5 | 60.0 | 68.3 | 100.9 | 74.7 | 78.2 | 101.7 | 30.1 | 39.6 | 88.1 | 118.8 | 114.8 |
| 0320 | Concrete Reinforcing | 105.8 | 94.9 | 100.1 | 100.4 | 74.9 | 87.3 | 105.8 | 95.1 | 100.3 | 98.9 | 76.1 | 87.1 | 100.7 | 75.2 | 87.5 | 99.6 | 104.8 | 102.2 |
| 0330 | Cast-in-Place Concrete | 101.2 | 88.7 | 96.4 | 89.1 | 53.0 | 75.3 | 105.2 | 90.0 | 99.3 | 96.2 | 76.3 | 88.6 | 98.7 | 40.3 | 76.4 | 99.7 | 114.5 | 105.4 |
| 03 | CONCRETE | 108.2 | 76.2 | 92.9 | 93.3 | 53.0 | 74.0 | 112.0 | 77.3 | 95.4 | 96.2 | 75.6 | 88.6 | 108.1 | 43.3 | 77.1 | 96.7 | 114.6 | 105.3 |
| 04 | MASONRY | 124.1 | 81.7 | 98.4 | 117.8 | 35.3 | 67.8 | 124.5 | 84.6 | 100.3 | 119.9 | 65.8 | 87.1 | 122.7 | 33.3 | 68.5 | 113.8 | 119.1 | 117.0 |
| 05 | METALS | 94.8 | 84.7 | 91.3 | 108.1 | 72.8 | 95.9 | 94.2 | 85.8 | 91.3 | 108.1 | 75.3 | 96.8 | 108.2 | 72.2 | 95.8 | 95.0 | 110.1 | 100.2 |
| 06 | WOOD, PLASTICS & COMPOSITES | 95.6 | 51.6 | 70.3 | 84.5 | 40.6 | 59.3 | 100.3 | 51.6 | 72.4 | 90.4 | 77.3 | 82.9 | 91.2 | 28.8 | 55.4 | 84.1 | 117.8 | 103.4 |
| 07 | THERMAL & MOISTURE PROTECTION | 148.4 | 74.6 | 119.1 | 95.8 | 45.4 | 75.8 | 148.6 | 75.3 | 119.5 | 96.2 | 62.5 | 82.8 | 97.0 | 39.4 | 74.1 | 98.4 | 113.3 | 104.3 |
| 08 | OPENINGS | 114.3 | 57.1 | 99.8 | 98.7 | 46.8 | 85.6 | 114.3 | 59.8 | 100.5 | 95.2 | 66.5 | 88.0 | 98.8 | 38.2 | 83.4 | 90.3 | 105.5 | 94.1 |
| 0920 | Plaster & Gypsum Board | 155.8 | 50.1 | 81.4 | 83.9 | 38.7 | 52.1 | 156.7 | 50.1 | 81.6 | 87.9 | 76.5 | 79.9 | 88.3 | 26.5 | 44.8 | 91.2 | 118.2 | 110.2 |
| 0950, 0980 | Ceilings & Acoustic Treatment | 160.9 | 50.1 | 87.5 | 119.1 | 38.7 | 65.8 | 160.9 | 50.1 | 87.5 | 128.5 | 76.5 | 94.0 | 122.0 | 26.5 | 58.7 | 89.5 | 118.2 | 108.5 |
| 0960 | Flooring | 136.5 | 48.5 | 111.2 | 97.1 | 45.5 | 82.3 | 139.5 | 90.3 | 125.3 | 100.4 | 67.1 | 90.8 | 101.6 | 45.5 | 85.5 | 95.1 | 113.2 | 100.3 |
| 0970, 0990 | Wall Finishes & Painting/Coating | 117.4 | 63.2 | 84.5 | 98.5 | 34.9 | 59.9 | 117.4 | 63.2 | 84.5 | 98.4 | 44.2 | 65.5 | 98.5 | 26.1 | 54.5 | 96.2 | 122.3 | 112.0 |
| 09 | FINISHES | 172.0 | 55.6 | 107.8 | 99.9 | 41.3 | 67.6 | 173.1 | 64.4 | 113.2 | 103.8 | 71.3 | 85.9 | 103.5 | 31.4 | 63.8 | 94.9 | 118.8 | 108.1 |
| COVERS | DIVS. 10 - 14, 25, 28, 41, 43, 44, 46 | 100.0 | 53.9 | 90.4 | 100.0 | 40.7 | 87.7 | 100.0 | 67.3 | 93.2 | 100.0 | 64.5 | 92.6 | 100.0 | 36.6 | 86.8 | 100.0 | 105.2 | 101.1 |
| 21, 22, 23 | FIRE SUPPRESSION, PLUMBING & HVAC | 99.8 | 77.4 | 90.5 | 100.7 | 60.8 | 84.1 | 100.8 | 78.8 | 91.6 | 99.9 | 67.2 | 86.3 | 99.9 | 56.9 | 82.0 | 94.7 | 106.5 | 99.6 |
| 26, 27, 3370 | ELECTRICAL, COMMUNICATIONS & UTIL. | 88.0 | 79.7 | 83.6 | 87.7 | 74.6 | 80.7 | 85.9 | 82.5 | 84.1 | 92.9 | 74.6 | 83.2 | 89.2 | 64.3 | 76.0 | 93.2 | 95.5 | 94.5 |
| MF2010 | WEIGHTED AVERAGE | 108.5 | 75.4 | 93.7 | 99.0 | 59.0 | 81.1 | 109.3 | 78.3 | 95.4 | 100.4 | 73.0 | 88.2 | 101.7 | 53.1 | 80.0 | 96.1 | 108.7 | 101.7 |

| | | ILLINOIS | | | | | | | | | | | | | | | | |
|---|---|---|---|---|---|---|---|---|---|---|---|---|---|---|---|---|---|---|
| | DIVISION | CARBONDALE | | | CENTRALIA | | | CHAMPAIGN | | | CHICAGO | | | DECATUR | | | EAST ST. LOUIS | | |
| | | 629 | | | 628 | | | 618 - 619 | | | 606 - 608 | | | 625 | | | 620 - 622 | | |
| | | MAT. | INST. | TOTAL | MAT. | INST. | TOTAL | MAT. | INST. | TOTAL | MAT. | INST. | TOTAL | MAT. | INST. | TOTAL | MAT. | INST. | TOTAL |
| 015433 | CONTRACTOR EQUIPMENT | | 107.4 | 107.4 | | 107.4 | 107.4 | | 101.7 | 101.7 | | 93.4 | 93.4 | | 101.7 | 101.7 | | 107.4 | 107.4 |
| 0241, 31 - 34 | SITE & INFRASTRUCTURE, DEMOLITION | 107.1 | 97.0 | 100.1 | 107.4 | 98.3 | 101.1 | 108.5 | 97.5 | 100.9 | 115.1 | 94.0 | 100.6 | 92.2 | 97.5 | 95.9 | 109.2 | 98.4 | 101.7 |
| 0310 | Concrete Forming & Accessories | 96.5 | 103.7 | 102.7 | 98.3 | 111.4 | 109.7 | 94.0 | 116.6 | 113.6 | 100.3 | 152.2 | 145.3 | 98.2 | 116.0 | 113.7 | 94.1 | 111.7 | 109.3 |
| 0320 | Concrete Reinforcing | 98.5 | 91.8 | 95.0 | 98.5 | 106.1 | 102.4 | 99.6 | 101.9 | 100.8 | 98.8 | 157.4 | 129.0 | 96.5 | 101.7 | 99.2 | 98.4 | 106.1 | 102.4 |
| 0330 | Cast-in-Place Concrete | 93.3 | 101.4 | 96.4 | 93.8 | 115.1 | 102.0 | 115.7 | 110.7 | 113.8 | 112.8 | 147.0 | 125.9 | 101.9 | 112.1 | 105.8 | 95.4 | 115.3 | 103.0 |
| 03 | CONCRETE | 87.8 | 101.5 | 94.3 | 88.2 | 112.3 | 99.8 | 109.3 | 111.7 | 110.5 | 104.4 | 150.3 | 126.3 | 98.0 | 111.9 | 104.7 | 89.2 | 112.5 | 100.4 |
| 04 | MASONRY | 76.0 | 103.0 | 92.4 | 76.1 | 117.2 | 101.0 | 136.9 | 114.1 | 123.1 | 96.0 | 154.3 | 131.3 | 72.8 | 113.2 | 97.3 | 76.2 | 117.3 | 101.1 |
| 05 | METALS | 94.3 | 110.4 | 99.8 | 94.3 | 119.0 | 102.8 | 95.0 | 107.1 | 99.2 | 95.6 | 133.4 | 108.6 | 97.9 | 107.4 | 101.1 | 95.5 | 119.4 | 103.7 |
| 06 | WOOD, PLASTICS & COMPOSITES | 98.0 | 102.2 | 100.4 | 100.5 | 108.8 | 105.3 | 90.7 | 116.0 | 105.2 | 100.2 | 149.9 | 128.7 | 97.9 | 116.0 | 108.3 | 95.3 | 108.8 | 103.1 |
| 07 | THERMAL & MOISTURE PROTECTION | 93.7 | 100.1 | 96.2 | 93.7 | 113.0 | 101.4 | 99.0 | 112.5 | 104.4 | 102.4 | 144.6 | 119.2 | 99.1 | 108.1 | 102.7 | 93.7 | 112.5 | 101.2 |
| 08 | OPENINGS | 86.4 | 107.1 | 91.7 | 86.4 | 114.4 | 93.5 | 91.0 | 111.2 | 96.1 | 100.9 | 152.9 | 114.1 | 96.8 | 111.1 | 100.4 | 86.5 | 114.4 | 93.6 |
| 0920 | Plaster & Gypsum Board | 95.6 | 102.2 | 100.3 | 97.0 | 109.0 | 105.4 | 94.3 | 116.4 | 109.9 | 96.3 | 151.3 | 135.0 | 98.5 | 116.4 | 111.1 | 94.3 | 109.0 | 104.7 |
| 0950, 0980 | Ceilings & Acoustic Treatment | 89.5 | 102.2 | 97.9 | 89.5 | 109.0 | 102.4 | 89.5 | 116.4 | 107.3 | 104.8 | 151.3 | 135.6 | 95.1 | 116.4 | 109.2 | 89.5 | 109.0 | 102.4 |
| 0960 | Flooring | 113.4 | 126.9 | 117.3 | 114.5 | 119.6 | 116.0 | 98.2 | 126.9 | 106.5 | 93.1 | 149.9 | 109.4 | 101.8 | 114.0 | 105.3 | 112.2 | 119.6 | 114.4 |
| 0970, 0990 | Wall Finishes & Painting/Coating | 104.0 | 98.4 | 100.6 | 104.0 | 108.0 | 106.4 | 96.2 | 114.7 | 107.4 | 94.9 | 156.2 | 132.1 | 96.2 | 110.7 | 105.0 | 104.0 | 105.8 | 105.1 |
| 09 | FINISHES | 98.4 | 106.5 | 102.8 | 98.9 | 112.3 | 106.3 | 97.0 | 118.8 | 109.0 | 98.0 | 152.8 | 128.2 | 99.0 | 116.0 | 108.4 | 97.9 | 112.3 | 105.9 |
| COVERS | DIVS. 10 - 14, 25, 28, 41, 43, 44, 46 | 100.0 | 104.0 | 100.8 | 100.0 | 106.9 | 101.4 | 100.0 | 104.1 | 100.9 | 100.0 | 125.5 | 105.3 | 100.0 | 103.8 | 100.8 | 100.0 | 106.9 | 101.4 |
| 21, 22, 23 | FIRE SUPPRESSION, PLUMBING & HVAC | 94.6 | 98.9 | 96.4 | 94.6 | 93.3 | 94.1 | 94.7 | 104.4 | 98.7 | 100.1 | 136.5 | 115.3 | 100.1 | 100.3 | 100.2 | 100.0 | 95.4 | 98.1 |
| 26, 27, 3370 | ELECTRICAL, COMMUNICATIONS & UTIL. | 92.0 | 100.6 | 96.6 | 93.5 | 103.9 | 99.0 | 96.8 | 95.8 | 96.3 | 95.3 | 134.8 | 116.3 | 96.3 | 91.3 | 93.6 | 93.0 | 103.9 | 98.8 |
| MF2010 | WEIGHTED AVERAGE | 92.9 | 102.3 | 97.1 | 93.2 | 106.9 | 99.3 | 99.8 | 107.3 | 103.1 | 99.7 | 138.9 | 117.2 | 97.0 | 105.3 | 100.7 | 94.6 | 107.5 | 100.4 |

| | | ILLINOIS | | | | | | | | | | | | | | | | |
|---|---|---|---|---|---|---|---|---|---|---|---|---|---|---|---|---|---|---|
| | DIVISION | EFFINGHAM | | | GALESBURG | | | JOLIET | | | KANKAKEE | | | LA SALLE | | | NORTH SUBURBAN | | |
| | | 624 | | | 614 | | | 604 | | | 609 | | | 613 | | | 600 - 603 | | |
| | | MAT. | INST. | TOTAL | MAT. | INST. | TOTAL | MAT. | INST. | TOTAL | MAT. | INST. | TOTAL | MAT. | INST. | TOTAL | MAT. | INST. | TOTAL |
| 015433 | CONTRACTOR EQUIPMENT | | 101.7 | 101.7 | | 101.0 | 101.0 | | 91.5 | 91.5 | | 91.5 | 91.5 | | 101.0 | 101.0 | | 91.5 | 91.5 |
| 0241, 31 - 34 | SITE & INFRASTRUCTURE, DEMOLITION | 100.1 | 97.1 | 98.0 | 101.4 | 96.7 | 98.1 | 114.8 | 93.2 | 99.9 | 107.9 | 92.3 | 97.2 | 100.8 | 97.8 | 98.7 | 113.8 | 92.5 | 99.1 |
| 0310 | Concrete Forming & Accessories | 103.1 | 110.2 | 109.2 | 93.9 | 114.5 | 111.8 | 102.4 | 154.7 | 147.8 | 96.0 | 130.8 | 126.2 | 107.2 | 119.9 | 118.2 | 101.8 | 148.3 | 142.1 |
| 0320 | Concrete Reinforcing | 99.6 | 97.0 | 98.3 | 99.1 | 104.5 | 101.9 | 98.8 | 145.2 | 122.7 | 99.5 | 138.3 | 119.5 | 99.2 | 135.4 | 117.9 | 98.8 | 152.0 | 126.2 |
| 0330 | Cast-in-Place Concrete | 101.6 | 107.0 | 103.6 | 102.7 | 105.5 | 103.8 | 112.7 | 144.7 | 125.0 | 105.1 | 127.3 | 113.6 | 102.6 | 117.5 | 108.3 | 112.8 | 142.1 | 124.0 |
| 03 | CONCRETE | 98.8 | 106.7 | 102.6 | 99.5 | 109.6 | 104.3 | 104.5 | 148.2 | 125.4 | 98.4 | 127.3 | 113.6 | 100.2 | 122.1 | 110.7 | 104.5 | 145.7 | 124.2 |
| 04 | MASONRY | 80.2 | 106.7 | 96.2 | 114.0 | 116.2 | 115.3 | 98.7 | 149.5 | 129.5 | 95.5 | 129.0 | 115.8 | 114.0 | 119.6 | 117.4 | 96.0 | 145.6 | 126.0 |
| 05 | METALS | 94.9 | 104.5 | 98.3 | 95.0 | 109.1 | 99.9 | 93.3 | 126.3 | 104.7 | 93.3 | 121.8 | 103.1 | 95.1 | 127.6 | 106.3 | 94.6 | 128.6 | 106.3 |
| 06 | WOOD, PLASTICS & COMPOSITES | 100.3 | 111.1 | 106.5 | 90.6 | 112.8 | 103.4 | 101.8 | 155.4 | 132.6 | 94.8 | 130.5 | 115.3 | 104.7 | 118.1 | 112.4 | 100.2 | 149.4 | 128.4 |
| 07 | THERMAL & MOISTURE PROTECTION | 98.7 | 106.7 | 101.8 | 98.5 | 107.4 | 102.1 | 102.2 | 141.5 | 117.8 | 101.4 | 130.1 | 112.8 | 98.7 | 114.4 | 104.9 | 102.7 | 135.8 | 115.8 |
| 08 | OPENINGS | 90.8 | 107.2 | 94.9 | 90.3 | 106.5 | 94.4 | 98.9 | 152.6 | 112.5 | 90.4 | 137.4 | 102.3 | 90.3 | 121.6 | 98.2 | 98.9 | 151.2 | 112.2 |
| 0920 | Plaster & Gypsum Board | 98.7 | 111.3 | 107.6 | 94.3 | 113.1 | 107.6 | 93.2 | 157.0 | 138.1 | 90.6 | 131.4 | 119.3 | 101.0 | 118.6 | 113.4 | 96.3 | 150.8 | 134.7 |
| 0950, 0980 | Ceilings & Acoustic Treatment | 89.5 | 111.3 | 103.9 | 89.5 | 113.1 | 105.2 | 104.8 | 157.0 | 139.4 | 104.8 | 131.4 | 122.4 | 89.5 | 118.6 | 108.8 | 104.8 | 150.8 | 135.3 |
| 0960 | Flooring | 103.0 | 126.9 | 109.9 | 98.1 | 113.2 | 102.5 | 92.7 | 139.3 | 106.1 | 89.8 | 139.3 | 104.1 | 105.2 | 117.4 | 108.7 | 93.1 | 139.3 | 106.4 |
| 0970, 0990 | Wall Finishes & Painting/Coating | 96.2 | 103.5 | 100.6 | 96.2 | 95.9 | 96.0 | 93.0 | 146.1 | 125.3 | 93.0 | 122.3 | 110.8 | 96.2 | 122.3 | 112.0 | 94.9 | 146.1 | 126.0 |
| 09 | FINISHES | 98.3 | 113.6 | 106.8 | 96.3 | 113.0 | 105.5 | 97.4 | 152.3 | 127.7 | 95.7 | 129.4 | 114.3 | 99.2 | 118.6 | 109.9 | 97.9 | 147.7 | 125.4 |
| COVERS | DIVS. 10 - 14, 25, 28, 41, 43, 44, 46 | 100.0 | 74.6 | 94.7 | 100.0 | 104.3 | 100.9 | 100.0 | 124.8 | 105.2 | 100.0 | 116.3 | 103.4 | 100.0 | 104.4 | 100.9 | 100.0 | 121.2 | 104.4 |
| 21, 22, 23 | FIRE SUPPRESSION, PLUMBING & HVAC | 94.7 | 100.9 | 97.3 | 94.7 | 101.9 | 97.7 | 100.2 | 136.2 | 115.2 | 94.8 | 122.1 | 106.2 | 94.7 | 122.9 | 106.4 | 100.1 | 123.0 | 109.6 |
| 26, 27, 3370 | ELECTRICAL, COMMUNICATIONS & UTIL. | 94.0 | 100.6 | 97.5 | 94.2 | 89.6 | 91.8 | 94.6 | 138.5 | 117.9 | 89.2 | 137.6 | 114.9 | 91.1 | 137.6 | 115.8 | 94.4 | 121.2 | 108.6 |
| MF2010 | WEIGHTED AVERAGE | 95.1 | 103.6 | 98.9 | 96.8 | 104.8 | 100.4 | 99.2 | 137.7 | 116.4 | 95.2 | 125.3 | 108.7 | 96.9 | 121.4 | 107.8 | 99.3 | 131.0 | 113.4 |

| | | ILLINOIS | | | | | | | | | | | | | | | | |
|---|---|---|---|---|---|---|---|---|---|---|---|---|---|---|---|---|---|---|
| | DIVISION | PEORIA | | | QUINCY | | | ROCK ISLAND | | | ROCKFORD | | | SOUTH SUBURBAN | | | SPRINGFIELD | | |
| | | 615 - 616 | | | 623 | | | 612 | | | 610 - 611 | | | 605 | | | 626 - 627 | | |
| | | MAT. | INST. | TOTAL | MAT. | INST. | TOTAL | MAT. | INST. | TOTAL | MAT. | INST. | TOTAL | MAT. | INST. | TOTAL | MAT. | INST. | TOTAL |
| 015433 | CONTRACTOR EQUIPMENT | | 101.0 | 101.0 | | 101.7 | 101.7 | | 101.0 | 101.0 | | 101.0 | 101.0 | | 91.5 | 91.5 | | 101.7 | 101.7 |
| 0241, 31 - 34 | SITE & INFRASTRUCTURE, DEMOLITION | 101.5 | 96.6 | 98.1 | 98.8 | 97.0 | 97.6 | 99.4 | 96.1 | 97.1 | 100.5 | 97.9 | 98.7 | 113.8 | 92.5 | 99.1 | 100.2 | 97.6 | 98.4 |
| 0310 | Concrete Forming & Accessories | 96.4 | 115.0 | 112.5 | 100.9 | 112.5 | 110.9 | 99.4 | 100.3 | 99.6 | 101.1 | 129.5 | 125.8 | 101.8 | 148.3 | 142.1 | 99.2 | 116.5 | 114.3 |
| 0320 | Concrete Reinforcing | 96.5 | 104.6 | 100.7 | 99.2 | 101.3 | 100.3 | 99.1 | 101.3 | 100.2 | 91.2 | 140.2 | 116.4 | 98.8 | 152.0 | 126.2 | 96.5 | 102.4 | 99.5 |
| 0330 | Cast-in-Place Concrete | 99.6 | 114.8 | 105.4 | 101.7 | 98.4 | 100.5 | 100.5 | 98.1 | 99.6 | 102.0 | 116.4 | 107.5 | 112.8 | 142.1 | 124.0 | 91.6 | 108.4 | 98.1 |
| 03 | CONCRETE | 96.7 | 113.0 | 104.5 | 98.5 | 105.6 | 101.9 | 97.6 | 100.1 | 98.8 | 97.4 | 126.9 | 111.5 | 104.5 | 145.7 | 124.2 | 92.9 | 111.0 | 101.6 |
| 04 | MASONRY | 114.1 | 115.4 | 114.9 | 99.9 | 100.7 | 100.4 | 113.8 | 97.6 | 104.0 | 88.9 | 139.1 | 119.3 | 96.0 | 145.6 | 126.0 | 77.6 | 114.3 | 99.9 |
| 05 | METALS | 97.9 | 110.0 | 102.0 | 95.0 | 106.2 | 98.8 | 95.1 | 106.9 | 99.2 | 97.8 | 129.4 | 108.6 | 94.6 | 128.6 | 106.3 | 100.2 | 107.7 | 102.8 |
| 06 | WOOD, PLASTICS & COMPOSITES | 97.9 | 112.8 | 106.5 | 97.7 | 116.0 | 108.2 | 92.1 | 99.1 | 96.1 | 97.9 | 127.5 | 114.8 | 100.2 | 149.4 | 128.4 | 98.9 | 116.0 | 108.7 |
| 07 | THERMAL & MOISTURE PROTECTION | 99.2 | 111.5 | 104.1 | 98.7 | 103.2 | 100.5 | 98.5 | 98.1 | 98.4 | 101.7 | 129.6 | 112.8 | 102.7 | 135.8 | 115.8 | 98.9 | 111.6 | 103.9 |
| 08 | OPENINGS | 96.9 | 110.5 | 100.4 | 91.6 | 111.1 | 96.6 | 90.3 | 98.4 | 92.3 | 96.9 | 130.6 | 105.5 | 98.9 | 151.2 | 112.2 | 97.9 | 111.3 | 101.3 |
| 0920 | Plaster & Gypsum Board | 98.5 | 113.1 | 108.8 | 97.0 | 116.4 | 110.7 | 94.3 | 99.0 | 97.6 | 98.5 | 128.2 | 119.4 | 96.3 | 150.8 | 134.7 | 100.3 | 116.4 | 111.6 |
| 0950, 0980 | Ceilings & Acoustic Treatment | 95.1 | 113.1 | 107.0 | 89.5 | 116.4 | 107.3 | 89.5 | 99.0 | 95.8 | 95.1 | 128.2 | 117.0 | 104.8 | 150.8 | 135.3 | 94.2 | 116.4 | 108.9 |
| 0960 | Flooring | 101.8 | 113.2 | 105.1 | 101.8 | 104.9 | 102.7 | 99.3 | 95.5 | 98.2 | 101.8 | 117.4 | 106.3 | 93.1 | 139.3 | 106.4 | 101.8 | 108.4 | 103.7 |
| 0970, 0990 | Wall Finishes & Painting/Coating | 96.2 | 122.3 | 112.0 | 96.2 | 105.6 | 101.9 | 96.2 | 99.0 | 97.9 | 96.2 | 129.6 | 116.4 | 94.9 | 146.1 | 126.0 | 96.2 | 110.2 | 104.7 |
| 09 | FINISHES | 99.2 | 115.7 | 108.3 | 97.7 | 111.8 | 105.4 | 96.6 | 99.2 | 98.0 | 99.2 | 127.6 | 114.8 | 97.9 | 147.7 | 125.4 | 98.9 | 115.1 | 107.8 |
| COVERS | DIVS. 10 - 14, 25, 28, 41, 43, 44, 46 | 100.0 | 104.3 | 100.9 | 100.0 | 76.8 | 95.2 | 100.0 | 98.2 | 99.6 | 100.0 | 96.8 | 99.3 | 100.0 | 121.2 | 104.4 | 100.0 | 104.2 | 100.9 |
| 21, 22, 23 | FIRE SUPPRESSION, PLUMBING & HVAC | 100.1 | 104.2 | 101.8 | 94.7 | 96.3 | 95.3 | 94.7 | 98.4 | 96.3 | 100.1 | 114.3 | 106.0 | 100.1 | 123.0 | 109.6 | 99.9 | 102.5 | 101.0 |
| 26, 27, 3370 | ELECTRICAL, COMMUNICATIONS & UTIL. | 95.4 | 95.7 | 95.5 | 91.2 | 83.4 | 87.0 | 85.7 | 96.5 | 91.5 | 95.7 | 127.1 | 112.4 | 94.4 | 121.2 | 108.6 | 99.9 | 94.4 | 96.5 |
| MF2010 | WEIGHTED AVERAGE | 99.2 | 107.3 | 102.8 | 95.8 | 99.6 | 97.5 | 95.7 | 99.0 | 97.2 | 98.1 | 122.7 | 109.1 | 99.3 | 131.0 | 113.4 | 97.5 | 106.2 | 101.4 |

## INDIANA

| DIVISION | | ANDERSON 460 | | | BLOOMINGTON 474 | | | COLUMBUS 472 | | | EVANSVILLE 476 - 477 | | | FORT WAYNE 467 - 468 | | | GARY 463 - 464 | | |
|---|---|---|---|---|---|---|---|---|---|---|---|---|---|---|---|---|---|---|---|
| | | MAT. | INST. | TOTAL | MAT. | INST. | TOTAL | MAT. | INST. | TOTAL | MAT. | INST. | TOTAL | MAT. | INST. | TOTAL | MAT. | INST. | TOTAL |
| 015433 | CONTRACTOR EQUIPMENT | | 96.3 | 96.3 | | 87.3 | 87.3 | | 87.3 | 87.3 | | 117.4 | 117.4 | | 96.3 | 96.3 | | 96.3 | 96.3 |
| 0241, 31 - 34 | SITE & INFRASTRUCTURE, DEMOLITION | 89.3 | 94.5 | 92.9 | 77.4 | 94.7 | 89.4 | 74.4 | 94.5 | 88.3 | 82.2 | 125.1 | 111.8 | 90.6 | 94.4 | 93.2 | 89.9 | 98.1 | 95.5 |
| 0310 | Concrete Forming & Accessories | 93.4 | 79.7 | 81.5 | 100.1 | 80.2 | 82.8 | 94.5 | 77.9 | 80.1 | 94.0 | 79.8 | 81.7 | 92.0 | 73.8 | 76.2 | 93.5 | 111.5 | 109.1 |
| 0320 | Concrete Reinforcing | 93.9 | 84.0 | 86.9 | 90.0 | 80.1 | 84.9 | 90.4 | 80.1 | 85.1 | 98.6 | 77.7 | 87.8 | 93.9 | 79.0 | 86.2 | 93.9 | 104.8 | 99.5 |
| 0330 | Cast-in-Place Concrete | 101.6 | 80.4 | 93.5 | 98.2 | 76.8 | 89.1 | 97.7 | 75.3 | 89.1 | 93.8 | 86.7 | 91.1 | 108.0 | 76.4 | 95.9 | 106.1 | 108.8 | 107.2 |
| 03 | CONCRETE | 94.1 | 80.6 | 87.6 | 100.8 | 78.8 | 90.3 | 100.2 | 77.3 | 89.2 | 101.0 | 82.0 | 91.9 | 97.2 | 76.3 | 87.2 | 96.4 | 109.2 | 102.5 |
| 04 | MASONRY | 84.2 | 82.2 | 83.0 | 88.4 | 78.5 | 82.4 | 88.3 | 78.4 | 82.3 | 84.2 | 83.2 | 83.6 | 87.7 | 78.9 | 82.4 | 85.5 | 112.7 | 102.0 |
| 05 | METALS | 93.4 | 88.3 | 91.7 | 94.5 | 77.4 | 88.6 | 94.6 | 76.6 | 88.4 | 88.5 | 84.6 | 87.2 | 93.4 | 87.4 | 91.4 | 93.4 | 105.6 | 97.6 |
| 06 | WOOD, PLASTICS & COMPOSITES | 109.6 | 79.7 | 92.4 | 120.4 | 80.1 | 97.2 | 115.1 | 77.1 | 93.3 | 98.9 | 78.0 | 86.9 | 109.3 | 72.6 | 88.3 | 107.0 | 110.7 | 109.1 |
| 07 | THERMAL & MOISTURE PROTECTION | 99.4 | 79.5 | 91.5 | 94.2 | 81.4 | 89.1 | 93.9 | 81.6 | 89.0 | 98.1 | 84.0 | 92.5 | 99.2 | 81.6 | 92.3 | 97.3 | 114.0 | 101.5 |
| 08 | OPENINGS | 97.3 | 80.6 | 93.1 | 105.3 | 80.9 | 99.1 | 100.5 | 79.2 | 95.1 | 97.3 | 79.0 | 92.6 | 97.3 | 73.8 | 91.4 | 97.3 | 114.0 | 101.5 |
| 0920 | Plaster & Gypsum Board | 99.5 | 79.5 | 85.4 | 94.0 | 80.1 | 84.2 | 90.0 | 77.0 | 80.9 | 88.5 | 76.8 | 80.2 | 98.6 | 72.2 | 80.1 | 92.0 | 111.5 | 105.7 |
| 0950, 0980 | Ceilings & Acoustic Treatment | 94.3 | 79.5 | 84.5 | 80.5 | 80.1 | 80.2 | 80.5 | 77.0 | 78.2 | 85.6 | 76.8 | 79.8 | 94.3 | 72.2 | 79.7 | 94.3 | 111.5 | 105.7 |
| 0960 | Flooring | 97.0 | 82.0 | 92.7 | 104.8 | 79.2 | 97.5 | 99.8 | 79.2 | 93.9 | 98.6 | 85.2 | 94.7 | 97.0 | 79.2 | 91.9 | 97.0 | 122.5 | 104.4 |
| 0970, 0990 | Wall Finishes & Painting/Coating | 102.9 | 71.5 | 83.9 | 92.2 | 85.1 | 87.9 | 92.2 | 85.1 | 87.9 | 97.6 | 84.5 | 89.7 | 102.9 | 76.0 | 86.6 | 102.9 | 118.6 | 112.4 |
| 09 | FINISHES | 97.6 | 79.3 | 87.5 | 96.3 | 80.5 | 87.6 | 94.3 | 78.7 | 85.7 | 94.8 | 81.3 | 87.3 | 97.3 | 75.0 | 85.0 | 96.6 | 114.4 | 106.4 |
| COVERS | DIVS. 10 - 14, 25, 28, 41, 43, 44, 46 | 100.0 | 91.9 | 98.3 | 100.0 | 89.8 | 97.9 | 100.0 | 89.5 | 97.8 | 100.0 | 77.8 | 95.4 | 100.0 | 90.3 | 98.0 | 100.0 | 106.4 | 101.3 |
| 21, 22, 23 | FIRE SUPPRESSION, PLUMBING & HVAC | 99.9 | 75.7 | 89.8 | 99.7 | 79.4 | 91.2 | 94.3 | 79.9 | 88.3 | 99.9 | 80.9 | 92.0 | 99.9 | 76.0 | 90.0 | 99.9 | 105.9 | 102.4 |
| 26, 27, 3370 | ELECTRICAL, COMMUNICATIONS & UTIL. | 85.2 | 88.2 | 86.8 | 100.0 | 83.8 | 91.4 | 99.1 | 88.1 | 93.3 | 96.1 | 86.3 | 90.9 | 85.9 | 75.7 | 80.5 | 97.5 | 107.4 | 102.8 |
| MF2010 | WEIGHTED AVERAGE | 95.3 | 82.8 | 89.7 | 98.1 | 81.5 | 90.7 | 95.9 | 81.6 | 89.5 | 95.9 | 86.0 | 91.5 | 95.9 | 79.2 | 88.5 | 96.7 | 108.1 | 101.8 |

## INDIANA

| DIVISION | | INDIANAPOLIS 461 - 462 | | | KOKOMO 469 | | | LAFAYETTE 479 | | | LAWRENCEBURG 470 | | | MUNCIE 473 | | | NEW ALBANY 471 | | |
|---|---|---|---|---|---|---|---|---|---|---|---|---|---|---|---|---|---|---|---|
| | | MAT. | INST. | TOTAL | MAT. | INST. | TOTAL | MAT. | INST. | TOTAL | MAT. | INST. | TOTAL | MAT. | INST. | TOTAL | MAT. | INST. | TOTAL |
| 015433 | CONTRACTOR EQUIPMENT | | 91.8 | 91.8 | | 96.3 | 96.3 | | 87.3 | 87.3 | | 106.1 | 106.1 | | 96.1 | 96.1 | | 95.8 | 95.8 |
| 0241, 31 - 34 | SITE & INFRASTRUCTURE, DEMOLITION | 88.8 | 97.8 | 95.0 | 86.6 | 94.5 | 92.0 | 75.4 | 94.5 | 88.6 | 73.4 | 111.9 | 100.0 | 77.5 | 94.9 | 89.5 | 69.6 | 98.0 | 89.2 |
| 0310 | Concrete Forming & Accessories | 94.1 | 86.5 | 87.5 | 96.5 | 76.2 | 78.9 | 92.4 | 81.2 | 82.6 | 91.0 | 74.9 | 77.0 | 92.6 | 79.5 | 81.2 | 88.2 | 72.7 | 74.8 |
| 0320 | Concrete Reinforcing | 93.3 | 83.6 | 88.3 | 85.0 | 80.2 | 82.5 | 90.0 | 80.3 | 85.0 | 89.3 | 79.5 | 84.2 | 99.6 | 80.3 | 89.7 | 90.7 | 77.7 | 84.0 |
| 0330 | Cast-in-Place Concrete | 97.7 | 82.9 | 92.1 | 100.6 | 83.3 | 94.0 | 98.3 | 82.7 | 92.3 | 91.9 | 74.3 | 85.2 | 103.2 | 79.6 | 94.1 | 94.9 | 73.4 | 86.7 |
| 03 | CONCRETE | 95.7 | 84.4 | 90.3 | 91.1 | 80.0 | 85.8 | 100.4 | 81.3 | 91.2 | 93.6 | 76.2 | 85.2 | 99.9 | 80.2 | 90.5 | 98.7 | 74.3 | 87.0 |
| 04 | MASONRY | 91.7 | 84.6 | 87.4 | 83.9 | 80.0 | 81.6 | 93.5 | 82.0 | 86.6 | 73.7 | 72.0 | 72.7 | 90.2 | 82.2 | 85.3 | 80.6 | 71.5 | 75.1 |
| 05 | METALS | 95.6 | 80.6 | 90.4 | 89.9 | 88.0 | 89.3 | 93.0 | 77.2 | 87.6 | 89.9 | 86.2 | 88.6 | 96.4 | 88.5 | 93.7 | 91.7 | 81.9 | 88.3 |
| 06 | WOOD, PLASTICS & COMPOSITES | 106.3 | 86.9 | 95.2 | 112.7 | 74.7 | 90.9 | 112.4 | 81.1 | 94.4 | 97.7 | 74.0 | 84.1 | 114.7 | 79.5 | 94.5 | 99.4 | 73.1 | 84.3 |
| 07 | THERMAL & MOISTURE PROTECTION | 97.4 | 84.7 | 92.4 | 99.1 | 78.2 | 90.7 | 93.8 | 83.1 | 89.6 | 97.4 | 78.2 | 89.8 | 96.8 | 79.5 | 89.9 | 85.9 | 71.1 | 80.0 |
| 08 | OPENINGS | 105.3 | 85.4 | 100.3 | 91.6 | 77.9 | 88.2 | 98.8 | 81.4 | 94.4 | 98.8 | 76.6 | 93.1 | 100.1 | 80.5 | 95.1 | 96.8 | 75.8 | 91.5 |
| 0920 | Plaster & Gypsum Board | 97.2 | 86.8 | 89.9 | 104.8 | 74.4 | 83.4 | 87.2 | 81.2 | 83.0 | 74.0 | 73.8 | 73.8 | 88.5 | 79.5 | 82.2 | 86.8 | 72.6 | 76.8 |
| 0950, 0980 | Ceilings & Acoustic Treatment | 98.0 | 86.8 | 90.6 | 94.3 | 74.4 | 81.1 | 76.8 | 81.2 | 79.7 | 91.3 | 73.8 | 79.7 | 81.5 | 79.5 | 80.2 | 85.6 | 72.6 | 77.0 |
| 0960 | Flooring | 96.8 | 88.2 | 94.4 | 101.4 | 89.6 | 98.0 | 98.7 | 89.9 | 96.2 | 73.4 | 88.2 | 77.7 | 98.3 | 82.0 | 93.6 | 96.7 | 66.4 | 88.0 |
| 0970, 0990 | Wall Finishes & Painting/Coating | 102.9 | 85.1 | 92.1 | 102.9 | 77.2 | 87.3 | 92.2 | 82.3 | 86.2 | 93.2 | 68.6 | 78.3 | 92.2 | 71.5 | 79.6 | 97.6 | 69.5 | 80.6 |
| 09 | FINISHES | 98.4 | 87.0 | 92.1 | 99.3 | 78.5 | 87.9 | 92.9 | 82.9 | 87.4 | 86.6 | 77.0 | 81.3 | 93.6 | 79.1 | 85.6 | 94.3 | 71.3 | 81.6 |
| COVERS | DIVS. 10 - 14, 25, 28, 41, 43, 44, 46 | 100.0 | 94.0 | 98.8 | 100.0 | 89.7 | 97.9 | 100.0 | 91.7 | 98.3 | 100.0 | 47.1 | 89.0 | 100.0 | 91.4 | 98.2 | 100.0 | 46.6 | 88.9 |
| 21, 22, 23 | FIRE SUPPRESSION, PLUMBING & HVAC | 99.9 | 81.3 | 92.1 | 94.5 | 80.0 | 88.5 | 94.3 | 79.7 | 88.2 | 95.0 | 75.8 | 87.0 | 99.7 | 75.7 | 89.7 | 94.5 | 76.1 | 86.8 |
| 26, 27, 3370 | ELECTRICAL, COMMUNICATIONS & UTIL. | 99.1 | 88.2 | 93.3 | 90.1 | 79.0 | 84.2 | 98.5 | 84.5 | 91.0 | 93.8 | 76.1 | 84.4 | 91.5 | 80.8 | 85.8 | 94.3 | 78.0 | 85.6 |
| MF2010 | WEIGHTED AVERAGE | 98.3 | 85.7 | 92.7 | 93.1 | 81.7 | 88.0 | 95.6 | 82.8 | 89.9 | 92.5 | 78.9 | 86.4 | 96.8 | 81.7 | 90.1 | 93.6 | 76.4 | 85.9 |

## INDIANA / IOWA

| DIVISION | | SOUTH BEND 465 - 466 | | | TERRE HAUTE 478 | | | WASHINGTON 475 | | | BURLINGTON 526 | | | CARROLL 514 | | | CEDAR RAPIDS 522 - 524 | | |
|---|---|---|---|---|---|---|---|---|---|---|---|---|---|---|---|---|---|---|---|
| | | MAT. | INST. | TOTAL | MAT. | INST. | TOTAL | MAT. | INST. | TOTAL | MAT. | INST. | TOTAL | MAT. | INST. | TOTAL | MAT. | INST. | TOTAL |
| 015433 | CONTRACTOR EQUIPMENT | | 105.0 | 105.0 | | 117.4 | 117.4 | | 117.4 | 117.4 | | 99.8 | 99.8 | | 99.8 | 99.8 | | 96.4 | 96.4 |
| 0241, 31 - 34 | SITE & INFRASTRUCTURE, DEMOLITION | 91.5 | 94.9 | 93.9 | 83.9 | 125.3 | 112.5 | 84.1 | 123.2 | 111.1 | 98.3 | 96.7 | 97.2 | 86.2 | 96.8 | 93.5 | 100.3 | 96.1 | 97.4 |
| 0310 | Concrete Forming & Accessories | 95.9 | 78.0 | 80.4 | 94.9 | 81.6 | 83.3 | 95.8 | 78.7 | 80.9 | 95.4 | 78.8 | 81.0 | 85.0 | 53.8 | 57.9 | 100.2 | 83.4 | 85.7 |
| 0320 | Concrete Reinforcing | 93.9 | 77.0 | 85.2 | 98.6 | 80.5 | 89.3 | 91.2 | 52.5 | 71.2 | 95.9 | 83.6 | 89.6 | 96.7 | 83.0 | 89.7 | 96.7 | 86.1 | 91.3 |
| 0330 | Cast-in-Place Concrete | 98.8 | 80.5 | 91.8 | 90.8 | 88.6 | 89.9 | 99.8 | 88.4 | 94.8 | 109.1 | 57.9 | 89.5 | 109.1 | 63.0 | 84.3 | 109.4 | 82.4 | 99.1 |
| 03 | CONCRETE | 89.0 | 80.0 | 84.7 | 103.5 | 84.0 | 94.1 | 109.2 | 77.1 | 93.8 | 104.0 | 73.3 | 89.3 | 103.0 | 63.8 | 84.3 | 104.1 | 84.2 | 94.6 |
| 04 | MASONRY | 84.0 | 81.1 | 82.2 | 91.4 | 82.2 | 85.8 | 84.4 | 84.5 | 84.4 | 103.2 | 69.8 | 83.0 | 105.6 | 76.6 | 88.0 | 109.6 | 82.1 | 92.9 |
| 05 | METALS | 93.4 | 99.2 | 95.4 | 89.1 | 87.1 | 88.4 | 83.8 | 70.5 | 79.2 | 89.2 | 92.5 | 90.3 | 89.2 | 92.1 | 90.2 | 91.6 | 94.6 | 92.6 |
| 06 | WOOD, PLASTICS & COMPOSITES | 106.8 | 77.4 | 89.9 | 100.9 | 81.1 | 89.5 | 101.3 | 77.9 | 87.9 | 97.4 | 79.6 | 87.2 | 85.5 | 48.6 | 64.4 | 103.0 | 82.4 | 91.2 |
| 07 | THERMAL & MOISTURE PROTECTION | 93.8 | 83.6 | 89.8 | 98.2 | 83.1 | 92.2 | 98.3 | 84.2 | 92.7 | 105.7 | 78.4 | 94.9 | 106.1 | 72.9 | 92.9 | 106.6 | 82.1 | 96.9 |
| 08 | OPENINGS | 90.6 | 77.6 | 87.3 | 97.8 | 81.4 | 93.7 | 94.4 | 69.0 | 87.9 | 90.7 | 72.6 | 86.1 | 95.9 | 57.3 | 86.1 | 96.8 | 83.4 | 93.4 |
| 0920 | Plaster & Gypsum Board | 90.8 | 77.1 | 81.2 | 88.5 | 79.9 | 82.5 | 88.3 | 76.7 | 80.1 | 99.2 | 79.0 | 85.0 | 104.7 | 47.2 | 61.1 | 103.3 | 82.2 | 88.4 |
| 0950, 0980 | Ceilings & Acoustic Treatment | 88.7 | 77.1 | 81.0 | 85.6 | 79.9 | 81.8 | 80.0 | 76.7 | 77.8 | 104.7 | 79.0 | 87.7 | 104.7 | 47.2 | 66.5 | 108.4 | 82.2 | 91.0 |
| 0960 | Flooring | 95.4 | 84.1 | 92.1 | 98.6 | 88.2 | 95.6 | 99.6 | 80.7 | 94.2 | 116.1 | 39.9 | 94.2 | 109.0 | 35.1 | 87.7 | 134.7 | 82.7 | 119.8 |
| 0970, 0990 | Wall Finishes & Painting/Coating | 102.4 | 85.7 | 92.3 | 97.6 | 87.0 | 91.2 | 97.6 | 85.1 | 90.0 | 101.4 | 84.5 | 91.1 | 101.4 | 80.0 | 88.4 | 104.3 | 73.9 | 85.9 |
| 09 | FINISHES | 94.8 | 79.9 | 86.6 | 94.8 | 83.3 | 88.5 | 94.2 | 80.4 | 86.6 | 107.0 | 72.2 | 87.8 | 102.4 | 51.2 | 74.2 | 113.9 | 82.2 | 96.4 |
| COVERS | DIVS. 10 - 14, 25, 28, 41, 43, 44, 46 | 100.0 | 73.1 | 94.4 | 100.0 | 94.5 | 98.9 | 100.0 | 77.7 | 95.4 | 100.0 | 90.6 | 98.1 | 100.0 | 72.4 | 94.3 | 100.0 | 91.7 | 98.3 |
| 21, 22, 23 | FIRE SUPPRESSION, PLUMBING & HVAC | 99.8 | 78.0 | 90.7 | 99.9 | 80.0 | 91.6 | 94.5 | 80.1 | 88.5 | 94.8 | 79.7 | 88.5 | 94.8 | 72.7 | 85.6 | 100.2 | 83.8 | 93.4 |
| 26, 27, 3370 | ELECTRICAL, COMMUNICATIONS & UTIL. | 102.7 | 86.0 | 93.8 | 94.0 | 83.8 | 88.6 | 94.6 | 86.3 | 90.2 | 98.6 | 73.4 | 85.2 | 99.4 | 84.2 | 91.3 | 96.0 | 82.7 | 89.0 |
| MF2010 | WEIGHTED AVERAGE | 95.3 | 83.3 | 89.9 | 96.6 | 86.7 | 92.2 | 94.6 | 83.3 | 89.5 | 97.5 | 78.6 | 89.1 | 97.3 | 73.6 | 86.7 | 100.5 | 85.5 | 93.8 |

# City Cost Indexes

## IOWA

| DIVISION | | COUNCIL BLUFFS 515 | | | CRESTON 508 | | | DAVENPORT 527 - 528 | | | DECORAH 521 | | | DES MOINES 500 - 503,509 | | | DUBUQUE 520 | | |
|---|---|---|---|---|---|---|---|---|---|---|---|---|---|---|---|---|---|---|---|
| | | MAT. | INST. | TOTAL | MAT. | INST. | TOTAL | MAT. | INST. | TOTAL | MAT. | INST. | TOTAL | MAT. | INST. | TOTAL | MAT. | INST. | TOTAL |
| 015433 | CONTRACTOR EQUIPMENT | | 96.1 | 96.1 | | 99.8 | 99.8 | | 99.8 | 99.8 | | 99.8 | 99.8 | | 101.5 | 101.5 | | 95.1 | 95.1 |
| 0241, 31 - 34 | SITE & INFRASTRUCTURE, DEMOLITION | 108.5 | 92.7 | 97.6 | 86.2 | 96.4 | 93.2 | 97.9 | 99.1 | 98.7 | 96.8 | 96.2 | 96.4 | 87.1 | 100.1 | 96.1 | 98.1 | 93.2 | 94.7 |
| 0310 | Concrete Forming & Accessories | 84.5 | 73.5 | 75.0 | 85.5 | 70.2 | 72.2 | 99.8 | 89.9 | 91.2 | 93.3 | 47.4 | 53.5 | 101.1 | 80.1 | 82.8 | 85.6 | 72.4 | 74.2 |
| 0320 | Concrete Reinforcing | 98.6 | 79.1 | 88.5 | 96.1 | 83.1 | 89.4 | 96.7 | 97.2 | 97.0 | 95.9 | 82.6 | 89.1 | 96.7 | 83.3 | 89.8 | 95.3 | 86.0 | 90.5 |
| 0330 | Cast-in-Place Concrete | 113.8 | 78.4 | 100.2 | 108.8 | 67.2 | 92.9 | 105.4 | 88.9 | 99.1 | 106.2 | 59.5 | 88.3 | 91.7 | 92.9 | 92.2 | 107.1 | 90.0 | 100.6 |
| 03 | CONCRETE | 106.8 | 77.1 | 92.5 | 102.8 | 72.5 | 88.3 | 102.0 | 91.5 | 97.0 | 101.9 | 59.7 | 81.7 | 94.1 | 85.8 | 90.1 | 100.9 | 81.8 | 91.8 |
| 04 | MASONRY | 110.9 | 77.5 | 90.7 | 105.6 | 83.8 | 92.4 | 106.0 | 87.7 | 94.9 | 125.3 | 72.7 | 93.4 | 99.8 | 84.9 | 90.8 | 110.6 | 78.3 | 91.0 |
| 05 | METALS | 96.7 | 90.9 | 94.7 | 89.1 | 92.4 | 90.2 | 91.6 | 102.4 | 95.3 | 89.3 | 90.2 | 89.6 | 91.7 | 94.7 | 92.8 | 90.2 | 94.0 | 91.5 |
| 06 | WOOD, PLASTICS & COMPOSITES | 84.5 | 73.0 | 77.9 | 86.0 | 66.9 | 75.0 | 103.0 | 89.0 | 94.9 | 94.7 | 40.1 | 63.4 | 102.8 | 77.8 | 88.4 | 86.1 | 68.6 | 76.1 |
| 07 | THERMAL & MOISTURE PROTECTION | 106.1 | 71.1 | 92.2 | 106.8 | 80.6 | 96.4 | 106.1 | 87.7 | 98.8 | 105.9 | 55.8 | 86.0 | 107.5 | 81.7 | 97.2 | 106.4 | 73.6 | 93.4 |
| 08 | OPENINGS | 95.9 | 78.2 | 91.4 | 108.5 | 67.5 | 98.1 | 96.8 | 90.8 | 95.3 | 94.2 | 52.6 | 83.7 | 96.8 | 83.8 | 93.5 | 95.9 | 77.7 | 91.3 |
| 0920 | Plaster & Gypsum Board | 94.3 | 72.4 | 78.9 | 94.3 | 66.0 | 74.4 | 103.3 | 88.7 | 93.0 | 97.8 | 38.4 | 56.0 | 95.4 | 77.2 | 82.6 | 94.7 | 68.0 | 75.9 |
| 0950, 0980 | Ceilings & Acoustic Treatment | 104.7 | 84.3 | 83.3 | 104.7 | 66.0 | 79.0 | 108.4 | 88.7 | 95.3 | 104.7 | 38.4 | 60.7 | 105.6 | 77.2 | 86.7 | 104.7 | 68.0 | 80.3 |
| 0960 | Flooring | 108.0 | 85.2 | 101.5 | 109.2 | 35.1 | 87.9 | 118.9 | 87.7 | 110.0 | 115.6 | 49.8 | 96.7 | 118.6 | 85.3 | 109.0 | 123.6 | 80.5 | 111.2 |
| 0970, 0990 | Wall Finishes & Painting/Coating | 98.4 | 69.0 | 80.5 | 101.4 | 80.0 | 88.4 | 101.4 | 90.5 | 94.7 | 101.4 | 35.4 | 61.4 | 101.4 | 84.2 | 91.0 | 103.3 | 68.6 | 82.3 |
| 09 | FINISHES | 103.6 | 74.9 | 87.8 | 102.5 | 63.8 | 81.1 | 109.1 | 89.4 | 98.2 | 106.6 | 44.8 | 72.5 | 106.0 | 81.0 | 92.2 | 108.5 | 72.6 | 88.7 |
| COVERS | DIVS. 10 - 14, 25, 28, 41, 43, 44, 46 | 100.0 | 84.8 | 96.8 | 100.0 | 77.0 | 95.2 | 100.0 | 93.7 | 98.7 | 100.0 | 85.0 | 96.9 | 100.0 | 91.6 | 98.3 | 100.0 | 89.6 | 97.8 |
| 21, 22, 23 | FIRE SUPPRESSION, PLUMBING & HVAC | 100.2 | 76.3 | 90.2 | 94.8 | 82.6 | 89.7 | 100.2 | 91.7 | 96.7 | 94.8 | 75.0 | 86.5 | 100.0 | 83.2 | 93.0 | 100.2 | 78.2 | 91.0 |
| 26, 27, 3370 | ELECTRICAL, COMMUNICATIONS & UTIL. | 101.7 | 83.8 | 92.2 | 93.8 | 84.2 | 88.7 | 93.8 | 88.5 | 91.0 | 96.0 | 48.1 | 70.6 | 107.0 | 84.2 | 94.9 | 100.2 | 80.8 | 89.9 |
| MF2010 | WEIGHTED AVERAGE | 101.4 | 80.3 | 91.9 | 98.0 | 80.2 | 90.0 | 99.4 | 92.0 | 96.1 | 98.4 | 66.7 | 84.3 | 98.9 | 86.2 | 93.2 | 99.7 | 81.2 | 91.4 |

## IOWA

| DIVISION | | FORT DODGE 505 | | | MASON CITY 504 | | | OTTUMWA 525 | | | SHENANDOAH 516 | | | SIBLEY 512 | | | SIOUX CITY 510 - 511 | | |
|---|---|---|---|---|---|---|---|---|---|---|---|---|---|---|---|---|---|---|---|
| | | MAT. | INST. | TOTAL | MAT. | INST. | TOTAL | MAT. | INST. | TOTAL | MAT. | INST. | TOTAL | MAT. | INST. | TOTAL | MAT. | INST. | TOTAL |
| 015433 | CONTRACTOR EQUIPMENT | | 99.8 | 99.8 | | 99.8 | 99.8 | | 95.1 | 95.1 | | 96.1 | 96.1 | | 99.8 | 99.8 | | 99.8 | 99.8 |
| 0241, 31 - 34 | SITE & INFRASTRUCTURE, DEMOLITION | 93.7 | 95.1 | 94.7 | 93.7 | 96.3 | 95.5 | 99.0 | 91.3 | 93.7 | 106.7 | 91.7 | 96.3 | 108.9 | 95.0 | 99.3 | 110.0 | 95.9 | 100.2 |
| 0310 | Concrete Forming & Accessories | 86.0 | 48.5 | 53.4 | 90.0 | 48.2 | 53.7 | 91.6 | 60.6 | 64.7 | 85.9 | 60.6 | 64.0 | 86.3 | 40.4 | 46.5 | 100.2 | 67.2 | 71.6 |
| 0320 | Concrete Reinforcing | 96.1 | 65.4 | 80.3 | 95.9 | 82.5 | 89.0 | 95.9 | 83.9 | 89.7 | 98.6 | 69.7 | 83.7 | 98.6 | 65.3 | 81.5 | 96.7 | 65.7 | 80.7 |
| 0330 | Cast-in-Place Concrete | 102.1 | 46.6 | 80.8 | 102.1 | 60.2 | 86.1 | 109.8 | 57.3 | 89.7 | 110.0 | 63.3 | 92.1 | 107.8 | 47.1 | 84.5 | 108.5 | 58.3 | 89.3 |
| 03 | CONCRETE | 98.0 | 52.5 | 76.2 | 98.3 | 60.2 | 80.1 | 103.7 | 65.1 | 85.2 | 104.1 | 64.4 | 85.1 | 103.0 | 49.0 | 77.2 | 103.4 | 64.9 | 84.9 |
| 04 | MASONRY | 103.8 | 40.9 | 65.7 | 117.9 | 71.5 | 89.7 | 106.9 | 60.7 | 78.9 | 110.5 | 77.5 | 90.5 | 131.1 | 41.2 | 76.6 | 103.0 | 57.6 | 75.5 |
| 05 | METALS | 89.2 | 81.7 | 86.6 | 89.2 | 90.6 | 89.7 | 89.1 | 92.1 | 90.1 | 95.7 | 85.5 | 92.2 | 89.4 | 80.4 | 86.3 | 91.6 | 83.2 | 88.7 |
| 06 | WOOD, PLASTICS & COMPOSITES | 86.5 | 48.9 | 65.0 | 90.6 | 40.1 | 61.6 | 92.5 | 62.1 | 75.0 | 86.1 | 57.2 | 69.6 | 86.8 | 38.8 | 59.3 | 103.0 | 67.3 | 82.5 |
| 07 | THERMAL & MOISTURE PROTECTION | 106.2 | 46.7 | 88.1 | 105.7 | 66.3 | 90.0 | 106.5 | 65.8 | 90.4 | 105.4 | 69.3 | 91.1 | 105.7 | 51.4 | 84.2 | 106.1 | 66.4 | 90.4 |
| 08 | OPENINGS | 101.1 | 51.4 | 88.5 | 91.7 | 52.5 | 81.8 | 95.8 | 67.6 | 88.7 | 85.6 | 58.1 | 78.6 | 92.0 | 45.9 | 80.3 | 96.8 | 63.2 | 88.3 |
| 0920 | Plaster & Gypsum Board | 94.7 | 47.5 | 61.5 | 95.2 | 38.4 | 55.2 | 96.1 | 61.2 | 71.5 | 94.7 | 56.3 | 67.6 | 94.7 | 37.0 | 54.1 | 103.3 | 66.4 | 77.3 |
| 0950, 0980 | Ceilings & Acoustic Treatment | 104.7 | 47.5 | 66.7 | 104.7 | 38.4 | 60.7 | 104.7 | 61.2 | 75.9 | 104.7 | 56.3 | 72.6 | 104.7 | 37.0 | 59.8 | 108.4 | 66.4 | 80.5 |
| 0960 | Flooring | 111.0 | 49.8 | 93.4 | 113.8 | 49.8 | 95.4 | 127.4 | 53.0 | 106.0 | 108.9 | 36.2 | 88.0 | 110.0 | 35.9 | 88.7 | 119.3 | 56.7 | 101.3 |
| 0970, 0990 | Wall Finishes & Painting/Coating | 101.4 | 63.3 | 78.3 | 101.4 | 32.4 | 59.5 | 103.3 | 82.9 | 90.9 | 98.4 | 69.0 | 80.5 | 101.4 | 63.3 | 78.3 | 101.8 | 63.3 | 78.5 |
| 09 | FINISHES | 104.6 | 49.3 | 74.1 | 105.5 | 44.6 | 71.9 | 109.9 | 60.6 | 82.7 | 103.8 | 55.8 | 77.3 | 106.4 | 40.6 | 70.1 | 111.0 | 64.7 | 85.5 |
| COVERS | DIVS. 10 - 14, 25, 28, 41, 43, 44, 46 | 100.0 | 80.7 | 96.0 | 100.0 | 85.4 | 97.0 | 100.0 | 83.1 | 96.5 | 100.0 | 73.7 | 94.5 | 100.0 | 79.7 | 95.8 | 100.0 | 87.4 | 97.4 |
| 21, 22, 23 | FIRE SUPPRESSION, PLUMBING & HVAC | 94.8 | 68.4 | 83.8 | 94.8 | 76.1 | 87.0 | 94.8 | 74.2 | 86.2 | 94.8 | 73.8 | 86.0 | 94.8 | 64.2 | 82.0 | 100.2 | 70.9 | 88.0 |
| 26, 27, 3370 | ELECTRICAL, COMMUNICATIONS & UTIL. | 101.2 | 45.5 | 71.6 | 100.2 | 60.4 | 79.0 | 98.4 | 69.4 | 83.0 | 96.0 | 83.8 | 89.5 | 96.0 | 45.0 | 68.9 | 96.0 | 76.0 | 85.4 |
| MF2010 | WEIGHTED AVERAGE | 97.7 | 60.6 | 81.1 | 97.5 | 68.9 | 84.7 | 98.4 | 71.9 | 86.6 | 97.9 | 73.6 | 87.1 | 98.9 | 57.3 | 80.3 | 100.1 | 71.9 | 87.5 |

## IOWA / KANSAS

| DIVISION | | IOWA SPENCER 513 | | | IOWA WATERLOO 506 - 507 | | | KANSAS BELLEVILLE 669 | | | KANSAS COLBY 677 | | | KANSAS DODGE CITY 678 | | | KANSAS EMPORIA 668 | | |
|---|---|---|---|---|---|---|---|---|---|---|---|---|---|---|---|---|---|---|---|
| | | MAT. | INST. | TOTAL | MAT. | INST. | TOTAL | MAT. | INST. | TOTAL | MAT. | INST. | TOTAL | MAT. | INST. | TOTAL | MAT. | INST. | TOTAL |
| 015433 | CONTRACTOR EQUIPMENT | | 99.8 | 99.8 | | 99.8 | 99.8 | | 102.7 | 102.7 | | 102.7 | 102.7 | | 102.7 | 102.7 | | 101.0 | 101.0 |
| 0241, 31 - 34 | SITE & INFRASTRUCTURE, DEMOLITION | 109.0 | 95.0 | 99.3 | 98.4 | 96.3 | 96.9 | 116.8 | 93.7 | 100.9 | 114.1 | 93.7 | 100.0 | 116.3 | 93.3 | 100.5 | 107.6 | 91.5 | 96.5 |
| 0310 | Concrete Forming & Accessories | 91.5 | 40.4 | 47.2 | 100.6 | 50.7 | 57.3 | 96.9 | 57.8 | 63.0 | 103.1 | 62.4 | 67.8 | 96.4 | 62.3 | 66.8 | 88.7 | 70.0 | 72.4 |
| 0320 | Concrete Reinforcing | 98.6 | 65.3 | 81.5 | 96.7 | 85.8 | 91.1 | 104.5 | 56.4 | 79.7 | 107.1 | 56.4 | 81.0 | 104.5 | 56.5 | 79.8 | 103.0 | 57.0 | 79.3 |
| 0330 | Cast-in-Place Concrete | 107.8 | 47.1 | 84.5 | 109.4 | 57.4 | 89.5 | 120.8 | 58.8 | 97.1 | 116.2 | 58.8 | 94.2 | 118.2 | 58.4 | 95.3 | 116.8 | 49.0 | 90.8 |
| 03 | CONCRETE | 103.3 | 49.0 | 77.3 | 104.1 | 61.1 | 83.5 | 116.9 | 59.2 | 89.3 | 114.1 | 61.2 | 88.8 | 115.3 | 61.0 | 89.3 | 109.6 | 61.4 | 86.5 |
| 04 | MASONRY | 131.1 | 41.2 | 76.6 | 104.5 | 77.5 | 88.2 | 97.9 | 59.7 | 74.8 | 99.3 | 59.7 | 75.3 | 106.3 | 62.3 | 79.6 | 103.6 | 70.2 | 83.4 |
| 05 | METALS | 89.3 | 80.4 | 86.6 | 91.6 | 93.5 | 92.2 | 93.2 | 79.0 | 88.3 | 93.5 | 79.0 | 88.5 | 95.1 | 77.9 | 89.2 | 92.9 | 80.3 | 88.5 |
| 06 | WOOD, PLASTICS & COMPOSITES | 92.3 | 38.8 | 61.6 | 103.5 | 40.1 | 67.1 | 92.3 | 56.6 | 71.8 | 99.2 | 62.8 | 78.3 | 91.7 | 62.8 | 75.1 | 84.3 | 72.4 | 77.5 |
| 07 | THERMAL & MOISTURE PROTECTION | 106.6 | 51.4 | 84.7 | 105.9 | 73.5 | 93.0 | 99.9 | 60.2 | 84.1 | 99.9 | 60.8 | 84.4 | 99.9 | 61.8 | 84.7 | 98.5 | 75.6 | 89.4 |
| 08 | OPENINGS | 105.2 | 45.9 | 90.2 | 92.6 | 55.2 | 83.1 | 96.6 | 51.6 | 85.2 | 96.7 | 55.0 | 86.2 | 96.7 | 55.0 | 86.1 | 94.3 | 60.2 | 85.6 |
| 0920 | Plaster & Gypsum Board | 96.1 | 37.0 | 54.5 | 103.3 | 38.4 | 57.6 | 96.5 | 55.2 | 67.4 | 101.0 | 61.6 | 73.3 | 95.6 | 61.6 | 71.7 | 92.5 | 71.5 | 77.7 |
| 0950, 0980 | Ceilings & Acoustic Treatment | 104.7 | 37.0 | 59.8 | 108.4 | 38.4 | 62.0 | 89.5 | 55.2 | 66.7 | 89.5 | 61.6 | 71.0 | 89.5 | 61.6 | 71.0 | 89.5 | 71.5 | 77.5 |
| 0960 | Flooring | 113.3 | 35.9 | 91.1 | 120.7 | 79.7 | 108.9 | 101.4 | 41.4 | 84.2 | 105.1 | 41.4 | 86.8 | 101.2 | 41.4 | 84.0 | 96.3 | 38.5 | 79.7 |
| 0970, 0990 | Wall Finishes & Painting/Coating | 101.4 | 63.3 | 78.3 | 101.8 | 83.7 | 90.8 | 96.2 | 41.9 | 63.2 | 96.2 | 41.9 | 63.2 | 96.2 | 41.9 | 63.2 | 96.2 | 41.9 | 63.2 |
| 09 | FINISHES | 107.5 | 40.6 | 70.6 | 109.6 | 57.4 | 80.8 | 98.9 | 52.8 | 73.5 | 100.4 | 56.5 | 76.2 | 98.7 | 56.5 | 75.4 | 95.9 | 61.5 | 76.9 |
| COVERS | DIVS. 10 - 14, 25, 28, 41, 43, 44, 46 | 100.0 | 79.7 | 95.8 | 100.0 | 86.8 | 97.3 | 100.0 | 44.9 | 88.6 | 100.0 | 45.7 | 88.7 | 100.0 | 45.7 | 88.7 | 100.0 | 44.6 | 88.5 |
| 21, 22, 23 | FIRE SUPPRESSION, PLUMBING & HVAC | 94.8 | 64.2 | 82.0 | 100.2 | 83.1 | 93.1 | 94.7 | 72.4 | 85.4 | 94.7 | 69.0 | 84.0 | 100.1 | 69.0 | 87.1 | 94.7 | 72.7 | 85.5 |
| 26, 27, 3370 | ELECTRICAL, COMMUNICATIONS & UTIL. | 97.9 | 45.0 | 69.7 | 96.0 | 60.4 | 77.1 | 98.8 | 68.7 | 82.8 | 98.8 | 68.7 | 82.8 | 95.3 | 75.3 | 84.7 | 95.7 | 75.2 | 84.8 |
| MF2010 | WEIGHTED AVERAGE | 100.6 | 57.3 | 81.3 | 99.4 | 73.4 | 87.8 | 99.6 | 66.5 | 84.8 | 99.5 | 66.8 | 84.9 | 101.0 | 67.8 | 86.2 | 97.8 | 70.8 | 85.7 |

## KANSAS

| DIVISION | | FORT SCOTT 667 | | | HAYS 676 | | | HUTCHINSON 675 | | | INDEPENDENCE 673 | | | KANSAS CITY 660 - 662 | | | LIBERAL 679 | | |
|---|---|---|---|---|---|---|---|---|---|---|---|---|---|---|---|---|---|---|---|
| | | MAT. | INST. | TOTAL | MAT. | INST. | TOTAL | MAT. | INST. | TOTAL | MAT. | INST. | TOTAL | MAT. | INST. | TOTAL | MAT. | INST. | TOTAL |
| 015433 | CONTRACTOR EQUIPMENT | | 101.9 | 101.9 | | 102.7 | 102.7 | | 102.7 | 102.7 | | 102.7 | 102.7 | | 99.5 | 99.5 | | 102.7 | 102.7 |
| 0241, 31 - 34 | SITE & INFRASTRUCTURE, DEMOLITION | 103.9 | 92.0 | 95.7 | 120.5 | 93.7 | 102.0 | 96.5 | 93.8 | 94.7 | 117.9 | 93.9 | 101.3 | 96.3 | 92.2 | 93.4 | 120.4 | 93.8 | 102.1 |
| 0310 | Concrete Forming & Accessories | 103.7 | 82.5 | 85.3 | 100.6 | 62.4 | 67.5 | 91.1 | 62.2 | 66.1 | 111.5 | 71.0 | 76.3 | 100.1 | 99.3 | 99.4 | 96.9 | 62.3 | 66.8 |
| 0320 | Concrete Reinforcing | 102.3 | 90.4 | 96.2 | 104.5 | 56.4 | 79.7 | 104.5 | 56.4 | 79.7 | 103.9 | 61.4 | 82.0 | 99.3 | 95.8 | 97.5 | 106.0 | 56.4 | 80.5 |
| 0330 | Cast-in-Place Concrete | 108.3 | 57.5 | 88.8 | 91.6 | 58.8 | 79.0 | 84.8 | 48.4 | 70.9 | 118.7 | 48.6 | 91.9 | 92.8 | 98.5 | 95.0 | 91.6 | 48.4 | 75.1 |
| 03 | CONCRETE | 104.5 | 76.1 | 90.9 | 104.5 | 61.2 | 83.8 | 88.0 | 57.6 | 73.5 | 116.5 | 62.5 | 90.7 | 96.0 | 98.6 | 97.2 | 106.5 | 57.6 | 83.1 |
| 04 | MASONRY | 104.3 | 60.7 | 77.9 | 107.0 | 59.7 | 78.3 | 98.6 | 61.9 | 76.4 | 96.4 | 64.6 | 77.1 | 106.8 | 99.6 | 102.5 | 104.1 | 61.9 | 78.5 |
| 05 | METALS | 92.8 | 92.0 | 92.5 | 93.1 | 79.0 | 88.2 | 92.9 | 77.8 | 87.7 | 92.8 | 80.4 | 88.6 | 100.6 | 99.9 | 100.4 | 93.4 | 77.8 | 88.0 |
| 06 | WOOD, PLASTICS & COMPOSITES | 101.2 | 90.8 | 95.2 | 96.5 | 62.8 | 77.2 | 86.8 | 62.8 | 73.0 | 108.8 | 73.9 | 88.8 | 97.4 | 99.8 | 98.8 | 92.3 | 62.8 | 75.4 |
| 07 | THERMAL & MOISTURE PROTECTION | 99.2 | 72.1 | 88.5 | 100.2 | 60.8 | 84.5 | 98.9 | 61.7 | 84.1 | 99.9 | 74.2 | 89.7 | 98.6 | 99.3 | 98.9 | 100.3 | 61.7 | 85.0 |
| 08 | OPENINGS | 94.3 | 82.2 | 91.2 | 96.6 | 55.0 | 86.1 | 96.6 | 55.0 | 86.1 | 94.3 | 62.2 | 86.1 | 95.7 | 92.6 | 94.9 | 96.7 | 55.0 | 86.1 |
| 0920 | Plaster & Gypsum Board | 99.2 | 90.4 | 93.0 | 99.2 | 61.6 | 72.7 | 94.3 | 61.6 | 71.3 | 107.6 | 73.0 | 83.2 | 91.2 | 99.7 | 97.2 | 96.5 | 61.6 | 71.9 |
| 0950, 0980 | Ceilings & Acoustic Treatment | 89.5 | 90.4 | 90.1 | 89.5 | 61.6 | 71.0 | 89.5 | 61.6 | 71.0 | 89.5 | 73.0 | 78.6 | 89.5 | 99.7 | 96.3 | 89.5 | 61.6 | 71.0 |
| 0960 | Flooring | 112.4 | 41.6 | 92.1 | 103.9 | 41.4 | 85.9 | 98.1 | 41.4 | 81.8 | 109.8 | 41.4 | 90.1 | 90.0 | 99.2 | 92.7 | 101.4 | 41.4 | 84.2 |
| 0970, 0990 | Wall Finishes & Painting/Coating | 98.1 | 41.9 | 64.0 | 96.2 | 41.9 | 63.2 | 96.2 | 41.9 | 63.2 | 96.2 | 41.9 | 63.2 | 104.0 | 72.9 | 85.1 | 96.2 | 41.9 | 63.2 |
| 09 | FINISHES | 101.7 | 72.1 | 85.3 | 100.4 | 56.5 | 76.2 | 95.9 | 56.5 | 74.2 | 102.7 | 63.0 | 80.8 | 96.2 | 95.6 | 95.9 | 99.6 | 56.5 | 75.8 |
| COVERS | DIVS. 10 - 14, 25, 28, 41, 43, 44, 46 | 100.0 | 51.4 | 89.9 | 100.0 | 45.7 | 88.7 | 100.0 | 45.7 | 88.7 | 100.0 | 47.0 | 89.0 | 100.0 | 68.8 | 93.5 | 100.0 | 45.7 | 88.7 |
| 21, 22, 23 | FIRE SUPPRESSION, PLUMBING & HVAC | 94.7 | 70.7 | 84.7 | 94.7 | 69.0 | 84.0 | 94.7 | 69.0 | 84.0 | 94.7 | 71.0 | 84.8 | 100.0 | 101.3 | 100.6 | 94.7 | 69.0 | 84.0 |
| 26, 27, 3370 | ELECTRICAL, COMMUNICATIONS & UTIL. | 94.9 | 75.2 | 84.4 | 97.6 | 68.7 | 82.3 | 92.0 | 68.7 | 79.6 | 94.5 | 75.1 | 84.2 | 101.4 | 98.0 | 99.6 | 95.3 | 71.1 | 82.5 |
| MF2010 | WEIGHTED AVERAGE | 97.6 | 75.2 | 87.6 | 98.7 | 66.8 | 84.4 | 94.5 | 66.4 | 82.0 | 99.2 | 70.6 | 86.4 | 99.2 | 97.3 | 98.3 | 98.5 | 66.8 | 84.3 |

## KANSAS / KENTUCKY

| DIVISION | | SALINA 674 | | | TOPEKA 664 - 666 | | | WICHITA 670 - 672 | | | ASHLAND 411 - 412 | | | BOWLING GREEN 421 - 422 | | | CAMPTON 413 - 414 | | |
|---|---|---|---|---|---|---|---|---|---|---|---|---|---|---|---|---|---|---|---|
| | | MAT. | INST. | TOTAL | MAT. | INST. | TOTAL | MAT. | INST. | TOTAL | MAT. | INST. | TOTAL | MAT. | INST. | TOTAL | MAT. | INST. | TOTAL |
| 015433 | CONTRACTOR EQUIPMENT | | 102.7 | 102.7 | | 101.0 | 101.0 | | 102.7 | 102.7 | | 98.4 | 98.4 | | 95.8 | 95.8 | | 102.3 | 102.3 |
| 0241, 31 - 34 | SITE & INFRASTRUCTURE, DEMOLITION | 105.8 | 93.8 | 97.5 | 97.8 | 90.5 | 92.8 | 97.6 | 92.4 | 94.0 | 106.2 | 87.0 | 93.0 | 69.6 | 98.8 | 89.8 | 78.9 | 99.0 | 92.8 |
| 0310 | Concrete Forming & Accessories | 93.1 | 58.2 | 62.8 | 100.3 | 43.5 | 51.1 | 99.4 | 47.4 | 54.3 | 88.7 | 102.1 | 100.4 | 84.6 | 81.1 | 81.5 | 89.3 | 48.6 | 54.0 |
| 0320 | Concrete Reinforcing | 103.9 | 65.3 | 84.0 | 96.5 | 94.7 | 95.6 | 96.5 | 77.0 | 86.4 | 92.1 | 100.7 | 96.6 | 89.4 | 81.4 | 85.2 | 90.2 | 65.1 | 77.3 |
| 0330 | Cast-in-Place Concrete | 102.6 | 52.1 | 83.2 | 92.4 | 48.1 | 75.4 | 90.2 | 51.2 | 75.3 | 86.3 | 102.2 | 92.4 | 85.6 | 101.7 | 91.8 | 95.5 | 60.0 | 81.9 |
| 03 | CONCRETE | 102.2 | 58.7 | 81.4 | 93.3 | 56.4 | 75.7 | 92.2 | 55.7 | 74.7 | 94.5 | 102.4 | 98.3 | 92.7 | 88.3 | 90.6 | 96.8 | 56.6 | 77.5 |
| 04 | MASONRY | 121.3 | 59.7 | 84.0 | 100.6 | 58.0 | 74.8 | 94.4 | 54.3 | 70.1 | 89.9 | 105.3 | 99.3 | 91.3 | 74.1 | 80.9 | 89.8 | 40.7 | 60.0 |
| 05 | METALS | 94.9 | 82.8 | 90.7 | 100.2 | 95.8 | 98.7 | 100.2 | 84.7 | 94.9 | 90.9 | 108.2 | 96.8 | 92.5 | 83.2 | 89.3 | 91.7 | 74.1 | 85.6 |
| 06 | WOOD, PLASTICS & COMPOSITES | 88.2 | 56.6 | 70.1 | 93.5 | 40.3 | 63.0 | 95.0 | 46.1 | 66.9 | 80.5 | 101.5 | 92.5 | 94.6 | 80.5 | 86.5 | 93.2 | 47.1 | 66.8 |
| 07 | THERMAL & MOISTURE PROTECTION | 99.4 | 60.2 | 83.8 | 100.4 | 64.2 | 86.1 | 98.8 | 55.4 | 81.5 | 90.4 | 98.8 | 93.7 | 85.8 | 82.5 | 84.5 | 98.5 | 51.5 | 79.8 |
| 08 | OPENINGS | 96.6 | 52.5 | 85.4 | 96.9 | 56.8 | 86.8 | 96.9 | 54.1 | 86.1 | 95.7 | 96.7 | 95.9 | 96.8 | 76.2 | 91.6 | 98.5 | 56.9 | 87.9 |
| 0920 | Plaster & Gypsum Board | 94.3 | 55.2 | 66.7 | 93.4 | 38.5 | 54.7 | 93.2 | 44.4 | 58.8 | 60.6 | 101.6 | 89.5 | 82.8 | 80.2 | 80.9 | 82.8 | 44.9 | 56.1 |
| 0950, 0980 | Ceilings & Acoustic Treatment | 89.5 | 55.2 | 66.7 | 90.4 | 38.5 | 56.0 | 89.5 | 44.4 | 59.6 | 80.4 | 101.6 | 94.5 | 85.6 | 80.2 | 82.0 | 85.6 | 44.9 | 58.6 |
| 0960 | Flooring | 99.4 | 65.4 | 89.6 | 102.8 | 77.8 | 95.6 | 101.6 | 66.6 | 91.5 | 78.5 | 105.9 | 86.4 | 94.6 | 80.1 | 90.4 | 96.1 | 43.6 | 81.0 |
| 0970, 0990 | Wall Finishes & Painting/Coating | 96.2 | 41.9 | 63.2 | 96.2 | 56.6 | 72.1 | 96.2 | 51.6 | 69.1 | 99.8 | 90.7 | 94.3 | 97.6 | 71.3 | 81.7 | 97.6 | 37.2 | 61.0 |
| 09 | FINISHES | 97.2 | 57.5 | 75.3 | 97.6 | 49.5 | 71.1 | 97.2 | 50.8 | 71.6 | 82.7 | 101.9 | 93.3 | 93.1 | 79.9 | 85.8 | 93.7 | 45.3 | 67.0 |
| COVERS | DIVS. 10 - 14, 25, 28, 41, 43, 44, 46 | 100.0 | 56.0 | 90.9 | 100.0 | 52.7 | 90.2 | 100.0 | 52.0 | 90.0 | 100.0 | 102.6 | 100.6 | 100.0 | 63.2 | 92.3 | 100.0 | 54.5 | 90.5 |
| 21, 22, 23 | FIRE SUPPRESSION, PLUMBING & HVAC | 100.1 | 69.1 | 87.2 | 100.1 | 58.5 | 82.8 | 99.9 | 65.1 | 85.4 | 94.4 | 90.0 | 92.6 | 99.9 | 83.0 | 92.9 | 94.6 | 66.3 | 82.8 |
| 26, 27, 3370 | ELECTRICAL, COMMUNICATIONS & UTIL. | 95.0 | 75.3 | 84.5 | 105.4 | 75.2 | 89.3 | 102.1 | 75.3 | 87.8 | 91.8 | 99.4 | 95.8 | 94.7 | 82.5 | 88.2 | 91.7 | 71.8 | 81.1 |
| MF2010 | WEIGHTED AVERAGE | 99.7 | 68.0 | 85.5 | 99.2 | 65.2 | 84.0 | 98.3 | 65.1 | 83.4 | 93.2 | 98.5 | 95.6 | 94.7 | 82.8 | 89.4 | 94.4 | 62.6 | 80.2 |

## KENTUCKY

| DIVISION | | CORBIN 407 - 409 | | | COVINGTON 410 | | | ELIZABETHTOWN 427 | | | FRANKFORT 406 | | | HAZARD 417 - 418 | | | HENDERSON 424 | | |
|---|---|---|---|---|---|---|---|---|---|---|---|---|---|---|---|---|---|---|---|
| | | MAT. | INST. | TOTAL | MAT. | INST. | TOTAL | MAT. | INST. | TOTAL | MAT. | INST. | TOTAL | MAT. | INST. | TOTAL | MAT. | INST. | TOTAL |
| 015433 | CONTRACTOR EQUIPMENT | | 102.3 | 102.3 | | 106.1 | 106.1 | | 95.8 | 95.8 | | 102.3 | 102.3 | | 102.3 | 102.3 | | 117.4 | 117.4 |
| 0241, 31 - 34 | SITE & INFRASTRUCTURE, DEMOLITION | 76.4 | 99.0 | 92.0 | 74.7 | 114.2 | 102.0 | 65.0 | 98.7 | 88.2 | 80.9 | 99.9 | 94.0 | 76.9 | 100.6 | 93.3 | 73.6 | 125.7 | 109.5 |
| 0310 | Concrete Forming & Accessories | 86.5 | 48.2 | 53.3 | 85.3 | 94.8 | 96.7 | 80.2 | 80.1 | 80.1 | 96.6 | 59.0 | 64.0 | 86.5 | 42.0 | 47.9 | 92.3 | 77.9 | 79.8 |
| 0320 | Concrete Reinforcing | 89.8 | 65.0 | 77.1 | 88.8 | 97.3 | 93.2 | 89.8 | 91.5 | 90.7 | 91.1 | 79.7 | 85.2 | 90.7 | 65.8 | 77.8 | 89.5 | 88.7 | 89.1 |
| 0330 | Cast-in-Place Concrete | 90.4 | 59.8 | 78.7 | 91.4 | 96.3 | 93.3 | 77.4 | 75.3 | 76.6 | 87.4 | 66.9 | 79.6 | 91.9 | 87.6 | 90.2 | 75.7 | 90.6 | 81.4 |
| 03 | CONCRETE | 92.0 | 56.3 | 74.9 | 95.1 | 97.4 | 96.2 | 85.1 | 80.8 | 83.0 | 91.4 | 66.3 | 79.4 | 93.6 | 63.1 | 79.0 | 89.8 | 84.5 | 87.2 |
| 04 | MASONRY | 89.5 | 41.4 | 60.4 | 103.2 | 108.8 | 106.6 | 76.0 | 80.1 | 78.5 | 87.1 | 73.2 | 78.6 | 86.7 | 50.1 | 64.6 | 95.3 | 93.1 | 94.0 |
| 05 | METALS | 91.6 | 73.7 | 85.4 | 89.9 | 94.1 | 91.3 | 91.6 | 87.9 | 90.3 | 98.4 | 83.3 | 93.2 | 91.7 | 73.9 | 85.6 | 83.6 | 89.0 | 85.4 |
| 06 | WOOD, PLASTICS & COMPOSITES | 90.2 | 47.1 | 65.5 | 91.1 | 91.4 | 91.3 | 89.7 | 80.5 | 84.4 | 102.0 | 56.9 | 76.2 | 90.2 | 34.6 | 58.3 | 96.9 | 73.7 | 83.6 |
| 07 | THERMAL & MOISTURE PROTECTION | 98.3 | 52.6 | 80.1 | 97.6 | 99.6 | 98.4 | 85.5 | 79.3 | 83.0 | 98.1 | 70.0 | 87.0 | 98.4 | 51.0 | 79.5 | 92.4 | 95.7 | 93.7 |
| 08 | OPENINGS | 97.7 | 49.6 | 85.5 | 99.9 | 90.2 | 97.4 | 96.9 | 83.8 | 93.6 | 97.0 | 67.8 | 89.6 | 98.9 | 40.6 | 84.1 | 94.9 | 79.3 | 90.9 |
| 0920 | Plaster & Gypsum Board | 81.4 | 44.9 | 55.7 | 70.1 | 91.7 | 85.3 | 81.0 | 80.2 | 80.4 | 93.1 | 55.0 | 66.3 | 81.4 | 32.0 | 46.6 | 85.2 | 72.3 | 76.1 |
| 0950, 0980 | Ceilings & Acoustic Treatment | 85.6 | 44.9 | 58.6 | 88.5 | 91.7 | 90.6 | 85.6 | 80.2 | 82.0 | 87.5 | 55.0 | 66.0 | 85.6 | 32.0 | 50.1 | 80.0 | 72.3 | 74.9 |
| 0960 | Flooring | 94.4 | 43.6 | 79.8 | 71.0 | 88.9 | 76.1 | 92.0 | 70.8 | 85.9 | 98.9 | 57.5 | 87.0 | 94.4 | 44.3 | 80.0 | 97.8 | 90.9 | 95.8 |
| 0970, 0990 | Wall Finishes & Painting/Coating | 97.6 | 37.2 | 61.0 | 93.2 | 83.1 | 87.1 | 97.6 | 78.7 | 86.1 | 97.6 | 50.6 | 69.1 | 97.6 | 30.5 | 56.9 | 97.6 | 95.8 | 96.5 |
| 09 | FINISHES | 92.8 | 45.3 | 66.6 | 85.0 | 94.9 | 90.5 | 91.7 | 78.0 | 84.1 | 96.2 | 56.7 | 74.5 | 92.9 | 39.7 | 63.6 | 92.5 | 81.6 | 86.5 |
| COVERS | DIVS. 10 - 14, 25, 28, 41, 43, 44, 46 | 100.0 | 49.1 | 89.4 | 100.0 | 105.0 | 101.0 | 100.0 | 93.4 | 98.6 | 100.0 | 67.0 | 93.1 | 100.0 | 56.2 | 90.9 | 100.0 | 70.6 | 93.9 |
| 21, 22, 23 | FIRE SUPPRESSION, PLUMBING & HVAC | 94.5 | 65.8 | 82.5 | 95.0 | 93.0 | 94.2 | 94.9 | 81.5 | 89.3 | 100.0 | 67.3 | 86.4 | 94.6 | 43.1 | 73.2 | 94.9 | 76.3 | 87.1 |
| 26, 27, 3370 | ELECTRICAL, COMMUNICATIONS & UTIL. | 92.0 | 71.8 | 81.2 | 96.1 | 80.5 | 87.8 | 91.5 | 82.5 | 86.7 | 103.1 | 94.6 | 98.6 | 91.7 | 54.5 | 71.9 | 94.0 | 82.4 | 87.8 |
| MF2010 | WEIGHTED AVERAGE | 93.5 | 62.0 | 79.4 | 94.4 | 96.0 | 95.1 | 91.0 | 83.4 | 87.6 | 97.1 | 74.3 | 86.9 | 93.7 | 55.7 | 76.7 | 92.3 | 86.3 | 89.6 |

## KENTUCKY

| DIVISION | | LEXINGTON 403 - 405 | | | LOUISVILLE 400 - 402 | | | OWENSBORO 423 | | | PADUCAH 420 | | | PIKEVILLE 415 - 416 | | | SOMERSET 425 - 426 | | |
|---|---|---|---|---|---|---|---|---|---|---|---|---|---|---|---|---|---|---|---|---|
| | | MAT. | INST. | TOTAL | MAT. | INST. | TOTAL | MAT. | INST. | TOTAL | MAT. | INST. | TOTAL | MAT. | INST. | TOTAL | MAT. | INST. | TOTAL |
| 015433 | CONTRACTOR EQUIPMENT | | 102.3 | 102.3 | | 95.8 | 95.8 | | 117.4 | 117.4 | | 117.4 | 117.4 | | 98.4 | 98.4 | | 102.3 | 102.3 |
| 0241, 31 - 34 | SITE & INFRASTRUCTURE, DEMOLITION | 78.0 | 101.1 | 93.9 | 67.8 | 98.7 | 89.1 | 82.2 | 125.3 | 112.0 | 76.2 | 125.1 | 109.9 | 116.2 | 85.6 | 95.1 | 70.7 | 99.5 | 90.6 |
| 0310 | Concrete Forming & Accessories | 97.6 | 74.2 | 77.3 | 93.0 | 80.1 | 81.8 | 90.9 | 69.8 | 72.6 | 88.8 | 80.3 | 81.4 | 96.8 | 67.4 | 71.3 | 87.2 | 46.7 | 52.1 |
| 0320 | Concrete Reinforcing | 98.6 | 91.0 | 94.7 | 98.6 | 91.5 | 95.0 | 89.5 | 90.3 | 89.9 | 90.0 | 85.5 | 87.7 | 92.6 | 89.6 | 91.1 | 89.8 | 78.7 | 84.1 |
| 0330 | Cast-in-Place Concrete | 92.5 | 84.1 | 89.3 | 94.4 | 75.3 | 87.1 | 88.3 | 97.4 | 91.8 | 80.6 | 88.7 | 83.7 | 94.9 | 92.4 | 94.0 | 75.7 | 101.9 | 85.7 |
| 03 | CONCRETE | 95.1 | 81.1 | 88.4 | 95.8 | 80.8 | 88.6 | 101.5 | 83.5 | 92.9 | 94.1 | 84.3 | 89.4 | 107.6 | 81.4 | 95.1 | 80.4 | 72.5 | 76.6 |
| 04 | MASONRY | 88.3 | 63.7 | 73.4 | 89.4 | 80.1 | 83.8 | 88.2 | 69.3 | 76.7 | 91.2 | 90.5 | 90.8 | 88.0 | 103.0 | 97.1 | 82.2 | 60.8 | 69.2 |
| 05 | METALS | 94.0 | 88.3 | 92.0 | 100.7 | 87.9 | 96.3 | 85.1 | 88.3 | 86.2 | 82.0 | 86.9 | 83.7 | 90.8 | 99.1 | 93.6 | 91.6 | 80.3 | 87.7 |
| 06 | WOOD, PLASTICS & COMPOSITES | 104.6 | 74.6 | 87.4 | 102.0 | 80.5 | 89.7 | 95.0 | 66.2 | 78.5 | 92.7 | 79.4 | 85.1 | 89.4 | 59.7 | 72.3 | 90.8 | 39.9 | 61.6 |
| 07 | THERMAL & MOISTURE PROTECTION | 98.5 | 84.6 | 93.0 | 92.5 | 79.3 | 87.3 | 98.2 | 74.7 | 88.8 | 97.9 | 82.4 | 91.7 | 91.0 | 69.5 | 82.5 | 97.8 | 57.0 | 81.6 |
| 08 | OPENINGS | 97.8 | 77.9 | 92.8 | 97.8 | 83.8 | 94.3 | 94.9 | 76.7 | 90.3 | 94.0 | 78.5 | 90.1 | 96.5 | 68.4 | 89.4 | 97.7 | 56.3 | 87.3 |
| 0920 | Plaster & Gypsum Board | 91.4 | 73.2 | 78.6 | 84.4 | 80.2 | 81.4 | 83.9 | 64.5 | 70.2 | 82.5 | 78.2 | 79.5 | 64.2 | 58.5 | 60.2 | 81.9 | 37.5 | 50.6 |
| 0950, 0980 | Ceilings & Acoustic Treatment | 91.3 | 73.2 | 79.3 | 87.5 | 80.2 | 82.6 | 80.0 | 64.5 | 69.7 | 80.0 | 78.2 | 78.8 | 80.4 | 58.5 | 65.9 | 85.6 | 37.5 | 53.7 |
| 0960 | Flooring | 99.3 | 58.6 | 87.6 | 97.8 | 70.8 | 90.0 | 97.0 | 80.1 | 92.2 | 95.8 | 62.7 | 86.3 | 82.9 | 51.8 | 74.0 | 94.6 | 43.6 | 79.9 |
| 0970, 0990 | Wall Finishes & Painting/Coating | 97.6 | 67.6 | 79.4 | 97.6 | 78.7 | 86.1 | 97.6 | 95.6 | 96.4 | 97.6 | 80.6 | 87.3 | 99.8 | 87.4 | 92.3 | 97.6 | 51.3 | 69.5 |
| 09 | FINISHES | 96.7 | 70.0 | 82.0 | 94.1 | 78.0 | 85.2 | 92.7 | 73.6 | 82.2 | 91.8 | 77.0 | 83.6 | 85.3 | 65.4 | 74.3 | 92.3 | 44.8 | 66.1 |
| COVERS | DIVS. 10 - 14, 25, 28, 41, 43, 44, 46 | 100.0 | 93.4 | 98.6 | 100.0 | 93.4 | 98.6 | 100.0 | 79.8 | 95.8 | 100.0 | 67.4 | 93.2 | 100.0 | 59.0 | 91.5 | 100.0 | 58.3 | 91.3 |
| 21, 22, 23 | FIRE SUPPRESSION, PLUMBING & HVAC | 99.9 | 62.9 | 84.5 | 100.0 | 81.5 | 92.3 | 99.9 | 66.6 | 86.0 | 94.9 | 82.0 | 89.5 | 94.4 | 76.1 | 86.8 | 94.9 | 62.4 | 81.4 |
| 26, 27, 3370 | ELECTRICAL, COMMUNICATIONS & UTIL. | 95.2 | 71.3 | 82.5 | 106.2 | 82.5 | 93.6 | 94.1 | 82.4 | 87.9 | 96.8 | 82.3 | 89.0 | 95.2 | 79.6 | 86.9 | 92.2 | 71.2 | 81.0 |
| MF2010 | WEIGHTED AVERAGE | 96.3 | 75.3 | 86.9 | 97.7 | 83.4 | 91.3 | 95.1 | 80.2 | 88.4 | 92.6 | 86.0 | 89.7 | 95.7 | 80.3 | 88.8 | 91.6 | 66.6 | 80.4 |

## LOUISIANA

| DIVISION | | ALEXANDRIA 713 - 714 | | | BATON ROUGE 707 - 708 | | | HAMMOND 704 | | | LAFAYETTE 705 | | | LAKE CHARLES 706 | | | MONROE 712 | | |
|---|---|---|---|---|---|---|---|---|---|---|---|---|---|---|---|---|---|---|---|---|
| | | MAT. | INST. | TOTAL | MAT. | INST. | TOTAL | MAT. | INST. | TOTAL | MAT. | INST. | TOTAL | MAT. | INST. | TOTAL | MAT. | INST. | TOTAL |
| 015433 | CONTRACTOR EQUIPMENT | | 88.5 | 88.5 | | 88.3 | 88.3 | | 89.0 | 89.0 | | 89.0 | 89.0 | | 88.3 | 88.3 | | 88.5 | 88.5 |
| 0241, 31 - 34 | SITE & INFRASTRUCTURE, DEMOLITION | 100.0 | 85.4 | 90.0 | 108.0 | 85.1 | 92.2 | 109.7 | 86.4 | 93.6 | 110.3 | 86.3 | 93.7 | 110.9 | 84.9 | 93.0 | 100.0 | 84.8 | 89.5 |
| 0310 | Concrete Forming & Accessories | 82.6 | 36.8 | 42.9 | 94.2 | 63.1 | 67.2 | 79.1 | 49.3 | 53.2 | 95.6 | 54.1 | 59.6 | 97.1 | 58.4 | 63.5 | 82.1 | 37.4 | 43.3 |
| 0320 | Concrete Reinforcing | 102.0 | 58.3 | 79.5 | 100.6 | 60.1 | 79.7 | 99.2 | 61.9 | 80.0 | 100.6 | 59.7 | 79.5 | 100.6 | 59.1 | 79.2 | 100.8 | 58.1 | 78.8 |
| 0330 | Cast-in-Place Concrete | 97.2 | 39.4 | 75.1 | 95.1 | 58.4 | 81.0 | 97.9 | 45.2 | 77.7 | 97.4 | 45.7 | 77.6 | 102.5 | 67.6 | 89.2 | 97.2 | 56.6 | 81.6 |
| 03 | CONCRETE | 94.0 | 43.1 | 69.6 | 92.4 | 61.4 | 77.6 | 92.7 | 51.2 | 72.8 | 93.7 | 53.0 | 74.2 | 96.3 | 62.3 | 80.0 | 93.8 | 49.1 | 72.4 |
| 04 | MASONRY | 117.5 | 48.8 | 75.9 | 105.9 | 50.6 | 72.4 | 105.5 | 52.3 | 73.3 | 105.5 | 43.9 | 68.2 | 104.4 | 57.9 | 76.2 | 113.0 | 46.8 | 72.9 |
| 05 | METALS | 90.2 | 71.7 | 83.9 | 102.7 | 70.0 | 91.5 | 98.5 | 70.5 | 88.8 | 97.5 | 69.3 | 87.8 | 97.5 | 70.2 | 88.1 | 90.2 | 69.7 | 83.1 |
| 06 | WOOD, PLASTICS & COMPOSITES | 86.3 | 35.8 | 57.3 | 100.7 | 69.3 | 82.7 | 85.8 | 50.9 | 65.8 | 107.1 | 58.2 | 79.0 | 105.2 | 62.4 | 80.6 | 85.7 | 36.1 | 57.2 |
| 07 | THERMAL & MOISTURE PROTECTION | 97.0 | 47.9 | 77.5 | 95.7 | 57.2 | 80.4 | 97.5 | 56.8 | 81.3 | 98.0 | 52.0 | 79.7 | 97.3 | 59.2 | 82.1 | 97.0 | 50.7 | 78.6 |
| 08 | OPENINGS | 99.0 | 42.3 | 84.6 | 102.2 | 60.8 | 91.7 | 97.7 | 55.4 | 87.0 | 102.2 | 54.8 | 90.2 | 102.2 | 57.9 | 91.0 | 98.9 | 46.3 | 85.6 |
| 0920 | Plaster & Gypsum Board | 74.5 | 34.4 | 46.2 | 94.6 | 68.7 | 76.4 | 88.1 | 49.8 | 61.2 | 98.2 | 57.3 | 69.4 | 98.2 | 61.6 | 72.4 | 74.0 | 34.7 | 46.4 |
| 0950, 0980 | Ceilings & Acoustic Treatment | 89.4 | 34.4 | 52.9 | 103.7 | 68.7 | 80.5 | 100.7 | 49.8 | 67.0 | 98.9 | 57.3 | 71.3 | 99.9 | 61.6 | 74.5 | 89.4 | 34.7 | 53.2 |
| 0960 | Flooring | 100.0 | 44.7 | 89.9 | 103.9 | 64.7 | 92.6 | 94.8 | 48.9 | 81.6 | 104.1 | 36.6 | 84.7 | 104.1 | 77.0 | 96.3 | 99.6 | 60.3 | 88.3 |
| 0970, 0990 | Wall Finishes & Painting/Coating | 97.1 | 34.3 | 59.3 | 100.7 | 39.3 | 63.4 | 100.7 | 52.0 | 71.1 | 100.7 | 39.3 | 63.4 | 100.7 | 36.2 | 61.6 | 97.1 | 41.0 | 63.1 |
| 09 | FINISHES | 89.8 | 40.8 | 62.8 | 98.9 | 61.7 | 78.4 | 94.8 | 49.4 | 69.8 | 98.4 | 49.2 | 71.2 | 98.6 | 59.8 | 77.2 | 89.6 | 41.0 | 62.8 |
| COVERS | DIVS. 10 - 14, 25, 28, 41, 43, 44, 46 | 100.0 | 52.7 | 90.2 | 100.0 | 79.2 | 95.7 | 100.0 | 50.6 | 89.7 | 100.0 | 77.6 | 95.4 | 100.0 | 78.5 | 95.5 | 100.0 | 48.3 | 89.3 |
| 21, 22, 23 | FIRE SUPPRESSION, PLUMBING & HVAC | 100.2 | 52.3 | 80.3 | 100.0 | 60.4 | 83.5 | 94.7 | 46.9 | 74.8 | 100.1 | 60.6 | 83.7 | 100.1 | 61.1 | 83.9 | 100.2 | 51.3 | 79.8 |
| 26, 27, 3370 | ELECTRICAL, COMMUNICATIONS & UTIL. | 95.2 | 57.7 | 75.3 | 114.6 | 62.7 | 87.0 | 100.3 | 57.8 | 77.7 | 101.7 | 68.4 | 84.0 | 101.2 | 68.4 | 83.8 | 97.3 | 58.9 | 76.9 |
| MF2010 | WEIGHTED AVERAGE | 97.2 | 53.8 | 77.8 | 101.5 | 63.6 | 84.5 | 97.4 | 56.1 | 78.9 | 99.7 | 60.5 | 82.2 | 100.0 | 65.0 | 84.3 | 97.2 | 54.3 | 78.0 |

## LOUISIANA / MAINE

| DIVISION | | LOUISIANA NEW ORLEANS 700 - 701 | | | LOUISIANA SHREVEPORT 710 - 711 | | | LOUISIANA THIBODAUX 703 | | | MAINE AUGUSTA 043 | | | MAINE BANGOR 044 | | | MAINE BATH 045 | | |
|---|---|---|---|---|---|---|---|---|---|---|---|---|---|---|---|---|---|---|---|---|
| | | MAT. | INST. | TOTAL | MAT. | INST. | TOTAL | MAT. | INST. | TOTAL | MAT. | INST. | TOTAL | MAT. | INST. | TOTAL | MAT. | INST. | TOTAL |
| 015433 | CONTRACTOR EQUIPMENT | | 89.2 | 89.2 | | 88.5 | 88.5 | | 89.0 | 89.0 | | 100.3 | 100.3 | | 100.3 | 100.3 | | 100.3 | 100.3 |
| 0241, 31 - 34 | SITE & INFRASTRUCTURE, DEMOLITION | 113.7 | 88.0 | 95.9 | 102.6 | 84.7 | 90.2 | 112.4 | 86.3 | 94.4 | 82.5 | 100.6 | 95.0 | 84.7 | 98.4 | 94.1 | 83.1 | 100.6 | 95.2 |
| 0310 | Concrete Forming & Accessories | 98.9 | 64.1 | 68.7 | 96.7 | 39.5 | 47.1 | 91.7 | 62.9 | 66.7 | 95.2 | 99.3 | 98.7 | 90.8 | 89.7 | 89.8 | 87.0 | 99.2 | 97.6 |
| 0320 | Concrete Reinforcing | 100.6 | 62.2 | 80.8 | 100.4 | 58.6 | 78.9 | 99.2 | 61.9 | 80.0 | 88.3 | 99.8 | 94.3 | 88.0 | 99.9 | 94.1 | 87.0 | 99.7 | 93.5 |
| 0330 | Cast-in-Place Concrete | 101.5 | 71.0 | 89.8 | 96.7 | 45.3 | 77.0 | 105.1 | 48.4 | 83.4 | 89.7 | 62.6 | 79.3 | 78.6 | 107.0 | 89.5 | 78.6 | 62.5 | 72.5 |
| 03 | CONCRETE | 95.9 | 66.7 | 81.9 | 93.8 | 46.2 | 71.0 | 98.0 | 58.3 | 79.0 | 105.1 | 86.2 | 96.1 | 98.8 | 96.8 | 97.9 | 98.9 | 86.1 | 92.8 |
| 04 | MASONRY | 105.7 | 61.8 | 79.1 | 102.7 | 48.4 | 69.8 | 133.7 | 49.5 | 82.7 | 99.2 | 61.2 | 76.2 | 113.7 | 61.0 | 81.7 | 120.3 | 58.1 | 82.6 |
| 05 | METALS | 110.6 | 73.9 | 98.0 | 87.0 | 69.7 | 81.1 | 98.5 | 70.7 | 88.9 | 87.4 | 83.9 | 86.2 | 87.0 | 80.0 | 84.6 | 85.4 | 83.6 | 84.8 |
| 06 | WOOD, PLASTICS & COMPOSITES | 101.6 | 64.1 | 80.1 | 101.9 | 39.2 | 65.9 | 94.2 | 68.5 | 79.5 | 96.7 | 110.4 | 104.6 | 91.6 | 96.8 | 94.6 | 87.1 | 110.4 | 100.5 |
| 07 | THERMAL & MOISTURE PROTECTION | 97.7 | 62.2 | 83.6 | 96.2 | 49.6 | 77.7 | 97.7 | 55.2 | 80.8 | 96.1 | 64.8 | 83.7 | 96.6 | 66.2 | 84.6 | 96.6 | 64.0 | 83.7 |
| 08 | OPENINGS | 103.0 | 64.9 | 93.3 | 96.7 | 44.4 | 83.5 | 103.2 | 64.9 | 93.5 | 102.3 | 90.0 | 99.2 | 102.3 | 82.4 | 97.2 | 102.2 | 89.7 | 99.1 |
| 0920 | Plaster & Gypsum Board | 96.4 | 63.4 | 73.2 | 81.4 | 37.9 | 50.8 | 92.1 | 67.9 | 75.1 | 96.8 | 110.0 | 106.1 | 94.2 | 96.0 | 95.5 | 91.5 | 110.0 | 104.6 |
| 0950, 0980 | Ceilings & Acoustic Treatment | 98.9 | 63.4 | 75.4 | 90.3 | 37.9 | 55.6 | 100.7 | 67.9 | 79.0 | 91.9 | 110.0 | 103.9 | 91.9 | 96.0 | 94.7 | 89.8 | 110.0 | 103.2 |
| 0960 | Flooring | 104.5 | 64.7 | 93.1 | 108.4 | 64.7 | 95.9 | 101.2 | 46.6 | 85.5 | 95.8 | 58.4 | 85.1 | 93.8 | 64.9 | 85.5 | 92.0 | 58.4 | 82.3 |
| 0970, 0990 | Wall Finishes & Painting/Coating | 102.1 | 65.9 | 80.1 | 97.1 | 34.3 | 59.0 | 102.1 | 52.0 | 71.7 | 92.3 | 37.1 | 58.8 | 92.3 | 38.6 | 59.7 | 92.3 | 37.2 | 58.8 |
| 09 | FINISHES | 98.8 | 64.3 | 79.8 | 93.5 | 42.8 | 65.6 | 97.5 | 59.3 | 76.5 | 96.4 | 86.8 | 91.1 | 95.8 | 80.7 | 87.5 | 94.5 | 86.8 | 90.2 |
| COVERS | DIVS. 10 - 14, 25, 28, 41, 43, 44, 46 | 100.0 | 82.5 | 96.4 | 100.0 | 67.2 | 93.2 | 100.0 | 79.7 | 95.8 | 100.0 | 96.7 | 99.3 | 100.0 | 106.3 | 101.3 | 100.0 | 96.7 | 99.3 |
| 21, 22, 23 | FIRE SUPPRESSION, PLUMBING & HVAC | 100.0 | 66.0 | 85.8 | 100.0 | 56.3 | 81.8 | 94.7 | 61.0 | 80.7 | 100.0 | 66.8 | 86.2 | 100.2 | 62.7 | 84.6 | 94.8 | 66.8 | 83.2 |
| 26, 27, 3370 | ELECTRICAL, COMMUNICATIONS & UTIL. | 101.7 | 72.1 | 86.0 | 103.0 | 68.0 | 84.4 | 98.7 | 70.8 | 83.9 | 99.0 | 79.9 | 88.8 | 98.6 | 79.9 | 88.7 | 96.4 | 79.9 | 87.6 |
| MF2010 | WEIGHTED AVERAGE | 102.0 | 69.2 | 87.3 | 96.9 | 57.1 | 79.1 | 100.3 | 64.3 | 84.2 | 97.9 | 80.0 | 89.9 | 97.9 | 79.0 | 89.4 | 96.3 | 79.6 | 88.8 |

## MAINE

| DIVISION | | HOULTON 047 | | | KITTERY 039 | | | LEWISTON 042 | | | MACHIAS 046 | | | PORTLAND 040 - 041 | | | ROCKLAND 048 | | |
|---|---|---|---|---|---|---|---|---|---|---|---|---|---|---|---|---|---|---|---|
| | | MAT. | INST. | TOTAL | MAT. | INST. | TOTAL | MAT. | INST. | TOTAL | MAT. | INST. | TOTAL | MAT. | INST. | TOTAL | MAT. | INST. | TOTAL |
| 015433 | CONTRACTOR EQUIPMENT | | 100.3 | 100.3 | | 100.3 | 100.3 | | 100.3 | 100.3 | | 100.3 | 100.3 | | 100.3 | 100.3 | | 100.3 | 100.3 |
| 0241, 31 - 34 | SITE & INFRASTRUCTURE, DEMOLITION | 84.8 | 100.5 | 95.6 | 79.3 | 100.7 | 94.1 | 81.9 | 98.4 | 93.3 | 84.2 | 100.5 | 95.4 | 80.2 | 98.4 | 92.7 | 80.7 | 100.5 | 94.4 |
| 0310 | Concrete Forming & Accessories | 93.8 | 98.3 | 97.7 | 87.5 | 99.7 | 98.1 | 95.5 | 89.7 | 90.5 | 91.3 | 98.3 | 97.4 | 94.4 | 89.7 | 90.3 | 92.2 | 98.7 | 97.8 |
| 0320 | Concrete Reinforcing | 88.0 | 99.8 | 94.1 | 86.5 | 99.8 | 93.3 | 108.1 | 99.9 | 103.9 | 88.0 | 99.8 | 94.1 | 108.1 | 99.9 | 103.9 | 88.0 | 99.9 | 94.1 |
| 0330 | Cast-in-Place Concrete | 78.6 | 60.7 | 71.8 | 78.6 | 63.5 | 72.8 | 80.2 | 107.0 | 90.5 | 78.6 | 60.8 | 71.8 | 99.5 | 107.0 | 102.3 | 80.3 | 61.9 | 73.2 |
| 03 | CONCRETE | 99.8 | 85.2 | 92.8 | 94.0 | 86.7 | 90.5 | 98.5 | 96.8 | 97.7 | 99.3 | 85.2 | 92.5 | 108.1 | 96.8 | 102.7 | 96.1 | 85.7 | 91.1 |
| 04 | MASONRY | 97.0 | 59.2 | 74.1 | 110.3 | 60.8 | 80.3 | 97.6 | 61.0 | 75.4 | 97.0 | 59.5 | 74.3 | 97.7 | 61.0 | 75.5 | 91.4 | 61.0 | 73.0 |
| 05 | METALS | 85.6 | 83.8 | 85.0 | 85.3 | 83.8 | 84.8 | 90.1 | 80.0 | 86.6 | 85.6 | 83.7 | 85.0 | 91.4 | 80.0 | 87.5 | 85.5 | 83.9 | 84.9 |
| 06 | WOOD, PLASTICS & COMPOSITES | 95.5 | 110.4 | 104.0 | 87.5 | 110.4 | 100.6 | 97.4 | 96.8 | 97.1 | 92.5 | 110.4 | 102.8 | 96.6 | 96.8 | 96.7 | 93.4 | 110.4 | 103.1 |
| 07 | THERMAL & MOISTURE PROTECTION | 96.7 | 62.3 | 83.0 | 96.3 | 65.0 | 83.9 | 96.5 | 66.2 | 84.5 | 96.7 | 62.3 | 83.0 | 99.6 | 66.2 | 86.3 | 96.4 | 63.2 | 83.2 |
| 08 | OPENINGS | 102.3 | 88.0 | 98.7 | 102.2 | 89.3 | 98.9 | 105.6 | 82.4 | 99.7 | 102.3 | 88.0 | 98.7 | 105.6 | 82.4 | 99.7 | 102.3 | 88.0 | 98.7 |
| 0920 | Plaster & Gypsum Board | 96.8 | 110.0 | 106.1 | 91.5 | 110.0 | 104.6 | 100.2 | 96.0 | 97.3 | 95.0 | 110.0 | 105.6 | 101.0 | 96.0 | 97.5 | 95.0 | 110.0 | 105.6 |
| 0950, 0980 | Ceilings & Acoustic Treatment | 89.8 | 110.0 | 103.2 | 89.8 | 110.0 | 103.2 | 99.4 | 96.0 | 97.2 | 89.8 | 110.0 | 103.2 | 96.6 | 96.0 | 96.2 | 89.8 | 110.0 | 103.2 |
| 0960 | Flooring | 94.9 | 54.7 | 83.3 | 92.1 | 60.9 | 83.1 | 96.5 | 64.9 | 87.4 | 94.1 | 54.6 | 82.8 | 95.8 | 64.9 | 87.0 | 94.3 | 54.7 | 82.9 |
| 0970, 0990 | Wall Finishes & Painting/Coating | 92.3 | 38.6 | 59.7 | 92.3 | 37.2 | 58.8 | 92.3 | 38.6 | 59.7 | 92.3 | 38.6 | 59.7 | 92.3 | 38.6 | 59.7 | 92.3 | 38.6 | 59.7 |
| 09 | FINISHES | 96.0 | 85.8 | 90.4 | 94.1 | 87.5 | 90.5 | 98.5 | 80.7 | 88.7 | 95.5 | 85.8 | 90.2 | 97.4 | 80.7 | 88.2 | 95.2 | 86.0 | 90.1 |
| COVERS | DIVS. 10 - 14, 25, 28, 41, 43, 44, 46 | 100.0 | 96.0 | 99.2 | 100.0 | 98.4 | 99.7 | 100.0 | 106.3 | 101.3 | 100.0 | 90.8 | 98.1 | 100.0 | 106.3 | 101.3 | 100.0 | 96.3 | 99.2 |
| 21, 22, 23 | FIRE SUPPRESSION, PLUMBING & HVAC | 94.8 | 65.7 | 82.7 | 94.8 | 67.4 | 83.4 | 100.2 | 62.7 | 84.6 | 94.8 | 65.8 | 82.7 | 100.0 | 62.7 | 84.5 | 94.8 | 66.2 | 82.9 |
| 26, 27, 3370 | ELECTRICAL, COMMUNICATIONS & UTIL. | 100.9 | 79.9 | 89.7 | 96.4 | 79.9 | 87.6 | 100.9 | 79.9 | 89.7 | 100.9 | 79.9 | 89.7 | 104.5 | 79.9 | 91.4 | 100.8 | 79.9 | 89.7 |
| MF2010 | WEIGHTED AVERAGE | 95.9 | 79.1 | 88.4 | 95.0 | 80.2 | 88.4 | 98.2 | 79.0 | 89.6 | 95.8 | 78.9 | 88.2 | 99.8 | 79.0 | 90.5 | 94.9 | 79.5 | 88.0 |

|  |  | MAINE | | | MARYLAND | | | | | | | | | | | | | | |
|---|---|---|---|---|---|---|---|---|---|---|---|---|---|---|---|---|---|---|---|
| DIVISION | | WATERVILLE 049 | | | ANNAPOLIS 214 | | | BALTIMORE 210 - 212 | | | COLLEGE PARK 207 - 208 | | | CUMBERLAND 215 | | | EASTON 216 | | |
| | | MAT. | INST. | TOTAL | MAT. | INST. | TOTAL | MAT. | INST. | TOTAL | MAT. | INST. | TOTAL | MAT. | INST. | TOTAL | MAT. | INST. | TOTAL |
| 015433 | CONTRACTOR EQUIPMENT | | 100.3 | 100.3 | | 99.1 | 99.1 | | 102.6 | 102.6 | | 103.5 | 103.5 | | 99.1 | 99.1 | | 99.1 | 99.1 |
| 0241, 31 - 34 | SITE & INFRASTRUCTURE, DEMOLITION | 84.8 | 100.6 | 95.7 | 98.4 | 90.4 | 92.9 | 98.8 | 94.4 | 95.8 | 111.1 | 91.4 | 97.5 | 90.3 | 89.7 | 89.9 | 97.1 | 85.9 | 89.4 |
| 0310 | Concrete Forming & Accessories | 86.5 | 99.3 | 97.6 | 92.5 | 51.6 | 57.0 | 95.2 | 76.8 | 79.2 | 86.7 | 72.4 | 74.3 | 86.6 | 78.7 | 79.7 | 84.7 | 29.7 | 37.0 |
| 0320 | Concrete Reinforcing | 88.0 | 99.8 | 94.1 | 96.5 | 83.3 | 89.7 | 96.5 | 83.4 | 89.7 | 94.6 | 81.9 | 88.1 | 85.2 | 73.4 | 79.1 | 84.5 | 17.2 | 49.9 |
| 0330 | Cast-in-Place Concrete | 78.6 | 62.6 | 72.5 | 128.8 | 77.7 | 109.2 | 114.7 | 78.5 | 100.8 | 130.9 | 78.3 | 110.8 | 102.4 | 79.8 | 93.7 | 113.7 | 39.7 | 85.3 |
| 03 | CONCRETE | 100.6 | 86.2 | 93.7 | 115.6 | 68.1 | 92.9 | 108.6 | 79.6 | 94.7 | 116.2 | 77.4 | 97.6 | 97.0 | 79.0 | 88.4 | 105.6 | 32.8 | 70.7 |
| 04 | MASONRY | 107.4 | 61.2 | 79.4 | 96.1 | 69.8 | 80.2 | 97.3 | 69.8 | 80.7 | 116.4 | 70.8 | 88.8 | 95.0 | 78.8 | 85.2 | 109.1 | 31.3 | 61.9 |
| 05 | METALS | 85.5 | 83.9 | 85.0 | 92.3 | 95.0 | 93.2 | 93.4 | 95.8 | 94.2 | 81.3 | 96.1 | 86.4 | 90.0 | 90.7 | 90.3 | 90.2 | 59.0 | 79.5 |
| 06 | WOOD, PLASTICS & COMPOSITES | 86.6 | 110.4 | 100.2 | 93.5 | 44.3 | 65.3 | 96.3 | 78.2 | 85.9 | 82.3 | 72.7 | 76.8 | 86.7 | 78.0 | 81.7 | 84.3 | 29.7 | 53.0 |
| 07 | THERMAL & MOISTURE PROTECTION | 96.7 | 64.8 | 84.1 | 94.9 | 76.6 | 87.6 | 95.2 | 80.4 | 89.3 | 99.4 | 78.4 | 91.0 | 94.6 | 77.4 | 87.8 | 94.7 | 37.1 | 71.8 |
| 08 | OPENINGS | 102.3 | 90.0 | 99.2 | 95.4 | 64.7 | 87.6 | 97.7 | 83.0 | 94.0 | 98.3 | 76.5 | 92.8 | 96.6 | 78.4 | 92.0 | 94.6 | 26.8 | 77.4 |
| 0920 | Plaster & Gypsum Board | 91.5 | 110.0 | 104.6 | 96.9 | 43.0 | 59.0 | 98.5 | 77.8 | 83.9 | 95.4 | 71.9 | 78.9 | 93.2 | 77.7 | 82.2 | 91.8 | 27.9 | 46.8 |
| 0950, 0980 | Ceilings & Acoustic Treatment | 89.8 | 110.0 | 103.2 | 93.2 | 43.0 | 60.0 | 94.2 | 77.8 | 83.3 | 98.3 | 71.9 | 80.8 | 94.2 | 77.7 | 83.2 | 94.2 | 27.9 | 50.3 |
| 0960 | Flooring | 91.8 | 58.4 | 82.2 | 94.2 | 81.6 | 90.6 | 94.2 | 81.6 | 90.6 | 111.9 | 54.5 | 95.4 | 90.1 | 82.9 | 88.0 | 89.3 | 17.2 | 68.6 |
| 0970, 0990 | Wall Finishes & Painting/Coating | 92.3 | 37.1 | 58.8 | 98.5 | 85.3 | 90.5 | 98.5 | 85.3 | 90.5 | 121.0 | 80.4 | 96.4 | 98.5 | 79.0 | 86.6 | 98.5 | 21.9 | 52.0 |
| 09 | FINISHES | 94.6 | 86.8 | 90.3 | 94.9 | 58.1 | 74.6 | 95.5 | 78.1 | 85.9 | 102.8 | 69.3 | 84.3 | 93.0 | 79.4 | 85.5 | 93.0 | 27.1 | 56.7 |
| COVERS | DIVS. 10 - 14, 25, 28, 41, 43, 44, 46 | 100.0 | 96.7 | 99.3 | 100.0 | 83.0 | 96.5 | 100.0 | 87.2 | 97.3 | 100.0 | 57.8 | 91.2 | 100.0 | 89.5 | 97.8 | 100.0 | 39.1 | 87.3 |
| 21, 22, 23 | FIRE SUPPRESSION, PLUMBING & HVAC | 94.8 | 66.8 | 83.2 | 100.0 | 85.7 | 94.1 | 100.1 | 85.7 | 94.1 | 94.9 | 88.1 | 92.0 | 94.6 | 71.3 | 84.9 | 94.6 | 27.8 | 66.8 |
| 26, 27, 3370 | ELECTRICAL, COMMUNICATIONS & UTIL. | 100.9 | 79.9 | 89.7 | 96.4 | 91.2 | 93.6 | 97.3 | 93.7 | 95.4 | 100.9 | 97.2 | 98.9 | 94.9 | 82.2 | 88.2 | 94.4 | 62.5 | 77.4 |
| MF2010 | WEIGHTED AVERAGE | 96.3 | 80.0 | 89.0 | 99.1 | 78.5 | 89.9 | 98.9 | 84.8 | 92.6 | 99.2 | 83.0 | 92.0 | 94.7 | 80.1 | 88.1 | 96.4 | 41.9 | 72.0 |

|  |  | MARYLAND | | | | | | | | | | | | | | | MASSACHUSETTS | | |
|---|---|---|---|---|---|---|---|---|---|---|---|---|---|---|---|---|---|---|---|
| DIVISION | | ELKTON 219 | | | HAGERSTOWN 217 | | | SALISBURY 218 | | | SILVER SPRING 209 | | | WALDORF 206 | | | BOSTON 020 - 022, 024 | | |
| | | MAT. | INST. | TOTAL | MAT. | INST. | TOTAL | MAT. | INST. | TOTAL | MAT. | INST. | TOTAL | MAT. | INST. | TOTAL | MAT. | INST. | TOTAL |
| 015433 | CONTRACTOR EQUIPMENT | | 99.1 | 99.1 | | 99.1 | 99.1 | | 99.1 | 99.1 | | 96.2 | 96.2 | | 96.2 | 96.2 | | 106.1 | 106.1 |
| 0241, 31 - 34 | SITE & INFRASTRUCTURE, DEMOLITION | 84.6 | 86.3 | 85.8 | 88.3 | 90.8 | 90.0 | 97.1 | 85.5 | 89.1 | 97.6 | 84.0 | 88.2 | 105.0 | 83.1 | 89.9 | 90.3 | 107.9 | 102.4 |
| 0310 | Concrete Forming & Accessories | 90.3 | 79.8 | 81.2 | 85.8 | 78.5 | 79.4 | 97.6 | 28.2 | 37.4 | 95.6 | 52.3 | 58.0 | 102.4 | 45.7 | 53.2 | 99.3 | 140.8 | 135.3 |
| 0320 | Concrete Reinforcing | 84.5 | 50.4 | 67.0 | 85.2 | 73.3 | 79.1 | 84.5 | 50.4 | 67.0 | 93.6 | 76.9 | 85.0 | 94.2 | 50.4 | 71.7 | 107.0 | 147.5 | 127.8 |
| 0330 | Cast-in-Place Concrete | 92.0 | 45.0 | 74.0 | 97.5 | 79.7 | 90.7 | 113.7 | 38.5 | 84.9 | 134.0 | 60.5 | 105.9 | 150.2 | 53.9 | 113.3 | 101.4 | 151.4 | 120.6 |
| 03 | CONCRETE | 89.0 | 62.8 | 76.4 | 93.0 | 78.8 | 86.2 | 106.4 | 37.3 | 73.3 | 114.7 | 61.4 | 89.2 | 126.6 | 50.9 | 90.4 | 112.3 | 144.7 | 127.8 |
| 04 | MASONRY | 95.3 | 34.1 | 58.2 | 101.0 | 78.8 | 87.5 | 108.7 | 33.1 | 62.9 | 115.5 | 61.3 | 82.7 | 98.7 | 57.9 | 74.0 | 116.6 | 159.3 | 142.5 |
| 05 | METALS | 90.3 | 67.5 | 82.4 | 90.1 | 90.6 | 90.3 | 90.3 | 60.6 | 80.1 | 85.3 | 90.8 | 87.2 | 85.3 | 74.1 | 81.4 | 102.0 | 128.6 | 111.2 |
| 06 | WOOD, PLASTICS & COMPOSITES | 91.0 | 94.4 | 93.0 | 85.5 | 78.1 | 81.2 | 100.2 | 26.3 | 57.8 | 88.5 | 51.2 | 67.1 | 95.7 | 45.9 | 67.1 | 100.7 | 140.2 | 123.4 |
| 07 | THERMAL & MOISTURE PROTECTION | 94.3 | 46.2 | 75.2 | 94.3 | 71.2 | 85.1 | 94.9 | 54.2 | 78.7 | 101.7 | 74.2 | 90.8 | 102.2 | 69.9 | 89.4 | 96.8 | 150.0 | 117.9 |
| 08 | OPENINGS | 94.6 | 67.4 | 87.7 | 94.6 | 77.2 | 90.2 | 94.9 | 22.5 | 76.5 | 91.4 | 63.7 | 84.4 | 92.2 | 44.5 | 80.2 | 101.6 | 142.8 | 112.0 |
| 0920 | Plaster & Gypsum Board | 95.4 | 94.6 | 94.8 | 91.8 | 77.8 | 81.9 | 101.1 | 24.5 | 47.2 | 101.2 | 50.6 | 65.5 | 104.7 | 45.1 | 62.7 | 103.1 | 140.8 | 129.7 |
| 0950, 0980 | Ceilings & Acoustic Treatment | 94.2 | 94.6 | 94.4 | 95.2 | 77.8 | 83.7 | 94.2 | 24.5 | 48.0 | 107.6 | 50.6 | 69.8 | 107.6 | 45.1 | 66.2 | 98.4 | 140.8 | 126.5 |
| 0960 | Flooring | 91.9 | 51.7 | 80.4 | 89.8 | 82.9 | 87.8 | 95.7 | 51.7 | 83.1 | 119.0 | 54.5 | 100.5 | 123.2 | 51.7 | 102.7 | 96.3 | 176.2 | 119.3 |
| 0970, 0990 | Wall Finishes & Painting/Coating | 98.5 | 80.4 | 87.5 | 98.5 | 33.5 | 59.0 | 98.5 | 24.7 | 53.7 | 127.9 | 80.4 | 99.0 | 127.9 | 57.6 | 85.3 | 94.9 | 157.1 | 132.6 |
| 09 | FINISHES | 93.5 | 77.8 | 84.8 | 92.8 | 74.4 | 82.7 | 96.1 | 31.0 | 60.2 | 102.7 | 54.1 | 75.9 | 104.8 | 47.1 | 73.0 | 98.4 | 149.6 | 126.7 |
| COVERS | DIVS. 10 - 14, 25, 28, 41, 43, 44, 46 | 100.0 | 47.5 | 89.1 | 100.0 | 89.6 | 97.8 | 100.0 | 38.3 | 87.2 | 100.0 | 50.1 | 89.6 | 100.0 | 41.0 | 87.7 | 100.0 | 120.3 | 104.2 |
| 21, 22, 23 | FIRE SUPPRESSION, PLUMBING & HVAC | 94.6 | 73.5 | 85.9 | 100.0 | 85.0 | 93.8 | 94.6 | 63.5 | 81.7 | 94.9 | 82.7 | 89.8 | 94.9 | 80.3 | 88.8 | 100.2 | 129.0 | 112.2 |
| 26, 27, 3370 | ELECTRICAL, COMMUNICATIONS & UTIL. | 96.3 | 53.5 | 73.5 | 94.7 | 82.2 | 88.1 | 93.1 | 62.5 | 76.8 | 98.1 | 97.2 | 97.6 | 95.3 | 97.2 | 96.3 | 100.0 | 134.8 | 118.5 |
| MF2010 | WEIGHTED AVERAGE | 93.6 | 64.9 | 80.7 | 95.5 | 82.2 | 89.5 | 96.8 | 51.0 | 76.3 | 98.3 | 74.5 | 87.7 | 99.1 | 68.5 | 85.4 | 102.3 | 136.8 | 117.7 |

# City Cost Indexes

## MASSACHUSETTS

| DIVISION | | BROCKTON 023 | | | BUZZARDS BAY 025 | | | FALL RIVER 027 | | | FITCHBURG 014 | | | FRAMINGHAM 017 | | | GREENFIELD 013 | | |
|---|---|---|---|---|---|---|---|---|---|---|---|---|---|---|---|---|---|---|---|
| | | MAT. | INST. | TOTAL | MAT. | INST. | TOTAL | MAT. | INST. | TOTAL | MAT. | INST. | TOTAL | MAT. | INST. | TOTAL | MAT. | INST. | TOTAL |
| 015433 | CONTRACTOR EQUIPMENT | | 102.2 | 102.2 | | 102.2 | 102.2 | | 103.2 | 103.2 | | 100.3 | 100.3 | | 101.4 | 101.4 | | 100.3 | 100.3 |
| 0241, 31 - 34 | SITE & INFRASTRUCTURE, DEMOLITION | 87.5 | 104.3 | 99.1 | 79.0 | 104.2 | 96.4 | 86.6 | 104.4 | 98.8 | 80.5 | 104.2 | 96.8 | 77.3 | 103.8 | 95.6 | 83.7 | 102.7 | 96.8 |
| 0310 | Concrete Forming & Accessories | 99.7 | 129.4 | 125.5 | 97.5 | 128.7 | 124.6 | 99.7 | 129.0 | 125.2 | 91.9 | 120.5 | 116.7 | 98.5 | 129.2 | 125.1 | 90.4 | 109.8 | 107.2 |
| 0320 | Concrete Reinforcing | 108.1 | 147.4 | 128.3 | 86.7 | 123.1 | 105.4 | 108.1 | 123.1 | 115.8 | 86.7 | 142.7 | 115.6 | 86.7 | 147.0 | 117.8 | 90.3 | 121.9 | 106.6 |
| 0330 | Cast-in-Place Concrete | 96.6 | 151.3 | 117.5 | 80.2 | 151.0 | 107.3 | 93.4 | 151.6 | 115.7 | 84.1 | 145.3 | 107.5 | 84.1 | 139.1 | 105.2 | 86.5 | 123.2 | 100.5 |
| 03 | CONCRETE | 109.6 | 139.4 | 123.8 | 92.5 | 134.4 | 112.6 | 108.0 | 134.8 | 120.8 | 90.6 | 132.4 | 110.6 | 93.7 | 135.1 | 113.5 | 94.6 | 116.0 | 104.9 |
| 04 | MASONRY | 112.6 | 155.6 | 138.6 | 103.2 | 155.6 | 134.9 | 113.0 | 155.5 | 138.8 | 99.5 | 149.3 | 129.7 | 106.1 | 149.4 | 132.4 | 103.7 | 123.4 | 115.6 |
| 05 | METALS | 98.7 | 126.2 | 108.1 | 93.5 | 115.7 | 101.1 | 98.7 | 116.1 | 104.7 | 93.4 | 120.8 | 102.8 | 93.4 | 125.3 | 104.4 | 95.6 | 107.8 | 99.8 |
| 06 | WOOD, PLASTICS & COMPOSITES | 99.6 | 127.6 | 115.6 | 96.7 | 127.6 | 114.4 | 99.6 | 127.8 | 115.8 | 93.2 | 116.5 | 106.6 | 99.2 | 127.3 | 115.3 | 91.4 | 108.4 | 101.1 |
| 07 | THERMAL & MOISTURE PROTECTION | 96.7 | 145.8 | 116.2 | 96.1 | 139.7 | 113.4 | 96.7 | 138.9 | 113.5 | 96.2 | 136.6 | 112.3 | 96.2 | 141.7 | 114.3 | 96.2 | 117.8 | 104.8 |
| 08 | OPENINGS | 99.6 | 133.1 | 108.1 | 94.9 | 122.4 | 101.9 | 99.6 | 122.6 | 105.4 | 104.2 | 126.0 | 109.7 | 95.4 | 132.9 | 104.9 | 104.4 | 112.1 | 106.3 |
| 0920 | Plaster & Gypsum Board | 97.3 | 127.7 | 118.7 | 93.0 | 127.7 | 117.4 | 97.3 | 127.7 | 118.7 | 94.2 | 116.3 | 109.8 | 97.3 | 127.7 | 118.7 | 95.4 | 108.0 | 104.2 |
| 0950, 0980 | Ceilings & Acoustic Treatment | 100.5 | 127.7 | 118.5 | 90.9 | 127.7 | 115.3 | 100.5 | 127.7 | 118.5 | 89.8 | 116.3 | 107.4 | 89.8 | 127.7 | 114.9 | 97.3 | 108.0 | 104.4 |
| 0960 | Flooring | 96.8 | 176.2 | 119.7 | 95.5 | 176.2 | 118.7 | 96.6 | 176.2 | 119.5 | 93.8 | 176.2 | 117.5 | 95.3 | 176.2 | 118.6 | 93.0 | 135.4 | 105.2 |
| 0970, 0990 | Wall Finishes & Painting/Coating | 94.4 | 142.1 | 123.3 | 94.4 | 142.1 | 123.3 | 94.4 | 142.1 | 123.3 | 92.3 | 142.1 | 122.5 | 93.4 | 142.1 | 122.9 | 92.3 | 111.7 | 104.1 |
| 09 | FINISHES | 98.4 | 139.6 | 121.1 | 94.8 | 139.6 | 119.5 | 98.3 | 139.8 | 121.2 | 94.1 | 133.1 | 115.6 | 94.8 | 139.4 | 119.4 | 96.0 | 114.7 | 106.3 |
| COVERS | DIVS. 10 - 14, 25, 28, 41, 43, 44, 46 | 100.0 | 117.5 | 103.6 | 100.0 | 117.5 | 103.6 | 100.0 | 118.1 | 103.8 | 100.0 | 92.9 | 98.5 | 100.0 | 117.0 | 103.5 | 100.0 | 105.6 | 101.2 |
| 21, 22, 23 | FIRE SUPPRESSION, PLUMBING & HVAC | 100.2 | 110.3 | 104.4 | 94.8 | 110.0 | 101.1 | 100.2 | 110.0 | 104.3 | 95.6 | 112.5 | 102.7 | 95.6 | 123.2 | 107.1 | 95.6 | 103.5 | 98.9 |
| 26, 27, 3370 | ELECTRICAL, COMMUNICATIONS & UTIL. | 99.1 | 101.0 | 100.1 | 96.0 | 101.0 | 98.6 | 99.1 | 101.0 | 100.1 | 101.4 | 104.0 | 102.8 | 97.5 | 126.9 | 113.1 | 101.4 | 95.0 | 98.0 |
| MF2010 | WEIGHTED AVERAGE | 100.9 | 124.5 | 111.5 | 95.0 | 122.2 | 107.1 | 100.7 | 122.3 | 110.4 | 96.2 | 121.1 | 107.3 | 95.6 | 129.4 | 110.7 | 97.4 | 108.6 | 102.4 |

## MASSACHUSETTS

| DIVISION | | HYANNIS 026 | | | LAWRENCE 019 | | | LOWELL 018 | | | NEW BEDFORD 027 | | | PITTSFIELD 012 | | | SPRINGFIELD 010 - 011 | | |
|---|---|---|---|---|---|---|---|---|---|---|---|---|---|---|---|---|---|---|---|
| | | MAT. | INST. | TOTAL | MAT. | INST. | TOTAL | MAT. | INST. | TOTAL | MAT. | INST. | TOTAL | MAT. | INST. | TOTAL | MAT. | INST. | TOTAL |
| 015433 | CONTRACTOR EQUIPMENT | | 102.2 | 102.2 | | 102.2 | 102.2 | | 100.3 | 100.3 | | 103.2 | 103.2 | | 100.3 | 100.3 | | 100.3 | 100.3 |
| 0241, 31 - 34 | SITE & INFRASTRUCTURE, DEMOLITION | 84.4 | 104.2 | 98.0 | 88.2 | 104.3 | 99.3 | 87.3 | 104.2 | 99.0 | 85.2 | 104.4 | 98.4 | 88.3 | 102.6 | 98.2 | 87.8 | 102.9 | 98.2 |
| 0310 | Concrete Forming & Accessories | 92.1 | 128.7 | 123.9 | 99.5 | 129.5 | 125.5 | 96.5 | 129.6 | 125.2 | 99.7 | 129.0 | 125.2 | 96.5 | 108.4 | 106.8 | 96.7 | 110.4 | 108.6 |
| 0320 | Concrete Reinforcing | 86.7 | 123.1 | 105.4 | 107.2 | 137.3 | 122.7 | 108.1 | 137.0 | 123.0 | 108.1 | 123.1 | 115.8 | 89.6 | 106.2 | 98.2 | 108.1 | 121.9 | 115.2 |
| 0330 | Cast-in-Place Concrete | 88.1 | 151.0 | 112.2 | 97.4 | 148.2 | 116.8 | 88.5 | 148.2 | 111.4 | 82.3 | 151.6 | 108.8 | 96.6 | 121.4 | 106.1 | 92.1 | 124.2 | 104.4 |
| 03 | CONCRETE | 99.6 | 134.4 | 116.3 | 109.8 | 136.5 | 122.6 | 100.3 | 136.3 | 117.5 | 102.4 | 134.8 | 117.9 | 101.6 | 111.8 | 106.5 | 102.1 | 116.7 | 109.1 |
| 04 | MASONRY | 111.5 | 155.6 | 138.2 | 111.7 | 155.6 | 138.3 | 98.2 | 149.3 | 129.2 | 111.4 | 155.5 | 138.1 | 98.9 | 120.3 | 111.9 | 98.5 | 125.1 | 114.6 |
| 05 | METALS | 95.0 | 115.7 | 102.1 | 96.0 | 122.1 | 105.0 | 96.0 | 119.4 | 104.0 | 98.7 | 116.1 | 104.7 | 95.8 | 101.2 | 97.6 | 98.6 | 107.8 | 101.8 |
| 06 | WOOD, PLASTICS & COMPOSITES | 90.7 | 127.6 | 111.8 | 99.6 | 127.6 | 115.6 | 99.0 | 127.6 | 115.4 | 99.6 | 127.8 | 115.8 | 99.0 | 108.4 | 104.4 | 99.0 | 108.4 | 104.4 |
| 07 | THERMAL & MOISTURE PROTECTION | 96.4 | 139.7 | 113.6 | 96.7 | 145.4 | 116.0 | 96.5 | 143.4 | 115.1 | 96.7 | 138.9 | 113.5 | 96.5 | 116.5 | 104.5 | 96.5 | 118.6 | 105.3 |
| 08 | OPENINGS | 95.6 | 122.4 | 102.4 | 99.6 | 130.4 | 107.4 | 105.6 | 130.4 | 111.9 | 99.6 | 122.6 | 105.4 | 105.6 | 108.0 | 106.2 | 105.6 | 112.1 | 107.3 |
| 0920 | Plaster & Gypsum Board | 89.5 | 127.7 | 116.4 | 100.2 | 127.7 | 119.6 | 100.2 | 127.7 | 119.6 | 97.3 | 127.7 | 118.7 | 100.2 | 108.0 | 105.7 | 100.2 | 108.0 | 105.7 |
| 0950, 0980 | Ceilings & Acoustic Treatment | 93.0 | 127.7 | 116.0 | 99.4 | 127.7 | 118.2 | 99.4 | 127.7 | 118.2 | 100.5 | 127.7 | 118.5 | 99.4 | 108.0 | 105.1 | 99.4 | 108.0 | 105.1 |
| 0960 | Flooring | 93.4 | 176.2 | 117.2 | 95.8 | 176.2 | 118.9 | 95.8 | 176.2 | 118.9 | 96.6 | 176.2 | 119.5 | 96.0 | 135.4 | 107.4 | 95.7 | 135.4 | 107.2 |
| 0970, 0990 | Wall Finishes & Painting/Coating | 94.4 | 142.1 | 123.3 | 92.4 | 142.1 | 122.5 | 92.3 | 142.1 | 122.5 | 94.4 | 142.1 | 123.3 | 92.3 | 111.7 | 104.1 | 93.8 | 111.7 | 104.7 |
| 09 | FINISHES | 94.6 | 139.6 | 119.4 | 98.1 | 139.6 | 120.9 | 98.0 | 139.6 | 120.9 | 98.2 | 139.8 | 121.1 | 98.1 | 114.0 | 106.8 | 98.1 | 115.1 | 107.5 |
| COVERS | DIVS. 10 - 14, 25, 28, 41, 43, 44, 46 | 100.0 | 117.5 | 103.6 | 100.0 | 117.6 | 103.7 | 100.0 | 117.6 | 103.7 | 100.0 | 118.1 | 103.8 | 100.0 | 104.5 | 100.9 | 100.0 | 106.2 | 101.3 |
| 21, 22, 23 | FIRE SUPPRESSION, PLUMBING & HVAC | 100.2 | 110.0 | 104.3 | 100.0 | 113.0 | 105.5 | 100.2 | 123.2 | 109.8 | 100.2 | 110.0 | 104.3 | 100.2 | 96.6 | 98.7 | 100.2 | 104.4 | 102.0 |
| 26, 27, 3370 | ELECTRICAL, COMMUNICATIONS & UTIL. | 96.5 | 101.0 | 98.9 | 100.3 | 126.9 | 114.4 | 100.9 | 126.8 | 114.7 | 99.9 | 101.0 | 100.5 | 100.9 | 95.0 | 97.7 | 100.9 | 96.9 | 98.8 |
| MF2010 | WEIGHTED AVERAGE | 98.0 | 122.2 | 108.8 | 100.6 | 127.8 | 112.8 | 99.4 | 129.0 | 112.6 | 100.0 | 122.3 | 110.0 | 99.6 | 105.3 | 102.2 | 100.0 | 109.5 | 104.2 |

## MASSACHUSETTS / MICHIGAN

| DIVISION | | MASSACHUSETTS WORCESTER 015 - 016 | | | ANN ARBOR 481 | | | BATTLE CREEK 490 | | | BAY CITY 487 | | | DEARBORN 481 | | | DETROIT 482 | | |
|---|---|---|---|---|---|---|---|---|---|---|---|---|---|---|---|---|---|---|---|
| | | MAT. | INST. | TOTAL | MAT. | INST. | TOTAL | MAT. | INST. | TOTAL | MAT. | INST. | TOTAL | MAT. | INST. | TOTAL | MAT. | INST. | TOTAL |
| 015433 | CONTRACTOR EQUIPMENT | | 100.3 | 100.3 | | 110.9 | 110.9 | | 103.6 | 103.6 | | 110.9 | 110.9 | | 110.9 | 110.9 | | 98.5 | 98.5 |
| 0241, 31 - 34 | SITE & INFRASTRUCTURE, DEMOLITION | 87.7 | 104.2 | 99.1 | 73.4 | 96.1 | 89.1 | 82.5 | 87.9 | 86.2 | 66.2 | 94.7 | 85.8 | 73.2 | 96.1 | 89.0 | 86.5 | 97.7 | 94.2 |
| 0310 | Concrete Forming & Accessories | 97.0 | 124.7 | 121.1 | 98.7 | 104.0 | 103.3 | 98.2 | 91.4 | 92.3 | 98.7 | 89.0 | 90.3 | 98.5 | 112.1 | 110.3 | 99.2 | 112.2 | 110.5 |
| 0320 | Concrete Reinforcing | 108.1 | 146.3 | 127.8 | 100.8 | 125.5 | 113.6 | 98.6 | 92.5 | 95.5 | 100.8 | 124.7 | 113.1 | 100.8 | 125.7 | 113.6 | 100.1 | 125.7 | 113.3 |
| 0330 | Cast-in-Place Concrete | 91.6 | 145.4 | 112.2 | 82.3 | 115.3 | 94.9 | 92.4 | 104.8 | 97.2 | 78.7 | 95.2 | 85.0 | 80.5 | 113.3 | 93.0 | 88.3 | 113.3 | 97.9 |
| 03 | CONCRETE | 101.9 | 135.0 | 117.7 | 90.3 | 112.3 | 100.9 | 95.0 | 95.5 | 95.3 | 88.5 | 98.6 | 93.4 | 89.4 | 115.3 | 101.8 | 93.3 | 114.2 | 103.3 |
| 04 | MASONRY | 98.0 | 149.3 | 129.1 | 96.8 | 110.4 | 105.0 | 95.3 | 88.1 | 90.9 | 96.4 | 89.6 | 92.3 | 96.6 | 113.7 | 107.0 | 95.8 | 113.7 | 106.7 |
| 05 | METALS | 98.7 | 122.4 | 106.8 | 93.1 | 120.2 | 102.4 | 92.9 | 87.7 | 91.1 | 93.7 | 116.9 | 101.7 | 93.1 | 120.6 | 102.6 | 97.0 | 102.8 | 99.0 |
| 06 | WOOD, PLASTICS & COMPOSITES | 99.4 | 122.0 | 112.4 | 100.4 | 101.7 | 101.2 | 100.9 | 91.6 | 95.6 | 100.4 | 88.3 | 93.5 | 100.4 | 112.2 | 107.2 | 100.8 | 112.2 | 107.3 |
| 07 | THERMAL & MOISTURE PROTECTION | 96.5 | 137.2 | 112.7 | 100.2 | 111.5 | 104.7 | 94.3 | 89.1 | 92.2 | 98.0 | 97.9 | 98.0 | 98.8 | 116.7 | 105.9 | 97.3 | 116.7 | 105.0 |
| 08 | OPENINGS | 105.6 | 129.9 | 111.8 | 92.1 | 105.6 | 95.5 | 92.0 | 85.5 | 90.4 | 92.1 | 95.9 | 93.1 | 92.1 | 110.6 | 96.8 | 93.6 | 111.3 | 98.1 |
| 0920 | Plaster & Gypsum Board | 100.2 | 122.0 | 115.5 | 99.1 | 100.7 | 100.3 | 90.1 | 87.4 | 88.2 | 99.1 | 86.9 | 90.5 | 99.1 | 111.5 | 107.8 | 99.1 | 111.5 | 107.8 |
| 0950, 0980 | Ceilings & Acoustic Treatment | 99.4 | 122.0 | 114.4 | 94.3 | 100.7 | 98.6 | 91.3 | 87.4 | 88.7 | 95.3 | 86.9 | 89.7 | 94.3 | 111.5 | 105.7 | 95.3 | 111.5 | 106.0 |
| 0960 | Flooring | 95.8 | 170.2 | 117.2 | 89.7 | 118.1 | 97.9 | 98.2 | 97.1 | 97.9 | 89.6 | 81.2 | 87.2 | 89.4 | 118.9 | 97.8 | 89.6 | 118.9 | 98.0 |
| 0970, 0990 | Wall Finishes & Painting/Coating | 92.3 | 142.1 | 122.5 | 93.0 | 104.2 | 99.8 | 97.6 | 82.8 | 88.6 | 93.0 | 85.1 | 88.2 | 93.0 | 104.3 | 99.9 | 94.5 | 104.3 | 100.5 |
| 09 | FINISHES | 98.0 | 135.1 | 118.4 | 92.6 | 106.2 | 100.1 | 96.1 | 91.6 | 93.6 | 92.3 | 86.4 | 89.1 | 92.4 | 112.7 | 103.6 | 93.8 | 112.7 | 104.2 |
| COVERS | DIVS. 10 - 14, 25, 28, 41, 43, 44, 46 | 100.0 | 109.9 | 102.1 | 100.0 | 106.1 | 101.3 | 100.0 | 106.8 | 101.4 | 100.0 | 100.2 | 100.0 | 100.0 | 107.5 | 101.6 | 100.0 | 107.5 | 101.6 |
| 21, 22, 23 | FIRE SUPPRESSION, PLUMBING & HVAC | 100.2 | 112.6 | 105.3 | 100.0 | 102.1 | 100.9 | 99.9 | 92.0 | 96.6 | 100.0 | 87.5 | 94.8 | 100.0 | 107.8 | 103.3 | 100.1 | 111.2 | 104.7 |
| 26, 27, 3370 | ELECTRICAL, COMMUNICATIONS & UTIL. | 100.9 | 104.8 | 103.0 | 96.9 | 110.3 | 104.1 | 94.7 | 85.2 | 89.7 | 95.9 | 90.5 | 93.0 | 96.9 | 111.6 | 104.7 | 97.9 | 111.5 | 105.2 |
| MF2010 | WEIGHTED AVERAGE | 100.0 | 122.7 | 110.1 | 95.2 | 107.7 | 100.8 | 95.8 | 90.4 | 93.4 | 94.6 | 93.9 | 94.3 | 95.0 | 111.2 | 102.2 | 96.7 | 110.2 | 102.8 |

717

# City Cost Indexes

## MICHIGAN

| DIVISION | | FLINT 484 - 485 | | | GAYLORD 497 | | | GRAND RAPIDS 493,495 | | | IRON MOUNTAIN 498 - 499 | | | JACKSON 492 | | | KALAMAZOO 491 | | |
|---|---|---|---|---|---|---|---|---|---|---|---|---|---|---|---|---|---|---|---|
| | | MAT. | INST. | TOTAL | MAT. | INST. | TOTAL | MAT. | INST. | TOTAL | MAT. | INST. | TOTAL | MAT. | INST. | TOTAL | MAT. | INST. | TOTAL |
| 015433 | CONTRACTOR EQUIPMENT | | 110.9 | 110.9 | | 105.9 | 105.9 | | 103.6 | 103.6 | | 94.4 | 94.4 | | 105.9 | 105.9 | | 103.6 | 103.6 |
| 0241, 31 - 34 | SITE & INFRASTRUCTURE, DEMOLITION | 64.2 | 95.4 | 85.8 | 78.2 | 86.1 | 83.7 | 83.9 | 87.7 | 86.6 | 85.4 | 94.5 | 91.7 | 98.0 | 87.4 | 90.7 | 82.8 | 87.9 | 86.3 |
| 0310 | Concrete Forming & Accessories | 101.6 | 90.9 | 92.3 | 95.4 | 56.9 | 62.0 | 98.4 | 80.8 | 83.1 | 87.3 | 88.0 | 87.9 | 92.9 | 88.5 | 89.1 | 98.2 | 91.1 | 92.0 |
| 0320 | Concrete Reinforcing | 100.8 | 125.0 | 113.3 | 91.9 | 94.1 | 93.1 | 98.6 | 91.8 | 95.1 | 91.7 | 95.8 | 93.8 | 89.3 | 124.8 | 107.6 | 98.6 | 92.4 | 95.4 |
| 0330 | Cast-in-Place Concrete | 82.9 | 97.1 | 88.3 | 92.2 | 72.5 | 84.6 | 98.4 | 96.9 | 97.8 | 109.1 | 91.6 | 102.4 | 92.0 | 95.5 | 93.4 | 94.3 | 104.7 | 98.3 |
| 03 | CONCRETE | 90.8 | 100.2 | 95.3 | 91.4 | 70.8 | 81.6 | 98.1 | 87.9 | 93.2 | 101.1 | 90.8 | 96.2 | 86.5 | 98.4 | 92.2 | 98.3 | 95.3 | 96.9 |
| 04 | MASONRY | 96.8 | 98.0 | 97.5 | 105.6 | 84.1 | 92.6 | 91.2 | 81.8 | 85.5 | 92.1 | 90.8 | 91.3 | 86.3 | 94.7 | 91.4 | 94.0 | 88.1 | 90.4 |
| 05 | METALS | 93.1 | 117.7 | 101.6 | 94.0 | 101.7 | 96.6 | 94.0 | 85.1 | 90.9 | 93.4 | 91.5 | 92.7 | 94.2 | 115.2 | 101.4 | 92.9 | 87.3 | 91.0 |
| 06 | WOOD, PLASTICS & COMPOSITES | 103.6 | 88.6 | 95.0 | 93.3 | 49.8 | 68.4 | 98.1 | 80.1 | 87.8 | 87.9 | 88.0 | 88.0 | 92.1 | 86.9 | 89.1 | 100.9 | 91.6 | 95.6 |
| 07 | THERMAL & MOISTURE PROTECTION | 98.0 | 101.5 | 99.4 | 92.8 | 71.9 | 84.5 | 96.1 | 69.6 | 85.6 | 96.0 | 85.8 | 92.0 | 92.2 | 98.1 | 94.5 | 94.3 | 89.0 | 92.2 |
| 08 | OPENINGS | 92.1 | 96.1 | 93.1 | 94.0 | 55.0 | 84.2 | 95.7 | 70.7 | 89.4 | 99.3 | 74.9 | 93.1 | 93.0 | 88.3 | 91.8 | 92.0 | 85.5 | 90.4 |
| 0920 | Plaster & Gypsum Board | 100.9 | 87.2 | 91.2 | 89.3 | 47.0 | 59.5 | 90.7 | 75.6 | 80.0 | 59.5 | 88.1 | 79.6 | 87.8 | 85.2 | 85.9 | 90.1 | 87.4 | 88.2 |
| 0950, 0980 | Ceilings & Acoustic Treatment | 94.3 | 87.2 | 89.6 | 88.5 | 47.0 | 61.0 | 89.4 | 75.6 | 80.2 | 89.2 | 88.1 | 88.5 | 89.4 | 85.2 | 86.6 | 91.3 | 87.4 | 88.7 |
| 0960 | Flooring | 89.6 | 83.9 | 87.9 | 91.1 | 50.9 | 79.5 | 98.4 | 35.8 | 80.4 | 112.3 | 97.7 | 108.1 | 89.6 | 70.0 | 84.0 | 98.2 | 76.5 | 91.9 |
| 0970, 0990 | Wall Finishes & Painting/Coating | 93.0 | 92.0 | 92.4 | 93.2 | 46.5 | 64.9 | 97.6 | 34.7 | 59.5 | 114.3 | 58.3 | 80.3 | 93.2 | 77.3 | 83.5 | 97.6 | 82.8 | 88.6 |
| 09 | FINISHES | 91.9 | 88.8 | 90.2 | 96.1 | 52.6 | 72.1 | 95.9 | 67.0 | 80.0 | 99.6 | 86.7 | 92.5 | 97.2 | 83.0 | 89.4 | 96.1 | 87.5 | 91.4 |
| COVERS | DIVS. 10 - 14, 25, 28, 41, 43, 44, 46 | 100.0 | 101.5 | 100.3 | 100.0 | 95.1 | 99.0 | 100.0 | 105.2 | 101.1 | 100.0 | 97.3 | 99.4 | 100.0 | 101.4 | 100.3 | 100.0 | 106.8 | 101.4 |
| 21, 22, 23 | FIRE SUPPRESSION, PLUMBING & HVAC | 100.0 | 96.1 | 98.4 | 94.9 | 82.6 | 89.7 | 99.9 | 80.6 | 91.8 | 94.8 | 87.4 | 91.7 | 94.9 | 93.1 | 94.1 | 99.9 | 83.0 | 92.9 |
| 26, 27, 3370 | ELECTRICAL, COMMUNICATIONS & UTIL. | 96.9 | 97.0 | 97.0 | 92.4 | 50.0 | 69.8 | 103.1 | 49.9 | 74.8 | 99.4 | 87.2 | 92.9 | 96.8 | 110.3 | 104.0 | 94.6 | 88.8 | 91.5 |
| MF2010 | WEIGHTED AVERAGE | 94.9 | 98.2 | 96.4 | 94.5 | 73.4 | 85.1 | 97.3 | 76.8 | 88.1 | 96.7 | 88.8 | 93.2 | 94.0 | 96.8 | 95.2 | 96.1 | 88.4 | 92.5 |

## MICHIGAN / MINNESOTA

| DIVISION | | LANSING 488 - 489 | | | MUSKEGON 494 | | | ROYAL OAK 480,483 | | | SAGINAW 486 | | | TRAVERSE CITY 496 | | | BEMIDJI 566 | | |
|---|---|---|---|---|---|---|---|---|---|---|---|---|---|---|---|---|---|---|---|
| | | MAT. | INST. | TOTAL | MAT. | INST. | TOTAL | MAT. | INST. | TOTAL | MAT. | INST. | TOTAL | MAT. | INST. | TOTAL | MAT. | INST. | TOTAL |
| 015433 | CONTRACTOR EQUIPMENT | | 110.9 | 110.9 | | 103.6 | 103.6 | | 95.8 | 95.8 | | 110.9 | 110.9 | | 94.4 | 94.4 | | 98.6 | 98.6 |
| 0241, 31 - 34 | SITE & INFRASTRUCTURE, DEMOLITION | 81.0 | 94.9 | 90.5 | 80.5 | 87.8 | 85.5 | 77.0 | 95.5 | 89.8 | 67.0 | 94.7 | 86.1 | 72.8 | 94.0 | 87.4 | 95.7 | 98.4 | 97.6 |
| 0310 | Concrete Forming & Accessories | 102.2 | 89.9 | 91.6 | 98.6 | 89.6 | 90.8 | 94.9 | 101.2 | 100.3 | 98.7 | 88.9 | 90.2 | 87.3 | 72.3 | 74.3 | 87.4 | 96.1 | 95.0 |
| 0320 | Concrete Reinforcing | 100.8 | 124.8 | 113.2 | 99.3 | 91.8 | 95.4 | 91.4 | 122.4 | 107.3 | 100.8 | 124.7 | 113.1 | 93.1 | 91.6 | 92.3 | 98.7 | 110.4 | 104.7 |
| 0330 | Cast-in-Place Concrete | 87.4 | 94.9 | 90.3 | 92.1 | 97.8 | 94.3 | 72.0 | 113.0 | 87.7 | 81.3 | 95.2 | 86.6 | 85.1 | 64.5 | 77.3 | 103.6 | 109.5 | 105.9 |
| 03 | CONCRETE | 93.1 | 99.0 | 95.9 | 93.4 | 92.2 | 92.8 | 77.9 | 108.4 | 92.5 | 89.8 | 98.6 | 94.0 | 83.1 | 73.9 | 78.7 | 99.1 | 104.4 | 101.6 |
| 04 | MASONRY | 90.5 | 95.2 | 93.3 | 92.5 | 77.3 | 83.3 | 91.9 | 113.5 | 105.0 | 98.3 | 89.6 | 93.0 | 90.3 | 84.0 | 86.5 | 103.6 | 107.6 | 106.0 |
| 05 | METALS | 91.7 | 117.1 | 100.4 | 90.6 | 86.1 | 89.1 | 96.1 | 99.1 | 97.2 | 93.2 | 116.8 | 101.3 | 93.3 | 89.3 | 91.9 | 92.5 | 123.9 | 103.3 |
| 06 | WOOD, PLASTICS & COMPOSITES | 103.0 | 88.4 | 94.6 | 97.0 | 90.4 | 93.2 | 95.8 | 97.8 | 96.9 | 96.1 | 88.3 | 91.6 | 87.9 | 71.0 | 78.2 | 73.2 | 93.0 | 84.6 |
| 07 | THERMAL & MOISTURE PROTECTION | 99.2 | 94.1 | 97.2 | 93.5 | 78.9 | 87.7 | 96.7 | 114.6 | 103.8 | 98.8 | 97.9 | 98.5 | 95.2 | 72.5 | 86.2 | 107.4 | 99.7 | 104.3 |
| 08 | OPENINGS | 92.1 | 96.0 | 93.1 | 91.3 | 89.0 | 90.7 | 92.2 | 102.5 | 94.8 | 90.0 | 95.9 | 91.5 | 99.2 | 66.2 | 90.9 | 96.6 | 113.3 | 100.8 |
| 0920 | Plaster & Gypsum Board | 91.9 | 87.0 | 88.5 | 76.9 | 86.1 | 83.4 | 95.9 | 96.7 | 96.4 | 99.1 | 86.9 | 90.5 | 59.5 | 70.6 | 67.3 | 100.6 | 93.1 | 95.3 |
| 0950, 0980 | Ceilings & Acoustic Treatment | 94.3 | 87.0 | 89.4 | 92.2 | 86.1 | 88.2 | 93.5 | 96.7 | 95.6 | 94.3 | 86.9 | 89.4 | 89.2 | 70.6 | 76.8 | 132.5 | 93.1 | 106.3 |
| 0960 | Flooring | 99.1 | 83.9 | 94.8 | 97.4 | 65.9 | 88.3 | 87.5 | 117.4 | 96.1 | 89.7 | 81.2 | 87.3 | 112.2 | 50.9 | 94.6 | 112.6 | 127.6 | 116.9 |
| 0970, 0990 | Wall Finishes & Painting/Coating | 103.9 | 85.9 | 93.0 | 96.9 | 83.1 | 88.5 | 94.5 | 104.2 | 100.4 | 93.0 | 85.1 | 88.2 | 114.3 | 46.5 | 73.2 | 98.4 | 102.4 | 100.8 |
| 09 | FINISHES | 96.0 | 87.7 | 91.4 | 93.8 | 84.5 | 88.6 | 91.6 | 103.9 | 98.4 | 92.2 | 86.4 | 89.0 | 98.6 | 64.8 | 80.0 | 111.4 | 102.1 | 106.3 |
| COVERS | DIVS. 10 - 14, 25, 28, 41, 43, 44, 46 | 100.0 | 101.1 | 100.2 | 100.0 | 106.4 | 101.3 | 100.0 | 104.5 | 100.9 | 100.0 | 100.2 | 100.0 | 100.0 | 93.6 | 98.7 | 100.0 | 99.4 | 99.9 |
| 21, 22, 23 | FIRE SUPPRESSION, PLUMBING & HVAC | 99.9 | 93.8 | 97.3 | 99.8 | 85.0 | 93.6 | 95.0 | 104.9 | 99.1 | 100.0 | 87.2 | 94.7 | 94.8 | 82.1 | 89.5 | 94.9 | 90.1 | 92.9 |
| 26, 27, 3370 | ELECTRICAL, COMMUNICATIONS & UTIL. | 99.5 | 95.0 | 97.1 | 95.0 | 78.8 | 86.4 | 99.3 | 96.1 | 97.6 | 94.5 | 92.1 | 93.2 | 94.1 | 78.3 | 85.7 | 102.4 | 101.4 | 101.9 |
| MF2010 | WEIGHTED AVERAGE | 95.7 | 96.5 | 96.1 | 94.7 | 85.3 | 90.5 | 92.8 | 103.6 | 97.6 | 94.5 | 94.0 | 94.3 | 93.4 | 79.4 | 87.2 | 98.5 | 102.2 | 100.2 |

## MINNESOTA

| DIVISION | | BRAINERD 564 | | | DETROIT LAKES 565 | | | DULUTH 556 - 558 | | | MANKATO 560 | | | MINNEAPOLIS 553 - 555 | | | ROCHESTER 559 | | |
|---|---|---|---|---|---|---|---|---|---|---|---|---|---|---|---|---|---|---|---|
| | | MAT. | INST. | TOTAL | MAT. | INST. | TOTAL | MAT. | INST. | TOTAL | MAT. | INST. | TOTAL | MAT. | INST. | TOTAL | MAT. | INST. | TOTAL |
| 015433 | CONTRACTOR EQUIPMENT | | 101.5 | 101.5 | | 98.6 | 98.6 | | 101.0 | 101.0 | | 101.5 | 101.5 | | 104.7 | 104.7 | | 101.0 | 101.0 |
| 0241, 31 - 34 | SITE & INFRASTRUCTURE, DEMOLITION | 93.8 | 103.5 | 100.5 | 93.7 | 98.9 | 97.3 | 96.5 | 101.9 | 100.2 | 90.1 | 104.2 | 99.8 | 99.1 | 107.5 | 104.9 | 98.0 | 101.8 | 100.6 |
| 0310 | Concrete Forming & Accessories | 90.7 | 99.1 | 98.0 | 84.8 | 99.0 | 97.1 | 100.6 | 118.1 | 115.8 | 97.4 | 107.0 | 105.7 | 102.6 | 132.1 | 128.2 | 103.3 | 113.0 | 111.7 |
| 0320 | Concrete Reinforcing | 97.6 | 110.6 | 104.3 | 98.7 | 110.4 | 104.7 | 93.1 | 110.8 | 102.2 | 97.4 | 121.2 | 109.7 | 93.3 | 121.6 | 107.9 | 93.1 | 121.3 | 107.6 |
| 0330 | Cast-in-Place Concrete | 112.6 | 114.6 | 113.4 | 100.6 | 113.9 | 105.7 | 105.0 | 109.4 | 106.7 | 103.7 | 113.7 | 107.5 | 117.5 | 122.1 | 119.3 | 115.2 | 105.7 | 111.6 |
| 03 | CONCRETE | 103.4 | 107.5 | 105.4 | 96.5 | 107.2 | 101.6 | 102.0 | 114.3 | 107.9 | 98.5 | 112.8 | 105.4 | 110.3 | 127.0 | 118.3 | 107.3 | 112.9 | 110.0 |
| 04 | MASONRY | 127.8 | 120.2 | 123.2 | 127.6 | 115.4 | 120.2 | 116.3 | 119.4 | 118.2 | 115.4 | 117.0 | 116.4 | 115.7 | 131.5 | 125.3 | 114.9 | 119.6 | 117.7 |
| 05 | METALS | 93.6 | 124.2 | 104.1 | 92.4 | 123.8 | 103.2 | 101.0 | 126.1 | 109.6 | 93.4 | 130.3 | 106.1 | 103.9 | 133.7 | 114.2 | 100.9 | 131.9 | 111.6 |
| 06 | WOOD, PLASTICS & COMPOSITES | 94.4 | 92.8 | 93.5 | 70.7 | 93.0 | 83.5 | 109.8 | 119.0 | 115.1 | 102.4 | 104.9 | 103.8 | 113.0 | 130.6 | 123.1 | 113.0 | 111.9 | 112.3 |
| 07 | THERMAL & MOISTURE PROTECTION | 107.0 | 113.6 | 109.6 | 107.2 | 113.9 | 109.9 | 107.9 | 119.4 | 112.5 | 107.3 | 104.8 | 106.3 | 107.7 | 133.6 | 118.0 | 108.0 | 106.5 | 107.4 |
| 08 | OPENINGS | 84.8 | 113.1 | 92.0 | 96.6 | 113.3 | 100.8 | 93.7 | 122.6 | 101.1 | 89.5 | 123.7 | 98.2 | 96.8 | 137.6 | 107.1 | 93.7 | 127.4 | 102.3 |
| 0920 | Plaster & Gypsum Board | 85.4 | 93.1 | 90.8 | 99.7 | 93.1 | 95.0 | 92.3 | 120.0 | 111.8 | 89.4 | 105.6 | 100.8 | 100.0 | 131.9 | 122.4 | 99.4 | 112.6 | 108.7 |
| 0950, 0980 | Ceilings & Acoustic Treatment | 65.5 | 93.1 | 83.8 | 132.5 | 93.1 | 106.3 | 88.3 | 120.0 | 109.3 | 65.5 | 105.6 | 92.0 | 90.2 | 131.9 | 117.8 | 88.3 | 112.6 | 104.5 |
| 0960 | Flooring | 111.4 | 127.6 | 116.0 | 111.2 | 127.6 | 115.9 | 108.3 | 129.0 | 114.2 | 113.5 | 127.6 | 117.6 | 106.1 | 127.6 | 112.3 | 108.5 | 127.6 | 114.0 |
| 0970, 0990 | Wall Finishes & Painting/Coating | 92.7 | 102.4 | 98.6 | 98.4 | 102.4 | 100.8 | 89.0 | 110.0 | 101.7 | 104.0 | 105.4 | 104.8 | 97.7 | 130.8 | 117.8 | 92.3 | 105.4 | 100.2 |
| 09 | FINISHES | 94.6 | 103.9 | 99.7 | 110.7 | 104.0 | 107.0 | 96.2 | 119.9 | 109.3 | 96.2 | 110.6 | 104.1 | 97.6 | 131.6 | 116.4 | 97.4 | 115.1 | 107.2 |
| COVERS | DIVS. 10 - 14, 25, 28, 41, 43, 44, 46 | 100.0 | 101.7 | 100.4 | 100.0 | 102.1 | 100.4 | 100.0 | 101.0 | 100.2 | 100.0 | 101.9 | 100.4 | 100.0 | 111.1 | 102.3 | 100.0 | 104.9 | 101.0 |
| 21, 22, 23 | FIRE SUPPRESSION, PLUMBING & HVAC | 94.3 | 94.1 | 94.2 | 94.9 | 93.8 | 94.5 | 100.0 | 99.7 | 99.9 | 94.3 | 92.9 | 93.7 | 99.9 | 124.0 | 109.9 | 99.9 | 100.3 | 100.1 |
| 26, 27, 3370 | ELECTRICAL, COMMUNICATIONS & UTIL. | 99.8 | 111.2 | 105.9 | 102.2 | 70.7 | 85.5 | 100.0 | 111.2 | 106.0 | 106.9 | 91.0 | 98.4 | 98.9 | 115.6 | 107.8 | 98.0 | 91.0 | 94.2 |
| MF2010 | WEIGHTED AVERAGE | 97.6 | 107.3 | 101.9 | 99.3 | 100.7 | 99.9 | 100.6 | 112.2 | 105.8 | 97.6 | 106.4 | 101.5 | 102.3 | 125.0 | 112.5 | 101.0 | 109.2 | 104.7 |

| DIVISION | | MINNESOTA | | | | | | | | | | | | | | | | MISSISSIPPI | | |
|---|---|---|---|---|---|---|---|---|---|---|---|---|---|---|---|---|---|---|---|---|
| | | SAINT PAUL | | | ST. CLOUD | | | THIEF RIVER FALLS | | | WILLMAR | | | WINDOM | | | BILOXI | | |
| | | 550 - 551 | | | 563 | | | 567 | | | 562 | | | 561 | | | 395 | | |
| | | MAT. | INST. | TOTAL | MAT. | INST. | TOTAL | MAT. | INST. | TOTAL | MAT. | INST. | TOTAL | MAT. | INST. | TOTAL | MAT. | INST. | TOTAL |
| 015433 | CONTRACTOR EQUIPMENT | | 101.0 | 101.0 | | 101.5 | 101.5 | | 98.6 | 98.6 | | 101.5 | 101.5 | | 101.5 | 101.5 | | 98.3 | 98.3 |
| 0241, 31 - 34 | SITE & INFRASTRUCTURE, DEMOLITION | 98.8 | 102.6 | 101.5 | 88.1 | 105.5 | 100.1 | 94.4 | 98.4 | 97.1 | 88.0 | 103.2 | 98.5 | 81.3 | 101.9 | 95.5 | 100.9 | 88.5 | 92.4 |
| 0310 | Concrete Forming & Accessories | 97.4 | 131.8 | 127.3 | 88.3 | 127.6 | 122.4 | 87.9 | 95.5 | 94.5 | 88.1 | 98.3 | 97.0 | 91.9 | 89.8 | 90.1 | 95.5 | 57.3 | 62.3 |
| 0320 | Concrete Reinforcing | 89.9 | 121.6 | 106.2 | 97.6 | 121.4 | 109.9 | 99.1 | 110.3 | 104.9 | 97.2 | 121.1 | 109.5 | 97.2 | 120.1 | 109.0 | 91.8 | 60.6 | 75.7 |
| 0330 | Cast-in-Place Concrete | 117.5 | 121.5 | 119.0 | 99.3 | 121.1 | 107.7 | 102.7 | 93.6 | 99.2 | 100.9 | 91.4 | 97.3 | 87.3 | 68.3 | 80.0 | 106.2 | 56.1 | 87.0 |
| 03 | CONCRETE | 112.0 | 126.7 | 119.0 | 94.3 | 124.6 | 108.8 | 97.7 | 98.7 | 98.2 | 94.3 | 101.3 | 97.7 | 84.8 | 89.5 | 87.0 | 95.3 | 59.1 | 78.0 |
| 04 | MASONRY | 125.3 | 131.5 | 129.1 | 112.2 | 129.3 | 122.6 | 103.6 | 107.6 | 106.0 | 116.7 | 120.2 | 118.8 | 127.2 | 92.1 | 105.9 | 89.1 | 46.7 | 63.4 |
| 05 | METALS | 100.4 | 133.4 | 111.7 | 94.2 | 131.6 | 107.1 | 92.6 | 123.2 | 103.1 | 93.3 | 129.1 | 105.7 | 93.3 | 126.6 | 104.7 | 92.7 | 84.0 | 89.7 |
| 06 | WOOD, PLASTICS & COMPOSITES | 106.1 | 130.5 | 120.1 | 91.9 | 125.0 | 110.9 | 74.1 | 93.0 | 84.9 | 91.7 | 93.0 | 92.4 | 95.5 | 90.1 | 92.4 | 99.8 | 60.9 | 77.5 |
| 07 | THERMAL & MOISTURE PROTECTION | 109.2 | 133.4 | 118.8 | 107.1 | 124.9 | 114.2 | 108.3 | 102.6 | 106.0 | 107.0 | 112.0 | 109.0 | 106.9 | 87.2 | 99.1 | 91.6 | 58.0 | 78.2 |
| 08 | OPENINGS | 91.3 | 137.5 | 103.0 | 89.5 | 134.6 | 100.9 | 96.6 | 113.3 | 100.8 | 87.1 | 98.0 | 89.9 | 91.3 | 96.4 | 92.6 | 97.9 | 61.8 | 88.8 |
| 0920 | Plaster & Gypsum Board | 89.6 | 131.9 | 119.3 | 85.0 | 126.3 | 114.1 | 100.1 | 93.1 | 95.2 | 85.0 | 93.3 | 90.9 | 85.4 | 90.3 | 88.8 | 98.6 | 60.2 | 71.5 |
| 0950, 0980 | Ceilings & Acoustic Treatment | 86.5 | 131.9 | 116.6 | 65.5 | 126.3 | 105.8 | 132.5 | 93.1 | 106.3 | 65.5 | 93.3 | 83.9 | 65.5 | 90.3 | 81.9 | 96.5 | 60.2 | 72.4 |
| 0960 | Flooring | 102.4 | 127.6 | 109.7 | 107.2 | 127.6 | 113.0 | 112.2 | 127.6 | 116.6 | 109.2 | 127.6 | 114.5 | 112.0 | 127.6 | 116.5 | 119.4 | 55.8 | 101.1 |
| 0970, 0990 | Wall Finishes & Painting/Coating | 97.7 | 128.2 | 116.2 | 104.0 | 130.8 | 120.3 | 98.4 | 102.4 | 100.8 | 98.4 | 102.4 | 100.8 | 98.4 | 105.4 | 102.6 | 105.9 | 46.8 | 70.0 |
| 09 | FINISHES | 94.5 | 131.2 | 114.8 | 93.6 | 128.3 | 112.7 | 111.2 | 102.1 | 106.2 | 93.8 | 101.5 | 98.0 | 94.1 | 96.7 | 95.5 | 105.4 | 56.1 | 78.2 |
| COVERS | DIVS. 10 - 14, 25, 28, 41, 43, 44, 46 | 100.0 | 110.8 | 102.2 | 100.0 | 108.7 | 101.8 | 100.0 | 99.3 | 99.9 | 100.0 | 101.7 | 100.3 | 100.0 | 96.5 | 99.3 | 100.0 | 62.0 | 92.1 |
| 21, 22, 23 | FIRE SUPPRESSION, PLUMBING & HVAC | 99.9 | 116.8 | 106.9 | 99.7 | 123.9 | 109.8 | 94.9 | 89.7 | 92.8 | 94.3 | 112.6 | 101.9 | 94.3 | 87.3 | 91.4 | 99.8 | 51.3 | 79.6 |
| 26, 27, 3370 | ELECTRICAL, COMMUNICATIONS & UTIL. | 98.5 | 115.6 | 107.6 | 99.8 | 115.6 | 108.2 | 99.3 | 70.7 | 84.1 | 99.3 | 92.5 | 95.9 | 106.9 | 90.9 | 98.4 | 98.7 | 59.4 | 77.8 |
| MF2010 | WEIGHTED AVERAGE | 101.6 | 123.0 | 111.2 | 97.3 | 123.2 | 108.9 | 98.1 | 97.1 | 97.7 | 95.9 | 107.2 | 100.9 | 96.2 | 95.3 | 95.8 | 97.6 | 60.8 | 81.2 |

| DIVISION | | MISSISSIPPI | | | | | | | | | | | | | | | | | | |
|---|---|---|---|---|---|---|---|---|---|---|---|---|---|---|---|---|---|---|---|---|
| | | CLARKSDALE | | | COLUMBUS | | | GREENVILLE | | | GREENWOOD | | | JACKSON | | | LAUREL | | |
| | | 386 | | | 397 | | | 387 | | | 389 | | | 390 - 392 | | | 394 | | |
| | | MAT. | INST. | TOTAL | MAT. | INST. | TOTAL | MAT. | INST. | TOTAL | MAT. | INST. | TOTAL | MAT. | INST. | TOTAL | MAT. | INST. | TOTAL |
| 015433 | CONTRACTOR EQUIPMENT | | 98.3 | 98.3 | | 98.3 | 98.3 | | 98.3 | 98.3 | | 98.3 | 98.3 | | 98.3 | 98.3 | | 98.3 | 98.3 |
| 0241, 31 - 34 | SITE & INFRASTRUCTURE, DEMOLITION | 98.6 | 88.3 | 91.5 | 99.7 | 88.6 | 92.1 | 104.6 | 88.6 | 93.6 | 101.6 | 88.0 | 92.3 | 96.2 | 88.6 | 90.9 | 106.0 | 88.4 | 93.9 |
| 0310 | Concrete Forming & Accessories | 86.4 | 53.4 | 57.8 | 84.5 | 55.6 | 59.4 | 83.0 | 69.2 | 71.0 | 95.4 | 53.4 | 59.0 | 94.8 | 62.3 | 66.6 | 84.5 | 57.4 | 61.0 |
| 0320 | Concrete Reinforcing | 98.9 | 44.8 | 71.0 | 98.3 | 45.9 | 71.3 | 99.5 | 65.0 | 81.7 | 98.9 | 56.6 | 77.1 | 91.8 | 58.6 | 74.7 | 98.9 | 42.1 | 69.7 |
| 0330 | Cast-in-Place Concrete | 103.6 | 57.0 | 85.7 | 108.3 | 62.5 | 90.8 | 106.7 | 59.3 | 88.5 | 111.4 | 56.6 | 90.4 | 90.4 | 60.1 | 78.8 | 105.9 | 58.8 | 87.9 |
| 03 | CONCRETE | 93.5 | 54.7 | 74.9 | 96.9 | 57.7 | 78.2 | 98.2 | 66.3 | 82.9 | 99.6 | 56.6 | 79.0 | 87.3 | 62.3 | 75.3 | 98.7 | 56.6 | 78.6 |
| 04 | MASONRY | 88.9 | 48.6 | 64.5 | 111.8 | 55.8 | 77.8 | 130.4 | 52.2 | 83.0 | 89.5 | 48.5 | 64.6 | 89.8 | 52.2 | 67.0 | 108.3 | 52.0 | 74.2 |
| 05 | METALS | 89.8 | 76.4 | 85.1 | 89.6 | 77.3 | 85.4 | 90.8 | 86.1 | 89.2 | 89.8 | 77.9 | 85.7 | 92.6 | 83.4 | 89.5 | 89.7 | 76.2 | 85.1 |
| 06 | WOOD, PLASTICS & COMPOSITES | 87.5 | 55.6 | 69.2 | 85.5 | 55.6 | 68.9 | 83.9 | 74.0 | 78.2 | 99.8 | 55.6 | 74.4 | 98.4 | 64.7 | 79.1 | 86.3 | 59.2 | 70.8 |
| 07 | THERMAL & MOISTURE PROTECTION | 91.4 | 50.6 | 75.2 | 91.5 | 55.2 | 77.1 | 91.7 | 64.2 | 80.8 | 91.7 | 50.5 | 75.4 | 93.4 | 63.2 | 81.4 | 91.7 | 57.0 | 77.9 |
| 08 | OPENINGS | 97.2 | 53.4 | 86.1 | 97.2 | 53.5 | 86.1 | 97.2 | 66.4 | 89.4 | 97.2 | 54.8 | 86.5 | 98.3 | 59.6 | 88.5 | 93.8 | 54.4 | 83.8 |
| 0920 | Plaster & Gypsum Board | 87.1 | 54.7 | 64.3 | 86.6 | 55.8 | 64.9 | 85.8 | 73.7 | 77.3 | 97.3 | 54.7 | 67.3 | 92.7 | 64.1 | 72.6 | 85.8 | 58.5 | 66.6 |
| 0950, 0980 | Ceilings & Acoustic Treatment | 89.7 | 54.7 | 66.5 | 89.7 | 55.8 | 67.2 | 91.8 | 73.7 | 79.8 | 89.7 | 54.7 | 66.5 | 93.7 | 64.1 | 74.1 | 89.7 | 58.5 | 69.0 |
| 0960 | Flooring | 113.0 | 55.8 | 96.6 | 111.8 | 62.6 | 97.7 | 110.9 | 55.8 | 95.1 | 119.4 | 55.8 | 101.1 | 119.8 | 55.8 | 101.4 | 109.9 | 55.8 | 94.4 |
| 0970, 0990 | Wall Finishes & Painting/Coating | 105.9 | 49.6 | 71.8 | 105.9 | 57.0 | 76.3 | 105.9 | 70.9 | 84.7 | 105.9 | 49.6 | 71.8 | 105.9 | 70.9 | 84.7 | 105.9 | 65.6 | 81.4 |
| 09 | FINISHES | 100.5 | 53.8 | 74.7 | 100.3 | 57.4 | 76.6 | 100.8 | 67.8 | 82.6 | 104.0 | 53.8 | 76.3 | 104.4 | 62.4 | 81.2 | 100.2 | 58.6 | 77.2 |
| COVERS | DIVS. 10 - 14, 25, 28, 41, 43, 44, 46 | 100.0 | 62.4 | 92.2 | 100.0 | 63.6 | 92.4 | 100.0 | 65.5 | 92.8 | 100.0 | 62.3 | 92.2 | 100.0 | 64.4 | 92.6 | 100.0 | 41.6 | 87.9 |
| 21, 22, 23 | FIRE SUPPRESSION, PLUMBING & HVAC | 97.5 | 45.3 | 75.8 | 96.9 | 41.2 | 73.7 | 99.8 | 52.2 | 80.0 | 97.5 | 43.2 | 74.9 | 99.8 | 57.8 | 82.3 | 96.9 | 50.3 | 77.5 |
| 26, 27, 3370 | ELECTRICAL, COMMUNICATIONS & UTIL. | 98.0 | 50.3 | 72.6 | 95.9 | 53.3 | 73.3 | 98.0 | 68.0 | 82.1 | 98.0 | 47.2 | 70.9 | 101.9 | 68.0 | 83.9 | 97.6 | 58.9 | 77.0 |
| MF2010 | WEIGHTED AVERAGE | 95.7 | 56.3 | 78.1 | 97.0 | 57.7 | 79.4 | 99.3 | 66.1 | 84.5 | 97.0 | 55.8 | 78.6 | 96.9 | 65.3 | 82.8 | 97.0 | 59.3 | 80.2 |

| DIVISION | | MISSISSIPPI | | | | | | | | | MISSOURI | | | | | | | | | |
|---|---|---|---|---|---|---|---|---|---|---|---|---|---|---|---|---|---|---|---|---|
| | | MCCOMB | | | MERIDIAN | | | TUPELO | | | BOWLING GREEN | | | CAPE GIRARDEAU | | | CHILLICOTHE | | |
| | | 396 | | | 393 | | | 388 | | | 633 | | | 637 | | | 646 | | |
| | | MAT. | INST. | TOTAL | MAT. | INST. | TOTAL | MAT. | INST. | TOTAL | MAT. | INST. | TOTAL | MAT. | INST. | TOTAL | MAT. | INST. | TOTAL |
| 015433 | CONTRACTOR EQUIPMENT | | 98.3 | 98.3 | | 98.3 | 98.3 | | 98.3 | 98.3 | | 106.5 | 106.5 | | 106.5 | 106.5 | | 101.6 | 101.6 |
| 0241, 31 - 34 | SITE & INFRASTRUCTURE, DEMOLITION | 92.3 | 88.2 | 89.5 | 96.2 | 88.9 | 91.1 | 96.1 | 88.2 | 90.7 | 92.4 | 93.4 | 93.1 | 93.6 | 93.4 | 93.4 | 98.8 | 90.4 | 93.0 |
| 0310 | Concrete Forming & Accessories | 84.5 | 55.3 | 59.2 | 81.9 | 68.6 | 70.4 | 83.5 | 55.6 | 59.3 | 96.2 | 89.0 | 90.0 | 89.7 | 76.6 | 78.3 | 91.2 | 91.9 | 91.8 |
| 0320 | Concrete Reinforcing | 99.5 | 44.2 | 71.0 | 98.3 | 58.7 | 77.9 | 96.8 | 56.7 | 76.1 | 103.7 | 98.4 | 101.0 | 105.0 | 84.6 | 94.5 | 97.9 | 101.7 | 99.9 |
| 0330 | Cast-in-Place Concrete | 94.0 | 56.4 | 79.6 | 100.6 | 62.7 | 86.1 | 103.6 | 58.5 | 86.3 | 94.0 | 77.2 | 87.6 | 93.0 | 84.3 | 89.7 | 94.2 | 66.7 | 83.7 |
| 03 | CONCRETE | 87.1 | 55.1 | 71.8 | 91.2 | 66.0 | 79.2 | 93.0 | 58.2 | 76.3 | 94.5 | 87.9 | 91.4 | 93.7 | 82.3 | 88.2 | 99.7 | 85.5 | 92.9 |
| 04 | MASONRY | 113.3 | 47.9 | 73.7 | 88.7 | 56.5 | 69.2 | 119.9 | 51.7 | 78.6 | 126.0 | 98.0 | 109.1 | 123.1 | 79.0 | 96.4 | 99.7 | 92.8 | 95.5 |
| 05 | METALS | 89.9 | 74.5 | 84.6 | 90.8 | 83.6 | 88.3 | 89.7 | 78.2 | 85.7 | 96.5 | 112.5 | 102.0 | 97.6 | 106.0 | 100.5 | 92.0 | 97.2 | 93.8 |
| 06 | WOOD, PLASTICS & COMPOSITES | 85.5 | 58.6 | 70.1 | 82.8 | 70.7 | 75.9 | 84.5 | 56.6 | 68.5 | 93.5 | 89.3 | 91.1 | 86.5 | 73.8 | 79.2 | 85.7 | 98.1 | 92.9 |
| 07 | THERMAL & MOISTURE PROTECTION | 91.2 | 49.7 | 74.7 | 91.3 | 65.7 | 81.1 | 91.4 | 56.8 | 77.7 | 99.0 | 100.4 | 99.6 | 98.9 | 82.5 | 92.4 | 97.6 | 86.2 | 93.1 |
| 08 | OPENINGS | 97.2 | 55.2 | 86.6 | 97.2 | 68.9 | 90.0 | 97.2 | 54.7 | 86.4 | 99.6 | 84.8 | 95.8 | 99.5 | 75.8 | 93.5 | 89.6 | 86.2 | 88.7 |
| 0920 | Plaster & Gypsum Board | 86.6 | 57.8 | 66.4 | 85.8 | 70.3 | 74.9 | 85.8 | 55.8 | 64.7 | 94.8 | 89.0 | 90.7 | 92.1 | 73.0 | 78.7 | 98.0 | 97.9 | 97.9 |
| 0950, 0980 | Ceilings & Acoustic Treatment | 89.7 | 57.8 | 68.6 | 91.8 | 70.3 | 77.6 | 89.7 | 55.8 | 67.2 | 85.9 | 89.0 | 87.9 | 85.9 | 73.0 | 77.3 | 86.8 | 97.9 | 94.2 |
| 0960 | Flooring | 111.8 | 55.8 | 95.7 | 109.9 | 55.8 | 94.3 | 111.3 | 55.8 | 95.3 | 97.2 | 96.6 | 97.0 | 93.6 | 88.8 | 92.2 | 94.7 | 45.4 | 80.6 |
| 0970, 0990 | Wall Finishes & Painting/Coating | 105.9 | 45.5 | 69.3 | 105.9 | 82.0 | 91.4 | 105.9 | 63.2 | 80.0 | 100.9 | 105.6 | 103.8 | 100.9 | 74.3 | 84.8 | 95.1 | 48.3 | 66.7 |
| 09 | FINISHES | 99.7 | 54.9 | 75.0 | 99.7 | 68.3 | 82.4 | 99.8 | 56.7 | 76.0 | 95.8 | 91.1 | 93.2 | 94.4 | 76.7 | 84.6 | 96.0 | 81.0 | 87.7 |
| COVERS | DIVS. 10 - 14, 25, 28, 41, 43, 44, 46 | 100.0 | 66.2 | 93.0 | 100.0 | 66.6 | 93.1 | 100.0 | 63.6 | 92.4 | 100.0 | 92.9 | 98.5 | 100.0 | 93.9 | 98.7 | 100.0 | 57.5 | 91.2 |
| 21, 22, 23 | FIRE SUPPRESSION, PLUMBING & HVAC | 96.9 | 37.9 | 72.3 | 99.8 | 60.1 | 83.3 | 97.8 | 47.5 | 76.9 | 94.5 | 88.7 | 92.1 | 99.9 | 97.1 | 98.7 | 94.9 | 45.0 | 74.1 |
| 26, 27, 3370 | ELECTRICAL, COMMUNICATIONS & UTIL. | 94.3 | 53.7 | 72.7 | 97.6 | 60.5 | 77.9 | 97.7 | 53.5 | 74.2 | 99.2 | 84.7 | 91.5 | 99.2 | 110.2 | 105.0 | 92.5 | 35.7 | 62.3 |
| MF2010 | WEIGHTED AVERAGE | 95.4 | 55.3 | 77.5 | 95.9 | 67.0 | 83.0 | 97.1 | 58.8 | 80.0 | 98.1 | 92.2 | 95.5 | 99.2 | 91.3 | 95.7 | 95.2 | 71.1 | 84.4 |

## MISSOURI

| DIVISION | | COLUMBIA 652 | | | FLAT RIVER 636 | | | HANNIBAL 634 | | | HARRISONVILLE 647 | | | JEFFERSON CITY 650 - 651 | | | JOPLIN 648 | | |
|---|---|---|---|---|---|---|---|---|---|---|---|---|---|---|---|---|---|---|---|
| | | MAT. | INST. | TOTAL | MAT. | INST. | TOTAL | MAT. | INST. | TOTAL | MAT. | INST. | TOTAL | MAT. | INST. | TOTAL | MAT. | INST. | TOTAL |
| 015433 | CONTRACTOR EQUIPMENT | | 107.4 | 107.4 | | 106.5 | 106.5 | | 106.5 | 106.5 | | 101.6 | 101.6 | | 107.4 | 107.4 | | 105.0 | 105.0 |
| 0241, 31 - 34 | SITE & INFRASTRUCTURE, DEMOLITION | 107.4 | 95.7 | 99.3 | 95.2 | 93.5 | 94.0 | 90.0 | 93.2 | 92.2 | 90.6 | 95.5 | 94.0 | 103.5 | 95.6 | 98.0 | 98.0 | 97.8 | 97.8 |
| 0310 | Concrete Forming & Accessories | 89.4 | 81.1 | 82.2 | 101.9 | 89.0 | 90.7 | 94.6 | 79.0 | 81.0 | 88.4 | 106.1 | 103.7 | 99.3 | 81.0 | 83.4 | 102.6 | 75.0 | 78.7 |
| 0320 | Concrete Reinforcing | 100.3 | 113.4 | 107.0 | 105.0 | 107.4 | 106.2 | 103.1 | 98.3 | 100.6 | 97.5 | 113.2 | 105.6 | 99.0 | 108.4 | 103.9 | 101.0 | 73.4 | 86.8 |
| 0330 | Cast-in-Place Concrete | 91.2 | 84.3 | 88.6 | 97.1 | 96.5 | 96.9 | 89.0 | 76.7 | 84.3 | 96.6 | 103.4 | 99.2 | 91.8 | 84.3 | 88.7 | 102.1 | 67.0 | 86.8 |
| 03 | CONCRETE | 86.3 | 89.7 | 88.0 | 97.5 | 96.3 | 96.9 | 90.7 | 83.3 | 87.2 | 96.2 | 106.7 | 101.2 | 87.8 | 88.7 | 88.2 | 99.5 | 72.9 | 86.7 |
| 04 | MASONRY | 133.3 | 87.4 | 105.5 | 123.6 | 80.5 | 97.5 | 117.1 | 98.0 | 105.5 | 94.5 | 104.6 | 100.6 | 96.6 | 87.4 | 91.0 | 92.6 | 86.6 | 89.0 |
| 05 | METALS | 92.9 | 119.5 | 102.1 | 96.4 | 115.7 | 103.1 | 96.5 | 111.3 | 101.6 | 92.4 | 112.7 | 99.4 | 92.3 | 117.3 | 100.9 | 95.1 | 88.8 | 92.9 |
| 06 | WOOD, PLASTICS & COMPOSITES | 90.7 | 78.8 | 83.9 | 100.6 | 89.3 | 94.2 | 91.3 | 76.8 | 83.0 | 82.3 | 106.0 | 95.9 | 101.2 | 78.8 | 88.3 | 96.5 | 76.1 | 84.8 |
| 07 | THERMAL & MOISTURE PROTECTION | 93.8 | 87.6 | 91.4 | 99.2 | 95.7 | 97.8 | 98.9 | 94.1 | 97.0 | 96.9 | 106.5 | 100.7 | 93.9 | 87.6 | 91.4 | 97.3 | 73.2 | 87.7 |
| 08 | OPENINGS | 91.3 | 95.3 | 92.3 | 99.5 | 87.5 | 96.5 | 99.6 | 78.0 | 94.1 | 89.9 | 108.2 | 94.5 | 88.6 | 94.0 | 89.9 | 90.8 | 73.8 | 86.5 |
| 0920 | Plaster & Gypsum Board | 93.9 | 78.1 | 82.8 | 100.1 | 89.0 | 92.3 | 92.6 | 76.0 | 80.9 | 91.2 | 106.0 | 101.7 | 98.9 | 78.1 | 84.3 | 105.1 | 75.2 | 84.0 |
| 0950, 0980 | Ceilings & Acoustic Treatment | 89.5 | 78.1 | 81.9 | 85.9 | 89.0 | 87.9 | 85.9 | 76.0 | 79.3 | 86.8 | 106.0 | 99.6 | 88.6 | 78.1 | 81.6 | 88.7 | 75.2 | 79.7 |
| 0960 | Flooring | 109.4 | 96.7 | 105.8 | 100.4 | 88.8 | 97.0 | 96.3 | 96.6 | 96.4 | 90.3 | 104.9 | 94.5 | 117.0 | 96.7 | 111.2 | 119.9 | 42.6 | 97.7 |
| 0970, 0990 | Wall Finishes & Painting/Coating | 104.0 | 81.1 | 90.1 | 100.9 | 63.9 | 78.5 | 100.9 | 105.6 | 103.8 | 99.8 | 107.9 | 104.7 | 104.0 | 81.1 | 90.1 | 94.6 | 37.8 | 60.1 |
| 09 | FINISHES | 96.9 | 82.6 | 89.0 | 97.6 | 85.1 | 90.7 | 95.1 | 83.7 | 88.8 | 93.6 | 106.0 | 100.4 | 99.8 | 82.6 | 90.3 | 102.8 | 65.6 | 82.3 |
| COVERS | DIVS. 10 - 14, 25, 28, 41, 43, 44, 46 | 100.0 | 95.6 | 99.1 | 100.0 | 96.2 | 99.2 | 100.0 | 56.6 | 91.0 | 100.0 | 99.9 | 100.0 | 100.0 | 95.6 | 99.1 | 100.0 | 64.1 | 92.5 |
| 21, 22, 23 | FIRE SUPPRESSION, PLUMBING & HVAC | 100.0 | 97.4 | 98.9 | 94.5 | 88.7 | 92.1 | 94.5 | 88.7 | 92.0 | 94.8 | 103.5 | 98.5 | 100.0 | 91.6 | 96.5 | 100.3 | 49.1 | 79.0 |
| 26, 27, 3370 | ELECTRICAL, COMMUNICATIONS & UTIL. | 94.3 | 89.9 | 91.9 | 104.2 | 110.2 | 107.4 | 97.8 | 84.7 | 90.8 | 99.7 | 103.8 | 101.9 | 99.1 | 89.9 | 94.2 | 90.2 | 70.5 | 79.8 |
| MF2010 | WEIGHTED AVERAGE | 97.4 | 93.7 | 95.7 | 99.1 | 94.7 | 97.1 | 96.9 | 88.8 | 93.3 | 94.8 | 104.8 | 99.3 | 95.9 | 92.1 | 94.2 | 97.1 | 71.4 | 85.6 |

## MISSOURI

| DIVISION | | KANSAS CITY 640 - 641 | | | KIRKSVILLE 635 | | | POPLAR BLUFF 639 | | | ROLLA 654 - 655 | | | SEDALIA 653 | | | SIKESTON 638 | | |
|---|---|---|---|---|---|---|---|---|---|---|---|---|---|---|---|---|---|---|---|
| | | MAT. | INST. | TOTAL | MAT. | INST. | TOTAL | MAT. | INST. | TOTAL | MAT. | INST. | TOTAL | MAT. | INST. | TOTAL | MAT. | INST. | TOTAL |
| 015433 | CONTRACTOR EQUIPMENT | | 102.8 | 102.8 | | 98.9 | 98.9 | | 101.1 | 101.1 | | 107.4 | 107.4 | | 99.5 | 99.5 | | 101.1 | 101.1 |
| 0241, 31 - 34 | SITE & INFRASTRUCTURE, DEMOLITION | 92.0 | 97.7 | 96.0 | 91.6 | 89.8 | 90.4 | 78.1 | 93.0 | 88.4 | 107.0 | 93.8 | 97.9 | 98.3 | 92.1 | 94.0 | 81.6 | 93.2 | 89.6 |
| 0310 | Concrete Forming & Accessories | 101.3 | 106.9 | 106.2 | 87.2 | 79.6 | 80.6 | 87.6 | 74.1 | 75.9 | 96.4 | 92.4 | 92.9 | 93.6 | 81.5 | 83.3 | 106.4 | 84.5 | 86.9 |
| 0320 | Concrete Reinforcing | 96.0 | 114.1 | 105.3 | 103.9 | 87.3 | 95.4 | 107.1 | 84.6 | 95.5 | 100.8 | 68.9 | 84.4 | 99.3 | 106.4 | 102.9 | 79.8 | 83.9 | 81.3 |
| 0330 | Cast-in-Place Concrete | 94.9 | 107.8 | 99.8 | 97.0 | 76.6 | 89.2 | 74.6 | 81.5 | 77.3 | 93.3 | 87.5 | 91.1 | 97.5 | 85.1 | 92.7 | 87.6 | 80.9 | 84.4 |
| 03 | CONCRETE | 95.5 | 108.8 | 101.8 | 107.6 | 80.9 | 94.9 | 84.0 | 79.7 | 81.9 | 88.1 | 87.6 | 87.9 | 99.7 | 88.3 | 94.2 | 87.6 | 80.9 | 84.4 |
| 04 | MASONRY | 96.5 | 108.4 | 103.7 | 129.6 | 89.3 | 105.2 | 121.5 | 73.1 | 92.2 | 110.0 | 80.5 | 92.1 | 116.1 | 87.8 | 99.0 | 121.2 | 77.9 | 95.0 |
| 05 | METALS | 102.7 | 113.7 | 106.5 | 96.1 | 97.6 | 96.6 | 96.7 | 97.7 | 97.0 | 92.2 | 103.6 | 96.2 | 90.9 | 108.2 | 96.8 | 97.0 | 96.9 | 97.0 |
| 06 | WOOD, PLASTICS & COMPOSITES | 95.8 | 106.4 | 101.9 | 79.6 | 78.8 | 79.1 | 78.8 | 73.8 | 76.0 | 97.8 | 95.2 | 96.3 | 90.8 | 78.8 | 83.9 | 80.2 | 73.8 | 76.6 |
| 07 | THERMAL & MOISTURE PROTECTION | 97.0 | 109.5 | 101.9 | 104.9 | 87.3 | 97.9 | 103.6 | 82.3 | 95.1 | 94.0 | 92.8 | 93.5 | 99.2 | 94.8 | 97.4 | 103.8 | 80.7 | 94.6 |
| 08 | OPENINGS | 96.9 | 108.8 | 99.9 | 103.9 | 76.5 | 96.9 | 104.9 | 75.1 | 97.3 | 91.3 | 89.1 | 90.7 | 94.7 | 88.9 | 93.2 | 104.9 | 68.5 | 95.7 |
| 0920 | Plaster & Gypsum Board | 100.9 | 106.5 | 104.8 | 87.2 | 78.1 | 80.8 | 87.7 | 73.0 | 77.3 | 95.6 | 95.0 | 95.2 | 88.5 | 78.1 | 81.2 | 89.5 | 73.0 | 77.9 |
| 0950, 0980 | Ceilings & Acoustic Treatment | 94.3 | 106.5 | 102.4 | 84.0 | 78.1 | 80.1 | 85.9 | 73.0 | 77.3 | 89.5 | 95.0 | 93.2 | 89.5 | 78.1 | 81.9 | 85.9 | 73.0 | 77.3 |
| 0960 | Flooring | 95.9 | 108.8 | 99.6 | 74.8 | 96.7 | 81.1 | 88.7 | 82.2 | 86.8 | 113.3 | 96.7 | 108.6 | 86.8 | 97.9 | 90.0 | 89.1 | 82.2 | 87.1 |
| 0970, 0990 | Wall Finishes & Painting/Coating | 99.8 | 115.5 | 109.4 | 96.3 | 81.1 | 87.0 | 95.8 | 73.7 | 82.4 | 104.0 | 88.2 | 94.4 | 104.0 | 107.9 | 106.4 | 95.8 | 58.9 | 73.4 |
| 09 | FINISHES | 97.9 | 108.0 | 103.5 | 94.3 | 81.7 | 87.3 | 94.4 | 74.1 | 83.2 | 98.3 | 92.2 | 94.9 | 95.2 | 86.1 | 90.2 | 95.0 | 73.0 | 82.8 |
| COVERS | DIVS. 10 - 14, 25, 28, 41, 43, 44, 46 | 100.0 | 101.3 | 100.3 | 100.0 | 54.9 | 90.6 | 100.0 | 91.8 | 98.3 | 100.0 | 59.1 | 91.5 | 100.0 | 93.1 | 98.6 | 100.0 | 93.5 | 98.7 |
| 21, 22, 23 | FIRE SUPPRESSION, PLUMBING & HVAC | 100.2 | 108.7 | 103.7 | 94.5 | 88.1 | 91.8 | 94.5 | 93.5 | 94.1 | 94.6 | 89.0 | 92.3 | 94.6 | 98.2 | 96.1 | 94.5 | 86.4 | 91.1 |
| 26, 27, 3370 | ELECTRICAL, COMMUNICATIONS & UTIL. | 101.3 | 104.2 | 102.8 | 98.0 | 84.6 | 90.9 | 98.3 | 110.2 | 104.6 | 92.6 | 89.8 | 91.1 | 94.0 | 103.8 | 99.2 | 97.3 | 110.2 | 104.1 |
| MF2010 | WEIGHTED AVERAGE | 99.0 | 107.3 | 102.7 | 100.1 | 85.4 | 93.5 | 96.6 | 88.4 | 92.9 | 95.0 | 89.5 | 92.5 | 96.5 | 94.6 | 95.7 | 97.1 | 87.1 | 92.6 |

## MISSOURI / MONTANA

| DIVISION | | SPRINGFIELD 656 - 658 | | | ST. JOSEPH 644 - 645 | | | ST. LOUIS 630 - 631 | | | BILLINGS 590 - 591 | | | BUTTE 597 | | | GREAT FALLS 594 | | |
|---|---|---|---|---|---|---|---|---|---|---|---|---|---|---|---|---|---|---|---|
| | | MAT. | INST. | TOTAL | MAT. | INST. | TOTAL | MAT. | INST. | TOTAL | MAT. | INST. | TOTAL | MAT. | INST. | TOTAL | MAT. | INST. | TOTAL |
| 015433 | CONTRACTOR EQUIPMENT | | 101.9 | 101.9 | | 101.6 | 101.6 | | 107.6 | 107.6 | | 98.9 | 98.9 | | 98.6 | 98.6 | | 98.6 | 98.6 |
| 0241, 31 - 34 | SITE & INFRASTRUCTURE, DEMOLITION | 100.1 | 93.7 | 95.6 | 93.5 | 93.7 | 93.6 | 93.4 | 97.2 | 96.1 | 95.9 | 96.8 | 96.5 | 106.4 | 96.1 | 99.3 | 110.4 | 96.8 | 101.0 |
| 0310 | Concrete Forming & Accessories | 101.7 | 75.7 | 79.1 | 101.2 | 89.4 | 91.0 | 101.9 | 102.7 | 102.6 | 98.0 | 67.3 | 71.3 | 87.2 | 65.3 | 68.2 | 97.8 | 68.8 | 72.6 |
| 0320 | Concrete Reinforcing | 96.5 | 112.9 | 104.9 | 94.8 | 112.8 | 104.1 | 96.0 | 109.2 | 102.8 | 96.6 | 81.5 | 88.8 | 104.9 | 81.5 | 92.8 | 96.7 | 81.5 | 88.8 |
| 0330 | Cast-in-Place Concrete | 99.1 | 73.9 | 89.4 | 94.8 | 104.5 | 98.5 | 93.0 | 108.3 | 98.9 | 113.0 | 67.6 | 95.6 | 124.3 | 69.3 | 103.3 | 131.4 | 65.6 | 106.2 |
| 03 | CONCRETE | 96.8 | 83.0 | 90.2 | 95.3 | 99.7 | 97.4 | 93.1 | 106.7 | 99.6 | 106.0 | 71.0 | 89.3 | 109.7 | 70.8 | 91.1 | 115.0 | 71.1 | 94.0 |
| 04 | MASONRY | 90.0 | 86.3 | 87.8 | 96.2 | 96.0 | 96.1 | 103.7 | 110.5 | 107.8 | 130.4 | 71.4 | 94.7 | 125.4 | 73.2 | 93.8 | 130.1 | 75.6 | 97.0 |
| 05 | METALS | 96.9 | 106.0 | 100.0 | 98.9 | 111.6 | 103.3 | 102.5 | 119.6 | 108.4 | 110.0 | 90.8 | 103.4 | 102.9 | 90.8 | 98.7 | 106.4 | 90.8 | 101.0 |
| 06 | WOOD, PLASTICS & COMPOSITES | 98.1 | 74.9 | 84.8 | 96.5 | 86.9 | 91.0 | 100.5 | 100.7 | 100.6 | 98.0 | 67.6 | 80.6 | 87.4 | 64.9 | 74.5 | 99.1 | 67.6 | 81.0 |
| 07 | THERMAL & MOISTURE PROTECTION | 98.0 | 81.6 | 91.5 | 97.4 | 97.0 | 97.3 | 99.0 | 107.1 | 102.2 | 106.3 | 70.2 | 92.0 | 106.2 | 70.1 | 91.9 | 106.7 | 69.0 | 91.7 |
| 08 | OPENINGS | 96.9 | 82.1 | 93.2 | 95.5 | 98.0 | 96.1 | 98.9 | 108.3 | 101.3 | 95.5 | 67.6 | 88.4 | 93.5 | 65.5 | 86.4 | 96.8 | 66.9 | 89.3 |
| 0920 | Plaster & Gypsum Board | 97.0 | 74.0 | 80.8 | 106.4 | 86.4 | 92.3 | 101.0 | 100.7 | 100.8 | 94.6 | 67.0 | 75.1 | 95.3 | 64.2 | 73.4 | 103.3 | 67.0 | 77.7 |
| 0950, 0980 | Ceilings & Acoustic Treatment | 89.5 | 74.0 | 79.2 | 93.4 | 86.4 | 88.7 | 90.5 | 100.7 | 97.3 | 102.1 | 67.0 | 78.8 | 106.5 | 64.2 | 78.5 | 108.4 | 67.0 | 80.9 |
| 0960 | Flooring | 111.2 | 42.6 | 91.5 | 99.1 | 97.9 | 98.7 | 100.1 | 100.1 | 100.1 | 111.9 | 74.2 | 101.1 | 109.1 | 78.4 | 100.3 | 117.8 | 78.4 | 106.3 |
| 0970, 0990 | Wall Finishes & Painting/Coating | 98.1 | 54.0 | 71.4 | 95.1 | 90.7 | 92.4 | 100.9 | 109.0 | 105.8 | 100.0 | 50.4 | 69.9 | 98.4 | 55.2 | 72.2 | 98.4 | 52.4 | 70.5 |
| 09 | FINISHES | 100.7 | 66.6 | 81.9 | 99.1 | 90.4 | 94.3 | 98.5 | 102.2 | 100.5 | 104.3 | 66.3 | 83.3 | 104.1 | 66.2 | 83.2 | 108.4 | 68.3 | 86.3 |
| COVERS | DIVS. 10 - 14, 25, 28, 41, 43, 44, 46 | 100.0 | 93.7 | 98.7 | 100.0 | 97.1 | 99.4 | 100.0 | 102.2 | 100.5 | 100.0 | 77.7 | 95.4 | 100.0 | 77.5 | 95.3 | 100.0 | 79.2 | 95.7 |
| 21, 22, 23 | FIRE SUPPRESSION, PLUMBING & HVAC | 100.1 | 72.3 | 88.5 | 100.3 | 89.2 | 95.7 | 99.9 | 103.3 | 101.3 | 100.1 | 73.1 | 88.8 | 100.2 | 70.0 | 87.6 | 100.2 | 75.2 | 89.8 |
| 26, 27, 3370 | ELECTRICAL, COMMUNICATIONS & UTIL. | 98.7 | 74.0 | 85.5 | 99.7 | 78.1 | 88.2 | 101.9 | 110.6 | 106.5 | 94.7 | 78.3 | 86.0 | 102.7 | 74.1 | 87.5 | 94.8 | 75.1 | 84.3 |
| MF2010 | WEIGHTED AVERAGE | 98.2 | 80.9 | 90.5 | 98.3 | 93.2 | 96.0 | 99.4 | 106.6 | 102.6 | 103.2 | 75.9 | 91.0 | 103.2 | 74.6 | 90.4 | 104.7 | 76.5 | 92.1 |

# City Cost Indexes

## MONTANA

| DIVISION | | HAVRE 595 MAT. | INST. | TOTAL | HELENA 596 MAT. | INST. | TOTAL | KALISPELL 599 MAT. | INST. | TOTAL | MILES CITY 593 MAT. | INST. | TOTAL | MISSOULA 598 MAT. | INST. | TOTAL | WOLF POINT 592 MAT. | INST. | TOTAL |
|---|---|---|---|---|---|---|---|---|---|---|---|---|---|---|---|---|---|---|---|
| 015433 | CONTRACTOR EQUIPMENT | | 98.6 | 98.6 | | 98.6 | 98.6 | | 98.6 | 98.6 | | 98.6 | 98.6 | | 98.6 | 98.6 | | 98.6 | 98.6 |
| 0241, 31 - 34 | SITE & INFRASTRUCTURE, DEMOLITION | 115.9 | 96.5 | 102.5 | 99.1 | 96.0 | 96.9 | 95.6 | 95.7 | 95.7 | 103.2 | 96.1 | 98.3 | 86.4 | 95.9 | 93.0 | 124.4 | 96.2 | 105.0 |
| 0310 | Concrete Forming & Accessories | 81.3 | 58.4 | 61.4 | 98.0 | 68.0 | 72.0 | 89.8 | 56.3 | 60.7 | 96.5 | 56.8 | 62.1 | 89.8 | 63.6 | 67.1 | 90.3 | 51.8 | 56.8 |
| 0320 | Concrete Reinforcing | 105.7 | 65.3 | 84.9 | 100.0 | 65.0 | 82.0 | 107.7 | 81.2 | 94.0 | 105.2 | 65.3 | 84.7 | 106.6 | 81.5 | 93.7 | 106.8 | 65.3 | 85.4 |
| 0330 | Cast-in-Place Concrete | 133.9 | 64.2 | 107.2 | 102.0 | 64.6 | 87.6 | 107.9 | 61.7 | 90.2 | 118.2 | 63.8 | 97.4 | 91.6 | 63.8 | 80.9 | 132.4 | 64.2 | 106.3 |
| 03 | CONCRETE | 118.4 | 62.8 | 91.8 | 100.7 | 67.2 | 84.6 | 98.9 | 64.1 | 82.2 | 106.3 | 62.0 | 85.1 | 86.8 | 68.1 | 77.8 | 122.3 | 59.9 | 92.4 |
| 04 | MASONRY | 126.4 | 73.6 | 94.4 | 121.2 | 70.0 | 90.2 | 124.3 | 68.3 | 90.4 | 132.3 | 68.7 | 93.7 | 150.9 | 71.8 | 102.9 | 133.5 | 69.4 | 94.7 |
| 05 | METALS | 99.1 | 82.6 | 93.5 | 105.5 | 82.0 | 97.4 | 99.0 | 89.2 | 95.6 | 98.3 | 82.5 | 92.9 | 99.6 | 89.9 | 96.2 | 98.4 | 82.5 | 92.9 |
| 06 | WOOD, PLASTICS & COMPOSITES | 80.0 | 55.2 | 64.8 | 99.1 | 67.6 | 81.1 | 90.3 | 53.5 | 69.2 | 96.6 | 55.7 | 73.1 | 90.3 | 63.7 | 75.1 | 89.4 | 48.5 | 65.9 |
| 07 | THERMAL & MOISTURE PROTECTION | 106.5 | 66.8 | 90.7 | 106.5 | 71.0 | 92.4 | 105.8 | 67.7 | 90.7 | 106.1 | 64.9 | 89.8 | 105.5 | 69.5 | 91.2 | 107.0 | 63.9 | 89.9 |
| 08 | OPENINGS | 93.6 | 55.6 | 83.9 | 96.3 | 61.2 | 87.4 | 93.5 | 59.3 | 84.9 | 93.0 | 55.8 | 83.6 | 93.5 | 64.6 | 86.2 | 93.0 | 51.7 | 82.6 |
| 0920 | Plaster & Gypsum Board | 90.4 | 54.2 | 64.9 | 95.4 | 67.0 | 75.4 | 95.7 | 52.4 | 65.2 | 102.3 | 54.6 | 68.7 | 95.7 | 62.9 | 72.6 | 97.4 | 47.2 | 62.1 |
| 0950, 0980 | Ceilings & Acoustic Treatment | 106.5 | 54.2 | 71.8 | 105.6 | 67.0 | 80.0 | 106.5 | 52.4 | 70.6 | 104.7 | 54.6 | 71.5 | 106.5 | 62.9 | 77.6 | 104.7 | 47.2 | 66.6 |
| 0960 | Flooring | 105.7 | 78.4 | 97.8 | 117.8 | 62.4 | 101.9 | 111.3 | 78.4 | 101.8 | 117.5 | 74.2 | 105.1 | 111.3 | 78.4 | 101.8 | 113.0 | 74.2 | 101.9 |
| 0970, 0990 | Wall Finishes & Painting/Coating | 98.4 | 52.4 | 70.5 | 98.4 | 46.6 | 66.9 | 98.4 | 40.3 | 63.1 | 98.4 | 48.3 | 68.0 | 98.4 | 51.2 | 69.8 | 98.4 | 52.4 | 70.5 |
| 09 | FINISHES | 103.3 | 60.5 | 79.7 | 106.3 | 64.9 | 83.5 | 104.2 | 57.7 | 78.6 | 106.9 | 58.4 | 80.1 | 103.5 | 64.7 | 82.1 | 106.5 | 54.7 | 77.9 |
| COVERS | DIVS. 10 - 14, 25, 28, 41, 43, 44, 46 | 100.0 | 77.1 | 95.2 | 100.0 | 79.2 | 95.7 | 100.0 | 76.1 | 95.0 | 100.0 | 75.4 | 94.9 | 100.0 | 76.9 | 95.2 | 100.0 | 74.8 | 94.8 |
| 21, 22, 23 | FIRE SUPPRESSION, PLUMBING & HVAC | 94.8 | 73.8 | 86.0 | 100.1 | 71.5 | 88.2 | 94.8 | 69.7 | 84.3 | 94.8 | 71.2 | 85.0 | 100.2 | 69.1 | 87.2 | 94.8 | 71.6 | 85.1 |
| 26, 27, 3370 | ELECTRICAL, COMMUNICATIONS & UTIL. | 94.8 | 67.4 | 80.2 | 101.0 | 75.1 | 87.2 | 99.3 | 73.5 | 85.6 | 94.8 | 70.2 | 81.7 | 100.4 | 73.5 | 86.1 | 94.8 | 65.4 | 79.2 |
| MF2010 | WEIGHTED AVERAGE | 101.9 | 71.3 | 88.2 | 102.5 | 73.2 | 89.4 | 99.4 | 71.3 | 86.8 | 100.6 | 70.2 | 87.0 | 100.5 | 73.4 | 88.4 | 103.2 | 68.6 | 87.7 |

## NEBRASKA

| DIVISION | | ALLIANCE 693 MAT. | INST. | TOTAL | COLUMBUS 686 MAT. | INST. | TOTAL | GRAND ISLAND 688 MAT. | INST. | TOTAL | HASTINGS 689 MAT. | INST. | TOTAL | LINCOLN 683 - 685 MAT. | INST. | TOTAL | MCCOOK 690 MAT. | INST. | TOTAL |
|---|---|---|---|---|---|---|---|---|---|---|---|---|---|---|---|---|---|---|---|
| 015433 | CONTRACTOR EQUIPMENT | | 96.8 | 96.8 | | 101.0 | 101.0 | | 101.0 | 101.0 | | 101.0 | 101.0 | | 101.0 | 101.0 | | 101.0 | 101.0 |
| 0241, 31 - 34 | SITE & INFRASTRUCTURE, DEMOLITION | 100.8 | 97.4 | 98.5 | 99.4 | 90.8 | 93.5 | 103.9 | 92.0 | 95.7 | 103.0 | 91.2 | 94.9 | 91.3 | 92.0 | 91.8 | 103.9 | 91.0 | 95.0 |
| 0310 | Concrete Forming & Accessories | 90.4 | 57.5 | 61.8 | 98.8 | 60.4 | 65.5 | 98.4 | 77.7 | 80.4 | 101.6 | 75.9 | 79.3 | 103.2 | 61.6 | 67.1 | 96.1 | 57.8 | 62.9 |
| 0320 | Concrete Reinforcing | 114.4 | 48.3 | 80.4 | 105.6 | 47.1 | 75.5 | 105.1 | 76.6 | 90.4 | 105.1 | 47.8 | 75.6 | 96.5 | 76.1 | 86.0 | 107.1 | 76.2 | 91.2 |
| 0330 | Cast-in-Place Concrete | 110.2 | 59.4 | 90.7 | 112.3 | 56.2 | 90.8 | 118.9 | 71.2 | 100.6 | 118.9 | 65.2 | 98.4 | 88.8 | 73.7 | 83.0 | 118.9 | 59.1 | 96.0 |
| 03 | CONCRETE | 118.3 | 56.9 | 88.9 | 104.6 | 57.5 | 82.1 | 109.3 | 75.9 | 93.3 | 109.5 | 67.6 | 89.5 | 91.7 | 69.5 | 81.1 | 109.4 | 62.8 | 87.1 |
| 04 | MASONRY | 110.8 | 68.3 | 85.1 | 115.9 | 73.2 | 90.0 | 108.9 | 80.8 | 91.8 | 117.8 | 90.5 | 101.3 | 97.2 | 71.2 | 81.4 | 106.0 | 68.4 | 83.2 |
| 05 | METALS | 98.5 | 65.2 | 87.0 | 91.1 | 74.9 | 85.5 | 92.9 | 87.0 | 90.9 | 93.4 | 76.5 | 87.6 | 97.9 | 86.1 | 93.8 | 93.7 | 84.9 | 90.6 |
| 06 | WOOD, PLASTICS & COMPOSITES | 84.3 | 56.6 | 68.4 | 95.5 | 61.6 | 76.0 | 94.7 | 78.5 | 85.4 | 98.4 | 78.5 | 87.0 | 97.6 | 57.0 | 74.3 | 92.7 | 56.7 | 72.1 |
| 07 | THERMAL & MOISTURE PROTECTION | 104.9 | 68.9 | 90.6 | 98.8 | 65.8 | 85.7 | 98.9 | 76.6 | 90.0 | 98.9 | 88.4 | 94.7 | 98.4 | 71.8 | 87.9 | 98.9 | 66.0 | 85.8 |
| 08 | OPENINGS | 90.9 | 52.7 | 81.2 | 90.6 | 57.3 | 82.2 | 90.7 | 74.2 | 86.5 | 90.7 | 66.5 | 84.5 | 96.1 | 58.4 | 86.6 | 90.7 | 59.9 | 82.9 |
| 0920 | Plaster & Gypsum Board | 86.4 | 55.3 | 64.5 | 95.2 | 60.4 | 70.7 | 94.3 | 77.8 | 82.7 | 96.5 | 77.8 | 83.3 | 96.4 | 55.6 | 67.7 | 94.3 | 55.3 | 66.8 |
| 0950, 0980 | Ceilings & Acoustic Treatment | 93.9 | 55.3 | 68.3 | 89.5 | 60.4 | 70.2 | 89.5 | 77.8 | 81.7 | 89.5 | 77.8 | 81.7 | 92.3 | 55.6 | 68.0 | 89.5 | 55.3 | 66.8 |
| 0960 | Flooring | 101.4 | 86.7 | 97.2 | 100.4 | 106.0 | 102.0 | 100.2 | 113.3 | 104.0 | 101.6 | 106.0 | 102.8 | 101.8 | 88.3 | 97.9 | 99.2 | 86.7 | 95.6 |
| 0970, 0990 | Wall Finishes & Painting/Coating | 172.0 | 59.0 | 103.4 | 96.2 | 68.6 | 79.4 | 96.2 | 72.8 | 82.0 | 96.2 | 68.6 | 79.4 | 96.2 | 83.8 | 88.7 | 96.2 | 51.4 | 69.0 |
| 09 | FINISHES | 100.6 | 62.2 | 79.4 | 97.6 | 68.6 | 81.6 | 97.7 | 82.9 | 89.6 | 98.4 | 80.1 | 88.3 | 98.5 | 68.0 | 81.7 | 97.4 | 61.3 | 77.5 |
| COVERS | DIVS. 10 - 14, 25, 28, 41, 43, 44, 46 | 100.0 | 70.5 | 93.9 | 100.0 | 80.1 | 95.9 | 100.0 | 85.7 | 97.0 | 100.0 | 84.3 | 96.7 | 100.0 | 83.2 | 96.5 | 100.0 | 69.7 | 93.7 |
| 21, 22, 23 | FIRE SUPPRESSION, PLUMBING & HVAC | 94.8 | 60.8 | 80.7 | 94.7 | 74.0 | 86.1 | 100.1 | 80.1 | 91.7 | 94.7 | 77.2 | 87.4 | 99.9 | 80.0 | 91.7 | 94.7 | 74.5 | 86.3 |
| 26, 27, 3370 | ELECTRICAL, COMMUNICATIONS & UTIL. | 91.2 | 75.3 | 82.7 | 90.8 | 83.7 | 87.0 | 89.4 | 68.8 | 78.4 | 88.7 | 90.7 | 89.8 | 100.4 | 68.8 | 83.6 | 93.5 | 75.3 | 83.8 |
| MF2010 | WEIGHTED AVERAGE | 99.6 | 66.8 | 85.0 | 96.7 | 72.9 | 86.1 | 98.5 | 79.8 | 90.1 | 97.7 | 80.7 | 90.1 | 97.7 | 74.9 | 87.5 | 97.5 | 71.9 | 86.0 |

## NEBRASKA / NEVADA

| DIVISION | | NORFOLK 687 MAT. | INST. | TOTAL | NORTH PLATTE 691 MAT. | INST. | TOTAL | OMAHA 680 - 681 MAT. | INST. | TOTAL | VALENTINE 692 MAT. | INST. | TOTAL | CARSON CITY 897 MAT. | INST. | TOTAL | ELKO 898 MAT. | INST. | TOTAL |
|---|---|---|---|---|---|---|---|---|---|---|---|---|---|---|---|---|---|---|---|
| 015433 | CONTRACTOR EQUIPMENT | | 91.8 | 91.8 | | 101.0 | 101.0 | | 91.8 | 91.8 | | 94.6 | 94.6 | | 100.6 | 100.6 | | 100.6 | 100.6 |
| 0241, 31 - 34 | SITE & INFRASTRUCTURE, DEMOLITION | 81.4 | 90.2 | 87.5 | 104.8 | 91.2 | 95.4 | 81.4 | 90.9 | 88.0 | 86.9 | 94.5 | 92.2 | 65.9 | 102.3 | 91.0 | 54.3 | 102.0 | 87.2 |
| 0310 | Concrete Forming & Accessories | 84.6 | 75.0 | 76.3 | 98.3 | 75.7 | 78.7 | 93.5 | 74.6 | 77.1 | 86.0 | 57.0 | 60.9 | 101.5 | 87.1 | 89.0 | 108.4 | 84.6 | 87.8 |
| 0320 | Concrete Reinforcing | 105.8 | 49.8 | 75.0 | 105.6 | 77.3 | 91.5 | 101.1 | 76.6 | 88.5 | 107.2 | 64.8 | 76.1 | 103.6 | 119.3 | 111.7 | 105.3 | 93.2 | 99.1 |
| 0330 | Cast-in-Place Concrete | 113.4 | 64.4 | 94.7 | 118.9 | 66.7 | 98.9 | 95.3 | 80.4 | 89.6 | 105.1 | 56.7 | 86.6 | 98.2 | 86.4 | 93.7 | 99.1 | 84.0 | 93.3 |
| 03 | CONCRETE | 103.5 | 66.9 | 86.0 | 109.5 | 73.5 | 92.3 | 94.3 | 77.2 | 86.1 | 105.9 | 55.7 | 81.9 | 102.3 | 93.1 | 97.9 | 100.0 | 86.2 | 93.4 |
| 04 | MASONRY | 122.7 | 71.0 | 91.4 | 93.1 | 90.5 | 91.6 | 102.7 | 81.6 | 89.9 | 105.8 | 68.4 | 83.1 | 116.9 | 75.8 | 92.0 | 119.3 | 68.5 | 88.5 |
| 05 | METALS | 94.3 | 68.1 | 85.3 | 93.1 | 86.9 | 91.0 | 97.2 | 79.1 | 91.0 | 104.3 | 66.3 | 91.2 | 94.7 | 104.2 | 98.0 | 93.7 | 91.4 | 92.9 |
| 06 | WOOD, PLASTICS & COMPOSITES | 79.1 | 78.0 | 78.4 | 94.6 | 78.5 | 85.4 | 89.2 | 74.1 | 80.5 | 79.0 | 56.2 | 65.9 | 90.2 | 87.0 | 88.3 | 99.6 | 86.3 | 92.0 |
| 07 | THERMAL & MOISTURE PROTECTION | 98.3 | 75.1 | 89.0 | 98.8 | 82.1 | 92.2 | 94.5 | 82.0 | 89.5 | 99.0 | 68.8 | 87.0 | 99.7 | 85.8 | 94.2 | 101.0 | 76.4 | 91.2 |
| 08 | OPENINGS | 92.3 | 67.0 | 85.9 | 89.9 | 73.5 | 85.8 | 99.0 | 73.9 | 92.6 | 92.4 | 53.9 | 82.7 | 95.2 | 106.4 | 98.0 | 98.0 | 83.9 | 94.4 |
| 0920 | Plaster & Gypsum Board | 94.7 | 77.8 | 82.8 | 94.3 | 77.8 | 82.7 | 99.3 | 73.8 | 81.4 | 96.7 | 55.3 | 67.5 | 94.0 | 86.4 | 88.7 | 98.9 | 85.8 | 89.6 |
| 0950, 0980 | Ceilings & Acoustic Treatment | 104.9 | 77.8 | 86.9 | 89.5 | 77.8 | 81.7 | 105.8 | 73.8 | 84.6 | 107.4 | 55.3 | 72.8 | 119.1 | 86.4 | 97.5 | 122.0 | 85.8 | 98.0 |
| 0960 | Flooring | 126.9 | 106.0 | 120.9 | 100.1 | 106.0 | 101.8 | 131.6 | 86.6 | 118.7 | 129.4 | 84.3 | 116.5 | 101.7 | 82.7 | 96.3 | 107.1 | 54.0 | 91.8 |
| 0970, 0990 | Wall Finishes & Painting/Coating | 170.4 | 68.6 | 108.6 | 96.2 | 68.6 | 79.4 | 170.4 | 70.4 | 109.7 | 170.4 | 71.1 | 110.2 | 99.5 | 85.0 | 90.7 | 98.5 | 85.0 | 90.3 |
| 09 | FINISHES | 118.9 | 80.1 | 97.5 | 97.7 | 80.1 | 88.0 | 121.1 | 76.1 | 96.3 | 120.9 | 62.6 | 88.8 | 103.2 | 85.3 | 93.3 | 104.6 | 79.1 | 90.5 |
| COVERS | DIVS. 10 - 14, 25, 28, 41, 43, 44, 46 | 100.0 | 83.0 | 96.5 | 100.0 | 73.4 | 94.5 | 100.0 | 84.0 | 96.7 | 100.0 | 67.6 | 93.3 | 100.0 | 98.3 | 99.6 | 100.0 | 69.3 | 93.6 |
| 21, 22, 23 | FIRE SUPPRESSION, PLUMBING & HVAC | 94.5 | 77.1 | 87.2 | 100.1 | 75.4 | 89.8 | 99.9 | 79.9 | 91.5 | 94.4 | 58.8 | 79.6 | 99.9 | 79.9 | 91.5 | 97.4 | 76.5 | 88.7 |
| 26, 27, 3370 | ELECTRICAL, COMMUNICATIONS & UTIL. | 89.7 | 84.3 | 86.8 | 91.5 | 83.7 | 87.3 | 100.1 | 84.6 | 91.9 | 88.5 | 94.0 | 91.4 | 95.8 | 100.6 | 98.3 | 93.7 | 100.6 | 97.4 |
| MF2010 | WEIGHTED AVERAGE | 98.4 | 76.4 | 88.6 | 97.8 | 80.9 | 90.3 | 99.8 | 80.6 | 91.2 | 99.5 | 68.7 | 85.7 | 98.6 | 90.8 | 95.1 | 97.7 | 84.4 | 91.8 |

## NEVADA / NEW HAMPSHIRE

| DIVISION | | ELY 893 MAT. | INST. | TOTAL | LAS VEGAS 889-891 MAT. | INST. | TOTAL | RENO 894-895 MAT. | INST. | TOTAL | CHARLESTON 036 MAT. | INST. | TOTAL | CLAREMONT 037 MAT. | INST. | TOTAL | CONCORD 032-033 MAT. | INST. | TOTAL |
|---|---|---|---|---|---|---|---|---|---|---|---|---|---|---|---|---|---|---|---|
| 015433 | CONTRACTOR EQUIPMENT | | 100.6 | 100.6 | | 100.6 | 100.6 | | 100.6 | 100.6 | | 100.3 | 100.3 | | 100.3 | 100.3 | | 100.3 | 100.3 |
| 0241, 31-34 | SITE & INFRASTRUCTURE, DEMOLITION | 58.9 | 102.5 | 89.0 | 60.7 | 104.9 | 91.2 | 58.8 | 102.3 | 88.8 | 81.6 | 96.4 | 91.8 | 75.8 | 96.4 | 90.0 | 86.8 | 98.3 | 94.8 |
| 0310 | Concrete Forming & Accessories | 102.0 | 106.9 | 106.2 | 103.7 | 117.3 | 115.5 | 98.8 | 87.0 | 88.6 | 85.3 | 44.2 | 49.7 | 90.9 | 44.2 | 50.4 | 92.4 | 86.2 | 87.0 |
| 0320 | Concrete Reinforcing | 104.1 | 86.3 | 94.9 | 96.0 | 127.3 | 112.1 | 98.4 | 126.0 | 112.6 | 86.5 | 47.4 | 66.4 | 86.5 | 47.4 | 66.4 | 86.5 | 90.8 | 88.7 |
| 0330 | Cast-in-Place Concrete | 106.4 | 106.3 | 106.3 | 103.0 | 114.1 | 107.3 | 112.5 | 86.3 | 102.5 | 94.1 | 50.6 | 77.4 | 86.5 | 50.6 | 72.7 | 102.4 | 88.0 | 96.9 |
| 03 | CONCRETE | 108.5 | 102.5 | 105.6 | 103.7 | 117.6 | 110.3 | 108.5 | 94.3 | 101.7 | 103.9 | 47.9 | 77.1 | 95.2 | 47.9 | 72.5 | 106.3 | 87.6 | 97.3 |
| 04 | MASONRY | 124.4 | 80.8 | 98.0 | 111.7 | 109.1 | 110.1 | 118.4 | 75.8 | 92.6 | 94.0 | 38.6 | 60.4 | 93.5 | 38.6 | 60.2 | 96.4 | 81.9 | 87.6 |
| 05 | METALS | 93.7 | 91.5 | 92.9 | 100.8 | 112.1 | 104.7 | 95.1 | 106.9 | 99.2 | 92.7 | 61.6 | 82.0 | 92.7 | 61.6 | 82.0 | 93.5 | 87.3 | 91.4 |
| 06 | WOOD, PLASTICS & COMPOSITES | 91.0 | 108.7 | 101.1 | 90.2 | 117.5 | 105.8 | 86.6 | 87.0 | 86.8 | 85.9 | 45.8 | 62.9 | 91.9 | 45.8 | 65.5 | 95.7 | 89.8 | 92.3 |
| 07 | THERMAL & MOISTURE PROTECTION | 101.7 | 88.7 | 96.5 | 115.2 | 102.8 | 110.3 | 101.2 | 85.8 | 95.1 | 95.9 | 41.8 | 74.4 | 95.8 | 41.8 | 74.3 | 94.7 | 84.9 | 90.8 |
| 08 | OPENINGS | 97.9 | 94.1 | 96.9 | 96.8 | 124.8 | 103.9 | 95.5 | 101.8 | 97.1 | 102.1 | 41.4 | 86.7 | 103.4 | 41.4 | 87.7 | 103.4 | 87.0 | 97.7 |
| 0920 | Plaster & Gypsum Board | 94.5 | 108.8 | 104.6 | 94.8 | 117.9 | 111.0 | 87.9 | 86.4 | 86.9 | 91.5 | 43.5 | 57.7 | 93.3 | 43.5 | 58.3 | 95.9 | 88.8 | 90.9 |
| 0950, 0980 | Ceilings & Acoustic Treatment | 122.0 | 108.8 | 113.3 | 132.3 | 117.9 | 122.7 | 128.5 | 86.4 | 100.6 | 89.8 | 43.5 | 59.1 | 89.8 | 43.5 | 59.1 | 88.9 | 88.8 | 88.8 |
| 0960 | Flooring | 104.6 | 60.4 | 91.9 | 95.7 | 104.3 | 98.2 | 101.3 | 82.7 | 96.0 | 90.7 | 33.9 | 74.4 | 93.1 | 33.9 | 76.1 | 90.8 | 94.6 | 91.9 |
| 0970, 0990 | Wall Finishes & Painting/Coating | 98.5 | 120.8 | 112.0 | 101.1 | 124.7 | 115.4 | 98.5 | 85.0 | 90.3 | 92.3 | 49.2 | 66.2 | 92.3 | 49.2 | 66.2 | 92.3 | 98.6 | 96.2 |
| 09 | FINISHES | 103.9 | 100.6 | 102.1 | 103.4 | 116.0 | 110.4 | 103.3 | 85.3 | 93.4 | 92.8 | 43.4 | 65.5 | 93.2 | 43.4 | 65.7 | 93.9 | 89.9 | 91.7 |
| COVERS | DIVS. 10-14, 25, 28, 41, 43, 44, 46 | 100.0 | 69.4 | 93.6 | 100.0 | 96.9 | 99.4 | 100.0 | 98.3 | 99.6 | 100.0 | 39.7 | 87.5 | 100.0 | 39.7 | 87.5 | 100.0 | 82.9 | 92.5 |
| 21, 22, 23 | FIRE SUPPRESSION, PLUMBING & HVAC | 97.4 | 107.2 | 101.5 | 100.0 | 110.0 | 104.2 | 99.9 | 79.9 | 91.6 | 94.8 | 40.2 | 72.0 | 94.8 | 40.2 | 72.1 | 100.0 | 83.2 | 93.0 |
| 26, 27, 3370 | ELECTRICAL, COMMUNICATIONS & UTIL. | 94.0 | 111.3 | 103.2 | 98.1 | 119.1 | 109.2 | 94.4 | 100.6 | 97.7 | 98.0 | 34.9 | 64.4 | 98.0 | 34.9 | 64.4 | 98.9 | 60.7 | 78.6 |
| MF2010 | WEIGHTED AVERAGE | 99.1 | 99.6 | 99.3 | 100.2 | 112.8 | 105.9 | 99.2 | 91.1 | 95.6 | 96.5 | 47.6 | 74.6 | 95.4 | 47.6 | 74.0 | 98.7 | 82.5 | 91.5 |

## NEW HAMPSHIRE / NEW JERSEY

| DIVISION | | KEENE 034 MAT. | INST. | TOTAL | LITTLETON 035 MAT. | INST. | TOTAL | MANCHESTER 031 MAT. | INST. | TOTAL | NASHUA 030 MAT. | INST. | TOTAL | PORTSMOUTH 038 MAT. | INST. | TOTAL | ATLANTIC CITY 082,084 MAT. | INST. | TOTAL |
|---|---|---|---|---|---|---|---|---|---|---|---|---|---|---|---|---|---|---|---|
| 015433 | CONTRACTOR EQUIPMENT | | 100.3 | 100.3 | | 100.3 | 100.3 | | 100.3 | 100.3 | | 100.3 | 100.3 | | 100.3 | 100.3 | | 98.4 | 98.4 |
| 0241, 31-34 | SITE & INFRASTRUCTURE, DEMOLITION | 88.6 | 96.5 | 94.1 | 75.9 | 96.1 | 89.9 | 85.1 | 98.3 | 94.3 | 89.3 | 98.3 | 95.6 | 83.5 | 99.7 | 94.7 | 92.1 | 103.6 | 100.0 |
| 0310 | Concrete Forming & Accessories | 89.6 | 42.3 | 48.5 | 99.7 | 47.2 | 54.1 | 94.8 | 86.8 | 87.8 | 96.7 | 86.8 | 88.1 | 86.6 | 90.7 | 90.2 | 104.5 | 126.8 | 123.8 |
| 0320 | Concrete Reinforcing | 86.5 | 56.4 | 71.0 | 87.3 | 34.3 | 60.0 | 108.1 | 90.9 | 99.3 | 108.1 | 90.9 | 99.3 | 86.5 | 91.0 | 88.8 | 82.4 | 116.4 | 99.9 |
| 0330 | Cast-in-Place Concrete | 94.5 | 52.1 | 78.3 | 84.8 | 52.8 | 72.6 | 105.3 | 110.8 | 107.4 | 89.5 | 110.8 | 97.6 | 84.9 | 117.0 | 97.2 | 80.2 | 130.2 | 99.4 |
| 03 | CONCRETE | 103.5 | 49.2 | 77.5 | 94.5 | 47.4 | 71.9 | 111.1 | 95.6 | 103.7 | 103.2 | 95.6 | 99.6 | 94.6 | 99.6 | 97.0 | 96.6 | 124.9 | 110.1 |
| 04 | MASONRY | 97.7 | 40.8 | 63.2 | 105.7 | 44.3 | 68.5 | 99.5 | 81.9 | 88.9 | 98.8 | 81.9 | 88.6 | 94.6 | 91.4 | 92.6 | 101.4 | 127.0 | 116.9 |
| 05 | METALS | 93.4 | 65.9 | 83.9 | 93.4 | 56.2 | 80.6 | 98.6 | 87.9 | 94.9 | 98.6 | 87.9 | 94.9 | 95.0 | 90.3 | 93.6 | 93.2 | 104.8 | 97.2 |
| 06 | WOOD, PLASTICS & COMPOSITES | 90.3 | 40.9 | 62.0 | 100.6 | 46.8 | 69.8 | 95.7 | 89.8 | 92.3 | 99.0 | 89.8 | 93.7 | 87.1 | 89.8 | 88.6 | 109.1 | 126.8 | 119.2 |
| 07 | THERMAL & MOISTURE PROTECTION | 96.3 | 43.8 | 75.5 | 95.9 | 47.7 | 76.7 | 95.9 | 88.4 | 92.9 | 96.7 | 88.4 | 93.4 | 96.3 | 112.0 | 102.5 | 96.4 | 126.6 | 108.4 |
| 08 | OPENINGS | 100.6 | 46.2 | 86.8 | 104.5 | 39.2 | 87.9 | 105.6 | 81.0 | 99.4 | 105.6 | 81.0 | 99.4 | 106.2 | 76.3 | 98.6 | 103.0 | 124.1 | 108.4 |
| 0920 | Plaster & Gypsum Board | 92.4 | 38.4 | 54.4 | 105.7 | 44.5 | 62.6 | 103.7 | 88.8 | 93.2 | 100.2 | 88.8 | 92.2 | 91.5 | 88.8 | 89.6 | 104.3 | 126.9 | 120.2 |
| 0950, 0980 | Ceilings & Acoustic Treatment | 89.8 | 38.4 | 55.8 | 89.8 | 44.5 | 59.8 | 95.7 | 88.8 | 91.1 | 99.4 | 88.8 | 92.4 | 91.9 | 88.8 | 89.9 | 89.8 | 126.9 | 114.4 |
| 0960 | Flooring | 92.6 | 55.6 | 82.0 | 101.6 | 33.9 | 82.2 | 96.9 | 94.6 | 96.2 | 95.8 | 94.6 | 95.5 | 91.2 | 94.6 | 92.2 | 99.5 | 134.3 | 109.5 |
| 0970, 0990 | Wall Finishes & Painting/Coating | 92.3 | 49.2 | 66.2 | 92.3 | 64.5 | 75.5 | 92.3 | 98.6 | 96.2 | 92.3 | 98.6 | 96.2 | 92.3 | 98.6 | 96.2 | 92.3 | 123.7 | 111.3 |
| 09 | FINISHES | 94.7 | 45.2 | 67.4 | 97.2 | 46.8 | 69.4 | 97.7 | 89.9 | 93.4 | 98.5 | 89.9 | 93.8 | 94.1 | 92.1 | 93.0 | 98.6 | 128.4 | 115.1 |
| COVERS | DIVS. 10-14, 25, 28, 41, 43, 44, 46 | 100.0 | 54.5 | 90.6 | 100.0 | 41.5 | 87.8 | 100.0 | 63.9 | 92.5 | 100.0 | 63.9 | 92.5 | 100.0 | 67.1 | 93.2 | 100.0 | 112.7 | 102.6 |
| 21, 22, 23 | FIRE SUPPRESSION, PLUMBING & HVAC | 94.8 | 43.5 | 73.4 | 94.8 | 67.1 | 83.3 | 100.0 | 83.2 | 93.0 | 100.2 | 83.2 | 93.1 | 100.2 | 88.0 | 95.1 | 99.8 | 123.7 | 109.7 |
| 26, 27, 3370 | ELECTRICAL, COMMUNICATIONS & UTIL. | 98.0 | 34.9 | 64.4 | 99.3 | 60.6 | 78.7 | 99.6 | 77.7 | 87.9 | 100.8 | 77.7 | 88.5 | 98.6 | 77.7 | 87.5 | 95.5 | 135.1 | 116.6 |
| MF2010 | WEIGHTED AVERAGE | 97.0 | 49.9 | 76.0 | 96.7 | 57.4 | 79.1 | 100.8 | 86.1 | 94.2 | 100.2 | 86.1 | 93.9 | 97.7 | 89.9 | 94.2 | 98.1 | 122.7 | 109.1 |

## NEW JERSEY

| DIVISION | | CAMDEN 081 MAT. | INST. | TOTAL | DOVER 078 MAT. | INST. | TOTAL | ELIZABETH 072 MAT. | INST. | TOTAL | HACKENSACK 076 MAT. | INST. | TOTAL | JERSEY CITY 073 MAT. | INST. | TOTAL | LONG BRANCH 077 MAT. | INST. | TOTAL |
|---|---|---|---|---|---|---|---|---|---|---|---|---|---|---|---|---|---|---|---|
| 015433 | CONTRACTOR EQUIPMENT | | 98.4 | 98.4 | | 100.3 | 100.3 | | 100.3 | 100.3 | | 100.3 | 100.3 | | 98.4 | 98.4 | | 98.0 | 98.0 |
| 0241, 31-34 | SITE & INFRASTRUCTURE, DEMOLITION | 92.8 | 104.1 | 100.6 | 103.9 | 104.7 | 104.5 | 108.0 | 104.7 | 105.7 | 104.6 | 105.2 | 105.0 | 92.8 | 105.2 | 101.3 | 97.8 | 105.0 | 102.8 |
| 0310 | Concrete Forming & Accessories | 96.7 | 126.9 | 122.9 | 93.2 | 127.5 | 123.0 | 104.0 | 127.5 | 124.4 | 93.2 | 127.5 | 123.0 | 96.7 | 127.6 | 123.5 | 97.2 | 127.3 | 123.3 |
| 0320 | Concrete Reinforcing | 108.1 | 118.2 | 113.3 | 83.3 | 125.3 | 104.9 | 83.3 | 125.3 | 104.9 | 83.3 | 125.3 | 104.9 | 108.1 | 125.3 | 117.0 | 83.3 | 125.3 | 104.9 |
| 0330 | Cast-in-Place Concrete | 77.8 | 132.0 | 98.6 | 99.8 | 129.2 | 111.0 | 85.7 | 129.2 | 102.4 | 97.5 | 129.2 | 109.6 | 77.8 | 129.3 | 97.5 | 86.5 | 130.1 | 103.2 |
| 03 | CONCRETE | 97.4 | 125.9 | 111.0 | 104.6 | 126.7 | 115.2 | 99.9 | 126.8 | 112.8 | 102.6 | 126.8 | 114.2 | 97.4 | 126.6 | 111.4 | 102.3 | 126.8 | 114.0 |
| 04 | MASONRY | 92.5 | 130.3 | 115.4 | 96.9 | 127.5 | 115.5 | 112.5 | 127.5 | 121.6 | 100.7 | 127.5 | 117.0 | 89.7 | 127.5 | 112.6 | 105.2 | 127.0 | 118.4 |
| 05 | METALS | 98.5 | 105.2 | 100.8 | 93.2 | 112.3 | 99.8 | 94.7 | 112.3 | 100.8 | 93.2 | 112.4 | 99.8 | 98.5 | 110.0 | 102.5 | 93.3 | 109.8 | 99.0 |
| 06 | WOOD, PLASTICS & COMPOSITES | 99.0 | 126.8 | 114.9 | 98.4 | 126.8 | 114.7 | 112.3 | 126.8 | 120.6 | 98.4 | 126.8 | 114.7 | 99.0 | 126.8 | 114.9 | 100.1 | 126.7 | 115.4 |
| 07 | THERMAL & MOISTURE PROTECTION | 96.3 | 126.9 | 108.4 | 96.7 | 127.4 | 108.9 | 96.9 | 127.4 | 109.0 | 96.5 | 124.9 | 107.8 | 96.2 | 127.4 | 108.6 | 96.4 | 126.8 | 108.5 |
| 08 | OPENINGS | 105.6 | 124.1 | 110.3 | 108.6 | 125.0 | 112.8 | 106.6 | 125.0 | 111.2 | 105.8 | 125.0 | 110.7 | 105.6 | 125.0 | 110.5 | 101.4 | 125.0 | 107.4 |
| 0920 | Plaster & Gypsum Board | 100.2 | 126.9 | 119.0 | 96.8 | 126.9 | 118.0 | 104.8 | 126.9 | 120.3 | 96.8 | 126.9 | 118.0 | 100.2 | 126.9 | 119.0 | 99.0 | 126.9 | 119.0 |
| 0950, 0980 | Ceilings & Acoustic Treatment | 99.4 | 126.9 | 117.6 | 89.8 | 126.9 | 114.4 | 91.9 | 126.9 | 115.1 | 89.8 | 126.9 | 114.4 | 99.4 | 126.9 | 117.6 | 89.8 | 126.9 | 114.4 |
| 0960 | Flooring | 95.8 | 134.3 | 106.9 | 94.8 | 165.7 | 115.1 | 99.9 | 165.7 | 118.8 | 94.8 | 165.7 | 115.1 | 95.8 | 165.7 | 115.9 | 96.1 | 165.7 | 116.1 |
| 0970, 0990 | Wall Finishes & Painting/Coating | 92.3 | 123.7 | 111.3 | 92.2 | 125.7 | 112.5 | 92.2 | 125.7 | 112.5 | 92.2 | 125.7 | 112.5 | 92.3 | 125.7 | 112.6 | 92.3 | 123.7 | 111.3 |
| 09 | FINISHES | 99.0 | 128.4 | 115.2 | 96.5 | 134.1 | 117.2 | 99.9 | 134.1 | 118.8 | 96.3 | 134.1 | 117.2 | 98.9 | 134.1 | 118.4 | 97.4 | 133.7 | 117.5 |
| COVERS | DIVS. 10-14, 25, 28, 41, 43, 44, 46 | 100.0 | 112.8 | 102.7 | 100.0 | 114.6 | 103.0 | 100.0 | 114.6 | 103.0 | 100.0 | 114.6 | 103.0 | 100.0 | 114.6 | 103.0 | 100.0 | 112.6 | 102.6 |
| 21, 22, 23 | FIRE SUPPRESSION, PLUMBING & HVAC | 100.2 | 123.7 | 110.0 | 99.8 | 127.1 | 111.2 | 100.2 | 127.6 | 111.6 | 99.8 | 127.1 | 111.2 | 100.2 | 127.1 | 111.4 | 99.8 | 127.5 | 111.4 |
| 26, 27, 3370 | ELECTRICAL, COMMUNICATIONS & UTIL. | 101.2 | 139.9 | 121.8 | 96.9 | 135.9 | 117.7 | 97.7 | 135.9 | 118.0 | 96.9 | 135.6 | 117.5 | 102.3 | 135.5 | 120.0 | 96.6 | 127.9 | 113.2 |
| MF2010 | WEIGHTED AVERAGE | 99.4 | 124.0 | 110.4 | 99.6 | 125.5 | 111.2 | 100.6 | 125.6 | 111.8 | 99.3 | 125.4 | 111.0 | 99.3 | 125.3 | 110.9 | 98.9 | 124.1 | 110.2 |

## NEW JERSEY

| DIVISION | | NEW BRUNSWICK 088 - 089 | | | NEWARK 070 - 071 | | | PATERSON 074 - 075 | | | POINT PLEASANT 087 | | | SUMMIT 079 | | | TRENTON 085 - 086 | | |
|---|---|---|---|---|---|---|---|---|---|---|---|---|---|---|---|---|---|---|---|---|
| | | MAT. | INST. | TOTAL | MAT. | INST. | TOTAL | MAT. | INST. | TOTAL | MAT. | INST. | TOTAL | MAT. | INST. | TOTAL | MAT. | INST. | TOTAL |
| 015433 | CONTRACTOR EQUIPMENT | | 98.0 | 98.0 | | 100.3 | 100.3 | | 100.3 | 100.3 | | 98.0 | 98.0 | | 100.3 | 100.3 | | 98.0 | 98.0 |
| 0241, 31 - 34 | SITE & INFRASTRUCTURE, DEMOLITION | 105.4 | 105.1 | 105.2 | 108.2 | 104.7 | 105.8 | 106.1 | 105.2 | 105.5 | 107.0 | 105.0 | 105.6 | 105.5 | 104.7 | 105.0 | 90.3 | 105.0 | 100.5 |
| 0310 | Concrete Forming & Accessories | 99.5 | 127.5 | 123.8 | 93.2 | 127.5 | 123.0 | 95.3 | 127.5 | 123.3 | 94.6 | 127.1 | 122.8 | 95.9 | 127.5 | 123.3 | 94.0 | 116.5 | 113.5 |
| 0320 | Concrete Reinforcing | 83.3 | 125.3 | 104.9 | 108.1 | 125.3 | 117.0 | 108.1 | 125.3 | 117.0 | 83.3 | 125.2 | 104.9 | 83.3 | 125.3 | 104.9 | 108.1 | 120.6 | 114.6 |
| 0330 | Cast-in-Place Concrete | 99.1 | 130.4 | 111.1 | 103.4 | 129.2 | 113.3 | 99.2 | 129.2 | 110.7 | 99.1 | 130.0 | 111.0 | 82.8 | 129.2 | 100.6 | 95.5 | 130.1 | 108.8 |
| 03 | CONCRETE | 114.8 | 127.0 | 120.7 | 110.0 | 126.8 | 118.0 | 108.0 | 126.8 | 117.0 | 114.5 | 126.6 | 120.3 | 96.8 | 126.8 | 111.1 | 106.1 | 121.1 | 113.3 |
| 04 | MASONRY | 99.0 | 127.5 | 116.3 | 102.1 | 127.5 | 117.5 | 97.3 | 127.5 | 115.6 | 90.0 | 127.0 | 112.4 | 99.3 | 127.5 | 116.4 | 88.9 | 127.0 | 112.0 |
| 05 | METALS | 93.2 | 109.9 | 99.0 | 98.5 | 112.3 | 103.3 | 93.8 | 112.4 | 100.2 | 93.2 | 109.6 | 98.9 | 93.2 | 112.3 | 99.8 | 93.7 | 107.4 | 98.4 |
| 06 | WOOD, PLASTICS & COMPOSITES | 103.2 | 126.7 | 116.7 | 97.7 | 126.8 | 114.4 | 100.8 | 126.8 | 115.7 | 96.7 | 126.7 | 113.9 | 101.9 | 126.8 | 116.2 | 96.1 | 112.2 | 105.4 |
| 07 | THERMAL & MOISTURE PROTECTION | 96.5 | 127.0 | 108.6 | 96.4 | 127.4 | 108.8 | 96.8 | 124.9 | 107.9 | 96.6 | 126.8 | 108.6 | 97.1 | 127.4 | 109.2 | 92.6 | 124.9 | 105.4 |
| 08 | OPENINGS | 97.4 | 125.0 | 104.4 | 111.7 | 125.0 | 115.0 | 111.7 | 125.0 | 115.0 | 99.5 | 127.1 | 106.5 | 113.9 | 125.0 | 116.7 | 106.2 | 115.6 | 108.6 |
| 0920 | Plaster & Gypsum Board | 101.2 | 126.9 | 119.3 | 97.9 | 126.9 | 118.3 | 100.2 | 126.9 | 119.0 | 96.8 | 126.9 | 118.0 | 99.0 | 126.9 | 118.6 | 98.1 | 112.0 | 107.9 |
| 0950, 0980 | Ceilings & Acoustic Treatment | 89.8 | 126.9 | 114.4 | 95.7 | 126.9 | 116.4 | 99.4 | 126.9 | 117.6 | 89.8 | 126.9 | 114.4 | 89.8 | 126.9 | 114.4 | 96.6 | 112.0 | 106.8 |
| 0960 | Flooring | 97.4 | 165.7 | 117.0 | 96.0 | 165.7 | 116.1 | 95.8 | 165.7 | 115.9 | 94.8 | 134.3 | 106.1 | 96.1 | 165.7 | 116.1 | 96.2 | 165.7 | 116.2 |
| 0970, 0990 | Wall Finishes & Painting/Coating | 92.3 | 125.7 | 112.6 | 92.2 | 125.7 | 112.5 | 92.2 | 125.7 | 112.5 | 92.3 | 123.7 | 111.3 | 92.2 | 125.7 | 112.5 | 92.3 | 123.7 | 111.3 |
| 09 | FINISHES | 98.9 | 134.1 | 118.3 | 97.8 | 134.1 | 117.9 | 99.1 | 134.1 | 118.4 | 97.6 | 128.4 | 114.6 | 97.5 | 134.1 | 117.7 | 97.7 | 125.2 | 112.9 |
| COVERS | DIVS. 10 - 14, 25, 28, 41, 43, 44, 46 | 100.0 | 114.5 | 103.0 | 100.0 | 114.6 | 103.0 | 100.0 | 114.6 | 103.0 | 100.0 | 108.1 | 101.7 | 100.0 | 114.6 | 103.0 | 100.0 | 111.0 | 102.3 |
| 21, 22, 23 | FIRE SUPPRESSION, PLUMBING & HVAC | 99.8 | 127.8 | 111.4 | 100.3 | 127.6 | 111.7 | 100.2 | 127.1 | 111.4 | 99.8 | 127.4 | 111.3 | 99.8 | 127.6 | 111.4 | 100.3 | 127.5 | 111.6 |
| 26, 27, 3370 | ELECTRICAL, COMMUNICATIONS & UTIL. | 96.3 | 132.8 | 115.7 | 100.6 | 135.6 | 119.2 | 102.3 | 135.9 | 120.2 | 95.5 | 127.9 | 112.7 | 97.7 | 135.9 | 118.0 | 102.4 | 137.7 | 121.2 |
| MF2010 | WEIGHTED AVERAGE | 100.1 | 125.0 | 111.2 | 102.3 | 125.6 | 112.7 | 101.4 | 125.5 | 112.2 | 99.6 | 123.3 | 110.2 | 99.6 | 125.6 | 111.2 | 99.4 | 122.7 | 109.8 |

| DIVISION | | NEW JERSEY VINELAND 080,083 | | | NEW MEXICO ALBUQUERQUE 870 - 872 | | | CARRIZOZO 883 | | | CLOVIS 881 | | | FARMINGTON 874 | | | GALLUP 873 | | |
|---|---|---|---|---|---|---|---|---|---|---|---|---|---|---|---|---|---|---|---|---|
| | | MAT. | INST. | TOTAL | MAT. | INST. | TOTAL | MAT. | INST. | TOTAL | MAT. | INST. | TOTAL | MAT. | INST. | TOTAL | MAT. | INST. | TOTAL |
| 015433 | CONTRACTOR EQUIPMENT | | 98.4 | 98.4 | | 113.0 | 113.0 | | 113.0 | 113.0 | | 113.0 | 113.0 | | 113.0 | 113.0 | | 113.0 | 113.0 |
| 0241, 31 - 34 | SITE & INFRASTRUCTURE, DEMOLITION | 96.6 | 104.1 | 101.8 | 75.1 | 108.5 | 98.1 | 94.7 | 108.5 | 104.2 | 83.6 | 108.4 | 100.7 | 81.1 | 108.5 | 100.0 | 88.4 | 108.5 | 102.2 |
| 0310 | Concrete Forming & Accessories | 92.3 | 126.9 | 122.3 | 99.2 | 67.2 | 71.5 | 99.2 | 67.2 | 71.5 | 99.2 | 67.1 | 71.3 | 99.2 | 67.2 | 71.5 | 99.2 | 67.2 | 71.5 |
| 0320 | Concrete Reinforcing | 82.4 | 116.4 | 99.9 | 99.1 | 72.2 | 85.2 | 106.2 | 72.2 | 88.7 | 107.4 | 72.1 | 89.7 | 108.2 | 72.2 | 89.7 | 103.6 | 72.2 | 87.4 |
| 0330 | Cast-in-Place Concrete | 86.5 | 130.3 | 103.3 | 105.2 | 75.3 | 93.8 | 97.9 | 75.3 | 89.3 | 97.9 | 75.3 | 89.2 | 106.2 | 75.3 | 94.4 | 99.9 | 75.3 | 90.5 |
| 03 | CONCRETE | 101.8 | 124.9 | 112.9 | 105.0 | 72.0 | 89.2 | 119.2 | 72.0 | 96.6 | 107.9 | 71.9 | 90.7 | 108.6 | 72.0 | 91.1 | 114.5 | 72.0 | 94.2 |
| 04 | MASONRY | 89.2 | 127.0 | 112.1 | 107.9 | 64.3 | 81.5 | 104.8 | 64.3 | 80.3 | 104.8 | 64.3 | 80.3 | 114.8 | 64.3 | 84.2 | 104.9 | 64.3 | 80.3 |
| 05 | METALS | 93.1 | 104.9 | 97.2 | 101.0 | 89.4 | 97.0 | 97.8 | 89.4 | 94.9 | 97.5 | 89.2 | 94.6 | 98.8 | 89.4 | 95.5 | 97.8 | 89.4 | 94.9 |
| 06 | WOOD, PLASTICS & COMPOSITES | 93.9 | 126.8 | 112.7 | 90.3 | 67.6 | 77.3 | 90.4 | 67.6 | 77.3 | 90.4 | 67.6 | 77.3 | 90.4 | 67.6 | 77.3 | 90.4 | 67.6 | 77.3 |
| 07 | THERMAL & MOISTURE PROTECTION | 96.2 | 126.6 | 108.2 | 100.7 | 75.0 | 90.5 | 102.5 | 75.0 | 91.6 | 101.2 | 75.0 | 90.8 | 101.0 | 75.0 | 90.7 | 102.1 | 75.0 | 91.3 |
| 08 | OPENINGS | 99.0 | 124.1 | 105.4 | 95.3 | 70.2 | 88.9 | 94.1 | 70.2 | 88.0 | 94.3 | 70.2 | 88.2 | 98.3 | 70.2 | 91.2 | 98.3 | 70.2 | 91.2 |
| 0920 | Plaster & Gypsum Board | 95.0 | 126.9 | 117.5 | 93.2 | 66.2 | 74.2 | 85.0 | 66.2 | 71.8 | 85.0 | 66.2 | 71.8 | 85.0 | 66.2 | 71.8 | 85.0 | 66.2 | 71.8 |
| 0950, 0980 | Ceilings & Acoustic Treatment | 89.8 | 126.9 | 114.4 | 122.4 | 66.2 | 85.2 | 118.2 | 66.2 | 83.7 | 118.2 | 66.2 | 83.7 | 118.2 | 66.2 | 83.7 | 118.2 | 66.2 | 83.7 |
| 0960 | Flooring | 94.0 | 134.3 | 105.6 | 99.2 | 69.7 | 90.7 | 101.3 | 69.7 | 92.2 | 101.3 | 69.7 | 92.2 | 101.3 | 69.7 | 92.2 | 101.3 | 69.7 | 92.2 |
| 0970, 0990 | Wall Finishes & Painting/Coating | 92.3 | 123.7 | 111.3 | 103.1 | 52.8 | 72.6 | 98.5 | 52.8 | 70.8 | 98.5 | 52.8 | 70.8 | 98.5 | 52.8 | 70.8 | 98.5 | 52.8 | 70.8 |
| 09 | FINISHES | 96.3 | 128.4 | 114.0 | 102.8 | 65.9 | 82.5 | 103.2 | 65.9 | 82.6 | 102.0 | 65.9 | 82.1 | 101.6 | 65.9 | 81.9 | 102.7 | 65.9 | 82.4 |
| COVERS | DIVS. 10 - 14, 25, 28, 41, 43, 44, 46 | 100.0 | 112.8 | 102.7 | 100.0 | 83.3 | 96.5 | 100.0 | 83.3 | 96.5 | 100.0 | 83.3 | 96.5 | 100.0 | 83.3 | 96.5 | 100.0 | 83.3 | 96.5 |
| 21, 22, 23 | FIRE SUPPRESSION, PLUMBING & HVAC | 99.8 | 123.7 | 109.7 | 100.0 | 69.8 | 87.4 | 96.9 | 69.8 | 85.6 | 96.9 | 69.4 | 85.4 | 99.9 | 69.8 | 87.4 | 96.7 | 69.8 | 85.5 |
| 26, 27, 3370 | ELECTRICAL, COMMUNICATIONS & UTIL. | 95.5 | 135.1 | 116.6 | 89.7 | 72.6 | 80.6 | 89.7 | 72.6 | 80.6 | 87.1 | 72.6 | 79.4 | 87.8 | 72.6 | 79.7 | 86.9 | 72.6 | 79.3 |
| MF2010 | WEIGHTED AVERAGE | 97.5 | 122.8 | 108.8 | 99.1 | 75.0 | 88.3 | 100.0 | 75.0 | 88.8 | 97.9 | 74.9 | 87.6 | 99.8 | 75.0 | 88.7 | 99.3 | 75.0 | 88.5 |

## NEW MEXICO

| DIVISION | | LAS CRUCES 880 | | | LAS VEGAS 877 | | | ROSWELL 882 | | | SANTA FE 875 | | | SOCORRO 878 | | | TRUTH/CONSEQUENCES 879 | | |
|---|---|---|---|---|---|---|---|---|---|---|---|---|---|---|---|---|---|---|---|---|
| | | MAT. | INST. | TOTAL | MAT. | INST. | TOTAL | MAT. | INST. | TOTAL | MAT. | INST. | TOTAL | MAT. | INST. | TOTAL | MAT. | INST. | TOTAL |
| 015433 | CONTRACTOR EQUIPMENT | | 87.9 | 87.9 | | 113.0 | 113.0 | | 113.0 | 113.0 | | 113.0 | 113.0 | | 113.0 | 113.0 | | 87.9 | 87.9 |
| 0241, 31 - 34 | SITE & INFRASTRUCTURE, DEMOLITION | 85.8 | 86.7 | 86.4 | 80.7 | 108.5 | 99.9 | 85.4 | 108.5 | 101.3 | 79.0 | 108.5 | 99.3 | 77.3 | 108.5 | 98.8 | 96.6 | 86.7 | 89.8 |
| 0310 | Concrete Forming & Accessories | 96.1 | 66.0 | 70.0 | 99.2 | 67.2 | 71.5 | 99.2 | 67.2 | 71.5 | 98.4 | 67.2 | 71.4 | 99.2 | 67.2 | 71.5 | 96.1 | 66.0 | 70.0 |
| 0320 | Concrete Reinforcing | 102.4 | 72.0 | 86.7 | 105.3 | 72.2 | 88.2 | 107.4 | 72.2 | 89.3 | 106.2 | 72.2 | 88.7 | 107.4 | 72.2 | 89.3 | 101.2 | 72.0 | 86.2 |
| 0330 | Cast-in-Place Concrete | 92.4 | 67.4 | 82.8 | 103.3 | 75.3 | 92.6 | 97.9 | 75.3 | 89.2 | 107.6 | 75.3 | 95.2 | 101.2 | 75.3 | 91.3 | 111.0 | 67.5 | 94.3 |
| 03 | CONCRETE | 86.6 | 68.4 | 77.9 | 105.0 | 72.0 | 89.7 | 108.7 | 72.0 | 91.1 | 107.2 | 72.0 | 90.3 | 104.5 | 72.0 | 89.0 | 99.0 | 68.4 | 84.4 |
| 04 | MASONRY | 100.8 | 64.0 | 78.5 | 105.2 | 64.3 | 80.4 | 115.5 | 64.3 | 84.5 | 109.1 | 64.3 | 82.0 | 105.1 | 64.3 | 80.4 | 102.8 | 64.0 | 79.3 |
| 05 | METALS | 98.7 | 82.6 | 93.1 | 97.6 | 89.4 | 94.8 | 98.7 | 89.4 | 95.5 | 98.7 | 89.4 | 95.5 | 97.8 | 89.4 | 94.9 | 97.4 | 82.7 | 92.3 |
| 06 | WOOD, PLASTICS & COMPOSITES | 81.6 | 66.4 | 72.9 | 90.4 | 67.6 | 77.3 | 90.4 | 67.6 | 77.3 | 89.9 | 67.6 | 77.1 | 90.4 | 67.6 | 77.3 | 81.6 | 66.4 | 72.9 |
| 07 | THERMAL & MOISTURE PROTECTION | 87.1 | 70.3 | 80.4 | 100.6 | 75.0 | 90.5 | 101.2 | 75.0 | 90.8 | 100.3 | 75.0 | 90.3 | 100.6 | 75.0 | 90.4 | 87.7 | 70.3 | 80.8 |
| 08 | OPENINGS | 88.3 | 69.5 | 83.6 | 94.3 | 70.2 | 88.2 | 94.1 | 70.2 | 88.0 | 94.2 | 70.2 | 88.1 | 94.1 | 70.2 | 88.0 | 88.2 | 69.5 | 83.5 |
| 0920 | Plaster & Gypsum Board | 87.4 | 66.2 | 72.5 | 85.0 | 66.2 | 71.8 | 85.0 | 66.2 | 71.8 | 92.9 | 66.2 | 74.1 | 85.0 | 66.2 | 71.8 | 87.4 | 66.2 | 72.5 |
| 0950, 0980 | Ceilings & Acoustic Treatment | 108.8 | 66.2 | 80.6 | 118.2 | 66.2 | 83.7 | 118.2 | 66.2 | 83.7 | 119.1 | 66.2 | 84.0 | 118.2 | 66.2 | 83.7 | 108.8 | 66.2 | 80.6 |
| 0960 | Flooring | 133.7 | 69.7 | 115.3 | 101.3 | 69.7 | 92.2 | 101.3 | 69.7 | 92.2 | 101.3 | 69.7 | 92.2 | 101.3 | 69.7 | 92.2 | 133.7 | 69.7 | 115.3 |
| 0970, 0990 | Wall Finishes & Painting/Coating | 90.2 | 52.8 | 67.5 | 98.5 | 52.8 | 70.8 | 98.5 | 52.8 | 70.8 | 98.5 | 52.8 | 70.8 | 98.5 | 52.8 | 70.8 | 90.2 | 52.8 | 67.5 |
| 09 | FINISHES | 115.9 | 65.0 | 87.8 | 101.4 | 65.9 | 81.8 | 102.1 | 65.9 | 82.1 | 102.8 | 65.9 | 82.4 | 101.4 | 65.9 | 81.8 | 116.5 | 65.0 | 88.1 |
| COVERS | DIVS. 10 - 14, 25, 28, 41, 43, 44, 46 | 100.0 | 80.4 | 95.9 | 100.0 | 83.3 | 96.5 | 100.0 | 83.3 | 96.5 | 100.0 | 83.3 | 96.5 | 100.0 | 83.3 | 96.5 | 100.0 | 80.4 | 95.9 |
| 21, 22, 23 | FIRE SUPPRESSION, PLUMBING & HVAC | 100.3 | 69.5 | 87.5 | 96.7 | 69.8 | 85.5 | 99.9 | 69.8 | 87.4 | 99.9 | 69.8 | 87.4 | 96.7 | 69.8 | 85.5 | 96.6 | 69.5 | 85.3 |
| 26, 27, 3370 | ELECTRICAL, COMMUNICATIONS & UTIL. | 88.3 | 67.8 | 77.4 | 89.7 | 72.6 | 80.6 | 88.6 | 72.6 | 80.1 | 100.0 | 72.6 | 85.4 | 87.5 | 72.6 | 79.6 | 91.4 | 72.6 | 81.4 |
| MF2010 | WEIGHTED AVERAGE | 96.3 | 70.9 | 85.0 | 97.7 | 75.0 | 87.6 | 99.6 | 75.0 | 88.6 | 100.1 | 75.0 | 88.9 | 97.2 | 75.0 | 87.3 | 97.5 | 71.6 | 85.9 |

| DIVISION | | NEW MEXICO TUCUMCARI 884 | | | NEW YORK ALBANY 120 - 122 | | | NEW YORK BINGHAMTON 137 - 139 | | | NEW YORK BRONX 104 | | | NEW YORK BROOKLYN 112 | | | NEW YORK BUFFALO 140 - 142 | | |
|---|---|---|---|---|---|---|---|---|---|---|---|---|---|---|---|---|---|---|---|
| | | MAT. | INST. | TOTAL | MAT. | INST. | TOTAL | MAT. | INST. | TOTAL | MAT. | INST. | TOTAL | MAT. | INST. | TOTAL | MAT. | INST. | TOTAL |
| 015433 | CONTRACTOR EQUIPMENT | | 113.0 | 113.0 | | 114.2 | 114.2 | | 114.5 | 114.5 | | 114.5 | 114.5 | | 115.4 | 115.4 | | 96.1 | 96.1 |
| 0241, 31 - 34 | SITE & INFRASTRUCTURE, DEMOLITION | 83.3 | 108.4 | 100.6 | 70.4 | 107.4 | 95.9 | 94.4 | 91.9 | 92.7 | 123.4 | 125.4 | 124.8 | 121.6 | 129.3 | 126.9 | 94.9 | 96.9 | 96.3 |
| 0310 | Concrete Forming & Accessories | 99.2 | 67.1 | 71.3 | 99.0 | 97.4 | 97.6 | 103.6 | 84.4 | 86.9 | 104.2 | 171.6 | 162.7 | 110.2 | 171.2 | 163.1 | 100.2 | 112.4 | 110.8 |
| 0320 | Concrete Reinforcing | 105.3 | 72.1 | 88.2 | 97.1 | 90.0 | 93.5 | 96.2 | 90.3 | 93.2 | 97.1 | 182.2 | 141.0 | 97.5 | 182.2 | 141.1 | 100.1 | 100.3 | 100.2 |
| 0330 | Cast-in-Place Concrete | 97.9 | 75.3 | 89.2 | 76.6 | 104.5 | 87.3 | 98.3 | 96.7 | 97.7 | 91.8 | 159.3 | 117.7 | 100.3 | 157.6 | 122.3 | 109.1 | 111.2 | 110.1 |
| 03 | CONCRETE | 107.1 | 71.9 | 90.3 | 96.0 | 99.1 | 97.5 | 98.0 | 91.4 | 94.8 | 99.8 | 167.6 | 132.3 | 110.7 | 166.8 | 137.6 | 109.1 | 112.2 | 110.6 |
| 04 | MASONRY | 117.0 | 64.3 | 85.1 | 92.0 | 103.1 | 98.7 | 107.2 | 91.4 | 97.6 | 94.4 | 165.1 | 137.2 | 117.4 | 165.0 | 146.3 | 107.8 | 116.4 | 113.0 |
| 05 | METALS | 97.5 | 89.2 | 94.6 | 100.0 | 104.1 | 101.4 | 96.6 | 113.6 | 102.5 | 102.2 | 143.1 | 116.2 | 104.5 | 140.8 | 117.0 | 101.7 | 94.2 | 99.1 |
| 06 | WOOD, PLASTICS & COMPOSITES | 90.4 | 67.6 | 77.3 | 99.1 | 96.2 | 97.4 | 109.6 | 83.0 | 94.3 | 103.3 | 175.2 | 144.6 | 111.3 | 174.9 | 147.8 | 103.1 | 112.3 | 108.4 |
| 07 | THERMAL & MOISTURE PROTECTION | 101.1 | 75.0 | 90.7 | 89.9 | 95.5 | 92.1 | 104.9 | 84.8 | 97.0 | 107.7 | 159.2 | 128.1 | 107.5 | 158.2 | 127.6 | 103.3 | 107.9 | 105.1 |
| 08 | OPENINGS | 94.0 | 70.2 | 88.0 | 94.5 | 88.6 | 93.0 | 89.6 | 79.9 | 87.1 | 87.3 | 171.6 | 108.6 | 89.4 | 170.0 | 109.8 | 90.6 | 100.4 | 93.1 |
| 0920 | Plaster & Gypsum Board | 85.0 | 66.2 | 71.8 | 103.6 | 95.8 | 98.1 | 114.4 | 82.0 | 91.6 | 109.4 | 177.1 | 157.1 | 112.0 | 177.1 | 157.8 | 100.8 | 112.4 | 109.0 |
| 0950, 0980 | Ceilings & Acoustic Treatment | 118.2 | 66.2 | 83.7 | 97.8 | 95.8 | 96.5 | 100.6 | 82.0 | 88.3 | 88.4 | 177.1 | 147.2 | 97.0 | 177.1 | 150.1 | 101.1 | 112.4 | 108.6 |
| 0960 | Flooring | 101.3 | 69.7 | 92.2 | 88.8 | 104.8 | 93.4 | 100.5 | 90.4 | 97.6 | 99.4 | 175.0 | 121.1 | 107.5 | 175.0 | 126.9 | 95.8 | 115.3 | 101.4 |
| 0970, 0990 | Wall Finishes & Painting/Coating | 98.5 | 52.8 | 70.8 | 85.8 | 80.1 | 82.4 | 87.9 | 83.5 | 85.3 | 99.8 | 156.0 | 133.9 | 116.0 | 156.0 | 140.3 | 94.5 | 111.6 | 104.9 |
| 09 | FINISHES | 101.9 | 65.9 | 82.0 | 95.7 | 97.1 | 96.4 | 98.0 | 85.1 | 90.9 | 104.9 | 171.5 | 141.6 | 114.4 | 171.2 | 145.7 | 97.3 | 113.7 | 106.3 |
| COVERS | DIVS. 10 - 14, 25, 28, 41, 43, 44, 46 | 100.0 | 83.3 | 96.5 | 100.0 | 93.9 | 98.7 | 100.0 | 94.6 | 98.9 | 100.0 | 127.5 | 105.7 | 100.0 | 126.6 | 105.5 | 100.0 | 105.9 | 101.2 |
| 21, 22, 23 | FIRE SUPPRESSION, PLUMBING & HVAC | 96.9 | 69.4 | 85.4 | 100.1 | 94.2 | 97.6 | 100.5 | 82.5 | 93.0 | 100.3 | 161.5 | 125.8 | 99.8 | 161.4 | 125.4 | 99.9 | 92.6 | 96.9 |
| 26, 27, 3370 | ELECTRICAL, COMMUNICATIONS & UTIL. | 89.7 | 72.6 | 80.6 | 102.9 | 93.6 | 98.0 | 101.4 | 92.2 | 96.5 | 97.0 | 172.8 | 137.3 | 100.8 | 172.8 | 139.1 | 99.9 | 98.7 | 99.3 |
| MF2010 | WEIGHTED AVERAGE | 98.6 | 74.9 | 88.0 | 97.2 | 97.9 | 97.5 | 98.8 | 90.2 | 94.6 | 100.9 | 160.3 | 126.9 | 104.0 | 160.1 | 129.1 | 100.5 | 102.9 | 101.6 |

| DIVISION | | NEW YORK ELMIRA 148 - 149 | | | NEW YORK FAR ROCKAWAY 116 | | | NEW YORK FLUSHING 113 | | | NEW YORK GLENS FALLS 128 | | | NEW YORK HICKSVILLE 115,117,118 | | | NEW YORK JAMAICA 114 | | |
|---|---|---|---|---|---|---|---|---|---|---|---|---|---|---|---|---|---|---|---|
| | | MAT. | INST. | TOTAL | MAT. | INST. | TOTAL | MAT. | INST. | TOTAL | MAT. | INST. | TOTAL | MAT. | INST. | TOTAL | MAT. | INST. | TOTAL |
| 015433 | CONTRACTOR EQUIPMENT | | 115.9 | 115.9 | | 115.4 | 115.4 | | 115.4 | 115.4 | | 114.2 | 114.2 | | 115.4 | 115.4 | | 115.4 | 115.4 |
| 0241, 31 - 34 | SITE & INFRASTRUCTURE, DEMOLITION | 90.8 | 93.3 | 92.5 | 125.7 | 129.3 | 128.2 | 125.7 | 129.3 | 128.2 | 62.6 | 106.7 | 93.0 | 114.0 | 128.1 | 123.8 | 119.5 | 129.3 | 126.2 |
| 0310 | Concrete Forming & Accessories | 85.2 | 87.1 | 86.8 | 96.6 | 171.2 | 161.3 | 100.5 | 171.2 | 161.8 | 87.5 | 88.6 | 88.5 | 93.3 | 142.7 | 136.2 | 100.5 | 171.2 | 161.8 |
| 0320 | Concrete Reinforcing | 101.4 | 87.2 | 94.1 | 97.5 | 182.2 | 141.1 | 99.2 | 182.2 | 141.9 | 96.8 | 90.0 | 93.3 | 97.5 | 182.0 | 141.0 | 100.3 | 182.2 | 141.1 |
| 0330 | Cast-in-Place Concrete | 97.5 | 96.7 | 97.2 | 108.5 | 157.6 | 127.3 | 108.5 | 157.6 | 127.3 | 74.8 | 102.1 | 85.2 | 92.1 | 153.1 | 115.5 | 100.3 | 157.6 | 122.3 |
| 03 | CONCRETE | 102.6 | 92.2 | 97.6 | 117.2 | 166.8 | 140.9 | 117.7 | 166.8 | 141.2 | 88.4 | 94.4 | 91.3 | 102.2 | 152.4 | 126.3 | 110.1 | 166.8 | 137.3 |
| 04 | MASONRY | 104.1 | 93.0 | 97.3 | 121.0 | 165.0 | 147.7 | 114.8 | 165.0 | 145.2 | 92.9 | 98.9 | 96.5 | 111.2 | 158.9 | 140.1 | 119.3 | 165.0 | 147.0 |
| 05 | METALS | 96.9 | 114.7 | 103.0 | 104.5 | 140.8 | 117.0 | 104.5 | 140.8 | 117.0 | 96.0 | 103.8 | 98.7 | 106.1 | 137.6 | 116.9 | 104.5 | 140.8 | 117.0 |
| 06 | WOOD, PLASTICS & COMPOSITES | 86.1 | 86.3 | 86.2 | 94.9 | 174.9 | 140.8 | 99.7 | 174.9 | 142.8 | 85.7 | 86.8 | 86.3 | 91.2 | 140.8 | 119.7 | 99.7 | 174.9 | 142.8 |
| 07 | THERMAL & MOISTURE PROTECTION | 103.9 | 86.1 | 96.9 | 107.4 | 158.2 | 127.6 | 107.5 | 158.2 | 127.6 | 89.4 | 92.0 | 90.4 | 107.1 | 150.0 | 124.1 | 107.3 | 158.2 | 127.5 |
| 08 | OPENINGS | 91.6 | 80.9 | 88.9 | 87.1 | 170.0 | 108.6 | 87.7 | 170.0 | 108.6 | 88.1 | 83.5 | 86.9 | 87.7 | 151.6 | 103.9 | 87.7 | 170.0 | 108.6 |
| 0920 | Plaster & Gypsum Board | 94.7 | 85.6 | 88.3 | 101.1 | 177.1 | 154.6 | 103.8 | 177.1 | 155.4 | 97.0 | 86.1 | 89.3 | 100.2 | 142.0 | 129.6 | 103.8 | 177.1 | 155.4 |
| 0950, 0980 | Ceilings & Acoustic Treatment | 97.3 | 85.6 | 89.6 | 86.7 | 177.1 | 146.6 | 86.7 | 177.1 | 146.6 | 95.1 | 86.1 | 89.1 | 85.6 | 142.0 | 123.0 | 86.7 | 177.1 | 146.6 |
| 0960 | Flooring | 85.7 | 90.4 | 87.1 | 102.3 | 175.0 | 123.2 | 103.8 | 175.0 | 124.3 | 80.8 | 104.8 | 87.7 | 101.2 | 150.7 | 115.4 | 103.8 | 175.0 | 124.3 |
| 0970, 0990 | Wall Finishes & Painting/Coating | 93.0 | 79.5 | 84.8 | 116.0 | 156.0 | 140.3 | 116.0 | 156.0 | 140.3 | 85.8 | 83.6 | 84.5 | 116.0 | 156.0 | 140.3 | 116.0 | 156.0 | 140.3 |
| 09 | FINISHES | 92.4 | 86.7 | 89.3 | 109.9 | 171.2 | 143.7 | 110.7 | 171.2 | 144.1 | 91.3 | 90.9 | 91.1 | 108.2 | 144.7 | 128.3 | 110.2 | 171.2 | 143.8 |
| COVERS | DIVS. 10 - 14, 25, 28, 41, 43, 44, 46 | 100.0 | 91.1 | 98.2 | 100.0 | 126.6 | 105.5 | 100.0 | 126.6 | 105.5 | 100.0 | 85.6 | 97.0 | 100.0 | 120.5 | 104.3 | 100.0 | 126.6 | 105.5 |
| 21, 22, 23 | FIRE SUPPRESSION, PLUMBING & HVAC | 94.7 | 85.4 | 90.9 | 94.4 | 161.4 | 122.3 | 94.4 | 161.4 | 122.3 | 95.0 | 86.5 | 91.5 | 99.8 | 146.9 | 119.4 | 94.4 | 161.4 | 122.3 |
| 26, 27, 3370 | ELECTRICAL, COMMUNICATIONS & UTIL. | 98.1 | 95.5 | 96.7 | 109.0 | 172.8 | 142.9 | 109.0 | 172.8 | 142.9 | 96.3 | 90.4 | 93.2 | 100.1 | 144.0 | 123.5 | 99.0 | 172.8 | 138.2 |
| MF2010 | WEIGHTED AVERAGE | 96.9 | 92.0 | 94.7 | 103.9 | 160.1 | 129.1 | 103.8 | 160.1 | 129.0 | 92.6 | 93.2 | 92.8 | 101.7 | 145.2 | 121.2 | 101.9 | 160.1 | 127.9 |

| DIVISION | | NEW YORK JAMESTOWN 147 | | | NEW YORK KINGSTON 124 | | | NEW YORK LONG ISLAND CITY 111 | | | NEW YORK MONTICELLO 127 | | | NEW YORK MOUNT VERNON 105 | | | NEW YORK NEW ROCHELLE 108 | | |
|---|---|---|---|---|---|---|---|---|---|---|---|---|---|---|---|---|---|---|---|
| | | MAT. | INST. | TOTAL | MAT. | INST. | TOTAL | MAT. | INST. | TOTAL | MAT. | INST. | TOTAL | MAT. | INST. | TOTAL | MAT. | INST. | TOTAL |
| 015433 | CONTRACTOR EQUIPMENT | | 92.9 | 92.9 | | 115.4 | 115.4 | | 115.4 | 115.4 | | 115.4 | 115.4 | | 114.5 | 114.5 | | 114.5 | 114.5 |
| 0241, 31 - 34 | SITE & INFRASTRUCTURE, DEMOLITION | 92.1 | 92.6 | 92.4 | 121.3 | 125.7 | 124.3 | 122.9 | 129.3 | 127.3 | 116.5 | 125.7 | 122.9 | 133.0 | 120.1 | 124.1 | 131.5 | 120.1 | 123.6 |
| 0310 | Concrete Forming & Accessories | 85.2 | 80.0 | 80.7 | 89.3 | 109.8 | 107.1 | 104.7 | 171.2 | 162.4 | 96.2 | 109.8 | 108.0 | 94.8 | 131.9 | 127.0 | 109.4 | 131.9 | 128.9 |
| 0320 | Concrete Reinforcing | 101.5 | 74.6 | 87.7 | 97.1 | 133.3 | 115.8 | 97.5 | 182.2 | 141.1 | 96.4 | 133.3 | 115.8 | 95.9 | 181.1 | 139.8 | 96.0 | 181.1 | 139.8 |
| 0330 | Cast-in-Place Concrete | 101.2 | 95.6 | 99.1 | 101.0 | 133.0 | 113.3 | 103.5 | 157.6 | 124.2 | 94.6 | 133.0 | 109.3 | 102.6 | 132.9 | 114.2 | 102.6 | 132.9 | 114.2 |
| 03 | CONCRETE | 105.9 | 84.4 | 95.6 | 111.1 | 121.8 | 116.2 | 113.4 | 166.8 | 138.9 | 105.7 | 121.8 | 113.4 | 110.1 | 140.5 | 124.6 | 109.3 | 140.5 | 124.3 |
| 04 | MASONRY | 113.1 | 91.0 | 99.7 | 108.4 | 136.2 | 125.2 | 114.3 | 165.0 | 145.0 | 101.7 | 136.2 | 122.6 | 100.0 | 130.1 | 118.2 | 100.0 | 130.1 | 118.2 |
| 05 | METALS | 94.2 | 80.9 | 89.6 | 104.5 | 114.5 | 107.9 | 104.5 | 140.8 | 117.0 | 104.5 | 114.6 | 107.9 | 101.9 | 134.0 | 112.9 | 102.2 | 134.0 | 113.1 |
| 06 | WOOD, PLASTICS & COMPOSITES | 84.6 | 76.6 | 80.0 | 87.0 | 104.9 | 97.2 | 105.3 | 174.9 | 145.2 | 95.4 | 104.9 | 100.8 | 92.6 | 135.5 | 117.2 | 110.6 | 135.5 | 124.9 |
| 07 | THERMAL & MOISTURE PROTECTION | 103.5 | 84.9 | 96.1 | 109.1 | 135.7 | 119.7 | 107.4 | 158.2 | 127.6 | 108.8 | 135.7 | 119.5 | 108.6 | 137.5 | 120.1 | 108.6 | 137.5 | 120.1 |
| 08 | OPENINGS | 91.5 | 74.7 | 87.2 | 93.0 | 121.3 | 100.2 | 87.7 | 170.0 | 108.6 | 87.8 | 121.3 | 96.3 | 87.3 | 148.7 | 102.8 | 87.3 | 148.7 | 102.9 |
| 0920 | Plaster & Gypsum Board | 89.3 | 75.7 | 79.7 | 95.8 | 105.0 | 102.2 | 107.3 | 177.1 | 156.4 | 99.3 | 105.0 | 103.3 | 103.5 | 136.1 | 126.5 | 116.8 | 136.1 | 130.4 |
| 0950, 0980 | Ceilings & Acoustic Treatment | 94.5 | 75.7 | 82.0 | 80.1 | 105.0 | 96.6 | 86.7 | 177.1 | 146.6 | 80.1 | 105.0 | 96.6 | 86.5 | 136.1 | 119.4 | 86.5 | 136.1 | 119.4 |
| 0960 | Flooring | 88.6 | 93.1 | 89.9 | 94.7 | 76.6 | 89.5 | 105.4 | 175.0 | 125.4 | 97.4 | 76.6 | 91.4 | 91.3 | 173.8 | 115.0 | 98.4 | 173.8 | 120.1 |
| 0970, 0990 | Wall Finishes & Painting/Coating | 94.5 | 79.9 | 85.6 | 113.6 | 105.0 | 108.4 | 116.0 | 156.0 | 140.3 | 113.6 | 103.9 | 107.7 | 97.8 | 156.0 | 133.1 | 97.8 | 156.0 | 133.1 |
| 09 | FINISHES | 92.2 | 81.8 | 86.5 | 105.1 | 101.0 | 102.8 | 111.3 | 171.2 | 144.3 | 105.9 | 100.9 | 103.1 | 102.3 | 141.9 | 124.1 | 105.7 | 141.9 | 125.7 |
| COVERS | DIVS. 10 - 14, 25, 28, 41, 43, 44, 46 | 100.0 | 93.2 | 98.6 | 100.0 | 109.4 | 101.9 | 100.0 | 126.6 | 105.5 | 100.0 | 109.4 | 101.9 | 100.0 | 115.3 | 103.2 | 100.0 | 115.3 | 103.2 |
| 21, 22, 23 | FIRE SUPPRESSION, PLUMBING & HVAC | 94.6 | 83.9 | 90.2 | 94.9 | 117.4 | 104.3 | 99.8 | 161.4 | 125.4 | 94.9 | 119.9 | 105.3 | 95.0 | 120.5 | 105.6 | 95.0 | 120.5 | 105.6 |
| 26, 27, 3370 | ELECTRICAL, COMMUNICATIONS & UTIL. | 97.0 | 81.8 | 88.9 | 97.9 | 118.2 | 108.7 | 99.5 | 172.8 | 138.5 | 97.9 | 118.2 | 108.7 | 95.2 | 153.6 | 126.3 | 95.2 | 153.6 | 126.3 |
| MF2010 | WEIGHTED AVERAGE | 97.2 | 84.4 | 91.5 | 101.6 | 118.7 | 109.2 | 103.6 | 160.1 | 128.9 | 100.0 | 119.2 | 108.6 | 100.0 | 134.3 | 115.3 | 100.3 | 134.3 | 115.5 |

# City Cost Indexes

## NEW YORK

| DIVISION | | NEW YORK 100 - 102 | | | NIAGARA FALLS 143 | | | PLATTSBURGH 129 | | | POUGHKEEPSIE 125 - 126 | | | QUEENS 110 | | | RIVERHEAD 119 | | |
|---|---|---|---|---|---|---|---|---|---|---|---|---|---|---|---|---|---|---|---|
| | | MAT. | INST. | TOTAL | MAT. | INST. | TOTAL | MAT. | INST. | TOTAL | MAT. | INST. | TOTAL | MAT. | INST. | TOTAL | MAT. | INST. | TOTAL |
| 015433 | CONTRACTOR EQUIPMENT | | 114.8 | 114.8 | | 92.9 | 92.9 | | 99.0 | 99.0 | | 115.4 | 115.4 | | 115.4 | 115.4 | | 115.4 | 115.4 |
| 0241, 31 - 34 | SITE & INFRASTRUCTURE, DEMOLITION | 133.1 | 126.2 | 128.4 | 94.2 | 93.4 | 93.6 | 87.3 | 102.5 | 97.8 | 117.3 | 125.6 | 123.0 | 117.6 | 129.3 | 125.6 | 115.1 | 128.1 | 124.1 |
| 0310 | Concrete Forming & Accessories | 108.7 | 187.6 | 177.1 | 85.2 | 108.1 | 105.0 | 91.9 | 82.4 | 83.6 | 89.3 | 156.6 | 147.7 | 93.5 | 171.2 | 160.9 | 97.7 | 142.6 | 136.7 |
| 0320 | Concrete Reinforcing | 102.8 | 182.3 | 143.7 | 100.1 | 92.7 | 96.3 | 101.2 | 89.4 | 95.1 | 97.1 | 133.6 | 115.9 | 99.2 | 182.2 | 141.9 | 99.3 | 179.0 | 140.4 |
| 0330 | Cast-in-Place Concrete | 103.4 | 169.5 | 128.7 | 104.7 | 102.4 | 103.8 | 91.5 | 94.3 | 92.6 | 97.8 | 131.5 | 110.7 | 95.3 | 157.6 | 119.2 | 93.6 | 153.1 | 116.4 |
| 03 | CONCRETE | 110.6 | 178.2 | 142.9 | 108.4 | 102.5 | 105.6 | 102.3 | 87.6 | 95.3 | 108.2 | 142.2 | 124.4 | 105.6 | 166.8 | 134.9 | 102.9 | 151.9 | 126.3 |
| 04 | MASONRY | 105.7 | 173.9 | 147.0 | 121.1 | 111.7 | 115.4 | 88.9 | 88.3 | 88.5 | 101.7 | 132.2 | 120.2 | 108.6 | 165.0 | 142.8 | 116.8 | 158.9 | 142.3 |
| 05 | METALS | 116.5 | 143.3 | 125.7 | 96.9 | 88.7 | 94.1 | 100.5 | 83.4 | 94.6 | 104.5 | 116.5 | 108.6 | 104.5 | 140.8 | 117.0 | 106.5 | 136.5 | 116.8 |
| 06 | WOOD, PLASTICS & COMPOSITES | 107.8 | 192.2 | 156.2 | 84.6 | 107.4 | 97.7 | 93.5 | 80.5 | 86.1 | 87.0 | 168.4 | 133.7 | 91.4 | 174.9 | 139.3 | 96.7 | 140.8 | 122.0 |
| 07 | THERMAL & MOISTURE PROTECTION | 107.7 | 165.0 | 130.4 | 103.6 | 102.1 | 103.0 | 102.8 | 87.1 | 96.6 | 109.1 | 141.9 | 122.1 | 107.1 | 158.2 | 127.4 | 107.7 | 150.0 | 124.5 |
| 08 | OPENINGS | 92.6 | 180.7 | 114.9 | 91.5 | 95.4 | 92.5 | 97.3 | 79.4 | 92.8 | 93.0 | 152.3 | 108.0 | 87.7 | 170.0 | 108.6 | 87.7 | 150.9 | 103.7 |
| 0920 | Plaster & Gypsum Board | 117.5 | 194.6 | 171.8 | 89.3 | 107.4 | 102.0 | 119.9 | 79.2 | 91.2 | 95.8 | 170.4 | 148.3 | 100.2 | 177.1 | 154.4 | 102.3 | 142.0 | 130.2 |
| 0950, 0980 | Ceilings & Acoustic Treatment | 108.1 | 194.6 | 165.4 | 94.5 | 107.4 | 103.0 | 109.9 | 79.2 | 89.5 | 80.1 | 170.4 | 140.0 | 86.7 | 177.1 | 146.6 | 86.6 | 142.0 | 123.3 |
| 0960 | Flooring | 100.7 | 175.0 | 122.0 | 88.4 | 107.6 | 94.0 | 104.3 | 100.7 | 103.3 | 94.7 | 147.1 | 109.8 | 101.2 | 175.0 | 122.4 | 102.3 | 150.7 | 116.2 |
| 0970, 0990 | Wall Finishes & Painting/Coating | 99.8 | 156.0 | 133.9 | 94.5 | 102.5 | 99.3 | 110.6 | 77.0 | 90.2 | 113.6 | 103.9 | 107.7 | 116.0 | 156.0 | 140.3 | 116.0 | 156.0 | 140.3 |
| 09 | FINISHES | 111.2 | 183.6 | 151.2 | 92.3 | 108.2 | 101.1 | 101.9 | 85.2 | 92.7 | 104.9 | 151.7 | 130.7 | 108.6 | 171.2 | 143.1 | 108.9 | 144.7 | 128.6 |
| COVERS | DIVS. 10 - 14, 25, 28, 41, 43, 44, 46 | 100.0 | 137.6 | 107.8 | 100.0 | 100.8 | 100.2 | 100.0 | 51.6 | 89.9 | 100.0 | 115.3 | 103.2 | 100.0 | 126.6 | 105.5 | 100.0 | 120.5 | 104.3 |
| 21, 22, 23 | FIRE SUPPRESSION, PLUMBING & HVAC | 100.2 | 166.0 | 127.6 | 94.6 | 92.4 | 93.7 | 94.9 | 81.4 | 89.2 | 94.9 | 119.7 | 105.2 | 99.8 | 161.4 | 126.4 | 99.9 | 145.6 | 119.0 |
| 26, 27, 3370 | ELECTRICAL, COMMUNICATIONS & UTIL. | 105.6 | 173.1 | 141.5 | 95.4 | 90.6 | 92.8 | 94.6 | 84.6 | 89.3 | 97.9 | 118.2 | 108.7 | 100.1 | 172.8 | 138.8 | 101.9 | 144.0 | 124.3 |
| MF2010 | WEIGHTED AVERAGE | 106.0 | 166.3 | 133.0 | 98.3 | 98.0 | 98.2 | 97.5 | 85.0 | 91.9 | 100.7 | 130.5 | 114.0 | 101.9 | 160.1 | 127.9 | 102.5 | 144.7 | 121.4 |

## NEW YORK

| DIVISION | | ROCHESTER 144 - 146 | | | SCHENECTADY 123 | | | STATEN ISLAND 103 | | | SUFFERN 109 | | | SYRACUSE 130 - 132 | | | UTICA 133 - 135 | | |
|---|---|---|---|---|---|---|---|---|---|---|---|---|---|---|---|---|---|---|---|
| | | MAT. | INST. | TOTAL | MAT. | INST. | TOTAL | MAT. | INST. | TOTAL | MAT. | INST. | TOTAL | MAT. | INST. | TOTAL | MAT. | INST. | TOTAL |
| 015433 | CONTRACTOR EQUIPMENT | | 116.6 | 116.6 | | 114.2 | 114.2 | | 114.5 | 114.5 | | 114.5 | 114.5 | | 114.2 | 114.2 | | 114.2 | 114.2 |
| 0241, 31 - 34 | SITE & INFRASTRUCTURE, DEMOLITION | 71.8 | 108.5 | 97.1 | 72.0 | 107.3 | 96.4 | 138.0 | 125.4 | 129.3 | 127.8 | 121.1 | 123.2 | 93.4 | 106.7 | 102.6 | 70.2 | 104.4 | 93.8 |
| 0310 | Concrete Forming & Accessories | 101.5 | 93.7 | 94.7 | 103.4 | 97.0 | 97.8 | 94.3 | 171.6 | 161.4 | 103.2 | 125.5 | 122.6 | 102.4 | 89.1 | 90.9 | 104.0 | 82.2 | 85.1 |
| 0320 | Concrete Reinforcing | 101.0 | 90.6 | 95.7 | 95.8 | 90.0 | 92.8 | 97.1 | 182.2 | 141.0 | 96.0 | 133.6 | 115.3 | 97.1 | 90.7 | 93.8 | 97.1 | 88.3 | 92.6 |
| 0330 | Cast-in-Place Concrete | 97.9 | 99.0 | 98.3 | 90.3 | 103.9 | 95.5 | 102.6 | 159.3 | 124.3 | 99.5 | 129.8 | 111.1 | 91.3 | 96.9 | 93.4 | 83.5 | 89.3 | 85.7 |
| 03 | CONCRETE | 116.3 | 95.8 | 106.5 | 103.0 | 98.7 | 100.9 | 111.9 | 167.6 | 138.6 | 106.2 | 127.9 | 116.6 | 101.6 | 92.9 | 97.5 | 99.8 | 86.8 | 93.6 |
| 04 | MASONRY | 104.2 | 98.4 | 100.7 | 91.1 | 102.0 | 97.7 | 108.2 | 165.1 | 142.7 | 100.1 | 125.6 | 115.6 | 98.5 | 100.1 | 99.5 | 89.8 | 87.8 | 88.6 |
| 05 | METALS | 98.2 | 105.4 | 100.7 | 100.1 | 104.1 | 101.5 | 100.1 | 143.1 | 114.9 | 100.1 | 118.5 | 106.4 | 100.0 | 103.2 | 101.1 | 97.9 | 102.0 | 99.3 |
| 06 | WOOD, PLASTICS & COMPOSITES | 100.0 | 93.7 | 96.4 | 105.8 | 96.2 | 100.3 | 91.1 | 175.2 | 139.4 | 102.9 | 131.1 | 119.1 | 105.8 | 87.4 | 95.2 | 105.8 | 84.2 | 93.4 |
| 07 | THERMAL & MOISTURE PROTECTION | 98.7 | 94.6 | 97.1 | 90.5 | 95.0 | 92.3 | 108.0 | 159.2 | 128.3 | 108.5 | 134.2 | 118.7 | 101.1 | 96.2 | 99.2 | 90.3 | 90.7 | 90.5 |
| 08 | OPENINGS | 93.8 | 87.3 | 92.2 | 94.5 | 88.6 | 93.0 | 87.3 | 171.6 | 108.6 | 87.3 | 137.3 | 100.0 | 92.7 | 80.1 | 89.5 | 94.5 | 79.9 | 90.8 |
| 0920 | Plaster & Gypsum Board | 106.2 | 93.5 | 97.3 | 108.6 | 95.8 | 99.6 | 103.6 | 177.1 | 155.3 | 108.4 | 131.7 | 124.8 | 108.6 | 86.8 | 93.3 | 108.6 | 83.5 | 90.9 |
| 0950, 0980 | Ceilings & Acoustic Treatment | 102.1 | 93.5 | 96.4 | 100.6 | 95.8 | 97.4 | 88.4 | 177.1 | 147.2 | 86.5 | 131.7 | 116.5 | 100.6 | 86.8 | 91.5 | 100.6 | 83.5 | 89.3 |
| 0960 | Flooring | 84.6 | 104.0 | 90.2 | 88.2 | 104.8 | 93.0 | 95.4 | 175.0 | 118.3 | 94.6 | 150.7 | 110.7 | 90.1 | 90.5 | 90.2 | 88.2 | 91.2 | 89.1 |
| 0970, 0990 | Wall Finishes & Painting/Coating | 94.3 | 97.6 | 96.3 | 85.8 | 80.1 | 82.4 | 99.8 | 156.0 | 133.9 | 97.8 | 112.3 | 106.6 | 92.5 | 95.4 | 94.3 | 85.8 | 95.4 | 91.6 |
| 09 | FINISHES | 98.5 | 96.3 | 97.3 | 96.7 | 96.8 | 96.8 | 104.5 | 171.5 | 141.4 | 103.4 | 126.3 | 116.0 | 98.5 | 89.8 | 93.7 | 96.8 | 85.2 | 90.4 |
| COVERS | DIVS. 10 - 14, 25, 28, 41, 43, 44, 46 | 100.0 | 94.5 | 98.8 | 100.0 | 93.5 | 98.7 | 100.0 | 127.5 | 105.7 | 100.0 | 113.2 | 102.7 | 100.0 | 95.7 | 99.1 | 100.0 | 87.2 | 97.3 |
| 21, 22, 23 | FIRE SUPPRESSION, PLUMBING & HVAC | 99.9 | 89.1 | 95.4 | 100.3 | 93.6 | 97.5 | 100.3 | 161.5 | 125.8 | 95.0 | 114.4 | 103.1 | 100.3 | 86.7 | 94.6 | 100.3 | 80.2 | 91.9 |
| 26, 27, 3370 | ELECTRICAL, COMMUNICATIONS & UTIL. | 100.1 | 91.6 | 95.6 | 101.5 | 93.6 | 97.3 | 97.0 | 172.8 | 137.3 | 103.7 | 126.4 | 115.8 | 101.5 | 92.4 | 96.6 | 99.6 | 92.4 | 95.8 |
| MF2010 | WEIGHTED AVERAGE | 100.3 | 95.6 | 98.2 | 98.2 | 97.5 | 97.9 | 102.1 | 160.3 | 128.1 | 100.1 | 123.1 | 110.4 | 99.4 | 93.5 | 96.7 | 97.1 | 88.7 | 93.4 |

## NEW YORK / NORTH CAROLINA

| DIVISION | | NEW YORK WATERTOWN 136 | | | NEW YORK WHITE PLAINS 106 | | | NEW YORK YONKERS 107 | | | NORTH CAROLINA ASHEVILLE 287 - 288 | | | NORTH CAROLINA CHARLOTTE 281 - 282 | | | NORTH CAROLINA DURHAM 277 | | |
|---|---|---|---|---|---|---|---|---|---|---|---|---|---|---|---|---|---|---|---|
| | | MAT. | INST. | TOTAL | MAT. | INST. | TOTAL | MAT. | INST. | TOTAL | MAT. | INST. | TOTAL | MAT. | INST. | TOTAL | MAT. | INST. | TOTAL |
| 015433 | CONTRACTOR EQUIPMENT | | 114.2 | 114.2 | | 114.5 | 114.5 | | 114.5 | 114.5 | | 94.9 | 94.9 | | 94.9 | 94.9 | | 100.4 | 100.4 |
| 0241, 31 - 34 | SITE & INFRASTRUCTURE, DEMOLITION | 78.2 | 106.8 | 97.9 | 123.0 | 120.1 | 121.0 | 132.5 | 119.9 | 123.8 | 109.5 | 74.8 | 85.6 | 110.7 | 74.9 | 86.0 | 109.8 | 83.6 | 91.7 |
| 0310 | Concrete Forming & Accessories | 89.2 | 92.0 | 91.6 | 108.1 | 131.9 | 128.8 | 108.4 | 131.9 | 128.8 | 92.0 | 38.1 | 45.3 | 96.5 | 40.0 | 47.4 | 94.6 | 38.7 | 46.1 |
| 0320 | Concrete Reinforcing | 97.7 | 90.7 | 94.1 | 96.0 | 181.1 | 139.8 | 99.8 | 181.2 | 141.7 | 90.2 | 41.6 | 65.2 | 90.6 | 38.8 | 63.9 | 91.6 | 48.7 | 69.5 |
| 0330 | Cast-in-Place Concrete | 97.2 | 96.7 | 97.0 | 91.1 | 132.9 | 107.1 | 101.9 | 132.9 | 113.8 | 113.2 | 45.8 | 87.4 | 119.5 | 47.7 | 92.0 | 115.1 | 46.0 | 88.6 |
| 03 | CONCRETE | 113.0 | 94.1 | 104.0 | 99.3 | 140.5 | 119.0 | 109.5 | 140.5 | 124.3 | 107.1 | 43.3 | 76.6 | 109.9 | 44.3 | 78.5 | 107.7 | 45.0 | 77.7 |
| 04 | MASONRY | 90.9 | 101.0 | 97.0 | 99.5 | 130.1 | 118.0 | 105.3 | 130.1 | 120.3 | 85.2 | 39.0 | 57.2 | 91.1 | 42.7 | 61.8 | 90.5 | 35.1 | 56.9 |
| 05 | METALS | 98.0 | 103.2 | 99.8 | 101.7 | 134.0 | 112.8 | 111.9 | 134.0 | 119.5 | 84.7 | 72.7 | 80.6 | 88.6 | 71.8 | 82.8 | 97.4 | 75.9 | 90.0 |
| 06 | WOOD, PLASTICS & COMPOSITES | 87.5 | 91.4 | 89.7 | 108.9 | 135.5 | 124.1 | 108.7 | 135.5 | 124.1 | 93.5 | 37.6 | 61.5 | 99.1 | 39.7 | 65.0 | 96.2 | 39.1 | 63.4 |
| 07 | THERMAL & MOISTURE PROTECTION | 90.6 | 97.0 | 93.1 | 108.3 | 137.5 | 119.9 | 108.6 | 137.5 | 120.1 | 98.2 | 42.1 | 75.9 | 95.7 | 43.6 | 75.0 | 98.9 | 44.0 | 77.1 |
| 08 | OPENINGS | 94.5 | 86.0 | 92.4 | 87.4 | 148.7 | 102.9 | 90.5 | 148.7 | 105.2 | 94.5 | 37.4 | 80.0 | 98.6 | 38.8 | 83.5 | 98.6 | 41.3 | 84.1 |
| 0920 | Plaster & Gypsum Board | 99.3 | 90.9 | 93.4 | 111.9 | 136.1 | 130.5 | 117.0 | 136.1 | 130.5 | 99.7 | 35.6 | 54.6 | 100.0 | 37.8 | 56.2 | 104.5 | 37.1 | 57.0 |
| 0950, 0980 | Ceilings & Acoustic Treatment | 100.6 | 90.9 | 94.2 | 86.5 | 136.1 | 119.4 | 106.2 | 136.1 | 126.0 | 92.0 | 35.6 | 54.6 | 94.8 | 37.8 | 57.0 | 95.7 | 37.1 | 56.8 |
| 0960 | Flooring | 81.3 | 92.0 | 84.4 | 96.7 | 173.8 | 118.8 | 96.3 | 173.8 | 118.5 | 106.2 | 44.0 | 88.3 | 109.3 | 44.2 | 90.6 | 109.9 | 44.0 | 91.0 |
| 0970, 0990 | Wall Finishes & Painting/Coating | 85.8 | 86.5 | 86.2 | 97.8 | 156.0 | 133.1 | 97.8 | 156.0 | 133.1 | 114.7 | 34.5 | 66.0 | 114.7 | 39.4 | 69.0 | 114.7 | 36.1 | 67.0 |
| 09 | FINISHES | 94.3 | 91.4 | 92.7 | 103.8 | 141.9 | 124.8 | 109.3 | 141.9 | 127.2 | 100.7 | 38.6 | 66.4 | 101.8 | 40.5 | 68.0 | 103.2 | 39.0 | 67.8 |
| COVERS | DIVS. 10 - 14, 25, 28, 41, 43, 44, 46 | 100.0 | 96.1 | 99.2 | 100.0 | 115.3 | 103.2 | 100.0 | 120.4 | 104.2 | 100.0 | 66.9 | 93.1 | 100.0 | 67.4 | 93.2 | 100.0 | 71.6 | 94.1 |
| 21, 22, 23 | FIRE SUPPRESSION, PLUMBING & HVAC | 100.3 | 81.8 | 92.6 | 100.5 | 120.5 | 108.8 | 100.5 | 120.5 | 108.8 | 100.3 | 37.8 | 74.3 | 100.0 | 37.9 | 74.2 | 100.4 | 36.6 | 73.8 |
| 26, 27, 3370 | ELECTRICAL, COMMUNICATIONS & UTIL. | 101.5 | 87.5 | 94.1 | 95.2 | 153.6 | 126.3 | 103.8 | 153.6 | 130.3 | 97.1 | 37.4 | 65.3 | 99.5 | 48.6 | 72.4 | 97.8 | 47.3 | 71.0 |
| MF2010 | WEIGHTED AVERAGE | 98.9 | 92.6 | 96.1 | 99.9 | 134.3 | 115.3 | 104.8 | 134.5 | 118.0 | 97.4 | 46.0 | 74.4 | 99.2 | 48.4 | 76.5 | 100.3 | 48.4 | 77.1 |

### NORTH CAROLINA

| DIVISION | | ELIZABETH CITY 279 MAT. | INST. | TOTAL | FAYETTEVILLE 283 MAT. | INST. | TOTAL | GASTONIA 280 MAT. | INST. | TOTAL | GREENSBORO 270,272-274 MAT. | INST. | TOTAL | HICKORY 286 MAT. | INST. | TOTAL | KINSTON 285 MAT. | INST. | TOTAL |
|---|---|---|---|---|---|---|---|---|---|---|---|---|---|---|---|---|---|---|---|
| 015433 | CONTRACTOR EQUIPMENT | | 104.7 | 104.7 | | 100.4 | 100.4 | | 94.9 | 94.9 | | 100.4 | 100.4 | | 100.4 | 100.4 | | 100.4 | 100.4 |
| 0241, 31 - 34 | SITE & INFRASTRUCTURE, DEMOLITION | 115.3 | 84.8 | 94.2 | 108.5 | 83.7 | 91.4 | 109.2 | 74.8 | 85.4 | 109.7 | 83.7 | 91.8 | 108.9 | 83.7 | 91.5 | 107.6 | 83.5 | 91.0 |
| 0310 | Concrete Forming & Accessories | 81.4 | 33.6 | 39.9 | 91.7 | 39.4 | 46.4 | 98.3 | 39.0 | 46.9 | 94.3 | 39.4 | 46.6 | 88.4 | 37.2 | 44.0 | 85.0 | 31.5 | 38.6 |
| 0320 | Concrete Reinforcing | 89.7 | 40.5 | 64.4 | 93.7 | 48.8 | 70.6 | 90.6 | 41.6 | 65.3 | 90.6 | 57.1 | 73.3 | 90.2 | 18.9 | 53.5 | 89.7 | 18.3 | 53.0 |
| 0330 | Cast-in-Place Concrete | 115.3 | 37.3 | 85.4 | 118.7 | 49.0 | 92.0 | 110.7 | 46.2 | 88.6 | 114.2 | 47.3 | 88.6 | 113.2 | 46.9 | 87.8 | 109.2 | 37.5 | 81.8 |
| 03 | CONCRETE | 108.1 | 38.2 | 74.6 | 109.3 | 46.4 | 79.2 | 105.6 | 46.2 | 77.2 | 107.1 | 47.3 | 78.5 | 106.9 | 39.2 | 74.5 | 103.5 | 33.3 | 69.9 |
| 04 | MASONRY | 102.9 | 26.3 | 56.4 | 88.5 | 40.0 | 59.1 | 89.7 | 36.9 | 57.7 | 88.4 | 35.8 | 56.5 | 74.9 | 40.2 | 53.9 | 81.0 | 28.2 | 49.0 |
| 05 | METALS | 86.7 | 72.2 | 81.7 | 101.1 | 76.0 | 92.5 | 85.6 | 72.7 | 81.2 | 91.7 | 79.1 | 87.4 | 84.7 | 66.1 | 78.3 | 83.8 | 65.0 | 77.3 |
| 06 | WOOD, PLASTICS & COMPOSITES | 80.5 | 34.8 | 54.3 | 92.5 | 39.0 | 61.8 | 101.8 | 39.0 | 65.8 | 95.9 | 39.0 | 63.2 | 88.3 | 37.4 | 59.1 | 85.0 | 31.7 | 54.5 |
| 07 | THERMAL & MOISTURE PROTECTION | 98.2 | 33.1 | 72.3 | 98.1 | 42.6 | 76.0 | 98.4 | 42.3 | 76.1 | 98.7 | 41.7 | 76.0 | 98.5 | 41.0 | 75.6 | 98.4 | 31.8 | 71.9 |
| 08 | OPENINGS | 95.5 | 32.7 | 79.6 | 94.6 | 41.3 | 81.1 | 98.6 | 38.9 | 83.5 | 98.6 | 43.6 | 84.7 | 94.6 | 30.3 | 78.3 | 94.7 | 27.3 | 77.6 |
| 0920 | Plaster & Gypsum Board | 96.3 | 32.0 | 51.0 | 104.2 | 37.1 | 56.9 | 106.5 | 37.0 | 57.6 | 106.5 | 37.0 | 57.6 | 99.7 | 35.4 | 54.4 | 99.4 | 29.5 | 50.2 |
| 0950, 0980 | Ceilings & Acoustic Treatment | 95.7 | 32.0 | 53.5 | 92.9 | 37.1 | 55.9 | 95.7 | 37.0 | 56.8 | 95.7 | 37.0 | 56.8 | 92.0 | 35.4 | 54.4 | 95.7 | 29.5 | 51.8 |
| 0960 | Flooring | 100.3 | 23.9 | 78.3 | 106.3 | 44.0 | 88.4 | 109.9 | 44.0 | 91.0 | 109.9 | 40.7 | 90.0 | 106.0 | 33.9 | 85.3 | 103.1 | 19.4 | 79.0 |
| 0970, 0990 | Wall Finishes & Painting/Coating | 114.7 | 44.5 | 72.1 | 114.7 | 34.5 | 66.0 | 114.7 | 39.5 | 69.1 | 114.7 | 33.5 | 65.5 | 114.7 | 33.5 | 65.5 | 114.7 | 30.5 | 63.6 |
| 09 | FINISHES | 99.7 | 32.8 | 62.8 | 101.4 | 39.4 | 67.2 | 103.3 | 39.8 | 68.3 | 103.4 | 38.7 | 67.7 | 100.8 | 35.9 | 65.0 | 100.5 | 29.1 | 61.1 |
| COVERS | DIVS. 10 - 14, 25, 28, 41, 43, 44, 46 | 100.0 | 75.5 | 94.9 | 100.0 | 72.4 | 94.3 | 100.0 | 66.8 | 93.1 | 100.0 | 67.1 | 93.2 | 100.0 | 66.8 | 93.1 | 100.0 | 70.5 | 93.9 |
| 21, 22, 23 | FIRE SUPPRESSION, PLUMBING & HVAC | 94.9 | 29.6 | 67.7 | 100.2 | 37.7 | 74.2 | 100.3 | 37.4 | 74.1 | 100.3 | 37.9 | 74.3 | 94.9 | 31.1 | 68.3 | 94.9 | 25.1 | 65.8 |
| 26, 27, 3370 | ELECTRICAL, COMMUNICATIONS & UTIL. | 97.6 | 29.7 | 61.5 | 95.7 | 51.6 | 72.2 | 96.4 | 34.7 | 63.6 | 96.8 | 37.6 | 65.3 | 94.4 | 31.4 | 60.9 | 94.1 | 49.0 | 70.1 |
| MF2010 | WEIGHTED AVERAGE | 97.5 | 41.0 | 72.3 | 100.0 | 50.0 | 77.6 | 98.2 | 46.0 | 74.8 | 99.2 | 47.9 | 76.2 | 95.2 | 42.8 | 71.8 | 94.9 | 40.6 | 70.6 |

### NORTH CAROLINA / NORTH DAKOTA

| DIVISION | | MURPHY 289 MAT. | INST. | TOTAL | RALEIGH 275 - 276 MAT. | INST. | TOTAL | ROCKY MOUNT 278 MAT. | INST. | TOTAL | WILMINGTON 284 MAT. | INST. | TOTAL | WINSTON-SALEM 271 MAT. | INST. | TOTAL | BISMARCK 585 MAT. | INST. | TOTAL |
|---|---|---|---|---|---|---|---|---|---|---|---|---|---|---|---|---|---|---|---|
| 015433 | CONTRACTOR EQUIPMENT | | 94.9 | 94.9 | | 100.4 | 100.4 | | 100.4 | 100.4 | | 94.9 | 94.9 | | 100.4 | 100.4 | | 98.6 | 98.6 |
| 0241, 31 - 34 | SITE & INFRASTRUCTURE, DEMOLITION | 111.5 | 74.5 | 86.0 | 110.2 | 83.8 | 92.0 | 113.2 | 83.4 | 92.6 | 110.8 | 74.8 | 85.9 | 110.0 | 83.7 | 91.8 | 97.2 | 97.2 | 97.2 |
| 0310 | Concrete Forming & Accessories | 98.8 | 24.3 | 34.1 | 94.8 | 40.2 | 47.4 | 87.3 | 29.3 | 36.9 | 93.2 | 38.7 | 45.9 | 95.7 | 38.4 | 45.9 | 95.7 | 43.1 | 50.1 |
| 0320 | Concrete Reinforcing | 89.8 | 18.2 | 52.9 | 90.6 | 48.8 | 69.1 | 89.7 | 26.7 | 57.3 | 90.9 | 47.0 | 68.3 | 90.6 | 38.6 | 63.8 | 104.9 | 86.9 | 95.6 |
| 0330 | Cast-in-Place Concrete | 117.1 | 35.5 | 85.9 | 122.4 | 52.4 | 95.6 | 112.8 | 34.7 | 82.9 | 112.8 | 47.0 | 87.6 | 117.1 | 44.7 | 89.4 | 99.2 | 49.9 | 80.3 |
| 03 | CONCRETE | 110.4 | 29.4 | 71.6 | 111.3 | 47.9 | 80.9 | 108.7 | 32.8 | 72.3 | 106.9 | 45.0 | 77.3 | 108.6 | 42.5 | 77.0 | 99.9 | 55.1 | 78.4 |
| 04 | MASONRY | 77.5 | 26.0 | 46.3 | 89.6 | 40.9 | 60.1 | 81.3 | 26.3 | 48.0 | 75.2 | 37.0 | 52.1 | 88.6 | 36.8 | 57.2 | 107.5 | 60.9 | 79.3 |
| 05 | METALS | 82.8 | 64.5 | 76.5 | 89.6 | 75.9 | 84.9 | 85.9 | 65.3 | 78.8 | 84.6 | 75.0 | 81.3 | 89.6 | 71.3 | 83.3 | 96.3 | 86.7 | 93.0 |
| 06 | WOOD, PLASTICS & COMPOSITES | 102.5 | 22.4 | 56.5 | 96.2 | 39.7 | 63.8 | 87.3 | 29.4 | 54.1 | 95.3 | 38.4 | 62.6 | 95.9 | 38.3 | 62.9 | 82.3 | 37.6 | 56.7 |
| 07 | THERMAL & MOISTURE PROTECTION | 98.4 | 31.1 | 71.6 | 97.9 | 42.9 | 76.0 | 98.7 | 29.1 | 71.1 | 98.2 | 42.0 | 75.9 | 98.7 | 41.3 | 75.9 | 106.3 | 53.5 | 85.3 |
| 08 | OPENINGS | 94.5 | 24.9 | 76.9 | 95.4 | 41.4 | 81.7 | 94.7 | 26.0 | 77.3 | 94.7 | 39.9 | 80.8 | 98.6 | 38.2 | 83.3 | 96.9 | 49.4 | 84.8 |
| 0920 | Plaster & Gypsum Board | 105.4 | 19.9 | 45.2 | 100.0 | 37.7 | 56.1 | 98.6 | 27.1 | 48.3 | 101.3 | 36.4 | 55.6 | 106.5 | 36.3 | 57.1 | 95.9 | 36.0 | 53.7 |
| 0950, 0980 | Ceilings & Acoustic Treatment | 92.0 | 19.9 | 44.2 | 94.8 | 37.7 | 57.0 | 92.9 | 27.1 | 49.3 | 92.9 | 36.4 | 55.4 | 95.7 | 36.3 | 56.4 | 139.9 | 36.0 | 71.0 |
| 0960 | Flooring | 110.2 | 24.8 | 85.7 | 109.9 | 44.0 | 91.0 | 104.4 | 22.5 | 80.9 | 107.0 | 44.0 | 88.9 | 109.9 | 44.0 | 91.0 | 118.2 | 76.2 | 106.2 |
| 0970, 0990 | Wall Finishes & Painting/Coating | 114.7 | 18.5 | 56.3 | 114.7 | 38.0 | 68.2 | 114.7 | 32.4 | 64.8 | 114.7 | 33.4 | 65.4 | 114.7 | 35.2 | 66.5 | 98.4 | 31.7 | 57.9 |
| 09 | FINISHES | 102.8 | 23.1 | 58.8 | 102.2 | 40.4 | 68.2 | 100.8 | 28.0 | 60.6 | 101.3 | 38.8 | 66.8 | 103.4 | 38.9 | 67.9 | 114.3 | 46.1 | 76.7 |
| COVERS | DIVS. 10 - 14, 25, 28, 41, 43, 44, 46 | 100.0 | 64.0 | 92.5 | 100.0 | 72.7 | 94.3 | 100.0 | 69.5 | 93.7 | 100.0 | 72.1 | 94.2 | 100.0 | 66.8 | 93.1 | 100.0 | 81.1 | 96.1 |
| 21, 22, 23 | FIRE SUPPRESSION, PLUMBING & HVAC | 94.9 | 18.8 | 63.2 | 100.1 | 36.1 | 73.5 | 94.9 | 32.7 | 69.0 | 100.3 | 37.5 | 74.2 | 100.3 | 36.5 | 73.7 | 100.2 | 56.4 | 82.0 |
| 26, 27, 3370 | ELECTRICAL, COMMUNICATIONS & UTIL. | 98.1 | 26.2 | 59.9 | 103.5 | 37.3 | 68.2 | 99.9 | 35.8 | 65.8 | 97.3 | 36.1 | 64.7 | 96.8 | 40.9 | 67.1 | 100.4 | 66.0 | 82.1 |
| MF2010 | WEIGHTED AVERAGE | 96.2 | 33.4 | 68.1 | 99.6 | 48.1 | 76.6 | 96.6 | 39.8 | 71.2 | 96.9 | 46.3 | 74.3 | 99.1 | 46.5 | 75.6 | 100.8 | 63.0 | 83.9 |

### NORTH DAKOTA

| DIVISION | | DEVILS LAKE 583 MAT. | INST. | TOTAL | DICKINSON 586 MAT. | INST. | TOTAL | FARGO 580 - 581 MAT. | INST. | TOTAL | GRAND FORKS 582 MAT. | INST. | TOTAL | JAMESTOWN 584 MAT. | INST. | TOTAL | MINOT 587 MAT. | INST. | TOTAL |
|---|---|---|---|---|---|---|---|---|---|---|---|---|---|---|---|---|---|---|---|
| 015433 | CONTRACTOR EQUIPMENT | | 98.6 | 98.6 | | 98.6 | 98.6 | | 98.6 | 98.6 | | 98.6 | 98.6 | | 98.6 | 98.6 | | 98.6 | 98.6 |
| 0241, 31 - 34 | SITE & INFRASTRUCTURE, DEMOLITION | 106.9 | 95.1 | 98.7 | 116.7 | 93.7 | 100.9 | 97.8 | 97.2 | 97.4 | 111.2 | 93.7 | 99.1 | 105.7 | 93.7 | 97.4 | 108.2 | 97.2 | 100.6 |
| 0310 | Concrete Forming & Accessories | 99.1 | 37.5 | 45.6 | 89.6 | 36.7 | 43.7 | 96.9 | 43.8 | 50.9 | 93.0 | 36.7 | 44.2 | 90.9 | 36.4 | 43.6 | 89.4 | 61.9 | 65.5 |
| 0320 | Concrete Reinforcing | 104.9 | 87.2 | 95.8 | 105.8 | 42.7 | 73.3 | 96.7 | 87.0 | 91.7 | 103.3 | 87.1 | 94.9 | 105.5 | 50.5 | 77.2 | 106.8 | 87.2 | 96.7 |
| 0330 | Cast-in-Place Concrete | 120.4 | 48.4 | 92.9 | 109.0 | 46.9 | 85.2 | 98.3 | 51.8 | 80.5 | 109.0 | 46.9 | 85.2 | 118.9 | 46.7 | 91.3 | 108.9 | 48.9 | 85.9 |
| 03 | CONCRETE | 113.7 | 52.1 | 84.2 | 113.4 | 42.7 | 79.5 | 108.5 | 56.1 | 83.4 | 110.2 | 51.0 | 81.9 | 112.3 | 44.0 | 79.6 | 110.0 | 63.2 | 86.9 |
| 04 | MASONRY | 116.3 | 66.7 | 86.2 | 119.4 | 60.3 | 83.6 | 109.4 | 60.9 | 80.0 | 110.6 | 66.2 | 83.7 | 130.1 | 35.4 | 72.7 | 110.0 | 65.6 | 83.1 |
| 05 | METALS | 96.3 | 85.8 | 92.7 | 96.2 | 64.7 | 85.4 | 98.7 | 87.1 | 94.7 | 96.3 | 81.1 | 91.1 | 96.3 | 66.0 | 85.8 | 96.6 | 87.4 | 93.4 |
| 06 | WOOD, PLASTICS & COMPOSITES | 85.7 | 34.6 | 56.4 | 75.6 | 34.6 | 52.1 | 82.2 | 38.1 | 56.9 | 79.2 | 34.6 | 53.6 | 77.2 | 34.6 | 52.8 | 75.3 | 62.7 | 68.1 |
| 07 | THERMAL & MOISTURE PROTECTION | 107.1 | 51.6 | 85.1 | 107.7 | 49.1 | 84.4 | 104.7 | 54.6 | 84.8 | 107.4 | 51.6 | 85.2 | 107.0 | 43.5 | 81.8 | 107.1 | 57.3 | 87.4 |
| 08 | OPENINGS | 96.9 | 43.1 | 83.3 | 96.9 | 32.5 | 80.6 | 96.8 | 49.7 | 84.9 | 96.9 | 43.1 | 83.3 | 96.9 | 34.7 | 81.1 | 97.0 | 62.9 | 88.4 |
| 0920 | Plaster & Gypsum Board | 113.6 | 32.9 | 56.8 | 104.7 | 32.9 | 54.2 | 95.9 | 36.6 | 54.1 | 106.5 | 32.9 | 54.7 | 106.0 | 32.9 | 54.6 | 104.7 | 61.8 | 74.5 |
| 0950, 0980 | Ceilings & Acoustic Treatment | 140.9 | 32.9 | 69.3 | 140.9 | 32.9 | 69.3 | 139.9 | 36.6 | 71.4 | 140.9 | 32.9 | 69.3 | 140.9 | 32.9 | 69.3 | 113.8 | 84.9 | 105.5 |
| 0960 | Flooring | 122.5 | 37.9 | 98.2 | 114.0 | 37.9 | 92.2 | 117.9 | 76.2 | 105.9 | 116.4 | 37.9 | 93.9 | 115.1 | 37.9 | 92.9 | 113.8 | 84.9 | 105.5 |
| 0970, 0990 | Wall Finishes & Painting/Coating | 98.4 | 24.2 | 53.4 | 98.4 | 24.2 | 53.4 | 98.4 | 75.3 | 84.4 | 98.4 | 24.2 | 53.4 | 98.4 | 24.2 | 53.4 | 98.4 | 26.5 | 54.8 |
| 09 | FINISHES | 118.6 | 34.6 | 72.3 | 116.0 | 34.6 | 71.1 | 114.3 | 51.3 | 79.6 | 116.3 | 34.6 | 71.3 | 115.4 | 34.6 | 70.8 | 115.1 | 62.1 | 85.8 |
| COVERS | DIVS. 10 - 14, 25, 28, 41, 43, 44, 46 | 100.0 | 35.9 | 86.7 | 100.0 | 36.1 | 86.7 | 100.0 | 81.1 | 96.1 | 100.0 | 36.0 | 86.7 | 100.0 | 78.9 | 95.6 | 100.0 | 84.0 | 96.7 |
| 21, 22, 23 | FIRE SUPPRESSION, PLUMBING & HVAC | 94.9 | 56.1 | 78.7 | 94.9 | 69.6 | 84.4 | 100.2 | 59.4 | 83.2 | 100.3 | 36.4 | 73.7 | 94.9 | 38.4 | 71.4 | 100.3 | 56.4 | 82.0 |
| 26, 27, 3370 | ELECTRICAL, COMMUNICATIONS & UTIL. | 94.1 | 38.8 | 64.7 | 103.5 | 66.5 | 83.8 | 102.1 | 64.0 | 81.9 | 97.9 | 58.1 | 76.7 | 94.1 | 38.8 | 64.7 | 101.3 | 82.1 | 91.1 |
| MF2010 | WEIGHTED AVERAGE | 101.7 | 56.0 | 81.3 | 102.8 | 58.2 | 82.9 | 102.4 | 64.3 | 85.4 | 102.6 | 53.8 | 80.8 | 101.9 | 46.8 | 77.3 | 102.5 | 70.0 | 88.0 |

# City Cost Indexes

## NORTH DAKOTA / OHIO

| DIVISION | | WILLISTON 588 | | | AKRON 442 - 443 | | | ATHENS 457 | | | CANTON 446 - 447 | | | CHILLICOTHE 456 | | | CINCINNATI 451 - 452 | | |
|---|---|---|---|---|---|---|---|---|---|---|---|---|---|---|---|---|---|---|---|
| | | MAT. | INST. | TOTAL | MAT. | INST. | TOTAL | MAT. | INST. | TOTAL | MAT. | INST. | TOTAL | MAT. | INST. | TOTAL | MAT. | INST. | TOTAL |
| 015433 | CONTRACTOR EQUIPMENT | | 98.6 | 98.6 | | 96.8 | 96.8 | | 91.3 | 91.3 | | 96.8 | 96.8 | | 101.6 | 101.6 | | 101.4 | 101.4 |
| 0241, 31 - 34 | SITE & INFRASTRUCTURE, DEMOLITION | 109.3 | 93.7 | 98.6 | 94.1 | 103.8 | 100.8 | 93.1 | 92.2 | 92.5 | 94.2 | 103.6 | 100.7 | 81.9 | 106.0 | 98.5 | 79.2 | 105.8 | 97.5 |
| 0310 | Concrete Forming & Accessories | 94.5 | 36.7 | 44.4 | 101.2 | 93.9 | 94.9 | 93.2 | 90.1 | 90.5 | 101.2 | 82.3 | 84.8 | 94.9 | 94.7 | 94.7 | 96.6 | 83.7 | 85.4 |
| 0320 | Concrete Reinforcing | 107.9 | 42.7 | 74.3 | 99.1 | 91.0 | 94.9 | 97.2 | 85.6 | 91.2 | 99.1 | 76.3 | 87.4 | 94.0 | 81.8 | 87.7 | 99.6 | 84.4 | 91.8 |
| 0330 | Cast-in-Place Concrete | 109.0 | 46.9 | 85.2 | 100.5 | 100.4 | 100.5 | 98.8 | 96.7 | 98.0 | 101.5 | 96.0 | 99.4 | 89.7 | 90.0 | 89.8 | 82.6 | 88.3 | 84.8 |
| 03 | CONCRETE | 110.1 | 42.7 | 77.9 | 98.8 | 94.8 | 96.9 | 98.4 | 90.5 | 94.6 | 99.3 | 85.3 | 92.6 | 93.0 | 90.5 | 91.8 | 88.0 | 85.5 | 86.8 |
| 04 | MASONRY | 104.5 | 60.3 | 77.7 | 95.5 | 97.6 | 96.8 | 72.4 | 91.0 | 83.7 | 96.2 | 84.1 | 88.9 | 80.5 | 93.5 | 88.4 | 80.1 | 87.9 | 84.8 |
| 05 | METALS | 96.4 | 64.7 | 85.5 | 93.0 | 80.2 | 88.6 | 100.8 | 70.2 | 90.3 | 93.0 | 73.5 | 86.3 | 93.3 | 84.8 | 90.4 | 95.7 | 86.0 | 92.4 |
| 06 | WOOD, PLASTICS & COMPOSITES | 80.4 | 34.6 | 54.1 | 99.9 | 93.4 | 96.2 | 81.1 | 92.3 | 87.5 | 100.4 | 82.7 | 90.2 | 93.3 | 96.3 | 95.0 | 95.6 | 81.7 | 87.6 |
| 07 | THERMAL & MOISTURE PROTECTION | 107.3 | 49.1 | 84.2 | 107.8 | 98.5 | 104.1 | 101.4 | 93.4 | 98.2 | 108.9 | 92.0 | 102.2 | 102.6 | 90.9 | 97.9 | 100.9 | 92.2 | 97.5 |
| 08 | OPENINGS | 97.0 | 32.5 | 80.6 | 111.6 | 92.8 | 106.8 | 101.3 | 82.6 | 96.5 | 105.4 | 79.1 | 98.7 | 92.6 | 87.8 | 91.4 | 101.1 | 81.8 | 96.2 |
| 0920 | Plaster & Gypsum Board | 106.5 | 32.9 | 54.7 | 92.0 | 92.9 | 92.6 | 91.8 | 91.7 | 91.8 | 93.4 | 81.8 | 85.2 | 93.3 | 96.4 | 95.5 | 95.3 | 81.4 | 85.5 |
| 0950, 0980 | Ceilings & Acoustic Treatment | 140.9 | 32.9 | 69.3 | 91.5 | 92.9 | 92.4 | 97.3 | 91.7 | 93.6 | 91.5 | 81.8 | 85.1 | 91.3 | 96.4 | 94.7 | 92.2 | 81.4 | 85.0 |
| 0960 | Flooring | 117.6 | 37.9 | 94.7 | 101.1 | 92.6 | 98.6 | 129.8 | 97.6 | 120.5 | 101.2 | 76.2 | 94.0 | 105.8 | 98.2 | 103.6 | 106.8 | 91.2 | 102.3 |
| 0970, 0990 | Wall Finishes & Painting/Coating | 98.4 | 24.2 | 53.4 | 101.3 | 101.2 | 101.2 | 105.9 | 44.9 | 68.9 | 101.3 | 83.6 | 90.6 | 102.9 | 86.6 | 93.0 | 102.9 | 85.1 | 92.1 |
| 09 | FINISHES | 116.6 | 34.6 | 71.4 | 99.5 | 94.5 | 96.8 | 102.0 | 87.9 | 94.2 | 99.8 | 81.3 | 89.6 | 99.0 | 95.5 | 97.0 | 99.4 | 84.8 | 91.4 |
| COVERS | DIVS. 10 - 14, 25, 28, 41, 43, 44, 46 | 100.0 | 36.1 | 86.7 | 100.0 | 81.4 | 96.1 | 100.0 | 58.0 | 91.3 | 100.0 | 61.3 | 91.9 | 100.0 | 77.7 | 95.4 | 100.0 | 92.5 | 98.5 |
| 21, 22, 23 | FIRE SUPPRESSION, PLUMBING & HVAC | 94.9 | 69.6 | 84.4 | 99.9 | 89.5 | 95.5 | 94.6 | 56.0 | 78.5 | 99.9 | 74.7 | 89.4 | 95.0 | 82.2 | 89.7 | 99.9 | 82.6 | 92.7 |
| 26, 27, 3370 | ELECTRICAL, COMMUNICATIONS & UTIL. | 98.3 | 66.5 | 81.4 | 99.9 | 92.5 | 96.0 | 98.3 | 98.0 | 98.1 | 99.1 | 88.8 | 93.6 | 98.4 | 79.5 | 88.3 | 96.9 | 80.6 | 88.2 |
| MF2010 | WEIGHTED AVERAGE | 101.0 | 58.2 | 81.9 | 99.8 | 92.6 | 96.6 | 97.0 | 81.0 | 89.8 | 99.2 | 82.5 | 91.8 | 94.5 | 88.5 | 91.8 | 96.0 | 86.3 | 91.7 |

## OHIO

| DIVISION | | CLEVELAND 441 | | | COLUMBUS 430 - 432 | | | DAYTON 453 - 454 | | | HAMILTON 450 | | | LIMA 458 | | | LORAIN 440 | | |
|---|---|---|---|---|---|---|---|---|---|---|---|---|---|---|---|---|---|---|---|
| | | MAT. | INST. | TOTAL | MAT. | INST. | TOTAL | MAT. | INST. | TOTAL | MAT. | INST. | TOTAL | MAT. | INST. | TOTAL | MAT. | INST. | TOTAL |
| 015433 | CONTRACTOR EQUIPMENT | | 97.1 | 97.1 | | 95.6 | 95.6 | | 96.4 | 96.4 | | 101.6 | 101.6 | | 93.8 | 93.8 | | 96.8 | 96.8 |
| 0241, 31 - 34 | SITE & INFRASTRUCTURE, DEMOLITION | 94.1 | 104.6 | 101.4 | 92.3 | 100.7 | 98.1 | 78.3 | 105.6 | 97.2 | 78.2 | 105.9 | 97.3 | 87.8 | 93.9 | 92.0 | 93.5 | 103.3 | 100.3 |
| 0310 | Concrete Forming & Accessories | 101.2 | 99.3 | 99.6 | 99.3 | 88.9 | 90.3 | 96.6 | 83.1 | 84.9 | 96.6 | 83.2 | 85.0 | 93.2 | 86.9 | 87.7 | 101.2 | 89.2 | 90.8 |
| 0320 | Concrete Reinforcing | 99.6 | 91.3 | 95.4 | 99.6 | 84.2 | 91.7 | 99.6 | 80.4 | 89.7 | 99.6 | 84.4 | 91.8 | 97.2 | 80.5 | 88.6 | 99.1 | 91.1 | 95.0 |
| 0330 | Cast-in-Place Concrete | 98.6 | 108.0 | 102.2 | 88.2 | 93.0 | 90.0 | 76.8 | 89.4 | 81.6 | 82.3 | 87.8 | 84.4 | 90.8 | 98.8 | 93.9 | 95.7 | 102.6 | 98.4 |
| 03 | CONCRETE | 97.9 | 100.0 | 98.9 | 93.4 | 89.0 | 91.3 | 85.1 | 84.4 | 84.8 | 87.9 | 85.1 | 86.5 | 91.9 | 89.5 | 90.7 | 96.3 | 93.5 | 95.0 |
| 04 | MASONRY | 100.3 | 104.6 | 102.9 | 91.0 | 97.1 | 94.7 | 79.7 | 85.0 | 82.9 | 80.0 | 86.9 | 84.2 | 98.8 | 84.5 | 90.1 | 91.8 | 97.5 | 95.3 |
| 05 | METALS | 94.7 | 83.9 | 91.0 | 94.7 | 81.2 | 90.1 | 94.4 | 77.3 | 88.8 | 95.0 | 85.9 | 91.8 | 100.8 | 80.4 | 93.8 | 93.6 | 81.8 | 89.6 |
| 06 | WOOD, PLASTICS & COMPOSITES | 98.9 | 97.0 | 97.8 | 103.4 | 86.7 | 93.8 | 96.9 | 82.1 | 88.4 | 95.6 | 81.7 | 87.6 | 81.0 | 87.0 | 84.5 | 99.9 | 87.2 | 92.6 |
| 07 | THERMAL & MOISTURE PROTECTION | 106.5 | 108.3 | 107.3 | 101.0 | 96.4 | 99.2 | 105.6 | 90.4 | 99.6 | 102.7 | 92.0 | 98.4 | 100.9 | 96.7 | 99.2 | 108.8 | 101.4 | 105.9 |
| 08 | OPENINGS | 101.3 | 94.8 | 99.6 | 103.5 | 83.2 | 98.4 | 101.1 | 81.0 | 96.0 | 98.5 | 81.8 | 94.2 | 101.3 | 81.1 | 96.2 | 105.4 | 89.4 | 101.4 |
| 0920 | Plaster & Gypsum Board | 91.1 | 96.6 | 95.0 | 97.1 | 86.3 | 89.5 | 95.3 | 81.8 | 85.8 | 95.3 | 81.4 | 85.5 | 91.8 | 86.3 | 87.9 | 92.0 | 86.5 | 88.1 |
| 0950, 0980 | Ceilings & Acoustic Treatment | 89.6 | 96.6 | 94.2 | 88.1 | 86.3 | 86.9 | 93.2 | 81.8 | 85.7 | 92.2 | 81.4 | 85.1 | 96.3 | 86.3 | 89.1 | 91.5 | 86.5 | 88.2 |
| 0960 | Flooring | 100.9 | 103.4 | 101.6 | 95.5 | 90.8 | 94.2 | 109.7 | 81.2 | 101.5 | 106.8 | 91.2 | 102.3 | 129.0 | 89.7 | 117.7 | 101.2 | 105.4 | 102.5 |
| 0970, 0990 | Wall Finishes & Painting/Coating | 101.3 | 106.6 | 104.5 | 97.8 | 93.3 | 95.1 | 102.9 | 80.1 | 89.1 | 102.9 | 81.4 | 89.9 | 105.9 | 85.9 | 93.8 | 101.3 | 106.6 | 104.5 |
| 09 | FINISHES | 99.1 | 100.7 | 100.0 | 97.5 | 89.5 | 93.1 | 100.5 | 82.2 | 90.4 | 99.4 | 84.2 | 91.0 | 101.2 | 87.2 | 93.5 | 99.6 | 93.8 | 96.4 |
| COVERS | DIVS. 10 - 14, 25, 28, 41, 43, 44, 46 | 100.0 | 98.6 | 99.7 | 100.0 | 95.0 | 99.0 | 100.0 | 92.1 | 98.4 | 100.0 | 92.1 | 98.4 | 100.0 | 79.9 | 95.8 | 100.0 | 95.4 | 99.0 |
| 21, 22, 23 | FIRE SUPPRESSION, PLUMBING & HVAC | 99.8 | 99.3 | 99.6 | 99.9 | 90.7 | 96.1 | 100.6 | 78.9 | 91.6 | 100.4 | 82.1 | 92.8 | 94.6 | 85.2 | 90.7 | 99.9 | 84.8 | 93.6 |
| 26, 27, 3370 | ELECTRICAL, COMMUNICATIONS & UTIL. | 99.4 | 108.9 | 104.5 | 99.3 | 86.7 | 92.6 | 95.6 | 78.8 | 86.6 | 95.9 | 65.1 | 79.5 | 98.7 | 76.0 | 86.6 | 99.2 | 91.1 | 94.9 |
| MF2010 | WEIGHTED AVERAGE | 99.0 | 100.5 | 99.7 | 97.9 | 90.3 | 94.5 | 95.8 | 83.6 | 90.4 | 95.7 | 83.8 | 90.4 | 97.4 | 85.0 | 91.9 | 98.7 | 91.5 | 95.5 |

## OHIO

| DIVISION | | MANSFIELD 448 - 449 | | | MARION 433 | | | SPRINGFIELD 455 | | | STEUBENVILLE 439 | | | TOLEDO 434 - 436 | | | YOUNGSTOWN 444 - 445 | | |
|---|---|---|---|---|---|---|---|---|---|---|---|---|---|---|---|---|---|---|---|
| | | MAT. | INST. | TOTAL | MAT. | INST. | TOTAL | MAT. | INST. | TOTAL | MAT. | INST. | TOTAL | MAT. | INST. | TOTAL | MAT. | INST. | TOTAL |
| 015433 | CONTRACTOR EQUIPMENT | | 96.8 | 96.8 | | 95.3 | 95.3 | | 96.4 | 96.4 | | 100.0 | 100.0 | | 98.2 | 98.2 | | 96.8 | 96.8 |
| 0241, 31 - 34 | SITE & INFRASTRUCTURE, DEMOLITION | 90.5 | 104.3 | 100.0 | 88.6 | 99.3 | 96.0 | 78.6 | 104.2 | 96.3 | 134.3 | 109.0 | 116.8 | 91.5 | 100.2 | 97.5 | 94.0 | 104.2 | 101.0 |
| 0310 | Concrete Forming & Accessories | 91.3 | 87.9 | 88.4 | 96.1 | 75.4 | 78.2 | 96.6 | 86.0 | 87.4 | 98.7 | 87.8 | 89.2 | 99.3 | 96.8 | 97.2 | 101.2 | 84.1 | 86.4 |
| 0320 | Concrete Reinforcing | 90.2 | 76.9 | 83.3 | 91.8 | 84.1 | 87.9 | 99.6 | 81.4 | 90.3 | 89.4 | 86.6 | 87.9 | 99.6 | 91.0 | 95.2 | 99.1 | 86.4 | 92.5 |
| 0330 | Cast-in-Place Concrete | 93.1 | 91.8 | 92.6 | 80.6 | 89.5 | 84.0 | 78.9 | 89.3 | 82.9 | 87.5 | 94.6 | 90.2 | 88.2 | 104.7 | 94.5 | 99.6 | 98.6 | 99.2 |
| 03 | CONCRETE | 91.1 | 86.6 | 88.9 | 85.7 | 81.7 | 83.8 | 86.1 | 85.9 | 86.0 | 89.6 | 89.3 | 89.4 | 93.4 | 98.0 | 95.6 | 98.3 | 88.9 | 93.8 |
| 04 | MASONRY | 94.6 | 97.2 | 96.2 | 92.9 | 96.1 | 94.8 | 79.8 | 85.0 | 83.0 | 79.8 | 92.5 | 87.5 | 99.6 | 101.0 | 100.5 | 95.8 | 90.0 | 92.3 |
| 05 | METALS | 93.8 | 75.6 | 87.5 | 93.9 | 77.6 | 88.3 | 94.8 | 77.5 | 88.9 | 90.6 | 79.0 | 86.6 | 94.4 | 86.8 | 91.8 | 93.0 | 78.4 | 88.0 |
| 06 | WOOD, PLASTICS & COMPOSITES | 88.4 | 87.2 | 87.7 | 99.6 | 72.9 | 84.3 | 98.3 | 86.4 | 91.5 | 95.4 | 87.0 | 90.6 | 103.4 | 96.7 | 99.5 | 99.9 | 82.3 | 89.8 |
| 07 | THERMAL & MOISTURE PROTECTION | 108.3 | 96.8 | 103.7 | 100.6 | 84.4 | 94.2 | 105.5 | 90.8 | 99.7 | 112.4 | 94.5 | 105.3 | 102.7 | 105.4 | 103.8 | 109.0 | 91.5 | 102.1 |
| 08 | OPENINGS | 104.5 | 80.2 | 98.3 | 97.5 | 74.3 | 91.6 | 98.9 | 81.7 | 94.5 | 98.3 | 84.1 | 94.7 | 101.3 | 92.8 | 99.1 | 105.4 | 84.2 | 100.0 |
| 0920 | Plaster & Gypsum Board | 85.2 | 86.5 | 86.1 | 94.8 | 72.1 | 78.8 | 95.3 | 86.3 | 88.9 | 95.4 | 86.1 | 88.8 | 97.1 | 96.5 | 96.7 | 92.0 | 81.5 | 84.6 |
| 0950, 0980 | Ceilings & Acoustic Treatment | 92.4 | 86.5 | 88.5 | 88.1 | 72.1 | 77.5 | 93.2 | 86.3 | 88.6 | 84.9 | 86.1 | 85.7 | 88.1 | 96.5 | 93.7 | 91.5 | 81.5 | 84.8 |
| 0960 | Flooring | 96.6 | 104.2 | 98.8 | 94.0 | 73.7 | 88.1 | 109.7 | 81.2 | 101.5 | 122.5 | 84.1 | 111.4 | 94.8 | 100.2 | 96.4 | 101.2 | 87.8 | 97.4 |
| 0970, 0990 | Wall Finishes & Painting/Coating | 101.3 | 93.1 | 96.3 | 97.8 | 51.1 | 69.5 | 102.9 | 80.1 | 89.1 | 113.1 | 91.4 | 99.9 | 97.8 | 105.1 | 102.2 | 101.3 | 90.9 | 95.0 |
| 09 | FINISHES | 97.3 | 91.7 | 94.2 | 96.4 | 72.2 | 83.0 | 100.5 | 84.7 | 91.8 | 115.8 | 87.2 | 100.0 | 97.3 | 98.6 | 98.0 | 99.7 | 85.0 | 91.6 |
| COVERS | DIVS. 10 - 14, 25, 28, 41, 43, 44, 46 | 100.0 | 83.3 | 96.5 | 100.0 | 58.3 | 91.3 | 100.0 | 92.6 | 98.5 | 100.0 | 83.5 | 96.6 | 100.0 | 82.4 | 96.3 | 100.0 | 78.7 | 95.6 |
| 21, 22, 23 | FIRE SUPPRESSION, PLUMBING & HVAC | 94.5 | 83.6 | 89.9 | 94.6 | 80.7 | 88.8 | 100.6 | 78.9 | 91.6 | 94.8 | 87.1 | 91.6 | 99.9 | 92.7 | 96.9 | 99.9 | 90.8 | 96.1 |
| 26, 27, 3370 | ELECTRICAL, COMMUNICATIONS & UTIL. | 96.5 | 78.9 | 87.1 | 92.7 | 74.3 | 82.9 | 95.6 | 76.5 | 85.4 | 87.3 | 98.2 | 93.1 | 99.6 | 97.6 | 98.5 | 99.2 | 80.0 | 89.0 |
| MF2010 | WEIGHTED AVERAGE | 96.2 | 87.0 | 92.1 | 94.1 | 80.8 | 88.2 | 95.7 | 83.8 | 90.4 | 96.4 | 90.5 | 93.8 | 98.1 | 95.9 | 97.1 | 99.1 | 87.5 | 93.9 |

# City Cost Indexes

## OHIO / OKLAHOMA

| DIVISION | | ZANESVILLE 437-438 MAT. | INST. | TOTAL | ARDMORE 734 MAT. | INST. | TOTAL | CLINTON 736 MAT. | INST. | TOTAL | DURANT 747 MAT. | INST. | TOTAL | ENID 737 MAT. | INST. | TOTAL | GUYMON 739 MAT. | INST. | TOTAL |
|---|---|---|---|---|---|---|---|---|---|---|---|---|---|---|---|---|---|---|---|
| 015433 | CONTRACTOR EQUIPMENT | | 95.3 | 95.3 | | 81.0 | 81.0 | | 80.1 | 80.1 | | 80.1 | 80.1 | | 80.1 | 80.1 | | 80.1 | 80.1 |
| 0241, 31 - 34 | SITE & INFRASTRUCTURE, DEMOLITION | 91.3 | 100.9 | 97.9 | 103.2 | 90.3 | 94.3 | 104.7 | 88.9 | 93.8 | 97.7 | 88.6 | 91.4 | 106.0 | 88.9 | 94.2 | 109.5 | 87.4 | 94.3 |
| 0310 | Concrete Forming & Accessories | 93.2 | 75.4 | 77.7 | 91.7 | 40.3 | 47.1 | 90.6 | 34.9 | 42.2 | 83.9 | 40.3 | 46.1 | 93.8 | 34.2 | 42.1 | 97.3 | 22.2 | 32.1 |
| 0320 | Concrete Reinforcing | 91.3 | 86.1 | 88.6 | 100.3 | 78.9 | 89.3 | 100.8 | 78.9 | 89.5 | 101.2 | 62.5 | 81.3 | 100.2 | 78.9 | 89.2 | 100.8 | 27.7 | 63.1 |
| 0330 | Cast-in-Place Concrete | 84.9 | 88.7 | 86.3 | 99.2 | 43.4 | 77.9 | 95.9 | 43.4 | 75.8 | 92.7 | 43.3 | 73.8 | 95.9 | 45.8 | 76.7 | 95.9 | 31.0 | 71.1 |
| 03 | CONCRETE | 89.1 | 81.8 | 85.6 | 95.7 | 49.2 | 73.5 | 95.0 | 46.8 | 71.9 | 90.4 | 46.0 | 69.1 | 95.5 | 47.3 | 72.4 | 98.1 | 27.0 | 64.1 |
| 04 | MASONRY | 89.9 | 84.0 | 86.3 | 96.8 | 57.0 | 72.7 | 125.2 | 57.0 | 83.9 | 95.4 | 65.0 | 77.0 | 106.0 | 57.0 | 76.3 | 99.9 | 22.6 | 53.1 |
| 05 | METALS | 95.2 | 79.9 | 89.9 | 91.5 | 69.8 | 84.1 | 91.6 | 69.7 | 84.1 | 91.6 | 63.1 | 81.8 | 93.1 | 69.7 | 85.1 | 92.1 | 39.8 | 74.1 |
| 06 | WOOD, PLASTICS & COMPOSITES | 95.8 | 72.9 | 82.7 | 97.2 | 38.9 | 63.8 | 96.2 | 31.6 | 59.1 | 88.1 | 38.9 | 59.9 | 99.9 | 30.7 | 60.2 | 103.6 | 20.3 | 55.8 |
| 07 | THERMAL & MOISTURE PROTECTION | 100.7 | 87.7 | 95.5 | 97.0 | 60.1 | 82.3 | 97.1 | 59.3 | 82.1 | 96.8 | 57.9 | 81.3 | 97.2 | 59.2 | 82.1 | 97.5 | 28.2 | 70.0 |
| 08 | OPENINGS | 97.5 | 76.3 | 92.1 | 95.1 | 49.6 | 83.5 | 95.1 | 45.6 | 82.6 | 95.1 | 45.5 | 82.5 | 95.1 | 45.0 | 82.4 | 95.2 | 21.7 | 76.6 |
| 0920 | Plaster & Gypsum Board | 92.2 | 72.1 | 78.0 | 78.4 | 37.8 | 49.8 | 77.9 | 30.2 | 44.3 | 73.5 | 37.8 | 48.4 | 79.3 | 29.3 | 44.1 | 79.8 | 18.6 | 36.7 |
| 0950, 0980 | Ceilings & Acoustic Treatment | 88.1 | 72.1 | 77.5 | 82.8 | 37.8 | 53.0 | 82.8 | 30.2 | 47.9 | 82.8 | 37.8 | 53.0 | 82.8 | 29.3 | 47.4 | 84.7 | 18.6 | 40.9 |
| 0960 | Flooring | 92.2 | 77.6 | 88.0 | 105.9 | 45.6 | 88.6 | 104.5 | 43.5 | 87.0 | 100.7 | 65.4 | 90.6 | 106.6 | 43.5 | 88.5 | 108.6 | 25.7 | 84.8 |
| 0970, 0990 | Wall Finishes & Painting/Coating | 97.8 | 72.8 | 82.7 | 97.1 | 55.2 | 71.7 | 97.1 | 55.2 | 71.7 | 97.1 | 55.2 | 71.7 | 97.1 | 55.2 | 71.7 | 97.1 | 19.7 | 50.1 |
| 09 | FINISHES | 95.8 | 74.7 | 84.2 | 91.3 | 41.0 | 63.6 | 91.0 | 36.3 | 60.9 | 88.8 | 45.2 | 64.7 | 91.9 | 35.8 | 60.9 | 93.4 | 22.2 | 54.1 |
| COVERS | DIVS. 10 - 14, 25, 28, 41, 43, 44, 46 | 100.0 | 73.7 | 94.5 | 100.0 | 75.3 | 94.9 | 100.0 | 74.5 | 94.7 | 100.0 | 75.4 | 94.9 | 100.0 | 74.4 | 94.7 | 100.0 | 72.3 | 94.2 |
| 21, 22, 23 | FIRE SUPPRESSION, PLUMBING & HVAC | 94.6 | 84.9 | 90.6 | 94.8 | 62.7 | 81.4 | 94.8 | 62.7 | 81.4 | 94.8 | 62.8 | 81.5 | 100.2 | 62.7 | 84.6 | 94.8 | 23.6 | 65.1 |
| 26, 27, 3370 | ELECTRICAL, COMMUNICATIONS & UTIL. | 92.8 | 69.5 | 80.4 | 95.6 | 70.0 | 82.0 | 96.7 | 70.0 | 82.5 | 98.6 | 52.8 | 74.2 | 96.7 | 70.0 | 82.5 | 98.6 | 16.7 | 55.0 |
| MF2010 | WEIGHTED AVERAGE | 94.6 | 81.1 | 88.6 | 95.2 | 61.0 | 79.9 | 96.7 | 59.6 | 80.2 | 94.3 | 58.5 | 78.3 | 97.4 | 59.6 | 80.5 | 96.5 | 31.1 | 67.3 |

## OKLAHOMA

| DIVISION | | LAWTON 735 MAT. | INST. | TOTAL | MCALESTER 745 MAT. | INST. | TOTAL | MIAMI 743 MAT. | INST. | TOTAL | MUSKOGEE 744 MAT. | INST. | TOTAL | OKLAHOMA CITY 730-731 MAT. | INST. | TOTAL | PONCA CITY 746 MAT. | INST. | TOTAL |
|---|---|---|---|---|---|---|---|---|---|---|---|---|---|---|---|---|---|---|---|
| 015433 | CONTRACTOR EQUIPMENT | | 81.0 | 81.0 | | 80.1 | 80.1 | | 88.5 | 88.5 | | 88.5 | 88.5 | | 81.2 | 81.2 | | 80.1 | 80.1 |
| 0241, 31 - 34 | SITE & INFRASTRUCTURE, DEMOLITION | 101.5 | 90.3 | 93.8 | 90.5 | 89.0 | 89.4 | 92.3 | 86.2 | 88.1 | 92.3 | 84.8 | 87.1 | 99.9 | 90.8 | 93.7 | 98.3 | 88.9 | 91.8 |
| 0310 | Concrete Forming & Accessories | 97.1 | 46.1 | 52.9 | 82.0 | 41.6 | 47.0 | 94.0 | 70.7 | 73.8 | 98.2 | 30.6 | 39.6 | 96.2 | 40.0 | 47.4 | 89.8 | 40.7 | 47.2 |
| 0320 | Concrete Reinforcing | 100.4 | 78.9 | 89.4 | 100.8 | 41.6 | 70.3 | 99.3 | 79.0 | 88.8 | 100.2 | 31.6 | 64.9 | 100.4 | 78.9 | 89.3 | 100.2 | 78.9 | 89.2 |
| 0330 | Cast-in-Place Concrete | 92.7 | 45.9 | 74.8 | 81.3 | 45.9 | 67.7 | 85.2 | 45.2 | 69.9 | 86.2 | 35.9 | 66.9 | 93.9 | 48.6 | 76.5 | 95.2 | 38.5 | 73.5 |
| 03 | CONCRETE | 91.8 | 52.6 | 73.1 | 81.3 | 43.5 | 63.2 | 85.6 | 64.1 | 75.3 | 87.3 | 33.9 | 61.7 | 92.4 | 50.8 | 72.5 | 92.4 | 47.7 | 71.0 |
| 04 | MASONRY | 99.8 | 57.0 | 73.9 | 114.6 | 57.6 | 80.1 | 98.1 | 58.2 | 73.9 | 117.4 | 46.2 | 74.3 | 99.3 | 58.8 | 74.8 | 90.7 | 57.6 | 70.6 |
| 05 | METALS | 96.7 | 69.8 | 87.5 | 91.5 | 53.5 | 78.4 | 91.5 | 82.3 | 88.3 | 93.1 | 55.9 | 80.3 | 98.6 | 69.8 | 88.7 | 91.5 | 69.5 | 83.9 |
| 06 | WOOD, PLASTICS & COMPOSITES | 102.6 | 46.7 | 70.6 | 85.7 | 42.2 | 60.7 | 99.8 | 79.2 | 88.0 | 104.0 | 30.4 | 61.8 | 101.9 | 37.6 | 65.0 | 95.5 | 39.5 | 63.4 |
| 07 | THERMAL & MOISTURE PROTECTION | 97.0 | 60.9 | 82.7 | 96.4 | 56.7 | 80.6 | 96.8 | 61.3 | 82.7 | 96.8 | 41.8 | 75.0 | 93.7 | 60.7 | 80.6 | 96.9 | 59.3 | 82.0 |
| 08 | OPENINGS | 96.7 | 53.7 | 85.8 | 95.1 | 40.9 | 81.4 | 95.1 | 71.9 | 89.2 | 95.0 | 29.5 | 78.4 | 96.7 | 48.7 | 84.6 | 95.1 | 49.9 | 83.6 |
| 0920 | Plaster & Gypsum Board | 81.4 | 45.8 | 56.3 | 72.2 | 41.2 | 50.3 | 79.3 | 79.1 | 79.2 | 81.4 | 28.8 | 44.4 | 83.4 | 36.4 | 50.3 | 77.9 | 38.4 | 50.1 |
| 0950, 0980 | Ceilings & Acoustic Treatment | 90.3 | 45.8 | 60.8 | 82.8 | 41.2 | 55.2 | 82.8 | 79.1 | 80.4 | 90.3 | 28.8 | 49.6 | 90.3 | 36.4 | 54.6 | 82.8 | 38.4 | 53.4 |
| 0960 | Flooring | 109.0 | 43.5 | 90.2 | 99.6 | 43.5 | 83.5 | 107.4 | 65.4 | 95.4 | 110.0 | 42.0 | 90.5 | 109.0 | 43.5 | 90.2 | 104.1 | 43.5 | 86.7 |
| 0970, 0990 | Wall Finishes & Painting/Coating | 97.1 | 55.2 | 71.7 | 97.1 | 40.7 | 62.9 | 97.1 | 83.4 | 88.8 | 97.1 | 29.6 | 56.1 | 97.1 | 55.2 | 71.7 | 97.1 | 55.2 | 71.7 |
| 09 | FINISHES | 93.9 | 45.2 | 67.0 | 87.7 | 41.1 | 62.0 | 90.7 | 72.2 | 80.5 | 93.4 | 32.5 | 59.8 | 93.8 | 40.2 | 64.3 | 90.4 | 41.1 | 63.3 |
| COVERS | DIVS. 10 - 14, 25, 28, 41, 43, 44, 46 | 100.0 | 76.2 | 95.1 | 100.0 | 75.9 | 95.0 | 100.0 | 80.7 | 96.0 | 100.0 | 74.0 | 94.6 | 100.0 | 75.8 | 95.0 | 100.0 | 75.5 | 94.9 |
| 21, 22, 23 | FIRE SUPPRESSION, PLUMBING & HVAC | 100.2 | 62.7 | 84.6 | 94.8 | 30.8 | 68.2 | 94.8 | 61.6 | 81.0 | 100.2 | 23.9 | 68.4 | 100.0 | 63.6 | 84.9 | 94.8 | 61.2 | 80.8 |
| 26, 27, 3370 | ELECTRICAL, COMMUNICATIONS & UTIL. | 98.6 | 70.0 | 83.4 | 96.8 | 67.8 | 81.4 | 98.5 | 67.9 | 82.2 | 96.3 | 28.6 | 60.3 | 104.0 | 70.0 | 85.9 | 96.3 | 64.0 | 79.1 |
| MF2010 | WEIGHTED AVERAGE | 97.6 | 62.3 | 81.8 | 93.7 | 51.2 | 74.7 | 94.0 | 69.0 | 82.8 | 96.7 | 39.6 | 71.2 | 98.2 | 61.5 | 81.8 | 94.3 | 59.6 | 78.8 |

## OKLAHOMA / OREGON

| DIVISION | | POTEAU 749 MAT. | INST. | TOTAL | SHAWNEE 748 MAT. | INST. | TOTAL | TULSA 740-741 MAT. | INST. | TOTAL | WOODWARD 738 MAT. | INST. | TOTAL | BEND 977 MAT. | INST. | TOTAL | EUGENE 974 MAT. | INST. | TOTAL |
|---|---|---|---|---|---|---|---|---|---|---|---|---|---|---|---|---|---|---|---|
| 015433 | CONTRACTOR EQUIPMENT | | 87.8 | 87.8 | | 80.1 | 80.1 | | 88.5 | 88.5 | | 80.1 | 80.1 | | 99.4 | 99.4 | | 99.4 | 99.4 |
| 0241, 31 - 34 | SITE & INFRASTRUCTURE, DEMOLITION | 76.2 | 84.8 | 82.2 | 101.6 | 88.8 | 92.8 | 98.6 | 86.2 | 90.0 | 105.1 | 88.9 | 93.9 | 115.3 | 104.6 | 107.9 | 103.3 | 104.6 | 104.2 |
| 0310 | Concrete Forming & Accessories | 87.8 | 43.0 | 48.9 | 83.8 | 39.3 | 45.2 | 98.1 | 39.8 | 47.5 | 90.7 | 34.8 | 42.2 | 109.7 | 95.3 | 97.2 | 106.4 | 95.2 | 96.7 |
| 0320 | Concrete Reinforcing | 101.3 | 78.9 | 89.8 | 100.2 | 45.7 | 72.1 | 100.4 | 78.8 | 89.3 | 100.2 | 78.9 | 89.2 | 104.3 | 102.9 | 103.6 | 108.8 | 102.9 | 105.8 |
| 0330 | Cast-in-Place Concrete | 85.2 | 44.7 | 69.7 | 98.3 | 42.9 | 77.1 | 94.0 | 44.7 | 75.1 | 95.9 | 43.4 | 75.8 | 100.2 | 106.2 | 102.5 | 97.0 | 106.2 | 100.5 |
| 03 | CONCRETE | 87.6 | 51.6 | 70.4 | 94.2 | 42.3 | 69.3 | 92.5 | 50.2 | 72.2 | 95.3 | 46.7 | 72.0 | 111.8 | 100.3 | 106.3 | 102.8 | 100.2 | 101.6 |
| 04 | MASONRY | 99.1 | 57.6 | 74.0 | 115.9 | 56.4 | 79.8 | 98.9 | 58.8 | 74.6 | 93.8 | 57.0 | 71.5 | 109.9 | 101.5 | 104.8 | 107.4 | 101.5 | 103.8 |
| 05 | METALS | 91.5 | 81.9 | 88.2 | 91.4 | 55.3 | 79.0 | 96.4 | 80.9 | 91.1 | 91.7 | 69.7 | 84.1 | 88.7 | 97.9 | 91.9 | 89.4 | 97.8 | 92.3 |
| 06 | WOOD, PLASTICS & COMPOSITES | 92.6 | 42.4 | 63.8 | 87.9 | 38.7 | 59.7 | 103.2 | 38.7 | 66.2 | 96.4 | 31.5 | 59.1 | 98.0 | 93.1 | 95.1 | 93.9 | 93.1 | 93.4 |
| 07 | THERMAL & MOISTURE PROTECTION | 96.9 | 57.0 | 81.0 | 96.9 | 59.9 | 82.2 | 96.8 | 55.4 | 80.4 | 97.2 | 59.3 | 82.1 | 106.3 | 94.2 | 101.5 | 105.6 | 91.2 | 99.9 |
| 08 | OPENINGS | 95.1 | 52.0 | 84.2 | 95.0 | 38.1 | 80.6 | 96.7 | 48.9 | 84.6 | 95.1 | 45.6 | 82.6 | 97.5 | 100.6 | 98.3 | 97.0 | 100.6 | 98.6 |
| 0920 | Plaster & Gypsum Board | 76.6 | 41.2 | 51.7 | 73.5 | 37.5 | 48.2 | 81.4 | 37.4 | 50.4 | 78.5 | 30.1 | 44.4 | 106.9 | 92.7 | 96.9 | 104.6 | 92.7 | 96.2 |
| 0950, 0980 | Ceilings & Acoustic Treatment | 82.8 | 41.2 | 55.2 | 82.8 | 37.5 | 52.8 | 90.3 | 37.4 | 55.3 | 84.7 | 30.1 | 48.5 | 101.1 | 92.7 | 95.5 | 102.1 | 92.7 | 95.8 |
| 0960 | Flooring | 103.4 | 65.4 | 92.5 | 100.6 | 34.6 | 81.7 | 108.8 | 45.2 | 90.5 | 104.5 | 45.6 | 87.6 | 110.6 | 100.7 | 107.8 | 108.8 | 100.7 | 106.5 |
| 0970, 0990 | Wall Finishes & Painting/Coating | 97.1 | 55.2 | 71.7 | 97.1 | 36.7 | 60.5 | 97.1 | 43.6 | 64.7 | 97.1 | 55.2 | 71.7 | 108.1 | 70.5 | 85.3 | 108.1 | 70.5 | 85.3 |
| 09 | FINISHES | 88.3 | 47.2 | 65.6 | 89.0 | 36.7 | 60.2 | 93.2 | 40.0 | 63.9 | 91.6 | 36.6 | 61.3 | 107.4 | 93.1 | 99.5 | 105.5 | 93.1 | 98.7 |
| COVERS | DIVS. 10 - 14, 25, 28, 41, 43, 44, 46 | 100.0 | 76.2 | 95.0 | 100.0 | 75.0 | 94.8 | 100.0 | 76.2 | 95.0 | 100.0 | 74.4 | 94.7 | 100.0 | 100.3 | 100.1 | 100.0 | 100.1 | 100.1 |
| 21, 22, 23 | FIRE SUPPRESSION, PLUMBING & HVAC | 94.8 | 61.3 | 80.8 | 94.8 | 62.0 | 81.1 | 100.2 | 48.4 | 78.6 | 94.8 | 62.7 | 81.4 | 94.7 | 98.7 | 96.4 | 100.1 | 98.7 | 99.5 |
| 26, 27, 3370 | ELECTRICAL, COMMUNICATIONS & UTIL. | 96.4 | 67.9 | 81.3 | 98.7 | 70.0 | 83.5 | 98.6 | 37.3 | 66.0 | 98.5 | 70.0 | 83.3 | 102.4 | 95.2 | 98.6 | 100.9 | 95.2 | 97.9 |
| MF2010 | WEIGHTED AVERAGE | 93.3 | 62.3 | 79.5 | 96.0 | 57.3 | 78.7 | 97.4 | 54.2 | 78.1 | 95.4 | 59.7 | 79.4 | 100.3 | 98.4 | 99.5 | 99.8 | 98.3 | 99.1 |

## OREGON

| DIVISION | | KLAMATH FALLS 976 | | | MEDFORD 975 | | | PENDLETON 978 | | | PORTLAND 970 - 972 | | | SALEM 973 | | | VALE 979 | | |
|---|---|---|---|---|---|---|---|---|---|---|---|---|---|---|---|---|---|---|---|
| | | MAT. | INST. | TOTAL | MAT. | INST. | TOTAL | MAT. | INST. | TOTAL | MAT. | INST. | TOTAL | MAT. | INST. | TOTAL | MAT. | INST. | TOTAL |
| 015433 | CONTRACTOR EQUIPMENT | | 99.4 | 99.4 | | 99.4 | 99.4 | | 96.7 | 96.7 | | 99.4 | 99.4 | | 99.4 | 99.4 | | 96.7 | 96.7 |
| 0241, 31 - 34 | SITE & INFRASTRUCTURE, DEMOLITION | 120.0 | 104.6 | 109.3 | 112.1 | 104.6 | 106.9 | 110.5 | 98.0 | 101.8 | 106.2 | 104.6 | 105.1 | 98.4 | 104.6 | 102.7 | 97.0 | 97.9 | 97.6 |
| 0310 | Concrete Forming & Accessories | 103.2 | 95.1 | 96.2 | 102.4 | 95.1 | 96.1 | 102.9 | 95.5 | 96.5 | 107.5 | 95.4 | 97.0 | 105.2 | 95.3 | 96.6 | 108.9 | 94.5 | 96.4 |
| 0320 | Concrete Reinforcing | 104.3 | 102.9 | 103.6 | 106.1 | 102.9 | 104.4 | 103.6 | 103.1 | 103.3 | 109.6 | 103.1 | 106.2 | 109.7 | 103.1 | 106.3 | 101.0 | 102.9 | 102.0 |
| 0330 | Cast-in-Place Concrete | 100.2 | 106.2 | 102.5 | 100.2 | 106.2 | 102.5 | 100.9 | 107.6 | 103.5 | 99.7 | 106.3 | 102.4 | 88.9 | 106.2 | 95.6 | 79.5 | 107.2 | 90.1 |
| 03 | CONCRETE | 114.8 | 100.2 | 107.8 | 109.2 | 100.2 | 104.9 | 94.6 | 100.9 | 97.6 | 104.3 | 100.4 | 102.4 | 98.8 | 100.4 | 99.5 | 79.2 | 100.3 | 89.3 |
| 04 | MASONRY | 124.5 | 101.5 | 110.6 | 104.3 | 101.5 | 102.6 | 114.1 | 103.7 | 107.8 | 108.0 | 103.6 | 105.3 | 112.3 | 103.6 | 107.0 | 112.4 | 103.7 | 107.1 |
| 05 | METALS | 88.7 | 97.7 | 91.8 | 89.0 | 97.7 | 92.0 | 94.5 | 98.6 | 95.9 | 90.2 | 98.2 | 93.0 | 89.8 | 98.1 | 92.6 | 94.4 | 97.6 | 95.5 |
| 06 | WOOD, PLASTICS & COMPOSITES | 89.2 | 93.1 | 91.4 | 88.1 | 93.1 | 90.9 | 91.0 | 93.2 | 92.2 | 94.8 | 93.1 | 93.8 | 91.1 | 93.1 | 92.2 | 99.2 | 93.2 | 95.8 |
| 07 | THERMAL & MOISTURE PROTECTION | 106.5 | 92.7 | 101.0 | 106.2 | 92.7 | 100.8 | 98.0 | 93.5 | 96.2 | 105.6 | 97.3 | 102.3 | 103.5 | 94.8 | 100.1 | 97.4 | 91.1 | 94.9 |
| 08 | OPENINGS | 97.5 | 100.6 | 98.3 | 100.5 | 100.6 | 100.5 | 93.1 | 100.7 | 95.0 | 95.5 | 100.6 | 96.8 | 97.3 | 100.6 | 98.2 | 93.1 | 93.2 | 93.1 |
| 0920 | Plaster & Gypsum Board | 102.1 | 92.7 | 95.5 | 101.2 | 92.7 | 95.2 | 82.4 | 92.7 | 89.6 | 104.9 | 92.7 | 96.3 | 99.2 | 92.7 | 94.6 | 89.1 | 92.7 | 91.6 |
| 0950, 0980 | Ceilings & Acoustic Treatment | 109.5 | 92.7 | 98.3 | 114.7 | 92.7 | 100.1 | 69.1 | 92.7 | 84.7 | 104.2 | 92.7 | 96.6 | 104.4 | 92.7 | 96.6 | 69.1 | 92.7 | 84.7 |
| 0960 | Flooring | 107.3 | 100.7 | 105.4 | 106.6 | 100.7 | 104.9 | 74.9 | 100.7 | 82.3 | 106.3 | 100.7 | 104.7 | 109.0 | 100.7 | 106.6 | 76.8 | 100.7 | 83.7 |
| 0970, 0990 | Wall Finishes & Painting/Coating | 108.1 | 62.6 | 80.5 | 108.1 | 62.6 | 80.5 | 96.2 | 70.5 | 80.6 | 107.7 | 70.5 | 85.1 | 108.1 | 70.5 | 85.3 | 96.2 | 70.5 | 80.6 |
| 09 | FINISHES | 108.1 | 92.3 | 99.4 | 108.0 | 92.3 | 99.3 | 73.1 | 93.2 | 84.2 | 105.3 | 93.1 | 98.6 | 104.9 | 93.1 | 98.4 | 73.4 | 93.2 | 84.3 |
| COVERS | DIVS. 10 - 14, 25, 28, 41, 43, 44, 46 | 100.0 | 100.3 | 100.1 | 100.0 | 100.3 | 100.1 | 100.0 | 98.8 | 99.8 | 100.0 | 100.4 | 100.1 | 100.0 | 100.3 | 100.1 | 100.0 | 100.6 | 100.1 |
| 21, 22, 23 | FIRE SUPPRESSION, PLUMBING & HVAC | 94.7 | 98.7 | 96.4 | 100.1 | 98.7 | 99.5 | 96.1 | 108.3 | 101.2 | 100.1 | 98.7 | 99.6 | 100.1 | 98.7 | 99.6 | 96.1 | 75.5 | 87.5 |
| 26, 27, 3370 | ELECTRICAL, COMMUNICATIONS & UTIL. | 101.0 | 84.7 | 92.4 | 104.8 | 84.7 | 94.1 | 92.6 | 96.4 | 94.6 | 101.1 | 102.4 | 101.8 | 106.3 | 95.2 | 100.4 | 92.6 | 78.4 | 85.0 |
| MF2010 | WEIGHTED AVERAGE | 101.5 | 96.7 | 99.3 | 101.5 | 96.7 | 99.3 | 94.9 | 100.3 | 97.3 | 100.0 | 99.7 | 99.9 | 99.8 | 98.6 | 99.3 | 92.6 | 90.4 | 91.6 |

## PENNSYLVANIA

| DIVISION | | ALLENTOWN 181 | | | ALTOONA 166 | | | BEDFORD 155 | | | BRADFORD 167 | | | BUTLER 160 | | | CHAMBERSBURG 172 | | |
|---|---|---|---|---|---|---|---|---|---|---|---|---|---|---|---|---|---|---|---|
| | | MAT. | INST. | TOTAL | MAT. | INST. | TOTAL | MAT. | INST. | TOTAL | MAT. | INST. | TOTAL | MAT. | INST. | TOTAL | MAT. | INST. | TOTAL |
| 015433 | CONTRACTOR EQUIPMENT | | 114.2 | 114.2 | | 114.2 | 114.2 | | 111.3 | 111.3 | | 114.2 | 114.2 | | 114.2 | 114.2 | | 113.5 | 113.5 |
| 0241, 31 - 34 | SITE & INFRASTRUCTURE, DEMOLITION | 92.3 | 105.9 | 101.7 | 96.3 | 105.9 | 103.0 | 112.5 | 102.8 | 105.8 | 91.8 | 105.3 | 101.1 | 87.5 | 107.3 | 101.2 | 86.2 | 103.5 | 98.1 |
| 0310 | Concrete Forming & Accessories | 101.8 | 114.3 | 112.6 | 87.2 | 79.4 | 80.4 | 87.6 | 81.8 | 82.6 | 89.4 | 79.7 | 81.0 | 88.7 | 93.7 | 93.0 | 86.7 | 80.9 | 81.7 |
| 0320 | Concrete Reinforcing | 97.1 | 113.3 | 105.5 | 94.2 | 91.5 | 92.8 | 95.9 | 80.5 | 88.0 | 96.2 | 91.5 | 93.8 | 94.8 | 103.7 | 99.4 | 94.6 | 99.9 | 97.3 |
| 0330 | Cast-in-Place Concrete | 82.7 | 102.5 | 90.3 | 92.2 | 86.6 | 90.0 | 106.3 | 74.0 | 93.9 | 88.1 | 89.0 | 88.4 | 81.4 | 97.6 | 87.6 | 89.4 | 73.7 | 83.4 |
| 03 | CONCRETE | 96.2 | 110.8 | 103.2 | 91.2 | 85.6 | 88.6 | 101.9 | 80.3 | 91.5 | 98.1 | 86.6 | 92.6 | 82.9 | 98.1 | 90.2 | 104.2 | 83.6 | 94.3 |
| 04 | MASONRY | 95.3 | 101.1 | 98.8 | 96.9 | 69.6 | 80.3 | 100.6 | 85.7 | 91.6 | 94.1 | 83.1 | 87.4 | 99.0 | 96.9 | 97.7 | 95.2 | 89.5 | 91.7 |
| 05 | METALS | 100.2 | 124.7 | 108.6 | 94.1 | 111.9 | 100.2 | 97.1 | 104.7 | 99.7 | 98.0 | 111.3 | 102.6 | 93.8 | 119.1 | 102.5 | 98.0 | 114.3 | 103.6 |
| 06 | WOOD, PLASTICS & COMPOSITES | 105.1 | 118.0 | 112.5 | 83.3 | 81.3 | 82.2 | 87.8 | 81.2 | 84.0 | 90.2 | 77.5 | 82.9 | 84.9 | 91.2 | 88.5 | 93.8 | 81.3 | 86.6 |
| 07 | THERMAL & MOISTURE PROTECTION | 101.1 | 117.1 | 107.4 | 100.5 | 91.4 | 96.9 | 105.5 | 89.0 | 98.9 | 101.1 | 90.2 | 96.8 | 100.2 | 100.2 | 100.2 | 100.2 | 75.9 | 90.5 |
| 08 | OPENINGS | 92.7 | 116.9 | 98.8 | 87.3 | 85.9 | 86.9 | 93.9 | 82.9 | 91.1 | 93.1 | 87.6 | 91.7 | 87.3 | 101.4 | 90.8 | 89.5 | 82.8 | 87.8 |
| 0920 | Plaster & Gypsum Board | 106.5 | 118.3 | 114.8 | 97.8 | 80.5 | 85.6 | 98.3 | 80.5 | 85.7 | 99.2 | 76.6 | 83.3 | 98.3 | 90.7 | 92.9 | 103.2 | 80.5 | 87.2 |
| 0950, 0980 | Ceilings & Acoustic Treatment | 93.1 | 118.3 | 109.8 | 96.9 | 80.5 | 86.0 | 98.7 | 80.5 | 86.6 | 95.1 | 76.6 | 82.9 | 97.9 | 90.7 | 93.1 | 95.1 | 80.5 | 85.4 |
| 0960 | Flooring | 90.1 | 97.0 | 92.1 | 84.0 | 93.0 | 86.6 | 89.4 | 93.1 | 90.5 | 84.2 | 97.3 | 87.9 | 84.8 | 85.6 | 85.0 | 87.8 | 48.8 | 76.6 |
| 0970, 0990 | Wall Finishes & Painting/Coating | 92.5 | 76.0 | 82.5 | 88.6 | 105.7 | 99.0 | 95.0 | 72.8 | 81.5 | 92.5 | 94.2 | 93.6 | 88.6 | 105.7 | 99.0 | 92.5 | 62.2 | 74.1 |
| 09 | FINISHES | 96.7 | 107.6 | 102.7 | 95.1 | 83.9 | 88.9 | 96.9 | 81.5 | 88.4 | 94.5 | 82.7 | 88.0 | 95.0 | 92.2 | 93.5 | 95.1 | 72.6 | 82.6 |
| COVERS | DIVS. 10 - 14, 25, 28, 41, 43, 44, 46 | 100.0 | 105.2 | 101.1 | 100.0 | 95.8 | 99.1 | 100.0 | 99.0 | 99.8 | 100.0 | 98.9 | 99.8 | 100.0 | 103.1 | 100.6 | 100.0 | 95.9 | 99.1 |
| 21, 22, 23 | FIRE SUPPRESSION, PLUMBING & HVAC | 100.3 | 110.3 | 104.5 | 99.8 | 82.9 | 92.8 | 94.7 | 87.6 | 91.8 | 94.9 | 89.2 | 92.5 | 94.5 | 94.5 | 94.5 | 95.0 | 89.0 | 92.5 |
| 26, 27, 3370 | ELECTRICAL, COMMUNICATIONS & UTIL. | 100.7 | 96.8 | 98.6 | 90.0 | 105.4 | 98.2 | 93.6 | 105.4 | 99.9 | 93.6 | 105.4 | 99.8 | 90.7 | 104.2 | 97.9 | 92.1 | 84.4 | 88.0 |
| MF2010 | WEIGHTED AVERAGE | 98.3 | 108.6 | 102.9 | 95.0 | 90.5 | 93.0 | 97.6 | 91.1 | 94.7 | 95.9 | 93.1 | 94.7 | 92.6 | 100.3 | 96.0 | 96.1 | 88.6 | 92.7 |

## PENNSYLVANIA

| DIVISION | | DOYLESTOWN 189 | | | DUBOIS 158 | | | ERIE 164 - 165 | | | GREENSBURG 156 | | | HARRISBURG 170 - 171 | | | HAZLETON 182 | | |
|---|---|---|---|---|---|---|---|---|---|---|---|---|---|---|---|---|---|---|---|
| | | MAT. | INST. | TOTAL | MAT. | INST. | TOTAL | MAT. | INST. | TOTAL | MAT. | INST. | TOTAL | MAT. | INST. | TOTAL | MAT. | INST. | TOTAL |
| 015433 | CONTRACTOR EQUIPMENT | | 94.4 | 94.4 | | 111.3 | 111.3 | | 114.2 | 114.2 | | 111.3 | 111.3 | | 113.5 | 113.5 | | 114.2 | 114.2 |
| 0241, 31 - 34 | SITE & INFRASTRUCTURE, DEMOLITION | 107.2 | 91.6 | 96.4 | 118.5 | 102.9 | 107.8 | 92.8 | 106.5 | 102.3 | 107.9 | 104.9 | 105.8 | 82.2 | 104.8 | 97.8 | 85.6 | 106.1 | 99.8 |
| 0310 | Concrete Forming & Accessories | 86.1 | 131.0 | 125.1 | 87.1 | 82.9 | 83.4 | 101.3 | 87.3 | 89.1 | 93.7 | 93.6 | 93.6 | 94.7 | 86.0 | 87.1 | 83.9 | 87.2 | 86.7 |
| 0320 | Concrete Reinforcing | 94.0 | 141.0 | 118.2 | 95.4 | 91.8 | 93.5 | 96.2 | 93.4 | 94.8 | 95.4 | 103.7 | 99.6 | 97.1 | 102.0 | 99.6 | 94.3 | 108.0 | 101.3 |
| 0330 | Cast-in-Place Concrete | 78.2 | 91.1 | 83.1 | 102.4 | 92.3 | 98.5 | 90.6 | 85.5 | 88.6 | 98.6 | 97.2 | 98.1 | 87.1 | 94.3 | 89.9 | 78.2 | 90.9 | 83.1 |
| 03 | CONCRETE | 91.9 | 119.0 | 104.8 | 103.0 | 89.1 | 96.3 | 90.0 | 89.1 | 89.6 | 96.7 | 97.9 | 97.3 | 96.1 | 93.3 | 94.7 | 88.7 | 93.8 | 91.1 |
| 04 | MASONRY | 98.5 | 130.7 | 118.0 | 100.6 | 86.6 | 92.1 | 87.6 | 91.2 | 89.7 | 110.9 | 96.9 | 102.4 | 95.1 | 89.6 | 91.7 | 107.4 | 96.1 | 100.6 |
| 05 | METALS | 97.7 | 129.7 | 108.7 | 97.1 | 110.8 | 101.8 | 94.3 | 112.2 | 100.5 | 97.0 | 117.7 | 104.2 | 102.3 | 118.6 | 107.9 | 99.9 | 121.2 | 107.2 |
| 06 | WOOD, PLASTICS & COMPOSITES | 85.4 | 133.2 | 112.8 | 86.7 | 81.2 | 83.5 | 101.1 | 85.7 | 92.3 | 94.4 | 91.2 | 92.5 | 98.9 | 85.0 | 90.9 | 84.0 | 85.0 | 84.5 |
| 07 | THERMAL & MOISTURE PROTECTION | 99.4 | 134.3 | 113.3 | 105.7 | 94.7 | 101.3 | 100.8 | 93.2 | 97.8 | 105.3 | 99.8 | 103.1 | 107.0 | 108.0 | 107.4 | 100.7 | 106.4 | 102.9 |
| 08 | OPENINGS | 94.3 | 142.9 | 106.6 | 93.9 | 89.6 | 92.8 | 87.4 | 88.4 | 87.7 | 93.9 | 101.4 | 95.8 | 92.7 | 91.6 | 92.4 | 93.4 | 92.8 | 93.2 |
| 0920 | Plaster & Gypsum Board | 96.2 | 133.9 | 122.8 | 97.0 | 80.5 | 85.4 | 106.5 | 85.1 | 91.4 | 99.4 | 90.7 | 93.3 | 103.3 | 84.3 | 89.9 | 96.8 | 84.3 | 88.0 |
| 0950, 0980 | Ceilings & Acoustic Treatment | 92.3 | 133.9 | 119.9 | 98.7 | 80.5 | 86.6 | 93.1 | 85.1 | 87.8 | 97.8 | 90.7 | 93.1 | 96.9 | 84.3 | 88.5 | 94.2 | 84.3 | 87.6 |
| 0960 | Flooring | 74.4 | 134.5 | 91.7 | 89.1 | 99.2 | 92.0 | 89.5 | 97.0 | 91.6 | 92.6 | 71.9 | 86.6 | 90.7 | 92.6 | 91.3 | 81.6 | 95.7 | 85.6 |
| 0970, 0990 | Wall Finishes & Painting/Coating | 92.0 | 74.3 | 81.3 | 95.0 | 105.7 | 101.5 | 97.0 | 89.4 | 92.4 | 95.0 | 105.7 | 101.5 | 92.5 | 87.8 | 89.7 | 92.5 | 94.6 | 93.8 |
| 09 | FINISHES | 87.5 | 124.8 | 108.1 | 97.2 | 87.5 | 91.9 | 97.4 | 88.9 | 92.7 | 97.4 | 90.6 | 93.6 | 95.8 | 86.9 | 90.9 | 92.7 | 87.4 | 89.8 |
| COVERS | DIVS. 10 - 14, 25, 28, 41, 43, 44, 46 | 100.0 | 76.9 | 95.2 | 100.0 | 99.0 | 99.8 | 100.0 | 100.9 | 100.2 | 100.0 | 103.0 | 100.6 | 100.0 | 96.3 | 99.2 | 100.0 | 99.8 | 100.0 |
| 21, 22, 23 | FIRE SUPPRESSION, PLUMBING & HVAC | 94.5 | 127.3 | 108.1 | 94.7 | 87.9 | 91.9 | 99.8 | 92.2 | 96.6 | 94.7 | 92.1 | 93.7 | 100.1 | 89.8 | 95.8 | 94.9 | 98.3 | 96.4 |
| 26, 27, 3370 | ELECTRICAL, COMMUNICATIONS & UTIL. | 93.0 | 120.5 | 107.6 | 94.3 | 105.4 | 100.2 | 92.0 | 86.0 | 88.8 | 94.3 | 105.4 | 100.2 | 97.3 | 84.5 | 90.5 | 94.5 | 87.6 | 90.8 |
| MF2010 | WEIGHTED AVERAGE | 95.1 | 121.9 | 107.1 | 98.0 | 94.2 | 96.3 | 94.8 | 93.5 | 94.2 | 97.5 | 99.4 | 98.3 | 98.0 | 93.8 | 96.1 | 95.5 | 97.3 | 96.3 |

## PENNSYLVANIA

| DIVISION | | INDIANA 157 MAT. | INST. | TOTAL | JOHNSTOWN 159 MAT. | INST. | TOTAL | KITTANNING 162 MAT. | INST. | TOTAL | LANCASTER 175-176 MAT. | INST. | TOTAL | LEHIGH VALLEY 180 MAT. | INST. | TOTAL | MONTROSE 188 MAT. | INST. | TOTAL |
|---|---|---|---|---|---|---|---|---|---|---|---|---|---|---|---|---|---|---|---|
| 015433 | CONTRACTOR EQUIPMENT | | 111.3 | 111.3 | | 111.3 | 111.3 | | | | | 114.2 | 114.2 | | 113.5 | 113.5 | | 114.2 | 114.2 |
| 0241, 31 - 34 | SITE & INFRASTRUCTURE, DEMOLITION | 105.8 | 103.7 | 104.4 | 113.2 | 104.1 | 106.9 | 90.2 | 107.3 | 102.0 | 78.2 | 104.8 | 96.6 | 89.7 | 105.9 | 100.9 | 88.6 | 104.1 | 99.3 |
| 0310 | Concrete Forming & Accessories | 88.2 | 86.0 | 86.3 | 87.1 | 83.1 | 83.6 | 88.7 | 93.8 | 93.1 | 88.6 | 85.7 | 86.1 | 95.4 | 114.1 | 111.7 | 84.7 | 86.3 | 86.1 |
| 0320 | Concrete Reinforcing | 94.6 | 103.7 | 99.3 | 95.9 | 103.5 | 99.8 | 94.8 | 103.8 | 99.4 | 94.3 | 101.9 | 98.2 | 94.3 | 113.3 | 104.1 | 98.8 | 107.8 | 103.4 |
| 0330 | Cast-in-Place Concrete | 96.7 | 97.1 | 96.8 | 102.7 | 91.8 | 101.3 | 84.6 | 97.7 | 89.6 | 76.1 | 94.5 | 83.1 | 84.6 | 102.4 | 91.4 | 83.0 | 89.3 | 85.4 |
| 03 | CONCRETE | 94.4 | 94.4 | 94.4 | 102.7 | 91.3 | 97.2 | 85.5 | 98.2 | 91.6 | 91.4 | 93.2 | 92.3 | 95.2 | 110.7 | 102.6 | 93.8 | 92.6 | 93.2 |
| 04 | MASONRY | 96.8 | 98.5 | 97.9 | 97.7 | 86.1 | 90.7 | 103.1 | 98.5 | 100.3 | 100.7 | 90.2 | 94.3 | 95.3 | 101.1 | 98.8 | 95.3 | 95.8 | 95.6 |
| 05 | METALS | 97.2 | 117.5 | 104.2 | 97.1 | 116.3 | 103.7 | 93.9 | 119.3 | 102.7 | 98.0 | 118.1 | 104.9 | 99.8 | 124.3 | 108.3 | 98.1 | 116.3 | 104.4 |
| 06 | WOOD, PLASTICS & COMPOSITES | 88.6 | 81.2 | 84.4 | 86.7 | 81.2 | 83.5 | 84.9 | 91.2 | 88.5 | 96.4 | 85.0 | 89.9 | 97.2 | 118.0 | 109.1 | 84.8 | 85.3 | 85.1 |
| 07 | THERMAL & MOISTURE PROTECTION | 105.2 | 97.8 | 102.3 | 105.5 | 93.2 | 100.6 | 100.3 | 100.6 | 100.4 | 99.8 | 97.1 | 98.7 | 101.0 | 105.6 | 102.8 | 100.7 | 94.4 | 98.2 |
| 08 | OPENINGS | 93.9 | 93.2 | 93.7 | 93.9 | 89.5 | 92.8 | 87.3 | 98.7 | 90.2 | 89.5 | 95.6 | 91.1 | 93.4 | 116.9 | 99.4 | 89.7 | 93.0 | 90.5 |
| 0920 | Plaster & Gypsum Board | 98.8 | 80.5 | 85.9 | 96.7 | 80.5 | 85.3 | 98.3 | 90.7 | 92.9 | 105.4 | 84.3 | 90.5 | 101.2 | 118.3 | 113.3 | 97.0 | 84.6 | 88.3 |
| 0950, 0980 | Ceilings & Acoustic Treatment | 98.7 | 80.5 | 86.6 | 97.8 | 80.5 | 86.3 | 97.9 | 90.7 | 93.1 | 95.1 | 84.3 | 87.9 | 94.2 | 118.3 | 110.2 | 95.1 | 84.6 | 88.2 |
| 0960 | Flooring | 89.8 | 99.2 | 92.5 | 89.1 | 93.1 | 90.3 | 84.8 | 103.3 | 90.1 | 88.8 | 92.6 | 89.9 | 86.9 | 97.0 | 89.8 | 82.2 | 61.2 | 76.2 |
| 0970, 0990 | Wall Finishes & Painting/Coating | 95.0 | 105.7 | 101.5 | 95.0 | 105.7 | 101.5 | 88.6 | 105.7 | 99.0 | 92.5 | 61.5 | 73.7 | 92.5 | 71.5 | 79.8 | 92.5 | 98.4 | 96.1 |
| 09 | FINISHES | 96.6 | 88.8 | 92.3 | 96.5 | 86.4 | 90.9 | 95.1 | 95.4 | 95.3 | 95.0 | 84.1 | 89.0 | 95.2 | 106.0 | 101.1 | 93.4 | 82.5 | 87.4 |
| COVERS | DIVS. 10 - 14, 25, 28, 41, 43, 44, 46 | 100.0 | 101.8 | 100.4 | 100.0 | 99.2 | 99.8 | 100.0 | 103.1 | 100.6 | 100.0 | 96.3 | 99.2 | 100.0 | 105.4 | 101.1 | 100.0 | 96.5 | 99.3 |
| 21, 22, 23 | FIRE SUPPRESSION, PLUMBING & HVAC | 94.7 | 92.1 | 93.6 | 94.7 | 88.2 | 92.0 | 94.5 | 96.7 | 95.4 | 95.0 | 90.0 | 92.9 | 94.9 | 110.3 | 101.3 | 94.9 | 97.8 | 96.1 |
| 26, 27, 3370 | ELECTRICAL, COMMUNICATIONS & UTIL. | 94.3 | 105.4 | 100.2 | 94.3 | 105.4 | 100.2 | 90.0 | 105.4 | 98.2 | 93.6 | 47.8 | 69.2 | 94.5 | 141.5 | 119.5 | 93.6 | 95.1 | 94.4 |
| MF2010 | WEIGHTED AVERAGE | 96.3 | 98.2 | 97.1 | 97.6 | 94.9 | 96.4 | 93.2 | 101.4 | 96.9 | 94.7 | 88.3 | 91.8 | 96.0 | 114.2 | 104.2 | 94.9 | 96.3 | 95.5 |

## PENNSYLVANIA

| DIVISION | | NEW CASTLE 161 MAT. | INST. | TOTAL | NORRISTOWN 194 MAT. | INST. | TOTAL | OIL CITY 163 MAT. | INST. | TOTAL | PHILADELPHIA 190-191 MAT. | INST. | TOTAL | PITTSBURGH 150-152 MAT. | INST. | TOTAL | POTTSVILLE 179 MAT. | INST. | TOTAL |
|---|---|---|---|---|---|---|---|---|---|---|---|---|---|---|---|---|---|---|---|
| 015433 | CONTRACTOR EQUIPMENT | | 114.2 | 114.2 | | 97.9 | 97.9 | | 114.2 | 114.2 | | 97.3 | 97.3 | | 112.6 | 112.6 | | 113.5 | 113.5 |
| 0241, 31 - 34 | SITE & INFRASTRUCTURE, DEMOLITION | 88.0 | 107.3 | 101.3 | 88.0 | 98.9 | 95.6 | 86.4 | 105.5 | 99.6 | 104.8 | 137.6 | 133.3 | 101.1 | 94.9 | 95.7 | 80.6 | 86.7 | 85.9 |
| 0310 | Concrete Forming & Accessories | 88.7 | 94.0 | 93.3 | 88.6 | 133.4 | 127.5 | 88.7 | 81.0 | 82.0 | 95.1 | 148.5 | 122.6 | 96.5 | 104.0 | 100.3 | 93.6 | 101.9 | 97.8 |
| 0320 | Concrete Reinforcing | 93.6 | 93.4 | 93.5 | 92.2 | 148.5 | 121.2 | 94.8 | 93.2 | 94.0 | 95.1 | 148.5 | 122.6 | 102.4 | 97.4 | 100.5 | 86.7 | 85.0 | 85.7 |
| 0330 | Cast-in-Place Concrete | 82.2 | 96.7 | 87.8 | 78.1 | 131.3 | 98.5 | 79.9 | 94.6 | 85.5 | 95.8 | 131.3 | 109.4 | 102.4 | 97.4 | 100.5 | 81.0 | 78.8 | 80.2 |
| 03 | CONCRETE | 83.3 | 96.0 | 89.4 | 87.5 | 135.2 | 110.3 | 81.7 | 89.4 | 85.4 | 99.7 | 137.1 | 117.6 | 100.1 | 98.6 | 99.3 | 95.4 | 88.3 | 92.0 |
| 04 | MASONRY | 98.3 | 96.9 | 97.4 | 110.0 | 130.8 | 122.6 | 99.2 | 90.5 | 94.0 | 96.4 | 131.0 | 117.4 | 92.0 | 101.9 | 98.0 | 95.0 | 89.7 | 91.8 |
| 05 | METALS | 93.9 | 113.8 | 100.8 | 100.4 | 135.2 | 112.4 | 93.9 | 112.2 | 100.2 | 104.3 | 135.3 | 115.0 | 98.6 | 118.4 | 105.4 | 86.7 | 85.0 | 85.7 |
| 06 | WOOD, PLASTICS & COMPOSITES | 84.9 | 92.5 | 89.3 | 86.8 | 133.1 | 113.4 | 84.8 | 77.5 | 80.7 | 106.0 | 138.7 | 124.7 | 104.1 | 92.4 | 97.4 | 99.9 | 105.6 | 102.2 |
| 07 | THERMAL & MOISTURE PROTECTION | 100.2 | 97.3 | 99.0 | 97.1 | 134.8 | 112.1 | 100.1 | 93.4 | 97.4 | 107.9 | 135.4 | 112.8 | 105.5 | 101.3 | 103.8 | 99.9 | 105.6 | 102.2 |
| 08 | OPENINGS | 87.3 | 96.0 | 89.5 | 89.7 | 145.0 | 103.7 | 87.3 | 80.0 | 85.4 | 98.2 | 148.0 | 110.8 | 96.8 | 102.2 | 98.1 | 89.6 | 91.0 | 89.9 |
| 0920 | Plaster & Gypsum Board | 98.3 | 92.0 | 93.8 | 89.7 | 133.9 | 120.8 | 98.3 | 76.6 | 83.0 | 98.1 | 139.7 | 127.4 | 106.7 | 92.0 | 96.4 | 99.7 | 84.3 | 88.8 |
| 0950, 0980 | Ceilings & Acoustic Treatment | 97.9 | 92.0 | 94.0 | 104.3 | 133.9 | 123.9 | 97.9 | 76.6 | 83.8 | 104.3 | 139.7 | 127.7 | 98.7 | 92.0 | 94.3 | 95.1 | 84.3 | 87.9 |
| 0960 | Flooring | 84.8 | 60.7 | 77.9 | 85.2 | 140.6 | 101.1 | 84.8 | 99.2 | 88.9 | 92.8 | 140.6 | 106.5 | 96.2 | 101.0 | 97.6 | 85.0 | 92.6 | 87.2 |
| 0970, 0990 | Wall Finishes & Painting/Coating | 88.6 | 105.7 | 99.0 | 96.2 | 145.4 | 126.0 | 88.6 | 105.7 | 99.0 | 96.2 | 142.3 | 124.2 | 95.0 | 114.8 | 107.0 | 92.5 | 99.1 | 96.5 |
| 09 | FINISHES | 95.0 | 88.4 | 91.4 | 95.3 | 135.8 | 117.7 | 94.8 | 85.1 | 89.5 | 98.9 | 138.8 | 120.9 | 99.7 | 97.0 | 98.2 | 93.4 | 88.7 | 90.8 |
| COVERS | DIVS. 10 - 14, 25, 28, 41, 43, 44, 46 | 100.0 | 103.2 | 100.7 | 100.0 | 119.3 | 104.0 | 100.0 | 100.1 | 100.0 | 100.0 | 121.5 | 104.5 | 100.0 | 103.1 | 100.7 | 100.0 | 97.1 | 99.4 |
| 21, 22, 23 | FIRE SUPPRESSION, PLUMBING & HVAC | 94.5 | 94.3 | 94.4 | 94.6 | 128.7 | 108.8 | 94.5 | 91.8 | 93.4 | 100.0 | 129.0 | 112.1 | 100.1 | 100.4 | 100.3 | 95.0 | 97.6 | 96.0 |
| 26, 27, 3370 | ELECTRICAL, COMMUNICATIONS & UTIL. | 90.7 | 93.3 | 92.1 | 93.4 | 141.8 | 119.1 | 92.7 | 104.2 | 98.8 | 98.3 | 141.9 | 121.5 | 96.9 | 105.4 | 101.4 | 91.6 | 89.6 | 90.5 |
| MF2010 | WEIGHTED AVERAGE | 92.6 | 97.2 | 94.7 | 95.1 | 131.2 | 111.3 | 92.6 | 94.9 | 93.6 | 99.7 | 132.2 | 114.3 | 99.3 | 102.9 | 100.9 | 94.6 | 95.6 | 95.1 |

## PENNSYLVANIA

| DIVISION | | READING 195-196 MAT. | INST. | TOTAL | SCRANTON 184-185 MAT. | INST. | TOTAL | STATE COLLEGE 168 MAT. | INST. | TOTAL | STROUDSBURG 183 MAT. | INST. | TOTAL | SUNBURY 178 MAT. | INST. | TOTAL | UNIONTOWN 154 MAT. | INST. | TOTAL |
|---|---|---|---|---|---|---|---|---|---|---|---|---|---|---|---|---|---|---|---|
| 015433 | CONTRACTOR EQUIPMENT | | 117.7 | 117.7 | | 114.2 | 114.2 | | 113.5 | 113.5 | | 114.2 | 114.2 | | 114.2 | 114.2 | | 111.3 | 111.3 |
| 0241, 31 - 34 | SITE & INFRASTRUCTURE, DEMOLITION | 92.0 | 111.1 | 105.2 | 92.7 | 106.2 | 102.0 | 83.3 | 104.8 | 98.2 | 87.3 | 104.2 | 99.0 | 91.4 | 105.9 | 101.4 | 106.4 | 104.9 | 105.4 |
| 0310 | Concrete Forming & Accessories | 104.3 | 87.0 | 89.3 | 101.9 | 87.7 | 89.6 | 85.8 | 79.3 | 80.2 | 90.0 | 87.4 | 87.7 | 93.3 | 85.8 | 86.8 | 81.7 | 93.6 | 92.0 |
| 0320 | Concrete Reinforcing | 93.5 | 102.1 | 97.9 | 97.1 | 108.0 | 102.7 | 95.4 | 92.0 | 93.7 | 97.4 | 120.3 | 109.2 | 96.2 | 100.2 | 98.2 | 95.4 | 103.7 | 99.6 |
| 0330 | Cast-in-Place Concrete | 70.0 | 95.8 | 79.9 | 86.4 | 91.1 | 88.2 | 83.3 | 69.4 | 78.0 | 81.4 | 75.4 | 79.1 | 88.6 | 94.2 | 90.7 | 96.7 | 97.2 | 96.9 |
| 03 | CONCRETE | 86.4 | 94.3 | 90.2 | 98.0 | 94.1 | 96.2 | 98.4 | 79.9 | 89.5 | 92.6 | 90.8 | 91.7 | 98.8 | 92.8 | 95.9 | 94.1 | 97.9 | 95.9 |
| 04 | MASONRY | 98.3 | 92.9 | 95.0 | 95.6 | 96.2 | 96.0 | 100.8 | 80.6 | 88.5 | 91.7 | 100.7 | 97.1 | 94.1 | 87.3 | 90.0 | 113.1 | 96.9 | 103.2 |
| 05 | METALS | 100.7 | 119.3 | 107.1 | 102.4 | 121.5 | 108.9 | 97.8 | 112.7 | 102.9 | 99.9 | 121.8 | 107.5 | 97.9 | 117.4 | 104.6 | 96.9 | 117.7 | 104.0 |
| 06 | WOOD, PLASTICS & COMPOSITES | 106.1 | 84.6 | 93.8 | 105.1 | 85.3 | 93.8 | 92.5 | 81.3 | 86.1 | 91.1 | 85.3 | 87.8 | 94.7 | 85.0 | 89.1 | 80.2 | 91.2 | 86.5 |
| 07 | THERMAL & MOISTURE PROTECTION | 97.5 | 109.8 | 102.4 | 101.0 | 97.9 | 99.8 | 100.0 | 92.3 | 96.9 | 100.8 | 87.0 | 95.4 | 101.0 | 104.6 | 102.4 | 105.1 | 99.8 | 103.0 |
| 08 | OPENINGS | 95.6 | 95.4 | 95.5 | 92.7 | 93.0 | 92.8 | 89.5 | 85.9 | 88.6 | 93.4 | 89.7 | 92.5 | 89.7 | 90.4 | 89.9 | 93.8 | 101.4 | 95.8 |
| 0920 | Plaster & Gypsum Board | 102.6 | 84.0 | 89.5 | 108.6 | 84.6 | 91.7 | 101.4 | 80.5 | 86.6 | 98.9 | 84.6 | 88.9 | 100.0 | 84.3 | 88.9 | 97.8 | 90.7 | 93.1 |
| 0950, 0980 | Ceilings & Acoustic Treatment | 93.9 | 84.0 | 87.3 | 100.6 | 84.6 | 90.0 | 93.2 | 80.5 | 84.8 | 92.3 | 84.6 | 87.2 | 93.2 | 84.3 | 87.3 | 86.5 | 99.2 | 90.1 |
| 0960 | Flooring | 90.5 | 92.6 | 91.1 | 90.1 | 107.6 | 95.2 | 87.4 | 103.0 | 91.9 | 84.6 | 54.2 | 75.9 | 85.9 | 92.6 | 87.8 | 86.5 | 99.2 | 90.1 |
| 0970, 0990 | Wall Finishes & Painting/Coating | 94.7 | 99.1 | 97.4 | 92.5 | 98.4 | 96.1 | 92.5 | 105.7 | 100.5 | 92.5 | 69.5 | 78.6 | 92.5 | 99.1 | 96.5 | 95.0 | 105.7 | 101.5 |
| 09 | FINISHES | 96.5 | 88.0 | 91.8 | 98.5 | 91.8 | 94.8 | 94.1 | 85.9 | 89.6 | 93.7 | 78.9 | 85.5 | 94.7 | 88.1 | 91.1 | 94.7 | 94.4 | 94.6 |
| COVERS | DIVS. 10 - 14, 25, 28, 41, 43, 44, 46 | 100.0 | 99.0 | 99.8 | 100.0 | 99.9 | 100.0 | 100.0 | 92.6 | 98.5 | 100.0 | 67.2 | 93.2 | 100.0 | 96.3 | 99.2 | 100.0 | 103.0 | 100.6 |
| 21, 22, 23 | FIRE SUPPRESSION, PLUMBING & HVAC | 100.1 | 106.9 | 102.9 | 100.3 | 98.4 | 99.5 | 95.0 | 84.7 | 90.7 | 94.9 | 100.1 | 97.1 | 94.9 | 89.6 | 92.7 | 94.7 | 92.1 | 93.7 |
| 26, 27, 3370 | ELECTRICAL, COMMUNICATIONS & UTIL. | 100.5 | 89.6 | 94.7 | 100.7 | 95.1 | 97.7 | 92.6 | 105.4 | 99.4 | 94.5 | 141.5 | 119.5 | 92.0 | 88.5 | 90.1 | 90.9 | 105.4 | 98.6 |
| MF2010 | WEIGHTED AVERAGE | 97.4 | 99.6 | 98.4 | 99.0 | 98.7 | 98.9 | 95.5 | 91.4 | 93.7 | 95.3 | 102.3 | 98.4 | 95.5 | 94.0 | 94.8 | 96.6 | 99.9 | 98.0 |

## PENNSYLVANIA

| DIVISION | | WASHINGTON 153 | | | WELLSBORO 169 | | | WESTCHESTER 193 | | | WILKES-BARRE 186 - 187 | | | WILLIAMSPORT 177 | | | YORK 173 - 174 | | |
|---|---|---|---|---|---|---|---|---|---|---|---|---|---|---|---|---|---|---|---|
| | | MAT. | INST. | TOTAL | MAT. | INST. | TOTAL | MAT. | INST. | TOTAL | MAT. | INST. | TOTAL | MAT. | INST. | TOTAL | MAT. | INST. | TOTAL |
| 015433 | CONTRACTOR EQUIPMENT | | 111.3 | 111.3 | | 114.2 | 114.2 | | 97.9 | 97.9 | | 114.2 | 114.2 | | 114.2 | 114.2 | | 113.5 | 113.5 |
| 0241, 31 - 34 | SITE & INFRASTRUCTURE, DEMOLITION | 106.4 | 104.8 | 105.3 | 95.9 | 104.0 | 101.5 | 93.7 | 96.7 | 95.8 | 85.4 | 106.1 | 99.7 | 82.6 | 104.7 | 97.8 | 81.5 | 104.8 | 97.6 |
| 0310 | Concrete Forming & Accessories | 88.4 | 93.7 | 93.0 | 88.7 | 84.2 | 84.8 | 95.3 | 132.1 | 127.2 | 92.5 | 87.6 | 88.3 | 90.0 | 62.9 | 66.5 | 83.9 | 86.2 | 85.9 |
| 0320 | Concrete Reinforcing | 95.4 | 103.8 | 99.7 | 95.4 | 107.6 | 101.7 | 91.3 | 114.6 | 103.3 | 96.2 | 108.0 | 102.3 | 95.4 | 69.4 | 82.0 | 96.2 | 102.0 | 99.2 |
| 0330 | Cast-in-Place Concrete | 96.7 | 97.3 | 96.9 | 87.3 | 87.0 | 87.2 | 86.5 | 130.7 | 103.4 | 78.2 | 90.9 | 83.1 | 75.0 | 75.3 | 75.1 | 81.4 | 94.7 | 86.5 |
| 03 | CONCRETE | 94.5 | 98.0 | 96.2 | 100.8 | 90.9 | 96.1 | 95.1 | 127.9 | 110.8 | 89.5 | 94.0 | 91.7 | 86.3 | 70.3 | 78.6 | 96.6 | 93.5 | 95.1 |
| 04 | MASONRY | 97.3 | 98.5 | 98.0 | 100.8 | 89.9 | 94.2 | 103.7 | 130.8 | 120.1 | 108.1 | 95.8 | 100.7 | 86.2 | 91.5 | 89.4 | 95.1 | 90.2 | 92.1 |
| 05 | METALS | 96.8 | 117.9 | 104.1 | 97.9 | 116.4 | 104.3 | 100.4 | 118.6 | 106.7 | 98.1 | 121.6 | 106.2 | 98.0 | 101.9 | 99.3 | 99.7 | 118.6 | 106.2 |
| 06 | WOOD, PLASTICS & COMPOSITES | 88.6 | 91.2 | 90.1 | 89.6 | 85.0 | 86.9 | 94.5 | 133.1 | 116.6 | 93.7 | 85.2 | 88.8 | 91.1 | 55.3 | 70.6 | 90.5 | 85.0 | 87.3 |
| 07 | THERMAL & MOISTURE PROTECTION | 105.2 | 100.2 | 103.3 | 101.3 | 92.1 | 97.7 | 97.4 | 134.4 | 112.1 | 100.6 | 106.3 | 102.9 | 100.5 | 102.7 | 101.4 | 99.9 | 108.3 | 103.3 |
| 08 | OPENINGS | 93.8 | 101.4 | 95.8 | 93.1 | 92.8 | 93.0 | 89.7 | 129.4 | 99.7 | 89.7 | 97.5 | 91.7 | 89.7 | 57.9 | 81.6 | 89.5 | 91.6 | 90.1 |
| 0920 | Plaster & Gypsum Board | 98.5 | 90.7 | 93.0 | 98.7 | 84.3 | 88.5 | 91.9 | 133.9 | 121.5 | 100.1 | 84.5 | 89.1 | 99.7 | 53.8 | 67.3 | 101.5 | 84.3 | 89.3 |
| 0950, 0980 | Ceilings & Acoustic Treatment | 97.8 | 90.7 | 93.1 | 93.2 | 84.3 | 87.3 | 104.3 | 133.9 | 123.9 | 95.1 | 84.5 | 88.1 | 95.1 | 53.8 | 67.7 | 94.1 | 84.3 | 87.6 |
| 0960 | Flooring | 89.9 | 99.2 | 92.6 | 83.9 | 53.7 | 75.2 | 88.6 | 140.1 | 103.5 | 85.6 | 95.7 | 88.5 | 84.6 | 52.5 | 75.4 | 86.4 | 92.6 | 88.2 |
| 0970, 0990 | Wall Finishes & Painting/Coating | 95.0 | 105.7 | 101.5 | 92.5 | 98.4 | 96.1 | 96.2 | 145.4 | 126.0 | 92.5 | 98.4 | 96.1 | 92.5 | 98.4 | 96.1 | 92.5 | 87.8 | 89.7 |
| 09 | FINISHES | 96.4 | 94.8 | 95.5 | 94.4 | 79.7 | 86.3 | 97.1 | 133.7 | 117.3 | 94.5 | 89.3 | 91.6 | 94.0 | 63.9 | 77.4 | 93.9 | 87.0 | 90.1 |
| COVERS | DIVS. 10 - 14, 25, 28, 41, 43, 44, 46 | 100.0 | 103.0 | 100.6 | 100.0 | 92.8 | 98.5 | 100.0 | 119.2 | 104.0 | 100.0 | 99.7 | 99.9 | 100.0 | 93.1 | 98.6 | 100.0 | 96.5 | 99.3 |
| 21, 22, 23 | FIRE SUPPRESSION, PLUMBING & HVAC | 94.7 | 95.5 | 95.1 | 94.9 | 89.0 | 92.5 | 94.6 | 127.2 | 108.1 | 94.9 | 98.2 | 96.3 | 94.9 | 89.7 | 92.8 | 100.3 | 90.1 | 96.0 |
| 26, 27, 3370 | ELECTRICAL, COMMUNICATIONS & UTIL. | 93.6 | 105.4 | 99.9 | 93.6 | 79.9 | 86.3 | 93.3 | 123.4 | 109.3 | 94.5 | 87.6 | 90.8 | 92.5 | 50.6 | 70.2 | 93.6 | 84.5 | 88.8 |
| MF2010 | WEIGHTED AVERAGE | 96.2 | 100.8 | 98.3 | 96.7 | 91.0 | 94.2 | 96.1 | 124.7 | 108.9 | 95.2 | 97.7 | 96.3 | 93.3 | 79.7 | 87.2 | 96.5 | 94.0 | 95.4 |

| DIVISION | | PUERTO RICO SAN JUAN 009 | | | RHODE ISLAND NEWPORT 028 | | | PROVIDENCE 029 | | | SOUTH CAROLINA AIKEN 298 | | | BEAUFORT 299 | | | CHARLESTON 294 | | |
|---|---|---|---|---|---|---|---|---|---|---|---|---|---|---|---|---|---|---|---|
| | | MAT. | INST. | TOTAL | MAT. | INST. | TOTAL | MAT. | INST. | TOTAL | MAT. | INST. | TOTAL | MAT. | INST. | TOTAL | MAT. | INST. | TOTAL |
| 015433 | CONTRACTOR EQUIPMENT | | 89.6 | 89.6 | | 101.9 | 101.9 | | 101.9 | 101.9 | | 100.0 | 100.0 | | 100.0 | 100.0 | | 100.0 | 100.0 |
| 0241, 31 - 34 | SITE & INFRASTRUCTURE, DEMOLITION | 125.4 | 90.5 | 101.3 | 82.5 | 103.5 | 97.0 | 83.8 | 103.5 | 97.4 | 127.9 | 86.0 | 99.0 | 122.3 | 84.2 | 96.0 | 103.5 | 83.8 | 89.9 |
| 0310 | Concrete Forming & Accessories | 92.1 | 17.6 | 27.4 | 99.7 | 115.6 | 113.5 | 98.4 | 115.6 | 113.3 | 96.0 | 71.0 | 74.3 | 95.0 | 40.1 | 47.4 | 94.1 | 41.9 | 48.8 |
| 0320 | Concrete Reinforcing | 194.4 | 12.4 | 100.7 | 108.1 | 114.4 | 111.3 | 108.1 | 114.4 | 111.3 | 91.7 | 68.8 | 79.9 | 90.8 | 27.7 | 58.3 | 90.6 | 59.8 | 74.7 |
| 0330 | Cast-in-Place Concrete | 101.6 | 30.6 | 74.4 | 78.6 | 120.3 | 94.5 | 96.5 | 120.3 | 105.6 | 89.3 | 72.9 | 83.0 | 89.3 | 49.5 | 74.1 | 104.8 | 50.2 | 83.9 |
| 03 | CONCRETE | 112.8 | 22.0 | 69.3 | 100.5 | 116.5 | 108.2 | 109.5 | 116.5 | 112.8 | 109.8 | 72.3 | 91.8 | 106.9 | 43.0 | 76.3 | 102.4 | 49.8 | 77.2 |
| 04 | MASONRY | 88.9 | 16.6 | 45.1 | 105.1 | 123.8 | 116.4 | 108.5 | 123.8 | 117.7 | 79.5 | 63.9 | 70.0 | 95.2 | 35.5 | 59.0 | 96.5 | 39.7 | 62.1 |
| 05 | METALS | 112.9 | 33.5 | 85.6 | 98.6 | 109.6 | 102.4 | 98.6 | 109.6 | 102.4 | 83.6 | 86.4 | 84.5 | 83.6 | 69.7 | 78.8 | 85.4 | 77.5 | 82.7 |
| 06 | WOOD, PLASTICS & COMPOSITES | 92.5 | 17.3 | 49.4 | 99.5 | 115.2 | 108.5 | 98.6 | 115.2 | 108.1 | 98.8 | 72.3 | 83.6 | 97.2 | 40.9 | 64.9 | 95.9 | 40.9 | 64.3 |
| 07 | THERMAL & MOISTURE PROTECTION | 125.6 | 21.1 | 84.0 | 96.6 | 115.4 | 104.1 | 94.3 | 115.4 | 102.7 | 99.5 | 69.8 | 87.7 | 99.2 | 38.5 | 75.1 | 98.3 | 43.1 | 76.4 |
| 08 | OPENINGS | 151.4 | 14.7 | 116.8 | 99.6 | 114.2 | 103.3 | 100.2 | 114.2 | 103.8 | 94.6 | 68.2 | 87.9 | 94.6 | 39.6 | 80.7 | 98.6 | 44.6 | 84.9 |
| 0920 | Plaster & Gypsum Board | 116.2 | 14.8 | 44.8 | 96.8 | 114.9 | 109.6 | 94.7 | 114.9 | 109.0 | 103.7 | 71.3 | 80.9 | 107.4 | 39.0 | 59.2 | 109.4 | 39.0 | 59.8 |
| 0950, 0980 | Ceilings & Acoustic Treatment | 234.7 | 14.8 | 88.9 | 96.5 | 114.9 | 108.7 | 95.8 | 114.9 | 108.5 | 92.0 | 71.3 | 78.3 | 95.7 | 39.0 | 58.1 | 95.7 | 39.0 | 58.1 |
| 0960 | Flooring | 200.5 | 13.9 | 146.9 | 96.6 | 131.6 | 106.7 | 96.3 | 131.6 | 106.4 | 108.7 | 69.9 | 97.5 | 110.4 | 54.1 | 94.2 | 109.9 | 61.5 | 96.0 |
| 0970, 0990 | Wall Finishes & Painting/Coating | 197.9 | 12.9 | 85.6 | 94.4 | 123.4 | 112.0 | 94.4 | 123.4 | 112.0 | 114.7 | 73.4 | 89.6 | 114.7 | 37.3 | 67.8 | 114.7 | 42.2 | 70.7 |
| 09 | FINISHES | 204.5 | 16.6 | 100.8 | 97.5 | 119.8 | 109.8 | 96.8 | 119.8 | 109.5 | 105.1 | 71.1 | 86.3 | 106.3 | 42.8 | 71.3 | 104.2 | 45.1 | 71.6 |
| COVERS | DIVS. 10 - 14, 25, 28, 41, 43, 44, 46 | 100.0 | 17.1 | 82.8 | 100.0 | 103.2 | 100.7 | 100.0 | 103.2 | 100.7 | 100.0 | 79.4 | 95.7 | 100.0 | 48.2 | 89.2 | 100.0 | 72.5 | 94.3 |
| 21, 22, 23 | FIRE SUPPRESSION, PLUMBING & HVAC | 103.4 | 13.6 | 66.0 | 100.2 | 108.0 | 103.4 | 100.0 | 108.0 | 103.3 | 94.9 | 65.6 | 82.7 | 94.9 | 25.9 | 66.2 | 100.3 | 40.8 | 75.5 |
| 26, 27, 3370 | ELECTRICAL, COMMUNICATIONS & UTIL. | 128.5 | 12.9 | 67.0 | 99.9 | 101.2 | 100.6 | 98.3 | 101.2 | 99.8 | 94.4 | 69.4 | 81.1 | 98.2 | 32.9 | 63.5 | 96.6 | 96.4 | 96.5 |
| MF2010 | WEIGHTED AVERAGE | 121.8 | 24.0 | 78.1 | 99.3 | 111.5 | 104.7 | 100.3 | 111.5 | 105.3 | 96.7 | 71.9 | 85.6 | 97.5 | 43.0 | 73.1 | 98.0 | 58.3 | 80.2 |

| DIVISION | | SOUTH CAROLINA COLUMBIA 290 - 292 | | | FLORENCE 295 | | | GREENVILLE 296 | | | ROCK HILL 297 | | | SPARTANBURG 293 | | | SOUTH DAKOTA ABERDEEN 574 | | |
|---|---|---|---|---|---|---|---|---|---|---|---|---|---|---|---|---|---|---|---|
| | | MAT. | INST. | TOTAL | MAT. | INST. | TOTAL | MAT. | INST. | TOTAL | MAT. | INST. | TOTAL | MAT. | INST. | TOTAL | MAT. | INST. | TOTAL |
| 015433 | CONTRACTOR EQUIPMENT | | 100.0 | 100.0 | | 100.0 | 100.0 | | 100.0 | 100.0 | | 100.0 | 100.0 | | 100.0 | 100.0 | | 98.6 | 98.6 |
| 0241, 31 - 34 | SITE & INFRASTRUCTURE, DEMOLITION | 101.3 | 83.7 | 89.2 | 115.0 | 83.8 | 93.4 | 109.3 | 83.4 | 91.4 | 107.1 | 83.3 | 90.6 | 109.1 | 83.4 | 91.4 | 96.1 | 94.6 | 95.0 |
| 0310 | Concrete Forming & Accessories | 95.8 | 41.0 | 48.2 | 82.3 | 41.1 | 46.6 | 93.5 | 41.0 | 47.9 | 92.1 | 38.6 | 45.7 | 96.6 | 41.0 | 48.3 | 96.3 | 39.7 | 47.2 |
| 0320 | Concrete Reinforcing | 90.6 | 59.6 | 74.6 | 90.3 | 59.7 | 74.5 | 90.2 | 47.6 | 68.3 | 91.0 | 47.6 | 68.6 | 90.2 | 47.6 | 68.3 | 103.4 | 41.0 | 71.3 |
| 0330 | Cast-in-Place Concrete | 107.6 | 50.8 | 85.8 | 89.3 | 50.0 | 74.2 | 89.3 | 49.9 | 74.2 | 89.3 | 44.8 | 72.2 | 89.3 | 49.9 | 74.2 | 103.8 | 46.2 | 81.8 |
| 03 | CONCRETE | 103.9 | 49.5 | 77.9 | 101.3 | 49.3 | 76.4 | 99.7 | 47.0 | 74.5 | 97.8 | 44.1 | 72.1 | 99.9 | 47.0 | 74.6 | 102.7 | 43.6 | 74.4 |
| 04 | MASONRY | 90.8 | 37.6 | 58.5 | 79.5 | 39.7 | 55.4 | 77.3 | 39.7 | 54.5 | 103.0 | 34.8 | 61.7 | 79.5 | 39.7 | 55.4 | 111.6 | 57.9 | 79.1 |
| 05 | METALS | 85.4 | 76.7 | 82.4 | 84.4 | 77.0 | 81.9 | 84.4 | 72.6 | 80.3 | 83.6 | 72.1 | 79.6 | 84.4 | 72.6 | 80.3 | 102.4 | 66.2 | 90.0 |
| 06 | WOOD, PLASTICS & COMPOSITES | 98.3 | 40.4 | 65.1 | 81.7 | 40.4 | 58.0 | 95.6 | 40.4 | 63.9 | 93.8 | 38.4 | 62.1 | 99.8 | 40.4 | 65.7 | 97.2 | 38.9 | 63.8 |
| 07 | THERMAL & MOISTURE PROTECTION | 97.2 | 41.8 | 75.2 | 98.5 | 43.1 | 76.6 | 98.5 | 43.1 | 76.5 | 98.4 | 40.6 | 75.5 | 98.6 | 43.1 | 76.5 | 106.1 | 51.0 | 84.2 |
| 08 | OPENINGS | 98.6 | 44.4 | 84.9 | 94.6 | 44.4 | 81.9 | 94.6 | 41.5 | 81.1 | 94.6 | 41.5 | 81.1 | 94.6 | 41.5 | 81.1 | 92.7 | 39.7 | 79.3 |
| 0920 | Plaster & Gypsum Board | 100.0 | 38.5 | 56.7 | 95.5 | 38.5 | 55.4 | 101.9 | 38.5 | 57.2 | 101.0 | 36.4 | 55.5 | 105.0 | 38.5 | 58.2 | 105.4 | 37.4 | 57.5 |
| 0950, 0980 | Ceilings & Acoustic Treatment | 94.8 | 38.5 | 57.5 | 92.9 | 38.5 | 56.8 | 92.0 | 38.5 | 56.5 | 92.0 | 36.4 | 55.2 | 92.0 | 38.5 | 56.5 | 108.1 | 37.4 | 61.2 |
| 0960 | Flooring | 106.5 | 46.1 | 89.1 | 100.8 | 46.1 | 85.1 | 107.4 | 60.4 | 93.9 | 106.4 | 47.0 | 89.3 | 108.9 | 60.4 | 95.0 | 118.6 | 53.2 | 99.8 |
| 0970, 0990 | Wall Finishes & Painting/Coating | 112.3 | 42.2 | 69.7 | 114.7 | 42.2 | 70.7 | 114.7 | 42.2 | 70.7 | 114.7 | 42.2 | 70.7 | 114.7 | 42.2 | 70.7 | 98.4 | 40.0 | 62.9 |
| 09 | FINISHES | 101.1 | 41.7 | 68.4 | 100.6 | 42.2 | 68.4 | 102.6 | 44.7 | 70.7 | 101.9 | 40.2 | 67.9 | 103.5 | 44.7 | 71.0 | 108.3 | 41.8 | 71.6 |
| COVERS | DIVS. 10 - 14, 25, 28, 41, 43, 44, 46 | 100.0 | 72.2 | 94.2 | 100.0 | 72.2 | 94.2 | 100.0 | 72.2 | 94.2 | 100.0 | 71.4 | 94.0 | 100.0 | 72.2 | 94.2 | 100.0 | 45.2 | 88.6 |
| 21, 22, 23 | FIRE SUPPRESSION, PLUMBING & HVAC | 100.0 | 36.1 | 73.4 | 100.3 | 36.1 | 73.6 | 100.3 | 34.6 | 73.0 | 94.9 | 33.7 | 69.4 | 100.3 | 34.6 | 73.0 | 100.1 | 43.7 | 76.6 |
| 26, 27, 3370 | ELECTRICAL, COMMUNICATIONS & UTIL. | 98.2 | 49.3 | 72.2 | 94.4 | 46.9 | 69.1 | 96.6 | 50.7 | 72.2 | 96.6 | 50.7 | 72.2 | 96.6 | 50.7 | 72.2 | 98.4 | 55.5 | 75.6 |
| MF2010 | WEIGHTED AVERAGE | 97.6 | 50.0 | 76.3 | 96.2 | 50.0 | 75.5 | 96.2 | 49.6 | 75.3 | 95.7 | 47.7 | 74.3 | 96.4 | 49.6 | 75.5 | 101.2 | 52.8 | 79.6 |

# City Cost Indexes

## SOUTH DAKOTA

| DIVISION | | MITCHELL 573 MAT. | INST. | TOTAL | MOBRIDGE 576 MAT. | INST. | TOTAL | PIERRE 575 MAT. | INST. | TOTAL | RAPID CITY 577 MAT. | INST. | TOTAL | SIOUX FALLS 570 - 571 MAT. | INST. | TOTAL | WATERTOWN 572 MAT. | INST. | TOTAL |
|---|---|---|---|---|---|---|---|---|---|---|---|---|---|---|---|---|---|---|---|
| 015433 | CONTRACTOR EQUIPMENT | | 98.6 | 98.6 | | 98.6 | 98.6 | | 98.6 | 98.6 | | 98.6 | 98.6 | | 99.7 | 99.7 | | 98.6 | 98.6 |
| 0241, 31 - 34 | SITE & INFRASTRUCTURE, DEMOLITION | 93.1 | 94.6 | 94.1 | 93.0 | 94.6 | 94.1 | 91.3 | 94.6 | 93.6 | 94.1 | 94.7 | 94.5 | 91.9 | 96.3 | 95.0 | 93.0 | 94.6 | 94.1 |
| 0310 | Concrete Forming & Accessories | 95.6 | 40.0 | 47.4 | 87.7 | 39.8 | 46.2 | 95.1 | 40.8 | 48.0 | 102.4 | 38.8 | 47.2 | 95.7 | 43.4 | 50.4 | 84.8 | 39.6 | 45.6 |
| 0320 | Concrete Reinforcing | 102.8 | 49.9 | 75.5 | 105.5 | 41.0 | 72.3 | 102.9 | 60.0 | 80.8 | 96.7 | 60.1 | 77.8 | 96.7 | 60.2 | 77.9 | 100.0 | 41.1 | 69.7 |
| 0330 | Cast-in-Place Concrete | 100.9 | 44.9 | 79.5 | 100.9 | 46.2 | 80.0 | 87.1 | 44.5 | 70.8 | 100.1 | 45.1 | 79.1 | 92.6 | 45.5 | 74.6 | 100.9 | 46.2 | 80.0 |
| 03 | CONCRETE | 100.4 | 44.8 | 73.8 | 100.3 | 43.6 | 73.2 | 93.5 | 47.1 | 71.3 | 99.6 | 46.5 | 74.2 | 94.2 | 48.7 | 72.4 | 99.4 | 43.6 | 72.7 |
| 04 | MASONRY | 99.8 | 55.7 | 73.1 | 109.0 | 57.9 | 78.1 | 107.1 | 56.0 | 76.1 | 109.0 | 57.1 | 77.5 | 104.6 | 57.8 | 76.2 | 136.1 | 60.2 | 90.1 |
| 05 | METALS | 101.4 | 66.3 | 89.4 | 101.5 | 66.1 | 89.3 | 102.4 | 75.0 | 93.0 | 104.3 | 75.5 | 94.4 | 104.5 | 75.8 | 94.6 | 101.4 | 66.7 | 89.5 |
| 06 | WOOD, PLASTICS & COMPOSITES | 96.2 | 39.5 | 63.7 | 86.9 | 39.2 | 59.6 | 95.1 | 39.8 | 63.4 | 100.3 | 36.0 | 63.4 | 95.9 | 42.5 | 65.3 | 84.0 | 38.9 | 58.1 |
| 07 | THERMAL & MOISTURE PROTECTION | 105.9 | 48.7 | 83.2 | 106.0 | 51.1 | 84.2 | 105.9 | 49.0 | 83.3 | 106.4 | 50.6 | 84.2 | 108.2 | 52.0 | 85.9 | 105.8 | 51.6 | 84.3 |
| 08 | OPENINGS | 91.2 | 40.3 | 78.3 | 94.1 | 39.3 | 80.2 | 95.8 | 45.8 | 83.2 | 96.8 | 43.8 | 83.4 | 97.1 | 47.3 | 84.5 | 91.2 | 39.4 | 78.1 |
| 0920 | Plaster & Gypsum Board | 103.8 | 38.0 | 57.5 | 97.9 | 37.7 | 55.5 | 95.9 | 38.3 | 55.3 | 104.6 | 34.4 | 55.2 | 102.5 | 41.1 | 59.3 | 96.2 | 37.4 | 54.8 |
| 0950, 0980 | Ceilings & Acoustic Treatment | 105.3 | 38.0 | 60.7 | 108.1 | 37.7 | 61.4 | 104.3 | 38.3 | 60.5 | 109.9 | 34.4 | 59.9 | 107.1 | 41.1 | 63.4 | 105.3 | 37.4 | 60.3 |
| 0960 | Flooring | 118.1 | 53.2 | 99.4 | 112.4 | 53.2 | 95.4 | 117.8 | 37.9 | 94.8 | 117.8 | 76.0 | 105.8 | 117.8 | 78.9 | 106.6 | 110.7 | 53.2 | 94.2 |
| 0970, 0990 | Wall Finishes & Painting/Coating | 98.4 | 43.7 | 65.2 | 98.4 | 44.8 | 65.9 | 98.4 | 48.0 | 67.8 | 98.4 | 48.0 | 67.8 | 98.4 | 48.0 | 67.8 | 98.4 | 40.0 | 62.9 |
| 09 | FINISHES | 107.2 | 42.5 | 71.5 | 105.4 | 42.5 | 70.7 | 105.8 | 40.2 | 69.6 | 108.3 | 45.8 | 73.9 | 107.3 | 50.0 | 75.7 | 104.1 | 41.8 | 69.9 |
| COVERS | DIVS. 10 - 14, 25, 28, 41, 43, 44, 46 | 100.0 | 41.9 | 87.9 | 100.0 | 45.2 | 88.6 | 100.0 | 76.9 | 95.2 | 100.0 | 76.8 | 95.2 | 100.0 | 77.3 | 95.3 | 100.0 | 45.2 | 88.6 |
| 21, 22, 23 | FIRE SUPPRESSION, PLUMBING & HVAC | 94.7 | 43.3 | 73.3 | 94.7 | 43.8 | 73.5 | 100.0 | 43.4 | 76.4 | 100.1 | 43.9 | 76.7 | 100.1 | 42.2 | 76.0 | 94.7 | 43.8 | 73.5 |
| 26, 27, 3370 | ELECTRICAL, COMMUNICATIONS & UTIL. | 96.7 | 47.7 | 70.7 | 98.4 | 48.7 | 72.0 | 98.4 | 53.2 | 74.4 | 94.8 | 53.2 | 72.7 | 100.7 | 64.5 | 81.4 | 95.8 | 48.7 | 70.7 |
| MF2010 | WEIGHTED AVERAGE | 98.3 | 51.6 | 77.4 | 99.1 | 52.0 | 78.0 | 99.7 | 54.5 | 79.5 | 101.0 | 55.3 | 80.6 | 100.6 | 57.9 | 81.5 | 99.7 | 52.2 | 78.4 |

## TENNESSEE

| DIVISION | | CHATTANOOGA 373 - 374 MAT. | INST. | TOTAL | COLUMBIA 384 MAT. | INST. | TOTAL | COOKEVILLE 385 MAT. | INST. | TOTAL | JACKSON 383 MAT. | INST. | TOTAL | JOHNSON CITY 376 MAT. | INST. | TOTAL | KNOXVILLE 377 - 379 MAT. | INST. | TOTAL |
|---|---|---|---|---|---|---|---|---|---|---|---|---|---|---|---|---|---|---|---|
| 015433 | CONTRACTOR EQUIPMENT | | 103.8 | 103.8 | | 97.8 | 97.8 | | 97.8 | 97.8 | | 104.0 | 104.0 | | 97.7 | 97.7 | | 97.7 | 97.7 |
| 0241, 31 - 34 | SITE & INFRASTRUCTURE, DEMOLITION | 101.5 | 98.4 | 99.3 | 89.5 | 86.9 | 87.7 | 95.5 | 86.3 | 89.2 | 98.0 | 96.9 | 97.2 | 110.0 | 87.1 | 94.2 | 87.7 | 87.6 | 87.6 |
| 0310 | Concrete Forming & Accessories | 97.2 | 58.2 | 63.4 | 83.5 | 62.3 | 65.1 | 83.7 | 38.0 | 44.0 | 90.2 | 48.3 | 53.8 | 84.5 | 41.5 | 47.1 | 96.0 | 44.6 | 51.4 |
| 0320 | Concrete Reinforcing | 83.5 | 63.1 | 73.0 | 86.6 | 63.1 | 74.5 | 86.6 | 63.0 | 74.5 | 86.6 | 59.9 | 72.8 | 84.1 | 61.9 | 72.7 | 83.5 | 62.0 | 72.4 |
| 0330 | Cast-in-Place Concrete | 97.6 | 63.3 | 84.5 | 92.3 | 54.0 | 77.7 | 104.6 | 45.6 | 82.1 | 102.2 | 52.2 | 83.0 | 78.5 | 64.0 | 72.9 | 91.7 | 50.8 | 76.0 |
| 03 | CONCRETE | 89.4 | 62.6 | 76.5 | 90.3 | 61.2 | 76.4 | 99.8 | 47.5 | 74.8 | 92.3 | 53.7 | 73.8 | 93.2 | 55.1 | 75.0 | 86.5 | 52.0 | 70.0 |
| 04 | MASONRY | 98.3 | 52.8 | 70.7 | 110.6 | 57.8 | 78.6 | 105.2 | 47.2 | 70.1 | 110.9 | 50.0 | 74.0 | 109.8 | 47.6 | 72.1 | 75.4 | 55.5 | 63.4 |
| 05 | METALS | 96.4 | 87.7 | 93.4 | 90.7 | 88.2 | 89.8 | 90.8 | 87.2 | 89.5 | 92.9 | 86.2 | 90.6 | 93.7 | 86.6 | 91.2 | 97.1 | 86.8 | 93.5 |
| 06 | WOOD, PLASTICS & COMPOSITES | 109.3 | 60.0 | 81.0 | 74.1 | 64.2 | 68.4 | 74.3 | 36.4 | 52.6 | 90.7 | 49.2 | 66.9 | 81.4 | 40.0 | 57.6 | 95.4 | 40.0 | 63.6 |
| 07 | THERMAL & MOISTURE PROTECTION | 93.7 | 60.5 | 80.5 | 89.2 | 60.0 | 77.6 | 89.6 | 49.0 | 73.5 | 91.4 | 52.0 | 75.7 | 89.3 | 56.4 | 76.2 | 87.5 | 60.0 | 76.6 |
| 08 | OPENINGS | 101.6 | 58.3 | 90.7 | 92.2 | 58.6 | 83.7 | 92.2 | 44.7 | 80.1 | 101.4 | 53.7 | 89.4 | 97.1 | 47.8 | 84.6 | 94.0 | 47.9 | 82.3 |
| 0920 | Plaster & Gypsum Board | 78.4 | 59.4 | 65.0 | 83.3 | 63.6 | 69.4 | 83.3 | 35.0 | 49.3 | 85.8 | 48.1 | 59.3 | 96.2 | 38.7 | 55.7 | 104.4 | 38.7 | 58.1 |
| 0950, 0980 | Ceilings & Acoustic Treatment | 97.4 | 59.4 | 72.2 | 84.1 | 63.6 | 70.5 | 84.1 | 35.0 | 51.5 | 93.3 | 48.1 | 63.4 | 93.3 | 38.7 | 57.1 | 94.2 | 38.7 | 57.4 |
| 0960 | Flooring | 100.0 | 59.8 | 88.5 | 94.0 | 21.7 | 73.2 | 94.1 | 61.9 | 84.9 | 93.0 | 43.1 | 78.7 | 96.2 | 42.9 | 80.9 | 101.2 | 57.7 | 88.7 |
| 0970, 0990 | Wall Finishes & Painting/Coating | 103.1 | 56.8 | 75.0 | 94.6 | 35.9 | 59.0 | 94.6 | 39.3 | 61.0 | 95.9 | 53.4 | 70.1 | 100.6 | 45.3 | 67.1 | 100.6 | 52.3 | 71.3 |
| 09 | FINISHES | 96.4 | 58.5 | 75.5 | 96.0 | 52.5 | 72.0 | 96.6 | 41.3 | 66.1 | 95.9 | 47.1 | 69.0 | 99.8 | 41.0 | 67.4 | 92.6 | 46.7 | 67.3 |
| COVERS | DIVS. 10 - 14, 25, 28, 41, 43, 44, 46 | 100.0 | 45.8 | 88.7 | 100.0 | 52.4 | 90.1 | 100.0 | 45.6 | 88.7 | 100.0 | 47.9 | 89.2 | 100.0 | 48.1 | 89.2 | 100.0 | 50.9 | 89.8 |
| 21, 22, 23 | FIRE SUPPRESSION, PLUMBING & HVAC | 100.0 | 63.1 | 84.7 | 97.1 | 76.1 | 88.3 | 97.1 | 70.7 | 86.1 | 99.9 | 70.5 | 87.7 | 99.8 | 57.6 | 82.2 | 99.8 | 66.5 | 85.9 |
| 26, 27, 3370 | ELECTRICAL, COMMUNICATIONS & UTIL. | 103.3 | 70.8 | 86.0 | 94.8 | 62.3 | 77.5 | 96.6 | 62.3 | 78.4 | 101.7 | 63.5 | 81.4 | 93.0 | 50.3 | 70.3 | 99.3 | 58.7 | 77.7 |
| MF2010 | WEIGHTED AVERAGE | 98.2 | 66.9 | 84.2 | 94.9 | 67.3 | 82.6 | 96.2 | 60.3 | 80.1 | 98.2 | 63.8 | 82.8 | 97.6 | 57.4 | 79.6 | 94.5 | 61.7 | 79.9 |

## TENNESSEE / TEXAS

| DIVISION | | MCKENZIE 382 MAT. | INST. | TOTAL | MEMPHIS 375,380 - 381 MAT. | INST. | TOTAL | NASHVILLE 370 - 372 MAT. | INST. | TOTAL | ABILENE 795 - 796 MAT. | INST. | TOTAL | AMARILLO 790 - 791 MAT. | INST. | TOTAL | AUSTIN 786 - 787 MAT. | INST. | TOTAL |
|---|---|---|---|---|---|---|---|---|---|---|---|---|---|---|---|---|---|---|---|
| 015433 | CONTRACTOR EQUIPMENT | | 97.8 | 97.8 | | 101.7 | 101.7 | | 104.8 | 104.8 | | 88.5 | 88.5 | | 88.5 | 88.5 | | 87.9 | 87.9 |
| 0241, 31 - 34 | SITE & INFRASTRUCTURE, DEMOLITION | 95.1 | 86.6 | 89.3 | 89.6 | 92.6 | 91.7 | 97.2 | 99.9 | 99.0 | 99.0 | 85.8 | 89.9 | 99.1 | 86.9 | 90.7 | 95.2 | 86.0 | 88.8 |
| 0310 | Concrete Forming & Accessories | 91.3 | 41.0 | 47.6 | 96.2 | 61.5 | 66.1 | 97.4 | 65.2 | 69.5 | 95.6 | 40.9 | 48.2 | 96.2 | 55.2 | 60.7 | 89.7 | 54.4 | 59.1 |
| 0320 | Concrete Reinforcing | 86.7 | 63.8 | 74.9 | 89.7 | 64.5 | 76.7 | 89.7 | 63.5 | 76.2 | 97.5 | 52.6 | 74.3 | 97.5 | 50.8 | 73.4 | 97.7 | 48.6 | 72.4 |
| 0330 | Cast-in-Place Concrete | 102.4 | 61.0 | 86.6 | 88.3 | 63.7 | 78.9 | 89.2 | 67.9 | 81.1 | 97.5 | 49.1 | 79.0 | 96.6 | 56.1 | 81.1 | 84.0 | 49.7 | 70.8 |
| 03 | CONCRETE | 98.4 | 54.3 | 77.3 | 85.9 | 64.4 | 75.6 | 86.5 | 67.2 | 77.3 | 93.7 | 47.1 | 71.4 | 93.3 | 55.5 | 75.2 | 82.4 | 52.4 | 68.0 |
| 04 | MASONRY | 109.2 | 49.9 | 73.2 | 91.1 | 65.1 | 75.3 | 80.9 | 60.5 | 68.5 | 102.7 | 58.8 | 76.1 | 102.7 | 59.6 | 76.5 | 96.1 | 48.6 | 67.3 |
| 05 | METALS | 90.8 | 88.5 | 90.0 | 98.1 | 90.0 | 95.3 | 98.1 | 89.2 | 95.0 | 99.9 | 69.0 | 89.3 | 99.9 | 68.8 | 89.2 | 95.0 | 64.2 | 84.4 |
| 06 | WOOD, PLASTICS & COMPOSITES | 83.2 | 39.2 | 58.0 | 100.8 | 62.5 | 78.8 | 106.1 | 66.4 | 83.3 | 99.7 | 39.4 | 65.1 | 98.1 | 57.7 | 75.0 | 94.1 | 57.8 | 73.3 |
| 07 | THERMAL & MOISTURE PROTECTION | 89.6 | 50.2 | 74.0 | 88.4 | 65.6 | 79.3 | 92.2 | 63.3 | 80.7 | 96.9 | 51.6 | 78.9 | 97.0 | 54.8 | 80.2 | 87.2 | 51.6 | 73.0 |
| 08 | OPENINGS | 92.2 | 46.7 | 80.7 | 100.7 | 62.9 | 91.1 | 99.5 | 65.2 | 90.8 | 93.1 | 42.9 | 80.4 | 93.1 | 54.8 | 83.4 | 98.4 | 55.0 | 87.4 |
| 0920 | Plaster & Gypsum Board | 87.3 | 37.9 | 52.5 | 96.2 | 61.8 | 72.0 | 90.8 | 65.8 | 73.2 | 81.4 | 38.2 | 50.9 | 82.9 | 57.0 | 64.7 | 81.3 | 57.1 | 64.2 |
| 0950, 0980 | Ceilings & Acoustic Treatment | 84.1 | 37.9 | 53.5 | 95.5 | 61.8 | 73.1 | 92.3 | 65.8 | 74.7 | 90.3 | 38.2 | 55.7 | 90.3 | 57.0 | 68.2 | 84.3 | 57.1 | 66.2 |
| 0960 | Flooring | 97.3 | 41.6 | 81.3 | 96.5 | 38.1 | 79.7 | 102.3 | 67.9 | 92.4 | 109.0 | 77.9 | 100.1 | 108.8 | 77.9 | 99.9 | 90.8 | 41.8 | 76.7 |
| 0970, 0990 | Wall Finishes & Painting/Coating | 94.6 | 53.4 | 69.6 | 96.7 | 56.4 | 72.3 | 102.6 | 70.4 | 83.0 | 95.7 | 52.9 | 69.7 | 95.7 | 52.2 | 69.3 | 93.0 | 43.0 | 62.7 |
| 09 | FINISHES | 97.9 | 41.1 | 66.6 | 96.6 | 56.1 | 74.3 | 98.4 | 66.1 | 80.6 | 93.2 | 48.3 | 68.5 | 93.5 | 59.2 | 74.6 | 85.4 | 50.8 | 66.3 |
| COVERS | DIVS. 10 - 14, 25, 28, 41, 43, 44, 46 | 100.0 | 30.3 | 85.5 | 100.0 | 80.8 | 96.0 | 100.0 | 81.6 | 96.2 | 100.0 | 75.7 | 94.9 | 100.0 | 73.9 | 94.6 | 100.0 | 77.6 | 95.3 |
| 21, 22, 23 | FIRE SUPPRESSION, PLUMBING & HVAC | 97.1 | 70.8 | 86.2 | 99.9 | 71.2 | 87.9 | 99.8 | 75.9 | 89.8 | 100.2 | 46.2 | 77.7 | 100.0 | 52.3 | 80.2 | 100.1 | 56.5 | 82.0 |
| 26, 27, 3370 | ELECTRICAL, COMMUNICATIONS & UTIL. | 96.4 | 68.3 | 81.4 | 100.3 | 66.4 | 82.3 | 100.9 | 62.3 | 80.9 | 98.8 | 51.6 | 73.7 | 101.5 | 68.5 | 84.0 | 95.1 | 65.9 | 82.2 |
| MF2010 | WEIGHTED AVERAGE | 96.3 | 62.1 | 81.1 | 96.6 | 70.3 | 84.9 | 96.7 | 72.7 | 85.9 | 97.9 | 54.8 | 78.6 | 98.1 | 61.9 | 81.9 | 95.1 | 59.4 | 79.1 |

# City Cost Indexes

## TEXAS

| DIVISION | | BEAUMONT 776-777 | | | BROWNWOOD 768 | | | BRYAN 778 | | | CHILDRESS 792 | | | CORPUS CHRISTI 783-784 | | | DALLAS 752-753 | | |
|---|---|---|---|---|---|---|---|---|---|---|---|---|---|---|---|---|---|---|---|
| | | MAT. | INST. | TOTAL | MAT. | INST. | TOTAL | MAT. | INST. | TOTAL | MAT. | INST. | TOTAL | MAT. | INST. | TOTAL | MAT. | INST. | TOTAL |
| 015433 | CONTRACTOR EQUIPMENT | | 89.5 | 89.5 | | 88.5 | 88.5 | | 89.5 | 89.5 | | 88.5 | 88.5 | | 95.4 | 95.4 | | 98.1 | 98.1 |
| 0241, 31 - 34 | SITE & INFRASTRUCTURE, DEMOLITION | 93.2 | 86.0 | 88.2 | 105.6 | 85.2 | 91.5 | 84.2 | 87.6 | 86.6 | 110.7 | 85.7 | 93.4 | 131.5 | 81.7 | 97.1 | 116.2 | 87.3 | 96.2 |
| 0310 | Concrete Forming & Accessories | 102.8 | 51.1 | 57.9 | 94.1 | 34.7 | 42.6 | 81.7 | 42.3 | 47.5 | 94.1 | 62.8 | 66.9 | 96.2 | 38.7 | 46.3 | 94.8 | 65.1 | 69.1 |
| 0320 | Concrete Reinforcing | 97.4 | 43.3 | 69.5 | 97.8 | 49.8 | 73.1 | 99.9 | 48.7 | 73.5 | 97.6 | 50.0 | 71.8 | 97.0 | 48.0 | 71.8 | 99.1 | 52.2 | 75.0 |
| 0330 | Cast-in-Place Concrete | 92.9 | 53.8 | 78.0 | 105.1 | 44.1 | 81.8 | 74.4 | 67.2 | 71.7 | 100.0 | 50.7 | 81.1 | 93.1 | 46.4 | 75.3 | 101.5 | 57.0 | 84.4 |
| 03 | CONCRETE | 95.2 | 51.5 | 74.3 | 101.0 | 41.9 | 72.7 | 79.7 | 53.2 | 67.0 | 101.8 | 56.7 | 80.2 | 89.5 | 45.0 | 68.2 | 97.5 | 61.2 | 80.1 |
| 04 | MASONRY | 106.7 | 59.8 | 78.3 | 132.1 | 51.3 | 83.1 | 144.8 | 58.4 | 92.5 | 107.5 | 52.6 | 74.2 | 85.1 | 50.3 | 64.0 | 103.3 | 59.6 | 76.8 |
| 05 | METALS | 98.0 | 65.4 | 86.8 | 97.5 | 62.4 | 85.4 | 97.5 | 68.0 | 87.3 | 97.3 | 64.4 | 86.0 | 94.5 | 75.4 | 88.0 | 100.9 | 80.0 | 93.7 |
| 06 | WOOD, PLASTICS & COMPOSITES | 112.5 | 51.3 | 77.4 | 98.3 | 30.8 | 59.6 | 78.5 | 35.5 | 53.9 | 99.1 | 68.0 | 81.3 | 113.6 | 38.0 | 70.2 | 98.4 | 68.1 | 81.0 |
| 07 | THERMAL & MOISTURE PROTECTION | 103.9 | 57.6 | 85.5 | 97.1 | 46.1 | 76.8 | 96.4 | 57.6 | 81.0 | 97.5 | 50.8 | 79.0 | 96.2 | 46.4 | 76.4 | 90.2 | 64.4 | 80.0 |
| 08 | OPENINGS | 96.5 | 47.8 | 84.2 | 89.0 | 38.2 | 76.1 | 97.9 | 43.6 | 84.1 | 90.1 | 57.0 | 81.7 | 105.5 | 39.3 | 88.7 | 101.9 | 59.0 | 91.0 |
| 0920 | Plaster & Gypsum Board | 92.1 | 50.4 | 62.7 | 80.3 | 29.2 | 44.4 | 80.0 | 34.2 | 47.8 | 80.3 | 67.6 | 71.4 | 86.0 | 36.6 | 51.2 | 95.1 | 67.6 | 75.7 |
| 0950, 0980 | Ceilings & Acoustic Treatment | 102.1 | 50.4 | 67.8 | 86.6 | 29.2 | 48.6 | 95.2 | 34.2 | 54.8 | 86.6 | 67.6 | 74.0 | 84.3 | 36.6 | 52.7 | 93.1 | 67.6 | 76.2 |
| 0960 | Flooring | 107.2 | 75.3 | 98.0 | 107.9 | 47.2 | 90.5 | 80.0 | 55.7 | 73.0 | 106.9 | 43.2 | 88.6 | 103.3 | 66.6 | 92.8 | 99.8 | 50.4 | 85.6 |
| 0970, 0990 | Wall Finishes & Painting/Coating | 92.3 | 51.2 | 67.4 | 95.7 | 30.8 | 56.3 | 88.1 | 63.8 | 73.3 | 95.7 | 40.7 | 62.3 | 105.3 | 60.8 | 78.3 | 103.8 | 54.3 | 73.7 |
| 09 | FINISHES | 90.2 | 55.3 | 70.9 | 92.6 | 35.8 | 61.3 | 78.4 | 45.3 | 60.1 | 93.0 | 57.8 | 73.6 | 93.2 | 45.7 | 67.0 | 99.0 | 61.6 | 78.4 |
| COVERS | DIVS. 10 - 14, 25, 28, 41, 43, 44, 46 | 100.0 | 79.2 | 95.7 | 100.0 | 40.4 | 87.6 | 100.0 | 78.4 | 95.5 | 100.0 | 76.1 | 95.0 | 100.0 | 76.5 | 95.1 | 100.0 | 81.8 | 96.2 |
| 21, 22, 23 | FIRE SUPPRESSION, PLUMBING & HVAC | 100.0 | 60.6 | 83.6 | 94.8 | 49.5 | 75.9 | 94.6 | 67.2 | 83.2 | 94.8 | 53.9 | 77.8 | 100.0 | 41.0 | 75.5 | 99.9 | 64.5 | 85.1 |
| 26, 27, 3370 | ELECTRICAL, COMMUNICATIONS & UTIL. | 97.7 | 74.2 | 85.2 | 94.2 | 46.4 | 68.8 | 95.2 | 67.0 | 80.2 | 98.7 | 46.0 | 70.7 | 95.0 | 55.9 | 74.2 | 97.4 | 69.1 | 82.3 |
| MF2010 | WEIGHTED AVERAGE | 98.2 | 62.9 | 82.4 | 98.0 | 49.6 | 76.3 | 95.1 | 62.1 | 80.3 | 97.5 | 58.1 | 79.9 | 97.6 | 52.8 | 77.6 | 99.9 | 67.5 | 85.4 |

## TEXAS

| DIVISION | | DEL RIO 788 | | | DENTON 762 | | | EASTLAND 764 | | | EL PASO 798 - 799,885 | | | FORT WORTH 760 - 761 | | | GALVESTON 775 | | |
|---|---|---|---|---|---|---|---|---|---|---|---|---|---|---|---|---|---|---|---|
| | | MAT. | INST. | TOTAL | MAT. | INST. | TOTAL | MAT. | INST. | TOTAL | MAT. | INST. | TOTAL | MAT. | INST. | TOTAL | MAT. | INST. | TOTAL |
| 015433 | CONTRACTOR EQUIPMENT | | 87.9 | 87.9 | | 94.7 | 94.7 | | 88.5 | 88.5 | | 88.5 | 88.5 | | 88.5 | 88.5 | | 98.8 | 98.8 |
| 0241, 31 - 34 | SITE & INFRASTRUCTURE, DEMOLITION | 115.1 | 84.5 | 94.0 | 105.7 | 78.4 | 86.8 | 108.6 | 85.2 | 92.4 | 96.3 | 85.2 | 88.6 | 101.0 | 86.2 | 90.8 | 116.8 | 85.5 | 95.2 |
| 0310 | Concrete Forming & Accessories | 90.8 | 38.4 | 45.3 | 99.5 | 34.1 | 42.8 | 94.7 | 37.7 | 45.3 | 94.9 | 45.8 | 52.3 | 91.8 | 64.5 | 68.1 | 90.6 | 63.1 | 66.8 |
| 0320 | Concrete Reinforcing | 97.7 | 45.0 | 70.5 | 99.2 | 51.1 | 74.4 | 98.0 | 49.8 | 73.2 | 97.5 | 47.9 | 71.9 | 97.5 | 52.1 | 74.1 | 99.3 | 65.0 | 81.6 |
| 0330 | Cast-in-Place Concrete | 100.2 | 43.9 | 78.7 | 81.3 | 46.0 | 67.8 | 111.4 | 44.8 | 85.7 | 91.1 | 42.0 | 72.3 | 100.3 | 52.7 | 82.1 | 98.8 | 68.3 | 87.1 |
| 03 | CONCRETE | 105.3 | 42.5 | 75.3 | 79.8 | 43.4 | 62.4 | 105.6 | 43.5 | 75.9 | 90.4 | 45.8 | 69.1 | 94.9 | 58.7 | 77.6 | 96.8 | 66.8 | 82.4 |
| 04 | MASONRY | 100.0 | 51.3 | 71.5 | 140.7 | 49.7 | 85.5 | 101.2 | 51.1 | 70.9 | 100.4 | 48.9 | 69.2 | 96.8 | 59.6 | 74.2 | 101.4 | 58.5 | 75.4 |
| 05 | METALS | 93.4 | 59.3 | 81.7 | 97.1 | 76.0 | 89.9 | 97.3 | 62.8 | 85.4 | 99.7 | 63.3 | 87.2 | 96.9 | 68.4 | 87.1 | 99.0 | 88.0 | 95.2 |
| 06 | WOOD, PLASTICS & COMPOSITES | 93.5 | 36.5 | 60.8 | 108.1 | 33.7 | 65.4 | 105.2 | 34.6 | 64.7 | 99.2 | 48.5 | 70.1 | 100.7 | 68.0 | 81.9 | 95.8 | 63.1 | 77.1 |
| 07 | THERMAL & MOISTURE PROTECTION | 93.0 | 45.6 | 74.1 | 95.5 | 45.9 | 75.8 | 97.4 | 45.3 | 76.7 | 93.1 | 54.4 | 77.7 | 97.8 | 54.8 | 80.7 | 95.7 | 66.2 | 83.9 |
| 08 | OPENINGS | 99.2 | 34.8 | 82.9 | 103.5 | 39.0 | 87.2 | 78.0 | 40.3 | 68.5 | 93.1 | 44.5 | 80.8 | 87.9 | 58.9 | 80.6 | 102.8 | 62.9 | 92.7 |
| 0920 | Plaster & Gypsum Board | 80.9 | 35.2 | 48.7 | 83.8 | 32.1 | 47.4 | 80.3 | 33.2 | 47.1 | 85.9 | 47.4 | 58.8 | 83.9 | 67.6 | 72.4 | 87.0 | 62.5 | 69.8 |
| 0950, 0980 | Ceilings & Acoustic Treatment | 80.5 | 35.2 | 50.5 | 91.3 | 32.1 | 52.1 | 86.6 | 33.2 | 51.2 | 90.3 | 47.4 | 61.9 | 92.2 | 67.6 | 75.9 | 98.0 | 62.5 | 74.5 |
| 0960 | Flooring | 87.9 | 36.0 | 73.0 | 100.5 | 45.2 | 84.6 | 140.7 | 47.2 | 113.9 | 109.0 | 65.9 | 96.6 | 142.0 | 43.0 | 113.6 | 93.8 | 55.7 | 82.8 |
| 0970, 0990 | Wall Finishes & Painting/Coating | 93.0 | 27.1 | 53.0 | 105.5 | 31.2 | 60.4 | 97.1 | 32.2 | 57.7 | 95.7 | 36.8 | 60.0 | 97.1 | 53.8 | 70.9 | 99.5 | 49.5 | 69.2 |
| 09 | FINISHES | 85.6 | 36.4 | 58.5 | 89.1 | 35.2 | 59.4 | 102.9 | 38.2 | 67.2 | 93.4 | 48.8 | 68.8 | 104.2 | 60.0 | 79.8 | 88.3 | 59.9 | 72.7 |
| COVERS | DIVS. 10 - 14, 25, 28, 41, 43, 44, 46 | 100.0 | 40.1 | 87.6 | 100.0 | 38.9 | 87.3 | 100.0 | 41.6 | 87.9 | 100.0 | 77.4 | 95.3 | 100.0 | 81.4 | 96.1 | 100.0 | 83.4 | 96.6 |
| 21, 22, 23 | FIRE SUPPRESSION, PLUMBING & HVAC | 94.6 | 47.3 | 74.9 | 94.8 | 40.7 | 72.3 | 94.8 | 49.8 | 76.1 | 100.0 | 36.2 | 73.5 | 100.0 | 58.9 | 82.9 | 94.6 | 67.2 | 83.2 |
| 26, 27, 3370 | ELECTRICAL, COMMUNICATIONS & UTIL. | 96.8 | 31.9 | 62.3 | 96.6 | 46.5 | 70.0 | 94.1 | 46.4 | 68.7 | 97.7 | 54.4 | 74.7 | 97.7 | 63.8 | 79.7 | 97.1 | 68.7 | 82.0 |
| MF2010 | WEIGHTED AVERAGE | 97.0 | 47.0 | 74.6 | 97.3 | 48.4 | 75.4 | 96.8 | 50.3 | 76.0 | 97.0 | 51.7 | 76.8 | 97.7 | 63.6 | 82.4 | 97.6 | 69.2 | 84.9 |

## TEXAS

| DIVISION | | GIDDINGS 789 | | | GREENVILLE 754 | | | HOUSTON 770 - 772 | | | HUNTSVILLE 773 | | | LAREDO 780 | | | LONGVIEW 756 | | |
|---|---|---|---|---|---|---|---|---|---|---|---|---|---|---|---|---|---|---|---|
| | | MAT. | INST. | TOTAL | MAT. | INST. | TOTAL | MAT. | INST. | TOTAL | MAT. | INST. | TOTAL | MAT. | INST. | TOTAL | MAT. | INST. | TOTAL |
| 015433 | CONTRACTOR EQUIPMENT | | 87.9 | 87.9 | | 95.7 | 95.7 | | 98.6 | 98.6 | | 89.5 | 89.5 | | 87.9 | 87.9 | | 90.3 | 90.3 |
| 0241, 31 - 34 | SITE & INFRASTRUCTURE, DEMOLITION | 101.3 | 84.8 | 89.9 | 108.6 | 82.1 | 90.3 | 114.5 | 85.2 | 94.3 | 101.9 | 85.6 | 90.7 | 95.1 | 85.7 | 88.6 | 103.6 | 89.4 | 93.8 |
| 0310 | Concrete Forming & Accessories | 89.0 | 58.8 | 62.8 | 87.2 | 28.6 | 36.4 | 92.5 | 63.3 | 67.1 | 88.3 | 58.3 | 62.3 | 90.8 | 38.9 | 45.8 | 84.3 | 29.5 | 36.8 |
| 0320 | Concrete Reinforcing | 98.3 | 46.9 | 71.8 | 99.6 | 26.0 | 61.7 | 99.1 | 65.5 | 81.8 | 100.2 | 49.0 | 73.8 | 97.7 | 47.9 | 72.0 | 98.6 | 23.7 | 60.0 |
| 0330 | Cast-in-Place Concrete | 85.0 | 44.4 | 69.4 | 92.4 | 38.2 | 71.7 | 95.8 | 66.3 | 85.3 | 102.4 | 49.1 | 82.0 | 71.8 | 60.0 | 67.2 | 107.4 | 35.8 | 80.0 |
| 03 | CONCRETE | 85.8 | 52.2 | 69.7 | 90.1 | 33.5 | 63.0 | 94.5 | 66.9 | 81.3 | 102.8 | 54.1 | 79.5 | 80.9 | 48.9 | 65.6 | 105.6 | 31.9 | 70.3 |
| 04 | MASONRY | 108.1 | 53.1 | 74.7 | 164.1 | 48.4 | 94.0 | 101.2 | 68.7 | 81.5 | 144.2 | 44.8 | 83.9 | 93.8 | 48.5 | 66.3 | 159.5 | 44.6 | 89.9 |
| 05 | METALS | 92.9 | 61.7 | 82.1 | 98.1 | 63.5 | 86.2 | 102.0 | 88.9 | 97.5 | 97.4 | 63.9 | 85.9 | 96.0 | 63.5 | 84.8 | 91.8 | 50.6 | 77.6 |
| 06 | WOOD, PLASTICS & COMPOSITES | 92.5 | 63.1 | 75.6 | 89.7 | 27.7 | 54.2 | 98.1 | 63.1 | 78.0 | 87.0 | 61.7 | 72.5 | 93.5 | 37.9 | 61.6 | 83.9 | 30.0 | 53.0 |
| 07 | THERMAL & MOISTURE PROTECTION | 93.6 | 49.9 | 76.2 | 90.3 | 41.1 | 70.7 | 95.3 | 68.5 | 84.6 | 97.4 | 47.9 | 77.7 | 92.0 | 49.4 | 75.1 | 91.5 | 36.4 | 69.6 |
| 08 | OPENINGS | 98.4 | 51.2 | 86.5 | 100.7 | 27.3 | 82.1 | 105.0 | 62.9 | 94.3 | 97.8 | 50.3 | 85.8 | 99.0 | 40.0 | 84.1 | 91.6 | 27.3 | 75.4 |
| 0920 | Plaster & Gypsum Board | 79.6 | 62.5 | 67.6 | 89.6 | 26.1 | 44.9 | 89.6 | 62.5 | 70.5 | 84.4 | 61.1 | 68.0 | 82.0 | 36.6 | 50.0 | 87.3 | 28.5 | 45.9 |
| 0950, 0980 | Ceilings & Acoustic Treatment | 80.5 | 62.5 | 68.6 | 89.4 | 26.1 | 47.4 | 102.7 | 62.5 | 76.1 | 95.2 | 61.1 | 72.6 | 84.3 | 36.6 | 52.7 | 85.6 | 28.5 | 47.8 |
| 0960 | Flooring | 88.0 | 36.0 | 73.0 | 95.7 | 49.6 | 82.5 | 95.0 | 55.7 | 83.7 | 83.5 | 36.0 | 69.9 | 87.8 | 66.6 | 81.7 | 101.1 | 36.4 | 82.5 |
| 0970, 0990 | Wall Finishes & Painting/Coating | 93.0 | 32.2 | 56.1 | 103.8 | 26.5 | 56.9 | 99.5 | 63.8 | 77.8 | 88.1 | 33.0 | 54.7 | 93.0 | 57.3 | 71.3 | 95.6 | 30.9 | 56.3 |
| 09 | FINISHES | 84.4 | 52.9 | 67.0 | 95.7 | 31.9 | 60.5 | 95.5 | 61.5 | 76.8 | 81.1 | 52.6 | 65.4 | 84.8 | 45.2 | 63.0 | 101.2 | 30.9 | 62.4 |
| COVERS | DIVS. 10 - 14, 25, 28, 41, 43, 44, 46 | 100.0 | 66.4 | 88.9 | 100.0 | 37.1 | 86.9 | 100.0 | 83.4 | 96.6 | 100.0 | 47.0 | 89.0 | 100.0 | 74.3 | 94.7 | 100.0 | 23.4 | 84.1 |
| 21, 22, 23 | FIRE SUPPRESSION, PLUMBING & HVAC | 94.6 | 61.5 | 80.9 | 94.6 | 46.7 | 74.6 | 99.9 | 67.4 | 86.4 | 94.6 | 62.4 | 81.2 | 100.0 | 39.7 | 74.9 | 94.5 | 29.5 | 67.4 |
| 26, 27, 3370 | ELECTRICAL, COMMUNICATIONS & UTIL. | 93.3 | 46.0 | 68.2 | 93.7 | 28.7 | 59.1 | 99.4 | 67.0 | 82.2 | 95.2 | 46.0 | 69.0 | 96.9 | 63.6 | 79.2 | 94.5 | 46.6 | 69.0 |
| MF2010 | WEIGHTED AVERAGE | 94.1 | 57.0 | 77.5 | 99.6 | 43.7 | 74.6 | 100.0 | 70.4 | 86.8 | 98.6 | 56.7 | 79.9 | 94.7 | 53.1 | 76.1 | 99.7 | 40.7 | 73.4 |

## TEXAS

| DIVISION | | LUBBOCK 793 - 794 | | | LUFKIN 759 | | | MCALLEN 785 | | | MCKINNEY 750 | | | MIDLAND 797 | | | ODESSA 797 | | |
|---|---|---|---|---|---|---|---|---|---|---|---|---|---|---|---|---|---|---|---|
| | | MAT. | INST. | TOTAL | MAT. | INST. | TOTAL | MAT. | INST. | TOTAL | MAT. | INST. | TOTAL | MAT. | INST. | TOTAL | MAT. | INST. | TOTAL |
| 015433 | CONTRACTOR EQUIPMENT | | 97.2 | 97.2 | | 90.3 | 90.3 | | 95.4 | 95.4 | | 95.7 | 95.7 | | 97.2 | 97.2 | | 88.5 | 88.5 |
| 0241, 31 - 34 | SITE & INFRASTRUCTURE, DEMOLITION | 125.6 | 84.1 | 97.0 | 98.0 | 90.5 | 92.8 | 136.2 | 81.6 | 98.5 | 104.2 | 82.7 | 89.4 | 129.4 | 84.1 | 98.1 | 99.2 | 86.1 | 90.2 |
| 0310 | Concrete Forming & Accessories | 93.6 | 53.6 | 58.9 | 87.3 | 59.9 | 63.5 | 95.6 | 36.3 | 44.1 | 86.4 | 38.3 | 44.7 | 97.0 | 40.3 | 47.8 | 95.6 | 40.2 | 47.5 |
| 0320 | Concrete Reinforcing | 98.7 | 52.6 | 74.9 | 100.2 | 61.7 | 80.4 | 97.2 | 47.9 | 71.8 | 99.6 | 51.1 | 74.6 | 99.7 | 51.9 | 75.1 | 97.5 | 51.8 | 74.0 |
| 0330 | Cast-in-Place Concrete | 97.6 | 49.4 | 79.2 | 96.1 | 45.6 | 76.8 | 101.2 | 44.7 | 79.6 | 86.6 | 46.8 | 71.4 | 103.8 | 46.6 | 81.9 | 97.5 | 45.6 | 77.6 |
| 03 | CONCRETE | 92.5 | 53.6 | 73.9 | 97.7 | 55.9 | 77.7 | 94.9 | 43.3 | 70.2 | 85.5 | 45.4 | 66.3 | 97.0 | 46.7 | 72.9 | 93.7 | 45.5 | 70.6 |
| 04 | MASONRY | 101.9 | 57.1 | 74.7 | 121.2 | 44.8 | 74.9 | 100.1 | 50.3 | 69.9 | 176.8 | 51.0 | 100.6 | 120.1 | 44.6 | 74.3 | 102.7 | 44.5 | 67.4 |
| 05 | METALS | 103.2 | 80.9 | 95.5 | 98.2 | 67.5 | 87.7 | 93.3 | 74.9 | 87.0 | 98.0 | 75.0 | 90.1 | 101.3 | 79.7 | 93.9 | 99.2 | 67.5 | 88.3 |
| 06 | WOOD, PLASTICS & COMPOSITES | 98.9 | 57.9 | 75.3 | 92.1 | 63.0 | 75.4 | 112.2 | 35.5 | 68.2 | 88.4 | 38.5 | 59.7 | 103.3 | 40.0 | 67.0 | 99.7 | 39.9 | 65.4 |
| 07 | THERMAL & MOISTURE PROTECTION | 87.7 | 53.8 | 74.2 | 91.3 | 47.6 | 73.9 | 96.2 | 44.7 | 75.7 | 90.2 | 47.9 | 73.4 | 87.9 | 45.2 | 70.9 | 96.9 | 44.4 | 76.1 |
| 08 | OPENINGS | 103.1 | 51.7 | 90.1 | 74.8 | 55.6 | 69.9 | 103.8 | 38.4 | 87.3 | 100.7 | 41.4 | 85.7 | 102.0 | 41.6 | 86.7 | 93.1 | 41.6 | 80.0 |
| 0920 | Plaster & Gypsum Board | 81.9 | 57.0 | 64.4 | 87.0 | 62.5 | 69.8 | 86.0 | 33.9 | 49.3 | 88.3 | 37.1 | 52.3 | 83.3 | 38.6 | 51.8 | 81.4 | 38.6 | 51.3 |
| 0950, 0980 | Ceilings & Acoustic Treatment | 92.2 | 57.0 | 68.9 | 80.0 | 62.5 | 68.4 | 84.3 | 33.9 | 50.9 | 89.4 | 37.1 | 54.7 | 89.4 | 38.6 | 55.7 | 90.3 | 38.6 | 56.1 |
| 0960 | Flooring | 101.5 | 40.8 | 84.1 | 135.2 | 39.6 | 107.7 | 102.8 | 66.6 | 92.4 | 95.4 | 36.9 | 78.6 | 103.0 | 50.4 | 87.9 | 109.0 | 50.4 | 92.2 |
| 0970, 0990 | Wall Finishes & Painting/Coating | 106.0 | 52.2 | 73.3 | 95.6 | 39.8 | 61.7 | 105.3 | 30.4 | 59.9 | 103.8 | 43.4 | 67.2 | 106.0 | 52.2 | 73.3 | 95.7 | 52.2 | 69.3 |
| 09 | FINISHES | 95.8 | 51.2 | 71.2 | 109.8 | 54.8 | 79.5 | 93.5 | 40.8 | 64.4 | 95.1 | 38.6 | 63.9 | 96.3 | 42.7 | 66.7 | 93.2 | 42.6 | 65.3 |
| COVERS | DIVS. 10 - 14, 25, 28, 41, 43, 44, 46 | 100.0 | 77.5 | 95.3 | 100.0 | 77.4 | 95.3 | 100.0 | 74.2 | 94.6 | 100.0 | 43.0 | 88.2 | 100.0 | 77.1 | 95.2 | 100.0 | 76.8 | 95.2 |
| 21, 22, 23 | FIRE SUPPRESSION, PLUMBING & HVAC | 99.8 | 45.3 | 77.1 | 94.5 | 62.5 | 81.2 | 94.6 | 31.1 | 68.2 | 94.6 | 44.4 | 73.7 | 94.4 | 45.5 | 74.0 | 100.2 | 45.4 | 77.4 |
| 26, 27, 3370 | ELECTRICAL, COMMUNICATIONS & UTIL. | 97.4 | 52.6 | 73.5 | 96.0 | 46.0 | 69.4 | 94.1 | 36.3 | 63.3 | 93.8 | 48.0 | 69.4 | 97.4 | 42.8 | 68.3 | 98.9 | 42.7 | 69.0 |
| MF2010 | WEIGHTED AVERAGE | 99.6 | 57.5 | 80.8 | 96.7 | 59.1 | 79.9 | 97.5 | 46.9 | 74.9 | 99.5 | 50.8 | 77.7 | 99.7 | 51.9 | 78.3 | 97.8 | 50.7 | 76.8 |

## TEXAS

| DIVISION | | PALESTINE 758 | | | SAN ANGELO 769 | | | SAN ANTONIO 781 - 782 | | | TEMPLE 765 | | | TEXARKANA 755 | | | TYLER 757 | | |
|---|---|---|---|---|---|---|---|---|---|---|---|---|---|---|---|---|---|---|---|
| | | MAT. | INST. | TOTAL | MAT. | INST. | TOTAL | MAT. | INST. | TOTAL | MAT. | INST. | TOTAL | MAT. | INST. | TOTAL | MAT. | INST. | TOTAL |
| 015433 | CONTRACTOR EQUIPMENT | | 90.3 | 90.3 | | 88.5 | 88.5 | | 90.4 | 90.4 | | 88.5 | 88.5 | | 90.3 | 90.3 | | 90.3 | 90.3 |
| 0241, 31 - 34 | SITE & INFRASTRUCTURE, DEMOLITION | 105.7 | 90.3 | 95.1 | 101.6 | 86.2 | 91.0 | 94.8 | 90.3 | 91.7 | 89.3 | 84.9 | 86.3 | 91.5 | 90.2 | 90.6 | 103.2 | 90.3 | 94.3 |
| 0310 | Concrete Forming & Accessories | 79.5 | 59.0 | 61.8 | 94.4 | 38.0 | 45.5 | 90.9 | 57.0 | 61.5 | 97.3 | 35.8 | 43.9 | 93.5 | 35.8 | 43.4 | 88.8 | 34.6 | 41.8 |
| 0320 | Concrete Reinforcing | 97.8 | 49.8 | 73.1 | 97.7 | 51.8 | 74.0 | 104.6 | 47.4 | 75.2 | 97.8 | 49.5 | 72.9 | 97.7 | 51.8 | 74.1 | 98.6 | 52.0 | 74.6 |
| 0330 | Cast-in-Place Concrete | 87.8 | 44.1 | 71.1 | 99.2 | 43.7 | 77.9 | 70.4 | 68.1 | 69.5 | 81.3 | 49.3 | 69.1 | 88.5 | 44.9 | 71.8 | 105.5 | 43.8 | 81.9 |
| 03 | CONCRETE | 98.5 | 52.7 | 76.6 | 96.3 | 43.7 | 71.1 | 81.3 | 59.7 | 71.0 | 82.4 | 44.2 | 64.1 | 90.8 | 43.2 | 68.0 | 105.0 | 42.4 | 75.0 |
| 04 | MASONRY | 116.5 | 51.5 | 77.1 | 128.6 | 43.9 | 77.2 | 93.7 | 60.5 | 73.6 | 140.1 | 49.8 | 85.4 | 181.5 | 47.4 | 100.2 | 170.5 | 50.3 | 97.7 |
| 05 | METALS | 98.0 | 62.3 | 85.7 | 97.7 | 64.5 | 86.3 | 96.7 | 66.2 | 86.2 | 97.4 | 63.4 | 85.7 | 91.7 | 66.8 | 83.1 | 97.8 | 67.3 | 87.3 |
| 06 | WOOD, PLASTICS & COMPOSITES | 82.9 | 63.0 | 71.5 | 98.6 | 38.1 | 63.9 | 93.5 | 55.2 | 71.6 | 107.6 | 34.6 | 65.7 | 94.4 | 34.1 | 59.8 | 93.5 | 32.2 | 58.3 |
| 07 | THERMAL & MOISTURE PROTECTION | 91.7 | 48.7 | 74.6 | 96.9 | 44.0 | 75.9 | 91.7 | 64.1 | 80.7 | 96.5 | 45.7 | 76.3 | 91.1 | 49.4 | 74.5 | 91.6 | 53.9 | 76.6 |
| 08 | OPENINGS | 74.7 | 52.6 | 69.1 | 89.0 | 40.6 | 76.8 | 100.9 | 52.9 | 88.8 | 75.9 | 36.9 | 66.0 | 91.6 | 39.2 | 78.3 | 74.7 | 38.2 | 65.4 |
| 0920 | Plaster & Gypsum Board | 82.2 | 62.5 | 68.3 | 80.3 | 36.8 | 49.7 | 82.0 | 54.4 | 62.5 | 80.3 | 33.2 | 47.1 | 91.3 | 32.7 | 50.1 | 87.0 | 30.7 | 47.4 |
| 0950, 0980 | Ceilings & Acoustic Treatment | 80.0 | 62.5 | 68.4 | 86.6 | 36.8 | 53.6 | 84.3 | 54.4 | 64.5 | 86.6 | 33.2 | 51.2 | 85.6 | 32.7 | 50.6 | 80.0 | 30.7 | 47.3 |
| 0960 | Flooring | 126.9 | 61.2 | 108.0 | 108.0 | 36.0 | 87.3 | 87.8 | 66.8 | 81.7 | 143.3 | 43.3 | 114.6 | 108.7 | 61.6 | 95.2 | 136.9 | 38.4 | 108.6 |
| 0970, 0990 | Wall Finishes & Painting/Coating | 95.6 | 32.2 | 57.1 | 95.7 | 52.2 | 69.3 | 93.0 | 57.3 | 71.3 | 97.1 | 35.4 | 59.7 | 95.6 | 41.5 | 62.8 | 95.6 | 41.5 | 62.8 |
| 09 | FINISHES | 107.5 | 57.4 | 79.9 | 92.4 | 38.3 | 62.6 | 84.9 | 58.0 | 70.0 | 102.2 | 36.6 | 66.1 | 103.2 | 40.1 | 68.4 | 110.7 | 34.3 | 68.6 |
| COVERS | DIVS. 10 - 14, 25, 28, 41, 43, 44, 46 | 100.0 | 44.2 | 88.4 | 100.0 | 74.2 | 94.6 | 100.0 | 80.6 | 96.0 | 100.0 | 38.4 | 87.2 | 100.0 | 74.3 | 94.7 | 100.0 | 74.0 | 94.6 |
| 21, 22, 23 | FIRE SUPPRESSION, PLUMBING & HVAC | 94.5 | 59.0 | 79.7 | 94.8 | 44.6 | 73.9 | 99.9 | 66.4 | 86.0 | 94.8 | 35.7 | 70.2 | 94.5 | 35.2 | 69.8 | 94.5 | 59.8 | 80.1 |
| 26, 27, 3370 | ELECTRICAL, COMMUNICATIONS & UTIL. | 91.6 | 46.0 | 67.4 | 98.7 | 48.6 | 72.1 | 97.3 | 63.6 | 79.4 | 95.5 | 68.8 | 81.3 | 95.8 | 64.8 | 79.3 | 94.5 | 55.1 | 73.5 |
| MF2010 | WEIGHTED AVERAGE | 96.1 | 57.4 | 78.8 | 97.6 | 50.1 | 76.3 | 95.0 | 65.1 | 81.6 | 95.3 | 50.0 | 75.1 | 99.1 | 51.5 | 77.8 | 100.2 | 54.9 | 80.0 |

## TEXAS / UTAH

| DIVISION | | VICTORIA 779 | | | WACO 766 - 767 | | | WAXAHACHIE 751 | | | WHARTON 774 | | | WICHITA FALLS 763 | | | LOGAN 843 | | |
|---|---|---|---|---|---|---|---|---|---|---|---|---|---|---|---|---|---|---|---|
| | | MAT. | INST. | TOTAL | MAT. | INST. | TOTAL | MAT. | INST. | TOTAL | MAT. | INST. | TOTAL | MAT. | INST. | TOTAL | MAT. | INST. | TOTAL |
| 015433 | CONTRACTOR EQUIPMENT | | 97.6 | 97.6 | | 88.5 | 88.5 | | 95.7 | 95.7 | | 98.8 | 98.8 | | 88.5 | 88.5 | | 99.8 | 99.8 |
| 0241, 31 - 34 | SITE & INFRASTRUCTURE, DEMOLITION | 122.3 | 80.7 | 93.6 | 97.8 | 86.2 | 89.8 | 106.6 | 83.2 | 90.5 | 129.0 | 83.6 | 97.7 | 98.5 | 85.8 | 89.7 | 85.2 | 99.5 | 95.1 |
| 0310 | Concrete Forming & Accessories | 90.9 | 37.1 | 44.2 | 96.0 | 41.3 | 48.5 | 86.4 | 60.4 | 63.8 | 85.6 | 38.6 | 44.8 | 96.0 | 39.2 | 46.7 | 105.4 | 61.4 | 67.2 |
| 0320 | Concrete Reinforcing | 95.4 | 31.6 | 62.6 | 97.5 | 48.6 | 72.3 | 99.6 | 51.1 | 74.7 | 99.3 | 53.9 | 75.9 | 97.5 | 51.8 | 73.9 | 98.9 | 77.0 | 87.6 |
| 0330 | Cast-in-Place Concrete | 111.0 | 42.2 | 84.7 | 88.0 | 53.9 | 74.9 | 91.4 | 42.6 | 72.7 | 114.1 | 52.7 | 90.6 | 94.0 | 46.2 | 75.7 | 89.2 | 73.9 | 83.4 |
| 03 | CONCRETE | 104.5 | 40.0 | 73.6 | 88.9 | 48.1 | 69.4 | 89.3 | 53.9 | 72.4 | 108.6 | 48.3 | 79.7 | 91.9 | 45.2 | 69.6 | 108.0 | 69.1 | 89.4 |
| 04 | MASONRY | 118.0 | 36.2 | 68.4 | 99.4 | 57.0 | 73.7 | 164.7 | 51.0 | 95.8 | 102.6 | 47.2 | 69.0 | 99.9 | 58.8 | 75.0 | 107.3 | 63.4 | 80.7 |
| 05 | METALS | 97.6 | 71.3 | 88.5 | 99.8 | 66.4 | 88.3 | 98.1 | 77.0 | 90.9 | 99.0 | 78.6 | 92.0 | 99.7 | 69.0 | 89.2 | 100.4 | 77.6 | 92.6 |
| 06 | WOOD, PLASTICS & COMPOSITES | 99.7 | 37.9 | 64.2 | 106.0 | 37.1 | 66.5 | 88.4 | 68.1 | 76.7 | 89.7 | 34.4 | 58.0 | 106.0 | 37.0 | 66.4 | 82.2 | 59.7 | 69.3 |
| 07 | THERMAL & MOISTURE PROTECTION | 99.2 | 41.1 | 76.1 | 97.4 | 49.2 | 78.3 | 90.3 | 48.0 | 73.4 | 96.0 | 51.2 | 78.2 | 97.4 | 52.6 | 79.6 | 99.8 | 65.1 | 86.0 |
| 08 | OPENINGS | 102.6 | 37.1 | 86.0 | 87.9 | 37.9 | 75.3 | 100.7 | 57.4 | 89.8 | 102.8 | 39.0 | 86.7 | 87.9 | 42.1 | 76.3 | 90.0 | 58.7 | 82.1 |
| 0920 | Plaster & Gypsum Board | 85.0 | 36.5 | 50.9 | 81.4 | 35.8 | 49.2 | 88.6 | 67.6 | 73.8 | 82.6 | 32.9 | 47.6 | 81.4 | 35.6 | 49.2 | 85.3 | 58.3 | 66.3 |
| 0950, 0980 | Ceilings & Acoustic Treatment | 99.0 | 36.5 | 57.6 | 90.3 | 35.8 | 54.2 | 90.3 | 67.6 | 75.3 | 98.0 | 32.9 | 54.9 | 90.3 | 35.6 | 54.1 | 119.1 | 58.3 | 78.8 |
| 0960 | Flooring | 92.9 | 39.6 | 77.6 | 142.0 | 46.8 | 114.7 | 95.4 | 39.6 | 79.3 | 91.1 | 47.2 | 78.5 | 142.8 | 77.9 | 124.2 | 101.3 | 50.4 | 86.7 |
| 0970, 0990 | Wall Finishes & Painting/Coating | 99.6 | 39.8 | 63.3 | 97.1 | 35.4 | 59.7 | 103.8 | 39.8 | 65.0 | 99.5 | 33.0 | 59.2 | 99.6 | 52.2 | 70.8 | 98.5 | 63.3 | 77.1 |
| 09 | FINISHES | 86.4 | 37.9 | 59.7 | 103.3 | 40.5 | 68.7 | 95.6 | 55.7 | 73.6 | 87.9 | 38.6 | 60.7 | 103.7 | 46.8 | 72.3 | 102.8 | 58.9 | 78.6 |
| COVERS | DIVS. 10 - 14, 25, 28, 41, 43, 44, 46 | 100.0 | 39.6 | 87.4 | 100.0 | 77.9 | 95.4 | 100.0 | 43.6 | 88.3 | 100.0 | 41.9 | 87.9 | 100.0 | 71.2 | 94.0 | 100.0 | 59.1 | 91.5 |
| 21, 22, 23 | FIRE SUPPRESSION, PLUMBING & HVAC | 94.6 | 33.5 | 69.2 | 100.2 | 52.4 | 80.3 | 94.6 | 43.4 | 73.3 | 94.6 | 62.5 | 81.2 | 100.2 | 47.7 | 78.3 | 99.9 | 70.2 | 87.6 |
| 26, 27, 3370 | ELECTRICAL, COMMUNICATIONS & UTIL. | 102.6 | 42.1 | 70.5 | 98.9 | 69.1 | 83.0 | 93.8 | 69.1 | 80.6 | 101.5 | 46.5 | 72.2 | 100.7 | 59.1 | 78.6 | 94.4 | 73.2 | 83.2 |
| MF2010 | WEIGHTED AVERAGE | 99.9 | 44.4 | 75.1 | 97.5 | 57.1 | 79.4 | 99.4 | 58.2 | 81.0 | 99.8 | 54.8 | 79.7 | 98.1 | 55.5 | 79.1 | 99.5 | 70.7 | 86.6 |

# City Cost Indexes

| | | UTAH | | | | | | | | | | | VERMONT | | | | | |
|---|---|---|---|---|---|---|---|---|---|---|---|---|---|---|---|---|---|---|
| | | OGDEN | | | PRICE | | | PROVO | | | SALT LAKE CITY | | | BELLOWS FALLS | | | BENNINGTON | | |
| DIVISION | | 842,844 | | | 845 | | | 846 - 847 | | | 840 - 841 | | | 051 | | | 052 | | |
| | | MAT. | INST. | TOTAL | MAT. | INST. | TOTAL | MAT. | INST. | TOTAL | MAT. | INST. | TOTAL | MAT. | INST. | TOTAL | MAT. | INST. | TOTAL |
| 015433 | CONTRACTOR EQUIPMENT | | 99.8 | 99.8 | | 98.7 | 98.7 | | 98.7 | 98.7 | | 99.7 | 99.7 | | 100.3 | 100.3 | | 100.3 | 100.3 |
| 0241, 31 - 34 | SITE & INFRASTRUCTURE, DEMOLITION | 74.6 | 99.5 | 91.8 | 83.1 | 97.8 | 93.2 | 81.9 | 97.8 | 92.8 | 74.4 | 99.4 | 91.7 | 78.4 | 98.2 | 92.1 | 77.7 | 98.0 | 91.7 |
| 0310 | Concrete Forming & Accessories | 105.5 | 61.4 | 67.3 | 107.5 | 52.1 | 59.4 | 106.8 | 61.3 | 67.3 | 108.0 | 61.3 | 67.5 | 96.5 | 49.7 | 55.9 | 94.4 | 54.1 | 59.4 |
| 0320 | Concrete Reinforcing | 98.5 | 77.0 | 87.5 | 106.0 | 76.9 | 91.0 | 106.9 | 77.0 | 91.5 | 100.8 | 77.0 | 88.6 | 86.5 | 64.6 | 75.2 | 86.5 | 64.6 | 75.2 |
| 0330 | Cast-in-Place Concrete | 90.6 | 73.9 | 84.2 | 89.3 | 63.1 | 79.3 | 89.3 | 73.8 | 83.4 | 99.0 | 73.8 | 89.4 | 91.2 | 72.4 | 84.0 | 91.2 | 69.0 | 82.7 |
| 03 | CONCRETE | 97.9 | 69.1 | 84.2 | 109.2 | 61.3 | 86.3 | 107.7 | 69.1 | 89.2 | 117.2 | 69.1 | 94.2 | 99.9 | 60.9 | 81.2 | 99.7 | 61.7 | 81.5 |
| 04 | MASONRY | 100.9 | 63.4 | 78.1 | 112.7 | 63.4 | 82.8 | 112.8 | 63.4 | 82.8 | 114.3 | 63.4 | 83.4 | 97.5 | 61.9 | 75.9 | 106.6 | 61.8 | 79.4 |
| 05 | METALS | 100.9 | 77.6 | 92.9 | 97.7 | 76.9 | 90.5 | 98.7 | 77.4 | 91.4 | 105.6 | 77.4 | 95.9 | 93.0 | 69.4 | 84.8 | 92.9 | 68.4 | 84.5 |
| 06 | WOOD, PLASTICS & COMPOSITES | 82.2 | 59.7 | 69.3 | 84.9 | 48.1 | 63.8 | 83.5 | 59.7 | 69.8 | 83.8 | 59.7 | 70.0 | 99.0 | 39.1 | 64.7 | 96.2 | 45.7 | 67.2 |
| 07 | THERMAL & MOISTURE PROTECTION | 98.6 | 65.1 | 85.3 | 101.6 | 60.7 | 85.4 | 101.6 | 65.1 | 87.1 | 105.4 | 65.1 | 89.4 | 95.9 | 60.3 | 81.8 | 95.8 | 55.3 | 79.7 |
| 08 | OPENINGS | 90.0 | 58.7 | 82.1 | 94.6 | 51.3 | 83.6 | 94.6 | 58.7 | 85.5 | 91.9 | 58.7 | 83.5 | 102.2 | 41.9 | 86.9 | 102.2 | 44.2 | 87.5 |
| 0920 | Plaster & Gypsum Board | 85.3 | 58.3 | 66.3 | 87.5 | 46.4 | 58.5 | 85.7 | 58.3 | 66.4 | 88.2 | 58.3 | 67.2 | 98.1 | 36.6 | 54.8 | 95.9 | 43.5 | 59.0 |
| 0950, 0980 | Ceilings & Acoustic Treatment | 119.1 | 58.3 | 78.8 | 119.1 | 46.4 | 70.9 | 119.1 | 58.3 | 78.8 | 112.9 | 58.3 | 76.7 | 89.8 | 36.6 | 54.6 | 89.8 | 43.5 | 59.1 |
| 0960 | Flooring | 99.1 | 50.4 | 85.1 | 102.5 | 37.5 | 83.8 | 102.1 | 50.4 | 87.3 | 102.5 | 50.4 | 87.5 | 95.8 | 32.2 | 77.6 | 94.7 | 32.2 | 76.7 |
| 0970, 0990 | Wall Finishes & Painting/Coating | 98.5 | 63.3 | 77.1 | 98.5 | 40.4 | 63.2 | 98.5 | 51.7 | 70.1 | 101.4 | 51.7 | 71.2 | 92.3 | 28.7 | 53.7 | 92.3 | 28.7 | 53.7 |
| 09 | FINISHES | 100.9 | 58.9 | 77.7 | 103.9 | 46.9 | 72.5 | 103.5 | 57.6 | 78.2 | 101.7 | 57.6 | 77.4 | 94.8 | 42.8 | 66.1 | 94.2 | 46.7 | 68.0 |
| COVERS | DIVS. 10 - 14, 25, 28, 41, 43, 44, 46 | 100.0 | 59.1 | 91.5 | 100.0 | 51.3 | 89.9 | 100.0 | 59.1 | 91.5 | 100.0 | 59.1 | 91.5 | 100.0 | 47.3 | 89.1 | 100.0 | 47.9 | 89.2 |
| 21, 22, 23 | FIRE SUPPRESSION, PLUMBING & HVAC | 99.9 | 70.2 | 87.6 | 97.2 | 70.4 | 86.1 | 99.9 | 70.2 | 87.5 | 100.1 | 70.2 | 87.6 | 94.8 | 96.9 | 95.7 | 94.8 | 91.5 | 93.4 |
| 26, 27, 3370 | ELECTRICAL, COMMUNICATIONS & UTIL. | 94.7 | 75.3 | 84.4 | 100.3 | 75.3 | 87.0 | 95.1 | 64.9 | 79.1 | 97.6 | 64.9 | 80.2 | 100.6 | 96.9 | 98.7 | 100.6 | 66.3 | 82.4 |
| MF2010 | WEIGHTED AVERAGE | 97.5 | 70.7 | 85.5 | 100.0 | 67.1 | 85.3 | 100.0 | 68.9 | 86.1 | 102.0 | 69.1 | 87.3 | 96.6 | 73.7 | 86.4 | 97.0 | 68.9 | 84.4 |

| | | VERMONT | | | | | | | | | | | | | | | | |
|---|---|---|---|---|---|---|---|---|---|---|---|---|---|---|---|---|---|---|
| | | BRATTLEBORO | | | BURLINGTON | | | GUILDHALL | | | MONTPELIER | | | RUTLAND | | | ST. JOHNSBURY | | |
| DIVISION | | 053 | | | 054 | | | 059 | | | 056 | | | 057 | | | 058 | | |
| | | MAT. | INST. | TOTAL | MAT. | INST. | TOTAL | MAT. | INST. | TOTAL | MAT. | INST. | TOTAL | MAT. | INST. | TOTAL | MAT. | INST. | TOTAL |
| 015433 | CONTRACTOR EQUIPMENT | | 100.3 | 100.3 | | 100.3 | 100.3 | | 100.3 | 100.3 | | 100.3 | 100.3 | | 100.3 | 100.3 | | 100.3 | 100.3 |
| 0241, 31 - 34 | SITE & INFRASTRUCTURE, DEMOLITION | 79.1 | 98.2 | 92.3 | 80.2 | 96.9 | 91.7 | 77.5 | 96.2 | 90.4 | 81.2 | 96.9 | 92.0 | 80.9 | 96.9 | 91.9 | 77.6 | 96.4 | 90.6 |
| 0310 | Concrete Forming & Accessories | 96.7 | 49.5 | 55.8 | 93.9 | 52.3 | 57.8 | 95.0 | 44.6 | 51.3 | 99.1 | 52.3 | 58.5 | 97.0 | 52.3 | 58.2 | 94.0 | 44.7 | 51.2 |
| 0320 | Concrete Reinforcing | 85.6 | 64.6 | 74.8 | 108.1 | 81.2 | 94.3 | 87.3 | 38.2 | 62.0 | 86.1 | 81.2 | 83.6 | 108.1 | 81.2 | 94.3 | 85.6 | 43.6 | 64.0 |
| 0330 | Cast-in-Place Concrete | 94.1 | 69.3 | 84.6 | 99.5 | 101.2 | 100.1 | 88.4 | 55.5 | 75.8 | 105.3 | 101.2 | 103.7 | 89.4 | 101.2 | 93.9 | 88.4 | 55.8 | 75.9 |
| 03 | CONCRETE | 102.3 | 59.8 | 81.9 | 108.1 | 75.0 | 92.2 | 97.3 | 47.9 | 73.7 | 108.1 | 75.0 | 92.3 | 103.2 | 75.0 | 89.7 | 97.0 | 49.0 | 74.0 |
| 04 | MASONRY | 105.7 | 61.9 | 79.1 | 109.1 | 50.1 | 73.3 | 106.0 | 39.9 | 65.9 | 92.6 | 50.1 | 66.9 | 87.9 | 50.1 | 65.0 | 131.4 | 39.9 | 76.0 |
| 05 | METALS | 92.9 | 69.2 | 84.8 | 100.4 | 77.9 | 92.7 | 93.0 | 57.9 | 80.9 | 92.9 | 77.9 | 87.8 | 98.6 | 77.9 | 91.5 | 93.0 | 58.3 | 81.0 |
| 06 | WOOD, PLASTICS & COMPOSITES | 99.3 | 39.1 | 64.8 | 94.2 | 51.7 | 69.8 | 95.7 | 45.7 | 67.0 | 98.4 | 51.7 | 71.7 | 99.4 | 51.7 | 72.1 | 92.0 | 45.7 | 65.5 |
| 07 | THERMAL & MOISTURE PROTECTION | 95.9 | 54.2 | 79.3 | 95.8 | 60.1 | 81.6 | 95.7 | 41.7 | 74.3 | 95.3 | 60.6 | 81.5 | 96.0 | 60.6 | 81.9 | 95.7 | 41.7 | 74.2 |
| 08 | OPENINGS | 102.2 | 41.9 | 86.9 | 105.6 | 52.8 | 92.3 | 102.2 | 38.9 | 86.2 | 102.2 | 52.8 | 89.7 | 105.6 | 52.8 | 92.3 | 102.2 | 38.9 | 86.2 |
| 0920 | Plaster & Gypsum Board | 98.1 | 36.6 | 54.8 | 99.7 | 49.6 | 64.5 | 103.9 | 43.5 | 61.3 | 107.4 | 49.6 | 66.7 | 98.9 | 49.6 | 64.2 | 107.4 | 43.5 | 62.4 |
| 0950, 0980 | Ceilings & Acoustic Treatment | 89.8 | 36.6 | 54.6 | 91.9 | 49.6 | 63.9 | 89.8 | 43.5 | 59.1 | 88.9 | 49.6 | 62.9 | 94.7 | 49.6 | 64.8 | 89.8 | 43.5 | 59.1 |
| 0960 | Flooring | 96.0 | 32.2 | 77.6 | 96.0 | 58.3 | 85.2 | 99.0 | 32.2 | 79.8 | 102.5 | 58.3 | 89.8 | 95.8 | 58.3 | 85.0 | 102.5 | 32.2 | 82.3 |
| 0970, 0990 | Wall Finishes & Painting/Coating | 92.3 | 28.7 | 53.7 | 92.3 | 33.4 | 56.6 | 92.3 | 22.3 | 49.8 | 92.3 | 33.4 | 56.6 | 92.3 | 33.4 | 56.6 | 92.3 | 23.4 | 50.5 |
| 09 | FINISHES | 94.9 | 42.8 | 66.2 | 95.6 | 50.2 | 70.6 | 96.3 | 40.1 | 65.3 | 97.8 | 50.2 | 71.5 | 96.0 | 50.2 | 70.7 | 97.8 | 40.2 | 66.0 |
| COVERS | DIVS. 10 - 14, 25, 28, 41, 43, 44, 46 | 100.0 | 47.3 | 89.1 | 100.0 | 73.1 | 94.4 | 100.0 | 39.7 | 87.5 | 100.0 | 73.1 | 94.4 | 100.0 | 73.1 | 94.4 | 100.0 | 39.7 | 87.5 |
| 21, 22, 23 | FIRE SUPPRESSION, PLUMBING & HVAC | 94.8 | 94.7 | 94.8 | 100.0 | 63.5 | 84.8 | 94.8 | 61.7 | 81.0 | 94.6 | 63.5 | 81.6 | 100.2 | 63.5 | 84.9 | 94.8 | 61.7 | 81.0 |
| 26, 27, 3370 | ELECTRICAL, COMMUNICATIONS & UTIL. | 100.6 | 96.9 | 98.7 | 99.8 | 66.3 | 82.0 | 100.6 | 66.3 | 82.4 | 101.2 | 66.3 | 82.6 | 100.7 | 66.3 | 82.4 | 100.6 | 66.3 | 82.4 |
| MF2010 | WEIGHTED AVERAGE | 97.4 | 72.9 | 86.5 | 100.9 | 66.1 | 85.4 | 96.9 | 55.7 | 78.5 | 97.7 | 66.2 | 83.6 | 99.2 | 66.2 | 84.4 | 98.3 | 55.9 | 79.3 |

| | | VERMONT | | | VIRGINIA | | | | | | | | | | | | | | |
|---|---|---|---|---|---|---|---|---|---|---|---|---|---|---|---|---|---|---|---|
| | | WHITE RIVER JCT. | | | ALEXANDRIA | | | ARLINGTON | | | BRISTOL | | | CHARLOTTESVILLE | | | CULPEPER | | |
| DIVISION | | 050 | | | 223 | | | 222 | | | 242 | | | 229 | | | 227 | | |
| | | MAT. | INST. | TOTAL | MAT. | INST. | TOTAL | MAT. | INST. | TOTAL | MAT. | INST. | TOTAL | MAT. | INST. | TOTAL | MAT. | INST. | TOTAL |
| 015433 | CONTRACTOR EQUIPMENT | | 100.3 | 100.3 | | 101.7 | 101.7 | | 100.4 | 100.4 | | 100.4 | 100.4 | | 104.8 | 104.8 | | 100.4 | 100.4 |
| 0241, 31 - 34 | SITE & INFRASTRUCTURE, DEMOLITION | 81.4 | 96.3 | 91.7 | 120.9 | 88.1 | 98.3 | 132.3 | 84.8 | 99.5 | 114.9 | 84.4 | 93.9 | 120.2 | 86.5 | 97.0 | 119.1 | 85.2 | 95.7 |
| 0310 | Concrete Forming & Accessories | 91.9 | 42.9 | 49.4 | 89.4 | 74.7 | 76.6 | 88.9 | 71.6 | 73.9 | 85.2 | 45.0 | 50.3 | 83.7 | 51.2 | 55.5 | 81.0 | 70.3 | 71.7 |
| 0320 | Concrete Reinforcing | 86.5 | 38.2 | 61.6 | 80.5 | 81.0 | 80.8 | 90.9 | 72.6 | 81.5 | 90.9 | 69.5 | 79.9 | 90.3 | 70.5 | 80.1 | 90.9 | 66.7 | 78.4 |
| 0330 | Cast-in-Place Concrete | 94.1 | 55.2 | 79.2 | 117.1 | 81.5 | 103.5 | 114.0 | 77.3 | 99.9 | 113.5 | 49.8 | 89.1 | 118.1 | 58.1 | 95.1 | 116.7 | 67.8 | 98.0 |
| 03 | CONCRETE | 104.3 | 47.1 | 76.9 | 110.0 | 79.2 | 95.3 | 114.4 | 74.4 | 95.3 | 110.6 | 53.1 | 83.1 | 111.6 | 58.9 | 86.4 | 108.3 | 69.6 | 89.8 |
| 04 | MASONRY | 118.4 | 38.6 | 70.0 | 91.8 | 71.6 | 79.5 | 105.7 | 66.3 | 81.8 | 93.9 | 42.5 | 62.7 | 119.7 | 49.1 | 76.9 | 107.6 | 63.2 | 80.7 |
| 05 | METALS | 93.0 | 58.0 | 80.9 | 93.0 | 95.2 | 93.7 | 91.7 | 82.3 | 88.5 | 90.5 | 86.2 | 89.0 | 90.8 | 88.6 | 90.1 | 90.9 | 81.2 | 87.5 |
| 06 | WOOD, PLASTICS & COMPOSITES | 93.2 | 44.0 | 65.0 | 95.0 | 73.8 | 82.8 | 91.9 | 73.8 | 81.5 | 84.6 | 43.2 | 60.9 | 83.0 | 49.3 | 63.7 | 81.9 | 73.8 | 77.3 |
| 07 | THERMAL & MOISTURE PROTECTION | 96.0 | 36.7 | 72.4 | 97.2 | 80.5 | 90.6 | 99.1 | 69.1 | 87.2 | 98.7 | 49.1 | 79.0 | 98.3 | 54.7 | 81.0 | 98.5 | 63.2 | 84.5 |
| 08 | OPENINGS | 102.2 | 40.2 | 86.5 | 98.6 | 75.7 | 92.8 | 96.5 | 69.6 | 89.7 | 99.9 | 49.4 | 87.1 | 97.9 | 54.3 | 86.9 | 98.3 | 65.1 | 89.9 |
| 0920 | Plaster & Gypsum Board | 94.2 | 41.6 | 57.2 | 106.5 | 72.8 | 82.8 | 103.0 | 72.8 | 81.8 | 98.4 | 41.4 | 58.2 | 97.9 | 46.9 | 62.0 | 98.6 | 72.8 | 80.5 |
| 0950, 0980 | Ceilings & Acoustic Treatment | 89.8 | 41.6 | 57.9 | 95.7 | 72.8 | 80.6 | 92.9 | 72.8 | 79.6 | 92.0 | 41.4 | 58.4 | 92.0 | 46.9 | 62.1 | 92.9 | 72.8 | 79.6 |
| 0960 | Flooring | 93.8 | 32.2 | 76.1 | 109.9 | 85.7 | 102.9 | 108.1 | 63.9 | 95.4 | 104.2 | 39.5 | 85.6 | 102.7 | 46.6 | 86.6 | 102.7 | 63.9 | 91.5 |
| 0970, 0990 | Wall Finishes & Painting/Coating | 92.3 | 19.5 | 48.2 | 129.1 | 108.3 | 116.5 | 129.1 | 82.8 | 101.0 | 114.7 | 44.8 | 72.3 | 114.7 | 63.3 | 83.5 | 129.1 | 56.3 | 84.9 |
| 09 | FINISHES | 94.0 | 38.6 | 63.4 | 104.4 | 79.8 | 90.8 | 104.2 | 71.9 | 86.4 | 100.5 | 43.1 | 68.9 | 100.1 | 50.8 | 72.9 | 100.8 | 68.8 | 83.2 |
| COVERS | DIVS. 10 - 14, 25, 28, 41, 43, 44, 46 | 100.0 | 39.2 | 87.4 | 100.0 | 88.2 | 97.5 | 100.0 | 72.5 | 94.3 | 100.0 | 75.9 | 95.0 | 100.0 | 75.8 | 95.0 | 100.0 | 61.6 | 92.0 |
| 21, 22, 23 | FIRE SUPPRESSION, PLUMBING & HVAC | 94.8 | 61.3 | 80.9 | 100.3 | 89.6 | 95.9 | 100.3 | 86.6 | 94.6 | 94.9 | 55.0 | 78.3 | 94.9 | 58.2 | 79.6 | 94.9 | 64.3 | 82.2 |
| 26, 27, 3370 | ELECTRICAL, COMMUNICATIONS & UTIL. | 100.6 | 66.3 | 82.4 | 96.8 | 96.9 | 96.9 | 94.2 | 96.7 | 95.5 | 96.5 | 38.4 | 65.6 | 96.4 | 70.1 | 82.5 | 99.2 | 56.6 | 76.5 |
| MF2010 | WEIGHTED AVERAGE | 98.3 | 55.1 | 79.0 | 100.2 | 85.5 | 93.6 | 101.2 | 80.1 | 91.8 | 98.4 | 55.1 | 79.0 | 99.8 | 63.4 | 83.5 | 99.1 | 67.8 | 85.1 |

735

# City Cost Indexes

## VIRGINIA

| DIVISION | | FAIRFAX 220 - 221 MAT. | INST. | TOTAL | FARMVILLE 239 MAT. | INST. | TOTAL | FREDERICKSBURG 224 - 225 MAT. | INST. | TOTAL | GRUNDY 246 MAT. | INST. | TOTAL | HARRISONBURG 228 MAT. | INST. | TOTAL | LYNCHBURG 245 MAT. | INST. | TOTAL |
|---|---|---|---|---|---|---|---|---|---|---|---|---|---|---|---|---|---|---|---|
| 015433 | CONTRACTOR EQUIPMENT | | 100.4 | 100.4 | | 104.8 | 104.8 | | 100.4 | 100.4 | | 100.4 | 100.4 | | 100.4 | 100.4 | | 100.4 | 100.4 |
| 0241, 31 - 34 | SITE & INFRASTRUCTURE, DEMOLITION | 131.4 | 85.7 | 99.8 | 115.9 | 85.8 | 95.1 | 118.5 | 85.4 | 95.6 | 112.5 | 83.2 | 92.3 | 128.0 | 84.5 | 98.0 | 113.4 | 84.6 | 93.5 |
| 0310 | Concrete Forming & Accessories | 83.6 | 72.1 | 73.7 | 95.0 | 46.6 | 53.0 | 83.6 | 50.9 | 55.2 | 88.1 | 38.8 | 45.3 | 79.9 | 45.1 | 49.7 | 85.2 | 64.5 | 67.2 |
| 0320 | Concrete Reinforcing | 90.9 | 80.9 | 85.7 | 90.9 | 53.7 | 71.7 | 91.6 | 80.0 | 85.6 | 89.7 | 54.3 | 71.4 | 90.9 | 48.9 | 69.2 | 90.3 | 70.2 | 79.9 |
| 0330 | Cast-in-Place Concrete | 114.0 | 78.3 | 100.3 | 118.1 | 56.1 | 94.3 | 115.7 | 54.7 | 92.3 | 113.5 | 50.1 | 89.3 | 114.0 | 61.3 | 93.8 | 113.5 | 61.2 | 93.5 |
| 03 | CONCRETE | 114.0 | 77.1 | 96.4 | 110.5 | 52.8 | 82.9 | 107.9 | 59.4 | 84.7 | 109.3 | 47.2 | 79.5 | 111.8 | 53.4 | 83.8 | 109.2 | 65.8 | 88.4 |
| 04 | MASONRY | 105.7 | 66.6 | 82.0 | 102.6 | 48.3 | 69.7 | 106.6 | 47.0 | 70.5 | 97.1 | 51.1 | 69.2 | 103.9 | 53.2 | 73.2 | 111.8 | 55.4 | 77.6 |
| 05 | METALS | 90.9 | 96.1 | 92.7 | 90.8 | 78.0 | 86.4 | 90.9 | 92.0 | 91.3 | 90.5 | 71.7 | 84.1 | 90.8 | 84.5 | 88.6 | 90.7 | 87.6 | 89.6 |
| 06 | WOOD, PLASTICS & COMPOSITES | 84.6 | 73.8 | 78.4 | 97.0 | 47.6 | 68.7 | 84.6 | 48.6 | 64.0 | 87.5 | 37.9 | 59.0 | 80.8 | 42.2 | 58.6 | 84.6 | 67.9 | 75.0 |
| 07 | THERMAL & MOISTURE PROTECTION | 99.1 | 69.5 | 87.3 | 98.3 | 49.5 | 78.9 | 98.5 | 52.7 | 80.3 | 98.6 | 48.8 | 78.9 | 98.9 | 62.8 | 84.6 | 98.5 | 56.9 | 82.0 |
| 08 | OPENINGS | 96.5 | 77.0 | 91.6 | 98.2 | 46.4 | 85.1 | 97.9 | 56.3 | 87.4 | 99.9 | 37.1 | 84.0 | 98.3 | 52.9 | 86.8 | 98.2 | 62.8 | 89.2 |
| 0920 | Plaster & Gypsum Board | 98.6 | 72.8 | 80.5 | 105.4 | 45.2 | 63.0 | 98.6 | 46.9 | 62.2 | 98.8 | 35.9 | 54.5 | 97.9 | 40.3 | 57.3 | 98.4 | 66.8 | 76.1 |
| 0950, 0980 | Ceilings & Acoustic Treatment | 92.9 | 72.8 | 79.6 | 92.0 | 45.2 | 61.0 | 92.9 | 46.9 | 62.4 | 92.0 | 35.9 | 54.8 | 92.0 | 40.3 | 57.7 | 92.0 | 66.8 | 75.3 |
| 0960 | Flooring | 104.5 | 66.8 | 93.7 | 109.9 | 63.7 | 96.6 | 104.5 | 78.2 | 96.9 | 105.7 | 35.6 | 85.6 | 102.3 | 78.2 | 95.4 | 104.2 | 44.1 | 86.9 |
| 0970, 0990 | Wall Finishes & Painting/Coating | 129.1 | 82.8 | 101.0 | 114.7 | 41.7 | 70.4 | 129.1 | 69.1 | 92.7 | 114.7 | 37.8 | 68.1 | 129.1 | 54.6 | 83.9 | 114.7 | 44.8 | 72.3 |
| 09 | FINISHES | 102.6 | 72.4 | 85.9 | 102.8 | 49.5 | 73.4 | 101.3 | 56.8 | 76.8 | 100.9 | 37.8 | 66.1 | 101.2 | 51.1 | 73.6 | 100.3 | 59.0 | 77.5 |
| COVERS | DIVS. 10 - 14, 25, 28, 41, 43, 44, 46 | 100.0 | 72.5 | 94.3 | 100.0 | 50.8 | 89.8 | 100.0 | 76.7 | 95.2 | 100.0 | 48.6 | 89.3 | 100.0 | 75.6 | 94.9 | 100.0 | 79.5 | 95.7 |
| 21, 22, 23 | FIRE SUPPRESSION, PLUMBING & HVAC | 94.9 | 86.9 | 91.6 | 94.9 | 37.5 | 71.0 | 94.9 | 80.4 | 88.9 | 94.9 | 42.1 | 72.9 | 94.9 | 54.9 | 78.3 | 94.9 | 57.7 | 79.4 |
| 26, 27, 3370 | ELECTRICAL, COMMUNICATIONS & UTIL. | 97.7 | 96.7 | 97.2 | 93.4 | 50.9 | 70.8 | 94.4 | 93.2 | 93.8 | 96.5 | 48.4 | 70.9 | 96.7 | 84.8 | 90.4 | 97.6 | 57.2 | 76.1 |
| MF2010 | WEIGHTED AVERAGE | 99.9 | 82.3 | 92.0 | 98.7 | 53.0 | 78.3 | 98.5 | 72.2 | 86.7 | 98.3 | 50.3 | 76.9 | 99.3 | 64.0 | 83.5 | 99.0 | 64.6 | 83.6 |

## VIRGINIA

| DIVISION | | NEWPORT NEWS 236 MAT. | INST. | TOTAL | NORFOLK 233 - 235 MAT. | INST. | TOTAL | PETERSBURG 238 MAT. | INST. | TOTAL | PORTSMOUTH 237 MAT. | INST. | TOTAL | PULASKI 243 MAT. | INST. | TOTAL | RICHMOND 230 - 232 MAT. | INST. | TOTAL |
|---|---|---|---|---|---|---|---|---|---|---|---|---|---|---|---|---|---|---|---|
| 015433 | CONTRACTOR EQUIPMENT | | 104.8 | 104.8 | | 105.4 | 105.4 | | 104.8 | 104.8 | | 104.7 | 104.7 | | 100.4 | 100.4 | | 104.8 | 104.8 |
| 0241, 31 - 34 | SITE & INFRASTRUCTURE, DEMOLITION | 114.0 | 87.2 | 95.5 | 112.1 | 88.2 | 95.6 | 118.7 | 87.5 | 97.2 | 112.3 | 86.6 | 94.5 | 111.8 | 83.2 | 92.1 | 111.3 | 87.5 | 94.9 |
| 0310 | Concrete Forming & Accessories | 94.3 | 67.4 | 71.0 | 95.8 | 67.5 | 71.3 | 88.1 | 60.8 | 64.4 | 84.3 | 55.4 | 59.3 | 88.1 | 39.0 | 45.5 | 93.8 | 60.8 | 65.2 |
| 0320 | Concrete Reinforcing | 90.6 | 70.9 | 80.4 | 90.6 | 70.9 | 80.4 | 90.3 | 71.2 | 80.4 | 90.3 | 70.8 | 80.3 | 89.7 | 54.3 | 71.5 | 90.6 | 71.2 | 80.6 |
| 0330 | Cast-in-Place Concrete | 114.7 | 61.4 | 94.3 | 125.4 | 61.5 | 100.9 | 121.9 | 56.2 | 96.7 | 113.6 | 61.0 | 93.5 | 113.5 | 50.2 | 89.3 | 110.6 | 56.2 | 89.8 |
| 03 | CONCRETE | 107.4 | 67.3 | 88.2 | 112.8 | 67.3 | 91.1 | 113.4 | 62.6 | 89.1 | 106.2 | 61.7 | 84.9 | 109.3 | 47.3 | 79.6 | 105.3 | 62.6 | 84.9 |
| 04 | MASONRY | 97.4 | 55.3 | 71.8 | 103.4 | 55.3 | 74.2 | 111.1 | 56.8 | 78.2 | 102.8 | 55.1 | 73.9 | 90.6 | 51.1 | 66.7 | 95.1 | 56.8 | 71.9 |
| 05 | METALS | 93.0 | 89.6 | 91.8 | 92.0 | 89.7 | 91.2 | 90.9 | 90.1 | 90.6 | 92.0 | 88.1 | 90.6 | 90.6 | 71.9 | 84.2 | 95.1 | 90.1 | 93.4 |
| 06 | WOOD, PLASTICS & COMPOSITES | 95.9 | 71.8 | 82.1 | 98.0 | 71.8 | 82.9 | 88.0 | 63.4 | 73.9 | 84.3 | 55.7 | 67.9 | 87.5 | 37.9 | 59.0 | 95.0 | 63.4 | 76.9 |
| 07 | THERMAL & MOISTURE PROTECTION | 98.3 | 57.3 | 82.0 | 95.9 | 57.3 | 80.6 | 98.3 | 57.6 | 82.2 | 98.3 | 55.0 | 81.1 | 98.6 | 48.8 | 78.9 | 97.1 | 57.6 | 81.4 |
| 08 | OPENINGS | 98.6 | 65.5 | 90.2 | 98.6 | 65.5 | 90.2 | 97.9 | 62.0 | 88.8 | 98.7 | 56.8 | 88.1 | 100.0 | 37.1 | 84.0 | 98.6 | 62.0 | 89.3 |
| 0920 | Plaster & Gypsum Board | 106.5 | 70.1 | 80.9 | 100.2 | 70.1 | 79.0 | 99.1 | 61.5 | 72.6 | 98.5 | 53.5 | 66.8 | 98.8 | 35.9 | 54.5 | 109.7 | 61.5 | 75.8 |
| 0950, 0980 | Ceilings & Acoustic Treatment | 95.7 | 70.1 | 78.7 | 95.7 | 70.1 | 78.7 | 92.9 | 61.5 | 72.1 | 95.7 | 53.5 | 67.7 | 92.0 | 35.9 | 54.8 | 98.5 | 61.5 | 74.0 |
| 0960 | Flooring | 109.9 | 53.4 | 93.7 | 109.3 | 53.4 | 93.2 | 105.2 | 75.2 | 96.6 | 101.8 | 67.5 | 92.0 | 105.7 | 35.6 | 85.6 | 109.3 | 75.2 | 99.5 |
| 0970, 0990 | Wall Finishes & Painting/Coating | 114.7 | 45.3 | 72.6 | 114.7 | 69.3 | 87.1 | 114.7 | 63.3 | 83.5 | 114.7 | 69.3 | 87.1 | 114.7 | 37.8 | 68.1 | 114.7 | 63.3 | 83.5 |
| 09 | FINISHES | 103.5 | 62.9 | 81.1 | 102.1 | 65.6 | 82.0 | 101.1 | 63.9 | 80.6 | 100.2 | 58.5 | 77.2 | 100.9 | 37.8 | 66.1 | 103.8 | 63.9 | 81.8 |
| COVERS | DIVS. 10 - 14, 25, 28, 41, 43, 44, 46 | 100.0 | 81.3 | 96.1 | 100.0 | 81.3 | 96.1 | 100.0 | 76.1 | 95.0 | 100.0 | 79.5 | 95.7 | 100.0 | 48.6 | 89.3 | 100.0 | 76.1 | 95.0 |
| 21, 22, 23 | FIRE SUPPRESSION, PLUMBING & HVAC | 100.3 | 64.2 | 85.3 | 100.1 | 63.5 | 84.9 | 94.9 | 56.3 | 78.8 | 100.3 | 63.5 | 85.0 | 94.9 | 42.4 | 73.0 | 100.1 | 56.3 | 81.9 |
| 26, 27, 3370 | ELECTRICAL, COMMUNICATIONS & UTIL. | 96.5 | 59.3 | 76.7 | 101.7 | 61.2 | 80.1 | 96.7 | 70.1 | 82.6 | 94.6 | 61.1 | 76.8 | 96.5 | 52.3 | 73.0 | 101.3 | 70.1 | 84.7 |
| MF2010 | WEIGHTED AVERAGE | 100.0 | 67.6 | 85.5 | 101.0 | 68.2 | 86.3 | 99.6 | 66.8 | 84.9 | 99.4 | 65.5 | 84.2 | 98.0 | 50.9 | 77.0 | 100.2 | 66.8 | 85.3 |

## VIRGINIA / WASHINGTON

| DIVISION | | ROANOKE 240 - 241 MAT. | INST. | TOTAL | STAUNTON 244 MAT. | INST. | TOTAL | WINCHESTER 226 MAT. | INST. | TOTAL | CLARKSTON 994 MAT. | INST. | TOTAL | EVERETT 982 MAT. | INST. | TOTAL | OLYMPIA 985 MAT. | INST. | TOTAL |
|---|---|---|---|---|---|---|---|---|---|---|---|---|---|---|---|---|---|---|---|
| 015433 | CONTRACTOR EQUIPMENT | | 100.4 | 100.4 | | 104.8 | 104.8 | | 100.4 | 100.4 | | 91.3 | 91.3 | | 103.9 | 103.9 | | 103.9 | 103.9 |
| 0241, 31 - 34 | SITE & INFRASTRUCTURE, DEMOLITION | 111.5 | 84.5 | 92.9 | 116.5 | 86.3 | 95.7 | 126.5 | 84.6 | 97.6 | 99.2 | 90.6 | 93.3 | 94.5 | 113.8 | 107.8 | 95.8 | 113.8 | 108.2 |
| 0310 | Concrete Forming & Accessories | 94.1 | 64.2 | 68.2 | 87.7 | 53.5 | 58.0 | 82.2 | 50.2 | 54.4 | 122.4 | 71.5 | 78.2 | 109.1 | 101.4 | 102.4 | 99.0 | 101.6 | 101.2 |
| 0320 | Concrete Reinforcing | 90.6 | 70.2 | 80.1 | 90.3 | 52.7 | 70.9 | 90.3 | 85.1 | 87.6 | 110.2 | 85.8 | 97.6 | 115.1 | 102.0 | 108.3 | 115.7 | 102.0 | 108.7 |
| 0330 | Cast-in-Place Concrete | 129.0 | 61.1 | 103.0 | 118.1 | 54.3 | 93.6 | 114.0 | 57.4 | 92.3 | 102.6 | 83.7 | 95.4 | 103.8 | 108.3 | 105.6 | 94.5 | 108.5 | 99.9 |
| 03 | CONCRETE | 114.5 | 65.6 | 91.1 | 110.7 | 55.3 | 84.2 | 111.1 | 60.8 | 87.0 | 100.7 | 78.5 | 90.1 | 99.6 | 103.3 | 101.4 | 95.5 | 103.5 | 99.3 |
| 04 | MASONRY | 98.4 | 57.1 | 73.3 | 108.1 | 55.5 | 76.2 | 101.5 | 51.0 | 70.9 | 108.3 | 80.5 | 91.5 | 132.1 | 101.4 | 113.5 | 121.0 | 101.5 | 109.2 |
| 05 | METALS | 92.8 | 87.4 | 90.9 | 90.8 | 83.0 | 88.1 | 90.9 | 91.2 | 91.0 | 89.0 | 81.2 | 86.3 | 101.8 | 92.4 | 98.6 | 97.0 | 92.5 | 95.4 |
| 06 | WOOD, PLASTICS & COMPOSITES | 95.9 | 67.9 | 79.9 | 87.5 | 54.0 | 68.3 | 83.2 | 48.6 | 63.4 | 96.8 | 69.3 | 81.0 | 105.4 | 101.2 | 103.0 | 92.0 | 101.2 | 97.3 |
| 07 | THERMAL & MOISTURE PROTECTION | 98.3 | 56.0 | 81.5 | 98.3 | 53.4 | 80.5 | 99.0 | 60.1 | 83.5 | 152.5 | 77.5 | 122.7 | 101.8 | 99.6 | 100.9 | 101.4 | 98.5 | 100.3 |
| 08 | OPENINGS | 98.6 | 62.8 | 89.5 | 98.2 | 52.0 | 86.5 | 100.0 | 57.6 | 89.3 | 106.6 | 72.4 | 97.9 | 102.3 | 100.7 | 101.9 | 102.2 | 100.7 | 101.8 |
| 0920 | Plaster & Gypsum Board | 105.6 | 66.8 | 78.6 | 98.8 | 51.8 | 65.7 | 98.6 | 46.9 | 62.2 | 134.2 | 68.2 | 87.7 | 119.5 | 101.2 | 106.6 | 108.7 | 101.2 | 103.4 |
| 0950, 0980 | Ceilings & Acoustic Treatment | 95.7 | 66.8 | 76.6 | 92.0 | 51.8 | 65.3 | 92.9 | 46.9 | 62.4 | 103.9 | 68.2 | 80.2 | 113.3 | 101.2 | 105.3 | 107.5 | 101.2 | 103.7 |
| 0960 | Flooring | 109.9 | 47.6 | 92.0 | 105.1 | 41.7 | 86.9 | 103.8 | 78.2 | 96.4 | 104.4 | 49.8 | 88.7 | 117.4 | 96.7 | 111.4 | 110.2 | 96.9 | 106.4 |
| 0970, 0990 | Wall Finishes & Painting/Coating | 114.7 | 44.8 | 72.3 | 114.7 | 31.2 | 64.1 | 129.1 | 48.1 | 79.9 | 108.3 | 63.0 | 80.8 | 104.7 | 76.4 | 87.6 | 104.7 | 85.9 | 93.3 |
| 09 | FINISHES | 103.3 | 59.8 | 79.3 | 100.6 | 49.5 | 72.4 | 101.8 | 54.4 | 75.7 | 118.5 | 65.5 | 89.3 | 110.8 | 97.9 | 103.7 | 106.3 | 99.0 | 102.3 |
| COVERS | DIVS. 10 - 14, 25, 28, 41, 43, 44, 46 | 100.0 | 79.4 | 95.7 | 100.0 | 79.1 | 95.7 | 100.0 | 57.1 | 91.1 | 100.0 | 83.8 | 96.6 | 100.0 | 100.3 | 100.1 | 100.0 | 101.7 | 100.3 |
| 21, 22, 23 | FIRE SUPPRESSION, PLUMBING & HVAC | 100.3 | 48.5 | 78.8 | 94.9 | 56.7 | 79.0 | 94.9 | 69.0 | 84.1 | 95.0 | 76.2 | 87.2 | 100.1 | 92.8 | 97.1 | 99.9 | 92.9 | 97.0 |
| 26, 27, 3370 | ELECTRICAL, COMMUNICATIONS & UTIL. | 96.5 | 41.3 | 67.1 | 95.2 | 44.9 | 68.4 | 94.9 | 46.6 | 69.2 | 101.8 | 94.2 | 97.8 | 104.4 | 84.1 | 93.6 | 105.8 | 95.8 | 100.5 |
| MF2010 | WEIGHTED AVERAGE | 100.7 | 60.7 | 82.8 | 98.9 | 59.1 | 81.1 | 99.2 | 63.2 | 83.1 | 101.9 | 79.8 | 92.0 | 103.4 | 97.1 | 100.6 | 101.3 | 98.9 | 100.2 |

## WASHINGTON

| DIVISION | | RICHLAND 993 | | | SEATTLE 980 - 981,987 | | | SPOKANE 990 - 992 | | | TACOMA 983 - 984 | | | VANCOUVER 986 | | | WENATCHEE 988 | | |
|---|---|---|---|---|---|---|---|---|---|---|---|---|---|---|---|---|---|---|---|
| | | MAT. | INST. | TOTAL | MAT. | INST. | TOTAL | MAT. | INST. | TOTAL | MAT. | INST. | TOTAL | MAT. | INST. | TOTAL | MAT. | INST. | TOTAL |
| 015433 | CONTRACTOR EQUIPMENT | | 91.3 | 91.3 | | 103.7 | 103.7 | | 91.3 | 91.3 | | 103.9 | 103.9 | | 98.2 | 98.2 | | 103.9 | 103.9 |
| 0241, 31 - 34 | SITE & INFRASTRUCTURE, DEMOLITION | 100.2 | 90.8 | 93.7 | 98.5 | 111.9 | 107.7 | 99.5 | 90.7 | 93.5 | 97.6 | 113.8 | 108.8 | 108.7 | 100.5 | 103.0 | 108.6 | 111.8 | 110.8 |
| 0310 | Concrete Forming & Accessories | 122.6 | 81.2 | 86.6 | 100.7 | 102.2 | 102.0 | 127.0 | 80.7 | 86.8 | 100.7 | 101.6 | 101.4 | 101.6 | 87.5 | 89.4 | 102.1 | 79.2 | 82.2 |
| 0320 | Concrete Reinforcing | 105.7 | 85.8 | 95.5 | 113.6 | 102.1 | 107.7 | 106.5 | 85.8 | 95.8 | 113.6 | 102.0 | 107.6 | 114.6 | 101.7 | 108.0 | 114.5 | 86.3 | 100.0 |
| 0330 | Cast-in-Place Concrete | 102.9 | 85.3 | 96.2 | 109.0 | 108.7 | 108.9 | 106.9 | 85.1 | 98.6 | 106.8 | 108.5 | 107.4 | 119.3 | 101.5 | 112.4 | 109.0 | 80.3 | 98.0 |
| 03 | CONCRETE | 100.3 | 83.3 | 92.2 | 102.6 | 104.0 | 103.2 | 102.7 | 83.1 | 93.3 | 101.5 | 103.5 | 102.4 | 111.8 | 94.9 | 103.7 | 109.0 | 80.9 | 95.6 |
| 04 | MASONRY | 108.1 | 83.0 | 92.9 | 125.5 | 101.5 | 111.0 | 108.8 | 83.0 | 93.1 | 125.4 | 101.5 | 110.9 | 126.5 | 96.1 | 108.1 | 128.5 | 82.9 | 100.9 |
| 05 | METALS | 89.4 | 81.7 | 86.8 | 103.7 | 94.5 | 100.5 | 91.8 | 81.3 | 88.2 | 103.7 | 92.5 | 99.8 | 101.0 | 93.0 | 98.3 | 101.0 | 82.8 | 94.7 |
| 06 | WOOD, PLASTICS & COMPOSITES | 97.0 | 80.6 | 87.7 | 96.5 | 101.2 | 99.2 | 105.6 | 80.4 | 91.2 | 95.2 | 101.2 | 98.7 | 87.5 | 85.2 | 86.2 | 96.3 | 78.2 | 85.9 |
| 07 | THERMAL & MOISTURE PROTECTION | 153.7 | 79.8 | 124.3 | 101.6 | 99.7 | 100.8 | 150.1 | 80.7 | 122.5 | 101.5 | 98.5 | 100.3 | 101.8 | 91.0 | 97.5 | 102.0 | 81.1 | 93.7 |
| 08 | OPENINGS | 109.1 | 74.0 | 100.2 | 104.6 | 100.7 | 103.6 | 109.7 | 74.0 | 100.6 | 102.9 | 100.7 | 102.3 | 100.5 | 91.1 | 98.1 | 102.5 | 72.8 | 94.9 |
| 0920 | Plaster & Gypsum Board | 134.2 | 79.7 | 95.8 | 112.8 | 101.2 | 104.6 | 134.5 | 79.7 | 95.9 | 115.6 | 101.2 | 105.5 | 113.6 | 85.0 | 93.5 | 117.4 | 77.4 | 89.3 |
| 0950, 0980 | Ceilings & Acoustic Treatment | 109.1 | 79.7 | 89.6 | 115.9 | 101.2 | 106.1 | 106.0 | 79.7 | 88.5 | 116.1 | 101.2 | 106.2 | 111.5 | 85.0 | 94.0 | 109.3 | 77.4 | 88.2 |
| 0960 | Flooring | 104.6 | 84.6 | 98.8 | 110.1 | 104.8 | 108.6 | 104.7 | 90.1 | 100.5 | 110.6 | 96.9 | 106.7 | 116.7 | 107.9 | 114.1 | 113.6 | 66.2 | 100.0 |
| 0970, 0990 | Wall Finishes & Painting/Coating | 108.3 | 63.0 | 80.8 | 104.7 | 85.9 | 93.3 | 108.3 | 63.0 | 80.8 | 104.7 | 85.9 | 93.3 | 109.6 | 66.6 | 83.6 | 104.7 | 63.0 | 79.4 |
| 09 | FINISHES | 119.9 | 79.7 | 97.7 | 108.7 | 100.6 | 104.2 | 119.4 | 80.8 | 98.1 | 109.1 | 99.0 | 103.5 | 107.3 | 88.7 | 97.0 | 109.8 | 74.6 | 90.4 |
| COVERS | DIVS. 10 - 14, 25, 28, 41, 43, 44, 46 | 100.0 | 69.2 | 93.6 | 100.0 | 101.7 | 100.4 | 100.0 | 69.1 | 93.6 | 100.0 | 101.7 | 100.3 | 100.0 | 65.7 | 92.9 | 100.0 | 85.3 | 96.9 |
| 21, 22, 23 | FIRE SUPPRESSION, PLUMBING & HVAC | 100.5 | 102.2 | 101.2 | 100.0 | 118.3 | 107.6 | 100.4 | 77.5 | 90.8 | 100.1 | 92.9 | 97.1 | 100.2 | 94.1 | 97.7 | 94.7 | 84.4 | 90.4 |
| 26, 27, 3370 | ELECTRICAL, COMMUNICATIONS & UTIL. | 99.5 | 95.1 | 97.2 | 104.2 | 110.2 | 107.4 | 96.5 | 80.8 | 88.1 | 104.2 | 95.8 | 99.7 | 109.5 | 100.1 | 104.5 | 105.1 | 81.9 | 92.8 |
| MF2010 | WEIGHTED AVERAGE | 103.4 | 88.0 | 96.5 | 103.8 | 106.6 | 105.0 | 103.7 | 80.8 | 93.5 | 103.5 | 98.9 | 101.4 | 104.8 | 93.8 | 99.9 | 103.3 | 83.7 | 94.6 |

| DIVISION | | WASHINGTON YAKIMA 989 | | | WEST VIRGINIA BECKLEY 258 - 259 | | | BLUEFIELD 247 - 248 | | | BUCKHANNON 262 | | | CHARLESTON 250 - 253 | | | CLARKSBURG 263 - 264 | | |
|---|---|---|---|---|---|---|---|---|---|---|---|---|---|---|---|---|---|---|---|
| | | MAT. | INST. | TOTAL | MAT. | INST. | TOTAL | MAT. | INST. | TOTAL | MAT. | INST. | TOTAL | MAT. | INST. | TOTAL | MAT. | INST. | TOTAL |
| 015433 | CONTRACTOR EQUIPMENT | | 103.9 | 103.9 | | 100.4 | 100.4 | | 100.4 | 100.4 | | 100.4 | 100.4 | | 100.4 | 100.4 | | 100.4 | 100.4 |
| 0241, 31 - 34 | SITE & INFRASTRUCTURE, DEMOLITION | 100.4 | 112.4 | 108.7 | 106.1 | 86.7 | 92.7 | 105.8 | 86.7 | 92.6 | 112.6 | 86.4 | 94.5 | 103.7 | 88.3 | 93.1 | 113.2 | 86.4 | 94.7 |
| 0310 | Concrete Forming & Accessories | 101.0 | 96.7 | 97.2 | 83.8 | 88.3 | 87.7 | 85.1 | 87.4 | 87.1 | 84.5 | 85.5 | 85.4 | 99.4 | 90.2 | 91.4 | 82.3 | 85.5 | 85.1 |
| 0320 | Concrete Reinforcing | 114.1 | 86.4 | 99.8 | 89.3 | 85.3 | 87.2 | 89.3 | 67.1 | 77.9 | 89.8 | 84.3 | 87.0 | 90.6 | 85.5 | 88.0 | 89.8 | 84.9 | 87.3 |
| 0330 | Cast-in-Place Concrete | 114.1 | 84.9 | 102.9 | 111.0 | 102.7 | 107.9 | 111.0 | 102.4 | 107.7 | 110.7 | 101.1 | 107.0 | 102.7 | 102.9 | 102.8 | 121.3 | 101.1 | 113.6 |
| 03 | CONCRETE | 106.5 | 90.2 | 98.7 | 104.7 | 93.3 | 99.2 | 104.8 | 89.4 | 97.4 | 107.9 | 91.3 | 100.0 | 101.6 | 94.2 | 98.1 | 112.7 | 91.4 | 102.5 |
| 04 | MASONRY | 118.3 | 82.8 | 96.8 | 97.5 | 88.9 | 92.3 | 92.3 | 88.9 | 90.2 | 103.7 | 84.7 | 92.2 | 94.0 | 90.8 | 92.1 | 106.2 | 84.7 | 93.2 |
| 05 | METALS | 101.7 | 82.5 | 95.1 | 90.8 | 96.0 | 92.6 | 90.8 | 88.7 | 90.1 | 90.9 | 96.5 | 92.9 | 93.0 | 97.2 | 94.5 | 90.9 | 96.7 | 92.9 |
| 06 | WOOD, PLASTICS & COMPOSITES | 95.7 | 101.2 | 98.9 | 84.6 | 89.8 | 87.6 | 86.3 | 89.8 | 88.3 | 85.6 | 85.2 | 85.4 | 100.2 | 89.8 | 95.0 | 82.4 | 85.2 | 84.0 |
| 07 | THERMAL & MOISTURE PROTECTION | 101.6 | 82.7 | 94.1 | 98.4 | 87.1 | 93.9 | 98.4 | 87.1 | 93.9 | 98.7 | 86.9 | 94.0 | 95.7 | 87.9 | 92.6 | 98.6 | 86.9 | 94.0 |
| 08 | OPENINGS | 102.4 | 81.8 | 97.2 | 99.3 | 82.9 | 95.1 | 100.2 | 78.7 | 94.7 | 100.2 | 80.1 | 95.1 | 99.9 | 82.9 | 95.5 | 100.2 | 82.1 | 95.6 |
| 0920 | Plaster & Gypsum Board | 115.7 | 101.2 | 105.5 | 96.1 | 89.3 | 91.3 | 96.9 | 89.3 | 91.6 | 97.5 | 84.6 | 88.4 | 99.7 | 89.3 | 92.4 | 95.3 | 84.6 | 87.7 |
| 0950, 0980 | Ceilings & Acoustic Treatment | 111.4 | 101.2 | 104.6 | 90.1 | 89.3 | 89.6 | 90.1 | 89.3 | 89.6 | 92.0 | 84.6 | 87.1 | 93.8 | 89.3 | 90.8 | 92.0 | 84.6 | 87.1 |
| 0960 | Flooring | 111.8 | 55.0 | 95.4 | 101.2 | 106.0 | 102.6 | 101.8 | 106.0 | 103.0 | 101.5 | 100.2 | 101.1 | 109.9 | 106.0 | 108.8 | 100.3 | 100.2 | 100.2 |
| 0970, 0990 | Wall Finishes & Painting/Coating | 104.7 | 63.0 | 79.4 | 114.7 | 37.7 | 68.0 | 114.7 | 37.7 | 68.0 | 114.7 | 85.5 | 97.0 | 114.7 | 84.9 | 96.7 | 114.7 | 85.5 | 97.0 |
| 09 | FINISHES | 108.7 | 85.9 | 96.1 | 98.3 | 87.0 | 92.1 | 98.6 | 87.0 | 92.2 | 99.5 | 88.1 | 93.2 | 101.6 | 92.7 | 96.7 | 98.8 | 88.1 | 92.9 |
| COVERS | DIVS. 10 - 14, 25, 28, 41, 43, 44, 46 | 100.0 | 98.7 | 99.7 | 100.0 | 59.3 | 91.5 | 100.0 | 59.2 | 91.5 | 100.0 | 97.7 | 99.5 | 100.0 | 99.3 | 99.9 | 100.0 | 97.7 | 99.5 |
| 21, 22, 23 | FIRE SUPPRESSION, PLUMBING & HVAC | 100.1 | 102.3 | 101.0 | 94.9 | 82.0 | 89.5 | 94.9 | 56.1 | 78.8 | 94.9 | 89.1 | 92.5 | 100.1 | 84.0 | 93.4 | 94.9 | 88.9 | 92.4 |
| 26, 27, 3370 | ELECTRICAL, COMMUNICATIONS & UTIL. | 107.5 | 95.1 | 100.9 | 92.4 | 89.4 | 90.8 | 95.3 | 70.4 | 82.0 | 96.8 | 93.5 | 95.0 | 99.8 | 89.5 | 94.3 | 96.8 | 93.5 | 95.0 |
| MF2010 | WEIGHTED AVERAGE | 103.7 | 93.1 | 99.0 | 97.0 | 87.2 | 92.6 | 97.1 | 77.7 | 88.4 | 98.5 | 89.7 | 94.6 | 99.0 | 90.1 | 95.0 | 99.2 | 89.7 | 94.9 |

## WEST VIRGINIA

| DIVISION | | GASSAWAY 266 | | | HUNTINGTON 255 - 257 | | | LEWISBURG 249 | | | MARTINSBURG 254 | | | MORGANTOWN 265 | | | PARKERSBURG 261 | | |
|---|---|---|---|---|---|---|---|---|---|---|---|---|---|---|---|---|---|---|---|
| | | MAT. | INST. | TOTAL | MAT. | INST. | TOTAL | MAT. | INST. | TOTAL | MAT. | INST. | TOTAL | MAT. | INST. | TOTAL | MAT. | INST. | TOTAL |
| 015433 | CONTRACTOR EQUIPMENT | | 100.4 | 100.4 | | 100.4 | 100.4 | | 100.4 | 100.4 | | 100.4 | 100.4 | | 100.4 | 100.4 | | 100.4 | 100.4 |
| 0241, 31 - 34 | SITE & INFRASTRUCTURE, DEMOLITION | 109.7 | 86.9 | 93.9 | 110.5 | 89.1 | 95.8 | 123.1 | 86.7 | 98.0 | 110.0 | 84.5 | 92.4 | 106.9 | 86.6 | 92.9 | 115.4 | 88.2 | 96.7 |
| 0310 | Concrete Forming & Accessories | 84.0 | 85.7 | 85.5 | 95.3 | 92.5 | 92.9 | 82.6 | 87.9 | 87.2 | 83.9 | 78.4 | 79.1 | 82.6 | 85.2 | 84.9 | 86.5 | 87.7 | 87.5 |
| 0320 | Concrete Reinforcing | 89.8 | 85.5 | 87.6 | 90.6 | 85.6 | 88.0 | 89.8 | 67.1 | 78.2 | 89.3 | 72.4 | 80.6 | 89.8 | 84.8 | 87.2 | 89.3 | 84.2 | 86.7 |
| 0330 | Cast-in-Place Concrete | 115.9 | 101.2 | 110.3 | 121.1 | 103.8 | 114.5 | 111.1 | 102.6 | 107.8 | 116.3 | 92.2 | 107.1 | 110.7 | 101.0 | 107.0 | 113.2 | 94.0 | 105.8 |
| 03 | CONCRETE | 108.5 | 91.6 | 100.5 | 110.6 | 95.6 | 103.4 | 114.7 | 89.7 | 102.7 | 108.7 | 82.6 | 96.2 | 104.5 | 91.3 | 98.2 | 110.1 | 89.9 | 100.4 |
| 04 | MASONRY | 107.3 | 87.7 | 95.4 | 97.8 | 92.4 | 94.5 | 96.5 | 88.9 | 91.9 | 99.3 | 72.1 | 82.8 | 126.4 | 84.7 | 101.1 | 82.4 | 86.6 | 84.9 |
| 05 | METALS | 90.9 | 97.2 | 93.1 | 93.1 | 97.4 | 94.5 | 90.9 | 89.2 | 90.3 | 91.1 | 84.5 | 88.8 | 91.0 | 96.4 | 92.9 | 91.7 | 96.5 | 93.3 |
| 06 | WOOD, PLASTICS & COMPOSITES | 84.7 | 85.2 | 85.0 | 95.9 | 91.9 | 93.6 | 82.7 | 89.8 | 86.8 | 84.6 | 79.9 | 81.9 | 82.7 | 85.2 | 84.1 | 85.9 | 86.8 | 86.4 |
| 07 | THERMAL & MOISTURE PROTECTION | 98.4 | 87.7 | 94.2 | 98.5 | 88.8 | 94.6 | 99.3 | 87.1 | 94.4 | 98.6 | 71.6 | 87.9 | 98.5 | 86.9 | 93.9 | 98.3 | 87.2 | 93.9 |
| 08 | OPENINGS | 98.1 | 80.4 | 93.6 | 98.6 | 84.1 | 94.9 | 100.2 | 78.7 | 94.7 | 101.6 | 67.7 | 93.0 | 101.6 | 82.1 | 96.7 | 99.2 | 81.0 | 94.6 |
| 0920 | Plaster & Gypsum Board | 96.6 | 84.6 | 88.1 | 105.4 | 91.5 | 95.7 | 95.3 | 89.3 | 91.1 | 96.6 | 79.1 | 84.3 | 95.3 | 84.6 | 87.7 | 98.4 | 86.2 | 89.8 |
| 0950, 0980 | Ceilings & Acoustic Treatment | 92.0 | 84.6 | 87.1 | 92.0 | 91.5 | 91.7 | 92.0 | 89.3 | 90.2 | 92.0 | 79.1 | 83.4 | 92.0 | 84.6 | 87.1 | 92.0 | 86.2 | 88.2 |
| 0960 | Flooring | 101.3 | 106.0 | 102.6 | 109.7 | 106.0 | 108.6 | 100.4 | 106.0 | 102.0 | 101.2 | 56.7 | 88.4 | 100.4 | 100.2 | 100.3 | 104.7 | 97.4 | 102.6 |
| 0970, 0990 | Wall Finishes & Painting/Coating | 114.7 | 84.9 | 96.7 | 114.7 | 84.9 | 96.7 | 114.7 | 37.7 | 68.0 | 114.7 | 44.5 | 72.1 | 114.7 | 85.5 | 97.0 | 114.7 | 85.0 | 96.7 |
| 09 | FINISHES | 99.0 | 89.2 | 93.6 | 102.3 | 94.4 | 97.9 | 100.0 | 87.0 | 92.8 | 99.0 | 71.4 | 83.8 | 98.4 | 88.1 | 92.7 | 100.7 | 89.1 | 94.3 |
| COVERS | DIVS. 10 - 14, 25, 28, 41, 43, 44, 46 | 100.0 | 97.7 | 99.5 | 100.0 | 100.2 | 100.0 | 100.0 | 59.2 | 91.5 | 100.0 | 57.3 | 91.1 | 100.0 | 68.5 | 93.4 | 100.0 | 98.8 | 99.8 |
| 21, 22, 23 | FIRE SUPPRESSION, PLUMBING & HVAC | 94.9 | 89.9 | 92.8 | 100.3 | 83.3 | 93.3 | 94.9 | 81.8 | 89.5 | 94.9 | 84.5 | 90.6 | 94.9 | 88.8 | 92.4 | 100.2 | 86.8 | 94.6 |
| 26, 27, 3370 | ELECTRICAL, COMMUNICATIONS & UTIL. | 96.8 | 89.5 | 92.9 | 96.5 | 95.9 | 96.2 | 92.4 | 70.4 | 80.7 | 98.8 | 82.1 | 89.9 | 99.8 | 93.5 | 95.1 | 97.0 | 92.2 | 94.4 |
| MF2010 | WEIGHTED AVERAGE | 98.4 | 89.9 | 94.6 | 100.2 | 91.6 | 96.3 | 98.9 | 83.2 | 91.9 | 98.6 | 79.1 | 89.9 | 99.2 | 88.8 | 94.6 | 99.1 | 89.4 | 94.8 |

| DIVISION | | WEST VIRGINIA | | | | | | | | | WISCONSIN | | | | | | | | |
|---|---|---|---|---|---|---|---|---|---|---|---|---|---|---|---|---|---|---|---|
| | | PETERSBURG 268 | | | ROMNEY 267 | | | WHEELING 260 | | | BELOIT 535 | | | EAU CLAIRE 547 | | | GREEN BAY 541 - 543 | | |
| | | MAT. | INST. | TOTAL | MAT. | INST. | TOTAL | MAT. | INST. | TOTAL | MAT. | INST. | TOTAL | MAT. | INST. | TOTAL | MAT. | INST. | TOTAL |
| 015433 | CONTRACTOR EQUIPMENT | | 100.4 | 100.4 | | 100.4 | 100.4 | | 100.4 | 100.4 | | 101.0 | 101.0 | | 100.7 | 100.7 | | 98.6 | 98.6 |
| 0241, 31 - 34 | SITE & INFRASTRUCTURE, DEMOLITION | 106.1 | 87.0 | 92.9 | 109.1 | 87.0 | 93.9 | 116.2 | 87.3 | 96.2 | 94.2 | 106.3 | 102.5 | 95.9 | 103.1 | 100.9 | 99.9 | 99.4 | 99.5 |
| 0310 | Concrete Forming & Accessories | 85.4 | 80.6 | 81.2 | 82.0 | 80.8 | 80.9 | 88.1 | 85.7 | 86.0 | 96.4 | 96.7 | 96.7 | 98.3 | 98.5 | 98.4 | 102.9 | 97.8 | 98.4 |
| 0320 | Concrete Reinforcing | 89.3 | 73.3 | 81.0 | 89.8 | 70.6 | 79.9 | 88.7 | 84.9 | 86.7 | 98.3 | 118.7 | 108.8 | 95.2 | 103.3 | 99.4 | 93.5 | 89.7 | 91.5 |
| 0330 | Cast-in-Place Concrete | 110.7 | 95.7 | 104.9 | 115.9 | 95.7 | 108.2 | 113.2 | 101.2 | 108.6 | 101.9 | 95.2 | 99.3 | 101.6 | 96.0 | 99.5 | 105.1 | 100.7 | 103.4 |
| 03 | CONCRETE | 104.6 | 85.2 | 95.3 | 108.4 | 84.8 | 97.1 | 110.1 | 91.5 | 101.2 | 99.5 | 100.5 | 100.0 | 101.0 | 98.8 | 99.9 | 104.0 | 97.5 | 100.9 |
| 04 | MASONRY | 99.3 | 84.7 | 90.5 | 96.9 | 84.0 | 89.0 | 106.9 | 84.7 | 93.5 | 104.7 | 95.9 | 99.4 | 95.6 | 100.1 | 98.3 | 129.4 | 99.3 | 111.2 |
| 05 | METALS | 91.0 | 91.6 | 91.2 | 91.1 | 90.8 | 91.0 | 91.8 | 96.8 | 93.5 | 94.0 | 109.5 | 99.3 | 95.5 | 104.3 | 98.5 | 97.7 | 98.2 | 97.9 |
| 06 | WOOD, PLASTICS & COMPOSITES | 86.5 | 79.8 | 82.7 | 81.9 | 79.8 | 80.7 | 87.5 | 85.2 | 86.2 | 102.8 | 99.2 | 100.7 | 108.7 | 99.3 | 103.3 | 109.0 | 99.3 | 103.4 |
| 07 | THERMAL & MOISTURE PROTECTION | 98.5 | 71.7 | 87.9 | 98.6 | 72.0 | 88.0 | 98.7 | 87.4 | 94.2 | 105.2 | 93.1 | 100.4 | 105.2 | 86.7 | 97.9 | 106.7 | 87.0 | 98.9 |
| 08 | OPENINGS | 101.6 | 74.7 | 94.8 | 101.5 | 74.1 | 94.6 | 100.1 | 82.1 | 95.6 | 100.0 | 107.0 | 101.8 | 98.2 | 97.9 | 98.1 | 96.6 | 96.6 | 96.6 |
| 0920 | Plaster & Gypsum Board | 97.5 | 79.1 | 84.5 | 94.8 | 79.1 | 83.7 | 98.8 | 84.6 | 88.8 | 95.7 | 99.5 | 98.4 | 100.4 | 99.5 | 99.8 | 94.5 | 99.5 | 98.0 |
| 0950, 0980 | Ceilings & Acoustic Treatment | 92.0 | 79.1 | 83.4 | 92.0 | 79.1 | 83.4 | 92.0 | 84.6 | 87.1 | 83.7 | 99.5 | 94.2 | 97.2 | 99.5 | 98.7 | 89.8 | 99.5 | 96.3 |
| 0960 | Flooring | 102.3 | 100.2 | 101.7 | 100.1 | 98.7 | 99.7 | 105.7 | 100.2 | 104.1 | 107.3 | 118.6 | 110.5 | 98.7 | 114.3 | 103.2 | 118.8 | 114.3 | 117.5 |
| 0970, 0990 | Wall Finishes & Painting/Coating | 114.7 | 84.9 | 96.7 | 114.7 | 84.9 | 96.7 | 114.7 | 85.5 | 97.0 | 95.9 | 99.5 | 98.0 | 90.0 | 93.1 | 91.9 | 100.4 | 82.0 | 89.2 |
| 09 | FINISHES | 99.2 | 84.9 | 91.3 | 98.5 | 84.6 | 90.8 | 101.0 | 88.1 | 93.9 | 100.6 | 101.3 | 101.0 | 99.0 | 101.2 | 100.2 | 104.0 | 99.8 | 101.7 |
| COVERS | DIVS. 10 - 14, 25, 28, 41, 43, 44, 46 | 100.0 | 67.8 | 93.3 | 100.0 | 97.1 | 99.4 | 100.0 | 99.4 | 99.9 | 100.0 | 99.2 | 99.8 | 100.0 | 95.3 | 99.0 | 100.0 | 95.0 | 99.0 |
| 21, 22, 23 | FIRE SUPPRESSION, PLUMBING & HVAC | 94.9 | 89.9 | 92.8 | 94.9 | 89.8 | 92.8 | 100.3 | 88.9 | 95.6 | 99.9 | 91.1 | 96.2 | 100.1 | 86.7 | 94.5 | 100.3 | 84.8 | 93.9 |
| 26, 27, 3370 | ELECTRICAL, COMMUNICATIONS & UTIL. | 100.5 | 82.2 | 90.8 | 99.7 | 82.2 | 90.4 | 93.9 | 93.5 | 93.7 | 99.1 | 89.4 | 94.0 | 102.1 | 92.2 | 96.8 | 96.1 | 84.6 | 90.0 |
| MF2010 | WEIGHTED AVERAGE | 98.2 | 85.0 | 92.3 | 98.4 | 85.6 | 92.7 | 100.3 | 89.9 | 95.7 | 99.3 | 97.8 | 98.6 | 99.3 | 96.1 | 97.9 | 101.7 | 93.3 | 98.0 |

| DIVISION | | WISCONSIN | | | | | | | | | | | | | | | | | |
|---|---|---|---|---|---|---|---|---|---|---|---|---|---|---|---|---|---|---|---|
| | | KENOSHA 531 | | | LA CROSSE 546 | | | LANCASTER 538 | | | MADISON 537 | | | MILWAUKEE 530,532 | | | NEW RICHMOND 540 | | |
| | | MAT. | INST. | TOTAL | MAT. | INST. | TOTAL | MAT. | INST. | TOTAL | MAT. | INST. | TOTAL | MAT. | INST. | TOTAL | MAT. | INST. | TOTAL |
| 015433 | CONTRACTOR EQUIPMENT | | 99.1 | 99.1 | | 100.7 | 100.7 | | 101.0 | 101.0 | | 101.0 | 101.0 | | 89.0 | 89.0 | | 101.5 | 101.5 |
| 0241, 31 - 34 | SITE & INFRASTRUCTURE, DEMOLITION | 100.6 | 103.8 | 102.8 | 89.1 | 103.2 | 98.8 | 93.9 | 106.6 | 102.7 | 89.9 | 106.7 | 101.5 | 95.0 | 96.7 | 96.2 | 90.9 | 103.8 | 99.8 |
| 0310 | Concrete Forming & Accessories | 99.8 | 105.5 | 104.8 | 87.1 | 98.6 | 97.1 | 95.8 | 97.6 | 97.4 | 95.8 | 99.4 | 98.9 | 98.3 | 117.3 | 114.8 | 94.5 | 105.4 | 104.0 |
| 0320 | Concrete Reinforcing | 98.1 | 101.9 | 100.1 | 95.0 | 92.0 | 93.4 | 99.6 | 91.8 | 95.6 | 98.3 | 92.2 | 95.2 | 98.3 | 102.3 | 100.4 | 92.5 | 103.2 | 98.0 |
| 0330 | Cast-in-Place Concrete | 110.9 | 99.8 | 106.7 | 91.4 | 93.4 | 92.1 | 101.3 | 99.0 | 100.4 | 93.2 | 99.7 | 95.7 | 99.6 | 111.8 | 104.2 | 105.8 | 82.9 | 97.0 |
| 03 | CONCRETE | 104.2 | 102.9 | 103.6 | 92.1 | 95.8 | 93.9 | 99.2 | 97.1 | 98.2 | 95.1 | 98.2 | 96.6 | 98.6 | 111.7 | 104.9 | 98.9 | 97.4 | 98.2 |
| 04 | MASONRY | 101.9 | 109.4 | 106.4 | 94.8 | 102.1 | 99.2 | 104.7 | 101.8 | 102.9 | 102.0 | 104.4 | 103.4 | 104.6 | 120.8 | 114.4 | 123.8 | 94.6 | 107.2 |
| 05 | METALS | 94.7 | 103.3 | 97.6 | 95.4 | 98.9 | 96.6 | 91.5 | 95.9 | 93.0 | 96.7 | 98.0 | 97.1 | 97.5 | 96.8 | 97.2 | 95.5 | 103.3 | 98.2 |
| 06 | WOOD, PLASTICS & COMPOSITES | 102.5 | 104.9 | 103.9 | 95.7 | 99.3 | 97.7 | 102.1 | 99.2 | 100.4 | 100.1 | 99.2 | 99.6 | 104.4 | 117.3 | 111.8 | 98.3 | 111.5 | 105.8 |
| 07 | THERMAL & MOISTURE PROTECTION | 105.0 | 98.1 | 102.3 | 104.8 | 86.3 | 97.4 | 105.0 | 81.2 | 95.6 | 101.8 | 92.5 | 98.1 | 103.2 | 113.8 | 107.4 | 107.2 | 93.9 | 101.9 |
| 08 | OPENINGS | 95.4 | 106.5 | 98.3 | 98.2 | 85.6 | 95.0 | 95.3 | 86.4 | 93.1 | 100.0 | 99.5 | 99.9 | 102.1 | 113.2 | 104.9 | 86.7 | 104.5 | 91.2 |
| 0920 | Plaster & Gypsum Board | 80.9 | 105.4 | 98.1 | 95.0 | 99.5 | 98.2 | 94.6 | 99.5 | 98.1 | 97.8 | 99.5 | 99.0 | 96.7 | 118.0 | 111.7 | 86.1 | 112.3 | 104.6 |
| 0950, 0980 | Ceilings & Acoustic Treatment | 83.7 | 105.4 | 98.1 | 95.3 | 99.5 | 98.1 | 79.9 | 99.5 | 92.9 | 87.6 | 99.5 | 95.5 | 87.4 | 118.0 | 107.7 | 64.5 | 112.3 | 96.2 |
| 0960 | Flooring | 125.9 | 113.2 | 122.2 | 91.3 | 114.3 | 97.9 | 106.4 | 107.1 | 106.6 | 101.3 | 107.1 | 103.0 | 110.3 | 120.4 | 113.2 | 111.2 | 114.3 | 112.1 |
| 0970, 0990 | Wall Finishes & Painting/Coating | 104.3 | 105.9 | 105.2 | 90.0 | 74.0 | 80.3 | 95.9 | 71.5 | 81.1 | 95.9 | 99.5 | 98.0 | 97.8 | 116.2 | 108.9 | 104.0 | 93.1 | 97.4 |
| 09 | FINISHES | 105.2 | 107.5 | 106.4 | 95.2 | 99.6 | 97.6 | 99.4 | 96.2 | 97.7 | 99.5 | 101.2 | 100.4 | 102.6 | 118.4 | 111.3 | 95.0 | 105.7 | 100.9 |
| COVERS | DIVS. 10 - 14, 25, 28, 41, 43, 44, 46 | 100.0 | 100.3 | 100.1 | 100.0 | 96.0 | 99.2 | 100.0 | 52.8 | 90.2 | 100.0 | 95.9 | 99.1 | 100.0 | 103.9 | 100.8 | 100.0 | 93.0 | 98.5 |
| 21, 22, 23 | FIRE SUPPRESSION, PLUMBING & HVAC | 100.1 | 97.0 | 98.8 | 100.1 | 87.5 | 94.9 | 94.5 | 87.3 | 91.5 | 99.8 | 96.8 | 98.6 | 99.9 | 104.0 | 101.6 | 94.3 | 84.2 | 90.1 |
| 26, 27, 3370 | ELECTRICAL, COMMUNICATIONS & UTIL. | 99.6 | 101.5 | 100.6 | 102.4 | 92.4 | 97.1 | 98.9 | 92.4 | 95.4 | 102.1 | 96.2 | 98.9 | 100.6 | 100.3 | 100.4 | 100.0 | 92.4 | 96.0 |
| MF2010 | WEIGHTED AVERAGE | 100.0 | 102.7 | 101.2 | 97.6 | 94.9 | 96.4 | 97.0 | 93.2 | 95.3 | 98.9 | 99.1 | 99.0 | 100.2 | 107.6 | 103.5 | 97.3 | 96.2 | 96.8 |

| DIVISION | | WISCONSIN | | | | | | | | | | | | | | | | | |
|---|---|---|---|---|---|---|---|---|---|---|---|---|---|---|---|---|---|---|---|
| | | OSHKOSH 549 | | | PORTAGE 539 | | | RACINE 534 | | | RHINELANDER 545 | | | SUPERIOR 548 | | | WAUSAU 544 | | |
| | | MAT. | INST. | TOTAL | MAT. | INST. | TOTAL | MAT. | INST. | TOTAL | MAT. | INST. | TOTAL | MAT. | INST. | TOTAL | MAT. | INST. | TOTAL |
| 015433 | CONTRACTOR EQUIPMENT | | 98.6 | 98.6 | | 101.0 | 101.0 | | 101.0 | 101.0 | | 98.6 | 98.6 | | 101.5 | 101.5 | | 98.6 | 98.6 |
| 0241, 31 - 34 | SITE & INFRASTRUCTURE, DEMOLITION | 91.3 | 98.3 | 96.1 | 84.0 | 106.7 | 99.7 | 94.0 | 107.3 | 103.2 | 105.1 | 99.5 | 101.2 | 87.3 | 104.1 | 98.9 | 86.7 | 99.4 | 95.5 |
| 0310 | Concrete Forming & Accessories | 88.6 | 86.5 | 86.8 | 89.9 | 98.5 | 97.4 | 96.7 | 105.5 | 104.4 | 86.6 | 97.7 | 96.2 | 93.1 | 94.7 | 94.5 | 88.2 | 97.7 | 96.5 |
| 0320 | Concrete Reinforcing | 93.6 | 78.9 | 86.0 | 99.7 | 91.9 | 95.7 | 98.3 | 101.9 | 100.1 | 93.7 | 94.1 | 93.9 | 92.5 | 94.6 | 93.6 | 93.7 | 92.0 | 92.8 |
| 0330 | Cast-in-Place Concrete | 97.5 | 90.5 | 94.8 | 87.0 | 101.3 | 92.5 | 90.6 | 99.8 | 94.8 | 110.5 | 101.4 | 107.0 | 99.7 | 99.3 | 99.5 | 90.8 | 100.7 | 94.6 |
| 03 | CONCRETE | 93.6 | 86.9 | 90.4 | 87.1 | 98.3 | 92.5 | 98.5 | 102.8 | 100.6 | 106.2 | 98.5 | 102.5 | 93.2 | 96.6 | 94.8 | 88.4 | 97.9 | 93.0 |
| 04 | MASONRY | 109.1 | 89.9 | 97.4 | 103.6 | 101.7 | 102.5 | 104.6 | 109.3 | 107.5 | 126.6 | 101.3 | 111.3 | 123.1 | 106.3 | 112.9 | 108.6 | 99.3 | 103.0 |
| 05 | METALS | 95.7 | 90.8 | 94.1 | 92.2 | 96.8 | 93.8 | 95.7 | 103.3 | 98.3 | 95.6 | 99.2 | 96.9 | 96.5 | 100.8 | 98.0 | 95.5 | 99.1 | 96.7 |
| 06 | WOOD, PLASTICS & COMPOSITES | 92.1 | 88.1 | 89.8 | 94.5 | 99.2 | 97.2 | 103.0 | 104.9 | 104.1 | 89.8 | 99.3 | 95.2 | 96.8 | 93.3 | 94.8 | 91.6 | 99.3 | 96.0 |
| 07 | THERMAL & MOISTURE PROTECTION | 105.9 | 79.9 | 95.5 | 104.5 | 91.0 | 99.1 | 105.3 | 97.7 | 102.3 | 106.7 | 85.3 | 98.2 | 106.9 | 97.0 | 103.0 | 105.7 | 86.0 | 97.9 |
| 08 | OPENINGS | 92.0 | 84.9 | 90.2 | 95.5 | 93.5 | 95.0 | 100.0 | 106.5 | 101.6 | 92.0 | 88.6 | 91.2 | 86.7 | 93.3 | 88.3 | 92.2 | 97.2 | 93.5 |
| 0920 | Plaster & Gypsum Board | 84.4 | 88.0 | 86.9 | 89.1 | 99.5 | 96.4 | 95.7 | 105.4 | 102.5 | 83.9 | 99.5 | 94.9 | 85.9 | 93.6 | 91.3 | 84.4 | 99.5 | 95.0 |
| 0950, 0980 | Ceilings & Acoustic Treatment | 89.8 | 88.0 | 88.6 | 80.9 | 99.5 | 93.2 | 83.7 | 105.4 | 98.1 | 89.8 | 99.5 | 96.3 | 65.5 | 93.6 | 84.1 | 89.8 | 99.5 | 96.3 |
| 0960 | Flooring | 109.2 | 100.6 | 106.7 | 102.1 | 113.7 | 105.4 | 107.3 | 113.2 | 109.0 | 108.2 | 114.3 | 109.9 | 112.7 | 123.6 | 115.8 | 109.0 | 114.3 | 110.5 |
| 0970, 0990 | Wall Finishes & Painting/Coating | 98.4 | 69.8 | 81.0 | 95.9 | 71.5 | 81.1 | 95.8 | 111.0 | 105.0 | 98.4 | 65.0 | 78.1 | 92.7 | 102.5 | 98.6 | 98.4 | 82.0 | 88.4 |
| 09 | FINISHES | 99.2 | 86.9 | 92.4 | 96.9 | 98.3 | 97.7 | 100.6 | 108.0 | 104.7 | 100.0 | 97.4 | 98.6 | 94.5 | 100.8 | 98.0 | 98.8 | 99.8 | 99.3 |
| COVERS | DIVS. 10 - 14, 25, 28, 41, 43, 44, 46 | 100.0 | 66.4 | 93.0 | 100.0 | 68.4 | 93.4 | 100.0 | 100.3 | 100.1 | 100.0 | 58.9 | 91.5 | 100.0 | 92.6 | 98.5 | 100.0 | 95.0 | 99.0 |
| 21, 22, 23 | FIRE SUPPRESSION, PLUMBING & HVAC | 94.9 | 76.3 | 87.2 | 94.5 | 92.3 | 93.6 | 99.9 | 97.0 | 98.7 | 94.9 | 86.8 | 91.5 | 94.3 | 91.8 | 93.2 | 94.9 | 85.5 | 91.0 |
| 26, 27, 3370 | ELECTRICAL, COMMUNICATIONS & UTIL. | 100.7 | 77.8 | 88.5 | 103.0 | 95.7 | 99.1 | 98.9 | 99.4 | 99.2 | 100.0 | 83.2 | 91.0 | 105.3 | 98.9 | 101.9 | 101.9 | 83.2 | 91.9 |
| MF2010 | WEIGHTED AVERAGE | 97.0 | 84.1 | 91.2 | 95.4 | 96.3 | 95.8 | 99.4 | 102.8 | 100.9 | 99.8 | 92.2 | 96.4 | 97.1 | 98.1 | 97.6 | 96.2 | 93.4 | 95.0 |

| DIVISION | | WYOMING | | | | | | | | | | | | | | | | |
|---|---|---|---|---|---|---|---|---|---|---|---|---|---|---|---|---|---|---|
| | | CASPER | | | CHEYENNE | | | NEWCASTLE | | | RAWLINS | | | RIVERTON | | | ROCK SPRINGS | | |
| | | 826 | | | 820 | | | 827 | | | 823 | | | 825 | | | 829 - 831 | | |
| | | MAT. | INST. | TOTAL | MAT. | INST. | TOTAL | MAT. | INST. | TOTAL | MAT. | INST. | TOTAL | MAT. | INST. | TOTAL | MAT. | INST. | TOTAL |
| 015433 | CONTRACTOR EQUIPMENT | | 100.6 | 100.6 | | 100.6 | 100.6 | | 100.6 | 100.6 | | 100.6 | 100.6 | | 100.6 | 100.6 | | 100.6 | 100.6 |
| 0241, 31 - 34 | SITE & INFRASTRUCTURE, DEMOLITION | 89.4 | 98.5 | 95.7 | 82.6 | 98.6 | 93.6 | 75.8 | 97.4 | 90.7 | 89.1 | 97.4 | 94.8 | 82.9 | 97.4 | 92.9 | 79.0 | 97.4 | 91.7 |
| 0310 | Concrete Forming & Accessories | 103.0 | 42.0 | 50.1 | 104.7 | 57.1 | 63.4 | 95.9 | 44.7 | 51.5 | 99.8 | 44.7 | 52.0 | 94.8 | 44.7 | 51.3 | 101.5 | 44.7 | 52.2 |
| 0320 | Concrete Reinforcing | 105.3 | 65.1 | 84.6 | 99.4 | 65.9 | 82.1 | 106.8 | 65.7 | 85.6 | 106.4 | 65.7 | 85.5 | 107.4 | 65.7 | 85.9 | 107.4 | 67.3 | 86.8 |
| 0330 | Cast-in-Place Concrete | 96.4 | 73.3 | 87.6 | 94.7 | 73.5 | 86.6 | 95.7 | 52.9 | 79.4 | 95.8 | 52.9 | 79.4 | 95.8 | 52.9 | 79.4 | 95.7 | 52.9 | 79.3 |
| 03 | CONCRETE | 101.7 | 58.0 | 80.8 | 100.6 | 64.9 | 83.5 | 101.1 | 52.3 | 77.7 | 114.2 | 52.3 | 84.6 | 109.0 | 52.3 | 81.8 | 101.6 | 52.6 | 78.1 |
| 04 | MASONRY | 101.5 | 36.6 | 62.2 | 98.3 | 50.5 | 69.4 | 95.1 | 45.3 | 64.9 | 95.1 | 45.3 | 64.9 | 95.1 | 45.3 | 64.9 | 153.6 | 45.3 | 88.0 |
| 05 | METALS | 98.1 | 68.7 | 88.0 | 99.3 | 70.3 | 89.3 | 94.4 | 68.1 | 85.3 | 94.4 | 68.1 | 85.4 | 94.5 | 68.1 | 85.4 | 95.3 | 68.7 | 86.1 |
| 06 | WOOD, PLASTICS & COMPOSITES | 94.3 | 39.2 | 62.7 | 96.0 | 59.0 | 74.8 | 86.3 | 45.7 | 63.0 | 90.3 | 45.7 | 64.7 | 85.3 | 45.7 | 62.6 | 93.7 | 45.7 | 66.2 |
| 07 | THERMAL & MOISTURE PROTECTION | 98.0 | 48.4 | 78.3 | 99.6 | 54.2 | 81.5 | 101.0 | 47.5 | 79.7 | 102.7 | 47.5 | 80.8 | 102.0 | 47.5 | 80.4 | 101.1 | 47.5 | 79.8 |
| 08 | OPENINGS | 94.7 | 44.1 | 81.9 | 96.8 | 54.9 | 86.2 | 101.1 | 47.6 | 87.5 | 100.7 | 47.6 | 87.3 | 100.9 | 47.6 | 87.4 | 101.5 | 48.0 | 87.9 |
| 0920 | Plaster & Gypsum Board | 101.1 | 37.2 | 56.1 | 92.6 | 57.7 | 68.0 | 86.6 | 43.9 | 56.6 | 87.9 | 43.9 | 57.0 | 86.6 | 43.9 | 56.6 | 96.3 | 43.9 | 59.4 |
| 0950, 0980 | Ceilings & Acoustic Treatment | 124.9 | 37.2 | 66.8 | 116.1 | 57.7 | 77.4 | 119.3 | 43.9 | 69.3 | 119.3 | 43.9 | 69.3 | 119.3 | 43.9 | 69.3 | 119.3 | 43.9 | 69.3 |
| 0960 | Flooring | 100.3 | 34.9 | 81.5 | 104.7 | 56.0 | 90.7 | 98.5 | 49.3 | 84.3 | 101.3 | 49.3 | 86.3 | 97.8 | 49.3 | 83.8 | 103.4 | 49.3 | 87.9 |
| 0970, 0990 | Wall Finishes & Painting/Coating | 98.5 | 46.3 | 66.8 | 101.8 | 46.3 | 68.2 | 97.7 | 29.7 | 56.4 | 97.7 | 29.7 | 56.4 | 97.7 | 29.7 | 56.4 | 97.7 | 29.7 | 56.4 |
| 09 | FINISHES | 106.9 | 39.8 | 69.9 | 106.2 | 55.8 | 78.4 | 101.0 | 43.2 | 69.1 | 103.6 | 43.2 | 70.3 | 101.8 | 43.2 | 69.5 | 103.7 | 43.2 | 70.3 |
| COVERS | DIVS. 10 - 14, 25, 28, 41, 43, 44, 46 | 100.0 | 77.1 | 95.2 | 100.0 | 79.4 | 95.7 | 100.0 | 59.3 | 91.5 | 100.0 | 59.3 | 91.5 | 100.0 | 59.3 | 91.5 | 100.0 | 59.3 | 91.5 |
| 21, 22, 23 | FIRE SUPPRESSION, PLUMBING & HVAC | 100.0 | 56.2 | 81.8 | 100.0 | 56.3 | 81.8 | 97.2 | 53.9 | 79.2 | 97.2 | 53.9 | 79.2 | 97.2 | 53.9 | 79.2 | 99.9 | 53.9 | 80.8 |
| 26, 27, 3370 | ELECTRICAL, COMMUNICATIONS & UTIL. | 98.9 | 68.3 | 82.6 | 97.3 | 74.4 | 85.2 | 95.2 | 68.3 | 80.9 | 95.2 | 68.3 | 80.9 | 95.2 | 68.3 | 80.9 | 93.3 | 68.3 | 80.0 |
| MF2010 | WEIGHTED AVERAGE | 99.5 | 58.5 | 81.2 | 99.2 | 64.7 | 83.8 | 97.3 | 58.0 | 79.7 | 99.6 | 58.0 | 81.0 | 98.6 | 58.0 | 80.4 | 101.5 | 58.1 | 82.1 |

| DIVISION | | WYOMING | | | | | | | | | | | | CANADA | | | | | |
|---|---|---|---|---|---|---|---|---|---|---|---|---|---|---|---|---|---|---|---|
| | | SHERIDAN | | | WHEATLAND | | | WORLAND | | | YELLOWSTONE NAT'L PA | | | BARRIE, ONTARIO | | | BATHURST, NEW BRUNSWICK | | |
| | | 828 | | | 822 | | | 824 | | | 821 | | | | | | | | |
| | | MAT. | INST. | TOTAL | MAT. | INST. | TOTAL | MAT. | INST. | TOTAL | MAT. | INST. | TOTAL | MAT. | INST. | TOTAL | MAT. | INST. | TOTAL |
| 015433 | CONTRACTOR EQUIPMENT | | 100.6 | 100.6 | | 100.6 | 100.6 | | 100.6 | 100.6 | | 100.6 | 100.6 | | 101.6 | 101.6 | | 101.4 | 101.4 |
| 0241, 31 - 34 | SITE & INFRASTRUCTURE, DEMOLITION | 83.0 | 98.6 | 93.7 | 79.8 | 97.5 | 92.0 | 77.5 | 97.4 | 91.2 | 77.5 | 97.8 | 91.5 | 115.0 | 101.5 | 105.7 | 94.2 | 97.3 | 96.3 |
| 0310 | Concrete Forming & Accessories | 102.6 | 47.1 | 54.4 | 97.8 | 45.1 | 52.1 | 97.9 | 44.7 | 51.7 | 97.9 | 47.0 | 53.7 | 123.1 | 95.7 | 99.3 | 102.2 | 68.0 | 72.5 |
| 0320 | Concrete Reinforcing | 107.4 | 65.8 | 86.0 | 106.8 | 65.7 | 85.6 | 107.4 | 65.7 | 85.9 | 109.3 | 66.8 | 87.4 | 180.1 | 89.0 | 133.2 | 148.4 | 60.0 | 102.9 |
| 0330 | Cast-in-Place Concrete | 99.0 | 73.4 | 89.2 | 99.9 | 53.5 | 82.1 | 95.7 | 52.9 | 79.3 | 95.7 | 56.4 | 80.6 | 167.6 | 94.5 | 139.6 | 138.2 | 65.1 | 110.2 |
| 03 | CONCRETE | 109.1 | 60.4 | 85.8 | 106.0 | 52.6 | 80.4 | 101.3 | 52.3 | 77.9 | 101.6 | 54.7 | 79.1 | 152.8 | 94.1 | 124.7 | 135.5 | 66.1 | 102.3 |
| 04 | MASONRY | 95.4 | 49.9 | 67.8 | 95.4 | 46.2 | 65.6 | 95.1 | 45.3 | 64.9 | 95.2 | 51.4 | 68.6 | 168.4 | 104.3 | 129.5 | 158.7 | 69.0 | 104.4 |
| 05 | METALS | 98.1 | 70.0 | 88.4 | 94.4 | 68.1 | 85.3 | 94.5 | 68.1 | 85.4 | 95.2 | 68.5 | 86.0 | 107.4 | 93.5 | 102.6 | 104.7 | 73.4 | 94.0 |
| 06 | WOOD, PLASTICS & COMPOSITES | 97.2 | 45.7 | 67.7 | 88.0 | 45.7 | 63.7 | 88.0 | 45.7 | 63.8 | 88.0 | 45.7 | 63.8 | 118.4 | 94.2 | 104.5 | 94.9 | 68.3 | 79.6 |
| 07 | THERMAL & MOISTURE PROTECTION | 102.1 | 52.6 | 82.4 | 101.3 | 48.5 | 80.3 | 101.0 | 47.5 | 79.7 | 100.6 | 50.7 | 80.8 | 111.4 | 94.9 | 104.8 | 104.8 | 66.1 | 89.4 |
| 08 | OPENINGS | 101.8 | 48.6 | 88.3 | 99.6 | 47.6 | 86.5 | 101.3 | 47.6 | 87.7 | 94.1 | 47.9 | 82.4 | 91.8 | 92.1 | 91.9 | 86.9 | 59.5 | 80.0 |
| 0920 | Plaster & Gypsum Board | 111.4 | 43.9 | 63.9 | 86.6 | 43.9 | 56.6 | 86.6 | 43.9 | 56.6 | 86.9 | 43.9 | 56.6 | 157.5 | 94.0 | 112.8 | 163.1 | 67.2 | 95.6 |
| 0950, 0980 | Ceilings & Acoustic Treatment | 122.4 | 43.9 | 70.4 | 119.3 | 43.9 | 69.3 | 119.3 | 43.9 | 69.3 | 120.2 | 43.9 | 69.6 | 95.2 | 94.0 | 94.4 | 96.2 | 67.2 | 77.0 |
| 0960 | Flooring | 101.3 | 49.3 | 86.4 | 99.8 | 49.3 | 85.3 | 99.8 | 49.3 | 85.3 | 99.8 | 49.3 | 85.3 | 132.0 | 102.7 | 123.5 | 115.2 | 49.0 | 96.2 |
| 0970, 0990 | Wall Finishes & Painting/Coating | 99.9 | 46.3 | 67.4 | 97.7 | 29.7 | 56.4 | 97.7 | 29.7 | 56.4 | 97.7 | 29.7 | 56.4 | 112.2 | 95.1 | 101.8 | 111.4 | 54.6 | 76.9 |
| 09 | FINISHES | 108.5 | 46.2 | 74.2 | 101.8 | 43.4 | 69.6 | 101.4 | 43.2 | 69.3 | 101.7 | 44.7 | 70.2 | 118.8 | 97.2 | 106.9 | 112.9 | 63.2 | 85.5 |
| COVERS | DIVS. 10 - 14, 25, 28, 41, 43, 44, 46 | 100.0 | 77.9 | 95.4 | 100.0 | 59.6 | 91.6 | 100.0 | 59.3 | 91.5 | 100.0 | 61.4 | 92.0 | 140.0 | 82.1 | 128.0 | 140.0 | 71.9 | 125.9 |
| 21, 22, 23 | FIRE SUPPRESSION, PLUMBING & HVAC | 97.2 | 56.2 | 80.2 | 97.2 | 54.3 | 79.4 | 97.2 | 53.9 | 79.2 | 97.2 | 57.0 | 80.5 | 102.4 | 107.3 | 104.5 | 102.6 | 75.4 | 91.3 |
| 26, 27, 3370 | ELECTRICAL, COMMUNICATIONS & UTIL. | 97.7 | 68.3 | 82.1 | 95.2 | 74.4 | 84.1 | 95.2 | 68.3 | 80.9 | 94.1 | 68.3 | 80.4 | 120.1 | 96.6 | 107.6 | 121.9 | 66.0 | 92.2 |
| MF2010 | WEIGHTED AVERAGE | 100.1 | 61.5 | 82.8 | 98.0 | 59.2 | 80.6 | 97.5 | 58.0 | 79.8 | 96.8 | 60.0 | 80.4 | 118.9 | 98.8 | 109.9 | 114.2 | 71.2 | 94.9 |

| DIVISION | | CANADA | | | | | | | | | | | | | | | | | |
|---|---|---|---|---|---|---|---|---|---|---|---|---|---|---|---|---|---|---|---|
| | | BRANDON, MANITOBA | | | BRANTFORD, ONTARIO | | | BRIDGEWATER, NOVA SCOTIA | | | CALGARY, ALBERTA | | | CAP-DE-LA-MADELEINE, QUEBEC | | | CHARLESBOURG, QUEBEC | | |
| | | MAT. | INST. | TOTAL | MAT. | INST. | TOTAL | MAT. | INST. | TOTAL | MAT. | INST. | TOTAL | MAT. | INST. | TOTAL | MAT. | INST. | TOTAL |
| 015433 | CONTRACTOR EQUIPMENT | | 103.8 | 103.8 | | 101.6 | 101.6 | | 101.2 | 101.2 | | 106.6 | 106.6 | | 102.0 | 102.0 | | 102.0 | 102.0 |
| 0241, 31 - 34 | SITE & INFRASTRUCTURE, DEMOLITION | 110.6 | 100.3 | 103.5 | 112.7 | 101.8 | 105.2 | 96.9 | 99.0 | 98.4 | 125.5 | 105.6 | 111.8 | 92.5 | 100.7 | 98.2 | 92.5 | 100.7 | 98.2 |
| 0310 | Concrete Forming & Accessories | 122.4 | 77.4 | 83.4 | 122.9 | 103.7 | 106.2 | 96.4 | 76.5 | 79.2 | 127.5 | 98.4 | 102.2 | 128.1 | 94.1 | 98.6 | 128.1 | 94.1 | 98.6 |
| 0320 | Concrete Reinforcing | 161.6 | 56.5 | 107.5 | 175.0 | 87.6 | 130.0 | 148.4 | 49.6 | 97.5 | 148.3 | 73.6 | 109.8 | 148.4 | 83.7 | 115.1 | 148.4 | 83.7 | 115.1 |
| 0330 | Cast-in-Place Concrete | 151.1 | 81.0 | 124.2 | 163.5 | 117.6 | 145.9 | 168.0 | 76.2 | 132.9 | 200.4 | 108.8 | 165.3 | 131.9 | 103.9 | 121.2 | 131.9 | 103.9 | 121.2 |
| 03 | CONCRETE | 135.2 | 75.3 | 106.5 | 150.0 | 105.2 | 128.6 | 148.0 | 72.0 | 111.6 | 167.1 | 97.3 | 133.7 | 130.6 | 95.5 | 113.8 | 130.6 | 95.5 | 113.8 |
| 04 | MASONRY | 162.8 | 71.9 | 107.7 | 164.7 | 109.1 | 131.0 | 159.7 | 76.5 | 109.3 | 215.0 | 92.7 | 140.9 | 159.7 | 92.6 | 119.0 | 159.7 | 92.6 | 119.0 |
| 05 | METALS | 107.6 | 79.0 | 97.8 | 106.8 | 94.2 | 102.5 | 105.9 | 76.3 | 95.7 | 148.0 | 89.8 | 128.0 | 105.0 | 90.5 | 100.0 | 105.0 | 90.5 | 100.0 |
| 06 | WOOD, PLASTICS & COMPOSITES | 114.0 | 78.4 | 93.6 | 116.3 | 102.3 | 108.3 | 87.1 | 75.9 | 80.7 | 114.5 | 98.1 | 105.1 | 127.7 | 94.0 | 108.4 | 127.7 | 94.0 | 108.4 |
| 07 | THERMAL & MOISTURE PROTECTION | 105.0 | 76.3 | 93.6 | 111.3 | 100.9 | 107.2 | 106.3 | 74.4 | 93.6 | 123.3 | 95.8 | 112.4 | 105.2 | 95.7 | 101.4 | 105.2 | 95.7 | 101.4 |
| 08 | OPENINGS | 92.9 | 69.0 | 86.8 | 91.7 | 98.7 | 93.4 | 83.5 | 69.0 | 79.8 | 92.9 | 87.5 | 91.5 | 92.9 | 86.3 | 91.2 | 92.9 | 86.3 | 91.2 |
| 0920 | Plaster & Gypsum Board | 130.2 | 77.4 | 93.0 | 157.5 | 102.4 | 118.7 | 156.2 | 75.1 | 99.1 | 173.7 | 97.6 | 120.1 | 188.5 | 93.7 | 121.7 | 188.5 | 93.7 | 121.7 |
| 0950, 0980 | Ceilings & Acoustic Treatment | 95.2 | 77.4 | 83.4 | 95.2 | 102.4 | 100.0 | 95.2 | 75.1 | 81.9 | 142.4 | 97.6 | 112.7 | 95.2 | 93.7 | 94.2 | 95.2 | 93.7 | 94.2 |
| 0960 | Flooring | 131.9 | 73.0 | 115.0 | 131.9 | 102.7 | 123.5 | 111.6 | 69.9 | 99.6 | 134.2 | 96.2 | 123.3 | 131.9 | 103.4 | 123.7 | 131.9 | 103.4 | 123.7 |
| 0970, 0990 | Wall Finishes & Painting/Coating | 111.4 | 62.1 | 81.5 | 111.4 | 104.4 | 107.2 | 111.4 | 67.9 | 85.0 | 111.3 | 112.0 | 111.7 | 111.4 | 97.1 | 102.7 | 111.4 | 97.1 | 102.7 |
| 09 | FINISHES | 114.6 | 75.3 | 92.9 | 117.5 | 104.0 | 110.1 | 110.5 | 74.6 | 90.7 | 131.9 | 99.9 | 114.3 | 120.3 | 96.4 | 107.2 | 120.3 | 96.4 | 107.2 |
| COVERS | DIVS. 10 - 14, 25, 28, 41, 43, 44, 46 | 140.0 | 74.4 | 126.4 | 140.0 | 84.5 | 128.5 | 140.0 | 73.3 | 126.1 | 140.0 | 105.4 | 132.8 | 140.0 | 94.0 | 130.5 | 140.0 | 94.0 | 130.5 |
| 21, 22, 23 | FIRE SUPPRESSION, PLUMBING & HVAC | 102.6 | 90.9 | 97.7 | 102.6 | 110.6 | 105.9 | 102.6 | 91.2 | 97.9 | 100.4 | 94.7 | 98.0 | 103.2 | 98.4 | 101.2 | 103.2 | 98.4 | 101.2 |
| 26, 27, 3370 | ELECTRICAL, COMMUNICATIONS & UTIL. | 121.8 | 74.3 | 96.6 | 119.7 | 96.0 | 107.1 | 125.8 | 68.7 | 95.4 | 127.0 | 86.0 | 105.2 | 119.0 | 77.5 | 96.9 | 119.0 | 77.5 | 96.9 |
| MF2010 | WEIGHTED AVERAGE | 116.1 | 80.3 | 100.1 | 118.1 | 103.0 | 111.3 | 115.8 | 79.1 | 99.4 | 130.9 | 94.8 | 114.8 | 114.9 | 93.0 | 105.1 | 114.9 | 93.0 | 105.1 |

## CANADA

| DIVISION | | CHARLOTTETOWN, PRINCE EDWARD ISLAND | | | CHICOUTIMI, QUEBEC | | | CORNER BROOK, NEWFOUNDLAND | | | CORNWALL, ONTARIO | | | DALHOUSIE, NEW BRUNSWICK | | | DARTMOUTH, NOVA SCOTIA | | |
|---|---|---|---|---|---|---|---|---|---|---|---|---|---|---|---|---|---|---|---|
| | | MAT. | INST. | TOTAL | MAT. | INST. | TOTAL | MAT. | INST. | TOTAL | MAT. | INST. | TOTAL | MAT. | INST. | TOTAL | MAT. | INST. | TOTAL |
| 015433 | CONTRACTOR EQUIPMENT | | 101.1 | 101.1 | | 102.0 | 102.0 | | 102.4 | 102.4 | | 101.6 | 101.6 | | 101.4 | 101.4 | | 101.2 | 101.2 |
| 0241, 31 - 34 | SITE & INFRASTRUCTURE, DEMOLITION | 104.7 | 96.4 | 99.0 | 92.0 | 99.4 | 97.1 | 112.7 | 97.7 | 102.4 | 110.7 | 101.3 | 104.2 | 94.1 | 97.3 | 96.3 | 104.6 | 99.0 | 100.7 |
| 0310 | Concrete Forming & Accessories | 96.6 | 62.1 | 66.7 | 128.1 | 91.7 | 96.5 | 103.8 | 65.8 | 70.8 | 120.9 | 95.8 | 99.2 | 102.2 | 68.2 | 72.7 | 96.4 | 76.5 | 79.2 |
| 0320 | Concrete Reinforcing | 158.7 | 49.6 | 102.5 | 111.9 | 93.5 | 102.4 | 148.4 | 51.5 | 98.5 | 175.0 | 87.3 | 129.8 | 148.4 | 60.0 | 102.9 | 155.0 | 49.6 | 100.7 |
| 0330 | Cast-in-Place Concrete | 159.3 | 64.3 | 122.9 | 130.1 | 99.0 | 118.2 | 172.3 | 74.5 | 134.8 | 147.0 | 107.1 | 131.7 | 138.2 | 65.2 | 110.2 | 169.8 | 76.2 | 134.0 |
| 03 | CONCRETE | 153.7 | 61.2 | 109.4 | 124.3 | 94.5 | 110.1 | 162.4 | 66.9 | 116.7 | 141.6 | 98.1 | 120.8 | 135.5 | 66.2 | 102.3 | 149.9 | 72.0 | 112.6 |
| 04 | MASONRY | 160.1 | 66.6 | 103.4 | 160.2 | 92.6 | 119.2 | 160.0 | 68.2 | 104.4 | 163.4 | 100.0 | 125.0 | 156.6 | 69.0 | 103.5 | 171.2 | 76.5 | 113.8 |
| 05 | METALS | 105.9 | 69.1 | 93.2 | 104.5 | 91.6 | 100.1 | 108.1 | 74.7 | 96.6 | 106.8 | 92.8 | 101.9 | 104.7 | 73.6 | 94.0 | 108.0 | 76.3 | 97.1 |
| 06 | WOOD, PLASTICS & COMPOSITES | 87.4 | 61.6 | 72.6 | 127.7 | 91.7 | 107.1 | 95.9 | 64.8 | 78.1 | 115.1 | 95.1 | 103.7 | 94.9 | 68.3 | 79.6 | 87.1 | 75.9 | 80.7 |
| 07 | THERMAL & MOISTURE PROTECTION | 105.8 | 63.4 | 88.9 | 105.1 | 94.8 | 101.0 | 108.3 | 65.7 | 91.4 | 111.1 | 95.1 | 104.7 | 108.8 | 66.1 | 91.8 | 106.3 | 74.4 | 93.6 |
| 08 | OPENINGS | 95.2 | 53.1 | 84.5 | 92.4 | 79.4 | 89.1 | 99.2 | 60.8 | 89.5 | 92.9 | 91.8 | 92.6 | 86.9 | 59.5 | 80.0 | 83.5 | 69.0 | 79.8 |
| 0920 | Plaster & Gypsum Board | 160.3 | 60.4 | 90.0 | 188.5 | 91.3 | 120.1 | 163.1 | 63.6 | 93.0 | 228.3 | 95.0 | 134.4 | 163.1 | 67.2 | 95.6 | 158.8 | 75.1 | 99.8 |
| 0950, 0980 | Ceilings & Acoustic Treatment | 108.3 | 60.4 | 76.6 | 95.2 | 91.3 | 92.6 | 96.2 | 63.6 | 74.6 | 98.0 | 95.0 | 96.0 | 96.2 | 67.2 | 77.0 | 104.6 | 75.1 | 85.0 |
| 0960 | Flooring | 111.6 | 65.3 | 98.3 | 131.9 | 103.4 | 123.7 | 115.9 | 59.5 | 99.7 | 131.9 | 101.2 | 123.1 | 115.2 | 74.9 | 103.7 | 111.6 | 69.9 | 99.6 |
| 0970, 0990 | Wall Finishes & Painting/Coating | 111.4 | 45.5 | 71.4 | 111.4 | 103.2 | 106.4 | 111.4 | 65.4 | 83.5 | 111.4 | 97.3 | 102.8 | 111.4 | 54.6 | 76.9 | 111.4 | 67.9 | 85.0 |
| 09 | FINISHES | 114.7 | 61.3 | 85.3 | 120.3 | 95.5 | 106.6 | 114.1 | 64.6 | 86.8 | 126.5 | 97.3 | 110.4 | 112.9 | 68.4 | 88.4 | 112.8 | 74.6 | 91.7 |
| COVERS | DIVS. 10 - 14, 25, 28, 41, 43, 44, 46 | 140.0 | 71.4 | 125.7 | 140.0 | 93.4 | 130.3 | 140.0 | 72.2 | 125.9 | 140.0 | 81.7 | 127.9 | 140.0 | 71.9 | 125.8 | 140.0 | 73.3 | 126.1 |
| 21, 22, 23 | FIRE SUPPRESSION, PLUMBING & HVAC | 102.6 | 68.9 | 88.6 | 102.6 | 97.8 | 100.6 | 102.6 | 77.6 | 92.2 | 103.2 | 108.2 | 105.3 | 102.6 | 75.4 | 91.3 | 102.6 | 91.2 | 97.9 |
| 26, 27, 3370 | ELECTRICAL, COMMUNICATIONS & UTIL. | 119.0 | 55.9 | 85.4 | 118.3 | 84.8 | 100.5 | 118.9 | 62.2 | 88.8 | 120.4 | 97.1 | 108.0 | 122.1 | 62.4 | 90.4 | 124.0 | 68.7 | 94.6 |
| MF2010 | WEIGHTED AVERAGE | 117.6 | 66.4 | 94.7 | 113.9 | 93.2 | 104.7 | 119.6 | 71.5 | 98.1 | 118.0 | 99.1 | 109.5 | 114.2 | 71.4 | 95.1 | 117.2 | 79.1 | 100.1 |

## CANADA

| DIVISION | | EDMONTON, ALBERTA | | | FORT MCMURRAY, ALBERTA | | | FREDERICTON, NEW BRUNSWICK | | | GATINEAU, QUEBEC | | | GRANBY, QUEBEC | | | HALIFAX, NOVA SCOTIA | | |
|---|---|---|---|---|---|---|---|---|---|---|---|---|---|---|---|---|---|---|---|
| | | MAT. | INST. | TOTAL | MAT. | INST. | TOTAL | MAT. | INST. | TOTAL | MAT. | INST. | TOTAL | MAT. | INST. | TOTAL | MAT. | INST. | TOTAL |
| 015433 | CONTRACTOR EQUIPMENT | | 106.6 | 106.6 | | 103.8 | 103.8 | | 101.4 | 101.4 | | 102.0 | 102.0 | | 102.0 | 102.0 | | 101.2 | 101.2 |
| 0241, 31 - 34 | SITE & INFRASTRUCTURE, DEMOLITION | 150.7 | 105.6 | 119.6 | 118.6 | 102.4 | 107.4 | 93.7 | 97.3 | 96.2 | 92.3 | 100.7 | 98.1 | 92.9 | 100.7 | 98.3 | 97.2 | 98.8 | 98.3 |
| 0310 | Concrete Forming & Accessories | 129.8 | 98.4 | 102.5 | 121.1 | 98.6 | 101.5 | 120.3 | 68.6 | 75.4 | 128.1 | 94.0 | 98.5 | 128.1 | 93.9 | 98.4 | 96.6 | 79.0 | 81.3 |
| 0320 | Concrete Reinforcing | 148.3 | 73.6 | 109.8 | 161.6 | 73.6 | 116.3 | 149.3 | 60.2 | 103.4 | 157.1 | 83.7 | 119.3 | 157.1 | 83.6 | 119.3 | 155.0 | 62.1 | 107.2 |
| 0330 | Cast-in-Place Concrete | 208.9 | 108.8 | 170.5 | 217.2 | 108.0 | 175.4 | 131.5 | 65.2 | 106.1 | 130.1 | 103.9 | 120.1 | 134.4 | 103.8 | 122.7 | 171.2 | 82.2 | 137.1 |
| 03 | CONCRETE | 171.5 | 97.3 | 136.0 | 174.9 | 97.2 | 137.7 | 133.6 | 66.4 | 101.4 | 130.9 | 95.4 | 113.9 | 133.1 | 95.4 | 115.1 | 150.6 | 77.5 | 115.6 |
| 04 | MASONRY | 202.5 | 92.7 | 136.0 | 204.3 | 92.7 | 136.6 | 177.3 | 70.7 | 112.7 | 159.6 | 92.6 | 119.0 | 160.0 | 92.6 | 119.1 | 172.8 | 84.2 | 119.1 |
| 05 | METALS | 148.6 | 89.8 | 128.4 | 132.1 | 89.9 | 117.6 | 104.6 | 74.2 | 94.2 | 105.0 | 90.3 | 100.0 | 105.0 | 90.3 | 100.0 | 119.0 | 81.8 | 106.2 |
| 06 | WOOD, PLASTICS & COMPOSITES | 111.0 | 98.1 | 103.6 | 110.8 | 98.1 | 103.5 | 116.9 | 68.3 | 89.0 | 127.7 | 94.0 | 108.4 | 127.7 | 94.0 | 108.4 | 87.3 | 77.9 | 81.9 |
| 07 | THERMAL & MOISTURE PROTECTION | 124.4 | 95.8 | 113.0 | 118.8 | 95.2 | 109.4 | 109.0 | 67.2 | 92.4 | 105.2 | 95.7 | 101.4 | 105.1 | 94.1 | 100.7 | 106.8 | 78.5 | 95.6 |
| 08 | OPENINGS | 92.9 | 87.5 | 91.5 | 92.9 | 87.5 | 91.5 | 86.8 | 58.3 | 79.6 | 92.9 | 81.3 | 89.9 | 92.9 | 81.3 | 89.9 | 83.5 | 72.8 | 80.8 |
| 0920 | Plaster & Gypsum Board | 164.5 | 97.6 | 117.4 | 156.8 | 97.7 | 115.2 | 181.0 | 67.2 | 100.9 | 147.7 | 93.7 | 109.7 | 150.6 | 93.7 | 110.5 | 159.2 | 77.1 | 101.4 |
| 0950, 0980 | Ceilings & Acoustic Treatment | 156.8 | 97.6 | 117.6 | 104.6 | 97.7 | 100.0 | 101.8 | 67.2 | 78.9 | 95.2 | 93.7 | 94.2 | 95.2 | 93.7 | 94.2 | 104.6 | 77.1 | 86.4 |
| 0960 | Flooring | 135.3 | 96.2 | 124.1 | 131.9 | 96.2 | 121.6 | 127.5 | 78.2 | 113.3 | 131.9 | 103.4 | 123.7 | 131.9 | 103.4 | 123.7 | 111.6 | 91.8 | 105.9 |
| 0970, 0990 | Wall Finishes & Painting/Coating | 111.4 | 112.0 | 111.8 | 111.4 | 101.1 | 105.1 | 111.4 | 69.6 | 86.0 | 111.4 | 97.1 | 102.7 | 111.4 | 97.1 | 102.7 | 111.4 | 81.5 | 93.2 |
| 09 | FINISHES | 136.0 | 99.9 | 116.1 | 120.9 | 98.7 | 108.6 | 119.9 | 70.7 | 92.8 | 115.5 | 96.4 | 105.0 | 115.8 | 96.4 | 105.1 | 112.8 | 91.9 | 95.8 |
| COVERS | DIVS. 10 - 14, 25, 28, 41, 43, 44, 46 | 140.0 | 105.4 | 132.8 | 140.0 | 105.2 | 132.8 | 140.0 | 71.9 | 125.9 | 140.0 | 94.0 | 130.5 | 140.0 | 94.0 | 130.5 | 140.0 | 74.0 | 126.3 |
| 21, 22, 23 | FIRE SUPPRESSION, PLUMBING & HVAC | 100.4 | 94.7 | 98.0 | 103.0 | 106.8 | 104.6 | 102.6 | 85.7 | 95.6 | 103.2 | 98.4 | 101.2 | 102.6 | 98.4 | 100.8 | 99.7 | 79.8 | 91.4 |
| 26, 27, 3370 | ELECTRICAL, COMMUNICATIONS & UTIL. | 117.8 | 86.0 | 100.9 | 113.0 | 91.2 | 101.4 | 132.0 | 82.3 | 105.5 | 119.0 | 77.5 | 96.9 | 119.6 | 77.5 | 97.2 | 124.9 | 81.5 | 101.8 |
| MF2010 | WEIGHTED AVERAGE | 131.1 | 94.8 | 114.9 | 127.0 | 97.7 | 113.9 | 116.8 | 76.8 | 98.9 | 114.6 | 92.7 | 104.8 | 114.8 | 92.7 | 104.9 | 118.1 | 81.7 | 101.8 |

## CANADA

| DIVISION | | HAMILTON, ONTARIO | | | HULL, QUEBEC | | | JOLIETTE, QUEBEC | | | KAMLOOPS, BRITISH COLUMBIA | | | KINGSTON, ONTARIO | | | KITCHENER, ONTARIO | | |
|---|---|---|---|---|---|---|---|---|---|---|---|---|---|---|---|---|---|---|---|
| | | MAT. | INST. | TOTAL | MAT. | INST. | TOTAL | MAT. | INST. | TOTAL | MAT. | INST. | TOTAL | MAT. | INST. | TOTAL | MAT. | INST. | TOTAL |
| 015433 | CONTRACTOR EQUIPMENT | | 108.5 | 108.5 | | 102.0 | 102.0 | | 102.0 | 102.0 | | 105.3 | 105.3 | | 103.9 | 103.9 | | 103.8 | 103.8 |
| 0241, 31 - 34 | SITE & INFRASTRUCTURE, DEMOLITION | 113.4 | 113.3 | 113.3 | 92.3 | 100.7 | 98.1 | 93.0 | 100.7 | 98.3 | 116.6 | 104.3 | 108.1 | 110.7 | 105.1 | 106.8 | 100.3 | 105.3 | 103.7 |
| 0310 | Concrete Forming & Accessories | 130.9 | 96.4 | 100.9 | 128.1 | 94.0 | 98.5 | 128.1 | 94.1 | 98.6 | 123.3 | 97.1 | 100.6 | 121.0 | 95.9 | 99.2 | 117.3 | 87.5 | 91.4 |
| 0320 | Concrete Reinforcing | 167.9 | 87.2 | 126.4 | 157.1 | 83.7 | 119.3 | 148.4 | 83.7 | 115.1 | 116.6 | 81.5 | 98.5 | 175.0 | 87.3 | 129.8 | 140.7 | 88.8 | 120.9 |
| 0330 | Cast-in-Place Concrete | 151.9 | 98.1 | 131.3 | 130.1 | 103.9 | 120.1 | 135.5 | 103.9 | 123.4 | 117.3 | 107.9 | 113.7 | 147.0 | 107.0 | 131.7 | 122.5 | 88.1 | 106.1 |
| 03 | CONCRETE | 143.6 | 95.3 | 120.5 | 130.9 | 95.4 | 113.9 | 132.4 | 95.5 | 114.7 | 138.4 | 97.8 | 119.0 | 141.6 | 98.1 | 120.8 | 122.5 | 88.1 | 106.1 |
| 04 | MASONRY | 180.7 | 102.7 | 133.4 | 159.6 | 92.6 | 119.0 | 160.0 | 92.6 | 119.2 | 166.6 | 101.0 | 126.8 | 169.9 | 100.0 | 127.5 | 161.4 | 94.6 | 120.9 |
| 05 | METALS | 126.0 | 93.2 | 114.7 | 105.0 | 90.3 | 100.0 | 105.0 | 90.5 | 100.0 | 107.4 | 90.4 | 101.6 | 108.5 | 92.7 | 103.1 | 115.4 | 93.0 | 107.7 |
| 06 | WOOD, PLASTICS & COMPOSITES | 115.6 | 95.5 | 104.1 | 127.7 | 94.0 | 108.4 | 127.7 | 94.0 | 108.4 | 98.6 | 95.5 | 96.8 | 115.1 | 95.3 | 103.7 | 109.6 | 86.7 | 96.5 |
| 07 | THERMAL & MOISTURE PROTECTION | 121.8 | 97.6 | 112.2 | 105.1 | 95.7 | 101.4 | 105.1 | 95.7 | 101.4 | 120.5 | 92.4 | 109.3 | 111.1 | 96.3 | 105.2 | 110.7 | 92.6 | 103.5 |
| 08 | OPENINGS | 92.9 | 93.8 | 93.1 | 92.9 | 81.3 | 89.9 | 92.9 | 86.3 | 91.2 | 88.7 | 91.2 | 89.3 | 92.9 | 91.5 | 92.5 | 84.0 | 87.2 | 84.8 |
| 0920 | Plaster & Gypsum Board | 193.9 | 95.4 | 124.5 | 147.7 | 93.7 | 109.7 | 188.5 | 93.7 | 121.7 | 137.8 | 94.9 | 107.6 | 232.2 | 95.1 | 135.7 | 155.2 | 86.3 | 106.7 |
| 0950, 0980 | Ceilings & Acoustic Treatment | 124.1 | 95.4 | 105.0 | 95.2 | 93.7 | 94.2 | 95.2 | 93.7 | 94.2 | 95.2 | 94.9 | 95.0 | 112.1 | 95.1 | 100.8 | 104.6 | 86.3 | 92.5 |
| 0960 | Flooring | 134.2 | 106.2 | 126.2 | 131.9 | 103.4 | 123.7 | 131.9 | 103.4 | 123.7 | 129.8 | 57.8 | 109.1 | 131.9 | 101.2 | 123.1 | 127.9 | 106.2 | 121.7 |
| 0970, 0990 | Wall Finishes & Painting/Coating | 111.4 | 103.1 | 106.3 | 111.4 | 97.1 | 102.7 | 111.4 | 97.1 | 102.7 | 111.4 | 88.9 | 97.7 | 111.4 | 89.9 | 98.3 | 111.4 | 93.0 | 100.2 |
| 09 | FINISHES | 129.2 | 99.1 | 112.6 | 115.5 | 96.4 | 105.0 | 120.3 | 96.4 | 107.2 | 116.7 | 89.4 | 101.7 | 130.0 | 96.5 | 111.5 | 117.2 | 91.2 | 102.9 |
| COVERS | DIVS. 10 - 14, 25, 28, 41, 43, 44, 46 | 140.0 | 108.0 | 133.4 | 140.0 | 94.0 | 130.5 | 140.0 | 94.0 | 130.5 | 140.0 | 105.5 | 132.8 | 140.0 | 81.7 | 127.9 | 140.0 | 104.8 | 132.7 |
| 21, 22, 23 | FIRE SUPPRESSION, PLUMBING & HVAC | 100.5 | 86.0 | 94.5 | 102.6 | 86.0 | 95.0 | 102.6 | 98.4 | 100.9 | 102.6 | 102.0 | 102.4 | 103.2 | 108.4 | 105.3 | 99.7 | 91.2 | 96.1 |
| 26, 27, 3370 | ELECTRICAL, COMMUNICATIONS & UTIL. | 127.5 | 89.4 | 107.2 | 121.1 | 77.5 | 97.9 | 119.6 | 77.5 | 97.2 | 123.5 | 87.5 | 104.3 | 120.4 | 95.7 | 107.2 | 120.9 | 86.9 | 102.8 |
| MF2010 | WEIGHTED AVERAGE | 122.6 | 95.4 | 110.5 | 114.6 | 92.7 | 104.8 | 115.1 | 93.0 | 105.2 | 117.2 | 96.2 | 107.8 | 118.9 | 99.2 | 110.1 | 114.0 | 92.1 | 104.2 |

**CANADA**

| DIVISION | | LAVAL, QUEBEC | | | LETHBRIDGE, ALBERTA | | | LLOYDMINSTER, ALBERTA | | | LONDON, ONTARIO | | | MEDICINE HAT, ALBERTA | | | MONCTON, NEW BRUNSWICK | | |
|---|---|---|---|---|---|---|---|---|---|---|---|---|---|---|---|---|---|---|---|
| | | MAT. | INST. | TOTAL | MAT. | INST. | TOTAL | MAT. | INST. | TOTAL | MAT. | INST. | TOTAL | MAT. | INST. | TOTAL | MAT. | INST. | TOTAL |
| 015433 | CONTRACTOR EQUIPMENT | | 102.0 | 102.0 | | 103.8 | 103.8 | | 103.8 | 103.8 | | 104.0 | 104.0 | | 103.8 | 103.8 | | 101.4 | 101.4 |
| 0241, 31 - 34 | SITE & INFRASTRUCTURE, DEMOLITION | 92.8 | 100.7 | 98.3 | 110.6 | 103.1 | 105.4 | 110.5 | 102.4 | 104.9 | 113.3 | 105.5 | 107.9 | 109.2 | 102.5 | 104.6 | 93.6 | 97.5 | 96.3 |
| 0310 | Concrete Forming & Accessories | 128.2 | 94.0 | 98.5 | 122.3 | 98.6 | 101.8 | 120.6 | 88.1 | 92.4 | 131.0 | 90.2 | 95.6 | 122.2 | 88.1 | 92.6 | 102.2 | 68.5 | 73.0 |
| 0320 | Concrete Reinforcing | 157.1 | 83.7 | 119.3 | 161.6 | 73.6 | 116.3 | 161.6 | 73.6 | 116.3 | 146.7 | 86.0 | 115.4 | 161.6 | 73.6 | 116.3 | 148.4 | 63.2 | 104.5 |
| 0330 | Cast-in-Place Concrete | 134.4 | 103.9 | 122.7 | 162.9 | 108.0 | 141.9 | 151.1 | 104.0 | 133.1 | 149.2 | 95.9 | 128.8 | 151.1 | 104.0 | 133.0 | 133.1 | 65.4 | 107.2 |
| 03 | CONCRETE | 133.1 | 95.4 | 115.1 | 147.7 | 97.2 | 123.5 | 141.7 | 91.2 | 117.5 | 139.2 | 91.5 | 116.4 | 141.8 | 91.1 | 117.5 | 133.0 | 67.2 | 101.5 |
| 04 | MASONRY | 159.9 | 92.6 | 120.9 | 177.7 | 92.7 | 126.2 | 159.9 | 85.7 | 114.9 | 182.1 | 99.8 | 132.2 | 159.9 | 85.7 | 114.9 | 158.3 | 69.1 | 104.2 |
| 05 | METALS | 105.1 | 90.3 | 100.0 | 125.5 | 89.9 | 113.3 | 107.6 | 89.8 | 101.5 | 123.6 | 92.4 | 112.9 | 107.6 | 89.7 | 101.5 | 104.7 | 79.4 | 96.0 |
| 06 | WOOD, PLASTICS & COMPOSITES | 127.8 | 94.0 | 108.4 | 114.1 | 98.1 | 104.9 | 110.8 | 87.6 | 97.5 | 115.6 | 88.3 | 99.9 | 114.1 | 87.6 | 98.9 | 94.9 | 68.3 | 79.6 |
| 07 | THERMAL & MOISTURE PROTECTION | 105.7 | 95.7 | 101.7 | 116.2 | 95.2 | 107.9 | 113.4 | 90.5 | 104.3 | 122.6 | 95.3 | 111.8 | 119.4 | 90.5 | 107.9 | 108.8 | 67.6 | 92.4 |
| 08 | OPENINGS | 92.9 | 81.3 | 89.9 | 92.9 | 87.5 | 91.5 | 92.9 | 81.8 | 90.1 | 94.0 | 88.7 | 92.7 | 92.9 | 81.8 | 90.1 | 86.9 | 62.8 | 80.8 |
| 0920 | Plaster & Gypsum Board | 151.1 | 93.7 | 110.6 | 138.8 | 97.7 | 109.9 | 133.8 | 86.9 | 100.8 | 197.3 | 88.0 | 120.3 | 136.5 | 86.9 | 101.5 | 163.1 | 67.2 | 95.6 |
| 0950, 0980 | Ceilings & Acoustic Treatment | 95.2 | 93.7 | 94.2 | 103.7 | 97.7 | 99.7 | 95.2 | 86.9 | 89.7 | 136.3 | 88.0 | 104.3 | 95.2 | 86.9 | 89.7 | 96.2 | 67.2 | 77.0 |
| 0960 | Flooring | 131.9 | 103.4 | 123.7 | 131.9 | 96.2 | 121.6 | 131.9 | 96.2 | 121.6 | 134.4 | 106.2 | 126.3 | 131.9 | 96.2 | 121.6 | 115.2 | 74.9 | 103.7 |
| 0970, 0990 | Wall Finishes & Painting/Coating | 111.4 | 97.1 | 102.7 | 111.3 | 110.3 | 110.7 | 111.4 | 86.0 | 96.0 | 111.4 | 99.8 | 104.3 | 111.3 | 86.0 | 95.9 | 111.4 | 54.6 | 76.9 |
| 09 | FINISHES | 115.9 | 96.4 | 105.2 | 117.6 | 99.7 | 107.7 | 115.3 | 89.0 | 100.8 | 132.3 | 94.0 | 111.2 | 115.4 | 89.0 | 100.9 | 112.9 | 68.4 | 88.4 |
| COVERS | DIVS. 10 - 14, 25, 28, 41, 43, 44, 46 | 140.0 | 94.0 | 130.5 | 140.0 | 105.2 | 132.8 | 140.0 | 101.5 | 132.0 | 140.0 | 106.6 | 133.1 | 140.0 | 101.5 | 132.0 | 140.0 | 71.9 | 125.9 |
| 21, 22, 23 | FIRE SUPPRESSION, PLUMBING & HVAC | 99.7 | 98.4 | 99.1 | 102.8 | 103.2 | 103.0 | 103.2 | 103.3 | 103.2 | 100.4 | 83.1 | 93.2 | 102.6 | 99.6 | 101.4 | 102.6 | 75.5 | 91.3 |
| 26, 27, 3370 | ELECTRICAL, COMMUNICATIONS & UTIL. | 120.8 | 77.5 | 97.8 | 114.5 | 91.2 | 102.1 | 112.0 | 91.2 | 100.9 | 119.5 | 86.9 | 102.2 | 112.0 | 91.2 | 100.9 | 127.2 | 66.0 | 94.7 |
| MF2010 | WEIGHTED AVERAGE | 114.3 | 92.7 | 104.6 | 120.9 | 97.1 | 110.3 | 116.2 | 93.6 | 106.1 | 121.4 | 91.9 | 108.2 | 116.3 | 92.8 | 105.8 | 114.5 | 72.8 | 95.8 |

**CANADA**

| DIVISION | | MONTREAL, QUEBEC | | | MOOSE JAW, SASKATCHEWAN | | | NEW GLASGOW, NOVA SCOTIA | | | NEWCASTLE, NEW BRUNSWICK | | | NORTH BAY, ONTARIO | | | OSHAWA, ONTARIO | | |
|---|---|---|---|---|---|---|---|---|---|---|---|---|---|---|---|---|---|---|---|
| | | MAT. | INST. | TOTAL | MAT. | INST. | TOTAL | MAT. | INST. | TOTAL | MAT. | INST. | TOTAL | MAT. | INST. | TOTAL | MAT. | INST. | TOTAL |
| 015433 | CONTRACTOR EQUIPMENT | | 103.6 | 103.6 | | 100.0 | 100.0 | | 101.2 | 101.2 | | 101.4 | 101.4 | | 101.6 | 101.6 | | 103.8 | 103.8 |
| 0241, 31 - 34 | SITE & INFRASTRUCTURE, DEMOLITION | 99.6 | 99.7 | 99.7 | 110.3 | 96.7 | 100.9 | 97.1 | 99.0 | 98.4 | 94.2 | 97.3 | 96.3 | 112.7 | 100.8 | 104.5 | 112.6 | 105.1 | 107.4 |
| 0310 | Concrete Forming & Accessories | 136.0 | 92.0 | 98.0 | 106.1 | 64.7 | 70.1 | 96.4 | 76.5 | 79.2 | 102.2 | 68.3 | 72.7 | 122.9 | 93.1 | 97.0 | 123.1 | 95.3 | 99.0 |
| 0320 | Concrete Reinforcing | 153.4 | 93.6 | 122.6 | 114.0 | 65.7 | 89.1 | 148.4 | 49.6 | 97.5 | 148.4 | 60.0 | 102.9 | 175.0 | 86.8 | 129.6 | 175.0 | 89.6 | 131.0 |
| 0330 | Cast-in-Place Concrete | 180.9 | 100.4 | 150.1 | 146.6 | 75.1 | 119.2 | 169.8 | 76.2 | 134.0 | 138.2 | 65.2 | 110.2 | 163.5 | 92.4 | 136.3 | 162.8 | 93.3 | 136.2 |
| 03 | CONCRETE | 156.4 | 95.1 | 127.1 | 125.2 | 69.0 | 98.3 | 148.9 | 72.0 | 112.1 | 135.5 | 66.2 | 102.3 | 150.0 | 91.8 | 122.1 | 149.7 | 93.6 | 122.8 |
| 04 | MASONRY | 167.3 | 92.6 | 122.1 | 158.5 | 68.4 | 103.9 | 159.8 | 76.5 | 109.3 | 158.7 | 69.0 | 104.4 | 164.7 | 95.6 | 122.8 | 164.3 | 100.4 | 125.6 |
| 05 | METALS | 122.2 | 92.1 | 111.8 | 103.9 | 76.1 | 94.3 | 105.9 | 76.3 | 95.7 | 104.7 | 73.6 | 94.0 | 106.8 | 92.4 | 101.8 | 106.8 | 94.7 | 102.7 |
| 06 | WOOD, PLASTICS & COMPOSITES | 127.8 | 92.0 | 107.3 | 96.4 | 63.1 | 77.3 | 87.1 | 75.9 | 80.7 | 94.9 | 68.3 | 79.6 | 116.3 | 93.6 | 103.3 | 116.5 | 94.2 | 103.7 |
| 07 | THERMAL & MOISTURE PROTECTION | 118.6 | 95.3 | 109.3 | 104.6 | 68.2 | 90.1 | 106.3 | 74.4 | 93.6 | 108.8 | 66.1 | 91.8 | 111.3 | 91.2 | 103.3 | 111.7 | 93.0 | 104.3 |
| 08 | OPENINGS | 92.9 | 81.4 | 90.0 | 87.7 | 60.4 | 80.8 | 83.5 | 69.0 | 79.8 | 86.9 | 59.5 | 80.0 | 91.7 | 89.1 | 91.0 | 91.7 | 93.6 | 92.2 |
| 0920 | Plaster & Gypsum Board | 168.2 | 91.3 | 114.1 | 130.8 | 61.9 | 82.3 | 156.2 | 75.1 | 99.1 | 163.1 | 67.2 | 95.6 | 157.5 | 93.4 | 112.4 | 159.0 | 94.0 | 113.3 |
| 0950, 0980 | Ceilings & Acoustic Treatment | 113.8 | 91.3 | 98.9 | 95.2 | 61.9 | 73.2 | 95.2 | 75.1 | 81.9 | 96.2 | 67.2 | 77.0 | 95.2 | 93.4 | 94.0 | 99.0 | 94.0 | 95.7 |
| 0960 | Flooring | 133.6 | 103.4 | 124.9 | 120.6 | 65.7 | 104.9 | 111.6 | 69.9 | 99.6 | 115.2 | 74.9 | 103.7 | 131.9 | 101.2 | 123.1 | 131.9 | 109.0 | 125.3 |
| 0970, 0990 | Wall Finishes & Painting/Coating | 111.4 | 103.2 | 106.4 | 111.4 | 70.7 | 86.7 | 111.4 | 67.9 | 85.0 | 111.4 | 54.6 | 76.9 | 111.4 | 96.5 | 102.4 | 111.4 | 107.2 | 108.8 |
| 09 | FINISHES | 122.6 | 95.7 | 107.7 | 111.4 | 65.2 | 85.9 | 110.5 | 74.6 | 90.7 | 112.9 | 68.4 | 88.4 | 117.5 | 95.4 | 105.3 | 118.4 | 99.1 | 107.7 |
| COVERS | DIVS. 10 - 14, 25, 28, 41, 43, 44, 46 | 140.0 | 94.0 | 130.4 | 140.0 | 71.1 | 125.7 | 140.0 | 73.3 | 126.1 | 140.0 | 71.9 | 125.9 | 140.0 | 80.2 | 127.6 | 140.0 | 107.5 | 133.3 |
| 21, 22, 23 | FIRE SUPPRESSION, PLUMBING & HVAC | 100.4 | 97.9 | 99.4 | 103.2 | 82.2 | 94.5 | 102.6 | 91.2 | 97.9 | 102.6 | 75.4 | 91.3 | 102.6 | 106.1 | 104.1 | 99.7 | 108.8 | 103.5 |
| 26, 27, 3370 | ELECTRICAL, COMMUNICATIONS & UTIL. | 128.3 | 84.8 | 105.1 | 122.1 | 67.0 | 92.8 | 119.3 | 68.7 | 92.4 | 121.1 | 66.0 | 91.8 | 119.7 | 97.0 | 107.6 | 122.2 | 96.6 | 108.6 |
| MF2010 | WEIGHTED AVERAGE | 122.0 | 93.5 | 109.3 | 113.4 | 73.6 | 95.6 | 115.3 | 79.1 | 99.1 | 114.2 | 71.9 | 95.3 | 118.1 | 96.7 | 108.5 | 117.7 | 100.0 | 109.8 |

**CANADA**

| DIVISION | | OTTAWA, ONTARIO | | | OWEN SOUND, ONTARIO | | | PETERBOROUGH, ONTARIO | | | PORTAGE LA PRAIRIE, MANITOBA | | | PRINCE ALBERT, SASKATCHEWAN | | | PRINCE GEORGE, BRITISH COLUMBIA | | |
|---|---|---|---|---|---|---|---|---|---|---|---|---|---|---|---|---|---|---|---|
| | | MAT. | INST. | TOTAL | MAT. | INST. | TOTAL | MAT. | INST. | TOTAL | MAT. | INST. | TOTAL | MAT. | INST. | TOTAL | MAT. | INST. | TOTAL |
| 015433 | CONTRACTOR EQUIPMENT | | 103.8 | 103.8 | | 101.6 | 101.6 | | 101.6 | 101.6 | | 103.8 | 103.8 | | 100.0 | 100.0 | | 105.3 | 105.3 |
| 0241, 31 - 34 | SITE & INFRASTRUCTURE, DEMOLITION | 111.4 | 105.4 | 107.2 | 115.0 | 101.3 | 105.5 | 112.7 | 101.2 | 104.8 | 111.0 | 100.2 | 103.6 | 105.0 | 96.9 | 99.4 | 120.3 | 104.3 | 109.3 |
| 0310 | Concrete Forming & Accessories | 131.9 | 92.8 | 97.9 | 123.1 | 91.5 | 95.7 | 122.9 | 94.2 | 98.0 | 122.4 | 76.9 | 82.9 | 106.1 | 64.4 | 70.0 | 111.4 | 91.5 | 94.1 |
| 0320 | Concrete Reinforcing | 160.4 | 85.9 | 122.0 | 180.1 | 89.0 | 133.2 | 175.0 | 87.4 | 129.9 | 161.6 | 65.6 | 91.6 | 119.2 | 65.6 | 91.6 | 116.6 | 81.5 | 98.5 |
| 0330 | Cast-in-Place Concrete | 167.3 | 95.1 | 139.6 | 167.6 | 88.1 | 137.1 | 163.5 | 94.2 | 137.0 | 151.1 | 80.3 | 124.0 | 132.8 | 75.0 | 110.7 | 147.1 | 107.9 | 132.1 |
| 03 | CONCRETE | 150.3 | 92.4 | 122.6 | 152.8 | 90.0 | 122.7 | 150.0 | 93.0 | 122.7 | 135.2 | 74.8 | 106.3 | 119.0 | 68.9 | 95.0 | 152.6 | 95.3 | 125.2 |
| 04 | MASONRY | 181.1 | 99.8 | 131.8 | 168.4 | 101.7 | 127.9 | 164.7 | 102.8 | 127.2 | 162.8 | 70.8 | 107.0 | 157.5 | 68.4 | 103.5 | 169.0 | 101.0 | 127.8 |
| 05 | METALS | 124.4 | 92.3 | 113.4 | 107.4 | 93.3 | 102.5 | 106.8 | 92.9 | 102.0 | 107.6 | 79.0 | 97.8 | 103.9 | 75.9 | 94.3 | 107.4 | 90.5 | 101.6 |
| 06 | WOOD, PLASTICS & COMPOSITES | 115.4 | 92.0 | 102.0 | 118.4 | 90.1 | 102.2 | 116.3 | 92.2 | 102.5 | 114.0 | 78.4 | 93.6 | 96.4 | 63.1 | 77.3 | 98.6 | 87.7 | 92.3 |
| 07 | THERMAL & MOISTURE PROTECTION | 124.8 | 94.8 | 112.9 | 111.4 | 92.0 | 103.7 | 111.3 | 97.1 | 105.7 | 105.0 | 75.8 | 93.4 | 114.7 | 66.9 | 89.6 | 114.7 | 91.6 | 105.5 |
| 08 | OPENINGS | 92.9 | 90.7 | 92.3 | 91.8 | 88.4 | 91.0 | 91.7 | 91.0 | 91.5 | 92.9 | 69.0 | 86.8 | 86.3 | 60.4 | 79.7 | 88.7 | 87.0 | 88.3 |
| 0920 | Plaster & Gypsum Board | 234.1 | 91.8 | 133.9 | 157.5 | 89.8 | 109.8 | 157.5 | 91.9 | 111.3 | 130.2 | 77.4 | 93.0 | 130.8 | 61.9 | 82.3 | 134.8 | 86.9 | 101.1 |
| 0950, 0980 | Ceilings & Acoustic Treatment | 133.5 | 91.8 | 105.8 | 95.2 | 89.8 | 91.6 | 95.2 | 91.9 | 93.0 | 95.2 | 77.4 | 83.4 | 95.2 | 61.9 | 73.2 | 95.2 | 86.9 | 89.7 |
| 0960 | Flooring | 134.2 | 101.2 | 124.7 | 132.0 | 102.7 | 123.5 | 131.9 | 101.2 | 123.1 | 131.9 | 73.0 | 115.0 | 120.6 | 65.7 | 104.9 | 126.8 | 79.2 | 113.1 |
| 0970, 0990 | Wall Finishes & Painting/Coating | 111.4 | 94.6 | 101.2 | 112.2 | 95.1 | 101.8 | 111.4 | 99.0 | 103.8 | 111.4 | 62.1 | 81.5 | 111.4 | 60.3 | 80.4 | 111.4 | 88.9 | 97.7 |
| 09 | FINISHES | 135.4 | 94.7 | 113.0 | 118.8 | 94.1 | 105.2 | 117.5 | 96.1 | 105.7 | 114.6 | 75.1 | 92.8 | 111.4 | 64.0 | 85.2 | 115.5 | 88.5 | 100.6 |
| COVERS | DIVS. 10 - 14, 25, 28, 41, 43, 44, 46 | 140.0 | 104.4 | 132.6 | 140.0 | 80.7 | 127.7 | 140.0 | 81.9 | 127.9 | 140.0 | 73.9 | 126.3 | 140.0 | 71.1 | 125.7 | 140.0 | 104.7 | 132.7 |
| 21, 22, 23 | FIRE SUPPRESSION, PLUMBING & HVAC | 100.5 | 83.6 | 93.4 | 102.4 | 106.0 | 103.9 | 102.6 | 109.9 | 105.7 | 102.6 | 90.3 | 97.5 | 103.2 | 74.1 | 91.1 | 102.6 | 102.1 | 102.4 |
| 26, 27, 3370 | ELECTRICAL, COMMUNICATIONS & UTIL. | 119.5 | 87.8 | 102.7 | 121.6 | 95.7 | 107.8 | 119.7 | 96.6 | 107.4 | 121.8 | 64.1 | 91.1 | 122.1 | 67.0 | 92.8 | 119.8 | 87.5 | 102.6 |
| MF2010 | WEIGHTED AVERAGE | 123.0 | 92.3 | 109.3 | 119.1 | 96.8 | 109.1 | 118.1 | 98.8 | 109.5 | 116.1 | 78.5 | 99.3 | 112.3 | 71.6 | 94.1 | 118.5 | 95.4 | 108.2 |

## CANADA

| DIVISION | | QUEBEC, QUEBEC | | | RED DEER, ALBERTA | | | REGINA, SASKATCHEWAN | | | RIMOUSKI, QUEBEC | | | ROUYN-NORANDA, QUEBEC | | | SAINT HYACINTHE, QUEBEC | | |
|---|---|---|---|---|---|---|---|---|---|---|---|---|---|---|---|---|---|---|---|
| | | MAT. | INST. | TOTAL | MAT. | INST. | TOTAL | MAT. | INST. | TOTAL | MAT. | INST. | TOTAL | MAT. | INST. | TOTAL | MAT. | INST. | TOTAL |
| 015433 | CONTRACTOR EQUIPMENT | | 104.0 | 104.0 | | 103.8 | 103.8 | | 100.0 | 100.0 | | 102.0 | 102.0 | | 102.0 | 102.0 | | 102.0 | 102.0 |
| 0241, 31 - 34 | SITE & INFRASTRUCTURE, DEMOLITION | 97.4 | 99.8 | 99.0 | 109.2 | 102.5 | 104.6 | 111.6 | 98.4 | 102.5 | 92.8 | 99.4 | 97.4 | 92.3 | 100.7 | 98.1 | 92.8 | 100.7 | 98.3 |
| 0310 | Concrete Forming & Accessories | 137.5 | 92.2 | 98.2 | 139.4 | 88.1 | 94.9 | 106.3 | 90.0 | 92.2 | 128.1 | 91.7 | 96.5 | 128.1 | 94.0 | 98.5 | 128.1 | 94.0 | 98.5 |
| 0320 | Concrete Reinforcing | 145.4 | 93.6 | 118.7 | 161.6 | 73.6 | 116.3 | 126.2 | 79.0 | 101.8 | 111.9 | 93.5 | 102.4 | 157.1 | 83.7 | 119.3 | 157.1 | 83.7 | 119.3 |
| 0330 | Cast-in-Place Concrete | 148.2 | 100.8 | 130.0 | 151.1 | 104.0 | 133.0 | 156.2 | 95.4 | 132.9 | 136.9 | 99.0 | 122.4 | 130.1 | 103.9 | 120.1 | 134.4 | 103.9 | 122.7 |
| 03 | CONCRETE | 138.9 | 95.4 | 118.1 | 142.8 | 91.1 | 118.1 | 131.8 | 89.7 | 111.7 | 127.7 | 94.5 | 111.8 | 130.9 | 95.4 | 113.9 | 133.1 | 95.4 | 115.1 |
| 04 | MASONRY | 174.5 | 92.6 | 124.9 | 159.9 | 85.7 | 114.9 | 165.6 | 93.1 | 121.7 | 159.4 | 92.6 | 118.9 | 159.6 | 92.6 | 119.0 | 159.9 | 92.6 | 119.1 |
| 05 | METALS | 123.8 | 92.4 | 113.0 | 107.6 | 89.7 | 101.5 | 104.8 | 84.2 | 97.7 | 104.5 | 91.6 | 100.1 | 105.0 | 90.3 | 100.0 | 105.0 | 90.3 | 100.0 |
| 06 | WOOD, PLASTICS & COMPOSITES | 128.6 | 92.0 | 107.6 | 114.1 | 87.6 | 98.9 | 96.6 | 90.8 | 93.3 | 127.7 | 91.7 | 107.1 | 127.7 | 94.0 | 108.4 | 127.7 | 94.0 | 108.4 |
| 07 | THERMAL & MOISTURE PROTECTION | 120.8 | 95.5 | 110.7 | 129.1 | 90.5 | 113.8 | 106.1 | 85.9 | 98.1 | 105.1 | 94.8 | 101.0 | 105.1 | 95.7 | 101.4 | 105.4 | 95.7 | 101.6 |
| 08 | OPENINGS | 92.9 | 89.6 | 92.1 | 92.9 | 81.8 | 90.1 | 87.7 | 79.2 | 85.6 | 92.4 | 79.4 | 89.1 | 92.9 | 81.3 | 89.9 | 92.9 | 81.3 | 89.9 |
| 0920 | Plaster & Gypsum Board | 196.6 | 91.3 | 122.5 | 136.5 | 86.9 | 101.5 | 166.4 | 90.5 | 113.0 | 188.0 | 91.3 | 119.9 | 147.2 | 93.7 | 109.5 | 150.1 | 93.7 | 110.4 |
| 0950, 0980 | Ceilings & Acoustic Treatment | 119.1 | 91.3 | 100.7 | 95.2 | 86.9 | 89.7 | 116.8 | 90.5 | 99.4 | 93.4 | 91.3 | 92.0 | 93.4 | 93.7 | 93.6 | 93.4 | 93.7 | 93.6 |
| 0960 | Flooring | 134.2 | 103.4 | 125.3 | 134.2 | 96.2 | 123.3 | 120.6 | 68.0 | 105.5 | 131.9 | 103.4 | 123.7 | 131.9 | 103.4 | 123.7 | 131.9 | 103.4 | 123.7 |
| 0970, 0990 | Wall Finishes & Painting/Coating | 111.8 | 103.2 | 106.6 | 111.3 | 86.0 | 95.9 | 111.4 | 89.2 | 97.9 | 111.4 | 103.2 | 106.4 | 111.4 | 97.1 | 102.7 | 111.4 | 97.1 | 102.7 |
| 09 | FINISHES | 127.5 | 95.7 | 110.0 | 116.1 | 89.0 | 101.2 | 120.5 | 86.4 | 101.7 | 119.9 | 95.5 | 106.4 | 115.0 | 96.4 | 104.8 | 115.4 | 96.4 | 104.9 |
| COVERS | DIVS. 10 - 14, 25, 28, 41, 43, 44, 46 | 140.0 | 94.1 | 130.5 | 140.0 | 101.5 | 132.0 | 140.0 | 77.9 | 127.1 | 140.0 | 93.4 | 130.3 | 140.0 | 94.0 | 130.5 | 140.0 | 94.0 | 130.5 |
| 21, 22, 23 | FIRE SUPPRESSION, PLUMBING & HVAC | 100.4 | 97.9 | 99.4 | 102.6 | 99.6 | 101.4 | 99.8 | 84.2 | 93.3 | 102.6 | 97.8 | 100.6 | 102.6 | 98.4 | 100.9 | 98.4 | 98.4 | 98.4 |
| 26, 27, 3370 | ELECTRICAL, COMMUNICATIONS & UTIL. | 119.6 | 84.8 | 101.1 | 112.0 | 91.2 | 100.9 | 123.6 | 89.2 | 105.3 | 119.6 | 84.8 | 101.1 | 119.6 | 77.5 | 97.2 | 120.4 | 77.5 | 97.6 |
| MF2010 | WEIGHTED AVERAGE | 120.1 | 94.0 | 108.4 | 116.8 | 92.8 | 106.1 | 114.9 | 87.7 | 102.7 | 114.4 | 93.2 | 104.9 | 114.5 | 92.7 | 104.7 | 113.9 | 92.7 | 104.4 |

## CANADA

| DIVISION | | SAINT JOHN, NEW BRUNSWICK | | | SARNIA, ONTARIO | | | SASKATOON, SASKATCHEWAN | | | SAULT STE MARIE, ONTARIO | | | SHERBROOKE, QUEBEC | | | SOREL, QUEBEC | | |
|---|---|---|---|---|---|---|---|---|---|---|---|---|---|---|---|---|---|---|---|
| | | MAT. | INST. | TOTAL | MAT. | INST. | TOTAL | MAT. | INST. | TOTAL | MAT. | INST. | TOTAL | MAT. | INST. | TOTAL | MAT. | INST. | TOTAL |
| 015433 | CONTRACTOR EQUIPMENT | | 101.4 | 101.4 | | 101.6 | 101.6 | | 100.0 | 100.0 | | 101.6 | 101.6 | | 102.0 | 102.0 | | 102.0 | 102.0 |
| 0241, 31 - 34 | SITE & INFRASTRUCTURE, DEMOLITION | 94.4 | 98.3 | 97.1 | 111.2 | 101.3 | 104.4 | 106.1 | 98.4 | 100.8 | 101.2 | 100.9 | 101.0 | 92.9 | 100.7 | 98.3 | 93.0 | 100.7 | 98.3 |
| 0310 | Concrete Forming & Accessories | 120.3 | 65.2 | 72.5 | 121.7 | 101.9 | 104.6 | 106.3 | 90.0 | 92.2 | 111.4 | 89.9 | 92.8 | 128.1 | 94.0 | 98.5 | 128.1 | 94.1 | 98.6 |
| 0320 | Concrete Reinforcing | 148.4 | 63.4 | 104.6 | 123.5 | 88.8 | 105.6 | 119.2 | 79.0 | 98.5 | 112.4 | 87.0 | 99.3 | 157.1 | 83.7 | 119.3 | 148.4 | 83.7 | 115.1 |
| 0330 | Cast-in-Place Concrete | 136.1 | 65.2 | 109.0 | 150.8 | 108.7 | 134.7 | 142.0 | 95.4 | 124.1 | 135.5 | 92.5 | 119.0 | 134.4 | 103.9 | 122.7 | 135.5 | 103.9 | 123.4 |
| 03 | CONCRETE | 135.8 | 65.7 | 102.3 | 136.0 | 101.6 | 119.5 | 123.7 | 89.7 | 107.4 | 119.8 | 90.4 | 105.7 | 133.1 | 95.4 | 115.1 | 132.4 | 95.5 | 114.7 |
| 04 | MASONRY | 176.7 | 71.3 | 112.8 | 175.2 | 105.8 | 133.1 | 165.3 | 93.1 | 121.6 | 161.2 | 97.2 | 122.4 | 159.9 | 92.6 | 119.1 | 160.0 | 92.6 | 119.1 |
| 05 | METALS | 104.6 | 80.0 | 96.2 | 106.7 | 93.3 | 102.1 | 104.9 | 84.2 | 97.7 | 105.9 | 91.5 | 101.0 | 105.0 | 90.3 | 100.0 | 105.0 | 90.5 | 100.0 |
| 06 | WOOD, PLASTICS & COMPOSITES | 116.9 | 63.7 | 86.4 | 115.4 | 101.2 | 107.2 | 94.9 | 90.8 | 92.6 | 103.0 | 88.9 | 94.9 | 127.7 | 94.0 | 108.4 | 127.7 | 94.0 | 108.4 |
| 07 | THERMAL & MOISTURE PROTECTION | 109.0 | 66.5 | 92.1 | 111.3 | 100.9 | 107.2 | 105.2 | 85.9 | 97.5 | 110.2 | 92.7 | 103.2 | 105.2 | 95.7 | 101.4 | 105.1 | 95.7 | 101.4 |
| 08 | OPENINGS | 86.8 | 59.1 | 79.8 | 94.0 | 95.2 | 94.3 | 86.8 | 79.2 | 84.9 | 84.4 | 87.2 | 85.1 | 92.9 | 81.3 | 89.9 | 92.9 | 86.3 | 91.2 |
| 0920 | Plaster & Gypsum Board | 181.0 | 62.5 | 97.6 | 179.8 | 101.2 | 124.5 | 146.0 | 90.5 | 106.9 | 141.9 | 88.6 | 104.4 | 150.1 | 93.7 | 110.4 | 188.0 | 93.7 | 121.6 |
| 0950, 0980 | Ceilings & Acoustic Treatment | 101.8 | 62.5 | 75.7 | 99.0 | 101.2 | 100.5 | 116.8 | 90.5 | 99.4 | 95.2 | 88.6 | 90.8 | 93.4 | 93.7 | 93.6 | 93.4 | 93.7 | 93.6 |
| 0960 | Flooring | 127.5 | 74.9 | 112.4 | 131.9 | 110.7 | 125.8 | 120.6 | 68.0 | 105.5 | 123.8 | 104.7 | 118.3 | 131.9 | 103.4 | 123.7 | 93.4 | 93.7 | 93.6 |
| 0970, 0990 | Wall Finishes & Painting/Coating | 111.4 | 73.2 | 88.2 | 111.4 | 111.9 | 111.7 | 111.4 | 89.2 | 97.9 | 111.4 | 103.6 | 106.7 | 111.4 | 97.1 | 102.7 | 111.4 | 97.1 | 102.7 |
| 09 | FINISHES | 119.9 | 67.6 | 91.1 | 121.0 | 105.1 | 112.2 | 117.8 | 86.4 | 100.5 | 112.4 | 94.0 | 102.3 | 115.4 | 96.4 | 104.9 | 119.9 | 96.4 | 107.0 |
| COVERS | DIVS. 10 - 14, 25, 28, 41, 43, 44, 46 | 140.0 | 71.3 | 125.7 | 140.0 | 83.5 | 128.3 | 140.0 | 77.9 | 127.1 | 140.0 | 106.1 | 132.9 | 140.0 | 94.0 | 130.5 | 140.0 | 94.0 | 130.5 |
| 21, 22, 23 | FIRE SUPPRESSION, PLUMBING & HVAC | 102.6 | 80.8 | 93.5 | 102.6 | 116.5 | 108.4 | 99.7 | 84.2 | 93.3 | 102.6 | 102.4 | 102.5 | 103.2 | 98.4 | 101.2 | 102.6 | 98.4 | 100.9 |
| 26, 27, 3370 | ELECTRICAL, COMMUNICATIONS & UTIL. | 130.6 | 86.4 | 107.1 | 122.9 | 99.0 | 110.2 | 123.9 | 89.2 | 105.4 | 121.5 | 97.0 | 108.4 | 119.7 | 77.5 | 97.2 | 119.6 | 77.5 | 97.2 |
| MF2010 | WEIGHTED AVERAGE | 116.9 | 76.5 | 98.8 | 117.8 | 103.6 | 111.4 | 113.4 | 87.7 | 101.9 | 112.7 | 96.3 | 105.4 | 114.9 | 92.7 | 105.0 | 115.1 | 93.0 | 105.2 |

## CANADA

| DIVISION | | ST CATHARINES, ONTARIO | | | ST JEROME, QUEBEC | | | ST JOHNS, NEWFOUNDLAND | | | SUDBURY, ONTARIO | | | SUMMERSIDE, PRINCE EDWARD ISLAND | | | SYDNEY, NOVA SCOTIA | | |
|---|---|---|---|---|---|---|---|---|---|---|---|---|---|---|---|---|---|---|---|
| | | MAT. | INST. | TOTAL | MAT. | INST. | TOTAL | MAT. | INST. | TOTAL | MAT. | INST. | TOTAL | MAT. | INST. | TOTAL | MAT. | INST. | TOTAL |
| 015433 | CONTRACTOR EQUIPMENT | | 101.6 | 101.6 | | 102.0 | 102.0 | | 102.4 | 102.4 | | 101.6 | 101.6 | | 101.1 | 101.1 | | 101.2 | 101.2 |
| 0241, 31 - 34 | SITE & INFRASTRUCTURE, DEMOLITION | 101.1 | 102.0 | 101.8 | 92.3 | 100.7 | 98.1 | 113.1 | 99.2 | 103.5 | 101.2 | 101.5 | 101.4 | 105.0 | 96.4 | 99.1 | 92.3 | 99.0 | 96.9 |
| 0310 | Concrete Forming & Accessories | 115.2 | 96.1 | 98.6 | 128.1 | 94.0 | 98.5 | 103.8 | 74.0 | 78.0 | 111.4 | 89.9 | 92.8 | 96.6 | 62.2 | 66.8 | 96.4 | 76.5 | 79.2 |
| 0320 | Concrete Reinforcing | 111.5 | 87.2 | 98.9 | 157.1 | 83.7 | 119.3 | 164.6 | 65.0 | 113.3 | 112.4 | 85.6 | 98.6 | 148.4 | 49.7 | 97.5 | 148.4 | 49.6 | 97.5 |
| 0330 | Cast-in-Place Concrete | 134.4 | 97.9 | 120.4 | 130.1 | 103.9 | 120.1 | 172.3 | 86.9 | 139.6 | 135.5 | 94.2 | 119.7 | 161.1 | 64.3 | 124.1 | 130.9 | 76.2 | 109.9 |
| 03 | CONCRETE | 119.3 | 95.0 | 107.7 | 130.9 | 95.4 | 113.9 | 164.7 | 77.3 | 122.9 | 119.8 | 90.8 | 105.9 | 153.1 | 61.3 | 109.2 | 129.4 | 72.0 | 101.9 |
| 04 | MASONRY | 160.8 | 102.7 | 125.6 | 159.6 | 92.6 | 119.0 | 162.9 | 79.0 | 112.0 | 160.9 | 97.7 | 122.6 | 159.1 | 66.6 | 103.0 | 156.7 | 76.5 | 108.1 |
| 05 | METALS | 105.9 | 92.8 | 101.4 | 105.0 | 90.3 | 100.0 | 108.1 | 79.8 | 98.3 | 105.9 | 92.4 | 101.2 | 105.9 | 69.3 | 93.3 | 105.9 | 76.3 | 95.7 |
| 06 | WOOD, PLASTICS & COMPOSITES | 107.1 | 95.5 | 100.5 | 127.7 | 94.0 | 108.4 | 96.0 | 72.7 | 82.6 | 96.0 | 88.9 | 94.9 | 87.4 | 61.6 | 72.6 | 87.1 | 75.9 | 80.7 |
| 07 | THERMAL & MOISTURE PROTECTION | 110.7 | 98.3 | 105.8 | 105.1 | 95.7 | 101.4 | 109.7 | 76.8 | 96.6 | 110.2 | 92.7 | 103.2 | 105.7 | 64.9 | 89.5 | 106.3 | 74.4 | 93.6 |
| 08 | OPENINGS | 83.4 | 92.3 | 85.7 | 92.9 | 81.3 | 89.9 | 99.2 | 67.5 | 91.2 | 84.4 | 87.2 | 85.1 | 105.7 | 64.9 | 89.5 | 83.5 | 69.0 | 79.8 |
| 0920 | Plaster & Gypsum Board | 136.9 | 95.4 | 107.7 | 147.2 | 93.7 | 109.5 | 165.6 | 71.7 | 99.5 | 141.9 | 88.6 | 104.4 | 156.6 | 60.4 | 88.9 | 156.2 | 75.1 | 99.1 |
| 0950, 0980 | Ceilings & Acoustic Treatment | 99.0 | 95.4 | 96.6 | 93.4 | 93.7 | 93.6 | 103.7 | 71.7 | 82.5 | 95.2 | 88.6 | 90.8 | 95.2 | 60.4 | 72.2 | 95.2 | 75.1 | 81.9 |
| 0960 | Flooring | 126.4 | 102.7 | 119.6 | 131.9 | 103.4 | 123.7 | 115.9 | 61.5 | 100.3 | 123.8 | 104.7 | 118.3 | 111.6 | 65.3 | 98.3 | 111.6 | 69.9 | 99.6 |
| 0970, 0990 | Wall Finishes & Painting/Coating | 111.4 | 103.1 | 106.3 | 111.4 | 97.1 | 102.7 | 111.4 | 76.3 | 90.1 | 111.4 | 94.0 | 100.9 | 111.4 | 45.5 | 71.4 | 111.4 | 67.9 | 85.0 |
| 09 | FINISHES | 113.3 | 98.5 | 105.2 | 115.0 | 96.4 | 104.8 | 116.0 | 71.9 | 91.7 | 112.4 | 93.1 | 101.7 | 111.5 | 61.3 | 83.8 | 110.5 | 74.6 | 90.7 |
| COVERS | DIVS. 10 - 14, 25, 28, 41, 43, 44, 46 | 140.0 | 81.8 | 127.9 | 140.0 | 94.0 | 130.5 | 140.0 | 74.7 | 126.4 | 140.0 | 106.1 | 133.0 | 140.0 | 71.4 | 125.7 | 140.0 | 73.3 | 126.1 |
| 21, 22, 23 | FIRE SUPPRESSION, PLUMBING & HVAC | 99.7 | 84.8 | 93.5 | 102.6 | 98.4 | 100.9 | 99.9 | 72.9 | 88.7 | 102.6 | 87.9 | 96.5 | 102.6 | 68.9 | 88.6 | 102.6 | 91.2 | 97.9 |
| 26, 27, 3370 | ELECTRICAL, COMMUNICATIONS & UTIL. | 122.7 | 87.9 | 104.2 | 120.3 | 77.5 | 97.6 | 119.7 | 73.8 | 95.3 | 119.7 | 87.9 | 102.8 | 118.2 | 55.9 | 85.1 | 119.3 | 68.7 | 92.4 |
| MF2010 | WEIGHTED AVERAGE | 112.1 | 93.1 | 103.6 | 114.5 | 92.7 | 104.8 | 119.7 | 76.9 | 100.6 | 112.5 | 92.1 | 103.4 | 117.1 | 66.4 | 94.5 | 112.6 | 79.1 | 97.6 |

## CANADA

| DIVISION | | THUNDER BAY, ONTARIO | | | TIMMINS, ONTARIO | | | TORONTO, ONTARIO | | | TROIS RIVIERES, QUEBEC | | | TRURO, NOVA SCOTIA | | | VANCOUVER, BRITISH COLUMBIA | | |
|---|---|---|---|---|---|---|---|---|---|---|---|---|---|---|---|---|---|---|---|
| | | MAT. | INST. | TOTAL | MAT. | INST. | TOTAL | MAT. | INST. | TOTAL | MAT. | INST. | TOTAL | MAT. | INST. | TOTAL | MAT. | INST. | TOTAL |
| 015433 | CONTRACTOR EQUIPMENT | | 101.6 | 101.6 | | 101.6 | 101.6 | | 104.0 | 104.0 | | 102.0 | 102.0 | | 101.2 | 101.2 | | 112.1 | 112.1 |
| 0241, 31 - 34 | SITE & INFRASTRUCTURE, DEMOLITION | 106.5 | 102.0 | 103.4 | 112.7 | 100.8 | 104.5 | 138.2 | 106.1 | 116.1 | 93.0 | 100.7 | 98.3 | 97.1 | 99.0 | 98.4 | 120.5 | 108.6 | 112.3 |
| 0310 | Concrete Forming & Accessories | 123.0 | 93.6 | 97.5 | 122.9 | 93.1 | 97.0 | 132.3 | 101.7 | 105.7 | 128.1 | 94.1 | 98.6 | 96.4 | 76.6 | 79.2 | 126.6 | 88.9 | 93.9 |
| 0320 | Concrete Reinforcing | 99.7 | 86.6 | 92.9 | 175.0 | 86.8 | 129.6 | 167.3 | 87.8 | 126.4 | 148.4 | 83.7 | 115.1 | 148.4 | 49.6 | 97.5 | 157.2 | 80.9 | 117.9 |
| 0330 | Cast-in-Place Concrete | 148.0 | 96.3 | 128.2 | 163.5 | 92.4 | 136.3 | 153.2 | 117.9 | 139.7 | 135.5 | 103.9 | 123.4 | 169.8 | 76.2 | 134.0 | 162.8 | 103.4 | 140.1 |
| 03 | CONCRETE | 128.3 | 93.3 | 111.5 | 150.0 | 91.8 | 122.1 | 144.3 | 104.4 | 125.2 | 132.4 | 95.5 | 114.7 | 148.9 | 72.0 | 112.1 | 153.2 | 92.6 | 124.2 |
| 04 | MASONRY | 161.7 | 101.9 | 125.4 | 164.7 | 95.6 | 122.8 | 194.1 | 107.3 | 141.5 | 160.0 | 92.6 | 119.1 | 159.8 | 76.5 | 109.3 | 176.7 | 95.9 | 127.7 |
| 05 | METALS | 105.7 | 91.7 | 100.9 | 106.8 | 92.4 | 101.8 | 126.1 | 93.8 | 115.0 | 105.0 | 90.5 | 100.0 | 105.9 | 76.3 | 95.7 | 162.8 | 91.8 | 138.3 |
| 06 | WOOD, PLASTICS & COMPOSITES | 116.5 | 92.3 | 102.6 | 116.3 | 93.6 | 103.3 | 116.5 | 100.2 | 107.2 | 127.7 | 94.0 | 108.4 | 87.1 | 75.9 | 80.7 | 114.9 | 87.0 | 98.9 |
| 07 | THERMAL & MOISTURE PROTECTION | 110.9 | 95.0 | 104.6 | 111.3 | 91.2 | 103.3 | 122.7 | 102.5 | 114.7 | 105.1 | 95.7 | 101.4 | 106.3 | 74.4 | 93.6 | 133.0 | 90.6 | 116.2 |
| 08 | OPENINGS | 82.4 | 89.9 | 84.3 | 91.7 | 89.1 | 91.0 | 91.7 | 98.4 | 93.4 | 92.9 | 86.3 | 91.2 | 83.5 | 69.0 | 79.8 | 94.6 | 86.0 | 92.4 |
| 0920 | Plaster & Gypsum Board | 172.5 | 92.0 | 115.8 | 157.5 | 93.4 | 112.4 | 180.3 | 100.2 | 123.9 | 188.0 | 93.7 | 121.6 | 156.2 | 75.1 | 99.1 | 155.1 | 86.1 | 106.5 |
| 0950, 0980 | Ceilings & Acoustic Treatment | 95.2 | 92.0 | 93.1 | 95.2 | 93.4 | 94.0 | 131.7 | 100.2 | 110.8 | 93.4 | 93.7 | 93.6 | 95.2 | 75.1 | 81.9 | 148.7 | 86.1 | 107.2 |
| 0960 | Flooring | 131.9 | 110.0 | 125.6 | 131.9 | 101.2 | 123.1 | 135.3 | 112.6 | 128.8 | 131.9 | 103.4 | 123.7 | 111.6 | 69.9 | 99.6 | 134.6 | 99.7 | 124.5 |
| 0970, 0990 | Wall Finishes & Painting/Coating | 111.4 | 91.4 | 99.3 | 111.4 | 96.5 | 102.4 | 113.4 | 107.2 | 109.6 | 111.4 | 97.1 | 102.7 | 111.4 | 67.9 | 85.0 | 111.3 | 83.1 | 94.2 |
| 09 | FINISHES | 118.9 | 96.7 | 106.6 | 117.5 | 95.4 | 105.3 | 131.6 | 104.4 | 116.6 | 119.9 | 96.4 | 107.0 | 110.5 | 74.6 | 90.7 | 130.8 | 89.9 | 108.2 |
| COVERS | DIVS. 10 - 14, 25, 28, 41, 43, 44, 46 | 140.0 | 81.6 | 127.9 | 140.0 | 80.2 | 127.6 | 140.0 | 109.8 | 133.7 | 140.0 | 94.0 | 130.5 | 140.0 | 73.3 | 126.1 | 140.0 | 103.3 | 132.4 |
| 21, 22, 23 | FIRE SUPPRESSION, PLUMBING & HVAC | 99.7 | 84.6 | 93.4 | 102.6 | 106.1 | 104.1 | 100.4 | 99.9 | 100.2 | 102.6 | 98.4 | 100.9 | 102.6 | 91.2 | 97.9 | 100.4 | 86.6 | 94.7 |
| 26, 27, 3370 | ELECTRICAL, COMMUNICATIONS & UTIL. | 120.9 | 86.3 | 102.5 | 121.5 | 97.0 | 108.4 | 121.1 | 90.1 | 104.6 | 119.6 | 77.5 | 97.2 | 119.3 | 68.7 | 92.4 | 121.8 | 79.4 | 99.3 |
| MF2010 | WEIGHTED AVERAGE | 113.6 | 91.9 | 103.9 | 118.3 | 96.7 | 108.6 | 123.6 | 100.7 | 113.4 | 115.1 | 93.0 | 105.2 | 115.3 | 79.1 | 99.1 | 129.1 | 90.7 | 111.9 |

## CANADA

| DIVISION | | VICTORIA, BRITISH COLUMBIA | | | WHITEHORSE, YUKON | | | WINDSOR, ONTARIO | | | WINNIPEG, MANITOBA | | | YARMOUTH, NOVA SCOTIA | | | YELLOWKNIFE, NWT | | |
|---|---|---|---|---|---|---|---|---|---|---|---|---|---|---|---|---|---|---|---|
| | | MAT. | INST. | TOTAL | MAT. | INST. | TOTAL | MAT. | INST. | TOTAL | MAT. | INST. | TOTAL | MAT. | INST. | TOTAL | MAT. | INST. | TOTAL |
| 015433 | CONTRACTOR EQUIPMENT | | 108.6 | 108.6 | | 101.4 | 101.4 | | 101.6 | 101.6 | | 105.9 | 105.9 | | 101.2 | 101.2 | | 101.2 | 101.2 |
| 0241, 31 - 34 | SITE & INFRASTRUCTURE, DEMOLITION | 120.3 | 108.1 | 111.9 | 106.7 | 97.5 | 100.3 | 96.2 | 101.9 | 100.1 | 119.0 | 101.7 | 107.1 | 96.9 | 99.0 | 98.4 | 128.7 | 102.2 | 110.4 |
| 0310 | Concrete Forming & Accessories | 111.4 | 88.3 | 91.3 | 106.2 | 63.6 | 69.2 | 123.0 | 91.9 | 96.0 | 131.5 | 71.2 | 79.2 | 96.4 | 76.5 | 79.2 | 106.3 | 87.2 | 89.7 |
| 0320 | Concrete Reinforcing | 116.6 | 80.8 | 98.2 | 119.2 | 63.7 | 90.6 | 109.2 | 85.9 | 97.2 | 148.3 | 59.0 | 102.3 | 148.4 | 49.6 | 97.5 | 119.2 | 67.6 | 92.6 |
| 0330 | Cast-in-Place Concrete | 147.1 | 102.7 | 130.1 | 146.6 | 74.2 | 118.8 | 137.7 | 97.2 | 122.2 | 190.9 | 76.8 | 147.2 | 168.0 | 76.2 | 132.9 | 243.2 | 94.7 | 186.4 |
| 03 | CONCRETE | 152.6 | 91.9 | 123.6 | 126.0 | 67.9 | 98.2 | 121.1 | 92.7 | 107.5 | 162.5 | 71.3 | 118.9 | 148.0 | 72.0 | 111.6 | 174.5 | 86.2 | 132.3 |
| 04 | MASONRY | 168.1 | 95.9 | 124.4 | 157.9 | 66.7 | 102.6 | 161.0 | 100.7 | 124.4 | 180.3 | 72.4 | 114.9 | 159.7 | 76.5 | 109.3 | 166.4 | 80.9 | 114.6 |
| 05 | METALS | 107.4 | 87.6 | 100.6 | 105.1 | 75.6 | 95.0 | 105.8 | 92.2 | 101.1 | 156.2 | 75.6 | 128.5 | 105.9 | 76.3 | 95.7 | 106.0 | 81.8 | 97.7 |
| 06 | WOOD, PLASTICS & COMPOSITES | 98.6 | 86.9 | 91.9 | 96.4 | 62.2 | 76.8 | 116.5 | 90.8 | 101.7 | 114.5 | 72.1 | 90.1 | 87.1 | 75.9 | 80.7 | 96.4 | 88.9 | 92.1 |
| 07 | THERMAL & MOISTURE PROTECTION | 114.7 | 90.4 | 105.1 | 104.5 | 65.9 | 89.1 | 110.7 | 94.0 | 104.1 | 116.0 | 73.6 | 99.1 | 106.3 | 74.4 | 93.6 | 107.4 | 84.7 | 98.4 |
| 08 | OPENINGS | 88.7 | 82.0 | 87.0 | 86.3 | 59.0 | 79.4 | 82.1 | 89.6 | 84.0 | 92.9 | 63.9 | 85.5 | 83.5 | 69.0 | 79.8 | 86.3 | 74.6 | 83.3 |
| 0920 | Plaster & Gypsum Board | 134.8 | 86.1 | 100.5 | 119.2 | 60.9 | 78.1 | 160.8 | 90.5 | 111.3 | 164.3 | 70.7 | 98.4 | 156.2 | 75.1 | 99.1 | 169.1 | 88.5 | 112.3 |
| 0950, 0980 | Ceilings & Acoustic Treatment | 95.2 | 86.1 | 89.2 | 95.2 | 60.9 | 72.5 | 95.2 | 90.5 | 92.1 | 136.4 | 70.7 | 92.9 | 95.2 | 75.1 | 81.9 | 128.0 | 88.5 | 101.8 |
| 0960 | Flooring | 126.8 | 79.2 | 113.1 | 120.6 | 63.7 | 104.3 | 131.9 | 107.1 | 124.8 | 133.0 | 75.5 | 116.5 | 111.6 | 69.9 | 99.6 | 120.6 | 97.8 | 114.1 |
| 0970, 0990 | Wall Finishes & Painting/Coating | 111.4 | 83.1 | 94.2 | 111.4 | 59.3 | 79.8 | 111.4 | 95.9 | 102.0 | 111.4 | 52.7 | 75.8 | 111.4 | 67.9 | 85.0 | 111.4 | 86.4 | 96.2 |
| 09 | FINISHES | 115.5 | 86.3 | 99.4 | 110.0 | 62.9 | 84.0 | 117.1 | 95.4 | 105.1 | 128.0 | 70.4 | 96.3 | 110.5 | 74.6 | 90.7 | 122.9 | 88.5 | 103.9 |
| COVERS | DIVS. 10 - 14, 25, 28, 41, 43, 44, 46 | 140.0 | 77.2 | 127.0 | 140.0 | 70.3 | 125.5 | 140.0 | 80.9 | 127.7 | 140.0 | 73.0 | 126.1 | 140.0 | 73.3 | 126.1 | 140.0 | 75.0 | 126.5 |
| 21, 22, 23 | FIRE SUPPRESSION, PLUMBING & HVAC | 102.6 | 86.6 | 95.9 | 102.6 | 80.8 | 93.5 | 99.7 | 84.2 | 93.3 | 100.5 | 69.5 | 87.6 | 102.6 | 91.2 | 97.9 | 102.9 | 103.2 | 103.1 |
| 26, 27, 3370 | ELECTRICAL, COMMUNICATIONS & UTIL. | 120.9 | 79.4 | 98.8 | 122.5 | 66.4 | 92.7 | 125.6 | 87.9 | 105.5 | 125.6 | 69.8 | 95.9 | 119.3 | 68.7 | 92.4 | 122.8 | 92.5 | 106.7 |
| MF2010 | WEIGHTED AVERAGE | 118.5 | 88.7 | 105.2 | 113.2 | 72.4 | 95.0 | 112.7 | 91.6 | 103.3 | 128.8 | 73.5 | 104.1 | 115.1 | 79.1 | 99.0 | 121.5 | 90.5 | 107.7 |

# Location Factors

Costs shown in RSMeans cost data publications are based on national averages for materials and installation. To adjust these costs to a specific location, simply multiply the base cost by the factor and divide by 100 for that city. The data is arranged alphabetically by state and postal zip code numbers. For a city not listed, use the factor for a nearby city with similar economic characteristics.

| STATE/ZIP | CITY | MAT. | INST. | TOTAL |
|---|---|---|---|---|
| **ALABAMA** | | | | |
| 350-352 | Birmingham | 97.0 | 75.6 | 87.4 |
| 354 | Tuscaloosa | 95.8 | 59.2 | 79.4 |
| 355 | Jasper | 96.0 | 58.2 | 79.1 |
| 356 | Decatur | 95.8 | 60.8 | 80.1 |
| 357-358 | Huntsville | 95.8 | 69.0 | 83.8 |
| 359 | Gadsden | 95.6 | 58.2 | 78.9 |
| 360-361 | Montgomery | 96.8 | 57.2 | 79.1 |
| 362 | Anniston | 95.1 | 65.5 | 81.9 |
| 363 | Dothan | 95.8 | 52.6 | 76.5 |
| 364 | Evergreen | 95.3 | 54.1 | 76.9 |
| 365-366 | Mobile | 96.8 | 67.9 | 83.9 |
| 367 | Selma | 95.5 | 52.5 | 76.3 |
| 368 | Phenix City | 96.2 | 56.3 | 78.4 |
| 369 | Butler | 95.7 | 52.6 | 76.4 |
| **ALASKA** | | | | |
| 995-996 | Anchorage | 127.2 | 114.5 | 121.5 |
| 997 | Fairbanks | 124.4 | 115.0 | 120.2 |
| 998 | Juneau | 122.8 | 112.6 | 118.3 |
| 999 | Ketchikan | 136.9 | 112.6 | 126.0 |
| **ARIZONA** | | | | |
| 850,853 | Phoenix | 98.6 | 75.9 | 88.5 |
| 851,852 | Mesa/Tempe | 98.2 | 72.8 | 86.8 |
| 855 | Globe | 98.5 | 62.0 | 82.2 |
| 856-857 | Tucson | 96.9 | 72.0 | 85.8 |
| 859 | Show Low | 98.6 | 63.5 | 82.9 |
| 860 | Flagstaff | 100.7 | 74.4 | 88.9 |
| 863 | Prescott | 98.4 | 63.0 | 82.6 |
| 864 | Kingman | 96.7 | 74.2 | 86.7 |
| 865 | Chambers | 96.8 | 63.5 | 81.9 |
| **ARKANSAS** | | | | |
| 716 | Pine Bluff | 96.7 | 64.8 | 82.4 |
| 717 | Camden | 94.2 | 50.7 | 74.8 |
| 718 | Texarkana | 95.9 | 51.9 | 76.3 |
| 719 | Hot Springs | 93.4 | 53.6 | 75.6 |
| 720-722 | Little Rock | 97.0 | 65.9 | 83.1 |
| 723 | West Memphis | 95.9 | 63.7 | 81.5 |
| 724 | Jonesboro | 96.7 | 61.6 | 81.0 |
| 725 | Batesville | 94.3 | 56.9 | 77.6 |
| 726 | Harrison | 95.7 | 53.0 | 76.6 |
| 727 | Fayetteville | 92.9 | 54.9 | 75.9 |
| 728 | Russellville | 94.2 | 56.1 | 77.2 |
| 729 | Fort Smith | 97.0 | 63.4 | 82.0 |
| **CALIFORNIA** | | | | |
| 900-902 | Los Angeles | 100.6 | 115.4 | 107.2 |
| 903-905 | Inglewood | 96.1 | 112.6 | 103.5 |
| 906-908 | Long Beach | 97.8 | 112.6 | 104.5 |
| 910-912 | Pasadena | 96.5 | 112.7 | 103.7 |
| 913-916 | Van Nuys | 99.7 | 112.7 | 105.5 |
| 917-918 | Alhambra | 98.6 | 112.7 | 104.9 |
| 919-921 | San Diego | 100.6 | 108.2 | 104.0 |
| 922 | Palm Springs | 97.5 | 111.5 | 103.8 |
| 923-924 | San Bernardino | 95.2 | 111.2 | 102.3 |
| 925 | Riverside | 100.0 | 113.3 | 105.9 |
| 926-927 | Santa Ana | 97.2 | 111.4 | 103.6 |
| 928 | Anaheim | 99.9 | 113.7 | 106.1 |
| 930 | Oxnard | 101.4 | 113.1 | 106.6 |
| 931 | Santa Barbara | 100.6 | 113.1 | 106.2 |
| 932-933 | Bakersfield | 101.4 | 110.4 | 105.4 |
| 934 | San Luis Obispo | 101.2 | 109.8 | 105.0 |
| 935 | Mojave | 98.2 | 107.7 | 102.4 |
| 936-938 | Fresno | 101.7 | 115.3 | 107.8 |
| 939 | Salinas | 102.0 | 123.0 | 111.4 |
| 940-941 | San Francisco | 109.6 | 140.8 | 123.6 |
| 942,956-958 | Sacramento | 103.9 | 117.2 | 109.8 |
| 943 | Palo Alto | 102.3 | 129.4 | 114.4 |
| 944 | San Mateo | 105.1 | 131.4 | 116.9 |
| 945 | Vallejo | 102.9 | 127.8 | 114.0 |
| 946 | Oakland | 107.5 | 133.7 | 119.2 |
| 947 | Berkeley | 106.8 | 133.6 | 118.8 |
| 948 | Richmond | 106.1 | 128.9 | 116.3 |
| 949 | San Rafael | 107.4 | 130.3 | 117.6 |
| 950 | Santa Cruz | 106.4 | 123.1 | 113.9 |

| STATE/ZIP | CITY | MAT. | INST. | TOTAL |
|---|---|---|---|---|
| **CALIFORNIA (CONT'D)** | | | | |
| 951 | San Jose | 104.6 | 134.3 | 117.9 |
| 952 | Stockton | 101.9 | 117.3 | 108.8 |
| 953 | Modesto | 101.9 | 116.6 | 108.5 |
| 954 | Santa Rosa | 101.7 | 134.9 | 116.5 |
| 955 | Eureka | 103.1 | 115.2 | 108.5 |
| 959 | Marysville | 102.4 | 113.8 | 107.5 |
| 960 | Redding | 104.1 | 113.8 | 108.4 |
| 961 | Susanville | 103.1 | 113.8 | 107.9 |
| **COLORADO** | | | | |
| 800-802 | Denver | 101.4 | 84.9 | 94.0 |
| 803 | Boulder | 98.5 | 83.7 | 91.8 |
| 804 | Golden | 100.6 | 82.5 | 92.5 |
| 805 | Fort Collins | 102.3 | 77.1 | 91.0 |
| 806 | Greeley | 99.6 | 73.6 | 88.0 |
| 807 | Fort Morgan | 99.1 | 82.3 | 91.6 |
| 808-809 | Colorado Springs | 101.3 | 82.4 | 92.8 |
| 810 | Pueblo | 100.8 | 80.2 | 91.6 |
| 811 | Alamosa | 102.1 | 76.1 | 90.5 |
| 812 | Salida | 102.0 | 77.8 | 91.2 |
| 813 | Durango | 102.6 | 75.3 | 90.4 |
| 814 | Montrose | 101.2 | 74.7 | 89.4 |
| 815 | Grand Junction | 104.7 | 75.2 | 91.5 |
| 816 | Glenwood Springs | 102.1 | 78.0 | 91.3 |
| **CONNECTICUT** | | | | |
| 060 | New Britain | 101.0 | 117.8 | 108.5 |
| 061 | Hartford | 102.1 | 117.8 | 109.1 |
| 062 | Willimantic | 101.6 | 117.8 | 108.9 |
| 063 | New London | 97.7 | 117.7 | 106.6 |
| 064 | Meriden | 99.7 | 117.7 | 107.8 |
| 065 | New Haven | 102.9 | 117.7 | 109.5 |
| 066 | Bridgeport | 102.2 | 116.4 | 108.6 |
| 067 | Waterbury | 101.7 | 117.7 | 108.9 |
| 068 | Norwalk | 101.6 | 116.5 | 108.3 |
| 069 | Stamford | 101.8 | 123.0 | 111.3 |
| **D.C.** | | | | |
| 200-205 | Washington | 102.8 | 91.9 | 97.9 |
| **DELAWARE** | | | | |
| 197 | Newark | 99.3 | 108.6 | 103.4 |
| 198 | Wilmington | 99.6 | 108.6 | 103.6 |
| 199 | Dover | 99.9 | 108.6 | 103.8 |
| **FLORIDA** | | | | |
| 320,322 | Jacksonville | 97.8 | 69.7 | 85.2 |
| 321 | Daytona Beach | 98.0 | 76.4 | 88.3 |
| 323 | Tallahassee | 98.6 | 59.7 | 81.2 |
| 324 | Panama City | 99.2 | 57.5 | 80.6 |
| 325 | Pensacola | 101.9 | 65.1 | 85.4 |
| 326,344 | Gainesville | 99.5 | 70.6 | 86.6 |
| 327-328,347 | Orlando | 101.2 | 74.2 | 89.1 |
| 329 | Melbourne | 100.4 | 79.6 | 91.1 |
| 330-332,340 | Miami | 99.1 | 78.1 | 89.7 |
| 333 | Fort Lauderdale | 97.3 | 76.7 | 88.1 |
| 334,349 | West Palm Beach | 96.0 | 75.7 | 86.9 |
| 335-336,346 | Tampa | 99.0 | 83.6 | 92.1 |
| 337 | St. Petersburg | 101.3 | 66.1 | 85.6 |
| 338 | Lakeland | 98.0 | 83.5 | 91.5 |
| 339,341 | Fort Myers | 97.3 | 75.4 | 87.5 |
| 342 | Sarasota | 99.2 | 77.3 | 89.4 |
| **GEORGIA** | | | | |
| 300-303,399 | Atlanta | 97.2 | 76.8 | 88.1 |
| 304 | Statesboro | 96.8 | 54.1 | 77.7 |
| 305 | Gainesville | 95.5 | 64.7 | 81.7 |
| 306 | Athens | 94.9 | 64.9 | 81.5 |
| 307 | Dalton | 96.8 | 60.5 | 80.6 |
| 308-309 | Augusta | 95.9 | 65.5 | 82.3 |
| 310-312 | Macon | 95.6 | 65.0 | 81.9 |
| 313-314 | Savannah | 97.7 | 60.8 | 81.2 |
| 315 | Waycross | 96.7 | 61.8 | 81.1 |
| 316 | Valdosta | 96.7 | 61.2 | 80.9 |
| 317,398 | Albany | 96.9 | 62.0 | 81.3 |
| 318-319 | Columbus | 96.8 | 65.3 | 82.7 |

| STATE/ZIP | CITY | MAT. | INST. | TOTAL |
|---|---|---|---|---|
| **HAWAII** | | | | |
| 967 | Hilo | 111.9 | 120.6 | 115.8 |
| 968 | Honolulu | 115.1 | 120.6 | 117.5 |
| **STATES & POSS.** | | | | |
| 969 | Guam | 133.3 | 61.7 | 101.3 |
| **IDAHO** | | | | |
| 832 | Pocatello | 100.4 | 73.0 | 88.2 |
| 833 | Twin Falls | 101.7 | 53.1 | 80.0 |
| 834 | Idaho Falls | 99.0 | 59.0 | 81.1 |
| 835 | Lewiston | 109.3 | 78.3 | 95.4 |
| 836-837 | Boise | 99.5 | 73.3 | 87.8 |
| 838 | Coeur d'Alene | 108.5 | 75.4 | 93.7 |
| **ILLINOIS** | | | | |
| 600-603 | North Suburban | 99.3 | 131.0 | 113.4 |
| 604 | Joliet | 99.2 | 137.7 | 116.4 |
| 605 | South Suburban | 99.3 | 131.0 | 113.4 |
| 606-608 | Chicago | 99.7 | 138.9 | 117.2 |
| 609 | Kankakee | 95.2 | 125.3 | 108.7 |
| 610-611 | Rockford | 98.1 | 122.7 | 109.1 |
| 612 | Rock Island | 95.7 | 99.0 | 97.2 |
| 613 | La Salle | 96.9 | 121.4 | 107.8 |
| 614 | Galesburg | 96.8 | 104.8 | 100.4 |
| 615-616 | Peoria | 99.2 | 107.3 | 102.8 |
| 617 | Bloomington | 96.1 | 108.7 | 101.7 |
| 618-619 | Champaign | 99.8 | 107.3 | 103.1 |
| 620-622 | East St. Louis | 94.6 | 107.5 | 100.4 |
| 623 | Quincy | 95.8 | 99.6 | 97.5 |
| 624 | Effingham | 95.1 | 103.6 | 98.9 |
| 625 | Decatur | 97.0 | 105.3 | 100.7 |
| 626-627 | Springfield | 97.5 | 106.2 | 101.4 |
| 628 | Centralia | 93.2 | 106.9 | 99.3 |
| 629 | Carbondale | 92.9 | 102.3 | 97.1 |
| **INDIANA** | | | | |
| 460 | Anderson | 95.3 | 82.8 | 89.7 |
| 461-462 | Indianapolis | 98.3 | 85.7 | 92.7 |
| 463-464 | Gary | 96.7 | 108.1 | 101.8 |
| 465-466 | South Bend | 95.3 | 83.3 | 89.9 |
| 467-468 | Fort Wayne | 95.9 | 79.2 | 88.5 |
| 469 | Kokomo | 93.1 | 81.7 | 88.0 |
| 470 | Lawrenceburg | 92.5 | 78.9 | 86.4 |
| 471 | New Albany | 93.6 | 76.4 | 85.9 |
| 472 | Columbus | 95.9 | 81.6 | 89.5 |
| 473 | Muncie | 96.8 | 81.7 | 90.1 |
| 474 | Bloomington | 98.1 | 81.5 | 90.7 |
| 475 | Washington | 94.6 | 83.3 | 89.5 |
| 476-477 | Evansville | 95.9 | 86.0 | 91.5 |
| 478 | Terre Haute | 96.6 | 86.7 | 92.2 |
| 479 | Lafayette | 95.6 | 82.8 | 89.9 |
| **IOWA** | | | | |
| 500-503,509 | Des Moines | 98.9 | 86.2 | 93.2 |
| 504 | Mason City | 97.5 | 68.9 | 84.7 |
| 505 | Fort Dodge | 97.7 | 60.6 | 81.1 |
| 506-507 | Waterloo | 99.4 | 73.4 | 87.8 |
| 508 | Creston | 98.0 | 80.2 | 90.0 |
| 510-511 | Sioux City | 100.1 | 71.9 | 87.5 |
| 512 | Sibley | 98.9 | 57.3 | 80.3 |
| 513 | Spencer | 100.6 | 57.3 | 81.3 |
| 514 | Carroll | 97.3 | 73.6 | 86.7 |
| 515 | Council Bluffs | 101.4 | 80.3 | 91.9 |
| 516 | Shenandoah | 97.9 | 73.6 | 87.1 |
| 520 | Dubuque | 99.7 | 81.2 | 91.4 |
| 521 | Decorah | 98.4 | 66.7 | 84.3 |
| 522-524 | Cedar Rapids | 100.5 | 85.5 | 93.8 |
| 525 | Ottumwa | 98.4 | 71.9 | 86.6 |
| 526 | Burlington | 97.5 | 78.6 | 89.1 |
| 527-528 | Davenport | 99.4 | 92.0 | 96.1 |
| **KANSAS** | | | | |
| 660-662 | Kansas City | 99.2 | 97.3 | 98.3 |
| 664-666 | Topeka | 99.2 | 65.2 | 84.0 |
| 667 | Fort Scott | 97.6 | 75.2 | 87.6 |
| 668 | Emporia | 97.8 | 70.8 | 85.7 |
| 669 | Belleville | 99.6 | 66.5 | 84.8 |
| 670-672 | Wichita | 98.3 | 65.1 | 83.4 |
| 673 | Independence | 99.2 | 70.6 | 86.4 |
| 674 | Salina | 99.7 | 68.0 | 85.5 |
| 675 | Hutchinson | 94.5 | 66.4 | 82.0 |
| 676 | Hays | 98.7 | 66.8 | 84.4 |
| 677 | Colby | 99.5 | 66.8 | 84.9 |

| STATE/ZIP | CITY | MAT. | INST. | TOTAL |
|---|---|---|---|---|
| **KANSAS (CONT'D)** | | | | |
| 678 | Dodge City | 101.0 | 67.8 | 86.2 |
| 679 | Liberal | 98.5 | 66.8 | 84.3 |
| **KENTUCKY** | | | | |
| 400-402 | Louisville | 97.7 | 83.4 | 91.3 |
| 403-405 | Lexington | 96.3 | 75.3 | 86.9 |
| 406 | Frankfort | 97.1 | 74.3 | 86.9 |
| 407-409 | Corbin | 93.5 | 62.0 | 79.4 |
| 410 | Covington | 94.4 | 96.0 | 95.1 |
| 411-412 | Ashland | 93.2 | 98.5 | 95.6 |
| 413-414 | Campton | 94.4 | 62.6 | 80.2 |
| 415-416 | Pikeville | 95.7 | 80.3 | 88.8 |
| 417-418 | Hazard | 93.7 | 55.7 | 76.7 |
| 420 | Paducah | 92.6 | 86.0 | 89.7 |
| 421-422 | Bowling Green | 94.7 | 82.8 | 89.4 |
| 423 | Owensboro | 95.1 | 80.2 | 88.4 |
| 424 | Henderson | 92.3 | 86.3 | 89.6 |
| 425-426 | Somerset | 91.6 | 66.6 | 80.4 |
| 427 | Elizabethtown | 91.0 | 83.4 | 87.6 |
| **LOUISIANA** | | | | |
| 700-701 | New Orleans | 102.0 | 69.2 | 87.3 |
| 703 | Thibodaux | 100.3 | 64.3 | 84.2 |
| 704 | Hammond | 97.4 | 56.1 | 78.9 |
| 705 | Lafayette | 99.7 | 60.5 | 82.2 |
| 706 | Lake Charles | 100.0 | 65.0 | 84.3 |
| 707-708 | Baton Rouge | 101.5 | 63.6 | 84.5 |
| 710-711 | Shreveport | 96.9 | 57.1 | 79.1 |
| 712 | Monroe | 97.2 | 54.3 | 78.0 |
| 713-714 | Alexandria | 97.2 | 53.8 | 77.8 |
| **MAINE** | | | | |
| 039 | Kittery | 95.0 | 80.2 | 88.4 |
| 040-041 | Portland | 99.8 | 79.0 | 90.5 |
| 042 | Lewiston | 98.2 | 79.0 | 89.6 |
| 043 | Augusta | 97.9 | 80.0 | 89.9 |
| 044 | Bangor | 97.9 | 79.0 | 89.4 |
| 045 | Bath | 96.3 | 79.6 | 88.8 |
| 046 | Machias | 95.8 | 78.9 | 88.2 |
| 047 | Houlton | 95.9 | 79.1 | 88.4 |
| 048 | Rockland | 94.9 | 79.5 | 88.0 |
| 049 | Waterville | 96.3 | 80.0 | 89.0 |
| **MARYLAND** | | | | |
| 206 | Waldorf | 99.1 | 68.5 | 85.4 |
| 207-208 | College Park | 99.2 | 83.0 | 92.0 |
| 209 | Silver Spring | 98.3 | 74.5 | 87.7 |
| 210-212 | Baltimore | 98.9 | 84.8 | 92.6 |
| 214 | Annapolis | 99.1 | 78.5 | 89.9 |
| 215 | Cumberland | 94.7 | 80.1 | 88.1 |
| 216 | Easton | 96.4 | 41.9 | 72.0 |
| 217 | Hagerstown | 95.5 | 82.2 | 89.5 |
| 218 | Salisbury | 96.8 | 51.0 | 76.3 |
| 219 | Elkton | 93.6 | 64.9 | 80.7 |
| **MASSACHUSETTS** | | | | |
| 010-011 | Springfield | 100.0 | 109.5 | 104.2 |
| 012 | Pittsfield | 99.6 | 105.3 | 102.2 |
| 013 | Greenfield | 97.4 | 108.6 | 102.4 |
| 014 | Fitchburg | 96.2 | 121.1 | 107.3 |
| 015-016 | Worcester | 100.0 | 122.7 | 110.1 |
| 017 | Framingham | 95.6 | 129.4 | 110.7 |
| 018 | Lowell | 99.4 | 129.0 | 112.6 |
| 019 | Lawrence | 100.6 | 127.8 | 112.8 |
| 020-022, 024 | Boston | 102.3 | 136.8 | 117.7 |
| 023 | Brockton | 100.9 | 124.5 | 111.5 |
| 025 | Buzzards Bay | 95.0 | 122.2 | 107.1 |
| 026 | Hyannis | 98.0 | 122.2 | 108.8 |
| 027 | New Bedford | 100.0 | 122.3 | 110.0 |
| **MICHIGAN** | | | | |
| 480,483 | Royal Oak | 92.8 | 103.6 | 97.6 |
| 481 | Ann Arbor | 95.2 | 107.7 | 100.8 |
| 482 | Detroit | 96.7 | 110.2 | 102.8 |
| 484-485 | Flint | 94.9 | 98.2 | 96.4 |
| 486 | Saginaw | 94.5 | 94.0 | 94.3 |
| 487 | Bay City | 94.6 | 93.9 | 94.3 |
| 488-489 | Lansing | 95.7 | 96.5 | 96.1 |
| 490 | Battle Creek | 95.8 | 90.4 | 93.4 |
| 491 | Kalamazoo | 96.1 | 88.4 | 92.7 |
| 492 | Jackson | 94.0 | 96.8 | 95.2 |
| 493,495 | Grand Rapids | 97.3 | 76.8 | 88.1 |
| 494 | Muskegon | 94.7 | 85.3 | 90.5 |

| STATE/ZIP | CITY | MAT. | INST. | TOTAL |
|---|---|---|---|---|
| **MICHIGAN (CONT'D)** | | | | |
| 496 | Traverse City | 93.4 | 79.4 | 87.2 |
| 497 | Gaylord | 94.5 | 73.4 | 85.1 |
| 498-499 | Iron mountain | 96.7 | 88.8 | 93.2 |
| **MINNESOTA** | | | | |
| 550-551 | Saint Paul | 101.6 | 123.0 | 111.2 |
| 553-555 | Minneapolis | 102.3 | 125.0 | 112.5 |
| 556-558 | Duluth | 100.6 | 112.2 | 105.8 |
| **MINNESOTA (CONT'd)** | | | | |
| 559 | Rochester | 101.0 | 109.2 | 104.7 |
| 560 | Mankato | 97.6 | 106.4 | 101.5 |
| 561 | Windom | 96.2 | 95.3 | 95.8 |
| 562 | Willmar | 95.9 | 107.2 | 100.9 |
| 563 | St. Cloud | 97.3 | 123.2 | 108.9 |
| 564 | Brainerd | 97.6 | 107.3 | 101.9 |
| 565 | Detroit Lakes | 99.3 | 100.7 | 99.9 |
| 566 | Bemidji | 98.5 | 102.2 | 100.2 |
| 567 | Thief River Falls | 98.1 | 97.1 | 97.7 |
| **MISSISSIPPI** | | | | |
| 386 | Clarksdale | 95.7 | 56.3 | 78.1 |
| 387 | Greenville | 99.3 | 66.1 | 84.5 |
| 388 | Tupelo | 97.1 | 58.8 | 80.0 |
| 389 | Greenwood | 97.0 | 55.8 | 78.6 |
| 390-392 | Jackson | 96.9 | 65.3 | 82.8 |
| 393 | Meridian | 95.9 | 67.0 | 83.0 |
| 394 | Laurel | 97.0 | 59.3 | 80.2 |
| 395 | Biloxi | 97.6 | 60.8 | 81.2 |
| 396 | Mccomb | 95.4 | 55.3 | 77.5 |
| 397 | Columbus | 97.0 | 57.7 | 79.4 |
| **MISSOURI** | | | | |
| 630-631 | St. Louis | 99.4 | 106.6 | 102.6 |
| 633 | Bowling Green | 98.1 | 92.2 | 95.5 |
| 634 | Hannibal | 96.9 | 88.8 | 93.3 |
| 635 | Kirksville | 100.1 | 85.4 | 93.5 |
| 636 | Flat River | 99.1 | 94.7 | 97.1 |
| 637 | Cape Girardeau | 99.2 | 91.3 | 95.7 |
| 638 | Sikeston | 97.1 | 87.1 | 92.6 |
| 639 | Poplar Bluff | 96.6 | 88.4 | 92.9 |
| 640-641 | Kansas City | 99.0 | 107.3 | 102.7 |
| 644-645 | St. Joseph | 98.3 | 93.2 | 96.0 |
| 646 | Chillicothe | 95.2 | 71.1 | 84.4 |
| 647 | Harrisonville | 94.8 | 104.8 | 99.3 |
| 648 | Joplin | 97.1 | 71.4 | 85.6 |
| 650-651 | Jefferson City | 95.9 | 92.1 | 94.2 |
| 652 | Columbia | 97.4 | 93.7 | 95.7 |
| 653 | Sedalia | 96.5 | 94.6 | 95.7 |
| 654-655 | Rolla | 95.0 | 89.5 | 92.5 |
| 656-658 | Springfield | 98.2 | 80.9 | 90.5 |
| **MONTANA** | | | | |
| 590-591 | Billings | 103.2 | 75.9 | 91.0 |
| 592 | Wolf Point | 103.2 | 68.6 | 87.7 |
| 593 | Miles City | 100.6 | 70.2 | 87.0 |
| 594 | Great Falls | 104.7 | 76.5 | 92.1 |
| 595 | Havre | 101.9 | 71.3 | 88.2 |
| 596 | Helena | 102.5 | 73.2 | 89.4 |
| 597 | Butte | 103.2 | 74.6 | 90.4 |
| 598 | Missoula | 100.5 | 73.4 | 88.4 |
| 599 | Kalispell | 99.4 | 71.3 | 86.8 |
| **NEBRASKA** | | | | |
| 680-681 | Omaha | 99.8 | 80.6 | 91.2 |
| 683-685 | Lincoln | 97.7 | 74.9 | 87.5 |
| 686 | Columbus | 96.7 | 72.9 | 86.1 |
| 687 | Norfolk | 98.4 | 76.4 | 88.6 |
| 688 | Grand Island | 98.5 | 79.8 | 90.1 |
| 689 | Hastings | 97.7 | 80.7 | 90.1 |
| 690 | Mccook | 97.5 | 71.9 | 86.0 |
| 691 | North Platte | 97.8 | 80.9 | 90.3 |
| 692 | Valentine | 99.5 | 68.7 | 85.7 |
| 693 | Alliance | 99.6 | 66.8 | 85.0 |
| **NEVADA** | | | | |
| 889-891 | Las Vegas | 100.2 | 112.8 | 105.9 |
| 893 | Ely | 99.1 | 99.6 | 99.3 |
| 894-895 | Reno | 99.2 | 91.1 | 95.6 |
| 897 | Carson City | 98.6 | 90.8 | 95.1 |
| 898 | Elko | 97.7 | 84.4 | 91.8 |
| **NEW HAMPSHIRE** | | | | |
| 030 | Nashua | 100.2 | 86.1 | 93.9 |
| 031 | Manchester | 100.8 | 86.1 | 94.2 |

| STATE/ZIP | CITY | MAT. | INST. | TOTAL |
|---|---|---|---|---|
| **NEW HAMPSHIRE (CONT'D)** | | | | |
| 032-033 | Concord | 98.7 | 82.5 | 91.5 |
| 034 | Keene | 97.0 | 49.9 | 76.0 |
| 035 | Littleton | 96.7 | 57.4 | 79.1 |
| 036 | Charleston | 96.5 | 47.6 | 74.6 |
| 037 | Claremont | 95.4 | 47.6 | 74.0 |
| 038 | Portsmouth | 97.7 | 89.9 | 94.2 |
| **NEW JERSEY** | | | | |
| 070-071 | Newark | 102.3 | 125.6 | 112.7 |
| 072 | Elizabeth | 100.6 | 125.6 | 111.8 |
| 073 | Jersey City | 99.3 | 125.3 | 110.9 |
| 074-075 | Paterson | 101.4 | 125.5 | 112.2 |
| 076 | Hackensack | 99.3 | 125.4 | 111.0 |
| 077 | Long Branch | 98.9 | 124.1 | 110.2 |
| 078 | Dover | 99.6 | 125.5 | 111.2 |
| 079 | Summit | 99.6 | 125.6 | 111.2 |
| 080,083 | Vineland | 97.5 | 122.8 | 108.8 |
| 081 | Camden | 99.4 | 124.0 | 110.4 |
| 082,084 | Atlantic City | 98.1 | 122.7 | 109.1 |
| 085-086 | Trenton | 99.4 | 122.7 | 109.8 |
| 087 | Point Pleasant | 99.6 | 123.3 | 110.2 |
| 088-089 | New Brunswick | 100.1 | 125.0 | 111.2 |
| **NEW MEXICO** | | | | |
| 870-872 | Albuquerque | 99.1 | 75.0 | 88.3 |
| 873 | Gallup | 99.3 | 75.0 | 88.5 |
| 874 | Farmington | 99.8 | 75.0 | 88.7 |
| 875 | Santa Fe | 100.1 | 75.0 | 88.9 |
| 877 | Las Vegas | 97.7 | 75.0 | 87.6 |
| 878 | Socorro | 97.2 | 75.0 | 87.3 |
| 879 | Truth/Consequences | 97.5 | 71.6 | 85.9 |
| 880 | Las Cruces | 96.3 | 70.9 | 85.0 |
| 881 | Clovis | 97.9 | 74.9 | 87.6 |
| 882 | Roswell | 99.6 | 75.0 | 88.6 |
| 883 | Carrizozo | 100.0 | 75.0 | 88.8 |
| 884 | Tucumcari | 98.6 | 74.9 | 88.0 |
| **NEW YORK** | | | | |
| 100-102 | New York | 106.0 | 166.3 | 133.0 |
| 103 | Staten Island | 102.1 | 160.3 | 128.1 |
| 104 | Bronx | 99.9 | 160.3 | 126.9 |
| 105 | Mount Vernon | 100.0 | 134.3 | 115.3 |
| 106 | White Plains | 99.9 | 134.3 | 115.3 |
| 107 | Yonkers | 104.8 | 134.5 | 118.0 |
| 108 | New Rochelle | 100.3 | 134.3 | 115.5 |
| 109 | Suffern | 100.1 | 123.1 | 110.4 |
| 110 | Queens | 101.9 | 160.1 | 127.9 |
| 111 | Long Island City | 103.6 | 160.1 | 128.9 |
| 112 | Brooklyn | 104.0 | 160.1 | 129.1 |
| 113 | Flushing | 103.8 | 160.1 | 129.0 |
| 114 | Jamaica | 101.9 | 160.1 | 127.9 |
| 115,117,118 | Hicksville | 101.7 | 145.2 | 121.2 |
| 116 | Far Rockaway | 103.9 | 160.1 | 129.1 |
| 119 | Riverhead | 102.5 | 144.7 | 121.4 |
| 120-122 | Albany | 97.2 | 97.9 | 97.5 |
| 123 | Schenectady | 98.2 | 97.5 | 97.9 |
| 124 | Kingston | 101.6 | 118.7 | 109.2 |
| 125-126 | Poughkeepsie | 100.7 | 130.5 | 114.0 |
| 127 | Monticello | 100.0 | 119.2 | 108.6 |
| 128 | Glens Falls | 92.6 | 93.2 | 92.8 |
| 129 | Plattsburgh | 97.5 | 85.0 | 91.9 |
| 130-132 | Syracuse | 99.4 | 93.5 | 96.7 |
| 133-135 | Utica | 97.1 | 88.7 | 93.4 |
| 136 | Watertown | 98.9 | 92.6 | 96.1 |
| 137-139 | Binghamton | 98.8 | 90.2 | 95.0 |
| 140-142 | Buffalo | 100.5 | 102.9 | 101.6 |
| 143 | Niagara Falls | 98.3 | 98.0 | 98.2 |
| 144-146 | Rochester | 100.3 | 95.6 | 98.2 |
| 147 | Jamestown | 97.2 | 84.4 | 91.5 |
| 148-149 | Elmira | 96.9 | 92.0 | 94.7 |
| **NORTH CAROLINA** | | | | |
| 270,272-274 | Greensboro | 99.2 | 47.9 | 76.2 |
| 271 | Winston-Salem | 99.1 | 46.5 | 75.6 |
| 275-276 | Raleigh | 99.6 | 48.1 | 76.6 |
| 277 | Durham | 100.3 | 48.4 | 77.1 |
| 278 | Rocky Mount | 96.6 | 39.8 | 71.2 |
| 279 | Elizabeth City | 97.5 | 41.0 | 72.3 |
| 280 | Gastonia | 98.2 | 46.0 | 74.8 |
| 281-282 | Charlotte | 99.2 | 48.4 | 76.5 |
| 283 | Fayetteville | 100.0 | 50.0 | 77.6 |
| 284 | Wilmington | 96.9 | 46.3 | 74.3 |
| 285 | Kinston | 94.9 | 40.6 | 70.6 |

# Location Factors

| STATE/ZIP | CITY | MAT. | INST. | TOTAL |
|---|---|---|---|---|
| **NORTH CAROLINA (CONT'D)** | | | | |
| 286 | Hickory | 95.2 | 42.8 | 71.8 |
| 287-288 | Asheville | 97.4 | 46.0 | 74.4 |
| 289 | Murphy | 96.2 | 33.4 | 68.1 |
| **NORTH DAKOTA** | | | | |
| 580-581 | Fargo | 102.4 | 64.3 | 85.4 |
| 582 | Grand Forks | 102.6 | 53.8 | 80.8 |
| 583 | Devils Lake | 101.7 | 56.0 | 81.3 |
| 584 | Jamestown | 101.9 | 46.8 | 77.3 |
| 585 | Bismarck | 100.8 | 63.0 | 83.9 |
| 586 | Dickinson | 102.8 | 58.2 | 82.9 |
| 587 | Minot | 102.5 | 70.0 | 88.0 |
| 588 | Williston | 101.0 | 58.2 | 81.9 |
| **OHIO** | | | | |
| 430-432 | Columbus | 97.9 | 90.3 | 94.5 |
| 433 | Marion | 94.1 | 80.8 | 88.2 |
| 434-436 | Toledo | 98.1 | 95.9 | 97.1 |
| 437-438 | Zanesville | 94.6 | 81.1 | 88.6 |
| 439 | Steubenville | 96.4 | 90.5 | 93.8 |
| 440 | Lorain | 98.7 | 91.5 | 95.5 |
| 441 | Cleveland | 99.0 | 100.5 | 99.7 |
| 442-443 | Akron | 99.8 | 92.6 | 96.6 |
| 444-445 | Youngstown | 99.1 | 87.5 | 93.9 |
| 446-447 | Canton | 99.2 | 82.5 | 91.8 |
| 448-449 | Mansfield | 96.2 | 87.0 | 92.1 |
| 450 | Hamilton | 95.7 | 83.8 | 90.4 |
| 451-452 | Cincinnati | 96.0 | 86.3 | 91.7 |
| 453-454 | Dayton | 95.8 | 83.6 | 90.3 |
| 455 | Springfield | 95.7 | 83.8 | 90.4 |
| 456 | Chillicothe | 94.5 | 88.5 | 91.8 |
| 457 | Athens | 97.0 | 81.0 | 89.8 |
| 458 | Lima | 97.4 | 85.0 | 91.9 |
| **OKLAHOMA** | | | | |
| 730-731 | Oklahoma City | 98.2 | 61.5 | 81.8 |
| 734 | Ardmore | 95.2 | 61.0 | 79.9 |
| 735 | Lawton | 97.6 | 62.3 | 81.8 |
| 736 | Clinton | 96.7 | 59.6 | 80.2 |
| 737 | Enid | 97.4 | 59.6 | 80.5 |
| 738 | Woodward | 95.4 | 59.7 | 79.4 |
| 739 | Guymon | 96.5 | 31.1 | 67.3 |
| 740-741 | Tulsa | 97.4 | 54.2 | 78.1 |
| 743 | Miami | 94.0 | 69.0 | 82.8 |
| 744 | Muskogee | 96.7 | 39.6 | 71.2 |
| 745 | Mcalester | 93.7 | 51.2 | 74.7 |
| 746 | Ponca City | 94.3 | 59.6 | 78.8 |
| 747 | Durant | 94.3 | 58.5 | 78.3 |
| 748 | Shawnee | 96.0 | 57.3 | 78.7 |
| 749 | Poteau | 93.3 | 62.3 | 79.5 |
| **OREGON** | | | | |
| 970-972 | Portland | 100.0 | 99.7 | 99.9 |
| 973 | Salem | 99.8 | 98.6 | 99.3 |
| 974 | Eugene | 99.8 | 98.3 | 99.1 |
| 975 | Medford | 101.5 | 96.7 | 99.3 |
| 976 | Klamath Falls | 101.5 | 96.7 | 99.3 |
| 977 | Bend | 100.3 | 98.4 | 99.5 |
| 978 | Pendleton | 94.9 | 100.3 | 97.3 |
| 979 | Vale | 92.6 | 90.4 | 91.6 |
| **PENNSYLVANIA** | | | | |
| 150-152 | Pittsburgh | 99.3 | 102.9 | 100.9 |
| 153 | Washington | 96.2 | 100.8 | 98.3 |
| 154 | Uniontown | 96.6 | 99.9 | 98.0 |
| 155 | Bedford | 97.6 | 91.1 | 94.7 |
| 156 | Greensburg | 97.5 | 99.4 | 98.3 |
| 157 | Indiana | 96.3 | 98.2 | 97.1 |
| 158 | Dubois | 98.0 | 94.2 | 96.3 |
| 159 | Johnstown | 97.6 | 94.9 | 96.4 |
| 160 | Butler | 92.6 | 100.3 | 96.0 |
| 161 | New Castle | 92.6 | 97.2 | 94.7 |
| 162 | Kittanning | 93.2 | 101.4 | 96.9 |
| 163 | Oil City | 92.6 | 94.9 | 93.6 |
| 164-165 | Erie | 94.8 | 93.5 | 94.2 |
| 166 | Altoona | 95.0 | 90.5 | 93.0 |
| 167 | Bradford | 95.9 | 93.1 | 94.7 |
| 168 | State College | 95.5 | 91.4 | 93.7 |
| 169 | Wellsboro | 96.7 | 91.0 | 94.2 |
| 170-171 | Harrisburg | 98.0 | 93.8 | 96.1 |
| 172 | Chambersburg | 96.1 | 88.6 | 92.7 |
| 173-174 | York | 96.5 | 94.0 | 95.4 |
| 175-176 | Lancaster | 94.7 | 88.3 | 91.8 |

| STATE/ZIP | CITY | MAT. | INST. | TOTAL |
|---|---|---|---|---|
| **PENNSYLVANIA (CONT'D)** | | | | |
| 177 | Williamsport | 93.3 | 79.7 | 87.2 |
| 178 | Sunbury | 95.5 | 94.0 | 94.8 |
| 179 | Pottsville | 94.6 | 95.6 | 95.1 |
| 180 | Lehigh Valley | 96.0 | 114.2 | 104.2 |
| 181 | Allentown | 98.3 | 108.6 | 102.9 |
| 182 | Hazleton | 95.5 | 97.3 | 96.3 |
| 183 | Stroudsburg | 95.3 | 102.3 | 98.4 |
| 184-185 | Scranton | 99.0 | 98.7 | 98.9 |
| 186-187 | Wilkes-Barre | 95.2 | 97.7 | 96.3 |
| 188 | Montrose | 94.9 | 96.3 | 95.5 |
| 189 | Doylestown | 95.1 | 121.9 | 107.1 |
| 190-191 | Philadelphia | 99.7 | 132.2 | 114.3 |
| 193 | Westchester | 96.1 | 124.7 | 108.9 |
| 194 | Norristown | 95.1 | 131.2 | 111.3 |
| 195-196 | Reading | 97.4 | 99.6 | 98.4 |
| **PUERTO RICO** | | | | |
| 009 | San Juan | 121.8 | 24.0 | 78.1 |
| **RHODE ISLAND** | | | | |
| 028 | Newport | 99.3 | 111.5 | 104.7 |
| 029 | Providence | 100.3 | 111.5 | 105.3 |
| **SOUTH CAROLINA** | | | | |
| 290-292 | Columbia | 97.6 | 50.0 | 76.3 |
| 293 | Spartanburg | 96.4 | 49.6 | 75.5 |
| 294 | Charleston | 98.0 | 58.3 | 80.2 |
| 295 | Florence | 96.2 | 50.0 | 75.5 |
| 296 | Greenville | 96.2 | 49.6 | 75.3 |
| 297 | Rock Hill | 95.7 | 47.7 | 74.3 |
| 298 | Aiken | 96.7 | 71.9 | 85.6 |
| 299 | Beaufort | 97.5 | 43.0 | 73.1 |
| **SOUTH DAKOTA** | | | | |
| 570-571 | Sioux Falls | 100.6 | 57.9 | 81.5 |
| 572 | Watertown | 99.7 | 52.2 | 78.4 |
| 573 | Mitchell | 98.3 | 51.6 | 77.4 |
| 574 | Aberdeen | 101.2 | 52.8 | 79.6 |
| 575 | Pierre | 99.7 | 54.5 | 79.5 |
| 576 | Mobridge | 99.1 | 52.0 | 78.0 |
| 577 | Rapid City | 101.0 | 55.3 | 80.6 |
| **TENNESSEE** | | | | |
| 370-372 | Nashville | 96.7 | 72.7 | 85.9 |
| 373-374 | Chattanooga | 98.2 | 66.9 | 84.2 |
| 375,380-381 | Memphis | 96.6 | 70.3 | 84.9 |
| 376 | Johnson City | 97.6 | 57.4 | 79.6 |
| 377-379 | Knoxville | 94.5 | 61.7 | 79.9 |
| 382 | Mckenzie | 96.3 | 62.1 | 81.1 |
| 383 | Jackson | 98.2 | 63.8 | 82.8 |
| 384 | Columbia | 94.9 | 67.3 | 82.6 |
| 385 | Cookeville | 96.2 | 60.3 | 80.1 |
| **TEXAS** | | | | |
| 750 | Mckinney | 99.5 | 50.8 | 77.7 |
| 751 | Waxahackie | 99.4 | 58.2 | 81.0 |
| 752-753 | Dallas | 99.9 | 67.5 | 85.4 |
| 754 | Greenville | 99.6 | 43.7 | 74.6 |
| 755 | Texarkana | 99.1 | 51.5 | 77.8 |
| 756 | Longview | 99.7 | 40.7 | 73.4 |
| 757 | Tyler | 100.2 | 54.9 | 80.0 |
| 758 | Palestine | 96.1 | 57.4 | 78.8 |
| 759 | Lufkin | 96.7 | 59.1 | 79.9 |
| 760-761 | Fort Worth | 97.7 | 63.6 | 82.4 |
| 762 | Denton | 97.3 | 48.4 | 75.4 |
| 763 | Wichita Falls | 98.1 | 55.5 | 79.1 |
| 764 | Eastland | 96.8 | 50.3 | 76.0 |
| 765 | Temple | 95.3 | 50.0 | 75.1 |
| 766-767 | Waco | 97.5 | 57.1 | 79.4 |
| 768 | Brownwood | 98.0 | 49.6 | 76.3 |
| 769 | San Angelo | 97.6 | 50.1 | 76.3 |
| 770-772 | Houston | 100.0 | 70.4 | 86.8 |
| 773 | Huntsville | 98.6 | 56.7 | 79.9 |
| 774 | Wharton | 99.8 | 54.8 | 79.7 |
| 775 | Galveston | 97.6 | 69.2 | 84.9 |
| 776-777 | Beaumont | 98.2 | 62.9 | 82.4 |
| 778 | Bryan | 95.1 | 62.1 | 80.3 |
| 779 | Victoria | 99.9 | 44.4 | 75.1 |
| 780 | Laredo | 94.7 | 53.1 | 76.1 |
| 781-782 | San Antonio | 95.0 | 65.1 | 81.6 |
| 783-784 | Corpus Christi | 97.6 | 52.8 | 77.6 |
| 785 | Mc Allen | 97.5 | 46.9 | 74.9 |
| 786-787 | Austin | 95.1 | 59.4 | 79.1 |

| STATE/ZIP | CITY | MAT. | INST. | TOTAL |
|---|---|---|---|---|
| **TEXAS (CONT'D)** | | | | |
| 788 | Del Rio | 97.0 | 47.0 | 74.6 |
| 789 | Giddings | 94.1 | 57.0 | 77.5 |
| 790-791 | Amarillo | 98.1 | 61.9 | 81.9 |
| 792 | Childress | 97.5 | 58.1 | 79.9 |
| 793-794 | Lubbock | 99.6 | 57.5 | 80.8 |
| 795-796 | Abilene | 97.9 | 54.8 | 78.6 |
| 797 | Midland | 99.7 | 51.9 | 78.3 |
| 798-799,885 | El Paso | 97.0 | 51.7 | 76.8 |
| **UTAH** | | | | |
| 840-841 | Salt Lake City | 102.0 | 69.1 | 87.3 |
| 842,844 | Ogden | 97.5 | 70.7 | 85.5 |
| 843 | Logan | 99.5 | 70.7 | 86.6 |
| 845 | Price | 100.0 | 67.1 | 85.3 |
| 846-847 | Provo | 100.0 | 68.9 | 86.1 |
| **VERMONT** | | | | |
| 050 | White River Jct. | 98.3 | 55.1 | 79.0 |
| 051 | Bellows Falls | 96.6 | 73.7 | 86.4 |
| 052 | Bennington | 97.0 | 68.9 | 84.4 |
| 053 | Brattleboro | 97.4 | 72.9 | 86.5 |
| 054 | Burlington | 100.9 | 66.1 | 85.4 |
| 056 | Montpelier | 97.7 | 66.2 | 83.6 |
| 057 | Rutland | 99.2 | 66.2 | 84.4 |
| 058 | St. Johnsbury | 98.3 | 55.9 | 79.3 |
| 059 | Guildhall | 96.9 | 55.7 | 78.5 |
| **VIRGINIA** | | | | |
| 220-221 | Fairfax | 99.9 | 82.3 | 92.0 |
| 222 | Arlington | 101.2 | 80.1 | 91.8 |
| 223 | Alexandria | 100.2 | 85.5 | 93.6 |
| 224-225 | Fredericksburg | 98.5 | 72.2 | 86.7 |
| 226 | Winchester | 99.2 | 63.2 | 83.1 |
| 227 | Culpeper | 99.1 | 67.8 | 85.1 |
| 228 | Harrisonburg | 99.3 | 64.0 | 83.5 |
| 229 | Charlottesville | 99.8 | 63.4 | 83.5 |
| 230-232 | Richmond | 100.2 | 66.8 | 85.3 |
| 233-235 | Norfolk | 101.0 | 68.2 | 86.3 |
| 236 | Newport News | 100.0 | 67.6 | 85.5 |
| 237 | Portsmouth | 99.4 | 65.5 | 84.2 |
| 238 | Petersburg | 99.6 | 66.8 | 84.9 |
| 239 | Farmville | 98.7 | 53.0 | 78.3 |
| 240-241 | Roanoke | 100.7 | 60.7 | 82.8 |
| 242 | Bristol | 98.4 | 55.1 | 79.0 |
| 243 | Pulaski | 98.0 | 50.9 | 77.0 |
| 244 | Staunton | 98.9 | 59.1 | 81.1 |
| 245 | Lynchburg | 99.0 | 64.6 | 83.6 |
| 246 | Grundy | 98.3 | 50.3 | 76.9 |
| **WASHINGTON** | | | | |
| 980-981,987 | Seattle | 103.8 | 106.6 | 105.0 |
| 982 | Everett | 103.4 | 97.1 | 100.6 |
| 983-984 | Tacoma | 103.5 | 98.9 | 101.4 |
| 985 | Olympia | 101.3 | 98.9 | 100.2 |
| 986 | Vancouver | 104.8 | 93.8 | 99.9 |
| 988 | Wenatchee | 103.3 | 83.7 | 94.6 |
| 989 | Yakima | 103.7 | 93.1 | 99.0 |
| 990-992 | Spokane | 103.7 | 80.8 | 93.5 |
| 993 | Richland | 103.4 | 88.0 | 96.5 |
| 994 | Clarkston | 101.9 | 79.8 | 92.0 |
| **WEST VIRGINIA** | | | | |
| 247-248 | Bluefield | 97.1 | 77.7 | 88.4 |
| 249 | Lewisburg | 98.9 | 83.2 | 91.9 |
| 250-253 | Charleston | 99.0 | 90.1 | 95.0 |
| 254 | Martinsburg | 98.6 | 79.1 | 89.9 |
| 255-257 | Huntington | 100.2 | 91.6 | 96.3 |
| 258-259 | Beckley | 97.0 | 87.2 | 92.6 |
| 260 | Wheeling | 100.3 | 89.9 | 95.7 |
| 261 | Parkersburg | 99.1 | 89.4 | 94.8 |
| 262 | Buckhannon | 98.5 | 89.7 | 94.6 |
| 263-264 | Clarksburg | 99.2 | 89.7 | 94.9 |
| 265 | Morgantown | 99.2 | 88.8 | 94.6 |
| 266 | Gassaway | 98.4 | 89.9 | 94.6 |
| 267 | Romney | 98.4 | 85.6 | 92.7 |
| 268 | Petersburg | 98.2 | 85.0 | 92.3 |
| **WISCONSIN** | | | | |
| 530,532 | Milwaukee | 100.2 | 107.6 | 103.5 |
| 531 | Kenosha | 100.0 | 102.7 | 101.2 |
| 534 | Racine | 99.4 | 102.8 | 100.9 |
| 535 | Beloit | 99.3 | 97.8 | 98.6 |
| 537 | Madison | 98.9 | 99.1 | 99.0 |

| STATE/ZIP | CITY | MAT. | INST. | TOTAL |
|---|---|---|---|---|
| **WISCONSIN (CONT'D)** | | | | |
| 538 | Lancaster | 97.0 | 93.2 | 95.3 |
| 539 | Portage | 95.4 | 96.3 | 95.8 |
| 540 | New Richmond | 97.3 | 96.2 | 96.8 |
| 541-543 | Green Bay | 101.7 | 93.3 | 98.0 |
| 544 | Wausau | 96.2 | 93.4 | 95.0 |
| 545 | Rhinelander | 99.8 | 92.2 | 96.4 |
| 546 | La Crosse | 97.6 | 94.9 | 96.4 |
| 547 | Eau Claire | 99.3 | 96.1 | 97.9 |
| 548 | Superior | 97.1 | 98.1 | 97.6 |
| 549 | Oshkosh | 97.0 | 84.1 | 91.2 |
| **WYOMING** | | | | |
| 820 | Cheyenne | 99.2 | 64.7 | 83.8 |
| 821 | Yellowstone Nat'l Park | 96.8 | 60.0 | 80.4 |
| 822 | Wheatland | 98.0 | 59.2 | 80.6 |
| 823 | Rawlins | 99.6 | 58.0 | 81.0 |
| 824 | Worland | 97.5 | 58.0 | 79.8 |
| 825 | Riverton | 98.6 | 58.0 | 80.4 |
| 826 | Casper | 99.5 | 58.5 | 81.2 |
| 827 | Newcastle | 97.3 | 58.0 | 79.7 |
| 828 | Sheridan | 100.1 | 61.5 | 82.8 |
| 829-831 | Rock Springs | 101.5 | 58.1 | 82.1 |
| **CANADIAN FACTORS (reflect Canadian currency)** | | | | |
| **ALBERTA** | | | | |
| | Calgary | 130.9 | 94.8 | 114.8 |
| | Edmonton | 131.1 | 94.8 | 114.9 |
| | Fort McMurray | 127.0 | 97.7 | 113.9 |
| | Lethbridge | 120.9 | 97.1 | 110.3 |
| | Lloydminster | 116.2 | 93.6 | 106.1 |
| | Medicine Hat | 116.3 | 92.8 | 105.8 |
| | Red Deer | 116.8 | 92.8 | 106.1 |
| **BRITISH COLUMBIA** | | | | |
| | Kamloops | 117.2 | 96.2 | 107.8 |
| | Prince George | 118.5 | 95.4 | 108.2 |
| | Vancouver | 129.1 | 90.7 | 111.9 |
| | Victoria | 118.5 | 88.7 | 105.2 |
| **MANITOBA** | | | | |
| | Brandon | 116.1 | 80.3 | 100.1 |
| | Portage la Prairie | 116.1 | 78.5 | 99.3 |
| | Winnipeg | 128.8 | 73.5 | 104.1 |
| **NEW BRUNSWICK** | | | | |
| | Bathurst | 114.2 | 71.2 | 94.9 |
| | Dalhousie | 114.2 | 71.4 | 95.1 |
| | Fredericton | 116.8 | 76.8 | 98.9 |
| | Moncton | 114.5 | 72.8 | 95.8 |
| | Newcastle | 114.2 | 71.9 | 95.3 |
| | St. John | 116.9 | 76.5 | 98.8 |
| **NEWFOUNDLAND** | | | | |
| | Corner Brook | 119.6 | 71.5 | 98.1 |
| | St Johns | 119.7 | 76.9 | 100.6 |
| **NORTHWEST TERRITORIES** | | | | |
| | Yellowknife | 121.5 | 90.5 | 107.7 |
| **NOVA SCOTIA** | | | | |
| | Bridgewater | 115.8 | 79.1 | 99.4 |
| | Dartmouth | 117.2 | 79.1 | 100.1 |
| | Halifax | 118.1 | 81.7 | 101.8 |
| | New Glasgow | 115.3 | 79.1 | 99.1 |
| | Sydney | 112.6 | 79.1 | 97.6 |
| | Truro | 115.3 | 79.1 | 99.1 |
| | Yarmouth | 115.1 | 79.1 | 99.0 |
| **ONTARIO** | | | | |
| | Barrie | 118.9 | 98.8 | 109.9 |
| | Brantford | 118.1 | 103.0 | 111.3 |
| | Cornwall | 118.0 | 99.1 | 109.5 |
| | Hamilton | 122.6 | 95.4 | 110.5 |
| | Kingston | 118.9 | 99.2 | 110.1 |
| | Kitchener | 114.0 | 92.1 | 104.2 |
| | London | 121.4 | 91.9 | 108.2 |
| | North Bay | 118.1 | 96.7 | 108.5 |
| | Oshawa | 117.7 | 100.0 | 109.8 |
| | Ottawa | 123.0 | 92.3 | 109.3 |
| | Owen Sound | 119.1 | 96.8 | 109.1 |
| | Peterborough | 118.1 | 98.8 | 109.5 |
| | Sarnia | 117.8 | 103.6 | 111.4 |

# Location Factors

| STATE/ZIP | CITY | MAT. | INST. | TOTAL |
|---|---|---|---|---|
| ONTARIO (CONT'D) | Sault Ste Marie | 112.7 | 96.3 | 105.4 |
| | St. Catharines | 112.1 | 93.1 | 103.6 |
| | Sudbury | 112.5 | 92.1 | 103.4 |
| | Thunder Bay | 113.6 | 91.9 | 103.9 |
| | Timmins | 118.3 | 96.7 | 108.6 |
| | Toronto | 123.6 | 100.7 | 113.4 |
| | Windsor | 112.7 | 91.6 | 103.3 |
| | | | | |
| PRINCE EDWARD ISLAND | | | | |
| | Charlottetown | 117.6 | 66.4 | 94.7 |
| | Summerside | 117.1 | 66.4 | 94.5 |
| | | | | |
| QUEBEC | | | | |
| | Cap-de-la-Madeleine | 114.9 | 93.0 | 105.1 |
| | Charlesbourg | 114.9 | 93.0 | 105.1 |
| | Chicoutimi | 113.9 | 93.2 | 104.7 |
| | Gatineau | 114.6 | 92.7 | 104.8 |
| | Granby | 114.8 | 92.7 | 104.9 |
| | Hull | 114.6 | 92.7 | 104.8 |
| | Joliette | 115.1 | 93.0 | 105.2 |
| | Laval | 114.3 | 92.7 | 104.6 |
| | Montreal | 122.0 | 93.5 | 109.3 |
| | Quebec | 120.1 | 94.0 | 108.4 |
| | Rimouski | 114.4 | 93.2 | 104.9 |
| | Rouyn-Noranda | 114.5 | 92.7 | 104.7 |
| | Saint Hyacinthe | 113.9 | 92.7 | 104.4 |
| | Sherbrooke | 114.9 | 92.7 | 105.0 |
| | Sorel | 115.1 | 93.0 | 105.2 |
| | St Jerome | 114.5 | 92.7 | 104.8 |
| | Trois Rivieres | 115.1 | 93.0 | 105.2 |
| | | | | |
| SASKATCHEWAN | | | | |
| | Moose Jaw | 113.4 | 73.6 | 95.6 |
| | Prince Albert | 112.3 | 71.6 | 94.1 |
| | Regina | 114.9 | 87.7 | 102.7 |
| | Saskatoon | 113.4 | 87.7 | 101.9 |
| | | | | |
| YUKON | | | | |
| | Whitehorse | 113.2 | 72.4 | 95.0 |

## R011105-05  Tips for Accurate Estimating

1. Use pre-printed or columnar forms for orderly sequence of dimensions and locations and for recording telephone quotations.

2. Use only the front side of each paper or form except for certain pre-printed summary forms.

3. Be consistent in listing dimensions: For example, length x width x height. This helps in rechecking to ensure that, the total length of partitions is appropriate for the building area.

4. Use printed (rather than measured) dimensions where given.

5. Add up multiple printed dimensions for a single entry where possible.

6. Measure all other dimensions carefully.

7. Use each set of dimensions to calculate multiple related quantities.

8. Convert foot and inch measurements to decimal feet when listing. Memorize decimal equivalents to .01 parts of a foot (1/8″ equals approximately .01′).

9. Do not "round off" quantities until the final summary.

10. Mark drawings with different colors as items are taken off.

11. Keep similar items together, different items separate.

12. Identify location and drawing numbers to aid in future checking for completeness.

13. Measure or list everything on the drawings or mentioned in the specifications.

14. It may be necessary to list items not called for to make the job complete.

15. Be alert for: Notes on plans such as N.T.S. (not to scale); changes in scale throughout the drawings; reduced size drawings; discrepancies between the specifications and the drawings.

16. Develop a consistent pattern of performing an estimate. For example:
    a. Start the quantity takeoff at the lower floor and move to the next higher floor.
    b. Proceed from the main section of the building to the wings.
    c. Proceed from south to north or vice versa, clockwise or counterclockwise.
    d. Take off floor plan quantities first, elevations next, then detail drawings.

17. List all gross dimensions that can be either used again for different quantities, or used as a rough check of other quantities for verification (exterior perimeter, gross floor area, individual floor areas, etc.).

18. Utilize design symmetry or repetition (repetitive floors, repetitive wings, symmetrical design around a center line, similar room layouts, etc.). Note: Extreme caution is needed here so as not to omit or duplicate an area.

19. Do not convert units until the final total is obtained. For instance, when estimating concrete work, keep all units to the nearest cubic foot, then summarize and convert to cubic yards.

20. When figuring alternatives, it is best to total all items involved in the basic system, then total all items involved in the alternates. Therefore you work with positive numbers in all cases. When adds and deducts are used, it is often confusing whether to add or subtract a portion of an item; especially on a complicated or involved alternate.

## R011110-10  Architectural Fees

Tabulated below are typical percentage fees by project size, for good professional architectural service. Fees may vary from those listed depending upon degree of design difficulty and economic conditions in any particular area.

Rates can be interpolated horizontally and vertically. Various portions of the same project requiring different rates should be adjusted proportionately. For alterations, add 50% to the fee for the first $500,000 of project cost and add 25% to the fee for project cost over $500,000.

Architectural fees tabulated below include Structural, Mechanical and Electrical Engineering Fees. They do not include the fees for special consultants such as kitchen planning, security, acoustical, interior design, etc.

Civil Engineering fees are included in the Architectural fee for project sites requiring minimal design such as city sites. However, separate Civil Engineering fees must be added when utility connections require design, drainage calculations are needed, stepped foundations are required, or provisions are required to protect adjacent wetlands.

| Building Types | Total Project Size in Thousands of Dollars | | | | | | |
|---|---|---|---|---|---|---|---|
| | 100 | 250 | 500 | 1,000 | 5,000 | 10,000 | 50,000 |
| Factories, garages, warehouses, repetitive housing | 9.0% | 8.0% | 7.0% | 6.2% | 5.3% | 4.9% | 4.5% |
| Apartments, banks, schools, libraries, offices, municipal buildings | 12.2 | 12.3 | 9.2 | 8.0 | 7.0 | 6.6 | 6.2 |
| Churches, hospitals, homes, laboratories, museums, research | 15.0 | 13.6 | 12.7 | 11.9 | 9.5 | 8.8 | 8.0 |
| Memorials, monumental work, decorative furnishings | — | 16.0 | 14.5 | 13.1 | 10.0 | 9.0 | 8.3 |

## R011110-30  Engineering Fees

Typical **Structural Engineering Fees** based on type of construction and total project size. These fees are included in Architectural Fees.

| Type of Construction | Total Project Size (in thousands of dollars) | | | |
|---|---|---|---|---|
| | $500 | $500-$1,000 | $1,000-$5,000 | Over $5000 |
| Industrial buildings, factories & warehouses | Technical payroll times 2.0 to 2.5 | 1.60% | 1.25% | 1.00% |
| Hotels, apartments, offices, dormitories, hospitals, public buildings, food stores | | 2.00% | 1.70% | 1.20% |
| Museums, banks, churches and cathedrals | | 2.00% | 1.75% | 1.25% |
| Thin shells, prestressed concrete, earthquake resistive | | 2.00% | 1.75% | 1.50% |
| Parking ramps, auditoriums, stadiums, convention halls, hangars & boiler houses | | 2.50% | 2.00% | 1.75% |
| Special buildings, major alterations, underpinning & future expansion | | Add to above 0.5% | Add to above 0.5% | Add to above 0.5% |

For complex reinforced concrete or unusually complicated structures, add 20% to 50%.

Typical **Mechanical and Electrical Engineering Fees** are based on the size of the subcontract. The fee structure for both are shown below. These fees are included in Architectural Fees.

| Type of Construction | Subcontract Size | | | | | | | |
|---|---|---|---|---|---|---|---|---|
| | $25,000 | $50,000 | $100,000 | $225,000 | $350,000 | $500,000 | $750,000 | $1,000,000 |
| Simple structures | 6.4% | 5.7% | 4.8% | 4.5% | 4.4% | 4.3% | 4.2% | 4.1% |
| Intermediate structures | 8.0 | 7.3 | 6.5 | 5.6 | 5.1 | 5.0 | 4.9 | 4.8 |
| Complex structures | 10.1 | 9.0 | 9.0 | 8.0 | 7.5 | 7.5 | 7.0 | 7.0 |

For renovations, add 15% to 25% to applicable fee.

---

**General Requirements** **R0121 Allowances**

## R012157-20  Construction Time Requirements

Table lists average construction time in months for different types, sizes, and values of building projects. Design time runs 25% to 40% of construction time.

| Building Type | Size S.F. | Project Value | Construction Duration |
|---|---|---|---|
| Industrial/Warehouse | 100,000 | $8,000,000 | 14 Months |
| | 500,000 | $32,000,000 | 19 Months |
| | 1,000,000 | $75,000,000 | 21 Months |
| Offices/Retail | 50,000 | $7,000,000 | 15 Months |
| | 250,000 | $28,000,000 | 23 months |
| | 500,000 | $58,000,000 | 34 months |
| Institutional/Hospitals/Laboratory | 200,000 | $45,000,000 | 31 months |
| | 500,000 | $110,000,000 | 52 months |
| | 750,000 | $160,000,000 | 55 months |
| | 1,000,000 | $210,000,000 | 60 months |

## R012909-80   Sales Tax by State

State sales tax on materials is tabulated below (5 states have no sales tax). Many states allow local jurisdictions, such as a county or city, to levy additional sales tax.

Some projects may be sales tax exempt, particularly those constructed with public funds.

| State | Tax (%) | State | Tax (%) | State | Tax (%) | State | Tax (%) |
|---|---|---|---|---|---|---|---|
| Alabama | 4 | Illinois | 6.25 | Montana | 0 | Rhode Island | 7 |
| Alaska | 0 | Indiana | 7 | Nebraska | 5.5 | South Carolina | 6 |
| Arizona | 5.6 | Iowa | 6 | Nevada | 6.85 | South Dakota | 4 |
| Arkansas | 6 | Kansas | 5.3 | New Hampshire | 0 | Tennessee | 7 |
| California | 8.25 | Kentucky | 6 | New Jersey | 7 | Texas | 6.25 |
| Colorado | 2.9 | Louisiana | 4 | New Mexico | 5 | Utah | 4.7 |
| Connecticut | 6 | Maine | 5 | New York | 4 | Vermont | 6 |
| Delaware | 0 | Maryland | 6 | North Carolina | 5.75 | Virginia | 5 |
| District of Columbia | 6 | Massachusetts | 6.25 | North Dakota | 5 | Washington | 6.5 |
| Florida | 6 | Michigan | 6 | Ohio | 5.5 | West Virginia | 6 |
| Georgia | 4 | Minnesota | 6.875 | Oklahoma | 4.5 | Wisconsin | 5 |
| Hawaii | 4 | Mississippi | 7 | Oregon | 0 | Wyoming | 4 |
| Idaho | 6 | Missouri | 4.225 | Pennsylvania | 6 | Average | 5.04 % |

## Sales Tax by Province (Canada)

GST - a value-added tax, which the government imposes on most goods and services provided in or imported into Canada. PST - a retail sales tax, which five of the provinces impose on the price of most goods and some services. QST - a value-added tax, similar to the federal GST, which Quebec imposes. HST - Three provinces have combined their retail sales tax with the federal GST into one harmonized tax.

| Province | PST (%) | QST (%) | GST(%) | HST(%) |
|---|---|---|---|---|
| Alberta | 0 | 0 | 5 | 0 |
| British Columbia | 0 | 0 | 0 | 12 |
| Manitoba | 7 | 0 | 5 | 0 |
| New Brunswick | 0 | 0 | 0 | 13 |
| Newfoundland | 0 | 0 | 0 | 13 |
| Northwest Territories | 0 | 0 | 5 | 0 |
| Nova Scotia | 0 | 0 | 0 | 15 |
| Ontario | 0 | 0 | 0 | 13 |
| Prince Edward Island | 10 | 0 | 5 | 0 |
| Quebec | 0 | 8.5 | 5 | 0 |
| Saskatchewan | 5 | 0 | 5 | 0 |
| Yukon | 0 | 0 | 5 | 0 |

## R012909-85 Unemployment Taxes and Social Security Taxes

State Unemployment Tax rates vary not only from state to state, but also with the experience rating of the contractor. The Federal Unemployment Tax rate is 6.2% of the first $7,000 of wages. This is reduced by a credit of up to 5.4% for timely payment to the state. The minimum Federal Unemployment Tax is 0.8% after all credits.

Social Security (FICA) for 2011 is estimated at time of publication to be 7.65% of wages up to $106,800.

## R012909-90 Overtime

One way to improve the completion date of a project or eliminate negative float from a schedule is to compress activity duration times. This can be achieved by increasing the crew size or working overtime with the proposed crew.

To determine the costs of working overtime to compress activity duration times, consider the following examples. Below is an overtime efficiency and cost chart based on a five, six, or seven day week with an eight through twelve hour day. Payroll percentage increases for time and one half and double time are shown for the various working days.

| Days per Week | Hours per Day | Production Efficiency | | | | | Payroll Cost Factors | |
|---|---|---|---|---|---|---|---|---|
| | | 1st Week | 2nd Week | 3rd Week | 4th Week | Average 4 Weeks | @ 1-1/2 Times | @ 2 Times |
| 5 | 8 | 100% | 100% | 100% | 100% | 100 % | 100 % | 100 % |
| | 9 | 100 | 100 | 95 | 90 | 96.25 | 105.6 | 111.1 |
| | 10 | 100 | 95 | 90 | 85 | 91.25 | 110.0 | 120.0 |
| | 11 | 95 | 90 | 75 | 65 | 81.25 | 113.6 | 127.3 |
| | 12 | 90 | 85 | 70 | 60 | 76.25 | 116.7 | 133.3 |
| 6 | 8 | 100 | 100 | 95 | 90 | 96.25 | 108.3 | 116.7 |
| | 9 | 100 | 95 | 90 | 85 | 92.50 | 113.0 | 125.9 |
| | 10 | 95 | 90 | 85 | 80 | 87.50 | 116.7 | 133.3 |
| | 11 | 95 | 85 | 70 | 65 | 78.75 | 119.7 | 139.4 |
| | 12 | 90 | 80 | 65 | 60 | 73.75 | 122.2 | 144.4 |
| 7 | 8 | 100 | 95 | 85 | 75 | 88.75 | 114.3 | 128.6 |
| | 9 | 95 | 90 | 80 | 70 | 83.75 | 118.3 | 136.5 |
| | 10 | 90 | 85 | 75 | 65 | 78.75 | 121.4 | 142.9 |
| | 11 | 85 | 80 | 65 | 60 | 72.50 | 124.0 | 148.1 |
| | 12 | 85 | 75 | 60 | 55 | 68.75 | 126.2 | 152.4 |

## R013113-40 Builder's Risk Insurance

Builder's Risk Insurance is insurance on a building during construction. Premiums are paid by the owner or the contractor. Blasting, collapse and underground insurance would raise total insurance costs above those listed. Floater policy for materials delivered to the job runs $.75 to $1.25 per $100 value. Contractor equipment insurance runs $.50 to $1.50 per $100 value. Insurance for miscellaneous tools to $1,500 value runs from $3.00 to $7.50 per $100 value.

Tabulated below are New England Builder's Risk insurance rates in dollars per $100 value for $1,000 deductible. For $25,000 deductible, rates can be reduced 13% to 34%. On contracts over $1,000,000, rates may be lower than those tabulated. Policies are written annually for the total completed value in place. For "all risk" insurance (excluding flood, earthquake and certain other perils) add $.025 to total rates below.

| Coverage | Frame Construction (Class 1) | | | Brick Construction (Class 4) | | | Fire Resistive (Class 6) | | |
|---|---|---|---|---|---|---|---|---|---|
| | Range | | Average | Range | | Average | Range | | Average |
| Fire Insurance | $.350 to | $ .850 | $.600 | $.158 to | $.189 | $.174 | $.052 to | $.080 | $.070 |
| Extended Coverage | .115 to | .200 | .158 | .080 to | .105 | .101 | .081 to | .105 | .100 |
| Vandalism | .012 to | .016 | .014 | .008 to | .011 | .011 | .008 to | .011 | .010 |
| Total Annual Rate | $.477 to | $1.066 | $.772 | $.246 to | $.305 | $.286 | $.141 to | $.196 | $.180 |

## R013113-50   General Contractor's Overhead

There are two distinct types of overhead on a construction project: Project Overhead and Main Office Overhead. Project Overhead includes those costs at a construction site not directly associated with the installation of construction materials. Examples of Project Overhead costs include the following:

1. Superintendent
2. Construction office and storage trailers
3. Temporary sanitary facilities
4. Temporary utilities
5. Security fencing
6. Photographs
7. Clean up
8. Performance and payment bonds

The above Project Overhead items are also referred to as General Requirements and therefore are estimated in Division 1. Division 1 is the first division listed in the CSI MasterFormat but it is usually the last division estimated. The sum of the costs in Divisions 1 through 49 is referred to as the sum of the direct costs.

All construction projects also include indirect costs. The primary components of indirect costs are the contractor's Main Office Overhead and profit. The amount of the Main Office Overhead expense varies depending on the following:

1. Owner's compensation
2. Project managers and estimator's wages
3. Clerical support wages
4. Office rent and utilities
5. Corporate legal and accounting costs
6. Advertising
7. Automobile expenses
8. Association dues
9. Travel and entertainment expenses

These costs are usually calculated as a percentage of annual sales volume. This percentage can range from 35% for a small contractor doing less than $500,000 to 5% for a large contractor with sales in excess of $100 million.

## R013113-60 Workers' Compensation Insurance Rates by Trade

The table below tabulates the national averages for Workers' Compensation insurance rates by trade and type of building. The average "Insurance Rate" is multiplied by the "% of Building Cost" for each trade. This produces the "Workers' Compensation Cost" by % of total labor cost, to be added for each trade by building type to determine the weighted average Workers' Compensation rate for the building types analyzed.

| Trade | Insurance Rate (% Labor Cost) Range | | Average | % of Building Cost Office Bldgs. | Schools & Apts. | Mfg. | Workers' Compensation Office Bldgs. | Schools & Apts. | Mfg. |
|---|---|---|---|---|---|---|---|---|---|
| Excavation, Grading, etc. | 3.5 % to | 18.2% | 9.1% | 4.8% | 4.9% | 4.5% | 0.44% | 0.45% | 0.41% |
| Piles & Foundations | 4.8 to | 51.8 | 18.0 | 7.1 | 5.2 | 8.7 | 1.28 | 0.94 | 1.57 |
| Concrete | 4.7 to | 28.9 | 13.3 | 5.0 | 14.8 | 3.7 | 0.67 | 1.97 | 0.49 |
| Masonry | 4 to | 28.5 | 12.7 | 6.9 | 7.5 | 1.9 | 0.88 | 0.95 | 0.24 |
| Structural Steel | 4.8 to | 155.9 | 36.3 | 10.7 | 3.9 | 17.6 | 3.88 | 1.42 | 6.39 |
| Miscellaneous & Ornamental Metals | 4 to | 21.2 | 9.9 | 2.8 | 4.0 | 3.6 | 0.28 | 0.40 | 0.36 |
| Carpentry & Millwork | 4.8 to | 53.2 | 16.1 | 3.7 | 4.0 | 0.5 | 0.60 | 0.64 | 0.08 |
| Metal or Composition Siding | 4.8 to | 40 | 14.5 | 2.3 | 0.3 | 4.3 | 0.33 | 0.04 | 0.62 |
| Roofing | 4.8 to | 77.1 | 29.2 | 2.3 | 2.6 | 3.1 | 0.67 | 0.76 | 0.91 |
| Doors & Hardware | 2.5 to | 40 | 10.9 | 0.9 | 1.4 | 0.4 | 0.10 | 0.15 | 0.04 |
| Sash & Glazing | 4.8 to | 34.6 | 12.9 | 3.5 | 4.0 | 1.0 | 0.45 | 0.52 | 0.13 |
| Lath & Plaster | 3.1 to | 29.9 | 11.4 | 3.3 | 6.9 | 0.8 | 0.38 | 0.79 | 0.09 |
| Tile, Marble & Floors | 2.8 to | 18.9 | 8.5 | 2.6 | 3.0 | 0.5 | 0.22 | 0.26 | 0.04 |
| Acoustical Ceilings | 3.1 to | 24.5 | 8.6 | 2.4 | 0.2 | 0.3 | 0.21 | 0.02 | 0.03 |
| Painting | 4.3 to | 29.6 | 11.1 | 1.5 | 1.6 | 1.6 | 0.17 | 0.18 | 0.18 |
| Interior Partitions | 4.8 to | 53.2 | 16.1 | 3.9 | 4.3 | 4.4 | 0.63 | 0.69 | 0.71 |
| Miscellaneous Items | 2.6 to | 109.2 | 14.7 | 5.2 | 3.7 | 9.7 | 0.77 | 0.55 | 1.43 |
| Elevators | 1.9 to | 11.9 | 6.1 | 2.1 | 1.1 | 2.2 | 0.13 | 0.07 | 0.13 |
| Sprinklers | 2.2 to | 17.2 | 7.1 | 0.5 | — | 2.0 | 0.04 | — | 0.14 |
| Plumbing | 2.6 to | 17.5 | 7.0 | 4.9 | 7.2 | 5.2 | 0.34 | 0.50 | 0.36 |
| Heat., Vent., Air Conditioning | 3.8 to | 22.7 | 8.9 | 13.5 | 11.0 | 12.9 | 1.20 | 0.98 | 1.15 |
| Electrical | 2.4 to | 11.7 | 5.8 | 10.1 | 8.4 | 11.1 | 0.59 | 0.49 | 0.64 |
| Total | 1.9 % to | 155.9% | — | 100.0% | 100.0% | 100.0% | 14.26% | 12.77% | 16.14% |
| Overall Weighted Average 14.39% | | | | | | | | | |

## Workers' Compensation Insurance Rates by States

The table below lists the weighted average Workers' Compensation base rate for each state with a factor comparing this with the national average of 14.1%.

| State | Weighted Average | Factor | State | Weighted Average | Factor | State | Weighted Average | Factor |
|---|---|---|---|---|---|---|---|---|
| Alabama | 20.0% | 142 | Kentucky | 16.9% | 120 | North Dakota | 11.2% | 79 |
| Alaska | 16.7 | 118 | Louisiana | 28.5 | 202 | Ohio | 8.6 | 61 |
| Arizona | 8.9 | 63 | Maine | 13.3 | 94 | Oklahoma | 13.7 | 97 |
| Arkansas | 11.3 | 80 | Maryland | 15.1 | 107 | Oregon | 12.4 | 88 |
| California | 28.0 | 199 | Massachusetts | 12.7 | 90 | Pennsylvania | 16.1 | 114 |
| Colorado | 7.7 | 55 | Michigan | 17.7 | 126 | Rhode Island | 11.7 | 83 |
| Connecticut | 21.3 | 151 | Minnesota | 26.2 | 186 | South Carolina | 20.0 | 142 |
| Delaware | 9.6 | 68 | Mississippi | 14.9 | 106 | South Dakota | 17.3 | 123 |
| District of Columbia | 11.1 | 79 | Missouri | 18.4 | 130 | Tennessee | 13.5 | 96 |
| Florida | 10.7 | 76 | Montana | 14.9 | 106 | Texas | 11.3 | 80 |
| Georgia | 27.1 | 192 | Nebraska | 19.4 | 138 | Utah | 7.8 | 55 |
| Hawaii | 9.0 | 64 | Nevada | 10.2 | 72 | Vermont | 14.8 | 105 |
| Idaho | 10.1 | 72 | New Hampshire | 20.1 | 143 | Virginia | 10.3 | 73 |
| Illinois | 22.9 | 162 | New Jersey | 12.8 | 91 | Washington | 9.6 | 68 |
| Indiana | 5.7 | 40 | New Mexico | 17.7 | 126 | West Virginia | 13.0 | 92 |
| Iowa | 10.9 | 77 | New York | 6.1 | 43 | Wisconsin | 15.3 | 109 |
| Kansas | 8.4 | 60 | North Carolina | 17.5 | 124 | Wyoming | 5.3 | 38 |
| Weighted Average for U.S. is 14.4% of payroll = 100% | | | | | | | | |

Rates in the following table are the base or manual costs per $100 of payroll for Workers' Compensation in each state. Rates are usually applied to straight time wages only and not to premium time wages and bonuses.
The weighted average skilled worker rate for 35 trades is 14.1%. For bidding purposes, apply the full value of Workers' Compensation directly to total labor costs, or if labor is 38%, materials 42% and overhead and profit 20% of total cost, carry 38/80 x 14.1% =6.7% of cost (before overhead and profit) into overhead. Rates vary not only from state to state but also with the experience rating of the contractor.

Rates are the most current available at the time of publication.

## R013113-60 Workers' Compensation Insurance Rates by Trade and State (cont.)

| State | Carpentry — 3 stories or less | Carpentry — interior cab. work | Carpentry — general | Concrete Work — NOC | Concrete Work — flat (flr, sdwk.) | Electrical Wiring — inside | Excavation — earth NOC | Excavation — rock | Glaziers | Insulation Work | Lathing | Masonry | Painting & Decorating | Pile Driving | Plastering | Plumbing | Roofing | Sheet Metal Work (HVAC) | Steel Erection — door & sash | Steel Erection — inter, ornam. | Steel Erection — structure | Steel Erection — NOC | Tile Work — (interior ceramic) | Waterproofing | Wrecking |
|---|---|---|---|---|---|---|---|---|---|---|---|---|---|---|---|---|---|---|---|---|---|---|---|---|---|
| | 5651 | 5437 | 5403 | 5213 | 5221 | 5190 | 6217 | 6217 | 5462 | 5479 | 5443 | 5022 | 5474 | 6003 | 5480 | 5183 | 5551 | 5538 | 5102 | 5102 | 5040 | 5057 | 5348 | 9014 | 5701 |
| AL | 25.95 | 16.64 | 27.92 | 13.20 | 9.70 | 9.29 | 11.19 | 11.19 | 22.26 | 17.03 | 10.11 | 25.50 | 17.74 | 21.90 | 15.41 | 8.77 | 47.76 | 9.10 | 9.40 | 9.40 | 45.83 | 21.35 | 15.04 | 7.80 | 45.83 |
| AK | 10.17 | 10.72 | 14.12 | 11.68 | 10.00 | 6.69 | 10.27 | 10.27 | 33.82 | 21.22 | 8.87 | 19.14 | 13.19 | 34.01 | 17.99 | 7.07 | 29.77 | 8.15 | 7.49 | 7.49 | 22.50 | 34.08 | 5.76 | 5.77 | 22.50 |
| AZ | 9.58 | 6.05 | 15.80 | 7.78 | 4.09 | 3.84 | 4.98 | 4.98 | 6.16 | 11.90 | 3.88 | 7.81 | 7.11 | 9.50 | 5.20 | 4.77 | 17.80 | 6.44 | 7.96 | 7.96 | 20.43 | 11.14 | 2.90 | 3.33 | 20.43 |
| AR | 10.25 | 6.32 | 13.03 | 7.90 | 6.10 | 5.13 | 7.72 | 7.72 | 8.76 | 9.71 | 5.62 | 8.38 | 8.83 | 12.60 | 11.45 | 5.58 | 21.23 | 5.31 | 6.19 | 6.19 | 27.01 | 32.27 | 6.75 | 3.52 | 27.01 |
| CA | 40.00 | 40.00 | 40.00 | 17.64 | 17.64 | 11.71 | 18.18 | 18.18 | 22.67 | 18.58 | 17.85 | 28.46 | 21.99 | 20.82 | 29.93 | 17.51 | 69.24 | 22.67 | 21.16 | 21.16 | 33.84 | 20.64 | 10.83 | 21.99 | 20.69 |
| CO | 7.64 | 4.80 | 6.96 | 6.20 | 4.69 | 2.91 | 5.75 | 5.75 | 5.42 | 9.91 | 3.93 | 7.92 | 5.68 | 8.44 | 5.60 | 4.11 | 14.33 | 3.85 | 5.41 | 5.41 | 24.62 | 9.37 | 4.39 | 3.11 | 9.37 |
| CT | 15.80 | 17.74 | 21.69 | 25.96 | 13.53 | 7.98 | 13.75 | 13.75 | 19.29 | 19.38 | 14.94 | 26.57 | 19.57 | 21.58 | 16.34 | 10.58 | 44.27 | 13.02 | 15.01 | 15.01 | 52.47 | 21.56 | 12.35 | 6.03 | 52.47 |
| DE | 9.85 | 9.85 | 7.39 | 7.74 | 6.33 | 3.39 | 5.87 | 5.87 | 8.38 | 7.39 | 8.38 | 8.64 | 10.01 | 11.74 | 8.38 | 5.02 | 17.66 | 6.50 | 8.74 | 8.74 | 17.91 | 8.74 | 5.92 | 8.64 | 17.91 |
| DC | 9.38 | 9.50 | 8.30 | 11.23 | 8.01 | 5.31 | 7.58 | 7.58 | 13.48 | 6.63 | 5.70 | 9.54 | 5.84 | 13.94 | 9.73 | 7.23 | 16.66 | 7.74 | 13.51 | 13.51 | 19.78 | 13.05 | 11.01 | 3.80 | 19.78 |
| FL | 9.80 | 6.83 | 10.73 | 10.69 | 5.31 | 4.72 | 5.78 | 5.78 | 8.37 | 7.60 | 4.04 | 8.78 | 7.50 | 38.24 | 12.58 | 4.75 | 17.85 | 6.49 | 8.06 | 8.06 | 20.73 | 10.19 | 4.07 | 4.15 | 20.73 |
| GA | 35.73 | 21.35 | 25.92 | 17.90 | 15.31 | 10.45 | 16.30 | 16.30 | 19.14 | 22.15 | 14.61 | 24.16 | 24.99 | 26.75 | 24.86 | 12.18 | 63.28 | 17.45 | 20.26 | 20.26 | 71.30 | 39.53 | 13.97 | 8.92 | 71.30 |
| HI | 7.56 | 7.43 | 13.92 | 8.59 | 4.90 | 4.26 | 5.38 | 5.38 | 8.46 | 6.34 | 6.06 | 9.00 | 7.28 | 11.84 | 10.47 | 4.00 | 12.79 | 4.33 | 6.28 | 6.28 | 29.40 | 10.64 | 5.80 | 4.77 | 29.40 |
| ID | 9.90 | 6.33 | 10.07 | 10.29 | 5.89 | 3.41 | 6.44 | 6.44 | 8.75 | 7.70 | 7.95 | 6.62 | 7.75 | 9.61 | 13.76 | 3.96 | 23.71 | 6.82 | 7.17 | 7.17 | 27.50 | 8.29 | 7.39 | 4.49 | 22.50 |
| IL | 23.76 | 14.07 | 22.20 | 28.86 | 11.72 | 9.02 | 10.65 | 10.65 | 18.69 | 19.31 | 15.58 | 20.36 | 12.84 | 27.34 | 20.79 | 11.11 | 35.16 | 13.05 | 18.56 | 18.56 | 84.61 | 24.43 | 16.31 | 5.51 | 84.61 |
| IN | 11.17 | 4.02 | 6.52 | 4.66 | 3.27 | 2.82 | 4.33 | 4.33 | 5.93 | 6.16 | 3.07 | 3.96 | 4.60 | 5.13 | 3.12 | 2.57 | 9.84 | 4.52 | 3.95 | 3.95 | 13.05 | 5.36 | 2.93 | 2.36 | 13.05 |
| IA | 13.13 | 9.09 | 11.58 | 14.33 | 7.45 | 4.42 | 7.61 | 7.61 | 8.93 | 6.14 | 5.23 | 8.37 | 6.58 | 9.12 | 8.83 | 5.86 | 20.63 | 4.44 | 5.40 | 5.40 | 33.37 | 19.27 | 7.35 | 3.98 | 33.37 |
| KS | 14.43 | 6.21 | 8.57 | 6.75 | 5.51 | 3.76 | 4.69 | 4.69 | 7.85 | 6.43 | 3.93 | 6.66 | 6.88 | 8.42 | 4.99 | 4.62 | 13.89 | 4.59 | 4.56 | 4.56 | 24.54 | 15.33 | 5.61 | 3.86 | 15.33 |
| KY | 12.63 | 12.30 | 24.88 | 12.96 | 7.51 | 5.36 | 11.65 | 11.65 | 16.68 | 13.04 | 8.83 | 7.70 | 11.00 | 21.98 | 12.55 | 5.87 | 34.41 | 11.42 | 10.07 | 10.07 | 60.00 | 17.23 | 18.88 | 4.22 | 60.00 |
| LA | 22.02 | 24.93 | 53.17 | 26.43 | 15.61 | 9.88 | 17.49 | 17.49 | 20.14 | 21.43 | 24.51 | 28.33 | 29.58 | 31.25 | 22.37 | 8.64 | 77.12 | 21.20 | 18.98 | 18.98 | 51.76 | 24.21 | 13.77 | 13.78 | 66.41 |
| ME | 14.70 | 10.85 | 21.69 | 14.37 | 9.79 | 5.55 | 9.56 | 9.56 | 13.86 | 11.69 | 6.39 | 11.50 | 12.62 | 13.12 | 11.21 | 8.38 | 21.49 | 8.22 | 6.97 | 6.97 | 30.84 | 16.36 | 5.67 | 4.87 | 30.84 |
| MD | 16.12 | 12.09 | 13.45 | 17.65 | 5.24 | 5.74 | 7.23 | 7.23 | 11.72 | 13.83 | 4.53 | 10.17 | 7.24 | 22.67 | 11.81 | 6.40 | 29.04 | 12.17 | 8.52 | 8.52 | 56.70 | 24.19 | 7.39 | 5.42 | 24.19 |
| MA | 7.50 | 5.93 | 11.92 | 21.45 | 6.62 | 3.17 | 4.17 | 4.17 | 8.93 | 8.94 | 6.53 | 11.62 | 5.01 | 14.99 | 4.89 | 3.88 | 30.35 | 5.74 | 6.76 | 6.76 | 44.61 | 39.97 | 6.45 | 2.50 | 29.46 |
| MI | 20.23 | 9.63 | 20.23 | 17.30 | 8.33 | 4.58 | 11.76 | 11.76 | 12.90 | 10.98 | 13.59 | 14.95 | 13.12 | 51.77 | 10.05 | 6.78 | 29.95 | 11.08 | 10.17 | 10.17 | 51.77 | 13.99 | 9.52 | 6.17 | 51.77 |
| MN | 21.58 | 21.93 | 31.55 | 10.75 | 14.50 | 5.25 | 11.40 | 11.40 | 34.58 | 20.50 | 12.03 | 17.40 | 14.38 | 28.68 | 12.03 | 9.30 | 59.58 | 18.78 | 11.58 | 11.58 | 155.90 | 9.80 | 13.50 | 6.93 | 9.80 |
| MS | 15.14 | 10.81 | 14.94 | 10.42 | 6.74 | 5.60 | 9.95 | 9.95 | 10.94 | 9.28 | 6.27 | 14.58 | 12.36 | 26.40 | 13.13 | 8.38 | 36.90 | 7.23 | 13.97 | 13.97 | 37.66 | 12.49 | 9.54 | 4.69 | 37.66 |
| MO | 18.48 | 13.84 | 14.93 | 21.56 | 11.45 | 8.02 | 12.16 | 12.16 | 11.41 | 15.50 | 10.00 | 16.97 | 12.30 | 21.47 | 15.98 | 11.55 | 34.76 | 9.09 | 12.35 | 12.35 | 34.33 | 46.40 | 12.73 | 6.91 | 34.33 |
| MT | 12.71 | 9.67 | 25.69 | 13.47 | 11.92 | 6.12 | 13.11 | 13.11 | 9.86 | 28.05 | 7.55 | 10.94 | 7.34 | 18.44 | 11.08 | 8.28 | 37.59 | 8.86 | 8.53 | 8.53 | 20.44 | 10.38 | 10.20 | 8.69 | 10.38 |
| NE | 22.48 | 16.10 | 22.18 | 21.90 | 12.55 | 8.83 | 14.33 | 14.33 | 13.63 | 21.55 | 9.85 | 18.65 | 13.40 | 17.78 | 15.03 | 10.45 | 37.18 | 11.60 | 13.93 | 13.93 | 45.53 | 36.08 | 9.05 | 6.43 | 64.50 |
| NV | 11.97 | 7.52 | 11.36 | 8.77 | 7.79 | 5.12 | 8.77 | 8.77 | 8.23 | 8.26 | 4.07 | 6.55 | 7.54 | 9.58 | 6.66 | 6.13 | 14.83 | 9.39 | 7.46 | 7.46 | 24.81 | 16.86 | 4.41 | 4.95 | 18.14 |
| NH | 19.09 | 11.63 | 18.72 | 25.24 | 11.74 | 6.65 | 11.22 | 11.22 | 10.69 | 13.49 | 8.48 | 21.03 | 25.47 | 16.39 | 10.11 | 9.96 | 43.20 | 9.01 | 12.97 | 12.97 | 79.22 | 23.21 | 11.47 | 6.64 | 79.22 |
| NJ | 14.86 | 9.38 | 14.81 | 14.44 | 9.32 | 4.32 | 8.00 | 8.00 | 7.69 | 10.21 | 10.65 | 12.88 | 10.30 | 16.15 | 10.65 | 5.67 | 38.90 | 6.72 | 9.23 | 9.23 | 16.02 | 16.09 | 7.69 | 5.87 | 16.52 |
| NM | 19.58 | 8.17 | 18.99 | 13.33 | 10.11 | 8.08 | 8.28 | 8.28 | 18.17 | 11.75 | 6.68 | 15.36 | 11.43 | 18.20 | 14.39 | 8.14 | 37.47 | 9.92 | 12.79 | 12.79 | 54.49 | 47.45 | 7.66 | 6.66 | 54.49 |
| NY | 5.97 | 2.54 | 5.59 | 6.71 | 5.39 | 2.37 | 3.49 | 3.49 | 15.45 | 2.98 | 5.00 | 6.56 | 4.25 | 6.03 | 3.16 | 3.16 | 12.06 | 4.13 | 5.04 | 5.04 | 10.02 | 7.90 | 2.79 | 2.53 | 4.13 |
| NC | 15.50 | 11.55 | 18.63 | 17.12 | 8.14 | 9.96 | 13.11 | 13.11 | 11.87 | 12.57 | 9.34 | 11.05 | 11.85 | 16.82 | 13.63 | 8.92 | 33.88 | 11.87 | 11.97 | 11.97 | 70.40 | 20.69 | 8.12 | 5.41 | 70.40 |
| ND | 10.56 | 10.56 | 10.56 | 5.17 | 5.17 | 3.66 | 4.44 | 4.44 | 10.56 | 10.56 | 7.94 | 6.35 | 6.94 | 14.58 | 7.94 | 5.56 | 20.76 | 5.56 | 14.58 | 14.58 | 14.58 | 19.14 | 10.56 | 20.76 | 9.31 |
| OH | 8.30 | 4.39 | 6.71 | 6.58 | 5.27 | 3.78 | 4.99 | 4.99 | 7.82 | 8.20 | 23.30 | 7.55 | 7.86 | 7.63 | 5.54 | 4.36 | 17.26 | 4.87 | 6.38 | 6.38 | 12.99 | 11.61 | 6.16 | 5.65 | 7.74 |
| OK | 11.94 | 8.70 | 12.16 | 13.07 | 6.43 | 6.00 | 10.98 | 10.98 | 12.76 | 12.68 | 8.07 | 9.29 | 7.99 | 15.81 | 16.64 | 7.06 | 21.53 | 9.05 | 11.30 | 11.30 | 36.18 | 23.97 | 6.37 | 6.21 | 36.18 |
| OR | 16.95 | 7.57 | 14.30 | 10.92 | 7.82 | 5.37 | 9.61 | 9.61 | 11.51 | 11.15 | 7.46 | 12.31 | 11.60 | 14.60 | 9.99 | 5.10 | 25.58 | 6.63 | 7.09 | 7.09 | 21.15 | 21.87 | 7.63 | 5.93 | 21.15 |
| PA | 16.22 | 16.22 | 13.12 | 16.61 | 12.01 | 6.79 | 9.95 | 9.95 | 12.89 | 13.12 | 12.89 | 14.10 | 14.78 | 17.69 | 12.89 | 8.58 | 32.97 | 8.95 | 16.30 | 16.30 | 24.57 | 15.11 | 9.36 | 14.00 | 24.57 |
| RI | 10.79 | 8.37 | 10.96 | 13.14 | 10.29 | 5.15 | 7.50 | 7.50 | 13.33 | 17.64 | 6.04 | 11.62 | 9.27 | 27.21 | 9.50 | 4.42 | 14.82 | 7.75 | 5.68 | 5.68 | 20.92 | 14.98 | 5.62 | 4.72 | 14.98 |
| SC | 21.43 | 18.13 | 22.49 | 16.89 | 9.59 | 11.45 | 11.78 | 11.78 | 16.99 | 18.58 | 11.06 | 13.93 | 15.50 | 20.61 | 15.75 | 12.36 | 50.08 | 13.06 | 13.91 | 13.91 | 41.22 | 32.53 | 10.50 | 6.33 | 41.22 |
| SD | 21.47 | 9.62 | 21.86 | 25.23 | 7.35 | 5.33 | 9.54 | 9.54 | 15.06 | 13.87 | 8.26 | 11.62 | 10.45 | 20.56 | 11.87 | 10.92 | 26.74 | 12.70 | 8.86 | 8.86 | 59.15 | 31.01 | 7.69 | 5.99 | 59.15 |
| TN | 14.09 | 10.66 | 11.66 | 12.50 | 6.18 | 6.09 | 12.14 | 12.14 | 12.52 | 9.27 | 8.02 | 12.10 | 9.94 | 24.60 | 10.80 | 6.06 | 25.65 | 8.80 | 8.58 | 8.58 | 25.35 | 25.53 | 8.11 | 4.08 | 25.35 |
| TX | 9.17 | 8.39 | 9.17 | 9.00 | 7.06 | 5.76 | 7.88 | 7.88 | 8.56 | 11.97 | 5.74 | 10.31 | 8.18 | 16.38 | 8.18 | 5.88 | 19.42 | 12.05 | 7.87 | 7.87 | 33.20 | 11.24 | 5.53 | 6.19 | 8.70 |
| UT | 11.11 | 6.27 | 8.58 | 8.13 | 6.04 | 2.96 | 6.70 | 6.70 | 7.35 | 5.44 | 3.32 | 7.07 | 6.19 | 8.25 | 4.53 | 3.79 | 21.11 | 5.03 | 5.21 | 5.21 | 14.21 | 11.11 | 4.24 | 3.59 | 18.87 |
| VT | 12.73 | 12.09 | 17.18 | 11.43 | 8.18 | 5.46 | 11.10 | 11.10 | 15.78 | 14.64 | 7.03 | 16.92 | 9.46 | 13.49 | 8.82 | 8.68 | 25.30 | 8.45 | 9.53 | 9.53 | 40.51 | 30.71 | 9.66 | 6.41 | 40.51 |
| VA | 10.44 | 8.34 | 8.71 | 9.83 | 5.37 | 4.39 | 7.46 | 7.46 | 8.80 | 7.36 | 8.95 | 7.50 | 8.31 | 10.07 | 7.92 | 4.93 | 23.55 | 6.37 | 6.66 | 6.66 | 27.15 | 20.56 | 4.98 | 3.09 | 27.15 |
| WA | 7.58 | 7.58 | 7.58 | 7.65 | 7.65 | 2.92 | 6.45 | 6.45 | 12.60 | 7.04 | 7.58 | 9.40 | 12.79 | 17.67 | 9.14 | 4.28 | 16.55 | 4.09 | 7.25 | 7.25 | 7.25 | 7.61 | 6.59 | 16.55 | 7.25 |
| WV | 14.41 | 10.17 | 14.80 | 12.02 | 6.15 | 6.42 | 8.69 | 8.69 | 11.48 | 9.73 | 8.06 | 10.16 | 10.75 | 17.16 | 10.53 | 6.79 | 24.97 | 8.18 | 7.97 | 7.97 | 34.22 | 7.91 | 8.31 | 3.83 | 17.17 |
| WI | 11.71 | 9.86 | 17.41 | 11.55 | 8.51 | 6.39 | 8.17 | 8.17 | 11.44 | 13.60 | 5.38 | 14.52 | 12.34 | 24.28 | 10.17 | 6.03 | 33.37 | 8.83 | 10.63 | 10.63 | 20.55 | 48.12 | 17.76 | 4.94 | 20.55 |
| WY | 4.77 | 4.77 | 4.77 | 4.77 | 4.77 | 4.77 | 4.77 | 4.77 | 4.77 | 4.77 | 4.77 | 4.77 | 4.77 | 4.77 | 4.77 | 4.77 | 4.77 | 4.77 | 4.77 | 4.77 | 4.77 | 4.77 | 4.77 | 4.77 | 4.77 |
| AVG. | 14.52 | 10.93 | 16.07 | 13.33 | 8.35 | 5.81 | 9.10 | 9.10 | 12.89 | 12.30 | 8.59 | 12.66 | 11.07 | 18.00 | 11.43 | 7.04 | 29.20 | 8.94 | 9.86 | 9.86 | 36.30 | 20.12 | 8.45 | 6.43 | 30.85 |

## R013113-60 Workers' Compensation (cont.) (Canada in Canadian dollars)

| Province | | Alberta | British Columbia | Manitoba | Ontario | New Brunswick | Newfndld. & Labrador | Northwest Territories | Nova Scotia | Prince Edward Island | Quebec | Saskatchewan | Yukon |
|---|---|---|---|---|---|---|---|---|---|---|---|---|---|
| Carpentry—3 stories or less | Rate | 6.42 | 3.64 | 4.23 | 4.35 | 5.20 | 10.67 | 3.54 | 6.91 | 5.61 | 13.18 | 4.07 | 9.22 |
| | Code | 42143 | 721028 | 40102 | 723 | 238130 | 4226 | 4-41 | 4226 | 401 | 80110 | 1226 | 202 |
| Carpentry—interior cab. work | Rate | 2.43 | 2.62 | 4.23 | 4.35 | 4.64 | 4.29 | 3.54 | 4.73 | 3.44 | 13.18 | 2.41 | 9.22 |
| | Code | 42133 | 721021 | 40102 | 723 | 238350 | 4279 | 4-41 | 4274 | 402 | 80110 | B1127 | 202 |
| CARPENTRY—general | Rate | 6.42 | 3.64 | 4.23 | 4.35 | 5.20 | 4.29 | 3.54 | 6.91 | 5.61 | 13.18 | 4.07 | 9.22 |
| | Code | 42143 | 721028 | 40102 | 723 | 238130 | 4299 | 4-41 | 4226 | 401 | 80110 | 12.02 | 202 |
| CONCRETE WORK—NOC | Rate | 4.11 | 3.87 | 6.79 | 16.50 | 5.20 | 10.67 | 3.54 | 4.85 | 5.61 | 14.82 | 4.80 | 7.12 |
| | Code | 42104 | 721010 | 40110 | 748 | 238110 | 4224 | 4-41 | 4224 | 401 | 80100 | B1314 | 203 |
| CONCRETE WORK—flat (flr. sidewalk) | Rate | 4.11 | 3.87 | 6.79 | 16.50 | 5.20 | 10.67 | 3.54 | 4.85 | 5.61 | 14.82 | 4.80 | 7.12 |
| | Code | 42104 | 721010 | 40110 | 748 | 238110 | 4224 | 4-41 | 4224 | 401 | 80100 | B1314 | 203 |
| ELECTRICAL Wiring—inside | Rate | 1.97 | 1.41 | 1.96 | 3.25 | 2.54 | 2.76 | 2.75 | 2.62 | 3.44 | 5.25 | 2.41 | 5.51 |
| | Code | 42124 | 721019 | 40203 | 704 | 238210 | 4261 | 4-46 | 4261 | 402 | 80170 | B1105 | 206 |
| EXCAVATION—earth NOC | Rate | 2.36 | 3.25 | 4.52 | 4.68 | 5.20 | 3.67 | 3.79 | 2.85 | 2.82 | 8.43 | 2.42 | 5.51 |
| | Code | 40604 | 721031 | 40706 | 711 | 238190 | 4214 | 4-43 | 4214 | 404 | 80030 | R1106 | 207 |
| EXCAVATION—rock | Rate | 2.36 | 3.25 | 4.52 | 4.68 | 5.20 | 3.67 | 3.79 | 2.85 | 2.82 | 8.43 | 2.42 | 5.51 |
| | Code | 40604 | 721031 | 40706 | 711 | 238190 | 4214 | 4-43 | 4214 | 404 | 80030 | R1106 | 207 |
| GLAZIERS | Rate | 2.75 | 3.05 | 4.23 | 9.25 | 5.20 | 5.40 | 3.54 | 5.76 | 3.44 | 12.63 | 4.80 | 5.51 |
| | Code | 42121 | 715020 | 40109 | 751 | 238150 | 4233 | 4-41 | 4233 | 402 | 80150 | B1304 | 212 |
| INSULATION WORK | Rate | 2.04 | 6.91 | 4.23 | 9.25 | 4.64 | 5.40 | 3.54 | 5.76 | 5.61 | 13.18 | 4.07 | 9.22 |
| | Code | 42184 | 721029 | 40102 | 751 | 238310 | 4234 | 4-41 | 4234 | 401 | 80110 | B1207 | 202 |
| LATHING | Rate | 4.15 | 6.27 | 4.23 | 4.35 | 4.64 | 4.29 | 3.54 | 4.73 | 3.44 | 13.18 | 4.80 | 9.22 |
| | Code | 42135 | 721042 | 40102 | 723 | 238390 | 4279 | 4-41 | 4271 | 402 | 80110 | B1316 | 202 |
| MASONRY | Rate | 4.11 | 3.50 | 4.23 | 11.15 | 5.20 | 5.40 | 3.54 | 5.76 | 5.61 | 14.82 | 4.80 | 9.22 |
| | Code | 42102 | 721037 | 40102 | 741 | 238140 | 4231 | 4-41 | 4231 | 401 | 80100 | B1318 | 202 |
| PAINTING & DECORATING | Rate | 3.39 | 3.36 | 3.49 | 6.75 | 4.64 | 4.29 | 3.54 | 4.73 | 3.44 | 13.18 | 4.07 | 9.22 |
| | Code | 42111 | 721041 | 40105 | 719 | 238320 | 4275 | 4-41 | 4275 | 402 | 80110 | B1201 | 202 |
| PILE DRIVING | Rate | 4.04 | 3.82 | 4.52 | 6.34 | 5.20 | 5.67 | 3.79 | 5.19 | 5.61 | 8.43 | 4.80 | 9.22 |
| | Code | 42159 | 722004 | 40706 | 732 | 238190 | 4129 | 4-43 | 4221 | 401 | 80030 | B1310 | 202 |
| PLASTERING | Rate | 4.15 | 6.27 | 4.46 | 6.75 | 4.64 | 4.29 | 3.54 | 4.73 | 3.44 | 13.18 | 4.07 | 9.22 |
| | Code | 42135 | 721042 | 40108 | 719 | 238390 | 4271 | 4-41 | 4271 | 402 | 80110 | B1221 | 202 |
| PLUMBING | Rate | 1.97 | 1.90 | 3.09 | 3.98 | 2.54 | 2.77 | 2.75 | 2.76 | 3.44 | 6.24 | 2.41 | 5.51 |
| | Code | 42122 | 721043 | 40204 | 707 | 238220 | 4241 | 4-46 | 4241 | 402 | 80160 | B1101 | 214 |
| ROOFING | Rate | 6.59 | 6.18 | 7.23 | 13.30 | 5.20 | 10.67 | 3.54 | 7.45 | 5.61 | 19.78 | 4.80 | 9.22 |
| | Code | 42118 | 721036 | 40403 | 728 | 238160 | 4236 | 4-41 | 4236 | 401 | 80130 | B1320 | 202 |
| SHEET METAL WORK (HVAC) | Rate | 1.97 | 1.90 | 7.23 | 3.98 | 2.54 | 2.77 | 2.75 | 2.76 | 3.44 | 6.24 | 2.41 | 5.51 |
| | Code | 42117 | 721043 | 40402 | 707 | 238220 | 4244 | 4-46 | 4244 | 402 | 80160 | B1107 | 208 |
| STEEL ERECTION—door & sash | Rate | 2.04 | 12.66 | 10.32 | 16.50 | 5.20 | 10.67 | 3.54 | 5.76 | 5.61 | 22.38 | 4.80 | 9.22 |
| | Code | 42106 | 722005 | 40502 | 748 | 238120 | 4227 | 4-41 | 4227 | 401 | 80080 | B1322 | 202 |
| STEEL ERECTION—inter., ornam. | Rate | 2.04 | 12.66 | 10.32 | 16.50 | 5.20 | 10.67 | 3.54 | 5.76 | 5.61 | 22.38 | 4.80 | 9.22 |
| | Code | 42106 | 722005 | 40502 | 748 | 238120 | 4227 | 4-41 | 4227 | 401 | 80080 | B1322 | 202 |
| STEEL ERECTION—structure | Rate | 2.04 | 12.66 | 10.32 | 16.50 | 5.20 | 10.67 | 3.54 | 5.76 | 5.61 | 22.38 | 4.80 | 9.22 |
| | Code | 42106 | 722005 | 40502 | 748 | 238120 | 4227 | 4-41 | 4227 | 401 | 80080 | B1322 | 202 |
| STEEL ERECTION—NOC | Rate | 2.04 | 12.66 | 10.32 | 16.50 | 5.20 | 10.67 | 3.54 | 5.76 | 5.61 | 22.38 | 4.80 | 9.22 |
| | Code | 42106 | 722005 | 40502 | 748 | 238120 | 4227 | 4-41 | 4227 | 401 | 80080 | B1322 | 202 |
| TILE WORK—inter. (ceramic) | Rate | 3.24 | 3.62 | 2.00 | 6.75 | 4.64 | 4.29 | 3.54 | 4.73 | 3.44 | 13.18 | 4.80 | 9.22 |
| | Code | 42113 | 721054 | 40103 | 719 | 238340 | 4276 | 4-41 | 4276 | 402 | 80110 | B1301 | 202 |
| WATERPROOFING | Rate | 3.39 | 3.35 | 4.23 | 4.35 | 5.20 | 4.29 | 3.54 | 5.76 | 3.44 | 19.78 | 4.07 | 9.22 |
| | Code | 42139 | 721016 | 40102 | 723 | 238190 | 4299 | 4-41 | 4239 | 402 | 80130 | B1217 | 202 |
| WRECKING | Rate | 2.36 | 4.44 | 7.34 | 16.50 | 5.20 | 3.67 | 3.79 | 2.85 | 5.61 | 13.18 | 2.42 | 9.22 |
| | Code | 40604 | 721005 | 40106 | 748 | 238190 | 4211 | 4-43 | 4211 | 401 | 80110 | R1125 | 202 |

## R013113-80    Performance Bond

This table shows the cost of a Performance Bond for a construction job scheduled to be completed in 12 months. Add 1% of the premium cost per month for jobs requiring more than 12 months to complete. The rates are "standard" rates offered to contractors that the bonding company considers financially sound and capable of doing the work. Preferred rates are offered by some bonding companies based upon financial strength of the contractor. Actual rates vary from contractor to contractor and from bonding company to bonding company. Contractors should prequalify through a bonding agency before submitting a bid on a contract that requires a bond.

| Contract Amount | Building Construction Class B Projects | | | Highways & Bridges | | | | | |
|---|---|---|---|---|---|---|---|---|---|
| | | | | Class A New Construction | | | Class A-1 Highway Resurfacing | | |
| First $ 100,000 bid | $25.00 per M | | | $15.00 per M | | | $9.40 per M | | |
| Next     400,000 bid | $ 2,500 | plus | $15.00 per M | $ 1,500 | plus | $10.00 per M | $    940 | plus | $7.20 per M |
| Next  2,000,000 bid | 8,500 | plus | 10.00 per M | 5,500 | plus | 7.00 per M | 3,820 | plus | 5.00 per M |
| Next  2,500,000 bid | 28,500 | plus | 7.50 per M | 19,500 | plus | 5.50 per M | 15,820 | plus | 4.50 per M |
| Next  2,500,000 bid | 47,250 | plus | 7.00 per M | 33,250 | plus | 5.00 per M | 28,320 | plus | 4.50 per M |
| Over  7,500,000 bid | 64,750 | plus | 6.00 per M | 45,750 | plus | 4.50 per M | 39,570 | plus | 4.00 per M |

## R015113-65    Temporary Power Equipment

Cost data for the temporary equipment was developed utilizing the following information.

1) Re-usable material-services, transformers, equipment and cords are based on new purchase and prorated to three projects.
2) PVC feeder includes trench and backfill.
3) Connections include disconnects and fuses.
4) Labor units include an allowance for removal.
5) No utility company charges or fees are included.
6) Concrete pads or vaults are not included.
7) Utility company conduits not included.

## R015423-10  Steel Tubular Scaffolding

On new construction, tubular scaffolding is efficient up to 60' high or five stories. Above this it is usually better to use a hung scaffolding if construction permits. Swing scaffolding operations may interfere with tenants. In this case, the tubular is more practical at all heights.

In repairing or cleaning the front of an existing building the cost of tubular scaffolding per S.F. of building front increases as the height increases above the first tier. The first tier cost is relatively high due to leveling and alignment.

The minimum efficient crew for erecting and dismantling is three workers. They can set up and remove 18 frame sections per day up to 5 stories high. For 6 to 12 stories high, a crew of four is most efficient. Use two or more on top and two on the bottom for handing up or hoisting. They can

also set up and remove 18 frame sections per day. At 7' horizontal spacing, this will run about 800 S.F. per day of erecting and dismantling. Time for placing and removing planks must be added to the above. A crew of three can place and remove 72 planks per day up to 5 stories. For over 5 stories, a crew of four can place and remove 80 planks per day.

The table below shows the number of pieces required to erect tubular steel scaffolding for 1000 S.F. of building frontage. This area is made up of a scaffolding system that is 12 frames (11 bays) long by 2 frames high.

For jobs under twenty-five frames, add 50% to rental cost. Rental rates will be lower for jobs over three months duration. Large quantities for long periods can reduce rental rates by 20%.

| Description of Component | Number of Pieces for 1000 S.F. of Building Front | Unit |
|---|:---:|:---:|
| 5' Wide Standard Frame, 6'-4" High | 24 | Ea. |
| Leveling Jack & Plate | 24 | |
| Cross Brace | 44 | |
| Side Arm Bracket, 21" | 12 | |
| Guardrail Post | 12 | |
| Guardrail, 7' section | 22 | |
| Stairway Section | 2 | |
| Stairway Starter Bar | 1 | |
| Stairway Inside Handrail | 2 | |
| Stairway Outside Handrail | 2 | |
| Walk-Thru Frame Guardrail | 2 | |

Scaffolding is often used as falsework over 15' high during construction of cast-in-place concrete beams and slabs. Two foot wide scaffolding is generally used for heavy beam construction. The span between frames depends upon the load to be carried with a maximum span of 5'.

Heavy duty shoring frames with a capacity of 10,000#/leg can be spaced up to 10' O.C. depending upon form support design and loading.

Scaffolding used as horizontal shoring requires less than half the material required with conventional shoring.

On new construction, erection is done by carpenters.

Rolling towers supporting horizontal shores can reduce labor and speed the job. For maintenance work, catwalks with spans up to 70' can be supported by the rolling towers.

## R015423-20  Pump Staging

Pump staging is generally not available for rent. The table below shows the number of pieces required to erect pump staging for 2400 S.F. of building

frontage. This area is made up of a pump jack system that is 3 poles (2 bays) wide by 2 poles high.

| Item | Number of Pieces for 2400 S.F. of Building Front | Unit |
|---|:---:|:---:|
| Aluminum pole section, 24' long | 6 | Ea. |
| Aluminum splice joint, 6' long | 3 | |
| Aluminum foldable brace | 3 | |
| Aluminum pump jack | 3 | |
| Aluminum support for workbench/back safety rail | 3 | |
| Aluminum scaffold plank/workbench, 14" wide x 24' long | 4 | |
| Safety net, 22' long | 2 | |
| Aluminum plank end safety rail | 2 | |

The cost in place for this 2400 S.F. will depend on how many uses are realized during the life of the equipment.

## R015433-10    Contractor Equipment

**Rental Rates** shown elsewhere in the book pertain to late model high quality machines in excellent working condition, rented from equipment dealers. Rental rates from contractors may be substantially lower than the rental rates from equipment dealers depending upon economic conditions; for older, less productive machines, reduce rates by a maximum of 15%. Any overtime must be added to the base rates. For shift work, rates are lower. Usual rule of thumb is 150% of one shift rate for two shifts; 200% for three shifts.

For periods of less than one week, operated equipment is usually more economical to rent than renting bare equipment and hiring an operator.

Costs to move equipment to a job site (mobilization) or from a job site (demobilization) are not included in rental rates, nor in any Equipment costs on any Unit Price line items or crew listings. These costs can be found elsewhere. If a piece of equipment is already at a job site, it is not appropriate to utilize mob/demob costs in an estimate again.

Rental rates vary throughout the country with larger cities generally having lower rates. Lease plans for new equipment are available for periods in excess of six months with a percentage of payments applying toward purchase.

Rental rates can also be treated as reimbursement costs for contractor-owned equipment. Owned equipment costs include depreciation, loan payments, interest, taxes, insurance, storage, and major repairs.

Monthly rental rates vary from 2% to 5% of the cost of the equipment depending on the anticipated life of the equipment and its wearing parts. Weekly rates are about 1/3 the monthly rates and daily rental rates about 1/3 the weekly rate.

The hourly operating costs for each piece of equipment include costs to the user such as fuel, oil, lubrication, normal expendables for the equipment, and a percentage of mechanic's wages chargeable to maintenance. The hourly operating costs listed do not include the operator's wages.

The daily cost for equipment used in the standard crews is figured by dividing the weekly rate by five, then adding eight times the hourly operating cost to give the total daily equipment cost, not including the operator. This figure is in the right hand column of the Equipment listings under Equipment Cost/Day.

**Pile Driving** rates shown for pile hammer and extractor do not include leads, crane, boiler or compressor. Vibratory pile driving requires an added field specialist during set-up and pile driving operation for the electric model. The hydraulic model requires a field specialist for set-up only. Up to 125 reuses of sheet piling are possible using vibratory drivers. For normal conditions, crane capacity for hammer type and size are as follows.

| Crane Capacity | Hammer Type and Size | | |
|---|---|---|---|
| | Air or Steam | Diesel | Vibratory |
| 25 ton | to 8,750 ft.-lb. | | 70 H.P. |
| 40 ton | 15,000 ft.-lb. | to 32,000 ft.-lb. | 170 H.P. |
| 60 ton | 25,000 ft.-lb. | | 300 H.P. |
| 100 ton | | 112,000 ft.-lb. | |

**Cranes** should be specified for the job by size, building and site characteristics, availability, performance characteristics, and duration of time required.

**Backhoes & Shovels** rent for about the same as equivalent size cranes but maintenance and operating expense is higher. Crane operators rate must be adjusted for high boom heights. Average adjustments: for 150' boom add 2% per hour; over 185', add 4% per hour; over 210', add 6% per hour; over 250', add 8% per hour and over 295', add 12% per hour.

**Tower Cranes** of the climbing or static type have jibs from 50' to 200' and capacities at maximum reach range from 4,000 to 14,000 pounds. Lifting capacities increase up to maximum load as the hook radius decreases.

Typical rental rates, based on purchase price are about 2% to 3% per month.

Erection and dismantling runs between 500 and 2000 labor hours. Climbing operation takes 10 labor hours per 20' climb. Crane dead time is about 5 hours per 40' climb. If crane is bolted to side of the building add cost of ties and extra mast sections. Climbing cranes have from 80' to 180' of mast while static cranes have 80' to 800' of mast.

**Truck Cranes** can be converted to tower cranes by using tower attachments. Mast heights over 400' have been used.

A single 100' high material **Hoist and Tower** can be erected and dismantled in about 400 labor hours; a double 100' high hoist and tower in about 600 labor hours. Erection times for additional heights are 3 and 4 labor hours

per vertical foot respectively up to 150', and 4 to 5 labor hours per vertical foot over 150' high. A 40' high portable Buck hoist takes about 160 labor hours to erect and dismantle. Additional heights take 2 labor hours per vertical foot to 80' and 3 labor hours per vertical foot for the next 100'. Most material hoists do not meet local code requirements for carrying personnel.

A 150' high **Personnel Hoist** requires about 500 to 800 labor hours to erect and dismantle. Budget erection time at 5 labor hours per vertical foot for all trades. Local code requirements or labor scarcity requiring overtime can add up to 50% to any of the above erection costs.

**Earthmoving Equipment**: The selection of earthmoving equipment depends upon the type and quantity of material, moisture content, haul distance, haul road, time available, and equipment available. Short haul cut and fill operations may require dozers only, while another operation may require excavators, a fleet of trucks, and spreading and compaction equipment. Stockpiled material and granular material are easily excavated with front end loaders. Scrapers are most economically used with hauls between 300' and 1-1/2 miles if adequate haul roads can be maintained. Shovels are often used for blasted rock and any material where a vertical face of 8' or more can be excavated. Special conditions may dictate the use of draglines, clamshells, or backhoes. Spreading and compaction equipment must be matched to the soil characteristics, the compaction required and the rate the fill is being supplied.

## R015433-15    Heavy Lifting

### Hydraulic Climbing Jacks

The use of hydraulic heavy lift systems is an alternative to conventional type crane equipment. The lifting, lowering, pushing, or pulling mechanism is a hydraulic climbing jack moving on a square steel jackrod from 1-5/8" to 4" square, or a steel cable. The jackrod or cable can be vertical or horizontal, stationary or movable, depending on the individual application. When the jackrod is stationary, the climbing jack will climb the rod and push or pull the load along with itself. When the climbing jack is stationary, the jackrod is movable with the load attached to the end and the climbing jack will lift or lower the jackrod with the attached load. The heavy lift system is normally operated by a single control lever located at the hydraulic pump.

The system is flexible in that one or more climbing jacks can be applied wherever a load support point is required, and the rate of lift synchronized.

Economic benefits have been demonstrated on projects such as: erection of ground assembled roofs and floors, complete bridge spans, girders and trusses, towers, chimney liners and steel vessels, storage tanks, and heavy machinery. Other uses are raising and lowering offshore work platforms, caissons, tunnel sections and pipelines.

## R024119-10 Demolition Defined

**Whole Building Demolition** - Demolition of the whole building with no concern for any particular building element, component, or material type being demolished. This type of demolition is accomplished with large pieces of construction equipment that break up the structure, load it into trucks and haul it to a disposal site, but disposal or dump fees are not included. Demolition of below-grade foundation elements, such as footings, foundation walls, grade beams, slabs on grade, etc., is not included. Certain mechanical equipment containing flammable liquids or ozone-depleting refrigerants, electric lighting elements, communication equipment components, and other building elements may contain hazardous waste, and must be removed, either selectively or carefully, as hazardous waste before the building can be demolished.

**Foundation Demolition** - Demolition of below-grade foundation footings, foundation walls, grade beams, and slabs on grade. This type of demolition is accomplished by hand or pneumatic hand tools, and does not include saw cutting, or handling, loading, hauling, or disposal of the debris.

**Gutting** - Removal of building interior finishes and electrical/mechanical systems down to the load-bearing and sub-floor elements of the rough building frame, with no concern for any particular building element, component, or material type being demolished. This type of demolition is accomplished by hand or pneumatic hand tools, and includes loading into trucks, but not hauling, disposal or dump fees, scaffolding, or shoring. Certain mechanical equipment containing flammable liquids or ozone-depleting refrigerants, electric lighting elements, communication equipment components, and other building elements may contain hazardous waste, and must be removed, either selectively or carefully, as hazardous waste, before the building is gutted.

**Selective Demolition** - Demolition of a selected building element, component, or finish, with some concern for surrounding or adjacent elements, components, or finishes (see the first Subdivision (s) at the beginning of appropriate Divisions). This type of demolition is accomplished by hand or pneumatic hand tools, and does not include handling, loading, storing, hauling, or disposal of the debris, scaffolding, or shoring. "Gutting" methods may be used in order to save time, but damage that is caused to surrounding or adjacent elements, components, or finishes may have to be repaired at a later time.

**Careful Removal** - Removal of a piece of service equipment, building element or component, or material type, with great concern for both the removed item and surrounding or adjacent elements, components or finishes. The purpose of careful removal may be to protect the removed item for later re-use, preserve a higher salvage value of the removed item, or replace an item while taking care to protect surrounding or adjacent elements, components, connections, or finishes from cosmetic and/or structural damage. An approximation of the time required to perform this type of removal is 1/3 to 1/2 the time it would take to install a new item of like kind (see Reference Number R220105-10). This type of removal is accomplished by hand or pneumatic hand tools, and does not include loading, hauling, or storing the removed item, scaffolding, shoring, or lifting equipment.

**Cutout Demolition** - Demolition of a small quantity of floor, wall, roof, or other assembly, with concern for the appearance and structural integrity of the surrounding materials. This type of demolition is accomplished by hand or pneumatic hand tools, and does not include saw cutting, handling, loading, hauling, or disposal of debris, scaffolding, or shoring.

**Rubbish Handling** - Work activities that involve handling, loading or hauling of debris. Generally, the cost of rubbish handling must be added to the cost of all types of demolition, with the exception of whole building demolition.

**Minor Site Demolition** - Demolition of site elements outside the footprint of a building. This type of demolition is accomplished by hand or pneumatic hand tools, or with larger pieces of construction equipment, and may include loading a removed item onto a truck (check the Crew for equipment used). It does not include saw cutting, hauling or disposal of debris, and, sometimes, handling or loading.

## R024119-20 Dumpsters

Dumpster rental costs on construction sites are presented in two ways.

The cost per week rental includes the delivery of the dumpster; its pulling or emptying once per week, and its final removal. The assumption is made that the dumpster contractor could choose to empty a dumpster by simply bringing in an empty unit and removing the full one. These costs also include the disposal of the materials in the dumpster.

The Alternate Pricing can be used when actual planned conditions are not approximated by the weekly numbers. For example, these lines can be used when a dumpster is needed for 4 weeks and will need to be emptied 2 or 3 times per week. Conversely the Alternate Pricing lines can be used when a dumpster will be rented for several weeks or months but needs to be emptied only a few times over this period.

## R026510-20 Underground Storage Tank Removal

Underground Storage Tank Removal can be divided into two categories: Non-Leaking and Leaking. Prior to removing an underground storage tank, tests should be made, with the proper authorities present, to determine whether a tank has been leaking or the surrounding soil has been contaminated.

To safely remove Liquid Underground Storage Tanks:
1. Excavate to the top of the tank.
2. Disconnect all piping.
3. Open all tank vents and access ports.
4. Remove all liquids and/or sludge.
5. Purge the tank with an inert gas.
6. Provide access to the inside of the tank and clean out the interior using proper personal protective equipment (PPE).
7. Excavate soil surrounding the tank using proper PPE for on-site personnel.
8. Pull and properly dispose of the tank.
9. Clean up the site of all contaminated material.
10. Install new tanks or close the excavation.

## Existing Conditions | R0282 Asbestos Remediation

### R028213-20　Asbestos Removal Process

Asbestos removal is accomplished by a specialty contractor who understands the federal and state regulations regarding the handling and disposal of the material. The process of asbestos removal is divided into many individual steps. An accurate estimate can be calculated only after all the steps have been priced.

The steps are generally as follows:

1. Obtain an asbestos abatement plan from an industrial hygienist.
2. Monitor the air quality in and around the removal area and along the path of travel between the removal area and transport area. This establishes the background contamination.
3. Construct a two part decontamination chamber at entrance to removal area.
4. Install a HEPA filter to create a negative pressure in the removal area.
5. Install wall, floor and ceiling protection as required by the plan, usually 2 layers of fireproof 6 mil polyethylene.
6. Industrial hygienist visually inspects work area to verify compliance with plan.
7. Provide temporary supports for conduit and piping affected by the removal process.
8. Proceed with asbestos removal and bagging process. Monitor air quality as described in Step #2. Discontinue operations when contaminate levels exceed applicable standards.
9. Document the legal disposal of materials in accordance with EPA standards.
10. Thoroughly clean removal area including all ledges, crevices and surfaces.
11. Post abatement inspection by industrial hygienist to verify plan compliance.
12. Provide a certificate from a licensed industrial hygienist attesting that contaminate levels are within acceptable standards before returning area to regular use.

## Existing Conditions | R0283 Lead Remediation

### R028319-60　Lead Paint Remediation Methods

Lead paint remediation can be accomplished by the following methods.

1. Abrasive blast
2. Chemical stripping
3. Power tool cleaning with vacuum collection system
4. Encapsulation
5. Remove and replace
6. Enclosure

Each of these methods has strengths and weakness depending on the specific circumstances of the project. The following is an overview of each method.

1. **Abrasive blasting** is usually accomplished with sand or recyclable metallic blast. Before work can begin, the area must be contained to ensure the blast material with lead does not escape to the atmosphere. The use of vacuum blast greatly reduces the containment requirements. Lead abatement equipment that may be associated with this work includes a negative air machine. In addition, it is necessary to have an industrial hygienist monitor the project on a continual basis. When the work is complete, the spent blast sand with lead must be disposed of as a hazardous material. If metallic shot was used, the lead is separated from the shot and disposed of as hazardous material. Worker protection includes disposable clothing and respiratory protection.

2. **Chemical stripping** requires strong chemicals be applied to the surface to remove the lead paint. Before the work can begin, the area under/adjacent to the work area must be covered to catch the chemical and removed lead. After the chemical is applied to the painted surface it is usually covered with paper. The chemical is left in place for the specified period, then the paper with lead paint is pulled or scraped off. The process may require several chemical applications. The paper with chemicals and lead paint adhered to it, plus the containment and loose scrapings collected by a HEPA (High Efficiency Particulate Air Filter) vac, must be disposed of as a hazardous material. The chemical stripping process usually requires a neutralizing agent and several wash downs after the paint is removed. Worker protection includes a neoprene or other compatible protective clothing and respiratory protection with face shield. An industrial hygienist is required intermittently during the process.

3. **Power tool cleaning** is accomplished using shrouded needle blasting guns. The shrouding with different end configurations is held up against the surface to be cleaned. The area is blasted with hardened needles and the shroud captures the lead with a HEPA vac and deposits it in a holding tank. An industrial hygienist monitors the project, protective clothing and a respirator is required until air samples prove otherwise. When the work is complete the lead must be disposed of as a hazardous material.

4. **Encapsulation** is a method that leaves the well bonded lead paint in place after the peeling paint has been removed. Before the work can begin, the area under/adjacent to the work must be covered to catch the scrapings. The scraped surface is then washed with a detergent and rinsed. The prepared surface is covered with approximately 10 mils of paint. A reinforcing fabric can also be embedded in the paint covering. The scraped paint and containment must be disposed of as a hazardous material. Workers must wear protective clothing and respirators.

5. **Remove and replace** is an effective way to remove lead paint from windows, gypsum walls and concrete masonry surfaces. The painted materials are removed and new materials are installed. Workers should wear a respirator and tyvek suit. The demolished materials must be disposed of as hazardous waste if it fails the TCLP (Toxicity Characteristic Leachate Process) test.

6. **Enclosure** is the process that permanently seals lead painted materials in place. This process has many applications such as covering lead painted drywall with new drywall, covering exterior construction with tyvek paper then residing, or covering lead painted structural members with aluminum or plastic. The seams on all enclosing materials must be securely sealed. An industrial hygienist monitors the project, and protective clothing and a respirator is required until air samples prove otherwise.

All the processes require clearance monitoring and wipe testing as required by the hygienist.

## R031113-10    Wall Form Materials

### Aluminum Forms

Approximate weight is 3 lbs. per S.F.C.A. Standard widths are available from 4″ to 36″ with 36″ most common. Standard lengths of 2′, 4′, 6′ to 8′ are available. Forms are lightweight and fewer ties are needed with the wider widths. The form face is either smooth or textured.

### Metal Framed Plywood Forms

Manufacturers claim over 75 reuses of plywood and over 300 reuses of steel frames. Many specials such as corners, fillers, pilasters, etc. are available. Monthly rental is generally about 15% of purchase price for first month and 9% per month thereafter with 90% of rental applied to purchase for the first month and decreasing percentages thereafter. Aluminum framed forms cost 25% to 30% more than steel framed.

After the first month, extra days may be prorated from the monthly charge. Rental rates do not include ties, accessories, cleaning, loss of hardware or freight in and out. Approximate weight is 5 lbs. per S.F. for steel; 3 lbs. per S.F. for aluminum.

Forms can be rented with option to buy.

### Plywood Forms, Job Fabricated

There are two types of plywood used for concrete forms.

1. Exterior plyform which is completely waterproof. This is face oiled to facilitate stripping. Ten reuses can be expected with this type with 25 reuses possible.
2. An overlaid type consists of a resin fiber fused to exterior plyform. No oiling is required except to facilitate cleaning. This is available in both high density (HDO) and medium density overlaid (MDO). Using HDO, 50 reuses can be expected with 200 possible.

Plyform is available in 5/8″ and 3/4″ thickness. High density overlaid is available in 3/8″, 1/2″, 5/8″ and 3/4″ thickness.

5/8″ thick is sufficient for most building forms, while 3/4″ is best on heavy construction.

### Plywood Forms, Modular, Prefabricated

There are many plywood forming systems without frames. Most of these are manufactured from 1-1/8″ (HDO) plywood and have some hardware attached. These are used principally for foundation walls 8′ or less high. With care and maintenance, 100 reuses can be attained with decreasing quality of surface finish.

### Steel Forms

Approximate weight is 6-1/2 lbs. per S.F.C.A. including accessories. Standard widths are available from 2″ to 24″, with 24″ most common. Standard lengths are from 2′ to 8′, with 4′ the most common. Forms are easily ganged into modular units.

Forms are usually leased for 15% of the purchase price per month prorated daily over 30 days.

Rental may be applied to sale price, and usually rental forms are bought. With careful handling and cleaning 200 to 400 reuses are possible.

Straight wall gang forms up to 12′ x 20′ or 8′ x 30′ can be fabricated. These crane handled forms usually lease for approx. 9% per month.

Individual job analysis is available from the manufacturer at no charge.

## R031113-30    Slipforms

The slipform method of forming may be used for forming circular silo and multi-celled storage bin type structures over 30′ high, and building core shear walls over eight stories high. The shear walls, usually enclose elevator shafts, stairwells, mechanical spaces, and toilet rooms. Reuse of the form on duplicate structures will reduce the height necessary and spread the cost of building the form. Slipform systems can be used to cast chimneys, towers, piers, dams, underground shafts or other structures capable of being extruded.

Slipforms are usually 4′ high and are raised semi-continuously by jacks climbing ·on rods which are embedded in the concrete. The jacks are powered by a hydraulic, pneumatic, or electric source and are available in 3, 6, and 22 ton capacities. Interior work decks and exterior scaffolds must be provided for placing inserts, embedded items, reinforcing steel, and concrete. Scaffolds below the form for finishers may be required. The interior work decks are often used as roof slab forms on silos and bin work. Form raising rates will range from 6″ to 20″ per hour for silos; 6″ to 30″ per hour for buildings; and 6″ to 48″ per hour for shaft work.

Reinforcing bars and stressing strands are usually hoisted by crane or gin pole, and the concrete material can be hoisted by crane, winch-powered skip, or pumps. The slipform system is operated on a continuous 24-hour day when a monolithic structure is desired. For least cost, the system is operated only during normal working hours.

Placing concrete will range from 0.5 to 1.5 labor-hours per C.Y. Bucks, blockouts, keyways, weldplates, etc. are extra.

## R031113-40 Forms for Reinforced Concrete

### Design Economy

Avoid many sizes in proportioning beams and columns.

From story to story avoid changing column dimensions. Gain strength by adding steel or using a richer mix. If a change in size of column is necessary, vary one dimension only to minimize form alterations. Keep beams and columns the same width.

From floor to floor in a multi-story building vary beam depth, not width, as that will leave slab panel form unchanged. It is cheaper to vary the strength of a beam from floor to floor by means of steel area than by 2″ changes in either width or depth.

### Cost Factors

Material includes the cost of lumber, cost of rent for metal pans or forms if used, nails, form ties, form oil, bolts and accessories.

Labor includes the cost of carpenters to make up, erect, remove and repair, plus common labor to clean and move. Having carpenters remove forms minimizes repairs.

Improper alignment and condition of forms will increase finishing cost. When forms are heavily oiled, concrete surfaces must be neutralized before finishing. Special curing compounds will cause spillages to spall off in first frost. Gang forming methods will reduce costs on large projects.

### Materials Used

Boards are seldom used unless their architectural finish is required. Generally, steel, fiberglass and plywood are used for contact surfaces. Labor on plywood is 10% less than with boards. The plywood is backed up with

2 x 4's at 12″ to 32″ O.C. Walers are generally 2 - 2 x 4's. Column forms are held together with steel yokes or bands. Shoring is with adjustable shoring or scaffolding for high ceilings.

### Reuse

Floor and column forms can be reused four or possibly five times without excessive repair. Remember to allow for 10% waste on each reuse.

When modular sized wall forms are made, up to twenty uses can be expected with exterior plyform.

When forms are reused, the cost to erect, strip, clean and move will not be affected. 10% replacement of lumber should be included and about one hour of carpenter time for repairs on each reuse per 100 S.F.

The reuse cost for certain accessory items normally rented on a monthly basis will be lower than the cost for the first use.

After fifth use, new material required plus time needed for repair prevent form cost from dropping further and it may go up. Much depends on care in stripping, the number of special bays, changes in beam or column sizes and other factors.

**Costs for multiple use of formwork may be developed as follows:**

| 2 Uses | 3 Uses | 4 Uses |
|---|---|---|
| $\dfrac{\text{(1st Use + Reuse)}}{2} = \text{avg. cost/2 uses}$ | $\dfrac{\text{(1st Use + 2 Reuse)}}{3} = \text{avg. cost/3 uses}$ | $\dfrac{\text{(1st use + 3 Reuse)}}{4} = \text{avg. cost/4 uses}$ |

### R031113-60  Formwork Labor-Hours

| Item | Unit | Hours Required | | | Total Hours | Multiple Use | | |
|---|---|---|---|---|---|---|---|---|
| | | Fabricate | Erect & Strip | Clean & Move | 1 Use | 2 Use | 3 Use | 4 Use |
| Beam and Girder, interior beams, 12" wide | 100 S.F. | 6.4 | 8.3 | 1.3 | 16.0 | 13.3 | 12.4 | 12.0 |
|   Hung from steel beams | | 5.8 | 7.7 | 1.3 | 14.8 | 12.4 | 11.6 | 11.2 |
|   Beam sides only, 36" high | | 5.8 | 7.2 | 1.3 | 14.3 | 11.9 | 11.1 | 10.7 |
|   Beam bottoms only, 24" wide | | 6.6 | 13.0 | 1.3 | 20.9 | 18.1 | 17.2 | 16.7 |
| Box out for openings | | 9.9 | 10.0 | 1.1 | 21.0 | 16.6 | 15.1 | 14.3 |
| Buttress forms, to 8' high | | 6.0 | 6.5 | 1.2 | 13.7 | 11.2 | 10.4 | 10.0 |
| Centering, steel, 3/4" rib lath | | | 1.0 | | 1.0 | | | |
|   3/8" rib lath or slab form | | | 0.9 | | 0.9 | | | |
| Chamfer strip or keyway | 100 L.F. | | 1.5 | | 1.5 | 1.5 | 1.5 | 1.5 |
| Columns, fiber tube 8" diameter | | | 20.6 | | 20.6 | | | |
|   12" | | | 21.3 | | 21.3 | | | |
|   16" | | | 22.9 | | 22.9 | | | |
|   20" | | | 23.7 | | 23.7 | | | |
|   24" | | | 24.6 | | 24.6 | | | |
|   30" | | | 25.6 | | 25.6 | | | |
| Columns, round steel, 12" diameter | | | 22.0 | | 22.0 | 22.0 | 22.0 | 22.0 |
|   16" | | | 25.6 | | 25.6 | 25.6 | 25.6 | 25.6 |
|   20" | | | 30.5 | | 30.5 | 30.5 | 30.5 | 30.5 |
|   24" | | | 37.7 | | 37.7 | 37.7 | 37.7 | 37.7 |
| Columns, plywood 8" x 8" | 100 S.F. | 7.0 | 11.0 | 1.2 | 19.2 | 16.2 | 15.2 | 14.7 |
|   12" x 12" | | 6.0 | 10.5 | 1.2 | 17.7 | 15.2 | 14.4 | 14.0 |
|   16" x 16" | | 5.9 | 10.0 | 1.2 | 17.1 | 14.7 | 13.8 | 13.4 |
|   24" x 24" | | 5.8 | 9.8 | 1.2 | 16.8 | 14.4 | 13.6 | 13.2 |
| Columns, steel framed plywood 8" x 8" | | | 10.0 | 1.0 | 11.0 | 11.0 | 11.0 | 11.0 |
|   12" x 12" | | | 9.3 | 1.0 | 10.3 | 10.3 | 10.3 | 10.3 |
|   16" x 16" | | | 8.5 | 1.0 | 9.5 | 9.5 | 9.5 | 9.5 |
|   24" x 24" | | | 7.8 | 1.0 | 8.8 | 8.8 | 8.8 | 8.8 |
| Drop head forms, plywood | | 9.0 | 12.5 | 1.5 | 23.0 | 19.0 | 17.7 | 17.0 |
| Coping forms | | 8.5 | 15.0 | 1.5 | 25.0 | 21.3 | 20.0 | 19.4 |
| Culvert, box | | | 14.5 | 4.3 | 18.8 | 18.8 | 18.8 | 18.8 |
| Curb forms, 6" to 12" high, on grade | | 5.0 | 8.5 | 1.2 | 14.7 | 12.7 | 12.1 | 11.7 |
|   On elevated slabs | | 6.0 | 10.8 | 1.2 | 18.0 | 15.5 | 14.7 | 14.3 |
| Edge forms to 6" high, on grade | 100 L.F. | 2.0 | 3.5 | 0.6 | 6.1 | 5.6 | 5.4 | 5.3 |
|   7" to 12" high | 100 S.F. | 2.5 | 5.0 | 1.0 | 8.5 | 7.8 | 7.5 | 7.4 |
| Equipment foundations | | 10.0 | 18.0 | 2.0 | 30.0 | 25.5 | 24.0 | 23.3 |
| Flat slabs, including drops | | 3.5 | 6.0 | 1.2 | 10.7 | 9.5 | 9.0 | 8.8 |
|   Hung from steel | | 3.0 | 5.5 | 1.2 | 9.7 | 8.7 | 8.4 | 8.2 |
|   Closed deck for domes | | 3.0 | 5.8 | 1.2 | 10.0 | 9.0 | 8.7 | 8.5 |
|   Open deck for pans | | 2.2 | 5.3 | 1.0 | 8.5 | 7.9 | 7.7 | 7.6 |
| Footings, continuous, 12" high | | 3.5 | 3.5 | 1.5 | 8.5 | 7.3 | 6.8 | 6.6 |
|   Spread, 12" high | | 4.7 | 4.2 | 1.6 | 10.5 | 8.7 | 8.0 | 7.7 |
|   Pile caps, square or rectangular | | 4.5 | 5.0 | 1.5 | 11.0 | 9.3 | 8.7 | 8.4 |
| Grade beams, 24" deep | | 2.5 | 5.3 | 1.2 | 9.0 | 8.3 | 8.0 | 7.9 |
| Lintel or Sill forms | | 8.0 | 17.0 | 2.0 | 27.0 | 23.5 | 22.3 | 21.8 |
| Spandrel beams, 12" wide | | 9.0 | 11.2 | 1.3 | 21.5 | 17.5 | 16.2 | 15.5 |
| Stairs | | | 25.0 | 4.0 | 29.0 | 29.0 | 29.0 | 29.0 |
| Trench forms in floor | | 4.5 | 14.0 | 1.5 | 20.0 | 18.3 | 17.7 | 17.4 |
| Walls, Plywood, at grade, to 8' high | | 5.0 | 6.5 | 1.5 | 13.0 | 11.0 | 9.7 | 9.5 |
|   8' to 16' | | 7.5 | 8.0 | 1.5 | 17.0 | 13.8 | 12.7 | 12.1 |
|   16' to 20' | | 9.0 | 10.0 | 1.5 | 20.5 | 16.5 | 15.2 | 14.5 |
| Foundation walls, to 8' high | | 4.5 | 6.5 | 1.0 | 12.0 | 10.3 | 9.7 | 9.4 |
|   8' to 16' high | | 5.5 | 7.5 | 1.0 | 14.0 | 11.8 | 11.0 | 10.6 |
| Retaining wall to 12' high, battered | | 6.0 | 8.5 | 1.5 | 16.0 | 13.5 | 12.7 | 12.3 |
| Radial walls to 12' high, smooth | | 8.0 | 9.5 | 2.0 | 19.5 | 16.0 | 14.8 | 14.3 |
|   2' chords | | 7.0 | 8.0 | 1.5 | 16.5 | 13.5 | 12.5 | 12.0 |
| Prefabricated modular, to 8' high | | — | 4.3 | 1.0 | 5.3 | 5.3 | 5.3 | 5.3 |
| Steel, to 8' high | | — | 6.8 | 1.2 | 8.0 | 8.0 | 8.0 | 8.0 |
|   8' to 16' high | | — | 9.1 | 1.5 | 10.6 | 10.3 | 10.2 | 10.2 |
| Steel framed plywood to 8' high | | — | 6.8 | 1.2 | 8.0 | 7.5 | 7.3 | 7.2 |
|   8' to 16' high | | — | 9.3 | 1.2 | 10.5 | 9.5 | 9.2 | 9.0 |

### R032110-10 Reinforcing Steel Weights and Measures

| Bar Designation No.** | Nominal Weight Lb./Ft. | U.S. Customary Units Nominal Dimensions* | | | SI Units Nominal Dimensions* | | | |
|---|---|---|---|---|---|---|---|---|
| | | Diameter in. | Cross Sectional Area, in.² | Perimeter in. | Nominal Weight kg/m | Diameter mm | Cross Sectional Area, cm² | Perimeter mm |
| 3 | .376 | .375 | .11 | 1.178 | .560 | 9.52 | .71 | 29.9 |
| 4 | .668 | .500 | .20 | 1.571 | .994 | 12.70 | 1.29 | 39.9 |
| 5 | 1.043 | .625 | .31 | 1.963 | 1.552 | 15.88 | 2.00 | 49.9 |
| 6 | 1.502 | .750 | .44 | 2.356 | 2.235 | 19.05 | 2.84 | 59.8 |
| 7 | 2.044 | .875 | .60 | 2.749 | 3.042 | 22.22 | 3.87 | 69.8 |
| 8 | 2.670 | 1.000 | .79 | 3.142 | 3.973 | 25.40 | 5.10 | 79.8 |
| 9 | 3.400 | 1.128 | 1.00 | 3.544 | 5.059 | 28.65 | 6.45 | 90.0 |
| 10 | 4.303 | 1.270 | 1.27 | 3.990 | 6.403 | 32.26 | 8.19 | 101.4 |
| 11 | 5.313 | 1.410 | 1.56 | 4.430 | 7.906 | 35.81 | 10.06 | 112.5 |
| 14 | 7.650 | 1.693 | 2.25 | 5.320 | 11.384 | 43.00 | 14.52 | 135.1 |
| 18 | 13.600 | 2.257 | 4.00 | 7.090 | 20.238 | 57.33 | 25.81 | 180.1 |

\* The nominal dimensions of a deformed bar are equivalent to those of a plain round bar having the same weight per foot as the deformed bar.

\*\* Bar numbers are based on the number of eighths of an inch included in the nominal diameter of the bars.

### R032110-20 Metric Rebar Specification - ASTM A615-81

| Grade 300 (300 MPa* = 43,560 psi; +8.7% vs. Grade 40) | | | | |
|---|---|---|---|---|
| Grade 400 (400 MPa* = 58,000 psi; −3.4% vs. Grade 60) | | | | |
| Bar No. | Diameter mm | Area mm² | Equivalent in.² | Comparison with U.S. Customary Bars |
| 10M | 11.3 | 100 | .16 | Between #3 & #4 |
| 15M | 16.0 | 200 | .31 | #5 (.31 in.²) |
| 20M | 19.5 | 300 | .47 | #6 (.44 in.²) |
| 25M | 25.2 | 500 | .78 | #8 (.79 in.²) |
| 30M | 29.9 | 700 | 1.09 | #9 (1.00 in.²) |
| 35M | 35.7 | 1000 | 1.55 | #11 (1.56 in.²) |
| 45M | 43.7 | 1500 | 2.33 | #14 (2.25 in.²) |
| 55M | 56.4 | 2500 | 3.88 | #18 (4.00 in.²) |

\* MPa = megapascals

## R032110-40  Weight of Steel Reinforcing Per Square Foot of Wall (PSF)

Reinforced Weights: The table below suggests the weights per square foot for reinforcing steel in walls. Weights are approximate and will be the same for all grades of steel bars. For bars in two directions, add weights for each size and spacing.

| C/C Spacing in Inches | #3 Wt. (PSF) | #4 Wt. (PSF) | #5 Wt. (PSF) | #6 Wt. (PSF) | #7 Wt. (PSF) | #8 Wt. (PSF) | #9 Wt. (PSF) | #10 Wt. (PSF) | #11 Wt. (PSF) |
|---|---|---|---|---|---|---|---|---|---|
| 2″ | 2.26 | 4.01 | 6.26 | 9.01 | 12.27 | | | | |
| 3″ | 1.50 | 2.67 | 4.17 | 6.01 | 8.18 | 10.68 | 13.60 | 17.21 | 21.25 |
| 4″ | 1.13 | 2.01 | 3.13 | 4.51 | 6.13 | 8.10 | 10.20 | 12.91 | 15.94 |
| 5″ | .90 | 1.60 | 2.50 | 3.60 | 4.91 | 6.41 | 8.16 | 10.33 | 12.75 |
| 6″ | .752 | 1.34 | 2.09 | 3.00 | 4.09 | 5.34 | 6.80 | 8.61 | 10.63 |
| 8″ | .564 | 1.00 | 1.57 | 2.25 | 3.07 | 4.01 | 5.10 | 6.46 | 7.97 |
| 10″ | .451 | .802 | 1.25 | 1.80 | 2.45 | 3.20 | 4.08 | 5.16 | 6.38 |
| 12″ | .376 | .668 | 1.04 | 1.50 | 2.04 | 2.67 | 3.40 | 4.30 | 5.31 |
| 18″ | .251 | .445 | .695 | 1.00 | 1.32 | 1.78 | 2.27 | 2.86 | 3.54 |
| 24″ | .188 | .334 | .522 | .751 | 1.02 | 1.34 | 1.70 | 2.15 | 2.66 |
| 30″ | .150 | .267 | .417 | .600 | .817 | 1.07 | 1.36 | 1.72 | 2.13 |
| 36″ | .125 | .223 | .348 | .501 | .681 | .890 | 1.13 | 1.43 | 1.77 |
| 42″ | .107 | .191 | .298 | .429 | .584 | .753 | .97 | 1.17 | 1.52 |
| 48″ | .094 | .167 | .261 | .376 | .511 | .668 | .85 | 1.08 | 1.33 |

## R032110-50  Minimum Wall Reinforcement Weight (PSF)

This table lists the approximate minimum wall reinforcement weights per S.F. according to the specification of .12% of gross area for vertical bars and .20% of gross area for horizontal bars.

| Location | Wall Thickness | Bar Size | Horizontal Steel Spacing C/C | Sq. In. Req'd per S.F. | Total Wt. per S.F. | Bar Size | Vertical Steel Spacing C/C | Sq. In. Req'd per S.F. | Total Wt. per S.F. | Horizontal & Vertical Steel Total Weight per S.F. |
|---|---|---|---|---|---|---|---|---|---|---|
| Both Faces | 10″ | #4 | 18″ | .24 | .89# | #3 | 18″ | .14 | .50# | 1.39# |
| | 12″ | #4 | 16″ | .29 | 1.00 | #3 | 16″ | .17 | .60 | 1.60 |
| | 14″ | #4 | 14″ | .34 | 1.14 | #3 | 13″ | .20 | .69 | 1.84 |
| | 16″ | #4 | 12″ | .38 | 1.34 | #3 | 11″ | .23 | .82 | 2.16 |
| | 18″ | #5 | 17″ | .43 | 1.47 | #4 | 18″ | .26 | .89 | 2.36 |
| One Face | 6″ | #3 | 9″ | .15 | .50 | #3 | 18″ | .09 | .25 | .75 |
| | 8″ | #4 | 12″ | .19 | .67 | #3 | 11″ | .12 | .41 | 1.08 |
| | 10″ | #5 | 15″ | .24 | .83 | #4 | 16″ | .14 | .50 | 1.34 |

## R032110-70  Bend, Place and Tie Reinforcing

Placing and tying by rodmen for footings and slabs runs from nine hrs. per ton for heavy bars to fifteen hrs. per ton for light bars. For beams, columns, and walls, production runs from eight hrs. per ton for heavy bars to twenty hrs. per ton for light bars. Overall average for typical reinforced concrete buildings is about fourteen hrs. per ton. These production figures include the time for placing of accessories and usual inserts, but not their material cost (allow 15% of the cost of delivered bent rods). Equipment handling is necessary for the larger-sized bars so that installation costs for the very heavy bars will not decrease proportionately.

Installation costs for splicing reinforcing bars include allowance for equipment to hold the bars in place while splicing as well as necessary scaffolding for iron workers.

## R032110-80  Shop-Fabricated Reinforcing Steel

The material prices for reinforcing, shown in the unit cost sections of the book, are for 50 tons or more of shop-fabricated reinforcing steel and include:
1. Mill base price of reinforcing steel
2. Mill grade/size/length extras
3. Mill delivery to the fabrication shop
4. Shop storage and handling
5. Shop drafting/detailing
6. Shop shearing and bending
7. Shop listing
8. Shop delivery to the job site

Both material and installation costs can be considerably higher for small jobs consisting primarily of smaller bars, while material costs may be slightly lower for larger jobs.

## Concrete | R0322 Welded Wire Fabric Reinforcing

### R032205-30  Common Stock Styles of Welded Wire Fabric

This table provides some of the basic specifications, sizes, and weights of welded wire fabric used for reinforcing concrete.

| | New Designation | Old Designation | | Steel Area per Foot | | | | Approximate Weight per 100 S.F. | |
| | Spacing — Cross Sectional Area (in.) — (Sq. in. 100) | Spacing — Wire Gauge (in.) — (AS & W) | | Longitudinal | | Transverse | | | |
| | | | | in. | cm | in. | cm | lbs | kg |
|---|---|---|---|---|---|---|---|---|---|
| Rolls | 6 x 6 — W1.4 x W1.4 | 6 x 6 — 10 x 10 | | .028 | .071 | .028 | .071 | 21 | 9.53 |
| | 6 x 6 — W2.0 x W2.0 | 6 x 6 — 8 x 8 | 1 | .040 | .102 | .040 | .102 | 29 | 13.15 |
| | 6 x 6 — W2.9 x W2.9 | 6 x 6 — 6 x 6 | | .058 | .147 | .058 | .147 | 42 | 19.05 |
| | 6 x 6 — W4.0 x W4.0 | 6 x 6 — 4 x 4 | | .080 | .203 | .080 | .203 | 58 | 26.91 |
| | 4 x 4 — W1.4 x W1.4 | 4 x 4 — 10 x 10 | | .042 | .107 | .042 | .107 | 31 | 14.06 |
| | 4 x 4 — W2.0 x W2.0 | 4 x 4 — 8 x 8 | 1 | .060 | .152 | .060 | .152 | 43 | 19.50 |
| | 4 x 4 — W2.9 x W2.9 | 4 x 4 — 6 x 6 | | .087 | .227 | .087 | .227 | 62 | 28.12 |
| | 4 x 4 — W4.0 x W4.0 | 4 x 4 — 4 x 4 | | .120 | .305 | .120 | .305 | 85 | 38.56 |
| Sheets | 6 x 6 — W2.9 x W2.9 | 6 x 6 — 6 x 6 | | .058 | .147 | .058 | .147 | 42 | 19.05 |
| | 6 x 6 — W4.0 x W4.0 | 6 x 6 — 4 x 4 | | .080 | .203 | .080 | .203 | 58 | 26.31 |
| | 6 x 6 — W5.5 x W5.5 | 6 x 6 — 2 x 2 | 2 | .110 | .279 | .110 | .279 | 80 | 36.29 |
| | 4 x 4 — W1.4 x W1.4 | 4 x 4 — 4 x 4 | | .120 | .305 | .120 | .305 | 85 | 38.56 |

NOTES: 1. Exact W—number size for 8 gauge is W2.1
2. Exact W—number size for 2 gauge is W5.4

## Concrete | R0330 Cast-In-Place Concrete

### R033053-50  Industrial Chimneys

Foundation requirements in C.Y. of concrete for various sized chimneys.

| Size Chimney | 2 Ton Soil | 3 Ton Soil | Size Chimney | 2 Ton Soil | 3 Ton Soil | Size Chimney | 2 Ton Soil | 3 Ton Soil |
|---|---|---|---|---|---|---|---|---|
| 75' x 3'-0" | 13  C.Y. | 11  C.Y. | 160' x 6'-6" | 86  C.Y. | 76  C.Y. | 300' x 10'-0" | 325  C.Y. | 245  C.Y. |
| 85' x 5'-6" | 19 | 16 | 175' x 7'-0" | 108 | 95 | 350' x 12'-0" | 422 | 320 |
| 100' x 5'-0" | 24 | 20 | 200' x 6'-0" | 125 | 105 | 400' x 14'-0" | 520 | 400 |
| 125' x 5'-6" | 43 | 36 | 250' x 8'-0" | 230 | 175 | 500' x 18'-0" | 725 | 575 |

## R033105-10 Proportionate Quantities

The tables below show both quantities per S.F. of floor areas as well as form and reinforcing quantities per C.Y. Unusual structural requirements would increase the ratios below. High strength reinforcing would reduce the steel weights. Figures are for 3000 psi concrete and 60,000 psi reinforcing unless specified otherwise.

| Type of Construction | Live Load | Span | Per S.F. of Floor Area | | | | Per C.Y. of Concrete | | |
|---|---|---|---|---|---|---|---|---|---|
| | | | Concrete | Forms | Reinf. | Pans | Forms | Reinf. | Pans |
| Flat Plate | 50 psf | 15 Ft. | .46 C.F. | 1.06 S.F. | 1.71lb. | | 62 S.F. | 101 lb. | |
| | | 20 | .63 | 1.02 | 2.40 | | 44 | 104 | |
| | | 25 | .79 | 1.02 | 3.03 | | 35 | 104 | |
| | 100 | 15 | .46 | 1.04 | 2.14 | | 61 | 126 | |
| | | 20 | .71 | 1.02 | 2.72 | | 39 | 104 | |
| | | 25 | .83 | 1.01 | 3.47 | | 33 | 113 | |
| Flat Plate (waffle construction) 20" domes | 50 | 20 | .43 | 1.00 | 2.10 | .84 S.F. | 63 | 135 | 53 S.F. |
| | | 25 | .52 | 1.00 | 2.90 | .89 | 52 | 150 | 46 |
| | | 30 | .64 | 1.00 | 3.70 | .87 | 42 | 155 | 37 |
| | 100 | 20 | .51 | 1.00 | 2.30 | .84 | 53 | 125 | 45 |
| | | 25 | .64 | 1.00 | 3.20 | .83 | 42 | 135 | 35 |
| | | 30 | .76 | 1.00 | 4.40 | .81 | 36 | 160 | 29 |
| Waffle Construction 30" domes | 50 | 25 | .69 | 1.06 | 1.83 | .68 | 42 | 72 | 40 |
| | | 30 | .74 | 1.06 | 2.39 | .69 | 39 | 87 | 39 |
| | | 35 | .86 | 1.05 | 2.71 | .69 | 33 | 85 | 39 |
| | | 40 | .78 | 1.00 | 4.80 | .68 | 35 | 165 | 40 |
| Flat Slab (two way with drop panels) | 50 | 20 | .62 | 1.03 | 2.34 | | 45 | 102 | |
| | | 25 | .77 | 1.03 | 2.99 | | 36 | 105 | |
| | | 30 | .95 | 1.03 | 4.09 | | 29 | 116 | |
| | 100 | 20 | .64 | 1.03 | 2.83 | | 43 | 119 | |
| | | 25 | .79 | 1.03 | 3.88 | | 35 | 133 | |
| | | 30 | .96 | 1.03 | 4.66 | | 29 | 131 | |
| | 200 | 20 | .73 | 1.03 | 3.03 | | 38 | 112 | |
| | | 25 | .86 | 1.03 | 4.23 | | 32 | 133 | |
| | | 30 | 1.06 | 1.03 | 5.30 | | 26 | 135 | |
| One Way Joists 20" Pans | 50 | 15 | .36 | 1.04 | 1.40 | .93 | 78 | 105 | 70 |
| | | 20 | .42 | 1.05 | 1.80 | .94 | 67 | 120 | 60 |
| | | 25 | .47 | 1.05 | 2.60 | .94 | 60 | 150 | 54 |
| | 100 | 15 | .38 | 1.07 | 1.90 | .93 | 77 | 140 | 66 |
| | | 20 | .44 | 1.08 | 2.40 | .94 | 67 | 150 | 58 |
| | | 25 | .52 | 1.07 | 3.50 | .94 | 55 | 185 | 49 |
| One Way Joists 8" x 16" filler blocks | 50 | 15 | .34 | 1.06 | 1.80 | .81 Ea. | 84 | 145 | 64 Ea. |
| | | 20 | .40 | 1.08 | 2.20 | .82 | 73 | 145 | 55 |
| | | 25 | .46 | 1.07 | 3.20 | .83 | 63 | 190 | 49 |
| | 100 | 15 | .39 | 1.07 | 1.90 | .81 | 74 | 130 | 56 |
| | | 20 | .46 | 1.09 | 2.80 | .82 | 64 | 160 | 48 |
| | | 25 | .53 | 1.10 | 3.60 | .83 | 56 | 190 | 42 |
| One Way Beam & Slab | 50 | 15 | .42 | 1.30 | 1.73 | | 84 | 111 | |
| | | 20 | .51 | 1.28 | 2.61 | | 68 | 138 | |
| | | 25 | .64 | 1.25 | 2.78 | | 53 | 117 | |
| | 100 | 15 | .42 | 1.30 | 1.90 | | 84 | 122 | |
| | | 20 | .54 | 1.35 | 2.69 | | 68 | 154 | |
| | | 25 | .69 | 1.37 | 3.93 | | 54 | 145 | |
| | 200 | 15 | .44 | 1.31 | 2.24 | | 80 | 137 | |
| | | 20 | .58 | 1.40 | 3.30 | | 65 | 163 | |
| | | 25 | .69 | 1.42 | 4.89 | | 53 | 183 | |
| Two Way Beam & Slab | 100 | 15 | .47 | 1.20 | 2.26 | | 69 | 130 | |
| | | 20 | .63 | 1.29 | 3.06 | | 55 | 131 | |
| | | 25 | .83 | 1.33 | 3.79 | | 43 | 123 | |
| | 200 | 15 | .49 | 1.25 | 2.70 | | 41 | 149 | |
| | | 20 | .66 | 1.32 | 4.04 | | 54 | 165 | |
| | | 25 | .88 | 1.32 | 6.08 | | 41 | 187 | |

## R033105-10 Proportionate Quantities (cont.)

| 4000 psi Concrete and 60,000 psi Reinforcing—Form and Reinforcing Quantities per C.Y. | | | | | |
|---|---|---|---|---|---|
| **Item** | **Size** | **Forms** | **Reinforcing** | **Minimum** | **Maximum** |
| Columns (square tied) | 10" x 10" | 130   S.F.C.A. | #5 to #11 | 220 lbs. | 875 lbs. |
| | 12" x 12" | 108 | #6 to #14 | 200 | 955 |
| | 14" x 14" | 92 | #7 to #14 | 190 | 900 |
| | 16" x 16" | 81 | #6 to #14 | 187 | 1082 |
| | 18" x 18" | 72 | #6 to #14 | 170 | 906 |
| | 20" x 20" | 65 | #7 to #18 | 150 | 1080 |
| | 22" x 22" | 59 | #8 to #18 | 153 | 902 |
| | 24" x 24" | 54 | #8 to #18 | 164 | 884 |
| | 26" x 26" | 50 | #9 to #18 | 169 | 994 |
| | 28" x 28" | 46 | #9 to #18 | 147 | 864 |
| | 30" x 30" | 43 | #10 to #18 | 146 | 983 |
| | 32" x 32" | 40 | #10 to #18 | 175 | 866 |
| | 34" x 34" | 38 | #10 to #18 | 157 | 772 |
| | 36" x 36" | 36 | #10 to #18 | 175 | 852 |
| | 38" x 38" | 34 | #10 to #18 | 158 | 765 |
| | 40" x 40" | 32 | #10 to #18 | 143 | 692 |

| **Item** | **Size** | **Form** | **Spiral** | **Reinforcing** | **Minimum** | **Maximum** |
|---|---|---|---|---|---|---|
| Columns (spirally reinforced) | 12" diameter | 34.5  L.F. | 190 lbs. | #4 to #11 | 165 lbs. | 1505 lb. |
| | | 34.5 | 190 | #14 & #18 | — | 1100 |
| | 14" | 25 | 170 | #4 to #11 | 150 | 970 |
| | | 25 | 170 | #14 & #18 | 800 | 1000 |
| | 16" | 19 | 160 | #4 to #11 | 160 | 950 |
| | | 19 | 160 | #14 & #18 | 605 | 1080 |
| | 18" | 15 | 150 | #4 to #11 | 160 | 915 |
| | | 15 | 150 | #14 & #18 | 480 | 1075 |
| | 20" | 12 | 130 | #4 to #11 | 155 | 865 |
| | | 12 | 130 | #14 & #18 | 385 | 1020 |
| | 22" | 10 | 125 | #4 to #11 | 165 | 775 |
| | | 10 | 125 | #14 & #18 | 320 | 995 |
| | 24" | 9 | 120 | #4 to #11 | 195 | 800 |
| | | 9 | 120 | #14 & #18 | 290 | 1150 |
| | 26" | 7.3 | 100 | #4 to #11 | 200 | 729 |
| | | 7.3 | 100 | #14 & #18 | 235 | 1035 |
| | 28" | 6.3 | 95 | #4 to #11 | 175 | 700 |
| | | 6.3 | 95 | #14 & #18 | 200 | 1075 |
| | 30" | 5.5 | 90 | #4 to #11 | 180 | 670 |
| | | 5.5 | 90 | #14 & #18 | 175 | 1015 |
| | 32" | 4.8 | 85 | #4 to #11 | 185 | 615 |
| | | 4.8 | 85 | #14 & #18 | 155 | 955 |
| | 34" | 4.3 | 80 | #4 to #11 | 180 | 600 |
| | | 4.3 | 80 | #14 & #18 | 170 | 855 |
| | 36" | 3.8 | 75 | #4 to #11 | 165 | 570 |
| | | 3.8 | 75 | #14 & #18 | 155 | 865 |
| | 40" | 3.0 | 70 | #4 to #11 | 165 | 500 |
| | | 3.0 | 70 | #14 & #18 | 145 | 765 |

## R033105-10  Proportionate Quantities (cont.)

| | | | | | | | |
|---|---|---|---|---|---|---|---|
| **3000 psi Concrete and 60,000 psi Reinforcing—Form and Reinforcing Quantities per C.Y.** | | | | | | | |
| Item | Type | Loading | Height | C.Y./L.F. | Forms/C.Y. | Reinf./C.Y. | |
| Retaining Walls | Cantilever | Level Backfill | 4 Ft. | 0.2 C.Y. | 49 S.F. | 35 lbs. | |
| | | | 8 | 0.5 | 42 | 45 | |
| | | | 12 | 0.8 | 35 | 70 | |
| | | | 16 | 1.1 | 32 | 85 | |
| | | | 20 | 1.6 | 28 | 105 | |
| | | Highway Surcharge | 4 | 0.3 | 41 | 35 | |
| | | | 8 | 0.5 | 36 | 55 | |
| | | | 12 | 0.8 | 33 | 90 | |
| | | | 16 | 1.2 | 30 | 120 | |
| | | | 20 | 1.7 | 27 | 155 | |
| | | Railroad Surcharge | 4 | 0.4 | 28 | 45 | |
| | | | 8 | 0.8 | 25 | 65 | |
| | | | 12 | 1.3 | 22 | 90 | |
| | | | 16 | 1.9 | 20 | 100 | |
| | | | 20 | 2.6 | 18 | 120 | |
| | Gravity, with Vertical Face | Level Backfill | 4 | 0.4 | 37 | None | |
| | | | 7 | 0.6 | 27 | | |
| | | | 10 | 1.2 | 20 | | |
| | | Sloping Backfill | 4 | 0.3 | 31 | | |
| | | | 7 | 0.8 | 21 | | |
| | | | 10 | 1.6 | 15 | ↓ | |

| | | Live Load in Kips per Linear Foot | | | | | | | |
|---|---|---|---|---|---|---|---|---|---|
| | Span | Under 1 Kip | | 2 to 3 Kips | | 4 to 5 Kips | | 6 to 7 Kips | |
| | | Forms | Reinf. | Forms | Reinf. | Forms | Reinf. | Forms | Reinf. |
| Beams | 10 Ft. | — | — | 90 S.F. | 170 # | 85 S.F. | 175 # | 75 S.F. | 185 # |
| | 16 | 130 S.F. | 165 # | 85 | 180 | 75 | 180 | 65 | 225 |
| | 20 | 110 | 170 | 75 | 185 | 62 | 200 | 51 | 200 |
| | 26 | 90 | 170 | 65 | 215 | 62 | 215 | — | — |
| | 30 | 85 | 175 | 60 | 200 | — | — | — | — |

| Item | Size | Type | Forms per C.Y. | Reinforcing per C.Y. |
|---|---|---|---|---|
| Spread Footings | Under 1 C.Y. | 1,000 psf soil | 24 S.F. | 44 lbs. |
| | | 5,000 | 24 | 42 |
| | | 10,000 | 24 | 52 |
| | 1 C.Y. to 5 C.Y. | 1,000 | 14 | 49 |
| | | 5,000 | 14 | 50 |
| | | 10,000 | 14 | 50 |
| | Over 5 C.Y. | 1,000 | 9 | 54 |
| | | 5,000 | 9 | 52 |
| | | 10,000 | 9 | 56 |
| Pile Caps (30 Ton Concrete Piles) | Under 5 C.Y. | shallow caps | 20 | 65 |
| | | medium | 20 | 50 |
| | | deep | 20 | 40 |
| | 5 C.Y. to 10 C.Y. | shallow | 14 | 55 |
| | | medium | 15 | 45 |
| | | deep | 15 | 40 |
| | 10 C.Y. to 20 C.Y. | shallow | 11 | 60 |
| | | medium | 11 | 45 |
| | | deep | 12 | 35 |
| | Over 20 C.Y. | shallow | 9 | 60 |
| | | medium | 9 | 45 |
| | | deep | 10 | 40 |

## R033105-10 Proportionate Quantities (cont.)

| 3000 psi Concrete and 60,000 psi Reinforcing — Form and Reinforcing Quantities per C.Y. | | | | | | |
|---|---|---|---|---|---|---|
| Item | Size | Pile Spacing | 50 T Pile | 100 T Pile | 50 T Pile | 100 T Pile |
| Pile Caps (Steel H Piles) | Under 5 C.Y. | 24" O.C. | 24 S.F. | 24 S.F. | 75 lbs. | 90 lbs. |
| | | 30" | 25 | 25 | 80 | 100 |
| | | 36" | 24 | 24 | 80 | 110 |
| | 5 C.Y. to 10 C.Y. | 24" | 15 | 15 | 80 | 110 |
| | | 30" | 15 | 15 | 85 | 110 |
| | | 36" | 15 | 15 | 75 | 90 |
| | Over 10 C.Y. | 24" | 13 | 13 | 85 | 90 |
| | | 30" | 11 | 11 | 85 | 95 |
| | | 36" | 10 | 10 | 85 | 90 |

| | | 8" Thick | | 10" Thick | | 12" Thick | | 15" Thick | |
|---|---|---|---|---|---|---|---|---|---|
| | Height | Forms | Reinf. | Forms | Reinf. | Forms | Reinf. | Forms | Reinf. |
| Basement Walls | 7 Ft. | 81 S.F. | 44 lbs. | 65 S.F. | 45 lbs. | 54 S.F. | 44 lbs. | 41 S.F. | 43 lbs. |
| | 8 | | 44 | | 45 | | 44 | | 43 |
| | 9 | | 46 | | 45 | | 44 | | 43 |
| | 10 | | 57 | | 45 | | 44 | | 43 |
| | 12 | | 83 | | 50 | | 52 | | 43 |
| | 14 | | 116 | | 65 | | 64 | | 51 |
| | 16 | | | | 86 | | 90 | | 65 |
| | 18 | ↓ | ↓ | ↓ | | ↓ | 106 | ↓ | 70 |

## R033105-20 Materials for One C.Y. of Concrete

This is an approximate method of figuring quantities of cement, sand and coarse aggregate for a field mix with waste allowance included.

With crushed gravel as coarse aggregate, to determine barrels of cement required, divide 10 by total mix; that is, for 1:2:4 mix, 10 divided by 7 = 1-3/7 barrels.

If the coarse aggregate is crushed stone, use 10-1/2 instead of 10 as given for gravel.

To determine tons of sand required, multiply barrels of cement by parts of sand and then by 0.2; that is, for the 1:2:4 mix, as above, 1-3/7 x 2 x .2 = .57 tons.

Tons of crushed gravel are in the same ratio to tons of sand as parts in the mix, or 4/2 x .57 = 1.14 tons.

| | | |
|---|---|---|
| 1 bag cement = 94# | 1 C.Y. sand or crushed gravel = 2700# | 1 C.Y. crushed stone = 2575# |
| 4 bags = 1 barrel | 1 ton sand or crushed gravel = 20 C.F. | 1 ton crushed stone = 21 C.F. |

Average carload of cement is 692 bags; of sand or gravel is 56 tons.

Do not stack stored cement over 10 bags high.

## R033105-30 Metric Equivalents of Cement Content for Concrete Mixes

| 94 Pound Bags per Cubic Yard | Kilograms per Cubic Meter | 94 Pound Bags per Cubic Yard | Kilograms per Cubic Meter |
|---|---|---|---|
| 1.0 | 55.77 | 7.0 | 390.4 |
| 1.5 | 83.65 | 7.5 | 418.3 |
| 2.0 | 111.5 | 8.0 | 446.2 |
| 2.5 | 139.4 | 8.5 | 474.0 |
| 3.0 | 167.3 | 9.0 | 501.9 |
| 3.5 | 195.2 | 9.5 | 529.8 |
| 4.0 | 223.1 | 10.0 | 557.7 |
| 4.5 | 251.0 | 10.5 | 585.6 |
| 5.0 | 278.8 | 11.0 | 613.5 |
| 5.5 | 306.7 | 11.5 | 641.3 |
| 6.0 | 334.6 | 12.0 | 669.2 |
| 6.5 | 362.5 | 12.5 | 697.1 |

a. If you know the cement content in pounds per cubic yard, multiply by .5933 to obtain kilograms per cubic meter.

b. If you know the cement content in 94 pound bags per cubic yard, multiply by 55.77 to obtain kilograms per cubic meter.

## R033105-40 Metric Equivalents of Common Concrete Strengths
### (to convert other psi values to megapascals, multiply by 0.006895)

| U.S. Values psi | SI Value Megapascals | Non-SI Metric Value kgf/cm²* |
|---|---|---|
| 2000 | 14 | 140 |
| 2500 | 17 | 175 |
| 3000 | 21 | 210 |
| 3500 | 24 | 245 |
| 4000 | 28 | 280 |
| 4500 | 31 | 315 |
| 5000 | 34 | 350 |
| 6000 | 41 | 420 |
| 7000 | 48 | 490 |
| 8000 | 55 | 560 |
| 9000 | 62 | 630 |
| 10,000 | 69 | 705 |

* kilograms force per square centimeter

## R033105-50 Quantities of Cement, Sand and Stone for One C.Y. of Concrete per Various Mixes

This table can be used to determine the quantities of the ingredients for smaller quantities of site mixed concrete.

| Concrete (C.Y.) | Mix = 1:1:1-3/4 | | | Mix = 1:2:2.25 | | | Mix = 1:2.25:3 | | | Mix = 1:3:4 | | |
|---|---|---|---|---|---|---|---|---|---|---|---|---|
| | Cement (sacks) | Sand (C.Y.) | Stone (C.Y.) | Cement (sacks) | Sand (C.Y.) | Stone (C.Y.) | Cement (sacks) | Sand (C.Y.) | Stone (C.Y.) | Cement (sacks) | Sand (C.Y.) | Stone (C.Y.) |
| 1 | 10 | .37 | .63 | 7.75 | .56 | .65 | 6.25 | .52 | .70 | 5 | .56 | .74 |
| 2 | 20 | .74 | 1.26 | 15.50 | 1.12 | 1.30 | 12.50 | 1.04 | 1.40 | 10 | 1.12 | 1.48 |
| 3 | 30 | 1.11 | 1.89 | 23.25 | 1.68 | 1.95 | 18.75 | 1.56 | 2.10 | 15 | 1.68 | 2.22 |
| 4 | 40 | 1.48 | 2.52 | 31.00 | 2.24 | 2.60 | 25.00 | 2.08 | 2.80 | 20 | 2.24 | 2.96 |
| 5 | 50 | 1.85 | 3.15 | 38.75 | 2.80 | 3.25 | 31.25 | 2.60 | 3.50 | 25 | 2.80 | 3.70 |
| 6 | 60 | 2.22 | 3.78 | 46.50 | 3.36 | 3.90 | 37.50 | 3.12 | 4.20 | 30 | 3.36 | 4.44 |
| 7 | 70 | 2.59 | 4.41 | 54.25 | 3.92 | 4.55 | 43.75 | 3.64 | 4.90 | 35 | 3.92 | 5.18 |
| 8 | 80 | 2.96 | 5.04 | 62.00 | 4.48 | 5.20 | 50.00 | 4.16 | 5.60 | 40 | 4.48 | 5.92 |
| 9 | 90 | 3.33 | 5.67 | 69.75 | 5.04 | 5.85 | 56.25 | 4.68 | 6.30 | 45 | 5.04 | 6.66 |
| 10 | 100 | 3.70 | 6.30 | 77.50 | 5.60 | 6.50 | 62.50 | 5.20 | 7.00 | 50 | 5.60 | 7.40 |
| 11 | 110 | 4.07 | 6.93 | 85.25 | 6.16 | 7.15 | 68.75 | 5.72 | 7.70 | 55 | 6.16 | 8.14 |
| 12 | 120 | 4.44 | 7.56 | 93.00 | 6.72 | 7.80 | 75.00 | 6.24 | 8.40 | 60 | 6.72 | 8.88 |
| 13 | 130 | 4.82 | 8.20 | 100.76 | 7.28 | 8.46 | 81.26 | 6.76 | 9.10 | 65 | 7.28 | 9.62 |
| 14 | 140 | 5.18 | 8.82 | 108.50 | 7.84 | 9.10 | 87.50 | 7.28 | 9.80 | 70 | 7.84 | 10.36 |
| 15 | 150 | 5.56 | 9.46 | 116.26 | 8.40 | 9.76 | 93.76 | 7.80 | 10.50 | 75 | 8.40 | 11.10 |
| 16 | 160 | 5.92 | 10.08 | 124.00 | 8.96 | 10.40 | 100.00 | 8.32 | 11.20 | 80 | 8.96 | 11.84 |
| 17 | 170 | 6.30 | 10.72 | 131.76 | 9.52 | 11.06 | 106.26 | 8.84 | 11.90 | 85 | 9.52 | 12.58 |
| 18 | 180 | 6.66 | 11.34 | 139.50 | 10.08 | 11.70 | 112.50 | 9.36 | 12.60 | 90 | 10.08 | 13.32 |
| 19 | 190 | 7.04 | 11.98 | 147.26 | 10.64 | 12.36 | 118.76 | 9.84 | 13.30 | 95 | 10.64 | 14.06 |
| 20 | 200 | 7.40 | 12.60 | 155.00 | 11.20 | 13.00 | 125.00 | 10.40 | 14.00 | 100 | 11.20 | 14.80 |
| 21 | 210 | 7.77 | 13.23 | 162.75 | 11.76 | 13.65 | 131.25 | 10.92 | 14.70 | 105 | 11.76 | 15.54 |
| 22 | 220 | 8.14 | 13.86 | 170.05 | 12.32 | 14.30 | 137.50 | 11.44 | 15.40 | 110 | 12.32 | 16.28 |
| 23 | 230 | 8.51 | 14.49 | 178.25 | 12.88 | 14.95 | 143.75 | 11.96 | 16.10 | 115 | 12.88 | 17.02 |
| 24 | 240 | 8.88 | 15.12 | 186.00 | 13.44 | 15.60 | 150.00 | 12.48 | 16.80 | 120 | 13.44 | 17.76 |
| 25 | 250 | 9.25 | 15.75 | 193.75 | 14.00 | 16.25 | 156.25 | 13.00 | 17.50 | 125 | 14.00 | 18.50 |
| 26 | 260 | 9.64 | 16.40 | 201.52 | 14.56 | 16.92 | 162.52 | 13.52 | 18.20 | 130 | 14.56 | 19.24 |
| 27 | 270 | 10.00 | 17.00 | 209.26 | 15.12 | 17.56 | 168.76 | 14.04 | 18.90 | 135 | 15.02 | 20.00 |
| 28 | 280 | 10.36 | 17.64 | 217.00 | 15.68 | 18.20 | 175.00 | 14.56 | 19.60 | 140 | 15.68 | 20.72 |
| 29 | 290 | 10.74 | 18.28 | 224.76 | 16.24 | 18.86 | 181.26 | 15.08 | 20.30 | 145 | 16.24 | 21.46 |

## R033105-65 Field-Mix Concrete

Presently most building jobs are built with ready-mixed concrete except at isolated locations and some larger jobs requiring over 10,000 C.Y. where land is readily available for setting up a temporary batch plant.

The most economical mix is a controlled mix using local aggregate proportioned by trial to give the required strength with the least cost of material.

## R033105-70 Placing Ready-Mixed Concrete

For ground pours allow for 5% waste when figuring quantities.

Prices in the front of the book assume normal deliveries. If deliveries are made before 8 A.M. or after 5 P.M. or on Saturday afternoons add 30%. Negotiated discounts for large volumes are not included in prices in front of book.

For the lower floors without truck access, concrete may be wheeled in rubber-tired buggies, conveyer handled, crane handled or pumped. Pumping is economical if there is top steel. Conveyers are more efficient for thick slabs.

At higher floors the rubber-tired buggies may be hoisted by a hoisting tower and wheeled to location. Placement by a conveyer is limited to three floors

and is best for high-volume pours. Pumped concrete is best when building has no crane access. Concrete may be pumped directly as high as thirty-six stories using special pumping techniques. Normal maximum height is about fifteen stories.

Best pumping aggregate is screened and graded bank gravel rather than crushed stone.

Pumping downward is more difficult than pumping upward. Horizontal distance from pump to pour may increase preparation time prior to pour. Placing by cranes, either mobile, climbing or tower types, continues as the most efficient method for high-rise concrete buildings.

## R033105-85 Lift Slabs

The cost advantage of the lift slab method is due to placing all concrete, reinforcing steel, inserts and electrical conduit at ground level and in reduction of formwork. Minimum economical project size is about 30,000 S.F. Slabs may be tilted for parking garage ramps.

It is now used in all types of buildings and has gone up to 22 stories high in apartment buildings. Current trend is to use post-tensioned flat plate slabs with spans from 22' to 35'. Cylindrical void forms are used when deep slabs are required. One pound of prestressing steel is about equal to seven pounds of conventional reinforcing.

To be considered cured for stressing and lifting, a slab must have attained 75% of design strength. Seven days are usually sufficient with four to five days possible if high early strength cement is used. Slabs can be stacked using two coats of a non-bonding agent to insure that slabs do not stick to each other. Lifting is done by companies specializing in this work. Lift rate is 5' to 15' per hour with an average of 10' per hour. Total areas up to 33,000 S.F. have been lifted at one time. 24 to 36 jacking columns are common. Most economical bay sizes are 24' to 28' with four to fourteen stories most efficient. Continuous design reduces reinforcing steel cost. Use of post-tensioned slabs allows larger bay sizes.

## R034105-30 Prestressed Precast Concrete Structural Units

| Type | Location | Depth | Span in Ft. | | Live Load Lb. per S.F. |
|---|---|---|---|---|---|
| Double Tee | Floor | 28" to 34" | 60 to 80 | | 50 to 80 |
| | Roof | 12" to 24" | 30 to 50 | | 40 |
| | Wall | Width 8' | Up to 55' high | | Wind |
| Multiple Tee | Roof | 8" to 12" | 15 to 40 | | 40 |
| | Floor | 8" to 12" | 15 to 30 | | 100 |
| Plank | Roof | | Roof | Floor | |
| | | 4" | 13 | 12 | 40 for Roof |
| | or | 6" | 22 | 18 | |
| or | | 8" | 26 | 25 | |
| | | 10" | 33 | 29 | 100 for Floor |
| | Floor | 12" | 42 | 32 | |
| Single Tee | Roof | 28" 32" 36" 48" | 40 80 100 120 | | 40 |
| AASHTO Girder | Bridges | Type 4 5 6 | 100 110 125 | | Highway |
| Box Beam | Bridges | 15" 27" 33" | 40 to 100 | | Highway |

The majority of precast projects today utilize double tees rather than single tees because of speed and ease of installation. As a result casting beds at manufacturing plants are normally formed for double tees. Single tee projects will therefore require an initial set up charge to be spread over the individual single tee costs.

For floors, a 2" to 3" topping is field cast over the shapes. For roofs, insulating concrete or rigid insulation is placed over the shapes.

Member lengths up to 40' are standard haul, 40' to 60' require special permits and lengths over 60' must be escorted. Over width and/or over length can add up to 100% on hauling costs.

Large heavy members may require two cranes for lifting which would increase erection costs by about 45%. An eight man crew can install 12 to 20 double tees, or 45 to 70 quad tees or planks per day.

Grouting of connections must also be included.

Several system buildings utilizing precast members are available. Heights can go up to 22 stories for apartment buildings. Optimum design ratio is 3 S.F. of surface to 1 S.F. of floor area.

## R034136-90　Prestressed Concrete, Post-Tensioned

In post-tensioned concrete the steel tendons are tensioned after the concrete has reached about 3/4 of its ultimate strength. The cableways are grouted after tensioning to provide bond between the steel and concrete. If bond is to be prevented, the tendons are coated with a corrosion-preventative grease and wrapped with waterproofed paper or plastic. Bonded tendons are usually used when ultimate strength (beams & girders) are controlling factors.

High strength concrete is used to fully utilize the steel, thereby reducing the size and weight of the member. A plasticizing agent may be added to reduce water content. Maximum size aggregate ranges from 1/2" to 1-1/2" depending on the spacing of the tendons.

The types of steel commonly used are bars and strands. Job conditions determine which is best suited. Bars are best for vertical prestresses since they are easy to support. The trend is for steel manufacturers to supply a finished package, cut to length, which reduces field preparation to a minimum.

Bars vary from 3/4" to 1-3/8" diameter. Table below gives time in labor-hours per tendon for placing, tensioning and grouting (if required) a 75' beam. Tendons used in buildings are not usually grouted; tendons for bridges usually are grouted. For strands the table indicates the labor-hours per pound for typical prestressed units 100' long. Simple span beams usually require one-end stressing regardless of lengths. Continuous beams are usually stressed from two ends. Long slabs are poured from the center outward and stressed in 75' increments after the initial 150' center pour.

| Labor Hours per Tendon and per Pound of Prestressed Steel | | | | | | |
|---|---|---|---|---|---|---|
| Length | 100' Beam | | 75' Beam | | 100' Slab | |
| Type Steel | Strand | | Bars | | Strand | |
| Diameter | 0.5" | | 3/4" | 1-3/8" | 0.5" | 0.6" |
| Number | 4 | 12 | 1 | 1 | 1 | 1 |
| Force in Kips | 100 | 300 | 42 | 143 | 25 | 35 |
| Preparation & Placing Cables | 3.6 | 7.4 | 0.9 | 2.9 | 0.9 | 1.1 |
| Stressing Cables | 2.0 | 2.4 | 0.8 | 1.6 | 0.5 | 0.5 |
| Grouting, if required | 2.5 | 3.0 | 0.6 | 1.3 | | |
| Total Labor Hours | 8.1 | 12.8 | 2.3 | 5.8 | 1.4 | 1.6 |
| Prestressing Steel Weights (Lbs.) | 215 | 640 | 115 | 380 | 53 | 74 |
| Labor-hours per Lb. Bonded | 0.038 | 0.020 | 0.020 | 0.015 | | |
| Non-bonded | | | | | 0.026 | 0.022 |

Flat Slab construction — 4000 psi concrete with span-to-depth ratio between 36 and 44. Two way post-tensioned steel averages 1.0 lb. per S.F. for 24' to 28' bays (usually strand) and additional reinforcing steel averages .5 lb. per S.F.

Pan and Joist construction — 4000 psi concrete with span-to-depth ratio 28 to 30. Post-tensioned steel averages .8 lb. per S.F. and reinforcing steel about 1.0 lb. per S.F. Placing and stressing averages 40 hours per ton of total material.

Beam construction — 4000 to 5000 psi concrete. Steel weights vary greatly.

Labor cost per pound goes down as the size and length of the tendon increase. The primary economic consideration is the cost per kip for the member.

Post-tensioning becomes feasible for beams and girders over 30' long; for continuous two-way slabs over 20' clear; also in transferring upper building loads over longer spans at lower levels. Post-tension suppliers will provide engineering services at no cost to the user. Substantial economies are possible by using post-tensioned Lift Slabs.

## R034513-10 Precast Concrete Wall Panels

Panels are either solid or insulated with plain, colored or textured finishes. Transportation is an important cost factor. Prices shown in the unit cost section of the book are based on delivery within 50 miles of a plant including fabricators' overhead and profit. Engineering data is available from fabricators to assist with construction details. Usual minimum job size for economical use of panels is about 5000 S.F. Small jobs can double the prices shown. For large, highly repetitive jobs, deduct up to 15% from the prices shown.

2" thick panels cost about the same as 3" thick panels, and maximum panel size is less. For building panels faced with granite, marble or stone, add the material prices from those unit cost sections to the plain panel price shown. There is a growing trend toward aggregate facings and broken rib finish rather than plain gray concrete panels.

No allowance has been made in the unit cost section for supporting steel framework. On one story buildings, panels may rest on grade beams and require only wind bracing and fasteners. On multi-story buildings panels can span from column to column and floor to floor. Plastic-designed steel-framed structures may have large deflections which slow down erection and raise costs.

Large panels are more economical than small panels on a S.F. basis. When figuring areas include all protrusions, returns, etc. Overhangs can triple erection costs. Panels over 45' have been produced. Larger flat units should be prestressed. Vacuum lifting of smooth finish panels eliminates inserts and can speed erection.

## R034713-20 Tilt Up Concrete Panels

The advantage of tilt up construction is in the low cost of forms and the placing of concrete and reinforcing. Panels up to 75' high and 5-1/2" thick have been tilted using strongbacks. Tilt up has been used for one to five story buildings and is well-suited for warehouses, stores, offices, schools and residences.

The panels are cast in forms on the floor slab. Most jobs use 5-1/2" thick solid reinforced concrete panels. Sandwich panels with a layer of insulating materials are also used. Where dampness is a factor, lightweight aggregate is used. Optimum panel size is 300 to 500 S.F.

Slabs are usually poured with 3000 psi concrete which permits tilting seven days after pouring. Slabs may be stacked on top of each other and are separated from each other by either two coats of bond breaker or a film of polyethylene. Use of high early-strength cement allows tilting two days after a pour. Tilting up is done with a roller outrigger crane with a capacity of at least 1-1/2 times the weight of the panel at the required reach. Exterior precast columns can be set at the same time as the panels; interior precast columns can be set first and the panels clipped directly to them. The use of cast-in-place concrete columns is diminishing due to shrinkage problems. Structural steel columns are sometimes used if crane rails are planned. Panels can be clipped to the columns or lowered between the flanges. Steel channels with anchors may be used as edge forms for the slab. When the panels are lifted the channels form an integral steel column to take structural loads. Roof loads can be carried directly by the panels for wall heights to 14'.

Requirements of local building codes may be a limiting factor and should be checked. Building floor slabs should be poured first and should be a minimum of 5" thick with 100% compaction of soil or 6" thick with less than 100% compaction.

Setting times as fast as nine minutes per panel have been observed, but a safer expectation would be four panels per hour with a crane and a four-man setting crew. If crane erects from inside building, some provision must be made to get crane out after walls are erected. Good yarding procedure is important to minimize delays. Equalizing three-point lifting beams and self-releasing pick-up hooks speed erection. If panels must be carried to their final location, setting time per panel will be increased and erection costs may approach the erection cost range of architectural precast wall panels. Placing panels into slots formed in continuous footers will speed erection.

Reinforcing should be with #5 bars with vertical bars on the bottom. If surface is to be sandblasted, stainless steel chairs should be used to prevent rust staining.

Use of a broom finish is popular since the unavoidable surface blemishes are concealed.

Precast columns run from three to five times the C.Y. price of the panels only.

## R035216-10 Lightweight Concrete

Lightweight aggregate concrete is usually purchased ready mixed, but it can also be field mixed.

Vermiculite or Perlite comes in bags of 4 C.F. under various trade names. Weight is about 8 lbs. per C.F. For insulating roof fill use 1:6 mix. For structural deck use 1:4 mix over gypsum boards, steeltex, steel centering, etc., supported by closely spaced joists or bulb trees. For structural slabs use 1:3:2 vermiculite sand concrete over steeltex, metal lath, steel centering, etc., on joists spaced 2'-0" O.C. for maximum L.L. of 80 P.S.F. Use same mix

for slab base fill over steel flooring or regular reinforced concrete slab when tile, terrazzo or other finish is to be laid over.

For slabs on grade use 1:3:2 mix when tile, etc., finish is to be laid over. If radiant heating units are installed use a 1:6 mix for a base. After coils are in place, cover with a regular granolithic finish (mix 1:3:2) to a minimum depth of 1-1/2" over top of units.

Reinforce all slabs with 6 x 6 or 10 x 10 welded wire mesh.

## R040130-10   Cleaning Face Brick

On smooth brick a person can clean 70 S.F. an hour; on rough brick 50 S.F. per hour. Use one gallon muriatic acid to 20 gallons of water for 1000

S.F. Do not use acid solution until wall is at least seven days old, but a mild soap solution may be used after two days.

Time has been allowed for clean-up in brick prices.

## R040513-10   Cement Mortar (material only)

Type N - 1:1:6 mix by volume. Use everywhere above grade except as noted below. - 1:3 mix using conventional masonry cement which saves handling two separate bagged materials.

Type M - 1:1/4:3 mix by volume, or 1 part cement, 1/4 (10% by wt.) lime, 3 parts sand. Use for heavy loads and where earthquakes or hurricanes may occur. Also for reinforced brick, sewers, manholes and everywhere below grade.

Mix Proportions by Volume and Compressive Strength of Mortar

| Where Used | Mortar Type | Allowable Proportions by Volume | | | | Compressive Strength @ 28 days |
|---|---|---|---|---|---|---|
| | | Portland Cement | Masonry Cement | Hydrated Lime | Masonry Sand | |
| Plain Masonry | M | 1 | 1 | — | 6 | |
| | | 1 | — | 1/4 | 3 | 2500 psi |
| | S | 1/2 | 1 | — | 4 | |
| | | 1 | — | 1/4 to 1/2 | 4 | 1800 psi |
| | N | — | 1 | — | 3 | |
| | | 1 | — | 1/2 to 1-1/4 | 6 | 750 psi |
| | O | — | 1 | — | 3 | |
| | | 1 | — | 1-1/4 to 2-1/2 | 9 | 350 psi |
| | K | 1 | — | 2-1/2 to 4 | 12 | 75 psi |
| Reinforced Masonry | PM | 1 | 1 | — | 6 | 2500 psi |
| | PL | 1 | — | 1/4 to 1/2 | 4 | 2500 psi |

Note: The total aggregate should be between 2.25 to 3 times the sum of the cement and lime used.

The labor cost to mix the mortar is included in the productivity and labor cost of unit price lines in unit cost sections for brickwork, blockwork and stonework.

The material cost of mixed mortar is included in the material cost of those same unit price lines and includes the cost of renting and operating a 10 C.F. mixer at the rate of 200 C.F. per day.

There are two types of mortar color used. One type is the inert additive type with about 100 lbs. per M brick as the typical quantity required. These colors are also available in smaller-batch-sized bags (1 lb. to 15 lb.) which can be placed directly into the mixer without measuring. The other type is premixed and replaces the masonry cement. Dark green color has the highest cost.

## R040519-50   Masonry Reinforcing

Horizontal joint reinforcing helps prevent wall cracks where wall movement may occur and in many locations is required by code. Horizontal joint reinforcing is generally not considered to be structural reinforcing and an unreinforced wall may still contain joint reinforcing.

Reinforcing strips come in 10′ and 12′ lengths and in truss and ladder shapes, with and without drips. Field labor runs between 2.7 to 5.3 hours per 1000 L.F. for wall thicknesses up to 12″.

The wire meets ASTM A82 for cold drawn steel wire and the typical size is 9 ga. sides and ties with 3/16″ diameter also available. Typical finish is mill galvanized with zinc coating at .10 oz. per S.F. Class I (.40 oz. per S.F.) and Class III (.80 oz per S.F.) are also available, as is hot dipped galvanizing at 1.50 oz. per S.F.

## R042110-10  Economy in Bricklaying

Have adequate supervision. Be sure bricklayers are always supplied with materials so there is no waiting. Place best bricklayers at corners and openings.

Use only screened sand for mortar. Otherwise, labor time will be wasted picking out pebbles. Use seamless metal tubs for mortar as they do not leak or catch the trowel. Locate stack and mortar for easy wheeling.

Have brick delivered for stacking. This makes for faster handling, reduces chipping and breakage, and requires less storage space. Many dealers will deliver select common in 2' x 3' x 4' pallets or face brick packaged. This affords quick handling with a crane or forklift and easy tonging in units of ten, which reduces waste.

Use wider bricks for one wythe wall construction. Keep scaffolding away from wall to allow mortar to fall clear and not stain wall.

On large jobs develop specialized crews for each type of masonry unit.

Consider designing for prefabricated panel construction on high rise projects.

Avoid excessive corners or openings. Each opening adds about 50% to labor cost for area of opening.

Bolting stone panels and using window frames as stops reduces labor costs and speeds up erection.

## R042110-20  Common and Face Brick

Common building brick manufactured according to ASTM C62 and facing brick manufactured according to ASTM C216 are the two standard bricks available for general building use.

Building brick is made in three grades; SW, where high resistance to damage caused by cyclic freezing is required; MW, where moderate resistance to cyclic freezing is needed; and NW, where little resistance to cyclic freezing is needed. Facing brick is made in only the two grades SW and MW. Additionally, facing brick is available in three types; FBS, for general use; FBX, for general use where a higher degree of precision and lower permissible variation in size than FBS is needed; and FBA, for general use to produce characteristic architectural effects resulting from non-uniformity in size and texture of the units.

In figuring the material cost of brickwork, an allowance of 25% mortar waste and 3% brick breakage was included. If bricks are delivered palletized with 280 to 300 per pallet, or packaged, allow only 1-1/2% for breakage. Packaged or palletized delivery is practical when a job is big enough to have a crane or other equipment available to handle a package of brick. This is so on all industrial work but not always true on small commercial buildings.

The use of buff and gray face is increasing, and there is a continuing trend to the Norman, Roman, Jumbo and SCR brick.

Common red clay brick for backup is not used that often. Concrete block is the most usual backup material with occasional use of sand lime or cement brick. Building brick is commonly used in solid walls for strength and as a fire stop.

Brick panels built on the ground and then crane erected to the upper floors have proven to be economical. This allows the work to be done under cover and without scaffolding.

## R042110-50  Brick, Block & Mortar Quantities

| Running Bond | | | | | | | For Other Bonds Standard Size Add to S.F. Quantities in Table to Left | | |
|---|---|---|---|---|---|---|---|---|---|
| Number of Brick per S.F. of Wall - Single Wythe with 3/8" Joints | | | | | C.F. of Mortar per M Bricks, Waste Included | | | | |
| Type Brick | Nominal Size (incl. mortar) L    H    W | Modular Coursing | Number of Brick per S.F. | | 3/8" Joint | 1/2" Joint | Bond Type | Description | Factor |
| Standard | 8 x 2-2/3 x 4 | 3C=8" | 6.75 | | 10.3 | 12.9 | Common | full header every fifth course | +20% |
| Economy | 8 x 4 x 4 | 1C=4" | 4.50 | | 11.4 | 14.6 | | full header every sixth course | +16.7% |
| Engineer | 8 x 3-1/5 x 4 | 5C=16" | 5.63 | | 10.6 | 13.6 | English | full header every second course | +50% |
| Fire | 9 x 2-1/2 x 4-1/2 | 2C=5" | 6.40 | | 550 # Fireclay | — | Flemish | alternate headers every course | +33.3% |
| Jumbo | 12 x 4 x 6 or 8 | 1C=4" | 3.00 | | 23.8 | 30.8 | | every sixth course | +5.6% |
| Norman | 12 x 2-2/3 x 4 | 3C=8" | 4.50 | | 14.0 | 17.9 | Header = W x H exposed | | +100% |
| Norwegian | 12 x 3-1/5 x 4 | 5C=16" | 3.75 | | 14.6 | 18.6 | Rowlock = H x W exposed | | +100% |
| Roman | 12 x 2 x 4 | 2C=4" | 6.00 | | 13.4 | 17.0 | Rowlock stretcher = L x W exposed | | +33.3% |
| SCR | 12 x 2-2/3 x 6 | 3C=8" | 4.50 | | 21.8 | 28.0 | Soldier = H x L exposed | | — |
| Utility | 12 x 4 x 4 | 1C=4" | 3.00 | | 15.4 | 19.6 | Sailor = W x L exposed | | -33.3% |

| Concrete Blocks Nominal Size | | Approximate Weight per S.F. | | Blocks per 100 S.F. | Mortar per M block, waste included | |
|---|---|---|---|---|---|---|
| | | Standard | Lightweight | | Partitions | Back up |
| 2" | x 8" x 16" | 20    PSF | 15   PSF | 113 | 27   C.F. | 36   C.F. |
| 4" | | 30 | 20 | | 41 | 51 |
| 6" | | 42 | 30 | | 56 | 66 |
| 8" | | 55 | 38 | | 72 | 82 |
| 10" | | 70 | 47 | | 87 | 97 |
| 12" | | 85 | 55 | | 102 | 112 |

## R042210-20 Concrete Block

The material cost of special block such as corner, jamb and head block can be figured at the same price as ordinary block of same size. Labor on specials is about the same as equal-sized regular block.

Bond beam and 16″ high lintel blocks are more expensive than regular units of equal size. Lintel blocks are 8″ long and either 8″ or 16″ high.

Use of motorized mortar spreader box will speed construction of continuous walls.

Hollow non-load-bearing units are made according to ASTM C129 and hollow load-bearing units according to ASTM C90.

## R050516-30 Coating Structural Steel

On field-welded jobs, the shop-applied primer coat is necessarily omitted. All painting must be done in the field and usually consists of red oxide rust inhibitive paint or an aluminum paint. The table below shows paint coverage and daily production for field painting.

See Division 09 97 13.23 for hot-dipped galvanizing and for field-applied cold galvanizing and other paints and protective coatings.

See Division 05 01 10.51 for steel surface preparation treatments such as wire brushing, pressure washing and sand blasting.

| Type Construction | Surface Area per Ton | Coat | One Gallon Covers | | In 8 Hrs. Person Covers | | Average per Ton Spray | |
|---|---|---|---|---|---|---|---|---|
| | | | Brush | Spray | Brush | Spray | Gallons | Labor-hours |
| Light Structural | 300 S.F. to 500 S.F. | 1st | 500 S.F. | 455 S.F. | 640 S.F. | 2000 S.F. | 0.9 gals. | 1.6 L.H. |
| | | 2nd | 450 | 410 | 800 | 2400 | 1.0 | 1.3 |
| | | 3rd | 450 | 410 | 960 | 3200 | 1.0 | 1.0 |
| Medium | 150 S.F. to 300 S.F. | All | 400 | 365 | 1600 | 3200 | 0.6 | 0.6 |
| Heavy Structural | 50 S.F. to 150 S.F. | 1st | 400 | 365 | 1920 | 4000 | 0.2 | 0.2 |
| | | 2nd | 400 | 365 | 2000 | 4000 | 0.2 | 0.2 |
| | | 3rd | 400 | 365 | 2000 | 4000 | 0.2 | 0.2 |
| Weighted Average | 225 S.F. | All | 400 | 365 | 1350 | 3000 | 0.6 | 0.6 |

## R050521-20 Welded Structural Steel

Usual weight reductions with welded design run 10% to 20% compared with bolted or riveted connections. This amounts to about the same total cost compared with bolted structures since field welding is more expensive than bolts. For normal spans of 18′ to 24′ figure 6 to 7 connections per ton.

Trusses — For welded trusses add 4% to weight of main members for connections. Up to 15% less steel can be expected in a welded truss compared to one that is shop bolted. Cost of erection is the same whether shop bolted or welded.

General — Typical electrodes for structural steel welding are E6010, E6011, E60T and E70T. Typical buildings vary between 2# to 8# of weld rod per

ton of steel. Buildings utilizing continuous design require about three times as much welding as conventional welded structures. In estimating field erection by welding, it is best to use the average linear feet of weld per ton to arrive at the welding cost per ton. The type, size and position of the weld will have a direct bearing on the cost per linear foot. A typical field welder will deposit 1.8# to 2# of weld rod per hour manually. Using semiautomatic methods can increase production by as much as 50% to 75%.

## R050523-10 High Strength Bolts

Common bolts (A307) are usually used in secondary connections (see Division 05 05 23.10).

High strength bolts (A325 and A490) are usually specified for primary connections such as column splices, beam and girder connections to columns, column bracing, connections for supports of operating equipment or of other live loads which produce impact or reversal of stress, and in structures carrying cranes of over 5-ton capacity.

Allow 20 field bolts per ton of steel for a 6 story office building, apartment house or light industrial building. For 6 to 12 stories allow 25 bolts per ton, and above 12 stories, 30 bolts per ton. On power stations, 20 to 25 bolts per ton are needed.

## R051223-10   Structural Steel

The bare material prices for structural steel, shown in the unit cost sections of the book, are for 100 tons of shop-fabricated structural steel and include:

1. Mill base price of structural steel
2. Mill scrap/grade/size/length extras
3. Mill delivery to a metals service center (warehouse)
4. Service center storage and handling
5. Service center delivery to a fabrication shop
6. Shop storage and handling
7. Shop drafting/detailing
8. Shop fabrication
9. Shop coat of primer paint
10. Shop listing
11. Shop delivery to the job site

In unit cost sections of the book that contain items for field fabrication of steel components, the bare material cost of steel includes:

1. Mill base price of structural steel
2. Mill scrap/grade/size/length extras
3. Mill delivery to a metals service center (warehouse)
4. Service center storage and handling
5. Service center delivery to the job site

## R051223-20   Steel Estimating Quantities

One estimate on erection is that a crane can handle 35 to 60 pieces per day. Say the average is 45. With usual sizes of beams, girders, and columns, this would amount to about 20 tons per day. The type of connection greatly affects the speed of erection. Moment connections for continuous design slow down production and increase erection costs.

Short open web bar joists can be set at the rate of 75 to 80 per day, with 50 per day being the average for setting long span joists.

After main members are calculated, add the following for usual allowances: base plates 2% to 3%; column splices 4% to 5%; and miscellaneous details 4% to 5%, for a total of 10% to 13% in addition to main members.

The ratio of column to beam tonnage varies depending on type of steels used, typical spans, story heights and live loads.

It is more economical to keep the column size constant and to vary the strength of the column by using high strength steels. This also saves floor space. Buildings have recently gone as high as ten stories with 8" high strength columns. For light columns under W8X31 lb. sections, concrete filled steel columns are economical.

High strength steels may be used in columns and beams to save floor space and to meet head room requirements. High strength steels in some sizes sometimes require long lead times.

Round, square and rectangular columns, both plain and concrete filled, are readily available and save floor area, but are higher in cost per pound than rolled columns. For high unbraced columns, tube columns may be less expensive.

Below are average minimum figures for the weights of the structural steel frame for different types of buildings using A36 steel, rolled shapes and simple joints. For economy in domes, rise to span ratio = .13. Open web joist framing systems will reduce weights by 10% to 40%. Composite design can reduce steel weight by up to 25% but additional concrete floor slab thickness may be required. Continuous design can reduce the weights up to 20%. There are many building codes with different live load requirements and different structural requirements, such as hurricane and earthquake loadings which can alter the figures.

| Structural Steel Weights per S.F. of Floor Area | | | | | | | | | |
|---|---|---|---|---|---|---|---|---|---|
| Type of Building | No. of Stories | Avg. Spans | L.L. #/S.F. | Lbs. Per S.F. | Type of Building | No. of Stories | Avg. Spans | L.L. #/S.F. | Lbs. Per S.F. |
| Steel Frame Mfg. | 1 | 20'x20'<br>30'x30'<br>40'x40' | 40 | 8<br>13<br>18 | Apartments | 2-8<br>9-25 | 20'x20' | 40 | 8<br>14 |
| Parking garage | 4 | Various | 80 | 8.5 | Office | to 10<br>20<br>30<br>over 50 | Various | 80 | 10<br>18<br>26<br>35 |
| Domes (Schwedler)* | 1 | 200'<br>300' | 30 | 10<br>15 | | | | | |

## R051223-25   Common Structural Steel Specifications

ASTM A992 (formerly A36, then A572 Grade 50) is the all-purpose carbon grade steel widely used in building and bridge construction.

The other high-strength steels listed below may each have certain advantages over ASTM A992 stuctural carbon steel, depending on the application. They have proven to be economical choices where, due to lighter members, the reduction of dead load and the associated savings in shipping cost can be significant.

ASTM A588 atmospheric weathering, high-strength low-alloy steels can be used in the bare (uncoated) condition, where exposure to normal atmosphere causes a tightly adherant oxide to form on the surface protecting the steel from further oxidation. ASTM A242 corrosion-resistant, high-strength low-alloy steels have enhanced atmospheric corrosion resistance of at least two times that of carbon structural steels with copper, or four times that of carbon structural steels without copper. The reduction or elimination of maintenance resulting from the use of these steels often offsets their higher initial cost.

| Steel Type | ASTM Designation | Minimum Yield Stress in KSI | Shapes Available |
|---|---|---|---|
| Carbon | A36 | 36 | All structural shape groups, and plates & bars up thru 8″ thick |
|  | A529 | 50 | Structural shape group 1, and plates & bars up thru 2″ thick |
| High-Strength Low-Alloy Quenched & Self-Tempered | A913 | 50 | All structural shape groups |
|  |  | 60 |  |
|  |  | 65 |  |
|  |  | 70 |  |
| High-Strength Low-Alloy Columbium-Vanadium | A572 | 42 | All structural shape groups, and plates & bars up thru 6″ thick |
|  |  | 50 | All structural shape groups, and plates & bars up thru 4″ thick |
|  |  | 55 | Structural shape groups 1 & 2, and plates & bars up thru 2″ thick |
|  |  | 60 | Structural shape groups 1 & 2, and plates & bars up thru 1-1/4″ thick |
|  |  | 65 | Structural shape group 1, and plates & bars up thru 1-1/4″ thick |
| High-Strength Low-Alloy Columbium-Vanadium | A992 | 50 | All structural shape groups |
| Weathering High-Strength Low-Alloy | A242 | 42 | Structural shape groups 4 & 5, and plates & bars over 1-1/2″ thru 4″ thick |
|  |  | 46 | Structural shape group 3, and plates & bars over 3/4″ up thru 1-1/2″ thick |
|  |  | 50 | Structural shape groups 1 & 2, and plates & bars up thru 3/4″ thick |
| Weathering High-Strength Low-Alloy | A588 | 42 | Plates & bars over 5″ up thru 8″ thick |
|  |  | 46 | Plates & bars over 4″ up thru 5″ thick |
|  |  | 50 | All structural shape groups, and plates & bars up thru 4″ thick |
| Quenched and Tempered Low-Alloy | A852 | 70 | Plates & bars up thru 4″ thick |
| Quenched and Tempered Alloy | A514 | 90 | Plates & bars over 2-1/2″ up thru 6″ thick |
|  |  | 100 | Plates & bars up thru 2-1/2″ thick |

## R051223-30   High Strength Steels

The mill price of high strength steels may be higher than A992 carbon steel but their proper use can achieve overall savings thru total reduced weights. For columns with L/r over 100, A992 steel is best; under 100, high strength steels are economical. For heavy columns, high strength steels are economical when cover plates are eliminated. There is no economy using high strength steels for clip angles or supports or for beams where deflection governs. Thinner members are more economical than thick.

The per ton erection and fabricating costs of the high strength steels will be higher than for A992 since the same number of pieces, but less weight, will be installed.

## R051223-35  Common Steel Sections

The upper portion of this table shows the name, shape, common designation and basic characteristics of commonly used steel sections. The lower portion explains how to read the designations used for the above illustrated common sections.

| Shape & Designation | Name & Characteristics | Shape & Designation | Name & Characteristics |
|---|---|---|---|
| W | W Shape — Parallel flange surfaces | MC | Miscellaneous Channel — Infrequently rolled by some producers |
| S | American Standard Beam (I Beam) — Sloped inner flange | L | Angle — Equal or unequal legs, constant thickness |
| M | Miscellaneous Beams — Cannot be classified as W, HP or S; infrequently rolled by some producers | T | Structural Tee — Cut from W, M or S on center of web |
| C | American Standard Channel — Sloped inner flange | HP | Bearing Pile — Parallel flanges and equal flange and web thickness |

Common drawing designations follow:

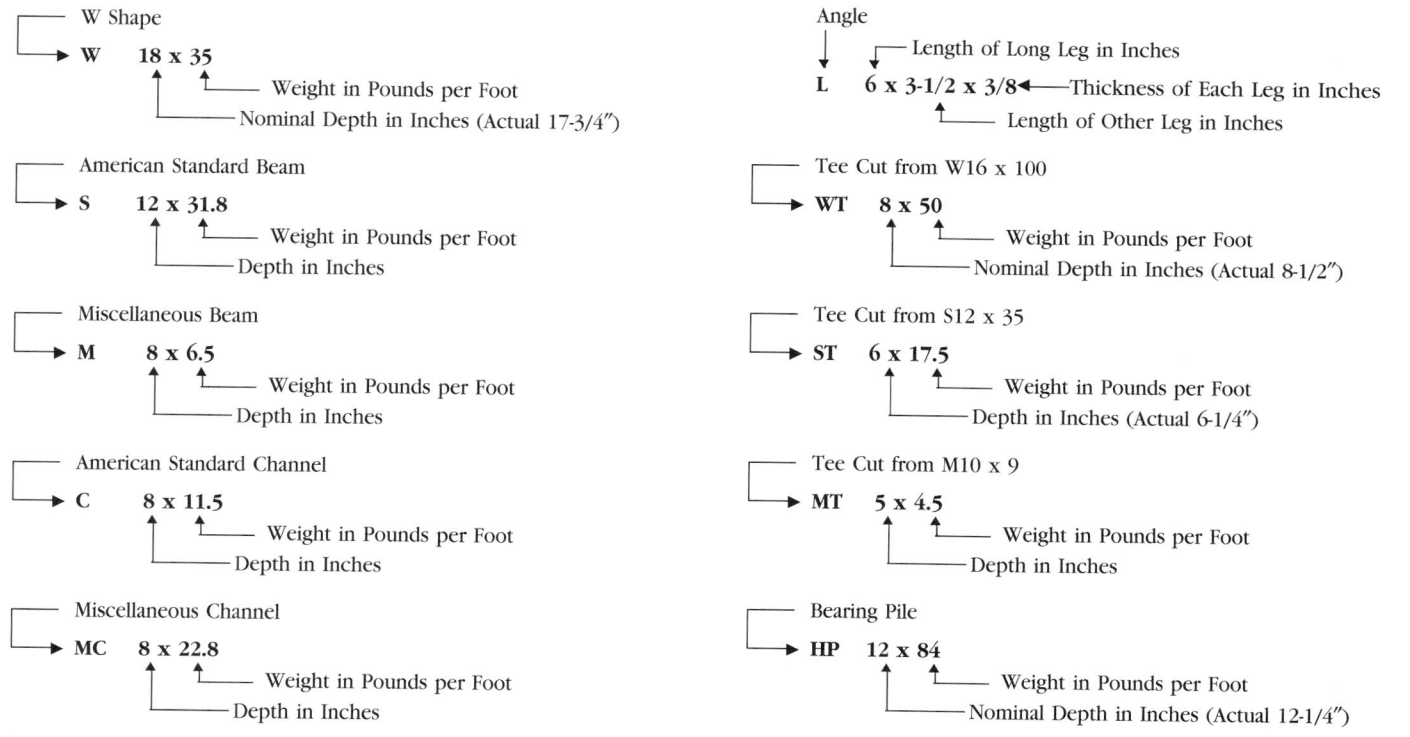

W Shape
W **18 x 35**
  Weight in Pounds per Foot
  Nominal Depth in Inches (Actual 17-3/4")

American Standard Beam
S **12 x 31.8**
  Weight in Pounds per Foot
  Depth in Inches

Miscellaneous Beam
M **8 x 6.5**
  Weight in Pounds per Foot
  Depth in Inches

American Standard Channel
C **8 x 11.5**
  Weight in Pounds per Foot
  Depth in Inches

Miscellaneous Channel
MC **8 x 22.8**
  Weight in Pounds per Foot
  Depth in Inches

Angle
Length of Long Leg in Inches
L **6 x 3-1/2 x 3/8** — Thickness of Each Leg in Inches
  Length of Other Leg in Inches

Tee Cut from W16 x 100
WT **8 x 50**
  Weight in Pounds per Foot
  Nominal Depth in Inches (Actual 8-1/2")

Tee Cut from S12 x 35
ST **6 x 17.5**
  Weight in Pounds per Foot
  Depth in Inches (Actual 6-1/4")

Tee Cut from M10 x 9
MT **5 x 4.5**
  Weight in Pounds per Foot
  Depth in Inches

Bearing Pile
HP **12 x 84**
  Weight in Pounds per Foot
  Nominal Depth in Inches (Actual 12-1/4")

## R051223-45    Installation Time for Structural Steel Building Components

The following tables show the expected average installation times for various structural steel shapes. Table A presents installation times for columns, Table B for beams, Table C for light framing and bolts, and Table D for structural steel for various project types.

### Table A

| Description | Labor-Hours | Unit |
|---|---|---|
| Columns | | |
|   Steel, Concrete Filled | | |
|     3-1/2″ Diameter | .933 | Ea. |
|     6-5/8″ Diameter | 1.120 | Ea. |
|   Steel Pipe | | |
|     3″ Diameter | .933 | Ea. |
|     8″ Diameter | 1.120 | Ea. |
|     12″ Diameter | 1.244 | Ea. |
|   Structural Tubing | | |
|     4″ x 4″ | .966 | Ea. |
|     8″ x 8″ | 1.120 | Ea. |
|     12″ x 8″ | 1.167 | Ea. |
|   W Shape 2 Tier | | |
|     W8 x 31 | .052 | L.F. |
|     W8 x 67 | .057 | L.F. |
|     W10 x 45 | .054 | L.F. |
|     W10 x 112 | .058 | L.F. |
|     W12 x 50 | .054 | L.F. |
|     W12 x 190 | .061 | L.F. |
|     W14 x 74 | .057 | L.F. |
|     W14 x 176 | .061 | L.F. |

### Table B

| Description | Labor-Hours | Unit | Labor-Hours | Unit |
|---|---|---|---|---|
| Beams, W Shape | | | | |
|   W6 x 9 | .949 | Ea. | .093 | L.F. |
|   W10 x 22 | 1.037 | Ea. | .085 | L.F. |
|   W12 x 26 | 1.037 | Ea. | .064 | L.F. |
|   W14 x 34 | 1.333 | Ea. | .069 | L.F. |
|   W16 x 31 | 1.333 | Ea. | .062 | L.F. |
|   W18 x 50 | 2.162 | Ea. | .088 | L.F. |
|   W21 x 62 | 2.222 | Ea. | .077 | L.F. |
|   W24 x 76 | 2.353 | Ea. | .072 | L.F. |
|   W27 x 94 | 2.581 | Ea. | .067 | L.F. |
|   W30 x 108 | 2.857 | Ea. | .067 | L.F. |
|   W33 x 130 | 3.200 | Ea. | .071 | L.F. |
|   W36 x 300 | 3.810 | Ea. | .077 | L.F. |

### Table C

| Description | Labor-Hours | Unit |
|---|---|---|
| Light Framing | | |
|   Angles 4″ and Larger | .055 | lbs. |
|   Less than 4″ | .091 | lbs. |
|   Channels 8″ and Larger | .048 | lbs. |
|   Less than 8″ | .072 | lbs. |
|   Cross Bracing Angles | .055 | lbs. |
|     Rods | .034 | lbs. |
|   Hanging Lintels | .069 | lbs. |
| High Strength Bolts in Place | | |
|   3/4″ Bolts | .070 | Ea. |
|   7/8″ Bolts | .076 | Ea. |

### Table D

| Description | Labor-Hours | Unit | Labor-Hours | Unit |
|---|---|---|---|---|
| Apartments, Nursing Homes, etc. | | | | |
|   1-2 Stories | 4.211 | Piece | 7.767 | Ton |
|   3-6 Stories | 4.444 | Piece | 7.921 | Ton |
|   7-15 Stories | 4.923 | Piece | 9.014 | Ton |
|   Over 15 Stories | 5.333 | Piece | 9.209 | Ton |
| Offices, Hospitals, etc. | | | | |
|   1-2 Stories | 4.211 | Piece | 7.767 | Ton |
|   3-6 Stories | 4.741 | Piece | 8.889 | Ton |
|   7-15 Stories | 4.923 | Piece | 9.014 | Ton |
|   Over 15 Stories | 5.120 | Piece | 9.209 | Ton |
| Industrial Buildings | | | | |
|   1 Story | 3.478 | Piece | 6.202 | Ton |

## R051223-50    Subpurlins

Bulb tee subpurlins are structural members designed to support and reinforce a variety of roof deck systems such as precast cement fiber roof deck tiles, monolithic roof deck systems, and gypsum or lightweight concrete over formboard. Other uses include interstitial service ceiling systems, wall panel systems, and joist anchoring in bond beams. See Unit Price section for pricing on a square foot basis at 32-5/8″ O.C. Maximum span is based on a 3-span condition with a total allowable vertical load of 40 psf.

## R051223-80    Dimensions and Weights of Sheet Steel

| Gauge No. | Approximate Thickness | | | | Weight | | |
|---|---|---|---|---|---|---|---|
| | Inches (in fractions) | Inches (in decimal parts) | | Millimeters | per S.F. in Ounces | per S.F. in Lbs. | per Square Meter in Kg. |
| | Wrought Iron | Wrought Iron | Steel | Steel | | | |
| 0000000 | 1/2" | .5 | .4782 | 12.146 | 320 | 20.000 | 97.650 |
| 000000 | 15/32" | .46875 | .4484 | 11.389 | 300 | 18.750 | 91.550 |
| 00000 | 7/16" | .4375 | .4185 | 10.630 | 280 | 17.500 | 85.440 |
| 0000 | 13/32" | .40625 | .3886 | 9.870 | 260 | 16.250 | 79.330 |
| 000 | 3/8" | .375 | .3587 | 9.111 | 240 | 15.000 | 73.240 |
| 00 | 11/32" | .34375 | .3288 | 8.352 | 220 | 13.750 | 67.130 |
| 0 | 5/16" | .3125 | .2989 | 7.592 | 200 | 12.500 | 61.030 |
| 1 | 9/32" | .28125 | .2690 | 6.833 | 180 | 11.250 | 54.930 |
| 2 | 17/64" | .265625 | .2541 | 6.454 | 170 | 10.625 | 51.880 |
| 3 | 1/4" | .25 | .2391 | 6.073 | 160 | 10.000 | 48.820 |
| 4 | 15/64" | .234375 | .2242 | 5.695 | 150 | 9.375 | 45.770 |
| 5 | 7/32" | .21875 | .2092 | 5.314 | 140 | 8.750 | 42.720 |
| 6 | 13/64" | .203125 | .1943 | 4.935 | 130 | 8.125 | 39.670 |
| 7 | 3/16" | .1875 | .1793 | 4.554 | 120 | 7.500 | 36.320 |
| 8 | 11/64" | .171875 | .1644 | 4.176 | 110 | 6.875 | 33.570 |
| 9 | 5/32" | .15625 | .1495 | 3.797 | 100 | 6.250 | 30.520 |
| 10 | 9/64" | .140625 | .1345 | 3.416 | 90 | 5.625 | 27.460 |
| 11 | 1/8" | .125 | .1196 | 3.038 | 80 | 5.000 | 24.410 |
| 12 | 7/64" | .109375 | .1046 | 2.657 | 70 | 4.375 | 21.360 |
| 13 | 3/32" | .09375 | .0897 | 2.278 | 60 | 3.750 | 18.310 |
| 14 | 5/64" | .078125 | .0747 | 1.897 | 50 | 3.125 | 15.260 |
| 15 | 9/128" | .0713125 | .0673 | 1.709 | 45 | 2.813 | 13.730 |
| 16 | 1/16" | .0625 | .0598 | 1.519 | 40 | 2.500 | 12.210 |
| 17 | 9/160" | .05625 | .0538 | 1.367 | 36 | 2.250 | 10.990 |
| 18 | 1/20" | .05 | .0478 | 1.214 | 32 | 2.000 | 9.765 |
| 19 | 7/160" | .04375 | .0418 | 1.062 | 28 | 1.750 | 8.544 |
| 20 | 3/80" | .0375 | .0359 | .912 | 24 | 1.500 | 7.324 |
| 21 | 11/320" | .034375 | .0329 | .836 | 22 | 1.375 | 6.713 |
| 22 | 1/32" | .03125 | .0299 | .759 | 20 | 1.250 | 6.103 |
| 23 | 9/320" | .028125 | .0269 | .683 | 18 | 1.125 | 5.490 |
| 24 | 1/40" | .025 | .0239 | .607 | 16 | 1.000 | 4.882 |
| 25 | 7/320" | .021875 | .0209 | .531 | 14 | .875 | 4.272 |
| 26 | 3/160" | .01875 | .0179 | .455 | 12 | .750 | 3.662 |
| 27 | 11/640" | .0171875 | .0164 | .417 | 11 | .688 | 3.357 |
| 28 | 1/64" | .015625 | .0149 | .378 | 10 | .625 | 3.052 |

## R053100-10 Decking Descriptions

### General - All Deck Products

Steel deck is made by cold forming structural grade sheet steel into a repeating pattern of parallel ribs. The strength and stiffness of the panels are the result of the ribs and the material properties of the steel. Deck lengths can be varied to suit job conditions, but because of shipping considerations, are usually less than 40 feet. Standard deck width varies with the product used but full sheets are usually 12″, 18″, 24″, 30″, or 36″. Deck is typically furnished in a standard width with the ends cut square. Any cutting for width, such as at openings or for angular fit, is done at the job site.

Deck is typically attached to the building frame with arc puddle welds, self-drilling screws, or powder or pneumatically driven pins. Sheet to sheet fastening is done with screws, button punching (crimping), or welds.

### Composite Floor Deck

After installation and adequate fastening, floor deck serves several purposes. It (a) acts as a working platform, (b) stabilizes the frame, (c) serves as a concrete form for the slab, and (d) reinforces the slab to carry the design loads applied during the life of the building. Composite decks are distinguished by the presence of shear connector devices as part of the deck. These devices are designed to mechanically lock the concrete and deck together so that the concrete and the deck work together to carry subsequent floor loads. These shear connector devices can be rolled-in embossments, lugs, holes, or wires welded to the panels. The deck profile can also be used to interlock concrete and steel.

Composite deck finishes are either galvanized (zinc coated) or phosphatized/painted. Galvanized deck has a zinc coating on both the top and bottom surfaces. The phosphatized/painted deck has a bare (phosphatized) top surface that will come into contact with the concrete. This bare top surface can be expected to develop rust before the concrete is placed. The bottom side of the deck has a primer coat of paint.

Composite floor deck is normally installed so the panel ends do not overlap on the supporting beams. Shear lugs or panel profile shape often prevent a tight metal to metal fit if the panel ends overlap; the air gap caused by overlapping will prevent proper fusion with the structural steel supports when the panel end laps are shear stud welded.

Adequate end bearing of the deck must be obtained as shown on the drawings. If bearing is actually less in the field than shown on the drawings, further investigation is required.

### Roof Deck

Roof deck is not designed to act compositely with other materials. Roof deck acts alone in transferring horizontal and vertical loads into the building frame. Roof deck rib openings are usually narrower than floor deck rib openings. This provides adequate support of rigid thermal insulation board.

Roof deck is typically installed to endlap approximately 2″ over supports. However, it can be butted (or lapped more than 2″) to solve field fit problems. Since designers frequently use the installed deck system as part of the horizontal bracing system (the deck as a diaphragm), any fastening substitution or change should be approved by the designer. Continuous perimeter support of the deck is necessary to limit edge deflection in the finished roof and may be required for diaphragm shear transfer.

Standard roof deck finishes are galvanized or primer painted. The standard factory applied paint for roof deck is a primer paint and is not intended to weather for extended periods of time. Field painting or touching up of abrasions and deterioration of the primer coat or other protective finishes is the responsibility of the contractor.

### Cellular Deck

Cellular deck is made by attaching a bottom steel sheet to a roof deck or composite floor deck panel. Cellular deck can be used in the same manner as floor deck. Electrical, telephone, and data wires are easily run through the chase created between the deck panel and the bottom sheet.

When used as part of the electrical distribution system, the cellular deck must be installed so that the ribs line up and create a smooth cell transition at abutting ends. The joint that occurs at butting cell ends must be taped or otherwise sealed to prevent wet concrete from seeping into the cell. Cell interiors must be free of welding burrs, or other sharp intrusions, to prevent damage to wires.

When used as a roof deck, the bottom flat plate is usually left exposed to view. Care must be maintained during erection to keep good alignment and prevent damage.

Cellular deck is sometimes used with the flat plate on the top side to provide a flat working surface. Installation of the deck for this purpose requires special methods for attachment to the frame because the flat plate, now on the top, can prevent direct access to the deck material that is bearing on the structural steel. It may be advisable to treat the flat top surface to prevent slipping.

Cellular deck is always furnished galvanized or painted over galvanized.

### Form Deck

Form deck can be any floor or roof deck product used as a concrete form. Connections to the frame are by the same methods used to anchor floor and roof deck. Welding washers are recommended when welding deck that is less than 20 gauge thickness.

Form deck is furnished galvanized, prime painted, or uncoated. Galvanized deck must be used for those roof deck systems where form deck is used to carry a lightweight insulating concrete fill.

## Wood, Plastics & Comp.  R0611 Wood Framing

### R061110-30  Lumber Product Material Prices

The price of forest products fluctuates widely from location to location and from season to season depending upon economic conditions. The bare material prices in the unit cost sections of the book show the National Average material prices in effect Jan. 1 of this book year. It must be noted that lumber prices in general may change significantly during the year.

Availability of certain items depends upon geographic location and must be checked prior to firm-price bidding.

## Wood, Plastics & Comp.  R0616 Sheathing

### R061636-20  Plywood

There are two types of plywood used in construction: interior, which is moisture-resistant but not waterproofed, and exterior, which is waterproofed.

The grade of the exterior surface of the plywood sheets is designated by the first letter: A, for smooth surface with patches allowed; B, for solid surface with patches and plugs allowed; C, which may be surface plugged or may have knot holes up to 1″ wide; and D, which is used only for interior type plywood and may have knot holes up to 2-1/2″ wide. "Structural Grade" is specifically designed for engineered applications such as box beams. All CC & DD grades have roof and floor spans marked on them.

Underlayment-grade plywood runs from 1/4″ to 1-1/4″ thick. Thicknesses 5/8″ and over have optional tongue and groove joints which eliminate the need for blocking the edges. Underlayment 19/32″ and over may be referred to as Sturd-i-Floor.

The price of plywood can fluctuate widely due to geographic and economic conditions.

Typical uses for various plywood grades are as follows:

AA-AD Interior — cupboards, shelving, paneling, furniture

BB Plyform — concrete form plywood

CDX — wall and roof sheathing

Structural — box beams, girders, stressed skin panels

AA-AC Exterior — fences, signs, siding, soffits, etc.

Underlayment — base for resilient floor coverings

Overlaid HDO — high density for concrete forms & highway signs

Overlaid MDO — medium density for painting, siding, soffits & signs

303 Siding — exterior siding, textured, striated, embossed, etc.

## Thermal & Moist. Protec.  R0731 Shingles & Shakes

### R073126-20  Roof Slate

16″, 18″ and 20″ are standard lengths, and slate usually comes in random widths. For standard 3/16″ thickness use 1-1/2″ copper nails. Allow for 3% breakage.

## Thermal & Moist. Protec.  R0751 Built-Up Bituminous Roofing

### R075113-20  Built-Up Roofing

Asphalt is available in kegs of 100 lbs. each; coal tar pitch in 560 lb. kegs. Prepared roofing felts are available in a wide range of sizes, weights and characteristics. However, the most commonly used are #15 (432 S.F. per roll, 13 lbs. per square) and #30 (216 S.F. per roll, 27 lbs. per square).

Inter-ply bitumen varies from 24 lbs. per sq. (asphalt) to 30 lbs. per sq. (coal tar) per ply, MF4@ 25%. Flood coat bitumen also varies from 60 lbs. per sq. (asphalt) to 75 lbs. per sq. (coal tar), MF4@ 25%. Expendable equipment (mops, brooms, screeds, etc.) runs about 16% of the bitumen cost. For new, inexperienced crews this factor may be much higher.

Rigid insulation board is typically applied in two layers. The first is mechanically attached to nailable decks or spot or solid mopped to non-nailable decks; the second layer is then spot or solid mopped to the first layer. Membrane application follows the insulation, except in protected membrane roofs, where the membrane goes down first and the insulation on top, followed with ballast (stone or concrete pavers). Insulation and related labor costs are NOT included in prices for built-up roofing.

## R075213-30 Modified Bitumen Roofing

The cost of modified bitumen roofing is highly dependent on the type of installation that is planned. Installation is based on the type of modifier used in the bitumen. The two most popular modifiers are atactic polypropylene (APP) and styrene butadiene styrene (SBS). The modifiers are added to heated bitumen during the manufacturing process to change its characteristics. A polyethylene, polyester or fiberglass reinforcing sheet is then sandwiched between layers of this bitumen. When completed, the result is a pre-assembled, built-up roof that has increased elasticity and weatherablility. Some manufacturers include a surfacing material such as ceramic or mineral granules, metal particles or sand.

The preferred method of adhering SBS-modified bitumen roofing to the substrate is with hot-mopped asphalt (much the same as built-up roofing). This installation method requires a tar kettle/pot to heat the asphalt, as well as the labor, tools and equipment necessary to distribute and spread the hot asphalt.

The alternative method for applying APP and SBS modified bitumen is as follows. A skilled installer uses a torch to melt a small pool of bitumen off the membrane. This pool must form across the entire roll for proper adhesion. The installer must unroll the roofing at a pace slow enough to melt the bitumen, but fast enough to prevent damage to the rest of the membrane.

Modified bitumen roofing provides the advantages of both built-up and single-ply roofing. Labor costs are reduced over those of built-up roofing because only a single ply is necessary. The elasticity of single-ply roofing is attained with the reinforcing sheet and polymer modifiers. Modifieds have some self-healing characteristics and because of their multi-layer construction, they offer the reliability and safety of built-up roofing.

## R078413-30 Firestopping

Firestopping is the sealing of structural, mechanical, electrical and other penetrations through fire-rated assemblies. The basic components of firestop systems are safing insulation and firestop sealant on both sides of wall penetrations and the top side of floor penetrations.

Pipe penetrations are assumed to be through concrete, grout, or joint compound and can be sleeved or unsleeved. Costs for the penetrations and sleeves are not included. An annular space of 1″ is assumed. Escutcheons are not included.

Metallic pipe is assumed to be copper, aluminum, cast iron or similar metallic material. Insulated metallic pipe is assumed to be covered with a thermal insulating jacket of varying thickness and materials.

Non-metallic pipe is assumed to be PVC, CPVC, FR Polypropylene or similar plastic piping material. Intumescent firestop sealant or wrap strips are included. Collars on both sides of wall penetrations and a sheet metal plate on the underside of floor penetrations are included.

Ductwork is assumed to be sheet metal, stainless steel or similar metallic material. Duct penetrations are assumed to be through concrete, grout or joint compound. Costs for penetrations and sleeves are not included. An annular space of 1/2″ is assumed.

Multi-trade openings include costs for sheet metal forms, firestop mortar, wrap strips, collars and sealants as necessary.

Structural penetrations joints are assumed to be 1/2″ or less. CMU walls are assumed to be within 1-1/2″ of metal deck. Drywall walls are assumed to be tight to the underside of metal decking.

Metal panel, glass or curtain wall systems include a spandrel area of 5′ filled with mineral wool foil-faced insulation. Fasteners and stiffeners are included.

## R081313-20    Steel Door Selection Guide

Standard steel doors are classified into four levels, as recommended by the Steel Door Institute in the chart below. Each of the four levels offers a range of construction models and designs, to meet architectural requirements for preference and appearance, including full flush, seamless, and stile & rail. Recommended minimum gauge requirements are also included.

For complete standard steel door construction specifications and available sizes, refer to the Steel Door Institute Technical Data Series, ANSI A250.8-98 (SDI-100), and ANSI A250.4-94 Test Procedure and Acceptance Criteria for Physical Endurance of Steel Door and Hardware Reinforcements.

| Level | | Model | Construction | For Full Flush or Seamless | | |
| --- | --- | --- | --- | --- | --- | --- |
| | | | | Min. Gauge | Thickness (in) | Thickness (mm) |
| I | Standard Duty | 1 | Full Flush | | | |
| | | 2 | Seamless | 20 | 0.032 | 0.8 |
| II | Heavy Duty | 1 | Full Flush | | | |
| | | 2 | Seamless | 18 | 0.042 | 1.0 |
| III | Extra Heavy Duty | 1 | Full Flush | | | |
| | | 2 | Seamless | | | |
| | | 3 | *Stile & Rail | 16 | 0.053 | 1.3 |
| IV | Maximum Duty | 1 | Full Flush | | | |
| | | 2 | Seamless | 14 | 0.067 | 1.6 |

*Stiles & rails are 16 gauge; flush panels, when specified, are 18 gauge

## R085123-10    Steel Sash

Ironworker crew will erect 25 S.F. or 1.3 sash unit per hour, whichever is less.

Mechanic will point 30 L.F. per hour.

Painter will paint 90 S.F. per coat per hour.

Glazier production depends on light size.

Allow 1 lb. special steel sash putty per 16" x 20" light.

## R085216-10    Window Estimates

To ensure a complete window estimate, be sure to include the material and labor costs for each window, as well as the material and labor costs for an interior wood trim set.

### R087120-10   Hinges

All closer equipped doors should have ball bearing hinges. Lead lined or extremely heavy doors require special strength hinges. Usually 1-1/2 pair of hinges are used per door up to 7'-6" high openings. Table below shows typical hinge requirements.

| Use Frequency | Type Hinge Required | Type of Opening | Type of Structure |
|---|---|---|---|
| High | Heavy weight | Entrances | Banks, Office buildings, Schools, Stores & Theaters |
|  | ball bearing | Toilet Rooms | Office buildings and Schools |
| Average | Standard | Entrances | Dwellings |
|  | weight | Corridors | Office buildings and Schools |
|  | ball bearing | Toilet Rooms | Stores |
| Low | Plain bearing | Interior | Dwellings |

| Door Thickness | Weight of Doors in Pounds per Square Foot | | | | |
|---|---|---|---|---|---|
|  | White Pine | Oak | Hollow Core | Solid Core | Hollow Metal |
| 1-3/8" | 3psf | 6psf | 1-1/2psf | 3-1/2 — 4psf | 6-1/2psf |
| 1-3/4" | 3-1/2 | 7 | 2-1/2 | 4-1/2 — 5-1/4 | 6-1/2 |
| 2-1/4" | 4-1/2 | 9 | — | 5-1/2 — 6-3/4 | 6-1/2 |

### R088110-10   Glazing Productivity

Some glass sizes are estimated by the "united inch" (height + width). The table below shows the number of lights glazed in an eight-hour period by the crew size indicated, for glass up to 1/4" thick. Square or nearly square lights are more economical on a S.F. basis. Long slender lights will have a high S.F. installation cost. For insulated glass reduce production by 33%. For 1/2" float glass reduce production by 50%. Production time for glazing with two glaziers per day averages: 1/4" float glass 120 S.F.; 1/2" float glass 55 S.F.; 1/2" insulated glass 95 S.F.; 3/4" insulated glass 75 S.F.

| Glazing Method | United Inches per Light | | | | | | | |
|---|---|---|---|---|---|---|---|---|
|  | 40" | 60" | 80" | 100" | 135" | 165" | 200" | 240" |
| Number of Men in Crew | 1 | 1 | 1 | 1 | 2 | 3 | 3 | 4 |
| Industrial sash, putty | 60 | 45 | 24 | 15 | 18 | — | — | — |
| With stops, putty bed | 50 | 36 | 21 | 12 | 16 | 8 | 4 | 3 |
| Wood stops, rubber | 40 | 27 | 15 | 9 | 11 | 6 | 3 | 2 |
| Metal stops, rubber | 30 | 24 | 14 | 9 | 9 | 6 | 3 | 2 |
| Structural glass | 10 | 7 | 4 | 3 | — | — | — | — |
| Corrugated glass | 12 | 9 | 7 | 4 | 4 | 4 | 3 | — |
| Storefronts | 16 | 15 | 13 | 11 | 7 | 6 | 4 | 4 |
| Skylights, putty glass | 60 | 36 | 21 | 12 | 16 | — | — | — |
| Thiokol set | 15 | 15 | 11 | 9 | 9 | 6 | 3 | 2 |
| Vinyl set, snap on | 18 | 18 | 13 | 12 | 12 | 7 | 5 | 4 |
| Maximum area per light | 2.8 S.F. | 6.3 S.F. | 11.1 S.F. | 17.4 S.F. | 31.6 S.F. | 47 S.F. | 69 S.F. | 100 S.F. |

**Reference Tables**

## R092000-50 Lath, Plaster and Gypsum Board

**Gypsum board lath** is available in 3/8″ thick x 16″ wide x 4′ long sheets as a base material for multi-layer plaster applications. It is also available as a base for either multi-layer or veneer plaster applications in 1/2″ and 5/8″ thick–4′ wide x 8′, 10′ or 12′ long sheets. Fasteners are screws or blued ring shank nails for wood framing and screws for metal framing.

**Metal lath** is available in diamond mesh pattern with flat or self-furring profiles. Paper backing is available for applications where excessive plaster waste needs to be avoided. A slotted mesh ribbed lath should be used in areas where the span between structural supports is greater than normal. Most metal lath comes in 27″ x 96″ sheets. Diamond mesh weighs 1.75, 2.5 or 3.4 pounds per square yard, slotted mesh lath weighs 2.75 or 3.4 pounds per square yard. Metal lath can be nailed, screwed or tied in place.

Many **accessories** are available. Corner beads, flat reinforcing strips, casing beads, control and expansion joints, furring brackets and channels are some examples. Note that accessories are not included in plaster or stucco line items.

**Plaster** is defined as a material or combination of materials that when mixed with a suitable amount of water, forms a plastic mass or paste. When applied to a surface, the paste adheres to it and subsequently hardens, preserving in a rigid state the form or texture imposed during the period of elasticity.

**Gypsum plaster** is made from ground calcined gypsum. It is mixed with aggregates and water for use as a base coat plaster.

**Vermiculite plaster** is a fire-retardant plaster covering used on steel beams, concrete slabs and other heavy construction materials. Vermiculite is a group name for certain clay minerals, hydrous silicates or aluminum, magnesium and iron that have been expanded by heat.

**Perlite plaster** is a plaster using perlite as an aggregate instead of sand. Perlite is a volcanic glass that has been expanded by heat.

**Gauging plaster** is a mix of gypsum plaster and lime putty that when applied produces a quick drying finish coat.

**Veneer plaster** is a one or two component gypsum plaster used as a thin finish coat over special gypsum board.

**Keenes cement** is a white cementitious material manufactured from gypsum that has been burned at a high temperature and ground to a fine powder. Alum is added to accelerate the set. The resulting plaster is hard and strong and accepts and maintains a high polish, hence it is used as a finishing plaster.

**Stucco** is a Portland cement based plaster used primarily as an exterior finish.

Plaster is used on both interior and exterior surfaces. Generally it is applied in multiple-coat systems. A three-coat system uses the terms scratch, brown and finish to identify each coat. A two-coat system uses base and finish to describe each coat. Each type of plaster and application system has attributes that are chosen by the designer to best fit the intended use.

| Gypsum Plaster Quantities for 100 S.Y. | 2 Coat, 5/8″ Thick | | 3 Coat, 3/4″ Thick | | |
|---|---|---|---|---|---|
| | Base | Finish | Scratch | Brown | Finish |
| | 1:3 Mix | 2:1 Mix | 1:2 Mix | 1:3 Mix | 2:1 Mix |
| Gypsum plaster | 1,300 lb. | | 1,350 lb. | 650 lb. | |
| Sand | 1.75 C.Y. | | 1.85 C.Y. | 1.35 C.Y. | |
| Finish hydrated lime | | 340 lb. | | | 340 lb. |
| Gauging plaster | | 170 lb. | | | 170 lb. |

| Vermiculite or Perlite Plaster Quantities for 100 S.Y. | 2 Coat, 5/8″ Thick | | 3 Coat, 3/4″ Thick | | |
|---|---|---|---|---|---|
| | Base | Finish | Scratch | Brown | Finish |
| Gypsum plaster | 1,250 lb. | | 1,450 lb. | 800 lb. | |
| Vermiculite or perlite | 7.8 bags | | 8.0 bags | 3.3 bags | |
| Finish hydrated lime | | 340 lb. | | | 340 lb. |
| Gauging plaster | | 170 lb. | | | 170 lb. |

| Stucco–Three-Coat System Quantities for 100 S.Y. | On Wood Frame | On Masonry |
|---|---|---|
| Portland cement | 29 bags | 21 bags |
| Sand | 2.6 C.Y. | 2.0 C.Y. |
| Hydrated lime | 180 lb. | 120 lb. |

## R092910-10 Levels of Gypsum Drywall Finish

In the past, contract documents often used phrases such as "industry standard" and "workmanlike finish" to specify the expected quality of gypsum board wall and ceiling installations. The vagueness of these descriptions led to unacceptable work and disputes.

In order to resolve this problem, four major trade associations concerned with the manufacture, erection, finish, and decoration of gypsum board wall and ceiling systems have developed an industry-wide *Recommended Levels of Gypsum Board Finish.*

The finish of gypsum board walls and ceilings for specific final decoration is dependent on a number of factors. A primary consideration is the location of the surface and the degree of decorative treatment desired. Painted and unpainted surfaces in warehouses and other areas where appearance is normally not critical may simply require the taping of wallboard joints and 'spotting' of fastener heads. Blemish-free, smooth, monolithic surfaces often intended for painted and decorated walls and ceilings in habitated structures, ranging from single-family dwellings through monumental buildings, require additional finishing prior to the application of the final decoration.

Other factors to be considered in determining the level of finish of the gypsum board surface are (1) the type of angle of surface illumination (both natural and artificial lighting), and (2) the paint and method of application or the type and finish of wallcovering specified as the final decoration. Critical lighting conditions, gloss paints, and thin wall coverings require a higher level of gypsum board finish than do heavily textured surfaces which are subsequently painted or surfaces which are to be decorated with heavy grade wall coverings.

The following descriptions were developed jointly by the Association of the Wall and Ceiling Industries-International (AWCI), Ceiling & Interior Systems Construction Association (CISCA), Gypsum Association (GA), and Painting and Decorating Contractors of America (PDCA) as a guide.

**Level 0: Used in temporary construction or wherever the final decoration has not been determined. Unfinished. No taping, finishing or corner beads are required.** Also could be used where non-predecorated panels will be used in demountable-type partitions that are to be painted as a final finish.

**Level 1: Frequently used in plenum areas above ceilings, in attics, in areas where the assembly would generally be concealed, or in building service corridors and other areas not normally open to public view. Some degree of sound and smoke control is provided; in some geographic areas, this level is referred to as "fire-taping," although this level of finish does not typically meet fire-resistant assembly requirements.** Where a fire resistance rating is required for the gypsum board assembly, details of construction should be in accordance with reports of fire tests of assemblies that have met the requirements of the fire rating acceptable.

All joints and interior angles shall have tape embedded in joint compound. Accessories are optional at specifier discretion in corridors and other areas with pedestrian traffic. Tape and fastener heads need not be covered with joint compound. Surface shall be free of excess joint compound. Tool marks and ridges are acceptable.

**Level 2: It may be specified for standard gypsum board surfaces in garages, warehouse storage, or other similar areas where surface appearance is not of primary importance.**

All joints and interior angles shall have tape embedded in joint compound and shall be immediately wiped with a joint knife or trowel, leaving a thin coating of joint compound over all joints and interior angles. Fastener heads and accessories shall be covered with a coat of joint compound. Surface shall be free of excess joint compound. Tool marks and ridges are acceptable.

**Level 3: Typically used in areas that are to receive heavy texture (spray or hand applied) finishes before final painting, or where commercial-grade (heavy duty) wall coverings are to be applied as the final decoration.** This level of finish should not be used where smooth painted surfaces or where lighter weight wall coverings are specified. The prepared surface shall be coated with a drywall primer prior to the application of final finishes.

All joints and interior angles shall have tape embedded in joint compound and shall be immediately wiped with a joint knife or trowel, leaving a thin coating of joint compound over all joints and interior angles. One additional coat of joint compound shall be applied over all joints and interior angles. Fastener heads and accessories shall be covered with two separate coats of joint compound. All joint compounds shall be smooth and free of tool marks and ridges. The prepared surface shall be covered with a drywall primer prior to the application of the final decoration.

**Level 4: This level should be used where residential grade (light duty) wall coverings, flat paints, or light textures are to be applied. The prepared surface shall be coated with a drywall primer prior to the application of final finishes. Release agents for wall coverings are specifically formulated to minimize damage if coverings are subsequently removed.**

The weight, texture, and sheen level of the wall covering material selected should be taken into consideration when specifying wall coverings over this level of drywall treatment. Joints and fasteners must be sufficiently concealed if the wall covering material is lightweight, contains limited pattern, has a glossy finish, or has any combination of these features. In critical lighting areas, flat paints applied over light textures tend to reduce joint photographing. Gloss, semi-gloss, and enamel paints are not recommended over this level of finish.

All joints and interior angles shall have tape embedded in joint compound and shall be immediately wiped with a joint knife or trowel, leaving a thin coating of joint compound over all joints and interior angles. In addition, two separate coats of joint compound shall be applied over all flat joints and one separate coat of joint compound applied over interior angles. Fastener heads and accessories shall be covered with three separate coats of joint compound. All joint compounds shall be smooth and free of tool marks and ridges. The prepared surface shall be covered with a drywall primer like Sheetrock® First Coat prior to the application of the final decoration.

**Level 5: The highest quality finish is the most effective method to provide a uniform surface and minimize the possibility of joint photographing and of fasteners showing through the final decoration.** This level of finish is required where gloss, semi-gloss, or enamel is specified; when flat joints are specified over an untextured surface; or where critical lighting conditions occur. The prepared surface shall be coated with a drywall primer prior to the application of final decoration.

All joints and interior angles shall have tape embedded in joint compound and be immediately wiped with a joint knife or trowel, leaving a thin coating of joint compound over all joints and interior angles. Two separate coats of joint compound shall be applied over all flat joints and one separate coat of joint compound applied over interior angles. Fastener heads and accessories shall be covered with three separate coats of joint compound.

A thin skim coat of joint compound shall be trowel applied to the entire surface. Excess compound is immediately troweled off, leaving a film or skim coating of compound completely covering the paper. As an alternative to a skim coat, a material manufactured especially for this purpose may be applied such as Sheetrock® Tuff-Hide primer surfacer. The surface must be smooth and free of tool marks and ridges. The prepared surface shall be covered with a drywall primer prior to the application of the final decoration.

## Finishes | R0966 Terrazzo Flooring

### R096613-10   Terrazzo Floor

The table below lists quantities required for 100 S.F. of 5/8″ terrazzo topping, either bonded or not bonded.

| Description | Bonded to Concrete 1-1/8″ Bed, 1:4 Mix | Not Bonded 2-1/8″ Bed and 1/4″ Sand |
|---|---|---|
| Portland cement, 94 lb. Bag | 6 bags | 8 bags |
| Sand | 10 C.F. | 20 C.F. |
| Divider strips, 4′ squares | 50 L.F. | 50 L.F. |
| Terrazzo fill, 50 lb. Bag | 12 bags | 12 bags |
| 15 Lb. tarred felt | | 1 C.S.F. |
| Mesh 2 x 2 #14 galvanized | | 1 C.S.F. |
| Crew J-3 | 0.77 days | 0.87 days |

2′ x 2′ panels require 1.00 L.F. divider strip per S.F.

3′ x 3′ panels require 0.67 L.F. divider strip per S.F.

4′ x 4′ panels require 0.50 L.F. divider strip per S.F.

5′ x 5′ panels require 0.40 L.F. divider strip per S.F.

6′ x 6′ panels require 0.33 L.F. divider strip per S.F.

## Finishes | R0972 Wall Coverings

### R097223-10   Wall Covering

The table below lists the quantities required for 100 S.F. of wall covering.

| Description | Medium-Priced Paper | Expensive Paper |
|---|---|---|
| Paper | 1.6 dbl. rolls | 1.6 dbl. rolls |
| Wall sizing | 0.25 gallon | 0.25 gallon |
| Vinyl wall paste | 0.6 gallon | 0.6 gallon |
| Apply sizing | 0.3 hour | 0.3 hour |
| Apply paper | 1.2 hours | 1.5 hours |

Most wallpapers now come in double rolls only.

To remove old paper, allow 1.3 hours per 100 S.F.

## R099100-10 Painting Estimating Techniques

Proper estimating methodology is needed to obtain an accurate painting estimate. There is no known reliable shortcut or square foot method. The following steps should be followed:

- List all surfaces to be painted, with an accurate quantity (area) of each. Items having similar surface condition, finish, application method and accessibility may be grouped together.
- List all the tasks required for each surface to be painted, including surface preparation, masking, and protection of adjacent surfaces. Surface preparation may include minor repairs, washing, sanding and puttying.
- Select the proper Means line for each task. Review and consider all adjustments to labor and materials for type of paint and location of work. Apply the height adjustment carefully. For instance, when applying the adjustment for work over 8' high to a wall that is 12' high, apply the adjustment only to the area between 8' and 12' high, and not to the entire wall.

When applying more than one percent (%) adjustment, apply each to the base cost of the data, rather than applying one percentage adjustment on top of the other.

When estimating the cost of painting walls and ceilings remember to add the brushwork for all cut-ins at inside corners and around windows and doors as a LF measure. One linear foot of cut-in with brush equals one square foot of painting.

All items for spray painting include the labor for roll-back.

Deduct for openings greater than 100 SF, or openings that extend from floor to ceiling and are greater than 5' wide. Do not deduct small openings.

The cost of brushes, rollers, ladders and spray equipment are considered to be part of a painting contractor's overhead, and should not be added to the estimate. The cost of rented equipment such as scaffolding and swing staging should be added to the estimate.

## R099100-20 Painting

| Item | Coat | One Gallon Covers | | | In 8 Hours a Laborer Covers | | | Labor-Hours per 100 S.F. | | |
|------|------|-------|--------|-------|-------|--------|-------|-------|--------|-------|
| | | Brush | Roller | Spray | Brush | Roller | Spray | Brush | Roller | Spray |
| Paint wood siding | prime | 250 S.F. | 225 S.F. | 290 S.F. | 1150 S.F. | 1300 S.F. | 2275 S.F. | .695 | .615 | .351 |
| | others | 270 | 250 | 290 | 1300 | 1625 | 2600 | .615 | .492 | .307 |
| Paint exterior trim | prime | 400 | — | — | 650 | — | — | 1.230 | — | — |
| | 1st | 475 | — | — | 800 | — | — | 1.000 | — | — |
| | 2nd | 520 | — | — | 975 | — | — | .820 | — | — |
| Paint shingle siding | prime | 270 | 255 | 300 | 650 | 975 | 1950 | 1.230 | .820 | .410 |
| | others | 360 | 340 | 380 | 800 | 1150 | 2275 | 1.000 | .695 | .351 |
| Stain shingle siding | 1st | 180 | 170 | 200 | 750 | 1125 | 2250 | 1.068 | .711 | .355 |
| | 2nd | 270 | 250 | 290 | 900 | 1325 | 2600 | .888 | .603 | .307 |
| Paint brick masonry | prime | 180 | 135 | 160 | 750 | 800 | 1800 | 1.066 | 1.000 | .444 |
| | 1st | 270 | 225 | 290 | 815 | 975 | 2275 | .981 | .820 | .351 |
| | 2nd | 340 | 305 | 360 | 815 | 1150 | 2925 | .981 | .695 | .273 |
| Paint interior plaster or drywall | prime | 400 | 380 | 495 | 1150 | 2000 | 3250 | .695 | .400 | .246 |
| | others | 450 | 425 | 495 | 1300 | 2300 | 4000 | .615 | .347 | .200 |
| Paint interior doors and windows | prime | 400 | — | — | 650 | — | — | 1.230 | — | — |
| | 1st | 425 | — | — | 800 | — | — | 1.000 | — | — |
| | 2nd | 450 | — | — | 975 | — | — | .820 | — | — |

**Special Construction** | **R1311 Swimming Pools**

## R131113-20 Swimming Pools

Pool prices given per square foot of surface area include pool structure, filter and chlorination equipment, pumps, related piping, ladders/steps, maintenance kit, skimmer and vacuum system. Decks and electrical service to equipment are not included.

Residential in-ground pool construction can be divided into two categories: vinyl lined and gunite. Vinyl lined pool walls are constructed of different materials including wood, concrete, plastic or metal. The bottom is often graded with sand over which the vinyl liner is installed. Vermiculite or soil cement bottoms may be substituted for an added cost.

Gunite pool construction is used both in residential and municipal installations. These structures are steel reinforced for strength and finished with a white cement limestone plaster.

Municipal pools will have a higher cost because plumbing codes require more expensive materials, chlorination equipment and higher filtration rates.

Municipal pools greater than 1,800 S.F. require gutter systems to control waves. This gutter may be formed into the concrete wall. Often a vinyl/stainless steel gutter or gutter/wall system is specified, which will raise the pool cost.

Competition pools usually require tile bottoms and sides with contrasting lane striping, which will also raise the pool cost.

# Special Construction — R1331 Fabric Structures

## R133113-10 Air Supported Structures

Air supported structures are made from fabrics that can be classified into two groups: temporary and permanent. Temporary fabrics include nylon, woven polyethylene, vinyl film, and vinyl coated dacron. These have lifespans that range from five to fifteen plus years. The cost per square foot includes a fabric shell, tension cables, primary and back-up inflation systems and doors. The lower cost structures are used for construction shelters, bulk storage and pond covers. The more expensive are used for recreational structures and warehouses.

Permanent fabrics are teflon coated fiberglass. The life of this structure is twenty plus years. The high cost limits its application to architectural designed structures which call for a clear span covered area, such as stadiums and convention centers. Both temporary and permanent structures are available in translucent fabrics which eliminates the need for daytime lighting.

Areas to be covered vary from 10,000 S.F. to any area up to 1000 foot wide by any length. Height restrictions range from a maximum of 1/2 of

width to a minimum of 1/6 of the width. Erection of even the largest of the temporary structures requires no more than a week.

Centrifugal fans provide the inflation necessary to support the structure during application of live loads. Airlocks are usually used at large entrances to prevent loss of static pressure. Some manufacturers employ propeller fans which generate sufficient airflow (30,000 CFM) to eliminate the need for airlocks. These fans may also be automatically controlled to resist high wind conditions, regulate humidity (air changes), and provide cooling and heat.

Insulation can be provided with the addition of a second or even third interior liner, creating a dead air space with an "R" value of four to nine. Some structures allow for the liner to be collapsed into the outer shell to enable the internal heat to melt accumulated snow. For cooling or air conditioning, the exterior face of the liner can be aluminized to reflect the sun's heat.

## R133113-90 Seismic Bracing

Sometimes referred to as anti-sway bracing, this support system is required in earthquake areas. The individual components must be assembled to

make a required system.

| Example | | | | Additionally, height factors must be taken into account. Add the following percentages to labor for elevated installations: | |
|---|---|---|---|---|---|
| | C-Clamp 3/8" rod | 2 ea. | | 15' to 20' high | 10% |
| | Rod, continuous thread 3/8" | 10 L.F. | | 21' to 25' high | 20% |
| | Field, weld 1" | 2 ea. | | 26' to 30' high | 30% |
| | | | | 31' to 35' high | 40% |
| | | | | 36' to 40' high | 50% |
| | | | | 41' to 50' high | 60% |

# Special Construction — R1334 Fabricated Engineered Structures

## R133419-10 Pre-Engineered Steel Buildings

These buildings are manufactured by many companies and normally erected by franchised dealers throughout the U.S. The four basic types are: Rigid Frames, Truss type, Post and Beam and the Sloped Beam type. Most popular roof slope is low pitch of 1" in 12". The minimum economical area of these buildings is about 3000 S.F. of floor area. Bay sizes are usually 20' to 24' but can go as high as 30' with heavier girts and purlins. Eave heights are usually 12' to 24' with 18' to 20' most typical.

Material prices shown in the Unit Price section are bare costs for the building shell only and do not include floors, foundations, anchor bolts, interior finishes or utilities. Costs assume at least three bays of 24' each, a 1" in 12" roof slope, and they are based on 30 psf roof load and 20 psf wind load

(wind load is a function of wind speed, building height, and terrain characteristics; this should be determined by a Registered Structural Engineer) and no unusual requirements. Costs include the structural frame, 26 ga. non-insulated colored corrugated or ribbed roofing and siding panels, fasteners, closures, trim and flashing but no allowance for insulation, doors, windows, skylights, gutters or downspouts. Very large projects would generally cost less for materials than the prices shown. For roof panel substitutions and wall panel substitutions, see appropriate Unit Price sections.

Conditions at the site, weather, shape and size of the building, and labor availability will affect the erection cost of the building.

## R133423-30 Dome Structures

**Steel** — The four types are Lamella, Schwedler, Arch and Geodesic. For maximum economy, rise should be about 15 to 20% of diameter. Most common diameters are in the 200' to 300' range. Lamella domes weigh about 5 P.S.F. of floor area less than Schwedler domes. Schwedler dome weight in lbs. per S.F. approaches .046 times the diameter. Domes below 125' diameter weigh .07 times diameter and the cost per ton of steel is higher. See R051223-20 for estimating weight.

**Wood** — Small domes are of sawn lumber, larger ones are laminated. In larger sizes, triaxial and triangular cost about the same; radial domes cost more. Radial domes are economical in the 60' to 70' diameter range. Most economical range of all types is 80' to 200' diameters. Diameters can

run over 400'. All costs are quoted above the foundation. Prices include 2" decking and a tension tie ring in place.

**Plywood** — Stock prefab geodesic domes are available with diameters from 24' to 60'.

**Fiberglass** — Aluminum framed translucent sandwich panels with spans from 5' to 45' are commercially available.

**Aluminum** — Stressed skin aluminum panels form geodesic domes with spans ranging from 82' to 232'. An aluminum space truss, triangulated or nontriangulated, with aluminum or clear acrylic closure panels can be used for clear spans of 40' to 415'.

## R142000-10   Freight Elevators

Capacities run from 2,000 lbs. to over 100,000 lbs. with 3,000 lbs. to 10,000 lbs. most common. Travel speeds are generally lower and control less intricate than on passenger elevators. Costs in the Unit Price sections are for hydraulic and geared elevators.

## R142000-20   Elevator Selective Costs   See R142000-40 for cost development.

| | Passenger | | Freight | | Hospital | |
|---|---|---|---|---|---|---|
| A. Base Unit | **Hydraulic** | **Electric** | **Hydraulic** | **Electric** | **Hydraulic** | **Electric** |
| Capacity | 1,500 lb. | 2,000 lb. | 2,000 lb. | 4,000 lb. | 4,000 lb. | 4,000 lb. |
| Speed | 100 F.P.M. | 200 F.P.M. | 100 F.P.M. | 200 F.P.M. | 100 F.P.M. | 200 F.P.M. |
| #Stops/Travel Ft. | 2/12 | 4/40 | 2/20 | 4/40 | 2/20 | 4/40 |
| Push Button Oper. | Yes | Yes | Yes | Yes | Yes | Yes |
| Telephone Box & Wire | " | " | " | " | " | " |
| Emergency Lighting | " | " | No | No | " | " |
| Cab | Plastic Lam. Walls | Plastic Lam. Walls | Painted Steel | Painted Steel | Plastic Lam. Walls | Plastic Lam. Walls |
| Cove Lighting | Yes | Yes | No | No | Yes | Yes |
| Floor | V.C.T. | V.C.T. | Wood w/Safety Treads | Wood w/Safety Treads | V.C.T. | V.C.T. |
| Doors, & Speedside Slide | Yes | Yes | Yes | Yes | Yes | Yes |
| Gates, Manual | No | No | No | No | No | No |
| Signals, Lighted Buttons | Car and Hall | Car and Hall | Car and Hall | Car and Hall | Car and Hall | Car and Hall |
| O.H. Geared Machine | N.A. | Yes | N.A. | Yes | N.A. | Yes |
| Variable Voltage Contr. | " | " | N.A. | " | " | " |
| Emergency Alarm | Yes | " | Yes | " | Yes | " |
| Class "A" Loading | N.A. | N.A. | " | " | N.A. | N.A. |

## R142000-30   Passenger Elevators

Electric elevators are used generally but hydraulic elevators can be used for lifts up to 70' and where large capacities are required. Hydraulic speeds are limited to 200 F.P.M. but cars are self leveling at the stops. On low rises, hydraulic installation runs about 15% less than standard electric types but on higher rises this installation cost advantage is reduced. Maintenance of hydraulic elevators is about the same as electric type but underground portion is not included in the maintenance contract.

In electric elevators there are several control systems available, the choice of which will be based upon elevator use, size, speed and cost criteria. The two types of drives are geared for low speeds and gearless for 450 F.P.M. and over.

The tables on the preceding pages illustrate typical installed costs of the various types of elevators available.

## R142000-40   Elevator Cost Development

To price a new car or truck from the factory, you must start with the manufacturer's basic model, then add or exchange optional equipment and features. The same is true for pricing elevators.

Requirement: One-passenger elevator, five-story hydraulic, 2,500 lb. capacity, 12' floor to floor, speed 150 F.P.M., emergency power switching and maintenance contract.

**Example:**

| Description | Adjustment |
|---|---|
| A. Base Elevator: Hydraulic Passenger, 1500 lb. Capacity, 100 fpm, 2 Stops, Standard Finish | 1 Ea. |
| B. Capacity Adjustment (2,500 lb.) | 1 Ea. |
| C. Excess Travel Adjustment: 48' Total Travel (4 x 12') minus 12' Base Unit Travel = | 36 V.L.F. |
| D. Stops Adjustment: 5 Total Stops minus 2 Stops (Base Unit) = | 3 Stops |
| E. Speed Adjustment (150 F.P.M.) | 1 Ea. |
| F. Options: | |
| 1. Intercom Service | 1 Ea. |
| 2. Emergency Power Switching, Automatic | 1 Ea. |
| 3. Stainless Steel Entrance Doors | 5 Ea. |
| 4. Maintenance Contract (12 Months) | 1 Ea. |
| 5. Position Indicator for main floor level (none indicated in Base Unit) | 1 Ea. |

## Conveying Equipment — R1431 Escalators

### R143110-10 Escalators

Moving stairs can be used for buildings where 600 or more people are to be carried to the second floor or beyond. Freight cannot be carried on escalators and at least one elevator must be available for this function.

Carrying capacity is 5,000 to 8,000 people per hour. Power requirement is 2 to 3 KW per hour and incline angle is 30°.

## Conveying Equipment — R1432 Moving Walks

### R143210-20 Moving Ramps and Walks

These are a specialized form of conveyor 3' to 6' wide with capacities of 3,600 to 18,000 persons per hour. Maximum speed is 140 F.P.M. and normal incline is 0° to 15°.

Local codes will determine the maximum angle. Outdoor units would require additional weather protection.

## Plumbing — R2201 Operation & Maintenance of Plumbing

### R220105-10 Demolition (Selective vs. Removal for Replacement)

Demolition can be divided into two basic categories.

One type of demolition involves the removal of material with no concern for its replacement. The labor-hours to estimate this work are found under "Selective Demolition" in the Fire Protection, Plumbing and HVAC Divisions. It is selective in that individual items or all the material installed as a system or trade grouping such as plumbing or heating systems are removed. This may be accomplished by the easiest way possible, such as sawing, torch cutting, or sledge hammer as well as simple unbolting.

The second type of demolition is the removal of some item for repair or replacement. This removal may involve careful draining, opening of unions,

disconnecting and tagging of electrical connections, capping of pipes/ducts to prevent entry of debris or leakage of the material contained as well as transport of the item away from its in-place location to a truck/dumpster. An approximation of the time required to accomplish this type of demolition is to use half of the time indicated as necessary to install a new unit. For example; installation of a new pump might be listed as requiring 6 labor-hours so if we had to estimate the removal of the old pump we would allow an additional 3 hours for a total of 9 hours. That is, the complete replacement of a defective pump with a new pump would be estimated to take 9 labor-hours.

## Plumbing — R2211 Facility Water Distribution

### R221113-50 Pipe Material Considerations

1. Malleable fittings should be used for gas service.
2. Malleable fittings are used where there are stresses/strains due to expansion and vibration.
3. Cast fittings may be broken as an aid to disassembling of heating lines frozen by long use, temperature and minerals.
4. Cast iron pipe is extensively used for underground and submerged service.
5. Type M (light wall) copper tubing is available in hard temper only and is used for nonpressure and less severe applications than K and L.
6. Type L (medium wall) copper tubing, available hard or soft for interior service.
7. Type K (heavy wall) copper tubing, available in hard or soft temper for use where conditions are severe. For underground and interior service.
8. Hard drawn tubing requires fewer hangers or supports but should not be bent. Silver brazed fittings are recommended, however soft solder is normally used.
9. Type DMV (very light wall) copper tubing designed for drainage, waste and vent plus other non-critical pressure services.

#### Domestic/Imported Pipe and Fittings Cost

The prices shown in this publication for steel/cast iron pipe and steel, cast iron, malleable iron fittings are based on domestic production sold at the normal trade discounts. The above listed items of foreign manufacture may be available at prices of 1/3 to 1/2 those shown. Some imported items after minor machining or finishing operations are being sold as domestic to further complicate the system.

**Caution:** Most pipe prices in this book also include a coupling and pipe hangers which for the larger sizes can add significantly to the per foot cost and should be taken into account when comparing "book cost" with quoted supplier's cost.

## R224000-40   Plumbing Fixture Installation Time

| Item | Rough-In | Set | Total Hours | Item | Rough-In | Set | Total Hours |
|---|---|---|---|---|---|---|---|
| Bathtub | 5 | 5 | 10 | Shower head only | 2 | 1 | 3 |
| Bathtub and shower, cast iron | 6 | 6 | 12 | Shower drain | 3 | 1 | 4 |
| Fire hose reel and cabinet | 4 | 2 | 6 | Shower stall, slate | | 15 | 15 |
| Floor drain to 4 inch diameter | 3 | 1 | 4 | Slop sink | 5 | 3 | 8 |
| Grease trap, single, cast iron | 5 | 3 | 8 | Test 6 fixtures | | | 14 |
| Kitchen gas range | | 4 | 4 | Urinal, wall | 6 | 2 | 8 |
| Kitchen sink, single | 4 | 4 | 8 | Urinal, pedestal or floor | 6 | 4 | 10 |
| Kitchen sink, double | 6 | 6 | 12 | Water closet and tank | 4 | 3 | 7 |
| Laundry tubs | 4 | 2 | 6 | Water closet and tank, wall hung | 5 | 3 | 8 |
| Lavatory wall hung | 5 | 3 | 8 | Water heater, 45 gals. gas, automatic | 5 | 2 | 7 |
| Lavatory pedestal | 5 | 3 | 8 | Water heaters, 65 gals. gas, automatic | 5 | 2 | 7 |
| Shower and stall | 6 | 4 | 10 | Water heaters, electric, plumbing only | 4 | 2 | 6 |

Fixture prices in front of book are based on the cost per fixture set in place. The rough-in cost, which must be added for each fixture, includes carrier, if required, some supply, waste and vent pipe connecting fittings and stops. The lengths of rough-in pipe are nominal runs which would connect to the larger runs and stacks. The supply runs and DWV runs and stacks must be accounted for in separate entries. In the eastern half of the United States it is common for the plumber to carry these to a point 5' outside the building.

## R224000-50   Water Cooler Application

| Type of Service | Requirement |
|---|---|
| Office, School or Hospital | 12 persons per gallon per hour |
| Office, Lobby or Department Store | 4 or 5 gallons per hour per fountain |
| Light manufacturing | 7 persons per gallon per hour |
| Heavy manufacturing | 5 persons per gallon per hour |
| Hot heavy manufacturing | 4 persons per gallon per hour |
| Hotel | .08 gallons per hour per room |
| Theatre | 1 gallon per hour per 100 seats |

### R233100-20 Ductwork

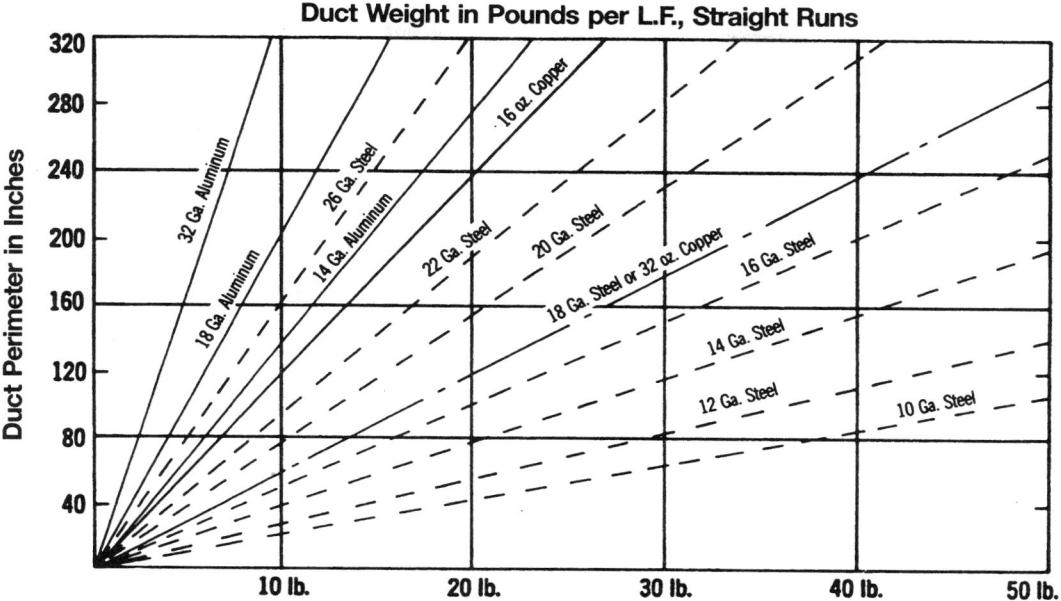

Duct Weight in Pounds per L.F., Straight Runs

Add to the above for fittings; 90° elbow is 3 L.F.; 45° elbow is 2.5 L.F.; offset is 4 L.F.; transition offset is 6 L.F.; square-to-round transition is 4 L.F.;

90° reducing elbow is 5 L.F. For bracing and waste, add 20% to aluminum and copper, 15% to steel.

## Heating, Ventilating & A.C. R2334 HVAC Fans

### R233400-10 Recommended Ventilation Air Changes

Table below lists range of time in minutes per change for various types of facilities.

| | | | | | |
|---|---|---|---|---|---|
| Assembly Halls | 2-10 | Dance Halls | 2-10 | Laundries | 1-3 |
| Auditoriums | 2-10 | Dining Rooms | 3-10 | Markets | 2-10 |
| Bakeries | 2-3 | Dry Cleaners | 1-5 | Offices | 2-10 |
| Banks | 3-10 | Factories | 2-5 | Pool Rooms | 2-5 |
| Bars | 2-5 | Garages | 2-10 | Recreation Rooms | 2-10 |
| Beauty Parlors | 2-5 | Generator Rooms | 2-5 | Sales Rooms | 2-10 |
| Boiler Rooms | 1-5 | Gymnasiums | 2-10 | Theaters | 2-8 |
| Bowling Alleys | 2-10 | Kitchens-Hospitals | 2-5 | Toilets | 2-5 |
| Churches | 5-10 | Kitchens-Restaurant | 1-3 | Transformer Rooms | 1-5 |

CFM air required for changes = Volume of room in cubic feet ÷ Minutes per change.

## Heating, Ventilating & A.C. R2350 Central Heating Equipment

### R235000-35 Heating (42° degrees latitude)

$$\text{Approximate S.F. radiation} = \frac{\text{S.F. sash}}{2} + \frac{\text{S.F. wall} + \text{roof} - \text{sash}}{20} + \frac{\text{C.F. building}}{200}$$

## R235000-50 Factor for Determining Heat Loss for Various Types of Buildings

**General:** While the most accurate estimates of heating requirements would naturally be based on detailed information about the building being considered, it is possible to arrive at a reasonable approximation using the following procedure:

1. Calculate the cubic volume of the room or building.
2. Select the appropriate factor from Table 1 below. Note that the factors apply only to inside temperatures listed in the first column and to 0° F outside temperature.
3. If the building has bad north and west exposures, multiply the heat loss factor by 1.1.
4. If the outside design temperature is other than 0° F, multiply the factor from Table 1 by the factor from Table 2.
5. Multiply the cubic volume by the factor selected from Table 1. This will give the estimated BTUH heat loss which must be made up to maintain inside temperature.

| Table 1 — Building Type | Conditions | Qualifications | Loss Factor* |
|---|---|---|---|
| Factories & Industrial Plants General Office Areas at 70° F | One Story | Skylight in Roof | 6.2 |
| | | No Skylight in Roof | 5.7 |
| | Multiple Story | Two Story | 4.6 |
| | | Three Story | 4.3 |
| | | Four Story | 4.1 |
| | | Five Story | 3.9 |
| | | Six Story | 3.6 |
| | All Walls Exposed | Flat Roof | 6.9 |
| | | Heated Space Above | 5.2 |
| | One Long Warm Common Wall | Flat Roof | 6.3 |
| | | Heated Space Above | 4.7 |
| | Warm Common Walls on Both Long Sides | Flat Roof | 5.8 |
| | | Heated Space Above | 4.1 |
| Warehouses at 60° F | All Walls Exposed | Skylights in Roof | 5.5 |
| | | No Skylight in Roof | 5.1 |
| | | Heated Space Above | 4.0 |
| | One Long Warm Common Wall | Skylight in Roof | 5.0 |
| | | No Skylight in Roof | 4.9 |
| | | Heated Space Above | 3.4 |
| | Warm Common Walls on Both Long Sides | Skylight in Roof | 4.7 |
| | | No Skylight in Roof | 4.4 |
| | | Heated Space Above | 3.0 |

*Note: This table tends to be conservative particularly for new buildings designed for minimum energy consumption.

| Table 2 — Outside Design Temperature Correction Factor (for Degrees Fahrenheit) | | | | | | | | | |
|---|---|---|---|---|---|---|---|---|---|
| Outside Design Temperature | 50 | 40 | 30 | 20 | 10 | 0 | -10 | -20 | -30 |
| Correction Factor | .29 | .43 | .57 | .72 | .86 | 1.00 | 1.14 | 1.28 | 1.43 |

## R235616-60 Solar Heating (Space and Hot Water)

Collectors should face as close to due South as possible, however, variations of up to 20 degrees on either side of true South are acceptable. Local climate and collector type may influence the choice between east or west deviations. Obviously they should be located so they are not shaded from the sun's rays. Incline collectors at a slope of latitude minus 5 degrees for domestic hot water and latitude plus 15 degrees for space heating.

Flat plate collectors consist of a number of components as follows: Insulation to reduce heat loss through the bottom and sides of the collector. The enclosure which contains all the components in this assembly is usually weatherproof and prevents dust, wind and water from coming in contact with the absorber plate. The cover plate usually consists of one or more layers of a variety of glass or plastic and reduces the reradiation by creating an air space which traps the heat between the cover and the absorber plates.

The absorber plate must have a good thermal bond with the fluid passages. The absorber plate is usually metallic and treated with a surface coating which improves absorptivity. Black or dark paints or selective coatings are used for this purpose, and the design of this passage and plate combination helps determine a solar system's effectiveness.

Heat transfer fluid passage tubes are attached above and below or integral with an absorber plate for the purpose of transferring thermal energy from the absorber plate to a heat transfer medium. The heat exchanger is a device for transferring thermal energy from one fluid to another.

Piping and storage tanks should be well insulated to minimize heat losses.

Size domestic water heating storage tanks to hold 20 gallons of water per user, minimum, plus 10 gallons per dishwasher or washing machine. For domestic water heating an optimum collector size is approximately 3/4 square foot of area per gallon of water storage. For space heating of residences and small commercial applications the collector is commonly sized between 30% and 50% of the internal floor area. For space heating of large commercial applications, collector areas less than 30% of the internal floor area can still provide significant heat reductions.

A supplementary heat source is recommended for Northern states for December through February.

The solar energy transmission per square foot of collector surface varies greatly with the material used. Initial cost, heat transmittance and useful life are obviously interrelated.

## **Heating, Ventilating & A.C.** R2360 Central Cooling Equipment

## R236000-20 Air Conditioning Requirements

BTUs per hour per S.F. of floor area and S.F. per ton of air conditioning.

| Type of Building | BTU/Hr per S.F. | S.F. per Ton | Type of Building | BTU/Hr per S.F. | S.F. per Ton | Type of Building | BTU/Hr per S.F. | S.F. per Ton |
|---|---|---|---|---|---|---|---|---|
| Apartments, Individual | 26 | 450 | Dormitory, Rooms | 40 | 300 | Libraries | 50 | 240 |
| Corridors | 22 | 550 | Corridors | 30 | 400 | Low Rise Office, Exterior | 38 | 320 |
| Auditoriums & Theaters | 40 | 300/18* | Dress Shops | 43 | 280 | Interior | 33 | 360 |
| Banks | 50 | 240 | Drug Stores | 80 | 150 | Medical Centers | 28 | 425 |
| Barber Shops | 48 | 250 | Factories | 40 | 300 | Motels | 28 | 425 |
| Bars & Taverns | 133 | 90 | High Rise Office—Ext. Rms. | 46 | 263 | Office (small suite) | 43 | 280 |
| Beauty Parlors | 66 | 180 | Interior Rooms | 37 | 325 | Post Office, Individual Office | 42 | 285 |
| Bowling Alleys | 68 | 175 | Hospitals, Core | 43 | 280 | Central Area | 46 | 260 |
| Churches | 36 | 330/20* | Perimeter | 46 | 260 | Residences | 20 | 600 |
| Cocktail Lounges | 68 | 175 | Hotel, Guest Rooms | 44 | 275 | Restaurants | 60 | 200 |
| Computer Rooms | 141 | 85 | Corridors | 30 | 400 | Schools & Colleges | 46 | 260 |
| Dental Offices | 52 | 230 | Public Spaces | 55 | 220 | Shoe Stores | 55 | 220 |
| Dept. Stores, Basement | 34 | 350 | Industrial Plants, Offices | 38 | 320 | Shop'g. Ctrs., Supermarkets | 34 | 350 |
| Main Floor | 40 | 300 | General Offices | 34 | 350 | Retail Stores | 48 | 250 |
| Upper Floor | 30 | 400 | Plant Areas | 40 | 300 | Specialty | 60 | 200 |

*Persons per ton
12,000 BTU = 1 ton of air conditioning

## R260519-92 Minimum Copper and Aluminum Wire Size Allowed for Various Types of Insulation

| Amperes | Copper | | Aluminum | | Amperes | Copper | | Aluminum | |
| | THW THWN or XHHW | THHN XHHW * | THW XHHW | THHN XHHW * | | THW THWN or XHHW | THHN XHHW * | THW XHHW | THHN XHHW * |
|---|---|---|---|---|---|---|---|---|---|
| 15A | #14 | #14 | #12 | #12 | 195 | 3/0 | 2/0 | 250kcmil | 4/0 |
| 20 | #12 | #12 | #10 | #10 | 200 | 3/0 | 3/0 | 250kcmil | 4/0 |
| 25 | #10 | #10 | #10 | #10 | 205 | 4/0 | 3/0 | 250kcmil | 4/0 |
| 30 | #10 | #10 | # 8 | # 8 | 225 | 4/0 | 3/0 | 300kcmil | 250kcmil |
| 40 | # 8 | # 8 | # 8 | # 8 | 230 | 4/0 | 4/0 | 300kcmil | 250kcmil |
| 45 | # 8 | # 8 | # 6 | # 8 | 250 | 250kcmil | 4/0 | 350kcmil | 300kcmil |
| 50 | # 8 | # 8 | # 6 | # 6 | 255 | 250kcmil | 4/0 | 400kcmil | 300kcmil |
| 55 | # 6 | # 8 | # 4 | # 6 | 260 | 300kcmil | 4/0 | 400kcmil | 350kcmil |
| 60 | # 6 | # 6 | # 4 | # 6 | 270 | 300kcmil | 250kcmil | 400kcmil | 350kcmil |
| 65 | # 6 | # 6 | # 4 | # 4 | 280 | 300kcmil | 250kcmil | 500kcmil | 350kcmil |
| 75 | # 4 | # 6 | # 3 | # 4 | 285 | 300kcmil | 250kcmil | 500kcmil | 400kcmil |
| 85 | # 4 | # 4 | # 2 | # 3 | 290 | 350kcmil | 250kcmil | 500kcmil | 400kcmil |
| 90 | # 3 | # 4 | # 2 | # 2 | 305 | 350kcmil | 300kcmil | 500kcmil | 400kcmil |
| 95 | # 3 | # 4 | # 1 | # 2 | 310 | 350kcmil | 300kcmil | 500kcmil | 500kcmil |
| 100 | # 3 | # 3 | # 1 | # 2 | 320 | 400kcmil | 300kcmil | 600kcmil | 500kcmil |
| 110 | # 2 | # 3 | 1/0 | # 1 | 335 | 400kcmil | 350kcmil | 600kcmil | 500kcmil |
| 115 | # 2 | # 2 | 1/0 | # 1 | 340 | 500kcmil | 350kcmil | 600kcmil | 500kcmil |
| 120 | # 1 | # 2 | 1/0 | 1/0 | 350 | 500kcmil | 350kcmil | 700kcmil | 500kcmil |
| 130 | # 1 | # 2 | 2/0 | 1/0 | 375 | 500kcmil | 400kcmil | 700kcmil | 600kcmil |
| 135 | 1/0 | # 1 | 2/0 | 1/0 | 380 | 500kcmil | 400kcmil | 750kcmil | 600kcmil |
| 150 | 1/0 | # 1 | 3/0 | 2/0 | 385 | 600kcmil | 500kcmil | 750kcmil | 600kcmil |
| 155 | 2/0 | 1/0 | 3/0 | 3/0 | 420 | 600kcmil | 500kcmil | | 700kcmil |
| 170 | 2/0 | 1/0 | 4/0 | 3/0 | 430 | | 500kcmil | | 750kcmil |
| 175 | 2/0 | 2/0 | 4/0 | 3/0 | 435 | | 600kcmil | | 750kcmil |
| 180 | 3/0 | 2/0 | 4/0 | 4/0 | 475 | | 600kcmil | | |

*Dry Locations Only

**Notes:**
1. Size #14 to 4/0 is in AWG units (American Wire Gauge).
2. Size 250 to 750 is in kcmil units (Thousand Circular Mils).
3. Use next higher ampere value if exact value is not listed in table.
4. For loads that operate continuously increase ampere value by 25% to obtain proper wire size.
5. Refer to Table R260519-91 for the maximum circuit length for the various size wires.
6. Table R260519-92 has been written for estimating purpose only, based on ambient temperature of 30°C (86° F); for ambient temperature other than 30°C (86° F), ampacity correction factors will be applied.

## R260533-22  Conductors in Conduit

Table below lists maximum number of conductors for various sized conduit using THW, TW or THWN insulations.

| Copper Wire Size | 1/2" | | | 3/4" | | | 1" | | | 1-1/4" | | | 1-1/2" | | | 2" | | | 2-1/2" | | | 3" | | 3-1/2" | | 4" | |
|---|---|---|---|---|---|---|---|---|---|---|---|---|---|---|---|---|---|---|---|---|---|---|---|---|---|---|---|
| | TW | THW | THWN | TW | THW | THWN | TW | THW | THWN | TW | THW | THWN | TW | THW | THWN | TW | THW | THWN | TW | THW | THWN | THW | THWN | THW | THWN | THW | THWN |
| #14 | 9 | 6 | 13 | 15 | 10 | 24 | 25 | 16 | 39 | 44 | 29 | 69 | 60 | 40 | 94 | 99 | 65 | 154 | 142 | 93 | | 143 | | 192 | | | |
| #12 | 7 | 4 | 10 | 12 | 8 | 18 | 19 | 13 | 29 | 35 | 24 | 51 | 47 | 32 | 70 | 78 | 53 | 114 | 111 | 76 | 164 | 117 | | 157 | | | |
| #10 | 5 | 4 | 6 | 9 | 6 | 11 | 15 | 11 | 18 | 26 | 19 | 32 | 36 | 26 | 44 | 60 | 43 | 73 | 85 | 61 | 104 | 95 | 160 | 127 | | 163 | |
| #8 | 2 | 1 | 3 | 4 | 3 | 5 | 7 | 5 | 9 | 12 | 10 | 16 | 17 | 13 | 22 | 28 | 22 | 36 | 40 | 32 | 51 | 49 | 79 | 66 | 106 | 85 | 136 |
| #6 | | 1 | 1 | | 2 | 4 | | 4 | 6 | | 7 | 11 | | 10 | 15 | | 16 | 26 | | 23 | 37 | 36 | 57 | 48 | 76 | 62 | 98 |
| #4 | | 1 | 1 | | 1 | 2 | | 3 | 4 | | 5 | 7 | | 7 | 9 | | 12 | 16 | | 17 | 22 | 27 | 35 | 36 | 47 | 47 | 60 |
| #3 | | 1 | 1 | | 1 | 1 | | 2 | 3 | | 4 | 6 | | 6 | 8 | | 10 | 13 | | 15 | 19 | 23 | 29 | 31 | 39 | 40 | 51 |
| #2 | | 1 | 1 | | 1 | 1 | | 2 | 3 | | 4 | 5 | | 5 | 7 | | 9 | 11 | | 13 | 16 | 20 | 25 | 27 | 33 | 34 | 43 |
| #1 | | | | | 1 | 1 | | 1 | 1 | | 3 | 3 | | 4 | 5 | | 6 | 8 | | 9 | 12 | 14 | 18 | 19 | 25 | 25 | 32 |
| 1/0 | | | | | 1 | 1 | | 1 | 1 | | 2 | 3 | | 3 | 4 | | 5 | 7 | | 8 | 10 | 12 | 15 | 16 | 21 | 21 | 27 |
| 2/0 | | | | | 1 | 1 | | 1 | 1 | | 1 | 2 | | 3 | 3 | | 5 | 6 | | 7 | 8 | 10 | 13 | 14 | 17 | 18 | 22 |
| 3/0 | | | | | 1 | 1 | | 1 | 1 | | 1 | 1 | | 2 | 3 | | 4 | 5 | | 6 | 7 | 9 | 11 | 12 | 14 | 15 | 18 |
| 4/0 | | | | | | 1 | | 1 | 1 | | 1 | 1 | | 1 | 2 | | 3 | 4 | | 5 | 6 | 7 | 9 | 10 | 12 | 13 | 15 |
| 250 kcmil | | | | | | | | 1 | 1 | | 1 | 1 | | 1 | 1 | | 2 | 3 | | 4 | 4 | 6 | 7 | 8 | 10 | 10 | 12 |
| 300 | | | | | | | | 1 | 1 | | 1 | 1 | | 1 | 1 | | 2 | 3 | | 3 | 4 | 5 | 6 | 7 | 8 | 9 | 11 |
| 350 | | | | | | | | | 1 | | 1 | 1 | | 1 | 1 | | 1 | 2 | | 3 | 3 | 4 | 5 | 6 | 7 | 8 | 9 |
| 400 | | | | | | | | | | | 1 | 1 | | 1 | 1 | | 1 | 1 | | 2 | 3 | 4 | 5 | 5 | 6 | 7 | 8 |
| 500 | | | | | | | | | | | 1 | 1 | | 1 | 1 | | 1 | 1 | | 1 | 2 | 3 | 4 | 4 | 5 | 6 | 7 |
| 600 | | | | | | | | | | | | 1 | | 1 | 1 | | 1 | 1 | | 1 | 1 | 3 | 3 | 4 | 4 | 5 | 5 |
| 700 | | | | | | | | | | | | | | 1 | 1 | | 1 | 1 | | 1 | 1 | 2 | 3 | 3 | 4 | 4 | 5 |
| 750 | | | | | | | | | | | | | | 1 | 1 | | 1 | 1 | | 1 | 1 | 2 | 2 | 3 | 3 | 4 | 4 |

## R312316-40    Excavating

The selection of equipment used for structural excavation and bulk excavation or for grading is determined by the following factors.

1. Quantity of material
2. Type of material
3. Depth or height of cut
4. Length of haul
5. Condition of haul road
6. Accessibility of site
7. Moisture content and dewatering requirements
8. Availability of excavating and hauling equipment

Some additional costs must be allowed for hand trimming the sides and bottom of concrete pours and other excavation below the general excavation.

Number of B.C.Y. per truck = 1.5 C.Y. bucket x 8 passes = 12 loose C.Y.

$$= 12 \text{ x } \frac{100}{118} = 10.2 \text{ B.C.Y. per truck}$$

Truck Haul Cycle:

| Load truck, 8 passes | = | 4 minutes |
|---|---|---|
| Haul distance, 1 mile | = | 9 minutes |
| Dump time | = | 2 minutes |
| Return, 1 mile | = | 7 minutes |
| Spot under machine | = | 1 minute |
| | | 23 minute cycle |

Add the mobilization and demobilization costs to the total excavation costs. When equipment is rented for more than three days, there is often no mobilization charge by the equipment dealer. On larger jobs outside of urban areas, scrapers can move earth economically provided a dump site or fill area and adequate haul roads are available. Excavation within sheeting bracing or cofferdam bracing is usually done with a clamshell and production

When planning excavation and fill, the following should also be considered.

1. Swell factor
2. Compaction factor
3. Moisture content
4. Density requirements

A typical example for scheduling and estimating the cost of excavation of a 15' deep basement on a dry site when the material must be hauled off the site is outlined below.

Assumptions:

1. Swell factor, 18%
2. No mobilization or demobilization
3. Allowance included for idle time and moving on job
4. No dewatering, sheeting, or bracing
5. No truck spotter or hand trimming

Fleet Haul Production per day in B.C.Y.

$$4 \text{ trucks x } \frac{50 \text{ min. hour}}{23 \text{ min. haul cycle}} \text{ x 8 hrs. x 10.2 B.C.Y.}$$

$$= 4 \text{ x } 2.2 \text{ x } 8 \text{ x } 10.2 = 718 \text{ B.C.Y./day}$$

is low, since the clamshell may have to be guided by hand between the bracing. When excavating or filling an area enclosed with a wellpoint system, add 10% to 15% to the cost to allow for restricted access. When estimating earth excavation quantities for structures, allow work space outside the building footprint for construction of the foundation and a slope of 1:1 unless sheeting is used.

## R312316-45   Excavating Equipment

The table below lists THEORETICAL hourly production in C.Y./hr. bank measure for some typical excavation equipment. Figures assume 50 minute hours, 83% job efficiency, 100% operator efficiency, 90° swing and properly sized hauling units, which must be modified for adverse digging and loading conditions. Actual production costs in the front of the book average about 50% of the theoretical values listed here.

| Equipment | Soil Type | B.C.Y. Weight | % Swell | 1 C.Y. | 1-1/2 C.Y. | 2 C.Y. | 2-1/2 C.Y. | 3 C.Y. | 3-1/2 C.Y. | 4 C.Y. |
|---|---|---|---|---|---|---|---|---|---|---|
| Hydraulic Excavator "Backhoe" 15' Deep Cut | Moist loam, sandy clay | 3400 lb. | 40% | 165 | 195 | 200 | 275 | 330 | 385 | 440 |
| | Sand and gravel | 3100 | 18 | 140 | 170 | 225 | 240 | 285 | 330 | 380 |
| | Common earth | 2800 | 30 | 150 | 180 | 230 | 250 | 300 | 350 | 400 |
| | Clay, hard, dense | 3000 | 33 | 120 | 140 | 190 | 200 | 240 | 260 | 320 |
| Power Shovel Optimum Cut (Ft.) | Moist loam, sandy clay | 3400 | 40 | 170 (6.0) | 245 (7.0) | 295 (7.8) | 335 (8.4) | 385 (8.8) | 435 (9.1) | 475 (9.4) |
| | Sand and gravel | 3100 | 18 | 165 (6.0) | 225 (7.0) | 275 (7.8) | 325 (8.4) | 375 (8.8) | 420 (9.1) | 460 (9.4) |
| | Common earth | 2800 | 30 | 145 (7.8) | 200 (9.2) | 250 (10.2) | 295 (11.2) | 335 (12.1) | 375 (13.0) | 425 (13.8) |
| | Clay, hard, dense | 3000 | 33 | 120 (9.0) | 175 (10.7) | 220 (12.2) | 255 (13.3) | 300 (14.2) | 335 (15.1) | 375 (16.0) |
| Drag Line Optimum Cut (Ft.) | Moist loam, sandy clay | 3400 | 40 | 130 (6.6) | 180 (7.4) | 220 (8.0) | 250 (8.5) | 290 (9.0) | 325 (9.5) | 385 (10.0) |
| | Sand and gravel | 3100 | 18 | 130 (6.6) | 175 (7.4) | 210 (8.0) | 245 (8.5) | 280 (9.0) | 315 (9.5) | 375 (10.0) |
| | Common earth | 2800 | 30 | 110 (8.0) | 160 (9.0) | 190 (9.9) | 220 (10.5) | 250 (11.0) | 280 (11.5) | 310 (12.0) |
| | Clay, hard, dense | 3000 | 33 | 90 (9.3) | 130 (10.7) | 160 (11.8) | 190 (12.3) | 225 (12.8) | 250 (13.3) | 280 (12.0) |

| | | | | Wheel Loaders | | | | Track Loaders | | |
|---|---|---|---|---|---|---|---|---|---|---|
| | | | | 3 C.Y. | 4 C.Y. | 6 C.Y. | 8 C.Y. | 2-1/4 C.Y. | 3 C.Y. | 4 C.Y. |
| Loading Tractors | Moist loam, sandy clay | 3400 | 40 | 260 | 340 | 510 | 690 | 135 | 180 | 250 |
| | Sand and gravel | 3100 | 18 | 245 | 320 | 480 | 650 | 130 | 170 | 235 |
| | Common earth | 2800 | 30 | 230 | 300 | 460 | 620 | 120 | 155 | 220 |
| | Clay, hard, dense | 3000 | 33 | 200 | 270 | 415 | 560 | 110 | 145 | 200 |
| | Rock, well-blasted | 4000 | 50 | 180 | 245 | 380 | 520 | 100 | 130 | 180 |

## R312319-90   Wellpoints

A single stage wellpoint system is usually limited to dewatering an average 15' depth below normal ground water level. Multi-stage systems are employed for greater depth with the pumping equipment installed only at the lowest header level. Ejectors with unlimited lift capacity can be economical when two or more stages of wellpoints can be replaced or when horizontal clearance is restricted, such as in deep trenches or tunneling projects, and where low water flows are expected. Wellpoints are usually spaced on 2-1/2' to 10' centers along a header pipe. Wellpoint spacing, header size, and pump size are all determined by the expected flow as dictated by soil conditions.

In almost all soils encountered in wellpoint dewatering, the wellpoints may be jetted into place. Cemented soils and stiff clays may require sand wicks about 12" in diameter around each wellpoint to increase efficiency and eliminate weeping into the excavation. These sand wicks require 1/2 to 3 C.Y. of washed filter sand and are installed by using a 12" diameter steel casing and hole puncher jetted into the ground 2' deeper than the wellpoint. Rock may require predrilled holes.

Labor required for the complete installation and removal of a single stage wellpoint system is in the range of 3/4 to 2 labor-hours per linear foot of header, depending upon jetting conditions, wellpoint spacing, etc.

Continuous pumping is necessary except in some free draining soil where temporary flooding is permissible (as in trenches which are backfilled after each day's work). Good practice requires provision of a stand-by pump during the continuous pumping operation.

Systems for continuous trenching below the water table should be installed three to four times the length of expected daily progress to ensure uninterrupted digging, and header pipe size should not be changed during the job.

For pervious free draining soils, deep wells in place of wellpoints may be economical because of lower installation and maintenance costs. Daily production ranges between two to three wells per day, for 25' to 40' depths, to one well per day for depths over 50'.

Detailed analysis and estimating for any dewatering problem is available at no cost from wellpoint manufacturers. Major firms will quote "sufficient equipment" quotes or their affiliates offer lump sum proposals to cover complete dewatering responsibility.

| Description for 200' System with 8" Header | | Quantities |
|---|---|---|
| Equipment & Material | Wellpoints 25' long, 2" diameter @ 5' O.C. | 40 Each |
| | Header pipe, 8" diameter | 200 L.F. |
| | Discharge pipe, 8" diameter | 100 L.F. |
| | 8" valves | 3 Each |
| | Combination jetting & wellpoint pump (standby) | 1 Each |
| | Wellpoint pump, 8" diameter | 1 Each |
| | Transportation to and from site | 1 Day |
| | Fuel for 30 days x 60 gal./day | 1800 Gallons |
| | Lubricants for 30 days x 16 lbs./day | 480 Lbs. |
| | Sand for points | 40 C.Y. |
| Labor | Technician to supervise installation | 1 Week |
| | Labor for installation and removal of system | 300 Labor-hours |
| | 4 Operators straight time 40 hrs./wk. for 4.33 wks. | 693 Hrs. |
| | 4 Operators overtime 2 hrs./wk. for 4.33 wks. | 35 Hrs. |

## R312323-30   Compacting Backfill

Compaction of fill in embankments, around structures, in trenches, and under slabs is important to control settlement. Factors affecting compaction are:

1. Soil gradation
2. Moisture content
3. Equipment used
4. Depth of fill per lift
5. Density required

Production Rate:

$$\frac{1.75' \text{ plate width x 50 F.P.M. x 50 min./hr. x .67' lift}}{27 \text{ C.F. per C.Y.}} = 108.5 \text{ C.Y./hr.}$$

Production Rate for 4 Passes:

$$\frac{108.5 \text{ C.Y.}}{4 \text{ passes}} = 27.125 \text{ C.Y./hr. x 8 hrs.} = 217 \text{ C.Y./day}$$

**Example:**

Compact granular fill around a building foundation using a 21" wide x 24" vibratory plate in 8" lifts. Operator moves at 50 F.P.M. working a 50 minute hour to develop 95% Modified Proctor Density with 4 passes.

## R314116-40   Wood Sheet Piling

Wood sheet piling may be used for depths to 20' where there is no ground water. If moderate ground water is encountered Tongue & Groove sheeting will help to keep it out. When considerable ground water is present, steel sheeting must be used.

### For estimating purposes on trench excavation, sizes are as follows:

| Depth | Sheeting | Wales | Braces | B.F. per S.F. |
|-------|----------|-------|--------|---------------|
| To 8' | 3 x 12's | 6 x 8's, 2 line | 6 x 8's, @ 10' | 4.0 @ 8' |
| 8' x 12' | 3 x 12's | 10 x 10's, 2 line | 10 x 10's, @ 9' | 5.0 average |
| 12' to 20' | 3 x 12's | 12 x 12's, 3 line | 12 x 12's, @ 8' | 7.0 average |

Sheeting to be toed in at least 2' depending upon soil conditions. A five person crew with an air compressor and sheeting driver can drive and brace 440 SF/day at 8' deep, 360 SF/day at 12' deep, and 320 SF/day at 16' deep.

For normal soils, piling can be pulled in 1/3 the time to install. Pulling difficulty increases with the time in the ground. Production can be increased by high pressure jetting.

## R314116-45   Steel Sheet Piling

Limiting weights are 22 to 38#/S.F. of wall surface with 27#/S.F. average for usual types and sizes. (Weights of piles themselves are from 30.7#/L.F. to 57#/L.F. but they are 15" to 21" wide.) Lightweight sections 12" to 28" wide from 3 ga. to 12 ga. thick are also available for shallow excavations. Piles may be driven two at a time with an impact or vibratory hammer (use vibratory to pull) hung from a crane without leads. A reasonable estimate of the life of steel sheet piling is 10 uses with up to 125 uses possible if a vibratory hammer is used. Used piling costs from 50% to 80% of new piling depending on location and market conditions. Sheet piling and H piles can be rented for about 30% of the delivered mill price for the first month and 5% per month thereafter. Allow 1 labor-hour per pile for cleaning and trimming after driving. These costs increase with depth and hydrostatic head. Vibratory drivers are faster in wet granular soils and are excellent for pile extraction. Pulling difficulty increases with the time in the ground and may cost more than driving. It is often economical to abandon the sheet piling, especially if it can be used as the outer wall form. Allow about 1/3 additional length or more for toeing into ground. Add bracing, waler and strut costs. Waler costs can equal the cost per ton of sheeting.

## R314513-90   Vibroflotation and Vibro Replacement Soil Compaction

**Vibroflotation** is a proprietary system of compacting sandy soils in place to increase relative density to about 70%. Typical bearing capacities attained will be 6000 psf for saturated sand and 12,000 psf for dry sand. Usual range is 4000 to 8000 psf capacity. Costs in the front of the book are for a vertical foot of compacted cylinder 6' to 10' in diameter.

**Vibro replacement** is a proprietary system of improving cohesive soils in place to increase bearing capacity. Most silts and clays above or below the water table can be strengthened by installation of stone columns.

The process consists of radial displacement of the soil by vibration. The created hole is then backfilled in stages with coarse granular fill which is thoroughly compacted and displaced into the surrounding soil in the form of a column.

The total project cost would depend on the number and depth of the compacted cylinders. The installing company guarantees relative soil density of the sand cylinders after compaction and the bearing capacity of the soil after the replacement process. Detailed estimating information is available from the installer at no cost.

## R316326-60 Caissons

The three principal types of cassions are:

**(1) Belled Caissons,** which except for shallow depths and poor soil conditions, are generally recommended. They provide more bearing than shaft area. Because of its conical shape, no horizontal reinforcement of the bell is required.

**(2) Straight Shaft Caissons** are used where relatively light loads are to be supported by caissons that rest on high value bearing strata. While the shaft is larger in diameter than for belled types this is more than offset by the saving in time and labor.

**(3) Keyed Caissons** are used when extremely heavy loads are to be carried. A keyed or socketed caisson transfers its load into rock by a combination of end-bearing and shear reinforcing of the shaft. The most economical shaft often consists of a steel casing, a steel wide flange core and concrete. Allowable compressive stresses of .225 f'c for concrete, 16,000 psi for the wide flange core, and 9,000 psi for the steel casing are commonly used. The usual range of shaft diameter is 18″ to 84″. The number of sizes specified for any one project should be limited due to the problems of casing and auger storage. When hand work is to be performed, shaft diameters should not be less than 32″. When inspection of borings is required a minimum shaft diameter of 30″ is recommended. Concrete caissons are intended to be poured against earth excavation so permanent forms which add to cost should not be used if the excavation is clean and the earth sufficiently impervious to prevent excessive loss of concrete.

| Soil Conditions for Belling | | |
|---|---|---|
| Good | Requires Handwork | Not Recommended |
| Clay | Hard Shale | Silt |
| Sandy Clay | Limestone | Sand |
| Silty Clay | Sandstone | Gravel |
| Clayey Silt | Weathered Mica | Igneous Rock |
| Hard-pan | | |
| Soft Shale | | |
| Decomposed Rock | | |

## R329219-50 Seeding

The type of grass is determined by light, shade and moisture content of soil plus intended use. Fertilizer should be disked 4″ before seeding. For steep slopes disk five tons of mulch and lay two tons of hay or straw on surface per acre after seeding. Surface mulch can be staked, lightly disked or tar emulsion sprayed. Material for mulch can be wood chips, peat moss, partially rotted hay or straw, wood fibers and sprayed emulsions. Hemp seed blankets with fertilizer are also available. For spring seeding, watering is necessary. Late fall seeding may have to be reseeded in the spring. Hydraulic seeding, power mulching, and aerial seeding can be used on large areas.

# R3311 Water Utility Distribution Piping

## R331113-80    Piping Designations

There are several systems currently in use to describe pipe and fittings. The following paragraphs will help to identify and clarify classifications of piping systems used for water distribution.

Piping may be classified by schedule. Piping schedules include 5S, 10S, 10, 20, 30, Standard, 40, 60, Extra Strong, 80, 100, 120, 140, 160 and Double Extra Strong. These schedules are dependent upon the pipe wall thickness. The wall thickness of a particular schedule may vary with pipe size.

**Ductile iron pipe** for water distribution is classified by Pressure Classes such as Class 150, 200, 250, 300 and 350. These classes are actually the rated water working pressure of the pipe in pounds per square inch (psi). The pipe in these pressure classes is designed to withstand the rated water working pressure plus a surge allowance of 100 psi.

The American Water Works Association (AWWA) provides standards for various types of **plastic pipe.** C-900 is the specification for polyvinyl chloride (PVC) piping used for water distribution in sizes ranging from 4″ through 12″. C-901 is the specification for polyethylene (PE) pressure pipe, tubing and fittings used for water distribution in sizes ranging from 1/2″ through 3″. C-905 is the specification for PVC piping sizes 14″ and greater.

**PVC pressure-rated pipe** is identified using the standard dimensional ratio (SDR) method. This method is defined by the American Society for Testing and Materials (ASTM) Standard D 2241. This pipe is available in SDR numbers 64, 41, 32.5, 26, 21, 17, and 13.5. Pipe with an SDR of 64 will have the thinnest wall while pipe with an SDR of 13.5 will have the thickest wall. When the pressure rating (PR) of a pipe is given in psi, it is based on a line supplying water at 73 degrees F.

The National Sanitation Foundation (NSF) seal of approval is applied to products that can be used with potable water. These products have been tested to ANSI/NSF Standard 14.

**Valves and strainers** are classified by American National Standards Institute (ANSI) Classes. These Classes are 125, 150, 200, 250, 300, 400, 600, 900, 1500 and 2500. Within each class there is an operating pressure range dependent upon temperature. Design parameters should be compared to the appropriate material dependent, pressure-temperature rating chart for accurate valve selection.

---

# Transportation    R3472 Railway Construction

## R347216-10    Single Track R.R. Siding

The costs for a single track RR siding in the Unit Price section include the components shown in the table below.

| Description of Component | Qty. per L.F. of Track | Unit |
|---|---|---|
| Ballast, 1-1/2″ crushed stone | .667 | C.Y. |
| 6″ x 8″ x 8′-6″ Treated timber ties, 22″ O.C. | .545 | Ea. |
| Tie plates, 2 per tie | 1.091 | Ea. |
| Track rail | 2.000 | L.F. |
| Spikes, 6″, 4 per tie | 2.182 | Ea. |
| Splice bars w/ bolts, lock washers & nuts, @ 33′ O.C. | .061 | Pair |
| Crew B-14 @ 57 L.F./Day | .018 | Day |

## R347216-20    Single Track, Steel Ties, Concrete Bed

The costs for a R.R. siding with steel ties and a concrete bed in the Unit Price section include the components shown in the table below.

| Description of Component | Qty. per L.F. of Track | Unit |
|---|---|---|
| Concrete bed, 9′ wide, 10″ thick | .278 | C.Y. |
| Ties, W6x16 x 6′-6″ long, @ 30″ O.C. | .400 | Ea. |
| Tie plates, 4 per tie | 1.600 | Ea. |
| Track rail | 2.000 | L.F. |
| Tie plate bolts, 1″, 8 per tie | 3.200 | Ea. |
| Splice bars w/bolts, lock washers & nuts, @ 33′ O.C. | .061 | Pair |
| Crew B-14 @ 22 L.F./Day | .045 | Day |

# Change Orders

## Change Order Considerations

A Change Order is a written document, usually prepared by the design professional, and signed by the owner, the architect/engineer and the contractor. A change order states the agreement of the parties to: an addition, deletion, or revision in the work; an adjustment in the contract sum, if any; or an adjustment in the contract time, if any. Change orders, or "extras" in the construction process occur after execution of the construction contract and impact architects/engineers, contractors and owners.

Change orders that are properly recognized and managed can ensure orderly, professional and profitable progress for all who are involved in the project. There are many causes for change orders and change order requests. In all cases, change orders or change order requests should be addressed promptly and in a precise and prescribed manner. The following paragraphs include information regarding change order pricing and procedures.

## The Causes of Change Orders

Reasons for issuing change orders include:

- Unforeseen field conditions that require a change in the work
- Correction of design discrepancies, errors or omissions in the contract documents
- Owner-requested changes, either by design criteria, scope of work, or project objectives
- Completion date changes for reasons unrelated to the construction process
- Changes in building code interpretations, or other public authority requirements that require a change in the work
- Changes in availability of existing or new materials and products

## Procedures

Properly written contract documents must include the correct change order procedures for all parties—owners, design professionals and contractors—to follow in order to avoid costly delays and litigation.

Being "in the right" is not always a sufficient or acceptable defense. The contract provisions requiring notification and documentation must be adhered to within a defined or reasonable time frame.

The appropriate method of handling change orders is by a written proposal and acceptance by all parties involved. Prior to starting work on a project, all parties should identify their authorized agents who may sign and accept change orders, as well as any limits placed on their authority.

Time may be a critical factor when the need for a change arises. For such cases, the contractor might be directed to proceed on a "time and materials" basis, rather than wait for all paperwork to be processed—a delay that could impede progress. In this situation, the contractor must still follow the prescribed change order procedures, including but not limited to, notification and documentation.

All forms used for change orders should be dated and signed by the proper authority. Lack of documentation can be very costly, especially if legal judgments are to be made and if certain field personnel are no longer available. For time and material change orders, the contractor should keep accurate daily records of all labor and material allocated to the change. Forms that can be used to document change order work are available in *Means Forms for Building Construction Professionals*.

Owners or awarding authorities who do considerable and continual building construction (such as the federal government) realize the inevitability of change orders for numerous reasons, both predictable and unpredictable. As a result, the federal government, the American Institute of Architects (AIA), the Engineers Joint Contract Documents Committee (EJCDC) and other contractor, legal and technical organizations have developed standards and procedures to be followed by all parties to achieve contract continuance and timely completion, while being financially fair to all concerned.

In addition to the change order standards put forth by industry associations, there are also many books available on the subject.

## Pricing Change Orders

When pricing change orders, regardless of their cause, the most significant factor is when the change occurs. The need for a change may be perceived in the field or requested by the architect/engineer *before* any of the actual installation has begun, or may evolve or appear *during* construction when the item of work in question is partially installed. In the latter cases, the original sequence of construction is disrupted, along with all contiguous and supporting systems. Change orders cause the greatest impact when they occur *after* the installation has been completed and must be uncovered, or even replaced. Post-completion changes may be caused by necessary design changes, product failure, or changes in the owner's requirements that are not discovered until the building or the systems begin to function.

Specified procedures of notification and record keeping must be adhered to and enforced regardless of the stage of construction: *before*, *during*, or *after* installation. Some bidding documents anticipate change orders by requiring that unit prices including overhead and profit percentages—for additional as well as deductible changes—be listed. Generally these unit prices do not fully take into account the ripple effect, or impact on other trades, and should be used for general guidance only.

When pricing change orders, it is important to classify the time frame in which the change occurs. There are two basic time frames for change orders: *pre-installation change orders*, which occur before the start of construction, and *post-installation change orders*, which involve reworking after the original installation. Change orders that occur between these stages may be priced according to the extent of work completed using a combination of techniques developed for pricing *pre-* and *post-installation* changes.

The following factors are the basis for a check list to use when preparing a change order estimate.

## Factors To Consider When Pricing Change Orders

As an estimator begins to prepare a change order, the following questions should be reviewed to determine their impact on the final price.

### General

- Is the change order work *pre-installation* or *post-installation?*

Change order work costs vary according to how much of the installation has been completed. Once workers have the project scoped in their mind, even though they have not started, it can be difficult to refocus.

Consequently they may spend more than the normal amount of time understanding the change. Also, modifications to work in place such as trimming or refitting usually take more time than was initially estimated. The greater the amount of work in place, the more reluctant workers are to change it. Psychologically they may resent the change and as a result the rework takes longer than normal. Post-installation change order estimates must include demolition of existing work as required to accomplish the change. If the work is performed at a later time, additional obstacles such as building finishes may be present which must be protected. Regardless of whether the change

occurs pre-installation or post-installation, attempt to isolate the identifiable factors and price them separately. For example, add shipping costs that may be required pre-installation or any demolition required post-installation. Then analyze the potential impact on productivity of psychological and/or learning curve factors and adjust the output rates accordingly. One approach is to break down the typical workday into segments and quantify the impact on each segment. The following chart may be useful as a guide:

| | Activities (Productivity) Expressed as Percentages of a Workday | | |
|---|---|---|---|
| Task | Means Mechanical Cost Data (for New Construction) | Pre-Installation Change Orders | Post-Installation Change Orders |
| 1. Study plans | 3% | 6% | 6% |
| 2. Material procurement | 3% | 3% | 3% |
| 3. Receiving and storing | 3% | 3% | 3% |
| 4. Mobilization | 5% | 5% | 5% |
| 5. Site movement | 5% | 5% | 8% |
| 6. Layout and marking | 8% | 10% | 12% |
| 7. Actual installation | 64% | 59% | 54% |
| 8. Clean-up | 3% | 3% | 3% |
| 9. Breaks—non-productive | 6% | 6% | 6% |
| Total | 100% | 100% | 100% |

### Change Order Installation Efficiency

The labor-hours expressed (for new construction) are based on average installation time, using an efficiency level of approximately 60-65%. For change order situations, adjustments to this efficiency level should reflect the daily labor-hour allocation for that particular occurrence.

If any of the specific percentages expressed in the above chart do not apply to a particular project situation, then those percentage points should be reallocated to the appropriate task(s). Example: Using data for new construction, assume there is no new material being utilized. The percentages for Tasks 2 and 3 would therefore be reallocated to other tasks. If the time required for Tasks 2 and 3 can now be applied to installation, we can add the time allocated for *Material Procurement* and *Receiving and Storing* to the *Actual Installation* time for new construction, thereby increasing the Actual Installation percentage.

This chart shows that, due to reduced productivity, labor costs will be higher than those for new construction by 5% to 15% for pre-installation change orders and by 15% to 25% for post-installation change orders. Each job and change order is unique and must be examined individually. Many factors, covered elsewhere in this section, can each have a significant impact on productivity and change order costs. All such factors should be considered in every case.

- Will the change substantially delay the original completion date?

A significant change in the project may cause the original completion date to be extended. The extended schedule may subject the contractor to new wage rates dictated by relevant labor contracts. Project supervision and other project overhead must also be extended beyond the original completion date. The schedule extension may also put installation into a new weather season. For example, underground piping scheduled for October installation was delayed until January. As a result, frost penetrated the trench area, thereby changing the degree of difficulty of the task. Changes and delays may have a ripple effect throughout the project. This effect must be analyzed and negotiated with the owner.

- What is the net effect of a deduct change order?

In most cases, change orders resulting in a deduction or credit reflect only bare costs. The contractor may retain the overhead and profit based on the original bid.

## Materials

- Will you have to pay more or less for the new material, required by the change order, than you paid for the original purchase?

The same material prices or discounts will usually apply to materials purchased for change orders as new construction. In some instances, however, the contractor may forfeit the advantages of competitive pricing for change orders. Consider the following example:

A contractor purchased over $20,000 worth of fan coil units for an installation, and obtained the maximum discount. Some time later it was determined the project required an additional matching unit. The contractor has to purchase this unit from the original supplier to ensure a match. The supplier at this time may not discount the unit because of the small quantity, and the fact that he is no longer in a competitive situation. The impact of quantity on purchase can add between 0% and 25% to material prices and/or subcontractor quotes.

- If materials have been ordered or delivered to the job site, will they be subject to a cancellation charge or restocking fee?

Check with the supplier to determine if ordered materials are subject to a cancellation charge. Delivered materials not used as result of a change order may be subject to a restocking fee if returned to the supplier. Common restocking charges run between 20% and 40%. Also, delivery charges to return the goods to the supplier must be added.

## Labor

- How efficient is the existing crew at the actual installation?

Is the same crew that performed the initial work going to do the change order? Possibly the change consists of the installation of a unit identical to one already installed; therefore the change should take less time. Be sure to consider this potential productivity increase and modify the productivity rates accordingly.

- If the crew size is increased, what impact will that have on supervision requirements?

Under most bargaining agreements or management practices, there is a point at which a working foreman is replaced by a nonworking foreman. This replacement increases project overhead by adding a nonproductive worker. If additional workers are added to accelerate the project or to perform changes while maintaining the schedule, be sure to add additional supervision time if warranted. Calculate the hours involved and the additional cost directly if possible.

- What are the other impacts of increased crew size?

The larger the crew, the greater the potential for productivity to decrease. Some of the factors that cause this productivity loss are: overcrowding (producing restrictive conditions in the working space), and possibly a shortage of any special tools and equipment required. Such factors affect not only the crew working on the elements directly involved in the change order, but other crews whose movement may also be hampered.

As the crew increases, check its basic composition for changes by the addition or deletion of apprentices or nonworking foreman and quantify the potential effects of equipment shortages or other logistical factors.

- As new crews, unfamiliar with the project, are brought onto the site, how long will it take them to become oriented to the project requirements?

The orientation time for a new crew to become 100% effective varies with the site and type of project. Orientation is easiest at a new construction site, and most difficult at existing, very restrictive renovation sites. The type of work also affects orientation time. When all elements of the work are exposed, such as concrete or masonry work, orientation is decreased. When the work is concealed or less visible, such as existing electrical systems, orientation takes longer. Usually orientation can be accomplished in one day or less. Costs for added orientation should be itemized and added to the total estimated cost.

- How much actual production can be gained by working overtime?

Short term overtime can be used effectively to accomplish more work in a day. However, as overtime is scheduled to run beyond several weeks, studies have shown marked decreases in output. The following chart shows the effect of long term overtime on worker efficiency. If the anticipated change requires extended overtime to keep the job on schedule, these factors can be used as a guide to predict the impact on time and cost. Add project overhead, particularly supervision, that may also be incurred.

| Days per Week | Hours per Day | Production Efficiency | | | | | Payroll Cost Factors | |
|---|---|---|---|---|---|---|---|---|
| | | 1 Week | 2 Weeks | 3 Weeks | 4 Weeks | Average 4 Weeks | @ 1-1/2 Times | @ 2 Times |
| 5 | 8 | 100% | 100% | 100% | 100% | 100% | 100% | 100% |
| | 9 | 100 | 100 | 95 | 90 | 96.25 | 105.6 | 111.1 |
| | 10 | 100 | 95 | 90 | 85 | 91.25 | 110.0 | 120.0 |
| | 11 | 95 | 90 | 75 | 65 | 81.25 | 113.6 | 127.3 |
| | 12 | 90 | 85 | 70 | 60 | 76.25 | 116.7 | 133.3 |
| 6 | 8 | 100 | 100 | 95 | 90 | 96.25 | 108.3 | 116.7 |
| | 9 | 100 | 95 | 90 | 85 | 92.50 | 113.0 | 125.9 |
| | 10 | 95 | 90 | 85 | 80 | 87.50 | 116.7 | 133.3 |
| | 11 | 95 | 85 | 70 | 65 | 78.75 | 119.7 | 139.4 |
| | 12 | 90 | 80 | 65 | 60 | 73.75 | 122.2 | 144.4 |
| 7 | 8 | 100 | 95 | 85 | 75 | 88.75 | 114.3 | 128.6 |
| | 9 | 95 | 90 | 80 | 70 | 83.75 | 118.3 | 136.5 |
| | 10 | 90 | 85 | 75 | 65 | 78.75 | 121.4 | 142.9 |
| | 11 | 85 | 80 | 65 | 60 | 72.50 | 124.0 | 148.1 |
| | 12 | 85 | 75 | 60 | 55 | 68.75 | 126.2 | 152.4 |

## Effects of Overtime

Caution: Under many labor agreements, Sundays and holidays are paid at a higher premium than the normal overtime rate.

The use of long-term overtime is counterproductive on almost any construction job; that is, the longer the period of overtime, the lower the actual production rate. Numerous studies have been conducted, and while they have resulted in slightly different numbers, all reach the same conclusion. The figure above tabulates the effects of overtime work on efficiency.

As illustrated, there can be a difference between the *actual* payroll cost per hour and the *effective* cost per hour for overtime work. This is due to the reduced production efficiency with the increase in weekly hours beyond 40. This difference between actual and effective cost results from overtime work over a prolonged period. Short-term overtime work does not result in as great a reduction in efficiency, and in such cases, effective cost may not vary significantly from the actual payroll cost. As the total hours per week are increased on a regular basis, more time is lost because of fatigue, lowered morale, and an increased accident rate.

As an example, assume a project where workers are working 6 days a week, 10 hours per day. From the figure above (based on productivity studies), the average effective productive hours over a four-week period are:

$$0.875 \times 60 = 52.5$$

Depending upon the locale and day of week, overtime hours may be paid at time and a half or double time. For time and a half, the overall (average) *actual* payroll cost (including regular and overtime hours) is determined as follows:

$$\frac{40 \text{ reg. hrs.} + (20 \text{ overtime hrs.} \times 1.5)}{60 \text{ hrs.}} = 1.167$$

Based on 60 hours, the payroll cost per hour will be 116.7% of the normal rate at 40 hours per week. However, because the effective production (efficiency) for 60 hours is reduced to the equivalent of 52.5 hours, the effective cost of overtime is calculated as follows:

For time and a half:

$$\frac{40 \text{ reg. hrs.} + (20 \text{ overtime hrs.} \times 1.5)}{52.5 \text{ hrs.}} = 1.33$$

Installed cost will be 133% of the normal rate (for labor).

Thus, when figuring overtime, the actual cost per unit of work will be higher than the apparent overtime payroll dollar increase, due to the reduced productivity of the longer workweek. These efficiency calculations are true only for those cost factors determined by hours worked. Costs that are applied weekly or monthly, such as equipment rentals, will not be similarly affected.

## Equipment

- What equipment is required to complete the change order?

Change orders may require extending the rental period of equipment already on the job site, or the addition of special equipment brought in to accomplish the change work. In either case, the additional rental charges and operator labor charges must be added.

## *Summary*

The preceding considerations and others you deem appropriate should be analyzed and applied to a change order estimate. The impact of each should be quantified and listed on the estimate to form an audit trail.

Change orders that are properly identified, documented, and managed help to ensure the orderly, professional and profitable progress of the work. They also minimize potential claims or disputes at the end of the project.

# Estimating Tips

- The cost figures in this Square Foot Cost section were derived from approximately 11,200 projects contained in the RSMeans database of completed construction projects. They include the contractor's overhead and profit, but do not generally include architectural fees or land costs. The figures have been adjusted to January of the current year. New projects are added to our files each year, and outdated projects are discarded. For this reason, certain costs may not show a uniform annual progression. In no case are all subdivisions of a project listed.

- These projects were located throughout the U.S. and reflect a tremendous variation in square foot (S.F.) and cubic foot (C.F.) costs. This is due to differences, not only in labor and material costs, but also in individual owners' requirements. For instance, a bank in a large city would have different features than one in a rural area. This is true of all the different types of buildings analyzed. Therefore, caution should be exercised when using these Square Foot costs. For example, for court houses, costs in the database are local court house costs and will not apply to the larger, more elaborate federal court houses. As a general rule, the projects in the 1/4 column do not include any site work or equipment, while the projects in the 3/4 column may include both equipment and site work. The median figures do not generally include site work.

- None of the figures "go with" any others. All individual cost items were computed and tabulated separately. Thus, the sum of the median figures for Plumbing, HVAC and Electrical will not normally total up to the total Mechanical and Electrical costs arrived at by separate analysis and tabulation of the projects.

- Each building was analyzed as to total and component costs and percentages. The figures were arranged in ascending order with the results tabulated as shown. The 1/4 column shows that 25% of the projects had lower costs and 75% had higher. The 3/4 column shows that 75% of the projects had lower costs and 25% had higher. The median column shows that 50% of the projects had lower costs and 50% had higher.

- There are two times when square foot costs are useful. The first is in the conceptual stage when no details are available. Then square foot costs make a useful starting point. The second is after the bids are in and the costs can be worked back into their appropriate units for information purposes. As soon as details become available in the project design, the square foot approach should be discontinued and the project priced as to its particular components. When more precision is required, or for estimating the replacement cost of specific buildings, the current edition of *RSMeans Square Foot Costs* should be used.

- In using the figures in this section, it is recommended that the median column be used for preliminary figures if no additional information is available. The median figures, when multiplied by the total city construction cost index figures (see City Cost Indexes) and then multiplied by the project size modifier at the end of this section, should present a fairly accurate base figure, which would then have to be adjusted in view of the estimator's experience, local economic conditions, code requirements, and the owner's particular requirements. There is no need to factor the percentage figures, as these should remain constant from city to city. All tabulations mentioning air conditioning had at least partial air conditioning.

- The editors of this book would greatly appreciate receiving cost figures on one or more of your recent projects, which would then be included in the averages for next year. All cost figures received will be kept confidential, except that they will be averaged with other similar projects to arrive at Square Foot cost figures for next year's book. See the last page of the book for details and the discount available for submitting one or more of your projects.

| 50 17 00 | S.F. Costs | | UNIT | UNIT COSTS | | | % OF TOTAL | | |
|---|---|---|---|---|---|---|---|---|---|
| | | | | 1/4 | MEDIAN | 3/4 | 1/4 | MEDIAN | 3/4 |
| **01** | 0010 | **APARTMENTS Low Rise (1 to 3 story)** | S.F. | 67 | 84 | 112 | | | | **01** |
| | 0020 | Total project cost | C.F. | 6 | 7.95 | 9.80 | | | |
| | 0100 | Site work | S.F. | 5.70 | 7.80 | 13.70 | 6.05% | 10.55% | 14.05% |
| | 0500 | Masonry | | 1.31 | 3.05 | 5.30 | 1.54% | 3.67% | 6.35% |
| | 1500 | Finishes | | 7.10 | 9.70 | 12 | 9.05% | 10.75% | 12.85% |
| | 1800 | Equipment | | 2.18 | 3.31 | 4.92 | 2.73% | 4.03% | 5.95% |
| | 2720 | Plumbing | | 5.20 | 6.70 | 8.50 | 6.65% | 8.95% | 10.05% |
| | 2770 | Heating, ventilating, air conditioning | | 3.31 | 4.08 | 6 | 4.20% | 5.60% | 7.60% |
| | 2900 | Electrical | | 3.86 | 5.15 | 6.90 | 5.20% | 6.65% | 8.40% |
| | 3100 | Total: Mechanical & Electrical | ↓ | 13.40 | 17.05 | 21.50 | 15.90% | 18.05% | 23% |
| | 9000 | Per apartment unit, total cost | Apt. | 62,000 | 95,000 | 140,000 | | | |
| | 9500 | Total: Mechanical & Electrical | " | 11,800 | 18,500 | 24,200 | | | |
| **02** | 0010 | **APARTMENTS Mid Rise (4 to 7 story)** | S.F. | 88.50 | 107 | 132 | | | | **02** |
| | 0020 | Total project costs | C.F. | 6.90 | 9.55 | 13.05 | | | |
| | 0100 | Site work | S.F. | 3.54 | 7 | 12.60 | 5.25% | 6.70% | 9.15% |
| | 0500 | Masonry | | 5.90 | 8.10 | 11.10 | 5.10% | 7.25% | 10.50% |
| | 1500 | Finishes | | 11.15 | 14.50 | 18.30 | 10.55% | 13.45% | 17.70% |
| | 1800 | Equipment | | 2.57 | 3.85 | 5.05 | 2.54% | 3.48% | 4.31% |
| | 2500 | Conveying equipment | | 1.91 | 2.43 | 2.94 | 1.94% | 2.27% | 2.69% |
| | 2720 | Plumbing | | 5.20 | 8.30 | 8.80 | 5.70% | 7.20% | 8.95% |
| | 2900 | Electrical | | 5.85 | 7.90 | 9.60 | 6.65% | 7.20% | 8.95% |
| | 3100 | Total: Mechanical & Electrical | ↓ | 18.70 | 23.50 | 28.50 | 18.50% | 21% | 23% |
| | 9000 | Per apartment unit, total cost | Apt. | 100,000 | 118,000 | 195,500 | | | |
| | 9500 | Total: Mechanical & Electrical | " | 18,900 | 21,900 | 27,600 | | | |
| **03** | 0010 | **APARTMENTS High Rise (8 to 24 story)** | S.F. | 100 | 116 | 139 | | | | **03** |
| | 0020 | Total project costs | C.F. | 9.75 | 11.35 | 14.50 | | | |
| | 0100 | Site work | S.F. | 3.64 | 5.90 | 8.25 | 2.58% | 4.84% | 6.15% |
| | 0500 | Masonry | | 5.80 | 10.55 | 13.15 | 4.74% | 9.65% | 11.05% |
| | 1500 | Finishes | | 11.15 | 13.90 | 16.45 | 9.75% | 11.80% | 13.70% |
| | 1800 | Equipment | | 3.23 | 3.97 | 5.25 | 2.78% | 3.49% | 4.35% |
| | 2500 | Conveying equipment | | 2.28 | 3.46 | 4.71 | 2.23% | 2.78% | 3.37% |
| | 2720 | Plumbing | | 7.40 | 8.70 | 12.15 | 6.80% | 7.20% | 10.45% |
| | 2900 | Electrical | | 6.90 | 8.70 | 11.75 | 6.45% | 7.65% | 8.80% |
| | 3100 | Total: Mechanical & Electrical | ↓ | 20.50 | 26.50 | 31.50 | 17.95% | 22.50% | 24.50% |
| | 9000 | Per apartment unit, total cost | Apt. | 104,500 | 115,000 | 159,500 | | | |
| | 9500 | Total: Mechanical & Electrical | " | 22,600 | 25,800 | 27,300 | | | |
| **04** | 0010 | **AUDITORIUMS** | S.F. | 104 | 142 | 204 | | | | **04** |
| | 0020 | Total project costs | C.F. | 6.55 | 9.10 | 13.05 | | | |
| | 2720 | Plumbing | S.F. | 6.30 | 9.15 | 11.10 | 5.85% | 7.20% | 8.70% |
| | 2900 | Electrical | | 8.05 | 11.75 | 19.30 | 6.80% | 9.05% | 11.35% |
| | 3100 | Total: Mechanical & Electrical | ↓ | 50.50 | 65.50 | 90.50 | 24.50% | 29% | 31.50% |
| **05** | 0010 | **AUTOMOTIVE SALES** | S.F. | 77 | 105 | 130 | | | | **05** |
| | 0020 | Total project costs | C.F. | 5.10 | 6.10 | 7.90 | | | |
| | 2720 | Plumbing | S.F. | 3.52 | 6.10 | 6.65 | 2.89% | 6.05% | 6.50% |
| | 2770 | Heating, ventilating, air conditioning | | 5.45 | 8.30 | 8.95 | 4.61% | 10% | 10.35% |
| | 2900 | Electrical | | 6.20 | 9.75 | 13.25 | 7.25% | 8.80% | 12.15% |
| | 3100 | Total: Mechanical & Electrical | ↓ | 19.50 | 27.50 | 33.50 | 18.90% | 20.50% | 22% |
| **06** | 0010 | **BANKS** | S.F. | 152 | 189 | 239 | | | | **06** |
| | 0020 | Total project costs | C.F. | 10.80 | 14.70 | 19.40 | | | |
| | 0100 | Site work | S.F. | 17.35 | 26.50 | 38.50 | 7.85% | 12.95% | 16.95% |
| | 0500 | Masonry | | 8.15 | 15.60 | 27.50 | 3.36% | 6.95% | 10.05% |
| | 1500 | Finishes | | 14 | 20.50 | 25.50 | 5.85% | 8.60% | 11.55% |
| | 1800 | Equipment | | 5.70 | 12.55 | 25.50 | 1% | 5.55% | 10.50% |
| | 2720 | Plumbing | | 4.74 | 6.80 | 9.90 | 2.82% | 3.90% | 4.93% |
| | 2770 | Heating, ventilating, air conditioning | | 9 | 12.05 | 16.05 | 4.86% | 7.15% | 8.50% |
| | 2900 | Electrical | | 14.40 | 19.15 | 25 | 8.20% | 10.20% | 12.20% |
| | 3100 | Total: Mechanical & Electrical | ↓ | 34.50 | 46.50 | 55.50 | 16.25% | 19.45% | 23% |
| | 3500 | See also division 11 16 00 & 11 17 00 | | | | | | | |

## 50 17 00 | S.F. Costs

| | | | UNIT | UNIT COSTS | | | % OF TOTAL | | |
|---|---|---|---|---|---|---|---|---|---|
| | | | | 1/4 | MEDIAN | 3/4 | 1/4 | MEDIAN | 3/4 |
| **13** | 0010 | **CHURCHES** | S.F. | 102 | 129 | 168 | | | |
| | 0020 | Total project costs | C.F. | 6.30 | 8 | 10.50 | | | |
| | 1800 | Equipment | S.F. | 1.22 | 2.91 | 6.20 | .95% | 2.11% | 4.31% |
| | 2720 | Plumbing | | 3.97 | 5.55 | 8.15 | 3.51% | 4.96% | 6.25% |
| | 2770 | Heating, ventilating, air conditioning | | 9.30 | 12.10 | 17.10 | 7.50% | 10% | 12% |
| | 2900 | Electrical | | 8.60 | 11.80 | 15.85 | 7.30% | 8.80% | 10.95% |
| | 3100 | Total: Mechanical & Electrical | | 27 | 35.50 | 47.50 | 18.30% | 22% | 25% |
| | 3500 | See also division 11 91 00 | | | | | | | |
| **15** | 0010 | **CLUBS, COUNTRY** | S.F. | 109 | 132 | 166 | | | |
| | 0020 | Total project costs | C.F. | 8.80 | 10.75 | 14.85 | | | |
| | 2720 | Plumbing | S.F. | 6.60 | 9.80 | 22.50 | 5.60% | 7.90% | 10% |
| | 2900 | Electrical | | 8.60 | 11.80 | 15.35 | 7% | 8.95% | 11% |
| | 3100 | Total: Mechanical & Electrical | | 45.50 | 57 | 60.50 | 19% | 26.50% | 29.50% |
| **17** | 0010 | **CLUBS, SOCIAL Fraternal** | S.F. | 87 | 125 | 168 | | | |
| | 0020 | Total project costs | C.F. | 5.45 | 8.25 | 9.85 | | | |
| | 2720 | Plumbing | S.F. | 5.50 | 6.85 | 10.35 | 5.60% | 6.90% | 8.55% |
| | 2770 | Heating, ventilating, air conditioning | | 7.90 | 9.55 | 12.30 | 8.20% | 9.25% | 14.40% |
| | 2900 | Electrical | | 6.55 | 10.80 | 12.35 | 6.50% | 9.50% | 10.55% |
| | 3100 | Total: Mechanical & Electrical | | 34 | 36.50 | 46.50 | 21% | 23% | 23.50% |
| **18** | 0010 | **CLUBS, Y.M.C.A.** | S.F. | 114 | 146 | 181 | | | |
| | 0020 | Total project costs | C.F. | 5.05 | 8.45 | 12.60 | | | |
| | 2720 | Plumbing | S.F. | 6.90 | 13.80 | 15.45 | 5.65% | 7.60% | 10.85% |
| | 2900 | Electrical | | 8.95 | 10.95 | 16.40 | 6.25% | 7.95% | 9.25% |
| | 3100 | Total: Mechanical & Electrical | | 33.50 | 40.50 | 68 | 20.50% | 22.50% | 28.50% |
| **19** | 0010 | **COLLEGES Classrooms & Administration** | S.F. | 111 | 151 | 201 | | | |
| | 0020 | Total project costs | C.F. | 8.10 | 11.75 | 18.10 | | | |
| | 0500 | Masonry | S.F. | 8.10 | 15.30 | 18.80 | 5.10% | 7.45% | 10.50% |
| | 2720 | Plumbing | | 5.65 | 11.55 | 21.50 | 5.10% | 6.60% | 8.95% |
| | 2900 | Electrical | | 9.20 | 14 | 19.45 | 7.80% | 9.85% | 12% |
| | 3100 | Total: Mechanical & Electrical | | 36.50 | 51.50 | 69 | 23.50% | 28% | 32% |
| **21** | 0010 | **COLLEGES Science, Engineering, Laboratories** | S.F. | 206 | 241 | 279 | | | |
| | 0020 | Total project costs | C.F. | 11.80 | 17.20 | 19.55 | | | |
| | 1800 | Equipment | S.F. | 11.45 | 26 | 28.50 | 2% | 6.45% | 12.65% |
| | 2900 | Electrical | | 16.95 | 24 | 37 | 7.10% | 9.40% | 12.10% |
| | 3100 | Total: Mechanical & Electrical | | 63 | 74.50 | 116 | 28.50% | 31.50% | 41% |
| | 3500 | See also division 11 53 00 | | | | | | | |
| **23** | 0010 | **COLLEGES Student Unions** | S.F. | 131 | 178 | 215 | | | |
| | 0020 | Total project costs | C.F. | 7.30 | 9.60 | 11.85 | | | |
| | 3100 | Total: Mechanical & Electrical | S.F. | 49.50 | 53.50 | 63 | 23.50% | 26% | 29% |
| **25** | 0010 | **COMMUNITY CENTERS** | S.F. | 108 | 134 | 181 | | | |
| | 0020 | Total project costs | C.F. | 7.05 | 10.10 | 13.10 | | | |
| | 1800 | Equipment | S.F. | 2.62 | 4.44 | 7.05 | 1.50% | 3.04% | 5.45% |
| | 2720 | Plumbing | | 5.15 | 9 | 12.25 | 4.85% | 7% | 8.95% |
| | 2770 | Heating, ventilating, air conditioning | | 8.20 | 12.05 | 17.15 | 6.80% | 10.35% | 12.90% |
| | 2900 | Electrical | | 8.85 | 11.80 | 17 | 7.20% | 8.95% | 10.45% |
| | 3100 | Total: Mechanical & Electrical | | 30.50 | 38 | 55 | 20% | 25% | 31% |
| **28** | 0010 | **COURT HOUSES** | S.F. | 156 | 184 | 254 | | | |
| | 0020 | Total project costs | C.F. | 12 | 14.35 | 18.05 | | | |
| | 2720 | Plumbing | S.F. | 7.45 | 10.40 | 14.90 | 5.95% | 7.45% | 8.20% |
| | 2900 | Electrical | | 16.60 | 18.45 | 27 | 8.90% | 10.45% | 11.85% |
| | 3100 | Total: Mechanical & Electrical | | 46.50 | 61 | 66.50 | 22.50% | 27.50% | 30% |
| **30** | 0010 | **DEPARTMENT STORES** | S.F. | 58 | 78 | 98.50 | | | |
| | 0020 | Total project costs | C.F. | 3.10 | 4.01 | 5.45 | | | |
| | 2720 | Plumbing | S.F. | 1.80 | 2.27 | 3.45 | 1.82% | 4.21% | 5.90% |
| | 2770 | Heating, ventilating, air conditioning | | 5.25 | 8.10 | 12.20 | 8.20% | 9.10% | 14.80% |

| | | 50 17 00 \| S.F. Costs | UNIT | UNIT COSTS | | | % OF TOTAL | | | |
|---|---|---|---|---|---|---|---|---|---|---|
| | | | | 1/4 | MEDIAN | 3/4 | 1/4 | MEDIAN | 3/4 | |
| 30 | 2900 | Electrical | S.F. | 6.65 | 9.10 | 10.75 | 9.05% | 12.15% | 14.95% | 30 |
| | 3100 | Total: Mechanical & Electrical | ↓ | 11.65 | 14.95 | 26 | 13.20% | 21.50% | 50% | |
| 31 | 0010 | **DORMITORIES Low Rise (1 to 3 story)** | S.F. | 110 | 143 | 190 | | | | 31 |
| | 0020 | Total project costs | C.F. | 6.20 | 10 | 15 | | | | |
| | 2720 | Plumbing | S.F. | 6.60 | 8.80 | 11.15 | 8.05% | 9% | 9.65% | |
| | 2770 | Heating, ventilating, air conditioning | | 6.95 | 8.35 | 11.10 | 4.61% | 8.05% | 10% | |
| | 2900 | Electrical | | 7.25 | 11.05 | 15.10 | 6.40% | 8.65% | 9.50% | |
| | 3100 | Total: Mechanical & Electrical | ↓ | 38 | 40.50 | 63 | 22% | 25% | 27% | |
| | 9000 | Per bed, total cost | Bed | 46,400 | 51,500 | 110,500 | | | | |
| 32 | 0010 | **DORMITORIES Mid Rise (4 to 8 story)** | S.F. | 135 | 176 | 217 | | | | 32 |
| | 0020 | Total project costs | C.F. | 14.85 | 16.30 | 19.50 | | | | |
| | 2900 | Electrical | S.F. | 14.30 | 16.25 | 22 | 8.20% | 10.20% | 11.95% | |
| | 3100 | Total: Mechanical & Electrical | " | 40 | 80 | 81 | 25.50% | 34.50% | 37.50% | |
| | 9000 | Per bed, total cost | Bed | 19,200 | 43,700 | 253,000 | | | | |
| 34 | 0010 | **FACTORIES** | S.F. | 51 | 76 | 117 | | | | 34 |
| | 0020 | Total project costs | C.F. | 3.27 | 4.87 | 8.10 | | | | |
| | 0100 | Site work | S.F. | 5.80 | 10.60 | 16.75 | 6.95% | 11.45% | 17.95% | |
| | 2720 | Plumbing | | 2.74 | 5.10 | 8.45 | 3.73% | 6.05% | 8.10% | |
| | 2770 | Heating, ventilating, air conditioning | | 5.35 | 7.65 | 10.35 | 5.25% | 8.45% | 11.35% | |
| | 2900 | Electrical | | 6.35 | 10 | 15.30 | 8.10% | 10.50% | 14.20% | |
| | 3100 | Total: Mechanical & Electrical | ↓ | 18.10 | 27.50 | 36.50 | 21% | 28.50% | 35.50% | |
| 36 | 0010 | **FIRE STATIONS** | S.F. | 101 | 139 | 187 | | | | 36 |
| | 0020 | Total project costs | C.F. | 5.90 | 8.10 | 10.80 | | | | |
| | 0500 | Masonry | S.F. | 14.95 | 26.50 | 35 | 8.35% | 11.55% | 16.15% | |
| | 1140 | Roofing | | 3.29 | 8.90 | 10.15 | 1.90% | 4.94% | 5.05% | |
| | 1580 | Painting | | 2.56 | 3.82 | 3.91 | 1.37% | 1.57% | 2.07% | |
| | 1800 | Equipment | | 1.26 | 2.41 | 4.45 | .62% | 1.86% | 3.54% | |
| | 2720 | Plumbing | | 5.65 | 9.10 | 13.15 | 5.85% | 7.35% | 9.45% | |
| | 2770 | Heating, ventilating, air conditioning | | 5.60 | 9.05 | 14 | 5.15% | 7.40% | 9.40% | |
| | 2900 | Electrical | | 7.20 | 12.65 | 17.15 | 6.80% | 8.60% | 10.60% | |
| | 3100 | Total: Mechanical & Electrical | ↓ | 37 | 47.50 | 53.50 | 18.40% | 23% | 26% | |
| 37 | 0010 | **FRATERNITY HOUSES & Sorority Houses** | S.F. | 101 | 130 | 178 | | | | 37 |
| | 0020 | Total project costs | C.F. | 10.05 | 10.50 | 12.65 | | | | |
| | 2720 | Plumbing | S.F. | 7.65 | 8.75 | 16 | 6.80% | 8% | 10.85% | |
| | 2900 | Electrical | | 6.65 | 14.40 | 17.65 | 6.60% | 9.90% | 10.65% | |
| | 3100 | Total: Mechanical & Electrical | ↓ | 7.85 | 25.50 | 30.50 | | 15.10% | 15.90% | |
| 38 | 0010 | **FUNERAL HOMES** | S.F. | 107 | 145 | 264 | | | | 38 |
| | 0020 | Total project costs | C.F. | 10.90 | 12.10 | 23.50 | | | | |
| | 2900 | Electrical | S.F. | 4.71 | 8.65 | 9.45 | 3.58% | 4.44% | 5.95% | |
| | 3100 | Total: Mechanical & Electrical | ↓ | 16.75 | 24.50 | 34 | 12.90% | 12.90% | 12.90% | |
| 39 | 0010 | **GARAGES, COMMERCIAL (Service)** | S.F. | 60.50 | 93.50 | 129 | | | | 39 |
| | 0020 | Total project costs | C.F. | 3.97 | 5.85 | 8.50 | | | | |
| | 1800 | Equipment | S.F. | 3.40 | 7.65 | 11.90 | 2.21% | 4.62% | 6.80% | |
| | 2720 | Plumbing | | 4.18 | 6.45 | 11.75 | 5.45% | 7.85% | 10.65% | |
| | 2730 | Heating & ventilating | | 5.50 | 7.30 | 9.85 | 5.25% | 6.85% | 8.20% | |
| | 2900 | Electrical | | 5.75 | 8.75 | 12.60 | 7.15% | 9.25% | 10.85% | |
| | 3100 | Total: Mechanical & Electrical | ↓ | 12.65 | 24.50 | 36 | 12.35% | 17.40% | 26% | |
| 40 | 0010 | **GARAGES, MUNICIPAL (Repair)** | S.F. | 94 | 119 | 169 | | | | 40 |
| | 0020 | Total project costs | C.F. | 5.55 | 7.05 | 12.10 | | | | |
| | 0500 | Masonry | S.F. | 8.30 | 16.25 | 25.50 | 5.60% | 9.15% | 12.50% | |
| | 2720 | Plumbing | | 3.98 | 7.65 | 14.40 | 3.59% | 6.70% | 7.95% | |
| | 2730 | Heating & ventilating | | 6.80 | 9.85 | 19 | 6.15% | 7.45% | 13.50% | |
| | 2900 | Electrical | | 6.55 | 10.50 | 14.95 | 6.65% | 9.25% | 11.15% | |
| | 3100 | Total: Mechanical & Electrical | ↓ | 30.50 | 42 | 61 | 21.50% | 23% | 28.50% | |

## 50 17 00 | S.F. Costs

| | | | UNIT | UNIT COSTS | | | % OF TOTAL | | |
|---|---|---|---|---|---|---|---|---|---|
| | | | | 1/4 | MEDIAN | 3/4 | 1/4 | MEDIAN | 3/4 |
| **41** | 0010 | **GARAGES, PARKING** | S.F. | 34.50 | 50.50 | 86.50 | | | | **41** |
| | 0020 | Total project costs | C.F. | 3.24 | 4.40 | 6.40 | | | |
| | 2720 | Plumbing | S.F. | .98 | 1.52 | 2.34 | 1.72% | 2.70% | 3.85% |
| | 2900 | Electrical | | 1.89 | 2.32 | 3.65 | 4.33% | 5.20% | 6.30% |
| | 3100 | Total: Mechanical & Electrical | ↓ | 3.87 | 5.40 | 6.70 | 7% | 8.90% | 11.05% |
| | 3200 | | | | | | | | |
| | 9000 | Per car, total cost | Car | 14,600 | 18,300 | 23,300 | | | |
| **43** | 0010 | **GYMNASIUMS** | S.F. | 96 | 128 | 173 | | | | **43** |
| | 0020 | Total project costs | C.F. | 4.79 | 6.50 | 8 | | | |
| | 1800 | Equipment | S.F. | 2.28 | 4.28 | 8.20 | 1.98% | 3.30% | 6.70% |
| | 2720 | Plumbing | | 6.05 | 7.20 | 9.30 | 4.65% | 6.40% | 7.75% |
| | 2770 | Heating, ventilating, air conditioning | | 6.55 | 9.95 | 19.95 | 5.15% | 9.05% | 11.10% |
| | 2900 | Electrical | | 7.40 | 10.05 | 13.25 | 6.60% | 8.30% | 10.30% |
| | 3100 | Total: Mechanical & Electrical | ↓ | 27 | 36.50 | 43.50 | 19.60% | 23.50% | 28% |
| | 3500 | See also division 11 67 00 | | | | | | | |
| **46** | 0010 | **HOSPITALS** | S.F. | 184 | 230 | 315 | | | | **46** |
| | 0020 | Total project costs | C.F. | 13.95 | 17.40 | 25 | | | |
| | 1800 | Equipment | S.F. | 4.68 | 9 | 15.50 | .80% | 2.53% | 4.80% |
| | 2720 | Plumbing | | 15.85 | 22.50 | 29 | 7.60% | 9.10% | 10.85% |
| | 2770 | Heating, ventilating, air conditioning | | 23.50 | 30 | 41.50 | 7.80% | 12.95% | 16.65% |
| | 2900 | Electrical | | 20.50 | 27.50 | 39 | 9.90% | 11.75% | 14.10% |
| | 3100 | Total: Mechanical & Electrical | ↓ | 59.50 | 82.50 | 123 | 28% | 33.50% | 37% |
| | 9000 | Per bed or person, total cost | Bed | 213,500 | 294,000 | 338,500 | | | |
| | 9900 | See also division 11 71 00 | | | | | | | |
| **48** | 0010 | **HOUSING For the Elderly** | S.F. | 90.50 | 115 | 141 | | | | **48** |
| | 0020 | Total project costs | C.F. | 6.45 | 8.95 | 11.45 | | | |
| | 0100 | Site work | S.F. | 6.30 | 9.80 | 14.35 | 5.05% | 7.90% | 12.10% |
| | 0500 | Masonry | | 2.76 | 10.30 | 15.05 | 1.30% | 6.05% | 11% |
| | 1800 | Equipment | | 2.19 | 3.02 | 4.80 | 1.88% | 3.23% | 4.43% |
| | 2510 | Conveying systems | | 2.20 | 2.96 | 4.02 | 1.78% | 2.20% | 2.81% |
| | 2720 | Plumbing | | 6.75 | 8.60 | 10.80 | 8.15% | 9.55% | 10.50% |
| | 2730 | Heating, ventilating, air conditioning | | 3.45 | 4.89 | 7.30 | 3.30% | 5.60% | 7.25% |
| | 2900 | Electrical | | 6.75 | 9.15 | 11.75 | 7.30% | 8.50% | 10.25% |
| | 3100 | Total: Mechanical & Electrical | ↓ | 24.50 | 29 | 37 | 18.10% | 22.50% | 29% |
| | 9000 | Per rental unit, total cost | Unit | 84,500 | 98,500 | 110,000 | | | |
| | 9500 | Total: Mechanical & Electrical | " | 18,800 | 21,600 | 25,200 | | | |
| **50** | 0010 | **HOUSING Public (Low Rise)** | S.F. | 76 | 106 | 138 | | | | **50** |
| | 0020 | Total project costs | C.F. | 6.80 | 8.45 | 10.50 | | | |
| | 0100 | Site work | S.F. | 9.70 | 13.95 | 22.50 | 8.35% | 11.75% | 16.50% |
| | 1800 | Equipment | | 2.07 | 3.38 | 5.15 | 2.26% | 3.03% | 4.24% |
| | 2720 | Plumbing | | 5.50 | 7.25 | 9.20 | 7.15% | 9.05% | 11.60% |
| | 2730 | Heating, ventilating, air conditioning | | 2.76 | 5.35 | 5.85 | 4.26% | 6.05% | 6.45% |
| | 2900 | Electrical | | 4.61 | 6.85 | 9.50 | 5.10% | 6.55% | 8.25% |
| | 3100 | Total: Mechanical & Electrical | ↓ | 22 | 28 | 31.50 | 14.50% | 17.55% | 26.50% |
| | 9000 | Per apartment, total cost | Apt. | 83,500 | 95,000 | 119,500 | | | |
| | 9500 | Total: Mechanical & Electrical | " | 17,800 | 22,000 | 24,400 | | | |
| **51** | 0010 | **ICE SKATING RINKS** | S.F. | 65 | 152 | 167 | | | | **51** |
| | 0020 | Total project costs | C.F. | 4.79 | 4.90 | 5.65 | | | |
| | 2720 | Plumbing | S.F. | 2.43 | 4.56 | 4.66 | 3.12% | 3.23% | 5.65% |
| | 2900 | Electrical | | 6.95 | 10.70 | 11.30 | 6.30% | 10.15% | 15.05% |
| | 3100 | Total: Mechanical & Electrical | ↓ | 11.55 | 16.35 | 20 | 18.95% | 18.95% | 18.95% |
| **52** | 0010 | **JAILS** | S.F. | 198 | 256 | 330 | | | | **52** |
| | 0020 | Total project costs | C.F. | 17.85 | 25 | 29 | | | |
| | 1800 | Equipment | S.F. | 7.75 | 23 | 39 | 2.77% | 5.55% | 10.35% |
| | 2720 | Plumbing | | 19.10 | 25.50 | 34 | 7% | 8.90% | 13.35% |
| | 2770 | Heating, ventilating, air conditioning | | 17.90 | 24 | 46 | 7.50% | 9.45% | 17.75% |
| | 2900 | Electrical | ↓ | 21 | 28 | 35 | 8.80% | 11.55% | 14.95% |

| | | 50 17 00 | S.F. Costs | UNIT | UNIT COSTS | | | % OF TOTAL | | | |
|---|---|---|---|---|---|---|---|---|---|---|
| | | | | | 1/4 | MEDIAN | 3/4 | 1/4 | MEDIAN | 3/4 | |
| 52 | 3100 | Total: Mechanical & Electrical | S.F. | 54 | 98.50 | 117 | 27.50% | 30% | 34% | 52 |
| 53 | 0010 | **LIBRARIES** | S.F. | 127 | 162 | 209 | | | | 53 |
| | 0020 | Total project costs | C.F. | 8.55 | 10.75 | 13.70 | | | | |
| | 0500 | Masonry | S.F. | 10 | 17.50 | 29.50 | 5.80% | 7.60% | 11.60% | |
| | 1800 | Equipment | | 1.71 | 4.60 | 6.95 | .37% | 1.50% | 4.07% | |
| | 2720 | Plumbing | | 4.63 | 6.75 | 9.10 | 3.38% | 4.60% | 5.70% | |
| | 2770 | Heating, ventilating, air conditioning | | 10.25 | 17.35 | 22.50 | 7.80% | 10.95% | 12.80% | |
| | 2900 | Electrical | | 12.85 | 16.65 | 21.50 | 8.35% | 10.30% | 11.95% | |
| | 3100 | Total: Mechanical & Electrical | | 37.50 | 48 | 59.50 | 20.50% | 23% | 26.50% | |
| 54 | 0010 | **LIVING, ASSISTED** | S.F. | 116 | 137 | 161 | | | | 54 |
| | 0020 | Total project costs | C.F. | 9.75 | 11.40 | 12.95 | | | | |
| | 0500 | Masonry | S.F. | 3.41 | 4.06 | 4.77 | 2.36% | 3.16% | 3.86% | |
| | 1800 | Equipment | | 2.63 | 3.06 | 3.92 | 2.12% | 2.47% | 2.87% | |
| | 2720 | Plumbing | | 9.70 | 13 | 13.45 | 6.05% | 8.15% | 10.60% | |
| | 2770 | Heating, ventilating, air conditioning | | 11.50 | 12.05 | 13.20 | 7.95% | 9.35% | 9.70% | |
| | 2900 | Electrical | | 11.35 | 12.55 | 14.50 | 9% | 10% | 10.70% | |
| | 3100 | Total: Mechanical & Electrical | | 31.50 | 37 | 43 | 25% | 28.50% | 31.50% | |
| 55 | 0010 | **MEDICAL CLINICS** | S.F. | 118 | 145 | 184 | | | | 55 |
| | 0020 | Total project costs | C.F. | 8.65 | 11.20 | 14.85 | | | | |
| | 1800 | Equipment | S.F. | 3.19 | 6.70 | 10.40 | 1.05% | 2.94% | 6.35% | |
| | 2720 | Plumbing | | 7.85 | 11.05 | 14.75 | 6.15% | 8.40% | 10.10% | |
| | 2770 | Heating, ventilating, air conditioning | | 9.35 | 12.25 | 18 | 6.65% | 8.85% | 11.35% | |
| | 2900 | Electrical | | 10.15 | 14.40 | 18.80 | 8.10% | 10% | 12.25% | |
| | 3100 | Total: Mechanical & Electrical | | 32.50 | 44 | 60 | 22.50% | 27% | 33.50% | |
| | 3500 | See also division 11 71 00 | | | | | | | | |
| 57 | 0010 | **MEDICAL OFFICES** | S.F. | 111 | 137 | 168 | | | | 57 |
| | 0020 | Total project costs | C.F. | 8.30 | 11.20 | 15.15 | | | | |
| | 1800 | Equipment | S.F. | 3.66 | 7.25 | 10.30 | .70% | 5.10% | 7.05% | |
| | 2720 | Plumbing | | 6.15 | 9.45 | 12.75 | 5.60% | 6.80% | 8.50% | |
| | 2770 | Heating, ventilating, air conditioning | | 7.40 | 10.70 | 14.10 | 6.10% | 8% | 9.70% | |
| | 2900 | Electrical | | 9.05 | 12.85 | 17.90 | 7.60% | 9.80% | 11.40% | |
| | 3100 | Total: Mechanical & Electrical | | 24 | 34.50 | 50 | 19.30% | 22.50% | 28.50% | |
| 59 | 0010 | **MOTELS** | S.F. | 70 | 101 | 132 | | | | 59 |
| | 0020 | Total project costs | C.F. | 6.25 | 8.35 | 13.65 | | | | |
| | 2720 | Plumbing | S.F. | 7.10 | 9.05 | 10.75 | 9.45% | 10.60% | 12.55% | |
| | 2770 | Heating, ventilating, air conditioning | | 4.32 | 6.45 | 11.55 | 5.60% | 5.60% | 10% | |
| | 2900 | Electrical | | 6.60 | 8.35 | 10.40 | 7.45% | 9.05% | 10.45% | |
| | 3100 | Total: Mechanical & Electrical | | 21 | 28 | 48.50 | 18.50% | 24% | 25.50% | |
| | 5000 | | | | | | | | | |
| | 9000 | Per rental unit, total cost | Unit | 35,600 | 67,500 | 73,000 | | | | |
| | 9500 | Total: Mechanical & Electrical | " | 6,950 | 10,500 | 12,200 | | | | |
| 60 | 0010 | **NURSING HOMES** | S.F. | 110 | 142 | 173 | | | | 60 |
| | 0020 | Total project costs | C.F. | 8.60 | 10.75 | 14.70 | | | | |
| | 1800 | Equipment | S.F. | 3.46 | 4.58 | 7.65 | 2.02% | 3.62% | 4.99% | |
| | 2720 | Plumbing | | 9.40 | 14.25 | 17.15 | 8.75% | 10.10% | 12.70% | |
| | 2770 | Heating, ventilating, air conditioning | | 9.90 | 15.05 | 19.90 | 9.70% | 11.45% | 11.80% | |
| | 2900 | Electrical | | 10.85 | 13.55 | 18.45 | 9.40% | 10.55% | 12.45% | |
| | 3100 | Total: Mechanical & Electrical | | 26 | 36.50 | 60.50 | 26% | 29.50% | 30.50% | |
| | 9000 | Per bed or person, total cost | Bed | 48,700 | 61,000 | 78,500 | | | | |
| 61 | 0010 | **OFFICES Low Rise (1 to 4 story)** | S.F. | 92.50 | 120 | 156 | | | | 61 |
| | 0020 | Total project costs | C.F. | 6.60 | 9.10 | 12 | | | | |
| | 0100 | Site work | S.F. | 7.40 | 13.15 | 19.35 | 6.20% | 9.70% | 13.55% | |
| | 0500 | Masonry | | 3.58 | 7.35 | 13.15 | 2.62% | 5.45% | 8.45% | |
| | 1800 | Equipment | | .98 | 1.92 | 5.25 | .60% | 1.50% | 3.50% | |
| | 2720 | Plumbing | | 3.31 | 5.10 | 7.50 | 3.66% | 4.50% | 6.10% | |
| | 2770 | Heating, ventilating, air conditioning | | 7.30 | 10.20 | 14.90 | 7.20% | 10.30% | 11.70% | |
| | 2900 | Electrical | | 7.55 | 10.75 | 15.25 | 7.45% | 9.65% | 11.35% | |

| | | 50 17 00 \| S.F. Costs | UNIT | UNIT COSTS | | | % OF TOTAL | | | |
|---|---|---|---|---|---|---|---|---|---|---|
| | | | | 1/4 | MEDIAN | 3/4 | 1/4 | MEDIAN | 3/4 | |
| 61 | 3100 | Total: Mechanical & Electrical | S.F. | 21 | 28.50 | 42.50 | 18.20% | 22% | 27% | 61 |
| 62 | 0010 | **OFFICES Mid Rise (5 to 10 story)** | S.F. | 98 | 119 | 161 | | | | 62 |
| | 0020 | Total project costs | C.F. | 6.95 | 8.85 | 12.55 | | | | |
| | 2720 | Plumbing | S.F. | 2.96 | 4.58 | 6.60 | 2.83% | 3.74% | 4.50% | |
| | 2770 | Heating, ventilating, air conditioning | | 7.45 | 10.60 | 16.95 | 7.65% | 9.40% | 11% | |
| | 2900 | Electrical | | 7.25 | 9.40 | 14.05 | 6.35% | 7.80% | 10% | |
| | 3100 | Total: Mechanical & Electrical | | 18.80 | 24.50 | 46 | 18.95% | 21% | 27.50% | |
| 63 | 0010 | **OFFICES High Rise (11 to 20 story)** | S.F. | 120 | 152 | 187 | | | | 63 |
| | 0020 | Total project costs | C.F. | 8.40 | 10.50 | 15.05 | | | | |
| | 2900 | Electrical | S.F. | 7.30 | 8.90 | 13.20 | 5.80% | 7.85% | 10.50% | |
| | 3100 | Total: Mechanical & Electrical | | 23.50 | 31.50 | 53 | 16.90% | 23.50% | 34% | |
| 64 | 0010 | **POLICE STATIONS** | S.F. | 144 | 189 | 239 | | | | 64 |
| | 0020 | Total project costs | C.F. | 11.50 | 14.05 | 19.25 | | | | |
| | 0500 | Masonry | S.F. | 13.55 | 24 | 30 | 6.70% | 9.10% | 11.35% | |
| | 1800 | Equipment | | 2.29 | 10.05 | 15.95 | .98% | 3.35% | 6.70% | |
| | 2720 | Plumbing | | 8.05 | 16.05 | 20 | 5.65% | 6.90% | 10.75% | |
| | 2770 | Heating, ventilating, air conditioning | | 12.60 | 16.75 | 23.50 | 5.85% | 10.55% | 11.70% | |
| | 2900 | Electrical | | 15.75 | 22.50 | 30 | 9.80% | 11.85% | 14.80% | |
| | 3100 | Total: Mechanical & Electrical | | 51.50 | 62 | 84 | 25% | 31.50% | 32.50% | |
| 65 | 0010 | **POST OFFICES** | S.F. | 114 | 141 | 179 | | | | 65 |
| | 0020 | Total project costs | C.F. | 6.85 | 8.65 | 9.85 | | | | |
| | 2720 | Plumbing | S.F. | 5.15 | 6.35 | 8 | 4.24% | 5.30% | 5.60% | |
| | 2770 | Heating, ventilating, air conditioning | | 8 | 9.90 | 11 | 6.65% | 7.15% | 9.35% | |
| | 2900 | Electrical | | 9.40 | 13.20 | 15.70 | 7.25% | 9% | 11% | |
| | 3100 | Total: Mechanical & Electrical | | 27 | 35.50 | 40 | 16.25% | 18.80% | 22% | |
| 66 | 0010 | **POWER PLANTS** | S.F. | 790 | 1,050 | 1,925 | | | | 66 |
| | 0020 | Total project costs | C.F. | 21.50 | 47.50 | 101 | | | | |
| | 2900 | Electrical | S.F. | 56 | 118 | 176 | 9.30% | 12.75% | 21.50% | |
| | 8100 | Total: Mechanical & Electrical | | 139 | 450 | 1,025 | 32.50% | 32.50% | 52.50% | |
| 67 | 0010 | **RELIGIOUS EDUCATION** | S.F. | 91.50 | 120 | 147 | | | | 67 |
| | 0020 | Total project costs | C.F. | 5.10 | 7.30 | 9.10 | | | | |
| | 2720 | Plumbing | S.F. | 3.83 | 5.45 | 7.70 | 4.40% | 5.30% | 7.10% | |
| | 2770 | Heating, ventilating, air conditioning | | 9.70 | 10.95 | 15.50 | 10.05% | 11.45% | 12.35% | |
| | 2900 | Electrical | | 7.25 | 10.25 | 13.55 | 7.60% | 9.05% | 10.35% | |
| | 3100 | Total: Mechanical & Electrical | | 29.50 | 39 | 47.50 | 22% | 23% | 27% | |
| 69 | 0010 | **RESEARCH Laboratories & Facilities** | S.F. | 137 | 197 | 287 | | | | 69 |
| | 0020 | Total project costs | C.F. | 10.35 | 20 | 24 | | | | |
| | 1800 | Equipment | S.F. | 6.10 | 12 | 29.50 | .94% | 4.58% | 9.10% | |
| | 2720 | Plumbing | | 13.75 | 17.60 | 28.50 | 6.15% | 8.30% | 10.80% | |
| | 2770 | Heating, ventilating, air conditioning | | 12.35 | 41.50 | 49 | 7.25% | 16.50% | 17.50% | |
| | 2900 | Electrical | | 16.05 | 26.50 | 43.50 | 9.45% | 11.15% | 15.40% | |
| | 3100 | Total: Mechanical & Electrical | | 52.50 | 91.50 | 132 | 31.50% | 37.50% | 42% | |
| 70 | 0010 | **RESTAURANTS** | S.F. | 133 | 172 | 223 | | | | 70 |
| | 0020 | Total project costs | C.F. | 11.20 | 14.70 | 19.25 | | | | |
| | 1800 | Equipment | S.F. | 8.60 | 21.50 | 32 | 6.10% | 13% | 15.65% | |
| | 2720 | Plumbing | | 10.55 | 12.80 | 16.80 | 6.10% | 8.15% | 9% | |
| | 2770 | Heating, ventilating, air conditioning | | 13.40 | 18.50 | 22.50 | 9.20% | 12% | 12.40% | |
| | 2900 | Electrical | | 14.10 | 17.35 | 22.50 | 8.35% | 10.55% | 11.55% | |
| | 3100 | Total: Mechanical & Electrical | | 43.50 | 46.50 | 60.50 | 21% | 24.50% | 29.50% | |
| | 9000 | Per seat unit, total cost | Seat | 4,875 | 6,500 | 7,700 | | | | |
| | 9500 | Total: Mechanical & Electrical | " | 1,225 | 1,625 | 1,925 | | | | |
| 72 | 0010 | **RETAIL STORES** | S.F. | 62 | 83.50 | 111 | | | | 72 |
| | 0020 | Total project costs | C.F. | 4.21 | 6 | 8.35 | | | | |
| | 2720 | Plumbing | S.F. | 2.25 | 3.76 | 6.40 | 3.26% | 4.60% | 6.80% | |
| | 2770 | Heating, ventilating, air conditioning | | 4.87 | 6.65 | 10 | 6.75% | 8.75% | 10.15% | |
| | 2900 | Electrical | | 5.60 | 7.65 | 11.05 | 7.25% | 9.90% | 11.60% | |
| | 3100 | Total: Mechanical & Electrical | | 14.90 | 19.05 | 26.50 | 17.05% | 21% | 23.50% | |

## 50 17 00 | S.F. Costs

| | | | UNIT | UNIT COSTS | | | % OF TOTAL | | |
|---|---|---|---|---|---|---|---|---|---|
| | | | | 1/4 | MEDIAN | 3/4 | 1/4 | MEDIAN | 3/4 |
| 74 | 0010 | **SCHOOLS Elementary** | S.F. | 101 | 125 | 154 | | | |
| | 0020 | Total project costs | C.F. | 6.65 | 8.50 | 11 | | | |
| | 0500 | Masonry | S.F. | 9.10 | 15.60 | 23 | 5.45% | 10.65% | 14.10% |
| | 1800 | Equipment | | 2.77 | 4.73 | 8.70 | 1.90% | 3.20% | 4.70% |
| | 2720 | Plumbing | | 5.85 | 8.25 | 11 | 5.70% | 7.15% | 9.35% |
| | 2730 | Heating, ventilating, air conditioning | | 8.75 | 13.90 | 19.45 | 8.15% | 10.80% | 14.90% |
| | 2900 | Electrical | | 9.55 | 12.65 | 16.10 | 8.40% | 10.05% | 11.70% |
| | 3100 | Total: Mechanical & Electrical | ↓ | 34.50 | 43 | 53 | 25% | 27.50% | 30% |
| | 9000 | Per pupil, total cost | Ea. | 11,600 | 17,300 | 38,700 | | | |
| | 9500 | Total: Mechanical & Electrical | " | 3,275 | 4,150 | 10,500 | | | |
| 76 | 0010 | **SCHOOLS Junior High & Middle** | S.F. | 105 | 128 | 157 | | | |
| | 0020 | Total project costs | C.F. | 6.65 | 8.60 | 9.65 | | | |
| | 0500 | Masonry | S.F. | 13.40 | 17.60 | 20.50 | 8.60% | 11.60% | 14.35% |
| | 1800 | Equipment | | 3.34 | 5.40 | 8.10 | 1.79% | 3.09% | 4.86% |
| | 2720 | Plumbing | | 6.10 | 7.50 | 9.30 | 5.30% | 6.80% | 7.25% |
| | 2770 | Heating, ventilating, air conditioning | | 12.15 | 14.80 | 26 | 8.90% | 11.55% | 14.20% |
| | 2900 | Electrical | | 10.25 | 12.35 | 15.95 | 7.90% | 9.35% | 10.60% |
| | 3100 | Total: Mechanical & Electrical | ↓ | 33.50 | 43 | 53 | 23.50% | 27% | 29.50% |
| | 9000 | Per pupil, total cost | Ea. | 13,300 | 17,400 | 23,400 | | | |
| 78 | 0010 | **SCHOOLS Senior High** | S.F. | 108 | 133 | 167 | | | |
| | 0020 | Total project costs | C.F. | 6.55 | 9.65 | 15.55 | | | |
| | 1800 | Equipment | S.F. | 2.86 | 6.75 | 9.50 | 1.88% | 2.91% | 4.56% |
| | 2720 | Plumbing | | 6.10 | 9.15 | 16.75 | 5.60% | 6.90% | 8.30% |
| | 2770 | Heating, ventilating, air conditioning | | 12.45 | 14.25 | 27 | 8.95% | 11.60% | 15% |
| | 2900 | Electrical | | 11 | 14.25 | 21.50 | 8.65% | 10.20% | 12.25% |
| | 3100 | Total: Mechanical & Electrical | ↓ | 36 | 42.50 | 71 | 23.50% | 26.50% | 28.50% |
| | 9000 | Per pupil, total cost | Ea. | 10,300 | 20,900 | 26,100 | | | |
| 80 | 0010 | **SCHOOLS Vocational** | S.F. | 88 | 128 | 158 | | | |
| | 0020 | Total project costs | C.F. | 5.45 | 7.85 | 10.85 | | | |
| | 0500 | Masonry | S.F. | 5.15 | 12.75 | 19.50 | 3.20% | 4.61% | 10.95% |
| | 1800 | Equipment | | 2.75 | 6.85 | 9.55 | 1.24% | 3.10% | 4.26% |
| | 2720 | Plumbing | | 5.65 | 8.40 | 12.35 | 5.40% | 6.90% | 8.55% |
| | 2770 | Heating, ventilating, air conditioning | | 7.90 | 14.65 | 24.50 | 8.60% | 11.90% | 14.65% |
| | 2900 | Electrical | | 9.15 | 12.55 | 17.15 | 8.45% | 11% | 13.20% |
| | 3100 | Total: Mechanical & Electrical | ↓ | 34.50 | 35 | 60.50 | 27.50% | 29.50% | 31% |
| | 9000 | Per pupil, total cost | Ea. | 12,300 | 32,800 | 49,000 | | | |
| 83 | 0010 | **SPORTS ARENAS** | S.F. | 77 | 103 | 158 | | | |
| | 0020 | Total project costs | C.F. | 4.18 | 7.50 | 9.65 | | | |
| | 2720 | Plumbing | S.F. | 4.46 | 6.75 | 14.30 | 4.35% | 6.35% | 9.40% |
| | 2770 | Heating, ventilating, air conditioning | | 9.60 | 11.35 | 15.75 | 8.80% | 10.20% | 13.55% |
| | 2900 | Electrical | | 8 | 10.90 | 14.05 | 8.60% | 9.90% | 12.25% |
| | 3100 | Total: Mechanical & Electrical | ↓ | 19.95 | 35 | 47 | 21.50% | 25% | 27.50% |
| 85 | 0010 | **SUPERMARKETS** | S.F. | 71 | 82.50 | 96.50 | | | |
| | 0020 | Total project costs | C.F. | 3.96 | 4.79 | 7.25 | | | |
| | 2720 | Plumbing | S.F. | 3.97 | 5 | 5.80 | 5.40% | 6% | 7.45% |
| | 2770 | Heating, ventilating, air conditioning | | 5.85 | 7.75 | 9.45 | 8.60% | 8.65% | 9.60% |
| | 2900 | Electrical | | 8.90 | 10.25 | 12.10 | 10.40% | 12.45% | 13.60% |
| | 3100 | Total: Mechanical & Electrical | ↓ | 23 | 25 | 32.50 | 20.50% | 26.50% | 31% |
| 86 | 0010 | **SWIMMING POOLS** | S.F. | 115 | 193 | 410 | | | |
| | 0020 | Total project costs | C.F. | 9.25 | 11.50 | 12.55 | | | |
| | 2720 | Plumbing | S.F. | 10.65 | 12.15 | 16.45 | 4.80% | 9.70% | 20.50% |
| | 2900 | Electrical | | 8.65 | 14 | 20.50 | 5.75% | 6.95% | 7.60% |
| | 3100 | Total: Mechanical & Electrical | ↓ | 21 | 54.50 | 73 | 11.15% | 14.10% | 23.50% |
| 87 | 0010 | **TELEPHONE EXCHANGES** | S.F. | 153 | 224 | 284 | | | |
| | 0020 | Total project costs | C.F. | 9.50 | 15.25 | 21 | | | |
| | 2720 | Plumbing | S.F. | 6.45 | 9.95 | 14.60 | 4.52% | 5.80% | 6.90% |
| | 2770 | Heating, ventilating, air conditioning | ↓ | 14.95 | 30 | 37.50 | 11.80% | 16.05% | 18.40% |

| | | 50 17 00 \| S.F. Costs | UNIT | UNIT COSTS | | | % OF TOTAL | | | |
|---|---|---|---|---|---|---|---|---|---|---|
| | | | | 1/4 | MEDIAN | 3/4 | 1/4 | MEDIAN | 3/4 | |
| 87 | 2900 | Electrical | S.F. | 15.55 | 24.50 | 43.50 | 10.90% | 14% | 17.85% | 87 |
| | 3100 | Total: Mechanical & Electrical | ↓ | 46 | 87 | 123 | 29.50% | 33.50% | 44.50% | |
| 91 | 0010 | **THEATERS** | S.F. | 96 | 120 | 182 | | | | 91 |
| | 0020 | Total project costs | C.F. | 4.44 | 6.55 | 9.65 | | | | |
| | 2720 | Plumbing | S.F. | 3.20 | 3.47 | 14.20 | 2.92% | 4.70% | 6.80% | |
| | 2770 | Heating, ventilating, air conditioning | | 9.35 | 11.30 | 14 | 8% | 12.25% | 13.40% | |
| | 2900 | Electrical | | 8.40 | 11.35 | 23 | 8.05% | 9.35% | 12.25% | |
| | 3100 | Total: Mechanical & Electrical | ↓ | 21.50 | 29 | 34.50 | 21.50% | 25.50% | 27.50% | |
| 94 | 0010 | **TOWN HALLS City Halls & Municipal Buildings** | S.F. | 107 | 136 | 178 | | | | 94 |
| | 0020 | Total project costs | C.F. | 9.80 | 11.75 | 16.50 | | | | |
| | 2720 | Plumbing | S.F. | 4.48 | 8.40 | 15.40 | 4.31% | 5.95% | 7.95% | |
| | 2770 | Heating, ventilating, air conditioning | | 8.10 | 16.10 | 23.50 | 7.05% | 9.05% | 13.45% | |
| | 2900 | Electrical | | 10.20 | 14.55 | 19.85 | 8.05% | 9.45% | 11.65% | |
| | 3100 | Total: Mechanical & Electrical | ↓ | 35.50 | 45 | 68.50 | 22% | 26.50% | 31% | |
| 97 | 0010 | **WAREHOUSES & Storage Buildings** | S.F. | 40 | 59.50 | 85.50 | | | | 97 |
| | 0020 | Total project costs | C.F. | 2.10 | 3.28 | 5.45 | | | | |
| | 0100 | Site work | S.F. | 4.13 | 8.20 | 12.40 | 6.05% | 12.95% | 19.85% | |
| | 0500 | Masonry | | 2.41 | 5.70 | 12.30 | 3.73% | 7.40% | 12.30% | |
| | 1800 | Equipment | | .64 | 1.39 | 7.80 | .91% | 1.82% | 5.55% | |
| | 2720 | Plumbing | | 1.34 | 2.39 | 4.48 | 2.90% | 4.80% | 6.55% | |
| | 2730 | Heating, ventilating, air conditioning | | 1.53 | 4.30 | 5.75 | 2.41% | 5% | 8.90% | |
| | 2900 | Electrical | | 2.37 | 4.46 | 7.40 | 5.15% | 7.20% | 10.10% | |
| | 3100 | Total: Mechanical & Electrical | ↓ | 6.60 | 10.15 | 20 | 12.75% | 18.90% | 26% | |
| 99 | 0010 | **WAREHOUSE & OFFICES Combination** | S.F. | 49 | 65.50 | 90 | | | | 99 |
| | 0020 | Total project costs | C.F. | 2.52 | 3.66 | 5.40 | | | | |
| | 1800 | Equipment | S.F. | .86 | 1.65 | 2.46 | .52% | 1.20% | 2.40% | |
| | 2720 | Plumbing | | 1.90 | 3.37 | 4.93 | 3.74% | 4.76% | 6.30% | |
| | 2770 | Heating, ventilating, air conditioning | | 3 | 4.69 | 6.55 | 5% | 5.65% | 10.05% | |
| | 2900 | Electrical | | 3.30 | 4.91 | 7.75 | 5.75% | 8% | 10% | |
| | 3100 | Total: Mechanical & Electrical | ↓ | 9.20 | 14.20 | 22.50 | 14.40% | 19.95% | 24.50% | |

# Square Foot Project Size Modifier

One factor that affects the S.F. cost of a particular building is the size. In general, for buildings built to the same specifications in the same locality, the larger building will have the lower S.F. cost. This is due mainly to the decreasing contribution of the exterior walls plus the economy of scale usually achievable in larger buildings. The Area Conversion Scale shown below will give a factor to convert costs for the typical size building to an adjusted cost for the particular project.

The Square Foot Base Size lists the median costs, most typical project size in our accumulated data, and the range in size of the projects.

The Size Factor for your project is determined by dividing your project area in S.F. by the typical project size for the particular Building Type. With this factor, enter the Area Conversion Scale at the appropriate Size Factor and determine the appropriate cost multiplier for your building size.

**Example:** Determine the cost per S.F. for a 100,000 S.F. Mid-rise apartment building.

$$\frac{\text{Proposed building area} = 100,000 \text{ S.F.}}{\text{Typical size from below} = 50,000 \text{ S.F.}} = 2.00$$

Enter Area Conversion scale at 2.0, intersect curve, read horizontally the appropriate cost multiplier of .94. Size adjusted cost becomes .94 x $107.00 = $101.00 based on national average costs.

Note: For Size Factors less than .50, the Cost Multiplier is 1.1
For Size Factors greater than 3.5, the Cost Multiplier is .90

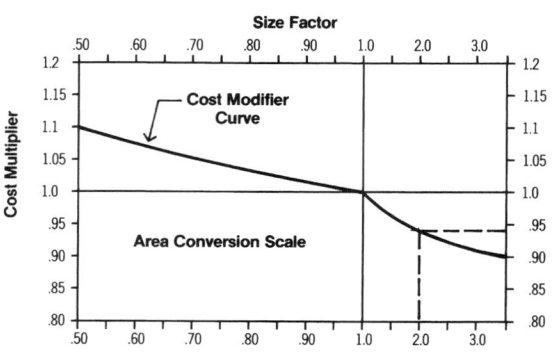

| Square Foot Base Size | | | | | | | |
|---|---|---|---|---|---|---|---|
| Building Type | Median Cost per S.F. | Typical Size Gross S.F. | Typical Range Gross S.F. | Building Type | Median Cost per S.F. | Typical Size Gross S.F. | Typical Range Gross S.F. |
| Apartments, Low Rise | $ 84.00 | 21,000 | 9,700 - 37,200 | Jails | $ 256.00 | 40,000 | 5,500 - 145,000 |
| Apartments, Mid Rise | 107.00 | 50,000 | 32,000 - 100,000 | Libraries | 162.00 | 12,000 | 7,000 - 31,000 |
| Apartments, High Rise | 116.00 | 145,000 | 95,000 - 600,000 | Living, Assisted | 137.00 | 32,300 | 23,500 - 50,300 |
| Auditoriums | 142.00 | 25,000 | 7,600 - 39,000 | Medical Clinics | 145.00 | 7,200 | 4,200 - 15,700 |
| Auto Sales | 105.00 | 20,000 | 10,800 - 28,600 | Medical Offices | 137.00 | 6,000 | 4,000 - 15,000 |
| Banks | 189.00 | 4,200 | 2,500 - 7,500 | Motels | 101.00 | 40,000 | 15,800 - 120,000 |
| Churches | 129.00 | 17,000 | 2,000 - 42,000 | Nursing Homes | 142.00 | 23,000 | 15,000 - 37,000 |
| Clubs, Country | 132.00 | 6,500 | 4,500 - 15,000 | Offices, Low Rise | 120.00 | 20,000 | 5,000 - 80,000 |
| Clubs, Social | 125.00 | 10,000 | 6,000 - 13,500 | Offices, Mid Rise | 119.00 | 120,000 | 20,000 - 300,000 |
| Clubs, YMCA | 146.00 | 28,300 | 12,800 - 39,400 | Offices, High Rise | 152.00 | 260,000 | 120,000 - 800,000 |
| Colleges (Class) | 151.00 | 50,000 | 15,000 - 150,000 | Police Stations | 189.00 | 10,500 | 4,000 - 19,000 |
| Colleges (Science Lab) | 241.00 | 45,600 | 16,600 - 80,000 | Post Offices | 141.00 | 12,400 | 6,800 - 30,000 |
| College (Student Union) | 178.00 | 33,400 | 16,000 - 85,000 | Power Plants | 1050.00 | 7,500 | 1,000 - 20,000 |
| Community Center | 134.00 | 9,400 | 5,300 - 16,700 | Religious Education | 120.00 | 9,000 | 6,000 - 12,000 |
| Court Houses | 184.00 | 32,400 | 17,800 - 106,000 | Research | 197.00 | 19,000 | 6,300 - 45,000 |
| Dept. Stores | 78.00 | 90,000 | 44,000 - 122,000 | Restaurants | 172.00 | 4,400 | 2,800 - 6,000 |
| Dormitories, Low Rise | 143.00 | 25,000 | 10,000 - 95,000 | Retail Stores | 83.50 | 7,200 | 4,000 - 17,600 |
| Dormitories, Mid Rise | 176.00 | 85,000 | 20,000 - 200,000 | Schools, Elementary | 125.00 | 41,000 | 24,500 - 55,000 |
| Factories | 76.00 | 26,400 | 12,900 - 50,000 | Schools, Jr. High | 128.00 | 92,000 | 52,000 - 119,000 |
| Fire Stations | 139.00 | 5,800 | 4,000 - 8,700 | Schools, Sr. High | 133.00 | 101,000 | 50,500 - 175,000 |
| Fraternity Houses | 130.00 | 12,500 | 8,200 - 14,800 | Schools, Vocational | 128.00 | 37,000 | 20,500 - 82,000 |
| Funeral Homes | 145.00 | 10,000 | 4,000 - 20,000 | Sports Arenas | 103.00 | 15,000 | 5,000 - 40,000 |
| Garages, Commercial | 93.50 | 9,300 | 5,000 - 13,600 | Supermarkets | 82.50 | 44,000 | 12,000 - 60,000 |
| Garages, Municipal | 119.00 | 8,300 | 4,500 - 12,600 | Swimming Pools | 193.00 | 20,000 | 10,000 - 32,000 |
| Garages, Parking | 50.50 | 163,000 | 76,400 - 225,300 | Telephone Exchange | 224.00 | 4,500 | 1,200 - 10,600 |
| Gymnasiums | 128.00 | 19,200 | 11,600 - 41,000 | Theaters | 120.00 | 10,500 | 8,800 - 17,500 |
| Hospitals | 230.00 | 55,000 | 27,200 - 125,000 | Town Halls | 136.00 | 10,800 | 4,800 - 23,400 |
| House (Elderly) | 115.00 | 37,000 | 21,000 - 66,000 | Warehouses | 59.50 | 25,000 | 8,000 - 72,000 |
| Housing (Public) | 106.00 | 36,000 | 14,400 - 74,400 | Warehouse & Office | 65.50 | 25,000 | 8,000 - 72,000 |
| Ice Rinks | 152.00 | 29,000 | 27,200 - 33,600 | | | | |

| | | | |
|---|---|---|---|
| A | Area Square Feet; Ampere | BTUH | BTU per Hour |
| ABS | Acrylonitrile Butadiene Stryrene; Asbestos Bonded Steel | B.U.R. | Built-up Roofing |
| A.C. | Alternating Current; Air-Conditioning; Asbestos Cement; Plywood Grade A & C | BX | Interlocked Armored Cable |
| | | °C | degree centegrade |
| | | c | Conductivity, Copper Sweat |
| | | C | Hundred; Centigrade |
| A.C.I. | American Concrete Institute | C/C | Center to Center, Cedar on Cedar |
| AD | Plywood, Grade A & D | C-C | Center to Center |
| Addit. | Additional | Cab. | Cabinet |
| Adj. | Adjustable | Cair. | Air Tool Laborer |
| af | Audio-frequency | Calc | Calculated |
| A.G.A. | American Gas Association | Cap. | Capacity |
| Agg. | Aggregate | Carp. | Carpenter |
| A.H. | Ampere Hours | C.B. | Circuit Breaker |
| A hr. | Ampere-hour | C.C.A. | Chromate Copper Arsenate |
| A.H.U. | Air Handling Unit | C.C.F. | Hundred Cubic Feet |
| A.I.A. | American Institute of Architects | cd | Candela |
| AIC | Ampere Interrupting Capacity | cd/sf | Candela per Square Foot |
| Allow. | Allowance | CD | Grade of Plywood Face & Back |
| alt. | Altitude | CDX | Plywood, Grade C & D, exterior glue |
| Alum. | Aluminum | | |
| a.m. | Ante Meridiem | Cefi. | Cement Finisher |
| Amp. | Ampere | Cem. | Cement |
| Anod. | Anodized | CF | Hundred Feet |
| Approx. | Approximate | C.F. | Cubic Feet |
| Apt. | Apartment | CFM | Cubic Feet per Minute |
| Asb. | Asbestos | c.g. | Center of Gravity |
| A.S.B.C. | American Standard Building Code | CHW | Chilled Water; Commercial Hot Water |
| Asbe. | Asbestos Worker | | |
| ASCE. | American Society of Civil Engineers | C.I. | Cast Iron |
| A.S.H.R.A.E. | American Society of Heating, Refrig. & AC Engineers | C.I.P. | Cast in Place |
| | | Circ. | Circuit |
| A.S.M.E. | American Society of Mechanical Engineers | C.L. | Carload Lot |
| | | Clab. | Common Laborer |
| A.S.T.M. | American Society for Testing and Materials | Clam. | Common maintenance laborer |
| | | C.L.F. | Hundred Linear Feet |
| Attchmt. | Attachment | CLF | Current Limiting Fuse |
| Avg., Ave. | Average | CLP | Cross Linked Polyethylene |
| A.W.G. | American Wire Gauge | cm | Centimeter |
| AWWA | American Water Works Assoc. | CMP | Corr. Metal Pipe |
| Bbl. | Barrel | C.M.U. | Concrete Masonry Unit |
| B&B | Grade B and Better; Balled & Burlapped | CN | Change Notice |
| | | Col. | Column |
| B.&S. | Bell and Spigot | CO₂ | Carbon Dioxide |
| B.&W. | Black and White | Comb. | Combination |
| b.c.c. | Body-centered Cubic | Compr. | Compressor |
| B.C.Y. | Bank Cubic Yards | Conc. | Concrete |
| BE | Bevel End | Cont. | Continuous; Continued, Container |
| B.F. | Board Feet | Corr. | Corrugated |
| Bg. cem. | Bag of Cement | Cos | Cosine |
| BHP | Boiler Horsepower; Brake Horsepower | Cot | Cotangent |
| | | Cov. | Cover |
| B.I. | Black Iron | C/P | Cedar on Paneling |
| Bit., Bitum. | Bituminous | CPA | Control Point Adjustment |
| Bit., Conc. | Bituminous Concrete | Cplg. | Coupling |
| Bk. | Backed | C.P.M. | Critical Path Method |
| Bkrs. | Breakers | CPVC | Chlorinated Polyvinyl Chloride |
| Bldg. | Building | C.Pr. | Hundred Pair |
| Blk. | Block | CRC | Cold Rolled Channel |
| Bm. | Beam | Creos. | Creosote |
| Boil. | Boilermaker | Crpt. | Carpet & Linoleum Layer |
| B.P.M. | Blows per Minute | CRT | Cathode-ray Tube |
| BR | Bedroom | CS | Carbon Steel, Constant Shear Bar Joist |
| Brg. | Bearing | | |
| Brhe. | Bricklayer Helper | Csc | Cosecant |
| Bric. | Bricklayer | C.S.F. | Hundred Square Feet |
| Brk. | Brick | CSI | Construction Specifications Institute |
| Brng. | Bearing | | |
| Brs. | Brass | C.T. | Current Transformer |
| Brz. | Bronze | CTS | Copper Tube Size |
| Bsn. | Basin | Cu | Copper, Cubic |
| Btr. | Better | Cu. Ft. | Cubic Foot |
| Btu | British Thermal Unit | cw | Continuous Wave |
| | | C.W. | Cool White; Cold Water |

| | | | |
|---|---|---|---|
| Cwt. | 100 Pounds | | |
| C.W.X. | Cool White Deluxe | | |
| C.Y. | Cubic Yard (27 cubic feet) | | |
| C.Y./Hr. | Cubic Yard per Hour | | |
| Cyl. | Cylinder | | |
| d | Penny (nail size) | | |
| D | Deep; Depth; Discharge | | |
| Dis., Disch. | Discharge | | |
| Db. | Decibel | | |
| Dbl. | Double | | |
| DC | Direct Current | | |
| DDC | Direct Digital Control | | |
| Demob. | Demobilization | | |
| d.f.u. | Drainage Fixture Units | | |
| D.H. | Double Hung | | |
| DHW | Domestic Hot Water | | |
| DI | Ductile Iron | | |
| Diag. | Diagonal | | |
| Diam., Dia | Diameter | | |
| Distrib. | Distribution | | |
| Div. | Division | | |
| Dk. | Deck | | |
| D.L. | Dead Load; Diesel | | |
| DLH | Deep Long Span Bar Joist | | |
| Do. | Ditto | | |
| Dp. | Depth | | |
| D.P.S.T. | Double Pole, Single Throw | | |
| Dr. | Drive | | |
| Drink. | Drinking | | |
| D.S. | Double Strength | | |
| D.S.A. | Double Strength A Grade | | |
| D.S.B. | Double Strength B Grade | | |
| Dty. | Duty | | |
| DWV | Drain Waste Vent | | |
| DX | Deluxe White, Direct Expansion | | |
| dyn | Dyne | | |
| e | Eccentricity | | |
| E | Equipment Only; East | | |
| Ea. | Each | | |
| E.B. | Encased Burial | | |
| Econ. | Economy | | |
| E.C.Y | Embankment Cubic Yards | | |
| EDP | Electronic Data Processing | | |
| EIFS | Exterior Insulation Finish System | | |
| E.D.R. | Equiv. Direct Radiation | | |
| Eq. | Equation | | |
| EL | elevation | | |
| Elec. | Electrician; Electrical | | |
| Elev. | Elevator; Elevating | | |
| EMT | Electrical Metallic Conduit; Thin Wall Conduit | | |
| Eng. | Engine, Engineered | | |
| EPDM | Ethylene Propylene Diene Monomer | | |
| EPS | Expanded Polystyrene | | |
| Eqhv. | Equip. Oper., Heavy | | |
| Eqlt. | Equip. Oper., Light | | |
| Eqmd. | Equip. Oper., Medium | | |
| Eqmm. | Equip. Oper., Master Mechanic | | |
| Eqol. | Equip. Oper., Oilers | | |
| Equip. | Equipment | | |
| ERW | Electric Resistance Welded | | |
| E.S. | Energy Saver | | |
| Est. | Estimated | | |
| esu | Electrostatic Units | | |
| E.W. | Each Way | | |
| EWT | Entering Water Temperature | | |
| Excav. | Excavation | | |
| Exp. | Expansion, Exposure | | |
| Ext. | Exterior | | |
| Extru. | Extrusion | | |
| f. | Fiber stress | | |
| F | Fahrenheit; Female; Fill | | |
| Fab. | Fabricated | | |

| | |
|---|---|
| FBGS | Fiberglass |
| F.C. | Footcandles |
| f.c.c. | Face-centered Cubic |
| f'c. | Compressive Stress in Concrete; Extreme Compressive Stress |
| F.E. | Front End |
| FEP | Fluorinated Ethylene Propylene (Teflon) |
| F.G. | Flat Grain |
| F.H.A. | Federal Housing Administration |
| Fig. | Figure |
| Fin. | Finished |
| Fixt. | Fixture |
| Fl. Oz. | Fluid Ounces |
| Flr. | Floor |
| F.M. | Frequency Modulation; Factory Mutual |
| Fmg. | Framing |
| Fdn. | Foundation |
| Fori. | Foreman, Inside |
| Foro. | Foreman, Outside |
| Fount. | Fountain |
| fpm | Feet per Minute |
| FPT | Female Pipe Thread |
| Fr. | Frame |
| F.R. | Fire Rating |
| FRK | Foil Reinforced Kraft |
| FRP | Fiberglass Reinforced Plastic |
| FS | Forged Steel |
| FSC | Cast Body; Cast Switch Box |
| Ft. | Foot; Feet |
| Ftng. | Fitting |
| Ftg. | Footing |
| Ft lb. | Foot Pound |
| Furn. | Furniture |
| FVNR | Full Voltage Non-Reversing |
| FXM | Female by Male |
| Fy. | Minimum Yield Stress of Steel |
| g | Gram |
| G | Gauss |
| Ga. | Gauge |
| Gal., gal. | Gallon |
| gpm, GPM | Gallon per Minute |
| Galv. | Galvanized |
| Gen. | General |
| G.F.I. | Ground Fault Interrupter |
| Glaz. | Glazier |
| GPD | Gallons per Day |
| GPH | Gallons per Hour |
| GPM | Gallons per Minute |
| GR | Grade |
| Gran. | Granular |
| Grnd. | Ground |
| H | High Henry |
| H.C. | High Capacity |
| H.D. | Heavy Duty; High Density |
| H.D.O. | High Density Overlaid |
| H.D.P.E. | high density polyethelene |
| Hdr. | Header |
| Hdwe. | Hardware |
| Help. | Helper Average |
| HEPA | High Efficiency Particulate Air Filter |
| Hg | Mercury |
| HIC | High Interrupting Capacity |
| HM | Hollow Metal |
| HMWPE | high molecular weight polyethylene |
| H.O. | High Output |
| Horiz. | Horizontal |
| H.P. | Horsepower; High Pressure |
| H.P.F. | High Power Factor |
| Hr. | Hour |
| Hrs./Day | Hours per Day |
| HSC | High Short Circuit |
| Ht. | Height |
| Htg. | Heating |
| Htrs. | Heaters |
| HVAC | Heating, Ventilation & Air-Conditioning |
| Hvy. | Heavy |
| HW | Hot Water |
| Hyd.;Hydr. | Hydraulic |
| Hz. | Hertz (cycles) |
| I. | Moment of Inertia |
| IBC | International Building Code |
| I.C. | Interrupting Capacity |
| ID | Inside Diameter |
| I.D. | Inside Dimension; Identification |
| I.F. | Inside Frosted |
| I.M.C. | Intermediate Metal Conduit |
| In. | Inch |
| Incan. | Incandescent |
| Incl. | Included; Including |
| Int. | Interior |
| Inst. | Installation |
| Insul. | Insulation/Insulated |
| I.P. | Iron Pipe |
| I.P.S. | Iron Pipe Size |
| I.P.T. | Iron Pipe Threaded |
| I.W. | Indirect Waste |
| J | Joule |
| J.I.C. | Joint Industrial Council |
| K | Thousand; Thousand Pounds; Heavy Wall Copper Tubing, Kelvin |
| K.A.H. | Thousand Amp. Hours |
| KCMIL | Thousand Circular Mils |
| KD | Knock Down |
| K.D.A.T. | Kiln Dried After Treatment |
| kg | Kilogram |
| kG | Kilogauss |
| kgf | Kilogram Force |
| kHz | Kilohertz |
| Kip. | 1000 Pounds |
| KJ | Kiljoule |
| K.L. | Effective Length Factor |
| K.L.F. | Kips per Linear Foot |
| Km | Kilometer |
| K.S.F. | Kips per Square Foot |
| K.S.I. | Kips per Square Inch |
| kV | Kilovolt |
| kVA | Kilovolt Ampere |
| K.V.A.R. | Kilovar (Reactance) |
| KW | Kilowatt |
| KWh | Kilowatt-hour |
| L | Labor Only; Length; Long; Medium Wall Copper Tubing |
| Lab. | Labor |
| lat | Latitude |
| Lath. | Lather |
| Lav. | Lavatory |
| lb.; # | Pound |
| L.B. | Load Bearing; L Conduit Body |
| L. & E. | Labor & Equipment |
| lb./hr. | Pounds per Hour |
| lb./L.F. | Pounds per Linear Foot |
| lbf/sq.in. | Pound-force per Square Inch |
| L.C.L. | Less than Carload Lot |
| L.C.Y. | Loose Cubic Yard |
| Ld. | Load |
| LE | Lead Equivalent |
| LED | Light Emitting Diode |
| L.F. | Linear Foot |
| L.F. Nose | Linear Foot of Stair Nosing |
| L.F. Rsr | Linear Foot of Stair Riser |
| Lg. | Long; Length; Large |
| L & H | Light and Heat |
| LH | Long Span Bar Joist |
| L.H. | Labor Hours |
| L.L. | Live Load |
| L.L.D. | Lamp Lumen Depreciation |
| lm | Lumen |
| lm/sf | Lumen per Square Foot |
| lm/W | Lumen per Watt |
| L.O.A. | Length Over All |
| log | Logarithm |
| L-O-L | Lateralolet |
| long. | longitude |
| L.P. | Liquefied Petroleum; Low Pressure |
| L.P.F. | Low Power Factor |
| LR | Long Radius |
| L.S. | Lump Sum |
| Lt. | Light |
| Lt. Ga. | Light Gauge |
| L.T.L. | Less than Truckload Lot |
| Lt. Wt. | Lightweight |
| L.V. | Low Voltage |
| M | Thousand; Material; Male; Light Wall Copper Tubing |
| M²CA | Meters Squared Contact Area |
| m/hr.; M.H. | Man-hour |
| mA | Milliampere |
| Mach. | Machine |
| Mag. Str. | Magnetic Starter |
| Maint. | Maintenance |
| Marb. | Marble Setter |
| Mat; Mat'l. | Material |
| Max. | Maximum |
| MBF | Thousand Board Feet |
| MBH | Thousand BTU's per hr. |
| MC | Metal Clad Cable |
| M.C.F. | Thousand Cubic Feet |
| M.C.F.M. | Thousand Cubic Feet per Minute |
| M.C.M. | Thousand Circular Mils |
| M.C.P. | Motor Circuit Protector |
| MD | Medium Duty |
| M.D.O. | Medium Density Overlaid |
| Med. | Medium |
| MF | Thousand Feet |
| M.F.B.M. | Thousand Feet Board Measure |
| Mfg. | Manufacturing |
| Mfrs. | Manufacturers |
| mg | Milligram |
| MGD | Million Gallons per Day |
| MGPH | Thousand Gallons per Hour |
| MH, M.H. | Manhole; Metal Halide; Man-Hour |
| MHz | Megahertz |
| Mi. | Mile |
| MI | Malleable Iron; Mineral Insulated |
| mm | Millimeter |
| Mill. | Millwright |
| Min., min. | Minimum, minute |
| Misc. | Miscellaneous |
| ml | Milliliter, Mainline |
| M.L.F. | Thousand Linear Feet |
| Mo. | Month |
| Mobil. | Mobilization |
| Mog. | Mogul Base |
| MPH | Miles per Hour |
| MPT | Male Pipe Thread |

| | |
|---|---|
| MRT | Mile Round Trip |
| ms | Millisecond |
| M.S.F. | Thousand Square Feet |
| Mstz. | Mosaic & Terrazzo Worker |
| M.S.Y. | Thousand Square Yards |
| Mtd., mtd. | Mounted |
| Mthe. | Mosaic & Terrazzo Helper |
| Mtng. | Mounting |
| Mult. | Multi; Multiply |
| M.V.A. | Million Volt Amperes |
| M.V.A.R. | Million Volt Amperes Reactance |
| MV | Megavolt |
| MW | Megawatt |
| MXM | Male by Male |
| MYD | Thousand Yards |
| N | Natural; North |
| nA | Nanoampere |
| NA | Not Available; Not Applicable |
| N.B.C. | National Building Code |
| NC | Normally Closed |
| N.E.M.A. | National Electrical Manufacturers Assoc. |
| NEHB | Bolted Circuit Breaker to 600V. |
| N.L.B. | Non-Load-Bearing |
| NM | Non-Metallic Cable |
| nm | Nanometer |
| No. | Number |
| NO | Normally Open |
| N.O.C. | Not Otherwise Classified |
| Nose. | Nosing |
| N.P.T. | National Pipe Thread |
| NQOD | Combination Plug-on/Bolt on Circuit Breaker to 240V. |
| N.R.C. | Noise Reduction Coefficient/ Nuclear Regulator Commission |
| N.R.S. | Non Rising Stem |
| ns | Nanosecond |
| nW | Nanowatt |
| OB | Opposing Blade |
| OC | On Center |
| OD | Outside Diameter |
| O.D. | Outside Dimension |
| ODS | Overhead Distribution System |
| O.G. | Ogee |
| O.H. | Overhead |
| O&P | Overhead and Profit |
| Oper. | Operator |
| Opng. | Opening |
| Orna. | Ornamental |
| OSB | Oriented Strand Board |
| O.S.&Y. | Outside Screw and Yoke |
| Ovhd. | Overhead |
| OWG | Oil, Water or Gas |
| Oz. | Ounce |
| P. | Pole; Applied Load; Projection |
| p. | Page |
| Pape. | Paperhanger |
| P.A.P.R. | Powered Air Purifying Respirator |
| PAR | Parabolic Reflector |
| Pc., Pcs. | Piece, Pieces |
| P.C. | Portland Cement; Power Connector |
| P.C.F. | Pounds per Cubic Foot |
| P.C.M. | Phase Contrast Microscopy |
| P.E. | Professional Engineer; Porcelain Enamel; Polyethylene; Plain End |
| Perf. | Perforated |
| PEX | Cross linked polyethylene |
| Ph. | Phase |
| P.I. | Pressure Injected |
| Pile. | Pile Driver |
| Pkg. | Package |

| | |
|---|---|
| Pl. | Plate |
| Plah. | Plasterer Helper |
| Plas. | Plasterer |
| Pluh. | Plumbers Helper |
| Plum. | Plumber |
| Ply. | Plywood |
| p.m. | Post Meridiem |
| Pntd. | Painted |
| Pord. | Painter, Ordinary |
| pp | Pages |
| PP, PPL | Polypropylene |
| P.P.M. | Parts per Million |
| Pr. | Pair |
| P.E.S.B. | Pre-engineered Steel Building |
| Prefab. | Prefabricated |
| Prefin. | Prefinished |
| Prop. | Propelled |
| PSF, psf | Pounds per Square Foot |
| PSI, psi | Pounds per Square Inch |
| PSIG | Pounds per Square Inch Gauge |
| PSP | Plastic Sewer Pipe |
| Pspr. | Painter, Spray |
| Psst. | Painter, Structural Steel |
| P.T. | Potential Transformer |
| P. & T. | Pressure & Temperature |
| Ptd. | Painted |
| Ptns. | Partitions |
| Pu | Ultimate Load |
| PVC | Polyvinyl Chloride |
| Pvmt. | Pavement |
| Pwr. | Power |
| Q | Quantity Heat Flow |
| Qt. | Quart |
| Quan., Qty. | Quantity |
| Q.C. | Quick Coupling |
| r | Radius of Gyration |
| R | Resistance |
| R.C.P. | Reinforced Concrete Pipe |
| Rect. | Rectangle |
| Reg. | Regular |
| Reinf. | Reinforced |
| Req'd. | Required |
| Res. | Resistant |
| Resi. | Residential |
| Rgh. | Rough |
| RGS | Rigid Galvanized Steel |
| R.H.W. | Rubber, Heat & Water Resistant; Residential Hot Water |
| rms | Root Mean Square |
| Rnd. | Round |
| Rodm. | Rodman |
| Rofc. | Roofer, Composition |
| Rofp. | Roofer, Precast |
| Rohe. | Roofer Helpers (Composition) |
| Rots. | Roofer, Tile & Slate |
| R.O.W. | Right of Way |
| RPM | Revolutions per Minute |
| R.S. | Rapid Start |
| Rsr | Riser |
| RT | Round Trip |
| S. | Suction; Single Entrance; South |
| SC | Screw Cover |
| SCFM | Standard Cubic Feet per Minute |
| Scaf. | Scaffold |
| Sch., Sched. | Schedule |
| S.C.R. | Modular Brick |
| S.D. | Sound Deadening |
| S.D.R. | Standard Dimension Ratio |
| S.E. | Surfaced Edge |
| Sel. | Select |
| S.E.R., S.E.U. | Service Entrance Cable |
| S.F. | Square Foot |

| | |
|---|---|
| S.F.C.A. | Square Foot Contact Area |
| S.F. Flr. | Square Foot of Floor |
| S.F.G. | Square Foot of Ground |
| S.F. Hor. | Square Foot Horizontal |
| S.F.R. | Square Feet of Radiation |
| S.F. Shlf. | Square Foot of Shelf |
| S4S | Surface 4 Sides |
| Shee. | Sheet Metal Worker |
| Sin. | Sine |
| Skwk. | Skilled Worker |
| SL | Saran Lined |
| S.L. | Slimline |
| Sldr. | Solder |
| SLH | Super Long Span Bar Joist |
| S.N. | Solid Neutral |
| S-O-L | Socketolet |
| sp | Standpipe |
| S.P. | Static Pressure; Single Pole; Self-Propelled |
| Spri. | Sprinkler Installer |
| spwg | Static Pressure Water Gauge |
| S.P.D.T. | Single Pole, Double Throw |
| SPF | Spruce Pine Fir |
| S.P.S.T. | Single Pole, Single Throw |
| SPT | Standard Pipe Thread |
| Sq. | Square; 100 Square Feet |
| Sq. Hd. | Square Head |
| Sq. In. | Square Inch |
| S.S. | Single Strength; Stainless Steel |
| S.S.B. | Single Strength B Grade |
| sst, ss | Stainless Steel |
| Sswk. | Structural Steel Worker |
| Sswl. | Structural Steel Welder |
| St.; Stl. | Steel |
| S.T.C. | Sound Transmission Coefficient |
| Std. | Standard |
| Stg. | Staging |
| STK | Select Tight Knot |
| STP | Standard Temperature & Pressure |
| Stpi. | Steamfitter, Pipefitter |
| Str. | Strength; Starter; Straight |
| Strd. | Stranded |
| Struct. | Structural |
| Sty. | Story |
| Subj. | Subject |
| Subs. | Subcontractors |
| Surf. | Surface |
| Sw. | Switch |
| Swbd. | Switchboard |
| S.Y. | Square Yard |
| Syn. | Synthetic |
| S.Y.P. | Southern Yellow Pine |
| Sys. | System |
| t. | Thickness |
| T | Temperature; Ton |
| Tan | Tangent |
| T.C. | Terra Cotta |
| T & C | Threaded and Coupled |
| T.D. | Temperature Difference |
| Tdd | Telecommunications Device for the Deaf |
| T.E.M. | Transmission Electron Microscopy |
| TFE | Tetrafluoroethylene (Teflon) |
| T. & G. | Tongue & Groove; Tar & Gravel |
| Th., Thk. | Thick |
| Thn. | Thin |
| Thrded | Threaded |
| Tilf. | Tile Layer, Floor |
| Tilh. | Tile Layer, Helper |
| THHN | Nylon Jacketed Wire |
| THW. | Insulated Strand Wire |

# Abbreviations

| | | | | | |
|---|---|---|---|---|---|
| THWN | Nylon Jacketed Wire | USP | United States Primed | Wrck. | Wrecker |
| T.L. | Truckload | UTP | Unshielded Twisted Pair | W.S.P. | Water, Steam, Petroleum |
| T.M. | Track Mounted | V | Volt | WT., Wt. | Weight |
| Tot. | Total | V.A. | Volt Amperes | WWF | Welded Wire Fabric |
| T-O-L | Threadolet | V.C.T. | Vinyl Composition Tile | XFER | Transfer |
| T.S. | Trigger Start | VAV | Variable Air Volume | XFMR | Transformer |
| Tr. | Trade | VC | Veneer Core | XHD | Extra Heavy Duty |
| Transf. | Transformer | Vent. | Ventilation | XHHW, | Cross-Linked Polyethylene Wire |
| Trhv. | Truck Driver, Heavy | Vert. | Vertical | XLPE | Insulation |
| Trlr | Trailer | V.F. | Vinyl Faced | XLP | Cross-linked Polyethylene |
| Trlt. | Truck Driver, Light | V.G. | Vertical Grain | Y | Wye |
| TTY | Teletypewriter | V.H.F. | Very High Frequency | yd | Yard |
| TV | Television | VHO | Very High Output | yr | Year |
| T.W. | Thermoplastic Water Resistant | Vib. | Vibrating | Δ | Delta |
| | Wire | V.L.F. | Vertical Linear Foot | % | Percent |
| UCI | Uniform Construction Index | Vol. | Volume | ~ | Approximately |
| UF | Underground Feeder | VRP | Vinyl Reinforced Polyester | Ø | Phase; diameter |
| UGND | Underground Feeder | W | Wire; Watt; Wide; West | @ | At |
| U.H.F. | Ultra High Frequency | w/ | With | # | Pound; Number |
| U.I. | United Inch | W.C. | Water Column; Water Closet | < | Less Than |
| U.L. | Underwriters Laboratory | W.F. | Wide Flange | > | Greater Than |
| Uld. | unloading | W.G. | Water Gauge | Z | zone |
| Unfin. | Unfinished | Wldg. | Welding | | |
| URD | Underground Residential | W. Mile | Wire Mile | | |
| | Distribution | W-O-L | Weldolet | | |
| US | United States | W.R. | Water Resistant | | |

# Index

# Index

# Index

# Index

# Index

# Index

# Index

# Division Notes

| | CREW | DAILY OUTPUT | LABOR-HOURS | UNIT | 2011 BARE COSTS | | | | TOTAL INCL O&P |
|---|---|---|---|---|---|---|---|---|---|
| | | | | | MAT. | LABOR | EQUIP. | TOTAL | |
| | | | | | | | | | |
| | | | | | | | | | |
| | | | | | | | | | |
| | | | | | | | | | |
| | | | | | | | | | |
| | | | | | | | | | |
| | | | | | | | | | |
| | | | | | | | | | |
| | | | | | | | | | |
| | | | | | | | | | |
| | | | | | | | | | |
| | | | | | | | | | |
| | | | | | | | | | |
| | | | | | | | | | |
| | | | | | | | | | |
| | | | | | | | | | |
| | | | | | | | | | |
| | | | | | | | | | |
| | | | | | | | | | |
| | | | | | | | | | |
| | | | | | | | | | |
| | | | | | | | | | |
| | | | | | | | | | |
| | | | | | | | | | |
| | | | | | | | | | |

# Division Notes

| | CREW | DAILY OUTPUT | LABOR-HOURS | UNIT | 2011 BARE COSTS | | | | TOTAL INCL O&P |
|---|---|---|---|---|---|---|---|---|---|
| | | | | | MAT. | LABOR | EQUIP. | TOTAL | |
| | | | | | | | | | |
| | | | | | | | | | |
| | | | | | | | | | |
| | | | | | | | | | |
| | | | | | | | | | |
| | | | | | | | | | |
| | | | | | | | | | |
| | | | | | | | | | |
| | | | | | | | | | |
| | | | | | | | | | |
| | | | | | | | | | |
| | | | | | | | | | |
| | | | | | | | | | |
| | | | | | | | | | |
| | | | | | | | | | |
| | | | | | | | | | |
| | | | | | | | | | |
| | | | | | | | | | |
| | | | | | | | | | |
| | | | | | | | | | |
| | | | | | | | | | |
| | | | | | | | | | |
| | | | | | | | | | |

# Division Notes

| | CREW | DAILY OUTPUT | LABOR-HOURS | UNIT | 2011 BARE COSTS | | | | TOTAL INCL O&P |
|---|---|---|---|---|---|---|---|---|---|
| | | | | | MAT. | LABOR | EQUIP. | TOTAL | |
| | | | | | | | | | |
| | | | | | | | | | |
| | | | | | | | | | |
| | | | | | | | | | |
| | | | | | | | | | |
| | | | | | | | | | |
| | | | | | | | | | |
| | | | | | | | | | |
| | | | | | | | | | |
| | | | | | | | | | |
| | | | | | | | | | |
| | | | | | | | | | |
| | | | | | | | | | |
| | | | | | | | | | |
| | | | | | | | | | |
| | | | | | | | | | |
| | | | | | | | | | |
| | | | | | | | | | |
| | | | | | | | | | |
| | | | | | | | | | |
| | | | | | | | | | |
| | | | | | | | | | |
| | | | | | | | | | |
| | | | | | | | | | |

# Division Notes

| | CREW | DAILY OUTPUT | LABOR-HOURS | UNIT | 2011 BARE COSTS | | | | TOTAL INCL O&P |
|---|---|---|---|---|---|---|---|---|---|
| | | | | | MAT. | LABOR | EQUIP. | TOTAL | |
| | | | | | | | | | |
| | | | | | | | | | |
| | | | | | | | | | |
| | | | | | | | | | |
| | | | | | | | | | |
| | | | | | | | | | |
| | | | | | | | | | |
| | | | | | | | | | |
| | | | | | | | | | |
| | | | | | | | | | |
| | | | | | | | | | |
| | | | | | | | | | |
| | | | | | | | | | |
| | | | | | | | | | |
| | | | | | | | | | |
| | | | | | | | | | |
| | | | | | | | | | |
| | | | | | | | | | |
| | | | | | | | | | |
| | | | | | | | | | |
| | | | | | | | | | |
| | | | | | | | | | |
| | | | | | | | | | |
| | | | | | | | | | |
| | | | | | | | | | |
| | | | | | | | | | |
| | | | | | | | | | |

# Division Notes

| | CREW | DAILY OUTPUT | LABOR-HOURS | UNIT | 2011 BARE COSTS | | | | TOTAL INCL O&P |
|---|---|---|---|---|---|---|---|---|---|
| | | | | | MAT. | LABOR | EQUIP. | TOTAL | |
| | | | | | | | | | |
| | | | | | | | | | |
| | | | | | | | | | |
| | | | | | | | | | |
| | | | | | | | | | |
| | | | | | | | | | |
| | | | | | | | | | |
| | | | | | | | | | |
| | | | | | | | | | |
| | | | | | | | | | |
| | | | | | | | | | |
| | | | | | | | | | |
| | | | | | | | | | |
| | | | | | | | | | |
| | | | | | | | | | |
| | | | | | | | | | |
| | | | | | | | | | |
| | | | | | | | | | |
| | | | | | | | | | |
| | | | | | | | | | |
| | | | | | | | | | |

# Reed Construction Data/RSMeans— a tradition of excellence in construction cost information and services since 1942

**For more information visit the RSMeans website at www.rsmeans.com**

## Book Selection Guide

The following table provides definitive information on the content of each cost data publication. The number of lines of data provided in each unit price or assemblies division, as well as the number of reference tables and crews, is listed for each book. The presence of other elements such as equipment rental costs, historical cost indexes, city cost indexes, square foot models, or cross-referenced indexes is also indicated. You can use the table to help select the RSMeans book that has the quantity and type of information you most need in your work.

| Unit Cost Divisions | Building Construction | Mechanical | Electrical | Commercial Renovation | Square Foot | Site Work Landsc. | Green Building | Interior | Concrete Masonry | Open Shop | Heavy Construction | Light Commercial | Facilities Construction | Plumbing | Residential |
|---|---|---|---|---|---|---|---|---|---|---|---|---|---|---|---|
| 1 | 571 | 392 | 401 | 492 | | 509 | 39 | 317 | 455 | 570 | 499 | 235 | 985 | 402 | 172 |
| 2 | 746 | 266 | 85 | 685 | | 967 | 168 | 364 | 221 | 743 | 706 | 440 | 1163 | 290 | 242 |
| 3 | 1526 | 216 | 206 | 887 | | 1329 | 859 | 304 | 1833 | 1526 | 1468 | 375 | 1552 | 183 | 359 |
| 4 | 898 | 24 | 0 | 698 | | 700 | 182 | 573 | 1133 | 874 | 593 | 490 | 1145 | 0 | 405 |
| 5 | 1916 | 263 | 167 | 1024 | | 870 | 1803 | 1025 | 833 | 1907 | 1117 | 895 | 1910 | 339 | 727 |
| 6 | 1855 | 67 | 69 | 1442 | | 450 | 537 | 1161 | 300 | 1834 | 381 | 1544 | 1466 | 22 | 2056 |
| 7 | 1437 | 209 | 125 | 1465 | | 534 | 766 | 544 | 485 | 1435 | 26 | 1143 | 1497 | 218 | 896 |
| 8 | 2093 | 61 | 10 | 2233 | | 297 | 1146 | 1731 | 676 | 2018 | 0 | 1622 | 2450 | 0 | 1397 |
| 9 | 1812 | 72 | 26 | 1627 | | 278 | 441 | 1910 | 386 | 1774 | 15 | 1542 | 2064 | 54 | 1354 |
| 10 | 988 | 17 | 10 | 608 | | 221 | 22 | 803 | 159 | 988 | 29 | 490 | 1074 | 232 | 221 |
| 11 | 1012 | 208 | 165 | 484 | | 126 | 54 | 862 | 28 | 997 | 0 | 231 | 1026 | 169 | 110 |
| 12 | 562 | 0 | 2 | 318 | | 248 | 150 | 1706 | 12 | 551 | 19 | 364 | 1756 | 23 | 317 |
| 13 | 738 | 119 | 111 | 247 | | 349 | 128 | 255 | 69 | 737 | 261 | 80 | 748 | 74 | 79 |
| 14 | 280 | 36 | 0 | 229 | | 28 | 0 | 259 | 0 | 279 | 0 | 12 | 299 | 21 | 6 |
| 21 | 78 | 0 | 16 | 30 | | 0 | 0 | 250 | 0 | 78 | 0 | 60 | 368 | 373 | 0 |
| 22 | 1141 | 7183 | 155 | 1105 | | 1510 | 1015 | 621 | 20 | 1103 | 1676 | 834 | 7114 | 8929 | 632 |
| 23 | 1186 | 7077 | 608 | 799 | | 157 | 894 | 677 | 38 | 1106 | 110 | 661 | 5154 | 1872 | 416 |
| 26 | 1249 | 496 | 9916 | 1006 | | 746 | 555 | 1068 | 55 | 1237 | 523 | 1095 | 9852 | 441 | 560 |
| 27 | 67 | 0 | 261 | 34 | | 9 | 0 | 59 | 0 | 67 | 35 | 48 | 263 | 0 | 4 |
| 28 | 96 | 58 | 124 | 71 | | 0 | 21 | 78 | 0 | 99 | 0 | 40 | 139 | 44 | 25 |
| 31 | 1477 | 735 | 608 | 798 | | 2794 | 38 | 7 | 1208 | 1421 | 2865 | 600 | 1540 | 660 | 606 |
| 32 | 759 | 54 | 0 | 840 | | 4195 | 343 | 359 | 274 | 730 | 1710 | 380 | 1611 | 166 | 429 |
| 33 | 502 | 1010 | 465 | 201 | | 1950 | 41 | 0 | 225 | 501 | 1873 | 113 | 1536 | 1181 | 111 |
| 34 | 109 | 0 | 20 | 4 | | 144 | 0 | 0 | 31 | 64 | 122 | 0 | 129 | 0 | 0 |
| 35 | 18 | 0 | 0 | 0 | | 166 | 0 | 0 | 0 | 18 | 281 | 0 | 83 | 0 | 0 |
| 41 | 52 | 0 | 0 | 24 | | 7 | 0 | 22 | 0 | 52 | 30 | 0 | 59 | 14 | 0 |
| 44 | 75 | 83 | 0 | 0 | | 0 | 0 | 0 | 0 | 0 | 0 | 0 | 75 | 75 | 0 |
| 46 | 23 | 6 | 0 | 0 | | 112 | 99 | 0 | 0 | 23 | 112 | 0 | 23 | 23 | 0 |
| 48 | 0 | 0 | 0 | 0 | | 0 | 19 | 0 | 0 | 0 | 0 | 0 | 0 | 0 | 0 |
| **Totals** | 23266 | 18652 | 13550 | 17351 | | 18696 | 9320 | 14955 | 8441 | 22732 | 14451 | 13294 | 47081 | 15805 | 11094 |

| Assem Div | Building Construction | Mechanical | Electrical | Commercial Renovation | Square Foot | Site Work Landscape | Assemblies | Green Building | Interior | Concrete Masonry | Heavy Construction | Light Commercial | Facilities Construction | Plumbing | Asm Div | Residential |
|---|---|---|---|---|---|---|---|---|---|---|---|---|---|---|---|---|
| A | | 15 | 0 | 188 | 150 | 575 | 596 | 0 | 0 | 534 | 569 | 154 | 24 | 0 | 1 | 374 |
| B | | 0 | 0 | 825 | 2465 | 0 | 5562 | 56 | 329 | 1911 | 368 | 2025 | 143 | 0 | 2 | 217 |
| C | | 0 | 0 | 647 | 866 | 0 | 1236 | 0 | 1561 | 146 | 0 | 766 | 248 | 0 | 3 | 588 |
| D | | 1064 | 811 | 712 | 1793 | 0 | 2423 | 233 | 765 | 0 | 0 | 1270 | 1026 | 1083 | 4 | 853 |
| E | | 0 | 0 | 86 | 260 | 0 | 297 | 0 | 5 | 0 | 0 | 257 | 5 | 0 | 5 | 392 |
| F | | 0 | 0 | 0 | 123 | 0 | 124 | 0 | 0 | 0 | 0 | 123 | 3 | 0 | 6 | 358 |
| G | | 527 | 294 | 317 | 202 | 2673 | 668 | 0 | 0 | 535 | 1189 | 200 | 293 | 685 | 7 | 300 |
| | | | | | | | | | | | | | | | 8 | 760 |
| | | | | | | | | | | | | | | | 9 | 80 |
| | | | | | | | | | | | | | | | 10 | 0 |
| | | | | | | | | | | | | | | | 11 | 0 |
| | | | | | | | | | | | | | | | 12 | 0 |
| **Totals** | | 1606 | 1105 | 2775 | 5859 | 3248 | 10906 | 289 | 2660 | 3126 | 2126 | 4795 | 1742 | 1768 | | 3922 |

| Reference Section | Building Construction Costs | Mechanical | Electrical | Commercial Renovation | Square Foot | Site Work Landscape | Assem. | Green Building | Interior | Concrete Masonry | Open Shop | Heavy Construction | Light Commercial | Facilities Construction | Plumbing | Resi. |
|---|---|---|---|---|---|---|---|---|---|---|---|---|---|---|---|---|
| Reference Tables | yes | yes | yes | yes | no | yes | yes | yes | yes | yes | yes | yes | yes | yes | yes | yes |
| Models | | | | | 107 | | | | | | | | 46 | | | 28 |
| Crews | 505 | 505 | 505 | 483 | | 505 | 505 | 505 | 505 | 505 | 481 | 505 | 481 | 483 | 505 | 481 |
| Equipment Rental Costs | yes | yes | yes | yes | | yes | | yes | yes | yes | yes | yes | yes | yes | yes | yes |
| Historical Cost Indexes | yes | yes | yes | yes | yes | yes | yes | yes | yes | yes | yes | yes | yes | yes | yes | no |
| City Cost Indexes | yes | yes | yes | yes | yes | yes | yes | yes | yes | yes | yes | yes | yes | yes | yes | yes |

# Annual Cost Guides

For more information
visit the RSMeans website
at www.rsmeans.com

## RSMeans Building Construction Cost Data 2011

Offers you unchallenged unit price reliability in an easy-to-use format. Whether used for verifying complete, finished estimates or for periodic checks, it supplies more cost facts better and faster than any comparable source. Over 21,000 unit prices have been updated for 2011. The City Cost Indexes and Location Factors cover over 930 areas, for indexing to any project location in North America. Order and get *RSMeans Quarterly Update Service* FREE. You'll have year-long access to the RSMeans Estimating **HOTLINE** FREE with your subscription. Expert assistance when using RSMeans data is just a phone call away.

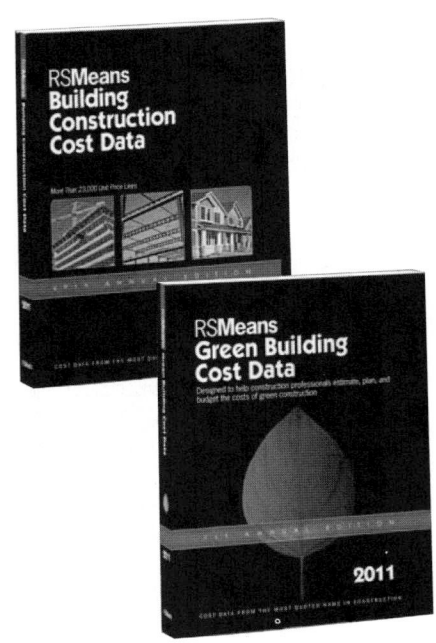

$174.95 per copy
Available Sept. 2010
Catalog no. 60011

**Unit prices organized according to the latest MasterFormat!**

## RSMeans Green Building Cost Data 2011

Estimate, plan, and budget the costs of green building for both new commercial construction and renovation work with this first edition of RSMeans Green Building Cost Data. More than 9,000 unit costs for a wide array of green building products plus assemblies costs. Easily identified cross references to LEED and Green Globes building rating systems criteria.

$144.95 per copy
Available Nov. 2010
Catalog no. 60551

## RSMeans Mechanical Cost Data 2011

- **HVAC**
- **Controls**

Total unit and systems price guidance for mechanical construction. . . materials, parts, fittings, and complete labor cost information. Includes prices for piping, heating, air conditioning, ventilation, and all related construction.

Plus new 2011 unit costs for:

- Thousands of installed HVAC/controls assemblies components
- "On-site" Location Factors for over 930 cities and towns in the U.S. and Canada
- Crews, labor, and equipment

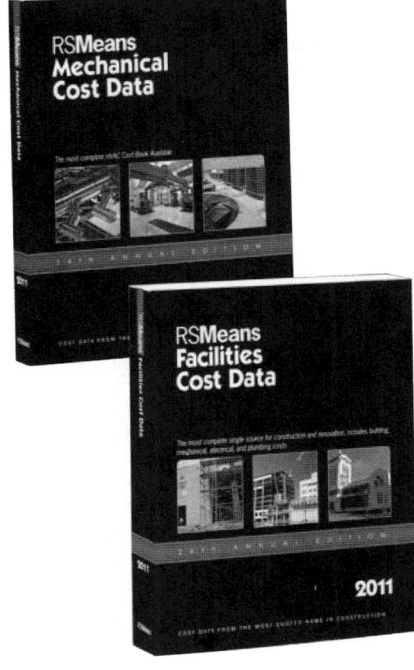

$174.95 per copy
Available Oct. 2010
Catalog no. 60021

**Unit prices organized according to the latest MasterFormat!**

## RSMeans Facilities Construction Cost Data 2011

For the maintenance and construction of commercial, industrial, municipal, and institutional properties. Costs are shown for new and remodeling construction and are broken down into materials, labor, equipment, and overhead and profit. Special emphasis is given to sections on mechanical, electrical, furnishings, site work, building maintenance, finish work, and demolition.

More than 45,000 unit costs, plus assemblies costs and a comprehensive Reference Section are included.

$439.95 per copy
Available Nov. 2010
Catalog no. 60201

For more information
visit the RSMeans website
at www.rsmeans.com

# Annual Cost Guides

## RSMeans Square Foot Costs 2011

*Accurate and Easy To Use!*

- **Updated price information** based on nationwide figures from suppliers, estimators, labor experts, and contractors

- "How-to-Use" sections, with **clear examples** of commercial, residential, industrial, and institutional structures

- Realistic graphics, offering true-to-life illustrations of building projects

- Extensive information on using square foot cost data, including sample estimates and alternate pricing methods

$184.95 per copy
Available Oct. 2010
Catalog no. 60051

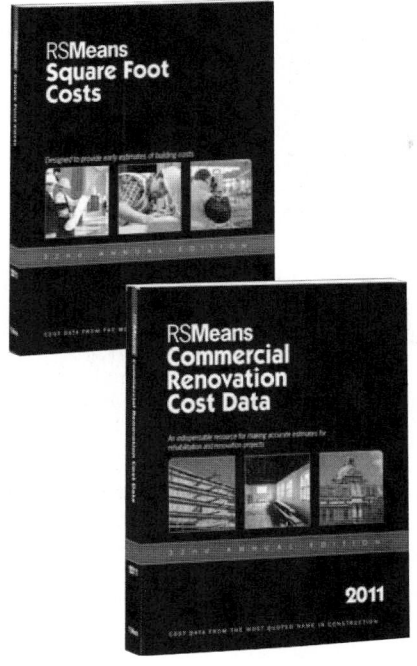

## RSMeans Commercial Renovation Cost Data 2011
### Commercial/Multifamily Residential

Use this valuable tool to estimate commercial and multifamily residential renovation and remodeling.

**Includes: New costs for hundreds of unique methods, materials, and conditions that only come up in repair and remodeling, PLUS:**

- Unit costs for more than 15,000 construction components
- Installed costs for more than 90 assemblies
- More than 930 "on-site" localization factors for the U.S. and Canada

**Unit prices organized according to the latest MasterFormat!**

$144.95 per copy
Available Oct. 2010
Catalog no. 60041

## RSMeans Electrical Cost Data 2011

Pricing information for every part of electrical cost planning. More than 13,000 unit and systems costs with design tables; clear specifications and drawings; engineering guides; illustrated estimating procedures; complete labor-hour and materials costs for better scheduling and procurement; and the latest electrical products and construction methods.

- A variety of special electrical systems including cathodic protection

- Costs for maintenance, demolition, HVAC/mechanical, specialties, equipment, and more

$174.95 per copy
Available Oct. 2010
Catalog no. 60031

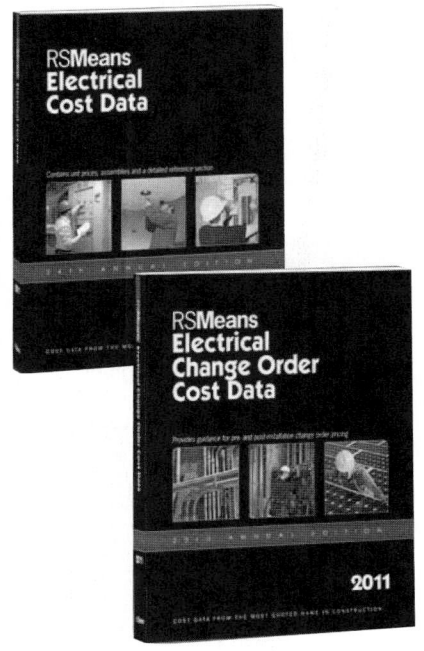

**Unit prices organized according to the latest MasterFormat!**

## RSMeans Electrical Change Order Cost Data 2011

*RSMeans Electrical Change Order Cost Data* provides you with electrical unit prices exclusively for pricing change orders—based on the recent, direct experience of contractors and suppliers. Analyze and check your own change order estimates against the experience others have had doing the same work. It also covers productivity analysis and change order cost justifications. With useful information for calculating the effects of change orders and dealing with their administration.

$169.95 per copy
Available Dec. 2010
Catalog no. 60231

## RSMeans Assemblies Cost Data 2011

*RSMeans Assemblies Cost Data* takes the guesswork out of preliminary or conceptual estimates. Now you don't have to try to calculate the assembled cost by working up individual component costs. We've done all the work for you.

Presents detailed illustrations, descriptions, specifications, and costs for every conceivable building assembly—240 types in all—arranged in the easy-to-use UNIFORMAT II system. Each illustrated "assembled" cost includes a complete grouping of materials and associated installation costs, including the installing contractor's overhead and profit.

**$285.95 per copy**
**Available Sept. 2010**
**Catalog no. 60061**

## RSMeans Open Shop Building Construction Cost Data 2011

The latest costs for accurate budgeting and estimating of new commercial and residential construction. . . renovation work. . . change orders. . . cost engineering.

*RSMeans Open Shop "BCCD"* will assist you to:
• Develop benchmark prices for change orders
• Plug gaps in preliminary estimates and budgets
• Estimate complex projects
• Substantiate invoices on contracts
• Price ADA-related renovations

**Unit prices organized according to the latest MasterFormat!**

**$134.95 per copy**
**Available Dec. 2010**
**Catalog no. 60151**

## RSMeans Residential Cost Data 2011

Contains square foot costs for 30 basic home models with the look of today, plus hundreds of custom additions and modifications you can quote right off the page. With costs for the 100 residential systems you're most likely to use in the year ahead. Complete with blank estimating forms, sample estimates, and step-by-step instructions.

Now contains line items for cultured stone and brick, PVC trim lumber, and TPO roofing.

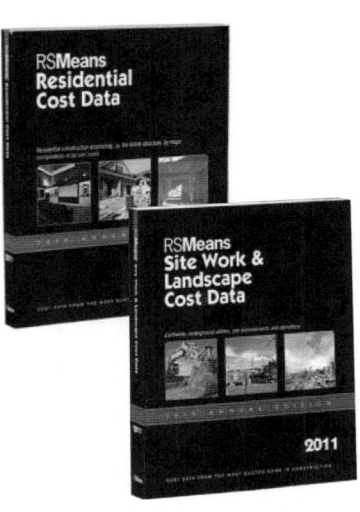

**Unit prices organized according to the latest MasterFormat!**

**$109.95 per copy**
**Available Oct. 2010**
**Catalog no. 60171**

## RSMeans Site Work & Landscape Cost Data 2011

Includes unit and assemblies costs for earthwork, sewerage, piped utilities, site improvements, drainage, paving, trees & shrubs, street openings/repairs, underground tanks, and more. Contains 57 tables of assemblies costs for accurate conceptual estimates.

Includes:
• Estimating for infrastructure improvements
• Environmentally-oriented construction
• ADA-mandated handicapped access
• Hazardous waste line items

**$169.95 per copy**
**Available Dec. 2010**
**Catalog no. 60281**

## RSMeans Facilities Maintenance & Repair Cost Data 2011

*RSMeans Facilities Maintenance & Repair Cost Data* gives you a complete system to manage and plan your facility repair and maintenance costs and budget efficiently. Guidelines for auditing a facility and developing an annual maintenance plan. Budgeting is included, along with reference tables on cost and management, and information on frequency and productivity of maintenance operations.

The only nationally recognized source of maintenance and repair costs. Developed in cooperation with the Civil Engineering Research Laboratory (CERL) of the Army Corps of Engineers.

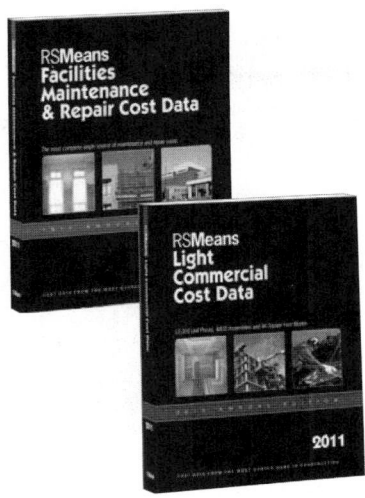

**$389.95 per copy**
**Available Nov. 2010**
**Catalog no. 60301**

## RSMeans Light Commercial Cost Data 2011

Specifically addresses the light commercial market, which is a specialized niche in the construction industry. Aids you, the owner/designer/contractor, in preparing all types of estimates—from budgets to detailed bids. Includes new advances in methods and materials.

Assemblies Section allows you to evaluate alternatives in the early stages of design/planning.

Over 11,000 unit costs ensure that you have the prices you need. . . when you need them.

**Unit prices organized according to the latest MasterFormat !**

**$119.95 per copy**
**Available Nov. 2010**
**Catalog no. 60181**

For more information
visit the RSMeans website
at www.rsmeans.com

# Annual Cost Guides

## RSMeans Concrete & Masonry Cost Data 2011

Provides you with cost facts for virtually all concrete/masonry estimating needs, from complicated formwork to various sizes and face finishes of brick and block—all in great detail. The comprehensive Unit Price Section contains more than 8,000 selected entries. Also contains an Assemblies [Cost] Section, and a detailed Reference Section that supplements the cost data.

**Unit prices organized according to the latest MasterFormat !**

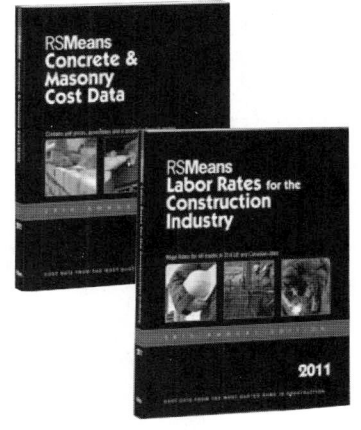

$159.95 per copy
Available Dec. 2010
Catalog no. 60111

## RSMeans Labor Rates for the Construction Industry 2011

Complete information for estimating labor costs, making comparisons, and negotiating wage rates by trade for more than 300 U.S. and Canadian cities. With 46 construction trades listed by local union number in each city, and historical wage rates included for comparison. Each city chart lists the county and is alphabetically arranged with handy visual flip tabs for quick reference.

$389.95 per copy
Available Dec. 2010
Catalog no. 60121

## RSMeans Construction Cost Indexes 2011

What materials and labor costs will change unexpectedly this year? By how much?

- Breakdowns for 318 major cities
- National averages for 30 key cities
- Expanded five major city indexes
- Historical construction cost indexes

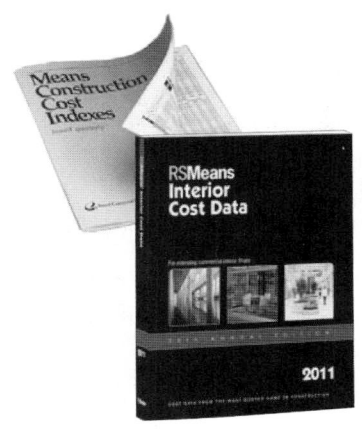

$362.00 per year (subscription)
Catalog no. 50141

$90.50 individual quarters
Catalog no. 60141 A,B,C,D

## RSMeans Interior Cost Data 2011

Provides you with prices and guidance needed to make accurate interior work estimates. Contains costs on materials, equipment, hardware, custom installations, furnishings, and labor costs. . . for new and remodel commercial and industrial interior construction, including updated information on office furnishings, and reference information.

**Unit prices organized according to the latest MasterFormat !**

$179.95 per copy
Available Nov. 2010
Catalog no. 60091

## RSMeans Heavy Construction Cost Data 2011

A comprehensive guide to heavy construction costs. Includes costs for highly specialized projects such as tunnels, dams, highways, airports, and waterways. Information on labor rates, equipment, and materials costs is included. Features unit price costs, systems costs, and numerous reference tables for costs and design.

**Unit prices organized according to the latest MasterFormat!**

$174.95 per copy
Available Dec. 2010
Catalog no. 60161

## RSMeans Plumbing Cost Data 2011

Comprehensive unit prices and assemblies for plumbing, irrigation systems, commercial and residential fire protection, point-of-use water heaters, and the latest approved materials. This publication and its companion, *RSMeans Mechanical Cost Data*, provide full-range cost estimating coverage for all the mechanical trades.

Now contains updated costs for potable water and radiant heat systems and high efficiency fixtures.

$174.95 per copy
Available Oct. 2010
Catalog no. 60211

# Reference Books

**For more information
visit the RSMeans website
at www.rsmeans.com**

## Value Engineering: Practical Applications

*For Design, Construction, Maintenance & Operations*

**by Alphonse Dell'Isola, PE**

A tool for immediate application—for engineers, architects, facility managers, owners, and contractors. Includes making the case for VE—the management briefing; integrating VE into planning, budgeting, and design; conducting life cycle costing; using VE methodology in design review and consultant selection; case studies; VE workbook; and a life cycle costing program on disk.

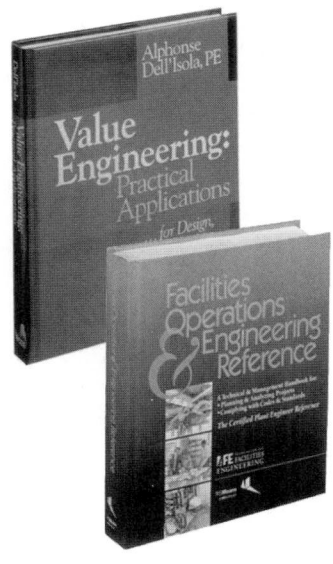

**$79.95 per copy
Over 450 pages, illustrated, softcover
Catalog no. 67319A**

## Facilities Operations & Engineering Reference

**by the Association for Facilities Engineering and RSMeans**

An all-in-one technical reference for planning and managing facility projects and solving day-to-day operations problems. Selected as the official certified plant engineer reference, this handbook covers financial analysis, maintenance, HVAC and energy efficiency, and more.

**$54.98 per copy
Over 700 pages, illustrated, hardcover
Catalog no. 67318**

## The Building Professional's Guide to Contract Documents

**3rd Edition**

**by Waller S. Poage, AIA, CSI, CVS**

A comprehensive reference for owners, design professionals, contractors, and students

- Structure your documents for maximum efficiency.
- Effectively communicate construction requirements.
- Understand the roles and responsibilities of construction professionals.
- Improve methods of project delivery.

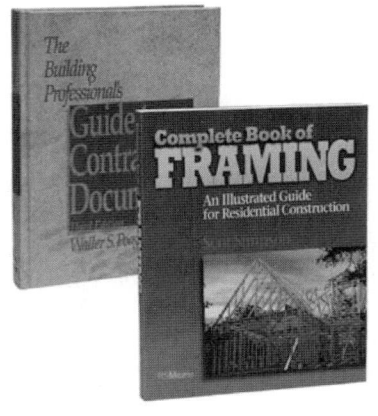

**$32.48 per copy, 400 pages
Diagrams and construction forms, hardcover
Catalog no. 67261A**

## Complete Book of Framing
**by Scot Simpson**

This straightforward, easy-to-learn method will help framers, carpenters, and handy homeowners build their skills in rough carpentry and framing. Shows how to frame all the parts of a house: floors, walls, roofs, door & window openings, and stairs—with hundreds of color photographs and drawings that show every detail.

The book gives beginners all the basics they need to go from zero framing knowledge to a journeyman level.

**$29.95 per copy
352 pages, softcover
Catalog no. 67353**

## Cost Planning & Estimating for Facilities Maintenance

In this unique book, a team of facilities management authorities shares their expertise on:

- Evaluating and budgeting maintenance operations
- Maintaining and repairing key building components
- Applying *RSMeans Facilities Maintenance & Repair Cost Data* to your estimating

Covers special maintenance requirements of the ten major building types.

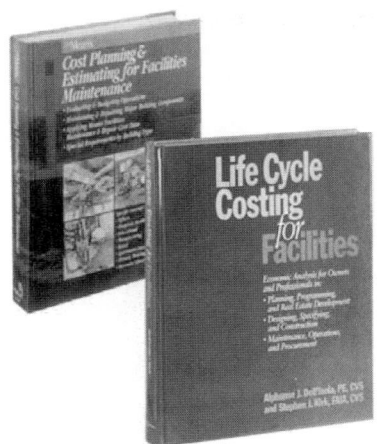

**$89.95 per copy
Over 475 pages, hardcover
Catalog no. 67314**

## Life Cycle Costing for Facilities

**by Alphonse Dell'Isola and Dr. Steven Kirk**

Guidance for achieving higher quality design and construction projects at lower costs! Cost-cutting efforts often sacrifice quality to yield the cheapest product. Life cycle costing enables building designers and owners to achieve both. The authors of this book show how LCC can work for a variety of projects — from roads to HVAC upgrades to different types of buildings.

**$99.95 per copy
450 pages, hardcover
Catalog no. 67341**

For more information
visit the RSMeans website
at www.rsmeans.com

# Reference Books

## Interior Home Improvement Costs 9th Edition

Updated estimates for the most popular remodeling and repair projects—from small, do-it-yourself jobs to major renovations and new construction. Includes: kitchens & baths; new living space from your attic, basement, or garage; new floors, paint, and wallpaper; tearing out or building new walls; closets, stairs, and fireplaces; new energy-saving improvements, home theaters, and more!

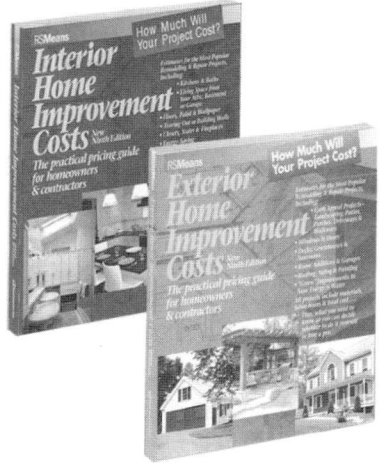

**$24.95 per copy**
**250 pages, illustrated, softcover**
**Catalog no. 67308E**

## Exterior Home Improvement Costs 9th Edition

Updated estimates for the most popular remodeling and repair projects—from small, do-it-yourself jobs, to major renovations and new construction. Includes: curb appeal projects—landscaping, patios, porches, driveways, and walkways; new windows and doors; decks, greenhouses, and sunrooms; room additions and garages; roofing, siding, and painting; "green" improvements to save energy & water.

**$24.95 per copy**
**Over 275 pages, illustrated, softcover**
**Catalog no. 67309E**

## Builder's Essentials: Plan Reading & Material Takeoff

*For Residential and Light Commercial Construction*

### by Wayne J. DelPico

A valuable tool for understanding plans and specs, and accurately calculating material quantities. Step-by-step instructions and takeoff procedures based on a full set of working drawings.

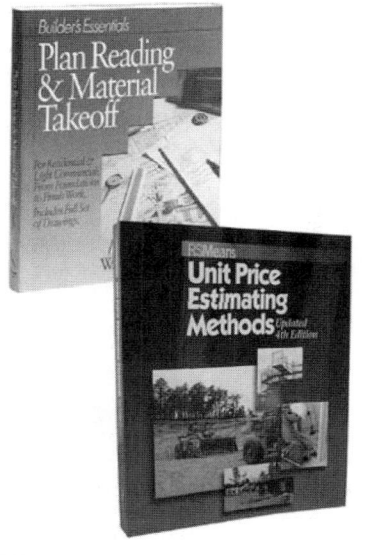

**$35.95 per copy**
**Over 420 pages, softcover**
**Catalog no. 67307**

## Means Unit Price Estimating Methods

### New 4th Edition

This new edition includes up-to-date cost data and estimating examples, updated to reflect changes to the CSI numbering system and new features of RSMeans cost data. It describes the most productive, universally accepted ways to estimate, and uses checklists and forms to illustrate shortcuts and timesavers. A model estimate demonstrates procedures. A new chapter explores computer estimating alternatives.

**$59.95 per copy**
**Over 350 pages, illustrated, softcover**
**Catalog no. 67303B**

## Concrete Repair and Maintenance Illustrated

### by Peter Emmons

Hundreds of illustrations show users how to analyze, repair, clean, and maintain concrete structures for optimal performance and cost effectiveness. From parking garages to roads and bridges to structural concrete, this comprehensive book describes the causes, effects, and remedies for concrete wear and failure. Invaluable for planning jobs, selecting materials, and training employees, this book is a must-have for concrete specialists, general contractors, facility managers, civil and structural engineers, and architects.

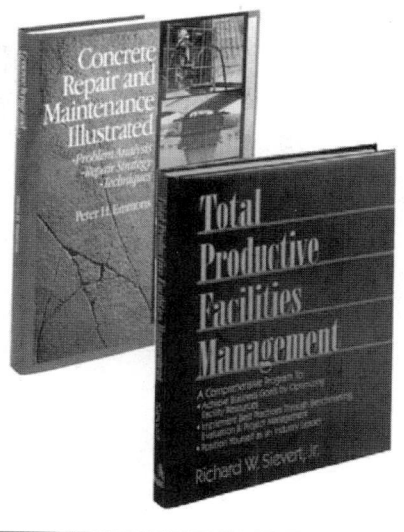

**$34.98 per copy**
**300 pages, illustrated, softcover**
**Catalog no. 67146**

## Total Productive Facilities Management

### by Richard W. Sievert, Jr.

Today, facilities are viewed as strategic resources. . . elevating the facility manager to the role of asset manager supporting the organization's overall business goals. Now, Richard Sievert Jr., in this well-articulated guidebook, sets forth a new operational standard for the facility manager's emerging role. . . a comprehensive program for managing facilities as a true profit center.

**$29.98 per copy**
**275 pages, softcover**
**Catalog no. 67321**

# Reference Books

**For more information
visit the RSMeans website
at www.rsmeans.com**

## Means Illustrated Construction
### Dictionary Condensed, 2nd Edition

The best portable dictionary for office or field use—an essential tool for contractors, architects, insurance and real estate personnel, facility managers, homeowners, and anyone who needs quick, clear definitions for construction terms. The second edition has been further enhanced with updates and hundreds of new terms and illustrations . . . in keeping with the most recent developments in the construction industry.

Now with a quick-reference Spanish section. Includes tools and equipment, materials, tasks, and more!

**$59.95 per copy
Over 500 pages, softcover
Catalog no. 67282A**

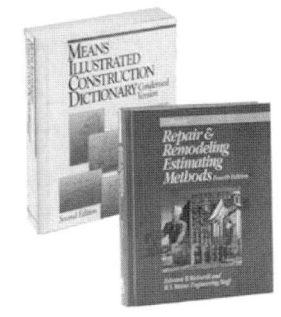

## Repair & Remodeling
### Estimating 4th Edition
### By Edward B. Wetherill and RSMeans

This important reference focuses on the unique problems of estimating renovations of existing structures, and helps you determine the true costs of remodeling through careful evaluation of architectural details and a site visit.

*New section on disaster restoration costs.*

**$69.95 per copy
Over 450 pages, illustrated, hardcover
Catalog no. 67265B**

## Facilities Planning &
## Relocation
### by David D. Owen

A complete system for planning space needs and managing relocations. Includes step-by-step manual, over 50 forms, and extensive reference section on materials and furnishings.

*New lower price and user-friendly format.*

**$49.98 per copy
Over 450 pages, softcover
Catalog no. 67301**

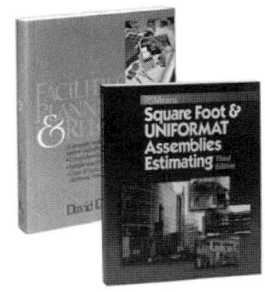

## Square Foot &
## UNIFORMAT Assemblies
### Estimating Methods 3rd Edition

Develop realistic square foot and assemblies costs for budgeting and construction funding. The new edition features updated guidance on square foot and assemblies estimating using UNIFORMAT II. An essential reference for anyone who performs conceptual estimates.

**$69.95 per copy
Over 300 pages, illustrated, softcover
Catalog no. 67145B**

## Electrical Estimating
### Methods 3rd Edition

Expanded edition includes sample estimates and cost information in keeping with the latest version of the CSI MasterFormat and UNIFORMAT II. Complete coverage of fiber optic and uninterruptible power supply electrical systems, broken down by components, and explained in detail. Includes a new chapter on computerized estimating methods. A practical companion to *RSMeans Electrical Cost Data.*

**$64.95 per copy
Over 325 pages, hardcover
Catalog no. 67230B**

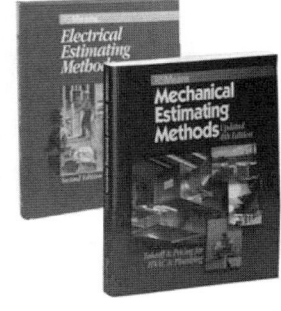

## Mechanical
### Estimating Methods 4th Edition

Completely updated, this guide assists you in making a review of plans, specs, and bid packages, with suggestions for takeoff procedures, listings, substitutions, and pre-bid scheduling for all components of HVAC. Includes suggestions for budgeting labor and equipment usage. Compares materials and construction methods to allow you to select the best option.

**$64.95 per copy
Over 350 pages, illustrated, softcover
Catalog no. 67294B**

## Construction Business
## Management
### by Nick Ganaway

Only 43% of construction firms stay in business after four years. Make sure your company thrives with valuable guidance from a pro with 25 years of success as a commercial contractor. Find out what it takes to build all aspects of a business that is profitable, enjoyable, and enduring. With a bonus chapter on retail construction.

**$49.95 per copy
200 pages, softcover
Catalog no. 67352**

## Project Scheduling &
## Management for Construction
### 3rd Edition
### by David R. Pierce, Jr.

A comprehensive yet easy-to-follow guide to construction project scheduling and control—from vital project management principles through the latest scheduling, tracking, and controlling techniques. The author is a leading authority on scheduling, with years of field and teaching experience at leading academic institutions. Spend a few hours with this book and come away with a solid understanding of this essential management topic.

**$64.95 per copy
Over 300 pages, illustrated, hardcover
Catalog no. 67247B**

For more information
visit the RSMeans website
at www.rsmeans.com

# Reference Books

# The Practice of Cost Segregation Analysis

## by Bruce A. Desrosiers and Wayne J. DelPico

This expert guide walks you through the practice of cost segregation analysis, which enables property owners to defer taxes and benefit from "accelerated cost recovery" through depreciation deductions on assets that are properly identified and classified.

With a glossary of terms, sample cost segregation estimates for various building types, key information resources, and updates via a dedicated website, this book is a critical resource for anyone involved in cost segregation analysis.

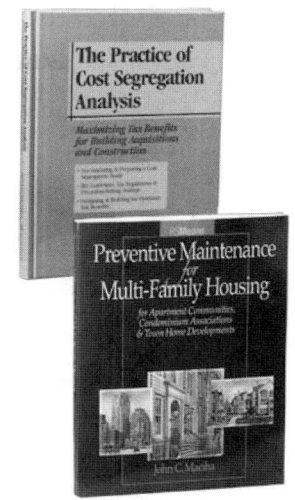

**$99.95 per copy**
**Over 225 pages**
**Catalog no. 67345**

# Preventive Maintenance for Multi-Family Housing

## by John C. Maciha

Prepared by one of the nation's leading experts on multi-family housing.

This complete PM system for apartment and condominium communities features expert guidance, checklists for buildings and grounds maintenance tasks and their frequencies, a reusable wall chart to track maintenance, and a dedicated website featuring customizable electronic forms. A must-have for anyone involved with multi-family housing maintenance and upkeep.

**$89.95 per copy**
**225 pages**
**Catalog no. 67346**

# How to Estimate with Means Data & CostWorks

**New 3rd Edition**

## by RSMeans and Saleh A. Mubarak, Ph.D.

New 3rd Edition—fully updated with new chapters, plus new CD with updated *CostWorks* cost data and MasterFormat organization. Includes all major construction items—with more than 300 exercises and two sets of plans that show how to estimate for a broad range of construction items and systems—including general conditions and equipment costs.

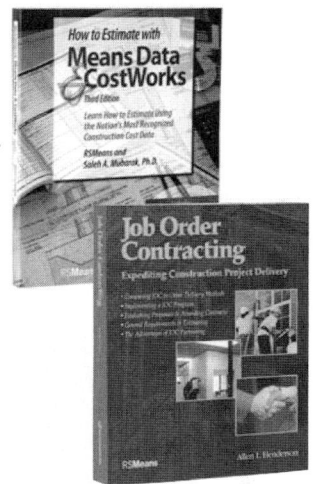

**$59.95 per copy**
**272 pages, softcover**
**Includes CostWorks CD**
**Catalog no. 67324B**

# Job Order Contracting

**Expediting Construction Project Delivery**

## by Allen Henderson

Expert guidance to help you implement JOC—fast becoming the preferred project delivery method for repair and renovation, minor new construction, and maintenance projects in the public sector and in many states and municipalities. The author, a leading JOC expert and practitioner, shows how to:

- Establish a JOC program
- Evaluate proposals and award contracts
- Handle general requirements and estimating
- Partner for maximum benefits

**$89.95 per copy**
**192 pages, illustrated, hardcover**
**Catalog no. 67348**

# Estimating Building Costs

**For the Residential & Light Commercial Contractor**

## by Wayne J. DelPico

Step-by-step estimating methods for residential and light commercial contractors. Includes a detailed look at every construction specialty—explaining all the components, takeoff units, and labor needed for well-organized, complete estimates. Covers correctly interpreting plans and specifications, and developing accurate and complete labor and material costs.

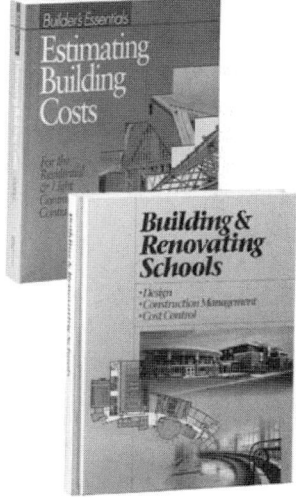

**$29.95 per copy**
**Over 400 pages, illustrated, softcover**
**Catalog no. 67343**

# Building & Renovating Schools

This all-inclusive guide covers every step of the school construction process—from initial planning, needs assessment, and design, right through moving into the new facility. A must-have resource for anyone concerned with new school construction or renovation. With square foot cost models for elementary, middle, and high school facilities, and real-life case studies of recently completed school projects.

The contributors to this book—architects, construction project managers, contractors, and estimators who specialize in school construction—provide start-to-finish, expert guidance on the process.

**$69.98 per copy**
**Over 425 pages, hardcover**
**Catalog no. 67342**

# Reference Books

**For more information
visit the RSMeans website
at www.rsmeans.com**

## The Homeowner's Guide to Mold By Michael Pugliese

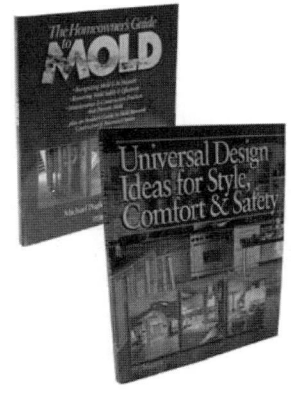

Expert guidance to protect your health and your home.

Mold, whether caused by leaks, humidity or flooding, is a real health and financial issue—for homeowners and contractors. This full-color book explains:
- Construction and maintenance practices to prevent mold
- How to inspect for and remove mold
- Mold remediation procedures and costs
- What to do after a flood
- How to deal with insurance companies

**$21.95 per copy
144 pages, softcover
Catalog no. 67344**

## Universal Design Ideas for Style, Comfort & Safety

### by RSMeans and Lexicon Consulting, Inc.

Incorporating universal design when building or remodeling helps people of any age and physical ability more fully and safely enjoy their living spaces. This book shows how universal design can be artfully blended into the most attractive homes. It discusses specialized products like adjustable countertops and chair lifts, as well as simple ways to enhance a home's safety and comfort. With color photos and expert guidance, every area of the home is covered. Includes budget estimates that give an idea how much projects will cost.

**$21.95
160 pages, illustrated, softcover
Catalog no. 67354**

## Landscape Estimating Methods

### 5th Edition

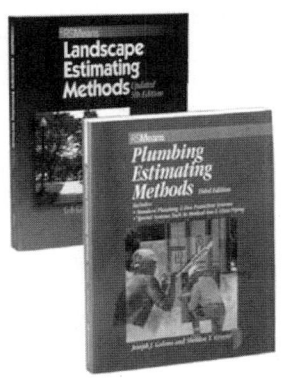

Answers questions about preparing competitive landscape construction estimates, with up-to-date cost estimates and the new MasterFormat classification system. Expanded and revised to address the latest materials and methods, including new coverage on approaches to green building. Includes:
- Step-by-step explanation of the estimating process
- Sample forms and worksheets that save time and prevent errors

**$64.95 per copy
Over 350 pages, softcover
Catalog no. 67295C**

## Plumbing Estimating Methods 3rd Edition

### by Joseph Galeno and Sheldon Greene

Updated and revised! This practical guide walks you through a plumbing estimate, from basic materials and installation methods through change order analysis. *Plumbing Estimating Methods* covers residential, commercial, industrial, and medical systems, and features sample takeoff and estimate forms and detailed illustrations of systems and components.

**$29.98 per copy
330+ pages, softcover
Catalog no. 67283B**

## Understanding & Negotiating Construction Contracts

### by Kit Werremeyer

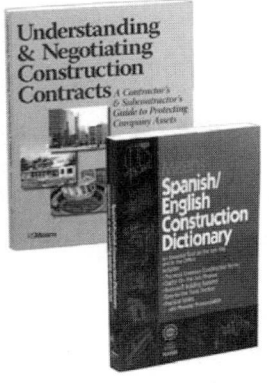

Take advantage of the author's 30 years' experience in small-to-large (including international) construction projects. Learn how to identify, understand, and evaluate high risk terms and conditions typically found in all construction contracts—then negotiate to lower or eliminate the risk, improve terms of payment, and reduce exposure to claims and disputes. The author avoids "legalese" and gives real-life examples from actual projects.

**$69.95 per copy
300 pages, softcover
Catalog no. 67350**

## Spanish/English Construction Dictionary 2nd Edition

### by RSMeans and the International Code Council

This expanded edition features thousands of the most common words and useful phrases in the construction industry with easy-to-follow pronunciations in both Spanish and English. Over 800 new terms, phrases, and illustrations have been added. It also features a new stand-alone "Safety & Emergencies" section, with colored pages for quick access.

Unique to this dictionary are the systems illustrations showing the relationship of components in the most common building systems for all major trades.

**$23.95 per copy
Over 400 pages
Catalog no. 67327A**

## How Your House Works
### by Charlie Wing

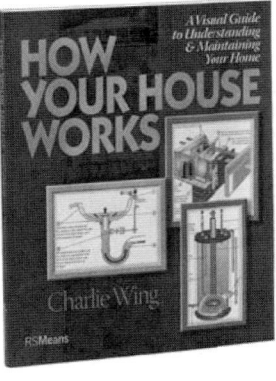

A must-have reference for every homeowner, handyman, and contractor—for repair, remodeling, and new construction. This book uncovers the mysteries behind just about every major appliance and building element in your house—from electrical, heating and AC, to plumbing, framing, foundations and appliances. Clear, "exploded" drawings show exactly how things should be put together and how they function—what to check if they don't work, and what you can do that might save you having to call in a professional.

**$21.95 per copy
160 pages, softcover
Catalog no. 67351**

For more information
visit the RSMeans website
at www.rsmeans.com

## Means CostWorks® CD Training

This one-day course helps users become more familiar with the functionality of *Means CostWorks* program. Each menu, icon, screen, and function found in the program is explained in depth. Time is devoted to hands-on estimating exercises.

**Some of what you'll learn:**
- Searching the database utilizing all navigation methods
- Exporting RSMeans Data to your preferred spreadsheet format
- Viewing crews, assembly components, and much more!
- Automatically regionalizing the database

You are required to bring your own laptop computer for this course.

When you register for this course you will receive an outline for your laptop requirements.

**Also offering web training for CostWorks CD!**

## Unit Price Estimating

This interactive *two-day* seminar teaches attendees how to interpret project information and process it into final, detailed estimates with the greatest accuracy level.

The most important credential an estimator can take to the job is the ability to visualize construction, and estimate accurately.

**Some of what you'll learn:**
- Interpreting the design in terms of cost
- The most detailed, time-tested methodology for accurate pricing
- Key cost drivers—material, labor, equipment, staging, and subcontracts
- Understanding direct and indirect costs for accurate job cost accounting and change order management

**Who should attend:** Corporate and government estimators and purchasers, architects, engineers... and others needing to produce accurate project estimates.

## Square Foot & Assemblies Estimating

This *two-day* course teaches attendees how to quickly deliver accurate square foot estimates using limited budget and design information.

**Some of what you'll learn:**
- How square foot costing gets the estimate done faster
- Taking advantage of a "systems" or "assemblies" format
- The RSMeans "building assemblies/square foot cost approach"
- How to create a reliable preliminary systems estimate using bare-bones design information

**Who should attend:** Facilities managers, facilities engineers, estimators, planners, developers, construction finance professionals... and others needing to make quick, accurate construction cost estimates at commercial, government, educational, and medical facilities.

## Facilities Maintenance & Repair Estimating

This *two-day* course teaches attendees how to plan, budget, and estimate the cost of ongoing and preventive maintenance and repair for existing buildings and grounds.

**Some of what you'll learn:**
- The most financially favorable maintenance, repair and replacement scheduling and estimating
- Auditing and value engineering facilities
- Preventive planning and facilities upgrading
- Determining both in-house and contract-out service costs
- Annual, asset-protecting M&R plan

**Who should attend:** Facility managers, maintenance supervisors, buildings and grounds superintendents, plant managers, planners, estimators... and others involved in facilities planning and budgeting.

## Mechanical & Electrical Estimating

This *two-day* course teaches attendees how to prepare more accurate and complete mechanical/electrical estimates, avoiding the pitfalls of omission and double-counting, while understanding the composition and rationale within the RSMeans Mechanical/Electrical database.

**Some of what you'll learn:**
- The unique way mechanical and electrical systems are interrelated
- M&E estimates–conceptual, planning, budgeting, and bidding stages
- Order of magnitude, square foot, assemblies, and unit price estimating
- Comparative cost analysis of equipment and design alternatives

**Who should attend:** Architects, engineers, facilities managers, mechanical and electrical contractors... and others needing a highly reliable method for developing, understanding, and evaluating mechanical and electrical contracts.

## Green Building: Planning & Construction

In this *two-day* course, learn about tailoring a building and its placement on the site to the local climate, site conditions, culture, and community. Includes information to reduce resource consumption and augment resource supply.

**Some of what you'll learn:**
- Green technologies, materials, systems, and standards
- Modeling tools and how to use them
- Cost vs. value of green products over their life cycle
- Low-cost green strategies, and economic incentives and funding
- Health, comfort, and productivity

**Who should attend:** Contractors, project managers, building owners, building officials, healthcare and insurance professionals.

## Practical Project Management for Construction Professionals

In this *two-day* course, acquire the essential knowledge and develop the skills to effectively and efficiently execute the day-to-day responsibilities of the construction project manager.

**Covers:**
- General conditions of the construction contract
- Contract modifications: change orders and construction change directives
- Negotiations with subcontractors and vendors
- Effective writing: notification and communications
- Dispute resolution: claims and liens

**Who should attend:** Architects, engineers, owner's representatives, project managers.

## Facilities Construction Estimating

In this *two-day* course, professionals working in facilities management can get help with their daily challenges to establish budgets for all phase of a project

**Some of what you'll learn:**
- Determine the full scope of a project
- Identify the scope of risks & opportunities
- Creative solutions to estimating issues
- Organizing estimates for presentation & discussion
- Special techniques for repair/remodel and maintenance projects
- Negotiating project change orders

**Who should attend:** Facility managers, engineers, contractors, facility trades-people, planners & project managers.

For more information
visit the RSMeans website
at www.rsmeans.com

## Assessing Scope of Work for Facility Construction Estimating

This *two-day* practical training program addresses the vital importance of understanding the SCOPE of projects in order to produce accurate cost estimates for facility repair and remodeling.

**Some of what you'll learn:**

- Discussions of site visits, plans/specs, record drawings of facilities, and site-specific lists
- Review of CSI divisions, including means, methods, materials, and the challenges of scoping each topic
- Exercises in SCOPE identification and SCOPE writing for accurate estimating of projects
- Hands-on exercises that require SCOPE, take-off, and pricing

**Who should attend:** Corporate and government estimators, planners, facility managers, and others who need to produce accurate project estimates.

## Site Work & Heavy Construction Estimating

This *two-day* course teaches attendees how to estimate earthwork, site utilities, foundations, and site improvements, using the assemblies and the unit price methods.

**Some of what you'll learn:**

- Basic site work and heavy construction estimating skills
- Estimating foundations, utilities, earthwork, and site improvements
- Correct equipment usage, quality control, and site investigation for estimating purposes

**Who should attend:** Project managers, design engineers, estimators, and contractors doing site work and heavy construction.

# 2011 RSMeans Seminar Schedule

Note: call for exact dates and details.

| Location | Dates |
|---|---|
| Las Vegas, NV | March |
| Washington, DC | April |
| Denver, CO | May |
| San Francisco, CA | June |
| Washington, DC | September |
| Dallas, TX | September |
| Las Vegas, NV | October |
| Atlantic City, NJ | October |
| Orlando, FL | November |
| San Diego, CA | December |

# 2011 Regional Seminar Schedule

| Location | Dates | Location | Dates |
|---|---|---|---|
| Seattle, WA | January and August | St. Louis, MO | August |
| Dallas/Ft. Worth, TX | January | Houston, TX | October |
| Austin, TX | February | Colorado Springs, CO | October |
| Anchorage, AK | March and September | Baltimore, MD | November |
| Oklahoma City, OK | March | Sacramento, CA | November |
| Phoenix, AZ | March | San Antonio, TX | December |
| Aiea, HI | May | Raleigh, NC | December |
| Kansas City, MO | May | | |
| Bethesda, MD | June | | |
| Columbus, GA | June | | |
| El Segundo, CA | August | | |

**1-800-334-3509, ext. 5115**

For more information
visit the RSMeans website
at www.rsmeans.com

# Professional Development

## Registration Information

**Register early... Save up to $100!** Register 30 days before the start date of a seminar and save $100 off your total fee. *Note: This discount can be applied only once per order. It cannot be applied to team discount registrations or any other special offer.*

**How to register** Register by phone today! The RSMeans toll-free number for making reservations is **1-800-334-3509, ext. 5115.**

**Individual seminar registration fee - $935.** *Means CostWorks®* **training registration fee - $375.** To register by mail, complete the registration form and return, with your full fee, to: RSMeans Seminars, 700 Longwater Drive, Norwell, MA 02061.

**Government pricing** All federal government employees save off the regular seminar price. Other promotional discounts cannot be combined with the government discount.

**Team discount program** for two to four seminar registrations. Call for pricing: 1-800-334-3509, ext. 5115

**Multiple course discounts** When signing up for two or more courses, call for pricing.

**Refund policy** Cancellations will be accepted up to ten business days prior to the seminar start. There are no refunds for cancellations received later than ten working days prior to the first day of the seminar. A $150 processing fee will be applied for all cancellations. Written notice of cancellation is required. Substitutions can be made at any time before the session starts. **No-shows are subject to the full seminar fee.**

**AACE approved courses** Many seminars described and offered here have been approved for 14 hours (1.4 recertification credits) of credit by the AACE International Certification Board toward meeting the continuing education requirements for recertification as a Certified Cost Engineer/Certified Cost Consultant.

**AIA Continuing Education** We are registered with the AIA Continuing Education System (AIA/CES) and are committed to developing quality learning activities in accordance with the CES criteria. Many seminars meet the AIA/CES criteria for Quality Level 2. AIA members may receive (14) learning units (LUs) for each two-day RSMeans course.

**Daily course schedule** The first day of each seminar session begins at 8:30 a.m. and ends at 4:30 p.m. The second day begins at 8:00 a.m. and ends at 4:00 p.m. Participants are urged to bring a hand-held calculator, since many actual problems will be worked out in each session.

**Continental breakfast** Your registration includes the cost of a continental breakfast, and a morning and afternoon refreshment break. These informal segments allow you to discuss topics of mutual interest with other seminar attendees. (You are free to make your own lunch and dinner arrangements.)

**Hotel/transportation arrangements** RSMeans arranges to hold a block of rooms at most host hotels. To take advantage of special group rates when making your reservation, be sure to mention that you are attending the RSMeans seminar. You are, of course, free to stay at the lodging place of your choice. (**Hotel reservations and transportation arrangements should be made directly by seminar attendees.**)

**Important** Class sizes are limited, so please register as soon as possible.

*Note: Pricing subject to change.*

---

## Registration Form

ADDS-1000

**Call 1-800-334-3509, ext. 5115 to register or FAX this form 1-800-632-6732. Visit our website: www.rsmeans.com**

Please register the following people for the RSMeans construction seminars as shown here. We understand that we must make our own hotel reservations if overnight stays are necessary.

☐ Full payment of $_____ enclosed.

☐ Bill me

Please print name of registrant(s)

(To appear on certificate of completion)

_____

_____

_____

_____

P.O. #:_____
GOVERNMENT AGENCIES MUST SUPPLY PURCHASE ORDER NUMBER OR TRAINING FORM.

Firm name_____

Address_____

City/State/Zip_____

Telephone no._____ Fax no._____

E-mail address_____

Charge registration(s) to: ☐ MasterCard ☐ VISA ☐ American Express

Account no._____ Exp. date_____

Cardholder's signature_____

Seminar name_____

Seminar City _____

**Please mail check to: RSMeans Seminars, 700 Longwater Drive, Norwell, MA 02061 USA**

For more information
visit the RSMeans website
at www.rsmeans.com

# New Titles

## The Gypsum Construction Handbook by USG

An invaluable reference of construction procedures for gypsum drywall, cement board, veneer plaster, and conventional plaster. This new edition includes the newest product developments, installation methods, fire- and sound-rated construction information, illustrated framing-to-finish application instructions, estimating and planning information, and more. Great for architects and engineers; contractors, builders, and dealers; apprentices and training programs; building inspectors and code officials; and anyone interested in gypsum construction. Features information on tools and safety practices, a glossary of construction terms, and a list of important agencies and associations. Also available in Spanish.

**$29.95**
**Over 575 pages, illustrated, softcover**
**Catalog no. 67357A**

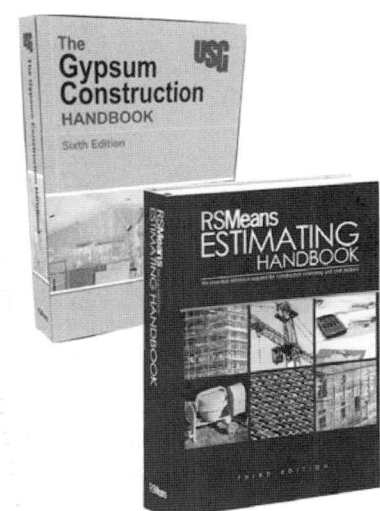

## RSMeans Estimating Handbook 3rd Edition

Widely used in the industry for tasks ranging from routine estimates to special cost analysis projects, this handbook has been completely updated and reorganized with new and expanded technical information.

*RSMeans Estimating Handbook* will help construction professionals:

• Evaluate architectural plans and specifications
• Prepare accurate quantity takeoffs
• Compare design alternatives and costs
• Perform value engineering
• Double-check estimates and quotes
• Estimate change orders

**$99.95 per copy**
**Over 900 pages, hardcover**
**Catalog No. 67276B**

## Means Illustrated Construction Dictionary
### Unabridged 4th Edition, with CD-ROM

Long regarded as the industry's finest, *Means Illustrated Construction Dictionary* is now even better. With nearly 20,000 terms and more than 1,400 illustrations and photos, it is the clear choice for the most comprehensive and current information. The companion CD-ROM that comes with this new edition adds extra features such as: larger graphics and expanded definitions.

**$99.95 per copy**
**Over 790 pages, illustrated, hardcover**
**Catalog no. 67292B**

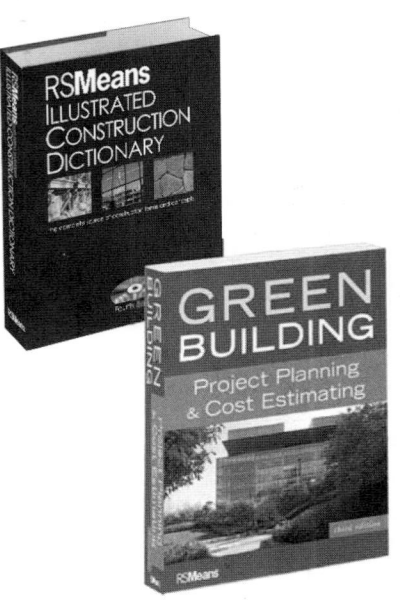

## Green Building: Project Planning & Cost Estimating 3rd Edition

Since the widely read first edition of this book, green building has gone from a growing trend to a major force in design and construction.

This new edition has been updated with the latest in green building technologies, design concepts, standards, and costs. Full-color with all new case studies—plus a new chapter on commercial real estate.

**$99.95 per copy**
**Over 450 pages, softcover**
**Catalog no. 67338B**

## Green Home Improvement
### by Daniel D. Chiras, PhD

With energy costs rising and environmental awareness increasing, many people are looking to make their homes greener. This book makes it easy, with 65 projects and actual costs and projected savings that will help homeowners prioritize their green improvements.

Projects range from simple water savers that cost only a few dollars, to bigger-ticket items such as efficient appliances and HVAC systems, green roofing, flooring, landscaping, and much more. With color photos and cost estimates, each project compares options and describes the work involved, the benefits, and the savings.

**$24.95**
**320 pages, illustrated, softcover**
**Catalog no. 67355**

## Residential & Light Commercial Construction Standards 3rd Edition, Updated by RSMeans and contributing authors

This book provides authoritative requirements and recommendations compiled from leading professional associations, industry publications, and building code organizations. This all-in-one reference helps establish a standard for workmanship, quickly resolve disputes, and avoid defect claims. Updated third edition includes new coverage of green building, seismic, hurricane, and mold-resistant construction.

**$59.95**
**Over 550 pages, illustrated, softcover**
**Catalog no. 67322B**

# 2011 Order Form

| Qty. | Book no. | COST ESTIMATING BOOKS | Unit Price | Total |
|------|----------|----------------------|-----------|-------|
| | 60061 | Assemblies Cost Data 2011 | $285.95 | |
| | 60011 | Building Construction Cost Data 2011 | 174.95 | |
| | 60111 | Concrete & Masonry Cost Data 2011 | 154.95 | |
| | 50141 | Construction Cost Indexes 2011 (subscription) | 362.00 | |
| | 60141A | Construction Cost Index–January 2011 | 90.50 | |
| | 60141B | Construction Cost Index–April 2011 | 90.50 | |
| | 60141C | Construction Cost Index–July 2011 | 90.50 | |
| | 60141D | Construction Cost Index–October 2011 | 90.50 | |
| | 60341 | Contr. Pricing Guide: Resid. R & R Costs 2011 | 39.95 | |
| | 60231 | Electrical Change Order Cost Data 2011 | 169.95 | |
| | 60031 | Electrical Cost Data 2011 | 174.95 | |
| | 60201 | Facilities Construction Cost Data 2011 | 439.95 | |
| | 60301 | Facilities Maintenance & Repair Cost Data 2011 | 389.95 | |
| | 60551 | Green Building Cost Data 2011 | 144.95 | |
| | 60161 | Heavy Construction Cost Data 2011 | 174.95 | |
| | 60091 | Interior Cost Data 2011 | 179.95 | |
| | 60121 | Labor Rates for the Const. Industry 2011 | 389.95 | |
| | 60181 | Light Commercial Cost Data 2011 | 119.95 | |
| | 60021 | Mechanical Cost Data 2011 | 174.95 | |
| | 60151 | Open Shop Building Const. Cost Data 2011 | 134.95 | |
| | 60211 | Plumbing Cost Data 2011 | 174.95 | |
| | 60041 | Commercial Renovation Cost Data 2011 | 144.95 | |
| | 60171 | Residential Cost Data 2011 | 109.95 | |
| | 60281 | Site Work & Landscape Cost Data 2011 | 169.95 | |
| | 60051 | Square Foot Costs 2011 | 184.95 | |
| | 62010 | Yardsticks for Costing (2010) | 174.95 | |
| | 62011 | Yardsticks for Costing (2011) | 174.95 | |
| | | **REFERENCE BOOKS** | | |
| | 67329 | Bldrs Essentials: Best Bus. Practices for Bldrs | 29.95 | |
| | 67298A | Bldrs Essentials: Framing/Carpentry 2nd Ed. | 15.98 | |
| | 67298AS | Bldrs Essentials: Framing/Carpentry Spanish | 15.98 | |
| | 67307 | Bldrs Essentials: Plan Reading & Takeoff | 35.95 | |
| | 67342 | Building & Renovating Schools | 69.98 | |
| | 67339 | Building Security: Strategies & Costs | 44.98 | |
| | 67353 | Complete Book of Framing | 29.95 | |
| | 67146 | Concrete Repair & Maintenance Illustrated | 34.98 | |
| | 67352 | Construction Business Management | 49.95 | |
| | 67314 | Cost Planning & Est. for Facil. Maint. | 89.95 | |
| | 67328A | Designing & Building with the IBC, 2nd Ed. | 99.95 | |
| | 67230B | Electrical Estimating Methods, 3rd Ed. | 64.95 | |
| | 64777A | Environmental Remediation Est. Methods, 2nd Ed. | 49.98 | |
| | 67343 | Estimating Bldg. Costs for Resi. & Lt. Comm. | 29.95 | |
| | 67276B | Estimating Handbook, 3rd Ed. | $99.95 | |
| | 67318 | Facilities Operations & Engineering Reference | 54.98 | |
| | 67301 | Facilities Planning & Relocation | 49.98 | |
| | 67338B | Green Building: Proj. Planning & Cost Est., 3rd Ed. | 99.95 | |
| | 67355 | Green Home Improvement | 24.95 | |
| | 67357A | The Gypsum Construction Handbook | 29.95 | |
| | 67357S | The Gypsum Construction Handbook (Spanish) | 29.95 | |
| | 67323 | Historic Preservation: Proj. Planning & Est. | 49.98 | |
| | 67308E | Home Improvement Costs–Int. Projects, 9th Ed. | 24.95 | |

| Qty. | Book no. | REFERENCE BOOKS (Cont.) | Unit Price | Total |
|------|----------|------------------------|-----------|-------|
| | 67309E | Home Improvement Costs–Ext. Projects, 9th Ed. | 24.95 | |
| | 67344 | Homeowner's Guide to Mold | 21.95 | |
| | 67324B | How to Est.w/Means Data & CostWorks, 3rd Ed. | 59.95 | |
| | 67351 | How Your House Works | 21.95 | |
| | 67282A | Illustrated Const. Dictionary, Condensed, 2nd Ed. | 59.95 | |
| | 67292B | Illustrated Const. Dictionary, w/CD-ROM, 4th Ed. | 99.95 | |
| | 67348 | Job Order Contracting | 89.95 | |
| | 67295C | Landscape Estimating Methods, 5th Ed. | 64.95 | |
| | 67341 | Life Cycle Costing for Facilities | 99.95 | |
| | 67294B | Mechanical Estimating Methods, 4th Ed. | 64.95 | |
| | 67283B | Plumbing Estimating Methods, 3rd Ed. | 29.98 | |
| | 67345 | Practice of Cost Segregation Analysis | 99.95 | |
| | 67337 | Preventive Maint. for Higher Education Facilities | 149.95 | |
| | 67346 | Preventive Maint. for Multi-Family Housing | 89.95 | |
| | 67326 | Preventive Maint. Guidelines for School Facil. | 149.95 | |
| | 67247B | Project Scheduling & Management for Constr. 3rd Ed. | 64.95 | |
| | 67265B | Repair & Remodeling Estimating Methods, 4th Ed. | 69.95 | |
| | 67322B | Resi. & Light Commercial Const. Stds., 3rd Ed. | 59.95 | |
| | 67327A | Spanish/English Construction Dictionary, 2nd Ed. | 23.95 | |
| | 67145B | Sq. Ft. & Assem. Estimating Methods, 3rd Ed. | 69.95 | |
| | 67321 | Total Productive Facilities Management | 29.98 | |
| | 67350 | Understanding and Negotiating Const. Contracts | 69.95 | |
| | 67303B | Unit Price Estimating Methods, 4th Ed. | 59.95 | |
| | 67354 | Universal Design | 21.95 | |
| | 67319A | Value Engineering: Practical Applications | 79.95 | |

MA residents add 6.25% state sales tax

Shipping & Handling**

Total (U.S. Funds)*

Prices are subject to change and are for U.S. delivery only. *Canadian customers may call for current prices. **Shipping & handling charges: Add 7% of total order for check & credit card payments. Add 9% of total order for invoiced orders.

**Send order:**                                                 ADDV-1000

Name (please print) _____

Company _____

☐ Comy

☐ Hon Address _____

_____

City/St/Zip _____

Phone _____ P.O. # _____

(Must accompany all orders being billed)

# RSMeans Project Cost Report

By filling out this report, your project data will contribute to the database that supports the RSMeans Project Cost Square Foot Data. When you fill out this form, RSMeans will provide a $30 discount off one of the RSMeans products advertised in the preceding pages. Please complete the form including all items where you have cost data, and all the items marked (✔).

---

### $30.00 Discount per product for each report you submit.

---

## Project Description (No remodeling projects, please.)

✔ Building Use (Office, School...) _____

✔ Address (City, State) _____

_____

✔ Total Building Area (SF) _____

✔ Ground Floor (SF) _____

✔ Frame (Wood, Steel...) _____

✔ Exterior Wall (Brick, Tilt-up...) _____

✔ Basement: (check one) ☐ Full ☐ Partial ☐ None

✔ Number Stories _____

✔ Floor-to-Floor Height _____

✔ Volume (C.F.) _____

% Air Conditioned _____ Tons _____

**Total Project Cost** $ _____

Owner _____

Architect _____

General Contractor _____

✔ Bid Date _____

✔ Typical Bay Size _____

✔ Capacity _____

✔ Labor Force: _____ % Union _____ % Non-Union

✔ Project Description (Circle one number in each line)

1. Economy 2. Average 3. Custom 4. Luxury
1. Square 2. Rectangular 3. Irregular 4. Very Irregular

Comments _____

| | | | |
|---|---|---|---|
| A | ✔ General Conditions | | |
| B | ✔ Site Work | | |
| C | ✔ Concrete | | |
| D | ✔ Masonry | | |
| E | ✔ Metals | | |
| F | ✔ Wood & Plastics | | |
| G | ✔ Thermal & Moisture Protection | $ | |
| GR | Roofing & Flashing | $ | |
| H | ✔ Doors and Windows | $ | |
| J | ✔ Finishes | $ | |
| JP | Painting & Wall Covering | $ | |

| | | | |
|---|---|---|---|
| K | ✔ Specialties | $ | |
| L | ✔ Equipment | $ | |
| M | ✔ Furnishings | $ | |
| N | ✔ Special Construction | $ | |
| P | ✔ Conveying Systems | $ | |
| Q | ✔ Mechanical | $ | |
| QP | Plumbing | $ | |
| QB | HVAC | $ | |
| R | ✔ Electrical | $ | |
| S | ✔ Mech./Elec. Combined | $ | |

Please specify the RSMeans product you wish to receive Complete the address information.

Product Name _____

Product Number _____

Your Name _____

Title _____

Company _____
☐ Company
☐ Home  Street Address _____

City, State, Zip _____

Return by mail or Fax 888-492-6770.

Method of Payment:

Credit Card # _____

Expiration Date _____

Check _____

Purchase Order _____

**RSMeans**
**a division of Reed Construction Data**
Square Foot Costs Department
700 Longwater Drive
Norwell, MA 02061